CLIMATE CHANGE 2001:
THE SCIENTIFIC BASIS

Climate Change 2001: The Scientific Basis is the most comprehensive and up-to-date scientific assessment of past, present and future climate change. The report:

- Analyses an enormous body of observations of all parts of the climate system.
- Catalogues increasing concentrations of atmospheric greenhouse gases.
- Assesses our understanding of the processes and feedbacks which govern the climate system.
- Projects scenarios of future climate change using a wide range of models of future emissions of greenhouse gases and aerosols.
- Makes a detailed study of whether a human influence on climate can be identified.
- Suggests gaps in information and understanding that remain in our knowledge of climate change and how these might be addressed.

Simply put, this latest assessment of the IPCC will again form the standard scientific reference for all those concerned with climate change and its consequences, including students and researchers in environmental science, meteorology, climatology, biology, ecology and atmospheric chemistry, and policymakers in governments and industry worldwide.

J.T. Houghton is Co-Chair of Working Group I, IPCC.

Y. Ding is Co-Chair of Working Group I, IPCC.

D.J. Griggs is the Head of the Technical Support Unit, Working Group I, IPCC.

M. Noguer is the Deputy Head of the Technical Support Unit, Working Group I, IPCC.

P.J. van der Linden is the Project Administrator, Technical Support Unit, Working Group I, IPCC.

X. Dai is a Visiting Scientist, Technical Support Unit, Working Group I, IPCC.

K. Maskell is a Climate Scientist, Technical Support Unit, Working Group I, IPCC.

C.A. Johnson is a Climate Scientist, Technical Support Unit, Working Group I, IPCC.

Climate Change 2001:
The Scientific Basis

Edited by

J.T. Houghton
Co-Chair of Working Group I, IPCC

Y. Ding
Co-Chair of Working Group I, IPCC

D.J. Griggs
Head of Technical Support Unit, Working Group I, IPCC

M. Noguer
Deputy Head of Technical Support Unit, Working Group I, IPCC

P.J. van der Linden
Project Administrator, Technical Support Unit, Working Group I, IPCC

X. Dai
Visiting Scientist, Technical Support Unit, Working Group I, IPCC

K. Maskell
Climate Scientist, Technical Support Unit, Working Group I, IPCC

C.A. Johnson
Climate Scientist, Technical Support Unit, Working Group I, IPCC

Contribution of Working Group I to the Third Assessment Report
of the Intergovernmental Panel on Climate Change

Published for the Intergovernmental Panel on Climate Change

CAMBRIDGE
UNIVERSITY PRESS

PUBLISHED BY THE PRESS SYNDICATE OF THE UNIVERSITY OF CAMBRIDGE
The Pitt Building, Trumpington Street, Cambridge, United Kingdom

CAMBRIDGE UNIVERSITY PRESS
The Edinburgh Building, Cambridge CB2 2RU, UK
40 West 20th Street, New York, NY 10011–4211, USA
10 Stamford Road, Oakleigh, Melbourne 3166, Australia
Ruiz de Alarcón 13, 28014 Madrid, Spain
Dock House, The Waterfront, Cape Town 8001, South Africa

http://www.cambridge.org

First published 2001

Printed in the United States of America

A catalogue record for this book is available from the British Library

Library of Congress cataloguing in publication data available

ISBN 0521 80767 0 hardback
ISBN 0521 01495 6 paperback

When citing chapters or the Technical Summary from this report, please use the authors in the order given on the chapter frontpage, for example, Chapter 2 is referenced as:
Folland, C.K., T.R. Karl, J.R. Christy, R.A. Clarke, G.V. Gruza, J. Jouzel, M.E. Mann, J. Oerlemans, M.J. Salinger and S.-W. Wang, 2001: Observed Climate Variability and Change. In: *Climate Change 2001: The Scientific Basis. Contribution of Working Group I to the Third Assessment Report of the Intergovernmental Panel on Climate Change* [Houghton, J.T., Y. Ding, D.J. Griggs, M. Noguer, P.J. van der Linden, X. Dai, K. Maskell, and C.A. Johnson (eds.)]. Cambridge University Press, Cambridge, United Kingdom and New York, NY, USA, 881pp.

Reference to the whole report is:
IPCC, 2001: *Climate Change 2001: The Scientific Basis. Contribution of Working Group I to the Third Assessment Report of the Intergovernmental Panel on Climate Change* [Houghton, J.T., Y. Ding, D.J. Griggs, M. Noguer, P.J. van der Linden, X. Dai, K. Maskell, and C.A. Johnson (eds.)]. Cambridge University Press, Cambridge, United Kingdom and New York, NY, USA, 881pp.

Cover photo © Science Photo Library

Contents

Foreword

The Intergovernmental Panel on Climate Change (IPCC) was jointly established by the World Meteorological Organization (WMO) and the United Nations Environment Programme (UNEP) in 1988. Its terms of reference include (i) to assess available scientific and socio-economic information on climate change and its impacts and on the options for mitigating climate change and adapting to it and (ii) to provide, on request, scientific/technical/socio-economic advice to the Conference of the Parties (COP) to the United Nations Framework Convention on Climate Change (UNFCCC). From 1990, the IPCC has produced a series of Assessment Reports, Special Reports, Technical Papers, methodologies and other products that have become standard works of reference, widely used by policy-makers, scientists and other experts.

This volume, which forms part of the Third Assessment Report (TAR), has been produced by Working Group I (WGI) of the IPCC and focuses on the science of climate change. It consists of 14 chapters covering the physical climate system, the factors that drive climate change, analyses of past climate and projections of future climate change, and detection and attribution of human influences on recent climate.

As is usual in the IPCC, success in producing this report has depended first and foremost on the knowledge, enthusiasm and co-operation of many hundreds of experts worldwide, in many related but different disciplines. We would like to express our gratitude to all the Co-ordinating Lead Authors, Lead Authors, Contributing Authors, Review Editors and Reviewers. These individuals have devoted enormous time and effort to produce this report and we are extremely grateful for their commitment to the IPCC process. We would like to thank the staff of the WGI Technical Support Unit and the IPCC Secretariat for their dedication in co-ordinating the production of another successful IPCC report. We are also grateful to the governments, who have supported their scientists' participation in the IPCC process and who have contributed to the IPCC Trust Fund to provide for the essential participation of experts from developing countries and countries with economies in transition. We would like to express our appreciation to the governments of France, Tanzania, New Zealand and Canada who hosted drafting sessions in their countries, to the government of China, who hosted the final session of Working Group I in Shanghai, and to the government of the United Kingdom, who funded the WGI Technical Support Unit.

We would particularly like to thank Dr Robert Watson, Chairman of the IPCC, for his sound direction and tireless and able guidance of the IPCC, and Sir John Houghton and Prof. Ding Yihui, the Co-Chairmen of Working Group I, for their skillful leadership of Working Group I through the production of this report.

G.O.P. Obasi
Secretary General
World Meteorological Organization

K. Töpfer
Executive Director
United Nations Environment Programme
and
Director-General
United Nations Office in Nairobi

Preface

This report is the first complete assessment of the science of climate change since Working Group I (WGI) of the IPCC produced its second report Climate Change 1995: The Science of Climate Change in 1996. It enlarges upon and updates the information contained in that, and previous, reports, but primarily it assesses new information and research, produced in the last five years. The report analyses the enormous body of observations of all parts of the climate system, concluding that this body of observations now gives a collective picture of a warming world. The report catalogues the increasing concentrations of atmospheric greenhouse gases and assesses the effects of these gases and atmospheric aerosols in altering the radiation balance of the Earth-atmosphere system. The report assesses the understanding of the processes that govern the climate system and by studying how well the new generation of climate models represent these processes, assesses the suitability of the models for projecting climate change into the future. A detailed study is made of human influence on climate and whether it can be identified with any more confidence than in 1996, concluding that there is new and stronger evidence that most of the observed warming observed over the last 50 years is attributable to human activities. Projections of future climate change are presented using a wide range of scenarios of future emissions of greenhouse gases and aerosols. Both temperature and sea level are projected to continue to rise throughout the 21st century for all scenarios studied. Finally, the report looks at the gaps in information and understanding that remain and how these might be addressed.

This report on the scientific basis of climate change is the first part of Climate Change 2001, the Third Assessment Report (TAR) of the IPCC. Other companion assessment volumes have been produced by Working Group II (Impacts, Adaptation and Vulnerability) and by Working Group III (Mitigation). An important aim of the TAR is to provide objective information on which to base climate change policies that will meet the Objective of the FCCC, expressed in Article 2, of stabilisation of greenhouse gas concentrations in the atmosphere at a level that would prevent dangerous anthropogenic interference with the climate system. To assist further in this aim, as part of the TAR a Synthesis Report is being produced that will draw from the Working Group Reports scientific and socio-economic information relevant to nine questions addressing particular policy issues raised by the FCCC objective.

This report was compiled between July 1998 and January 2001, by 122 Lead Authors. In addition, 515 Contributing Authors submitted draft text and information to the Lead Authors. The draft report was circulated for review by experts, with 420 reviewers submitting valuable suggestions for improvement. This was followed by review by governments and experts, through which several hundred more reviewers participated. All the comments received were carefully analysed and assimilated into a revised document for consideration at the session of Working Group I held in Shanghai, 17 to 20 January 2001. There the Summary for Policymakers was approved in detail and the underlying report accepted.

Strenuous efforts have also been made to maximise the ease of utility of the report. As in 1996 the report contains a Summary for Policymakers (SPM) and a Technical Summary (TS), in addition to the main chapters in the report. The SPM and the TS follow the same structure, so that more information on items of interest in the SPM can easily be found in the TS. In turn, each section of the SPM and TS has been referenced to the appropriate section of the relevant chapter by the use of Source Information, so that material in the SPM and TS can easily be followed up in further detail in the chapters. The report also contains an index at Appendix VIII, which although not comprehensive allows for a search of the report at relatively top-level broad categories. By the end of 2001 a more in-depth search will be possible on an electronic version of the report, which will be found on the web at http://www.ipcc.ch.

We wish to express our sincere appreciation to all the Co-ordinating Lead Authors, Lead Authors and Review Editors whose expertise, diligence and patience have underpinned the successful completion of this report, and to the many contributors and reviewers for their valuable and painstaking dedication and work. We are grateful to Jean Jouzel, Hervé Le Treut, Buruhani Nyenzi, Jim Salinger, John Stone and Francis Zwiers for helping to organise drafting meetings; and to Wang Caifang for helping to organise the session of Working Group I held in Shanghai, 17 to 20 January 2001.

We would also like to thank members of the Working Group I Bureau, Buruhani Nyenzi, Armando Ramirez-Rojas, John Stone, John Zillman and Fortunat Joos for their wise counsel and guidance throughout the preparation of the report.

We would particularly like to thank Dave Griggs, Maria
Noguer, Paul van der Linden, Kathy Maskell, Xiaosu Dai,
Cathy Johnson, Anne Murrill and David Hall in the Working
Group I Technical Support Unit, with added assistance from
Alison Renshaw, for their tireless and good humoured
support throughout the preparation of the report. Thanks go
to Christoph Ritz and Bettina Buechler of ProClim (Forum
for Climate and Global Change), a programme of the Swiss
Academy of Sciences in Bern for their assistance in
producing the index to this report. We would also like to
thank Narasimhan Sundararaman, the Secretary of IPCC,
Renate Christ, Deputy Secretary, and the staff of the IPCC
Secretariat, Rudie Bourgeois, Chantal Ettori and Annie
Courtin who provided logistical support for government
liaison and travel of experts from the developing and transitional economy countries.

Robert Watson
IPCC Chairman

John Houghton
Co-chair IPCC WGI

Ding Yihui
Co-chair IPCC WGI

Summary for Policymakers

A Report of Working Group I of the Intergovernmental Panel on Climate Change

Based on a draft prepared by:

Daniel L. Albritton, Myles R. Allen, Alfons P. M. Baede, John A. Church, Ulrich Cubasch, Dai Xiaosu, Ding Yihui, Dieter H. Ehhalt, Christopher K. Folland, Filippo Giorgi, Jonathan M. Gregory, David J. Griggs, Jim M. Haywood, Bruce Hewitson, John T. Houghton, Joanna I. House, Michael Hulme, Ivar Isaksen, Victor J. Jaramillo, Achuthan Jayaraman, Catherine A. Johnson, Fortunat Joos, Sylvie Joussaume, Thomas Karl, David J. Karoly, Haroon S. Kheshgi, Corrine Le Quéré, Kathy Maskell, Luis J. Mata, Bryant J. McAvaney, Mack McFarland, Linda O. Mearns, Gerald A. Meehl, L. Gylvan Meira-Filho, Valentin P. Meleshko, John F. B. Mitchell, Berrien Moore, Richard K. Mugara, Maria Noguer, Buruhani S. Nyenzi, Michael Oppenheimer, Joyce E. Penner, Steven Pollonais, Michael Prather, I. Colin Prentice, Venkatchalam Ramaswamy, Armando Ramirez-Rojas, Sarah C. B. Raper, M. Jim Salinger, Robert J. Scholes, Susan Solomon, Thomas F. Stocker, John M. R. Stone, Ronald J. Stouffer, Kevin E. Trenberth, Ming-Xing Wang, Robert T. Watson, Kok S. Yap, John Zillman

with contributions from many authors and reviewers.

Summary for Policymakers

The Third Assessment Report of Working Group I of the Intergovernmental Panel on Climate Change (IPCC) builds upon past assessments and incorporates new results from the past five years of research on climate change[1]. Many hundreds of scientists[2] from many countries participated in its preparation and review.

This Summary for Policymakers (SPM), which was approved by IPCC member governments in Shanghai in January 2001[3], describes the current state of understanding of the climate system and provides estimates of its projected future evolution and their uncertainties. Further details can be found in the underlying report, and the appended Source Information provides cross references to the report's chapters.

An increasing body of observations gives a collective picture of a warming world and other changes in the climate system.

Since the release of the Second Assessment Report (SAR[4]), additional data from new studies of current and palaeoclimates, improved analysis of data sets, more rigorous evaluation of their quality, and comparisons among data from different sources have led to greater understanding of climate change.

The global average surface temperature has increased over the 20th century by about 0.6°C.

● The global average surface temperature (the average of near surface air temperature over land, and sea surface temperature) has increased since 1861. Over the 20th century the increase has been $0.6 \pm 0.2°C$[5,6] (Figure 1a). This value is about 0.15°C larger than that estimated by the SAR for the period up to 1994, owing to the relatively high temperatures of the additional years (1995 to 2000) and improved methods of processing the data. These numbers take into account various adjustments, including urban heat island effects. The record shows a great deal of variability; for example, most of the warming occurred during the 20th century, during two periods, 1910 to 1945 and 1976 to 2000.

● Globally, it is very likely[7] that the 1990s was the warmest decade and 1998 the warmest year in the instrumental record, since 1861 (see Figure 1a).

● New analyses of proxy data for the Northern Hemisphere indicate that the increase in temperature in the 20th century is likely[7] to have been the largest of any century during the past 1,000 years. It is also likely[7] that, in the Northern Hemisphere, the 1990s was the warmest decade and 1998 the warmest year (Figure 1b). Because less data are available, less is known about annual averages prior to 1,000 years before present and for conditions prevailing in most of the Southern Hemisphere prior to 1861.

● On average, between 1950 and 1993, night-time daily minimum air temperatures over land increased by about 0.2°C per decade. This is about twice the rate of increase in daytime daily maximum air temperatures (0.1°C per decade). This has lengthened the freeze-free season in many mid- and high latitude regions. The increase in sea surface temperature over this period is about half that of the mean land surface air temperature.

---

[1] *Climate change* in IPCC usage refers to any change in climate over time, whether due to natural variability or as a result of human activity. This usage differs from that in the Framework Convention on Climate Change, where *climate change* refers to a change of climate that is attributed directly or indirectly to human activity that alters the composition of the global atmosphere and that is in addition to natural climate variability observed over comparable time periods.

[2] In total 122 Co-ordinating Lead Authors and Lead Authors, 515 Contributing Authors, 21 Review Editors and 420 Expert Reviewers.

[3] Delegations of 99 IPCC member countries participated in the Eighth Session of Working Group I in Shanghai on 17 to 20 January 2001.

[4] The IPCC Second Assessment Report is referred to in this Summary for Policymakers as the SAR.

[5] Generally temperature trends are rounded to the nearest 0.05°C per unit time, the periods often being limited by data availability.

[6] In general, a 5% statistical significance level is used, and a 95% confidence level.

[7] In this Summary for Policymakers and in the Technical Summary, the following words have been used where appropriate to indicate judgmental estimates of confidence: *virtually certain* (greater than 99% chance that a result is true); *very likely* (90–99% chance); *likely* (66–90% chance); *medium likelihood* (33–66% chance); *unlikely* (10–33% chance); *very unlikely* (1–10% chance); *exceptionally unlikely* (less than 1% chance). The reader is referred to individual chapters for more details.

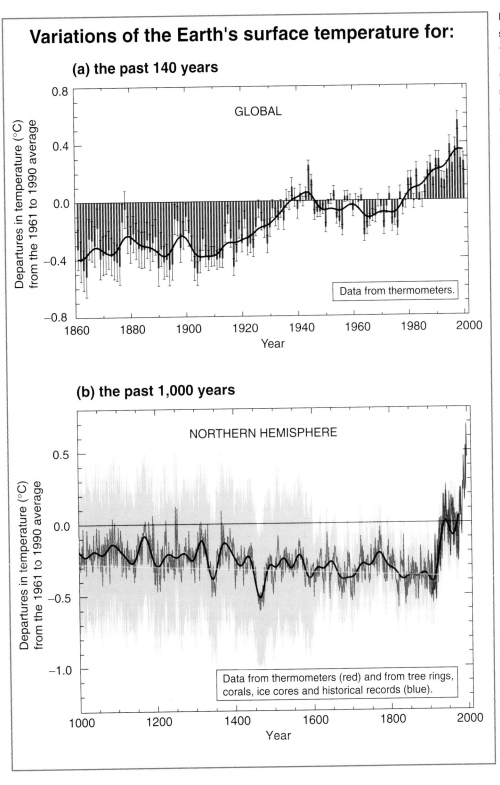

Variations of the Earth's surface temperature for:

(a) the past 140 years

GLOBAL

Departures in temperature (°C) from the 1961 to 1990 average

Data from thermometers.

Year

(b) the past 1,000 years

NORTHERN HEMISPHERE

Departures in temperature (°C) from the 1961 to 1990 average

Data from thermometers (red) and from tree rings, corals, ice cores and historical records (blue).

Year

Figure 1: Variations of the Earth's surface temperature over the last 140 years and the last millennium.

(a) The Earth's surface temperature is shown year by year (red bars) and approximately decade by decade (black line, a filtered annual curve suppressing fluctuations below near decadal time-scales). There are uncertainties in the annual data (thin black whisker bars represent the 95% confidence range) due to data gaps, random instrumental errors and uncertainties, uncertainties in bias corrections in the ocean surface temperature data and also in adjustments for urbanisation over the land. Over both the last 140 years and 100 years, the best estimate is that the global average surface temperature has increased by 0.6 ± 0.2°C.

(b) Additionally, the year by year (blue curve) and 50 year average (black curve) variations of the average surface temperature of the Northern Hemisphere for the past 1000 years have been reconstructed from "proxy" data calibrated against thermometer data (see list of the main proxy data in the diagram). The 95% confidence range in the annual data is represented by the grey region. These uncertainties increase in more distant times and are always much larger than in the instrumental record due to the use of relatively sparse proxy data. Nevertheless the rate and duration of warming of the 20th century has been much greater than in any of the previous nine centuries. Similarly, it is likely[7] that the 1990s have been the warmest decade and 1998 the warmest year of the millennium.

[Based upon (a) Chapter 2, Figure 2.7c and (b) Chapter 2, Figure 2.20]

Temperatures have risen during the past four decades in the lowest 8 kilometres of the atmosphere.

- Since the late 1950s (the period of adequate observations from weather balloons), the overall global temperature increases in the lowest 8 kilometres of the atmosphere and in surface temperature have been similar at 0.1°C per decade.

- Since the start of the satellite record in 1979, both satellite and weather balloon measurements show that the global average temperature of the lowest 8 kilometres of the atmosphere has changed by +0.05 ± 0.10°C per decade, but the global average surface temperature has increased significantly by +0.15 ± 0.05°C per decade. The difference in the warming rates is statistically significant. This difference occurs primarily over the tropical and sub-tropical regions.

- The lowest 8 kilometres of the atmosphere and the surface are influenced differently by factors such as stratospheric ozone depletion, atmospheric aerosols, and the El Niño phenomenon. Hence, it is physically plausible to expect that over a short time period (e.g., 20 years) there may be differences in temperature trends. In addition, spatial sampling techniques can also explain some of the differences in trends, but these differences are not fully resolved.

Snow cover and ice extent have decreased.

- Satellite data show that there are very likely[7] to have been decreases of about 10% in the extent of snow cover since the late 1960s, and ground-based observations show that there is very likely[7] to have been a reduction of about two weeks in the annual duration of lake and river ice cover in the mid- and high latitudes of the Northern Hemisphere, over the 20th century.

- There has been a widespread retreat of mountain glaciers in non-polar regions during the 20th century.

- Northern Hemisphere spring and summer sea-ice extent has decreased by about 10 to 15% since the 1950s. It is likely[7] that there has been about a 40% decline in Arctic sea-ice thickness during late summer to early autumn in recent decades and a considerably slower decline in winter sea-ice thickness.

Global average sea level has risen and ocean heat content has increased.

- Tide gauge data show that global average sea level rose between 0.1 and 0.2 metres during the 20th century.

- Global ocean heat content has increased since the late 1950s, the period for which adequate observations of sub-surface ocean temperatures have been available.

Changes have also occurred in other important aspects of climate.

- It is very likely[7] that precipitation has increased by 0.5 to 1% per decade in the 20th century over most mid- and high latitudes of the Northern Hemisphere continents, and it is likely[7] that rainfall has increased by 0.2 to 0.3% per decade over the tropical (10°N to 10°S) land areas. Increases in the tropics are not evident over the past few decades. It is also likely[7] that rainfall has decreased over much of the Northern Hemisphere sub-tropical (10°N to 30°N) land areas during the 20th century by about 0.3% per decade. In contrast to the Northern Hemisphere, no comparable systematic changes have been detected in broad latitudinal averages over the Southern Hemisphere. There are insufficient data to establish trends in precipitation over the oceans.

- In the mid- and high latitudes of the Northern Hemisphere over the latter half of the 20th century, it is likely[7] that there has been a 2 to 4% increase in the frequency of heavy precipitation events. Increases in heavy precipitation events can arise from a number of causes, e.g., changes in atmospheric moisture, thunderstorm activity and large-scale storm activity.

- It is likely[7] that there has been a 2% increase in cloud cover over mid- to high latitude land areas during the 20th century. In most areas the trends relate well to the observed decrease in daily temperature range.

- Since 1950 it is very likely[7] that there has been a reduction in the frequency of extreme low temperatures, with a smaller increase in the frequency of extreme high temperatures.

- Warm episodes of the El Niño-Southern Oscillation (ENSO) phenomenon (which consistently affects regional variations of precipitation and temperature over much of the tropics, sub-tropics and some mid-latitude areas) have been more frequent, persistent and intense since the mid-1970s, compared with the previous 100 years.

- Over the 20th century (1900 to 1995), there were relatively small increases in global land areas experiencing severe drought or severe wetness. In many regions, these changes are dominated by inter-decadal and multi-decadal climate variability, such as the shift in ENSO towards more warm events.

- In some regions, such as parts of Asia and Africa, the frequency and intensity of droughts have been observed to increase in recent decades.

Some important aspects of climate appear not to have changed.

- A few areas of the globe have not warmed in recent decades, mainly over some parts of the Southern Hemisphere oceans and parts of Antarctica.

- No significant trends of Antarctic sea-ice extent are apparent since 1978, the period of reliable satellite measurements.

- Changes globally in tropical and extra-tropical storm intensity and frequency are dominated by inter-decadal to multi-decadal variations, with no significant trends evident over the 20th century. Conflicting analyses make it difficult to draw definitive conclusions about changes in storm activity, especially in the extra-tropics.

- No systematic changes in the frequency of tornadoes, thunder days, or hail events are evident in the limited areas analysed.

Emissions of greenhouse gases and aerosols due to human activities continue to alter the atmosphere in ways that are expected to affect the climate.

Changes in climate occur as a result of both internal variability within the climate system and external factors (both natural and anthropogenic). The influence of external factors on climate can be broadly compared using the concept of radiative forcing[8]. A positive radiative forcing, such as that produced by increasing concentrations of greenhouse gases, tends to warm the surface. A negative radiative forcing, which can arise from an increase in some types of aerosols (microscopic airborne particles) tends to cool the surface. Natural factors, such as changes in solar output or explosive volcanic activity, can also cause radiative forcing. Characterisation of these climate forcing agents and their changes over time (see Figure 2) is required to understand past climate changes in the context of natural variations and to project what climate changes could lie ahead. Figure 3 shows current estimates of the radiative forcing due to increased concentrations of atmospheric constituents and other mechanisms.

---

[8] *Radiative forcing* is a measure of the influence a factor has in altering the balance of incoming and outgoing energy in the Earth-atmosphere system, and is an index of the importance of the factor as a potential climate change mechanism. It is expressed in Watts per square metre (Wm^{-2}).

Indicators of the human influence on the atmosphere during the Industrial Era

(a) Global atmospheric concentrations of three well mixed greenhouse gases

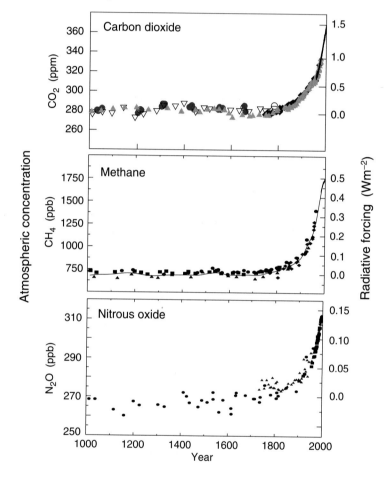

(b) Sulphate aerosols deposited in Greenland ice

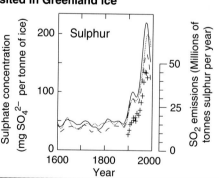

Figure 2: Long records of past changes in atmospheric composition provide the context for the influence of anthropogenic emissions.

(a) shows changes in the atmospheric concentrations of carbon dioxide (CO_2), methane (CH_4), and nitrous oxide (N_2O) over the past 1000 years. The ice core and firn data for several sites in Antarctica and Greenland (shown by different symbols) are supplemented with the data from direct atmospheric samples over the past few decades (shown by the line for CO_2 and incorporated in the curve representing the global average of CH_4). The estimated positive radiative forcing of the climate system from these gases is indicated on the right-hand scale. Since these gases have atmospheric lifetimes of a decade or more, they are well mixed, and their concentrations reflect emissions from sources throughout the globe. All three records show effects of the large and increasing growth in anthropogenic emissions during the Industrial Era.

(b) illustrates the influence of industrial emissions on atmospheric sulphate concentrations, which produce negative radiative forcing. Shown is the time history of the concentrations of sulphate, not in the atmosphere but in ice cores in Greenland (shown by lines; from which the episodic effects of volcanic eruptions have been removed). Such data indicate the local deposition of sulphate aerosols at the site, reflecting sulphur dioxide (SO_2) emissions at mid-latitudes in the Northern Hemisphere. This record, albeit more regional than that of the globally-mixed greenhouse gases, demonstrates the large growth in anthropogenic SO_2 emissions during the Industrial Era. The pluses denote the relevant regional estimated SO_2 emissions (right-hand scale).

[Based upon (a) Chapter 3, Figure 3.2b (CO_2); Chapter 4, Figure 4.1a and b (CH_4) and Chapter 4, Figure 4.2 (N_2O) and (b) Chapter 5, Figure 5.4a]

Concentrations of atmospheric greenhouse gases and their radiative forcing have continued to increase as a result of human activities.

- The atmospheric concentration of carbon dioxide (CO_2) has increased by 31% since 1750. The present CO_2 concentration has not been exceeded during the past 420,000 years and likely[7] not during the past 20 million years. The current rate of increase is unprecedented during at least the past 20,000 years.

- About three-quarters of the anthropogenic emissions of CO_2 to the atmosphere during the past 20 years is due to fossil fuel burning. The rest is predominantly due to land-use change, especially deforestation.

- Currently the ocean and the land together are taking up about half of the anthropogenic CO_2 emissions. On land, the uptake of anthropogenic CO_2 very likely[7] exceeded the release of CO_2 by deforestation during the 1990s.

- The rate of increase of atmospheric CO_2 concentration has been about 1.5 ppm[9] (0.4%) per year over the past two decades. During the 1990s the year to year increase varied from 0.9 ppm (0.2%) to 2.8 ppm (0.8%). A large part of this variability is due to the effect of climate variability (e.g., El Niño events) on CO_2 uptake and release by land and oceans.

- The atmospheric concentration of methane (CH_4) has increased by 1060 ppb[9] (151%) since 1750 and continues to increase. The present CH_4 concentration has not been exceeded during the past 420,000 years. The annual growth in CH_4 concentration slowed and became more variable in the 1990s, compared with the 1980s. Slightly more than half of current CH_4 emissions are anthropogenic (e.g., use of fossil fuels, cattle, rice agriculture and landfills). In addition, carbon monoxide (CO) emissions have recently been identified as a cause of increasing CH_4 concentration.

- The atmospheric concentration of nitrous oxide (N_2O) has increased by 46 ppb (17%) since 1750 and continues to increase. The present N_2O concentration has not been exceeded during at least the past thousand years. About a third of current N_2O emissions are anthropogenic (e.g., agricultural soils, cattle feed lots and chemical industry).

- Since 1995, the atmospheric concentrations of many of those halocarbon gases that are both ozone-depleting and greenhouse gases (e.g., $CFCl_3$ and CF_2Cl_2), are either increasing more slowly or decreasing, both in response to reduced emissions under the regulations of the Montreal Protocol and its Amendments. Their substitute compounds (e.g., CHF_2Cl and CF_3CH_2F) and some other synthetic compounds (e.g., perfluorocarbons (PFCs) and sulphur hexafluoride (SF_6)) are also greenhouse gases, and their concentrations are currently increasing.

- The radiative forcing due to increases of the well-mixed greenhouse gases from 1750 to 2000 is estimated to be 2.43 Wm^{-2}: 1.46 Wm^{-2} from CO_2; 0.48 Wm^{-2} from CH_4; 0.34 Wm^{-2} from the halocarbons; and 0.15 Wm^{-2} from N_2O. (See Figure 3, where the uncertainties are also illustrated.)

- The observed depletion of the stratospheric ozone (O_3) layer from 1979 to 2000 is estimated to have caused a negative radiative forcing (-0.15 Wm^{-2}). Assuming full compliance with current halocarbon regulations, the positive forcing of the halocarbons will be reduced as will the magnitude of the negative forcing from stratospheric ozone depletion as the ozone layer recovers over the 21st century.

- The total amount of O_3 in the troposphere is estimated to have increased by 36% since 1750, due primarily to anthropogenic emissions of several O_3-forming gases. This corresponds to a positive radiative forcing of 0.35 Wm^{-2}. O_3 forcing varies considerably by region and responds much more quickly to changes in emissions than the long-lived greenhouse gases, such as CO_2.

---

[9] ppm (parts per million) or ppb (parts per billion, 1 billion = 1,000 million) is the ratio of the number of greenhouse gas molecules to the total number of molecules of dry air. For example: 300 ppm means 300 molecules of a greenhouse gas per million molecules of dry air.

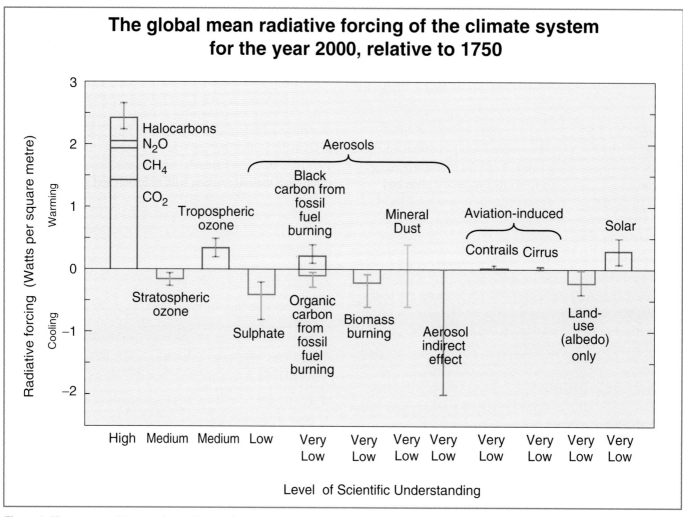

Figure 3: Many external factors force climate change.

These radiative forcings arise from changes in the atmospheric composition, alteration of surface reflectance by land use, and variation in the output of the sun. Except for solar variation, some form of human activity is linked to each. The rectangular bars represent estimates of the contributions of these forcings – some of which yield warming, and some cooling. Forcing due to episodic volcanic events, which lead to a negative forcing lasting only for a few years, is not shown. The indirect effect of aerosols shown is their effect on the size and number of cloud droplets. A second indirect effect of aerosols on clouds, namely their effect on cloud lifetime, which would also lead to a negative forcing, is not shown. Effects of aviation on greenhouse gases are included in the individual bars. The vertical line about the rectangular bars indicates a range of estimates, guided by the spread in the published values of the forcings and physical understanding. Some of the forcings possess a much greater degree of certainty than others. A vertical line without a rectangular bar denotes a forcing for which no best estimate can be given owing to large uncertainties. The overall level of scientific understanding for each forcing varies considerably, as noted. Some of the radiative forcing agents are well mixed over the globe, such as CO_2, thereby perturbing the global heat balance. Others represent perturbations with stronger regional signatures because of their spatial distribution, such as aerosols. For this and other reasons, a simple sum of the positive and negative bars cannot be expected to yield the net effect on the climate system. The simulations of this assessment report (for example, Figure 5) indicate that the estimated net effect of these perturbations is to have warmed the global climate since 1750. [Based upon Chapter 6, Figure 6.6]

Anthropogenic aerosols are short-lived and mostly produce negative radiative forcing.

● The major sources of anthropogenic aerosols are fossil fuel and biomass burning. These sources are also linked to degradation of air quality and acid deposition.

● Since the SAR, significant progress has been achieved in better characterising the direct radiative roles of different types of aerosols. Direct radiative forcing is estimated to be -0.4 Wm^{-2} for sulphate, -0.2 Wm^{-2} for biomass burning aerosols, -0.1 Wm^{-2} for fossil fuel organic carbon and $+0.2$ Wm^{-2} for fossil fuel black carbon aerosols. There is much less confidence in the ability to quantify the total aerosol direct effect, and its evolution over time, than for the gases listed above. Aerosols also vary considerably by region and respond quickly to changes in emissions.

● In addition to their direct radiative forcing, aerosols have an indirect radiative forcing through their effects on clouds. There is now more evidence for this indirect effect, which is negative, although of very uncertain magnitude.

Natural factors have made small contributions to radiative forcing over the past century.

● The radiative forcing due to changes in solar irradiance for the period since 1750 is estimated to be about $+0.3$ Wm^{-2}, most of which occurred during the first half of the 20th century. Since the late 1970s, satellite instruments have observed small oscillations due to the 11-year solar cycle. Mechanisms for the amplification of solar effects on climate have been proposed, but currently lack a rigorous theoretical or observational basis.

● Stratospheric aerosols from explosive volcanic eruptions lead to negative forcing, which lasts a few years. Several major eruptions occurred in the periods 1880 to 1920 and 1960 to 1991.

● The combined change in radiative forcing of the two major natural factors (solar variation and volcanic aerosols) is estimated to be negative for the past two, and possibly the past four, decades.

Confidence in the ability of models to project future climate has increased.

Complex physically-based climate models are required to provide detailed estimates of feedbacks and of regional features. Such models cannot yet simulate all aspects of climate (e.g., they still cannot account fully for the observed trend in the surface-troposphere temperature difference since 1979) and there are particular uncertainties associated with clouds and their interaction with radiation and aerosols. Nevertheless, confidence in the ability of these models to provide useful projections of future climate has improved due to their demonstrated performance on a range of space and time-scales.

● Understanding of climate processes and their incorporation in climate models have improved, including water vapour, sea-ice dynamics, and ocean heat transport.

● Some recent models produce satisfactory simulations of current climate without the need for non-physical adjustments of heat and water fluxes at the ocean-atmosphere interface used in earlier models.

● Simulations that include estimates of natural and anthropogenic forcing reproduce the observed large-scale changes in surface temperature over the 20th century (Figure 4). However, contributions from some additional processes and forcings may not have been included in the models. Nevertheless, the large-scale consistency between models and observations can be used to provide an independent check on projected warming rates over the next few decades under a given emissions scenario.

● Some aspects of model simulations of ENSO, monsoons and the North Atlantic Oscillation, as well as selected periods of past climate, have improved.

There is new and stronger evidence that most of the warming observed over the last 50 years is attributable to human activities.

The SAR concluded: "The balance of evidence suggests a discernible human influence on global climate". That report also noted that the anthropogenic signal was still emerging from the background of natural climate variability. Since the SAR, progress has been made in reducing uncertainty, particularly with respect to distinguishing and quantifying the magnitude of responses to different external influences. Although many of the sources of uncertainty identified in the SAR still remain to some degree, new evidence and improved understanding support an updated conclusion.

• There is a longer and more closely scrutinised temperature record and new model estimates of variability. The warming over the past 100 years is very unlikely[7] to be due to internal variability alone, as estimated by current models. Reconstructions of climate data for the past 1,000 years (Figure 1b) also indicate that this warming was unusual and is unlikely[7] to be entirely natural in origin.

• There are new estimates of the climate response to natural and anthropogenic forcing, and new detection techniques have been applied. Detection and attribution studies consistently find evidence for an anthropogenic signal in the climate record of the last 35 to 50 years.

• Simulations of the response to natural forcings alone (i.e., the response to variability in solar irradiance and volcanic eruptions) do not explain the warming in the second half of the 20th century (see for example Figure 4a). However, they indicate that natural forcings may have contributed to the observed warming in the first half of the 20th century.

• The warming over the last 50 years due to anthropogenic greenhouse gases can be identified despite uncertainties in forcing due to anthropogenic sulphate aerosol and natural factors (volcanoes and solar irradiance). The anthropogenic sulphate aerosol forcing, while uncertain, is negative over this period and therefore cannot explain the warming. Changes in natural forcing during most of this period are also estimated to be negative and are unlikely[7] to explain the warming.

• Detection and attribution studies comparing model simulated changes with the observed record can now take into account uncertainty in the magnitude of modelled response to external forcing, in particular that due to uncertainty in climate sensitivity.

• Most of these studies find that, over the last 50 years, the estimated rate and magnitude of warming due to increasing concentrations of greenhouse gases alone are comparable with, or larger than, the observed warming. Furthermore, most model estimates that take into account both greenhouse gases and sulphate aerosols are consistent with observations over this period.

• The best agreement between model simulations and observations over the last 140 years has been found when all the above anthropogenic and natural forcing factors are combined, as shown in Figure 4c. These results show that the forcings included are sufficient to explain the observed changes, but do not exclude the possibility that other forcings may also have contributed.

In the light of new evidence and taking into account the remaining uncertainties, most of the observed warming over the last 50 years is likely[7] to have been due to the increase in greenhouse gas concentrations.

Furthermore, it is very likely[7] that the 20th century warming has contributed significantly to the observed sea level rise, through thermal expansion of sea water and widespread loss of land ice. Within present uncertainties, observations and models are both consistent with a lack of significant acceleration of sea level rise during the 20th century.

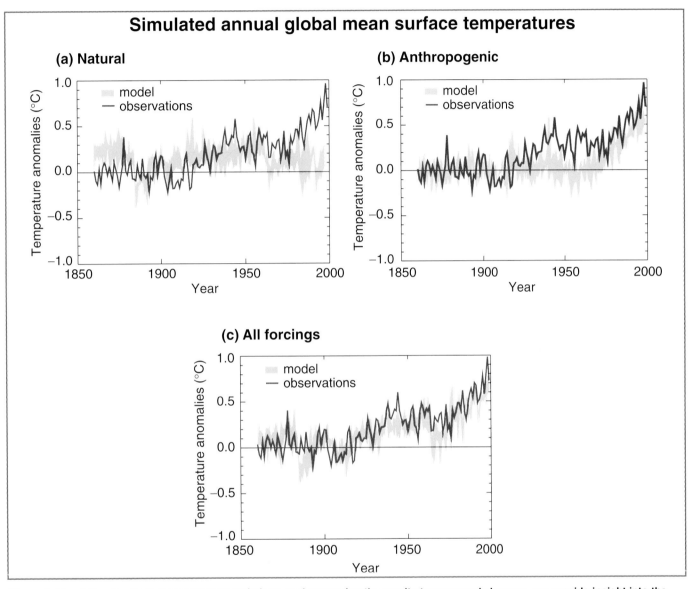

Figure 4: Simulating the Earth's temperature variations, and comparing the results to measured changes, can provide insight into the underlying causes of the major changes.

A climate model can be used to simulate the temperature changes that occur both from natural and anthropogenic causes. The simulations represented by the band in (a) were done with only natural forcings: solar variation and volcanic activity. Those encompassed by the band in (b) were done with anthropogenic forcings: greenhouse gases and an estimate of sulphate aerosols, and those encompassed by the band in (c) were done with both natural and anthropogenic forcings included. From (b), it can be seen that inclusion of anthropogenic forcings provides a plausible explanation for a substantial part of the observed temperature changes over the past century, but the best match with observations is obtained in (c) when both natural and anthropogenic factors are included. These results show that the forcings included are sufficient to explain the observed changes, but do not exclude the possibility that other forcings may also have contributed. The bands of model results presented here are for four runs from the same model. Similar results to those in (b) are obtained with other models with anthropogenic forcing. [Based upon Chapter 12, Figure 12.7]

Human influences will continue to change atmospheric composition throughout the 21st century.

Models have been used to make projections of atmospheric concentrations of greenhouse gases and aerosols, and hence of future climate, based upon emissions scenarios from the IPCC Special Report on Emission Scenarios (SRES) (Figure 5). These scenarios were developed to update the IS92 series, which were used in the SAR and are shown for comparison here in some cases.

Greenhouse gases

- Emissions of CO_2 due to fossil fuel burning are virtually certain[7] to be the dominant influence on the trends in atmospheric CO_2 concentration during the 21st century.

- As the CO_2 concentration of the atmosphere increases, ocean and land will take up a decreasing fraction of anthropogenic CO_2 emissions. The net effect of land and ocean climate feedbacks as indicated by models is to further increase projected atmospheric CO_2 concentrations, by reducing both the ocean and land uptake of CO_2.

- By 2100, carbon cycle models project atmospheric CO_2 concentrations of 540 to 970 ppm for the illustrative SRES scenarios (90 to 250% above the concentration of 280 ppm in the year 1750), Figure 5b. These projections include the land and ocean climate feedbacks. Uncertainties, especially about the magnitude of the climate feedback from the terrestrial biosphere, cause a variation of about −10 to +30% around each scenario. The total range is 490 to 1260 ppm (75 to 350% above the 1750 concentration).

- Changing land use could influence atmospheric CO_2 concentration. Hypothetically, if all of the carbon released by historical land-use changes could be restored to the terrestrial biosphere over the course of the century (e.g., by reforestation), CO_2 concentration would be reduced by 40 to 70 ppm.

- Model calculations of the concentrations of the non-CO_2 greenhouse gases by 2100 vary considerably across the SRES illustrative scenarios, with CH_4 changing by −190 to +1,970 ppb (present concentration 1,760 ppb), N_2O changing by +38 to +144 ppb (present concentration 316 ppb), total tropospheric O_3 changing by −12 to +62%, and a wide range of changes in concentrations of HFCs, PFCs and SF_6, all relative to the year 2000. In some scenarios, total tropospheric O_3 would become as important a radiative forcing agent as CH_4 and, over much of the Northern Hemisphere, would threaten the attainment of current air quality targets.

- Reductions in greenhouse gas emissions and the gases that control their concentration would be necessary to stabilise radiative forcing. For example, for the most important anthropogenic greenhouse gas, carbon cycle models indicate that stabilisation of atmospheric CO_2 concentrations at 450, 650 or 1,000 ppm would require global anthropogenic CO_2 emissions to drop below 1990 levels, within a few decades, about a century, or about two centuries, respectively, and continue to decrease steadily thereafter. Eventually CO_2 emissions would need to decline to a very small fraction of current emissions.

Aerosols

- The SRES scenarios include the possibility of either increases or decreases in anthropogenic aerosols (e.g., sulphate aerosols (Figure 5c), biomass aerosols, black and organic carbon aerosols) depending on the extent of fossil fuel use and policies to abate polluting emissions. In addition, natural aerosols (e.g., sea salt, dust and emissions leading to the production of sulphate and carbon aerosols) are projected to increase as a result of changes in climate.

Radiative forcing over the 21st century

- For the SRES illustrative scenarios, relative to the year 2000, the global mean radiative forcing due to greenhouse gases continues to increase through the 21st century, with the fraction due to CO_2 projected to increase from slightly more than half to about three quarters. The change in the direct plus indirect aerosol radiative forcing is projected to be smaller in magnitude than that of CO_2.

Global average temperature and sea level are projected to rise under all IPCC SRES scenarios.

In order to make projections of future climate, models incorporate past, as well as future emissions of greenhouse gases and aerosols. Hence, they include estimates of warming to date and the commitment to future warming from past emissions.

Temperature

- The globally averaged surface temperature is projected to increase by 1.4 to 5.8°C (Figure 5d) over the period 1990 to 2100. These results are for the full range of 35 SRES scenarios, based on a number of climate models[10,11].

- Temperature increases are projected to be greater than those in the SAR, which were about 1.0 to 3.5°C based on the six IS92 scenarios. The higher projected temperatures and the wider range are due primarily to the lower projected sulphur dioxide emissions in the SRES scenarios relative to the IS92 scenarios.

- The projected rate of warming is much larger than the observed changes during the 20th century and is very likely[7] to be without precedent during at least the last 10,000 years, based on palaeoclimate data.

- By 2100, the range in the surface temperature response across the group of climate models run with a given scenario is comparable to the range obtained from a single model run with the different SRES scenarios.

- On timescales of a few decades, the current observed rate of warming can be used to constrain the projected response to a given emissions scenario despite uncertainty in climate sensitivity. This approach suggests that anthropogenic warming is likely[7] to lie in the range of 0.1 to 0.2°C per decade over the next few decades under the IS92a scenario, similar to the corresponding range of projections of the simple model used in Figure 5d.

- Based on recent global model simulations, it is very likely[7] that nearly all land areas will warm more rapidly than the global average, particularly those at northern high latitudes in the cold season. Most notable of these is the warming in the northern regions of North America, and northern and central Asia, which exceeds global mean warming in each model by more than 40%. In contrast, the warming is less than the global mean change in south and southeast Asia in summer and in southern South America in winter.

- Recent trends for surface temperature to become more El Niño-like in the tropical Pacific, with the eastern tropical Pacific warming more than the western tropical Pacific, with a corresponding eastward shift of precipitation, are projected to continue in many models.

Precipitation

- Based on global model simulations and for a wide range of scenarios, global average water vapour concentration and precipitation are projected to increase during the 21st century. By the second half of the 21st century, it is likely[7] that precipitation will have increased over northern mid- to high latitudes and Antarctica in winter. At low latitudes there are both regional increases and decreases over land areas. Larger year to year variations in precipitation are very likely[7] over most areas where an increase in mean precipitation is projected.

---

[10] Complex physically based climate models are the main tool for projecting future climate change. In order to explore the full range of scenarios, these are complemented by simple climate models calibrated to yield an equivalent response in temperature and sea level to complex climate models. These projections are obtained using a simple climate model whose climate sensitivity and ocean heat uptake are calibrated to each of seven complex climate models. The climate sensitivity used in the simple model ranges from 1.7 to 4.2°C, which is comparable to the commonly accepted range of 1.5 to 4.5°C.

[11] This range does not include uncertainties in the modelling of radiative forcing, e.g. aerosol forcing uncertainties. A small carbon-cycle climate feedback is included.

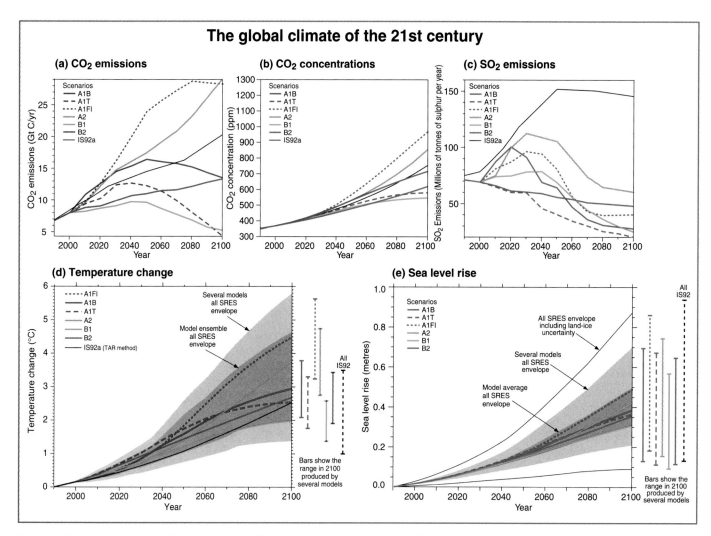

Figure 5: The global climate of the 21st century will depend on natural changes and the response of the climate system to human activities.

Climate models project the response of many climate variables – such as increases in global surface temperature and sea level – to various scenarios of greenhouse gas and other human-related emissions. (a) shows the CO_2 emissions of the six illustrative SRES scenarios, which are summarised in the box on page 18, along with IS92a for comparison purposes with the SAR. (b) shows projected CO_2 concentrations. (c) shows anthropogenic SO_2 emissions. Emissions of other gases and other aerosols were included in the model but are not shown in the figure. (d) and (e) show the projected temperature and sea level responses, respectively. The "several models all SRES envelope" in (d) and (e) shows the temperature and sea level rise, respectively, for the simple model when tuned to a number of complex models with a range of climate sensitivities. All SRES envelopes refer to the full range of 35 SRES scenarios. The "model average all SRES envelope" shows the average from these models for the range of scenarios. Note that the warming and sea level rise from these emissions would continue well beyond 2100. Also note that this range does not allow for uncertainty relating to ice dynamical changes in the West Antarctic ice sheet, nor does it account for uncertainties in projecting non-sulphate aerosols and greenhouse gas concentrations. [Based upon (a) Chapter 3, Figure 3.12, (b) Chapter 3, Figure 3.12, (c) Chapter 5, Figure 5.13, (d) Chapter 9, Figure 9.14, (e) Chapter 11, Figure 11.12, Appendix II]

Extreme Events

Table 1 depicts an assessment of confidence in observed changes in extremes of weather and climate during the latter half of the 20th century (left column) and in projected changes during the 21st century (right column)[a]. This assessment relies on observational and modelling studies, as well as the physical plausibility of future projections across all commonly-used scenarios and is based on expert judgement[7].

- For some other extreme phenomena, many of which may have important impacts on the environment and society, there is currently insufficient information to assess recent trends, and climate models currently lack the spatial detail required to make confident projections. For example, very small-scale phenomena, such as thunderstorms, tornadoes, hail and lightning, are not simulated in climate models.

Table 1: Estimates of confidence in observed and projected changes in extreme weather and climate events.

Confidence in observed changes (latter half of the 20th century)	Changes in Phenomenon	Confidence in projected changes (during the 21st century)
Likely[7]	**Higher maximum temperatures and more hot days over nearly all land areas**	Very likely[7]
Very likely[7]	**Higher minimum temperatures, fewer cold days and frost days over nearly all land areas**	Very likely[7]
Very likely[7]	**Reduced diurnal temperature range over most land areas**	Very likely[7]
Likely[7], over many areas	**Increase of heat index[12] over land areas**	Very likely[7], over most areas
Likely[7], over many Northern Hemisphere mid- to high latitude land areas	**More intense precipitation events[b]**	Very likely[7], over many areas
Likely[7], in a few areas	**Increased summer continental drying and associated risk of drought**	Likely[7], over most mid-latitude continental interiors. (Lack of consistent projections in other areas)
Not observed in the few analyses available	**Increase in tropical cyclone peak wind intensities[c]**	Likely[7], over some areas
Insufficient data for assessment	**Increase in tropical cyclone mean and peak precipitation intensities[c]**	Likely[7], over some areas

[a] For more details see Chapter 2 (observations) and Chapter 9, 10 (projections).

[b] For other areas, there are either insufficient data or conflicting analyses.

[c] Past and future changes in tropical cyclone location and frequency are uncertain.

---

[12] Heat index: A combination of temperature and humidity that measures effects on human comfort.

El Niño

- Confidence in projections of changes in future frequency, amplitude, and spatial pattern of El Niño events in the tropical Pacific is tempered by some shortcomings in how well El Niño is simulated in complex models. Current projections show little change or a small increase in amplitude for El Niño events over the next 100 years.

- Even with little or no change in El Niño amplitude, global warming is likely[7] to lead to greater extremes of drying and heavy rainfall and increase the risk of droughts and floods that occur with El Niño events in many different regions.

Monsoons

- It is likely[7] that warming associated with increasing greenhouse gas concentrations will cause an increase of Asian summer monsoon precipitation variability. Changes in monsoon mean duration and strength depend on the details of the emission scenario. The confidence in such projections is also limited by how well the climate models simulate the detailed seasonal evolution of the monsoons.

Thermohaline circulation

- Most models show weakening of the ocean thermohaline circulation which leads to a reduction of the heat transport into high latitudes of the Northern Hemisphere. However, even in models where the thermohaline circulation weakens, there is still a warming over Europe due to increased greenhouse gases. The current projections using climate models do not exhibit a complete shut-down of the thermohaline circulation by 2100. Beyond 2100, the thermohaline circulation could completely, and possibly irreversibly, shut-down in either hemisphere if the change in radiative forcing is large enough and applied long enough.

Snow and ice

- Northern Hemisphere snow cover and sea-ice extent are projected to decrease further.

- Glaciers and ice caps are projected to continue their widespread retreat during the 21st century.

- The Antarctic ice sheet is likely[7] to gain mass because of greater precipitation, while the Greenland ice sheet is likely[7] to lose mass because the increase in runoff will exceed the precipitation increase.

- Concerns have been expressed about the stability of the West Antarctic ice sheet because it is grounded below sea level. However, loss of grounded ice leading to substantial sea level rise from this source is now widely agreed to be very unlikely[7] during the 21st century, although its dynamics are still inadequately understood, especially for projections on longer time-scales.

Sea level

- Global mean sea level is projected to rise by 0.09 to 0.88 metres between 1990 and 2100, for the full range of SRES scenarios. This is due primarily to thermal expansion and loss of mass from glaciers and ice caps (Figure 5e). The range of sea level rise presented in the SAR was 0.13 to 0.94 metres based on the IS92 scenarios. Despite the higher temperature change projections in this assessment, the sea level projections are slightly lower, primarily due to the use of improved models, which give a smaller contribution from glaciers and ice sheets.

Anthropogenic climate change will persist for many centuries.

- Emissions of long-lived greenhouse gases (i.e., CO_2, N_2O, PFCs, SF_6) have a lasting effect on atmospheric composition, radiative forcing and climate. For example, several centuries after CO_2 emissions occur, about a quarter of the increase in CO_2 concentration caused by these emissions is still present in the atmosphere.

- After greenhouse gas concentrations have stabilised, global average surface temperatures would rise at a rate of only a few tenths of a degree per century rather than several degrees per century as projected for the 21st century without stabilisation. The lower the level at which concentrations are stabilised, the smaller the total temperature change.

- Global mean surface temperature increases and rising sea level from thermal expansion of the ocean are projected to continue for hundreds of years after stabilisation of greenhouse gas concentrations (even at present levels), owing to the long timescales on which the deep ocean adjusts to climate change.

- Ice sheets will continue to react to climate warming and contribute to sea level rise for thousands of years after climate has been stabilised. Climate models indicate that the local warming over Greenland is likely[7] to be one to three times the global average. Ice sheet models project that a local warming of larger than 3°C, if sustained for millennia, would lead to virtually a complete melting of the Greenland ice sheet with a resulting sea level rise of about 7 metres. A local warming of 5.5°C, if sustained for 1,000 years, would be likely[7] to result in a contribution from Greenland of about 3 metres to sea level rise.

- Current ice dynamic models suggest that the West Antarctic ice sheet could contribute up to 3 metres to sea level rise over the next 1,000 years, but such results are strongly dependent on model assumptions regarding climate change scenarios, ice dynamics and other factors.

Further action is required to address remaining gaps in information and understanding.

Further research is required to improve the ability to detect, attribute and understand climate change, to reduce uncertainties and to project future climate changes. In particular, there is a need for additional systematic and sustained observations, modelling and process studies. A serious concern is the decline of observational networks. The following are high priority areas for action.

- Systematic observations and reconstructions:

  - Reverse the decline of observational networks in many parts of the world.

  - Sustain and expand the observational foundation for climate studies by providing accurate, long-term, consistent data including implementation of a strategy for integrated global observations.

  - Enhance the development of reconstructions of past climate periods.

  - Improve the observations of the spatial distribution of greenhouse gases and aerosols.

- Modelling and process studies:

  - Improve understanding of the mechanisms and factors leading to changes in radiative forcing.

  - Understand and characterise the important unresolved processes and feedbacks, both physical and biogeochemical, in the climate system.

  - Improve methods to quantify uncertainties of climate projections and scenarios, including long-term ensemble simulations using complex models.

  - Improve the integrated hierarchy of global and regional climate models with a focus on the simulation of climate variability, regional climate changes and extreme events.

  - Link more effectively models of the physical climate and the biogeochemical system, and in turn improve coupling with descriptions of human activities.

Cutting across these foci are crucial needs associated with strengthening international co-operation and co-ordination in order to better utilise scientific, computational and observational resources. This should also promote the free exchange of data among scientists. A special need is to increase the observational and research capacities in many regions, particularly in developing countries. Finally, as is the goal of this assessment, there is a continuing imperative to communicate research advances in terms that are relevant to decision making.

The Emissions Scenarios of the Special Report on Emissions Scenarios (SRES)

A1. The A1 storyline and scenario family describes a future world of very rapid economic growth, global population that peaks in mid-century and declines thereafter, and the rapid introduction of new and more efficient technologies. Major underlying themes are convergence among regions, capacity building and increased cultural and social interactions, with a substantial reduction in regional differences in per capita income. The A1 scenario family develops into three groups that describe alternative directions of technological change in the energy system. The three A1 groups are distinguished by their technological emphasis: fossil intensive (A1FI), non-fossil energy sources (A1T), or a balance across all sources (A1B) (where balanced is defined as not relying too heavily on one particular energy source, on the assumption that similar improvement rates apply to all energy supply and end use technologies).

A2. The A2 storyline and scenario family describes a very heterogeneous world. The underlying theme is self-reliance and preservation of local identities. Fertility patterns across regions converge very slowly, which results in continuously increasing population. Economic development is primarily regionally oriented and per capita economic growth and technological change more fragmented and slower than other storylines.

B1. The B1 storyline and scenario family describes a convergent world with the same global population, that peaks in mid-century and declines thereafter, as in the A1 storyline, but with rapid change in economic structures toward a service and information economy, with reductions in material intensity and the introduction of clean and resource-efficient technologies. The emphasis is on global solutions to economic, social and environmental sustainability, including improved equity, but without additional climate initiatives.

B2. The B2 storyline and scenario family describes a world in which the emphasis is on local solutions to economic, social and environmental sustainability. It is a world with continuously increasing global population, at a rate lower than A2, intermediate levels of economic development, and less rapid and more diverse technological change than in the B1 and A1 storylines. While the scenario is also oriented towards environmental protection and social equity, it focuses on local and regional levels.

An illustrative scenario was chosen for each of the six scenario groups A1B, A1FI, A1T, A2, B1 and B2. All should be considered equally sound.

The SRES scenarios do not include additional climate initiatives, which means that no scenarios are included that explicitly assume implementation of the United Nations Framework Convention on Climate Change or the emissions targets of the Kyoto Protocol.

Source Information: Summary for Policymakers

This appendix provides the cross-reference of the topics in the Summary for Policymakers (page and bullet point topic) to the sections of the chapters of the full report that contain expanded information about the topic.

An increasing body of observations gives a collective picture of a warming world and other changes in the climate system.

SPM Page	Cross-Reference: SPM Topic – Chapter Section
2	*The global average surface temperature has increased over the 20th century by about 0.6°C.* ● Chapter 2.2.2 ● Chapter 2.2.2 ● Chapter 2.3 ● Chapter 2.2.2
4	*Temperatures have risen during the past four decades in the lowest 8 kilometres of the atmosphere.* ● Chapter 2.2.3 and 2.2.4 ● Chapter 2.2.3 and 2.2.4 ● Chapter 2.2.3, 2.2.4 and Chapter 12.3.2
4	*Snow cover and ice extent have decreased.* All three bullet points: Chapter 2.2.5 and 2.2.6
4	*Global average sea level has risen and ocean heat content has increased.* ● Chapter 11.3.2 ● Chapter 2.2.2 and Chapter 11.2.1
4 – 5	*Changes have also occurred in other important aspects of climate.* ● Chapter 2.5.2 ● Chapter 2.7.2 ● Chapter 2.2.2 and 2.5.5 ● Chapter 2.7.2 ● Chapter 2.6.2 and 2.6.3 ● Chapter 2.7.3 ● Chapter 2.7.3
5	*Some important aspects of climate appear not to have changed.* ● Chapter 2.2.2 ● Chapter 2.2.5 ● Chapter 2.7.3 ● Chapter 2.7.3

Emissions of greenhouse gases and aerosols due to human activities continue to alter the atmosphere in ways that are expected to affect the climate system.

SPM Page	Cross-Reference: SPM Topic – Chapter Section
5	Chapeau: "Changes in climate occur ..." Chapter 1, Chapter 3.1, Chapter 4.1, Chapter 5.1, Chapter 6.1, 6.2, 6.9, 6.11 and 6.13
7	*Concentrations of atmospheric greenhouse gases and their radiative forcing have continued to increase as a result of human activities.* Carbon dioxide: ● Chapter 3.3.1, 3.3.2, 3.3.3 and 3.5.1 ● Chapter 3.5.1 ● Chapter 3.2.2, 3.2.3, 3.5.1 and Table 3.1 ● Chapter 3.5.1 and 3.5.2 Methane: ● Chapter 4.2.1 Nitrous oxide: ● Chapter 4.2.1 Halocarbons: ● Chapter 4.2.2 Radiative forcing of well-mixed gases: ● Chapter 4.2.1 and Chapter 6.3 Stratospheric ozone: ● Chapter 4.2.2 and Chapter 6.4 Tropospheric ozone: ● Chapter 4.2.4 and Chapter 6.5
9	*Anthropogenic aerosols are short-lived and mostly produce negative radiative forcing.* ● Chapter 5.2 and 5.5.4 ● Chapter 5.1, 5.2 and Chapter 6.7 ● Chapter 5.3.2, 5.4.3 and Chapter 6.8
9	*Natural factors have made small contributions to radiative forcing over the past century.* ● Chapter 6.11 and 6.15.1 ● Chapter 6.9 and 6.15.1 ● Chapter 6.15.1

Confidence in the ability of models to project future climate has increased.

SPM Page	Cross-Reference: SPM Topic – Chapter Section
9	Chapeau: "Complex physically-based ..." Chapter 8.3.2, 8.5.1, 8.6.1, 8.10.3 and Chapter 12.3.2
9	● Chapter 7.2.1, 7.5.2 and 7.6.1 ● Chapter 8.4.2 ● Chapter 8.6.3 and Chapter 12.3.2 ● Chapter 8.5.5, 8.7.1 and 8.7.5

There is new and stronger evidence that most of the warming observed over the last 50 years is attributable to human activities.

SPM Page	Cross-Reference: SPM Topic – Chapter Section
10	Chapeau: "The SAR concluded: The balance of evidence suggests ..." Chapter 12.1.2 and 12.6
10	● Chapter 12.2.2, 12.4.3 and 12.6 ● Chapter 12.4.1, 12.4.2, 12.4.3 and 12.6 ● Chapter 12.2.3, 12.4.1, 12.4.2, 12.4.3 and 12.6 ● Chapter 12.4.3 and 12.6. ● Chapter 12.6 ● Chapter 12.4.3 ● Chapter 12.4.3 and 12.6
10	"In the light of new evidence and taking into account the ..." Chapter 12.4 and 12.6
10	"Furthermore, it is very likely that the 20th century warming has ..." Chapter 11.4

Human influences will continue to change atmospheric composition throughout the 21st century.

SPM Page	Cross-Reference: SPM Topic – Chapter Section
12	Chapeau: "Models have been used to make projections ..." Chapter 4.4.5 and Appendix II
12	*Greenhouse gases* ● Chapter 3.7.3 and Appendix II ● Chapter 3.7.1, 3.7.2, 3.7.3 and Appendix II ● Chapter 3.7.3 and Appendix II ● Chapter 3.2.2 and Appendix II ● Chapter 4.4.5, 4.5, 4.6 and Appendix II ● Chapter 3.7.3
12	*Aerosols* ● Chapter 5.5.2, 5.5.3 and Appendix II
12	*Radiative forcing over the 21st century* ● Chapter 6.15.2 and Appendix II

Global average temperature and sea level are projected to rise under all IPCC SRES scenarios.

SPM Page	Cross-Reference: SPM Topic – Chapter Section
13	*Temperature* ● Chapter 9.3.3 ● Chapter 9.3.3 ● Chapter 2.2.2, 2.3.2 and 2.4 ● Chapter 9.3.3 and Chapter 10.3.2 ● Chapter 8.6.1, Chapter 12.4.3, Chapter 13.5.1 and 13.5.2 ● Chapter 10.3.2 and Box 10.1 ● Chapter 9.3.2
13	*Precipitation* ● Chapter 9.3.1, 9.3.6, Chapter 10.3.2 and Box 10.1
15	*Extreme events* Table 1: Chapter 2.1, 2.2, 2.5, 2.7.2, 2.7.3, Chapter 9.3.6 and Chapter 10.3.2 ● Chapter 2.7.3 and Chapter 9.3.6
16	*El Niño* ● Chapter 9.3.5 ● Chapter 9.3.5
16	*Monsoons* ● Chapter 9.3.5
16	*Thermohaline circulation* ● Chapter 9.3.4
16	*Snow and ice* ● Chapter 9.3.2 ● Chapter 11.5.1 ● Chapter 11.5.1 ● Chapter 11.5.4
16	*Sea level* ● Chapter 11.5.1

Anthropogenic climate change will persist for many centuries.

SPM Page	Cross-Reference: SPM Topic – Chapter Section
17	● Chapter 3.2.3, Chapter 4.4 and Chapter 6.15 ● Chapter 9.3.3 and 9.3.4 ● Chapter 11.5.4 ● Chapter 11.5.4 ● Chapter 11.5.4

Further work is required to address remaining gaps in information and understanding.

SPM Page	Cross-Reference: SPM Topic – Chapter Section
17 – 18	All bullet points: Chapter 14, Executive Summary

Technical Summary

A report accepted by Working Group I of the IPCC but not approved in detail

"Acceptance" of IPCC Reports at a Session of the Working Group or Panel signifies that the material has not been subject to line by line discussion and agreement, but nevertheless presents a comprehensive, objective and balanced view of the subject matter.

Co-ordinating Lead Authors

D.L. Albritton (USA), L.G. Meira Filho (Brazil)

Lead Authors

U. Cubasch (Germany), X. Dai (China), Y. Ding (China), D.J. Griggs (UK), B. Hewitson (South Africa), J.T. Houghton (UK), I. Isaksen (Norway), T. Karl (USA), M. McFarland (USA), V.P. Meleshko (Russia), J.F.B. Mitchell (UK), M. Noguer (UK), B.S. Nyenzi (Tanzania), M. Oppenheimer (USA), J.E. Penner (USA), S. Pollonais (Trinidad and Tobago), T. Stocker (Switzerland), K.E. Trenberth (USA)

Contributing Authors

M.R. Allen, (UK), A.P.M. Baede (Netherlands), J.A. Church (Australia), D.H. Ehhalt (Germany), C.K. Folland (UK), F. Giorgi (Italy), J.M. Gregory (UK), J.M. Haywood (UK), J.I. House (Germany), M. Hulme (UK), V.J. Jaramillo (Mexico), A. Jayaraman (India), C.A. Johnson (UK), S. Joussaume (France), D.J. Karoly (Australia), H. Kheshgi (USA), C. Le Quéré (France), L.J. Mata (Germany), B.J. McAvaney (Australia), L.O. Mearns (USA), G.A. Meehl (USA), B. Moore III (USA), R.K. Mugara (Zambia), M. Prather (USA), C. Prentice (Germany), V. Ramaswamy (USA), S.C.B. Raper (UK), M.J. Salinger (New Zealand), R. Scholes (S. Africa), S. Solomon (USA), R. Stouffer (USA), M-X. Wang (China), R.T. Watson (USA), K-S. Yap (Malaysia)

Review Editors

F. Joos (Switzerland), A. Ramirez-Rojas (Venzuela), J.M.R. Stone (Canada), J. Zillman (Australia)

Technical Summary of the Working Group I Report

A. Introduction

A.1 The IPCC and its Working Groups

The Intergovernmental Panel on Climate Change (IPCC) was established by the World Meteorological Organisation (WMO) and the United Nations Environment Programme (UNEP) in 1988. The aim was, and remains, to provide an assessment of the understanding of all aspects of climate change[1], including how human activities can cause such changes and can be impacted by them. It had become widely recognised that human-influenced emissions of greenhouse gases have the potential to alter the climate system (see Box 1), with possible deleterious or beneficial effects. It was also recognised that addressing such global issues required organisation on a global scale, including assessment of the understanding of the issue by the worldwide expert communities.

At its first session, the IPCC was organised into three Working Groups. The current remits of the Working Groups are for Working Group I to address the scientific aspects of the climate system and climate change, Working Group II to address the impacts of and adaptations to climate change, and Working Group III to address the options for the mitigation of climate change. The IPCC provided its first major assessment report in 1990 and its second major assessment report in 1996.

The IPCC reports are (i) up-to-date descriptions of the knowns and unknowns of the climate system and related factors, (ii) based on the knowledge of the international expert communities, (iii) produced by an open and peer-reviewed professional process, and (iv) based upon scientific publications whose findings are summarised in terms useful to decision makers. While the assessed information is policy relevant, the IPCC does not establish or advocate public policy.

The scope of the assessments of Working Group I includes observations of the current changes and trends in the climate system, a reconstruction of past changes and trends, an understanding of the processes involved in those changes, and the incorporation of this knowledge into models that can attribute the causes of changes and that can provide simulation of natural and human-induced future changes in the climate system.

A.2 The First and Second Assessment Reports of Working Group I

In the First Assessment Report in 1990, Working Group I broadly described the status of the understanding of the climate system and climate change that had been gained over the preceding decades of research. Several major points were emphasised. The greenhouse effect is a natural feature of the planet, and its fundamental physics is well understood. The atmospheric abundances of greenhouse gases were increasing, due largely to human activities. Continued future growth in greenhouse gas emissions was predicted to lead to significant increases in the average surface temperature of the planet, increases that would exceed the natural variation of the past several millennia and that could be reversed only slowly. The past century had, at that time, seen a surface warming of nearly 0.5°C, which was broadly consistent with that predicted by climate models for the greenhouse gas increases, but was also comparable to what was then known about natural variation. Lastly, it was pointed out that the current level of understanding at that time and the existing capabilities of climate models limited the prediction of changes in the climate of specific regions.

Based on the results of additional research and Special Reports produced in the interim, IPCC Working Group I assessed the new state of understanding in its Second Assessment Report (SAR[2]) in 1996. The report underscored that greenhouse gas abundances continued to increase in the atmosphere and that very substantial cuts in emissions would be required for stabilisation of greenhouse gas concentrations in the atmosphere (which is the ultimate goal of Article 2 of the Framework Convention on Climate Change). Further, the general increase in

---

[1] *Climate change* in IPCC usage refers to any change in climate over time, whether due to natural variability or as a result of human activity. This usage differs from that in the Framework Convention on Climate Change, where *climate change* refers to a change of climate that is attributed directly or indirectly to human activity that alters the composition of the global atmosphere and that is in addition to natural climate variability observed over comparable time periods. For a definition of scientific and technical terms: see the Glossary in Appendix I.

[2] The IPCC Second Assessment Report is referred to in this Technical Summary as the SAR.

global temperature continued, with recent years being the warmest since at least 1860. The ability of climate models to simulate observed events and trends had improved, particularly with the inclusion of sulphate aerosols and stratospheric ozone as radiative forcing agents in climate models. Utilising this simulative capability to compare to the observed patterns of regional temperature changes, the report concluded that the ability to quantify the human influence on global climate was limited. The limitations arose because the expected signal was still emerging from the noise of natural variability and because of uncertainties in other key factors. Nevertheless, the report also concluded that "the balance of evidence suggests a discernible human influence on global climate". Lastly, based on a range of scenarios of future greenhouse gas abundances, a set of responses of the climate system was simulated.

A.3 The Third Assessment Report: This Technical Summary

The third major assessment report of IPCC Working Group I builds upon these past assessments and incorporates the results of the past five years of climate research. This Technical Summary is based on the underlying information of the chapters, which is cross-referenced in the Source Notes in the Appendix. This Summary aims to describe the major features (see Figure 1) of the understanding of the climate system and climate change at the outset of the 21st century. Specifically:

● What does the observational record show with regard to past climate changes, both globally and regionally and both on the average and in the extremes? (Section B)

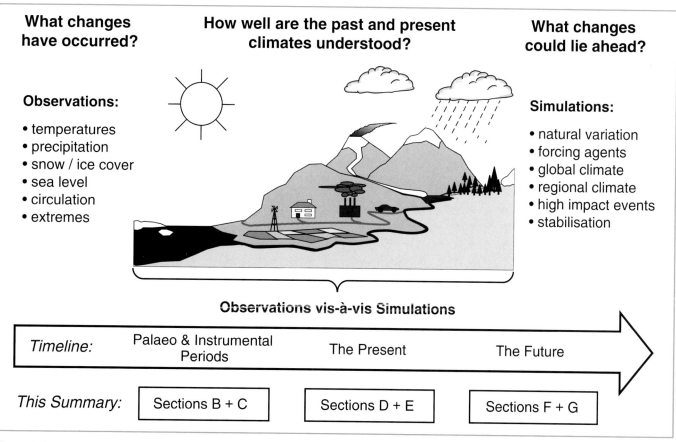

Figure 1: Key questions about the climate system and its relation to humankind. This Technical Summary, which is based on the underlying information in the chapters, is a status report on the answers, presented in the structure indicated.

- How quantitative is the understanding of the agents that cause climate to change, including both those that are natural (e.g., solar variation) and human-related (e.g., greenhouse gases) phenomena? (Section C)

- What is the current ability to simulate the responses of the climate system to these forcing agents? In particular, how well are key physical and biogeochemical processes described by present global climate models? (Section D)

- Based on today's observational data and today's climate predictive capabilities, what does the comparison show regarding a human influence on today's climate? (Section E)

- Further, using current predictive tools, what could the possible climate future be? Namely, for a wide spectrum of projections for several climate-forcing agents, what does current understanding project for global temperatures, regional patterns of precipitation, sea levels, and changes in extremes? (Section F)

Finally, what are the most urgent research activities that need to be addressed to improve our understanding of the climate system and to reduce our uncertainty regarding future climate change?

The Third Assessment Report of IPCC Working Group I is the product of hundreds of scientists from the developed and developing world who contributed to its preparation and review. What follows is a summary of their understanding of the climate system.

Box 1: What drives changes in climate?

The Earth absorbs radiation from the Sun, mainly at the surface. This energy is then redistributed by the atmospheric and oceanic circulations and radiated back to space at longer (infrared) wavelengths. For the annual mean and for the Earth as a whole, the incoming solar radiation energy is balanced approximately by the outgoing terrestrial radiation. Any factor that alters the radiation received from the Sun or lost to space, or that alters the redistribution of energy within the atmosphere and between the atmosphere, land, and ocean, can affect climate. A change in the net radiative energy available to the global Earth-atmosphere system is termed here, and in previous IPCC reports, a radiative forcing. Positive radiative forcings tend to warm the Earth's surface and lower atmosphere. Negative radiative forcings tend to cool them.

Increases in the concentrations of greenhouse gases will reduce the efficiency with which the Earth's surface radiates to space. More of the outgoing terrestrial radiation from the surface is absorbed by the atmosphere and re-emitted at higher altitudes and lower temperatures. This results in a positive radiative forcing that tends to warm the lower atmosphere and surface. Because less heat escapes to space, this is the enhanced greenhouse effect – an enhancement of an effect that has operated in the Earth's atmosphere for billions of years due to the presence of naturally occurring greenhouse gases: water vapour, carbon dioxide, ozone, methane and nitrous oxide. The amount of radiative forcing depends on the size of the increase in concentration of each greenhouse gas, the radiative properties of the gases involved, and the concentrations of other greenhouse gases already present in the atmosphere. Further, many greenhouse gases reside in the atmosphere for centuries after being emitted, thereby introducing a long-term commitment to positive radiative forcing.

Anthropogenic aerosols (microscopic airborne particles or droplets) in the troposphere, such as those derived from fossil fuel and biomass burning, can reflect solar radiation, which leads to a cooling tendency in the climate system. Because it can absorb solar radiation, black carbon (soot) aerosol tends to warm the climate system. In addition, changes in aerosol concentrations can alter cloud amount and cloud reflectivity through their effect on cloud properties and lifetimes. In most cases, tropospheric aerosols tend to produce a negative radiative forcing and a cooler climate. They have a much shorter lifetime (days to weeks) than most greenhouse gases (decades to centuries), and, as a result, their concentrations respond much more quickly to changes in emissions.

Volcanic activity can inject large amounts of sulphur-containing gases (primarily sulphur dioxide) into the stratosphere, which are transformed into sulphate aerosols. Individual eruptions can produce a large, but transitory, negative radiative forcing, tending to cool the Earth's surface and lower atmosphere over periods of a few years.

The Sun's output of energy varies by small amounts (0.1%) over an 11-year cycle and, in addition, variations over longer periods may occur. On time-scales of tens to thousands of years, slow variations in the Earth's orbit, which are well understood, have led to changes in the seasonal and latitudinal distribution of solar radiation. These changes have played an important part in controlling the variations of climate in the distant past, such as the glacial and inter-glacial cycles.

When radiative forcing changes, the climate system responds on various time-scales. The longest of these are due to the large heat capacity of the deep ocean and dynamic adjustment of the ice sheets. This means that the transient response to a change (either positive or negative) may last for thousands of years. Any changes in the radiative balance of the Earth, including those due to an increase in greenhouse gases or in aerosols, will alter the global hydrological cycle and atmospheric and oceanic circulation, thereby affecting weather patterns and regional temperatures and precipitation.

Any human-induced changes in climate will be embedded in a background of natural climatic variations that occur on a whole range of time- and space-scales. Climate variability can occur as a result of natural changes in the forcing of the climate system, for example variations in the strength of the incoming solar radiation and changes in the concentrations of aerosols arising from volcanic eruptions. Natural climate variations can also occur in the absence of a change in external forcing, as a result of complex interactions between components of the climate system, such as the coupling between the atmosphere and ocean. The El Niño-Southern Oscillation (ENSO) phenomenon is an example of such natural "internal" variability on interannual time-scales. To distinguish anthropogenic climate changes from natural variations, it is necessary to identify the anthropogenic "signal" against the background "noise" of natural climate variability.

B. The Observed Changes in the Climate System

Is the Earth's climate changing? The answer is unequivocally "Yes". A suite of observations supports this conclusion and provides insight about the rapidity of those changes. These data are also the bedrock upon which to construct the answer to the more difficult question: "*Why* is it changing?", which is addressed in later Sections.

This Section provides an updated summary of the observations that delineate how the climate system has changed in the past. Many of the variables of the climate system have been measured directly, i.e., the "instrumental record". For example, widespread direct measurements of surface temperature began around the middle of the 19th century. Near global observations of other surface "weather" variables, such as precipitation and winds, have been made for about a hundred years. Sea level measurements have been made for over 100 years in some places, but the network of tide gauges with long records provides only limited global coverage. Upper air observations have been made systematically only since the late 1940s. There are also long records of surface oceanic observations made from ships since the mid-19th century and by dedicated buoys since about the late 1970s. Sub-surface oceanic temperature measurements with near global coverage are now available from the late 1940s. Since the late 1970s, other data from Earth-observation satellites have been used to provide a wide range of global observations of various components of the climate system. In addition, a growing set of palaeoclimatic data, e.g., from trees, corals, sediments, and ice, are giving information about the Earth's climate of centuries and millennia before the present.

This Section places particular emphasis on current knowledge of past changes in key climate variables: temperature, precipitation and atmospheric moisture, snow cover, extent of land and sea ice, sea level, patterns in atmospheric and oceanic circulation, extreme weather and climate events, and overall features of the climate variability. The concluding part of this Section compares the observed trends in these various climate indicators to see if a collective picture emerges. The degree of this internal consistency is a critical factor in assessing the level of confidence in the current understanding of the climate system.

B.1 Observed Changes in Temperature

Temperatures in the instrumental record for land and oceans

The global average surface temperature has increased by 0.6 ± 0.2°C[3] since the late 19th century. It is very likely that the 1990s was the warmest decade and 1998 the warmest year in the instrumental record since 1861 (see Figure 2). The main cause of the increased estimate of global warming of 0.15°C since the SAR is related to the record warmth of the additional six years (1995 to 2000) of data. A secondary reason is related to improved methods of estimating change. The current, slightly larger uncertainty range (±0.2°C, 95% confidence interval) is also more objectively based. Further, the scientific basis for confidence in the estimates of the increase in global

1910 to 1945 and since 1976. The rate of increase of temperature for both periods is about 0.15°C/decade. Recent warming has been greater over land compared to oceans; the increase in sea surface temperature over the period 1950 to 1993 is about half that of the mean land-surface air temperature. The high global temperature associated with the 1997 to 1998 El Niño event stands out as an extreme event, even taking into account the recent rate of warming.

The regional patterns of the warming that occurred in the early part of the 20th century were different than those that occurred in the latter part. Figure 3 shows the regional patterns of the warming that have occurred over the full 20th century, as well as for three component time periods. The most recent period of warming (1976 to 1999) has been almost global, but the largest increases in temperature have occurred over the mid- and high latitudes of the continents in the Northern Hemisphere. Year-round cooling is evident in the north-western North Atlantic and the central North Pacific Oceans, but the North Atlantic cooling trend has recently reversed. The recent regional patterns of temperature change have been shown to be related, in part, to various phases of atmospheric-oceanic oscillations, such as the North Atlantic-Arctic Oscillation and possibly the Pacific Decadal Oscillation. Therefore, regional temperature trends over a few decades can be strongly influenced by regional variability in the climate system and can depart appreciably from a global average. The 1910 to 1945 warming was initially concentrated in the North Atlantic. By contrast, the period 1946 to 1975 showed significant cooling in the North Atlantic, as well as much of the Northern Hemisphere, and warming in much of the Southern Hemisphere.

New analyses indicate that global ocean heat content has increased significantly since the late 1950s. More than half of the increase in heat content has occurred in the upper 300 m

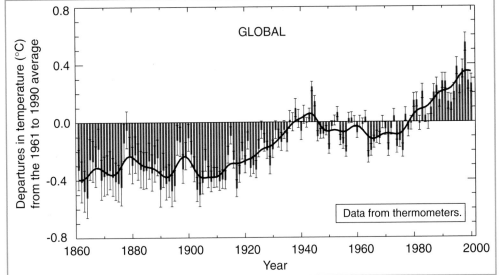

Figure 2: Combined annual land-surface air and sea surface temperature anomalies (°C) 1861 to 2000, relative to 1961 to 1990. Two standard error uncertainties are shown as bars on the annual number. [Based on Figure 2.7c]

temperature since the late 19th century has been strengthened since the SAR. This is due to the improvements derived from several new studies. These include an independent test of the corrections used for time-dependent biases in the sea surface temperature data and new analyses of the effect of urban "heat island" influences on global land-temperature trends. As indicated in Figure 2, most of the increase in global temperature since the late 19th century has occurred in two distinct periods:

---

[3] Generally, temperature trends are rounded to the nearest 0.05°C per unit of time, the periods often being limited by data availability.

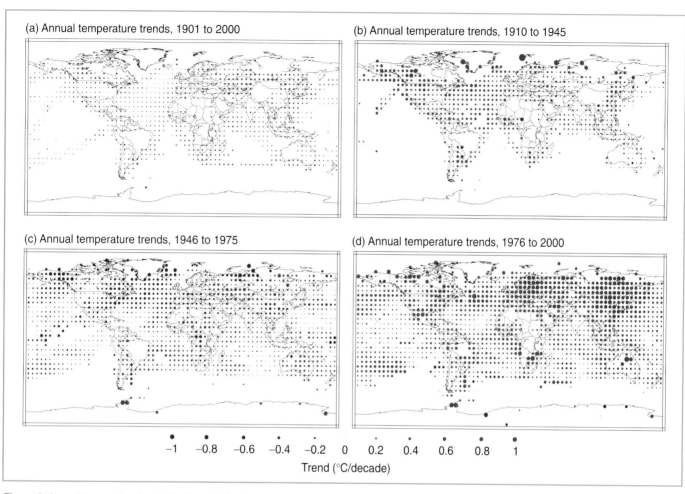

Figure 3: Annual temperature trends for the periods 1901 to 1999, 1910 to 1945, 1946 to 1975 and 1976 to 1999 respectively. Trends are represented by the area of the circle with red representing increases, blue representing decreases, and green little or no change. Trends were calculated from annually averaged gridded anomalies with the requirement that the calculation of annual anomalies include a minimum of 10 months of data. For the period 1901 to 1999, trends were calculated only for those grid boxes containing annual anomalies in at least 66 of the 100 years. The minimum number of years required for the shorter time periods (1910 to 1945, 1946 to 1975, and 1976 to 1999) was 24, 20, and 16 years respectively. [Based on Figure 2.9]

of the ocean, equivalent to a rate of temperature increase in this layer of about 0.04°C/decade.

New analyses of daily maximum and minimum land-surface temperatures for 1950 to 1993 continue to show that this measure of diurnal temperature range is decreasing very widely, although not everywhere. On average, minimum temperatures are increasing at about twice the rate of maximum temperatures (0.2 versus 0.1°C/decade).

Temperatures above the surface layer from satellite and weather balloon records
Surface, balloon and satellite temperature measurements show that the troposphere and Earth's surface have warmed and that the stratosphere has cooled. Over the shorter time period for which there have been both satellite and weather balloon data (since 1979), the balloon and satellite records show significantly less lower-tropospheric warming than observed at the surface. Analyses of temperature trends since 1958 for the lowest 8 km of the atmosphere and at the surface are in

good agreement, as shown in Figure 4a, with a warming of about 0.1°C per decade. However, since the beginning of the satellite record in 1979, the temperature data from both satellites and weather balloons show a warming in the global middle-to-lower troposphere at a rate of approximately 0.05 ± 0.10°C per decade. The global average surface temperature has increased significantly by 0.15 ± 0.05°C/decade. The difference in the warming rates is statistically significant. By contrast, during the period 1958 to 1978, surface temperature trends were near zero, while trends for the lowest 8 km of the atmosphere were near 0.2°C/decade. About half of the observed difference in warming since 1979 is likely[4] to be due to the combination of the differences in spatial coverage of the surface and tropospheric observations and the physical effects of the sequence of volcanic eruptions and a substantial El Niño (see Box 4 for a general description of ENSO) that occurred within this period. The remaining difference is very likely real and not an observing bias. It arises primarily due to differences in the rate of temperature change over the tropical and sub-tropical regions, which were faster in the lowest 8 km of the atmosphere before about 1979, but which have been slower since then. There are no significant differences in warming rates over mid-latitude continental regions in the Northern Hemisphere. In the upper troposphere, no significant global temperature trends have been detected since the early 1960s. In the stratosphere, as shown in Figure 4b, both satellites and balloons show substantial cooling, punctuated by sharp warming episodes of one to two years long that are due to volcanic eruptions.

Surface temperatures during the pre-instrumental period from the proxy record

It is likely that the rate and duration of the warming of the 20th century is larger than any other time during the last 1,000 years. The 1990s are likely to have been the warmest decade of the millennium in the Northern Hemisphere, and 1998 is likely to have been the warmest year. There has been a considerable advance in understanding of temperature change that occurred over the last millennium, especially from the synthesis of individual temperature reconstructions. This new detailed temperature record for the Northern Hemisphere is shown in

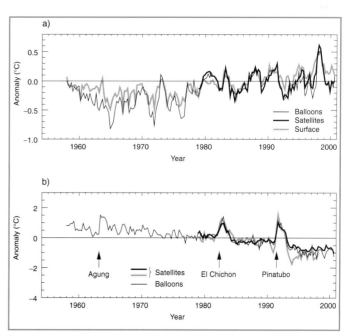

Figure 4: (a) Time-series of seasonal temperature anomalies of the troposphere based on balloons and satellites in addition to the surface. (b) Time-series of seasonal temperature anomalies of the lower stratosphere from balloons and satellites. [Based on Figure 2.12]

Figure 5. The data show a relatively warm period associated with the 11th to 14th centuries and a relatively cool period associated with the 15th to 19th centuries in the Northern Hemisphere. However, evidence does not support these "Medieval Warm Period" and "Little Ice Age" periods, respectively, as being globally synchronous. As Figure 5 indicates, the rate and duration of warming of the Northern Hemisphere in the 20th century appears to have been unprecedented during the millennium, and it cannot simply be considered as a recovery from the "Little Ice Age" of the 15th to 19th centuries. These analyses are complemented by sensitivity analysis of the spatial representativeness of available palaeoclimatic data, indicating that the warmth of the recent decade is outside the 95% confidence interval of temperature uncertainty, even during the warmest periods of the last millennium. Moreover, several different analyses have now been completed, each suggesting

---

[4] In this Technical Summary and in the Summary for Policymakers, the following words have been used to indicate approximate judgmental estimates of confidence: *virtually certain* (greater than 99% chance that a result is true); *very likely* (90–99% chance); *likely* (66–90% chance); *medium likelihood* (33–66% chance); *unlikely* (10–33% chance); *very unlikely* (1–10% chance); *exceptionally unlikely* (less than 1% chance). The reader is referred to individual chapters for more details.

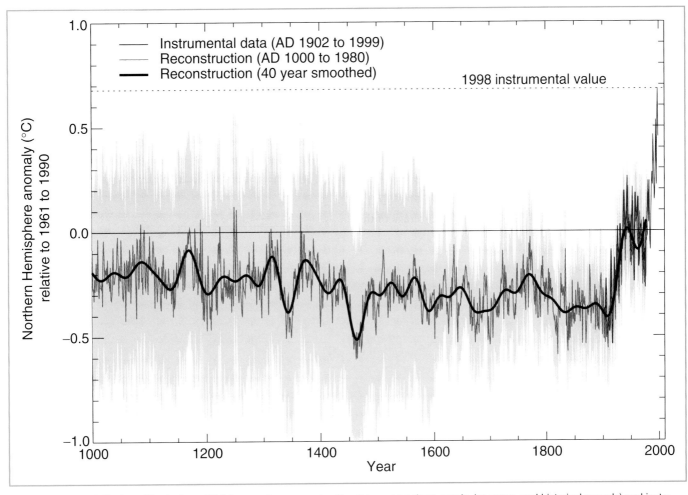

Figure 5: Millennial Northern Hemisphere (NH) temperature reconstruction (blue – tree rings, corals, ice cores, and historical records) and instrumental data (red) from AD 1000 to 1999. Smoother version of NH series (black), and two standard error limits (gray shaded) are shown. [Based on Figure 2.20]

that the Northern Hemisphere temperatures of the past decade have been warmer than any other time in the past six to ten centuries. This is the time-span over which temperatures with annual resolution can be calculated using hemispheric-wide tree-ring, ice-cores, corals, and and other annually-resolved proxy data. Because less data are available, less is known about annual averages prior to 1,000 years before the present and for conditions prevailing in most of the Southern Hemisphere prior to 1861.

It is likely that large rapid decadal temperature changes occurred during the last glacial and its deglaciation (between about 100,000 and 10,000 years ago), particularly in high latitudes of the Northern Hemisphere. In a few places during the deglaciation, local increases in temperature of 5 to 10°C are likely to have occurred over periods as short as a few decades. During the last 10,000 years, there is emerging evidence of significant rapid regional temperature changes, which are part of the natural variability of climate.

B.2 Observed Changes in Precipitation and Atmospheric Moisture

Since the time of the SAR, annual land precipitation has continued to increase in the middle and high latitudes of the Northern Hemisphere (very likely to be 0.5 to 1%/decade), except over Eastern Asia. Over the sub-tropics (10°N to 30°N), land-surface rainfall has decreased on average (likely to be about 0.3%/decade), although this has shown signs of recovery in recent years. Tropical land-surface precipitation measurements indicate that precipitation likely has increased by about 0.2 to 0.3%/decade over the 20th century, but increases are not evident over the past few decades and the amount of tropical land (versus ocean) area for the latitudes 10°N to 10°S is relatively small. Nonetheless, direct measurements of precipitation and model reanalyses of inferred precipitation indicate that rainfall has also increased over large parts of the tropical oceans. Where and when available, changes in annual streamflow often relate well to changes in total precipitation. The increases in precipitation over Northern Hemisphere mid- and high latitude land areas have a strong correlation to long-term increases in total cloud amount. In contrast to the Northern Hemisphere, no comparable systematic changes in precipitation have been detected in broad latitudinal averages over the Southern Hemisphere.

It is likely that total atmospheric water vapour has increased several per cent per decade over many regions of the Northern Hemisphere. Changes in water vapour over approximately the past 25 years have been analysed for selected regions using *in situ* surface observations, as well as lower-tropospheric measurements from satellites and weather balloons. A pattern of overall surface and lower-tropospheric water vapour increases over the past few decades is emerging from the most reliable data sets, although there are likely to be time-dependent biases in these data and regional variations in the trends. Water vapour in the lower stratosphere is also likely to have increased by about 10% per decade since the beginning of the observational record (1980).

Changes in total cloud amounts over Northern Hemisphere mid- and high latitude continental regions indicate a likely increase in cloud cover of about 2% since the beginning of the 20th century, which has now been shown to be positively correlated with decreases in the diurnal temperature range. Similar changes have been shown over Australia, the only Southern Hemisphere continent where such an analysis has been completed. Changes in total cloud amount are uncertain both over sub-tropical and tropical land areas, as well as over the oceans.

B.3 Observed Changes in Snow Cover and Land- and Sea-Ice Extent

Decreasing snow cover and land-ice extent continue to be positively correlated with increasing land-surface temperatures. Satellite data show that there are very likely to have been decreases of about 10% in the extent of snow cover since the late 1960s. There is a highly significant correlation between increases in Northern Hemisphere land temperatures and the decreases. There is now ample evidence to support a major retreat of alpine and continental glaciers in response to 20th century warming. In a few maritime regions, increases in precipitation due to regional atmospheric circulation changes have overshadowed increases in temperature in the past two decades, and glaciers have re-advanced. Over the past 100 to 150 years, ground-based observations show that there is very likely to have been a reduction of about two weeks in the annual duration of lake and river ice in the mid- to high latitudes of the Northern Hemisphere.

Northern Hemisphere sea-ice amounts are decreasing, but no significant trends in Antarctic sea-ice extent are apparent. A retreat of sea-ice extent in the Arctic spring and summer of 10 to 15% since the 1950s is consistent with an increase in spring temperatures and, to a lesser extent, summer temperatures in the high latitudes. There is little indication of reduced Arctic sea-ice extent during winter when temperatures have increased in the surrounding region. By contrast, there is no readily apparent relationship between decadal changes of Antarctic temperatures and sea-ice extent since 1973. After an initial decrease in the mid-1970s, Antarctic sea-ice extent has remained stable, or even slightly increased.

New data indicate that there likely has been an approximately 40% decline in Arctic sea-ice thickness in late summer to early autumn between the period of 1958 to 1976 and the mid-1990s, and a substantially smaller decline in winter. The relatively short record length and incomplete sampling limit the interpretation of these data. Interannual variability and inter-decadal variability could be influencing these changes.

B.4 Observed Changes in Sea Level

Changes during the instrumental record

Based on tide gauge data, the rate of global mean sea level rise during the 20th century is in the range 1.0 to 2.0 mm/yr, with a central value of 1.5 mm/yr (the central value should not be interpreted as a best estimate). (See Box 2 for the factors that influence sea level.) As Figure 6 indicates, the longest instrumental records (two or three centuries at most) of local sea level come from tide gauges. Based on the very few long tide-gauge records, the average rate of sea level rise has been larger during the 20th century than during the 19th century. No significant acceleration in the rate of sea level rise during the 20th century has been detected. This is not inconsistent with model results due to the possibility of compensating factors and the limited data.

Changes during the pre-instrumental record

Since the last glacial maximum about 20,000 years ago, the sea level in locations far from present and former ice sheets has risen by over 120 m as a result of loss of mass from these ice sheets. Vertical land movements, both upward and downward, are still occurring in response to these large transfers of mass from ice sheets to oceans. The most rapid rise in global sea level was between 15,000 and 6,000 years

---

Box 2: What causes sea level to change?

The level of the sea at the shoreline is determined by many factors in the global environment that operate on a great range of time-scales, from hours (tidal) to millions of years (ocean basin changes due to tectonics and sedimentation). On the time-scale of decades to centuries, some of the largest influences on the average levels of the sea are linked to climate and climate change processes.

Firstly, as ocean water warms, it expands. On the basis of observations of ocean temperatures and model results, thermal expansion is believed to be one of the major contributors to historical sea level changes. Further, thermal expansion is expected to contribute the largest component to sea level rise over the next hundred years. Deep ocean temperatures change only slowly; therefore, thermal expansion would continue for many centuries even if the atmospheric concentrations of greenhouse gases were to stabilise.

The amount of warming and the depth of water affected vary with location. In addition, warmer water expands more than colder water for a given change in temperature. The geographical distribution of sea level change results from the geographical variation of thermal expansion, changes in salinity, winds, and ocean circulation. The range of regional variation is substantial compared with the global average sea level rise.

Sea level also changes when the mass of water in the ocean increases or decreases. This occurs when ocean water is exchanged with the water stored on land. The major land store is the water frozen in glaciers or ice sheets. Indeed, the main reason for the lower sea level during the last glacial period was the amount of water stored in the large extension of the ice sheets on the continents of the Northern Hemisphere. After thermal expansion, the melting of mountain glaciers and ice caps is expected to make the largest contribution to the rise of sea level over the next hundred years. These glaciers and ice caps make up only a few per cent of the world's land-ice area, but they are more sensitive to climate change than the larger ice sheets in Greenland and Antarctica, because the ice sheets are in colder climates with low precipitation and low melting rates. Consequently, the large ice sheets are expected to make only a small net contribution to sea level change in the coming decades.

Sea level is also influenced by processes that are not explicitly related to climate change. Terrestrial water storage (and hence, sea level) can be altered by extraction of ground water, building of reservoirs, changes in surface runoff, and seepage into deep aquifers from reservoirs and irrigation. These factors may be offsetting a significant fraction of the expected acceleration in sea level rise from thermal expansion and glacial melting. In addition, coastal subsidence in river delta regions can also influence local sea level. Vertical land movements caused by natural geological processes, such as slow movements in the Earth's mantle and tectonic displacements of the crust, can have effects on local sea level that are comparable to climate-related impacts. Lastly, on seasonal, interannual, and decadal time-scales, sea level responds to changes in atmospheric and ocean dynamics, with the most striking example occurring during El Niño events.

ago, with an average rate of about 10 mm/yr. Based on geological data, eustatic sea level (i.e., corresponding to a change in ocean volume) may have risen at an average rate of 0.5 mm/yr over the past 6,000 years and at an average rate of 0.1 to 0.2 mm/yr over the last 3,000 years. This rate is about one tenth of that occurring during the 20th century. Over the past 3,000 to 5,000 years, oscillations in global sea level on time-scales of 100 to 1,000 years are unlikely to have exceeded 0.3 to 0.5 m.

B.5 Observed Changes in Atmospheric and Oceanic Circulation Patterns

The behaviour of ENSO (see Box 4 for a general description), has been unusual since the mid-1970s compared with the previous 100 years, with warm phase ENSO episodes being relatively more frequent, persistent, and intense than the opposite cool phase. This recent behaviour of ENSO is reflected in variations in precipitation and temperature over much of the global tropics and sub-tropics. The overall effect

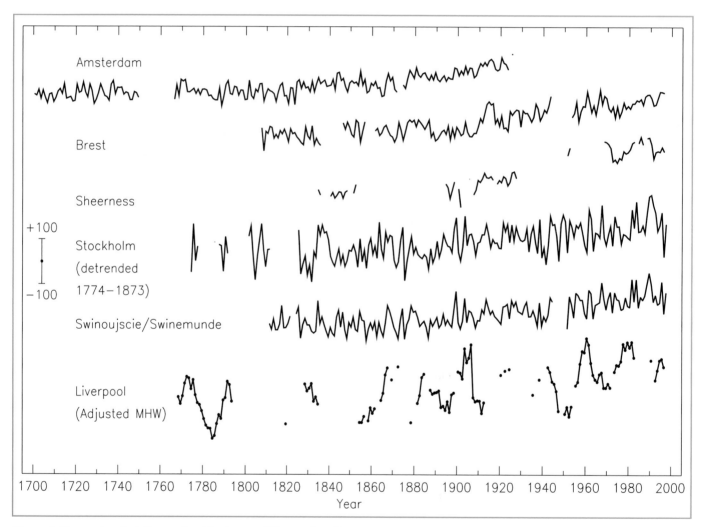

Figure 6: Time-series of relative sea level for the past 300 years from Northern Europe: Amsterdam, Netherlands; Brest, France; Sheerness, UK; Stockholm, Sweden (detrended over the period 1774 to 1873 to remove to first order the contribution of post-glacial rebound); Swinoujscie, Poland (formerly Swinemunde, Germany); and Liverpool, UK. Data for the latter are of "Adjusted Mean High Water" rather than Mean Sea Level and include a nodal (18.6 year) term. The scale bar indicates ±100 mm. [Based on Figure 11.7]

is likely to have been a small contribution to the increase in global temperatures during the last few decades. The Inter-decadal Pacific Oscillation and the Pacific Decadal Oscillation are associated with decadal to multidecadal climate variability over the Pacific basin. It is likely that these oscillations modulate ENSO-related climate variability.

Other important circulation features that affect the climate in large regions of the globe are being characterised. The North Atlantic Oscillation (NAO) is linked to the strength of the westerlies over the Atlantic and extra-tropical Eurasia. During winter the NAO displays irregular oscillations on interannual to multi-decadal time-scales. Since the 1970s, the winter NAO has often been in a phase that contributes to stronger westerlies, which correlate with cold season warming over Eurasia. New evidence indicates that the NAO and changes in Arctic sea ice are likely to be closely coupled. The NAO is now believed to be part of a wider scale atmospheric Arctic Oscillation that affects much of the extratropical Northern Hemisphere. A similar Antarctic Oscillation has been in an enhanced positive phase during the last 15 years, with stronger westerlies over the Southern Oceans.

B.6 Observed Changes in Climate Variability and Extreme Weather and Climate Events

New analyses show that in regions where total precipitation has increased, it is very likely that there have been even more pronounced increases in heavy and extreme precipitation events. The converse is also true. In some regions, however, heavy and extreme events (i.e., defined to be within the upper or lower ten percentiles) have increased despite the fact that total precipitation has decreased or remained constant. This is attributed to a decrease in the frequency of precipitation events. Overall, it is likely that for many mid- and high latitude areas, primarily in the Northern Hemisphere, statistically significant increases have occurred in the proportion of total annual precipitation derived from heavy and extreme precipitation events; it is likely that there has been a 2 to 4% increase in the frequency of heavy precipitation events over the latter half of the 20th century. Over the 20th century (1900 to 1995), there were relatively small increases in global land areas experiencing severe drought or severe wetness. In some regions, such as parts of

Asia and Africa, the frequency and intensity of drought have been observed to increase in recent decades. In many regions, these changes are dominated by inter-decadal and multi-decadal climate variability, such as the shift in ENSO towards more warm events. In many regions, inter-daily temperature variability has decreased, and increases in the daily minimum temperature are lengthening the freeze-free period in most mid- and high latitude regions. Since 1950 it is very likely that there has been a significant reduction in the frequency of much-below-normal seasonal mean temperatures across much of the globe, but there has been a smaller increase in the frequency of much-above-normal seasonal temperatures.

There is no compelling evidence to indicate that the characteristics of tropical and extratropical storms have changed. Changes in tropical storm intensity and frequency are dominated by interdecadal to multidecadal variations, which may be substantial, e.g., in the tropical North Atlantic. Owing to incomplete data and limited and conflicting analyses, it is uncertain as to whether there have been any long-term and large-scale increases in the intensity and frequency of extra-tropical cyclones in the Northern Hemisphere. Regional increases have been identified in the North Pacific, parts of North America, and Europe over the past several decades. In the Southern Hemisphere, fewer analyses have been completed, but they suggest a decrease in extra-tropical cyclone activity since the 1970s. Recent analyses of changes in severe local weather (e.g., tornadoes, thunderstorm days, and hail) in a few selected regions do not provide compelling evidence to suggest long-term changes. In general, trends in severe weather events are notoriously difficult to detect because of their relatively rare occurrence and large spatial variability.

B.7 The Collective Picture: A Warming World and Other Changes in the Climate System

As summarised above, a suite of climate changes is now well-documented, particularly over the recent decades to century time period, with its growing set of direct measurements. Figure 7 illustrates these trends in temperature indicators (Figure 7a) and hydrological and storm-related indicators (Figure 7b), as well as also providing an indication of certainty about the changes.

Taken together, these trends illustrate a collective picture of a warming world:

- Surface temperature measurements over the land and oceans (with two separate estimates over the latter) have been measured and adjusted independently. All data sets show quite similar upward trends globally, with two major warming periods globally: 1910 to 1945 and since 1976. There is an emerging tendency for global land-surface air temperatures to warm faster than the global ocean-surface temperatures.

- Weather balloon measurements show that lower-tropospheric temperatures have been increasing since 1958, though only slightly since 1979. Since 1979, satellite data are available and show similar trends to balloon data.

- The decrease in the continental diurnal temperature range coincides with increases in cloud amount, precipitation, and increases in total water vapour.

- The nearly worldwide decrease in mountain glacier extent and ice mass is consistent with worldwide surface temperature increases. A few recent exceptions in coastal regions are consistent with atmospheric circulation variations and related precipitation increases.

- The decreases in snow cover and the shortening seasons of lake and river ice relate well to increases in Northern Hemispheric land-surface temperatures.

- The systematic decrease in spring and summer sea-ice extent and thickness in the Arctic is consistent with increases in temperature over most of the adjacent land and ocean.

- Ocean heat content has increased, and global average sea level has risen.

- The increases in total tropospheric water vapour in the last 25 years are qualitatively consistent with increases in tropospheric temperatures and an enhanced hydrologic cycle, resulting in more extreme and heavier precipitation events in many areas with increasing precipitation, e.g., middle and high latitudes of the Northern Hemisphere.

Some important aspects of climate appear not to have changed.

- A few areas of the globe have not warmed in recent decades, mainly over some parts of the Southern Hemisphere oceans and parts of Antarctica.

- No significant trends in Antarctic sea-ice extent are apparent over the period of systematic satellite measurements (since 1978).

- Based on limited data, the observed variations in the intensity and frequency of tropical and extra-tropical cyclones and severe local storms show no clear trends in the last half of the 20th century, although multi-decadal fluctuations are sometimes apparent.

The variations and trends in the examined indicators imply that it is virtually certain that there has been a generally increasing trend in global surface temperature over the 20th century, although short-term and regional deviations from this trend occur.

(a) Temperature Indicators

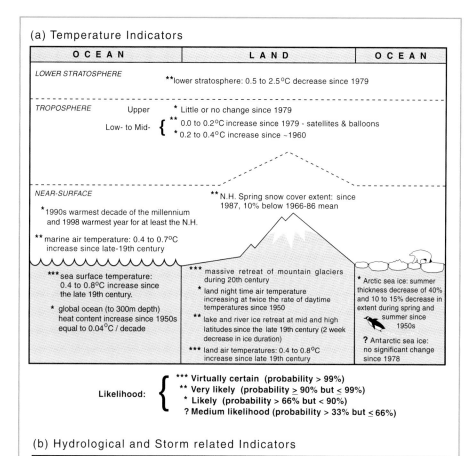

Figure 7a: Schematic of observed variations of the temperature indicators. [Based on Figure 2.39a]

(b) Hydrological and Storm related Indicators

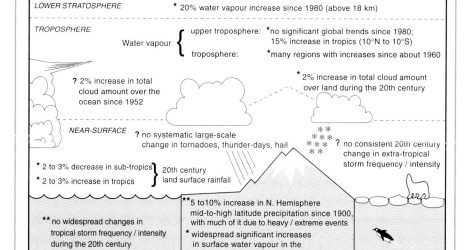

Figure 7b: Schematic of observed variations of the hydrological and storm-related indicators. [Based on Figure 2.39b]

C. The Forcing Agents That Cause Climate Change

In addition to the past variations and changes in the Earth's climate, observations have also documented the changes that have occurred in agents that can cause climate change. Most notable among these are increases in the atmospheric concentrations of greenhouse gases and aerosols (microscopic airborne particles or droplets) and variations in solar activity, both of which can alter the Earth's radiation budget and hence climate. These observational records of climate-forcing agents are part of the input needed to understand the past climate changes noted in the preceding Section and, very importantly, to predict what climate changes could lie ahead (see Section F).

Like the record of past climate changes, the data sets for forcing agents are of varying length and quality. Direct measurements of solar irradiance exist for only about two decades. The sustained direct monitoring of the atmospheric concentrations of carbon dioxide (CO_2) began about the middle of the 20th century and, in later years, for other long-lived, well-mixed gases such as methane. Palaeo-atmospheric data from ice cores reveal the concentration changes occurring in earlier millennia for some greenhouse gases. In contrast, the time-series measurements for the forcing agents that have relatively short residence times in the atmosphere (e.g., aerosols) are more recent and are far less complete, because they are harder to measure and are spatially heterogeneous. Current data sets show the human influence on atmospheric concentrations of both the long-lived greenhouse gases and short-lived forcing agents during the last part of the past millennium. Figure 8 illustrates the effects of the large growth over the Industrial Era in the anthropogenic emissions of greenhouse gases and sulphur dioxide, the latter being a precursor of aerosols.

A change in the energy available to the global Earth-atmosphere system due to changes in these forcing agents is termed radiative forcing (Wm^{-2}) of the climate system (see Box 1). Defined in this manner, radiative forcing of climate change constitutes an index of the relative global mean impacts on the surface-troposphere system due to different natural and anthropogenic causes. This Section updates the knowledge of the radiative forcing of climate change that has occurred from pre-industrial times to the present. Figure 9 shows the estimated radiative forcings from the beginning of the Industrial Era (1750) to 1999 for the quantifiable natural and anthropogenic

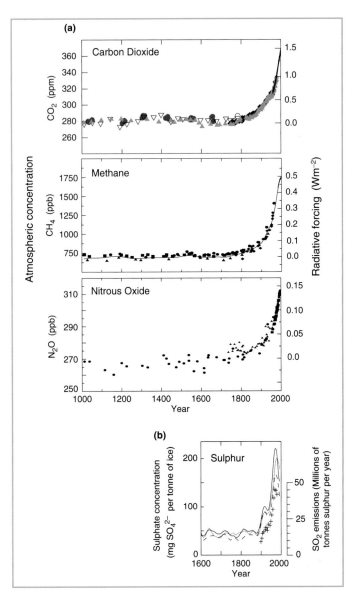

Figure 8: Records of changes in atmospheric composition. (a) Atmospheric concentrations of CO_2, CH_4 and N_2O over the past 1,000 years. Ice core and firn data for several sites in Antarctica and Greenland (shown by different symbols) are supplemented with the data from direct atmospheric samples over the past few decades (shown by the line for CO_2 and incorporated in the curve representing the global average of CH_4). The estimated radiative forcing from these gases is indicated on the right-hand scale. (b) Sulphate concentration in several Greenland ice cores with the episodic effects of volcanic eruptions removed (lines) and total SO_2 emissions from sources in the US and Europe (crosses). [Based on (a) Figure 3.2b (CO_2), Figure 4.1a and b (CH_4) and Figure 4.2 (N_2O) and (b) Figure 5.4a]

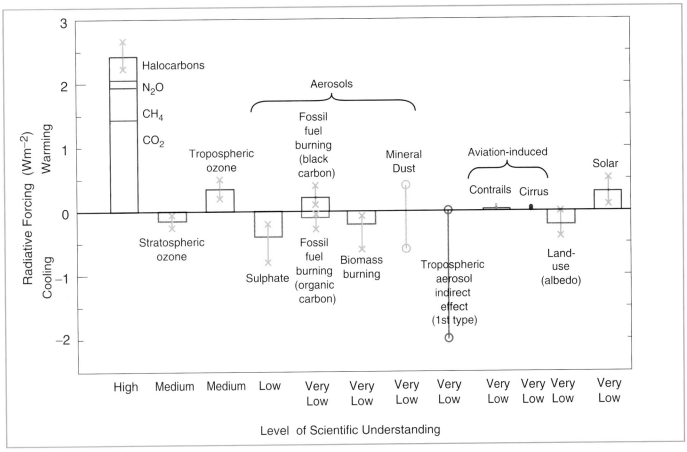

Figure 9: Global, annual-mean radiative forcings (Wm^{-2}) due to a number of agents for the period from pre-industrial (1750) to present (late 1990s; about 2000) (numerical values are also listed in Table 6.11 of Chapter 6). For detailed explanations, see Chapter 6.13. The height of the rectangular bar denotes a central or best estimate value, while its absence denotes no best estimate is possible. The vertical line about the rectangular bar with "x" delimiters indicates an estimate of the uncertainty range, for the most part guided by the spread in the published values of the forcing. A vertical line without a rectangular bar and with "o" delimiters denotes a forcing for which no central estimate can be given owing to large uncertainties. The uncertainty range specified here has no statistical basis and therefore differs from the use of the term elsewhere in this document. A "level of scientific understanding" index is accorded to each forcing, with high, medium, low and very low levels, respectively. This represents the subjective judgement about the reliability of the forcing estimate, involving factors such as the assumptions necessary to evaluate the forcing, the degree of knowledge of the physical/chemical mechanisms determining the forcing, and the uncertainties surrounding the quantitative estimate of the forcing (see Table 6.12). The well-mixed greenhouse gases are grouped together into a single rectangular bar with the individual mean contributions due to CO_2, CH_4, N_2O and halocarbons shown (see Tables 6.1 and 6.11). Fossil fuel burning is separated into the "black carbon" and "organic carbon" components with its separate best estimate and range. The sign of the effects due to mineral dust is itself an uncertainty. The indirect forcing due to tropospheric aerosols is poorly understood. The same is true for the forcing due to aviation via its effects on contrails and cirrus clouds. Only the "first" type of indirect effect due to aerosols as applicable in the context of liquid clouds is considered here. The "second" type of effect is conceptually important, but there exists very little confidence in the simulated quantitative estimates. The forcing associated with stratospheric aerosols from volcanic eruptions is highly variable over the period and is not considered for this plot (however, see Figure 6.8). All the forcings shown have distinct spatial and seasonal features (Figure 6.7) such that the global, annual means appearing on this plot do not yield a complete picture of the radiative perturbation. They are only intended to give, in a relative sense, a first-order perspective on a global, annual mean scale and cannot be readily employed to obtain the climate response to the total natural and/or anthropogenic forcings. As in the SAR, it is emphasised that the positive and negative global mean forcings cannot be added up and viewed *a priori* as providing offsets in terms of the complete global climate impact. [Based on Figure 6.6]

forcing agents. Although not included in the figure due to their episodic nature, volcanic eruptions are the source of another important natural forcing. Summaries of the information about each forcing agent follow in the sub-sections below.

The forcing agents included in Figure 9 vary greatly in their form, magnitude and spatial distribution. Some of the greenhouse gases are emitted directly into the atmosphere; some are chemical products from other emissions. Some greenhouse gases have long atmospheric residence times and, as a result, are well-mixed throughout the atmosphere. Others are short-lived and have heterogeneous regional concentrations. Most of the gases originate from both natural and anthropogenic sources. Lastly, as shown in Figure 9, the radiative forcings of individual agents can be positive (i.e., a tendency to warm the Earth's surface) or negative (i.e., a tendency to cool the Earth's surface).

C.1 Observed Changes in Globally Well-Mixed Greenhouse Gas Concentrations and Radiative Forcing

Over the millennium before the Industrial Era, the atmospheric concentrations of greenhouse gases remained relatively constant. Since then, however, the concentrations of many greenhouse gases have increased directly or indirectly because of human activities.

Table 1 provides examples of several greenhouse gases and summarises their 1750 and 1998 concentrations, their change during the 1990s, and their atmospheric lifetimes. The contribution of a species to radiative forcing of climate change depends on the molecular radiative properties of the gas, the size of the increase in atmospheric concentration, and the residence time of the species in the atmosphere, once emitted. *The latter – the atmospheric residence time of the greenhouse gas – is a highly policy relevant characteristic. Namely, emissions of a greenhouse gas that has a long atmospheric residence time is a quasi-irreversible commitment to sustained radiative forcing over decades, centuries, or millennia, before natural processes can remove the quantities emitted.*

Table 1: Examples of greenhouse gases that are affected by human activities. [Based upon Chapter 3 and Table 4.1]

	CO_2 (Carbon Dioxide)	CH_4 (Methane)	N_2O (Nitrous Oxide)	CFC-11 (Chlorofluoro-carbon-11)	HFC-23 (Hydrofluoro-carbon-23)	CF_4 (Perfluoro-methane)
Pre-industrial concentration	about 280 ppm	about 700 ppb	about 270 ppb	zero	zero	40 ppt
Concentration in 1998	365 ppm	1745 ppb	314 ppb	268 ppt	14 ppt	80 ppt
Rate of concentration change[b]	1.5 ppm/yr[a]	7.0 ppb/yr[a]	0.8 ppb/yr	−1.4 ppt/yr	0.55 ppt/yr	1 ppt/yr
Atmospheric lifetime	5 to 200 yr[c]	12 yr[d]	114 yr[d]	45 yr	260 yr	>50,000 yr

[a] Rate has fluctuated between 0.9 ppm/yr and 2.8 ppm/yr for CO_2 and between 0 and 13 ppb/yr for CH_4 over the period 1990 to 1999.

[b] Rate is calculated over the period 1990 to 1999.

[c] No single lifetime can be defined for CO_2 because of the different rates of uptake by different removal processes.

[d] This lifetime has been defined as an "adjustment time" that takes into account the indirect effect of the gas on its own residence time.

Carbon dioxide (CO₂)

The atmospheric concentration of CO_2 has increased from 280 ppm[5] in 1750 to 367 ppm in 1999 (31%, Table 1). Today's CO_2 concentration has not been exceeded during the past 420,000 years and likely not during the past 20 million years. The rate of increase over the past century is unprecedented, at least during the past 20,000 years (Figure 10).

The CO_2 isotopic composition and the observed decrease in Oxygen (O_2) demonstrates that the observed increase in CO_2 is predominately due to the oxidation of organic carbon by fossil-fuel combustion and deforestation. An expanding set of palaeo-atmospheric data from air trapped in ice over hundreds of millennia provide a context for the increase in CO_2 concentrations during the Industrial Era (Figure 10). Compared to the relatively stable CO_2 concentrations (280 ± 10 ppm) of the preceding several thousand years, the increase during the Industrial Era is dramatic. The average rate of increase since 1980 is 0.4%/yr. The increase is a consequence of CO_2 emissions. Most of the emissions during the past 20 years are due to fossil fuel burning, the rest (10 to 30%) is predominantly due to land-use change, especially deforestation. As shown in Figure 9, CO_2 is the dominant human-influenced greenhouse gas, with a current radiative forcing of 1.46 Wm⁻², being 60% of the total from the changes in concentrations of all of the long-lived and globally mixed greenhouse gases.

Direct atmospheric measurements of CO_2 concentrations made over the past 40 years show that year to year fluctuations in the rate of increase of atmospheric CO_2 are large. In the 1990s, the annual rates of CO_2 increase in the atmosphere varied from 0.9 to 2.8 ppm/yr, equivalent to 1.9 to 6.0 PgC/yr. Such annual changes can be related statistically to short-term climate variability, which alters the rate at which atmospheric CO_2 is taken up and released by the oceans and land. The highest rates of increase in atmospheric CO_2 have typically been in strong El Niño years (Box 4). These higher rates of increase can be plausibly explained by reduced terrestrial uptake (or terrestrial outgassing) of CO_2 during El Niño years, overwhelming the tendency of the ocean to take up more CO_2 than usual.

Partitioning of anthropogenic CO_2 between atmospheric increases and land and ocean uptake for the past two decades can now be calculated from atmospheric observations. Table 2 presents a global CO_2 budget for the 1980s (which proves to be similar to the one constructed with the help of ocean model results in the SAR) and for the 1990s. Measurements of the decrease in atmospheric oxygen (O_2) as well as the increase in CO_2 were used in the construction of these new budgets. Results from this approach are consistent with other analyses based on the isotopic composition of atmospheric CO_2 and with independent estimates based on measurements of CO_2 and ¹³CO_2 in seawater. The 1990s budget is based on newly available measurements and updates the budget for

Table 2: Global CO_2 budgets (in PgC/yr) based on measurements of atmospheric CO_2 and O_2. Positive values are fluxes to the atmosphere; negative values represent uptake from the atmosphere. [Based upon Tables 3.1 and 3.3]

	SAR[a,b] 1980 to 1989	This Report[a] 1980 to 1989	This Report[a] 1990 to 1999
Atmospheric increase	3.3 ± 0.1	3.3 ± 0.1	3.2 ± 0.1
Emissions (fossil fuel, cement)[c]	5.5 ± 0.3	5.4 ± 0.3	6.3 ± 0.4
Ocean-atmosphere flux	−2.0 ± 0.5	−1.9 ± 0.6	−1.7 ± 0.5
Land-atmosphere flux[d]	−0.2 ± 0.6	−0.2 ± 0.7	−1.4 ± 0.7

[a] Note that the uncertainties cited in this table are ±1 standard error. The uncertainties cited in the SAR were ±1.6 standard error (i.e., approximately 90% confidence interval). Uncertainties cited from the SAR were adjusted to ±1 standard error. Error bars denote uncertainty, not interannual variability, which is substantially greater.

[b] Previous IPCC carbon budgets calculated ocean uptake from models and the land-atmosphere flux was inferred by difference.

[c] The fossil fuel emissions term for the 1980s has been revised slightly downward since the SAR.

[d] The land-atmosphere flux represents the balance of a positive term due to land-use change and a residual terrestrial sink. The two terms cannot be separated on the basis of current atmospheric measurements. Using independent analyses to estimate the land-use change component for 1980 to 1989, the residual terrestrial sink can be inferred as follows: Land-use change 1.7 PgC/yr (0.6 to 2.5); Residual terrestrial sink −1.9 PgC/yr (−3.8 to 0.3). Comparable data for the 1990s are not yet available.

[5] Atmospheric abundances of trace gases are reported here as the mole fraction (molar mixing ratio) of the gas relative to dry air (ppm = 10⁻⁶, ppb = 10⁻⁹, ppt = 10⁻¹²). Atmospheric burden is reported as the total mass of the gas (e.g., Mt = Tg = 10¹² g). The global carbon cycle is expressed in PgC = GtC.

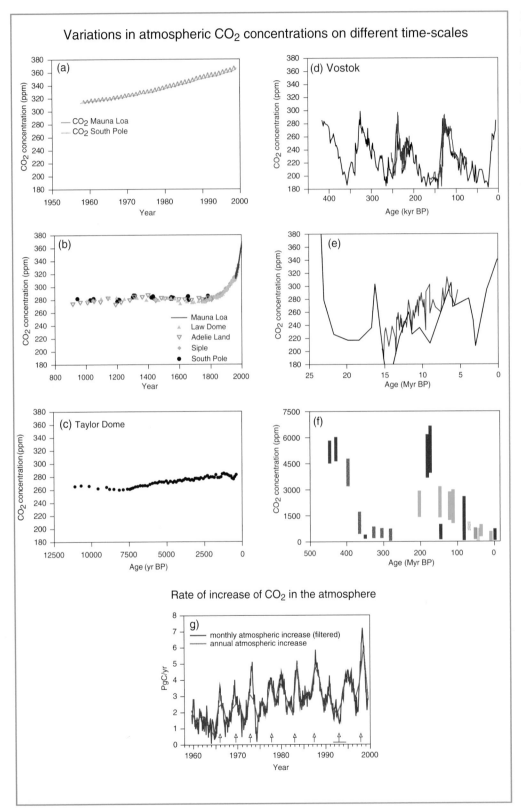

Figure 10: Variations in atmospheric CO_2 concentration on different time-scales. (a) Direct measurements of atmospheric CO_2. (b) CO_2 concentration in Antarctic ice cores for the past millenium. Recent atmospheric measurements (Mauna Loa) are shown for comparison. (c) CO_2 concentration in the Taylor Dome Antarctic ice core. (d) CO_2 concentration in the Vostok Antarctic ice core. (Different colours represent results from different studies.) (e to f) Geochemically inferred CO_2 concentrations. (Coloured bars and lines represent different published studies) (g) Annual atmospheric increases in CO_2. Monthly atmospheric increases have been filtered to remove the seasonal cycle. Vertical arrows denote El Niño events. A horizontal line defines the extended El Niño of 1991 to 1994. [Based on Figures 3.2 and 3.3]

1989 to 1998 derived using SAR methodology for the IPCC Special Report on Land Use, Land-Use Change and Forestry (2000). The terrestrial biosphere as a whole has gained carbon during the 1980s and 1990s; i.e., the CO_2 released by land-use change (mainly tropical deforestation) was more than compensated by other terrestrial sinks, which are likely located in both the northern extra-tropics and in the tropics. There remain large uncertainties associated with estimating the CO_2 release due to land-use change (and, therefore, with the magnitude of the residual terrestrial sink).

Process-based modelling (terrestrial and ocean carbon models) has allowed preliminary quantification of mechanisms in the global carbon cycle. Terrestrial model results indicate that enhanced plant growth due to higher CO_2 (CO_2 fertilisation) and anthropogenic nitrogen deposition contribute significantly to CO_2 uptake, i.e., are potentially responsible for the residual terrestrial sink described above, along with other proposed mechanisms, such as changes in land-management practices. The modelled effects of climate change during the 1980s on the terrestrial sink are small and of uncertain sign.

Methane (CH_4)

Atmospheric methane (CH_4) concentrations have increased by about 150% (1,060 ppb) since 1750. The present CH_4 concentration has not been exceeded during the past 420,000 years. Methane (CH_4) is a greenhouse gas with both natural (e.g., wetlands) and human-influenced sources (e.g., agriculture, natural gas activities, and landfills). Slightly more than half of current CH_4 emissions are anthropogenic. It is removed from the atmosphere by chemical reactions. As Figure 11 shows, systematic, globally representative measurements of the concentration of CH_4 in the atmosphere have been made since 1983, and the record of atmospheric concentrations has been extended to earlier times from air extracted from ice cores and firn layers. The current direct radiative forcing of 0.48 Wm^{-2} from CH_4 is 20% of the total from all of the long-lived and globally mixed greenhouse gases (see Figure 9).

The atmospheric abundance of CH_4 continues to increase, from about 1,610 ppb in 1983 to 1,745 ppb in 1998, but the observed annual increase has declined during this period. The increase was highly variable during the 1990s; it was near zero in 1992 and as large as 13 ppb during 1998. There is no clear quantitative explanation for this variability. Since

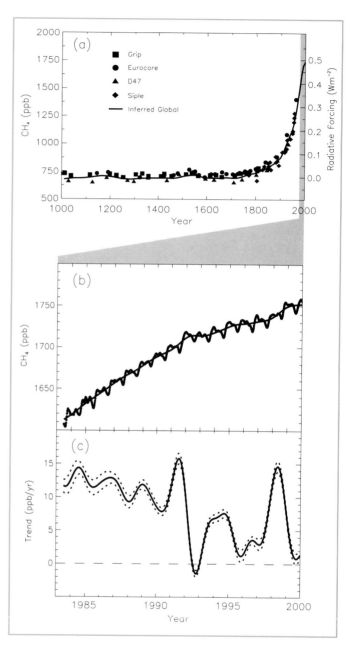

Figure 11: (a) Change in CH_4 abundance (mole fraction, in ppb = 10^{-9}) determined from ice cores, firn, and whole air samples plotted for the last 1000 years. Radiative forcing, approximated by a linear scale since the pre-industrial era, is plotted on the right axis. (b) Globally averaged CH_4 (monthly varying) and deseasonalised CH_4 (smooth line) abundance plotted for 1983 to 1999. (c) Instantaneous annual growth rate (ppb/yr) in global atmospheric CH_4 abundance from 1983 through 1999 calculated as the derivative of the deseasonalised trend curve above. Uncertainties (dotted lines) are ±1 standard deviation. [Based on Figure 4.1]

the SAR, quantification of certain anthropogenic sources of CH_4, such as that from rice production, has improved.

The rate of increase in atmospheric CH_4 is due to a small imbalance between poorly characterised sources and sinks, which makes the prediction of future concentrations problematic. Although the major contributors to the global CH_4 budget likely have been identified, most of them are quite uncertain quantitatively because of the difficulty in assessing emission rates of highly variable biospheric sources. The limitations of poorly quantified and characterised CH_4 source strengths inhibit the prediction of future CH_4 atmospheric concentrations (and hence its contribution to radiative forcing) for any given anthropogenic emission scenario, particularly since both natural emissions and the removal of CH_4 can be influenced substantially by climate change.

Nitrous oxide (N_2O)

The atmospheric concentration of nitrous oxide (N_2O) has steadily increased during the Industrial Era and is now 16% (46 ppb) larger than in 1750. The present N_2O concentration has not been exceeded during at least the past thousand years. Nitrous oxide is another greenhouse gas with both natural and anthropogenic sources, and it is removed from the atmosphere by chemical reactions. Atmospheric concentrations of N_2O continue to increase at a rate of 0.25%/yr (1980 to 1998). Significant interannual variations in the upward trend of N_2O concentrations are observed, e.g., a 50% reduction in annual growth rate from 1991 to 1993. Suggested causes are several-fold: a decrease in use of nitrogen-based fertiliser, lower biogenic emissions, and larger stratospheric losses due to volcanic-induced circulation changes. Since 1993, the growth of N_2O concentrations has returned to rates closer to those observed during the 1980s. While this observed multi-year variance has provided some potential insight into what processes control the behaviour of atmospheric N_2O, the multi-year trends of this greenhouse gas remain largely unexplained.

The global budget of nitrous oxide is in better balance than in the SAR, but uncertainties in the emissions from individual sources are still quite large. Natural sources of N_2O are estimated to be approximately 10 TgN/yr (1990), with soils being about 65% of the sources and oceans about 30%. New, higher estimates of the emissions from anthropogenic sources (agriculture, biomass burning, industrial activities, and livestock management) of approximately 7 TgN/yr have

brought the source/sink estimates closer in balance, compared with the SAR. However, the predictive understanding associated with this significant, long-lived greenhouse gas has not improved significantly since the last assessment. The radiative forcing is estimated at 0.15 Wm^{-2}, which is 6% of the total from all of the long-lived and globally mixed greenhouse gases (see Figure 9).

Halocarbons and related compounds

The atmospheric concentrations of many of those gases that are both ozone-depleting and greenhouse gases are either decreasing (CFC-11, CFC-113, CH_3CCl_3 and CCl_4) or increasing more slowly (CFC-12) in response to reduced emissions under the regulations of the Montreal Protocol and its Amendments. Many of these halocarbons are also radiatively effective, long-lived greenhouse gases. Halocarbons are carbon compounds that contain fluorine, chlorine, bromine or iodine. For most of these compounds, human activities are the sole source. Halocarbons that contain chlorine (e.g., chlorofluorocarbons - CFCs) and bromine (e.g., halons) cause depletion of the stratospheric ozone layer and are controlled under the Montreal Protocol. The combined tropospheric abundance of ozone-depleting gases peaked in 1994 and is slowly declining. The atmospheric abundances of some of the major greenhouse halocarbons have peaked, as shown for CFC-11 in Figure 12. The concentrations of CFCs and chlorocarbons in the troposphere are consistent with reported emissions. Halocarbons contribute a radiative forcing of 0.34 Wm^{-2}, which is 14% of the radiative forcing from all of the globally mixed greenhouse gases (Figure 9).

The observed atmospheric concentrations of the substitutes for the CFCs are increasing, and some of these compounds are greenhouse gases. The abundances of the hydrochlorofluoro-carbons (HCFCs) and hydrofluorocarbons (HFCs) are increasing as a result of continuation of earlier uses and of their use as substitutes for the CFCs. For example, the concen-tration of HFC-23 has increased by more than a factor of three between 1978 and 1995. Because current concentrations are relatively low, the present contribution of HFCs to radiative forcing is relatively small. The present contribution of HCFCs to radiative forcing is also relatively small, and future emissions of these gases are limited by the Montreal Protocol.

The perfluorocarbons (PFCs, e.g., CF_4 and C_2F_6) and sulphur hexafluoride (SF_6) have anthropogenic sources, have extremely

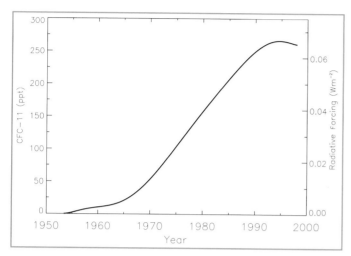

Figure 12: Global mean CFC-11 (CFCl$_3$) tropospheric abundance (ppt) from 1950 to 1998 based on smoothed measurements and emission models. CFC-11's radiative forcing is shown on the right axis. [Based on Figure 4.6]

long atmospheric residence times, and are strong absorbers of infrared radiation. Therefore, these compounds, even with relatively small emissions, have the potential to influence climate far into the future. Perfluoromethane (CF$_4$) resides in the atmosphere for at least 50,000 years. It has a natural background; however, current anthropogenic emissions exceed natural ones by a factor of 1,000 or more and are responsible for the observed increase. Sulphur hexafluoride (SF$_6$) is 22,200 times more effective a greenhouse gas than CO$_2$ on a per-kg basis. The current atmospheric concentrations are very small (4.2 ppt), but have a significant growth rate (0.24 ppt/yr). There is good agreement between the observed atmospheric growth rate of SF$_6$ and the emissions based on revised sales and storage data.

C.2 Observed Changes in Other Radiatively Important Gases

Atmospheric ozone (O$_3$)

Ozone (O$_3$) is an important greenhouse gas present in both the stratosphere and troposphere. The role of ozone in the atmospheric radiation budget is strongly dependent on the altitude at which changes in ozone concentrations occur. The changes in ozone concentrations are also spatially variable.

Further, ozone is not a directly emitted species, but rather it is formed in the atmosphere from photochemical processes involving both natural and human-influenced precursor species. Once formed, the residence time of ozone in the atmosphere is relatively short, varying from weeks to months. As a result, estimation of ozone's radiative role is more complex and much less certain than for the above long-lived and globally well-mixed greenhouse gases.

The observed losses of stratospheric ozone layer over the past two decades have caused a negative forcing of 0.15 ± 0.1 Wm^{-2} (i.e., a tendency toward cooling) of the surface troposphere system. It was reported in Climate Change 1992: The Supplementary Report to the IPCC Scientific Assessment, that depletion of the ozone layer by anthropogenic halocarbons introduces a negative radiative forcing. The estimate shown in Figure 9 is slightly larger in magnitude than that given in the SAR, owing to the ozone depletion that has continued over the past five years, and it is more certain as a result of an increased number of modelling studies. Studies with General Circulation Models indicate that, despite the inhomogeneity in ozone loss (i.e., lower stratosphere at high latitudes), such a negative forcing does relate to a surface temperature decrease in proportion to the magnitude of the negative forcing. Therefore, this negative forcing over the past two decades has offset some of the positive forcing that is occurring from the long-lived and globally well-mixed greenhouse gases (Figure 9). A major source of uncertainty in the estimation of the negative forcing is due to incomplete knowledge of ozone depletion near the tropopause. Model calculations indicate that increased penetration of ultraviolet radiation to the troposphere, as a result of stratospheric ozone depletion, leads to enhanced removal rates of gases like CH$_4$, thus amplifying the negative forcing due to ozone depletion. As the ozone layer recovers in future decades because of the effects of the Montreal Protocol, relative to the present, future radiative forcing associated with stratospheric ozone is projected to become positive.

The global average radiative forcing due to increases in tropospheric ozone since pre-industrial times is estimated to have enhanced the anthropogenic greenhouse gas forcing by 0.35 ± 0.2 Wm^{-2}. This makes tropospheric ozone the third most important greenhouse gas after CO$_2$ and CH$_4$. Ozone is formed by photochemical reactions and its future change will be determined by, among other things, emissions of CH$_4$ and pollutants (as noted below). Ozone concentrations respond relatively quickly to changes in the emissions of pollutants. On the basis of limited observations and several modelling studies, tropospheric ozone is estimated to have increased by about 35% since the Pre-industrial Era, with some regions experiencing larger and some with smaller increases. There have been few observed increases in ozone concentrations in the global troposphere since the mid-1980s at most of the few remote locations where it is regularly measured. The lack of observed increase over North America and Europe is related to the lack of a sustained increase in ozone-precursor emissions from those continents. However, some Asian stations indicate a possible rise in tropospheric ozone, which could be related to the increase in East Asian emissions. As a result of more modelling studies than before, there is now an increased confidence in the estimates of tropospheric ozone forcing. The confidence, however, is still much less than that for the well-mixed greenhouse gases, but more so than that for aerosol forcing. Uncertainties arise because of limited information on pre-industrial ozone distributions and limited information to evaluate modelled global trends in the modern era (i.e., post-1960).

Gases with only indirect radiative influences

Several chemically reactive gases, including reactive nitrogen species (NO_x), carbon monoxide (CO), and the volatile organic compounds (VOCs), control, in part, the oxidising capacity of the troposphere, as well as the abundance of ozone. These pollutants act as indirect greenhouse gases through their influence not only on ozone, but also on the lifetimes of CH_4 and other greenhouse gases. The emissions of NO_x and CO are dominated by human activities.

Carbon monoxide is identified as an important indirect greenhouse gas. Model calculations indicate that emission of 100 Mt of CO is equivalent in terms of greenhouse gas perturbations to the emission of about 5 Mt of CH_4. The abundance of CO in the Northern Hemisphere is about twice that in the Southern Hemisphere and has increased in the second half of the 20th century along with industrialisation and population.

The reactive nitrogen species NO and NO_2, (whose sum is denoted NO_x), are key compounds in the chemistry of the troposphere, but their overall radiative impact remains difficult to quantify. The importance of NO_x in the radiation budget is because increases in NO_x concentrations perturb several greenhouse gases; for example, decreases in methane and the HFCs and increases in tropospheric ozone. Deposition

of the reaction products of NO_x fertilises the biosphere, thereby decreasing atmospheric CO_2. While difficult to quantify, increases in NO_x that are projected to the year 2100 would cause significant changes in greenhouse gases.

C.3 Observed and Modelled Changes in Aerosols

Aerosols (very small airborne particles and droplets) are known to influence significantly the radiative budget of the Earth/atmosphere. Aerosol radiative effects occur in two distinct ways: (i) the direct effect, whereby aerosols themselves scatter and absorb solar and thermal infrared radiation, and (ii) the indirect effect, whereby aerosols modify the microphysical and hence the radiative properties and amount of clouds. Aerosols are produced by a variety of processes, both natural (including dust storms and volcanic activity) and anthropogenic (including fossil fuel and biomass burning). The atmospheric concentrations of tropospheric aerosols are thought to have increased over recent years due to increased anthropogenic emissions of particles and their precursor gases, hence giving rise to radiative forcing. Most aerosols are found in the lower troposphere (below a few kilometres), but the radiative effect of many aerosols is sensitive to the vertical distribution. Aerosols undergo chemical and physical changes while in the atmosphere, notably within clouds, and are removed largely and relatively rapidly by precipitation (typically within a week). Because of this short residence time and the inhomogeneity of sources, aerosols are distributed inhomogeneously in the troposphere, with maxima near the sources. The radiative forcing due to aerosols depends not only on these spatial distributions, but also on the size, shape, and chemical composition of the particles and various aspects (e.g., cloud formation) of the hydrological cycle as well. As a result of all of these factors, obtaining accurate estimates of this forcing has been very challenging, from both the observational and theoretical standpoints.

*Nevertheless, substantial progress has been achieved in better defining the **direct effect** of a wider set of different aerosols.* The SAR considered the direct effects of only three anthropogenic aerosol species: sulphate aerosols, biomass-burning aerosols, and fossil fuel black carbon (or soot). Observations have now shown the importance of organic materials in both fossil fuel carbon aerosols and biomass-burning carbon aerosols. Since the SAR, the inclusion of estimates for the abundance of fossil fuel organic carbon aerosols has led to an increase in the predicted total optical depth (and consequent negative forcing) associated with industrial aerosols. Advances in observations and in aerosol and radiative models have allowed quantitative estimates of these separate components, as well as an estimate for the range of radiative forcing associated with mineral dust, as shown in Figure 9. Direct radiative forcing is estimated to be -0.4 Wm^{-2} for sulphate, -0.2 Wm^{-2} for biomass-burning aerosols, -0.1 Wm^{-2} for fossil fuel organic carbon, and $+0.2$ Wm^{-2} for fossil fuel black carbon aerosols. Uncertainties remain relatively large, however. These arise from difficulties in determining the concentration and radiative characteristics of atmospheric aerosols and the fraction of the aerosols that are of anthropogenic origin, particularly the knowledge of the sources of carbonaceous aerosols. This leads to considerable differences (i.e., factor of two to three range) in the burden and substantial differences in the vertical distribution (factor of ten). Anthropogenic dust aerosol is also poorly quantified. Satellite observations, combined with model calculations, are enabling the identification of the spatial signature of the total aerosol radiative effect in clear skies; however, the quantitative amount is still uncertain.

Estimates of the **indirect radiative** *forcing by anthropogenic aerosols remain problematic, although observational evidence points to a negative aerosol-induced indirect forcing in warm clouds.* Two different approaches exist for estimating the indirect effect of aerosols: empirical methods and mechanistic methods. The former have been applied to estimate the effects of industrial aerosols, while the latter have been applied to estimate the effects of sulphate, fossil fuel carbonaceous aerosols, and biomass aerosols. In addition, models for the indirect effect have been used to estimate the effects of the initial change in droplet size and concentrations (a first indirect effect), as well as the effects of the subsequent change in precipitation efficiency (a second indirect effect). The studies represented in Figure 9 provide an expert judgement for the range of the first of these; the range is now slightly wider than in the SAR; the radiative perturbation associated with the second indirect effect is of the same sign and could be of similar magnitude compared to the first effect.

The indirect radiative effect of aerosols is now understood to also encompass effects on ice and mixed-phase clouds, but the magnitude of any such indirect effect is not known, although it is likely to be positive. It is not possible to estimate the number of anthropogenic ice nuclei at the present time. Except at cold temperatures (below $-45°C$) where homogeneous nucleation is expected to dominate, the mechanisms of ice formation in these clouds are not yet known.

C.4 Observed Changes in Other Anthropogenic Forcing Agents

Land-use (albedo) change

Changes in land use, deforestation being the major factor, appear to have produced a negative radiative forcing of -0.2 ± 0.2 Wm^{-2} (Figure 8). The largest effect is estimated to be at the high latitudes. This is because deforestation has caused snow-covered forests with relatively low albedo to be replaced with open, snow-covered areas with higher albedo. The estimate given above is based on simulations in which pre-industrial vegetation is replaced by current land-use patterns. However, the level of understanding is very low for this forcing, and there have been far fewer investigations of this forcing compared to investigations of other factors considered in this report.

C.5 Observed and Modelled Changes in Solar and Volcanic Activity

Radiative forcing of the climate system due to solar irradiance change is estimated to be 0.3 ± 0.2 Wm^{-2} for the period 1750 to the present (Figure 8), and most of the change is estimated to have occurred during the first half of the 20th century. The fundamental source of all energy in the Earth's climate system is radiation from the Sun. Therefore, variation in solar output is a radiative forcing agent. The absolute value of the spectrally integrated total solar irradiance (TSI) incident on the Earth is not known to better than about 4 Wm^{-2}, but satellite observations since the late 1970s show relative variations over the past two solar 11-year activity cycles of about 0.1%, which is equivalent to a variation in radiative forcing of about 0.2 Wm^{-2}. Prior to these satellite observations, reliable direct measurements of solar irradiance are not available. Variations over longer periods may have been larger, but the techniques used to reconstruct historical values of TSI from proxy observations (e.g., sunspots) have not been adequately verified. Solar variation varies more substantially in the ultraviolet region, and studies with climate models suggest that inclusion of spectrally resolved solar irradiance variations and solar-induced stratospheric ozone changes may improve the realism of model simulations of the impact of solar variability on climate. Other mechanisms for the amplification of solar effects on climate have been proposed, but do not have a rigorous theoretical or observational basis.

Stratospheric aerosols from explosive volcanic eruptions lead to negative forcing that lasts a few years. Several explosive eruptions occurred in the periods 1880 to 1920 and 1960 to 1991, and no explosive eruptions since 1991. Enhanced stratospheric aerosol content due to volcanic eruptions, together with the small solar irradiance variations, result in a net negative natural radiative forcing over the past two, and possibly even the past four, decades.

C.6 Global Warming Potentials

Radiative forcings and Global Warming Potentials (GWPs) are presented in Table 3 for an expanded set of gases. GWPs are a measure of the relative radiative effect of a given substance compared to CO_2, integrated over a chosen time horizon. New categories of gases in Table 3 include fluorinated organic molecules, many of which are ethers that are proposed as halocarbon substitutes. Some of the GWPs have larger uncertainties than that of others, particularly for those gases where detailed laboratory data on lifetimes are not yet available. The direct GWPs have been calculated relative to CO_2 using an improved calculation of the CO_2 radiative forcing, the SAR response function for a CO_2 pulse, and new values for the radiative forcing and lifetimes for a number of halocarbons. Indirect GWPs, resulting from indirect radiative forcing effects, are also estimated for some new gases, including carbon monoxide. The direct GWPs for those species whose lifetimes are well characterised are estimated to be accurate within ±35%, but the indirect GWPs are less certain.

D. The Simulation of the Climate System and its Changes

The preceding two Sections reported on the climate from the distant past to the present day through the observations of climate variables and the forcing agents that cause climate to change. This Section bridges to the climate of the future by describing the only tool that provides quantitative estimates of future climate changes, namely, numerical models. The basic understanding of the energy balance of the Earth system means that quite simple models can provide a broad quantitative estimate of some globally averaged variables, but more accurate estimates of feedbacks and of regional detail can only come from more elaborate climate models. The complexity of the processes in the climate system prevents the use of extrapolation of past trends or statistical and other purely empirical techniques for projections. Climate models can be used to simulate the climate responses to different input scenarios of future forcing agents (Section F). Similarly, projection of the fate of emitted CO_2 (i.e., the relative sequestration into the various reservoirs) and other greenhouse gases requires an understanding of the biogeo-chemical processes involved and incorporating these into a numerical carbon cycle model.

A climate model is a simplified mathematical representation of the Earth's climate system (see Box 3). The degree to which the model can simulate the responses of the climate system hinges to a very large degree on the level of understanding of the physical, geophysical, chemical and biological processes that govern the climate system. Since the SAR, researchers have made substantial improvements in the simulation of the Earth's climate system with models. First, the current understanding of some of the most important processes that govern the climate system and how well they are represented in present climate models are summarised here. Then, this Section presents an assessment of the overall ability of present models to make useful projections of future climate.

D.1 Climate Processes and Feedbacks

Processes in the climate system determine the natural variability of the climate system and its response to perturbations, such as the increase in the atmospheric concentrations of greenhouse gases. Many basic climate processes of importance are well-known and modelled exceedingly well. Feedback processes amplify (a positive feedback) or reduce (a negative

Table 3: Direct Global Warming Potentials (GWPs) relative to carbon dioxide (for gases for which the lifetimes have been adequately characterised). GWPs are an index for estimating relative global warming contribution due to atmospheric emission of a kg of a particular greenhouse gas compared to emission of a kg of carbon dioxide. GWPs calculated for different time horizons show the effects of atmospheric lifetimes of the different gases. [Based upon Table 6.7]

Gas		Lifetime (years)	Global Warming Potential (Time Horizon in years)		
			20 yrs	100 yrs	500 yrs
Carbon dioxide	CO_2		1	1	1
Methane[a]	CH_4	12.0^b	62	23	7
Nitrous oxide	N_2O	$114^{\,b}$	275	296	156
Hydrofluorocarbons					
HFC-23	CHF_3	260	9400	12000	10000
HFC-32	CH_2F_2	5.0	1800	550	170
HFC-41	CH_3F	2.6	330	97	30
HFC-125	CHF_2CF_3	29	5900	3400	1100
HFC-134	CHF_2CHF_2	9.6	3200	1100	330
HFC-134a	CH_2FCF_3	13.8	3300	1300	400
HFC-143	CHF_2CH_2F	3.4	1100	330	100
HFC-143a	CF_3CH_3	52	5500	4300	1600
HFC-152	CH_2FCH_2F	0.5	140	43	13
HFC-152a	CH_3CHF_2	1.4	410	120	37
HFC-161	CH_3CH_2F	0.3	40	12	4
HFC-227ea	CF_3CHFCF_3	33	5600	3500	1100
HFC-236cb	$CH_2FCF_2CF_3$	13.2	3300	1300	390
HFC-236ea	CHF_2CHFCF_3	10	3600	1200	390
HFC-236fa	$CF_3CH_2CF_3$	220	7500	9400	7100
HFC-245ca	$CH_2FCF_2CHF_2$	5.9	2100	640	200
HFC-245fa	$CHF_2CH_2CF_3$	7.2	3000	950	300
HFC-365mfc	$CF_3CH_2CF_2CH_3$	9.9	2600	890	280
HFC-43-10mee	$CF_3CHFCHFCF_2CF_3$	15	3700	1500	470
Fully fluorinated species					
SF_6		3200	15100	22200	32400
CF_4		50000	3900	5700	8900
C_2F_6		10000	8000	11900	18000
C_3F_8		2600	5900	8600	12400
C_4F_{10}		2600	5900	8600	12400
$c-C_4F_8$		3200	6800	10000	14500
C_5F_{12}		4100	6000	8900	13200
C_6F_{14}		3200	6100	9000	13200
Ethers and Halogenated Ethers					
CH_3OCH_3		0.015	1	1	<<1
HFE-125	CF_3OCHF_2	150	12900	14900	9200
HFE-134	CHF_2OCHF_2	26.2	10500	6100	2000
HFE-143a	CH_3OCF_3	4.4	2500	750	230
HCFE-235da2	$CF_3CHClOCHF_2$	2.6	1100	340	110
HFE-245fa2	$CF_3CH_2OCHF_2$	4.4	1900	570	180
HFE-254cb2	$CHF_2CF_2OCH_3$	0.22	99	30	9
HFE-7100	$C_4F_9OCH_3$	5.0	1300	390	120
HFE-7200	$C_4F_9OC_2H_5$	0.77	190	55	17
H-Galden 1040x	$CHF_2OCF_2OC_2F_4OCHF_2$	6.3	5900	1800	560
HG-10	$CHF_2OCF_2OCHF_2$	12.1	7500	2700	850
HG-01	$CHF_2OCF_2CF_2OCHF_2$	6.2	4700	1500	450

[a] The methane GWPs include an indirect contribution from stratospheric H_2O and O_3 production.

[b] The values for methane and nitrous oxide are adjustment times, which incorporate the indirect effects of emission of each gas on its own lifetime.

Box 3: Climate Models: How are they built and how are they applied?

Comprehensive climate models are based on physical laws represented by mathematical equations that are solved using a three-dimensional grid over the globe. For climate simulation, the major components of the climate system must be represented in sub-models (atmosphere, ocean, land surface, cryosphere and biosphere), along with the processes that go on within and between them. Most results in this report are derived from the results of models, which include some representation of all these components. Global climate models in which the atmosphere and ocean components have been coupled together are also known as Atmosphere-Ocean General Circulation Models (AOGCMs). In the atmospheric module, for example, equations are solved that describe the large-scale evolution of momentum, heat and moisture. Similar equations are solved for the ocean. Currently, the resolution of the atmospheric part of a typical model is about 250 km in the horizontal and about 1 km in the vertical above the boundary layer. The resolution of a typical ocean model is about 200 to 400 m in the vertical, with a horizontal resolution of about 125 to 250 km. Equations are typically solved for every half hour of a model integration. Many physical processes, such as those related to clouds or ocean convection, take place on much smaller spatial scales than the model grid and therefore cannot be modelled and resolved explicitly. Their average effects are approximately included in a simple way by taking advantage of

The Development of Climate models, Past, Present and Future

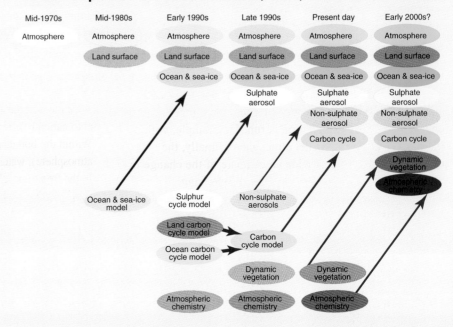

Box 3, Figure 1: The development of climate models over the last 25 years showing how the different components are first developed separately and later coupled into comprehensive climate models.

physically based relationships with the larger-scale variables. This technique is known as parametrization.

In order to make quantitative projections of future climate change, it is necessary to use climate models that simulate all the important processes governing the future evolution of the climate. Climate models have developed over the past few decades as computing power has increased. During that time, models of the main components, atmosphere, land, ocean and sea ice have been developed separately and then gradually integrated. This coupling of the various components is a difficult process. Most recently, sulphur cycle components

have been incorporated to represent the emissions of sulphur and how they are oxidised to form aerosol particles. Currently in progress, in a few models, is the coupling of the land carbon cycle and the ocean carbon cycle. The atmospheric chemistry component currently is modelled outside the main climate model. The ultimate aim is, of course, to model as much as possible of the whole of the Earth's climate system so that all the components can interact and, thus, the predictions of climate change will continuously take into account the effect of feedbacks among components. The Figure above shows the past, present and possible future evolution of climate models.

Some models offset errors and surface flux imbalances through "flux adjustments", which are empirically determined systematic adjustments at the atmosphere-ocean interface held fixed in time in order to bring the simulated climate closer to the observed state. A strategy has been designed for carrying out climate experiments that removes much of the effects of some model errors on results. What is often done is that first a "control" climate simulation is run with the model. Then, the climate change experiment simulation is run, for example, with increased CO_2 in the model atmosphere. Finally, the difference is taken to provide an estimate of the change in climate due to the perturbation. The differencing technique removes most of the effects of any artificial adjustments in the model, as well as systematic errors that are common to both runs. However, a comparison of different model results makes it apparent that the nature of some errors still influences the outcome.

Many aspects of the Earth's climate system are chaotic – its evolution is sensitive to small perturbations in initial conditions. This sensitivity limits predictability of the detailed evolution of weather to about two weeks. However, predictability of climate is not so limited because of the systematic influences on the atmosphere of the more slowly varying components of the climate system. Nevertheless, to be able to make reliable forecasts in the presence of both initial condition and model uncertainty, it is desirable to repeat the prediction many times from different perturbed initial states and using different global models. These ensembles are the basis of probability forecasts of the climate state.

Comprehensive AOGCMs are very complex and take large computer resources to run. To explore different scenarios of emissions of greenhouse gases and the effects of assumptions or approximations in parameters in the model more thoroughly, simpler models are also widely used. The simplifications may include coarser resolution and simplified dynamics and physical processes. Together, simple, intermediate, and comprehensive models form a "hierarchy of climate models", all of which are necessary to explore choices made in parametrizations and assess the robustness of climate changes.

feedback) changes in response to an initial perturbation and hence are very important for accurate simulation of the evolution of climate.

Water vapour

A major feedback accounting for the large warming predicted by climate models in response to an increase in CO_2 is the increase in atmospheric water vapour. An increase in the temperature of the atmosphere increases its water-holding capacity; however, since most of the atmosphere is undersaturated, this does not automatically mean that water vapour, itself, must increase. Within the boundary layer (roughly the lowest 1 to 2 km of the atmosphere), water vapour increases with increasing temperature. In the free troposphere above the boundary layer, where the water vapour greenhouse effect is most important, the situation is harder to quantify. Water vapour feedback, as derived from current models, approximately doubles the warming from what it would be for fixed water vapour. Since the SAR, major improvements have occurred in the treatment of water vapour in models, although detrainment of moisture from clouds remains quite uncertain and discrepancies exist between model water vapour distributions and those observed. Models are capable of simulating the moist and very dry regions observed in the tropics and sub-tropics and how they evolve with the seasons and from year to year. While reassuring, this does not provide a check of the feedbacks, although the balance of evidence favours a positive clear-sky water vapour feedback of the magnitude comparable to that found in simulations.

Clouds

As has been the case since the first IPCC Assessment Report in 1990, probably the greatest uncertainty in future projections of climate arises from clouds and their interactions with radiation. Clouds can both absorb and reflect solar radiation (thereby cooling the surface) and absorb and emit long wave radiation (thereby warming the surface). The competition between these effects depends on cloud height, thickness and radiative properties. The radiative properties and evolution of clouds depend on the distribution of atmospheric water vapour, water drops, ice particles, atmospheric aerosols and cloud thickness. The physical basis of cloud parametrizations is greatly improved in models through inclusion of bulk representation of cloud microphysical properties in a cloud water budget equation, although considerable uncertainty remains. Clouds represent a significant source of potential error in climate simulations. The possibility that models underestimate systematically solar absorption in clouds remains a controversial matter. The sign of the net cloud feedback is still a matter of uncertainty, and the various models exhibit a large spread. Further uncertainties arise from precipitation processes and the difficulty in correctly simulating the diurnal cycle and precipitation amounts and frequencies.

Stratosphere

There has been a growing appreciation of the importance of the stratosphere in the climate system because of changes in its structure and recognition of the vital role of both radiative and dynamical processes. The vertical profile of temperature change in the atmosphere, including the stratosphere, is an important indicator in detection and attribution studies. Most of the observed decreases in lower-stratospheric temperatures have been due to ozone decreases, of which the Antarctic "ozone hole" is a part, rather than increased CO_2 concentrations. Waves generated in the troposphere can propagate into the stratosphere where they are absorbed. As a result, stratospheric changes alter where and how these waves are absorbed, and the effects can extend downward into the troposphere. Changes in solar irradiance, mainly in the ultraviolet (UV), lead to photochemically-induced ozone changes and, hence, alter the stratospheric heating rates, which can alter the tropospheric circulation. Limitations in resolution and relatively poor representation of some stratospheric processes adds uncertainty to model results.

Ocean

Major improvements have taken place in modelling ocean processes, in particular heat transport. These improvements, in conjunction with an increase in resolution, have been important in reducing the need for flux adjustment in models and in producing realistic simulations of natural large-scale circulation patterns and improvements in simulating El Niño (see Box 4). Ocean currents carry heat from the tropics to higher latitudes. The ocean exchanges heat, water (through evaporation and precipitation) and CO_2 with the atmosphere. Because of its huge mass and high heat capacity, the ocean slows climate change and influences the time-scales of variability in the ocean-atmosphere system. Considerable progress has been made in the understanding of ocean processes relevant for climate change. Increases in resolution, as well as improved representation (parametrization) of important sub-grid scale processes (e.g., mesoscale eddies), have increased the realism of simulations. Major uncertainties

still exist with the representation of small-scale processes, such as overflows (flow through narrow channels, e.g., between Greenland and Iceland), western boundary currents (i.e., large-scale narrow currents along coastlines), convection and mixing. Boundary currents in climate simulations are weaker and wider than in nature, although the consequences of this for climate are not clear.

Cryosphere

The representation of sea-ice processes continues to improve, with several climate models now incorporating physically based treatments of ice dynamics. The representation of land-ice processes in global climate models remains rudimentary. The cryosphere consists of those regions of Earth that are seasonally or perennially covered by snow and ice. Sea ice is important because it reflects more incoming solar radiation than the sea surface (i.e., it has a higher albedo), and it insulates the sea from heat loss during the winter. Therefore, reduction of sea ice gives a positive feedback on climate warming at high latitudes. Furthermore, because sea ice contains less salt than sea water, when sea ice is formed the salt content (salinity) and density of the surface layer of the ocean is increased. This promotes an exchange of water with deeper layers of the ocean, affecting ocean circulation. The formation of icebergs and the melting of ice shelves returns fresh water from the land to the ocean, so that changes in the rates of these processes could affect ocean circulation by changing the surface salinity. Snow has a higher albedo than the land surface; hence, reductions in snow cover lead to a similar positive albedo feedback, although weaker than for sea ice. Increasingly complex snow schemes and sub-grid scale variability in ice cover and thickness, which can significantly influence albedo and atmosphere-ocean exchanges, are being introduced in some climate models.

Land surface

Research with models containing the latest representations of the land surface indicates that the direct effects of increased CO_2 on the physiology of plants could lead to a relative reduction in evapotranspiration over the tropical continents, with associated regional warming and drying over that predicted for conventional greenhouse warming effects. Land surface changes provide important feedbacks as anthropogenic climate changes (e.g., increased temperature, changes in precipitation, changes in net radiative heating, and the direct effects of CO_2) will influence the state of the land surface (e.g., soil moisture, albedo, roughness and vegetation). Exchanges of energy, momentum, water, heat and carbon between the land surface and the atmosphere can be defined in models as functions of the type and density of the local vegetation and the depth and physical properties of the soil, all based on land-surface data bases that have been improved using satellite observations. Recent advances in the understanding of vegetation photosynthesis and water use have been used to couple the terrestrial energy, water and carbon cycles within a new generation of land surface parametrizations, which have been tested against field observations and implemented in a few GCMs, with demonstrable improvements in the simulation of land-atmosphere fluxes. However, significant problems remain to be solved in the areas of soil moisture processes, runoff prediction, land-use change and the treatment of snow and sub-grid scale heterogeneity.

Changes in land-surface cover can affect global climate in several ways. Large-scale deforestation in the humid tropics (e.g., South America, Africa, and Southeast Asia) has been identified as the most important ongoing land-surface process, because it reduces evaporation and increases surface temperature. These effects are qualitatively reproduced by most models. However, large uncertainties still persist on the quantitative impact of large-scale deforestation on the hydrological cycle, particularly over Amazonia.

Carbon cycle

Recent improvements in process-based terrestrial and ocean carbon cycle models and their evaluation against observations have given more confidence in their use for future scenario studies. CO_2 naturally cycles rapidly among the atmosphere, oceans and land. However, the removal of the CO_2 perturbation added by human activities from the atmosphere takes far longer. This is because of processes that limit the rate at which ocean and terrestrial carbon stocks can increase. Anthropogenic CO_2 is taken up by the ocean because of its high solubility (caused by the nature of carbonate chemistry), but the rate of uptake is limited by the finite speed of vertical mixing. Anthropogenic CO_2 is taken up by terrestrial ecosystems through several possible mechanisms, for example, land management, CO_2 fertil-isation (the enhancement of plant growth as a result of increased atmospheric CO_2 concentration) and increasing anthropogenic inputs of nitrogen. This uptake is limited by the relatively small fraction of plant carbon that can enter long-term storage (wood and humus). The fraction of emitted CO_2 that can be taken up by the oceans and land is expected to decline with increasing CO_2

concentrations. Process-based models of the ocean and land carbon cycles (including representations of physical, chemical and biological processes) have been developed and evaluated against measurements pertinent to the natural carbon cycle. Such models have also been set up to mimic the human perturbation of the carbon cycle and have been able to generate time-series of ocean and land carbon uptake that are broadly consistent with observed global trends. There are still substantial differences among models, especially in how they treat the physical ocean circulation and in regional responses of terrestrial ecosystem processes to climate. Nevertheless, current models consistently indicate that when the effects of climate change are considered, CO_2 uptake by oceans and land becomes smaller.

D.2 The Coupled Systems

As noted in Section D.1, many feedbacks operate within the individual components of the climate system (atmosphere, ocean, cryosphere and land surface). However, many important processes and feedbacks occur through the coupling of the climate system components. Their representation is important to the prediction of large-scale responses.

Modes of natural variability

There is an increasing realisation that natural circulation patterns, such as ENSO and NAO, play a fundamental role in global climate and its interannual and longer-term variability. The strongest natural fluctuation of climate on interannual time-scales is the ENSO phenomenon (see Box 4). It is an inherently coupled atmosphere-ocean mode with its core activity in the tropical Pacific, but with important regional climate impacts throughout the world. Global climate models are only now beginning to exhibit variability in the tropical Pacific that resembles ENSO, mainly through increased meridional resolution at the equator. Patterns of sea surface temperature and atmospheric circulation similar to those occurring during ENSO on interannual time-scales also occur on decadal and longer time-scales.

The North Atlantic Oscillation (NAO) is the dominant pattern of northern wintertime atmospheric circulation variability and is increasingly being simulated realistically. The NAO is closely related to the Arctic Oscillation (AO), which has an additional annular component around the Arctic. There is strong evidence that the NAO arises mainly from internal atmospheric processes involving the entire troposphere-stratosphere system.

Box 4: The El Niño-Southern Oscillation (ENSO)

The strongest natural fluctuation of climate on interannual time-scales is the El Niño-Southern Oscillation (ENSO) phenomenon. The term "El Niño" originally applied to an annual weak warm ocean current that ran southwards along the coast of Peru about Christmas-time and only subsequently became associated with the unusually large warmings. The coastal warming, however, is often associated with a much more extensive anomalous ocean warming to the International Dateline, and it is this Pacific basinwide phenomenon that forms the link with the anomalous global climate patterns. The atmospheric component tied to "El Niño" is termed the "Southern Oscillation". Scientists often call this phenomenon, where the atmosphere and ocean collaborate together, ENSO (El Niño-Southern Oscillation).

ENSO is a natural phenomenon, and there is good evidence from cores of coral and glacial ice in the Andes that it has been going on for millennia. The ocean and atmospheric conditions in the tropical Pacific are seldom average, but instead fluctuate somewhat irregularly between El Niño events and the opposite "La Niña" phase, consisting of a basinwide cooling of the tropical Pacific, with a preferred period of about three to six years. The most intense phase of each event usually lasts about a year.

A distinctive pattern of sea surface temperatures in the Pacific Ocean sets the stage for ENSO events. Key features are the "warm pool" in the tropical western Pacific, where the warmest ocean waters in the world reside, much colder waters in the eastern Pacific, and a cold tongue along the equator that is most pronounced about October and weakest in March. The atmospheric easterly trade winds in the tropics pile up the warm waters in the west, producing an upward slope of sea level along the equator of 0.60 m from east to west. The winds drive the surface ocean currents, which determine where the surface waters flow and diverge. Thus, cooler nutrient-rich waters upwell from below along the equator and western coasts of the Americas, favouring development of phytoplankton, zooplankton, and hence fish. Because convection and thunderstorms preferentially occur over warmer waters, the pattern of sea surface temperatures determines the distribution of rainfall in the tropics, and this in turn determines the atmospheric heating patterns through the release of latent heat. The heating drives the large-scale monsoonal-type circulations in the tropics, and consequently determines the winds. This strong coupling between the atmosphere and ocean in the tropics gives rise to the El Niño phenomenon.

During El Niño, the warm waters from the western tropical Pacific migrate eastward as the trade winds weaken, shifting the pattern of tropical rainstorms, further weakening the trade winds, and thus reinforcing the changes in sea temperatures. Sea level drops in the west, but rises in the east by as much as 0.25 m, as warm waters surge eastward along the equator. However, the changes in atmospheric circulation are not confined to the tropics, but extend globally and influence the jet streams and storm tracks in mid-latitudes. Approximately reverse patterns occur during the opposite La Niña phase of the phenomenon.

Changes associated with ENSO produce large variations in weather and climate around the world from year to year. These often have a profound impact on humanity and society because of associated droughts, floods, heat waves and other changes that can severely disrupt agriculture, fisheries, the environment, health, energy demand, air quality and also change the risks of fire. ENSO also plays a prominent role in modulating exchanges of CO_2 with the atmosphere. The normal upwelling of cold nutrient-rich and CO_2-rich waters in the tropical Pacific is suppressed during El Niño.

Fluctuations in Atlantic Sea Surface Temperatures (SSTs) are related to the strength of the NAO, and a modest two-way interaction between the NAO and the Atlantic Ocean, leading to decadal variability, is emerging as important in projecting climate change.

Climate change may manifest itself both as shifting means, as well as changing preference of specific climate regimes, as evidenced by the observed trend toward positive values for the last 30 years in the NAO index and the climate "shift" in the tropical Pacific about 1976. While coupled models simulate features of observed natural climate variability, such as the NAO and ENSO, which suggests that many of the relevant processes are included in the models, further progress is needed to depict these natural modes accurately. Moreover, because ENSO and NAO are key determinants of regional climate change and can possibly result in abrupt and counter intuitive changes, there has been an increase in uncertainty in those aspects of climate change that critically depend on regional changes.

The thermohaline circulation (THC)

The thermohaline circulation (THC) is responsible for the major part of the meridional heat transport in the Atlantic Ocean. The THC is a global-scale overturning in the ocean driven by density differences arising from temperature and salinity effects. In the Atlantic, heat is transported by warm surface waters flowing northward and cold saline waters from the North Atlantic returning at depth. Reorganisations in the Atlantic THC can be triggered by perturbations in the surface buoyancy, which is influenced by precipitation, evaporation, continental runoff, sea-ice formation, and the exchange of heat, processes that could all change with consequences for regional and global climate. Interactions between the atmosphere and the ocean are also likely to be of considerable importance on decadal and longer time-scales, where the THC is involved. The interplay between the large-scale atmospheric forcing, with warming and evaporation in low latitudes and cooling and increased precipitation at high latitudes, forms the basis of a potential instability of the present Atlantic THC. ENSO may also influence the Atlantic THC by altering the fresh water balance of the tropical Atlantic, therefore providing a coupling between low and high latitudes. Uncertainties in the representation of small-scale flows over sills and through narrow straits and of ocean convection limit the ability of models to simulate situations involving substantial changes in the THC. The less saline North Pacific means that a deep THC does not occur in the Pacific.

Non-linear events and rapid climate change

The possibility for rapid and irreversible changes in the climate system exists, but there is a large degree of uncertainty about the mechanisms involved and hence also about the likelihood or time-scales of such transitions. The climate system involves many processes and feedbacks that interact in complex non-linear ways. This interaction can give rise to thresholds in the climate system that can be crossed if the system is perturbed sufficiently. There is evidence from polar ice cores suggesting that atmospheric regimes can change within a few years and that large-scale hemispheric changes can evolve as fast as a few decades. For example, the possibility of a threshold for a rapid transition of the Atlantic THC to a collapsed state has been

demonstrated with a hierarchy of models. It is not yet clear what this threshold is and how likely it is that human activity would lead it to being exceeded (see Section F.6). Atmospheric circulation can be characterised by different preferred patterns; e.g., arising from ENSO and the NAO/AO, and changes in their phase can occur rapidly. Basic theory and models suggest that climate change may be first expressed in changes in the frequency of occurrence of these patterns. Changes in vegetation, through either direct anthropogenic deforestation or those caused by global warming, could occur rapidly and could induce further climate change. It is supposed that the rapid creation of the Sahara about 5,500 years ago represents an example of such a non-linear change in land cover.

D.3 Regionalisation Techniques

Regional climate information was only addressed to a limited degree in the SAR. Techniques used to enhance regional detail have been substantially improved since the SAR and have become more widely applied. They fall into three categories: high and variable resolution AOGCMs; regional (or nested limited area) climate models (RCMs); and empirical/statistical and statistical/dynamical methods. The techniques exhibit different strengths and weaknesses and their use at the continental scale strongly depends on the needs of specific applications.

Coarse resolution AOGCMs simulate atmospheric general circulation features well in general. At the regional scale, the models display area-average biases that are highly variable from region to region and among models, with sub-continental area averaged seasonal temperature biases typically ±4°C and precipitation biases between −40 and +80%. These represent an important improvement compared to AOGCMs evaluated in the SAR.

The development of high resolution/variable resolution Atmospheric General Circulation Models (AGCMs) since the SAR generally shows that the dynamics and large-scale flow in the models improves as resolution increases. In some cases, however, systematic errors are worsened compared to coarser resolution models, although only very few results have been documented.

High resolution RCMs have matured considerably since the SAR. Regional models consistently improve the spatial detail of simulated climate compared to AGCMs. RCMs driven by observed boundary conditions evidence area-averaged temperature biases (regional scales of 10^5 to 10^6 km^2) generally below 2°C, while precipitation biases are below 50%. Regionalisation work indicates at finer scales that the changes can be substantially different in magnitude or sign from the large area-average results. A relatively large spread exists among models, although attribution of the cause of these differences is unclear.

D.4 Overall Assessment of Abilities

Coupled models have evolved and improved significantly since the SAR. In general, they provide credible simulations of climate, at least down to sub-continental scales and over temporal scales from seasonal to decadal. Coupled models, as a class, are considered to be suitable tools to provide useful projections of future climates. These models cannot yet simulate all aspects of climate (e.g., they still cannot account fully for the observed trend in the surface-troposphere temperature differences since 1979). Clouds and humidity also remain sources of significant uncertainty, but there have been incremental improvements in simulations of these quantities. No single model can be considered "best", and it is important to utilise results from a range of carefully evaluated coupled models to explore effects of different formulations. The rationale for increased confidence in models arises from model performance in the following areas.

Flux adjustment

The overall confidence in model projections is increased by the improved performance of several models that do not use flux adjustment. These models now maintain stable, multi-century simulations of surface climate that are considered to be of sufficient quality to allow their use for climate change projections. The changes whereby many models can now run without flux adjustment have come from improvements in both the atmospheric and oceanic components. In the model atmosphere, improvements in convection, the boundary layer, clouds, and surface latent heat fluxes are most notable. In the model ocean, the improvements are in resolution, boundary layer mixing, and in the representation of eddies. The results from climate change studies with flux adjusted and non-flux adjusted models are broadly in agreement; nonetheless, the development of stable non-flux adjusted models increases confidence in their ability to simulate future climates.

Climate of the 20th century

Confidence in the ability of models to project future climates is increased by the ability of several models to reproduce warming trends in the 20th century surface air temperature when driven by increased greenhouse gases and sulphate aerosols. This is illustrated in Figure 13. However, only idealized scenarios of sulphate aerosols have been used and contributions from some additional processes and forcings may not have been included in the models. Some modelling studies suggest that inclusion of additional forcings like solar variability and volcanic aerosols may improve some aspects of the simulated climate variability of the 20th century.

Extreme events

Analysis of and confidence in extreme events simulated within climate models are still emerging, particularly for storm tracks and storm frequency. "Tropical-cyclone-like" vortices are being simulated in climate models, although enough uncertainty remains over their interpretation to warrant caution in projections of tropical cyclone changes. However, in general, the analysis of extreme events in both observations (see Section B.6) and coupled models is underdeveloped.

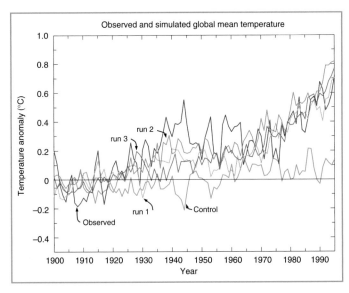

Figure 13: Observed and modelled global annual mean temperature anomalies (°C) relative to the average of the observations over the period 1900 to 1930. The control and three independent simulations with the same greenhouse gas plus aerosol forcing and slightly different initial conditions are shown from an AOGCM. The three greenhouse gas plus aerosol simulations are labeled 'run 1', 'run 2', and 'run 3' respectively. [Based on Figure 8.15]

Interannual variability

The performance of coupled models in simulating ENSO has improved; however, its variability is displaced westward and its strength is generally underestimated. When suitably initialised with surface wind and sub-surface ocean data, some coupled models have had a degree of success in predicting ENSO events.

Model intercomparisons

The growth in systematic intercomparisons of models provides the core evidence for the growing capabilities of climate models. For example, the Coupled Model Intercomparison Project (CMIP) is enabling a more comprehensive and systematic evaluation and intercomparison of coupled models run in a standardised configuration and responding to standardised forcing. Some degree of quantification of improvements in coupled model performance has now been demonstrated. The Palaeoclimate Model Intercomparison Project (PMIP) provides intercomparisons of models for the mid-Holocene (6,000 years before present) and for the Last Glacial Maximum (21,000 years before present). The ability of these models to simulate some aspects of palaeoclimates, compared to a range of palaeoclimate proxy data, gives confidence in models (at least the atmospheric component) over a range of difference forcings.

E. The Identification of a Human Influence on Climate Change

Sections B and C characterised the observed past changes in climate and in forcing agents, respectively. Section D examined the capabilities of climate models to predict the response of the climate system to such changes in forcing. This Section uses that information to examine the question of whether a human influence on climate change to date can be identified.

This is an important point to address. The SAR concluded that "the balance of evidence suggests that there is a discernible human influence on global climate". It noted that the detection and attribution of anthropogenic climate change signals will be accomplished through a gradual accumulation of evidence. The SAR also noted uncertainties in a number of factors, including internal variability and the magnitude and patterns of forcing and response, which prevented them from drawing a stronger conclusion.

E.1 The Meaning of Detection and Attribution

***Detection** is the process of demonstrating that an observed change is significantly different (in a statistical sense) than can be explained by natural variability.* ***Attribution** is the process of establishing cause and effect with some defined level of confidence, including the assessment of competing hypotheses.* The response to anthropogenic changes in climate forcing occurs against a backdrop of natural internal and externally forced climate variability. Internal climate variability, i.e., climate variability not forced by external agents, occurs on all time-scales from weeks to centuries and even millennia. Slow climate components, such as the ocean, have particularly important roles on decadal and century time-scales because they integrate weather variability. Thus, the climate is capable of producing long time-scale variations of considerable magnitude without external influences. Externally forced climate variations (signals) may be due to changes in natural forcing factors, such as solar radiation or volcanic aerosols, or to changes in anthropogenic forcing factors, such as increasing concentrations of greenhouse gases or aerosols. The presence of this natural climate variability means that the detection and attribution of anthropogenic climate change is a statistical "signal to noise" problem. *Detection* studies demonstrate whether or not an observed change is highly unusual in a statistical sense, but this does not necessarily imply that we understand its causes. The *attribution* of climate change to anthropogenic causes involves statistical analysis and the careful assessment of multiple lines of evidence to demonstrate, within a pre-specified margin of error, that the observed changes are:

- unlikely to be due entirely to internal variability;

- consistent with the estimated responses to the given combination of anthropogenic and natural forcing; and

- not consistent with alternative, physically plausible explanations of recent climate change that exclude important elements of the given combination of forcings.

E.2 A Longer and More Closely Scrutinised Observational Record

Three of the last five years (1995, 1997 and 1998) were the warmest globally in the instrumental record. The impact of observational sampling errors has been estimated for the global and hemispheric mean temperature record. There is also a better understanding of the errors and uncertainties in the satellite-based (Microwave Sounding Unit, MSU) temperature record. Discrepancies between MSU and radiosonde data have largely been resolved, although the observed trend in the difference between the surface and lower tropospheric temperatures cannot fully be accounted for (see Section B). New reconstructions of temperature over the last 1,000 years indicate that the temperature changes over the last hundred years are unlikely to be entirely natural in origin, even taking into account the large uncertainties in palaeo-reconstructions (see Section B).

E.3 New Model Estimates of Internal Variability

The warming over the past 100 years is very unlikely to be due to internal variability alone, as estimated by current models. The instrumental record is short and covers the period of human influence and palaeo-records include natural forced variations, such as those due to variations in solar irradiance and in the frequency of major volcanic eruptions. These limitations leave few alternatives to using long "control" simulations with coupled models for the estimation of internal climate variability. Since the SAR, more models have been used to estimate the magnitude of internal climate variability, a representative sample of which is given in Figure 14. As can be seen, there is a wide range of global scale internal

variability in these models. Estimates of the longer time-scale variability relevant to detection and attribution studies is uncertain, but, on interannual and decadal time-scales, some models show similar or larger variability than observed, even though models do not include variance from external sources. Conclusions on detection of an anthropogenic signal are insensitive to the model used to estimate internal variability, and recent changes cannot be accounted for as pure internal variability, even if the amplitude of simulated internal variations is increased by a factor of two or perhaps more. Most recent detection and attribution studies find no evidence that model-estimated internal variability at the surface is inconsistent with the residual variability that remains in the observations after removal of the estimated anthropogenic signals on the large spatial and long time-scales used in detection and attribution studies. Note, however, the ability to detect inconsistencies is limited. As Figure 14 indicates, no model control simulation shows a trend in surface air temperature as large as the observed trend over the last 1,000 years.

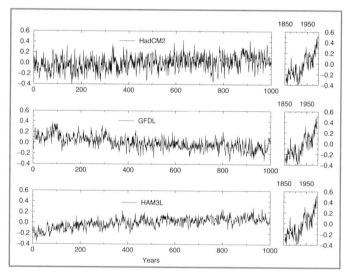

Figure 14: Global mean surface air temperature anomalies from 1,000 year control simulations with three different climate models, – Hadley, Geophysical Fluid Dynamics Laboratory and Hamburg, compared to the recent instrumental record. No model control simulation shows a trend in surface air temperature as large as the observed trend. If internal variability is correct in these models, the recent warming is likely not due to variability produced within the climate system alone. [Based on Figure 12.1]

E.4 New Estimates of Responses to Natural Forcing

Assessments based on physical principles and model simulations indicate that natural forcing alone is unlikely to explain the recent observed global warming or the observed changes in vertical temperature structure of the atmosphere. Fully coupled ocean-atmosphere models have used reconstructions of solar and volcanic forcings over the last one to three centuries to estimate the contribution of natural forcing to climate variability and change. Although the reconstruction of natural forcings is uncertain, including their effects produces an increase in variance at longer (multi-decadal) time-scales. This brings the low-frequency variability closer to that deduced from palaeo-reconstructions. It is likely that the net natural forcing (i.e., solar plus volcanic) has been negative over the past two decades, and possibly even the past four decades. Statistical assessments confirm that simulated natural variability, both internal and naturally forced, is unlikely to explain the warming in the latter half of the 20th century (see Figure 15). However, there is evidence for a detectable volcanic influence on climate and evidence that suggests a detectable solar influence, especially in the early part of the 20th century. Even if the models underestimate the magnitude of the response to solar or volcanic forcing, the spatial and temporal patterns are such that these effects alone cannot explain the observed temperature changes over the 20th century.

E.5 Sensitivity to Estimates of Climate Change Signals

There is a wide range of evidence of qualitative consistencies between observed climate changes and model responses to anthropogenic forcing. Models and observations show increasing global temperature, increasing land-ocean temperature contrast, diminishing sea-ice extent, glacial retreat, and increases in precipitation at high latitudes in the Northern Hemisphere. Some qualitative inconsistencies remain, including the fact that models predict a faster rate of warming in the mid- to upper troposphere than is observed in either satellite or radiosonde tropospheric temperature records.

All simulations with greenhouse gases and sulphate aerosols that have been used in detection studies have found that a significant anthropogenic contribution is required to account for surface and tropospheric trends over at least the last 30 years. Since the SAR, more simulations with increases in greenhouse gases and some representation of aerosol effects have become available. Several studies have included an explicit representation of greenhouse gases (as opposed to an equivalent increase in CO_2). Some have also included tropospheric ozone changes, an interactive sulphur cycle, an explicit radiative treatment of the scattering of sulphate aerosols, and improved estimates of the changes in stratospheric ozone. Overall, while detection of the climate response to these other anthropogenic factors is often ambiguous, detection of the influence of greenhouse gases on the surface temperature changes over the past 50 years is robust. In some cases, ensembles of simulations have been run to reduce noise in the estimates of the time-dependent response. Some studies have evaluated seasonal variation of the response. Uncertainties in the estimated climate change signals have made it difficult to attribute the observed climate change to one specific combination of anthropogenic and natural influences, but all studies have found a significant anthropogenic contribution is required to account for surface and tropospheric trends over at least the last thirty years.

E.6 A Wider Range of Detection Techniques

Temperature

Evidence of a human influence on climate is obtained over a substantially wider range of detection techniques. A major advance since the SAR is the increase in the range of techniques used and the evaluation of the degree to which the results are independent of the assumptions made in applying those techniques. There have been studies using pattern correlations, optimal detection studies using one or more fixed patterns and time-varying patterns, and a number of other techniques. The increase in the number of studies, breadth of techniques, increased rigour in the assessment of the role of anthropogenic forcing in climate, and the robustness of results to the assumptions made using those techniques, has increased the confidence in these aspects of detection and attribution.

Results are sensitive to the range of temporal and spatial scales that are considered. Several decades of data are necessary to separate forced signals from internal variability. Idealised studies have demonstrated that surface temperature changes are detectable only on scales in the order of 5,000 km. Such studies show that the level of agreement found between simulations and observations in pattern correlation studies is close to what one would expect in theory.

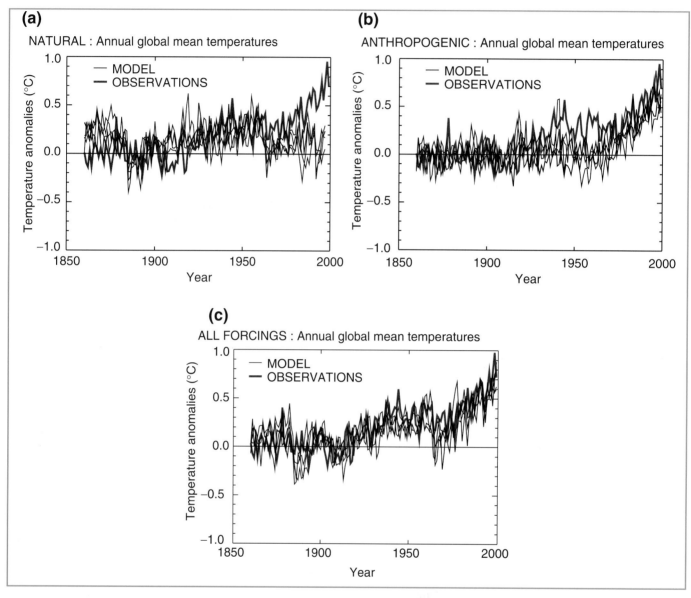

Figure 15: Global mean surface temperature anomalies relative to the 1880 to 1920 mean from the instrumental record compared with ensembles of four simulations with a coupled ocean-atmosphere climate model forced (a) with solar and volcanic forcing only, (b) with anthropogenic forcing including well mixed greenhouse gases, changes in stratospheric and tropospheric ozone and the direct and indirect effects of sulphate aerosols, and (c) with all forcings, both natural and anthropogenic. The thick line shows the instrumental data while the thin lines show the individual model simulations in the ensemble of four members. Note that the data are annual mean values. The model data are only sampled at the locations where there are observations. The changes in sulphate aerosol are calculated interactively, and changes in tropospheric ozone were calculated offline using a chemical transport model. Changes in cloud brightness (the first indirect effect of sulphate aerosols) were calculated by an off line simulation and included in the model. The changes in stratospheric ozone were based on observations. The volcanic and solar forcing were based on published combinations of measured and proxy data. The net anthropogenic forcing at 1990 was 1.0 Wm^{-2} including a net cooling of 1.0 Wm^{-2} due to sulphate aerosols. The net natural forcing for 1990 relative to 1860 was 0.5 Wm^{-2}, and for 1992 was a net cooling of 2.0 Wm^{-2} due to Mount Pinatubo. Other models forced with anthropogenic forcing give similar results to those shown in (b). [Based on Figure 12.7]

Most attribution studies find that, over the last 50 years, the estimated rate and magnitude of global warming due to increasing concentrations of greenhouse gases alone are comparable with or larger than the observed warming. Attribution studies address the question of "whether the magnitude of the simulated response to a particular forcing agent is consistent with observations". The use of multi-signal techniques has enabled studies that discriminate between the effects of different factors on climate. The inclusion of the time dependence of signals has helped to distinguish between natural and anthropogenic forcings. As more response patterns are included, the problem of degeneracy (different combinations of patterns yielding near identical fits to the observations) inevitably arises. Nevertheless, even with all the major responses that have been included in the analysis, a distinct greenhouse gas signal remains detectable. Furthermore, most model estimates that take into account both greenhouse gases and sulphate aerosols are consistent with observations over this period. The best agreement between model simulations and observations over the last 140 years is found when both anthropogenic and natural factors are included (see Figure 15). These results show that the forcings included are sufficient to explain the observed changes, but do not exclude the possibility that other forcings have also contributed. Overall, the magnitude of the temperature response to increasing concentrations of greenhouse gases is found to be consistent with observations on the scales considered (see Figure 16), but there remain discrepies between modelled and observed response to other natural and anthropogenic factors.

Uncertainties in other forcings that have been included do not prevent identification of the effect of anthropogenic greenhouse gases over the last 50 years. The sulphate forcing, while uncertain, is negative over this period. Changes in natural forcing during most of this period are also estimated to be negative. Detection of the influence of anthropogenic greenhouse gases therefore cannot be eliminated either by the uncertainty in sulphate aerosol forcing or because natural forcing has not been included in all model simulations. Studies that distinguish the separate responses to greenhouse gas, sulphate aerosol and natural forcing produce uncertain estimates of the amplitude of the sulphate aerosol and natural signals, but almost all studies are nevertheless able to detect the presence of the anthropogenic greenhouse gas signal in the recent climate record.

The detection and attribution methods used should not be sensitive to errors in the amplitude of the global mean response to individual forcings. In the signal-estimation methods used in this report, the amplitude of the signal is estimated from the observations and not the amplitude of the simulated response. Hence the estimates are independent of those factors determining the simulated amplitude of the response, such as the climate sensitivity of the model used. In addition, if the signal due to a given forcing is estimated individually, the amplitude is largely independent of the magnitude of the forcing used to derive the response. Uncertainty in the amplitude of the solar and indirect sulphate aerosol forcing should not affect the magnitude of the estimated signal.

Sea level

It is very likely that the 20th century warming has contributed significantly to the observed sea level rise, through thermal expansion of sea water and widespread loss of land ice. Within present uncertainties, observations and models are both consistent with a lack of significant acceleration of sea level rise during the 20th century.

E.7 Remaining Uncertainties in Detection and Attribution

Some progress has been made in reducing uncertainty, though many of the sources of uncertainty identified in the SAR still exist. These include:

● *Discrepancies between the vertical profile of temperature change in the troposphere seen in observations and models.* These have been reduced as more realistic forcing histories have been used in models, although not fully resolved. Also, the difference between observed surface and lower-tropospheric trends over the last two decades cannot be fully reproduced by model simulations.

● *Large uncertainties in estimates of internal climate variability from models and observations.* Although as noted above, these are unlikely (bordering on very unlikely) to be large enough to nullify the claim that a detectable climate change has taken place.

● *Considerable uncertainty in the reconstructions of solar and volcanic forcing which are based on proxy or limited observational data for all but the last two decades.* Detection of the influence of greenhouse gases on climate

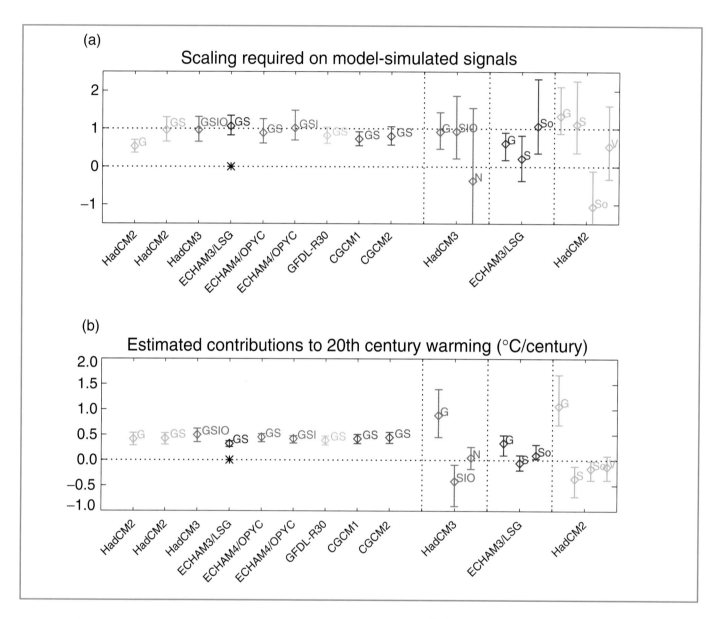

Figure 16: (a) Estimates of the "scaling factors" by which the amplitude of several model-simulated signals must be multiplied to reproduce the corresponding changes in the observed record. The vertical bars indicate the 5 to 95% uncertainty range due to internal variability. A range encompassing unity implies that this combination of forcing amplitude and model-simulated response is consistent with the corresponding observed change, while a range encompassing zero implies that this model-simulated signal is not detectable. Signals are defined as the ensemble mean response to external forcing expressed in large-scale (>5,000 km) near-surface temperatures over the 1946 to 1996 period relative to the 1896 to 1996 mean. The first entry (G) shows the scaling factor and 5 to 95% confidence interval obtained with the assumption that the observations consist only of a response to greenhouse gases plus internal variability. The range is significantly less than one (consistent with results from other models), meaning that models forced with greenhouse gases alone significantly over predict the observed warming signal. The next eight entries show scaling factors for model-simulated responses to greenhouse and sulphate forcing (GS), with two cases including indirect sulphate and tropospheric ozone forcing, one of these also including stratospheric ozone depletion (GSI and GSIO, respectively). All but one (CGCM1) of these ranges is consistent with unity. Hence there is little evidence that models are

appears to be robust to possible amplification of the solar forcing by ozone-solar or solar-cloud interactions, provided these do not alter the pattern or time-dependence of the response to solar forcing. Amplification of the solar signal by these processes, which are not yet included in models, remains speculative.

● *Large uncertainties in anthropogenic forcing are associated with the effects of aerosols.* The effects of some anthropogenic factors, including organic carbon, black carbon, biomass aerosols, and changes in land use, have not been included in detection and attribution studies. Estimates of the size and geographic pattern of the effects of these forcings vary considerably, although individually their global effects are estimated to be relatively small.

● Large differences in the response of different models to the same forcing. These differences, which are often greater than the difference in response in the same model with and without aerosol effects, highlight the large uncertainties in climate change prediction and the need to quantify uncertainty and reduce it through better observational data sets and model improvement.

E.8 Synopsis

In the light of new evidence and taking into account the remaining uncertainties, most of the observed warming over the last 50 years is likely to have been due to the increase in greenhouse gas concentrations.

systematically over- or under predicting the amplitude of the observed response under the assumption that model-simulated GS signals and internal variability are an adequate representation (i.e., that natural forcing has had little net impact on this diagnostic). Observed residual variability is consistent with this assumption in all but one case (ECHAM3, indicated by the asterisk). One is obliged to make this assumption to include models for which only a simulation of the anthropogenic response is available, but uncertainty estimates in these single signal cases are incomplete since they do not account for uncertainty in the naturally forced response. These ranges indicate, however, the high level of confidence with which internal variability, as simulated by these various models, can be rejected as an explanation of recent near-surface temperature change. A more complete uncertainty analysis is provided by the next three entries, which show corresponding scaling factors on individual greenhouse (G), sulphate (S), solar-plus-volcanic (N), solar-only (So) and volcanic-only (V) signals for those cases in which the relevant simulations have been performed. In these cases, multiple factors are estimated simultaneously to account for uncertainty in the amplitude of the naturally forced response. The uncertainties increase but the greenhouse signal remains consistently detectable. In one case (ECHAM3) the model appears to be overestimating the greenhouse response (scaling range in the G signal inconsistent with unity), but this result is sensitive to which component of the control is used to define the detection space. It is also not known how it would respond to the inclusion of a volcanic signal. In cases where both solar and volcanic forcing is included (HadCM2 and HadCM3), G and S signals remain detectable and consistent with unity independent of whether natural signals are estimated jointly or separately (allowing for different errors in S and V responses).

 (b) Estimated contributions to global mean warming over the 20th century, based on the results shown in (a), with 5 to 95% confidence intervals. Although the estimates vary depending on which model's signal and what forcing is assumed, and are less certain if more than one signal is estimated, all show a significant contribution from anthropogenic climate change to 20th century warming. [Based on Figure 12.12]

F. The Projections of the Earth's Future Climate

The tools of climate models are used with future scenarios of forcing agents (e.g., greenhouse gases and aerosols) as input to make a suite of projected future climate changes that illustrates the possibilities that could lie ahead. Section F.1 provides a description of the future scenarios of forcing agents given in the IPCC Special Report on Emission Scenarios (SRES) on which, wherever possible, the future changes presented in this section are based. Sections F.2 to F.9 present the resulting projections of changes to the future climate. Finally, Section F.10 presents the results of future projections based on scenarios of a future where greenhouse gas concentrations are stabilised.

F.1 The IPCC Special Report on Emissions Scenarios (SRES)

In 1996, the IPCC began the development of a new set of emissions scenarios, effectively to update and replace the well-known IS92 scenarios. The approved new set of scenarios is described in the IPCC Special Report on Emission Scenarios (SRES). Four different narrative storylines were developed to describe consistently the relationships between the forces driving emissions and their evolution and to add context for the scenario quantification. The resulting set of 40 scenarios (35 of which contain data on the full range of gases required to force climate models) cover a wide range of the main demographic, economic and technological driving forces of future greenhouse gas and sulphur emissions. Each scenario represents a specific quantification of one of the four storylines. All the scenarios based on the same storyline constitute a scenario "family" (See Box 5, which briefly describes the main characteristics of the four SRES storylines and scenario families). The SRES scenarios do not include additional climate initiatives, which means that no scenarios are included that explicitly assume implementation of the United Nations Framework Convention on Climate Change or the emissions targets of the Kyoto Protocol. However, greenhouse gas emissions are directly affected by non-climate change policies designed for a wide range of other purposes (e.g., air quality). Furthermore, government policies can, to varying degrees, influence the greenhouse gas emission drivers, such as

demographic change, social and economic development, technological change, resource use, and pollution management. This influence is broadly reflected in the storylines and resulting scenarios.

Since the SRES was not approved until 15 March 2000, it was too late for the modelling community to incorporate the final approved scenarios in their models and have the results available in time for this Third Assessment Report. However, draft scenarios were released to climate modellers earlier to facilitate their input to the Third Assessment Report, in accordance with a decision of the IPCC Bureau in 1998. At that time, one marker scenario was chosen from each of four of the scenario groups based directly on the storylines (A1B, A2, B1, and B2). The choice of the markers was based on which of the initial quantifications best reflected the storyline and features of specific models. Marker scenarios are no more or less likely than any other scenarios, but are considered illustrative of a particular storyline. Scenarios were also selected later to illustrate the other two scenario groups (A1FI and A1T) within the A1 family, which specifically explore alternative technology developments, holding the other driving forces constant. Hence there is an illustrative scenario for each of the six scenario groups, and all are equally plausible. Since the latter two illustrative scenarios were selected at a late stage in the process, the AOGCM modelling results presented in this report only use two of the four draft marker scenarios. At present, only scenarios A2 and B2 have been integrated by more than one AOGCM. The AOGCM results have been augmented by results from simple climate models that cover all six illustrative scenarios. The IS92a scenario is also presented in a number of cases to provide direct comparison with the results presented in the SAR.

The final four marker scenarios contained in the SRES differ in minor ways from the draft scenarios used for the AOGCM experiments described in this report. In order to ascertain the likely effect of differences in the draft and final SRES scenarios, each of the four draft and final marker scenarios were studied using a simple climate model. For three of the four marker scenarios (A1B, A2, and B2) temperature change from the draft and marker scenarios are very similar. The primary difference is a change to the standardised values for 1990 to 2000, which is common to all these scenarios. This results in a higher forcing early in the period.

Box 5: The Emissions Scenarios of the Special Report on Emissions Scenarios (SRES)

A1. The A1 storyline and scenario family describes a future world of very rapid economic growth, global population that peaks in mid-century and declines thereafter, and the rapid introduction of new and more efficient technologies. Major underlying themes are convergence among regions, capacity building and increased cultural and social interactions, with a substantial reduction in regional differences in per capita income. The A1 scenario family develops into three groups that describe alternative directions of technological change in the energy system. The three A1 groups are distinguished by their technological emphasis: fossil intensive (A1FI), non-fossil energy sources (A1T), or a balance across all sources (A1B) (where balanced is defined as not relying too heavily on one particular energy source, on the assumption that similar improvement rates apply to all energy supply and end-use technologies).

A2. The A2 storyline and scenario family describes a very heterogeneous world. The underlying theme is self-reliance and preservation of local identities. Fertility patterns across regions converge very slowly, which results in continuously increasing population. Economic development is primarily regionally oriented and per capita economic growth and techno-logical change more fragmented and slower than other storylines.

B1. The B1 storyline and scenario family describes a convergent world with the same global population, that peaks in mid-century and declines thereafter, as in the A1 storyline, but with rapid change in economic structures toward a service and information economy, with reductions in material intensity and the introduction of clean and resource-efficient technologies. The emphasis is on global solutions to economic, social and environmental sustainability, including improved equity, but without additional climate initiatives.

B2. The B2 storyline and scenario family describes a world in which the emphasis is on local solutions to economic, social and environmental sustainability. It is a world with continuously increasing global population, at a rate lower than A2, intermediate levels of economic development, and less rapid and more diverse technological change than in the A1 and B1 storylines. While the scenario is also oriented towards environmental protection and social equity, it focuses on local and regional levels.

There are further small differences in net forcing, but these decrease until, by 2100, differences in temperature change in the two versions of these scenarios are in the range 1 to 2%. For the B1 scenario, however, temperature change is significantly lower in the final version, leading to a difference in the temperature change in 2100 of almost 20%, as a result of generally lower emissions across the full range of greenhouse gases.

Anthropogenic emissions of the three main greenhouse gases, CO_2, CH_4 and N_2O, together with anthropogenic sulphur dioxide emissions, are shown for the six illustrative SRES scenarios in Figure 17. It is evident that these scenarios encompass a wide range of emissions. For comparison, emissions are also shown for IS92a. Particularly noteworthy are the much lower future sulphur dioxide emissions for the six SRES scenarios, compared to the IS92 scenarios, due to structural changes in the energy system as well as concerns about local and regional air pollution.

F.2 Projections of Future Changes in Greenhouse Gases and Aerosols

Models indicate that the illustrative SRES scenarios lead to very different CO_2 concentration trajectories (see Figure 18). By 2100, carbon cycle models project atmospheric CO_2 concentrations of 540 to 970 ppm for the illustrative SRES scenarios (90 to 250% above the concentration of 280 ppm in 1750). The net effect of land and ocean climate feedbacks as indicated by models is to further increase projected atmospheric CO_2 concentrations by reducing both the ocean and land uptake of CO_2. These projections include the land and ocean climate feedbacks. Uncertainties, especially about the magnitude of the climate feedback from the terrestrial biosphere, cause a variation of about −10 to +30% around each scenario. The total range is 490 to 1260 ppm (75 to 350% above the 1750 concentration).

Measures to enhance carbon storage in terrestrial ecosystems could influence atmospheric CO_2 concentration, but the upper bound for reduction of CO_2 concentration by such means is 40

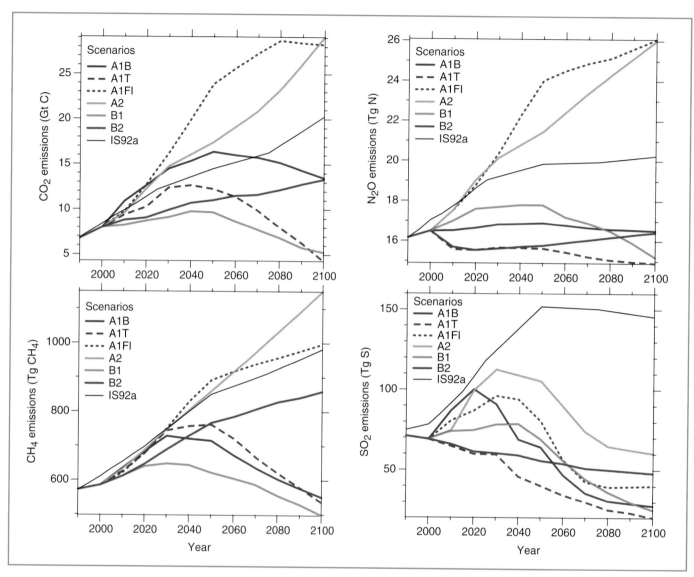

Figure 17: Anthropogenic emissions of CO_2, CH_4, N_2O and sulphur dioxide for the six illustrative SRES scenarios, A1B, A2, B1 and B2, A1FI and A1T. For comparison the IS92a scenario is also shown. [Based on IPCC Special Report on Emissions Scenarios.]

to 70 ppm. If all the carbon released by historic land-use changes could be restored to the terrestrial biosphere over the course of the century (e.g., by reforestation), CO_2 concentration would be reduced by 40 to 70 ppm. Thus, fossil fuel CO_2 emissions are virtually certain to remain the dominant control over trends in atmospheric CO_2 concentration during this century.

Model calculations of the abundances of the primary non-CO_2 greenhouse gases by the year 2100 vary considerably across the six illustrative SRES scenarios. In general A1B, A1T and B1 have the smallest increases, and A1FI and A2, the largest. The CH_4 changes from 1998 to 2100 range from −190 to +1970 ppb (−11 to +112%), and N_2O increases from +38 to +144 ppb (+12 to +46%) (see Figures 17b and c). The HFCs (134a, 143a, and 125) reach abundances of a few hundred to a thousand ppt from negligible levels today. The PFC CF4 is projected to increase to 200 to 400 ppt, and SF_6 is projected to increase to 35 to 65 ppt.

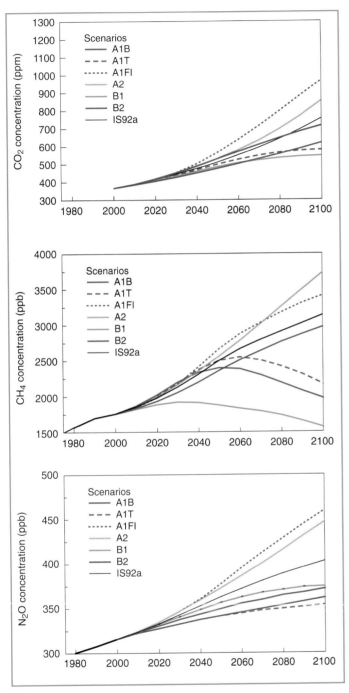

Figure 18: Atmospheric concentrations of CO_2, CH_4 and N_2O resulting from the six SRES scenarios and from the IS92a scenario computed with current methodology. [Based on Figures 3.12 and 4.14]

For the six illustrative SRES emissions scenarios, projected emissions of indirect greenhouse gases (NO_x, CO, VOC), together with changes in CH_4, are projected to change the global mean abundance of the tropospheric hydroxyl radical (OH), by –20% to +6% over the next century. Because of the importance of OH in tropospheric chemistry, comparable, but opposite sign, changes occur in the atmospheric lifetimes of the greenhouse gases CH_4 and HFCs. This impact depends in large part on the magnitude of and the balance between NO_x and CO emissions. Changes in tropospheric O_3 of –12 to +62% are calculated from 2000 until 2100. The largest increase predicted for the 21st century is for scenarios A1FI and A2 and would be more than twice as large as that experienced since the Pre-industrial Era. These O_3 increases are attributable to the concurrent and large increases in anthropogenic NO_x and CH_4 emissions.

The large growth in emissions of greenhouse gases and other pollutants as projected in some of the six illustrative SRES scenarios for the 21st century will degrade the global environment in ways beyond climate change. Changes projected in the SRES A2 and A1FI scenarios would degrade air quality over much of the globe by increasing background levels of tropospheric O_3. In northern mid-latitudes during summer, the zonal average of O_3 increases near the surface are about 30 ppb or more, raising background levels to about 80 ppb, threatening the attainment of current air quality standards over most metropolitan and even rural regions and compromising crop and forest productivity. This problem reaches across continental boundaries and couples emissions of NO_x on a hemispheric scale.

Except for sulphate and black carbon, models show an approximately linear dependence of the abundance of aerosols on emissions. The processes that determine the removal rate for black carbon differ substantially between the models, leading to major uncertainty in the future projections of black carbon. Emissions of natural aerosols such as sea salt, dust, and gas phase precursors of aerosols such as terpenes, sulphur dioxide (SO_2), and dimethyl sulphide oxidation may increase as a result of changes in climate and atmospheric chemistry.

The six illustrative SRES scenarios cover nearly the full range of forcing that results from the full set of SRES scenarios. Estimated total historical anthropogenic radiative forcing from 1765 to 1990 followed by forcing resulting from the six

SRES scenarios are shown in Figure 19. The forcing from the full range of 35 SRES scenarios is shown on the figure as a shaded envelope, since the forcings resulting from individual scenarios cross with time. The direct forcing from biomass-burning aerosols is scaled with deforestation rates. The SRES scenarios include the possibility of either increases or decreases in anthropogenic aerosols (e.g., sulphate aerosols, biomass aerosols, and black and organic carbon aerosols), depending on the extent of fossil fuel use and policies to abate polluting emissions. The SRES scenarios do not include emissions estimates for non-sulphate aerosols. Two methods for projecting these emissions were considered in this report: the first scales the emissions of fossil fuel and biomass aerosols with CO while the second scales the emissions with SO_2 and deforestation. Only the second method was used for climate projections. For comparison, radiative forcing is also shown for the IS92a scenario. It is evident that the range for the new SRES scenarios is shifted higher compared to the IS92 scenarios. This is mainly due to the reduced future SO_2 emissions of the SRES scenarios compared to the IS92 scenarios, but also to the slightly larger cumulative carbon emissions featured in some SRES scenarios.

In almost all SRES scenarios, the radiative forcing due to CO_2, CH_4, N_2O and tropospheric O_3 continue to increase, with the fraction of the total radiative forcing due to CO_2 projected to increase from slightly more than half to about

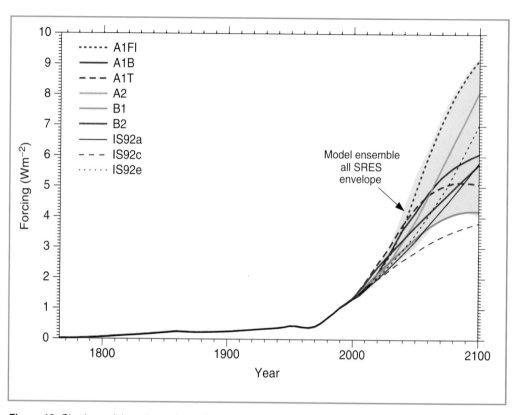

Figure 19: Simple model results: estimated historical anthropogenic radiative forcing up to the year 2000 followed by radiative forcing for the six illustrative SRES scenarios. The shading shows the envelope of forcing that encompasses the full set of thirty five SRES scenarios. The method of calculation closely follows that explained in the chapters. The values are based on the radiative forcing for a doubling of CO_2 from seven AOGCMs. The IS92a, IS92c, and IS92e forcing is also shown following the same method of calculation. [Based on Figure 9.13a]

three-quarters of the total. The radiative forcing due to O_3-depleting gases decreases due to the introduction of emission controls aimed at curbing stratospheric ozone depletion. The direct aerosol (sulphate and black and organic carbon components taken together) radiative forcing (evaluated relative to present day, 2000) varies in sign for the different scenarios. The direct plus indirect aerosol effects are projected to be smaller in magnitude than that of CO_2. No estimates are made for the spatial aspects of the future forcings. The indirect effect of aerosols on clouds is included in simple climate model calculations and scaled non-linearly with SO_2 emissions, assuming a present day value of -0.8 Wm^{-2}, as in the SAR.

F.3 Projections of Future Changes in Temperature

AOGCM results

Climate sensitivity is likely to be in the range of 1.5 to 4.5°C. This estimate is unchanged from the first IPCC Assessment Report in 1990 and the SAR. The climate sensitivity is the equilibrium response of global surface temperature to a doubling of equivalent CO_2 concentration. The range of estimates arises from uncertainties in the climate models and their internal feedbacks, particularly those related to clouds and related processes. Used for the first time in this IPCC report is the Transient Climate Response (TCR). The TCR is defined as the globally averaged surface air temperature change, at the time of doubling of CO_2, in a 1%/yr CO_2-increase experiment. This rate of CO_2 increase is assumed to represent the radiative forcing from all greenhouse gases. The TCR combines elements of model sensitivity and factors that affect response (e.g., ocean heat uptake). The range of the TCR for current AOGCMs is 1.1 to 3.1°C.

Including the direct effect of sulphate aerosols reduces global mean mid-21st century warming. The surface temperature response pattern for a given model, with and without sulphate aerosols, is more similar than the pattern between two models using the same forcing.

Models project changes in several broad-scale climate variables. As the radiative forcing of the climate system changes, the land warms faster and more than the ocean, and there is greater relative warming at high latitudes. Models project a smaller surface air temperature increase in the North Atlantic and circumpolar southern ocean regions relative to the global mean. There is projected to be a decrease in diurnal temperature range in many areas, with night-time lows increasing more than daytime highs. A number of models show a general decrease of daily variability of surface air temperature in winter and increased daily variability in summer in the Northern Hemisphere land areas. As the climate warms, the Northern Hemisphere snow cover and sea-ice extent are projected to decrease. Many of these changes are consistent with recent observational trends, as noted in Section B.

Multi-model ensembles of AOGCM simulations for a range of scenarios are being used to quantify the mean climate change and uncertainty based on the range of model results. For the end of the 21st century (2071 to 2100), the mean change in global average surface air temperature, relative to the period 1961 to 1990, is 3.0°C (with a range of 1.3 to 4.5°C) for the A2 draft marker scenario and 2.2°C (with a range of 0.9 to 3.4°C) for the B2 draft marker scenario. The B2 scenario produces a smaller warming that is consistent with its lower rate of increased CO_2 concentration.

On time-scales of a few decades, the current observed rate of warming can be used to constrain the projected response to a given emissions scenario despite uncertainty in climate sensitivity. Analysis of simple models and intercomparisons of AOGCM responses to idealised forcing scenarios suggest that, for most scenarios over the coming decades, errors in large-scale temperature projections are likely to increase in proportion to the magnitude of the overall response. The estimated size of and uncertainty in current observed warming rates attributable to human influence thus provides a relatively model-independent estimate of uncertainty in multi-decade projections under most scenarios. To be consistent with recent observations, anthropogenic warming is likely to lie in the range 0.1 to 0.2°C/decade over the next few decades under the IS92a scenario. This is similar to the range of responses to this scenario based on the seven versions of the simple model used in Figure 22.

Most of the features of the geographical response in the SRES scenario experiments are similar for different scenarios (see Figure 20) and are similar to those for idealised 1% CO_2-increase integrations. The biggest difference between the 1% CO_2-increase experiments, which have no sulphate aerosol, and the SRES experiments is the regional moderating of the warming over industrialised areas, in the SRES experiments, where the negative forcing from sulphate aerosols is greatest. This regional effect was noted in the SAR for only two models, but this has now been shown to be a consistent response across the greater number of more recent models.

It is very likely that nearly all land areas will warm more rapidly than the global average, particularly those at northern high latitudes in the cold season. Results (see Figure 21) from recent AOGCM simulations forced with SRES A2 and B2 emissions scenarios indicate that in winter the warming for all high-latitude northern regions exceeds the global mean warming in each model by more than 40% (1.3 to 6.3°C for the range of models and scenarios considered). In summer, warming is in excess of 40% above the global mean change in central and northern Asia. Only in south Asia and southern South America in June/July/August, and Southeast Asia for both seasons, do the models consistently show warming less than the global average.

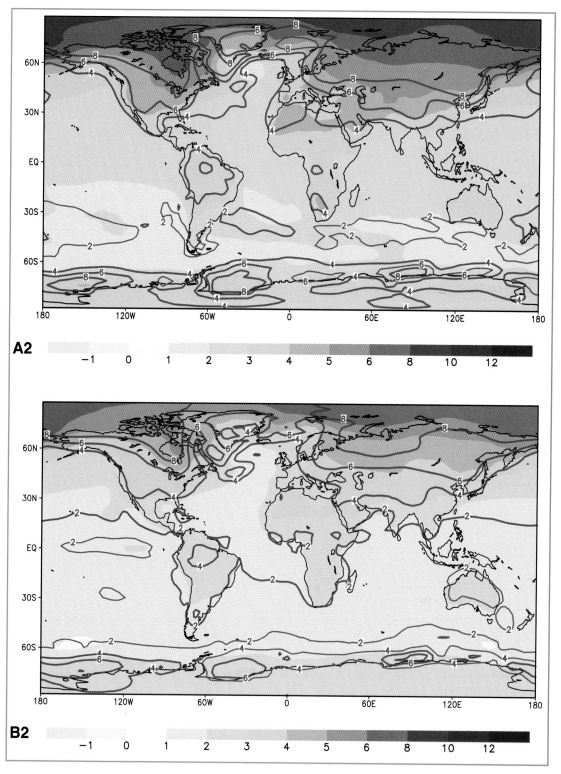

A2

B2

Figure 20: The annual mean change of the temperature (colour shading) and its range (isolines) (Unit: °C) for the SRES scenario A2 (upper panel) and the SRES scenario B2 (lower panel). Both SRES scenarios show the period 2071 to 2100 relative to the period 1961 to 1990 and were performed by OAGCMs. [Based on Figures 9.10d and 9.10e]

Simple climate model results

Due to computational expense, AOGCMs can only be run for a limited number of scenarios. A simple model can be calibrated to represent globally averaged AOGCM responses and run for a much larger number of scenarios.

The globally averaged surface temperature is projected to increase by 1.4 to 5.8°C (Figure 22(a)) over the period 1990 to 2100. These results are for the full range of 35 SRES scenarios, based on a number of climate models.[6,7] Temperature increases are projected to be greater than those in the SAR, which were about 1.0 to 3.5°C based on six IS92 scenarios. The higher

projected temperatures and the wider range are due primarily to the lower projected SO_2 emissions in the SRES scenarios relative to the IS92 scenarios. The projected rate of warming is much larger than the observed changes during the 20th century and is very likely to be without precedent during at least the last 10,000 years, based on palaeoclimate data.

The relative ranking of the SRES scenarios in terms of global mean temperature changes with time. In particular, for scenarios with higher fossil fuel use (hence, higher carbon dioxide emissions, e.g., A2), the SO_2 emissions are also higher. In the near term (to around 2050), the cooling effect of

---

[6] Complex physically based climate models are the main tool for projecting future climate change. In order to explore the range of scenarios, these are complemented by simple climate models calibrated to yield an equivalent response in temperature and sea level to complex climate models. These projections are obtained using a simple climate model whose climate sensitivity and ocean heat uptake are calibrated to each of 7 complex climate models. The climate sensitivity used in the simple model ranges from 1.7 to 4.2°C, which is comparable to the commonly accepted range of 1.5 to 4.5°C.

[7] This range does not include uncertainties in the modelling of radiative forcing, e.g. aerosol forcing uncertainties. A small carbon cycle climate feedback is included.

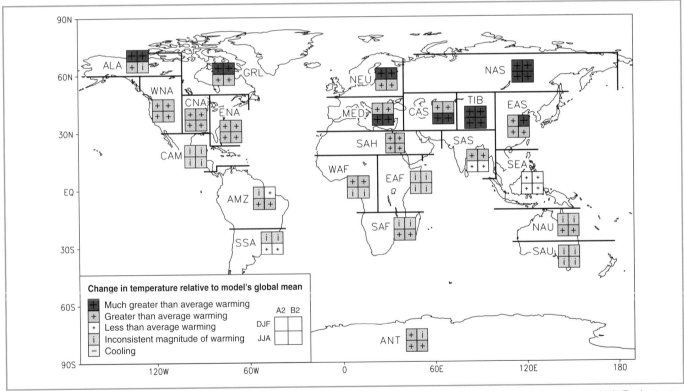

Figure 21: Analysis of inter-model consistency in regional relative warming (warming relative to each model's global average warming). Regions are classified as showing either agreement on warming in excess of 40% above the global average ('Much greater than average warming'), agreement on warming greater than the global average ('Greater than average warming'), agreement on warming less than the global average ('Less than average warming'), or disagreement amongst models on the magnitude of regional relative warming ('Inconsistent magnitude of warming'). There is also a category for agreement on cooling (which never occurs). A consistent result from at least seven of the nine models is deemed necessary for agreement. The global annual average warming of the models used span 1.2 to 4.5°C for A2 and 0.9 to 3.4°C for B2, and therefore a regional 40% amplification represents warming ranges of 1.7 to 6.3°C for A2 and 1.3 to 4.7°C for B2. [Based on Chapter 10, Box 1, Figure 1]

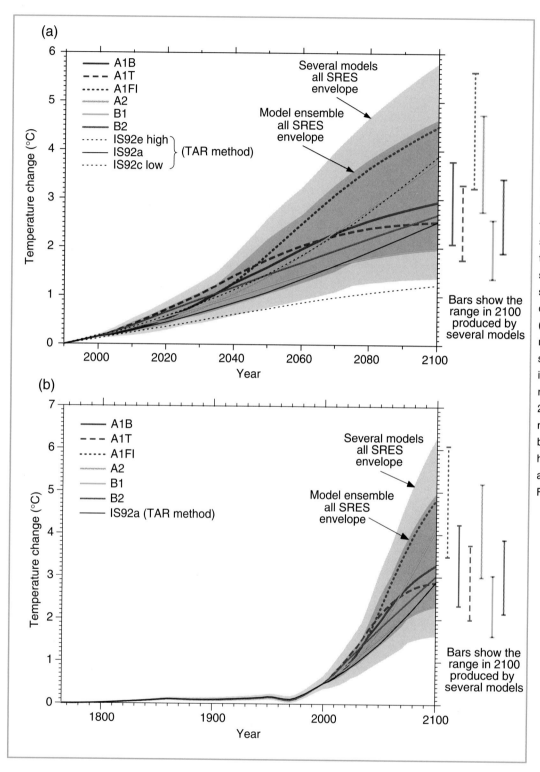

Figure 22: Simple model results: (a) global mean temperature projections for the six illustrative SRES scenarios using a simple climate model tuned to a number of complex models with a range of climate sensitivities. Also for comparison, following the same method, results are shown for IS92a. The darker shading represents the envelope of the full set of thirty-five SRES scenarios using the average of the model results (mean climate sensitivity is 2.8°C). The lighter shading is the envelope based on all seven model projections (with climate sensitivity in the range 1.7 to 4.2°C). The bars show, for each of the six illustrative SRES scenarios, the range of simple model results in 2100 for the seven AOGCM model tunings. (b) Same as (a) but results using estimated historical anthropogenic forcing are also used. [Based on Figures 9.14 and 9.13b]

higher sulphur dioxide emissions significantly reduces the warming caused by increased emissions of greenhouse gases in scenarios such as A2. The opposite effect is seen for scenarios B1 and B2, which have lower fossil fuel emissions as well as lower SO_2 emissions, and lead to a larger near-term warming. In the longer term, however, the level of emissions of long-lived greenhouse gases such as CO_2 and N_2O become the dominant determinants of the resulting climate changes.

By 2100, differences in emissions in the SRES scenarios and different climate model responses contribute similar uncertainty to the range of global temperature change. Further uncertainties arise due to uncertainties in the radiative forcing. The largest forcing uncertainty is that due to the sulphate aerosols.

F.4 Projections of Future Changes in Precipitation

Globally averaged water vapour, evaporation and precipitation are projected to increase. At the regional scale both increases and decreases in precipitation are seen. Results (see Figure 23) from recent AOGCM simulations forced with SRES A2 and B2 emissions scenarios indicate that it is likely for precipitation to increase in both summer and winter over high-latitude regions. In winter, increases are also seen over northern mid-latitudes, tropical Africa and Antarctica, and in summer in southern and eastern Asia. Australia, central America, and southern Africa show consistent decreases in winter rainfall.

Based on patterns emerging from a limited number of studies with current AOGCMs, older GCMs, and regionalisation studies, there is a strong correlation between precipitation

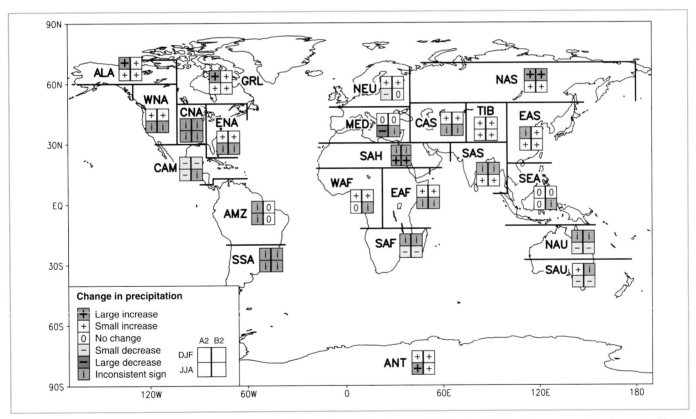

Figure 23: Analysis of inter-model consistency in regional precipitation change. Regions are classified as showing either agreement on increase with an average change of greater than 20% ('Large increase'), agreement on increase with an average change between 5 and 20% ('Small increase'), agreement on a change between −5 and +5% or agreement with an average change between −5 and 5% ('No change'), agreement on decrease with an average change between −5 and −20% ('Small decrease'), agreement on decrease with an average change of less than −20% ('Large decrease'), or disagreement ('Inconsistent sign'). A consistent result from at least seven of the nine models is deemed necessary for agreement. [Based on Chapter 10, Box 1, Figure 2]

interannual variability and mean precipitation. Future increases in mean precipitation will likely lead to increases in variability. Conversely, precipitation variability will likely decrease only in areas of reduced mean precipitation.

F.5 Projections of Future Changes in Extreme Events

It is only recently that changes in extremes of weather and climate observed to date have been compared to changes projected by models (Table 4). More hot days and heat waves are very likely over nearly all land areas. These increases are projected to be largest mainly in areas where soil moisture decreases occur. Increases in daily minimum temperature are

projected to occur over nearly all land areas and are generally larger where snow and ice retreat. Frost days and cold waves are very likely to become fewer. The changes in surface air temperature and surface absolute humidity are projected to result in increases in the heat index (which is a measure of the combined effects of temperature and moisture). The increases in surface air temperature are also projected to result in an increase in the "cooling degree days" (which is a measure of the amount of cooling required on a given day once the temperature exceeds a given threshold) and a decrease in "heating degree days". Precipitation extremes are projected to increase more than the mean and the intensity of precipitation events are projected to increase. The frequency of extreme

Table 4: Estimates of confidence in observed and projected changes in extreme weather and climate events. The table depicts an assessment of confidence in observed changes in extremes of weather and climate during the latter half of the 20th century (left column) and in projected changes during the 21st century (right column)[a]. This assessment relies on observational and modelling studies, as well as physical plausibility of future projections across all commonly used scenarios and is based on expert judgement (see Footnote 4). [Based upon Table 9.6]

Confidence in observed changes (latter half of the 20th century)	Changes in Phenomenon	Confidence in projected changes (during the 21st century)
Likely	**Higher maximum temperatures and more hot days over nearly all land areas**	Very likely
Very likely	**Higher minimum temperatures, fewer cold days and frost days over nearly all land areas**	Very likely
Very likely	**Reduced diurnal temperature range over most land areas**	Very likely
Likely, over many areas	**Increase of heat index[8] over land areas**	Very likely, over most areas
Likely, over many Northern Hemisphere mid- to high latitude land areas	**More intense precipitation events[b]**	Very likely, over many areas
Likely, in a few areas	**Increased summer continental drying and associated risk of drought**	Likely, over most mid-latitude continental interiors (Lack of consistent projections in other areas)
Not observed in the few analyses available	**Increase in tropical cyclone peak wind intensities[c]**	Likely, over some areas
Insufficient data for assessment	**Increase in tropical cyclone mean and peak precipitation intensities[c]**	Likely, over some areas

[a] For more details see Chapter 2 (observations) and Chapters 9, 10 (projections).

[b] For other areas there are either insufficient data or conflicting analyses.

[c] Past and future changes in tropical cyclone location and frequency are uncertain.

[8] Heat index: A combination of temperature and humidity that measures effects on human comfort

precipitation events is projected to increase almost everywhere. There is projected to be a general drying of the mid-continental areas during summer. This is ascribed to a combination of increased temperature and potential evaporation that is not balanced by increases of precipitation. There is little agreement yet among models concerning future changes in mid-latitude storm intensity, frequency, and variability. There is little consistent evidence that shows changes in the projected frequency of tropical cyclones and areas of formation. However, some measures of intensities show projected increases, and some theoretical and modelling studies suggest that the upper limit of these intensities could increase. Mean and peak precipitation intensities from tropical cyclones are likely to increase appreciably.

For some other extreme phenomena, many of which may have important impacts on the environment and society, there is currently insufficient information to assess recent trends, and confidence in models and understanding is inadequate to make firm projections. In particular, very small-scale phenomena such as thunderstorms, tornadoes, hail, and lightning are not simulated in global models. Insufficient analysis has occurred of how extra-tropical cyclones may change.

F.6 Projections of Future Changes in Thermohaline Circulation

Most models show weakening of the Northern Hemisphere Thermohaline Circulation (THC), which contributes to a reduction of the surface warming in the northern North Atlantic. Even in models where the THC weakens, there is still a warming over Europe due to increased greenhouse gases. In experiments where the atmospheric greenhouse gas concentration is stabilised at twice its present day value, the North Atlantic THC is projected to recover from initial weakening within one to several centuries. The THC could collapse entirely in either hemisphere if the rate of change in radiative forcing is large enough and applied long enough. Models indicate that a decrease of the THC reduces its resilience to perturbations, i.e., a once reduced THC appears to be less stable and a shut-down can become more likely. However, it is too early to say with confidence whether an irreversible collapse in the THC is likely or not, or at what threshold it might occur and what the climate implications could be. None of the current projections with coupled models exhibits a complete shut-down of the THC by 2100. Although the North

Atlantic THC weakens in most models, the relative roles of surface heat and fresh water fluxes vary from model to model. Wind stress changes appear to play only a minor role in the transient response.

F.7 Projections of Future Changes in Modes of Natural Variability

Many models show a mean El Niño-like response in the tropical Pacific, with the central and eastern equatorial Pacific sea surface temperatures projected to warm more than the western equatorial Pacific and with a corresponding mean eastward shift of precipitation. Although many models show an El Niño-like change of the mean state of tropical Pacific sea surface temperatures, the cause is uncertain. It has been related to changes in the cloud radiative forcing and/or evaporative damping of the east-west sea surface temperature gradient in some models. Confidence in projections of changes in future frequency, amplitude, and spatial pattern of El Niño events in the tropical Pacific is tempered by some shortcomings in how well El Niño is simulated in complex models. Current projections show little change or a small increase in amplitude for El Niño events over the next 100 years. However, even with little or no change in El Niño amplitude, global warming is likely to lead to greater extremes of drying and heavy rainfall and increase the risk of droughts and floods that occur with El Niño events in many regions. It also is likely that warming associated with increasing greenhouse gas concentrations will cause an increase of Asian summer monsoon precipitation variability. Changes in monsoon mean duration and strength depend on the details of the emission scenario. The confidence in such projections is limited by how well the climate models simulate the detailed seasonal evolution of the monsoons. There is no clear agreement on changes in frequency or structure of naturally occurring modes of variability, such as the North Atlantic Oscillation, i.e., the magnitude and character of the changes vary across the models.

F.8 Projections of Future Changes in Land Ice (Glaciers, Ice Caps and Ice Sheets), Sea Ice and Snow Cover

Glaciers and ice caps will continue their widespread retreat during the 21st century and Northern Hemisphere snow cover and sea ice are projected to decrease further. Methods have been developed recently for estimating glacier melt from seasonally and geographically dependent patterns of surface air temperature change, that are obtained from AOGCM experiments. Modelling studies suggest that the evolution of glacial mass is controlled principally by temperature changes, rather than precipitation changes, on the global average.

The Antarctic ice sheet is likely to gain mass because of greater precipitation, while the Greenland ice sheet is likely to lose mass because the increase in runoff will exceed the precipitation increase. The West Antarctic Ice Sheet (WAIS)

has attracted special attention because it contains enough ice to raise sea level by 6 m and because of suggestions that instabilities associated with its being grounded below sea level may result in rapid ice discharge when the surrounding ice shelves are weakened. However, loss of grounded ice leading to substantial sea level rise from this source is now widely agreed to be very unlikely during the 21st century, although its dynamics are still inadequately understood, especially for projections on longer time-scales.

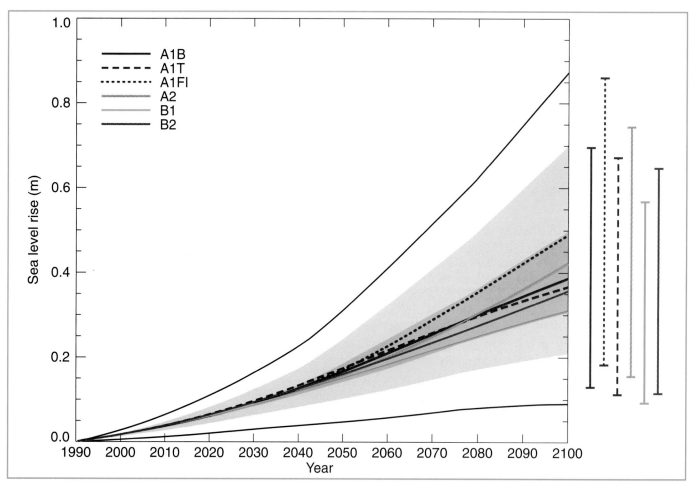

Figure 24: Global average sea level rise 1990 to 2100 for the SRES scenarios. Thermal expansion and land ice changes were calculated using a simple climate model calibrated separately for each of seven AOGCMs, and contributions from changes in permafrost, the effect of sediment deposition and the long-term adjustment of the ice sheets to past climate change were added. Each of the six lines appearing in the key is the average of AOGCMs for one of the six illustrative scenarios. The region in dark shading shows the range of the average of AOGCMs for all thirty five SRES scenarios. The region in light shading shows the range of all AOGCMs for all thirty five scenarios. The region delimited by the outermost lines shows the range of all AOGCMs and scenarios including uncertainty in land-ice changes, permafrost changes and sediment deposition. Note that this range does not allow for uncertainty relating to ice-dynamic changes in the West Antarctic ice sheet. [Based on Figure 11.12]

F.9 Projections of Future Changes in Sea Level

Projections of global average sea level rise from 1990 to 2100, using a range of AOGCMs following the IS92a scenario (including the direct effect of sulphate aerosol emissions), lie in the range 0.11 to 0.77 m. This range reflects the systematic uncertainty of modelling. The main contributions to this sea level rise are:

- a thermal expansion of 0.11 to 0.43 m, accelerating through the 21st century;

- a glacier contribution of 0.01 to 0.23 m;

- a Greenland contribution of −0.02 to 0.09 m; and

- an Antarctic contribution of −0.17 to +0.02 m.

Also included in the computation of the total change are smaller contributions from thawing of permafrost, deposition of sediment, and the ongoing contributions from ice sheets as a result of climate change since the Last Glacial Maximum. To establish the range of sea level rise resulting from the choice of different SRES scenarios, results for thermal expansion and land-ice change from simple models tuned to several AOGCMs are used (as in Section F.3 for temperature).

For the full set of SRES scenarios, a sea level rise of 0.09 to 0.88 m is projected for 1990 to 2100 (see Figure 24), primarily from thermal expansion and loss of mass from glaciers and ice caps. The central value is 0.48 m, which corresponds to an average rate of about two to four times the rate over the 20th century. The range of sea level rise presented in the SAR was 0.13 to 0.94 m based on the IS92 scenarios. Despite higher temperature change projections in this assessment, the sea level projections are slightly lower, primarily due to the use of improved models which give a smaller contribution from glaciers and ice sheets. If terrestrial storage continues at its current rates, the projections could be changed by −0.21 to 0.11 m. For an average of the AOGCMs, the SRES scenarios give results that differ by 0.02 m or less for the first half of the 21st century. By 2100, they vary over a range amounting to about 50% of the central value. Beyond the 21st century, sea level rise depends strongly on the emissions scenario.

Models agree on the qualitative conclusion that the range of regional variation in sea level change is substantial compared to global average sea level rise. However, confidence in the regional distribution of sea level change from AOGCMs is low because there is little similarity between models, although nearly all models project greater than average rise in the Arctic Ocean and less than average rise in the Southern Ocean. Further, land movements, both isostatic and tectonic, will continue through the 21st century at rates that are unaffected by climate change. It can be expected that by 2100, many regions currently experiencing relative sea level fall will instead have a rising relative sea level. Lastly, extreme high water levels will occur with increasing frequency as a result of mean sea level rise. Their frequency may be further increased if storms become more frequent or severe as a result of climate change.

F.10 Projections of Future Changes in Response to CO_2 Concentration Stabilisation Profiles

Greenhouse gases and aerosols

All of the stabilisation profiles studied require CO_2 emissions to eventually drop well below current levels. Anthropogenic CO_2 emission rates that arrive at stable CO_2 concentration levels from 450 to 1,000 ppm were deduced from the prescribed CO_2 profiles (Figure 25a). The results (Figure 25b) are not substantially different from those presented in the SAR; however, the range is larger, mainly due to the range of future terrestrial carbon uptake caused by different assumptions in the models. Stabilisation at 450, 650 or 1,000 ppm would require global anthropogenic emissions to drop below 1990 levels within a few decades, about a century, or about two centuries, respectively, and continue to steadily decrease thereafter. Although there is sufficient uptake capacity in the ocean to incorporate 70 to 80% of foreseeable anthropogenic CO_2 emissions to the atmosphere, this process takes centuries due to the rate of ocean mixing. As a result, even several centuries after emissions occurred, about a quarter of the increase in concentration caused by these emissions is still present in the atmosphere. To maintain constant CO_2 concentration beyond 2300 requires emissions to drop to match the rate of carbon sinks at that time. Natural land and ocean sinks with the capacity to persist for hundreds or thousands of years are small (<0.2 PgC/yr).

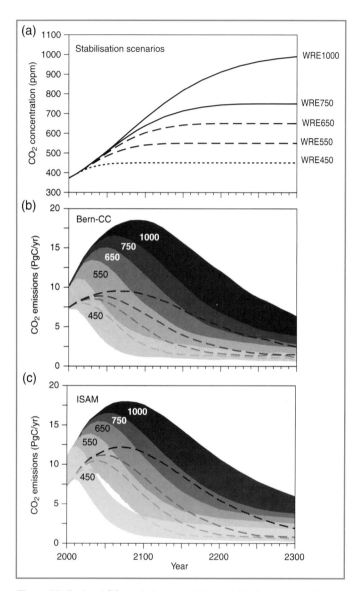

Figure 25: Projcted CO_2 emissions permitting stabilisation of atmospheric CO_2 concentrations at different final values. Panel (a) shows the assumed trajectories of CO_2 concentration (WRE scenarios) and panels (b) and (c) show the implied CO_2 emissions, as projected with two fast carbon cycle models, Bern-CC and ISAM. The model ranges for ISAM were obtained by tuning the model to approximate the range of responses to CO_2 and climate from model intercomparisons. This approach yields a lower bound on uncertainties in the carbon cycle response. The model ranges for Bern-CC were obtained by combining different bounding assumptions about the behaviour of the CO_2 fertilisation effect, the response of heterotrophic respiration to temperature and the turnover time of the ocean, thus approaching an upper bound on uncertainties in the carbon cycle response. For each model, the upper and lower bounds are indicated by the top and bottom of the shaded area. Alternatively, the lower bound (where hidden) is indicated by a hatched line. [Based on Figure 3.13]

Temperature

Global mean temperature continues to increase for hundreds of years at a rate of a few tenths of a degree per century after concentrations of CO_2 have been stabilised, due to long time-scales in the ocean. The temperature implications of CO_2 concentration profiles leading to stabilisation from 450 ppm to 1,000 ppm were studied using a simple climate model tuned to seven AOGCMs with a mean climate sensitivity of 2.8°C. For all the pathways leading to stabilisation, the climate system shows considerable warming during the 21st century and beyond (see Figure 26). The lower the level at which concentrations stabilise, the smaller the total temperature change.

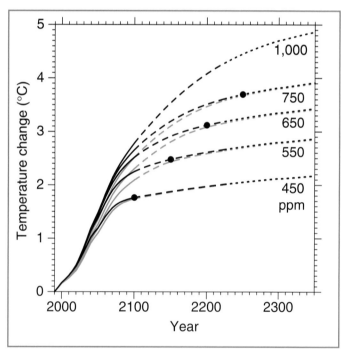

Figure 26: Simple model results: Projected global mean temperature changes when the concentration of CO_2 is stabilised following the WRE profiles (see Chapter 9 Section 9.3.3). For comparison, results based on the S profiles in the SAR are also shown in green (S1000 not available). The results are the average produced by a simple climate model tuned to seven AOGCMs. The baseline scenario is scenario A1B, this is specified only to 2100. After 2100, the emissions of gases other than CO_2 are assumed to remain constant at their A1B 2100 values. The projections are labelled according to the level of CO_2 stabilisation. The broken lines after 2100 indicate increased uncertainty in the simple climate model results beyond 2100. The black dots indicate the time of CO_2 stabilisation. The stabilisation year for the WRE1000 profile is 2375. [Based on Figure 9.16]

Sea level

If greenhouse gas concentrations were stabilised (even at present levels), sea level would nonetheless continue to rise for hundreds of years. After 500 years, sea level rise from thermal expansion may have reached only half of its eventual level, which models suggest may lie within a range of 0.5 to 2.0 m and 1 to 4 m for CO_2 levels of twice and four times pre-industrial, respectively. The long time-scale is characteristic of the weak diffusion and slow circulation processes that transport heat into the deep ocean.

The loss of a substantial fraction of the total glacier mass is likely. Areas that are currently marginally glaciated are most likely to become ice-free.

Ice sheets will continue to react to climatic change during the next several thousand years, even if the climate is stabilised. Together, the present Antarctic and Greenland ice sheets contain enough water to raise sea level by almost 70 m if they were to melt, so that only a small fractional change in their volume would have a significant effect.

Models project that a local annual average warming of larger than 3°C, sustained for millennia, would lead to virtually a complete melting of the Greenland ice sheet with a reulting sea level rise of about 7 m. Projected temperatures over Greenland are generally greater than globally averaged temperatures by a factor of 1.2 to 3.1 for the range of models

used in Chapter 11. For a warming over Greenland of 5.5°C, consistent with mid-range stabilisation scenarios (see Figure 26), the Greenland ice sheet is likely to contribute about 3 m in 1,000 years. For a warming of 8°C, the contribution is about 6 m, the ice sheet being largely eliminated. For smaller warmings, the decay of the ice sheet would be substantially slower (see Figure 27).

Current ice dynamic models project that the West Antarctic ice sheet (WAIS) will contribute no more than 3 mm/yr to sea level rise over the next thousand years, even if significant changes were to occur in the ice shelves. Such results are strongly dependent on model assumptions regarding climate change scenarios, ice dynamics and other factors. Apart from the possibility of an internal ice dynamic instability, surface melting will affect the long-term viability of the Antarctic ice sheet. For warmings of more than 10°C, simple runoff models predict that a zone of net mass loss would develop on the ice sheet surface. Irreversible disintegration of the WAIS would result because the WAIS cannot retreat to higher ground once its margins are subjected to surface melting and begin to recede. Such a disintegration would take at least a few millennia. Thresholds for total disintegration of the East Antarctic ice sheet by surface melting involve warmings above 20°C, a situation that has not occurred for at least 15 million years and which is far more than predicted by any scenario of climate change currently under consideration.

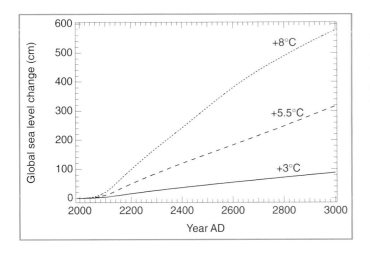

Figure 27: Response of the Greenland ice sheet to three climatic warming scenarios during the third millennium expressed in equivaient changes of global sea level. The curve labels refer to the mean annual temperature rise over Greenland by 3000 AD as predicted by a two-dimensional climate and ocean model forced by greenhouse gas concen-tration rises until 2130 AD and kept constant after that. Note that projected temperatures over Greenland are generally greater than globally averaged temperatures by a factor of 1.2 to 3.1 for the range of models used in Chapter 11. [Based on Figure 11.16]

G. Advancing Understanding

The previous sections have contained descriptions of the current state of knowledge of the climate of the past and present, the current understanding of the forcing agents and processes in the climate system and how well they can be represented in climate models. Given the knowledge possessed today, the best assessment was given whether climate change can be detected and whether that change can be attributed to human influence. With the best tools available today, projections were made of how the climate could change in the future for different scenarios of emissions of greenhouse gases.

This Section looks into the future in a different way. Uncertainties are present in each step of the chain from emissions of greenhouse gases and aerosols, through to the impacts that they have on the climate system and society (see Figure 28). Many factors continue to limit the ability to detect, attribute, and understand current climate change and to project what future climate changes may be. Further work is needed in nine broad areas.

G.1 Data

Arrest the decline of observational networks in many parts of the world. Unless networks are significantly improved, it may be difficult or impossible to detect climate change in many areas of the globe.

Expand the observational foundation for climate studies to provide accurate, long-term data with expanded temporal and spatial coverage. Given the complexity of the climate system and the inherent multi-decadal time-scale, there is a need for long-term consistent data to support climate and environmental change investigations and projections. Data from the present and recent past, climate-relevant data for the last few centuries, and for the last several millennia are all needed. There is a particular shortage of data in polar regions and data for the quantitative assessment of extremes on the global scale.

G.2 Climate Processes and Modelling

Estimate better future emissions and concentrations of greenhouse gases and aerosols. It is particularly important that improvements are realised in deriving concentrations from emissions of gases and particularly aerosols, in addressing biogeochemical sequestration and cycling, and specifically, and in determining the spatial-temporal distribution of CO_2 sources and sinks, currently and in the future.

Understand and characterise more completely dominant processes (e.g., ocean mixing) and feedbacks (e.g., from clouds and sea ice) in the atmosphere, biota, land and ocean surfaces, and deep oceans. These sub-systems, phenomena, and processes are important and merit increased attention to improve prognostic capabilities generally. The interplay of observation and models will be the key for progress. The rapid forcing of a non-linear system has a high prospect of producing surprises.

Address more completely patterns of long-term climate variability. This topic arises both in model calculations and in the climate system. In simulations, the issue of climate drift within model calculations needs to be clarified better in part because it compounds the difficulty of distinguishing signal and noise. With respect to the long-term natural variability in the climate system per se, it is important to understand this variability and to expand the emerging capability of predicting patterns of organised variability such as ENSO.

Explore more fully the probabilistic character of future climate states by developing multiple ensembles of model calculations. The climate system is a coupled non-linear chaotic system, and therefore the long-term prediction of future exact climate states is not possible. Rather the focus must be upon the prediction of the probability distribution of the system's future possible states by the generation of ensembles of model solutions.

Improve the integrated hierarchy of global and regional climate models with emphasis on improving the simulation of regional impacts and extreme weather events. This will require improvements in the understanding of the coupling between the major atmospheric, oceanic, and terrestrial systems, and extensive diagnostic modelling and observational studies that evaluate and improve simulative performance. A particularly important issue is the adequacy of data needed to attack the question of changes in extreme events.

G.3 Human Aspects

Link more formally physical climate-biogeochemical models with models of the human system and thereby provide the basis for expanded exploration of possible cause-effect-cause patterns linking human and non-human components of the Earth system. At present, human influences generally are treated only through emission scenarios that provide external forcings to the climate system. In future more comprehensive models are required in which human activities need to begin to interact with the dynamics of physical, chemical, and biological sub-systems through a diverse set of contributing activities, feedbacks and responses.

G.4 International Framework

Accelerate internationally progress in understanding climate change by strengthening the international framework that is needed to co-ordinate national and institutional efforts so that research, computational, and observational resources may be used to the greatest overall advantage. Elements of this framework exist in the international programmes supported by the International Council of Scientific Unions (ICSU), the World Meteorological Organization (WMO), the United Nations Environment Programme (UNEP), and the United Nations Education, Scientific and Cultural Organisation (UNESCO). There is a corresponding need for strengthening the co-operation within the international research community, building research capacity in many regions and, as is the goal of this assessment, effectively describing research advances in terms that are relevant to decision making.

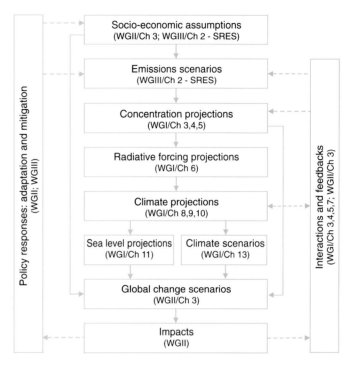

Figure 28: The cascade of uncertainties in projections to be considered in developing climate and related scenarios for climate change impact, adaptation, and mitigation assessment. [Based on Figure 13.2]

Source Information: Technical Summary

This Appendix provides the cross-reference of the topics in the Technical Summary (page and section) to the sections of the chapters that contain expanded information about the topic.

Section A: Introduction

Section B: The Observed Changes in the Climate System

Section E: The Identification of a Human Influence on Climate Change

Section F: The Projections of the Earth's Future Climate

1

The Climate System: an Overview

Co-ordinating Lead Author
A.P.M. Baede

Lead Authors
E. Ahlonsou, Y. Ding, D. Schimel

Review Editors
B. Bolin, S. Pollonais

Contents

1.1 Introduction to the Climate System

1.1.1 Climate

Weather and climate

Weather and climate have a profound influence on life on Earth. They are part of the daily experience of human beings and are essential for health, food production and well-being. Many consider the prospect of human-induced climate change as a matter of concern. The IPCC Second Assessment Report (IPCC, 1996) (hereafter SAR) presented scientific evidence that human activities may already be influencing the climate. If one wishes to understand, detect and eventually predict the human influence on climate, one needs to understand the system that determines the climate of the Earth and of the processes that lead to climate change.

In common parlance the notions "weather" and "climate" are loosely defined[1]. The "weather", as we experience it, is the fluctuating state of the atmosphere around us, characterised by the temperature, wind, precipitation, clouds and other weather elements. This weather is the result of rapidly developing and decaying weather systems such as mid-latitude low and high pressure systems with their associated frontal zones, showers and tropical cyclones. Weather has only limited predictability. Mesoscale convective systems are predictable over a period of hours only; synoptic scale cyclones may be predictable over a period of several days to a week. Beyond a week or two individual weather systems are unpredictable. "Climate" refers to the average weather in terms of the mean and its variability over a certain time-span and a certain area. Classical climatology provides a classification and description of the various climate regimes found on Earth. Climate varies from place to place, depending on latitude, distance to the sea, vegetation, presence or absence of mountains or other geographical factors. Climate varies also in time; from season to season, year to year, decade to decade or on much longer time-scales, such as the Ice Ages. Statistically significant variations of the mean state of the climate or of its variability, typically persisting for decades or longer, are referred to as "climate change". The Glossary gives definitions of these important and central notions of "climate variability" and "climate change".

Climate variations and change, caused by external forcings, may be partly predictable, particularly on the larger, continental and global, spatial scales. Because human activities, such as the emission of greenhouse gases or land-use change, do result in external forcing, it is believed that the large-scale aspects of human-induced climate change are also partly predictable. However the ability to actually do so is limited because we cannot accurately predict population change, economic change, technological development, and other relevant characteristics of future human activity. In practice, therefore, one has to rely on carefully constructed scenarios of human behaviour and determine climate projections on the basis of such scenarios.

---

[1] For a definition of scientific and technical terms used in this Report: see Appendix I: Glossary.

Climate variables

The traditional knowledge of weather and climate focuses on those variables that affect daily life most directly: average, maximum and minimum temperature, wind near the surface of the Earth, precipitation in its various forms, humidity, cloud type and amount, and solar radiation. These are the variables observed hourly by a large number of weather stations around the globe.

However this is only part of the reality that determines weather and climate. The growth, movement and decay of weather systems depend also on the vertical structure of the atmosphere, the influence of the underlying land and sea and many other factors not directly experienced by human beings. Climate is determined by the atmospheric circulation and by its interactions with the large-scale ocean currents and the land with its features such as albedo, vegetation and soil moisture. The climate of the Earth as a whole depends on factors that influence the radiative balance, such as for example, the atmospheric composition, solar radiation or volcanic eruptions. To understand the climate of our planet Earth and its variations and to understand and possibly predict the changes of the climate brought about by human activities, one cannot ignore any of these many factors and components that determine the climate. We must understand the *climate system*, the complicated system consisting of various components, including the dynamics and composition of the atmosphere, the ocean, the ice and snow cover, the land surface and its features, the many mutual interactions between them, and the large variety of physical, chemical and biological processes taking place in and among these components. "Climate" in a wider sense refers to the state of the climate system as a whole, including a statistical description of its variations. This chapter provides the reader with an overview of the climate system and the climate in this wider sense, and acts as an introduction to the Report.

1.1.2 The Climate System

Its components

The climate system, as defined in this Report, is an interactive system consisting of five major components: the atmosphere, the hydrosphere, the cryosphere, the land surface and the biosphere, forced or influenced by various external forcing mechanisms, the most important of which is the Sun (see Figure 1.1). Also the direct effect of human activities on the climate system is considered an external forcing.

The *atmosphere* is the most unstable and rapidly changing part of the system. Its composition, which has changed with the evolution of the Earth, is of central importance to the problem assessed in this Report. The Earth's dry atmosphere is composed mainly of nitrogen (N_2, 78.1% volume mixing ratio), oxygen (O_2, 20.9% volume mixing ratio, and argon (Ar, 0.93% volume mixing ratio). These gases have only limited interaction with the incoming solar radiation and they do not interact with the infrared radiation emitted by the Earth. However there are a number of trace gases, such as carbon dioxide (CO_2), methane (CH_4), nitrous oxide (N_2O) and ozone (O_3), which do absorb and emit infrared radiation. These so called greenhouse gases, with a total volume mixing ratio in dry air of less than 0.1% by volume, play an essential role in the Earth's energy budget. Moreover the

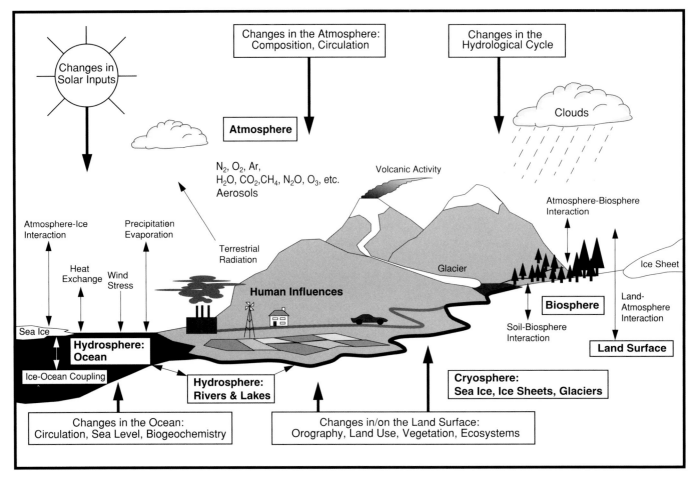

Figure 1.1: Schematic view of the components of the global climate system (bold), their processes and interactions (thin arrows) and some aspects that may change (bold arrows).

atmosphere contains water vapour (H_2O), which is also a natural greenhouse gas. Its volume mixing ratio is highly variable, but it is typically in the order of 1%. Because these greenhouse gases absorb the infrared radiation emitted by the Earth and emit infrared radiation up- and downward, they tend to raise the temperature near the Earth's surface. Water vapour, CO_2 and O_3 also absorb solar short-wave radiation.

The atmospheric distribution of ozone and its role in the Earth's energy budget is unique. Ozone in the lower part of the atmosphere, the troposphere and lower stratosphere, acts as a greenhouse gas. Higher up in the stratosphere there is a natural layer of high ozone concentration, which absorbs solar ultra-violet radiation. In this way this so-called ozone layer plays an essential role in the stratosphere's radiative balance, at the same time filtering out this potentially damaging form of radiation.

Beside these gases, the atmosphere also contains solid and liquid particles (aerosols) and clouds, which interact with the incoming and outgoing radiation in a complex and spatially very variable manner. The most variable component of the atmosphere is water in its various phases such as vapour, cloud droplets, and ice crystals. Water vapour is the strongest greenhouse gas. For these reasons and because the transition between the various phases absorb and release much energy, water vapour is central to the climate and its variability and change.

The *hydrosphere* is the component comprising all liquid surface and subterranean water, both fresh water, including rivers, lakes and aquifers, and saline water of the oceans and seas. Fresh water runoff from the land returning to the oceans in rivers influences the ocean's composition and circulation. The oceans cover approximately 70% of the Earth's surface. They store and transport a large amount of energy and dissolve and store great quantities of carbon dioxide. Their circulation, driven by the wind and by density contrasts caused by salinity and thermal gradients (the so-called thermohaline circulation), is much slower than the atmospheric circulation. Mainly due to the large thermal inertia of the oceans, they damp vast and strong temperature changes and function as a regulator of the Earth's climate and as a source of natural climate variability, in particular on the longer time-scales.

The *cryosphere*, including the ice sheets of Greenland and Antarctica, continental glaciers and snow fields, sea ice and permafrost, derives its importance to the climate system from its high reflectivity (albedo) for solar radiation, its low thermal conductivity, its large thermal inertia and, especially, its critical role in driving deep ocean water circulation. Because the ice sheets store a large amount of water, variations in their volume are a potential source of sea level variations (Chapter 11).

Vegetation and soils at the *land surface* control how energy received from the Sun is returned to the atmosphere. Some is returned as long-wave (infrared) radiation, heating the atmosphere as the land surface warms. Some serves to evaporate water, either in the soil or in the leaves of plants, bringing water back into the atmosphere. Because the evaporation of soil moisture requires energy, soil moisture has a strong influence on the surface temperature. The texture of the land surface (its roughness) influences the atmosphere dynamically as winds blow over the land's surface. Roughness is determined by both topography and vegetation. Wind also blows dust from the surface into the atmosphere, which interacts with the atmospheric radiation.

The marine and terrestrial *biospheres* have a major impact on the atmosphere's composition. The biota influence the uptake and release of greenhouse gases. Through the photosynthetic process, both marine and terrestrial plants (especially forests) store significant amounts of carbon from carbon dioxide. Thus, the biosphere plays a central role in the carbon cycle, as well as in the budgets of many other gases, such as methane and nitrous oxide. Other biospheric emissions are the so-called volatile organic compounds (VOC) which may have important effects on atmospheric chemistry, on aerosol formation and therefore on climate. Because the storage of carbon and the exchange of trace gases are influenced by climate, feedbacks between climate change and atmospheric concentrations of trace gases can occur. The influence of climate on the biosphere is preserved as fossils, tree rings, pollen and other records, so that much of what is known of past climates comes from such biotic indicators.

Interactions among the components
Many physical, chemical and biological interaction processes occur among the various components of the climate system on a wide range of space and time scales, making the system extremely complex. Although the components of the climate system are very different in their composition, physical and chemical properties, structure and behaviour, they are all linked by fluxes of mass, heat and momentum: all subsystems are open and interrelated.

As an example, the atmosphere and the oceans are strongly coupled and exchange, among others, water vapour and heat through evaporation. This is part of the hydrological cycle and leads to condensation, cloud formation, precipitation and runoff, and supplies energy to weather systems. On the other hand, precipitation has an influence on salinity, its distribution and the thermohaline circulation. Atmosphere and oceans also exchange, among other gases, carbon dioxide, maintaining a balance by dissolving it in cold polar water which sinks into the deep ocean and by outgassing in relatively warm upwelling water near the equator.

Some other examples: sea ice hinders the exchanges between atmosphere and oceans; the biosphere influences the carbon dioxide concentration by photosynthesis and respiration, which in turn is influenced by climate change. The biosphere also affects the input of water in the atmosphere through evapotranspiration, and the atmosphere's radiative balance through the amount of sunlight reflected back to the sky (albedo).

These are just a few examples from a virtually inexhaustible list of complex interactions some of which are poorly known or perhaps even unknown. Chapter 7 provides an assessment of the present knowledge of physical climate processes and feedbacks, whilst Chapter 3 deals with biological feedbacks.

Any change, whether natural or anthropogenic, in the components of the climate system and their interactions, or in the external forcing, may result in climate variations. The following sections introduce various aspects of natural climate variations, followed by an introduction to the human influence on the climate system.

1.2 Natural Climate Variations

1.2.1 Natural Forcing of the Climate System

The Sun and the global energy balance
The ultimate source of energy that drives the climate system is radiation from the Sun. About half of the radiation is in the visible short-wave part of the electromagnetic spectrum. The other half is mostly in the near-infrared part, with some in the ultraviolet part of the spectrum. Each square metre of the Earth's spherical surface outside the atmosphere receives an average throughout the year of 342 Watts of solar radiation, 31% of which is immediately reflected back into space by clouds, by the atmosphere, and by the Earth's surface. The remaining 235 Wm^{-2} is partly absorbed by the atmosphere but most (168 Wm^{-2}) warms the Earth's surface: the land and the ocean. The Earth's surface returns that heat to the atmosphere, partly as infrared radiation, partly as sensible heat and as water vapour which releases its heat when it condenses higher up in the atmosphere. This exchange of energy between surface and atmosphere maintains under present conditions a global mean temperature near the surface of 14°C, decreasing rapidly with height and reaching a mean temperature of −58°C at the top of the troposphere.

For a stable climate, a balance is required between incoming solar radiation and the outgoing radiation emitted by the climate system. Therefore the climate system itself must radiate on average 235 Wm^{-2} back into space. Details of this energy balance can be seen in Figure 1.2, which shows on the left hand side what happens with the incoming solar radiation, and on the right hand side how the atmosphere emits the outgoing infrared radiation. Any physical object radiates energy of an amount and at wavelengths typical for the temperature of the object: at higher temperatures more energy is radiated at shorter wavelengths. For the Earth to radiate 235 Wm^{-2}, it should radiate at an effective emission temperature of −19°C with typical wavelengths in the infrared part of the spectrum. This is 33°C lower than the average temperature of 14°C at the Earth's surface. To understand why this is so, one must take into account the radiative properties of the atmosphere in the infrared part of the spectrum.

The natural greenhouse effect
The atmosphere contains several trace gases which absorb and emit infrared radiation. These so-called greenhouse gases absorb infrared radiation, emitted by the Earth's surface, the atmosphere

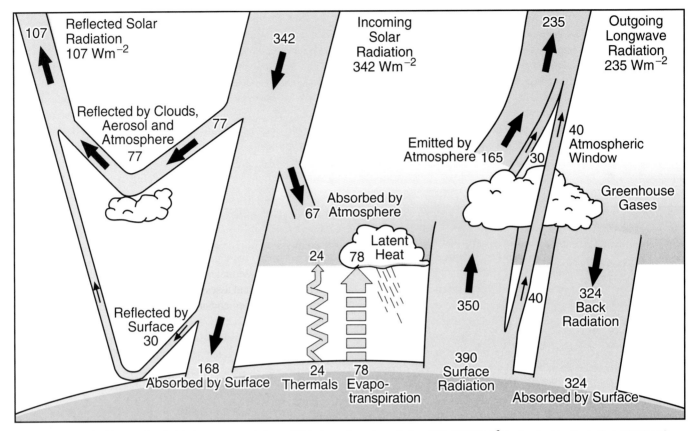

Figure 1.2: The Earth's annual and global mean energy balance. Of the incoming solar radiation, 49% (168 Wm^{-2}) is absorbed by the surface. That heat is returned to the atmosphere as sensible heat, as evapotranspiration (latent heat) and as thermal infrared radiation. Most of this radiation is absorbed by the atmosphere, which in turn emits radiation both up and down. The radiation lost to space comes from cloud tops and atmospheric regions much colder than the surface. This causes a greenhouse effect. Source: Kiehl and Trenberth, 1997: Earth's Annual Global Mean Energy Budget, *Bull. Am. Met. Soc.* 78, 197-208.

and clouds, except in a transparent part of the spectrum called the "atmospheric window", as shown in Figure 1.2. They emit in turn infrared radiation in all directions including downward to the Earth's surface. Thus greenhouse gases trap heat within the atmosphere. This mechanism is called the natural greenhouse effect. The net result is an upward transfer of infrared radiation from warmer levels near the Earth's surface to colder levels at higher altitudes. The infrared radiation is effectively radiated back into space from an altitude with a temperature of, on average, −19°C, in balance with the incoming radiation, whereas the Earth's surface is kept at a much higher temperature of on average 14°C. This effective emission temperature of −19°C corresponds in mid-latitudes with a height of approximately 5 km. Note that it is essential for the greenhouse effect that the temperature of the lower atmosphere is not constant (isothermal) but decreases with height. The natural greenhouse effect is part of the energy balance of the Earth, as can be seen schematically in Figure 1.2.

Clouds also play an important role in the Earth's energy balance and in particular in the natural greenhouse effect. Clouds absorb and emit infrared radiation and thus contribute to warming the Earth's surface, just like the greenhouse gases. On the other hand, most clouds are bright reflectors of solar radiation and tend to cool the climate system. The net average effect of the Earth's cloud cover in the present climate is a slight cooling: the

reflection of radiation more than compensates for the greenhouse effect of clouds. However this effect is highly variable, depending on height, type and optical properties of clouds.

This introduction to the global energy balance and the natural greenhouse effect is entirely in terms of the global mean and in radiative terms. However, for a full understanding of the greenhouse effect and of its impact on the climate system, dynamical feedbacks and energy transfer processes should also be taken into account. Chapter 7 presents a more detailed analysis and assessment.

Radiative forcing and forcing variability

In an equilibrium climate state the average net radiation at the top of the atmosphere is zero. A change in either the solar radiation or the infrared radiation changes the net radiation. The corresponding imbalance is called "radiative forcing". In practice, for this purpose, the top of the troposphere (the tropopause) is taken as the top of the atmosphere, because the stratosphere adjusts in a matter of months to changes in the radiative balance, whereas the surface-troposphere system adjusts much more slowly, owing principally to the large thermal inertia of the oceans. The radiative forcing of the surface troposphere system is then the change in net irradiance at the tropopause after allowing for stratospheric temperatures to re-adjust to radiative equilibrium, but with surface and tropospheric temperatures and state held fixed at the unperturbed values. A detailed explanation and discussion of the radiative forcing concept may be found in Appendix 6.1 to Chapter 6.

External forcings, such as the solar radiation or the large amounts of aerosols ejected by volcanic eruption into the atmosphere, may vary on widely different time-scales, causing natural variations in the radiative forcing. These variations may be negative or positive. In either case the climate system must react to restore the balance. A positive radiative forcing tends to warm the surface on average, whereas a negative radiative forcing tends to cool it. Internal climate processes and feedbacks may also cause variations in the radiative balance by their impact on the reflected solar radiation or emitted infrared radiation, but such variations are not considered part of radiative forcing. Chapter 6 assesses the present knowledge of radiative forcing and its variations, including the anthropogenic change of the atmospheric composition.

1.2.2 Natural Variability of Climate

Internally and externally induced climate variability
Climate variations, both in the mean state and in other statistics such as, for example, the occurrence of extreme events, may result from radiative forcing, but also from internal interactions between components of the climate system. A distinction can therefore be made between externally and internally induced natural climate variability and change.

When variations in the external forcing occur, the response time of the various components of the climate system is very different. With regard to the atmosphere, the response time of the troposphere is relatively short, from days to weeks, whereas the stratosphere comes into equilibrium on a time-scale of typically a few months. Due to their large heat capacity, the oceans have a much longer response time, typically decades but up to centuries or millennia. The response time of the strongly coupled surface-troposphere system is therefore slow compared with that of the stratosphere, and is mainly determined by the oceans. The biosphere may respond fast, e.g. to droughts, but also very slowly to imposed changes. Therefore the system may respond to variations in external forcing on a wide range of space- and time-scales. The impact of solar variations on the climate provides an example of such externally induced climate variations.

But even without changes in external forcing, the climate may vary naturally, because, in a system of components with very different response times and non-linear interactions, the components are never in equilibrium and are constantly varying. An example of such internal climate variation is the El Niño-Southern Oscillation (ENSO), resulting from the interaction between atmosphere and ocean in the tropical Pacific.

Feedbacks and non-linearities
The response of the climate to the internal variability of the climate system and to external forcings is further complicated by feedbacks and non-linear responses of the components. A process is called a feedback when the result of the process affects its origin thereby intensifying (positive feedback) or reducing (negative feedback) the original effect. An important example of a positive feedback is the water vapour feedback in which the amount of water vapour in the atmosphere increases as the Earth warms. This increase in turn may amplify the warming because water vapour is a strong greenhouse gas. A strong and very basic negative feedback is radiative damping: an increase in temperature strongly increases the amount of emitted infrared radiation. This limits and controls the original temperature increase.

A distinction is made between physical feedbacks involving physical climate processes, and biogeochemical feedbacks often involving coupled biological, geological and chemical processes. An example of a physical feedback is the complicated interaction between clouds and the radiative balance. Chapter 7 provides an overview and assessment of the present knowledge of such feedbacks. An important example of a biogeochemical feedback is the interaction between the atmospheric CO_2 concentration and the carbon uptake by the land surface and the oceans. Understanding this feedback is essential for an understanding of the carbon cycle. This is discussed and assessed in detail in Chapter 3.

Many processes and interactions in the climate system are non-linear. That means that there is no simple proportional relation between cause and effect. A complex, non-linear system may display what is technically called chaotic behaviour. This means that the behaviour of the system is critically dependent on very small changes of the initial conditions. This does not imply, however, that the behaviour of non-linear chaotic systems is entirely unpredictable, contrary to what is meant by "chaotic" in colloquial language. It has, however, consequences for the nature of its variability and the predictability of its variations. The daily weather is a good example. The evolution of weather systems responsible for the daily weather is governed by such non-linear chaotic dynamics. This does not preclude successful weather prediction, but its predictability is limited to a period of at most two weeks. Similarly, although the climate system is highly non-linear, the quasi-linear response of many models to present and predicted levels of external radiative forcing suggests that the large-scale aspects of human-induced climate change may be predictable, although as discussed in Section 1.3.2 below, unpredictable behaviour of non-linear systems can never be ruled out. The predictability of the climate system is discussed in Chapter 7.

Global and hemispheric variability
Climate varies naturally on all time-scales. During the last million years or so, glacial periods and interglacials have alternated as a result of variations in the Earth's orbital parameters. Based on Antarctic ice cores, more detailed information is available now about the four full glacial cycles during the last 500,000 years. In recent years it was discovered that during the last glacial period large and very rapid temperature variations took place over large parts of the globe, in particular in the higher latitudes of the Northern Hemisphere. These abrupt events saw temperature changes of many degrees within a human lifetime. In contrast, the last 10,000 years appear to have been relatively more stable, though locally quite large changes have occurred.

Recent analyses suggest that the Northern Hemisphere climate of the past 1,000 years was characterised by an irregular but steady cooling, followed by a strong warming during the 20th century. Temperatures were relatively warm during the 11th to 13th centuries and relatively cool during the 16th to 19th centuries. These periods coincide with what are traditionally known as the medieval Climate Optimum and the Little Ice Age, although these anomalies appear to have been most distinct only in and around the North Atlantic region. Based on these analyses, the warmth of the late 20th century appears to have been unprecedented during the millennium. A comprehensive review and assessment of observed global and hemispheric variability may be found in Chapter 2.

The scarce data from the Southern Hemisphere suggest temperature changes in past centuries markedly different from those in the Northern Hemisphere, the only obvious similarity being the strong warming during the 20th century.

Regional patterns of climate variability
Regional or local climate is generally much more variable than climate on a hemispheric or global scale because regional or local variations in one region are compensated for by opposite variations elsewhere. Indeed a closer inspection of the spatial structure of climate variability, in particular on seasonal and longer time-scales, shows that it occurs predominantly in preferred large-scale and geographically anchored spatial patterns. Such patterns result from interactions between the atmospheric circulation and the land and ocean surfaces. Though geographically anchored, their amplitude can change in time as, for example, the heat exchange with the underlying ocean changes.

A well-known example is the quasi-periodically varying ENSO phenomenon, caused by atmosphere-ocean interaction in the tropical Pacific. The resulting El Niño and La Niña events have a worldwide impact on weather and climate.

Another example is the North Atlantic Oscillation (NAO), which has a strong influence on the climate of Europe and part of Asia. This pattern consists of opposing variations of barometric pressure near Iceland and near the Azores. On average, a westerly current, between the Icelandic low pressure area and the Azores high-pressure area, carries cyclones with their associated frontal systems towards Europe. However the pressure difference between Iceland and the Azores fluctuates on time-scales of days to decades, and can be reversed at times. The variability of NAO has considerable influence on the regional climate variability in Europe, in particular in wintertime. Chapter 7 discusses the internal processes involved in NAO variability.

Similarly, although data are scarcer, leading modes of variability have been identified over the Southern Hemisphere. Examples are a North-South dipole structure over the Southern Pacific, whose variability is strongly related to ENSO variability, and the Antarctic Oscillation, a zonal pressure fluctuation between middle and high latitudes of the Southern Hemisphere. A detailed account of regional climate variability may be found in Chapter 2.

1.2.3 Extreme Events

Climate as defined is associated with a certain probability distribution of weather events. Weather events with values far away from the mean (such as heat waves, droughts and flooding) are by definition less likely to occur. The least likely events in a statistical sense are called "extreme events". Extreme weather in one region (e.g. a heat wave) may be normal in another. In both regions nature and society are adapted to the regional weather averaged over longer periods, but much less to extremes. For example, tropical African temperatures could severely damage vegetation or human health if they occurred in Northern Europe. Impacts of extreme events are felt strongly by ecosystems and society and may be destructive.

Small changes in climate may, but will not necessarily, have a large impact on the probability distribution of weather events in space and time, and on the intensity of extremes. Nature and society are often particularly ill prepared and vulnerable for such changes. This is the reason why since the SAR much more attention has been paid to observed and projected variations of extremes. Chapter 2 gives an assessment of the present knowledge.

1.3 Human-induced Climate Variations

1.3.1 Human Influence on the Climate System

Human beings, like other living organisms, have always influenced their environment. It is only since the beginning of the Industrial Revolution, mid-18th century, that the impact of human activities has begun to extend to a much larger scale, continental or even global. Human activities, in particular those involving the combustion of fossil fuels for industrial or domestic usage, and biomass burning, produce greenhouse gases and aerosols which affect the composition of the atmosphere. The emission of chlorofluorocarbons (CFCs) and other chlorine and bromine compounds has not only an impact on the radiative forcing, but has also led to the depletion of the stratospheric ozone layer. Land-use change, due to urbanisation and human forestry and agricultural practices, affect the physical and biological properties of the Earth's surface. Such effects change the radiative forcing and have a potential impact on regional and global climate.

Anthropogenic perturbation of the atmospheric composition
For about a thousand years before the Industrial Revolution, the amount of greenhouse gases in the atmosphere remained relatively constant. Since then, the concentration of various greenhouse gases has increased. The amount of carbon dioxide, for example, has increased by more than 30% since pre-industrial times and is still increasing at an unprecedented rate of on average 0.4% per year, mainly due to the combustion of fossil fuels and deforestation. We know that this increase is anthropogenic because the changing isotopic composition of the atmospheric CO_2 betrays the fossil origin of the increase. The concentration of other natural radiatively active atmospheric components, such as methane and nitrous oxide, is increasing as well due to agricultural, industrial and other activities. The concentration of the nitrogen oxides (NO and NO_2) and of carbon monoxide (CO) are also increasing. Although these gases are not greenhouse gases, they play a role in the atmospheric chemistry and have led to an increase in tropospheric ozone, a greenhouse gas, by 40% since pre-industrial times (Chapter 4). Moreover,

NO_2 is an important absorber of visible solar radiation. Chlorofluorocarbons and some other halogen compounds do not occur naturally in the atmosphere but have been introduced by human activities. Beside their depleting effect on the stratospheric ozone layer, they are strong greenhouse gases. Their greenhouse effect is only partly compensated for by the depletion of the ozone layer which causes a negative forcing of the surface-troposphere system. All these gases, except tropospheric ozone and its precursors, have long to very long atmospheric lifetimes and therefore become well-mixed throughout the atmosphere.

Human industrial, energy related, and land-use activities also increase the amount of aerosol in the atmosphere, in the form of mineral dust, sulphates and nitrates and soot. Their atmospheric lifetime is short because they are removed by rain. As a result their concentrations are highest near their sources and vary substantially regionally, with global consequences. The increases in greenhouse gas concentrations and aerosol content in the atmosphere result in a change in the radiative forcing to which the climate system must act to restore the radiative balance.

The enhanced greenhouse effect
The increased concentration of greenhouse gases in the atmosphere enhances the absorption and emission of infrared radiation. The atmosphere's opacity increases so that the altitude from which the Earth's radiation is effectively emitted into space becomes higher. Because the temperature is lower at higher altitudes, less energy is emitted, causing a positive radiative forcing. This effect is called the enhanced greenhouse effect, which is discussed in detail in Chapter 6.

If the amount of carbon dioxide were doubled instantaneously, with everything else remaining the same, the outgoing infrared radiation would be reduced by about 4 Wm^{-2}. In other words, the radiative forcing corresponding to a doubling of the CO_2 concentration would be 4 Wm^{-2}. To counteract this imbalance, the temperature of the surface-troposphere system would have to increase by 1.2°C (with an accuracy of ±10%), in the absence of other changes. In reality, due to feedbacks, the response of the climate system is much more complex. It is believed that the overall effect of the feedbacks amplifies the temperature increase to 1.5 to 4.5°C. A significant part of this uncertainty range arises from our limited knowledge of clouds and their interactions with radiation. To appreciate the magnitude of this temperature increase, it should be compared with the global mean temperature difference of perhaps 5 or 6°C from the middle of the last Ice Age to the present interglacial.

The so-called water vapour feedback, caused by an increase in atmospheric water vapour due to a temperature increase, is the most important feedback responsible for the amplification of the temperature increase. Concern has been expressed about the strength of this feedback, in particular in relation to the role of upper tropospheric humidity. Since the SAR, thinking about this feedback has become increasingly sophisticated thanks both to modelling and to observational studies. Feedbacks are discussed and assessed in Chapter 7. In particular, the present state of knowledge of the water vapour feedback is examined in Section 7.2.1.

It has been suggested that the absorption by CO_2 is already saturated so that an increase would have no effect. This, however, is not the case. Carbon dioxide absorbs infrared radiation in the middle of its 15 μm band to the extent that radiation in the middle of this band cannot escape unimpeded: this absorption is saturated. This, however, is not the case for the band's wings. It is because of these effects of partial saturation that the radiative forcing is not proportional to the increase in the carbon dioxide concentration but shows a logarithmic dependence. Every further doubling adds an additional 4 Wm^{-2} to the radiative forcing.

The other human-made greenhouse gases add to the effect of increased carbon dioxide. Their total effect at the surface is often expressed in terms of the effect of an equivalent increase in carbon dioxide.

The effect of aerosols
The effect of the increasing amount of aerosols on the radiative forcing is complex and not yet well known. The direct effect is the scattering of part of the incoming solar radiation back into space. This causes a negative radiative forcing which may partly, and locally even completely, offset the enhanced greenhouse effect. However, due to their short atmospheric lifetime, the radiative forcing is very inhomogeneous in space and in time. This complicates their effect on the highly non-linear climate system. Some aerosols, such as soot, absorb solar radiation directly, leading to local heating of the atmosphere, or absorb and emit infrared radiation, adding to the enhanced greenhouse effect.

Aerosols may also affect the number, density and size of cloud droplets. This may change the amount and optical properties of clouds, and hence their reflection and absorption. It may also have an impact on the formation of precipitation. As discussed in Chapter 5, these are potentially important indirect effects of aerosols, resulting probably in a negative radiative forcing of as yet very uncertain magnitude.

Land-use change
The term "land-use change" refers to a change in the use or management of land. Such change may result from various human activities such as changes in agriculture and irrigation, deforestation, reforestation and afforestation, but also from urbanisation or traffic. Land-use change results in changing the physical and biological properties of the land surface and thus the climate system.

It is now recognized that land-use change on the present scale may contribute significantly to changing the local, regional or even global climate and moreover has an important impact on the carbon cycle. Physical processes and feedbacks caused by land-use change, that may have an impact on the climate, include changes in albedo and surface roughness, and the exchange between land and atmosphere of water vapour and greenhouse gases. These climatic consequences of land-use change are discussed and evaluated in Section 4 of Chapter 7. Land-use change may also affect the climate system through biological processes and feedbacks involving the terrestrial vegetation, which may lead to changes in the sources and sinks of carbon in its various forms. Chapter 3 reviews the consequences for the carbon cycle. Obviously the combined effect of these physical and biogeochemical processes and feedbacks is complex, but new data sets and models start to shed light on this.

Urbanisation is another kind of land-use change. This may affect the local wind climate through its influence on the surface roughness. It may also create a local climate substantially warmer than the surrounding countryside by the heat released by densely populated human settlements, by changes in evaporation characteristics and by modifying the outgoing long-wave radiation through interception by tall buildings etc. This is known as an "urban heat island". The influence on the regional climate may be noticeable but small. It may however have a significant influence on long instrumental temperature records from stations affected by expanding urbanisation. The consequences of this urbanisation effect for the global surface temperature record has been the subject of debate. It is discussed in Section 2.2.2 of Chapter 2.

Climate response

The increase in greenhouse gas and aerosol concentrations in the atmosphere and also land-use change produces a radiative forcing or affects processes and feedbacks in the climate system. As discussed in Chapter 7, the response of the climate to these human-induced forcings is complicated by such feedbacks, by the strong non-linearity of many processes and by the fact that the various coupled components of the climate system have very different response times to perturbations. Qualitatively, an increase of atmospheric greenhouse gas concentrations leads to an average increase of the temperature of the surface-troposphere system. The response of the stratosphere is entirely different. The stratosphere is characterised by a radiative balance between absorption of solar radiation, mainly by ozone, and emission of infrared radiation mainly by carbon dioxide. An increase in the carbon dioxide concentration therefore leads to an increase of the emission and thus to a cooling of the stratosphere.

The only means available to quantify the non-linear climate response is by using numerical models of the climate system based on well-established physical, chemical and biological principles, possibly combined with empirical and statistical methods.

1.3.2 Modelling and Projection of Anthropogenic Climate Change

Climate models

The behaviour of the climate system, its components and their interactions, can be studied and simulated using tools known as climate models. These are designed mainly for studying climate processes and natural climate variability, and for projecting the response of the climate to human-induced forcing. Each component or coupled combination of components of the climate system can be represented by models of varying complexity.

The nucleus of the most complex atmosphere and ocean models, called General Circulation Models (Atmospheric General Circulation Models (AGCMs) and Ocean General Circulation Models (OGCMs)) is based upon physical laws describing the dynamics of atmosphere and ocean, expressed by mathematical equations. Since these equations are non-linear, they need to be solved numerically by means of well-established mathematical techniques. Current atmosphere models are solved spatially on a three-dimensional grid of points on the globe with

a horizontal resolution typically of 250 km and some 10 to 30 levels in the vertical. A typical ocean model has a horizontal resolution of 125 to 250 km and a resolution of 200 to 400 m in the vertical. Their time-dependent behaviour is computed by taking time steps typically of 30 minutes. The impact of the spatial resolution on the model simulations is discussed in Section 8.9 of Chapter 8.

Models of the various components of the climate system may be coupled to produce increasingly complex models. The historical development of such coupled climate models is shown in Box 3 of the Technical Summary. Processes taking place on spatial and temporal scales smaller than the model's resolution, such as individual clouds or convection in atmosphere models, or heat transport through boundary currents or mesoscale eddies in ocean models, are included through a parametric representation in terms of the resolved basic quantities of the model. Coupled atmosphere-ocean models, including such parametrized physical processes, are called Atmosphere-Ocean General Circulation Models (AOGCMs). They are combined with mathematical representations of other components of the climate system, sometimes based on empirical relations, such as the land surface and the cryosphere. The most recent models may include representations of aerosol processes and the carbon cycle, and in the near future perhaps also the atmospheric chemistry. The development of these very complex coupled models goes hand in hand with the availability of ever larger and faster computers to run the models. Climate simulations require the largest, most capable computers available.

A realistic representation of the coupling between the various components of the climate system is essential. In particular the coupling between the atmosphere and the oceans is of central importance. The oceans have a huge heat capacity and a decisive influence on the hydrological cycle of the climate system, and store and exchange large quantities of carbon dioxide. To a large degree the coupling between oceans and atmosphere determines the energy budget of the climate system. There have been difficulties modelling this coupling with enough accuracy to prevent the model climate unrealistically drifting away from the observed climate. Such climate drift may be avoided by adding an artificial correction to the coupling, the so-called "flux adjustment". The evaluation in Chapter 8 of recent model results identifies improvements since the SAR, to the point that there is a reduced reliance on such corrections, with some recent models operating with minimal or no adjustment.

For various reasons, discussed in Section 8.3 of Chapter 8, it is important to also develop and use simpler models than the fully coupled comprehensive AOGCMs, for example to study only one or a specific combination of components of the climate system or even single processes, or to study many different alternatives, which is not possible or is impractical with comprehensive models. In IPCC (1997) a hierarchy of models used in the IPCC assessment process was identified and described, differing in such aspects as the number of spatial dimensions, the extent to which physical processes are explicitly represented, the level to which empirical parametrization is involved, and the computational costs of running the models. In the IPCC context, simple models are also used to compute the consequences of greenhouse gas emission scenarios. Such models are tuned to the AOGCMs to give similar results when globally averaged.

Projections of climate change

Climate models are used to simulate and quantify the climate response to present and future human activities. The first step is to simulate the present climate for extended simulation periods, typically many decades, under present conditions without any change in external climate forcing.

The quality of these simulations is assessed by systematically comparing the simulated climate with observations of the present climate. In this way the model is evaluated and its quality established. A range of diagnostic tools has been developed to assist the scientists in carrying out the evaluation. This step is essential to gain confidence in and provide a baseline for projections of human-induced climate change. Models may also be evaluated by running them under different palaeoclimate (e.g. Ice Age) conditions. Chapter 8 of this report presents a detailed assessment of the latest climate models of various complexity, in particular the AOGCMs. Once the quality of the model is established, two different strategies have been applied to make projections of future climate change.

The first, so-called equilibrium method is to change, e.g. double, the carbon dioxide concentration and to run the model again to a new equilibrium. The differences between the climate statistics of the two simulations provide an estimate of the climate change corresponding to the doubling of carbon dioxide, and of the sensitivity of the climate to a change in the radiative forcing. This method reduces systematic errors present in both simulations. If combined with simple slab ocean models, this strategy is relatively cheap because it does not require long runs to reach equilibrium. However it does not provide insight in to the time dependence of climate change.

The second, so-called transient, method, common nowadays with improved computer resources, is to force the model with a greenhouse gas and aerosol scenario. The difference between such simulation and the original baseline simulation provides a time-dependent projection of climate change.

This transient method requires a time-dependent profile of greenhouse gas and aerosol concentrations. These may be derived from so-called emission scenarios. Such scenarios have been developed, among others by IPCC, on the basis of various internally coherent assumptions concerning future socio-economic and demographic developments. In the SAR the IPCC Scenarios IS92 were used (IPCC, 1994). The most recent IPCC emission scenarios are described in the IPCC Special Report on Emission Scenarios (Nakićenović *et al*., 2000). Different assumptions concerning e.g. the growth of the world population, energy intensity and efficiency, and economic growth, lead to considerably different emission scenarios. For example the two extreme estimates in the IPCC IS92 scenarios of the carbon dioxide emission by 2100 differ by a factor of 7. Because scenarios by their very nature should not be used and regarded as predictions, the term "climate projections" is used in this Report.

Transient simulations may also be based on artificially constructed, so-called idealised, scenarios. For example, scenarios have been constructed, assuming a gradual increase of greenhouse gas concentrations followed by stabilisation at various levels. Climate simulations based on such idealised scenarios may provide insight in to the climate response to potential policy measures leading to a stabilisation of the GHG concentrations, which is the ultimate objective of the United Nations Framework Convention on Climate Change (UNFCCC) as formulated in its Article 2. See Section 3 of Chapter 9 for an assessment.

Projections from present models show substantial agreement, but at the same time there is still a considerable degree of ambiguity and difference between the various models. All models show an increase in the globally averaged equilibrium surface temperature and global mean precipitation. In Chapter 9 the results of various models and intercomparison projects are assessed. Model results are more ambiguous about the spatial patterns of climate change than about the global response. Regional patterns depend significantly on the time dependence of the forcing, the spatial distribution of aerosol concentrations and details of the modelled climate processes. Research tools have been developed to generate more reliable regional climate information. These tools and their results are presented and assessed in Chapter 10.

To study the impact of climate change, a plausible and consistent description of a possible future climate is required. The construction of such climate change scenarios relies mainly on results from model projections, although sometimes information from past climates is used. The basis for and development of such scenarios is assessed in Chapter 13. Global and regional sea-level change scenarios are reviewed in Chapter 11.

Predictability, global and regional

In trying to quantify climate change, there is a fundamental question to be answered: is the evolution of the state of the climate system predictable? Since the pioneering work by Lorenz in the 1960s, it is well known that complex non-linear systems have limited predictability, even though the mathematical equations describing the time evolution of the system are perfectly deterministic.

The climate system is, as we have seen, such a non-linear complex system with many inherent time scales. Its predictability may depend on the type of climate event considered, the time and space scales involved and whether internal variability of the system or variability from changes in external forcing is involved. Internal variations caused by the chaotic dynamics of the climate system may be predictable to some extent. Recent experience has shown that the ENSO phenomenon may possess a fair degree of predictability for several months or even a year ahead. The same may be true for other events dominated by the long oceanic time-scales, such as perhaps the NAO. On the other hand, it is not known, for example, whether the rapid climate changes observed during the last glacial period are at all predictable or are unpredictable consequences of small changes resulting in major climatic shifts.

There is evidence to suggest that climate variations on a global scale resulting from variations in external forcing are partly predictable. Examples are the mean annual cycle and short-term climate variations from individual volcanic eruptions, which models simulate well. Regularities in past climates, in particular the cyclic succession of warm and glacial periods forced by geometrical changes in the Sun-Earth orbit, are simulated by simple models with a certain degree of success. The global and continental scale aspects of human-induced climate change, as simulated by the models forced by increasing greenhouse gas concentration, are largely reproducible. Although this is not an absolute proof, it provides evidence that such externally forced climate change may be predictable, if their forcing mechanisms are known or can be predicted.

Finally, global or continental scale climate change and variability may be more predictable than regional or local scale change, because the climate on very large spatial scales is less influenced by internal dynamics, whereas regional and local climate is much more variable under the influence of the internal chaotic dynamics of the system. See Chapter 7 for an assessment of the predictability of the climate system.

Rapid climate change
A non-linear system such as the climate system may exhibit rapid climate change as a response to internal processes or rapidly changing external forcing. Because the probability of their occurrence may be small and their predictability limited, they are colloquially referred to as "unexpected events" or "surprises". The abrupt events that took place during the last glacial cycle are often cited as an example to demonstrate the possibility of such rapid climate change. Certain possible abrupt events as a result of the rapidly increasing anthropogenic forcing could be envisioned. Examples are a possible reorganization of the thermohaline ocean circulation in the North Atlantic resulting in a more southerly course of the Gulf Stream, which would have a profound influence on the climate of Western Europe, a possible reduction of upper-level ocean cycling in the Southern Ocean, or a possible but unlikely rapid disintegration of part of the Antarctic ice sheet with dramatic consequences for the global sea level.

More generally, with a rapidly changing external forcing, the non-linear climate system may experience as yet unenvisionable, unexpected, rapid change. Chapter 7, in particular Section 7.7, of this Report reviews and assesses the present knowledge of non-linear events and rapid climate change. Potential rapid changes in sea level are assessed in Chapter 11.

1.3.3 Observing Anthropogenic Climate Change

Observing the climate
The question naturally arises whether the system has already undergone human-induced climate change. To answer this question, accurate and detailed observations of climate and climate variability are required. Instrumental observations of land and ocean surface weather variables and sea surface temperature have been made increasingly widely since the mid-19th century. Recently, ships' observations have been supplemented by data from dedicated buoys. The network of upper-air observations, however, only became widespread in the late 1950s. The density of observing stations always has been and still is extremely inhomogeneous, with many stations in densely populated areas and virtually none in huge oceanic areas. In recent times special

earth-observation satellites have been launched, providing a wide range of observations of various components of the climate system all over the globe. The correct interpretation of such data still requires high quality *in situ* and surface data. The longer observational records suffer from changes in instrumentation, measurement techniques, exposure and gaps due to political circumstances or wars. Satellite data also require compensation for orbital and atmospheric transmission effects and for instrumental biases and instabilities. Earlier the problems related to urbanisation were mentioned. To be useful for the detection of climate change, observational records have to be adjusted carefully for all these effects.

Concern has been expressed about the present condition of the observational networks. The number of upper-air observations, surface stations and observations from ships is declining, partly compensated for by an increasing number of satellite observations. An increasing number of stations are being automated, which may have an impact on the quality and homogeneity of the observations. Maintaining and improving the quality and density of existing observational networks is essential for necessary high standard information. In order to implement and improve systematic observations of all components of the climate system, the World Meteorological Organization and the International Oceanographic Commission have established a Global Climate Observing System (GCOS). Initially GCOS uses existing atmospheric, oceanic and terrestrial networks. Later GCOS will aim to amplify and improve the observational networks where needed and possible.

Observations alone are not sufficient to produce a coherent and global picture of the state of the climate system. So-called data assimilation systems have been developed, which combine observations and their temporal and spatial statistics with model information to provide a coherent quantitative estimate in space and time of the state of the climate system. Data assimilation also allows the estimation of properties which cannot easily be observed directly but which are linked to the observations through physical laws. Some institutions have recently reanalysed several decades of data by means of the most recent and most sophisticated version of their data assimilation system, avoiding in this way inhomogenities due to changes in their system. However inhomogeneities in these reanalyses may still exist due to changing sources of information, such as the introduction of new satellite systems.

The 20th century
Historically, human activities such as deforestation may have had a local or regional impact, but there is no reason to expect any large human influence on the global climate before the 20th century. Observations of the global climate system during the 20th century are therefore of particular importance. Chapter 2 presents evidence that there has been a mean global warming of 0.4 to 0.8°C of the atmosphere at the surface since the late 19th century. Figure 2.1 of Chapter 2 shows that this increase took place in two distinct phases, the first one between 1910 and 1945, and recently since 1976. Recent years have been exceptionally warm, with a larger increase in minimum than in maximum temperatures possibly related, among other factors, to an increase

in cloud cover. Surface temperature records indicate that the 1990s are likely to have been the warmest decade of the millennium in the Northern hemisphere, and 1998 is likely to have been the warmest year. For instrumentally recorded history, 1998 has been the warmest year globally. Concomitant with this temperature increase, sea level has risen during the 20th century by 10 to 20 cm and there has been a general retreat of glaciers worldwide, except in a few maritime regions, e.g. Norway and New Zealand (Chapter 11).

Regional changes are also apparent. The observed warming has been largest over the mid- and high-latitude continents in winter and spring. Precipitation trends vary considerably geographically and, moreover, data in most of the Southern Hemisphere and over the oceans are scarce. From the data available, it appears that precipitation has increased over land in mid- and high latitudes of the Northern Hemisphere, especially during winter and early spring, and over most Southern Hemisphere land areas. Over the tropical and the Northern Hemisphere subtropical land areas, particularly over the Mediterranean region during winter, conditions have become drier. In contrast, over large parts of the tropical oceans rainfall has increased.

There is considerable variability of the atmospheric circulation at long time-scales. The NAO for example, with its strong influence on the weather and climate of extratropical Eurasia, fluctuates on multi-annual and multi-decadal time-scales, perhaps influenced by varying temperature patterns in the Atlantic Ocean. Since the 1970s the NAO has been in a phase that gives stronger westerly winds in winter. Recent ENSO behaviour seems to have been unusual compared to that of previous decades: there is evidence that El Niño episodes since the mid-1970s have been relatively more frequent than the opposite La Niña episodes.

There are suggestions that the occurrence of extreme weather events has changed in certain areas, but a global pattern is not yet apparent. For example, it is likely that in many regions of the world, both in the Northern and Southern Hemisphere, there has been a disproportionate increase in heavy and extreme precipitation rates in areas where the total precipitation has increased. Across most of the globe there has been a decrease in the frequency of much below-normal seasonal temperatures.

A detailed assessment of observed climate variability and change may be found in Chapter 2, and of observed sea level change in Chapter 11. Figure 2.39 of Chapter 2 summarises observed variations in temperature and the hydrological cycle.

Detection and attribution
The fact that the global mean temperature has increased since the late 19th century and that other trends have been observed does not necessarily mean that an anthropogenic effect on the climate system has been identified. Climate has always varied on all time-scales, so the observed change may be natural. A more detailed analysis is required to provide evidence of a human impact.

Identifying human-induced climate change requires two steps. First it must be demonstrated that an observed climate change is unusual in a statistical sense. This is the detection problem. For this to be successful one has to know quantitatively how climate varies naturally. Although estimates have improved since the SAR, there is still considerable uncertainty in the magnitude of this natural climate variability. The SAR concluded nevertheless, on the basis of careful analyses, that "the observed change in global mean, annually averaged temperature over the last century is unlikely to be due entirely to natural fluctuations of the climate system".

Having detected a climatic change, the most likely cause of that change has to be established. This is the attribution problem. Can one attribute the detected change to human activities, or could it also be due to natural causes? Also attribution is a statistical process. Neither detection nor attribution can ever be "certain", but only probable in a statistical sense. The attribution problem has been addressed by comparing the temporal and spatial patterns of the observed temperature increase with model calculations based on anthropogenic forcing by greenhouse gases and aerosols, on the assumption that these patterns carry a fingerprint of their cause. In this way the SAR found that "there is evidence of an emerging pattern of climate response to forcing by greenhouse gases and sulphate aerosols in the observed climate record". Since the SAR new results have become available which tend to support this conclusion. The present status of the detection of climate change and attribution of its causes is assessed in Chapter 12.

1.4 A 'Road-map' to this Report

This Report, the third IPCC Working Group I Assessment Report since 1990, assesses the state of scientific understanding of the climate system and its variability and change, in particular human-induced climate change. This section provides a 'road map' to the 14 chapters of this report and the major issues they are designed to address. Each chapter provides an initial summary of the Working Group I Second Assessment Report (IPCC, 1996) and then goes on to emphasise the progress made since then. The chapters can be viewed as covering the following three broad areas: *past* changes and the factors that can force climate change (Chapters 2 to 6), our *present* understanding and ability to model the climate system (Chapters 7, 8 and 14) and possible *future* climate change (Chapters 9 to 13).

In order to understand, assess and quantify the possible human influence on climate, an analysis of past climate variability and change is required (Chapter 2). The chapter tackles such questions as: how much is the world warming and is the recent global warming unusual? It looks in detail at trends and variability during the recent instrumental period (the last 100 years or so) and draws on palaeo-data to put them into the context of climate over much longer periods.

There are many factors that are known to influence climate, both natural and human-induced. The increase in concentrations of greenhouse gases and aerosols through human activity is of particular concern. Chapters 3 to 5 examine how well the three most important human contributions to the changing composition of the atmosphere; carbon dioxide, other greenhouse gases and aerosols, are understood, including the physical, chemical and biological processes which determine the atmospheric concentrations of these components. The next step, taken in Chapter 6, is to evaluate how this change in atmospheric composition has affected radiative forcing within the context of other factors such as land-use change, volcanic eruptions and solar variations.

Understanding the climate response to these various radiative forcings and projecting how they could affect future climate requires an understanding of the physical processes and feedbacks in the climate system and an ability to model them (Chapter 7). The only tools available for such projections of future climate are numerical models of the climate system of various complexity. An evaluation of such models against observations of the present and past climate and model intercomparisons provide the basis for confidence in such tools (Chapter 8).

Climate models together with scenarios of future emissions of radiatively active atmospheric components, as for example the SRES scenarios (Nakićenović *et al.*, 2000), recently developed by IPCC specifically for this purpose, are used to project future climate change. State-of-the-art projections for the next 100 years are assessed in Chapter 9, mainly at a global level, but also including large-scale patterns, their spatial and temporal variability and extreme events. Partly in response to the need for more details of climate change at a regional level, research in this area has been particularly active over the last 5 years. A new chapter, compared to previous assessments, has been included which examines the various techniques available to derive regional climate projections and, as far as is currently possible, assesses regional climate change information (Chapter 10). Chapter 11 assesses the current state of knowledge of the rate of change of global average and regional sea level in response to climate change.

A key conclusion from the SAR was that "the balance of evidence suggests that there is a discernible human influence on global climate". Chapter 12 assesses research over the last 5 years on the detection and attribution of climate change drawing on the developments in observational research (Chapters 2 to 6) and modelling (Chapters 7 to 10) to consider how this conclusion has changed.

Data derived directly from projections with climate models are often inappropriate for assessing the impacts of climate change which can require detailed, regional or local information as well as observational data describing current (or baseline) climate. Climate change scenarios are plausible representations of future climate constructed explicitly for impact assessment and form a key link between IPCC Working Groups I and II. For the first time, Working Group I have included a chapter dedicated to climate scenarios (Chapter 13) – this is intended to provide an assessment of scenario generation techniques, rather than to present scenarios themselves.

All chapters of the report highlight areas of certainty and uncertainty, and gaps in current knowledge. Chapter 14 draws together this information to present key areas that need to be addressed to advance understanding and reduce uncertainty in the science of climate change.

A comprehensive and integrated summary of all results of this assessment report may be found in the Technical Summary in this volume. A brief summary highlighting points of particular policy relevance is presented in the Summary for Policymakers.

References

IPCC, 1994: *Climate Change 1994: Radiative Forcing of Climate Change and an Evaluation of the IPCC IS92 Emission Scenarios*, [J.T. Houghton, L.G. Meira Filho, J. Bruce, Hoesung Lee, B.A. Callander, E. Haites, N. Harris and K. Maskell (eds.)]. Cambridge University Press, Cambridge, UK and New York, NY, USA, 339 pp.

IPCC, 1996: *Climate Change 1995: The Science of Climate Change. Contribution of Working Group I to the Second Assessment Report of the Intergovernmental Panel on Climate Change* [Houghton, J.T., L.G. Meira Filho, B.A. Callander, N. Harris, A. Kattenberg, and K. Maskell (eds.)]. Cambridge University Press, Cambridge, United Kingdom and New York, NY, USA, 572 pp.

IPCC, 1997: *IPCC Technical Paper II: An introduction to simple climate models used in the IPCC Second Assessment Report*, [J.T. Houghton, L.G. Meira Filho, D.J. Griggs and K. Maskell (eds.)].

Nakićenović, N., J. Alcamo, G. Davis, B. de Vries, J. Fenhann, S. Gaffin, K. Gregory, A. Grübler, T. Y. Jung, T. Kram, E. L. La Rovere, L. Michaelis, S. Mori, T. Morita, W. Pepper, H. Pitcher, L. Price, K. Raihi, A. Roehrl, H-H. Rogner, A. Sankovski, M. Schlesinger, P. Shukla, S. Smith, R. Swart, S. van Rooijen, N. Victor, Z. Dadi, 2000: *IPCC Special Report on Emissions Scenarios*, Cambridge University Press, Cambridge, United Kingdom and New York, NY, USA, 599 pp.

2

Observed Climate Variability and Change

Co-ordinating Lead Authors
C.K. Folland, T.R. Karl

Lead Authors
J.R. Christy, R.A. Clarke, G.V. Gruza, J. Jouzel, M.E. Mann, J. Oerlemans, M.J. Salinger, S.-W. Wang

Contributing Authors
J. Bates, M. Crowe, P. Frich, P. Groisman, J. Hurrell, P. Jones, D. Parker, T. Peterson, D. Robinson, J. Walsh, M. Abbott, L. Alexander, H. Alexandersson, R. Allan, R. Alley, P. Ambenje, P. Arkin, L. Bajuk, R. Balling, M.Y. Bardin, R. Bradley, R. Brázdil, K.R. Briffa, H. Brooks, R.D. Brown, S. Brown, M. Brunet-India, M. Cane, D. Changnon, S. Changnon, J. Cole, D. Collins, E. Cook, A. Dai, A. Douglas, B. Douglas, J.C. Duplessy, D. Easterling, P. Englehart, R.E. Eskridge, D. Etheridge, D. Fisher, D. Gaffen, K. Gallo, E. Genikhovich, D. Gong, G. Gutman, W. Haeberli, J. Haigh, J. Hansen, D. Hardy, S. Harrison, R. Heino, K. Hennessy, W. Hogg, S. Huang, K. Hughen, M.K. Hughes, M. Hulme, H. Iskenderian, O.M. Johannessen, D. Kaiser, D. Karoly, D. Kley, R. Knight, K.R. Kumar, K. Kunkel, M. Lal, C. Landsea, J. Lawrimore, J. Lean, C. Leovy, H. Lins, R. Livezey, K.M. Lugina, I. Macadam, J.A. Majorowicz, B. Manighetti, J. Marengo, E. Mekis, M.W. Miles, A. Moberg, I. Mokhov, V. Morgan, L. Mysak, M. New, J. Norris, L. Ogallo, J. Overpeck, T. Owen, D. Paillard, T. Palmer, C. Parkinson, C.R. Pfister, N. Plummer, H. Pollack, C. Prentice, R. Quayle, E.Y. Rankova, N. Rayner, V.N. Razuvaev, G. Ren, J. Renwick, R. Reynolds, D. Rind, A. Robock, R. Rosen, S. Rösner, R. Ross, D. Rothrock, J.M. Russell, M. Serreze, W.R. Skinner, J. Slack, D.M. Smith, D. Stahle, M. Stendel, A. Sterin, T. Stocker, B. Sun, V. Swail, V. Thapliyal, L. Thompson, W.J. Thompson, A. Timmermann, R. Toumi, K. Trenberth, H. Tuomenvirta, T. van Ommen, D. Vaughan, K.Y. Vinnikov, U. von Grafenstein, H. von Storch, M. Vuille, P. Wadhams, J.M. Wallace, S. Warren, W. White, P. Xie, P. Zhai

Review Editors
R. Hallgren, B. Nyenzi

Contents

Executive Summary

Overview

The best estimate of global surface temperature change is a 0.6°C increase since the late 19th century with a 95% confidence interval of 0.4 to 0.8°C. The increase in temperature of 0.15°C compared to that assessed in the IPCC WGI Second Assessment Report (IPCC, 1996) (hereafter SAR) is partly due to the additional data for the last five years, together with improved methods of analysis and the fact that the SAR decided not to update the value in the First Assessment Report, despite slight additional warming. It is likely that there have been real differences between the rate of warming in the troposphere and the surface over the last twenty years, which are not fully understood. New palaeoclimate analyses for the last 1,000 years over the Northern Hemisphere indicate that the magnitude of 20th century warming is likely to have been the largest of any century during this period. In addition, the 1990s are likely to have been the warmest decade of the millennium. New analyses indicate that the global ocean has warmed significantly since the late 1940s: more than half of the increase in heat content has occurred in the upper 300 m, mainly since the late 1950s. The warming is superimposed on strong global decadal variability. Night minimum temperatures are continuing to increase, lengthening the freeze-free season in many mid- and high latitude regions. There has been a reduction in the frequency of extreme low temperatures, without an equivalent increase in the frequency of extreme high temperatures. Over the last twenty-five years, it is likely that atmospheric water vapour has increased over the Northern Hemisphere in many regions. There has been quite a widespread reduction in daily and other sub-monthly time-scales of temperature variability during the 20th century. New evidence shows a decline in Arctic sea-ice extent, particularly in spring and summer. Consistent with this finding are analyses showing a near 40% decrease in the average thickness of summer Arctic sea ice over approximately the last thirty years, though uncertainties are difficult to estimate and the influence of multi-decadal variability cannot yet be assessed. Widespread increases are likely to have occurred in the proportion of total precipitation derived from heavy and extreme precipitation events over land in the mid- and high latitudes of the Northern Hemisphere.

Changes in Temperature and Related Variables

Changes in near-surface temperature from the instrumental record

- Average global surface temperature has increased by approximately 0.6°C since the late 19th century, with 95% confidence limits of close to 0.4 and 0.8°C. Most of this increase has occurred in two periods, from about 1910 to 1945 and since 1976, and the largest recent warming is in the winter extra-tropical Northern Hemisphere. The warming rate since 1976, 0.17°C/decade, has been slightly larger than the rate of warming during the 1910 to 1945 period (0.14°C/decade), although the total increase in temperature is larger for the 1910 to 1945 period. The most recent warming period also has a faster rate of warming over land compared with the oceans. The high global

temperature associated with the 1997/98 El Niño event stands out in both surface and tropospheric temperatures as an extreme event, even after consideration of the recent rat· of warming.

- Confidence in the magnitude of global warming since the late 19th century has increased since the SAR due to new analyses, including model simulations of land-surface temperature changes and new studies of the effect of urbanisation on global land temperature trends. There is a high level of consistency between changes in sea surface temperatures (SSTs) and near-surface land air temperatures across the land-ocean boundary over the 20th century, despite independent observing systems and independent bias correction factors for SSTs before 1942. The assessed warming is considerably larger than the total contributions of the plausible sources of error.

- Twentieth century temperature trends show a broad pattern of tropical warming, while extra-tropical trends have been more variable. Warming from 1910 to 1945 was initially concentrated in the North Atlantic and nearby regions. The Northern Hemisphere shows cooling during the period 1946 to 1975 while the Southern Hemisphere shows warming. The recent 1976 to 2000 warming was largely globally synchronous, but emphasised in the Northern Hemisphere continents during winter and spring, with year-round cooling in parts of the Southern Hemisphere oceans and Antarctica. North Atlantic cooling between about 1960 and 1985 has recently reversed. Overall, warming over the Southern Hemisphere has been more uniform during the instrumental record than that over the Northern Hemisphere.

- The patterns of global temperature change since the 1970s are related in part to the positive westerly phase of the North Atlantic/Arctic Oscillation and possibly to decadal to multi-decadal variability in the Pacific.

- A multi-decadal fluctuation of SST in the North Atlantic has been in a rising phase since about the mid-1980s. Warming in many regions of this ocean has accelerated over the last five years and is likely to have contributed to quite rapid parallel increases of near-surface air temperature in much of Europe.

- New analysis shows that the global ocean heat content has increased since the late 1950s. This increase is superimposed on substantial global decadal variability. More than half the heating is contained in the uppermost 300 m where it is equivalent to an average temperature increase of 0.037°C/decade.

- Analyses of mean daily maximum and minimum land surface air temperatures continue to support a reduction in the diurnal temperature range in many parts of the world, with, globally, minimum temperatures increasing at nearly twice the rate of maximum temperatures between about 1950 and 1993. The rate of temperature increase during this time has been 0.1°C and 0.2°C for the maximum and minimum, respectively. This is about half of the rate of temperature increase over the oceans during this time.

Changes in temperature-related variables

- Alpine and continental glaciers have extensively retreated in response to 20th century warming. Glaciers in a few maritime regions are advancing, mainly due to increases in precipitation related to atmospheric circulation changes, e.g., Norway, New Zealand.

- The duration of Northern Hemisphere lake-ice and river-ice cover over the past century, or more, shows widespread decreases averaging to about two fewer weeks of ice cover.

- There is a highly significant interannual (+0.6) and multi-decadal correlation between increases in the Northern Hemisphere spring land temperature and a reduction in the Northern Hemisphere spring snow cover since data have been available (1966). Snow cover extent has decreased by about 10% since 1966.

- A 10 to 15% reduction in sea-ice extent in the Arctic spring and summer since the 1950s is consistent with an increase in spring, and to a lesser extent, summer temperatures in the high latitudes. There is little indication of reduced Arctic sea-ice extent during winter when temperatures have increased in the surrounding region.

- New data from submarines indicate that there has been about a 40% decline in Arctic sea-ice thickness in summer or early autumn between the period 1958 to 1976 and the mid-1990s, an average of near 4 cm per year. Other independent observations show a much slower decrease in winter sea-ice thickness of about 1 cm per year. The influence of substantial interannual and inter-decadal variability on these changes cannot be assessed because of restricted sampling.

- By contrast, there is no readily apparent relationship between decadal changes in Antarctic temperatures and sea-ice extent since 1973. Satellite data indicate that after a possible initial decrease in the mid-1970s, Antarctic sea-ice extent has stayed almost stable or even increased since 1978.

Changes in temperature above the surface layer

- Analysis of global temperature trends since 1958 in the low to mid-troposphere from balloons shows a warming of about +0.1°C/decade, which is similar to the average rate of warming at the surface. Since the early 1960s no significant trends have been detected for the global mean temperature in the uppermost troposphere.

- Satellites have only been available since 1979. Between 1979 and 2000, based on satellites and balloons, the lower-tropospheric trend has been +0.04 ± 0.11°C/decade and 0.03 ± 0.10°C/decade, respectively. By contrast, surface temperature trends for 1979 to 2000 were greater, at 0.16 ± 0.06°C/decade. The trend in the difference of the surface and lower-tropospheric series of 0.13 ± 0.06°C/decade is clearly statistically significant. This is in contrast to near zero surface temperature trends over 1958 to 1978 when the global lower-tropospheric temperature warmed by 0.03°C/decade relative to the surface.

- It is very likely that these significant differences in trends between the surface and lower troposphere are real and not solely an artefact of measurement bias, though differences in spatial and temporal sampling are likely to contribute. The differences are particularly apparent in many parts of the tropics and sub-tropics where the surface has warmed faster than the lower troposphere. In some other regions, e.g., North America, Europe and Australia, lower-tropospheric and surface trends are very similar.

- Throughout the stratosphere, negative temperature trends have been observed since 1979, ranging from a decrease of 0.5 or 0.6°C/decade in the lower stratosphere to 2.5°C/decade in the upper stratosphere.

Changes in temperature during the pre-instrumental period

The past millennium

- New analyses indicate that the magnitude of Northern Hempher warming over the 20th century is likely to have been the largest of any century in the last 1,000 years.

- The 1990s are likely to have been the warmest decade of the millennium in the Northern Hemisphere and 1998 is likely to have been the warmest year. Because less data are available, less is known about annual averages prior to 1,000 years before the present and for conditions prevailing in most of the Southern Hemisphere prior to 1861.

- Evidence does not support the existence of globally synchronous periods of cooling or warming associated with the 'Little Ice Age' and 'Medieval Warm Period'. However, reconstructed Northern Hemisphere temperatures do show a cooling during the 15th to 19th centuries and a relatively warm period during the 11th to 14th centuries, though the latter period is still cooler than the late 20th century.

- Analyses of borehole temperatures indicate a non-linear increase in global average ground surface temperature over land of 1.0 ± 0.3°C over the last 500 years, with most of the increase occurring since the late 19th century. There may be additional uncertainties due to the assumptions used in this technique, and decreasing resolution back in time limits confidence in the exact timing of the warming.

Changes across the last 500,000 years

- It is very likely that large and rapid decadal temperature changes occurred during the last glacial and its deglaciation (between about 100,000 and 10,000 years ago), particularly in higher latitudes of the Northern Hemisphere. During the last deglaciation, local increases in temperature are likely to have been as large as 5 to 10°C over a few decades. Over the same period there is evidence of less pronounced but nearly synchronous changes worldwide, except in high southern latitudes.

- Antarctic ice cores have provided new evidence of almost in-phase changes of temperature, carbon dioxide and methane through the ice age cycles over the past 420,000 years.

- There is emerging evidence for significant, rapid (time-scales of several decades or more), regional temperature changes during the last 10,000 years. However, the evidence does not indicate that any such events were global in scale.

Changes in Precipitation and Related Variables

Precipitation

- Instrumental records of land-surface precipitation continue to show an increase of 0.5 to 1%/decade in much of the Northern Hemisphere mid- and high latitudes. A notable exception includes parts of eastern Russia. In contrast, over much of the sub-tropical land areas rainfall has decreased during the 20th century (by −0.3%/decade), but this trend has weakened in recent decades. Other precipitation indicators suggest that large parts of the tropical oceans have had more precipitation in recent decades, and that precipitation has significantly increased over tropical land areas during the 20th century (2.4%/century). The increase in precipitation over the tropics is not evident during the past few decades.

- In the Southern Hemisphere, the pattern of island rainfall in parts of the South Pacific has changed since the mid-1970s, associated with the more frequent occurrence of the warm phase of the El Niño-Southern Oscillation (ENSO).

- Where data are available, changes in annual streamflow usually relate well to changes in total precipitation.

Water vapour

- Changes in water vapour mixing ratio have been analysed for selected regions using *in situ* surface observations as well as lower-tropospheric measurements based on satellites and weather balloons. A pattern of overall surface and lower-tropospheric water vapour mixing ratio increases over the past few decades is emerging, although there are likely to be some time-dependent biases in these data and regional variations in trends. The more reliable data sets show that it is likely that total atmospheric water vapour has increased several per cent per decade over many regions of the Northern Hemisphere since the early 1970s. Changes over the Southern Hemisphere cannot yet be assessed.

- Satellite observations of upper-tropospheric humidity from 1980 to 1997 show statistically significant positive trends of 0.1%/year for the zone 10°N to 10°S. Other trends are not statistically significant, but include a 0.04%/year positive trend for the zone 60°N to 60°S but a negative trend of −0.1%/year over the region 30°S to 60°S.

- Balloon observations of stratospheric water vapour above 18 km show an increase of about 1%/year for the period from 1981 to 2000. Shorter satellite records show a similar positive trend, suggesting that the change is global in character, but they also indicate a slowing of the positive trend after 1996.

Clouds

- It is likely that there has been an increase in total cloud cover of about 2% over many mid- to high latitude land areas since the beginning of the 20th century. The increases in total cloud amount are positively correlated with decreases in the diurnal temperature range. Changes in total cloud amount are uncertain both over sub-tropical and tropical land areas as well as over the oceans.

Changes in Atmospheric/Oceanic Circulation

El Niño-Southern Oscillation (ENSO)

- The frequency and intensity of ENSO has been unusual since the mid-1970s compared with the previous 100 years. Warm phase ENSO episodes have been relatively more frequent, persistent, or intense than the opposite cold phase during this period.

- This recent behaviour of ENSO is related to variations in precipitation and temperature over much of the global tropics and sub-tropics and some mid-latitude areas. The overall effect is likely to have made a small contribution to the increase in global surface temperature during the last few decades.

Other Oscillations

- The Inter-decadal Pacific Oscillation is likely to be a Pacific-wide manifestation of the Pacific Decadal Oscillation. Both are associated with decadal climate variability over the Pacific basin. It is likely that these related phenomena modulate ENSO-related climate variability.

- The winter North Atlantic Oscillation (NAO) and the associated Arctic Oscillation (AO), which appear to be largely the same phenomenon, show decadal to multi-decadal variability. Since the 1970s these oscillations have been in a phase that gives stronger westerly winds over much of extra-tropical Eurasia in the winter half year. This is associated with cold season warming over extra-tropical Eurasia, but cooling in some regions further south.

- The High Latitude Mode (HLM) or Antarctic Oscillation (AAO) in the Southern Hemisphere has been in an enhanced positive phase in the last fifteen years, with stronger westerly winds over the Southern Ocean.

- It is likely that rapid (time-scales of several decades or more) changes of atmospheric and ocean circulation occurred during inter-glacial periods, affecting regional climate, without human interference.

Changes in Extreme (within the upper or lower ten percentiles) Weather and Climate Events

Precipitation

- New analyses show that in regions where total precipitation has increased it is very likely that there have been even more pronounced increases in heavy and extreme precipitation events. The converse is also true.

- In some regions, heavy and extreme precipitation events have increased despite the fact that total precipitation has decreased or remained constant. This is attributed to a decrease in the frequency of precipitation events. Changes in the frequency of heavy precipitation events can arise from several causes, e.g., changes in atmospheric moisture or circulation.

- Over the latter half of the 20th century it is likely that there has been a 2 to 4% increase in the frequency of heavy precipitation events reported by the available observing stations in the mid- and high latitudes of the Northern Hemisphere.

- Trends for severe drought and wet area statistics for 1900 to 1995 are relatively small over global land areas. However, during the last two or three decades there have been some increases in the globally combined severe dry and wet areas.

Temperature
- In many regions inter-daily temperature variability has decreased. Increases in the daily minimum temperatures are lengthening the freeze-free season in most mid- and high latitude regions.

- A significant reduction in the frequency of extreme low monthly and seasonal average temperatures across much of the globe has occurred since the late 19th century. However, a relatively smaller increase in the frequency of extreme high monthly and seasonal average temperatures has been observed.

Storms
- Changes in tropical and extra-tropical storm intensity and frequency are dominated by inter-decadal to multi-decadal variations, with no significant trends over the 20th century evident. Conflicting analyses make it difficult to draw definitive conclusions about changes in storm activity, especially in the extra-tropics.

- No systematic changes in the frequency of tornadoes, thunder days, or hail events are evident in the limited areas analysed.

2.1 Introduction

Observed climate change and variability (for definitions, see the IPCC Glossary, Appendix I) are considered in this chapter by addressing seven commonly asked questions related to the detection of climate change and sensitivity of the climate to anthropogenic activity. The questions are:

How much is the world warming?
Is the recent warming unusual?
How rapidly did climate change in the distant past?
Have precipitation and atmospheric moisture changed?
Are the atmospheric/oceanic circulations changing?
Has climate variability, or have climate extremes, changed?
Are the observed trends internally consistent?

This chapter emphasises change against a background of variability. The certainty of conclusions that can be drawn about climate from observations depends critically on the availability of accurate, complete and consistent series of observations. For many variables important in documenting, detecting, and attributing climate change, Karl *et al.* (1995a) demonstrate that the data are still not good enough for really firm conclusions to be reached, as noted in the IPCC WGI Second Assessment Report (IPCC, 1996) (hereafter SAR). This especially applies to global trends in variables that have large regional variations, such as precipitation, whereas conclusions about temperature changes are often considerably more firmly based. The recently designated Global Climate Observing System (GCOS) upper air network (Wallis, 1998) and a GCOS surface network (Peterson *et al.*, 1997), maintained and reporting to higher standards, may have had a limited positive impact on the quality and availability of some of our results. New data sets e.g., on surface humidity, sea-ice thickness and sub-surface ocean temperature, have widened the range of conclusions than can be drawn since the SAR, albeit tentatively. However, a wider range of analytical techniques and tests of the data have increased our confidence in areas such as surface temperature changes.

Throughout the chapter we try to consistently indicate the degree of our confidence in trends and other results. Sometimes we provide quantitative estimates of uncertainty, as far as possible the value of twice the standard error, or we estimate statistical significance at the 0.05 (5%) level. This is the appropriate terminology and implies that what we see is very unusual, given the null hypothesis. We use the word "trend" to designate a generally progressive change in the level of a variable. Where numerical values are given, they are equivalent linear trends, though more complex changes in the variable will often be clear from the description. We use the word "consistent" to imply similarity between results or data sets that are expected to be related on physical grounds. Where this is not possible, we use the following words to indicate judgmental estimates of confidence: virtually certain (>99% chance that a result is true); very likely (≥90% but ≤99% chance); likely (>66% but <90% chance); medium likelihood (>33% but ≤66% chance), unlikely (>10% but ≤33% chance); very unlikely (≥1% but ≤10% chance) and exceptionally unlikely (<1% chance).

2.2 How Much is the World Warming?

2.2.1 Background

The SAR concluded that, on a global average, land-surface air and sea surface temperature rose by between 0.3°C and 0.6°C between the late 19th century and 1994. In this section, the recent warming is re-examined, using updated data. We include recent analyses of the diurnal asymmetry of the warming and its geographical structure. Conventional temperature observations are supplemented by indirect evidence and by satellite-based data. For the first time, we make objective estimates of uncertainties in the surface temperature data, though these are preliminary. We also assess recent work in compiling hemispheric and global temperature records from palaeoclimatic data, especially for the most recent millennium.

2.2.2 Temperature in the Instrumental Record for Land and Oceans

Note that all data sets are adjusted to have zero anomaly when averaged over the period 1961 to 1990.

2.2.2.1 Land-surface air temperature

The SAR reviewed the three databases of land-surface air temperature due to Jones (1994), Hansen and Lebedeff (1988) and Vinnikov *et al.* (1990). The first and second databases have been updated by Jones *et al.* (2001) and Hansen *et al.* (1999), respectively, and a further analysis has become available (Peterson and Vose, 1997; Peterson *et al.*, 1998a, 1999). The last paper also separates rural temperature stations in the Global Historical Climatology Network (GHCN) (Peterson and Vose, 1997) from the full set of stations which, in common with the other three analyses, have been screened for urbanisation effects. While there is little difference in the long-term (1880 to 1998) rural (0.70°C/century) and full set of station temperature trends (actually less at 0.65°C/century), more recent data (1951 to 1989), as cited in Peterson *et al.* (1999), do suggest a slight divergence in the rural (0.80°C/century) and full set of station trends (0.92°C/century). However, neither pair of differences is statistically significant. In addition, while not reported in Peterson *et al.*, the 1951 to 1989 trend for urban stations alone was 0.10°C/decade. We conclude that estimates of long-term (1880 to 1998) global land-surface air temperature variations and trends are relatively little affected by whether the station distribution typically used by the four global analyses is used, or whether a special effort is made to concentrate on rural stations using elaborate criteria to identify them. Part of the reason for this lack of sensitivity is that the average trends in available worldwide *urban* stations for 1951 to 1989 are not greatly more than those for all land stations (0.09°C/decade). The differences in trend between rural and all stations are also virtually unaffected by elimination of areas of largest temperature change, like Siberia, because such areas are well represented in both sets of stations.

These results confirm the conclusions of Jones *et al.* (1990) and Easterling *et al.* (1997) that urban effects on 20th century globally and hemispherically averaged land air temperature time-

Box 2.1: Urban Heat Island and the Observed Increases in Land Air Temperature.

There are two primary reasons why urban heat islands have been suspected as being partially responsible for the observed increases in land air temperatures over the last few decades. The first is related to the observed decrease in the diurnal temperature range and the second is related to a lower rate of warming observed over the past twenty years in the lower troposphere compared with the surface.

Since the 1950s both daily maximum and minimum temperatures are available over more than 50% of the global land area. These data indicate that on average the mean minimum temperature has increased at nearly twice the rate of the maximum temperature, reducing the daily temperature range by about 0.8°C over these areas. This has raised questions related to whether the growth of urban heat islands may be responsible for a substantial portion of the observed mean temperature increase, because it is well-known that compared to non-urban areas urban heat islands raise night-time temperatures more than daytime temperatures. Nonetheless, the relatively strong correlation between observed decreases in the daily temperature range with increases of both precipitation (leading to more moist surface conditions) and total cloud amount support the notion that the reduction in diurnal temperature range is in response to these physical changes.

Since 1979 satellite observations and weather balloons (which generally agree well) show substantially less warming of the global lower troposphere (around 2 km) than surface temperatures (0.03 and 0.04°C/decade, respectively, compared to 0.16°C/decade at the surface). However, over the Northern Hemisphere land areas where urban heat islands are most apparent, both the trends of lower-tropospheric temperature and surface air temperature show no significant differences. In fact, the lower-tropospheric temperatures warm at a slightly greater rate over North America (about 0.28°C/decade using satellite data) than do the surface temperatures (0.27°C/decade), although again the difference is not statistically significant. In the global average, the trend differences arise largely from the tropical and sub-tropical oceans. In many such regions, the near-surface marine air temperatures tend to be cool and dense compared with conditions aloft, allowing for the lapse rate with height, disconnecting near-surface (up to about 1 km) conditions from higher layers in the atmosphere. Thus the surface marine layer and the troposphere above can have differing variations and trends.

Clearly, the urban heat island effect is a real climate change in urban areas, but is not representative of larger areas. Extensive tests have shown that the urban heat island effects are no more than about 0.05°C up to 1990 in the global temperature records used in this chapter to depict climate change. Thus we have assumed an uncertainty of zero in global land-surface air temperature in 1900 due to urbanisation, linearly increasing to 0.06°C (two standard deviations 0.12°C) in 2000.

series do not exceed about 0.05°C over the period 1900 to 1990 (assumed here to represent one standard error in the assessed non-urban trends). However, greater urbanisation influences in future cannot be discounted. Note that changes in borehole temperatures (Section 2.3.2), the recession of the glaciers (Section 2.2.5.4), and changes in marine temperature (Section 2.2.2.2), which are not subject to urbanisation, agree well with the instrumental estimates of surface warming over the last century. Reviews of the homogeneity and construction of current surface air temperature databases appear in Peterson *et al.* (1998b) and Jones *et al.* (1999a). The latter shows that global temperature anomalies can be converted into absolute temperature values with only a small extra uncertainty.

Figure 2.1a shows the Jones *et al.* (2001) CRU (Climatic Research Unit) annual averages, together with an approximately decadally smoothed curve, to highlight decadal and longer changes. This is compared with smoothed curves from the other three analyses in Figure 2.1b. We do not show standard errors for the CRU land data using the Jones *et al.* (1997b) method as tests suggest that these may not be reliable for land data on its own. Instead we use an optimum averaging method (Folland *et al.*, 2001) where the calculated uncertainties are centred on the simple CRU average. We have added an estimate of the additional, independent, uncertainty (twice the standard error)

due to urbanisation increasing from zero in 1900 to 0.12°C in 2000. (The Jones *et al.* (1990) estimates can be interpreted as one standard error equal to 10% of the global warming to that time of about 0.05°C, see also Box 2.1 on urbanisation.) Note that the warming substantially exceeds the calculated uncertainties. (We have not included the possible refinement of assuming urbanisation uncertainties to apply to the cold side of the trend line only, which would reduce the total uncertainty range in Figure 2.1.)

Over global land, a further warming of surface air temperature has occurred since the SAR. The Peterson and Vose (1997) NCDC (National Climate Data Center) series gives distinctly more warming than does the CRU series since the mid-1980s. The former series is a straightforward average of local land areas, weighted according to their size, whereas the CRU series is a simple average of the two hemispheres which gives more weight to the relatively small area of the Southern Hemisphere land. Because the Northern Hemisphere land has warmed considerably faster than the Southern Hemisphere land since the mid-1980s (reflected in Table 2.1), the simple average results in less warming. The Hansen *et al.* (1999) GISS (Goddard Institute for Space Studies) series has recently been revised and shows a little less warming than the CRU series since the late 1980s. One reason for this behaviour lies in the way that the Hansen series is constructed. Among other differences, this series gives much more weight to oceanic islands and Antarctica. Because the oceans and Antarctica have warmed less than the rest of the global land in the last fifteen years (see below), the Hansen series can be expected to show less warming. Some of these considerations apply to the Vinnikov *et al.* (1990) SHI (State Hydrological Institute) series, though this excludes areas south of 60°S.

A new record was set in all four series in 1998 (anomalies relative to 1961 to 1990 of CRU, 0.68°C; NCDC, 0.87°C; GISS, 0.58°C; and SHI, 0.58°C). 1998 was influenced by the strong 1997/98 El Niño; the warming influence of El Niño on global temperature is empirically well attested (e.g., Jones, 1994) and the physical causes are starting to be uncovered (Meehl *et al.*, 1998). However, 1998 was considerably warmer than 1983, a year warmed by the comparable 1982/83 El Niño. In fact 1998 was between 0.34 and 0.54°C warmer than 1983 over land, depending on the temperature series used, though there was some offsetting cooling from volcanic aerosols from the 1982 El Chichon eruption in 1983. 1999 was globally much cooler than 1998, with an anomaly of 0.40°C in the CRU series, as it was cooled by the strongest La Niña since 1988/89. Despite its relative coolness, 1999 was still the fifth warmest year in the CRU record. Depending on the record used, 1999 was between 0.11°C and 0.33°C warmer than the last comparable La Niña year, 1989. It is noteworthy, however, that north of 20°N, 1999 was nearly as warm as 1998. Mitigation of the warming trend in the early 1990s was short-lived and was mainly due to the cooling influence of the eruption of Mount Pinatubo in 1991 (Parker *et al.*, 1996), highlighted in the SAR. The ten warmest years in all four records have occurred after 1980, six or seven of them in the 1990s in each series.

Based on the CRU series, equivalent linear trends in global, Northern and Southern Hemisphere land-surface air temperature are shown in Table 2.1. Because warming may not persist at the rates shown, all trends are shown in °C/decade. The two main periods of warming in all three series are between about 1910 to 1945 and between 1976 to 2000 (updated from Karl *et al.*, 2000). Trends have been calculated using a restricted maximum likelihood method (Diggle *et al.*, 1999) that allows for serial correlation in the data. It gives larger standard errors than ordinary least squares methods when data have a complex temporal structure, as is true here. Table 2.1 and Figure 2.1 show that the rate of global and hemispheric warming in land-surface air temperature from 1976 to 2000 was about twice as fast (but interannually more variable) than that for the period 1910 to 1945. However, trends over such short periods are very susceptible to end effects so the values in Table 2.1, and Table 2.2 below, should be viewed with caution for these periods. Both periods of warming are statistically significant, as is (easily) the warming since 1861 or 1901. Uncertainties in the annual values due to data gaps, including an additional estimate of uncertainties due to urbanisation, are included for land-surface air temperature but equivalent uncertainties are not currently available for the marine data alone. Thus estimates in Table 2.1 for the marine data may be conservative, though the effect of adding the influence of annual uncertainties to the land-surface air temperature data trends was small. The period 1946 to 1975 had no significant change of temperature, though there was a small non-significant, but regionally more marked, cooling over the Northern Hemisphere, as discussed by Parker *et al.* (1994).

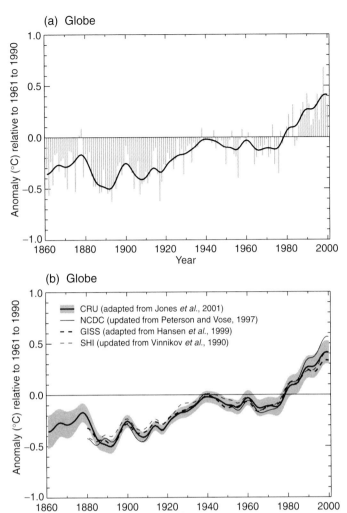

Figure 2.1: (a) Annual anomalies of global average land-surface air temperature (°C), 1861 to 2000, relative to 1961 to 1990 values. Bars and solid curve are from CRU (Jones *et al.*, 2001). Values are the simple average of the anomalies for the two hemispheres. The smoothed curve was created using a 21-point binomial filter giving near decadal averages. (b) As (a) but smoothed curves only from NCDC (updated from Peterson and Vose, 1997) – thin solid curve; GISS (adapted from Hansen *et al.*, 1999) – thick dashed curve; SHI (updated from Vinnikov *et al.*, 1990) – thin dashed curve to 1999 only; Peterson and Vose (1997) – thin solid curve. Thick solid curve – as in (a). Two standard error uncertainties are centred on the CRU curve and are estimated using an optimum averaging method (Folland *et al.*, 2001) and include uncertainties due to urbanisation but not due to uncertainties in thermometer exposures. The NCDC curve is the weighted average of the two hemispheres according to the area sampled, which accounts for most of the differences from the CRU curve.

The equivalent linear changes in global average CRU land-surface air temperature over 1861 to 2000 and 1901 to 2000 that take into account annual sampling errors and uncertainties due to urbanisation are 0.63 ± 0.24°C and 0.61 ± 0.18°C respectively. Corresponding Northern and Southern Hemisphere changes for 1901 to 2000 are 0.71 ± 0.31°C and 0.52 ± 0.13°C, respectively. Marine surface temperatures are discussed further in Section 2.2.2.2.

Table 2.1: *Restricted maximum likelihood linear trends in annual average land-surface air temperature (LSAT) anomalies from CRU and sea surface temperature (SST) and night marine air temperature (NMAT) anomalies from the UK Met Office (UKMO). Twice the standard errors of the trends are shown in brackets. Trends significant at the 5% level or better, according to calculations made using an appropriate form of the t test, are shown in bold type. The significances of the trends are indicated beneath their twice standard errors. The method for calculating the trends, standard errors and significances allows for serial correlation and can result in a trend for the globe that is not exactly equal to the average of the trends for the hemispheres, consistent with uncertainties in the trends. The estimates of trends and errors for the land data account for uncertainties in the annual anomalies due to data gaps and urbanisation. Uncertainties in annual marine anomalies are not available. Trends are given in °C/decade.*

	1861 to 2000	1901 to 2000	1910 to 1945	1946 to 1975	1976 to 2000
Northern Hemisphere CRU LSAT (Jones *et al.*, 2001)	**0.06** **(0.02)** 1%	**0.07** **(0.03)** 1%	**0.14** **(0.05)** 1%	−0.04 (0.06)	**0.31** **(0.11)** 1%
Southern Hemisphere CRU LSAT (Jones *et al.*, 2001)	**0.03** **(0.01)** 1%	**0.05** **(0.01)** 1%	**0.08** **(0.04)** 1%	0.02 (0.05)	**0.13** **(0.08)** 1%
Global CRU LSAT (Jones *et al.*, 2001)	**0.05** **(0.02)** 1%	**0.06** **(0.02)** 1%	**0.11** **(0.03)** 1%	−0.01 (0.05)	**0.22** **(0.08)** 1%
Northern Hemisphere UKMO SST (Jones *et al.*, 2001)	**0.03** **(0.01)** 1%	**0.05** **(0.02)** 1%	**0.15** **(0.04)** 1%	−0.05 (0.10)	**0.18** **(0.05)** 1%
Southern Hemisphere UKMO SST (Jones *et al.*, 2001)	**0.04** **(0.01)** 1%	**0.06** **(0.01)** 1%	**0.13** **(0.05)** 1%	0.06 (0.07)	**0.10** **(0.05)** 1%
Global UKMO SST (Jones *et al.*, 2001)	**0.04** **(0.01)** 1%	**0.06** **(0.01)** 1%	**0.15** **(0.04)** 1%	0.01 (0.06)	**0.14** **(0.04)** 1%
Global UKMO NMAT (Parker *et al.*, 1995)		**0.05** **(0.02)** 1%	**0.14** **(0.04)** 1%	−0.01 (0.06)	**0.11** **(0.05)** 1%

Maximum and minimum temperature

As reported in the SAR, and updated by Easterling *et al.* (1997), the increase in temperature in recent decades has involved a faster rise in daily minimum than daily maximum temperature in many continental regions. This gives a decrease in the diurnal temperature range (DTR) in many parts of the world. The analysis by Easterling *et al.* (1997) increased total global coverage from 37% to 54% of global land area. Large parts of the world have still not been analysed due to a lack of observations or inaccessible data, particularly in the tropics. Updating all the data remains a problem, so the analysis ends in 1993.

The overall global trend for the maximum temperature during 1950 to 1993 is approximately 0.1°C/decade and the trend for the minimum temperature is about 0.2°C/decade. Consequently, the trend in the DTR is about −0.1°C/decade. The rate of temperature increase for both maximum and minimum temperature over this period is greater than for the mean temperature over the entire 20th century, reflecting the strong warming in recent decades. Note that these trends for 1950 to 1993 will differ from the global trends due to the restricted data coverage so we only quote trends to 0.1°C.

Since the DTR is the maximum temperature minus the minimum temperature, the DTR can decrease when the trend in the maximum or minimum temperature is downward, upward, or unchanging. This contributes to less spatial coherence on the DTR map than on maps of mean temperature trend. Maximum temperatures have increased over most areas with the notable exception of eastern Canada, the southern United States, portions of Eastern

and southern Europe (Brunetti *et al.*, 2000a), southern China, and parts of southern South America. Minimum temperatures, however, increased almost everywhere except in eastern Canada and small areas of Eastern Europe and the Middle East. The DTR decreased in most areas, except over middle Canada, and parts of southern Africa, south-west Asia, Europe, and the western tropical Pacific Islands. In some areas the pattern of temperature change has been different. In both New Zealand (Salinger, 1995) and central Europe (Weber *et al.*, 1994; Brázdil *et al.*, 1996) maximum and minimum temperatures have increased at similar rates. In India the DTR has increased due to a decrease in the minimum temperature (Kumar *et al.*, 1994). Eastern Canada also shows a slight increase in DTR due to a stronger cooling in maximum temperatures relative to minimum temperatures (Easterling *et al.*, 1997). However, recently annual mean maximum and minimum temperatures for Canada have been analysed using newly homogenised data (Vincent, 1998; Vincent and Gullet, 1999); these have increased by 0.3 and 0.4°C, respectively, over the last fifty years (Zhang *et al.*, 1999). Central England temperature also shows no decrease in DTR since 1878 (Parker and Horton, 1999). Similarly, a new temperature data set for north-east Spain (not available on Figure 2.2 below, Brunet-India *et al.*, 1999a,b), shows an increase in maximum temperature over 1913 to 1998 to be about twice as fast as that of minimum temperature. Recent analyses by Quintana-Gomez (1999) reveal a large reduction in the DTR over Venezuela and Colombia, primarily due to increasing minimum temperatures (up to 0.5°C/decade). In northern China, the decrease in DTR is due to a stronger warming

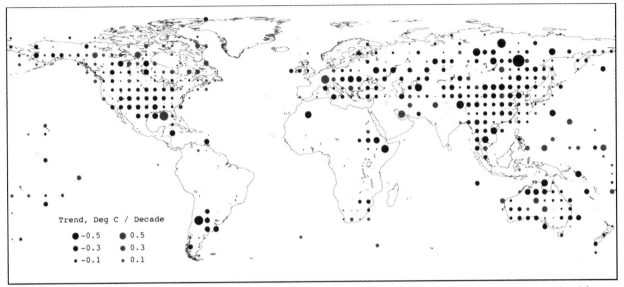

Figure. 2.2: Trends in annual diurnal temperature range (DTR, °C/decade), from 1950 to 1993, for non-urban stations only, updated from Easterling *et al.* (1997). Decreases are in blue and increases in red. This data set of maximum and minimum temperature differs from and has more restricted coverage than those of mean temperature used elsewhere in Section 2.2.

in minimum temperature compared with maximum temperatures. However, in southern China the decreased DTR is due to a cooling in maximum with a slight warming in minimum temperature (Zhai and Ren, 1999).

The DTR is particularly susceptible to urban effects. Gallo *et al.* (1996) examined differences in DTR between stations based on predominant land use in the vicinity of the observing site. Results show statistically significant differences in DTR between stations associated with predominantly rural land use/land cover and those associated with more urban land use/land cover, with rural settings generally having larger DTR than urban settings. Although this shows that the distinction between urban and rural land use is important as one of the factors that can influence the trends observed in temperatures, Figure 2.2 shows annual mean trends in diurnal temperature range in worldwide non-urban stations over the period 1950 to 1993 (from Easterling *et al.,* 1997). The trends for both the maximum and minimum temperatures are about 0.005°C/decade smaller than the trends for the full network including urban sites, which is consistent with earlier estimated urban effects on global temperature anomaly time-series (Jones *et al.*, 1990).

Minimum temperature for both hemispheres increased abruptly in the late 1970s, coincident with an apparent change in the character of the El Niño-Southern Oscillation (ENSO) phenomenon, giving persistently warmer sea temperatures in the tropical central and east Pacific (see Section 2.6.2). Seasonally, the strongest changes in the DTR were in the boreal winter (–0.13°C/decade for rural stations) and the smallest changes were during boreal summer (–0.065°C/decade), indicating some seasonality in the changes. Preliminary extensions of the Easterling *et al.* (1997) analysis to 1997 show that the declining trends in DTR have continued in much of North America and Asia.

Figure 2.3 shows the relationship between cloudiness and the DTR for a number of regions where long-term cloud cover data are available (Dai *et al.*, 1997a). For each region there was an increase in cloud cover over the 20th century and generally a

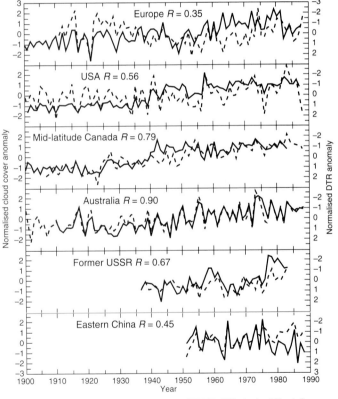

Figure 2.3: Cloud cover (solid line) and DTR (°C, dashed line) for Europe, USA, Canada, Australia, the former Soviet Union, and eastern China (from Dai *et al.*, 1997a). Note that the axis for DTR has been inverted. Therefore, a positive correlation of cloud cover with inverted DTR indicates a negative cloud cover/DTR correlation.

decrease in DTR. In some instances the correlation between annual cloud cover and annual DTR is remarkably strong, suggesting a distinct relationship between cloud cover and DTR. This would be expected since cloud dampens the diurnal cycle of radiation balance at the surface. Anthropogenically-caused increases in tropospheric aerosol loadings have been implicated in some of these cloud cover changes, while the aerosols themselves can cause small changes in DTR without cloud changes (Hansen *et al.*, 1998 and Chapter 6).

2.2.2.2 *Sea surface temperature and ocean air temperature*

The analyses of SST described here all estimate the sub-surface bulk temperature, (i.e. the temperature in the first few metres of the ocean) not the skin temperature. Thus the Reynolds and Smith (1994) and Smith *et al.* (1996) data, which incorporate polar orbiting satellite temperatures, utilise skin temperatures that have been adjusted to estimate bulk SST values through a calibration procedure.

Many historical *in situ* marine data still remain to be digitised and incorporated into the database, to improve coverage and reduce the uncertainties in our estimates of marine climatic variations. A combined physical-empirical method (Folland and Parker, 1995) is used, as in the SAR, to estimate adjustments to ships' SST data obtained up to 1941 to compensate for heat losses from uninsulated (mainly canvas) or partly-insulated (mainly wooden) buckets (see Box 2.2). The corrections are independent of the land-surface air temperature data. Confirmation that these spatially and temporally complex adjustments are quite realistic globally is emerging from simulations of the Jones (1994) land-surface air temperature anomalies using the Hadley Centre atmospheric climate model HadAM3 forced with observed SST and sea-ice extents since 1871, updated from Rayner *et al.* (1996). Figure 2.4 (Folland *et al.*, 2001) shows simulations of global land-surface air temperature anomalies in model runs forced with SST, with and without bias adjustments to the SST data before 1942. All runs with uncorrected SST (only the average is shown) give too cold a simulation of land-surface air temperature for much of the period before 1941 relative to the 1946 to 1965 base period, with a dramatic increase in 1942. All

six individual runs with bias-adjusted SST (only the average is shown) give simulated land air temperatures close to those observed so that internal model variability is small on decadal time-scales compared to the signal being sought. These global results are mostly confirmed by ten similar large regional land-surface air temperature analyses (not shown). Hanawa *et al.* (2000) have provided independent confirmation of the SST bias corrections around Japan. Therefore, our confidence in the SST data sets has increased. Marine data issues are discussed further in Box 2.2, in Trenberth *et al.* (1992) and Folland *et al.* (1993).

Figure 2.5a shows annual values of global SST, using a recently improved UKMO analysis that does not fill regions of missing data (Jones *et al.*, 2001), together with decadally smoothed values of SST from the same analysis. NMAT is also shown. These generally agree well after 1900, but NMAT data are warmer before that time with a slow cooling trend from 1860 not seen in the SSTs, though the minimum around 1910 is seen in both series. The SST analysis from the SAR is also shown. The changes in SST since the SAR are generally fairly small, though the peak warmth in the early 1940s is more evident in the more recent analysis, supported by the NMAT analysis. A contribution to decadally averaged global warmth at that time is likely to have arisen from closely spaced multiple El Niño events centred near 1939 to 1941 and perhaps 1942 to 1944 (Bigg and Inonue, 1992; and Figure 2.29). The NMAT data largely avoid daytime heating of ships' decks (Bottomley *et al.*, 1990; Folland and Parker, 1995). Although NMAT data have been corrected for warm biases in World War II they may still be too warm in the Northern Hemisphere at that time (Figure 2.5c), though there is good agreement in the Southern Hemisphere (Figure 2.5d). The NMAT analysis is based on that in Parker *et al.* (1995) but differs from that used in the SAR in that it incorporates optimal inter-polated data using orthogonal spatial patterns (eigenvectors). This is similar to the technique described by Kaplan *et al.* (1997, 1998) but with additional allowance for non-stationarity of the data (Parker *et al*, 1995). Great care is needed in making these reconstructions in a changing climate, as pointed out by Hurrell and Trenberth (1999). This NMAT analysis has been chosen because of the often very sparse data. NMAT confirms the SST trends in the 20th century until 1991 (see also Table 2.1). After 1991, NMAT warmed at a slower rate than SST in parts of the Southern Hemisphere, notably the South Indian and the tropical South Pacific Oceans. Overall, however, the SST data should be regarded as more reliable, though the relative changes in NMAT since 1991 may be partly real (Christy *et al.*, 2001). The similar trends in SST and island air temperature found by Folland *et al.* (1997) for four regions of the tropical and extra-tropical South Pacific over much of the last century support the generally greater reliability of the SST data.

Figure 2.5b shows three time-series of changes in global SST. The UKMO series (as in Figure 2.5a) does not include polar orbiting satellite data because of possible time-varying biases in them that remain difficult to correct fully (Reynolds, 1993) though the NCEP (National Centers for Environmental Prediction) data (adapted from Smith *et al.*, 1996 and Reynolds and Smith, 1994), starting in 1950, do include satellite data after 1981. The NCDC series (updated from Quayle *et al.*, 1999) starts in 1880 and

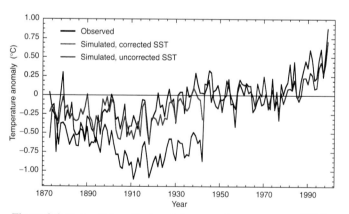

Figure 2.4: Tests of bias adjustments to sea surface temperature (SST) using a climate model (Folland *et al.*, 2001). Black line - annual mean observed land surface air temperature (SAT) anomaly (°C) from a 1946 to 1965 average (Jones, 1994), a period before major anthropogenic warming. Red line – annual averages of four simulations of SAT anomalies using uncorrected SST data, 1872 to 1941, and an average of six simulations for 1941 to 1998. Blue line – average of six simulations of SAT, forced with SST data corrected up to 1941 (Folland and Parker, 1995). Simulated data are collocated with available observations.

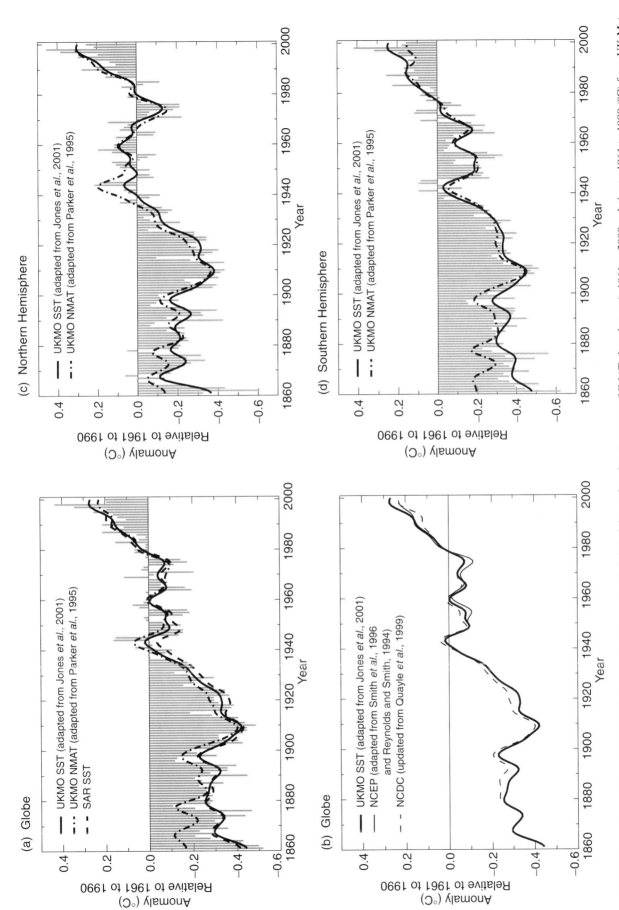

Figure 2.5: (a) Annual anomalies of global SST (bars and solid curve) and global night marine air temperature (NMAT, dotted curve), 1861 to 2000, relative to 1961 to 1990 (°C) from UK Met Office analyses (NMAT updated from Parker *et al.*, 1995). Smoothed curves were created using a 21-point binomial filter to give near-decadal averages. Also shown are the equivalent SST anomalies from the SAR – dashed curve. (b) Smoothed annual global SST (°C), 1861 to 2000, relative to 1961 to 1990, from USA National Climate Data Centre, Quayle *et al.* (1999) (thin dashed line, includes satellite data); USA National Centres for Environmental Prediction, Reynolds and Smith (1994) and Smith *et al.* (1996) (thin solid line, includes satellite data, to 1999 only), and UK Met Office (Jones *et al.*, 2001) (thick line). (c) UKMO SST and NMAT anomaly time-series from a 1961 to 1990 average for the Northern Hemisphere. (d) As (c) but for the Southern Hemisphere. Both for 1861 to 2000.

Box 2.2: Adjustments and Corrections to Marine Observations.

The SST data used here comprise over 80 million observations from the UK Main Marine Data Bank, the United States Comprehensive Ocean Atmosphere Data Set (COADS) and recent information telecommunicated from ships and buoys from the World Weather Watch. These observations have been carefully checked for homogeneity and carefully corrected for the use of uninsulated wooden and canvas buckets for collecting seawater prior to 1942. However, corrections prior to about 1900 are less well known because of uncertainties in the mix of wooden and canvas buckets. Nevertheless, Figure 2.4 provides good evidence that even in the 1870s, SST was little biased relative to land-surface air temperatures globally. Since 1941, observations mainly come from ship engine intake measurements, better insulated buckets and, latterly, from buoys. SST anomalies (from a 1961 to 1990 average) are first averaged into 1° latitude by 1° longitude boxes for five-day periods; the anomaly for a given observation is calculated from a 1° box climatology that changes each day throughout the year. The five-day 1° box anomalies are then aggregated into 5° boxes for the whole month with outlying values rejected, and monthly average anomalies calculated. Further adjustments are made to monthly SST anomalies for the varying numbers of observations in each 5° box because when observations are few, random errors tend to increase the variance of the monthly mean. NMAT data are treated similarly and have quite similar characteristics. However, a variance adjustment to NMAT data is not yet made. NMAT data are also corrected for the progressive increase in the height of thermometer screens on ships above the ocean surface, though no corrections have been made since 1930. Because there are only about half as many NMAT as SST data and NMAT have smaller temporal persistence, monthly NMAT anomalies may be less representative than SST anomalies even on quite large space scales. On longer time-scales, and over the majority of large ocean regions in the 20th century, there is good agreement between NMAT and SST. 19th century NMAT anomaly time-series should be viewed cautiously because of the sparse character of the constituent observations, and regionally varying biases, only some of which have been corrected.

includes satellite data to provide nearly complete global coverage. Up to 1981, the Quayle *et al* series is based on the UKMO series, adjusted by linear regression to match the NCEP series after 1981. It has a truly global coverage based on the optimally interpolated Reynolds and Smith data. The Kaplan *et al.* (1998) global analysis is not shown because it makes no allowance for non-stationarity (here the existence of global warming) in its optimum interpolation procedures, as noted by Hurrell and Trenberth (1999). The warmest year globally in each SST record was 1998 (UKMO, 0.43°C, NCDC, 0.39°C, and NCEP, 0.34°C, above the 1961 to 1990 average). The latter two analyses are in principle affected by artificially reduced trends in the satellite data (Hurrell and Trenberth, 1999), though the data we show include recent attempts to reduce this. The global SST show mostly similar trends to those of the land-surface air temperature until 1976, but the trend since 1976 is markedly less (Table 2.1). NMAT trends are not calculated from 1861, as they are too unreliable. The difference in trend between global SST and global land air temperature since 1976 does not appear to be significant, but the trend in NMAT (despite any residual data problems) does appear to be less than that in the land air temperature since 1976. Figures 2.5c and d show that NMAT and SST trends remain very similar in the Northern Hemisphere to the end of the record, but diverge rather suddenly in the Southern Hemisphere from about 1991, as mentioned above. The five warmest years in each of the UKMO, NCDC and NCEP SST analyses have occurred after 1986, four of them in the 1990s in the UKMO analysis.

Particularly strong warming has occurred in the extra-tropical North Atlantic since the mid-1980s (approximately 35° to 65°N, 0° to 35°W not shown). This warming appears to be related in part to the warming phase of a multi-decadal fluctuation (Folland *et al.*, 1986, 1999a; Delworth and Mann, 2000; see Section 2.6), perhaps not confined to the North Atlantic (Minobe, 1997; Chao *et al.*, 2000), though global warming is likely to be

contributing too. In addition, the cooling in the north-western North Atlantic south of Greenland, reported in the SAR, has ceased. These features were noted by Hansen *et al.* (1999).

2.2.2.3 *Land and sea combined*

Figure 2.6 summarises the relative changes of UKMO SST, UKMO NMAT and CRU land-surface air temperature. The greater warming of the land in recent years is clear, but otherwise all three curves have a generally similar shape except that modest cooling of NMAT in the late 19th century is not seen in the SST data as noted for Figure 2.5. The relative coldness of the land around 1885 to 1895 comes from the Northern Hemisphere continental interiors, particularly in winter, as global coastal land air temperature and adjacent SST anomalies agree well at this time (Parker *et al.*, 1995), confirmed by the Jones *et al.* (2001) data. Note that there are some systematic compensating differences between the land and SST in the late 19th century in both hemispheres (not shown). The CRU land data are generally about 0.1 to 0.2°C colder in the Northern Hemisphere except at the beginning of the record (early 1860s), when they agree, and rather colder than this in 1885 to 1890. The opposite is seen in the Southern Hemisphere before 1885 when SST is generally 0.1 to 0.2°C colder and 0.3°C colder around 1875. Overall the SST data are less variable in each hemisphere in these rather poorly observed periods. The Southern Hemisphere land temperature at this time can actually represent a very small observed area of the hemisphere while the SST data, though sparse, are generally considerably more widespread. The sharp cooling in SST around 1903/4 in Figures 2.5 and 2.6, seen in the land as well as the two ocean surface data sets, was discussed for the North Atlantic and Indian Oceans by Helland-Hansen and Nansen (1920) not long after the event. The reduced warming of the NMAT in the last decade reflects differences in the Southern Hemisphere discussed above. Slightly greater warming of the global ocean than the

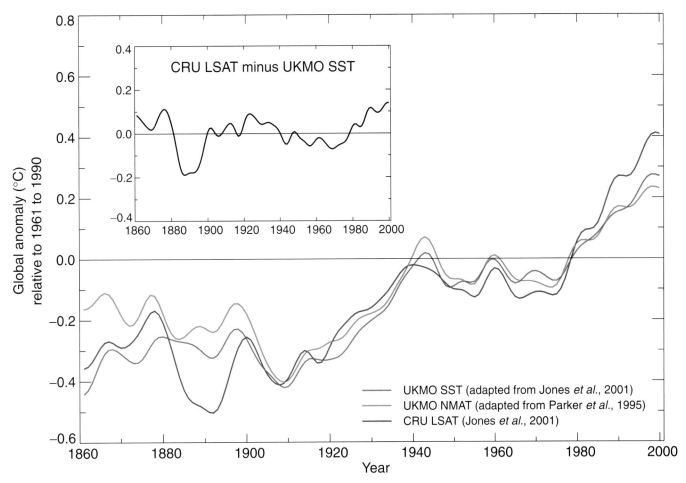

Figure 2.6: Smoothed annual anomalies of global average sea surface temperature (°C) 1861 to 2000, relative to 1961 to 1990 (blue curve), night marine air temperature (green curve), and land-surface air temperature (red curve). The data are from UK Met Office and CRU analyses (adapted from Jones *et al.*, 2001, and Parker *et al.*, 1995). The smoothed curves were created using a 21-point binomial filter giving near-decadal averages. Also shown (inset) are the smoothed differences between the land-surface air and sea surface temperature anomalies.

global land in 1910 to 1945 (seen in Table 2.1) is within the uncertainties of either data set, as a slightly slower warming of the ocean might be expected on physical grounds.

Figures 2.7a to c show annual time-series of anomalies of combined land-surface air temperature and SST for the hemispheres and globe since 1861, based on the latest CRU land air temperature data and the UKMO SST data. Jones *et al.* (2001) temperature data have been averaged by both a standard weighting method, used in the SAR, as shown by the dashed smoothed curves, and by an optimum averaging method (Shen *et al.*, 1998; Folland *et al.*, 2001) as shown by the bars and solid smoothed curves. The latter method uses the variance-covariance matrix instead of correlation functions (Kagan, 1997). The calculated uncertainties (twice the standard error) in the annual values are also shown (including the independent urbanisation and SST bias correction uncertainties). Optimum averaging gives less weight to areas of high data uncertainty than do ordinary averaging methods, and it takes much better account of data gaps. It also gives more weight to Antarctica, the great bulk of which (away from the Antarctic Peninsula) has warmed little in the last two decades (Comiso, 2000). Optimum averages can affect individual

years markedly when data are sparse. Thus extra warmth of the warm year 1878 (strongly affected by the 1877/78 El Niño) in the Northern relative to the Southern Hemisphere in the area weighted average (not shown) disappears when optimum averages are used. In the Northern Hemisphere, the optimum averages are little different from area weighted averages, but they are consistently warmer in the sparsely sampled Southern Hemisphere before 1940, often by more than one tenth of a degree. The overall effect on global temperature is small, however (Figure 2.7c)

The five warmest global optimally averaged years since the beginning of the record in 1861 all occurred in the 1990s with 1998 having the warmest anomaly (0.55°C). This year was significantly warmer than the second warmest year, 1995 (0.38°C), while 1999 was fourth warmest year, despite the strong La Niña event. The remarkably consistent monthly global warmth of 1998 is discussed in Karl *et al.* (2000).

Table 2.2 shows linear trends of the annual optimum averages, and twice their standard errors, for the globe and hemispheres using the restricted maximum likelihood method as in Table 2.1 and allowing for the annual uncertainties due to data gaps, urbanisation over land, and bias corrections to SST. Since

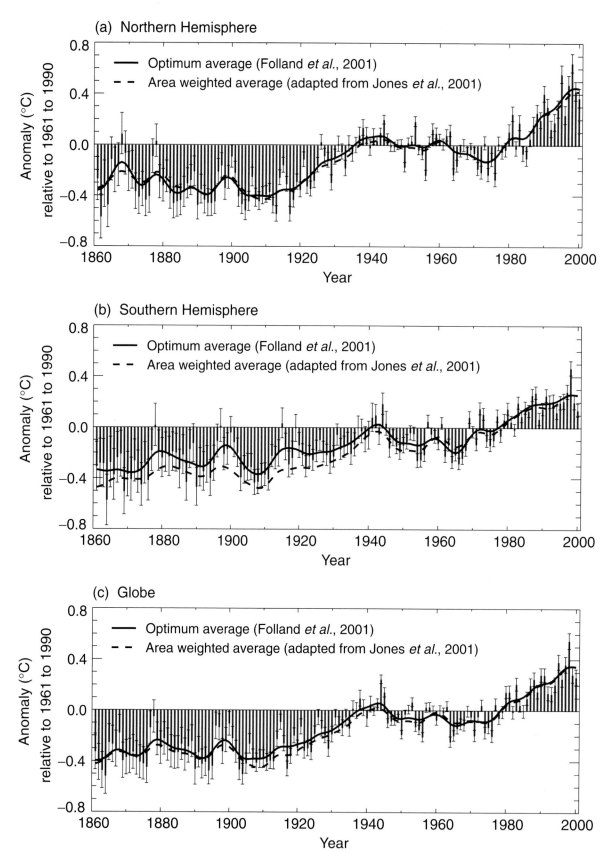

Figure 2.7: Smoothed annual anomalies of combined land-surface air and sea surface temperatures (°C), 1861 to 2000, relative to 1961 to 1990, for (a) Northern Hemisphere; (b) Southern Hemisphere; and (c) Globe. The smoothed curves were created using a 21-point binomial filter giving near-decadal averages. Optimally averaged anomalies (Folland *et al.*, 2001) – solid curves; standard area weighted anomalies (adapted from Jones *et al.*, 2001) – dashed curves. Also shown are the unsmoothed optimum averages – red bars, and twice their standard errors – width denoted by black "I". Note that optimum averages for the Southern Hemisphere are a little warmer before 1940, when the data are sparse, than the area-weighted averages. However, the two types of averaging give similar results in the Northern Hemisphere.

Table 2.2: *As Table 2.1 but for annual optimally averaged combined CRU land-surface air temperature anomalies and UKMO sea surface temperature anomalies (CRU LSAT + UKMO SST). All of the estimates of trends and errors in the table account for uncertainties in the annual anomalies due to data gaps, urbanisation over land, and bias corrections to SST.*

	1861 to 2000	1901 to 2000	1910 to 1945	1946 to 1975	1976 to 2000
Northern Hemisphere CRU LSAT + UKMO SST (Folland *et al.*, 2001)	**0.05** (0.02) 1%	**0.06** (0.02) 1%	**0.17** (0.03) 1%	−0.05 (0.05)	**0.24** (0.07) 1%
Southern Hemisphere CRU LSAT + UKMO SST (Folland *et al.*, 2001)	**0.04** (0.01) 1%	**0.05** (0.02) 1%	**0.09** (0.05) 1%	0.03 (0.07)	**0.11** (0.05) 1%
Global CRU LSAT + UKMO SST (Folland *et al.*, 2001)	**0.04** (0.01) 1%	**0.06** (0.02) 1%	**0.14** (0.04) 1%	−0.01 (0.04)	**0.17** (0.05) 1%

1861 the hemispheres have warmed by approximately the same amount. However both the earlier period of warming (1910 to 1945) and the more recent one (1976 to 1999) saw rates of warming about twice as great in the Northern Hemisphere. There was continued (non-significant) warming in the Southern Hemisphere, though at a reduced rate, in 1946 to 1975, which partially offset (non-significant) cooling in the Northern Hemisphere over the same period to give a (non-significant) 0.03°C cooling globally. The global trend from 1861 to 2000 can be cautiously interpreted as an equivalent linear warming of 0.61°C over the 140-year period, with a 95% confidence level uncertainty of ± 0.16°C. From 1901 an equivalent warming of 0.57°C has occurred, with an uncertainty of ± 0.17°C.

Figure 2.8 shows a smoothed optimally averaged annual global time-series with estimates of uncertainty at ± twice the standard error of the smoothed (near decadal) estimate. Note that the optimum average uncertainties increase in earlier years mainly because of the much larger data gaps. Also shown are uncertainties estimated by Jones *et al.* (1997b) using a different method centred on the Jones *et al.* (2001) land and sea surface temperature series. This series uses the average of anomalies from all available grid boxes, weighted according to grid box area. Therefore, in contrast to the Jones *et al.* (2001) global land-surface air temperature data, the global land and sea surface temperature data are not a simple average of the hemispheres. The optimally averaged uncertainties vary from about 15 to 65% less than those given by Jones *et al.* (1997b). This is reasonable as optimum averages have minimum variance amongst the range of unbiased estimates of the average. Not surprisingly, there is relatively little difference in the decadal averages themselves. However unlike the Jones *et al.* estimates of uncertainty, the optimum average also includes uncertainties in bias corrections to SST up to 1941 (Folland and Parker, 1995) and the uncertainties (as included in Figure 2.1) in the land data component that are due to urbanisation. Cessation of the SST component of uncertainty after 1941 is the reason for a lack of increase in uncertainties in the fairly poorly observed period 1942 to 1945. Uncertainties due to changes in thermometer screens are poorly known but could be 0.1°C globally in the 19th and early 20th centuries (Parker, 1994); they are not included here, but a preliminary analysis appears in Folland *et al.* (2001). For further discussion of changes in land and ocean surface temperature, see Jones *et al.* (1999a).

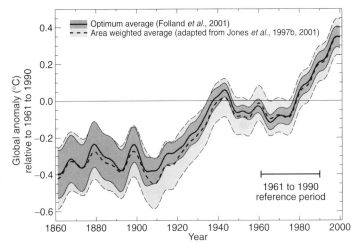

Figure 2.8: Smoothed annual anomalies of global combined land-surface air and sea surface temperatures (°C), 1861 to 2000, relative to 1961 to 1990, and twice their standard errors. The smoothed curves and shaded areas were created using a 21-point binomial filter giving near-decadal averages, with appropriate errors. Optimally averaged anomalies and uncertainties (Folland *et al.*, 2001) – solid curve and dark shading; standard area weighted anomalies and uncertainties (adapted from Jones *et al.*, 1997b, 2001) – dashed curve and light shading. Note that uncertainties decrease after 1941 due to the cessation of uncertainties due to bias corrections in sea surface temperature. On the other hand, uncertainties due to urbanisation of the land component, assessed as zero in 1900, continue to increase after 1941 to a maximum in 2000.

Referring back to Table 2.2 and including the second decimal place, our best estimate of the equivalent linear rate of global land and ocean surface warming between 1861 to 2000 is 0.044°C/decade, or a warming of 0.61 ± 0.16°C. Over the period 1901 to 2000, the equivalent values are 0.058°C/decade or a warming of 0.57 ± 0.17°C. These values include the modifying effects of the annual uncertainties. So we calculate that since the late 19th or the beginning of the 20th century, up to 2000, global warming has been 0.6 ± 0.2°C. This is 0.15°C more warming than the 0.3 to 0.6°C estimated more subjectively up to 1994 by the SAR. This relatively large increase is explained by the increase in temperature since the SAR was completed, improved

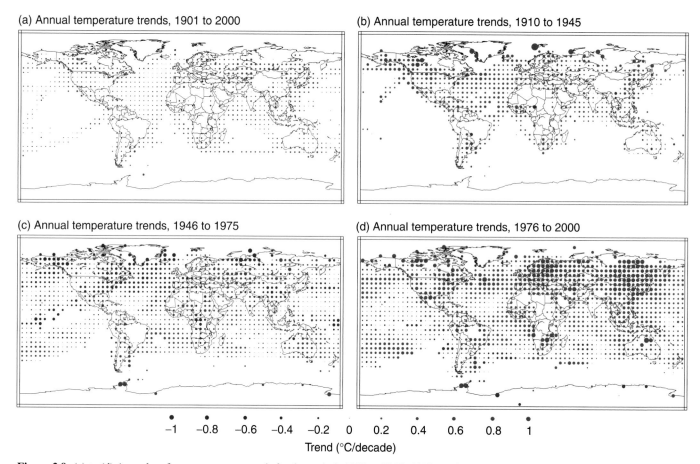

(a) Annual temperature trends, 1901 to 2000

(b) Annual temperature trends, 1910 to 1945

(c) Annual temperature trends, 1946 to 1975

(d) Annual temperature trends, 1976 to 2000

Trend (°C/decade)

Figure 2.9: (a) to (d) Annual surface temperature trends for the periods 1901 to 2000, 1910 to 1945, 1946 to 1975, and 1976 to 2000, respectively (°C/decade), calculated from combined land-surface air and sea surface temperatures adapted from Jones *et al.* (2001). The red, blue and green circles indicate areas with positive trends, negative trends and little or no trend respectively. The size of each circle reflects the size of the trend that it represents. Trends were calculated from annually averaged gridded anomalies with the requirement that annual anomalies include a minimum of 10 months of data. For the period 1901 to 2000, trends were calculated only for those grid boxes containing annual anomalies in at least 66 of the 100 years. The minimum number of years required for the shorter time periods (1910 to 1945, 1946 to 1975, and 1976 to 2000) was 24, 20, and 16 years, respectively.

methods of analysis and the fact that the SAR decided not update the value in the First Assessment Report, despite slight additional warming. The latter decision was likely to have been due to a cautious interpretation of overall uncertainties which had at that time to be subjectively assessed.

2.2.2.4 Are the land and ocean surface temperature changes mutually consistent?
Most of the warming in the 20th century occurred in two distinct periods separated by several decades of little overall globally averaged change, as objectively identified by Karl *et al.* (2000) and discussed in IPCC (1990, 1992, 1996) and several references quoted therein. Figures 2.9 and 2.10 highlight the worldwide behaviour of temperature change in the three periods. These linear trends have been calculated from the Jones *et al.* (2001) gridded combination of UKMO SST and CRU land-surface air tempera-ture, from which the trends in Table 2.2 were calculated. Optimum averaging has not been used for Figures 2.9 and 2.10, and only trends for grid boxes where reasonably complete time-series of data exist are shown. The periods chosen are 1910 to 1945 (first

warming period), 1946 to 1975 (period of little global temperature change), 1976 to 2000 (second warming period, where all four seasons are shown in Figure 2.10) and the 20th century, 1901 to 2000. It can be seen that there is a high degree of local consistency between the SST and land air temperature across the land-ocean boundary, noting that the corrections to SST (Folland and Parker, 1995) are independent of the land data. The consistency with which this should be true locally is not known physically, but is consistent with the similarity of larger-scale coastal land and ocean surface temperature anomalies on decadal time-scales found by Parker *et al.* (1995). The warming observed in the period from 1910 to 1945 was greatest in the Northern Hemisphere high latitudes, as discussed in Parker *et al.* (1994). By contrast, the period from 1946 to 1975 shows widespread cooling in the Northern Hemisphere relative to much of the Southern, consistent with Tables 2.1 and 2.2 and Parker *et al.* (1994). Much of the cooling was seen in the Northern Hemisphere regions that showed most warming in 1910 to 1945 (Figure 2.9 and Parker *et al.*, 1994). In accord with the results in the SAR, recent warming (1976 to 2000) has been greatest over the mid-latitude Northern

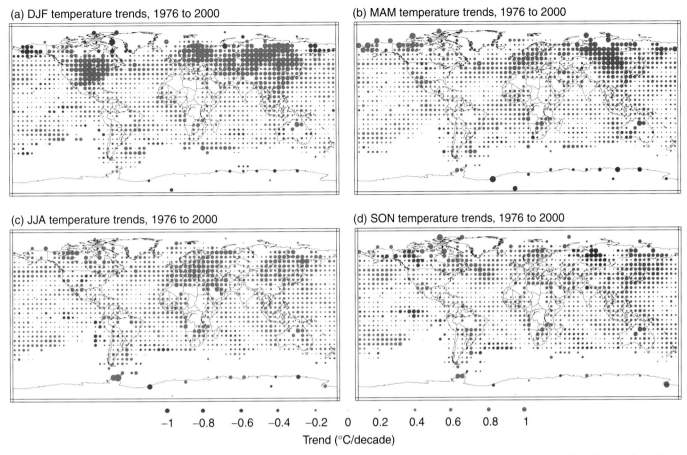

Figure 2.10: (a) to (d) Seasonal surface temperature trends for the period 1976 to 2000 (°C/decade), calculated from combined land-surface air and sea surface temperatures adapted from Jones *et al.* (2001). The red, blue and green circles indicate areas with positive trends, negative trends and little or no trend respectively. The size of each circle reflects the size of the trend that it represents. Trends were calculated from seasonally averaged gridded anomalies with the requirement that the calculation of seasonal anomalies should include all three months. Trends were calculated only for those grid boxes containing seasonal anomalies in at least 16 of the 24 years.

Hemisphere continents in winter. However, the updated data shows only very limited areas of year-round cooling in the north-west North Atlantic and mid-latitude North Pacific. Over 1901 to 2000 as a whole, noting the strong consistency across the land-ocean boundary, most warming is observed over mid- and high latitude Asia and parts of western Canada. The only large areas of observed cooling are just south and east of Greenland and in a few scattered continental regions in the tropics and sub-tropics.

Faster warming of the land-surface temperature than the ocean surface temperature in the last two decades, evident in Figure 2.6, could in part be a signal of anthropogenic warming (see Chapters 9 and 12). However, a component, at least in the Northern Hemisphere north of 40 to 45°N, may result from the sharp increase in the positive phase of the winter half year North Atlantic Oscillation (NAO)/Arctic Oscillation (AO) since about 1970 (Section 2.6.5), though this itself might have an anthropogenic component (Chapter 7). There has also been a strong bias to the warm phase of El Niño since about 1976 (Section 2.6.2). In particular, Hurrell and van Loon (1997) and Thompson *et al.* (2000a) show that the positive phase of the NAO advects additional warm air over extra-tropical Eurasia north of about 45°N. The positive phase of the NAO or AO is therefore likely to be a major cause of

the winter half-year warming in Siberia and northern Europe in Figure 2.10, as also quantified by Hurrell (1996). Cooling over the western North Atlantic Ocean also occurs, partly due to advection of cold air in an enhanced north to north-west airflow. Hurrell (1996) also shows that the warm phase of El Niño is associated with widespread extra-tropical continental warming, particularly over North America and parts of Siberia, with cooling over the North Pacific Ocean. Both effects are consistent with the strong warming over Siberia in winter in 1976 to 2000 (Figure 2.10), warming over much of North America and cooling over the Davis Strait region. Note that some regional details of the seasonal trends for 1976 to 2000 in Figure 2.10 may be sensitive to small changes in record length. A test for the shorter period 1980 to 1997 showed the same general worldwide pattern of (generally somewhat reduced) seasonal warming trends as in Figure 2.10, but with some regional changes, particularly over North America, almost certainly related to atmospheric circulation fluctuations. However, Siberian trends were considerably more robust.

We conclude that in the 20th century we have seen a consistent large-scale warming of the land and ocean surface. Some regional details can be explained from accompanying atmospheric circulation changes.

2.2.2.5 Sub-surface ocean temperatures and salinities

While the upper ocean temperature and salinity are coupled to the atmosphere on diurnal and seasonal time-scales, the deep ocean responds on much longer time-scales. During the last decade, data set development, rescue, declassification and new global surveys have made temperature and salinity profile data more readily available (Levitus *et al.*, 1994, 2000a).

Global

Levitus *et al.* (1997, 2000b) made annual estimates of the heat content of the upper 300 m of the world ocean from 1948 through to 1998 (Figure 2.11). The Atlantic and Indian Oceans each show a similar change from relatively cold to relatively warm conditions around 1976. The Pacific Ocean exhibits more of a bidecadal signal in heat storage. In 1998, the upper 300 m of the world ocean contained $(1.0 \pm 0.5) \times 10^{23}$ Joules more heat than it did in the mid-1950s, which represents a warming of $0.3 \pm 0.15°C$. A least squares linear regression to the annual temperature anomalies from 1958 to 1998 gives a warming of $0.037°C$/decade. White *et al.* (1997, 1998b) computed changes in diabatic heat storage within the seasonal mixed layer from 1955 to 1996 between 20°S and 60°N and observed a warming of $0.15 \pm 0.02°C$ or $0.036°C$/decade.

Extension of the analysis to the upper 3,000 m shows that similar changes in heat content have occurred over intermediate and deep waters in all the basins, especially in the North and South Atlantic and the South Indian Oceans. The change in global ocean heat content from the 1950s to the 1990s is equivalent to a net downwards surface heat flux of 0.3 Wm^{-2} over the whole period.

Pacific

The winter and spring mixed-layer depths over the sub-tropical gyre of the North Pacific deepened 30 to 80% over the period 1960 to 1988 (Polovina *et al.*, 1995). Over the sub-polar gyre, mixed-layer depths reduced by 20 to 30% over the same period. The surface layer of the sub-polar gyre in the north-east Pacific has both warmed and freshened, resulting in a lower surface density (Freeland *et al.*, 1997). Wong *et al.* (1999) compared trans-Pacific data from the early 1990s to historical data collected about twenty years earlier. The changes in temperature and salinity are consistent with surface warming and freshening at mid- and higher latitudes and the subsequent subduction (downward advection) of these changes into the thermocline. From 1968/69 and 1990/91, the South Pacific waters beneath the base of the thermocline have cooled and freshened (Johnson and Orsi, 1997); the greatest cooling and freshening of $-1.0°C$ and 0.25, respectively, occurred near 48°S and were still observed at 20°S. All the deep water masses show a cooling and freshening at these high southern latitudes.

Arctic

Recent surveys of the Arctic Ocean (Quadfasel *et al.*, 1993; Carmack *et al.*, 1995; Jones *et al.*, 1996) have revealed a sub-surface Atlantic-derived warm water layer that is up to 1°C warmer and whose temperature maximum is up to 100 dbar shallower than observed from ice camps from the 1950s to the

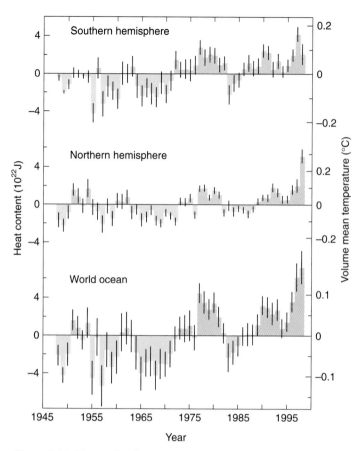

Figure 2.11: Time-series for 1948 to 1998 of ocean heat content anomalies in the upper 300 m for the two hemispheres and the global ocean. Note that 1.5×10^{22} J equals 1 watt-year-m^{-2} averaged over the entire surface of the earth. Vertical lines through each yearly estimate are $\pm$ one standard error (Levitus *et al.*, 2000b).

1980s, as well as from ice-breaker data in the late 1980s and early 1990s. Warming is greatest in the Eurasian Basin. Annual surveys of the southern Canada Basin since 1979 (Melling, 1998), have shown a warming and deepening lower Atlantic layer, the lower halocline layer cooling by 0.12°C and the upper halocline layer warming by 0.15°C. Steele and Boyd (1998) compared winter temperature and salinity profiles obtained over the central and eastern Arctic Basins from submarine transects in 1995 and 1993 with Soviet data collected over the period 1950 to 1989 (Environmental Working Group, 1997). They showed that the cold halocline waters cover significantly less area in the newer data. This is consistent with a decreased supply of cold, fresh halocline waters from the Pacific Shelf areas.

Atlantic

The sub-arctic North Atlantic exhibits decadal variability in both temperature and salinity (Belkin *et al.*, 1998). Reverdin *et al.* (1997) found that the variability of salinity around the entire subarctic gyre for the period 1948 to 1990 was most prominent at periods of 10 years and longer, and extended from the surface to below the base of the winter mixed layer. This salinity signal was only coherent with elsewhere in the north-western Atlantic. A single spatial pattern explains 70% of the variance of the upper ocean salt content of the subarctic gyre, corresponding to a signal propagating from the west to the north-east. Reverdin *et al.* also found that fluctuations in the outflow of fresh water from the Arctic are associated with periods of greater or fewer than usual northerly winds east of Greenland or off the Canadian Archipelago.

North Atlantic deep waters begin as intermediate waters in the Nordic seas. These waters have freshened over the 1980s and 1990s (Bönisch *et al.*, 1997). In addition, the absence of deep convection over the same period has caused Nordic Sea bottom waters to become warmer, saltier and less dense. The Faroes-Shetland Channel is the principal pathway between the north-east Atlantic and the Norwegian Sea and has been surveyed regularly since 1893 (Turrell *et al.*, 1999). Unfortunately, the quality of the salinity measurements was poor from 1930 through to 1960. Since the mid-1970s, the intermediate and bottom waters entering the North Atlantic through the channel have freshened at rates of 0.02/decade and 0.01/decade, respectively. The decreased salinities have resulted in decreased water densities and a decrease of between 1 and 7%/decade in the transport of deep water into the North Atlantic.

In the Labrador Sea, winter oceanic deep convection was intense during the earlier 1990s, extending to deeper than 2,400 m in 1992 to 1994. This produced a Labrador Sea water mass colder, denser and fresher than has been observed over at least the last five decades (Lazier, 1995; Dickson *et al.*, 1996).

Within the tropical and sub-tropical gyres of the North Atlantic, the deep and intermediate water masses are warming. Ocean station S (south-east of Bermuda, 32°17′N, 64°50′W) has been sampled bi-weekly since 1954. Joyce and Robins (1996) extended the hydrographic record from ocean station S back from 1954 to 1922 using nearby observations. They find an almost constant rate of warming over the 1,500 to 2,500 dbar layer of 0.05°C/decade over the 73-year period 1922 to 1995. This corresponds to a net downward heat flux of 0.7 Wm^{-2}. Sections completed in 1958, 1985 and 1997 along 52°W and 66°W between 20°N to 35°N (Joyce *et al.*, 1999) show a rate of warming of 0.06°C/decade, similar to that seen at Bermuda but averaged over a larger 1,700 m depth interval. Trans-Atlantic sections along 24°N in 1957, 1981 and 1992 show a similar warming between 800 and 2,500 m (Parrilla *et al.*, 1994; Bryden *et al.*, 1996). The maximum warming at 1,100 m is occurring at a rate of 0.1°C/decade. At 8°N between 1957 and 1993, Arhan *et al.* (1998) showed warming from 1,150 and 2,800 m with the maximum warming of 0.15°C at 1,660 m.

The Antarctic bottom water in the Argentine Basin of the South Atlantic experienced a marked cooling (0.05°C) and freshening (0.008) during the 1980s (Coles *et al.*, 1996). The bottom waters of the Vema Channel at the northern end of the Argentine basin did not change significantly during the 1980s but warmed steadily during a 700-day set of current meter deployments from 1992 to 1994 (Zenk and Hogg, 1996).

The Indian Ocean

Bindoff and Mcdougall (2000) have examined changes between historical data collected mostly in the period 1959 to 1966 with WOCE data collected in 1987 in the southern Indian Ocean at latitudes 30 to 35°S. They found warming throughout the upper 900 m of the water column (maximum average warming over this section of 0.5°C at 220 dbar).

2.2.3 Temperature of the Upper Air

Uncertainties in discerning changes

Several measuring systems are available to estimate the temperature variations and trends of the air above the surface, though all contain significant time-varying biases as outlined below.

Weather balloons

The longest data sets of upper air temperature are derived from instruments carried aloft by balloons (radiosondes). Changes in balloon instrumentation and data processing over the years have been pervasive, however, resulting in discontinuities in these temperature records (Gaffen, 1994; Parker and Cox, 1995; Parker *et al.*, 1997). Gaffen *et al.* (2000b) attempted to identify these biases by using statistical tests to determine "change-points" – sudden temperature shifts not likely to be of natural origin (e.g., instrument changes). However, they found that alternative methods for identifying change points yield different trend estimates and that the analysis was hampered by the lack of complete documentation of instrument and data processing changes for many stations. This study, however, only analysed change points in the time-series of individual stations in isolation. Another technique, used successfully with surface data, relies on differences produced from comparisons among several stations in close proximity. In addition, Santer *et al.* (1999) noted that temperature trends estimated from radiosonde data sets are sensitive to how temperature shifts are dealt with, which stations are utilised, and the method used for areal averaging.

Worldwide temperatures from the Microwave Sounding Unit (MSU) data (Christy *et al.*, 2000) have been available from the beginning of 1979 for intercomparison studies. Parker *et al.* (1997) used the lower-stratospheric and lower-tropospheric MSU products to adjust monthly radiosonde reports for stations in Australia and New Zealand at times when instrumental or data-processing changes were documented. Some individual stratospheric corrections were as much as 3°C due to radiosonde instrument changes. The main disadvantage of the Parker *et al.* technique is that the raw MSU record has time-varying biases which must first be estimated and eliminated (Christy *et al.*, 2000).

Gaffen *et al.* (2000b) compared trends for 1959 to 1995, calculated using linear regression, for twenty-two stations with nearly complete data records at levels between 850 and 30 hPa. Each of these stations is included in two data sets created since the SAR: (a) monthly mean temperatures reported by the weather balloon station operators (Parker *et al.*, 1997; CLIMAT TEMP data) and (b) monthly mean temperatures calculated from archived daily weather balloon releases (Eskridge *et al.*, 1995; CARDS data). Decadal trends at individual sites differed randomly between the two data sets by typically 0.1°C/decade, with the largest differences at highest altitudes. In a few cases the differences were larger and statistically significant at the 1% level. The discrepancies were sometimes traceable to time-of-observation differences of the data used to calculate the averages.

The analysis of trends requires long station data records with minimal missing data. The records for 180 stations in the combined Global Climate Observing System Upper Air Network (GUAN) and the Angell (1988, 2000) network do not generally meet this standard, as only 74 of the GUAN stations, for instance, have at least 85% of tropospheric monthly means available for 1958 to 1998. In the lower stratosphere (up to 30 hPa), only twenty-two stations meet this requirement (Gaffen *et al.*, 2000b). These deficiencies present the dilemma of using either relatively small networks of stations with adequate data (the Southern Hemisphere, in particular, is poorly sampled) or larger networks with poorer quality data (adding uncertainty to the resulting trend estimates).

Characteristics, such as spatial coverage, of each data set derived from the weather balloon data are different. For example, Sterin (1999) used data from over 800 stations from the CARDS and telecommunicated data sets, with only gross spatial and temporal consistency checks. The data were objectively interpolated to all unobserved regions, introducing extra uncertainty. Parker *et al.* (1997) placed CLIMAT TEMP data into 5° latitude × 10° longitude grid boxes from about 400 sites, leaving unobserved boxes missing. Further data sets were created employing limited spatial interpolation and bias-adjustments, but uncertainties related to spatial under-sampling remain (Hurrell *et al.*, 2000). Angell (1988) placed observations from 63 stations into seven broad latitudinal bands, calculated the simple average for each band and produced global, hemispheric and zonal mean anomalies.

Satellites

Radiosondes measure temperatures at discrete levels, but satellite instruments observe the intensity of radiation from deep atmospheric layers. The advantage of satellites is the essentially uniform, global, coverage. The three temperature products that are commonly available from MSU are: the low to mid-troposphere (MSU 2LT, surface to about 8 km), mid-troposphere (MSU 2, surface to about 18 km, hence including some stratospheric emissions) and the lower stratosphere (MSU 4, 15 to 23 km, hence including some tropical tropospheric emissions) (Christy *et al.*, 2000). No other data, such as from radiosondes, are used to construct these MSU data sets. It is important to note that the troposphere and stratosphere are two distinct layers of the atmosphere with substantially different temperature variations and trends. The altitude of the troposphere/stratosphere boundary varies with latitude, being about 16 to 17 km in the tropics but only 8 to 10 km at high latitudes.

Since the SAR, several issues have emerged regarding MSU temperatures. Mo (1995) reported that for one of the longest-lived satellites (NOAA-12, 1991 to 1998) the non-linear calibration coefficients were erroneous, affecting MSU 2 and MSU 2LT. Wentz and Schabel (1998) discovered that satellite orbit decay introduces gradual, spurious cooling in MSU 2LT. Christy *et al.* (1998, 2000) found that instrument responses often differ between the laboratory assessments and on-orbit performance, requiring further corrections. Additional adjustments were also made by recalculating and removing spurious temperature trends due to diurnal effects induced by the east-west drift of the spacecraft (Christy *et al.*, 2000). The magnitude of the spurious trends (1979 to 1998) removed from version D compared to version C were:

orbit decay, −0.11; instrument response, +0.04 and diurnal drift, +0.03°C/decade.

Version D of the MSU data is used in Figure 2.12. The SAR presented version B that for the low to mid-troposphere indicated a global trend about 0.05°C/decade more negative than version D (for 1979 to 1995). Quite separately, Prabhakara *et al.* (1998) generated a version of MSU 2 without corrections for satellite drift or instrument body effects, in many ways similar to MSU 2 version A of Spencer and Christy (1992a, 1992b).

The Stratospheric Sounding Unit (SSU) detects the intensity of thermal emissions and measures deep layer temperatures at altitudes above 20 km (Nash and Forrester, 1986). As with the MSU products, adjustments are required for radiometer biases, diurnal sampling and orbital drift (Chanin and Ramaswamy, 1999).

Rocketsondes and lidar

Data sets generated from rocketsondes have been updated (Golitsyn *et al.*, 1996; Lysenko *et al.*, 1997), providing temperature information to as high as 75 km. Important difficulties arise with these data due to different types of instrumentation, tidal cycles (amplitude 2°C) and to assumed corrections for aerodynamic heating. The last set of adjustments has the most significant impact on trends. The approximately 11-year solar cycle forces a temperature perturbation of >1°C in the mid- to upper stratosphere (30 to 50 km). Keckhut *et al.* (1999) and Dunkerton *et al.* (1998) created a quality-controlled data set of these measurements, which is used in Chanin and Ramaswamy (1999). The very limited number of launch sites leads to some uncertainty in deduced temperatures, and most launches were terminated in the mid-1990s.

Rayleigh lidar measurements began in 1979 at the Haute Provence Observatory in southern France and have spread to locations around the world. Lidar techniques generate the vertical profile of temperature from 30 to 90 km, providing absolute temperatures within 2.5°C accuracy. Chanin and Ramaswamy (1999) have combined MSU 4, SSU, radiosonde, lidar and rocketsonde data to estimate 5-km thick layer temperature variations for altitudes of 15 to 50 km, generally limited to the Northern Hemisphere mid-latitudes.

Reanalyses

The principle of reanalysis is to use observations in the data assimilation scheme of a fixed global weather forecasting model to create a dynamically consistent set of historical atmospheric analyses (Kalnay *et al.*, 1996). Within the assimilation scheme, potentially errant data are amended or excluded using comparisons with neighbours and/or calculated conditions. However, small, time-dependent biases in the observations, of magnitudes important for climate change, are virtually impossible to detect in the model, even in areas of adequate *in situ* data. Furthermore, in areas with few *in situ* data the reanalyses are often affected by inadequate model physics or satellite data for which time-varying biases have not been removed.

Though interannual variability is reproduced well, known discontinuities in reanalysed data sets indicate that further research is required to reduce time-dependent errors to a level suitable for climate change studies (Basist and Chelliah, 1997;

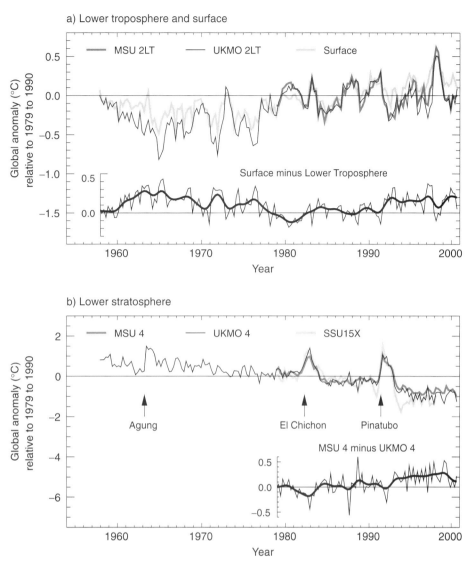

Figure 2.12: (a) Seasonal anomalies of global average temperature (°C), 1958 to 2000, relative to 1979 to 1990 for the lower troposphere, as observed from satellites (MSU 2LT) and balloons (UKMO 2LT), and for the surface (adapted from Jones *et al.*, 2001). Also shown (bottom graph) are the differences between the surface temperature anomalies and the averages of the satellite and balloon-based observations of the lower-tropospheric temperature anomalies. (b) As (a) but for the temperature of the lower stratosphere, as observed from satellites (MSU 4 and SSU 15X) and balloons (UKMO 4). The times of the major explosive eruptions of the Agung, El Chichon and Mt. Pinatubo volcanoes are marked. Also shown (bottom graph) are the differences between the MSU 4 and UKMO 4 based temperature anomalies.

Hurrell and Trenberth, 1998; Santer *et al.*, 1999, 2000; Stendel *et al.*, 2000). It is anticipated that future assessments of climate change will utilise reanalysis products to which substantial improvements will have been made. Data from the NCEP reanalysis are included below for comparison purposes, but longer-term stratospheric trends from NCEP are especially suspect due to a large shift in temperature when satellite data were incorporated for the first time in 1978 (Santer *et al.*, 1999).

2.2.4 How do Surface and Upper Air Temperature Variations Compare?

In Figure 2.12 we display the surface, tropospheric and stratospheric temperature variations using representative data sets from those described above. Trend values (°C/decade) are shown in

Table 2.3 with 95% confidence intervals, which in part represent uncertainties due to temporal sampling, not those due to measurement error (see below). The effect of explosive volcanic events (Agung, 1963; El Chichon, 1982; and Mt. Pinatubo, 1991) is evident in Figure 2.12, as is a relative shift to warmer temperatures in the lower troposphere compared to the surface in the late 1970s, followed by large variations in both due to ENSO (particularly in 1998). After the shift in the late 1970s, the overall tropospheric temperature trend is near zero but the surface has warmed (see Figure 2.12a and Table 2.3).

Global variations and trends in the lower stratosphere are temporally more coherent than in the troposphere (Figure 2.12b), though the warming effects due to the volcanic eruptions are clearly evident. For the period 1958 to 2000, all stratospheric data sets except NCEP 4, which contains erroneous trends, show signif-

Table 2.3: As Table 2.1 but for annual average surface and upper air temperature anomalies from various data sets. The surface temperature trends are of combined land-surface air temperature (LSAT) and sea surface temperature (SST) or sea ice and sea surface temperature (ISST) anomalies. The upper air trends are of temperature anomalies corresponding to or approximately corresponding to temperature anomalies from MSU channels 2LT and 4. The tropical region is defined as the latitude band 20°S to 20°N for all the data sets except for the GISS LSAT + UKMO ISST data set where the region is defined as the latitude band 23.6°S to 23.6°N. The last line of the table shows trends in the differences between temperature anomalies for the surface, from the UKMO LSAT + UKMO SST data set, and for the lower troposphere, taken as the average of the UKMO 2LT and MSU 2LT anomalies for 1979 to 2000 and as the UKMO 2LT anomalies alone before 1979. None of the estimates of trends and errors account for uncertainties in the annual anomalies as these are not available. All calculations use data to the end of 2000 except for those for the NOAA data sets, which include data up to August 2000 only.

	1958 to 2000		1958 to 1978		1979 to 2000	
	Globe	Tropics	Globe	Tropics	Globe	Tropics
Surface						
UKMO LSAT + UKMO SST (Jones *et al.*, 2001)	**0.10** **(0.05)**	**0.08** **(0.06)**	−0.05 (0.07)	−0.09 (0.12)	**0.16** **(0.06)**	0.10 (0.10)
GISS LSAT + UKMO ISST (Hansen *et al.*, 1999; Rayner *et al.*, 2000)	**0.09** **(0.04)**	**0.09** **(0.06)**	−0.03 (0.07)	−0.09 (0.11)	**0.13** **(0.07)**	0.09 (0.10)
NCDC LSAT + NCEP SST (Quayle *et al.*, 1999; Reynolds and Smith, 1994)	**0.09** **(0.05)**	**0.09** **(0.06)**	−0.05 (0.06)	−0.08 (0.11)	**0.14** **(0.06)**	0.10 (0.11)
Lower troposphere						
UKMO 2LT (Parker *et al.*, 1997)	**0.11** **(0.07)**	**0.13** **(0.08)**	−0.03 (0.12)	0.07 (0.16)	0.03 (0.10)	−0.08 (0.12)
MSU 2LT (Christy *et al.*, 2000)					0.04 (0.11)	−0.06 (0.16)
NCEP 2LT (Stendel *et al.*, 2000)	**0.13** **(0.07)**	**0.08** **(0.08)**	0.02 (0.18)	−0.05 (0.17)	0.01 (0.11)	−0.07 (0.14)
NOAA 850−300hPa (Angell, 2000)	0.07 (0.08)	**0.07** **(0.07)**	−0.08 (0.15)	0.04 (0.20)	−0.03 (0.15)	−0.11 (0.19)
RIHMI 850−300hPa (Sterin, 1999)	**0.04** **(0.04)**	**0.07** **(0.05)**	−0.03 (0.06)	0.07 (0.08)	0.00 (0.07)	−0.06 (0.09)
Lower stratosphere						
UKMO 4 (Parker *et al.*, 1997)	**−0.39** **(0.15)**	**−0.31** **(0.19)**	**−0.37** **(0.21)**	−0.07 (0.51)	**−0.64** **(0.47)**	−0.50 (0.54)
MSU 4 (Christy *et al.*, 2000)					**−0.52** **(0.48)**	−0.29 (0.51)
NCEP 4 (Stendel *et al.*, 2000)	−0.25 (0.62)	−0.04 (0.32)	**−0.36** **(0.33)**	**−0.46** **(0.29)**	−0.61 (1.21)	−0.57 (0.77)
NOAA 100−50hPa (Angell, 2000)	**−0.64** **(0.30)**	**−0.58** **(0.39)**	−0.23 (0.22)	0.20 (0.43)	**−1.10** **(0.58)**	−0.68 (2.08)
RIHMI 100−50hPa (Sterin, 1999)	**−0.25** **(0.12)**	**−0.22** **(0.12)**	−0.20 (0.27)	−0.08 (0.10)	**−0.43** **(0.24)**	**−0.45** **(0.28)**
Surface minus lower troposphere						
	−0.01 (0.05)	−0.05 (0.07)	−0.03 (0.08)	**−0.16** **(0.10)**	**0.13** **(0.06)**	**0.17** **(0.06)**

icant negative trends (Table 2.3). Note that MSU 4, and simulations of MSU 4 (UKMO 4 and NCEP 4), include a portion of the upper troposphere below 100 hPa and so are expected to show less negative trends than those measuring at higher altitudes (e.g., the 100 to 50 hPa layers in Table 2.3 and the SSU in Figure 2.12b).

Blended information for 5 km thick levels in the stratosphere at 45°N compiled by Chanin and Ramaswamy (1999) show a negative trend in temperature increasing with height: −0.5°C/decade at 15 km, −0.8°C/decade at 20 to 35 km, and −2.5°C/decade at 50 km. These large, negative trends are consistent with models of the combined effects of ozone depletion and increased concentrations of infrared radiating gases, mainly water vapour and carbon dioxide (Chapters 6 and 12).

The vertical profile of temperature trends based on surface data and radiosondes is consistent with the satellite temperatures. Global trends since 1979 are most positive at the surface, though less positive for night marine air temperatures in the Southern Hemisphere (see Section 2.2.2.2), near zero for levels between 850 to 300 hPa (1.5 to 8 km) and negative at 200 hPa (11 km) and above. Thus during the past two decades, the surface, most of the troposphere, and the stratosphere have responded differently to climate forcings because different physical processes have dominated in each of these regions during that time (Trenberth *et al.*, 1992; Christy and McNider, 1994; NRC, 2000 and Chapter 12). On a longer time-scale, the tropospheric temperature trend since 1958, estimated from a sparser radiosonde network, is closer to that of the surface, about +0.10°C/decade (Figure 2.12a and Table 2.3) (Angell, 1999, 2000; Brown *et al.*, 2000; Gaffen *et al.*, 2000a). Gaffen *et al.* (2000a) and Brown *et al.* (2000) noted a decreasing lower-tropospheric lapse rate from 1958 to 1980, and an increasing lower-tropospheric lapse rate after 1980 (Figure 2.12a). However, Folland *et al.* (1998) showed that global upper-tropospheric temperature has changed little since the late 1960s because the observed stratospheric cooling extended into the uppermost regions of the troposphere.

Between 1979 and 2000, the magnitude of trends between the surface and MSU 2LT is generally most similar in many parts of the Northern Hemisphere extra-tropics (20°N to pole) where deep vertical mixing is often a characteristic of the troposphere. For example, the northern extra-tropical trends for the surface and MSU 2LT were 0.28 and 0.21°C/decade, respectively, and over the North American continent trends were 0.27 ± 0.24 and 0.28 ± 0.23°C/decade, respectively, with an annual correlation of 0.92. Over Europe the rates were 0.38 ± 0.36 and 0.38 ± 0.30°C/decade, respectively. Some additional warming of the surface relative to the lower troposphere would be expected in the winter half year over extra-tropical Asia (whole year warming rates of 0.35 ± 0.20 and 0.18 ± 0.18°C/decade, respectively), consistent with the vertical temperature structure of the increased positive phase of the Arctic Oscillation (Thompson *et al.*, 2000a, Figure 2.30). The vertical structure of the atmosphere in marine environments, however, generally reveals a relatively shallow inversion layer (surface up to 0.7 to 2 km) which is somewhat decoupled from the deep troposphere above (Trenberth *et al.*, 1992; Christy, 1995; Hurrell and Trenberth, 1996). Not only are local surface versus tropospheric correlations often near zero in these regions, but surface and tropospheric trends can be quite different (Chase *et al.*, 2000). This is seen in the different trends for the period 1979 to 2000 in the tropical band, 0.10 ± 0.10 and −0.06 ± 0.16°C/decade, respectively (Table 2.3) and also in the southern extra-tropics where the trends are 0.08 ± 0.06 and −0.05 ± 0.08°C/decade, respectively. Trends calculated for the differences between the surface and the troposphere for 1979 to 2000 are statistically significant globally at 0.13 ± 0.06°C/decade, and even more so in the tropics at 0.17 ± 0.06°C/decade. Statistical significance arises because large interannual variations in the parent time-series are strongly correlated and so largely disappear in the difference time-series (Santer *et al.*, 2000; Christy *et al.*, 2001). However, as implied above, they are not significant over many extra-tropical regions of the Northern Hemisphere such as North America and Europe and they are also insignificant in some Southern Hemisphere areas. The sequence of volcanic eruption, ENSO events, and the trends in the Arctic Oscillation have all been linked to some of this difference in warming rates (Michaels and Knappenburg, 2000; Santer *et al.*, 2000; Thompson *et al.*, 2000a; Wigley, 2000) and do explain a part of the difference in the rates of warming (see Chapter 12).

The linear trend is a simple measure of the overall tendency of a time-series and has several types of uncertainty; temporal sampling uncertainty owing to short data sets, spatial sampling errors owing to incomplete spatial sampling, and various other forms of measurement error, such as instrument or calibration errors. Temporal sampling uncertainties are present even when the data are perfectly known because trends calculated for short periods are unrepresentative of other short periods, or of the longer term, due to large interannual to decadal variations. Thus confidence intervals for estimates of trend since 1979 due to temporal sampling uncertainty can be relatively large, as high as ± 0.2°C/decade below 300 hPa (Table 2.3, Santer *et al.*, 2000). Accordingly, the period from 1979 to 2000 provides limited information on long-term trends, or trends for other 22-year periods.

Uncertainties arising from measurement errors due to the factors discussed in Section 2.2.3, including incomplete spatial sampling, can be substantial. One estimate of this uncertainty can be made from comparisons between the various analyses in Table 2.3. For trends below 300 hPa, this uncertainty may be as large as ± 0.10°C/decade since 1979, though Christy *et al.* (2000) estimate the 95% confidence interval as ± 0.06°C for the MSU 2LT layer average. For example, Santer *et al.* (2000) find that when the satellite observations from MSU 2LT are masked to match the less than complete global coverage of the surface observations during the past few decades, the differences in the trends between the surface and the troposphere are reduced by about one third.

Summarising, it is very likely that the surface has warmed in the global average relative to the troposphere, and the troposphere has warmed relative to the stratosphere since 1979 (Figure 2.12a,b; Pielke *et al.*, 1998a,b; Angell, 1999, 2000; Brown *et al.*, 2000; Christy *et al.*, 2000; Gaffen *et al.*, 2000a; Hurrell *et al.*, 2000; NRC, 2000; Stendel *et al.*, 2000). However, the relative warming is spatially very variable and most significant in the tropics and sub-tropics. There is evidence that the troposphere warmed relative to the surface in the pre-satellite era (1958 to 1979, see Brown *et al.*, 2000; Gaffen *et al.*, 2000a), though confidence in this finding is lower. Uncertainties due to limited temporal sampling prevent confident extrapolation of these trends to other or longer time periods (Christy *et al.*, 2000; Hurrell *et al.*, 2000; NRC, 2000; Santer *et al.*, 2000). Some physical explanations for changes in the vertical profile of global temperature trends are discussed in Chapter 12 but a full explanation of the lower-tropospheric lapse rate changes since 1958 requires further research.

2.2.5 Changes in the Cryosphere

This chapter does not describe changes in the major ice sheets as this is dealt with in detail in Chapter 11.

2.2.5.1 Snow cover, including snowfall

Satellite records indicate that the Northern Hemisphere annual snow-cover extent (SCE) has decreased by about 10% since 1966 largely due to decreases in spring and summer since the mid-1980s over both the Eurasian and American continents (Figure 2.13a; Robinson, 1997, 1999). Winter and autumn SCE show no statistically significant change. Reduction in snow cover during the mid- to late 1980s was strongly related to temperature

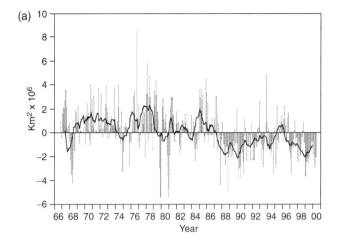

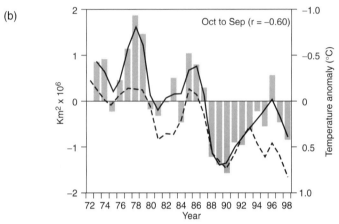

Figure 2.13: (a) Anomalies of monthly snow cover over the Northern Hemisphere lands (including Greenland) between November 1966 and May 2000. Also shown are twelve-month running anomalies of hemispheric snow extent, plotted on the seventh month of a given interval. Anomalies are calculated from NOAA/NESDIS snow maps. Mean hemispheric snow extent is 25.2 million km^2 for the full period of record. the curve of running means is extrapolated by using period of record monthly means for 12 months in the late 1960s in order to create a continuous curve of running means. Missing months fell between May and October, and no winter months are missing. June 1999 to May 2000 values are based on preliminary analyses. (b) Seasonal snowcover anomalies (in million km^2) versus temperature anomalies (in °C). Both snow and temperature anomalies are area averages over the region for which climatological values of seasonal snow-cover frequency (based on the 1973 to 1998 period) are between 10 and 90%. Season is indicated at the top of each panel. Axis for snow anomaly on the left-hand-side y axis, axis for temperature anomaly is on the right-hand-side y axis. Bar plot indicates time-series of snow cover anomalies. Continuous colour curve indicates nine-point weighted average of snow-cover anomaly. Dashed black curve indicates time-series of nine-point weighted average of area average temperature anomaly. Snow-cover calculations are based on the NOAA/NESDIS snow cover data for the period 1973 to 1998 (updated from Robinson *et al.*, 1993). Temperature calculations are based on the Jones data set, hence anomalies are with respect to the time period 1961 to 1990. Snow anomalies are with respect to the time period 1973 to 1998. Correlation coefficient (r) between seasonal snow cover anomalies and temperature anomalies is indicated in parentheses. (Figure contributed by David A. Robinson and Anjuli Bamzai, Rutgers University.)

increases in snow covered areas (Figure 2.13b). There is a highly significant interannual (+0.6) and multi-decadal correlation between increases in the Northern Hemisphere spring land temperature and a reduction in the Northern Hemisphere spring snow cover since data have been available (1966). Snow cover extent has decreased about 10% since 1966. The improvements in the quantity and quality of the visible satellite imagery used to produce the operational snow-cover product cannot account for the observed changes in snow cover.

Longer regional time-series based on station records and reconstructions suggest that Northern Hemisphere spring and summer SCEs in the past decade have been at their lowest values in the past 100 years. In the other seasons, it is likely that extents in the latter portion of the 20th century exceeded those of earlier years (Brown, 2000).

Reconstructions for North America suggest that while there has been a general decrease in spring SCE since 1915, it is likely that winter SCE has increased (Brown and Goodison, 1996; Frei *et al.*, 1999; Hughes and Robinson, 1996; Hughes *et al.*, 1996). Similar to the results in North America, in Eurasia April SCE has significantly decreased; but lack of data has prevented an analysis of winter trends (Brown, 2000). Over Canada, there has been a general decrease in snow depth since 1946, especially during spring, in agreement with decreases in SCE (Brown and Braaten, 1998). Winter depths have declined over European Russia since 1900 (Meshcherskaya *et al.*, 1995), but have increased elsewhere over Russia in the past few decades (Fallot *et al.*, 1997). The common thread between studies that have examined seasonality is an overall reduction in spring snow cover in the latter half of the 20th century.

There have been relatively few studies of snowfall trends across the globe. Statistically significant increases in seasonal snowfall have been observed over the central USA in the 20th century (Hughes and Robinson, 1996). In recent decades, snowfall has also been heavier to the lee of the North American Great Lakes than earlier in the century (Leathers and Ellis, 1996). These findings are in line with observations from Canada and the former Soviet Union, reflecting a trend towards increased precipitation over the mid-latitude lands in the Northern Hemisphere (Groisman and Easterling, 1994; Brown and Goodison, 1996; Ye *et al.*, 1998).

2.2.5.2 *Sea-ice extent and thickness*
Sea-ice extent
Sea-ice extent is expected to become a sensitive indicator of a warming climate, although only recently have long records become available in the Arctic, and our knowledge of Antarctic sea-ice extent before the 1970s is very limited.

Sea-ice extent (the area within the ice-ocean margin) was observed from space from 1973 to 1976 using the ESMR (Electrically Scanning Microwave Radiometer) satellite-based instrument, and then continuously from 1978 using the SSMR (Scanning Multichannel Microwave Radiometer) (1978 to 1987) and SSM/I (Special Sensor Microwave/Imager) (1987 to present). By inter-calibrating data from different satellites, Bjørgo *et al.* (1997) and subsequently Cavalieri *et al.* (1997) obtained uniform monthly estimates of sea-ice extent for both hemispheres from

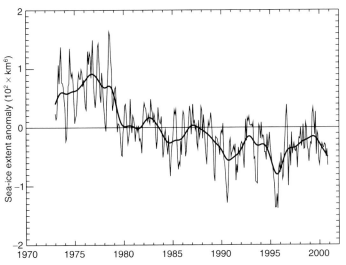

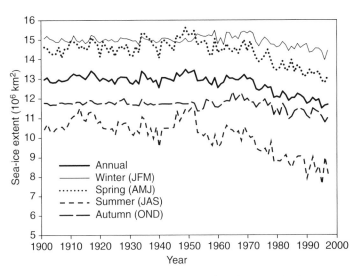

Figure 2.14: Monthly Arctic sea-ice extent anomalies, 1973 to 2000, relative to 1973 to 1996. The data are a blend of updated Walsh (Walsh, 1978), Goddard Space Flight Center satellite passive microwave (Scanning Multichannel Microwave Radiometer (SMMR) and Special Sensor Microwave/Imager (SSM/I)) derived data (Cavalieri *et al.*, 1997) and National Centers for Environmental Prediction satellite passive microwave derived data (Grumbine, 1996). Updated digitised ice data for the Great Lakes are also included (Assel, 1983).

Fig 2.15: Time-series of annual and seasonal sea-ice extent in the Northern Hemisphere, 1901 to 1999, (Annual values from Vinnikov *et al.*, 1999b; seasonal values updated from Chapman and Walsh, 1993.)

November 1978 through to December 1996. Over this period, the sea-ice extent over the Northern Hemisphere showed a decrease of −2.8 ± 0.3%/decade (Parkinson *et al.*, 1999), consistent with Johannessen *et al.* (1995) (Figure 2.14). The Arctic decrease was strongest in the Eastern Hemisphere and most apparent in summer (Maslanik *et al.*, 1996; Parkinson *et al.*, 1999).

Hemispheric and regional data sets for the Arctic enable the satellite-derived trends in Figure 2.14 to be placed into a century scale context. Figure 2.15 shows annual time-series of the Northern Hemisphere ice extent by season from 1901 to 1999 using *in situ* data before the satellite era (Vinnikov *et al.*, 1999a). It should be emphasised that the spatial coverage of earlier data is not complete, with the largest data voids in the autumn and winter. Because few data were available, the variability of the autumn and wintertime series in Figure 2.15 is smaller during the early decades of the century. Essentially no data for summer and autumn are available for the World War II period. The summer decrease that is largely responsible for the overall downward trend during the satellite era is present during the entire second half of the 20th century (Figure 2.15). This decrease represents about 15% of the average summer extent in the first half of the 20th century. Spring values show a somewhat weaker negative trend over the same period with a total reduction of near 8%, but there is only a slight and uncertain downward trend in autumn and winter since about 1970.

The overall recent decrease of Arctic ice extent is, at first sight, consistent with the recent pattern of high latitude temperature change, which includes a warming over most of the sub-arctic land areas (Section 2.2.2.1). Some of this pattern of warming has been attributed to recent trends in the atmospheric circulation of the North Atlantic Oscillation and its Arctic-wide manifestation, the Arctic Oscillation (Section 2.6).

Related to the decline in sea-ice extent is a decrease in the length of the sea-ice season (Parkinson, 2000) and an increase in the length of the Arctic summer melting season between 1979 and 1998, also derived from satellite data. The shortest season was 1979 (57 days) and the longest was in 1998 (81 days) with an increasing trend of 5 days per decade (Smith, 1998, updated). The 7% per decade reduction in the multi-year ice area during 1978 to 1998 is relatively large compared with an approximately 2%/decade decrease in the total ice area in winter (Johannessen *et al.*, 1999). This reflects greater summer melting, consistent with the results of Smith (1998).

Over the period 1979 to 1996, the Antarctic (Cavalieri *et al.*, 1997; Parkinson *et al.*, 1999) shows a weak increase of 1.3 ± 0.2%/decade. Figure 2.16 (for 1973 to 1998) shows a new integrated data set of Antarctic sea-ice extent that was put together for the new European Centre for Medium-range Weather Forecasts (ECMWF) 40-year reanalysis that extends the record back to 1973. While showing the same weak increase after 1979, it also suggests greater ice extents in the mid-1970s. Although century scale time-series cannot be constructed for the Antarctic, de la Mare (1997) has used whaling ship logs to infer significantly greater ice extent in the Southern Ocean during the 1930s and 1940s than during recent decades. The indirect nature of the earlier evidence, however, introduces substantial uncertainty into this conclusion.

Antarctic Peninsula ice shelves
Although warming over Antarctica as a whole appears to have been perhaps half of a degree in the last half century (Jacka and Budd, 1998), the Antarctic Peninsula has warmed more rapidly, by more than 2°C since the 1940s (King, 1994). This regional

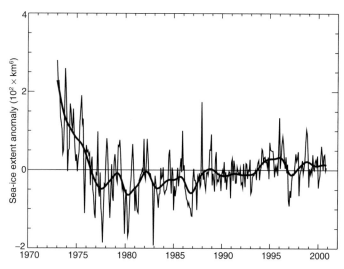

Figure 2.16: Monthly Antarctic sea-ice extent anomalies, 1973 to 2000, relative to 1973 to 1996. The data are a blend of National Ice Center (NIC) chart-derived data (Knight, 1984), Goddard Space Flight Center satellite passive-microwave (Scanning Multichannel Microwave Radiometer (SMMR) and Special Sensor Microwave/Imager (SSM/I)) derived data (Cavalieri *et al.*, 1997) and National Centers for Environmental Prediction satellite passive-microwave derived data (Grumbine, 1996). It is uncertain as to whether the decrease in interannual variability of sea ice after about 1988 is real or an observing bias.

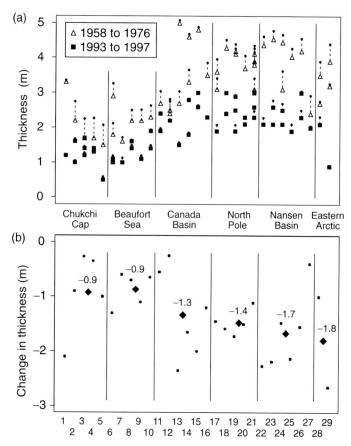

Figure 2.17: Mean ice thickness at places where early cruises were (nearly) collocated with cruises in the 1990s. Early data (1958 to 1976) are shown by open triangles, and those from the 1990s by solid squares, both seasonally adjusted to September 15. The small dots show the original data before the seasonal adjustment. The crossings are grouped into six regions separated by the solid lines. From Rothrock *et al.* (1999).

warming, whose cause has yet to be fully discovered (but see Section 2.6.6), has led to a southerly migration of the climatic limit of ice shelves so that five ice shelves have retreated over the last century (Vaughan and Doake, 1996). The progressive retreat of ice shelves eventually resulted in the spectacular final-stage collapse of the Prince Gustav and parts of the Larsen ice shelves in 1995. Each left only a small residual shelf. After the collapse, James Ross Island, situated off the northern end of the Antarctic Peninsula, is now circumnavigable by ship for the first time since it was discovered in the early 19th century (Vaughan and Lachlan-Cope, 1995).

Sea-ice thickness

Our knowledge of sea-ice thickness in the Arctic comes largely from upward sonar profiling by USA and British submarines since 1958 and 1971, respectively. Rothrock *et al.* (1999) compared late summer September to October data from 1993, 1996 and 1997 from an USA civilian submarine research programme with data from six summer cruises from the period 1958 to 1976. Thickness was adjusted to mid-September values to account for seasonal variability. The significant decline in mean ice thickness was observed for all regions, increasing from the Canada Basin towards Europe (Figure 2.17). Overall, there was a mean reduction in thickness of 42% from 3.1 to 1.8 m the earlier period to the present.

Wadhams and Davis (2000) have compared ice thickness changes between October 1976 and September 1996 between 81°N and 90°N near the 0° meridian. The overall decline in mean sea-ice thickness between 1976 and 1996 was 43%. Over every one degree of latitude, both a significant decline in ice thickness

and some completely open water were observed. Despite these dramatic results, it is not known whether these changes reflect long-term change or a major mode of multi-decadal variability. Vinje *et al.* (1998) measured the thickness of ice exiting the Arctic Ocean through Fram Strait from 1990 to 1996 using moored upward looking sonars and reported a rather different result. The mean annual ice thickness in Fram Strait varied from 2.64 to 3.41 m. These observations were consistent with Arctic Ocean-wide ice thickness estimates made by drilling from Soviet Ice Stations from 1972 to 1981 and from submarine transects from 1960 to 1982, suggesting little change in ice thickness from the 1960s and 1970s to the 1990s.

Nagurnyi *et al.* (1994, 1999) used measurements of long surface gravity waves in the Arctic ice pack to estimate the mean ice thickness from wave attenuation. These measurements are available for the winters of 1978/79 to 1990/91. Johannessen *et al.* (1999) demonstrated a strong correlation between these ice thickness estimates and the area of multi-year (MY) ice in the Arctic Ocean as obtained from the SSMR and SSM/I. Both the area of MY ice and the ice thickness (winter) estimates show a decrease of 5 to 7%/decade, considerably less than the submarine estimates (late summer). Even though the satellite measurements have continued for more than twenty years, they are inadequate to distinguish between changes due to long-term trends or interannual/inter-decadal variability (Johannessen *et al.*, 1999).

2.2.5.3 *Permafrost*

About 25% of the land mass of the Northern Hemisphere is underlain by permafrost, including large regions of Canada, China, Russia and Alaska, with smaller permafrost areas in mountain chains of many other countries in both the Northern and Southern Hemisphere (Brown *et al.*, 1997; Zhang *et al.*, 1999). Permafrost in large part depends on climate. Over half of the world's permafrost is at temperatures a few degrees below 0°C. Temperature variations in near-surface permafrost (20 to 200 m depth) can be used as a sensitive indicator of the inter-annual and decade-to-century climatic variability and long-term changes in the surface energy balance (Lachenbruch and Marshall, 1986; Lachenbruch *et al.*, 1988; Clow *et al.*, 1991; Beltrami and Taylor, 1994; Majorowicz and Judge, 1994). Very small changes in surface climate can produce important changes in permafrost temperatures. Lachenbruch and Marshall (1986) used climate reconstructions from deep (>125 m depth) temperature measurements in permafrost to show that there has been a general warming of the permafrost in the Alaskan Arctic of 2 to 4°C over the last century.

Evidence of change in the southern extent of the discontinuous permafrost zone in the last century has also been recorded. In North America, the southern boundary of the discontinuous permafrost zone has migrated northward in response to warming after the Little Ice Age, and continues to do so today (Thie, 1974; Vitt *et al.*, 1994; Halsey *et al.*, 1995; Laberge and Payette, 1995; French and Egorov, 1998). In China both an increase in the lower altitudinal limit of mountain permafrost and a decrease in areal extent have been observed (Wang *et al.*, 2000).

Long-term monitoring of shallow permafrost began in earnest in the last few decades. Recent analyses indicate that permafrost in many regions of the earth is currently warming (Gravis *et al.*, 1988; Haeberli *et al.*, 1993; Osterkamp, 1994; Pavlov, 1994; Wang and French, 1994; Ding, 1998; Sharkhuu, 1998; Vonder Mühll *et al.*, 1998; Weller and Anderson, 1998; Osterkamp and Romanovsky, 1999; Romanovsky and Osterkamp, 1999). However, the onset, magnitude (from a few tenths to a few degrees centigrade) and rate of warming varies regionally, and not all sites in a given region show the same trend (Osterkamp and Romanovsky, 1999). This variability, as well as short-term (decadal or less) trends superimposed on long-term (century) trends, is briefly discussed in Serreze *et al.* (2000). There has also been evidence of recent permafrost cooling into the mid-1990s in parts of north-eastern and north-western Canada (Allard *et al.*, 1995; Burn, 1998). However, there are regional data gaps, such as in the central and high Arctic in North America. A new international permafrost thermal monitoring network (Burgess *et al.*, 2000) is being developed to help address these gaps and document the spatial and temporal variability across the globe.

Properties of the surface and the active layer (that having seasonal freezing and thawing) affect surface heat exchanges in permafrost regions. Other conditions remaining constant, the thickness of the active layer could be expected to increase in response to warming of the climate. A circumpolar network to monitor active-layer thickness at representative locations was developed in the 1990s to track long-term trends in active layer thickness (Nelson and Brown, 1997). Active layer thickness time-series are becoming available (Nelson *et al.*, 1998; Nixon and Taylor, 1998), and evidence of increasing thaw depths is starting to be reported (Pavlov, 1998; Wolfe *et al.*, 2000).

2.2.5.4 *Mountain glaciers*

The recession of mountain glaciers was used in IPCC (1990) to provide qualitative support to the rise in global temperatures since the late 19th century. Work on glacier recession has considerable potential to support or qualify the instrumental record of temperature change and to cast further light on regional or worldwide temperature changes before the instrumental era. Two types of data from glaciers contain climatic information: (i) mass balance observations and (ii) data on the geometry of glaciers, notably glacier length. More comprehensive information is now becoming available and worldwide glacier inventories have been updated (e.g., IAHS (ICSI)/UNEP/UNESCO, 1999). Note that changes in the Greenland and Antarctic ice sheets are discussed in Chapter 11.

We first discuss mass balance observations. The specific mass balance is defined as the net annual gain or loss of mass at the glacier surface, per unit area of the surface. The mass balance averaged over an entire glacier is denoted by B_m. Systematic investigations of glacier mass balance started after 1945, so these records are shorter than the instrumental climate records normally available in the vicinity. In contrast to frequently made statements, B_m is not necessarily a more precise indicator of climate change than is glacier length. Time-series of B_m do contain year-to-year variability reflecting short-term fluctuations in meteorological quantities but of concern on longer time-scales is the effect of changing glacier geometry. A steadily retreating glacier will get thinner and the mass balance will become more negative because of a slowly increasing surface air temperature due to a lowering surface that is not reflected in a large-scale temperature signal. Climatic interpretation of long-term trends in of mass balance data requires the use of coupled mass balance-ice flow models to separate the climatic and geometric parts of the signal. Such studies have only just begun. However, mass balance observations are needed for estimating the contribution of glacier melt to sea level rise, so are discussed further in Chapter 11.

A wealth of information exists on the geometry of valley glaciers. Glacier records are very useful for studies of Holocene climate variability (e.g., Haeberli *et al.*, 1998; and Section 2.4). Written documents going back to the 16th century exist that describe catastrophic floods caused by the bursting of glacier-dammed lakes or arable land and farms destroyed by advancing glaciers, e.g., in 18th century Norway (Østrem *et al.*, 1977). A large amount of information is available from sketches, etchings, paintings and old photographs of glaciers, though many show the same glaciers (Holzhauser and Zumbühl, 1996). About fifty glaciers have two or more useful pictures from distinctly different times. In many cases geomorphologic evidence in the form of terminal moraines and trimlines can be used as reliable complementary information to construct the history of a glacier over the last few centuries. Systematic mapping of glaciers started only 100 years ago and has been limited to a few glaciers. The most comprehensive data are of length variations. Glacier length records complement the instrumental meteorological record because (i) some extend further back in time; (ii) some records are from remote regions where few meteorological observations exist; (iii) on average, glaciers exist at a significantly higher altitude than meteorological stations.

The last point is of particular interest in the light of the discrepancy between recent tropical glacier length reduction and lack of warming in the lower troposphere since 1979 indicated by satellites and radiosondes in the tropics (Section 2.2.3). Long-term monitoring of glacier extent provides abundant evidence that tropical glaciers are receding at an increasing rate in all tropical mountain areas. This applies to the tropical Andes (Brecher and Thompson, 1993; Hastenrath and Ames, 1995; Ames, 1998), Mount Kenya and Kilimanjaro (Hastenrath and Kruss, 1992; Hastenrath and Greischar, 1997) and to the glaciers in Irian Jaya (Peterson and Peterson, 1994).

Relating mass balance fluctuations to meteorological conditions is more complicated for tropical glaciers than for mid- and high latitude glaciers, and it has not been demonstrated that temperature is the most important factor. Nevertheless, the fast glacier recession in the tropics seems at first sight to be consistent with an increase in tropical freezing heights of 100 m over the period 1970 to 1986 as reported by Diaz and Graham (1996), corresponding to an increase of 0.5°C at tropical high mountain levels, which they also link to increases in tropical SST since the mid-1970s (Figure 2.10). However, although Gaffen *et al.* (2000) found a similar increase over 1960 to 1997, they found a lowering of freezing level over 1979 to 1997 which, at least superficially, is not consistent with glacier recession.

Figure 2.18 shows a representative selection of glacier length records from different parts of the world and updates the diagram in IPCC (1990). It is clear from Figure 2.18 that glacier retreat on the century time-scale is worldwide. The available data suggest that this retreat generally started later at high latitudes but in low and mid-latitudes the retreat generally started in the mid-19th century.

On the global scale, air temperature is considered by most glaciologists to be the most important factor reflecting glacier retreat. This is based on calculations with mass balance models (Greuell and Oerlemans, 1987; Oerlemans, 1992; Fleming *et al.*, 1997; Jóhannesson, 1997). For a typical mid-latitude glacier, a 30% decrease in cloudiness or a 25% decrease in precipitation would have the same effect as a 1°C temperature rise. Such changes in cloudiness or precipitation can occur locally or even regionally on a decadal time-scale associated with changes in circulation, but global trends of this size on a century time-scale are very unlikely. As mentioned in the SAR, Oerlemans (1994) concluded that a warming rate of 0.66 ± 0.20°C per century at the mean glacier altitude could explain the linear part of the observed retreat of 48 widely distributed glaciers.

Glaciers are generally not in equilibrium with the prevailing climatic conditions and a more refined analysis should deal with the different response times of glaciers which involves modelling (Oerlemans *et al.*, 1998). It will take some time before a large

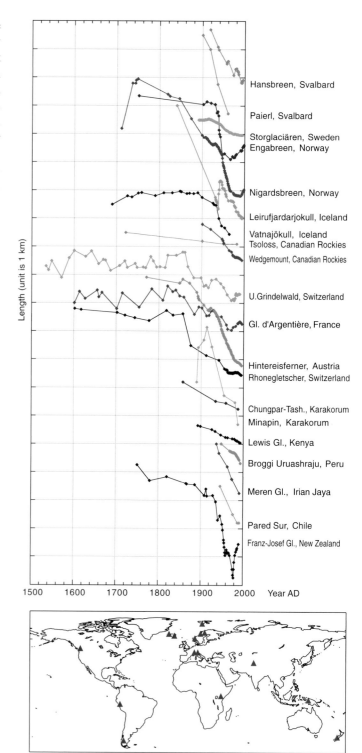

Figure 2.18: A collection of twenty glacier length records from different parts of the world. Curves have been translated along the vertical axis to make them fit in one frame. The geographical distribution of the data is also shown, though a single triangle may represent more than one glacier. Data are from the World Glacier Monitoring Service (http://www.geo.unizh.ch/wgms/) with some additions from various unpublished sources.

number of glaciers are modelled. Nevertheless, work done so far indicates that the response times of glacier lengths shown in Figure 2.18 are in the 10 to 70 year range. Therefore the timing of the onset of glacier retreat implies that a significant global warming is likely to have started not later than the mid-19th century. This conflicts with the Jones *et al.* (2001) global land instrumental temperature data (Figure 2.1), and the combined hemispheric and global land and marine data (Figure 2.7), where clear warming is not seen until the beginning of the 20th century. This conclusion also conflicts with some (but not all) of the palaeo-temperature reconstructions in Figure 2.21, Section 2.3 , where clear warming, e.g., in the Mann *et al.* (1999) Northern Hemisphere series, starts at about the same time as in the Jones *et al.* (2001) data. These discrepancies are currently unexplained.

For the last two to three decades, far more records have been available than are shown in Figure 2.18. Many are documented at the World Glacier Monitoring Service in Zürich, Switzerland (e.g., IAHS (ICSI)/UNEP/UNESCO, 1998) The general picture is one of widespread retreat, notably in Alaska, Franz-Josef Land, Asia, the Alps, Indonesia and Africa, and tropical and sub-tropical regions of South America. In a few regions a considerable number of glaciers are currently advancing (e.g., Western Norway, New Zealand). In Norway this is very likely to be due to increases in precipitation owing to the positive phase of the North Atlantic Oscillation (Section 2.6), and in the Southern Alps of New Zealandand due to wetter conditions with little warming since about 1980. Finally, indications in the European Alps that current glacier recession is reaching levels not seen for perhaps a few thousand years comes from the exposure of radiocarbon-dated ancient remains in high glacial saddles. Here there is no significant ice flow and melting is assumed to have taken place *in situ* for the first time in millennia (e.g., the finding of the 5,000-year-old Oetzal "ice man").

2.2.5.5 *Lake and river ice*

Numerous studies suggest the importance of lake and river ice break-up as an index of climate variability and change, especially as related to temperature and snow cover (Palecki and Barry, 1986; Schindler *et al.*, 1990; Robertson *et al.*, 1992; Assel and Robertson, 1995; Anderson *et al.*, 1996; Wynne *et al.*, 1998; Magnuson *et al.*, 2000). Records of lake and river ice can be used to independently evaluate changes of temperature and, to some extent, snow cover. Like other proxy measurements they have limitations, and are subject to their own time-dependent biases such as changes in observers and protocols related to the identification of "ice on" and "ice off" conditions. Larger lakes often have the best records, but are often located near human settlements which can affect the homogeneity of the record, e.g., associated cooling water discharges and urban heat islands, so care is needed to select suitable lakes.

A recent analysis has been made of trends in 39 extensive Northern Hemisphere lake and river ice records over the 150-year period from 1846 to 1995. Ice break-up dates now occur on average about nine days earlier in the spring than at the beginning of the record, and autumn freeze-up occurs on average about ten days later (Magnuson *et al.*, 2000). Only one of the 39 records, in Japan, showed changes that indicate a slight cooling.

2.2.6 *Are the Retreat of Glaciers, Sea Ice, and Snow Cover Consistent with the Surface Temperature Trends?*

A significant relationship has been found between interannual variations (correlation = −0.60) of the Northern Hemisphere snow-cover extent and land-surface air temperature in spring since the 1960s. However, the observed increase in temperature during the winter is not reflected in a reduced snow-cover extent. Reduced ice cover on the Northern Hemisphere lakes and rivers, primarily due to earlier onset in spring of ice-free conditions during the 20th century, is consistent with reduced snow cover extent in that season. Sea-ice retreat in the Arctic spring and summer is also consistent with an increase in spring, and to a lesser extent, summer temperatures in the high northern latitudes. Summer temperature increases have been less than in spring in nearby land areas, but Arctic sea-ice extent and especially thickness have markedly decreased. Nevertheless, there is only a small indication of reduced Arctic sea ice during winter when temperatures have also increased. Antarctic sea-ice extent has not decreased since the late 1970s, possibly related to recent indications of little change in Antarctic temperatures over much of the continent in that period . There is now ample evidence to support a major retreat of most mountain glaciers during the last 100 years in response to widespread increases in temperature. There has been especially fast glacial recession in the tropics in recent decades, although tropical temperatures in the free atmosphere near glacier levels have increased little since 1980 according to radiosonde and MSU data.

2.2.7 *Summary*

Global surface temperatures have increased between 0.4 and 0.8°C since the late 19th century, but most of this increase has occurred in two distinct periods, 1910 to 1945 and since 1976. The rate of temperature increase since 1976 has been over 0.15°C/decade. Our confidence in the rate of warming has increased since the SAR due to new analyses including: model simulations using observed SSTs with and without corrections for time-dependent biases, new studies of the effect of urbanisation on global land temperature trends, new evidence for mass ablation of glaciers, continued reductions in snow-cover extent, and a significant reduction in Arctic sea-ice extent in spring and summer, and in thickness. However, there is some disagreement between warming rates in the various land and ocean-based data sets in the 1990s, though all agree on appreciable warming.

New analyses of mean daily maximum and minimum temperatures continue to support a reduction in the diurnal temperature range with minimum temperatures increasing at about twice the rate of maximum temperatures over the second half of the 20th century. Seasonally, the greatest warming since 1976 over land has occurred during the Northern Hemisphere winter and spring, but significant warming has also occurred in the Northern Hemisphere summer. Southern Hemisphere warming has also been strongest during the winter over land, but little difference between the seasons is apparent when both land and oceans are considered. The largest rates of warming continue to be found in the mid- and high latitude continental regions of the Northern Hemisphere.

Analyses of overall temperature trends in the low to mid-troposphere and near the surface since 1958 are in good agreement, with a warming of about 0.1°C per decade. Since the beginning of the satellite record (1979), however, low to mid-troposphere temperatures have warmed in both satellite and weather balloon records at a global rate of only 0.04 and 0.03°C/decade respectively. This is about 0.12°C/decade less than the rate of temperature increase near the surface since 1979. About half of this difference in warming rate is very likely to be due to the combination of differences in spatial coverage and the real physical affects of volcanoes and ENSO (Santer *et al.*, 2000), see also Chapter 12. The remaining difference remains unexplained, but is likely to be real. In the stratosphere, both satellites and weather balloons continue to show substantial cooling. The faster rate of recession of tropical mountain glaciers in the last twenty years than might have been expected from the MSU and radiosonde records remains unexplained, though some glaciers may still be responding to the warming indicated by radiosondes that occurred around 1976 to 1981.

2.3 Is the Recent Warming Unusual?

2.3.1 Background

To determine whether 20th century warming is unusual, it is essential to place it in the context of longer-term climate variability. Owing to the sparseness of instrumental climate records prior to the 20th century (especially prior to the mid-19th century), estimates of global climate variability during past centuries must often rely upon indirect "proxy" indicators – natural or human documentary archives that record past climate variations, but must be calibrated against instrumental data for a meaningful climate interpretation (Bradley, 1999, gives a review). Coarsely resolved climate trends over several centuries are evident in many regions e.g., from the recession of glaciers (Grove and Switsur, 1994; and Section 2.2.5.4) or the geothermal information provided by borehole measurements (Pollack *et al.*, 1998). Large-scale estimates of decadal, annual or seasonal climate variations in past centuries, however, must rely upon sources that resolve annual or seasonal climatic variations. Such proxy information includes width and density measurements from tree rings (e.g., Cook, 1995; see Fritts, 1991, for a review), layer thickness from laminated sediment cores (e.g., Hughen *et al.*, 1996; Lamoureux and Bradley, 1996), isotopes, chemistry, and accumulation from annually resolved ice cores (e.g., Claussen *et al.*, 1995; Fisher *et al.*, 1998), isotopes from corals (e.g., Tudhope *et al.*, 1995; Dunbar and Cole, 1999), and the sparse historical documentary evidence available over the globe during the past few centuries (see e.g., Bradley and Jones, 1995; Pfister *et al.*, 1998). Taken as a whole, such proxy climate data can provide global scale sampling of climate variations several centuries into the past, with the potential to resolve large-scale patterns of climate change prior to the instrumental period, albeit with important limitations and uncertainties.

The SAR examined evidence for climate change in the past, on time-scales of centuries to millennia. Based on information from a variety of proxy climate indicators, reconstructions of

mountain glacier mass and extent, and geothermal sub-surface information from boreholes, it was concluded that summer temperatures in the Northern Hemisphere during recent decades are the warmest in at least six centuries. While data prior to AD 1400 were considered too sparse for reliable inferences regarding hemispheric or global mean temperatures, regional inferences were nonetheless made about climate changes further back in time.

Since the SAR, a number of studies based on considerably expanded databases of palaeoclimate information have allowed more decisive conclusions about the spatial and temporal patterns of climate change in past centuries. A number of important advances have been in key areas such as ice core palaeoclimatology (e.g., White *et al.*, 1998a), dendroclimatology (e.g., Cook, 1995; Briffa *et al.*, 1998b), and geothermal palaeo-temperature estimation (e.g., Pollack *et al.*, 1998). Moreover, the latest studies based on global networks of "multi-proxy" data have proved particularly useful for describing global or hemispheric patterns of climate variability in past centuries (e.g., Bradley and Jones, 1993; Hughes and Diaz, 1994; Mann *et al.*, 1995; Fisher, 1997; Overpeck *et al.*, 1997; Mann *et al.*, 1998, 1999). Such estimates allow the observed trends of the 20th century to be put in a longer-term perspective. These have also allowed better comparisons with possible physical influences on climate forcings (Lean *et al.*, 1995; Crowley and Kim, 1996, 1999; Overpeck *et al.*, 1997; Mann *et al.*, 1998; Waple *et al.*, 2001), and for new evaluations of the low-frequency climate variability exhibited by numerical climate models (Barnett *et al.*, 1996; Jones *et al.*, 1998; Crowley and Kim, 1999; Delworth and Mann, 2000).

2.3.2 Temperature of the Past 1,000 Years

The past 1,000 years are a particularly important time-frame for assessing the background natural variability of the climate for climate change detection. Astronomical boundary conditions have strayed relatively little from their modern-day values over this interval (but see Section 2.3.4 for a possible caveat) and, with the latest evidence, the spatial extent of large-scale climate change during the past millennium can now be meaningfully characterised (Briffa *et al.*, 1998b; Jones *et al.*, 1998; Mann *et al.*, 1998; 1999; 2000a; 2000b). Moreover, estimates of volcanic and solar climate forcings are also possible over this period, allowing model-based estimates of their climate effects (Crowley and Kim, 1999; Free and Robock, 1999).

2.3.2.1 Palaeoclimate proxy indicators

A "proxy" climate indicator is a local record that is interpreted using physical or biophysical principles to represent some combination of climate-related variations back in time. Palaeoclimate proxy indicators have the potential to provide evidence for large-scale climatic changes prior to the existence of widespread instrumental or historical documentary records. Typically, the interpretation of a proxy climate record is complicated by the presence of "noise" in which climate information is immersed, and a variety of possible distortions of the underlying climate information (e.g., Bradley, 1999; Ren, 1999a,b). Careful calibration and cross-validation procedures are necessary to establish a reliable relationship between a proxy indicator and the climatic variable or variables it is assumed to represent, providing a "transfer" function through which past climatic conditions can be estimated. High-resolution proxy climate indicators, including tree rings, corals, ice cores, and laminated lake/ocean sediments, can be used to provide detailed information on annual or near-annual climate variations back in time. Certain coarser resolution proxy information (from e.g., boreholes, glacial moraines, and non-laminated ocean sediment records) can usefully supplement this high-resolution information. Important recent advances in the development and interpretation of proxy climate indicators are described below.

Tree rings
Tree-ring records of past climate are precisely dated, annually resolved, and can be well calibrated and verified (Fritts, 1976). They typically extend from the present to several centuries or more into the past, and so are useful for documenting climate change in terrestrial regions of the globe. Many recent studies have sought to reconstruct warm-season and annual temperatures several centuries or more ago from either the width or the density of annual growth rings (Briffa *et al.*, 1995; D'Arrigo *et al.*, 1996; Jacoby *et al.*, 1996; D'Arrigo *et al.*, 1998; Wiles *et al.*, 1998; Hughes *et al.*, 1999; Cook *et al.*, 2000). Recently, there has been a concerted effort to develop spatial reconstructions of past temperature variations (e.g., Briffa *et al.*, 1996) and estimates of hemispheric and global temperature change (e.g., Briffa *et al.*, 1998b; Briffa, 2000). Tree-ring networks are also now being used to reconstruct important indices of climate variability over several centuries such as the Southern Oscillation Index (Stahle *et al.*, 1998), the North Atlantic Oscillation (Cook *et al.*, 1998; Cullen *et al.*, 2001) and the Antarctic Oscillation Index (Villalba *et al.*, 1997) (see also Section 2.6), as well as patterns of pre-instrumental precipitation and drought (Section 2.5.2.2).

Several important caveats must be borne in mind when using tree-ring data for palaeoclimate reconstructions. Not least is the intrinsic sampling bias. Tree-ring information is available only in terrestrial regions, so is not available over substantial regions of the globe, and the climate signals contained in tree-ring density or width data reflect a complex biological response to climate forcing. Non-climatic growth trends must be removed from the tree-ring chronology, making it difficult to resolve time-scales longer than the lengths of the constituent chronologies (Briffa, 2000). Furthermore, the biological response to climate forcing may change over time. There is evidence, for example, that high latitude tree-ring density variations have changed in their response to temperature in recent decades, associated with possible non-climatic factors (Briffa *et al.*, 1998a). By contrast, Vaganov *et al.* (1999) have presented evidence that such changes may actually be climatic and result from the effects of increasing winter precipitation on the starting date of the growing season (see Section 2.7.2.2). Carbon dioxide fertilization may also have an influence, particularly on high-elevation drought-sensitive tree species, although attempts have been made to correct for this effect where appropriate (Mann *et al.*, 1999). Thus climate reconstructions based entirely on tree-ring data are susceptible to several sources of contamination or non-stationarity of response. For these reasons, investigators have increasingly found tree-ring data most

useful when supplemented by other types of proxy information in "multi-proxy" estimates of past temperature change (Overpeck *et al.*, 1997; Jones *et al.*, 1998; Mann *et al.*, 1998; 1999; 2000a; 2000b; Crowley and Lowery, 2000).

Corals
Palaeoclimate reconstructions from corals provide insights into the past variability of the tropical and sub-tropical oceans and atmosphere, prior to the instrumental period, at annual or seasonal resolutions, making them a key addition to terrestrial information. Because of their potential to sample climate variations in ENSO-sensitive regions, a modest network of high-quality coral site records can resolve key large-scale patterns of climate variability (Evans *et al.*, 1998). The corals used for palaeoclimate reconstruction grow throughout the tropics in relatively shallow waters, often living for several centuries. Accurate annual age estimates are possible for most sites using a combination of annual variations in skeletal density and geochemical parameters. Palaeoclimate reconstructions from corals generally rely on geochemical characteristics of the coral skeleton such as temporal variations in trace elements or stable isotopes or, less frequently, on density or variations in fluorescence. Dunbar and Cole (1999) review the use of coral records for palaeoclimatic reconstruction.

Ice cores
Ice cores from polar regions of northern Greenland, Canada and the islands of the North Atlantic and Arctic Oceans, Antarctica, and alpine, tropical and sub-tropical locations (e.g., Thompson, 1996) can provide several climate-related indicators. These indicators include stable isotopes (e.g., ^{18}O), the fraction of melting ice, the rate of accumulation of precipitation, concentrations of various salts and acids, the implied atmospheric loading of dust pollen, and trace gases such as CH_4 and CO_2.

Recently, there has been increased activity in creating high-resolution Antarctic ice core series e.g., for the past millennium (Peel *et al.*, 1996; Mayewski and Goodwin, 1997; Morgan and van Ommen, 1997). In certain regions, isotope information from ice cores shows the late 20th century temperatures as the warmest few decades in the last 1,000 years (Thompson *et al.*, 2000a). Key strengths of ice core information are their high resolution (annual or even seasonal where accumulations rates are particularly high – see van Ommen and Morgan, 1996, 1997), availability in polar and high-elevation regions where other types of proxy climate information like tree-ring data are not available, and their provision of multiple climate- and atmosphere-related variables from the same reasonably well dated physical location (e.g., the GISP2 core; White *et al.*, 1998a). A weakness of ice core data is regional sampling bias (high elevation or high latitude) and melt water and precipitation accumulation data are not easy to date accurately.

The best dated series are based on sub-annual sampling of cores and the counting of seasonal ice layers. Such series may have absolute dating errors as small as a few years in a millennium (Fisher *et al.*, 1996). Dating is sometimes performed using volcanic acid layers with assumed dates (e.g., Clausen *et al.*, 1995) but uncertainties in the volcanic dates can result in dating uncertainties throughout the core (Fisher *et al.*, 1998).

Lake and ocean sediments

Annually laminated (varved) lake sediments offer considerable potential as high-resolution archives of palaeo-environmental conditions where other high-resolution proxy indicators are not available (e.g., arid terrestrial regions), and latitudes poleward of the treeline (Lamoureux and Bradley, 1996; Wohlfarth *et al.*, 1998; Hughen *et al.*, 2000). When annual deposition of the varves can be independently confirmed (e.g., through radiometric dating), they provide seasonal to interannual resolution over centuries to millennia. Varved sediments can be formed from biological processes or from the deposition of inorganic sediments, both of which are often influenced by climate variations. Three primary climate variables may influence lake varves: (a) summer temperature, serving as an index of the energy available to melt the seasonal snowpack, or snow and ice on glaciers; (b) winter snowfall, which governs the volume of discharge capable of mobilising sediments when melting; and (c) rainfall. Laminated lake sediments dominated by (a) can be used for inferences about past high latitude summer temperature changes (e.g., Overpeck *et al.*, 1997), while sediments dominated by the latter two influences can be used to estimate past drought and precipitation patterns (Section 2.5.2.2).

Ocean sediments may also be useful for high-resolution climate reconstructions. In rare examples, annually laminated sediments can be found (e.g., Hughen *et al.*, 1996; Black *et al.*, 1999) and it is possible to incorporate isotope and other information in climate reconstructions, much as varved lake sediments are used. Otherwise, sedimentation rates may sometimes still be sufficiently high that century-scale variability is resolvable (e.g., the Bermuda rise ocean sediment oxygen isotope record of Keigwin, 1996). Dating in such cases, however, must rely on radiometric methods with relatively poor age control.

Borehole measurements

Borehole measurements attempt to relate profiles of temperature with depth to the history of temperature change at the ground surface. The present global database of more than 600 borehole temperature-depth profiles has the densest geographic coverage in North America and Europe, but sparser data are available in other regions (e.g., Australia, Asia, Africa and South America). The depths of the temperature profiles range from about 200 to greater than 1,000 m, allowing palaeo-temperature reconstructions back several hundred to a thousand years. Although large-scale temperature reconstructions have been made to more than a millennium ago (Huang *et al.*, 1997), they show substantial sensitivity to assumptions that are needed to convert the temperature profiles to ground surface temperature changes. Borehole data are probably most useful for climate reconstructions over the last five centuries (Pollack *et al.*, 1998).

Figure 2.19 shows a reconstructed global ground surface temperature history (Pollack *et al.*, 1998; see also Huang *et al.*, 2000) from an average of the 358 individual sites, most located in North America and Eurasia, but some located in Africa, South America and Australia (similar results are obtained by Huang *et al.*, 2000, using an updated network of 616 sites). Superimposed is an instrumental estimate of global surface air temperature (Jones and Briffa, 1992). The ensemble of reconstructions shows

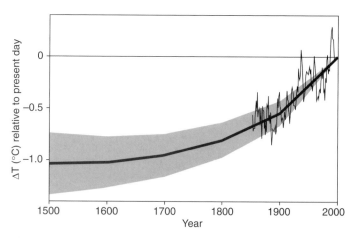

Figure 2.19: Reconstructed global ground temperature estimate from borehole data over the past five centuries, relative to present day. Shaded areas represent ± two standard errors about the mean history (Pollack *et al.*, 1998). Superimposed is a smoothed (five-year running average) of the global surface air temperature instrumental record since 1860 (Jones and Briffa, 1992).

that the average ground temperature of the Earth has increased by about 0.5°C during the 20th century, and that this was the warmest of the past five centuries. About 80% of the sites experienced a net warming over this period. The estimated mean cumulative ground surface temperature change since 1500 is close to 1.0 ± 0.3°C. Uncertainties due to spatial sampling (see Pollack *et al.*, 1998 and Huang *et al.*, 2000) are also shown. It should be noted that the temporal resolution of the borehole estimates decreases sharply back in time, making it perilous to compare the shape of the trend shown in Figure 2.19 with better-resolved trends determined from higher-resolution climate proxy data discussed below.

While borehole data provide a direct estimate of ground surface temperatures under certain simplifying assumptions about the geothermal properties of the earth near the borehole, a number of factors complicate their interpretation. Non-temperature-related factors such as land-use changes, natural land cover variations, long-term variations in winter snow cover and soil moisture change the sub-surface thermal properties and weaken the interpretation of the reconstructions as estimates of surface air temperature change. In central England, where seasonal snow cover is not significant, and major land-use changes occurred many centuries ago, borehole ground surface temperature trends do tend to be similar to those in long instrumental records (Jones, 1999). In contrast, Skinner and Majorowicz (1999) show that borehole estimates of ground surface temperature warming during the 20th century in north-western North America are 1 to 2°C greater than in corresponding instrumental estimates of surface air temperature. They suggest that this discrepancy may be due to land-use changes that can enhance warming of the ground surface relative to that of the overlying atmospheric boundary layer (see also Lewis, 1998). Such factors need to be better understood before borehole temperature measurements can be confidently interpreted.

Documentary evidence

Historical documentary data are valuable sources of information about past climate (e.g., Brown and Issar, 1998; Bradley, 1999). However, their use requires great care, as such documents may be biased towards describing only the more extreme events, and are, in certain cases, prone to the use of inconsistent language between different writers and different epochs, and to errors in dating. As for all proxy information, historical documents require careful calibration and verification against modern instrumental data. Two areas particularly strong in historical documents describing climate are Europe and China. In Europe, attempts have been made to extend long climate series back in time using a combination of documentary evidence and fragmentary instrumental records (e.g., Pfister, 1995; Pfister *et al.*, 1998). Additional information about past climate change has also been obtained purely from documentary records in Europe (e.g., Martin-Vide and Barriendos, 1995; Brázdil, 1996; Pfister *et al.*, 1996, 1998, 1999; Pfister and Brázdil, 1999; Rodrigo *et al.*, 1999). In China, regional instrumental temperature series have been extended back over much of the past millennium using documentary data combined with inferences from ice cores and tree rings (Wang *et al.*, 1998a, 1998b; Wang and Gong, 2000).

Mountain glacier moraines

The position of moraines or till left behind by receding glaciers can provide information on the advances (and, less accurately, the retreats) of mountain glaciers. Owing to the complex balance between local changes in melting and ice accumulation, and the effects of topography which influence mountain glaciers (see Section 2.2.5.4), it is difficult to reconstruct regional (as opposed to global) climate changes from the extent of mountain glaciers alone (Oerlemans, 1989). For example, both increased winter precipitation (through greater accumulation) and lower summer temperatures (through decreased melting or "ablation") can lead to more positive glacial mass balances. The inertia of large glaciers dictates that they respond to climate change relatively slowly, with delays of decades or occasionally centuries. For smaller, fast moving glaciers in regions where precipitation and accumulation are moderate, temperature changes are usually the dominant factor influencing mountain glacier masses and lengths. Here glacier moraine evidence in combination with other lines of evidence can provide reliable information on past regional temperature changes (Salinger, 1995; Holzhauser and Zumbühl, 1996; Raper *et al.*, 1996; Salinger *et al.*, 1996).

2.3.2.2 Multi-proxy synthesis of recent temperature change

Since the SAR there have been several attempts to combine various types of high-resolution proxy climate indicators to create large-scale palaeoclimate reconstructions that build on earlier work by e.g., Bradley and Jones (1993); Hughes and Diaz (1994) and Mann *et al.* (1995). Overpeck *et al.* (1997) and Fisher (1997) have sought to combine information from ice cores, varved lake sediment cores, and tree rings to reconstruct high latitude climate trends for past centuries. Jones *et al.* (1998) estimated extra-tropical Northern and Southern Hemisphere warm-season temperature changes during the past millennium using a sparse set of extra-tropical warm-season temperature proxy indicators (10 and

8 respectively). Mann *et al.* (1998) reconstructed global patterns of annual surface temperature several centuries back in time. They calibrated a combined terrestrial (tree ring, ice core and historical documentary indicator) and marine (coral) multi-proxy climate network against dominant patterns of 20th century global surface temperature. Averaging the reconstructed temperature patterns over the far more data-rich Northern Hemisphere half of the global domain, they estimated the Northern Hemisphere mean temperature back to AD 1400, a reconstruction which had significant skill in independent cross-validation tests. Self-consistent estimates were also made of the uncertainties. This work has now been extended back to AD 1000 (Figure 2.20, based on Mann *et al.*, 1999). The uncertainties (the shaded region in Figure 2.20) expand considerably in earlier centuries because of the sparse network of proxy data. Taking into account these substantial uncertainties, Mann *et al.* (1999) concluded that the 1990s were likely to have been the warmest decade, and 1998 the warmest year, of the past millennium for at least the Northern Hemisphere. Jones *et al.* (1998) came to a similar conclusion from largely independent data and an entirely independent methodology. Crowley and Lowery (2000) reached the similar conclusion that medieval temperatures were no warmer than mid-20th century temperatures. Borehole data (Pollack *et al.*, 1998) independently support this conclusion for the past 500 years although, as discussed earlier (Section 2.3.2.1), detailed interpretations comparison with long-term trends from such of such data are perilous owing to loss of temporal resolution back in time.

The largely independent multi-proxy Northern Hemisphere temperature reconstructions of Jones *et al.* (1998) and Mann *et al.* (1999) are compared in Figure 2.21, together with an independent (extra-tropical, warm-season) Northern Hemisphere temperature estimate by Briffa (2000) based on tree-ring density data. The estimated uncertainties shown are those for the smoothed Mann *et al.* series. Significant differences between the three reconstructions are evident during the 17th and early 19th centuries where either the Briffa *et al.* or Jones *et al.* series lie outside the estimated uncertainties in the Mann *et al.* series. Much of these differences appear to result from the different latitudinal and seasonal emphases of the temperature estimates. This conclusion is supported by the observation that the Mann *et al.* hemispheric temperature average, when restricted to just the extra-tropical (30 to 70°N band) region of the Northern Hemisphere, shows greater similarity in its trend over the past few centuries to the Jones *et al.* reconstruction. The differences between these reconstructions emphasise the importance of regional and seasonal variations in climate change. These are discussed in the next section.

2.3.3 Was there a "Little Ice Age" and a "Medieval Warm Period"?

The terms "Little Ice Age" and "Medieval Warm Period" have been used to describe two past climate epochs in Europe and neighbouring regions during roughly the 17th to 19th and 11th to 14th centuries, respectively. The timing, however, of these cold and warm periods has recently been demonstrated to vary geographically over the globe in a considerable way (Bradley and

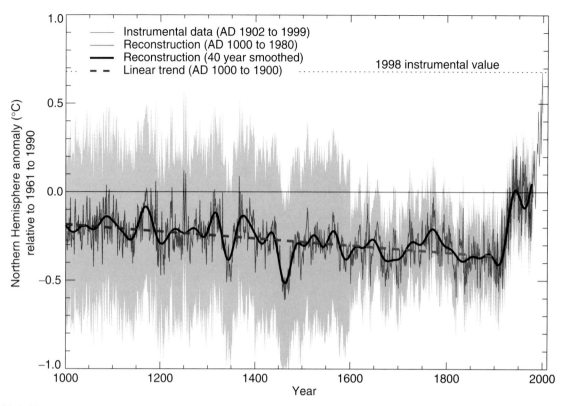

Figure 2.20: Millennial Northern Hemisphere (NH) temperature reconstruction (blue) and instrumental data (red) from AD 1000 to 1999, adapted from Mann *et al.* (1999). Smoother version of NH series (black), linear trend from AD 1000 to 1850 (purple-dashed) and two standard error limits (grey shaded) are shown.

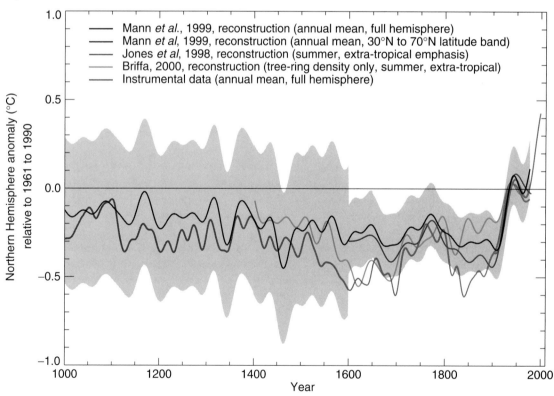

Figure 2.21: Comparison of warm-season (Jones *et al.*, 1998) and annual mean (Mann *et al.*, 1998, 1999) multi-proxy-based and warm season tree-ring-based (Briffa, 2000) millennial Northern Hemisphere temperature reconstructions. The recent instrumental annual mean Northern Hemisphere temperature record to 1999 is shown for comparison. Also shown is an extra-tropical sampling of the Mann *et al.* (1999) temperature pattern reconstructions more directly comparable in its latitudinal sampling to the Jones *et al.* series. The self-consistently estimated two standard error limits (shaded region) for the smoothed Mann *et al.* (1999) series are shown. The horizontal zero line denotes the 1961 to 1990 reference period mean temperature. All series were smoothed with a 40-year Hamming-weights lowpass filter, with boundary constraints imposed by padding the series with its mean values during the first and last 25 years.

Jones, 1993; Hughes and Diaz, 1994; Crowley and Lowery, 2000). Evidence from mountain glaciers does suggest increased glaciation in a number of widely spread regions outside Europe prior to the 20th century, including Alaska, New Zealand and Patagonia (Grove and Switsur, 1994). However, the timing of maximum glacial advances in these regions differs considerably, suggesting that they may represent largely independent regional climate changes, not a globally-synchronous increased glaciation (see Bradley, 1999). Thus current evidence does not support globally synchronous periods of anomalous cold or warmth over this timeframe, and the conventional terms of "Little Ice Age" and "Medieval Warm Period" appear to have limited utility in describing trends in hemispheric or global mean temperature changes in past centuries. With the more widespread proxy data and multi-proxy reconstructions of temperature change now available, the spatial and temporal character of these putative climate epochs can be reassessed.

Mann *et al.* (1998) and Jones *et al.* (1998) support the idea that the 15th to 19th centuries were the coldest of the millennium over the Northern Hemisphere overall. However, viewed hemispherically, the "Little Ice Age" can only be considered as a modest cooling of the Northern Hemisphere during this period of less than 1°C relative to late 20th century levels (Bradley and Jones, 1993; Jones *et al.*, 1998; Mann *et al.*, 1998; 1999; Crowley and Lowery, 2000). Cold conditions appear, however, to have been considerably more pronounced in particular regions. Such regional variability can be understood in part as reflecting accompanying changes in atmospheric circulation. The "Little Ice Age" appears to have been most clearly expressed in the North Atlantic region as altered patterns of atmospheric circulation (O'Brien *et al.*, 1995). Unusually cold, dry winters in central Europe (e.g., 1 to 2°C below normal during the late 17th century) were very likely to have been associated with more frequent flows of continental air from the north-east (Wanner *et al.*, 1995; Pfister, 1999). Such conditions are consistent (Luterbacher *et al.*, 1999) with the negative or enhanced easterly wind phase of the NAO (Sections 2.2.2.3 and 2.6.5), which implies both warm and cold anomalies over different regions in the North Atlantic sector. Such strong influences on European temperature demonstrate the difficulty in extrapolating the sparse early information about European climate change to the hemispheric, let alone global, scale. While past changes in the NAO have likely had an influence in eastern North America, changes in the El Niño phenomenon (see also Section 2.6), are likely to have had a particularly significant influence on regional temperature patterns over North America.

The hemispherically averaged coldness of the 17th century largely reflected cold conditions in Eurasia, while cold hemispheric conditions in the 19th century were more associated with cold conditions in North America (Jones *et al.*, 1998; Mann *et al.*, 2000b). So, while the coldest decades of the 19th century appear to have been approximately 0.6 to 0.7°C colder than the latter decades of the 20th century in the hemispheric mean (Mann *et al.*, 1998), the coldest decades for the North American continent were closer to 1.5°C colder (Mann *et al.*, 2000b). In addition, the timing of peak coldness was often specific to particular seasons. In Switzerland, for example, the first particularly cold winters appear to have been in the 1560s, with cold springs beginning around 1568, and with 1573 the first unusually cold summer (Pfister, 1995).

The evidence for temperature changes in past centuries in the Southern Hemisphere is quite sparse. What evidence is available at the hemispheric scale for summer (Jones *et al.*, 1998) and annual mean conditions (Mann *et al.*, 2000b) suggests markedly different behaviour from the Northern Hemisphere. The only obvious similarity is the unprecedented warmth of the late 20th century. Speleothem evidence (isotopic evidence from calcite deposition in stalagmites and stalactites) from South Africa indicates anomalously cold conditions only prior to the 19th century, while speleothem (records derived from analysing stalagmites and stalagtites) and glacier evidence from the Southern Alps of New Zealand suggests cold conditions during the mid-17th and mid-19th centuries (Salinger, 1995). Dendroclimatic evidence from nearby Tasmania (Cook *et al.*, 2000) shows no evidence of unusual coldness at these times. Differences in the seasons most represented by this proxy information prevent a more direct comparison.

As with the "Little Ice Age", the posited "Medieval Warm Period" appears to have been less distinct, more moderate in amplitude, and somewhat different in timing at the hemispheric scale than is typically inferred for the conventionally-defined European epoch. The Northern Hemisphere mean temperature estimates of Jones *et al.* (1998), Mann *et al.* (1999), and Crowley and Lowery (2000) show temperatures from the 11th to 14th centuries to be about 0.2°C warmer than those from the 15th to 19th centuries, but rather below mid-20th century temperatures. The long-term hemispheric trend is best described as a modest and irregular cooling from AD 1000 to around 1850 to 1900, followed by an abrupt 20th century warming. Regional evidence is, however, quite variable. Crowley and Lowery (2000) show that western Greenland exhibited anomalous warmth locally only around AD 1000 (and to a lesser extent, around AD 1400), with quite cold conditions during the latter part of the 11th century, while Scandinavian summer temperatures appeared relatively warm only during the 11th and early 12th centuries. Crowley and Lowery (2000) find no evidence for warmth in the tropics. Regional evidence for medieval warmth elsewhere in the Northern Hemisphere is so variable that eastern, yet not western, China appears to have been warm by 20th century standards from the 9th to 13th centuries. The 12th and 14th centuries appear to have been mainly cold in China (Wang *et al.*, 1998a,b; Wang and Gong, 2000). The restricted evidence from the Southern Hemisphere, e.g., the Tasmanian tree-ring temperature reconstruction of Cook *et al.* (1999), shows no evidence for a distinct Medieval Warm Period.

Medieval warmth appears, in large part, to have been restricted to areas in and neighbouring the North Atlantic. This may implicate the role of ocean circulation-related climate variability. The Bermuda rise sediment record of Keigwin (1996) suggests warm medieval conditions and cold 17th to 19th century conditions in the Sargasso Sea of the tropical North Atlantic. A sediment record just south of Newfoundland (Keigwin and Pickart, 1999), in contrast, indicates cold medieval and warm 16th to 19th century upper ocean temperatures. Keigwin and Pickart (1999) suggest that these temperature contrasts were associated with changes in ocean currents in the North Atlantic. They argue that the "Little Ice Age" and "Medieval Warm Period" in the Atlantic region may in large measure reflect century-scale changes in the North Atlantic Oscillation (see Section 2.6). Such regional changes in oceanic and atmospheric processes, which are also relevant to the natural variability of the climate on millennial and longer time-scales (see Section 2.4.2), are greatly diminished or absent in their influence on hemispheric or global mean temperatures.

2.3.4 Volcanic and Solar Effects in the Recent Record

Recent studies comparing reconstructions of surface temperature and natural (solar and volcanic) radiative forcing (e.g., Lean *et al.*, 1995; Crowley and Kim, 1996, 1999; Overpeck *et al.*, 1997; Mann *et al.*, 1998; Damon and Peristykh, 1999; Free and Robock, 1999; Waple *et al.*, 2001) suggest that a combination of solar and volcanic influences have affected large-scale temperature in past centuries. The primary features of the Northern Hemisphere mean annual temperature histories of Mann *et al.* (1999a) and Crowley and Lowery (2000) from AD 1000 to 1900 have been largely reproduced based on experiments using an Energy Balance Model forced by estimates of these natural radiative forcings (Crowley, 2000; Mann, 2000) making the argument that the "Little Ice Age" and "Medieval Warm Period", at the hemispheric mean scale, are consistent with estimates of naturally-forced climate variability. Several studies indicate that the combined effect of these influences has contributed a small component to the warming of the 20th century. Most of these studies isolate greenhouse radiative forcing as being dominant during late 20th century warming (see Crowley, 2000). This argues against a close empirical relationship between certain sun-climate parameters and large-scale temperature that has been claimed for the 20th century (Hoyt and Schatten, 1997). The reader is referred to Chapter 6 for a detailed discussion of these radiative forcings, and to Chapter 12 for comparisons of observed and model simulations of recent climate change.

2.3.5 Summary

Since the SAR there have been considerable advances in our knowledge of temperature change over the last millennium. It is likely that temperatures were relatively warm in the Northern Hemisphere as a whole during the earlier centuries of the millennium, but it is much less likely that a globally-synchronous, well defined interval of "Medieval warmth" existed, comparable to the near global warmth of the late 20th century. Marked warmth seems to have been confined to Europe and regions neighbouring the North Atlantic. Relatively colder hemispheric or global-scale conditions did appear to set in after about AD 1400 and persist through the 19th century, but peak coldness is observed during substantially different epochs in different regions. By contrast, the warming of the 20th century has had a much more convincing global signature (see Figure 2.9). This is consistent with the palaeoclimate evidence that the rate and magnitude of global or hemispheric surface 20th century warming is likely to have been the largest of the millennium, with the 1990s and 1998 likely to have been the warmest decade and year, respectively, in the Northern Hemisphere. Independent estimates of hemispheric and global ground temperature trends over the past five centuries from sub-surface information contained in borehole data confirm the conclusion that late 20th century warmth is anomalous in a long-term context. Decreasing temporal resolution back in time of these estimates and potential complications in inferring surface air temperature trends from sub-surface ground temperature measurements precludes, however, a meaningful direct comparison of the borehole estimates with high-resolution temperature estimates based on other proxy climate data. Because less data are available, less is known about annual averages prior to 1,000 years before the present and for conditions prevailing in most of the Southern Hemisphere prior to 1861.

2.4 How Rapidly did Climate Change in the Distant Past?

2.4.1 Background

Only during the 1980s was the possibility of rapid climatic changes occurring at the time-scale of human life more or less fully recognised, largely due to the Greenland ice core drilled at Dye 3 in Southern Greenland (Dansgaard *et al.*, 1982, 1989). A possible link between such events and the mode of operation of the ocean was then subsequently suggested (Oeschger *et al.*, 1984; Broecker *et al.*, 1985; see Broecker, 1997, for a recent review). The SAR reviewed the evidence of such changes since the peak of the last inter-glacial period about 120 ky BP (thousands of years Before Present). It concluded that: (1) large and rapid climatic changes occurred during the last Ice Age and during the transition towards the present Holocene; (2) temperatures were far less variable during this latter period; and (3) suggestions that rapid changes may have also occurred during the last inter-glacial required confirmation.

These changes are now best documented from ice core, deep-sea sediment and continental records. Complementary and generally discontinuous information comes from coral and lake level data. The time-scale for the Pleistocene deep-sea core record is based on the orbitally tuned oxygen isotope record from marine sediments (Martinson *et al.*, 1987), constrained by two radiometrically dated horizons, the peak of the last inter-glacial (about 124 ky BP) and the Brunhes/Matuyama reversal of the Earth's magnetic field at about 780 ky BP. ^{14}C-dating is also used in the upper 50 ky BP; the result is a deep-sea core chronology believed to be accurate to within a few per cent for the last million years. ^{14}C-dating is also used for dating continental records as well as the counting of annual layers in tree rings and varved lake records, whereas ice-core chronologies are obtained by combining layer counting, glaciological models and comparison with other dated records. The use of globally representative records, such as changes in continental ice volume recorded in the isotopic composition of deep-sea sediments, or changes in atmospheric composition recorded in air bubbles trapped in ice cores, now allow such local records to be put into a global perspective. Studies still largely focus on the more recent glacial-interglacial cycle (the last 120 to 130 ky). Table 2.4 is a guide to terminology.

Table 2.4: *Guide to terminology used in palaeoclimate studies of the last 150,000 years.*

"Event", Stage	Estimated age (calendar years)
Holocene	~10 ky BP to present
Holocene maximum warming (also referred to as "climatic optimum")	Variable? ~4.5 to 6 ky BP (Europe) 10 to 6 ky BP (SH)
Last deglaciation	~18 to 10 ky BP
Termination 1	~14 ky BP
Younger Dryas	~12.7 to 11.5 ky BP
Antarctic cold reversal	14 to 13 ky BP
Bölling-Alleröd warm period	14.5 to 13 ky BP (Europe)
Last glacial	~74 to 14 ky BP
LGM (last glacial maximum)	~25 to 18 ky BP
Last interglacial peak	~124 ky BP
Termination 2	~130 ky BP
Eemian/MIS stage 5e	~128 to 118 ky BP
Heinrich events	Peaks of ice-rafted detritus in marine sediments, ~7 to 10 ky time-scale.
Dansgaard-Oeschger events	Warm-cold oscillations determined from ice cores with duration ~2 to 3 ky.
Bond cycles	A quasi-cycle during the last Ice Age whose period is equal to the time between successive Heinrich events.
Terminations	Periods of rapid deglaciation.

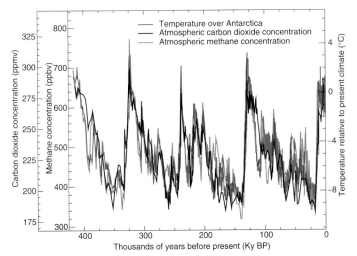

Figure 2.22: Variations of temperature, methane, and atmospheric carbon dioxide concentrations derived from air trapped within ice cores from Antarctica (adapted from Sowers and Bender, 1995; Blunier *et al.*, 1997; Fischer *et al.*, 1999; Petit *et al.*, 1999).

Before reviewing important recent information about rapid changes, we briefly mention progress made on two aspects of the palaeoclimate record of relevance for future climate. The first deals with the relationship between modern and past terrestrial data and SSTs around the time of the Last Glacial Maximum (about 20 ky BP); this is important because of the use of glacial data to validate climate models. New results obtained since the SAR both from marine and terrestrial sources (reviewed in Chapter 8), agree on a tropical cooling of about 3°C. The second concerns the greenhouse gas record (CO$_2$ and CH$_4$) which has now been considerably extended due to the recent completion of drilling of the Vostok ice

core in central East Antarctica. The strong relationship between CO$_2$ and CH$_4$ and Antarctic climate documented over the last climatic cycle has been remarkably confirmed over four climatic cycles, spanning about 420 ky (Figure 2.22). Present day levels of these two important greenhouse gases appear unprecedented during this entire interval (Petit *et al.*, 1999; and Figure 2.22). From a detailed study of the last three glacial terminations in the Vostok ice core, Fischer *et al.* (1999) conclude that CO$_2$ increases started 600 ± 400 years after the Antarctic warming. However, considering the large uncertainty in the ages of the CO$_2$ and ice (1,000 years or more if we consider the ice accumulation rate uncertainty), Petit *et al.* (1999) felt it premature to ascertain the sign of the phase relationship between CO$_2$ and Antarctic temperature at the initiation of the terminations. In any event, CO$_2$ changes parallel Antarctic temperature changes during deglaciations (Sowers and Bender, 1995; Blunier *et al.*, 1997; Petit *et al.*, 1999). This is consistent with a significant contribution of these greenhouse gases to the glacial-interglacial changes by amplifying the initial orbital forcing (Petit *et al.*, 1999).

We also now have a better knowledge of climate variability over the last few climatic cycles as illustrated by selected palaeo-temperature records back to about 400 ky (Figure 2.23). The amplitude of the glacial-interglacial temperature change was lower in tropical and equatorial regions (e.g., curve c) than in mid- and high latitudes (other curves). During glacial periods, the climate of the North Atlantic and adjacent regions (curves a and b) was more variable than in the Southern Hemisphere (curve d). Also (not shown), full glacial periods were characterised by very high fluxes of dust (seen in ice-core records and in continental and marine records). A combination of increased dust source area, stronger atmospheric transport and a weaker hydrological cycle (Yung *et al.*, 1996; Mahowald *et al.*, 1999; Petit *et al.*, 1999) probably generated these changes.

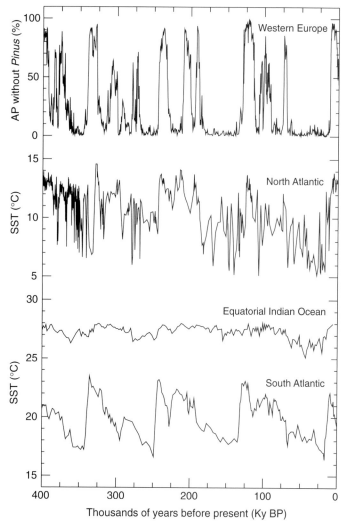

Figure 2.23: Time-series illustrating temperature variability over the last about 400 ky (updated from Rostek *et al.*, 1993; Schneider *et al.*, 1996; MacManus *et al.*, 1999; Reille *et al.*, 2000). The uppermost time-series describes the percentage of tree pollen that excludes pollen from pine tree species. The higher this percentage, the warmer was the climate.

2.4.2 How Stable was the Holocene Climate?

Ice core, marine and terrestrial records show that the Holocene was marked by a millennial-scale mode of variability (Meese *et al.*, 1994; O'Brien *et al.*, 1995; Bond *et al.*, 1997; Yiou *et al.*, 1997a,b). These variations affect both atmospheric (Mayewski *et al.*, 1997) and oceanic (Bianchi and McCave, 1999) indicators. The occurrence of very large floods in the south-western United States also reflects substantial low-frequency variability (Ely *et al.*, 1993). SSTs reconstructed from analyses of a sub-tropical, high sedimentation rate site off West Africa might indicate a remarkably high amplitude Holocene variability of 5 to 8°C on a time-scale about 1,500 years (deMenocal, 1998). During the later Holocene, New Zealand speleothems indicate a lowering of temperature after about 7 ky BP, with small advances of the mountain glaciers in the Southern Alps near about 4 and 2.5 ky BP (Salinger and McGlone, 1989). Speleothem records also

indicate a temperature decrease of about 1.5°C some 2 to 3 ky ago (Williams *et al.*, 1999). These indications are consistent with cooler periods at these times shown by South African speleothems (Partridge, 1997). By contrast, temperature peaks appeared in China at about 7 ky BP and at 5.5 to 6 ky BP (Wang and Gong, 2000).

Central Greenland ice cores and European lake isotopic records show correlated temperature variations within the Holocene, with a roughly 50% higher amplitude at Summit Greenland, compared to Europe (Figure 2.24). The most prominent event in both records occurred about 8,200 years BP (Alley *et al.*, 1997; von Grafenstein *et al.*, 1998; Barber *et al.*, 1999) when annual mean temperatures dropped by as much as 2°C in mid-Europe and the European alpine timberline fell by about 200 m (Wick and Tinner, 1999). The event may be related to a significant decrease of SST in the Norwegian Sea (Klitgaard-Kristensen *et al.*, 1998). Lake records from the southern border of the Sahara indicate extremely dry conditions during this time, and probably also during other cool but less dramatic events of this kind (Street-Perrot and Perrot, 1990 ; Gasse and Van Campo, 1994). The about 8,200 year cooling may also have been worldwide (Stager and Mayewski, 1997), although abrupt early Holocene climate changes recorded in a North American lake are thought to reflect a different event (Hu *et al.*, 1999). Thus cooling is indicated in the New Zealand Southern Alps, with small advances of the mountain glaciers at about 8,000 years BP (Salinger and McGlone, 1989).

Further abrupt climatic changes and reversals on millennial time-scales during the Holocene are documented from pollen and lake level records e.g., in Europe (Magny, 1995; Pazdur *et al.*, 1995; Combourieu-Nebout *et al.*, 1998), North Africa (Gasse *et al.*, 1990; Lamb *et al.*, 1995), North America (Jacobson *et al.*, 1987; Overpeck *et al.*, 1991) and Australia (Kershaw *et al.*, 1991). Holocene lake level changes in Europe have been shown to correlate (Magny, 1995; Yu and Harrisson, 1996) with millennial-scale changes in North Atlantic SST and salinity records (Duplessy *et al.*, 1992; Gasse and van Campo, 1994), suggesting a possible link between millennial thermohaline circulation variability and atmospheric circulation over Europe.

The early Holocene was generally warmer than the 20th century but the period of maximum warmth depends on the region considered. It is seen at the beginning of the Holocene (about 11 to 10 ky BP) in most ice cores from high latitude regions e.g., north-west Canada (Ritchie *et al.*, 1989), central Antarctica (Ciais *et al.*, 1992; Masson *et al.*, 2000) and in some tropical ice cores such as Huascaran in Peru (Thompson *et al.*, 1995). It is also seen during the early Holocene in the Guliya ice core in China (Thompson *et al.*, 1998) but not in two other Chinese cores (Dunde, Thompson *et al.*, 1989; and Dasuopu, to be published). North Africa experienced a greatly expanded monsoon in the early and mid-Holocene, starting at 11 ky BP (Petit-Maire and Guo, 1996), and declining thereafter. In New Zealand the warmest conditions occurred between about 10 to 8 ky BP, when there was a more complete forest cover than at any other time. Glacial activity was at a minimal level in the Southern Alps and speleothem analyses indicate temperatures were about 2°C warmer than present (Salinger and McGlone, 1989; Williams *et al.*, 1999).

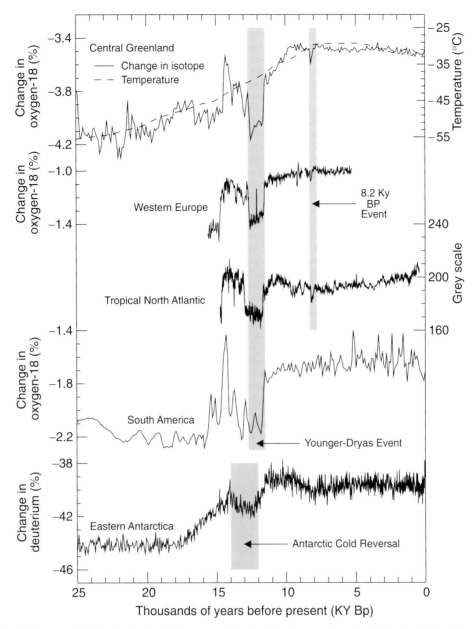

Figure 2.24: Records of climate variability during the Holocene and the last climatic transition, including the 8.2 ky BP event (adapted from Johnsen *et al.*, 1992; Hughen *et al.*, 1996; Thompson *et al.*, 1998; von Grafenstein *et al.*, 1999; Jouzel *et al.*, 2001). The shaded areas show the 8.2 ky BP event, the Younger Dryas event and the Antarctic Cold Reversal. The grey scale used in the Tropical North Atlantic record is a measure of sea surface temperature, deduced from the colour of plankton rich layers within an ocean sediment core.

By contrast, central Greenland (Dahl-Jensen *et al.*, 1998), and regions downstream of the Laurentide ice sheet, did not warm up until after 8 ky BP (including Europe: COHMAP Members, 1988; eastern North America: Webb *et al.*, 1993). The East Asian monsoon did not commence its expanded phase until after 8 ky BP (Sun and Chen, 1991; Harrison *et al.*, 1996; Yu and Qin, 1997; Ren and Zhang, 1998). A more detailed description of the climate at 6 ky BP as well as of the mechanisms involved is given in Chapter 8. Long-term climate changes during the Holocene are consistent with the effects of orbital forcing, modified by the persistence of the Laurentide ice sheet (which finally disappeared around 6 ky BP).

Seasonal to interannual climate variability may also have varied its character during the Holocene. This is a period for

which a variety of palaeo-proxies and archaeological investigations (e.g., Sandweiss *et al.*, 1996; Rodbell *et al.*, 1999) provide evidence for past variations in the strength and frequency of ENSO extremes. A 16-year long time-series of temperature and hydrological balance from a coral dated at 5,370 years BP from the Great Barrier Reef (Gagan *et al.*, 1998) implies that ENSO, or its teleconnections to Australia, were substantially different in the mid-Holocene than today. Mid-Holocene changes in the spectrum of ENSO variability have also been implicated by sedimentary palaeoclimatic records in Australasia (McGlone *et al.*, 1992; Shulmeister and Lees, 1995) and South America (Sandweiss *et al.*, 1996; Rodbell *et al.*, 1999).

To sum up, the Holocene shows both long-term trends (including changes in the nature of ENSO) and millennial time-scale variability although the amplitude of the variability is small compared with that characteristic of Ice Ages. As more detailed information becomes available, the timing of the Holocene maximum warmth is seen to differ across the globe. There appears to be a south to north pattern, with southern latitudes displaying maximum warming a few millennia before the Northern Hemisphere regions. Interestingly, the Holocene appears by far the longest warm "stable" period (as far as seen from the Antarctic climate record) over the last 400 ky, with profound implications for the development of civilisation (Petit *et al.*, 1999).

2.4.3 How Fast did Climate Change during the Glacial Period?

The most extreme manifestation of climate change in the geological record is the transition from full glacial to full inter-glacial conditions. During the most recent glacial cycle, peak glacial conditions prevailed from about 25 to 18 ky BP. Temperatures close to those of today were restored by approximately 10 ky BP. However, warming was not continuous. The deglaciation was accomplished in two main stages, with a return to colder conditions (Younger Dryas/Antarctic Cold Reversal) or, at the least, a pause in the deglaciation.

The central Greenland ice core record (GRIP and GISP2) has a near annual resolution across the entire glacial to Holocene transition, and reveals episodes of very rapid change. The return to the cold conditions of the Younger Dryas from the incipient inter-glacial warming 13,000 years ago took place within a few decades or less (Alley *et al.*, 1993). The warming phase, that took place about 11,500 years ago, at the end of the Younger Dryas was also very abrupt and central Greenland temperatures increased by 7°C or more in a few decades (Johnsen *et al.*, 1992; Grootes *et al.*, 1993; Severinghaus *et al.*, 1998). Most of the changes in wind-blown materials and some other climate indicators were accomplished in a few years (Alley *et al.*, 1993; Taylor *et al.*, 1993; Hammer *et al.*, 1997). Broad regions of the Earth experienced almost synchronous changes over periods of 0 to 30 years (Severinghaus *et al.*, 1998), and changes were very abrupt in at least some regions (Bard *et al.*, 1987), e.g. requiring as little as 10 years off Venezuela (Hughen *et al.*, 1996). Fluctuations in ice conductivity indicate that atmospheric circulation was reorganised extremely rapidly (Taylor *et al.*, 1993). A similar, correlated sequence of abrupt deglacial events also occurred in the tropical and temperate North Atlantic (Bard *et al.*, 1987; Hughen *et al.*, 1996) and in Western Europe (von Grafenstein *et al.*, 1999).

A Younger-Dryas type event is also recorded in a Bolivian ice core (Thompson *et al.*, 1998; Sajama, South America in Figure 2.24) and in a major advance of a mountain glacier in the Southern Alps of New Zealand (Denton and Hendy, 1994). However there is recent evidence against a significant Younger Dryas cooling here (Singer *et al.*, 1998) and at other sites of the Southern Hemisphere (reviewed by Alley and Clarke, 1999). Instead, the Antarctic (and Southern Ocean) climate was characterised by a less pronounced cooling (the Antarctic Cold Reversal: Jouzel *et al.*, 1987) which preceded the Younger Dryas by more than 1 ky (Jouzel *et al.*, 1995; Sowers and Bender, 1995; Blunier *et al.*, 1997). Curiously, one coastal site in Antarctica,

Taylor Dome (Steig *et al.*, 1998) exhibited cooling in phase with the North Atlantic. Recent series obtained at Law Dome, another coastal site of East Antarctica, show instead a cold reversal preceding the Younger Dryas as in other Antarctic records. This suggests that the Taylor Dome record is of limited geographical significance but it also suggests that there is more to be discovered about this cooling event in the Southern Hemisphere.

The inception of deglacial warming about 14.5 ky BP was also very rapid, leading to the Bölling-Alleröd warm period in less than twenty years (Severinghaus and Brook, 1999). Almost synchronously, major vegetation changes occurred in Europe and North America with a rise in African lake levels (Gasse and van Campo, 1994). There was also a pronounced warming of the North Atlantic and North Pacific (Koç and Janssen, 1994; Sarnthein *et al.*, 1994; Kotilainen and Shackleton, 1995; Thunnell and Mortyn, 1995; Wansaard, 1996; Watts *et al.*, 1996; Webb *et al.*, 1998).

The rate of temperature change during the recovery phase from the last glacial maximum provides a benchmark against which to assess warming rates in the late 20th century. Available data indicate an average warming rate of about 2°C/millennium between about 20 and 10 ky BP in Greenland, with lower rates for other regions. Speleothem data from New Zealand, and positions of mountain glacier moraine termini suggest warming rates of 2°C/millennium from 15 to 13 ky BP (Salinger and McGlone, 1989). Speleothem data for South Africa suggest a warming rate of 1.5°C/millennium (Partridge, 1997) over the same time period. On the other hand, very rapid warming at the start of the Bölling-Alleröd period, or at the end of the Younger Dryas may have occurred at rates as large as 10°C/50 years for a significant part of the Northern Hemisphere.

Oxygen isotope measurements in Greenland ice cores demonstrate that a series of rapid warm and cold oscillations called Dansgaard-Oeschger events punctuated the last glaciation (Figure 2.23, see North Atlantic SST panel, and Dansgaard *et al.*, 1993). Associated temperature changes may be as high as 16°C (Lang *et al.*, 1999). These oscillations are correlated with SST variations in several North Atlantic deep-sea cores (Bond *et al.*, 1993). There was clearly a close relation between these ice core temperature cycles and another prominent feature of North Atlantic deep-sea core records, the Heinrich events. Heinrich events occurred every 7,000 to 10,000 years during times of sea surface cooling in the form of brief, exceptionally large, discharges of icebergs from the Laurentide and European ice sheets which left conspicuous layers of detrital rocks in deep-sea sediments. Accompanying the Heinrich events were large decreases in the oxygen isotope ratio of planktonic foraminifera, providing evidence of lowered surface salinity probably caused by melting of drifting ice (Bond *et al.*, 1993). Heinrich events appear at the end of a series of saw-toothed shaped, near millennial temperature cycles. Each set of millennial cycles is known as a Bond cycle. Each cycle was characterised by a succession of progressively cooler relatively warm periods (interstadials) during the Ice Age period. Each cooling trend ended with a very rapid, high amplitude, warming and a massive discharge of icebergs. The impact of these Heinrich events on the climate system extended far beyond the northern North Atlantic. At the time of major iceberg discharges, strong vegetation changes have been detected in Florida (Grimm *et al.*, 1993; Watts *et al.*, 1996), oceanic changes occurred in the Santa Barbara Basin off California (Behl and Kennet, 1996) and changes in loess grain-size, associated with atmospheric circulation changes, have been detected in China (Porter and An, 1995; Ding *et al.*, 1998).

Deep-sea cores also show the presence of ice rafting cycles in the intervals between Heinrich events (Bond and Lotti, 1995). Their duration varies between 2,000 and 3,000 years and they closely coincide with the Dansgaard-Oeschger events of the last glaciation. A study of the ice-rafted material suggests that, coincident with the Dansgaard-Oeschger cooling, ice within the Icelandic ice cap and within or near the Gulf of Saint Lawrence underwent nearly synchronous increases in rates of calving. The Heinrich events reflect a slower rhythm of iceberg discharges, probably from the Hudson Strait.

Air temperature, SST and salinity variations in the North Atlantic are associated with major changes in the thermohaline circulation. A core from the margin of the Faeroe-Shetland channel covering the last glacial period reveals numerous oscillations in benthic and planktonic foraminifera, oxygen isotopes and ice-rafted detritus (Rasmussen *et al.*, 1996a). These oscillations correlate with the Dansgaard-Oeschger cycles, showing a close relationship between the deep ocean circulation and the abrupt climatic changes of the last glaciation. Warm episodes were associated with higher SST and the presence of oceanic convection in the Norwegian Greenland Sea. Cold episodes were associated with low SST and salinity and no convection in the Norwegian Greenland Sea (Rasmussen *et al.*, 1996b). Cores from the mid-latitudes of the North Atlantic show that the iceberg discharges in Heinrich events resulted in both low salinity and a reduced thermohaline circulation (Cortijo *et al.*, 1997; Vidal *et al.*, 1997).

These rapid climatic events of the last glacial period, best documented in Greenland and the North Atlantic, have smoothed counterparts in Antarctica (Bender *et al.*, 1994; Jouzel *et al.*, 1994). A peak in the concentration of the isotope beryllium-10 in ice cores (Yiou *et al.*, 1997a), changes in the concentration of atmospheric methane (Blunier *et al.*, 1998) and in the isotopic content of oxygen in ice cores (Bender *et al.*, 1999) indicate links between the Northern and Southern Hemisphere climates over this period. Large Greenland warming events around 36 and 45 ky BP lag their Antarctic counterparts by more than 1,000 years. This argues against coupling between northern and southern polar regions via the atmosphere but favours a connection via the ocean (Blunier *et al.*, 1998).

New evidence suggests that the North Atlantic has three modes of operation. These are: deep-water sinking in the GIN (Greenland-Iceland-Norwegian) Seas and the Labrador Sea, deep-water sinking in the North Atlantic or in the Labrador Sea but not the GIN Seas (Duplessy *et al.*, 1991; Labeyrie *et al.*, 1992) in the cold phase of Dansgaard-Oeschger events and at glacial maximum, and little deep-water sinking in the GIN or Labrador Seas (Heinrich events) (Sarnthein *et al.*, 1994; Vidal *et al.*, 1997, 1998; Alley and Clark, 1999; Stocker, 2000). The first type corresponds to modern, warm conditions. Shut-down of convection in the GIN Seas has a strong effect on the high latitude Atlantic atmosphere and on areas that respond to it such as the monsoon regions of north Africa (Street-Perrott and Perrott, 1990). However, cross-equatorial Atlantic ocean surface transport that supplies the water for the formation of the Labrador Sea deep-water continues to remove heat from the South Atlantic under these conditions. The additional "Heinrich shut-down" of the North Atlantic and Labrador Sea deep-water formation allows this heat to remain in the South Atlantic (Crowley, 1992), and may increase deep-water formation either south of the area affected by melt-water injection (Vidal *et al.*, 1997, 1998) or in the Southern Ocean (Broecker, 1998). This reorganisation could cause warming of regions of the South Atlantic and downwind of it (Charles *et al.*, 1996; Blunier *et al.*, 1998) through a seesaw relationship with the North Atlantic. However, the behaviour of Taylor Dome in the Antarctic and several other southern sites (see above) which exhibit cooling in phase with the North Atlantic argue for an additional atmospheric link to some southern regions.

2.4.4 How Stable was the Previous Inter-glacial?

Assessment of present day climate variability benefits from comparison with conditions during inter-glacial periods that are broadly comparable with the Holocene. The most recent such inter-glacial began about 130 ky BP, lasting until about 71 ky BP when final deterioration into the last glacial began. However, only the Eemian interval, from about 130 to 120 ky BP corresponds to a climate as warm as, or warmer than, today e.g., Figure 2.22.

The study of atmospheric composition changes has revealed that rapid changes of properties observed for the lowest part of the Greenland cores (GRIP Project Members, 1993; Grootes *et al.*, 1993) do not correspond to climatic instabilities during the last inter-glacial (Chappellaz *et al.*, 1997). The extent to which climate was more or less stable during this last inter-glacial than during the Holocene is unclear. Early evidence from marine cores (CLIMAP, 1984; McManus *et al.*, 1994) and other ice cores (Jouzel *et al.*, 1993) indicated that the Eemian climate was rather stable. A high resolution North Atlantic record shows a lack of substantial fluctuations during the last inter-glacial but also indicates that the Eemian began and ended with abrupt changes in deep-water flow, with transitions occurring in less than 400 years (Adkins *et al.*, 1997). In New Zealand, there were at least three periods of milder climate than typical of the Holocene during the last inter-glacial (Salinger and McGlone, 1989). Study of an Indonesian fossil coral indicates that ENSO was robust during the last glacial period (Hughen *et al.*, 1999).

A rapid and significant cooling event within the Eemian period has been detected from European continental pollen records (Cheddadi *et al.*, 1998). High winter temperatures prevailed for 3.5 to 4 ky after the deglaciation, but then dropped by as much as 6 to 10°C in mid-Eemian times, accompanied by a decrease in precipitation. In Antarctica, the last inter-glacial is also marked by a short (about 5 ky) period of warm temperatures followed by a slightly cooler interval (Petit *et al.*, 1999). Further evidence for Eemian climate variability is found in marine records. An invasion of cold, low salinity water in the Norwegian

Sea (Cortijo *et al.*, 1994) was probably associated with a reduction in warm water transport by the North Atlantic Drift and the thermohaline circulation. Overall, the last inter-glacial appears, at least during its first part, warmer than present day climates by at least 2°C in many sites, i.e., comparable to anthropogenic warming expected by the year 2100. However, the geographical coverage of reliable and well-dated temperature time-series is too sparse to provide a global estimate.

2.4.5 Summary

Current evidence indicates that very rapid and large temperature changes, generally associated with changes in oceanic and atmospheric circulation, occurred during the last glacial period and during the last deglaciation, particularly in higher latitudes of the Northern Hemisphere. During the warming phases, and the Younger Dryas pause, there is evidence of almost worldwide, nearly synchronous events. However, as with the Holocene maximum warming and the Last Glacial Maximum, these changes appear to have occurred asynchronously between the Northern Hemisphere and at least part of the Southern Hemisphere. During the Holocene smaller but locally quite large climate changes occurred sporadically; similar changes may have occurred in the last inter-glacial. Evidence is increasing, therefore, that a rapid reorganisation of atmospheric and ocean circulation (time-scales of several decades or more) can occur during inter-glacial periods without human interference.

2.5 How have Precipitation and Atmospheric Moisture Changed?

2.5.1 Background

Increasing global surface temperatures are very likely to lead to changes in precipitation and atmospheric moisture, because of changes in atmospheric circulation, a more active hydrological cycle, and increases in the water holding capacity throughout the atmosphere. Atmospheric water vapour is also a climatically critical greenhouse gas, and an important chemical constituent in the troposphere and stratosphere.

Precipitation measurement and analysis are made more difficult by accompanying natural phenomena such as wind and the use of different instruments and techniques (Arkin and Ardanuy, 1989). Because of the substantial under-catch of precipitation gauges during solid precipitation, frequent light rainfall events, or windy conditions, the true precipitation in the Arctic is more than 50% higher than the measured values (Førland and Hanssen-Bauer, 2000). Gauge under-catch is substantially less in warmer, less windy climates with heavier rainfall. New, satellite-derived precipitation estimates offer the prospect of near-global climatologies covering at least one or two decades, but multi-decadal global changes cannot be estimated with high confidence.

For all these reasons it is useful to compare changes in many of the moisture-related variables, such as streamflow and soil moisture, with precipitation to help validate long-term precipitation trends.

2.5.2 Changes in Precipitation and Related Variables

2.5.2.1 Land

Overall, global land precipitation has increased by about 2% since the beginning of the 20th century (Jones and Hulme, 1996; Hulme *et al.*, 1998). The increase is statistically significant but has been neither spatially nor temporally uniform (Karl and Knight, 1998; Doherty *et al.*, 1999). Dai *et al.* (1997b) found a global secular increase in precipitation separate from ENSO and other modes of variability. Data from over 20,000 stations contributed to the changes since 1900 shown in Figure 2.25. The effects of changes in windshields on winter precipitation measurements were taken into account for most mid- and high latitude observations. Dai *et al.* (1997b) indicated that instrumental discontinuities are unlikely to significantly impact other observations.

Mid- and high latitudes
Over the 20th century, annual zonally averaged precipitation increased by between 7 to 12% for the zones 30°N to 85°N and by about 2% between 0°S to 55°S (Figure 2.25(ii)). The increase in the Northern Hemisphere is likely to be slightly biased because adjustments have not been made for the increasing fraction of precipitation falling in liquid as opposed to frozen form. The exact rate of precipitation increase depends on the method of calculating the changes, but the bias is expected to be small because the amount of annual precipitation affected by this trend is generally only about a few per cent. Nevertheless, this unsteady, but highly statistically significant trend toward more precipitation in many of these regions is continuing. For example, in 1998 the Northern Hemisphere high latitudes (55°N and higher) had their wettest year on record and the mid-latitudes have had precipitation totals exceeding the 1961 to 1990 mean every year since 1995.

Figure 2.25(i) shows mostly increasing precipitation in the Northern Hemisphere mid- and high latitudes, especially during the autumn and winter, but these increases vary both spatially and temporally. For example, precipitation over the United States has increased by between 5 to 10% since 1900 (Figure 2.25(ii)) but this increase has been interrupted by multi-year anomalies such as the drought years of the 1930s and early 1950s (Karl and Knight, 1998; Groisman *et al.*, 1999). The increase is most pronounced during the warm seasons. Using data selected to be relatively free of anthropogenic influences such as ground water pumpage or land use changes, several recent analyses (Lettenmaier *et al.*, 1999; Lins and Slack, 1999; Groisman *et al.*, 2001) have detected increases in streamflow across much of the contiguous United States, confirming the general tendency to increasing precipitation. However, Lins and Michaels (1994) found in some regions that increased streamflow did not relate well to an increase in rainfall. This has been further evaluated by Groisman *et al.* (2001) who show that changes in snow-cover extent also influence the timing and volume of streamflow.

Regionally, Mekis and Hogg (1999) showed that precipitation in Canada has increased by an average of more than 10% over the 20th century. Zhang *et al.* (2000) report an increase in Canadian heavy snowfall amounts north of 55°N and Akinremi *et al.* (1999) found rainfall significantly increasing in the Canadian prairies from 1956 to 1995. Multi-decadal streamflow data in Canada are not extensive, but there are no apparent inconsistencies between observed changes in streamflow or precipitation (Zhang *et al.*, 2000).

Over the last 50 years there has been a slight decrease in annual precipitation over China (Zhai *et al.*, 1999a), which is supported by a significant (5% confidence level) decrease in the number of rainy days (3.9%/decade; Figure 2.25 (ii)). In contrast, the area affected by the upper 10% of heaviest precipitation has significantly increased. Zhai *et al.* (1999b) show a significant increase in precipitation over the middle and lower reaches of the Yangtze River and west China during the latter part of the 20th century, while also detecting a declining trend in precipitation over northern China.

There have been marked increases in precipitation in the latter part of the 20th century over northern Europe, with a general decrease southward to the Mediterranean (Schönwiese and Rapp, 1997; Figure 2.25(i)). Dry wintertime conditions over southern Europe and the Mediterranean (Piervitali *et al.*, 1998; Romero *et al.*, 1998) and wetter than normal conditions over many parts of northern Europe and Scandinavia (Hanssen-Bauer and Førland, 2000) are linked to strong positive values of the North Atlantic Oscillation, with more anticyclonic conditions over southern Europe and stronger westerlies over northern Europe (Section 2.6.5).

Based on recent research (Bogdanova and Mescherskaya, 1998; Groisman and Rankova, 2001), the precipitation trend for the last century over the former USSR as reported by the SAR was slightly overestimated. The new results indicate that precipitation has increased since 1891 by about 5% west of 90°E for both warm and cold seasons. Georgievsky *et al.* (1996) also noted increases in precipitation over the last several decades over western Russia, accompanied by increases in streamflow and a rise in the level of the Caspian Sea. In eastern Russia a negative precipitation trend since 1945 is embedded in the century-long positive precipitation trend (Figure 2.25(ii); Gruza *et al.*, 1999). Soil moisture data for large regions of Eurasia (Robock *et al.*, 2000) show large upward trends. The rate of increase is more than 1 cm/decade in the available soil moisture in the top 1 m of soil. These large positive trends occur simultaneously with positive trends in temperature that would normally reduce soil moisture. Increases in precipitation (and cloud cover, Section 2.5.5) are believed to have more than compensated for the increased losses due to evapotranspiration.

An analysis of rainfall data since 1910 by Haylock and Nicholls (2000) reveals a large decrease in total precipitation and related rain days in south-western Australia. Annual total rainfall has increased over much of Australia with significant increases of 15 to 20% in large areas. The increase in total rainfall has been accompanied by a significant 10% rise in the average number of rain days over Australia (Hennessy *et al.*, 1999). Elsewhere in the Southern Hemisphere, a long-term increase in precipitation in Argentina has been observed for the period 1900 to 1998 (Figure 2.25(i); Dai *et al.*, 1997b).

Tropics and sub-tropics
The increase in precipitation in the mid- and high latitudes contrasts with decreases in the northern sub-tropics (with

marginal statistical significance) which were largely responsible for the decade-long reduction in global land precipitation from the mid-1980s through the mid-1990s. Since the SAR, record low precipitation has been observed in equatorial regions, while the sub-tropics have recovered from their anomalously low values of the 1980s.

Regionally positive but non-significant trends have occurred in the rainy season rainfall in north-east Brazil and northern Amazonia (Marengo *et al.*, 1998). River data from northern Amazonia indicate wetter periods in the mid-1970s, and in 1990, as well as drier periods between 1980 to 1990, consistent with rainfall anomalies. Northern Amazonian rainfall appears to be modulated by multi-decadal climate variations.

There is little evidence for a long-term trend in Indian monsoonal rainfall but there are multi-decadal variations (Kumar *et al.*, 1999a,b). From 1906 to about 1960, monsoonal rainfall increased then decreased through 1974 and has increased since (see Section 2.6). In central America for much of the period from the early 1940s to present, western Mexico has experienced an increasingly erratic monsoonal rainfall (Douglas and Englehart, 1999).

Since 1976, increases in precipitation in the South Pacific have occurred to the north-east of the South Pacific Convergence Zone (SPCZ) while decreases have occurred to its south-west (Salinger *et al.*, 1996). Manton *et al.* (2001) found significant decreases in rain days since 1961 throughout Southeast Asia and western and central South Pacific, but increases in the north of French Polynesia and Fiji.

Streamflow data for major rivers in south-eastern South America for the period 1901 to 1995 show that streamflow has increased since the mid-1960s, and was accompanied by a significant decrease in the amplitude of the seasonal cycle of most of those rivers (Garcia and Vargas, 1998; Genta *et al.*, 1998). Figure 2.25(i) shows increases in precipitation since 1900 along the South American eastern coastal areas, with less extensive increases since 1976.

There has been a pattern of continued aridity since the late 1960s throughout North Africa south of the Sahara. This pattern is most persistent in the western region. The driest period was in the 1980s with some recovery occurring during the 1990s, particularly in the easternmost sectors where rainfall in some years was near or just above the long-term mean (Nicholson *et al.*, 2000). Southern Africa was relatively moist in the 1950s and 1970s (Nicholson *et al.*, 2000); but Hulme (1996) found significant decreases in precipitation being observed since the late 1970s. Early 2000, however, has seen flood-producing rains in the eastern part of southern Africa.

2.5.2.2 Palaeo-drought

Palaeoclimate proxy evidence (tree rings, lake sediments and pollen evidence) has been used to estimate variability in drought and precipitation patterns in past centuries. Much of the recent research has emphasised the North American region (e.g., Cook *et al.*, 1999a), where a key conclusion is that the range of regional drought variability observed during the 20th century may not be representative of the larger range of

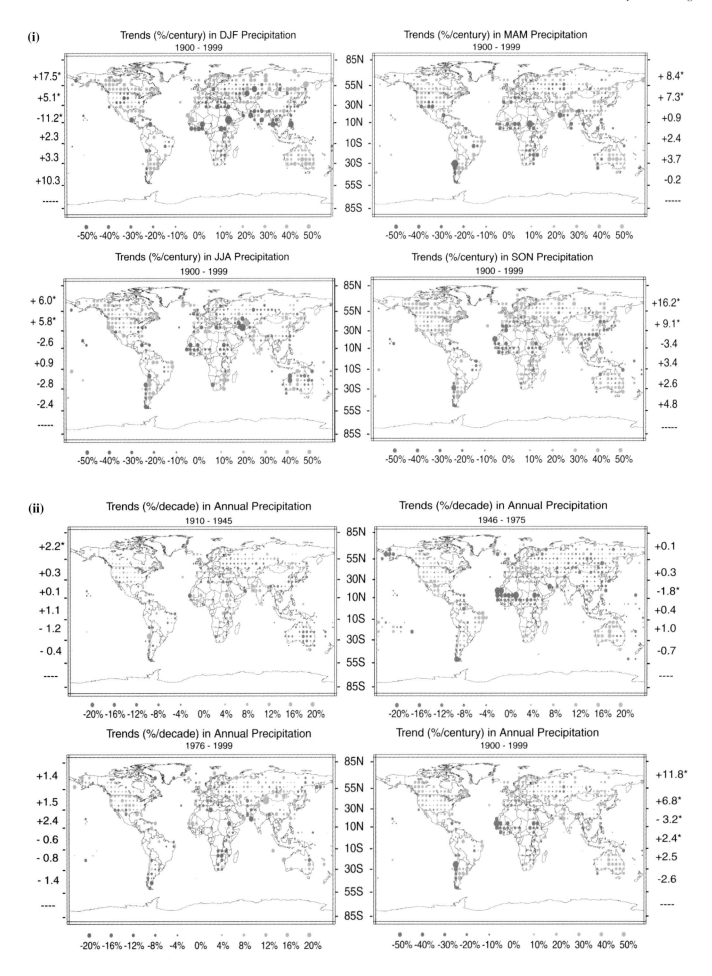

Figure 2.25(i): Trends for 1900 to 1999 for the four seasons. Precipitation trends are represented by the area of the circle with green representing increases and brown representing decreases. Annual and seasonal trends were calculated using the following method. Precipitation anomalies in physical units were calculated for each station based on 1961 to 1990 normals and averaged into $5° \times 5°$ grid cells on a monthly basis. The 1961 to 1990 monthly mean precipitation for each grid cell was added to the monthly anomalies and the resulting grid cell values summed into annual and seasonal totals. This series was converted into percentages of normal precipitation, and trends calculated from the percentages. Average trends within six latitude bands (85°N to 55°N, 55°N to 30°N, 30°N to 10°N, 10°N to 10°S, 10°S to 30°S, 30°S to 55°S) are shown in the legend of each map. The 1961 to 1990 monthly mean precipitation for the latitude band was added to the anomaly time-series and the resulting values totalled across all months within the season or year. The significance of each trend (based on a 5% level) was determined using a t-test and a non-parametric test statistic. Trends found to be significant under both tests are indicated with an asterisk.

Figure 2.25(ii): As in Figure 2.25(i) except annual trends for the three periods of changing rates of global temperature (shown in Figure 2.9) and the full period, 1900 to 1999. During the 100-year periods, calculation of grid cell trends required at least 66% of the years without missing data and at least three years of data within each decade except the first and last. During the shorter periods, calculation of grid cell trends required at least 75% of the years without missing data. Stations with more than one sixth of their data missing during the normal period and grid cells with more than one season or year without any measurable precipitation during the normal period were excluded from consideration. Due to the nature of trend estimation, it is not possible to cumulatively sum the trends for each of the three periods to obtain an overall trend.

drought evident in past centuries (Laird *et al.*, 1996; Woodhouse and Overpeck, 1998). Hughes and Graumlich (1996) and Hughes and Funkhouser (1999) provide evidence of multi-decadal mega-droughts in the western Great Basin of North America in the 10th to 14th centuries. Nonetheless, the 20th century dust bowl still stands out as the most extreme drought of the past several centuries, the period when North American continental scale reconstruction is possible. Swetnam and Betancourt (1998) argue that recent spring wetness in the American south-west is greater than that observed in at least the last thousand years. Evidence of significant changes in regional hydroclimatic patterns is not limited, however, to North America. Stine (1994) argues that enhanced drought conditions occurred synchronously in South America. Ice accumulation at Quelccaya in the Andes, and on the Dunde Ice Cap on the Tibetan Plateau (Thompson, 1996) was slower in the first half of the last millennium than the last 500 years, but 500-year averages are not easily related to the palaeo-temperature data (Figure 2.21). Pollen evidence indicates significant changes in summer rainfall patterns in China in the earlier centuries of the past millennium (Ren, 1998). The relationship between such past changes in regional drought and precipitation patterns, and large-scale atmospheric circulation patterns associated with ENSO, for example, is an area of active current research (e.g., Cole and Cook, 1998).

2.5.2.3 Ocean

The strong spatial variability inherent in precipitation requires the use of estimates based on satellite observations for many regions. Thus satellite data are essential to infer global changes in precipitation, as the oceans account for 70% of the global surface area. Since adequate observations were not made until the early 1970s, no satellite-based record is sufficiently long to permit estimates of century-long changes. The first satellite instrument specifically designed to make estimates of precipitation did not begin operation until 1987. At this time three data sets are available: (a) the Global Precipitation Climatology Project (GPCP) product, which spans the period from 1987 to the present (Huffman *et al.*, 1997); (b) the CPC Merged Analysis of Precipitation (CMAP) product, covering the period from 1979 to 1998 (Xie and Arkin, 1997); and (c) MSU-derived precipitation estimates since 1979 (Spencer, 1993). While the period from 1987 appears to be well observed, it is too short to draw conclusions regarding decadal-scale variations. The longer CMAP data set assumes that the various satellite-derived estimates have no trend over the period, and hence no longer time-scale conclusions are possible. Nonetheless, analyses of the CMAP product and associated data from the NCEP reanalysis project indicate that there have been substantial average increases in precipitation over the tropical oceans during the last twenty years, related to increased frequency and intensity of ENSO (Trenberth *et al.*, 2001). ENSO conditions are not related to positive precipitation anomalies everywhere over the tropical oceans (e.g., south-western Tropical Pacific).

2.5.3 Water Vapour

Although measurement problems hinder the analysis of long-term water vapour changes (Elliott, 1995; Rind, 1998), several recent studies tend to confirm and extend the findings of lower tropospheric water vapour increases reported in the SAR. Furthermore, new analyses indicate upward trends in near-surface humidity. Knowledge about changes in water vapour at upper tropospheric and lower stratospheric levels is of great importance because strong alterations in radiative forcing can result from small absolute changes in water vapour at these levels (Chapters 6 and 7). New data presented here from the SPARC WAVAS (Stratospheric Processes and their Role in Climate / Water Vapour Assessment) project (Kley *et al.*, 2000) are starting to cast light on changes at these levels. Note that water vapour pressure, and specific humidity (for a constant relative humidity) increase non-linearly with increasing temperature.

2.5.3.1 Surface water vapour

Water vapour pressure, dew-point or relative humidity at the surface is conventionally measured using wet and dry bulb thermometers exposed in thermometer screens at climate stations. The quality of these data has been little studied. Wet bulb thermometers are not usually aspirated, so that the cooling of the wet bulb, and therefore the deduced specific or relative humidity, depends on the flow rate of air within the screen. This may often differ from the assumed airflow. Occasionally wet bulbs may dry out. Thus it is not possible to judge fully the accuracy of surface vapour pressure trends presented here.

Schönwiese *et al.* (1994) and Schönwiese and Rapp (1997) found small increases in surface vapour pressure over most of Europe from 1961 to 1990. The annual trends are weak. Statistically significant changes are confined to increases of about 0.5 to 1.5 hPa (relative to mean values of 12 to 15 hPa) in the southern and eastern Mediterranean region (with the largest increase in summer) and decreases of about 0.5 hPa over parts of Turkey (mainly in springtime).

Specific humidity trends over the United States were overwhelmingly positive for the period 1961 to 1995, with magnitudes of several per cent per decade, and with the largest and most statistically significant trends in spring and summer (Gaffen and Ross, 1999). Night-time specific humidity trends were generally stronger than daytime trends. Relative humidity showed smaller increases, especially in winter and spring. The specific humidity and derived dew point trends are broadly consistent, both spatially and in their day-night differences, with temperature trends. Schwartzman *et al.* (1998) found that the diurnal dewpoint cycle is changing over North America, with a relative decline in late afternoon and a small rise at midday.

Increases in water vapour over the former Soviet Union, Eastern China, the United States and tropical Western Pacific islands have been found in some seasons by Sun *et al.* (2000) in the second half of the 20th century, but with decreases in Canada in autumn. The selective character of the findings prevents any assessment of statistical significance. Wang and Gaffen (2001) found that specific humidity trends over China were overwhelmingly positive over 1951 to 1994, with the largest and most statistically significant trends in north-west China north of 35°N and

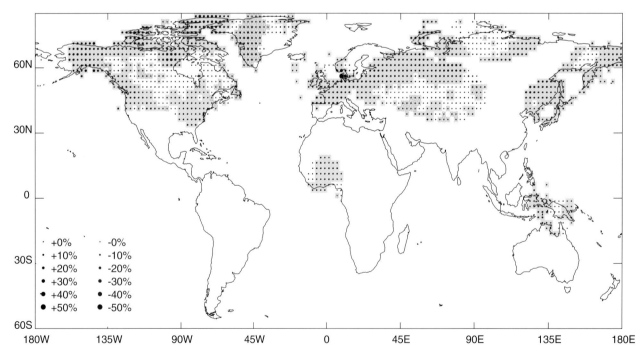

Figure 2.26: Trends in annual mean surface water vapour pressure, 1975 to 1995, expressed as a percentage of the 1975 to 1995 mean. Areas without dots have no data. Blue shaded areas have nominally significant increasing trends and brown shaded areas have significant decreasing trends, both at the 5% significance level. Biases in these data have been little studied so the level of significance may be overstated. From New *et al.* (2000).

west of 105°E. Trends were larger in summer and night-time trends were generally larger than daytime ones.

Recently New *et al.* (2000) have estimated linear trends for annual and seasonal values of surface vapour pressure over land using calculated monthly vapour pressure data from climate stations. Figure 2.26 shows trends for the 21 years from 1975 to 1995, corresponding to much of the recent period of global warming described in Section 2.2.2.3. Although the uncertain quality of the data prevents any definitive conclusions about statistical significance, nominal significance of trends at the 5% level was estimated after smoothing the annual data to reduce the influence of outliers at the beginning and end of this short series. Few Southern Hemisphere data have been analysed, but Figure 2.26 shows that there have been widespread nominally significant increases in annual mean water vapour in the Northern Hemisphere. These increases are reflected in the individual seasons, although nominally significant annual mean increases are more extensive. Regional decreases over eastern Canada are explained by colder conditions in the winter half year associated with the increasingly positive phase of the North Atlantic Oscillation (Section 2.6.5).

2.5.3.2 Lower-tropospheric water vapour

Radiosonde and satellite observations of water vapour above the surface have been analysed for evidence of long-term change. Both data sources have had serious data quality and temporal homogeneity problems (Elliott, 1995), although recent work to determine trends in water vapour from the surface to 500 hPa since 1973 has been based on radiosonde data judged to be largely unaffected by these problems (Ross and Elliott, 2001). Published satellite data are insufficiently homogeneous or too short in length to deduce reliable trends or low-frequency variations.

Radiosonde observations

Ross and Elliott (1996, 1998) analysed surface-to-500 hPa precipitable water over the Northern Hemisphere for 1973 to 1995 using quality-controlled data. Increases in precipitable water were found over North America except for north-east Canada. Over Eurasia,

only China and the Pacific islands show coherent regional increases. The remainder of Eurasia shows a mixture of positive and negative trends, with a tendency for negative trends over Eastern Europe and western Russia. Mid-tropospheric water vapour trends tend to be of the same sign as temperature trends over North America, China, and the Pacific, but elsewhere the temperature trends are more consistently positive than the water vapour trends. Figure 2.27 summarises the results. Lower-tropospheric dew-point data for the period 1961 to 1995 also show increases, though smaller than those for the 1973 to 1995 period, and few are statistically significant (Ross and Elliott, 2001).

Zhai and Eskridge (1997) found increases of about 1 to 3%/decade in surface-to-200 hPa precipitable water over China for 1970 to 1990. Increases were most significant in spring. Percentage trends were larger over the 700 to 400 hPa layer than the surface-700 hPa layer. Gutzler (1996) found that specific humidity data at 1,000, 700, and 300 hPa at four western tropical Pacific radiosonde stations from 1973 to 1993 gave increases of 3 to 9%/decade, with larger percentage increases at increasing height above the surface. In contrast, Peixoto and Oort (1996) found decreases in zonal mean relative humidity between 1974 and 1988. The decreases are more marked at 300 hPa, where they are more likely to be associated with instrument changes than at lower levels, and are more pronounced at higher latitudes than in the tropics.

2.5.3.3 Upper-tropospheric and lower-stratospheric water vapour

Recently assessed increases in lower stratospheric water vapour mixing ratio over the last few decades are likely to have caused a decrease in stratospheric temperatures by an amount comparable to that produced by ozone decreases (Forster and Shine, 1999; Smith *et al.*, 2001) (see lower-stratospheric temperature trends in Section 2.2.3). These changes also impact on ozone chemistry (Chapter 4) and on radiative forcing of the atmosphere (Chapters 6 and 7). Data from over twenty-five instruments that measure water vapour concentration and relative humidity in the upper troposphere and stratosphere were recently compared and

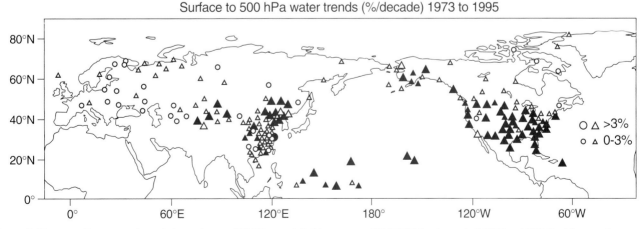

Figure 2.27: Annually averaged trends in surface to 500 hPa precipitable water at 0000UTC for the period 1973 to 1995. Positive trends are indicated by triangles and negative trends by circles. Filled symbols indicate the trends were statistically significant at the 5% level according to the Spearman test. The two sizes of symbols give an indication of the magnitude of the trend. From Ross and Elliott (2001).

assessed in the international SPARC study (Kley *et al.*, 2000). The purpose of the study, which included measurements made by both *in situ* and remote sensing techniques utilising balloons, aircraft and satellites, was to determine the data quality and to estimate the magnitude of any trends. The study showed that some stratospheric instruments have sampled over a long enough period that several overlapping time-series of intermediate length (8 to 15 years) can be used to help evaluate stratospheric changes. A reasonable degree of consistency was found among stratospheric measurements made from near the tropopause up to as high as 50 km (about 1 hPa). Most observations were within ±10% of the grand mean of all measurements to which they were compared.

Accurate balloon observations of lower-stratospheric water vapour are available from 1964 to 1976 over Washington, D.C. and from 1980 to present over Boulder, Colorado, USA (e.g., Mastenbrook, 1968; Harries, 1976; Mastenbrook and Oltmans, 1983; Oltmans and Hofmann, 1995). The SPARC study shows that these point measurements are nevertheless representative of global stratospheric conditions above about 18 to 20 km, but not of the lowest stratosphere where there can be significant regional and seasonal changes. A positive lower stratosphere trend of about 1 to 1.5%/year in specific humidity (about 0.04 ppm/year) since the mid-1960s is indicated by the balloon data (Oltmans *et al.*, 2000). The increase was not monotonic but showed several rapid rises with plateaux in between. Even though the recent satellite record is relatively short, these measurements have revealed changes of the same character. The satellite results show a spatial pattern of trends in the lower stratosphere, and suggest a slowing in the positive trend after 1996 (Smith *et al.*, 2000). Although not definitive, these observations are consistent in suggesting that lower-stratospheric water vapour has increased globally on average at about 1%/year over at least the past forty years, but at a variable rate.

Although radiosondes have made observations of water vapour in the upper troposphere (i.e., above 500 hPa) since the 1950s, these observations have suffered from instrumental errors (Elliott and Gaffen, 1991). Peixoto and Oort (1996) have re-examined these observations for the period 1974 to 1988 and found large trends in upper-tropospheric humidity at the 300 hPa level. They concluded that these trends were unrealistically large and were likely to be due to instrument changes. Satellite observations of upper-tropospheric humidity (UTH) measurements made by TOVS (Television infrared observation satellite Operational Vertical Sounder) since 1979, and representative of a deep layer between 500 to 200 hPa, show very large interannual variability (Bates *et al.*, 1996). The SPARC assessment of these observations (Kley *et al.*, 2000) indicated that they were of sufficient quality for trend analyses. The SPARC study and an analysis by Bates and Jackson (2001) show large regional trends that are attributed to circulation changes associated with ENSO, decadal variability over equatorial Africa, and decadal variability of the Arctic Oscillation (see Section 2.6). Statistically significant positive trends of 0.1%/year are found for 10°N to 10°S, and a non-significant trend of 0.04%/year for 60°N to 60°S, but this includes a component negative trend of −0.1%/year for 30°S to 60°S. The trends in large zonal bands

tend to be residuals from cancellations in sign and magnitude of much larger regional trends. These UTH trends should be treated with caution especially in the deep tropics because of significant interannual variability and persistence, both of which hamper trend detection.

In summary, *in situ* and radiosonde measurements tend to show increasing water vapour in the lower troposphere and near the surface, though this is not seen everywhere, and data quality is still an issue. The longer, more reliable data sets suggest multi-decadal increases in atmospheric water vapour of several per cent per decade over regions of the Northern Hemisphere. New analyses of balloon and satellite records indicate that stratospheric water vapour above 18 km shows an increase of about 1%/year for the period 1981 to 2000 but with a slowing of the positive trend after 1996. Satellite observations of upper-tropospheric humidity from 1980 to 1997 show statistically significant positive trends of 0.1%/year for the zone 10°N to 10°S.

2.5.4 Evaporation

Only evaporation from the land surface is discussed, as nothing new since the SAR has emerged on oceanic evaporation changes.

2.5.4.1 Land

The SAR reported widespread decreases of pan evaporation over the USA and Russia during the 20th century. Pan evaporation measurements are an index of evaporation from a surface with an unlimited supply of water (potential evaporation). Interpretation of this result involving potential evaporation as a decrease in actual land surface evaporation is contradictory to the temperature and precipitation increase reported in these areas, and the general intensification of the hydrological cycle over northern extra-tropical land areas (Brutsaert and Parlange, 1998). Further analysis by Lawrimore and Peterson (2000) supports Brutsaert and Parlange's (1998) interpretation, as does Golubev *et al.* (2001). Using parallel observations of actual evaporation and pan evaporation at five Russian experimental sites, Golubev *et al.* (2001) developed a method to estimate actual land surface evaporation from the pan evaporation measurements. They showed that using this method, actual evaporation is shown to have increased during the second half of the 20th century over most dry regions of the United States and Russia. Similarly, over humid maritime regions of the eastern United States (and north-eastern Washington state) actual evaporation during the warm season was also found to increase. Only over the heavily forested regions of Russia and the northern United States did actual evaporation decrease. The increase in actual evaporation is related to the greater availability of moisture at the surface, due to increases in precipitation and the higher temperatures.

2.5.5 Clouds

Clouds are important in the Earth's climate system because of their effects on solar radiation, terrestrial radiation and precipitation. Different cloud types contribute to total cloud amount and are associated with a wide variety of thermal and dynamic processes in the climate system (see Chapter 7, Section 7.2.2).

Therefore knowing the variations in total cloud amount and different cloud types would significantly contribute to improving our understanding of the role of clouds in contemporary climate change. Several analyses of cloud amounts for regions of the world have been performed since the SAR. Problems with data homogeneity, particularly concerning biases with changing times of observation (Sun and Groisman, 2000; Sun *et al.*, 2001) have been addressed in several studies, but other issues continue to be a source of uncertainty.

2.5.5.1 Land

Dai *et al.* (1997a, 1999) and Kaiser (1998) examined cloud cover changes over the former USSR and China during the last four to five decades, to add to earlier analyses for Europe, the United States, Canada, and Australia by Henderson-Sellers (1992) and Karl and Steurer (1990). These studies show 20th century increases in cloud cover over much of the United States (mostly confined to the first 80 years) and the former USSR, which are significantly negatively correlated with changes in the diurnal range of surface air temperature (DTR) (as shown earlier in Figure 2.3). Sun and Groisman (2000) showed that in the former USSR low-level cloud cover significantly decreased during the period 1936 to 1990. However, this was more than offset by a significant increase in cumulus and cirrus clouds during the past several decades. Over much of China, however, daytime and night-time total cloud cover exhibited significant decreasing trends of 1 to 2% sky cover/decade for both day and night observations between 1951 and 1994 (Kaiser, 1998, 2000), which the DTR failed to follow (Figure 2.3). This discrepancy may result from the increasing effect of industrial aerosols on the DTR since the late 1970s (Dai *et al.*, 1999). Tuomenvirta *et al.* (2000) show increasing trends in cloud cover during the period 1910 to 1995 for northern Europe, which are consistent with decreases in the DTR. A new analysis (Neff, 1999) reveals a dramatic increase (15 to 20%) of spring and summer cloud amount at the South Pole during the past four decades in this region. This appears to be related to the observed delay in the breakdown of the spring polar vortex and is believed to be related to decreases in stratospheric temperatures.

There are few analyses of the amounts of various cloud types or changes over the tropics and sub-tropics. Correlations with observed precipitation and clouds observed by satellites suggest that much of the increase in the total cloud amount is likely to have resulted from increases in thick, precipitating clouds (Dai *et al.*, 1997a). Hahn *et al.* (1996) show decreasing decadal scale trends in cloud cover over much of China, as well as over most of South America and Africa for the period 1971 to 1991. The latter two areas have little surface-based information.

2.5.5.2 Ocean

The SAR presented analyses of inter-decadal changes in marine cloud coverage. The data have now been re-examined and doubt has been cast on some of the previous findings (Bajuk and Leovy, 1998a; Norris, 1999). Additional data have also reversed some of the previous trends. In the SAR a 3% increase in cumulonimbus clouds was reported for the period 1952 to 1981. An update of this analysis showed a gradual rise in cumulonimbus cloud amount from the mid-1950s to the mid-1970s, with a gradual decline thereafter (Bajuk and Leovy, 1998a). Bajuk and Leovy (1998b) cast doubt on the homogeneity of the cloud amounts derived from ship data. They find that inter-decadal variations of the frequency of occurrence of cloud amount for a given cloud type are generally unrelated to similar time-scale variations in SST and large-scale divergence of the surface winds. Nonetheless, some regional changes and variations based on ship reports of low and middle clouds are likely to be rather robust. Variations in these categories of cloud are consistent with variations of other climate system variables. Examples include: (1) a long-term upward trend in altostratus and nimbostratus across the mid-latitude North Pacific and North Atlantic Oceans (Parungo *et al.*, 1994; Norris and Leovy, 1995); (2) ENSO related variations in the frequency of low cloud types across the Pacific and Indian Oceans (Bajuk and Leovy, 1998b); and (3) interannual variations in summer season stratiform clouds across the North Pacific (Norris *et al.*, 1998). Norris (1999) found an increase in total sky cover of approximately 2%, and an increase of approximately 4% in low cloud cover in his analyses of ship reports between 1952 and 1995. He finds no evidence for changes in observation practices that may have affected these trends. The trends are dominated by a globally consistent mode and are as large or larger in the tropics and Southern Hemisphere as in the Northern Hemisphere. This argues against attribution to increased anthropogenic aerosol amounts.

2.5.5.3 Global

Although satellite estimates of changes and variations in cloud amount and type contain systematic biases, Rossow and Schiffer (1999) showed improved calibration and cloud detection sensitivities for the International Satellite Cloud Climatology Program (ISCCP) data set. Using data from 1983 to 1994, a globally increasing trend in monthly mean cloudiness reversed during the late 1980s and early 1990s. There now appears to be an overall trend toward reduced total cloud amounts over both land and ocean during this period. An estimate for aircraft-induced cirrus cover for the late 1990s ranges from 0 to 0.2% of the surface of the Earth (IPCC, 1999).

2.5.6 *Summary*

Since the SAR, land surface precipitation has continued to increase in the Northern Hemisphere mid- and high latitudes; over the sub-tropics, the drying trend has been ameliorated somewhat. Where data are available, changes in annual stream-flow relate well to changes in total precipitation. Over the Southern Hemisphere land areas no pronounced changes in total precipitation are evident since the SAR. The changes in precipitation in mid- and high latitudes over land have a strong correlation with long-term changes in total cloud amount. Little can be said about changes in ocean precipitation as satellite data sets have not yet been adequately tested for time-dependent biases. Changes in water vapour have been analysed most for selected Northern Hemisphere regions, and show an emerging pattern of surface and tropospheric water vapour increases over the past few decades, although there are still untested or uncorrected biases in these data. Limited data from the stratosphere also suggest increases in water vapour but this result must be viewed with great caution. Over land, an increase in cloud cover of a few per cent since the turn of the century is observed, which is shown to closely relate to changes in the diurnal temperature range. Changes in ocean cloud amount and type show systematic increases of a few per cent since the 1950s, but these relate poorly to SST or surface wind divergence changes, casting some doubt on the integrity of the trends. No changes in observing practices can be identified, however, that might have led to time-dependent biases in the ocean cloud amount and frequency statistics.

2.6 Are the Atmospheric/Oceanic Circulations Changing?

2.6.1 Background

Changes or fluctuations in atmospheric and oceanic circulation are important elements of climate. Such circulation changes are the main cause of variations in climate elements on a regional scale, sometimes mediated by parallel changes in the land surface (IPCC, 1990, 1996). ENSO and NAO are such examples. On decadal time-scales, the Pacific Decadal Oscillation (PDO) and the related Inter-decadal Pacific Oscillation (IPO) may account for approximately half the global mean variation in surface temperatures. They are also prominently linked to regional variations in temperature and precipitation (Higgins *et al.*, 2000). This section documents regional changes and slow fluctuations in atmospheric circulation over past decades, and demonstrates that these are consistent with large-scale changes in other variables, especially temperature and precipitation. Note that there is much evidence that many of the atmospheric circulation changes we observe, particularly in the extra-tropics, are the net result of irregular fluctuations between preferred states of the atmosphere (Palmer, 1993, 1999) that last for much shorter times. Thus changes in circulation on decadal time-scales involve changes in the frequency of such states. Chapter 7 discusses this in more detail. The focus of this section is on long-term variation and change, rather than on shorter-term variability.

2.6.2 El Niño-Southern Oscillation and Tropical/Extra-tropical Interaction

ENSO is the primary global mode of climate variability in the 2 to 7 year time band. El Niño is defined by SST anomalies in the eastern tropical Pacific while the Southern Oscillation Index (SOI) is a measure of the atmospheric circulation response in the Pacific-Indian Ocean region. This sub-section assesses the variability of ENSO over the past few centuries.

Multiproxy-based reconstructions of the behaviour of ENSO have recently been attempted for the past few centuries, including a boreal winter season SOI reconstruction based on highly ENSO-sensitive tree-ring indicators (Stahle *et al.*, 1998). A multiproxy-based reconstruction of the boreal cold-season (Oct-Mar) NINO 3 (SST anomalies in the tropical Pacific from 5°N to 5°S, 150°W to 90°W) index (Mann *et al.*, 2000b) has also been made. Figure 2.28 compares the behaviour of these

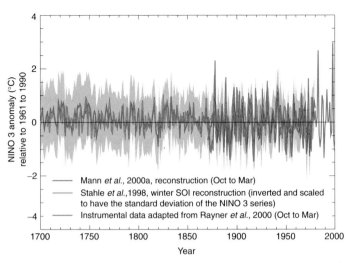

Figure 2.28: Reconstructions since 1700 of proxy-based ENSO indices. Shown are the Northern Hemisphere cold-season (Oct-Mar) mean NINO 3 index of Mann *et al.* (2000a) and the Northern Hemisphere winter SOI index of Stahle *et al.* (1998). The SOI series is scaled to have the same standard deviation as the NINO 3 index, and is reversed in sign to be positively correlated with the NINO 3 series. An instrumental NINO 3 index from 1871 to 2000 is shown for comparison (Rayner *et al.*, 2000; see also Figure 2.29), with two standard error limits (grey shaded) of the proxy NINO 3 reconstruction.

two series with recent ENSO behaviour. The SOI reconstruction has been rescaled to have the sign and variance of the NINO 3 reconstruction; the two reconstructions, based on independent methods and partially independent data, have a linear correlation (r=0.64) during the pre-calibration interval. While the estimated uncertainties in these reconstructed series are substantial, they suggest that the very large 1982/83 and 1997/98 warm events might be outside the range of variability of the past few centuries. However, the reconstructions tend to underestimate the amplitude of ENSO events, as is clearly evident for the large 1877/78 event. Only a richer network of ENSO-sensitive proxy indicators can improve this situation, such as the new long tropical coral series becoming available (see Dunbar and Cole, 1999).

Instrumental records have been examined to search for possible changes in ENSO over the past 120 years. Three new reconstructions of SST in the eastern Equatorial Pacific (Figure 2.29) that use optimum interpolation methods exhibit strong similarities. The dominant 2 to 6 year time-scale in ENSO is apparent. Both the activity and periodicity of ENSO have varied considerably since 1871 with considerable irregularity in time. There was an apparent "shift" in the temperature of the tropical Pacific around 1976 to warmer conditions, discussed in the SAR, which appeared to continue until at least 1998. During this period ENSO events were more frequent, intense or persistent. It is unclear whether this warm state continues, with the persistence of the long La Niña from late 1998 until early 2001. ENSO has been related to variations of precipitation and temperature over much of the tropics and sub-tropics, and some mid-latitude areas.

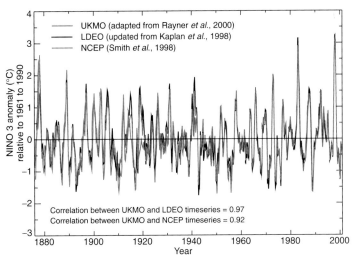

Figure 2.29: El Niño-La Niña variations from 1876 to 2000 measured by sea surface temperature in the region 5°N to 5°S, 150 to 90°W. Reconstructions using pattern analysis methods from (a) red: UK Met Office (UKMO) Hadley Centre sea ice and sea surface temperature data set version 1 (Rayner *et al.*, 2000); (b) black: from the Lamont-Doherty Earth Observatory (LDEO) (Kaplan *et al.*, 1998); (c) blue: the National Centers for Environmental Prediction (NCEP) analysis (Smith *et al.*, 1998). 1876 is close to the earliest date for which reasonably reliable reconstructions can be made.

A number of recent studies have found changes in the interannual variability of ENSO over the last century, related in part to an observed reduction in ENSO variability between about 1920 and 1960. Various studies (Wang and Wang, 1996; Torrence and Compo, 1998; Torrence and Webster, 1998; Kestin *et al.*, 1999) show more robust signals in the quasi-biennial and 'classical' 3 to 4 year ENSO bands (3.4 and 7 years) during the first and last 40 to 50 years of the instrumental record. A period of very weak signal strength (with a near 5-year periodicity) occurs in much of the intervening epoch.

The 1990s have received considerable attention, as the recent behaviour of ENSO seems unusual relative to that of previous decades. A protracted period of low SOI from 1990 to 1995, during which several weak to moderate El Niño events occurred with no intervening La Niña events (Goddard and Graham, 1997) was found by some studies (e.g., Trenberth and Hoar, 1996) to be statistically very rare. Whether global warming is influencing El Niño, especially given the remarkable El Niño of 1997/98, is a key question (Trenberth, 1998b), especially as El Niño affects global temperature itself (Section 2.2 and Chapter 7).

2.6.3 *Decadal to Inter-decadal Pacific Oscillation, and the North Pacific Oscillation*

Recently, 'ENSO-like' spatial patterns in the climate system, which operate on decadal to multi-decadal time-scales, have been identified. This lower-frequency SST variability is less equatorially confined in the central and eastern Pacific, and relatively more prominent over the extra-tropics, especially the north-west

Pacific, and has a similar counterpart in night marine air temperatures (Tanimoto *et al.*, 1993; Folland *et al.*, 1999a; Allan, 2000). The corresponding sea level pressure (SLP) signature is also strongest over the North Pacific, and its December-February counterpart in the mid-troposphere more closely resembles the Pacific-North America (PNA) pattern (Zhang *et al.*, 1997b; Livezey and Smith, 1999). There is ambiguity about whether inter-decadal Pacific-wide features are independent of global warming. In the longer Folland *et al.* (1999) analyses since 1911 they appear to be largely independent, but in the Livezey and Smith analysis of more recent SST data they are an integral part of a global warming signal. Using a different method of analysis of data since 1901, Moron *et al.* (1998) find a global warming signal whose pattern in the Pacific is intermediate between these two analyses.

The PDO of Mantua *et al.* (1997), with lower-frequency variations in the leading North Pacific SST pattern, may be related to the same Pacific-wide features, and parallels the dominant pattern of North Pacific SLP variability. The relationship is such that cooler than average SSTs occur during periods of lower than average SLP over the central North Pacific and *vice versa*. Recently, the IPO, a Pacific basin-wide feature, has been described, which includes low-frequency variations in climate over the North Pacific (Power *et al.*, 1998, 1999; Folland *et al.*, 1999a). The time-series of this feature is broadly similar to the inter-decadal part of the North Pacific PDO index of Mantua *et al.* (1997). The IPO may be a Pacific-wide manifestation of the PDO and seems to be part of a continuous spectrum of low-frequency modulation of ENSO, and so may be partly stochastic. When the IPO is in a positive phase, SST over a large area of the south-west Pacific is cold, as is SST over the extra-tropical north-west Pacific. SST over the central tropical Pacific is warm but less obviously warm over the equatorial far eastern Pacific unlike ENSO. Warmth also extends into the tropical west Pacific, unlike the situation on the ENSO time-scale.

The IPO shows three major phases this century: positive from 1922 to 1946 and from 1978 to at least 1998, with a negative phase between 1947 and 1976. Arguably, the structure of this pattern, nearly symmetrical about the equator and only subtly different from ENSO, is a strong indication of the importance of the tropical Pacific for many remote climates on all time-scales (Garreaud and Battisti, 1999). Power *et al.* (1999) showed that the two phases of the IPO appear to modulate year-to-year ENSO precipitation variability over Australia. Salinger and Mullan (1999) showed that prominent sub-bidecadal climate variations in New Zealand, identified in the temperature signal by Folland and Salinger (1995), are related to a SST pattern like the IPO. The IPO is a significant source of decadal climate variation throughout the South Pacific, and modulates ENSO climate variability in this region (Salinger *et al.*, 2001). Similarly, the PDO (and likely the IPO) may play a key role in modulating ENSO teleconnections across North America on inter-decadal time-scales (Gershunov and Barnett, 1998; Livezey and Smith, 1999).

A simple and robust index of climate variability over the North Pacific is the area-weighted mean SLP, averaged over most of the extra-tropical North Pacific Ocean, of Trenberth and Hurrell

(1994). A general reduction in SLP after about 1976 has been particularly evident during the winter half (November to March) of many of these years. This is characterised by a deeper-than-normal Aleutian low pressure system, accompanied by stronger-than-normal westerly winds across the central North Pacific and enhanced southerly to south-westerly flow along the west coast of North America, as reviewed in the SAR (Figure 3.17). Consequently, there have been increases in surface air temperature and SST over much of western North America and the eastern North Pacific, respectively, over the past two decades, especially in winter, but decreases in SST, or only modest warming, over parts of the central extra-tropical North Pacific (Figure 2.10). Numerous studies have suggested that the mid-1970s changes in the atmospheric and oceanic circulation may reflect one or more low-frequency variations over the North Pacific, one being the PDO (Kawamura, 1994; Latif and Barnett, 1994; Mann and Park, 1994, 1996; Deser and Blackmon 1995; Zhang et al., 1997b; White and Cayan, 1998; Enfield and Mestas-Nuñez, 1999).

2.6.4 Monsoons

Variations in the behaviour of the North African summer monsoon were highlighted in IPCC (1990). Moron (1997) demonstrated that long-term variations of Sahel annual rainfall, particularly the wet 1950s and the dry 1970 to 1980s, are seen over the Guinea coast area, although trends are strongest in the Sahel. The significant decrease in Guinea coast rainfall (Ward, 1998) is present in both the first and second rainy seasons, but is strongest in the second. Janicot et al. (1996) and Moron (1997) demonstrated that the moderate influence of ENSO (towards drier conditions) has increased since 1960, with warm events associated more strongly with large-scale anomalous dry conditions over the Guinea and Sahel belts. Ward et al. (1999) show that the Sahel has become moderately wetter since 1987, despite the increased drying influence of ENSO events, a trend that continued to 1999 (Parker and Horton, 2000). This recent behaviour may be related to a quasi-hemispheric variation of SST (e.g., Enfield and Mestas-Nuñez, 1999) shown to be related to Sahel rainfall by Folland et al. (1986), and which may be related to the recent strong increase in North Atlantic SST mentioned in Section 2.2.2.2. Many other parts of tropical Africa are influenced by ENSO towards either drier or wetter conditions than normal, sometimes modulated by regional SST anomalies near Africa (e.g., Nicholson and Kim, 1997; Nicholson, 1997; Indeje et al., 2000), but few trends can be discerned.

Multi-decadal and decadal variations of the Indian monsoon have been widely noted (e.g., Pant and Kumar, 1997) but links with El Niño do not now seem straightforward (Slingo et al., 1999). However, despite the recent strong El Niño episodes, the inverse relationship between the ENSO and the Indian summer monsoon (weak monsoon arising from an ENSO event) has broken down in the recent two decades (Kumar et al., 1999a). This link operated on multi-decadal time-scales with NINO 3 SST until at least 1970. Kumar et al. suggest that persistently increased surface temperatures over Eurasia in winter and spring (Figure 2.10) have favoured an enhanced land-ocean thermal

gradient conducive to stronger monsoons; they also observe a shift away from India in the sinking node of the Walker circulation in El Niño. Changes have also occurred in relationships with Indian monsoon precursors (Kumar et al., 1999b). One possibility is that warming over the Indian Ocean (Figures 2.9, 2.10) may have increased moisture and rainfall for a given state of the atmospheric circulation (Kitoh et al., 1997). There may be a link to multi-decadal variations in Pacific SST, but this remains to be investigated, together with other monsoon indices (e.g., Goswami et al., 1997).

It has been known for some time that the position of the western North Pacific sub-tropical high affects the East Asian monsoon. Gong and Wang (1999a) showed that summer (June to August) precipitation over central and eastern China near 30°N is positively correlated with the intensity of the high, with negative correlations to the north and south. A location of the sub-tropical high further south than normal is conducive to heavy summer rainfall in this region. Time-series of the sub-tropical high show an increase in areal extent in the 1920s, then another increase from the mid-1970s to 1998, giving frequent wet summers in this region in recent years. The north-east winter monsoon has also shown low-frequency variations. Thus the strength of the Siberian high increased to a peak around 1968, and then weakened to a minimum around 1990 (Gong and Wang, 1999b), in phase with the increased frequency of the positive phase of the NAO (Wallace, 2000 and next section). This is likely to have contributed to strong recent winter warming in China shown in Figure 2.10.

2.6.5 The Northern Hemisphere excluding the North Pacific Ocean

The atmospheric circulation over the Northern Hemisphere has exhibited anomalous behaviour over the past several decades. In particular, the dominant patterns of atmospheric variability in the winter half-year have tended to be strongly biased to one phase. Thus SLP has been lower than average over the mid- and high latitudes of the North Atlantic Ocean, as well as over much of the Arctic, while it has been higher than average over the sub-tropical oceans, especially the Atlantic. Moreover, in the past thirty years, changes in these leading patterns of natural atmospheric variability appear to be unusual in the context of the observational record.

The dominant pattern of atmospheric circulation variability over the North Atlantic is known as the NAO, and its wintertime index is shown in Figure 2.30 (updated from Hurrell, 1995). As discussed in the SAR, positive values of the NAO give stronger than average westerlies over the mid-latitudes of the Atlantic with low SLP anomalies in the Icelandic region and over much of the Arctic and high SLP anomalies across the sub-tropical Atlantic and into southern Europe. The positive, enhanced westerly, phase of the NAO is associated with cold winters over the north-west Atlantic and warm winters over Europe, Siberia and eastern Asia (Thompson and Wallace, 2001) as well as wet conditions from Iceland to Scandinavia and dry winters over southern Europe. A sharp reversal is evident in the NAO index starting around 1970 from a negative towards a positive phase.

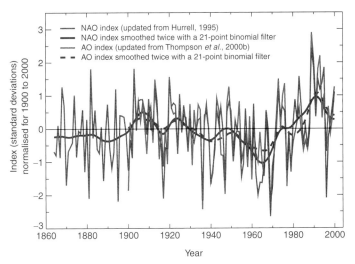

Figure 2.30: December to March North Atlantic Oscillation (NAO) indices, 1864 to 2000, and Arctic Oscillation (AO) indices, 1900 to 2000, updated from Hurrell (1995) and updated from Thompson and Wallace (2000) and Thompson *et al.* (2000b), respectively. The indices were normalised using the means and standard deviations from their common period, 1900 to 2000, smoothed twice using a 21-point binomial filter where indicated and then plotted according to the years of their Januarys.

Since about 1985, the NAO has tended to remain in a strong positive phase, though with substantial interannual variability. Hurrell (1996) and Thompson *et al.* (2000a) showed that the recent upward trend in the NAO accounts for much of the regional surface winter half-year warming over northern Europe and Asia north of about 40°N over the past thirty years, as well as the cooling over the north-west Atlantic (see Section 2.2.2.3). Moreover, when circulation changes over the North Pacific are also considered, much of the pattern of the Northern Hemisphere winter half-year surface temperature changes since the mid-1970s can be explained. This can be associated with changes in the NAO, and in the PNA atmospheric pattern related to ENSO or the PDO (Graf *et al.*, 1995; Wallace *et al.*, 1995; Shabbar *et al.*, 1997; Thompson and Wallace, 1998, 2000).

The changes in atmospheric circulation over the Atlantic are also connected with much of the observed pressure fall over the Arctic in recent years (Walsh *et al.*, 1996). Other features related to the circulation changes include the strengthening of sub-polar westerlies from the surface of the North Atlantic up, in winter as high as the lower stratosphere (Thompson *et al.*, 2000a) and pronounced regional changes in precipitation patterns (Hurrell, 1995; Dai *et al.*, 1997b; Hurrell and van Loon 1997; Section 2.5.2.1). Associated precipitation increases have resulted in the notable advance of some Scandinavian glaciers (Hagen *et al.*, 1995), while decreases to the south of about 50°N have contributed to the further retreat of Alpine glaciers (Frank, 1997; see also Section 2.2.5.3)

The NAO is regarded (largely) by some as the regional expression of a zonally symmetrical hemispheric mode of variability characterised by a seesaw of atmospheric mass between the polar cap and the mid-latitudes in both the Atlantic

and Pacific Ocean basins (Thompson and Wallace, 1998, 2001). This mode has been named the AO (Figure 2.30). The time-series of the NAO and AO are quite similar: the correlation of monthly anomalies of station data SLP series of NAO and AO is about 0.7 (depending on their exact definitions and epochs) while seasonal variations shown in Figure 2.30 have even higher correlations. The NAO and AO can be viewed as manifestations of the same basic phenomenon (Wallace, 2000).

Changes and decadal fluctuations in sea-ice cover in the Labrador and Greenland Seas, as well as over the Arctic, appear well correlated with the NAO (Chapman and Walsh, 1993; Maslanik *et al.*, 1996; McPhee *et al.*, 1998; Mysak and Venegas, 1998; Parkinson *et al.*, 1999; Deser *et al.*, 2000). The relationship between the SLP and ice anomaly fields is consistent with the idea that atmospheric circulation anomalies force the sea-ice variations (Prisenberg *et al.*, 1997). Feedbacks or other influences of winter ice anomalies on the atmosphere have been more difficult to detect, although Deser *et al.* (2000) suggest that a local response of the atmospheric circulation to the reduced sea-ice cover east of Greenland in recent years is also apparent (see also Section 2.2.5.2).

A number of studies have placed the recent positive values of the NAO into a longer-term perspective (Jones *et al.*, 1997a; Appenzeller *et al.*, 1998; Cook *et al.*, 1998; Luterbacher *et al.*, 1999; Osborn *et al.*, 1999) back to the 1700s. The recent strength of the positive phase of the NAO seems unusual from these reconstructions but, as in Figure 2.28, these proxy data reconstructions may underestimate variability. An extended positive phase occurred in the early 20th century (Figure 2.30), particularly pronounced in January (Parker and Folland, 1988), comparable in length to the recent positive phase. Higher-frequency variability of the NAO also appears to have varied. Hurrell and van Loon (1997) showed that quasi-decadal (6 to 10 year) variability has become more pronounced over the latter half of the 20th century, while quasi-biennial variability dominated in the early instrumental record.

2.6.6 The Southern Hemisphere

Since the SAR there has been more emphasis on analysis of decadal variability over the Southern Hemisphere. The Southern Hemisphere gridded SLP data for the period 1950 to 1994 show two dominant modes in annual average values, similar to those identified by Karoly *et al.* (1996) using station data. The first mode unambiguously represents the Southern Oscillation and reflects the tendency towards more frequent and intense negative phases over the past several decades. The second mode represents anomalies throughout the mid-latitude regions across the Indian Ocean and western Pacific, which contrast with anomalies elsewhere.

The Trans-Polar Index (TPI) is the only large-scale station pressure-based extra-tropical Southern Hemisphere circulation index in regular use. It is based on the normalised pressure difference between New Zealand and South America and has been recalculated and extended by Jones *et al.* (1999b). On decadal and longer time-scales the TPI reflects movement in the phase of wave number one around the Southern Hemisphere.

Troughing (low pressure) was more frequent in the New Zealand region in the 1920s, and at a maximum in the 1940s. Anticyclonicity was favoured from the late 1950s to 1976, with troughing in the South American sector. Troughing was again apparent in the New Zealand sector in the 1990s (Salinger *et al.*, 1996).

A leading mode of variability in the extra-tropical Southern Hemisphere circulation on interannual to multi-decadal time-scales is a zonally elongated north-south dipole structure over the Pacific, stretching from the sub-tropics to the Antarctic coast (Mo and Higgins, 1998; Kidson, 1999; Kiladis and Mo, 1999). It is strongly related to ENSO variability. The lower-frequency dipole structure contributes to variability in blocking frequency across the far south Pacific (Renwick, 1998; Renwick and Revell, 1999).

ENSO variability is also implicated in the modulation of a "High Latitude Mode" (HLM) (Kidson, 1988; Karoly, 1990), especially over the austral summer. The HLM is now also called the "Antarctic Oscillation" (AAO); they appear to be the same phenomenon with the same structure (Thompson and Wallace, 2000). The AAO is a zonal pressure fluctuation between mid- and high latitudes of the Southern Hemisphere, centred on 55 to 60°S. It has recently been further studied (Gong and Wang, 1999c; Kidson, 1999; Thompson and Wallace, 2001; Figure 2.31) and shown to extend into the lower stratosphere between the Antarctic and the sub-tropical latitudes of the Southern Hemisphere. The AAO appears to persist all year but may be most active from mid-October to mid-December when it extends into the stratosphere (Thompson and Wallace, 2001). In its high index phase, it consists of low pressure or heights above the Antarctic and the near Southern Ocean with high heights north of about 50°S. Although the data are sparse, there is evidence that, like the NAO, the AAO has tended to move more towards a positive index phase, despite lower pressures being observed over the New Zealand region during the 1990s. This change is also associated with with increasing westerly winds in mid-latitudes. Thompson and Wallace (2001) show that most of Antarctica is rather cold in this phase, except for the Antarctic Peninsula which is warm due to additional advection of relatively warm air from seas to the west. This may explain some of the behaviour of Antarctic temperatures in the last two decades (Figure 2.10; Comiso, 2000).

Other work has identified the likely existence of an Antarctic Circumpolar Wave (ACW) (Jacobs and Mitchell, 1996; White and Peterson, 1996), a multi-annual climate signal in the Southern Ocean, with co-varying and perhaps coupled SST and SLP anomalies that move around the Southern Ocean. Its long-term variability is not yet known.

2.6.7 *Summary*

The interannual variability of ENSO has fluctuated substantially over the last century, with notably reduced variability during the period 1920 to 1960, compared with adjacent periods. It remains unclear whether global warming has influenced the shift towards less frequent La Niña episodes from 1976 to 1998, including the abnormally protracted ENSO

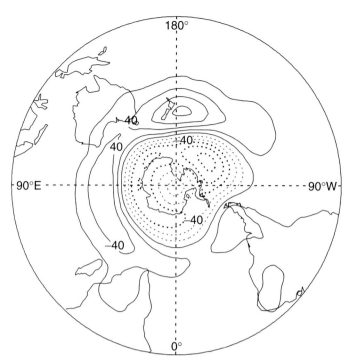

Figure 2.31: The High Latitude Mode (Kidson, 1988) or Antarctic Oscillation (AAO), defined as the first orthogonal pattern (covariance eigenvector of the Southern Hemisphere monthly surface pressure, January 1958 to December 1997) (Gong and Wang, 1999c; Kiladis and Mo, 1999). Data from NCAR/NCEP Reanalysis (Kalnay *et al.*, 1996). Note that Thompson and Wallace (2000) use 850 hPa height to define their AAO.

1990 to 1995 event and the exceptionally strong 1982/83 and 1997/98 events. Analysis of SST patterns indicates that a global warming pattern may have increased the background temperature in the region most affected by ENSO, but there is some ambiguity in the details of this pattern.

Since the SAR, 'ENSO-like' features operating on decadal to multi-decadal time-scales have been identified, such as the PDO and IPO. They appear to be part of a continuous spectrum of ENSO variability that has subtly changing SST patterns as time-scales increase and which may have distinctive effects on regional climate around the Pacific basin. For the period since 1900, El Niño (La Niña) events are more prevalent during positive (negative) phases of the IPO.

In the Northern Hemisphere, pronounced changes in winter atmospheric and oceanic circulations over the North Pacific in the 1970s (the North Pacific Oscillation) have been paralleled by wintertime circulation changes over the North Atlantic, recorded by the NAO. In the North Pacific, spatially coherent changes have occurred in surface temperature across the North Pacific and western North America, while the enhanced westerly phase of the NAO has caused considerable winter half-year temperature and precipitation changes over a vast area of extra-tropical Eurasia. In the Southern Hemisphere, a feature quite like the NAO, the HLM or the AAO, also appears to have moved into an enhanced westerly phase in middle latitudes.

2.7 Has Climate Variability, or have Climate Extremes, Changed?

2.7.1 Background

Changes in climate variability and extremes of weather and climate events have received increased attention in the last few years. Understanding changes in climate variability and climate extremes is made difficult by interactions between the changes in the mean and variability (Meehl *et al.*, 2000). Such interactions vary from variable to variable depending on their statistical distribution. For example, the distribution of temperatures often resembles a normal distribution where non-stationarity of the distribution implies changes in the mean or variance. In such a distribution, an increase in the

mean leads to new record high temperatures (Figure 2.32a), but a change in the mean does not imply any change in variability. For example, in Figure 2.32a, the range between the hottest and coldest temperatures does not change. An increase in variability without a change in the mean implies an increase in the probability of both hot and cold extremes as well as the absolute value of the extremes (Figure 2.32b). Increases in both the mean and the variability are also possible (Figure 2.32c), which affects (in this example) the probability of hot and cold extremes, with more frequent hot events with more extreme high temperatures and fewer cold events. Other combinations of changes in both mean and variability would lead to different results.

Consequently, even when changes in extremes can be documented, unless a specific analysis has been completed, it is often uncertain whether the changes are caused by a change in the mean, variance, or both. In addition, uncertainties in the rate of change of the mean confound interpretation of changes in variance since all variance statistics are dependent on a reference level, i.e., the mean.

For variables that are not well approximated by normal distributions, like precipitation, the situation is even more complex, especially for dry climates. For precipitation, for example, changes in the mean total precipitation can be accompanied by other changes like the frequency of precipitation or the shape of the distribution including its variability. All these changes can affect the various aspects of precipitation extremes including the intensity of precipitation (amount per unit time).

This section considers the changes in variability and extremes simultaneously for two variables, temperature and precipitation. We include new analyses and additional data compiled since the SAR which provide new insights. We also assess new information related to changes in extreme weather and climate phenomena, e.g., tropical cyclones, tornadoes, etc. In these analyses, the primary focus is on assessing the stationarity (e.g., the null hypothesis of no change) of these events, given numerous inhomogeneities in monitoring.

2.7.2 Is There Evidence for Changes in Variability or Extremes?

The issues involved in measuring and assessing changes in extremes have recently been comprehensively reviewed by Trenberth and Owen (1999), Nicholls and Murray (1999), and Folland *et al.* (1999b). Despite some progress described below, there remains a lack of accessible daily climate data sets which can be intercompared over large regions (Folland *et al.*, 2000). Extremes are a key aspect of climate change. Changes in the frequency of many extremes (increases or decreases) can be surprisingly large for seemingly modest mean changes in climate (Katz, 1999) and are often the most sensitive aspects of climate change for ecosystem and societal responses. Moreover, changes in extremes are often most sensitive to inhomogeneous climate monitoring practices, making assessment of change more difficult than assessing the change in the mean.

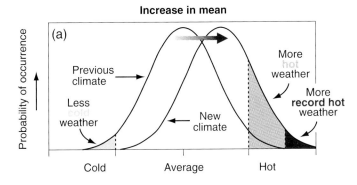

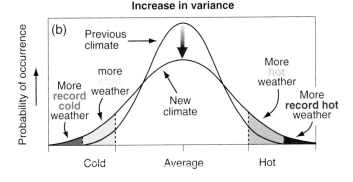

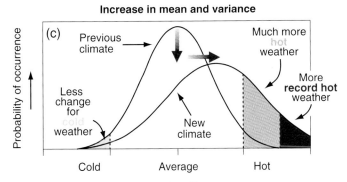

Figure 2.32: Schematic showing the effect on extreme temperatures when (a) the mean temperature increases, (b) the variance increases, and (c) when both the mean and variance increase for a normal distribution of temperature.

2.7.2.1 *Temperature*

Given the number of ways in which extreme climate events and variability about the mean can be defined, (e.g., extreme daily temperatures, large areas experiencing unusual temperatures, severity of heat waves, number of frosts or freezes, changes in interannual variability of large area temperatures, etc.) extreme care must be exercised in generalising results. Here we assess the evidence for changes in temperature extremes or variability, first based on global analyses and then on more detailed regional analyses.

Parker *et al.* (1994) compared the interannual variability of seasonal temperature anomalies from the 1954 to 1973 period to the 1974 to 1993 period for most of the globe. They found a small increase in variability overall with an especially large increase in central North America. By restricting the analyses to the latter half of the 20th century, Parker *et al.* (1994) minimised the potential biases due to an increasing number of observations in this period. Several other studies found a reduction in other aspects of variability over longer time periods. Jones (1999) also analysed global data and found no change in variability, but since 1951 the rise in global mean temperatures can be attributed to an increase (decrease) in areas with much above (below) normal temperatures. They also analysed the change in the aggregated total of much below and much above normal temperatures (upper and lower ten percentiles). They found little overall change, except for a reduced number of much above or below normal temperatures during the 1960s and 1970s. Michaels *et al.* (1998) examined 5° latitude × 5° longitude monthly temperature anomalies for many grid cells around the world and found an overall decrease in intra-annual variance over the past 50 to 100 years. They also examined the daily maximum and minimum temperatures from the United States, China, and the former Soviet Union and found a general decline in the intra-monthly temperature variability. As reported in the SAR, a related analysis by Karl *et al.* (1995b) found reduced day-to-day variability during the 20th century in the Northern Hemisphere, particularly in the United States and China. Recently, Collins *et al.* (2000) has identified similar trends in Australia. By analysing a long homogenised daily temperature index for four stations in Northern Europe, Moberg *et al.* (2000) also found a progressive reduction in all-seasons inter-daily variability of about 7% between 1880 and 1998. Balling (1998) found an overall decrease in the spatial variance of both satellite-based lower-tropospheric measurements from 1979 to 1996 and in near-surface air temperatures from 1897 to 1996.

Consequently, there is now little evidence to suggest that the interannual variability of global temperatures has increased over the past few decades, but there is some evidence to suggest that the variability of intra-annual temperatures has actually quite widely decreased. Several analyses find a decrease in spatial and temporal variability of temperatures on these shorter time-scales.

There have been a number of new regional studies related to changes in extreme temperature events during the 20th century. Gruza *et al.* (1999) found statistically significant increases in the number of days with extreme high temperatures across Russia using data back to 1961 and on a monthly basis back to 1900. Frich *et al.* (2001) analysed data spanning the last half of the 20th century across most of the Northern Hemisphere mid- and high

latitudes and found a statistically significant increase (5 to >15%) in the growing season length in many regions. Heino *et al.* (1999) also found that there has been a reduction in the number of days with frost (the number of days with minimum temperature ≤0°C) in northern and central Europe. Thus, some stations now have as many as 50 fewer days of frost per year compared with earlier in the 20th century. Easterling *et al.* (2000) found there has been a significant decrease in the number of days below freezing over the central United States (about seven per year). For Canada, Bonsal *et al.* (2001) also found fewer days with extreme low temperatures during winter, spring and summer, and more days with extreme high temperatures during winter and spring. This has led to a significant increase in the frost-free period. Decreasing numbers of days with freezing temperatures have been found in Australia and New Zealand over recent decades (Plummer *et al.*, 1999; Collins *et al.*, 2000). In addition, while increases in the frequency of warm days have been observed, decreases in the number of cool nights have been stronger. Frich *et al.* (2001) show a reduced number of days with frost across much of the globe (Figure 2.33) while Michaels *et al.* (2000) find that much of the warming during the 20th century has been during the cold season in the mid- to high latitudes, consistent with the reduction of extremely low temperatures. Frich *et al.* (2001) have also found a statistically significant reduction in the difference between the annual extremes of daily maximum and minimum temperatures during the latter half of the 20th century. In China, strong increases in the absolute minimum temperature have been observed, with decreases in the 1-day seasonal extreme maximum temperature (Zhai *et al.*, 1999a) since the 1950s. Wang and Gaffen (2001), however, for a similar period, found an increase in "hot" days in China. Hot days were defined as those days above the 85th percentile during July and August based on an "apparent temperature" index related to human discomfort in China (Steadman, 1984). The number of extremely cold days has also been shown to be decreasing in China (Zhai *et al.*, 1999a). Manton *et al.* (2001) found significant increases in hot days and warm nights, and decreases in cool days and cold nights since 1961 across the Southeast Asia and South Pacific Region. Jones *et al.* (1999c) have analysed the 230-year-long daily central England data set that has been adjusted for observing inhomogeneities. They found that the increase in temperature observed in central England corresponds mainly to a reduction in the frequency of much below normal daily temperatures. An increase of the frequency of much above normal temperatures was less apparent.

Analyses of 20th century trends in the United States of short-duration episodes (a few days) of extreme hot or cold weather did not show any significant changes in frequency or intensity (Kunkel *et al.*, 1996, 1999; Karl and Knight, 1997). For Australia, Collins *et al.* (2000) found higher frequencies of multi-day warm nights and days, and decreases in the frequency of cool days and nights. In an extensive assessment of the change in frequency of heat waves during the latter half of the 20th century, Frich *et al.* (2001) find some evidence for an increase in heat-wave frequency, but several regions have opposite trends (Figure 2.33c). The extreme heat in the United States during several years in the 1930s dominates the time-series of heat waves in that region. On the other hand, trends in the frequency of extreme apparent tempera-

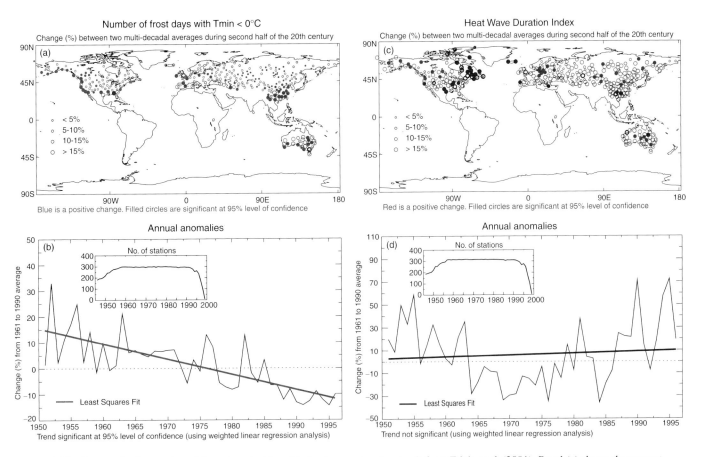

Figure 2.33: Changes in the number of frost days (a, b) and in heat-wave duration (c, d) from Frich *et al.* (2001). Panel (a) shows the percent changes in the total number of days with a minimum temperature of less than 0°C between the first and last half of the period, approximately 1946 to 1999. The red circles indicate negative changes and the blue circles indicate positive changes. Panel (c) shows percentage changes in the maximum number of consecutive days (for periods with >5 such days) with maximum temperatures >5°C above the 1961 to 1990 daily normal. The changes are for the first and second half of the period, approximately 1946 to 1999. The red circles indicate positive changes and the blue circles indicate negative changes. In both panels, the size of each circle reflects the size of the change and solid circles represent statistically significant changes. Panels (b) and (d) show the average annual values of these quantities expressed as percentage differences from their 1961 to 1990 average values. The trend shown in panel (b) is statistically significant at the 5% level.

tures are significantly larger for 1949 to 1995 during summer over most of the USA (Gaffen and Ross, 1998). Warm humid nights more than doubled in number over 1949 to 1995 at some locations. Trends in nocturnal apparent temperature in the USA, however, are likely to be associated, in part, with increased urbanisation. Nevertheless, using methods and data sets to minimise urban heat island effects and instrument changes, Easterling *et al.* (2000) arrived at conclusions similar to those of Gaffen and Ross (1998).

During the 1997/98 El Niño event, global temperature records were broken for sixteen consecutive months from May 1997 through to August 1998. Karl *et al.* (2000) describe this as an unusual event and such a monthly sequence is unprecedented in the observational record. More recently, Wigley (2000) argues that if it were not for the eruption of Mt. Pinatubo, an approximately equal number of record-breaking temperatures would have been set during the El Niño of 1990/91. As temperatures continue to warm, more events like these are likely, especially when enhanced by other factors, such as El Niño.

2.7.2.2 Precipitation

A better understanding of the relationship between changes in total precipitation and intense precipitation events has been achieved since the SAR. Although many areas of the globe have not been analysed, and considerable data remain inaccessible, enough data have been analysed to confirm some basic properties of the changes in extreme precipitation. Groisman *et al.* (1999) developed a simple statistical model of the frequency of daily precipitation based on the gamma distribution. They applied this model to a variety of regions around the world (40% of the global land area) during the season of greatest precipitation. Although Wilks (1999) shows that the gamma distribution under some circumstances can underestimate the probability of the highest rainfall amounts, Groisman *et al.* (1999) applied the distribution to the upper 5 and 10 percentiles of the distribution which are less subject to underestimation. Their analysis period varied from region to region, but within each region it generally spanned at least the last several decades, and for some regions much of the 20th century (Australia, United States, Norway, and South

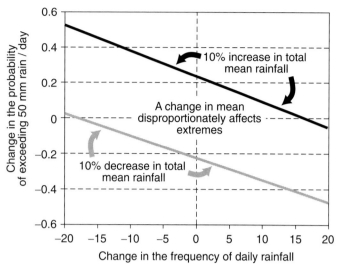

Figure 2.34: An example (from Groisman *et al.*, 1999) of the sensitivity of the frequency of heavy daily rainfall to a shift in the mean total rainfall, based on station data from Guangzhou, China. This example uses a threshold of 50 mm of precipitation per day. It shows the effects of a 10% increase and a 10% decrease in mean total summer rainfall, based on a gamma distribution of the rainfall with a constant shape parameter.

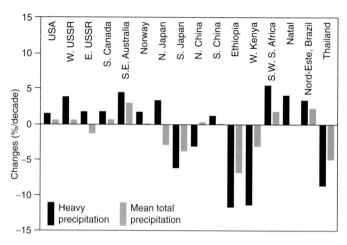

Figure 2.35: Linear trends (%/decade) of heavy precipitation (above the 90th percentile) and total precipitation during the rainy season over various regions of the globe. Seasons for each region usually span at least 50 years. Note that the magnitudes of the changes in heavy precipitation frequencies are always higher than changes in mean precipitation totals (Easterling *et al.*, 2000).

Africa). In the model used by Groisman *et al.* (1999), the mean total precipitation is also proportional to the shape and scale parameters of the gamma distribution as well as to the probability of precipitation on any given day. The shape parameter of the gamma distribution tends to be relatively stable across a wide range of precipitation regimes, in contrast to the scale parameter. Given the conservative nature of the shape parameter, it is possible to illustrate the relationships between changes in the mean total precipitation, the probability of precipitation (which is proportional to the number of days with precipitation), and changes in heavy precipitation (Figure 2.34). Given no change in the frequency (number of days) of precipitation, a 10% change in the mean total precipitation is amplified to a larger percentage change in heavy precipitation rates compared to the change in the mean. Using the statistical theory of extremes, Katz (1999) obtained results consistent with those of Groisman *et al.* (1999). For many regions of the world it appears that the changes in the frequency or probability of precipitation events are either small enough, or well enough expressed in the high rainfall rates (Karl and Knight, 1998; Gruza *et al.*, 1999; Haylock and Nicholls, 2000) that an increase in the mean total precipitation is disproportionately reflected in increased heavy precipitation rates (Figure 2.35).

Given the patterns of mean total precipitation changes (Section 2.5.2) during the 20th century, it could be anticipated that, in general, for those areas with increased mean total precipitation, the percentage increase in heavy precipitation rates should be significantly larger, and *vice versa* for total precipitation decreases. Regional analyses of annual precipitation in the United States (Karl and Knight, 1998; Trenberth, 1998a; Kunkel *et al.*, 1999); Canada (Stone *et al.*, 1999); Switzerland (Frei and Schär, 2001); Japan (Iwashima and Yamamoto, 1993; Yamamoto and Sakurai, 1999); wintertime precipitation in the UK (Osborn *et al.*, 2000); and rainy season precipitation in Norway, South

Africa, the Nord Este of Brazil, and the former USSR (Groisman *et al.*, 1999; Gruza *et al.*, 1999; Easterling *et al.*, 2000) confirm this characteristic of an amplified response for the heavy and extreme events.

Increases in heavy precipitation have also been documented even when mean total precipitation decreases (for example, see Northern Japan in Figure 2.35, or Manton *et al.*, 2001). This can occur when the probability of precipitation (the number of events) decreases, or if the shape of the precipitation distribution changes, but this latter situation is less likely (Buffoni *et al.*, 1999; Groisman *et al.*, 1999; Brunetti *et al.*, 2000a,b). For example, in Siberia for the summer season during the years 1936 to 1994 there was a statistically significant decrease in total precipitation of 1.3%/decade, but the number of days with precipitation also decreased. This resulted in an increase (1.9%/decade) in the frequency of heavy rainfall above 25 mm. The opposite can also occur when the number of rainfall events increases; thus Førland *et al.* (1998) found no trends in 1-day annual maximum precipitation in the Nordic countries, even when mean total precipitation increased.

There has also been a 10 to 45% increase in heavy rainfall, as defined by the 99th percentile of daily totals, over many regions of Australia from 1910 to 1995, but few individual trends were statistically significant (Hennessy *et al.*, 1999). In southwest Australia, however, a 15% decrease has been observed in winter rainfall on very wet days (Hennessy *et al.*, 1999; Haylock and Nicholls, 2000).

In Niger, a recent analysis of hourly rainfall data (Shinoda *et al.*, 1999) reveals that the droughts in the 1970s and 1980s were characterised primarily by a reduced frequency of heavy rainfall events (those exceeding 30 mm/day) rather than by a reduction in rainfall amount within heavy events. Such a result is still consistent with the model of Groisman *et al.* (1999), as a decrease in the

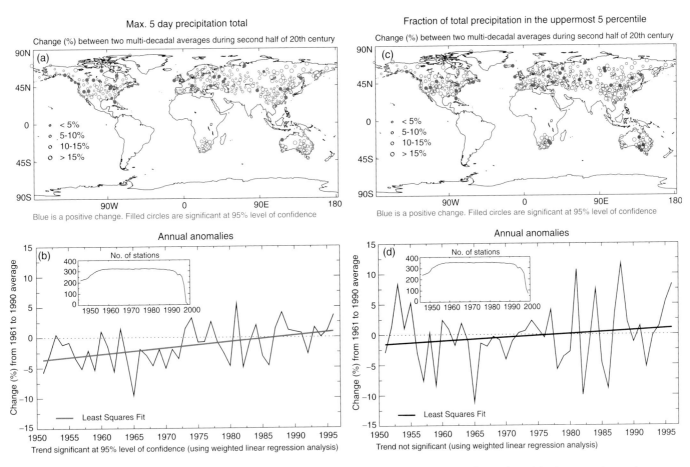

Figure 2.36: Changes in the maximum annual 5-day precipitation total (a, b) and in the proportion of annual precipitation occurring on days on which the 95th percentile of daily precipitation, defined over the period 1961 to 1990, was exceeded (c, d). The analysis shown is from Frich *et al.* (2001). Panels (a) and (c) show percentage changes in these quantities between the first and last half of the period, approximately 1946 to 1999. In both panels, the red circles indicate negative changes and the blue circles indicate positive changes. The size of each circle reflects the size of the change and solid circles represent statistically significant changes. Panels (b) and (d) show the average annual values of the quantities expressed as percentage differences from their 1961 to 1990 average values. The trend shown in panel (b) is statistically significant at the 5% level.

frequency of rainfall events has been responsible for the decrease in total rainfall. In the Sahel region of Nigeria, however, there has been a decrease in the heaviest daily precipitation amounts, coincident with an overall decrease in annual rainfall. This pattern is apparent throughout the Sudano-Sahel Zone, including the Ethiopian plateau (Nicholson, 1993; Tarhule and Woo, 1998; Easterling *et al.*, 2000). Again, it is apparent that there has been an amplified response of the heaviest precipitation rates relative to the percentage change in total precipitation.

Since large portions of the mid- and high latitude land areas have had increasing precipitation during the last half the 20th century, the question arises as to how much of this area is affected by increases in heavy and extreme precipitation rates. The Frich *et al.* (2001) analysis suggests an overall increase in the area affected by more intense daily rainfall. Figure 2.36 shows that widely distributed parts of the mid- and high latitudes have locally statistically significant increases in both the proportion of mean annual total precipitation falling into the upper five percentiles and in the annual maximum consecutive 5-day precipitation total. However, for the regions of the globe sampled taken as a whole, only the latter statistic shows a significant increase. Regional analyses in

Russia (Gruza *et al.*, 1999), the United States (Karl and Knight, 1998) and elsewhere (Groisman *et al.*, 1999; Easterling *et al.*, 2000) confirm this trend. Although the trends are by no means uniform, as would be anticipated with the relatively high spatial and interannual variability of precipitation, about 10% of the stations analysed show statistically significant increases at the 5% level. This equates to about a 4% increase in the annual maximum 5-day precipitation total (Figure 2.36b). The number of stations reflecting a locally significant increase in the proportion of total annual precipitation occurring in the upper five percentiles of daily precipitation totals outweighs the number of stations with significantly decreasing trends by more than 3 to 1 (Figure 2.36c). Although not statistically significant when averaging over all stations, there is about a 1% increase in the proportion of daily precipitation events occurring in the upper five percentiles (Figure 2.36d). Overall, it is likely that there has been a 2 to 4% increase in the number of heavy precipitation events when averaged across the mid- and high latitudes.

It has been noted that an increase (or decrease) in heavy precipitation events may not necessarily translate into annual peak (or low) river levels. For example, in the United States, Lins and Slack (1999) could not detect an increase in the upper quantiles of streamflow, despite the documented increase in heavy and extreme precipitation events. It is possible that this null result is partly due to the method of analysis, but it is also attributable to the timing of the annual peak streamflow discharge, which in the United States is usually in late winter or early spring. A reduced snow cover extent in the mountainous West changes the peak river flow, as does timing of increases in heavy and extreme precipitation reported in the United States, which is best reflected during the warm season. Groisman *et al.* (2001) and Zhang *et al.* (2000) also show reduced peak streamflow in areas with reduced spring snow cover extent. Nonetheless, in much of the United States where spring snow melt does not dominate peak or normal flow, Groisman *et al.* (2001) show increasing high streamflow related to increasing heavy precipitation.

It is noteworthy that the influence of warmer temperatures and increased water vapour in the atmosphere (Section 2.5.3) are not independent events, and are likely to be jointly related to increases in heavy and extreme precipitation events.

2.7.3 *Is There Evidence for Changes in Extreme Weather or Climate Events?*

In this section we assess changes in the intensity and frequency of various weather phenomena. One aspect of change that is important, but which is beyond the analysis of present records, relates to changes in the tracks of storms. Severe storms are often rare, so the analysis of large areas and long lengths of homogeneous storm records are required to assess changes. So far this combination of data is not available.

2.7.3.1 *Tropical cyclones*

This section updates the information provided in the SAR regarding changes in tropical cyclones across various ocean basins and those affecting the nearby continents. As reported in the SAR, a part of the multi-decadal trend of tropical cyclones occurring in the Australian region (105° to 160°E) is likely to be artificial, as the forecasters in the region no longer classify some weak (>990 hPa central pressure) systems as "cyclones" (Nicholls *et al.*, 1998). By considering only the moderate and intense tropical cyclones (central pressure ≤990 hPa), this artificial trend is eliminated. The remaining moderate and strong tropical cyclones reveal a numerical decline since the late 1980s, but the trend is not statistically significant. Similarly, the trend in intense tropical cyclones (minimum central pressure below 970 hPa) is not significantly different from zero. Nicholls *et al.* (1998) attributed the decrease in moderate cyclones to more frequent occurrences of El Niño during the 1980s and 1990s. However, a weak trend in the intense tropical cyclones implies that while ENSO modulates the total frequency of cyclones in the region, other factors must be more important in regulating their intensity. For example, new work by Higgins and Shi (2000) and Maloney and Hartmann (2001) show that 30 to 80 day Madden-Julian oscillations modulate tropical cyclone activity.

As reported in the SAR, the north-east sub-tropical Pacific has experienced a significant upward trend in tropical cyclone frequency in the short period examined, but additional data since

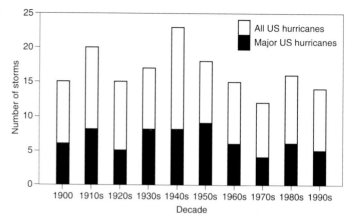

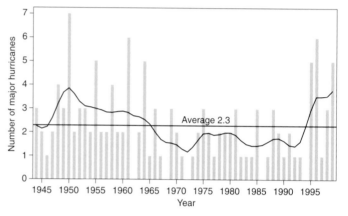

Figure 2.37: Top figure, decadal variations in hurricanes making landfall in the USA (updated from Karl *et al.*, 1995). Bottom figure, interannual variability in the number of major hurricanes (Saffir-Simpson categories 3, 4, and 5) and the long-term average across the North Atlantic (from Landsea *et al.*, 1999).

that time show no appreciable trend. There is no appreciable long-term variation of the total number of tropical storm strength cyclones observed in the north Indian, south-west Indian and south-west Pacific Oceans east of 160°E. (Neumann, 1993; Lander and Guard, 1998). For the north-west sub-tropical Pacific basin, Chan and Shi (1996) found that the frequency of typhoons and the total number of tropical storms and typhoons have been more variable since about 1980. There was an increase from 1981 to 1994, which was preceded by a nearly identical magnitude of decrease from about 1960 to 1980. No analysis has been done on the frequency of intense typhoons (having winds of at least 50 m/s) due to an overestimation of the intensity of such storms in the 1950s and 1960s (Black, 1993).

There has been an extensive analysis of the North Atlantic basin for the entire basin back to 1944, and also for the United States landfall tropical storms and hurricanes back to 1899. The all-basin data, however, have been affected by a bias in the measurement of strong hurricanes. This bias has been removed in an approximate way to provide estimates of the true occurrence of intense (or major) hurricanes since 1944 in the North Atlantic (Landsea, 1993). Earlier events lack reliable data on the strong inner core of the hurricanes. The United States record of landfall frequency and intensity of hurricanes is very reliable because of the availability of central pressure measurements at landfall (Jarrell *et al.*, 1992). Both of these data sets continue to show considerable inter-decadal variability, but no significant long-term trends (Figure 2.37, from Landsea *et al.*, 1999). Active years occurred from the late 1940s to the mid-1960s, quiet years occurred from the 1970s to the early 1990s, and then there was a shift again to active conditions from 1995 to 1999. Concurrent with these frequency changes, there have been periods with a strong mean intensity of the North Atlantic tropical cyclones (mid-1940s to the 1960s and 1995 to 1999) and a weak intensity (1970s to early 1990s). There has been no significant change in the peak intensity reached by the strongest hurricane each year (Landsea *et al.*, 1996). As might be anticipated, there is a close correspondence between the intensity of hurricanes in the North Atlantic and those making landfall in the United States (Figure 2.37).

Using historical records, Fernandez-Partagas and Diaz (1996) estimated that overall Atlantic tropical storm and hurricane activity for the years 1851 to 1890 was 12% lower than the corresponding forty year period of 1951 to 1990, although little can be said regarding the intense hurricanes. They based this assessment upon a constant ratio of USA landfalling tropical cyclones to all-basin activity, which is likely to be valid for multi-decadal time-scales. However, this also assumes that Fernandez-Partagas and Diaz were able to uncover all USA landfalling tropical cyclones back to 1851, which may be more questionable.

2.7.3.2 Extra-tropical cyclones

Extra-tropical cyclones are baroclinic low pressure systems that occur throughout the mid-latitudes of both hemispheres. Their potential for causing property damage, particularly as winter storms, is well documented, where the main interest is in wind and wind-generated waves. In place of direct wind measurements, which suffer from lack of consistency of instrumentation, methodology and exposure, values based on SLP gradients have been derived which are more reliable for discerning long-term changes. Over the oceans, the additional measurements of wave heights and tide gauge measurements provide additional ways of indirectly evaluating changes in extra-tropical storm strength and frequency (see Chapter 11, Section 11.3.3). Global analyses of changes in extra-tropical storm frequency and intensity have not been attempted, but there have been several large-scale studies. Jones *et al.* (1999c) developed a gale index of geostrophic flow and vorticity over the UK for the period 1881 to 1997. This revealed an increase in the number of severe gale days over the UK since the 1960s, but no long-term increase when considering the century period. Serreze *et al.* (1997) found increases in cold season cyclones in the Arctic region for the period 1966 to 1993. Angel and Isard (1998) found significant increases in strong cyclones (<993 mb) in the Great Lakes region from 1900 to 1990 during the cold season. Graham and Diaz (2001) find evidence for increases in strong cyclones over the Pacific Ocean between 25 and 40°N since 1948 and link the increase to increasing sea surface temperatures in the western Tropical Pacific. Alexandersson *et al.* (1998, 2000) similarly studied extreme geostrophic wind events in the north-western European area based on homogenised observations during the period 1881 to 1998. These studies revealed an increase in the number of

extreme wind events around and to the north of the North Sea. The WASA group (1998) similarly investigated the storm related sea level variations at gauge stations in the south-eastern part of the North Sea. They found no long-term trend during the last 100 years, but a clear rise since a minimum of storminess in the 1960s, which is consistent with the rise in extreme geostrophic wind found by Jones *et al.* (1999c). This increase is also consistent with changes in the NAO (Figure 2.30). Some analyses have focused on hemispheric changes in cyclone activity. Lambert (1996) analysed gridded SLP over both the North Atlantic and North Pacific Oceans for the period 1891 to 1991. He found a significant increase in intense extra-tropical storms, especially over the last two decades of his analysis, but the data were not completely homogenised. Simmonds and Keay (2000) used data from 1958 to 1997 in the Southern Hemisphere and found an increase in cyclone activity through 1972 before decreasing through 1997 with strong decreases during the 1990s.

Hourly values of water levels provide a unique record of tropical and extra-tropical storms where stations exist. Zhang *et al.* (1997a) have analysed century-long records along the East Coast of the United States. They calculated several different measures of storm severity, but did not find any long-term trends. On the other hand, they did find that the effect of sea level rise over the last century has exacerbated the beach erosion and flooding from modern storms that would have been less damaging a century ago.

Another proxy for cyclone intensity is wave height (see Chapter 11, Section 11.3.3). Several studies report increased wave height over the past three decades in the North Atlantic (approximately 2.5 cm/yr) and in coastal areas, though no longer-term trends were evident (Carter and Draper, 1988; Bacon and Carter, 1991; Bouws *et al.*, 1996; Kushnir *et al.*, 1997; WASA Group, 1998).

It appears that recent work points towards increases over time in extra-tropical cyclone activity during the latter half of the 20th century in the Northern Hemisphere, and decreased activity in the Southern Hemisphere. However, the mechanisms involved are not clear, and it is not certain whether the trends are multi-decadal fluctuations, or rather part of a longer-term trend. Furthermore decreased cyclone activity in higher latitudes of the Southern Hemisphere is not obviously consistent with an increase in the positive phase of the Antarctic Oscillation in the last fifteen years or so (Section 2.6.6). A more fundamental question is whether we would expect more or fewer extra-tropical cyclones with increased warming. As pointed out by Simmonds and Keay (2000), the specific humidity increases as temperatures increase, and this increased moisture should enhance extra-tropical cyclones, but Zhang and Wang (1997) suggest that cyclones transport energy more efficiently in a more moist atmosphere, therefore requiring fewer extra-tropical cyclones (see Chapters 7 and 10 for more discussion).

2.7.3.3 Droughts and wet spells

In the SAR, an intensification of the hydrological cycle was projected to occur as the globe warms. One measure of such intensification is to examine whether the frequency of droughts and wet spells are increasing. Karl *et al.* (1995c) examined the proportion of land areas having a severe drought and a severe moisture surplus over the United States. Dai *et al.* (1998) extended this analysis to global land areas using the water balance approach of the Palmer Drought Severity Index. Long-term global trends for 1900 to 1995 are relatively small for both severe drought and wet area statistics. However, during the last two to three decades, there have been some increases in the globally combined severe dry and wet areas, resulting from increases in either the dry area, e.g., over the Sahel, eastern Asia and southern Africa or the wet areas, e.g., over the United States and Europe. Most of the increases occurred after 1970. Except for the Sahel, however, the magnitude of dry and wet areas of the recent decades is not unprecedented during this century, but it should be noted that rainfall in the Sahel since the height of the drought has substantially increased. In related work, Frich *et al.* (2001) found that in much of the mid- and high latitudes, there has been a statistically significant increase in both the number of days with precipitation exceeding 10 mm per day and in the number of consecutive days with precipitation during the second half of the 20th century.

Recent changes in the areas experiencing severe drought or wet spells are closely related to the shift in ENSO towards more warm events since the late 1970s, and coincide with record high global mean temperatures. Dai *et al.* (1998) found that for a given value of ENSO intensity, the response in areas affected by drought or excessive wetness since the 1970s is more extreme than prior to the 1970s, also suggesting an intensification of the hydrological cycle.

2.7.3.4 Tornadoes, hail and other severe local weather

Small-scale severe weather phenomena (SCSWP) are primarily characterised by quasi-random temporal and spatial events. These events, in turn, have local and regional impacts, often with significant damage and sometimes loss of life. Tornadoes and thunderstorms and related phenomena such as lightning, hail, wind, dust, water spouts, downpours and cloudbursts belong to this group. In the light of the very strong spatial variability of SCSWP, the density of surface meteorological observing stations is too coarse to measure all such events. Moreover, areally consistent values of SCSWP are inherently elusive. Statistics of relatively rare events are not stable at single stations, observational practices can be subjective and change over time, and the metadata outlining these practices are often not readily available to researchers. For these reasons, monitoring the occurrence of local maxima and minima in smoothed SCSWP series, as well as checking for trends of the same sign for different but related SCSWP (e.g., thunderstorms, hail, cloud bursts), are important for checking inconsistencies. Because of the inherent difficulty in working with these data, there have been relatively few large-scale analyses of changes and variations in these events. Nonetheless, a few new regional analyses have been completed since the SAR.

A regional analysis by Dessens (1995) and more recent global analysis by Reeve and Toumi (1999) show that there is a significant interannual correlation between hail and lightning and mean minimum temperature and wet bulb temperatures. Using a three-year data set, Reeve and Toumi (1999) found a statistically

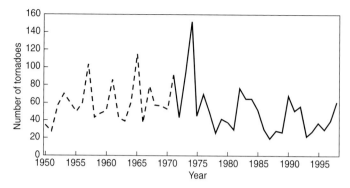

Figure 2.38: Annual total number of very strong through violent (F3-F5) tornadoes reported in the USA, which are defined as having estimated wind speeds from approximately 70 to 164 ms[-1]. The Fujita tornado classification scale was implemented in 1971. Prior to 1971, these data are based on storm damage reports (National Climatic Data Center, NOAA).

significant relationship between lightning frequency and wet bulb temperature. They show that with a 1°C increase in global wet-bulb temperature there is a 40% increase in lightning activity, with larger increases over the Northern Hemisphere land areas (56%). Unfortunately, there are few long-term data sets that have been analysed for lightning and related phenomena such as hail or thunderstorms, to calculate multi-decadal hemispheric or global trends.

A regional analysis assessed the temporal fluctuations and trends in hail-day and thunder-day occurrences during a 100-year period, from 1896 to 1995, derived from carefully screened records of 67 stations distributed across the United States. Upward hail day trends were found in the High Plains-Rockies and the south-east, contrasting with areas with no trend in the northern Midwest and along the East Coast, and with downward trends elsewhere (Changnon and Changnon, 2000). The major regions of decrease and increase in hail activity match regions of increased and decreased thunder activity for 1901 to 1980 well (Changnon, 1985; Gabriel and Changnon, 1990) and also crop-hail insurance losses (Changnon *et al.*, 1996; Changnon and Changnon, 1997). In general, hail frequency shows a general decrease for most of the United States over the last century, with increases over the High Plains, the region where most of the crop-hail damage occurs in the United States. So, despite an increase in minimum temperature of more than 1°C since 1900 and an increase in tropospheric water vapour over the United States since 1973 (when records are deemed reliable), no systematic increase in hail or thunder days was found.

In south Moravia, Czech Republic, a decreasing linear trend in the frequency of thunderstorms, hailstorms and heavy rain from 1946 to 1995 was related to a significant decrease in the occurrence of these phenomena during cyclonic situations, when 90% of these phenomena occur in that region (Brázdil and Vais, 1997). Temperatures have increased in this area since 1946.

Since 1920, the number of tornadoes reported annually in the United States has increased by an order of magnitude, but this increase reflects greater effectiveness in collecting tornado reports (Doswell and Burgess, 1988; Grazulis, 1993; Grazulis *et al.*, 1998). On the other hand, severe tornadoes are not easily overlooked. Restricting the analysis to very strong and violent tornadoes results in a much different assessment (Figure 2.38) showing little long-term change, though some years like 1974 show a very large number of tornadoes. Furthermore, consideration of the number of days with tornadoes, rather than number of tornadoes, reduces the artificial changes that result from modern, more detailed damage surveys (e.g., Doswell and Burgess, 1988). The data set of "significant" tornado days developed by Grazulis (1993) shows a slow increase in number of days with significant tornadoes from the early 1920s through the 1960s, followed by a decrease since that time.

2.7.4 Summary

Based on new analyses since the SAR, it is likely that there has been a widespread increase in heavy and extreme precipitation events in regions where total precipitation has increased, e.g., the mid- and high latitudes of the Northern Hemisphere. Increases in the mean have often been found to be amplified in the highest precipitation rates total. In some regions, increases in heavy rainfall have been identified where the total precipitation has decreased or remained constant, such as eastern Asia. This is attributed to a decrease in the frequency of precipitation. Fewer areas have been identified where decreases in total annual precipitation have been associated with decreases in the highest precipitation rates, but some have been found. Temperature variability has decreased on intra-seasonal and daily time-scales in limited regional studies. New record high night-time minimum temperatures are lengthening the freeze and frost season in many mid- and high latitude regions. The increase in global temperatures has resulted mainly from a significant reduction in the frequency of much below normal seasonal mean temperatures across much of the globe, with a corresponding smaller increase in the frequency of much above normal temperatures. There is little sign of long-term changes in tropical storm intensity and frequency, but inter-decadal variations are pronounced. Owing to incomplete data and relatively few analyses, we are uncertain as to whether there has been any large-scale, long-term increase in the Northern Hemisphere extra-tropical cyclone intensity and frequency though some, sometimes strong, multi-decadal variations and recent increases were identified in several regions. Limited evidence exists for a decrease in cyclone frequency in the Southern Hemisphere since the early 1970s, but there has been a paucity of analyses and data. Recent analyses of changes in severe local weather (tornadoes, thunder days, lightning and hail) in a few selected regions provide no compelling evidence for widespread systematic long-term changes.

2.8 Are the Observed Trends Internally Consistent?

It is very important to compare trends in the various indicators to see if a physically consistent picture emerges, as this will critically affect the final assessment of our confidence in any such changes. A number of qualitative consistencies among the various indicators of climate change have increased our confidence in our analyses of the historical climate record: Figure 2.39a and b summarises the changes in various temperature and hydrological indicators, respectively, and provides a measure of confidence about each change. Of particular relevance are the changes identified below:

- Temperature over the global land and oceans, with two estimates for the latter, are measured and adjusted independently, yet all three show quite consistent increasing trends (0.52 to 0.61°C/century) over the 20th century.

- The nearly worldwide decrease in mountain glacier extent and mass is consistent with 20th century global temperature increases. A few recent exceptions in maritime areas have been affected by atmospheric circulation variations and related precipitation increases.

- Though less certain, substantial proxy evidence points to the exceptional warmth of the late 20th century relative to the last 1,000 years. The 1990s are likely to have been the warmest decade of the past 1,000 years over the Northern Hemisphere as a whole.

- Satellite and balloon measurements agree that lower-tropospheric temperatures have increased only slightly since 1979, though there has been a faster rate of global surface temperature increase. Balloon measurements indicate a larger lower-tropospheric temperature increase since 1958, similar to that shown by global surface temperature measurements over the same period. Balloon and satellite measurements agree that lower-stratospheric temperatures have declined significantly since 1979.

- Since 1979, trends in worldwide land-surface air temperature derived from weather stations in the Northern Hemisphere, in regions where urbanisation is likely to have been strong, agree closely with satellite derived temperature trends in the lower troposphere above the same regions. This suggests that urban heat island biases have not significantly affected surface temperature over the period.

- The decrease in the continental diurnal temperature range since around 1950 coincides with increases in cloud amount and, at least since the mid-1970s in the Northern Hemisphere, increases in water vapour.

- Decreases in spring snow cover extent since the 1960s, and in the duration of lake and river ice over at least the last century, relate well to increases in Northern Hemispheric surface air temperatures.

- The systematic decrease in spring and summer Arctic sea-ice extent in recent decades is broadly consistent with increases of temperature over most of the adjacent land and ocean. The large reduction in the thickness of summer and early autumn Arctic sea ice over the last thirty to forty years is consistent with this decrease in spatial extent, but we are unsure to what extent poor temporal sampling and multi-decadal variability are affecting the conclusions.

- The increases in lower-tropospheric water vapour and temperature since the mid-1970s are qualitatively consistent with an enhanced hydrological cycle. This is in turn consistent with a greater fraction of precipitation being delivered from extreme and heavy precipitation events, primarily in areas with increasing precipitation, e.g., mid- and high latitudes of the Northern Hemisphere.

- Where data are available, changes in precipitation generally correspond with consistent changes in streamflow and soil moisture.

We conclude that the variations and trends of the examined indicators consistently and very strongly support an increasing global surface temperature over at least the last century, although substantial shorter-term global and regional deviations from this warming trend are very likely to have occurred.

Temperature Indicators

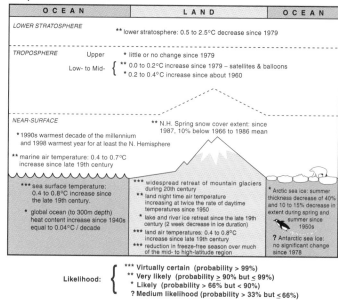

Figure 2.39a: Schematic of observed variations of various temperature indicators.

Hydrological and Storm-Related Indicators

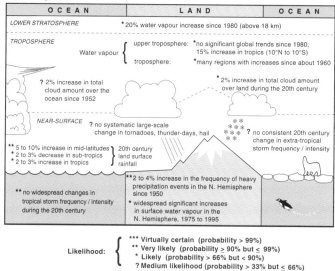

Figure 2.39b: Schematic of observed variations of various hydrological and storm-related indicators.

References

Adkins, J.F., E.W. Boyle, L. Keigwin and E. Cortijo, 1997: Variability of the North Atlantic thermohaline circulation during the Last interglacial period. *Nature*, **390**, 154-156.

Akinremi, O.O., S.M. McGinn and H.W. Cutforth, 1999: Precipitation trends on the Canadian Prairies. *J. Climate*, **12**, 2996-3003.

Alexandersson, H., T. Schmith, K. Iden and H. Tuomenvirta, 1998: Long-term variations of the storm climate over NW Europe. *Global Atmosphere and Ocean Systems*, **6**, 97-120.

Alexandersson, H., T. Schmith, K. Iden and H. Tuomenvirta, 2000: Trends in storms in NW Europe derived from an updated pressure data set. *Clim. Res.*, **14**, 71-73.

Allan, R.J., 2000: ENSO and climatic variability in the last 150 years. In: *El Niño and the Southern Oscillation: Multiscale Variability, Global and Regional Impacts*, edited by Diaz, H.F. and V. Markgraf, Cambridge University Press, Cambridge, UK, pp. 3-56.

Allard, M., B. Wang and J.A. Pilon, 1995: Recent cooling along the southern shore of Hudson Strait Quebec, Canada, documented from permafrost temperature measurements. *Arctic and Alpine Research*, **27**, 157-166.

Alley, R.B. and P.U. Clark, 1999: The deglaciation of the Northern Hemisphere: a global perspective. *Ann. Rev. Earth Planet. Sci.*, 149-182.

Alley, R.B., D.A. Meese, C.A. Shuman, A.J. Gow, K.C. Taylor, P.M. Grootes, J.W.C. White, M. Ram, E.D. Waddington, P.A. Mayewski and G.A. Zielinski, 1993: Abrupt increase in Greenland snow accumulation at the end of the Younger Dryas event. *Nature*, **362**, 527-529.

Alley, R.B., P.A. Mayewski, T. Sowers, M. Stuiver, K.C. Taylor and P.U. Clark, 1997: Holocene climatic instability: A prominent, widespread event 8200 years ago. *Geology*, **25**, 483-486.

Ames, A., 1998: A documentation of glacier tongue variations and lake development in the Cordillera Blanca, Peru. *Z. Gletscherkd. Glazialgeol.*, **34**(1), 1-36.

Anderson, W.L., D.M. Robertson and J.J. Magnuson, 1996: Evidence of recent warming and El Nino-related variations in ice breakup of Wisconsin Lakes. *Limnol. Oceanogr.*, **41**(5), 815-821.

Angel, J.R. and S.A. Isard, 1998: The frequency and intensity of Great Lake cyclones. *J. Climate*, **11**, 61-71.

Angell, J.K., 1988: Variations and trends in tropospheric and stratospheric global temperatures, 1958-87. *J. Climate*, **1**, 1296-1313.

Angell, J.K., 1999: Comparison of surface and tropospheric temperature trends estimated from a 63-station radiosonde network, 1958-1998. *Geophys. Res. Lett.*, **26**, 2761-2764.

Angell, J.K, 2000: Difference in radiosonde temperature trend for the period 1979-1998 of MSU data and the period 1959-1998 twice as long. *Geophys. Res. Lett.*, **27**, 2177-2180.

Appenzeller, C., T.F. Stocker and M. Anklin, 1998: North Atlantic oscillation dynamics recorded in Greenland ice cores. *Science*, **282**, 446-449.

Arhan, M., H. Mescier, B. Bourles and Y. Gouriou, 1998: Hydrographic sections across the Atlantic at 7°30'N and 4°30'S. *Deep Sea Res., Part I*, **45**, 829-872.

Arkin, P.A. and P.E. Ardanuy, 1989: Estimating climatic-scale precipitation from space: a review. *J. Climate*, **2**, 1229-1238.

Assel, R.A., 1983: Description and analysis of a 20-year (1960-79) digital ice-concentration database for the Great Lakes of North America. *Ann. Glaciol.*, **4**, 14-18.

Assel, R.A. and D.M. Robertson, 1995: Changes in winter air temperatures near Lake Michigan during 1851-1993 as determined from regional lake-ice records. *Limnol. Oceanogr.*, **40**, 165-176.

Bacon, S. and D.J.T. Carter, 1991: Wave Climate Changes in the North Atlantic and the North Sea. *Int. J. Climatol.*, **11**, 545-558.

Bajuk, L.J. and C.B. Leovy, 1998a: Are there real interdecadal variations in marine low clouds? *J. Climate*, **11**, 2910-2921.

Bajuk, L.J. and C.B. Leovy, 1998b: Seasonal and interannual variations in stratiform and convective clouds over the tropical Pacific and Indian Oceans from ship observations. *J. Climate*, **11**, 2922-2941.

Balling, R.C., Jr., 1998: Analysis of daily and monthly spatial variance components in historical temperature records. *Physical Geography*, **18**, 544-552.

Barber, D.C., A. Dyke, C. Hillaire-Marcel, A.E. Jennings, J.T. Andrews, M.W. Kerwin, G. Bilodeau, R. McNeely, J. Southon, M.D. Morehead and J.M. Gagnon, 1999: Forcing of the cold event of 8,200 years ago by catastrophic drainage of Laurentide lakes. *Nature*, **400**, 344-347.

Bard, E., M. Arnold, P. Maurice, J. Duprat, J. Moyes and J.C. Duplessy, 1987: Retreat velocity of the North Atlantic polar front during the last deglaciation determined by 14C accelerator mass spectrometry. *Science*, **328**, 791-794.

Barnett, T.P., B. Santer, P.D. Jones and R.S. Bradley, 1996: Estimates of low frequency natural variability in near-surface air temperature. *The Holocene*, **6**, 255-263.

Basist, A.N. and M. Chelliah, 1997: Comparison of tropospheric temperatures derived from the NCEP/NCAR reanalysis, NCEP operational analysis and the Microwave Sounding Unit. *Bull. Am. Met. Soc.*, **78**, 1431-1447.

Bates, J.J. and D.L. Jackson, 2001: Trends in upper tropospheric humidity, *Geophys. Res Lett.*, in press.

Bates, J., X. Wu and D. Jackson, 1996: Interannual variability of upper-tropospheric water vapor band brightness temperature. *J. Climate*, **9**, 427-438.

Behl, R.J. and J.P. Kennet, 1996: Brief interstadial events in the Santa Barbara basin, NE Pacific, during the past 60 kyr. *Nature*, **379**, 243-246.

Belkin, I.M., S. Levitus, J. Anotonov and S.-A. Malmberg, 1998: "Great Salinity Anomalies" in the North Atlantic. *Progress in Oceanography*, **41**, 1-68.

Beltrami, H. and A.E. Taylor, 1994: Records of climatic change in the Canadian Arctic:Combination of geothermal and oxygen isotope data yields high resolution ground temperature histories. *EOS, Transactions*, American Geophysical Union, **75**(44), 75.

Bender, M., T. Sowers, M.L. Dickson, J. Orchado, P. Grootes, P.A. Mayewski and D.A. Meese, 1994: Climate connection between Greenland and Antarctica during the last 100,000 years. *Nature*, **372**, 663-666.

Bender, M., B. Malaize, J. Orchado, T. Sowers and J. Jouzel, 1999: High precision correlations of Greenland and Antarctic ice core records over the last 100 kyr. *Geophysical Monograph*, **112**, Mechanisms of global climate change at millenial timescales, edited by P.U. Clark, R.S. Webb and L.D. Keigwin, 149-164.

Bianchi, G.G. and N.I. McCave, 1999: Holocene periodicity in North Atlantic climate and deep-ocean flow South of Iceland. *Nature*, **397**, 515-517.

Bigg, G.R. and M. Inonue, 1992: Rossby waves and El Nino 1935-1946. *Quart. J. R. Met. Soc.*, **118**, 125-152.

Bindoff, N.L. and T.J. McDougall, 2000: Decadal changes along an Indian Ocean section at 32 S. and their interpretation. *J. Phys.Oceanogr.*, **30**, 1207-1222.

Bjørgo, E., O.M. Johannessen and M.W. Miles, 1997: Analysis of merged SMMR/SSMI time series of Arctic and Antarctic sea ice parameters, *Geophys. Res. Lett.*, **24**, 413-416.

Black, D.E., L.C. Peterson, J.T. Overpeck, A. Kaptan, M.N. Evans and M. Kashgarian, 1999: Eight centuries of North Atlantic Ocean Atmosphere Variability. *Science*, **286**, 1709-1713.

Black, P.G., 1993: Evolution of maximum wind estimates in typhoons. In: *Tropical Cyclone Disasters*, edited by J. Lighthill, Z. Zhemin, G. Holland and K. Emanuel, Peking University Press, Beijing, 104-115.

Blunier, T., J. Schwander, B. Stauffer, T. Stocker, A. Dällenbach, A. Indermühle, J. Tschumi, J. Chappellaz, D. Raynaud, J.M. Barnola,

1997: Timing of the Antarctic Cold Reversal and the atmospheric CO_2 increase with respect to the Younger Dryas event. *Geophys. Res. Lett.*, **24**(21), 2683-2686.

Blunier, T., J. Chappellaz, J. Schwander, A. Dällenbach, B. Stauffer, T. Stocker, D. Raynaud, J. Jouzel, H.B. Clausen, C.U. Hammer and S.J. Johnsen, 1998: Asynchrony of Antarctic and Greenland climate change during the last glacial period. *Nature*, **394**, 739-743.

Bogdanova, E.G. and A.V. Mestcherskaya, 1998: Influence of moistening losses on the homogeneity of annual precipitation time series. *Russian Meteorol. Hydrol.*, **11**, 88-99.

Bond, G. and R. Lotti, 1995: Iceberg discharges into the North Atlantic on millennial time scales during the last glaciation. *Science*, **267**, 1005-1010.

Bond, G., W.S. Broecker, S.J. Johnsen, J. McManus, L.D. Labeyrie, J. Jouzel and G. Bonani, 1993: Correlations between climate records from North Atlantic sediments and Greenland ice. *Nature*, **365**, 143-147.

Bond, G., Showers, W. Cheseby, M. Lotti, R. Almasi, P. deMenocal, P. Priore, P. Cullen, H.I. Hajdas and G. Bonani, 1997: A pervasive Millennial-scale cycle in North Atlantic Holocene and glacial climates. *Science*, **278**, 1257-1266.

Bönisch, G., J. Blindheim, J.L. Bullister, P. Schlosser and D.W.R. Wallace, 1997: Long-term trends of temperature, salinity, density, and transient tracers in the central Greenland Sea. *J. Geophys. Res.*, **102**(C8), 18553-18571.

Bonsal, B.R., X. Zhang, L.A.Vincent and W.D. Hogg, 2001: Characteristics of daily and extreme temperatures over Canada. *J. Climate*, in press.

Bottomley, M., C.K. Folland, J. Hsiung, R.E. Newell and D.E. Parker, 1990: Global Ocean Surface Temperature Atlas "GOSTA". HMSO, London, 20pp+iv, 313 plates.

Bouws, E., D.Jannink and G.J. Komen, 1996: The Increasing Wave Height in the North Atlantic Ocean. *Bull. Am. Met. Soc.*, **77**, 2275-2277.

Bradley, R.S., 1999: Paleoclimatology: reconstructing climates of the Quaternary Harcourt. Academic Press, San Diego, 610 pp.

Bradley, R.S. and P.D. Jones, 1993: `Little Ice Age' summer temperature variations: their nature and relevance to recent global warming trends. *The Holocene*, **3**, 367-376.

Bradley, R.S. and P.D. Jones (eds.), 1995: Climate Since A.D. 1500. (Revised edition) Routledge, London, 706 pp.

Brázdil, R., 1996: Reconstructions of past climate from historical sources in the Czechs Lands. In: *Climatic Variations and Forcing Mechanisms of the Last 2000 Years*, P.D. Jones, R.S. Bradley and J. Jouzel (eds.), NATO ASI Series, Springer Verlag, Berlin, Heidelberg, **41**, 409-431.

Brázdil, R. and T. Vais, 1997: Thunderstorms and related weather extremes in south Moravia, Czech Republic in 1946-1995: Data, results, impacts. *Preprints of the Workshop on Indices and Indicators for Climate Extremes, NOAA/NCDC, Asheville NC USA, 3-6 June 1997*, 4 pp.

Brázdil, R., M. Budykov, I. Auer, R. Böhm, T. Cegnar, P. Fasko, M Lapin, M. Gajic-Capka, K. Zaninovic, E. Koleva, T. Niedzwiedz, S. Szalai, Z. Ustrnul and R.O. Weber, 1996: Trends of maximum and minimum daily temperatures in central and southeastern Europe. *Int. J. Climatol.*, **16**, 765-782.

Brecher, H.H. and L.G. Thompson, 1993: Measurement of the retreat of Qori Kalis glacier in the tropical Andes of Peru by terrestrial photogrammetry. *Photogrammetric Engineering and Remote Sensing*, **59**(6), 1017-1022.

Briffa, K.R., 2000: Annual climate variability in the Holocene: interpreting the message of ancient trees. *Quat. Sci. Rev.*, **19**, 87-105.

Briffa, K.R., P.D. Jones, F.H. Schweingruber, S.G. Shiyatov, and E.R. Cook, 1995: Unusual twentieth-century summer warmth in a 1,000-year temperature record from Siberia. *Nature*, **376**, 156-159.

Briffa, K.R., P.D. Jones, F.H. Schweingruber, S.G. Shiyatov and E.A. Vaganov, 1996: Development of a North Eurasian chronology network: Rationale and preliminary results of comparative ring-width and densitometric analyses in northern Russia. In: *Tree Rings, Environment, and Humanity. Radiocarbon 1996*, J.S. Dean, D.M. Meko and T.W. Swetnam (eds.), Department of Geosciences, The University of Arizona, Tucson, pp. 25-41.

Briffa, K.R., F.H. Schweingruber, P.D. Jones, T.J. Osborn, S.G. Shiyatov and E.A. Vaganov, 1998a: Reduced sensitivity of recent tree-growth to temperature at high northern latitudes. *Nature*, **391**, 678-682.

Briffa, K.R., P.D. Jones, F.H. Schweingruber and T.J. Osborn, 1998b: Influence of volcanic eruptions on Northern Hemisphere summer temperature over the past 600 years. *Nature*, **393**, 450-455.

Broecker, W.S., 1997: Thermohaline circulation, the Achilles heel of our climate system: Will man-made CO_2 upset the current balance? *Science*, **278**, 1582-1588.

Broecker, W.S., 1998: Paleocean circulation during the last deglaciation A bipolar seasaw? *Paleoceanography*, **13**, 119-121.

Broecker, W.S., D.M. Peteet and D. Rind, 1985: Does the ocean-atmosphere system have more than one mode of operation? *Nature*, **315**, 21-26.

Brown, J., O.J. Ferrians, Jr., J.A. Heginbottom and E.S. Melnikov, 1997: Circum-Arctic map of permafrost and ground-ice conditions. *U.S. Geological Survey Circum-Pacific Map* CP- 45, 1:10,000,000, Reston, Virginia.

Brown, N. and A. Issar (eds.), 1998: Water, Environment and Society in Times of Climatic Change. Kluwer, pp. 241-271.

Brown, R.D., 2000: Northern Hemisphere snow cover variability and change, 1915-1997. *J. Climate*, **13**, 2339-2355.

Brown, R.D. and B.E. Goodison, 1996: Interannual variability in reconstructed Canadian snow cover, 1915-1992. *J. Climate*, **9**, 1299-1318.

Brown, R.D. and R.O. Braaten, 1998: Spatial and temporal variability of Canadian monthly snow depths, 1946-1995. *Atmosphere-Ocean*, **36**, 37-45.

Brown, S.J., D.E. Parker, C.K. Folland and I. Macadam, 2000: Decadal variability in the lower-tropospheric lapse rate. *Geophys. Res. Lett.*, **27**, 997-1000.

Brunet-India, M., E. Aguilar, O. Saladie, J. Sigro and D. Lopez, 1999a: Evolución térmica reciente de la región catalana a partir de la construccion de series climáticas regionales. In: Raso Nadal, J. M. and Martin-Vide, J.: *La Climatología española en los albores del siglo XXI*, Barcelona: Publicaciones de la A.E.C., Serie A, **1**, 91-101.

Brunet-India, M., E. Aguilar, O. Saladie, J. Sigro and D. Lopez, 1999b: Variaciones y tendencias contemporaneas de la temperatura máxima, mínima y amplitud térmica diaria en el NE de España, In: Raso Nadal, J. M. Martin-Vide, J.: *La Climatología española en los albores del siglo XXI*, Barcelona: Publicaciones de la A.E.C., Serie A, **1**, 103-112.

Brunetti, M., L. Buffoni, M. Maugeri and T. Nanni, 2000a: Trends of minimum and maximum daily temperatures in Italy from 1865 to 1996. *Theoretical and Applied Climatology*, **66**, 49-60.

Brunetti, M., S. Cecchini, M. Maugeri and T. Nanni, 2000b: Solar and terrestrial signals in precipitation and temperature in Italy from 1865 to 1996. *Advances in Geosciences*, W. Schroeder (Editor), IAGA, pp. 124-133.

Brutsaert, W. and M.B. Parlange, 1998: Hydrological cycle explains the evaporation paradox. *Nature*, **396**, 30.

Bryden, H.L., M.J. Griffiths, A.M. Lavin, R.C. Millard, G. Parrilla and W.M. Smethie, 1996: Decadal changes in water masses characteristics at 24°N in the subtropical Atlantic Ocean. *J. Climate*, **9**, 3162-3186.

Buffoni, L., M. Maugeri and T. Nanni, 1999: Precipitation in Italy from 1833 to 1996. *Theoretical and Applied Climatology*, **63**, 33-40.

Burgess, M.M., S.L. Smith, J. Brown, V. Romanovsky and K. Hinkel,

2000: The Global Terrestrial Network for Permafrost (GTNet-P): permafrost monitoring contributing to global climate observations. *Geological Survey of Canada*, Current Research 2000E-14, 8 pp. (online, http://www.nrcan.gc.ca/gsc/bookstore)

Burn, C.R., 1998: Field investigations of permafrost and climatic change in northwest North America; *Proceedings of Seventh International Conference on Permafrost*, Yellowknife, Canada, June 1998, Université Laval, Quebec, Collection Nordicana No. 57, pp. 107-120

Carmack, E.C., R.W. MacDonald, R.W. Perkin, F.A. McLaughlin and R.J. Pearson, 1995: Evidence for warming of Atlantic water in the southern Canadian Basin of the Arctic Ocean: Results from the Larsen-93 expedition. *Geophys. Res. Lett.*, **22**, 1061-1064.

Carter, D.J.T. and L. Draper, 1988: Has the Northeast Atlantic Become Rougher? *Nature*, **332**, 494.

Cavalieri, D.J., P. Gloersen, C.L. Parkinson, J.C. Comiso and H.J. Zwally, 1997: Observed hemispheric asymmetry in global sea ice changes. *Science*, **278**, 1104-1106.

Chan, J.C.L. and J. Shi, 1996: Long-term trends and interannual variability in tropical cyclone activity over the western North Pacific. *Geophys. Res. Lett.*, **23**, 2765-2767.

Changnon, D. and S.A. Changnon, 1997: Surrogate data to estimate crop-hail loss. *J. Appl. Met.*, **36**, 1202-1210.

Changnon, S.A., 1985: Secular variations in thunder-day frequencies in the twentieth century. *J. Geophys. Res.*, **90**, 6181-6194.

Changnon, S.A. and D. Changnon, 2000: Long-term fluctuations in hail incidences in the United States. *J. Climate*, **13**, 658-664.

Changnon, S.A., D. Changnon, E.R. Fosse, D.C. Hoganson, R.J. Roth and J. Totsch, 1996: Impacts and Responses of the Weather Insurance Industry to Recent Weather Extremes. *Final Report to UCAR from Changnon Climatologist*, CRR-41, Mahomet, IL, 166 pp.

Chanin, M.L. and V. Ramaswamy, 1999: "Trends in Stratospheric Temperatures" in WMO (World Meteorological Organization), Scientific Assessment of Ozone Depletion: 1998, *Global Ozone Research and Monitoring Project* - Report No. 44, Geneva, pp. 5.1-5.59.

Chao, Y., M. Ghil and J.C. McWilliams, 2000: Pacific Interdecadal variability in this century's sea surface temperatures. *Geophys. Res. Lett.*, **27**, 2261-2264.

Chapman, W.L. and J.E. Walsh, 1993: Recent variations of sea ice and air temperature in high latitudes. *Bull. Am. Met. Soc.*, **74**, 33-47.

Chappellaz, J., E. Brook, T. Blunier and B. Malaizé, 1997: CH_4 and $\delta^{18}O$ of O_2 records from Greenland ice: A clue for stratigraphic disturbance in the bottom part of the Greeland . Ice Core Project and the Greeland Ice Sheet Project 2 ice-cores. *J. Geophys. Res.*, **102**, 26547-26557.

Charles, C.D., J. Lynch-Stieglitz, U.S. Niennemann and R.G. Fairbanks, 1996: Climate connections between the two hemispheres revealed by deep sea sediment core/ice core correlations. *Earth Planet Sci. Lett.*, **142**, 19-27.

Chase, T.N., R.A. Pielke Sr., J.A. Knaff, T.G.F. Kittel and J.L. Eastman, 2000: A comparison of regional trends in 1979-1997 depth-averaged tropospheric temperatures. *Int. J. Climatol.*, **20**, 503-518.

Cheddadi, R., K. Mamakowa, J. Guiot, J.L. de Beaulieu, M. Reille, V. Andrieu, W. Grasnoszewki and O. Peyron, 1998: Was the climate of the Eemian stable ? A quntitative climate reconstruction from seven European climate records. *Paleogeography, Paleoclimatology, Paleoecology*, **143**, 73-85.

Christy, J.R., 1995: Temperature above the surface. *Clim. Change*, **31**, 455-474.

Christy, J.R. and R. T. McNider, 1994: Satellite greenhouse warming. *Nature*, **367**, 325.

Christy, J.R., R.W. Spencer and E. Lobl, 1998: Analysis of the merging procedure for the MSU daily temperature time series. *J. Climate*, **5**, 2016-2041.

Christy, J.R., R.W. Spencer and W.D. Braswell, 2000: MSU tropospheric temperatures:Dataset construction and radiosonde comparisons. *J. Atmos. Oceanic Tech.*, **17**, 1153-1170.

Christy, J.R., D.E. Parker, S.J. Brown, I. Macadam, M. Stendel and W.B. Norris, 2001: Differential trends in tropical sea surface and atmospheric temperatures. *Geophys. Res. Lett.*, **28**, 183-186.

Ciais, P., J.R. Petit, J. Jouzel, C. Lorius, N.I. Barkov, V. Lipenkov and V. Nicolaïev, 1992: Evidence for an Early Holocene climatic optimum in the Antarctic deep ice core record. *Clim. Dyn.*, **6**, 169-177.

Clausen, H.B., C.U. Hammer, J. Christensen, C.S. Schott Hvidberg, D. Dahl-Jensen, M. Legrand and J.P. Steffensen, 1995: 1250 years of global volcanism as revealed by central Greenland ice cores. In: *Ice Core Studies of Global Biogeochemical Cycles*, Nato ASI, Series I, vol 30., edited by R.J. Delmas, Springer-Verlag, New York, pp. 517-532.

CLIMAP Project Members, 1984: The last interglacial ocean. *Quat. Res.*, **21**, 123-224.

Clow, G.D., A.H. Lachenbruch and C.P. McKay, 1991: Investigation of borehole temperature data for recent climate changes: Examples from Alaskan Arctic and Antarctica. In: *Proceedings of the International Conference on the Role of Polar Regions in Global Change*, June 11-15, 1990, Geophysical Institute, University of Alaska, Fairbanks, vol. 2, 533.

COHMAP Members, 1988: Climatic changes of the last 18,000 years: observations and model simulations. *Science*, **241**, 1043-1052.

Cole, J.E. and E.R. Cook, 1998: The changing relationship between ENSO variability and moisture balance in the continental United States. *Geophys. Res. Lett.*, **25**, 4529-4532.

Coles, V.J., M.S. McCartney, D.B. Olson and W.M. Smethie Jr., 1996: Changes in the Antarctic bottom water properties in the western South Atlantic in the late 1980's. *J. Geophys. Res.*, **101**(C4), 8957-8970.

Collins, D.A., P.M. Della-Marta, N. Plummer and B.C. Trewin, 2000: Trends in annual frequencies of extreme temperature events in Australia. *Australian Meteorological Magazine*, **49**, 277-292.

Combourieu-Nebout, N., M. Paterne and J.L. Turon, 1998: A high-resolution record of the last deglaciation in the Central Mediterranean Sea: Palaeovegetation and palaeohydrological evolution. *Quat. Sci. Rev.*, **17**, 303-317.

Comiso, F., 2000: Variability and trends in Antarctic temperatures from in situ and satellite infrared measurements. *J. Climate*, **13**, 1674-1696.

Cook, E.R., 1995: Temperature histories in tree rings and corals. *Clim. Dyn.*, **11**, 211-222.

Cook, E.R., R.D. D'Arrigo and K.R. Briffa, 1998: A reconstruction of the North Atlantic Oscillation using tree-ring chronologies from North America and Europe. *The Holocene*, **8**, 9-17.

Cook, E.R., D.M. Meko, D.W. Stahle and M.K. Cleaveland, 1999: Drought reconstructions for the continental United States. *J. Climate*, **12**, 1145-1162.

Cook, E.R., B.M. Buckley and R.D. D'Arrigo, 2000: Warm-Season Temperatures since 1600 B.C. Reconstructed from Tasmanian Tree Rings and Their Relationship to Large-Scale Sea Surface Temperature Anomalies. *Clim. Dyn.*, **16**, 79-91.

Cortijo, E., J. Duplessy, L. Labeyrie, H. Leclaire, J. Duprat and T. van Weering, 1994: Eemian cooling in the Norwegian Sea and North Atlantic ocean preceding ice-sheet growth. *Nature*, **372**, 446-449.

Cortijo, E., L.D. Labeyrie, L. Vidal, M. Vautravers, M. Chapman, J.C. Duplessy, M. Elliot, M. Arnold and G. Auffret, 1997: Changes in the sea surface hydrology associated with Heinrich event 4 in the North Atlantic Ocean (40-60°N). *Earth Planet. Sci. Lett.*, **146**, 29-45.

Crowley, T.J., 1992: North Alantic Deep Water Cools The Southern Hemisphere. *Paleoceanography*, **7**, 489-497.

Crowley, T.J., 2000: Causes of Climate Change Over the Past 1000 Years. *Science*, **289**, 270-277.

Crowley, T.J. and K.Y. Kim, 1996: Comparison of proxy records of climate change and solar forcing. *Geophys. Res. Lett.*, **23**, 359-362.

Crowley, T.J. and K.Y. Kim, 1999: Modeling the temperature response to forced climate change over the last six centuries. *Geophys. Res. Lett.*, **26**, 1901-1904.

Crowley, T.J. and T. Lowery, 2000: How warm was the Medieval warm period? *Ambio*, **29**, 51-54.

Cullen, H., R. D'Arrigo, E. Cook and M.E. Mann, 2001: Multiproxy-based reconstructions of the North Atlantic Oscillation over the past three centuries. *Paleoceanography*, **16**, 27-39.

Dahl-Jensen, D., K. Mosegaard, N. Gundestrup, G.D. Clow, S.J. Johnsen, A.W. Hansen and N. Balling, 1998: Past temperatures directly from the Greenland ice sheet. *Science*, **282**, 268-271.

Dai, A., A.D. DelGenio and I.Y. Fung, 1997a: Clouds, precipitation, and temperature range. *Nature*, **386**, 665-666.

Dai, A., I.Y. Fung and A.D. Del Genio, 1997b: Surface observed global land precipitation variations during 1900-88. *J. Climate*, **10**, 2943-2962.

Dai, A., K.E. Trenberth and T.R. Karl, 1998: Global variations in droughts and wet spells: 1900-1995. *Geophys. Res. Lett.*, **25**, 3367-3370.

Dai, A., K.E. Trenberth and T.R. Karl, 1999: Effects of clouds, soil moisture, precipitation and water vapor on diurnal temperature range. *J. Climate*, **12**, 2452-2473.

Damon, P.E. and A.N. Peristykh, 1999: Solar cycle length and 20th century Northern Hemisphere Warming. *Geophys. Res. Lett.*, **26**, 2469-2472.

Dansgaard, W., H.B. Clausen, N. Gundestrup, C.U. Hammer, S.J. Johnsen, P. Krinstindottir and N. Reeh, 1982: A new Greenland deep ice core. *Science*, **218**, 1273-1277.

Dansgaard, W., J.W. White and S.J. Johnsen, 1989: The abrupt termination of the Younger Dryas climate event. *Nature*, **339**, 532-534.

Dansgaard, W., S.J. Johnsen, H.B. Clausen, D. Dahl-Jensen, N.S. Gundestrup, C.U. Hammer, C.S. Hvidberg, J.P. Steffensen, A.E. Sveinbjörnsdottir, J. Jouzel and G. Bond, 1993: Evidence for general instability of past climate from a 250 kyr ice core. *Nature*, **364**, 218-219.

D'Arrigo, R.D., E.R. Cook and G.C. Jacoby, 1996: Annual to decadal-scale variations in northwest Atlantic sector temperatures inferred from Labrador tree rings. *Canadian Journal of Forest Research*, **26**, 143-148.

D'Arrigo, R.D., E.R. Cook, M.J. Salinger, J. Palmer, P.J. Krusic, B.M. Buckley and R. Villalba, 1998: Tree-ring records from New Zealand: long-term context for recent warming trend. *Clim. Dyn.*, **14**, 191-199.

De la Mare, W.K., 1997: Abrupt mid-twentieth century decline in Antarctic sea-ice extent from whaling records. *Nature*, **389**, 57-60.

Delworth, T.L. and M.H. Mann, 2000: Observed and simulated multidecadal variability in the Northern Hemisphere. *Clim. Dyn.*, **16**, 661-676.

de Menocal, P., 1998: Subtropical signatures of millenial-scale Holocene climate variability. International Conference of Paleoceanography.

Denton, G. and C.H. Hendy, 1994: Younger Dryas age advance of the Franz Josef glacier in the Southern Alps of New Zealand. *Science*, **264**, 1434-1437.

Deser, C. and M.L. Blackmon, 1995: On the relationship between tropical and North Pacific sea surface temperature variations. *J. Climate*, **8**, 1677-1680.

Deser, C., J.E. Walsh and M.S. Timlin, 2000: Arctic sea ice variability in the context of recent wintertime atmospheric circulation trends. *J. Climate*, **13**, 617-633.

Dessens, J., 1995: Severe convective weather in the context of a nighttime global warming. *Geophys. Res. Lett.*, **22**, 1241-1244.

Diaz, H.F. and H.F. Graham, 1996: Recent changes in tropical freezing heights and the role of sea surface temperature. *Nature*, **383**, 152-155.

Dickson, R.R., J. Lazier, J. Meincke, P. Rhines and J. Swift, 1996: Long-term co-ordinated changes in the convective activity of the North Atlantic. *Prog. Oceanogr.*, **38**, 241-295.

Diggle, P.J., K.Y. Liang and S.L. Zeger, 1999: Analysis of longitudinal data. Clarendon Press, Oxford, 253 pp.

Ding, Y., 1998: Recent degradation of permafrost in China and the response to climatic warming. In: *Proceedings of the Seventh International Conference on Permafrost*, Yellowknife, Canada, June 1998, Université Laval, Quebec, Collection Nordicana No. 57, pp. 221-224.

Ding, Z.L., N.W. Rutter, T.S. Liu, J.M. Sun, J.Z. Ren, D. Rokosh and S.F. Xiong, 1998: Correlation of Dansgaard-Oeschger cycles between Greenland ice and Chinese loess. *Paleoclimates*, **2**, 281-291.

Doherty, R.M., M. Hulme and C.G. Jones, 1999: A gridded reconstruction of land and ocean precipitation for the extended Tropics from 1974-1994. *Int. J. Climatol.*, **19**, 119-142.

Doswell, C.A. III and D.W. Burgess, 1988: On some issues of United States tornado climatology. *Mon. Wea. Rev.*, **116**, 495-501.

Douglas, A.V. and P.J. Englehart, 1999: Inter-monthly variability of the Mexican summer monsoon. *Proceedings of the Twenty-Second Annual Climate Diagnostics and Prediction Workshop*, Berkeley, CA, October 6-10, 1997, Washington, D.C.: U.S. Department of Commerce, NOM, NTIS #PB97-159164, pp. 246-249.

Dunbar, R.B. and J.E.Cole, 1999: Annual Records of Tropical Systems, (ARTS): A PAGES Report 99-1/CLIVAR Initiative: Recommendations for Research. Summary of scientific priorities and implementation strategies: *ARTS Planning Workshop*, Kauai, Hawaii, PAGES Report 99-1.

Dunkerton, T., D. Delisi and M. Baldwin, 1998: Middle atmosphere cooling trend in historical rocketsonde data. *Geophys. Res. Lett.*, **25**, 3371-3374.

Duplessy, J.C., L. Labeyrie, A. Juillet-Leclerc, F. Maitre, J. Duprat and M. Sarnthein, 1991: Surface salinity reconstruction of the North Atlantic Ocean during the last glacial maximum. *Oceanologica Acta*, **14**, 311-324.

Duplessy, J.C., L. Labeyrie, M. Arnold, M. Paterne, J. Duprat and T.C.E. van Weering, 1992: Changes in surface salinity of the North Atlantic Ocean during the last deglaciation. *Nature*, **358**, 485-487.

Easterling, D.R., B. Horton, P.D. Jones, T.C. Peterson, T.R. Karl, D.E. Parker, M.J. Salinger, V. Razuvayev, N. Plummer, P. Jamason and C.K. Folland, 1997: Maximum and minimum temperature trends for the globe. *Science*, **277**, 364-367.

Easterling, D.R., J.L. Evans, P.Ya. Groisman, T.R. Karl, K.E. Kunkel and P. Ambenje, 2000: Observed variability and trends in extreme climate events. *Bull. Am. Met. Soc.*, **81**, 417-425.

Elliott, W.P., 1995: On detecting long-term changes in atmospheric moisture. *Clim. Change*, **31**, 349-367.

Elliott, W. and D. Gaffen, 1991: On the utility of radiosonde humidity archives for climate studies. *Bull. Am. Met. Soc.*, **72**, 1507-1520.

Ely, L.L., E. Yehouda, V.R. Baker and D.R. Cayan, 1993: A 5000-year record of extreme floods and climate change in the Southwestern United States. *Science*, **262**, 410-412.

Enfield, D.B. and A.M. Mestas-Nuñez, 1999: Multiscale variabilities in global sea surface temperatures and their relationships with tropospheric climate patterns. *J. Climate*, **12**, 2719-2733.

Environmental Working Group (EWG), 1997: *Joint U.S.-Russian Atlas of the Arctic Ocean [CD-ROM]*, Natl. Snow and Ice Data Centre, Boulder, Colorado, USA.

Eskridge, R.E., O.A. Alduchov, I.V. Chernykh, Z. Panmao, A.C. Polansky and S.R. Doty, 1995: A Comprehensive aerological reference data set (CARDS): Rough and systematic errors. *Bull. Am. Met. Soc.*, **76**, 1759-1775.

Evans, J.S., R. Toumi, J.E. Harries, M.P. Chipperfield and J.R. Russell III, 1998: Trends in stratospheric humidity and the sensitivity of ozone to those trends. *J. Geophys. Res.*, **103**, 8715-8725.

Fallot, J.-M., R.G. Barry and D. Hoogstrate, 1997: Variations of mean cold season temperature, precipitation and snow depths during the last 100 years in the Former Soviet Union (FSU). *Hydrol. Sci. J.,* **42**, 301-327.

Fernandez-Partagas, J. and H.F. Diaz, 1996: Atlantic hurricanes in the second half of the 19th Century. *Bull. Am. Met. Soc.,* **77**, 2899-2906.

Fischer, H., M. Wahlen, J. Smith, D. Mastroiani and B. Deck, 1999: Ice core records of atmospheric CO_2 around the last three glacial terminations. *Science,* **283**, 1712-1714.

Fisher, D.A. 1997: High resolution reconstructed Northern Hemisphere temperatures for the last few centuries: using regional average tree ring, ice core and historical annual time series. Paper U32C-7 in *Supplement to EOS, Transactions,* American Geophysical Union Vol. 78 No. 46, abstract.

Fisher, D.A., R.M. Koerner, K. Kuivinen, H.B. Clausen, S.J. Johnsen, J.P. Steffensen, N. Gundestrup and C.U. Hammer, 1996: Inter-comparison of ice core (O-18) and precipitation records from sites in Canada and Greenland over the last 3500 years and over the last few centuries in detail using EOF techniques. In: *Climate Variations and Forcing Mechanisms of the Last 2000 Years,* edited by P.D. Jones, R.S. Bradley and J. Jouzel, NATO ASI Series I, Vol. 41, pp. 297-328.

Fisher, D.A., R.M. Koerner, J.C. Bourgeois, G. Zielinski, C. Wake, C.U. Hammer, H.B. Clausen, N. Gundestrup, S.J. Johnsen, K. Goto-Azuma, T. Hondoh, E. Blake and M. Gerasimoff, 1998: Penny Ice Cap, Baffin Island, Canada, and the Wisconsinan Foxe Dome Connection: two states of Hudson Bay ice cover. *Science,* **279**, 692-695.

Fleming, K.M., J.A. Dowdeswell and J. Oerlemans, 1997: Modelling the mass balance of northwest Spitsbergen glaciers and response to climate change. *Ann. Glaciol.,* **24**, 203-210.

Folland, C.K. and D.E. Parker, 1995: Correction of instrumental biases in historical sea surface temperature data. *Quart. J. R. Met. Soc.,* **121**, 319-367.

Folland, C.K. and M.J. Salinger, 1995: Surface temperature trends in New Zealand and the surrounding ocean, 1871-1993. *Int. J. Climatol.,* **15**, 1195-1218.

Folland, C.K., D.E. Parker and T.N. Palmer, 1986: Sahel rainfall and worldwide sea temperatures 1901-85. *Nature,* **320**, 602-607.

Folland, C.K., R.W. Reynolds, M. Gordon and D.E. Parker, 1993: A study of six operational sea surface temperature analyses. *J. Climate,* **6**, 96-113.

Folland, C.K., M.J. Salinger and N. Rayner, 1997: A comparison of annual South Pacific island and ocean surface temperatures. *Weather and Climate,* **17**, 23-42.

Folland, C.K., D.M.H. Sexton, D.J. Karoly, C.E. Johnson, D.P. Rowell and D.E. Parker, 1998: Influences of anthropogenic and oceanic forcing on recent climate change. *Geophys. Res. Lett.,* **25**, 353-356.

Folland, C.K., D.E. Parker, A.W. Colman and R.Washington, 1999a: Large scale modes of ocean surface temperature since the late nineteenth century. In: *Beyond El Niño: Decadal and Interdecadal Climate Variability,* A. Navarra (ed.), Springer-Verlag, Berlin, pp. 73-102.

Folland, C.K., C. Miller, D. Bader, M. Crowe, P. Jones, N. Plummer, D.E. Parker, J. Rogers and P. Scholefield, 1999b: Workshop on Indices and Indicators for climate extremes, Asheville, NC, USA, 3-6 June 1999: Breakout Group C: Temperature indices for Climate Extremes. *Clim. Change,* **42**, 31-43.

Folland, C.K., N. Rayner, P. Frich, T. Basnett, D. Parker and B. Horton, 2000: Uncertainties in climate data sets – a challenge for WMO. *WMO Bull.,* **49**, 59-68.

Folland, C.K., N.A. Rayner, S.J. Brown, T.M. Smith, S.S. Shen, D.E. Parker, I. Macadam, P.D. Jones, R.N. Jones, N. Nicholls and D.M.H. Sexton, 2001: Global temperature change and it uncertainties since 1861. *Geophys. Res. Lett.,* in press.

Førland, E.J. and I. Hassen-Bauer, 2000: Increased precipitation in the Norwegian Arctic: True or false? *Clim. Change,* **46**, 485-509.

Førland, E.J., H. Alexandersson, A. Drebs, I. Hassen-Bauer, H. Vedin and O.E. Tveito, 1998: Trends in maximum 1-day precipitation in the Nordic region, *DNMI-KLIMA* 14/98, pp. 55, Norwegian Meteorological Institute, N-0313 Oslo, Norway.

Forster, P.M. and K.P. Shine, 1999: Stratospheric water vapour changes as a possible contributor to observed stratospheric cooling. *Geophys. Res. Lett.,* **26**, 3309-3312.

Frank, P., 1997: Changes in the glacier area in the Austrian Alps between 1973 and 1992 derived from LANDSAT data. *MPI report* 242, 21 pp.

Free, M. and A. Robock, 1999: Global Warming in the Context of the Little Ice Age. *J. Geophys. Res.,* **104** (D16), 19057-19070.

Freeland, H., K. Denman, C.S. Wong, F. Whitney and R. Jacques, 1997: Evidence of change in the winter mixed layer in the Northeast Pacific Ocean. *Deep Sea Res., Part I,* **44**(12), 2117-2129.

Frei, A., D.A. Robinson and M.G. Hughes, 1999: North American snow extent: 1900-1994. *Int. J. Climatol.,* **19**, 1517-1534.

Frei, C. and C. Schär, 2001: Detection probability of trends in rare events: Theory and application to heavy precipitation in the Alpine Region. *J. Climate,* **14**, 1568-1584.

French, H.M. and I.E. Egorov, 1998: 20th century variations in the southern limit of permafrost near Thompson, northern Manitoba, Canada. In: *Proceedings of Seventh International Conference on Permafrost,* Yellowknife, Canada, June 1998, Université Laval, Quebec, Collection Nordicana No. 57, pp. 297-304.

Frich, P., L.V. Alexander, P. Della-Marta, B. Gleason, M. Haylock, A. Klein-Tank and T. Peterson, 2001: Observed coherent changes in climatic extremes during the second half of the 20th Century. *Clim. Res.,* in press.

Fritts, H.C., 1976: Tree Rings and Climate. Academic Press, London.

Fritts, H.C., 1991: Reconstructing large-scale climatic patterns from Tree Ring Data. The University of Arizona Press, Tucson.

Gabriel, K.R. and S.A. Changnon, 1990: Temporal features in thunder days in the United States. *Clim. Change,* **15**, 455-477.

Gaffen, D.J., 1994: Temporal inhomogeneities in radiosonde temperature records. *J. Geophys. Res.,* **99**, 3667-3676.

Gaffen, D.J. and R.J. Ross, 1998: Increased summertime heat stress in the U.S. *Nature,* **396**, 529-530.

Gaffen, D.J. and R.J. Ross, 1999: Climatology and trends of U.S. surface humidity and temperature. *J. Climate,* **13**, 811-828.

Gaffen, D.J., B.D. Santer, J.S. Boyle, J.R. Christy, N.E. Graham and R. J. Ross, 2000a: Multidecadal changes in the vertical structure of the tropical troposphere. *Science,* **287**, 1242-1245.

Gaffen, D.J., M.A. Sargent, R.E. Habermann and J.R. Lazante, 2000b: Sensitivity of tropospheric and stratospheric temperature trends to radiosonde data quality. *J. Climate,* **13**, 1776-1796.

Gagan, M.K., L.K. Ayliffe, D. Hopley, J.A. Cali, G.E. Mortimer, J. Chappell, M.T. McCulloch and M.J. 1998: Heat, temperature and surface-ocean water balance of the mid-Holocene tropical western Pacific. *Science,* **279**, 1014-1018.

Gallo, K.P., D.R. Easterling and T.C. Peterson, 1996: The influence of land use/land cover on climatological values of the diurnal temperature range. *J. Climate,* **9**, 2941-2944.

Garcia, N.O. and W.M. Vargas, 1998: The temporal climatic variabilitiy in the 'Rio de la Plata' Basin displayed by the river discharges. *Clim. Change,* **38**, 359-379.

Garreaud, R.D. and D.S. Battisti, 1999: Interannual (ENSO) and interdecadal variability in the Southern Hemisphere tropospheric circulation. *J. Climate,* **12**, 2113-2123.

Gasse, F. and E. Van Campo, 1994: Abrupt post-glacials eveents in West asia and north Africa. *Earth Planet Sci. Lett.,* **126**, 453-456.

Gasse, F., R. Tehet and A. Durand, 1990: The arid-humid transition in the Sahara and the Sahel during the last deglaciation. *Nature,* **346**, 141-146.

Genta, J.L., G. Perez-Iribarren and C.R. Mechoso, 1998: A recent increasing trend in the streamflow of rivers in southeastern South America. *J. Climate,* **11**, 2858-2862.

Georgievsky, V.Yu., A.V. Ezhov, A.L. Shalygin, I.A. Shiklomanov and A.I. Shiklomanov, 1996: Assessment of the effect of possible climate changes on hydrological regime and water resources of rivers in the former USSR. *Russian Meteorol. And Hydrol.*, **11**, 66-74.

Gershunov, A. and T.P. Barnett, 1998: Interdecadal modulation of ENSO teleconnections. *Bull. Am. Met. Soc.*, **79**, 2715-2725.

Goddard, L. and N.E. Graham, 1997: El Niño in the 1990s. *J. Geophys. Res.*, **102**, 10423-10436.

Golitsyn, G.S., A.I. Semenov, N.N. Shefov, L.M. Fishkova, E.V. Lysenko and S.P. Perov, 1996: Long-term temperature trends in the middle and upper atmosphere. *Geophys. Res. Lett.*, **23**, 1741-1744.

Golubev, V.S., J.H. Lawrimore, P.Ya. Groisman, N.A. Speranskaya, S.A. Zhuravin, M.J. Menne, T.C. Peterson and R.W. Malone, 2001: Evaporation changes over the contiguous United States and the Former USSR: The re-assessment. *Geophys. Res. Lett.*, in press.

Gong, D.Y. and S.W. Wang, 1999a: Experiments on the reconstruction of historical monthly mean northern hemispheric 500hPa heights from surface data. *Report on the Department of Geophysics, Peking University*.

Gong, D.Y. and S.W. Wang, 1999b: Variability of the Siberian High and the possible connection to global warming. *Acta Geographica Sinica*, **54** (2), 142-150 (in Chinese).

Gong, D.Y. and S.W. Wang, 1999c: Definition of Antarctic Oscillation index. *Geophys. Res. Lett.*, **26**, 459-462.

Goswami, B.N., V. Krishnamurthy and H. Annamalai, 1997: A broad scale circulation index for the interannual variability of the Indian summer monsoon. *Report No. 46, COLA*, 4041 Powder Mill Road, Suite 302, Calverton, MD, 20705, USA.

Graf, H.F., J. Perlwitz, I. Kirchner and I. Schult, 1995: Recent northern winter climate trends, ozone changes and increased greenhouse gas forcing. *Contrib. Phys. Atmos.*, **68**, 233-248.

Graham, N.E. and H.F. Diaz, 2001: Evidence for intensification of North Pacific Winter Cyclones since 1948. *J. Climate*, in press.

Gravis, G.F., N.G. Moskalenko and A.V. Pavlov, 1988: Perennial changes in natural complexes of the cryolithozone. In: *Proceedings of the Fifth International Conference on Permafrost*, Trondheim, Norway, vol. 1, 165-169.

Grazulis, T.P., 1993: Significant Tornadoes, 1680-1991. *Environmental Films*, St. Johnsbury, VT, 1326 pp.

Grazulis, T.P., C.A. Doswell III, H.E. Brooks and M. Biddle, 1998: A new perspective of the societal impacts of North American tornadoes covering two centuries. Preprints, *19th Conference on Severe Local Storms*. American Meteorological Society, Minneapolis, MN, 196-199.

Greuell, J.W. and J. Oerlemans, 1987: Sensitivity studies with a mass balance model including temperature profile calculations inside the glacier. *Zeits. Gletscherk. Glaziageol.*, **22**, 101-124.

Grimm, E.C., G.L. Jacobson, W.A. Watts, B.C.S. Hansen and K.A. Maasch, 1993: A 50000-year record of climate oscillations from Florida and its temporal correlation with the Heinrich events. *Science*, **261**, 198-200.

GRIP project members, 1993: Climatic instability during the last interglacial period revealed in the Greenland summit ice-core. *Nature*, **364**, 203-207.

Groisman, P.Ya. and D. Easterling, 1994: Variability and trends of total precipitation and snowfall over the United States and Canada. *J. Climate*, **7**, 184-205.

Groisman, P.Ya. and E.Ya. Rankova, 2001: Precipitation trends over the Russian permafrost-free zone: Removing the artifacts of pre-processing. *Int. J. Climatol.*, 21, 657-678.

Groisman, P.Ya., T.R. Karl, D.R. Easterling, R.W. Knight, P.B. Jamason, K.J. Hennessy, R. Suppiah, C.M. Page, J. Wibig, K. Fortuniak, V.N. Razuvaev, A. Douglas, E. Førland and P.M. Zhai, 1999: Changes in the probability of heavy precipitation: Important indicators of climatic change. *Clim. Change*, **42**, 243-283.

Groisman, P.Ya., R.W. Knight and T.R. Karl, 2001: Heavy precipitation and high streamflow in the United States: Trends in the 20th century. *Bull. Am. Met. Soc.*, **82**, 219-246.

Grootes, P.M., M. Stuiver, J.W.C. White, S.J. Johnsen and J. Jouzel, 1993: Comparison of the oxygen isotope records from the GISP2 and GRIP Greenland ice cores. *Nature*, **366**, 552-554.

Grove, J.M. and R. Switsur, 1994: Glacial geological evidence for the`Medieval Warm Period. *Clim. Change*, **26**, 143-169.

Grumbine, R.W., 1996: Automated Passive Microwave Sea Ice Concentration Analysis at NCEP. US Department of Commerce, National Ocean And Atmospheric Administration, National Weather Service, National Centers for Environmental Prediction, Technical Note, OMB contribution 120, March, 1996, 13 pp. Also: ⟨http://polar.wwb.noaa.gov/seaice/docs/ssmi.auto/ssmi120.html⟩

Gruza, G., E. Rankova, V. Razuvaev and O. Bulygina, 1999: Indicators of climate change for the Russian Federation. *Clim. Change*, **42**, 219-242.

Gutzler, D., 1996: Low-frequency ocean-atmosphere variability across the tropical western Pacific. *J. Atmos. Sci.*, **53**, 2773-2785.

Haeberli, W., G. Cheng, A.P. Gorbunov and S.A. Harris, 1993: Mountain permafrost and climatic change. *Permafrost and Periglacial Processes*, **4**, 165-174.

Haeberli, W., M. Hoelzle and S. Suter (eds.), 1998: Into the second century of worldwide glacier monitoring: prospects and strategies, A contribution to the International Hydrological Programme (IHP) and the Global Environment Monitoring System (GEMS). *UNESCO Studies and Reports in Hydrology, 56*, Paris.

Hagen, J.O., K. Melvold, T. Eiken, E. Isaksson and B. Lefauconnier, 1995: Recent trends in the mass balance of glaciers in Scandinavia and Svalbard. *Proceedings of the international symposium on environmental research in the Arctic*. Watanabe, Okitsugu (Eds.), Tokyo, Japan, 19-21 July, 1995, National Institute of Polar Research, 343-354.

Hahn, C.J., S.G. Warren and J. London, 1996: Edited synoptic cloud reports from ships and land stations over the globe, 1982-1991. Rep#NDP026B, 45 pp. [*Available from Carbon Dioxide Information Analysis Center*, Oak Ridge National Laboratory, P.O. Box 2008, Oak Ridge, TN 37831-6050.]

Halsey, L.A., D.H. Vitt and S.C. Zoltai, 1995: Disequilibrium response of permafrost in boreal continental western Canada to climate change. *Clim. Change*, **30**, 57-73.

Hammer, C.U., H.B. Clausen and C.C. Langway, 1997: 50,000 years of recorded global volcanism. *Clim. Change*, **35**, 1-15.

Hanawa, K., S.Yasunaka, T. Manabe and N. Iwasaka, 2000: Examination of correction to historical SST data using long-term coastal SST data taken around Japan. *J. Met. Soc. Japan*, **78**, 187-195.

Hansen, J. and S. Lebedeff, 1988: Global surface temperatures: update through 1987. *Geophys. Res. Lett.*, **15**, 323-326.

Hansen, J.E., M. Sato, A. Lacis, R. Ruedy, I. Tegen and E. Matthews, 1998: Climate forcings in the Industrial era. *Proc. Natl. Acad. Sci., USA*, **95**, 12753-12758.

Hansen, J., R. Ruedy, J. Glascoe and M. Sato, 1999: GISS analysis of surface temperature change. *J. Geophys. Res.*, **104**(D24), 30997-31022.

Hanssen-Bauer, I. and E.J. Førland, 2000: Temperature and precipitation variations in Norway 1900-1994 and their links to atmospheric circulation. *Int. J. Climatol.*, **20**, 1693-1708.

Harries, J.E., 1976: The distribution of water vapor in the stratosphere. *Rev. Geophys. Space Phys.*, **14**, 565-575.

Harrison, S.P., G. Yu and P.E. Tarasov, 1996: Late quaternary lake-level record from Northern Eurasia. *Quat. Res.*, **45**, 138-159.

Hastenrath, S. and P.D. Kruss, 1992: The dramatic retreat of Mount Kenya's glaciers between 1963 and 1987: greenhouse forcing. *Ann. Glaciol.*, **16**, 127-133.

Hastenrath, S. and A. Ames, 1995: Recession of Yanamarey glacier in

Cordillera Blanca, Peru during the 20th century. *J. Glaciol.*, **41**(137), 191-196.

Hastenrath, S. and L. Greischar, 1997: Glacier recession on Kilimanjaro, East Africa, 1912-89. *J. Glaciol.*, **43**, 455-459.

Haylock, M. and N. Nicholls, 2000: Trends in extreme rainfall indices for an updated high quality data set for Australia, 1910-1998. *Int. J. Climatol.*, **20**, 1533-1541.

Heino, R., R. Brázdil, E. Forland, H. Tuomenvirta, H. Alexandersson, M. Beniston, C. Pfister, M. Rebetez, G. Rosenhagen, S. Rösner and J. Wibig, 1999: Progress in the study of climatic extremes in Northern and Central Europe. *Clim. Change*, **42**, 151-181.

Helland-Hansen, B. and F. Nansen, 1920: Temperature variations in the North Atlantic Ocean and in the atmosphere. Introductory studies on the cause of climatological variations. *Smithsonian Miscellaneous Collections*, 70(4), publication 2537, Washington, DC.

Henderson-Sellers, A., 1992: Continental cloudiness changes this century. *Geo Journal*, **27**, 255-262.

Hennessy, K.J., R. Suppiah and C.M. Page, 1999: Australian rainfall changes, 1910-1995. *Australian Meteorological Magazine*, **48**, 1-13.

Higgins, R.W. and W. Shi, 2000: Dominant factors responsible for interannual variability of the summer, monsoon in the southwestern United States. *J. Climate*, **13**, 759-776.

Higgins, R.W., A. Leetmaa, Y. Xue and A. Barnston, 2000: Dominant factors influencing the seasonal predictability of US precipitation and surface air temperature. *J. Climate*, **13**, 3994-4017.

Holzhauser, H. and H.J. Zumbühl, 1996: To the history of the Lower Grindelwald Glacier during the last 2800 years - paleosols, fossil wood and historical pictorial records - new results. *Z. Geomorph. N. F.*, **104**, 95-127.

Hoyt, D.V. and K.H. Schatten, 1997: The role of the sun in climatic change. Oxford University Press, Oxford, 279 pp.

Hu, F.S., D. Slawinski, H.E.J. Wright, E. Ito, R.G. Johnson, K.R. Kelts, R.F. McEwan and A. Boedigheimer, 1999: Abrupt changes in North American climate during early Holocene times. *Nature*, **400**, 437-440.

Huang, S., H.N. Pollack and P.Y. Shen, 1997: Late quaternary temperature changes seen in world-wide continental heat flow measurements. *Geophys. Res. Lett.*, **24**, 1947-1950.

Huang, S., H.N. Pollack and P.Y. Shen, 2000: Temperature trends over the past five centuries reconstructed from borehole temperatures. *Nature*, **403**, 756-758.

Huffman, G., R.F. Adler, P.A. Arkin, J. Janowiak, P. Xie, R. Joyce, R. Ferraro, A. Chang, A. McNab, A. Gruber and B. Rudolf, 1997: The Global Precipitation Climatology Project (GPCP) merged precipitation data sets. *Bull. Am. Met. Soc.*, **78**, 5-20.

Hughen, K.A., J.T. Overpeck, L.C. Peterson and S. Trumbore, 1996: Rapid climate changes in the tropical Atlantic region during the last deglaciation. *Nature*, **380**, 51-54.

Hughen, K.H., D. P. Schrag, S.B. Jacobsen and W. Hantor, 1999: El Niño during the last interglacial period recorded by a fossil coral from Indonesia. *Geophys. Res. Lett.*, **26**, 3129-3132.

Hughen, K.A., J.T. Overpeck and R. Anderson, 2000: Recent warming in a 500-year paleoclimate record from Upper Soper Lake, Baffin Island, Canada. *The Holocene*, **10**, 9-19.

Hughes, M.G. and D.A. Robinson, 1996: Historical snow cover variability in the Great Plains region of the USA: 1910 through to 1993. *Int. J. Climatol.*, **16**, 1005-1018.

Hughes, M.G., A. Frei and D.A. Robinson, 1996: Historical analysis of North American snow cover extent: merging satellite and station derived snow cover observations. *Proc.1996 Eastern Snow Conf.*, Williamsburg, VA, 21-32.

Hughes, M.K. and H.F. Diaz, 1994: Was there a "Medieval Warm Period" and if so, where and when? *Clim. Change*, **26**, 109-142.

Hughes, M.K. and L.J. Graumlich, 1996: Multimillennial dendroclimatic records from Western North America. In: *Climatic Variations and Forcing Mechanisms of the Last 2000 Years*, R.S. Bradley, P.D. Jones and J. Jouzel (eds.), Springer Verlag, Berlin, pp. 109-124.

Hughes, M.K. and G. Funkhouser, 1999: Extremes of moisture availability reconstructed from tree rings for recent millennia in the Great Basin of Western North America. In: *The Impacts of Climatic Variability on Forests*, Beinston, M. and J. Innes (eds.), Springer-Verlag, Berlin, pp. 99-107.

Hughes, M.K., E.A. Vaganov, S. Shiyatov, R.Touchan and G. Funkhouser, 1999: Twentieth century summer warmth in northern Yakutia in a 600 year context. *The Holocene*, **9**, 603-308.

Hulme, M., 1996: Recent climatic change in the world's drylands. *Geophys. Res. Lett.*, **23**, 61-64.

Hulme, M., T.J. Osborn and T.C. Johns, 1998: Precipitation sensitivity to global warming: Comparison of observations with HadCM2 simulations. *Geophys. Res. Lett.*, **25**, 3379-3382.

Hurrell, J.W., 1995: Decadal trends in the North Atlantic Oscillation regional temperatures and precipitation. *Science*, **269**, 676-679.

Hurrell, J.W., 1996: Influence of variations in extratropical wintertime teleconnections on Northern Hemisphere temperatures. *Geophys. Res. Lett.*, **23**, 665-668.

Hurrell, J.W. and K.E. Trenberth, 1996: Satellite versus surface estimates of air temperature since 1979. *J. Climate*, **9**, 2222-2232.

Hurrell, J.W. and H. van Loon, 1997: Decadal variations in climate associated with the North Atlantic Oscillation. *Clim. Change*, **36**, 301-326.

Hurrell, J.W. and K.E. Trenberth, 1998: Difficulties in obtaining reliable temperature trends: reconciling the surface and satellite Microwave Sounding Unit records. *J. Climate*, **11**, 945-967.

Hurrell, J.W. and K.E. Trenberth, 1999: Global sea surface temperature analyses: multiple problems and their implications for climate analysis, modeling and reanalysis, *Bull. Am. Met. Soc.*, **80**, 2661-2678.

Hurrell, J.W., S.J. Brown, K.E. Trenberth and J.R. Christy, 2000: Comparison of tropospheric temperatures from radiosondes and satellites: 1979-1998. *Bull. Am. Met. Soc.*, **81**, 2165-2177.

IAHS(ICSI)/UNEP/UNESCO, 1998: Fluctuations of the Glaciers, 1990-95. W. Haeberli, M. Hoelzle, S. Suter and R. Frauenfelder (eds.), *World Glacier Monitoring Service*, University and ETH, Zurich.

IAHS(ICSI)/UNEP/UNESCO, 1999: Glacier mass balance bulletin no. 5, W. Haeberli, M. Hoelzle and R. Frauenfelder (eds.), *World Glacier Monitoring Service*, University and ETH, Zurich.

Indeje, M., H.M. Semazzi and L.J. Ogallo, 2000: ENSO signals in East African rainfall seasons. *Int. J. Climatol.*, **20**, 19-46.

IPCC, 1990: Climate Change, The IPPC Scientific Assessment. J.T. Houghton, G.J. Jenkins and J.J. Ephraums (eds.), Cambridge University Press, Cambridge, UK, 365 pp.

IPCC, 1992: Climate Change 1992: The Supplementary Report to the IPCC Scientific Assessment. J.T. Houghton, B.A. Callander and S.K. Varney (eds.), Cambridge University Press, Cambridge, UK, 198 pp.

IPCC, 1996: Climate Change 1995: The Science of Climate Change. Contribution of Working Group I to the Second Assessment Report of the Intergovernmental Panel on Climate Change [Houghton, J.T., L.G. Meira Filho, B.A. Callander, N. Harris, A. Kattenberg, and K. Maskell (eds.)]. Cambridge University Press, Cambridge, United Kingdom and New York, NY, USA, 572 pp.

IPCC, 1999: IPCC Special Report Aviation and the Global Atmosphere. Cambridge University Press, Cambridge, UK, 373 pp.

Iwashima, T. and R. Yamamoto, 1993: A statistical analysis of the extreme events: Long-term trend of heavy daily precipitation. *J. Met. Soc. Japan*, **71**, 637-640.

Jacka, T.H. and W.F. Budd, 1998: Detection of temperature and sea-ice-extent change in the Antarctic and Southern Ocean, 1949-96. *Ann. Glaciol.*, **27**, 553-559.

Jacobs, G.A. and J.L. Mitchell, 1996: Ocean circulation variations associated with the Antarctic Circumpolar Wave. *Geophys. Res. Lett.,* **23**, 2947-2950.

Jacobson, G.L., T. Webb III and E.C. Grimm, 1987: Patterns and rates of vegetation change during deglaciation of eastern North America. In: *North American and Adjacent Oceans during the Last Deglaciation* (eds. W.F. Ruddiman and H.E. Wright), Decade of North American Geology, G.S.A., Boulder, CO, pp. 277-288.

Jacoby, G.C., R.D. D'Arrigo and T. Davaajamts, 1996: Mongolian tree rings and 20th century warming. *Science,* **273**, 771-773.

Janicot, S., V. Moron and B. Fontaine, 1996: Sahel droughts and ENSO dynamics. *Geophys. Res. Lett.,* **23**, 551-554.

Jarrell, J.D., P.J. Hebert and M. Mayfield, 1992: Hurricane experience levels of coastal county populations from Texas to Maine. *NOAA Tech. Memo,* NWS NHC 46, Coral Gables, Florida, USA, 152 pp.

Johannessen, O.M., M.W. Miles and E. Bjørgo, 1995: The Arctic's shrinking sea ice. *Nature,* **376**, 126-127.

Johannessen, O.M., E.V. Shalina and M.W. Miles, 1999: Satellite evidence for an Arctic Sea Ice Cover in Transformation. *Science,* **286**, 1937-1939.

Jóhannesson, T., 1997: The response of two Icelandic glaciers to climate warming computed with a degree-day glacier mass balance model coupled to a dynamic glacier model. *J. Glaciol.,* **43**, 321-327.

Johnsen, S.J., H.B. Clausen, W. Dansgaard, N. Fuhrer, N. Gundestrup, C.U. Hammer, P. Iversen, J. Jouzel, B. Stauffer and J.P. Steffensen, 1992: Irregular glacial interstadials recorded in a new Greenland ice core. *Nature,* **359**, 311-313.

Johnson, G.C. and A.H. Orsi, 1997: Southwest Pacific Ocean water-mass changes between 1968/69 and 1990/91. *J. Climate,* **10**, 306–316.

Jones, E.P., K. Aagaard, E.C. Carmack, R.W. MacDonals, F.A. McLaughlin, R.G. Perkin and J.H. Swift, 1996: Recent Changes in Arctic Ocean Thermohaline Structure: Results from the Canada/US 1994 Arctic Ocean Section. *Mem. Natl. Inst. Polar Res. Special Issue,* **51**, 307-315.

Jones, P.D., 1994: Hemispheric surface air temperature variations: a reanalysis and an update to 1993. *J. Climate,* **7**, 1794-1802.

Jones, P.D., 1999: Classics in physical geography revisited – Manley's CET series. *Progress in Physical Geography,* **23**, 425-428.

Jones, P.D. and K.R. Briffa, 1992: Global surface air temperature variations during the twentieth century: Part 1, spatial temporal and seasonal details. *The Holocene,* **2**, 165-179.

Jones, P.D. and M. Hulme, 1996: Calculating regional climatic time series for temperature and precipitation: methods and illustrations. *Int. J. Climatol.,* **16**, 361-377.

Jones, P.D., P.Ya. Groisman, M. Coughlan, N. Plummer, W.C Wang, and T.R. Karl, 1990: Assessment of urbanization effects in time series of surface air temperature over land. *Nature,* **347**, 169-172.

Jones, P.D., T. Jónsson and D. Wheeler, 1997a: Extension of the North Atlantic Oscillation using early instrumental pressure observations from Gibraltar and south-west Iceland. *Int. J. Climatol.,* **17**, 1433-1450.

Jones, P.D., T.J. Osborn and K.R. Briffa, 1997b: Estimating sampling errors in large-scale temperature averages. *J. Climate,* **10**, 2548-2568.

Jones, P.D., K.R. Briffa, T.P., Barnett and S.F.B. Tett, 1998: High-resolution palaeoclimatic records for the last millennium: interpretation, integration and comparison with General Circulation Model control run temperatures. *The Holocene,* **8**, 455-471.

Jones, P.D., M. New, D.E. Parker, S. Martin and I.G. Rigor, 1999a: Surface air temperature and its changes over the past 150 years. *Rev. Geophys.,* **37**, 173-199.

Jones, P.D., M.J. Salinger and A.B. Mullan, 1999b: Extratropical circulation indices in the Southern Hemisphere. *Int. J. Climatol.,* **19**, 1301-1317.

Jones, P.D., E.B. Horton, C.K. Folland, M. Hulme, D.E. Parker and T.A. Basnett, 1999c: The use of indices to identify changes in climatic extremes. *Clim. Change,* **42**, 131-149.

Jones, P.D., T.J. Osborn, K.R. Briffa, C.K. Folland, E.B. Horton, L.V. Alexander, D.E. Parker and N.A. Rayner, 2001: Adjusting for sampling density in grid box land and ocean surface temperature time series. *J. Geophys. Res.,* **106**, 3371-3380.

Jouzel, J., C. Lorius, J.R. Petit, C. Genthon, N.I. Barkov, V.M. Kotlyakov and V.M. Petrov, 1987: Vostok ice core: a continuous isotope temperature record over the last climatic cycle (160,000 years). *Nature,* **329**, 402-408.

Jouzel, J., N.I. Barkov, J.M. Barnola, M. Bender, J. Bender, J. Chappelaz, C. Genthron, V.M. Kotlyakov, V. Lipenkiv, C. Lorius, J.R. Petit, D. Raynaud, G. Raisbeck, C. Ritz, T. Sowers, M. Stievenard, F. Yiou and P. Yiou, 1993: Extending the Vostok ice core record of paleoclimate to the penultimate glacial period. *Nature,* **364**, 407-412.

Jouzel, J., C. Lorius, S.J. Johnsen and P. Grootes, 1994: Climate instabilities: Greenland and Antarctic records. *C.R. Acad. Sci.* Paris, t 319, série II, pp. 65-77.

Jouzel, J., R. Vaikmae, J.R. Petit, M. Martin, Y. Duclos, M. Stievenard, C. Lorius, M. Toots, M.A. Mélières, L.H. Burckle, N.I. Barkov and V.M. Kotlyakov, 1995: The two-step shape and timing of the last deglaciation in Antarctica. *Clim. Dyn.,* **11**, 151-161.

Jouzel, J., V. Masson, O. Cattani, S. Falourd, M. Stievenard, B. Stenni, A. Longinelli, S.J. Johnson, J.P. Steffenssen, J.R. Petit, J. Schwander and R. Souchez, 2001: A new 27 kyr high resolution East Antarctic climate record. *Geophys. Res. Lett.,* in press.

Joyce, T.M. and P. Robbins, 1996: The long-term hydrographic record at Bermuda. *J. Climate,* **9**, 3121-3131.

Joyce, T.M., R.S. Pickart and R.C. Millard, 1999: Long-term hydrographic changes at 52° and 66°W in the North Atlantic Subtropical Gyre and Caribbean. *Deep-Sea Res., Part II,* **46**, 245-278.

Kagan, R.L., 1997: Averaging of Meteorological Fields. *Translation by UK Ministry of Defence Linguistic Services of original Russian 1979 text.* Eds: L.S. Gandon and T.M. Smith, Kluwer, London, 279 pp.

Kaiser, D.P., 1998: Analysis of total cloud amount over China, 1951-1994. *Geophys. Res. Lett.,* **25**, 3599-3602.

Kaiser, D.P., 2000: Decreasing cloudiness over China! An updated analysis examining additional variables. *Geophys. Res. Lett.,* **27**, 2193-2196.

Kalnay, E., M. Kanamitsu, R. Kistler, W Collins, D. Deaven, I. Gandin, M. Iredell, S. Saha, G. White, J. Woollen, Y. Zhu, M. Chelliah, W. Ebisuzaki, W. Higgins, J. Janowiak, K.C. Mo, C. Ropelewski, J. Wang, A. Leetmaa, R. Reynolds, R. Jenne and D. Joseph, 1996: The NCEP/NCAR 40-year Reanalysis Project. *Bull. Am. Met. Soc.,* **77**, 437-471.

Kaplan, A., Y. Kushnir, M.A. Cane and M. Benno Blumenthal, 1997: Reduced space optimal analysis for historical data sets: 136 years of Atlantic sea surface temperatures. *J. Geophys. Res.,* **102**(C13), 27835-27860.

Kaplan, A., M.A. Cane, Y.A. Kushnir and A.C. Clement, 1998: Analyses of global sea surface temperature, 1856-1991. *J. Geophys. Res.,* **103**(C9), 18567-18589.

Karl, T.R. and P.M. Steurer, 1990: Increased cloudiness in the United States during the first half of the twentieth century: fact or fiction? *Geophys. Res. Lett.,* **17**, 1925-1928.

Karl, T.R. and R.W. Knight, 1997: The 1995 Chicago heat wave: How likely is a recurrence? *Bull. Am. Met. Soc.,* **78**, 1107-1119.

Karl, T.R. and R.W. Knight, 1998: Secular trends of precipitation amount, frequency, and intensity in the USA. *Bull. Am. Met. Soc.,* **79**, 231-241.

Karl, T.R., V.E. Derr, D.R. Easterling, C.K. Folland, D.J. Hofmann, S. Levitus, N. Nicholls, D.E. Parker and G.W. Withee, 1995a: Critical issues for long-term climate monitoring. *Clim. Change,* **31**, 185-221, 1995a and In: *Long-term climate Monitoring by the Global Climate Observing System,* T. Karl (ed.), Kluwer, Dordrecht, pp. 55-91.

Karl, T.R., R.W. Knight and N. Plummer, 1995b: Trends in high-frequency climate variability in the twentieth century. *Nature*, **377**, 217-220.

Karl, T.R., R.W. Knight, D.R. Easterling and R.G. Quayle, 1995c: Trends in U.S. climate during the Twentieth Century. *Consequences*, **1**, 3-12.

Karl, T.R., R.W. Knight, and B. Baker, 2000: The record breaking global temperatures of 1997 and 1998: evidence for an increase in the rate of global warming? *Geophys. Res. Lett.*, **27**, 719-722.

Karoly, D.J., 1990: The role of transient eddies in low-frequency zonal variations of the Southern Hemisphere circulation. *Tellus,* **42A**, 41-50.

Karoly, D.J., P. Hope and P.D. Jones, 1996: Decadal variations of the Southern Hemisphere circulation. *Int. J. Climatol.*, **16**, 723-738.

Katz, R.W., 1999: Extreme value theory for precipitation: Sensitivity analysis for climate change. *Advances in Water Resources*, **23**, 133-139.

Kawamura, R., 1994: A rotated EOF analysis of global sea surface temperature variability with interannual and interdecadal scales. *J. Phys. Oceanogr.*, **24**, 707-715.

Keckhut, P., F.J. Schmidlin, A. Hauchecorne and M.-L. Chanin, 1999: Stratospheric and mesospheric cooling trend estimates from US rocketsondes at low latitude stations (8°S-34°N), taking into account instrumental changes and natural variability. *J. Atmos. And Solar-Terr. Phys.*, **61**, 447-459.

Keigwin, L., 1996: The Little Ice Age and Medieval Warm Period in the Sargasso Sea. *Science*, **274**, 1504-1508.

Keigwin, L.D. and R.S. Pickart, 1999: Slope water current over the Laurentian Fan on Interannual to Millennial Time Scales. *Science*, **286**, 520-523.

Kershaw, A.P., D.M. D'Costa, J.R.C.M. Mason and B.E. Wagstaff, 1991: Palynological evidence for Quaternary vegetation and environments of Mainland Southeastern Australia. *Quat. Sci. Rev.*, **10**, 391-404.

Kestin, T.S., D.J. Karoly, J.I. Jano and N.A. Rayner, 1999: Time-frequency variability of ENSO and stochastic simulations. *J. Climate*, **11**, 2258-2272.

Kidson, J.W., 1988: Interannual variations in the Southern Hemisphere circulation. *J. Climate*, **1**, 1177-1198.

Kidson, J.W., 1999: Principal modes of Southern Hemisphere low frequency variability obtained from NCEP/NCAR reanalyses. *J. Climate*, **12**, 2808-2830.

Kiladis, G.N. and K.C. Mo, 1999: Interannual and intraseasonal variability in the Southern Hemisphere, Chapter 8. In: *Meteorology of the Southern Hemisphere*, American Meteorological Society, Boston.

King, J.C., 1994: Recent climate variability in the vicinity of the Antarctic Peninsula. *J. Climate*, **14**, 357-361.

Kley, D., J.M. Russell and C. Phillips (eds.), 2000: SPARC Assessment of upper tropospheric and tratospheric water vapour. WCRP-No. 113, WMO/TD-No. 1043, SPARC Report No.2, 325 pp.

Klitgaard-Kristensen, D., H.P. Sejrup, H. Haflidason, S. Johnsen and M. Spurk, 1998: The short cold period 8,200 years ago documented in oxygen isotope records of precipitation in Europe and Greenland. *J. Quaternary Sciences*, **13**, 165-169.

Knight, R.W., 1984: Introduction to a new sea-ice database. *Ann. Glaciol.*, **5**, 81-84.

Koç, N. and E. Jansen, 1994: Response of the high-latitude Northern-Hemisphere to orbital climate forcing-evidence from the Nordic seas. *Geology*, **22**, 523-526.

Kotilainin, A.T. and N.J. Shackleton, 1995: Rapid climate variability in the North Pacific Ocean during the past 95,000 years. *Nature*, **377**, 323-326.

Kumar, K., K. Rupa, K.K. Kumar and G.B. Pant, 1994: Diurnal asymmetry of surface temperature trends over India. *Geophys. Res. Lett.*, **21**, 677-680.

Kumar, K.K., R. Kleeman, M.A. Crane and B. Rajaopalan, 1999a: Epochal changes in Indian monsoon-ENSO precursors. *Geophys. Res. Lett.*, **26**, 75-78.

Kumar, K.K., B. Rajaopalan and M.A. Crane, 1999b: On the weakening relationship between the Indian monsoon and ENSO. *Science*, **284**, 2156-2159.

Kunkel, K.E., S.A. Changnon, B.C. Reinke and R.W. Arritt, 1996: The July 1995 heat wave in the Midwest: A climatic perspective and critical weather factors. *Bull. Am. Met. Soc.*, **77**, 1507-1518.

Kunkel, K.E., K. Andsager and D.R. Easterling, 1999: Long-term trends in extreme precipitation events over the conterminous United States and Canada. *J. Climate*, **12**, 2515-2527.

Kushnir, Y., V.J. Cardon, J.G. Greenwood and M.A. Cane, 1997: The recent increase in North Atlantic wave heights. *J. Climate*, **10**, 2107-2113.

Laberge, M.J. and S. Payette, 1995: Long-term monitoring of permafrost change in a palsa peatland in northern Quebec, Canada: 1983-1993. *Arctic and Alpine Research*, **27**, 167-171.

Labeyrie, L., J.C. Duplessy, J. Duprat, A. Juillet-Leclerc, J. Moyes, E. Michel, N. Kallel and N.J. Shackleton, 1992: Changes in vertical structure of the North Atlantic Ocean between glacial and modern times. *Quat. Sci. Rev.*, **11**, 401-413.

Lachenbruch, A.H. and B.V. Marshall, 1986: Changing climate: geothermal evidence from permafrost in the Alaskan Arctic. *Science*, **234**, 689-696.

Lachenbruch, A.H., T.T Cladouhos and R.W. Saltus, 1988: Permafrost temperature and the changing climate. In: *Proceedings of the Fifth International Conference on Permafrost*, Trondheim, Norway, 3, 9-17.

Laird, K.R., S.C. Fritz, K.A. Maasch and B.F. Cumming, 1996: Greater Drought Intensity and frequency before AD 1200 in the Northern Great Plains. *Nature*, **384**, 552-554.

Lamb, H.F., F. Gasse, A. Bekaddour, N. El Hamouti, S. van der Kaars, W.T. Perkins, N.J. Pearce and C.N. Roberts, 1995: Relation between century-scale Holocene arid intervals in temperate and tropical zones. *Nature*, **373**, 134-137.

Lambert, S.J., 1996: Intense extratopical Northern Hemisphere winter cyclone events: 1189-1991. *J. Geophys. Res.*, **101**, 21319-21325.

Lamoureux, S.F. and R.S. Bradley, 1996: A 3300 year varved sediment record of environmental change from northern Ellesmere Island, Canada. J. *Paleolimnology*, **16**, 239-255.

Lander, M.A. and C.P. Guard, 1998: A look at global tropical cyclone activity during 1995: Contrasting high Atlantic activity with low activity in other basins. *Mon. Wea. Rev.*, **126**, 1163-1173.

Landsea, C.W., 1993: A climatology of intense (or major) Atlantic hurricanes. *Mon. Wea. Rev.*, **121**, 1703-1713.

Landsea, C.W., N. Nicholls, W.M. Gray and L.A. Avila, 1996: Downward trends in the frequency of intense Atlantic hurricanes during the past five decades. *Geophys. Res. Lett.*, **23**, 1697-1700.

Landsea, C.W., R.A. Pielke, Jr., A.M. Mestas-Nunez and J.A. Knaff, 1999: Atlantic basin hurricanes: Indices of climatic changes. *Clim. Change*, **42**, 89-129.

Lang, C., M. Leuenberger, J. Schwander and J. Johnsen, 1999: 16°C rapid temperature variation in central Greenland 70000 years ago. *Science*, **286**, 934-937.

Latif, M. and T.P. Barnett, 1994: Causes of decadal climate variability over the North Pacific and North America. *Science*, **266**, 634-637.

Lawrimore, J.H. and T.C. Peterson, 2000: Pan evaporation trends in dry and humid regions of the United States. *J. Hydrometeor.*, **1**, 543-546.

Lazier, J.R.N., 1995: The salinity decrease in the Labrador Sea over the past thirty years. In: *Natural Climate Variability on Decade-to-Century Time Scales*, D.G. Martinson, K. Bryan, M. Ghil, M.M. Hall, T. Karl, E.S. Sarachik, S. Sorooshian, and L.D. Talley (eds.), National Academy Press, Washington, D.C., pp. 295-305.

Lean, J., J. Beer and R.S. Bradley, 1995: Reconstruction of solar irradiance since 1610: Implications for climatic change. *Geophys. Res. Lett.*, **22**, 3195-3198.

Leathers, D.J. and A.W. Ellis, 1996: Synoptic mechanisms associated with snowfall increases to the lee of Lakes Erie and Ontario. *Int. J. Climatol.*, **16**, 1117-1135.

Lettenmaier, D.P., A.W. Wood, R.N. Palmer, E.F. Wood and E.Z. Stakhiv, 1999: Water resources implications of global warming: A U.S. regional perspective. *Clim. Change*, **43**, 537-579.

Levitus, S. and J. Antonov, 1997: Variability of heat storage of and the rate of heat storage of the world ocean. NOAA NESDIS Atlas 16, US Government Printing Office, Washington, D.C., 6 pp., 186 figures.

Levitus, S., R. Gelfeld, T. Boyer and D. Johnson, 1994: Results of the NODC and IOC Data Archaeology and Rescue projects In: *Key to Oceanographic Records Documentation No. 19*, National Oceanographic Data Center, Washington, D.C., 67 pp.

Levitus, S., R. Gelfeld, M. E. Conkright, T. Boyer, D. Johnson, T. O'Brien, C. Stephens, C. Forgy, O. Baranova, I. Smolyar, G. Trammell and R. Moffatt, 2000a: Results of the NODC and IOC Data Archaeology and Rescue projects. In: *Key to Oceanographic Records Documentation No. 19*, National Oceanographic Data Center, Washington, D.C., 19 pp.

Levitus, S., J.Antonov, T.P. Boyer and C. Stephens, 2000b: Warming of the World Ocean. *Science*, **287**, 2225-2229.

Lewis, T., 1998: The effect of deforestation on ground surface temperatures. *Global and Planetary Change*, **18**, 1-13.

Lins, H.F. and P.J. Michaels, 1994: Increasing U.S. streamflow linked to greenhouse forcing. *Eos Trans. AGU*, **75**, 281, 284-285.

Lins, H F. and J.R. Slack, 1999: Streamflow trends in the United States. *Geophys. Res. Lett.*, **26**, 227-230.

Livezey, R.E. and T.M. Smith, 1999: Covariability of aspects of North American climate with global sea surface temperatures on interannual to interdecadal timescales. *J. Climate*, **12**, 289-302.

Luterbacher, J., C. Schmutz, D. Gyalistras, E. Xoplaki and H. Wanner, 1999: Reconstruction of monthly NAO and EU indices back to A.D. 1675. *Geophys. Res. Lett.*, **26**, 759-762.

Lysenko, E.V., G. Nelidova and A. Prostova, 1997: Changes in the stratospheric and mesospheric thermal conditions during the last 3 decades: 1. The evolution of a temperature trend. *Isvestia, Atmos. and Oceanic Physics*, **33**(2), 218-225.

MacManus, J., D.W. Oppo and J.L.Cullen, 1999: A 0.5 Million-Year Record of Millenial scale climate variability in the North Atlantic. *Science*, **283**, 971-975.

Magnuson, J.J., D.M. Robertson, B.J. Benson, R.H. Wynne, D.M. Livingston, T. Arai, R.A. Assel, R.G. Barry, V. Card, E. Kuusisto, N.G. Granin, T.D. Prowse, K.M. Stewart and V.S. Vuglinski, 2000: Historical trends in lake and river ice cover in the Northern Hemisphere. *Science*, **289**, 1743-1746.

Magny, M., 1995: Successive oceanic and solar forcing indicated by Younger Dryas and early Holocene climatic oscillations in the Jura. *Quat. Res.*, **43**, 279-285.

Mahowald, N., K.E. Kohfeld, M. Hansson, Y. Balkanski, S.P. Harrison, I.C. Prentice, M. Schulz and H. Rodhe, 1999: Dust sources and deposition during the last glacial maximum and current climate: a comparison of model results with paleodata from ice cores and marine sediments. *J. Geophys. Res.*, **104**, 15895-15916.

Majorowicz, J.A. and A. Judge, 1994: Climate induced ground warming at the southern margins of permafrost. *EOS, Transactions*, American Geophysical Union, **75**(44), 84.

Maloney, E.D. and D.L. Hartmann, 2001: The Madden-julian Oscillation, Barotropic Dynamics, and North Pacific Tropical Cyclone Formation, Part I: Observations. *J. Atmos. Sci.*, in press.

Mann, M.E., 2000: Lessons for a New Millennium. *Science*, **289**(14), 253-254.

Mann, M.E. and J. Park, 1994: Global-scale modes of surface temperature variability on interannual to century timescales. *J. Geophys. Res.*, **99**, 25819-25833.

Mann, M.E. and J. Park, 1996: Joint spatiotemporal modes of surface temperature and sea level pressure variability in the Northern Hemisphere during the last century. *J. Climate*, **9**, 2137-2162.

Mann, M.E., J. Park and R.S. Bradley, 1995: Global interdecadal and century-scale oscillations during the past five centuries. *Nature*, **378**, 266-270.

Mann, M.E., R.S. Bradley and M.K. Hughes, 1998: Global-scale temperature patterns and climate forcing over the past six centuries. *Nature*, **392**, 779-787.

Mann, M.E., R.S. Bradley, and M.K. Hughes, 1999: Northern Hemisphere Temperatures During the Past Millennium: Inferences, Uncertainties, and Limitations. *Geophys. Res. Lett.*, **26**, 759-762.

Mann, M.E., R.S. Bradley and M.K. Hughes, 2000a: Long-term variability in the El Nino Southern Oscillation and associated teleconnections. In: *El Nino and the Southern Oscillation: Multiscale Variability and its Impacts on Natural Ecosystems and Society*, H.F. Diaz and V. Markgraf (eds.), Cambridge University Press, Cambridge, UK, 357-412.

Mann, M.E., E. Gille, R.S. Bradley, M.K. Hughes, J.T. Overpeck, F.T. Keimig and W. Gross, 2000b: Global temperature patterns in past centuries: An interactive presentation. *Earth Interactions*, **4/4**, 1-29.

Manton, M.J., P.M. Della-Marta, M.R. Haylock, K.J. Hennessy, N. Nicholls, L.E. Chambers, D.A. Collins, G. Daw, A. Finet, D. Gunawan, K. Inape, H. Isobe, T.S. Kestin, P. Lafale, C.H. Leyu, T. Lwin, L. Maitrepierre, N. Ouprasitwong, C.M. Page, J. Pahalad, N. Plummer, M.J. Salinger, R. Suppiah, V.L. Tran, B. Trewin, I. Tibig and D. Yee, 2001: Trends in extreme daily rainfall and temperature in Southeast Asia and the South Pacific: 1961-1998. *Int. J. Climatol.*, **21**, 269-284.

Mantua, N.J., S.R. Hare, Y. Zhang, J.M. Wallace and R.C. Francis, 1997: A Pacific interdecadal climate oscillation with impacts on salmon production. *Bull. Am. Met. Soc.*, **78**, 1069-1079.

Marengo, J.A., J. Tomasella and C.R. Uvo, 1998: Trends in streamflow and rainfall in tropical South America: Amazonia, Eastern Brazil and Northwestern Peru. *J. Geophys. Res.*, **103**, 1775-1783.

Martinson, D.G., N.G. Pisias, J.D. Hays, J. Imbrie, T.C. Moore and N.J. Shackleton, 1987: Age Dating and the Orbital Theory of the Ice Ages: Development of a High-Resolution 0-300,000 Years Chronostratigraphy. *Quat. Res.*, **27**, 1-30.

Martin-Vide, J. and M. Barriendos, 1995: The use of rogation ceremony records in climatic reconstruction: A case study from Catalonia (Spain). *Clim. Change*, **30**, 201-221.

Maslanik, J.A., M.C. Serreze and R.G. Barry, 1996: Recent decreases in Arctic summer ice cover and linkages to atmospheric circulation anomalies. *Geophys. Res. Lett.*, **23**, 1677-1680.

Masson, V., F. Vimeux, J. Jouzel, V. Morgan, M. Delmotte, C. Hammer, S.J. Johnsen, V. Lipenkov, J.R. Petit, E. Steig, M. Stievenard and R. Sousmis Vaikmae, 2000: Holocene climate variability in Antarctica based on 11 ice-core isotopic records. *Quat. Res.*, **54**, 348-358.

Mastenbrook, H.J., 1968: Water vapor distribution in the stratosphere and high troposphere. *J. Atmos. Sci.*, **25**, 299-311.

Mastenbrook, H.J. and S. Oltmans, 1983: Stratospheric water vapor variability for Washington D.C./Boulder, CO.: 1964-1982. *J. Atmos. Sci.*, **40**, 2157-2165.

Mayewski, P.A. and I.D. Goodwin, 1997: International Trans-Antarctic Scientific Expedition (ITASE). *PAGES/SCAR Workshop Report Series, 97-1*.Bern Switzerland, 48 pp.

Mayewski, P.A., L.D. Meeker, M.S. Twickler, S. Whitlow, Y. Qinzhao, W.B. Lyons and M. Prentice, 1997: Major features and forcing of high-latitude northern hemisphere atmospheric circulation using a 110,000-year-long glaciochemical series. *J. Goephys. Res.*, **102**, 26345-26366.

McGlone, M.S., A.P. Kershaw and V. Markgraf, 1992: El Niño/Southern Oscillation and climatic variability in Australasian and South American paleoenvironmental records. In: *El Niño: Historical and paleoclimatic aspects of the Southern Oscillation*, H.F. Diaz and V. Markgraf (eds.), Cambridge, Cambridge University Press, pp. 435-

462.

McManus, J.F., G.C. Bond, W.S. Broecker, S. Johnsen, L. Labeyrie and S. Higgins, 1994: High-resolution climate records from the North Atlantic during the last interglacial. *Nature*, **317**, 326-329.

McPhee, M.G., T.P. Stanton, J.H. Morison and D.G. Martinson, 1998: Freshening of the upper ocean in the Arctic: Is perennial sea ice disappearing? *Geophys. Res. Lett.*, **25**, 1729-1732.

Meehl, G.A., J. Arblaster and W. Strand, 1998: Global decadal scale climate variability. *Geophys. Res. Lett.*, **25**, 3983-3986.

Meehl, G.A., T. Karl, D.R. Easterling, S. Changnon, R. Pielke, Jr., D. Changnon, J. Evans, P.Ya. Groisman, T.R. Knutson, K.E. Knukel, L.O. Mearns, C. Parmesan, R. Pulwarty, T. Root, R.T. Sylves, P. Whetton and F. Zwiers, 2000: An introduction to trends in extreme weather and climate events: Observations, socioeconomic impacts, terrestrial ecological impacts, and model projections. *Bull. Am. Met. Soc.*, **81**, 413-416.

Meese, D.A. and 13 others, 1994: The accumulation record from the GISP2 core as an indicator of climate change throughout the Holocene. *Science*, **266**, 1680-1682.

Mekis, E. and W.D. Hogg, 1999: Rehabilitation and analysis of Canadian daily precipitation time series. *Atmosphere-Ocean*, **37**(1), 53-85.

Melling, H., 1998: Hydrographic changes in the Canada Basin of the Arctic Ocean, 1979-1996. *J. Geophys. Res.*, **103**(C4), 7637-7645.

Meshcherskaya, A.V., I.G. Belyankina and M.P. Golod, 1995: Monitoring tolshching cnozhnogo pokprova v osnovioi zerno proizvodyashchei zone Byvshego SSSR za period instrumental'nykh nablyugenii. Izvestiya Akad. Nauk SSR, *Sser. Geograf.*, pp. 101-110.

Michaels, P.J. and P.C. Knappenberger, 2000: Natural signals in the MSU lower tropospheric temperature record. *Geophys. Res. Lett.*, **27**, 2905-2908.

Michaels, P.J., R.C. Balling, Jr. , R.S. Vose and P.C. Knappenberger, 1998: Analysis of trends in the variability of daily and monthly historical temperature measurements. *Clim. Res.*, **10**, 27-33.

Michaels, P.J., P.C. Knappenberger, R.C. Balling Jr. and R.E. Davis, 2000: Observed warming in cold anticyclones. *Clim. Res.*, **14**, 1-6.

Minobe, S., 1997: A 50-70 year climatic oscillation over the North Pacific and North America. *Geophys. Res. Lett.*, **24**, 683-686.

Mo, K.C. and R.W. Higgins, 1998: The Pacific South American modes and tropical convection during the Southern Hemisphere winter. *Mon. Wea. Rev.*, **126**, 1581-1596.

Mo, T., 1995: A study of the Microwave Sounding Unit on the NOAA-12 satellite. *IEEE Trans. Geoscience and Remote Sensing*, **33**, 1141-1152.

Moberg, A., P.D. Jones, M. Barriendos, H. BergstrØm. D. Camuffo, C. Cocheo, T.D. Davies, G. Demar?e, J. Martin-Vide, M. Maugeri, R. Rodriquez and T. Verhoeve, 2000: Day-to-day temperature variability trends in 160-275-year long European instrumental records. *J. Geophys. Res.*, **105**(D18), 22849-22868.

Morgan, V.I. and T.D. van Ommen, 1997: Seasonality in late-Holocene climate from ice core records. *The Holocene*, **7**, 351-354.

Moron, V., 1997: Trend, decadal and interannual variability in annual rainfall in subequatorial and tropical North Africa (1900-1994). *Int. J. Climatol.*, **17**, 785-806.

Moron, V., R. Vautard and M. Ghil, 1998: Trends, Interdecadal and interannual oscillations in global sea-surface temepratures. *Clim. Dyn.*, **14**, 545-569.

Mysak, L.A. and S.A. Venegas, 1998: Decadal climate oscillations in the Arctic: a new feedback loop for atmospheric-ice-ocean interactions. *Geophys. Res. Lett.*, **25**, 3607-3610.

Nagurnyi, A.P., V.G. Korostelev and P.A. Abaza, 1994: Wave method for evaluating the effective thickness of sea ice in climate monitoring. *Bulletin of the Russian Academy of Sciences, Physics Supplement, Physics of Vibrations*, pp. 168-241.

Nagurnyi, A.P., V.G. Korostelev and V.V. Ivanov, 1999: Multiyear variability of sea ice thickness in the Arctic Basin measured by

elastic-gravity oscillation of the ice cover. *Meteorologiya I gidrologiya*, **3**, 72-78.

Nash, J. and G.F. Forrester, 1986: Long-term monitoring of stratospheric temperature trends using radiance measurements obtained by the TIROS-N series of NOAA spacecraft. *Adv. Space Res.*, **6**, 37-44.

National Climatic Data Center (NCDC), 1997: Products and Services Guide, Asheville, NC: US Department of Commerce, NOAA, 60 pp.

Neff, W.D., 1999: Decadal time scale trends and variability in the tropospheric circulation over the South Pole. *J. Geophys. Res.*, **104**(D22), 27217-27251.

Nelson, F.E. and J. Brown, 1997: Global change and permafrost. *Frozen Ground*, **21**, 21-24.

Nelson, F.E., K.M. Hinkel, N.I. Shiklomanov, G.R. Mueller, L.L. Miller and D.A. Walker, 1998: Active-layer thickness in north-central Alaska: systematic sampling, scale, and spatial autocorrelation. *J. Geophys. Res.*, **103**, 28963-28973.

Neumann, C.J., 1993: Global Overview, Global Guide to Tropical Cyclone Forecasting. WMO/TC-No. 560, Report No. TCP-31, World Meteorological Organization, Geneva, pp. 1.1-1.43.

New, M., M. Hulme and P.D. Jones, 2000: Representing twentieth century space-time climate variability, II: Development of 1901-1996 monthly grids of terrestrial surface climate. *J. Climate*, **13**, 2217-2238.

Nicholls, N. and W. Murray, 1999: Workshop on Indices and Indicators for climate extremes, Asheville, NC, USA, 3-6 June 1999. Breakout Group B: Precipitation. *Clim. Change*, **42**, 23-29.

Nicholls, N., C.W. Landsea and J. Gill, 1998: Recent trends in Australian region tropical cyclone activity. *Met. Atmos. Phys.*, **65**, 197-205.

Nicholson, S.E., 1993: An overview of African rainfall fluctuations of the last decade. *J. Climate*, **6**, 1463-1466.

Nicholson, S.E., 1997: An analysis of the ENSO signal in the tropical Atlantic and western Indian oceans. *Int. J. Climatol.*, **17**, 345-375.

Nicholson, S.E. and J. Kim, 1997: The relationship of the El Nino-Southern Oscillation to African rainfall. *Int. J. Climatol.*, **17**, 117-135.

Nicholson, S.E., B. Some and B. Kane, 2000: An analysis of recent rainfall conditions in west Arica, including the rainy seasons of the 1997 El Niño and the 1998 La Niña years. *J. Climate*, **13**, 2628-2640.

Nixon, F.M. and A.E. Taylor, 1998: Regional active layer monitoring across the sporadic, discontinuous and continuous permafrost zones, Mackenzie Valley, northwestern Canada. *In: Proceedings of the Seventh International Conference on Permafrost* (Lewcowicz, A.G. and M. Allard (eds.)) Centre d'Etudes Nordiques, Université Laval, Québec, pp. 815-820.

Norris, J.R., 1999: On trends and possible artifacts in global ocean cloud cover between 1952 and 1995. *J. Climate*, **12**, 1864-1870.

Norris, J.R. and C.B. Leovy, 1995: Comments on "Trends in global marine cloudiness and anthropogenic sulphur". *J. Climate*, **8**, 2109-2110.

Norris, J.R., Y. Zhang and J.M. Wallace, 1998: Role of low clouds in summertime atmosphere-ocean interactions over the North Pacific. *J. Climate*, **11**, 2482-2490.

NRC (National Research Council), 2000: Reconciling Observations of Global Temperature Change. National Academy Press, Washington D.C., 85 pp.

O'Brien, S., P.A. Mayewski, L.D. Meeker, D.A. Meese, M.S. Twickler and S.I. Whitlow, 1995: Complexity of Holocene climate as reconstructed from a Greenland ice core. *Science*, **270**, 1962-1964.

Oerlemans, J., 1989: On the response of valley glaciers to climatic change. In: *Glacier fluctuations and climatic change*, J. Oerlemans, (ed.), Dordrecht, Kluwer Academic, pp. 353-372.

Oerlemans, J., 1992: Climate sensitivity of glaciers in southern Norway: application of an energy-balance model to Nigardsbreen, Hellstugubreen and Alfotbreen. *J. Glaciol.*, **38**, 223-232.

Oerlemans, J., 1994: Quantifying global warming from the retreat of glaciers. *Science*, **264**, 243-245.

Oerlemans, J., B. Anderson, A. Hubbard, P. Huybrechts, T. Jóhannesson, W.H. Knap, M. Schmeits, A.P. Stroeven, R.S.W. van de Wal, J. Wallinga and Z. Zuo, 1998: Modelling the response of glaciers to climate warming. *Clim. Dyn.*, **14**, 267-274.

Oeschger, H., J. Beer, U. Siegenthaler, B. Stauffer, W. Dansgaard and C.C. Langway, 1984: Late glacial climate history from ice cores. In: *Climate processes and climate sensitivity*, J.E. Hansen and T. Takahashi (eds.), American Geophysical Union, Washington.

Oltmans, S.J. and D.J. Hofmann, 1995: Increase in lower-stratospheric water vapour at a mid-latitude Northern Hemisphere site from 1981-1994. *Nature*, **374**, 146-149.

Oltmans, S.J., S.H. Voemel, D. Hofmann, K. Rosenlof and D. Kley, 2000: The increase in stratospheric water vapor from balloon-borne, frostpoint hygrometer measurements at Washington, DC and Boulder, Colorado. *Geophys. Res. Lett.*, **27**, 3453-3456.

Osborn, T.J., K.R. Briffa, S.F.B. Tett and P.D. Jones, 1999: Evaluation of the North Atlantic Oscillation as simulated by a coupled climate model. *Clim. Dyn.*, **15**, 685-702.

Osborn, T.J., M. Hulme, P.D. Jones and T.A, Basnett, 2000: Observed trends in the daily intensity of United Kingdom precipitation. *Int. J. Climatol.*, **20**, 347-364.

Osterkamp, T.E., 1994: Evidence for warming and thawing of discontinuous permafrost in Alaska. *EOS, Transactions*, American Geophysical Union, **75**, 85.

Osterkamp, T.E. and V.E. Romanovsky, 1999: Evidence for warming and thawing of discontinuous permafrost in Alaska. *Permafrost and Periglacial Processes*, **10**(1), 17-37.

Østrem, G., O. Liestøl and B. Wold, 1977: Glaciological investigations at Nigardsbreen, Norway. *Norsk Geogr. Tidsskr.*, **30**, 187-209.

Overpeck, J.T., P.J. Bartlein and T. Webb III, 1991: Potential magnitude of future vegetation change in eastern North America: Comparisons with the past. *Science*, **252**, 692-695.

Overpeck, J., K. Hughen, D. Hardy, R. Bradley, R. Case, M. Douglas, B. Finney, K. Gajewski, G. Jacoby, A. Jennings, S. Lamoureux, A. Lasca, G. MacDonald, J. Moore, M. Retelle, S. Smith, A. Wolfe and G. Zielinski, 1997: Arctic environmental change of the last four centuries. *Science*, **278**, 1251-1256.

Palecki, M.A. and R.G. Barry, 1986: Freeze-up and break-up of lakes as an index of temperature changes during the transition seasons: A case study for Finland. *J. Clim. Appl. Met.*, **25**, 893-902.

Palmer, T.N., 1993: A nonlinear dynamical perspective on climate change. *Weather*, **48**, 313-326.

Palmer, T.N., 1999: A nonlinear dynamical perspective on climate prediction. *J. Climate*, **12**, 575-591.

Pant, G.B. and K.R. Kumar, 1997: Climates of South Asia, John Wiley, Chichester, 320pp.

Parker, D.E., 1994: Effects of changing exposures of thermometers at land stations. *Int. J. Climatol.*, **14**, 102-113.

Parker, D.E. and C.K. Folland, 1988: The nature of climatic variability. *Met. Mag.*, **117**, 201-210.

Parker, D.E. and D.I. Cox, 1995: Towards a consistent global climatological rawinsonde data-base. *Int. J. Climatol.*, **15**, 473-496.

Parker, D.E. and E.B. Horton, 1999: Global and regional climate in 1998. *Weather*, **54**, 173-184.

Parker, D.E. and E.B. Horton, 2000: Global and regional climate in 1999. *Weather*, **55**, 188-199.

Parker, D.E., P.D. Jones, C.K. Folland and A. Bevan, 1994: Interdecadal changes of surface temperature since the late nineteenth century. *J. Geophys. Res.*, **99**, 14373-14399.

Parker, D.E., C.K. Folland and M. Jackson, 1995: Marine surface temperature: observed variations and data requirements. *Clim. Change*, **31**, 559-600.

Parker, D.E., H. Wilson, P.D. Jones, J. Christy and C.K. Folland, 1996: The impact of Mount Pinatubo on climate. *Int. J. Climatol.*, **16**, 487-

497.

Parker, D.E., M. Gordon, D.P.N. Cullum, D.M.H. Sexton, C.K. Folland and N. Rayner, 1997: A new global gridded radiosonde temperature data base and recent temperature trends. *Geophys. Res. Lett.*, **24**, 1499-1502.

Parkinson, C.L., 2000: Variability of Arctic sea ice. The view from space, an 18-year record. *Arctic*, **53**, 341-358.

Parkinson, C.L., D.J. Cavalieri, P. Gloersen, H.J. Zwally and J.C. Comiso, 1999: Arctic sea ice extents, areas, and trends, 1978-1996. *J. Geophys. Res.*, **104**(C9), 20837-20856.

Parrilla, G., A. Lavín, H.L. Bryden, M.J. García and R. Millard, 1994: Rising temperatures in the subtropical North Atlantic Ocean over the past 35 years. *Nature,* **369**, 48-51.

Partridge, T.C., 1997: Cainozoic environmental change in southern Africa, with special emhpasis on the last 200,000 years. *Progress in Physical Geography*, **21**, 3-22.

Parungo, F., J.F. Boatman, H. Sievering, S.W. Wilkison and B.B. Hicks, 1994: Trends in global marine cloudiness and anthropogenic sulphur. *J. Climate*, **7**, 434-440.

Pavlov, A.V., 1994: Current changes of climate and permafrost in the Arctic and Sub-Arctic of Russia. *Permafrost and Periglacial Processes*, **5**, 101-110.

Pavlov, A.V., 1998: Active layer monitoring in Northern West Siberia. *Proceedings of the Seventh International Conference on Permafrost*, Yellowknife, Canada, June 1998, Université Laval, Quebec, Collection Nordicana No. 57, pp. 875-881.

Pazdur, A., M.R. Fontugne and T. Goslar, 1995: Late glacial and Holocene water-level changes of the Gosciaz Lake, central Poland, derived from carbon-isotope studies of laminated sediment. *Quat. Sci. Rev.*, **14**, 125-135.

Peel, D.A., R. Mulvaney, E.C. Pasteur and C. Chenery, 1996: Climate changes in the Atlantic Sector of Antarctica over the past 500 years from ice-core and other evidence. In: *Climate Variations and Forcing Mechanisms of the Last 2000 Years*. NATO ASI Series I vol 41, P.D. Jones, R.S. Bradley and J. Jouzel (eds.), pp. 243-262.

Peixoto, J.P. and A.H. Oort, 1996: The climatology of relative humidity in the atmosphere. *J. Climate*, **9**, 3443-3463.

Peterson, J.A. and L.F. Peterson, 1994: Ice retreat from the neoglacial maxima in the Puncak Jayakesuma area, Republic of Indonesia. *Z. Gletscherkd. Glazialgeol.*, **30**, 1-9.

Peterson, T.C. and R.S.Vose, 1997: An overview of the global historical climatology network temperature data base. *Bull. Am. Met. Soc.*, **78**, 2837-2849.

Peterson, T.C., H. Daan, and P.D. Jones, 1997: Initial selection of a GCOS surface network. *Bull. Am. Met. Soc.*, **78**, 2145-2152.

Peterson, T.C., T.R. Karl, P.F. Jamason, R. Knight and D.R. Easterling, 1998a: The first difference method: maximizing station density for the calculation of long-term temperature change. *J. Geophys. Res. - Atmos.*, **103**, 25967-25974.

Peterson, T.C., D.R. Easterling, T.R. Karl, P. Groisman, N. Nicholls, N. Plummer, S. Torok, I. Auer, R. Boehm, D. Gullett, L. Vincent, R. Heino, H. Tuomenvirta, O. Mestre, T. Szentimrey, J. Salinger, E.J. Førland, I. Hanssen-Bauer, H. Alexandersson, P. Jones and D. Parker, 1998b: Homogeneity adjustments of in situ atmospheric climate data: a review. *Int. J. Climatol.*, **18**, 1495-1517.

Peterson, T.C., K.P. Gallo, J. Livermore, T.W. Owen, A. Huang and D.A. McKittrick, 1999: Global rural temperature trends. *Geophys. Res. Lett.*, **26**, 329-332.

Petit, J.R., J. Jouzel, D. Raynaud, N.I. Barkov, J.M. Barnola, I. Basile, M. Bender, J. Chappellaz, J. Davis, G. Delaygue, M. Delmotte, V.M. Kotyakov, M. Legrand, V.Y. Lipenkov, C. Lorius, L. Pepin, C. Ritz, E. Saltzman and M. Stievenard, 1999: Climate and Atmospheric History of the Past 420,000 years from the Vostok Ice Core, Antarctica. *Nature*, **399**, 429-436.

Petit-Maire and Z.T. Guo, 1996: Mise en evidence de variations cimatiques, holocenes rapides, en phase dans les deserts actuels de Chine du nor et due Nord de l'Afrique, *C.R. Acad. Sci.,* Paris, 322, Serie Iia, pp. 847-851.

Pfister, C., 1995: Monthly temperature and precipitation in central Europe from 1525-1979: quantifying documentary evidence on weather and its effects. In: *Climate since A.D. 1500,* R.S. Bradley and P.D.Jones (eds.), Routledge, London, pp. 118-142.

Pfister, C., 1999: Wetternachhersage: 500 Jahre Klimavariationen und Naturkatastrophen 1496-1995. Paul Haupt, Bern, 304 pp.

Pfister, C. and R. Brázdil, 1999: Climatic Variability in Sixteenth-Century Europe and its Social Dimension: A Synthesis. In: Climatic Variability in Sixteenth-Century Europe and its Social Dimension, C. Pfister, R. Brázdil and R. Glaser (eds.), *Special Issue of Clim. Change,* **43,** 5-54.

Pfister, C., G. Kleinlogel, G. Schwarz-Zanetti and M. Wegmann, 1996: Winters in Europe: The fourteenth century. *Clim. Change,* **34,** 91-108.

Pfister, C., J. Luterbacher, G. Schwarz-Zanetti and M. Wegmann, 1998: Winter air temperature variations in Central Europe during the Early and High Middle Ages (A.D. 750-1300). *Holocene,* **8,** 547-564.

Pfister, C., R. Brázdil, R. Glaser, M. Barriendos Vallvé, D. Camuffo, M. Deutsch, P. Dobrovoln?, S. Enzi, E. Guidoboni, O. Kotyza, S. Militzer, L. Rácz, and F.S. Rodrigo, 1999: Documentary Evidence on Climate in Sixteenth-Century Europe. In: *Climatic Variability in Sixteenth-Century Europe and its Social Dimension,* C. Pfister, R. Brázdil and R. Glaser (eds.), Kluwer, Dordrech,*Special Issue of Clim. Change,* **43,** 55-110.

Pielke, Sr. , R.A., J. Eastman, T.N. Chase, J. Knaff and T.G.F. Kittel, 1998a: Errata to 1973-1996 Trends in depth-averaged tropospheric temperature. *J. Geophys. Res.,* **103**(D14), 16927-16933.

Pielke, Sr. , R.A., J. Eastman, T.N. Chase, J. Knaff and T.G.F. Kittel, 1998b: 1973-1996 Trends in depth-averaged tropospheric temperature. *J. Geophys. Res.,* **103**(D22), 28909-28912.

Piervitali, E., M. Colacino and M. Conte, 1998: Rainfall over the Central-Western Mediterranean basin in the period 1951-1995. Part I: Precipitation trends. *Geophysics and Space Physics,* **21**C(3), 331-344.

Plummer, N., M.J. Salinger, N. Nicholls, R. Suppiah, K.J. Hennessy, R.M. Leighton, B.C. Trewin, C.M. Page and J.M. Lough, 1999: Changes in climate extremes over the Australian region and New Zealand during the twentieth century. *Clim. Change,* **42,** 183-202.

Pollack, H., S. Huang and P.Y. Shen, 1998: Climate change revealed by subsurface temperatures: A global perspective. *Science ,* **282,** 279-281.

Polovina, J.J., G.T. Mitchum and G.T. Evans, 1995: Decadal and basin-scale variation in mixed layer depth and the impact on biological production in the Central and North Pacific, 1960-88. *Deep Sea Res., Part I,* **42**(10), 1701-1716.

Porter, S.C. and Z. An, 1995: Correlation between climate events in the North Atlantic and China during the last glaciation. *Nature,* **375,** 305-308.

Power, S., F. Tseitkin, S. Torok, B. Lavery, R. Dahni and B. McAvaney, 1998: Australian temperature, Australian rainfall and the Southern Oscillation, 1910-1992: coherent variability and recent changes. *Australian Met. Mag.,* **47,** 85-101.

Power, S., T. Casey, C.K. Folland, A. Colman and V. Mehta, 1999: Inter-decadal modulation of the impact of ENSO on Australia. *Clim. Dyn.,* **15,** 319-323.

Prabhakara, C., R. Iacovassi Jr. and J.-M. Yoo, 1998: Global warming deduced from MSU. *Geophys. Res. Lett.,* **25,** 1927-1930.

Prisenberg, S.J., I.K. Peterson, S. Narayanan and J.U. Umoh, 1997: Interaction between atmosphere, ice cover, and ocean off Labrador and Newfoundland from 1962-1992. *Can. J. Aquat. Sci.,* **54,** 30-39.

Quadfasel, D., A. Sy and B. Rudels, 1993: A ship of opportunity section to the North Pole: Upper ocean temperature observations. *Deep Sea Res.,* **40,** 777-789.

Quayle, R.G., T.C. Peterson, A.N. Basist and C.S. Godfrey, 1999: An operational near-real-time global temperature index. *Geophys. Res. Lett.,* **26,** 333-335.

Quintana-Gomez, R.A., 1999: Trends of maximum and minimum temperatures in northern South America. *J. Climate,* **12,** 2104-2112.

Raper, S.C.B., K.R. Briffa and T.M.L. Wigley, 1996: Glacier change in northern Sweden from AD 500: a simple geometric model of Storglaciären. *J. Glaciol.,* **42,** 341-351.

Rasmussen, T.L., E. Thomsen, L.D. Labeyrie and T.C.E. van Weering, 1996a: Circulation changes in the Faeroe-Shetland Channel correlating with cold events during the last glacial period (58-10 ka). *Geology,* **24,** 937-940.

Rasmussen, T.L., T.C.E. van Weering and L.D. Labeyrie, 1996b: Climatic instability, ice sheets and ocean dynamics at high northern latitudes during the last glacial period (58-10 ka). *Quaternary Science Reviews,* **15,** 1-10.

Rayner, N.A., E.B. Horton, D.E. Parker, C.K. Folland and R.B. Hackett, 1996: Version 2.2 of the global sea-ice and sea surface temperature data set, 1903-1994. *Climate Research Technical Note 74,* 43pp. (Available from National Meteorological Library, London Road, Bracknell, UK, RG12 2SZ).

Rayner, N.A., D.E. Parker, P. Frich, E.B. Horton, C.K. Folland and L.V. Alexander, 2000: SST and sea-ice fields for ERA40. In *Proc. Second Int. WCRP Conf. On Reanalyses,* Wokefield Park, Reading, UK, 23-27 August 1999. WCRP-109, WMO/TD-NO. 985.

Reeve, N. and R. Toumi, 1999: Lightning activity as an indicator of climate change. *Quart. J. R. Met. Soc.,* **125,** 893-903.

Reille, M., J.L. de Beaulieu, H. Svobodova, V. Andrieu-Ponel and C. Goeury, 2000: Pollen biostratigraphy of the last five climatic cycles from a long continental sequence from the Velay region (Massif Central, France). *J. Quat. Sci.,* **15,** 665-685.

Ren, G., 1998: Pollen evidence for increased summer rainfall in the Medieval warm period at Maili, Northeast China. *Geophys. Res. Lett.,* **25,** 1931-1934.

Ren, G., 1999a: Some paleoclimatological problems associated with the present global warming. *J. Appl. Met.,* **7**(3), 361-370 (in Chinese with English abstract).

Ren, G., 1999b: Some progresses and problems in Paloeclimatology. *Scientia Geographic Sinica,* **19,** 368-378.

Ren, G. and L. Zhang, 1998: A preliminary mapped summary of Holocene pollen data for Northeast China. *Quat. Sci. Rev.,* **17,** 669-688.

Renwick, J.A., 1998: ENSO-related variability in the frequency of South Pacific blocking. *Mon. Wea. Rev.,* **126,** 3117-3123.

Renwick, J.A. and M.J. Revell, 1999: Blocking over the South Pacific and Rossby Wave Propagation. *Mon. Wea. Rev.,* **127,** 2233-2247.

Reverdin, G., D.R. Cayan and Y. Kushnir, 1997: Decadal variability of hydrography in the upper northern North Atlantic in 1948-1990. *J. Geophys. Res.,* **102**(C4), 8505-8531.

Reynolds, R.W., 1993: Impact of Mount Pinatubo aerosols on satellite-derived sea surface temperatures. *J. Climate,* **6,** 768-774.

Reynolds, R.W. and T.M. Smith, 1994: Improved global sea surface temperature analyses using optimum interpolation. *J. Climate,* **7,** 929-948.

Rind, D., 1998: Just add water vapor. *Science,* **281,** 1152-1153.

Ritchie, J.C., L.C. Cwynar and R.W. Spear, 1983: Evidence from North-West Canada for an early Holocene Milankovitch thermal maximum. *Nature,* **305,** 126-128.

Robertson, D.M., R.R. Ragotzkie and J.J. Magnuson, 1992: Lake ice records used to detect historical and future climatic changes. *Clim. Change,* **21,** 407-427.

Robinson, D.A., 1997: Hemispheric snow cover and surface albedo for model validation. *Ann. Glaciol.,* **25,** 241-245.

Robinson, D.A., 1999: Northern Hemisphere snow cover during the satellite era. *Proc. 5th Conf. Polar Met. and Ocean.*, Dallas, TX, American Meteorological Society, Boston, MA, pp. 255-260.

Robinson, D.A., K.F. Dewey and R.R. Heim, 1993: Global snow cover monitoring: An update. *Bull. Am. Met. Soc.*, **74**, 1689-1696.

Robock, A., Y.V. Konstantin, G. Srinivasan, J.K. Entin, S.E. Hollinger, N.A. Speranskaya, S. Liu and A. Namkhai, 2000: The global soil moisture data bank. *Bull. Am. Met. Soc.*, **81**, 1281-1299.

Rodbell, D., G.O. Seltzer, D.M. Anderson, D.B. Enfield, M.B. Abbott and J.H. Newman, 1999: A high-resolution 15000 year record of El Nino driven alluviation in southwestern Ecuador. *Science*, **283**, 516-520.

Rodrigo, F.S., M.J. Esteban-Parra, D. Pozo-Vazquez and Y. Castro-Diez, 1999: A 500-year precipitation record in Southern Spain. *Int. J. Climatol.*, **19**, 1233-1253.

Romanovsky, V.E. and T.E. Osterkamp, 1999: Permafrost Temperature Dynamics in Alaska and East Siberia During the Last 50 years. 11th Arctic Forum, ARCUS, Washington, DC, March 22-23.

Romero, R., J.A. Guijarro, C. Ramis and S. Alonso, 1998: A 30-year (1964-1993) daily rainfall data base for the Spanish Mediterranean regions: First exploratory study. *Int. J. Climatol.*, **18**, 541-560.

Ross, R.J. and W.P. Elliott, 1996: Tropospheric water vapor climatology and trends over North America: 1973-93. *J. Climate*, **9**, 3561-3574.

Ross, R.J. and W.P. Elliott, 1998: Northern hemisphere water vapor trends. Ninth Symposium on Global Change Studies, *Amer. Meteor. Soc., Preprints*, pp. 39-41.

Ross, R.J. and W.P. Elliott, 2001: Radiosonde-based Northern Hemisphere tropospheric water vapour trends. *J. Climate*, **14**, 1602-1612.

Rossow, W.B. and R.A. Schiffer, 1999: Advances in understanding clouds from ISCCP. *Bull. Am. Met. Soc.*, **80**, 2261-2287.

Rostek, F., G. Ruhland, F. Bassinot, P.J. Müller, L. Labeyrie, Y. Lancelot and E. Bard, 1993: Reconstructing sea surface temperature and salinity using $\delta^{18}O$ and alkenone records. *Nature*, **364**, 319-321.

Rothrock, D.A., Y. Yu and G.A. Maykut, 1999: Thinning of the Arctic Sea-Ice Cover. *Geophys. Res. Lett.*, **26**, 3469-3472.

Salinger, M.J., 1995: Southwest Pacific temperature: trends in maximum and minimum temperatures. *Atmos. Res.*, **37**, 87-100.

Salinger, M.J. and M.S. McGlone, 1989: New Zealand Climate – The past two million years. The New Zealand Climate report 1990, Royal Society of New Zealand, Wellington, 13-17.

Salinger, M.J. and A.B. Mullan, 1999: New Zealand climate: temperature and precipitation variations and their links with atmospheric circulation. *Int. J. Climatol.*, **19**, 1049-1071.

Salinger, M.J., R.J. Allan, N. Bindoff, J. Hannah, B. Lavery, Z. Lin, J. Lindesay, N. Nicholls, N. Plummer and S. Torok, 1996: Observed variability and change in climate and sea level in Australia, New Zealand and the South Pacific. In: *Greenhouse: Coping with Climate Change*, W.J. Bouma, G.I. Pearman and M.R. Manning (eds.), CSIRO, Melbourne, Australia, pp. 100-126.

Salinger, M.J., J.A. Renwick and A.B. Mullan, 2001: Interdecadal Pacific Oscillation and South Pacific climate. *Int. J. Climatol.*, accepted.

Sandweiss, D.H., J.B. Richardson III, E.J. Reitz, H.B.R. Rollins and K.A. Maasch, 1996: Geoarcheological evidence from Paru for a 5000 years B.P. onset of El Nino. *Science*, **273**, 1531-1533.

Santer, B.D., J.J. Hnilo, T.M.L. Wrigley, J.S. Boyle, C. Doutriaux, M. Fiorino, D.E. Parker and K.E. Taylor, 1999: Uncertainties in observational based estimates of temperature change in the free atmosphere. *J. Geophys. Res.*, **104**, 6305-6333.

Santer, B.D., T.M.L. Wigley, J.S. Boyle, D.J. Gaffen, J.J. Hnilo, D. Nychka, D.E. Parker and K.E. Taylor, 2000: Statistical significance of trend differences in layer-average temperature time series. *J. Geophys. Res.*, **105**, 7337-7356.

SAR, see IPCC, 1996.

Sarnthein, M., K. Winn, S.J.A. Jung, J.C. Duplessy, L. Labeyrie, H.

Erlenkeuser and G. Ganssen, 1994: Changes in east Atlantic deep water circulation over the last 30,000 years. *Paleoceanography*, **9**, 209-267.

Schindler, D.W., K.G. Beaty, E.J. Fee, D.R. Cruikshank, E.R. Devruyn, D.L. Findlay, G.A. Linsey, J.A. Shearer, M.P. Stainton and M.A. Turner, 1990: Effects of climatic warming on lakes of the central boreal forest. *Science*, **250**, 967-970.

Schneider, R.R., P.J. Müller, G. Ruhland, G. Meinecke, H. Schmidt and G. Wefer, 1996: Late Quaternary surface temperatures and productivity in east-equatoricl South Atlantic: Response to changes in trade/monsoon wind forcing and surface water advection, 1996. In: *The South Atlantic: Present and Past Circulation*, G. Wefer, W.H. Berger, G. Siedler and D. Webb (eds.), Springer-Verlag, Berlin , pp. 527-551.

Schönwiese, C.D. and J. Rapp, 1997: Climate Trend Atlas of Europe Based on Observations 1891-1990. Kluwer Academic Publishers, Dordrecht, 228 pp.

Schönwiese, C.D., J. Rapp, T. Fuchs, and M. Denhard, 1994: Observed climate trends in Europe 1891-1990. *Meteorol. Zeitschrift*, **3**, 22-28.

Schwartzman, P.D., P.J. Michaels and P.C. Knappenberger, 1998: Observed changes in the diurnal dewpoint cycles across North America. *Geophys. Res. Lett.*, **25**, 2265-2268.

Serreze, M.C., F. Carse and R.G. Barry, 1997: Icelandic low cyclone activity climatological features, linkages with the NAO, and relationships with recent changes in the Northern Hemisphere circulation. *J. Climate*, **10**, 453-464.

Serreze, M.C., J.E. Walsh, F.S. Chapin III, T. Osterkamp, M. Dyurgerov, V. Romanovsky, W.C. Oechel, J. Morison, T. Zhang and R.G. Barry, 2000: Observational evidence of recent change in the northern high-latitude environment. *Clim. Change*, **46**, 159-207.

Severinghaus, J.P. and E. Brook, 1999: Abrupt climate change at the End of the last glacial period inferred from trapped air in polar ice. *Science*, **286**, 930-934.

Severinghaus, J.P., T. Sowers, E. Brook, R.B. Alley and M.L. Bender, 1998: Timing of abrupt climate change at the end of the Younger Dryas interval from thermally fractionated gases in polar ice. *Nature*, **391**, 141-146.

Shabbar, A., K. Higuchi, W. Skinner and J.L. Knox, 1997: The association between the BWA index and winter surface temperature variability over eastern Canada and west Greenland. *Int. J. Climatol.*, **17**, 1195-1210.

Sharkhuu, N., 1998: Trends of permafrost development in the Selenge River Basin, Mongolia. *Proceedings of the Seventh International Conference on Permafrost*, Yellowknife, Canada, June 1998, Université Laval, Quebec, Collection Nordicana No. 57, pp. 979-986.

Shen, S.S., M. Thomas, C.F. Ropelewski and R.E. Livezey, 1998: An optimal regional averaging method with error estimates and a test using tropical Pacific SST data. *J. Climate*, **11**, 2340-2350.

Shinoda, M., T. Okatani and M. Saloum, 1999: Diurnal variations of rainfall over Niger in the West African Sahel: A comparison between wet and drought years. *Int. J. Climatol.*, **19**, 81-94.

Shulmeister, J. and B.G. Lees, 1995: Pollen evidence form tropical Australia for the onset of an ENSO-dominated climate at c. 4000 BP. *The Holocene*, **5**, 10-18.

Simmonds, I. and K. Keay, 2000: Variability of Southern Hemisphere extratropical cyclone behavior, 1958-97. *J. Climate*, **13**(3), 550–561.

Singer, C., J. Shulmeister and B. McLea, 1998: Evidence against a significant Younger Dryas cooling event in New Zealand. *Science*, **281**, 812-814.

Skinner, W.R. and J.A. Majorowicz, 1999: Regional climatic warming and associated twentieth century land-cover changes in north-western North America. *Clim. Res.*, **12**, 39-52.

Slingo, J.M., D.P. Rowell, K.R. Sperber, and F. Nortley, 1999: On the predictability of the interannual behaviour of the Madden-Julian Oscillation and its relationship with El Nino. *Quart. J. R. Met. Soc.*,

125, 583-609.

Smith, C.A., R. Toumi and J.D. Haigh, 2000: Seasonal trends in stratospheric water vapor. *Geophys. Res. Lett.*, **27**, 1687-1690.

Smith, C.A., J.D. Haigh and R. Toumi, 2001: Radiative forcing due to trends in stratospheric water vapor. *Geophys. Res. Lett.*, **28**, 179-182.

Smith, D.M., 1998: Recent increase in the length of the melt season of perennial Arctic sea ice. *Geophys. Res. Lett.*, **25**, 655-658.

Smith, T.M., R.W. Reynolds, R.E. Livezey and D.C. Stokes, 1996: Reconstruction of historical sea surface temperatures using empirical orthogonal functions. *J. Climate*, **9**, 1403-1420.

Smith, T.M., R.E. Livezey and S.S. Shen, 1998: An improved method for analyzing sparce and irregularly distributed SST data on a regular grid: The tropical Pacific Ocean. *J. Climate*, **11**, 1717-1729.

Sowers, T. and M. Bender, 1995: Climate records covering the last deglaciation. *Science,* **269**, 210-214.

Spencer, R.W., 1993: Global oceanic precipitation from the MSU during 1979-92 and comparisons to other climatologies. *J. Climate*, **6**, 1301-1326.

Spencer, R.W. and J.R. Christy, 1992a: Precision and radiosonde validation of satellite gridpoint temperature anomalies, Part I: MSU channel 2. *J. Climate*, **5**, 847-857.

Spencer, R.W. and J.R. Christy, 1992b: Precision and radiosonde validation of satellite gridpoint temperature anomalies, Part II: A tropospheric retrieval and trends 1979-90. *J. Climate*, **5**, 858-866.

Stager, J.C. and P.A. Mayewski, 1997: Abrupt Early to Mid-Holocene Climatic transition registered at the Equator and the Poles. *Science*, **276**, 1834-1836.

Stahle, D.W., M.K. Cleaveland, M.D. Therrell, D.A. Gay, R.D. D`Arrigo, P.J. Krusic, E.R. Cook, R.J. Allan, J.E. Cole, R.B. Dunbar, M.D. Moore, M.A. Stokes, B.T. Burns, J. Villanueva-Diaz and L.G. Thompson, 1998: Experimental Dendroclimatic Reconstruction of the Southern Oscillation. *Bull. Am. Met. Soc.*, **79**, 2137-2152.

Steadman, R.G., 1984: A universal scale of apparent temperature. *J. Clim. Appl. Met.,* **23**, 1674-1687.

Steele, M. and T. Boyd, 1998: Retreat of the cold halocline layer in the Arctic Ocean. *J. Geophys. Res.*, **103**(C5), 10419-10435.

Steig, E., E.J. Brook, J.W.C. White, C.M. Sucher, M.L. Bender, S.J. Lehman, D.L. Morse, E.D. Waddigton and G.D. Clow, 1998: Synchronous climate changes in Antarctica and the North Atlantic. *Science*, **282**, 92-95.

Stendel, M., J.R. Christy and L. Bengtsson, 2000: Assessing levels of uncertainty in recent temperature time series. *Clim. Dyn.*, **16**(8), 587-601.

Sterin, A.M., 1999: An analysis of linear trends in the free atmosphere temperature series for 1958-1997. *Meteorologiai Gidrologia*, **5**, 52-68.

Stine, S., 1994: Extreme and persistent drought in California and Patagonia during medieval time. *Nature*, **369**, 546-549.

Stocker, T.F., 2000: Past and further reorganization in the climate system. *Quat. Sci. Rev.*, **19**, 301-319.

Stone, D.A., A.J. Weaver and F.W. Zwiers, 1999: Trends in Canadian precipitation intensity. *Atmos. Ocean*, **2**, 321-347.

Street-Perrott, F.A. and R.A. Perrott, 1990: Abrupt climate fluctuations in the tropics: the influence of Atlantic ocean circulation. *Nature*, **343**, 607-612.

Sun, B. and P.Ya. Groisman, 2000: Cloudiness variations over the former Soviet Union. *Int. J. Climatol.*, **20**, 1097-1111.

Sun, B., P.Ya. Groisman, R.S. Bradley, and F.T. Keimig, 2000: Temporal changes in the observed relationship between cloud cover and surface air temperature. *J. Climate*, **13**, 4341-4357.

Sun, B., P.Ya. Groisman and I.I. Mokhov, 2001: Recent changes in cloud type frequency and inferred increases in convection over the United States and the Former USSR. *J. Climate*, **14**, 1864-1880.

Sun, X.J. and Y.S. Chen, 1991: Palynological records of the last 11,000 years in China. *Quat. Sci. Rev.*, **10**, 537-544.

Swetnam, T.W. and J. L. Betancourt, 1998: Mesoscale disturbance amd ecological response to decadal climate variability in the American Southwest. *J. Climate*, **11**, 3128-3147.

Tanimoto, Y., N. Iwasaka, K. Hanawa and Y. Toba, 1993: Characteristic variations of sea surface temperature with multiple time scales in the North Pacific. *J. Climate*, **6**, 1153-1160.

Tarhule, A. and M. Woo, 1998: Changes in rainfall characteristics in northern Nigeria. *Int. J. Climatol.*, **18**, 1261-1271.

Taylor, K.C., G.W. Lamorey, G.A. Doyle, R.B. Alley, P.M. Grootes, P.A. Mayewski, J.W.C. White and L.K. Barlow, 1993: The "flickering switch" of late Pleistocene climate change. *Nature,* **361**, 432-436.

Thie, J., 1974: Distribution and thawing of permafrost in the southern part of the discontinuous permafrost zone in Manitoba. *Arctic*, **27**, 189-200.

Thompson, D.W.J. and J.M. Wallace, 1998: The Arctic oscillation signature in the wintertime geopotential height and temperature fields. *Geophys. Res. Lett.*, **25**, 1297-1300.

Thompson, D.W.J. and J.M. Wallace, 2000: Annual modes in the extratropical circulation Part I: month-to-month variability. *J. Climate*, **13**, 1000-1016.

Thompson, D.W.J. and J.M. Wallace, 2001: Regional climate impacts of the Northern Hemisphere annular mode and associated climate trends. *Nature*, in press.

Thompson, D.W.J., J.M. Wallace and G.C. Hegerl, 2000b: Annual modes in the extratropical circulation Part II: trends. *J. Climate*, **13**, 1018-1036.

Thompson, L.G., 1996: Climate changes for the last 2000 years inferred from ice core evidence in tropical ice cores. In: *Climate Variations and Forcing Mechanisms of the Last 2000 Years, NATO ASI Series I,* P.D. Jones, R.S. Bradley and J. Jouzel (eds.), **41**, 281-297.

Thompson, L.G. and 13 others, 1989: Pleistocene climate record from Qinghai-Tibetan Plateau ice cores. *Science*, **246**, 474-477.

Thompson, L.G., E. Mosley-Thompson, M.E. Davis, P.N. Lin, K.A. Henderson, J. Cole-Dai, J.F. Bolzan and K.B. Liu, 1995: Late Glacial Stage and Holocene Tropical Ice Core Records from Huascaran, Peru. *Science*, **269**, 46-50.

Thompson, L.G., M.E. Davis, E. Mosley-Thompson, T.A. Sowers, K.A. Henderson, V.S. Zagorodnov, P.N. Lin, V.N. Mikhalenko, R.K. Campen, J.F. Bolzan, J. Cole-Dai and B. Francou, 1998: A 25,000-year tropical climate history from Bolivian ice cores. *Science*, **282**, 1858-1864.

Thompson, L.G., T. Yao, E. Mosley-Thompson, M.E. Davis, K.A. Henderson and P.N. Lin, 2000a: A high resolution millennial record of the South Asian Monsoon from Himalayan ice cores. *Science*, **289**, 1916-1919.

Thunnell, R.C. and P.G. Mortyn, 1995: Glacial climate instability in the northeast Pacific Ocean. *Nature*, **376**, 504-506.

Torrence, C. and G.P. Compo, 1998: A practical guide to wavelet analysis. *Bull. Am. Met. Soc.*, **79**, 61-78.

Torrence, C. and P.J. Webster, 1998: The annual cycle of persistence in the El Niño/Southern Oscillation. *Quart. J. R. Met. Soc.*, **124**, 1985-2004.

Torrence, C. and P.J. Webster, 1999: Interdecadal changes in the ENSO-monsoon system. *J. Climate,* **12**, 2679-2690.

Trenberth, K.E., 1998a: Atmospheric moisture residence times and cycling: Implications for rainfall rates with climate change. *Clim. Change*, **39**, 667-694.

Trenberth, K.E., 1998b: El Nino and global warming. *J. Marine Education*, **15**, 12-18.

Trenberth, K.E. and J.W. Hurrell, 1994: Decadal atmosphere-ocean variations in the Pacific. *Clim. Dyn.*, **9**, 303-319.

Trenberth, K.E. and T.J. Hoar, 1996: The 1990-1995 El Niño-Southern Oscillation event: longest on record. *Geophys. Res. Lett.*, **23**, 57-60.

Trenberth, K.E. and T.W. Owen, 1999: Workshop on Indices and Indicators for climate extremes, Asheville, NC, USA, 3-6 June 1999:

Breakout Group A: Storms. *Clim. Change*, **42**, 9-21.

Trenberth, K.E., J.R. Christy and J.W. Hurrell, 1992: Monitoring global monthly mean surface temperatures. *J. Climate*, **5**, 1405-1423.

Trenberth, K.E., J.M. Caron and D.P. Stepaniak, 2001: The Atmospheric Energy Budget and Implications for Surface Fluxes and Ocean Heat Transports. *Clim. Dyn.*, **17**, 259-276.

Tudhope, A.W., G.B. Shimmield, C.P. Chilcott, M. Jebb, A.E. Fallick and A.N. Dalgleish, 1995: Recent changes in climate in the far western equatorial Pacific and their relationship to the Southern Oscillation: oxygen isotope records from massive corals, Papua, New Guinea. *Earth and Planetary Science Letters*, **136**, 575-590.

Tuomenvirta, H., H. Alexandersson, A. Drebs, P. Frich and P.O. Nordli, 2000: Trends in Nordic and Arctic temperature extremes and ranges. *J. Climate*, **13**, 977-990.

Turrell, W.R., G. Slesser, R.D. Adams, R. Payne and P.A. Gillibrand, 1999: Decadal variability in the composition of Faroe Shetland Channel bottom water. *Deep Sea Res. Part I*, **46**, 1-25.

Vaganov, E.A., M.K. Hughes, A.V. Kirdyanov, F.H. Schweingruber and P.P. Silkin, 1999: Influence of snowfall and melt timing on tree growth in subarctic Eurasia. *Nature*, **400**, 149-151.

Van Ommen, T.D. and V. Morgan, 1996: Peroxide concentrations in the DSS ice core, Law Dome, Antarctica. *J. Geophys. Res.*, **101**(D10), 15147-15152.

Van Ommen, T.D. and V. Morgan, 1997: Calibrating the ice core paleothermometer using seasonality. *J. Geophys. Res.*, **102**(D8), 9351-9357.

Vaughan, D.G. and T. Lachlan-Cope, 1995: Recent retreat of ice shelves on the Antarctic Peninsula. *Weather*, **50** , 374-376.

Vaughan, D.G. and C.S.M. Doake, 1996: Recent atmospheric warming and retreat of ice shelves on the Antarctic Peninsula. *Nature*, **379**, 328-331

Vidal, L., L.D. Labeyrie, E. Cortijo, M. Arnold, J.C. Duplessy, E. Michel, S. Becque and T.C.E. van Weering, 1997: Evidence for changes in the North Atlantic Deep Water linked to meltwater surges during the Heinrich events. *Earth Planet. Sci. Lett.*, **146**, 13-27.

Vidal, L., L.D. Labeyrie and T.C.E. van Weering, 1998: Benthic δ^{18}O records in the North Atlantic over the last glacial period (60-10 kyr): Evidence for brine formation. *Paleoceanography*, **13**, 245-251.

Villalba, R., E.R. Cook, R. D'Arrigo, G.C. Jacoby, P.D. Jones, J.M. Salinger and J. Palmer, 1997: Sea-level pressure variability around Antarctica since A.D. 1750 inferred from subantarctic tree-ring records. *Clim. Dyn.*, **13**, 375-390.

Vincent, L.A., 1998: A technique for the identification of inhomogeneities in Canadian temperature series. *J. Climate*, **11**, 1094-1104.

Vincent, L.A. and D.W. Gullett, 1999: Canadian historical and homogeneous temperature datasets for climate change analysis. *Int. J. Climatol.*, **19**, 1375-1388.

Vinje, T., N. Nordlund and Å. Kvambekk, 1998: Monitoring ice thickness in Fram Strait. *J. Geophys. Res.*, **103**(C5), 10437-10449.

Vinnikov, K.Ya., P.Ya. Groisman and K.M. Lugina, 1990: Empirical data on contemporary global climate changes (temperature and precipitation). *J. Climate*, **3**, 662-677.

Vinnikov, K.Y., A. Robock, S. Qiu and J.K. Entin, 1999a: Optimal design of surface networks for observation of soil moisture. *J. Geophys. Res.*, **104**, 19743-19749.

Vinnikov, K.Y., A. Robock, R.J. Stouffer, J.E. Walsh, C.L. Parkinson, D.J. Cavalieri, J.F.B. Mitchell, D. Garrett and V.F. Zakharov, 1999b: Global warming and Northern Hemisphere sea ice extent. *Science*, **286**, 1934-1937.

Vitt, D.H., L.A. Halsey and S.C. Zoltai, 1994: The bog landforms of continental western Canada in relation to climate and permafrost patterns. *Artic and Alpine Res.*, **26**, 1-13.

Vonder Mühll, D., T. Stucki and W. Haeberly, 1998: Borehole temperatures in Alpine permafrost: a ten years series. *Proceedings of the Seventh International Conference on Permafrost*, Yellowknife, Canada, June 1998, Université Laval, Quebec, Collection Nordicana No. 57, pp. 1089-1096.

von Grafenstein, U., H. Erlenkeuser, J. Muller, J. Jouzel and S.J. Johnsen, 1998: The cold event 8,200 years ago documented in oxygen isotope records of precipitation in Europe and Greenland. *Clim. Dyn.*, **14**, 73-81.

von Grafenstein, U., H. Erlenkeuser, A. Brauer, J. Jouzel and S.J. Johnsen, 1999: A mid-European decadal isotope-climate record from 15,500 to 5,000 years B.P. *Science*, **284**, 1654-1657.

Wadhams, P. and N.R. Davis, 2000: Further evidence of sea ice thinning in the Arctic Ocean. *Geophys. Res. Lett.*, **27**, 3973-3976.

Wallace, J.M., 2000: North Atlantic Oscillation / Northern Hemisphere annular mode: Two paradigms - One phenomenon. *Quart. J. R. Met. Soc.*, **126**, 791-805.

Wallace, J.M., Y. Zhang and J.A. Renwick, 1995: Dynamic contribution to hemispheric mean temperature trends. *Science*, **270**, 780-783.

Wallis, T.W.R., 1998: A subset of core stations from the Comprehensive Aerological Data Set (CARDS). *J. Climate*, **12**, 272-282.

Walsh, J.E., 1978: Data set on Northern Hemisphere sea-ice extent. *Glaciological Data, Report GD-2*, World Data Center-A for Glaciology (Snow and Ice), part 1, pp. 49-51.

Walsh, J.M., W.L. Chapman and T.L. Shy, 1996: Recent decrease of sea level pressure in the central Arctic. *J. Climate*, **9**, 480-486.

Wang, B. and H.M. French, 1994: Climate controls and high-altitude permafrost, Qinghai-Xizang (Tibet) Plateau, China. *Permafrost and Periglacial Processes*, **5**, 87-100.

Wang, B. and Y. Wang, 1996: Temporal structure of the Southern Oscillation as revealed by waveform and wavelet analysis. *J. Climate*, **9**, 1586-1598.

Wang, S.L., H.J. Jin, S. Li and L. Zhao, 2000: Permafrost Degradation on the Qinghai-Tibet Plateau and its Environmental Impacts. *Permafrost and Periglacial Processes*, **11**, 43-53.

Wang, S.W. and D.Y. Gong, 2000: Climate in China during the four special periods in Holocene. *Progress in Nature Science*, **10**(5), 379-386.

Wang, S.W., J. Ye, D. Gong and J. Zhu, 1998a: Construction of mean annual temperature series for the last one hundred years in China, *Quart. J. Appl. Met.*, **9**(4), 392-401 (in Chinese).

Wang, S.W., J. Ye and D. Gong, 1998b: Climate in China during the Little Ice Age. *Quaternary Sciences*, **1**, 54-64 (in Chinese).

Wang, X.L. and D.J. Gaffen, 2001: Late twentieth century climatology and trends of surface humidity and temperature in China. *J. Climate*, in press.

Wanner, H., C. Pfister, R. Bràzdil, P. Frich, K. Fruydendahl, T. Jonsson, J. Kington, H.H. Lamb, S. Rosenorn and E. Wishman, 1995: Wintertime European circulation patterns during the Late Maunder Minimum Cooling Period (1675-1704). *Theor. Appl. Climatol.*, **51**, 167-175.

Wansard, G., 1996: Quantification of paleotemperature changes during isotopic stage 2 in the La Draga continental sequence (NE Spain) based on the Mg/Ca ratio of freshwater ostracods. *Quarternary Science Review*, **15**, 237-245.

Waple, A., M.E. Mann and R.S. Bradley, 2001: Long-term Patterns of Solar Irradiance Forcing in Model Experiments and Proxy-based Surface Temperature Reconstructions. *Clim. Dyn.*, in press.

Ward, M.N., 1998: Diagnosis and short-lead time prediction of summer rainfall in tropical North Africa and interannual and multi-decadal timescales. *J. Climate*, **11**, 3167-3191.

Ward, M.N., P.J. Lamb, D.H. Portis, M. El Hamly, and R. Sebbari, 1999: Climate Variability in Northern Africa: Understanding Droughts in the Sahel and the Mahgreb. In: *Beyond El Nino: Decadal and Interdecadal Climate Variability*, A. Navarra (ed.), Springer, Berlin, pp. 119-140.

WASA Group (von Storch et al.), 1998: Changing waves and storms in the Northeast Atlantic? *Bull. Am. Met. Soc.*, **79**, 741-760.

Watts, W.A., J.R.M. Allen and B. Huntley, 1996: Vegetation history and palaeoclimate of the last glacial period at Lago Grande Di Monticchio, Southern Italy. *Quat. Sci. Rev.*, **15**, 133-151.

Webb, I., Thompson and J.E. Kutzbach, 1998: An introduction to Late Quaternary Climates: Data Syntheses and Model Experiments. *Quat. Sci. Rev.*, **17**, 465-471.

Weber, R.O., P. Talkner and G. Stefanicki, 1994: Asymmetric diurnal temperature change in the Alpine region. *Geophys. Res. Lett.*, **21**, 673-676.

Weller, G. and P.A. Anderson (eds.), 1998: Implications of Global Change in Alaska and the Bering Sea Region. *Proceedings of a Workshop, June 1997, Centre for Global Change and Arctic System Research*, University of Alaska Fairbanks, Fairbanks, Alaska, 157 pp.

Wentz, F.J. and M. Schabel, 1998: Effects of orbital decay on satellite-derived lower-tropospheric temperature trends. *Nature*, **394**, 661-664.

White, J.W.C., L.K. Barlow, D.A. Fisher, P. Grootes, J. Jouzel, S. Johnsen, and P.A. Mayewski, 1998a: The climate signal in the stable isotopes of snow from Summit Greenland: results of comparisons with modern climate observations. *Special Issue J. Geophys. Res.*, American Geophysical Union, 26425-26440.

White, W.B. and R. Peterson, 1996: An Antarctic circumpolar wave in surface pressure, wind, temperature, and sea ice extent. *Nature*, **380**, 699-702.

White, W.B. and D.R. Cayan, 1998: Quasi-periodicity and global symmetries in interdecadal upper ocean temperature variability. *J. Geophys. Res.*, **103**(C10), 21335-21354.

White, W.B., J. Lean, D.R. Cayan and M.D. Dettinger, 1997: A response of global upper ocean temperature to changing solar irradiance. *J. Geophys. Res.*, **102**, 3255-3266.

White, W.B., D.R. Cayan and J. Lean, 1998b: Global upper ocean heat storage response to radiative forcing from changing solar irradiance and increasing greenhouse gas/aerosol concentrations. *J. Geophys. Res.*, **103**, 21355-21366.

Wick, L. and W. Tinner, 1999: Vegetation changes and timberline fluctuations in the Central Alps as indicators of Holocene climate oscillations. *Arctic and Alpine Research*, **29**, 445-458.

Wigley, T.M.L., 2000: ENSO, volcanoes and record-breaking temperatures. *Geophys. Res. Lett.*, **27**, 4101-4104.

Wiles, G.C., R.D. D'Arrigo and G.C. Jacoby, 1998: Gulf of Alaska atmosphere-ocean variability over recent centuries inferred from coastal tree-ring records. *Clim. Change*, **38**, 289-306.

Wilks, D.S., 1999: Interannual variability and extreme-value characteristics of several stochastic daily precipitation models. *Agric. For. Meteorol.*, **93**, 153-169.

Williams, P.W., A. Marshall, D.C. Ford and A.V. Jenkinson, 1999: Palaeoclimatic interpretation of stable isotope data from Holocene speleotherms of the Waitomo district, North Island, New Zealand. *Holocene*, **9**, 649-657.

Wohlfarth, B., H. Linderson, B. Holmquist and I. Cato, 1998: The climatic significance of clastic varves in the Angermanalven Estuary, northern Sweden, AD 1860-1950. *The Holocene*, **8**, 525-534.

Wolfe, S.A., E. Kotler and F.M. Nixon, 2000: Recent warming impacts in the Mackenzie Delta, Northwest Territories, and northern Yukon Territory coastal areas. *Geological Survey of Canada*, Current Research 2000-B1, 9 pp.

Wong, A.P.S., N.L. Bindoff and J.A. Church, 1999: Large-scale freshening of intermediate waters in the Pacific and Indian Oceans. *Nature*, **400**, 440-443.

Woodhouse, C.A. and J.T. Overpeck, 1998: 2000 years of drought variability in the central United States. *Bull. Am. Met. Soc.*, **79**, 2693-2714.

Wynne, R.H., T.M. Lilles, M.K. Clayton and J.J. Magnuson, 1998: The predominant spatial trends of mean ice breakup dates can be attributed to latitude and snowfall. *Photogrammetric Engineering and Remote Sensing*, ISSN: 0099-1112 (Falls Church, VA), 64, 607-618.

Xie, P. and P.A. Arkin, 1997: Global precipitation: A 17-year monthly analysis based on gauge observations, satellite estimates and numerical model outputs. *Bull. Am. Met. Soc.*, **78**, 2539-2558.

Yamamoto, R. and Y. Sakurai, 1999: Long-term intensification of extremely heavy rainfall intensity in recent 100 years. *World Resource Rev.*, **11**, 271-281.

Ye, H., H.R. Cho and P.E. Gustafson, 1998: The changes in Russian winter snow accumulation during 1936-83 and its spatial patterns. *J. Climate*, **11**, 856-863.

Yiou, F., G.M. Raisbeck., S. Baumgartner, J. Beer, C. Hammer, S. Johnsen, J. Jouzel, P.W. Kubik, J. Lestringuez, M. Stievenard, M. Suter and P. Yiou, 1997a: Beryllium 10 in the Greenland Ice Core Project ice core at Summit Greenland. *J. Geophys. Res.*, **102**, 26783-26794.

Yiou, P., K. Fuhrer, L.D. Meeker, J. Jouzel, S.J. Johnsen and P.A. Mayewski, 1997b: Paleoclimatic variability inferred from the spectral analysis of Greenland and Antarctic ice core data. *J. Geophys. Res.*, **102**, 26441-26454.

Yu, G. and S.P. Harrison, 1996: An evaluation of the simulated water balance of Eurasia and northern Africa at 6000 y BP using lake status data. *Clim. Dyn.*, **12**, 723-735.

Yu, G. and B. Qin, 1997: Holocene temperature and precipitation reconstructions and monsoonal climates in eastern China using pollen data. *Paleoclimates*, **2**, 1-32.

Yung, Y.L., T. Lee, C.H. Wang and Y.T. Shieh, 1996: Dust: A diagnostic of the hydrologic cycle during the Last Glacial Maximum. *Science*, **271**, 962-963.

Zenk, W. and N. Hogg, 1996: Warming trend in Antarctic Bottom Water flowing into the Brazil Basin. *Deep Sea Res., Part I*, **43**, 1461-1473.

Zhai, P.M. and R.E. Eskridge, 1997: Atmospheric water vapor over China. *J. Climate*, **10**, 2643-2652.

Zhai, P.M. and F.M. Ren, 1999: Changes of China's maximum and minimum temperatures in 1951-1990. *Acta Meteor. Sinica*, **13**, 278-290.

Zhai, P.M., A. Sun, F. Ren, X. Liu, B. Gao and Q. Zhang, 1999a: Changes of climate extremes in China. *Clim. Change*, **42**, 203-218.

Zhai, P.M., F.M. Ren and Q. Zhang, 1999b: Detection of trends in China's precipitation extremes. *Acta Meteorologica Sinica*, **57**, 208-216.

Zhang, K., B.C. Douglas and S.P. Leatherman, 1997a: East Coast storm surges provide unique climate record. *EOS Trans. American Geophysical Union*, **78**(37).

Zhang, T., R.G. Barry, K. Knowles, J.A. Heginbottom and J. Brown, 1999: Statistics and characteristics of permafrost and ground-ice distribution in the Northern Hemisphere. *Polar Geography*, **2**, 132-154.

Zhang, X., L.A. Vincent, W.D. Hogg and A. Niitsoo, 2000: Temperature and precipitation trends in Canada during the 20th Century. *Atmosphere-Ocean*, **38**, 395-429.

Zhang, Y. and W.C. Wang, 1997: Model-simulated northern winter cyclone and anticyclone activity under a greenhouse warming scenario. *J. Climate*, **10**, 1616-1634.

Zhang, Y., J.M. Wallace and D.S. Battisti, 1997b: ENSO-like interdecadal variability: 1900-93. *J. Climate*, **10**, 1004-1020.

3

The Carbon Cycle and Atmospheric Carbon Dioxide

Co-ordinating Lead Author
I.C. Prentice

Lead Authors
G.D. Farquhar, M.J.R. Fasham, M.L. Goulden, M. Heimann, V.J. Jaramillo, H.S. Kheshgi, C. Le Quéré,
R.J. Scholes, D.W.R. Wallace

Contributing Authors
D. Archer, M.R. Ashmore, O. Aumont, D. Baker, M. Battle, M. Bender, L.P. Bopp, P. Bousquet, K. Caldeira,
P. Ciais, P.M. Cox, W. Cramer, F. Dentener, I.G. Enting, C.B. Field, P. Friedlingstein, E.A. Holland,
R.A. Houghton, J.I. House, A. Ishida, A.K. Jain, I.A. Janssens, F. Joos, T. Kaminski, C.D. Keeling,
R.F. Keeling, D.W. Kicklighter, K.E. Kohfeld, W. Knorr, R. Law, T. Lenton, K. Lindsay, E. Maier-Reimer,
A.C. Manning, R.J. Matear, A.D. McGuire, J.M. Melillo, R. Meyer, M. Mund, J.C. Orr, S. Piper, K. Plattner,
P.J. Rayner, S. Sitch, R. Slater, S. Taguchi, P.P. Tans, H.Q. Tian, M.F. Weirig, T. Whorf, A. Yool

Review Editors
L. Pitelka, A. Ramirez Rojas

Contents

Executive Summary

CO$_2$ concentration trends and budgets

Before the Industrial Era, circa 1750, atmospheric carbon dioxide (CO$_2$) concentration was 280 ± 10 ppm for several thousand years. It has risen continuously since then, reaching 367 ppm in 1999.

The present atmospheric CO$_2$ concentration has not been exceeded during the past 420,000 years, and likely not during the past 20 million years. The rate of increase over the past century is unprecedented, at least during the past 20,000 years.

The present atmospheric CO$_2$ increase is caused by anthropogenic emissions of CO$_2$. About three-quarters of these emissions are due to fossil fuel burning. Fossil fuel burning (plus a small contribution from cement production) released on average 5.4 ± 0.3 PgC/yr during 1980 to 1989, and 6.3 ± 0.4 PgC/yr during 1990 to 1999. Land use change is responsible for the rest of the emissions.

The rate of increase of atmospheric CO$_2$ content was 3.3 ± 0.1 PgC/yr during 1980 to 1989 and 3.2 ± 0.1 PgC/yr during 1990 to 1999. These rates are less than the emissions, because some of the emitted CO$_2$ dissolves in the oceans, and some is taken up by terrestrial ecosystems. Individual years show different rates of increase. For example, 1992 was low (1.9 PgC/yr), and 1998 was the highest (6.0 PgC/yr) since direct measurements began in 1957. This variability is mainly caused by variations in land and ocean uptake.

Statistically, high rates of increase in atmospheric CO$_2$ have occurred in most El Niño years, although low rates occurred during the extended El Niño of 1991 to 1994. Surface water CO$_2$ measurements from the equatorial Pacific show that the natural source of CO$_2$ from this region is reduced by between 0.2 and 1.0 PgC/yr during El Niño events, counter to the atmospheric increase. It is likely that the high rates of CO$_2$ increase during most El Niño events are explained by reductions in land uptake, caused in part by the effects of high temperatures, drought and fire on terrestrial ecosystems in the tropics.

Land and ocean uptake of CO$_2$ can now be separated using atmospheric measurements (CO$_2$, oxygen (O$_2$) and ^{13}CO$_2$). For 1980 to 1989, the ocean-atmosphere flux is estimated as −1.9 ± 0.6 PgC/yr and the land-atmosphere flux as −0.2 ± 0.7 PgC/yr based on CO$_2$ and O$_2$ measurements (negative signs denote net uptake). For 1990 to 1999, the ocean-atmosphere flux is estimated as −1.7 ± 0.5 PgC/yr and the land-atmosphere flux as −1.4 ± 0.7 PgC/yr. These figures are consistent with alternative budgets based on CO$_2$ and ^{13}CO$_2$ measurements, and with independent estimates based on measurements of CO$_2$ and ^{13}CO$_2$ in sea water. The new 1980s estimates are also consistent with the ocean-model based carbon budget of the IPCC WGI Second Assessment Report (IPCC, 1996a) (hereafter SAR). The new 1990s estimates update the budget derived using SAR methodologies for the IPCC Special Report on Land Use, Land Use Change and Forestry (IPCC, 2000a).

The net CO$_2$ release due to land-use change during the 1980s has been estimated as 0.6 to 2.5 PgC/yr (central estimate 1.7 PgC/yr). This net CO$_2$ release is mainly due to deforestation in the tropics. Uncertainties about land-use changes limit the accuracy of these estimates. Comparable data for the 1990s are not yet available.

The land-atmosphere flux estimated from atmospheric observations comprises the *balance* of net CO$_2$ release due to land-use changes and CO$_2$ uptake by terrestrial systems (the "residual terrestrial sink"). The residual terrestrial sink is estimated as −1.9 PgC/yr (range −3.8 to +0.3 PgC/yr) during the 1980s. It has several likely causes, including changes in land management practices and fertilisation effects of increased atmospheric CO$_2$ and nitrogen (N) deposition, leading to increased vegetation and soil carbon.

Modelling based on atmospheric observations (inverse modelling) enables the land-atmosphere and ocean-atmosphere fluxes to be partitioned between broad latitudinal bands. The sites of anthropogenic CO$_2$ uptake in the ocean are not resolved by inverse modelling because of the large, natural background air-sea fluxes (outgassing in the tropics and uptake in high latitudes). Estimates of the land-atmosphere flux north of 30°N during 1980 to 1989 range from −2.3 to −0.6 PgC/yr; for the tropics, −1.0 to +1.5 PgC/yr. These results imply substantial terrestrial sinks for anthropogenic CO$_2$ in the northern extra-tropics, and in the tropics (to balance deforestation). The pattern for the 1980s persisted into the 1990s.

Terrestrial carbon inventory data indicate carbon sinks in northern and tropical forests, consistent with the results of inverse modelling.

East-west gradients of atmospheric CO$_2$ concentration are an order of magnitude smaller than north-south gradients. Estimates of continental-scale CO$_2$ balance are possible in principle but are poorly constrained because there are too few well-calibrated CO$_2$ monitoring sites, especially in the interior of continents, and insufficient data on air-sea fluxes and vertical transport in the atmosphere.

The global carbon cycle and anthropogenic CO$_2$

The global carbon cycle operates through a variety of response and feedback mechanisms. The most relevant for decade to century time-scales are listed here.

Responses of the carbon cycle to changing CO$_2$ concentrations

- Uptake of anthropogenic CO$_2$ by the ocean is primarily governed by ocean circulation and carbonate chemistry. So long as atmospheric CO$_2$ concentration is increasing there is net uptake of carbon by the ocean, driven by the atmosphere-ocean difference in partial pressure of CO$_2$. The fraction of anthropogenic CO$_2$ that is taken up by the ocean declines with increasing CO$_2$ concentration, due to reduced buffer capacity of the carbonate system. The fraction taken up by the ocean also declines with the rate of increase of atmospheric CO$_2$, because the rate of mixing between deep water and surface water limits CO$_2$ uptake.

- Increasing atmospheric CO$_2$ has no significant fertilisation effect on marine biological productivity, but it decreases pH. Over a century, changes in marine biology brought about by changes in calcification at low pH could increase the ocean uptake of CO$_2$ by a few percentage points.

- Terrestrial uptake of CO_2 is governed by net biome production (NBP), which is the balance of net primary production (NPP) and carbon losses due to heterotrophic respiration (decomposition and herbivory) and fire, including the fate of harvested biomass. NPP increases when atmospheric CO_2 concentration is increased above present levels (the "fertilisation" effect occurs directly through enhanced photosynthesis, and indirectly through effects such as increased water use efficiency). At high CO_2 concentration (800 to 1,000 ppm) any further direct CO_2 fertilisation effect is likely to be small. The effectiveness of terrestrial uptake as a carbon sink depends on the transfer of carbon to forms with long residence times (wood or modified soil organic matter). Management practices can enhance the carbon sink because of the inertia of these "slow" carbon pools.

Feedbacks in the carbon cycle due to climate change
- Warming reduces the solubility of CO_2 and therefore reduces uptake of CO_2 by the ocean.
- Increased vertical stratification in the ocean is likely to accompany increasing global temperature. The likely consequences include reduced outgassing of upwelled CO_2, reduced transport of excess carbon to the deep ocean, and changes in biological productivity.
- On short time-scales, warming increases the rate of heterotrophic respiration on land, but the extent to which this effect can alter land-atmosphere fluxes over longer time-scales is not yet clear. Warming, and regional changes in precipitation patterns and cloudiness, are also likely to bring about changes in terrestrial ecosystem structure, geographic distribution and primary production. The net effect of climate on NBP depends on regional patterns of climate change.

Other impacts on the carbon cycle
- Changes in management practices are very likely to have significant effects on the terrestrial carbon cycle. In addition to deforestation and afforestation/reforestation, more subtle management effects can be important. For example, fire suppression (e.g., in savannas) reduces CO_2 emissions from burning, and encourages woody plant biomass to increase. On agricultural lands, some of the soil carbon lost when land was cleared and tilled can be regained through adoption of low-tillage agriculture.
- Anthropogenic N deposition is increasing terrestrial NPP in some regions; excess tropospheric ozone (O_3) is likely to be reducing NPP.
- Anthropogenic inputs of nutrients to the oceans by rivers and atmospheric dust may influence marine biological productivity, although such effects are poorly quantified.

Modelling and projection of CO_2 concentration

Process-based models of oceanic and terrestrial carbon cycling have been developed, compared and tested against *in situ* measurements and atmospheric measurements. The following are consistent results based on several models.
- Modelled ocean-atmosphere flux during 1980 to 1989 was in the range −1.5 to −2.2 PgC/yr for the 1980s, consistent with earlier model estimates and consistent with the atmospheric budget.
- Modelled land-atmosphere flux during 1980 to 1989 was in the range −0.3 to −1.5 PgC/yr, consistent with or slightly more negative than the land-atmosphere flux as indicated by the atmospheric budget. CO_2 fertilisation and anthropogenic N deposition effects contributed significantly: their combined effect was estimated as −1.5 to −3.1 PgC/yr. Effects of climate change during the 1980s were small, and of uncertain sign.
- In future projections with ocean models, driven by CO_2 concentrations derived from the IS92a scenario (for illustration and comparison with earlier work), ocean uptake becomes progressively larger towards the end of the century, but represents a smaller fraction of emissions than today. When climate change feedbacks are included, ocean uptake becomes less in all models, when compared with the situation without climate feedbacks.
- In analogous projections with terrestrial models, the rate of uptake by the land due to CO_2 fertilisation increases until mid-century, but the models project smaller increases, or no increase, after that time. When climate change feedbacks are included, land uptake becomes less in all models, when compared with the situation without climate feedbacks. Some models have shown a rapid decline in carbon uptake after the mid-century.

Two simplified, fast models (ISAM and Bern-CC) were used to project future CO_2 concentrations under IS92a and six SRES scenarios, and to project future emissions under five CO_2 stabilisation scenarios. Both models represent ocean and terrestrial climate feedbacks, in a way consistent with process-based models, and allow for uncertainties in climate sensitivity and in ocean and terrestrial responses to CO_2 and climate.

- The reference case projections (which include climate feedbacks) of both models under IS92a are, by coincidence, close to those made in the SAR (which neglected feedbacks).
- The SRES scenarios lead to divergent CO_2 concentration trajectories. Among the six emissions scenarios considered, the projected range of CO_2 concentrations at the end of the century is 550 to 970 ppm (ISAM model) or 540 to 960 ppm (Bern-CC model).
- Variations in climate sensitivity and ocean and terrestrial model responses add at least −10 to +30% uncertainty to these values, and to the emissions implied by the stabilisation scenarios.
- The net effect of land and ocean climate feedbacks is always to increase projected atmospheric CO_2 concentrations. This is equivalent to reducing the allowable emissions for stabilisation at any one CO_2 concentration.
- New studies with general circulation models including interactive land and ocean carbon cycle components also indicate that climate feedbacks have the potential to increase atmospheric CO_2 but with large uncertainty about the magnitude of the terrestrial biosphere feedback.

Implications

CO_2 emissions from fossil fuel burning are virtually certain to be the dominant factor determining CO_2 concentrations during the 21st century. There is scope for land-use changes to increase or decrease CO_2 concentrations on this time-scale. If all of the carbon so far released by land-use changes could be restored to the terrestrial biosphere, CO_2 at the end of the century would be 40 to 70 ppm less than it would be if no such intervention had occurred. By comparison, global deforestation would add two to four times more CO_2 to the atmosphere than reforestation of all cleared areas would subtract.

There is sufficient uptake capacity in the ocean to incorporate 70 to 80% of foreseeable anthropogenic CO_2 emissions to the atmosphere, this process takes centuries due to the rate of ocean mixing. As a result, even several centuries after emissions occurred, about a quarter of the increase in concentration caused by these emissions is still present in the atmosphere.

CO_2 stabilisation at 450, 650 or 1,000 ppm would require global anthropogenic CO_2 emissions to drop below 1990 levels, within a few decades, about a century, or about two centuries respectively, and continue to steadily decrease thereafter. Stabilisation requires that net anthropogenic CO_2 emissions ultimately decline to the level of persistent natural land and ocean sinks, which are expected to be small (<0.2 PgC/yr).

3.1 Introduction

The concentration of CO_2 in the atmosphere has risen from close to 280 parts per million (ppm) in 1800, at first slowly and then progressively faster to a value of 367 ppm in 1999, echoing the increasing pace of global agricultural and industrial development. This is known from numerous, well-replicated measurements of the composition of air bubbles trapped in Antarctic ice. Atmospheric CO_2 concentrations have been measured directly with high precision since 1957; these measurements agree with ice-core measurements, and show a continuation of the increasing trend up to the present.

Several additional lines of evidence confirm that the recent and continuing increase of atmospheric CO_2 content is caused by anthropogenic CO_2 emissions – most importantly fossil fuel burning. First, atmospheric O_2 is declining at a rate comparable with fossil fuel emissions of CO_2 (combustion consumes O_2). Second, the characteristic isotopic signatures of fossil fuel (its lack of ^{14}C, and depleted content of ^{13}C) leave their mark in the atmosphere. Third, the increase in observed CO_2 concentration has been faster in the northern hemisphere, where most fossil fuel burning occurs.

Atmospheric CO_2 is, however, increasing only at about half the rate of fossil fuel emissions; the rest of the CO_2 emitted either dissolves in sea water and mixes into the deep ocean, or is taken up by terrestrial ecosystems. Uptake by terrestrial ecosystems is due to an excess of primary production (photosynthesis) over respiration and other oxidative processes (decomposition or combustion of organic material). Terrestrial systems are also an anthropogenic source of CO_2 when land-use changes (particularly deforestation) lead to loss of carbon from plants and soils. Nonetheless, the global balance in terrestrial systems is currently a net uptake of CO_2.

The part of fossil fuel CO_2 that is taken up by the ocean and the part that is taken up by the land can be calculated from the changes in atmospheric CO_2 and O_2 content because terrestrial processes of CO_2 exchange involve exchange of oxygen whereas dissolution in the ocean does not. Global carbon budgets based on CO_2 and O_2 measurements for the 1980s and 1990s are shown in Table 3.1. The human influence on the fluxes of carbon among the three "reservoirs" (atmosphere, ocean, and terrestrial biosphere) represent a small but significant perturbation of a huge global cycle (Figure 3.1).

This chapter summarises current knowledge of the global carbon cycle, with special reference to the fate of fossil fuel CO_2 and the factors that influence the uptake or release of CO_2 by the oceans and land. These factors include atmospheric CO_2 concentration itself, the naturally variable climate, likely climate changes caused by increasing CO_2 and other greenhouse gases, changes in ocean circulation and biology, fertilising effects of atmospheric CO_2 and nitrogen deposition, and direct human actions such as land conversion (from native vegetation to agriculture and vice versa), fire suppression and land management for carbon storage as provided for by the Kyoto Protocol (IPCC, 2000a). Any changes in the function of either the terrestrial biosphere or the ocean – whether intended or not – could potentially have significant effects, manifested within years to decades, on the fraction of fossil fuel CO_2 that stays in the atmosphere. This perspective has

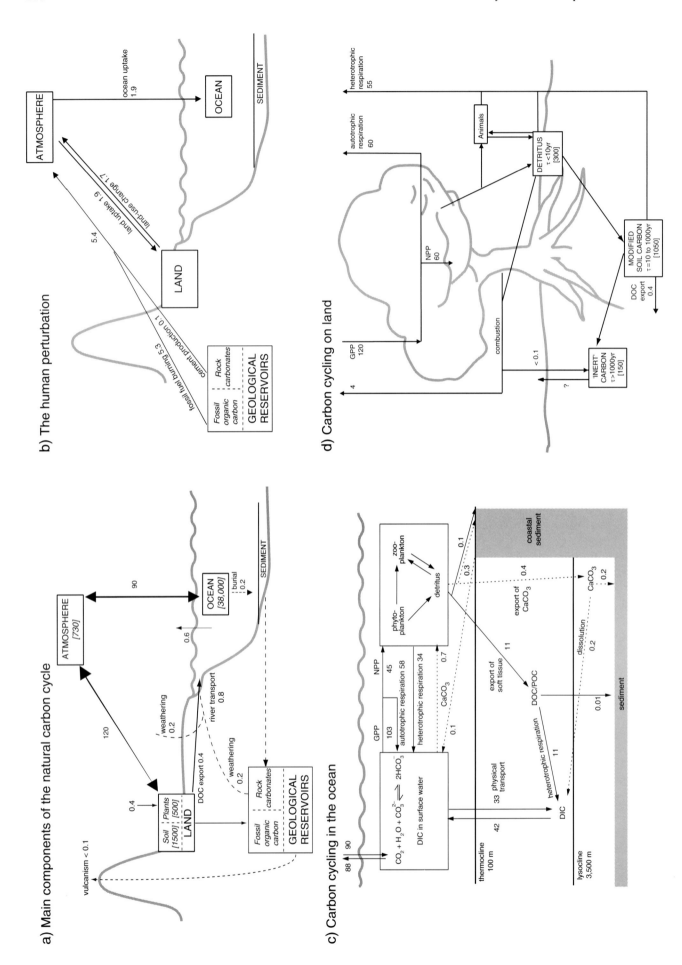

a) Main components of the natural carbon cycle

b) The human perturbation

c) Carbon cycling in the ocean

d) Carbon cycling on land

driven a great deal of research during the years since the IPCC WGI Second Assessment report (IPCC, 1996) (hereafter SAR) (Schimel *et al.,* 1996; Melillo *et al.,* 1996; Denman *et al.,* 1996). Some major areas where advances have been made since the SAR are as follows:

- Observational research (atmospheric, marine and terrestrial) aimed at a better quantification of carbon fluxes on local, regional and global scales. For example, improved precision and repeatability in atmospheric CO_2 and stable isotope measurements; the development of highly precise methods to measure changes in atmospheric O_2 concentrations; local terrestrial CO_2 flux measurements from towers, which are now being performed continuously in many terrestrial ecosystems; satellite observations of global land cover and change; and enhanced monitoring of geographical, seasonal and interannual variations of biogeochemical parameters in the sea, including measurements of the partial pressure of CO_2 (pCO_2) in surface waters.

- Experimental manipulations, for example: laboratory and greenhouse experiments with raised and lowered CO_2 concentrations; field experiments on ecosystems using free-air carbon dioxide enrichment (FACE) and open-top chamber studies of raised CO_2 effects, studies of soil warming and nutrient enrichment effects; and *in situ* fertilisation experiments on marine ecosystems and associated pCO_2 measurements.

- Theory and modelling, especially applications of atmospheric transport models to link atmospheric observations to surface fluxes (inverse modelling); the development of process-based models of terrestrial and marine carbon cycling and programmes to compare and test these models against observations; and the use of such models to project climate feedbacks on the uptake of CO_2 by the oceans and land.

As a result of this research, there is now a more firmly based knowledge of several central features of the carbon cycle. For example:

- Time series of atmospheric CO_2, O_2 and $^{13}CO_2$ measurements have made it possible to observationally constrain the partitioning of CO_2 between terrestrial and oceanic uptake and to confirm earlier budgets, which were partly based on model results.

- *In situ* experiments have explored the nature and extent of CO_2 responses in a variety of terrestrial ecosystems (including forests), and have confirmed the existence of iron limitations on marine productivity.

- Process-based models of terrestrial and marine biogeochemical processes have been used to represent a complex array of feedbacks in the carbon cycle, allowing the net effects of these processes to be estimated for the recent past and for future scenarios.

Figure 3.1: The global carbon cycle: storages (PgC) and fluxes (PgC/yr) estimated for the 1980s. (a) *Main components of the natural cycle.* The thick arrows denote the most important fluxes from the point of view of the contemporary CO_2 balance of the atmosphere: gross primary production and respiration by the land biosphere, and physical air-sea exchange. These fluxes are approximately balanced each year, but imbalances can affect atmospheric CO_2 concentration significantly over years to centuries. The thin arrows denote additional natural fluxes (dashed lines for fluxes of carbon as $CaCO_3$), which are important on longer time-scales. The flux of 0.4 PgC/yr from atmospheric CO_2 via plants to inert soil carbon is approximately balanced on a time-scale of several millenia by export of dissolved organic carbon (DOC) in rivers (Schlesinger, 1990). A further 0.4 PgC/yr flux of dissolved inorganic carbon (DIC) is derived from the weathering of $CaCO_3$, which takes up CO_2 from the atmosphere in a 1:1 ratio. These fluxes of DOC and DIC together comprise the river transport of 0.8 PgC/yr. In the ocean, the DOC from rivers is respired and released to the atmosphere, while $CaCO_3$ production by marine organisms results in half of the DIC from rivers being returned to the atmosphere and half being buried in deep-sea sediments – which are the precursor of carbonate rocks. Also shown are processes with even longer time-scales: burial of organic matter as fossil organic carbon (including fossil fuels), and outgassing of CO_2 through tectonic processes (vulcanism). Emissions due to vulcanism are estimated as 0.02 to 0.05 PgC/yr (Williams *et al.,* 1992; Bickle, 1994). (b) *The human perturbation* (data from Table 3.1). Fossil fuel burning and land-use change are the main anthropogenic processes that release CO_2 to the atmosphere. Only a part of this CO_2 stays in the atmosphere; the rest is taken up by the land (plants and soil) or by the ocean. These uptake components represent imbalances in the large natural two-way fluxes between atmosphere and ocean and between atmosphere and land. (c) *Carbon cycling in the ocean.* CO_2 that dissolves in the ocean is found in three main forms (CO_2, CO_3^{2-}, HCO_3^-, the sum of which is DIC). DIC is transported in the ocean by physical and biological processes. Gross primary production (GPP) is the total amount of organic carbon produced by photosynthesis (estimate from Bender *et al.,* 1994); net primary production (NPP) is what is what remains after autotrophic respiration, i.e., respiration by photosynthetic organisms (estimate from Falkowski *et al.,* 1998). Sinking of DOC and particulate organic matter (POC) of biological origin results in a downward flux known as export production (estimate from Schlitzer, 2000). This organic matter is tranported and respired by non-photosynthetic organisms (heterotrophic respiration) and ultimately upwelled and returned to the atmosphere. Only a tiny fraction is buried in deep-sea sediments. Export of $CaCO_3$ to the deep ocean is a smaller flux than total export production (0.4 PgC/yr) but about half of this carbon is buried as $CaCO_3$ in sediments; the other half is dissolved at depth, and joins the pool of DIC (Milliman, 1993). Also shown are approximate fluxes for the shorter-term burial of organic carbon and $CaCO_3$ in coastal sediments and the re-dissolution of a part of the buried $CaCO_3$ from these sediments. (d) *Carbon cycling on land.* By contrast with the ocean, most carbon cycling through the land takes place locally within ecosystems. About half of GPP is respired by plants. The remainer (NPP) is approximately balanced by heterotrophic respiration with a smaller component of direct oxidation in fires (combustion). Through senescence of plant tissues, most of NPP joins the detritus pool; some detritus decomposes (i.e., is respired and returned to the atmosphere as CO_2) quickly while some is converted to modified soil carbon, which decomposes more slowly. The small fraction of modified soil carbon that is further converted to compounds resistant to decomposition, and the small amount of black carbon produced in fires, constitute the "inert" carbon pool. It is likely that biological processes also consume much of the "inert" carbon as well but little is currently known about these processes. Estimates for soil carbon amounts are from Batjes (1996) and partitioning from Schimel *et al.* (1994) and Falloon *et al.* (1998). The estimate for the combustion flux is from Scholes and Andreae (2000). 'τ' denotes the turnover time for different components of soil organic matter.

Box 3.1: Measuring terrestrial carbon stocks and fluxes.

Estimating the carbon stocks in terrestrial ecosystems and accounting for changes in these stocks requires adequate information on land cover, carbon density in vegetation and soils, and the fate of carbon (burning, removals, decomposition). Accounting for changes in all carbon stocks in all areas would yield the net carbon exchange between terrestrial ecosystems and the atmosphere (NBP).

Global land cover maps show poor agreement due to different definitions of cover types and inconsistent sources of data (de Fries and Townshend, 1994). Land cover changes are difficult to document, uncertainties are large, and historical data are sparse. Satellite imagery is a valuable tool for estimating land cover, despite problems with cloud cover, changes at fine spatial scales, and interpretation (for example, difficulties in distinguishing primary and secondary forest). Aerial photography and ground measurements can be used to validate satellite-based observations.

The carbon density of vegetation and soils has been measured in numerous ecological field studies that have been aggregated to a global scale to assess carbon stocks and NPP (e.g., Atjay *et al.*, 1979; Olson *et al.*, 1983; Saugier and Roy, 2001; Table 3.2), although high spatial and temporal heterogeneity and methodological differences introduce large uncertainties. Land inventory studies tend to measure the carbon stocks in vegetation and soils over larger areas and/or longer time periods. For example, the United Nations Food and Agricultural Organisation (FAO) has been compiling forest inventories since 1946 providing detailed data on carbon stocks, often based on commercial wood production data. Inventory studies include managed forests with mixed age stands, thus average carbon stock values are often lower than those based on ecological site studies, which have generally been carried out in relatively undisturbed, mature ecosystems. Fluxes of carbon can be estimated from changes in inventoried carbon stocks (e.g., UN-ECE/FAO, 2000), or from combining data on land-use change with methods to calculate changes in carbon stock (e.g., Houghton, 1999). The greatest uncertainty in both methods is in estimating the fate of the carbon: the fraction which is burned, rates of decomposition, the effect of burning and harvesting on soil carbon, and subsequent land management.

Ecosystem-atmosphere CO_2 exchange on short time-scales can be measured using micrometeorological techniques such as eddy covariance, which relies on rapidly responding sensors mounted on towers to resolve the net flux of CO_2 between a patch of land and the atmosphere (Baldocchi *et al.*, 1988). The annual integral of the measured CO_2 exchange is approximately equivalent to NEP (Wofsy *et al.*, 1993; Goulden *et. al*, 1996; Aubinet *et al.*, 2000). This innovation has led to the establishment of a rapidly expanding network of long-term monitoring sites (FLUXNET) with many sites now operating for several years, improving the understanding of the physiological and ecological processes that control NEP (e.g., Valentini *et al.*, 2000). The distribution of sites is currently biased toward regrowing forests in the Northern Hemisphere, and there are still technical problems and uncertainties, although these are being tackled. Current flux measurement techniques typically integrate processes at a scale less than 1 km^2.

Table 3.1: *Global CO_2 budgets (in PgC/yr) based on intra-decadal trends in atmospheric CO_2 and O_2. Positive values are fluxes to the atmosphere; negative values represent uptake from the atmosphere. The fossil fuel emissions term for the 1980s (Marland et al., 2000) has been slightly revised downward since the SAR. Error bars denote uncertainty ($\pm 1\sigma$), not interannual variability, which is substantially greater.*

	1980s	**1990s**
Atmospheric increase	3.3 ± 0.1	3.2 ± 0.1
Emissions (fossil fuel, cement)	5.4 ± 0.3	6.3 ± 0.4
Ocean-atmosphere flux	-1.9 ± 0.6	-1.7 ± 0.5
Land-atmosphere flux*	-0.2 ± 0.7	-1.4 ± 0.7
partitioned as follows		
Land-use change	1.7 (0.6 to 2.5)	NA
Residual terrestrial sink	-1.9 (-3.8 to 0.3)	NA

* The land-atmosphere flux represents the balance of a positive term due to land-use change and a residual terrestrial sink. The two terms cannot be separated on the basis of current atmospheric measurements. Using independent analyses to estimate the land-use change component for the 1980s based on Houghton (1999), Houghton and Hackler (1999), Houghton *et al.* (2000), and the CCMLP (McGuire *et al.*, 2001) the residual terrestrial sink can be inferred for the 1980s. Comparable global data on land-use changes through the 1990s are not yet available.

3.2 Terrestrial and Ocean Biogeochemistry: Update on Processes

3.2.1 Overview of the Carbon Cycle

The first panel of Figure 3.1 shows the major components of the carbon cycle, estimates of the current storage in the active compartments, and estimates of the gross fluxes between compartments. The second panel shows best estimates of the additional flux (release to the atmosphere – positive; uptake – negative) associated with the human perturbation of the carbon cycle during the 1980s. Note that the gross amounts of carbon annually exchanged between the ocean and atmosphere, and between the land and atmosphere, represent a sizeable fraction of the atmospheric CO_2 content – and are many times larger than the total anthropogenic CO_2 input. In consequence, an imbalance in these exchanges could easily lead to an anomaly of comparable magnitude to the direct anthropogenic perturbation. This implies that it is important to consider how these fluxes may be changing in response to human activities.

To understand how the changing global environment may alter the carbon cycle, it is necessary to further analyse the fluxes and examine the physicochemical and biological processes that determine them. The remaining two panels of Figure 3.1 indicate the main constituent fluxes in the terrestrial and marine systems, with current estimates of their magnitude. The following sections explain the controls on these fluxes, with special reference to processes by which anthropogenic changes may influence the overall carbon balance of the land and oceans on time-scales from years to centuries.

3.2.2 Terrestrial Carbon Processes

3.2.2.1 Background

Higher plants acquire CO_2 by diffusion through tiny pores (stomata) into leaves and thus to the sites of photosynthesis. The total amount of CO_2 that dissolves in leaf water amounts to about 270 PgC/yr, i.e., more than one-third of all the CO_2 in the atmosphere (Farquhar *et al.*, 1993; Ciais *et al.*, 1997). This quantity is measurable because this CO_2 has time to exchange oxygen atoms with the leaf water and is imprinted with the corresponding ^{18}O "signature" (Francey and Tans, 1987; Farquhar *et al.*, 1993). Most of this CO_2 diffuses out again without participating in photosynthesis. The amount that is "fixed" from the atmosphere, i.e., converted from CO_2 to carbohydrate during photosynthesis, is known as gross primary production (GPP). Terrestrial GPP has been estimated as about 120 PgC/yr based on ^{18}O measurements of atmospheric CO_2 (Ciais *et al.*, 1997). This is also the approximate value necessary to support observed plant growth, assuming that about half of GPP is incorporated into new plant tissues such as leaves, roots and wood, and the other half is converted back to atmospheric CO_2 by autotrophic respiration (respiration by plant tissues) (Lloyd and Farquhar, 1996; Waring *et al.*, 1998).

Annual plant growth is the difference between photosynthesis and autotrophic respiration, and is referred to as net primary production (NPP). NPP has been measured in all major

ecosystem types by sequential harvesting or by measuring plant biomass (Hall *et al.*, 1993). Global terrestrial NPP has been estimated at about 60 PgC/yr through integration of field measurements (Table 3.2) (Atjay *et al.*, 1979; Saugier and Roy, 2001). Estimates from remote sensing and atmospheric CO_2 data (Ruimy *et al.*, 1994; Knorr and Heimann, 1995) concur with this value, although there are large uncertainties in all methods. Eventually, virtually all of the carbon fixed in NPP is returned to the atmospheric CO_2 pool through two processes: heterotrophic respiration (Rh) by decomposers (bacteria and fungi feeding on dead tissue and exudates) and herbivores; and combustion in natural or human-set fires (Figure 3.1d).

Most dead biomass enters the detritus and soil organic matter pools where it is respired at a rate that depends on the chemical composition of the dead tissues and on environmental conditions (for example, low temperatures, dry conditions and flooding slow down decomposition). Conceptually, several soil carbon pools are distinguished. Detritus and microbial biomass have a short turnover time (<10 yr). Modified soil organic carbon has decadal to centennial turnover time. Inert (stable or recalcitrant) soil organic carbon is composed of molecules more or less resistant to further decomposition. A very small fraction of soil organic matter, and a small fraction of burnt biomass, are converted into inert forms (Schlesinger, 1990; Kuhlbusch *et al.*, 1996). Natural processes and management regimes may reduce or increase the amount of carbon stored in pools with turnover times on the order of tens to hundreds of years (living wood, wood products and modified soil organic matter) and thus influence the time evolution of atmospheric CO_2 over the century.

The difference between NPP and Rh determines how much carbon is lost or gained by the ecosystem in the absence of disturbances that remove carbon from the ecosystem (such as harvest or fire). This carbon balance, or net ecosystem production (NEP), can be estimated from changes in carbon stocks, or by measuring the fluxes of CO_2 between patches of land and the atmosphere (see Box 3.1). Annual NEP flux measurements are in the range 0.7 to 5.9 MgC/ha/yr for tropical forests and 0.8 to 7.0 MgC/ha/yr for temperate forests; boreal forests can reach up to 2.5 MgC/ha/yr although they have been shown to be carbon-neutral or to release carbon in warm and/or cloudy years (Valentini *et al.*, 2000). Integration of these and other results leads to an estimated global NEP of about 10 PgC/yr, although this is likely to be an overestimate because of the current biased distribution of flux measuring sites (Bolin *et al.*, 2000).

When other losses of carbon are accounted for, including fires, harvesting/removals (eventually combusted or decomposed), erosion and export of dissolved or suspended organic carbon (DOC) by rivers to the oceans (Schlesinger and Melack, 1981; Sarmiento and Sundquist; 1992), what remains is the net biome production (NBP), i.e., the carbon accumulated by the terrestrial biosphere (Schulze and Heimann, 1998). This is what the atmosphere ultimately "sees" as the net land uptake on a global scale over periods of a year or more. NBP is estimated in this chapter to have averaged -0.2 ± 0.7 PgC/yr during the 1980s and -1.4 ± 0.7 PgC/yr during the 1990s, based on atmospheric measurements of CO_2 and O_2 (Section 3.5.1 and Table 3.1).

Box 3.2: Maximum impacts of reforestation and deforestation on atmospheric CO_2.

Rough upper bounds for the impact of reforestation on atmospheric CO_2 concentration over a century time-scale can be calculated as follows. Cumulative carbon losses to the atmosphere due to land-use change during the past 1 to 2 centuries are estimated as 180 to 200 PgC (de Fries *et al.*, 1999) and cumulative fossil fuel emissions to year 2000 as 280 PgC (Marland *et al.*, 2000), giving cumulative anthropogenic emissions of 480 to 500 PgC. Atmospheric CO_2 content has increased by 90 ppm (190 PgC). Approximately 40% of anthropogenic CO_2 emissions has thus remained in the atmosphere; the rest has been taken up by the land and oceans in roughly equal proportions (see main text). Conversely, if land-use change were completely reversed over the 21st century, a CO_2 reduction of $0.40 \times 200 = 80$ PgC (about 40 ppm) might be expected. This calculation assumes that future ecosystems will not store more carbon than pre-industrial ecosystems, and that ocean uptake will be less because of lower CO_2 concentration in the atmosphere (see Section 3.2.3.1) .

A higher bound can be obtained by assuming that the carbon taken up by the land during the past 1 to 2 centuries, i.e. about half of the carbon taken up by the land and ocean combined, will be retained there. This calculation yields a CO_2 reduction of $0.70 \times 200 = 140$ PgC (about 70 ppm). These calculations are not greatly influenced by the choice of reference period. Both calculations require the extreme assumption that a large proportion of today's agricultural land is returned to forest.

The maximum impact of total deforestation can be calculated in a similar way. Depending on different assumptions about vegetation and soil carbon density in different ecosystem types (Table 3.2) and the proportion of soil carbon lost during deforestation (20 to 50%; IPCC, 1997), complete conversion of forests to climatically equivalent grasslands would add 400 to 800 PgC to the atmosphere. Thus, global deforestation could theoretically add two to four times more CO_2 to the atmosphere than could be subtracted by reforestation of cleared areas.

Table 3.2: Estimates of terrestrial carbon stocks and NPP (global aggregated values by biome).

Biome	Area (10⁹ ha)		Global Carbon Stocks (PgC)[f]						Carbon density (MgC/ha)				NPP (PgC/yr)	
	WBGU[a]	MRS[b]	WBGU[a]			MRS[b]		IGBP[c]	WBGU[a]		MRS[b]	IGBP[c]	Atjay[a]	MRS[b]
			Plants	Soil	Total	Plants	Soil	Total	Plants	Soil	Plants	Soil		
Tropical forests	1.76	1.75	212	216	428	340	213	553	120	123	194	122	13.7	21.9
Temperate forests	1.04	1.04	59	100	159	139[e]	153	292	57	96	134	147	6.5	8.1
Boreal forests	1.37	1.37	88[d]	471	559	57	338	395	64	344	42	247	3.2	2.6
Tropical savannas & grasslands	2.25	2.76	66	264	330	79	247	326	29	117	29	90	17.7	14.9
Temperate grasslands & shrublands	1.25	1.78	9	295	304	23	176	199	7	236	13	99	5.3	7.0
Deserts and semi deserts	4.55[h]	2.77	8	191	199	10	159	169	2	42	4	57	1.4	3.5
Tundra	0.95	0.56	6	121	127	2	115	117	6	127	4	206	1.0	0.5
Croplands	1.60	1.35	3	128	131	4	165	169	2	80	3	122	6.8	4.1
Wetlands[g]	0.35	–	15	225	240	–	–	–	43	643	–	–	4.3	–
Total	15.12	14.93[h]	466	2011	2477	654	1567	2221					59.9	62.6

[a] WBGU (1988): forest data from Dixon *et al.* (1994); other data from Atjay *et al.* (1979).

[b] MRS: Mooney, Roy and Saugier (MRS) (2001). Temperate grassland and Mediterranean shrubland categories combined.

[c] IGBP-DIS (International Geosphere-Biosphere Programme – Data Information Service) soil carbon layer (Carter and Scholes, 2000) overlaid with De Fries *et al.* (1999) current vegetation map to give average ecosystem soil carbon.

[d] WBGU boreal forest vegetation estimate is likely to be to high, due to high Russian forest density estimates including standing dead biomass.

[e] MRS temperate forest estimate is likely to be too high, being based on mature stand density.

[f] Soil carbon values are for the top 1 m, although stores are also high below this depth in peatlands and tropical forests.

[g] Variations in classification of ecosystems can lead to inconsistencies. In particular, wetlands are not recognised in the MRS classification.

[h] Total land area of 14.93×10^9 ha in MRS includes 1.55×10^9 ha ice cover not listed in this table. InWBGU, ice is included in deserts and semi-deserts category.

By definition, for an ecosystem in steady state, Rh and other carbon losses would just balance NPP, and NBP would be zero. In reality, human activities, natural disturbances and climate variability alter NPP and Rh, causing transient changes in the terrestrial carbon pool and thus non-zero NBP. If the rate of carbon input (NPP) changes, the rate of carbon output (Rh) also changes, in proportion to the altered carbon content; but there is a time lag between changes in NPP and changes in the slower responding carbon pools. For a *step* increase in NPP, NBP is expected to increase at first but to relax towards zero over a period of years to decades as the respiring pool "catches up". The globally averaged lag required for Rh to catch up with a change in NPP has been estimated to be of the order of 10 to 30 years (Raich and Schlesinger, 1992). A *continuous* increase in NPP is expected to produce a sustained positive NBP, so long as NPP is still increasing, so that the increased terrestrial carbon has not been processed through the respiring carbon pools (Taylor and Lloyd, 1992; Friedlingstein *et al.,* 1995a; Thompson *et al.,* 1996; Kicklighter *et al.,* 1999), and provided that the increase is not outweighed by compensating increases in mortality or disturbance.

The terrestrial system is currently acting as a global sink for carbon (Table 3.1) despite large releases of carbon due to deforestation in some regions. Likely mechanisms for the sink are known, but their relative contribution is uncertain. Natural climate variability and disturbance regimes (including fire and herbivory) affect NBP through their impacts on NPP, allocation to long- versus short-lived tissues, chemical and physical properties of litter, stocks of living biomass, stocks of detritus and soil carbon, environmental controls on decomposition and rates of biomass removal. Human impacts occur through changes in land use and land management, and through indirect mechanisms including climate change, and fertilisation due to elevated CO_2 and deposition of nutrients (most importantly, reactive nitrogen). These mechanisms are discussed individually in the following sections.

3.2.2.2 Effects of changes in land use and land management
Changes in land use and management affect the amount of carbon in plant biomass and soils. Historical cumulative carbon losses due to changes in land use have been estimated to be 180 to 200 PgC by comparing maps of "natural" vegetation in the absence of human disturbance (derived from ground-based information (Matthews, 1983) or from modelled potential vegetation based on climate (Leemans, 1990)) to a map of current vegetation derived from 1987 satellite data (de Fries *et al.,* 1999). Houghton (1999, 2000) estimated emissions of 121 PgC (approximately 60% in tropical areas and 40% in temperate areas) for the period 1850 to 1990 from statistics on land-use change, and a simple model tracking rates of decomposition from different pools and rates of regrowth on abandoned or reforested land. There was substantial deforestation in temperate areas prior to 1850, and this may be partially reflected in the difference between these two analyses. The estimated land-use emissions during 1850 to 1990 of 121 PgC (Houghton, 1999, 2000) can be compared to estimated net terrestrial flux of 39 PgC to the atmosphere over the same period inferred from an atmospheric

increase of 144 PgC (Etheridge *et al.,* 1996; Keeling and Whorf, 2000), a release of 212 PgC due to fossil fuel burning (Marland *et al.,* 2000), and a modelled ocean-atmosphere flux of about −107 PgC (Gruber, 1998, Sabine *et al.,* 1999, Feely *et al.,* 1999a). The difference between the net terrestrial flux and estimated land-use change emissions implies a residual land-atmosphere flux of −82 PgC (i.e., a terrestrial sink) over the same period. Box 3.2 indicates the theoretical upper bounds for additional carbon storage due to land-use change, similar bounds for carbon loss by continuing deforestation, and the implications of these calculations for atmospheric CO_2.

Land use responds to social and economic pressures to provide food, fuel and wood products, for subsistence use or for export. Land clearing can lead to soil degradation, erosion and leaching of nutrients, and may therefore reduce the subsequent ability of the ecosystem to act as a carbon sink (Taylor and Lloyd, 1992). Ecosystem conservation and management practices can restore, maintain and enlarge carbon stocks (IPCC, 2000a). Fire is important in the carbon budget of some ecosystems (e.g., boreal forests, grasslands, tropical savannas and woodlands) and is affected directly by management and indirectly by land-use change (Apps *et al.,* 1993). Fire is a major short-term source of carbon, but adds to a small longer-term sink (<0.1 PgC/yr) through production of slowly decomposing and inert black carbon.

Forests
Deforestation has been responsible for almost 90% of the estimated emissions due to land-use change since 1850, with a 20% decrease of the global forest area (Houghton, 1999). Deforestation appears to be slowing slightly in tropical countries (FAO, 1997; Houghton, 2000), and some deforested areas in Europe and North America have been reforested in recent decades (FAO, 1997). Managed or regenerated forests generally store less carbon than natural forests, even at maturity. New trees take up carbon rapidly, but this slows down towards maturity when forests can be slight sources or sinks (Buchmann and Schulze, 1999). To use land continuously in order to take up carbon, the wood must be harvested and turned into long-lived products and trees must be re-planted. The trees may also be used for biomass energy to avoid future fossil fuel emissions (Hall *et al.,* 2000). Analysis of scenarios for future development show that expanded use of biomass energy could reduce the rate of atmospheric CO_2 increase (IPCC 1996b; Leemans *et al.,* 1996; Edmonds *et al.,* 1996; Ishitani *et al.,* 1996; IPCC, 2000a). IPCC (1996b) estimated that slowing deforestation and promoting natural forest regeneration and afforestation could increase carbon stocks by about 60 to 87 PgC over the period 1995 to 2050, mostly in the tropics (Brown *et al.,* 1996).

Savannas and grasslands – fire and grazing
Grasslands and mixed tree-grass systems are vulnerable to subtle environmental and management changes that can lead to shifts in vegetation state (Scholes and Archer, 1997; House and Hall, 2001). Livestock grazing on these lands is the land use with the largest global areal extent (FAO, 1993a). Extensive clearing of trees (for agricultural expansion) has occurred in some areas. In

other areas, fire suppression, eradication of indigenous browsers and the introduction of intensive grazing and exotic trees and shrubs have caused an increase in woody plant density known as woody encroachment or tree thickening (Archer *et al.*, 2001). This process has been estimated to result in a CO_2 sink of up to 0.17 PgC/yr in the USA during the 1980s (Houghton *et al.*, 1999) and at least 0.03 PgC/yr in Australia (Burrows, 1998). Grassland ecosystems have high root production and store most of their carbon in soils where turnover is relatively slow, allowing the possibility of enhancement through management (e.g., Fisher *et al.*, 1994).

Peatlands/wetlands
Peatlands/wetlands are large reserves of carbon, because anaerobic soil conditions and (in northern peatlands) low temperatures reduce decomposition and promote accumulation of organic matter. Total carbon stored in northern peatlands has been estimated as about 455 PgC (Gorham, 1991) with a current uptake rate in extant northern peatlands of 0.07 PgC/yr (Clymo *et al.*, 1998). Anaerobic decomposition releases methane (CH_4) which has a global warming potential (GWP) about 23 times that of CO_2 (Chapter 6). The balance between CH_4 release and CO_2 uptake and release is highly variable and poorly understood. Draining peatlands for agriculture increases total carbon released by decomposition, although less is in the form of CH_4. Forests grown on drained peatlands may be sources or sinks of CO_2 depending on the balance of decomposition and tree growth (Minkkinen and Laine, 1998).

Agricultural land
Conversion of natural vegetation to agriculture is a major source of CO_2, not only due to losses of plant biomass but also, increased decomposition of soil organic matter caused by disturbance and energy costs of various agricultural practices (e.g., fertilisation and irrigation; Schlesinger, 2000). Conversely, the use of high-yielding plant varieties, fertilisers, irrigation, residue management and reduced tillage can reduce losses and enhance uptake within managed areas (Cole *et al.*, 1996; Blume *et al.*, 1998). These processes have led to an estimated increase of soil carbon in agricultural soils in the USA of 0.14 PgC/yr during the 1980s (Houghton *et al.*, 1999). IPCC (1996b) estimated that appropriate management practices could increase carbon sinks by 0.4 to 0.9 PgC/yr , or a cumulative carbon storage of 24 to 43 PgC over 50 years; energy efficiency improvements and production of energy from dedicated crops and residues would result in a further mitigation potential of 0.3 to 1.4 PgC/yr, or a cumulative carbon storage of 16 to 68 PgC over 50 years (Cole *et al.*, 1996).

Scenarios
The IPCC Special Report on Land Use, Land-Use Change and Forestry (IPCC, 2000a) (hereafter SRLULUCF) derived scenarios of land-use emissions for the period 2008 to 2012. It was estimated that a deforestation flux of 1.79 PgC/yr is likely to be offset by reforestation and afforestation flux of −0.20 to −0.58 PgC/yr, yielding a net release of 1.59 to 1.20 PgC/yr (Schlamadinger *et al.*, 2000). The potential for net carbon storage from several "additional activities" such as improved land management and other land-use changes was estimated to amount to a global land-atmosphere flux in the region of −1.3 PgC/yr in 2010 and −2.5 PgC/yr in 2040, not including wood products and bioenergy (Sampson *et al.*, 2000).

3.2.2.3 Effects of climate
Solar radiation, temperature and available water affect photosynthesis, plant respiration and decomposition, thus climate change can lead to changes in NEP. A substantial part of the interannual variability in the rate of increase of CO_2 is likely to reflect terrestrial biosphere responses to climate variability (Section 3.5.3). Warming may increase NPP in temperate and arctic ecosystems where it can increase the length of the seasonal and daily growing cycles, but it may decrease NPP in water-stressed ecosystems as it increases water loss. Respiratory processes are sensitive to temperature; soil and root respiration have generally been shown to increase with warming in the short term (Lloyd and Taylor, 1994; Boone *et al.*, 1998) although evidence on longer-term impacts is conflicting (Trumbore, 2000; Giardina and Ryan, 2000; Jarvis and Linder, 2000). Changes in rainfall pattern affect plant water availability and the length of the growing season, particularly in arid and semi-arid regions. Cloud cover can be beneficial to NPP in dry areas with high solar radiation, but detrimental in areas with low solar radiation. Changing climate can also affect the distribution of plants and the incidence of disturbances such as fire (which could increase or decrease depending on warming and precipitation patterns, possibly resulting under some circumstances in rapid losses of carbon), wind, and insect and pathogen attacks, leading to changes in NBP. The global balance of these positive and negative effects of climate on NBP depends strongly on regional aspects of climate change.

The climatic sensitivity of high northern latitude ecosystems (tundra and taiga) has received particular attention as a consequence of their expanse, high carbon density, and observations of disproportionate warming in these regions (Chapman and Walsh, 1993; Overpeck *et al.*, 1997). High-latitude ecosystems contain about 25% of the total world soil carbon pool in the permafrost and the seasonally-thawed soil layer. This carbon storage may be affected by changes in temperature and water table depth. High latitude ecosystems have low NPP, in part due to short growing seasons, and slow nutrient cycling because of low rates of decomposition in waterlogged and cold soils. Remotely sensed data (Myneni *et al.*, 1997) and phenological observations (Menzel and Fabian, 1999) independently indicate a recent trend to longer growing seasons in the boreal zone and temperate Europe. Such a trend might be expected to have increased annual NPP. A shift towards earlier and stronger spring depletion of atmospheric CO_2 has also been observed at northern stations, consistent with earlier onset of growth at mid- to high northern latitudes (Manning, 1992; Keeling *et al.*, 1996a; Randerson, 1999). However, recent flux measurements at individual high-latitude sites have generally failed to find appreciable NEP (Oechel *et al.*, 1993; Goulden *et al.*, 1998; Schulze *et al.*, 1999; Oechel *et al.*, 2000). These studies suggest that, at least in the short term, any direct effect of warming on NPP may be more than offset by an increased respiration of soil carbon caused by the effects of increased depth of soil thaw. Increased decomposition, may, however also increase nutrient mineralisation and thereby indirectly stimulate NPP (Melillo *et al.*, 1993; Jarvis and Linder, 2000; Oechel *et al.*, 2000).

Large areas of the tropics are arid and semi-arid, and plant production is limited by water availability. There is evidence that even evergreen tropical moist forests show reduced GPP during the dry season (Malhi *et al.*, 1998) and may become a carbon source under the hot, dry conditions of typical El Niño years. With a warmer ocean surface, and consequently generally increased precipitation, the global trend in the tropics might be expected to be towards increased NPP, but changing precipitation patterns could lead to drought, reducing NPP and increasing fire frequency in the affected regions.

3.2.2.4 Effects of increasing atmospheric CO_2

CO_2 and O_2 compete for the reaction sites on the photosynthetic carbon-fixing enzyme, Rubisco. Increasing the concentration of CO_2 in the atmosphere has two effects on the Rubisco reactions: increasing the rate of reaction with CO_2 (carboxylation) and decreasing the rate of oxygenation. Both effects increase the rate of photosynthesis, since oxygenation is followed by photorespiration which releases CO_2 (Farquhar *et al.*, 1980). With increased photsynthesis, plants can develop faster, attaining the same final size in less time, or can increase their final mass. In the first case, the overall rate of litter production increases and so the soil carbon stock increases; in the second case, both the below-ground and above-ground carbon stocks increase. Both types of growth response to elevated CO_2 have been observed (Masle, 2000).

The strength of the response of photosynthesis to an increase in CO_2 concentration depends on the photosynthetic pathway used by the plant. Plants with a photosynthetic pathway known as C_3 (all trees, nearly all plants of cold climates, and most agricultural crops including wheat and rice) generally show an increased rate of photosynthesis in response to increases in CO_2 concentration above the present level (Koch and Mooney, 1996; Curtis, 1996; Mooney *et al.*, 1999). Plants with the C_4 photosynthetic pathway (tropical and many temperate grasses, some desert shrubs, and some crops including maize and sugar cane) already have a mechanism to concentrate CO_2 and therefore show either no direct photosynthetic response, or less response than C_3 plants (Wand *et al.*, 1999). Increased CO_2 has also been reported to reduce plant respiration under some conditions (Drake *et al.*, 1999), although this effect has been questioned.

Increased CO_2 concentration allows the partial closure of stomata, restricting water loss during transpiration and producing an increase in the ratio of carbon gain to water loss ("water-use efficiency", WUE) (Field *et al.*, 1995a; Drake *et al.*, 1997; Farquhar, 1997; Körner, 2000). This effect can lengthen the duration of the growing season in seasonally dry ecosystems and can increase NPP in both C_3 and C_4 plants.

Nitrogen-use efficiency also generally improves as carbon input increases, because plants can vary the ratio between carbon and nitrogen in tissues and require lower concentrations of photosynthetic enzymes in order to carry out photosynthesis at a given rate; for this reason, low nitrogen availability does not consistently limit plant responses to increased atmospheric CO_2 (McGuire *et al.*, 1995; Lloyd and Farquhar, 1996; Curtis and Wang, 1998; Norby *et al.*, 1999; Körner, 2000). Increased CO_2 concentration may also stimulate nitrogen fixation (Hungate *et al.*, 1999; Vitousek and Field, 1999). Changes in tissue nutrient concentration may affect herbivory and decomposition, although long-term decomposition studies have shown that the effect of elevated CO_2 in this respect is likely to be small (Norby and Cortufo, 1998) because changes in the C:N ratio of leaves are not consistently reflected in the C:N ratio of leaf litter due to nitrogen retranslocation (Norby *et al.*, 1999).

The process of CO_2 "fertilisation" thus involves direct effects on carbon assimilation and indirect effects such as those via water saving and interactions between the carbon and nitrogen cycles. Increasing CO_2 can therefore lead to structural and physiological changes in plants (Pritchard *et al.*, 1999) and can further affect plant competition and distribution patterns due to responses of different species. Field studies show that the relative stimulation of NPP tends to be greater in low-productivity years, suggesting that improvements in water- and nutrient-use efficiency can be more important than direct NPP stimulation (Luo *et al.*, 1999).

Although NPP stimulation is not automatically reflected in increased plant biomass, additional carbon is expected to enter the soil, via accelerated ontogeny, which reduces lifespan and results in more rapid shoot death, or by enhanced root turnover or exudation (Koch and Mooney, 1996; Allen *et al.*, 2000). Because the soil microbial community is generally limited by the availability of organic substrates, enhanced addition of labile carbon to the soil tends to increase heterotrophic respiration unless inhibited by other factors such as low temperature (Hungate *et al.*, 1997; Schlesinger and Andrews, 2000). Field studies have indicated increases in soil organic matter, and increases in soil respiration of about 30%, under elevated CO_2 (Schlesinger and Andrews, 2000). The potential role of the soil as a carbon sink under elevated CO_2 is crucial to understanding NEP and long-term carbon dynamics, but remains insufficiently well understood (Trumbore, 2000).

C_3 crops show an average increase in NPP of around 33% for a doubling of atmospheric CO_2 (Koch and Mooney, 1996). Grassland and crop studies combined show an average biomass increase of 14%, with a wide range of responses among individual studies (Mooney *et al.*, 1999). In cold climates, low temperatures restrict the photosynthetic response to elevated CO_2. In tropical grasslands and savannas, C_4 grasses are dominant, so it has been assumed that trees and C_3 grasses would gain a competitive advantage at high CO_2 (Gifford, 1992; Collatz *et al.*, 1998). This is supported by carbon isotope evidence from the last glacial maximum, which suggests that low CO_2 favours C_4 plants (Street-Perrott *et al.*, 1998). However, field experiments suggest a more complex picture with C_4 plants sometimes doing better than C_3 under elevated CO_2 due to improved WUE at the ecosystem level (Owensby *et al.*, 1993; Polley *et al.*, 1996). Highly productive forest ecosystems have the greatest potential for *absolute* increases in productivity due to CO_2 effects. Long-term field studies on young trees have typically shown a stimulation of photosynthesis of about 60% for a doubling of CO_2 (Saxe *et al.*, 1998; Norby *et al.*, 1999). A FACE experiment in a fast growing young pine forest showed an increase of 25% in NPP for an increase in atmospheric CO_2 to 560 ppm (DeLucia *et al.*, 1999). Some of this additional NPP is allocated to root metabolism and associated microbes; soil CO_2 efflux increases, returning a part (but not all) of the extra NPP to the atmosphere (Allen *et al.*, 2000). The response of mature forests to increases in atmospheric CO_2 concentration has not been shown experimentally; it may be different from that of young forests for various reasons, including changes in leaf C:N ratios and stomatal responses to water vapour deficits as trees mature (Curtis and Wang, 1998; Norby *et al.*, 1999).

At high CO_2 concentrations there can be no further increase in photosynthesis with increasing CO_2 (Farquhar *et al.*, 1980), except through further stomatal closure, which may produce continued increases in WUE in water-limited environments. The shape of the response curve of global NPP at higher CO_2 concentrations than present is uncertain because the response at the level of gas exchange is modified by incompletely understood plant- and ecosystem-level processes (Luo *et al.*, 1999). Based on photosynthetic physiology, it is likely that the additional carbon that could be taken up globally by enhanced photosynthesis as a direct consequence of rising atmospheric CO_2 concentration is small at atmospheric concentrations above 800 to 1,000 ppm. Experimental studies indicate that some ecosystems show greatly reduced CO_2 fertilisation at lower concentrations than this (Körner, 2000).

3.2.2.5 Effects of anthropogenic nitrogen deposition
Nitrogen availability is an important constraint on NPP (Vitousek *et al.*, 1997), although phosphorus and calcium may be more important limiting nutrients in many tropical and sub-tropical regions (Matson, 1999). Reactive nitrogen is released into the atmosphere in the form of nitrogen oxides (NO_x) during fossil fuel and biomass combustion and ammonia emitted by industrial regions, animal husbandry and fertiliser use (Chapter 4). This nitrogen is then deposited fairly near to the source, and can act as a fertiliser for terrestrial plants. There has been a rapid increase in reactive nitrogen deposition over the last 150 years (Vitousek *et al.*, 1997; Holland *et al.*, 1999). Much field evidence on nitrogen fertilisation effects on plants (e.g., Chapin, 1980; Vitousek and Howarth, 1991; Bergh *et al.*, 1999) supports the hypothesis that additional nitrogen deposition will result in increased NPP, including the growth of trees in Europe (Spiecker *et al.*, 1996). There is also evidence (Fog, 1988; Bryant *et al.*, 1998) that N fertilisation enhances the formation of modified soil organic matter and thus increases the residence time of carbon in soils.

Tracer experiments with addition of the stable isotope ^{15}N provide insight into the short-term fate of deposited reactive nitrogen (Gundersen *et al.*, 1998). It is clear from these experiments that most of the added N added to the soil surface is retained in the ecosystem rather than being leached out via water transport or returned to the atmosphere in gaseous form (as N_2, NO, N_2O or NH_3). Studies have also shown that the tracer is found initially in the soil (Nadelhoffer *et al.*, 1999), but that it enters the vegetation after a few years (Clark 1977; Schimel and Chapin, 1996; Delgado *et al.*, 1996; Schulze, 2000).

There is an upper limit to the amount of added N that can fertilise plant growth. This limit is thought to have been reached in highly polluted regions of Europe. With nitrogen saturation, ecosystems are no longer able to process the incoming nitrogen deposition, and may also suffer from deleterious effects of associated pollutants such as ozone (O_3), nutrient imbalance, and aluminium toxicity (Schulze *et al.*, 1989; Aber *et al.*, 1998).

3.2.2.6 Additional impacts of changing atmospheric chemistry
Current tropospheric O_3 concentrations in Europe and North America cause visible leaf injury on a range of crop and tree species and have been shown to reduce the growth and yield of crops and young trees in experimental studies. The longer-term effects of O_3 on forest productivity are less certain, although significant negative associations between ozone exposure and forest growth have been reported in North America (Mclaughlin and Percy, 2000) and in central Europe (Braun *et al.*, 2000). O_3 is taken up through stomata, so decreased stomatal conductance at elevated CO_2 may reduce the effects of O_3 (Semenov *et al.*, 1998, 1999). There is also evidence of significant interactions between O_3 and soil water availability in effects on stem growth or NPP from field studies (e.g., Mclaughlin and Downing, 1995) and from modelling studies (e.g., Ollinger *et al.*, 1997). The regional impacts of O_3 on NPP elsewhere in the world are uncertain, although significant impacts on forests have been reported close to major cities. Fowler *et al.* (2000) estimate that the proportion of global forests exposed to potentially damaging ozone concentrations will increase from about 25% in 1990 to about 50% in 2100.

Other possible negative effects of industrially generated pollution on plant growth include effects of soil acidification due to deposition of NO_3^- and SO_4^{2-}. Severe forest decline has been observed in regions with high sulphate deposition, for instance in parts of eastern Europe and southern China. The wider effects are less certain and depend on soil sensitivity. Fowler *et al.* (2000) estimate that 8% of global forest cover received an annual sulphate deposition above an estimated threshold for effects on acid sensitive soils, and that this will increase to 17% in 2050. The most significant long-term effect of continued acid deposition for forest productivity may be through depletion of base cations, with evidence of both increased leaching rates and decreased foliar concentrations (Mclaughlin and Percy, 2000), although the link between these changes in nutrient cycles and NPP needs to be quantified.

3.2.2.7 Additional constraints on terrestrial CO_2 uptake
It is very likely that there are upper limits to carbon storage in ecosystems due to mechanical and resource constraints on the amount of above ground biomass and physical limits to the amount of organic carbon that can be held in soils (Scholes *et al.*, 1999). It is also generally expected that increased above-ground NPP (production of leaves and stem) will to some extent be counterbalanced by an increased rate of turnover of the biomass as upper limits are approached.

3.2.3 Ocean Carbon Processes

3.2.3.1 Background

The total amount of carbon in the ocean is about 50 times greater than the amount in the atmosphere, and is exchanged with the atmosphere on a time-scale of several hundred years. Dissolution in the oceans provides a large sink for anthropogenic CO_2, due in part to its high solubility, but above all because of its dissociation into ions and interactions with sea water constituents (see Box 3.3).

The annual two-way gross exchange of CO_2 between the atmosphere and surface ocean is about 90 PgC/yr, mediated by molecular diffusion across the air-sea interface. Net CO_2 transfer can occur whenever there is a partial pressure difference of CO_2 across this interface. The flux can be estimated as the product of a gas transfer coefficient, the solubility of CO_2, and the partial pressure difference of CO_2 between air and water. The gas transfer coefficient incorporates effects of many physical factors but is usually expressed as a non-linear function of wind speed alone. There is considerable uncertainty about this function (Liss and Merlivat, 1986; Wanninkhof, 1992; Watson *et al.*, 1995). Improvements in the ability to measure CO_2 transfer directly (e.g., Wanninkhof and McGillis, 1999) may lead to a better knowledge of gas transfer coefficients.

Despite extensive global measurements conducted during the 1990s, measurements of surface water pCO_2 remain sparse, and extensive spatial and temporal interpolation is required in order to produce global fields. Takahashi *et al.* (1999) interpolated data collected over three decades in order to derive monthly values of surface water pCO_2 over the globe for a single "virtual" calendar year (1995). A wind speed dependent gas transfer coefficient was used to calculate monthly net CO_2 fluxes. The resulting estimates, although subject to large uncertainty, revealed clear regional and seasonal patterns in net fluxes.

Regional net CO_2 transfers estimated from contemporary surface water pCO_2 data should not be confused with the uptake of anthropogenic CO_2. The uptake of anthropogenic CO_2 is the *increase* in net transfer over the pre-industrial net transfer, and is therefore superimposed on a globally varying pattern of relatively large natural transfers. The natural transfers result from heating and cooling, and biological production and respiration. Carbon is transferred within the ocean from natural sink regions to natural source regions via ocean circulation and the sinking of carbon rich particles. This spatial separation of natural sources and sinks dominates the regional distribution of net annual air-sea fluxes.

CO_2 solubility is temperature dependent, hence air-sea heat transfer contributes to seasonal and regional patterns of air-sea CO_2 transfer (Watson *et al.*, 1995). Net cooling of surface waters tends to drive CO_2 uptake; net warming drives outgassing. Regions of cooling and heating are linked via circulation, producing vertical gradients and north-south transports of carbon within the ocean (e.g., of the order 0.5 to 1 PgC/yr southward transport in the Atlantic Basin; Broecker and Peng, 1992; Keeling and Peng, 1995; Watson *et al.*, 1995; Holfort *et al.*, 1998).

Biological processes also drive seasonal and regional distributions of CO_2 fluxes (Figure 1c). The gross primary production by ocean phytoplankton has been estimated by Bender *et al.* (1994) to be 103 PgC/yr. Part of this is returned to DIC through autotrophic respiration, with the remainder being net primary production, estimated on the basis of global remote sensing data

---

Box 3.3: The varying CO_2 uptake capacity of the ocean.

Because of its solubility and chemical reactivity, CO_2 is taken up by the ocean much more effectively than other anthropogenic gases (e.g., chlorofluorocarbons (CFCs) and CH_4). CO_2 that dissolves in seawater is found in three main forms. The sum of these forms constitutes dissolved inorganic carbon (DIC). The three forms are: (1) dissolved CO_2 (non-ionic, about 1% of the total) which can be exchanged with the atmosphere until the partial pressure in surface water and air are equal, (2) bicarbonate ion (HCO_3^-, about 91%); and (3) carbonate ion (CO_3^{2-}, about 8%). As atmospheric CO_2 increases, the dissolved CO_2 content of surface seawater increases at a similar rate, but most of the added CO_2 ends up as HCO_3^-. Meanwhile, the CO_3^{2-} content decreases, since the net effect of adding CO_2 is a reaction with CO_3^{2-} to form HCO_3^- (Figure 3.1). There is therefore less available CO_3^{2-} to react with further CO_2 additions, causing an increasing proportion of the added CO_2 to remain in its dissolved form. This restricts further uptake, so that the overall ability of surface sea water to take up CO_2 decreases at higher atmospheric CO_2 levels. The effect is large. For a 100 ppm increase in atmospheric CO_2 above today's level (i.e., from 370 to 470 ppm) the DIC concentration increase of surface sea water is already about 40% smaller than would have been caused by a similar 100 ppm increase relative to pre-industrial levels (i.e., from 280 to 380 ppm). The contemporary DIC increase is about 60% greater than would result if atmospheric CO_2 were to increase from 750 to 850 ppm.

The uptake capacity for CO_2 also varies significantly due to additional factors, most importantly seawater temperature, salinity and alkalinity (the latter being a measurable quantity approximately equal to $[HCO_3^-] + 2 \times [CO_3^{2-}]$). Alkalinity is influenced primarily by the cycle of $CaCO_3$ formation (in shells and corals) and dissolution (see Figure 3.1c).

to be about 45 PgC/yr (Longhurst *et al.*, 1995; Antoine *et al.*, 1996; Falkowski *et al.*, 1998; Field *et al.*, 1998; Balkanski *et al.*, 1999). About 14 to 30% of the total NPP occurs in coastal areas (Gattuso *et al.*, 1998). The resulting organic carbon is consumed by zooplankton (a quantitatively more important process than herbivory on land) or becomes detritus. Some organic carbon is released in dissolved form (DOC) and oxidised by bacteria (Ducklow, 1999) with a fraction entering the ocean reservoir as net DOC production (Hansell and Carlson, 1998). Sinking of particulate organic carbon (POC) composed of dead organisms and detritus together with vertical transfer of DOC create a downward flux of organic carbon from the upper ocean known as "export production". Recent estimates for global export production range from roughly 10 to 20 PgC/yr (Falkowski *et al.*, 1998; Laws *et al.*, 2000). An alternative estimate for global export production of 11 PgC/yr has been derived using an inverse model of physical and chemical data from the world's oceans (Schlitzer, 2000). Only a small fraction (about 0.1 PgC) of the export production sinks in sediments, mostly in the coastal ocean (Gattuso *et al.*, 1998). Heterotrophic respiration at depth converts the remaining organic carbon back to DIC. Eventually, and usually at another location, this DIC is upwelled into the ocean's surface layer again and may re-equilibrate with the atmospheric CO_2. These mechanisms, often referred to as the biological pump, maintain higher DIC concentrations at depth and cause atmospheric CO_2 concentrations to be about 200 ppm lower than would be the case in the absence of such mechanisms (Sarmiento and Toggweiler, 1984; Maier-Reimer *et al.*, 1996).

Marine organisms also form shells of solid calcium carbonate ($CaCO_3$) that sink vertically or accumulate in sediments, coral reefs and sands. This process depletes surface CO_3^{2-}, reduces alkalinity, and tends to increase pCO_2 and drive more outgassing of CO_2 (see Box 3.3 and Figure 3.1). The effect of $CaCO_3$ formation on surface water pCO_2 and air-sea fluxes is therefore counter to the effect of organic carbon production. For the surface ocean globally, the ratio between the export of organic carbon and the export of calcium carbonate (the "rain ratio") is a critical factor controlling the overall effect of biological activity on surface ocean pCO_2 (Figure 3.1; Archer and Maier-Reimer, 1994). Milliman (1993) estimated a global production of $CaCO_3$ of 0.7 PgC/yr, with roughly equivalent amounts produced in shallow water and surface waters of the deep ocean. Of this total, approximately 60% accumulates in sediments. The rest re-dissolves either in the water column or within the sediment. An estimate of $CaCO_3$ flux analogous to the export production of organic carbon, however, should include sinking out of the upper layers of the open ocean, net accumulation in shallow sediments and reefs, and export of material from shallow systems into deep sea environments. Based on Milliman's (1993) budget, this quantity is about 0.6 PgC/yr (± 25 to 50 % at least). The global average rain ratio has been variously estimated from models of varying complexity to be 4 (Broecker and Peng, 1982), 3.5 to 7.5 (Shaffer, 1993), and 11 (Yamanaka and Tajika, 1996). (It should be noted that rain ratios are highly depth dependent due to rapid oxidation of organic carbon at shallow depth compared to the

depths at which sinking $CaCO_3$ starts to dissolve.) If one accepts an organic carbon export production value of 11 PgC/yr (Schlitzer, 2000), then only Yamanaka and Tajika's (1996) value for the rain ratio approaches consistency with the observation-based estimates of the export of $CaCO_3$ and organic carbon from the ocean surface layer.

The overall productivity of the ocean is determined largely by nutrient supply from deep water. There are multiple potentially limiting nutrients: in practice nitrate and/or phosphate are commonly limiting (Falkowski *et al.*, 1998; Tyrell, 1999). Silicate plays a role in limiting specific types of phytoplankton and hence in determining the qualitative nature of primary production, and potentially the depth to which organic carbon sinks. A role for iron in limiting primary productivity in regions with detectable phosphate and nitrate but low productivity (HNLC or "high nutrient, low chlorophyll regions") has been experimentally demonstrated in the equatorial Pacific (Coale *et al.*, 1996) and the Southern Ocean (Boyd *et al.*, 2000). In both regions artificial addition of iron stimulated phytoplankton growth, resulting in decreased surface-water pCO_2. In HLNC regions, the supply of iron from deep water, while an important source, is generally insufficient to meet the requirements of phytoplankton. An important additional supply of iron to surface waters far removed from sediment and riverine sources is aeolian transport and deposition (Duce and Tindale, 1991; Fung *et al.*, 2000; Martin, 1990). This aeolian supply of iron may limit primary production in HNLC regions, although the effect is ultimately constrained by the availability of nitrate and phosphate. Iron has been hypothesised to play an indirect role over longer time-scales (e.g., glacial-interglacial) through limitation of oceanic nitrogen fixation and, consequently, the oceanic content of nitrate (Falkowski *et al.*, 1998; Broecker and Henderson, 1998; Box 3.4). The regional variability of oceanic nitrogen fixation (Gruber and Sarmiento, 1997) and its temporal variability and potential climate-sensitivity have recently become apparent based on results from long time-series and global surveys (Karl *et al.*, 1997; Hansell and Feely, 2000).

Carbon (organic and inorganic) derived from land also enters the ocean via rivers as well as to some extent via groundwater. This transport comprises a natural carbon transport together with a significant anthropogenic perturbation. The global natural transport from rivers to the ocean is about 0.8 PgC/yr, half of which is organic and half inorganic (Meybeck 1982, 1993; Sarmiento and Sundquist 1992; Figure 3.1). Additional fluxes due to human activity have been estimated (Meybeck, 1993) to be about 0.1 PgC/yr (mainly organic carbon). Much of the organic carbon is deposited and/or respired and outgassed close to land, mostly within estuaries (Smith and Hollibaugh, 1993). The outgassing of anthropogenic carbon from estuaries can be a significant term in comparison with regional CO_2 emissions estimates (e.g., 5 to 10% for Western Europe; Frankignoulle *et al.*, 1998). The natural DIC transport via rivers, however, is part of a large-scale cycling of carbon between the open ocean and land associated with dissolution and precipitation of carbonate minerals. This natural cycle drives net outgassing from the ocean of the order 0.6 PgC/yr globally, which should be included in any assessment of net air-sea and atmosphere-terrestrial biosphere transfers (Sarmiento and Sundquist, 1992) and ocean transports (e.g., Holfort *et al.*, 1998).

3.2.3.2 Uptake of anthropogenic CO_2

Despite the importance of biological processes for the ocean's natural carbon cycle, current thinking maintains that the oceanic uptake of anthropogenic CO_2 is primarily a physically and chemically controlled process superimposed on a biologically driven carbon cycle that is close to steady state. This differs from the situation on land because of the different factors which control marine and terrestrial primary productivity. On land, experiments have repeatedly shown that current CO_2 concentrations are limiting to plant growth (Section 3.2.2.4). In the ocean, experimental evidence is against control of productivity by CO_2 concentrations, except for certain species at lower than contemporary CO_2 concentrations (Riebesell *et al.*, 1993; Falkowski, 1994). Further, deep ocean concentrations of major nutrients and DIC are tightly correlated, with the existing ratios closely (but not exactly, see Section 3.2.3.3) matching the nutritional requirements of marine organisms (the "Redfield ratios": Redfield *et al.*, 1963). This implies that as long as nutrients that are mixed into the ocean surface layer are largely removed by organic carbon production and export, then there is little potential to drive a net air-sea carbon transfer simply through alteration of the global rate of production. Terrestrial ecosystems show greater variability in this respect because land plants have multiple ways to acquire nutrients, and have greater plasticity in their chemical composition (Melillo and Gosz, 1983). There are, however, extensive regions of the ocean surface where major nutrients are not fully depleted, and changes in these regions may play a significant role in altering atmosphere-ocean carbon partitioning (see Section 3.2.3.3).

The increase of atmospheric pCO$_2$ over pre-industrial levels has tended to increase uptake into natural CO_2 sink regions and decreased release from natural outgassing regions. Contemporary net air-sea fluxes comprise spatially-varying mixtures of natural and anthropogenic CO_2 flux components and cannot be equated with anthropogenic CO_2 uptake, except on a global scale. Uptake of anthropogenic CO_2 is strongest in regions where "old" waters, which have spent many years in the ocean interior since their last contact with the atmosphere, are re-exposed at the sea surface to a contemporary atmosphere which now contains anthropogenic CO_2 (e.g., Sarmiento *et al.*, 1992; Doney, 1999). In an upwelling region, for example, the natural component of the air-sea flux may be to outgas CO_2 to the atmosphere. The higher atmospheric pCO$_2$ of the contemporary atmosphere acts to reduce this outgassing relative to the natural state, implying that more carbon remains in the ocean. This represents uptake of anthropogenic CO_2 by a region which is a source of CO_2 to the atmosphere. The additional carbon in the ocean resulting from such uptake is then transported by the surface ocean circulation, and eventually stored as surface waters sink, or are mixed, into the deep ocean interior. Whereas upwelling into the surface layer is quantitatively balanced on a global scale by sinking, the locations where deep waters rise and sink can be separated by large horizontal distances.

Air-sea gas transfer allows older waters to approach a new steady state with higher atmospheric CO_2 levels after about a year at the sea surface. This is fast relative to the rate of ocean mixing, implying that anthropogenic CO_2 uptake is limited by the rate at which "older" waters are mixed towards the air-sea interface. The rate of exposure of older, deeper waters is therefore a critical factor limiting the uptake of anthropogenic CO_2. In principle, there is sufficient uptake capacity (see Box 3.3) in the ocean to incorporate 70 to 80% of anthropogenic CO_2 emissions to the atmosphere, even when total emissions of up to 4,500 PgC are considered (Archer *et al.*, 1997). The finite rate of ocean mixing, however, means that it takes several hundred years to access this capacity (Maier-Reimer and Hasselmann, 1987; Enting *et al.*, 1994; Archer *et al.*, 1997). Chemical neutralisation of added CO_2 through reaction with CaCO$_3$ contained in deep ocean sediments could potentially absorb a further 9 to 15% of the total emitted amount, reducing the airborne fraction of cumulative emissions by about a factor of 2; however the response time of deep ocean sediments is in the order of 5,000 years (Archer *et al.*, 1997).

Using time-series and global survey data, the increasing oceanic carbon content has been directly observed, although the signal is small compared to natural variability and requires extremely accurate measurements (Sabine *et al.*, 1997). A long-term increase of surface water CO_2 levels tracking the mean atmospheric CO_2 increase has been observed in the ocean's subtropical gyres (Bates *et al.*, 1996; Winn *et al.*, 1998) and the equatorial Pacific (Feely *et al.*, 1999b). However, very few such time-series exist and the response of other important oceanic regions to the atmospheric pCO$_2$ increase cannot yet be assessed. Inter-decadal increases in DIC concentrations at depth have been resolved from direct measurements (Wallace, 1995; Peng *et al.*, 1998; Ono *et al.*, 1998; Sabine *et al.*, 1999). The total amounts of anthropogenic CO_2 accumulated in the ocean since the pre-industrial era can also be estimated from measurements using recent refinements (Gruber *et al.*, 1996) of long-standing methods for separating the natural and anthropogenic components of oceanic DIC. A comparison of such analyses with ocean model results is presented in Section 3.6.3.

3.2.3.3 Future changes in ocean CO_2 uptake

This section lists processes that may be important for the future uptake of anthropogenic CO_2. These changes can represent changes in anthropogenic CO_2 uptake itself (mainly physical and chemical processes), or changes in the natural biologically-linked cycling of carbon between the atmosphere and ocean.

Physical and chemical processes

Buffering changes. The capacity of surface waters to take up anthropogenic CO_2 is decreasing as CO_2 levels increase (see Box 3.3). The magnitude of this effect is substantial. This decrease in uptake capacity of the ocean makes atmospheric CO_2 more sensitive to anthropogenic emissions and other changes in the natural cycling of carbon.

Emissions rate. Even assuming no other changes to the carbon cycle, the proportion of emitted CO_2 that can be taken up by the ocean decreases as the rate of emission increases. This is due to the finite rate of exposure of 'older', deeper waters to the anthropogenic CO_2 contained in the atmosphere.

Warming. CO_2 is less soluble in warmer water, and the equilibrium pCO_2 in seawater increases by about 10 to 20 ppm per °C temperature increase. Warming of surface water would therefore tend to increase surface water pCO_2, driving CO_2 from the surface ocean to the atmosphere. The expected effect of such warming on atmospheric CO_2 may be smaller, depending on the rate of exchange between ocean surface waters and the deep ocean at high latitudes (e.g., Bacastow, 1993).

Vertical mixing and stratification. Several coupled atmosphere-ocean models have shown global warming to be accompanied by an increase in vertical stratification (see Chapter 7). Such a change would reduce the rate of mixing between surface and deep waters, and therefore reduce the effective volume of the ocean that is exposed to high atmospheric CO_2. On its own, this effect would tend to reduce the ocean CO_2 uptake. However, changes in stratification may also drive changes in the natural carbon cycle. The magnitude and even the sign of changes in the natural cycle are much more difficult to predict because of the complexity of ocean biological processes (Sarmiento *et al.*, 1998; Matear and Hirst, 1999).

Biologically-linked processes

Qualitative and quantitative changes in carbon uptake arising from changes in marine ecosystems are more speculative (Denman *et al.*, 1996; Falkowski *et al.*, 1998; Watson and Liss, 1998), but are likely to have occurred over glacial-interglacial time-scales (Section 3.3). Falkowski *et al.* (1998) listed three major classes of biologically linked factors that can in principal alter the air-sea partitioning of CO_2: (1) changes in surface nutrient utilisation (e.g., in HNLC areas); (2) changes in total ocean content of major nutrients; (3) changes in the elemental composition of biogenic material (including the rain ratio). Our incomplete understanding of present day nutrient controls on productivity limits our ability to predict future changes in ocean biology and their effect on CO_2 levels. For example, the possible identification of changes in deep ocean C:N:P ratios (Pahlow and Riebesell, 2000) leaves open the question of the extent to which ocean biological carbon cycling is in steady state, or is likely to remain so in the future.

Changes in surface nutrient utilisation. Changes in the utilisation of surface nutrients in HNLC regions have the potential to alter export production and carbon storage in the ocean interior. Most attention focuses on the role of inadvertent or deliberate changes in the external supply of iron to such regions. The sign of possible future responses of ocean biota due to iron supply changes is difficult to assess. Future iron supply may increase due to erosion (enhanced by agriculture and urbanisation) which tends to increase dust export and aeolian iron deposition (Tegen and Fung, 1995). Conversely, a globally enhanced hydrological cycle and increased water-use efficiency of terrestrial plants may tend to reduce future dust export (Harrison *et al.*, 2001). The delivery of dust to the HNLC regions will be sensitive to regional changes in erosion and the hydrological cycle, affecting the important regions of dust export, rather than to global scale changes (Dai *et al.*, 1998).

Surface nutrient supply could be reduced if ocean stratification reduces the supply of major nutrients carried to the surface waters from the deep ocean (Sarmiento *et al.*, 1998). The impact of strati-fication on marine productivity depends on the limiting factor. In regions limited by deep ocean nutrients, stratification would reduce marine productivity and the strength of the export of carbon by biological processes. Conversely, stratification also increases the light exposure of marine organisms, which would increase productivity in regions where light is limiting.

Changes in total ocean content of major nutrients. Changes in the delivery of the major biologically limiting nutrients (N, P, Fe, Si) from riverine, atmospheric or sedimentary sources, or changes in removal rates (e.g., denitrification), could alter oceanic nutrient inventories and hence export production and ocean carbon storage. On the global scale, the upward fluxes of major nutrients are slightly depleted in N relative to P with respect to the nutrient requirements of phytoplankton (Fanning, 1992). This relative supply of N versus P may be sensitive to climate and circulation related changes in the rate of fixed-nitrogen removal by denitrification (Ganeshram *et al.*, 1995) or via changes in the rate of nitrogen fixation. Changes in river flow and composition are also affecting the supply of nutrients (Frankignoulle *et al.*, 1998). The hypothesised link between nitrogen fixation in certain ocean regions and the external iron supply (Falkowski, 1997; Wu *et al.*, 2000) could play a role in future nutrient and carbon budgets. Nitrogen fixation rates may also be affected by changes in stratification and mixing. For example, Karl *et al.* (1997) have identified interannual variability in nitrogen fixation rates in the subtropical Pacific which are apparently linked to ENSO variability in upper ocean dynamics.

Changes in the elemental composition of biogenic material. The structure and biogeochemistry of marine ecosystems can be affected by numerous climate-related factors including temperature, cloudiness, nutrient availability, mixed-layer physics and sea-ice extent. In turn the structure of marine ecosystems, and particularly the species composition of phytoplankton, exert a control on the partitioning of carbon between the ocean and the atmosphere. For example, a change in distribution of calcareous versus siliceous planktonic organisms could affect CO_2 uptake in the future, as it may have done in the past (Archer and Maier-Reimer, 1994). Precipitation of $CaCO_3$ by marine organisms (calcification) removes dissolved CO_3^{2-}, thus decreasing surface water alkalinity and reducing the capacity of sea water to dissolve atmospheric CO_2 (see Box 3.3). Recent experimental evidence suggests that as a direct result of increasing atmospheric and surface water pCO_2 levels, oceanic calcification will decrease significantly over the next 100 years. Model-based calculations suggest that decreases in coral reef calcification rates of the order 17 to 35% relative to pre-industrial rates are possible (Kleypas *et al.*, 1999). Experimental studies with corals have confirmed such effects (Langdon *et al.*, 2000). Field and laboratory studies have shown that planktonic calcification is also highly sensitive to pCO_2 levels. The calcification rate of coccolithophorids decreases by 16 to 83% at pCO_2 levels of 750 ppm (Riebesell *et al.*, 2000). Such an effect would tend to favour CO_2 storage in the upper ocean and act as a negative feedback on atmospheric growth rates of CO_2. However, long-term predictions of such biological responses are hampered by a lack of understanding concerning physiological acclimation and genetic adaptations of species to increasing pCO_2.

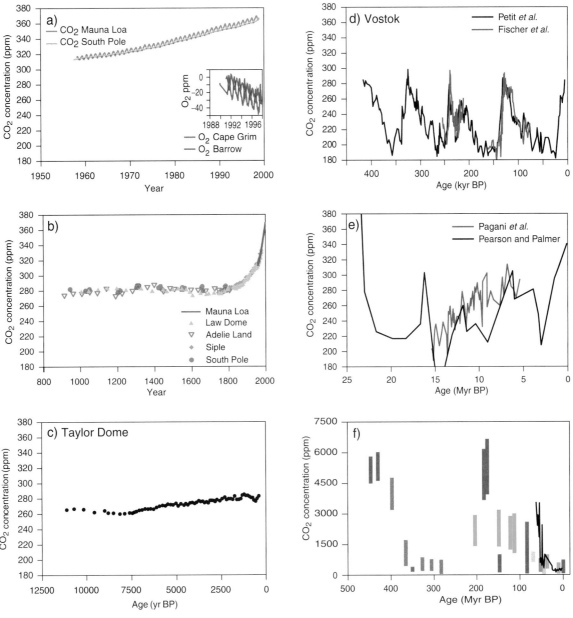

Figure 3.2: Variations in atmospheric CO_2 concentration on different time-scales. (a) Direct measurements of atmospheric CO_2 concentration (Keeling and Whorf, 2000), and O_2 from 1990 onwards (Battle *et al.*, 2000). O_2 concentration is expressed as the change from an arbitrary standard. (b) CO_2 concentration in Antarctic ice cores for the past millenium (Siegenthaler *et al.*, 1988; Neftel *et al.*, 1994; Barnola *et al.*, 1995; Etheridge *et al.*, 1996). Recent atmospheric measurements at Mauna Loa (Keeling and Whorf, 2000) are shown for comparison. (c) CO_2 concentration in the Taylor Dome Antarctic ice core (Indermühle *et al.*, 1999). (d) CO_2 concentration in the Vostok Antarctic ice core (Petit *et al.*, 1999; Fischer *et al.*, 1999). (e) Geochemically inferred CO_2 concentrations, from Pagani *et al.* (1999a) and Pearson and Palmer (2000). (f) Geochemically inferred CO_2 concentrations: coloured bars represent different published studies cited by Berner (1997). The data from Pearson and Palmer (2000) are shown by a black line. (BP = before present.)

3.3 Palaeo CO_2 and Natural Changes in the Carbon Cycle

3.3.1 Geological History of Atmospheric CO_2

Atmospheric CO_2 concentration has varied on all time-scales during the Earth's history (Figure 3.2). There is evidence for very high CO_2 concentrations (>3,000 ppm) between 600 and 400 Myr BP and between 200 and 150 Myr BP (Figure 3.2f). On long time-scales, atmospheric CO_2 content is determined by the balance among geochemical processes including organic carbon burial in sediments, silicate rock weathering, and vulcanism (Berner, 1993, 1997). In particular, terrestrial vegetation has enhanced the rate of silicate weathering, which consumes CO_2 while releasing base cations that end up in the ocean. Subsequent deep-sea burial of Ca and Mg (as carbonates, for example $CaCO_3$) in the shells of marine organisms removes CO_2. The net effect of slight imbalances in the carbon cycle over tens to hundreds of millions of years has been to reduce

Box 3.4: Causes of glacial/inter-glacial changes in atmospheric CO_2.

One family of hypotheses to explain glacial/inter-glacial variations of atmospheric CO_2 relies on physical mechanisms that could change the dissolution and outgassing of CO_2 in the ocean. The solubility of CO_2 is increased at low temperature, but reduced at high salinity. These effects nearly cancel out over the glacial/inter-glacial cycle, so simple solubility changes are not the answer. Stephens and Keeling (2000) have proposed that extended winter sea ice prevented outgassing of upwelled, CO_2-rich water around the Antarctic continent during glacial times. A melt-water "cap" may have further restricted outgassing of CO_2 during summer (François *et al.,* 1997). These mechanisms could explain the parallel increases of Antarctic temperature and CO_2 during deglaciation. However, they require less vertical mixing to occur at low latitudes than is normally assumed. The relative importance of high and low latitudes for the transport of CO_2 by physical processes is not well known, and may be poorly represented in most ocean carbon models (Toggweiler, 1999; Broecker *et al.,* 1999).

Several authors have hypothesised increased utilisation of surface nutrients by marine ecosystems in high latitudes, leading to stronger vertical gradients of DIC and thus reduced atmospheric CO_2 during glacial times (Sarmiento and Toggweiler, 1984; Siegenthaler and Wenk, 1984; Knox and McElroy, 1984). Other hypotheses call for an increased external supply of nutrients to the ocean (McElroy, 1983; Martin *et al.,* 1990; Broecker and Henderson, 1998). The supply of iron-rich dust to the Southern Ocean is increased during glacial periods, due to expanded deserts in the Patagonian source region (Andersen *et al.,* 1998; Mahowald *et al.,* 1999; Petit *et al.,* 1999); dust-borne iron concentration in Antarctic ice is also increased (Edwards *et al.,* 1998). Fertilisation of marine productivity by iron from this source could have influenced atmospheric CO_2. Most of these mechanisms, however, can only account for about 30 ppm, or less, of the change (Lefèvre and Watson, 1999; Archer and Johnson, 2000). Palaeo-nutrient proxies have also been used to argue against large changes in total high latitude productivity (Boyle, 1988; Rickaby and Elderfield, 1999; Elderfield and Rickaby, 2000), even if the region of high productivity in the Southern Ocean may have been shifted to the north (Kumar *et al.,* 1995; François *et al.,* 1997). Increased productivity over larger regions might have been caused by decreased denitrification (Altabet *et al.,* 1995; Ganeshram *et al.,* 1995) or iron stimulated N_2 fixation (Broecker and Henderson, 1998) leading to an increase in the total ocean content of reactive nitrogen.

Another family of hypotheses invokes ocean alkalinity changes by a variety of mechanisms (Opdyke and Walker, 1992; Archer and Maier-Reimer, 1994; Kleypas, 1997), including increased silica supply through dust, promoting export production by siliceous rather than calcareous phytoplankton (Harrison, 2000). Although there is geochemical evidence for higher ocean pH during glacial times (Sanyal *et al.,* 1995), a large increase in alkalinity would result in a much deeper lysocline, implying an increase in $CaCO_3$ preservation that is not observed in deep-sea sediments (Catubig *et al.,* 1998; Sigman *et al.,* 1998; Archer *et al.,* 2000).

Given the complex timing of changes between climate changes and atmospheric CO_2 on glacial-interglacial time-scales, it is plausible that more than one mechanism has been in operation; and indeed most or all of the hypotheses encounter difficulties if called upon individually to explain the full magnitude of the change.

---

atmospheric CO_2. The rates of these processes are extremely slow, hence they are of limited relevance to the atmospheric CO_2 response to emissions over the next hundred years.

It is pertinent, however, that photosynthesis evolved at a time when O_2 concentrations were far less than at present. O_2 has accumulated in the atmosphere over geological time because photosynthesis results in the burial of reduced chemical species: pyrite (FeS_2) derived from sulphur-reducing bacteria, and organic carbon. This accumulation has consequences for terrestrial and marine ecosystems today. Primary production is carbon limited in terrestrial ecosystems in part because of (geologically speaking) low CO_2 concentrations, and in part because Rubisco (the enzyme that fixes CO_2 in all plants) also has an affinity for O_2 that reduces its efficiency in photosynthesis (see Section 3.2.2.4). Primary production is iron limited in some marine ecosystems mainly because of the extreme insolubility of Fe(III), the predominant form of iron in the present, O_2-rich environment. These difficulties faced by contemporary organisms represent a legacy of earlier evolution under very different biogeochemical conditions.

In more recent times, atmospheric CO_2 concentration continued to fall after about 60 Myr BP and there is geochemical evidence that concentrations were <300 ppm by about 20 Myr BP

(Pagani *et al.,* 1999a; Pearson and Palmer, 1999, 2000; Figure 3.2e). Low CO_2 concentrations may have been the stimulus that favoured the evolution of C_4 plants, which increased greatly in abundance between 7 and 5 Myr BP (Cerling *et al.,* 1993, 1997; Pagani *et al.,* 1999b). Although contemporary CO_2 concentrations were exceeded during earlier geological epochs, they are likely higher now than at any time during the past 20 million years.

3.3.2 Variations in Atmospheric CO_2 during Glacial/inter-glacial Cycles

The purity of Antarctic ice allows the CO_2 concentration in trapped air bubbles to be accurately measured (Tschumi and Stauffer, 2000). The CO_2 record from the Vostok ice core is the best available for the glacial/inter-glacial time-scale and covers the past four glacial/inter-glacial cycles (420 kyr) with a resolution of 1 to 2 kyr (Petit *et al.,* 1999; Fischer *et al.,* 1999). The general pattern is clear (Figure 3.2d): atmospheric CO_2 has been low (but $\geq$ 180 ppm) during glacial periods, and higher (but $\leq$300 ppm) during interglacials. Natural processes during the glacial-interglacial cycles have maintained CO_2 concentrations within these bounds, despite considerable variability on multi-millenial time-scales. The present CO_2 concentration is higher than at any time during the 420 kyr period covered by the Vostok record.

The terrestrial biosphere stores 300 to 700 Pg *more* carbon during interglacial periods than during glacial periods, based on a widely accepted interpretation of the $\delta^{13}C$ record in deep-sea sediments (Shackleton, 1977; Bird *et al.*, 1994; Crowley, 1995). Terrestrial modelling studies (e.g., Friedlingstein *et al.*, 1995b; Peng *et al.*, 1998) have reached the same conclusion. Thus, the terrestrial biosphere does not cause the difference in atmospheric CO_2 between glacial and interglacial periods. The cause must lie in the ocean, and indeed the amount of atmospheric change to be accounted for must be augmented to account for a fraction of the carbon transferred between the land and ocean. The mechanism remains controversial (see Box 3.4). In part this is because a variety of processes that could be effective in altering CO_2 levels on a century time-scale can be largely cancelled on multi-millenial time-scales by changes in $CaCO_3$ sedimentation or dissolution, as discussed in Section 3.2.3.1.

Orbital variations (Berger, 1978) are the pacemaker of climate change on multi-millenial time-scales (Hays *et al.*, 1976). Atmospheric CO_2 is one of many Earth system variables that show the characteristic "Milankovitch" periodicities, and has been implicated as a key factor in locking natural climate changes to the 100 kyr eccentricity cycle (Shackleton, 2000). Whatever the mechanisms involved, lags of up to 2,000 to 4,000 years in the drawdown of CO_2 at the start of glacial periods suggests that the low CO_2 concentrations during glacial periods amplify the climate change but do not initiate glaciations (Lorius and Oeschger, 1994; Fischer *et al.*, 1999). Once established, the low CO_2 concentration is likely to have enhanced global cooling (Hewitt and Mitchell, 1997). During the last deglaciation, rising CO_2 paralleled Southern Hemisphere warming and was ahead of Northern Hemisphere warming (Chapter 2).

During glacial periods, the atmospheric CO_2 concentration does not track the "fast" changes in climate (e.g., decade to century scale warming events) associated with Dansgaard-Oeschger events, although there are CO_2 fluctuations of up to 20 ppm associated with the longer-lived events (Stauffer *et al.*, 1998; Indermühle *et al.*, 2000) (see Chapter 2 for explanations of these terms). During the last deglaciation, atmospheric CO_2 concentration continued to increase, by about 12 ppm, through the Younger Dryas cold reversal (12.7 to 11.6 kyr BP) seen in Northern Hemisphere palaeoclimate records (Fischer *et al.*, 1999; Smith *et al.*, 1999). Palaeo-oceanographic evidence shows that the Younger Dryas event was marked by a prolonged shut-down of the thermo-haline circulation, which is likely to have been triggered by the release of melt water into the North Atlantic. Similar behaviour, with a slight rise in CO_2 accompanying a major Northern Hemisphere cooling and shutdown of North Atlantic Deep Water production, has been produced in a coupled atmosphere-ocean model (Marchal *et al.*, 1998). The observed CO_2 rise during the Younger Dryas period was modest, suggesting that atmospheric CO_2 has, under natural conditions, been well buffered against abrupt changes in climate, including thermohaline collapse. This buffering is a direct consequence of the large reservoir of DIC in the ocean.

3.3.3 Variations in Atmospheric CO_2 during the Past 11,000 Years

Natural variations in CO_2 during the past 11,000 years (Figure 3.2c) have been small (about 20 ppm) according to the best available measurements, which are from the Taylor Dome ice core (Smith *et al.*, 1999; Indermühle *et al.*, 1999). These measurements show a short-lived maximum around 11 kyr BP, followed by a slight fall, which may have been caused by increasing carbon storage in the terrestrial biosphere. Atmospheric CO_2 concentration was about 260 ppm at its Holocene minimum around 8 kyr BP and increased towards about 280 ppm in the pre-industrial period. The same pattern and the same CO_2 concentration levels over the past 8 kyr have also been shown in another ice core, BH7 near Vostok (Peybernès *et al.*, 2000). The causes of these changes are not known. Preliminary $\delta^{13}C$ measurements (see Box 3.6) suggest that this increase may have been due to a gradual reduction in terrestrial carbon storage (Indermühle *et al.*, 1999; Smith *et al.*, 1999) but others have considered an oceanic explanation more likely.

Atmospheric CO_2 concentrations have also been reconstructed indirectly, from stomatal index measurements on sub-fossil leaves (Van de Water *et al.*, 1994; Beerling *et al.*, 1995; Rundgren and Beerling, 1999; Wagner *et al.*, 1999). Stomatal density and stomatal index of many species respond to atmospheric CO_2 (Woodward, 1987; Woodward and Bazzaz, 1988) but are influenced by other environmental variables as well (Poole *et al.*, 1996). One recent stomatal index record, interpreted as implying high (up to 350 ppm) and rapidly fluctuating CO_2 concentrations in the early Holocene (Wagner *et al.*, 1999), is clearly incompatible with the ice core record of Indermühle *et al.* (1999), whereas a continuous stomatal index record from 9 kyr BP onwards (Rundgren and Beerling, 1999) has shown concentration trends consistent with the ice-core records.

Figure 3.2b shows the excellent agreement among different high-resolution Antarctic ice cores covering the past 1,000 years. Atmospheric CO_2 concentration fell by about 8 to 10 ppm during the Little Ice Age (from 1280 to 1860, see Chapter 2) (Figure. 3.2b, c; Barnola *et al.*, 1995; Etheridge *et al.*, 1996; Indermühle *et al.*, 1999; Rundgren and Beerling, 1999). A slight contemporaneous increase in $\delta^{13}C$ of atmospheric CO_2 has led to the suggestion that this effect was caused by enhanced carbon storage on land (Francey *et al.*, 1999b; Trudinger *et al.*, 1999).

3.3.4 Implications

The Vostok record of atmospheric CO_2 and Antarctic climate is consistent with a view of the climate system in which CO_2 concentration changes amplify orbitally-induced climate changes on glacial/inter-glacial time-scales (Shackleton, 2000). Changes during the present inter-glacial (until the start of the anthropogenic CO_2 rise) have been small by comparison. Although complete explanations for these changes in the past are lacking, high-resolution ice core records establish that the human-induced increase of atmospheric CO_2 over the past century is at least an order of magnitude faster than has occurred during the preceeding 20,000 years.

3.4 Anthropogenic Sources of CO₂

3.4.1 Emissions from Fossil Fuel Burning and Cement Production

Current anthropogenic emissions of CO₂ are primarily the result of the consumption of energy from fossil fuels. Estimates of annual global emissions from fossil fuel burning and cement production have been made for the period from 1751 through 1999. Figure 3.3 summarises emissions over the period from 1959 to 1999 (Keeling and Whorf, 2000).

Estimates of annual global emissions from fossil fuel burning and cement production by Marland *et al.* (2000) span the period from 1751 through to 1997, reaching a maximum in 1997 of 6.6 PgC/yr (0.2 PgC/yr of this was from cement production). The primary data for these estimates are annual energy statistics compiled by the United Nations (2000). Emissions for 1998 and 1999 have been estimated based on energy statistics compiled by British Petroleum (2000). Emission factors (IPCC, 1997) were applied to consumption statistics (British Petroleum, 2000) to calculate emissions over the period 1990 to 1999. Emissions were then scaled to match the estimates for emissions from fossil fuel burning and cement production from Marland *et al.* (2000) over the overlap period from 1990 to 1997. The scaled emission estimates, therefore, implicitly include emissions from cement production.

The average value of emissions for the 1980s given by Marland *et al.* (2000) is 5.44 ± 0.3 PgC/yr, revised from the earlier estimate (Marland *et al.* 1994; Andres *et al.* 2000) of 5.46 ± 0.3 PgC/yr used in the SAR and in the Special Report on Radiative Forcing (IPCC, 1994) (hereafter SRRF). Estimated emissions rose from 6.1 PgC/yr in 1990 to 6.5 PgC/yr in 1999. The average value of emissions in the 1990s was 6.3 ± 0.4 PgC/yr.

3.4.2 Consequences of Land-use Change

About 10 to 30% of the current total anthropogenic emissions of CO₂ are estimated to be caused by land-use conversion. Such estimates rely on land cover data sets which are highly variable, and estimates of average carbon density of vegetation types, which are also highly variable with stand age and local conditions (see Box 3.1). Hence they cannot be specified as accurately as is possible for fossil fuel emissions. Historical emissions are treated in Section 3.2.2.2; this section focuses on the contemporary situation.

Net land-use flux, comprising the balance of positive terms due to deforestation and negative terms due to regrowth on abandoned agricultural land, has been estimated based on land-use statistics and simple models of rates of decomposition and regrowth, excluding possible climate, CO₂ and N fertilisation effects (Houghton, 1999). Not all land-use emissions are included, for example mining of peatlands. The analysis of Houghton (1999) indicated that the net flux due to land-use change was 2.0 ± 0.8 PgC/yr during the 1980s, almost entirely due to deforestation of tropical regions. Temperate forests were found to show an approximate balance between carbon uptake in regrowing forests and carbon lost in oxidation of wood

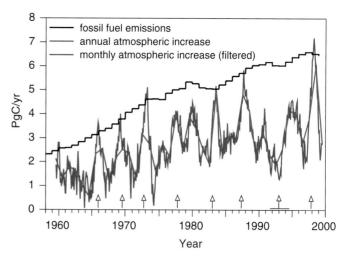

Figure 3.3: Fossil fuel emissions and the rate of increase of CO₂ concentration in the atmosphere. The annual atmospheric increase is the measured increase during a calendar year. The monthly atmospheric increases have been filtered to remove the seasonal cycle. Vertical arrows denote El Niño events. A horizontal line defines the extended El Niño of 1991 to 1994. Atmospheric data are from Keeling and Whorf (2000), fossil fuel emissions data are from Marland *et al.* (2000) and British Petroleum (2000), see explanations in text.

products, except in Europe, which showed a small net accumulation. The estimate of 2.0 PgC/yr is somewhat higher than Houghton and Hackler's (1995) earlier estimate of 1.6 PgC/yr for the same period, which was used in the SAR, because of a reanalysis of data from tropical Asia (Houghton and Hackler, 1999). However, other recent analyses by the same authors reduce the estimated emissions from the Brazilian Amazon by half (Houghton *et al.*, 2000), and point to other previously unaccounted for sinks of carbon in the USA such as fire suppression and woody encroachment, and changes in the management of agricultural soils (Houghton *et al.*, 1999). Consideration of these additional studies brings the overall total back down to 1.7 ± 0.8 PgC/yr (Houghton, 2000), as given in the SRLULUCF.

An independent analysis (see Section 3.6.2.2) by the Carbon Cycle Model Linkage Project (CCMLP) also calculated the marginal effects of land-use changes on the global terrestrial carbon budget (McGuire *et al.*, 2001). Land-use change data (conversions between native vegetation and crops) were derived from Ramankutty and Foley (2000). The estimates obtained for net land-use flux during the 1980s were between 0.6 and 1.0 PgC/yr, i.e., substantially smaller than the fluxes calculated by Houghton (1999). The reasons for this discrepancy are unclear. The CCMLP estimates may be too low because they neglected conversions to pasture. However, data presented in Houghton (1999) indicate that the main changes during recent decades were due to land conversion for crops. A more important difference may lie in the timing of deforestation in different regions in the tropics, where Ramankutty and Foley (2000) show higher overall rates in the 1970s and lower rates in the 1980s than Houghton does (1999).

Box 3.5: The use of O_2 measurements to assess the fate of fossil fuel CO_2.

The amount of CO_2 that remains in the atmosphere each year has been consistently less than the amount emitted by fossil fuel burning. This is because some CO_2 dissolves and mixes in the ocean, and some is taken up by the land. These two modes of uptake have different effects on the concentration of O_2 in the atmosphere. Fossil fuel burning consumes O_2 and causes a decline in atmospheric O_2 concentration (Figure 3.4). Dissolution of CO_2 in the ocean has no effect on atmospheric O_2. Terrestrial uptake of CO_2, by contrast, implies that photosynthesis (which releases O_2) is exceeding respiration and other oxidation processes, including fire (which consume O_2). Thus, net terrestrial uptake of CO_2 implies a net release of O_2, in a known stochiometric ratio. This difference can be used to partition the total CO_2 uptake into land and ocean components, as shown graphically in Figure 3.4. Strictly speaking, the atmospheric $O_2 - CO_2$ budget method can only distinguish between net non-biological ocean uptake and net biospheric uptake, which in principle includes both the terrestrial and the marine biosphere. However, since biological oxygen uptake is not expected to have changed significantly during recent decades because of nutrient limitations in most parts of the ocean (see Section 3.2.3.2), this inferred biospheric uptake is attributed to the land.

Measurement of changes in O_2 presents a technical challenge because changes of a few ppm caused by fossil fuel burning have to be determined against a background concentration of 209,000 ppm (about 21%). For technical reasons, O_2 is measured relative to N_2, the main constituent of the atmosphere, as a reference gas. For simplicity this chapter refers to O_2 concentrations, although strictly it is $O_2 : N_2$ ratios that are measured. The impact of nitrification-denitrification changes on atmospheric N_2 content are assumed not to be problematic because they are small and the inventory of N_2 is very large. Increases in ocean temperatures (Levitus *et al.*, 2000), because of their effect on the temperature dependent solubility, induce small outgassing fluxes of O_2 and N_2 (Keeling *et al.*, 1993) that have to be taken into account (see Figure 3.4) although their magnitude is only approximately known. Impacts on atmospheric O_2 caused by changes in the ventilation of deeper, oxygen depleted waters have been observed on interannual time-scales (Keeling *et al.*, 1993, Bender *et al.*, 1996). They could also occur on longer time-scales, e.g., through increased ocean stratification induced by ocean warming.

Another analysis calculated a substantially higher net source due to land-use change in the tropics of 2.4 ± 1.0 PgC/yr during the 1980s (Fearnside, 2000). This analysis did not deal with temperate regions, and is not used in the global budget estimates.

No complete global assessment of deforestation effects covering the 1990s is available. Rates of deforestation appear to be declining. The FAO (1997) tropical forest assessment reported annual losses of 15.5×10^6 ha in the 1980s, and 13.7×10^6 ha in 1990 to 1995. Independent studies show a significant decline in deforestation rates in the Amazon region (Skole and Tucker, 1993; Fearnside, 2000). The annual flux of carbon from land-use change for the period from 1990 to 1995 has been estimated to be 1.6 PgC/yr from 1990 to 1995, consisting of a source of 1.7 PgC/yr in the tropics and a small sink in temperate and boreal areas (Houghton, 2000).

3.5 Observations, Trends and Budgets

3.5.1 Atmospheric Measurements and Global CO_2 Budgets

Continuous time-series of highly precise measurements of the atmospheric composition are of central importance to current understanding of the contemporary carbon cycle. CO_2 has been measured at the Mauna Loa and South Pole stations since 1957 (Keeling *et al.*, 1995; Figure 3.2a), and through a global surface sampling network developed in the 1970s that is becoming progressively more extensive and better inter-calibrated (Conway *et al.*, 1994; Keeling *et al.*, 1995). Associated measurements of $\delta^{13}C$ in atmospheric CO_2 began in 1977 (Francey *et al.*, 1995, Keeling *et al.*, 1995, Trolier *et al.*, 1996). More recently, comple-

mentary information has been available from O_2 concentrations (measured as ratios of $O_2:N_2$, see Box 3.5), which have been regularly measured since the early 1990s (Keeling and Shertz, 1992; Keeling *et al.*, 1993; Bender *et al.*, 1996; Keeling *et al.*, 1996b; Battle *et al.*, 2000; Manning, 2001; Figure 3.2a). O_2 concentration data for the 1980s have been gleaned by two methods: sampling of archived air flasks that were collected during the 1980s (Langenfelds *et al.*, 1999), and measuring the air trapped in Antarctic firn (Battle *et al.*, 1996).

In addition to fossil fuel CO_2 emissions, Figure 3.3 shows the observed seasonally corrected growth rate of the atmospheric CO_2 concentrations, based on the two longest running atmospheric CO_2 recording stations (Keeling and Whorf, 2000). It is evident from this comparison that a part of the anthropogenic CO_2 has not remained in the atmosphere; in other words, CO_2 has been taken up by the land or the ocean or both. This comparison also shows that there is considerable interannual variability in the total rate of uptake.

O_2 and CO_2 measurements are used here to provide observationally-based budgets of atmospheric CO_2 (Table 3.1). CO_2 budgets are presented here (Table 3.1) for the 1980s (for comparison with previous work; Table 3.3), and for the 1990s. The reported error ranges are based on uncertainties of global fossil fuel emissions, determination of the decadal average changes in the atmospheric CO_2 concentration, and $O_2:N_2$ ratio; and uncertainties in the assumed $O_2:CO_2$ stoichiometric ratios in the combustion of fossil fuels and in photosynthesis and respiration. The error ranges reflect *uncertainties* of the decadal mean averaged values; they do not reflect interannual variability in annual values, which far exceeds uncertainty in

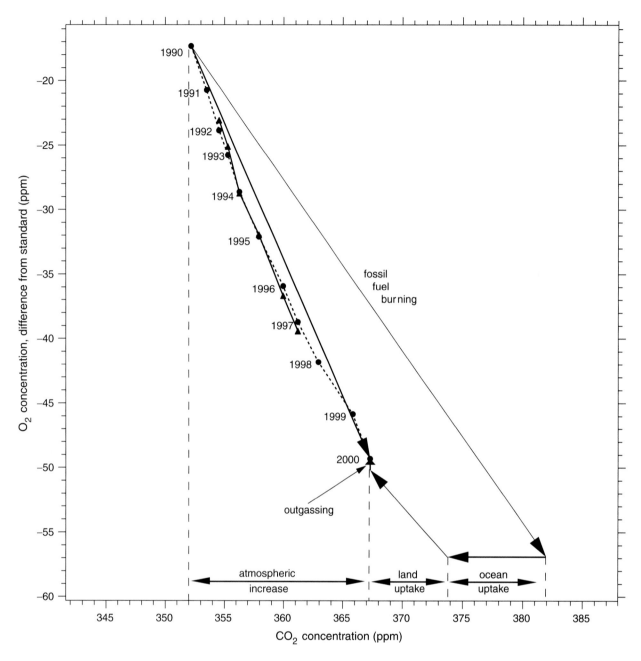

Figure 3.4: Partitioning of fossil fuel CO_2 uptake using O_2 measurements (Keeling and Shertz, 1992; Keeling *et al.*, 1993; Battle *et al.*, 1996, 2000; Bender *et al.*, 1996; Keeling *et al.*, 1996b; Manning, 2001). The graph shows the relationship between changes in CO_2 (horizontal axis) and O_2 (vertical axis). Observations of annual mean concentrations of O_2, centred on January 1, are shown from the average of the Alert and La Jolla monitoring stations (Keeling *et al.*, 1996b; Manning, 2001; solid circles) and from the average of the Cape Grim and Point Barrow monitoring stations (Battle *et al.*, 2000; solid triangles). The records from the two laboratories, which use different reference standards, have been shifted to optimally match during the mutually overlapping period. The CO_2 observations represent global averages compiled from the stations of the NOAA network (Conway *et al.*, 1994) with the methods of Tans *et al.* (1989). The arrow labelled "fossil fuel burning" denotes the effect of the combustion of fossil fuels (Marland *et al.*, 2000; British Petroleum, 2000) based on the relatively well known O_2:CO_2 stoichiometric relation of the different fuel types (Keeling, 1988). Uptake by land and ocean is constrained by the known O_2:CO_2 stoichiometric ratio of these processes, defining the slopes of the respective arrows. A small correction is made for differential outgassing of O_2 and N_2 with the increased temperature of the ocean as estimated by Levitus *et al.* (2000).

Box 3.6: Stable carbon isotopes in atmospheric CO_2.

$\delta^{13}C$, a measure of the relative abundance of the two stable carbon isotopes, ^{13}C and ^{12}C, in atmospheric CO_2 gives in principle similar possibilities to O_2 for the partitioning of atmospheric CO_2 uptake (Keeling *et al.*, 1979, 1980; Mook *et al.*, 1983; Keeling *et al.*, 1989; Francey *et al.*, 1995; Keeling *et al.*, 1995). The principle of using $\delta^{13}C$ to separate land and ocean components of the carbon budget relies on the fractionation during photosynthesis by C_3 plants, which discriminates against ^{13}C. This fractionation leads to biospheric carbon being depleted in ^{13}C by about 18‰ relative to the atmosphere. In contrast, exchanges with the ocean involve relatively small fractionation effects. Changes in the $^{13}C/^{12}C$ ratio of atmospheric CO_2 thus indicate the extent to which concurrent CO_2 variations can be ascribed to variations in biospheric uptake. The calculation also requires specification of the turnover times of carbon in the ocean and on land, because fossil fuel burning implies a continuous release of isotopically light carbon to the atmosphere. This leads to a lowering of the atmospheric $^{13}C/^{12}C$ isotope ratio, which takes years to centuries to work its way through the carbon cycle (Keeling *et al.*, 1980; Tans *et al.*, 1993; Ciais *et al.*, 1995a,b).

There are some complications. C_3 plants discriminate against ^{13}C more strongly than C_4 plants (Lloyd and Farquhar, 1994), thus the distributions of C_3 and C_4 photosynthesis need to be known. The oceanic disequilibrium can in principle be estimated observationally (Tans *et al.*, 1993; Heimannn and Maier-Reimer, 1996; Bacastow *et al.*, 1996; Gruber *et al.*, 1999), while the terrestrial disequilibrium has to be estimated by means of models (e.g., Ciais *et al.*, 1999). Langenfelds *et al.* (1999) and Battle *et al.* (2000) have shown that recently estimated values for the disequilibrium terms lead to consistency between the partitioning of CO_2 uptake into land and ocean uptake based on O_2 and on $\delta^{13}C$ measurements.

---

the decadal mean rate of increase, as is further discussed in Section 3.5.2. The salient facts are as follows:

- During the 1980s, fossil fuel emissions were on average 5.4 ± 0.3 PgC/yr and atmospheric CO_2 content increased on average by 3.3 ± 0.1 PgC/yr. Partitioning of CO_2 uptake was estimated based on archived flask O_2 measurements (Langenfelds *et al.*, 1999) for the 1979 to 1997 period, taking the O_2 trend during 1991 to 1997 (Battle *et al.*, 2000) into account. The resulting estimate of the ocean-atmosphere flux was -1.9 ± 0.6 PgC/yr and of the land-atmosphere flux -0.2 ± 0.7 PgC/yr. This partitioning is adopted here in Table 3.1. It is corroborated by independent O_2 measurements in Antarctic firn (Battle *et al.*, 1996). Restricting the analysis to the Battle *et al.* (1996) data for the 1980 to 1989 period, an ocean-atmosphere flux of -1.8 ± 1.0 PgC/yr and a land-atmosphere flux of -0.4 ± 1.0 PgC/yr were obtained, i.e., results indistinguishable from the values in Table 3.1.

- Despite a greater emission rate of 6.3 ± 0.4 PgC/yr (see Section 3.4.1), the average atmospheric increase during the 1990s was 3.2 ± 0.1 PgC/yr, i.e., about the same as during the 1980s. An exceptionally low rate of increase during the early 1990s was balanced by a high rate during the late 1990s. Based on the longest existing O_2 records from La Jolla (California, USA) and Alert (northern Canada) (Keeling *et al.*, 1996b; Manning, 2001; see Figure 3.4), the ocean-atmosphere flux during the 1990s was -1.7 ± 0.5 PgC/yr and the land-atmosphere flux was -1.4 ± 0.7 PgC/yr.

Ocean uptake in the 1980s as estimated from O_2 and CO_2 measurements thus agrees with the estimates in the SRRF (Schimel *et al.*, 1995) and the SAR (Schimel *et al.*, 1996) (although these were model-based estimates; this section presents only observationally-based estimates (Table 3.3)). Considering the uncertainties, the ocean sink in the 1990s was not significantly different from that in

the 1980s. The land-atmosphere flux was close to zero in the 1980s, as also implied by the SAR budget. The land appears to have taken up more carbon during the 1990s than during the 1980s. The causes cannot yet be reliably quantified, but possible mechanisms include a slow down in deforestation (Section 3.4.2), and climate variability that resulted in temporarily increased land and/or ocean uptake in the early 1990s (Section 3.5.2). These budgets are consistent with information from atmospheric $\delta^{13}C$ measurements (see Box 3.6 and Table 3.4) and with budgets presented in the SRLULUCF (Bolin *et al.* 2000) except that estimated ocean uptake is smaller, and land uptake accordingly larger, than given in the SRLULUCF (see Table 3.3, footnote *i*).

Several alternative approaches to estimating the ocean-atmosphere and land-atmosphere fluxes of CO_2 are summarised in Table 3.4. Alternative methods for estimating the global ocean-atmosphere flux, based on surface-water pCO_2 measurements and ocean $\delta^{13}C$ changes (Quay *et al.*, 1992; Tans *et al.*, 1993, Heimann and Maier-Reimer, 1996; Sonnerup *et al.*, 1999), respectively, have yielded a range of -1.5 to -2.8 PgC/yr (for various recent periods). The total anthropogenic CO_2 added to the ocean since pre-industrial times can also be estimated indirectly using oceanic observations (Gruber *et al.*, 1996). A global value of 107 ± 27 PgC by 1990 can be estimated from the basin-scale values of 40 ± 9 PgC for the Atlantic in the 1980s (Gruber, 1998), 20 ± 3 PgC for the Indian Ocean in 1995 (Sabine *et al.*, 1999), and the preliminary value of 45 PgC for the Pacific Ocean in 1990 to 1996 (Feely *et al.*, 1999a) with a large uncertainty of the order of ± 15 PgC. Assuming that accumulation of CO_2 in the ocean follows a curve similar to the (better known) accumulation in the atmosphere, the value for the ocean-atmosphere flux for 1980 to 1989 would be between -1.6 and -2.7 PgC/yr. Although each individual method has large uncertainty, all of these ocean-based measurements give results comparable with the fluxes presented in Table 3.1. Consideration of model-based estimates of ocean uptake in Table 3.4 is deferred to Section 3.6.2.2.

Table 3.3: *Comparison of the global CO_2 budgets from Table 3.1 with previous IPCC estimates[a,b,c] (units are PgC/yr).*

	1980s				1990s	1989 to 1998
	This chapter	**SRLULUCF**[d]	**SAR**[e]	**SRRF**[f]	**This chapter**	**SRLULUCF**[d]
Atmospheric increase	3.3 ± 0.1	3.3 ± 0.1	3.3 ± 0.1	3.2 ± 0.1	3.2 ± 0.1	3.3 ± 0.1
Emissions (fossil fuel, cement)	5.4 ± 0.3	5.5 ± 0.3	5.5 ± 0.3	5.5 ± 0.3	6.4 ± 0.4	6.3 ± 0.4
Ocean-atmosphere flux	-1.9 ± 0.6	-2.0 ± 0.5[i]	-2.0 ± 0.5	-2.0 ± 0.5	-1.7 ± 0.5	-2.3 ± 0.5[i]
Land-atmosphere flux* *partitioned as follows*	-0.2 ± 0.7[g]	-0.2 ± 0.6	-0.2 ± 0.6	-0.3 ± 0.6	-1.4 ± 0.7	-0.7 ± 0.6
Land-use change	$1.7\ (0.6\ \text{to}\ 2.5)$[g]	1.7 ± 0.8	1.6 ± 1.0	1.6 ± 1.0	*insufficient*	1.6 ± 0.8[j]
Residual terrestrial sink	$-1.9\ (-3.8\ \text{to}\ 0.3)$	-1.9 ± 1.3	-1.8 ± 1.6[h]	-1.9 ± 1.6	*data*	-2.3 ± 1.3

[a] Positive values are fluxes to the atmosphere; negative values represent uptake from the atmosphere.

[b] Previous IPCC carbon budgets calculated ocean uptake and land-use change from models. The residual terrestrial sink was inferred. Here the implied land-atmosphere flux (with its error) is derived from these previous budgets as required for comparison with Table 3.1.

[c] Error ranges are expressed in this book as 67% confidence intervals ($\pm 1\sigma$). Previous IPCC estimates have used 90% confidence intervals ($\pm 1.6\sigma$). These error ranges have been scaled down as required for comparison with Table 3.1. Uncertainty ranges for land-use change emissions have not been altered in this way.

[d] IPCC Special Report on Land Use, Land-use Change and Forestry (SRLULUCF) (IPCC, 2000a; Bolin *et al.*, 2000).

[e] IPCC Second Assessment Report (SAR) (IPCC, 1996a; Schimel *et al.*, 1996).

[f] IPCC Special Report on Radiative Forcing (SRRF) (Schimel *et al.*, 1995).

[g] Ranges based on Houghton (1999, 2000), Houghton and Hackler (1999), and CCMLP model results (McGuire *et al.*, 2001).

[h] The sink of 0.5 ± 0.5 PgC/yr in "northern forest regrowth" cited in the SAR budget is assigned here to be part of the residual terrestrial sink, following Bolin *et al.* (2000).

[i] Based on an ocean carbon cycle model (Jain *et al.*, 1995) used in the IPCC SAR (IPCC, 1996; Harvey *et al.*, 1997), tuned to yield an ocean-atmosphere flux of 2.0 PgC/yr in the 1980s for consistency with the SAR. After re-calibration to match the mean behaviour of OCMIP models and taking account of the effect of observed changes in temperature aon CO_2 and solubility, the same model yields an ocean-atmosphere flux of -1.7 PgC/yr for the 1980s and -1.9 PgC/yr for 1989 to 1998.

[j] Based on annual average estimated emissions for 1989 to 1995 (Houghton, 2000).

The land-atmosphere flux based on atmospheric measurements represents the *balance* of a net land-use flux (currently a positive flux, or carbon source, dominated by tropical deforestation) and a residual component which is, by inference, a negative flux or carbon sink. Using the land-atmosphere flux estimates from Table 3.1, assuming that land-use change contributed +1.7 PgC/yr to the atmosphere during the 1980s (Section 3.4.2), then a residual terrestrial flux of -1.9 PgC/yr (i.e., a residual sink of similar magnitude to the total ocean uptake) is required for mass balance. This is the term popularly (and misleadingly) known as the "missing sink". The central estimate of its magnitude agrees with previous analyses, e.g., in the SAR (if "northern forest regrowth" is combined with "residual terrestrial sink" terms in the SAR budget; Schimel *et al.*, 1996) and the SRLULUCF (Bolin *et al.*, 2000) (Table 3.3). The uncertainty around this number is rather large, however, because it compounds the uncertainty in the atmospheric budget with a major uncertainty about changes in land use. Using an error range corresponding to 90% confidence intervals around the atmospheric estimate of -0.2 PgC/yr (i.e., 1.6σ, giving confidence intervals of ± 1.1 PgC/yr), and taking the range of estimates for CO_2 released due to land-use change during the 1980s from Section 3.4.2, the residual terrestrial sink is estimated to range from -3.8 to $+0.3$ PgC/yr for the 1980s. Model-based analysis of the components of the residual terrestrial sink (Table 3.4) is discussed in Section 3.6.2.2.

3.5.2 Interannual Variability in the Rate of Atmospheric CO_2 Increase

The rate of increase in the globally averaged atmospheric concentration of CO_2 varies greatly from year to year. "Fast" and "slow" years have differed by 3 to 4 PgC/yr within a decade (Figure 3.3). This variability cannot be accounted for by fossil fuel emissions, which do not show short-term variability of this magnitude. The explanation must lie in variability of the land-atmosphere flux or the ocean-atmosphere flux or both. Variability in both systems could be induced by climate variability.

An association between CO_2 variability and El Niño in particular has been reported for over twenty years and has been confirmed by recent statistical analyses (Bacastow, 1976; Keeling and Revelle, 1985; Thompson *et al.*, 1986; Siegenthaler, 1990; Elliott *et al.*, 1991; Braswell *et al.*, 1997; Feely *et al.*, 1997; Dettinger and Ghil, 1998; Rayner *et al.*, 1999b). During most of the observational record, El Niño events have been marked by high rates of increase in atmospheric CO_2 concentration compared with surrounding years, in the order of > 1 PgC/yr higher during most El Niño events (Figure 3.3). Direct measurements of oceanic CO_2 in the equatorial Pacific over the last 20 years have shown that the natural efflux of CO_2 from this region is reduced by between 0.2 to 1.0 PgC/yr during El Niño (Keeling and Revelle, 1985; Smethie *et al.*, 1985; Takahashi *et al.*, 1986; Inoue and Sugimura, 1992; Wong *et al.*, 1993; Feely *et al.*, 1997;

Table 3.4: *Alternative estimates of ocean-atmosphere and land-atmosphere fluxes.*

	Ocean-atmosphere flux	Land-atmosphere flux
Oceanic observations		
1970 to 1990 Ocean ^{13}C inventory	-2.1 ± 0.8 [a] -2.1 ± 0.9 [b]	
1985 to 1995 Ocean ^{13}C inventory	-1.5 ± 0.9 [c]	
1995 Surface-water pCO$_2$	-2.8 ± 1.5 [d]	
1990 Inventory of anthropogenic CO$_2$	-1.6 to -2.7 [e]	
Atmospheric observations		
1980 to 1989 O$_2$ in Antarctic firn	-1.8 [f]	-0.4 [f]
1990 to 1999 Atmospheric CO$_2$ and δ^{13}C	-1.8 [g] -2.4 [h]	-1.4 [g] -0.8 [h]
Models		
1980 to 1989 OCMIP	-1.5 to -2.2 [i]	
CCMLP * *partitioned as follows:*		-0.3 to -1.5 [j] * *partitioned as follows:*
Land-use change		0.6 to 1.0
CO$_2$ and N fertilisation		-1.5 to -3.1
Climate variability		-0.2 to $+0.9$

Sources of data:

[a] Quay *et al.* (1992).

[b] Heimann and Maier-Reimer (1996).

[c] Gruber and Keeling (2001).

[d] Takahashi *et al.* (1999) with -0.6 PgC/yr correction for land-ocean river flux.

[e] Gruber (1998), Sabine *et al.* (1999); Feely *et al.* (1999a), assuming that the ocean and atmospheric CO$_2$ increase follow a similar curve.

[f] This chapter, from data of Battle *et al.* (1996).

[g] Updated calculation of Ciais *et al.* (1995b); Tans *et al.* (1989); Trolier *et al.* (1996) (no error bars given).

[h] Keeling and Piper (2000) (no error bars given).

[i] Orr *et al.* (2000), Orr and Dutay (1999).

[j] McGuire *et al.* (2001).

1999b), mainly due to the reduced upwelling of CO$_2$-rich waters (Archer *et al.*, 1996). The ocean response to El Niño in the most active region thus tends to increase global CO$_2$ uptake, counter to the increasing atmospheric concentration. Although it cannot be ruled out that other ocean basins may play a significant role for global interannual variability in ocean-atmosphere flux, the existing oceanic measurements suggest (by default) that the response of the terrestrial biosphere is the cause of the typically high rates of CO$_2$ increase during El Niño.

Associated variations in the north-south gradient of CO$_2$ indicate that the El Niño CO$_2$ anomalies originate in the tropics (Conway *et al.*, 1994; Keeling and Piper, 2000). Typical El Niño events are characterised by changed atmospheric circula-

tion and precipitation patterns (Zeng, 1999) that give rise to high tropical land temperatures (which would be expected to increase Rh and reduce NPP); concurrent droughts which reduce NPP, especially in the most productive regions such as the Amazon rain forest; and increased incidence of fires in tropical regions. Increased cloudiness associated with enhanced south-east Asian monsoons during the late phase of El Niño has also been suggested as a factor reducing global NPP (Yang and Wang, 2000). Typically, although not invariably, the rate of atmospheric CO$_2$ increase declines around the start of an El Niño, then rapidly rises during the late stages (Elliott *et al.*, 1991; Conway *et al.*, 1994). It has been suggested that this pattern represents early onset of enhanced ocean CO$_2$ uptake, followed by reduced terrestrial CO_2 uptake or terrestrial CO_2 release (Feely *et al.*, 1987, 1999b; Rayner *et al.*, 1999b; Yang and Wang, 2000).

Atmospheric $\delta^{13}C$ and, more recently, O_2 measurements have been used to partition the interannual variability of the atmospheric CO_2 increase into oceanic and terrestrial components. Analyses based on $\delta^{13}C$ by Keeling *et al.* (1995) and Francey *et al.* (1995) reached contradictory conclusions, but the discrepancies are now thought to be due at least in part to $\delta^{13}C$ measurement calibration problems during the 1980s, which have largely been resolved during the 1990s (Francey *et al.*, 1999a). For the 1990s, a range of analyses using different atmospheric observations and/or data analysis techniques estimate that the amplitude of annual peak to peak variation associated with the ocean is about 2 to 3 PgC/yr and the amplitude associated with the terrestrial biosphere is about 4 to 5 PgC/yr (Rayner *et al.*, 1999a; Joos *et al.*, 1999a; Battle *et al.*, 2000 (O_2-based analysis); Keeling and Piper, 2000; Manning, 2001). A similar partitioning was estimated by Bousquet *et al.* (2000) based on the spatial pattern of CO_2 measurements using the approach described in the next section (3.5.3). However, the various reconstructed time sequences of terrestrial and ocean uptake differ in many details and do not provide conclusive evidence of the mechanisms involved.

The early 1990s were unusual in that the growth rate in atmospheric CO_2 was low (1.9 PgC/yr in 1992), especially in the Northern Hemisphere (Conway *et al.*, 1994), while an extended El Niño event occurred in the equatorial Pacific. Various mechanisms have been suggested, but none fully explain this unusual behavior of the carbon cycle. The slow down in the CO_2 increase has been linked to the predominantly mid- to high latitude cooling caused by the Pinatubo eruption (Conway *et al.*, 1994; Ciais *et al.*, 1995a,b; Schimel *et al.*, 1996), but there is no proof of any connection between these events. Other partial explanations could come from a temporary slow down of tropical deforestation (Houghton *et al.*, 2000), or natural decadal variability in the ocean-atmosphere or land-atmosphere fluxes (Keeling *et al.*, 1995). In any case, the slowdown proved to be temporary, and the El Niño of 1998 was marked by the highest rate of CO_2 increase on record, 6.0 PgC/yr.

3.5.3 Inverse Modelling of Carbon Sources and Sinks

Inverse modelling attempts to resolve regional patterns of CO_2 uptake and release from observed spatial and temporal patterns in atmospheric CO_2 concentrations, sometimes also taking into consideration O_2 and/or $\delta^{13}C$ measurements. The most robust results are for the latitudinal partitioning of sources and sinks between northern and southern mid- to high latitudes and the tropics. The observed annual mean latitudinal gradient of atmospheric CO_2 concentration during the last 20 years is relatively large (about 3 to 4 ppm) compared with current measurement accuracy. It is however not as large as would be predicted from the geographical distribution of fossil fuel burning – a fact that suggests the existence of a northern sink for CO_2, as already recognised a decade ago (Keeling *et al.*, 1989; Tans *et al.*, 1990; Enting and Mansbridge, 1991).

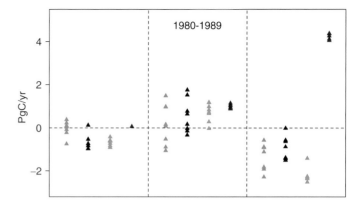

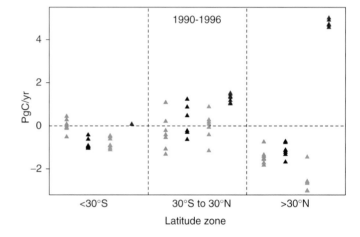

Figure 3.5: Inverse model estimates of fossil fuel CO_2 uptake by latitude bands according to eight models using different techniques and sets of atmospheric observations (results summarised by Heimann, 2001). Positive numbers denote fluxes to the atmosphere; negative numbers denote uptake from the atmosphere. The ocean-atmosphere fluxes represent mainly the natural carbon cycle; the land-atmosphere fluxes may be considered as estimates of the uptake of anthropogenic CO_2 by the land (with some caveats as discussed in the text). The sum of land-atmosphere and ocean-atmosphere fluxes is shown because it is somewhat better constrained by observations than the separate fluxes, especially for the 1980s when the measurement network was less extensive than it is today. The 1990s are represented by the period 1990 to 1996 only, because when this exercise was carried out the modelling groups did not have access to all of the necessary data for more recent years.

The nature of this sink, however, cannot be determined from atmospheric CO_2 concentration measurements alone. It might reflect, at least in part, a natural source-sink pattern of oceanic CO_2 fluxes (Keeling *et al.*, 1989; Broecker and Peng, 1992). This view is supported by the early atmospheric CO_2 data from the 1960s (Bolin and Keeling, 1963) which do not show a clear latitudinal gradient, despite the fact that at that time the fossil

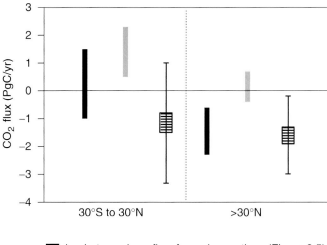

Figure 3.6: Partitioning the 1980s land-atmosphere flux for the tropics and the northern extratropics. The residual terrestrial sink in different latitude bands can be inferred by subtracting the land-use change flux for the 1980s (estimated by modelling studies: Houghton, 1999; Houghton and Hackler, 1999; Houghton *et al.*, 2000; McGuire *et al.*, 2001) from the net land-atmosphere flux as obtained from atmospheric observations by inverse modelling for the same period (Heimann, 2001; results from Figure 3.5). Positive numbers denote fluxes to the atmosphere; negative numbers denote uptake from the atmosphere. This calculation is analogous to the global budget calculation in Table 3.1, but now the model results are broken down geographically and the land-atmosphere fluxes are obtained by inverse modelling. The upper and lower bounds on the residual sink are obtained by pairing opposite extremes of the ranges of values accepted for the two terms in this calculation (for example, by subtracting the bottom of the range of values for land-use change with the top of the range for the land-atmosphere flux). The mid-ranges are obtained by combining similar extremes (for example, subtracting the bottom of the range for land-use change emissions from the bottom of the range land-atmosphere flux).

emissions were already at least half as large as in the 1990s. Quantitative analysis shows that the Northern Hemisphere sink has not changed much in magnitude since the 1960s (Keeling *et al.*, 1989; Fan *et al.*, 1999). On the other hand, the existing air-sea flux measurements do not support the idea of a large oceanic uptake of CO_2 in the Northern Hemisphere (Tans *et al.*, 1990; Takahashi, 1999). An alternative view, therefore, locates a significant fraction of this Northern Hemisphere sink on land. This view is corroborated, at least for the 1990s, by analyses of the concurrent latitudinal gradients of $\delta^{13}C$ (Ciais *et al.*, 1995a,b) and O_2 (Keeling *et al.*, 1996b).

Results of analyses for the 1980s and 1990 to 1996, carried out by eight modelling groups using different atmospheric transport models, observational data, constraints and mathematical procedures, are summarised in Figure 3.5. Only the most robust findings, i.e., estimates of the mean carbon balance for three latitude bands averaged over the two time periods, are

shown. The latitude bands are: "southern extratropics" (>30°S), "tropics" (30°S to 30°N) and "northern extratropics" (>30°N). The carbon balance estimates are broken down into land and ocean compartments within each latitude band (Heimann, 2001).

Although the ranges of the estimates in Figure 3.5 limit the precision of any inference from these analyses, some clear features emerge. The inferred ocean uptake pattern shows the sum of two components: the natural carbon cycle in which CO_2 is outgassed in the tropics and taken up in the extratropics, and the perturbation uptake of anthropogenic CO_2. Separation of these two components cannot be achieved from atmospheric measurements alone.

The estimates for the land, on the other hand, in principle indicate the locations of terrestrial anthropogenic CO_2 uptake (albeit with caveats listed below). For 1980 to 1989, the inverse-model estimates of the land-atmosphere flux are −2.3 to −0.6 PgC/yr in the northern extratropics and −1.0 to +1.5 PgC/yr in the tropics. These estimates imply that anthropogenic CO_2 was taken up *both* in the northern extratropics and in the tropics (balancing deforestation), as illustrated in Figure 3.6. The estimated land-atmosphere flux in the southern extratropics is estimated as close to zero, which is expected given the small land area involved. Estimates of CO_2 fluxes for the period 1990 to 1996 show a general resemblance to those for the 1980s. For 1990 to 1996, the inverse-model estimates of the land-atmosphere flux are −1.8 to −0.7 PgC/yr in the northern extratropics and −1.3 to +1.1 PgC/yr in the tropics. These results suggest a tendency towards a reduced land-atmosphere flux in the tropics, compared to the 1980s. Such a trend could be produced by reduced deforestation, increased CO_2 uptake or a combination of these.

Inverse modelling studies usually attempt greater spatial resolution of sources and sinks than is presented in this section. However, there are large unresolved differences in longitudinal patterns obtained by inverse modelling, especially in the northern hemisphere and in the tropics (Enting *et al.*, 1995, Law *et al.*, 1996; Fan *et al.*, 1998; Rayner *et al.*, 1999a; Bousquet *et al.*, 1999; Kaminski *et al.*, 1999). These differences may be traced to different approaches and several difficulties in inverse modelling of atmospheric CO_2 (Heimann and Kaminski, 1999):

• The longitudinal variations in CO_2 concentration reflecting net surface sources and sinks are on annual average typically <1 ppm. Resolution of such a small signal (against a background of seasonal variations up to 15 ppm in the Northern Hemisphere) requires high quality atmospheric measurements, measurement protocols and calibration procedures within and between monitoring networks (Keeling *et al.*, 1989; Conway *et al.*, 1994).

• Inverse modelling results depend on the properties of the atmospheric transport models used. The north-south transport of the models can be checked by comparing simulations of the relatively well-known inert anthropogenic tracer SF_6 with measured atmospheric concentrations of this tracer, as recently investigated in the TRANSCOM intercomparison project (Denning *et al.*, 1999). Unfortunately there is no currently measured tracer that can be used to evaluate the models' representation of longitudinal transport. Furthermore, the strong seasonality of the terrestrial CO_2 flux in the Northern Hemisphere together with covarying seasonal variations in atmospheric transport may induce significant mean annual gradients in concentration which do not reflect net annual sources and sinks, but which nevertheless have to be modelled correctly if inverse model calculations are to be reliable (Bolin and Keeling, 1963; Heimann *et al.*, 1986; Keeling *et al.*, 1989; Denning *et al.*, 1995; Law *et al.*, 1996). Even the sign of this so-called "rectifier effect" is uncertain. Some scientists believe that it may be responsible for a part of the apparent Northern Hemisphere uptake of CO_2 implied by inverse modelling results (Taylor, 1989; Taylor and Orr, 2000).

- The spatial partitioning of CO_2 uptake could also be distorted by a few tenths of 1 PgC/yr because the atmospheric concentration gradients also reflect the natural fluxes induced by weathering, transport of carbon by rivers and subsequent outgassing from the ocean (see Figure 3.1) (Sarmiento and Sundquist, 1992; Aumont *et al.*, 2001b). Furthermore, the effects of atmospheric transport of carbon released as CO and CH_4 (especially from incomplete fossil fuel burning, tropical biomass burning, and CH_4 from tropical wetlands) with subsequent oxidation to CO_2 is generally neglected. Their inclusion in the inversion leads to corrections of the latitudinal partitioning of up to 0.1 PgC/yr (Enting and Mansbridge, 1991).

- The distribution of atmospheric CO_2 measurement stations (Figure 3.7) is uneven, and severely underrepresents the continents. This underrepresentation is due in part to the problem of finding continental locations where measurements will not be overwhelmed by local sources and sinks.

- Because of the finite number of monitoring stations, the mathematical inversion problem is highly underdetermined. In principle a multitude of different surface source/sink configurations are compatible with the atmospheric data, within their measurement accuracy. Therefore, in order to extract a meaningful solution, additional information on the sources and sinks has to be introduced into the calculation. Examples of this additional information include maps of air-sea fluxes from observations or ocean models, patterns of terrestrial CO_2 exchanges inferred by terrestrial models, and remote sensing data. Thus, many methodological choices about the use of auxiliary data can influence the outcome of the analysis.

Interannual variability of climate is likely to strongly influence the spatial distribution of CO_2 sources and sinks, so that analyses based on a few years of data are insufficient to establish a long-term trend.

In conclusion, the present atmospheric measurement network, current information on air-sea fluxes and current understanding of vertical atmospheric transport are not sufficient to allow full use of the potential of inverse modelling techniques to infer geographically detailed source-sink distributions of anthropogenic CO_2.

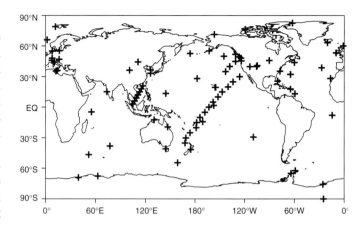

Figure 3.7: The atmospheric CO_2 measuring station network as represented by GLOBAL VIEW–CO_2 (Comparative Atmosphere Data Integration Project – Carbon Dioxide, NOAA/CMDL, http://www.cmdl.noaa.gov/ccg/co2).

3.5.4 Terrestrial Biomass Inventories

Inventory studies measure changes in carbon stocks over large areas, and can thus provide spatially aggregated estimates of large-scale fluxes of CO_2 over multi-annual time-scales (Box 3.1). Mid- and high latitude forests are covered by extensive national inventories based on repeated measurements of many thousands of plots. Inventories in the tropics are by comparison generally inadequate, particularly in view of the high rates of land-use change and extremely heterogeneous carbon density in many tropical ecosystems. There are still therefore large uncertainties in attempting to balance the terrestrial carbon budget on a global scale using inventory data.

The FAO Temperate and Boreal Forest Resource Assessment (TBFRA-2000) is a recent synthesis of inventories of forests and other wooded lands in Annex I (developed) countries for the early 1990s (UN-ECE/FAO, 2000). Many countries reported substantial increases in forest areas in recent years, as well as increasing carbon density in existing forests. According to TBFRA-2000, the land-atmosphere flux was −0.9 PgC/yr for all Annex I countries combined (the net annual increment of trees accounted for −1.5 PgC/yr, while losses due to fellings were 0.6 PgC/yr). Of this flux, −0.8 PgC/yr was due to uptake in "northern" forests (Europe, CIS, Japan and North America). An earlier review of individual regional and national studies by Dixon *et al.* (1994), highlighted in the IPCC WGII Second Assessment Report (IPCC, 1996b; Brown *et al.*, 1996), gave a range of −0.6 to −0.9 PgC/yr for the land atmosphere flux in northern forests. While TBFRA-2000 estimated biomass of woody vegetation only, the analyses reviewed in Dixon *et al.* (1994) included other vegetation, soils, litter and wood products. Under the United Nations Framework Convention for Climate Change (UNFCCC) signatory countries are required to report greenhouse gas emissions, including those from land-use change and forestry. Compilation of these data implies a land-atmosphere flux of −0.6 PgC/yr for all Annex I countries, and −0.6 PgC/yr for Annex I countries in the northern latitudes only (UNFCCC, 2000). While the TBFRA synthesised country statistics and adjusted data to fit FAO definitions and methodologies for calculating carbon stocks, the UNFCCC report summarises emissions data reported by each country according to IPCC guidelines; interpretation of guidelines is variable, and not all countries had reported data on land use. The implications of definitions and methodologies in calculating carbon fluxes, particularly in relation to implementation of the Kyoto Protocol, is discussed in detail in the SRLULUCF (IPCC, 2000a).

A recent compilation of data from 478 permanent plots in mature tropical moist forests throughout the tropics over at least two decades found these were taking up carbon due to increasing rates of tree growth. Extrapolation from these plots led to an estimated land-atmosphere flux of (0.6 ± 0.3 PgC/yr in Latin America; growth trends in African and Asian forests were not significantly different from zero (Phillips *et al.*, 1998). This net uptake is offset by emissions due to deforestation. Dixon *et al.* (1994) estimated tropical forests overall to be a net source of carbon with a land-atmosphere flux 1.7 ± 0.4, based mostly on FAO (1993b) inventory data and simple models of the effect of land-use change (Houghton, 1995). It will not be possible to assess trends and fluxes for the 1990s in the tropics from inventory data until a full data set is available from the FAO Global Forest Resources Assessment 2000. Among those countries that have reported land-use emissions data to the UNFCCC, there are significant discrepancies between the primary data used in emissions inventories and the data available in international surveys; for example, rates of deforestation differ from rates reported by FAO (1993b) by as much as a factor of six (Houghton and Ramakrishna, 1999).

The results of globally aggregated forest inventories show a greater uptake of carbon in forest growth than model-based calculations of the marginal effects of land-use change (e.g., Houghton, 2000). Thus, inventory studies provide independent evidence for the existence of a residual terrestrial sink; and they show that a substantial part of this sink, at least, is located in northern extratropical and tropical forests. Additional evidence from individual inventory studies in mature forests that have not undergone land-use changes shows that carbon stocks in such forests are increasing (e.g., Lugo and Brown, 1993; Phillips *et al.*, 1998; Schulze *et al.*, 1999). The difference between the northern extra-tropical land-atmosphere flux of around −0.8 PgC/yr calculated by inventories (TBFRA-2000) and that of −0.1 PgC/yr from land-use statistics (Houghton, 2000), both for the early 1990s, implies a residual terrestrial sink on the order of −0.7 PgC/yr in northern mid- and high latitudes. Combining this with the estimated sink of −0.6 PgC/yr in mature tropical moist forests (Phillips *et al.,* 1998) makes it plausible that at least a significant fraction of the current global terrestrial sink (Table 3.1) could be explained by an increase of carbon stocks in extant forests. The inventory-based estimate of land-atmosphere flux in northern forests (−0.8 PgC/yr) is at the positive end of the range calculated by inverse modelling studies for the >30°N latitude band from 1990 to 1996 (−1.8 to −0.7 PgC/yr, Section 3.5.3), either because of biases in inverse modelling that might tend to increase apparent uptake in the north (Section 3.5.3), or because possible sinks in other ecosys-

tems (e.g., temperate grassland soils) have not been considered in the inventories. In the tropics, the difference between the uptake of carbon estimated by inventory studies in mature forests of Latin America (−0.6 PgC/yr) (Phillips *et al.*, 1998) and the estimated emissions due to deforestation in the tropics of 1.7 PgC/yr (Houghton, 2000) yields an estimated land-atmosphere flux of 1.1 PgC/yr, which is at the positive end of the range calculated by inverse modelling studies for 30°S to 30°N (−1.3 to +1.1 PgC/yr, Section 3.5.3). Again, it should be noted that possible additional sinks (e.g., in savannas) are neglected by the land-use and inventory-based calculations.

3.6 Carbon Cycle Model Evaluation

3.6.1 Terrestrial and Ocean Biogeochemistry Models

The interactions of complex processes as discussed in Section 3.2 can be analysed with models that incorporate current knowledge at the process level, including syntheses of experimental results. Process-based models make it possible to explore the potential consequences of climate variability for the global carbon cycle, and to project possible future changes in carbon cycling associated with changes in atmospheric and ocean circulation. Models can be run with prescribed inputs such as observations of surface climate and CO_2 or the output of climate models. They can also be coupled to atmospheric general circulation models (Cox *et al.*, 2000; Friedlingstein *et al.*, 2000), to allow simulation of a wider range of interactions between climate and the carbon cycle.

Process-based terrestrial models used in carbon cycle studies are (a) terrestrial biogeochemical models (TBMs), which simulate fluxes of carbon, water and nitrogen coupled within terrestrial ecosystems, and (b) dynamic global vegetation models (DGVMs), which further couple these processes interactively with changes in ecosystem structure and composition (competition among different plant functional types; Prentice *et al.*, 2000). The treatment of carbon-nutrient interaction varies widely; for example, some models treat nitrogen supply explicitly as a constraint on NPP, while others do not. There are currently about 30 TBMs and <10 DGVMs. Cramer and Field (1999) and Cramer *et al.* (2001) reported results from intercomparisons of TBMs and DGVMs respectively. A current international project, Ecosystem Model/Data Intercomparison (EMDI), aims to test models of both types against a large set of terrestrial measurements, in order to better constrain the modelled responses of terrestrial carbon cycling to changes in CO_2 and climate.

Process-based ocean models used in carbon cycle studies include surface exchange of CO_2 with the atmosphere, carbon chemistry, transport by physical processes in the ocean, and transport by marine biology. The parametrization of marine biology can be classified as (a) nutrient-based models where the export of carbon below the surface ocean (approximately the top 50 m) is a function of surface nutrient concentration, (b) nutrient-restoring models in which biological carbon fluxes are set to the rates required for maintaining observed nutrient concentration gradients against dissipation by ocean mixing, and (c) models that explicitly represent the food chain involving nutrients, phytoplankton, zooplankton and detritus (NPZD models). In current models, the uptake of anthropogenic CO_2 is controlled mainly by physical transport and surface carbon chemistry, whereas the natural carbon cycle is controlled by physical, chemical and biological processes. The Ocean Carbon Cycle Model Intercomparison Project (OCMIP) compared the performance of four ocean models with respect to natural and anthropogenic tracers (Sarmiento *et al.*, 2000; Orr *et al.*, 2001), and is currently undergoing a similar comparison with 13 models and an extended data set (Orr and Dutay, 1999).

3.6.2 Evaluation of Terrestrial Models

Evaluation of terrestrial carbon cycle models requires different types of data to test processes operating on a range of time-scales from hours to centuries (see Section 3.2.2), including short-term environmental responses of CO_2 and water fluxes between vegetation canopies and the atmosphere (e.g., Cienciala *et al.*, 1998), responses of ecosystem carbon balance to interannual climate variability (e.g. Kindermann *et al.*, 1996; Heimann *et al.*, 1997; Gérard *et al.*, 1999; Knorr, 2000; Prentice *et al.*, 2000), and longer-term consequences of historical land-use change (McGuire *et al.*, 2001). Differences and uncertainties in model behaviour have been evaluated through model intercomparison (Cramer *et al.*, 1999, 2001) and sensitivity analyses (Knorr, 2000; Knorr and Heimann, 2001a).

3.6.2.1 Natural carbon cycling on land
Terrestrial model evaluation has traditionally been carried out as comparisons with *in situ* field observations of ecosystem variables (e.g., Raich *et al.*, 1991; Foley, 1994; Haxeltine and Prentice, 1996). The largest data sets of relevant field measurements are for NPP and soil carbon content. Other "target" variables include soil moisture, nitrogen mineralisation rate, and the amounts of carbon and nitrogen in different compartments of the ecosystem. Such comparisons have generally shown reasonable agreement between observed and modelled geographic patterns of these variables, but they do not test the time-dependent response of models to environmental variability.

Time-dependent data sets for *in situ* comparisons are now becoming available, thanks to eddy-covariance measurements of CO_2 fluxes (Section 3.2.2.1; Box 3.1). Daily and seasonal cycles of CO_2 and water fluxes provide a test of the coupling between the carbon and hydrological cycles as simulated by terrestrial models (Cienciala *et al.*, 1998). Flux measurements are now being carried out on a multi-annual basis at an increasing number of stations, although global coverage remains uneven, with the greatest concentration in Europe and North America and few measurements from the tropics (see Box 3.1). Field campaigns have started to retrieve flux data from more remote regions (e.g., Schulze *et al.*, 1999). The Large-scale Biosphere Atmosphere Experiment in Amazonia, LBA, will yield more comprehensive data on the carbon, water and energy exchanges of tropical terrestrial ecosystems and will allow a more rigorous evaluation of the performance of models in the tropics than has been possible up until now (e.g., Tian *et al.*, 1998). As current models show conflicting

responses of global NPP to climate (Cramer *et al.*, 1999), systematic comparisons with seasonal and interannual flux measurements are a priority to reduce uncertainties in terrestrial carbon modelling.

Terrestrial models have also been evaluated at a global scale by comparing simulated ecosystem water balance with river runoff (e.g., Neilson and Marks, 1994; Foley *et al.*, 1996; Kucharik *et al.*, 2000), and simulated seasonal leaf area with satellite observations of "greenness", often based on the normalised difference vegetation index (NDVI) (Field *et al.*, 1995b; de Fries *et al.*, 1995). NDVI data can be translated into estimates of the plant-absorbed fraction of incoming photosynthetically active radiation (FPAR) (Asrar *et al*, 1992), which is related to leaf area index (LAI). The first terrestrial model intercomparison showed differences among model simulations of LAI and its seasonality (Bondeau *et al.*, 1999). More recently, it has been shown that constraining a terrestrial model with remotely sensed spatial patterns of FPAR can lead to a reduction of uncertainty in NPP simulations by about one third (Knorr and Heimann, 2001b). Agreement with patterns of remotely sensed FPAR has thus become a standard benchmark for terrestrial models (Haxeltine and Prentice, 1996; Kucharik *et al.*, 2000) and attention has been focused on improving the simulation of LAI and its seasonal variations.

A more direct test of the simulated net exchange of CO_2 between the terrestrial biosphere and the atmosphere is provided by comparison with atmospheric CO_2 measurements at remote monitoring sites. The comparison requires the use of an atmospheric transport model to simulate CO_2 as a passive tracer (Kaminski *et al.*, 1996). The seasonal cycle of atmospheric CO_2 shows a strong latitudinal pattern in amplitude and phase, and is dominated by the terrestrial biosphere (Heimann *et al.*, 1998). The ability to simulate this seasonal cycle thus constitutes a benchmark for terrestrial models' response to climate (Denning *et al.*, 1996; Hunt *et al.*, 1996; Heimann *et al.*, 1998; Nemry *et al.*, 1999). Generally, the observed seasonal cycles of CO_2 in northern and tropical latitudes can be well simulated, with terrestrial models using NDVI data as input (Knorr and Heimann, 1995), or by fully prognostic models, including DGVMs (Prentice *et al.*, 2000).

Major features of interannual variability of the CO_2 increase are also simulated by terrestrial models (Kindermann *et al.*, 1996; Heimann *et al.*, 1997; Gérard *et al.*, 1999; Ito and Oikawa, 2000; Knorr, 2000; Prentice *et al.*, 2000). This finding supports the hypothesis (Section 3.5.2) that terrestrial effects are important in determining the interannual variability of CO_2 uptake. During typical El Niño events, terrestrial model results consistently show strongly reduced CO_2 uptake or CO_2 release by the land. This result has been obtained with a range of models, even when the models differ substantially in the relative sensitivitivities of NPP and heterotrophic respiration to temperature (Heimann *et al.*, 1997; Knorr, 2000). The low CO_2 growth rate during the early 1990s has been simulated by some terrestrial models (Prentice *et al.*, 2000; Knorr, 2000).

At the longest time-scales of interest, spanning the industrial period, models of the natural terrestrial carbon cycle show a pronounced response to rising atmospheric CO_2 levels

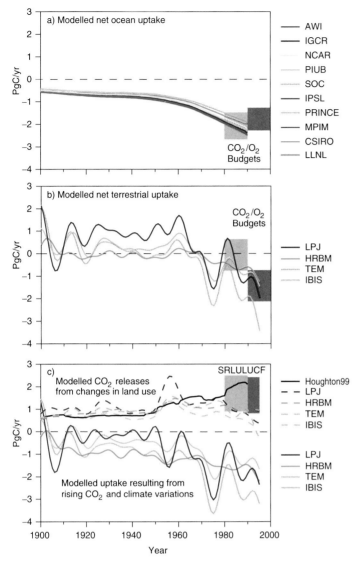

Figure 3.8: Modelled fluxes of anthropogenic CO_2 over the past century. (a) Ocean model results from OCMIP (Orr and Dutay, 1999; Orr *et al.*, 2000); (b), (c) terrestrial model results from CCMLP (McGuire *et al.*, 2001). Positive numbers denote fluxes to the atmosphere; negative numbers denote uptake from the atmosphere. The ocean model results appear smooth because they contain no interannual variability, being forced only by historical changes in atmospheric CO_2. The results are truncated at 1990 because subsequent years were simulated using a CO_2 concentration scenario rather than actual measurements, leading to a likely overestimate of uptake for the 1990s. The terrestrial model results include effects of historical CO_2 concentrations, climate variations, and land-use changes based on Ramankutty and Foley (2000). The results were smoothed using a 10-year running mean to remove short-term variability. For comparison, grey boxes denote observational estimates of CO_2 uptake by the ocean in panel (a) and by the land in panel (b) (from Table 3.1). Land-use change flux estimates from Houghton *et al.* (1999) are shown by the black line in panel (c). The grey boxes in panel (c) indicate the range of decadal average values for the land-use change flux accepted by the SRLULUCF (Bolin *et al.*, 2000) for the 1980s and for 1990 to 1995.

as a result of CO_2 fertilisation, generally larger than the NPP response to the climate change over this period (Kicklighter *et al.*, 1999). According to CCMLP results, the CO_2 increase maintains a lead of NPP over Rh and an increase of the amplitude of the seasonal CO_2 cycle (McGuire *et al.*, 2001), consistent with long-term observations (Keeling *et al.*, 1996a), which indicate an increase in amplitude of about 20% since accurate atmospheric measurements began. However, the magnitude of this effect was greatly over- or under estimated by some models, reflecting unresolved differences in the parametrization of the CO_2 fertilisation response.

3.6.2.2 Uptake and release of anthropogenic CO_2 by the land

The most comprehensive model-based estimates of the terrestrial components of the anthropogenic CO_2 budget are those that have been produced by the CCMLP. McGuire *et al.* (2001) used two TBMs and two DGVMs driven by changes in atmospheric CO_2, then changes in CO_2 with historical changes in climate (from observations), and finally changes in CO_2 and climate with land-use change from Ramankutty and Foley (2000) (Figure 3.8; Table 3.4). In these simulations, CO_2 fertilisation accounted for a land-atmosphere flux of −0.9 to −3.1 PgC/yr, land-use change a positive flux of 0.6 to 1.0 PgC/yr, and climate variability a small additional effect of uncertain sign, −0.2 to 0.9 PgC/yr during the 1980s. The total land-atmosphere flux simulated for the 1980s amounted to −0.3 to −1.5 PgC/yr, which is consistent with or slightly more negative than the observationally-based estimate of −0.2 ± 0.7 PgC/yr (Table 3.1). Net uptake by all models reported in McGuire *et al.* (2001) is shown to be occurring mainly in tropical, temperate and boreal forests − consistent with forest inventory data (Section 3.5.4) − while some regions (notably semi-arid tropical and sub-tropical regions) show net carbon loss. The model estimates of the CO_2 source due to land-use change are substantially smaller than the estimate of Houghton (1999) (Section 3.4.2). This divergence primarily reflects disagreements between the Houghton (1999) and Ramankutty and Foley (2000) data sets as to the timing of tropical deforestation in different regions (see Section 3.4.2).

There is no general agreement on how to model the linkage between reactive nitrogen deposition and vegetation productivity, and recent model estimates of the additional effect of anthropogenic nitrogen fertilisation on the global carbon cycle vary widely. The anthropogenic nitrogen input itself (Holland *et al.*, 1999), the fate of anthropogenic nitrogen in the ecosystem (Nadelhoffer *et al.*, 1999; Jenkinson *et al.*, 1999), and changes in ecosystem nitrogen fixation (Vitousek and Field, 1999) represent major sources of uncertainty. Estimates of the anthropogenic nitrogen effect range from −0.2 PgC/yr (Nadelhoffer *et al.*, 1999) to −1.1 or −1.4 PgC/yr (Holland *et al.*, 1997). The model with the smallest CO_2 fertilisation effect (−0.9 PgC/yr) in the McGuire *et al.* (2001) study has been shown to respond strongly to anthropogenic nitrogen input, yielding a combined (CO_2 and nitrogen) fertilisation effect of −1.5 PgC/yr. A modelling study by Lloyd (1999) suggests that CO_2 and nitrogen fertilisation effects may by synergistic. Evaluation of model results on carbon-nitrogen coupling against experimental results is a current research focus.

3.6.3 Evaluation of Ocean Models

Natural and anthropogenic tracers have been extensively measured, most recently as part of the Joint Global Ocean Flux Study (JGOFS) and World Ocean Circulation Experiment (WOCE). Because of these measurement campaigns, such tracers provide important opportunities to evaluate representations of ocean physics and biogeochemistry in models.

3.6.3.1 Natural carbon cycling in the ocean

Most global ocean models of the carbon cycle are successful in reproducing the main vertical and horizontal features of ocean carbon content (Maier-Reimer, 1993; Aumont, 1998; Murnane *et al.*, 1999). The observed features reasonably reproduced by all ocean models are the mean vertical gradient in DIC, with enriched deep ocean concentrations (Goyet and Davies, 1997), and the spatial patterns of surface pCO_2 with outgassing in the tropics and uptake at higher latitudes (Takahashi *et al.*, 1999). Furthermore, models which incorporate marine biology (including DOC and plankton dynamics) roughly reproduce the seasonal cycle of surface ocean pCO_2, atmospheric O_2 after it has been corrected for seasonal land variability, and surface chlorophyll (Six and Maier-Reimer, 1996; Stephens *et al.*, 1998; Aumont *et al.*, 2001a). Ocean carbon models can also roughly reproduce the phase and amplitude of interannual variability of ocean pCO_2 in the equatorial Pacific (Winguth *et al.*, 1994; Le Quéré *et al.*, 2000) in agreement with available observations (Feely *et al.*, 1997; 1999b; Boutin *et al.*, 1999).

Although many first-order features can be reproduced by global models, there are still important aspects of the ocean carbon cycle that are not well simulated, because either marine biology or ocean physics are imperfectly reproduced. Ocean carbon models have difficulties in reproducing the spatial structure of the deep ocean ^{14}C (Orr *et al.*, 2001), which suggests problems in simulating the physical exchange of carbon between surface and the deep ocean. Models display their largest disagreements where fewest observations exist, in particular in the important region of the Southern Ocean where the mixing of tracers is subject to large uncertainties (Caldeira and Duffy, 2000; Sarmiento *et al.*, 2000; Orr *et al.*, 2001). In spite of these differences, all ocean carbon models estimate zero interhemispheric transport of carbon (Sarmiento *et al.*, 2000) whereas a transport as large as 1 PgC/yr has been inferred from atmospheric CO_2 measurements (Keeling *et al.*, 1989). Consideration of the global transport of carbon by rivers reduces the discrepancy but does not remove it (Sarmiento and Sundquist, 1992; Aumont *et al.*, 2001b). Atmospheric CO_2 and O_2 measurements suggest that interhemispheric transport may be incorrectly simulated by ocean models (Stephens *et al.*, 1998), and could hint at difficulties in modelling heat transport (Murnane *et al.*, 1999). Recent data from the Southern Ocean, however, seem closer to model results (Stephens, 1999) and the question about interhemispheric transport thus remains open. These problems could partly be resolved by a better representation of the physical transport of carbon in the ocean, especially isopycnal diffusion, sub-grid eddy mixing, and sea-ice formation (Stephens *et al.*, 1999).

Three common problems related to marine biology in global ocean models are discussed here. First, most models poorly represent the formation and dissolution of $CaCO_3$, which controls alkalinity. This process is often parameterized as a function of direct or indirect observations (salinity, temperature, nutrients). Although correct for the present day ocean, this parametrisation may not hold for past or future conditions with different ocean circulation and surface water fluxes. The alkalinity cycle is difficult to represent because the rate of $CaCO_3$ formation derived from observations is consistently larger than the one required by models for reproducing observed deep ocean alkalinity (Maier-Reimer, 1993; Yamanaka and Tajika, 1996). Second, marine productivity tends to be underestimated by models in sub-tropical regions and overestimated in the equatorial oceans and at high latitudes in the North Pacific and Southern Oceans. The overestimation may be caused by limitation in plankton growth by iron (Coale *et al.*, 1996; Boyd *et al.*, 2000; Archer and Johnson, 2000), while underestimation in the sub-tropics partly stems from neglecting mesoscale variability (McGillicuddy and Robinson, 1997; Oschlies and Garçon, 1998). The remaining discrepancies might be attributed in part to more complex processes involving nitrogen fixation (Karl *et al.*, 1997). Finally, the tight coupling between carbon and either nitrogen or phosphate, which is generally implicit in ocean carbon models, precludes the simulation of past or future marine biological feedback mechanisms that involve a partial decoupling between carbon and nutrients (see Section 3.2.3).

3.6.3.2 Uptake of anthropogenic CO₂ by the ocean

Ocean uptake is constrained to some degree by observations of anthropogenic tracers. Three transient tracers are commonly used. First, anthropogenic CO_2 itself gives a direct benchmark for model estimates of the quantity and distribution of anthropogenic CO_2 that has penetrated the ocean since the pre-industrial era. Anthropogenic CO_2 can be inventoried by an indirect method whereby carbon concentration is compared to what would be expected from water exposed to pre-industrial air (Gruber *et al.*, 1996). The $^{14}CO_2$ released in the early 1960s by atmospheric nuclear testing (commonly called bomb ^{14}C) provides a second tracer; the content of bomb ^{14}C in the ocean is used to constrain global air-sea CO_2 exchange (Wanninkhof *et al.*, 1992), and ocean model results can be compared with its penetration depth as a benchmark for vertical transport (Broecker *et al.*, 1995). Bomb ^{14}C is computed by subtracting the observed ^{14}C concentration from an estimate of its pre-industrial value (Broecker *et al.*, 1995). Finally, CFCs also constrain the downward transport of tracers in ocean models. No natural background needs to be subtracted from CFCs. None of these three tracers provide a perfect indicator of anthropogenic CO_2 uptake: CO_2 equilibrates with the atmosphere ten times faster than ^{14}C and ten times slower than CFCs; anthropogenic CO_2 and ^{14}C are indirectly estimated. As part of the Ocean Carbon-Cycle Model Intercomparison Project (OCMIP), a comparison of carbon models with respect to all three anthropogenic tracers is in progress (Orr and Dutay, 1999; Orr *et al.*, 2001).

Although regional estimates show discrepancies, modelled estimates of anthropogenic tracers agree reasonably well with observations when integrated globally. The mean value of the penetration depth of bomb ^{14}C for all observational sites during

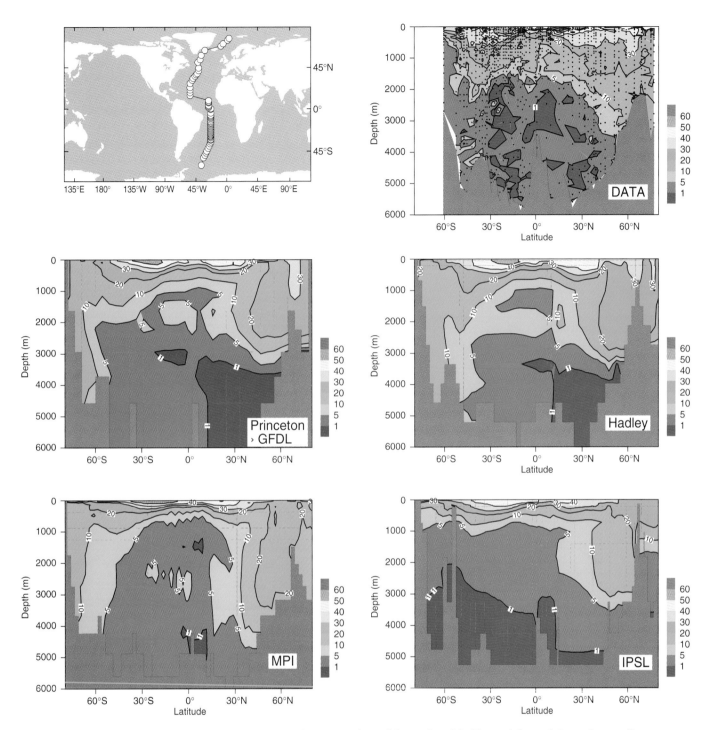

Figure 3.9: Anthropogenic CO_2 in the Atlantic Ocean (μmol/kg): comparison of data and models. The top left panel shows the sampling transect; the top right panel shows estimates of anthropogenic CO_2 content along this transect using observations from several cruises between 1981 and 1989 (Gruber, 1998). Anthropogenic CO_2 is not measured directly but is separated from the large background of oceanic carbon by an indirect method based on observations (Gruber *et al.*, 1996). The remaining panels show simulations of anthropogenic CO_2 content made with four ocean carbon models forced by the same atmospheric CO_2 concentration history (Orr *et al.*, 2000).

the late 1970s is 390 ± 39 m (Broecker *et al.*, 1995). For the same years and stations, modelled estimates range between 283 and 376 m (Orr *et al.*, 2001). Modelled and observed CFC concentrations have been compared locally but not yet globally (England 1995; Robitaille and Weaver, 1995; Orr and Dutay, 1999). Modelled anthropogenic CO_2 inventory since 1800 is

comparable to the estimate of 40 ± 9 PgC for the Atlantic Ocean (Gruber, 1998) and 20 ± 3 PgC for the Indian Ocean (Sabine *et al.*, 1999; Orr *et al.*, 2001). Latitude-depth profiles of anthropogenic CO_2 in the Atlantic, extracted from data and from models, are shown in Figure 3.9. Modelled CO_2 uptake for the global ocean between 1800 and 1990 ranges between 100 and

133 PgC (Figure 3.8), comparable to the preliminary data-based estimate of 107 ± 27 PgC for the global ocean, which includes the Pacific value of 45 ± 15 PgC (Feely *et al.*, 1999a). Although in reasonable agreement with basin and global estimates of anthropogenic CO_2, modelled inventories exhibit large differences at the regional scale: models tend to underestimate the inventory of anthropogenic CO_2 between 50°S and 50°N in the Atlantic and Indian Oceans, and to overestimate it at high latitudes (Sabine *et al.*, 1999; Orr *et al.*, 2001). In the Southern Ocean the uptake of anthropogenic CO_2 varies by a factor of two among models (Orr *et al.*, 2001). The difficulty for models in reproducing the spatial structure of anthropogenic tracers may be indicative of problems in ocean physics mentioned earlier, and may be responsible for the increasing range of model estimates when future CO_2 uptake is projected by the same models (Figure 3.10c).

The most recent model estimates of the ocean-atmosphere flux obtained with process-based models are −1.5 to −2.2 for 1980 to 1989 (Table 3.4), in agreement with earlier model estimates for the same period (Enting *et al.*, 1994; Orr *et al.*, 2001). These estimates are fully consistent with the budget based on atmospheric observations alone (Table 3.1), with estimates based on pCO_2 and $\delta^{13}C$ observations (Table 3.4), and with the SAR estimate of −2.0 ± 0.8 PgC/yr. Figure 3.8 shows modelled ocean CO_2 uptake for 1900 to 2000. (These results do not include natural variability and therefore appear smoother than in reality.) The oceanic regions absorbing the largest quantities of anthropogenic CO_2 according to models are those where older waters come in contact with the atmosphere, such as high latitudes and upwelling regions of the equator. In contrast, modelled sub-tropical regions rapidly saturate at atmospheric CO_2 level and do not absorb large quantities of anthropogenic CO_2 (Sarmiento *et al.*, 1992; Orr *et al.*, 2001).

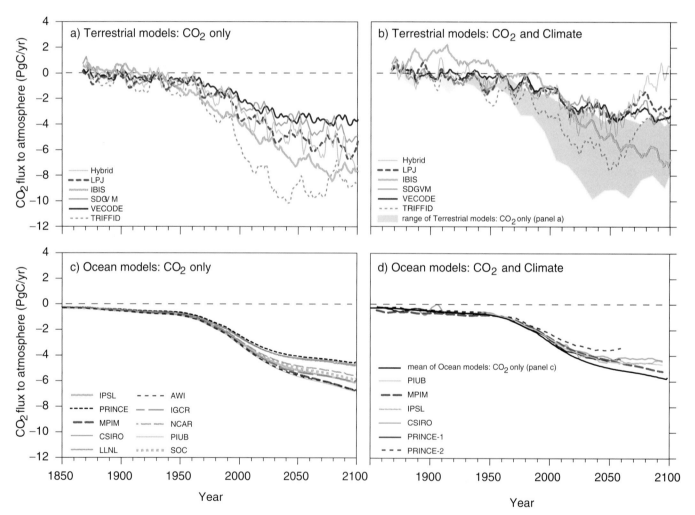

Figure 3.10: Projections of anthropogenic CO_2 uptake by process-based models. Six dynamic global vegetation models were run with IS92a CO_2 concentrations as given in the SAR: (a) CO_2 only, and (b) with these CO_2 concentrations plus simulated climate changes obtained from the Hadley Centre climate model with CO_2 and sulphate aerosol forcing from IS92a (Cramer *et al.*, 2000). Panel (b) also shows the envelope of the results from panel (a) (in grey). (c) Ten process-based ocean carbon models were run with the same CO_2 concentrations, assuming a constant climate (Orr and Dutay, 1999; Orr *et al.*, 2000). A further six models were used to estimate the climate change impact on ocean CO_2 uptake as a proportional change from the CO_2-only case. The resulting changes were imposed on the mean trajectory of the simulations shown in panel (c), shown by the black line in panel (d), yielding the remaining trajectories in panel (d). The range of model results in panel (d) thus represents only the climate change impact on CO_2 uptake; the range does not include the range of representations of ocean physical transport, which is depicted in panel (c).

3.7 Projections of CO₂ Concentration and their Implications

3.7.1 Terrestrial Carbon Model Responses to Scenarios of Change in CO₂ and Climate

Possible feedbacks from terrestrial carbon cycling to atmospheric CO_2 were assessed using multiple models by Cramer *et al.* (2001). Six DGVMs (Figure 3.10a) (Foley *et al.*, 1996; Brovkin *et al.*, 1997; Friend *et al.*, 1997; Woodward *et al.*, 1998; Huntingford *et al.*, 2000; Sitch, 2000) were driven first by CO_2 concentrations derived from the IS92a emissions scenario as in the SAR, and then with CO_2 changes plus climate changes derived from the HadCM2 coupled ocean-atmosphere general circulation model simulation including sulphate aerosol effects as described by Mitchell *et al.* (1995). Except for one empirical model (VECODE; Brovkin *et al.*, 1997), the models included explicit representation of all the following processes: the CO_2 fertilisation effect on NPP (modelled explicitly in terms of photosynthesis, respiration, and feedbacks associated with carbon allocation); responses of NPP to climate specific to each plant functional type (PFT); competition among PFTs for light and water; dynamic shifts in vegetation structure due to climate and CO_2 effects; competitive limits to above-ground biomass; natural disturbance regimes and their interaction with PFT composition; soil temperature and moisture effects on heterotrophic respiration. Two models include an interactive N cycle. Land use and anthropogenic N deposition were not considered.

Driven by increases in CO_2 beyond the present day, the modelled sink due to CO_2 fertilisation continued to increase. By the middle of the 21st century the simulated land-atmosphere flux due to CO_2 was in the range −8.7 to −3.6 PgC/yr. Beyond mid-century the rate of increase became less, due to the declining photosynthetic response to CO_2. When the climate change scenario was included as well as the CO_2 increase, modelled uptake was reduced compared with the CO_2-only analysis. At mid-century, climate change reduced the uptake by 21 to 43%. A marked decline in terrestrial uptake after the mid-century was seen in two models, and one model had zero terrestrial uptake by 2100. By 2100 the range of model estimates of the land-atmosphere flux had widened to −6.7 to +0.4 PgC/yr. Increasing heterotrophic respiration in response to warming (Cao and Woodward, 1998a,b; Cramer *et al.*, 2001) was a common factor (but not the only one) leading to reduced land uptake. The differences among the modelled climate responses were largely due to unresolved discrepancies in the response of global NPP to temperature. The balance of positive versus negative regional effects of climate change on NPP was estimated differently by these models, to the extent that the sign of the global response of NPP to climate change alone was not consistent. In addition, one model simulated a partial replacement of the Amazon rainforest by C_4 grassland. This response was not shown, or occurred on a much smaller scale, in the other models. The details of this modelling exercise are presumably dependent on sensitivity of the particular climate model, and regional aspects of the simulated climate change (Cramer *et al.*, 2001).

3.7.2 Ocean Carbon Model Responses to Scenarios of Change in CO₂ and Climate

Analogous simulations have been performed with several ocean carbon models (Figure 3.10c,d). To compute the impact of increasing CO_2 alone (no climate change), OCMIP models were forced to follow the atmospheric CO_2 concentration derived from the IS92a scenario as in the DGVM experiment (Figure 3.10a,b) (Orr and Dutay, 1999). All models agreed in projecting that the annual ocean-atmosphere flux of CO_2 continues to become larger, reaching −6.7 to −4.5 PgC/yr by 2100 (Figure 3.10c). Since surface conditions (temperature, wind speed, alkalinity) were prescribed, the range in model estimates stems only from different representations of physical transport processes.

Several atmosphere-ocean models were used to project the effect of climate change (Maier-Reimer *et al.*, 1996; Sarmiento *et al.*, 1998; Matear and Hirst, 1999; Joos *et al.*, 1999b; Bopp *et al.*, 2001). These models include most processes previously discussed, including all processes associated with carbonate chemistry and gas exchange, physical and biological transport of CO_2, and changes in temperature, salinity, wind speed, and ice cover. They account for simple changes in biological productivity, but not for changes in external nutrient supply, species composition, pH, or Redfield ratios, all of which could be involved in more complex biological feedbacks. Coupled models estimate the impact of climate change as a departure, reported in per cent, from a "control" experiment modelling the effect of increasing atmospheric CO_2 alone.

In climate change simulations, warming of surface waters and increased stratification of the upper ocean produced an overall positive feedback that reduced the accumulated ocean uptake of CO_2 by 6 to 25% between 1990 and the middle of the 21st century, as compared with the CO_2-only case. In the first part of the simulation, the climate-mediated feedback is mainly due to the temperature effect on CO_2 solubility (Sarmiento and Le Quéré, 1996; Matear and Hirst, 1999). Towards the mid-century, the impact of circulation changes becomes significant in most models, with the net effect of further reducing ocean CO_2 uptake. To investigate the effect of climate change on the IS92a scenario, the average of the OCMIP CO_2-only projections (mean of results in Figure 3.10c) was used as a baseline and the reduction in atmosphere-ocean CO_2 flux caused by climate change (in per cent since the beginning of the simulation) was applied to this curve (Figure 3.10d). The range in model results (Figure 3.10d) must be attributed to uncertainties related to climate change feedback, and not to uncertainties in the modelling of physical transport as shown in Figure 3.10c.

The range of model estimates of the climate change impact is dependent on the choice of scenario for atmospheric CO_2 and on assumptions concerning marine biology (Joos *et al.*, 1999b). At high CO_2 concentrations, marine biology can have a greater impact on atmospheric CO_2 than at low concentrations because the buffering capacity of the ocean is reduced (see Box 3.3) (Sarmiento and Le Quéré, 1996). Although the impact of changes in marine biology is highly uncertain and many key processes discussed in Section 3.2.3.1 are not included in current models, sensitivity studies can provide approximate upper and lower bounds for the potential impact of marine biology on future ocean CO_2 uptake. A sensitivity study of two extreme scenarios for nutrient supply to marine biology gave a range of 8 to 25% for the reduction of CO_2 uptake by mid-century (Sarmiento *et al.*, 1998). This range is comparable to other uncertainties, including those stemming from physical transport (Figure 3.10c).

3.7.3 Coupled Model Responses and Implications for Future CO_2 Concentrations

Carbon cycle models have indicated the potential for climate change to influence the rate of CO_2 uptake by both land (Section 3.7.1) and oceans (Section 3.7.2) and thereby influence the time course of atmospheric CO_2 concentration for any given emissions scenario. Coupled models are required to quantify these effects.

Two general circulation model simulations have included interactive land and ocean carbon cycle components (Cox *et al.*, 2000; Friedlingstein *et al.*, 2001). The Cox *et al.* (2000) model was driven by CO_2 emissions from the IS92a scenario (Legget *et al.*, 1992) and the Friedlingstein *et al.* (2001) model was driven by CO_2 emissions from the SRES A2 scenario (IPCC, 2000b). Both simulations indicate a positive feedback, i.e., both CO_2 concentrations and climate change at the end of the 21st century are increased due to the coupling. The simulated magnitudes of the effect differ (+70 ppm, Friedlingstein *et al.*, 2001; +270 ppm, Cox *et al.*, 2000). In the Cox *et al.* (2000) simulation, which included a DGVM, the increased atmospheric CO_2 is caused mainly by loss of soil carbon and in part by tropical forest die back. The magnitude of the climate-carbon cycle feedback still has large uncertainties associated with the response of the terrestrial biosphere to climate change, especially the response of heterotrophic respiration and tropical forest NPP to temperature (Cox *et al.*, 2000; see Sections 3.2.2.3 and 3.7.1). In the following section, simplified models are used to assess these uncertainties.

Figure 3.11: Projected CO_2 concentrations resulting from the IS92a emissions scenario. For a strict comparison with previous work, IS92a-based projections were made with two fast carbon cycle models, Bern-CC and ISAM (see Box 3.7), based on CO_2 changes only, and on CO_2 changes plus land and ocean climate feedbacks. Panel (a) shows the CO_2 emissions prescribed by IS92a; the panels (b) and (c) show projected CO_2 concentrations for the Bern-CC and ISAM models, respectively. Results obtained for the SAR, using earlier versions of the same models, are also shown. The model ranges for ISAM were obtained by tuning the model to approximate the range of responses to CO_2 and climate shown by the models in Figure 3.10, combined with a range of climate sensitivities from 1.5 to 4.5°C rise for a doubling of CO_2. This approach yields a lower bound on uncertainties in the carbon cycle and climate. The model ranges for Bern-CC were obtained by combining different bounding assumptions about the behaviour of the CO_2 fertilisation effect, the response of heterotrophic respiration to temperature and the turnover time of the ocean, thus approaching an upper bound on uncertainties in the carbon cycle. The effect of varying climate sensitivity from 1.5 to 4.5°C is shown separately for Bern-CC. Both models adopted a "reference case" with mid-range behaviour of the carbon cycle and climate sensitivity of 2.5°C.

3.7.3.1 Methods for assessing the response of atmospheric CO_2 to different emissions pathways and model sensitivities

This section follows the approach of previous IPCC reports in using simplified, fast models (sometimes known as reduced-form models) to assess the relationship between CO_2 emissions and concentrations, under various assumptions about their future time course. Results are shown from two models, whose salient features are summarised in Box 3.7. The models lend themselves to somewhat different approaches to estimating uncertainties. In the ISAM model, "high-CO_2" and "low-CO_2" alternatives are calculated for every emissions scenario, based on tuning the model to match the range of responses included in the model

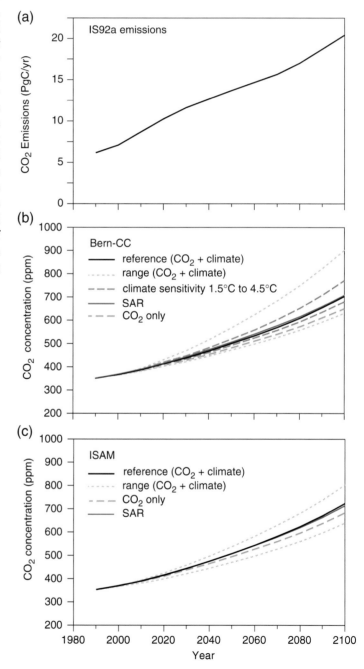

Box 3.7: Fast, simplified models used in this assessment.

The ***Bern-CC*** model comprises:

- A box-diffusion type ocean carbon model, (HILDA version K(z); Siegenthaler and Joos, 1992; Joos *et al.*, 1996), already used in the SAR. In addition to the SAR version, the effect of sea surface warming on carbonate chemistry is included (Joos *et al.*, 1999b).

- An impulse-response climate model (Hooss *et al.*, 1999), which converts radiative forcing into spatial patterns of changes in temperature, precipitation and cloud cover on a global grid. The patterns of the climate anomalies are derived from the first principal component of the climate response shown by the full three-dimensional atmosphere-ocean GCM, ECHAM-3/LSG (Voss and Mikolajewicz, 1999). Their magnitude is scaled according to the prescribed climate sensitivity.

- The terrestrial carbon model LPJ, as described in Sitch *et al.* (2000) and Cramer *et al.* (2001). LPJ is a process-based DGVM that falls in the mid-range of CO_2 and climate responses as shown in Cramer *et al.* (2001). It is used here at $3.75° \times 2.5°$ resolution, as in Cramer *et al.* (2001).

- A radiative forcing module. The radiative forcing of CO_2, the concentration increase of non-CO_2 greenhouse gases and their radiative forcing, direct forcing due to sulphate, black carbon and organic aerosols, and indirect forcing due to sulphate aerosols are projected using a variant of SAR models (Harvey *et al.*, 1997; Fuglestvedt and Berntsen, 1999) updated with information summarised in Chapters 4, 5 and 6. The concentrations of non-CO_2 greenhouse gases, aerosol loadings, and radiative forcings are consistent with those given in Appendix II.

Sensitivities of projected CO_2 concentrations to model assumptions were assessed as follows. Rh was assumed either to be independent of global warming (Giardina and Ryan, 2000; Jarvis and Linder 2000), or to increase with temperature according to Lloyd and Taylor (1994). CO_2 fertilisation was either capped after year 2000 by keeping CO_2 at the year 2000 value in the photosynthesis module, or increased asymptotically following Haxeltine and Prentice (1996). (Although apparently unrealistic, capping the CO_2 fertilisation in the model is designed to mimic the possibility that other, transient factors such as land management changes might be largely responsible for current terrestrial carbon uptake.) Transport parameters of the ocean model (including gas exchange) were scaled by a factor of 1.5 and 0.5. Average ocean uptake for the 1980s is 2.0 PgC/yr in the reference case, 1.46 PgC/yr for the "slow ocean" and 2.54 PgC/yr for the "fast ocean", roughly in accord with the range of observational estimates (Table 3.1, Section 3.2.3.2). A "low-CO_2" parametrization was obtained by combining the fast ocean and no response of Rh to tempera- ture. A "high-CO_2" parametrization was obtained by combining the slow ocean and capping CO_2 fertilisation. Climate sensitivity was set at 2.5 °C for a doubling of CO_2. Effects of varying climate sensitivity from 1.5°C to 4.5°C are also shown for one case.

The ***ISAM*** model was described by Jain *et al.* (1994) and used in the SAR for CO_2-only analyses, with a different set of model parameters from those used here (Jain, 2000). The full configuration of ISAM comprises:

- A globally aggregated upwelling-diffusion ocean model including the effects of temperature on CO_2 solubility and carbonate chemistry (Jain *et al.*, 1995).

- An energy balance climate model of the type used in the IPCC 1990 assessment (Hoffert *et al.*, 1980; Bretherton *et al.*, 1990). In this model, heat is transported as a tracer in the ocean and shares the same transport parameters as DIC.

- A six-box globally aggregated terrestrial carbon model including empirical parametrizations of CO_2 fertilisation and temperature effects on productivity and respiration (Harvey, 1989; Kheshgi *et al.*, 1996).

- The radiative forcing of CO_2 projected using a SAR model (Harvey *et al.*, 1997) modified with information summarised in Chapter 6. Radiative forcing from agents other than CO_2 are identical to that used in the Bern-CC model.

In addition to varying the climate sensitivity (1.5 to 4.5°C), parameters of the terrestrial and ocean components (strength of CO_2 fertilisation, temperature response of NPP and heterotrophic respiration; ocean heat and DIC transport) were adjusted to mimic the ranges of CO_2 and climate responses as shown by existing process-based models (Figure 3.10). A reference case was defined with climate sensitivity 2.5°C, ocean uptake corresponding to the mean of the ocean model results in Figure 3.10, and terrestrial uptake corresponding to the mean of the responses of the mid-range models LPJ, IBIS and SDGVM (Figure 3.10). A "low CO_2" parametri- sation was chosen with climate sensitivity 1.5°C, and maximal CO_2 uptake by oceans and land; and a "high-CO_2" parametrization with climate sensitivity 4.5°C, and minimal CO_2 uptake by oceans and land.

intercomparisons shown in Figure 3.10. Uncertainties cited from the ISAM model can be regarded as providing a lower bound on uncertainty since they do not admit possible behaviours outside the range considered in recent modelling studies. In the Bern-CC model, "high-CO_2" and "low-CO_2" alternatives are calculated by making bounding assumptions about carbon cycle processes (for example, in the high-CO_2 parametrization CO_2 fertilisation is capped at year 2000; in the low-CO_2 parametrization Rh does not increase with warming). This approach yields generally larger ranges of projected CO_2 concentrations than the ISAM approach. The ranges cited from the Bern-CC model can be regarded as approaching an upper bound on uncertainty, since the true system response is likely to be less extreme than the bounding assumptions, and because the combination of "best" and "worst" case assumptions for every process is intrinsically unlikely.

3.7.3.2 Concentration projections based on IS92a, for comparison with previous studies

Illustrative model runs (Figure 3.11) based on the IS92a scenario (Leggett *et al.*, 1992) are shown first so as to allow comparison with earlier model results presented in the SAR and the SRRF (Schimel *et al.*, 1995). In the SRRF comparison of eighteen global carbon cycle models (Enting *et al.*, 1994; Schimel *et al.*, 1995) the CO_2 fertilisation response of the land was calibrated to match the central estimate of the global carbon budget for the 1980s, assuming a land-use source of 1.6 PgC/yr in the 1980s and attributing the residual terrestrial sink to CO_2 fertilisation. This intercomparison yielded CO_2 concentrations in 2100 of 668 to 734 ppm; results presented in Schimel *et al.* (1996) (from the Bern model) gave 688 ppm. After recalibrating to match a presumed land-use source of 1.1 PgC/yr, implying a weaker CO_2 response, the 2100 CO_2 concentration was given as 712 ppm in the SAR (Schimel *et al.*, 1996). An IPCC Technical Paper (Wigley *et al.*, 1997) evaluated the sensitivity of IS92a results to this calibration procedure. Wigley *et al.*, (1997) found that a range of assumed values from 0.4 to 1.8 PgC/yr for the land-use source during the 1980s gave rise to a range of 2100 CO_2 concentrations from 667 to 766 ppm.

In contrast with the SAR, the results presented here are based on approximating the behaviour of spatially resolved process-based models in which CO_2 and climate responses are not constrained by prior assumptions about the global carbon budget. The CO_2-only response of both models' reference cases (Figure 3.11) leads to a 2100 CO_2 concentration of 682 ppm (ISAM) or 651 ppm (Bern-CC). These values are slightly lower than projected in the SAR – 715 ppm (ISAM; SAR version) and 712 ppm (Bern Model; SAR version) – because current process-based terrestrial models typically yield a stronger CO_2 response than was assumed in the SAR. With climate feedbacks included, the 2100 CO_2 concentration in the reference case becomes, by coincidence, effectively indistinguishable from that given in the SAR: 723 ppm (ISAM) and 706 ppm (Bern-CC). The ranges of 164 ppm or −12% / +11% (about the reference case) (ISAM) and 273 ppm or −10% / +28% (Bern-CC) in the 2100 CO_2 concentration indicate that there is significant uncertainty about the future CO_2 concentrations due to any one pathway of changes in emissions. Separate calculations with the Bern-CC model (Figure 3.11) show that the effect of changing climate sensitivity alone is less important than

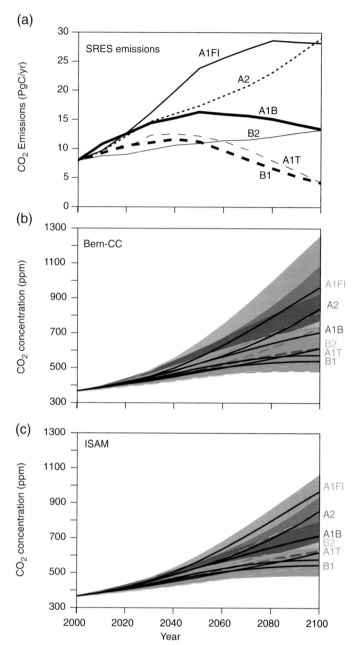

Figure 3.12: Projected CO_2 concentrations resulting from six SRES scenarios. The SRES scenarios represent the outcome of different assumptions about the future course of economic development, demography and technological change (see Appendix II). Panel (a) shows CO_2 emissions for the selected scenarios and panels (b) and (c) show resulting CO_2 concentrations as projected by two fast carbon cycle models, Bern-CC and ISAM (see Box 3.7 and Figure 3.11). The ranges represent effects of different model parametrizations and assumptions as indicated in the text and in the caption to Figure 3.11. For each model, and each scenario the reference case is shown by a black line, the upper bound (high-CO_2 parametrization) is indicated by the top of the coloured area, and the lower bound (low-CO_2 parametrization) by the bottom of the coloured area or (where hidden) by a dashed coloured line.

the effect of varying assumptions in the carbon cycle model's components. The effect of increasing climate sensitivity to 4.5 °C (increasing the climate feedback) is much larger than the effect of reducing climate sensitivity to 1.5 °C. The "low-CO_2" parametrization of Bern-CC yields CO_2 concentrations closer to the reference case than the "high-CO_2" parametrization, in which the terrestrial sink is forced to approach zero during the first few decades of the century due to the capping of the CO_2 fertilisation effect.

The reference simulations with ISAM yielded an implied average land-use source during the 1980s of 0.9 PgC/yr. The range was 0.2 to 2.0 PgC/yr. Corresponding values for Bern-CC were 0.6 PgC/yr and a range of 0.0 to 1.5 PgC/yr. These ranges broadly overlap the range estimates of the 1980s land-use source given in Table 3.1. Present knowledge of the carbon budget is therefore not precise enough to allow much narrowing of the uncertainty associated with future land and ocean uptake as expressed in these projections. However, the lowest implied land-use source values fall below the range given in Table 3.1.

3.7.3.3 SRES scenarios and their implications for future CO_2 concentration

The Special Report on Emissions Scenarios (SRES) (IPCC, 2000b) produced a series of scenarios, of which six are used here, representing outcomes of distinct narratives of economic development and demographic and technological change. In ISAM model runs with these scenarios, past fossil emissions (see Section 3.4.1), CO_2 concentrations (Enting *et al.*, 1994; Keeling and Whorf, 2000) and mean global temperatures (Jones *et al.*, 2000) were specified up to and including 1999; scenario-based analyses started in 2000. In the Bern-CC model runs, observed CO_2 (Etheridge, *et al.*, 1996, Keeling and Whorf, 2000) and past fossil emissions (Marland *et al.*, 1999) were prescribed, and historical temperature changes were modelled, based on radiative forcing from greenhouse gases and aerosols; again, scenario-based analyses started in 2000. Past emissions from changing land use were calculated in order to balance the carbon budget.

The six scenarios lead to substantial differences in projected CO_2 concentration trajectories (Figure 3.12). Significant uncertainties are introduced by the range of model parametrizations considered, so that the trajectories calculated for "adjacent" scenarios overlap, especially during the first half-century. The reference cases of the six scenarios account for a range of 2100 CO_2 concentrations from 541 to 963 ppm in the Bern-CC model and 549 to 970 ppm in the ISAM model. The uncertainties around the 2100 values due to model parametrizations are −12 to +10 % (ISAM) and −14 to +31 % (Bern-CC).

These uncertainties reflect incomplete understanding of climate sensitivity and the carbon cycle. They substantially limit our current ability to make quantitative predictions about the future consequences of a given emissions trajectory. Nevertheless, the results show that higher emissions are always expected to lead to higher projected atmospheric concentrations. They also show that the range of emissions scenarios currently accepted as plausible leads to a range of CO_2 concentrations that exceeds the likely upper bound of uncertainties due to differences among model parameterizations and assumptions.

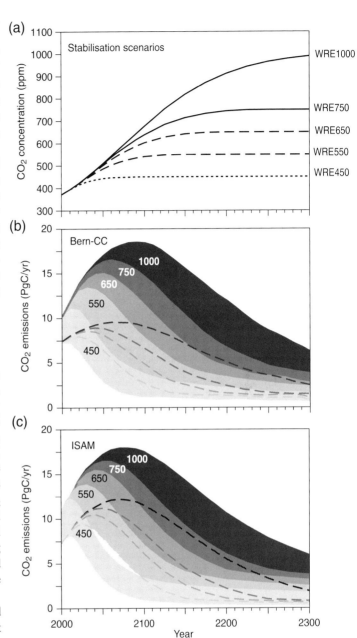

Figure 3.13: Projected CO_2 emissions leading to stabilisation of atmospheric CO_2 concentrations at different final values. Panel (a) shows the assumed trajectories of CO_2 concentration (WRE scenarios; Wigley *et al.*, 1996) and panels (b) and (c) show the implied CO_2 emissions, as projected with two fast carbon cycle models, Bern-CC and ISAM (see Box 3.7 and Figure 3.11). The ranges represent effects of different model parametrizations and assumptions as indicated in the text and in the caption to Figure 3.11. For each model, the upper and lower bounds (corresponding to low- and high-CO_2 parametrizations, respectively) are indicated by the top and bottom of the shaded area. Alternatively, the lower bound (where hidden) is indicated by a dashed line.

3.7.3.4 *Stabilisation scenarios and their implications for future CO$_2$ emissions*

Stabilisation scenarios illustrate implied rates of CO$_2$ emission that would arrive at various stable CO$_2$ concentration levels. These have been projected using a similar methodology to that applied in the analysis of emissions scenarios. The WRE trajectories follow CO$_2$ concentrations consistent with the IS92a scenario beginning in 1990 and branch off to reach constant CO$_2$ concentrations of 450, 550, 650, 750 and 1,000 ppm (Wigley *et al.*, 1996). The rationale for various alternative time trajectories and stabilisation levels is discussed in Chapter 2 of the IPCC WGIII Third Assessment Report (Morita *et al.*, 2001). Differences in emissions pathways for different time trajectories leading to a certain stabilisation target (e.g., S versus WRE profiles) are discussed in Schimel *et al.*, (1997). Here, we have calculated emissions for one set of emission profiles to illustrate differences in implied emissions that arise from updating models since the SAR.

As in Section 3.7.3.2, the models were initialised up to present. Then anthropogenic emissions for the prescribed CO$_2$ stabilization profiles were calculated; deduced emissions equal the change in modeled ocean and terrestrial carbon inventories plus the prescribed change in atmospheric CO$_2$ content. To estimate the strength of carbon cycle-climate feedbacks, global temperature (ISAM) and changes in the fields of temperature, precipitation and cloud cover (Bern-CC) were projected from CO$_2$ radiative forcing only, neglecting effects of other greenhouse gases and aerosols which are not specified in the WRE profiles. The results for the reference cases are not substantially different from those presented in the SAR (Figure 3.13). However, the range based on alternative model parametrizations is larger than presented in the SAR, mainly due to the range of simulated terrestrial CO$_2$ uptake. CO$_2$ stabilisation at 450, 650 or 1,000ppm would require global anthropogenic CO$_2$ emissions to drop below 1990 levels, within a few decades, about a century, or about two centuries, respectively.

In all cases, once CO$_2$ concentration becomes constant, the implied anthropogenic emission declines steadily. This result was expected. It highlights the fact that to maintain a constant future CO$_2$ concentration, anthropogenic CO$_2$ emissions would ultimately have to be reduced to the level of persistent natural sinks. Persistent terrestrial sinks are not well quantified; peatlands may be a candidate, but the gradual rise in atmospheric CO$_2$ concentration during the present interglacial (Figure 3.2) argues against any such sink. Estimates of current uptake by peatlands are <0.1 PgC/yr (Clymo *et al.*, 1998). Mixing of ocean DIC between surface and deep waters should continue to produce ocean uptake for several centuries after an input of anthropogenic atmospheric CO$_2$ (Siegenthaler and Oeschger, 1978; Maier-Reimer and Hasselmann, 1987; Sarmiento *et al.*, 1992). This mixing is the main reason for continued uptake (and therefore positive calculated emissions) after stabilisation. However, the main, known natural sink expected to persist longer than a few centuries is that due to dissolution of CaCO$_3$ in ocean sediments, which increases ocean alkalinity and thereby allows additional CO$_2$ to dissolve in the ocean. For CO$_2$ concentrations about 1,000 ppm, this sink is estimated to be smaller than about −0.1 PgC/yr (Archer *et al.*, 1998). Thus, for any significant CO$_2$ emissions to persist over centuries without continuing to increase atmospheric CO$_2$ would require some method of producing an artificial carbon sink.

3.7.4 *Conclusions*

The differences among the CO$_2$ concentrations projected with the various SRES scenarios considered are larger than the differences caused by inclusion or omission of climate-mediated feedbacks. The range of uptake rates projected by process-based models for any one scenario is, however, considerable, due to uncertainties about (especially) terrestrial ecosystem responses to high CO$_2$ concentrations, which have not yet been resolved experimentally, and uncertainties about the response of global NPP to changes in climate (Cramer *et al.*, 1999). A smaller feedback would be implied if, as some models indicate, global NPP increases with warming throughout the relevant range of climates and no forest die back occurs. Larger positive feedbacks would be implied if regional drying caused partial die back of tropical forests, as some of the DGVMs in Cramer *et al.* (2001), and one coupled climate-carbon model study of Cox *et al.* (2000), suggest; however, another coupled climate-carbon model study (Friedlingstein *et al.*, 2001) suggests a smaller feedback. Uncertainty also arises due to differences in the climate responses of ocean models, especially as regards the extent and effects (biological as well as physical) of increased stratification in a warmer climate (Joos *et al.*, 1999b).

In conclusion, *anthropogenic CO$_2$ emissions are virtually certain to be the dominant factor determining CO$_2$ concentrations throughout the 21st century.* The importance of anthropogenic emissions is underlined by the expectation that the proportion of emissions taken up by both ocean and land will decline at high atmospheric CO$_2$ concentrations (even if absolute uptake by the ocean continues to rise). There is considerable uncertainty in projections of future CO$_2$ concentration, because of uncertainty about the effects of climate change on the processes determining ocean and land uptake of CO$_2$. These uncertainties do not negate the main finding that anthropogenic emissions will be the main control.

Large-scale manipulations of terrestrial ecosystems have been proposed as a means of slowing the increase of atmospheric CO$_2$ during the 21st century in support of the aims of the Kyoto Protocol (Tans and Wallace, 1999; IPCC, 2000a). Based on current understanding of land use in the carbon cycle, the impacts of future land use on terrestrial biosphere-atmosphere exchanges have the potential to modify atmospheric CO$_2$ concentrations on this time-scale. Direct effects of land-use changes are thought to represent about 10 to 30% of total anthropogenic CO$_2$ emissions (Table 3.1), so there is scope for either intended or unintended changes in land use to reduce or increase total anthropogenic emissions. But the possibilities for enhancing natural sinks have to be placed in perspective: a rough upper bound for the reduction in CO$_2$ concentration that could be achieved by enhancing terrestrial carbon uptake through land-use change over the coming century is 40 to 70 ppm (Section 3.2.2.2), to be considered against a two to four times larger potential for increasing CO$_2$ concentraion by deforestation, and a >400 ppm range among the SRES scenarios (Figure 3.12).

References

Aber, J., W. McDowell, K. Nadelhoffer, A. Magill, G. Bernstson, M. Kamakea, S. McNulty, W. Currie, L. Rustad, I. Fernandez, 1998: Nitrogen saturation in temperate forest ecosystems. *BioScience*, **48**, 921-934.

Allen, A.S., J.A. Andrews, A.C. Finzi, R. Matamala, D.D. Richter and W.H. Schlesinger, 2000: Effects of free-air CO_2 enrichment (FACE) on belowground processes in a *Pinus taeda* forest. *Ecological Applications*, **10**, 437-448.

Altabet, M.A., R. Francois, D.W. Murray, and W.L. Prell, 1995: Climate-Related Variations in Denitrification in the Arabian Sea From Sediment $^{15}N/^{14}N$ Ratios. *Nature*, **373**, 506-509.

Andersen, K. K., A. Armengaud, and C. Genthon, 1998: Atmospheric dust under glacial and interglacial conditions, *Geophysical Research Letters*, **25**, 2281-2284, 1998.

Andres, R. J., Marland, G., Boden, T., and Bischof, S., 2000: Carbon dioxide emissions from fossil fuel consumption and cement manufacture, 1751-1991, and an estimate of their isotopic composition and latitudinal distribution. In: *The Carbon Cycle*, [Wigley, T.M.L. and D.S. Schimel (eds.)]. Cambridge University Press, New York, pp. 53-62.

Antoine, D., J.M. Andre, and A. Morel, 1996: Oceanic primary production. 2. Estimation at global scale from satellite (coastal zone color scanner) chlorophyll. *Global Biogeochemical Cycles*, **10**, 57-69.

Apps, M.J., W.A. Kurz, R.J. Luxmoore, L.O. Nilsson, R.A. Sedjo, R. Schmidt, L.G. Simpson, and T.S. Vinson, 1993: Boreal Forests and Tundra. *Water Air Soil Pollution*, **70**, 39-53.

Archer, D.E, and E. Maier-Reimer, 1994: Effect of deep-sea sedimentary calcite preservation on atmospheric CO_2 concentration. *Nature*, **367**, 260-263.

Archer, D.E., and K. Johnson, 2000: A Model of the iron cycle in the ocean. *Global Biogeochemical Cycles*, **14**, 269-279.

Archer, D.E., T. Takahashi, S. Sutherland, J. Goddard, D. Chipman, K. Rodgers, and H. Ogura, 1996: Daily, seasonal and interannual variability of sea-surface carbon and nutrient concentration in the equatorial Pacific Ocean. *Deep-Sea Research Part II-Topical Studies in Oceanography*, **43**, 779-808.

Archer, D.E, H. Kheshgi, and E. Maier-Reimer, 1997: Multiple timescales for neutralization of fossil fuel CO_2. *Geophysical Research Letters*, **24**, 405-408.

Archer, D.E, H. Kheshgi, and E. Maier-Reimer, 1998: Dynamics of fossil fuel CO_2 neutralization by marine $CaCO_3$. *Global Biogeochemical Cycles*, **12**, 259-276.

Archer, D.E, A. Winguth, D. Lea, and N. Mahowald, 2000: What caused the glacial/interglacial atmospheric pCO_2 cycles? *Reviews of Geophysics*, **38**, 159-189.

Archer, S., T.W. Boutton, and K.A. Hibbard, 2001: Trees in grasslands: biogeochemical consequences of woody plant expansion. In: *Global Biogeochemical Cycles and their Interrelationship with Climate* [Schulze, E.-A., S.P. Harrison, M. Heimann, E.A. Holland, J. Lloyd, I.C. Prentice, and D.S. Schimel (eds.)], Academic Press.

Asrar, G., Myneni R.B., and Choudhury, B.J., 1992: Spatial heterogeneity in vegetation canopies and remote sensing of absorbed photosynthetically active radiation: a modeling study. *Remote Sensing Environment*, **41**, 85-103.

Atjay, G.L., P. Ketner, and P. Duvigneaud, 1979: Terrestrial primary production and phytomass. In: *The Global Carbon Cycle* [Bolin, B., E.T. Degens, S. Kempe, and P. Ketner (eds.)], John Wiley & Sons, Chichester, pp. 129-181.

Aubinet, M., A. Grelle, A. Ibrom, U. Rannik, J. Moncrieff, T. Foken, A.S. Kowalski, P.H. Martin, P. Berbigier, C. Bernhofer, R. Clement, J. Elbers, A. Granier, T. Grunwald, K. Morgenstern, K. Pilegaard, C. Rebmann, W. Snijders, R. Valentini and T. Vesala, 2000: Estimates of the annual net carbon and water exchange of forests: The EUROFLUX methodology. *Advances in Ecological Research*, **30**, 113-175.

Aumont, O., 1998: Étude du cycle naturel du carbone dans un modèle 3D de l'océan mondial. Ph.D. thesis, Université Pierre et Marie Curie, 4 place Jussieu, Paris 95005.

Aumont, O., S. Belviso and P. Monfray, 2001a: Dimethylsulfoniopropionate (DMSP) and dimethylsulfide (DMS) sea surface distributions simulated from a global 3-D ocean carbon cycle model. *Journal of Geophysical Research*, (in press)

Aumont, O., J.C. Orr, P. Monfray, W. Ludwig, P. Amiotte-Suchet and J.L. Probst, 2001b: Riverine-driven interhemispheric transport of carbon. *Global Biogeochemical Cycles*, (in press)

Bacastow, R.B., 1976: Modulation of atmospheric carbon dioxide by the southern oscillation. *Nature*, **261**, 116-118.

Bacastow, R.B., 1993: The effect of temperature change of the warm surface waters of the oceans on atmospheric CO_2. *Global Biogeochemical Cycles*, **10**, 319-333.

Bacastow, R.B., C.D. Keeling, T.J. Lueker, M. Wahlen, and W.G. Mook, 1996: The ^{13}C Suess effect in the world surface oceans and its implications for oceanic uptake of CO_2 – Analysis of observations at Bermuda. *Global Biogeochemical Cycles*, **10**, 335-346.

Baldocchi, D.D., B.B. Hicks, T.P. Meyers, 1988: Measuring biosphere-atmosphere exchanges of biologically related gases with micrometeorological methods. *Ecology*, **69**, 1331-1340.

Balkanski, Y., P. Monfray, M. Battle, and M. Heimann, 1999: Ocean primary production derived from satellite data: An evaluation with atmospheric oxygen measurements. *Global Biogeochemical Cycles*, **13**, 257-271.

Barnola, J.M., M. Anklin, J. Porcheron, D. Raynaud, J. Schwander, and B. Stauffer, 1995: CO_2 evolution during the last millennium as recorded by Antarctic and Greenland ice. *Tellus Series B-Chemical and Physical Meteorology*, **47**, 264-272.

Bates, N.R., A.F. Michaels, and A.H. Knap, 1996: Seasonal and interannual variability of oceanic carbon dioxide species at the US JGOFS Bermuda Atlantic Time-series Study (BATS) site. *Deep-Sea Research Part II-Topical Studies in Oceanography*, **43**, 347-383.

Batjes, N.H., 1996: Total carbon and nitrogen in the soils of the world. *European Journal of Soil Science* **47**: 151-163.

Battle, M., M. Bender, T. Sowers, P.P. Tans, J.H. Butler, J.W. Elkins, J.T. Ellis, T. Conway, N. Zhang, P. Lang, and A.D. Clarke, 1996: Atmospheric gas concentrations over the past century measured in air from firn at the South Pole. *Nature*, **383**, 231-235.

Battle, M., M. Bender, P.P. Tans, J.W.C. White, J.T. Ellis, T. Conway, and R.J. Francey, 2000: Global carbon sinks and their variability, inferred from atmospheric O_2 and $\delta^{13}C$. *Science*, **287**, 2467-2470.

Beerling, D.J., H.H. Birks, and F.I. Woodward, 1995: Rapid late-glacial atmospheric CO_2 changes reconstructed from the stomatal density record of fossil leaves. *Journal of Quaternary Science*, **10**, 379-384.

Bender, M., T. Sowers, and L. Labeyrie, 1994: The Dole effect and its variations during the last 130,000 years as measured in the VOSTOK ice core. *Global Biogeochemical Cycles*, **8**, 363-376.

Bender, M., T. Ellis, P. Tans, R. Francey, and D. Lowe, 1996: Variability in the O_2/N_2 ratio of southern hemisphere air, 1991-1994 - Implications for the carbon cycle. *Global Biogeochemical Cycles*, **10**, 9-21.

Berger, A.L., 1978: Long-term variations of caloric insolation resulting from the Earth's orbital elements. *Quaternary Research*, **9**, 139-167.

Bergh, J., S. Linder, T. Lundmark, B. Elfving, 1999: The effect of water and nutrient availability on the productivity of Norway spruce in northern and southern Sweden. *Forest Ecology and Management*, **119**, 51-62.

Berner, R.A., 1993: Weathering and its effect on atmospheric CO_2 over phanerozoic time. *Chemical Geology*, **107**, 373-374.

Berner, R.A., 1997: The rise of plants and their effect on weathering and atmospheric CO_2. *Science*, **276**, 544-546.

Bickle, M.J., 1994: The role of metamorphic decarbonation reactions in returning strontium to the silicate sediment mass. *Nature*, **367**, 699-704.

Bird, M.I., J. Lloyd, and G.D. Farquhar, 1994: Terrestrial carbon storage at the LGM. Nature, 371, 585.

Blume, H.P., H. Eger, E. Fleischhauer, A. Hebel, C. Reij, and K.G. Steiner, 1998: Towards sustainable land use. *Advances in Geoecology*, **31**, 1625pp.

Bolin, B. and Keeling C.D., 1963: Large-scale atmospheric mixing as deduced from the seasonal and meridional variations of carbon dioxide. *Journal of Geophysical Research*, **68**, 3899-3920.

Bolin, B., Sukumar, R., P. Ciais, W. Cramer, P. Jarvis, H. Kheshgi, C. Nobre, S. Semenov, W. Steffen, 2000: Global Perspective. In: *IPCC, Land Use, Land-Use Change, and Forestry. A Special Report of the IPCC* [Watson, R.T., I.R. Noble, B. Bolin, N.H. Ravindranath, D.J. Verardo and D.J. Dokken (eds.)]. Cambridge University Press, Cambridge, UK, pp. 23-51.

Bondeau, A., D.W. Kicklighter, and J. Kaduk, 1999: Comparing global models of terrestrial net primary productivity (NPP): importance of vegetation structure on seasonal NPP estimates. *Global Change Biology*, **5**, 35-45.

Boone, R.D., K.J. Nadelhoffer, J.D. Canary, and J.P. Kaye, 1998: Roots exert a strong influence on the temperature sensitivity of soil respiration. *Nature*, **396**, 570-572.

Bopp, L., P. Monfray, O. Aumont, J.-L. Dufresne, H. LeTreut, G. Madec, L. Terray, and J. Orr, 2001: Potential impact of climate change on marine export production. *Global Biogeochemical Cycles* **15(1)**, 81-100.

Bousquet, P., P. Ciais, P. Peylin, M. Ramonet, and P. Monfray, 1999: Inverse modeling of annual atmospheric CO_2 sources and sinks 1. Method and control inversion. *Journal of Geophysical Research-Atmospheres*, **104**, 26161-26178.

Bousquet, P., P. Peylin, P. Ciais, C. Le Quéré, P. Friedlingstein and P. P. Tans, 2000: Regional changes of CO_2 fluxes over land and oceans since 1980. *Science*, **290**, 1342-1346.

Boutin, J., J. Etcheto, Y. Dandonneau, D.C.E. Bakker, R.A. Feely, H.Y. Inoue, M. Ishii, R.D. Ling, P.D. Nightingale, N. Metzl, and R. Wanninkhof, 1999: Satellite sea surface temperature: a powerful tool for interpreting in situ pCO_2 measurements in the equatorial Pacific Ocean. *Tellus Series B-Chemical and Physical Meteorology*, **51**, 490-508.

Boyd, P.W., A. Watson, C.S. Law, E. Abraham, T. Trull, R. Murdoch, D.C.E. Bakker, A.R. Bowie, K. Buesseler, H. Chang, M. Charette, P. Croot, K. Downing, R. Frew, M. Gall, M. Hadfield, J. Hall, M. Harvey, G. Jameson, J. La Roche, M. Liddicoat, R. Ling, M. Maldonado, R.M. McKay, S. Nodder, S. Pickmere, R. Pridmore, S. Rintoul, K. Safi, P. Sutton, R. Strzepek, K. Tanneberger, S. Turner, A. Waite, and J. Zeldis, 2000: A mesoscale pytoplankton bloom in the polar Southern Ocean stimulated by iron fertilization. *Nature*, **407**, 695-702.

Boyle, E.A., 1988: The role of vertical chemical fractionation in controlling late Quaternary atmospheric carbon dioxide. *Journal of Geophysical Research*, **93**, 15701-15714.

Braun, S., Rihm, B., Schindler, C., and Fluckiger, W., 2000: Growth of mature beech in relation to ozone and nitrogen deposition: an epidemiological approach. *Water Air and Soil Pollution*, **116**, 357-364.

Braswell, B.H., D.S. Schimel, E. Linder, and B. Moore, 1997: The response of global terrestrial ecosystems to interannual temperature variability. *Science*, **278**, 870-872.

Bretherton, F.P., K. Bryan, and J.D. Woods, 1990: Time-dependent greenhouse-gas-induced climate change. In: *Climate Change, The IPCC Scientific Assessment* [Houghton, J.T., G.J. Jenkins and J.J. Ephraums (eds.)]. Cambridge University Press, Cambridge, pp. 173-194.

British Petroleum Company, 2000: BP Statistical Review of World Energy 1999. British Petroleum Company, London, UK.

Broecker, W.S., and T.H. Peng, 1982: Tracers in the sea, Eldigio, Palisades, pp. 691.

Broecker, W.S. and T.H. Peng, 1992: Interhemispheric transport of carbon dioxide by ocean circulation. *Nature*, **356**, 587-589.

Broecker, W.S., and G.M. Henderson, 1998: The sequence of events surrounding Termination II and their implications for the cause of glacial-interglacial CO_2 changes. *Paleoceanography*, **13**, 352-364.

Broecker, W.S., S. Sutherland, W. Smethie, T.H. Peng, and G. Ostlund, 1995: Oceanic radiocarbon: separation of the natural and bomb components. *Global Biogeochemical Cycles*, **9**, 263-288.

Broecker, W., J. Lynch-Stieglitz, D. Archer, M. Hofmann, E. Maier-Reimer, O. Marchal, T. Stocker, and N. Gruber, 1999: How strong is the Harvardton-Bear constraint? *Global Biogeochemical Cycles*, **13**, 817-820.

Brovkin, V., A. Ganopolski, and Y. Svirezhev, 1997: A continuous climate-vegetation classification for use in climate-biosphere studies. *Ecological Modelling*, **101**, 251-261.

Brown, S., J. Sathaye, M. Cannell, P. Kauppi, P. Burschel, A. Grainger, J. Heuveldop, R. Leemans, P. Moura Costa, M. Pinard, S. Nilsson, W. Schopfhauser, R. Sedjo, N. Singh, M. Trexler, J. van Minnen, S. Weyers, 1996: Management of forests for mitigation of greenhouse gas emissions. In: *IPCC Climate Change 1995 - Impacts, Adaptations and Mitigation of Climate Change: Scientific-Technical Analyses, Contribution of Working Group II to the Second Assessment Report of the Intergovernmental Panel on Climate Change* [Watson, R.T., M.C. Zinyowera, R.H. Moss and D.J. Dokken (eds.)], Cambridge University Press, Cambridge, pp. 773-797.

Bryant, D.M., E.A. Holland, T.R. Seastedt and M.D. Walker, 1998: Analysis of litter decomposition in an alpine tundra. *Canadian Journal of Botany-Revue Canadienne De Botanique*, **76**, 1295-1304.

Buchmann, N., and E.D. Schulze, 1999: Net CO_2 and H_2O fluxes of terrestrial ecosystems. *Global Biogeochemical Cycles*, **13**, 751-760.

Burrows, W.H., J.F. Compton, and M.B. Hoffmann, 1998: Vegetation thickening and carbon sinks in the grazed woodlands of north-east Australia. In: *Proceedings Australian Forest Growers Conference*, Lismore, NSW, pp. 305-316.

Caldeira, K., and P.B. Duffy, 2000: The role of the Southern Ocean in uptake and storage of anthropogenic carbon dioxide. *Science*, **287**, 620-622.

Cao, M., and F.I. Woodward, 1998a: Dynamic responses of terrestrial ecosystem carbon cycling to global climate change. *Nature*, **393**, 249-252.

Cao, M.K., and F.I. Woodward, 1998b: Net primary and ecosystem production and carbon stocks of terrestrial ecosystems and their responses to climate change. *Global Change Biology*, **4**, 185-198.

Carter, A.J. and R.J. Scholes, 2000: Spatial Global Database of Soil Properties. IGBP Global Soil Data Task CD-ROM. International Geosphere-Biosphere Programme (IGBP) Data Information Systems. Toulouse, France.

Catubig, N.R., D.E. Archer, R. Francois, P. deMenocal, W. Howard and E.-F. Yu, 1998: Global deep-sea burial rate of calcium carbonate during the last glacial maximum. *Paleoceanography*, **13**, 298-310.

Cerling, T.E., Y. Wang, and J. Quade, 1993: Expansion of C_4 ecosystems as an indicator of global ecological change in the late miocene. *Nature*, **361**, 344-345.

Cerling, T.E., J.M. Harris, B.J. MacFadden, M.G. Leakey, J. Quade, V. Eisenmann, 1997: Global vegetation change through the Miocene-Pliocene boundary. *Nature*, **389**, 153-158.

Chapin, F.S., 1980: The mineral-nutrition of wild plants. *Annual Review of Ecology and Systematics*, **11**, 233-260.

Chapman, W.L., and J.E. Walsh, 1993: Recent variations of sea ice and air-temperature in high-latitudes. *Bulletin of the American Meteorological Society*, **74**, 33-47.

Ciais, P., P.P. Tans, and M. Trolier, 1995a: A large northern-hemisphere terrestrial CO_2 sink indicated by the $^{13}C/^{12}C$ ratio of atmospheric CO_2. *Science,* **269**, 1098-1102.

Ciais, P., P.P. Tans, J.W.C. White, M. Trolier, R.J. Francey, J.A. Berry, D.R. Randall, P.J. Sellers, J.G. Collatz, and D.S. Schimel, 1995b: Partitioning of ocean and land uptake of CO_2 as inferred by $\delta^{13}C$ measurements from the NOAA Climate Monitoring and Diagnostics Laboratory Global Air Sampling Network. *Journal of Geophysical Research-Atmospheres,* **100**, 5051-5070.

Ciais, P., A.S. Denning, P.P. Tans, J.A. Berry, D.A. Randall, G.J. Collatz, P.J. Sellers, J.W.C. White, M. Trolier, H.A.J. Meijer, R.J. Francey, P. Monfray, and M. Heimann, 1997: A three-dimensional synthesis study of $\delta^{18}O$ in atmospheric CO_2. 1. Surface Fluxes. *Journal of Geophysical Research - Atmosphere,* **102**, 5857-5872.

Ciais, P., P. Friedlingstein, D.S. Schimel, and P.P. Tans, 1999: A global calculation of the delta ^{13}C of soil respired carbon: Implications for the biospheric uptake of anthropogenic CO_2. *Global Biogeochemical Cycles,* **13**, 519-530.

Cienciala, E., S. W. Running, A. Lindroth, A. Grelle, and M.G. Ryan, 1998: Analysis of carbon and water fluxes from the NOPEX boreal forest: comparison of measurments with Forest-BGC simulations. *Journal of Hydrology,* **212-213**, 62-78.

Clark, F.E., 1977: Internal cycling of ^{15}N in shortgrass prairie. *Ecology,* **58**, 1322-1333.

Clymo, R.S., J. Turunen, and K. Tolonen, 1998: Carbon accumulation in peatland. *Oikos,* **81**, 368-388.

Coale, K.H., K.S. Johnson, S.E. Fitzwater, R.M. Gordon, S. Tanner, F.P. Chavez, L. Ferioli, C. Sakamoto, P. Rogers, F. Millero, P. Steinberg, P. Nightingale, D. Cooper, W.P. Cochlan, M.R. Landry, J. Constantinou, G. Rollwagen, A. Trasvina, and R. Kudela, 1996: A massive phytoplankton bloom induced by an ecosystem-scale iron fertilization experiment in the equatorial Pacific Ocean. *Nature,* **383**, 495-501.

Cole, C.V., C. Cerri, K. Minami, A. Mosier, N. Rosenberg, D. Sauerbeck, J. Dumanski, J. Duxbury, J. Freney, R. Gupta, O. Heinemeyer, T. Kolchugina, J. Lee, K. Paustian, D. Powlson, N. Sampson, H. Tiessen, M. Van Noordwijk, Q. Zhao, I.P. Abrol, T. Barnwell, C. Campbell, R.L Desjardin, C. Feller, P. Garin, M.J. Glendining, E.G. Gregorich, D. Johnson, J. Kimble, R. Lal, C. Monreal, D.S. Ojima, M. Padgett, W. Post, W. Sombroek, C. Tarnocai, T. Vinson, S. Vogel, and G. Ward, 1996: Agricultural options for mitigation of greenhouse gas emissions. In: *Climate Change 1995 – Impacts, Adaptations and Mitigation of Climate Change: Scientific–Technical Analyses* [Watson, R.T., M.C. Zinyowera and R.H. Moss (eds.)], Cambridge University Press, Cambridge, pp. 745-771.

Collatz, G.J., J.A. Berry, and J.S. Clark, 1998: Effects of climate and atmospheric CO_2 partial pressure on the global distribution of C_4 grasses: present, past, and future. *Oecologia,* **114**, 441-454.

Conway, T.J., P.P. Tans, L.S. Waterman, K.W. Thoning, D.R. Kitzis, K.A. Masarie, and N. Zhang, 1994: Evidence for interannual variability of the carbon cycle from the National Oceanic and Atmospheric Administration/Climate Monitoring and Diagnostics Laboratory global air sampling network. *Journal of Geophysical Research,* **99**, 22831-22855.

Cox, P.M., R.A. Betts, C.D. Jones, S.A. Spall, and I.J. Totterdell, 2000: Acceleration of global warming due to carbon-cycle feedbacks in a coupled model. *Nature, 408, 184-187.*

Cramer, W. and C.B. Field, 1999: Comparing global models of terrestrial net primary productivity (NPP): introduction. *Global Change Biology,* **5**, III-IV.

Cramer, W., D.W. Kicklighter, A. Bondeau, B. Moore III, G. Churkina, B. Nemry, A. Ruimy, A.L. Schloss and the participants of the Potsdam NPP Model Intercomparison, 1999: Comparing global models of terrestrial net primary productivity (NPP) overview and key results. *Global Change Biology,* **5**, 1-16.

Cramer, W., A. Bondeau, F.I. Woodward, I.C. Prentice, R.A. Betts, V.

Brovkin, P.M. Cox, V. Fisher, J.A. Foley, A.D. Friend, C. Kucharik, M.R. Lomas, N. Ramankutty, S. Sitch, B. Smith, A. White, and C. Young-Molling, 2001: Global response of terrestrial ecosystem structure and function to CO_2 and climate change: results from six dynamic global vegetation models. *Global Change Biology* (in press).

Crowley, T.J., 1995: Ice-age terrestrial carbon changes revisited. *Global Biogeochemical Cycles,* **9**, 377-389.

Curtis, P.S., 1996: A meta-analysis of leaf gas exchange and nitrogen in trees grown under elevated carbon dioxide. *Plant Cell and Environment,* **19**, 127-137.

Curtis, P.S., and X.Z. Wang, 1998: A meta-analysis of elevated CO_2 effects on woody plant mass, form, and physiology. *Oecologia,* **113**, 299-313.

Dai, A., and I.Y. Fung, 1993: Can climate variability contribute to the "missing" CO_2 sink? *Global Biogeochemical Cycles,* **7**, 599-609.

Dai, A., K.E. Trenberth and T.R. Karl, 1998: Global variations in droughts and wet spells: 1900-1995. *Geophysical Research Letters,* **25**, 3367-3370.

De Fries, R.S. and J.R.G. Townshend, 1994: NDVI-derived land-cover classifications at a global-scale. *International Journal of Remote Sensing,* **15**, 3567-3586.

De Fries, R.S., C.B. Field, I. Fung, C.O. Justice, S. Los, P.A. Matson, E. Matthews, H.A. Mooney, C.S. Potter, K. Prentice, P.J. Sellers, J.R.G. Townshend, C.J. Tucker, S.L. Ustin and P.M. Vitousek, 1995: Mapping the land-surface for global atmosphere-biosphere models - toward continuous distributions of vegetations functional- properties. *Journal of Geophysical Research-Atmospheres,* **100**, 20867-20882.

De Fries, R.S., C.B. Field, I. Fung, G.J. Collatz, and L. Bounoua, 1999: Combining satellite data and biogeochemical models to estimate global effects of human-induced land cover change on carbon emissions and primary productivity. *Global Biogeochemical Cycles,* **13**, 803-815.

Delgado, J.A., A.R. Mosier, D.W. Valentine, D.S. Schimel, and W.J. Parton, 1996: Long term ^{15}N studies in a catena of the shortgrass steppe. *Biogeochemistry,* **32**, 41-52.

DeLucia, E.H., J.G. Hamilton, S.L. Naidu, R.B. Thomas, J.A. Andrews, A. Finzi, R. Lavine, R. Matamala, J.E. Mohan, G.R. Hendrey, and W.H. Schlesinger, 1999: Net primary production of a forest ecosystem with experimental CO_2 enrichment. *Science,* **284**, 1177-1179.

Denman, K., E. Hofmann, and H. Marchant, 1996: Marine biotic responses to environmental change and feedbacks to climate, In: *Climate Change 1995. The Science of Climate Change.* [Houghton, J.T., L.G.M. Filho, B.A. Callander, N. Harris, A. Kattenberg, and K. Maskell (eds.)]. University Press, Cambridge, pp. 449-481.

Denning, A.S., I.Y. Fung, and D. Randall, 1995: Latitudinal gradient of atmospheric CO_2 due to seasonal exchange with land biota. *Nature,* **376**, 240-243.

Denning, A.S., D.A. Randall, G.J. Collatz, and P.J. Sellers, 1996: Simulations of terrestrial carbon metabolism and atmospheric CO_2 in a general circulation model. Part 2: Spatial and temporal variations of atmospheric CO_2. *Tellus,* 48B, 543-567.

Denning, A. S., M. Holzer, K. R. Gurney, M. Heimann, R. M. Law, P. J. Rayner, I. Y. Fung, S.-M. Fan, S. Taguchi, P. Friedlingstein, Y. Balkanski, J. Taylor, M. Maiss, and I. Levin, 1999. Three-dimensional transport and concentration of SF_6: A model intercomparison study (TransCom 2). *Tellus,* 51B, 266-297.

Dettinger, M.D. and M. Ghil, 1998: Seasonal and interannual variations of atmospheric CO_2 and climate. *Tellus Series B-Chemical and Physical Meteorology,* **50**, 1-24.

Dixon, R.K., S. Brown, R.A. Houghton, A.M. Solomon, M.C. Trexler and J. Wisniewski, 1994: Carbon Pools and Flux of Global Forest Ecosystems. *Science, 263,* 185-190.

Doney, S.C., 1999: Major challenges confronting marine biogeochemical modeling. *Global Biogeochemical Cycles,* **13,** 705-714.

Drake, B.G., M.A. Gonzalez-Meler, and S.P. Long, 1997: More efficient plants: a consequence of rising atmospheric CO_2? *Annual Review of Plant Physiology and Plant Molecular Biology,* **48,** 609-639.

Drake, B.G., J. Azcon-Bieto, J. Berry, J. Bunce, P. Dijkstra, J. Farrar, R.M. Gifford, M.A. Gonzalez-Meler, G. Koch, H. Lambers, J. Siedow and S. Wullschleger, 1999: Does elevated atmospheric CO_2 concentration inhibit mitochondrial respiration in green plants? *Plant Cell and Environment,* **22,** 649-657.

Duce, R.A., and N.W. Tindale, 1991: Atmospheric transport of iron and its deposition in the ocean. *Limnology and Oceanography,* **36,** 1715-1726.

Ducklow, H.W., 1999: The bacterial content of the oceanic euphotic zone. *FEMS Microbiology-Ecology,* **30,** 1-10.

Edmonds, J.A., M.A. Wise, R.D. Sands, R.A. Brown, and H.S. Kheshgi, 1996: Agriculture, Land Use, and Commercial Biomass Energy: A Preliminary Integrated Analysis of the Potential Role of Biomass Energy for Reducing Future Greenhouse Related Emissions. PNNL–111555. *Pacific Northwest National Laboratory, Washington, DC, USA.*

Edwards, R., P.N. Sedwick, V. Morgan, C.F. Boutron and S. Hong, 1998: Iron in ice cores from Law Dome, East Antarctica: implications for past deposition of aerosol iron. *Annals of Glaciology,* **27,** 365-370.

Elderfield, H. and R.E.M. Rickaby, 2000: Oceanic Cd/P ratio and nutrient utilization in the glacial Southern Ocean. *Nature,* **405,** 305-310.

Elliott, W.P., J.K. Angell and K.W. Thoning, 1991: Relation of atmospheric CO_2 to tropical sea and air temperatures and precipitation. *Tellus Series B-Chemical and Physical Meteorology,* **43,** 144-155.

England, M.H., 1995: Using chlorofluorocarbons to assess ocean climate models. *Geophysical Research Letters,* **22,** 3051-3054.

Enting, I.G. and J.V. Mansbridge, 1991: Latitudinal distribution of sources and sinks of CO_2 - results of an inversion study. *Tellus Series B-Chemical and Physical Meteorology,* **43,** 156-170

Enting, I.G., T.M.L. Wigley, and M. Heimann, 1994: Future emissions and concentrations of carbon dioxide: Key ocean/atmosphere/land analyses. Division of Atmospheric Research, Commonw. Science and Ind. Research Organisation, Melbourne, Victoria.

Enting, I.G., C.M. Trudinger, and R.J. Francey, 1995: A Synthesis Inversion of the Concentration and $\delta^{13}C$ of Atmospheric CO_2. *Tellus Series B-Chemical and Physical Meteorology*, **47,** 35-52.

Eppley, R.W., and B.J. Peterson, 1979: Particulate organic-matter flux and planktonic new production in the deep ocean. *Nature,* **282,** 677-680.

Etheridge, D.M., L.P. Steele, R.L. Langenfelds, R.J. Francey, J.M. Barnola, and V.I. Morgan, 1996: Natural and anthropogenic changes in atmospheric CO_2 over the last 1000 years from air in Antarctic ice and firn. *Journal of Geophysical Research - Atmosphere,* **101,** 4115-4128.

Falkowski, P.G., 1994: The role of phytoplankton photosynthesis in global biogeochemical cycles. *Photosynthesis Research,* **39,** 235-258.

Falkowski, P.G., 1997: Evolution of the nitrogen cycle and its influence on the biological CO_2 pump in the ocean. *Nature,* **387,** 272-275.

Falkowski, P.G., R.T. Barber, and V. Smetacek, 1998: Biogeochemical controls and feedbacks on ocean primary production. *Science,* **281,** 200-206.

Falloon, P., Smith, P., Coleman, K. and Marshall, S. 1998: Estimating the size of the inert organic matter pool from total soil organic carbon content for use in the Rothamsted Carbon Model. *Soil Biology and Biochemistry,* **30,** 1207-1211.

Fan, S., M. Gloor, and J. Mahlman, S. Pacala, J. Sarmiento, T. Takahashi and P. Tans, 1998: A large terrestrial carbon sink in North America implied by atmospheric and oceanic carbon dioxide data and models. *Science,* **282,** 442-446.

Fan, S.M., Blaine T.L., and Sarmiento J.L., 1999: Terrestrial carbon sink in the Northern Hemisphere estimated from the atmospheric CO_2 difference between Manna Loa and the South Pole since 1959. *Tellus Series B-Chemical and Physical Meteorology,* **51,** 863-870.

Fanning, K.A., 1992: Nutrient provinces in the sea: concentration ratios, reaction rate ratios, and ideal covariation. *Journal of Geophysical Research,* **97,** 5693-5712.

FAO, 1993a: 1992 Production Yearbook. FAO, Rome, Italy

FAO, 1993b: Forest Resources Assessment 1990. *Tropical Countries. FAO Forestry Pap. No. 112.* UN Food Agric. Org., Rome, Italy.

FAO, 1997: State of the world's forests. FAO, Rome, Italy.

Farquhar, G.D., 1997: Carbon dioxide and vegetation. *Science,* **278,** 1411.

Farquhar, G.D., S.V. von Caemmerer, and J.A. Berry, 1980: A biochemical-model of photosynthetic CO_2 assimilation in leaves of C_3 plants. *Planta,* **149,** 78-90.

Farquhar, G.D., J. Lloyd, J.A. Taylor, L.B. Flanagan, J.P. Syvertsen, K.T. Hubick, S.C. Wong, and J.R. Ehleringer, 1993: Vegetation effects on the isotope composition of oxygen in atmospheric CO_2. *Nature,* **363,** 439-443.

Fearnside, P.M., 2000: Global warming and tropical land-use change: greenhouse gas emissions from biomass burning, decomposition and soils in forest conversion, shifting cultivation and secondary vegetation. *Climatic Change,* **46,** 115-158.

Feely, R.A., R.H. Gammon, B.A. Taft, P.E. Pullen, L.S. Waterman, T.J. Conway, J.F. Gendron and D.P. Wisegarver, 1987: Distribution of chemical tracers in the eastern equatorial Pacific during and after the 1982-1983 El-Niño/Southern Oscillation event. *Journal of Geophysical Research,* **92,** 6545-6558.

Feely, R.A., R. Wanninkhof, C. Goyet, D.E. Archer, and T. Takahashi, 1997: Variability of CO_2 distributions and sea-air fluxes in the central and eastern equatorial Pacific during the 1991-1994 El Niño. *Deep-Sea Research Part II-Topical Studies in Oceanography,* **44,** 1851-1867.

Feely, R.A., C.L. Sabine, R.M. Key, and T.H. Peng, 1999a: CO_2 synthesis results: estimating the anthropogenic carbon dioxide sink in the Pacific ocean. *US. JGOFS News,* **9,** 1-5.

Feely, R.A., R. Wanninkhof, T. Takahashi, and P. Tans, 1999b: Influence of El Niño on the equatorial Pacific contribution to atmospheric CO_2 accumulation. *Nature,* **398,** 597-601.

Fearnside, P.M., 2000: Global warming and tropical land-use change: Greenhouse gas emissions from biomass burning, decomposition and soils in forest conversion, shifting cultivation and secondary vegetation. *Climatic Change,* **46,** 115-158.

Field, C.B., R.B. Jackson, and H.A. Mooney, 1995a: Stomatal responses to increased CO_2: implications from the plant to the global scale. *Plant Cell and Environment,* **18,** 1214-1225.

Field, C.B., J.T. Randerson, and C.M. Malmstrom, 1995b: Global net primary production - combining ecology and remote- sensing. *Remote Sensing of Environment,* **51,** 74-88.

Field, C.B., M.J. Behrenfeld, J.T. Randerson, and P. Falkowski, 1998: Primary production of the biosphere: Integrating terrestrial and oceanic components. *Science,* **281,** 237-240.

Fischer, H., M. Whalen, J. Smith, D. Mastroianni, and B. Deck, 1999: Ice core records of atmospheric CO_2 around the last three glacial terminations. *Science,* **283,** 1712-1714.

Fisher, M.J., I.M. Rao, M.A. Ayarza, C.E. Lascano, J.I. Sanz, R.J. Thomas, and R.R. Vera, 1994: Carbon storage by introduced deep-rooted grasses in the South- American Savannas. *Nature,* **371,** 236-238.

Fog, K., 1988: The effect of added nitrogen on the rate of decomposition of organic -matter. *Biological Reviews of the Cambridge Philosophical Society,* **63,** 433-462.

Foley, J.A., 1994: The sensitivity of the terrestrial biosphere to climatic change: A simulation of the middle Holocene. *Global Biogeochemical Cycles,* **8,** 505-525.

Foley, J.A., I.C. Prentice, N. Ramankutty, S. Levis, D. Pollard, S. Sitch, and A. Haxeltine, 1996: An integrated biosphere model of land surface processes, terrestrial carbon balance, and vegetation dynamics. *Global Biogeochemical Cycles,* **10,** 603-628.

Foley, N.A., S. Levis, and I.C. Prentice, 1998: Coupling dynamic models of climate and vegetation. *Global Change Biology,* **4,** 561-579.

Fowler, D., J.N. Cape, M. Coyle, C. Flechard, J. Kuylenstierna, K. Hicks, D. Derwent, C. Johnson, and D. Stevenson, 2000: The global exposure of forests to air pollutants. *Water Air and Soil Pollution,* **116,** 5-32.

François, R., M.A. Altabet, E.F. Yu, D.M. Sigman, M.P. Bacon, M. Frank, G. Bohrmann, G. Bareille and L.D. Labeyrie, 1997: Contribution of Southern Ocean surface-water stratification to low atmospheric CO_2 concentrations during the last glacial period. *Nature,* **389,** 929-935.

Francey, R.J., and P.P. Tans, 1987: Latitudinal variation in O^{18} of atmospheric CO_2. *Nature,* **327,** 495-497.

Francey, R.J., P.P. Tans, and C.E. Allison, 1995: Changes in oceanic and terrestrial carbon uptake since 1982. *Nature,* **373,** 326-330.

Francey, R., P. Rayner, R. Langenfelds, and C. Trudinger, 1999a: The inversion of atmospheric CO_2 mixing ratios and isotopic composition to constrain large-scale air-sea fluxes. In: *2nd international symposium CO_2 in the oceans,* Center for Global Environmental Research, National Institute for Environmental Study, Tsukuba, Japan, 237-243.

Francey, R.J., C.E. Allison, D.M. Etheridge, C.M. Trudinger, I.G. Enting, M. Leuenberger, R.L. Langenfelds, E. Michel, and L.P. Steele, 1999b: A 1000-year high precision record of delta ^{13}C in atmospheric CO_2. *Tellus Series B-Chemical and Physical Meteorology,* **51,** 170-193.

Frankignoulle, M., G. Abril, A. Borges, I. Bourge, C. Canon, B. Delille, E. Libert, and J.-M. Théate, 1998: Carbon dioxide emission from European estuaries. *Science,* **282,** 434-436.

Friedlingstein, P., I. Fung, E. Holland, J. John, G. Brasseur, D. Erickson and D. Schimel, 1995a: On the contribution of CO_2 fertilization to the missing biospheric sink. *Global Biogeochemical Cycles,* **9,** 541-556.

Friedlingstein, P., K.C. Prentice, I.Y. Fung, J.G. John, and G.P. Brasseur, 1995b: Carbon-biosphere-climate interactions in the last glacial maximum climate. *Journal of Geophysical Research - Atmosphere,* **100,** 7203-7221.

Friedlingstein, P., L. Bopp, P. Ciais, J.-L. Dufresne, L. Fairhead, H. LeTreut, P. Monfray, and J. Orr, 2001: Positive feedback between future climate change and the carbon cycle. Note du Pole de Modelisation, *Geophys. Res. Lett.,* **28,** 1543-1546.

Friend, A.D., A.K. Stevens, R.G. Knox, and M.G.R. Cannell, 1997: A process-based, terrestrial biosphere model of ecosystem dynamics (Hybrid v3.0). *Ecological Modelling,* **95,** 249-287.

Fuglestvedt, J.S., and T. Berntsen, 1999: A simple model for scenario studies of changes in global climate. *Technical Report # 1999:2,* Center for International Climate and Environmental Research, Oslo.

Fung, I.Y., S.K. Meyn, I. Tegen, S.C. Doney, J. John, and J.K.B. Bishop, 2000: Iron supply and demand in the upper ocean. *Global Biogeochemical Cycles,* **14,** 281-295.

Gattuso, J.-P., M. Frankignoulle and R. Wollast, 1998: Carbon and carbonate metabolism in coastal aquatic ecosystems. *Annual Review of Ecology and Systematics,* **29,** 405-434.

Ganeshram, R.S., T.F. Pedersen, S.E. Calvert, and J.W. Murray, 1995: Large changes in oceanic nutrient inventories from glacial to interglacial periods. *Nature,* **376,** 755-758.

Gérard, J.C., B. Nemry, L.M. Francois and P. Warnant, 1999: The interannual change of atmospheric CO_2: contribution of subtropical ecosystems? *Geophys. Res. Lett.,* **26,** 243-246.

Giardina, C.P. and M.G. Ryan, 2000: Evidence that decomposition rates of organic carbon in mineral soil do not vary with temperature. *Nature,* **404,** 858-861.

Gifford, R.M., 1992: Implications of the globally increasing atmospheric CO_2 concentration and temperature for the Australian terrestrial carbon budget - integration using a simple-model. *Australian Journal of Botany,* **40,** 527-543.

Gorham, E., 1991: Northern peatlands - role in the carbon-cycle and probable responses to climatic warming. *Ecological Applications,* **1,** 182-195.

Goulden, M.L., J.W. Munger, S.M. Fan, B.C. Daube, and S.C. Wofsy, 1996: Exchange of carbon dioxide by a deciduous forest: Response to interannual climate variability. *Science,* **271,** 1576-1578.

Goulden, M.L., S.C. Wofsy, J.W. Harden, S.E. Trumbore, P.M. Crill, S.T. Gower, T. Fries, B.C. Daube, S.M. Fan, D.J. Sutton, A. Bazzaz, and J.W. Munger, 1998: Sensitivity of boreal forest carbon balance to soil thaw. *Science,* **279,** 214-217.

Goyet, C., and D. Davis, 1997: Estimation of total CO_2 concentration throughout the water column. *Deep-Sea Research Part I-Oceanographic Research Papers,* **44,** 859-877.

Gruber, N., 1998: Anthropogenic CO_2 in the Atlantic Ocean. *Global Biogeochemical Cycles,* **12,** 165-191.

Gruber, N., and J. L. Sarmiento, 1997: Global patterns of marine nitrogen fixation and denitrification. *Global Biogeochemical Cycles,* **11,** 235-266.

Gruber, N. and C. D. Keeling, 2001: An improved estimate of the isotopic air-sea disequilibrium of CO_2: Implications for the oceanic uptake of anthropogenic CO_2, GRL, vol. 28, pg 555-558.

Gruber, N., J.L. Sarmiento, and T.F. Stocker, 1996: An improved method for detecting anthropogenic CO_2 in the oceans. *Global Biogeochemical Cycles,* **10,** 809-837.

Gruber, N., C.D. Keeling, R.B. Bacastow, P.R. Guenther, T.J. Lueker, M. Wahlen, H.A.J. Meijer, W.G. Mook, and T.F. Stocker, 1999: Spatiotemporal patterns of carbon-13 in the global surface oceans and the oceanic Suess effect. *Global Biogeochemical Cycles,* **13,** 307-335.

Gunderson, P., B.A. Emmett, O.J. Kjønaas, C.J. Koopmans, A. Tietema, 1998: Impact of nitrogen deposition on nitrogen cycling in forests: a synthesis of NITREX data. *Forest Ecology and Management,* **101,** 37-55

Hall, D.O., J.M.O. Scurlock, H.R. Bolhar-Nordenkampfe, P.C. Leegood, and S.P. Long, 1993: Photosynthesis and Production in a Changing Environment: A Field and Laboratory Manual, (eds.)]. Chapman & Hall, London, 464pp.

Hall, D.O., J. House, and I. Scrase, 2000: An overview of biomass energy. In: *Industrial Uses of Biomass Energy: the Example of Brazil* [Rosillo-Calle, F., H. Rothman and S.V. Bajay (eds.)]. Taylor & Francis, London.

Hansell, D.A., and C.A. Carlson, 1998: Deep-ocean gradients in the concentration of dissolved organic carbon. *Nature,* **395,** 263-266.

Hansell, D.A., and R.A. Feely, 2000: Atmospheric intertropical convergence impacts surface ocean carbon and nitrogen biogeochemistry in the western tropical Pacific. *Geophys. Res. Lett.,* **27,** 1013-1016.

Harrison, K.G., 2000: Role of increased marine silica input on paleo-pCO_2 levels. *Paleoceanography,* **15,** 292-298.

Harrison, S.P., K.E. Kohfeld, C. Roelandt, and T. Claquin, 2001: The role of dust in climate changes today, at the last glacial maximum and in the future. *Earth Science Reviews* (in press).

Harvey, L.D.D., 1989: Effect of model structure on the response of terrestrial biosphere models to CO_2 and temperature increases. *Global Biogeochemical Cycles,* **3,** 137-153.

Harvey, L. D. D., Gregory, J., Hoffert, M., Jain, A., Lal, M., Leemans, R., Raper, S. C. B., Wigley, T. M. L., and de Wolde, J. R., 1997: *An introduction to simple climate models used in the IPCC Second Assessment Report. IPCC Technical Paper II,* [Houghton, J.T., L.G.

Meira Filho, D.J. Griggs and K. Maskell (eds)]. IPCC, Geneva, Switzerland. 50 pp.

Haxeltine, A., and I.C. Prentice, 1996: BIOME3: An equilibrium terrestrial biosphere model based on ecophysiological constraints, resource availability, and competition among plant functional types. *Global Biogeochemical Cycles,* **10,** 693 - 709.

Hays, J.D., J. Imbrie, and N.J. Shackleton, 1976: Variations in Earths Orbit - Pacemaker of Ice Ages. *Science,* **194,** 1121-1132.

Heimann, M., 2001: Atmospheric Inversion Calculations Performed for IPCC Third Assesment Report Chapter 3 (The Carbon Cycle and Atmospheric CO_2), Technical Reports - Max-Planck-Institute für Biogeochemie No. 2.

Heimann, M. and Maier-Reimer, E., 1996: On the relations between the oceanic uptake of CO_2 and its carbon isotopes. *Global Biogeochemical Cycles,* **10,** 89-110.

Heimann, M., and T. Kaminski, 1999: Inverse modelling approaches to infer surface trace gas fluxes from observed atmospheric mixing ratios. In: *Approaches to scaling a trace gas fluxes in ecosystems* [Bowman A.F. (ed.)]. Elsevier Science, pp. 277-295.

Heimann, M., C.D. Keeling, and I.Y. Fung, 1986: Simulating the atmospheric carbon dioxide distribution with a three dimensional tracer model. In: *The Changing Carbon Cycle: A Global Analysis* [J.R. Trabalka and D. E. Reichle (eds.)]. Springer-Verlag, New York.

Heimann, M., G. Esser, J. Kaduk, D. Kicklighter, G. Kohlmaier, D. McGuire, B. Morre III, C. Prentice, W. Sauf, A. Schloss, U. Wittenberg, and G. Würth, 1997: Interannual variability of CO_2 exchange fluxes as simulated by four terrestrial biogeochemical models. In: *Fifth International Carbon Dioxide Conference*, Cairns, Queensland, Australia.

Heimann, M., G. Esser, A. Haxeltine, J. Kaduk, D.W. Kicklighter, W. Knorr, G.H. Kohlmaier, A.D. McGuire, J. Melillo, B. Moore, R.D. Otto, I.C. Prentice, W. Sauf, A. Schloss, S. Sitch, U. Wittenberg, and G. Wurth, 1998: Evaluation of Terrestrial Carbon Cycle Models Through Simulations of the Seasonal Cycle of Atmospheric CO_2 - First Results of a Model Intercomparison Study. *Global Biogeochemical Cycles,* **12,** 1-24.

Hewitt, C.D., and J.F.B. Mitchell, 1997: Radiative forcing and response of a GCM to ice age boundary conditions: cloud feedback and climate sensitivity. *Climate Dynamics,* **13,** 821-834.

Hoffert, M.I., A.J. Callegari, and C.T. Hsieh, 1980: The role of deep-sea heat-storage in the secular response to climatic forcing. *Journal of Geophysical Research-Oceans and Atmospheres,* **85,** 6667-6679.

Holfort, J., K.M. Johnson, B. Schneider, G. Siedler, and D.W.R. Wallace, 1998: Meridional transport of dissolved inorganic carbon in the South Atlantic Ocean. *Global Biogeochemical Cycles,* **12,** 479-499.

Holland, E.A., B.H. Braswell, J.F. Lamarque, A. Townsend, J. Sulzman, J.F. Muller, F. Dentener, G. Brasseur, H. Levy, J.E. Penner, and G.J. Roelofs, 1997: Variations in the predicted spatial distribution of atmospheric nitrogen depostion and their impact on carbon uptake by terrestrial ecosystems. *Journal of Geophysical Research - Atmosphere,* **102,** 15849-15866.

Holland, E.A., F.J. Dentener, B.H. Braswell, and J.M. Sulzman, 1999: Contemporary and pre-industrial global reactive nitrogen budgets. *Biogeochemistry,* **46,** 7-43.

Hooss, G., R. Voss, K. Hasselmann, E. Meier-Reimer, and F. Joss, 1999: A nonlinear impulse response model of the coupled carbon cycle-ocean-atmosphere climate system. Max-Planck-Institut für Meteorologie, Hamburg.

Houghton, R.A., 1995: Effects of land-use change, surface temperature, and CO_2 concentration on terrestrial stores of carbon. In: *Biotic Feedbacks in teh Global Climate System* [Woodwell, G.M. (ed.)]. Oxford University Press, London, pp. 333-350.

Houghton, R.A., 1999: The annual net flux of carbon to the atmosphere from cahgnes in land use 1850-1990. *Tellus,* **51B,** 298-313.

Houghton, R.A., 2000: A new estimate of global sources and sinks of

carbon from land-use change. *EOS,* **81,** supplement s281.

Houghton, R.A., and J.L. Hackler, 1995: Continental scale estimates of the biotic carbon flux from land cover change: 1850-1980. ORNL/CDIAC-79, NDP-050. Oak Ridge National Laboratory, Oak Ridge, Tennessee, pp. 144.

Houghton, R.A., and J.L. Hackler, 1999: Emissions of carbon from forestry and land-use change in tropical Asia. *Global Change Biology,* **5,** 481-492.

Houghton, R.A., and K. Ramakrishna, 1999: A review of national emissions inventories from select non-annex I countries: Implications for counting sources and sinks of carbon. *Annual Review of Energy Environment,* **24,** 571-605.

Houghton, R.A., J.L. Hackler, and K.T. Lawrence, 1999: The US carbon budget: Contributions from land-use change. *Science,* **285,** 574-578.

Houghton, R.A., D.L. Skole, C.A. Nobre, J.L. Hackler, K.T. Lawerence and W.H. Chomentowski, 2000: Annual fluxes of carbon from deforestation and regrowth in the Brazilian Amazon. *Nature,* **403,** 301-304.

House, J.I., and D.O. Hall, 2001: Net primary production of savannas and tropical grasslands, In: *Terrestrial Global Productivity: Past, Present and Future* [Mooney, H., J. Roy and B. Saugier (eds.)], Academic Press, San Diego (in press).

Hungate, B.A., E.A. Holland, R.B. Jackson, F.S. Chapin, H.A. Mooney, and C.B. Field, 1997: The fate of carbon in grasslands under carbon dioxide enrichment. *Nature,* **388,** 576-579.

Hungate, B.A., P. Dijkstra, D.W. Johnson, C.R. Hinkle and B.G. Drake, 1999: Elevated CO_2 increases nitrogen fixation and decreases soil nitrogen mineralization in Florida scrub oak. *Global Change Biology,* **5,** 781-789.

Hunt, E.R., S.C. Piper, R. Nemani, C.D. Keeling, R.D. Otto, and S.W. Running, 1996: Global net carbon exchange and intra-annual atmospheric CO_2 concentrations predicted by an ecosystem process model and three dimensional atmospheric transport model. *Global Biogeochemical Cycles,* **10,** 431-456.

Huntingford, C., P.M. Cox, and T.M. Lenton, 2000. Contrasting responses of a simple terrestrial ecosystem model to global change. *Ecological Modelling,* **134**(1), 41-58.

Indermühle, A., T.F. Stocker, F. Joss, H. Fischer, H.J. Smith, M. Wahlen, B. Deck, D. Mastroianni, J. Tschumi, T. Blunier, R. Meyer, and B. Stauffer, 1999: Holocene carbon-cycle dynamics based on CO_2 trapped in ice at Taylor Dome, Antarctica. *Nature,* **398,** 121-126.

Indermühle, A., E. Monnin B Stauffer, T. Stocker, and M. Wahlen, 2000. Atmospheric CO_2 concentration from 60 to 20 kyr BP from Taylor Dome ice core, Antarctica, Geophys. Res. Lett. **29,** 753-758.

Inoue, H.Y., and Y. Sugimura, 1992: Variations and distributions of CO_2 in and over the equatorial pacific during the period from the 1986/88 El-Niño event to the 1988/89 La-Niña event. *Tellus Series B,* **44,** 1-22.

Ishitani, H., T.B. Johansson, and S. Al–Khouli, 1996: Energy Supply Mitigation Options. In: *Climate Change 1995. Impacts, Adaptations and Mitigation of Climate Change: Scientific-Technical Analyses. Contribution of Working Group II to the Second Assessment Report of the Intergovernmental Panel on Climate Change.* [Watson, R.T.. M.C. Zinyowera and R.H. Moss (eds.)]. Cambridge University Press, Cambridge, pp. 587-648.

IPCC, 1994: Radiative Forcing of Climate Change and An Evaluation of the IPCC IS92 Emissions Scenarios [Houghton, J.T., L.G. Meira Filho, J. Bruce, Hoesung Lee, B.A. Callander, E. Haites, N. Harris and K. Maskell (eds)]. Cambridge University Press, Cambridge, pp. 339.

IPCC, 1996a: Climate Change 1995. The Science of Climate Change. Contribution of Working Group I to the Second Assessment Report of the Intergovernmental Panel on Climate Change. [Houghton, J.T. , L.G. Meira Filho, B.A. Callander, N. Harris, A. Kattenberg and K. Maskell (eds.)]. Cambridge University Press, Cambridge, UK.

IPCC, 1996b: Climate Change 1995. Impacts, Adaptations and Mitigation of Climate Change: Scientific-Technical Analyses. Contribution of Working Group II to the Second Assessment Report of the Intergovernmental Panel on Climate Change. [Watson, R.T., M.C. Zinyowera, R.H. Moss and D.J. Dokken (eds.)]. Cambridge University Press, Cambridge, UK

IPCC, 1997: Revised 1996 IPCC Guidelines for National Greenhouse Gas Inventories. Reference Manual, Intergovernmental Panel on Climate Change. [Houghton, J.T., L.G. Meira Filho, B. Lim, K. Treanton, I. Mamaty, Y. Bonduki, D.J. Griggs and B.A. Callender (eds.)]. Cambridge University Press, Cambridge, UK.

IPCC, 2000a: Land Use, Land-Use Change, and Forestry. A Special Report of the IPCC. [Watson, R.T., I.R. Noble, B. Bolin, N.H. Ravindranath, D.J. Verardo and D.J. Dokken (eds.)]. Cambridge University Press, Cambridge, UK.

IPCC, 2000b: Special Report on Emissions Scenarios. A Special Report of Working Group III of the intergovernmental Panel on Climate Change. [Nakićenović, N., J. Alcamo, G. Davis, B. de Vries, J. Fenhann, S. Gaffin, K. Gregory, A. Grübler, T. Yong Jung, T. Kram, E.L. La Rovere, L. Michaelis, S. Mori, T. Morita, W. Pepper, H. Pitcher, L. Price, K. Riahi, A. Roehrl, H.-H. Rogner, A. Sankovski, M. Schlesinger, P. Shukla, S. Smith, R. Swart, S. van Rooijen, N. Victor and Z. Dadi (eds.)]. Cambridge University Press, Cambridge, UK.

Ito A. and T. Oikawa, 2000: The large carbon emission from terrestrial ecosystem in 1998: A model simulation, *Journal of the Meterological Society of Japan*, **78,** 103-110

Jain A.K., 2000: The Web Interface of Integrated Science Assessment Model (ISAM). http://frodo.atmos.uiuc.edu/isam.

Jain, A.K., H.S. Kheshgi, and D.J. Wuebbles, 1994: Integrated Science Model for Assessment of Climate Change. Lawrence Livermore National Laboratory, UCRL-JC-116526.

Jain, A.K., H.S. Kheshgi, M.I. Hoffert, and D.J. Wuebbles, 1995: Distribution of radiocarbon as a test of global carbon-cycle models. *Global Biogeochemical Cycles,* **9,** 153-166.

Jarvis, P. and S. Linder, 2000: Botany - Constraints to growth of boreal forests. *Nature,* **405,** 904-905.

Jenkinson, D.S., K. Goulding, and D.S. Powlson, 1999: Nitrogen deposition and carbon sequestration. *Nature,* **400,** 629-629.

Joos, F., M. Bruno, R. Fink, U. Siegenthaler, T.F. Stocker, and C. Le Quéré, 1996: An efficient and accurate representation of complex oceanic and biospheric models of anthropogenic carbon uptake. *Tellus Series B,* **48,** 397-417.

Joos, F., R. Meyer, M. Bruno, and M. Leuenberger, 1999a: The variability in the carbon sinks as reconstructed for the last 1000 years. *Geophys. Res. Lett.,* **26,** 1437-1441.

Joos, F., G.-K. Plattner, T.F. Stocker, O. Marchal, and A. Schmittner, 1999b: Global warming and marine carbon cycle feedbacks on future atmospheric CO_2. *Science,* **284,** 464-467.

Jones, P.D., Parker, D.E., Osborn, T.J., and Briffa, K.R., 2000: Global and hemispheric temperature anomalies - land and marine instrumental records. In: *Trends: A Compendium of Data on Global Change,* Oak Ridge National Laboratory, Oak Ridge, Tenn., U.S.A.

Kaminski, T., R. Giering, and M. Heimann, 1996: Sensitivity of the seasonal cycle of CO_2 at remote monitoring stations with respect to seasonal surface exchange fluxes determined with the adjoint of an atmospheric transport model. *Physics and Chemistry of the Earth,* **21,** 457-462.

Kaminski, T., M. Heimann, and R. Giering, 1999: A coarse grid three-dimensional global inverse model of the atmospheric transport - 1. Adjoint model and Jacobian matrix. *Journal of Geophysical Research-Atmospheres,* **104,** 18535-18553.

Karl, D., R. Letelier, L. Tupas, J. Dore, J. Christian, and D. Hebel, 1997: The role of nitrogen fixation in biogeochemical cycling in the subtropical North Pacific Ocean. *Nature,* **388,** 533-538.

Keeling, C.D., and R. Revelle, 1985: Effects of El-Niño southern oscillation on the atmospheric content of carbon-dioxide. *Meteoritics,* **20,** 437-450.

Keeling, C.D., and T.P. Whorf, 2000: Atmospheric CO_2 records from sites in the SIO air sampling network. In: *Trends: A compendium of data on global change.* Carbon Dioxide Information Analysis Center, Oak Ridge National Laboratory, Oak Ridge, Tenn., USA.

Keeling, C.D., and Piper S.C., 2000: Interannual variations of exchanges of atmospheric CO_2 and $^{13}CO_2$ with the terrestrial biosphere and oceans from 1978 to 2000: III. Simulated sources and sinks. Scripps Institution of Oceanography Reference No. 00-14, University of California, San Diego, 68 pp.

Keeling, C.D., W.G. Mook, and P.P. Tans, 1979: Recent trends in the ^{13}C-^{12}C ratio of atmospheric carbon-dioxide. *Nature,* **277,** 121-123.

Keeling, C.D., R.B. Bacastow, and P.P. Tans, 1980: Predicted shift in the ^{13}C-^{12}C ratio of atmospheric carbon-dioxide. *Geophys. Res. Lett.,* **7,** 505-508.

Keeling, C.D., R.B. Bacastow, A.l. Carter, S.C. Piper, T.P. Whorf, M. Heimann, W.G. Mook, and H. Roeloffzen, 1989: A three dimensional model of atmospheric CO_2 transport based on observed winds: 1. Analysis of observational data. In: *Aspects of Climate Variability in the Pacific and the Western Americas* [Peterson D.H. (ed.)]. American Geophysical Union, Washington, DC, pp. 165-236.

Keeling, C.D., T.P. Whorf, M. Wahlen, and J. Vanderplicht, 1995: Interannual extremes in the rate of rise of atmospheric carbon dioxide since 1980. *Nature,* **375,** 666-670.

Keeling, C.D., J.F.S. Chin, and T.P. Whorf, 1996a: Increased activity of northern vegetation inferred from atmospheric CO_2 measurements. *Nature,* **382,** 146-149.

Keeling, R.F., 1988: Measuring correlations between atmospheric oxygen and carbon-dioxide mole fractions - a preliminary-study in urban air. *Journal Of Atmospheric Chemistry,* **7,** 153-176.

Keeling, R.F., and S.R. Shertz, 1992: Seasonal and interannual variations in atmospheric oxygen and implications for the global carbon cycle. *Nature,* **358,** 723-727.

Keeling, R.F., and T.H. Peng, 1995: Transport of Heat, CO_2 and O_2 by the Atlantic thermohaline circulation. *Philosophical Transactions of the Royal Society of London Series B,* **348,** 133-142.

Keeling, R.F., R.P. Najjar, and M.L. Bender, 1993: What atmospheric oxygen measurements can tell us about the global carbon-cycle. *Global Biogeochemical Cycles,* **7,** 37-67.

Keeling, R.F., S.C. Piper, and M. Heimann, 1996b: Global and hemispheric CO_2 sinks deduced from changes in atmospheric O_2 concentration. *Nature,* **381,** 218-221.

Kheshgi, H.S., A.K. Jain, and D.J. Wuebbles, 1996: Accounting for the missing carbon sink with the CO_2 fertilization effect. *Climatic Change,* **33,** 31-62.

Kicklighter, D.W., M. Bruno, S. Dönges, G. Esser, M. Heimann, J. Helfrich, F. Ift, F. Joos, J. Kadku, G.H. Kohlmaier, A.D. McGuire, J.M. Melillo, R. Meyer, B. Moore III, A. Nadler, I.C. Prentice, W. Sauf, A.L. Schloss, S. Sitch, U. Wittenberg, and G. Würth, 1999: A first order analysis of the potential of CO_2 fertilization to affect the global carbon budget: A comparison of four terrestrial biosphere models. *Tellus,* **51B,** 343-366.

Kindermann, J., G. Wurth, G.H. Kohlmaier, and F.W. Badeck, 1996: Interannual variation of carbon exchange fluxes in terrestrial ecosystems. *Global Biogeochemical Cycles,* **10,** 737-755.

Klepper, O., and B.J. De Haan, 1995: A sensitivity study of the effect of global change on ocean carbon uptake. *Tellus Series B,* **47,** 490-500.

Kleypas, J.A., 1997: Modeled estimates of global reef habitat and carbonate production since the last glacial maximum. *Paleoceanography,* **12,** 533-545.

Kleypas, J.A., R.W. Buddemeier, D. Archer, J.P. Gattuso, C. Langdon, and B.N. Opdyke, 1999: Geochemical consequences of increased atmospheric carbon dioxide on coral reefs. *Science,* **284,** 118-120.

Knorr, W., 2000: Annual and interannual CO_2 exchanges of the terrestrial biosphere: process-based simulations and uncertainties. *Global Ecology and Biogeography,* **9,** 225-252.

Knorr, W., and M. Heimann, 1995: Impact of drought stress and other factors on seasonal land biosphere CO_2 exchange studied through an atmospheric tracer transport model. *Tellus,* **47B,** 471-489.

Knorr, W. and M. Heimann, 2001a: Uncertainties in global terrestrial biosphere modeling, Part I: a comprehensive sensitivity analysis with a new photosynthesis and energy balance scheme. *Global Biogeochemical Cycles,* **15(1),** 207-225.

Knorr, W., and M. Heimann, 2001b: Uncertainties in global terrestrial biosphere modeling, Part II: global constraints for a process-based vegetation model. *Global Biogeochemical Cycles,* **15(1),** 227-246.

Knox, F. and M.B. McElroy, 1984: Changes in atmospheric CO_2: influence of the marine biota at high latitude. *Journal of Geophysical Research,* **89,** 4629-4637.

Koch, G.W., and H.A. Mooney, 1996: Response of terrestrial ecosystems to elevated CO_2: a synthesis and summary. In: *Carbon Dioxide and Terrestrial Ecosystems* [Koch, G.W. and H.A. Mooney (eds.)]. Academic Press, San Diego, pp. 415-429.

Körner, C., 2000: Biosphere responses to CO_2-enrichment. *Ecological Applications, 10, 1590-1619.*

Kucharik, C.J., J.A. Foley, C. Delire, V.A. Fisher, M.T. Coe, J.D. Lenters, C. Young-Molling, N. Ramankutty, J.M. Norman and S.T. Gower, 2000: Testing the performance of a Dynamic Global Ecosystem Model: Water balance, carbon balance, and vegetation structure. *Global Biogeochemical Cycles,* **14,** 795-825.

Kuhlbusch, T.A.J., M.O. Andeae, H. Cahier, J.G. Goldammer, J.P. Lacaux, R. Shea, and P.J. Crutzen, 1996: Black carbon formation by savanna fires: Measurements and implications for the global carbon cycle. *Journal of Geophysical Research,* **101,** 23651-23665.

Kumar, N., R.F. Anderson, R.A. Mortlock, P.N. Froelich, P. Kubik, B. Dittrich-Hannen, and M. Suter, 1995: Increased biological productivity and export production in the glacial Southern Ocean. *Nature,* **378,** 675-680.

Langenfelds, R.L., R.J. Francey, and L.P. Steele, 1999: Partitioning of the global fossil CO_2 sink using a 19-year trend in atmospheric O_2. *Geophys. Res. Lett.* **26,** 1897-1900.

Langdon, C., T. Takahashi, C. Sweeney, D. Chipman, J. Goddard, F. Marubini, H. Aceves, H. Barnett and M. Atkinson, 2000: Effect of calcium carbonate saturation state on the calcification rate of an experimental coral reef. *Global Biogeochemical Cycles,* **14,** 639-654.

Law, R.M., P.J. Rayner, A.S. Denning, D. Erickson, I.Y. Fung, M. Heimann, S.C. Piper, M. Ramonet, S. Taguchi, J.A. Taylor, C.M. Trudinger, and I.G. Watterson, 1996: Variations in modeled atmospheric transport of carbon dioxide and the consequences for CO_2 inversions. *Global Biogeochemical Cycles,* **10,** 783-796.

Laws, E.A., P.G. Falkowski, W.O. Smith Jr., H. Ducklow, and J.J. McCarthy, 2000: Temperature effects on export production in the open ocean. *Global Biogeochemical Cycles,* 14(4), 1231-1246.

Le Quéré, C., J. C. Orr, P. Monfray, O. Aumont, and G. Madec, 2000: Interannual variability of the oceanic sink of CO_2 from 1979 through 1997. *Global Biogeochemical Cycles, 14(4), 1247-1265.*

Leemans, R., 1990: Global data sets collected and compiled by the Biosphere Project, Working Paper, International Institute for Applied Analysis (IIASA), Laxenburg, Austria.

Leemans, R., A. vanAmstel, C. Battjes, E. Kreileman, and S. Toet, 1996: The land cover and carbon cycle consequences of large-scale utilizations of biomass as an energy source. *Global Environmental Change-Human and Policy Dimensions,* **6,** 335-357.

Lefèvre, N., and A.J. Watson, 1999: Modeling the geochemical cycle of iron in the oceans and its impact on atmospheric CO_2 concentrations. *Global Biogeochemical Cycles,* **13,** 727-736.

Leggett, J.A., W.J. Pepper, and R.J. Swart, 1992: Emissions scenarios for the IPCC: an update. In: *Climate Change 1992. The Supplementary Report to the IPCC Scientific assessment.* [Houghton, J.T., B.A. Callander and S.K. Varney (eds.)]. Cambridge University Press, Cambridge, pp. 69-95.

Levitus, S., J.I. Antonov, T.P. Boyer, and C. Stephens, 2000: Warming of the world ocean. *Science,* **287,** 2225-2229.

Liss, P.S., and L. Merlivat, 1986: Air-sea gas exchange: Introduction and synthesis. In: *The Role of Air-Sea Exchange in Geochemical Cycling* [Buat-Ménard, P. (ed.)]. D. Reidel, Hingham, Massachusetts, 113-127.

Lloyd, J., 1999: The CO_2 dependence of photosynthesis, plant growth responses to elevated CO_2 concentrations and their interaction with soil nutrient status, II. Temperate and boreal forest productivity and the combined effects of increasing CO_2 concentrations and increased nitrogen deposition at a global scale. *Functional Ecology,* **13,** 439-459.

Lloyd, J., and G.D. Farquhar, 1994: ^{13}C discrimination during CO_2 assimilation by the terrestrial biosphere. *Oecologia,* **99,** 201-215.

Lloyd, J., and J.A. Taylor, 1994: On the temperature-dependence of soil respiration. *Functinal Ecolology,* **8,** 315-323.

Lloyd, J., and G.D. Farquhar, 1996: The CO_2 dependence of photosynthesis, plant growth responses to elevated atmospheric CO_2 concentrations, and their interaction with soil nutrient status. I. General principles and forest ecosystems. *Functional Ecology,* **10,** 4-32.

Longhurst, A., S. Sathyendranath, T. Platt, and C. Caverhill, 1995: An Estimate of global primary production in the ocean from satellite radiometer data. *Journal of Plankton Research,* **17,** 1245-1271.

Lorius, C., and H. Oeschger, 1994: Palaeo-perspectives - reducing uncertainties in global change. *Ambio,* **23,** 30-36.

Lugo, A.E. and S. Brown, 1993: Management of tropical soils as sinks or sources of atmospheric carbon. *Plant and Soil,* **149,** 27-41.

Luo, Y.Q., J. Reynolds, and Y.P. Wang, 1999: A search for predictive understanding of plant responses to elevated $[CO_2]$. *Global Change Biology,* **5,** 143-156.

Mahowald, N., K.E. Kohfeld, M. Hansson, Y. Balkanski, S.P. Harrison, I.C. Prentice, H. Rodhe, and M. Schulz, 1999: Dust sources and deposition during the Last Glacial Maximum and current climate: a comparison of model results with paleodata from ice cores and marine sediments. *Journal of Geophysical Research,* **104,** 15,895-16,436.

Maier-Reimer, E., 1993: Geochemical cycles in an ocean general-circulation model - preindustrial tracer distributions. *Global Biogeochemical Cycles,* **7,** 645-677.

Maier-Reimer, E., and K. Hasselmann, 1987: Transport and storage of CO_2 in the ocean - An inorganic ocean-circulation carbon cycle model. *Climate Dynamics,* **2,** 63-90.

Maier-Reimer, E., U. Mikolajewicz, and A. Winguth, 1996: Future ocean uptake of CO_2 - Interaction between ocean circulation and biology. *Climate Dynamics,* **12,** 711-721.

Malhi, Y., A.D. Nobre, J. Grace, B. Kruijt, M.G.P. Pereira, A. Culf, S. Scott, 1998: Carbon dioxide transfer over a Central Amazonian rain forest. *Journal of Geophysical Research-Atmospheres,* **103,** 31593-31612.

Manning, A.C., 2001. *Temporal variability of atmospheric oxygen from both continuous measurements and a flask sampling network: Tools for studying the global carbon cycle.* Ph. D. thesis, University of California, San Diego, La Jolla, California, U.S.A., 190 pp.

Manning, M.R., 1992: Seasonal cycles in atmospheric CO_2 concentrations. In: *The Global Carbon Cycle: Proceedings of the NATO.* [M. Heimann, (ed.)]. Advanced Study Institute I15, Il Ciocco, September 8-20, 1991, Springer Verlag, Berlin.

Marchal, O., T.F. Stocker, and F. Joos, 1998: A latitude-depth, circulation biogeochemical ocean model for paleoclimate studies - development and sensitivities. *Tellus,* **50B,** 290-316.

Marland, G., R.J. Andres, and T.A. Boden, 1994: Global, regional, and national CO_2 emissions. In: *Trends 93: A Compendium of Data on*

Global Change [T. A. Boden, D. P. Kaiser, R. J. Sepanski, and F. W. Stoss (eds.)], Oak Ridge National Laboratory, ORNL/CDIAC-65, Oak Ridge, Tenn., USA, 505-584.

Marland, G., R.J. Andres, T.A. Boden, C. Hohnston, and A. Brenkert, 1999: Global, Regional, and National CO_2 Emission Estimates from Fossil Fuel Burning, Cement Production, and Gas Flaring: 1751-1996. Carbon Dioxide Information Analysis Center, Oak Ridge National Laboratory, Oak Ridge, Tenn., USA.

Marland, G., T.A. Boden, and R.J. Andres, 2000: Global, regional, and national CO_2 emissions. In: *Trends: A Compendium of Data on Global Change*. Carbon Dioxide Information Analysis Center, Oak Ridge National Laboratory, U. S. Department of Energy, Oak Ridge, Tenn., USA.

Martin, J., 1990: Glacial-interglacial CO_2 change: the iron hypothesis. *Paleoceanography,* **5**, 1-13.

Martin, J.H., S.E. Fitzwater, and R.M. Gordon, 1990: Iron deficiency limits phytoplankton growth in Antarctic waters. *Global Biogeochemical Cycles,* **4**, 5-12.

Masle, J. 2000: The effects of elevated CO_2 concentrations on cell division rates, growth patterns, and blade anatomy in young wheat plants are modulated by factors related to leaf position, vernalization, and genotype. *Plant Physiology,* **122**, 1399-1415.

Matear, R.J., and A.C. Hirst, 1999: Climate change feedback on the future oceanic CO_2 uptake. *Tellus Series B,* **51**, 722-733.

Matson, P.A., W.H. McDowell, A.R. Townsend and P.M. Vitousek, 1999: The globalization of N deposition: ecosystem consequences in tropical environments. *Biogeochemistry,* **46**, 67-83.

Matthews, E., 1983: Global vegetation and land-use - new high-resolution data-bases for climate studies. *Journal of Climate and Applied Meteorology,* **22**, 474-487.

McElroy, M.B., 1983: Marine biological-controls on atmospheric CO_2 and climate. *Nature,* **302**, 328-329.

McGillicuddy, D.J., and A.R. Robinson, 1997: Eddy-induced nutrient supply and new production in the Sargasso Sea. *Deep-Sea Research Part I-Oceanographic Research Papers,* **44**, 1427-1450.

McGuire, A.D., J.M. Melillo, and L.A. Joyce, 1995: The role of nitrogen in the response of forest net primary production to elevated atmospheric carbon-dioxide. *Annual Review of Ecology and Systematics,* **26**, 473-503.

McGuire, A.D., S. Sitch, J.S. Clein, R. Dargaville, G. Esser, J. Foley, M. Heimann, F. Joos, J. Kaplan, D.W. Kicklighter, R.A. Meier, J.M. Melillo, B. Moore III, I.C. Prentice, N. Ramankutty, T. Reichenau, A. Schloss, H. Tian, L.J. Williams, U. Wittenberg, 2001 Carbon balance of the terrestrial biosphere in the twentieth century: Analyses of CO_2, climate and land-use effects with four process-based ecosystem models. *Global Biogeochemical Cycles,* **15(1)**, 183-206.

Mclaughlin, S., and Downing, D.J., 1995: Interactive effects of ambient ozone and climate measured on growth of mature forest trees. *Nature,* **374**, 252-254.

Mclaughlin, S., and Percy, K., 2000: Forest health in North America: some perspectives on actual and potential roles of climate and air pollution. *Water Air and Soil Pollution,* **116**, 151-197.

Melillo, J.M., and J.R. Gosz, 1983: Interactions of biogeochemical cycles in forest ecosystems. In: *The major biogeochemical cycles and their interactions* [Bolin, B. and R.B. Cook (eds.)]. John Wiley and Sons, New York, pp. 177-222.

Melillo, J.M., A.D. McGuire, D.W. Kicklighter, B.M. III, C.J. Vörösmarty and A.L. Schloss, 1993: Global climate change and terrestrial net primary production. *Nature,* **363**, 234-240.

Melillo, J.M., I.C. Prentice, G.D. Farquhar, E.-D. Schulze, and O.E. Sala, 1996: Terrestrial biotic response to environmental change and feedbacks to climate, In:*Climate Change 1995. The Science of Climate Change.*, [J.T. Houghton, L.G.M. Filho, B.A. Callander, N. Harris, A. Kattenberg and K. Maskell (eds.)]. University Press, Cambridge, pp. 449-481.

Menzel, A., and P. Fabian, 1999: Growing season extended in Europe. *Nature,* **397**, 659.

Meybeck, M., 1982: Carbon, nitrogen, and phosphorus transport by world rivers. *American Journal of Science,* **282**, 401-450.

Meybeck, M., 1993: Riverine transport of atmospheric carbon – sources, global typology and budget. *Water, Air and Soil Pollution,* **70**, 443-463

Milliman, J.D., 1993: Production and accumulation of calcium-carbonate in the ocean - budget of a nonsteady state. *Global Biogeochemical Cycles,* **7**, 927-957.

Minkkinen, K., and J. Laine, 1998: Long-term effect of forest drainage on the peat carbon stores of pine mires in Finland. *Canadian Journal of Forest Research-Revue Canadienne De Recherche Forestiere,* **28**, 1267-1275.

Mitchell, J.F.B., T.C. Johns, J.M. Gregory, and S.F.B. Tett, 1995: Climate response to increasing levels of greenhouse gases and sulphate aerosols. *Nature,* **376**, 501-504.

Mook, W.G., M. Koopmans, A.F. Carter, and C.D. Keeling, 1983: Seasonal, latitudinal, and secular variations in the abundance and isotopic-ratios of atmospheric carbon-dioxide. 1. Results from land stations. *Journal of Geophysical Research-Oceans and Atmospheres,* **88**, 915-933.

Mooney, H.A., J. Canadell, F.S. Chapin III, J.R. Ehleringer, Ch. Körner, R.E. McMurtrie, W.J. Parton, L.F. Pitelka, and E.D. Schulze, 1999: Ecosystem physiology responses to global change. In: *Implications of Global Change for Natural and Managed Ecosystems. A Synthesis of GCTE and Related Research. IGBP Book Series No. 4,* [Walker, B.H., W.L. Steffen, J. Canadell and J.S.I. Ingram (eds.)]. Cambridge University Press, Cambridge, pp. 141-189.

Mooney, H., J. Roy and B. Saugier (eds.) 2001. *Terrestrial Global Productivity: Past, Present and Future,* Academic Press, San Diego (in press).

Morita, T., J. Robinson, A. Adegbulugbe, J. Alcamo, E. Lebre La Rovere, N. Nakićenović, H. Pitcher, P. Raskin, V. Sokolov, B. de Vries, D. Zhou, 2001. Greenhouse gas emission mitigation scenarios and implications. In: *Climate Change: Mitigation. Contribution of Working Group III to the Third Assessment Report of the Intergovernmental Panel on Climate Change* [O. Davidson, B. Metz and R. Swart (eds)]. Cambridge University Press, Cambridge, United Kingdom and New York, NY, USA, in press.

Murnane, R.J., J.L. Sarmiento, and C. Le Quéré, 1999: Spatial distribution of air-sea CO_2 fluxes and the interhemispheric transport of carbon by the oceans. *Global Biogeochemical Cycles,* **13**, 287-305.

Myneni, R.B., C.D. Keeling, C.J. Tucker, G. Asrar, and R.R. Nemani, 1997: Increased plant growth in the northern high latitudes from 1981 to 1991. *Nature,* **386**, 698-702.

Nadelhoffer, K.J., B.A. Emmett, and P. Gundersen, 1999: Nitrogen deposition makes a minor contribution to carbon sequestration in temperate forests. *Nature,* **398**, 145-148.

Neftel, A., H. Friedli, E. Moor, H. Lötscher, H. Oeschger, U. Siegenthaler, and B. Stauffer, 1994: Historical CO_2 record from the Siple station ice core. In: *Trends '93: A Compendium of Data on Global Change.* [T.A. Boden, D.P. Kaiser, R.J. Sepanski, and F.W. Stoss (eds.)], Carbon Dioxide Inf. Anal. Cent., Oak Ridge., pp. 11-14.

Neilson, R.P., and D. Marks, 1994: A global perspective of regional vegetation and hydrologic sensitivities from climatic change. *Journal of Vegetation Science,* **5**, 715-730.

Nemry, B., L. Francois, J.C. Gerard, A. Bondeau, and M. Heimann, 1999: Comparing global models of terrestrial net primary productivity (NPP): analysis of the seasonal atmospheric CO_2 signal. *Global Change Biology,* **5**, 65-76.

Norby, R.J. and M.F. Cotrufo, 1988: Global change - A question of litter quality. *Nature,* **396**, 17-18.

Norby, R.J., S.D. Wullschleger, C.A. Gunderson, D.W. Johnson, and R. Ceulemans, 1999: Tree responses to rising CO_2 in field experiments:

implications for the future forest. *Plant Cell and Environment,* **22,** 683-714.

Oechel, W.C., S.J. Hastings, G. Vourlitis, M. Jenkins, G. Riechers, and N. Grulke, 1993: Recent change of arctic tundra ecosystems from a net carbon- dioxide sink to a source. *Nature,* **361,** 520-523.

Oechel, W.C., G.L. Vourlitis, S.J. Hastings, R.C. Zulueta, L. Hinzman, and D. Kane, 2000: Acclimation of ecosystem CO_2 exchange in the Alaskan Arctic in response to decadal climate warming. *Nature,* **406,** 978-981.

Ollinger, S.V., J.S. Aber, and P.B. Reich, 1997: Simulating ozone effects on forest productivity: interactions among leaf-, canopy-, and stand-level processes. *Ecological Applications,* **7,** 1237-1251.

Olson, J.S., J.A. Watts, and L.J. Allison, 1983: Carbon in Live Vegetation of Major World Ecosystems. Oak Ridge, Tennessee:Oak Ridge National Laboratory. ORNL-5862.

Ono, T., S. Watanabe, K. Okuda, and M. Fukasawa, 1998: Distribution of total carbonate and related properties in the North Pacific along 30 degrees N. *Journal of Geophysical Research-Oceans,* **103,** 30873-30883.

Opdyke, B.N., and J.C.G. Walker, 1992: Return of the coral-reef hypothesis - basin to shelf partitioning of $CaCO_3$ and Its Effect On Atmospheric CO_2. *Geology,* **20,** 733-736.

Orr, J.C., and J.-C. Dutay, 1999: OCMIP mid-project workshop. *Research GAIM Newsletter,* **3,** 4-5.

Orr, J., E. Maier-Reimer, U. Mikolajewicz, P. Monfray, J. L. Sarmiento, J. R. Toggweiler, N. K. Taylor, J. Palmer, N. Gruber, C. L. Sabine, C. Le Quéré, R. M. Key, and J. Boutin, 2001: Estimates of anthropogenic carbon uptake from four 3-D global ocean models. *Global Biogeochemical Cycles,* **15,** 43-60.

Oschlies, A., and V. Garçon, 1998: Eddy-induced enhancement of primary production in a model of the north Atlantic Ocean. *Nature,* **394,** 266-269.

Overpeck, J., K. Hughen, D. Hardy, R. Bradley, R. Case, M. Douglas, B. Finney, K. Gajewski, G. Jacoby, A. Jennings, S. Lamoureux, A. Lasca, G. MacDonald, J. Moore, M. Retelle, S. Smith, A. Wolfe, and G. Zielinski, 1997: Arctic environmental change of the last four centuries. *Science,* **278,** 1251-1256.

Owensby, C.E., P.I. Coyne, and J.M. Ham, 1993: Biomass production in a tallgrass prairie ecosystem exposed to ambient and elevated CO_2. *Ecological Applications,* **3,** 644-653.

Pagani, M., M.A. Arthur and K.H. Freeman, 1999a: Miocene evolution of atmospheric carbon dioxide. *Paleoceanography,* **14,** 273-292.

Pagani, M., K. Freeman, and M.A. Arthur, 1999b: Late Miocene atmospheric CO_2 concentrations and the expansion of C_4 grasses. *Science,* **285,** 876-879.

Pahlow, M., and U. Riebesell, 2000: Temporal trends in deep ocean Redfield ratios. *Science,* **287,** 831-833.

Pearson, P.N. and M.R. Palmer, 1999: Middle eocene seawater pH and atmospheric carbon dioxide concentrations. *Science,* **284,** 1824-1826.

Pearson, P.N. and M.R. Palmer, 2000: Atmospheric carbon dioxide concentrations over the past 60 million years. *Nature,* **406,** 695-699.

Peng, C.H., J. Guiot, and E.V. Campo, 1998: Past and future carbon balance of European ecosystems from pollen data and climatic models simulations. *Global Planetary Change,* **18,** 189-200.

Petit, J.R., J. Jouzel, D. Raynaud, N.I. Barkov, J.M. Barnola, I. Basile, M. Bender, J. Chappellaz, M. Davis, G. Delaygue, M. Delmotte, V.M. Kotlyakov, M. Legrand, V.Y. Lipenkov, C. Lorius, L. Pepin, C. Ritz, E. Saltzman, and M. Stievenard, 1999: Climate and atmospheric history of the past 420,000 years from the Vostok ice core, Antarctica. *Nature,* **399,** 429-436.

Peybernès, N., E. Michel, J.-M. Barnola, M. Delmotte, J. Chappellaz, and D. Raynaud, 2000: Information on carbon cycle during the last 8,000 years deduced from CO_2, $\delta^{13}CO_2$ and CH_4 profiles obtained on a Vostok Ice core. Presented at: *EGS, Nice, France, 25-26 April 2000.*

Phillips, O.L., Y. Malhi, N. Higuchi, W.F. Laurance, P.V. Nunez, R.M. Vasquez, S.G. Laurance, L.V. Ferreira, M. Stern, S. Brown and J. Grace, 1998: Changes in the carbon balance of tropical forests: Evidence from long-term plots. *Science,* **282,** 439-442.

Polley, H.W., H.B. Johnson, H.S. Mayeux, and C.R. Tischler, 1996: Are some of the recent changes in grassland communities a response to rising CO_2 concentrations? In: *Carbon Dioxide, Populations, and Communities* [Körner, C. and F.A. Bazzaz (eds.)]. Academic Press, San Diego, pp. 177-195.

Prentice, I.C., M. Heimann, and S. Sitch, 2000: The carbon balance of the terrestrial biosphere: Ecosystem models and atmospheric observations. *Ecological Applications,* **10,** 1553-1573.

Pritchard, S.G., H.H. Rogers, S.A. Prior, and C.M. Peterson, 1999: Elevated CO_2 and plant structure: a review. *Global Change Biology,* **5,** 807-837.

Poole, I., J.D.B. Weyers, T. Lawson and J.A. Raven, 1996: Variations in stomatal density and index: Implications for palaeoclimatic reconstructions. *Plant Cell and Environment,* **19,** 705-712.

Quay, P.D., B. Tilbrook, C.S. Won, 1992: Oceanic uptake of fossil fuel CO_2: Carbon-13 evidence. *Science,* **256,** 74-79.

Raich, J.W., E.B. Rastetter, J.M. Melillo, D.W. Kicklighter, P.A. Steudler, B.J. Peterson, A.L. Grace, B. Moore and C.J. Vorosmarty, 1991: Potential net primary productivity in South-America - application of a global-model. *Ecological Applications,* **1,** 399-429.

Raich, J.W., and W.H. Schlesinger, 1992: The global carbon-dioxide flux in soil respiration and its relationship to vegetation and climate. *Tellus Series B-Chemical and Physical Meteorology,* **44,** 81-99

Ramankutty, N., and J.A. Foley, 2000: Estimating historical changes in global land cover: Croplands from 1700 to 1992. *Global Biogeochemical Cycles,* **13,** 997-1027.

Randerson, J.T., C.B. Field, I.Y. Fung, and P.P. Tans, 1999: Increases in early season ecosystem uptake explain recent changes in the seasonal cycle of atmospheric CO_2 at high northern latitudes. *Geophys. Res. Lett.,* **26,** 2765-2768.

Rayner, P.J., I.G. Enting, R.J. Francey, and R. Langenfelds, 1999a: Reconstructing the recent carbon cycle from atmospheric CO_2, $\delta^{13}C$ and O_2/N_2 observations. *Tellus,* **51B,** 213-232.

Rayner, P.J., R.M. Law, and R. Dargaville, 1999b: The relationship between tropical CO_2 fluxes and the El Niño - Southern Oscillation. *Geophys. Res. Lett.,* **26,** 493-496.

Redfield, A.C., B.H. Ketchum, and F.A. Richards, 1963: The influence of organisms on the composition of seawater. In: *The Sea* [Hill, M.N. (ed.)]. Wiley, New York, NY, pp. 26-77.

Rickaby, R.E.M., and H. Elderfield, 1999: Planktonic foraminiferal Cd/Ca: Paleonutrients or paleotemperature? *Paleoceanography,* **14,** 293-303.

Riebesell, U., D.A. Wolf-Gladrow, and V. Smetacek, 1993: Carbon-dioxide limitation of marine-phytoplankton growth-rates. *Nature,* **361,** 249-251.

Riebesell, U., I. Zondervan, B. Rost, P.D. Tortell, R.E. Zeebe, and F.M. Morel, 2000: Reduced calcification of marine phytoplankton in response to increased atmospheric CO_2. *Nature,* **407,** 364-367.

Robitaille, D.Y., and A.J. Weaver, 1995: Validation of sub-grid-scale mixing schemes using CFCs in a global ocean model. *Geophys. Res. Lett.,* **22,** 2917-2920.

Ruimy, A., B. Saugier, and G. Dedieu, 1994: Methodology for the estimation of terrestrial net primary production from remotely sensed data. *Journal of Geophysical Research-Atmosphere,* **99,** 5263-5283.

Rundgren, M., and D. Beerling, 1999: A Holocene CO_2 record from the stomatal index of subfossil Salix herbacea L. leaves from northern Sweden. *Holocene,* **9,** 509-513.

Sabine, C.L., D.W.R. Wallace, and F.J. Millero, 1997: Survey of CO_2 in the Oceans reveals clues about global carbon cycle. *EOS, Transaction of the American Geophysical Union,* **78,** 54-55.

Sabine, C.L., R.M. Key, K.M. Johnson, F.J. Millero, A. Poisson, J.L.

Sarmiento, D.W.R. Wallace, and C.D. Winn, 1999: Anthropogenic CO_2 inventory of the Indian Ocean. *Global Biogeochemical Cycles,* **13,** 179-198.

Sampson, R.N. and R.J. Scholes, 2000: Additional Human-Induced Activities-Article 3.4. In: *Land Use, Land-Use Change, and Forestry* [Watson, R.T., I.R. Noble, B. Bolin, N.H. Ravindranath, D.J. Verardo and D.J. Dokken (eds.)]. Cambridge University Press, Cambridge, UK, pp. 181-282.

Sanyal, A., N.G. Hemming, G.N. Hanson, and W.S. Broecker, 1995: Evidence for a higher pH in the glacial ocean from boron isotopes in foraminifera. *Nature,* **373,** 234-236.

Sarmiento, J.L., and J.R. Toggweiler, 1984: A new model for the role of the oceans in determining atmospheric pCO_2. *Nature,* **308,** 621-624.

SAR, see IPCC (1996a).

Sarmiento, J.L., J.C. Orr, and U. Siegenthaler, 1992: A perturbation simulation of CO_2 uptake in an ocean general- circulation model. *Journal of Geophysical Research-Oceans,* **97,** 3621-3645.

Sarmiento, J.L., and E.T. Sundquist, 1992: Revised budget for the oceanic uptake of anthropogenic carbon- dioxide. *Nature,* **356,** 589-593.

Sarmiento, J.L., and C. Le Quéré, 1996: Oceanic carbon dioxide uptake in a model of century-scale global warming. *Science,* **274,** 1346-1350.

Sarmiento, J.L., T.M.C. Hughes, R.J. Stouffer, and S. Manabe, 1998: Simulated response of the ocean carbon cycle to anthropogenic climate warming. *Nature,* **393,** 245-249.

Sarmiento, J.L., P. Monfray, E. Maier-Reimer, O. Aumont, R. Murnane, and J. C. Orr, 2000: Sea-air CO_2 fluxes and carbon transport: a comparison of three ocean general circulation models. *Global Biogeochemical Cycles, 14(4), 1267-1281.*

Saugier, B., and J. Roy, 2001: Estimations of global terrestrial productivity: converging towards a single number? In: *Global terrestrial productivity: past, present and future* [Roy, J., B. Saugier and H.A. Mooney (eds.)]. Academic Press, (in press).

Saxe, H., D.S. Ellsworth, and J. Heath, 1998: Tree and forest functioning in an enriched CO_2 atmosphere. *New Phytologist,* **139,** 395-436.

Schimel, J.P., and F.S. Chapin, 1996: Tundra plant uptake of amino acid and $NH4^+$ nitrogen in situ: Plants compete well for amino acid N. *Ecology,* **77,** 2142-2147.

Schimel, D.S., B.H. Braswell, E.A. Holland, R. McKeown, D.S. Ojima, T.H. Painter, W.J. Parton, and A.R. Townsend, 1994: Climatic, edaphic and biotic controls over carbon and turnover of carbon in soils. *Global Biogeochemical Cycles,* **8,** 279-293.

Schimel, D., I.G. Enting, M. Heimann, T.M.L. Wigley, D. Raynaud, D. Alves, and U. Siegenthaler, 1995: CO_2 and the carbon cycle. In: *Climate Change 1994: Radiative Forcing of Climate Change and an Evaluation of the IPCC IS92 Emission Scenarios* [Houghton, J.T., L.G. Meira Filho, J. Bruce, Hoesung Lee, B.A. Callander, E. Haites, E. Harris, K. Maskell (eds.)]. Cambridge University Press, Cambridge, pp. 35-71.

Schimel, D., D. Alves, I. Enting, M. Heimann, F. Joos, D. Raynaud, and T. Wigley, 1996: CO_2 and the carbon cycle. In: *Climate Change 1995: The Science of Climate Change: Contribution of WGI to the Second Assessment Report of the IPCC* [Houghton, J.T., L.G. Meira Filho, B.A. Callander, N. Harris, A. Kattenberg and K. Maskell (eds.)]. Cambridge University Press, Cambridge, pp. 65-86.

Schimel, D., M. Grubb, F. Joos, R. Kaufmann, R. Moss, W. Ogana, R. Richels, and T.M.L. Wigley, 1997: *Stabilization of Atmospheric Greenhouse Gases: Physical, Biological and Socio-economic Implications — IPCC Technical Paper 3* [Houghton, J.T.,, L.G. Meira Filho, D.J. Griggs and K. Maskell (eds)]. Intergovernmental Panel on Climate Change, Bracknell, UK , 53 pp.

Schlamadinger, B. and T. Karjalainen, 2000: Afforestation, Reforestation, and Deforestation (ARD) Activites. In: *Land Use, Land-Use Change, and Forestry* [Watson, R.T., I.R. Noble, B. Bolin, N.H. Ravindranath, D.J. Verardo and D.J. Dokken (eds.)]. Cambridge

University Press, Cambridge, UK, pp. 127-181.

Schlesinger, W.H., 1990: Evidence from chronosequence studies for a low carbon-storage potential of soils. *Nature,* **348,** 233-234.

Schlesinger, W.H., 2000: Carbon sequestration in soils. *Science,* **284,** 2095.

Schlesinger, W.H. and J.M. Melack, 1981: Transport of organic-carbon in the worlds rivers. *Tellus,* **33,** 172-187

Schlesinger, W.H., and J.A. Andrews, 2000: Soil respiration and the global carbon cycle. *Biogeochemistry,* **48,** 7-20.

Schlitzer, R. 2000: Applying the adjoint method for biogeochemical modeling: export of particulate organic matter in the world ocean. In: *Inverse methods in global biogeochemical cycles* [Kasibhatla, P., M. Heimann, P. Rayner, N. Mahowald, R.G. Prinn, and D.E. Hartley (eds)], Geophysical Monograph Series, **114,** 107-124.

Scholes, R.J., and S.R. Archer, 1997: Tree-grass interactions in savannas. *Annual Review of Ecology and Systematics,* **28,** 517-544.

Scholes, M., and M.O. Andreae, 2000: Biogenic and pyrogenic emissions from Africa and their impact on the global atmosphere. *Ambio,* **29,** 23-29.

Scholes, R.J., E.D. Schulze, L.F. Pitelka, and D.O. Hall, 1999: Biogeochemistry of terrestrial ecosystems. In: *The terrestrial biosphere and global change* [Walker B., Steffen W., Canadell J. and Ingram J. (eds.)]. Cambridge University Press., Cambridge, pp. 271-303.

Schulze, E.D., 2000: Carbon and Nitrogen Cycling in European Forest Ecosystems. Ecological Studies Vol. 142., Springer Verlag.

Schulze, E.D., and M. Heimann, 1998: Carbon and water exchange of terrestrial ecosystems. In: *Asian change in the context of global change* [Galloway, J.N. and J. Melillo (eds.)]. Cambridge University Press, Cambridge.

Schulze, E.D., W. Devries, and M. Hauhs, 1989: Critical loads for nitrogen deposition on forest ecosystems. *Water Air and Soil Pollution,* **48,** 451-456.

Schulze, E.D., J. Lloyd, F.M. Kelliher, C. Wirth, C. Rebmann, B. Luhker, M. Mund, A. Knohl, I.M. Milyukova, W. Schulze, W. Ziegler, A.B. Varlagin, A.F. Sogachev, R. Valentini, S. Dore, S. Grigoriev, O. Kolle, M.I. Panfyorov, N. Tchebakova, and N.N. Vygodskaya, 1999: Productivity of forests in the Eurosiberian boreal region and their potential to act as a carbon sink - a synthesis. *Global Change Biology,* **5,** 703-722.

Semenov, S.M., I.M. Kounina, and B.A. Koukhta, 1998: An ecological analysis of anthropogenic changes in ground–level concentrations of O_3, SO_2, and CO_2 in Europe. *Doklady Biological Sciences,* **361,** 344-347.

Semenov, S.M., I.M. Kounina, and B.A. Koukhta, 1999: Tropospheric ozone and plant growth in Europe., Publishing Centre, 'Meteorology and Hydrology', Moscow, 208 pp.

Shackleton, N.J., 1977: Carbon-13 in Uvigerina: Tropical rainforest history and the equatorial Pacific carbonate dissolution cycles. In: *The Fate of Fossil Fuel CO_2 in the Oceans* [Anderson, N.R. and A. Malahoff (eds.)], 401-427.

Shackleton, N.J., 2000: The 100,000-year ice-age cycle identified and found to lag temperatrue, carb on dioxide, and orbital eccentricity. *Science,* **289,** 1897-1902.

Shaffer, G., 1993: Effects of the marine carbon biota on global carbon cycling. In: *The Global Carbon Cycle* [Heimann, M. (ed.)]. **I 15,** Springer Verlag, Berlin, pp. 431-455.

Siegenthaler, U., 1990: Biogeochemical Cycles - El-Niño and Atmospheric CO_2. *Nature,* **345,** 295-296.

Siegenthaler, U. and H. Oeschger, 1978: Predicting future atmospheric carbon-dioxide levels. *Science,* **199,** 388-395

Siegenthaler, U., and T. Wenk, 1984: Rapid atmospheric CO_2 variations and ocean circulation. *Nature,* **308,** 624-627.

Siegenthaler, U. and F. Joos, 1992: Use of a simple-model for studying oceanic tracer distributions and the global carbon-cycle. *Tellus Series*

B-Chemical and Physical Meteorology, **44,** 186-207.

Siegenthaler, U., H. Friedli, H. Loetscher, E. Moor, A. Neftel, H. Oeschger and B. Stauffer, 1988: Stable-isotope ratios and concentration of CO_2 in air from polar ice cores. *Annals of Glaciology, 10,* 1-6.

Sigman, D.M., D.C. McCorkle, and W.R. Martin, 1998: The calcite lysocline as a constraint on glacial/interglacial low-latitude production changes. *Global Biogeochemical Cycles, 12,* 409-427.

Sitch, S., 2000: *The role of vegetation dynamics in the control of atmospheric CO_2 content.* PhD thesis, University of Lund, Sweden, 213 pp.

Six, K.D., and E. Maier-Reimer, 1996: Effects of plankton dynamics on seasonal carbon fluxes in an ocean general circulation model. *Global Biogeochemical Cycles, 10,* 559-583.

Skole, D. and C. Tucker, 1993: Tropical deforestation and habitat fragmentation in the Amazon - satellite data from 1978 to 1988. *Science, 260,* 1905-1910.

Smethie, W.M., T. Takahashi, and D.W. Chipman, 1985: Gas-exchange and CO_2 flux in the tropical atlantic-ocean determined from ^{222}Rn and pCO_2 measurements. *Journal of Geophysical Research-Oceans, 90,* 7005-7022.

Smith, H.J., H. Fischer, M. Wahlen, D. Mastroianni, and B. Deck, 1999: Dual modes of the carbon cycle since the Last Glacial Maximum. *Nature, 400,* 248-250.

Smith, S.V., and J.T. Hollibaugh, 1993: Coastal metabolism and the oceanic organic-carbon balance. *Reviews of Geophysics, 31,* 75-89.

Sonnerup, R. E., Quay P. D., McNichol A. P., Bullister J. L., Westby T. A., and Anderson H. L., 1999: Reconstructing the oceanic C-13 Suess effect. Global Biogeochemical Cycles, 13, 857-872

Spiecker, H., K. Mielikäinen, M. Köhl, J.P. Skovsgaard, 1996: Growth trends in European forests. *European Forest Institute Research Report No. 5,* Springer Verlag, Berlin-Heidelberg.

Stauffer, B., T. Blunier, A. Dällenbach, A. Indermühle, J. Schwander, T.F. Stocker, J. Tschumi, J. Chappellaz, D. Raynaud, C.U. Hammer and H.B. Clausen,1998: Atmospheric CO_2 concentration and millennial-scale climate change during the last glacial period. *Nature, 392,* 59-62.

Stephens, B.B., 1999: *Field-based atmospheric oxygen measurements and the ocean carbon cycle.* Unpublished Ph.D. thesis, Univ. of California, USA.

Stephens, B.B., and R.F. Keeling, 2000: The influence of Antarctic sea ice on glacial-interglacial CO_2 variations. *Nature, 404,* 171-174.

Stephens, B.B., R.F. Keeling, and M. Heimann, 1998: Testing global ocean carbon cycle models using measurements of atmospheric O_2 and CO_2 concentration. *Global Biogeochemical Cycles, 12,* 213-230.

Street-Perrott, F.A., Y. Huang, R.A. Perrott, and G. Eglinton, 1998: Carbon isotopes in lake sediments and peats of last glacial age: implications for the global carbon cycle, In: *Stable Isotopes* [Griffiths H. (ed.)]. BIOS Scientific Publishers Ltd, pp. 381-396.

Takahashi, T., J. Goddard, S. Sutherland, D.W. Chipman, and C. C. Breeze, 1986: Seasonal and geographic variability of carbon dioxide sink/source in the ocean areas, Lamont Doherty Geol. Obs., Palisades, New York, 66 pp.

Takahashi, T., R. H. Wanninkhof, R. A. Feely, R. F. Weiss, D. W. Chipman, N. Bates, J. Olafson, C. Sabine, and S.C. Sutherland, 1999: Net sea-air CO_2 flux over the global oceans: An improved estimate based on the sea-air pCO_2 difference. In: *Proceedings of the 2nd International Symposium CO_2 in the Oceans,* Center for Global Environmental Research, National Institute for Environmental Studies, Tsukuba, Japan, 9-15.

Tans, P.P., and D.W.R. Wallace, 1999: Carbon cycle research after Kyoto. *Tellus, 51B,* 562-571.

Tans, P.P., T.J. Conway, and T. Nakazawa, 1989: Latitudinal distribution of the sources and sinks of atmospheric carbon dioxide derived from surface observations and atmospheric transport model. *Journal of Geophysical Research, 94,* 5151-5172.

Tans, P.P., I.Y. Fung, and T. Takahashi, 1990: Observational constraints on the global atmospheric CO_2 budget. *Science, 247,* 1431-1438.

Tans, P.P., J.A. Berry, and R.F. Keeling, 1993: Oceanic $^{13}C/^{12}C$ observations - a new window on ocean CO_2 uptake. *Global Biogeochemical Cycles, 7,* 353-368.

Taylor, J. A., 1989. "A stochastic Lagrangian atmospheric transport model to determine global CO_2 sources and sinks: a preliminary discussion." *Tellus B,* 41(3), 272-285.

Taylor, J.A., and J. Lloyd, 1992: Sources and sinks of atmospheric CO_2. *Australian Journal of Botany, 40,* 407-418.

Taylor, J.A., and J.C. Orr, 2000: The natural latitudinal distribution of atmospheric CO_2, *Global and Planetary Change, 26,* 375-386.

Tegen, I., and I. Fung, 1995: Contribution to the atmospheric mineral aerosol load from land surface modification. *Journal of Geophysical Research, 100,* 18707-18726.

Thompson, M.L., I.G. Enting, G.I. Pearman, and P. Hyson, 1986. Interannual variation of atmospheric CO_2 concentration. *Journal of Atmospheric Chemistry, 4,* 125-155.

Thompson, M.V., J.T. Randerson, C.M. Malmstrom, and C.B. Field, 1996: Change in net primary production and heterotrophic respiration: How much is necessary to sustain the terrestrial carbon sink? *Global Biogeochemical Cycles, 10,* 711-726.

Tian, H.Q., J.M. Melillo, and D.W. Kicklighter, 1998: Effect of interannual climate variability on carbon storage in Amazonian ecosystems. *Nature, 396,* 664-667.

Toggweiler, J.R., 1999: Variation of atmospheric CO_2 by ventilation of the ocean's deepest water. *Paleoceanography, 14,* 571-588.

Trolier, M., J.W.C. White, P.P. Tans, K.A. Masarie, and P.A. Gemery, 1996: Monitoring the isotopic composition of atmospheric CO_2 - measurements from the NOAA Global Air Sampling Network. *Journal of Geophysical Research - Atmosphere, 101,* 25897-25916.

Trudinger, C.M., I.G. Enting, R.J. Francey, D.M. Etheridge, and P.J. Rayner, 1999: Long-term variability in the global carbon cycle inferred from a high-precision CO_2 and delta ^{13}C ice-core record. *Tellus, 51B,* 233-248.

Trumbore, S., 2000: Age of soil organic matter and soil respiration: radiocarbon constraints on belowground C dynamics. *Ecological Applications, 10,* 399-411.

Tschumi, J., and B. Stauffer, 2000. Reconstructing past atmospheric CO_2 concentrations based on ice-core analyses: open questions due to in situ production of CO_2 in ice. *Journal of Glaciology, 46,* 45-53.

Tyrell, T., 1999: The relative influences of nitrogen and phosphorus on oceanic primary production. *Nature, 400,* 525-531.

UNFCCC (United Nations Framework Convention on Climate Change), 2000: Methodological Issue. Land-use, land-use change and forestry. Synthesis report on national greenhouse gas information reported by Annex I Parties for the land-use change and forestry sector and agricultural soils category. Note by the Secretariat. Subsidiary Body for Scientific and Technological Advice. FCCC/SBSTA/2000/3. Bonn, Germany. 48pp.

United Nations, 2000: The United Nations Energy Statistics Database, United Nations Statistical Division, New York.

UN-ECE/FAO (United Nations Economics Commision for Europe/Food and Agriculture Organization of the United Nations), 2000: Forest Resources of Europe, CIS, North America, Australia, Japan and New Zealand (industrialized temperate/boreal countries) UN-ECE/FAO Contribution to the Global Forest Resources Assessment 2000. In: *Geneva Timber and Forest Study Papers, No. 17,* United Nations, New York, Geneva, pp. 445.

Valentini, R., G. Matteucci, A.J. Dolman, E.D. Schulze, C. Rebmann, E.J. Moors, A. Granier, P. Gross, N.O. Jensen, K. Pilegaard, A. Lindroth, A. Grelle, C. Bernhofer, T. Grunwald, M. Aubinet, R. Ceulemans, A.S. Kowalski, T. Vesala, U. Rannik, P. Berbigier, D. Loustau, J. Guomundsson, H. Thorgeirsson, A. Ibrom, K. Morgenstern, R. Clement, J. Moncrieff, L. Montagnani, S. Minerbi and P.G. Jarvis,

2000: Respiration as the main determinant of carbon balance in European forests. *Nature,* **404,** 861-865.

Van de Water, P.K., S.W. Leavitt and J.L. Betancourt, 1994: Trends in stomatal density and $^{13}C/^{12}C$ ratios of pinus- flexilis needles during last glacial-interglacial cycle. *Science,* **264,** 239-243.

Vitousek, P.M. and R.W. Howarth, 1991: Nitrogen limitation on land and in the sea - how can it occur. *Biogeochemistry,* **13,** 87-115.

Vitousek, P.M. and C.B. Field, 1999: Ecosystem constraints to symbiotic nitrogen fixers: a simple model and its implications. *Biogeochemistry,* **46,** 179-202.

Vitousek, P.M., J.D. Aber, R.W. Howarth, G.E. Likens, P.A. Matson, D.W. Schindler, W.H. Schlesinger, D.G. Tilman, 1997: Human alteration of the global nitrogen cycle: sources and consequesnces. *Ecological Applications,* **7,** 737-750.

Voss, R., and U. Mikolajewicz, 1999: Long-term climate changes due to increased CO_2 concentration in the coupled atmosphere-ocean general circulation model ECHAM3/LSG. *Technical Report # 298,* Max-Planck-Institut für Meteorologie, Hamburg.

Wagner, F., S.J.P. Bohncke, D.L. Dilcher, W.M. Kurschner, B. van Geel, and H. Visscher, 1999: Century-scale shifts in early Holocene atmospheric CO_2 concentration. *Science,* **284,** 1971-1973.

Wallace, D.W.R., 1995: Monitoring global ocean carbon inventories. Ocean observing system development panel background., Texas A&M University, College Station, TX, 54pp.

Wand, S.J.E., G.F. Midgley, M.H. Jones, and P.S. Curtis, 1999: Responses of wild C_4 and C_3 grass (Poaceae) species to elevated atmospheric CO_2 concentration: a meta-analytic test of current theories and perceptions. *Global Change Biology,* **5,** 723-741.

Wanninkhof, R., 1992: Relationship between wind-speed and gas-exchange over the ocean. *Journal of Geophysical Research-Oceans,* **97,** 7373-7382.

Wanninkhof, R., and W.R. McGillis, 1999: A cubic relationship between air-sea CO_2 exchange and wind speed. *Geophys. Res. Lett.,* **26,** 1889-1892.

Waring, R.H., J.J. Landsberg, and M. Williams, 1998: Net primary production of forests: a constant fraction of gross primary production? *Tree Physiology,* **18,** 129-134.

Watson, A.J., and P.S. Liss, 1998: Marine biological controls on climate via the carbon and sulphur geochemical cycles. *Philosophical Transactions of the Royal Society of London Series B-Biological Sciences,* **353,** 41-51.

Watson, A.J., P.D. Nightingale, and D.J. Cooper, 1995: Modeling atmosphere ocean CO_2 transfer. *Philosophical Transactions of the Royal Society of London Series B-Biological Sciences,* **348,** 125-132.

WBGU, (Wissenschaftlicher Beirat der Bundesregierung Globale Umweltveränerungen) 1988. Die Anrechnung biolischer Quellen und Senken in Kyoto-Protokoll: Fortschritt oder Rückschlag für den globalen Umweltschutz Sondergutachten 1988. Bremerhaven, Germany, 76pp, (available in English).

Wigley, T.M.L., Richels, R., and Edmonds, J.A., 1996: Economic and environmental choices in the stabilization of CO_2 concentrations: choosing the "right" emissions pathway. *Nature,* **379,** 240-243.

Wigley, T.M.L., A. K. Jain, F. Joos, B. S. Nyenzi, and P.R. Shukla, 1997: *Implications of Proposed CO_2 Emissions Limitations. IPCC Technical Paper 4.* [Houghton, J.T., L.G. Meira Filho, D.J. Griggs and M. Noguer (eds)]. Intergovernmental Panel on Climate Change, Bracknell, UK, 41 pp.

Williams, S.N., S.J. Schaefer, M.L. Calvache, and D. Lopez, 1992: Global carbon dioxide emission to the atmosphere by volcanoes. *Geochimica et Cosmochimica Acta,* **56,** 1765-1770.

Winguth, A.M.E., M. Heimann, K.D. Kurz, E. Maier-Reimer, U. Mikolajewicz, and J. Segschneider, 1994: El-Niño-Southern oscillation related fluctuations of the marine carbon cycle. *Global Biogeochemical Cycles,* **8,** 39-63.

Winn, C.D., Y.H. Li, F.T. Mackenzie, and D.M. Karl, 1998: Rising surface ocean dissolved inorganic carbon at the Hawaii Ocean Time-series site. *Marine Chemistry,* **60,** 33-47.

Wofsy, S.C., Goulden, M.L., Munger, J.W., Fan, S.M., Bakwin, P.S., Daube, B.C., Bassow, S.L., Bazzaz, F.A., 1993: Net exchange of CO_2 in a mid-latitude forest. *Science,* **260,** 1314-1317.

Wong, C.S., Y.H. Chan, J.S. Page, G.E. Smith, and R.D. Bellegay, 1993: Changes in equatorial CO_2 flux and new production estimated from CO_2 and nutrient levels in Pacific surface waters during the 1986/87 El-Niño. *Tellus Series B,* **45,** 64-79.

Woodward, F.I., 1987: Stomatal Numbers Are Sensitive to Increases in CO_2 From Preindustrial Levels. *Nature,* **327,** 617-618.

Woodward, F.I., and F.A. Bazzaz, 1988: The Responses of Stomatal Density to CO_2 Partial-Pressure. *Journal of Experimental Botany,* **39,** 1771-1781.

Woodward, F.I., M.R. Lomas, and R.A. Betts, 1998: Vegetation-climate feedbacks in a greenhouse world. *Philosophical Transactions of the Royal Society of London B,* **353,** 29-38.

Wu, J., W. Sunda, E.A. Boyle and D.M. Karl, 2000: Phosphate depletion in the western North Atlantic Ocean. *Science,* **289,** 759-762.

Yang, X. and M.X. Wang, 2000: Monsoon ecosystems control on atmospheric CO_2 interannual variability: inferred from a significant positive correlation between year-to-year changes in land precipitation and atmospheric CO_2 growth rate. *Geophys. Res. Lett.,* **27,** 1671-1674

Yamanaka, Y., and E. Tajika, 1996: The role of the vertical fluxes of particulate organic matter and calcite in the oceanic carbon cycle: Studies using an ocean biogeochemical general circulation model. *Global Biogeochemical Cycles,* **10,** 361-382

Zeng, X.B., 1999: The relationship among precipitation, cloud-top temperature, and precipitable water over the tropics. *Journal of Climate,* **12,** 2503-2514.

4

Atmospheric Chemistry and Greenhouse Gases

Co-ordinating Lead Authors
M. Prather, D. Ehhalt

Lead Authors
F. Dentener, R. Derwent, E. Dlugokencky, E. Holland, I. Isaksen, J. Katima, V. Kirchhoff, P. Matson,
P. Midgley, M. Wang

Contributing Authors
T. Berntsen, I. Bey, G. Brasseur, L. Buja, W.J. Collins, J. Daniel, W.B. DeMore, N. Derek, R. Dickerson,
D. Etheridge, J. Feichter, P. Fraser, R. Friedl, J. Fuglestvedt, M. Gauss, L. Grenfell, A. Grübler, N. Harris,
D. Hauglustaine, L. Horowitz, C. Jackman, D. Jacob, L. Jaeglé, A. Jain, M. Kanakidou, S. Karlsdottir,
M. Ko, M. Kurylo, M. Lawrence, J.A. Logan, M. Manning, D. Mauzerall, J. McConnell, L. Mickley,
S. Montzka, J.F. Müller, J. Olivier, K. Pickering, G. Pitari, G.J. Roelofs, H. Rogers, B. Rognerud, S. Smith,
S. Solomon, J. Staehelin, P. Steele, D. Stevenson, J. Sundet, A. Thompson, M. van Weele,
R. von Kuhlmann, Y. Wang, D. Weisenstein, T. Wigley, O. Wild, D. Wuebbles, R. Yantosca

Review Editors
F. Joos, M. McFarland

Contents

Executive Summary

Two important new findings since the IPCC WGI Second Assessment Report (IPCC, 1996) (hereafter SAR) demonstrate the importance of atmospheric chemistry in controlling greenhouse gases:

Currently, tropospheric ozone (O_3) is the third most important greenhouse gas after carbon dioxide (CO_2) and methane (CH_4). It is a product of photochemistry, and its future abundance is controlled primarily by emissions of CH_4, carbon monoxide (CO), nitrogen oxides (NO_x), and volatile organic compounds (VOC). There is now greater confidence in the model assessment of the increase in tropospheric O_3 since the pre-industrial period, which amounts to 30% when globally averaged, as well as the response to future emissions. For scenarios in which the CH_4 abundance doubles and anthropogenic CO and NO_x emissions triple, the tropospheric O_3 abundance is predicted to increase by an additional 50% above today's abundance.

CO is identified as an important indirect greenhouse gas. An addition of CO to the atmosphere perturbs the OH-CH_4-O_3 chemistry. Model calculations indicate that the emission of 100 Mt of CO stimulates an atmospheric chemistry perturbation that is equivalent to direct emission of about 5 Mt of CH_4.

A major conclusion of this report is that atmospheric abundances of almost all greenhouse gases reached the highest values in their measurement records during the 1990s:

The atmospheric abundance of CH_4 continues to increase, from about 1,520 ppb in 1978 to 1,745 ppb in 1998. However, the observed annual increase in CH_4 has declined during the last two decades. This increase is highly variable; it was near zero in 1992 and as large as +13 ppb during 1998. There is no clear, quantitative explanation for this variability. Since the SAR, quantification of certain anthropogenic sources of CH_4, such as that from rice production, has improved.

The atmospheric burden of nitrous oxide (N_2O) continues to increase by about 0.25%/yr. New, higher estimates of emissions from agricultural sources improve our understanding of the global N_2O budget.

The atmospheric abundances of major greenhouse gases that deplete stratospheric ozone are decreasing (CFC-11, CFC-113, CH_3CCl_3, CCl_4), or increasing more slowly (CFC-12), in response to the phase-out in their production agreed to under the Montreal Protocol and its Amendments.

HFC-152a and HFC-134a are increasing in the atmosphere. This growth is consistent with the rise in their industrial use. HFC-23, an unintended by-product of HCFC-22 production, is also increasing.

Perfluorocarbon (PFC) e.g., CF_4 (perfluoromethane) appears to have a natural background; however, current anthropogenic emissions exceed natural ones by a factor of 1,000 or more and are responsible for the observed increase.

There is good agreement between the increase in atmospheric abundances of sulphur hexafluoride (SF_6) and emissions estimates based on revised sales and storage data.

There has been little increase in global tropospheric O_3 since the 1980s at the few remote locations where it is regularly measured. Only two of the fourteen stations, one in Japan and one in Europe, had statistically significant increases in tropospheric O_3 between 1980 and 1995. By contrast, the four Canadian stations, all at high latitudes, had significant decreases in tropospheric O_3 for the same time period. However, limited observations from the late 19th and early 20th centuries combined with models suggest that tropospheric O_3 has increased from a global mean value of 25 DU (where 1 DU = 2.7×10^{16} O_3 molecules/cm^2) in the pre-industrial era to 34 DU today. While the SAR estimated similar values, the new analysis provides more confidence in this increase of 9 DU.

Changes in atmospheric composition and chemistry over the past century have affected, and those projected into the future will affect, the lifetimes of many greenhouse gases and thus alter the climate forcing of anthropogenic emissions:

The atmospheric lifetime relates emissions of a component to its atmospheric burden. In some cases, for instance for methane, a change in emissions perturbs the chemistry and thus the corresponding lifetime. The CH_4 feedback effect amplifies the climate forcing of an addition of CH_4 to the current atmosphere by lengthening the perturbation lifetime relative to the global atmospheric lifetime of CH_4 by a factor of 1.4. This earlier finding is corroborated here by new model studies that also predict only small changes in this CH_4 feedback for the different scenarios projected to year 2100. Another feedback has been identified for the addition of N_2O to the atmosphere; it is associated with stratospheric O_3 chemistry and shortens the perturbation lifetime relative to the global atmospheric lifetime of N_2O by about 5%.

Several chemically reactive gases – CO, NO_x (=NO+NO_2), and VOC – control in part the abundance of O_3 and the oxidising capacity (OH) of the troposphere. These pollutants act as indirect greenhouse gases through their influence on atmospheric chemistry, e.g., formation of tropospheric O_3 or changing the lifetime of CH_4. The emissions of NO_x and CO are dominated by human activities. The abundance of CO in the Northern Hemisphere is about twice that in the Southern Hemisphere and has increased in the second half of the 20th century along with industrialisation and population. The urban and regional abundance of NO_x has generally increased with industrialisation, but the global abundance of this short-lived, highly variable pollutant cannot be derived from measurements. Increased NO_x abundances will in general increase tropospheric O_3 and decrease CH_4. Deposition of NO_x reaction products fertilises the biosphere, stimulates CO_2 uptake, but also provides an input of acidic precipitation.

The IPCC Special Report on Emission Scenarios (SRES) generated six marker/illustrative scenarios (labelled A1B, A1T, A1FI, A2, B1, B2) plus four preliminary marker scenarios (labelled here A1p, A2p, B1p, and B2p). These projected changes in anthropogenic emissions of trace gases from year 2000 to year 2100, making different assumptions on population development, energy use, and technology. Results from both sets of scenarios are discussed here since the preliminary marker scenarios (December 1998) were used in this report:

Model calculations of the abundances of the primary greenhouse gases by year 2100 vary considerably across the SRES scenarios: in general A1B, A1T, and B1 have the smallest increases of emissions and burdens; and A1FI and A2 the largest. CH_4 changes from 1998 to 2100 range from -10 to $+115\%$; and N_2O increases from 13 to 47%. The HFCs – 134a, 143a, and 125 – reach abundances of a few hundred to nearly a thousand ppt from negligible levels today. The PFC CF_4 is projected to increase to between 200 and 400 ppt; and SF_6 to between 35 and 65 ppt.

SRES projected anthropogenic emissions of the indirect greenhouse gases (NO_x, CO and VOC) together with changes in CH_4 are expected to change the global mean abundance of tropospheric OH by -20 to $+6\%$ over the next century. Comparable, but opposite sign, changes occur in the atmospheric lifetimes of the greenhouse gases, CH_4 and HFCs. This impact depends in large part on the magnitude of, and the balance between, NO_x and CO emissions.

For the SRES scenarios, changes in tropospheric O_3 between years 2000 and 2100 range from -4 to $+21$ DU. The largest increase predicted for the 21st century (scenarios A1FI and A2) would be more than twice as large as that experienced since the pre-industrial era. These O_3 increases are attributable to the concurrent, large (almost factor of 3) increases in anthropogenic NO_x and CH_4 emissions.

The large growth in emissions of greenhouse gases and other pollutants as projected in some SRES scenarios for the 21st century will degrade the global environment in ways beyond climate change:

Changes projected in the SRES A2 and A1FI scenarios would degrade air quality over much of the globe by increasing background levels of O_3. In northern mid-latitudes during summer, the zonal average increases near the surface are about 30 ppb or more, raising background levels to nearly 80 ppb, threatening attainment of air quality standards over most metropolitan and even rural regions, and compromising crop and forest productivity. This problem reaches across continental boundaries since emissions of NO_x influence photochemistry on a hemispheric scale.

A more complete and accurate assessment of the human impact on greenhouse gases requires greater understanding of sources, processes, and coupling between different parts of the climate system:

The current assessment is notably incomplete in calculating the total impact of individual industrial / agricultural sectors on greenhouse gases and aerosols. The IPCC Special Report on Aviation demonstrates that the total impact of a sector is not represented by (nor scalable to) the direct emissions of primary greenhouse gases alone, but needs to consider a wide range of atmospheric changes.

The ability to hindcast the detailed changes in atmospheric composition over the past decade, particularly the variability of tropospheric O_3 and CO, is limited by the availability of measurements and their integration with models and emissions data. Nevertheless, since the SAR there have been substantial advances in measurement techniques, field campaigns, laboratory studies, global networks, satellite observations, and coupled models that have improved the level of scientific understanding of this assessment. Better simulation of the past two decades, and in due course the upcoming one, would reduce uncertainty ranges and improve the confidence level of our projections of greenhouse gases.

Feedbacks between atmospheric chemistry, climate, and the biosphere were not developed to the stage that they could be included in the projected numbers here. Failure to include such coupling is likely to lead to systematic errors and may substantially alter the projected increases in the major greenhouse gases.

4.1 Introduction

This chapter investigates greenhouse gases whose atmospheric burdens[1] and climate impacts generally depend on atmospheric chemistry. These greenhouse gases include those listed in the Kyoto Protocol – methane (CH_4), nitrous oxide (N_2O), hydrofluorocarbons (HFCs), perfluorocarbons (PFCs), sulphur hexafluoride (SF_6) – and those listed under the Montreal Protocol and its Amendments – the chlorofluorocarbons (CFCs), the hydrochlorofluorocarbons (HCFCs), and the halons. A major focus of this assessment is the change in tropospheric ozone (O_3). Stratospheric water vapour (H_2O) is also treated here, but tropospheric H_2O, which is part of the hydrological cycle and calculated within climate models, is not discussed. This chapter also treats the reactive gases carbon monoxide (CO), volatile organic compounds (VOC), and nitrogen oxides (NO_x = $NO+NO_2$), termed indirect greenhouse gases. These pollutants are not significant direct greenhouse gases, but through atmospheric chemistry they control the abundances[1] of direct greenhouse gases. This chapter reviews the factors controlling the current atmospheric abundances of direct and indirect greenhouse gases; it looks at the changes since the pre-industrial era and their attribution to anthropogenic activities; and it calculates atmospheric abundances to the year 2100 based on projected emissions of greenhouse gases and pollutants. Carbon dioxide (CO_2) is treated in Chapter 3; and aerosols in Chapter 5. The atmospheric abundances of greenhouse gases and aerosols from all chapters are combined in Chapter 6 to calculate current and future radiative forcing. This chapter is an update of the IPCC WGI Second Assessment Report (IPCC, 1996) (hereafter SAR). For a review of the chemical processes controlling the abundance of greenhouse gases see the SAR (Prather *et al.*, 1995) or Ehhalt (1999). More recent assessments of changing atmospheric chemistry and composition include the IPCC Special Report on Aviation and the Global Atmosphere (Penner *et al.*, 1999) and the World Meteorological Organization / United Nations Environmental Programme (WMO/UNEP) Scientific Assessment of Ozone Depletion (WMO, 1999).

4.1.1 Sources of Greenhouse Gases

Substantial, pre-industrial abundances for CH_4 and N_2O are found in the tiny bubbles of ancient air trapped in ice cores. Both gases have large, natural emission rates, which have varied over past climatic changes but have sustained a stable atmospheric abundance for the centuries prior to the Industrial Revolution (see Figures 4.1 and 4.2). Emissions of CH_4 and N_2O due to human activities are also substantial and have caused large relative increases in their respective burdens over the last century. The atmospheric burdens of CH_4 and N_2O over the next century will likely be driven by changes in both anthropogenic and natural sources. A second class of greenhouse gases – the synthetic HFCs, PFCs, SF_6, CFCs, and halons – did not exist in the atmosphere before the 20th century (Butler *et al.*, 1999). CF_4, a PFC, is detected in ice cores and appears to have an extremely small natural source (Harnisch and Eisenhauer, 1998). The current burdens of these latter gases are derived from atmospheric observations and represent accumulations of past anthropogenic releases; their future burdens depend almost solely on industrial production and release to the atmosphere. Stratospheric H_2O could increase, driven by *in situ* sources, such as the oxidation of CH_4 and exhaust from aviation, or by a changing climate.

Tropospheric O_3 is both generated and destroyed by photochemistry within the atmosphere. Its *in situ* sources are expected to have grown with the increasing industrial emissions of its precursors: CH_4, NO_x, CO and VOC. In addition, there is substantial transport of ozone from the stratosphere to the troposphere (see also Section 4.2.4). The effects of stratospheric O_3 depletion over the past three decades and the projections of its recovery, following cessation of emissions of the Montreal Protocol gases, was recently assessed (WMO, 1999).

The current global emissions, mean abundances, and trends of the gases mentioned above are summarised in Table 4.1a. Table 4.1b lists additional synthetic greenhouse gases without established atmospheric abundances. For the Montreal Protocol gases, political regulation has led to a phase-out of emissions that has slowed their atmospheric increases, or turned them into decreases, such as for CFC-11. For other greenhouse gases, the anthropogenic emissions are projected to increase or remain high in the absence of climate-policy regulations. Projections of future emissions for this assessment, i.e., the IPCC Special Report on Emission Scenarios (SRES) (Nakićenović *et al.*, 2000) anticipate future development of industries and agriculture that represent major sources of greenhouse gases in the absence of climate-policy regulations. The first draft of this chapter and many of the climate studies in this report used the greenhouse gas concentrations derived from the SRES preliminary marker scenarios (i.e., the SRES database as of January 1999 and labelled 'p' here). The scenario IS92a has been carried along in many tables to provide a reference of the changes since the SAR. The projections of greenhouse gases and aerosols for the six new SRES marker/illustrative scenarios are discussed here and tabulated in Appendix II.

An important policy issue is the complete impact of different industrial or agricultural sectors on climate. This requires aggregation of the SRES scenarios by sector (e.g., transportation) or sub-sector (e.g., aviation; Penner *et al.*, 1999), including not only emissions but also changes in land use or natural ecosystems. Due to chemical coupling, correlated emissions can have synergistic effects; for instance NO_x and CO from transportation produce regional O_3 increases. Thus a given sector may act through several channels on the future trends of greenhouse gases. In this chapter we will evaluate the data available on this subject in the current literature and in the SRES scenarios.

---

[1] Atmospheric *abundances* for trace gases are reported here as the mole fraction (molar mixing ratio) of the gas relative to dry air (ppm = 10^{-6}, ppb = 10^{-9}, ppt = 10^{-12}); whereas the *burden* is reported as the total mass of that gas in the atmosphere (e.g., Mt = Tg = 10^{12} g). For most trace gases in this chapter, the burden is based on the total weight of the molecule; for the N-containing gases, it includes only the mass of the N; and for some VOC budgets where noted, it includes only the mass of the C.

Table 4.1(a): *Chemically reactive greenhouse gases and their precursors: abundances, trends, budgets, lifetimes, and GWPs.*

Chemical species	Formula	Abundance[a] ppt		Trend ppt/yr[a]	Annual emission	Life-time	100-yr GWP[b]
		1998	1750	1990s	late 90s	(yr)	
Methane	CH_4(ppb)	1745	700	7.0	600 Tg	8.4/12[c]	23
Nitrous oxide	N_2O (ppb)	314	270	0.8	16.4 TgN	120/114[c]	296
Perfluoromethane	CF_4	80	40	1.0	~15 Gg	>50000	5700
Perfluoroethane	C_2F_6	3.0	0	0.08	~2 Gg	10000	11900
Sulphur hexafluoride	SF_6	4.2	0	0.24	~6 Gg	3200	22200
HFC-23	CHF_3	14	0	0.55	~7 Gg	260	12000
HFC-134a	CF_3CH_2F	7.5	0	2.0	~25 Gg	13.8	1300
HFC-152a	CH_3CHF_2	0.5	0	0.1	~4 Gg	1.40	120

Important greenhouse halocarbons under Montreal Protocol and its Amendments

CFC-11	$CFCl_3$	268	0	−1.4		45	4600
CFC-12	CF_2Cl_2	533	0	4.4		100	10600
CFC-13	CF_3Cl	4	0	0.1		640	14000
CFC-113	$CF_2ClCFCl_2$	84	0	0.0		85	6000
CFC-114	CF_2ClCF_2Cl	15	0	<0.5		300	9800
CFC-115	CF_3CF_2Cl	7	0	0.4		1700	7200
Carbon tetrachloride	CCl_4	102	0	−1.0		35	1800
Methyl chloroform	CH_3CCl_3	69	0	−14		4.8	140
HCFC-22	CHF_2Cl	132	0	5		11.9	1700
HCFC-141b	CH_3CFCl_2	10	0	2		9.3	700
HCFC-142b	CH_3CF_2Cl	11	0	1		19	2400
Halon-1211	CF_2ClBr	3.8	0	0.2		11	1300
Halon-1301	CF_3Br	2.5	0	0.1		65	6900
Halon-2402	CF_2BrCF_2Br	0.45	0	~ 0		<20	

Other chemically active gases dirctly or indirectly affecting radiative forcing

Tropospheric ozone	O_3 (DU)	34	25	?	see text	0.01-0.05	−
Tropospheric NO_x	$NO + NO_2$	5-999	?	?	~52 TgN	<0.01-0.03	−
Carbon monoxide	CO (ppb)[d]	80	?	6	~2800 Tg	0.08 - 0.25	[d]
Stratospheric water	H_2O (ppm)	3-6	3-5	?	see text	1-6	−

[a] All abundances are tropospheric molar mixing ratios in ppt (10^{-12})and trends are in ppt/yr unless superseded by units on line (ppb = 10^{-9}, ppm = 10^{-6}). Where possible, the 1998 values are global, annual averages and the trends are calculated for 1996 to 1998.

[b] GWPs are from Chapter 6 of this report and refer to the 100-year horizon values.

[c] Species with chemical feedbacks that change the duration of the atmospheric response; global mean atmospheric lifetime (LT) is given first followed by perturbation lifetime (PT). Values are taken from the SAR (Prather *et al.*, 1995; Schimel *et al.*, 1996) updated with WMO98 (Kurylo and Rodriguez, 1999; Prinn and Zander, 1999) and new OH-scaling, see text. Uncertainties in lifetimes have not changed substantially since the SAR.

[d] CO trend is very sensitive to the time period chosen. The value listed for 1996 to 1998, +6 ppb/yr, is driven by a large increase during 1998. For the period 1991 to 1999, the CO trend was −0.6 ppb/yr. CO is an indirect greenhouse gas: for comparison with CH_4 see this chapter; for GWP, see Chapter 6.

Table 4.1(b): *Additional synthetic greenhouse gases.*

Chemical species	Formula	Lifetime (yr)	GWP[b]
Perfluoropropane	C_3F_8	2600	8600
Perfluorobutane	C_4F_{10}	2600	8600
Perfluorocyclobutane	C_4F_8	3200	10000
Perfluoropentane	C_5F_{12}	4100	8900
Perfluorohexane	C_6F_{14}	3200	9000
Trifluoromethyl-sulphur pentafluoride	SF_5CF_3	1000	17500
Nitrogen trifluoride	NF_3	>500	10800
Trifluoroiodomethane	CF_3I	<0.005	1
HFC-32	CH_2F_2	5.0	550
HFC-41	CH_3F	2.6	97
HFC-125	CHF_2CF_3	29	3400
HFC-134	CHF_2CHF_2	9.6	1100
HFC-143	CH_2FCHF_2	3.4	330
HFC-143a	CH_3CF_3	52	4300
HFC-152	CH_2FCH_2F	0.5	43
HFC-161	CH_3CH_2F	0.3	12
HFC-227ea	CF_3CHFCF_3	33	3500
HFC-236cb	$CF_3CF_2CH_2F$	13.2	1300
HFC-236ea	$CF_3CHFCHF_2$	10.0	1200
HFC-236fa	$CF_3CH_2CF_3$	220	9400
HFC-245ca	$CH_2FCF_2CHF_2$	5.9	640
HFC-245ea	$CHF_2CHFCHF_2$	4.0	
HFC-245eb	CF_3CHFCH_2F	4.2	
HFC-245fa	$CHF_2CH_2CF_3$	7.2	950
HFC-263fb	$CF_3CH_2CH_3$	1.6	
HFC-338pcc	$CHF_2CF_2CF_2CF_2H$	11.4	
HFC-356mcf	$CF_3CF_2CH_2CH_2F$	1.2	
HFC-356mff	$CF_3CH_2CH_2CF_3$	7.9	
HFC-365mfc	$CF_3CH_2CF_2CH_3$	9.9	890
HFC-43-10mee	$CF_3CHFCHFCF_2CF_3$	15	1500
HFC-458mfcf	$CF_3CH_2CF_2CH_2CF_3$	22	
HFC-55-10mcff	$CF_3CF_2CH_2CH_2CF_2CF_3$	7.7	
HFE-125	CF_3OCHF_2	150	14900
HFE-134	CF_2HOCF_2H	26	2400
HFE-143a	CF_3OCH_3	4.4	750
HFE-152a	CH_3OCHF_2	1.5	
HFE-245fa2	$CHF_2OCH_2CF_3$	4.6	570
HFE-356mff2	$CF_3CH_2OCH_2CF_3$	0.4	

4.1.2 Atmospheric Chemistry and Feedbacks

All greenhouse gases except CO_2 and H_2O are removed from the atmosphere primarily by chemical processes within the atmosphere. Greenhouse gases containing one or more H atoms (e.g., CH_4, HFCs and HCFCs), as well as other pollutants, are removed primarily by the reaction with hydroxyl radicals (OH). This removal takes place in the troposphere, the lowermost part of the atmosphere, ranging from the surface up to 7 to 16 km depending on latitude and season and containing 80% of the mass of the atmosphere. The greenhouse gases N_2O, PFCs, SF_6, CFCs and halons do not react with OH in the troposphere. These gases are destroyed in the stratosphere or above, mainly by solar ultraviolet radiation (UV) at short wavelengths (<240 nm), and are long-lived. Because of the time required to transport these gases to the region of chemical loss, they have a minimum lifetime of about 20 years. CO_2 is practically inert in the atmosphere and does not directly influence the chemistry, but it has a small *in situ* source from the oxidation of CH_4, CO and VOC.

Tropospheric OH abundances depend on abundances of NO_x, CH_4, CO, VOC, O_3 and H_2O plus the amount of solar UV (>300 nm) that reaches the troposphere. As a consequence, OH varies widely with geographical location, time of day, and season. Likewise the local loss rates of all those gases reacting with OH also vary. Because of its dependence on CH_4 and other pollutants, tropospheric OH is expected to have changed since the pre-industrial era and to change again for future emission scenarios. For some of these gases other removal processes, such as photolysis or surface uptake, are also important; and the total sink of the gas is obtained by integrating over all such processes. The chemistry of tropospheric O_3 is closely tied to that of OH, and its abundance also varies with changing precursor emissions. The chemistry of the troposphere is also directly influenced by the stratospheric burden of O_3, climatic changes in temperature (T) and humidity (H_2O), as well as by interactions between tropospheric aerosols and trace gases. Such couplings provide a "feedback" between the climate change induced by increasing greenhouse gases and the concentration of these gases. Another feedback, internal to the chemistry, is the impact of CH_4 on OH and hence its own loss. These feedbacks are expected to be important for tropospheric O_3 and OH. Such chemistry-chemistry or climate-chemistry coupling has been listed under "indirect effects" in the SAR (Prather *et al.*, 1995; Schimel *et al.*, 1996).

This chapter uses 3-D chemistry-transport models (CTMs) to integrate the varying chemical processes over global conditions, to estimate their significance, and to translate the emission scenarios into abundance changes in the greenhouse gases CH_4, HFCs, and O_3. An extensive modelling exercise called OxComp (tropospheric oxidant model comparison) – involving model comparisons, sensitivity studies, and investigation of the IPCC SRES scenarios – was organised to support this report.

Stratospheric circulation and distribution of O_3 control the transport of the long-lived greenhouse gases to regions of photochemical loss as well as the penetration of solar UV into the atmosphere. At the same time, many of these gases (e.g., N_2O and CFCs) supply ozone-depleting radicals (e.g., nitric oxide (NO) and Cl) to the stratosphere, providing a feedback between the gas and its loss rate. Another consequence of the observed stratospheric ozone depletion is that tropospheric photochemical activity is expected to have increased, altering tropospheric OH and O_3. Climate change in the 21st century, including the radiative cooling of the stratosphere by increased levels of CO_2, is expected to alter stratospheric circulation and O_3, and, hence, the global mean loss rates of the long-lived gases. Some of these effects are discussed in WMO (1999) and are briefly considered here.

The biosphere's response to global change will impact the atmospheric composition of the 21st century. The anticipated changes in climate (e.g., temperature, precipitation) and in chemistry will alter ecosystems and thus the "natural", background emissions of trace gases. There is accumulating evidence that increased N deposition (the result of NO_x and ammonia (NH_3) emissions) and elevated surface O_3

abundances have opposite influences on plant CO_2 uptake: O_3 (>40 ppb) inhibits CO_2 uptake; while N deposition enhances it up to a threshold, above which the effects are detrimental. In addition, the increased N availability from atmospheric deposition and direct fertilisation accelerates the emission of N-containing trace gases (NO, N_2O and NH_3) and CH_4, as well as altering species diversity and biospheric functioning. These complex interactions represent a chemistry-biosphere feedback that may alter greenhouse forcing.

4.1.3 Trace Gas Budgets and Trends

The "budget" of a trace gas consists of three quantities: its global source, global sink and atmospheric burden. The burden is defined as the total mass of the gas integrated over the atmosphere and related reservoirs, which usually include just the troposphere and stratosphere. The global burden (in Tg) and its trend (i.e., the difference between sources and sinks, in Tg/yr) can be determined from atmospheric measurements and, for the long-lived gases, are usually the best-known quantities in the budgets. For short-lived, highly variable gases such as tropospheric O_3 and NO_x, the atmospheric burden cannot be measured with great accuracy. The global source strength is the sum of all sources, including emissions and *in situ* chemical production. Likewise, the sink strength (or global loss rate) can have several independent components.

The source strength (Tg/yr) for most greenhouse gases is comprised of surface emissions. For synthetic gases where industrial production and emissions are well documented, the source strengths may be accurately known. For CH_4 and N_2O, however, there are large, not well-quantified, natural emissions. Further, the anthropogenic emissions of these gases are primarily associated with agricultural sources that are difficult to quantify accurately. Considerable research has gone into identifying and quantifying the emissions from individual sources for CH_4 and N_2O, as discussed below. Such uncertainty in source strength also holds for synthetic gases with undocumented emissions. The source strength for tropospheric O_3 includes both a stratospheric influx and *in situ* production and is thus derived primarily from global chemical models.

The sink strength (Tg/yr) of long-lived greenhouse gases can be derived from a combination of atmospheric observations, laboratory experiments, and models. The atmospheric chemistry models are based on physical principles and laboratory data, and include as constraints the observed chemistry of the atmosphere over the past two decades. For example, stratospheric loss rates are derived from models either by combining observed trace gas distributions with theoretically calculated loss frequencies or from the measured correlation of the respective gas with a trace gas of known vertical flux. In such analyses there are a wide range of self-consistency checks. Mean global loss rates based on *a priori* modelling (e.g., the CH_4-lifetime studies from OxComp described later) can be compared with empirically-based loss rates that are scaled from a gas with similar loss processes that has well-known emissions and atmospheric burden (e.g., the AGAGE (Advanced Global Atmospheric Gases Experiment) calibration of mean tropospheric OH using methyl chloroform (CH_3CCl_3); Prinn *et al.*, 1995). Our knowledge of the current budget of a greenhouse gas provides a key constraint in modelling its future abundance. For example, in both the IS92a and SRES projected emissions of CH_4 and N_2O, we apply a constant offset to each set of emissions so that our calculated burden is consistent with the observed budget and trend during the 1990s.

4.1.4 Atmospheric Lifetimes and Time-Scales

The global atmospheric lifetime (yr) characterises the time required to turn over the global atmospheric burden. It is defined as the burden (Tg) divided by the mean global sink (Tg/yr) for a gas in steady state (i.e., with unchanging burden). This quantity was defined as both "lifetime" and "turnover time" in the SAR (see also Bolin and Rodhe, 1973). Lifetimes calculated in this manner are listed in Table 4.1. A corollary of this definition is that, when in steady state (i.e., source strength = sink strength), the atmospheric burden of a gas equals the product of its lifetime and its emissions. A further corollary is that the integrated atmospheric abundance following a single emission is equal to the product of its steady-state lifetime for that emission pattern and the amount emitted (Prather, 1996). This latter, new result since the SAR supports the market-basket approach of aggregating the direct emissions of different greenhouse gases with a GWP (Global Warming Potential) weighting.

The atmospheric lifetime is basically a scale factor relating (i) constant emissions (Tg/yr) to a steady-state burden (Tg), or (ii) an emission pulse (Tg) to the time-integrated burden of that pulse (Tg/yr). The lifetime is often additionally assumed to be a constant, independent of the sources; and it is also taken to represent the decay time (e-fold) of a perturbation. These latter assumptions apply rigorously only for a gas whose local chemical lifetime is constant in space and time, such as for the noble gas radon, whose lifetime is fixed by the rate of its radioactive decay. In such a case the mean atmospheric lifetime equals the local lifetime: the lifetime that relates source strength to global burden is exactly the decay time of a perturbation.

This general applicability of the atmospheric lifetime breaks down for those greenhouse gases and pollutants whose chemical losses vary in space and time. NO_x, for instance, has a local lifetime of <1 day in the lower troposphere, but >5 days in the upper troposphere; and both times are less than the time required for vertical mixing of the troposphere. In this case emission of NO_x into the upper troposphere will produce a larger atmospheric burden than the same emission into the lower troposphere. Consequently, the definition of the atmospheric lifetime of NO_x is not unique and depends on the location (and season) of its emissions. The same is true for any gas whose local lifetime is variable and on average shorter than about 0.5 year, i.e., the decay time of a north-south difference between hemispheres representing one of the longer time-scales for tropospheric mixing. The majority of greenhouse gases considered here have atmospheric lifetimes greater than 2 years, much longer than tropospheric mixing times; and hence their lifetimes are not significantly altered by the location of sources

within the troposphere. When lifetimes are reported for gases in Table 4.1, it is assumed that the gases are uniformly mixed throughout the troposphere. This assumption is unlikely for gases with lifetimes <1 year, and reported values must be viewed only as approximations.

Some gases have chemical feedbacks that change their lifetimes. For example, the increasing CH_4 abundance leads to a longer lifetime for CH_4 (Prather *et al.*, 1995; Schimel *et al.*, 1996). A chemical feedback with opposite effect has been identified for N_2O where a greater N_2O burden leads to increases in stratospheric NO_x which in turn depletes mid-stratospheric ozone. This ozone loss enhances the UV, and as a consequence N_2O is photolysed more rapidly (Prather, 1998). Such feedbacks cause the time-scale of a perturbation, henceforth called perturbation lifetime (*PT*), to differ from the global atmospheric lifetime (*LT*). In the limit of small perturbations, the relation between the perturbation lifetime of a gas and its global atmospheric lifetime can be derived from a simple budget relationship as $PT = LT / (1 - s)$, where the sensitivity coefficient $s = \partial ln(LT) / \partial ln(B)$ and B = burden. Without a feedback on lifetime, $s = 0$, and *PT* is identical to *LT*. The product, *PT* times a sustained change in emission, gives the resulting change in the burden. The ratio of *PT/LT* adopted here for CH_4, 1.4, is based on recent model studies (see Section 4.4) and is consistent with the SAR results.

To evaluate the total greenhouse effect of a given gas molecule, one needs to know, first, how long it remains in the atmosphere and, second, how it interacts chemically with other molecules. This effect is calculated by injecting a pulse of that gas (e.g., 1 Tg) into the atmosphere and watching the added abundance decay as simulated in a CTM. This decay is represented by a sum of exponential functions, each with its own decay time. These exponential functions are the chemical modes of the linearised chemistry-transport equations of the CTM (Prather, 1996). In the case of a CH_4 addition, the longest-lived mode has an e-fold time of 12 years, close to the perturbation lifetime (*PT*) of CH_4, and carries most of the added burden. (This e-fold time was called the adjustment time in the SAR.) In the case of a CO addition, this same mode is also excited, but at a reduced amplitude (Prather, 1996; Daniel and Solomon, 1998). The pulse of added CO, by causing the concentration of OH to decrease and thus the lifetime of CH_4 to increase temporarily, causes a build-up of CH_4 while the added burden of CO persists. After the initial period of a few months defined by the CO photochemical lifetime, this built-up CH_4 then decays in the same manner as would a direct pulse of CH_4. Similarly, an addition of NO_x (e.g., from aviation; see Isaksen and Jackman, 1999) will excite this mode, but with a negative amplitude. Thus, changes in the emissions of short-lived gases can generate long-lived perturbations as shown in 3-D CTMs (Wild and Prather, 2000; Derwent *et al.*, 2001). Changes in tropospheric O_3 accompany the CH_4 decay on a 12 year time-scale as an inherent component of this mode, a key example of chemical coupling in the troposphere. Thus, any chemically reactive gas, whether a greenhouse gas or not, will produce some level of indirect greenhouse effect through its impact on atmospheric chemistry.

4.2 Trace Gases: Current Observations, Trends, and Budgets

4.2.1 Non-CO$_2$ Kyoto Gases

4.2.1.1 Methane (CH$_4$)

Methane's globally averaged atmospheric surface abundance in 1998 was 1,745 ppb (see Figure 4.1), corresponding to a total burden of about 4,850 Tg(CH$_4$). The uncertainty in the burden is small ($\pm$5%) because the spatial and temporal distributions of tropospheric and stratospheric CH$_4$ have been determined by extensive high-precision measurements and the tropospheric variability is relatively small. For example, the Northern Hemisphere CH$_4$ abundances average about 5% higher than those in the Southern Hemisphere. Seasonal variations, with a minimum in late summer, are observed with peak-to-peak amplitudes of about 2% at mid-latitudes. The average vertical gradient in the troposphere is negligible, but CH$_4$ abundances in the stratosphere decrease rapidly with altitude, e.g., to 1,400 ppb at 30 km altitude in the tropics and to 500 ppb at 30 km in high latitude northern winter.

The most important known sources of atmospheric methane are listed in Table 4.2. Although the major source terms of atmospheric CH$_4$ have probably been identified, many of the source strengths are still uncertain due to the difficulty in assessing the global emission rates of the biospheric sources, whose strengths are highly variable in space and time: e.g., local emissions from most types of natural wetland can vary by a few orders of magnitude over a few metres. Nevertheless, new approaches have led to improved estimates of the global emissions rates from some source types. For instance, intensive studies on emissions from rice agriculture have substantially improved these emissions estimates (Ding and Wang, 1996; Wang and Shangguan, 1996). Further, integration of emissions over a whole growth period (rather than looking at the emissions on individual days with different ambient temperatures) has lowered the estimates of CH$_4$ emissions from rice agriculture from about 80 Tg/yr to about 40 Tg/yr (Neue and Sass, 1998; Sass *et al.*, 1999). There have also been attempts to deduce emission rates from observed spatial and temporal distributions of atmospheric CH$_4$ through inverse modelling (e.g., Hein *et al.*, 1997; Houweling *et al.*, 1999). The emissions so derived depend on the precise knowledge of the mean global loss rate and represent a relative attribution into aggregated sources of similar properties. The results of some of these studies have been included in Table 4.2. The global CH$_4$ budget can also be constrained by measurements of stable isotopes (δ^{13}C and δD) and radiocarbon (^{14}CH$_4$) in atmospheric CH$_4$ and in CH$_4$ from the major sources (e.g., Stevens and Engelkemeir, 1988; Wahlen *et al.*, 1989; Quay *et al.*, 1991, 1999; Lassey *et al.*, 1993; Lowe *et al.*, 1994). So far the measurements of isotopic composition of CH$_4$ have served mainly to constrain the contribution from fossil fuel related sources. The emissions from the various sources sum up to a global total of about 600 Tg/yr, of which about 60% are related to human activities such as agriculture, fossil fuel use and waste disposal. This is consistent with the SRES estimate of 347 Tg/yr for anthropogenic CH$_4$ emissions in the year 2000.

The current emissions from CH$_4$ hydrate deposits appear small, about 10 Tg/yr. However, these deposits are enormous, about 10^7 TgC (Suess *et al.*, 1999), and there is an indication of a catastrophic release of a gaseous carbon compound about 55 million years ago, which has been attributed to a large-scale perturbation of CH$_4$ hydrate deposits (Dickens, 1999; Norris and Röhl, 1999). Recent research points to regional releases of CH$_4$ from clathrates in ocean sediments during the last 60,000 years (Kennett *et al.*, 2000), but much of this CH$_4$ is likely to be oxidised by bacteria before reaching the atmosphere (Dickens, 2001). This evidence adds to the concern that the expected global warming may lead to an increase in these emissions and thus to another positive feedback in the climate system. So far, the size of that feedback has not been quantified. On the other hand, the historic record of atmospheric CH$_4$ derived from ice cores (Petit *et al.*, 1999), which spans several large temperature swings plus glaciations, constrains the possible past releases from methane hydrates to the atmosphere. Indeed, Brook *et al.* (2000) find little evidence for rapid, massive CH$_4$ excursions that might be associated with large-scale decomposition of methane hydrates in sediments during the past 50,000 years.

The mean global loss rate of atmospheric CH$_4$ is dominated by its reaction with OH in the troposphere.

$$OH + CH_4 \quad \rightarrow \quad CH_3 + H_2O$$

This loss term can be quantified with relatively good accuracy based on the mean global OH concentration derived from the methyl chloroform (CH$_3$CCL$_3$) budget described in Section 4.2.6 on OH. In that way we obtain a mean global loss rate of 507 Tg(CH$_4$)/yr for the current tropospheric removal of CH$_4$ by OH. In addition there are other minor removal processes for atmospheric CH$_4$. Reaction with Cl atoms in the marine boundary layer probably constitutes less than 2% of the total sink (Singh *et al.*, 1996). A recent process model study (Ridgwell *et al.*, 1999) suggested a soil sink of 38 Tg/yr, and this can be compared to SAR estimates of 30 Tg/yr. Minor amounts of CH$_4$ are also destroyed in the stratosphere by reactions with OH, Cl, and O(^{1}D), resulting in a combined loss rate of 40 Tg/yr. Summing these, our best estimate of the current global loss rate of atmospheric CH$_4$ totals 576 Tg/yr (see Table 4.2), which agrees reasonably with the total sources derived from process models. The atmospheric lifetime of CH$_4$ derived from this loss rate and the global burden is 8.4 years. Attributing individual lifetimes to the different components of CH$_4$ loss results in 9.6 years for loss due to tropospheric OH, 120 years for stratospheric loss, and 160 years for the soil sink (i.e., 1/8.4 yr = 1/9.6 yr + 1/120 yr + 1/160 yr).

The atmospheric abundance of CH$_4$ has increased by about a factor of 2.5 since the pre-industrial era (see Figure 4.1a) as evidenced by measurements of CH$_4$ in air extracted from ice cores and firn (Etheridge *et al.*, 1998). This increase still continues, albeit at a declining rate (see Figure 4.1b). The global tropospheric CH$_4$ growth rate averaged over the period 1992 through 1998 is about 4.9 ppb/yr, corresponding to an average annual increase in atmospheric burden of 14 Tg. Superimposed on this long-term decline in growth rate are interannual variations in the trend (Figure 4.1c). There are no clear quantitative explanations for this variability, but understanding these variations in

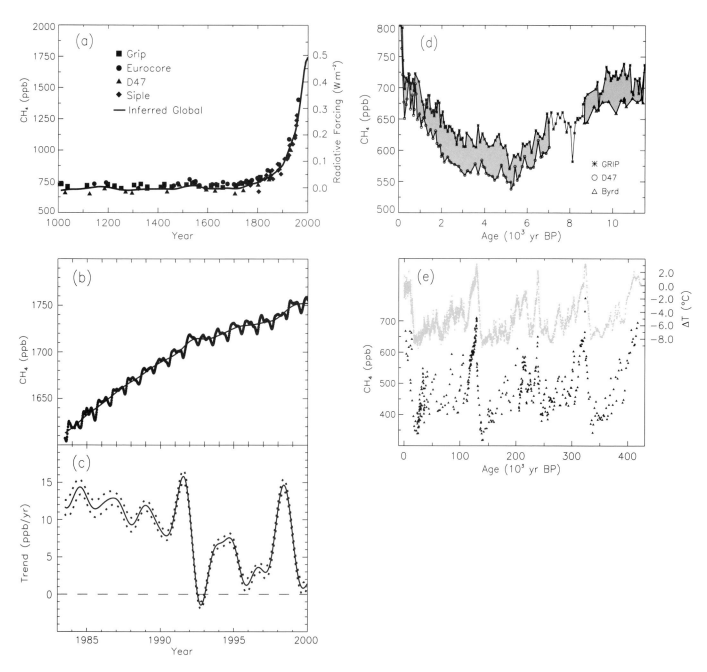

Figure 4.1: (a) Change in CH$_4$ abundance (mole fraction, in ppb = 10^{-9}) determined from ice cores, firn, and whole air samples plotted for the last 1,000 years. Data sets are as follows: Grip, Blunier *et al.* (1995) and Chappellaz *et al.* (1997); Eurocore, Blunier *et al.* (1993); D47, Chappellaz *et al.* (1997); Siple, Stauffer *et al.* (1985); Global (inferred from Antarctic and Greenland ice cores, firn air, and modern measurements), Etheridge *et al.* (1998) and Dlugokencky *et al.* (1998). Radiative forcing, approximated by a linear scale since the pre-industrial era, is plotted on the right axis. (b) Globally averaged CH$_4$ (monthly varying) and deseasonalised CH$_4$ (smooth line) abundance plotted for 1983 to 1999 (Dlugokencky *et al.*, 1998). (c) Instantaneous annual growth rate (ppb/yr) in global atmospheric CH$_4$ abundance from 1983 through 1999 calculated as the derivative of the deseasonalised trend curve above (Dlugokencky *et al.*, 1998). Uncertainties (dotted lines) are ±1 standard deviation. (d) Comparison of Greenland (GRIP) and Antarctic (D47 and Byrd) CH$_4$ abundances for the past 11.5 kyr (Chappellaz *et al.*, 1997). The shaded area is the pole-to-pole difference where Antarctic data exist. (e) Atmospheric CH$_4$ abundances (black triangles) and temperature anomalies with respect to mean recent temperature (grey diamonds) determined for the past 420 kyr from an ice core drilled at Vostok Station in East Antarctica (Petit *et al.*, 1999).

trend will ultimately help constrain specific budget terms. After the eruption of Mt. Pinatubo, a large positive anomaly in growth rate was observed at tropical latitudes. It has been attributed to short-term decreases in solar UV in the tropics immediately following the eruption that decreased OH formation rates in the troposphere (Dlugokencky *et al.*, 1996). A large decrease in growth was observed, particularly in high northern latitudes, in 1992. This feature has been attributed in part to decreased northern wetland emission rates resulting from anomalously low surface temperatures (Hogan and Harriss, 1994) and in part to

Table 4.2: *Estimates of the global methane budget (in Tg(CH_4/yr) from different sources compared with the values adopted for this report (TAR).*

Reference:	Fung et al. (1991)	Hein et al. (1997)	Lelieveld et al. (1998)	Houweling et al. (1999)	Mosier et al. (1998a)	Olivier et al. (1999)	Cao et al. (1998)	SAR	TAR[a]
Base year:	1980s	–	1992	–	1994	1990	–	1980s	1998
Natural sources									
Wetlands	115	237	225[c]	145			92		
Termites	20	–	20	20					
Ocean	10	–	15	15					
Hydrates	5	–	10	–					
Anthropogenic sources									
Energy	75	97	110	89		109			
Landfills	40	35	40	73		36			
Ruminants	80	90[b]	115	93	80	93[b]			
Waste treatment	–	[b]	25	–	14	[b]			
Rice agriculture	100	88	[c]	–	25-54	60	53		
Biomass burning	55	40	40	40	34	23			
Other	–	–	–	20	15				
Total source	**500**	**587**	**600**					**597**	**598**
Imbalance (trend)								+37	+22
Sinks									
Soils	10	–	30	30	44			30	30
Tropospheric OH	450	489	510					490	506
Stratospheric loss	–	46	40					40	40
Total sink	**460**	**535**	**580**					**560**	**576**

[a] TAR budget based on 1,745 ppb, 2.78 Tg/ppb, lifetime of 8.4 yr, and an imbalance of +8 ppb/yr.

[b] Waste treatment included under ruminants.

[c] Rice included under wetlands.

stratospheric ozone depletion that increased tropospheric OH (Bekki *et al.*, 1994; Fuglestvedt *et al.*, 1994). Records of changes in the $^{13}C/^{12}C$ ratios in atmospheric CH_4 during this period suggest the existence of an anomaly in the sources or sinks involving more than one causal factor (Lowe *et al.*, 1997; Mak *et al.*, 2000).

There is no consensus on the causes of the long-term decline in the annual growth rate. Assuming a constant mean atmospheric lifetime of CH_4 of 8.9 years as derived by Prinn *et al.* (1995), Dlugokencky *et al.* (1998) suggest that during the period 1984 to 1997 global emissions were essentially constant and that the decline in annual growth rate was caused by an approach to steady state between global emissions and atmospheric loss rate. Their estimated average source strength was about 550 Tg/yr. (Inclusion of a soil sink term of 30 Tg/yr would decrease the lifetime to 8.6 years and suggest an average source strength of about 570 Tg/yr.) Francey *et al.* (1999), using measurements of $^{13}CH_4$ from Antarctic firn air samples and archived air from Cape Grim, Tasmania, also concluded that the decreased CH_4 growth rate was consistent with constant OH and constant or very slowly increasing CH_4 sources after 1982. However, other analyses of the global methyl chloroform (CH_3CCl_3) budget (Krol *et al.*, 1998) and the changing chemistry of the atmosphere (Karlsdottir and Isaksen, 2000) argue for an increase in globally averaged OH of +0.5%/yr over the last two decades (see Section 4.2.6 below) and hence a parallel increase in global CH_4 emissions by +0.5%/yr.

The historic record of atmospheric CH_4 obtained from ice cores has been extended to 420,000 years before present (BP) (Petit *et al.*, 1999). As Figure 4.1e demonstrates, at no time during this record have atmospheric CH_4 mixing ratios approached today's values. CH_4 varies with climate as does CO_2. High values are observed during interglacial periods, but these maxima barely exceed the immediate pre-industrial CH_4 mixing ratio of 700 ppb. At the same time, ice core measurements from Greenland and Antarctica indicate that during the Holocene CH_4 had a pole-to-pole difference of about 44 ± 7 ppb with higher values in the Arctic as today, but long before humans influenced atmospheric methane concentrations (Chappelaz *et al.*, 1997; Figure 4.1d). Finally, study of CH_4 ice-core records at high time resolution reveals no evidence for rapid, massive CH_4 excursions that might be associated with large-scale decomposition of methane hydrates in sediments (Brook *et al.*, 2000).

The feedback of CH_4 on tropospheric OH and its own lifetime is re-evaluated with contemporary CTMs as part of OxComp, and results are summarised in Table 4.3. The calculated OH feedback, $\partial \ln(OH) / \partial \ln(CH_4)$, is consistent between the models, indicating that tropospheric OH abundances decline by 0.32% for every 1% increase in CH_4. The TAR value for the sensitivity coefficient $s = \partial \ln(LT) / \partial \ln(CH_4)$ is then 0.28 and the ratio PT/LT is 1.4. This 40% increase in the integrated effect of a CH_4 perturbation does not appear as a 40% larger amplitude in the perturbation but rather as a lengthening of the duration of the perturbation to 12 years. This feedback is difficult

Table 4.3: *Methane lifetime and feedback on tropospheric OH[a] for the 1990s.*

CTM	lifetime vs. OH(yr)[b]	$\delta\ln(OH)/$ $\delta\ln(CH_4)$	s= $\delta\ln(LT)/$ $\delta\ln(CH_4)$	PT/LT
IASB	8.1	−0.31	+0.27	1.37
KNMI	9.8	−0.35	+0.31	1.45
MPIC	8.5	−0.29	+0.25	1.33
UCI	9.0	−0.34 (−0.38)[c]	+0.30	1.43
UIO1	6.5	−0.33	+0.29	1.41
UKMO	8.3	−0.31 (−0.34)[c]	+0.27	1.37
ULAQ	13.8	−0.29	+0.25	1.33
TAR value[d]	9.6	−0.32		1.4

[a] Global mean tropospheric OH is weighted by the CH_4 loss rate.

[b] Lifetime against tropospheric OH loss at 1,745 ppb.

[c] Evaluated at 4,300 ppb CH_4 plus emissions for Y2100/draft-A2 scenario.

[d] TAR recommended OH lifetime for CH_4, 9.6 yr, is scaled from a CH_3CCl_3 OH lifetime of 5.7 yr (WMO, 1999; based on Prinn *et al.*, 1995) using a temperature of 272K (Spivakovsky *et al.*, 2000). Stratospheric (120 yr) and soil-loss (160 yr) lifetimes are added (inversely) to give mean atmospheric lifetime of 8.4 yr. Only the OH lifetime is diagnosed and is subject to chemical feedback factor, and thus the total atmospheric lifetime for a CH_4 perturbation is 12 yr. In the SAR, the feedback factor referred only to the increase in the lifetime against tropospheric OH, and hence was larger. For Chemistry Transport Model (CTM) code see Table 4.10.

to observe, since it would require knowledge of the *increase* in CH_4 sources plus other factors affecting OH over the past two decades. Unlike for the global mean tropospheric OH abundance, there is also no synthetic compound that can calibrate this feedback; but it is possible that an analysis of the budgets of $^{13}CH_4$ and $^{12}CH_4$ separately may lead to an observational constraint (Manning, 1999).

4.2.1.2 Nitrous oxide (N₂O)

The globally averaged surface abundance of N_2O was 314 ppb in 1998, corresponding to a global burden of 1510 TgN. N_2O abundances are about 0.8 ppb greater in the Northern Hemisphere than in the Southern Hemisphere, consistent with about 60% of emissions occurring in the Northern Hemisphere. Almost no vertical gradient is observed in the troposphere, but N_2O abundances decrease in the stratosphere, for example, falling to about 120 ppb by 30 km at mid-latitudes.

The known sources of N_2O are listed in Table 4.4 with estimates of their emission rates and ranges. As with methane, it remains difficult to assess global emission rates from individual sources that vary greatly over small spatial and temporal scales. Total N_2O emissions of 16.4 TgN/yr can be inferred from the N_2O global sink strength (burden/lifetime) plus the rate of increase in the burden. In the SAR the sum of N_2O emissions from specific sources was notably less than that inferred from the loss rate. The recent estimates of global N_2O emissions from Mosier *et al.* (1998b) and Kroeze *et al.* (1999) match the global loss rate and underline the progress that has been made on quantification of natural and agricultural sources. The former study calculated new values for N_2O agricultural emissions that

include the full impact of agriculture on the global nitrogen cycle and show that N_2O emissions from soils are the largest term in the budget (Table 4.4). The latter study combined these with emissions from other anthropogenic and natural sources to calculate a total emission of 17.7 TgN/yr for 1994.

The enhanced N_2O emissions from agricultural and natural ecosystems are believed to be caused by increasing soil N availability driven by increased fertilizer use, agricultural nitrogen (N_2) fixation, and N deposition; and this model can explain the increase in atmospheric N_2O abundances over the last 150 years (Nevison and Holland, 1997). Recent discovery of a faster-than-linear feedback in the emission of N_2O and NO from soils in response to external N inputs is important, given the projected increases of N fertilisation and deposition increases in tropical countries (Matson *et al.*, 1999). Tropical ecosystems, currently an important source of N_2O (and NO) are often phosphorus-limited rather than being N-limited like the Northern Hemispheric terrestrial ecosystems. Nitrogen fertiliser inputs into these phosphorus-limited ecosystems generate NO and N_2O fluxes that are 10 to 100 times greater than the same fertiliser addition to nearby N-limited ecosystems (Hall and Matson, 1999). In addition to N availability, soil N_2O emissions are regulated by temperature and soil moisture and so are likely to respond to climate changes (Frolking *et al.*, 1998; Parton *et al.*, 1998). The magnitude of this response will be affected by feedbacks operating through the biospheric carbon cycle (Li *et al.*, 1992, 1996).

The industrial sources of N_2O include nylon production, nitric acid production, fossil fuel fired power plants, and vehicular emissions. It was once thought that emission from

Table 4.4: *Estimates of the global nitrous oxide budget (in TgN/yr) from different sources compared with the values adopted for this report (TAR).*

Reference:	Mosier *et al.* (1998b) Kroeze *et al.* (1999)		Olivier *et al.* (1998)		SAR	TAR
Base year:	1994	range	1990	range	1980s	1990s
Sources						
Ocean	3.0	1 – 5	3.6	2.8 – 5.7	3	
Atmosphere (NH_3 oxidation)	0.6	0.3 – 1.2	0.6	0.3 – 1.2		
Tropical soils						
Wet forest	3.0	2.2 – 3.7			3	
Dry savannas	1.0	0.5 – 2.0			1	
Temperate soils						
Forests	1.0	0.1 – 2.0			1	
Grasslands	1.0	0.5 – 2.0			1	
All soils			6.6	3.3 – 9.9		
Natural sub-total	9.6	4.6 – 15.9	10.8	6.4 – 16.8	9	
Agricultural soils	4.2	0.6 – 14.8	1.9	0.7 – 4.3	3.5	
Biomass burning	0.5	0.2 – 1.0	0.5	0.2 – 0.8	0.5	
Industrial sources	1.3	0.7 – 1.8	0.7	0.2 – 1.1	1.3	
Cattle and feedlots	2.1	0.6 – 3.1	1.0	0.2 – 2.0	0.4	
Anthropogenic Sub-total	8.1	2.1 – 20.7	4.1	1.3 – 7.7	5.7	6.9[a]
Total sources	**17.7**	**6.7 – 36.6**	**14.9**	**7.7 – 24.5**	**14.7**[b]	
Imbalance (trend)	3.9	3.1 – 4.7			3.9	3.8
Total sinks (stratospheric)	**12.3**	**9 – 16**			**12.3**	**12.6**
Implied total source	16.2				16.2	16.4

[a] SRES 2000 anthropogenic N_2O emissions.

[b] N.B. total sources do not equal sink + imbalance.

automobile catalytic converters were a potential source of N_2O, but extrapolating measurements of N_2O emissions from automobiles in roadway tunnels in Stockholm and Hamburg during 1992 to the global fleet gives a source of only 0.24 ± 0.14 TgN/yr (Berges *et al.*, 1993). More recent measurements suggest even smaller global emissions from automobiles, 0.11 ± 0.04 TgN/yr (Becker *et al.*, 1999; Jiménez *et al.*, 2000).

The identified sinks for N_2O are photodissociation (90%) and reaction with electronically excited oxygen atoms ($O(^1D)$); they occur in the stratosphere and lead to an atmospheric lifetime of 120 years (SAR; Volk *et al.*, 1997; Prinn and Zander, 1999). The small uptake of N_2O by soils is not included in this lifetime, but is rather incorporated into the net emission of N_2O from soils because it is coupled to the overall N-partitioning.

Isotopic ($\delta^{15}N$ and $\delta^{18}O$) N_2O measurements are also used to constrain the N_2O budget. The isotopic composition of tropospheric N_2O derives from the flux-weighted isotopic composition of sources corrected for fractionation during destruction in the stratosphere. Typical observed values are $\delta^{15}N = 7 \permil$ and $\delta^{18}O = 20.7 \permil$ relative to atmospheric N_2 and oxygen (O_2) (Kim and Craig, 1990). Most surface sources are depleted in ^{15}N and ^{18}O relative to tropospheric N_2O (e.g., Kim and Craig, 1993), and so

other processes (sources or sinks) must lead to isotopic enrichment. Rahn and Wahlen (1997) use stratospheric air samples to show that the tropospheric isotope signature of N_2O can be explained by a return flux of isotopically enriched N_2O from the stratosphere, and no exotic sources of N_2O are needed. Yung and Miller (1997) point out that large isotopic fractionation can occur in the stratosphere during photolysis due to small differences in the zero point energies of the different isotopic species, and Rahn *et al.* (1998) have verified this latter effect with laboratory measurements. Wingen and Finlayson-Pitts (1998) failed to find evidence that reaction of CO_3 with N_2 (McElroy and Jones, 1996) is an atmospheric source of N_2O. The use of isotopes has not yet conclusively identified new sources nor constrained the N_2O budget better than other approaches, but the emerging data set of isotopic measurements, including measurements of the intra-molecular position of ^{15}N in N_2O isotopomers (Yoshida and Toyoda, 2000) will provide better constraints in the future.

Tropospheric N_2O abundances have increased from pre-industrial values of about 270 ppb (Machida *et al.*, 1995; Battle *et al.*, 1996; Flückiger *et al.*, 1999) to a globally averaged value of 314 ppb in 1998 (Prinn *et al.*, 1990, 1998; Elkins *et al.*, 1998) as shown in Figure 4.2. The pre-industrial source is estimated to

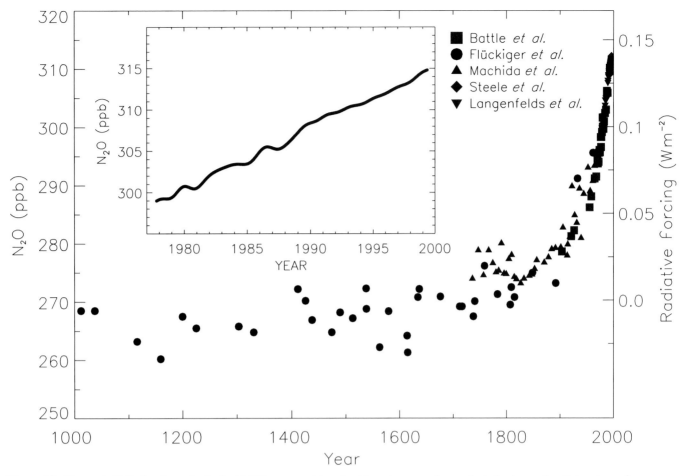

Figure 4.2: Change in N$_2$O abundance for the last 1,000 years as determined from ice cores, firn, and whole air samples. Data sets are from: Machida *et al.* (1995); Battle *et al.* (1996); Langenfelds *et al.* (1996); Steele *et al.* (1996); Flückiger *et al.* (1999). Radiative forcing, approximated by a linear scale, is plotted on the right axis. Deseasonalised global averages are plotted in the inset (Butler *et al.*, 1998b).

be 10.7 TgN/yr, which implies that current anthropogenic emissions are about 5.7 TgN/yr assuming no change in the natural emissions over this period. The average rate of increase during the period 1980 to 1998 determined from surface measurements was +0.8 ± 0.2 ppb/yr (+0.25 ± 0.05 %/yr) and is in reasonable agreement with measurements of the N$_2$O vertical column density above Jungfraujoch Station, +0.36 ± 0.06%/yr between 1984 and 1992 (Zander *et al.*, 1994). Large interannual variations in this trend are also observed. Thompson *et al.* (1994) report that the N$_2$O growth rate decreased from 1 ppb/yr in 1991 to 0.5 ppb/yr in 1993 and suggest that decreased use of nitrogen-containing fertiliser and lower temperatures in the Northern Hemisphere may have been in part responsible for lower biogenic soil emissions. Schauffler and Daniel (1994) suggest that the N$_2$O trend was affected by stratospheric circulation changes induced by massive increase in stratospheric aerosols following the eruption of Mt. Pinatubo. Since 1993, the N$_2$O increase has returned to rates closer to those observed during the 1980s.

The feedback of N$_2$O on its own lifetime (Prather, 1998) has been examined for this assessment with additional studies from established 2-D stratospheric chemical models. All models give similar results, see Table 4.5. The global mean atmospheric

lifetime of N$_2$O decreases about 0.5% for every 10% increase in N$_2$O ($s = -0.05$). This shift is small but systematic, and it is included in Table 4.1a as a shorter perturbation lifetime for N$_2$O, 114 years instead of 120 years. For N$_2$O (unlike for CH$_4$) the time to mix the gas into the middle stratosphere where it is destroyed, about 3 years, causes a separation between PT (about 114 years) and the e-fold of the long-lived mode (about 110 years).

4.2.1.3 Hydrofluorocarbons (HFCs)

The HFCs with the largest measured atmospheric abundances are (in order), HFC-23 (CHF$_3$), HFC-134a (CF$_3$CH$_2$F), and HFC-152a (CH$_3$CHF$_2$). The recent rises in these HFCs are shown in Figure 4.3 along with some major HCFCs, the latter being controlled under the Montreal Protocol and its Amendments. HFC-23 is a by-product of HCFC-22 production. It has a long atmospheric lifetime of 260 years, so that most emissions, which have occurred over the past two decades, will have accumulated in the atmosphere. Between 1978 and 1995, HFC-23 increased from about 3 to 10 ppt; and it continues to rise even more rapidly (Oram *et al.*, 1996). HFC-134a is used primarily as a refrigerant, especially in car air conditioners. It has an atmospheric lifetime of 13.8 years, and its annual emissions have grown from near

Table 4.5: *Nitrous oxide lifetime feedback and residence time.*

Models	Contributor	Lifetime LT (yr)	Sensitivity, $s = \partial \ln(LT)/\partial \ln(B)$	Decay Time of mode (yr)
AER 2D	Ko and Weisenstein	111	−0.062	102
GSFC 2D	Jackman	137	−0.052	127
UCI 1D	Prather	119	−0.046	110
Oslo 2D	Rognerud	97	−0.061	

Lifetime (LT_B) is calculated at steady-state for an N_2O burden (B) corresponding to a tropospheric abundance of 330 ppb. The sensitivity coefficient (s) is calculated by increasing the N_2O burden approximately 10% to $B+\Delta B$, calculating the new steady state atmospheric lifetime ($LT_{B+\Delta B}$), and then using a finite difference approximation for s, $\ln(LT_{B+\Delta B}/LT_B)/\ln(1+\Delta B/B)$. The perturbation lifetime (PT), i.e., the effective duration of an N_2O addition, can be derived as $PT = LT/(1-s)$ or equivalently from the simple budget-balance equation: $(B+\Delta B)/LT_{B+\Delta B} = B/LT_B + \Delta B/PT$.

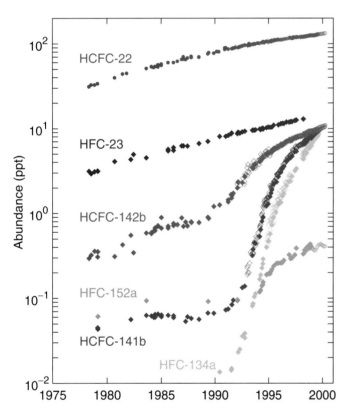

Figure 4.3: HFC-23 (blue, UEA scale), -152a (green, UEA scale), -134a (orange, NOAA scale), and HCFC-22 (magenta, SIO scale), -142b (red, NOAA scale), and -141b (purple, NOAA scale) abundances (ppt) at Cape Grim, Tasmania for the period 1978 to 1999. Different symbols are data from different measurement networks: SIO (filled circles), NOAA-CMDL (open diamonds, Montzka *et al.*, 1994, 1996a,b, 1999), UEA (filled diamonds, Oram *et al.*, 1995, 1996, 1998, 1999) and AGAGE (open circles, only for 1998 to 2000, all gases but HFC-23, Miller *et al.*, 1998; Sturrock *et al.*, 1999; Prinn *et al.*, 2000). Southern Hemisphere values (Cape Grim) are slightly lower than global averages.

zero in 1990 to an estimated 0.032 Tg/yr in 1996. The abundance continues to rise almost exponentially as the use of this HFC increases (Montzka *et al.*, 1996b; Oram *et al.*, 1996; Simmonds *et al.*, 1998). HFC-152a is a short-lived gas with a mean atmospheric lifetime of 1.4 years. Its rise has been steady, but its low emissions and a short lifetime have kept its abundance below 1 ppt.

4.2.1.4 Perfluorocarbons (PFCs) and sulphur hexafluoride (SF_6)
PFCs, in particular CF_4 and C_2F_6, as well as SF_6 have sources predominantly in the Northern Hemisphere, atmospheric lifetimes longer than 1,000 years, and large absorption cross-sections for terrestrial infra-red radiation. These compounds are far from a steady state between sources and sinks, and even small emissions will contribute to radiative forcing over the next several millennia. Current emissions of C_2F_6 and SF_6 are clearly anthropogenic and well quantified by the accumulating atmospheric burden. Harnisch and Eisenhauer (1998) have shown that CF_4 and SF_6 are naturally present in fluorites, and out-gassing from these materials leads to natural background abundances of 40 ppt for CF_4 and 0.01 ppt for SF_6. However, at present the anthropogenic emissions of CF_4 exceed the natural ones by a factor of 1,000 or more and are responsible for the rapid rise in atmospheric abundance. Atmospheric burdens of CF_4 and SF_6 are increasing as shown in Figures 4.4 and 4.5, respectively. Surface measurements show that SF_6 has increased by about 7%/yr during the 1980s and 1990s (Geller *et al.*, 1997; Maiss and Brenninkmeijer, 1998). Recent relative rates of increase are 1.3%/yr for CF_4 and 3.2%/yr for C_2F_6 (Harnisch *et al.*, 1996). The only important sinks for PFCs and SF_6 are photolysis or ion reactions in the mesosphere. These gases provide useful tracers of atmospheric transport in both troposphere and stratosphere.

A new, long-lived, anthropogenic greenhouse gas has recently been found in the atmosphere (Sturges *et al.*, 2000). Trifluoromethyl sulphur pentafluoride (SF_5CF_3) – a hybrid of PFCs and SF_6 not specifically addressed in Annex A of the Kyoto Protocol – has the largest radiative forcing, on a per molecule basis, of any gas found in the atmosphere to date. Its abundance has grown from near zero in the late 1960s to about 0.12 ppt in 1999.

4.2.2 Montreal Protocol Gases and Stratospheric Ozone (O₃)

The Montreal Protocol is an internationally accepted agreement whereby nations agree to control the production of ozone-depleting substances. Many of the chemicals that release chlorine atoms into the stratosphere, and deplete stratospheric O_3, are also greenhouse gases, so they are discussed briefly here. Detailed assessment of the current observations, trends, lifetimes, and emissions for substances covered by the protocol are in WMO (Kurylo and Rodriguez, 1999; Prinn and Zander, 1999). The ozone-depleting gases with the largest potential to influence climate are CFC-11 ($CFCl_3$), CFC-12 (CF_2Cl_2), and CFC-113 ($CF_2ClCFCl_2$). It is now clear from measurements in polar firn air that there are no natural sources of these compounds (Butler *et al.*, 1999). Surface measurements of these compounds show that their growth rates continue to

decline. Growth rates are slightly negative for CFC-11 and CFC-113 (Montzka *et al.*, 1996a, 1999; Prinn *et al.*, 2000); see Figure 4.6. CFC-12 increased by 4 ppt/yr during 1995 to 1996, down from about 12 ppt/yr in the late 1980s, see Figure 4.7). Methyl chloroform (CH_3CCl_3) has decreased dramatically since the Montreal Protocol was invoked, due to its relatively short lifetime (about 5 years) and the rapidity with which emissions were phased out. Its decline was 13 ppt/yr during the period 1995 to 1996 (Prinn *et al.*, 1998, 2000). The halon abundances are small relative to the CFCs, and will never become large if the Montreal Protocol is adhered to. Atmospheric measurements show that growth rates of halon-1301 and halon-2402 decreased in response to the Montreal Protocol, but halon-1211 continues to increase at rates larger than expected based on industrial emissions data (Butler *et al.*, 1998a; Fraser *et al.*, 1999; Montzka *et al.*, 1999).

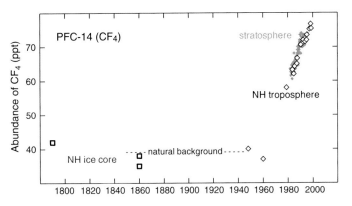

Figure 4.4: Abundance of CF_4 (ppt) over the last 200 years as measured in tropospheric air (open diamonds), stratospheric air (small filled diamonds), and ice cores (open squares) (Harnisch *et al.*, 1996; 1999).

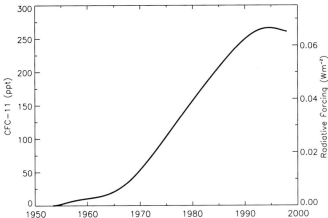

Figure 4.6: Global mean CFC-11 ($CFCl_3$) tropospheric abundance (ppt) from 1950 to 1998 based on smoothed measurements and emission models (Prinn *et al.*, 2000). CFC-11's radiative forcing is shown on the right axis.

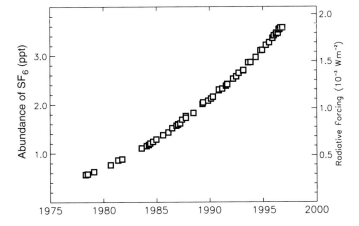

Figure 4.5: Abundance of SF_6 (ppt) measured at Cape Grim, Tasmania since 1978 (Maiss *et al.*, 1996; Maiss and Brenninkmeijer, 1998). Cape Grim values are about 3% lower than global averages.

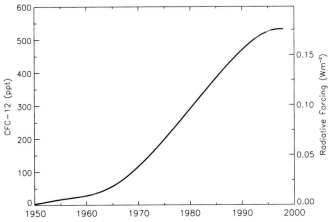

Figure 4.7: Global mean CFC-12 (CF_2Cl_2) tropospheric abundance (ppt) from 1950 to 1998 based on smoothed measurements and emission models (Prinn *et al.*, 2000). CFC-12's radiative forcing is shown on the right axis.

The depletion of stratospheric ozone over the past three decades has been substantial. Between 60°S and 60°N it averaged about 2%/decade. A thorough review of the direct and possible indirect effects of stratospheric ozone depletion are given in WMO (Granier and Shine, 1999). The depletion of O_3 (and its radiative forcing) is expected to follow the weighted halogen abundance in the stratosphere. Therefore, both will reach a maximum in about 2000 before starting to recover; however, detection of stratospheric O_3 recovery is not expected much before 2010 (Jackman *et al.*, 1996; Hofmann and Pyle, 1999). Methyl chloroform has been the main driver of the rapid turnaround in stratospheric chlorine during the late 1990s (Montzka *et al.*, 1999; Prinn *et al.*, 2000), and further recovery will rely on the more slowly declining abundances of CFC-11 and -12, and halons (Fraser *et al.*, 1999; Montzka *et al.*, 1999). It is expected that stratospheric ozone depletion due to halogens will recover during the next 50 to 100 years (Hofmann and Pyle, 1999). In the short run, climatic changes, such as cooling in the northern winter stratosphere, may enhance ozone depletion, but over the next century, the major uncertainties in stratospheric ozone lie with (i) the magnitude of future consumption of ozone-depleting substances by developing countries (Fraser and Prather, 1999; Montzka *et al.*, 1999), (ii) the projected abundances of CH_4 and N_2O, and (iii) the projected climate change impacts on stratospheric temperatures and circulation.

4.2.3 Reactive Gases

4.2.3.1 Carbon monoxide (CO) and hydrogen (H_2)

Carbon monoxide (CO) does not absorb terrestrial infrared radiation strongly enough to be counted as a direct greenhouse gas, but its role in determining tropospheric OH indirectly affects the atmospheric burden of CH_4 (Isaksen and Hov, 1987) and can lead to the formation of O_3. More than half of atmospheric CO emissions today are caused by human activities, and as a result the Northern Hemisphere contains about twice as much CO as the Southern Hemisphere. Because of its relatively short lifetime and distinct emission patterns, CO has large gradients in the atmosphere, and its global burden of about 360 Tg is more uncertain than those of CH_4 or N_2O. In the high northern latitudes, CO abundances vary from about 60 ppb during summer to 200 ppb during winter. At the South Pole, CO varies between about 30 ppb in summer and 65 ppb in winter. Observed abundances, supported by column density measurements, suggest that globally, CO was slowly increasing until the late 1980s, but has started to decrease since then (Zander *et al.* 1989; Khalil and Rasmussen, 1994), possibly due to decreased automobile emissions as a result of catalytic converters (Bakwin *et al.*, 1994). Measurements from a globally distributed network of sampling sites indicate that CO decreased globally by about 2 %/yr from 1991 to 1997 (Novelli *et al.*, 1998) but then increased in 1998. In the Southern Hemisphere, no long-term trend has been detected in CO measurements from Cape Point, South Africa for the period 1978 to 1998 (Labuschagne *et al.*, 1999).

Some recent evaluations of the global CO budget are presented in Table 4.6. The emissions presented by Hauglustaine *et al.* (1998) were used in a forward, i.e., top-down, modelling study of the CO budget; whereas Bergamasschi *et al.* (2000) used a model inversion to derive CO sources. These varied approaches do not yet lead to a consistent picture of the CO budget. Anthropogenic sources (deforestation, savanna and waste burning, fossil and domestic fuel use) dominate the direct emissions of CO, emitting 1,350 out of 1,550 Tg(CO)/yr. A source of 1,230 Tg(CO)/yr is estimated from *in situ* oxidation of CH_4 and other hydrocarbons, and about half of this source can be attributed to anthropogenic emissions. Fossil sources of CO have already been accounted for as release of fossil C in the CO_2 budget, and thus we do not double-count this CO as a source of CO_2.

It has been proposed that CO emissions should have a GWP because of their effects on the lifetimes of other greenhouse gases (Shine *et al.*, 1990; Fuglesvedt *et al.*, 1996; Prather, 1996). Daniel and Solomon (1998) estimate that the cumulative indirect radiative forcing due to anthropogenic CO emissions may be larger than that of N_2O. Combining these early box models with 3-D global CTM studies using models from OxComp (Wild and Prather, 2000; Derwent *et al.*, 2001) suggests that emitting 100 Tg(CO) is equivalent to emitting 5 Tg(CH_4): the resulting CH_4 perturbation appears after a few months and lasts 12 years as would a CH_4 perturbation; and further, the resulting tropospheric O_3 increase is global, the same as for a direct CH_4 perturbation. Effectively the CO emission excites the global 12-year chemical mode that is associated with CH_4 perturbations. This equivalency is not unique as the impact of CO appears to vary by as much as 20% with latitude of emission. Further, this equivalency systematically underestimates the impact of CO on greenhouse gases because it does not include the short-term tropospheric O_3 increase during the early period of very high CO abundances (< 6 months). Such O_3 increases are regional, however, and their magnitude depends on local conditions.

Molecular hydrogen (H_2) is not a direct greenhouse gas. But it can reduce OH and thus indirectly increase CH_4 and HFCs. Its atmospheric abundance is about 500 ppb. Simmonds *et al.* (2000) report a trend of +1.2 ± 0.8 ppb/yr for background air at Mace Head, Ireland between 1994 and 1998; but, in contrast, Novelli *et al.* (1999) report a trend of −2.3 ± 0.1 ppb/yr based on a global network of sampling sites. H_2 is produced in many of the same processes that produce CO (e.g., combustion of fossil fuel and atmospheric oxidation of CH_4), and its atmospheric measurements can be used to constrain CO and CH_4 budgets. Ehhalt (1999) estimates global annual emissions of about 70 Tg(H_2)/yr, of which half are anthropogenic. About one third of atmospheric H_2 is removed by reaction with tropospheric OH, and the remainder, by microbial uptake in soils. Due to the larger land area in the Northern Hemisphere than in the Southern Hemisphere, most H_2 is lost in the Northern Hemisphere. As a result, H_2 abundances are on average greater in the Southern Hemisphere despite 70% of emissions being in the Northern Hemisphere (Novelli *et al.*, 1999; Simmonds *et al.*, 2000). Currently the impact of H_2 on tropospheric OH is small, comparable to some of the VOC. No scenarios for changing H_2 emissions are considered here; however, in a possible fuel-cell economy, future atmospheric emissions may need to be considered as a potential climate perturbation.

Table 4.6: *Estimates of the global tropospheric carbon monoxide budget (in Tg(CO)/yr) from different sources compared with the values adopted for this report (TAR).*

Reference:	Hauglustaine *et al.* (1998)	Bergamasschi *et al.* (2000)	WMO (1999)	SAR (1996)	TAR[a]
Sources					
Oxidation of CH_4		795		400 – 1000	800
Oxidation of Isoprene		268		200 – 600[b]	270
Oxidation of Terpene		136			~0
Oxidation of industrial NMHC		203			110
Oxidation of biomass NMHC		–			30
Oxidation of Acetone		–			20
Sub-total *in situ* oxidation	**881**	**1402**			**1230**
Vegetation		–	100	60 – 160	150
Oceans		49	50	20 – 200	50
Biomass burning[c]		768	500	300 – 700	700
Fossil & domestic fuel		641	500	300 – 550	650
Sub-total direct emissions	**1219**	**1458**	**1150**		**1550**
Total sources	**2100**	**2860**		**1800 – 2700**	**2780**
Sinks					
Surface deposition	190			250 – 640	
OH reaction	1920			1500 – 2700	

Anthropogenic emissions by continent/region	Y2000	Y2100(A2p)			
Africa	80	480			
South America	36	233			
Southeast Asia	44	203			
India	64	282			
North America	137	218			
Europe	109	217			
East Asia	158	424			
Australia	8	20			
Other	400	407			
Sum	1036	2484			

[a] Recommended for OxComp model calculations for year 2000.

[b] Includes all VOC oxidation.

[c] From deforestation, savannah and waste burning.

4.2.3.2 *Volatile organic compounds (VOC)*

Volatile organic compounds (VOC), which include non-methane hydrocarbons (NMHC) and oxygenated NMHC (e.g., alcohols, aldehydes and organic acids), have short atmospheric lifetimes (fractions of a day to months) and small direct impact on radiative forcing. VOC influence climate through their production of organic aerosols and their involvement in photochemistry, i.e., production of O_3 in the presence of NO_x and light. The largest source, by far, is natural emission from vegetation. Isoprene, with the largest emission rate, is not stored in plants and is only emitted during photosynthesis (Lerdau and Keller, 1997). Isoprene emission is an important component in tropospheric photochemistry (Guenther *et al.*, 1995, 1999) and is included in

the OxComp simulations. Monoterpenes are stored in plant reservoirs, so they are emitted throughout the day and night. The monoterpenes play an important role in aerosol formation and are discussed in Chapter 5. Vegetation also releases other VOC at relatively small rates, and small amounts of NMHC are emitted naturally by the oceans. Anthropogenic sources of VOC include fuel production, distribution, and combustion, with the largest source being emissions (i) from motor vehicles due to either evaporation or incomplete combustion of fuel, and (ii) from biomass burning. Thousands of different compounds with varying lifetimes and chemical behaviour have been observed in the atmosphere, so most models of tropospheric chemistry include some chemical speciation of the VOC. Generally, fossil

Table 4.7(a): *Estimates of global VOC emissions (in TgC/yr) from different sources compared with the values adopted for this report (TAR).*

Ehhalt (1999)	Isoprene (C_5H_8)	Terpene ($C_{10}H_{16}$)	C_2H_6	C_3H_8	C_4H_{10}	C_2H_4	C_3H_6	C_2H_2	Benzene (C_6H_6)	Toluene (C_7H_8)
Fossil fuel [a]	–	–	4.8	4.9	8.3	8.6	8.6	2.3	4.6	13.7
Biomass burning	–	–	5.6	3.3	1.7	8.6	4.3	1.8	2.8	1.8
Vegetation	503	124	4.0	4.1	2.5	8.6	8.6	–	–	–
Oceans	–	–	0.8	1.1	–	1.6	1.4	–	–	–

TAR [b]	Total	Isoprene	Terpene	Acetone
Fossil fuel [a]	161			
Biomass burning	33			
Vegetation	377	220	127	30

[a] Fossil includes domestic fuel.
[b] TAR refers to recommended values for OxComp model calculations for the year 2000.

Table 4.7(b): *Detailed breakdown of VOC emissions by species adopted for this report (TAR).*

Species	Industrial wt%	Industrial #C atoms	Biomass burning wt%	Biomass burning #C atoms
Alcohols	3.2	2.5	8.1	1.5
Ethane	4.7	2.0	7.0	2.0
Propane	5.5	3.0	2.0	3.0
Butanes	10.9	4.0	0.6	4.0
Pentanes	9.4	5.0	1.4	5.0
Higher alkanes	18.2	7.5	1.3	8.0
Ethene	5.2	2.0	14.6	2.0
Propene	2.4	3.0	7.0	3.0
Ethyne	2.2	2.0	6.0	2.0
Other alkenes, alkynes, dienes	3.8	4.8	7.6	4.6
Benzene	3.0	6.0	9.5	6.0
Toluene	4.9	7.0	4.1	7.0
Xylene	3.6	8.0	1.2	8.0
Trimethylbenzene	0.7	9.0	–	–
Other aromatics	3.1	9.6	1.0	8.0
Esters	1.4	5.2	–	–
Ethers	1.7	4.7	5.5	5.0
Chlorinated HC's	0.5	2.6	–	–
Formaldehyde	0.5	1.0	1.2	1.0
Other aldehydes	1.6	3.7	6.1	3.7
Ketones	1.9	4.6	0.8	3.6
Acids	3.6	1.9	15.1	1.9
Others	8.1	4.9	–	–

wt% values are given for the individual VOC with the sums being: industrial, 210 Tg(VOC)/yr, corresponding to 161 TgC/yr; and biomass burning, 42 Tg(VOC)/yr, corresonding to 33 TgC/yr.

VOC sources have already been accounted for as release of fossil C in the CO_2 budgets and thus we do not count VOC as a source of CO_2.

Given their short lifetimes and geographically varying sources, it is not possible to derive a global atmospheric burden or mean abundance for most VOC from current measurements. VOC abundances are generally concentrated very near their sources. Natural emissions occur predominantly in the tropics (23°S to 23°N) with smaller amounts emitted in the northern mid-latitudes and boreal regions mainly in the warmer seasons. Anthropogenic emissions occur in heavily populated, industrialised regions (95% in the Northern Hemisphere peaking at 40°N to 50°N), where natural emissions are relatively low, so they have significant impacts on regional chemistry despite small global emissions. A few VOC, such as ethane and acetone, are longer-lived and impact tropospheric chemistry on hemispheric scales. Two independent estimates of global emissions (Ehhalt, 1999; and TAR/OxComp budget based on the Emission Database for Global Atmospheric Research (EDGAR)) are summarised in Table 4.7a. The OxComp specification of the hydrocarbon mixture for both industrial and biomass-burning emissions is given in Table 4.7b.

One of the NMHC with systematic global measurements is ethane (C_2H_6). Rudolph (1995) have used measurements from five surface stations and many ship and aircraft campaigns during 1980 to 1990 to derive the average seasonal cycle for ethane as a function of latitude. Ehhalt *et al.* (1991) report a trend of +0.8%/yr in the column density above Jungfraujoch, Switzerland for the period 1951 to 1988, but in the following years, the trend turned negative. Mahieu *et al.* (1997) report a trend in C_2H_6 of -2.7 ± 0.3%/yr at Jungfraujoch, Switzerland for 1985 to 1993; Rinsland *et al.* (1998) report a trend of -1.2 ± 0.4%/yr at Kitt Peak, Arizona for 1977 to 1997 and -0.6 ± 0.8%/yr at Lauder, New Zealand for 1993 to 1997. It is expected that anthropogenic emissions of most VOC have risen since pre-industrial times due to increased use of gasoline and other hydrocarbon products. Due to the importance of VOC abundance in determining tropospheric O_3 and OH, systematic measurements and analyses of their budgets will remain important in understanding the chemistry-climate coupling.

There is a serious discrepancy between the isoprene emissions derived by Guenther *et al.* (1995) based on a global scaling of emission from different biomes, about 500 TgC/yr, and those used in OxComp for global chemistry-transport modelling, about 200 TgC/yr. When the larger isoprene fluxes are used in the CTMs, many observational constraints on CO and even isoprene itself are poorly matched. This highlights a key uncertainty in global modelling of highly reactive trace gases: namely, what fraction of primary emissions escapes immediate reaction/removal in the vegetation canopy or immediate boundary layer and participates in the chemistry on the scales represented by global models? For the isoprene budget, there are no measurements of the deposition of reaction products within the canopy. More detail on the scaling of isoprene and monoterpene emissions is provided in Chapter 5. Although isoprene emissions are likely to change in response to evolving chemical and climate environment over the next century, this assessment was unable to include a projection of such changes.

4.2.3.3 Nitrogen oxides (NO_x)

Nitrogen oxides ($NO_x = NO + NO_2$) do not directly affect Earth's radiative balance, but they catalyse tropospheric O_3 formation through a sequence of reactions, e.g.,

$$
\begin{array}{lll}
OH + CO + O_2 & \rightarrow & CO_2 + HO_2 \\
HO_2 + NO & \rightarrow & NO_2 + OH \\
NO_2 + h\nu & \rightarrow & NO + O(^3P) \\
O(^3P) + O_2 + M & \rightarrow & O_3 + M \\
\end{array}
$$

$$
\text{net:} \quad CO + 2O_2 + h\nu \quad \rightarrow \quad CO_2 + O_3
$$

By rapidly converting HO_2 to OH, NO enhances tropospheric OH abundances and thus indirectly reduces the atmospheric burdens of CO, CH_4, and HFCs. Much of recent understanding of the role of NO_x in producing tropospheric O_3 and changing OH abundances is derived from *in situ* measurement campaigns that sample over a wide range of chemical conditions in the upper troposphere or at the surface (see Section 4.2.6 on tropospheric OH). These atmospheric measurements generally support the current photochemical models. There is substantial spatial and temporal variability in the measured abundance of NO_x, which ranges from a few ppt near the surface over the remote tropical Pacific Ocean to >100 ppb in urban regions. The local chemical lifetime of NO_x is always short, but varies widely throughout the troposphere, being 1 day or less in the polluted boundary layer, day or night, and 5 to 10 days in the upper troposphere. As with VOC, it is not possible to derive a global burden or average abundance for NO_x from measurements of atmospheric abundances.

Most tropospheric NO_x are emitted as NO, which photochemically equilibrates with nitrogen dioxide (NO_2) within a few minutes. Significant sources, summarised in Table 4.8, include both surface and *in situ* emissions, and only a small amount is transported down from the stratosphere. NO_x emitted within polluted regions are more rapidly removed than those in remote regions. Emissions directly into the free troposphere have a disproportionately large impact on global greenhouse gases. The major source of NO_x is fossil fuel combustion, with 40% coming from the transportation sector. Benkovitz *et al.* (1996) estimated global emissions at 21 TgN/yr for 1985. The NO_x emissions from fossil fuel use used in model studies here for year 2000 are considerably higher, namely 33 TgN/yr. The large American and European emissions are relatively stable, but emissions from East Asia are increasing by about +4%/yr (Kato and Akimoto, 1992). Other important, but more uncertain surface sources are biomass burning and soil emissions. The soil source recently derived from a bottom-up compilation of over 100 measurements from various ecosystems is 21 TgN/yr (Davidson and Kingerlee, 1997), much higher than earlier estimates. Part of the discrepancy can be explained by the trapping of soil-emitted NO in the vegetation canopy. Inclusion of canopy scavenging reduces the NO_x flux to the free troposphere to 13 TgN/yr, which is still twice the flux estimated by another recent study (Yienger and Levy, 1995). Emissions of NO_x in the free troposphere include NO_x from aircraft (8 to 12 km), ammonia oxidation, and lightning (Lee *et al.*, 1997). Estimates of the lightning NO_x source are quite variable; some recent global estimates are 12 TgN/yr (Price *et al.*, 1997a,b), while other studies recommend 3 to 5 TgN/yr (e.g., Huntrieser *et al.*, 1998; Wang *et al.*, 1998a). Recent studies indicate that the global lightning frequency may be lower than previously estimated (Christian *et al.*, 1999) but that intra-cloud lightning may be much more effective at producing NO (DeCaria *et al.*, 2000). In total, anthropogenic NO_x emissions dominate natural sources, with fossil fuel combustion concentrated in northern industrial regions. However, natural sources may control a larger fraction of the globe. Overall, anthropogenic NO_x emissions are expected to undergo a fundamental shift from the current dominance of the

Table 4.8: *Estimates of the global tropospheric NO$_x$ budget (in TgN/yr) from different sources compared with the values adopted for this report.*

Reference:	TAR	Ehhalt (1999)	Holland *et al.* (1999)	Penner *et al.* (1999)	Lee *et al.* (1997)
Base year	2000	~1985	~1985	1992	
Fossil fuel	33.0	21.0	20 – 24	21.0	22.0
Aircraft	0.7	0.45	0.23 – 0.6	0.5	0.85
Biomass burning	7.1	7.5	3 – 13	5 – 12	7.9
Soils	5.6	5.5	4 – 21	4 – 6	7.0
NH$_3$ oxidation	–	3.0	0.5 – 3	–	0.9
Lightning	5.0	7.0	3 – 13	3 – 5	5.0
Stratosphere	<0.5	0.15	0.1 –0.6	–	0.6
Total	**51.9**	**44.6**			**44.3**

Anthropogenic emissions by continent/region	Y2000	Y2100(A2p)
Africa	2.5	21.8
South America	1.4	10.8
Southeast Asia	1.2	6.8
India	1.7	10.0
North America	10.1	18.5
Europe	7.3	14.3
East Asia	5.6	24.1
Australia	0.5	1.1
Other	2.3	2.6
Sum	**32.6**	**110.0**

The TAR column was used in OxComp model calculations for year 2000; fossil fuel includes bio-fuels, but surface sources only; stratospheric source in TAR is upper limit and includes HNO$_3$; the range of values used in modelling for IPCC aviation assessment (Penner *et al.* 1999) is given.

Northern Hemisphere to a more tropical distribution of emissions. Asian emissions from fossil fuel are expected to drive an overall increase in NO$_x$ emissions in the 21st century (Logan, 1994; Van Aardenne *et al.*, 1999).

The dominant sink of NO$_x$ is atmospheric oxidation of NO$_2$ by OH to form nitric acid (HNO$_3$), which then collects on aerosols or dissolves in precipitation and is subsequently scavenged by rainfall. Other pathways for direct NO$_x$ removal occur through canopy scavenging of NO$_x$ and direct, dry deposition of NO$_x$, HNO$_3$, and particulate nitrates to the land surface and the ocean. Dry deposition can influence the surface exchanges and can thus alter the release of NO$_x$ and N$_2$O to the atmosphere. Peroxyacetyl nitrate (PAN), formed by the reaction of CH$_3$C(O)O$_2$ with NO$_2$, can transport HO$_x$ and NO$_x$ to remote regions of the atmosphere due to its stability at the cold temperatures of the upper troposphere. In addition tropospheric aerosols provide surfaces on which reactive nitrogen, in the form of NO$_3$ (nitrate radical) or N$_2$O$_5$, is converted to HNO$_3$ (Dentener and Crutzen, 1993; Jacob, 2000).

Some CTM studies argue against either the large soil source or the large lightning source of NO$_x$. A climatology of NO$_x$ measurements from aircraft was prepared by Emmons *et al.*

(1997) and compared with six chemical transport models. They found that the processes controlling NO$_x$ in the remote troposphere are not well modelled and that, of course, there is a paucity of global NO$_x$ measurements. For short-lived gases like NO$_x$, resolution of budget discrepancies is even more challenging than for the long-lived species, because the limited atmospheric measurements offer few real constraints on the global budget. However, an additional constraint on the NO$_x$ budget is emerging as the extensive measurements of wet deposition of nitrate over Northern Hemisphere continents are compiled and increasing numbers of surface measurements of dry deposition of HNO$_3$, NO$_2$, and particulate nitrate become available, and thus allow a much better estimate of the NO$_x$ sink.

4.2.4 Tropospheric O$_3$

Tropospheric O$_3$ is a direct greenhouse gas. The past increase in tropospheric O$_3$ is estimated to provide the third largest increase in direct radiative forcing since the pre-industrial era. In addition, through its chemical impact on OH, it modifies the lifetimes of other greenhouse gases, such as CH$_4$. Its budget, however, is much more difficult to derive than that of a long-lived gas for

Table 4.9: *Estimates of the change in tropospheric ozone since the pre-industrial era from various sources compared with the values recommended in this report.*

Current climatology of tropospheric ozone (Park *et al.*, 1999):

Global mean tropospheric O_3: 34 DU = 370 Tg(O)$_3$ content in the Northern Hemisphere = 36 DU, in the Southern Hemisphere = 32 DU.

SAR recommendation:

"50% of current Northern Hemisphere is anthropogenic" gives pre-industrial global mean content = 25 DU.

Increase = +9 DU

19th & early 20th century observations:[a]

Assume Northern Hemisphere tropospheric ozone has increased uniformly by >30 ppb.

Increase = +10 to +13 DU

Survey of CTM simulated change from pre-industrial:[b]

DU increase	Model	Reference
9.6	UIO	Berntsen *et al.* (1999)
7.9	GFDL	Haywood *et al.* (1998)
8.9	MOZART-1	Hauglustaine *et al.* (1998)
8.4	NCAR/2D	Kiehl *et al.* (1999)
9.5	GFDL-scaled	Levy *et al.* (1997)
12.0	Harvard/GISS	Mickley *et al.* (1999)
7.2	ECHAM4	Roelofs *et al.* (1997)
8.7	UKMO	Stevenson *et al.* (2000)
8.0	MOGUNTIA	VanDorland *et al.* (1997)

Increase = +7 to +12 DU (model range)

TAR recommendation:

Pre-industrial era global mean tropospheric O_3 has increased from 25 DU to 34 DU.

This increase, +9 DU, has a 67% likely range of 6 to 13 DU.

Increase = +9 DU (+6 to +13 DU)

The troposphere is defined as air with O_3 <150 ppb, see Logan (1999). The Dobson Unit is 1 DU = 2.687×10^{16} molecules of O_3 per square centimetre; globally 1 DU = 10.9 Tg(O_3) and 1 ppb of tropospheric O_3 = 0.65 DU. The change in CH_4 alone since pre-industrial conditions would give about +4 DU global increase in tropospheric O_3 alone (see Table 4.11).

[a] Early observations are difficult to interpret and do not provide coverage needed to derive the tropospheric burden of O_3 (see Harris *et al.*, 1997). The change in burden is derived here by shifting tropospheric O_3 uniformly in altitude to give 10 ppb at the surface in Northern Hemisphere mid-latitudes and 20 ppb at surface in Northern Hemisphere tropics (implies 10 DU), or by additionally reducing Southern Hemisphere tropics to 20 ppb and Southern Hemisphere mid-latitudes to 25 ppb at the surface (13 DU).

[b] From a survey of models by Hauglustaine and Solomon and Chapter 4. Except for Kiehl *et al.*, these were all CTMs; they used widely varying assumptions about pre-industrial conditions for CH , CO, NO , and biomass burning and they did not all report consistent diagnostics.

---

several reasons. Ozone abundances in the troposphere typically vary from less than 10 ppb over remote tropical oceans up to about 100 ppb in the upper troposphere, and often exceed 100 ppb downwind of polluted metropolitan regions. This variability, reflecting its rapid chemical turnover, makes it impossible to determine the tropospheric burden from the available surface sites, and we must rely on infrequent and sparsely sited profiles from ozone sondes (e.g., Logan, 1999). The total column of ozone is measured from satellites, and these observations have been used to infer the tropospheric ozone column after removing the much larger stratospheric column (e.g., Fishman and Brackett, 1997; Hudson and Thompson, 1998; Ziemke *et al.*, 1998). The current burden of tropospheric O_3 is about 370 Tg(O_3), which is equivalent to a globally averaged column density of 34 DU (Dobson Units, 1 DU = 2.687×10^{16} molecules/cm^{-2}) or a mean abundance of about 50 ppb, see Table 4.9.

The sources and sinks of tropospheric ozone are even more difficult to quantify than the burden. Influx of stratospheric air is a source of about 475 Tg(O_3)/yr based on observed correlations with other gases (Murphy and Fahey, 1994; McLinden *et al.*, 2000). The *in situ* photochemical sources are predicted to be many times larger, but are nearly balanced by equally large *in situ* chemical sinks (see discussion on CTM modelling of tropospheric O_3 in Sections 4.4 and 4.5, Table 4.12). Photochemical production of ozone is tied to the abundance of pollutants and thus varies widely over a range of spatial scales, the most important of which (e.g., biomass burning plumes, urban plumes, aircraft corridors, and convective outflows) are not well represented in most global CTMs and cannot be quantified globally with regional models. The dominant photochemical sinks for tropospheric O_3 are the catalytic destruction cycle involving the $HO_2 + O_3$ reaction and photolytic destruction by pathways involving the reaction of $O(^1D)$, a product of O_3 photodissociation. The other large sink, comparable in magnitude to the stratospheric source, is surface loss mainly to vegetation. Another loss of O_3 is observed under certain conditions in the polar marine boundary layer, notably at the end of Arctic winter. It indicates reactions involving halogen radicals and aerosols (Oum *et al.*, 1998; Dickerson *et al.*, 1999; Impey *et al.*, 1999; Platt and Moortgat, 1999; Prados *et al.*, 1999; Vogt *et al.*, 1999). The contribution of these processes to the global budget is not yet quantified, but is probably small.

Atmospheric measurement campaigns, both at surface sites and with aircraft, have focused on simultaneous observations of the many chemical species involved in tropospheric O_3 production. Primary areas of O_3 production are the mid-latitude industrialised and tropical biomass burning regions. For example, the North Atlantic Regional Experiment (NARE) and the Atmosphere Ocean Chemistry Experiment (AEROCE) showed that the prevailing westerly winds typically carry large quantities of ozone and precursors from the eastern USA over the North Atlantic, reaching Bermuda and beyond (e.g., Dickerson *et al.*, 1995; Penkett *et al.*, 1998; Prados *et al.*, 1999). The Pacific Exploratory Missions (PEM: Hoell *et al.*, 1997, 1999) measured extensive plumes of pollution including ozone and its precursors downwind of eastern Asia. Convective transport of emissions from biomass burning affect the abundance of O_3 in the mid- and upper troposphere (Pickering *et al.*, 1996). Emissions by tropical fires in South America and southern Africa have been identified as the cause of enhanced O_3 over the South Atlantic (Thompson *et al.*, 1996), and the effects of biomass burning were seen in the remote South Pacific in PEM Tropics A (Schultz *et al.*, 1999; Talbot *et al*., 1999). Due to the widely varying chemical environments, these extensive studies provide a statistical sampling of conditions along with a critical test of the photochemistry in CTM simulations, but they do not provide an integrated budget for tropospheric O_3. An example of such model-and-measurements study is given in the Section 4.2.6 discussion of tropospheric OH.

Recent trends in global tropospheric O_3 are extremely difficult to infer from the available measurements, while trends in the stratosphere are readily identified (Randel *et al.*, 1999; WMO, 1999). With photochemistry producing local lifetimes as short as

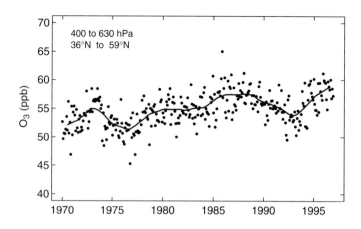

Figure 4.8: Mid-tropospheric O_3 abundance (ppb) in northern mid-latitudes (36°N-59°N) for the years 1970 to 1996. Observations between 630 and 400 hPa are averaged from nine ozone sonde stations (four in North America, three in Europe, two in Japan), following the data analysis of Logan *et al.* (1999). Values are derived from the residuals of the trend fit with the trend added back to allow for discontinuities in the instruments. Monthly data (points) are shown with a smoothed 12-month-running mean (line).

a few days in the boundary layer, the local measurement of tropospheric O_3 does not reflect the abundance over the surrounding continent, and a surface measurement is not representative of the bulk of the troposphere above. Thus it is not contradictory for decadal trends in different atmospheric regions to be different, e.g., driven by the regional changes in pollutants, particularly NO_x. Ozone sondes offer the best record of O_3 throughout the troposphere, although measurements at many stations are made only weekly (infrequently for a variable gas like O_3). Weekly continuous data since 1970 are available from only nine stations in the latitude range 36°N to 59°N (Logan *et al.*, 1999; WMO, 1999). Different trends are seen at different locations for different periods. Most stations show an increase from 1970 to 1980, but no clear trend from 1980 to 1996. A composite record of the mid-tropospheric O_3 abundance from 1970 to 1996 from the nine stations is taken from the analysis of Logan *et al.* (1999) and presented in Figure 4.8. There is no obvious linear increase in O_3 abundance over this period, although the second half of this record (about 57 ppb) is clearly greater than the first half (about 53 ppb). Of the fourteen stations with records since 1980, only two, one in Japan and one in Europe, had statistically significant increases in mid-tropospheric O_3 between 1980 and 1995. By contrast, the four Canadian stations, all at high latitudes, had significant decreases for the same time period. Surface O_3 measurements from seventeen background stations also show no clear trend, even in the northern mid-latitudes (Oltmans *et al.*, 1998; WMO, 1999). The largest negative trend in surface O_3 was −0.7 ± 0.2%/yr at the South Pole (1975 to 1997), while the largest positive trend was +1.5 ± 0.5%/yr at Zugspitze, Germany (1978 to 1995). This ambiguous record of change over the past two decades may possibly be reconciled with the model predictions (see Section 4.4) of increasing tropospheric O_3 driven regionally by increasing emissions of pollutants: the growth in NO_x emissions is expected to have shifted from North America and Europe to Asia.

The change in tropospheric O_3 since the pre-industrial era is even more difficult to evaluate on the basis of measurements alone. Since O_3 is reactive, atmospheric abundances cannot be retrieved from ice cores. Recent evaluations of surface measurements in the 19th and early 20th century in Europe (Volz and Kley, 1988; Staehelin *et al.*, 1994, 1998; Harris *et al.*, 1997) indicate much lower O_3 abundances than today, yet the scaling of these data to a tropospheric O_3 burden, even for northern mid-latitudes, is not obvious. In the SAR, these data were used to make a rough estimate that O_3 abundances in the Northern Hemispheric troposphere have doubled since the pre-industrial era. A similar difference, of 10 to 13 DU when globally averaged, is obtained using the climatology given by Park *et al.* (1999) for tropospheric O_3 today and a parallel one with abundances adjusted to match the 19th century measurements in the Northern Hemisphere. CTM calculations predict that current anthropogenic emissions of NO_x, CO, and VOC, as well as the increase in CH_4 should have increased tropospheric O_3 by a similar amount, primarily in the Northern Hemisphere. A recent survey of CTM studies gives global average increases ranging from 8 to 12 DU, although this small range does not adequately represent the uncertainty. These results are summarised in Table 4.9. Based on measurements, analyses, and models, the most likely increase in tropospheric O_3 was 9 DU globally averaged, with a 67% confidence range of 6 to 13 DU. For some of the emissions scenarios considered here, tropospheric O_3 is expected to increase even more in the 21st century as emissions of its precursors – NO_x, CO and VOC – continue to grow (see Section 4.4).

4.2.5 Stratospheric H_2O

Water vapour in the lower stratosphere is a very effective greenhouse gas. Baseline levels of stratospheric H_2O are controlled by the temperature of the tropical tropopause, a parameter that changes with climate (Moyer *et al.*, 1996; Rosenlof *et al.*, 1997; Dessler, 1998; Mote *et al.*, 1998). The oxidation of CH_4 is a source of mid-stratospheric H_2O and currently causes its abundance to increase from about 3 ppm at the tropopause to about 6 ppm in the upper stratosphere. In addition, future direct injections of H_2O from high-flying aircraft may add H_2O to the lower stratosphere (Penner *et al.*, 1999). Oltmans and Hofmann (1995) report statistically significant increases in lower stratospheric H_2O above Boulder, Colorado between 1981 and 1994. The vertical profile and amplitude of these changes do not correspond quantitatively with that expected from the recognised anthropogenic sources (CH_4 oxidation). Analyses of satellite and ground-based measurements (Nedoluha *et al.*, 1998; Michelsen *et al.*, 2000) find increases in upper stratospheric H_2O from 1985 to 1997, but at rates (>1%/yr) that exceed those from identified anthropogenic sources (i.e., aviation and methane increases) and that obviously could not have been maintained over many decades. In principle such a temporary trend could be caused by a warming tropopause, but a recent analysis indicates instead a cooling tropopause (Simmons *et al.*, 1999). It is important to resolve these apparent discrepancies; since, without a physical basis for this recent trend, no recommendation can be made here for projecting changes in lower stratospheric H_2O over the 21st century.

4.2.6 Tropospheric OH and Photochemical Modelling

The hydroxyl radical (OH) is the primary cleansing agent of the lower atmosphere, in particular, it provides the dominant sink for CH_4 and HFCs as well as the pollutants NO_x, CO and VOC. Once formed, tropospheric OH reacts with CH_4 or CO within a second. The local abundance of OH is controlled by the local abundances of NO_x, CO, VOC, CH_4, O_3, and H_2O as well as the intensity of solar UV; and thus it varies greatly with time of day, season, and geographic location.

The primary source of tropospheric OH is a pair of reactions that start with the photodissociation of O_3 by solar UV.

$$O_3 + h\nu \quad \rightarrow \quad O(^1D) + O_2$$
$$O(^1D) + H_2O \quad \rightarrow \quad OH + OH$$

Although in polluted regions and in the upper troposphere, photodissociation of other trace gases such as peroxides, acetone and formaldehyde (Singh *et al.*, 1995; Arnold *et al.*, 1997) may provide the dominant source (e.g., Folkins *et al.*, 1997; Prather and Jacob, 1997; Wennberg *et al.*, 1998; Müller and Brasseur, 1999). OH reacts with many atmospheric trace gases, in most cases as the first and rate-determining step of a reaction chain that leads to more or less complete oxidation of the compound. These chains often lead to formation of an HO_2 radical, which then reacts with O_3 or NO to recycle back to OH. Tropospheric OH is lost through reactions with other radicals, e.g., the reaction with HO_2 to form H_2O or with NO_2 to form HNO_3. In addition to providing the primary loss for CH_4 and other pollutants, HO_x radicals (OH and HO_2) together with NO_x are key catalysts in the production of tropospheric O_3 (see Section 4.2.3.3). The sources and sinks of OH involve most of the fast photochemistry of the troposphere.

Pre-industrial OH is likely to have been different than today, but because of the counteracting effects of lower CO and CH_4 (increasing OH) and reduced NO_x and O_3 (decreasing OH), there is no consensus on the magnitude of this change (e.g., Wang and Jacob, 1998). Trends in the current OH burden appear to be <1%/yr. Separate analyses of the CH_3CCl_3 observations for the period 1978 to 1994 report two different but overlapping trends in global OH: no trend within the uncertainty range (Prinn *et al.*, 1995), and $0.5 \pm 0.6\%$/yr (Krol *et al.*, 1998). Based on the OxComp workshop, the SRES projected emissions would lead to future changes in tropospheric OH that ranging from +5% to −20% (see Section 4.4).

4.2.6.1 Laboratory data and the OH lifetime of greenhouse gases

Laboratory data on the rates of chemical reactions and photodissociation provide a cornerstone for the chemical models used here. Subsequent to the SAR there have been a number of updates to the recommended chemical rate databases of the International Union of Pure and Applied Chemistry (IUPAC 1997a,b, 1999) and the Jet Propulsion Laboratory (JPL) (DeMore *et al.*, 1997; Sander *et al.*, 2000). The CTMs in the OxComp workshop generally used the JPL-1997 database (JPL, 1997) with some updated rates similar to JPL-2000 (JPL, 2000). The most significant changes or additions to the databases include: (i) revision of the low temperature reaction rate coefficients for OH + NO_2 leading to enhancement of HO_x

and NO_x abundances in the lower stratosphere and upper troposphere; (ii) extension of the production of $O(^1D)$ from O_3 photodissociation to longer wavelengths resulting in enhanced OH production in the upper troposphere; and (iii) identification of a new heterogeneous reaction involving hydrolysis of $BrONO_2$ which serves to enhance HO_x and suppress NO_x in the lower stratosphere. These database improvements, along with many other smaller refinements, do not change the overall understanding of atmospheric chemical processes but do impact the modelled tropospheric OH abundances and the magnitude of calculated O_3 changes by as much as 20% under certain conditions.

Reaction rate coefficients used in this chapter to calculate atmospheric lifetimes for gases destroyed by tropospheric OH are from the 1997 and 2000 NASA/JPL evaluations (DeMore *et al.*, 1997; Sander *et al.*, 2000) and from Orkin *et al.* (1999) for HFE-356mff2. These rate coefficients are sensitive to atmospheric temperature and can be ten times faster near the surface than in the upper troposphere. The global mean abundance of OH cannot be directly measured, but a weighted average of the OH sink for certain synthetic trace gases (whose budgets are well established and whose total atmospheric sinks are essentially controlled by OH) can be derived. The ratio of the atmospheric lifetimes against tropospheric OH loss for a gas is scaled to that of CH_3CCl_3 by the inverse ratio of their OH-reaction rate coefficients at an appropriate scaling temperature. A new analysis of the modelled global OH distribution predicts relatively greater abundances at mid-levels in the troposphere (where it is colder) and results in a new scaling temperature for the rate coefficients of 272K (Spivakovsky *et al.*, 2000), instead of 277K (Prather and Spivakovsky, 1990; SAR). The atmospheric lifetimes reported in Table 4.1 use this approach, adopting an "OH lifetime" of 5.7 years for CH_3CCl_3 (Prinn *et al.*, 1995; WMO, 1999). Stratospheric losses for all gases are taken from published values (Ko *et al.*, 1999; WMO, 1999) or calculated as 8% of the tropospheric loss (with a minimum lifetime of 30 years). The only gases in Table 4.1 with surface losses are CH_4 (a soil-sink lifetime of 160 years) and CH_3CCl_3 (an ocean-sink lifetime of 85 years). The lifetime for nitrogen trifluoride (NF_3) is taken from Molina *et al.* (1995). These lifetimes agree with the recent compendium of Naik *et al.* (2000).

Analysis of the CH_3CCl_3 burden and trend (Prinn *et al.*, 1995; Krol *et al.*, 1998; Montzka *et al.*, 2000) has provided a cornerstone of our empirical derivations of the OH lifetimes of most gases. Quantification of the "OH-lifetime" of CH_3CCl_3 has evolved over the past decade. The SAR adopted a value of 5.9 ± 0.7 years in calculating the lifetimes of the greenhouse gases. This range covered the updated analysis of Prinn *et al.* (1995), 5.7 years, which was used in WMO (1999) and adopted for this report. Montzka *et al.* (2000) extend the atmospheric record of CH_3CCl_3 to include the rapid decay over the last five years following cessation of emissions and derive an OH lifetime of 6.3 years. The new information on the CH_3CCl_3 lifetime by Montzka *et al.* (2000) has not been incorporated into this report, but it falls within the ±15% uncertainty for these lifetimes. If the new value of 6.3 years were adopted, then the lifetime of CH_4

would increase to 9.2 yr, and all lifetimes, perturbation lifetimes, and GWPs for gases controlled by tropospheric OH would be about 10% greater.

4.2.6.2 *Atmospheric measurements and modelling of photochemistry*

Atmospheric measurements provide another cornerstone for the numerical modelling of photochemistry. Over the last five years direct atmospheric measurements of HO_x radicals, made simultaneously with the other key species that control HO_x, have been conducted over a wide range of conditions: the upper troposphere and lower stratosphere (e.g., SPADE, ASHOE/MAESA, STRAT; SUCCESS, SONEX, PEM-TROPICS A & B), the remote Pacific (MLOPEX), and the polluted boundary layer and its outflow (POPCORN, NARE, SOS). These intensive measurement campaigns provide the first thorough tests of tropospheric OH chemistry and production of O_3 for a range of global conditions. As an example here, we present an analysis of the 1997 SONEX (Subsonic assessment program Ozone and Nitrogen oxide EXperiment) aircraft campaign over the North Atlantic that tests one of the chemical models from the OxComp workshop (HGIS).

The 1997 SONEX aircraft campaign over the North Atlantic provided the first airborne measurements of HO_x abundances concurrent with the controlling chemical background: H_2O_2, CH_3OOH, CH_2O, O_3, NO_x, H_2O, acetone and hydrocarbons. These observations allowed a detailed evaluation of our understanding of HO_x chemistry and O_3 production in the upper troposphere. Figure 4.9 (panels 1-3) shows a comparison between SONEX measurements and model calculations (Jaeglé *et al.*, 1999) for OH and HO_2 abundances and the ratio HO_2/OH. At each point the model used the local, simultaneously observed chemical abundances. The cycling between OH and HO_2 takes place on a time-scale of a few seconds and is mainly controlled by reaction of OH with CO producing HO_2, followed by reaction of HO_2 with NO producing OH. This cycle also leads to the production of ozone. As seen in Figure 4.9, the HO_2/OH ratio is reproduced by model calculations to within the combined uncertainties of observations (±20%) and those from propagation of rate coefficient errors in the model (±100%), implying that the photochemical processes driving the cycling between OH and HO_2 appear to be understood (Wennberg *et al.*, 1998; Brune *et al.*, 1999). The absolute abundances of OH and HO_2 are matched by model calculations to within 40% (the reported accuracy of the HO_x observations) and the median model-to-observed ratio for HO_2 is 1.12. The model captures 80% of the observed variance in HO_x, which is driven by the local variations in NO_x and the HO_x sources (Faloona *et al.*, 2000, Jaeglé *et al.*, 2000;). The predominant sources of HO_x during SONEX were reaction of $O(^1D)$ with H_2O and photodissociation of acetone; the role of H_2O_2 and CH_3OOH as HO_x sources was small. This was not necessarily the case in some of the other airborne campaigns, where large differences between measured and modelled OH, up to a factor of 5, were observed in the upper troposphere. In these campaigns the larger measured OH concentrations were tentatively ascribed to enhanced levels of OH precursors, such as H_2O_2, CH_3OOH, or CH_2O, whose concentrations had not been measured.

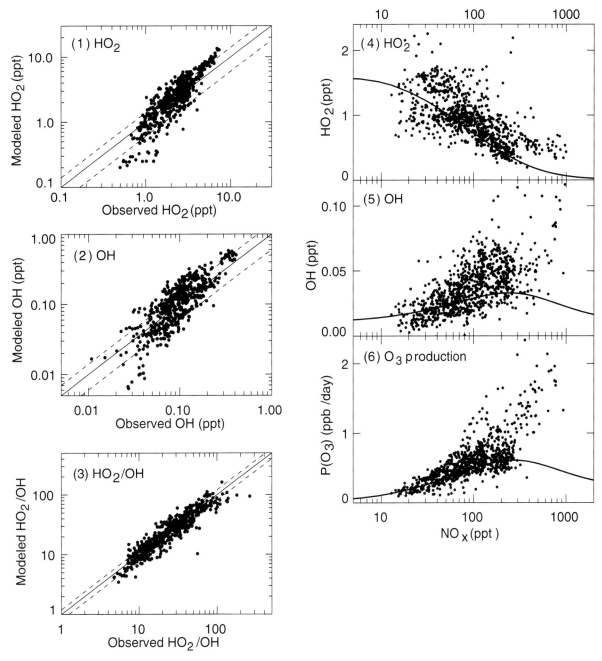

Figure 4.9: (left panel) Observed versus modelled (1) HO_2 abundance (ppt), (2) OH abundance (ppt), and (3) HO_2/OH ratio in the upper troposphere (8 to 12 km altitude) during SONEX. Observations are for cloud-free, daytime conditions. Model calculations are constrained with local observations of the photochemical background (H_2O_2, CH_3OOH, NO, O_3, H_2O, CO, CH_4, ethane, propane, acetone, temperature, pressure, aerosol surface area and actinic flux). The 1:1 line (solid) and instrumental accuracy range (dashed) are shown. Adapted from Brune *et al.* (1999). (right panel) Observed (4) HO_2 abundance (ppt), (5) OH abundance (ppt), and (6) derived O_3 production rate (ppb/day) as a function of the NO_x (NO+NO_2) abundance (ppt). Data taken from SONEX (8 to 12 km altitude, 40° to 60°N latitude) and adapted from Jaeglé *et al.* (1999). All values are 24-hour averages. The lines correspond to model-calculated values as a function of NO_x using the median photochemical background during SONEX rather than the instantaneous values (points).

Tropospheric O_3 production is tightly linked to the abundance of NO_x, and Figure 4.9 (panel 6) shows this production rate (calculated as the rate of the reaction of HO_2 with NO) for each set of observations as function of NO_x during the SONEX mission. Also shown in Figure 4.9 (panels 4-5) are the measured abundances of OH and HO_2 as a function of NO_x. The smooth curve on each panel 4-6 is a model simulation of the expected relationship if the chemical background except for NO_x remained unchanged at the observed median abundances. This curve shows the "expected" behaviour of tropospheric chemistry when only NO_x is increased: OH increases with NO_x abundances up to 300 ppt because HO_2 is shifted into OH; it decreases with increasing NO_x at higher NO_x abundances because the OH reaction with NO_2 forming HNO_3 becomes the dominant sink for HO_x radicals.

Production of O_3 is expected to follow a similar pattern with rates suppressed at NO_x abundances greater than 300 ppt under these atmospheric conditions (e.g., Ehhalt, 1998). These SONEX observations indicate, however, that both OH abundance and O_3 production may continue to increase with NO_x concentrations up to 1,000 ppt because the high NO_x abundances were often associated with convection and lightning events and occurred simultaneously with high HO_x sources. By segregating observations according to HO_x source strengths, Jaeglé *et al.* (1999) identified the approach to NO_x-saturated conditions predicted by the chemical models when HO_x sources remain constant. A NO_x-saturated environment was clearly found for the POPCORN (Photo-Oxidant formation by Plant emitted Compounds and OH Radicals in north-eastern Germany) boundary layer measurements in Germany (Rohrer *et al.*, 1998; Ehhalt and Rohrer, 2000). The impact of NO_x-saturated conditions on the production of O_3 is large in the boundary layer, where much of the NO_x is removed within a day, but may be less important in the upper troposphere, where the local lifetime of NO_x is several days and the elevated abundances of NO_x are likely to be transported and diluted to below saturation levels. This effective reduction of the NO_x-saturation effect due to 3-D atmospheric mixing is seen in the CTM modelling of aviation NO_x emissions where a linear increase in tropospheric O_3 is found, even with large NO_x emissions in the upper troposphere (Isaksen and Jackman, 1999).

4.3 Projections of Future Emissions

The IPCC SRES (Nakićenović *et al.*, 2000) developed 40 future scenarios that are characterised by distinctly different levels of population, economic, and technological development. Six of these scenarios were identified as illustrative scenarios and these were used for the analyses presented in this chapter. The SRES scenarios define only the changes in anthropogenic emissions and not the concurrent changes in natural emissions due either to direct human activities such as land-use change or to the indirect impacts of climate change. The annual anthropogenic emissions for all greenhouse gases, NO_x, CO, VOC and SO_2 (sulphur dioxide) are given in the SRES for the preliminary marker scenarios (Nakićenović *et al.*, 2000, Appendix VI) and the final marker/illustrative scenarios (Nakićenović *et al.*, 2000, Appendix VII). Much of these data is also tabulated in Appendix II to this report. There are insufficient data in the published SRES (Nakićenović *et al.*, 2000) to break down the individual contributions to HFCs, PFCs, and SF_6, but these emissions were supplied by Lead Authors of the SRES (available at sres.ciesin.org) and are also reproduced in this Appendix. The geographic distribution of emissions of the short-lived compounds – NO_x, CO, VOC, and SO_2 – is an important factor in their greenhouse forcing, and the preliminary gridded emissions were likewise supplied by the SRES Lead Authors (Tom Kram and Steven Smith, December 1998) and used in the OxComp model studies. A synopsis of the regional shift in CO and NO_x emissions projected by 2100 is given in Tables 4.6 and 4.8.

This chapter evaluates the SRES emissions from year 2000 to year 2100 in terms of their impact on the abundances of non-CO_2 greenhouse gases. A new feature of this report, i.e., use of NO_x,

CO and VOC emissions to project changes in tropospheric O_3 and OH, represents a significant advance over the level-of-science in the SAR. The original four preliminary marker scenarios (December 1998) are included here because they have been used in preliminary model studies for the TAR and are designated A1p, A2p, B1p, B2p. In January 1999, these emissions were converted into greenhouse gas abundances using the level-of-science and methodology in the SAR, and the radiative forcings from these greenhouse gas abundances were used in this report for some climate model simulations.

The recently approved six marker/illustrative scenarios (March 2000) are also evaluated and are designated A1B-AIM, A1T-MESSAGE, A1FI-MiniCAM, A2-ASF, B1-IMAGE, B2-MESSAGE (hereafter abbreviated as A1B, A1T, A1FI, A2, B1, B2). For comparison with the previous assessment, we also evaluate the IPCC emissions scenario IS92a used in the SAR; for the full range of IS92 scenarios, see the SAR. An agreed-upon property of all SRES scenarios is that there is no intervention to reduce greenhouse gases; but, in contrast, regional controls on SO_2 emissions across the illustrative SRES scenarios lead to emissions in the last two decades of the century that are well below those of 1990 levels. There appear to be few controls on NO_x, CO and VOC emissions across all scenarios; however, the large increases in surface O_3 abundances implied by these results may be inconsistent with the SRES storylines that underpin the emissions scenarios. As understanding of the relationship between emissions and tropospheric O_3 abundances improves, particularly on regional scales, more consistent emissions scenarios can be developed. The SRES scenarios project substantial emissions of HFC-134a as in IS92a, but only half as much HFC-125, and no emissions of HFC-152a. The SRES emissions scenarios do include a much larger suite of HFCs plus SF_6 and PFCs, which are not included in IS92a. The emissions of greenhouse gases under the Montreal Protocol and its Amendments (CFCs, HCFCs, halons) have been evaluated in WMO (Madronich and Velders, 1999). This report adopts the single WMO baseline Montreal Protocol Scenario A1 (no relation to SRES A1) for emissions and concentrations of these gases, while the SRES adopted a similar WMO Scenario A3 (maximum production); however, the differences between scenarios in terms of climate forcing is inconsequential. The resulting abundances of greenhouse gases are given in Appendix II and discussed in Section 4.4.5.

4.3.1 The Adjusted/Augmented IPCC/SRES Emission Scenarios

Among the four SRES preliminary marker scenarios, A2p has overall the highest emissions. For model simulations of future atmospheric chemistry in the OxComp workshop, we needed to focus on a single test case and chose scenario A2p in the year 2100 since it represents the largest increase in emissions of CH_4, CO, NO_x, and VOC. Once the response of O_3 and OH to these extreme emissions is understood, other scenarios and intermediate years can be interpolated with some confidence.

Y2000

For the OxComp workshop, we adopt Y2000 emissions that

include both natural and anthropogenic sources. The OxComp Y2000 anthropogenic emissions are roughly consistent with, but different in detail from, the anthropogenic emissions provided by SRES. These adjustments were necessary to be consistent with current budgets, to include natural sources as discussed previously, and to provide more detailed information on source categories, including temporal and spatial distribution of emissions that are not specified by SRES. Emissions of NO_x, CO and VOC for the year 2000 are based on GEIA(Global Emissions Inventory Activity)/EDGAR emissions for 1990 (Graedel *et al.*, 1993; Olivier *et al.*, 1999) projected to year 2000. Tropospheric abundances of long-lived gases such as CH_4 were fixed from recent observations. The difference between SRES and OxComp Y2000 emissions are nominally within the range of uncertainty for these quantities. The OxComp Y2000 simulations provide a "current" atmosphere to compare with observations.

Y2100(A2p)

Since the OxComp Y2000 emissions differ somewhat from the A2p emissions for the year 2000, we define Y2100(A2p) emissions by the sum of our adjusted Y2000 emissions plus the difference between the SRES-A2p emissions for the years 2100 and 2000. Thus our absolute increase in emissions matches that of SRES-A2p. In these Y2100(A2p) simulations, natural emissions were not changed.

4.3.2 Shifting Regional Emissions of NO_x CO, and VOC in 2100

A shift of the growth of anthropogenic emissions of NO_x, CO and VOC, such as that from North America and Europe to Southern and Eastern Asia over the past decades, is changing the geographic pattern of emissions, which in turn will change the distribution of the O_3 increases in the troposphere predicted for the year 2100. In contrast, for long-lived greenhouse gases, shifting the location of emissions has little impact. We use the SRES emission maps, to take into account such changes in emissions patterns. For Y2000 and Y2100(A2p) the emissions of CO and NO_x, broken down by continents, are given in Tables 4.6 and 4.8, respectively. In terms of assessing future changes in tropospheric OH and O_3, it is essential to have a coherent model for emissions scenarios that consistently projects the spatial patterns of the emissions along with the accompanying changes in urbanisation and land use.

4.3.3 Projections of Natural Emissions in 2100

SRES scenarios do not consider the changes in natural emissions and sinks of reactive gases that are induced by alterations in land use and agriculture or land-cover characteristics. (Land-use change statistics, however, are reported, and these could, in principle, be used to estimate such changes.) In some sense these altered emissions must be considered as anthropogenic changes. Examples of such changes may be increased NO_x, N_2O and NH_3 emissions from natural waters and ecosystems near agricultural areas with intensified use of N-fertiliser. A change of land cover, such as deforestation, may lead to reduced isoprene emissions but

to increases in soil emissions of NO_x. At present we can only point out the lack of projecting these parallel changes in once natural emissions as an uncertainty in this assessment.

4.4 Projections of Atmospheric Composition for the 21st Century

4.4.1 Introduction

Calculating the abundances of chemically reactive greenhouse gases in response to projected emissions requires models that can predict how the lifetimes of these gases are changed by an evolving atmospheric chemistry. This assessment focuses on predicting changes in the oxidative state of the troposphere, specifically O_3 (a greenhouse gas) and OH (the sink for many greenhouse gases). Many research groups have studied and predicted changes in global tropospheric chemistry, and we seek to establish a consensus in these predictions, using a standardised set of scenarios in a workshop organised for this report. The projection of stratospheric O_3 recovery in the 21st century – also a factor in radiative forcing and the oxidative state of the atmosphere – is reviewed extensively in WMO (Hofmann and Pyle, 1999), and no new evaluation is made here. The only stratospheric change included implicitly is the N_2O feedback on its lifetime. Overall, these projections of atmospheric composition for the 21st century include the most extensive set of trace gas emissions for IPCC assessments to date: greenhouse gases (N_2O, CH_4, HFCs, PFCs, SF_6) plus pollutants (NO_x, CO, VOC).

4.4.2 The OxComp Workshop

In the SAR, the chapter on atmospheric chemistry included two modelling studies: PhotoComp (comparison of ozone photochemistry in box models) and Delta-CH_4 (methane feedbacks in 2-D and 3-D tropospheric chemistry models). These model studies established standard model tests for participation in IPCC. They resulted in a consensus regarding the CH_4 feedback and identified the importance (and lack of uniform treatment) of NMHC chemistry on tropospheric O_3 production. This synthesis allowed for the SAR to use the CH_4-lifetime feedback and a simple estimate of tropospheric O_3 increase due solely to CH_4. The SAR noted that individual CTMs had calculated an impact of changing NO_x and CO emissions on global OH and CH_4 abundances, but that a consensus on predicting future changes in O_3 and OH did not exist.

Since 1995, considerable research has gone into the development and validation of tropospheric CTMs. The IPCC Special Report on Aviation and the Global Atmosphere (Derwent and Friedl, 1999) used a wide range of global CTMs to predict the enhancement of tropospheric O_3 due to aircraft NO_x emissions. The results were surprisingly robust, not only for the hemispheric mean O_3 increase, but also for the increase in global mean OH reported as a decrease in the CH_4 lifetime. The current state-of-modelling in global tropospheric chemistry has advanced since PhotoComp and Delta-CH_4 in the SAR and now includes as standard a three-dimensional synoptic meteorology and treatment of non-methane hydrocarbon chemistry. A survey of

Table 4.10: *Chemistry-Transport Models (CTM) contributing to the OxComp evaluation of predicting tropospheric O_3 and OH.*

CTM	Institute	Contributing authors	References
GISS	GISS	Shindell /Grenfell	Hansen *et al.* (1997b)
HGEO	Harvard U.	Bey / Jacob	Bey *et al.* (1999)
HGIS	Harvard U.	Mickley / Jacob	Mickley *et al.* (1999)
IASB	IAS/Belg.	Mülller	Müller and Brasseur (1995, 1999)
KNMI	KNMI/Utrecht	van Weele	Jeuken *et al.* (1999), Houweling *et al.* (2000)
MOZ1	NCAR/CNRS	Hauglustaine / Brasseur	Brasseur *et al.* (1998b), Hauglustaine *et al.* (1998)
MOZ2	NCAR	Horowitz/ Brasseur	Brasseur *et al.* (1998b), Hauglustaine *et al.* (1998)
MPIC	MPI/Chem	Kuhlmann / Lawrence	Crutzen *et al.* (1999), Lawrence *et al.* (1999)
UCI	UC Irvine	Wild	Hannegan *et al.* (1998), Wild and Prather (2000)
UIO	U. Oslo	Berntsen	Berntsen and Isaksen (1997), Fuglestvedt *et al.* (1999)
UIO2	U. Oslo	Sundet	Sundet (1997)
UKMO	UK Met Office	Stevenson	Collins *et al.* (1997), Johnson *et al.* (1999)
ULAQ	U. L. Aquila	Pitari	Pitari *et al.* (1997)
UCAM	U. Cambridge	Plantevin /Johnson	Law *et al.* (1998, 2000)(TOMCAT)

recent CTM-based publications on the tropospheric O_3 budget, collected for this report, is discussed in Section 4.5.

This assessment, building on these developments, organised a workshop to compare CTM results for a few, well-constrained atmospheric simulations. An open invitation, sent out in March 1999 to research groups involved in 3-D global tropospheric chemistry modelling, invited participation in this report's assessment of change in tropospheric oxidative state through a model intercomparison and workshop (OxComp). This workshop is an IPCC-focused follow-on to the Global Integration and Modelling (GIM) study (Kanakidou *et al.*, 1999). The infrastructure for OxComp (ftp site, database, graphics, and scientific support) was provided by the University of Oslo group, and the workshop meeting in July 1999 was hosted by the Max Planck Institute for Meteorology (MPI) Hamburg. Participating models are described by publications in peer-reviewed literature as summarised in Table 4.10; all include 3-D global tropospheric chemistry including NMHC; and assessment results are based on models returning a sufficient number of OxComp cases. The two goals of OxComp are (i) to build a consensus on current modelling capability to predict changes in tropospheric OH and O_3 and (ii) to develop a useful parametrization to calculate the greenhouse gases (including tropospheric O_3 but not CO_2) using the IPCC emissions scenarios.

4.4.3 Testing CTM Simulation of the Current (Y2000) Atmosphere

The OxComp workshop defined a series of atmospheres and emission scenarios. These included Y2000, a new reference atmosphere meant to represent year 2000 that provides a baseline from which all changes in greenhouse gases were calculated. For Y2000, abundances of long-lived gases were prescribed by 1998 measurements (Table 4.1a), and emissions of short-lived pollutants, NO_x, CO and VOC, were based primarily on projections to the year 2000 of GEIA/EDGAR emissions for 1990 (Olivier *et al.*, 1998, 1999), see Section 4.3.1. Stratospheric O_3

was calculated in some models and prescribed by current observation in others. The predicted atmospheric quantities in all these simulations are therefore short-lived tropospheric gases: O_3, CO, NO_x, VOC, OH and other radicals. Following the GIM model study (Kanakidou *et al.*, 1999), we use atmospheric measurement of O_3 and CO to test the model simulations of the current atmosphere. The Y2000 atmosphere was chosen because of the need for an IPCC baseline, and it does not try to match conditions over the 1980s and 1990s from which the measurements come. Although the observed trends in tropospheric O_3 and CO are not particularly large over this period and thus justify the present approach, a more thorough comparison of model results and measurements would need to use the regional distribution of the pollutant emissions for the observation period.

The seasonal cycle of O_3 in the free troposphere (700, 500, and 300 hPa) has been observed over the past decade from more than thirty ozone sonde stations (Logan, 1999). These measurements are compared with the OxComp Y2000 simulations for Resolute (75°N), Hohenpeissenberg (48°N), Boulder (40°N), Tateno (36°N), and Hilo (20°N) in Figure 4.10. Surface measurements from Cape Grim (40°S), representative of the marine boundary layer in southern mid-latitudes, are also compared with the models in Figure 4.10. With the exception of a few outliers, the model simulations are within ±30% of observed tropospheric O_3 abundance, and they generally show a maximum in spring to early summer as observed, although they often miss the month of maximum O_3. At 300 hPa the large springtime variation at many stations is due to the influence of stratospheric air that is approximately simulated at Resolute, but, usually overestimated at the other stations. The CTM simulations in the tropics (Hilo) at 700 to 500 hPa show much greater spread and hence generally worse agreement with observations. The mean concentration of surface O_3 observed at Cape Grim is well matched by most models, but the seasonality is underestimated.

Observed CO abundances are compared with the Y2000 model simulations in Figure 4.11 for surface sites at various altitudes and latitudes: Cape Grim (CGA, 94 m), Tae Ahn (KOR,

Table 4.11: *Changes in tropospheric O_3 (DU) and OH (%) relative to year 2000 for various perturbations to the atmosphere. Individual values calculated with chemistry transport-models (CTMs) plus the average values adopted for this report (TAR).*

CTM	Y2000 +10% CH_4 Case A	All A2x Case B	A2x: Y2100 − Y2000 −NO_x Case C	−NO_x−VOC−CH_4 Case D
Effective[a] tropospheric O_3 change (DU):				
HGIS		26.5		
GISS		25.2		
IASB	0.66	18.9	9.2	0.4
KNMI	0.63	18.0	9.0	
MOZ1		16.6		
MOZ2		22.4		
MPIC	0.40			
UCI	0.69	23.3	10.2	2.8
UIO	0.51	26.0	6.0	2.1
UKMO		18.9	4.6	3.1
ULAQ	0.85	22.2	14.5	5.9
TAR[b]	**0.64**	**22.0**	**8.9**	**2.0**
Tropospheric OH change (%)				
IASB	−2.9%	−7%		
KNMI	−3.3%	−25%	−41%	
MOZ1		−21%		
MOZ2		−18%		
MPIC	−2.7%			
UCI	−3.2%	−15%	−39%	−16.0%
UIO	−3.1%	−6%	−37%	−12.3%
UKMO	−2.9%	−12%	−37%	−10.8%
ULAQ	−2.7%	−17%	−43%	−22.0%
TAR[b]	**−3.0%**	**−16%**	**−40%**	**−14%**

Model results from OxComp workshop; all changes (DU for O_3 and % for OH) are relative to the year Y2000. Tropospheric mean OH is weighted by CH_4 loss rate. Mean O_3 changes (all positive) are derived from the standard reporting grid on which the CTMs interpolated their results. See Table 4.10 for the model key. The different cases include (A) a 10% increase in CH_4 to 1,920 ppb and (B) a full 2100 simulation following SRES draft marker scenario A2 (based on February 1999 calculations for preliminary work of this report). Case C drops the NO_x emissions back to Y2000 values; and case D drops NO_x, VOC, and CH_4 likewise.
Adopted CH_4 abundances and pollutant emissions from Y2000 to Y2100 are:
     Y2000: CH_4=1,745 ppb, e−NO_x=32.5 TgN/yr, e−CO=1,050 Tg/yr, e−VOC=150 Tg/yr.
     Y2100: CH_4=4,300 ppb, e−NO_x=110.0 TgN/yr, e−CO=2,500 Tg/yr, e−VOC=350 Tg/yr.

[a] N.B. Unfortunately, after the government review it was discovered that the method of integrating O_3 changes on the reporting grid was not well defined and resulted in some unintentional errors in the values reported above. Thus, the values here include in effect the O_3 increases predicted/expected in the lower stratosphere in addition to the troposphere. In terms of climate change, use of these values may not be unreasonable since O_3 changes in the lower stratosphere do contribute to radiative forcing. Nevertheless, the troposphere-only changes are about 25 to 33% less than the values above.
$$\delta(\text{tropospheric } O_3) = +5.0 \times \delta\ln(CH_4) + 0.125 \times \delta(e-NO_x) + 0.0011 \times \delta(e-CO) + 0.0033 \times \delta(e-VOC) \text{ in DU}.$$

[b] TAR adopts the weighted average for cases A to D as shown, where the weighting includes factors about model formulation and comparison with observations. A linear interpolation is derived from these results and used in the scenarios:
$$\delta\ln(\text{tropospheric OH}) = -0.32 \times \delta\ln(CH_4) + 0.0042 \times \delta(e-NO_x) - 1.05e-4 \times \delta(e-CO) - 3.15e-4 \times \delta(e-VOC),$$
$$\delta(\text{effective } O_3) = +6.7 \times \delta\ln(CH_4) + 0.17 \times \delta(e-NO_x) + 0.0014 \times \delta(e-CO) + 0.0042 \times \delta(e-VOC) \text{ in DU}.$$

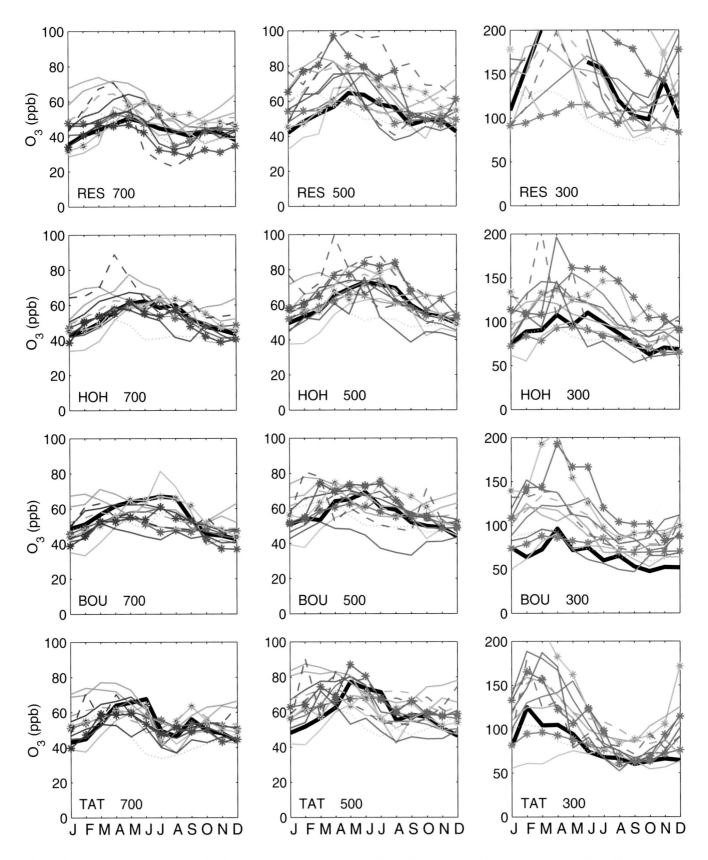

Figure 4.10: Observed monthly mean O₃ abundance (ppb) from sondes at 700 hPa (left column), 500 hPa (centre) and 300 hPa (right) from a sample of stations (thick black line) compared with Y2000 model simulations from OxComp (thin coloured lines, see model key in legend and Table 4.10). The sonde stations include RESolute (75°N, 95°W), HOHenpeissenberg (48°N, 11°E), BOUlder (40°N, 105°W), TATeno (36°N, 140°E), and HILo (20°N, 155°W). Surface monthly O₃ observations (thick black line) at Cape Grim Observatory (CGA, 40°S, 144°E, 94 m above mean sea level) are also compared with the models. (Continues opposite.)

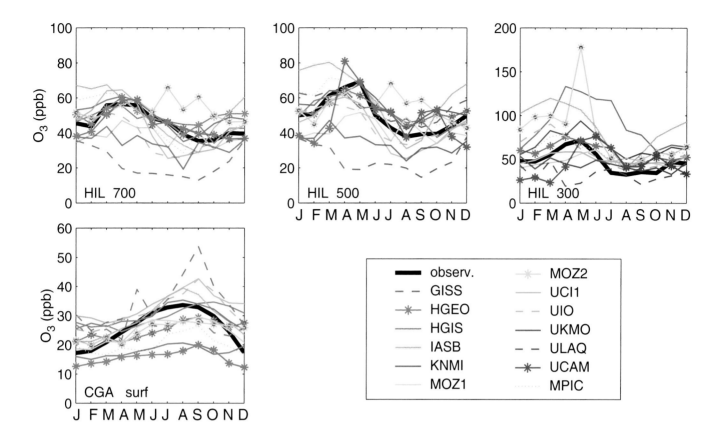

20 m), Mauna Loa (MLH, 3397 m), Alert (ALT, 210 m), and Niwot Ridge (NWR, 3475 m). The Alert abundances are well matched by most but not all models. Niwot Ridge and Mauna Loa are reasonably well modelled except for the February to March maximum. At Tae Ahn, the models miss the deep minimum in late summer, but do predict the much larger abundances downwind of Asian sources. At Cape Grim the seasonal cycle is matched, but the CO abundance is uniformly overestimated (30 to 50%) by all the models, probably indicating an error in Southern Hemisphere emissions of CO.

Overall, this comparison with CO and O_3 observations shows good simulations by the OxComp models of the global scale chemical features of the current troposphere as evidenced by CO and O_3; however, the critical NO_x chemistry emphasises variability on much smaller scales, such as biomass burning plumes and lightning storms, that are not well represented by the global models. With this large variability and small scales, the database of NO_x measurements needed to provide a test for the global models, equivalent to CO and O_3, would need to be much larger.

The current NO_x database (e.g., Emmons *et al.*, 1997; Thakur *et al.*, 1999) does not provide critical tests of CTM treatment of these sub-grid scales.

4.4.4 Model Simulations of Perturbed and Y2100 Atmospheres

The OxComp workshop also defined a series of perturbations to the Y2000 atmosphere for which the models reported the monthly averaged 3-D distribution of O_3 abundances and the budget for CH_4, specifically the loss due to reaction with tropospheric OH. From these diagnostics, the research group at Oslo calculated the change in global mean tropospheric O_3 (DU) and in OH (%) relative to Y2000, as shown in Table 4.11. For each model at every month, the "troposphere" was defined as where O_3 abundances were less than 150 ppb in the Y2000 simulation, a reasonably conservative diagnostic of the tropopause (see Logan, 1999). Because O_3 is more effective as a greenhouse gas when it lies above the surface boundary layer (SAR; Hansen *et al.*, 1997a; Prather and Sausen, 1999; Chapter 6 of this report), the model study diagnosed the O_3 change occurring in the 0 to 2 km layers of the model. This amount is typically 20 to 25% of the total change and is consistent across models and types of perturbations here.

Case A, a +10% increase in CH_4 abundance for Y2000, had consistent results across reporting models that differed little from the SAR's Delta-CH_4 model study. The adopted values for this report are −3% change in OH and +0.64 DU increase in O_3, as listed under the "TAR" row in Table 4.11.

The Y2100 atmosphere in OxComp mimics the increases in pollutant emissions in SRES A2p scenario from year 2000 to year 2100 with the year 2100 abundance of CH_4, 4,300 ppb, calculated with the SAR technology and named here A2x. (See discussion in section 4.4.5; for the SAR, only the CH_4-OH feedback is included.) The long-lived gases CO_2 and N_2O have no impact on these tropospheric chemistry calculations as specified.

Cases B-C-D are a sequence of three Y2100 atmospheres based on A2x: Case B is the full Y2100-A2x scenario; Case C is the same Y2100-A2x scenario but with unchanged (Y2000) NO_x

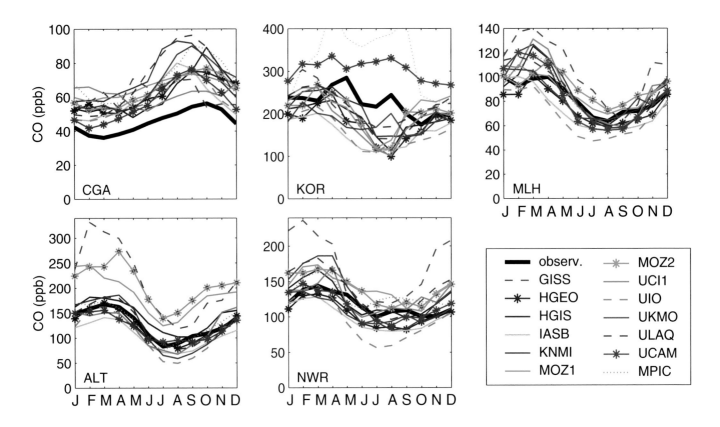

Figure 4.11: Observed seasonal surface CO abundance (ppb, thick black lines) at Cape Grim (CGA: 40°S, 144°E, 94 m above mean sea level), Tae Ahn (KOR: 36°N, 126°E, 20 m), Mauna Loa (MLH: 19°N, 155°W, 3397 m), Alert (ALT: 82°N, 62°W, 210 m), and Niwot Ridge (NWR: 40°N, 105°W, 3475 m) are compared with the OxComp model simulations from Y2000, see Figure 4.10.

emissions; and Case D is the same but with NO_x, VOC and CH_4 unchanged since Y2000 (i.e., only CO emissions change). Case B (Y2100-A2x) results are available from most OxComp participants. All models predict a decrease in OH, but with a wide range from −6 to −25%, and here we adopt a decrease of −16%. Given the different distributions of the O_3 increase from the OxComp models (Figures 4.12-13), the increases in globally integrated O_3 were remarkably consistent, ranging from +16.6 to +26.5 DU, and we adopt +22 DU. Without the increase in NO_x emissions (Case C) the O_3 increase drops substantially, ranging from +4.6 to +14.5 DU; and the OH decrease is large, −37 to −43%. With only CO emissions (Case D) the O_3 increase is smallest in all models, +0.4 to +5.9 DU.

This report adopts a weighted, rounded average of the changes in OH and O_3 for cases A-D as shown in the bold rows in Table 4.11. The weighting includes factors about model formulation and comparison with observations. This sequence of calculations (Y2000 plus Cases A-B-C-D) allows us to define a simple linear relationship for the absolute change in tropospheric O_3 and the relative change in OH as a function of the CH_4 abundance and the emission rates for NO_x, for CO, and for VOC. These two relationships are given in Table 4.11. Since the change in CH_4 abundance and other pollutant emissions for Y2100-A2x are among the largest in the SRES scenarios, we believe that interpolation of the O_3 and OH changes for different emission scenarios and years introduces little additional uncertainty.

The possibility that future emissions of CH_4 and CO overwhelm the oxidative capacity of the troposphere is tested (Case E, see Table 4.3 footnote &) with a +10% increase in CH_4 on top of Y2100-A2x (Case B). Even at 4,300 ppb CH_4, the decrease in OH calculated by two CTMs is only slightly larger than in Case A, and thus, at least for SRES A2p, the CH_4-feedback factor does not become as large as in the runaway case (Prather, 1996). This report assumes that the CH_4 feedback remains constant over the next century; however, equivalent studies for the low-NO_x future scenarios are not assessed.

The apparent agreement on predicting the single global, annual mean tropospheric O_3 increase, e.g., Case B in Table 4.11, belies the large differences as to where this increase occurs and what is its peak magnitude. The spatial distributions of the tropospheric O_3 increases in July for Case B are shown in Figure 4.12 (latitude by altitude zonal average abundance, ppb) and Figure 4.13 (latitude by longitude column density, DU) for nine CTMs. The largest increase in abundance occurs near the tropopause at 40°N latitude; yet some models concentrate this increase in the tropics and others push it to high latitudes. In terms of column density, models generally predict large increases along the southern edge of Asia from Arabia to eastern China; although the increases in tropical, biomass-burning regions varies widely from model to model.

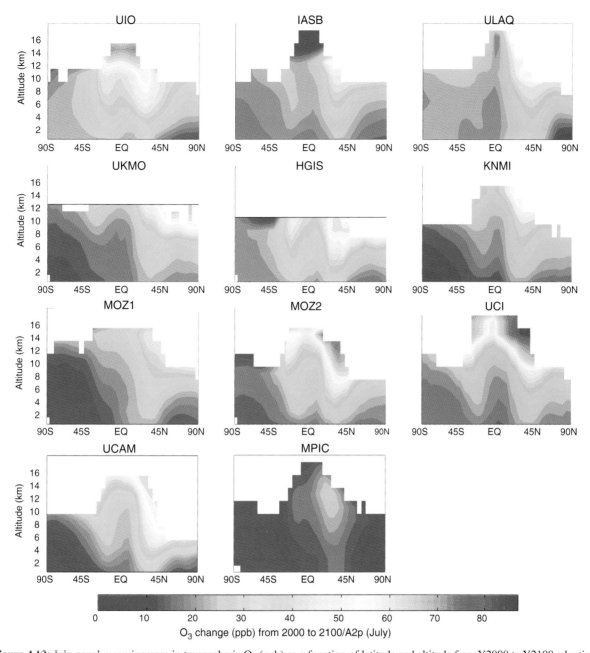

Figure 4.12: July zonal mean increase in tropospheric O_3 (ppb) as a function of latitude and altitude from Y2000 to Y2100 adopting SRES A2p projections for CH_4, CO, VOC, and NO_x. Results are shown for a sample of the chemistry-transport models (CTM) participating in IPCC OxComp workshop. Increases range from 0 to more than 80 ppb. Changes in the stratosphere (defined as $O_3 > 150$ ppb in that model's Y2000 simulation) are masked off, as are also regions in the upper troposphere for some CTMs (UKMO, HGIS) where O_3 is not explicitly calculated. See Table 4.10 for participating models.

This similarity in the total, but difference in the location, of the predicted O_3 increases is noted in Isaksen and Jackman (1999) and is probably due to the different transport formulations of the models as documented in previous CTM intercomparisons (Jacob *et al.*, 1997). Possibly, the agreement on the average O_3 increase may reflect a more uniform production of O_3 molecules as a function of NO_x emissions and CH_4 abundance across all models. Nevertheless, the large model range in the predicted patterns of O_3 perturbations leads to a larger uncertainty in climate impact than is indicated by Table 4.11.

The projected increases in tropospheric O_3 under SRES A2 and A1FI will have serious consequences on the air quality of most of the Northern Hemisphere by year 2100. Taking only the global numbers from Figure 4.14, the mean abundance of tropospheric O_3 will increase from about 52 ppb (typical mid-tropospheric abundances) to about 84 ppb in year 2100. Similar increases of about +30 ppb are seen near the surface at 40°N on a zonal average in Figure 4.12. Such increases will raise the "background" levels of O_3 in the northern mid-latitudes to close to the current clean-air standard.

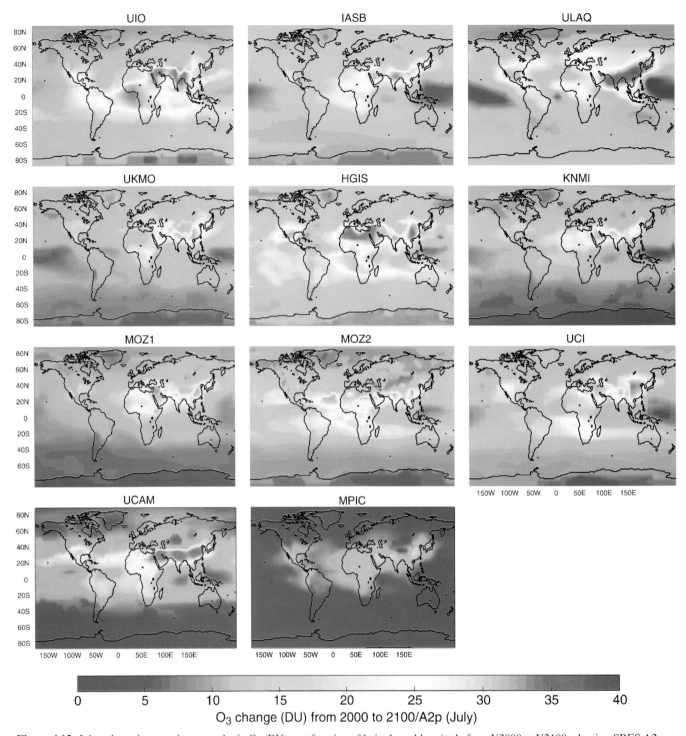

Figure 4.13: July column increase in tropospheric O_3 (DU) as a function of latitude and longitude from Y2000 to Y2100 adopting SRES A2p projections for CH_4, CO, VOC, and NO_x is shown for some OxComp simulations. See Figure 4.12.

4.4.5 Atmospheric Composition for the IPCC Scenarios to 2100

Mean tropospheric abundances of greenhouse gases and other chemical changes in the atmosphere are calculated by this chapter for years 2000 to 2100 from the SRES scenarios for anthropogenic emissions of CH_4, N_2O, HFCs, PFCs, SF_6, NO_x, CO, and VOC (corresponding emissions of CO_2 and aerosol precursors are not used). The emissions from the six SRES marker/illustrative scenarios (A1B, A1T, A1FI, A2, B1, B2) are tabulated in Appendix II, as are the resulting greenhouse gas abundances, including CO_2 and aerosol burdens. Chlorine- and bromine-containing greenhouse gases are not calculated here, and we adopt the single baseline scenario from the WMO assessment (Montreal Protocol Scenario A1 of Madronich and Velders, 1999), which is reproduced in Appendix II. Also given in Appendix II are the parallel data for the SRES preliminary marker scenarios (A1p, A2p, B1p, B2p) and, in many cases, the SAR scenario IS92a as a comparison with the previous assessment.

Greenhouse gas abundances are calculated using a methodology similar to the SAR: (1) The troposphere is treated as a single box with a fill-factor for each gas that relates the burden to the tropospheric mean abundance (e.g., Tg/ppb). (2) The atmospheric lifetime for each gas is recalculated each year based on conditions at the beginning of the year and the formulae in Table 4.11. (Changes in tropospheric OH are used to scale the lifetimes of CH_4 and HFCs, and the abundance of N_2O is used to calculate its new lifetime.) (3) The abundance of a gas is integrated exactly over the year assuming that emissions remain constant for 12 months. (4) Abundances are annual means, reported at the beginning of each year (e.g., year 2100 = 1 January 2100).

In the SAR, the only OH feedback considered was that of CH_4 on its own lifetime. For this report, we calculate the change in tropospheric OH due to CH_4 abundance as well as the immediate emissions of NO_x, CO and VOC. Likewise, the increase in tropospheric O_3 projected in the SAR considered only increases in CH_4; whereas now it includes the emissions of NO_x, CO and VOC. Thus the difference between IS92a in the SAR and in this report is similar to that noted by Kheshgi *et al.* (1999). Also, the feedback of N_2O on its lifetime is included here for the first time and shows up as reduction of 14 ppb by year 2100 in this report's IS92a scenario as compared to the SAR.

The 21st century abundances of CH_4, N_2O, tropospheric O_3, HFC-134a, CF_4, and SF_6 for the SRES scenarios are shown in Figure 4.14. Historical data are plotted before year 2000; and the SRES projections, thereafter to year 2100. CH_4 continues to rise in B2, A1FI, and A2 (like IS92a), with abundances reaching 2,970 to 3,730 ppb, in order. For A1B and A1T, CH_4 peaks in mid-century at about 2,500 ppb and then falls. For B1, CH_4 levels off and eventually falls to 1980-levels by year 2100. N_2O continues to rise in all scenarios, reflecting in part its long lifetime, and abundances by the end of the century range from 350 to 460 ppb. Most scenarios lead to increases in tropospheric O_3, with scenarios A1FI and A2 projecting the maximum tropospheric O_3 burdens of 55 DU by year 2100. This increase of about 60% from today is more than twice the change from pre-industrial to present. Scenario B1 is alone in projecting an overall decline in tropospheric O_3 over most of the century: the drop to 30 DU is about halfway back to pre-industrial values. HFC-134a, the HFC with the largest projected abundance, is expected to reach about 900 ppt by year 2100 for all scenarios except B1. Likewise by 2100, the abundance of CF_4 rises to 340 to 400 ppt in all scenarios except B1. The projected increase in SF_6 is much smaller in absolute abundance, reaching about 60 ppt in scenarios A1 and A2. For the major non-CO_2 greenhouse gases, the SRES A2 and A1FI increases are similar to, but slightly larger than, those of IS92a. The SRES mix of lesser greenhouse gases (HFCs, PFCs, SF_6) and their abundances are increased substantially relative to IS92a. The summed radiative forcings from these gases plus CO_2 and aerosols are given in Chapter 6.

The chemistry of the troposphere is changing notably in these scenarios, and this is illustrated in Figure 4.14 with the lifetime (LT) of CH_4 and the change in mean tropospheric OH relative to year 2000. In all scenarios except B1, OH decreases 10% or more by the end of the century, pushing the lifetime of CH_4 up from 8.4 years, to 9.2 to 10.0 years. While increasing emissions of NO_x in most of these scenarios increases O_3 and would tend to increase OH (see notes to Table 4.11), the increase in CH_4 abundance and the greater CO emissions appear to dominate, driving OH down. In such an atmosphere, emissions of CH_4 and HFCs persist longer with greater greenhouse impact. In contrast the B1 atmosphere is more readily able to oxidise these compounds and reduce their impact.

4.4.6 Gaps in These Projections – the Need for Coupled Models

There are some obvious gaps in these projections where processes influencing the greenhouse gas abundances have been omitted. One involves coupling of tropospheric chemistry with the stratosphere. For one, we did not include the recovery of stratospheric ozone expected over the next century. The slow recovery of stratospheric ozone depletion from the halogens will lead to an increase in the flux of ozone into the troposphere and also to reduced solar UV in the troposphere, effectively reversing over the next century what has occurred over the past two decades. A more important impact on the Y2100 stratosphere, however, is the response to increases in CH_4 and N_2O projected by most scenarios (see Hofmann and Pyle, 1999), which in terms of coupled stratosphere-troposphere chemistry models could be evaluated in only one of the OxComp models (ULAQ, Université degli studi dell' Aguila) and is not included here.

Another major gap in these projections is the lack of global models coupling the atmospheric changes with biogeochemical models. There have been studies that tackled individual parts of the problem, e.g., deposition of reactive N (Holland *et al.*, 1997), crop damage from O_3 (Chameides *et al.*, 1994). Integrated assessment studies have coupled N_2O and CH_4 emission models with lower dimension or parametrized climate and chemistry models (e.g., Alcamo, 1994; Holmes and Ellis, 1999; Prinn *et al.*, 1999). However, the inherent local nature of this coupling, along with the possible feedbacks through, for example NO and VOC emissions, point to the need for coupled 3-D global chemistry and ecosystem models in these assessments.

Finally, there is an obvious need to couple the physical changes in the climate system (water vapour, temperature, winds, convection) with the global chemical models. This has been partially accomplished for some cases that are highlighted here (Section 4.5.2), but like other gaps presents a major challenge for the next assessment.

4.4.7 Sensitivity Analysis for Individual Sectors

In order to assess the overall impact of changing industry or agriculture, it would be necessary to combine all emissions from a specific sector or sub-sector as has been done with the IPCC assessment of aviation (Penner *et al.*, 1999). Further, the impact on natural emissions and land-use change (e.g., albedo, aerosols) would also need to be included. Such a sector analysis would cut across Chapters 3, 4, 5 and 6 of this report (e.g., as in Prather and

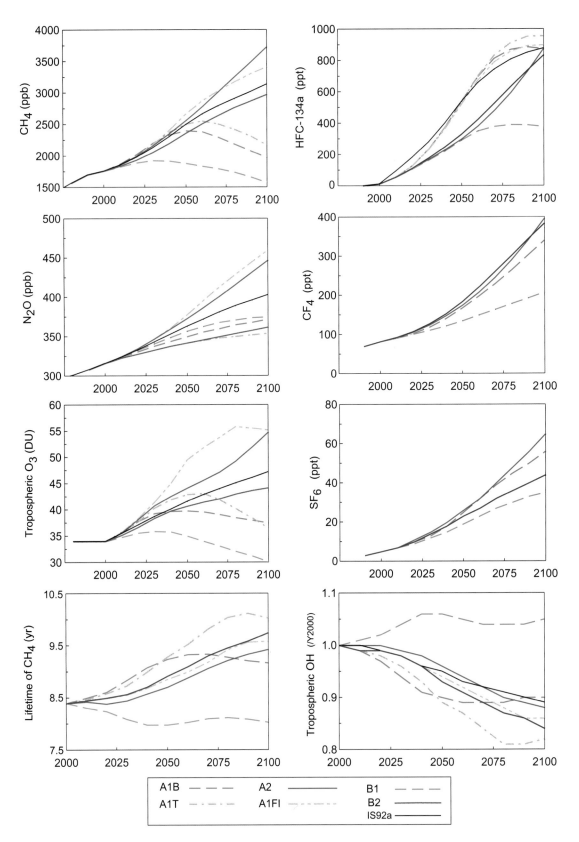

Figure 4.14: Atmospheric composition and properties predicted using the six SRES Marker-Illustrative scenarios for anthropogenic emissions: A1B (green dashed line), A1T (yellow dash-dotted), A1FI (orange dash-dot-dotted), A2 (red solid), B1 (cyan dashed), B2 (solid dark blue). Abundances prior to year 2000 are taken from observations, and the IS92a scenario computed with current methodology is shown for reference (thin black line). Results are shown for CH_4 (ppb), N_2O (ppb), tropospheric O_3 (DU), HFC-134a (ppt), CF_4 (ppt), SF_6 (ppt), the lifetime of CH_4 (yr), and the global annual mean abundance of tropospheric OH (scaled to year 2000 value). All SRES A1-type scenarios have the same emissions for HFCs, PFCs, and SF_6 (appearing a A1B), but the HFC-134a abundances vary because the tropospheric OH values differ affecting its lifetime. The IS92a scenario did not include emissions of PFCs and SF_6. For details, see chapter text and tables in Appendix II.

Sausen, 1999). Such an analysis cannot be done for the SRES emissions scenarios, which lack a breakdown by sector and also lack numbers for the changes in the land area of agriculture or urbanisation.

4.5 Open Questions

Many processes involving atmospheric chemistry, and the coupling of atmospheric chemistry with other elements of global change, have been proposed in the scientific literature. These are generally based on sound physical and chemical principles, but unfortunately, there is no consensus on their quantitative role in atmospheric chemistry on a global scale (e.g., the effects of clouds on tropospheric ozone: Lelieveld and Crutzen (1990) vs. Liang and Jacob (1997)), on the magnitude of possible compensating effects (e.g., net settling of HNO_3 on cloud particles: Lawrence and Crutzen (1999) vs. full cloud-scale dynamics), or even on how to implement them or whether these are already effectively included in many of the model calculations. While many of these processes may be important, there is inadequate information or consensus to make a quantitative evaluation in this assessment. This assessment is not a review, and so this section presents only a few examples of recent publications studying feedbacks or chemical processes, which are not included, but which are potentially important in this assessment.

4.5.1 Chemical Processes Important on the Global Scale

4.5.1.1 Missing chemistry, representation of small scales, and changing emission patterns

Analyses and observations (see Section 4.2.6) continue to test and improve the chemistry and transport used in the global CTMs. In terms of the chemistry, recent studies have looked, for example, at the representation of NMHC chemistry (Houweling *et al.* 1998; Wang *et al.* 1998b), the role of halogens in the O_3 budget of the remote marine troposphere, and the acetone source of upper tropospheric OH (see Sections 4.2.4 and 4.2.6). Most of these improvements in understanding will eventually become adopted as standard in the global CTMs, but at this stage, for example, the role of tropospheric halogen chemistry on the Y2100 predictions has not been evaluated in the CTMs.

Convection, as well as urban pollution and biomass burning plumes, occur on horizontal scales not resolved in global CTMs. These sub-grid features appear to be important in calculating OH abundances and O_3 production for biomass burning emissions (Pickering *et al.*, 1996; Folkins *et al.*, 1997), for the remote upper troposphere (Jaeglé *et al.*, 1997; Prather and Jacob, 1997; Wennberg *et al.*, 1998), and in urban plumes (e.g., Sillman *et al.*, 1990). Convection is represented in all CTMs here (e.g., Collins *et al.*, 1999; Müller and Brasseur, 1999) but in quite different ways, and it still involves parametrization of processes occurring on a sub-grid scale. A substantial element of the differences in CTM simulations appears to lie with the different representations of convection and boundary layer transport, particularly for the short-lived gases such as NO_x.

A change in the geographic emission pattern of the pollutants (NO_x, CO and VOC) can by itself alter tropospheric O_3

and OH abundances and in turn the abundances of CH_4 and HFCs. In one study of regional NO_x emissions and control strategy, Fuglestvedt *et al.* (1999) find that upper tropospheric O_3 is most sensitive to NO_x reductions in Southeast Asia and Australia and least to those in Scandinavia. Understanding trends in CO requires knowledge not only of the *in situ* chemistry of CO (e.g., Granier *et al.*, 1996; Kanakidou and Crutzen, 1999), but also of how local pollution control has altered the global pattern of emissions (e.g., Hallock-Waters *et al.*, 1999). These shifts have been included to some extent in the SRES emissions for year 2100 used here; however, the projected change in emission patterns have not been formally evaluated within the atmospheric chemistry community in terms of uncertainty in the Y2100 global atmosphere.

4.5.1.2 Aerosol interactions with tropospheric O_3 and OH

Over the past decade of assessments, stratospheric O_3 chemistry has been closely linked with aerosols, and global models in the recent WMO assessments have included some treatment of the stratospheric sulphate layer and polar stratospheric clouds. In the troposphere, studies have identified mechanisms that couple gas-phase and aerosol chemistry (Jacob, 2000). Many aerosols are photochemically formed from trace gases, and at rates that depend on the oxidative state of the atmosphere. Such processes are often included in global aerosol models (see Chapter 5). The feedback of the aerosols on the trace gas chemistry includes a wide range of processes: conversion of NO_x to nitrates, removal of HO_x, altering the UV flux and hence photodissociation rates (e.g., Dickerson *et al.*, 1997; Jacobson, 1998), and catalysing more exotic reactions leading to release of NO_x or halogen radicals. These processes are highly sensitive to the properties of the aerosol and the local chemical environment, and their importance on a global scale is not yet established. Only the first example above of aerosol chemistry is generally included in many of the CTMs represented here; however, the surface area of wet aerosols (that converts NO_x to HNO_3 via the intermediate species NO_3 and N_2O_5) is usually specified and not interactively calculated. More laboratory and field research is needed to define the processes so that implementation in global scale models can evaluate their quantitative impact on these calculations of greenhouse gases.

4.5.1.3 Stratosphere-troposphere coupling

The observed depletion of stratospheric ozone over the past three decades, which can be attributed in large part but not in total to the rise in stratospheric chlorine levels, has been reviewed extensively in WMO (1999). This depletion has lead to increases in tropospheric UV and hence forces tropospheric OH abundances upward (Bekki *et al.*, 1994). The total effect of such a change is not simple and involves the coupled stratosphere-troposphere chemical system; for example, ozone depletion may also have reduced the influx of O_3 from the stratosphere, which would reduce tropospheric O_3 (Karlsdottir *et al.*, 2000) and tend to reverse the OH trend. Such chemical feedbacks are reviewed as "climate-chemistry" feedbacks in WMO 1999 (Granier and Shine, 1999). There is insufficient understanding or quantitative consensus on these effects to be included in this assessment. While chlorine-driven O_3 depletion becomes much less of an issue in the latter half of the 21st century, the projected increases in CO_2, CH_4, and N_2O may cause even larger changes in stratospheric O_3. The lack of coupled CTMs that include stratospheric changes adds uncertainty to these projections.

4.5.1.4 Uncertainties in the tropospheric O_3 budget

An updated survey of global tropospheric CTM studies since the SAR focuses on the tropospheric O_3 budget and is reported in Table 4.12. In this case authors were asked for diagnostics that did not always appear in publication. The modelled tropospheric O_3 abundances generally agree with observations; in most cases the net budgets are in balance; and yet the individual components vary greatly. For example, the stratospheric source ranges from 400 to 1,400 Tg/yr, while the surface sink is only slightly more constrained, 500 to 1,200 Tg/yr. If absolute production is diagnosed as the reactions of HO_2 and other peroxy radicals with NO, then the globally integrated production is calculated to be very large, 2,300 to 4,300 Tg/yr and is matched by an equally large sink (see Sections 4.2.3.3 and 4.2.6). The differences between the flux from the stratosphere and the destruction at the surface is balanced by the net *in situ* photochemical production. In this survey, the net production varies widely, from −800 to +500 Tg/yr, indicating that in some CTMs the troposphere is a large chemical source and in others a large sink. Nevertheless, the large differences in the stratospheric source are apparently the driving force behind whether a model calculates a chemical source or sink of tropospheric O_3. Individual CTM studies of the relative roles of stratospheric influx versus tropospheric chemistry in determining the tropospheric O_3 abundance (e.g., Roelofs and Lelieveld, 1997; Wang et al., 1998a; Yienger et al., 1999) will not represent a consensus until all CTMs develop a more accurate representation of the stratospheric source consistent with observations (Murphy and Fahey, 1994).

4.5.2 Impacts of Physical Climate Change on Atmospheric Chemistry

As global warming increases in the next century, the first-order atmospheric changes that impact tropospheric chemistry are the anticipated rise in temperature and water vapour. For example, an early 2-D model study (Fuglestvedt et al., 1995) reports that tropospheric O_3 decreases by about 10% in response to a warmer, more humid climate projected for year 2050 as compared to an atmosphere with current temperature and H_2O. A recent study based on NCAR (National Center for Atmospheric Research) CCM (Community Climate Model) projected year 2050 changes in tropospheric temperature and H_2O (Brasseur et al., 1998a) finds a global mean 7% increase in the OH abundance and a 5% decrease in tropospheric O_3, again relative to the same calculation with the current physical climate.

A 3-D tropospheric chemistry model has been coupled to the Hadley Centre Atmosphere-Ocean General Circulation Model (AOGCM) and experiments performed using the SRES preliminary marker A2p emissions (i) as annual snapshots (Stevenson et al., 2000) and (ii) as a 110-year, fully coupled experiment (Johnson et al., 1999) for the period 1990 to 2100. By 2100, the experiments with coupled climate change have increases in CH_4 which are only

about three-quarters those of the simulation without climate change and increases in Northern Hemisphere mid-latitude O_3 which are reduced by half. The two major climate-chemistry feedback mechanisms identified in these and previous studies were (1) the change of chemical reaction rates with the average 3°C increase in tropospheric temperatures and (2) the enhanced photochemical destruction of tropospheric O_3 with the approximately 20% increase in water vapour. The role of changes in the circulation and convection appeared to play a lesser role but have not been fully evaluated. These studies clearly point out the importance of including the climate-chemistry feedbacks, but are just the beginning of the research that is needed for adequate assessment.

Thunderstorms, and their associated lightning, are a component of the physical climate system that provides a direct source of a key chemical species, NO_x. The magnitude and distribution of this lightning NO_x source controls the magnitude of the anthropogenic perturbations, e.g., that of aviation NO_x emissions on upper tropospheric O_3 (Berntsen and Isaksen, 1999). In spite of thorough investigations of the vertical distribution of lightning NO_x (Huntrieser et al., 1998; Pickering et al. 1998), uncertainty in the source strength of lightning NO_x cannot be easily derived from observations (Thakur et al., 1999; Thompson et al., 1999). The link of lightning with deep convection (Price and Rind, 1992) opens up the possibility that this source of NO_x would vary with climate change, however, no quantitative evaluation can yet be made.

4.5.3 Feedbacks through Natural Emissions

Natural emissions of N_2O and CH_4 are currently the dominant contributors to their respective atmospheric burdens, with terrestrial emissions greatest in the tropics. Emissions of both of these gases are clearly driven by changes in physical climate as seen in the ice-core record (Figure 4.1e). Soil N_2O emissions are sensitive to temperature and soil moisture and changes in rates of carbon and nitrogen cycling (Prinn et al., 1999). Similarly, methane emissions from wetlands are sensitive to the extent of inundation, temperature rise, and changes in rates of carbon and nitrogen cycling. Natural emissions of the pollutants NO_x, CO, and VOC play an important role in production of tropospheric O_3 and the abundance of OH; and these emissions are subject to similar forcings by both the physical and chemical climates. Terrestrial and aquatic ecosystems in turn respond to near-surface pollution (O_3, NO_2, acidic gases and aerosols) and to inadvertent fertilisation through deposition of reactive nitrogen (often emitted from the biosphere as NO or NH_3). This response can take the form of die back, reduced growth, or changed species composition competition that may alter trace gas surface exchange and ecosystem health and function. The coupling of this feedback system – between build-up of greenhouse gases, human-induced climate change, ecosystem responses, trace gas exchange at the surface, and back to atmospheric composition – has not been evaluated in this assessment. The variety and complexity of these feedbacks relating to ecosystems, beyond simple increases with rising temperatures and changing precipitation, argues strongly for the full interactive coupling of biogeochemical models of trace gas emissions with chemistry and climate models.

Table 4.12: Tropospheric ozone budgets for circa 1990 conditions from a sample of global 3-D CTMs since the SAR.

CTM	STE	Prod	Loss	P–L	SURF	Burden	Reference
			(Tg/yr)			(Tg)	
MATCH	1440	2490	3300	−810	620		Crutzen *et al.* (1999)
MATCH-MPIC	1103	2334	2812	−478	621		Lawrence *et al.* (1999)
ECHAM/TM3	768	3979	4065	−86	681	311	Houweling *et al.* (1998)
ECHAM/TM3[a]	740	2894	3149	−255	533	266	Houweling *et al.* (1998)
HARVARD	400	4100	3680	+420	820	310	Wang *et al.* (1998a)
GCTM	696			+128	825	298	Levy *et al.* (1997)
UIO	846			+295	1178	370	Berntsen *et al.* (1996)
ECHAM4	459	3425	3350	+75	534	271	Roelofs and Lelieveld (1997)
MOZART[b]	391	3018	2511	+507	898	193	Hauglustaine *et al.* (1998)
STOCHEM	432	4320	3890	+430	862	316	Stevenson *et al.* (2000)
KNMI	1429	2864	3719	−855	574		Wauben *et al.* (1998)
UCI	473	4229	3884	+345	812	288	Wild and Prather (2000)

STE = stratosphere-troposphere exchange (net flux from stratosphere) (Tg/yr).
Prod & Loss = *in situ* tropospheric chemical terms, P–L = net. (Tg/yr).
SURF = surface deposition (Tg/yr). Burden = total content (Tg, 34DU = 372Tg).
Budgets should balance exactly (STE+P–L=SURF), but may not due to roundoff.
[a] Results using CH_4-only chemistry without NMHC.
[b] Budget/burden calculated from surface to 250 hPa (missing part of upper troposphere).

4.6 Overall Impact of Global Atmospheric Chemistry Change

The projected growth in emissions of greenhouse gases and other pollutants in the IPCC SRES scenarios for the 21st century is expected to increase the atmospheric burden of non-CO_2 greenhouse gases substantially and contribute a sizable fraction to the overall increase in radiative forcing of the climate. These changes in atmospheric composition may, however, degrade the global environment in ways beyond climate change.

The impact of metropolitan pollution, specifically O_3 and CO, on the background air of the Atlantic and Pacific Oceans has been highlighted by many studies over the past decade. These have ranged from observations of anthropogenic pollution reaching across the Northern Hemisphere (e.g., Parrish *et al.*, 1993; Jaffe *et al.*, 1999) to analyses of rapidly increasing emissions of pollutants (NO_x, CO, VOC) in, for example, East Asia (Kato and Akimoto 1992; Elliott *et al.*, 1997). CTM studies have tried to quantify some of these projections for the near term: Berntsen *et al.* (1999) predict notable increases in CO and O_3 coming into the north-west USA from a doubling of current Asian emissions; Jacob *et al.* (1999) calculate that monthly mean O_3 abundances over the USA will increase by 1 to 6 ppb from a tripling of these emissions between 1985 and 2010; and Collins *et al.* (2000) project a 3 ppb increase from 1990 to 2015 in monthly mean O_3 over north-west Europe due to rising North American emissions. The impact of metropolitan pollution will expand over the coming decades as urban areas grow and use of resources intensifies.

What is new in this IPCC assessment is the extension of these projections to the year 2100, whereupon the cumulative impact of all Northern Hemisphere emissions, not just those immediately upwind, may for some scenarios double O_3 abundances over the northern mid-latitudes. Surface O_3 abundances during July over the industrialised continents of the Northern Hemisphere are about 40 ppb with 2000 emissions; and under SRES scenarios A2 and A1FI they would reach 45 to 50 ppb with 2030 emissions, 60 ppb with 2060 emissions, and >70 ppb with 2100 emissions. Since regional ozone episodes start with these background levels and build upon them with local smog production, it may be impossible under these circumstances to achieve a clean-air standard of <80 ppb over most populated regions. This problem reaches across continental boundaries and couples emissions of NO_x on a hemispheric scale. In the 21st century a global perspective will be needed to meet regional air quality objectives. The impact of this threatened degradation of air quality upon societal behaviour and policy decisions will possibly change the balance of future emissions impacting climate change (e.g., more fuel burn (CO_2) to achieve lower NO_x as in aviation; Penner *et al.*, 1999).

Under some emission scenarios, the large increases in tropospheric O_3 combined with the decreases in OH may alter the oxidation rate and the degradation paths for hydrocarbons and other hazardous substances. The damage caused by higher O_3 levels to both crops and natural systems needs to be assessed, and societal responses to this threat would likely change the emissions scenarios evaluated here (e.g., the current SRES scenarios anticipate the societal demand to control urban aerosols and acid rain by substantially cutting sulphur emissions) .

Coupling between atmospheric chemistry, the biosphere, and the climate are not at the stage that these feedbacks can be included in this assessment. There are indications, however, that the evolution of natural emissions and physical climate projected over the next century will change the baseline atmospheric chemistry and lead to altered biosphere-atmosphere exchanges and continued atmospheric change independent of anthropogenic emissions.

References

Alcamo, J. (ed.), 1994: IMAGE 2.0: Integrated Modeling of Global Climate Change. Special issue of *Water, Air and Soil Pollution*, **76**(1-2).

Arnold, F., J. Schneider, K. Gollinger, H. Schlager, P. Schulte, D.E. Hagen, P.D. Whitefield and P. van Velthoven, 1997: Observation of upper tropospheric sulfur dioxide and acetone pollution: potential implications for hydroxyl radical and aerosol formation. *Geophys. Res. Lett.*, **24**, 57-60.

Bakwin, P.S., P.P. Tans and P.C. Novelli, 1994: Carbon monoxide budget in the northern hemisphere. *Geophys. Res. Lett.*, **21**, 433-436.

Battle, M., M. Bender, T. Sowers, P.P. Tans, J.H. Butler, J.W. Elkins, T. Conway, N. Zhang, P. Lang and A.D. Clarke, 1996: Atmospheric gas concentrations over the past century measured in air from firn at South Pole. *Nature*, **383**, 231-235.

Becker, K.H., J.C. Lörzer, R. Kurtenbach, P. Wiesen, T.E. Jensen and T.J. Wallington, 1999: Nitrous oxide emissions from vehicles. *Envir. Sci. Tech.*, **33**, 4134-4139.

Bekki, S., K.S. Law and J.A. Pyle, 1994: Effect of ozone depletion on atmospheric CH_4 and CO concentrations. *Nature*, **371**, 595-597.

Benkovitz, C.M., M.T. Scholtz, J. Pacyna, L. Tawasou, J. Dignon, E.C. Voldner, P.A. Spiro, J.A. Logan and T.E. Graedel, 1996: Global gridded inventories of anthropogenic emissions of sulfur and nitrogen. *J. Geophys. Res.*, **101**, 29239-29252.

Bergamaschi, P., R. Hein, M. Heimann and P.J. Crutzen, 2000: Inverse modeling of the global CO cycle 1. Inversion of CO mixing ratios. *J. Geophys. Res.*, **105**, 1909-1927.

Berges, M.G.M., R.M. Hofmann, D. Schafre and P.J. Crutzen, 1993: Nitrous oxide emissions from motor vehicles in tunnels and their global extrapolation. *J. Geophys. Res.*, **98**, 18527-18531.

Berntsen, T. and I.S.A. Isaksen, 1997: A global three-dimensional chemical transport model for the troposphere. 1. Model description and CO and ozone results. *J. Geophys. Res.*, **102**, 21239-21280.

Berntsen, T. and I.S.A. Isaksen, 1999: Effects of lightning and convection on changes in tropospheric ozone due to NOx emissions from aircraft. *Tellus*, **51B**, 766-788.

Berntsen, T., I.S.A. Isaksen, W.-C. Wang and X.-Z. Liang, 1996: Impacts of increased anthropogenic emissions in Asia on tropospheric ozone and climate. A global 3-D model study. *Tellus*, **48B**, 13-32.

Berntsen, T.K., S. Karlsdottir and D.A. Jaffe, 1999: Influence of Asian emissions on the composition of air reaching the North Western United States. *Geophys. Res. Lett.*, **26**, 2171-2174.

Bey, I., R.M. Yantosca and D.J. Jacob, 1999: Export of pollutants from eastern Asia: a simulation of the PEM - West (B) aircraft mission using a 3-D model driven by assimilated meteorological fields. Presented at AGU spring meeting, 1999.

Bolin, B. and H. Rodhe, 1973: A note on the concepts of age distribution and transit time in natural reservoirs. *Tellus*, **25**, 58-62.

Blunier, T., J. Chappellaz, J. Schwander, J.-M. Barnola, T. Desperts, B. Stauffer and D. Raynaud, 1993: Atmospheric methane, record from a Greenland ice core over the last 1000 years. *J. Geophys. Res.*, **20**, 2219-2222.

Blunier, T., J. Chappellaz, J. Schwander, B. Stauffer and D. Raynaud, 1995: Variations in atmospheric methane concentration during the Holocene epoch. *Nature*, **374**, 46-49.

Brasseur, G.P., J.T. Kiehl, J.-F. Muller, T. Schneider, C. Granier, X.X. Tie and D. Hauglustaine, 1998a: Past and future changes in global tropospheric ozone: impact on radiative forcing. *Geophys. Res. Lett.*, **25**, 3807-3810.

Brasseur, G.P., D.A. Hauglustaine, S. Walters, P.J. Rasch, J.-F. Müller, C. Granier and X.X. Tie, 1998b: MOZART, a global chemical transport model for ozone and related chemical tracers: 1. Model description. *J. Geophys. Res.*, **103**, 28265-28289.

Brook, E.J., S. Harder, J. Severinghaus, E.J. Steig and C.M. Sucher, 2000: On the origin and timing of rapid changes in atmospheric methane during the last glacial period. *Global Biogechem. Cycles*, **14**, 559-572.

Brune, W.H., D. Tan, I.F. Faloona, L. Jaeglé, D.J. Jacob, B.G. Heikes, J. Snow, Y. Kondo, R. Shetter, G.W. Sachse, B. Anderson, G.L. Gregory, S. Vay, H.B. Singh, D.D. Davis, J.H. Crawford and D.R. Blake, 1999: OH and HO_2 chemistry in the north Atlantic free troposphere. *Geophys. Res. Lett.*, **26**, 3077-3080.

Butler, J.H., S.A. Montzka, A.D. Clarke, J.M. Lobert and J.W. Elkins, 1998a: Growth and distribution of halons in the atmosphere. *J. Geophys. Res.*, **103**, 1503-1511.

Butler, J.H. et al., 1998b: Nitrous oxide and halocompounds. In "Climate Monitoring and Diagnostics Laborartory Summary Report No. 24, 1996-1997," edited by D.J. Hofmann, J.T. Peterson and R.M. Rosson, pp. 91-121.

Butler, J.H., M. Battle, M.L. Bender, S.A. Montzka, A.D. Clark, E.S. Saltzman, C.M. Sucher, J.P. Severinghaus and J.W. Elkins, 1999: A record of atmospheric halocarbons during the twentieth century from polar firn air. *Nature*, **399**, 749-755.

Cao, M., K. Gregson and S. Marshall, 1998: Global methane emission from wetlands and its sensitivity to climate change. *Atmos. Env.*, **32**, 3293-3299.

Chameides, W.L., P.S. Kasibhatla, J. Yienger and H. Levy, 1994: Growth of continental-scale metro-agro-plexes, regional ozone pollution, and world food-production. *Science*, **264**, 74-77.

Chappellaz, J., T. Blunier, S. Kints, A. Dällenbach, J.-M. Barnola, J. Schwander, D. Raynaud and B. Stauffer, 1997: Changes in the atmospheric CH_4 gradient between Greenland and Antarctica during the Holocene. *J. Geophys. Res.*, **102**, 15987-15999.

Christian, H.J. et al., 1999: Global frequency and distribution of lightning as observed by the optical transient detector (OTD). 11th International Conference on Atmospheric Electricity, Guntersville, AL NASA/CP-1999-209261.

Collins, W.J., D.S. Stevenson, C.E. Johnson and R.G. Derwent, 1997: Tropospheric ozone in a global-scale three-dimensional Lagrangian model and its response to NOx emission controls. *J. Atmos. Chem.*, **26**, 223-274.

Collins, W.J., D.S. Stevenson, C.E. Johnson and R.G. Derwent, 1999: The role of convection in determining the budget of odd hydrogen in the upper troposphere. *J. Geophys. Res.*, **104**, 26927-26941.

Collins, W.J., D.S. Stevenson, C.E. Johnson and R.G. Derwent, 2000: The European regional ozone distribution and its links with the global scale for the years 1992 and 2015. *Atmos. Env.*, **34**, 255-267.

Crutzen, P.J., M.G. Lawrence and U. Poschl, 1999: On the background photochemistry of tropospheric ozone. *Tellus*, **51A-B**, 123-146.

Daniel, J.S. and S. Solomon, 1998: On the climate forcing of carbon monoxide. *J. Geophys. Res.*, **103**, 13249-13260.

Davidson, E.A and W. Kingerlee, 1997: A global inventory of nitric oxide emissions form soils. **48**, 37-50.

DeCaria, A., K.E. Pickering, G.L. Stenchikov, J.R. Scala, J.E. Dye, B.A. Ridley and P. Laroche, 2000: A cloud-scale model study of lightning-generated NOx in an individual thunderstorm during STERAO-A. *J. Geophys. Res.*, **105**, 11601-11616.

DeMore, W.B., S.P. Sander, D.M. Goldan, R.F. Hampson, M.J. Kurylo, C.J. Howard, A.R. Ravishankara, C.E. Kolb and M.J. Molina, 1997: Chemical kinetics and photochemical data for use in stratospheric modeling. Evaluation No. 12, JPL Publication 97-4, Jet Propulsion Laboratory, California Institute of Technology, Pasadena, CA.

Dentener, F.J. and P.J. Crutzen, 1993: Reaction of N_2O_5 on tropospheric aerosols: Impact on the global distribution of NOx, O_3, and OH. *J. Geophys. Res.*, **98**, 7149-7163.

Derwent, R. and R. Friedl, 1999: Chapter 2. Impacts of Aircraft Emissions on Atmospheric Ozone. In: Aviation and the Global Atmosphere, edited by J.E. Penner *et al.*, Cambridge University Press, pp 29-64.

Derwent, R.G., W.J. Collins, C.E. Johnson and D.S. Stevenson, 2001: Transient behaviour of tropospheric ozone precursors in a global 3-D CTM and their indirect greenhouse effects. *Clim Change*, **49:4**, in press.

Dessler, A.E., 1998: A reexamination of the 'stratospheric fountain' hypothesis. *Geophys. Res. Lett.*, **25**, 4165-4168.

Dickens, G.R., 1999: The blast in the past. *Nature*, **401**, 752-755.

Dickens, G., 2001: On the fate of past gas: what happens to methane released from a bacterially mediated gas hydrate capacitor? Geochem. Geophys. Geosyst., **2**, 2000GC000131.

Dickerson, R.R., B.G. Doddridge, P.K. Kelley and K.P. Rhoads, 1995: Large-scale pollution of the atmosphere over the North Atlantic Ocean: Evidence from Bermuda. *J. Geophys Res.*, **100**, 8945-8952.

Dickerson, R.R., S. Kondragunta, G. Stenchikov, K.L. Civerolo, B.G. Doddridge and B.N. Holben, 1997: The impact of aerosols on solar radiation and photochemical smog. *Science*, **278**, 827-830.

Dickerson, R.R., K.P. Rhoads, T.P. Carsey, S.J. Oltmans, J.P. Burrows and P.J, Crutzen, 1999: Ozone in the remote marine boundary layer: A possible role for halogens. *J. Geophys. Res.*, **104**, 21385-21395.

Ding, A. and M.X. Wang, 1996: A Model for methane emission from rice field and its application in southern China. *Advances in Atmospheric Sciences*, **13**, 159-168.

Dlugokencky, E.J., E.G. Dutton, P.C. Novelli, P.P. Tans, K.A. Masarie, K.O. Lantz and S. Madronich, 1996: Changes in CH_4 and CO growth rates after the eruption of Mt. Pinatubo and their link with changes in tropical tropospheric UV flux. *Geophys. Res. Lett.*, **23**, 2761-2764.

Dlugokencky, E.J., K.A. Masarie, P.M. Lang and P.P. Tans, 1998: Continuing decline in the growth rate of the atmospheric methane burden. *Nature*, **393**, 447-450.

Ehhalt, D.H., 1998: Radical ideas. *Science*, **279**, 1002-1003.

Ehhalt, D.H., 1999: Gas phase chemistry of the troposphere. Chapter 2 in "Global Aspects of Atmospheric Chemistry", edited by Deutsche Bunsen-Gesellschaft für Physikalische Chemie e.V., R. Zellner guest editor, Darmstadt Steinkopff, New York, Springer, pp. 21-109.

Ehhalt, D.H. and F. Rohrer, 2000: Dependence of the OH concentration on solar UV. *J. Geophys. Res.*, **105**, 3565-3571.

Ehhalt, D.H., U. Schmidt, R. Zander, P. Demoulin and C.P. Rinsland, 1991: Seasonal cycle and secular trend of the total and tropospheric column abundance of ethane above Jungfraujoch. *J. Geophys. Res.*, **96**, 4985-4994.

Elkins, J.W., J.H. Butler, D.F. Hurst, S.A. Montzka, F.L. Moore, T.M. Thompson and B.D. Hall, 1998: Halocarbons and other Atmospheric Trace Species Group/Climate Monitoring and Diagnostics Laboratory (HATS/CMDL), http://www.cmdl.noaa.gov/hats or anonymous ftp://ftp.cmdl.noaa.gov/hats, Boulder, Colorado, USA.

Elliott, S., D.R. Blake, R.A. Duce, C.A. Lai, I. McCreary, L.A. McNair, F.S. Rowland, A.G. Russell, G.E. Streit and R.P. Turco, 1997: Motorization of China implies changes in Pacific air chemistry and primary production. *Geophys. Res. Lett.*, **24**, 2671-2674.

Emmons, L.K., M.A. Carroll, D.A. Hauglustaine, G.P. Brasseur, C. Atherton, J. Penner, S. Sillman, H. Levy II, F. Rohrer, W.M.F. Wauben, P.F.J. Van Velthoven, Y. Wang, D. Jacob, P. Bakwin, R. Dickerson, B. Doddridge, C. Gerbig, R. Honrath, G. Hûbler, D. Jaffee, Y. Kondo, J.W. Munger, A. Torres and A. Volz-Thomas, 1997: Climatologies of NOx and NOy: A comparison of data and models. *Atmos. Env.*, **31**, 1851-1904.

Etheridge, D.M., L.P. Steele, R.J. Francey and R.L. Langenfelds, 1998: Atmospheric methane between 1000 A.D. and present: Evidence of anthropogenic emissions and climatic variability. *J. Geophys. Res.*, **103**, 15979-15993.

Faloona, I., D. Tan, W.H. Brune, L. Jaeglé, D.J. Jacob, Y. Kondo, M. Koike, R. Chatfield, R. Pueschel, G. Ferry, G. Sachse, S. Vay, B. Anderson, J. Hannon and H. Fuelberg, 2000: Observations of HOx and its relationship with NOx in the upper troposphere during SONEX. *J. Geophys. Res.*, **105**, 3771-3783.

Fishman, J. and V.G. Brackett, 1997: The climatological distribution of tropospheric ozone derived from satellite measurements using version 7 Total Ozone Mapping Spectrometer and Stratospheric Aerosol and Gas Experiment data sets. *J. Geophys. Res.*, **102**: 19275-19278.

Flückiger, J., A. Dällenbach, T. Blunier, B. Stauffer, T.F. Stocker, D. Raynaud and J.-M. Barnola, 1999: Variations in atmospheric N_2O concentration during abrupt climatic changes. *Science*, **285**, 227-230.

Folkins, I., P.O. Wennberg, T.F. Hanisco, J.G. Anderson and R.J. Salawitch, 1997: OH, HO_2 and NO in two biomass burning plumes: Sources of HOx and implications for ozone production. *Geophys. Res. Lett.*, **24**, 3185-3188.

Francey, R.J., M.R. Manning, C.E. Allison, S.A. Coram, D.M. Etheridge, R.L. Langenfelds, D.C. Lowe and L.P. Steele, 1999: A history of $\delta^{13}C$ in atmospheric CH_4 from the Cape Grim air archive and Antarctic firn air. *J. Geophys. Res.*, **104**, 23631-23643.

Fraser, P.J. and M.J. Prather, 1999: Uncertain road to ozone recovery. *Nature*, **398**, 663-664.

Fraser, P.J., D.E. Oram, C.E. Reeves, S.A. Penkett and A. McCulloch, 1999: Southern Hemisphere halon trends (1978-1998) and global halon emissions. *J. Geophys. Res.*, **104**, 15985-15999.

Frolking, S.E., A.R. Mosier, D.S. Ojima, C. Li, W.J. Parton, C.S. Potter, E. Priesack, R. Stenger, C. Haberbosch, P. Dorsch, H. Flessa, K.A. Smith, 1998: Comparison of N_2O emissions from soils at three temperate agricultural sites: simulations of year-round measurements by four models. *Nutr. Cycl. Agroecosys.*, **52**, 77-105.

Fuglestvedt, J.S., J.E. Jonson and I.S.A. Isaksen, 1994: Effects of reduction in stratospheric ozone on tropospheric chemistry through changes in photolysis rates. *Tellus*, **46B**, 172-192.

Fuglestvedt, J.S., J.E. Jonson, W.-C. Wang and I.S.A. Isaksen, 1995: Climate change and its effect on tropospheric ozone. In: Atmospheric Ozone as a Climate Gas, edited by W.C. Wang and I.S.A. Isaksen, pp. 145-162, NATO ASI Series vol. 132, Springer-Verlag, Berlin.

Fuglestvedt, J.S., I.S.A. Isaksen and W.-C. Wang, 1996: Estimates of indirect global warming potentials for CH_4, CO, and NOx. *Clim. Change*, **34**, 405-437.

Fuglestvedt, J.S., T. Berntsen, I.S.A. Isaksen, M. Liang and W.-C. Wang, 1999: Climatic forcing of nitrogen oxides through changes in tropospheric ozone and methane; global 3-D model studies. *Atmos. Env.*, **33**, 961-977.

Fung, I., J. John, J. Lerner, E. Matthews, M. Prather, L.P. Steele and P.J. Fraser, 1991: Three-dimensional model synthesis of the global methane cycle. *J. Geophys. Res.*, **96**, 13033-13065.

Geller, L.S., J.W. Elkins, J.M. Lobert, A.D. Clarke, D.F. Hurst, J.H. Butler and R.C. Myers, 1997: Tropospheric SF_6: observed latitudinal distribution and trends, derived emissions, and interhemispheric exchange time. *Geophys. Res. Lett.*, **24**, 675-678.

Graedel, T.E., T.S. Bates, A.F. Bouwman, D. Cunnold, J. Dignon, I. Fung, D.J. Jacob, B.K. Lamb, J.A. Logan, G. Marland, P. Middleton, J.M. Pacyna, M. Placet and C. Veldt , 1993: A compilation of inventories of emissions to the atmosphere. *Global Biogeochemical Cycles*, **7**, 1-26.

Granier, C. and K.P. Shine, 1999: Chapter 10. Climate Effects of Ozone and Halocarbon Changes. In: Scientific Assessment of Ozone Depletion: 1998. Global Ozone Research and Monitoring Project - Report No. 44, World Meteorological Organization, Geneva, Switzerland, pp. 10.1-10.38.

Granier, C., J.-F. Müller, S. Madronich and G.P. Brasseur, 1996: Possible causes of the 1990-1993 decrease in the global tropospheric CO abundances: A three-dimensional sensitivity study. *Atmos. Env.*, **30**, 1673-1682.

Guenther, A., C. Hewitt, D. Erickson, R. Fall, C. Geron, T. Graedel, P. Harley, L. Klinger, M. Lerdau, W. McKay, T. Pierce, B. Scholes, R. Steinbrecher, R. Tallamraju, J. Taylor and P. Zimmerman, 1995: A global model of natural volatile organic compound emissions. *J. Geophys. Res.*, **100**, 8873-8892.

Guenther, A., B. Baugh, G. Brasseur, J. Greenberg, P. Harley, L. Klinger, D. Serca and L. Vierling, 1999: Isoprene emission estimates and uncertainties for the Central African EXPRESSO study domain. *J. Geophys. Res.*, **104**, 30625-30639.

Hall, S. J. and P.A. Matson, 1999: Nitrogen oxide emissions after N additions in tropical forests. *Nature*, **400**, 152-155.

Hallock-Waters, K.A., B.G. Doddridge, R.R. Dickerson, S. Spitzer and J.D. Ray, 1999: Carbon monoxide in the U.S. mid-Atlantic troposphere: Evidence for a decreasing trend. *Geophys. Res. Lett.*, **26**, 2816-2864.

Hannegan, B., S. Olsen, M. Prather, X. Zhu, D. Rind and J. Lerner, 1998: The dry stratosphere: A limit on cometary water influx. *Geophys.Res.Lett.*, **25**, 1649-1652.

Hansen, J., M. Sato and R. Ruedy, 1997a: Radiative forcing and climate response. *J. Geophys. Res.*, **102**, 6831-6864.

Hansen, J., M. Sato, R. Ruedy, A. Lacis, K. Asamoah and 38 others, 1997b: Forcings and chaos in interannual to decadal climate change. *J. Geophys. Res.*, **102**, 25679-25720.

Harnisch, J. and A. Eisenhauer, 1998: Natural CF_4 and SF_6 on Earth. *Geophys. Res. Lett.*, **25**, 2401-2404.

Harnisch, J., R. Borchers, P. Fabian and M. Maiss, 1996: Tropospheric trends for CF_4 and CF_3CF_3 since 1982 derived from SF_6 dated stratospheric air. *Geophys. Res. Lett.*, **23**, 1099-1102.

Harnisch, J., R. Borchers, P. Fabian and M. Maiss, 1999: CF_4 and the age of mesospheric and polar vortex air. *Geophys. Res. Lett.*, **26**, 295-298.

Harris, N.R.P., G. Ancellet, L. Bishop, D.J. Hofmann, J.B. Kerr, R.D. McPeters, M. Prendez, W.J. Randel, J. Staehelin, B.H. Subbaraya, A. Volz-Thomas, J. Zawodny and C.S. Zerefos, 1997: Trends in stratospheric and free tropospheric ozone. *J. Geophys. Res.*, **102**, 1571-1590.

Hauglustaine, D.A., G.P. Brasseur, S. Walters, P.J. Rasch, J.-F. Müller, L.K. Emmons and M.A. Carroll, 1998: MOZART, a global chemical transport model for ozone and related chemical tracers: 2. Model results and evaluation. *J. Geophys. Res.*, **103**, 28291-28335.

Haywood, J.M., M.D. Schawrzkopf and V. Ramaswamy, 1998: Estimates of radiative forcing due to modeled increases in tropospheric ozone. *J. Geophys. Res.*, **103**, 16999-17007.

Hein, R., P.J. Crutzen and M. Heinmann, 1997: An inverse modeling approach to investigate the global atmospheric methane cycle. *Global Biogeochem. Cycles*, **11**, 43-76.

Hoell, J.M., D.D. Davis, S.C. Liu, R.E. Newell, H. Akimoto, R.J. McNeal and R.J. Bendura, 1997: The Pacific Exploratory Mission-West, Phase B: Feb-March, 1994. *J. Geophys. Res.*, **102**, 28223-28239.

Hoell, J.M., D.D. Davis, D.J. Jacob, M.O. Rogers, R.E. Newell, H.E. Fuelberg, R.J. McNeal, J.L. Raper and R.J. Bendura, 1999: Pacific Exploratory Mission in the tropical Pacific: PEM-Tropics A, August-September 1996. *J. Geophys. Res.*, **104**, 5567-5583.

Hofmann, D.J. and J.A. Pyle, 1999: Chapter 12. Predicting Future Ozone Changes and Detection of Recovery. In: Scientific Assessment of Ozone Depletion: 1998. Global Ozone Research and Monitoring Project - Report No. 44, World Meteorological Organization, Geneva, Switzerland, pp. 12.1-57.

Hogan, K.B. and R.C. Harriss, 1994: Comment on 'A dramatic decrease in the growth rate of atmospheric methane in the northern hemisphere during 1992' by Dlugokencky et al. *Geophys. Res. Lett.*, **21**, 2445-2446.

Holland, E.A., B.H. Braswell, J.F. Lamarque, A. Townsend, J. Sulzman, J.F. Muller, F. Dentener, G. Brasseur, H. Levy, J.E. Penner and G.J. Roelofs, 1997: Variations in the predicted spatial distribution of atmospheric nitrogen deposition and their impact on carbon uptake by terrestrial ecosystems. *J. Geophys. Res.*, **102**, 15849-15866.

Holland, E.A., F.J. Dentener, B.H. Braswell and J.M. Sulzman, 1999: Contemporary and pre-industrial reactive nitrogen budgets. *Biogeochemistry*, **46**, 7-43.

Holmes, K.J. and J.H. Ellis, 1999: An integrated assessment modeling framework for assessing primary and secondary impacts from carbon dioxide stabilization scenarios. *Env. Modeling and Assessment*, **4**, 45-63.

Houweling, S., F. Dentener and J. Lelieveld, 1998: The impact of non-methane hydrocarbon compounds on tropospheric photochemistry. *J. Geophys. Res.*, **103**, 10673-10696.

Houweling, S., T. Kaminski, F. Dentener, J. Lelieveld and M. Heimann, 1999: Inverse modeling of methane sources and sinks using the adjoint of a global transport model. *J. Geophys. Res.*, **104**, 26137-26160.

Houweling, S., F. Dentener, J. Lelieveld, B. Walter and E. Dlugokencky, 2000: The modeling of tropospheric methane: how well can point measurements be reproduced by a global model? *J. Geophys. Res.*, **105**, 8981-9002.

Hudson, R.D. and A.M. Thompson. 1998: Tropical tropospheric ozone from total ozone mapping spectrometer by a modified residual method. *J. Geophys. Res.*, **103**, 22129-22145.

Huntrieser, H., H. Schlager, C. Feigl and H. Höller, 1998: Transport and production of NOx in electrified thunderstorms: survey of previous studies and new observations at midlatitudes. *J. Geophys. Res.*, **103**, 28247-28264.

Impey, G.A., C.M. Mihele, K.G. Anlauf, L.A. Barrie, D.R. Hastie and P.B. Shepson, 1999: Measurements of photolyzable halogen compounds and bromine radicals during the Polar Sunrise Experiment 1997. *J. Atmos. Chem.*, **34**, 21-37.

IPCC, 1996: Climate Change 1995: The Science of Climate Change. Contribution of Working Group I to the Second Assessment Report of the Intergovernmental Panel on Climate Change [Houghton, J.T., L.G. Meira Filho, B.A. Callander, N. Harris, A. Kattenberg, and K. Maskell (eds.)]. Cambridge University Press, Cambridge, United Kingdom and New York, NY, USA, 572 pp.

Isaksen, I.S.A. and O. Hov, 1987: Calculations of trends in the tropospheric concentrations of O_3, OH, CO, CH_4 and NOx. *Tellus*, **39B**, 271-283.

Isaksen, I. and C. Jackman, 1999: Chapter 4. Modeling the Chemical Composition of the Future Atmosphere. In: Aviation and the Global Atmosphere, edited by J.E. Penner et al., Cambridge University Press, pp 121-183.

IUPAC, 1997a: Evaluated kinetic and photochemical data for atmospheric chemistry: Supplement V, IUPAC subcommittee on gas kinetic data evaluation for atmospheric chemistry. *Journal of Physical Chemistry Reference Data*, **26**, 512-1011.

IUPAC, 1997b: Evaluated kinetic and photochemical data for atmospheric chemistry: Supplement VI, IUPAC subcommittee on gas kinetic data evaluation for atmospheric chemistry. *Journal of Physical Chemistry Reference Data*, **26**, 1329-1499.

IUPAC, 1999: Evaluated kinetic and photochemical data for atmospheric chemistry: Supplement VII, IUPAC subcommittee on gas kinetic data evaluation for atmospheric chemistry. *Journal of Physical Chemistry Reference Data*, **28**, 191-393.

Jackman, C.H., E.L. Fleming, S. Chandra, D.B. Considine and J.E. Rosenfield, 1996: Past, present, and future modeled ozone trends with comparisons to observed trends. *J. Geophys. Res.*, **101**, 28753-28767.

Jacob, D.J., 2000: Heterogeneous chemistry and tropospheric ozone. *Atmos. Env.*, **34**, 2131-2159.

Jacob, D.J. et al. (30 authors), 1997: Evaluation and intercomparison of global atmospheric transport models using 222Rn and other short-lived tracers. *J. Geophys. Res.*, **102**, 5953-5970.

Jacob, D.J., J.A. Logan and P.P. Murti, 1999: Effect of rising Asian emissions on surface ozone in the United States. *Geophys. Res. Lett.*, **26**, 2175-2178.

Jacobson, M.Z., 1998: Studying the effects of aerosols on vertical photolysis rate coefficient and temperature profiles over an urban airshed. *J. Geophys. Res.*, **103**, 10593-10604.

Jaeglé, L., D.J. Jacob, P.O. Wennberg, C.M. Spivakovsky, T.F. Hanisco, E.L. Lanzendorf, E.J. Hintsa, D.W. Fahey, E.R. Keim, M.H. Proffitt, E. Atlas, F. Flocke, S. Schauffler, C.T. McElroy, C. Midwinter, L. Pfister and J.C. Wilson, 1997: Observed OH and HO2 in the upper troposphere suggest a major source from convective injection of peroxides. *Geophys. Res. Lett.*, **24**, 3181-3184.

Jaeglé, L., D.J. Jacob, W.H. Brune, I. Faloona, D. Tan, Y. Kondo, G. Sachse, B. Anderson, G.L. Gregory, S. Vay, H.B. Singh, D.R. Blake and R. Shetter, 1999: Ozone production in the upper troposphere and the influence of aircraft during SONEX: approach of NOx-saturated conditions. *Geophys. Res. Lett.*, **26**, 3081-3084.

Jaeglé, L., D.J. Jacob, W.H. Brune, I. Faloona, D. Tan, B.G. Heikes, Y. Kondo, G. W. Sachse, B. Anderson, G.L. Gregory, H.B. Singh, R. Pueschel, G. Ferry, D.R. Blake and R.Shetter, 2000: Photochemistry of HOx in the upper troposphere at northern midlatitudes. *J. Geophys. Res.*, **105**, 3877-3892.

Jaffe, D.A., T. Anderson, D. Covert, R. Kotchenruther, B.Trost J. Danielson, W. Simpson, T. Berntsen, S. Karlsdottir, D. Blake, J. Harris and G. Carmichael, 1999: Transport of Asian air pollution to North America. *Geophys. Res. Lett.*, **26**, 711-714.

Jeuken, A.B.M., H.J. Eskes, P.F.J. van Velthoven, H.M. Kelder and E.V. Hólm, 1999: Assimilation of total ozone satellite measurements in a three-dimensional tracer transport model. *J. Geophys. Res.*, **104**, 5551-5563.

Jiménez, J.L., J.B. McManus, J.H. Shorter, D.D. Nelson, M.S. Zahniser, M. Koplow, G.J. McRae and C.E. Kolb, 2000: Cross road and mobile tunable infrared laser measurements of nitrous oxide emissions from motor vehicles. *Chemosphere: Global Change Science*, (in press).

Johnson, C.E., W.J. Collins, D.S. Stevenson and R.G. Derwent, 1999: The relative roles of climate and emissions changes on future oxidant concentrations. *J. Geophys. Res.*, **104**, 18631-18645.

JPL, 1997: Chemical Kinetics and Photochemical Data for Use in Stratospheric Modeling: Evaluation 12. JPL-97-4, NASA Panel for Data Evaluation, Jet Propulsion Laboratory, NASA Aeronautics and Space Administration, Pasadena, CA, USA, 266 pp.

JPL, 2000: Chemical Kinetics and Photochemical Data for Use in Stratospheric Modeling: Evaluation 13. JPL-00-3, NASA Panel for Data Evaluation, Jet Propulsion Laboratory, NASA Aeronautics and Space Administration, Pasadena, CA, USA, 73 pp.

Kanakidou, M. and P.J. Crutzen, 1999: The photochemical source of carbon monoxide: importance, uncertainties and feedbacks. *Chemosphere: Global Change Science*, **1**, 91-109.

Kanakidou, M., F.J. Dentener, G.P. Brasseur, T.K. Berntsen, W.J. Collins, D.A. Hauglustaine, S. Houweling, I.S.A. Isaksen, M. Krol, M.G. Lawrence, J.-F. Müller, N. Poisson, G.J. Roelofs, Y. Wang and W.M.F. Wauben, 1999: 3-D global simulations of tropospheric CO distributions - results of the GIM/IGAC intercomparison 1997 exercise. *Chemosphere: Global Change Science*, **1**, 263-282.

Karlsdottir, S. and I.S.A. Isaksen, 2000: Changing methane lifetime: Possible cause for reduced growth. *Geophys. Res. Lett.*, **27**, 93-96.

Karlsdottir, S., I.S.A. Isaksen, G. Myhre and T.K. Berntsen, 2000: Trend analysis of O3 and CO in the period 1980 to 1996: A 3-D model study. *J. Geophys. Res.*, **105**, 28907-28933.

Kato, N. and H. Akimoto, 1992: Anthropogenic emissions of SO2 and NOx in Asia: emission inventories. *Atmos. Env.*, **26A**, 2997-3017.

Kennett, J.P., K.G. Cannariato, I.L. Hendy and R.J. Behl, 2000: Carbon isotopic evidence for methane hydrate instability during quaternary interstadials. *Science*, **288**, 128-133.

Khalil, M.A.K. and R. Rasmussen, 1994: Global decrease in atmospheric carbon monoxide. *Nature*, **370**, 639-641.

Kheshgi, H.S., A.K. Jain, R. Kotamarthi and D.J. Wuebbles, 1999: Future atmospheric methane concentrations in the context of the stabilization of greenhouse gas concentrations. *J. Geophys. Res.*, **104**, 19183-19190.

Kiehl, J.T., T.L. Schneider, R.W. Portmann and S. Solomon, 1999: Climate forcing due to tropospheric and stratospheric ozone. *J. Geophys. Res.*, **104**, 31239-31254.

Kim, K.-R. and H. Craig, 1990: Two-isotope characterization of N2O in the Pacific Ocean and constraints on its origin in deep water. *Nature*, **347**, 58-61.

Kim, K.-R. and H. Craig, 1993: Nitrogen-15 and oxygen-18 characteristics of nitrous oxide: A global perspective. *Science*, **262**, 1855-1857.

Ko, M.K.W., R.-L. Shia, N.-D. Sze, H. Magid and R.G. Bray, 1999: Atmospheric lifetime and global warming potential of HFC-245fa. *J. Geophys. Res.*, **104**, 8173-8181.

Kroeze, C., A. Mozier and L. Bouwman, 1999: Closing the N2O Budget: A retrospective analysis. *Global Biogeochem. Cycles*, **13**, 1-8.

Krol, M., P.J. Van Leeuwen and J. Lelieveld, 1998: Global OH trend inferred from methylchloroform measurements. *J. Geophys. Res.*, **103**, 10697-10711.

Kurylo, M.J. and J.M. Rodriguez, 1999: Chapter 2, Short-lived Ozone-Related Compounds. In: Scientific Assessment of Ozone Depletion: 1998. Global Ozone Research and Monitoring Project - Report No. 44, World Meteorological Organization, Geneva, Switzerland, pp 2.1-56.

Labuschagne, C., E.-G. Brunke, B. Parker and H.E. Scheel, 1999: Cape Point Trace Gas Observations Under Baseline and Non-Baseline Conditions. Poster Presentation at NOAA CMDL Annual Meeting, 12-13 May, 1999.

Langenfelds, R.L., P.J. Fraser, R.J. Francey, L.P. Steele, L.W. Porter and C.E. Allison, 1996: The Cape Grim Air Archive: The first seventeen years. In "Baseline Atmospheric Program Australia, 1994-95," edited by R.J. Francey, A.L. Dick and N. Derek, pp. 53-70.

Lassey, K.R., D.C. Lowe, C.A.M. Brenninkmeijer and A.J. Gomez, 1993: Atmospheric methane and its carbon isotopes in the southern hemisphere: Their time series and an instructive model. *Chemosphere: Global Change Science*, **26**, 95-100.

Law, K.S., P.-H. Plantevin, D.E. Shallcross, H.L.Rogers, J.A. Pyle, C.Grouhel, V. Thouret and A. Marenco, 1998: Evaluation of modelled O3 using MOZAIC data. *J. Geophys. Res*, **103**, 25721-25740.

Law, K.S., P.-H. Plantevin, V. Thouret, A. Marenco, W.A.H. Asman, M. Lawrence, P.J. Crutzen, J.F. Muller, D.A. Hauglustaine and M. Kanakidou, 2000: Comparison between global chemistry transport model results and MOZAIC data. *J. Geophys. Res.*, **105**, 1503-1525.

Lawrence, M. and P.J. Crutzen, 1999: The impact of cloud particle gravitational settling on soluble trace gas distributions. *Tellus*, **50B**, 263-289.

Lawrence, M.G., P.J. Crutzen, P.J. Rasch, B.E. Eaton and N.M. Mahowald, 1999: A model for studies of tropospheric photochemistry: description, global distributions and evaluation. *J. Geophys. Res.*, **104**, 26245-26277.

Lee, D.S., I. Köhler, E. Grobler, F. Rohrer, R. Sauen, L. Gallardo-Klenner, J.J.G. Olivier, F.J. Dentener and A.F. Bouwman, 1997: Estimates of global NOx emissions and their uncertainties. *Atmos. Env.*, **31**, 1735-1749.

Lelieveld, J. and P.J. Crutzen, 1990: Influences of cloud photochemical processes on tropospheric ozone. *Nature*, **343**, 227-233.

Lelieveld, J., P. Crutzen and F.J. Dentener, 1998: Changing concentration, lifetime and climate forcing of atmospheric methane. *Tellus*, **50B**, 128-150.

Lerdau, M. and M. Keller, 1997: Controls over isoprene emission from trees in a sub-tropical dry forest. *Plant, Cell, and Environment*, **20**, 569-578.

Levy II, H., P.S. Kasibhatla, W.J. Moxim, A.A. Klonecki, A.I. Hirsch, S.J. Oltmans and W.L. Chameides, 1997: The global impact of human activity on tropospheric ozone. *Geophys. Res. Lett.*, **24**, 791-794.

Li, C., S. Frolking and T.A. Frolking, 1992: A Model of nitrous oxide evolution from soil dirvben by rainfall events: I. Model structure and sensitivity. *J. Geophys. Res.*, **97**, 9759-9776.

Li, C., V. Narayanan and R. Harriss, 1996: Model estimates of nitrous oxide evolution from soil agricultural lands in the United States. *Global Biogeochem. Cycles*, **10**, 297-306.

Liang, J. and D.J. Jacob, 1997: Effect of aqueous phase cloud chemistry on tropospheric ozone. *J. Geophys. Res.*, **102**, 5993-6001.

Logan, J.A., 1994: Trends in the vertical distribution of ozone: an analysis of ozone sonde data. *J. Geophys. Res.*, **99**, 25553-25585.

Logan, J.A., 1999: An analysis of ozonesonde data for the troposphere: recommendations for testing 3-D models and development of a gridded climatology for tropospheric ozone. *J. Geophys. Res.*, **104**, 16115-16149.

Logan, J.A., I.A. Megretskaia, A.J. Miller, G.C. Tiao, D. Choi, L. Zhang, R.S. Stolarski, G.J. Labow, S.M. Hollandsworth, G.E. Bodeker, H. Claude, D. DeMuer, J.B. Kerr, D.W. Tarasick, S.J. Oltmans, B. Johnson, F. Schmidlin, J. Staehelin, P. Viatte and O. Uchino, 1999: Trends in the vertical distribution of ozone: A comparison of two analyses of ozonesonde data. *J. Geophys. Res.*, **104**, 26373-26399.

Lowe, D.C., C.A.M. Brenninkmeijer, G.W. Brailsford, K.R. Lassey and A.J. Gomez, 1994: Concentration and ^{13}C records of atmospheric methane in New Zealand and Antarctica: Evidence for changes in methane sources. *J. Geophys. Res.*, **99**, 16913-16925.

Lowe, D.C., M.R. Manning, G.W. Brailsford and A.M. Bromley, 1997: The 1991-1992 atmospheric methane anomaly: Southern Hemisphere C-13 decrease and growth rate fluctuations. *Geophys. Res. Lett.*, **24**, 857-860.

Machida, T, T. Nakazawa, Y. Fujii, S. Aoki and O. Watanabe, 1995: Increase in the atmospheric nitrous oxide concentration during the last 250 years. *Geophys. Res. Lett.*, **22**, 2921-2924.

Madronich, S. and G.J.M. Velders, 1999: Chapter 11, Halocarbon Scenarios for the Future Ozone Layer and Related Consequences. In: Scientific Assessment of Ozone Depletion: 1998. Global Ozone Research and Monitoring Project - Report No. 44, World Meteorological Organization, Geneva, Switzerland, pp 11.1-38.

Mahieu, E., R. Zander, L. Delbouille, P. Demoulin, G. Roland and C. Servais, 1997: Observed trends in total column abundances of atmospheric gases from IR solar spectra recorded at the Jungfraujoch. *J. Atmos. Chem.*, **28**, 227-243.

Maiss, M. and C.A.M. Brenninkmeijer, 1998: Atmospheric SF$_6$: trends, sources, and prospects. *Envir. Sci. Tech.*, **32**, 3077-3086.

Maiss, M., L.P. Steele, R.J. Francey, P.J. Fraser, R.L. Langenfelds, N.B.A. Trivett and I. Levin, 1996: Sulfur hexafluoride - A powerful new atmospheric tracer. *Atmos. Env.*, **30**, 1621-1629.

Mak, J.E., M.R. Manning and D.C. Lowe, 2000: Aircraft observations of delta C-13 of atmospheric methane over the Pacific in August 1991 and 1993: Evidence of an enrichment in (CH$_4$)-C-13 in the Southern Hemisphere. *J. Geophys. Res.*, **105**, 1329-1335.

Manning, M.R., 1999: Characteristic modes of isotopic variations in atmospheric chemistry. *Geophys. Res. Lett.*, **26**, 1263-1266.

Matson, P.A., W.H. McDowell, A.R. Townsend and P.M. Vitousek, 1999: The globalization of N deposition: ecosystem consequences in tropical environments. *Biogeochemistry*, **46**, 67-83.

McElroy, M.B. and D.B.A. Jones, 1996: Evidence for an additional source of atmospheric N$_2$O. *Global Biogeochem. Cycles*, **10**, 651-659.

McLinden, C., S. Olsen, B. Hannegan, O. Wild, M. Prather and J. Sundet, 2000: Stratospheric ozone in 3-D models: a simple chemistry and the cross-tropopause flux. *J. Geophys. Res.*, **105**, 14653-14665.

Michelsen, H.A., F.W. Irion, G.L. Manney, G.C. Toon and M.R. Gunson, 2000: Features and trends in ATMOS version 3 stratospheric water vapor and methane measurements. J. Geophys. Res., **105**, 22713-22724.

Mickley, L.J., P.P. Murti, D.J. Jacob, J.A. Logan, D. Rind and D. Koch, 1999: Radiative forcing from tropospheric ozone calculated with a unified chemistry-climate model. *J. Geophys. Res.*, **104**, 30153-30172.

Miller, B., J. Huang, R. Weiss, R. Prinn and P. Fraser, 1998: Atmospheric trend and lifetime of chlorodifluoromethane (HCFC-22) and the global tropospheric OH concentration. *J. Geophys. Res.*, **103**, 13237-13248.

Molina, L.T., P.J. Woodbridge and M.J. Molina, 1995: Atmospheric reactions and ultraviolet and infrared absorptivities of nitrogen trifluoride. *Geophys. Res. Lett.*, **22**, 1873-1876.

Montzka, S.A., R.C. Myers, J.H. Butler and J.W. Elkins, 1994: Early trends in the global tropospheric abundance of hydrochlorofluorocarbon -141b and -142b. *Geophys. Res. Lett.*, **21**, 2483-2486.

Montzka, S.A., J.H. Butler, R.C. Myers, T.M. Thompson. T.H. Swanson, A.D. Clarke, L.T. Lock and J.W. Elkins, 1996a: Decline in the tropospheric abundance of halogen from halocarbons: implications for stratospheric ozone depletion. *Science*, **272**, 1318-1322.

Montzka, S.A., R.C. Myers, J.H. Butler, J.W. Elkins, L.T. Lock, A.D. Clarke and A.H. Goldstein, 1996b: Observations of HFC-134a in the remote troposphere. *Geophys. Res. Lett.*, **23**, 169-172.

Montzka, S.A., J.H. Butler, J.W. Elkins, T.M. Thompson, A.D. Clarke and L.T. Lock, 1999: Present and future trends in the atmospheric burden of ozone-depleting halogens. *Nature*, **398**, 690-694.

Montzka, S.A., C.M. Spivakovsky, J.H. Butler, J.W. Elkins, L.T. Lock and D.J. Mondeel, 2000: New observational constraints for atmospheric hydroxyl on global and hemispheric scales. *Science*, **288**, 500-503.

Mosier, A.R., J.M. Duxbury, J.R. Freney, O. Heinemeyer, K. Minami and D.E. Johnson, 1998a: Mitigating agricultural emissions of methane. *Clim. Change*, **40**, 39-80.

Mosier, A, C. Kroeze, C. Nevison, O. Oenema, S. Seitzinger and O. van Cleemput, 1998b: Closing the global N$_2$O budget: nitrous oxide emissions through the agricultural nitrogen cycle - OECD/IPCC/IEA phase II development of IPCC guidelines for national greenhouse gas inventory methodology. *Nutrient Cycling in Agroecosystems*, **52**, 225-248.

Mote, P.W., T.J. Dunkerton, M.E. McIntyre, E.A. Ray, P.H. Haynes and J.M. Russell, 1998: Vertical velocity, vertical diffusion, and dilution by midlatitude air in the tropical lower stratosphere. *J. Geophys. Res.*, **103**, 8651-8666.

Moyer, E.J., F.W. Irion, Y.L. Yung and M.R. Gunson, 1996: ATMOS stratospheric deuterated water and implications for troposphere-stratosphere transport. *Geophys. Res. Lett.*, **23**, 2385-2388.

Müller, J.-F. and G. Brasseur, 1995: IMAGES: A three-dimensional chemical transport model of the global troposphere. *J. Geophys. Res.*, **100**, 16445-16490.

Müller, J.-F. and G. Brasseur, 1999: Sources of upper tropospheric HOx: A three-dimensional study. *J. Geophys. Res.*, **104**, 1705-1715.

Murphy, D.M. and D.W. Fahey, 1994: An estimate of the flux of stratospheric reactive nitrogen and ozone into the troposphere. *J. Geophys. Res.*, **99**, 5325-5332.

Naik, V., A.K. Jain, K.O. Patten, and D.J. Wuebbles, Consistent sets of atmospheric lifetimes and radiative forcing on climate for CFC replacements: HCFCs and HFCs, J. Geophys. Res., **105**, 6903-6914, 2000.

Nakićenović, N., J. Alcamo, G. Davis, B. de Vries, J. Fenhann, S. Gaffin, K. Gregory, A. Grübler, T. Y. Jung, T. Kram, E. L. La Rovere, L. Michaelis, S. Mori, T. Morita, W. Pepper, H. Pitcher, L. Price, K. Raihi, A. Roehrl, H-H. Rogner, A. Sankovski, M. Schlesinger, P. Shukla, S. Smith, R. Swart, S. van Rooijen, N. Victor, Z. Dadi, 2000: *IPCC Special Report on Emissions Scenarios*, Cambridge University Press, Cambridge, United Kingdom and New York, NY, USA, 599 pp.

Nedoluha, G.E., R.M. Bevilacqua, R.M. Gomez, D.E. Siskind, B.C. Hicks, J.M. Russell and B.J. Connor, 1998: Increases in middle atmospheric water vapor as observed by the Halogen Occultation Experiment and the ground-based Water Vapor Millimeter-wave Spectrometer from 1991 to 1997. *J. Geophys. Res.*, **103**, 3531-3543.

Neue, H.-U. and R. Sass, 1998: The budget of methane from rice fields. *IGACtivities Newsletter*, **12**, 3-11.

Nevison C.D. and E.A. Holland, 1997: A reexamination of the impact of anthropogenically fixed nitrogen on atmospheric N_2O and the stratospheric O_3 layer. *J. Geophys. Res.*, **102**, 25519-25536.

Norris, R.D. and U. Röhl, 1999: Carbon cycling and chronology of climate warming during the Palaeocene/Eocene transition. *Nature*, **401**, 775-778.

Novelli, P.C., K.A. Masarie and P.M. Lang, 1998: Distribution and recent changes of carbon monoxide in the lower troposphere. *J. Geophys. Res.*, **103**, 19015-19033.

Novelli, P.C., P.M. Lang, K.A. Masarie, D.F. Hurst, R. Myers and J.W. Elkins, 1999: Molecular hydrogen in the troposphere: Global distribution and budget. *J. Geophys. Res.*, **104**, 30427-30444.

Oltmans, S.J. and D.J. Hofmann, 1995: Increase in lower stratospheric water vapor at midlatitude northern hemisphere. *Nature*, **374**, 146-149.

Oltmans, S.J., A.S. Lefohn, H.E. Scheel, J.M. Harris, H. Levy, I.E. Galbally, E.G. Brunke, C.P. Meyer, J.A. Lathrop, B.J. Johnson, D.S. Shadwick, E. Cuevas, F.J. Schmidlin, D.W. Tarasick, H. Claude, J.B. Kerr, O. Uchino and V. Mohnen, 1998: Trends of ozone in the troposphere. *Geophys. Res. Lett.*, **25**, 139-142.

Olivier, J.G.J., A.F. Bouwman, K.W. van der Hoek and J.J.M. Berdowski, 1998: Global Air Emission Inventories for Anthropogenic Sources of NOx, NH_3 and N_2O in 1990. *Environmental Poll.*, **102**, 135-148.

Olivier, J.G.J., A.F. Bouwman, J.J.M. Berdowski, C. Veldt, J.P.J. Bloos, A.J.H. Visschedijk, C.W.M. van der Maas and P.Y.J. Zasndveld, 1999: Sectoral emission inventories of greenhouse gases for 1990 on a per country basis as well as on 1×1. *Envir. Sci. Policy*, **2**, 241-263.

Oram, D.E., C.E. Reeves, S.A. Penkett and P.J. Fraser, 1995: Measurements of HCFC-142b and HCFC-141b in the Cape Grim air archive: 1978-1993. *Geophys. Res. Lett.*, **22**, 2741-2744.

Oram, D.E., C.E. Reeves, W.T. Sturges, S.A. Penkett, P.J. Fraser and R.L. Langenfelds, 1996: Recent tropospheric growth rate and distribution of HFC-134a (CF_3CH_2F). *Geophys. Res. Lett.*, **23**, 1949-1952.

Oram, D.E., W.T. Sturges, S.A. Penkett, J.M. Lee, P.J. Fraser, A. McCulloch and A. Engel, 1998: Atmospheric measurements and emissions of HFC-23 (CHF_3). *Geophys. Res. Lett.*, **25**, 35-38.

Oram, D.E., W.T. Sturges, S.A. Penkett and P.J. Fraser, 1999: Tropospheric abundance and growth rates of radiatively-active halocarbon trace gases and estimates of global emissions. In *IUGG 99: abstracts, Birmingham.* [England]: International Union of Geodesy and Geophysics, 213 pp, abstract MI02/W/12-A4.

Orkin, V.L., E. Villenave, R.E. Huie and M.J. Kurylo, 1999: Atmospheric lifetimes and global warming potentials of hydrofluoroethers: Reactivity toward OH, UV spectra, and IR absorption cross sections. *J. Phys. Chem.*, **103**, 9770-9779.

Oum, K.W., M.J. Lakin and B.J. Finlayson-Pitts, 1998: Bromine activation in the troposphere by the dark reaction of O-3 with seawater ice. *Geophys. Res. Lett.*, **25**, 3923-3926.

Park, J.H., M.K.W. Ko, C.H. Jackman, R.A. Plumb, J.A. Kaye and K.H. Sage (eds.), 1999: M&M-2, NASA: Models and Measurements Intercomparison II. TM_1999_209554, September 1999, 502 pp.

Parrish, D.D., J.S. Holloway, M. Trainer, P.C. Murphy, G.L. Forbes and F.C. Fehsenfeld, 1993: Export of North-American ozone pollution to the North-Atlantic ocean. *Science*, **259**, 1436-1439.

Parton, W.J., M. Hartman, D. Ojima and D. Schimel, 1998: DAYCENT and its land surface submodel: description and testing. *Global Planet. Change*, **19**, 35-48.

Penkett, S.A., A. Volz-Thomas, D.D. Parrish, R.E. Honrath and F.C. Fehsenfeld, 1998: The North Atlantic Regional Experiment (NARE II), Preface. *J. Geophys. Res.*, **103**, 13353-13356.

Penner, J.E., D.H. Lister, D.J. Griggs, D.J. Dokken and M. McFarland (eds.), 1999: Aviation and the Global Atmosphere. A Special Report of IPCC Working Groups I and III, Cambridge University Press, Cambridge, UK, 373 pp.

Petit, J.R., J. Jouzel, D. Raynaud, N.I. Barkov, J.M. Barnola, I. Basile, M. Bender, J. Chappellaz, Davis, G. Delaygue, M. Delmotte, V.M. Kotlyakov, M. Legrand, V.Y. Lipenkov, C. Lorius, L. Pepin, C. Ritz, E. Saltzman and M. Stievenard, 1999: Climate and atmospheric history of the past 420,000 years from the Vostok ice core, Antarctica. *Nature*, **399**, 429-436.

Pickering, K.E., A.M. Thompson, Y. Wang, W.-K Tao, D.P. McNamara, V.W.J.H. Kirchoff, B.G. Heikes, G.W. Sachse, J.D. Bradshaw, G.L. Gregory and D.R. Blake, 1996: Convective transport of biomass burning emissions over Brazil during TRACE a. *J. Geophys. Res.*, **101**, 23993-24012.

Pickering, K.E., Y. Wang, W.-K. Tao, C. Price and J.-F. Müller, 1998: Vertical distribution of lightning NOx for use in regional and global chemical transport models. *J. Geophys. Res.*, **103**, 31203-31216.

Pitari, G., B. Grassi and G. Visconti, 1997: Results of a chemical-transport model with interactive aerosol microphysics. Proc. XVIII Quadrennial Ozone Symposium, R. Bojkov and G. Visconti (eds.), pp 759-762.

Platt, U. and G.K. Moortgat, 1999: Heterogeneous and homogeneous chemistry of reactive halogen compounds in the lower troposphere - XXIII General Assembly of the EGS. *J. Atmos. Chem.*, **34**, 1-8.

Prados, A.I., R.R. Dickerson, B.G. Doddridge, P.A. Milne, J.L. Moody and J.T. Merrill, 1999: Transport of ozone and pollutants from North America to the North Atlantic Ocean during the 1996 Atmosphere/Ocean Chemistry Experiment (AEROCE) intensive. *J. Geophys. Res.*, **104**, 26219-26233.

Prather, M.J., 1996: Natural modes and time scales in atmospheric chemistry: theory, GWPs for CH_4 and CO, and runaway growth. *Geophys. Res. Lett.*, **23**, 2597-2600.

Prather, M.J., 1998: Time scales in atmospheric chemistry: coupled perturbations to N_2O, NOy, and O_3. *Science*, **279**, 1339-1341.

Prather, M.J. and C.M. Spivakovsky, 1990: Tropospheric OH and the lifetimes of hydrochlorofluorocarbons (HCFCs). *J. Geophys. Res.*, **95**, 18723-18729.

Prather, M.J. and D.J. Jacob, 1997: A persistent imbalance in HOx and NOx photochemistry of the upper troposphere driven by deep tropical convection. *Geophys. Res. Lett.*, **24**, 3189-3192.

Prather, M. and R. Sausen, 1999: Chapter 6. Potential Climate Change from Aviation. In: Aviation and the Global Atmosphere, edited by J.E. Penner et al., Cambridge University Press, pp 185-215.

Prather, M., R. Derwent, D. Ehhalt, P. Fraser, E Sanhueza and X. Zhou, 1995: Other tracer gases and atmospheric chemistry. In: *Climate Change 1994*, [Houghton, J.T., L.G. Meira Filho, J. Bruce, Hoesung lee, B.A. Callander, E. Haites, N. Harris and K. Maskell (eds.)], Cambridge University Press, Cambridge, UK, pp 73-126.

Price, C. and D. Rind, 1992: A simple lightning parameterization for calculating global lightning distributions. *J. Geophys. Res.*, **97**, 9919-9933.

Price, C., J. Penner and M. Prather, 1997a: NOx from lightning 1. Global distribution based on lightning physics. *J. Geophys. Res.*, **102**, 5929-5941.

Price, C., J. Penner and M. Prather, 1997b: NOx from lightning 2. Constraints from the global electric circuit. *J. Geophys. Res.*, **102**, 5943-5951.

Prinn, R.G. and R. Zander, 1999: Chapter 1, Long-lived Ozone Related Compounds. In: Scientific Assessment of Ozone Depletion: 1998. Global Ozone Research and Monitoring Project - Report No. 44, World Meteorological Organization, Geneva, Switzerland, pp 1.1-54.

Prinn, R.G., D.M. Cunnold, R. Rasmussen, P.G. Simmonds, F.N. Alyea, A.J. Crawford, P.J. Fraser and R.D. Rosen, 1990: Atmospheric emissions and trends of nitrous oxide deduced from 10 years of ALE/GAGE data. *J. Geophys. Res.*, **95**, 18369-18385.

Prinn, R.G., R.F. Weiss, B.R. Miller, J. Huang, F.N. Alyea, D.M. Cunnold, P.J. Fraser, D.E. Hartley and P.G. Simmonds, 1995:

Atmospheric trend and lifetime of CH_3CCl_3 and global OH concentrations. *Science*, **269**, 187-192.

Prinn, R.G., R.F. Weiss, P.J. Fraser, P.G. Simmonds, F.N. Alyea and D.M. Cunnold, 1998: The ALE/GAGE/AGAGE database. DOE-CDIAC World Data Center (e-mail to: cpd@ornl.gov), Dataset No. DB-1001.

Prinn, R., H. Jacoby, A. Sokolov, C. Wang, X. Xiao, Z. Yang, R. Eckhaus, P. Stone, D. Ellerman, J. Melillo, J. Fitzmaurice, D. Kicklighter, G. Holian and Y. Liu, 1999: Integrated global system model for climate policy assessment: Feedbacks and sensitivity studies. *Clim. Change*, **41**, 469-546.

Prinn, R.G., R.F. Weiss, P.J. Fraser, P.G. Simmonds, D.M. Cunnold, F.N. Alyea, S. O'Doherty, P. Salameh, B.R. Miller, J. Huang, R.H.J. Wang, D.E. Hartley, C. Harth, L.P. Steele, G. Sturrock, P.M. Midgley and A. McCulloch, 2000: A history of chemically and radiatively important gases in air deduced from ALE/GAGE/AGAGE. *J. Geophys. Res.*, **105**, 17751-17792.

Quay, P.D., S.L. King, J. Stutsman, D.O. Wilbur, L.P. Steele, I. Fung, R.H. Gammon, T.A. Brown, G.W. Farwell, P.M. Grootes and F.H. Schmidt, 1991: Carbon isotopic composition of atmospheric CH_4: Fossil and biomass burning source components. *Global Biogeochem. Cycles*, **5**, 25-47.

Quay, P., J. Stutsman, D. Wilbur, A. Stover, E. Dlugokencky and T. Brown, 1999: The isotopic composition of atmospheric methane. *Global Biogeochem. Cycles*, **13**, 445-461.

Rahn, T. and M. Whalen, 1997: Stable isotope enrichment in stratospheric nitrous oxide. *Science*, **278**, 1776-1778.

Rahn, T., H. Zhang, M. Whalen and G.A. Blake, 1998: Stable isotope fractionation during ultraviolet photolysis of N_2O. *Geophys. Res. Lett.*, **25**, 4489-4492.

Randel, W.J., R.S. Stolarski, D.M. Cunnold, J.A. Logan, M.J. Newchurch and J.M. Zawodny, 1999: Trends in the vertical distribution of ozone. *Science*, **285**, 1689-1692.

Ridgwell, A.J., S.J. Marshall and K. Gregson, 1999: Consumption of methane by soils: A process-based model. *Global Biogeochem. Cycles*, **13**, 59-70.

Rinsland, C.P., N.B. Jones, B.J. Connor, J.A. Logan, N.S. Pougatchev, A. Goldman, F.J. Murcray, T.M. Stephen, A.S. Pine, R. Zander, E. Mahieu and P. Demoulin, 1998: Northern and southern hemisphere ground-based infrared spectroscopic measurements of tropospheric carbon monoxide and ethane. *J. Geophys. Res.*, **103**, 28197-28217.

Roelofs, G.-J. and J.S. Lelieveld, 1997: Model study of cross-tropopause O_3 transports on tropospheric O_3 levels. *Tellus*, **49B**, 38-55.

Roelofs, G.-J., J.S. Lelieveld and R. van Dorland, 1997: A three dimensional chemistry/general circulation model simulation of anthropogenically derived ozone in the troposphere and its radiative climate forcing. *J. Geophys. Res.*, **102**, 23389-23401.

Rohrer, F., D. Brüning, E. Gobler, M. Weber, D. Ehhalt, R. Neubert, W. Schüßler, and I. Levine, 1998: Mixing ratios and photostationary state of NO and NO_2 observed during the POPCORN field campaign at a rural site in Germany. *J. Atmos. Chem.*, **31**, 119-137.

Rosenlof, K.H., A.F. Tuck , K.K. Kelly, Kelly, J.M. Russell and M.P. McCormick, 1997: Hemispheric asymmetries in water vapor and inferences about transport in the lower stratosphere. *J. Geophys. Res.*, **102**, 13213-13234.

Rudolph, J., 1995: The tropospheric distribution and budget of ethane. *J. Geophys. Res.*, **100**, 11369-11381.

Sander, S.P., R.R. Friedl, W.B. DeMore, A.R. Ravishankara, D.M. Golden, C.E. Kolb, M.J. Kurylo, R.F. Hampson, R.E. Huie, M.J. Molina and G.K. Moortgat, 2000: Chemical Kinetics and Photochemical Data for Use in Stratospheric Modeling. Supplement to Evaluation No. 12 - Update of Key Reactions and Evaluation No.13, JPL Publication 00-3, Jet Propulsion Laboratory, California Institute of Technology, Pasadena, CA.

SAR, see IPCC, 1996.

Sass, R.L., F.M. Fisher Jr., A. Ding and Y. Huang, 1999: Exchange of methane from rice fields: national, regional, and global budgets. *J. Geophys. Res.*, **104**, 26943-26951.

Schauffler, S.M. and J.S. Daniel, 1994: On the effects of stratospheric circulation changes on trace gas trends. *J. Geophys. Res.*, **99**, 25747-25754.

Schimel, D., D. Alves, I. Enting, M. Heimann, F. Joos, D. Raynaud, T. Wigley, M. Prather, R. Derwent, D. Ehhalt, P. Fraser, E. Sanhueza, X. Zhou, P. Jonas, R. Charlson, H. Rodhe, S. Sadasivan, K.P. Shine, Y. Fouquart, V. Ramaswamy, S. Solomon, J. Srivinasan, D. Albritton, R. Derwent, I. Isaksen, M. Lal and D. Wuebbles, 1996: Chapter 2, Radiative Forcing of Climate Change. In: Climate Change 1995: The Science of Climate Change. Contribution of Working Group I to the Second Assessment Report of the Intergovernmental Panel on Climate Change [Houghton, J.T., L.G. Meira Filho, B.A. Callander, N. Harris, A. Kattenberg, and K. Maskell (eds.)]. Cambridge University Press, pp 65-131.

Schultz, M.G., D.J. Jacob, Y.H. Wang, J.A. Logan, E.L. Atlas, D.R. Blake, N.J. Blake, J.D. Bradshaw, E.V. Browell, M.A. Fenn, F. Flocke, G.L. Gregory, B.G. Heikes, G.W. Sachse, S.T. Sandholm, R.E. Shetter, H.B. Singh and R.W. Talbot, 1999: On the origin of tropospheric ozone and NOx over the tropical South Pacific. *J. Geophys. Res.*, **104**, 5829-5843.

Shine, K.P., R.G. Derwent, D.J. Wuebbles and J.-J. Morcrette, 1990: Chapter 2. Radiative Forcing of Climate. In: Climate Change: The IPCC Scientific Assessment, edited by J.T. Houghton, G.J. Jenkins, J.J. Ephraums, Cambridge University Press, Cambridge, UK, pp 41-68.

Sillman, S., J.A. Logan and S.C. Wofsy, 1990: A regional scale-model for ozone in the united-states with subgrid representation of urban and power-plant plumes. *J. Geophys. Res.*, **95**, 5731-5748.

Simmonds, P.G., S. O'Doherty, J. Huang, R. Prinn, R.G. Derwent, D. Ryall, G. Nickless and D. Cunnold, 1998: Calculated trends and the atmospheric abundance of 1,1,1,2-tetrafluoroethane, 1,1-dichloro-1-fluoroethane, and 1-chloro-1,1-difluoroethane using automated in-situ gas chromatography mass spectrometry measurements recorded at Mace Head, Ireland, from October 1994 to March 1997. *J. Geophys. Res.*, **103**, 16029-16037.

Simmonds, P.G., R.G. Derwent, S. O'Doherty, D.B. Ryall, L.P. Steele, R.L. Langenfelds, P. Salameh, H.J. Wang, C.H. Dimmer and L.E. Hudson, 2000: Continuous high-frequency observations of hydrogen at the Mace Head baseline atmospheric monitoring station over the 1994-1998 period. *J. Geophys. Res.*, **105**, 12105-12121.

Simmons, A.J., A. Untch, C. Jakob, P. Kallberg and P. Unden, 1999: Stratospheric water vapour and tropical tropopause temperatures in ECMWF analyses and multi-year simulations. *Quart. J. R. Met. Soc.*, Part A, **125**, 353-386.

Singh, H.B., M. Kanakidou, P.J. Crutzen and D.J. Jacob, 1995: High concentrations and photochemical fate of oxygenated hydrocarbons in the global troposphere. *Nature*, **378**, 50-54.

Singh, H.B., A.N. Thakur, Y.E. Chen and M. Kanakidou, 1996: Tetrachloroethylene as an indicator of low Cl atom concentrations in the troposphere. *Geophys. Res. Lett.*, **23**, 1529-1532.

Spivakovsky, C.M., J.A. Logan, S.A. Montzka, Y.J. Balkanski, M. Foreman-Fowler, D.B.A. Jones, L.W. Horowitz, A.C. Fusco, C.A.M. Brenninkmeijer, M.J. Prather, S.C. Wofsy and M.B. McElroy, 2000: Three-dimensional climatological distribution of tropospheric OH: update and evaluation. *J. Geophys. Res.*, **105**, 8931-8980.

Staehelin, J., J. Thudium, R. Buehler, A. Volz-Thomas and W. Graber, 1994: Trends in surface ozone concentrations at Arosa (Switzerland). *Atmos. Env.*, **28**, 75-87.

Staehelin, J., R. Kegel and N.R.P. Harris, 1998: Trend analysis of the homogenized total ozone series of Arosa (Switzerland), 1926-1996. *J. Geophys. Res.*, **103**, 8389-8399.

Stauffer, B., G. Fischer, A. Neftel and H. Oeschger, 1985: Increase of atmospheric methane recorded in Antarctic ice core. *Science*, **229**,

1386-1388.

Steele, L.P., R.L. Langenfelds, M.P. Lucarelli, P.J. Fraser, L.N. Cooper, D.A. Spenser, S. Chea and K. Broadhurst, 1996: Atmospheric methane, carbon dioxide, carbon monoxide, hydrogen, and nitrous oxide from Cape Grim air samples analysed by gas chromatography. In: Baseline Atmospheric Program Australia, 1994-95, edited by R.J. Francey, A.L. Dick and N. Derek, pp 107-110.

Stevens, C.M. and A. Engelkemeir, 1988: Stable carbon isotopic composition of methane from some natural and anthropogenic sources. *J. Geophys. Res.*, **93**, 725-733.

Stevenson, D.S., C.E. Johnson, W.J. Collins, R.G. Derwent and J.M. Edwards, 2000: Future tropospheric ozone radiative forcing and methane turnover - the impact of climate change. *Geophys. Res. Lett.*, **27**, 2073-2076.

Sturges, W.T., T.J. Wallington, M.D. Hurley, K.P. Shine, K. Sihra, A. Engel, D.E. Oram, S.A. Penkett, R. Mulvaney and C.A.M. Brenninkmeijer, 2000: A potent greenhouse gas identified in the atmosphere: SF_5CF_3. *Science*, **289**, 611-613.

Sturrock, G.A., S. O'Doherty, P.G. Simmonds and P.J. Fraser, 1999: *In situ* GC-MS measurements of the CFC replacement chemicals and other halocarbon species: The AGAGE program at Cape Grim, Tasmania. Proceedings of the Australian Symposium on Analytical Science, Melbourne, July 1999, Royal Australian Chemical Institute, 45-48.

Suess, E., G. Bohrmann, J. Greinert and E. Lausch, 1999: Flammable ice. Scientific American, 52-59/76-83.

Sundet, J.K., 1997: Model Studies with a 3-D Global CTM using ECMWF data. Ph.D. thesis, Dept. of Geophysics, University of Oslo, Norway.

Talbot, R.W., J.E. Dibb, E.M. Scheuer, D.R. Blake, N.J. Blake, G.L. Gregory, G.W. Sachse, J.D. Bradshaw, S.T. Sandholm and H.B. Singh, 1999: Influence of biomass combustion emissions on the distribution of acidic trace gases over the southern Pacific basin during austral springtime. *J. Geophys. Res.*, **104**, 5623-5634,.

Thakur, A.N., H.B. Singh, P. Mariani, Y. Chen, Y. Wang, D.J. Jacob, G. Brasseur, J.-F. Müller and M. Lawrence, 1999: Distribution or reactive nitrogen species in the remote free troposphere: data and model comparisons. *Atmos. Env.*, **33**, 1403-1422.

Thompson, T.M., J.W. Elkins, J.H. Butler, S.A. Montzka, R.C. Myers, T.H. Swanson, T.J. Baring, A.D. Clarke, G.S. Dutton, A.H. Hayden, J.M. Lobert, J.M. Gilligan and C.M. Volk, 1994: 5. Nitrous oxide and Halocarbons Division. In: Climate Monitoring and Diagnostics Laboratory - No. 22 Summary Report 1993, edited by J.T. Peterson and R.M. Rosson, US Department of Commerce, NOAA/ERL, Boulder, CO, pp 72-91.

Thompson, A.M., K.E. Pickering, D.P. McNamara, M.R. Schoeberl, R.D. Hudson, J.H. Kim, E.V. Browell, V.W.J.H. Kirchhoff and D. Nganga, 1996: Where did tropospheric ozone over southern Africa and the tropical Atlantic come from in October 1992? Insights from TOMS, GTE TRACE A and SAFARI 1992. *J. Geophys. Res.*, **101**, 24251-24278.

Thompson, A.M., L.C. Sparling, Y. Kondo, B.E. Anderson, G.L. Gregory and G.W. Sachse, 1999: Perspectives on NO, NOy and fine aerosol sources and variability during SONEX. *Geophys. Res. Lett.*, **26**, 3073-3076.

van Aardenne, J.A., G.R. Carmichael, H. Levy, D. Streets and L. Hordijk, 1999: Anthropogenic NOx emissions in Asia in the period 1990-2020. Atmos. Env., 33, 633-646.

Van Dorland, R., F.J. Dentener and J. Lelieveld, 1997: Radiative forcing due to tropospheric ozone and sulfate aerosols. *J. Geophys. Res.*, **102**, 28079-28100.

Vogt, R., R. Sander, R. Von Glasow and P.J. Crutzen, 1999: Iodine chemistry and its role in halogen activation and ozone loss in the marine boundary layer: A model study. *J. Atmos. Chem.*, **32**, 375-395.

Volk, C.M., J.W. Elkins, D.W. Fahey, G.S. Dutton, J.M. Gilligan, M. Lowenstein, J.R. Podolske, K.R. Chan and M.R. Gunson, 1997: Evaluation of source gas lifetimes from stratospheric observations. *J. Geophys. Res.*, **102**, 25543-25564.

Volz, A. and D. Kley, 1988: Evaluation of the Montsouris series of ozone measurements made in the nineteenth century. *Nature*, **332**, 240-242.

Wahlen, M., N. Tanaka, R. Henery, B. Deck, J. Zeglen, J.S. Vogel, J. Southon, A. Shemesh, R. Fairbanks and W. Broecker, 1989: Carbon-14 in methane sources and in atmospheric methane: The contribution of fossil carbon. *Science*, **245**, 286-290.

Wang, M.X. and X. Shangguan, 1996: CH_4 emission from various rice fields in P.R. China. *Theoretical and Applied Climatology*, **55**, 129-138.

Wang, Y.H. and D.J. Jacob, 1998: Anthropogenic forcing on tropospheric ozone and OH since preindustrial times. *J. Geophys. Res.*, **103**, 31123-31135.

Wang, Y., J.A. Logan and D.J. Jacob, 1998a: Global simulation of tropospheric O_3-NOx-hydrocarbon chemistry. 2. Model evaluation and global ozone budget. *J. Geophys. Res.*, **103**, 10727-10755.

Wang, Y., D.J. Jacob and J.A. Logan, 1998b: Global simulation of tropospheric O_3-NOx-hydrocarbon chemistry. 3. Origin of tropospheric ozone and effects of non-methane hydrocarbons. *J. Geophys. Res.*, **103**, 10757-10767.

Wauben, W.M.F., J.P.F. Fortuin, P.F.J. van Velthoven and H. Kelder, 1998: Comparison of modelled ozone distributions with sonde and satellite observations. *J. Geophys. Res.*, **103**, 3511-3530.

Wennberg, P.O., T.F. Hanisco, L. Jaegle, D.J. Jacob, E.J. Hintsa, E.J. Lanzendorf, J.G. Anderson, R.-S. Gao, E.R. Kein, S.G. Donnelly, L.A. Del Negro, D.W. Fahey, S.A. McKeen, R.J. Salawitch, C.R. Webster, R.D. May, R.L. Herman, M.H. Profitt, J.J. Margitan, E.L. Atlas, S.M. Schauffer, F. Flocke, C.T. McElroy and T.P. Bui, 1998: Hydrogen radicals, nitrogen radicals and the production of O_3 in the upper troposphere. *Science*, **279**, 49-53.

Wild, O. and M.J. Prather, 2000: Excitation of the primary tropospheric chemical mode in a global CTM. *J. Geophys. Res.*, **105**, 24647-24660.

Wingen, L.M. and B.J. Finlayson-Pitts, 1998: An upper limit on the production of N_2O from the reaction of $O(^1D)$ with CO_2 in the presence of N_2. *Geophys. Res. Lett.*, **25**, 517-520.

WMO, 1999: Scientific Assessment of Ozone Depletion: 1998. Global Ozone Research and Monitoring Project - Report No. 44, World Meteorological Organization, Geneva, Switzerland, 732 pp.

Yienger, J.J. and H. Levy II, 1995: Empirical model of global soil-biogenic NOx emissions. *J. Geophys. Res.*, **100**, 11447-11464.

Yienger, J.J., A.A. Klonecki, H. Levy II, W.J. Moxim and G.R. Carmichael, 1999: An evaluation of chemistry's role in the winter-spring ozone maximum found in the northern midlatitude free troposphere. *J.Geophys. Res.*, **104**, 3655-3667.

Yoshida, N. and S. Toyoda, 2000: Constraining the atmospheric N_2O budget from intramolecular site preference in N_2O isotopomers. *Nature*, **405**, 330-334.

Yung, Y.L. and C.E. Miller, 1997: Isotopic fractionation of stratospheric nitrous oxide. *Science*, **278**, 1778-1780.

Zander, R., P. Demoulin, D.H. Ehhalt, U. Schmidt and C.P. Rinsland, 1989: Secular increase of total vertical column abundance of carbon monoxide above central Europe since 1950. *J. Geophys., Res.*, **94**, 11021-11028.

Zander, R., D.H. Ehhalt, C.P. Rinsland, U. Schmidt, E. Mahieu, J. Rudolph, P. Demoulin, G. Roland, L. Delbouille and A.J. Sauval, 1994: Secular trend and seasonal variability of the column abundance of N_2O above Jungfraujoch station determined from IR solar spectra. *J. Geophys. Res.*, **99**, 16745-16756.

Ziemke, J.R., S. Chandra and P.K. Bhartia, 1998: Two new methods for deriving tropospheric column ozone from TOMS measurements: Assimilated UARS MLS/HALOE and convective-cloud differential techniques. *J. Geophys. Res.*, **103**, 22115-22127.

5

Aerosols, their Direct and Indirect Effects

Co-ordinating Lead Author
J.E. Penner

Lead Authors
M. Andreae, H. Annegarn, L. Barrie, J. Feichter, D. Hegg, A. Jayaraman, R. Leaitch, D. Murphy, J. Nganga, G. Pitari

Contributing Authors
A. Ackerman, P. Adams, P. Austin, R. Boers, O. Boucher, M. Chin, C. Chuang, B. Collins, W. Cooke, P. DeMott, Y. Feng, H. Fischer, I. Fung, S. Ghan, P. Ginoux, S.-L. Gong, A. Guenther, M. Herzog, A. Higurashi, Y. Kaufman, A. Kettle, J. Kiehl, D. Koch, G. Lammel, C. Land, U. Lohmann, S. Madronich, E. Mancini, M. Mishchenko, T. Nakajima, P. Quinn, P. Rasch, D.L. Roberts, D. Savoie, S. Schwartz, J. Seinfeld, B. Soden, D. Tanré, K. Taylor, I. Tegen, X. Tie, G. Vali, R. Van Dingenen, M. van Weele, Y. Zhang

Review Editors
B. Nyenzi, J. Prospero

Contents

Executive Summary

This chapter provides a synopsis of aerosol observations, source inventories, and the theoretical understanding required to enable an assessment of radiative forcing from aerosols and its uncertainty.

- The chemical and physical properties of aerosols are needed to estimate and predict direct and indirect climate forcing.

Aerosols are liquid or solid particles suspended in the air. They have a direct radiative forcing because they scatter and absorb solar and infrared radiation in the atmosphere. Aerosols also alter warm, ice and mixed-phase cloud formation processes by increasing droplet number concentrations and ice particle concentrations. They decrease the precipitation efficiency of warm clouds and thereby cause an indirect radiative forcing associated with these changes in cloud properties. Aerosols have most likely made a significant negative contribution to the overall radiative forcing. An important characteristic of aerosols is that they have short atmospheric lifetimes and therefore cannot be considered simply as a long-term offset to the warming influence of greenhouse gases.

The size distribution of aerosols is critical to all climate influences. Sub-micrometre aerosols scatter more light per unit mass and have a longer atmospheric lifetime than larger aerosols. The number of cloud condensation nuclei per mass of aerosol also depends on the chemical composition of aerosols as a function of size. Therefore, it is essential to understand the processes that determine these properties.

- Since the last IPCC report, there has been a greater appreciation of aerosol species other than sulphate, including sea salt, dust, and carbonaceous material. Regionally resolved emissions have been estimated for these species.

For sulphate, uncertainties in the atmospheric transformation of anthropogenic sulphur dioxide (SO_2) emissions to sulphate are larger than the 20 to 30% uncertainties in the emissions themselves. SO_2 from volcanoes has a disproportionate impact on sulphate aerosols due to the high altitude of the emissions, resulting in low SO_2 losses to dry deposition and a long aerosol lifetime. Modelled dust concentrations are systematically too high in the Southern Hemisphere, indicating that source strengths developed for the Sahara do not accurately predict dust uplift in other arid areas. Owing to a sensitive, non-linear dependence on wind speed of the flux of sea salt from ocean to atmosphere, estimates of global sea salt emissions from two present day estimates of wind speed differed by 55%. The two available inventories of black carbon emissions agree to 25% but the uncertainty is certainly greater than that and is subjectively estimated as a factor of two. The accuracy of source estimates for organic aerosol species has not been assessed, but organic species are believed to contribute significantly to both direct and indirect radiative forcing. Aerosol nitrate is regionally important but its global impact is uncertain.

- There is a great spatial and temporal variability in aerosol concentrations. Global measurements are not available for many aerosol properties, so models must be used to interpolate and extrapolate the available data. Such models now include the types of aerosols most important for climate change.

A model intercomparison was carried out as part of preparation for this assessment. All participating models simulated surface mass concentrations of non-sea-salt sulphate to within 50% of observations at most locations. Whereas sulphate aerosol models are now commonplace and reasonably well-tested, models of both organic and black carbon aerosol species are in early stages of development. They are not well-tested because there are few reliable measurements of black carbon or organic aerosols.

The vertical distribution of aerosol concentrations differs substantially from one model to the next, especially for components other than sulphate. For summertime tropopause conditions the range of model predictions is a factor of five for sulphate. The range of predicted concentrations is even larger for some of the other aerosol species. However, there are insufficient data to evaluate this aspect of the models. It will be important to narrow the uncertainties associated with this aspect of models in order to improve the assessment of aircraft effects, for example.

Although there are quite large spreads between the individual short-term observed and model-predicted concentrations at individual surface stations (in particular for carbonaceous aerosols), the calculated global burdens for most models agree to within a factor of 2.5 for sulphate, organics, and black carbon. The model-calculated range increases to three and to five for dust and sea salt with diameters less than 2 μm, respectively. The range for sea salt increases to a factor of six when different present day surface wind data sets are used.

- An analysis of the contributions of the uncertainties in the different factors needed to estimate direct forcing to the overall uncertainty in the direct forcing estimates can be made. This analysis leads to an overall uncertainty estimate for fossil fuel aerosols of 89% (or a range from –0.1 to –1.0 Wm^{-2}) while that for biomass aerosols is 85% (or a range from –0.1 to –0.5 Wm^{-2}).

For this analysis the central value for the forcing was estimated using the two-stream radiative transfer equation for a simple box model. Central values for all parameters were used and error propagation was handled by a standard Taylor expansion. Estimates of uncertainty associated with each parameter were developed from a combination of literature estimates for emissions, model results for determination of burdens, and atmospheric measurements for determination of mass scattering and absorption coefficients and water uptake effects. While such a simple approach has shortcomings (e.g., it tacitly assumes both horizontal and vertical homogeneity in such quantities as relative humidity, or at least that mean values can well represent the actual distributions), it allows a specific association of parameter uncertainties with their effects on forcing. With this approach, the most important uncertainties for fossil fuel aerosols are the upscatter fraction (or asymmetry parameter), the burden (which includes propagated uncertainties in emissions), and the mass scattering efficiency. The most important uncertainties for biomass aerosols are the single scattering albedo, the upscatter fraction, and the burden (including the propagated uncertainties in emissions).

• Preliminary estimates of aerosol concentrations have been made for future scenarios.

Several scenarios were explored which included future changes in anthropogenic emissions, temperature, and wind speed. Changes in the biosphere were not considered. Most models using present day meteorology show an approximate linear dependence of aerosol abundance on emissions. Sulphate and black carbon aerosols can respond in a non-linear fashion depending on the chemical parametrization used in the model. Projected changes in emissions may increase the relative importance of nitrate aerosols. If wind speeds increase in a future climate, as predicted by several GCMs, then increased emissions of sea salt aerosols may represent a significant negative climate feedback.

• There is now clear experimental evidence for the existence of a warm cloud aerosol indirect effect.

The radiative forcing of aerosols through their effect on liquid-water clouds consists of two parts: the 1st indirect effect (increase in droplet number associated with increases in aerosols) and the 2nd indirect effect (decrease in precipitation efficiency associated with increases in aerosols). The 1st indirect effect has strong observational support. This includes a recent study that established a link between changes in aerosols, cloud droplet number and cloud albedo (optical depth). There is also clear observational evidence for an effect of aerosols on precipitation efficiency. However, there is only limited support for an effect of changes in precipitation efficiency on cloud albedo. Models which include the second indirect effect find that it increases the overall indirect forcing by a factor of from 1.25 to more than a factor of two. Precipitation changes could be important to climate change even if their net radiative effect is small, but our ability to assess the changes in precipitation patterns due to aerosols is limited.

The response of droplet number concentration to increasing aerosols is largest when aerosol concentrations are small. Therefore, uncertainties in the concentrations of natural aerosols add an additional uncertainty of at least a factor of ± 1.5 to calculations of indirect forcing. Because the pre-existing, natural size distribution modulates the size distribution of anthropogenic mass there is further uncertainty associated with estimates of indirect forcing.

A major challenge is to develop and validate, through observations and small-scale modelling, parametrizations for GCMs of the microphysics of clouds and their interactions with aerosols. Projections of a future indirect effect are especially uncertain because empirical relationships between cloud droplet number and aerosol mass may not remain valid for possible future changes in aerosol size distributions. Mechanistic parametrizations have been developed but these are not fully validated.

• An analysis of the contributions of the uncertainties in the different factors needed to estimate indirect forcing of the first kind can be made. This analysis leads to an overall uncertainty estimate for indirect forcing over Northern Hemisphere marine regions by fossil fuel aerosols of 100% (or a range from 0 to -2.8 Wm^{-2}).

For this analysis the central value for the forcing was estimated using the two stream radiative transfer equation for a simple box model. Central values for all parameters were used and error propagation was handled as for the estimate of direct forcing uncertainty. This estimate is less quantitative than that for direct forcing because it is still difficult to estimate many of the parameters and the 2/3 uncertainty ranges are not firm. Moreover, several sources of uncertainty could not be estimated or included in the analysis. With this approach, the most important uncertainties are the determination of cloud liquid-water content and vertical extent. The relationship between sulphate aerosol concentration (with the propagated uncertainties from the burden calculation and emissions) and cloud droplet number concentration is of near equal importance in determining the forcing.

• The indirect radiative effect of aerosols also includes effects on ice and mixed phase clouds, but the magnitude of any indirect effect associated with the ice phase is not known.

It is not possible to estimate the number of anthropogenic ice nuclei at the present time. Except at very low temperatures ($<-40°C$), the mechanism of ice formation in clouds is not understood. Anthropogenic ice nuclei may have a large (probably positive) impact on forcing.

• There are linkages between policy on national air quality standards and climate change.

Policies and management techniques introduced to protect human health, improve visibility, and reduce acid rain will also affect the concentrations of aerosols relevant to climate.

5.1 Introduction

Aerosols have a direct radiative forcing because they scatter and absorb solar and infrared radiation in the atmosphere. Aerosols also alter the formation and precipitation efficiency of liquid-water, ice and mixed-phase clouds, thereby causing an indirect radiative forcing associated with these changes in cloud properties.

The quantification of aerosol radiative forcing is more complex than the quantification of radiative forcing by greenhouse gases because aerosol mass and particle number concentrations are highly variable in space and time. This variability is largely due to the much shorter atmospheric lifetime of aerosols compared with the important greenhouse gases. Spatially and temporally resolved information on the atmospheric burden and radiative properties of aerosols is needed to estimate radiative forcing. Important parameters are size distribution, change in size with relative humidity, complex refractive index, and solubility of aerosol particles. Estimating radiative forcing also requires an ability to distinguish natural and anthropogenic aerosols.

The quantification of indirect radiative forcing by aerosols is especially difficult. In addition to the variability in aerosol concentrations, some quite complicated aerosol influences on cloud processes must be accurately modelled. The warm (liquid-water) cloud indirect forcing may be divided into two components. The first indirect forcing is associated with the change in droplet concentration caused by increases in aerosol cloud condensation nuclei. The second indirect forcing is associated with the change in precipitation efficiency that results from a change in droplet number concentration. Quantification of the latter forcing necessitates understanding of a change in cloud liquid-water content and cloud amount. In addition to warm clouds, ice clouds may also be affected by aerosols.

5.1.1 Advances since the Second Assessment Report

Considerable progress in understanding the effects of aerosols on radiative balances in the atmosphere has been made since the IPCC WGI Second Assessment Report (IPCC, 1996) (hereafter SAR). A variety of field studies have taken place, providing both process-level understanding and a descriptive understanding of the aerosols in different regions. In addition, a variety of aerosol networks and satellite analyses have provided observations of regional differences in aerosol characteristics. Improved instrumentation is available for measurements of the chemical composition of single particles.

Models of aerosols have significantly improved since the SAR. Because global scale observations are not available for many aerosol properties, models are essential for interpolating and extrapolating available data to the global scale. Although there is a high degree of uncertainty associated with their use, models are presently the only tools with which to study past or future aerosol distributions and properties.

The very simplest models represent the global atmosphere as a single box in steady state for which the burden can be derived if estimates of sources and lifetimes are available. This approach was used in early assessments of the climatic effect of aerosols (e.g., Charlson *et al.*, 1987, 1992; Penner *et al.*, 1992; Andreae, 1995) since the information and modelling tools to provide a spatially- and temporally-resolved analysis were not available at the time. At the time of the SAR, three-dimensional models were only available for sulphate aerosols and soot. Since then, three-dimensional aerosol models have been developed for carbonaceous aerosols from biomass burning and from fossil fuels (Liousse *et al.*, 1996; Cooke and Wilson, 1996; Cooke *et al.*, 1999), dust aerosols (Tegen and Fung, 1994; Tegen *et al.*, 1996), sea salt aerosol (Gong *et al.*, 1998) and nitrate and ammonia in aerosols (Adams *et al.*, 1999, 2001; Penner *et al.*, 1999a). In this report, the focus is on a temporally and spatially resolved analysis of the atmospheric concentrations of aerosols and their radiative properties.

5.1.2 Aerosol Properties Relevant to Radiative Forcing

The radiatively important properties of atmospheric aerosols (both direct and indirect) are determined at the most fundamental level by the aerosol composition and size distribution. However, for purposes of the direct radiative forcing calculation and for assessment of uncertainties, these properties can be subsumed into a small set of parameters. Knowledge of a set of four quantities as a function of wavelength is necessary to translate aerosol burdens into first aerosol optical depths, and then a radiative perturbation: the mass light-scattering efficiency α_{sp}, the functional dependence of light-scattering on relative humidity $f(RH)$, the single-scattering albedo ω_0, and the asymmetry parameter g (cf., Charlson *et al.*, 1992; Penner *et al.*, 1994a).

Light scattering by aerosols is measurable as well as calculable from measured aerosol size and composition. This permits comparisons, called closure studies, of the different measurements for consistency. An example is the comparison of the derived optical depth with directly measured or inferred optical depths from sunphotometers or satellite radiometers. Indeed, various sorts of closure studies have been successfully conducted and lend added credibility to the measurements of the individual quantities (e.g., McMurry *et al.*, 1996; Clarke *et al.*, 1996; Hegg *et al.*, 1997; Quinn and Coffman, 1998; Wenny *et al.*, 1998; Raes *et al.*, 2000). Closure studies can also provide objective estimates of the uncertainty in calculating radiative quantities such as optical depth.

Aerosols in the accumulation mode, i.e., those with dry diameters between 0.1 and 1 μm (Schwartz, 1996) are of most importance. These aerosols can hydrate to diameters between 0.1 and 2 μm where their mass extinction efficiency is largest (see Figure 5.1). Accumulation mode aerosols not only have high scattering efficiency, they also have the longest atmospheric lifetime: smaller particles coagulate more quickly while nucleation to cloud drops or impaction onto the surface removes larger particles efficiently. Accumulation mode aerosols form the majority of cloud condensation nuclei (CCN). Hence, anthropogenic aerosol perturbations such as sulphur emissions have the greatest climate impact when, as is often the case, they produce or affect accumulation mode aerosols (Jones *et al.*, 1994).

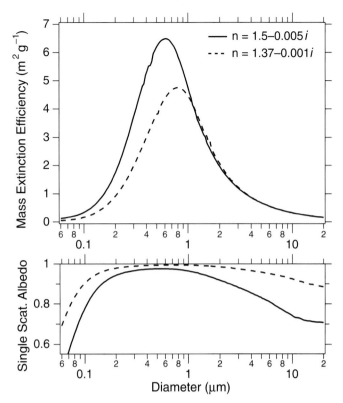

Figure 5.1: Extinction efficiency (per unit total aerosol mass) and single scattering albedo of aerosols. The calculations are integrated over a typical solar spectrum rather than using a single wavelength. Aerosols with diameters between about 0.1 and 2 μm scatter the most light per unit mass. Coarse mode aerosols (i.e., those larger than accumulation mode) have a smaller single scattering albedo even if they are made of the same material (i.e., refractive index) as accumulation mode aerosols. If the refractive index 1.37–0.001i is viewed as that of a hydrated aerosol then the curve represents the wet extinction efficiency. The dry extinction efficiency would be larger and shifted to slightly smaller diameters.

The direct radiative effect of aerosols is also very sensitive to the single scattering albedo ω_o. For example, a change in ω_o from 0.9 to 0.8 can often change the sign of the direct effect, depending on the albedo of the underlying surface and the altitude of the aerosols (Hansen *et al.*, 1997). Unfortunately, it is difficult to measure ω_o accurately. The mass of black carbon on a filter can be converted to light absorption, but the conversion depends on the size and mixing state of the black carbon with the rest of the aerosols. The mass measurements are themselves difficult, as discussed in Section 5.2.2.4. Aerosol light absorption can also be measured as the difference in light extinction and scattering. Very careful calibrations are required because the absorption is often a difference between two large numbers. As discussed in Section 5.2.4, it is difficult to retrieve ω_o from satellite data. Well-calibrated sunphotometers can derive ω_o by comparing light scattering measured away from the Sun with direct Sun extinction measurements (Dubovik *et al.*, 1998).

Some encouraging comparisons have been made between different techniques for measurements of ω_o and related quantities. Direct measurements of light absorption near Denver, Colorado using photo-acoustic spectroscopy were highly

correlated with a filter technique (Moosmüller *et al.*, 1998). However, these results also pointed to a possible strong wavelength dependence in the light absorption. An airborne comparison of six techniques (extinction cell, three filter techniques, irradiance measurements, and black carbon mass by thermal evolution) in biomass burning plumes and hazes was reported by Reid *et al.* (1998a). Regional averages of ω_o derived from all techniques except thermal evolution agreed within about 0.02 (ω_o is dimensionless), but individual data points were only moderately correlated (regression coefficient values of about 0.6).

Another complication comes from the way in which different chemical species are mixed in aerosols (e.g., Li and Okada, 1999). Radiative properties can change depending on whether different chemicals are in the same particles (internal mixtures) or different particles (external mixtures). Also, combining species may produce different aerosol size distributions than would be the case if the species were assumed to act independently. One example is the interaction of sulphate with sea salt or dust discussed in Section 5.2.2.6.

Fortunately, studies of the effects of mixing different refractive indices have yielded a fairly straightforward message: the type of mixing is usually significant only for absorbing material (Tang, 1996; Abdou *et al.*, 1997; Fassi-Fihri *et al.*, 1997). For non-absorbing aerosols, an average refractive index appropriate to the chemical composition at a given place and time is adequate. On the other hand, black carbon can absorb up to twice as much light when present as inclusions in scattering particles such as ammonium sulphate compared with separate particles (Ackerman and Toon, 1981; Horvath, 1993; Fuller *et al.*, 1999). Models of present day aerosols often implicitly include this effect by using empirically determined light absorption coefficients but future efforts will need to explicitly consider how black carbon is mixed with other aerosols. Uncertainties in the way absorbing aerosols are mixed may introduce a range of a factor of two in the magnitude of forcing by black carbon (Haywood and Shine, 1995; Jacobson, 2000).

To assess uncertainties associated with the basic aerosol parameters, a compilation is given in Table 5.1, stratified by a crude geographic/aerosol type differentiation. The values for the size distribution parameters given in the table were derived from the references to the table. The mass scattering efficiency and upscatter fraction shown in the table are derived from Mie calculations for spherical particles using these size distributions and a constant index of refraction for the accumulation mode. The scattering efficiency dependence on relative humidity (RH) and the single-scattering albedo were derived from the literature review of measurements.

The uncertainties given in the table for the central values of number modal diameter and geometric standard deviation (D_g and σ_g) are based on the ranges of values surveyed in the literature, as are those for f(RH) and ω_o. Those for the derivative quantities α_{sp} and $\bar{\beta}$, however, are based on Mie calculations using the upper and lower uncertainty limits for the central values of the size parameters, i.e., the propagation of errors is based on the functional relationships of Mie theory. The two calculations with coarse modes require some explanation.

Table 5.1: *Variation in dry aerosol optical properties at 550 nm by region/type.*

Aerosol type	D_g (mode) (μm)	Geometric standard dev. σ_g	α_{sp} (m^2 g^{-1})	Asymmetry parameter g	$\bar{\beta}$	f(RH) (at RH = 80%)	f_b(RH) (at RH = 80%)	Single-scattering albedo ω_o (dry)	Single-scattering albedo ω_o (80%)
Pacific marine									
w/single mode	0.19 ± 0.03	1.5 ± 0.15	3.7 ± 1.1	0.616 ± 0.11	0.23 ± 0.05	2.2 ± 0.3		0.99 ± 0.01	1.0 ± 0.01
w/accum & coarse	(0.46)	(2.1)	1.8 ± 0.5	0.661 ± 0.01	0.21 ± 0.003	2.2 ± 0.3		0.99 ± 0.01	1.0 ± 0.01
Atlantic marine	0.15 ± 0.05	1.9 ± 0.6	3.8 ± 1.0	0.664 ± 0.25	0.21 ± 0.11	2.2 ± 0.3		0.97 ± 0.03	1.0 ± 0.03
Fine soil dust	0.10 ± 0.26	2.8 ± 1.2	1.8 ± 0.1	0.682 ± 0.16	0.20 ± 0.07	1.3 ± 0.2		0.82 ± 0.06	0.83 ± 0.06
Polluted continental	0.10 ± 0.08	1.9 ± 0.3	3.5 ± 1.2	0.638 ± 0.28	0.22 ± 0.12	2.0 ± 0.3	0.81 ± 0.08	0.92 ± 0.05	0.95 ± 0.05
Background Continental									
w/single mode	0.08 ± 0.03	1.75 ± 0.34	2.2 ± 0.9	0.537 ± 0.26	0.27 ± 0.11	2.3 ± 0.4		0.97 ± 0.03	1.0 ± 0.03
w/accum. & coarse	(1.02)	(2.2)	1.0 ± 0.08	0.664 ± 0.07	0.21 ± 0.03	2.3 ± 0.4		0.97 ± 0.03	1.0 ± 0.03
Free troposphere	0.072 ± 0.03	2.2 ± 0.7	3.4 ± 0.7	0.649 ± 0.22	0.22 ± 0.10	—		—	
Biomass plumes	0.13 ± 0.02	1.75 ± 0.25	3.6 ± 1.0	0.628 ± 0.12	0.23 ± 0.05	1.1 ± 0.1		0.87 ± 0.06	0.87 ± 0.06
Biomass regional haze	0.16 ± 0.04	1.65 ± 0.25	3.6 ± 1.1	0.631 ± 0.16	0.22 ± 0.07	1.2 ± 0.2	0.81 ± 0.07	0.89 ± 0.05	0.90 ± 0.05

Literature references: Anderson *et al.* (1996); Bodhaine (1995); Carrico *et al.* (1998, 2000); Charlson *et al.* (1984); Clarke *et al.* (1999); Collins *et al.* (2000); Covert *et al.* (1996); Eccleston *et al.* (1974); Eck *et al.* (1999); Einfeld *et al.* (1991); Fitzgerald (1991); Fitzgerald and Hoppel (1982), Frick and Hoppel (1993); Gasso *et al.* (1999); Hegg *et al.* (1993, 1996a,b); Hobbs *et al.* (1997); Jaenicke (1993); Jennings *et al.* (1991); Kaufman (1987); Kotchenruther and Hobbs (1998); Kotchenruther *et al.* (1999); Leaitch and Isaac (1991); Le Canut *et al.* (1992, 1996); Lippman (1980); McInnes *et al.* (1997; 1998); Meszaros (1981); Nyeki *et al.* (1998); O'Dowd and Smith (1993); Quinn and Coffman (1998); Quinn *et al.* (1990; 1993, 1996); Radke *et al.* (1991); Raes *et al.* (1997); Reid and Hobbs (1998); Reid *et al.* (1998b); Remer *et al.* (1997); Saxena *et al.* (1995); Seinfeld and Pandis (1998); Sokolik and Golitsyn (1993); Takeda *et al.* (1987); Tang (1996); Tangren (1982); Torres *et al.* (1998); Waggoner *et al.* (1983); Whitby (1978); Zhang *et al.* (1993).

While the accumulation mode is generally thought to dominate light scattering, recent studies – as discussed below – have suggested that sea salt and dust can play a large role under certain conditions. To include this possibility, a sea salt mode has been added to the Pacific marine accumulation mode. The salt mode extends well into the accumulation size range and is consistent with O'Dowd and Smith (1993). It is optically very important at wind speeds above 7 to 10 ms^{-1}. For the case shown in the table, the sea salt mode accounts for about 50% of the local light scattering and could contribute over a third of the column optical depth, depending on assumptions about the scale height of the salt. Similarly, a soil dust coarse mode based on work by Whitby (1978) was added to the continental background accumulation mode. When present, this mode often dominates light scattering but, except for regions dominated by frequent dust outbreaks, is usually present over so small a vertical depth that its contribution to the column optical depth is generally slight. The importance of these coarse modes points to the importance of using size-resolved salt and dust fluxes such as those given in this report.

5.2 Sources and Production Mechanisms of Atmospheric Aerosols

5.2.1 Introduction

The concept of a "source strength" is much more difficult to define for aerosols than for most greenhouse gases. First, many aerosol species (e.g., sulphates, secondary organics) are not directly emitted, but are formed in the atmosphere from gaseous precursors. Second, some aerosol types (e.g., dust, sea salt) consist of particles whose physical properties, such as size and refractive index, have wide ranges. Since the atmospheric lifetimes and radiative effect of particles strongly depend on these properties, it makes little sense to provide a single value for the source strength of such aerosols. Third, aerosol species often combine to form mixed particles with optical properties and atmospheric lifetimes different from those of their components. Finally, clouds affect aerosols in a very complex way by scavenging aerosols, by adding mass through liquid phase chemistry, and through the formation of new aerosol particles in and near clouds. With regard to aerosol sources, we can report substantial progress over the previous IPCC assessment:

1) There are now better inventories of aerosol precursor emissions for many species (e.g., dimethylsulphide (DMS) and SO$_2$), including estimates of source fields for future scenarios. The present-day estimates on which this report is based are summarised in Table 5.2, see also Figure 5.2.

2) Emphasis is now on spatiotemporally resolved source and distribution fields.

3) There is now a better understanding of the conversion mechanisms that transform precursors into aerosol particles.

4) There is substantial progress towards the explicit representation of number/size and mass/size distributions and the specification of optical and hydration properties in models.

Table 5.2: *Annual source strength for present day emissions of aerosol precursors (Tg N, S or C /year). The reference year is indicated in parentheses behind individual sources, where applicable.*

	Northern Hemisphere	Southern Hemisphere	Global[a]	Range	Source
NO_x (as TgN/yr)	32	9	41		(see also Chapter 4).
Fossil fuel (1985)	20	1.1	21		Benkovitz *et al.* (1996)
Aircraft (1992)	0.54	0.04	0.58	0.4–0.9	Penner *et al.* (1999b); Daggett *et al.* (1999)
Biomass burning (ca. 1990)	3.3	3.1	6.4	2–12	Liousse *et al.* (1996); Atherton (1996)
Soils (ca. 1990)	3.5	2.0	5.5	3–12	Yienger and Levy (1995)
Agricultural soils			2.2	0–4	"
Natural soils			3.2	3–8	"
Lightning	4.4	2.6	7.0	2–12	Price *et al.* (1997); Lawrence *et al.* (1995)
NH_3 (as TgN/yr)	41	13	54	40–70	Bouwman *et al.* (1997)
Domestic animals (1990)	18	4.1	21.6	10–30	"
Agriculture (1990)	12	1.1	12.6	6–18	"
Human (1990)	2.3	0.3	2.6	1.3–3.9	"
Biomass burning (1990)	3.5	2.2	5.7	3–8	"
Fossil fuel and industry (1990)	0.29	0.01	0.3	0.1–0.5	"
Natural soils (1990)	1.4	1.1	2.4	1–10	"
Wild animals (1990)	0.10	0.02	0.1	0–1	"
Oceans	3.6	4.5	8.2	3–16	"
SO_2 (as TgS/yr)	76	12	88	67–130	
Fossil fuel and industry (1985)	68	8	76	60–100	Benkovitz *et al.* (1996)
Aircraft (1992)	0.06	0.004	0.06	0.03–1.0	Penner *et al.* (1998a); Penner *et al.* (1999b); Fahey *et al.* (1999)
Biomass burning (ca. 1990)	1.2	1.0	2.2	1–6	Spiro *et al.* (1992)
Volcanoes	6.3	3.0	9.3	6–20	Andres and Kasgnoc (1998) (incl. H_2S)
DMS or H_2S (as TgS/yr)	11.6	13.4	25.0	12–42	
Oceans	11	13	24	13–36	Kettle and Andreae (2000)
Land biota and soils	0.6	0.4	1.0	0.4–5.6	Bates *et al.* (1992); Andreae and Jaeschke (1992)
Volatile organic emissions (as TgC/yr)	171	65	236	100–560	
Anthropogenic (1985)	104	5	109	60–160	Piccot *et al.* (1992)
Terpenes (1990)	67	60	127	40–400	Guenther *et al.* (1995)

[a] The global figure may not equal the sum of the N. hemisphere and S. Hemisphere totals due to rounding.

5.2.2 Primary and Secondary Sources of Aerosols

5.2.2.1 Soil dust

Soil dust is a major contributor to aerosol loading and optical thickness, especially in sub-tropical and tropical regions. Estimates of its global source strength range from 1,000 to 5,000 Mt/yr (Duce, 1995; see Table 5.3), with very high spatial and temporal variability. Dust source regions are mainly deserts, dry lake beds, and semi-arid desert fringes, but also areas in drier regions where vegetation has been reduced or soil surfaces have been disturbed by human activities. Major dust sources are found in the desert regions of the Northern Hemisphere, while dust emissions in the Southern Hemisphere are relatively small. Unfortunately, this is not reflected in the source distribution shown in Figure 5.2(f), and represents a probable shortcoming of the dust mobilisation model used. Dust deflation occurs in a source region when the surface wind speed exceeds a threshold velocity, which is a function of surface roughness elements, grain size, and soil moisture. Crusting of soil surfaces and limitation of particle availability can reduce the dust release from a source region (Gillette, 1978). On the other hand, the disturbance of such surfaces by human activities can strongly enhance dust

Table 5.3: *Primary particle emissions for the year 2000 (Tg/yr)[a].*

	Northern Hemisphere	Southern Hemisphere	Global	Low	High	Source
Carbonaceous aerosols Organic Matter (0–2 μm)						
Biomass burning	28	26	54	45	80	Liousse *et al.* (1996), Scholes and Andreae (2000)
Fossil fuel	28	0.4	28	10	30	Cook *et al.* (1999), Penner *et al.* (1993)
Biogenic (>1μm)	—	—	56	0	90	Penner (1995)
Black Carbon (0–2 μm)						
Biomass burning	2.9	2.7	5.7	5	9	Liousse *et al.* (1996); Scholes and Andreae (2000)
Fossil fuel	6.5	0.1	6.6	6	8	Cooke *et al.* (1999); Penner *et al.* (1993)
Aircraft	0.005	0.0004	0.006			
Industrial Dust, etc. (> 1 μm)			100	40	130	Wolf and Hidy (1997); Andreae (1995) Gong *et al.* (1998)
Sea Salt						
d< 1 μm	23	31	54	18	100	
d=1–16μm	1,420	1,870	3,290	1,000	6,000	
Total	1,440	1,900	3,340	1,000	6,000	
Mineral (Soil) Dust [b]						
d< 1 μm	90	17	110	—	—	
d=1–2μm	240	50	290	—	—	
d=2–20μm	1,470	282	1,750	—	—	
Total	1,800	349	2,150	1,000	3,000	

[a] Range reflects estimates reported in the literature. The actual range of uncertainty may encompass values larger and smaller than those reported here.
[b] Source inventory prepared by P. Ginoux for the IPCC Model Intercomparison Workshop.

mobilisation. It has been estimated that up to 50% of the current atmospheric dust load originates from disturbed soil surfaces, and should therefore be considered anthropogenic in origin (Tegen and Fung, 1995), but this estimate must be considered highly uncertain. Furthermore, dust deflation can change in response to naturally occurring climate modes. For example, Saharan dust transport to Barbados increases during El Niño years (Prospero and Nees, 1986), and dust export to the Mediterranean and the North Atlantic is correlated with the North Atlantic Oscillation (Moulin *et al.*, 1997). Analysis of dust storm records shows regions with both increases and decreases in dust storm frequency over the last several decades (Goudie and Middleton, 1992).

The atmospheric lifetime of dust depends on particle size; large particles are quickly removed from the atmosphere by gravitational settling, while sub-micron sized particles can have atmospheric lifetimes of several weeks. A number of models of dust mobilisation and transport have been developed for regional to global scales (Marticorena *et al.*, 1997; Miller and Tegen, 1998; Tegen and Miller, 1998).

To estimate the radiative effects of dust aerosol, information is required about particle size, refractive index, and whether the minerals are mixed externally or as aggregates (Tegen *et al.*, 1996; Schulz *et al.*, 1998; Sokolik and Toon, 1999; Jacobson, 2001). Typical volume median diameters of dust particles are of the order of 2 to 4 μm. Refractive indices measured on Saharan dust have often been used to estimate the global dust radiative forcing (Tegen *et al.*, 1996). Since this dust has a single scattering

albedo significantly below one, the resulting forcing is small due to partial cancellation of solar and thermal forcing, as well as cancellation of positive and negative forcing over different geographic regions (Tegen and Lacis, 1996). However, different refractive indices of dust from different regions as well as regional differences in surface albedo lead to a large uncertainty in the resulting top-of-atmosphere dust forcing (Sokolik and Toon, 1996; Claquin *et al.*, 1998, 1999).

5.2.2.2 Sea salt

Sea salt aerosols are generated by various physical processes, especially the bursting of entrained air bubbles during whitecap formation (Blanchard, 1983; Monahan *et al.*, 1986), resulting in a strong dependence on wind speed. This aerosol may be the dominant contributor to both light scattering and cloud nuclei in those regions of the marine atmosphere where wind speeds are high and/or other aerosol sources are weak (O'Dowd *et al.*, 1997; Murphy *et al.*, 1998a; Quinn *et al.*, 1998). Sea salt particles are very efficient CCN, and therefore characterisation of their surface production is of major importance for aerosol indirect effects. For example, Feingold *et al.* (1999a) showed that in concentrations of 1 particle per litre, giant salt particles are able to modify stratocumulus drizzle production and cloud albedo significantly.

Sea salt particles cover a wide size range (about 0.05 to 10 μm diameter), and have a correspondingly wide range of atmospheric lifetimes. Thus, as for dust, it is necessary to analyse their emissions and atmospheric distribution in a size-resolved model. A semi-empirical formulation was used by Gong *et al.*

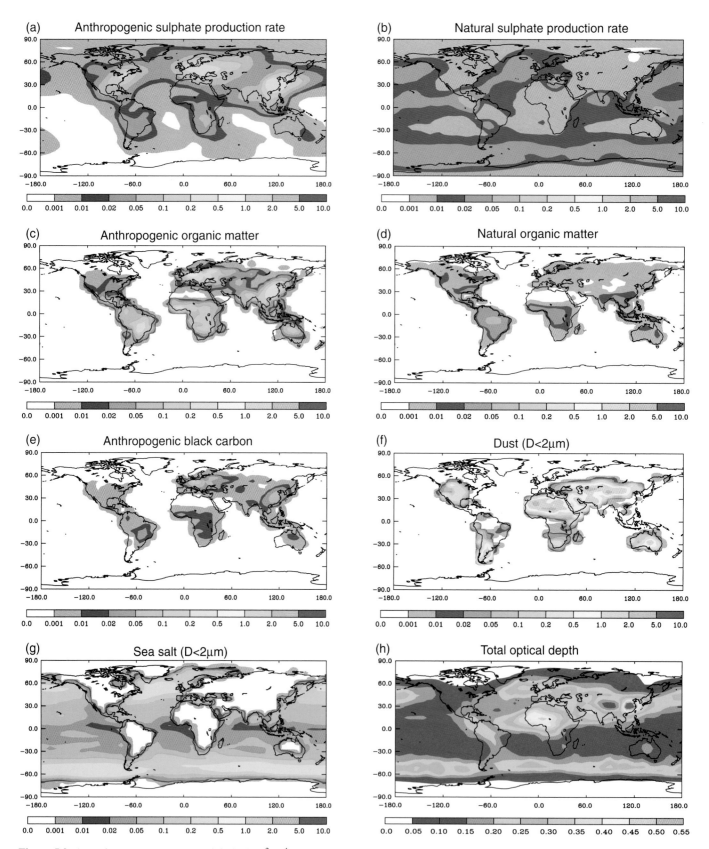

Figure 5.2: Annual average source strength in kg km^{-2} hr^{-1} for each of the aerosol types considered here (a to g) with total aerosol optical depth (h). Shown are (a) the column average H$_2$SO$_4$ production rate from anthropogenic sources, (b) the column average H$_2$SO$_4$ production rate from natural sources (DMS and SO$_2$ from volcanoes), (c) anthropogenic sources of organic matter, (d) natural sources of organic matter, (e) anthropogenic sources of black carbon, (f) dust sources for dust with diameters less than 2 μm, (g) sea salt sources for sea salt with diameters less than 2 μm, and (h) total optical depth for the sensitivity case ECHAM/GRANTOUR model (see Section 5.4.1.4).

(1998) to relate the size-segregated surface emission rates of sea salt aerosols to the wind field and produce global monthly sea salt fluxes for eight size intervals between 0.06 and 16 μm dry diameter (Figure 5.2g and Table 5.3). For the present-day climate, the total sea salt flux from ocean to atmosphere is estimated to be 3,300 Tg/yr, within the range of previous estimates (1,000 to 3,000 Tg/yr, Erickson and Duce, 1988; 5,900 Tg/yr, Tegen *et al.*, 1997).

5.2.2.3 Industrial dust, primary anthropogenic aerosols

Transportation, coal combustion, cement manufacturing, metallurgy, and waste incineration are among the industrial and technical activities that produce primary aerosol particles. Recent estimates for the current emission of these aerosols range from about 100 Tg/yr (Andreae, 1995) to about 200 Tg/yr (Wolf and Hidy, 1997). These aerosol sources are responsible for the most conspicuous impact of anthropogenic aerosols on environmental quality, and have been widely monitored and regulated. As a result, the emission of industrial dust aerosols has been reduced significantly, particularly in developed countries. Considering the source strength and the fact that much industrial dust is present in a size fraction that is not optically very active (>1 μm diameter), it is probably not of climatic importance at present. On the other hand, growing industrialisation without stringent emission controls, especially in Asia, may lead to increases in this source to values above 300 Tg/yr by 2040 (Wolf and Hidy, 1997).

5.2.2.4 Carbonaceous aerosols (organic and black carbon)

Carbonaceous compounds make up a large but highly variable fraction of the atmospheric aerosol (for definitions see Glossary). Organics are the largest single component of biomass burning aerosols (Andreae *et al.*, 1988; Cachier *et al.*, 1995; Artaxo *et al.*, 1998a). Measurements over the Atlantic in the haze plume from the United States indicated that aerosol organics scattered at least as much light as sulphate (Hegg *et al.*, 1997; Novakov *et al.*, 1997). Organics are also important constituents, perhaps even a majority, of upper-tropospheric aerosols (Murphy *et al.*, 1998b). The presence of polar functional groups, particularly carboxylic and dicarboxylic acids, makes many of the organic compounds in aerosols water-soluble and allows them to participate in cloud droplet nucleation (Saxena *et al.*, 1995; Saxena and Hildemann, 1996; Sempéré and Kawamura, 1996). Recent field measurements have confirmed that organic aerosols may be efficient cloud nuclei and consequently play an important role for the indirect climate effect as well (Rivera-Carpio *et al.*, 1996).

There are significant analytical difficulties in making valid measurements of the various organic carbon species in aerosols. Large artefacts can be produced by both adsorption of organics from the gas phase onto aerosol collection media, as well as evaporation of volatile organics from aerosol samples (Appel *et al.*, 1983; Turpin *et al.*, 1994; McMurry *et al.*, 1996). The magnitude of these artefacts can be comparable to the amount of organic aerosol in unpolluted locations. Progress has been made on minimising and correcting for these artefacts through several techniques: diffusion denuders to remove gas phase organics (Eatough *et al.*, 1996), impactors with relatively inert

surfaces and low pressure drops (Saxena *et al.*, 1995), and thermal desorption analysis to improve the accuracy of corrections from back-up filters (Novakov *et al.*, 1997). No rigorous comparisons of different techniques are available to constrain measurement errors.

Of particular importance for the direct effect is the light-absorbing character of some carbonaceous species, such as soot and tarry substances. Modelling studies suggest that the abundance of "black carbon" relative to non-absorbing constituents has a strong influence on the magnitude of the direct effect (e.g., Hansen *et al.*, 1997; Schult *et al.*, 1997; Haywood and Ramaswamy, 1998; Myhre *et al.*, 1998; Penner *et al.*, 1998b).

Given their importance, measurements of black carbon, and the differentiation between black and organic carbon, still require improvement (Heintzenberg *et al.*, 1997). Thermal methods measure the amount of carbon evolved from a filter sample as a function of temperature. Care must be taken to avoid errors due to pyrolysis of organics and interference from other species in the aerosol (Reid *et al.*, 1998a; Martins *et al.*, 1998). Other black carbon measurements use the light absorption of aerosol on a filter measured either in transmission or reflection. However, calibrations for converting the change in absorption to black carbon are not universally applicable (Liousse *et al.*, 1993). In part because of these issues, considerable uncertainties persist regarding the source strengths of light-absorbing aerosols (Bond *et al.*, 1998).

Carbonaceous aerosols from fossil fuel and biomass combustion

The main sources for carbonaceous aerosols are biomass and fossil fuel burning, and the atmospheric oxidation of biogenic and anthropogenic volatile organic compounds (VOC). In this section, we discuss that fraction of the carbonaceous aerosol which originates from biomass or fossil fuel combustion and is present predominantly in the sub-micron size fraction (Echalar *et al.*, 1998; Cooke, *et al.*, 1999). The global emission of organic aerosol from biomass and fossil fuel burning has been estimated at 45 to 80 and 10 to 30 Tg/yr, respectively (Liousse, *et al.*, 1996; Cooke, *et al.*, 1999; Scholes and Andreae, 2000). Combustion processes are the dominant source for black carbon; recent estimates place the global emissions from biomass burning at 6 to 9 Tg/yr and from fossil fuel burning at 6 to 8 Tg/yr (Penner *et al.*, 1993; Cooke and Wilson, 1996; Liousse *et al.*, 1996; Cooke *et al.*, 1999, Scholes and Andreae, 2000; see Table 5.3). A recent study by Bond *et al.* (1998), in which a different technique for the determination of black carbon emissions was used, suggests significantly lower emissions. Not enough measurements are available at the present time, however, to provide an independent estimate based on this technique. The source distributions are shown in Figures 5.2(c) and 5.2(e) for organic and black carbon, respectively.

The relatively close agreement between the current estimates of aerosol emission from biomass burning may underestimate the true uncertainty. Substantial progress has been made in recent years with regard to the emission factors, i.e., the amount of aerosol emitted per amount of biomass burned. In contrast, the estimation of the amounts of biomass combusted per unit area and time is still based on rather crude assessments and has not yet benefited significantly from the remote sensing tools becoming available. Where comparisons between different approaches to combustion estimates have been made, they have shown differences of almost an order of magnitude for specific regions (Scholes *et al.*, 1996; Scholes and Andreae, 2000). Extra-tropical fires were not included in the analysis by Liousse *et al.* (1996) and domestic biofuel use may have been underrepresented in most of the presently available studies. A recent analysis suggests that up to 3,000 Tg of biofuel may be burned worldwide (Ludwig *et al.*, 2001). This source may increase in the coming decades because it is mainly used in regions that are experiencing rapid population growth.

Organic aerosols from the atmospheric oxidation of hydrocarbons
Atmospheric oxidation of biogenic hydrocarbons yields compounds of low volatility that readily form aerosols. Because it is formed by gas-to-particle conversion, this secondary organic aerosol (SOA) is present in the sub-micron size fraction. Liousse *et al.* (1996) included SOA formation from biogenic precursors in their global study of carbonaceous aerosols; they employed a constant aerosol yield of 5% for all terpenes. Based on smog chamber data and an aerosol-producing VOC emissions rate of 300 to 500 TgC/yr, Andreae and Crutzen (1997) provided an estimate of the global aerosol production from biogenic precursors of 30 to 270 Tg/yr.

Recent analyses based on improved knowledge of reaction pathways and non-methane hydrocarbon source inventories have led to substantial downward revisions of this estimate. The total global emissions of monoterpenes and other reactive volatile organic compounds (ORVOC) have been estimated by ecosystem (Guenther *et al.*, 1995). By determining the predominant plant types associated with these ecosystems and identifying and quantifying the specific monoterpene and ORVOC emissions from these plants, the contributions of individual compounds to emissions of monoterpenes or ORVOC on a global scale can be inferred (Griffin, *et al.*, 1999b; Penner *et al.*, 1999a).

Experiments investigating the aerosol-forming potentials of biogenic compounds have shown that aerosol production yields depend on the oxidation mechanism. In general, oxidation by O_3 or NO_3 individually yields more aerosol than oxidation by OH (Hoffmann, *et al.*, 1997; Griffin, *et al.*, 1999a). However, because of the low concentrations of NO_3 and O_3 outside of polluted areas, on a global scale most VOC oxidation occurs through reaction with OH. The subsequent condensation of organic compounds onto aerosols is a function not only of the vapour pressure of the various molecules and the ambient temperature, but also the presence of other aerosol organics that can absorb products from gas-phase hydrocarbon oxidation (Odum *et al.*, 1996; Hoffmann *et al.*, 1997; Griffin *et al.*, 1999a).

When combined with appropriate transport and reaction mechanisms in global chemistry transport models, these hydrocarbon emissions yield estimated ranges of global biogenically derived SOA of 13 to 24 Tg/yr (Griffin *et al.*, 1999b) and 8 to 40 Tg/yr (Penner *et al.*, 1999a). Figure 5.2(d) shows the global distribution of SOA production from biogenic precursors derived from the terpene sources from Guenther *et al.* (1995) for a total source strength of 14 Tg/yr (see Table 5.3).

It should be noted that while the precursors of this aerosol are indeed of natural origin, the dependence of aerosol yield on the oxidation mechanism implies that aerosol production from biogenic emissions might be influenced by human activities. Anthropogenic emissions, especially of NO_x, are causing an increase in the amounts of O_3 and NO_3, resulting in a possible 3- to 4-fold increase of biogenic organic aerosol production since pre-industrial times (Kanakidou *et al.*, 2000). Recent studies in Amazonia confirm low aerosol yields and little production of new particles from VOC oxidation under unpolluted conditions (Artaxo *et al.*, 1998b; Roberts *et al.*, 1998). Given the vast amount of VOC emitted in the humid tropics, a large increase in SOA production could be expected from increasing development and anthropogenic emissions in this region.

Anthropogenic VOC can also be oxidised to organic particulate matter. Only the oxidation of aromatic compounds, however, yields significant amounts of aerosol, typically about 30 g of particulate matter for 1 kg of aromatic compounds oxidised under urban conditions (Odum *et al.*, 1996). The global emission of anthropogenic VOC has been estimated at 109 ± 27 Tg/yr, of which about 60% is attributable to fossil fuel use and the rest to biomass burning (Piccot *et al.*, 1992). The emission of aromatics amounts to about 19 ± 5 Tg/yr, of which 12 ± 3 Tg/yr is related to fossil fuel use. Using these data, we obtain a very small source strength for this aerosol type, about 0.6 ± 0.3 Tg/yr.

5.2.2.5 Primary biogenic aerosols

Primary biogenic aerosol consists of plant debris (cuticular waxes, leaf fragments, etc.), humic matter, and microbial particles (bacteria, fungi, viruses, algae, pollen, spores, etc.). Unfortunately, little information is available that would allow a reliable estimate of the contribution of primary biogenic particles to the atmospheric aerosol. In an urban, temperate setting, Matthias-Maser and Jaenicke (1995) have found concentrations of 10 to 30% of the total aerosol volume in both the sub-micron and super-micron size fractions. Their contribution in densely vegetated regions, particularly the moist tropics, could be even more significant. This view is supported by analyses of the lipid fraction in Amazonian aerosols (Simoneit *et al.*, 1990).

The presence of humic-like substances makes this aerosol light-absorbing, especially in the UV-B region (Havers *et al.*, 1998), and there is evidence that primary biogenic particles may be able to act both as cloud droplet and ice nuclei (Schnell and Vali, 1976). They may, therefore, be of importance for both direct and indirect climatic effects, but not enough is known at this time to assess their role with any confidence. Since their atmospheric abundance may undergo large changes as a result of land-use change, they deserve more scientific study.

5.2.2.6 Sulphates

Sulphate aerosols are produced by chemical reactions in the atmosphere from gaseous precursors (with the exception of sea salt sulphate and gypsum dust particles). The key controlling variables for the production of sulphate aerosol from its precursors are:

Table 5.4: *Estimates for secondary aerosol sources (in Tg substance/yr[a]).*

	Northern Hemisphere	Southern Hemisphere	Global	Low	High	Source
Sulphate (as NH_4HSO_4)	145	55	200	107	374	from Table 5.5
Anthropogenic	106	15	122	69	214	
Biogenic	25	32	57	28	118	
Volcanic	14	7	21	9	48	
Nitrate (as NO_3^-)[b]						
Anthropogenic	12.4	1.8	14.2	9.6	19.2	
Natural	2.2	1.7	3.9	1.9	7.6	
Organic compounds						
Anthropogenic	0.15	0.45	0.6	0.3	1.8	see text
VOC						
Biogenic VOC	8.2	7.4	16	8	40	Griffin *et al.* (1999b); Penner *et al.* (1999a)

[a] Total sulphate production calculated from data in Table 5.5, disaggregated into anthropogenic, biogenic and volcanic fluxes using the precursor data in Table 5.2 and the ECHAM/GRANTOUR model (see Table 5.8).

[b] Total net chemical tendency for HNO_3 from UCI model (Chapter 4) apportioned as NO_3^- according to the model of Penner *et al.* (1999a). Range corresponds to range from NO_x sources in Table 5.2.

(1) the source strength of the precursor substances,

(2) the fraction of the precursors removed before conversion to sulphate,

(3) the chemical transformation rates along with the gas-phase and aqueous chemical pathways for sulphate formation from SO_2.

The atmospheric burden of the sulphate aerosol is then regulated by the interplay of production, transport and deposition (wet and dry).

The two main sulphate precursors are SO_2 from anthropogenic sources and volcanoes, and DMS from biogenic sources, especially marine plankton (Table 5.2). Since SO_2 emissions are mostly related to fossil fuel burning, the source distribution and magnitude for this trace gas are fairly well-known, and recent estimates differ by no more than about 20 to 30% (Lelieveld *et al.*, 1997). Volcanic emissions will be addressed in Section 5.2.2.8.

Estimating the emission of marine biogenic DMS requires a gridded database on its concentration in surface sea water and a parametrization of the sea/air gas transfer process. A 1°×1° monthly data set of DMS in surface water has been obtained from some 16,000 observations using a heuristic interpolation scheme (Kettle *et al.*, 1999). Estimates for data-sparse regions are generated by assuming similarity to comparable biogeographic regions with adequate data coverage. Consequently, while the global mean surface DMS concentration is quite robust because of the large data set used (error estimate ± 50%), the estimates for specific regions and seasons remain highly uncertain in many ocean regions where sampling has been sparse (error up to factor of 5). These uncertainties are compounded with those resulting from the lack of a generally accepted air/sea flux parametrization. The approach of Liss and Merlivat (1986) and that of Wanninkhof (1992) yield fluxes differing by a factor of two (Kettle and Andreae, 2000). In Table 5.2, we use the mean of these two estimates (24 Tg S(DMS)/yr).

The chemical pathway of conversion of precursors to sulphate is important because it changes the radiative effects. Most SO_2 is converted to sulphate either in the gas phase or in cloud droplets that later evaporate. Model calculations suggest that aqueous phase oxidation is dominant globally (Table 5.5). Both processes produce sulphate mostly in sub-micron aerosols that are efficient light scatterers, but the precise size distribution of sulphate in aerosols is different for gas phase and aqueous production. The size distribution of the sulphate formed in the gas phase process also depends on the interplay between nucleation, condensation and coagulation. Models that describe this interplay are in an early stage of development, and, unfortunately, there are substantial inconsistencies between our theoretical description of nucleation and condensation and the rates of these processes inferred from atmospheric measurements (Eisele and McMurry, 1997; Weber *et al.*, 1999). Thus, most models of sulphate aerosol have simply assumed a size distribution based on present day measurements. Because there is no general reason that this same size should have applied in the past or will in the future, this lends considerable uncertainty to calculations of forcing. Many of the same issues about nucleation and condensation also apply to secondary organic aerosols.

Two types of chemical interaction have recently been recognised that can reduce the radiative impact of sulphate by causing some of it to condense onto larger particles with lower scattering efficiencies and shorter atmospheric lifetimes. The first is heterogeneous reactions of SO_2 on mineral aerosols (Andreae and Crutzen, 1997; Li-Jones and Prospero, 1998; Zhang and Carmichael, 1999). The second is oxidation of SO_2 to sulphate in sea salt-containing cloud droplets and deliquesced sea salt aerosols. This process can result in a substantial fraction of non-sea-salt sulphate to be present on large sea salt particles, especially under conditions where the rate of photochemical H_2SO_4 production is low and the amount of sea salt aerosol surface available is high (Sievering *et al.*, 1992; O'Dowd, *et al.*, 1997; Andreae *et al.*, 1999).

Because the models used to estimate sulphate aerosol production differ in the resolution and representation of physical processes and in the complexity of the chemical schemes, estimates of the amount of sulphate aerosol produced and its

Table 5.5: *Production parameters and burdens of SO_2 and aerosol sulphate as predicted by eleven different models.*

Model	Sulphur source Tg S/yr	Precursor deposition %	Gas phase oxidation %	Aqueous oxidation %	SO_2 burden Tg S	$\tau(SO_2)$ days	Sulphate dry deposition %	Sulphate wet deposition %	SO_4^{2-} burden Tg S	$\tau(SO_4^{2-})$ days	P days
A	94.5	47	8	45	0.30	1.1	16	84	0.77	5.0	2.9
B	122.8	49	5	46	0.20	0.6	27	73	0.80	4.6	2.3
C	100.7	49	17	34	0.43	1.5	13	87	0.63	4.4	2.2
D	80.4	44	16	39	0.56	2.6	20	80	0.73	5.7	3.3
E	106.0	54	6	40	0.36	1.2	11	89	0.55	4.1	1.9
F	90.0	18	18	64	0.61	2.4	22	78	0.96	4.7	3.8
G	82.5	33	12	56	0.40	1.9	7	93	0.57	3.8	2.5
H	95.7	45	13	42	0.54	2.4	18	82	1.03	7.2	3.9
I	125.6	47	9	44	0.63	2.0	16	84	0.74	3.6	2.2
J	90.0	24	15	59	0.60	2.3	25	75	1.10	5.3	4.5
K	92.5	56	15	27	0.43	1.8	13	87	0.63	5.8	2.5
Average	98.2	42	12	45	0.46	1.8	17	83	0.77	4.9	2.9
Standard deviation	14.7	12	5	11	0.14	0.6	6	6	0.19	1.0	0.8

Model/Reference: A MOGUNTIA/Langner and Rodhe, 1991; B: IMAGES/Pham *et al.*, 1996; C: ECHAM3/Feichter *et al.*, 1996; D: Harvard-GISS / Koch *et al.*, 1999; E: CCM1-GRANTOUR/Chuang *et al.*, 1997; F:ECHAM4/Roelofs *et al.*, 1998; G: CCM3/Barth *et al.*, 2000 and Rasch *et al.*, 2000a; H: CCC/Lohmann *et al.*, 1999a.; I: Iversen *et al.*, 2000; J: Lelieveld *et al.*, 1997; K: GOCART/Chin *et al.*, 2000.

atmospheric burden are highly model-dependent. Table 5.4 provides an overall model-based estimate of sulphate production and Table 5.5 emphasises the differences between different models. All the models shown in Table 5.5 include anthropogenic and natural sources and consider at least three species, DMS, SO_2 and SO_4^{2-}, B and D consider more species and have a more detailed representation of the gas-phase chemistry. C, F and G include a more detailed representation of the aqueous phase processes. The calculated residence times of SO_2, defined as the global burden divided by the global emission flux, range between 0.6 and 2.6 days as a result of different deposition parametrizations. Because of losses due to SO_2 deposition, only 46 to 82% of the SO_2 emitted undergoes chemical transformations and forms sulphate. The global turnover time of sulphate is mainly determined by wet removal and is estimated to be between 4 and 7 days. Because of the critical role that precipitation scavenging plays in controlling sulphate lifetime, it is important how well models predict vertical profiles.

The various models start with gaseous sulphur sources ranging from 80 to 130 TgS/yr, and arrive at SO_2 and SO_4^{2-} burdens of 0.2 to 0.6 and 0.6 to 1.1 TgS, respectively. It is noteworthy that there is little correlation between source strength and the resulting burden between models. In fact, the model with the second-highest precursor source (B) has the lowest SO_2 burden, and the model with the highest sulphate burden (J) starts with a much lower precursor source than the model with the lowest sulphate burden (E). Figures 5.2(a) and (b) show the global distribution of sulphate aerosol production from anthropogenic SO_2 and from natural sources (primarily DMS), respectively (see also Table 5.4).

The modelled production efficiency of atmospheric sulphate aerosol burden from a given amount of precursors is expressed as P, the ratio between the global sulphate burden to the global

sulphur emissions per day. At the global scale, this parameter varies between the models listed in Table 5.5 by more than a factor of two, from 1.9 to 4.5 days. Within a given model, the potential of a specific sulphur source to contribute to the global sulphate burden varies strongly as a function of where and in what form sulphur is introduced into the atmosphere. SO_2 from volcanoes (P=6.0 days) is injected at higher altitudes, and DMS (P=3.1 days) is not subject to dry deposition and can therefore be converted to SO_2 far enough from the ground to avoid large deposition losses. In contrast, most anthropogenic SO_2 (P=0.8 to 2.9 days) is released near the ground and therefore much of it is lost by deposition before oxidation can occur (Feichter *et al.*, 1997; Graf *et al.*, 1997). Regional differences in the conversion potential of anthropogenic emissions may be caused by the latitude-dependent oxidation capacity and by differences in the precipitation regime. For the same reasons P exhibits a distinct seasonality in mid- and high latitudes.

This comparison indicates that in addition to uncertainties in precursor source strengths, which may be ranging from factors of about 1.3 (SO_2) to 2 (DMS), the estimation of the production and deposition terms of sulphate aerosol introduces an additional uncertainty of at least a factor of 2 into the prediction of the sulphate burden. As the relationship between sulphur sources and resulting sulphate load depends on numerous parameters, the conversion efficiency must be expected to change with changing source patterns and with changing climate.

Sulphate in aerosol particles is present as sulphuric acid, ammonium sulphate, and intermediate compounds, depending on the availability of gaseous ammonia to neutralise the sulphuric acid formed from SO_2. In a recent modelling study, Adams *et al.* (1999) estimate that the global mean NH_4^+/SO_4^{2-} mole ratio is about one, in good agreement with available measurements. This increases the mass of sulphate aerosol by some 17%, but also changes the hydration behaviour and refractive index of the aerosol. The overall effects are of the order of 10%, relatively minor compared with the uncertainties discussed above (Howell and Huebert, 1998).

5.2.2.7 Nitrates

Aerosol nitrate is closely tied to the relative abundances of ammonium and sulphate. If ammonia is available in excess of the amount required to neutralise sulphuric acid, nitrate can form small, radiatively efficient aerosols. In the presence of accumulation-mode sulphuric acid containing aerosols, however, nitric acid deposits on larger, alkaline mineral or salt particles (Bassett and Seinfeld, 1984; Murphy and Thomson, 1997; Gard *et al.*, 1998). Because coarse mode particles are less efficient per unit mass at scattering light, this process reduces the radiative impact of nitrate (Yang *et al.*, 1994; Li-Jones and Prospero, 1998).

Until recently, nitrate has not been considered in assessments of the radiative effects of aerosols. Andreae (1995) estimated that the global burden of ammonium nitrate aerosol from natural and anthropogenic sources is 0.24 and 0.4 Tg (as NH_4NO_3), respectively, and that anthropogenic nitrates cause only 2% of the total direct forcing. Jacobson (2001) derived similar burdens, and estimated forcing by anthropogenic nitrate to be -0.024 Wm^{-2}. Adams *et al.* (1999) obtained an even lower value of 0.17 Tg (as NO_3^-) for the global nitrate burden. Part of this difference may be due to the fact that the latter model does not include nitrate deposition on sea salt aerosols. Another estimate (van Dorland *et al.*, 1997) suggested that forcing due to ammonium nitrate is about one tenth of the sulphate forcing. The importance of aerosol nitrate could increase substantially over the next century, however. For example, the SRES A2 emissions scenario projects that NO_x emissions will more than triple in that time period while SO_2 emissions decline slightly. Assuming increasing agricultural emissions of ammonia, it is conceivable that direct forcing by ammonium nitrate could become comparable in magnitude to that due to sulphate (Adams *et al.*, 2001).

Forcing due to nitrate aerosol is already important at the regional scale (ten Brink *et al.*, 1996). Observations and model results both show that in regions of elevated NO_x and NH_3 emissions, such as Europe, India, and parts of North America, NH_4NO_3 aerosol concentrations may be quite high and actually exceed those of sulphate. This is particularly evident when aerosol sampling techniques are used that avoid nitrate evaporation from the sampling substrate (Slanina *et al.*, 1999). Substantial amounts of NH_4NO_3 have also been observed in the European plume during ACE-2 (Andreae *et al.*, 2000).

5.2.2.8 Volcanoes

Two components of volcanic emissions are of most significance for aerosols: primary dust and gaseous sulphur. The estimated dust flux reported in Jones *et al.*, (1994a) for the1980s ranges from 4 to 10,000 Tg/yr, with a "best" estimate of 33 Tg/yr (Andreae, 1995). The lower limit represents continuous eruptive activity, and is about two orders of magnitude smaller than soil dust emission. The upper value, on the other hand, is the order of magnitude of volcanic dust mass emitted during large explosive eruptions. However, the stratospheric lifetime of these coarse particles is only about 1 to 2 months (NASA, 1992), due to the efficient removal by settling.

Sulphur emissions occur mainly in the form of SO_2, even though other sulphur species may be present in the volcanic plume, predominantly SO_4^{2-} aerosols and H_2S. Stoiber *et al.* (1987) have estimated that the amount of SO_4^{2-} and H_2S is commonly less than 1% of the total, although it may in some cases reach 10%. Graf *et al.* (1998), on the other hand, have estimated the fraction of H_2S and SO_4^{2-} to be about 20% of the total. Nevertheless, the error made in considering all the emitted sulphur as SO_2 is likely to be a small one, since H_2S oxidises to SO_2 in about 2 days in the troposphere or 10 days in the stratosphere. Estimates of the emission of sulphur containing species from quiescent degassing and eruptions range from 7.2 TgS/yr to 14 ± 6 TgS/yr (Stoiber *et al.*, 1987; Spiro *et al.*, 1992; Graf *et al.*, 1997; Andres and Kasgnoc, 1998). These estimates are highly uncertain because only very few of the potential sources have ever been measured and the variability between sources and between different stages of activity of the sources is considerable.

Volcanic aerosols in the troposphere

Graf *et al.* (1997) suggest that volcanic sources are important to the sulphate aerosol burden in the upper troposphere, where they might contribute to the formation of ice particles and thus represent a potential for a large indirect radiative effect (see Section 5.3.6). Sassen (1992) and Sassen *et al.* (1995) have presented evidence of cirrus cloud formation from volcanic aerosols and Song *et al.* (1996) suggest that the interannual variability of high level clouds is associated with explosive volcanoes.

Calculations using a global climate model (Graf *et al.*, 1997) have reached the "surprising" conclusion that the radiative effect of volcanic sulphate is only slightly smaller than that of anthropogenic sulphate, even though the anthropogenic SO_2 source strength is about five times larger. Table 5.6 shows that the calculated efficiency of volcanic sulphur in producing

Table 5.6: *Global annual mean sulphur budget (from Graf et al., 1997) and top-of-atmosphere forcing in percentage of the total (102 TgS/yr emission, about 1 TgS burden, -0.65 Wm^{-2} forcing). Efficiency is relative sulphate burden divided by relative source strength (i.e. column 3 / column 1).*

Source	Sulphur emission	SO_2 burden	SO_4^{2-} burden	Efficiency	Direct forcing TOA %
Anthropogenic	66	46	37	0.56	40
Biomass burning	2.5	1.2	1.6	0.64	2
DMS	18	18	25	1.39	26
Volcanoes	14	35	36	2.63	33

sulphate aerosols is about 4.5 times larger than that of anthropogenic sulphur. The main reason is that SO_2 released from volcanoes at higher altitudes has a longer residence time, mainly due to lower dry deposition rates than those calculated for surface emissions of SO_2 (cf.B,enkovitz *et al.*, 1994). On the other hand, because different models show major discrepancies in vertical sulphur transport and in upper tropospheric aerosol concentrations, the above result could be very model-dependent.

Volcanic aerosols emitted into the stratosphere
Volcanic emissions sufficiently cataclysmic to penetrate the stratosphere are rare. Nevertheless, the associated transient climatic effects are large and trends in the frequency of volcanic eruptions could lead to important trends in average surface temperature. The well-documented evolution of the Pinatubo plume illustrates the climate effects of a large eruption (Stenchikov *et al.*, 1998).

About three months of post-eruptive aging are needed for chemical and microphysical processes to produce the stratospheric peak of sulphate aerosol mass and optical thickness (Stowe *et al.*, 1992; McCormick *et al.*, 1995). Assuming at this stage a global stratospheric optical depth of the order of 0.1 at 0.55 μm, a total time of about 4 years is needed to return to the background value of 0.003 (WMO/UNEP, 1992; McCormick and Veiga, 1992) using one year as e-folding time for volcanic aerosol decay. This is, of course, important in terms of climate forcing: in the case of Pinatubo, a radiative forcing of about -4 Wm^{-2} was reached at the beginning of 1992, decaying exponentially to about -0.1 Wm^{-2} in a time frame of 4 years (Minnis *et al.*, 1993; McCormick *et al.*, 1995). This direct forcing is augmented by an indirect forcing associated with O_3 depletion that is much smaller than the direct forcing (about -0.1 Wm^{-2}).

The background amount of stratospheric sulphate is mainly produced by UV photolysis of organic carbonyl sulphide forming SO_2, although the direct contribution of tropospheric SO_2 injected in the stratosphere in the tropical tropopause region is significant for particle formation in the lower stratosphere and accounts for about one third of the total stratospheric sulphate mass (Weisenstein *et al.*, 1997). The observed sulphate load in the stratosphere is about 0.15 TgS (Kent and McCormick, 1984) during volcanically quiet periods, and this accounts for about 15% of the total sulphate (i.e., troposphere + stratosphere) (see Table 5.6).

The historical record of SO_2 emissions by erupting volcanoes shows that over 100 Tg of SO_2 can be emitted in a single event, such as the Tambora volcano eruption of 1815 (Stoiber *et al.*, 1987). Such large eruptions have led to strong transient cooling effects (-0.14 to $-0.31°C$ for eruptions in the 19th and 20th centuries), but historical and instrumental observations do not indicate any significant trend in the frequency of highly explosive volcanoes (Robertson *et al.*, 2001). Thus, while variations in volcanic activity may have influenced climate at decadal and shorter scales, it seems unlikely that trends in volcanic emissions could have played any role in establishing a longer-term temperature trend.

5.2.3 Summary of Main Uncertainties Associated with Aerosol Sources and Properties

In the case of primary aerosols, the largest uncertainties often lie in the extrapolation of experimentally determined source strengths to other regions and seasons. This is especially true for dust, for which many of the observations are for a Saharan source. The spatial and temporal distribution of biomass fires also remains uncertain. The non-linear dependence of sea salt aerosol formation on wind speed creates difficulties in parametrizations in large-scale models and the vertical profile of sea salt aerosols needs to be better defined.

Secondary aerosol species have uncertainties both in the sources of the precursor gases and in the atmospheric processes that convert some of those gases to aerosols. For sulphate, the uncertainties in the conversion from SO_2 (factor of 2) are larger than the uncertainties in anthropogenic sources (20 to 30%). For hydrocarbons there are large uncertainties both in the emissions of key precursor gases as a function of space and time as well as the fractional yield of aerosols as those gases are oxidised. Taken at face value, the combined uncertainties can be a factor of three for sulphate and more for organics. On the other hand, the success some models have had in predicting aerosol concentrations at observation sites (see Section 5.4) as well as wet deposition suggests that at least for sulphate the models have more skill than suggested by a worst-case propagation of errors. Nevertheless, we cannot be sure that these models achieve reasonable success for the right reasons.

Besides the problem of predicting the mass of aerosol species produced, there is the more complex issue of adequately describing their physical properties relevant to climate forcing. Here we would like to highlight that the situation is much better for models of present day aerosols, which can rely on empirical data for optical properties, than for predictions of future aerosol effects. Another issue for optical properties is that the quantity and sometimes the quality of observational data on single scattering albedo do not match those available for optical depth. Perhaps the most important uncertainty in aerosol properties is the production of cloud condensation nuclei (Section 5.3.3).

5.2.4 Global Observations and Field Campaigns

Satellite observations (reviewed by King *et al.*, 1999) are naturally suited to the global coverage demanded by regional variations in aerosols. Aerosol optical depth can be retrieved over the ocean in clear-sky conditions from satellite measurements of irradiance and a model of aerosol properties (Mishchenko *et al.*, 1999). These have been retrieved from satellite instruments such as AVHRR (Husar *et al.*, 1997; Higurashi and Nakajima, 1999), METEOSAT (e.g., Jankowiak and Tanré, 1992; Moulin *et al.*, 1997), ATSR (Veefkind *et al.*, 1999), and OCTS (Nakajima *et al.*, 1999). More recently-dedicated instruments such as MODIS and POLDER have been designed to monitor aerosol properties (Deuzé *et al.*, 1999; Tanré *et al.*, 1999; Boucher and Tanré, 2000). Aerosol retrievals over land, especially over low albedo regions, are developing rapidly but are complicated by the spectral and angular dependence of the surface reflectivity (e.g., Leroy *et al.*,

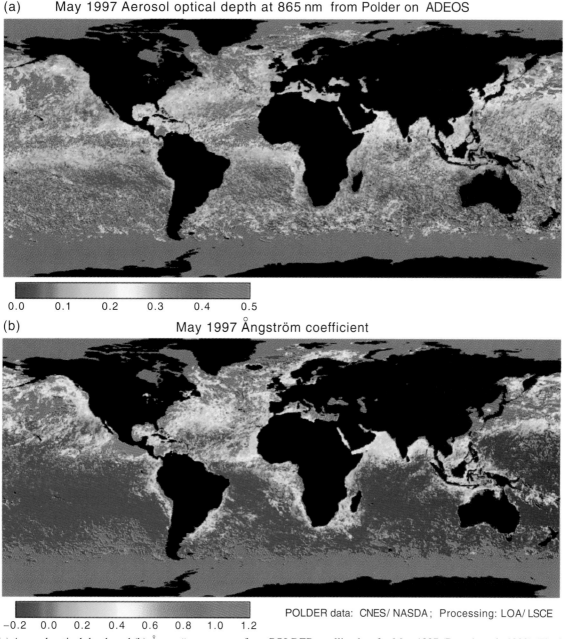

(a) May 1997 Aerosol optical depth at 865 nm from Polder on ADEOS

0.0   0.1   0.2   0.3   0.4   0.5

(b) May 1997 Ångström coefficient

−0.2   0.0   0.2   0.4   0.6   0.8   1.0   1.2

POLDER data: CNES/ NASDA; Processing: LOA/ LSCE

Figure 5.3: (a) Aerosol optical depth and (b) Ångström exponent from POLDER satellite data for May 1997 (Deuzé *et al.*, 1999). The largest optical depths over the Atlantic Ocean are from north African dust. The Ångström exponent expresses the wavelength dependence of scattered light between 670 and 865 nm. The African dust plume has a small Ångström exponent due to the importance of coarse mode aerosols whereas the larger Ångström exponents around the continents show the importance of accumulation mode aerosols in those locations.

1997; Wanner *et al.*, 1997; Soufflet *et al.*, 1997). The TOMS instrument has the capability to detect partially absorbing aerosols over land and ocean but the retrievals are only semi-quantitative (Hsu *et al.*, 1999). Comparisons of ERBE and SCARAB data with radiative transfer models show that aerosols must be included to accurately model the radiation budget (Cusack *et al.*, 1998; Haywood *et al.*, 1999).

There is not enough information content in a single observed quantity (scattered light) to retrieve the aerosol size distribution and the vertical profile in addition to the optical depth. Light scattering can be measured at more than one wavelength, but in most cases no more than two or three independent parameters can be derived even from observations at a large number of wavelengths (Tanré *et al.*, 1996; Kaufman *et al.*, 1997). Observations of the polarisation of backscattered light have the potential to add more information content (Herman *et al.*, 1997), as do observations at multiple angles of the same point in the atmosphere as a satellite moves overhead (Flowerdew and Haigh, 1996; Kahn *et al.*, 1997; Veefkind *et al.*, 1998).

In addition to aerosol optical depth, the vertically averaged Ångström exponent (which is related to aerosol size), can also be retrieved with reasonable agreement when compared to ground-based sunphotometer data (Goloub *et al.*, 1999; Higurashi and Nakajima, 1999). Vertical profiles of aerosols are available in the upper troposphere and stratosphere from the SAGE instrument but its limb scanning technique cannot be extended downward because of interference from clouds. Active sensing from space shows promise in retrieving vertical profiles of aerosols (Osborn *et al.*, 1998).

The single scattering albedo ω_0, which is important in determining the direct radiative effect, is difficult to retrieve from satellites, especially over oceans (Kaufman *et al.*, 1997). This is due to satellites' viewing geometry, which restricts measurements to light scattering rather than extinction for most tropospheric aerosols. Accurate values of ω_0 can be retrieved from satellite data for some special conditions, especially if combined with ground-based data (Fraser and Kaufman, 1985; Nakajima and Higurashi, 1997).

At this time there are no reliable methods for determining from satellite data the fraction of atmospheric aerosols that are anthropogenic. This is a major limitation in determining the radiative forcing in addition to the overall radiative effect of aerosols (Boucher and Tanré, 2000).

Field campaigns such as the Tropospheric Aerosol Radiative Forcing Observational Experiment (TARFOX), the Aerosol Characterisation Experiment (ACE-1, ACE-2), the Indian Ocean Experiment (INDOEX), SCAR-B, and monitoring networks such as AERONET provide essential information about the chemical and physical properties of aerosols (Novakov *et al.*, 1997; Hegg *et al.*, 1997; Hobbs *et al.*, 1997; Ross *et al.*, 1998; Holben *et al.*, 1998). These studies have also shown the importance of mixing and entrainment between the boundary layer and the free troposphere in determining aerosol properties over the oceans (e.g. Bates *et al.*, 1998; Johnson *et al.*, 2000). INDOEX found long-range transport of highly absorbing aerosol. The importance of absorption was shown by a change in irradiance at the surface that was two to three times that at the top of the atmosphere (Satheesh *et al.*, 1999; Podgorny *et al.*, 2000). TARFOX data also show the importance of black carbon in calculating broad-band irradiances, and confirm theoretical calculations that the strongest radiative effect occurred not at noon, but rather when the solar zenith angle was approximately 60 to 70 degrees (Russell *et al.*, 1999; Hignett *et al.*, 1999).

Local closure studies can compare size-resolved aerosol chemistry data with both total mass measurements and light scattering. Both comparisons have been made to within analytical uncertainties of 20 to 40% (Quinn and Coffman, 1998; Neusüß *et al.*, 2000; Putaud *et al.*, 2000). Local closure studies have also successfully compared predicted and measured hygroscopic growth (Carrico *et al.*, 1998; Swietlicki *et al.*, 1999). However, achieving closure between measured aerosol properties and observed cloud nucleation is more difficult (Brenguier *et al.*, 2000).

Column closure studies compare different ways of obtaining a vertically integrated optical property, usually optical depth. Comparisons were made during TARFOX between aerosol optical depths derived from satellites looking down and an aircraft sunphotometer looking up (Veefkind *et al.*, 1998). Optical depths from ATSR-2 retrievals were within 0.03 of

sunphotometer data and optical depths from AVHRR were systematically lower but highly correlated with sunphotometer data. Except in some dust layers, aerosol optical depths computed during ACE-2 from *in situ* data, measured from sunphotometers, and retrieved from satellites agreed within about 20% (Collins *et al.*, 2000; Schmid *et al.*, 2000). Optical depths computed from lidar profiles have agreed with sunphotometer measurements within 40% at several sites (Ferrare *et al.*, 1998, 2000; Flamant *et al.*, 2000).

There is considerable common ground between closure studies (Russell and Heintzenberg, 2000). An important uncertainty in both mass closure and hygroscopic growth is often the treatment of organics (e.g. McMurry *et al.*, 1996; Raes *et al.*, 2000). The sampling of coarse aerosols is often a limitation in computing scattering from in-situ data (e.g. Ferrare *et al.*, 1998; Quinn and Coffman, 1998). Black carbon and other light absorbing aerosols are difficult to treat because of difficulties both in measuring them (Section 5.2.2.4) and computing their effects (Section 5.1.2). Layers of dust aerosol pose special problems because they combine coarse particles, uncertainties in optical parameters, non-sphericity, and spatial inhomogeneity (Clarke *et al.*, 1996; Russell and Heintzenberg, 2000). For example, averaged discrepancies during ACE-2 for derived and measured scattering coefficients were less than 20% except in dust layers, where they were about 40% (Collins *et al.*, 2000). There has been a strong consensus from field studies that surface measurements of aerosol properties are rarely sufficient for computations of column properties such as optical depth. Some common reasons are strong vertical gradients in coarse aerosols and the transport of continental aerosols in the free troposphere above a marine boundary layer.

5.2.5 Trends in Aerosols

The regional nature of aerosols makes tropospheric aerosol trends more difficult to determine than trends in long-lived trace gases. Moreover, there are few long-term records of tropospheric aerosols (SAR). Ice cores provide records of species relevant to aerosols at a few locations. As shown in Figure 5.4, ice cores from both Greenland and the Alps display the strong anthropogenic influence of sulphate deposited during this century (Döscher *et al.*, 1995; Fischer *et al.*, 1998). Carbonaceous aerosols also show long-term trends (Lavanchy *et al.*, 1999). Sulphate in Antarctic ice cores shows no such trend (Dai *et al.*, 1995) since its source in the Southern Hemisphere is primarily natural. Aerosols have been measured from balloon sondes at Wyoming since 1971. For the number of aerosols larger than 0.15 μm, decreasing trends of -1.8 ± 1.4%/yr and -1.6 ± 1.8%/yr (90% confidence limits) were found in the 2.5 to 5 and 5 to 10 km altitude ranges, respectively (Hofmann, 1993). The total number of particles increased by 0.7 ± 0.1%/yr at Cape Grim from 1977 to 1991 but the number of particles large enough to nucleate cloud droplets (CCN) decreased by 1.5 ± 0.3%/yr from 1981 to 1991 (Gras, 1995). There are also some long-term data on visibility and turbidity. For example, summertime visibility in the eastern United States was worst in the 1970s, which was also a time of maximum SO_2 emissions (Husar and Wilson, 1993).

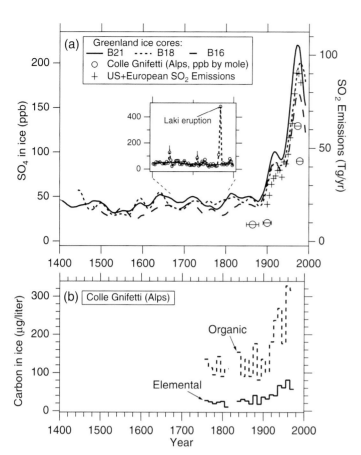

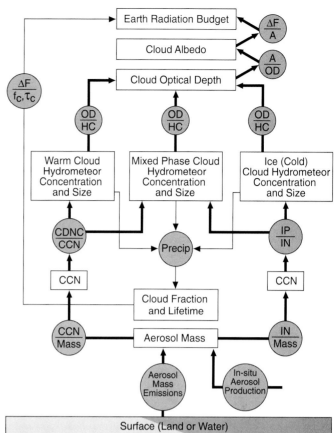

Figure 5.4: (a) Sulphate concentrations in several Greenland ice cores and an Alpine ice core (Fischer *et al.*, 1998; Döscher *et al.*, 1995). Also shown are the total SO_2 emissions from sources from the US and Europe (Gschwandtner *et al.*, 1986; Mylona, 1996). The inset shows how peaks due to major volcanic eruptions have been removed by a robust running median method followed by singular spectrum analysis. (b) Black carbon and organic carbon concentrations in alpine ice cores (Lavanchy *et al.*, 1999).

Figure 5.5: Flow chart showing the processes linking aerosol emissions or production with changes in cloud optical depth and radiative forcing. Bars indicate functional dependence of the quantity on top of the bar to that under the bar. Symbols: CCN (Cloud condensation nuclei); CDNC (Cloud droplet number concentration); IN (Ice nuclei); IP (Ice particles); OD (Optical depth); HC (Hydrometeor concentration); A (Albedo); f_c (Cloud fraction); τ_c (Cloud optical depth); ΔF (Radiative forcing).

5.3 Indirect Forcing Associated with Aerosols

5.3.1 Introduction

Indirect forcing by aerosols is broadly defined as the overall process by which aerosols perturb the Earth-atmosphere radiation balance by modulation of cloud albedo and cloud amount. It can be viewed as a series of processes linking various intermediate variables such as aerosol mass, cloud condensation nuclei (CCN) concentration, ice nuclei (IN) concentration, water phase partitioning, cloud optical depth, etc., which connect emissions of aerosols (or their precursors) to the top of the atmosphere radiative forcing due to clouds. A schematic of the processes involved in indirect forcing from this perspective is shown in Figure 5.5. Rather than attempt to discuss fully all of the processes shown in Figure 5.5, we concentrate here on a selected suite of linkages, selected either because significant progress towards quantification has been made in the last five years, or because they are vitally important. However, before delving into these relationships, we present a brief review of the observational evidence for indirect forcing.

5.3.2 Observational Support for Indirect Forcing

Observational support for indirect forcing by aerosols derives from several sources. Considering first remote sensing, satellite studies of clouds near regions of high SO_2 emissions have shown that polluted clouds have higher reflectivity on average than background clouds (Kuang and Yung, 2000). A study by Han *et al.* (1998a) has shown that satellite-retrieved column cloud drop concentrations in low-level clouds increase substantially from marine to continental clouds. They are also high in tropical areas where biomass burning is prevalent. Wetzel and Stowe (1999) showed that there is a statistically significant correlation of aerosol optical depth with cloud drop effective radius(r_{eff}) (negative correlation) and of aerosol optical depth with cloud optical depths (positive correlation) for clouds with optical depths less than 15. Han *et al.* (1998b), analysing ISCCP data, found an expected increase in cloud albedo with decreasing droplet size for all optically thick clouds but an unexpected decrease in albedo with decreasing droplet size in optically thinner clouds ($\tau_c<15$) over marine locations. This latter relationship may arise because of the modulation of the liquid-water path by cloud dynamics associated with absorption of solar radiation (Boers and Mitchell, 1994) but may also arise from the generally large spatial scale of some satellite retrievals which can yield misleading correlations. For example, Szczodrak *et al.* (2001), using 1 km resolution AVHRR data, do not see the increase in liquid-water path (LWP) with increasing effective radius for all clouds seen by Han *et al.* (1998b), who utilised 4 km resolution pixels. In any case, a relationship similar to that found by Han *et al.* (1998b) was found in the model of Lohmann *et al.* (1999b,c) and that model supports the finding of a significant indirect forcing with increases in aerosol concentrations (Lohmann *et al.*, 2000). Further evidence for an indirect forcing associated with increases in aerosol concentrations comes from the study by Nakajima *et al.* (2001). They found increases in cloud albedo, decreases in cloud droplet r_{eff}, and increases in cloud droplet number associated with increases in aerosol column number concentration.

In-situ measurement programmes have found linkages between CCN concentrations and both drizzle and cloud droplet r_{eff} in marine stratocumulus (cf., Hudson and Yum, 1997; Yum and Hudson, 1998). Moreover, several studies have contrasted the microstructure of polluted and clean clouds in the same airshed (e.g., Twohy *et al.*, 1995; Garrett and Hobbs, 1995) while others have linked seasonal variations in drop concentrations and r_{eff} with seasonal variations in CCN (Boers *et al.*, 1998). Indeed, recent studies by Rosenfeld (1999, 2000) show a dramatic impact of aerosols on cloud precipitation efficiency. Until recently, evidence of the impact of aerosols on cloud albedo itself has been confined primarily to studies of ship tracks (e.g., Coakley *et al.*, 1987; Ferek *et al.*, 1998; Ackerman *et al.*, 2000) although some other types of studies have indeed been done (e.g., Boers *et al.*, 1998). However, new analyses have not only identified changes in cloud microstructure due to aerosols but have associated these changes with increases in cloud albedo as well. An example of this, from Brenguier *et al.* (2000), can be seen in Figure 5.6, which displays measured cloud reflectances at two wavelengths for clean and polluted clouds examined during ACE-2. Thus, our understanding has advanced appreciably since the time of the SAR and these studies leave little doubt that anthropogenic aerosols have a non-zero impact on warm cloud radiative forcing. An estimate of the forcing over oceans (from satellite studies) ranged from −0.7 to −1.7 Wm^{-2} (Nakajima *et al.*, 2001).

For cirrus clouds, much of both the observational and theoretical support for indirect forcing by anthropogenic aerosols has been recently summarised in the report on the impact of aviation-derived aerosols on clouds (Fahey *et al.*, 1999). There is a remarkably consistent upward trend in cirrus fractional cloudiness in areas of high air traffic over the last two decades observable, in at least broad outline, in both ground-based and satellite databases (Fahey *et al.*, 1999). The suggested increase in cirrus cloud cover of about 4% since 1971 is much larger than can be attributed to linear-shaped contrails, with the resulting implication that aviation emissions are perturbing high altitude cloud amount. While purely meteorological trends (i.e., increasing upper atmospheric water vapour, changing circulation patterns) and/or increases associated with surface-based emissions cannot be excluded as a possible cause for these trends, these observations are very suggestive of a perturbation from aviation emissions.

5.3.3 Factors Controlling Cloud Condensation Nuclei

The effectiveness of an aerosol particle as a CCN depends on its size and response to water. Atmospheric aerosol particles are either hydrophobic (i.e. will not activate in cloud under any circumstances), water-insoluble but possess hydrophyllic sites that allow the particles to wet and activate at higher supersaturations, or have some water-soluble component and will activate at lower supersaturations given sufficient time to achieve their critical radius. Only particles with some water-soluble species are significant for the indirect forcing (e.g., Kulmala *et al.*, 1996; Eichel *et al.*, 1996). However, there are many water-soluble compounds in the atmospheric aerosol with widely varying degrees of solubility.

Sulphates, sodium chloride, and other water-soluble salts and inorganic acids are common to the atmospheric aerosol (Section 5.2) and to CCN (e.g., Hudson and Da, 1996). The abilities of these species to serve as CCN are relatively well

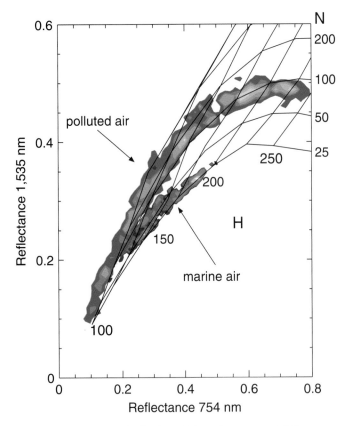

Figure 5.6: Frequency distributions of the reflectances at 1,535 nm versus reflectances at 754 nm determined during the ACE-2 experiment. Isolines of geometrical thickness (H) and droplet number concentration (N) demonstrate the higher reflectance in polluted cloud if normalised by a similar geometrical thickness (Brenguier *et al.* 2000).

known, whereas our understanding of the ability of organic species to act as CCN is relatively poor. This is a critical area of uncertainty for the global-scale modelling of cloud droplet nucleation.

The water-soluble fraction of organic species in aerosols can be high relative to sulphate (e.g., Li *et al.*, 1996; Zappoli *et al.*, 1999), and organics may be important sources of CCN in at least some circumstances (Novakov and Penner, 1993; Rivera-Carpio *et al.*, 1996). Aerosols from biomass burning are primarily composed of organics and may act as CCN, but some of their CCN activity may actually be due to co-resident inorganic constituents (Van-Dinh *et al.* 1994; Novakov and Corrigan, 1996; Leaitch *et al.*, 1996a). Measurements from a forested area suggest that some of the products of terpene oxidation may serve as CCN (Leaitch *et al.*, 1999), and Virkkula *et al.* (1999) found that particles from pinene oxidation absorbed some water at an RH of about 84%. Cruz and Pandis (1997) examined the CCN activity of particles of adipic acid and of glutaric acid (products of alkene oxidation) and found reasonable agreement with Köhler theory, whereas Corrigan and Novakov (1999) measured much higher activation diameters than predicted by theory. Volatile organic acids (formic, acetic, pyruvic, oxalic) may also contribute to the formation of CCN in areas covered with vegetation and in plumes from biomass burning (Yu, 2000). Pósfai *et al.* (1998) suggest that organic films can be responsible for relatively large water uptake at low RH.

The wide range of solubilities of the organic species may help to explain why observations do not always provide a consistent picture of the uptake of water. The water solubility of the oxygenated organic species tends to decrease with increasing the carbon number (Saxena and Hildemann, 1996). Shulman *et al.* (1996) showed that the cloud activation of organic species with lower solubilties might be delayed due to the increased time required for dissolution, and delays of 1−3 seconds have been observed (Shantz *et al.*, 1999). For a mixture of an inorganic salt or acid with an organic that is at least slightly soluble, the presence of the organic may contribute to some reduction in the critical supersaturation for activation (Corrigan and Novakov, 1999) especially if the organics reduce the surface tension (Shulman *et al.* 1996) as demonstrated in natural cloud water (Facchini *et al.*, 1999).

Historically, much of the interest in organics has focused on the inhibition of CCN activation in cloud by surface-active organics. Recently, Hansson *et al.* (1998) found that thick coatings of either of two insoluble high molecular weight organics (50 to 100% by mass of tetracosane or of lauric acid) reduced the hygroscopic growth factor for particles of NaCl. On the other hand, studies by Shulman *et al.* (1996), Cruz and Pandis (1998), and Virkkula *et al.* (1999) suggest that coatings of organic acids do not necessarily inhibit the effect of the other species on the condensation of water. The ability of the hydrophobic coating to form a complete barrier to the water vapour is a critical aspect of this issue. It is unlikely that such coatings are common in the atmosphere.

Although the CCN activities of inorganic components of the aerosol are well known, there are other aspects to the water activity of these species that need to be considered. Highly

soluble gases, such as HNO_3, can dissolve into a growing solution droplet prior to activation in cloud. The addition of this inorganic substance to the solution can decrease the critical supersaturation for activation (Kulmula *et al.*, 1993; Laaksonen *et al.*, 1998). The result is an increase in cloud droplet number but this is tempered somewhat by an enhanced condensation rate that contributes to a slight reduction in the cloud supersaturation. The importance of this effect will depend on the mixing ratio of such gases relative to the CCN solute concentrations and this has not been properly evaluated. An important consideration for the development of the CCN spectrum is the in-cloud oxidation of SO_2. Current models indicate that the fraction of secondary sulphate that is due to SO_2 oxidation in cloud can be in the range of 60 to 80% (Table 5.5). While the uncertainty in this estimate does not greatly impact the model sulphur budgets, it has significant consequences for the magnitude of the predicted indirect forcing (Chuang *et al.*, 1997; Zhang *et al.*, 1999).

Large-scale models must be able to represent several factors related to CCN in order to better assess the indirect effect: the size distribution of the mass of water-soluble species, the degree of solubility of the represented species, and the amount of mixing of individual species within a given size fraction. The most critical species are sulphates, organics, sea salt and nitrates.

5.3.4 Determination of Cloud Droplet Number Concentration

The impact of CCN on the cloud droplet number concentration (N_d) can be non-linear. One consequence of this is that the number of natural CCN can strongly influence the way that CCN from anthropogenic emissions affect the indirect radiative forcing. For example, Ghan *et al.* (1998) have shown that the presence of relatively high concentrations of sea salt particles can lead to increased N_d at lower sulphate concentrations and higher updraught speeds. Conversely, the N_d are lowered by high salt concentrations if sulphate concentrations are higher and updraught speeds are weaker (see also O'Dowd *et al.* (1999)). However, it is not clear whether these processes significantly affect the radiative forcing.

There are two general methods that have been used to relate changes in N_d to changes in aerosol concentrations. The first and simplest approach uses an empirical relationship that directly connects some aerosol quantity to N_d. Two such empirical treatments have been derived. Jones *et al.* (1994b) used a relationship between N_d and the number concentration of aerosol particles (N_a) above a certain size. This method is appropriate for the particles that serve as nuclei for cloud droplets in stratiform cloud, but it can be ambiguous for cumuliform cloud because of the activation of particles smaller than the threshold of the N_a (Isaac *et al.*, 1990; Gillani *et al.*, 1995). Boucher and Lohmann (1995) used observations of N_d and of CCN versus particulate or cloudwater sulphate to devise relationships between N_d and particulate sulphate. This approach has the advantage that it circumvents the assumptions required in deriving the aerosol number concentration N_a from sulphate mass. The use of sulphate as a surrogate for N_d implicitly accounts for other particulate species in the aerosol, but only as long as relationships are used that take into account the potential regional and seasonal differences in the chemical mixture of the aerosol (e.g., Van Dingenen *et al.*, 1995; Menon and Saxena, 1998). Thus, the empirical relationships derived for regions with high industrial sulphate loading may not be appropriate for biomass aerosols. A new empirical approach relates N_d to the particle mass concentration and mean volume diameter in the accumulation mode aerosol. It is based on mass scavenging efficiencies of the accumulation-mode aerosol by stratiform clouds (Glantz and Noone, 2000). When deriving empirical relationships, it is important to be aware of the effects of the averaging scale (Gultepe and Isaac, 1998). An advantage of the empirical methods is that they may account for the effects on N_d that are associated with cloud dynamics in an average sense. Variance in the cloud updraught velocities is one of the main reasons for the large scatter in the observations from which these empirical relationships are derived (Leaitch *et al.*, 1996b; Feingold *et al.*, 1999b).

The second method that has been used to relate changes in N_d to changes in aerosol concentrations is based on a prognostic parametrization of the cloud droplet formation process (Ghan *et al.*, 1993, 1995, 1997; Chuang and Penner, 1995; Abdul-Razzak *et al.*, 1998; Abdul-Razzak and Ghan, 2000). This type of approach requires a representation of the CCN activity of the particles and a representation of the dynamic and thermodynamic properties of the cloud. At present, some of the aerosol properties necessary to describe the CCN spectrum must be assumed in order to apply this approach.

A comparison of the empirical and prognostic methods currently in use for determining the N_d is shown in Figure 5.7. The empirical relationships are taken from Jones *et al.* (1994b) and Boucher and Lohmann (1995) and represent stratiform cloud. The curves labelled PROG follow the prognostic parametrization used by Chuang and Penner (1995). There is some relative general agreement between the empirical curves and the prognostic curves for low updraught velocity. The prognostic curves for the higher updraught velocity, i.e. more convective cloud, give much higher N_d than the empirical results for stratiform cloud. The relatively close agreement of the 10 cms^{-1} updraught curve with the empirical scheme for the ocean is somewhat fortuitous. Further work comparing both the empirical and prognostic schemes with observations is needed, especially for the climatologically important marine stratocumulus, to better understand the reasons for the agreements and differences in Figure 5.7.

There are also atmospheric dynamic factors that affect the prediction of N_d. Trajectories of air parcels through stratocumulus are highly variable (e.g., Feingold *et al.*, 1998). For the prognostic methods with their explicit representation of the updraught velocity, there is a need to understand not just the probability distribution function (PDF) of updraughts in these clouds, but the PDF of those that actually nucleate droplets. No satisfactory method of parametrizing the local updraught has yet been devised. The prognostic methods also produce an adiabatic N_d, which leads to the problem of how to represent the mean N_d in the cloud in the presence of the entrainment of dry air. Entrainment can even result from changes in N_d (Boers and Mitchell, 1994). These issues are critical for the prognostic prediction of N_d and deserve continued study to determine how best to take their effects into account.

5.3.5 Aerosol Impact on Liquid-Water Content and Cloud Amount

The indirect effect of aerosols on clouds is determined not only by the instantaneous mean droplet concentration change (i.e., the first indirect or Twomey effect; Twomey, 1977) but is also strongly associated with the development of precipitation and thus the cloud liquid-water path, lifetime of individual clouds and the consequent geographic extent of cloudiness (second indirect effect). These processes are well illustrated by marine stratiform clouds.

The longevity of marine stratiform clouds, a key cloud type for climate forcing in the lower troposphere, is dictated by a delicate balance between a number of source and sink terms for condensed water, including turbulent latent and sensible heat fluxes from the ocean surface, radiative cooling and heating rates, entrainment of dry air from above the cloud top inversion, and the precipitation flux out of the cloud. Unfortunately, changes to cloud extent induced by plausible changes in cloud droplet number concentration due to aerosol modulation can be slight and difficult to characterise (cf., Hignett, 1991). For example, Pincus and Baker (1994) point out that changes in short-wave absorption induced by changes in drop number act primarily to change cloud thickness – and cloud thickness is also strongly modulated by non-radiative processes. Depending on the cloud type, feedbacks involving cloud thickness can substantially reduce or enhance changes in cloud albedo due to change in droplet concentration (Boers and Mitchell 1994; Pincus and Baker, 1994). Feingold *et al.* (1997) have more recently examined the impact of drizzle modulation by aerosol on cloud optical depth.

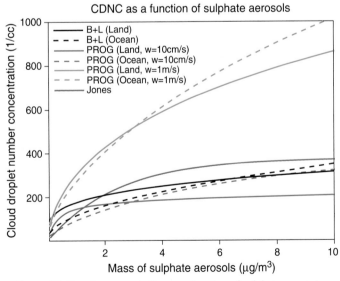

Figure 5.7. Droplet concentration as a function of sulphate concentration for 3 different treatments: the empirical treatment of Jones *et al.* (1994b), the empirical treatment of Boucher and Lohmann (1995) (denoted B+L), and the mechanistic treatment of Chuang and Penner (1995) (denoted PROG).

Nevertheless, precipitation processes are extremely important to marine cloud fraction, with varying precipitation efficiency leading to varying fractional cloudiness and liquid water content (Albrecht, 1989). The postulated mechanism is as follows. The activation of a larger number of aerosol particles limits the size to which drops can grow for an available cooling rate. Hence, the number of drops which grow large enough to initiate the collision-coalescence process (the dominant precipitation process in warm clouds) is decreased and precipitation rates are attenuated. With precipitation attenuated, a major sink for cloud drops is removed and cloud lifetime is enhanced. Liou and Cheng (1989) first estimated the potential global significance of this process. Further studies with more sophisticated models have supported the significance of the modulation of precipitation by aerosols and led to the consideration of several different processes that contribute to the effects of aerosols on clouds.

One such process is the modulation of cloud optical depth by precipitation. Pincus and Baker (1994) showed, using a simple mixed-layer model, that the cloud optical depth in marine stratiform clouds was a strong function of the initial aerosol concentration upon which the cloud formed, the dependence being close to exponential. Boers (1995) subsequently demonstrated, through modelling calculations coupled with field observations, that a substantial part of the seasonal cycle in cloud albedo at Cape Grim could be due to the modulation of cloud optical depth by aerosols and their effects on the efficiency of precipitation.

A second important process which may be affected by onset of precipitation is that of decoupling of the cloud layer from the surface. Precipitation may sometimes produce a sub-cloud stable layer that cuts off the moisture flux to the cloud. However, decoupling is not an inevitable consequence of precipitation formation and, under some circumstances, a balance between the moisture flux from the surface and precipitation sinks determine the cloud extent. Pincus *et al.* (1997) observed no difference between precipitating and non-precipitating stratocumulus with respect to cloud fraction and both Austin *et al.* (1995) and Stevens *et al.* (1998) found that observed stratocumulus seemed able to maintain themselves despite a considerable precipitation rate.

While the effect of precipitation modulation on cloud amount is supported by a number of studies, several others have argued that external thermodynamic factors such as sea surface temperature (SST) are the main factors determining the formation and dissipation of marine stratocumulus (cf., Wyant *et al.*, 1997). Such an analysis is also supported by the relationship of satellite-derived aerosol number concentration with cloud droplet number concentration and with liquid-water path. The former shows a positive correlation while the latter shows no particular relationship (Nakajima *et al.*, 2001). Thus, the climatological significance of this aspect of the indirect effect needs a great deal more investigation.

Finally, it is important to note that the impact of anthropogenic aerosols on precipitation modulation will be dependent on both the natural and anthropogenic aerosol size distributions. For example, a number of studies have suggested that both natural (e.g., Feingold *et al.*, 1999a) and anthropogenic (Eagen *et al.*, 1974) giant CCN have a great impact on precipitation and will influence the effect of smaller, anthropogenic CCN on precipitation.

5.3.6 Ice Formation and Indirect Forcing

Formation of ice in the atmosphere has long been recognised to be a topic of great importance due to its key role in the precipitation process. However, progress in elucidating this role has been plagued by a host of complex issues such as the precise mode of action of ice nuclei (IN) (e.g., Cooper, 1980), *in situ* modification of IN activity by various substrate coatings, including residual ice (Borys, 1989; Curry *et al.*, 1990; Rosinski and Morgan, 1991; Beard, 1992; Vali, 1992), secondary ice production (e.g., Beard, 1992) and a lack of consistency in measurement techniques (cf., Bigg, 1990; Vali, 1991; Rogers, 1993; Pruppacher and Klett, 1997).

Because of these issues, it is premature to quantitatively assess the impact of ice formation on indirect forcing. Instead, the potential importance of such formation is given a preliminary assessment by addressing a set of four fundamental and serial questions whose answers would, in principle, yield the desired assessment. A summary of what is known with respect to these questions is given below.

Does ice formation have an impact on radiative forcing?
In principle the phase partitioning of water in clouds should have a substantial impact on cloud radiative forcing, first because the ice hydrometeors will tend to be much larger than cloud drops and thus increase precipitation, and second because the size of hydrometeors (determined by both ice/vapour and ice/liquid partitioning) can have a significant impact on radiative balances. Several GCM sensitivity studies have supported these expectations (e.g., Senior and Mitchell, 1993; Fowler and Randall, 1996) by demonstrating that the fraction of supercooled water in the models which is converted to ice has a significant impact on the global radiative balance. A simple sensitivity study with the ECHAM model (cf., Lohmann and Feichter, 1997) conducted for this assessment revealed a very large difference in globally averaged cloud forcing of $+16.9\,\mathrm{Wm^{-2}}$ induced by allowing only ice in clouds with temperatures below $0°C$ as compared with allowing only water in clouds with temperatures above $-35°C$. The liquid-water path change in these experiments ($160\,\mathrm{gm^{-2}}$ to $54.9\,\mathrm{gm^{-2}}$) was larger than the 60% uncertainty in this quantity from measurements (Greenwald *et al.*, 1993; Weng and Grody, 1994). However, this certainly demonstrates that even small changes in ice formation could have a significant impact on the indirect climate forcing due to aerosols.

Is formation of the ice phase modulated by aerosols?
The relative roles of different types of ice nucleation in cirrus clouds is very complex (e.g., Sassen and Dodd, 1988; Heymsfield and Miloshevich, 1993, 1995; DeMott *et al.*, 1997, 1998; Strom *et al.*, 1997; Xu *et al.*, 1998; Martin, 1998; Koop *et al.*, 1999). Presumably there is a temperature-dependent transition from predominantly heterogeneous nucleation (i.e., initiated at a phase boundary with a substrate – the heterogeneous nucleus) to homogeneous nucleation (i.e., within the liquid phase alone – no substrate required) that depends on the chemistry and size of the precursor haze drops of the homogeneous process as well as on the chemistry and concentration of the heterogeneous nuclei.

Supersaturations with respect to ice in excess of 40 to 50% are necessary to freeze sulphate haze drops, even at quite low temperatures. Far lower supersaturations will be adequate if heterogeneous IN are present. There is thus a large supersaturation range in which heterogeneous IN could have a significant impact (Fahey *et al.*, 1999). However, in both heterogeneous and homogeneous cases, aerosols play an important role in glaciation.

For lower level, warmer (though still supercooled) clouds, in principle the answer to our query is necessarily positive. In this vast, liquid-water reservoir, temperatures are simply too warm for homogeneous freezing of cloud drops to occur and aerosol surfaces of some sort must provide the substrate for ice initiation. However, prolific secondary ice formation due to such processes as the Hallett-Mossop mechanism renders the establishment of a clear relationship between measured IN concentrations and ice particle concentrations quite difficult (cf., Beard, 1992; Rangno and Hobbs, 1994). Nevertheless, in some instances relationships between IN concentrations and ice formation in lower-tropospheric clouds have been found (e.g., Stith *et al.*, 1994; DeMott *et al.*, 1996). This lends credibility to the view that the actual ice initiation process must be modulated by aerosols.

For both upper and lower level clouds in the troposphere, it seems clear that the ice initiation process is dependent on aerosols, though the nucleation process can proceed via different pathways and from a variety of different nucleating chemical species.

Are a significant fraction of ice nucleating aerosols anthropogenic?

For cirrus cloud formed by homogeneous nucleation on haze particles, there are clearly solid grounds for asserting that there is a substantive anthropogenic component to the associated IN. Most chemical transport models (e.g., Penner *et al.*, 1994b; Feichter *et al.*, 1996; Graf *et al.*, 1997; Koch *et al.*, 1999; Rasch *et al.*, 2000a) as well as sulphur isotope studies (e.g., Norman *et al.*, 1999) suggest that a substantial fraction of the upper-tropospheric sulphate burden (and thus the sulphate haze droplets involved in homogeneous nucleation) is anthropogenic although quantitative estimates vary from model to model. Furthermore, for wintertime polar clouds, a special case of low-level cirrus, there are observations and modelling results to indicate that, through suppression of the freezing point, the acidic aerosols of Arctic haze favour the formation of large ice crystals rather than the smaller particles in unpolluted ice fog during the Arctic airmass cooling process. The larger particles increase the sedimentation flux and deplete water vapor from the atmosphere, providing a negative forcing (Blanchet and Girard, 1995). There are also grounds for asserting that changes in the sulphate abundance will lead to increases in ice nucleation (cf., Jensen and Toon, 1992, 1994; Kärcher *et al.*, 1998; Lin *et al.*, 1998; Fahey *et al.* 1999) so that a positive forcing is plausible. In any case, anthropogenic modulation of ice particles should certainly be considered.

For cirrus clouds in the upper troposphere formed on insoluble, heterogeneous IN, the situation is much less clear. Recent observations of heterogeneous IN in both the upper and lower troposphere are suggestive of a large role for crustally derived aerosols (e.g., Kumai, 1976; Hagen *et al.*, 1995; Heintzenberg *et al.*, 1996; Kriedenweis *et al.*, 1998). On the other hand, there is some evidence of a carbonaceous component as well (Kärcher *et al.*, 1996), quite likely coming from aircraft exhaust (Jensen and Toon, 1997; Chen *et al.*, 1998; Strom and Ohlsson, 1998; Petzold *et al.*, 1998). Hence, while the precise partitioning of heterogeneous IN amount between anthropogenic and natural is not currently feasible, a significant fraction might be anthropogenic.

For ice formation in other lower-tropospheric clouds, the situation is also unclear. All of the IN must necessarily be heterogeneous nuclei, with all of the ambiguities as to source that this implies. Natural ice nuclei are certainly not lacking. In addition to the soil source alluded to above, biogenic IN have been measured at significant levels, in both laboratory and field studies (cf., Schnell and Vali, 1976; Levin and Yankofsky, 1983). Much of this work has been recently reviewed in Szyrmer and Zawadzki (1997).

Numerous aerosol species of anthropogenic origin have been identified, both in the laboratory and in the field, as effective ice initiators (e.g., Hogan, 1967; Langer *et al.*, 1967; Van Valin *et al.*, 1976). However, measurement of enhanced IN concentrations from industrial sources has produced contradictory results. In some cases (e.g., Hobbs and Locatelli, 1970; Al-Naime and Saunders, 1985), above background IN concentrations have been found in urban plumes while in other instances below background levels have been found (e.g., Braham and Spyers-Duran, 1974; Perez *et al.*, 1985). Some of the discrepancies may be attributable to differing measurement techniques (Szyrmer and Zawadzki, 1997). However, it is possible that a good deal of it is due to differing degrees of deactivation of IN by pollutants such as sulphate, which readily forms in the atmosphere and can coat IN surfaces thereby deactivating them (particularly if the mechanism for ice formation involves deposition from the gas phase). Indeed, Bigg (1990) has presented evidence, albeit inconclusive, of a long-term decrease in IN associated with increasing pollution, perhaps acting through such a mechanism. There is also the possibility that secondary aerosol constituents such as aliphatic alcohols or dicarboxylic acids could coat inactive particles and thus transform them into efficient IN. If so, an incubation period after primary emission would be necessary and measurements could differ depending on time from emission. Although variability in natural heterogeneous IN sources may produce frequent circumstances where natural IN sources exceed anthropogenic IN in importance, it is difficult to see how the anthropogenic impact, both positive and negative, can be neglected.

How might the anthropogenic IN component vary with time?

This is the least tractable question of all, with all of the uncertainties in the answers to the previous questions propagating into any proposed answer. For homogeneous nucleation in upper-level, cirriform clouds, an assertion that the number of ice forming particles will be related to sulphur emissions is at least tenable. However, for the lower troposphere, little can be said at present. The impact of anthropogenic emissions could be negative, due to "poisoning" of natural IN; positive, due perhaps, to increased organic and inorganic carbon compared to inorganic non-carbon pollution emissions; or simply have little impact due to a small net source strength compared to those of natural IN.

5.4 Global Models and Calculation of Direct and Indirect Climate Forcing

5.4.1 Summary of Current Model Capabilities

In the past, many climate models used prescribed climatologies (Tanré *et al.*, 1984: d'Almeida, 1991) or precalculated monthly or annual mean column aerosol to describe the geographical distribution of aerosols and aerosol types. Optical properties were calculated offline by Mie-calculations assuming a uniform particle size, density and particle composition for each of the aerosol components (Shettle and Fenn, 1976; Krekov, 1993).

Most current GCMs are beginning to incorporate the calculation of aerosol mass interactively, taking account of the effect of aerosols on meteorology (Taylor and Penner, 1994; Roeckner *et al.*, 1999). In addition, models (GCMs and Chemistry Transport Models) are now available which directly use information on cloud formation and removal from the GCMs to account for the complex interactions between cloud processes, heterogeneous chemistry and wet removal (e.g. Feichter *et al.*, 1997; Roelofs *et al.*, 1998; Koch *et al.*, 1999; Rasch *et al.*, 2000a). These models are able to represent the high temporal and spatial variability of the aerosol particle mass distribution but must assume a size distribution for the aerosol to calculate their radiative effects.

The number and size of primary aerosols depend on the initial size distribution attributed to their source profiles together with the main growth and removal process. Models which represent number concentration have been developed for mineral dust studies (Tegen *et al.*, 1996; Schulz *et al.*, 1998) and for sea salt aerosols (Gong *et al.*, 1998). Representation of aerosol number is far more difficult for sulphate and secondary organics because the size distributions of condensing species depend on the size distribution of aerosols which are present before condensation and on cloud processes, but attempts to include these processes have been initiated (Herzog *et al.*, 2000; Ghan *et al.*, 2001a,b,c).

Because the processes for treating aerosol removal associated with precipitation and deposition and in-cloud conversion of SO_2 to sulphate are represented in climate models as sub-grid scale processes, there are significant variations in the efficiency of these processes between different models (see Section 5.2.2.6). In the past, model intercomparisons sponsored by the World Climate Research Programme (WCRP) have focussed on ^{222}Rn, ^{210}Pb, SO_2, and SO_4^{2-} (Jacob *et al.*, 1997; Rasch *et al.*, 2000b; Barrie *et al.*, 2001). These comparisons provide a "snapshot" in time of the relative performance of a major fraction of available large-scale sulphur models. They have shown that most of the models are able to simulate monthly average concentrations of species near the surface over continental sites to within a factor of two. Models are less sensitive to changes in removal rates near source regions, and they tend to agree more closely with observations over source regions than over remote regions. Comparison of models with observations at remote receptor sites can indicate whether transport and wet removal is well simulated but, for sulphate, may also be an indication of whether local source strengths are correctly estimated. The WCRP-sponsored model intercomparison in 1995 showed that model simulations differed significantly in the upper troposphere for species undergoing wet scavenging processes (Rasch *et al.*, 2000b) and the IPCC workshop (Section 5.4.1.2) demonstrated a similar sensitivity. Unfortunately, observations to characterise particle concentrations in remote regions and in the upper troposphere are limited. The vertical particle distribution affects aerosol forcing because scattering particles exhibit a greater forcing when they are located in the lower part of the troposphere, due to the effects of humidity on their size. Also, absorbing aerosols yield a greater forcing when the underlying surface albedo is high or when the aerosol mass is above low clouds (Haywood and Ramaswamy, 1998).

5.4.1.1 Comparison of large-scale sulphate models (COSAM)

One measure of our knowledge comes from the convergence in predictions made by a variety of models. Difficulties in the analysis and evaluation of such comparisons can result from models employing different emissions, meteorological fields etc. A set of standardised input was provided for the Comparison of Large Scale Sulphate Aerosol Models (COSAM) workshop which took place in 1998 and 1999 (see Barrie *et al.*, 2001). Ten models participated in this comparison. As noted above in Section 5.2.2.6, the simulation of the processes determining the sulphate concentration differed considerably between models. The fraction of sulphur removed by precipitation ranged from 50 to 80% of the total source of sulphur. The fraction of total chemical production of sulphate from SO_2 that took place in clear air (in contrast to in-cloud) ranged from 10 to 50%. This latter variability points to important uncertainties in current model capability to predict the indirect forcing by anthropogenic sulphate, because the mechanism of sulphate production determines the number of CCN produced (Section 5.3.3).

The ability of the models to predict the vertical distributions of aerosols was examined by comparing model predictions of the vertical distribution of SO_4^{2-}, SO_2 and related parameters such as ozone, hydrogen peroxide and cloud liquid-water content to mean profiles taken during aircraft campaigns at North Bay, a remote forested location (in southern Nova Scotia) 500 km north of the city of Toronto (Lohmann *et al.*, 2001). For SO_4^{2-}, the models were within a factor of 2 of the observed mean profiles (which were averages of 64 and 46 profiles for North Bay and Novia Scotia, respectively) but there was a tendency for the models to be higher than observations. The sulphur dioxide concentration was generally within a factor of two of the observations. Those models that overpredicted SO_4^{2-} also underpredicted SO_2, demonstrating that the model treatment for the chemical transformation of SO_2 to SO_4^{2-} is a source of uncertainty in the prediction of the vertical distribution of SO_4^{2-}.

A comparison of modelled and observed ground level sulphate at 25 remote sites showed that on average, most models predict surface level seasonal mean SO_4^{2-} aerosol mixing ratios to within 20%, but that surface SO_2 was overestimated by 100% or more. A high resolution limited area model performed best by matching both parameters within 20%. This was consistent with the large variation in the ability of models to transport and disperse sulphur in the vertical.

Both regional source budget analyses (Roelofs *et al.*, 2001) as well as long-range transport model tests (Barrie *et al.*, 2001) suggest that the dominant cause of model-to-model differences is the representation of cloud processes (e.g. aqueous-phase sulphate production rates, wet deposition efficiency and vertical transport efficiencies) and horizontal advection to remote regions.

In summary, the COSAM model comparison showed that both convective transport and oxidation of sulphur dioxide to sulphate are processes that are not well simulated in large-scale models. In addition, the treatment of dry and wet deposition of aerosols and aerosol precursors continues to lead to important variations among the models. These variations lead to uncertainties in atmospheric sulphate concentrations of up to a factor of 2. The uncertainties in SO_2 are larger than those for SO_4^{2-}.

5.4.1.2 The IPCC model comparison workshop: sulphate, organic carbon, black carbon, dust, and sea salt

Comparisons of models with observations for sulphate aerosols and other sulphur compounds are particularly relevant for assessing model capabilities because the emissions of sulphur bearing compounds are better known than the emissions of other aerosol compounds (Section 5.2). Thus, comparison can focus on the capabilities of the models to treat transport and oxidation processes. Recent field studies, however, have pointed out the importance of organic aerosol compounds (Hegg *et al.*, 1997), dust aerosols (Li-Jones and Prospero, 1998; Prospero, 1999), and sea salt aerosols (Murphy *et al.*, 1998a). Also, soot is important because it decreases the reflection and increases the absorption of solar radiation (Haywood and Shine, 1995). Furthermore, the magnitude of the indirect effect is sensitive to the abundance of natural aerosols (Penner *et al.* 1996; O'Dowd *et al.*, 1999; Chuang *et al.*, 2000). Therefore, an examination of model capability to represent this entire suite of aerosol components was undertaken as part of this report.

Emissions for this model comparison were specified by the most recently available emissions inventories for each component (see Tables 5.2, 5.3, Section 5.2 and Table 5.7). Eleven aerosol models participated in the model intercomparison of sulphate, and of these, nine treated black carbon, eight treated organic carbon, seven treated dust, and six treated sea salt. Eight scenarios were defined (see Table 5.7). The first, SC1, was selected to provide good estimates of present day aerosol emissions. SC2 was defined to simulate aerosol concentrations in 2030 according to preliminary estimates from the IPCC SRES A2 scenario (Nakićenović, *et al.*, 2000). SC3 was defined to simulate the draft A2 scenario in 2100 and SC4 to simulate the draft B1 scenario in 2100. SC1-SC4 used present day chemistry and natural emissions. In addition, we estimated possible future changes in emissions of the natural components DMS, terpenes, dust and sea salt in 2100 in SC5 for the A2 scenario and in SC8 for the B1 scenario. Scenario SC6 also estimated changes in emissions of other gas phase components associated with the production of sulphate in the A2 scenario in 2100 (see Chapter 4) and SC7 estimated changes in climate (temperature, winds and precipitation patterns) as well. Table 5.7 also shows the estimates of anthropogenic emissions in 2100 associated with the IS92a scenario. Some of the participants also provided estimates of direct and indirect forcing. These estimates, together with the

range of predicted concentrations among the models, help to define the uncertainty due to different model approaches in aerosol forcing for future scenarios. The models, participants, scenarios they provided, and the aerosol components treated are summarised in Table 5.8.

5.4.1.3 Comparison of modelled and observed aerosol concentrations

Like previous model intercomparisons, the IPCC comparison showed large differences (factor of 2) in model predictions of the vertical distribution of aerosols. The model simulations of surface sulphate concentrations (Figure 5.8) indicate that much of the difference in sulphate radiative forcing reported in the literature is most likely to be associated with either variations in the vertical distribution or with the response of sulphate aerosols to variations in relative humidity (Penner *et al.*, 1998b).

The IPCC comparison showed that the capability of models to simulate other aerosol components is inferior to their capability to simulate sulphate aerosol. For example, sea salt in the North and South Pacific shows poorer agreement, with an average absoute error of 8 μgm^{-3} (Figure 5.9) than the corresponding sulphate comparison which is less than 1 μgm^{-3} (Figure 5.8) (see Table 5.9 also).

For dust the model-observation comparison showed a better agreement with surface observations in the Northern than in the Southern Hemisphere. For example, the average absolute error in the Northern Pacific was 179%, while it was 268% in the Southern Pacific. In the Southern Hemisphere, almost all models predict concentrations higher than the observations at all stations poleward of 22°S. Thus, it appears that dust mobilisation estimates may be too high, particularly those for Australia and South America. The paucity of dust from these regions relative to other arid dust source areas has been noted previously (Prospero *et al.*, 1989; Tegen and Fung, 1994; Rea, 1994), and may reflect the relative tectonic stability, low weathering rates, duration of land-surface exposure, and low human impacts in this area.

The interpretation of the comparison of observed and model-predicted concentrations for both organic carbon and black carbon is more difficult because of both inaccuracies in the observations (Section 5.1.2) and the fact that most measured concentrations are only available on a campaign basis. In addition, the source strength and atmospheric removal processes of carbonaceous aerosols are poorly known. Most models were able to reproduce the observed concentrations of BC to within a factor of 10 (see Figure 5.10) and some models were consistently better than this. Both modelled and observed concentrations varied by a factor of about 1,000 between different sites, so agreement to within a factor of 10 demonstrates predictive capability. However, there are still large uncertainties remaining in modelling carbonaceous aerosols.

Table 5.9 presents an overview of the comparison between observed and calculated surface mixing ratios. Table 5.9a gives the comparison in terms of absolute mass concentrations while Table 5.9b gives the comparison in terms of average differences of percents. The average absolute error for sulphate surface concentrations is 26% (eleven models) and the agreement

Table 5.7: *Global emissions specified for the IPCC model intercomparison workshop.*

	SC1	SC2	SC3	SC4	SC5	
		A2	A2	B1	A2	IS92a
Year	2000	2030	2100	2100	2100+ natural	2100
Sulphur (as Tg S)[a]						
Anthropogenic SO_2	69.0	111.9	60.3	28.6	60.3	147.0
Ocean DMS	25.3	25.3	25.3	25.3	27.0	
Volcanic SO_2	9.6	9.6	9.6	9.6	9.6	
Organic Carbon (as OM)[b]						
Anthropogenic	81.4	108.6	189.5	75.6	189.5	126.5
Natural	14.4	14.4	14.4	14.4	20.7	
Black Carbon[b]						
Anthropogenic	12.4	16.2	28.8	12.0	28.8	19.3
Dust (<2 μm diameter)[c]	400	400	400	400	418.3	
Dust (>2 μm diameter)[c]	1,750	1,750	1,750	1,750	1,898	
Sea Salt (as Na) (<2 μm diameter)[d]	88.5	88.5	88.5	88.5	155.0	
Sea Salt (as Na) (>2 μm diameter)[d]	1,066	1,066	1,066	1,066	1,866	

[a] Anthropogenic SO_2 emissions were the preliminary emissions from Nakicenovic *et al.* (2000). DMS emissions were the average of the emissions based on Wanninkhof (1992) and those based on Liss and Merlivat (1986) using methods described in Kettle *et al.* (1999). Volcanic SO_2 emissions were those available from the IGAC Global Emissions Inventory Activity (http://www.geiacenter.org) described by Andres and Kasgnoc (1998).

[b] Organic carbon is given as Tg of organic matter (OM) while black carbon is Tg C. For anthropogenic organic matter, the black carbon inventory of Liousse *et al.* (1996) was scaled up by a factor of 4. This scaling approximately accounts for the production of secondary organic aerosols consistent with the analysis of Cooke *et al.* (1999). For 2030 and 2100, the ratio of the source strengths for CO in 2030 and 2100 to that in 2000 was used to scale the source of organic carbon and black carbon at each grid location. For natural organic aerosols, the terpene emissions from Guenther *et al.* (1995) were assumed to rapidly undergo oxidation yielding a source of aerosol organic matter of 11% by mass per unit C of the emitted terpenes.

[c] The dust emission inventory was prepared by P. Ginoux.

[d] The sea salt emissions were those developed by Gong *et al.* (1997a,b) based on the Canadian Climate Model winds.

Table 5.8: *Aerosol models participating in IPCC model intercomparison workshop.*[a]

Code	Model	Contributor	Aerosol components treated	Scenarios	Forcing provided
1	ULAQ	Pitari	Sulphate, OC, BC, dust, sea salt	SC1, SC2, SC3, SC4, SC5, SC7	
2	GISS	Koch, Tegen	Sulphate, OC, BC, dust, sea salt	SC1, SC2, SC3, SC4, SC8	Direct, Indirect
3	Georgia Tech./GSFC GOCART	Chin, Ginoux	Sulphate, OC, BC, dust, sea salt	SC1, SC2, SC3, SC4, SC5	
4	Hadley Center, UK Met Office	Roberts, Woodage, Woodward, Robinson	Sulphate, BC, dust	SC1, SC2, SC3, SC4, SC5	Direct
5	LLNL/U. Mich. (CCM1/ GRANTOUR)	Chuang Penner Zhang	Sulphate, OC, BC, dust, sea salt	SC1, SC2, SC3, SC4, SC5, SC8	Indirect
6	Max Planck/ Dalhousie U. (ECHAM4.0)	Feichter Land Lohmann	Sulphate, OC, BC	SC1, SC2, SC3	Indirect
7	U. Michigan (ECHAM3.6/ GRANTOUR)	Herzog, Penner Zhang	Sulphate, OC, BC, dust, sea salt	SC1, SC2, SC3, SC4, SC5, SC8	Direct
8	UKMO/Stochem	Collins	Sulphate	SC1,SC3	
9	NCAR/Mozart	Tie	Sulphate, BC	SC1	
10	KNMI/TM3	van Weele	Sulphate	SC1	
11	PNNL	Ghan, Easter	Sulphate, OC, BC, dust, sea salt	SC1, SC3	Direct, indirect

[a] References to the models: (1) Pitari *et al.*, (2001); Pitari and Mancini (2001); (2) Koch *et al.* (1999); Tegen and Miller (1998); (3) Chin *et al.* (2000); (4) Jones *et al.* (1994b); (5) Chuang *et al.* (2000); (6) Feichter *et al.* (1996); Lohmann *et al.* (1999b); (7) This model is similar to those described by Chuang *et al.* (1997) and Chuang *et al.* (2001), but with updated cloud scavenging and convective processes; (8) Stevenson *et al.* (1998); (9) Brasseur *et al.* (1998); (10) Houweling *et al.* (1998); Jeuken *et al.* (2001); (11) Ghan *et al.* (2001a,b,c).

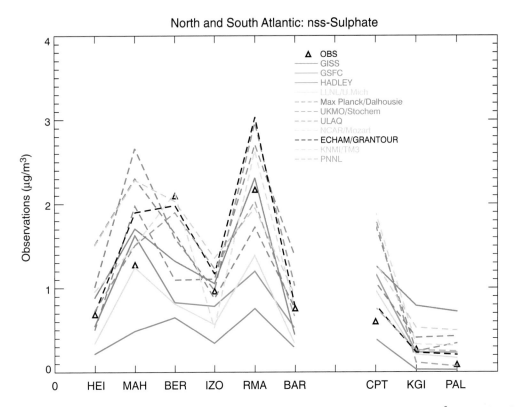

Figure 5.8: Observed and model-predicted annual average concentrations of non-sea salt sulphate (in μg m^{-3}) at a series of stations in the North and South Atlantic. The models are listed in Table 5.8. Data were provided by D. Savoie and J. Prospero (University of Miami). Stations refer to: Heimaey, Iceland (HEI); Mace Head, Ireland (MAH); Bermuda (BER); Izania (IZO); Miami, Florida (RMA); Ragged Point, Barbados (BAR); Cape Point, South Africa (CPT); King George Island (KGI); and Palmer Station, Antarctica (PAL).

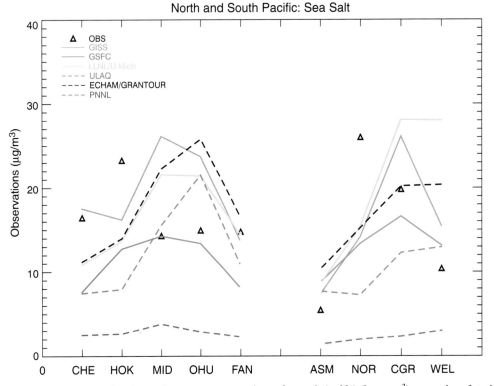

Figure 5.9: Observed and model-predicted annual average concentrations of sea salt (as Na) (in μgm^{-3}) at a series of stations in the North and South Pacific. The models are listed in Table 5.8. Data were provided by D. Savoie and J. Prospero (University of Miami). Stations refer to: Cheju, Korea (CHE); Hedo, Okinawa, Japan (HOK); Midway Island (MID); Oahu, Hawaii (OHU); Fanning Island (FAN); American Samoa (ASM); Norfolk Island (NOR); Cape Grim, Tasmania (CGR); and Wellington/Baring Head, New Zealand (WEL).

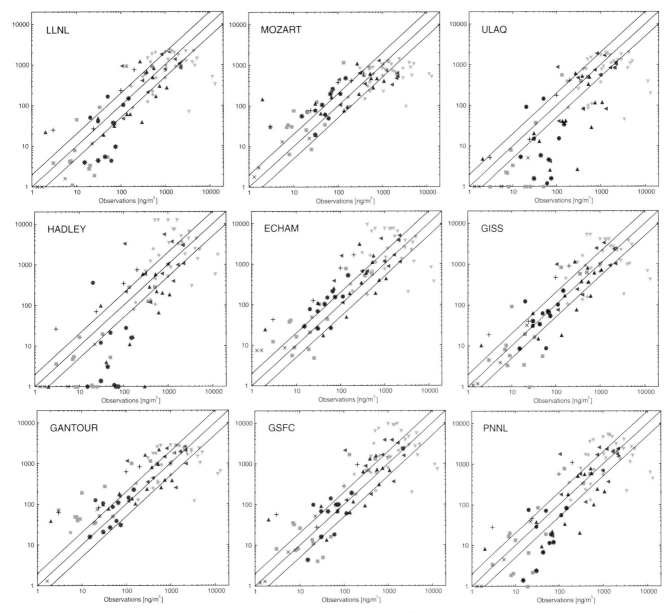

Figure 5.10: Observed and model-predicted concentrations of black carbon (in ng C m^{-3}) at a number of locations. The models are listed in Table 5.8. Observations refer to those summarised by Liousse *et al.* (1996) and Cooke *et al.* (1999). Symbols refer to: circle, Liousse Atlantic; square, Liousse Pacific; diamond, Liousse Northern Hemisphere rural; plus, Liousse Southern Hemisphere rural; asterisk, Liousse Northern Hemisphere remote; cross, Liousse Southern Hemisphere remote; upward triangle, Cooke remote; left triangle, Cooke rural; downward triangle, Cooke urban.

between modelled concentrations and observations is better for sulphate than for any other species. The largest difference with observed values is that of carbonaceous aerosols with an average absolute error (BC: nine models, OC: eight models) of about 179%. This may be partly due to the large uncertainties in the estimated strength of biomass burning and biogenic sources. The average absolute error for the dust (six models) and sea salt (five models) simulations is 70 and 46%, respectively.

In addition to the model comparison with observations, an analysis of the variation in aerosol burden among the models was considered. After throwing out the burdens from models that were outliers in terms of their comparison with observations, the model results still differed by a factor of 2.2 for sulphate, by a

factor of 2.2 for organic carbon (seven models), and by a factor of 2.5 for black carbon (eight models). In contrast, the range of total burdens for dust was a factor of 2.8 and 4.5 for aerosols with diameter less than and greater than 2 μm, respectively, while that for sea salt was a factor of 4.9 and 5.3, respectively. The range for sea salt with D<2 μm increases to 6 if the GSFC Gocart model is considered (with its higher sea salt flux). These differences were also evident in comparing concentrations in the upper troposphere. For example, in the upper troposphere (8 to 12 km) between 30 and 60°N, the range of predicted concentrations of sulphate, organic carbon and black carbon was about a factor of 5 to 10. This range increased to as much as a factor of 20 or more in the case of sea salt and dust.

Table 5.9a: *Comparison of models and observations of aerosol species at selected surface locations (µg/m³)[a,b].*

Model	Sulphate		Black carbon		Organic carbon		Dust		Sea salt	
	Average bias (µg/m³)	Average absolute error (µg/m³)	Average bias (µg/m³)	Average absolute error (µg/m³)	Average bias (µg/m³)	Average absolute error (µg/m³)	Average bias (µg/m³)	Average absolute error (µg/m³)	Average bias (µg/m³)	Average absolute error (µg/m³)
GISS	0.15	0.33	0.16	0.61	0.69	1.52	5.37	5.37	3.90	11.94
GSFC	−0.10	0.28	0.71	1.00	0.71	1.57	−0.5	1.98	−3.02	9.23
Hadley	−0.54	0.55	0.74	1.18			−2.47	3.48		
CCM/Grantour	−0.31	0.40	−0.18	0.50	−0.84	1.20	1.77	2.99	5.26	14.48
ECHAM	0.09	0.42	0.74	1.07	1.52	2.09				
Stochem	0.34	0.40								
ULAQ	0.18	0.34	−0.30	0.48	−0.47	1.43	1.82	3.69	0.81	12.57
Mozart	0.05	0.39	−0.34	0.51						
ECHAM/Grantour	0.26	0.28	0.07	0.55	−0.57	1.40	5.2	5.27	2.07	10.55
TM3	0.27	0.47								
PNNL	−0.04	0.28	0.16	0.64	0.79	1.50	−2.48	2.64	−13.46	13.74
Average of all models	0.03	0.38	0.20	0.73	0.26	1.53	2.73	3.86	−0.74	12.09

[a] Aerosol sulphate, dust and sea salt were compared to observations at a selection of marine locations. The observations for organic carbon and black carbon were those compiled by Liousse *et al.* (1996) and Cooke *et al.* (1999).

[b] The average bias and the average absolute error is the average differences between each model result and the observations over all stations.

Table 5.9b: *Comparison of models and observations of aerosol species at selected surface locations (%)[a,b].*

Model	Sulphate		Black carbon		Organic carbon		Dust		Sea salt	
	Average bias (%)	Average absolute error (%)	Average bias (%)	Average absolute error (%)	Average bias (%)	Average absolute error (%)	Average bias (%)	Average absolute error (%)	Average bias (%)	Average absolute error (%)
GISS	26	31	85	127	91	121	121	121	37	40
GSFC	7	15	189	219	109	134	39	42	21	30
Hadley	−11	16	140	220						
CCM/Grantour	1	15	43	111	13	85	78	80	63	68
ECHAM	32	35	253	276	273	285				
Stochem	30	34								
ULAQ	10	17	−10	84	23	100	21	35	81	88
Mozart	28	31	164	211						
ECHAM/Grantour	31	31	204	230	88	135	70	70	29	33
TM3	43	46								
PNNL	17	21	75	133	189	220			−12	16
Average of all models	19	26	127	179	112	154	66	70	36	46

[a] Aerosol sulphate, dust and sea salt were compared to observations at a selection of marine locations. The observations for organic carbon and black carbon were those compiled by Liousse *et al.* (1996) and Cooke *et al.* (1999).

[b] The average bias and the average absolute error is the average percentage difference between each model result and the observations over all stations.

Based on this model comparison study, the ability of the global models to reproduce the aerosol mixing ratios at the surface can be described as acceptable for sulphate. However, improvement is needed for the other aerosol species. The large differences in predicted atmospheric aerosol burden which is the relevant parameter in determining the forcing, impede an accurate estimate of the aerosol climate effect. Improvement of the assessment of aerosol burden requires more measurements within the free troposphere.

5.4.1.4 Comparison of modelled and satellite-derived aerosol optical depth

Comparison of model results with remote surface observations of the major components making up the composition of the atmospheric aerosol provide a test of whether the models treat transport and removal of individual species adequately. But as noted above, the difference between the vertical distribution of species among the models is significant (i.e. a difference of more

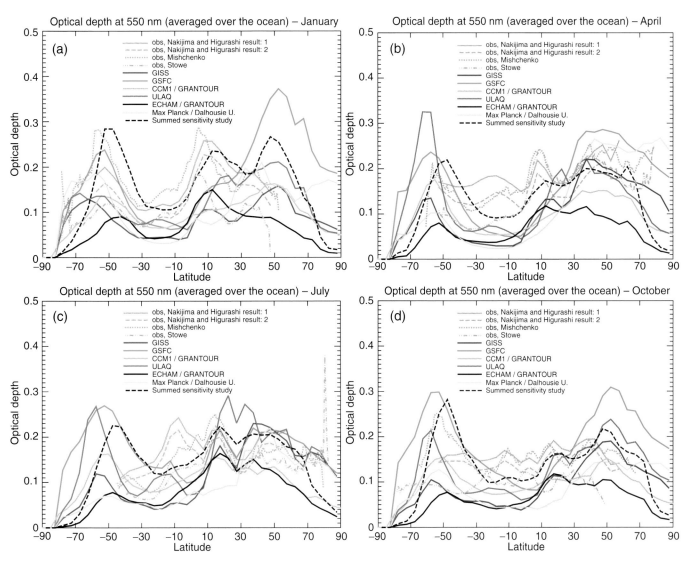

Figure 5.11: Aerosol optical depth derived from AVHRR satellite analysis following Nakajima *et al.* (1999) (labelled result: 1 and result: 2), Mishchenko *et al.* (1999) and Stowe *et al.* (1997) for January, April, July and October. The results from Nakajima refer to 1990, while those from Mishchenko and Stowe refer to an average over the years 1985 to 1988. The results derived from the models which participated in the IPCC-sponsored workshop are also shown (see Table 5.8). The case labelled "summed sensitivity study" shows the derived optical depth for the ECHAM/GRANTOUR model using a factor of two increase in the DMS flux and the monthly average sea salt fluxes derived using the SSM/I wind fields.

than a factor of 2 in the upper troposphere) and the global average abundance of individual components varies significantly (more than a factor of 2) between the models, especially for components such as dust and sea salt. Two methods have been used to try to understand whether the models adequately treat the total aerosol abundance. The first, comparison of total optical depth with satellite measurements, was employed by Tegen *et al.* (1997), while the second, comparison of total reflected short-wave radiation with satellite observations, was employed by Haywood *et al.* (1999). We examined both measures in an effort to understand whether the model-predicted forcing associated with aerosols is adequate.

Figure 5.11 shows the zonal average optical depth deduced from AVHRR data for 1990 by Nakajima and Higurashi (Nakajima *et al.*, 1999) and for an average of the time period

February 1985 to October 1988 by Mishchenko *et al.* (1999) and by Stowe *et al.* (1997). The two results from Nakajima *et al.* (1999) demonstrate the sensitivity of the retrieved optical depth to the assumed particle size distribution. Results from the models which participated in the intercomparison workshop are also included. Because the GISS, CCM1, ECHAM/GRANTOUR and ULAQ models all used the same sources, the differences between these models are due to model parametrization procedures. The GOCART (GSFC) model used a source distribution for sea salt that was derived from daily varying special sensor microwave imager (SSM/I; Atlas *et al.*, 1996) winds for 1990 and was, on average, 55% larger than the baseline sea salt source specified for the model workshop. The MPI/Dalhousie model used monthly average dust and sea salt distributions from prior CCM1 model simulations (cf., Lohmann *et al.*, 1999b,c).

Workshop participants were asked to report their derived optical depths. However, these varied widely and were often much smaller than that derived here. Therefore, we constructed the optical depths shown in Figure 5.11 from the frequency distribution of relative humidity from the T21 version of the ECHAM 3.6 general circulation model, together with the reported monthly average distributions of aerosol mixing ratios. The model optical depths were derived using extinction coefficients at 0.55 μm for dry sea salt of 3.45 m^2g^{-1}, 0.69 m^2g^{-1}, and 0.20 m^2g^{-1} for diameters in the size range from 0.2 to 2 μm, 2.0 to 8 μm and 8 to 20 μm, respectively, and an extinction coefficient of 9.94 m^2g^{-1} for sulphate. The humidity dependence of the extinction for sea salt and dust was determined using the model described in Penner *et al.* (1999a). The dust extinction coefficients were from Tegen *et al.* (1997) and organic and black carbon extinction coefficients for 80% relative humidity were from Haywood *et al.* (1999). We also examined the sensitivity of the modelled optical depths to a factor of two increase in the DMS flux and to the use of monthly average sea salt fluxes derived using the SSM/I wind fields. The OC and BC extinction coefficients were also varied, adding the humidity dependence determined by Penner *et al.* (1998b). Finally, the extinction coefficient for sulphate was altered to that calculated for an assumed ratio of total NH_3 and HNO_3 to H_2SO_4 of four. The line in the graph labelled "summed sensitivity study" shows the results for the ECHAM/GRANTOUR model when these parameters were varied. Most of the difference between the summed sensitivity case and that for the baseline ECHAM/GRANTOUR model is due to the use of larger DMS and sea salt fluxes.

The satellite-derived optical depths from Stowe *et al.* (1997) are lower on average by 0.05 and by 0.03 than those from Mischenko *et al.* (1999) and result 2 from Nakajima *et al.* (1999), respectively. The latter two retrievals make use of a two-wavelength technique which is thought to be more accurate than the one-wavelenth technique of Stowe *et al.* (1997). However, it is worth bearing in mind that most of the difference in retrieved aerosol optical depth may be related to cloud screening techniques (Mischenko *et al.*, 1999) or to assumed size distribution (Nakajima *et al.*, 1999).

Modelled optical depths north of 30°N are sometimes higher and sometimes lower than those of the retrieved AVHRR optical depths. For example, there is an average difference of 0.13 in July for the ULAQ model in comparison with result 2 for Nakajima *et al.* (1999) while the average difference is −0.09 in January for the ECHAM/GRANTOUR model in comparison with the retrieved optical depths from Mischenko *et al.* (1999). The modelled optical depths in the latitude band from 30°N and 40°N are systematically too high in July. For example, the average of the modelled optical depths is larger than the satellite-derived optical depth of Nakajima, Mischenko, and Stowe on average by 0.06, 0.05 and 0.04, respectively. We note that sulphate and dust provide the largest components of optical depth in this region with sea salt providing the third most important component. Since the sources represent the year 2000, while the measured optical depths refer to an average of the years 1985 to 1988, some of the overprediction of optical depth may be associated with higher sources than the time period of the measurements. The black dashed line shows the estimated optical depths from the

ECHAM/GRANTOUR model with the larger sea salt fluxes deduced from the SSM/I winds, with doubled DMS flux, and with optical properties for an assumed ratio of total NO_3 to H_2SO_4 of 4:1. Comparison of these results with those of the retrieved optical depths shows that the uncertainties in these parameters lead to changes in optical depth that are of the order of 0.05 or more.

Modelled aerosol optical depths near 10°N are dominated by dust with some contribution from organic carbon and sulphate (especially in January and April). They are systematically lower (by, on average, 0.08) than the average retrieved optical depth. The discrepancy between modelled and retrieved optical depths in this region, however, would be reduced if the sea salt fluxes derived from SSM/I winds and larger DMS fluxes had been used.

The modelled aerosol optical depths from 10°S to 30°S are due to a combination of different aerosol types. They are systematically lower than the average of the retrieved optical depths by an average of 0.06 with biases ranging from −0.14 to 0.01 in January, from −0.12 to −0.02 in April, from −0.13 to 0.07 in July and from −0.11 to 0.06 in October. As shown by the sensitivity study, much of the difference between the modelled and retrieved zonal average optical depths could be removed by using higher sea salt and DMS fluxes. However, the spatial character of the differences reveals that the cause of the discrepancies probably cannot be attributed to any single source. For example, Figure 5.12 shows the difference between the base case and sensitivity case for the ECHAM/GRANTOUR model and the optical depths retrieved by Mischenko *et al.* (1999). In January the differences are largest in the central Pacific Ocean. In April the differences appear largest in the Pacific Ocean and in the Indian Ocean west of Australia. In July and October the differences are mainly to the west and east of the African continent and also in the mid-Pacific Ocean and west of South America in October. We note that the large overprediction of optical depth in both the base case and the sensitivity case off the coast of Asia in July may be due to the difference in simulated year and the year of the optical depth retrievals. The model simulations used anthropogenic sulphate emissions appropriate to 2000 while the retrievals refer to an average of 1985 to 1988 data.

Modelled aerosol optical depths near 60°S are dominated by sea salt. This component appears to be reasonably well represented by the models, especially for the optical depths predicted using the SSM/I sea salt fluxes. However, if some of the other models had used the higher fluxes used in the GSFC model, the optical depths would be overpredicted in this region.

Haywood *et al.* (1999) found a large and significant difference between modelled and observed reflectivity at northern latitudes when higher sea salt fluxes based on the algorithm used here were used together with the SSM/I-derived winds. The use of this algorithm corrected a significant underprediction of reflected radiation in mid-ocean regions when the sea salt parametrization of Lovett (1978) was used. We used the aerosol burdens from the models in the IPCC intercomparison to scale the optical depths in the Haywood *et al.* study in order to compare the average reflected radiation from each of the models in the intercomparison with that from the ERBE satellite. For the GSFC model which also used winds derived from the SSM/I, results similar to those reported by Haywood *et al.* (1999) were found (i.e. an overpredicted flux at

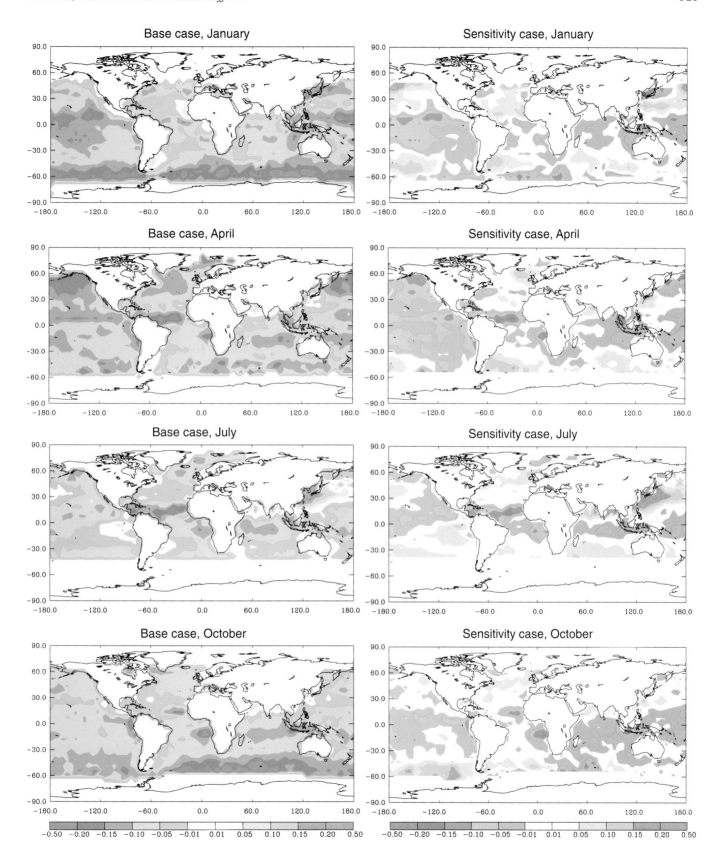

Figure 5.12: Difference between the ECHAM/GRANTOUR computation of optical depth and the satellite-retrieved optical depths from Mishchenko *et al.* (1999). The left column shows the optical depths derived for the standard set of sources, while the right panel shows the derived optical depths for the sensitivity study using a factor of two increase in the DMS flux and the monthly average sea salt fluxes derived using the SSM/I wind fields (see text). Note that the anthropogenic sulphate sources were for the year 2000 while the satellite analysis covers the time period 1985 to 1988. This may explain the systematic overestimate of optical depth off the coast of Asia in July.

northern latitudes, but no large biases elsewhere). For the other models, the reflected radiation in mid-ocean regions was under-predicted relative to ERBE, but the area of underprediction was not as large as the area of underprediction for the GFDL model with the low sea salt option. For these models, the predicted reflected radiation for the region north of 30°N was sometimes higher than that from the ERBE satellite, but this was not consistent across all models. In general, this analysis showed that the comparison of model results with the reflected radiation from ERBE is broadly consistent with that developed above for the comparison between modelled and AVHRR-retrieved aerosol optical depth.

In summary, analysis of the AVHRR comparisons indicates that significant uncertainties remain in both our ability to retrieve aerosol optical thickness from satellites and in our ability to model aerosol effects on the radiation budget. For example, there is a global average difference between the optical depth retrieved by Mishchenko *et al.* (1999) and that retrieved by Stowe *et al.* (1997) of 0.05 and a difference between Nakajima *et al.* (1999) and Stowe *et al.* (1997) of 0.03. The global average difference between the average optical depth from the models and the average optical depth from the satellite retrieval is of the same magnitude, namely −0.04. In the region 10°S to 30°S, the analysis indicates that modelled optical depths are consistently too low. Such a model underestimate might indicate that the retrieved optical depth from satellites is too high, for example, because the cloud screening algorithm is not adequate in this region. Alternatively, there may be a need for a larger source of aerosols in this region or for smaller modelled removal rates.

5.4.2 Overall Uncertainty in Direct Forcing Estimates

To illustrate how the uncertainty associated with each of the factors determining aerosol direct forcing contribute to the overall uncertainty in forcing, we adopt a simple box-model of the overall change in planetary albedo. This approach follows that presented in Penner *et al.* (1994a), but is updated in several respects. For this model, the change in global average planetary albedo associated with anthropogenic aerosols is described as (Chylek and Wong, 1995):

$$\Delta\alpha_p = [T_a^2(1-A_c)][2(1-R_s)^2\bar{\beta}f_b M\alpha_s f(RH) - 4R_s M\alpha_s f(RH)((1-\omega_0)/\omega_0)]$$

where,

T_a = atmospheric transmissivity above the main aerosol layer
A_c = global cloud fraction
R_s = global average surface albedo
$\bar{\beta}$ = upscatter fraction for isotropic incoming radiation
f_b = hygroscopic growth factor for upscatter fraction
M = global mean column burden for aerosol constituent, (gm^{-2})
α_s = aerosol mass scattering efficiency, (m^2g^{-1})
$f(RH)$ = hygroscopic growth factor for total particle scattering
ω_0 = single scattering albedo at ambient RH (assumed to be 80% for this analysis).

The key quantities that enter into the calculation of uncertainties are listed in Table 5.10, together with estimates of their 2/3 uncertainty range. Given these central values and

uncertainties, the variance (and thus uncertainty) associated with the planetary albedo change ($\Delta\alpha_p$) is determined by standard Taylor expansion of the function around the central values for the change in planetary albedo. Thus, with variances given by S_{xi}^2:

$$S_{\Delta\alpha p}^2 = \left[\frac{\partial\Delta\alpha_p}{\partial x_i}\right]^2 S_{xi}^2 + cov(x_i, x_j)\left[\frac{\partial\Delta\alpha_p}{\partial x_i}\right]\left[\frac{\partial\Delta\alpha_p}{\partial x_j}\right]$$

where the function $cov(x_i, x_j)$ is the covariance of the variables in the argument and a variable subscript (i.e., i or j) implicitly requires summation from 1 to n where n is the number of variables. Significant covariances are found between $\bar{\beta}$ and α_s, $\bar{\beta}$ and f_b, α_s and $f(RH)$, and ω_0 and $f(RH)$. For these variable pairs, Bravais-Pearson (linear) correlation coefficients were found to be −0.9, −0.9, 0.9 and 0.9, respectively. These were determined by sampling the probability distribution associated with each pair of variables to generate a large set of corresponding pairs of values. Linear regression analysis was then performed on these corresponding pairs to determine the linear correlation coefficient between the paired variables. The burden estimates in Table 5.10 are based on model calculations from the IPCC workshop. The uncertainty range was taken from the range in burdens from the models (assumed here to be a 2/3 uncertainty range) which may be a reasonable estimate for some of the "structural uncertainty" associated with different parametrization choices in the models (Pan *et al.*, 1997). It includes both "chemical" quantities such as the fraction of emitted SO_2 that is converted to sulphate aerosol and the mean residence time of the aerosol. The central emissions estimates are those specified in the model workshop (except where noted) and the uncertainties are those estimated in Section 5.2. The uncertainty range for the burden, M, used in evaluating the change in planetary albedo was calculated from the uncertainty in emissions together with the uncertainty in burden utilising the geometric concatenation procedure of Penner *et al.* (1994a).

The resulting overall uncertainty in the forcing by aerosols associated with fossil fuels and other industrial emissions is ± 0.42 Wm^{-2}; that is, if the global mean forcing is evaluated as −0.6 Wm^{-2} (the row value calculated by using the central values in the forcing equation), then the range of estimated global mean forcing is −0.1 to −1.0 Wm^{-2}. This range encompasses values for recent evaluations of the forcing for the individual components comprising the industrial aerosols (see Chapter 6). The main uncertainties are those for the upscatter fraction, the burden (which includes the propagated uncertainties in emissions), and the mass scattering efficiency, in that order. If the mode of mixing of black carbon were assumed to be external, instead of internal as used here, then the uncertainty range would be similar, but the central value for fossil fuel aerosols might be −0.7 Wm^{-2} instead of −0.6 Wm^{-2}.

The overall uncertainty in the forcing by biomass aerosols is ± 0.24 Wm^{-2}; thus, if the global mean forcing is −0.3 Wm^{-2}, then the 2/3 uncertainty range of estimated global mean forcing is −0.1 to −0.5 Wm^{-2}. The main uncertainties are the single scattering albedo, the upscatter fraction, and the burden.

Several assumptions are implicit in the above approach. First, there is a specific assumption concerning the determination of the single scattering albedo at 80% RH since most measure-

Table 5.10a: *Factors contributing to uncertainties in the estimates of the direct forcing by aerosols associated with fossil fuels and other industrial processes and their estimated range. Note that optical parameters are for a wavelength of 550 nm and are for dry aerosol.*

Quantity	Central Value	2/3 Uncertainty Range
Total emission of anthropogenic OC from fossil fuel burning (Tg/yr) [a]	20	10 to 30
Atmospheric burden of OC from fossil fuels (Tg) [b]	0.48	0.33 to 0.70
Total emission of anthropogenic BC from fossil fuel burning (Tg/yr)	7	4.67 to 10.5
Atmospheric burden of BC from fossil fuel burning (Tg) [b]	0.133	0.11 to 0.16
Total emission of anthropogenic sulfate from fossil fuel burning (Tg/yr)	69	57.5 to 82.8
Atmospheric burden of sulphate from fossil fuel burning (Tg S)	0.525	0.35 to 0.79
Fraction of light scattered into upward hemisphere, $\bar{\beta}$ [c]	0.23	0.17 to 0.29
Aerosol mass scattering efficiency (m^2g^{-1}), α_s [c]	3.5	2.3 to 4.7
Aerosol single scattering albedo (dry), ω_0 [c]	0.92	0.85 to 0.97
T_a, atmospheric transmittance above aerosol layer [d]	0.87	0.72 to 1.00
Fractional increase in aerosol scattering efficiency due to hygroscopic growth at RH=80%	2.0	1.7 to 2.3
Fraction of Earth not covered by cloud [d]	0.39	0.35 to 0.43
Mean surface albedo [d]	0.15	0.08 to 0.22
Result: If central value is −0.6 Wm^{-2} the 2/3 uncertainty range is from −0.1 to −1.0 Wm^{-2}		

[a] The central value estimated here was taken from Table 5.3, while the value used in intercomparison was 29.

[b] The burden was estimated from the model intercomparison for total anthropogenic carbon taking the fraction associated with fossil fuels from the fraction of emissions associated with fossil fuels.

[c] The central estimate and uncertainty range were calculated from the size distribution for polluted continental aerosols in Table 5.1.

[d] Central estimate and uncertainty range adapted from values used by Penner *et al.* (1994a).

Table 5.10b: *Factors contributing to uncertainties in the estimates of the direct forcing by biomass burning aerosols and their estimated range. Note that optical parameters are for a wavelength of 550 nm and are for dry aerosol.*

Quantity	Central Value	2/3 Uncertainty Range
Total emission of anthropogenic OC from biomass burning (Tg/yr)	62.5	45 to 80
Atmospheric burden of OC from smoke (Tg) [a]	1.04	0.75 to 1.51
Total emission of anthropogenic BC from biomass burning (Tg/yr)	7	5.0 to 9.8
Atmospheric burden of BC from biomass burning (Tg) [a]	0.133	0.11 to 0.16
Fraction of light scattered into upward hemisphere, $\bar{\beta}$ [b]	0.23	0.17 to 0.29
Aerosol mass scattering efficiency (m^2g^{-1}), α_s [b]	3.6	2.5 to 4.7
Aerosol single scattering albedo (dry), ω_0 [b]	0.89	0.83 to 0.95
T_a, atmospheric transmittance above aerosol layer [c]	0.87	0.72 to 1.00
Fractional increase in aerosol scattering efficiency due to hygroscopic growth at RH=80%	1.2	1.0 to 1.4
Fraction of Earth not covered by cloud [c]	0.39	0.35 to 0.43
Mean surface albedo [c]	0.15	0.08 to 0.22
Result: If central value is −0.3 Wm^{-2} the range is from −0.1 to −0.5 Wm^{-2}		

[a] The burden was estimated from the model intercomparison for total anthropogenic carbon taking the fraction associated with biomass aerosols from the fraction of emissions associated with biomass aerosols.

[b] The central estimate and uncertainty range were calculated from the size distribution for biomass regional haze in Table 5.1.

[c] Central estimate and uncertainty range adapted from values used by Penner *et al.* (1994a).

ments of this parameter are for low (dry) RH. Second, it was assumed that the specific aerosol absorption did not change with humidification while that of light scattering followed f(RH). While reasonable at this time, this issue needs further investigation. Another assumption implicit in the formula for planetary albedo change is that both absorption and scattering by aerosols are not significant compared to cloud effects when clouds are present. As Haywood and Ramaswamy (1998) have pointed out, this assumption is reasonable for scattering aerosols but not for strongly absorbing aerosols. The overall effect of including the effects of absorbing aerosols in the presence of clouds would be to reduce the net forcing relative to that calculated here. Turning to other issues, only second order terms were used in the Taylor expansion. While certainly valid for cases in which the uncertainties in the independent parameters do not exceed 20% of the central values, this neglect becomes somewhat problematic for larger uncertainties. Since there are several variables whose uncertainties are of the order of 50%, an assessment was made of possible underestimation of error by full functional evaluation, i.e., a population of values for the forcing function was generated from the population of the most uncertain variables (all other variables held constant) and the variance in the forcing function generated in this way compared with that derived from Taylor expansion. It was found that the error in variance propagation was second order (<10%) and has therefore been neglected here. Finally, there is an implicit assumption that all of the variables are normally distributed.

There are, of course, several sources of uncertainty that can not be evaluated using this approach. For example, uncertainties associated with the vertical distribution of the aerosol and with potential correlations between clouds and aerosol abundances are not evaluated. Furthermore, uncertainties associated with the radiative transfer treatment (which might be of order 20% (Boucher *et al.*, 1998)) are neglected. A further caveat is that the uncertainty estimates of Table 5.9 depend on the assumption that the data chosen for the size distribution are representative throughout those areas where these aerosol types contribute to forcing. Because we used observations from continental polluted regions for fossil fuel aerosols and from regional measurements for biomass aerosols, they may not encompass the full set of size distributions that may occur in regions outside of those where these measurements are valid.

5.4.3 Modelling the Indirect Effect of Aerosols on Global Climate Forcing

Published estimates of the indirect forcing by anthropogenic aerosols are summarised in Table 5.11. The results in Table 5.11a are for the assumption that changes in initial droplet concentrations do not lead to any changes in cloud properties. These results, therefore, did not include any feedbacks to liquid-water content or cloud amount as a result of changes to precipitation efficiency. The results in Table 5.11b include a calculation of the decrease in precipitation efficiency that results from an increase in N_d. Therefore, these results do include feedbacks to changes in liquid-water and cloud amount from changes in cloud microphysics.

The forcing calculations estimated from the models described by Boucher and Lohmann (1995), Jones and Slingo (1996), Feichter *et al.* (1997), Kiehl *et al.* (2000) and Rotstayn (1999) were performed as the difference between two simulations with fixed sea surface temperatures. These calculations, therefore, allow some feedback of the perturbation in the cloud albedo to both the dynamic fields and to water vapour and cloud amount. By definition, these GCM forcing estimates are therefore not strict radiative forcings, however, Rotstayn and Penner (2001) have shown that for the first indirect effect the inclusion of the feedback changes their forcing estimate by only about 10%. The other results are from chemical transport models/GCMs that do not include any feedbacks.

All models in Table 5.11a explicitly consider only sulphate, except those of Chuang *et al.* (1997, 2000) and Iversen *et al.* (2000). Five of the models in 5.11a use an empirical method to relate N_d to sulphate mass concentration, which implicitly includes some effect of other industrial aerosol components (e.g., OC), and their forcing estimates range from −0.4 to −1.8 Wm^{-2} with a median of −0.9 Wm^{-2}. Chuang *et al.* (2000) and Iversen *et al.* (2000) use mechanistic parametrizations to calculate N_d. Chuang *et al.* (2000) consider organic carbon, black carbon, sea salt and dust as well as sulphate. The estimate of indirect forcing from Chuang *et al.* (1997) is for sulphate only and that from Iversen *et al.* (2000) is for sulphate and BC. The range of the forcing estimates for industrial aerosols from the models using the mechanistic approaches for N_d (−0.82 to −1.36 Wm^{-2}) is similar to that of the empirical methods. The inclusion of forcing by biomass aerosols increases the total forcing to −1.85 Wm^{-2} in the model of Chuang *et al.* (2000).

The estimates of indirect forcing that include feedback to the liquid-water path and cloud amount from changes in cloud microphysics and precipitation efficiency range from −1.1 Wm^{-2} to −4.8 Wm^{-2} (Table 5.11b). Due to the nature of the calculation, these are not strict estimates of radiative forcing, but as noted above, Rotstayn and Penner (2001) have shown that the feedback effects are probably not significant so that these may be viewed as estimates of forcing. The largest absolute value, that obtained by Lohmann and Feichter (1997) using the Xu and Randall and the Beheng parametrization schemes, is considered less realistic because the comparison of the short- and long-wave cloud radiative fluxes with those from the Earth Radiation Budget Experiment (ERBE) was relatively poor. The next highest value (−3.2 Wm^{-2} from Ghan *et al.*, 2001b) was obtained with a model using a very simple cloud cover parametrization (full coverage if saturated and no coverage if unsaturated). Thus, based on the present modelling, a more likely range for the combination of the first and second indirect effects is −1.1 to −2.2 Wm^{-2}. However, this range only includes one model estimate for the effects of aerosols from biomass burning as well as industrial aerosols (i.e. −1.1 to −1.9 from Lohmann *et al.*, 2000) and the assessment of carbon aerosol forcing from this model (−0.9 Wm^{-2}) is less than that for carbon aerosols from the first indirect effect from the Chuang *et al.* (2000a) model in Table 5.11a (i.e. −1.51 Wm^{-2}). In view of this, the range of forcing estimates for the combination of first and second indirect effects is −1.1 to −3.7 Wm^{-2}. Clearly, the uncertainty is relatively large, and considerable effort will be required to improve indirect forcing estimates.

A few of the factors that may contribute to this uncertainty are:

• The pre-industrial sulphate used in these studies varies by a factor of just over two, and the anthropogenic indirect forcing is calculated from the difference between the pre-industrial and industrial scenarios. Considering the non-linearities in the parametrizations of N_d, the pre-industrial sulphate concentration can be a significant factor.

• The variance in the industrial sulphate concentrations is about 50%.

• The parametrizations or methods of relating droplet number concentrations to aerosol concentrations have been discussed in Section 5.3.4. Using Figure 5.7, we can estimate the range of uncertainty attached to simply the differences in the parametrizations shown in the figure. For stratiform cloud, assuming a mean sulphate concentration of 0.5 μg m^{-3} over the oceans and 2.0 μg m^{-3} over the land, the differences among the parametrizations in Figure 5.7 amount to about 30%. Neglecting that there is a serious question about the ability of the empirical methods to represent the global aerosol, this is a reasonable approximation based on current knowledge. As a couple of footnotes, Rotstayn (1999) compared the forcings using the Boucher and Lohmann (1995) parametrization with a similar parametrization from Roelofs *et al.* (1998), and found the result with the Roelofs *et al.* method was 25% higher. And the relatively close agreement of the Boucher and Lohmann (1995) method with the prognostic method (Figure 5.7) has also been seen in the simulated results from MIRAGE model (Ghan *et al.*, 2001a).

• Changes in the vertical distribution of the aerosol will affect the indirect forcing because of changes in its spatial relationship to cloud. Rotstayn (1999) found a 24% change in the radiative forcing depending on whether the aerosol concentration decreased exponentially or was uniform with height through the lower 10 km. Jones and Slingo (1996) distributed the aerosol mass uniformly, half in the lower 2 km, and half above, compared with Jones *et al.* (1994b) who confined half the aerosol mass to the lower 1.5 km. Jones and Slingo (1996) obtained an indirect forcing 15% higher than Jones *et al.* (1994b).

It is important to draw attention to some other factors concerning the comparison of the forcing estimates from the various models.

• Boucher and Lohmann (1995) found very little difference between the LMD (French) and ECHAM (German) models. Jones and Slingo (1996) found that the Hadley (UK Met Office) model produced a lower forcing in the Northern Hemisphere relative to the Boucher and Lohmann results, however, this may have been due to differences in the distribution of sulphate mass as well as the GCM.

• The effect of horizontal resolution of the GCM was considered by Ghan *et al.* (2001b). They found an increase in the indirect forcing of about 40% when the horizontal resolution was degraded from 2.8°×2.8° to 4.5°×7.5°.

• There have been no studies that have explicitly considered the effect of the cloud cover parametrization on the first indirect

effect alone. Lohmann and Feichter (1997) examined the effect of two different cloud cover parametrizations (Sundquist *et al.*, 1989 and Xu and Randall, 1996) on the combined first and second indirect effects, which gave a difference of more than a factor of 3 in the indirect forcing. Indeed, the effect of uncertainties in the GCM predictions of cloud cover (which are known to be inaccurate) is not known.

• The autoconversion parametrization is a critical step in the simulation of the second indirect effect. Lohmann and Feichter (1997) compared two such parametrizations and found a difference in the total indirect forcing of a factor of two. Ghan *et al.* (2001b) compared two other methods for autoconversion and found a difference of about 30%. Rotstayn (1999) performed a sensitivity experiment in which the droplet size threshold for autoconversion was increased. This resulted in a reduction of the second indirect effect, but there was a slightly compensating increase in the first indirect effect. This is an area of high sensitivity and uncertainty.

5.4.4 Model Validation of Indirect Effects

Validation of the simulations from global models is an essential component of estimating and reducing the uncertainties in the indirect forcing. Comparisons of observations and modelled concentrations of chemical species have been discussed in Section 5.4.1.3 while comparisons of modelled and satellite-derived aerosol optical depths were discussed in Section 5.4.1.4. Here, comparisons with observations of several other model products important for indirect forcing are examined. Unfortunately, there is only a very small set of observations of the physical, chemical, and radiative properties of clouds from *in situ* methods available. Thus, validations with these types of datasets are left to limited temporal and spatial scales and to comparing relationships among various quantities. Lohmann *et al.* (2001), for example, compared prognostic simulations with observations of the relationships between particulate sulphate and total particle mass, between particle number concentration and sulphate mass, and between N_d and sulphate mass. The relationships between sulphate mass and total particle number concentrations was larger than observations in the case of internal mixtures but was smaller than observations in the case of external mixtures. Ghan *et al.* (2001a) found that the results of their determination of N_d using their mechanistic parametrization were comparable to the results using the empirical parametrization of Boucher and Lohmann (1995). Such tests are important for large-scale model parametrizations because comparisons of absolute concentrations on the scale of the model grid size are difficult.

Satellites offer observations over large temporal and spatial scales; however, for the derived parameters of interest, they are much less accurate than *in situ* observations. Han *et al.* (1994) retrieved an r_{eff} for liquid-water clouds from ISCCP satellite data that showed a significant land/sea contrast. Smaller droplets were found over the continents, and there was a systematic difference between the two Hemispheres with larger droplets in the Southern Hemisphere clouds. The r_{eff} calculated by different models and the observations from Han *et al.* are shown in Table 5.12. Since the r_{eff} tends to increase with increasing height above

Table 5.11a: *Comparison of model-predicted indirect forcing without cloud amount and liquid-water path feedback.*

Model	Pre-industrial aerosol (Tg)	Industrial aerosol (Tg)	N_d parametrization	Cloud cover parametrization	Forcing (Wm^{-2})
B&L (1995)	0.34 Tg S	Sulphate: 0.44 Tg S	Various empirical results	Le Treut and Li (1991); Roeckner *et al.* (1991)	−0.5 to −1.4; −0.5 to −1.5
Jones and Slingo (1996)	0.16 Tg S	Sulphate: 0.3 Tg S	Jones *et al.* (1994b); Hegg (1994). B&L (1995)	Smith (1990)	−1.5; −0.5; −0.6
Chuang *et al.* (1997)	Sulphate: 0.25 Tg S, Carbon aerosols: 1.72 Tg	Sulphate: 0.30 Tg S	Chuang and Penner (1995)	NCAR-CCM1	−0.62 to −1.24 (internal mix); −1.64 (external mix)
Feichter *et al.* (1997)	Sulphate 0.3 Tg S	Sulphate: 0.38 Tg	B&L (1995)	Sundqvist *et al.* (1989)	−0.76
Lohmann and Feichter (1997)			B&L (1995)	Sundqvist *et al.* (1989)	−1.0
Chuang *et al.* (2000)	Sea salt: 0.79Tg, Dust: 4.93 Tg, Sulphate: 0.25 TgS, Organic matter: 0.12 Tg	Sulphate: 0.30 TgS, Organic matter: 1.4 Tg BC: 0.19 Tg	Chuang and Penner (1995)		−1.85 (all aerosols) −1.51 (all carbon aerosols) −1.16 (biomass aerosols only); − 0.30 (sulphate only)
Kiehl *et al.* (2000)			Martin *et al.* (1994); Martin *et al.* (1994) with N_d minima; Jones *et al.* (1994b); B&L (1995)	Rasch and Kristjansson (1998)	−0.68; −0.40; −0.80; −1.78
Rotstayn (1999)	Sulphate: 0.21 TgS	Sulphate: 0.30 TgS	B&L (1995); Roelofs *et al.* (1998)	Smith, (1990)	−1.2; −1.7
Iversen *et al.* (2000)	Sulphate: 0.14 TgS, BC: 0.01 TgC, Sea salt and dust included, but not quantified	Sulphate: 0.60 TgS, BC: 0.25 TgC	Similar to Chuang and Penner (1995)	Rasch and Kristjansson (1998)	−1.36

cloud base and the satellite observations of r_{eff} are weighted for cloud top, the satellite observations will tend to overestimate the overall r_{eff} compared with that determined from *in situ* studies. *In situ* data sets against which to make absolute comparisons are few in number (e.g. Boers and Kummel, 1998). However, for now model evaluations are better done using the contrasts in r_{eff} between the land and ocean and between the Southern Hemisphere and the Northern Hemisphere. Most of the models listed in Table 5.12, with the exception of Roelofs *et al.* (1998), approximate the difference between r_{eff} over the Southern Hemisphere ocean vs the Northern Hemisphere ocean. Over land, the Southern Hemisphere *vs* Northern Hemisphere difference from Roelofs *et al.* (1998) is closest to the observed difference. For r_{eff} over Southern Hemisphere land vs Southern Hemisphere ocean, several models are relatively close to the difference from the observations (Boucher and Lohmann, 1995; Chuang *et al.*, 1997; Roelofs *et al.*, 1998; Lohmann *et al.*, 1999b,c). Some models compare with observations better than others, but there is no model that is able to reproduce all the observed differences. The r_{eff} calculated with the

same parametrization but using different GCM meteorologies are quite different (compare Jones and Slingo (1996) vs Boucher and Lohmann (1995)). As noted above, Jones and Slingo determined the "cloud top" r_{eff} by assuming a LWC profile that increased with height from cloud base to cloud top. Such a profile is more similar to observed profiles and might be expected to produce a better comparison with the observations. While the Jones and Slingo (1996) model does reasonably well in terms of hemispheric contrasts, their results indicate a land-ocean contrast in the opposite direction to that from the other models and the observations. We note that many factors may affect the results of this type of comparison. For example, Roelofs *et al.* (1998) estimate the sensitivity of the r_{eff} calculations to uncertainties in the sulphate concentration field, cloud cover and cloud liquid-water content to be of the order of a few micrometres. Moreover, the satellite determination of r_{eff} is probably no more accurate than a few micrometres (Han *et al.*, 1994).

Column mean N_d has been retrieved from satellite observations (Han *et al.*, 1998a), and this offers another validation

Table 5.11b: Comparison of indirect forcing by different models with cloud amount and liquid-water path feedback.

Model	Pre-industrial aerosol (Tg)	Industrial aerosol (Tg)	Nucleation parametrization	Cloud cover; autoconversion parametrization	Forcing (Wm^{-2})
Lohmann and Feichter (1997)	Sulphate: 0.3 TgS (interactive)	Sulphate: 0.38 Tg S (interactive)	Boucher and Lohmann (1995)	Sundqvist et al. (1989); Beheng (1994)	−1.4
Lohmann and Feichter (1997)	Sulphate: 0.3 TgS (interactive)	Sulphate: 0.38 Tg S (interactive)	Boucher and Lohmann (1995)	Xu and Randall (1996); Berry (1967)	−2.2
Lohmann and Feichter (1997)	Sulphate: 0.3 TgS (interactive)	Sulphate: 0.38 Tg S (interactive)	Boucher and Lohmann (1995)	Xu and Randall (1996); Beheng (1994)	−4.8
Lohmann et al., (2000a)	Monthly average Sea salt: 0.79 Tg Monthly average Dust: 5.23 Tg Interactive Organic matter: 0.12 Tg Interactive sulfate	Total interactive sulphate: 1.04 Tg Interactive organic matter:1.69 Tg (tot) Interactive black carbon: 0.24 Tg	Chuang and Penner (1995)	Sundqvist et al. (1989); Beheng (1994)	Total forcing: −1.1 to −1.9 Carbon only: −0.9 Sulfate only: −0.4
Rotstayn (1999)	Monthly average Sulphate: 0.21 Tg S	Monthly average sulphate: 0.30 Tg S	Boucher and Lohmann (1995); Roelofs et al. (1998)	Cloud cover; Smith (1990)	−2.1; −3.2
Iversen et al., (2000)	Sulphatee: 0.14 TgS, BC: 0.01 TgC, Sea salt and dust included, but not quantified	Sulphate: 0.60 TgS, BC: 0.25 TgC	Similar to Chuang and Penner (1995)	Rasch and Kristjansson (1998)	−1.9
Ghan et al., (2001b)	Sulphate: 0.42 TgS, OC: 1.35 Tg, BC: 0.22 Tg, Sea salt: 4.3 Tg, Dust: 4.3 Tg	Sulphate: 0.68 TgS	Abdul-Razzak and Ghan (2000)	0 if unsaturated, 1 if saturated; Ziegler (1985)	−1.7 Sulphate; ($2.8°×2.8°$)
Ghan et al., (2001b)	Sulphate: 0.39 TgS, OC: 1.14 Tg, BC: 0.18 Tg, Sea salt: 3.8 Tg, Dust: 4.6 Tg	Sulphate: 0.58 TgS	Abdul-Razzak and Ghan (2000)	0 or 1; Ziegler (1985)	−2.4 ($4.5°×7.5°$)
Ghan et al., (2001b)	As above	As above	Abdul-Razzak and Ghan (2000)	0 or 1; modified Tripoli and Cotton (1980)	−3.2 ($4.5°×7.5°$)

parameter over the global scale, but the derivation of r_{eff} is implicit in the determination of the column N_d. Because of the inverse relationship between r_{eff} and N_d for a given optical depth, the column N_d will be lower than *in situ* observations, since r_{eff} is higher than *in situ* observations (Han *et al.*, 1998a). Lohmann *et al.* (1999b,c) were able to simulate the general pattern of column N_d found by Han *et al.*, but their absolute values were higher.

As discussed in Section 5.3.2, Han *et al.* (1998b) found an increase in cloud albedo with decreasing droplet size for all optically thick clouds ($\tau_c > 15$) and optically thin clouds ($\tau_c < 15$) over land. Such an observation is consistent with expectations associated with the first indirect effect. However, for optically thinner clouds over marine locations, they found that the albedo decreased with decreasing droplet size, which might seem to be in conflict with an indirect effect. Using the ECHAM GCM, Lohmann *et al.* (1999b,c), were able to simulate the observed pattern for optically thick clouds. In addition, optically thin clouds over the oceans conformed to the observations. The ability of this model, which predicts an indirect effect (Table 5.11b; Lohmann *et al.*, 2000), to reproduce this satellite observation shows that this observation does not necessarily negate the indirect effect. The reasons for optically thin clouds to display a reduction in albedo with decreasing r_{eff} have not been elucidated (LWP was found to increase with increasing r_{eff} for all water clouds), and this is a reminder that interpretations must be considered carefully.

Table 5.12: *Cloud droplet effective radius of warm clouds (in μm).*

All results for 45°S to 45°N	Ocean Southern Hemisphere	Ocean Northern Hemisphere	Land Southern Hemisphere	Land Northern Hemisphere	Total
Han *et al.* (1994)	11.9	11.1	9.0	7.4	10.7
Boucher and Lohmann (1995)	8.9 to 10.1	8.3 to 9.3	5.4 to 8.7	4.9 to 8.0	
Jones and Slingo (1996)	9.6 to 10.8	9.0 to 10.4	10.2 to 11.8	9.9 to 10.8	9.5 to 11.1
Roelofs *et al.* (1998)	12.2	10.3	8.8	6.9	10.4
Chuang *et al.* (1997)	11.6 to 12.0	10.7 to 11.4	8.8 to 9.1	8.6 to 9.0	10.7 to 11.2
Lohmann *et al.* (1999b,c)	10.7	10.2	8.3	4.9	
Rotstayn (1999)	11.2	10.9	9.8	9.5	10.7
Ghan *et al.* (2001a,b)					11.0 to 11.7

Nakajima *et al.* (2001) examined the variation of cloud albedo, droplet concentration, and effective radius with column aerosol concentration. They found that the cloud droplet column concentration increased while r_{eff} decreased over a range of values of column aerosol number concentration. They did not find a significant increase in liquid-water path as aerosol number concentration increased. This observation may indicate that the 2nd indirect effect is not important on a global scale, however, further work is needed to confirm this.

5.4.5 Assessment of the Uncertainty in Indirect Forcing of the First Kind

Estimation of the uncertainty in the complete (i.e. the first and second indirect effects) radiative forcing is not currently feasible due to a lack of analytical relationships to treat the indirect forcing of the second kind. However, the indirect forcing of the first kind can be treated if we adopt a simple box-model approach such as those used in early assessments of indirect forcing (e.g., Charlson *et al.*, 1992; Schwartz and Slingo, 1996). This evaluation of uncertainty is only illustrative both because of the box-model nature of the estimate and because our assessment of the uncertainty in the parameters is only first order. Nevertheless, it is useful to make such calculations since they can yield valuable information both on our current state of knowledge with regard to the indirect forcing and can help guide efforts to reduce uncertainty. Moreover they illustrate a rigorous method that could allow a more quantitative estimate of uncertainty than the methods followed in Chapter 6.

We adopt a functional relationship between sulphate concentrations and cloud droplet number concentration based on empirical relationships in order to render the calculations more tractable. Therefore, this analysis is only applicable to the Northern Hemisphere, since data from this region were used to derive the empirical relationship. The analysis is further restricted to the marine atmosphere and excludes any consideration of indirect forcing by biomass aerosols.

The indirect forcing of the first kind (hereafter called simply the indirect forcing) can be expressed as:

$$\Delta F = F_d T_a^2 f_c \Delta A_p \qquad (1)$$

where F_d is the average downward flux at the top of the atmosphere, T_a is the atmospheric transmission above the cloud layer, f_c is the fractional cloud cover of those clouds susceptible to aerosol modulation and ΔA_p is the change in planetary albedo (equivalent here to the above cloud albedo) associated with an increase in the cloud droplet number concentration (CDNC). To take into account multiple reflections between the cloud layer and the surface, the expression of Liou (1980) is used with an assumption of no absorption within the cloud. This assumption is very reasonable for the bulk of the incoming radiation which will be scattered by cloud drops, i.e., subject to the indirect effect, but does restrict the radiation band to a range from 0.3 to 0.7 μm. Thus:

$$A_p = A_c + R_s \left[(1 - A_c)^2 / (1 - R_s A_c) \right] \qquad (2)$$

where A_c is the cloud albedo and R_s is the albedo of the underlying surface. ΔA_p is then calculated as the difference between this function evaluated for A_c, the background cloud albedo, and A_c', the anthropogenically perturbed albedo (note that "primed" quantities will always refer to anthropogenically perturbed values of the quantity). Cloud albedo is, in turn, evaluated using the relationship:

$$A_c = \tau_c / (\tau_c + 7.7) \qquad (3)$$

which is an approximation of the two stream evaluation of cloud albedo for conservative scattering assuming an asymmetry parameter of 0.85 (Lacis and Hansen, 1974). Here, τ_c is the cloud optical depth, given by the expression of Twomey (1977):

$$\tau_c = h \left(9 \pi LWC^2 N_d / 2\rho^2 \right)^{1/3} \qquad (4)$$

where h is the cloud layer thickness, LWC is the layer mean liquid-water content, N_d is the CDNC, and ρ is the density of water. To relate the CDNC to anthropogenic emissions, we use the empirical expression of Boucher and Lohmann (1995) which has the form:

$$N_d = A(SO_4^{2-})^B \qquad (5)$$

where SO_4^{2-} is the mean concentration of sulphate aerosol at cloud base in μg m^{-3}, and A and B are empirical constants. We adopt the values of A=115 and B=0.48 which are appropriate for marine air (Boucher and Lohmann, 1995).

Using the same procedures as with the assessment of uncertainties in the direct forcing (e.g. Section 5.4.2), we first determine the uncertainties in the most fundamental parameters. We then use Taylor expansions of the various equations given above to determine the uncertainty in the forcing associated with the central values of the parameters used in the calculations. Thus the uncertainty involves the uncertainties in the concentration of SO_4^{2-} and in the empirical coefficients used to relate the concentration of SO_4^{2-} to N_d. Moreover, the calculation of uncertainty necessarily involves an evaluation of A_c for both the background and anthropogenically perturbed values. These quantities, together with corresponding values for LWC and h, which are based on available observations, are then used to generate uncertainties in the cloud optical depth. This hierarchical evaluation proceeds "upward" from the uncertainty in the primary variables until the uncertainty in the forcing itself, together with the associated central value can be assessed. Such an uncertainty estimate is different in philosophy from that used in Chapter 6 which only assesses the range of estimates in the literature.

Several issues arise that deserve some discussion. The first such issue is the concatenation of uncertainties in the empirical relationship between the concentration of SO_4^{2-} and N_d. The uncertainty in SO_4^{2-} is straightforward and is based on the assessment of uncertainty in burden and the uncertainty in emissions as used in Section 5.4.2. The uncertainty in the relationship between SO_4^{2-} and N_d, equation (3), requires an evaluation of the uncertainties in the coefficients A and B. Based on a comparison of the parametrization of Boucher and Lohmann (1995) with that of Jones *et al.* (1994b) (Figure 5.7), we assign an *ad hoc* contribution of the functional relationship to the uncertainty in N_d of 40% of the central value. We then further assume that the uncertainties in A and B contribute equally to the total uncertainty (i.e., 40%) and, with these constraints, derived the uncertainties in A and B. This allows us to use Taylor expansions for both the perturbed and unperturbed atmosphere. Due largely to the form of the functional relationship, this procedure yields uncertainties whose major components in both the perturbed and unperturbed

cases are attributable to the uncertainty in the parametrization (61% for the unperturbed case and 64% for the perturbed case) rather than the uncertainty in burden or emissions.

In order to assess the overall uncertainty in forcing, the covariance between the base parameters N_d, LWC and h must be evaluated. But both LWC and h are assumed to be constant in the first indirect effect, and therefore the covariance is assumed to be zero. Certainly, the limited observations available (cf., Hegg *et al.*, 1996b) do not show any covariance. Nevertheless, it is important to note this potential effect and the possible impact of the second indirect effect (precipitation modulation effects by aerosols) on the uncertainty in the first.

Another covariance which cannot be neglected arises in the evaluation of the uncertainty in ΔA_p from the possible covariance between A_c and A_c'. This covariance arises because of the dependence of the perturbation in cloud albedo on both the unperturbed aerosol concentration and the unperturbed albedo. We assessed this covariance by using equations (4) and (3) to generate a set of corresponding values of the perturbed and unperturbed albedos for different values of N_d. Then a linear fit to the parameters ΔN_d and A_c' was used to derive the correlation coefficient and then the covariance. Various sample sizes and incremental values of the aerosol perturbation were generated to test the stability of the correlation. This procedure resulted in a stable correlation coefficient between A_c and A_c' of 0.74.

A final issue arises as to the choice of susceptible cloud fraction (f_c) in the basic forcing equation. Here, we follow the analysis of Charlson *et al.* (1987) and use the estimates in the cloud atlas of Warren *et al.* (1988). The total fractional cover of low and mid level clouds is not used since mid-level clouds can be mixed phase and the relationship of these clouds to anthropogenic aerosol is still unclear (Section 5.3.6). Instead, we use the estimates of Charlson *et al.* (1987) for non-overlapped low marine cloud as a lower bound (2/3 bound) for the susceptible cloud fraction and the sum (correcting for overlap) of the low and

Table 5.13: *Factors contributing to uncertainties in the estimates of the first indirect forcing over Northern Hemisphere marine locations by aerosols associated with fossil fuels and other industrial processes and their estimated range.*

Quantity	Central Value	2/3 Uncertainty Range
Background N_d for Northern Hemisphere marine locations (cm^{-3})	140	66 to 214
Perturbed N_d for Northern Hemisphere marine location (cm^{-3})	217	124 to 310
Cloud mean liquid-water content (LWC) ($g\ m^{-3}$)	0.225	0.125 to 0.325
Background sulphate concentration ($\mu g\ m^{-3}$)[a]	1.5	0.85 to 2.15
Cloud layer thickness (m)	200	100 to 300
Perturbed sulphate concentration ($\mu g\ m^{-3}$)[b]	3.6	2.4 to 4.8
Susceptible cloud fraction, f_c	0.24	0.19 to 0.29
Atmospheric transmission above cloud layer, T_a	0.92	0.78 to 1.00
Mean surface albedo	0.06	0.03 to 0.09
Result: If central value is $-1.4\ Wm^{-2}$ the 2/3 uncertainty range is from 0 to $-2.8\ Wm^{-2}$		

[a] Calculated from a central value for the Northern Hemisphere marine burden of 0.12 TgS with an uncertainty range from 0.07 to 0.17 TgS. The central estimate for sulphur emissions in the Northern Hemisphere was 12 TgS/yr with an uncertainty range from 9 to 17 TgS/yr.

[b] Calculated from a central value for the Northern Hemisphere marine perturbed burden of 0.28 TgS with an uncertainty range from 0.15 to 0.41 TgS. The central value for the Northern Hemisphere emissions of natural and anthropogenic sulphur was 74 TgS/yr with a 2/3 uncertainty range of 57 to 91 Tg S/yr.

middle marine stratiform cloud as an upper limit for f_c. The central value is taken as midway between these extremes.

The above assumptions, together with the parameter values given in Table 5.13, yield a central value for the indirect forcing over marine areas of $-1.4\,\mathrm{Wm^{-2}}$ together with an uncertainty of $\pm1.4\,\mathrm{Wm^{-2}}$. Hence, the forcing lies in the range between zero and $-2.8\,\mathrm{Wm^{-2}}$. This range is in reasonable agreement with that given in Chapter 6, based on GCM assessments. Nearly all of the uncertainty in the forcing is associated with that in the planetary albedo which, in turn, is dominated by the uncertainties in the perturbed and unperturbed cloud albedos, and thus in the cloud optical depths. However, it is interesting to note that most of the uncertainty in the optical depths for both perturbed and unperturbed clouds arises from the uncertainties in the cloud LWC and cloud thickness, and not from the uncertainties in the CDNC. This is not actually particularly surprising and is simply due to the stronger functional dependence of the optical depth on the LWC and thickness, h. This allocation of uncertainty may be slightly misleading because the 1st indirect effect depends on an assumption of fixed h and LWC. Yet here we are concerned with evaluating both the central estimate as well as its uncertainty. Hence, knowledge of h and LWC are needed. Indeed, consideration of these parameters and the accuracy of their representation in GCMs must certainly contribute to any uncertainty in the estimates of forcing from GCMs. Thus, it seems clear that progress in reducing the uncertainty in the indirect forcing of the first kind will be at least as dependent on acquiring better data on cloud LWC and thickness as on better quantifying anthropogenic aerosol concentrations and their effects on N_d. At a somewhat lower priority, it is clearly quite important to better understand the relationship between cloud drop number concentration and sulphate concentration. The total uncertainty is clearly dependent on this relationship. If it were significantly different in form, a somewhat different order of contribution to total uncertainty might have arisen. On the other hand, the uncertainty in the magnitude of the emissions of sulphur gases as well as the uncertainty in the burden of sulphate played only a minor role in the determination of the overall uncertainty.

5.5 Aerosol Effects in Future Scenarios

5.5.1 Introduction

Aerosol concentrations and forcing will change in the future, both as a result of changing emissions and as a result of changing climate. The uncertainties associated with our knowledge of the present day distribution of aerosols noted in previous sections will carry over into uncertainties in analysis of future scenarios. Nonetheless, models are the best available tool for making an assessment of what changes might follow. To estimate these future changes, we specified a set of emissions for the IPCC model intercomparison workshop (Section 5.4.1) based on the draft scenarios developed for the IPCC Special Report on Emissions Scenarios (SRES) (Nakićenović *et al.*, 2000). The results from the workshop form the basis of the future aerosol forcing reported in Chapter 6 and contribute to the climate change scenarios reported in Chapter 9.

Separate estimates for the amount of biomass-burning activity were not available for the SRES scenarios (though growth in biomass burning was included in the SRES analysis). Also, estimates of emissions of organic carbon and black carbon aerosols from fossil fuels and industrial activity were not available. Therefore, these were constructed using the ratio of source strengths for CO in 2030 and 2100 to that in 2000, respectively. This ratio was then used to scale the emissions for organic carbon and black carbon from fossil fuel and biomass burning. Because the scenarios do not provide a breakdown of emissions for CO by source category, this scaling implicitly assumes that as a given region develops, the ratio of emissions of CO by biomass burning and by fossil fuel burning remains roughly constant. We note that our projected carbon particle emissions may be too large if countries choose to target particle emissions for reduction to a greater extent than they target emissions of CO.

In addition to the emissions of carbon particles, we constructed emissions for NH_3 in order to examine possible changes in the emissions of NH_3 and HNO_3 to aerosol abundance and forcing. Only the A2 scenario in 2100 was considered. For this simulation, the growth in anthropogenic NH_3 emissions was assumed to follow the growth in anthropogenic N_2O emissions. Anthropogenic emissions grew from 46.9 TgN/yr in 2000 to 111.5 TgN/yr in 2100. The anthropogenic NO_x emissions were 39.5 TgN/yr in 2000 and grew to 109.7 TgN/yr in 2100.

The emissions for 2030 and 2100 from the draft A2 and B1 scenarios that were considered in the IPCC workshop are shown in Table 5.7 (see the Appendix to this volume for the final SRES emissions). As noted there, SO_2 emissions in 2030 are about a factor of 1.6 higher than those in 2000 in the draft A2 scenario, but decrease thereafter to global average levels that are less than the present-day estimates in 2100. Carbon aerosol emissions grow by a factor of 1.3 in 2030 in the A2 scenario and continue to grow to 2100 by an additional factor of 1.8. In the B1 scenario, both SO_2 and carbon aerosol emissions are controlled by 2100, falling by 60 and 3%, respectively, compared to 2000. For comparison, the table also shows emissions for the IS92a scenario, with carbon aerosol emissions constructed as above for the SRES scenarios. In the IS92a scenario, growth continues throughout the time period for both SO_2 and carbon emissions: SO_2 and CO emissions are a factor of 1.8 and 1.6 larger in 2100, respectively, than the same emissions in 2000.

5.5.2 Climate Change and Natural Aerosol Emissions

Aerosols that originate from natural emissions may also be expected to change in future scenarios. For example, terpene emissions depend on temperature, precipitation and light levels, and sea salt emissions depend on wind speed and temperature. DMS emissions depend on wind speed and temperature, and dust emissions depend on soil moisture and wind speed. Changes to natural emissions associated with changes to these factors were considered in scenarios SC5 and SC8.

To construct future changes in natural emissions, the NCAR CSM simulations (Dai *et al.*, 2001) were used to estimate possible changes in climate. This climate simulation was one formulated to treat a "business as usual" (IS92a) scenario and resulted in a global average surface temperature change for the decade prior to 2100 relative to that for the decade prior to 2000 of 1.76°K. While methods are available to estimate the effect of changes in climate on natural emissions, given current vegetation cover, tools to define the impact of future changes in land use and land cover on spatially disaggregated emissions are not well developed. Thus, even though changes in land use would be expected to affect vegetation cover and this is responsible for emissions of terpenes and determines which areas are subject to dust uplift, these effects could not be included. Furthermore, the understanding of how phytoplankton populations that produce DMS may change with changes in climate is poor. Therefore, in the following, only the effects of changes in temperature, precipitation, light levels, and wind speeds on future emissions were considered. The changes in wind speed were estimated from the ratio of monthly average wind speeds for the years 2090 to 2100 to that for the years 1990 to 2000 associated with the IS92a scenario (Dai *et al.*, 2001). This ratio was used to adjust the wind speeds that were used to generate current (2000) emissions to obtain future (2100) emissions of both DMS and dust. This method gives only a first-order estimate of possible changes, since it assumes that the distribution of wind speeds within a month remains constant with time. For wind speeds associated with sea salt, a somewhat different method was used (see below). The terpene fluxes were estimated directly from the daily data that were projected by the CSM simulation for 2100.

5.5.2.1 Projection of DMS emissions in 2100
Emissions of DMS were projected using the procedures described in Kettle *et al.* (1999). Since the atmospheric concentration of DMS is negligible in relation to what it would be if the atmospheric and oceanic concentrations were in equilibrium, the DMS flux is assumed to depend only on the sea surface concentrations and the wind speed. The projected DMS flux (as well as that for 2000) used the average of the parametrizations of Liss and Merlivat (1986) and Wanninkhof (1992). A correction of the piston velocity for sea surface temperature was made using the Schmidt number dependence of Saltzman *et al.* (1993).

The DMS flux for 2000 was calculated from the monthly average sea surface temperature data of Levitus and Boyer (1994) and the sea surface DMS concentration data of Kettle *et al.* (1999). In addition, climatological wind speeds from Trenberth *et al.* (1989) and climatological sea ice cover fields from Chapman and Walsh (1993) were used. The global flux for the year 2100 was calculated as the product of this initial field and the ratio of the piston velocity field in 2100 to that in 2000 using the sea surface temperature and wind speed information from the NCAR CSM.

There are several possible sources of error in these calculations. The most serious assumption is that the DMS concentration fields do not change between the years 2000 to 2100. DMS is produced as part of phytoplankton bloom cycles, especially in high latitude areas. It is likely that the mean distribution of phytoplankton blooms in the upper ocean would change between 2000 and 2100 given any perturbation of the sea surface temperature, wind speed, and sunlight. The other major assumption is that the monthly climatological ice cover does not change

between 2000 and 2100. Ice acts as a lid on the ocean in upper latitudes through which DMS cannot pass.

Overall, the calculations suggest a small increase in global DMS flux between the year 2000 (with a global DMS flux of 26.0 TgS/yr) and the year 2100 (with a global DMS flux of 27.7 TgS/yr). The most noticeable features in the 2100 fields are the predicted increases in DMS fluxes in some areas of the North Atlantic, North Pacific, and some areas of the Southern Ocean immediately adjacent to the Antarctic continent. There are some localised increases predicted in the tropical and sub-tropical Pacific Ocean.

5.5.2.2 Projection of VOC emissions in 2100
Emissions of isoprene, monoterpenes, and other VOC were calculated using the GLOBEIS model (Guenther *et al.*, 1999) which estimates biogenic VOC emissions as a function of foliar density, an emission capacity (the emission at specified environmental conditions), and an emission activity factor that accounts for variations due to environmental conditions. The foliar densities, emission capacities, and algorithms used to determine emissions activities for both 2000 and 2100 are the same as those described by Guenther *et al.* (1995). One difference between this work and that of Guenther *et al.* (1995) is that here we used hourly temperatures for each hour of a month to determine the monthly average emission rate whereas Guenther *et al.* (1995) used monthly average temperatures to drive emission algorithms. This results in about a 20% increase in isoprene emissions in 2000 and 10% increase in emissions of other biogenic VOC.

Global annual emissions of monoterpenes for 2000 were 146 Tg (compared to the estimate of 127 Tg in Guenther *et al.* (1995) which was used for the workshop 2000 emissions). These increase by 23% for 2100 relative to the 2000 scenario. The changes are much higher at certain seasons and locations. The spatial distribution of changes in total monoterpene emissions range from a 17% decrease to a 200% increase.

The simple model used for this analysis does not consider a number of other factors that could significantly influence long-term trends in biogenic VOC emissions. For example, we did not consider changes in soil moisture which could significantly impact emission rates. In addition, we did not consider changes in future concentrations of OH, O_3, and NO_3 in determining the yield of aerosol products. Instead, a constant yield of 11% of the terpene emissions was assumed for all future emissions.

5.5.2.3 Projection of dust emissions in 2100
The meteorological variables used to compute the dust emission for 2000 corresponded to those computed for daily meteorological data from the Data Assimilation Office analysis for 1990 (the GEOS-1 DAS, see Schubert *et al.*, 1993) using algorithms outlined by P. Ginoux. For this computation, the total dust flux was scaled to yield a total emission of about 2000 Tg/yr. Emissions in four size categories were specified (diameter 0.2 to 12.0 μm).

In order to calculate dust flux in 2100, the monthly mean variables were computed by averaging the wind speed at the lowest level (100 m) and the soil moisture content over the ten years ending in 1999 and 2099. The fluxes for the 21st century were calculated using the same meteorological variables, corresponding to 1990, scaled by the monthly mean ratio of the equivalent variables computed by the NCAR CSM model.

Overall, predicted dust emissions increase by approximately 10%. There are substantial increases in some seasons and locations (e.g. increases in Australia (85%) in winter and in Europe (86%) and East Asia (42%) in summer). Because dust may be especially important as an ice-nucleating agent in clouds, these possible changes add substantial uncertainty to the projected future indirect forcing.

5.5.2.4 Projection of sea salt emissions in 2100

The production of sea salt aerosol is also a strong function of wind speed. The semi-empirical formulation of Monahan *et al.* (1986) was used to produce global monthly sea salt fluxes for eight size intervals (dry diameter of 0.06 to 16 μm) using procedures discussed in Gong *et al.* (1997a,b). In order to project sea salt emissions for the workshop, the ratio of daily average wind speed in the ten years prior to 2100 and 2000 was used to scale the 2000 daily average sea salt flux to the one in 2100. Because this method may overestimate emissions if the product of the daily average ratio with the daily average winds in 2000 produces high wind speeds with a high frequency, the calculation was checked using the ratio of the monthly average wind speeds. This produced a total sea salt flux that was 13% smaller than the projections given in the workshop specifications.

Predicted sea salt emissions were 3,340 Tg in 2000 and increased to 5,880 Tg in 2100. These increases point to a potentially important negative climate feedback. For example, the present day direct radiative impact of sea salt is estimated to be between -0.75 and -2.5 Wm^{-2} using the clear-sky estimates from Haywood *et al.* (1999) or -0.34 Wm^{-2} using the whole-sky estimates from Jacobson (2001). Assuming the ratio of whole-sky to clear-sky forcing from Jacobson (2000b), we project that these changes in sea salt emissions might lead to a radiative feedback in 2100 of up to -0.8 Wm^{-2}. If we assume that the near-doubling of the sea salt mass flux would result in a proportional increase in the number flux, a significant increase in reflected radiation may also result from the indirect effect. The LLNL/Umich model was used to evaluate the possible impact of these emissions. It was found that the changes in natural emissions in 2100 result in a radiative feedback of -1.16 Wm^{-2} in the 2100 A2 scenario. These projected climate changes rely on the projected wind speed changes from the NCAR CSM model. Because projections from other models may not be as large, we also calculated the projected sea salt emissions from three other climate models using the ratio of monthly average winds for the time period from 2090 to 2100 to that for the time period 1990 to 2000 to scale the 2000 sea salt fluxes. Compared to the projections from the NCAR CSM, these models resulted in annual average fluxes that were 37% higher (GFDL model), 13% smaller (Max Planck model), and 9% smaller (Hadley Centre UK Met Office model). The monthly average temperature change associated with these wind speed and sea salt projections was 2.8°K (GFDL model; Knutson *et al.* 1999), 2.8°K (Max Planck model; Roeckner *et al.* 1999) and 2.15°K (Hadley Centre UK Met Office model; Gordon *et al.*, 2000) compared to the temperature projection of 1.8°K from the NCAR CSM model (Dai *et al.*, 2001).

5.5.3 Simulation of Future Aerosol Concentrations

In order to project future aerosol concentrations, we formed the average burdens from the models that gave reasonable agreement with observations for the 2000 scenario. As noted above, the differences in the burdens calculated by the different models point to substantial uncertainties in the prediction of current burdens and these translate into similar uncertainties in projecting future burdens. Except for sulphate and black carbon, however, the future projected global average burdens scaled approximately linearly with emissions. Thus, we may assume that the projected uncertainty in future burdens is mainly determined by the uncertainty in the emissions themselves together with the uncertainty in the burdens associated with different model treatments. For SO_4^{2-}, future anthropogenic concentrations were not linear in the emissions. For example, some models projected increases in burden relative to emissions while others projected decreases. The range of projected changes in anthropogenic burden relative to emissions was -14% to $+25\%$ depending on the scenario. Use of the average of the models for the projection of anthropogenic SO_4^{2-} and total SO_4^{2-} may therefore bias the results somewhat, but the uncertainties in the projected SO_4^{2-} concentrations are smaller than those introduced from the range of estimates for the 2000 scenario itself. Table 5.14 gives our projected average burdens for each draft SRES scenario. Results for the burdens associated with the final SRES scenarios are reproduced in Appendix II and are shown in Figure 5.13 (see Chapter 9 and Nakićenović *et al.* (2000) for scenario definitions).

One issue of importance for future scenarios concerns the treatment of black carbon removal rates. Whereas most models projected changes for BC to be approximately linear, the ULAQ and GISS models had somewhat different treatments. In the ULAQ model, the reaction of O_3 with BC leads to additional losses for BC. This caused a change of -30% for BC in the A2 2100 scenario in the simulation that included consideration of chemistry feedbacks (SC6) relative to the scenario without changes in chemistry (SC3). The GISS model, on the other hand, did not consider any heterogeneous loss, but did assume that the loss of BC associated with wet scavenging depended on the interaction of BC with SO_4^{2-}. In this model, as SO_2 emissions decrease, the lifetime of BC increased. This model projects a 50% increase in BC lifetime (and thus in the concentrations relative to emissions) in 2100 associated with the A2 scenario. Thus, the uncertainty associated with the projection of BC concentrations is larger than that for the other aerosol types.

In addition to the studies outlined above, one additional study was performed. In this study, the A2 2100 emission scenario was used to simulate HNO_3 concentrations in 2100 using the Harvard University model (see Chapter 4) and the present day anthropogenic NH_3 emissions were scaled by the increase in N_2O emissions in 2100 from the draft SRES A2 scenario. Then the model described by Adams *et al.* (1999) was used to estimate direct forcing after condensation of the additional HNO_3 and NH_3 onto the calculated sulphate aerosol. The nitrate and ammonium burdens increased by about a factor of 4.7 and a factor of 1.9, respectively, in the A2 2100 scenario. The estimated forcing associated with anthropogenic SO_4^{2-},

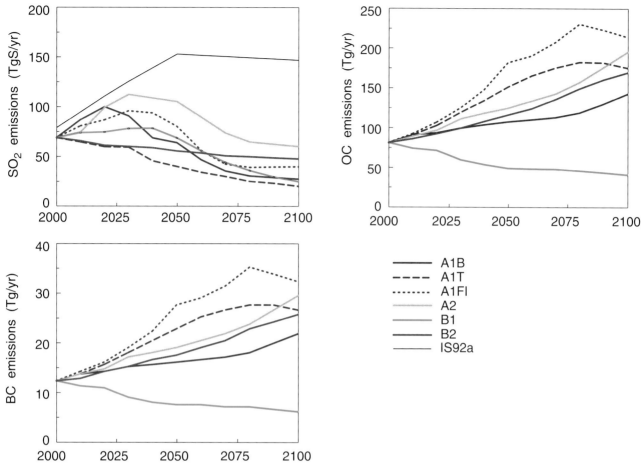

Figure 5.13: *Anthropogenic aerosol emissions projected for the SRES scenarios (Nakicenovic et al., 2000).*

Table 5.14: *Projected future aerosol burden for draft SRES scenarios. The range predicted from the models that participated in the workshop are given for 2000.*

	2000 (SC1)	A2 2030 (SC2)	A2 2100 (SC3)	B1 2100 (SC4)	A2 2100 with natural aerosols (SC5)	B1 2100 with natural aerosols (SC8)
Sulphate Natural (TgS)	0.26 0.15 - 0.36	0.26	0.26	0.26	0.28	0.28
Sulphate Anthr. (TgS)	0.52 0.35 - 0.75	0.90	0.55	0.28	0.54	0.26
Nitrate Natural (TgN)	0.02		0.02			
Nitrate Anthr. (TgN)	0.07		0.38			
Ammonium Nat. (TgN)	0.09		0.09			
Ammonium Anthr. (TgN)	0.33		0.72			
BC (Tg)	0.26 0.22 - 0.32	0.33	0.61	0.25	0.61	0.25
OC Natural (Tg)	0.15 0.08 - 0.28	0.15	0.15	0.15	0.22	0.22
OC Anthr. (Tg)	1.52 1.05 - 2.21	1.95	2.30	0.90	2.30	0.90
Dust (D<2 μm) (Tg)	12.98 6.24 - 17.73	13.52	13.52	13.52	13.54	13.54
Dust (D>2 μm) (Tg)	19.58 7.39 - 33.15	19.58	19.58	19.58	20.91	20.91
Sea salt (D<2 μm) (Tg-Na)	2.74 1.29 - 7.81	2.74	2.74	2.74	4.77	4.77
Sea salt (D>2 μm) (Tg-Na)	3.86 1.22 - 6.51	3.86	3.86	3.86	6.68	6.68

nitrate, and ammonium aerosols increased from -1.78 Wm^{-2} to -2.77 Wm^{-2}. Thus, the control of sulphate aerosol in future scenarios may not necessarily lead to decreases in forcing, if the levels of ammonium nitrate in aerosol increase.

Two final considerations include the possible impact of chemistry and climate changes on future concentrations. These were examined by the ULAQ model for the A2 2100 scenario. Future concentration changes were small for the simulation that included changes in chemistry only (scenario SC6). Climate feedbacks, however, were significant. Aerosol concentrations changed by as much as -20% relative to the model simulations that did not include climate change.

5.5.4 *Linkage to Other Issues and Summary*

As we have seen, anthropogenic aerosols may have a substantial effect on the present day aerosol abundance, optical depth and thus forcing of climate. While we have made substantial progress in defining the role of anthropogenic aerosols on direct forcing, significant uncertainties remain, particularly with the role of anthropogenic organic and black carbon aerosols in detemining this forcing. Our ability to assess the indirect forcing by aerosols has a much larger uncertainty associated with it. The largest estimates of negative forcing due to the warm-cloud indirect effect may approach or exceed the positive forcing due to long-lived greenhouse gases. On the other hand, there is sufficient uncertainty in the calculation of indirect forcing to allow values that are substantially smaller than the positive forcing by greenhouse gases. Other factors which have not been assessed include possible anthropogenic perturbations to high level cirrus clouds as well as to clouds in the reservoir between $0°C$ and $-35°C$. As we have discussed, significant positive forcing is possible from aerosol-induced increased formation of cirrus and/or from increased glaciation of clouds in the region between $0°C$ and $-35°C$.

Concerns about aerosols derive from a number of other considerations. These include visibility, toxic effects and human health, interactions of aerosols with chemical processes in the troposphere and stratosphere, acid deposition, and air pollution. Of these concerns, those associated with toxic effects and visibility have led the industrialised countries to promulgate standards to reduce the concentrations of aerosols in urban and also more pristine locations. Also, concerns about the effects of acid rain have led to increased controls over the emissions of SO_2. As shown above, the SRES scenarios for the future (Nakićenović *et al.*, 2000) have all assumed that emissions of SO_2 will eventually decrease, and the A1T and B2 scenarios predict that sulphur emissions start to decrease on a global average basis almost immediately.

Emissions of carbon aerosols in the future may not necessarily follow the scenarios outlined for SO_2 within the SRES scenarios. Furthermore, older scenarios, developed by IPCC (1992), do not presume that SO_2 emissions would necessarily decrease. We have evaluated a set of scenarios for future aerosols that account for a range of possible future emissions. We noted that one of the future effects that need to be evaluated includes the perturbation to the cycles of natural aerosols. Another includes the interaction of chemical cycles with aerosols and the prediction of how changes in other gas-phase emissions may be perturbing aerosols. We must also consider how changes in aerosols may perturb gas phase cycles. Our evaluation of these interactions has been necessarily limited. Both the interaction of atmospheric chemical cycles with aerosols and the perturbation to natural aerosol components through changing climate patterns need to be better understood.

5.6 Investigations Needed to Improve Confidence in Estimates of Aerosol Forcing and the Role of Aerosols in Climate Processes

Atmospheric measurements have lagged behind awareness of the importance of aerosols in climate. There is a great challenge in adequately characterising the nature and occurrence of atmospheric aerosols and in including their effects in models to reduce uncertainties in climate prediction. Because aerosols: (i) originate from a variety of sources, (ii) are distributed across a wide spectrum of particle sizes and (iii) have atmospheric lifetimes that are much shorter than those of most greenhouse gases, their concentrations and composition have great spatial and temporal variability. Satellite-based measurements of aerosols are a necessary but not sufficient component of an approach needed to acquire an adequate information base upon which progress in understanding the role of aerosols in climate can be built. Below we outline measurements and process level studies that are necessary to reduce uncertainties in both direct and indirect forcing. These studies and observations are needed both to improve the models on which climate forcing rely and to check our understanding of this forcing as aerosol concentrations change in the future.

I. Systematic Ground-Based Measurements
There is a need for countries of the world to develop and support a network of systematic ground-based observations of aerosol properties in the atmosphere that include a variety of physical and chemical measurements ranging from local *in situ* to remotely sensed total column or vertical profile properties. The Global Atmospheric Watch programme of the World Meterological Organization is but one player in organising routine aerosols measurements on a stage that includes other international organisations (e.g. International Atmoic Energy Agency, IAEA) as well as national research programmes (e.g. national environmental agencies, national atmospheric research agencies).

It is recommended that, at all levels, emphasis be placed in developing a *common strategy* for aerosol and gas measurements at a selected set of regionally representative sites. One possible model is to develop a set of primary, secondary and tertiary aerosol networks around the world. At the primary stations, a comprehensive suite of aerosol and gas measurements should be taken that are long-term in scope (gaseous precursors are an essential part of the aerosol story since much aerosol mass is formed in the atmosphere from gas-to-particle conversion). At secondary stations, a less comprehensive set of observations would be taken that would provide background information for intensive shorter term process-oriented studies. It would be desirable to co-locate vertical profiling networks that involve complex instrumentation such as lidars with these baseline stations. Tertiary stations may include stations operated by national research programmes that are related to urban aerosol issues and human health.

These measurements should be closely co-ordinated with satellite observations of aerosols. The types of measurements should include *in situ* size-segregated concentrations of aerosol physical properties such as number and mass but also chemical properties such as composition and optical properties. Total column properties such as aerosol optical depth, Angstrom coefficient, CO and O_3 add value to these data sets in evaluating the simulation of aerosols as active constitutents in climate models.

II. Systematic Vertical Profile Measurements

There is a paucity of systematic vertical profile measurements of size-segregated or even total atmospheric aerosol physical, chemical and optical properties. For these parameters, no climatological database exists that can be used to evaluate the performance of climate models that include aerosols as active constituents. The COSAM model comparison (Barrie *et al.*, 2001) had to use vertical profile observations from a few intensive aircraft campaigns of only a few months duration to evaluate climate model aerosol predictions. Such measurments would be best co-located with the ground-based network stations. Since they involve routine aircraft surveillance missions and are costly, the development of robust, sensitive lightweight instrument packages for deployment in small aircraft or on commercial airliners is a high priority. Both continuous real time measurements and collection of aerosols for post-flight analysis are needed.

The network design needs to be systematically developed and implemented. One possible model is to conduct observations at pairs of stations around – and downwind of – major aerosol sources types such as industrial (Europe, North America, Asia), soil dust (Sahara or Asian), biomass burning (Amazon or southern Africa) and sea salt (roaring forties of the southern Pacific Ocean region).

III. Characterisation of Aerosol Processes in Selected Regions

There is a need for integrated measurements to be undertaken in a number of situations to enhance the capability to quantitatively simulate the processes that influence the size-segregated concentration and composition of aerosols and their gaseous precursors. The situations need to be carefully selected and the observations sufficiently comprehensive that they constrain models of aerosol dynamics and chemistry. The International Global Atmospheric Chemistry (IGAC) programme in its series of Atmospheric

Chemistry Experiments in the roaring forties of the southern Pacific (ACE-1), the outflow from North Africa and Europe to the eastern North Atlantic and the 2001 study in Southeast Asia and downwind in the Pacific (ACE-Asia) are examples of attempts to do this that require support and continued adjustment of experimental design to match outstanding questions. Such studies need to be conducted in industrial continental and neighboring marine, upper-tropospheric, Arctic, remote oceanic and dust-dominated air masses. Closure of aerosol transport and transportation models as well as direct forcing closure studies should be an integral part of these studies.

IV. Indirect Forcing Studies

There is a need for several carefully designed multi-platform (surface-based boat, aircraft and satellite) closure studies that elucidate the processes that determine cloud microphysical (e.g., size-distributed droplet number concentration and chemical composition, hydrometeor type) and macrophysical properties (e.g. cloud thickness, cloud liquid-water content, precipitation rate, total column cloud, albedo). A second goal would be to understand how aerosols influence the interaction of clouds with solar radiation and precipitation formation. These studies should take place in a variety of regions so that a range of aerosol types as well as cloud types can be explored. Emphasis should be placed on reducing uncertainties related to scaling-up of the processes of aerosol-cloud interactions from individual clouds (about 1 to 10 km) to the typical resolution of a climate model (about 100 to 500 km). Can sub-grid parametrizations of cloud processes accurately represent cloud-radiation interactions and the role that aerosols play in that interaction? Answering this question will require that process studies be performed in conjunction with a range of model types such as models that include a detailed microphysical representation of clouds to models that include the parametrizations in climate models.

V. Measurements of Aerosol Characteristics from Space

An integrated strategy for reducing uncertainties should include high quality measurements of aerosols from space. At the time of this report, only measurements from AVHRR and POLDER were available. The latter instrument may yield measurements of aerosol optical depth over land, but it was operational for less than a year. High quality satellite measurements together with systematic comparisons with models and the process-level studies noted above should allow us to reduce the uncertainties in current aerosol models. Systematic comparison of models that include an analysis of the indirect effect with satellite measurements of clouds and with data gathered from process-level studies will also reduce uncertainties in indirect effects and in projected climate change.

References

Abdou, W.A., J.V. Martonchik, R.A. Kahn, R.A. West, and D.J. Diner, 1997: A modified linear-mixing method for calculating atmospheric path radiances of aerosol mixtures. *J. Geophys. Res.*, **102**, 16883-16888.

Abdul-Razzak, H. and S. Ghan, 2000: A parametrization of aerosol activation 2. Multiple aerosol types. *J. Geophys. Res.*, **105**, 6837-6844.

Abdul-Razzak, H., S. Ghan and C. Rivera-Carpio, 1998: A parametrization of aerosol activation 1. Single aerosol type. *J. Geophys. Res.*, **103**, 6123-6131.

Ackerman, T.P, and O.B. Toon, 1981: Absorption of visible radiation in atmosphere containing mixtures of absorbing and nonabsorbing particles. *Appl. Opt.*, **20**, 3661-3668.

Ackerman, A.S., O.B. Toon, J.P. Taylor, D.W. Johnson, P.V. Hobbs and R.J. Ferek, 2000: Effects of aerosols on cloud albedo: Evaluation of Twomey's parametrization of cloud susceptibility using measurements of ship tracks. *J. Atmos. Sci.*, **57** 2684-2695.

Adams, P.J., J.H. Seinfeld and D.M. Koch, 1999: Global concentrations of tropospheric sulphate, nitrate and ammonium aerosol simulated in a general circulation model. *J. Geophys. Res.*, **104**, 13791–13823.

Adams, P.J. J.H. Seinfeld, D. Koch, L. Mickley, and D. Jacob, 2001: General circulation model assessment of direct radiative forcing by the sulphate-nitrat-ammonium-water inorganic aerosol system. *J. Geophys. Res.*, **106**, 1097-1111.

Albrecht, B., 1989: Aerosols, cloud microphysics and fractional cloudiness. *Science*, **245**, 1227–1230.

Al-Naime, R. and C.P.R. Saunders, 1985: Ice nucleus measurements: Effects of site location and weather. *Tellus*, **37B**, 296–303.

Anderson, B.E., W.B. Grant, G.L. Gregory, E.V. Browell, J.E. Collins Jr., D.W. Sachse, D.R. Bagwell, C.H. Hudgins, D.R. Blake and N.J. Blake, 1996: Aerosols from biomass burning over the tropical South Atlantic region: Distributions and impacts. *J. Geophys. Res.*, **101**, 24,117-24,137.

Andreae M.O. and W.A. Jaeschke, 1992: Exchange of Sulphur between biosphere and atmosphere over temperate and tropical regions, in: Sulphur cycling on the continents: *Wetlands, Terrestrial Ecosystems, and Associated Water Bodies. SCOPE* **48**, R.W. Howarth, J.W.B. Stewart and M.V. Ivanov (eds), Wiley, Chichester, 27-61.

Andreae, M.O. 1995: Climatic effects of changing atmospheric aerosol levels. In: *World Survey of Climatology. Vol. 16: Future Climates of the World*, A. Henderson-Sellers (ed). Elsevier, Amsterdam, pp. 341–392.

Andreae, M.O. and P.J. Crutzen, 1997: Atmospheric aerosols: Biogeochemical sources and role in atmospheric chemistry. *Science*, **276**, 1052–1056.

Andreae, M.O., E.V. Browell, M. Garstang, G.L. Gregory, R.C. Harriss, G.F. Hill, D.J. Jacob, M.C. Pereira, G.W. Sachse, A.W. Setzer, P.L.S. Dias, R.W. Talbot, A.L. Torres and S.C. Wofsy, 1988: Biomass-burning emissions and associated haze layers over Amazonia. *J. Geophys. Res.*, **93**, 1509–1527.

Andreae, M.O., W. Elbert, R. Gabriel, D.W. Johnson, S. Osborne and R. Wood, 2000: Soluble ion chemistry of the atmospheric aerosol and SO_2 concentrations over the eastern North Atlantic during ACE-2. *Tellus*, **52**, 1066-1087.

Andreae, M.O., W. Elbert, Y. Cai, T.W. Andreae and J. Gras, 1999: Non-seasalt sulphate, methanesulfonate, and nitrate aerosol concentrations and size distributions at Cape Grim, Tasmania. *J. Geophys. Res.*, **104**, 21,695-21,706.

Andres, R.J. and A.D. Kasgnoc, 1998: A time-averaged inventory of sub-aerial volcanic sulphur emissions. *J. Geophys. Res.* Atmos., **103**, 25251–25261.

Appel, B.R., Y. Tokiwa, and E.L. Kothny, 1983: Sampling of carbonaceous particles in the atmosphere. *Atmos. Env.*, **17**, 1787-1796.

Artaxo, P., E.T. Fernandes, J.V. Martins, M.A. Yamasoe, P.V. Hobbs, W. Maenhaut, K.M. Longo and A. Castanho, 1998a: Large-scale aerosol source apportionment in Amazonia. *J. Geophys. Res.*, **103**, 31837–31847.

Artaxo, P., E. Swietlicki, J. Zhou, H.-C. Hansson, W. Maenhaut, M. Claeys, M.O. Andreae, J. Ström, J.V. Martins, M.A. Yamasoe and R. van Grieken, 1998b: Aerosol properties in the central Amazon Basin during the wet season during the LBA/CLAIRE experiment. *Eos Trans. AGU*, **79**, F155.

Atherton, C.A., 1996: Biomass burning sources of nitrogen oxides, carbon monoxide, and non-methane hydrocarbons, *Lawrence Livermore National Laboratory Report* UCRL-ID-122583, 1996.

Atlas, R.M., R.N. Hoffman, S.C. Bloom, J.C. Jusem, and J. Ardizzone, 1996: A multiyear global surface wind velocity dataset using SSM/I wind obsevations, *Bull. Am. Meteorol. Soc.*, **77**, 869-882.

Austin, P., Y. Wang,R. Pincus and V. Kujala, 1995: Precipitation in stratocumulus clouds: observational and modeling results, *J. Atmos. Sci.*, **52**, 2329-2352.

Barrie, L.A., Y. Yi, W.R. Leaitch, U. Lohmann, P. Kasibhatla, G.-J. Roelofs, J. Wilson, F. McGovern *et al.*, 2001: A comparison of large scale atmospheric sulphate aerosol models (COSAM): Overview and highlights. *Tellus B*, in press.

Barth, M.C., P.J. Rasch, J.T. Kiehl, C.M. Benkowitz, and S.E. Schwartz, 2000: Sulphur chemistry in the NCAR CCM: description, evaluation, features and sensitivity to aqueous chemistry. *J. Geophys. Res.*, **105**, 1387-1416.

Bassett, M., and J.H. Seinfeld, 1984: Atmospheric equilibrium model of sulphate and nitrate aerosols. II. Particle size analysis. *Atmos. Env.*, **18**, 1163.

Bates, T.S., B.K. Lamb, A. Guenther, J. Dignon, and R.E. Stoiber, 1992: Sulphur emissions to the atmosphere from natural sources. *J. Atmos. Chem.*, **14**, 315-337.

Bates, T.S., V.N. Kapustin, P.K. Quinn, D.S. Covert, D.J. Coffman, C. Mari, P.A. Durkee, W.J. De Bruyn, and E.S. Saltman, 1998: Processes controlling the distribution of aerosol particles in the lower marine boundary layer during the First Aerosol Characterization Experiment (ACE 1). *J. Geophys. Res.*, **103**, 16369-16383.

Beard, K.V., 1992: Ice nucleation in warm-base convective clouds: An assessment of micro-physical mechanisms. *Atmos. Res.*, 28, 125-152.

Beheng, K.D., 1994: A parametrization of warm cloud microphysical conversion processes. *Atmos. Res.*, **33**, 193-206.

Benkovitz, C.M., C.M. Berkowitz, R.C. Easter, S. Nemesure, R. Wagner, and S.E. Schwartz, 1994: Sulphate over the North Atlantic and adjacent regions: Evaluation for October and November 1986 using a three-dimensional model driven by observation-derived meteorology. *J. Geophys. Res.*, **99**, 20,725-20,756.

Benkovitz, C.M., M.T. Scholtz, J. Pacyna, L. Tarrason, J. Dignon, E.C.Voldner, P.A.Spiro, J.A. Logan, T.E. Graedel, 1996: Global gridded inventories of anthropogenic emissions of sulphur and nitrogen. *J. Geophys. Res.*, **101**, 29,239-29,253.

Benkovitz, C.M., S.E. Schwartz, 1997: Evaluation of modelled sulphate and SO2 over North America and Europe for four seasonal months in 1986-1987. *J. Geophys. Res.*, **102**, 25,305-25,338.

Berry, E.X., 1967: Cloud drop growth by collection. *J. Atmos. Sci.*, **24**, 688-701.

Bigg, E.K., 1990: Long-term trends in ice nucleus concentrations. *Atmos. Res.*, **25**, 409–425.

Blanchard, D.C., 1983: The production, distribution and bacterial enrichment of the sea-salt aerosol. In: *Air-Sea Exchange of Gases and Particles*, P.S. Liss and W.G.N. Slinn (eds), Reidel, Boston, USA, pp. 407–454.

Blanchet J.-P. and E. Girard, 1995: Water vapor and temperature feedback in the formation of continental Arctic air: Its implications for climate, *Sci. Total Environ.*, 160-161, 793-802.

Bodhaine, B.A., 1995: Aerosol absorption measurements at Barrow,

Mauna Loa and the South Pole. *J. Geophys. Res.*, **100**, 8967-8975.

Boers, R. and R.M. Mitchell, 1994: Absorption feedback in stratocumulus clouds − influence on cloud top albedo. *Tellus*, **46**, 229−241.

Boers, R., 1995: Influence of seasonal variation in cloud condensation nuclei, drizzle, and solar radiation, on marine stratocumulus optical depth. *Tellus*, **47B**, 578−586.

Boers, R. and P. Krummel, 1998: Microphysical properties of boundary layer clouds over the Southern Ocean during ACE I. *J. Geophys. Res.*, **103**, 16651-16663.

Boers, R., J. B. Jensen, P. B. Krummel 1998: Microphysical and radiative structure of marine stratocumulus clouds over the Southern Ocean: Summer results and seasonal differences. *Quart. J. Royal Meteor. Soc.*, **124**, 151-168.

Bond, T.C., R.J. Charlson and J. Heintzenberg, 1998: Quantifying the emission of light-absorbing particles: Measurements tailored to climate studies. *Geophys. Res. Lett.*, **25**, 337−340, 1998.

Borys, R. D., 1989: Studies of ice nucleation by arctic aerosol on AGASP-II. *J. Atmos. Chem.*, **9**, 169−185.

Boucher, O. and U. Lohmann, 1995: The sulphate-CCN-cloud albedo effect - A sensitivity study with two general circulation models. *Tellus*, **47B**, 281−300.

Boucher, O., and D. Tanré, 2000: Estimation of the aerosol perturbation to the Earth's radiative budget over oceans using POLDER satellite aerosol retrievals, *Geophys. Res. Lett.*, **27**, 1103-1106.

Boucher, O., S.E. Schwartz, T.P. Ackerman, T.L Anderson, B. Bergstrom, B. Bonnel, P. Chylek, A. Dahlback, Y. Fouquart, Q. Fu, R.N. Halthore, J.M. Haywood, T. Iversen, S. Kato, S. Kinne, A. Kirkevag, K.R. Knapp, A. Lacis, I. Laszlo, M.I. Mishchenko, S. Nemesure, V. Ramaswamy, D.L. Roberts, P. Russell, M.E. Schlesinger, G.L. Stephens, R. Wagener, M. Wang, J. Wong, F. Yang, 1998: Intercomparison of models representing direct shortwave radiative forcing by sulphate aerosols, *Geophys. Res*, **103**, 16,979-16,998.

Bouwman, A.F., D.S. Lee, W.A.H. Asman, F.J. Dentener, K.W. Van Der Hoek, and J.G.J. Olivier, 1997: A global high-resolution emission inventory for ammonia. *Global Biochem. Cycles*, **11**, 561-588.

Braham, R. R. Jr. and P. Spyers-Duran, 1974: Ice nuclei measurements in an urban atmosphere. *J. Appl. Meteor.*, **13**, 940−945.

Brasseur, G. P., D. A. Hauglustaine, S. Walters, P. J. Rasch, J.-F. Muller, C. Granier, X. Tie, 1998: MOZART, a global chemical transport model for ozone and related chemical tracers. Part 1: Model description . *J. Geophys. Res.*, **103**, 28,265-28,289.

Brenguier, J.-L., H. Pawlowska, L. Schuller, R. Preusker, J. Fischer and Y. Fouquart, 2000: Radiative properties of boundary layer clouds: droplet effective radius versus number concentration. *J. Atmos. Sci.*, **57**, 803-821.

Cachier, H., C. Liousse, P. Buat-Menard and A. Gaudichet, 1995: Particulate content of savanna fire emissions. *J. Atmos. Chem.*, **22**, 123−148.

Carrico, C.M., M.J. Rood and J.A. Ogren, 1998: Aerosol light scattering properties at Cape Grim, Tasmania during the First Aerosol Characterization Experiment (ACE-1). *J. Geophys. Res.*, **103**, 16,565-16,574.

Carrico, C.M., M.J. Rood, J.A. Ogren, C. Neuses, A. Wiedensolhler and J. Heintzenberg, 2000: Aerosol light scattering properties at Sagres, Portugal, during ACE-2, *Tellus*, **52**, 694-715.

Chapman, W. L. and J. E.,Walsh, 1993: Recent variations of sea ice and air temperature in high latitudes, *Bull. Am. Met. Soc.*, **74**, 33-47.

Charlson, R. J., D. S. Covert and T. V. Larson, 1984: Observation of the effect of humidity on light scattering by aerosols. In *Hygroscopic Aerosols*, ed. T. H. Rukube and A. Deepak, pp. 35-44, A. Deepak, Hampton, VA.

Charlson, R. J., J. E. Lovelock, M. O. Andreae, and S. G. Warren, 1987: Ocean phytoplankton, atmospheric sulphur, cloud albedo and climate. *Nature*, **326**, 655-661.

Charlson, R.J., S.E. Schwartz, J.M. Hales, R.D. Cess, J.A. Coakley, J.E. Hansen and D.J. Hofmann, 1992: Climate forcing by anthropogenic aerosols. *Science*, **255**, 423−430.

Chen, Y., S. M. Kreidenweis, L. M. McInnes, D. C. Rogers, and P. J. DeMott, 1998, Single particle analyses of ice nucleating aerosols in the upper troposphere and lower stratosphere. *Geophys. Res. Lett.*, **25**, 1391-1394.

Chin, M., R.B. Rood, S.-J. Lin, J.-F. Muller, and A. M. Thompson, 2000: Atmospheric sulphur cycle simulated in the global model GOCART: Model description and global properties. *J. Geophys. Res.*, **105**, 24671-24687.

Chuang, C.C. and J.E. Penner, 1995: Effects of anthropogenic sulphate on cloud drop nucleation and optical properties. *Tellus*, **47**, 566−577.

Chuang, C.C., J.E. Penner, K.E. Taylor, A.S. Grossman, and J.J. Walton, 1997: An assessment of the radiative effects of anthropogenic sulphate. *J. Geophys. Res.*, **102**, 3761-3778.

Chuang, C. C., J. E. Penner, K. E. Grant, J. M. Prospero, G. H. Rau and K. Kawamoto, 2000: Cloud susceptibility and the first aerosol indirect forcing: sensitivity to black carbon and aerosol concentrations. *J. Geophys. Res.* Submitted. Also: Lawrence Livermore National Laboratory Report, UCRL-JC-139097 Rev 1.

Chylek, P. and J. Wong, 1995: Effects of absorbing aerosols on the global radiation budget. *J. Geophys. Res. Lett.*, **22**, 929-931.

Claquin, T., M. Schulz, Y. Balkanski and O. Boucher, 1998: Uncertainties in assessing radiative forcing by mineral dust. *Tellus*, **50B**, 491−505.

Claquin, T., M. Schulz, and Y. J. Balkanski, 1999: Modeling the mineralogy of atmospheric dust sources. *J. Geophys. Res.*, **104**, 22,243-22,256.

Clarke, A. D. J. N. Porter, F. P. J. Valero, and P. Pilewskie, 1996: Vertical profiles, aerosol microphysics, and optical closure during the Atlantic Stratocumulus Transition Experiment: measured and modelled column optical properties. *J. Geophys. Res.*, **101**, 4443-4453.

Clarke, A. P., J. Eisele, V. N. Kapustin, K. Moore, D. Tanner, L. Mauldin, M. Litchy, B. Lienert, M. A. Carroll and G. Albercook, 1999: Nucleation in the equatorial free troposphere: Favorable environments during PEM-Tropics. *J. Geophys. Res.*, **104**, 5735-5744.

Coakley, J. A., Jr., R. L. Bernstein and P. A. Durkee, 1987, Effect of ship-stack effluents on cloud reflectivity. *Science*, **237**, 1020−1022.

Collins, D. R., H. H. Jonsson, R. C. Flagan, J. H. Seinfeld, K. J. Noone, E. Ostrom, D. A. Hegg, S. Gasso, P. B. Russell, J. M. Livingston, B. Schmid and L. M. Russell, 2000: In situ aerosol size distributions and clear column radiative closure during ACE-2. *Tellus*, **52B**, 498 525.

Cooke, W.F. and J.J.N. Wilson, 1996: A global black carbon aerosol model. *J. Geophys. Res. Atmos.*, **101**, 19395−19409.

Cooke, W.F., C. Liousse, H. Cachier, and J. Feichter, 1999: Construction of a 1° × 1° degree fossil fuel emission data set for carbonaceous aerosol and implementation and radiative impact in the ECHAM4 model. *J. Geophys. Res*, **104**, 22,137-22,162.

Cooper, W. A., 1980: A method of detecting contact ice nuclei using filter samples. Preprints Cloud Phys. Conf., Clermont-Ferrand, France, pp. 605−669.

Corrigan, C.E., and T. Novakov, 1999: Cloud condensation nucleus activity of organic compounds: A laboratory study. *Atmos. Env.*, **33**, 2661-2668.

Covert, D. S., V. N. Kapustin, T. S. Bates and P. K. Quinn, 1996: Physical properties of marine boundary layer aerosol particles of the mid-Pacific in relation to sources and meteorological transport. *J. Geophys. Res.*, **101**, 6919-6930.

Cruz, C.N. and S.N. Pandis, 1997: A study of the ability of secondary organic aerosol to act as cloud condensation nuclei. *Atmos. Environ.*, **31**, 2205−2214.

Cruz, C.N. and S.N. Pandis, 1998: The effect of organic coatings on the cloud condensation nuclei activation of inorganic atmospheric aerosol. *J. Geophys. Res. Atmos.*, **103**, 13111−13123.

Curry, J. A., F. G. Meyer, L. F. Radke, C. A. Brock and E. E. Ebert, 1990: The occurrence and characteristics of lower tropospheric ice crystals in the Arctic. *Int. J. Climate*, **10**, 749–764.

Cusack, S., A. Slingo, J. M. Edwards, and M. Wild, 1998: The radiative impact of a simple aerosol climatology on the Hadley Centre atmospheric GCM. *Q. J. R. Met. Soc*, **124**, 2517-2526.

Daggett, D. L., D. J. Sutkus Jr., D. P. DuPois, and S. L. Baughcum, 1999: An evaluation of aircraft emissions inventory methodology by comparisons with reported airline data. NASA/CR-1999-209480, 75 pp.

Dai, A., T.M.L. Wigley, B.A. Boville, J.T. Kiehl, and L.E. Buja, 2001: Climates of the 20th and 21st centuries simulated by the NCAR Climate System Model. *J. Climate*, in press.

Dai, J. C., L. G. Thompson, and E. Mosley-Thompson, 1995: A 485 year record of atmospheric chloride, nitrate and sulphate: results of chemical analysis of ice cores from Dyer Plateau, Antarctic Peninsula. *Annals of Glaciology*, **21**, 182-188.

D'Almeida, G.A., 1991: Atmospheric aerosols : Global climatology and radiative characteristics. A. Deepak Pub., Hampton, Va., USA.

DeMott, P.J., J.L. Stith, R.J. Zen and D.C. Rogers, 1996: Relations between aerosol and cloud properties in North Dakota cumulus clouds. Preprints, 12th Int. Conf. on Clouds and Precipitation, Zurich, Switzerland, 19-23 August, pp. 320–323.

DeMott, P. J., D.C. Rogers and S.M. Kreidenweis, 1997: The susceptibility of ice formation in upper tropospheric clouds to insoluble aerosol components. *J. Geophys. Res.*, **102**, 19575–19584.

DeMott, P.J., D.C. Rogers and S.M. Kreidenweis, Y. Chen, C.H. Twohy, D. Baumgardner, A.J. Heymsfield and K.R. Chan, 1998: The role of heterogeneous freezing nucleation in upper tropospheric clouds: inferences from SUCCESS. *J. Geophys. Res. Lett.*, **25**, 1387–1390.

Deuzé, J.L., M. Herman, P. Goloub, D. Tanré, and A. Marchand, 1999: Characterization of aerosols over ocean from POLDER/ADEOS-1. *Geophys. Res. Lett.*, **26**, 1421-1424.

Döscher, A., H., W. Gäggeler, U. Schotterer, and M. Schwikowski, 1995: A 130 years deposition record of sulphate, nitrate, and chloride from a high-alpine glacier. *Water Air Soil Pollut.*, **85**, 603-609.

Dubovik, O., B.N. Holben, Y.J. Kaufman, M. Yamasoe, A. Smirnov, D. Tanré and I. Slutsker, 1998: Single-scattering albedo of smoke retrieved from the sky radiance and solar transmittance measured from ground. *J. Geophys. Res.*, **103**, 31,903–31,923.

Duce, R., 1995: Distributions and fluxes of mineral aerosol. In *Aerosol Forcing of Climate*, R.J. Charlson and J. Heintzenberg (eds), John Wiley, Chichester, UK, pp. 43–72.

Eagen, R. C., P.V. Hobbs and L.F. Radke, 1974: Particle emissions from a large craft paper mill and their effects on the microstructure of warm clouds,, *J. Applied Meteorl.*, **13**, 535-557.

Eatough, D.J., D.A. Eatough, L. Lewis and E.A. Lewis, 1996: Fine particulate chemical composition and light extinction at Canyonlands National Park using organic particulate material concentrations obtained with a multi-system, multichannel diffusion denuder sampler. *J. Geophys. Res.*, **101**, 19515–19531.

Eccleston, A. J., N. K. King and D. R. Packham, 1974: The scattering coefficient and mass concentration of smoke from some Australian forest fires. J. Air. Poll. Cont. Assoc., **24**, 1047-1050.

Echalar, F., P. Artaxo, J.V. Martins, M.A. Yamasoe, F. Gerab, W. Maenhaut and B. Holben, 1998: Long-term monitoring of atmospheric aerosols in the Amazon Basin: Source identification and apportionment. *J. Geophys. Res.*, **103**, 31,849-31,864.

Eck, T. J., B. N. Holben, J. S. Reid, O. Duborik, S. Kinne, A. Smirnov, N. T. O'Neill and I. Slutsken, 1999: The wavelength dependence of the optical depth of biomass burning, urban, and desert dust aerosols. *J. Geophys. Res.*, **104**, 31,333-31,349.

Eichel, C., M. Kramer, L. Schutz, and S. Wurzler, 1996: The water-soluble fraction of atmospheric aerosol particles and its influence on cloud microphysics. *J. Geophys. Res.* Atmos., **101**, 29499–29510.

Einfeld, W., D. E. Ward and C. Hardy, 1991: Effects of fire behavior on prescribed fire smoke characteristics: A case study. In Global Biomass Burning: Atmospheric, Climate, and Biospheric Implications, ed. J. S. Levine, pp. 209-224, MIT Press, Cambridge, MA.

Eisele, F.L. and P.H. McMurry, 1997: Recent progress in understanding particle nucleation and growth, *Phil. Trans. R. Soc. Lond.* B, **352**, 191-201.

Erickson, D.J. III and R.A. Duce, 1988: On the global flux of atmospheric sea salt. *J. Geophys. Res.*, **93**, 14079–14088.

Facchini, M.C., M. Mircea, S. Fuzzi, and R.J. Charlson, , 1999: Cloud albedo enhancement by surface-active organic solutes in growing droplets. *Nature*, **401**, 257-259.

Fahey, D. W., U. Schumann, S. Ackerman, P. Artaxo, O. Boucher, M.Y. Danilin, B. Kärcher, P. Minnis, T. Nakajima, and O.B. Toon, 1999: Aviation-produced aerosols and cloudiness. In Aviation and the Global Atmosphere, J.E. Penner, D.H. Lister, D.J. Griggs, D.J. Dokken, and M. McFarland (Eds.), Cambridge University Press, Cambridge, U.K., 65-120.

Fassi-Fihri, A., K. Suhre, R. Rosset, 1997: Internal and external mixing in atmospheric aerosols by coagulation: impact on the optical and hygroscopic properties of the sulphate-soot system. *Atmos. Env.*, **31**, 1393-1402.

Feichter, J., E. Kjellstrom, H. Rodhe, F. Dentener, J. Lelieveld and G.-J. Roelofs. 1996: Simulation of the tropospheric sulphur cycle in a global climate model. *Atmos. Env.*, **30**, 1693–1707.

Feichter, J., U. Lohmann and I. Schult, 1997: The atmospheric sulphur cycle in ECHAM-4 and its impact on the shortwave radiation. *Climate Dynamics*, **13**, 235–246.

Feingold, G., R. Boers, B. Stevens and W.R. Cotton, 1997: A modelling study of the effect of drizzle on cloud optical depth and susceptibility. *J. Geophys. Res.*, **102**, 13527-13534.

Feingold, G., S.M. Kreidenweis, and Y. Zhang, 1998: Stratocumulus processing of gases and cloud condensation nuclei: Part I: trajectory ensemble model. *J. Geophys. Res.*, **103**, 19527-19542.

Feingold, G., W. R. Cotton, S. M. Kreidenweis, and J. T. Davis, 1999a: Impact of giant cloud condensation nuclei on drizzle formation in marine stratocumulus: Implications for cloud radiative properties. *J. Atmos. Sci.*, **56**, 4100-4117.

Feingold, G., A.S. Frisch, B. Stevens, and W.R. Cotton, 1999b: The stratocumulus boundary layer as viewed by K-band radar, microwave radiometer and lidar. *J. Geophys. Res.*, **104**, 22195-22203.

Ferek, R. J., D. A. Hegg, P. V. Hobbs, P. Durkee and K. Nielson, 1998: Measurements of ship-induced tracks in clouds off the Washington Coast. *J. Geophys. Res.* Atmos., **103**, 23199–23206.

Ferrare, R. A., Melfi, S. H. Whiteman, D. N., Evans, K. D., Leifer, R., 1998: Raman lidar measurements of aerosol extinction and backscattering, 1, Methods and comparisons. *J. Geophys. Res.*, **103**, 19,663-19,672.

Ferrare, R., S. Ismail, E. Browell, V. Brakett, . Clayton, S. Kooi, S. H. Melfi, D. Whiteman, G. Schwemmer, K. Evans, P. Russell, J. Livingston, B. Schmid, B. Holben, L. Remer, A. Smirnov, and P. V. Hobbs, 2000: Comparison of aerosol optical properties and water vapor among ground and airborne lidars and Sun photometers during TARFOX. *J. Geophys. Res.*, **105**, 9917-9933.

Fischer, H., D. Wagenbach, and J. Kipfstuhl, 1998: Sulphate and nitrate firn concentrations on the Greenland ice sheet 2. Temporal anthropogenic deposition changes. *J. Geophys. Res.*, **103**, 21935-21942.

Fitzgerald, J. W., 1991: Marine aerosols: A review. *Atmos. Environ.*, **25A**, 533-545.

Fitzgerald, J. W. and W. A. Hoppel, 1982: The size and scattering coefficient of urban aerosol particles at Washington D.C. as a function of relative humidity. *J. Atmos. Sci.*, **39**, 1838-1852.

Flamant, C., J. Pelon, P. Chazette, V. Trouillet, K. Quinn, R. Frouin, D. Bruneau, J. F. Leon, T. S. Bates, J. Johnson, and J. Livingston, 2000:

Airborne lidar measurements of aerosol spatial distribution and optical properties over the Atlantic Ocean during a European pollution outbreak of ACE-2, *Tellus*, **52B**, 662-677.

Flowerdew, R. J., and J. D. Haigh, 1996: Retrieval of aerosol optical thickness over land using the ATSR-2 dual-look satellite radiometer. *Geophys. Res. Lett.*, **23**, 351-354.

Fowler, L. D. and D. A. Randall, 1996: Liquid and ice cloud microphysics in the CSU general circulation model. Part III: Sensitivity to model assumption. *J. Climate*, **9**, 561–586.

Fraser, R.S. and Y.J. Kaufman, 1985: The relative importance of aerosol scattering and absorption in remote sensing. IEEE Trans Geoscience and Remote Sensing, **23**, 625–633.

Frick, G. M. and W. A. Hoppel, 1993: Airship measurements of aerosol size distributions, cloud droplet spectra, and trace gas concentrations in the marine boundary layer. *Bull. Amer. Met. Soc.*, **74**, 2195-2202.

Fuller, K.A., W.C. Malm, and S.M. Kreidenweis, 1999: Effects of mixing on extinction by carbonaceous particles. *J. Geophys. Res.*, **104**, 15,941-15,954.

Gard, E. E., M. J. Kleeman, D. S. Gross, L. S. Hughes, J. O. Allen, B. D. Morrical, D. P. Fergenson, T. Dienes, M. E. Galli, R. J. Johnson, G. R. Cass, Glen R, and K. A. Prather, 1998: Direct observation of heterogeneous chemistry in the atmosphere. *Science*, **279**, 1184-1187.

Garrett, T. J. and P. V. Hobbs, 1995: Long-range transport of continental aerosols over the Atlantic Ocean and their effects on cloud structures. *J. Atmos. Sci.*, **52**, 2977–2984.

Gasso, S., D.A. Hegg, D.S. Covert, D. Collins, K.J. Noone, E. Ostrom, B. Schmid, P.B. Russell, J.M. Livingston, P.A. Durkee and H. Jonsson, 2000: Influence of humidity on the aerosol scattering coefficient and its effect on the upwelling radiance during ACE-2. *Tellus*, **52B**, 546-567.

Ghan, S.J., C. Chuang and J.E. Penner, 1993: A parametrization of cloud droplet nucleation part I: single aerosol type. *Atmos. Res.*, **30**, 197–211.

Ghan, S.J., C.C. Chuang, R.C. Easter, and J.E. Penner, 1995: A parametrization of cloud droplet nucleation. 2. Multiple aerosol types. *Atmos. Res.*, **36**, 39-54.

Ghan, S., L.R. Leung, R.C. Easter, and H.Abdul-Razzak, 1997: Prediction of cloud droplet number in a general circulation model. *J. Geophys. Res. Atmos.*, **102**, 21777–21794.

Ghan, S.J., G. Guzman and H. Abdul-Razzak, 1998: Competition between sea salt and sulphate particles as cloud condensation nuclei. *J. Atmos. Sci.*, **55**, 3340–3347.

Ghan, S., R. Easter, J. Hudson, and F.-M. Bréon, 2001a: Evaluation of aerosol indirect radiative forcing in MIRAGE. *J. Geophysical Research*, in press.

Ghan, S., R. Easter, E. Chapman, H. Abdul-Razzak, Y. Zhang, R. Leung, N. Laulainen, R. D. Saylor, and R. Zaveri, 2001b: A physically-based estimate of radiative forcing by anthropogenic sulphate aerosol. *J. Geophysical Research*, in press.

Ghan, S., N. Laulainen, R. Easter, R. Wagener, S. Nemesure, E. Chapman, Y. Zhang, and R. Leung, 2001c: Evaluation of aerosol direct radiative forcing in MIRAGE, *J. Geophys. Res.*, in press.

Gillani, N., S.E. Schwartz, W.R. Leaitch, J.W. Strapp and G.A. Isaac, 1995: Field observations in continental stratiform clouds: partitioning of cloud droplets between droplets and unactivated interstitial aerosols. *J. Geophys. Res. Atmos.*, **100**, 18687–18706.

Gillette, D., 1978: Wind-tunnel simulation of erosion of soil: Effect of soil texture, sandblasting, wind speed and soil consolidation on dust production. *Atmos. Environ.*, **12**, 1735–1743.

Glantz, P. and K.J. Noone, 2000: A physically-based algorithm for estimating the relationship between aerosol mass to cloud droplet number. *Tellus*, **52**, 1216-1231.

Goloub, P., D. Tanré, J.-L. Deuzé, M. German, A. Marchand, and F.-M. Bréon, 1999: Valication of the first algorithm applied for deriving the aerosol properties over the ocean using the POLER/ADEOS measurements. IEEE.*Trans. Geosc. Rem. Sens.*, **27**, 1586-1596.

Gong, S.L., L.A. Barrie, and J.-P. Blanchet, 1997a: Modeling sea-salt aerosols in the atmosphere. Part 1: Model development, *J. Geophys. Res.*, **102**, 3805-3818.

Gong, S. L., L. A. Barrie, J. Prospero, D. L. Savoie, G. P. Ayers, J.-P. Blanchet, and L. Spacek, 1997b: Modeling sea-salt aerosols in the atmosphere, Part 2: Atmospheric concentrations and fluxes. *J. Geophys. Res.* **102**, 3819-3830.

Gong, S.L., L.A. Barrie, J.-P. Blanchet and L. Spacek, 1998: Modeling size-distributed sea salt aerosols in the atmosphere: An application using Canadian climate models. In: Air Pollution Modeling and Its Applications XII, S.-E. Gryning and N. Chaumerliac (eds), Plenum Press, New York.

Gordon, C., C. Cooper, C. A. Senior, H. Banks, J. M. Gregory, T. C. Johns, J. F. B. Mitchell, and R. A. Wood, 2000: The simulation of SST, sea-ice extents and ocean heat transport in a version of the Hadley Centre coupled model without fluxx adjustments. *Climate Dynamics*, **16**, 147-168.

Goudie, A.S. and N.J. Middleton, 1992: The changing frequency of dust storms through time. *Clim. Change*, **20**, 197-225.

Graf, H.-F., J. Feichter and B. Langmann, 1997: Volcanic sulphur emissions: Estimates of source strength and its contribution to the global sulphate distribution. *J. Geophys. Res.*, **102**, 10727–10738.

Graf, H.-F., B. Langmann and J. Feichter, 1998: The contribution of Earth degassing to the atmospheric sulphur budget. Chem. Geol., **147**, 131–145.

Gras, J. L., 1995: CN, CCN, and.particle size in Southern Ocean air at Cape Grim, *Atmos. Res.*, **35**, 233-251.

Greenwald, T. J., G. L. Stephens, T. H. Vonder Haar, and D. L. Jackson, 1993: A physical retrieval of cloud liquid water over the global oceans using special sensor microwave/imager (SSM/I) observations. *J. Geophys. Res.*, **98**, 18,471-18,488.

Griffin, R.J., D.R. Cocker, III, R.C. Flagan and J.H. Seinfeld, 1999a: Organic aerosol formation from the oxidation of biogenic hydrocarbons. *J. Geophys. Res.*, **104**, 3555–3567.

Griffin, R.J., D.R. Cocker, III, J.H. Seinfeld and D. Dabdub 1999b: Estimate of global atmospheric organic aerosols from oxidation of biogenic hydrocarbons. *Geophys. Res. Lett.*, **26**, 2721–2724.

Gschwandtner, G., K. Gschwandtner, K. Eldridge, C. Mann, and D. Mobley, 1986: Historic emissions of sulphur and nitrogen oxides in the United States from 1900 to 1980. *J. Air Pollut. Control Assoc.*, **36**, 139-149.

Guenther, A., C. Hewitt, D. Erickson, R. Fall, C. Geron, T. Graedel, P. Harley, L. Klinger, M. Lerdau, W. McKay, T. Pierce, B. Scholes, R. Steinbrecher, R. Tallamraju, J. Taylor and P. Zimmerman, 1995: A global model of natural volatile organic compound emissions. *J. Geophys. Res. Atmos.*, **100**, 8873–8892.

Guenther, A., B. Baugh, G. Brasseur, J. Greenberg, P. Harley, L. Klinger, D. Serca, and L. Vierling, 1999: Isoprene emission estimates and uncertainties for the Central African EXPRESSO study domain, *J. Geophys. Res.*, **104**, 30,625-30,639.

Gultepe, I. and G.A. Isaac, 1998: Scale effects on averaging of cloud droplet and aerosol number concentrations: observations and models. *J. Climate*, **12**, 1268-1279.

Hagen, D. E., J. Podzinsek and M. B. Trueblood, 1995: Upper-tropospheric aerosol sampled during FIRE IFO II. *J. Atmos. Sci.*, **52**, 4196–4209.

Han, Q., W. B. Rossow, and A. A. Lacis, 1994: Near-global survey of effective droplet radii in liquid water clouds using ISCCP data. *J. Climate*, **7**, 465-497.

Han, Q, W.B. Rossow, J. Chou and R.M. Welch 1998a: Global variation of column droplet concentration in low-level clouds, *Geophys. Res. Lett*, **25**, 1419-1422.

Han, Q., Rossow, W.B., Chou, J. and Welch, R.M., 1998b: Global survey of the relationships of cloud albedo and liquid water path with droplet size using ISCCP. *J. Climate*, **11**, 1516–1528.

Hansen, J.E., M. Sato and R. Ruedy, 1997: Radiative forcing and climate response. *J. Geophys. Res.*, **102**, 6831–6864.

Hansson, H.-C., M.J. Rood, S. Koloutsou-vakakis, K. Hameri, D. Orsini, and A. Wiedensohler, 1998: NaCl aerosol particle hygroscopicity dependence on mixing with organic compounds. *J. Atmos. Chem.*, **31**, 321–346.

Havers, N., Burba, P., Lambert, J. and Klockow, D., 1998: Spectroscopic characterization of humic-like substances in airborne particulate matter. *J. Atmos. Chem.*, **29**, 45-54.

Haywood, J.M. and K.P. Shine, 1995: The effect of anthropogenic sulphate and soot aerosol on the clear sky planetary radiation budget. *Geophys. Res. Lett.*, **22**, 603–606.

Haywood, J., and V. Ramaswamy, 1998: Global sensitivity studies of the direct radiative forcing due to anthropogenic sulphate and black carbon aerosols. *J. Geophys. Res.*, **103**, 6043-6058.

Haywood, J., V. Ramaswamy, and B. Soden, 1999: Tropospheric aerosol climate forcing in clear-sky satellite observations over the oceans, *Science*, **283**, 1299-1303.

Hegg, D. A., L. F. Radke and P. V. Hobbs, 1993: Aerosol size distributions in the cloudy atmospheric boundary layer of the North Atlantic Ocean. *J. Geophys. Res.*, **98**, 8841-8846.

Hegg, D.A., 1994: Cloud condensation nucleus-sulphate mass relationship and cloud albedo. *J. Geophys. Res.*, **99**, 25,903-25,907.

Hegg, D. A., D. S. Covert, M. J. Rood and P. V. Hobbs, 1996a: Measurement of the aerosol optical properties in marine air. *J. Geophys. Res.*, **96**, 12893-12903.

Hegg, D.A., P.V. Hobbs, S. Gasso, J.D. Nance and A.L. Rango, 1996b: Aerosol measurements in the Arctic relevant to direct and indirect radiative forcing. *J. Geophys. Res.*, **101**, 23349–23363.

Hegg, D.A., J. Livingston, P.V. Hobbs, T. Novakov and P. Russell, 1997: Chemical apportionment and aerosol column optical depths off the mid-Atlantic coast of the United States. *J. Geophys. Res.*, **102**, 25,293–25,303.

Heintzenberg, J., K. Okada and J. Strom, 1996: On the composition of non-volatile material in upper tropospheric aerosols and cirrus crystals. *Atmos. Res.*, **41**, 81–88.

Heintzenberg, J., R.J. Charlson, A.D. Clarke, C. Liousse, V. Ramaswamy, K.P. Shine, M. Wendisch and G. Helas, 1997: Measurements and modelling of aerosol single-scattering albedo: Progress, problems and prospects. *Contr. Atmosph. Phys.*, **70**, 249–263.

Herman, M., J. L. Deuzé, P. Goloub, F. M. Bréon, and D. Tanré, 1997: Remote sensing of aerosols over land surfaces including polarization measurements and application to POLDER measurements. *J. Geophys. Res.*, **102**, 17039-17049.

Herzog, M., J. E. Penner, J. J. Walton, S. M. Kreidenweis, D. Y. Harrington, and D. K. Weisenstein, 2000: Modeling of global sulphate aerosol number concentrations. In Nucleation and Atmospheric aerosols 2000: 15th International Conference, B. N. Hale and M. Kulmala, Eds., AIP Conference Proceedings, **534**, 677-679.

Heymsfield, A. J. and L. M. Miloshevich, 1993: Homogeneous ice nucleation and supercooled liquid water in orographic wave clouds. *J. Atmos. Sci.*, **50**, 2335–2353.

Heymsfield A,J,, and L.M. Miloshevich 1995: Relative humidity and temperature influences on cirrus formation and evolution -observations from wave clouds and FIRE II. *J. Atmos. Sci.*, **52**, 4302-4326.

Hignett, P., 1991: Observations of diurnal variation in a cloud-capped marine boundary layer. *J. Atmos. Sci.*, **48**, 1474-1482.

Hignett, P., J. T. Taylor, P. N. Francis, and M. D. Glew, 1999: Comparison of observed and modelled direct forcing during TARFOX, *J. Geophys. Res.*, **104**, 2279-2287.

Higurashi, A., and T. Nakajima, 1999: Development of a two channel aerosol algorithm on a global scale using NOAA/AVHRR. *J. Atmos. Sci*, **56**, 924-941.

Hobbs, P. V. and J. D. Locatelli, 1970: Ice nucleus measurements at three sites in western Washington. *J. Atmos. Sci.*, **27**, 90–100.

Hobbs, P. V., J. S. Reid, J. D. Herring, J. D. Nance, R. E. Weiss, J. L. Ross, D. A. Hegg, R. D. Ottmar and C. A. Liousse, 1997: Particle and trace-gas measurements in the smoke from prescribed burns of forest products in the Pacific Northwest. In Biomass Burning and Global Change, ed., J. S. Levine, pp. 697-715, MIT Press, Cambridge, MA.

Hoffmann, T., J.R. Odum, F. Bowman, D. Collins, D. Klockow, R.C. Flagan and J.H. Seinfeld, 1997: Formation of organic aerosols from the oxidation of biogenic hydrocarbons. *J. Atmos. Chem.*, **26**, 189–222.

Hofmann, D. J., 1993: Twenty years of balloon-borne tropospheric aerosol measurements at Laramie, Wyoming. *J. Geophys. Res.*, **98**, 12753-12766.

Hogan, A. W., 1967: Ice nuclei from direct reaction of iodine vapor with vapors from leaded gasoline. *Science*, **158**, 158.

Holben, B. N., T. F. Eck, I. Slutsker, D. Tanré, J. P. Buis, A. Stezer, E. Vermote, J. A. reagan, U. J. Kaufman, T. Nakajima, F. Lavenu, I. Jankowiak, and A. Smirnov, 1998: AERONET–A federated instrument network and data archive for aerosol characterization. *Remote Sens. Environ.*, **66**, 1-16.

Horvath, H., 1993: Atmospheric light absorption-a review. *Atmos. Env.*, **27A**, 293-317.

Houweling, S., F. Dentener, and J. Lelieveld, 1998: The impact of non-methane hydrocarbon compounds on tropospheric chemistry. *J. Geophys. Res.*, **103**, 10,673-106,96.

Howell, S.G. and B.J. Huebert, 1998: Determining marine aerosol scattering characteristics at ambient humidity from size-resolved chemical composition. *J. Geophys. Res. Atmos.*, **103**, 1391–1404.

Hsu, N. C., J. R. Herman, O. Torres, B. N. Holben, D. Tanré, T. F. Eck, A. Smirnov, B. Chatenet, and F. Lavenu, 1999: Comparisons of the TOMS aerosol index with Sun-photometer aerosol optical thickness: Results and applications. *J. Geophys. Res.*, **104**, 6269-6279.

Hudson, J. G. and S. S. Yum, 1997: Droplet spectral broadening in marine stratus. *J. Atmos. Sci.*, **54**, 2642–2654.

Hudson, J., and X. Da, 1996: Volatility and size of cloud condensation nuclei. *J. Geophys. Res.*, **101**, 4435–4442.

Husar, R. B., and W. E. Wilson, 1993: Haze and sulphur emission trends in the Eastern United States. *Envir. Sci. Tech.* **27**, 12-16.

Husar, R. B., J. M. Prospero, and L. L. Stowe, 1997: Characterization of tropospheric aerosols over the oceans with the NOAA advanced very high resolution radiometer optical thickness operational product, *J. Geophys. Res.*, **102**, 16889-16909.

IPCC, 1992: Climate Change 1992: The Supplementary Report to the IPCC Scientific Assessment. Houghton, J.T., B.A. Callander, and S.K.Varney (eds.). Cambridge University Press, Cambridge, UK, 200 pp.

IPCC, 1996: Climate Change 1995: The Science of Climate Change. Contribution of Working Group I to the Second Assessment Report of the Intergovernmental Panel on Climate Change. Houghton, J.T., L.G. Meira Filho, B.A. Callander, N. Harris, A. Kattenberg, and K. Maskell (eds.). Cambridge University Press, Cambridge, United Kingdom and New York, NY, USA, 572 pp.

Isaac, G.A., W.R. Leaitch and J.W. Strapp, 1990: The vertical distribution of aerosols and acid related compounds in air and cloudwater. *Atmos. Environ.*, **24A**, 3033–3046.

Iversen, T., A. Kirkevåg, J. E. Kristjansson, and Ø. Seland, 2000: Climate effects of sulphate and black carbon estimated in a global climate model. In Air Pollution Modeling and its Application XIV, S.-E. Gryning and F.A. Schiermeier, Eds. Kluwer/Plenum Publishers, New York, 335-342.

Jacob, D.J., M.J. Prather, P.J. Rasch, R.-L. Shia, Y.J. Balkanski, S.R. Beagley, D.J. Bergmann, W.T. Blackshear, M. Brown, M. Chiba,

M.P. Chipperfield, J. de Grandpré, J.E. Dignon, J. Feichter, C. Genthon, W.L. Grose, P.S. Kasibhatla, I. Köhler, M.A. Kritz, K. Law, J.E. Penner, M. Ramonet, C.E. Reeves, D.A. Rotman, D.Z. Stockwell, P.F.J. Van Velthoven, G. Verver, O. Wild, H. Yang, and P. Zimmermann, 1997: Evaluation and intercomparison of global atmospheric transport models using 222Rn and other short-lived tracers. *J. Geophys. Res.*, **102**, 5953-5970.

Jacobson, M. Z., 2000: A physically-based treatment of elemental carbon optics: Implications for global direct forcing of aerosols, *Geophys. Res. Lett.*, **27**, 217-220.

Jacobson, M.Z., 2001: Global direct radiative forcing due to multicomponent anthropogenic and natural aerosols. *J. Geophys. Res.*, in press.

Jaenicke, R., 1993: Tropospheric aerosols. In Aerosol-Cloud-Climate Interactions, ed. P. V. Hobbs, Academic Press, San Diego, CA, pp. 1-31.

Jankowiak, I., and D. Tanré, 1992: Satellite climatology of Saharan dust outbreaks: Method and preliminary results. *J. Clim.*, **4**, 646-656.

Jennings, S. G., C. D. O'Dowd, T. C. O'Conner and J. M. McGovern, 1991: Physical characteristics of the ambient aerosol at Mace Head. *Atmos. Environ.*, **25A**, 557-562.

Jensen, E.J., and O.B. Toon, 1992: The potential effects of volcanic aerosols on cirrus cloud microphysics. *Geophys. Res. Lett.*, **19**, 1759-1762.

Jensen, E.J., and O.B. Toon, 1994: Ice nucleation in the upper troposphere - sensitivity to aerosol number density, temperature, and cooling rate. *Geophys. Res. Lett.*, **21**, 2019-2022.

Jensen, E.J. and O.B. Toon, 1997: The potential impact of soot particles from aircraft exhaust on cirrus clouds. *Geophys. Res. Lett.*, **24**, 249–252.

Jeuken, A., P. Veefkind, F. Dentener, S. Metzger, 2001: Simulation of the aerosol optical depth over Europe for August 1997 and a comparison with observations. *J. Geophys. Res.*, in press.

Johnson, D.W., S. Osborne, R. Wood, K. Suhre, R. Johnson, S. Businger, P.K. Quinn, A. Weidensohler, P.A. Durkee, L.M. Russell, M.O. Andreae, C. O'Dowd, K.J. Noone, B.Bandy, J. Rudolph and S. Rapsomanikis, 2000: An overview of the Langrangian experiments undetaken during the North Atlantic regional Aerosol Characterisation Experiment (ACE-2), *Tellus*, **52B**, 290-320.

Jones, P.R., R.J. Charlson and H. Rodhe, 1994a: Aerosols. In: Climate Change 1994: Radiative Forcing of Climate Change and an evaluation of the IPCC IS92a Emission Scenarios. A special Report of IPCC Working Groups I and III. Houghton J.T., L.G. Meira Filho, J. Bruce, Hoesung Lee, B.A. Callander, E. Haites, N. Harris, and K. Maskell (eds.), Cambridge University Press, Cambridge, UK, pp. 131–162.

Jones, A., D.L. Roberts, and A. Slingo, 1994b: A climate model study of indirect radiative forcing by anthropogenic aerosols. *Nature*, **370**, 450–453.

Jones, A. and A. Slingo, 1996: Predicting cloud-droplet effective radius and indirect sulphate aerosol forcing using a general circulation model. *Quart. J. R. Met. Soc.*, **122**, 1573–1595.

Kahn, R. A., West, D. McDonald, B. Rheingans, and M. I. Mishchenko, 1997: Sensitivity of multiangle remote sensing observations to aerosol sphericity. *J. Geophys. Res.*, **102**, 16861-16870.

Kanakidou, M., K. Tsigaridis, F. J. Dentener, and P. J. Crutzen, 2000: Human activity enhances the formation of organic aerosols by biogenic hydrocarbon oxidation. *J. Geophys. Res.*, **1-5**, 9243-9254.

Kärcher, B., Th. Peter, U. M. Biermann, and U. Schumann, 1996: The initial composition of jet condensation trails, *J. Atmos. Sci.*, **53**, 3066-3083.

Kärcher, B., R. Busen, A. Petzold, F.P. Schröder, U. Schumann, and E.J. Jensen, 1998: Physicochemistry of aircraft-generated liquid aerosols, soot, and ice particles, 2, Comparison with observations and sensitivity studies. *J. Geophys. Res.*, **103**, 17129-17148.

Kaufman, Y. J. , 1987: Satellite sensing of aerosol absorption. *J. Geophys. Res.*, **92**, 4307-4317.

Kaufman, Y.J., D. Tanré, H.R. Gordon, T. Nakajima, J. Lenoble, R. Frouin, H. Grassl, B.M. Herman, M.D. King and P.M. Teillet, 1997: Passive remote sensing of tropospheric aerosol and atmospheric correction for the aerosol effect. *J. Geophys. Res.*, **102**, 16,815–16,830.

Kent, G.S., and M.P. McCormick, 1984: SAGE and SAM II measurements of global stratospheric aerosol optical depth and mass loading. *J. Geophys. Res.*, **89**, 5303-5314.

Kettle, A.J., M.O. Andreae, D. Amouroux, T.W. Andreae, T.S. Bates, V. Berrresheim, H. Bingemer, R. Boniforti, M.A.J. Curran, G.R. DiTullio, G. Helas, G.B. Jones, M.D. Keller, R.P. Kiene, C. Leck, M. Levasseur, G. Malin, M. Maspero, P. Matrai, A.R. McTaggart, N. Mihalopoulos, B.C. Nguyen, A. Novo, J.P. Putaud, S. Rapsomanikis, G. Roberts, G. Schebeske, S. Sharma, R. Simo, R. Staubes, S. Turner and G. Uher, 1999: A global database of sea surface dimethylsulfide (DMS) measurements and a procedure to predict sea surface DMS as a function of latitude, longitude and month. *Global Biogeochemical Cycles*, **13**, pp. 399–444.

Kettle, A.J., and M.O. Andreae, 2000: Flux of dimethylsulfide from the oceans: A comparison of updated data sets and flux models. *J. Geophys. Res.*, **105**, 26793-26808.

Kiehl, J.T., T.L. Schneider, P.J. Rasch, M.C. Barth and J. Wong, 2000: Radiative forcing due to sulphate aerosols from simulations with the NCAR community model (CCM3). *J. Geophys. Res.*, **105**, 1441-1458.

King, M.D., Y.J. Kaufman, D. Tanré, and T. Nakajima, 1999: Remote sensing of tropospheric aerosols from space: Past, present and future. *Bull.Amer. Meteor. Soc.*, 2222-2259.

Knutson, T.R., T.L. Delworth, K.W. Dixon and R.J. Stouffer, 1999: Model assessment of regional surface temperature trends (1949-97). *J. Geophys. Res.*, **104**, 30,981-30,996.

Koch, D., D. Jacob, I. Tegen, D. Rind and M. Chin, 1999: Tropospheric sulphur simulation and sulphate direct radiative forcing in the Goddard Institute for Space Studies general circulation model. *J. Geophys. Res.*, **104**, 23,799–23,822.

Koop, T., A.K. Bertram, L.T. Molina and M.J. Molina, 1999: Phase transitions in aqueous NH_4HSO_4 solutions. *J. Phys. Chem.*, **103**, 9042-9048.

Kotchenruther, R. A. and P. V. Hobbs, 1998: Humidification factors of aerosols from biomass burning in Brazil. *J. Geophys. Res.*, **103**, 32081-32090.

Kotchenruther, R. A., P. V. Hobbs and D. A. Hegg, 1999: Humidification factors for atmospheric aerosols off the mid-Atlantic coast of the United States. *J. Geophys. Res.*, **104**, 2239-2251.

Krekov, G.M., 1993: Models of atmospheric aerosols. In Aerosol Effects on Climate, S.G. Jennings, ed., U. of Arizona Press, Tucson, Ariz.

Kriedenweis, S.M., Y. Chen, D.C. Rogers and P.J. DeMott, 1998: Isolating and identifying atmospheric ice-nucleating aerosols: A new technique. *Atmos. Res.*, **46**, 263–279.

Kuang, Z. and Y.L. Yung, 2000: Reflectivity variations off the Peru coast: Evidence for indirect effect of anthropogenic sulphate aerosols on clouds. *Geophys. Res. Lett.*, **16**, 2501-2504.

Kulmula, M., A. Laaksonen, P. Korhonen, T. Vesala and T. Ahonen, 1993: The effect of atmospheric nitric acid vapor on cloud condensation nucleus activation. *J. Geophys. Res.*, **98**, 22949–22958.

Kulmala, M., P. Korhonen, T. Vesala, H.-C. Hansson, K. Noone and B. Svenningsson, 1996: The effect of hygroscopicity on cloud droplet formation. *Tellus*, **48B**, 347–360.

Kumai, M., 1976: Identification of nuclei and concentrations of chemical species in snow crystals sampled at the South Pole. *J. Atmos. Sci.*, **33**, 833–841.

Laaksonen, A., P. Korhonen, M. Kulmala and R.J. Charlson, 1998: Modification of the Kohler equation to include soluble trace gases and slightly soluble substances, *J. Atmos. Sci.*, **55**, 853-862.

Lacis, A. A. and J. E. Hansen, 1974: A parametrization for the absorption of solar radiation in the Earth's atmosphere. *J. Atmos. Sci.*, **31**, 118-133.

Langer, G., J. Posenski and C. P. Edwards, 1967: A continuous ice nucleus counter and its application to tracking in the troposphere. *J. Appl. Meteor.*, **6**, 114–125.

Langner, J. and H. Rodhe, 1991: A global three-dimensional model of the tropospheric sulphur cycle. *J. Atmos. Chem.*, **13**, 225–263.

Lavanchy, V. M. H., H. W. Gäggeler, U. Schotterer, M. Schwikowski, and U. Baltensperger, 1999: Historical record of carbonaceous particle concentrations from a European high-alpine glacier (Colle Gnifetti, Switzerland). *J. Geophys. Res.*, **104**, 21227-21236.

Lawrence, M.G., W.L. Chameides, P.S. Kasibhatla, H. Levy II, and W. Moxim, 1995: Lightning and atmospheric chemistry: The rate of atmospheric NO production, in Handbook of Atmospheric Electrodynamics, 1, H. Volland, Editor, 189-202, CRC Press, Boca Raton, Florida.

Leaitch, W. R. and G. A. Isaac, 1991: Tropospheric aerosol size distribution from 1982 to 1988 over Eastern North America. *Atmos. Environ.*, **25A**, 601-619.

Leaitch, W.R., S.-M., Li, P.S.K. Liu, C.M. Banic, A.M. Macdonald, G.A. Isaac, M.D. Couture, and J.W. Strapp, 1996a: Relationships among CCN, aerosol size distribution and ion chemistry from airborne measurements over the Bay of Fundy in August-September, 1995. In: Nucleation and Atmospheric Aerosols, M. Kulmala and P. Wagner (eds). Elsevier Science Inc., pp. 840–843.

Leaitch, W.R., C.M. Banic, G.A. Isaac, M.D. Couture, P.S.K. Liu, I. Gultepe, S.-M. Li, L.I. Kleinman, P.H. Daum and J.I. MacPherson, 1996b: Physical and chemical observations in marine stratus during 1993 NARE: Factors controlling cloud droplet number concentrations. *J. Geophys. Res. Atmos.*, **101**, 29123–29135.

Leaitch, W.R., J.W. Bottenheim, T.A. Biesenthal, S.-M. Li, P.S.K. Liu, K. Asalian, H. Dryfhout-Clark, F. Hopper and F.J. Brechtel, 1999: A case study of gas-to-particle conversion in an eastern Canadian forest. *J. Geophys. Res. Atmos.*, **104**, 8095–8111.

Le Canut, P., M. O. Andreae, G. W. Harris, J. G. Wienhold and T. Zenker, 1996: Airborne studies of emissions from savanna fires in southern Africa. 1. Aerosol emissions measured with a laser optical particle counter. *J. Geophys. Res.*, **101**, 23615-23630.

Le Canut, P., M. O. Andreae, G. W. Harris, J. G. Wienhold and T. Zenker, 1992: Aerosol optical properties over southern Africa during SAFARI-92. In Biomass Burning and Global Change, ed. J. S. Levine, pp. 441-459, MIT Press, Cambridge, MA.

Le Treut, H., and Z.X. Li, 1991: Sensitivity of an atmospheric general circulation model to prescribed SST changes: Feedback effects associated with the simulation of cloud optical properties, *Climate Dynam.*, **5**, 175-187.

Lelieveld, J., G.J. Roelofs, L. Ganzeveld, J. Feichter, and H. Rodhe, 1997: Terrestrial sources and distribution of atmospheric sulphur, *Phil. Trans. R. Soc. Lond. B.*, **352**, 149-158.

Leroy, M., J. -L. Deuzé, F. -M. Bréon, O. Hautecoeur, M. Herman, J.-C. Buriez, D. Tanré, S. Bouffiés, P. Chazette, and J.-L. Roujean, 1997: Retrieval of atmospheric properties and surface bidirectional reflectances over land from POLDER/ADEOS. *J. Geophys. Res.*, **102**, 17023-17037.

Levin, Z. and S. A. Yankofsky, 1983: Contact versus immersion freezing of freely suspended droplets by bacterial ice nuclei. *J. Climate* Appl. Meteor., **22**, 1964–1966.

Levitus, S. and T.P. Boyer, 1994: NOAA Atlas NESDIS 4, World Ocean Atlas 1994, vol. 4: Temperature. Nat. Environ. Satellite, Data, and Inf. Serv., Nat. Oceanic and Atmos. Admin., U.S. Dep. Of Comm. Washington, D.C.

Li, S.-M., C.M. Banic, W.R. Leaitch, P.S.K. Liu, G.A. Isaac, X.-L. Zhou, and Y.-N. Lee 1996: Water-soluble fractions of aerosol and the relation to size distributions based on aircraft measurements from the

North Atlantic Regional Experiment. *J. Geophys. Res. Atmos.*, **101**, 29,111-29,121.

Li, F, and K. Okada, Diffusion and modification of marine aerosol particles over the coastal areas in China, A case study using a single particle analysis, 1999: *J. Atmos. Sci.*, **56**, 241-248.

Li-Jones, X. and J.M. Prospero, 1998: Variations in the size distribution of non-sea-salt sulphate aerosol in the marine boundary layer at Barbados: Impact of African dust. *J. Geophys. Res.*, **103**, 16073–16084.

Lin, H., K.J. Noone, J. Strom, and A.J. Heymsfield, 1998: Small ice crystals in cirrus clouds: A model study and comparison with in-situ observations. *J. Atmos. Sci.*, **55**, 1928-1939.

Liou, K.-N., 1980: An Introduction to Atmospheric Radiation, Academic Press, New York.

Liou, K.-N. and S.-C. Cheng, 1989: Role of cloud microphysical processes in climate: an assessment from a one-dimensional prespective. *J. Geophy. Res.*, **94**, 8599-9607.

Liousse, C., H. Cachier and S.G. Jennings, 1993: Optical and thermal measurements of black carbon aerosol content in different environments: Variation of the specific attenuation cross-section, sigma (s). *Atmos. Environ.*, **27A**, 1203–1211.

Liousse, C., J.E. Penner, C. Chuang, J.J. Walton, H. Eddleman and H. Cachier, 1996: A global three-dimensional model study of carbonaceous aerosols. *J. Geophys. Res. Atmos.*, **101**, 19411–19432.

Lippman, M. , 1980: Size distribution in urban aerosols. Am. N. Y. Acad. Sci., **328**, 1-12.

Liss, P.S. and L. Merlivat, 1986: Air-sea gas exchange rates: Introduction and synthesis. In: The Role of Air-Sea Exchange in Geochemical Cycling. P. Buat Menard (ed), D. Reidel Publishing, Berlin, 113–127.

Lohmann, U. and J. Feichter, 1997: Impact of sulphate aerosols on albedo and lifetime of clouds: A sensitivity study with the ECHAM4 GCM. *J. Geophys. Res. Atmos.*, **102**, 13685–13700.

Lohmann, U., K. Von Salzen, and N. McFarlane, H.G. Leighton, and J. Feichter, 1999a: The tropospheric sulphur cycle in the Canadian general circulation model. *J. Geophys. Res.*, **26**,833-26,858.

Lohmann, U., J. Feichter, C.C. Chuang and J.E. Penner, 1999b: Prediction of the number of cloud droplets in the ECHAM GCM. *J. Geophys. Res.*, **104**, 9169–9198.

Lohmann, U., J. Feichter, C.C. Chuang, and J.E. Penner, 1999c: Erratum, *J. Geophys. Res.*, **104**, 24,557-24,563.

Lohmann, U., J. Feichter, J.E. Penner, and R. Leaitch, 2000: Indirect effect of sulphate and carbonaceous aerosols: A mechanistic treatment, *J. Geophys. Res.*, **105**. 12,193-12,206.

Lohmann, U., R. Leaitch, K. Law, L. Barrie, Y. Yi, D. Bergmann, M. Chin, R. Easter, J. Feichter, A. Jeukin, E. Kjellstroem, D. Koch, C. Land, P. Rasch, G.-J. Roelofs, 2001: Comparison of the vertical distributions of sulphur species from models which participated in the COSAM excercise with observations. *Tellus*, in press.

Lovett, R.F., 1978: Quantitative measurement of airborne sea salt in the North Atlantic. *Tellus*, **30**, 358-364.

Ludwig, J., L.T. Marufu, B. Huber, M.O. Andreae and G. Helas, 2001: Combustion of biomass fuels in developing countries - A major source of atmospheric pollutants. *Environmental Pollution*, in press.

Marticorena, B., G. Bergametti, B. Aumont, Y. Callot, C. Ndoume and M. Legrand, 1997: Modeling the atmospheric dust cycle. 2. Simulation of Saharan dust sources. *J. Geophys. Res. Atmos.*, **102**, 4387–4404.

Martin, S.T., 1998: Phase transformations of the ternary system (NH4)2SO4-H2SO4-H2O and the implications for cirrus cloud formation. *Geophys. Res. Lett.*, **25**, 1657–1660.

Martin, G.M., D.W. Johnson, and A. Spice, 1994: The measurement and parametrization of effective radius of droplets in warm stratocumulus clouds, *J. Atmos. Sci.* **51**, 1823-1842.

Martins, J.V., P. Artaxo, C. Liousse, J.S. Reid, P.V. Hobbs and Y.J. Kaufman, 1998: Effects of black carbon content, particle size, and mixing on light absorption by aerosols from biomass burning in Brazil. *J. Geophys. Res. Atmos.*, **103**, 32041–32050.

Matthias-Maser, S. and R. Jaenicke, 1995: The size distribution of primary biological aerosol particles with radii >0.2 μm in an urban-rural influenced region. *Atmospheric Research*, **39**, 279–286.

McCormick, M.P., and R.E. Veiga, 1992: SAGE II measurements of early Pinatubo aerosols. *Geophys. Res. Lett.*, **19**, 155-158.

McCormick, M.P., L.W. Thomason and C.R. Trepte, 1995: Atmospheric effects of the Mt. Pinatubo eruption. *Nature*, **373**, 399–404.

McInnes, L.,D. Covert and B. Baker, 1997: The number of sea-salt , sulphate, and carbonaaceous particles in the marine atmosphere: EM measurements consistent with the ambient size distribution, *Tellus*, **49B**, 300-313.

McInnes, L., M. Bergin, J. Ogren and S. Schwartz, 1998: Apportionment of light scattering and hydroscopic growth to aerosol composition. *Geophys. Res. Lett.*, **25**, 513-516.

McMurry, P.H., X.Q. Zhang and C.T. Lee, 1996: Issues in aerosol measurement for optics assessments. *J. Geophys. Res. Atmos.*, **101**, 19189–19197.

Menon, S., and V.K. Saxena, 1998: Role of sulphates in regional cloud-climate interactions. *Atmos. Res.*, **47-48**, 299–315.

Meszaros, E., 1981: Atmospheric Chemistry. Elsevier, New York, pp. 1-201.

Miller, R.L. and I. Tegen, 1998: Climate response to soil dust aerosols. *J. Climate*, **11**, 3247–3267.

Minnis, P., E. F. Harrison, L. L. Stowe, G. G. Gibson, F. M. Denn, D. R. Doelling, W. L. Smith, Jr., 1993: Radiative climate forcing by the Mt. Pinatubo eruption. *Science*, **259**, 1411-1415.

Mishchenko, M.K., I.V. Geogdzhayev, B. Cairns, W.B. Rossow, and A.A. Lacis, 1999: Aerosol retrievals over the ocean using channel 1 and 2 AVHRR data: A sensitivity analysis and preliminary results, *Applied Optics*, **38**, 7325-7341.

Monahan, E.C., D.E. Spiel and K.L. Davidson, 1986: A model of marine aerosol generation via whitecaps and wave disruption in oceanic whitecaps. In: Oceanic whitecaps and their role in air-sea exchange processes, E.C. Monahan and G.M. Niocaill (eds), D. Reidel Publishing, Dordrecht, Holland, pp. 167–174.

Moosmüller, H., W.P. Arnott, C.F. Rogers, J.C. Chow, C.A. Frazier, L.E. Sherman and D.L. Dietrich, 1998: Photo-acoustic and filter measure-ments related to aerosol light absorption during the Northern Front Range Air Quality Study (Colorado 1996/1997). *J. Geophys. Res. Atmos.*, **103**, 28149–28157.

Moulin, C., C.E. Lambert, F. Dulac and U. Dayan, 1997: Control of atmospheric export of dust from North Africa by the North Atlantic oscillation. *Nature*, **387**, 691–694.

Murphy, D. M., and D. S. Thomson, 1997: Chemical composition of single aerosol particles at Idaho Hill: Negative ion measurements. *J. Geophys. Res.* **102**, 6353-6368.

Murphy, D.M., J.R. Anderson, P.K. Quinn, L.M. McInnes, F.J. Brechtel, S.M. Kreidenweis, A.M. Middlebrook, M. Posfai, D.S. Thomson and P.R. Buseck, 1998a: Influence of sea-salt on aerosol radiative proper-ties in the Southern Ocean marine boundary layer. *Nature*, **392**, 62–65.

Murphy, D.M., D.S. Thomson and T.M.J. Mahoney, 1998b: In situ measurements of organics, meteoritic material, mercury, and other elements in aerosols at 5 to 19 kilometers. *Science*, **282**, 1664–1669.

Myhre, G., F. Stordal, K. Restad and I.S.A. Isaksen, 1998: Estimation of the direct radiative forcing due to sulphate and soot aerosols. *Tellus*, **50B**, 463–477.

Mylona, S., 1996: Sulphur dioxide emissions in Europe 1980-1991 and their effect on sulphur concentrations and depositions. *Tellus*, **48B**, 662-689.

Nakajima, T. and A. Higurashi, 1997: AVHRR remote sensing of aerosol optical properties in the Persian Gulf region, summer 1991. *J. Geophys. Res.*, **102**, 16,935–16,946.

Nakajima, T., A. Higurashi, K. Aoki, T. Endoh, H. Fukushima, M. Toratani, Y. Mitomi, B. G. Mitchell, and R. Frouin, 1999: Early phase analysis of OCTS radiance data for aerosol remote sensing, IEEE Trans. Geo. Remote Sens., **37**, 1575-1585.

Nakajima, T., A. Higurashi , K. Kawamoto, and J. E. Penner, 2001: A possible correlation between satellite-derived cloud and aerosol microphysical parameters. *Geophys. Res. Lett.*, in press.

Nakićenović, N., J. Alcamo, G. Davis, B. de Vries, J. Fenhann, S. Gaffin, K. Gregory, A. Grubler, T.Y. Jung, T. Kram, E.L. La Rovere, L. Michaelis, S. Mori, T. Morita, W. Pepper, H. Pitcher, L. Price, K. Raihi, A. Roehrl, H.-H. Rogner, A. Sankovski, M. Schlesinger, P. Shukla, S. Smith, R. Swart, S. van Rooijen, N. Victor, and Z. Dadi, 2000: Emissions Scenarios. A Special Report of Working Group III of the Intergovernmental Panel on Climate Change. Cambridge University Press, Cambridge, United Kingdom and New York, NY, USA, 599 pp.

NASA, 1992: The atmospheric effects of stratospheric aircraft: A first program report, M.J. Prather *et al.* (eds.), NASA Ref. Publ. 1272, pp. 64–91.

Neusüß , C., D. Weise, W. Birmili, H. Wex, A. Wiedensohler, and D. S. Covert, 2000: Size segregated chemical, gravimetric and number distribution-derived mass closure of the aerosol in Sagres, Portugal during ACE-2. *Tellus*, **52B**, 169-184.

Norman, A. L., L. A. Barrie, D. Toom-Sauntry, A. Sirois, H. R. Krouse, S. M. Li, and S. Sharma, 1999: Sources of aerosol sulphate at Alert: Apportionment using stable isotopes. *J. Geophys. Res.*, **104**, 11,619-11,631.

Novakov, T. and J.E. Penner, 1993: Large contribution of organic aerosols to cloud-condensation-nuclei concentrations. *Nature*, **365**, 823–826.

Novakov, T. and C. E. Corrigan, 1996: Cloud condensation nucleus activity of the organic component of biomass smoke particles. *Geophys. Res. Lett.*, **23**, 2141–2144.

Novakov, T., D.A. Hegg and P.V. Hobbs, 1997: Airborne measurements of carbonaceous aerosols on the East Coast of the United States. *J. Geophys. Res. Atmos.*, **102**, 30023–30030.

Nyeki, S., F. Li, E. Weingartner, N. Streit, L. Colbeck, H. W. Gaggeler and U. Boltensperger, 1998: The background aerosol size distribution in the free troposphere: An analysis of the annual cycle at a high-alpine site. *J. Geophys. Res.*, **103**, 31749-31761.

O'Dowd, C.D. and M.H. Smith, 1993: Physico-chemical properties of aerosols over the Northeast Atlantic: Evidence for wind-speed-related sub-micron sea-salt aerosol production. *J.Geophys. Res.*, **98**, 1137–1149.

O'Dowd,C., M.H.Smith, I.E.Consterdine, and J. A. Lowe, 1997: Marine aerosol, sea-salt, and the marine sulphur cycle: A short review. *Atmos. Environ.*, **31**, 73-80.

O'Dowd, C.D., J.A. Lowe, M.H. Smith, and A.D. Kaye, 1999: The relative importance of Nss-sulphate and sea-salt aerosol to the marine CCN population: an improved multi-component aerosol-cloud droplet parametrization. *Quart. J. Roy. Meteorol. Soc.*, **125**, 1295-1313.

Odum, J.R., T. Hoffmann, F. Bowman, D. Collins, R.C. Flagan and J.H. Seinfeld, 1996: Gas/particle partitioning and secondary organic aerosol yields. *Environ. Sci. Tech.*, **30**, 2580–2585.

Osborn, M. T., G. S. Kent, and C. R. Trepte, 1998: Stratospheric aerosol measurements by the Lidar in Space Technology Experiment. *J. Geophys. Res.*, **103**, 11447-11453.

Pan, W., M.A. Tatang, G.J. McRae and R.G. Prinn, 1997: Uncertainty analysis of direct radiative forcing by anthropogenic sulphate aerosols, *J. Geophys. Res.*, **102**, 21915–21924.

Penner, J.E., R.E. Dickinson and C.A. O'Neill, 1992: Effects of aerosol from biomass burning on the global radiation budget. *Science*, **256**, 1432–1434.

Penner, J.E., H. Eddleman and T. Novakov, 1993: Towards the development of a global inventory of black carbon emissions, *Atmos. Environ.*, **27A**, 1277–1295.

Penner, J.E., R.J. Charlson, J.M. Hales, N. Laulainen, R. Leifer, T. Novakov, J. Ogren, L.F. Radke, S.E. Schwartz, and L. Travis, 1994a: Quantifying and minimizing uncertainty of climate forcing by anthropogenic aerosols, *Bull. Am. Met. Soc.*, **75**, 375–400.

Penner, J.E., C.A. Atherton, and T.E. Graedel, 1994b: Global emissions and models of photochemically active compounds. In Global Atmospheric-Biospheric Chemistry, ed. R. Prinn, Plenum Publishing, N.Y., 223-248.

Penner, J.E., 1995: Carbonaceous aerosols influencing atmospheric radiation: black and organic carbon, in Aerosol Forcing of Climate, ed. R.J. Charlson and J. Heintzenberg, John Wiley and Sons, Chichester, 91-108.

Penner, J.E., C.C. Chuang, and C. Liousse, 1996: The contribution of carbonaceous aerosols to climate change. In Nucleation and Atmospheric Aerosols 1996, M. Kulmala and P.E. Wagner, eds., Elsevier Science, Ltd., 759-769.

Penner, J.E., D. Bergmann, J.J. Walton, D. Kinnison, M.J. Prather, D. Rotman, C. Price, K.E. Pickering, S.L. Baughcum, 1998a: An evaluation of upper tropospheric NOx with two models. *J. Geophys. Res.*, **103**, 22,097-22,114.

Penner, J.E., C.C. Chuang and K. Grant, 1998b: Climate forcing by carbonaceous and sulphate aerosols. *Clim. Dyn.*, **14**, 839–851.

Penner, J.E., C.C. Chuang and K. Grant, 1999a: Climate change and radiative forcing by anthropogenic aerosols: A review of research during the last five years. Paper presented at the La Jolla International School of Science, The Institute for Advanced Physics Studies, La Jolla, CA, U.S.A.

Penner, J.E., D. Lister, D. Griggs, D. Docken M. MacFarland (Eds.), 1999b: Aviation and the Global Atmosphere, Intergovernmental Panel on Climate Change Special Report, Cambridge University Press, Cambridge, U.K., 373 pp.

Perez, P. J., J. A. Garcia, and J. Casanova, 1985: Ice nuclei concentrations in Vallodolid, Spain and their relationship to meteorological parameters. *J. Res. Atmos.*, **19**, 153–158.

Petzold, A., J. Ström, S. Ohlsson, and F.P. Schröder, 1998: Elemental composition and morphology of ice-crystal residual particles in cirrus clouds and contrails. *Atmos. Res.* **49**, 21-34.

Pham, M., J.-F. Müller, G. P. Brasseur, C. Granier, and G. Mägie, 1996: A 3D model study of the global sulphur cycle: Contributions of anthropogenic and biogenic sources. *Atmos. Env.*, **30**, 1815-1822.

Piccot, S.D., J.J. Watson and J.W. Jones, 1992: A global inventory of volatile organic compound emissions from anthropogenic sources. *J. Geophys. Res.*, **97**, 9897–9912.

Pincus, R. and M.A. Baker, 1994: Effect of precipitation on the albedo susceptibility of clouds in the marine boundary layer. *Nature*, **372**, 250–252.

Pincus, R., M.A. Baker and C.S. Bretherton, 1997: What controls stratocumulus radiation properties? Lagrangian observations of cloud evolution. *J. Atmos. Sci.*, **54**, 2215–2236.

Pitari, G., E. Mancini, A. Bregman, H.L. Rogers, J.K. Sundet, V. Grewe, and O. Dessens, 2001: Sulphate particles from subsonic aviation: Impact on upper tropospheric and lower stratospheric ozone, Physics and Chemistry of the Earth, European Geophysical Society, in press.

Pitari, G., and E. Mancini, 2001: Climatic impact of future supersonic aircraft: Role of water vapour and ozone feedback on circulation, Physics and Chemistry of the Earth, European Geophysical Society, in press.

Podgorny, I.A., W. Conant, V. Ramanathan, and S. K. Sateesh, 2000: Aerosol modulation of atmospheric and surface solar heating over the tropical Indian Ocean. *Tellus B.*, **52**, 947-958.

Posfai, M., H. Xu, J.R. Anderson, and P.R. Buseck, 1998: Wet and dry sizes of atmopsheric aerosol particles: an AFM-TEM study. *Geophys. Res. Lett.*, **25**, 1907–1910.

Price, C., J.E. Penner, and M.J. Prather, 1997: NOx from lightning, Part I: Global distribution based on lightning physics. *J. Geophys. Res.*, **102**, 5929-2941, 1997.

Prospero, J.M., M. Uematsu, and D.L. Savoie, 1989: Mineral aerosol transport to the Pacific Ocean. In Chemical Oceanography, J.P. Riley, and R. Chester, Eds., Academic, London, 188-218.

Prospero, J.M. and R.T. Nees, 1986: Impact of the North African drought and El Nino on mineral dust in the Barbados trade winds. *Nature*, **320**, 735–738.

Prospero, J.M., 1999: Long-term measurements of the transport of African mineral dust to the southeastern United States: Implications for regional air quality. *J. Geophys. Res.*, **104**, 15,917–15,928.

Pruppacher, H.R. and J.D. Klett, 1997: Microphysics of Clouds and Precipitation. Reidel, Dordrecht, 954 pp.

Putaud, J.-P., R. Van Digenen, M. Mangoni, A. Virkkula, F. Raes, H. Maring, J.M. Prospero, E. Swietlicki, O.H. Berg, R. Hillami, and T. Mäkelä, 2000: Chemical mass closure and assessment of the origin of the submicron aersol in the marine boundary layer and the free troposphere at Tenerife during ACE-2. *Tellus*, **52B**, 141-168.

Quinn, P.K., T.S. Bates, J.E. Johnson, D.S. Covert and R.J. Charlson, 1990: Interactions between the sulphur and reduced nitrogen cycles over the central Pacific Ocean. *J. Geophys. Res.*, **95**, 16405-16416.

Quinn, P.K., D.S. Covert, T.S. Bates, V.N. Kapustin, D.C. Ramsey-Bell and L.M. McInnes, 1993: Dimethylsulfide/cloud condensation nuclei/climate system: Relevant size-resolved measurements of the chemical and physical properties of atmospheric aerosol particles. *J. Geophys. Res.*, **98**, 10411-10427.

Quinn, P.K., V.N. Kapustin, T.S. Bates and D.S. Covert, 1996: Chemical and optical properties of marine boundary layer aerosol particles of the mid-Pacific in relation to sources and meteorological transport. *J. Geophys. Res.*, **101**, 6931-6951.

Quinn, P.K. and D.J. Coffman, 1998: Local closure during the First Aerosol Characterization Experiment (ACE 1): Aerosol mass concentration and scattering and backscattering coefficients. *J. Geophys. Res.*, **103**, 15575-15596.

Quinn, P.K., D.J. Coffman, V.N. Kapustin, T.S. Bates and D.S. Covert, 1998: Aerosol optical properties in the marine boundary layer during the First Aerosol Characterization Experiment (ACE 1) and the underlying chemical and physical aerosol properties. *J. Geophys. Res.*, **103**, 16,547–16,563.

Radke, L. F., D. A. Hegg, P. V. Hobbs, J. D. Nance, J. H. Lyons, K. K. Laursen, R. E. Weiss, P. J. Riggan and D. E. Ward, 1991: Particulate and trace gas emissions from large biomass fires in North America. In Global Biomass Burning: Atmospheric, Climate and Biospheric Implications, ed. J. S. Levine, pp. 209-224, MIT Press, Cambridge, MA.

Raes, F., R. Van Dingenen, E. Cuevas, P.F.J. Van Velthoven and J.M. Prospero, 1997: Observations of aerosols in the free troposphere and marine boundary layer of the subtropical Northeast Atlantic: Discussion of processes determining their size distribution. *J. Geophys. Res.*, **102**, 21,315-21,328.

Raes, F., T.F. Bates, F.M. McGovern and M. van Liedekerke, 2000: The second Aerosol Characterization Experiment: General overview and main results. *Tellus*, **52**, 111-125.

Rangno, A. and P. V. Hobbs, 1994: Ice particle concentrations and precipitation development in small continental cumuliform clouds. *Q. J. Roy. Met. Soc.*, **120**, 573–601.

Rasch, P.J. and J.E. Kristjansson, 1998: A comparison of the CCM3 model climate using diagnosed and predicted condensate parametrizations. *J. Climate*, **11**, 1587-1614.

Rasch, P.J, M.C. Barth, J.T. Kiehl, S.E. Schwartz and C.M. Benkowitz, 2000a: A description of the global sulphur cycle and its controlling processes in the NCAR CCM3. *J. Geophys. Res.*, **105**, 1367-1386.

Rasch, P.J., H. Feichter, K. Law, N. Mahowald, J. Penner, C. Benkovitz, C. Genthon, C. Giannakopoulos, P. Kasibhatla, D. Koch, H. Levy, T.

Maki, M. Prather, D.L. Roberts, G.-J. Roelofs, D. Stevenson, Z. Stockwell, S. Taguchi, M. Kritz, M. Chipperfield, M. Baldocchi, P. McMurry, L. Barrie, Y. Balkanski, R. Chatfield, E. Kjellstrom, M. Lawrence, H.N. Lee, J. Lelieveld, K.J. Noone, J. Seinfeld, G. Stenchikov, S. Schwarz, C. Walcek, D. Williamson, 2000b: A comparison of scavenging and deposition processes in global models: Results from the WCRP Cambridge Workshop of 1995. *Tellus*, **52B**, 1025-1056.

Rea, D. K., 1994: The paleoclimatic record provided by eolian deposition in the deep sea: The geologic history of wind. *Rev. Geophys.*, **i32**, 159-195.

Reid, J. S. and P. V. Hobbs, 1998: Physical and optical properties of smoke from individual biomass fires in Brazil. *J. Geophys. Res.*, **103**, 32013-32031.

Reid, J.S., P.V. Hobbs, C. Liousse, J. V. Martins, R.E. Weiss, and T.F. Eck, 1998a: Comparisons of techniques for measuring shortwave absorption and black carbon content of aerosols from biomass burning in Brazil. *J. Geophys. Res.*, **103**, 32,031-32,040.

Reid, J.S., P.V. Hobbs, R.J. Ferek, D.R. Blake, J.V. Martins, M.R. Dunlap and C. Liousse, 1998b: Physical, chemical, and optical properties of regional hazes dominated by smoke in Brazil. *J. Geophys. Res.*, **103**, 32,059–32,080.

Remer, L. A., S. Gasso, D. A. Hegg, Y.J. Kaufman and B. N. Holben, 1997: Urban/industrial aerosol: Ground-based sun/sky radiometer and airborne in situ measurements. *J. Geophys. Res.*, **102**, 16849-16859.

Rivera-Carpio, C.A., C.E. Corrigan, T. Novakov, J.E. Penner, C.F. Rogers and J.C. Chow, 1996: Derivation of contributions of sulphate and carbonaceous aerosols to cloud condensation nuclei from mass size distributions. *J. Geophys. Res. Atmos.*, **101**, 19483–19493.

Roberts, G., M.O. Andreae, W. Maenhaut, P. Artaxo, J.V. Martins, J. Zhou and E. Swietlicki, 1998: Relationships of cloud condensation nuclei to size distribution and aerosol composition in the Amazon Basin. *Eos Trans. AGU*, **79**, F159.

Robertson, A.D., J.T. Overpeck, D. Rind, E. Mosley-Thompson, G.A. Zielinski, J.L. Lean, D. Koch, J.E. Penner, I. Tegen and R. Healy, 2001: Hypothesized climate forcing time series for the last 500 years. *J. Geophys. Res.*, in press.

Roeckner, E., Rieland, M. and Keup, E., 1991: Modelling of clouds and radiation in the ECHAM model. ECMWF/WCRP Workshop on Clouds, Radiative Transfer and the Hydrological Cycle, 199-222. ECMWF, Reading, U.K.

Roeckner, E., L. Bengtsson, J. Feichter, J. Lelieveld and H. Rodhe, 1999: Transient climate change simulations with a coupled atmosphere-ocean GCM including the tropospheric sulphur cycle. *J. Climate*, **12**, 3004–3032.

Roelofs, G.-J., J. Lelieveld and L. Ganzeveld, 1998: Simulation of global sulphate distribution and the influence on effective cloud drop radii with a coupled photochemistry-sulphur cycle model. *Tellus*, **50B**, 224–242.

Roelofs, G.J. , P. Kasibhatla, L. Barrie, D. Bergmann, C, Bridgeman, M. Chin, J. Christensen, R. Easter, J. Feichter, A. Jeuken, E. Kjellström, D. Koch, C. Land, U. Lohmann, P. Rasch, 2001: Analysis of regional budgets of sulphur species modelled for the COSAM exercise. *Tellus* B, in press.

Rogers, D., 1993: Measurements of natural ice nuclei with a continuous flow diffusion chamber. *Atmos. Res.*, **29**, 209–228.

Rosenfeld, D. 1999: TRMM observed first direct evidence of smoke from forest fires inhibiting rainfall. *Geophys. Res. Lett.*, **26**, 3105-3108.

Rosenfeld, D. 2000: Suppression of rain and snow by urban and industrial air pollution. *Science*. **287**, 1793-1796.

Rosinski, J. and G. Morgan, 1991: Cloud condensation nuclei as sources of ice-forming nuclei in clouds. *J. Aerosol Sci.*, **22**, 123–133.

Ross, J. I., P. V. Hobbs, and B. Holden, 1998: Radiative characteristics of regional hazes dominated by smoke from biomass burning in Brazil: Closure tests and direct radiative forcing, *J. Geophys. Res.*, **103**, 31925-31941.

Rotstayn, L. D., 1999: Indirect forcing by anthropogenic aerosols: A global climate model calculation of effective-radius and cloud-lifetime effects. *J. Geophys. Res. Atmos.*, **104**, 9369–9380.

Rotstayn, L. D. and J.E. Penner, 2001: Forcing, quasi-forcing and climate response. *J. Climate*, in press.

Russell, P. B., J. M. Livingston, P. Hignett, S. Kinne, J. Wong, A. Chien, R. Bergstrom, P. Durkee, and P. V. Hobbs, 1999: Aerosol-induced radiative flux changes off the United States mid-Atlantic coast: Comparison of values calculated from sunphotometer and in situ data with those measured by airborne pyranometer. *J. Geophys. Res.*, **104**, 2289-2307.

Russell, P. B., and J. Heintzenberg, 2000: An overview of the ACE-2 clear sky column closure experiment (CLEARCOLUMN). *Tellus*, **52B**, 463-483.

Saltzman, E. S., King, D. B., Holmen, K. and Leck, C., 1993: Experimental determination of the diffusion coefficient of dimethyl-sulfide in water. *J. Geophys. Res.*, **98**, 16481-16486.

SAR, see IPCC, 1996.

Sassen, K. and G. C. Dodd, 1988: Homogeneous nucleation rate for highly supercooled cirrus cloud droplets. *J. Atmos. Sci.*, **45**, 1357–1369.

Sassen, K., 1992: Evidence for liquid-phase cirrus cloud formation from volcanic aerosols: Climatic implications. *Science*, **257**, 516–519.

Sassen, K., D.O.C. Starr, G.G. Mace, M.R. Poellot, and others, 1995: The 5-6 December 1991 FIRE IFO II jet stream cirrus case study: Possible influences of volcanic aerosols. *J. Atmos. Sci.*, **52**, 97–123.

Satheesh, S. K., V. Ramanathon, X. Li-Jnes, J. M. Lobert, I. A. Podgorny, J. M. Prosper, B. N. Holben, and N. G. Loev, 1999: A model for the natural and anthropogenic aerosols over the tropical Indian Ocean derived from INDOEX data. *J. Geophys. Res.*, **104**, 27,421-27,440.

Saxena, P., L.M. Hildemann, P.H. McMurry and J.H. Seinfeld, 1995: Organics alter hygroscopic behavior of atmospheric particles. *J. Geophys. Res. Atmos.*, **100**, 18755–18770.

Saxena, P. and L.M. Hildemann, 1996: Water soluble organics in atmospheric particles: A critical review of the literature and application of thermodynamics to identify candidate compounds. *J. Atmos. Chem.*, **24**, 57–109.

Schnell, R. C. and G. Vali, 1976: Biogenic ice nuclei: Part I. Terrestrial and marine sources. *J. Atmos. Sci.*, **33**, 1554–1564.

Schmid, B. J. M. Livingston, P. B. Russell, P. A. Durkee, H. H. Jonsson, D. R. Collins, R. D. Flagan, J. H. Seinfeld, S. Gassó, D. A. Hegg, E. Öström, K. J. Noone, E. J. Welton, K. J. Voss, H. R. Gordon, P. Formenti, and M. O. Andreae, 2000: Clear-sky closure studies of lower tropospheric aerosol and water vapor during ACE-2 using airborne sunphotometer, airborne in situ, space-borne, and ground-based measurements. *Tellus*, **52B**, 568-593.

Scholes, M. and M.O. Andreae, 2000: Biogenic and pyrogenic emissions from Africa and their impact on the global atmosphere. *Ambio*, **29**, 23-29.

Scholes, R.J., D. Ward and C.O. Justice, 1996: Emissions of trace gases and aerosol particles due to vegetation burning in southern-hemisphere Africa. *J. Geophys. Res. Atmos.*, **101**, 23677–23682.

Schubert, S.D., R.B. Rood, and J. Pfaendtner, 1993: An assimilated data set for Earth science applications, *Bull. Amer. Met. Soc.*, **24**, 2331-2342.

Schult, I., J. Feichter and W.F. Cooke, 1997: Effect of black carbon and sulphate aerosols on the Global Radiation Budget. *J. Geophys. Res. Atmos.*, **102**, 30107–30117.

Schulz, M., Y.J. Balkanski, W. Guelle and F. Dulac, 1998: Role of aerosol size distribution and source location in a three-dimensional simulation of a Saharan dust episode tested against satellite-derived optical thickness. *J. Geophys. Res. Atmos.*, **103**, 10579–10592.

Schwartz, S. E., 1996: The whitehouse effect - Shortwave radiative forcing of climate by anthropogenic aerosols: An overview, *J. Aer. Sci.*, **27**, 359-383.

Schwartz, S.E., and A. Slingo, 1996: Enhanced shortwave cloud radiative forcing due to anthropogenic aerosols. In Clouds, Chemistry and Climate, P. J. Crutzen and V. Ramanathan, Eds., NATO ASI Series, Vol. I 35, Springer-Verlag, Berlin, 191-236.

Seinfeld, J. H. and S. N. Pandis, 1998: Atmospheric Chemistry and Physics From Air Pollution to Climate Change. Wiley Interscience, New York, 1326 pp.

Seinfeld, J.H. and S.N. Pandis, 1998: Atmospheric Chemistry and Physics: From Air Pollution to Climate Change, John Wiley and Sons, New York, 1326pp.

Sempéré, R. and K. Kawamura, 1996: Low molecular weight dicarboxylic acids and related polar compounds in the remote marine rain samples collected from western Pacific. *Atmos. Environ.*, **30**, 1609–1619.

Senior, C. A. and J.L.B. Mitchell, 1993: Carbon dioxide and climate: The impact of cloud parametrization. *J. Climate*, **6**, 393–418.

Shantz, N., R. Leaitch, S.-M. Li, W. Hoppel, G. Frick, P. Caffrey, D. Hegg, S. Gao, T. Albrechcinski, 1999: Controlled studies and field measurements of organic cloud condensation nuclei. Paper Number A22B-11, American Geophysical Union Fall Meeting, San Francisco.

Shettle, E.P. and R. Fenn, 1976: Models of the atmospheric aerosols and their optical properties, AGARD Conference Proceedings, No. 183, AGARD-CP-183.

Shulman, M.L., M.C. Jacobson, R.J. Charlson, R.E. Synovec and T.E. Young, 1996: Dissolution behaviour and surface tension effects of organic compounds in nucleating droplets. *Geophy. Res. Lett.*, **23**, 277–280.

Sievering, H., J. Boatman, E. Gorman, Y. Kim, L. Anderson, G. Ennis, M. Luria and S. Pandis, 1992: Removal of sulphur from the marine boundary layer by ozone oxidation in sea- salt aerosols. *Nature*, **360**, 571–573.

Simoneit, B.R.T., J.N. Cardoso and N. Robinson, 1990: An assessment of the origin and composition of higher molecular weight organic matter in aerosols over Amazonia. *Chemosphere*, **21**, 1285–1301.

Slanina, J., H.M. ten Brink and A. Khlystov, 1999: Fate of products of degradation processes: Consequences for climatic change. *Chemosphere*, **38**, 1429–1444.

Smith, R.N.B., 1990: A scheme for predicting layer clouds and their water content in a general circulation model. *Q.J.R. Meteorol. Soc.*, **116**, 435-460.

Sokolik, I. and G. Golitsyn, 1993: Investigation of optical and radiative properties of atmospheric dust aerosols. *Atmos. Environ.*, **27A**, 2509-2517.

Sokolik, I.N. and O.B. Toon, 1996: Direct radiative forcing by anthropogenic airborne mineral aerosols. *Nature*, **381**, 681–683.

Sokolik, I.N. and O.B. Toon, 1999: Incorporation of mineralogical composition into models of the radiative properties of mineral aerosol from UV to IR wavelengths. *J. Geophys. Res. Atmos.*, **104**, 9423–9444.

Song, N., D.O'C. Starr, D.J. Wuebbles, A. Williams and S.M. Larson, 1996: Volcanic aerosols and inter-annual variation of high clouds. *Geophys. Res. Lett.*, **23**, 2657–2660.

Soufflet, V., D. Tanré, A. Royer, and N. T. O'Neill, 1997: Remote sensing of aerosols over boreal forest and lake water from AVHRR data. *Rem. Sensing Env.*, **60**, 22-34.

Spiro, P.A., D.J. Jacob and J.A. Logan, 1992: Global inventory of sulphur emissions with 1x1 resolution. *J. Geophys. Res.*, **97**, 6023–6036.

Stenchikov, G.L., I. Kirchner, A. Robock, H.-F. Graf, J.C. Antuna, R. Grainger, A. Lambert, and L. Thomason, 1998: Radiative forcing from the 1991 Mt. Pinatubovolcanic eruption. *J. Geophys. Res.*, 103,
13837-13858.

Stevens, B., W.R. Cotton, G. Feingold, and C.-H. Moeng, 1998: Large eddy simulations of strongly precipitating, shallow, stratocumulus topped boundary layers. *J. Atmos. Sci.*, **55**, 3616-3638.

Stevenson, D.S., W.J. Collins, C.E. Johnson, R.G. Derwent, 1998: Inter-comparison and evaluation of atmospheric transport in a Lagrangian model (STOCHEM), and an Eulerian model (UM) using ^{222}Rn as a short-lived tracer. *Quart. J. Roy. Meteorol. Soc.*, **124**, 2477-2491.

Stith, J. L., D. A. Burrows and P. J. DeMott, 1994: Initiation of ice: Comparison of numerical model results with observations of ice development in a cumulus cloud. *Atmos. Res.*, **32**, 13–30.

Stoiber, R.E., S.N. Williams and B. Huebert, 1987: Annual contribution of sulphur dioxide to the atmosphere by volcanoes. *J. Volcanol. Geotherm. Res.*, **33**, 1–8.

Stowe, L.L., R.M. Carey and P.P. Pellegrino, 1992: Monitoring the Mt. Pinatubo aerosol layer with NOAA/11 AVHRR data. *Geophys. Res. Lett.*, **19**, 159–162.

Stowe, L.L., A.M. Ignatov, and R.R. Singh, 1997: Development, validation, and potential enhancements to the second-generation operational aerosol product at the National Environmental Satellite, Data, and Information Service of the National Oceanic and Atmospheric Administration. *J. Geophys. Res.*, **102**, 16923-16934.

Strom, J. and S. Ohlsson, 1998: In situ measurements of enhanced crystal number densities in cirrus clouds caused by aircraft exhaust. *J. Geophys. Res. Atmos.*, **103**, 11355–11361.

Strom, J., B. Strauss, T. Anderson, F. Schroder, J. Heintzenberg and R. Wiedding, 1997: In situ observations of the microphysical properties of young cirrus clouds. *J. Atmos. Sci.*, **54**, 2542–2553.

Sundqvist, H., E. Berge, and J.E. Kristjansson, 1989: Condensation and cloud parametrization studies with a mesoscale numerical weather prediction model. *Mon. Weather Rev.*, **117**, 1641-1657.

Swietlicki, E., J. Zhou, O.H. Berg, B.G. Martinsson, G.Frank, S.-I. Cederfelt, U. Dusek, A. Berner, W. Birmili, A. Wiedensohler, B. Yuskiewicz, and K.N. Bower, 1999: A closure study of sub-micrometer aerosol particle hygroscopic behaviour. *Atmos. Res.*, 205-240.

Szczodrak, M., P.H. Austin, and P. Krummel, 2001: Variability of optical depth and effective radius in marine stratocumulus clouds. *J. Atmos. Sci.*, in press.

Szyrmer, W. and I. Zawadski, 1997: Biogenic and anthropogenic sources of ice-forming nuclei: A review. *Bull. Amer. Meteor. Soc.*, **78**, 209–229.

Takeda, T., P.-M. Wu and K. Okada, 1987: Dependence of light scattering coefficient of aerosols on relative humidity in the atmosphere of Nagoya. *J. Met. Soc. Japan*, **64**, 957-966.

Tang, I.N., 1996: Chemical and size effects of hygroscopic aerosols on light scattering coefficients. *J. Geophys. Res.*, **101**, 19245–19250.

Tangren, C. D., 1982: Scattering coefficient and particulate matter concentration in forest fire smoke. *J. Air Poll. Control Assoc.*, **32**, 729-732.

Tanré, D., J.F. Geleyn, and J. Slingo, 1984: First results of the introduction of an advanced aerosol-radiation interaction in the ECMWF low resolution global model. In: *Aerosols and their Climate Effects.* H.F. Gerber and A. Deepak (Eds.), Deepak Publishing, 133-177.

Tanré, D., M. Herman, and Y. J. Kaufman, 1996: Information on aerosol size distribution contained in solar reflected spectral radiances. *J. Geophys. Res.*, **101**, 19043-19060.

Tanré, D., L. A. Remer, Y. J. Kaufman, S. Mattoo, P. V. Hobbs, J. M. Livingston, P. B. Russell, and A. Smirnov, 1999: Retrieval of aerosol optical thickness and size distribution over ocean from the MODIS airborne simulator during TARFOX. *J. Geophys. Res.*, **104**, 2261-2278.

Taylor, K.E. and J.E. Penner, 1994: Response of the climate system to atmospheric aerosols and greenhouse gases. *Nature*, **369**, 734-737.

Tegen, I., and I. Fung, 1994: Modeling of mineral dust in the atmosphere: Sources, transport, and optical thickness. *J. Geophys. Res.*, **99D**,

22,897-22,914.

Tegen, I., and I. Fung, 1995: Contribution to the atmospheric mineral aerosol load from land surface modification. *J. Geophys. Res.*, **100**, 18,707-18,726.

Tegen, I. and A.A. Lacis, 1996: Modelling of particle size distribution and its influence on the radiative properties of mineral dust aerosol. *J. Geophys. Res. Atmos.*, **101**, 19237–19244.

Tegen, I., A.A. Lacis and I. Fung, 1996: The influence on climate forcing of mineral aerosols from disturbed soils *Nature*, **380**, 419–422.

Tegen, I. and R. Miller, 1998: A general circulation model study of the inter-annual variability of soil dust aerosol. *J. Geophys. Res. Atmos.*, **103**, 25975–25995.

Tegen, I., P. Hollrig, M. Chin, I. Fung, D. Jacob and J.E. Penner, 1997: Contribution of different aerosol species to the global aerosol extinction optical thickness: Estimates from model results. *J. Geophys. Res. Atmos.*, **102**, 23895–23915.

ten Brink, H. M., J. P. Veefkind, A. Waijers-Ijpelaan, and J. C. van der Hage, 1996: Aerosol light-scattering in the Netherlands, *Atmos. Environ.*, **30**, 4251-4261.

Torres, O., P. K. Bhartia, J. R. Herman, Z. Ahmad and J. Gleason, 1998: Derivation of aerosol properties from satellite measurements of backscattered ultraviolet radiation: Theoretical basis. *J. Geophys. Res.*, **103**, 17099-17110.

Trenberth, Kevin E., Olson, Jerry G. and Large, William G., 1989: A Global Ocean Wind Stress Climatology Based on ECMWF Analysis. National Center for Atmospheric Research, Climate and Global Dynamics Division, Boulder, Colorado.

Tripoli, G.J. and W.R. Cotton, 1980: A numerical investigation of several factors contributing to the observed variable intensity of deep convection over south Florida. *J. Appl. Meteorol.*, **19**, 1037-1063.

Turpin, B.J., J.J. Huntzicker and S.V. Hering, 1994: Investigation of organic aerosol sampling artifacts in the Los Angeles basin. *Atmos. Environ.*, **28**, 3061–3071.

Twohy, C. H., P. A. Durkee, B. J. Huebert and R. J. Charlson, 1995: Effects of aerosol particles on the microphysics of coastal stratiform clouds. *J. Climate*, **8**, 773–783.

Twomey, S., 1977, Influence of pollution on the short-wave albedo of clouds. *J. Atmos. Sci.*, **34**, 1149–1152.

Vali, G., 1991: Nucleation of ice. In: Atmospheric Particles and Nuclei, G. Götz, E. Mészaros and G. Vali (eds), Akadémiai Kiado, 131-132.

Vali, G., 1992: Memory effect in the nucleation of ice on mercuric iodide. In: Nucleation and Atmospheric Aerosols, N. Fukuta and P. E. Wagner (eds), A. Deepak Publishing, Hampton, VA, USA, 259–262.

Van Dingenen, R.V., F. Raes and N.R. Jensen, 1995: Evidence for anthropogenic impact on number concentration and sulphate content of cloud-processed aerosol particles over the North Atlantic. *J. Geophys. Res. Atmos.*, **100**, 21057–21067.

Van Dinh, P., J.-P. Lacaux, and R. Serpolay, 1994: Cloud-active particles from African savanna combustion experiments. *Atmos. Res.*, **31**, 41-58.

Van Dorland, R., F.J. Dentener and J. Lelieveld, 1997: Radiative forcing due to tropospheric ozone and sulphate aerosols. *J. Geophys. Res.*, **102**, 28,079–28,100.

Van Valin, C. C., R. J. Pueschel, J. P. Parungo and R. A. Proulx, 1976: Cloud and ice nuclei from human activities. *Atmos. Environ.*, **10**, 27–31.

Veefkind, J. P., G. de Leeuw, and P. A. Durkee, 1998: Retrieval of aerosol optical depth over land using two-angle view satellite radiometry during TARFOX. *Geophys. Res. Lett.*, **25**, 3135-3138.

Veefkind, J.P., G. de Leeuw, P.B. Russell, P.V. Hobbs, and J.M Livingston, 1999: Aerosol optical depth retrieval using ATSR-2 and AVHRR data during TARFOX. *J. Geophys. Res.*, **194**, 2253-2260.

Virkkula, A., R.V. Dingenen, F. Raes and J. Hjorth, 1999: Hygroscopic properties of aerosol formed by oxidation of limonene, a-pinene, and

b-pinene. *J. Geophys. Res.*, **104**, 3569–3579.

Waggoner, A. P., R. E. Weiss and T. V. Larson, 1983: In-situ, rapid response measurement of $H_2SO_4/(NH_4)_2SO_4$ aerosols in urban Houston: A comparison with rural Virginia. *Atmos. Environ.*, **17**, 1723-1731.

Wanner, W., A.H. Srahler, B. Hu, P. Lewis, J.-P. Muller, X. Li, C.L. Barker Schaaf, and M.J. Barnsley, 1997: Global retrieval of bidirectional reflectance and albedo over land from EOS MODIS and MISR data: Theory and algorithm. *J. Geophys. Res.*, **102**, 17143-17161.

Wanninkhof, R., 1992: Relationship between wind speed and gas exchange over the ocean. *J. Geophys. Res.*, **97**, 7373–7382.

Warren, S. G., C. J. Hahn, J. London, R. M. Chervin, R. L. Jenne, 1988: Global Distribution of Total Cloud cover and Cloud Type Amounts over the Ocean, NCAR Technical Note, TN-317+STR (National Center for Atmospheric Research, Boulder, CO).

Weber, R.J., P.H. McMurry, R.L. Mauldin III, D.J. Tanner, F.L. Eisele, A.D. Clarke and V.N. Kapustin, 1999: New particle formation in the remote troposphere: A comparison of observations at various sites, *Geophys. Res. Lett.*, **26**, 307-310.

Weisenstein, D.K., G.K. Yue, M.K.W. Ko, N.-D. Sze, J.M. Rodriguez and C.J. Stott, 1997: A tw0 dimensional model of sulphur species and aerosols. *J. Geophys. Res.*, **102**, 13019-13035.

Weng, F. and N.C. Grody, 1994: Retrieval of cloud liquid water using the special sensor microwave imager (SSM/I). *J. Geophys. Res.*, **99**, 25,535-25,551.

Wenny, B.N., J.S. Schafer, J.J. DeLuisi, V.K. Saxena, W.F. Barnard, I.V. Petropavlovskikh, and A.J. Vergamini, 1998: A study of regional aerosol radiative properties and effects on ultraviolet-B radiation. *J. Geophys. Res.*, **103**, 17,083-17,097.

Wetzel, M. and L.L. Stowe, 1999: Satellite-observed patterns in the relationship of aerosol optical thickness to stratus cloud microphysics and shortwave radiative forcing. *J. Geophys. Res. Atmos.* **104**, 31,287-31,299.

Whitby, K.T., 1978: The physical characteristics of sulphur aerosols. *Atmos. Environ.*, **12**, 135-159.

WMO/UNEP, 1992: Scientific assessment of ozone depletion: 1991. Global ozoneresearch and monitoring project, report # 25, Geneva.

Wolf, M.E. and G.M. Hidy, 1997: Aerosols and climate: Anthropogenic emissions and trends for 50 years. *J. Geophys. Res. Atmos.*, **102**, 11113–11121.

Wyant, M.C., C.S. Bretherton, H.A. Rand and D.E. Stevens, 1997: Numerical simulations and a conceptual model of the stratocumulus to trade cumulus transition. *J. Atmos. Sci.*, **54**, 168–192.

Xu, J., D. Imre, R. McGraw and I. Tang, 1998: Ammonium sulphate: Equilibrium and metastability phase diagrams from 40 to 50 degrees C. *J. Phys. Chem.*, B **102**, 7462–7469.

Xu, K.M., and D.A. Randall, 1996: A semiempirical cloudiness parametrization for use in climate models. *J.Atmos. Sci.*, **53**, 3084-3102.

Yang, Z., S. Young, V. Kotamarthi, and G.R. Carmichael, 1994: Photochemical oxidant processes in the presence of dust: an evaluation of the impact of dust on particulate nitrate and ozone formation. *J. Appl. Meteorol.*, **33**, 813-824.

Yienger, J.J. and H. Levy, Empirical-model of global soil-biogenic NOx emissions. 1995: *J. Geophys. Res.*, **100**, 11,447-11,464.

Yu, S., 2000: Role of organic acids (formic, acetic, pyruvic and oxalic) in the formation of cloud condensation nuclei (CCN): A review. *Atmos. Res.*, **53**, 185-217.

Yum, S.S. and J.G. Hudson, 1998: Comparisons of cloud microphysics with cloud condensation nuclei spectra over the summertime Southern. *J. Geophys. Res. Atmos.*, **103**, 16625–16636.

Zappoli, S., A. Andracchio, S. Fuzzi, M.C. Facchini, A. Geleneser, G. Kiss, Z. Krivacsy, A. Molnar, E. Mozaros, H.-C. Hansson, K. Rosman, and Y. Zebuhr, 1999: Inorganic, organic and macromolecular components of fine aerosol in different areas of Europe in relation to their water solubility. *Atmos. Env.*, **33**, 2733-2743.

Zhang, X. Q., P.H. McMurry, S.V. Herring and G.S. Casuccio, 1993: Mixing characteristics and water content of submicron aerosols measured in Los Angeles and at the Grand Canyon. *Atmos. Environ.*, **27A**, 1593-1607.

Zhang, Y. and G.R. Carmichael, 1999: The role of mineral aerosol in tropospheric chemistry in East Asia-A model study, *J. App. Met.*, **38**, 353-366.

Zhang, Y., S.M. Kreidenweis and G. Feingold, 1999: Stratocumulus processing of gases and cloud condensation nuclei: Part II: chemistry sensitivity analysis. *J. Geophys. Res.*, **104**, 16,061-16,080.

Ziegler, C.L., 1985: Retrieval of thermal microphysical variables in observed convective storms, 1, Model development and preliminary testing. *J. Atmos. Sci.*, **42**, 1487-1509.

6

Radiative Forcing of Climate Change

Co-ordinating Lead Author
V. Ramaswamy

Lead Authors
O. Boucher, J. Haigh, D. Hauglustaine, J. Haywood, G. Myhre, T. Nakajima, G.Y. Shi, S. Solomon

Contributing Authors
R. Betts, R. Charlson, C. Chuang, J.S. Daniel, A. Del Genio, R. van Dorland, J. Feichter, J. Fuglestvedt, P.M. de F. Forster, S.J. Ghan, A. Jones, J.T. Kiehl, D. Koch, C. Land, J. Lean, U. Lohmann, K. Minschwaner, J.E. Penner, D.L. Roberts, H. Rodhe, G.J. Roelofs, L.D. Rotstayn, T.L. Schneider, U. Schumann, S.E. Schwartz, M.D. Schwarzkopf, K.P. Shine, S. Smith, D.S. Stevenson, F. Stordal, I. Tegen, Y. Zhang

Review Editors
F. Joos, J. Srinivasan

Contents

Executive Summary

- Radiative forcing continues to be a useful tool to estimate, to a first order, the relative climate impacts (viz., relative global mean surface temperature responses) due to radiatively induced perturbations. The practical appeal of the radiative forcing concept is due, in the main, to the assumption that there exists a general relationship between the global mean forcing and the global mean equilibrium surface temperature response (i.e., the global mean climate sensitivity parameter, λ) which is similar for all the different types of forcings. Model investigations of responses to many of the relevant forcings indicate an approximate near invariance of λ (to about 25%). There is some evidence from model studies, however, that λ can be substantially different for certain forcing types. Reiterating the IPCC WGI Second Assessment Report (IPCC, 1996a) (hereafter SAR), the global mean forcing estimates are not necessarily indicators of the detailed aspects of the potential climate responses (e.g., regional climate change).

- The simple formulae used by the IPCC to calculate the radiative forcing due to well-mixed greenhouse gases have been improved, leading to a slight change in the forcing estimates. Compared to the use of the earlier expressions, the improved formulae, for fixed changes in gas concentrations, decrease the carbon dioxide (CO_2) and nitrous oxide (N_2O) radiative forcing by 15%, increase the CFC-11 and CFC-12 radiative forcing by 10 to 15%, while yielding no change in the case of methane (CH_4). Using the new expressions, the radiative forcing due to the increases in the well-mixed greenhouse gases from the pre-industrial (1750) to present time (1998) is now estimated to be $+2.43$ Wm^{-2} (comprising CO_2 (1.46 Wm^{-2}), CH_4 (0.48 Wm^{-2}), N_2O (0.15 Wm^{-2}) and halocarbons (halogen-containing compounds) (0.34 Wm^{-2})), with an uncertainty[1] of 10% and a high level of scientific understanding (LOSU).

- The forcing due to the loss of stratospheric ozone (O_3) between 1979 and 1997 is estimated to be -0.15 Wm^{-2} (range: -0.05 to -0.25 Wm^{-2}). The magnitude is slightly larger than in the SAR owing to the longer period now considered. Incomplete knowledge of the O_3 losses near the tropopause continues to be the main source of uncertainty. The LOSU of this forcing is assigned a medium rank.

- The global average radiative forcing due to increases in tropospheric O_3 since pre-industrial times is estimated to be $+0.35 \pm 0.15$ Wm^{-2}. This estimate is consistent with the SAR estimate, but is based on a much wider range of model studies and a single analysis that is constrained by observations; there are uncertainties because of the inter-model

differences, the limited information on pre-industrial O_3 distributions, and the limited data that are available to evaluate the model trends for modern (post-1960) conditions. A rank of medium is assigned for the LOSU of this forcing.

- The changes in tropospheric O_3 are mainly driven by increased emissions of CH_4, carbon monoxide (CO), non-methane hydrocarbons (NMHCs) and nitrogen oxides (NO$_x$), but the specific contributions of each are not yet well quantified. Tropospheric and stratospheric photochemical processes lead to other indirect radiative forcings through, for instance, changes in the hydroxyl radical (OH) distribution and increase in stratospheric water vapour concentrations.

- Models have been used to estimate the direct radiative forcing for five distinct aerosol species of anthropogenic origin. The global, annual mean radiative forcing is estimated as -0.4 Wm^{-2} (-0.2 to -0.8 Wm^{-2}) for sulphate aerosols; -0.2 Wm^{-2} (-0.07 to -0.6 Wm^{-2}) for biomass burning aerosols; -0.10 Wm^{-2} (-0.03 to -0.30 Wm^{-2}) for fossil fuel organic carbon aerosols; $+0.2$ Wm^{-2} ($+0.1$ to $+0.4$ Wm^{-2}) for fossil fuel black carbon aerosols; and in the range -0.6 to $+0.4$ Wm^{-2} for mineral dust aerosols. The LOSU for sulphate aerosols is low while for biomass burning, fossil fuel organic carbon, fossil fuel black carbon, and mineral dust aerosols the LOSU is very low.

- Models have been used to estimate the "first" indirect effect of anthropogenic sulphate and carbonaceous aerosols (namely, a reduction in the cloud droplet size at constant liquid water content) as applicable in the context of liquid clouds, yielding global mean radiative forcings ranging from -0.3 to -1.8 Wm^{-2}. Because of the large uncertainties in aerosol and cloud processes and their parametrizations in general circulation models (GCMs), the potentially incomplete knowledge of the radiative effect of black carbon in clouds, and the possibility that the forcings for individual aerosol types may not be additive, a range of radiative forcing from 0 to -2 Wm^{-2} is adopted considering all aerosol types, with no best estimate. The LOSU for this forcing is very low.

- The "second" indirect effect of aerosols (a decrease in the precipitation efficiency, increase in cloud water content and cloud lifetime) is another potentially important mechanism for climate change. It is difficult to define and quantify in the context of current radiative forcing of climate change evaluations and current model simulations. No estimate is therefore given. Nevertheless, present GCM calculations suggest that the radiative flux perturbation associated with the second aerosol indirect effect is of the same sign and could be of similar magnitude compared to the first effect.

- Aerosol levels in the stratosphere have now fallen to well below the peak values seen in 1991 to 1993 in the wake of the Mt. Pinatubo eruption, and are comparable to the low values seen in about 1979, a quiescent period for volcanic activity. Although episodic in nature and transient in duration, stratospheric

---

[1] The "uncertainty range" for the global mean estimates of the various forcings in this chapter is guided, for the most part, by the spread in the published estimates. It is not statistically based and differs in this respect from the manner "uncertainty range" is treated elsewhere in this document.

aerosols from explosive volcanic eruptions can exert a significant influence on the time history of radiative forcing of climate.

- Owing to an increase in land-surface albedo during snow cover in deforested mid-latitude areas, changes in land use are estimated to yield a forcing of -0.2 Wm^{-2} (range: 0 to -0.4 Wm^{-2}). However, the LOSU is very low and there have been much less intensive investigations compared with other anthropogenic forcings.

- Radiative forcing due to changes in total solar irradiance (TSI) is estimated to be $+0.3 \pm 0.2$ Wm^{-2} for the period 1750 to the present. The wide range given, and the very low LOSU, are largely due to uncertainties in past values of TSI. Satellite observations, which now extend for two decades, are of sufficient precision to show variations in TSI over the solar 11-year activity cycle of about 0.08%. Variations over longer periods may have been larger but the techniques used to reconstruct historical values of TSI from proxy observations (e.g., sunspots) have not been adequately verified. Solar radiation varies more substantially in the ultraviolet region and GCM studies suggest that inclusion of spectrally resolved solar irradiance variations and solar-induced stratospheric O$_3$ changes may improve the realism of model simulations of the impact of solar variability on climate. Other mechanisms for the amplification of solar effects on climate, such as enhancement of the Earth's electric field causing electrofreezing of cloud particles, may exist but do not yet have a rigorous theoretical or observational basis.

- Radiative forcings and Global Warming Potentials (GWPs) are presented for an expanded set of gases. New categories of gases in the radiative forcing set include fluorinated organic molecules, many of which are ethers that may be considered as halocarbon substitutes. Some of the GWPs have larger uncertainties than others, particularly for those gases where detailed laboratory data on lifetimes are not yet available. The direct GWPs have been calculated relative to CO$_2$ using an improved calculation of the CO$_2$ radiative forcing, the SAR response function for a CO$_2$ pulse, and new estimates for the radiative forcing and lifetimes for a number of gases. As a consequence of changes in the radiative forcing for CO$_2$ and CFC-11, the revised GWPs are typically 20% higher than listed in the SAR. Indirect GWPs are also discussed for some new gases, including CO. The direct GWPs for those species whose lifetimes are well characterised are estimated to be accurate (relative to one another) to within ± 35%, but the indirect GWPs are less certain.

- The geographical distributions of each of the forcing mechanisms vary considerably. While well-mixed greenhouse gases exert a significant radiative forcing everywhere on the globe, the forcings due to the short-lived species (e.g., direct and indirect aerosol effects, tropospheric and stratospheric O$_3$) are not global in extent and can be highly spatially inhomogeneous. Furthermore, different radiative forcing mechanisms lead to differences in the partitioning of the perturbation between the atmosphere and surface. While the Northern to Southern Hemisphere ratio of the solar and well-mixed greenhouse gas forcings is very nearly 1, that for the fossil fuel generated sulphate and carbonaceous aerosols and tropospheric O$_3$ is substantially greater than 1 (i.e., primarily in the Northern Hemisphere), and that for stratospheric O$_3$ and biomass burning aerosol is less than 1 (i.e., primarily in the Southern Hemisphere).

- The global mean radiative forcing evolution comprises of a steadily increasing contribution due to the well-mixed greenhouse gases. Other greenhouse gas contributions are due to stratospheric O$_3$ from the late 1970s to the present, and tropospheric O$_3$ whose precise evolution over the past century is uncertain. The evolution of the direct aerosol forcing due to sulphates parallels approximately the secular changes in the sulphur emissions, but it is more difficult to estimate the temporal evolution due to the other aerosol components, while estimates for the indirect forcings are even more problematic. The temporal evolution estimates indicate that the net natural forcing (solar plus stratospheric aerosols from volcanic eruptions) has been negative over the past two and possibly even the past four decades. In contrast, the positive forcing by well-mixed greenhouse gases has increased rapidly over the past four decades.

- Estimates of the global mean radiative forcing due to different future scenarios (up to 2100) of the emissions of trace gases and aerosols have been performed (Nakićenović *et al.*, 2000; see also Chapters 3, 4 and 5). Although there is a large variation in the estimates from the different scenarios, the results indicate that the forcing (evaluated relative to pre-industrial times, 1750) due to the trace gases taken together is projected to increase, with the fraction of the total due to CO$_2$ becoming even greater than for the present day. The direct aerosol (sulphate, black and organic carbon components taken together) radiative forcing (evaluated relative to the present day, 2000) varies in sign for the different scenarios. The direct aerosol effects are estimated to be substantially smaller in magnitude than that of CO$_2$. No estimates are made for the spatial aspects of the future forcings. Relative to 2000, the change in the direct plus indirect aerosol radiative forcing is projected to be smaller in magnitude than that of CO$_2$.

6.1 Radiative Forcing

6.1.1 Definition

The term "radiative forcing" has been employed in the IPCC Assessments to denote an externally imposed perturbation in the radiative energy budget of the Earth's climate system. Such a perturbation can be brought about by secular changes in the concentrations of radiatively active species (e.g., CO_2, aerosols), changes in the solar irradiance incident upon the planet, or other changes that affect the radiative energy absorbed by the surface (e.g., changes in surface reflection properties). This imbalance in the radiation budget has the potential to lead to changes in climate parameters and thus result in a new equilibrium state of the climate system. In particular, IPCC (1990, 1992, 1994) and the Second Assessment Report (IPCC, 1996) (hereafter SAR) used the following definition for the radiative forcing of the climate system: "The radiative forcing of the surface-troposphere system due to the perturbation in or the introduction of an agent (say, a change in greenhouse gas concentrations) is the change in net (down minus up) irradiance (solar plus long-wave; in Wm^{-2}) at the tropopause AFTER allowing for stratospheric temperatures to readjust to radiative equilibrium, but with surface and tropospheric temperatures and state held fixed at the unperturbed values". In the context of climate change, the term forcing is restricted to changes in the radiation balance of the surface-troposphere system imposed by external factors, with no changes in stratospheric dynamics, without any surface and tropospheric feedbacks in operation (i.e., no secondary effects induced because of changes in tropospheric motions or its thermodynamic state), and with no dynamically-induced changes in the amount and distribution of atmospheric water (vapour, liquid, and solid forms). Note that one potential forcing type, the second indirect effect of aerosols (Chapter 5 and Section 6.8), comprises microphysically-induced changes in the water substance. The IPCC usage of the "global mean" forcing refers to the globally and annually averaged estimate of the forcing.

The prior IPCC Assessments as well as other recent studies (notably the SAR; see also Hansen *et al.* (1997a) and Shine and Forster (1999)) have discussed the rationale for this definition and its application to the issue of forcing of climate change. The salient elements of the radiative forcing concept that characterise its eventual applicability as a tool are summarised in Appendix 6.1 (see also WMO, 1986; SAR). Defined in the above manner, radiative forcing of climate change is a modelling concept that constitutes a simple but important means of estimating the relative impacts due to different natural and anthropogenic radiative causes upon the surface-troposphere system (see Section 6.2.1). The IPCC Assessments have, in particular, focused on the forcings between pre-industrial times (taken here to be 1750) and the present (1990s, and approaching 2000). Another period of interest in recent literature has been the 1980 to 2000 period, which corresponds to a time frame when a global coverage of the climate system from satellites has become possible.

We find no reason to alter our view of any aspect of the basis, concept, formulation, and application of radiative forcing, as laid down in the IPCC Assessments to date and as applicable to the forcing of climate change. Indeed, we reiterate the view of previous IPCC reports and recommend a continued usage of the forcing concept to gauge the relative strengths of various perturbation agents, but, as discussed below in Section 6.2, urge that the constraints on the interpretation of the forcing estimates and the limitations in its utility be noted.

6.1.2 Evolution of Knowledge on Forcing Agents

The first IPCC Assessment (IPCC, 1990) recognised the existence of a host of agents that can cause climate change including greenhouse gases, tropospheric aerosols, land-use change, solar irradiance and stratospheric aerosols from volcanic eruptions, and provided firm quantitative estimates of the well-mixed greenhouse gas forcing since pre-industrial times. Since that Assessment, the number of agents identified as potential climate changing entities has increased, along with knowledge on the space-time aspects of their operation and magnitudes. This has prompted the radiative forcing concept to be extended, and the evaluation to be performed for spatial scales less than global, and for seasonal time-scales.

IPCC (1992) recognised the importance of the forcing due to anthropogenic sulphate aerosols and assessed quantitative estimates for the first time. IPCC (1992) also recognised the forcing due to the observed loss of stratospheric O_3 and that due to an increase in tropospheric O_3. Subsequent assessments (IPCC, 1994; SAR) have performed better evaluations of the estimates of the forcings due to agents having a space-time dependence such as aerosols and O_3, besides strengthening further the confidence in the well-mixed greenhouse gas forcing estimates. More information on changes in solar irradiance have also become available since 1990. The status of knowledge on forcing arising due to changes in land use has remained somewhat shallow.

For the well-mixed greenhouse gases (CO_2, CH_4, N_2O and halocarbons), their long lifetimes and near uniform spatial distributions imply that a few observations coupled with a good knowledge of their radiative properties will suffice to yield a reasonably accurate estimate of the radiative forcing, accompanied by a high degree of confidence (SAR; Shine and Forster, 1999). But, in the case of short-lived species, notably aerosols, observations of the concentrations over wide spatial regions and over long time periods are needed. Such global observations are not yet in place. Thus, estimates are drawn from model simulations of their three-dimensional distributions. This poses an uncertainty in the computation of forcing which is sensitive to the space-time distribution of the atmospheric concentrations and chemical composition of the species.

6.2 Forcing-Response Relationship

6.2.1 Characteristics

As discussed in the SAR, the change in the net irradiance at the tropopause, as defined in Section 6.1.1, is, to a first order, a good indicator of the equilibrium global mean (understood to be globally and annually averaged) surface temperature change. The climate sensitivity parameter (global mean surface temperature response ΔT_s to the radiative forcing ΔF) is defined as:

$$\Delta T_s / \Delta F = \lambda \qquad (6.1)$$

(Dickinson, 1982; WMO, 1986; Cess *et al.*, 1993). Equation (6.1) is defined for the transition of the surface-troposphere system from one equilibrium state to another in response to an externally imposed radiative perturbation. In the one-dimensional radiative-convective models, wherein the concept was first initiated, λ is a nearly invariant parameter (typically, about 0.5 K/(Wm^{-2}); Ramanathan *et al.*, 1985) for a variety of radiative forcings, thus introducing the notion of a possible universality of the relationship between forcing and response. It is this feature which has enabled the radiative forcing to be perceived as a useful tool for obtaining first-order estimates of the relative climate impacts of different imposed radiative perturbations. Although the value of the parameter "λ" can vary from one model to another, within each model it is found to be remarkably constant for a wide range of radiative perturbations (WMO, 1986). The invariance of λ has made the radiative forcing concept appealing as a convenient measure to estimate the global, annual mean surface temperature response, without taking the recourse to actually run and analyse, say, a three-dimensional atmosphere-ocean general circulation model (AOGCM) simulation.

In the context of the three-dimensional AOGCMs, too, the applicability of a general global mean climate sensitivity parameter (i.e., global mean surface temperature response to global mean radiative forcing) has been explored. The GCM investigations include studies of (i) the responses to short-wave forcing such as a change in the solar constant or cloud albedo or doubling of CO_2, both forcing types being approximately spatially homogeneous (e.g., Manabe and Wetherald, 1980; Hansen *et al.*, 1984, 1997a; Chen and Ramaswamy, 1996a; Le Treut *et al.*, 1998), (ii) responses due to different considered mixtures of greenhouse gases, with the forcings again being globally homogeneous (Wang *et al.*, 1991, 1992), (iii) responses to the spatially homogeneous greenhouse gas and the spatially inhomogeneous sulphate aerosol direct forcings (Cox *et al.*, 1995), (iv) responses to different assumed profiles of spatially inhomogeneous species, e.g., aerosols and O_3 (Hansen *et al.*, 1997a), and (v) present-day versus palaeoclimate (e.g., last glacial maximum) simulations (Manabe and Broccoli, 1985; Rind *et al.*, 1989; Berger *et al.*, 1993; Hewitt and Mitchell, 1997).

Overall, the three-dimensional AOGCM experiments performed thus far show that the radiative forcing continues to serve as a good estimator for the global mean surface temperature response but not to a quantitatively rigorous extent as in the case of the one-dimensional radiative-convective models. Several GCM studies suggest a similar global mean climate sensitivity for the spatially homogeneous and for many but not all of the spatially inhomogeneous forcings of relevance for climate change in the industrial era (Wang *et al.*, 1992; Roeckner *et al.*, 1994; Taylor and Penner, 1994; Cox *et al.*, 1995; Hansen *et al.*, 1997a). Paleoclimate simulations (Manabe and Broccoli, 1985; Rind *et al.*, 1989) also suggest the idea of similarities in climate sensitivity for

a spatially homogeneous and an inhomogeneous forcing (arising due to the presence of continental ice sheets at mid- to high northern latitudes during the last glacial maximum). However, different values of climate sensitivity can result from the different GCMs which, in turn, are different from the λ values obtained with the radiative-convective models. Hansen *et al.* (1997a) show that the variation in λ for most of the globally distributed forcings suspected of influencing climate over the past century is typically within about 20%. Extending considerations to some of the spatially confined forcings yields a range of about 25 to 30% around a central estimate (see also Forster *et al.*, 2001). This is to be contrasted with the variation of 15% obtained in a smaller number of experiments (all with fixed clouds) by Ramaswamy and Chen (1997b). However, in a general sense and considering arbitrary forcing types, the variation in λ could be substantially higher (50% or more) and the climate response much more complex (Hansen *et al.*, 1997a). It is noted that the climate sensitivity for some of the forcings that have potentially occurred in the industrial era have yet to be comprehensively investigated.

While the total climate feedback for the spatially homogeneous and the considered inhomogeneous forcings does not differ significantly, leading to a near-invariant climate sensitivity, the individual feedback mechanisms (water vapour, ice albedo, lapse rate, clouds) can have different strengths (Chen and Ramaswamy, 1996a,b). The feedback effects can be of considerably larger magnitude than the initial forcing and govern the magnitude of the global mean response (Ramanathan, 1981; Wetherald and Manabe, 1988; Hansen *et al.*, 1997a). For different types of perturbations, the relative magnitudes of the feedbacks can vary substantially.

For spatially homogeneous forcings of opposite signs, the responses are somewhat similar in magnitude, although the ice albedo feedback mechanism can yield an asymmetry in the high latitude response with respect to the sign of the forcing (Chen and Ramaswamy, 1996a). Even if the forcings are spatially homogeneous, there could be changes in land surface energy budgets that depend on the manner of the perturbation (Chen and Ramaswamy, 1996a). Furthermore, for the same global mean forcing, dynamic feedbacks involving changes in convective heating and precipitation can be initiated in the spatially inhomogeneous perturbation cases that differ from those in the spatially homogeneous perturbation cases.

The nature of the response and the forcing-response relation (Equation 6.1) could depend critically on the vertical structure of the forcing (see WMO, 1999). A case in point is O_3 changes, since this initiates a vertically inhomogeneous forcing owing to differing characteristics of the solar and long-wave components (WMO, 1992). Another type of forcing is that due to absorbing aerosols in the troposphere (Kondratyev, 1999). In this instance, the surface experiences a deficit while the atmosphere gains short-wave radiative energy. Hansen *et al.* (1997a) show that, for both these special types of forcing, if the perturbation occurs close to the surface, complex feedbacks involving lapse rate and cloudiness could alter the climate sensitivity substantially from that prevailing for a similar magnitude of perturbation imposed at other altitudes. A different kind of example is illustrated by model experiments indicating that the climate sensitivity is considerably different for

O_3 losses occurring in the upper rather than lower stratosphere (Hansen *et al.*, 1997a; Christiansen, 1999). Yet another example is stratospheric aerosols in the aftermath of volcanic eruptions. In this case, the lower stratosphere is radiatively warmed while the surface-troposphere cools (Stenchikov *et al.*, 1998) so that the climate sensitivity parameter does not convey a complete picture of the climatic perturbations. Note that this contrasts with the effects due to CO_2 increases, wherein the surface-troposphere experiences a radiative heating and the stratosphere a cooling. The vertical partitioning of forcing between atmosphere and surface could also affect the manner of changes of parameters other than surface temperature, e.g., evaporation, soil moisture.

Zonal mean and regional scale responses for spatially inhomogeneous forcings can differ considerably from those for homogeneous forcings. Cox *et al.* (1995) and Taylor and Penner (1994) conclude that the spatially inhomogeneous sulphate aerosol direct forcing in the northern mid-latitudes tends to yield a significant response there that is absent in the spatially homogeneous case. Using a series of idealised perturbations, Ramaswamy and Chen (1997b) show that the gradient of the equator-to-pole surface temperature response to spatially homogeneous and inhomogeneous forcings is significantly different when scaled with respect to the global mean forcing, indicating that the more spatially confined the forcing, the greater the meridional gradient of the temperature response. In the context of the additive nature of the regional temperature change signature, Penner *et al.* (1997) suggest that there may be some limit to the magnitude of the forcings that yield a linear signal.

A related issue is whether responses to individual forcings can be linearly added to obtain the total response to the sum of the forcings. Indications from experiments that have attempted a very limited number of combinations are that the forcings can indeed be added (Cox *et al.*, 1995; Roeckner *et al.*, 1994; Taylor and Penner, 1994). These investigations have been carried out in the context of equilibrium simulations and have essentially dealt with the CO_2 and sulphate aerosol direct forcing. There tends to be a linear additivity not only for the global mean temperature, but also for the zonal mean temperature and precipitation (Ramaswamy and Chen, 1997a). Haywood *et al.* (1997c) have extended the study to transient simulations involving greenhouse gases and sulphate aerosol forcings in a GCM. They find the linear additivity to approximately hold for both the surface temperature and precipitation, even on regional scales. Parameters other than surface temperature and precipitation have not been tested extensively. Owing to the limited sets of forcings examined thus far, it is not possible as yet to generalise to all natural and anthropogenic forcings discussed in subsequent sections of this chapter.

One caveat that needs to be reiterated (see IPCC, 1994 and SAR) regarding forcing-response relationships is that, even if there is a cancellation in the global mean forcing due to forcings that are of opposite signs and distributed spatially in a different manner, and even if the responses are linearly additive, there could be spatial aspects of the responses that are not necessarily null. In particular, circulation changes could result in a distinct regional response even under conditions of a null global mean forcing and a null global mean surface temperature response (Ramaswamy

and Chen, 1997a). Sinha and Harries (1997) suggest that there can be characteristic vertical responses even if the net radiative forcing is zero.

6.2.2 Strengths and Limitations of the Forcing Concept

Radiative forcing continues to be a useful concept, providing a convenient first-order measure of the relative climatic importance of different agents (SAR; Shine and Forster, 1999). It is computationally much more efficient than a GCM calculation of the climate response to a specific forcing; the simplicity of the calculation allows for sophisticated, highly accurate radiation schemes, yielding accurate forcing estimates; the simplicity also allows for a relative ease in conducting model intercomparisons; it yields a first-order perspective that can then be used as a basis for more elaborate GCM investigations; it potentially bypasses the complex tasks of running and analysing equilibrium-response GCM integrations; it is useful for isolating errors and uncertainties due to radiative aspects of the problem.

In gauging the relative climatic significance of different forcings, an important question is whether they have similar climate sensitivities. As discussed in Section 6.2.1, while models indicate a reasonable similarity of climate sensitivities for spatially homogeneous forcings (e.g., CO_2 changes, solar irradiance changes), it is not possible as yet to make a generalisation applicable to all the spatially inhomogeneous forcing types. In some cases, the climate sensitivity differs significantly from that for CO_2 changes while, for some other cases, detailed studies have yet to be conducted. A related question is whether the linear additivity concept mentioned above can be extended to include all of the relevant forcings, such that the sum of the responses to the individual forcings yields the correct total climate response. As stated above, such tests have been conducted only for limited subsets of the relevant forcings.

Another important limitation of the concept is that there are parameters other than global mean surface temperature that need to be determined, and that are as important from a climate and societal impacts perspective; the forcing concept cannot provide estimates for such climate parameters as directly as for the global mean surface temperature response. There has been considerably less research on the relationship of the equilibrium response in such parameters as precipitation, ice extent, sea level, etc., to the imposed radiative forcing.

Although the radiative forcing concept was originally formulated for the global, annual mean climate system, over the past decade, it has been extended to smaller spatial domains (zonal mean), and smaller time-averaging periods (seasons) in order to deal with short-lived species that have a distinct geographical and seasonal character, e.g., aerosols and O_3 (see also the SAR). The global, annual average forcing estimate for these species masks the inhomogeneity in the problem such that the anticipated global mean response (via Equation 6.1) may not be adequate for gauging the spatial pattern of the actual climate change. For these classes of radiative perturbations, it is incorrect to assume that the characteristics of the responses would be necessarily co-located with the forcing, or that the magnitudes would follow the forcing patterns exactly (e.g., Cox *et al.*, 1995; Ramaswamy and Chen, 1997b).

6.3 Well-mixed Greenhouse Gases

The well-mixed greenhouse gases have lifetimes long enough to be relatively homogeneously mixed in the troposphere. In contrast, O_3 (Section 6.5) and the NMHCs (Section 6.6) are gases with relatively short lifetimes and are therefore not homogeneously distributed in the troposphere.

Spectroscopic data on the gaseous species have been improved with successive versions of the HITRAN (Rothman *et al.*, 1992, 1998) and GEISA databases (Jacquinet-Husson *et al.*, 1999). Pinnock and Shine (1998) investigated the effect of the additional hundred thousands of new lines in the 1996 edition of the HITRAN database (relative to the 1986 and the 1992 editions) on the infrared radiative forcing due to CO_2, CH_4, N_2O and O_3. They found a rather small effect due to the additional lines, less than a 5% effect for the radiative forcing of the cited gases and less than 1.5% for a doubling of CO_2. For the chlorofluorocarbons (CFCs) and their replacements, the uncertainties in the spectroscopic data are much larger than for CO_2, CH_4, N_2O and O_3, and differ more among the various laboratory studies. Christidis *et al.* (1997) found a range of 20% between ten different spectroscopic studies of CFC-11. Ballard *et al.* (2000) performed an intercomparison of laboratory data from five groups and found the range in the measured absorption cross-section of HCFC-22 to be about 10%.

Several previous studies of radiative forcing due to well-mixed greenhouse gases have been performed using single, mostly global mean, vertical profiles. Myhre and Stordal (1997) investigated the effects of spatial and temporal averaging on the globally and annually averaged radiative forcing due to the well-mixed greenhouse gases. The use of a single global mean vertical profile to represent the global domain, instead of the more rigorous latitudinally varying profiles, can lead to errors of about 5 to 10%; errors arising due to the temporal averaging process are much less (~1%). Freckleton *et al.* (1998) found similar effects and suggested three vertical profiles which could represent global atmospheric conditions satisfactorily in radiative transfer calculations. In the above two studies as well as in Forster *et al.* (1997), it is the dependence of the radiative forcing on the tropopause height and thereby also the vertical temperature profile, that constitutes the main reason for the need of a latitudinal resolution in radiative forcing calculations. The radiative forcing due to halocarbons depends on the tropopause height more than is the case for CO_2 (Forster *et al.*, 1997; Myhre and Stordal, 1997).

Not all greenhouse gases are well mixed vertically and horizontally in the troposphere. Freckleton *et al.* (1998) have investigated the effects of inhomogeneities in the concentrations of the greenhouse gases on the radiative forcing. For CH_4 (a well-mixed greenhouse gas), the assumption that it is well-mixed horizontally in the troposphere introduces an error much less than 1% relative to a calculation in which a chemistry-transport model predicted distribution of CH_4 was used. For most halocarbons, and to a lesser extent for CH_4 and N_2O, the mixing ratio decays with altitude in the stratosphere. For CH_4 and N_2O, this implies a reduction in the radiative forcing of up to about 3% (Freckleton *et al.*, 1998; Myhre *et al.*, 1998b). For most halocarbons, this implies a reduction in the radiative forcing up to about 10%

(Christidis *et al.*, 1997; Hansen *et al.*, 1997a; Minschwaner *et al.*, 1998; Myhre *et al.*, 1998b) while it is found to be up to 40% for a short-lived component found in Jain *et al.* (2000).

Trapping of the long-wave radiation due to the presence of clouds reduces the radiative forcing of the greenhouse gases compared to the clear-sky forcing. However, the magnitude of the effect due to clouds varies for different greenhouse gases. Relative to clear skies, clouds reduce the global mean radiative forcing due to CO_2 by about 15% (Pinnock *et al.*, 1995; Myhre and Stordal, 1997), that due to CH_4 and N_2O is reduced by about 20% (derived from Myhre *et al.*, 1998b), and that due to the halocarbons is reduced by up to 30% (Pinnock *et al.*, 1995; Christidis *et al.*, 1997; Myhre *et al.*, 1998b).

The effect of stratospheric temperature adjustment also differs between the various well-mixed greenhouse gases, owing to different gas optical depths, spectral overlap with other gases, and the vertical profiles in the stratosphere. The stratospheric temperature adjustment reduces the radiative forcing due to CO_2 by about 15% (Hansen *et al.*, 1997a). CH_4 and N_2O estimates are slightly modified by the stratospheric temperature adjustment, whereas the radiative forcing due to halocarbons can increase by up to 10% depending on the spectral overlap with O_3 (IPCC, 1994).

Radiative transfer calculations are performed with different types of radiative transfer schemes ranging from line-by-line models to band models (IPCC, 1994). Evans and Puckrin (1999) have performed surface measurements of downward spectral radiances which reveal the optical characteristics of individual greenhouse gases. These measurements are compared with line-by-line calculations. The agreement between the surface measurements and the line-by-line model is within 10% for the most important of the greenhouse gases: CO_2, CH_4, N_2O, CFC-11 and CFC-12. This is not a direct test of the irradiance change at the tropopause and thus of the radiative forcing, but the good agreement does offer verification of fundamental radiative transfer knowledge as represented by the line-by-line (LBL) model. This aspect concerning the LBL calculation is reassuring as several radiative forcing determinations which employ coarser spectral resolution models use the LBL as a benchmark tool (Freckleton *et al.*, 1996; Christidis *et al.*, 1997; Minschwaner *et al.*, 1998; Myhre *et al.*, 1998b; Shira *et al.*, 2001). Satellite observations can also be useful in estimates of radiative forcing and in the intercomparison of radiative transfer codes (Chazette *et al.*, 1998).

6.3.1 Carbon Dioxide

IPCC (1990) and the SAR used a radiative forcing of 4.37 Wm^{-2} for a doubling of CO_2 calculated with a simplified expression. Since then several studies, including some using GCMs (Mitchell and Johns, 1997; Ramaswamy and Chen, 1997b; Hansen *et al.*, 1998), have calculated a lower radiative forcing due to CO_2 (Pinnock *et al.*, 1995; Roehl *et al.*, 1995; Myhre and Stordal, 1997; Myhre *et al.*, 1998b; Jain *et al.*, 2000). The newer estimates of radiative forcing due to a doubling of CO_2 are between 3.5 and 4.1 Wm^{-2} with the relevant species and various overlaps between greenhouse gases included. The lower forcing in the cited newer studies is due to an accounting of the stratospheric temperature adjustment which was not properly taken into account in the simplified expression used in IPCC (1990) and the SAR (Myhre *et al.*, 1998b). In Myhre *et al.* (1998b) and Jain *et al.* (2000), the short-wave forcing due to CO_2 is also included, an effect not taken into account in the SAR. The short-wave effect results in a negative forcing contribution for the surface-troposphere system owing to the extra absorption due to CO_2 in the stratosphere; however, this effect is relatively small compared to the total radiative forcing (< 5%).

The new best estimate based on the published results for the radiative forcing due to a doubling of CO_2 is 3.7 Wm^{-2}, which is a reduction of 15% compared to the SAR. The forcing since pre-industrial times in the SAR was estimated to be 1.56 Wm^{-2}; this is now altered to 1.46 Wm^{-2} in accordance with the discussion above. The overall decrease of about 6% (from 1.56 to 1.46) accounts for the above effect and also accounts for the increase in CO_2 concentration since the time period considered in the SAR (the latter effect, by itself, yields an increase in the forcing of about 10%).

While an updating of the simplified expressions to account for the stratospheric adjustment becomes necessary for radiative forcing estimates, it is noted that GCM simulations of CO_2-induced climate effects already account for this physical effect implicitly (see also Chapter 9). In some climate studies, the sum of the non-CO_2 well-mixed greenhouse gases forcing is represented by that due to an equivalent amount of CO_2. Because the CO_2 forcing in the SAR was higher than the new estimate, the use of the equivalent CO_2 concept would underestimate the impact of the non-CO_2 well-mixed gases, if the IPCC values of radiative forcing were used in the scaling operation.

6.3.2 Methane and Nitrous Oxide

The SAR reported that several studies found a higher forcing due to CH_4 than IPCC (1990), up to 20%; however the recommendation was to use the same value as in IPCC (1990). The higher radiative forcing estimates were obtained using band models. Recent calculations using LBL and band models confirm these results (Lelieveld *et al.*, 1998; Minschwaner *et al.*, 1998; Jain *et al.*, 2000). Using two band models, Myhre *et al.* (1998b) found the computed radiative forcing to differ by almost 10%. This was attributed to difficulties in the treatment of CH_4 in band models since, given its present abundance, the CH_4 absorption lies between the weak line and the strong line limits (Ramanathan *et al.*, 1987). After updating for a small increase in concentration since the SAR, the radiative forcing due to CH_4 is 0.48 Wm^{-2} since pre-industrial times. This estimate for forcing due to CH_4 is only for the direct effect of CH_4; for radiative forcing of the indirect effect of CH_4, see Sections 6.5 and 6.6.

The problem mentioned above with the band models for CH_4 does not occur to the same degree in the case of N_2O, given the latter's present concentrations. Three recent studies, Myhre *et al.* (1998b) (two models), Minschwaner *et al.* (1998) (one model), and Jain *et al.* (2000) (one model), calculated lower radiative forcing for N_2O than reported in previous IPCC assessments, viz., 0.13, 0.12, 0.11, and 0.12 Wm^{-2}, respectively,

compared to 0.14 Wm^{-2} in the SAR. For N_2O, effects of change in spectroscopic data, stratospheric adjustment, and decay of the mixing ratio in the stratosphere are all found to be small effects. However, effects of clouds and different radiation schemes are potential sources for the difference between the newer estimates and the SAR. A value of 0.15 Wm^{-2} is now suggested for the radiative forcing due to N_2O, taking into account an increase in the concentration since the SAR, together with a smaller pre-industrial concentration than assumed in IPCC (1996a; Table 2.2) (see Chapter 4).

6.3.3 Halocarbons

The SAR referred to Pinnock *et al.* (1995), who obtained a higher radiative forcing for CFC-11 than used in previous IPCC reports, but refrained from changing the recommended value pending further investigations. Since then several papers have investigated CFC-11, confirming the higher forcing value (Christidis *et al.*, 1997; Hansen *et al.*, 1997a; Myhre and Stordal, 1997; Good *et al.*, 1998; Myhre *et al.*, 1998b; Jain *et al.*, 2000) with a range from 0.24 to 0.29 Wm^{-2} $ppbv^{-1}$. As mentioned above, Christidis *et al.* (1997) found a large discrepancy in the absorption data for CFC-11 in the literature. Other causes for the difference in the radiative forcing are different treatments of the decrease in mixing ratio in the stratosphere and the fact that some estimates are performed with a single global mean column atmospheric profile. Taking these effects into account, a radiative efficiency due to CFC-11 of 0.25 Wm^{-2} $ppbv^{-1}$ is used, the same value as in WMO (1999). For the present concentration of CFC-11, this yields a forcing of 0.07 Wm^{-2} since pre-industrial times. In previous IPCC reports, radiative forcing due to CFCs and their replacements have been given relative to CFC-11. CFC-11 is now revised and this introduces a complicating factor since the radiative forcing for the CFCs and CFC replacements are given as absolute values in some studies, but relative to CFC-11 in others. WMO (1999) updated several of the halocarbons giving radiative forcing in absolute values (in Wm^{-2} $ppbv^{-1}$).

CFC-12 is investigated in Hansen *et al.* (1997a), Myhre *et al.* (1998b), Minschwaner *et al.* (1998), Good *et al.* (1998) and Jain *et al.* (2000). The difference in the results is up to 20% which is due to differing impact of clouds, absorption cross-section data, and the vertical profile of decay of the mixing ratio in the stratosphere. The radiative forcing due to CFC-12 of 0.32 Wm^{-2} $ppbv^{-1}$ used in WMO (1999) is retained, which is slightly higher than the SAR value. The present radiative forcing due to CFC-12 is therefore 0.17 Wm^{-2}, which is the third highest forcing among the well-mixed greenhouse gases.

Radiative forcing values for well-mixed greenhouse gases with non-negligible contributions at present are included in Table 6.1. Several recent studies have investigated various CFC replacements (Imasu *et al.*, 1995; Gierczak *et al.*, 1996; Barry *et al.*, 1997; Christidis *et al.*, 1997; Grossman *et al.*, 1997; Papasavva *et al.*, 1997; Good *et al.*, 1998; Heathfield *et al.*, 1998b; Highwood and Shine, 2000; Ko *et al.*, 1999; Myhre *et al.*, 1999; Jain *et al.*, 2000; Li *et al.*, 2000; Naik *et al.*, 2000; Shira *et al.*, 2001). For some CFC replacements not included in Table 6.1, the radiative forcings are shown in Tables 6.7 and 6.8 (Section 6.12).

Table 6.1: *Pre-industrial (1750) and present (1998) abundances of well-mixed greenhouse gases and the radiative forcing due to the change in abundance. Volume mixing ratios for CO_2 are in ppm, for CH_4 and N_2O in ppb, and for the rest in ppt.*

Gas	Abundance (Year 1750)	Abundance (Year 1998)	Radiative forcing (Wm^{-2})
Gases relevant to radiative forcing only			
CO_2	278	365	1.46
CH_4	700	1745	0.48
N_2O	270	314	0.15
CF_4	40	80	0.003
C_2F_6	0	3	0.001
SF_6	0	4.2	0.002
HFC-23	0	14	0.002
HFC-134a	0	7.5	0.001
HFC-152a	0	0.5	0.000
Gases relevant to radiative forcing and ozone depletion			
CFC-11	0	268	0.07
CFC-12	0	533	0.17
CFC-13	0	4	0.001
CFC-113	0	84	0.03
CFC-114	0	15	0.005
CFC-115	0	7	0.001
CCl_4	0	102	0.01
CH_3CCl_3	0	69	0.004
HCFC-22	0	132	0.03
HCFC-141b	0	10	0.001
HCFC-142b	0	11	0.002
Halon-1211	0	3.8	0.001
Halon-1301	0	2.5	0.001

Table 6.2: *Simplified expressions for calculation of radiative forcing due to CO_2, CH_4, N_2O, and halocarbons. The first row for CO_2 lists an expression with a form similar to IPCC (1990) but with newer values of the constants. The second row for CO_2 is a more complete and updated expression similar in form to that of Shi (1992). The third row expression for CO_2 is from WMO (1999), based in turn on Hansen et al. (1988).*

Trace gas	Simplified expression Radiative forcing, ΔF (Wm^{-2})	Constants
CO_2	$\Delta F = \alpha \ln(C/C_0)$ $\Delta F = \alpha \ln(C/C_0) + \beta(\sqrt{C} - \sqrt{C_0})$ $\Delta F = \alpha(g(C) - g(C_0))$ where $g(C) = \ln(1 + 1.2C + 0.005C^2 + 1.4 \times 10^{-6}C^3)$	$\alpha = 5.35$ $\alpha = 4.841,\ \beta = 0.0906$ $\alpha = 3.35$
CH_4	$\Delta F = \alpha(\sqrt{M} - \sqrt{M_0}) - (f(M, N_0) - f(M_0, N_0))$	$\alpha = 0.036$
N_2O	$\Delta F = \alpha(\sqrt{N} - \sqrt{N_0}) - (f(M_0, N) - f(M_0, N_0))$	$\alpha = 0.12$
CFC-11[a]	$\Delta F = \alpha(X - X_0)$	$\alpha = 0.25$
CFC-12	$\Delta F = \alpha(X - X_0)$	$\alpha = 0.32$

$f(M, N) = 0.47 \ln[1 + 2.01 \times 10^{-5} (MN)^{0.75} + 5.31 \times 10^{-15} M(MN)^{1.52}]$

C is CO_2 in ppm

M is CH_4 in ppb

N is N_2O in ppb

X is CFC in ppb

The constant in the simplified expression for CO_2 for the first row is based on radiative transfer calculations with three-dimensional climatological meteorological input data (Myhre *et al.*, 1998b). For the second and third rows, constants are derived with radiative transfer calculations using one-dimensional global average meteorological input data from Shi (1992) and Hansen *et al.* (1988), respectively.

The subscript 0 denotes the unperturbed concentration.

[a] The same expression is used for all CFCs and CFC replacements, but with different values for α (i.e., the radiative efficiencies in Table 6.7).

The values of CFC-115 and CCl_4 have been substantially revised since the IPCC (1994) report, with a lower and higher radiative forcing estimate, respectively. Highwood and Shine (2000) calculated a radiative forcing due to chloroform ($CHCl_3$) which is much stronger than the SAR value. They suggest that this is due to the neglect of bands outside 800 to 1,200 cm^{-1} in previous studies of chloroform. Highwood and Shine (2000) found a radiative forcing due to HFC-23 which is substantially lower than the value given in the SAR.

6.3.4 Total Well-Mixed Greenhouse Gas Forcing Estimate

The radiative forcing due to all well-mixed greenhouse gases since pre-industrial times was estimated to be 2.45 Wm^{-2} in the SAR with an uncertainty of 15%. This is now altered to a radiative forcing of 2.43 Wm^{-2} with an uncertainty of 10%, based on the range of model results and the discussion of factors leading to uncertainties in the radiative forcing due to these greenhouse gases. The uncertainty in the radiative forcing due to CO_2 is estimated to be smaller than for the other well-mixed greenhouse gases; less than 10% (Section 6.3.1). For the CH_4 forcing the main uncertainty is connected to the radiative transfer

code itself and is estimated to be about 15% (Section 6.3.2). The uncertainty in N_2O (Section 6.3.2) is similar to that for CO_2, whereas the main uncertainties for halocarbons arise from the spectroscopic data. The estimated uncertainty for halocarbons is 10 to 15% for the most frequently studied species, but higher for some of the less investigated molecules (Section 6.3.3). A small increase in the concentrations of the well-mixed greenhouse gases since the SAR has compensated for the reduction in radiative forcing resulting from improved radiative transfer calculations. The rate of increase in the well-mixed greenhouse gas concentrations, and thereby the radiative forcing, has been smaller over the first half of the 1990s compared to previous decades (see also Hansen *et al.*, 1998). This is mainly a result of reduced growth in CO_2 and CH_4 concentrations and smaller increase or even reduction in the concentration of some of the halocarbons.

6.3.5 Simplified Expressions

IPCC (1990) used simplified analytical expressions for the well-mixed greenhouse gases based in part on Hansen *et al.* (1988). With updates of the radiative forcing, the simplified expressions need to be reconsidered, especially for CO_2 and N_2O. Shi (1992) investigated simplified expressions for the well-mixed greenhouse gases and Hansen *et al.* (1988, 1998) presented a simplified expression for CO_2. Myhre *et al.* (1998b) used the previous IPCC expressions with new constants, finding good agreement (within 5%) with high spectral resolution radiative transfer calculations. The already well established and simple functional forms of the expressions used in IPCC (1990), and their excellent agreement with explicit radiative transfer calculations, are strong bases for their continued usage, albeit with revised values of the constants, as listed in Table 6.2. Shi (1992) has suggested more physically based and accurate expressions which account for (i) additional absorption bands that could yield a separate functional form besides the one in IPCC (1990), and (ii) a better treatment of the overlap between gases. WMO (1999) used a simplified expression for CO_2 based on Hansen *et al.* (1988) and this simplified expression is used in the calculations of GWP in Section 6.12. For CO_2 the simplified expressions from Shi (1992) and Hansen *et al.* (1988) are also listed alongside the IPCC (1990)-like expression for CO_2 in Table 6.2. Compared to IPCC (1990) and the SAR and for similar changes in the concentrations of well-mixed greenhouse gases, the improved simplified expressions result in a 15% decrease in the estimate of the radiative forcing by CO_2 (first row in Table 6.2), a 15% decrease in the case of N_2O, an increase of 10 to 15% in the case of CFC-11 and CFC-12, and no change in the case of CH_4.

6.4 Stratospheric Ozone

6.4.1 Introduction

The observed stratospheric O_3 losses over the past two decades have caused a negative forcing of the surface-troposphere system (IPCC, 1992, 1994; SAR). In general, the sign and magnitude of the forcing due to stratospheric O_3 loss are governed by the vertical profile of the O_3 loss from the lower through to the upper stratosphere (WMO, 1999). Ozone depletion in the lower stratosphere, which occurs mainly in the mid- to high latitudes is the principal component of the forcing. It causes an increase in the solar forcing of the surface-troposphere system. However, the long-wave effects consist of a reduction of the emission from the stratosphere to the troposphere. This comes about due to the O_3 loss, coupled with a cooling of the stratospheric temperatures in the stratospheric adjustment process, with a colder stratosphere emitting less radiation. The long-wave effects, after adjustment of the stratospheric temperatures to the imposed perturbation, overwhelm the solar effect i.e., the negative long-wave forcing prevails over the positive solar to lead to a net negative radiative forcing of the surface-troposphere system (IPCC, 1992). The magnitude of the forcing is dependent on the loss in the lower stratosphere, with the estimates subject to some uncertainties in view of the fact that detailed observations on the vertical profile in this region of the atmosphere are difficult to obtain.

Typically, model-based estimates involve a local (i.e., over the grid box of the model) adjustment of the stratosphere (Section 6.1) assuming the dynamical heating to be fixed (FDH approximation; see also Appendix 6.1). An improved version of this scheme is the so-called seasonally evolving fixed dynamical heating (SEFDH; Forster *et al.*, 1997; Kiehl *et al.*, 1999). The adjustment of the stratosphere to a new thermal equilibrium state is a critical element for estimating the sign and magnitude of the forcing due to stratospheric O_3 loss (WMO, 1992, 1995). While the computational procedures are well established for the FDH and SEFDH approximations in the context of the surface-troposphere forcing, one test of the approximations lies in the comparison of the computed with observed temperature changes, since it is this factor that plays a large role in the estimate of the forcing. While the temperature changes going into the determination of the forcing are broadly consistent with the observations, there are challenges in comparing quantitatively the actual temperature changes (which undoubtedly are affected by other influences and may even contain feedbacks due to O_3 and other forcings) with the FDH or SEFDH model simulations (which necessarily do not contain feedback effects other than the stratospheric temperature response due to the essentially radiative adjustment process).

We reiterate both the concept of the forcing for stratospheric O_3 changes and the fact that this has led to a negative radiative forcing since the late 1970s. Further, the model-based estimates that necessarily rely on satellite observations of O_3 losses are likely the most reliable means to derive the forcing, notwithstanding the uncertainty in the vertical profile of loss in the vicinity of the tropopause. Since several model estimates have employed the Total Ozone Mapping Spectrometer (TOMS) observations as one of the inputs for the calculations, there is the likelihood of a small tropospheric O_3 change component contaminating the stratospheric O_3 loss amounts, especially for the lowermost regions of the lower stratosphere (Hansen *et al.*, 1997a; Shine and Forster, 1999). Both the estimates derived in the earlier IPCC assessments and the studies since the SAR show that the forcing pattern increases from the mid- to high latitudes consistent with the O_3 loss amounts. Seasonally, the winter/springtime forcings are the largest, again consistent with the temporal nature of the observed O_3 depletion.

It is logical to enquire into the realism of the computed coolings with the available observations using models more realistic than FDH/SEFDH, namely GCMs. Furthermore, comparison of the FDH and SEFDH derived temperature changes with those from a GCM constitutes another test of the approximations. WMO (1999) concluded, on the basis of intercomparisons of the temperature records as measured by different instruments, that there has been a distinct cooling of the global mean temperature of the lower stratosphere over the past two decades, with a value of about 0.5°C/decade. Model simulations from GCMs using the observed O_3 losses yield global mean temperature changes that are approximately consistent with the observations. Such a cooling is also much larger than that due to the well-mixed greenhouse gases taken together over the same time period. Although the possibility of other trace species also contributing to this cooling cannot be ruled out, the consistency between observations and model simulations enhances the general principle of an O_3-induced cooling of the lower stratosphere, and thus the negativity of the radiative forcing due to the O_3 loss. Going from global, annual mean to zonal, seasonal mean changes in the lower stratosphere, the agreement between models and observations tends to be less strong than for the global mean values, but the suggestion of an O_3-induced signal exists. Note though that water vapour changes could also be contributing (see Section 6.6.4; Forster and Shine, 1999), complicating the quantitative attribution of the cooling solely due to O_3. As far as the FDH models that have been employed to derive the forcing are concerned, their temperature changes are broadly consistent with the GCMs and the observed cooling. However, the mid- to high latitude cooling in FDH tends to be stronger than in the GCMs and is more than that observed. The SEFDH approximation tends to do better than the FDH calculation when compared against observations (Forster *et al.*, 1997).

6.4.2 Forcing Estimates

Earlier IPCC reports had quoted a value of about -0.1 Wm^{-2}/decade with a factor of two uncertainty. There have been revisions in this estimate based on new data available on the O_3 trends (Harris *et al.*, 1998; WMO, 1999), and an extension of the period over which the forcing is computed. Models using observed O_3 changes but with varied methods to derive the temperature changes in the stratosphere have obtained -0.05 to -0.19 Wm^{-2}/decade (WMO, 1999).

Hansen *et al.* (1997a) have extended the calculations to include the O_3 loss up to the mid-1990s and performed a variety of O_3 loss experiments to investigate the forcing and response. In particular, they obtained forcings of -0.2 and -0.28 Wm^{-2} for the period 1979 to 1994 using SAGE/TOMS and SAGE/SBUV satellite data, respectively. Hansen *et al.* (1998) updated their forcing to -0.2 Wm^{-2} with an uncertainty of 0.1 Wm^{-2} for the period 1970 to present. Forster and Shine (1997) obtained forcings of -0.17 Wm^{-2} and -0.22 Wm^{-2} for the period 1979 to 1996 using SAGE and SBUV observations, respectively. The WMO (1999) assessment gave a value of -0.2 Wm^{-2} with an uncertainty of ± 0.15 Wm^{-2} for the period from late 1970s to mid-1990s. Forster and Shine (1997) have also extended the computations back to 1964 using O_3 changes deduced from surface-based observations; combining these with an assumption that the decadal rate of change of forcing from 1979 to 1991 was sustained to the mid-1990s yielded a total stratospheric O_3 forcing of about -0.23 Wm^{-2}. Shine and Forster (1999) have revised this value to -0.15 Wm^{-2} for the period 1979 to 1997, choosing not to include the values prior to 1979 in view of the lack of knowledge on the vertical profile which makes the sign of the change also uncertain. They also revised the uncertainty to ± 0.12 Wm^{-2} around the central estimate. A more recent estimate by Forster (1999) yields -0.10 ± 0.02 Wm^{-2} for the 1979 to 1997 period using the SPARC O_3 profile (Harris *et al.*, 1998).

There have been attempts to use satellite-observed O_3 and temperature changes to gauge the forcing. Thus, Zhong *et al.* (1996, 1998) obtained a small value of -0.02 Wm^{-2}/decade; and with inclusion of the 14 micron band, a value of -0.05 Wm^{-2}/decade. It has been noted that the poor vertical resolution of the satellite temperature retrievals makes it difficult to estimate the forcing; in fact, a similar calculation using radiosonde-based temperatures yields a value of -0.1 Wm^{-2}/decade (Shine *et al.*,

1998). The main difficulty is that the temperature change in the vicinity of the lower stratosphere critically affects the emission from the stratosphere into the troposphere. Thus any uncertainty in the MSU satellite retrieval induced by the broad altitude weighting function (see WMO, 1999) becomes an important factor in the estimation of the forcing. Further, the degree of response of the climate system, embedded in the observed temperature change (i.e., feedbacks), is not resolved in an easy manner. This makes it difficult to distinguish quantitatively the part of temperature change that is a consequence of the stratospheric adjustment process (which would be, by definition, a legitimate component of the forcing estimate) and that which is due to mechanisms other than O_3 loss. Thus, using observed temperatures to estimate the forcing may be more uncertain than the model-based estimates. It must be noted though that both methods share the difficulty of quantifying the vertical and geographical distributions of the O_3 changes near the tropopause, and the rigorous association of this to the observed temperature changes. In an overall sense, it is a difficult task to verify the radiative forcing in cases where the stratospheric adjustment yields a dramatically different result than the instantaneous forcing i.e., where the species changes affect stratospheric temperatures and alter substantially the long-wave radiative effects at the tropopause. A related point is the possible upward movement of the tropopause which could explain in part the observed negative trends in O_3 and temperature (Fortuin and Kelder, 1996).

Kiehl *et al.* (1999) obtained a radiative forcing of -0.187 Wm^{-2} using the O_3 profile data set describing changes since the late 1970s due to stratospheric depletion alone, consistent with the range of other models (see Shine *et al.*, 1995). Kiehl *et al.* (1999) also present results using a very different set of O_3 change profiles deduced from satellite-derived total column O_3 and satellite-inferred tropospheric O_3 measurements to arrive at an implied O_3 forcing, considering changes at and above the tropopause, of -0.01 Wm^{-2}. The reason for the considerably weaker estimate reflects the increased O_3 in the tropopause region that is believed to have occurred since pre-industrial times (largely before 1970) in many polluted areas. How the changes in O_3 at the tropopause are prescribed is hence an important factor for the difference between this calculation and those from the other estimates.

Clearly, since WMO (1992), this forcing has been investigated in an intensive manner using different approaches, and the observational evidence of the O_3 losses, including the spatial and seasonal characteristics, are now on a firmer footing. In arriving at a best estimate for the forcing, we rely essentially on the studies that have made use of stratospheric O_3 observations directly. Based on this consideration, we adopt here a forcing of -0.15 ± 0.1 Wm^{-2} for the 1979 to 1997 period. However, it is cautioned that the small values obtained by the two specific studies mentioned above inhibit the placement of a high confidence in the estimate quoted.

In general, the reliability of the estimates above is affected by the fact that the O_3 changes in the lower stratosphere, tropopause, and upper troposphere are all poorly quantified, around the globe in general, such that the entire global domain from 200 to 50 hPa becomes crucial for the temperature change and the adjusted forcing. Forster and Shine (1997) note that the sensitivity of forcing to percentage of O_3 loss near the tropopause is more than when the changes occur lower in the atmosphere.

Myhre *et al.* (1998a) derived O_3 changes using a chemical model in contrast to observations. As the loss of O_3 in the upper stratosphere in the simulations was large, a positive forcing of 0.02 Wm^{-2}/decade was obtained (see Ramanathan and Dickinson (1979) for an explanation of the change of sign for a O_3 loss in the lower stratosphere versus the upper stratosphere). While there are difficulties in modelling the O_3 depletion in the global strato-sphere (WMO, 1999), this study reiterates the need to be cognisant of the role played by the vertical profile of O_3 loss amounts in the entire stratosphere, i.e., middle and upper strato-sphere as well, besides the lower stratosphere.

An important issue is whether the actual surface temperature responses to the forcing by stratospheric O_3 has the same relationship with forcing as obtained for, say, CO_2 or solar constant changes. Hansen *et al.* (1997a) and Christiansen (1999) have performed a host of GCM experiments to test this concept. The forcing by lower stratospheric O_3 is an unusual one in that it has a positive short-wave and a negative long-wave radiative forcing. Moreover, it has a unique vertical structure owing to the fact that the short-wave effects are felt at the surface while the long-wave is felt only initially at the upper troposphere (Ramanathan and Dickinson, 1979; WMO, 1992). Compared to, say, CO_2 change, the stratospheric O_3 forcing is not global in extent, being very small in the tropics and increasing from mid- to high latitudes; the O_3 forcing also differs in its vertical structure, since the radiative forcings for CO_2 change in both the troposphere and surface are of the same sign (WMO, 1986). The relationship between the global mean forcing and response differs by less than 20% for O_3 profiles, resembling somewhat the actual losses (Hansen *et al.*, 1997a). However, serious departures occur if the O_3 changes are introduced near surface layers when the lapse rate change, together with cloud feedbacks, make the climate sensitivity quite different from the nominal values. There also occur substantial differences in the climate sensitivity parameter for O_3 losses in the upper stratosphere. This is further substantiated by Christiansen (1999) who shows that the higher climate sensitivity for upper stratospheric O_3 losses relative to lower stratospheric depletion is related to the vertical partitioning of the forcing, in particular the relative roles of short-wave and long-wave radiation in the surface-troposphere system. It is encouraging that the global mean climate sensitivity parameter for cases involving lower stratospheric O_3 changes and that for CO_2 changes (viz., doubling) are reasonably similar in Christiansen (1999) while being within about 25% of a central value in Hansen *et al.* (1997a). An energy balance model study (Bintanja *et al.*, 1997) suggests a stronger albedo feedback for O_3 changes than for CO_2 perturbations (see also WMO, 1999).

The evolution of the forcing due to stratospheric O_3 loss hinges on the rate of recovery of the ozone layer, with special regards to the spatial structure of such a recovery in the mid- to high latitudes. If the O_3 losses are at their maximum or will reach a maximum within the next decade, then the forcing may not become much more negative (i.e., it was −0.1 Wm^{-2} for just the

1980s; inclusion of the 1990s increases the magnitude by about 50%). And, as the O_3 layer recovers, the forcing may remain static, eventually tending to become less negative. At this time, there will be a lesser offset of the positive greenhouse effects of the halocarbons and the other well-mixed greenhouse gases (WMO, 1999). Solomon and Daniel (1996) point out that the global mean stratospheric O_3 forcing can be expected to scale down substantially in importance relative to the well-mixed greenhouse gases, in view of the former's decline and the latter's sustained increase in concentrations. Note, however, that the evolution of the negativity of the stratospheric O_3 forcing may vary considerably with latitude and season i.e., the recovery may not occur at all locations and seasons at the same rate. Thus, the spatial and seasonal evolution of forcing in the future requires as much scrutiny as the global mean estimate.

6.5 Radiative Forcing by Tropospheric Ozone

6.5.1 Introduction

Human activities have long been known to influence tropospheric O_3, not only in urban areas where O_3 is a major component of 'smog', but also in the remote atmosphere (e.g., IPCC, 1990, 1994; the SAR and references therein). The current state of scientific understanding of tropospheric O_3 chemistry and trends is reviewed in Chapter 4 of this report, where it is emphasised that tropospheric O_3 has an average lifetime of the order of weeks. This relatively short lifetime implies that the distribution of tropospheric O_3, as well as the trends in that distribution (which in turn lead to radiative forcing) are highly variable in space and time. Studies relating to the evaluation of radiative forcing due to estimated tropospheric O_3 increases since pre-industrial times are discussed here. While there are a number of sites where high quality surface measurements of O_3 have been obtained for a few decades, there are fewer locations where ozonesonde data allow study of the vertical distribution of the trends, and fewer still with records prior to about 1970. A limited number of surface measure-ments in Europe date back to the late 19th century. These suggest that O_3 has more than tripled in the 20th century there (Marenco *et al.*, 1994). The lack of global information on pre-industrial tropospheric O_3 distributions is, however, a major uncertainty in the evaluation of the forcing of this key gas (see Chapter 4).

Biomass burning plays a significant role in tropospheric O_3 production and hence in tropical radiative forcing over large spatial scales, particularly in the tropical Atlantic west of the coast of Africa (e.g., Fishman, 1991; Fishman and Brackett, 1997; Portmann *et al.*, 1997; Hudson and Thompson, 1998) and in Indonesia (Hauglustaine *et al.*, 1999). Export of industrial pollution to the Arctic can lead to increased O_3 over a highly reflective snow or ice surface, and correspondingly large local radiative forcings (Hauglustaine *et al.*, 1998; Mickley *et al.*, 1999).

Chapter 4 and Sections 6.6.2 and 6.6.3 discuss the chemistry responsible for the tropospheric O_3 forcing; here we emphasise that the tropospheric O_3 forcing is driven by and broadly attribut-able to emissions of other gases. The observed regional variability of O_3 trends is related to the transport of key precursors, particularly reactive nitrogen, CO, and NMHCs (see Chapter 4). However, the chemistry of O_3 production can be non-linear, so that increased emissions of, for example, the nitric oxide precursor do not necessarily lead to linear responses in O_3 concentrations over all ranges of likely values (e.g, Kleinman, 1994; Klonecki and Levy, 1997). Further, the relationship of precursor emissions to O_3 trends may also vary in time. One study suggests that the O_3 production efficiency per mole of nitrogen oxide emitted has decreased globally by a factor of two since pre-industrial times (Wang and Jacob, 1998). Because of these complex and poorly understood interactions, the forcing due to tropospheric O_3 trends cannot be reliably and uniquely attributed in a quantitative fashion to the emissions of specific precursors.

For the purposes of this report, several evaluations of the global radiative forcing due to tropospheric O_3 changes since pre-industrial times have been intercompared. It will be shown that the uncertainties in radiative forcing can be better understood when both the absolute radiative forcing (Wm^{-2}) and normalised forcing (Wm^{-2} per Dobson Unit of tropospheric O_3 change) are considered. The results of this intercomparison and the availability of numerous models using different approaches suggest reduced uncertainties in the radiative forcing estimates compared to those of the SAR. Furthermore, recent work has shown that the dependence of the forcing on the altitude where the O_3 changes occur within the troposphere is less pronounced than previously thought, providing improved scientific understanding. Finally, some estimates of the likely magnitude of future tropospheric O_3 radiative forcing are presented and discussed.

6.5.2 Estimates of Tropospheric Ozone Radiative Forcing since Pre-Industrial Times

6.5.2.1 Ozone radiative forcing: process studies
The dependence of surface temperature response on the height of an imposed constant change in O_3 amount (often 10 Dobson Units, or DU) can be estimated using radiative convective models (e.g., Wang *et al.*, 1980; Lacis *et al.*, 1990; Forster and Shine, 1997). Figure 6.1a shows the results of such a calculation. This figure suggests that surface temperature is particularly sensitive to O_3 trends near 8 to 15 km, in the vicinity of the tropopause. However, Forster and Shine (1997) made the important point that since there is far less O_3 in the troposphere than near the tropopause and in the stratosphere, the use of a constant perturbation is unlikely to provide a realistic measure of the sensitivity profile. For example, 10 DU corresponds to roughly a 400% increase in mid-tropospheric O_3, but much less at higher levels. Figure 6.1b shows how a 10% (rather than 10 DU) local O_3 perturbation affects the calculated surface temperature as a function of the altitude where the O_3 change is imposed (from Forster and Shine, 1997). This figure suggests that the sensitivity of surface temperature to the altitude of O_3 perturbations is considerably smaller than suggested by earlier studies that employed constant absolute changes to probe these effects. While the study of Forster and Shine (1997) employed a simple radiative/convective model, Hansen *et al.* (1997a) carried out similar calculations using a GCM. Their study suggests that cloud feedbacks could further lower the altitude at which surface temperature is most sensitive to O_3 perturbations (Hansen *et al.*, 1997a; WMO, 1999), but this work employed perturbations of 10

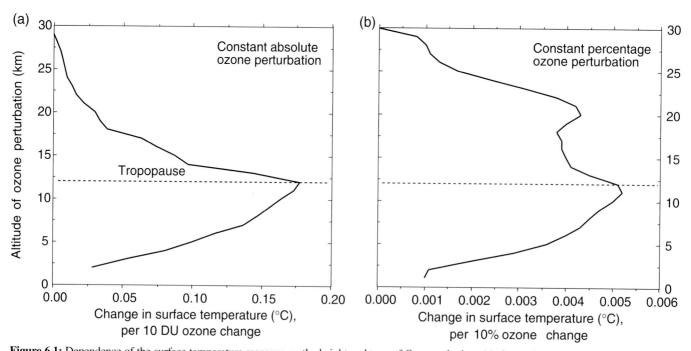

Figure 6.1: Dependence of the surface temperature response on the height and type of O_3 perturbation; (a) shows the sensitivity to a constant absolute change (10 DU), while (b) shows the sensitivity to a constant percentage change (10%). The model tropopause is at 12 km. From Forster and Shine (1997).

DU in each layer (which is necessary to obtain a significant signal in the GCM but is, as noted above, an unrealistically large value for the tropospheric levels in particular).

Portmann *et al.* (1997) and Kiehl *et al.* (1999) combined an O_3 climatology based upon satellite measurements of the tropospheric column content (from Fishman and Brackett, 1997) with a model calculation to derive estimates of the O_3 radiative forcing for the tropics and for the globe, respectively. They showed that the dependence of the normalised O_3 forcing (Wm^{-2} per DU of integrated O_3 column change) upon uncertainties in the vertical distribution of the perturbation was less than previously thought, about 0.03 to 0.06 Wm^{-2} per DU based upon their radiative code and considering a range of profile shapes. This rather limited sensitivity to the altitude distribution of the imposed perturbation is broadly consistent with the work of Forster and Shine (1997) highlighted above and in Figure 6.1.

6.5.2.2 Model estimates

Table 6.3 presents a comparison of estimates of globally averaged O_3 change since pre-industrial times and its corresponding radiative forcing based on published literature and including stratospheric temperature adjustment. Results from several studies are presented, one constrained by the climatology derived from observations discussed in the previous section, and ten from global chemistry/transport models. Results are presented for both clear sky and for total sky conditions, providing an indication of the role of clouds, and long-wave and short-wave contributions are shown separately. In addition, the globally averaged integrated tropospheric O_3 change (DU) is also indicated, to give a sense of the range in published model estimates of this key

factor. Finally, the normalised O_3 forcings are also presented (Wm^{-2} per DU).

Table 6.3 summarises the ranges obtained in these model studies. For total sky conditions, the range in globally and annual averaged tropospheric O_3 forcing from all of these models is from 0.28 to 0.43 Wm^{-2}, while the normalised forcing is 0.033 to 0.056 Wm^{-2} per DU. The tropospheric O_3 forcing constrained by the observational climatology is 0.32 Wm^{-2} for globally averaged, total sky conditions. These estimates are comparable in magnitude (and qualitatively similar in pattern, see Section 6.14) but opposite in sign to that believed to be due to the direct effect of sulphate aerosols.

Figure 6.2 presents the latitudinal distributions of absolute and normalised forcings for both January and July, from several chemistry/transport models and from the observationally constrained evaluation. The figure shows that the tropics and northern mid-latitude regions account for the bulk of the global forcing. The contribution from northern mid-latitudes is particularly large in summer, when it can reach as much as 1 Wm^{-2} locally near 40 to 50°N. These latitudinal gradients and seasonal changes in the zonally-averaged forcing are likely to be reflected at least in part in patterns of response (Hansen *et al.*, 1997a) and hence are a needed element in the detection and attribution of climate change.

Figure 6.2 shows that the large seasonal and inter-model ranges in absolute forcings compress somewhat when normalised forcings are considered. At 40°N for example, the absolute forcing differs over all models and seasons by more than a factor of 5 (Figure 6.2a), while the differences in normalised forcings are about a factor of 2.5 (Figure 6.2b). Mickley *et al.* (1999)

Table 6.3: *Tropospheric O_3 change (ΔO_3) in Dobson Units (DU) since pre-industrial times, and the accompanying short-wave (SW), long-wave (LW), and net (SW plus LW) radiative forcings (Wm^{-2}), after accounting for stratospheric temperature adjustment (using the Fixed Dynamical Heating method). Estimates are taken from the published literature. Normalised forcings (norm.) refer to radiative forcing per O_3 change (Wm^{-2} per DU).*

	Clear sky conditions						Total sky conditions			
Reference	ΔO_3	LW	SW	SW (norm.)	Net	Net (norm.)	LW	SW	Net	Net (norm.)
Berntsen *et al.* (1997) – [Reading model]	7.600	0.260	0.050	0.007	0.310	0.041	0.210	0.070	0.280	0.037
Stevenson *et al.* (1998)	8.700	0.326	0.065	0.007	0.391	0.045	0.201	0.088	0.289	0.033
Berntsen *et al.* (1997) – [Oslo model]	7.600	0.330	0.060	0.008	0.390	0.051	0.230	0.080	0.310	0.041
Haywood *et al.* (1998a)	7.900	0.330	0.050	0.006	0.380	0.048	0.230	0.080	0.310	0.039
Kiehl *et al.* (1999)	8.400	0.316	0.063	0.008	0.379	0.045	0.251	0.069	0.320	0.038
Berntsen *et al.* (2000)	9.600	0.357	0.071	0.007	0.428	0.045	0.246	0.096	0.342	0.036
Brasseur *et al.* (1998)	–	–	–	–	–	–	–	–	0.370	–
van Dorland *et al.* (1997)	8.070	0.390	0.054	0.007	0.443	0.055	0.304	0.076	0.380	0.047
Roelofs *et al.* (1997)	7.200	0.350	0.047	0.007	0.397	0.055	0.287	0.117	0.404	0.056
Lelieveld and Dentener (2000)	–	–	–	–	–	–	–	–	0.420	–
Hauglustaine *et al.* (1998)	8.940	0.448	0.063	0.007	0.511	0.057	0.338	0.088	0.426	0.048
Mean	8.224	0.345	0.058	0.007	0.403	0.049	0.255	0.085	0.343	0.042

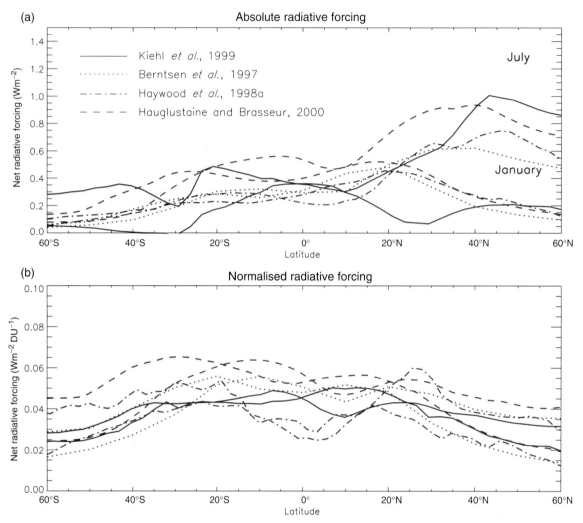

Figure 6.2: Latitudinal distribution of absolute (Wm^{-2}) and normalised (Wm^{-2} DU^{-1}) tropospheric O$_3$ radiative forcing for summer and winter conditions calculated by several models.

demonstrate that the normalised forcing at extra-tropical latitudes is largely determined by the temperature difference between the tropopause and the Earth's surface, while cloud cover plays a dominant role in the tropics. This suggests that the differences in physical climatology between models, and their comparison to the real atmosphere, may be a key factor in improving understanding of the normalised forcing. The difference between the normalised tropical and mid-latitude forcings also emphasises the need to quantify both the O$_3$ produced by biomass burning (and other tropical processes), and that produced by industrial practices, especially in the northern hemisphere. Some of the studies in Figure 6.2 suggest, for example, that a model producing larger O$_3$ changes in the tropics could produce a greater globally averaged tropospheric O$_3$ forcing than one producing larger changes in northern mid-latitudes, even if the globally averaged O$_3$ increases in the two studies were identical. Overall, this Figure suggests that the differences between model estimates of tropospheric O$_3$ forcing since pre-industrial times are likely to be dominated by chemistry (e.g., emission inventories, chemical processes, and transport) rather than by factors relating to radiative transfer.

6.5.3 *Future Tropospheric Ozone Forcing*

As noted in Harris *et al.* (1998) and Oltmans *et al.* (1998), the observed upward trends in surface O$_3$ in Europe and North America appear to be less steep in the past decade or two than in earlier periods (e.g., before about 1980), perhaps because of control measures designed to reduce emissions of O$_3$ precursors and mitigate urban pollution problems. Non-linear chemical feedbacks may also contribute to damping the recent past and future trends (Wang and Jacob, 1998). Theoretical studies using chemistry/transport models have attempted to prescribe likely future emissions of precursors and predict future tropospheric O$_3$ abundances. The largest future increases in O$_3$ forcings may occur in Asia in association with projected population growth and future development (van Dorland *et al.*, 1997; Brasseur *et al.*, 1998). Chalita *et al.* (1996) calculated a globally averaged radiative forcing from pre-industrial times to 2050 of 0.43 Wm^{-2}. The models of van Dorland *et al.* (1997) and Brasseur *et al.* (1998) suggest a higher globally averaged total radiative forcing from pre-industrial times to 2050 of 0.66 Wm^{-2} and 0.63 Wm^{-2}, respectively, while that of Stevenson *et al.* (1998) yields a forcing of 0.48 Wm^{-2} in 2100. As the analysis presented above shows, these differences are likely to be due to modelled differences in the latitudinal distributions or magnitudes of the projected O$_3$ change due for example to different emission inventories or model processes such as transport, and much less likely to be due to differences in radiative codes.

6.6 Indirect Forcings due to Chemistry

In addition to the direct forcings caused by injection of radiatively active gases to the atmosphere, some compounds or processes can also modify the radiative balance through indirect effects relating to chemical transformation or change in the distribution of radiatively active species. As previously indicated (IPCC, 1992, 1994; SAR), the tropospheric chemical processes determining the indirect greenhouse effects are highly complex and not fully understood. The uncertainties connected with estimates of the indirect effects are larger than the uncertainties of those connected to estimates of the direct effects. Because of the central role that O$_3$ and OH play in tropospheric chemistry, the chemistry of CH$_4$, CO, NMHC, and NO$_x$ is strongly intertwined, making the interpretation of the effects associated with emission changes rather complex. It should be noted that indirect effects involving OH feedback on the lifetime of well-mixed greenhouse gases, and tropospheric O$_3$ concentration changes since the pre-industrial are implicitly accounted for in Sections 6.3 and 6.5.

6.6.1 Effects of Stratospheric Ozone Changes on Radiatively Active Species

Increased penetration of ultraviolet radiation into the troposphere as a result of stratospheric ozone depletion leads to changes in the photodissociation rates of some key chemical species. One of the primary species affected by possible changes in photodissociation rates is the hydroxyl radical OH, which regulates the tropospheric lifetime of a large number of trace gases such as CH$_4$, CO, hydrochlorofluorocarbons (HCFCs), hydrofluorocarbons (HFCs), NO$_x$, and to a lesser extent, sulphur dioxide (SO$_2$) (see Chapter 4). The impact of stratospheric O$_3$ changes on the fate of tropospheric species has been discussed by Ma and van Weele (2000), Fuglestvedt *et al.* (1994), Bekki *et al.* (1994); and Fuglestvedt *et al.* (1995); on the basis of two-dimensional model simulations, and by Madronich and Granier (1994), Granier *et al.* (1996), Van Dop and Krol (1996) and Krol *et al.* (1998) using three-dimensional models. These studies and the updated calculations presented by WMO (1999) estimated that a 1% decrease in global total O$_3$ leads to a global increase in O$_3$ photolysis rate of 1.4%, resulting in a 0.7 to 0.8% increase in global OH. Changes in photolysis rates from reduced stratospheric O$_3$ also have the potential to alter tropospheric O$_3$ production and destruction rates. Based on stratospheric O$_3$ evolution over the period 1980 to 1996, Myhre *et al.* (2000) have calculated a reduction in tropospheric O$_3$ associated with increased ultraviolet penetration and a corresponding negative radiative forcing reaching −0.01 Wm^{-2} over that period of time. Another indirect impact of stratospheric O$_3$ depletion on

climate forcing has been proposed by Toumi *et al.* (1994, 1995). The hydroxyl radical oxidises SO$_2$ to gaseous sulphuric acid, which is a source of H$_2$SO$_4$ particles. Changes in H$_2$SO$_4$ formation resulting from OH changes might affect the number of particles which act as condensation nuclei. Rodhe and Crutzen (1995) challenged whether this mechanism was of importance.

6.6.2 Indirect Forcings of Methane, Carbon Monoxide and Non-Methane Hydrocarbons

CH$_4$ is oxidised primarily (>90%) in the troposphere through reaction with OH. Since the CH$_4$ oxidation cycle provides a substantial fraction of the OH loss in the troposphere, there is strong interaction between OH and CH$_4$. This causes the OH to decrease when CH$_4$ increases, leading to a further increase in CH$_4$. This chemical feedback is examined in Chapter 4. Lelieveld and Crutzen (1992), Brühl (1993), Lelieveld *et al.* (1993, 1998), Hauglustaine *et al.* (1994) and Fuglestvedt *et al.* (1996) estimated that this feedback added 25 to 35% to the direct CH$_4$ forcing depending on the initial CH$_4$ perturbation and the model used. These values are in line with the 30% contribution estimated by IPCC (1992) and the SAR. The tropospheric O$_3$ increase associated with photochemical production from the CH$_4$ oxidation cycle also contributes to enhancing the total CH$_4$ radiative forcing. The forcing from enhanced levels of tropospheric O$_3$ estimated by Lelieveld and Crutzen (1992), Lelieveld *et al.* (1993, 1998), Hauglustaine *et al.* (1994) and Fuglestvedt *et al.* (1996) contributes a further 30 to 40% on top of the CH$_4$ forcing, due directly to CH$_4$ emissions. IPCC (1992) estimated a lower contribution of O$_3$ of approximately 20%. Lelieveld *et al.* (1998) have estimated a direct radiative forcing associated with CH$_4$ increase since the pre-industrial era (1850 to 1992) of 0.33 Wm^{-2} and an additional forcing of 0.11 Wm^{-2} associated with OH feedback. The increased tropospheric O$_3$ contributes an additional 0.11 Wm^{-2} and stratospheric H$_2$O another 0.02 Wm^{-2}. These authors found that the total CH$_4$ forcing (0.57 Wm^{-2}) is higher by 73% than the direct forcing. The CH$_4$ radiative forcing given in Section 6.3 (0.5 Wm^{-2}) and based on recorded CH$_4$ concentration increase includes both the direct and OH indirect contributions. This updated forcing is higher by 14% than the 0.44 Wm^{-2} obtained by Lelieveld *et al.* (1998) on the basis of calculated present day and pre-industrial CH$_4$ distributions. In addition to that, oxidation of CH$_4$ also leads to the formation of CO$_2$, providing a further indirect effect.

In contrast to CH$_4$, the direct radiative forcing of CO is relatively small (Evans and Puckrin, 1995; Sinha and Toumi, 1996). Sinha and Toumi (1996) have estimated a clear sky radiative forcing of 0.024 Wm^{-2} for a uniform increase in CO from 25 ppbv to 100 ppbv. However, CO plays a primary role in governing OH abundances in the troposphere. As indicated by Prather (1994, 1996) and Daniel and Solomon (1998), CO emissions into the atmosphere may have a significant impact on climate forcing due to chemical impact on CH$_4$ lifetime, and tropospheric O$_3$ and CO$_2$ photochemical production (see Chapter 4). The contribution of these indirect effects to the GWPs based on box model calculations are presented in Section 6.12.

Similarly, NMHCs have a small direct radiative forcing. Highwood *et al.* (1999) estimated an upper limit of 0.015 Wm^{-2} (1% of the present day forcing due to other greenhouse gases) on the globally averaged direct forcing of sixteen NMHCs. As indicated by Johnson and Derwent (1996), the indirect forcing through changes in OH and tropospheric O_3 is also small for each NMHC taken individually but can be significant taken as a family. The indirect forcings of NMHCs are still poorly quantified and require the use of global three-dimensional chemical transport models. Accurate calculations of these effects are a notoriously difficult problem in atmospheric chemistry.

6.6.3 Indirect Forcing by NO$_x$ Emissions

Through production of tropospheric O_3, emissions of nitrogen oxides (NO$_x$ = NO + NO$_2$) lead to a positive radiative forcing of climate (warming), but by affecting the concentration of OH they reduce the levels of CH_4, providing a negative forcing (cooling) which partly offsets the O_3 forcing. Due to non-linearities in O_3 photochemical production together with differences in mixing regimes and removal processes, the O_3 and OH changes strongly depend on the localisation of the NO$_x$ surface emission perturbation, as calculated by Hauglustaine and Granier (1995), Johnson and Derwent (1996), Berntsen *et al.* (1996), Fuglestvedt *et al.* (1996, 1999) and Gupta *et al.* (1998). The CH_4 and O_3 forcings are similar in magnitude, but opposite in sign, as calculated by Fuglestvedt *et al.* (1999). Due to differences in CH_4 and O_3 lifetimes, the NO$_x$ perturbation on the CH_4 forcing acts on a global scale over a period of approximately a decade, while the O_3 forcing is of regional character and occurs over a period of weeks. Based on three-dimensional model results, Fuglestvedt *et al.* (1999) have calculated that the O_3 radiative forcing per change in NO$_x$ emission (10^{-2} Wm^{-2} per TgN/yr) is 0.35 and 0.29 for the USA and Scandinavia, respectively, and reaches 2.4 for Southeast Asia. The CH_4 forcing per change in NO$_x$ emission ranges from −0.37 (Scandinavia) and −0.5 (USA) to −2.3 (Southeast Asia) in the same units. Additional work is required to assess the impact of NO$_x$ on the radiative forcing of climate.

Deep convection can remove pollutants from the lower atmosphere and inject them rapidly into the middle and upper troposphere and, occasionally, into the stratosphere. Changes in convective regimes associated with climate changes have therefore the potential to significantly modify the distribution and the photochemistry of O_3 in a region where its impact on the radiative forcing is the largest. Based on the tropospheric O_3 column derived from satellite during the 1997 to 1998 El Niño, Chandra *et al.* (1998) reported a decrease in O_3 column of 4 to 8 Dobson Units (DU) in the eastern Pacific and an increase of 10 to 20 DU in the western Pacific, largely as a result of the eastward shift of the tropical convective activity. Lelieveld and Crutzen (1994) showed that convective transport can change the budget of O_3 in the troposphere. Berntsen and Isaksen (1999) have also indicated that changes in convection can modify the sensitivity of the atmosphere to anthropogenic perturbations as aircraft emissions. Their study indicates that reduced convective activity leads to a 40% increase in the O_3 response to aircraft NO$_x$ emissions due to modified background atmospheric concentrations. In addition to that, lightning is a major source of NO$_x$ in the troposphere and thus contributes to the photochemical production of O_3 (Huntrieser *et al.*, 1998; Wang *et al.*, 1998; Hauglustaine *et al.*, 2001; Lelieveld and Dentener, 2000). In the tropical mid- and upper troposphere, modelling studies by Lamarque *et al.* (1996), Levy *et al.* (1996), Penner *et al.* (1998a), and Allen *et al.* (2000) have calculated a contribution of lightning to the NO$_x$ levels of 60 to 90% during summer. Direct observations by Solomon *et al.* (1999) of absorption of visible radiation indicate that nitrogen dioxide can lead to local instantaneous radiative forcing exceeding 1 Wm^{-2}. These enhancements of NO$_2$ absorption are likely to be due both to pollution and to production by lightning in convective clouds. Further measurements are required to bracket the direct radiative forcing by NO$_2$ under a variety of storm and pollution conditions. On the basis of two-dimensional model calculations, Toumi *et al.* (1996) calculated that for a 20% increase of lightning the global mean radiative forcing by enhanced tropospheric O_3 production is about 0.1 Wm^{-2}. Based on an apparent correlation between lightning strike rates and surface temperatures, Sinha and Toumi (1997) have suggested a positive climate feedback through O_3 production from lightning NO$_x$ in a warmer climate. Improved modelling and observations are required to confirm this hypothesis.

Aircraft emissions also have the potential to alter the composition of the atmosphere and induce a radiative forcing of climate. According to IPCC (1999), in 1992 the NO$_x$ emissions from subsonic aircraft are estimated to have increased O_3 concentrations at cruise altitudes in the Northern Hemisphere mid-latitude summer by about 6% compared with an atmosphere without aircraft. The associated global mean radiative forcing is 0.023 Wm^{-2}. In addition to increasing tropospheric O_3, aircraft NO$_x$ are expected to decrease the concentration of CH_4 inducing a negative radiative forcing of −0.014 Wm^{-2} for the 1992 aircraft fleet. Again, due to different O_3 and CH_4 lifetimes, the two forcings show very different spatial distribution and seasonal evolution (IPCC, 1999). The O_3 forcing shows a marked maximum at northern mid-latitudes, reaching 0.06 Wm^{-2}, whilst the CH_4 forcing exhibits a more uniform distribution, with a maximum of about −0.02 Wm^{-2} in the tropical regions. Based on IPCC (1999), the O_3 forcing calculated for 2050 conditions is 0.060 Wm^{-2} and the CH_4 forcing is −0.045 Wm^{-2}.

6.6.4 Stratospheric Water Vapour

The oxidation of CH_4 produces water vapour that can contribute significantly to the levels of H_2O in the stratosphere. Water vapour is also directly emitted by aircraft in the lower stratosphere. Lelieveld and Crutzen (1992), Lelieveld *et al.* (1993, 1998), Hauglustaine *et al.* (1994) and Fuglestvedt *et al.* (1996) have estimated a contribution of stratospheric water vapour of 2 to 5% of the total CH_4 forcing. Oltmans and Hofmann (1995) have reported an increase in lower stratospheric H_2O (18 to 20 km) measured at Boulder of 0.8% per year for the period 1981 to 1994. Measurements of stratospheric H_2O from the HALOE instrument on board the UARS satellite also show an increase in stratospheric H_2O between 30 and 60 km of 40 to 100 ppbv/year for the period 1992 to 1997 (Evans *et al.*, 1998; Nedoluha *et al.*, 1998). A similar trend has also been reported by Randel *et al.* (1999) based on HALOE data for the period 1993 to 1997. In this study, a time variation and a flattening of the trend has been determined after 1996. This stratospheric H_2O increase is well above that expected from the rising CH_4 levels in the atmosphere. Based on these observed stratospheric H_2O trends, Forster and Shine (1999) have estimated a radiative forcing of 0.2 Wm^{-2} since 1980. Note that, just as for stratospheric O_3, there exists considerable uncertainty concerning the trend near the tropopause region globally. It should also be noted that if the changes in water vapour were a result of CH_4 oxidation, the changes in H_2O would be a forcing. However, if they result from changes in tropical tropopause temperature change or in dynamics, then they should be viewed as a feedback (as defined in Section 6.2). Additional measurements and analyses are clearly needed to explain the observed trends.

6.7 The Direct Radiative Forcing of Tropospheric Aerosols

Anthropogenic aerosols scatter and absorb short-wave and long-wave radiation, thereby perturbing the energy budget of the Earth/atmosphere system and exerting a direct radiative forcing. This section concentrates on estimates of the global mean direct effect of anthropogenic tropospheric aerosols and is necessarily dependent upon global models. Field campaigns which provide essential input parameters for the models, and satellite observational studies of the direct effect of tropospheric aerosols, which provide useful validation data for the models, are considered in detail in Chapter 5.

6.7.1 Summary of the IPCC WGI Second Assessment Report and Areas of Development

The SAR considered three anthropogenic aerosol species; sulphate, biomass burning aerosols, and fossil fuel black carbon (or soot). The SAR suggested a radiative forcing of −0.4 Wm^{-2} with a factor of two uncertainty for sulphate aerosols, −0.2 Wm^{-2} with a factor of three uncertainty for biomass burning, and +0.1 Wm^{-2} with a factor of three uncertainty for fossil fuel black carbon aerosols. The level of scientific understanding (referred to as a confidence level in the SAR, see Section 6.13) was classified as "low" for sulphate aerosol and "very low" for both fossil fuel black carbon and biomass burning aerosols. Since the SAR, there have been advances in both modelling and observational studies of the direct effect of tropospheric aerosols (see reviews by Shine and Forster (1999) and Haywood and Boucher (2000)). Global chemical transport modelling studies encompass a greater number of aerosol species and continue to improve the representation of the physical and chemical processes (see Chapters 4 and 5). Global models are more numerous and include more accurate radiative transfer codes, more sophisticated treatments of the effects of relative humidity for hygroscopic aerosols, better treatment of clouds, and better spatial and temporal resolution than some earlier studies. The present day direct radiative forcing due to aircraft emissions of sulphate and black carbon aerosol have been calculated to be insignificant (IPCC,

1999) and are not considered further. Spatial patterns of the calculated radiative forcings are not discussed in detail here but are presented in Section 6.14.

6.7.2 Sulphate Aerosol

Early one-dimensional box-model estimates of the radiative forcing (e.g., Charlson *et al.*, 1992) using simplified expressions for radiative forcing have been superseded by global calculations using prescribed aerosol concentrations from chemical transport models (CTMs). These studies use either three-dimensional observed fields of for example, clouds, relative humidity and surface reflectance (e.g., Kiehl and Briegleb, 1993; Myhre *et al.*, 1998c), or GCM generated fields (e.g., Boucher and Anderson, 1995; Haywood *et al.*, 1997a) together with the prescribed aerosol distributions from CTMs and detailed radiative transfer codes in calculating the radiative forcing. A growing number of studies perform both the chemical production, transformation, and transportation of aerosols and the radiative forcing calculations (see Chapter 5) with the advantage of correlating predicted aerosol distributions precisely with fields determining aerosol production and deposition such as clouds (e.g., Penner *et al.*, 1998b). Table 6.4 summarises estimates of the radiative forcing from global modelling studies.

The calculated global mean radiative forcing ranges from −0.26 to −0.82 Wm^{-2}, although most lie in the range −0.26 to −0.4 Wm^{-2}. The spatial distribution of the forcings is similar in the studies showing strongest radiative forcings over industrial regions of the Northern Hemisphere although the ratio of the annual mean Northern Hemisphere/Southern Hemisphere radiative forcing varies from 2.0 (Graf *et al.*, 1997) to 6.9 (Myhre *et al.*, 1998c) (see Section 6.14.2 for further details). The ratio of the annual mean radiative forcing over land to that over ocean also varies considerably, ranging from 1.3 (Kiehl *et al.*, 2000) to 3.4 (Boucher and Anderson, 1995). The direct radiative forcing (DRF) is strongest in the Northern Hemisphere summer when the insolation is the highest although different seasonal cycles of the sulphate burden from the chemical transport models result in maximum global mean radiative forcings ranging from May to August (e.g., Haywood and Ramaswamy, 1998), the ratio of the June-July-August/December-January-February radiative forcing being estimated to lie in the range less than 2 (e.g., van Dorland *et al.*, 1997) to > 5 (e.g., Penner *et al.*, 1998b; Grant *et al.*, 1999) with a mean of approximately 3.3. The range of uncertainty in the radiative forcings can be isolated from the uncertainties in the simulated sulphate loadings by considering the range in the normalised radiative forcing i.e., the radiative forcing per unit mass of sulphate aerosol (e.g., Nemesure *et al.*, 1995; Pilinis *et al.*, 1995). Table 6.4 shows that this is substantial, indicating that differences in the radiative forcing are not due solely to different mass loading.

The optical parameters for sulphate aerosol in each of the global studies vary. Although the single scattering albedo of pure sulphate and sulphate mixed with water is close to unity throughout most of the solar spectrum, some of the studies (e.g., Feichter *et al.*, 1997; van Dorland *et al.*, 1997; Hansen *et al.*,

Table 6.4: *The global and annual mean direct radiative forcing (DRF) for the period from pre-industrial (1750) to present day (2000) due to sulphate aerosols from different global studies. The anthropogenic column burden of sulphate and the source of the sulphate data are also shown together with the normalised radiative forcing. "Frac" indicates a cloud scheme with fractional grid box cloud amount, and "On/off" indicates that a grid box becomes overcast once a certain relative humidity threshold is reached. An asterisk indicates that the maximum hygroscopic growth of the aerosols was restricted to a relative humidity of 90%. The ratio of the Northern to the Southern Hemisphere (NH/SH) DRF and the ratio of the mean radiative forcing over land to the mean radiative forcing over oceans is also shown. NA indicates data is not available. "Langner and Rodhe (1991)s" indicates that the slow oxidation case was used in the calculations.*

Study	DRF (Wm^{-2})	Column burden (mgm^{-2})	Normalised DRF (Wg^{-1})	Cloud scheme	DRF NH/SH	DRF land/ ocean	Source of sulphate data
Ghan *et al.* (2001a)	−0.44	4.0	−110	On/off	5.2	1.9	Ghan *et al.* (2001a)
Jacobson (2001)	−0.32	2.55	−125	Frac	2.7	1.2	Jacobson (2001)
Boucher and Anderson (1995)	−0.29	2.32	−125	Frac	4.3	3.4	Langner and Rodhe (1991)
Graf *et al.* (1997)	−0.26	1.70	−153	Frac	2.0	NA	Graf *et al.* (1997)
Feichter *et al.* (1997)	−0.35	2.23	−157	Frac	4.2	1.4	Feichter *et al.* (1997)
Kiehl and Briegleb (1993)	−0.28	1.76	−159	Frac	3.3	NA	Langner and Rodhe (1991)s
Iversen *et al.* (2000)	−0.41	2.40	−167	Frac	4.1	NA	Iversen *et al.* (2000)
Myhre *et al.* (1998c)	−0.32	1.90	−169	Frac	6.9	NA	Restad *et al.* (1998)
van Dorland *et al.* (1997)	−0.36	2.11	−171	Frac	5.0	NA	van Dorland *et al.* (1997)
Koch *et al.* (1999)	−0.68	3.3	−200	Frac	NA	NA	Koch *et al.* (1999)
Kiehl and Rodhe (1995)	−0.66	3.23	−204	Frac	NA	NA	Pham *et al.* (1995)
	−0.29	1.76	−165	Frac	NA	NA	Langner and Rodhe (1991)s
Chuang *et al.* (1997)	−0.43	2.10	−205	On/off*	4.7	2.4	Chuang *et al.* (1997)
Haywood *et al.* (1997a)	−0.38	1.76	−215	Frac	4.0	NA	Langner and Rodhe (1991)s
Hansen *et al.* (1998)	−0.28	1.14	−246	Frac	NA	NA	Chin and Jacob (1996)
Kiehl *et al.* (2000)	−0.56	2.23	−251	Frac	2.7	1.3	Kiehl *et al.* (2000)
Haywood and Ramaswamy (1998)	−0.63	1.76	−358	On/off	3.6	2.6	Langner and Rodhe (1991)s
	−0.82	1.76	−460	On/off	5.8	2.7	Kasibhatla *et al.* (1997)
Penner *et al.* (1998b) and Grant *et al.* (1999)	−0.81	1.82	−445	On/off	4.5	2.3	Penner *et al.* (1998b)

1998) include some absorption. Charlson *et al.* (1999) show considerable variation in the specific extinction coefficient used in different studies, particularly when accounting for relative humidity effects. The treatment of the effects of relative humidity and clouds appear to be particularly important in determining the radiative forcing. The studies of Haywood and Ramaswamy (1998), Penner *et al.* (1998b) and Grant *et al.* (1999) produce normalised radiative forcings a factor of two to three higher than the other studies. Both Haywood and Ramaswamy (1998) and Penner *et al.* (1998b) acknowledge that their use of on/off cloud schemes where cloud fills an entire grid box once a threshold relative humidity is exceeded may lead to strong radiative

forcings due to strong non-linear relative humidity effects. Chuang *et al.* (1997) use an on/off cloud scheme and report a radiative forcing lower than these two studies, but the hygroscopic growth is rather suppressed above a relative humidity of 90%. The use of monthly mean relative humidity fields in some of the calculations leads to lower radiative forcings as temporal variations in relative humidity and associated non-linear effects are not accounted for (e.g., Kiehl and Briegleb, 1993; Myhre *et al.*, 1998c). Kiehl *et al.* (2000) improve the treatment of relative humidity compared to Kiehl and Briegleb (1993) and Kiehl and Rodhe (1995) by improving the relative humidity dependence of the aerosol optical properties and by

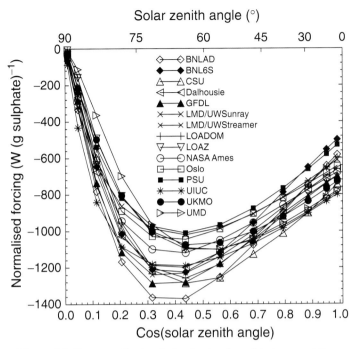

Figure 6.3: The normalised radiative forcing for sulphate aerosol from the intercomparison study of Boucher *et al.* (1998). A log-normal distribution with a geometric mean diameter of 0.170 μm and a geometric standard deviation of 1.105 together with an aerosol optical depth of 0.20 at 0.55 μm and a Lambertian surface reflectance of 0.15 was assumed. For details of the acronyms and the radiation codes used in the calculations see Boucher *et al.* (1998).

sulphate occurs in cloud-free regions. The global mean long-wave radiative forcing has been estimated to be less than 0.01 Wm⁻² and is insignificant (Haywood *et al.*, 1997a).

Boucher and Anderson (1995) investigate the effects of different size distributions, finding a 20 to 30% variation in the radiative forcing for reasonable size distributions. Nemesure *et al.* (1995) and Boucher *et al.* (1998) find a much larger sensitivity to the assumed size distribution as sulphate is modelled by much narrower size distributions. Column radiative forcing calculations by fifteen radiative transfer codes of varying complexity (Boucher *et al.*, 1998) show that, for well constrained input data, differences in the computed radiative forcing when clouds are excluded are relatively modest at approximately 20% (see Figure 6.3). This indicates that uncertainties in the input parameters and in the implementation of the radiative transfer codes and the inclusion of clouds lead to the large spread in estimates as suggested by Penner *et al.* (1994).

Additional column calculations show a weakened radiative forcing when the cloud optical depth is much greater than the aerosol optical depth (Haywood and Shine, 1997; Liao and Seinfeld, 1998) and show that the forcing is insensitive to the relative vertical position of the cloud and aerosol. Haywood *et al.* (1997b, 1998b) and Ghan and Easter (1998) used a cloud resolving model to investigate effects of sub-grid scale variations in relative humidity and cloud. For their case study, the optical depth and radiative forcing in a GCM sized grid box were underestimated by 60%. Effects of sub-grid scale variations in relative humidity and cloud on a global scale have not been rigorously investigated.

Until differences in estimates of radiative forcing due to sulphate aerosol can be reconciled, a radiative forcing of −0.4 Wm⁻² with a range of −0.2 to −0.8 Wm⁻² is retained. This estimate is based on the range of radiative forcings provided by the model estimates shown in Table 6.4. It is not possible to perform standard statistical procedures because of the limited number of studies and the fact that the resulting DRFs are not normally distributed. However, these results are broadly consistent with the estimates of the uncertainties derived in Chapter 5 (see Sections 5.4.2, and Sections 6.7.8).

6.7.3 Fossil Fuel Black Carbon Aerosol

Since the SAR there have been a number of more refined three-dimensional global model estimates of the radiative forcing due to black carbon (BC) aerosol from fossil fuel burning which have superseded calculations using a simple expression for the radiative forcing where the contribution from cloudy regions was not included (e.g., Chylek and Wong, 1995; Haywood and Shine, 1995). These estimates now include the contribution to the total radiative forcing from areas where BC exists either above or within clouds, although the treatment of BC within clouds remains crude. Table 6.5 includes recent global annual mean estimates of the radiative forcing due to BC aerosol from fossil fuels. Haywood *et al.* (1997a) and Myhre *et al.* (1998c) assumed that fossil fuel BC was directly proportional to the mass of sulphate from Langner and Rodhe (1991) and Restad *et al.* (1998), respectively, by applying a 7.5% mass scaling (equivalent

using interactive GCM relative humidities rather than monthly mean ECMWF analyses, resulting in a larger normalised radiative forcing. Ghan *et al.* (2001a) perform a sensitivity study and find that application of GCM relative humidities changes the direct radiative forcing by −0.2 Wm⁻² compared to the use of ECMWF analyses because the frequency of occurrence of relative humidities over 90% is higher when using GCM relative humidities. Haywood and Ramaswamy's (1998) GCM study indicates a stronger radiative forcing when sulphate resides near the surface because the relative humidity and subsequent hygroscopic growth of sulphate particles is higher. GCM sensitivity studies (Boucher and Anderson, 1995) and column calculations (Nemesure *et al.*, 1995) show that the radiative forcing is a strong function of relative humidity but relatively insensitive to chemical composition. Ghan *et al.* (2001a) suggest a number of reasons for the relatively low DRF from their study, including the relatively large fraction of anthropogenic sulphate in winter when the insolation is lowest, and the fact that sulphate aerosols are smaller in this study and therefore have smaller scattering efficiencies (see also Boucher and Anderson, 1995). The contribution to the global forcing from cloudy regions is predicted to be 4% (Haywood *et al.*, 1997a), 11% (Haywood and Ramaswamy, 1998), 22% (Boucher and Anderson, 1995) and 27% (Myhre *et al.*, 1998c) and hence remains uncertain. However, it should be noted that all of these studies suggest the majority of the global annual mean direct radiative forcing due to to a global mean burden of approximately 0.13 mgm^{-2} to 0.14 mgm^{-2}). Global annual mean radiative forcings of +0.20 Wm^{-2} and +0.16 Wm^{-2} are calculated for external mixtures. Both studies suggest that BC internally mixed with sulphate will exert a stronger forcing although the method of mixing was fairly crude in both of these studies. These estimates of the radiative forcing were performed before global modelling studies for BC were generally available.

Penner *et al.* (1998b) and Grant *et al.* (1999) used a chemical transport model in conjunction with a GCM to estimate a global column burden of BC from fossil fuel emissions of 0.16 mgm^{-2} and a radiative forcing of +0.20 Wm^{-2} for an external mixture. Cooke *et al.* (1999) estimated the global burden of optically active BC aerosol from fossil fuel burning from a 1° by 1° inventory of emissions to be 0.14 mgm^{-2} (note the good agreement with the assumed column burdens of Haywood *et al.* (1997a) and Myhre *et al.* (1998c)) and a subsequent radiative forcing of +0.17 Wm^{-2}. Haywood and Ramaswamy (1998) estimated the BC radiative forcing due to fossil fuel and biomass burning to be approximately +0.4 Wm^{-2} (see also Section 6.7.5), approximately half of which (+0.2 Wm^{-2}) is due to fossil fuel sources. Jacobson (2000, 2001) modelled BC from both fossil fuels and biomass burning sources and investigated the effects of different mixing schemes, finding a direct radiative forcing of +0.54 Wm^{-2} and a normalised direct radiative forcing of 1,200 Wg^{-1} when BC was modelled as an absorbing spherical core. These studies suggest that the global mean radiative forcing due to fossil fuel BC is strongest in June/July/August owing to the larger insolation coupled with higher atmospheric concentrations in the Northern Hemisphere. From these studies the normalised radiative forcing for an external mixture of BC aerosol appears better defined than for sulphate aerosol ranging from +1,123 Wg^{-1} to +1,525 Wg^{-1}. However, all these studies used the same size distribution and exclude effects of relative humidity, thus the modelled specific extinction is independent of the relative humidity at approximately 10 m^{2}g^{-1} at 0.55 µm. Observational studies show a wide range of specific extinction coefficients, α_{sp}, (see Section 5.1.2 and Table 5.1) of approximately 5 to 20 m^{2}g^{-1} at 0.55 µm, thus the uncertainty in the associated radiative forcing is likely to be higher than the global model results suggest. Additionally, if BC were modelled as an internal mixture, Haywood *et al.* (1997a) and Myhre *et al.* (1998c) suggest the degree of absorption may be considerably enhanced, the radiative forcing being estimated as +0.36 Wm^{-2} and +0.44 Wm^{-2}, respectively. Both of these studies use relatively simple effective medium mixing rules for determining the composite refractive index of internally mixed BC with sulphate and water which may overestimate the degree of absorption (e.g., Jacobson, 2000). Detailed scattering studies including a randomly positioned black carbon sphere in a scattering droplet show that the absorption is relatively well represented by effective medium approximations (Chylek *et al.*, 1996b). Column studies by Haywood and Shine (1997) and Liao and Seinfeld (1998) and the global studies by Haywood and Ramaswamy (1998) and Penner *et al.* (1998b) suggest that the radiative forcing due to BC will be enhanced if BC exists within or above the cloud, but reduced if the BC is below the cloud; thus the vertical profile of BC aerosol must be determined from observations and modelled accurately.

On the basis of the calculations summarised above, the estimate of the global mean radiative forcing for BC aerosols from fossil fuels is revised to +0.2 Wm^{-2} (from +0.1 Wm^{-2}) with a range +0.1 to +0.4 Wm^{-2}. The significant contribution to the global annual mean radiative forcing from cloudy regions is the main reason for the increase in the radiative forcing since the SAR. Uncertainties in determining the radiative forcing lie in modelling the mixing of BC with hygroscopic aerosols and the subsequent wet deposition processes which consequently influence the modelled atmospheric lifetime and burden of BC aerosol. Additionally, mixing of BC with other aerosol types and cloud droplets influences the optical parameters.

6.7.4 Fossil Fuel Organic Carbon Aerosol

Anthropogenic organic aerosols are a by-product of fossil fuel and biomass combustion and they consist of many complex chemical compounds and are released either as primary aerosol particles or as volatile organic gases (see Chapter 5). Studies that investigate the radiative forcing due to organic carbon (OC) from fossil fuels are included in Table 6.5. Penner *et al.* (1998b) and Grant *et al.* (1999) found a DRF of +0.16 Wm^{-2} when modelling the direct radiative forcing due to an internal mixture of fossil fuel BC and OC, and +0.2 Wm^{-2} when modelling the radiative forcing due to externally mixed fossil fuel BC. From these results, an annual global mean radiative forcing of −0.04 Wm^{-2} for fossil fuel OC from a global mean burden of approximately 0.7 mgm^{-2} may be derived. However, if OC were modelled as an external mixture with BC and/or if the effects of relative humidity are included, the radiative forcing due to OC from fossil fuels is likely to be more negative, thus this represents an approximate weakest limit. An alternative method for calculating the DRF due to fossil fuel OC from the results of Penner *et al.* (1998b) is to note that the absorption is approximately doubled when BC is modelled as an internal mixture rather than an external mixture (e.g., Haywood *et al.*, 1997a). Thus, for an external mixture of fossil fuel OC a radiative forcing of −0.24 Wm^{-2} may be more appropriate. Cooke *et al.* (1999) performed GCM calculations for externally mixed fossil fuel OC, finding a radiative forcing of −0.02 Wm^{-2} from a global mean burden of 0.34 mgm^{-2}. Myhre *et al.* (2001) scale the atmospheric concentrations of fossil fuel OC to modelled sulphate aerosol concentrations and include the effects of relative humidity to estimate a radiative forcing of −0.09 Wm^{-2} from a global mean burden of 0.66 mgm^{-2}. Thus, modelling estimates suggest that the normalised radiative forcing for OC is in the range −60 to −340 Wg^{-1}, which is smaller in magnitude than that due to BC due to the larger specific extinction coefficient for BC and the fact that BC may exert a significant radiative forcing in cloudy regions. Cooke *et al.* (1999) assume that OC is partially absorbing with a modelled single scattering albedo of approximately 0.97 at a wavelength of 0.55 µm. Hansen *et al.* (1998) use a three-dimensional GCM and the OC distribution from Liousse *et al.* (1996) and estimate the radiative forcing due to combined fossil fuel and biomass sources to be −0.41 Wm^{-2}. The approximate fraction of the atmospheric burden of the fossil fuel component may be estimated from the emission inventory of Liousse *et al.* (1996)

Table 6.5: *The global and annual mean direct radiative forcing for the period from pre-industrial (1750) to present day (2000) due to black carbon (BC) and organic carbon (OC) aerosols from different global studies. The anthropogenic column burden of BC and OC is shown, where appropriate, together with the normalised direct radiative forcing (DRF).*

Aerosol	Author and type of study	Mixing or optical parameters	DRF (Wm^{-2})	Column burden (mgm^{-2})	Normalised DRF (Wg^{-1})	Remarks
Fossil fuel BC	Haywood *et al.* (1997a)	External	+0.20	0.13	1525	GCM study. 7.5% mass scaling of BC to SO_4 assumed. SO_4 from Langner and Rodhe (1991) (slow oxidation case).
		Internal with sulphate	+0.36		2770	Internal mixing approximated by volume weighting the refractive indices of BC and SO_4
	Myhre *et al.* (1998c)	External	+0.16	0.14	1123	Three-dimensional study using global climatologies for cloud, surface reflectance etc. 7.5% mass scaling of BC to SO_4 assumed. SO_4 from Restad *et al.* (1998).
		Internal with sulphate	+0.42		3000	Internal mixing approximated by volume weighting the refractive indices of BC and SO_4
	Penner *et al.* (1998b) and Grant *et al.* (1999)	Internal with fossil fuel OC	+0.20	0.16	1287	BC modelled using chemical transport model and GCM.
	Cooke *et al.* (1999)	External	+0.17	0.14	1210	BC modelled using chemical transport model and GCM.
	Haywood and Ramaswamy (1998)	External mixture	+0.2	0.13	1500	Three-dimensional GCM study using Cooke and Wilson (1996) BC data scaled to Liousse *et al.* (1996) total BC mass. 50% of the BC mass assumed to be from fossil fuels.
Fossil fuel OC	Penner *et al.* (1998b)	Internal/external with fossil fuel BC	−0.04 to −0.24	−0.7	−60 to −340	OC modelled using chemical transport model and GCM. Weakest estimate corresponds to internal mixing with BC, strongest estimate corresponds to external mixing with BC.
	Cooke *et al.* (1999)	External mixture	−0.02	0.34	−70	OC modelled using chemical transport model and GCM. May be more negative due to effects of RH and assumption of partial absorption of OC.
	Myhre *et al.* (2001)	External mixture	−0.09	0.66	−135	Three-dimensional study using global climatologies for cloud, surface reflectance etc. OC scaled linearly to SO_4. SO_4 from Restad *et al.* (1998).
Biomass burning (BC+OC)	Hobbs *et al.* (1997)	Optical parameters of biomass smoke	−0.3	3.7	−80	Uses simplified expression from Chylek and Wong (1995). Neglects radiative forcing from cloudy areas. Other parameters including estimated column burden from Penner *et al.* (1992).
	Iacobellis *et al.* (1999)	Optical parameters of biomass smoke	−0.74	3.5	−210	Three-dimensional chemical transport model and GCM. Simplified expressions also examined.
	Penner *et al.* (1998b) and Grant *et al.* (1999)	Internal/external mixture of OC and BC	−0.14 to −0.23	1.76	−80 to −120	Three-dimensional chemical transport model and GCM using biomass optical parameters modelled from two observational studies.
Fossil fuel and biomass burning BC	Haywood and Ramaswamy (1998)	External mixture	+0.4	0.27	1500	Three-dimensional GCM study using Cooke and Wilson (1996) BC data scaled to Liousse *et al.* (1996) total BC mass.
	Hansen *et al.* (1998)	Observed single scattering albedo	+0.27	NA	NA	Adjustment of modelled single scattering albedo from 1.0 to 0.92-0.95 to account for the absorption properties of BC.
	Jacobson (2001)	Internal with BC core.	+0.54	0.45	1200	GCM study using data from Cooke and Wilson (1996) scaled by a factor of 0.85.
Fossil fuel and biomass burning OC	Hansen *et al.* (1998)	External mixture	−0.41	NA	NA	Three-dimensional GCM study using Liousse *et al.* (1996) OC data. OC modelled as scattering.
	Jacobson (2001)	Internal with BC core.	−0.04 to −0.06	1.8	−17	OC treated as shell with BC core. Strongest forcing for non-absorbing OC.

who estimate that fossil fuels contribute 38% to the total OC emissions. Thus the radiative forcing due to fossil fuel OC may be inferred to be approximately -0.16 Wm^{-2}. This may constitute an approximate upper estimate as the majority of fossil fuel OC occurs over mid-latitude land areas where the surface reflectance is generally higher and insolation is lower than in the equatorial regions where biomass burning is the major source of OC. From these calculations, the radiative forcing due to fossil fuel OC is estimated to be -0.10 Wm^{-2}. The uncertainty associated with this estimate is necessarily high due to the limited number of detailed studies and is estimated to be at least a factor of three.

6.7.5 Biomass Burning Aerosol

Biomass burning aerosol consists of two major chemical components: black carbon (BC), which primarily absorbs solar radiation, and organic carbon (OC), which primarily scatters solar radiation. Sources of biomass burning aerosol include burning of forests and savanna for colonisation and agriculture, burning of agricultural waste, and substances burned for fuel such as wood, dung and peat. Not all biomass aerosol comes from anthropogenic activities, as naturally occurring vegetation fires regularly occur. The fraction of anthropogenic biomass aerosol remains difficult to deduce. As for sulphate and fossil fuel BC aerosol, three-dimensional GCM estimates (e.g., Penner *et al.*, 1998b) have superseded earlier simple box-model calculations (e.g., Penner *et al.*, 1992) (see Table 6.5). Penner *et al.* (1998b) and Grant *et al.* (1999) used a GCM to model the radiative forcing from biomass aerosol, finding a combined total radiative forcing of -0.14 to -0.23 Wm^{-2} depending upon the mixing assumptions and size distributions used. The radiative forcing is calculated to be negative over the majority of the globe, but some limited areas of positive forcing exist over areas with a high surface reflectance (see Section 6.14.2). The seasonal cycle is strongly influenced by the seasonal cycle of biomass burning emissions (Grant *et al.*, 1999; Iacobellis *et al.*, 1999), the global mean being estimated to be a maximum during June/July/August. Hobbs *et al.* (1997) performed aircraft measurements of smoke from biomass burning during the Smoke Cloud and Radiation-Brazil (SCAR-B) experiment and used the model of Penner *et al.* (1992) to estimate a global mean radiative forcing of -0.3 Wm^{-2} as an approximate upper limit due to modified optical parameters. Ross *et al.* (1998) performed a similar measurement and modelling study estimating a local annual mean radiative forcing of -2 to -3 Wm^{-2} in intensive biomass burning regions, indicating that the global mean radiative forcing is likely to be significantly smaller than in Penner *et al.* (1992). Iacobellis *et al.* (1999) model the global radiative forcing as -0.7 Wm^{-2} but use an emission factor for biomass aerosols that Liousse *et al.* (1996) suggest is a factor of three too high, thus a radiative forcing of -0.25 Wm^{-2} is more likely. These estimates neglect any long-wave radiative forcing although Christopher *et al.* (1996) found a discernible signal in AVHRR data from Brazilian forest fires that opposes the short-wave radiative forcing; the magnitude of the long-wave signal will depend upon the size, optical parameters, and altitude of the aerosol.

The radiative effects of the individual BC and OC components from biomass burning have also received attention. Haywood and Ramaswamy (1998) re-scaled the global column burden of BC from Cooke and Wilson (1996) to that of Liousse *et al.* (1996) which is thought to be more representative of optically active BC and estimated the radiative forcing due to fossil fuel and biomass burning to be approximately $+0.4$ Wm^{-2}, approximately half of which, or $+0.2$ Wm^{-2}, is due to biomass sources. Hansen *et al.* (1998) adjusted the single scattering albedo of background aerosols from unity to 0.92 to 0.95 to account for the absorbing properties of BC from combined fossil fuel and biomass emissions and found a radiative forcing of $+0.27$ Wm^{-2}.

The radiative forcing due to the purely scattering OC component from combined fossil fuel and biomass emissions is estimated by Hansen *et al.* (1998) to be -0.41 Wm^{-2} with a approximate upper limit for the radiative forcing due to fossil fuel OC from this study of approximately -0.16 Wm^{-2} (see Section 6.7.4). Thus an approximate weakest limit for the radiative forcing due to biomass burning OC is -0.25 Wm^{-2}. Jacobson (2001) used a multi-component aerosol model to investigate the radiative forcing due to OC from combined fossil fuel and biomass aerosols using emissions from Liousse *et al.* (1996) finding a resultant radiative forcing of between -0.04 and -0.06 Wm^{-2}, the strongest forcing being for purely scattering aerosol. The small radiative forcings of Jacobson (2001) may be due to the aerosol being modelled as tri-modal with less aerosol in the optically active region of the spectrum, or due to the fact that absorption by BC may be enhanced in the modelled multi-component mixture.

On the basis of these studies, the estimate of the radiative forcing due to biomass burning aerosols remains the same at -0.2 Wm^{-2}. The uncertainty associated with the radiative forcing is very difficult to estimate due to the limited number of studies available and is estimated as at least a factor of three, leading to a range of -0.07 to -0.6 Wm^{-2}.

6.7.6 Mineral Dust Aerosol

Recent studies have suggested that 20% (Sokolik and Toon, 1996) and up to 30 to 50% (Tegen and Fung, 1995) of the total mineral dust in the atmosphere originates from anthropogenic activities, the precise fraction of mineral dust of anthropogenic origin being extremely difficult to determine. Only the radiative forcing from this anthropogenic component is considered as there is no evidence that the naturally occurring component has changed since 1750, although ice core measurements suggest that atmospheric concentrations of dust have varied substantially over longer time-scales (e.g., Petit, 1990). Because mineral dust particles are of a relatively large size and because it becomes lofted to high altitudes in the troposphere, in addition to the short-wave radiative forcing, it may exert a significant long-wave radiative forcing. The global mean short-wave radiative forcing will be negative due to the predominantly scattering nature in the solar spectrum (although partial absorption may lead to a local positive radiative forcing over high surface albedos and clouds) and the global mean long-wave forcing will be positive.

Sokolik and Toon (1996) used a simple box model and neglected forcing in cloudy regions to estimate a short-wave radiative forcing of −0.25 Wm^{-2} over ocean and −0.6 Wm^{-2} over land, leading to a global forcing of approximately −0.46 Wm^{-2}. They point out that this is offset to some extent by a positive long-wave forcing. Tegen and Fung (1995) performed a more detailed three-dimensional GCM modelling study of dust aerosol and estimated that approximately 30 to 50% of the total dust burden is due to changes in land use associated with anthropogenic activity. The radiative forcing using this data was estimated by Tegen *et al.* (1996) to be −0.25 Wm^{-2} in the short-wave and +0.34 Wm^{-2} in the long-wave, resulting in a net radiative forcing of +0.09 Wm^{-2}. Updated calculations of the net radiative forcing based on Miller and Tegen (1998) estimate the radiative forcing to be −0.22 Wm^{-2} in the short-wave and +0.16 Wm^{-2} in the long-wave, resulting in a net radiative forcing of −0.06 Wm^{-2}. Hansen *et al.* (1998) perform similar calculations and calculate a net radiative forcing of −0.12 Wm^{-2} by assuming a different vertical distribution, different optical parameters and using a different global model. Jacobson (2001) used a multi-component global aerosol model to estimate the direct radiative forcing to be −0.062 Wm^{-2} in the short-wave and +0.05 Wm^{-2} in the long-wave, resulting in a net radiative forcing of −0.012 Wm^{-2}. The effects of non-sphericity of the mineral dust are not accounted for in these calculations. Mishchenko *et al.* (1997) suggest that differences in the optical parameters between model spheroids and actual dust particles do not exceed 10 to 15%, although changes of this magnitude may have a large effect on the radiative forcing (Miller and Tegen, 1998). An example of the geographical distribution of the radiative forcing is shown in Figure 6.7g (data from Tegen *et al.*, 1996) which shows regions of positive and negative forcing. Positive forcing tends to exist over regions of high surface reflectance and negative radiative forcings tend to exist over areas of low surface reflectance. This is due to the dependency of the forcing on surface reflectance and the additional effects of the long-wave radiative forcing.

One problem that needs to be solved is uncertainty in representative refractive indices (Claquin *et al.*, 1998), and how they vary geographically due to different mineral composition of different source regions (e.g., Lindberg *et al.*, 1976). Sokolik *et al.* (1993) summarise the imaginary part of the refractive index for different geographic regions finding a range (−0.003i to approximately −0.02i) at a wavelength of 0.55 μm, and differences in refractive index in the long-wave from different geographical sources are also reported by Sokolik *et al.* (1998). Kaufman *et al.* (2001) use observations from the Landsat satellite coupled with ground-based sun photometer measurements and suggest that Saharan dust has a smaller imaginary refractive index (−0.001i) at 0.55 μm and absorbs less solar radiation than that used in the above modelling studies leading to a much enhanced shortwave radiative forcing. However, the increase is much less in the modelling study of Hansen *et al.* (1998) who find the net radiative forcing changes from −0.12 Wm^{-2} to −0.53 Wm^{-2} when dust is treated as conservatively scattering. von Hoyningen-Huene *et al.* (1999) determine the imaginary part of the refractive index from surface based absorption and scattering measurements and find that a refrac-

tive index of −0.005i best fits the observations for Saharan dust, which is in agreement with the values reported by Sokolik *et al.* (1993). The refractive indices together with the assumed size distributions determine the optical parameters. The radiative forcing is particularly sensitive to the single scattering albedo (Miller and Tegen, 1999). Additional uncertainties lie in modelling the size distributions (Tegen and Lacis, 1996; Claquin *et al.*, 1998) which, together with the refractive indices, determine the optical parameters. Measurements made by the Advanced Very High Resolution Radiometer (AVHRR) by Ackerman and Chung (1992), showed a local short-wave radiative perturbation off the west coast of Africa of −40 to −90 Wm^{-2} and a corresponding long-wave perturbation of +5 to +20 Wm^{-2} at the top of the atmosphere. Relating instantaneous observational measurements that do not account for the effects of clouds, diurnal averaging of the radiation, the seasonal signal associated with emissions and the fraction of mineral dust that is anthropogenic to the global mean radiative forcing is very difficult. Because the resultant global mean net radiative forcing is a residual obtained by summing the short-wave and the long-wave radiative forcings which are of roughly comparable magnitudes, the uncertainty in the radiative forcing is large and even the sign is in doubt due to the competing nature of the short-wave and long-wave effects. The studies above suggest, on balance, that the shortwave radiative forcing is likely to be of a larger magnitude than the long-wave radiative forcing, which indicates that the net radiative forcing is likely to be negative, although a net positive radiative forcing cannot be ruled out. Therefore a tentative range of −0.6 to +0.4 Wm^{-2} is adopted; a best estimate cannot be assigned as yet.

6.7.7 Nitrate Aerosol

Although IPCC (1994) identified nitrate aerosol as a significant anthropogenic source of aerosol, only three estimates of the radiative forcing are available. Van Dorland *et al.* (1997) has produced a very speculative radiative forcing estimate of approximately −0.03 Wm^{-2} for ammonium nitrate, while Adams *et al.* (2001) derived a radiative forcing of −0.22 Wm^{-2} from an anthropogenic burden of 0.62 mg(NO_3)m^{-2}, and Jacobson (2001) derived a radiative forcing of approximately −0.02 Wm^{-2} for a column burden of 0.7 mg(NO_3)m^{-2}. It appears that the large discrepancy between the results of Jacobson (2001) and Adams *et al.* (2001) is that 90% of the nitrate is in the coarse mode in Jacobson (2001), which reduces the scattering efficiency. Recent measurement studies by Veefkind *et al.* (1996) and ten Brink *et al.* (1997) in the Netherlands have shown that nitrate aerosol in the form of ammonium nitrate is a locally important aerosol species in terms of aerosol mass in the optically active sub-micron size range and hence the associated radiative forcing. They also emphasise the problems in measuring the concentrations and size distributions of nitrate which is a semi-volatile substance. The contradictory nature of the global studies means that no "best estimate" or range for the radiative forcing due to anthropogenic nitrate aerosol is presented in this report, though future studies may prove that it exerts a significant radiative forcing.

6.7.8 *Discussion of Uncertainties*

While the radiative forcing due to greenhouse gases may be determined to a reasonably high degree of accuracy (Section 6.3), the uncertainties relating to aerosol radiative forcings remain large, and rely to a large extent on the estimates from global modelling studies that are difficult to verify at the present time. The range of estimates presented in this section represents mainly the structural uncertainties (i.e., differences in the model structures and assumptions) rather than the parametric uncertainties (i.e., the uncertainties in the key parameters) (see Pan *et al.*, 1997). This is because many of the model calculations for sulphate aerosol use identical size distributions and refractive indices which lead to identical optical parameters. Thus the model results are not necessarily independent, and certainly do not include the full range of parametric uncertainties. The response of the direct radiative forcing to parametric uncertainties is investigated in sensitivity studies for different aerosol species (e.g., Boucher and Anderson, 1995; Haywood and Ramaswamy, 1998). Three major areas of uncertainty exist; uncertainties in the atmospheric burden and the anthropogenic contribution to it, uncertainties in the optical parameters, and uncertainties in implementation of the optical parameters and burden to give a radiative forcing. The atmospheric burden of an anthropogenic aerosol species is determined by factors such as emission, aging, convective transport, scavenging and deposition processes, each of which have an associated uncertainty (see Chapter 5). The optical parameters have uncertainties associated with uncertainties in size distribution, chemical composition, state of mixing, method of mixing, and asphericity. The problem of the degree of external/internal mixing in the atmosphere deserves highlighting, as global modelling studies tend to assume external mixtures which make modelling the sources, atmospheric transport and radiative properties simpler (e.g., Tegen *et al.*, 1997). However, single particle analysis of particles containing mineral dust (e.g., Levin *et al.*, 1996) and sea salt (Murphy *et al.*, 1998) have often shown them to be internally mixed with sulphate and other aerosols of anthropogenic origin. Thus, heterogeneous conversion of SO_2 to sulphate aerosol on dust or sea salt particles may effectively lead to sulphate becoming internally mixed with larger super-micron particles (e.g., Dentener *et al.*, 1996) leading to a reduction in extinction efficiency, an effect that has not yet been accounted for in global modelling studies. Studies that model internal mixtures of absorbing and scattering aerosols necessarily apply simplifying assumptions such as volume weighting the refractive indices (e.g., Haywood *et al.*, 1997a; Myhre *et al.*, 1998c) which may overestimate the degree of absorption (e.g., Jacobson, 2000). Modelling studies that examine the effects of internal mixing of multi-component aerosols are only just becoming available (Jacobson, 2001). Uncertainties in calculating the radiative forcing from specified burdens and optical parameters arise from uncertainties in the parametrization of relative humidity effects, the horizontal and vertical distributions of the aerosol, the uncertainties and sub-grid scale effects in other model fields such as clouds, humidity, temperature and surface reflectance, the representation of the diurnal cycle, and the accuracy of the radiation code used in the calculations. The short atmospheric lifetime of aerosols and the resultant large spatial variability imply a strong requirement for global observational data. Until a reliable global observational method for verifying the radiative effects of anthropogenic aerosols is available, it is likely that the radiative forcing of any aerosol species will remain difficult to quantify. Although satellite retrievals of aerosol optical properties have advanced substantially since the SAR, the difficult problem of separating anthropogenic from natural aerosol still remains (Chapter 5). Nevertheless, new analyses reiterate that global satellite measurements contain tropospheric aerosol signatures that include those due to anthropogenic aerosols (Haywood *et al.*, 1999; Boucher and Tanré, 2000). While the general spatial distribution of the radiative forcing for sulphate aerosol appears to be similar among the studies listed in Table 6.4, some important features, such as the seasonal cycle in the radiative forcing, remain highly uncertain which may have important consequences in terms of the detection and attribution of climate change (Chapter 12).

Here we examine the consistency of the ranges derived in this section with the ranges obtained using the approach of Chapter 5, Section 5.4.2. For the industrial sulphate, fossil fuel BC, and fossil fuel OC aerosols, the "best estimates" of the direct radiative forcing are -0.4 Wm^{-2} with a factor of 2 uncertainty, $+0.2$ Wm^{-2} with a factor of 2 uncertainty and -0.1 Wm^{-2} with a factor of 3 uncertainty, respectively. An estimate of the total direct radiative forcing from industrial aerosols (using RMS errors) leads to a range -0.07 to -1.24 Wm^{-2} which is reasonably consistent with -0.1 to -1.0 Wm^{-2} (one standard deviation) from Chapter 5, Section 5.4.2. For biomass burning aerosols the "best estimate" of the direct radiative forcing is -0.2 Wm^{-2} with a range -0.07 to -0.6 Wm^{-2}, which is reasonably consistent with -0.1 to -0.5 Wm^{-2} (one standard deviation) obtained from Section 5.4.2.

Additionally, while this section has concentrated upon the radiative forcing at the top of the atmosphere, the effects of anthropogenic aerosols upon the radiative budget at the surface of the Earth has not been considered in detail. For purely scattering aerosols in cloud-free conditions, the radiative effect at the surface is within a few per cent of that at the top of the atmosphere (e.g., Haywood and Shine, 1997). However, for partially absorbing aerosols, the radiative effect at the surface may be many times that at the top of the atmosphere, as evidenced by modelling and measurement studies (e.g., Haywood and Shine, 1997; Haywood *et al.*, 1999; Chapter 5). This is because, for partially absorbing aerosols, energy is transferred directly to the atmospheric column. Ackerman *et al.* (2000) point out that this can warm the atmosphere and "burn off" clouds. They conclude that during the northeast monsoon (dry season over India) daytime trade cumulus cloud cover over the northern Indian Ocean can be reduced by nearly half, although these results depend strongly upon the meteorological conditions and modelling assumptions. This process may also be important over the global domain as indicated by Hansen *et al.* (1997a), as the climate sensitivity parameter (Section 6.2) may differ significantly for absorbing aerosols due to diabatic heating in the aerosol layer modifying the temperature structure of the atmosphere, which affects the formation of clouds. The vertical partitioning of the forcing by absorbing aerosols is also a potentially important factor in determining climatic changes at the surface (e.g., evaporation, soil moisture).

6.8 The Indirect Radiative Forcing of Tropospheric Aerosols

6.8.1 Introduction

Aerosols serve as cloud condensation and ice nuclei, thereby modifying the microphysics, the radiative properties, and the lifetime of clouds. The physics and chemistry of the indirect effect of aerosols is discussed in detail in Chapter 5. Only aspects directly relevant to quantifying the indirect radiative forcing by aerosols are presented here. The aerosol indirect effect is usually split into two effects: the first indirect effect, whereby an increase in aerosols causes an increase in droplet concentration and a decrease in droplet size for fixed liquid water content (Twomey, 1974), and the second indirect effect, whereby the reduction in cloud droplet size affects the precipitation efficiency, tending to increase the liquid water content, the cloud lifetime (Albrecht, 1989), and the cloud thickness (Pincus and Baker, 1994). Until recently, the first indirect effect has received much more attention than the second. IPCC (1994) and the SAR only considered the first indirect effect. Shine *et al.* (1996) retained a range of radiative forcing from 0 to -1.5 Wm^{-2} with no best estimate, although a value of -0.8 Wm^{-2} was used for the year 1990 in the IS92a scenario (Kattenberg *et al.*, 1996). Here we review and discuss the various estimates for the globally averaged aerosol indirect forcing available in the literature. Because of the inherent complexity of the aerosol indirect effect, GCM studies dealing with its quantification necessarily include an important level of simplification. While this represents a legitimate approach, it should be clear that the GCM estimates of the aerosol indirect effect are very uncertain. Section 6.8.2 investigates the indirect radiative forcing due to sulphate aerosols, on which most efforts have concentrated, while other aerosol types are treated in Section 6.8.3. Section 6.8.4 is devoted to alternative approaches, while Section 6.8.6 describes the aerosol indirect effects on ice clouds.

6.8.2 Indirect Radiative Forcing by Sulphate Aerosols

6.8.2.1 Estimates of the first indirect effect

The studies reported in Table 6.6 use different GCMs and methods for computing the droplet number concentration (i.e., empirical relationships between the sulphate mass and the cloud droplet number concentration, empirical relationships between the sulphate aerosol number concentration and the cloud droplet number concentration, or parametrization of cloud nucleation processes). The forcing estimates for the first indirect effect from sulphate aerosols range from -0.3 to -1.8 Wm^{-2}, which is close to the range of 0 to -1.5 Wm^{-2} given in the SAR when only a few estimates were available.

There is a tendency for more and more studies to use interactive (on-line) rather than prescribed (monthly or annual mean) sulphate concentrations. Feichter *et al.* (1997) pointed out that the first indirect effect calculated from monthly mean sulphate concentrations is 20% larger than calculated from interactive sulphate concentrations. Jones *et al.* (1999) found that the total indirect effect was overesti-

mated by about 60% when they used seasonal or annual mean sulphate concentrations.

The various GCM studies show some disagreement on the spatial distribution of the forcing, an example of which is shown in Figure 6.7h. The Northern to Southern Hemisphere ratio varies from 1.4 to 4 depending on the models. It is generally smaller than the Northern to Southern Hemisphere ratio of anthropogenic sulphate aerosol concentrations because of the higher susceptibility of the clouds in the Southern Hemisphere (Platnick and Twomey, 1994; Taylor and McHaffie, 1994). The ocean to land ratio depends very much on the method used to relate the concentration of sulphate mass to the cloud droplet number concentration and on the natural background aerosol concentrations. It was generally found to be smaller than unity (Boucher and Lohmann, 1995; Jones and Slingo, 1997; Kiehl *et al.*, 2000). Larger ratios, such as 1.6 (Chuang *et al.*, 1997) and 5 (Jones and Slingo, 1997), are reported in some of the sensitivity experiments. Using a detailed inventory of ship sulphur emissions and a simple calculation of the aerosol indirect effect, Capaldo *et al.* (1999) suggested that a significant fraction of the effect over the oceans (-0.11 Wm^{-2}, averaged globally) could be due to ship-emitted particulate matter (sulphate plus organic material). So far this source of aerosols has not been included in the GCM studies.

Kogan *et al.* (1996, 1997) used the Warren *et al.* (1988) cloud climatology over the oceans rather than a GCM to predict the indirect effect by sulphate aerosols on cloud albedo. The cloud albedo susceptibility was evaluated from a large eddy simulation model applied to stratocumulus clouds. They found an indirect short-wave forcing of -1.1 Wm^{-2} over the oceans with a small hemispheric difference of 0.4 Wm^{-2} (i.e., a Northern to Southern Hemisphere ratio of about 1.4). In their study, the forcing had a strong seasonal cycle, with the Southern Hemisphere forcing prevailing in some seasons.

6.8.2.2 Estimates of the second indirect effect and of the combined effect

Whereas the first indirect effect can be computed diagnostically, assessment of the second indirect effect implies that two independent (i.e., with the same fixed sea surface temperatures, but with different meteorologies) GCM simulations be made, a first one with pre-industrial and a second one with present day aerosols. The difference in top of the atmosphere fluxes or cloud radiative forcings between two such simulations is used as a proxy for the forcings due to the second indirect and the combined effects. The simulations need to be sufficiently long (usually 5 years) so that the effects of natural variability are expected to average out. However, as a consequence, the estimates of the aerosol indirect effect may include some undesirable feedbacks involving climate parameters which would violate the definition adopted for radiative forcing of climate change by IPCC (see also Section 6.1.1). Rotstayn and Penner (2001) showed that, at least in their model, the difference in top of the atmosphere fluxes between two simulations did not differ by more than 10% from the radiative forcing for the first indirect effect computed diagnostically. Caution should nevertheless be exercised before this result can be generalised to other models and to the second indirect effect.

Table 6.6: *The global mean annual average aerosol indirect radiative forcing from different global studies. Letters P (prescribed) and C (computed) refer to off-line and on-line sulphate aerosol calculations, respectively. CCN and CDN stand for cloud condensation nuclei and cloud droplet number, respectively. In studies indicated by an asterisk, the estimate in flux change due to the indirect effect of aerosols was computed as the difference in top of atmosphere fluxes between two distinct simulations and therefore does not represent a forcing in the strict sense (see text). When several simulations are performed in the same study, "base" indicates the baseline calculation, while the range of estimates is given in parenthesis.*

Reference	Aerosol Type	Forcing estimate (Wm^{-2})				Remarks	
		First effect	**Second effect**	**Both effects**			
Boucher and Rodhe (1994)*	Sulphate			−0.65 to −1.35	P	Uses 3 relationships between sulphate mass and CCN/CDN concentrations.	
Chuang *et al.* (1994)	Sulphate	−0.47			C	Includes a parametrization of cloud nucleation processes.	
Jones *et al.* (1994)	Sulphate	−1.3			P	Uses a relationship between aerosol and droplet number concentrations.	
Boucher and Lohmann (1995)	Sulphate	−0.5 to −1.4			P	LMD GCM	Uses 4 different relationships between sulphate mass and CCN/CDN concentrations (A, B, C, and D).
Boucher and Lohmann (1995)	Sulphate	−0.45 to −1.5			P	ECHAM	
Jones and Slingo (1996)	Sulphate	−0.3 to −1.5			P	Uses 2 different sulphate distributions. Follows Jones *et al.* (1994), Hegg (1994), Boucher and Lohmann (1995) 'D'.	
Kogan *et al.* (1996) Kogan *et al.* (1997)	Sulphate	−1.1				Uses a cloud climatology rather than GCM-simulated clouds.	
Chuang *et al.* (1997)	Sulphate	−0.4 to −1.6			C	Includes a parametrization of cloud nucleation processes. Uses a mixture of pre-existing aerosols.	
Feichter *et al.* (1997)	Sulphate	−0.76			C	Uses Boucher and Lohmann (1995) 'A' relationship.	
Jones and Slingo (1997)	Sulphate	−0.55 to −1.50			P	Uses 2 different versions of the Hadley Centre model.	
Lohmann and Feichter (1997)*	Sulphate	−1		−1.4 to −4.8	C	Uses Boucher and Lohmann (1995) 'A' relationship.	
Rotstayn (1999)*	Sulphate	base −1.2 (−1.1 to −1.7)	base −1.0 (−0.4 to −1.0)	base −2.1 (−1.6 to −3.2)	P	Includes a (small) long-wave radiative forcing.	
Jones *et al.* (1999)* [a]	Sulphate	−0.91	base −0.66	−1.18	C	Includes a (small) long-wave radiative forcing. The two effects add non-linearly.	
Kiehl *et al.* (2000)	Sulphate	−0.40 to −1.78			C		
Ghan *et al.* (2001a)*	Sulphate	~50% for base	~50% for base	base −1.7 (−1.6 to −3.2)	C	Includes a parametrization of cloud nucleation. Predicted aerosol size distribution.	
Lohmann *et al.* (2000)*	Sulphate			base −0.4 (0 to −0.4)	C	Includes a parametrization of cloud nucleation processes.	
	Carb.			base −0.9 (−0.9 to −1.3)	C		
	Sulphate and Carb.	−40% for base	−60% for base	base −1.1 (−1.1 to −1.9)	C		
Chuang *et al.* (2000b)	Sulphate	−0.30			C	Includes a parametrization of cloud nucleation processes. Includes the effect of BC absorption in clouds.	
	Carb.	base −1.51 (−1.27 to −1.67)			C		
	Sulphate and Carb.	−1.85			C		

[a] This model predicts too low sulphate concentrations on average.

Jones *et al.* (1999) and Rotstayn (1999) provide estimates for the second indirect effect alone with ranges of −0.53 to −2.29 Wm⁻² and −0.4 to −1.0 Wm⁻², respectively. For the combined effect (first and second effects estimated together), Jones *et al.* (1999) and Rotstayn (1999) give best estimates of −1.18 and −2.1 Wm⁻². Larger radiative impacts are found in sensitivity tests by Rotstayn (1999) and Lohmann and Feichter (1997), with values of −3.2 and −4.8 Wm⁻², respectively. Rotstayn (1999) tried an alternative parametrization for cloud droplet concentration (Roelofs *et al.*, 1998) and Lohmann and Feichter (1997) tried an alternative cloud scheme. In contrast to these studies, Lohmann *et al.* (2000) adopted a "mechanistic" approach, where they introduced a prognostic equation for the cloud droplet number concentration. They predict a much smaller combined effect with radiative impact of 0.0 and −0.4 Wm⁻², assuming externally and internally mixed sulphate aerosols, respectively. The authors attribute this rather small radiative impact to the small increase in anthropogenic sulphate aerosol number concentrations. Ghan *et al.* (2001a) estimate a combined effect of −1.7 Wm⁻² using a mechanistic approach similar to that of Lohmann *et al.*, but with aerosol size distribution predicted rather than prescribed. The larger forcing is due to the absence of lower bounds on droplet and aerosol number concentrations in the simulations by Ghan *et al.* (2001a). Studies by Rotstayn (1999) and Jones *et al.* (1999) both indicate that the positive long-wave forcing associated with the indirect aerosol effect is small (between 0.0 and 0.1 Wm⁻² for each of the effects).

The partitioning of the total indirect radiative impact between the first and second effect is uncertain. Jones *et al.* (1999) found that the ratio between their best estimates of the first and second indirect effects, taken separately, was 1.38, while Lohmann *et al.* (2000) predicted a ratio of 0.71 (for sulphate and carbonaceous aerosols taken together). Rotstayn (1999) and Lohmann *et al.* (2000) estimated that the radiative impact for the combined effects was of similar magnitude than the sum of the two effects considered separately. In contrast, Jones *et al.* (1999) found that the combined radiative impact (−1.18 Wm⁻²) was less than the sum of the two effects (estimated as −0.91 and −0.66 Wm⁻², respectively, yielding a sum of −1.57 Wm⁻²).

6.8.2.3 Further discussion of uncertainties

Some authors have argued that sea salt particles may compete with sulphate aerosols as cloud condensation nuclei, thereby reducing the importance of anthropogenic sulphate in droplet nucleation (Ghan *et al.*, 1998; O'Dowd *et al.*, 1999). While this process is empirically accounted for in some of the above mentioned estimates (e.g., Jones *et al.*, 1999), it certainly adds further uncertainty to the estimates. Considerable sensitivity is found to the parametrization of the autoconversion process (Boucher *et al.*, 1995; Lohmann and Feichter, 1997; Delobbe and Gallée, 1998; Jones *et al.*, 1999) which complicates matters because there is a need to "tune" the autoconversion onset in GCMs (Boucher *et al.*, 1995; Fowler *et al.*, 1996; Rotstayn, 1999, 2000) to which the indirect aerosol forcing is sensitive (Jones *et al.*, 1999; Rotstayn, 1999; Ghan *et al.*, 2001a). The indirect aerosol forcing is also sensitive to the treatment of the pre-industrial aerosol concentration and properties (Jones *et al.*, 1999; Lohmann *et al.*, 2000) which

remain poorly characterised, the representation of the microphysics of mid-level clouds (Lohmann *et al.*, 2000), the representation of aerosol size distribution (Ghan *et al.*, 2001a), the parametrization of sub-grid scale clouds, the horizontal resolution of the GCM (Ghan *et al.*, 2001a), and the ability of GCMs to simulate the stratocumulus cloud fields. Finally, it should be noted that all the studies discussed above cannot be considered as truly "independent" because many of them (with the exceptions of Lohmann *et al.* (2000) and Ghan *et al.* (2001a)) use similar methodologies and similar relationships between sulphate mass and cloud droplet number concentration. Therefore it is suspected that the range of model results does not encompass the total range of uncertainties. In an overall sense, it can be concluded that the considerable sensitivities to the treatment of microphysical details associated with aerosol-cloud interactions, and their linkages with macroscopic cloud and circulation parameters, remain to be thoroughly explored.

6.8.3 Indirect Radiative Forcing by Other Species

6.8.3.1 Carbonaceous aerosols

In this section, carbonaceous aerosols refer to the mixture (internal or external) of OC and BC aerosols. Carbonaceous aerosols (and in particular biomass burning aerosols) are efficient cloud condensation nuclei (see Chapter 5 and e.g., Novakov and Penner, 1993; Novakov and Corrigan, 1996). There have been few GCM studies estimating the indirect forcing from carbonaceous aerosols. Penner *et al.* (1996) reported a range of forcing from −2.5 to −4.5 Wm⁻² (not included in Table 6.6), which is probably an overestimate because of neglect of other aerosol types such as dust and sea salt and underestimated natural emissions of organic aerosols. In a new set of simulations using an updated model accounting for dust and sea salt aerosols, Chuang *et al.* (2000b) obtained a forcing of −1.51 Wm⁻² for the first indirect effect from carbonaceous aerosols (−0.52 and −1.16 Wm⁻² for fossil fuel and biomass burning aerosols, respectively). This estimate includes the effect of black carbon absorption in cloud droplets. Lohmann *et al.* (2000) predicted a radiative impact for the combined effect (i.e., first and second effects) of −1.3 and −0.9 Wm⁻² for externally and internally mixed carbonaceous aerosols, respectively. These estimates do not include the effects of secondary organic aerosols, nor the effects of absorption of solar radiation by black carbon within the cloud.

Kaufman and Nakajima (1993) used AVHRR data to analyse bright warm clouds over Brazil during the biomass burning season. They found a decrease in the cloud reflectance when the smoke optical thickness increased. Kaufman and Fraser (1997) used a similar approach to observe thinner and less reflective clouds. They showed that smoke from biomass burning aerosols reduced the cloud droplet size and increased the cloud reflectance for smoke optical depth up to 0.8. They estimated the indirect radiative forcing by smoke to be −2 Wm⁻² over this region for the three months when biomass burning takes place, which would suggest a much smaller global average. However, using a combination of satellite observations and a global chemistry model, Remer *et al.* (1999) estimated that 50% of the cumulative biomass burning aerosol indirect forcing occurs for smoke optical depth smaller than 0.1, that is well away from the source regions.

6.8.3.2 Combination of sulphate and carbonaceous aerosols

Lohmann *et al.* (2000) found that the radiative impact of sulphate and carbonaceous aerosols considered simultaneously (−1.5 and −1.1 Wm^{-2} for the externally and internally mixed assumptions, respectively) is comparable to the sum of the radiative impacts calculated separately (see Table 6.6). Chuang *et al.* (2000b) reached a similar conclusion in their experiments where they considered only the first indirect effect, with a total radiative forcing of −1.85 Wm^{-2}. It is noteworthy, however, that in these two studies, the predicted radiative forcings for sulphate aerosols are on the low side. This indicates that it is not wise to add the radiative forcings due to sulphate and carbonaceous aerosols obtained separately from two different models. In fact, most of the GCM studies of the indirect aerosol effect used sulphate as a surrogate for the total anthropogenic fraction of the aerosol (e.g., Boucher and Lohmann, 1995; Feichter *et al.*, 1997; Lohmann and Feichter, 1997). In this case the computed forcings incorporate the effects of other aerosol types which have a similar spatial distribution to sulphate aerosols, such as nitrate aerosols or carbonaceous aerosols from fossil fuel combustion. It will not include, however, the effects of biomass burning aerosols which have a different spatial distribution from sulphate aerosols.

Another issue is the potential for light-absorbing aerosols to increase in-cloud absorption of solar radiation – and correspondingly decrease the cloud albedo – when incorporated inside cloud droplets. Twohy *et al.* (1989) concluded from measurements off the coast of California and from simple radiative calculations that the observed levels of soot would not lead to a significant impact on the cloud albedo. Chylek *et al.* (1996a) estimated an upper bound for increased absorption of solar radiation of 1 to 3 Wm^{-2} (global and annual average) for a black carbon concentration of 0.5 μgm^{-3}. Considering the modelled atmospheric concentrations of soot (Chapter 5 and Sections 6.7.3 and 6.7.5) and the fact that only a fraction of the soot is incorporated in the cloud droplets, the effect is probably smaller by one to two orders of magnitude. Heintzenberg and Wendisch (1996) showed that the decrease in radiative forcing due to a decrease in soot concentrations with increasing distances from the pollution sources could be compensated by a concurrent increase in the fraction of soot which is incorporated in the cloud droplets. Only one GCM study to date has considered the in-cloud absorption of soot. Chuang *et al.* (2000b) estimated a radiative forcing for in-cloud BC of +0.07 Wm^{-2} for the soot concentrations predicted by their model and using an effective medium approximation. More studies are needed to confirm these results.

6.8.3.3 Mineral dust aerosols

Levin *et al.* (1996) observed desert dust particles coated with sulphate. Such particles may originate from in-cloud scavenging of interstitial dust particles followed by evaporation of the cloud droplets, condensation of SO_2 onto dust followed by oxidation, or even coagulation of dust and sulphate particles. The presence of soluble material (which may be of anthropogenic origin) on the desert dust particles converts them into large and effective CCN which may affect the cloud microphysics. Whether this effect results in a significant climate forcing has not been investigated and cannot presently be quantified.

6.8.3.4 Effect of gas-phase nitric acid

Kulmala *et al.* (1993, 1995, 1998) argued that enhanced concentrations of condensable vapours (such as HNO_3 and HCl) in the atmosphere could affect cloud properties by facilitating the activation of cloud condensation nuclei. The impact of such an effect on the planetary cloud albedo has not been assessed.

6.8.4 Indirect Methods for Estimating the Indirect Aerosol Effect

6.8.4.1 The "missing" climate forcing

Hansen and colleagues have used two alternative approaches to characterise and quantify any "missing" climate forcing besides that due to greenhouse gases, solar constant, O_3, and aerosol direct effect. Hansen *et al.* (1995) used a simplified GCM to investigate the impacts of various climate forcings on the diurnal cycle of surface air temperature and compared them with observations. They found that, although the aerosol direct effect or an increase in continental cloud albedo could contribute to damp the surface temperature diurnal cycle, only an increase in continental cloud cover would be consistent with observations (Karl *et al.*, 1993). The required cloud increase depends on cloud height and would be of the order of 1% global coverage for low clouds (i.e., 2 to 5% over land). We cannot rule out that such a change is an unidentified cloud feedback rather than a forcing. Hansen *et al.* (1997b) also argued that agreement between observed and computed temperature trends requires the presence of another forcing of at least −1 Wm^{-2} which is inferred as being due to the *indirect* effect. In their calculations, the *direct* tropospheric aerosol effect does not play a large net role, because the moderately absorbing aerosol assumption leads to an offset between its sunlight reflecting and absorbing properties insofar as the top of the atmosphere irradiance change is concerned. However, this method assumes that the observed change in temperature since pre-industrial times is primarily a response to anthropogenic forcings, that all the other anthropogenic forcings are well quantified, and that the climate sensitivity parameter (Section 6.1) predicted by the GCM is correct (Rodhe *et al.*, 2000). Therefore it may simply be a coincidence that the estimate of Hansen *et al.* (1997b) is consistent with the GCM estimates discussed above.

6.8.4.2 Remote sensing of the indirect effect of aerosols

Han *et al.* (1994) analysed AVHRR satellite radiances to retrieve the cloud droplet size of low-level clouds. They reported significant inter-hemispheric differences for both maritime and continental clouds. Boucher (1995) showed that, if this difference is to be attributed to anthropogenic aerosols, it implies a differential forcing of about −1 Wm^{-2} between the two hemispheres. Assuming a Northern Hemisphere to Southern Hemisphere ratio of 2:1 for the aerosol indirect effect, this would imply a globally-averaged forcing of −1.5 Wm^{-2}. It is not clear, however, to what extent changes in cloud droplet size are related to change in aerosol concentrations. For instance, Han *et al.* (1998) showed that cloud albedo decreases with decreasing droplet size for the optically thinner clouds over the oceans. While this does not invalidate the aerosol indirect effect at all, it underlines the limitations in using satellite observed changes in droplet size to compute the aerosol indirect forcing. Therefore it seems difficult at present to use satellite observations to estimate the first aerosol indirect forcing unless some changes in cloud albedo could be tied to changes in aerosol concentrations under the assumption of constant liquid water content. Satellite observations do play, however, a key role for evaluating models of the indirect aerosol radiative effect (Ghan *et al.*, 2001b).

6.8.5 Forcing Estimates for This Report

Shine and Forster (1999) proposed a value of the aerosol indirect forcing due to all aerosols of -1 Wm^{-2} with at least a factor of two uncertainty. Several observational studies (see Chapter 5) support the existence of the first aerosol indirect effect on low-level clouds and a negative sign for the associated radiative forcing, but these studies do not give indications on what a (negative) upper bound of the forcing would be. GCM studies predict radiative forcing for the first indirect effect of industrial aerosols in the range of -0.3 to -1.8 Wm^{-2}. However, because of the uncertainties in the estimates discussed above, the limited validation of GCM parametrizations and results, and because in-cloud absorption by black carbon aerosols was not considered in all but one of the GCM studies, we retain 0 Wm^{-2} as an upper bound for the first aerosol indirect effect. A lower bound of -2 Wm^{-2} is selected on the basis of available GCM studies for the first indirect effect. Not too much emphasis should be given to the exact bounds of this interval because they do not carry any statistical meaning (Section 6.13.1) and because of the very low level of scientific understanding associated to this forcing (see Section 6.13.1 where this concept is defined).

Available GCM studies suggest that the radiative flux perturbations associated with the second effect could be of similar magnitude to that of the first effect. There are no studies yet to confirm unambiguously that the GCM estimates of the radiative impact associated with the second indirect effect can be interpreted in the strict sense of a radiative forcing (see Sections 6.1 and 6.8.2.2), and very few observations exist as yet to support the existence of a significant effect. Therefore we refrain from giving any estimate or range of estimates for the second aerosol indirect effect. However, this does not minimise the potential importance of this effect.

6.8.6 Aerosol Indirect Effect on Ice Clouds

6.8.6.1 Contrails and contrail-induced cloudiness
Using meteorological and air traffic data scaled to regional observations of contrail cover, Sausen *et al.* (1998) estimated the present day global mean cover by line-shaped contrails to be about 0.1%. This results in a global and annual mean radiative forcing by line-shaped contrails of 0.02 Wm^{-2} (Minnis *et al.*, 1999), subject to uncertainties in the contrail cover, optical depth, ice particle size and shape (Meerkötter *et al.*, 1999). We follow Fahey *et al.* (1999) and retain a range of 0.005 to 0.06 Wm^{-2} for the present day forcing, around the best estimate of 0.02 Wm^{-2}.

Contrails can evolve into extended cirrus clouds. Boucher (1999) and Fahey *et al.* (1999) have shown evidences that cirrus

occurrence and coverage may have increased in regions of high air traffic compared with the rest of the globe. Smith *et al.* (1998) reported the existence of nearly invisible layers of small ice crystals, which cause absorption of infrared radiation, and could be due to remnant contrail particles. From consideration of the spatial distribution of cirrus trends during the last 25 years, Fahey *et al.* (1999) gave a range of possible best estimates of 0 to 0.04 Wm^{-2} for the radiative forcing due to aviation-induced cirrus. The available information on cirrus clouds was deemed insufficient to determine a single best estimate or an uncertainty range.

6.8.6.2 Impact of anthropogenic aerosols on cirrus cloud microphysics
Measurements by Ström and Ohlsson (1998) in a region of high air traffic revealed higher crystal number concentrations in areas of the cloud affected by soot emissions from aircraft. If the observed enhancement in crystal number density (which is about a factor of 2) is associated with a reduction in the mean crystal size, as confirmed by the measurements of Kristensson *et al.* (2000), a change in cloud radiative forcing may result. Wyser and Ström (1998) estimated the forcing, although very uncertain, to be in the order of 0.3 Wm^{-2} in regions of dense air traffic under the assumption of a 20% decrease of the mean crystal size. No globally averaged radiative forcing is available.

The sedimentation of ice particles from contrails may remove water vapour from the upper troposphere. This effect is expected to be more important in strongly supersaturated air when ice particles can fall without evaporating (Fahey *et al.*, 1999). The impacts of such an effect on cirrus formation, vertical profile of humidity and the subsequent radiative forcing have not been assessed.

Aerosols also serve as ice nuclei although it is well recognised that there are fewer ice nuclei than cloud condensation nuclei. It is conceivable that anthropogenic aerosols emitted at the surface and transported to the upper troposphere affect the formation and properties of ice clouds. Jensen and Toon (1997) suggested that insoluble particles from the surface or soot particles emitted by aircraft, if they serve as effective ice nuclei, can result in an increase in the cirrus cloud coverage. Laaksonen *et al.* (1997) argued that nitric acid pollution is able to cause an increase in supercooled cirrus cloud droplet concentrations, and thereby influence climate (see Chapter 5, Section 5.3.6). Such effects, if significant at all, are not quantified at present.

6.9 Stratospheric Aerosols

IPCC (1992, 1994) and the SAR have dealt with the climatic effects of episodic, explosive volcanic events which result in significant enhancements of aerosol concentrations in the stratosphere. The most dramatic of these in recent times has been the 1991 eruption of Mt. Pinatubo. The radiative, chemical, dynamical and climatic consequences accompanying the transient duration of sulphuric acid aerosols in the stratosphere have been discussed in previous IPCC assessments. The eruption of Mt. Pinatubo reached a peak forcing of about -3 Wm^{-2} (uncertainty of 20%) in 1991 (Hansen *et al.*, 1998; Stenchikov *et al.*, 1998), and was plausibly the largest volcanic aerosol forcing of this century, perturbing the stratospheric and surface climate significantly (SAR). However, aerosol levels in the stratosphere have now fallen to well below the peak values seen in 1991 to 1993, and are comparable to the very low value seen in about 1979, which was a quiescent time as far as volcanic activity was concerned (WMO, 1999). It is likely that even the heterogeneous chemical effects initiated by aerosols upon the O_3 chemistry and its destruction (Solomon *et al.*, 1996) have diminished. One consequence of the aerosol-induced ozone depletion may have been a prolonged cooling of the lower stratosphere to abnormally low values through the mid-1990s, as estimated from satellite and radiosonde observations (WMO, 1999, Chapter 5). Although episodic in nature and transient in duration, volcanic events can exert a significant influence on the time history of the radiative forcing evolution and thereby on the long-term (interannual to decadal scale) temperature record (see Chapter 9).

As noted in previous IPCC Assessments, there are difficulties in compiling a good quantitative record of the episodic volcanic events (see also Rowntree, 1998), in particular the intensity of their forcings prior to the 1960s. Efforts have been made to compile the optical depths of the past volcanoes (SAR; Robock and Free, 1995, 1996; Andronova *et al.*, 1999); however, the estimated global forcings probably have an uncertainty of a factor of two or more. Several major volcanic eruptions occurred between 1880 and 1920, and between 1960 and 1991.

6.10 Land-use Change (Surface Albedo Effect)

Changes in land surface albedo can result from land-use changes (Henderson-Sellers, 1995) and thus be tied to an anthropogenic cause. Hansen *et al.* (1997b) estimate that a forcing of $-0.4 \, \mathrm{Wm^{-2}}$ has resulted, about half of which is estimated to have occurred in the Industrial Era. The largest effect is estimated to be at the high latitudes where snow-covered forests that have a lower albedo have been replaced by snow-covered deforested areas. Hansen *et al.* (1998) point out that the albedo of a cultivated field is affected more by a given snowfall than the albedo of an evergreen forest. They performed a simulation with pre-industrial vegetation replaced by current land-use patterns and found the global mean forcing to be $-0.21 \, \mathrm{Wm^{-2}}$, with the largest contributions coming from deforested areas in Eurasia and North America. In a similar study, Betts (2001) estimates an instantaneous radiative forcing of $-0.20 \, \mathrm{Wm^{-2}}$ by surface albedo change due to present day land use relative to natural vegetation. In agreement with Hansen *et al.* (1997b), the greatest effect is seen in the high latitude agricultural regions. If, as above, half of the land clearance is assumed to have taken place since the industrial revolution, this suggests a forcing of $-0.10 \, \mathrm{Wm^{-2}}$ by land use over this period. In a parallel simulation with the climate free to respond to the change in albedo and other vegetation characteristics, lower temperatures are simulated in the northern mid-latitudes. These are mainly attributed to the increased surface albedo, although increases in cloud cover cause further localised reductions in the net surface short-wave radiation in some regions. However, some areas exhibit higher temperatures in their dry season, consistent with a decrease in evapotranspiration due to reduced access of soil moisture by the shallower roots of the crops compared with forest.

Following Hansen *et al.* (1997b), Shine and Forster (1999) recommended in their review a value of $-0.2 \, \mathrm{Wm^{-2}}$ with at least a $0.2 \, \mathrm{Wm^{-2}}$ uncertainty. We adopt those values here for the best estimate and range, respectively; however, in view of the small number of investigations and uncertainty in historical land cover changes, there is very low confidence in these values at present.

Changes in land use can also exert other kinds of climatic impacts, e.g., changes in roughness length, turbulent fluxes, and soil moisture and heat budgets (see also Chapters 7 and 8). Further, there are a host of factors that are potentially affected by land-use change and that could have an impact on the atmospheric concentrations of radiatively active trace gases and aerosols. For instance, the dry deposition rates of species could be affected owing to the surface roughness change. Precipitation changes induced by deforestation etc. could affect the wet deposition of species and thereby bring about biogeochemical changes, leading to changes in lifetimes. The impacts due to such changes have not been comprehensively investigated.

6.11 Solar Forcing of Climate

In this section variations in total solar irradiance and how these translate into radiative forcing are described and potential mechanisms for amplification of solar effects are discussed. The detection of solar effects in observational records is covered in Chapters 2 and 12.

6.11.1 Total Solar Irradiance

6.11.1.1 The observational record

The fundamental source of all energy in the climate system is the Sun so that variation in solar output provides a means for radiative forcing of climate change. It is only since the late 1970s, however, and the advent of space-borne measurements of total solar irradiance (TSI), that it has been clear that the solar "constant" does, in fact, vary. These satellite instruments suggest a variation in annual mean TSI of the order 0.08% (or about $1.1 \, \mathrm{Wm^{-2}}$) between minimum and maximum of the 11-year solar cycle. While the instruments are capable of such precision their absolute calibration is much poorer such that, for example, TSI values for solar minimum 1986 to 1987 from the ERB radiometer on Nimbus 7 and the ERBE experiment on NOAA-9 disagree by about $7 \, \mathrm{Wm^{-2}}$ (Lean and Rind, 1998). More recent data from ACRIM on UARS, EURECA and VIRGO on SOHO cluster around the ERBE value (see Figure 6.4) so absolute uncertainty may be estimated at around $4 \, \mathrm{Wm^{-2}}$. Although individual instrument records last for a number of years, each sensor suffers degradation on orbit so that construction of a composite series of TSI from overlapping records becomes a complex task. Figure 6.4 shows TSI measurements made from satellites, rockets, and balloons since 1979.

Willson (1997) used ERB data to provide cross-calibration between the non-overlapping records of ACRIM-I and ACRIM-II and deduced that TSI was $0.5 \, \mathrm{Wm^{-2}}$ higher during the solar minimum of 1996 than during solar minimum in 1986. If this reflects an underlying trend in solar irradiance it would

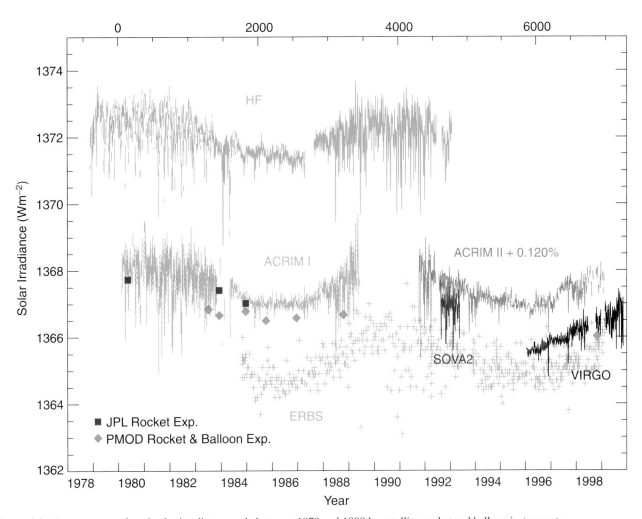

Figure 6.4: Measurements of total solar irradiance made between 1979 and 1999 by satellite, rocket and balloon instruments (http://www.pmodwrc.ch/solar_const/solar_const.html).

represent a radiative forcing[2] of 0.09 Wm^{-2} over that decade compared with about 0.4 Wm^{-2} due to well-mixed greenhouse gases. The factors used to correct ACRIM-I and ACRIM-II by Willson (1997) agree with those derived independently by Crommelynk *et al.* (1995) who derived a Space Absolute Radiometric Reference of TSI reportedly accurate to ± 0.15%. Fröhlich and Lean (1998), however, derived a composite TSI series which shows almost identical values in 1986 and 1996, in good agreement with a model of the TSI variability based on independent observations of sunspots and bright areas (faculae). The difference between these two assessments depends critically on the corrections necessary to compensate for problems of unexplained drift and uncalibrated degradation in both the Nimbus 7/ERB and ERBS time series. Thus, longer-

term and more accurate measurements are required before trends in TSI can be monitored to sufficient accuracy for application to studies of the radiative forcing of climate.

6.11.1.2 Reconstructions of past variations of total solar irradiance

As direct measurements of TSI are only available over the past two decades it is necessary to use other proxy measures of solar output to deduce variations at earlier dates. In the simplest type of reconstruction a proxy measure, such as sunspot number (Stevens and North, 1996) or solar diameter (Nesme-Ribes *et al.*, 1993), is calibrated against recent TSI measurements and extrapolated backwards using a linear relationship. The various proxies vary markedly as indicators of solar activity. For example, over the past century, sunspot number and 10.7 cm flux showed highest values at the solar maximum of 1958, whereas the aa index, which gives a measure of the magnitude of the solar magnetic field at the Earth, peaked during 1990. This is because whereas sunspot numbers return to essentially zero at each solar minimum the aa index shows 11-year cycles imposed on a longer-term modulation (Lean and Rind, 1998). Other terrestrially based indicators of solar activity recorded by cosmogenic isotopes in tree-rings and ice-

---

[2] Geometric factors affect the conversion from change in TSI to radiative forcing. It is necessary to divide by a factor of 4, representing the ratio of the area of the Earth's disc projected towards the Sun to the total surface area of the Earth, and to multiply by a factor of 0.69, to take account of the Earth's albedo of 31%. Thus a variation of 0.5 Wm^{-2} in TSI represents a variation in global average instantaneous (i.e. neglecting stratospheric adjustment) radiative forcing of about 0.09 Wm^{-2}.

cores also show longer term modulation. However, direct solar proxies other than the sunspot number cover too short a period to reliably detect such a trend. Thus, it is not clear which proxy, if any, can be satisfactorily used to indicate past values of TSI.

A more fundamental approach recognises that solar radiative output is determined by a balance between increases due to the development of faculae and decreases due to the presence of sunspots. Longer-term changes are also speculated to be occurring in the quiet Sun against which these variable active regions are set. The sunspot darkening depends on the area of the solar disc covered by the sunspots while the facular brightening has been related to a variety of indices. These include sunspot number (Lean *et al.*, 1995), emission of singly ionised calcium (Ca II at 393.4 nm) (Lean *et al.*, 1992), solar cycle length, solar cycle decay rate, solar rotation rate and various empirical combinations of all of these (Hoyt and Schatten, 1993; Solanki and Fligge, 1998).

In addition to estimates of the impact of active regions on TSI potential contributions due to the variation in brightness of the quiet Sun have been estimated (Lean *et al.*, 1992; White *et al.*, 1992). These were largely based on observations of the behaviour of Sun-like stars (Baliunas and Jastrow, 1990) and the assumption that during the Maunder Minimum (an extended period during the late 17th century during which no sunspots were observed) the Sun was in a non-cycling state. The various reconstructions vary widely in the values deduced for TSI during the Maunder Minimum relative to the present. Mendoza (1997) has pointed out that uncertainties in the assumptions made about the state of the Sun during that period could imply a range of between 1 and 15 Wm^{-2} reduction in TSI less than present mean values although most estimates lie in the 3 to 5.5 Wm^{-2} range. Figure 6.5 shows group sunspot numbers from 1610 to 1996 (Hoyt and Schatten, 1998) together with five TSI reconstructions. The sunspot numbers (grey curve, scaled to correspond to Nimbus-7 TSI observations for 1979 to 1993) show little long-term trend. Lean *et al.* (1995, solid red curve) determine long-term variability from sunspot cycle amplitude; Hoyt and Schatten (1993, black solid curve) use mainly the length of the sunspot cycle; the two Solanki and Fligge (1998) blue curves (dotted) are similar in derivation to the Lean *et al.* and Hoyt and Schatten methods. Lockwood and Stamper (1999, heavy dash-dot green curve) use an entirely different approach, based not on sunspot numbers but on the aa geomagnetic index, and predict somewhat larger variation over individual cycles but less on the longer term. Clearly, even disregarding the shifts due to absolute scaling, there are large differences between the TSI reconstructions. Thus knowledge of solar radiative forcing is uncertain, even over the 20th century and certainly over longer periods.

An alternative approach which has been used to reconstruct TSI (Reid, 1997; Soon *et al.*, 1996) is to assume that time variations in global surface temperature are due to a combination of the effects of solar variability and enhanced greenhouse gas concentrations and to find that combination of these two forcings which best combine to simulate surface temperature measurements. However, these authors did not take natural climatic variability into account and a TSI series derived by such methods could not be used as an independent measure of radiative forcing of climate.

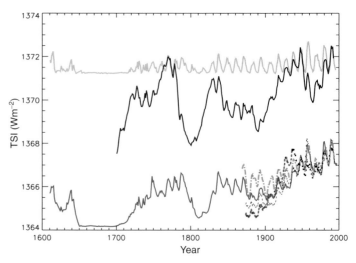

Figure 6.5: Reconstructions of total solar irradiance (TSI) by Lean *et al.* (1995, solid red curve), Hoyt and Schatten (1993, data updated by the authors to 1999, solid black curve), Solanki and Fligge (1998, dotted blue curves), and Lockwood and Stamper (1999, heavy dash-dot green curve); the grey curve shows group sunspot numbers (Hoyt and Schatten, 1998) scaled to Nimbus-7 observations for 1979 to 1993.

The estimate for solar radiative forcing since 1750 of 0.3 Wm^{-2}, shown in Figure 6.6, is based on the values in Figure 6.5 (taking the 11-year cycle minimum values in 1744 and 1996). Clearly the starting date of 1750 (chosen for the date of the pre-industrial atmosphere in Figure 6.6) is crucial here: a choice of 1700 would give values about twice as large; a choice of 1776 would give smaller values (particularly using the Hoyt and Schatten series). The range of 0.1 to 0.5 Wm^{-2} given in Figure 6.6 is based on the variability of the series, the differences between the reconstructions and uncertainties concerning stratospheric adjustment (see Section 6.11.2.1). However, because of the large uncertainty in the absolute value of TSI and the reconstruction methods our assessment of the "level of scientific understanding" is "very low".

6.11.2 Mechanisms for Amplification of Solar Forcing

6.11.2.1 Solar ultraviolet variation
The Sun emits radiation over the entire electromagnetic spectrum but with most energy in the near ultraviolet, visible, and near infrared regions: 80% of TSI lies between 400 and 1,600 nm. The variations in TSI, discussed above, of a few tenths of one per cent thus represent the integrated change across ultraviolet, visible, and near infrared wavelengths. Most of this radiation passes through the atmosphere unhindered to the tropopause but shorter wavelengths are absorbed in the middle atmosphere where they result in O_3 formation and local heating. In the ultraviolet the amplitude of variability is much higher. Since 1978 (see Cebula *et al.* (1998) for a review) satellite data have shown variations over the 27-day solar rotation period of e.g., 6% at 200 nm and 2.5% at 250 nm. Problems with sensor drift and calibration preclude direct detection of variability on the 11-year time-scale at wavelengths longer than about 250 nm. Instead ultraviolet irradiance cycles are

deduced from observations by scaling the 27-day variations to selected solar activity indices and assuming that the same scaling applies over longer time-scales (Lean *et al.*, 1997). With the launch of the SOLSTICE (Rottman *et al.*, 1993) and SUSIM (Brueckner *et al.*, 1993) instruments on UARS in 1991, measurements have now been made from near solar maximum to solar minimum with long-term precisions that approach 1% at some wavelengths (Rottman *et al.*, 1993; Floyd *et al.*, 1998). Careful cross-calibration of NOAA-11 SBUV/2 with Shuttle SBUV observations (Cebula *et al.*, 1998) have also produced spectral variations 1989 to 1994, but also with uncertainties of a few per cent, which exceeds the actual irradiance variability at the longer ultraviolet wavelengths. Comparison of the SOLSTICE, SUSIM, and SBUV/2 data show reasonable agreement during their 2 to 3 year overlap period (DeLand and Cebula, 1998) and suggest a decline of about 7% at 200 to 208 nm and of about 3.5% at 250 nm from solar maximum in 1989 to near solar minimum in 1994. These estimates agree well with the modelled scalings deduced from the 27-day variations (Lean *et al.*, 1997).

Variations in stratospheric composition and thermal structure resulting from ultraviolet irradiance changes may have an impact on tropospheric climate. The first mechanism whereby this might happen is through changes in radiative forcing (Haigh, 1994). Thus, in addition to a direct increase in downward short-wave irradiance at the tropopause, higher solar activity can cause an increase in downward infrared flux by heating the stratosphere and also radiative forcing due to O_3 changes. However, the sign of the O_3 effect is not well established. Haigh (1994) found that O_3 increases reduced the solar radiative forcing by about 0.1 Wm^{-2} at solar maximum, Wuebbles *et al.* (1998) computed a value of $-0.13\,Wm^{-2}$ due to O_3 increases since the Maunder Minimum and Myhre *et al.* (1998a) about $-0.02\,Wm^{-2}$ from minimum to maximum of the solar cycle. Hansen *et al.* (1997a) showed an additional forcing of about $+0.05\,Wm^{-2}$ from minimum to maximum of a solar cycle due to O_3 increases and lower stratospheric warming. Haigh (1999) and Larkin *et al.* (2000) suggest that the O_3 effect has little effect on radiative forcing at the tropopause but significant effect on where the additional radiation is absorbed (more within the troposphere rather than at the surface). These disparities may represent the different approaches used. Haigh (1994), Wuebbles *et al.* (1998), and Myhre *et al.* (1998a) calculated the O_3 response using two-dimensional chemical-transport models in which temperature changes are estimated using the fixed dynamic heating approximation. The Hansen *et al.* (1997a) value was deduced from studies with a simplified GCM of sensitivity to slab O_3 changes (of unspecified cause) and the assumption that the solar-induced O_3 change is all within the 10 to 150 hPa region. Haigh (1999) and Larkin *et al.* (2000) specified solar irradiance and O_3 changes and calculated the stratospheric temperature response in GCMs. The negative radiative forcing values probably correspond to O_3 change profiles which peak at higher altitudes, and thus have less impact on lower stratospheric temperature and long-wave radiative forcing, although the different methods for calculating temperature change may also be important.

The response of O_3 to solar variability is not well established. Two-dimensional models (e.g., Haigh, 1994; Fleming *et al.*,

1995; Wuebbles *et al.*, 1998) predict the largest fractional changes in the middle-upper stratosphere with monotonically decreasing effects towards the tropopause. Multiple regression analysis of satellite data as carried out with SBUV data by McCormack and Hood (1996) and SAGE data by Wang *et al.* (1996) suggest the largest changes in the upper and lower stratosphere and zero, or even slightly negative, changes in the middle stratosphere. However, as the data are only available over about one and a half solar cycles, have large uncertainties, especially in the lower stratosphere, and may not properly have accounted for the effects of volcanic aerosol (Solomon *et al.*, 1996), the true nature of solar-induced changes in stratospheric O_3 remains uncertain.

Chandra *et al.* (1999) have estimated tropical tropospheric O_3 column amounts by taking the difference between TOMS total columns in clear-sky areas and those with deep convective cloud. They deduced O_3 changes of the order 10% of the tropospheric column over the eleven year cycle occurring out of phase with the solar forcing. This they ascribed to a self-feedback effect on O_3 production with enhanced levels of O_3 aloft resulting in less ultraviolet reaching the troposphere. However, it is also possible that the O_3 changes reflect a response to solar effects in tropospheric dynamics. If the changes are real then solar radiative forcing, as represented in Figure 6.6, should be reduced by approximately 30% due to solar-induced decreases in tropospheric O_3

Changes in stratospheric thermal structure may also affect the troposphere through dynamical interactions rather than through radiative forcing. Kodera (1995) suggested that changes in stratospheric zonal wind structure, brought about by enhanced solar heating, could interact with vertically propagating planetary waves in the winter hemisphere to produce a particular mode of response. This mode, also seen in response to heating in the lower stratosphere caused by injection of volcanic aerosol, shows dipole anomalies in zonal wind structure which propagate down, over the winter period, into the troposphere. Rind and Balachandran (1995) investigated the impact of large increases in solar ultraviolet on the troposphere with a GCM and confirmed that altered refraction characteristics affect wave propagation in winter high latitudes. Solar cycle changes to wave propagation in the middle atmosphere were also investigated by Arnold and Robinson (1998) who used a three-dimensional model of the atmosphere between 10 and 140 km to study the effects of thermospheric heating. They found that non-linear interactions in the winter hemisphere resulted in changes to the stratospheric circulation. It is not clear that the signals discussed above are statistically robust in any of these studies.

Haigh (1996, 1999), Shindell *et al.* (1999), and Larkin *et al.* (2000) have introduced realistic changes in ultraviolet and O_3 into GCMs and found that the inclusion of the O_3 has a significant effect on simulated climate. Haigh (1999) using a GCM with a lid at 10 hPa and few stratospheric levels, showed a pattern of change in zonal mean temperature which was consistent over a range of assumptions concerning the magnitude of the ultraviolet and O_3 changes. This pattern consisted of warming in the stratosphere (except in winter high latitudes) and a vertical banding structure in the troposphere due to shifts in the positions of the sub-tropical jets. The predicted changes in 30 hPa geopotential heights were of similar form to those observed by Labitzke and van Loon (1997) but of smaller magnitude (by about a factor 3). Larkin *et al.* (2000) found very similar patterns of change to those of Haigh (1999) using the same solar irradiance/O_3 changes but an entirely different GCM with a lid at 0.1 hPa. Shindell *et al.* (1999) used a model with a detailed representation of the middle atmosphere and a parametrized photochemical scheme which allows O_3 to respond to changes in ultraviolet flux but not to changes in wind fields. They showed larger amplitude changes in 30 hPa heights in the winter hemisphere, more like those of Labitzke and van Loon (1997) but with little summer hemisphere response. These experiments suggest that further work is needed to establish the response of O_3 to solar variability and to how this affects climate.

6.11.2.2 Cosmic rays and clouds

Svensmark and Friis-Christensen (1997) demonstrated a high degree of correlation between total cloud cover, from the ISCCP C2 data set, and cosmic ray flux between 1984 and 1991. Changes in the heliosphere arising from fluctuations in the Sun's magnetic field mean that galactic cosmic rays (GCRs) are less able to reach the Earth when the Sun is more active so the cosmic ray flux is inversely related to solar activity. Svensmark and Friis-Christensen analysed monthly mean data of total cloud using only data over the oceans between 60°S and 60°N from geostationary satellites. They found an increase in cloudiness of 3 to 4% from solar maximum to minimum and speculated that (a) increased GCR flux causes an increase in total cloud and that (b) the increase in total cloud causes a cooling of climate. Svensmark and Friis-Christensen (1997) also extended this analysis to cover the years 1980 to 1996 using cloud data from the DMSP and Nimbus-7 satellites and showed that the high correlation with GCR flux is maintained. However, it was not possible to intercalibrate the different data sets so the validity of the extended data set as a measure of variations in absolute total cloudiness is open to question.

Svensmark (1998) showed that, at least for the limited period of this study, total cloud varies more closely with GCRs than with the 10.7 cm solar activity index over the past solar cycle. On longer time-scales he also demonstrated that Northern Hemisphere surface temperatures between 1937 and 1994 follow variations in cosmic ray flux and solar cycle length more closely than total irradiance or sunspot number. There has been a long-term decrease in cosmic ray flux since the late 17th century, as evidenced by the ^{10}Be and ^{14}C cosmogenic isotope records (Stuiver and Reimer, 1993; Beer *et al.*, 1994), and this mirrors the long-term increase in TSI. However, the TSI reconstruction of Hoyt and Schatten (1993), which is based on solar cycle lengths, does not appear to track the cosmogenic isotope records any more closely than that of Lean *et al.* (1995), which is based on sunspot cycle amplitude (Lean and Rind, 1998). Such use of different solar indices may help to identify which physical mechanisms, if any, are responsible for the apparent meteorological responses to solar activity.

Kuang *et al.* (1998) have repeated Svensmark and Friis-Christensen's analysis of ISCCP data and showed high correlations with an El Niño-Southern Osciallation (ENSO) index

difficult to distinguish from the GCR flux. Farrar (2000) showed that the pattern of change in cloudiness over that period, particularly in the Pacific Ocean, corresponds to what would be expected for the atmospheric circulation changes characteristic of El Niño. Kernthaler *et al.* (1999) have also studied the ISCCP dataset, using both geostationary and polar orbiter data and suggested that the correlation with cosmic ray flux is reduced if high latitude data are included. This would not be expected if cosmic rays were directly inducing increases in cloudiness, as cosmic ray flux is greatest at high latitudes. Kernthaler *et al.* (1999), Jørgensen and Hansen (2000), and Gierens and Ponater (1999), also noted that a mechanism whereby cosmic rays resulted in greater cloud cover would be most likely to affect high cloud as ionisation is greatest at these altitudes. Even if high cloud did respond to cosmic rays, it is not clear that this would cause global cooling as for thin high cloud the long-wave warming effects dominate the short-wave cooling effect. Kristjánsson and Kristiansen (2000) have additionally analysed the ISCCP D2 dataset, 1989 to 1993, and found little statistical evidence of a relationship between GCRs and cloud cover with the possible exception of low marine clouds in mid-latitudes. They also noted that there was no correlation between outgoing long-wave radiation, as represented in ERBE data, and GCRs. Thus the evidence for a cosmic ray impact on cloudiness remains unproven.

A further consideration must be potential physical mechanisms whereby cosmic rays might enhance cloudiness. Cosmic rays are the principal source of ionisation in the free troposphere. Furthermore, ionisation rates and atmospheric conductivity are observed to vary with solar activity. Svensmark and Friis-Christensen (1997) propose that the correlation between cosmic rays and cloud cover that they observed is due to an increase in efficiency of charged particles, over uncharged ones, in acting as cloud condensation nuclei. There is evidence for this occurring in thunderstorms (Pruppacher and Klett, 1997) but it is not clear to what extent this affects cloud development. There is also evidence that ions are sometimes critical in gas-to-particle conversion but again there is no evidence that this has any impact on cloud formation.

In a series of papers, Brian Tinsley has developed a more detailed mechanism for a link between cosmic rays and cloudiness (e.g., Tinsley, 1996). This is based on the premise that aerosols ionised by cosmic rays are more effective as ice nuclei and cause freezing of supercooled water in clouds. In clouds that are likely to cause precipitation the latent heat thus released then causes enhanced convection which promotes cyclonic development and hence increased storminess. There is some laboratory evidence to suggest that charging increases ice nucleation efficiency (Pruppacher, 1973) although there is no observational evidence of this process taking place in the atmosphere. Furthermore, only a small proportion of aerosol particles are capable of acting as ice nuclei, depending on chemical composition or shape. There are also laboratory studies (Abbas and Latham, 1969) which indicate the existence of "electrofreezing", but again no evidence in the real atmosphere. Thus Tinsley's mechanism is plausible but requires further observational and modelling studies to establish whether or not it could be of sufficient magnitude to result in the claimed effects (Harrison and Shine, 1999).

We conclude that mechanisms for the amplification of solar forcing are not well established. Variations in ultraviolet and solar-induced changes in O_3 may have a small effect on radiative forcing but additionally may affect climate through changing the distribution of solar heating and thus indirectly through a dynamical response. At present there is insufficient evidence to confirm that cloud cover responds to solar variability.

6.12 Global Warming Potentials

6.12.1 Introduction

Just as radiative forcing provides a simplified means of comparing the various factors that are believed to influence the climate system to one another, Global Warming Potentials (GWPs) are one type of simplified index based upon radiative properties that can be used to estimate the potential future impacts of emissions of different gases upon the climate system in a relative sense. The formulation of GWPs, reasons for the choice of various time horizons, and the effects of clouds, scenarios, and many other factors upon GWP values were discussed in detail in IPCC (1994). That discussion will not be repeated here. Section 6.2 discusses the relationship between radiative forcing and climate response and describes recent studies that have supported the view that many different kinds of forcing agents (e.g., various greenhouse gases, sulphate aerosols, solar activity, etc.) yield similar globally averaged climate responses per Wm^{-2} of forcing (albeit with different spatial patterns in some important cases). These parallels in global mean climate responses have motivated the use of simplified measures to estimate in an approximate fashion the relative effects of emissions of different gases on climate. The emphasis on relative rather than absolute effects (such as computed temperature change) avoids dependence upon any particular model of climate response.

The impact of greenhouse gas emissions upon the atmosphere is related not only to radiative properties, but also to the time-scale characterising the removal of the substance from the atmosphere. Radiative properties control the absorption of radiation per kilogram of gas present at any instant, but the lifetime (or adjustment time, see Chapter 4) controls how long an emitted kilogram is retained in the atmosphere and hence is able to influence the thermal budget. The climate system responds to changes in the thermal budget on time-scales ranging from the order of months to millennia depending upon processes within the atmosphere, ocean, cryosphere, etc.

GWPs are a measure of the relative radiative effect of a given substance compared to another, integrated over a chosen time horizon. The choice of the time horizon depends in part upon whether the user wishes to emphasise shorter-term processes (e.g., responses of cloud cover to surface temperature changes) or longer-term phenomena (such as sea level rise) that are linked to sustained alterations of the thermal budget (e.g., the slow transfer of heat between, for example, the atmosphere and ocean). In addition, if the speed of potential climate change is of greatest interest (rather than the

eventual magnitude), then a focus on shorter time horizons can be useful (IPCC, 1994; Skodvin and Fuglestvedt, 1997; Fuglestvedt *et al.*, 2000; Smith and Wigley, 2000a,b).

As in previous reports, here we present GWPs for 20, 100, and 500 year time horizons. The most recent GWP evaluations are those of WMO (1999) and the SAR, and the results presented here are drawn in large part from those assessments, with updates for those cases where significantly different new laboratory or radiative transfer results have been published. The sources used for input variables for the GWP calculations are indicated in this section and in the headers and footnotes to the tables, where sources of new estimates since the SAR are identified.

The GWP has been defined as the ratio of the time-integrated radiative forcing from the instantaneous release of 1 kg of a trace substance relative to that of 1 kg of a reference gas (IPCC, 1990):

$$ GWP(x) = \frac{\int_0^{TH} a_x \cdot [x(t)]\, dt}{\int_0^{TH} a_r \cdot [r(t)]\, dt} \qquad (6.2) $$

where *TH* is the time horizon over which the calculation is considered, a_x is the radiative efficiency due to a unit increase in atmospheric abundance of the substance in question (i.e., $Wm^{-2}\ kg^{-1}$), $[x(t)]$ is the time-dependent decay in abundance of the instantaneous release of the substance, and the corresponding quantities for the reference gas are in the denominator. The GWP of any substance therefore expresses the integrated forcing of a pulse (of given small mass) of that substance relative to the integrated forcing of a pulse (of the same mass) of the reference gas over some time horizon. The numerator of Equation 6.2 is the absolute (rather than relative) GWP of a given substance, referred to as the AGWP. The GWPs of various greenhouse gases can then be easily compared to determine which will cause the greatest integrated radiative forcing over the time horizon of interest. The direct relative radiative forcings per ppbv are derived from infrared radiative transfer models based on laboratory measurements of the molecular properties of each substance and considering the molecular weights. Updated information since the SAR is presented for many gases in Section 6.3. Many important changes in these quantities were recently reviewed in WMO (1999) and will be briefly summarised here. In addition, some gases can indirectly affect radiative forcing, mainly through chemical processes. For example, tropospheric O_3 provides a significant radiative forcing of the climate system, but its production occurs indirectly, as a result of atmospheric chemistry following emissions of precursors such as NO_x, CO, and NMHCs (see Section 6.6 and Chapter 4). Indirect effects will be described below for a number of key gases.

It is important to distinguish between the integrated relative effect of an emitted kilogram of gas which is represented by a GWP and the actual radiative forcings for specific gas amounts presented, for example, in Section 6.3 and in Figure 6.6. GWPs are intended for use in studying relative rather than absolute impacts of emissions, and pertain to specific time horizons.

The radiative efficiencies a_r and a_x are not necessarily constant over time. While the absorption of infrared radiation by many greenhouse gases varies linearly with their abundance, a few important ones display non-linear behaviour for current and likely future abundances (e.g., CO_2, CH_4, and N_2O). For those gases, the relative radiative forcing will depend upon abundance and hence upon the future scenario adopted. These issues were discussed in detail and some sensitivities to chosen scenarios were presented in IPCC (1994).

A key aspect of GWP calculations is the choice of the reference gas, taken here to be CO_2. In IPCC (1994), it was shown, for example, that a particular scenario for future growth of CO_2 (S650, see Chapter 3) would change the denominator of Equation 6.2 by as much as 15% compared to a calculation employing constant pre-industrial CO_2 mixing ratios.

The atmospheric response time of CO_2 is subject to substantial scientific uncertainties, due to limitations in our knowledge of key processes including its uptake by the biosphere and ocean. When CO_2 is used as the reference, the numerical values of the GWPs of all greenhouse gases can change substantially as research improves the understanding of the removal processes of CO_2. The removal function for CO_2 used for the GWPs presented here is based upon carbon cycle models such as those discussed in Chapter 3. The CO_2 radiative efficiency (a_r) used in this report has been updated since the SAR, as discussed in Section 6.3 (see below).

The lifetimes of non-CO_2 greenhouse gases are dependent largely on atmospheric photochemistry, which controls photolysis and related removal processes as discussed in Chapter 4. When the lifetime of the gas in question is comparable to the response time of CO_2 (nominally about 150 years, although it is clear that the removal of CO_2 cannot be adequately described by a single, simple exponential lifetime; see IPCC (1994) and the discussion below), the GWP is relatively insensitive to choice of time horizon, i.e., for N_2O. When the lifetime of the gas in question differs substantially from the response time of the reference gas, the GWP becomes sensitive to the choice of time horizon, which in turn implies a decision regarding the climate processes and impacts of interest, as noted above. For longer time horizons, those species that decay more rapidly than the reference gas display decreasing GWPs, with the slope of the decay being dependent mainly on the lifetime of the gas in question. Gases with lifetimes much longer than that of the reference gas (e.g., C_2F_6) display increasing GWPs over long time horizons (i.e., greater than a hundred years). We emphasise that the GWP is an integral from zero to the chosen time horizon; hence the values presented in the table for 25, 100, and 500 years are not additive.

A number of studies have suggested modified or different indices for evaluating relative future climate effects. Here we provide only an indication of the kinds of issues that are being considered in these alternative indices. Wuebbles and Calm (1997) emphasised the fact that some halocarbon substitutes used, for example, in refrigeration are less efficient than the CFCs they are replacing, so that more energy is consumed through their use and hence more CO_2 emitted per hour of operation. Evaluation of these technological factors would only be possible through detailed emission inventories coupled with scenarios, and are not included here. Some studies have argued for use of "discount rates" to reflect increasing uncertainty and

changing policies with time (e.g., to account for the possibility that new technologies will emerge to solve problems; see for example Lashof and Ahuja, 1990; Reilly *et al.*, 1999). Economic factors could be considered along with the technological ones mentioned above, adding another aspect to any scenario. These are not included here. Smith and Wigley (2000a,b), Fuglestvedt *et al.* (2000), and Reilly *et al.* (1999) have examined the relationship between GWPs and climate response using simple energy balance models. These studies emphasised the point made above regarding the links between choice of time horizon, lifetime of a particular substance, and the climate response of interest, noting for example that while the 100-year GWP for N_2O represented the model-calculated temperature responses (both instantaneous and integrated over time), and the calculated sea level rise to good accuracy, that for CH_4 (a short-lived gas with a lifetime of about 10 years) represented sea level rise far better than it did the instantaneous temperature change for that time horizon. For very short-lived gases, GWPs are often calculated using a 'slab' (continuous) rather than pulse emission. This approach involves the assumption of specific scenarios for the magnitudes of the slabs. For some very short-lived gases such as NMHCs, models must employ 'slab' emissions in GWP analyses since their impact depends critically on non-linear coupled chemical processes. Some examples are given in Sections 6.12.3.2 and 6.12.3.4.

6.12.2 Direct GWPs

The CO_2 response function used in this report is the same as that in WMO (1999) and the SAR and is based on the "Bern" carbon cycle model (see Siegenthaler and Joos, 1992; Joos *et al.*, 1996) run for a constant future mixing ratio of CO_2 over a period of 500 years. The Bern carbon cycle model was compared to others in IPCC (1994), where it was shown that different models gave a range of as much as 20% in the CO_2 response, with the greatest differences occurring over time-scales greater than 20 years.

The radiative efficiency per kilogram of CO_2 has been updated compared to previous IPCC assessments (IPCC, 1994; SAR). Here we employ the approach discussed in WMO (1999) using the simplified formula presented in Table 6.2. We assume a background CO_2 mixing ratio of 364 ppmv, close to the present day value (WMO (1995) used 354 ppmv). For this assumption, this expression agrees well with the adjusted total-sky radiative forcing calculations of Myhre and Stordal (1997); see also Myhre *et al.* (1998b). The revised forcing is about 12% lower than that in the SAR. For a small perturbation in CO_2 from 364 ppmv, the radiative efficiency is 0.01548 Wm^{-2}/ppmv. This value is used in the GWP calculations presented here. We emphasise that it applies only to GWP calculations and cannot be used to obtain the total radiative forcing for this key gas since pre-industrial times, due to time-dependent changes in mixing ratio as noted above. Because of this change in the CO_2 forcing per mass, the CO_2 AGWPs are 0.207, 0.696, and 2.241 Wm^{-2}/yr/ppmv for 20, 100, and 500 year time horizons, respectively. These are smaller than the values used in the SAR by 13%. AGWPs for any gas can be obtained from the GWP values given in the tables presented here by multiplying by these numbers.

The decreases in the CO_2 AGWPs will lead to proportionately larger GWPs for other gases compared to previous IPCC assessments, in the absence of other changes. In Tables 6.7 and 6.8, GWPs (on a mass basis) for 93 gases are tabulated for time horizons of 20, 100, and 500 years. The list includes CH_4, N_2O, CFCs, HCFCs, HFCs, hydrochlorocarbons, bromocarbons, iodocarbons, fully fluorinated species, fluoroalcohols, and fluoroethers. The radiative efficiencies per kilogram were derived from the values given per ppbv in Section 6.3. Several of these have been updated since the SAR, most notably that of CFC-11. Since the radiative forcings of the halocarbon replacement gases are scaled relative to CFC-11 in GWP calculations, the GWPs of those gases are also affected by this change. As discussed further in Section 6.3, the change in radiative forcing for CFC-11 reflects new studies (Pinnock *et al.,* 1995; Christidis *et al.,* l997; Hansen *et al.,* l997a; Myhre and Stordal, 1997; Good *et al.,* 1998) suggesting that the radiative forcing of this gas is about 0.25 Wm^{-2} $ppbv^{-1}$, an increase of about 14% compared to the value adopted in earlier IPCC reports, which was based on the study of Hansen *et al.* (l988).

The lifetimes and adjustment times used in Tables 6.7 and 6.8 come from Chapter 4 except where noted. For some gases (including several of the fluoroethers), lifetimes have not been derived from laboratory measurements, but have been estimated by various other means. For this reason, the lifetimes for these gases, and hence the GWPs, are considered to be much less reliable, and so these gases are listed separately in Table 6.8. NF_3 is listed in Table 6.8 because, although its photolytic destruction has been characterised, other loss processes may be significant but have not yet been characterised (Molina *et al.,* 1995). Note also that some gases, for example, trifluoromethyl iodide (CF_3I) and dimethyl ether (CH_3OCH_3) have very short lifetimes (less than a few months); GWPs for such very short-lived gases may need to be treated with caution, because the gases are unlikely to be evenly distributed globally, and hence estimates of, for example, their radiative forcing using global mean conditions may be subject to error.

Uncertainties in the lifetimes of CFC-11 and CH_3CCl_3 are thought to be about 10%, while uncertainties in the lifetimes of gases obtained relative to CFC-11 or CH_3CCl_3 are somewhat larger (20 to 30%) (SAR; WMO, l999). Uncertainties in the radiative forcing per unit mass of the majority of the gases considered in Table 6.7 are approximately ± 10%. The SAR suggested typical uncertainties of ± 35% (relative to the reference gas) for the GWPs, and we retain this uncertainty estimate for gases listed in Table 6.7. In addition to uncertainties in the CO_2 radiative efficiency per kilogram and in the response function, AGWPs of CO_2 are affected by assumptions concerning future CO_2 abundances as noted above. Furthermore, as the CO_2 mixing ratios and climate change, the pulse response function changes as well. In spite of these dependencies on the choice of future emission scenarios, it remains likely that the error introduced by these assumptions is smaller than the uncertainties introduced by our imperfect understanding of the carbon cycle (see Chapter 3). Finally, although any induced error in the CO_2 AGWPs will certainly affect the non-CO_2 GWPs, it will not affect intercomparisons among non-CO_2 GWPs.

6.12.3 Indirect GWPs

We next consider discuss indirect effects in more detail, and present GWPs for other gases, including estimates of their impacts. While direct GWPs are usually believed to be known reasonably accurately (±35%), indirect GWPs can be highly uncertain. A number of different processes contribute to indirect effects for various molecules; many of these are also discussed in Section 6.6.

6.12.3.1 Methane
Four types of indirect effects due to the presence of atmospheric CH_4 have been identified (see Chapter 4 and Section 6.6). The largest effect is potentially the production of O_3 (25% of the direct effect, or 19% of the total, as in the SAR). This effect is difficult to quantify, however, because the magnitude of O_3 production is highly dependent on the abundance and distribution of NO_x (IPCC, 1994; SAR). Other indirect effects include the production of stratospheric water vapour (assumed here to represent 5% of the direct effect, or 4% of the total, as in the SAR), the production of CO_2 (from certain CH_4 sources), and the temporal changes in the CH_4 adjustment time resulting from its coupling with OH (Lelieveld and Crutzen, 1992; Brühl, 1993; Prather, 1994, 1996; SAR; Fuglestvedt *et al.,* 1996). Here we adopt the values for each of these terms as given in the SAR, with a correction for the updated CO_2 AGWPs and adopting the perturbation lifetime given in Chapter 4. It should be noted that the climate forcing caused by CO_2 produced from the oxidation of CH_4 is not included in these GWP estimates. As discussed in the SAR, it is often the case that this CO_2 is included in national carbon production inventories. Therefore, depending on how the inventories are combined, including the CO_2 production from CH_4 could result in double counting this CO_2.

6.12.3.2 Carbon monoxide
CO has a small direct GWP but leads to indirect radiative effects that are similar to those of CH_4. As in the case of CH_4, the production of CO_2 from oxidised CO can lead to double counting of this CO_2 and is therefore not considered here. The emission of CO perturbs OH, which in turn can then lead to an increase in the CH_4 lifetime (Fuglestvedt *et al.,* l996; Prather, 1996; Daniel and Solomon, 1998). This term involves the same processes whereby CH_4 itself influences its own lifetime and hence GWP values (Prather, l996) and is subject to similar uncertainty. This term can be evaluated with reasonable accuracy using a box model, as shown by Prather (l996) and Daniel and Solomon (1998). Emissions of CO can also lead to the production of O_3 (see Chapter 4), with the magnitude of O_3 formation dependent on the amount of NO_x present. As with CH_4, this effect is quite difficult to quantify due to the highly variable and uncertain NO_x distribution (e.g., Emmons *et al.,* 1997). Because of the difficulty in accurately calculating the amount of O_3 produced by CO emissions, an accurate estimate of the entire indirect forcing of CO requires a three-dimensional chemical model. Table 6.9 presents estimates of the CO GWP due to O_3 production and to feedbacks on the CH_4 cycle from two recent multi-dimensional model studies (in which a "slab" emission of CO was imposed),

Table 6.7: *Direct Global Warming Potentials (mass basis) relative to carbon dioxide (for gases for which the lifetimes have been adequately characterised).*

Gas		Radiative efficiency $(Wm^{-2}ppb^{-1})$ (from (b) unless indicated)	Lifetime (years) (from Chapter 4 unless indicated)	Global Warming Potential Time horizon		
				20 years	100 years	500 years
Carbon dioxide	CO_2	See Section 6.12.2	See Section 6.12.2	1	1	1
Methane	CH_4	$3.7 \times 10^{-4\,\Sigma}$	12.0*	62	23	7
Nitrous oxide	N_2O	$3.1 \times 10^{-3\,\Sigma}$	114*	275	296	156
Chlorofluorocarbons						
CFC-11	CCl_3F	0.25	45	6300	4600	1600
CFC-12	CCl_2F_2	0.32	100	10200	10600	5200
CFC-13	$CClF_3$	0.25	640 (c)	10000	14000	16300
CFC-113	CCl_2FCClF_2	0.30	85	6100	6000	2700
CFC-114	$CClF_2CClF_2$	0.31	300	7500	9800	8700
CFC-115	CF_3CClF_2	$0.18^\dagger$	1700	4900	7200	9900
Hydrochlorofluorocarbons						
HCFC-21	$CHCl_2F$	0.17	2.0 (d)	700	210	65
HCFC-22	$CHClF_2$	$0.20^\S$	11.9	4800	1700	540
HCFC-123	CF_3CHCl_2	0.20	1.4 (a)	390	120	36
HCFC-124	CF_3CHClF	0.22	6.1 (a)	2000	620	190
HCFC-141b	CH_3CCl_2F	0.14	9.3	2100	700	220
HCFC-142b	CH_3CClF_2	0.20	19	5200	2400	740
HCFC-225ca	$CF_3CF_2CHCl_2$	0.27	2.1 (a)	590	180	55
HCFC-225cb	$CClF_2CF_2CHClF$	0.32	6.2 (a)	2000	620	190
Hydrofluorocarbons						
HFC-23	CHF_3	$0.16^\S$	260	9400	12000	10000
HFC-32	CH_2F_2	$0.09^\S$	5.0	1800	550	170
HFC-41	CH_3F	0.02	2.6	330	97	30
HFC-125	CHF_2CF_3	$0.23^\S$	29	5900	3400	1100
HFC-134	CHF_2CHF_2	0.18	9.6	3200	1100	330
HFC-134a	CH_2FCF_3	$0.15^\S$	13.8	3300	1300	400
HFC-143	CHF_2CH_2F	0.13	3.4	1100	330	100
HFC-143a	CF_3CH_3	$0.13^\S$	52	5500	4300	1600
HFC-152	CH_2FCH_2F	0.09	0.5	140	43	13
HFC-152a	CH_3CHF_2	$0.09^\S$	1.4	410	120	37
HFC-161	CH_3CH_2F	0.03	0.3	40	12	4
HFC-227ea	CF_3CHFCF_3	0.30	33.0	5600	3500	1100
HFC-236cb	$CH_2FCF_2CF_3$	0.23	13.2	3300	1300	390
HFC-236ea	CHF_2CHFCF_3	0.30	10.0	3600	1200	390
HFC-236fa	$CF_3CH_2CF_3$	0.28	220	7500	9400	7100
HFC-245ca	$CH_2FCF_2CHF_2$	0.23	5.9	2100	640	200
HFC-245fa	$CHF_2CH_2CF_3$	$0.28^\&$	7.2	3000	950	300
HFC-365mfc	$CF_3CH_2CF_2CH_3$	0.21 (k)	9.9	2600	890	280
HFC-43-10mee	$CF_3CHFCHFCF_2CF_3$	0.40	15	3700	1500	470
Chlorocarbons						
CH_3CCl_3		0.06	4.8	450	140	42
CCl_4		$0.13^{\dagger\dagger}$	35	2700	1800	580
$CHCl_3$		$0.11^\S$	0.51 (a)	100	30	9
CH_3Cl		0.01	1.3 (b)	55	16	5
CH_2Cl_2		0.03	0.46 (a)	35	10	3

Gas		Radiative Efficiency (Wm^{-2} ppb^{-1}) (from (b) unless indicated)	Lifetime (years) (from Chapter 4 unless indicated)	Global Warming Potential Time horizon		
				20 years	100 years	500 years
Bromocarbons						
CH$_3$Br		0.01	0.7 (b)	16	5	1
CH$_2$Br$_2$		0.01	0.41 (i)	5	1	<<1
CHBrF$_2$		0.14	7.0 (i)	1500	470	150
Halon-1211	CBrClF$_2$	0.30	11	3600	1300	390
Halon-1301	CBrF$_3$	0.32	65	7900	6900	2700
Iodocarbons						
FIC-13I1	CF$_3$I	0.23	0.005 (a)	1	1	<<1
Fully fluorinated species						
SF$_6$		0.52	3200	15100	22200	32400
CF$_4$		0.08	50000	3900	5700	8900
C$_2$F$_6$		0.26§	10000	8000	11900	18000
C$_3$F$_8$		0.26	2600	5900	8600	12400
C$_4$F$_{10}$		0.33	2600	5900	8600	12400
c-C$_4$F$_8$		0.32§	3200	6800	10000	14500
C$_5$F$_{12}$		0.41	4100	6000	8900	13200
C$_6$F$_{14}$		0.49	3200	6100	9000	13200
Ethers and Halogenated Ethers						
CH$_3$OCH$_3$		0.02	0.015 (e)	1	1	<<1
(CF$_3$)$_2$CFOCH$_3$		0.31	3.4 (l)	1100	330	100
(CF$_3$)CH$_2$OH		0.18	0.5 (m)	190	57	18
CF$_3$CF$_2$CH$_2$OH		0.24	0.4 (m)	140	40	13
(CF$_3$)$_2$CHOH		0.28	1.8 (m)	640	190	59
HFE-125	CF$_3$OCHF$_2$	0.44	150	12900	14900	9200
HFE-134	CHF$_2$OCHF$_2$	0.45	26.2	10500	6100	2000
HFE-143a	CH$_3$OCF$_3$	0.27	4.4	2500	750	230
HCFE-235da2	CF$_3$CHClOCHF$_2$	0.38	2.6 (i)	1100	340	110
HFE-245cb2	CF$_3$CF$_2$OCH$_3$	0.32	4.3 (l)	1900	580	180
HFE-245fa2	CF$_3$CH$_2$OCHF$_2$	0.31	4.4 (i)	1900	570	180
HFE-254cb2	CHF$_2$CF$_2$OCH$_3$	0.28	0.22 (h)	99	30	9
HFE-347mcc3	CF$_3$CF$_2$CF$_2$OCH$_3$	0.34	4.5 (l)	1600	480	150
HFE-356pcf3	CHF$_2$CF$_2$CH$_2$OCHF$_2$	0.39	3.2 (n)	1500	430	130
HFE-374pcf2	CHF$_2$CF$_2$OCH$_2$CH$_3$	0.25	5.0 (n)	1800	540	170
HFE-7100	C$_4$F$_9$OCH$_3$	0.31	5.0 (f)	1300	390	120
HFE-7200	C$_4$F$_9$OC$_2$H$_5$	0.30$^\Omega$	0.77 (g)	190	55	17
H-Galden 1040x	CHF$_2$OCF$_2$OC$_2$F$_4$OCHF$_2$	1.37(j)$^\Omega$	6.3$^\Omega$	5900	1800	560
HG-10	CHF$_2$OCF$_2$OCHF$_2$	0.66$^\Omega$	12.1$^\Omega$	7500	2700	850
HG-01	CHF$_2$OCF$_2$CF$_2$OCHF$_2$	0.87$^\Omega$	6.2$^\Omega$	4700	1500	450

* The values for CH$_4$ and N$_2$O are adjustment times including feedbacks of emission on lifetimes (see Chapter 4).

Σ From the formulas given in Table 6.2, with updated constants based on the IPCC (1990) expressions.

Note: For all gases destroyed by reaction with OH, updated lifetimes include scaling to CH$_3$CCl$_3$ lifetimes, as well as an estimate of the stratospheric destruction. See references below for rates along with Chapter 4 and WMO (1999).

(a) Taken from the SAR (b) Taken from WMO (1999) (c) Taken from WMO (1995) (d) DeMore *et al.* (1997)
(e) Good *et al.* (1998) (f) Wallington *et al.* (1997) (g) Christensen *et al.* (1998) (h) Heathfield *et al.* (1998a)
(i) Christidis *et al.* (1997) (j) Gierczak *et al.* (1996) (k) Barry *et al.* (1997) (l) Tokuhashi *et al.* (1999a)
(m) Tokuhashi *et al.* (1999b) (n) Tokuhashi *et al.* (2000)

† Myhre *et al.* (1998b) †† Jain *et al.* (2000) § Highwood and Shine (2000) $^\&$ Ko *et al.* (1999)
$^\Omega$ See Cavalli *et al.* (1998) and Myhre *et al.* (1999)

Table 6.8: *Direct Global Warming Potentials (mass basis) relative to carbon dioxide (for gases for whose lifetime has been determined only via indirect means, rather than laboratory measurements, or for whom there is uncertainty over the loss processes). Radiative efficiency is defined with respect to all sky.*

Gas		Radiative efficiency $(Wm^{-2}ppb^{-1})$ (from (b) unless indicated)	Estimated lifetime (years)	Global Warming Potential Time horizon		
				20 years	100 years	500 years
NF_3		0.13	740 (a)	7700	10800	13100
SF_5CF_3		0.57 (d)	>1000*	>12200	>17500	>22500
$c-C_3F_6$		0.42	>1000*	>11800	>16800	>21600
HFE-227ea	$CF_3CHFOCF_3$	0.40	11 (c)	4200	1500	460
HFE-236ea2	$CF_3CHFOCHF_2$	0.44	5.8 (c)	3100	960	300
HFE-236fa	$CF_3CH_2OCF_3$	0.34	3.7 (c)	1600	470	150
HFE-245fa1	$CHF_2CH_2OCF_3$	0.30	2.2 (c)	940	280	86
HFE-263fb2	$CF_3CH_2OCH_3$	0.20	0.1 (c)	37	11	3
HFE-329mcc2	$CF_3CF_2OCF_2CHF_2$	0.49	6.8 (c)	2800	890	280
HFE-338mcf2	$CF_3CF_2OCH_2CF_3$	0.43	4.3 (c)	1800	540	170
HFE-347mcf2	$CF_3CF_2OCH_2CHF_2$	0.41	2.8 (c)	1200	360	110
HFE-356mec3	$CF_3CHFCF_2OCH_3$	0.30	0.94 (c)	330	98	30
HFE-356pcc3	$CHF_2CF_2CF_2OCH_3$	0.33	0.93 (c)	360	110	33
HFE-356pcf2	$CHF_2CF_2OCH_2CHF_2$	0.37	2.0 (c)	860	260	80
HFE-365mcf3	$CF_3CF_2CH_2OCH_3$	0.27	0.11 (c)	38	11	4
$(CF_3)_2CHOCHF_2$		0.41	3.1 (c)	1200	370	110
$(CF_3)_2CHOCH_3$		0.30	0.25 (c)	88	26	8
$-(CF_2)_4CH(OH)-$		0.30	0.85 (c)	240	70	22

(a) Molina *et al.* (1995).
(b) WMO (1999).
(c) Imasu *et al.* (1995).
(d) Sturges *et al.* (2000).
* Estimated lower limit based upon perfluorinated structure.

along with the box-model estimate for the latter term alone from Daniel and Solomon (1998), which is based on the analytical formalism developed by Prather (1996). Table 6.9 shows that the 100-year GWP for CO is likely to be 1.0 to 3.0, while that for shorter time horizons is estimated at 2.8 to 10. These estimates are subject to large uncertainties, as discussed further in Chapter 4.

6.12.3.3 Halocarbons

In addition to their direct radiative forcing, chlorinated and brominated halocarbons can lead to a significant indirect forcing through their destruction of stratospheric O_3 (Section 6.4). By destroying stratospheric O_3, itself a greenhouse gas, halocarbons induce a negative indirect forcing that counteracts some or perhaps all (in certain cases) of their direct forcing. Furthermore, decreases in stratospheric O_3 act to increase the ultraviolet field of the troposphere and hence can increase OH and deplete those gases destroyed by reaction with the OH radical (particularly CH_4); this provides an additional negative forcing. Quantifying the magnitude of the negative indirect forcing is quite difficult for several reasons. As discussed in Section 6.4, the negative forcing

arising from the O_3 destruction is highly dependent on the altitude profile of the O_3 loss. The additional radiative effect due to enhanced tropospheric OH is similarly difficult to quantify (see e.g., WMO, 1999). While recognising these uncertainties, estimates have been made of the net radiative forcing due to particular halocarbons, which can then be used to determine net GWPs (including both direct and indirect effects). This was done by Daniel *et al.* (1995), where it was shown that if the enhanced tropospheric OH effect were ignored, and the negative forcing due to O_3 loss during the 1980s was -0.08 Wm^{-2}, the net GWPs for the bromocarbons were significantly negative, illustrating the impact of the negative forcing arising from the bromocarbon-induced ozone depletion. While the effect on the chlorocarbon GWPs was less pronounced, it was significant as well. Table 6.10 updates the results from Daniel *et al.*'s "constant-alpha" case A as in WMO (1999), where the effectiveness of bromine for O_3 loss relative to chlorine (called alpha) has been increased from 40 to 60. The updated radiative efficiency of CO_2 has also been included. An uncertainty in the 1980 to 1990 O_3 radiative forcing of -0.03 to -0.15 Wm^{-2} has been adopted based upon Section

Table 6.9: *Estimated indirect Global Warming Potentials for CO for time horizons of 20, 100, and 500 years.*

	Indirect Global Warming Potentials Time horizon		
	20 years	100 years	500 years
Daniel and Solomon (1998): box model considering CH_4 feedbacks only	2.8	1.0	0.3
Fuglestvedt *et al.* (1996): two-dimensional model including CH_4 feedbacks and tropospheric O_3 production by CO itself	10	3.0	1.0
Johnson and Derwent (1996): two-dimensional model including CH_4 feedbacks and tropospheric O_3 production by CO itself	—	2.1	—

Table 6.10: *Net Global Warming Potentials (mass basis) of selected halocarbons (updated from Daniel et al., 1995; based upon updated stratospheric O_3 forcing estimates, lifetimes, and radiative data from this report).*

Species	Time horizon = 2010 (20 years)			Time horizon = 2090 (100 years)		
	Direct	Min	Max	Direct	Min	Max
CFC-11	6300	100	5000	4600	−600	3600
CFC-12	10200	7100	9600	10600	7300	9900
CFC-113	6100	2400	5300	6000	2200	5200
HCFC-22	4800	4100	4700	1700	1400	1700
HCFC-123	390	100	330	120	20	100
HCFC-124	2000	1600	1900	620	480	590
HCFC-141b	2100	180	1700	700	−5	570
HCFC-142b	5200	4400	5100	2400	1900	2300
$CHCl_3$	450	−1800	10	140	−560	0
CCl_4	2700	−4700	1300	1800	−3900	660
CH_3Br	16	−8900	−1700	5	−2600	−500
Halon-1211	3600	−58000	−8600	1300	−24000	−3600
Halon-1301	7900	−79000	−9100	6900	−76000	−9300

6.4, and these correspond (respectively) to the maximum and minimum GWP estimates given in Table 6.10.

6.12.3.4 NO_x and non-methane hydrocarbons

The short lifetimes and complex non-linear chemistries of NO_x and NMHC make calculation of their indirect GWPs a challenging task subject to very large uncertainties (see Chapter 4). However, IPCC (1999) has probed in detail the issue of the relative differences in the impacts of NO_x upon O_3 depending on where it is emitted (in particular, surface emissions versus those from aircraft). Higher altitude emissions have greater impacts both because of longer NO_x residence times and more efficient tropospheric O_3 production, as well as enhanced radiative forcing sensitivity (see Section 6.5). Two recent two-dimensional model studies (Fuglestvedt *et al.*, 1996; Johnson and Derwent, 1996) have presented estimates of the GWPs for NO_x emitted from aircraft. These studies suggest GWPs of the order of 450 for aircraft NO_x emissions considering a 100-year time horizon, while those for surface emissions are likely to be much smaller, of the order of 5. While such numerical values are subject to very large quantitative uncertainties, they illustrate that the emissions of NO_x from aircraft are characterised by far greater GWPs than those of surface sources, due mainly to the longer lifetime of the emitted NO_x at higher altitudes.

6.13 Global Mean Radiative Forcings

6.13.1 Estimates

The global, annual mean radiative forcing estimates from 1750 to the present (late 1990s; about 2000) for the different agents are plotted in Figure 6.6, based on the discussions in the foregoing sections. As in the SAR, the height of the rectangular bar denotes a central or best estimate of the forcing, while the vertical line about the bar is an estimate of the uncertainty range, guided by the spread in the published results and physical understanding, and with no statistical connotation. The uncertainty range, as employed in this chapter, is not the product of systematic quantitative analyses of the various factors associated with the forcing, and thus lacks a rigorous statistical basis. The usage here is different from the manner "uncertainty range" is defined and addressed elsewhere in this document. The SAR had also stated a "confidence level" which represented a subjective judgement that the actual forcing would lie within the specified uncertainty range. In order to avoid the confusion over the use of the term "confidence level", we introduce in this assessment a "level of scientific understanding" (LOSU) that represents, again, a subjective judgement and expresses somewhat similar notions as in the SAR (refer also to IPCC, 1999). The LOSU index for each

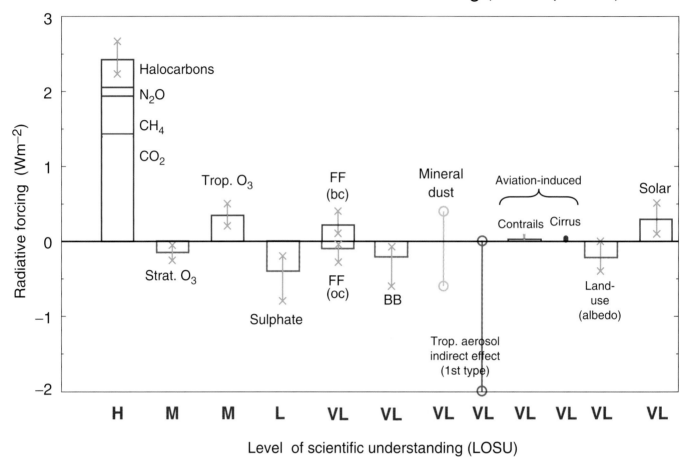

Figure 6.6: Global, annual mean radiative forcings (Wm^{-2}) due to a number of agents for the period from pre-industrial (1750) to present (late 1990s; about 2000) (numerical values are also listed in Table 6.11). For detailed explanations see Section 6.13. The height of the rectangular bar denotes a central or best estimate value while its absence denotes no best estimate is possible. The vertical line about the rectangular bar with "x" delimiters indicates an estimate of the uncertainty range, guided by the spread in the published values of the forcing and physical understanding. A vertical line without a rectangular bar and with "o" delimiters denotes a forcing for which no central estimate can be given owing to large uncertainties. The uncertainty range specified here has no statistical basis and therefore differs from the use of the term elsewhere in this document. A "level of scientific understanding" (LOSU) index is accorded to each forcing, with H, M, L and VL denoting high, medium, low and very low levels, respectively. This represents our subjective judgement about the reliability of the forcing estimate, involving factors such as the assumptions necessary to evaluate the forcing, the degree of our knowledge of the physical/chemical mechanisms determining the forcing, and the uncertainties surrounding the quantitative estimate of the forcing (see Table 6.12). The well-mixed greenhouse gases are grouped together into a single rectangular bar with the individual mean contributions due to CO_2, CH_4, N_2O, and halocarbons (see Tables 6.1 and 6.11) shown; halocarbons refers to all halogen-containing compounds listed in Table 6.1. "FF" denotes fossil fuel burning while "BB" denotes biomass burning aerosol. Fossil fuel burning is separated into the "black carbon" (bc) and "organic carbon" (oc) components with its separate best estimate and range. The sign of the effects due to mineral dust is itself an uncertainty. The indirect forcing due to tropospheric aerosols is poorly understood. The same is true for the forcing due to aviation via their effects on contrails and cirrus clouds. Only the *first* type of indirect effect due to aerosols as applicable in the context of liquid clouds is considered here. The *second* type of effect is conceptually important but there exists very little confidence in the simulated quantitative estimates. The forcing associated with stratospheric aerosols from volcanic eruptions is highly variable over the period and is not considered for this plot (however, see Figure 6.8d). All the forcings shown have distinct spatial and seasonal features (Figure 6.7) such that the global, annual means appearing on this plot do not yield a complete picture of the radiative perturbation. They are only intended to give, in a relative sense, a first-order perspective on a global, annual mean scale, and cannot be readily employed to obtain the climate response to the total natural and/or anthropogenic forcings. As in the SAR, it is emphasised that the positive and negative global mean forcings cannot be added up and viewed a priori as providing offsets in terms of the complete global climate impact.

Table 6.11: *Numerical values of the global and annual mean forcings from 1850 to about the early 1990s as presented in the SAR, and from 1750 to present (about 2000) as presented in this report. The estimate for the well-mixed greenhouse gases is partitioned into the contributions from CO₂, CH₄, N₂O, and halocarbons. An approximate estimate of the Northern Hemisphere (NH) to Southern Hemisphere (SH) ratio is also given for the present report (see also Figure 6.6). The uncertainty about the central estimate (if applicable) is listed in square brackets. No uncertainty is estimated for the NH/SH ratio.*

	Global mean radiative forcing (Wm^{-2}) [Uncertainty]		NH/SH ratio
	SAR	**This Report**	**This Report**
Well-mixed greenhouse gases {Comprising CO_2, CH_4, N_2O, and halocarbons}	+2.45 [15%] {CO_2 (1.56); CH_4 (0.47); N_2O (0.14); Halocarbons (0.28)}	+2.43 [10%] {CO_2 (1.46); CH_4 (0.48); N_2O (0.15); Halocarbons (0.34)}	1
Stratospheric O_3	−0.10 [2X]	−0.15 [67%]	<1
Tropospheric O_3	+0.40 [50%]	+0.35 [43%]	>1
Direct sulphate aerosols	−0.40 [2X]	−0.40 [2X]	>>1
Direct biomass burning aerosols	−0.20 [3X]	−0.20 [3X]	<1
Direct FF aerosols (BC)	+0.10 [3X]	+0.20 [2X]	>>1
Direct FF aerosols (OC)	*	−0.10 [3X]	>>1
Direct mineral dust aerosols	*	−0.60 to +0.40	*
Indirect aerosol effect	0 to −1.5 {sulphate aerosols}	0 to −2.0 {1st effect only; all aerosols}	>1
Contrails Aviation-induced cirrus	* *	0.02 [~3.5 X] 0 to 0.04	>>1 *
Land-use (albedo)	*	−0.20 [100%]	>>1
Solar	+0.30 [67%]	+0.30 [67%]	1

* No estimate given.

forcing agent is based on an assessment of the nature of assumptions involved, the uncertainties prevailing about the processes that govern the forcing, and the resulting confidence in the numerical value of the estimate. The subjectivity reflected in the LOSU index is unavoidable and is necessitated by the lack of sufficient quantitative information on the uncertainties, especially for the non-well-mixed greenhouse gas forcing mechanisms. In the case of some forcings, this is in part due to a lack of enough investigations. Thus, the application of rigorous statistical methods to quantify the uncertainties of all of the forcing agents in a uniform manner is not possible at present.

The discussions below relate to the changes with respect to the SAR estimates. In many respects, there is a similarity between the estimates, range and understanding levels listed here, and those stated in the recent studies of Hansen *et al.* (1998) and Shine and Forster (1999). Table 6.11 compares the numerical values with the estimates in the SAR. Also, the Northern to Southern Hemisphere ratio is shown for the present estimates (see also Section 6.14). Table 6.12 summarises the principal aspects known regarding the forcings, along with a brief listing of the key uncertainties in the processes which, in turn, lead to uncertainties in and affect the reliability of the quantitative estimates.

The total forcing estimate for well-mixed greenhouse gases is slightly less now, by about 1% (see Section 6.3) compared to the estimate given in the SAR. The uncertainty range remains

quite small and these estimates retain a "high" LOSU. This forcing continues to enjoy the highest confidence amongst the different natural and anthropogenic forcings.

The estimate for stratospheric O_3 has increased in magnitude, owing mainly to the inclusion of observed ozone depletions through mid-1995 and beyond. It is an encouraging feature that several different model calculations yield similar estimates for the forcing. The uncertainty range remains similar to that given in the SAR. These arguments suggest an elevation of the confidence in the forcing estimate relative to the SAR. Accordingly, a "medium" LOSU is assigned here. A still higher elevation of the rank is precluded because the O_3 loss profile near tropopause continues to be an uncertainty that is significant and that has not been adequately resolved. Also, the global stratospheric temperature change calculations involved in the forcing determination are not quantitatively identical to the observed changes, which in turn, affects, the precision of the forcing estimate.

The estimate for tropospheric O_3 (0.35 ± 0.15 Wm^{-2}) is on firmer grounds now than in the SAR. Since that assessment, many different models have been employed to compute the forcing, including one analysis constrained by observations. These have resulted in a narrowing of the uncertainty range and increased the confidence with regards to the central estimate. The preceding argument strongly suggests an advancement of the confidence in this forcing estimate. Hence, a "medium" LOSU is

Table 6.12: *Summary of the known inputs (i.e., very good knowledge available), and major uncertainties (i.e., key limitations) that affect the quantitative estimates of the global and annual mean radiative forcing of climate change due to the agents listed in Figure 6.6. The summary forms a basis for assigning a level of scientific understanding (LOSU) rank to each agent (H=High, M=Medium, L=Low, and VL=Very low; see Section 6.13).*

	Known inputs	**Major uncertainties**	**Overall rank**
Well-mixed greenhouse gases	Concentrations; spectroscopy		H
Stratospheric O$_3$	Global observations of column change; observations of profiles information at many sites; spectroscopy; qualitative observational evidence of global stratospheric cooling	Change before 1970s; profile of change near tropopause; quantitative attribution of observed stratospheric temperature change	M
Tropospheric O$_3$	Surface observations from many sites since about 1960s; near-global data on present day column; limited local data on vertical distribution; spectroscopy	Emissions, chemistry and transport of precursors and O$_3$; profile near tropopause; lack of global data on pre-industrial levels	M
Sulphate aerosols (direct effect)	Pre-industrial and present source regions and strengths; chemical transformation and water uptake; deposition in some regions; observational evidence of aerosol presence	Transport and chemistry of precursors; aerosol microphysics; optical properties and vertical distribution; cloud distributions; lack of quantitative global observations of distributions and/or forcing	L
Other aerosols (direct effect)	Source regions; some observational evidence of aerosol presence	Pre-industrial and present source strengths; in-cloud chemistry and water uptake; aerosol microphysics; optical properties and vertical distributions; cloud distributions; lack of quantitative global data on forcing	VL
Aerosols (indirect; 1st type)	Evidence for phenomenon in ship tracks; measurement of variations in cloud droplet size from satellite and field observations	Quantification of aerosol-cloud interactions; model simulation of aerosol and cloud distributions; difficulty in evaluation from observations; lack of global measurements; optical properties of mixtures; pre-industrial aerosol concentration and properties	VL
Contrails and aviation-induced cirrus	Air traffic patterns; contrail formation; cirrus clouds presence	Ice microphysics and optics; geographical distributions; quantification of induced-cirrus cloudiness	VL
Land use (albedo)	Present day limited observations of deforestation	Human and natural effects on vegetation since 1750; lack of quantitative information, including separation of natural and anthropogenic changes	VL
Solar	Variations over last 20 years; information on Sun-like stars; proxy indicators of solar activity	Relation between proxies and total solar irradiance; induced changes in O$_3$; effects in troposphere; lack of quantitative information going back more than 20 years; cosmic rays and atmospheric feedbacks	VL

accorded for tropospheric O_3 forcing. Key uncertainties remain concerning the pre-industrial distributions, the effects of stratospheric-tropospheric exchange and the manner of its evolution over time, as well as the seasonal cycle in some regions of the globe.

As the LOSU rankings are subjective and reflect qualitative considerations, the fact that tropospheric and stratospheric O_3 have the same ranks does not imply that the degree of confidence in their respective estimates is identical. In fact, from the observational standpoint, stratospheric O_3 forcing, which has occurred only since 1970s and is better documented, is on relatively firmer ground. Nevertheless, both O_3 components are less certain relative to the well-mixed greenhouse gases, but more so compared with the agents discussed below.

The estimate for the direct sulphate aerosol forcing has also seen multiple model investigations since the SAR, resulting in more estimates being available for this assessment. It is striking that consideration of all of the estimates available since 1996 lead to the same best estimate (-0.4 Wm^{-2}) and uncertainty (-0.2 to -0.8 Wm^{-2}) range as in the previous assessment. As in the case of O_3, that could be a motivation for elevating the status of knowledge of this forcing to a higher confidence level. However, there remain critical areas of uncertainty concerning the modelling of the geographical distribution of sulphate aerosols, spatial cloud distributions, effects due to relative humidity etc. Hence, we retain a "low" LOSU for this forcing.

The SAR stated a radiative forcing of $+0.1$ Wm^{-2} for fossil fuel (FF) black carbon aerosols with a range $+0.03$ to $+0.3$ Wm^{-2}, and a "very low" level of confidence. For biomass burning (BB) aerosols, the SAR stated a radiative forcing of -0.2 Wm^{-2} with a range -0.07 to -0.6 Wm^{-2}, and a "very low" level of confidence. In the present assessment, the radiative forcing of the black carbon component from FF is estimated to be $+0.2$ Wm^{-2} with a range from $+0.1$ to $+0.4$ Wm^{-2} based on studies since the SAR. A "very low" LOSU is accorded in view of the differences in the estimates from the various models. The organic carbon component from FF is estimated to yield a forcing of -0.1 Wm^{-2} with a range from -0.03 to -0.30 Wm^{-2}; this has a "very low" LOSU. Note that extreme caution must be exercised in adding the uncertainties of the organic and black carbon components to get the uncertainty for FF as a whole. For BB aerosols, no attempt is made to separate into black and organic carbon components, in view of considerable uncertainties. The central estimate and range for BB aerosols remains the same as in the SAR; this has a "very low" LOSU in view of the several uncertainties in the calculations (Section 6.7).

Mineral dust is a new component in the current assessment. The studies on the "disturbed" soils suggest an anthropogenic influence, with a range from $+0.4$ to -0.6 Wm^{-2}. In general, the evaluation for dust aerosol is complicated by the fact that the short-wave consists of a significant reflection and absorption component, and the long-wave also exerts a substantial contribution by way of a trapping of the infrared radiation. Thus, the net radiative energy gained or lost by the system is the difference between non-negligible positive and negative radiative flux changes operating simultaneously. Because of this complexity, we refrain from giving a best estimate and accord this component a "very low" LOSU.

As explained in Section 6.8, the "indirect" forcing due to all tropospheric aerosols can be thought of as comprising two effects. Only the first type of effect as applicable in the context of liquid clouds is considered here. As in the SAR, no best estimate is given in view of the large uncertainties prevailing in this problem (Section 6.8). The range (0 to -2 Wm^{-2}) is based on published estimates and subjective assessment of the uncertainties. Although several model studies suggest a non-zero, negative value as the upper bound (about -0.3 Wm^{-2}), substantial gaps in the knowledge remain which affect the confidence in the model simulations of this forcing (e.g., uncertainties in aerosol and cloud processes and their representations in GCMs, the potentially incomplete knowledge of the radiative effect of black carbon in clouds, and the possibility that the forcings for individual aerosol types may not be additive), such that the possibility of a very small negative value cannot be excluded; thus zero is retained as an upper bound as in the SAR. In view of the large uncertainties in the processes and the quantification, a "very low" LOSU is assigned to this forcing. Inclusion of the second indirect effect (Chapter 5) is fraught with even more uncertainties and, despite being conceptually valid as an anthropogenic perturbation, raises the question of whether the model estimates to-date can be unambiguously characterised as an aerosol radiative forcing.

Aviation introduces two distinct types of perturbation (Section 6.8). Contrails produced by aircraft constitute an anthropogenic perturbation. This is estimated to contribute 0.02 Wm^{-2} with an uncertainty of a factor of 3 or 4 (IPCC, 1999); the uncertainty factor is assumed to be 3.5 in Figure 6.6. This has an extremely low level of confidence associated with it. Additionally, aviation-produced cirrus is estimated by IPCC (1999) to yield a forcing of 0 to 0.04 Wm^{-2}, but no central estimate or uncertainty range was estimated in that report. Both components have a "very low" LOSU.

Volcanic aerosols that represent a transient forcing of the climate system following an eruption are not plotted since they are episodic events and cannot be categorised as a century-scale secular forcing, unlike the others. However, they can have substantial impacts on interannual to decadal scale temperature changes and hence are important factors in the time evolution of the forcing (see Section 6.15). Some studies (Hansen *et al.*, 1998; Shine and Forster, 1999) have attempted to scale the volcanic forcings in a particular decade with respect to that in a quiescent decade.

Land-use change was dealt with in IPCC (1990) but was not considered in the SAR. However, recent studies (e.g., Hansen *et al.*, 1998) have raised the possibility of a negative forcing due to deforestation and the ensuing effects of snow-covered land albedo changes in mid-latitudes. There are not many studies on this subject and rigorous investigations are lacking such that this forcing has a "very low" LOSU, with the range in the estimate being 0 to -0.4 Wm^{-2} (central estimate: -0.2 Wm^{-2}). Note that the land-use forcing here is restricted to that due to albedo change.

Solar forcing remains the same as in the SAR, in terms of best estimate, the uncertainty range and the confidence level. Thus, the range is 0.1 to 0.5 Wm^{-2} with a best estimate of 0.3 Wm^{-2}, and with a "very low" LOSU.

6.13.2 Limitations

It is important that the global mean forcing estimates be interpreted in a proper manner. Recall that the utility of the forcing concept is to afford a first-order perspective into the relative climatic impacts (viz., global-mean surface temperature change) of the different forcings. As stated in Section 6.2, for many of the relevant forcings (e.g., well-mixed greenhouse gases, solar, certain aerosol and O_3 profile cases), model studies suggest a reasonable similarity of the climate sensitivity factor (Equation 6.1), such that a comparison of these forcings is meaningful for assessing their relative effects on the global mean surface temperature. However, as mentioned earlier, the climate sensitivity factor for some of the spatially inhomogeneous forcings has yet to be fully explored. For some of the forcings (e.g., involving some absorbing aerosol and O_3 profile cases; see Hansen *et al.*, 1997a), the climate sensitivity is markedly different than for, say, the well-mixed greenhouse gases, while, for other forcings (e.g., indirect aerosol effect), more comprehensive studies are needed before a generalisation can become possible.

It is also cautioned that it may be inappropriate to perform a sum of the forcings to derive a best estimate "total" radiative forcing. Such an operation has the limitation that there are differing degrees of reliability of the global mean estimates of the various forcings, which do not necessarily lend themselves to a well-justified quantitative manipulation. For some forcings, there is not even a central or best estimate given at present (e.g., indirect aerosol forcing), essentially due to the substantial uncertainties.

The ranges given for the various forcings in Figure 6.6, as already pointed out, do not have a statistical basis and are guided mostly by the estimates from published model studies. Performing mathematical manipulations using these ranges to obtain a "net uncertainty range" for the total forcing, therefore, lacks a rigorous basis. Adding to the complexity is the fact that each forcing has associated with it an assessment of the level of knowledge that is subjective in nature viz., LOSU (Table 6.12). The LOSU index is not a quantitative indicator and, at best, yields a qualitative sense about the reliability of the estimates, with the well-mixed greenhouse gases having the highest reliability, those with "medium" rank having lesser reliability, and with even less reliability for the "low" and "very low" rankings. To some extent, the relatively lower ranking of the non-well-mixed greenhouse gases (e.g., aerosols, O_3) is associated with the fact that the forcing estimates for these agents depend on model simulations of species' concentrations, in contrast to the well-mixed greenhouse gases whose global concentrations are well quantified.

In a general sense, the strategy and usefulness of combining global mean estimates of forcings that have different signs, spatial patterns, vertical structures, uncertainties, and LOSUs, and the resulting significance in the context of the global climate response are yet to be fully explored. For some combinations of forcing agents (e.g., well-mixed greenhouse gases and sulphate aerosol; see Section 6.2), it is apparent from model tests that the global mean responses to the individual forcings can be added to yield the total global mean response. Because linear additivity

tests have yet to be performed for the complete set of agents shown in Figure 6.6, it is not possible to state with absolute certainty that the additivity concept will necessarily hold for the entire set of forcings.

Figure 6.6 depicts the uncertainties and LOSUs only for the global mean estimates. No attempt is made here to extend these subjective characterisations to the spatial domains associated with each of the forcings (see Figure 6.7, and Table 6.11 for the Northern to Southern Hemisphere ratios). As in the SAR, we reiterate that, in view of the spatial character of several of the forcing agents, the global mean estimates do not necessarily describe the complete spatial (horizontal and vertical dimensions) and seasonal climate responses to a particular radiative perturbation. Nor do they yield quantitative information about changes in parameters other than the global mean surface temperature response.

One diagnostic constraint on the total global mean forcing since pre-industrial times is likely to be provided by comparisons of model-simulated (driven by the combination of forcings) and observed climate changes, including spatially-based detection-attribution analyses (Chapter 12). However, the *a posteriori* inference involves a number of crucial assumptions, including the uncertainties associated with the forcings, the representativeness of the climate models' sensitivity to the forcings, and the model's representation of the real world's "natural" variations.

Overall, the net forcing comprises of a large positive value due to well-mixed greenhouse gases, followed by a number of other agents that have smaller positive or negative values. Thus, relative to IPCC (1990) and over this past decade, there are now more forcing agents to be accounted for, each with a sizeable uncertainty that can affect the estimated climate response. In this regard, consideration of the "newer" forcing agents brings on an additional element of uncertainty in climate change analyses, over and above those concerning climate feedbacks and natural variability (IPCC, 1990). Both the spatial character of the forcing and doubts about the magnitudes (and, in some cases, even the sign) add to the complexity of the climate change problem. However, this does not necessarily imply that the uncertainty associated with the forcings is now of much greater importance than the issue of climate sensitivity of models.

6.14 The Geographical Distribution of the Radiative Forcings

While previous sections have concentrated upon estimates of the global annual mean of the radiative forcing of particular mechanisms, this section presents the geographical distribution of the present day radiative forcings. Although the exact spatial distribution of the radiative forcing may differ between studies, many of them show similar features that are highlighted in this section. With the exception of well-mixed greenhouse gases, different studies calculate different magnitudes of the radiative forcing. In these cases, the spatial distributions are discussed, not the absolute values of the radiative forcing. It should be stressed that the Figure 6.7 represents plausible examples of geographical distributions only – significant differences may occur in other studies. In addition to Figure 6.7, Table 6.11 lists Northern to Southern Hemisphere ratio for forcings.

6.14.1 Gaseous Species

An example of the radiative forcing due to the combined effects of present day concentrations of CO_2, CH_4, N_2O, CFC-11 and CFC-12 is shown in Figure 6.7a (see Shine and Forster, 1999, for further details). The zonal nature of the radiative forcing is apparent and is similar to the radiative forcing due to well-mixed greenhouse gases from Kiehl and Briegleb (1993) presented in IPCC (1994). The radiative forcing ranges from approximately $+1\ Wm^{-2}$ at the polar regions to $+3\ Wm^{-2}$ in the sub-tropics. The pattern of the radiative forcing is governed mainly by variations of surface temperature and water vapour and the occurrence of high level cloud (Section 6.3).

An example of the radiative forcing due to stratospheric ozone depletion is shown in Figure 6.7b which was calculated using zonal mean stratospheric ozone depletions from 1979 to 1994 (WMO, 1995) by Shine and Forster (1999). The zonal nature of the radiative forcing is apparent with strongest radiative forcings occurring in polar regions which are areas of maximum ozone depletion (Section 6.4). The gradient of the radiative forcing tends to enhance the zonal gradient of the radiative forcing due to gaseous species shown in Figure 6.7a.

An example of the radiative forcing due to modelled increases in tropospheric O_3 is shown in Figure 6.7c (Berntsen *et al.*, 1997; Shine and Forster, 1999). While the exact spatial distribution of the radiative forcing may differ in other studies, the general pattern showing a maximum radiative forcing over North Africa and the Middle East is common to many other studies (Section 6.5). However, observational evidence presented by Kiehl *et al.* (1999) suggests that this might be an artefact introduced by the chemical transport models. The radiative forcing is much less homogeneous than for well-mixed greenhouse gases, the maximum radiative forcing being due to the coincidence of a relatively large O_3 change, warm surface temperatures, high surface reflectance, and cloud-free conditions.

6.14.2 Aerosol Species

An example of the direct radiative forcing due to sulphate aerosol is shown in Figure 6.7d (Haywood *et al.*, 1997a). In common with many other studies (see Section 6.7), the direct radiative forcing is negative everywhere, and there are three main areas where the radiative forcing is strongest in the Northern Hemisphere corresponding to the main industrialised regions of North America, Europe, and Southeast Asia. In the Southern Hemisphere, two less strong regions are seen. The ratio of the radiative forcing in the Northern Hemisphere to the Southern Hemisphere has been reported by many studies and varies from 2 (Graf *et al.*, 1997) to approximately 7 (Myhre *et al.*, 1998c). Generally, the strongest sulphate direct radiative forcing occurs over land areas although the low surface reflectance means that areas of water close to heavily industrialised regions such as the Mediterranean Sea, the Black Sea and the Baltic Sea result in strong local radiative forcings. Due to the large areal extent of ocean regions, the contribution to the total annual mean radiative forcing from ocean regions is significant. The ratio of the annual mean radiative forcing over land to that over oceans varies from

approximately 1.3 (Kiehl *et al.*, 2000) to 3.4 (Boucher and Anderson, 1995) (see Table 6.4).

An example of the direct radiative forcing due to organic carbon and black carbon from biomass burning is shown in Figure 6.7e (Penner *et al.*, 1998b; Grant *et al.*, 1999). While the radiative forcing is generally negative, positive forcing occurs in areas with a very high surface reflectance such as desert regions in North Africa, and the snow fields of the Himalayas. This is because biomass burning aerosols contain black carbon and are partially absorbing. The dependency of the sign of the radiative forcing from partially absorbing aerosols upon the surface reflectance has been investigated by a number of recent studies (e.g., Chylek and Wong, 1995; Chylek *et al.*, 1995; Haywood and Shine, 1995; Hansen *et al.*, 1997a). The strongest negative radiative forcing is associated with regions of intense biomass burning activity namely, South America, Africa, and Southern Asia and Indonesia and differ from the regions where the sulphate radiative forcing is strongest (Figure 6.7d), being confined to approximately 30°N to 30°S.

An example of the direct radiative forcing due to organic and black carbon from fossil fuel burning is shown in Figure 6.7f (Penner *et al.*, 1998b; Grant *et al.*, 1999). In contrast to the direct radiative forcing from biomass burning (Figure 6.7e), the modelled direct radiative forcing is generally positive except over some oceanic regions near industrialised regions such as the Mediterranean Sea and Black Sea. This is because, on average, aerosols emitted from fossil fuels contain a higher black/organic carbon ratio than biomass aerosols (Penner *et al.*, 1998b; Grant *et al.*, 1999) and are thus more absorbing. Comparison of the radiative forcing due to sulphate aerosols reveals that the areas of strongest sulphate direct radiative forcing are offset to some degree by the radiative forcing due to fossil fuel emissions of black carbon as shown in calculations by Haywood *et al.* (1997a) and Myhre *et al.* (1998c). Additional regions of moderate positive radiative forcing are present over areas of high surface reflectance such as northern polar regions and the North African deserts.

An example of the direct radiative forcing due to anthropogenic emissions of mineral dust is shown in Figure 6.7g (Tegen *et al.*, 1996). Areas of strong positive forcing are shown over regions with high surface reflectance such as desert regions in Africa and over the snow surfaces of the Himalayas and areas of strong negative forcing are apparent over ocean areas close to mineral dust sources such as off the coasts of Arabia and North Africa. The exact switchover between areas of positive and negative radiative forcing are not well established owing to uncertainties in the modelled mineral aerosol optical properties and depends upon the assumed single scattering albedo (Miller and Tegen, 1998), the long-wave properties and altitude of the aerosol (Section 6.7.6).

An example of the "first" indirect radiative effect (i.e., changes in the cloud reflectivity only) due to anthropogenic industrial aerosols is shown in Figure 6.7h. The forcing is calculated diagnostically in a similar way to Jones and Slingo (1997), but is based on a more recent version of the Hadley Centre model (HadAM3; Pope *et al.*, 2000), uses updated sulphur emission scenarios from the SRES scenario for the year 2000 (Johns *et al.*, 2001) and also includes a simple parametrization of

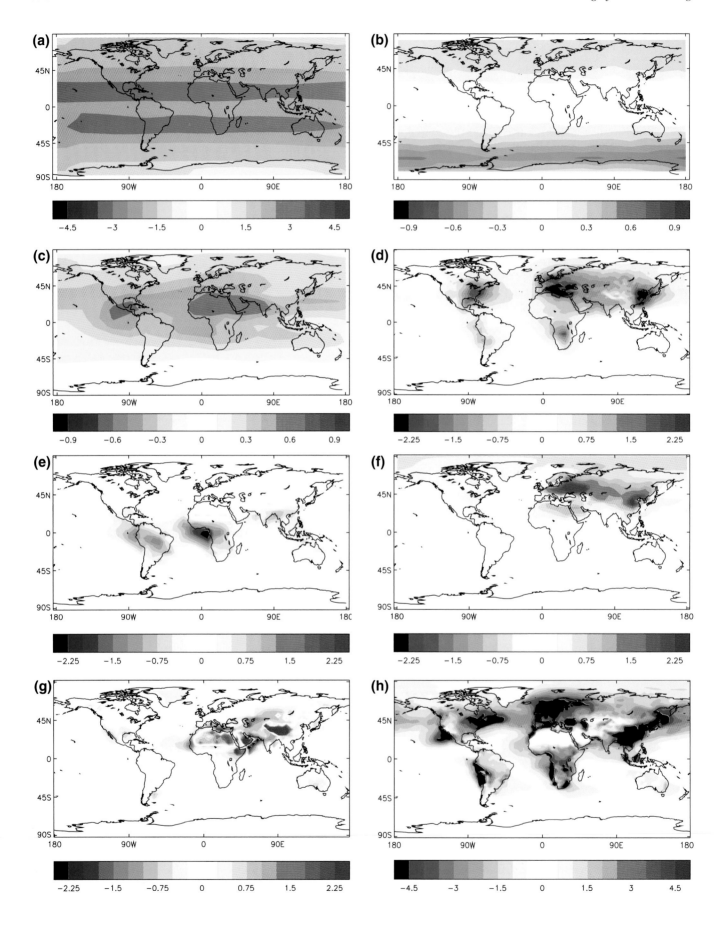

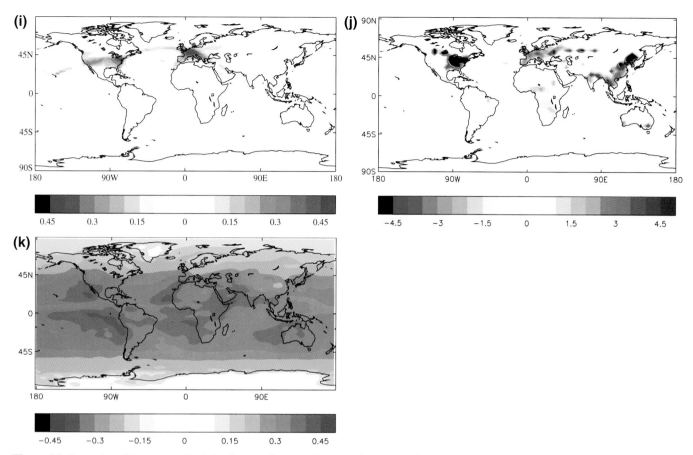

Figure 6.7: Examples of the geographical distribution of present-day annual-average radiative forcing (1750 to 2000) due to (a) well-mixed greenhouse gases including CO_2, CH_4, N_2O, CFC-11 and CFC-12 (Shine and Forster, 1999); (b) stratospheric ozone depletion over the period 1979 to 1994 given by WMO, 1995 (Shine and Forster, 1999); (c) increases in tropospheric O_3 (Berntsen *et al.*, 1997; Shine and Forster, 1999); (d) the direct effect of sulphate aerosol (Haywood *et al.*, 1997a); (e) the direct effect of organic carbon and black carbon from biomass burning (Penner *et al.*, 1998b; Grant *et al.*, 1999); (f) the direct effect of organic carbon and black carbon from fossil fuel burning (Penner *et al.*, 1998b; Grant *et al.*, 1999), (g) the direct effect of anthropogenic emissions of mineral dust (Tegen *et al.*, 1996); (h) the "first" indirect effect of sulphate aerosol calculated diagnostically in a similar way to Jones and Slingo (1997), but based on a more recent version of the Hadley Centre model (HadAM3; Pope *et al.*, 2000), using sulphur emission scenarios for year 2000 from the SRES scenario (Johns *et al.*, 2001) and including a simple parametrization of sea salt aerosol (Jones *et al.*, 1999); (i) contrails (Minnis *et al.*, 1999); (j) surface albedo change due to changes in land use (Hansen *et al.*, 1998), (k) solar variability (Haigh, 1996). Note that the scale differs for the various panels. Different modelling studies may show considerably different spatial patterns as described in the text. (Units: Wm^{-2})

sea salt aerosol (Jones *et al.*, 1999). The spatial distribution of the indirect radiative forcing is quite different from the direct radiative forcing with strong areas of forcing off the coasts of industrialised regions (note the change in scale of Figure 6.7h). There is a significant radiative forcing over land regions such as Europe and the Eastern coast of North America, and Southeast Asia. The spatial distribution of the indirect radiative forcing will depend critically upon the assumed spatial distribution of the background aerosol field and the applied anthropogenic perturbation and differs substantially between studies (see Section 6.8.5). It would have a very different spatial distribution if the effect of biomass burning aerosols were included. The "second" indirect effect whereby inclusion of aerosols influences the lifetime of clouds is not considered here due to the complications of necessarily including some cloud feedback processes in the estimates (Section 6.8.5), but may well resemble the spatial distribution of the "first" indirect effect.

6.14.3 Other Radiative Forcing Mechanisms

The spatial distribution of three other radiative forcing mechanisms are considered in this section: the radiative forcing due to contrails, land-use change, and solar variability. The radiative forcing due to other constituents such as nitrate aerosol and aviation-induced cirrus that are very difficult to quantify at present are not presented as geographic distributions of the radiative forcing are currently considered to be speculative.

An example of the present day radiative forcing due to the effect of contrails is shown in Figure 6.7i (Minnis *et al.*, 1999). The radiative forcing is very inhomogeneous, being confined to air-traffic corridors (IPCC, 1999). Future scenarios for aircraft emissions may shift the current geographical pattern of the radiative forcing as discussed in IPCC (1999).

An example of an estimate of the radiative forcing due to changes in land use is shown in Figure 6.7j (Hansen *et al.*, 1998).

The areas of strongest negative forcing occur at northern latitudes of the Northern Hemisphere due to the felling of forests which have a lower albedo when snow is present (see Section 6.13). Additional effects are due to the change in albedo between crop lands and naturally occurring vegetation. Examples where the radiative forcing is positive include areas where irrigation has enabled crop-growing on previously barren land.

An example of the present day radiative forcing due to solar variability is shown in Figure 6.7k. The solar radiative forcing was calculated by scaling the top of the atmosphere net solar radiation such that the global average is +0.3 Wm^{-2} (as deduced for global average radiative forcing since 1750, see Section 6.13.1). Thus it assumes a 0.125% increase in solar constant and no change in any other parameter (e.g., O_3, cloud). The cloud and radiation fields were calculated within a run of the UGAMP GCM (Haigh, 1996). The strongest radiative forcings exist where the surface reflectance is low (i.e., oceanic regions) and the insolation is highest (i.e., equatorial regions). The solar radiative forcing is also modulated by cloud amount, areas with low cloud amount showing the strongest radiative forcing. The solar radiative forcing is more inhomogeneous than the radiative forcing due to gaseous species (Section 6.14.1), but more homogeneous than the radiative forcing due to aerosol species (Section 6.14.2).

While the preceding sections have shown that the radiative forcing due to the different forcing mechanisms have very different spatial distributions, it is essential to note that the forcing/response relationship given in Section 6.2 relates global mean radiative forcings to global mean temperature response. Thus, it is not possible to simply map the geographical radiative forcing mechanisms by assuming a globally invariant climate sensitivity parameter to predict a geographic temperature response, due to the complex nature of the atmosphere-ocean system. Rather, the effects of spatial inhomogeneity in the distribution of the radiative forcing may lead to locally different responses in surface temperature (Section 6.2) indicating that the spatial distributions of the radiative forcing need to be accurately represented to improve regional estimates of surface temperature response and other physical parameters.

6.15 Time Evolution of Radiative Forcings

6.15.1 Past to Present

IPCC (1990) showed time evolution of the radiative forcing due to the well-mixed greenhouse gases. For the other radiative forcing mechanisms, the previous IPCC reports (IPCC, 1990; SAR) did not consider the evolution of the radiative forcing, and assessed mainly the radiative fluxes in the pre-industrial and present epochs. However, more recent studies have considered the time evolution of several forcing mechanisms (Hansen *et al.*, 1993; Wigley *et al.*, 1997; Myhre *et al.*, 2001). The information on the time evolution of the radiative forcing from pre-industrial times to present illustrates the differing importance of the various radiative forcing mechanisms over the various time periods, as well as the different start times of their perturbation of the radiative balance. It also gives useful information in the form of

inputs to climate models viz., as driving mechanisms to investigate potential causes of climate changes. Studies of the evolution of various anthropogenic as well as natural forcing mechanisms may then be used to explain potential causes of past climate change, e.g. since pre-industrial times (Chapter 12).

The knowledge of the various forcing mechanisms varies substantially (see Section 6.13) and, for some, the knowledge of their time evolution is more problematic than for others. For example, well-mixed greenhouse gas concentrations are observed very accurately from about 1950, with observations even further back in time, whereas data for most aerosol components are much more scarce and uncertain. Uncertainties in the evolution of many of the radiative forcing mechanisms shown in Figure 6.8 have not been assessed yet, and the presented time history should be regarded as an example of a possible evolution.

The forcing mechanisms, considered, where information on their time evolution is available, are plotted in Figure 6.8, with the present forcing corresponding to the best estimates given in Section 6.13. The time evolution differs considerably among the forcing mechanisms as different processes controlling the emissions and lifetimes are involved. Concentrations of the well-mixed greenhouse gases are taken from Chapters 3 and 4 and the simplified expressions for the radiative forcing in Table 6.2 (first row for CO_2) are used. The evolution of the radiative forcing due to tropospheric O_3 is taken from Berntsen *et al.* (2000). For the radiative forcing due to stratospheric O_3, information on the evolution is taken from the SAR and scaled to the present forcing of -0.15 Wm^{-2} (Section 6.4).

For sulphate, the time evolution for SO_2 emission (Schlesinger *et al.*, 1992) is used, updated with values after 1990 (Stern and Kaufmann, 1996). For black and organic carbon aerosols, the fossil fuel component is scaled to coal, diesel, and oil use and fossil fuel emission (IPCC, 1996b), respectively. Altered emission coefficients as a result of improved technology have not been taken into account and is a substantial uncertainty for the time evolution of the black carbon emission. The biomass component is scaled to the gross deforestation (see SAR). In addition to different forcing mechanisms of anthropogenic origin, two natural forcings have also been considered: solar irradiance variations and stratospheric aerosols of volcanic origin. The Lean *et al.* (1995) and Hoyt and Schatten (1993) estimates of direct solar forcing due to variation in total solar irradiance are shown. Differences between the solar irradiance constructions are due to use of different proxy parameters for the solar irradiance variations (Section 6.11). The optical depth of the stratospheric aerosols of volcanic origin is taken from Sato *et al.* (1993) and Robock and Free (1996). The data from Sato *et al.* (1993) (updated from www.giss.nasa.gov) are given for the period 1850 to 1998, whereas those from Robock and Free (1996) are for the period 1750 to 1988. In Robock and Free (1996) the optical depth is given for the Northern Hemisphere. The relationship between optical depth and radiative forcing from Lacis *et al.* (1992) is used. The stratospheric aerosols yield a very strong forcing immediately after major eruptions (Section 6.9); however, the lifetime of the stratospheric aerosols is only a few years. Therefore, the transient response due to the forcing by stratospheric aerosols

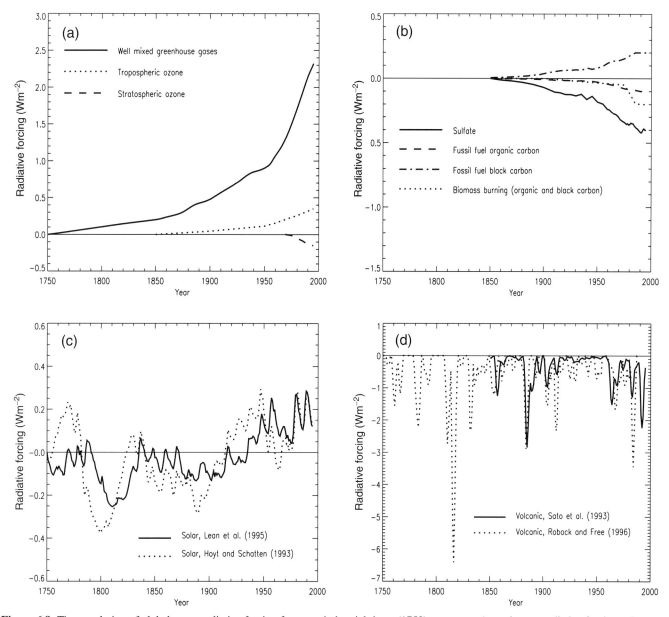

Figure 6.8: Time evolution of global mean radiative forcing from pre-industrial times (1750) to present. Annual mean radiative forcing values are shown, except for stratospheric aerosols of volcanic origin where a three-year running mean is used. Radiative forcings due to (a) well-mixed greenhouse gases, tropospheric O₃, and stratospheric O₃, (b) direct effect of sulphate aerosols, fossil fuel organic and black carbon aerosols, and biomass burning aerosols, (c) solar irradiance variations from Lean *et al.* (1995) and Hoyt and Schatten (1993), respectively, and (d) stratospheric aerosols of volcanic origin from Sato *et al.* (1993) (and updated with data from www.giss.nasa.gov) for the period 1850 to 1998 and from Robock and Free (1996) for the period 1750 to 1988 (for the Northern Hemisphere only). Note the change of scales in the different panels.

cannot readily be compared to that due to the more sustained or steadily increasing forcings.

The relative evolution of the strengths of the different forcing mechanisms presented above are seen to be very different. The forcing due to well-mixed greenhouse gases (consisting of the components listed in Table 6.11) is the dominant forcing mechanism over the century time period. For tropospheric O₃, the relative radiative forcing compared to the well-mixed greenhouse gases has slightly increased; it was 10% of the well-mixed greenhouse gas forcing in 1900 and is 15% at present. The stratospheric O₃ forcing is significant only over the last two

decades. The aerosols have short lifetimes compared with the well-mixed greenhouse gases, and their forcings at any given time depend on the current emissions. This is not the case for the well-mixed greenhouse gases for which both current and previous emissions are relevant. Table 6.13 shows five year averages for the radiative forcings denoted in Figure 6.8 over the period 1960 to 1995. For some of the forcing mechanisms a large increase in the radiative forcing (computed since pre-industrial times) is estimated to have occurred over the period 1960 to 1995. This is the case for the well-mixed greenhouse gases, tropospheric O₃, biomass burning aerosols and organic carbon

Table 6.13: *Radiative forcings (5-year averages) due to change in well-mixed greenhouse gases (WMGG), stratospheric O_3, tropospheric O_3, direct effect of sulphate aerosols, fossil fuel organic carbon aerosols (FF OC), fossil fuel black carbon aerosols (FF BC), aerosols from biomass burning (organic carbon and black carbon) (BB), stratospheric aerosols of volcanic origin, and changes in Total Solar Irradiance. All values are evaluated with respect to pre-industrial times (1750).*

Time period	WMGG	Strat O_3	Trop O_3	Sulphate	FF OC	FF BC	BB	Volcanic	Solar[a]	Solar[b]
1961 to 1965	1.14	0.00	0.17	−0.26	−0.05	0.14	−0.04	−1.00	0.11	−0.07
1966 to 1970	1.27	0.00	0.20	−0.29	−0.06	0.14	−0.04	−0.77	0.11	0.04
1971 to 1975	1.44	−0.01	0.22	−0.33	−0.07	0.15	−0.05	−0.28	0.06	0.03
1976 to 1980	1.64	−0.04	0.25	−0.35	−0.08	0.16	−0.09	−0.15	0.17	0.14
1981 to 1985	1.85	−0.07	0.28	−0.36	−0.09	0.19	−0.16	−0.88	0.17	0.20
1986 to 1990	2.07	−0.11	0.31	−0.40	−0.10	0.20	−0.20	−0.35	0.21	0.19
1991 to 1995	2.26	−0.14	0.34	−0.40	−0.10	0.20	−0.20	−1.42	0.18	0.18

[a] From Lean *et al.* (1995).

[b] From Hoyt and Schatten (1993).

aerosols from fossil fuel consumption, whereas, for sulphate aerosols and black carbon aerosols from fossil fuel consumption, a smaller increase is estimated.

As seen from Figure 6.8d the radiative forcing due to stratospheric aerosols of volcanic activity has very large year to year variations. The solar irradiance, according to the two reconstructions, generally increases and may have contributed in an important manner to the warming in the 20th century, particularly in the period from 1900 to 1950. Volcanic activity was particularly strong around 1900 and at different times since 1963. Table 6.13 shows a strong radiative forcing due to the temporal evolution of the stratospheric aerosols of volcanic origin during the period 1961 to 1965; however, the strongest forcing in the course of the past four decades has occurred over the period from 1991 to 1995. The temporal evolution of the stratospheric aerosol content together with the small solar irradiance variations during the last few (two to four) decades indicates that the natural forcing has been negative over the past two and possibly even the past four decades. In contrast, the positive forcing due to well-mixed greenhouse gases has increased rapidly over the past four decades.

6.15.2 SRES Scenarios

6.15.2.1 Well-mixed greenhouse gases

Emissions from the SRES scenarios (Nakićenović *et al.*, 2000) were used in Chapter 4 to simulate the concentrations of well-mixed greenhouse gases and O_3 in the atmosphere for the period 2000 to 2100. Here we compute the associated radiative forcing due to well-mixed greenhouse gases using the simplified expressions given in Table 6.2 (for CO_2, the expression in the first row of that Table is employed). The numbers displayed in Table 6.14 represent the radiative forcings in 2050 and 2100 with respect to the year 2000. For the gases relevant to radiative forcing only, the SRES scenarios give a wide range of radiative forcings for the various compounds. Note that the CO_2 forcing increases substantially in the future. All the SRES scenarios give positive radiative forcing values for the well-mixed greenhouse gases (except for the radiative forcing by CH_4 in the B1 scenario for 2100). The scenario for gases relevant to radiative forcing and ozone

depletion yields negative values of the radiative forcing, reflecting the reductions in atmospheric concentrations of ozone depleting gases due to emission control.

From Table 6.11, the radiative forcing due to CO_2 is at present slightly larger than 50% of the total greenhouse gas forcing (including tropospheric O_3). In 2050 and 2100 the SRES scenarios indicate a stronger dominance of CO_2 on the total greenhouse gas forcing (including tropospheric O_3), at about 70 to 80% of the total forcing. The SRES scenarios indicate that HFC-134a will give the largest radiative forcing amongst the HFCs.

6.15.2.2 Tropospheric ozone

The radiative forcings associated with future tropospheric O_3 increases are calculated on the basis of the O_3 changes calculated by Chapter 4 and presented in Appendix II, Table II.2.5 for the various SRES scenarios. The mean forcing per DU estimated from the various models and given in Table 6.3 (i.e., 0.042 Wm^{-2}/DU) is used to derive these future forcings. Most scenarios lead to increases in the abundances of tropospheric O_3 and consequently to positive radiative forcings in 2050 and 2100. Scenarios A1fi, A2, and A2p provide the maximum tropospheric O_3 forcings reaching 0.89 Wm^{-2} in 2100. Only scenario B1 predicts a decrease in tropospheric O_3 and a negative forcing of −0.16 Wm^{-2} in 2100.

6.15.2.3 Aerosol direct effect

The direct radiative forcing due to anthropogenic sulphate, black carbon (BC), and organic carbon (OC) aerosols are assessed using the scenarios described in Chapter 5, Section 5.5.3 and the column burdens presented in Appendix II, Tables II.2.7, II.2.8 and II.2.9 respectively. Scenarios for mineral dust are not considered here, as there is no "best estimate" available (Sections 6.7.6 and 6.13). For each aerosol species, the ratio of the column burdens for the particular scenario to that of the year 2000 is multiplied by the "best estimate" of the present day radiative forcing. The year 2000 "best estimate" radiative forcing is then subtracted to give the radiative forcing for the period from the year 2000 to the date of the scenario. Estimates of the direct radiative forcing for each of the aerosol species are given in Table 6.15.

Table 6.14: *Future (2050 and 2100) radiative forcings due to gases that are relevant for radiative forcing only (based on SRES scenarios), and due to gases that are relevant for radiative forcing and ozone depletion (based on WMO (1999) scenarios). For some gases the results for the IS92a emission scenario, with updated model calculations, are also shown (see Chapter 4 for further details). The radiative forcings presented here are with respect to the year 2000. They may be added to the approximately present day (1998) total radiative forcing given in Table 6.1 to obtain the approximate radiative forcings with respect to pre-industrial times (1750).*

Gas	Radiative forcing (Wm^{-2}) 2050										
	A1b	A1t	A1fi	A2	B1	B2	A1p	A2p	B1p	B2p	IS92a
Gases relevant to radiative forcing only											
CO_2	1.90	1.62	2.24	1.90	1.46	1.37	1.97	1.93	1.30	1.45	1.66
CH_4	0.22	0.25	0.30	0.27	0.04	0.20	0.31	0.26	0.09	0.24	0.25
N_2O	0.10	0.08	0.18	0.17	0.12	0.08	0.10	0.17	0.12	0.09	0.14
Tropospheric O_3	0.24	0.37	0.66	0.43	0.04	0.28	0.32	0.42	0.15	0.28	0.32
HFC-23	0.003	0.003	0.003	0.003	0.003	0.003	0.003	0.003	0.003	0.003	
HFC-125	0.013	0.013	0.013	0.008	0.008	0.008	0.013	0.007	0.008	0.009	0.020
HFC-32	0.001	0.001	0.001	0.001	0.001	0.001	0.001	0.001	0.001	0.001	
HFC-134a	0.079	0.078	0.076	0.043	0.042	0.048	0.111	0.054	0.057	0.070	0.079
HFC-143a	0.009	0.009	0.009	0.006	0.006	0.006	0.010	0.006	0.006	0.007	
HFC-152a	0.000	0.000	0.000	0.000	0.000	0.000	0.000	0.000	0.000	0.000	0.005
HFC-227ea	0.010	0.010	0.010	0.006	0.006	0.007	0.010	0.005	0.006	0.007	
HFC-245ca	0.017	0.017	0.016	0.008	0.009	0.010	0.017	0.008	0.009	0.011	
HFC-43-10mee	0.002	0.002	0.002	0.001	0.001	0.001	0.002	0.001	0.001	0.001	
CF_4	0.007	0.007	0.007	0.007	0.004	0.008	0.013	0.010	0.008	0.012	
C_2F_6	0.002	0.002	0.002	0.002	0.001	0.002	0.003	0.002	0.002	0.002	
C_4F_{10}	0.000	0.000	0.000	0.000	0.000	0.000	0.007	0.005	0.005	0.005	
SF_6	0.010	0.010	0.010	0.011	0.007	0.009	0.011	0.011	0.008	0.011	
Gases relevant to radiative forcing and ozone depletion											
CFC-11	−0.04										
CFC-12	−0.06										
CFC-113	−0.01										
CFC-114	−0.002										
CFC-115	0.000										
CCl_4	−0.009										
CH_3CCl_3	−0.004										
HCFC-22	−0.020										
HCFC-141b	−0.001										
HCFC-142b	0.000										
HCFC-123	0.000										
Halon-1211	−0.001										
Halon-1301	0.000										

Gas	Radiative forcing (Wm^{-2}) 2100										
	A1b	A1t	A1fi	A2	B1	B2	A1p	A2p	B1p	B2p	IS92a
Gases relevant to radiative forcing only											
CO_2	3.48	2.39	5.15	4.42	2.06	2.73	3.46	4.42	2.34	2.79	3.48
CH_4	0.08	0.14	0.51	0.59	−0.07	0.39	0.15	0.58	0.10	0.27	0.43
N_2O	0.16	0.11	0.40	0.36	0.17	0.14	0.13	0.36	0.21	0.12	0.25
Tropospheric O_3	0.15	0.11	0.89	0.87	−0.16	0.43	0.21	0.87	0.05	0.37	0.55
HFC-23	0.003	0.003	0.003	0.003	0.003	0.003	0.002	0.002	0.002	0.002	
HFC-125	0.031	0.032	0.031	0.025	0.013	0.023	0.030	0.023	0.013	0.023	
HFC-32	0.002	0.002	0.002	0.002	0.001	0.002	0.002	0.002	0.001	0.001	
HFC-134a	0.129	0.142	0.133	0.130	0.055	0.123	0.172	0.168	0.076	0.154	
HFC-143a	0.026	0.027	0.026	0.020	0.012	0.018	0.026	0.020	0.012	0.019	
HFC-152a	0.000	0.000	0.000	0.000	0.000	0.000	0.000	0.000	0.000	0.000	0.007
HFC-227ea	0.021	0.022	0.021	0.018	0.011	0.020	0.020	0.016	0.010	0.018	
HFC-245ca	0.021	0.023	0.022	0.024	0.009	0.024	0.020	0.023	0.009	0.020	
HFC-43-10mee	0.004	0.005	0.004	0.003	0.002	0.003	0.004	0.002	0.002	0.002	
CF_4	0.021	0.021	0.021	0.025	0.010	0.024	0.028	0.034	0.015	0.033	
C_2F_6	0.004	0.004	0.004	0.005	0.002	0.005	0.006	0.007	0.003	0.007	
C_4F_{10}	0.000	0.000	0.000	0.000	0.000	0.000	0.022	0.016	0.013	0.017	
SF_6	0.027	0.027	0.027	0.031	0.016	0.020	0.027	0.032	0.017	0.034	
Gases relevant to radiative forcing and ozone depletion											
CFC-11	−0.06										
CFC-12	−0.10										
CFC-113	−0.02										
CFC-114	−0.003										
CFC-115	0.000										
CCl_4	−0.010										
CH_3CCl_3	−0.004										
HCFC-22	−0.030										
HCFC-141b	−0.001										
HCFC-142b	−0.002										
HCFC-123	0.000										
Halon-1211	−0.001										
Halon-1301	−0.001										

Table 6.15. Direct aerosol radiative forcings (Wm⁻²) estimated as an average of different models for the IPCC SRES scenarios described in Chapter 5. The burdens used in calculating the radiative forcings are given in Appendix II, Tables II.2.7, II.2.8 and II.2.9, the radiative forcings presented here are from the year 2000 to the date of the scenario. They may be added to the present day forcings given in Section 6.7 (−0.4 Wm⁻² for sulphate aerosols, +0.4 Wm⁻² assumed for BC aerosols (from fossil fuel and biomass burning), and −0.5 Wm⁻² assumed for OC aerosols (from fossil fuel and biomass burning)) to obtain the radiative forcings from pre-industrial times.

Aerosol	Radiative forcing (Wm⁻²) 2050										
	A1b	A1t	A1fi	A2	B1	B2	A1p	A2p	B1p	B2p	IS92a
Sulphate	+0.03	+0.17	−0.07	−0.21	0.0	+0.08	+0.03	−0.21	+0.10	+0.07	−0.43
BC	+0.13	+0.35	+0.49	+0.21	−0.16	+0.17	+0.37	+0.21	−0.04	+0.12	+0.12
OC	−0.16	−0.43	−0.61	−0.27	+0.20	−0.21	−0.46	−0.27	+0.05	−0.15	−0.15

Aerosol	Radiative forcing (Wm⁻²) 2100										
	A1b	A1t	A1fi	A2	B1	B2	A1p	A2p	B1p	B2p	IS92a
Sulphate	+0.24	+0.28	+0.17	+0.05	+0.25	+0.12	+0.24	+0.05	+0.23	+0.12	−0.39
BC	+0.30	+0.46	+0.65	+0.56	−0.20	+0.44	+0.55	+0.56	−0.03	+0.40	+0.26
OC	−0.38	−0.58	−0.82	−0.70	+0.25	−0.54	−0.69	−0.70	+0.04	−0.50	−0.33

The uncertainty associated with these estimates of the direct radiative forcing is necessarily higher than the estimates for the year 2000. This is because estimates for the year 2000 are constrained as best as possible to observations. Additionally, the simplified approach applied here does not account for changes in the spatial pattern of the global distribution of the aerosol species that may arise from changes in the geographic distribution of emissions. The uncertainty is therefore estimated as a factor of three for the radiative forcing by sulphate aerosols, and a factor of four for the radiative forcing by BC and OC aerosols.

For sulphate aerosols, only scenarios A1fi, A2, and A2p show a negative direct radiative forcing for the period 2000 to 2050, with all scenarios showing positive radiative forcings over the period 2000 to 2100. This contrasts with the IS92a scenario which predicted an increasingly negative radiative forcing due to the higher column burden of sulphate. For BC aerosols the majority of the scenarios show an increasingly positive direct radiative forcing in 2050 and 2100 (except for B1 and B1p). For OC aerosols, the opposite is true, with the majority of scenarios showing an increasingly negative direct radiative forcing in 2050 and 2100 (except for B1 and B1p). The direct aerosol radiative forcing evolution due to sulphate, black and organic carbon aerosols taken together varies in sign for the different scenarios. The magnitudes are much smaller than that for CO_2. As with the trace gases, there is considerable variation amongst the different scenarios.

6.15.2.4 Aerosol indirect effect

The indirect radiative forcing was calculated by the LLNL/Umich model (Chuang *et al.*, 2000a), the GISS model (Koch *et al.*, 1999), and the Max Planck/Dalhousie (Lohmann *et al.*, 1999a,b, 2000) model. The LLNL/Umich model uses the mechanistic formulation for the determination of droplet concentration described by Chuang *et al.* (1997) but has been updated to include interactive dust and sea salt. This model provides an estimate of the "first" indirect effect. The GISS model used an empirical formulation for relating droplet concentrations and aerosol concentrations and

also provides an estimate of the "first" indirect effect. The Max-Planck/Dalhousie model used the mechanistic formulation of Chuang *et al.* (1997) as described in Lohmann *et al.* (1999a,b, 2000) and includes both the "first" and "second" indirect effects. In addition, the Max-Planck/Dalhousie model used monthly averaged dust and sea salt fields from the LLNL/Umich model. Further details of the models are provided in Chapter 5. The indirect radiative forcing scenarios are summarised in Table 6.16, and represent the radiative forcing from the year 2000 due to sulphate and carbonaceous aerosols. Some simulations suggest that changes in the concentrations of natural aerosols due to changes in the climate could also contribute to a negative aerosol indirect effect (see Chapter 5).

As for the aerosol direct effect, the uncertainty associated with these estimates is higher than for the estimates for the present day. Considering the very low LOSU of the indirect aerosol effect and the fact that no best estimate was recommended for the present day (Sections 6.8 and 6.13), the numbers provided in Table 6.16 give only a rough indication of how the forcings could change between the present and the future (according to the SRES scenarios) in three different models. Relative to 2000, the change in the direct plus indirect aerosol radiative forcing is projected to be smaller in magnitude than that of CO_2.

Table 6.16: Indirect aerosol radiative forcing (Wm⁻²) estimated by different models for the IPCC SRES scenarios described in Chapter 5. The sulphate burdens in each case are given in Chapter 5, Table 5.14. These estimates are from the average difference in cloud forcing between two simulations. The numbers represent the radiative forcing from 2000. No numbers are given for the year 2000 as no best estimate of the radiative forcing is suggested in Section 6.8.

	LLNL/Umich	GISS	Max Planck/ Dalhousie
A2 2030	−0.24	−0.29	−0.05
A2 2100	−0.47	−0.36	−0.32
B1 2100	+0.10		

Appendix 6.1 Elements of Radiative Forcing Concept

The principal elements of the radiative forcing concept are summarised below.

(A) The concept was developed first in the context of the one-dimensional radiative-convective models that investigated the equilibrated global, annual mean surface temperature responses to radiative perturbations caused by changes in the concentrations of radiatively active species (Manabe and Wetherald, 1967; Ramanathan and Coakley, 1978; WMO, 1986). Over the past decade, the concept has been extended to cover different spatial dimensions and seasonal time-scales (IPCC, 1992, 1994).

(B) In the one-dimensional radiative-convective model framework, the surface and troposphere are closely coupled, behaving as a single thermodynamic system under the joint control of radiative and convective processes, with a specified lapse rate determining the thermal structure. The stratospheric state is determined by the radiative equilibrium condition. The stratosphere and troposphere irradiances are together constrained by the requirement that the top of the atmosphere net total irradiance (i.e., radiative energy absorbed minus that emitted by the Earth's entire climate system) must be zero at equilibrium. In applying the forcing concept to arbitrary spatial and seasonal time-scales, as opposed to the global annual mean, it has been assumed (WMO, 1992; SAR) that the stratosphere is in radiative-dynamical (rather than radiative) equilibrium (see (D) below).

(C) The stratosphere is a *fast response* system which, in response to an imposed radiative perturbation, comes into equilibrium on a time-scale (about a few months) that is much more rapid than the surface-troposphere system (typically decades) (Hansen *et al.*, 1997a; Shine and Forster, 1999). The latter is a *slow response* system owing principally to the thermal inertia of the oceans.

(D) When a perturbation is applied (such as increases in well-mixed greenhouse gases), there is an *instantaneous* change in irradiances that is manifest in general as a radiative imbalance (forcing) at the surface, tropopause and the top of the atmosphere. The rapid thermal re-equilibration of the stratosphere leads to an alteration of the radiative imbalance imposed on the surface-troposphere system (WMO, 1992), thereby yielding an *adjusted* forcing (SAR). The surface and troposphere, operating in a *slow response* mode, are still in a process of adjustment while the stratosphere has already reached its new equilibrium state. The SAR points out the clear distinction existing between the *instantaneous* and *adjusted* forcings.

For the arbitrary space and time mean stratosphere, there arises the need to evaluate the radiative flux changes with the stratosphere in a radiative-dynamical equilibrium. A classical method to determine this is the "Fixed Dynamical Heating" (FDH; WMO, 1995) in which it is assumed that the dynamical heating rate in the stratosphere is unchanged and that the stratosphere comes to a new thermal equilibrium in response to the perturbation through adjustments in the temperature profile, such that a

new radiative-dynamical equilibrium is attained (*radiative response*; see Ramanathan and Dickinson (1979) and Fels *et al.* (1980)). The resulting adjustment process in the stratosphere makes an additional contribution to the forcing of the surface-troposphere system. When the stratosphere has adjusted to a new radiative-dynamical equilibrium with resultant changes to its thermal state, the change in flux at the tropopause and at the top of the atmosphere become identical. It is important that the stratosphere be in radiative-dynamical equilibrium and, as shown by Hansen *et al.* (1997a), it is the *adjusted* rather than the *instantaneous* forcing that is a more relevant indicator of the surface temperature response. The adjustment of stratosphere is crucial for some of the forcings, but not for all of them (Shine and Forster, 1999). In the case of some radiative perturbations, the stratosphere is hardly perturbed and the instantaneous and adjusted radiative flux changes thus tend to be similar (SAR). In other cases, there is only a small (20% or less) influence due to the stratospheric adjustment process. However, for the case of ozone depletion in the lower stratosphere, the effect of a stratospheric adjustment could even yield a change in the sign of the forcing.

(E) As a direct consequence of the above, the forcing definition most appropriate for the response of the surface-troposphere system to a radiative perturbation is the net (down minus up) radiative (solar plus long-wave) change defined at the tropopause after the stratosphere has come to a new thermal equilibrium state. Thus, the level at which the tropopause is assumed in the models is an important aspect for the quantitative determination of the forcing (Forster *et al.*, 1997), as is the model vertical resolution used to resolve the vicinity of the tropopause region (Shine *et al.*, 1995). The classical radiative-convective model definition of the tropopause considers it as a boundary between a region where radiative-convective equilibrium prevails (i.e., troposphere) and a region that is in radiative (or radiative-dynamical, i.e., stratosphere) equilibrium. Such a distinction could be ambiguous in the case of the three-dimensional GCMs or the real world. However, radiative forcing appears to be relatively robust to changes in the definition of the tropopause in a GCM (Christiansen, 1999).

(F) A major motivation for the radiative forcing concept is the ease of climate change analysis when radiative forcing, feedback, and climate response are distinguished from one another. Such a separation is possible in the modelling framework where forcing and feedback can be evaluated separately e.g., for the case of CO_2 doubling effects (see Dickinson, 1982; Dickinson and Cicerone, 1986; Cess and Potter, 1988; Cess *et al.*, 1993). The consideration of forcing, feedback and response as three distinct entities in the modelling framework (see also Charlson, 2000) while originating from the one-dimensional radiative/convective models, has made the transition to GCM studies of climate (Hansen *et al.*, 1981; Wetherald and Manabe, 1988; Chen and Ramaswamy, 1996a).

(G) A critical aspect of the separability mentioned in (F) is holding the surface and troposphere state fixed in the evaluation of the radiative forcing. For example, in the case of a change in the concentration of a radiatively active species, the term "state" implies that all parameters are held at the unperturbed values with only the concerned species' concentration being changed. Thus, in the strictest sense, temperature in the troposphere, water vapour and clouds in the entire atmosphere, and circulation, are held fixed in the computation of the irradiance changes at the tropopause, with only the stratosphere adjusted to a new thermal equilibrium state via the radiative response. In contrast to this prescription for calculating the forcing, the resulting changes in the meteorological and climatic parameters (e.g., tropospheric temperature and water vapour) constitute responses to the imposed perturbation (WMO, 1986; Charlson, 2000). In the IPCC context, the change in a radiative agent's characteristic has involved anthropogenic (e.g., CO_2) and natural (e.g., aerosols after a volcanic eruption) perturbations.

It is important to emphasise that changes in water vapour in the troposphere are viewed as a feedback variable rather than a forcing agent. However, in the case of the second indirect aerosol forcing, the separation is less distinct. Anthropogenic emissions (e.g., aircraft, fossil fuel) or precursors to water vapour are negligible. The same is true for changes in water vapour in the stratosphere, except in the instance that the oxidation of CH_4 provides an input. Changes in the condensed liquid and solid phases of water (i.e., clouds) are also considered as part of the climate feedback. The strict requirement of no feedbacks in the surface and troposphere demands that no secondary effects such as changes in troposphere motions or its thermodynamic state, or dynamically-induced changes in water substance in the surface and atmosphere, be included in the evaluation of the net irradiance change at the tropopause. (Note: the second indirect effect of aerosols consists of microphysically-induced changes in the water substance.)

(H) The foregoing governing factors reflect the fundamental recognition that, in response to an externally imposed radiative forcing, there is a shift in the equilibrium state of the climate system. This forcing of the climate change in the IPCC parlance is to be distinguished from *forcing* definitions initiated for other purposes, e.g., cloud forcing (Ramanathan *et al.*, 1989), sea surface temperature related forcing during ENSO periods, etc.

References

Abbas, M.A. and J. Latham, 1969: The electrofreezing of supercooled water drops. *J. Met. Soc. Japan*, **47**, 65-74.

Ackerman, S.A. and H. Chung, 1992: Radiative effects of airborne dust on regional energy budgets at the top of the atmosphere. *J. Appl. Meteor.*, **31**, 223-233.

Ackerman, A.S., O.B. Toon, D.E. Stevens, A.J. Heymsfield, V. Ramanathan, and E.J. Welton, 2000: Reduction of tropical cloudiness by soot. *Science*, **288**, 1042-1047.

Adams, P.J., J.H. Seinfeld, D.M. Koch, L Mickley, and D. Jacob, 2001: General circulation model assessment of direct radiative forcing by the sulfate-nitrate-ammonium-water inorganic aerosol system. *J. Geophys. Res.*, **106**, 1097-1111.

Albrecht, B.A., 1989: Aerosols, cloud microphysics, and fractional cloudiness. *Science*, **245**, 1227-1230.

Allen, D., K. Pickering, G. Stenchikov, A. Thompson, and Y. Kondo, 2000: A three-dimensional total odd nitrogen (NOy) simulation during SONEX using a stretched-grid chemical transport model. *J. Geophys. Res.*, **105**, 3851-3876.

Andronova, N.G., E.V. Rozanov, F. Yang. M.E. Schlesinger, and G.L. Stenchikov, 1999: Radiative forcing by volcanic aerosols from 1850 through 1994. *J. Geophys. Res.*, **104**, 16807-16826.

Arnold, N.F. and T.R. Robinson, 1998: Solar cycle changes to planetary wave propagation and their influence on the middle atmosphere circulation. *Annal. Geophys.*, **16**, 69-76.

Baliunas, S. and R. Jastrow, 1990: Evidence for long-term brightness changes of solar-type stars. *Nature*, **348**, 520-523.

Ballard, J., R.J. Knight, D.A. Newnham, J. Vander Auwera, M. Herman, G. Di Lonardo, G. Masciarelli, F.M. Nicolaisen, J.A. Beukes, L.K. Christensen, R. McPheat, G. Duxbury, R. Freckleton, and K.P. Shine, 2000: An intercomparison of laboratory measurements of absorption cross-sections and integrated absorption intensities for HCFC-22. *J. Quant. Spectrosc. Radiat. Transfer.*, **66**, 109-128.

Barry, J., G. Locke, D. Scollard, H. Sidebottom, J. Treacy, C. Clerbaux, R. Colin, and J. Franklin, 1997: 1,1,1,3,3,-pentafluorobutane (HFC-365mfc): Atmospheric degradation and contribution to radiative forcing. *Int. J. Chem. Kinet.*, **29**, 607-617.

Beer, J., S.T. Baumgartner, B. Dittrich-Hannen, J. Hauenstein, P. Kubik, C. Lukasczyk, W. Mende, R. Stellmacher, and M. Suter, 1994: Solar variability traced by cosmogenic isotopes. In: *The Sun as a Variable Star* [Pap, J.M., C. Fröhlich, H.S. Hudson, and S.K. Solanki (eds.)]. CUP 291-300.

Bekki, S., K.S. Law, and J.A. Pyle, 1994: Effect of ozone depletion on atmospheric CH_4 and CO concentrations. *Nature*, **371**, 595-597.

Berger, A., C. Tricot, H. Gallée, and M.F. Loutre, 1993: Water vapor, CO_2 and insolation over the last glacial-interglacial cycles. *Phil. Trans. Roy. Soc. London. B*, **341**, 253-261.

Berntsen, T.K. and I.S.A. Isaksen, 1999: Effects of lightning and convection on changes in tropospheric ozone due to NO_x emissions from aircraft. *Tellus*, **51B**, 766-788.

Berntsen, T., I.S.A. Isaksen, W.-C. Wang, and X-Z. Liang, 1996: Impacts of increased anthropogenic emissions in Asia on tropospheric ozone and climate. *Tellus*, **48B**, 13-32.

Berntsen, T.K., I.S.A. Isaksen, G. Myhre, J.S. Fuglestvedt, F. Stordal, T. A. Larsen, R.S. Freckleton, and K.P. Shine, 1997: Effects of anthropogenic emissions on tropospheric ozone and its radiative forcing. *J. Geophys. Res.*, **102**, 28101-28126.

Berntsen, T.K., G. Myhre, F. Stordal, and I.S.A. Isaksen, 2000: Time evolution of tropospheric ozone and its radiative forcing. *J. Geophys. Res.*, **105**, 8915-8930.

Betts, R., 2001: Biogeophysical impacts of land use on present-day climate: near-surface temperature and radiative forcing. *Atmos. Sci. Lett.*, doi:10.1006/asle.2000.0023.

Bintanja, R., J.P.F. Fortuin, and H. Kelder, 1997: Simulation of the meridionally and seasonally varying climate response caused by changes in ozone concentration. *J. Climate*, **10**, 1288-1311.

Boucher, O., 1995: GCM estimate of the indirect aerosol forcing using satellite-retrieved cloud effective droplet radii. *J. Climate*, **8**, 1403-1409.

Boucher, O., 1999: Aircraft can increase cirrus cloudiness. *Nature*, **397**, 30-31.

Boucher, O. and H. Rodhe, 1994: The sulfate-CCN-cloud albedo effect: A sensitivity study. Report CM-83, Department of Meteorology, Stockholm University, 20 pp.

Boucher, O. and T.L. Anderson, 1995: GCM assessment of the sensitivity of direct climate forcing by anthropogenic sulfate aerosols to aerosol size and chemistry. *J. Geophys. Res.*, **100**, 26117-26134.

Boucher, O. and U. Lohmann, 1995: The sulfate-CCN-cloud albedo effect: A sensitivity study using two general circulation models. *Tellus*, **47B**, 281-300.

Boucher, O. and D. Tanré, 2000: Estimation of the aerosol perturbation to the Earth's radiative budget over oceans using POLDER satellite aerosol retrievals. *Geophys. Res. Lett.*, **27**, 1103-1106.

Boucher, O., H. Le Treut, and M.B. Baker, 1995: Precipitation and radiation modelling in a GCM: Introduction of cloud microphysics. *J. Geophys. Res.*, **100**, 16395-16414.

Boucher, O., S.E. Schwartz, T.P. Ackerman, T.L. Anderson, B. Bergstrom, B. Bonnel, P. Chylek, A. Dahlback, Y. Fouquart, Q. Fu, R.N. Halthore, J.M. Haywood, T. Iversen, S. Kato, S. Kinne, A. Kirkevåg, K.R. Knapp, A. Lacis, I. Laszlo, M.I. Mishchenko, S. Nemesure, V. Ramaswamy, D.L. Roberts, P. Russell, M.E. Schlesinger, G.L. Stephens, R. Wagener, M. Wang, J. Wong, and F. Yang, 1998: Intercomparison of models representing direct shortwave radiative forcing by sulfate aerosols. *J. Geophys. Res.*, **103**, 16979-16998.

Brasseur, G.P., J.T. Kiehl, J.-F. Muller, T. Schneider, C. Granier, X.X. Tie, and D. Hauglustaine, 1998: Past and future changes in global tropospheric ozone: impact on radiative forcing. *Geophys. Res. Lett.*, **25**, 3807-3810.

Brueckner, G.E., K.L. Edlow, L.E. Floyd, J.L. Lean, and M.E. VanHoosier, 1993: The Solar Ultraviolet Spectral Irradiance Monitor (SUSIM) experiment on board the Upper Atmosphere Research Satellite (UARS). *J. Geophys. Res.*, **98**, 10695-10711.

Brühl, C., 1993: The impact of the future scenarios for methane and other chemically active gases on the GWP of methane. *Chemosphere*, **26**, 731-738.

Capaldo, K., J.J. Corbett, P. Kasibhatla, P. Fischbeck, and S.N. Pandis, 1999: Effects of ship emissions on sulphur cycling and radiative climate forcing over the ocean. *Nature*, **400**, 743-746.

Cavalli, F., M. Glasius, S. Hjorth, B. Rindone, and N. R. Jensen, 1998: Atmospheric lifetimes, infrared spectra, and degradation products of a series of hydrofluoroethers. *Atmos. Env.*, **32**, 3767-3773.

Cebula, R. P., M.T. DeLand, and E. Hilsenrath, 1998: NOAA 11 Solar Backscattered Ultraviolet, model 2 (SBUV/2) instrument solar spectral irradiance measurements in 1989-1994. 1. Observations and long-term calibration. *J. Geophys. Res.*, **103**, 16235-16249.

Cess, R.D. and G.L. Potter, 1988: A methodology for understanding and intercomparing atmospheric climate feedback processes in GCMs. *J. Geophys. Res.*, **93**, 8305-8314.

Cess, R.D., M.-H. Zhang, G.L. Potter, H.W. Barker, R.A. Colman, D.A. Dazlich, A.D. Del Genio, M. Esch, J.R. Fraser, V. Galin, W.L. Gates, J.J. Hack, W.J. Ingram, J.T. Kiehl, A.A. Lacis, H. Le Treut, Z.-X. Li, X.Z. Liang, J.-F. Mahfouf, B.J. McAvaney, K.P. Meleshko, J.-J. Morcrette, D.A. Randall, E. Roeckner, J.-F. Royer, A.P. Sokolov, P.V. Sporyshev, K.E. Taylor, W.-C. Wang, and R.T. Wetherald, 1993: Uncertainties in CO_2 radiative forcing in atmospheric general circulation models. *Science*, **262**, 1252-1255.

Chalita, S., D.A. Hauglustaine, H. Le Treut, and J.-F. Müller, 1996: Radiative forcing due to increased tropospheric ozone concentra-tions. *Atmos. Env.*, **30**, 1641-1646.

Chandra, S., J.R. Ziemke, W. Min, and W.G. Read, 1998: Effects of 1997-1998 El Niño on tropospheric ozone and water vapor. *Geophys. Res. Lett.*, **25**, 3867-3870.

Chandra, S., J.R. Ziemke, and R.W. Stewart, 1999: An 11-year solar cycle in tropospheric ozone from TOMS measurements. *Geophys. Res. Lett.*, **26**, 185-188.

Charlson, R.J., 2000: The coupling of biogeochemical cycles and climate: Forcings, feedbacks and responses. In "Earth System Sciences: from Biogeochemical Cycles to Global Change", Academic Press.

Charlson, R.J., S.E. Schwartz, J.M. Hales, R.D. Cess, J.A. Coakley, J.E. Hansen, and D.J. Hofmann, 1992: Climate forcing by anthropogenic aerosols. *Science*, **255**, 423-430.

Charlson, R.J., T.L. Anderson, and H. Rodhe, 1999: Direct climate forcing by anthropogenic aerosols: quantifying the link between sulfate and radiation. *Contrib. Atmos. Phys.*, **72**, 79-94.

Chazette, P., C. Clerbaux, and G. Mégie, 1998: Direct estimate of methane radiative forcing by use of nadir spectral radiances. *Applied Optics*, **37**, 3113-3120.

Chen, C-T. and V. Ramaswamy, 1996a: Sensitivity of simulated global climate to perturbations in low cloud microphysical properties. I. Globally uniform perturbations. *J. Climate*, **9**, 1385-1402.

Chen, C-T. and V. Ramaswamy, 1996b: Sensitivity of simulated global climate to perturbations in low cloud microphysical properties. II. Spatially localized perturbations. *J. Climate*, **9**, 2788-2801.

Chin, M. and D.J. Jacob, 1996: Anthropogenic and natural contributions to tropospheric sulfate: a global model analysis. *J. Geophys. Res.*, **101**, 18691-18699.

Christiansen, B., 1999: Radiative forcing and climate sensitivity: The ozone experience. *Quart. J. R. Meteorol. Soc.*, **125**, 3011-3035.

Christensen, L.K., J. Sehested, O.J. Nielsen, M. Bilde, T.J. Wallington, A. Guschin, L.T. Molina, and M.J. Molina, 1998: Atmospheric chemistry of HFE-7200 ($C_4F_9OC_2H_5$): Reaction with OH radicals, fate of $C_4F_9OCH_2CH_2O$ and C_4F_9OCHO CH_3 radicals. *J. Phys. Chem.* A, **102**, 4839-4845.

Christidis, N., M.D. Hurley, S. Pinnock, K.P. Shine, and T.J. Wallington, 1997: Radiative forcing of climate change by CFC-11 and possible CFC replacements. *J. Geophys. Res.*, **102**, 19597-19609.

Christopher, S.A., D.V., Kliche, J. Chou, and R.M. Welch, 1996: First estimates of the radiative forcing of aerosols generated from biomass burning using satellite data. *J. Geophys. Res.*, **101**, 21265-21273.

Chuang, C.C., J.E. Penner, K.E. Taylor, and J.J. Walton, 1994: Climate effects of anthropogenic sulfate: Simulations from a coupled chemistry/climate model. In: Preprints of the Conference on Atmospheric Chemistry, Nashville, Tennessee, January 1994. American Meteorological Society, Boston, USA, pp. 170-174.

Chuang, C.C., J.E. Penner, K.E. Taylor, A.S. Grossman, and J.J. Walton, 1997: An assessment of the radiative effects of anthropogenic sulfate. *J. Geophys. Res.*, **102**, 3761-3778.

Chuang, C.C., J.E. Penner, and Y. Zhang, 2000a: Simulations of aerosol indirect effect for IPCC emissions scenarios. In: Proceedings of the 11th Symposium on Global Change Studies, January 9-14, 2000, Long Beach, California. American Meteorological Society, Boston, USA, pp. 320-323.

Chuang, C.C., J.E. Penner, J.M. Prospero, K.E. Grant and G.H. Rau, 2000b: Effects of anthropogenic aerosols on cloud susceptibility: a sensitivity study of radiative forcing to aerosol characteristics and global concentration. Lawrence Livermore National Laboratory Internal Report, No. UCRL-JC-139097 Rev. 1., Lawrence Livermore National Laboratory, CA, USA.

Chylek, P. and J. Wong, 1995: Effect of absorbing aerosols on the global radiation budget. *Geophys. Res. Lett.*, **22**, 929-931.

Chylek, P., G. Videen, and D. Ngo, 1995: Effect of black carbon on the optical properties and climate forcing of sulfate aerosols. *J. Geophys.*

Res., **100**, 16325-16332.

Chylek, P., C.M. Banic, B. Johnson, P.A. Damiano, G.A. Isaac, W.R. Leaitch, P.S.K. Liu, F.S. Boudala, B. Winter, and D. Ngo, 1996a: Black carbon: Atmospheric concentrations and cloud water content measurements over southern Nova Scotia. *J. Geophys. Res.*, **101**, 29105-29110.

Chylek, P., G.B. Lesins, G. Videen, J.G.D. Wong, R.G. Pinnick, D. Ngo, and J.D. Klett, 1996b: Black carbon and absorption of solar radiation by clouds. *J. Geophys. Res.*, **101**, 23365-23371.

Claquin, T., M. Schulz, Y. Balkanski, and O. Boucher, 1998: Uncertainties in assessing radiative forcing by mineral dust. *Tellus*, **50B**, 491-505.

Cooke, W.F. and J.J.N. Wilson, 1996: A global black carbon model. *J. Geophys. Res.*, **101**, 19395-19409.

Cooke, W.F., C. Liousse, H. Cachier, and J. Feichter, 1999: Construction of a 1°x1° fossil-fuel emission dataset for carbonaceous aerosol and implementation and radiative impact in the ECHAM-4 model. *J. Geophys. Res.*, **104**, 22137-22162.

Cox, S.J., W.-C. Wang, and S.E. Schwartz, 1995: Climate response to forcings by sulfate aerosols and greenhouse gases. *Geophys. Res. Lett.*, **18**, 2509-2512.

Crommelynk, D., A. Fichot, R.B. Lee III, and J. Romero, 1995: First realisation of the space absolute radiometric reference (SARR) during the ATLAS 2 flight period. *Adv. Spac. Res.*, **16**, 17-23.

Daniel, J.S. and S. Solomon, 1998. On the climate forcing of carbon monoxide. *J. Geophys. Res.*, **103**, 13249-13260.

Daniel, J.S., S. Solomon, and D. Albritton, 1995: On the evaluation of halocarbon radiative forcing and global warming potentials. *J. Geophys. Res.*, **100**, 1271-1285.

DeLand, M.T. and R.P. Cebula, 1998: NOAA 11 Solar Backscattered Ultraviolet, model 2 (SBUV/2) instrument solar spectral irradiance measurements in 1989-1994. 2. Results, validation, and comparisons. *J. Geophys. Res.*, **103**, 16251-16273.

Delobbe, L. and H. Gallée, 1998: Simulation of marine stratocumulus: Effect of precipitation parameterization and sensitivity to droplet number concentration. *Boundary-Layer Meteorol.*, **89**, 75-107.

DeMore, W.B., S.P. Sander, D.M. Golden, R.F. Hampson, M.J. Kurylo, C.J. Howard, A.R. Ravishankara, C.E. Kolb, and M.J. Molina, 1997: Chemical kinetics and photochemical data for use in stratospheric modeling, Evaluation Number **12**, JPL Publ. 97-4, Jet Propulsion Laboratory, Pasadena, Calif.

Dentener, F.J., G.R. Carmichael, Y. Zhang, J. Lelieveld, and P. Crutzen, 1996: Role of mineral aerosol as a reactive surface in the global troposphere. *J. Geophys. Res.*, **101**, 22869-22889.

Dickinson, R.E., 1982: In: Carbon Dioxide Review [Clark, W.C. (ed.)]. Clarendon, New York, NY, USA, pp. 101-133.

Dickinson, R.E., and R.J. Cicerone, 1986: Future global warming from atmospheric trace gases. *Nature*, **319**, 109-115.

van Dorland, R., F.J. Dentener, and J. Lelieveld, 1997: Radiative forcing due to tropospheric ozone and sulfate aerosols. *J. Geophys. Res.*, **102**, 28079-28100.

Emmons, L.K., M.A. Carroll, D.A. Hauglustaine, G.P. Brasseur, C. Atherton, J. Penner, S. Sillman, H. Levy, F. Rohrer, W.M.F. Wauben, P.F.J. Van Velthoven, Y. Wang, D. Jacob, P. Bakwin, R. Dickerson, B. Doddridge, C. Gerbig, R. Honrath, G. Hübler, D. Jaffe, Y. Kondo, J.W. Munger, A. Torres, and A. Volz-Thomas, 1997: Climatologies of NOx and NOy: a comparison of data and models. *Atmos. Env.*, **31**, 1851-1904.

Evans, W.F.J. and E. Puckrin, 1995: An observation of the greenhouse radiation associated with carbon monoxide. *Geophys. Res. Lett.*, **22**, 925-928.

Evans, W.F.J. and E. Puckrin, 1999: A comparison of GCM models with experimental measurements of surface radiative forcing by greenhouse gases. 10th Symposium on Global Change Studies, American Meteorological Society, pp. 378-381, 10-15 January 1999,

Dallas, Texas.

Evans, S.J., R. Toumi, J.E. Harries, M.P. Chipperfield, and J.M. Russell III, 1998: Trends in stratospheric humidities and the sensitivity of ozone to these trends. *J. Geophys. Res.*, **103**, 8715-8725.

Fahey, D.W., U. Schumann, S. Ackerman, P. Artaxo, O. Boucher, M.Y. Danilin, B. Kärcher, P. Minnis, T. Nakajima, and O.B. Toon, 1999: Aviation-produced aerosols and cloudiness. In: Intergovernmental Panel on Climate Change Special Report on Aviation and the Global Atmosphere [Penner, J.E., D.H. Lister, D.J. Griggs, D.J. Dokken, and M. McFarland (eds.)]. Cambridge University Press, Cambridge, United Kingdom and New York, NY, USA, pp. 65-120.

Farrar, P.D., 2000: Are cosmic rays influencing oceanic cloud coverage – Or is it only El Niño? *Clim. Change*, **47**, 7-15.

Feichter, J., U. Lohmann, and I. Schult, 1997: The atmospheric sulfur cycle in ECHAM-4 and its impact on the shortwave radiation. *Clim. Dyn.*, **13**, 235-246.

Fels, S.B., J.D. Mahlman, M.D. Schwarzkopf, and R.W. Sinclair, 1980: Stratospheric sensitivity to perturbations in ozone and carbon dioxide: Radiative and dynamical response. *J. Atmos. Sci.*, **37**, 2265-2297.

Fishman, J., 1991: The global consequences of increasing tropospheric ozone concentrations. *Chemosphere*, **22**, 685-695.

Fishman, J. and V.G. Brackett, 1997: The climatological distribution of tropospheric ozone derived from satellite measurements using version 7 Total Ozone Mapping Spectrometer and Stratospheric Aerosol and Gas Experiment data sets. *J. Geophys. Res.*, **102**, 19275-19278.

Fleming, E.L., S. Chandra, C.H. Jackman, D.B. Considine, and A.R. Douglass, 1995: The middle atmosphere response to short and long term UV variations: An analysis of observations and 2D model results. *J. Atmos. Terr. Phys.*, **57**, 333-365.

Floyd, L.E., P.A. Reiser, P.C. Crane, L.C. Herring, D.K. Prinz, and G.E. Brueckner, 1998: Solar cycle 22 UV spectral irradiance variability: Current measurements by SUSIM UARS. *Solar Phys.*, **177**, 79-87.

Forster, P.M. de F., 1999: Radiative forcing due to stratospheric ozone changes 1979-1997, using updated trend estimates, *J. Geophys. Res.*, **104**, 24,395-24,399.

Forster, P. M. de F. and K.P. Shine, 1999: Stratospheric water vapour changes as a possible contributor to observed stratospheric cooling, *Geophys. Res. Lett.*, **26**, 3309-3312.

Forster, P.M. de F. and K.P. Shine, 1997: Radiative forcing and temperature trends from stratospheric ozone changes. *J. Geophys. Res.*, **102**, 10841-10857.

Forster, P.M. de F., R.S. Freckleton, and K.P. Shine, 1997: On aspects of the concept of radiative forcing. *Clim. Dyn.*, **13**, 547-560.

Forster, P.M. de F., M. Blackburn, R. Glover, and K.P. Shine, 2001: An examination of climate sensitivity for idealized climate change experiments in an intermediate general circulation model. *Clim. Dyn.*, **16**, 833-849.

Fortuin, J.P.F. and H. Kelder, 1996: Possible links between ozone and temperature profiles. *Geophys. Res. Lett.*, **23**, 1517-1520.

Fowler, L.D., D.A. Randall, and S.A. Rutledge, 1996: Liquid and ice cloud microphysics in the CSU general circulation model. Part I: Model description and simulated microphysical processes. *J. Climate*, **9**, 489-529.

Freckleton, R.S., S. Pinnock, and K.P. Shine, 1996: Radiative forcing of halocarbons: A comparison of line-by-line and narrow-band models using CF4 as an example. *J. Quant. Spectrosc. Radiat. Transfer*, **55**, 763-769.

Freckleton, R.S., E.J. Highwood, K.P. Shine, O. Wild, K.S. Law, and M.G. Sanderson, 1998: Greenhouse gas radiative forcing: Effects of averaging and inhomogeneities in trace gas distribution. *Q. J. R. Meteorol. Soc.*, **124**, 2099-2127.

Fröhlich, C. and J. Lean, 1998: The Sun's total irradiance: Cycles, trends and related climate change uncertainties since 1976. *Geophys. Res.*

Lett., **25**, 4377-4380.

Fuglestvedt, J.S., J.E. Jonson, and I.S.A. Isaksen, 1994: Effects of reduction in stratospheric ozone on tropospheric chemistry through changes in photolysis rates. *Tellus*, **46B**, 172-192.

Fuglestvedt, J.S., J.E. Jonson, W.-C. Wang, and I.S.A. Isaksen, 1995: Response in tropospheric chemistry to changes in UV fluxes, temperatures and water vapor densities. In: Atmospheric Ozone as a Climate Gas [Wang, W.-C. and I.S.A. Isaksen (eds.)]. NATO Series I 32, Springer-Verlag, Berlin, Germany, pp. 143-162.

Fuglestvedt, J.S., I.S.A. Isaksen, and W.-C. Wang, 1996: Estimates of indirect global warming potentials for CH_4, CO, and NO_x. *Clim. Change*, **34**, 405-437.

Fuglestvedt, J.S., T.K. Berntsen, I.S.A. Isaksen, H. Mao, X.-Z. Liang, and W.-C. Wang, 1999: Climate forcing of nitrogen oxides through changes in tropospheric ozone and methane: global 3D model studies. *Atmos. Env.*, **33**, 961-977.

Fuglestvedt, J. S., T. K. Berntsen, O. Godal, and T. Skodvin, 2000: Climate implications of GWP-based reductions in greenhouse gas emissions. *Geophys. Res. Lett.*, **27**, 409-412.

Ghan, S. and R. Easter, 1998: Comments: A limited-area-model case study of the effects of sub-grid scale variations in relative humidity and cloud upon the direct radiative forcing of sulfate aerosol. *Geophys. Res. Lett.*, **25**, 1039-1040.

Ghan, S.J., G. Guzman, and H. Abdul-Razzak, 1998: Competition between sea salt and sulfate particles as cloud condensation nuclei. *J. Atmos. Sci.*, **55**, 3340-3347.

Ghan, S.J., R.C. Easter, E. Chapman, H. Abdul-Razzak, Y. Zhang, R. Leung, N. Laulainen, R. Saylor, and R. Zaveri, 2001a: A physically-based estimate of radiative forcing by anthropogenic sulfate aerosol. *J. Geophys. Res.*, in press.

Ghan, S., R. Easter, J. Hudson, and F.-M. Bréon, 2001b: Evaluation of aerosol indirect radiative forcing in MIRAGE. *J. Geophys. Res.*, in press.

Gierczak, T., R.K. Talukdar, J.B. Burdkholder, R.W. Portmann, J.S. Daniel, S. Solomon, and A.R. Ravishankara, 1996: Atmospheric fate and greenhouse warming potentials of HFC-236fa and HFC-236ea. *J. Geophys. Res.*, **101**, 12905-12911.

Gierens, K. and M. Ponater, 1999: Comment on "Variation of cosmic ray flux and global cloud coverage - a missing link in solar-climate relationships" by H. Svensmark and E. Friis-Christensen. *J. Atmos. Solar-Terrest. Phys.*, **61**, 795-797.

Good, D.A., J.S. Francisco, A.K. Jain, D.J. Wuebbles, 1998: Lifetimes and global warming potentials for dimethyl ether and for fluorinated ethers: CH_3OCF_3 (E143a), CHF_2OCHF_2 (E134), CHF_2OCF_3 (E125). *J. Geophys. Res.*, **103**, 28181-28186.

Graf, H.-F., J. Feichter, and B. Langmann, 1997: Volcanic sulfur emissions: Estimates of source strength and its contribution to the global sulfate distribution. *J. Geophys. Res.*, **102**, 10727-10738.

Granier, C., J.-F. Müller, S. Madronich, and G. P. Brasseur, 1996: Possible causes for the 1990-1993 decrease in global tropospheric CO abundances: a three-dimensional sensitivity study. *Atmos. Env.*, **30**, 1673-1682.

Grant, K.E., C.C. Chuang, A.S. Grossman, and J.E. Penner: 1999: Modeling the spectral optical properties of ammonium sulfate and biomass aerosols: Parameterization of relative humidity effects and model results. *Atmos. Env.*, **33**, 2603-2620.

Grossman, A.S., K.E. Grant, W.E. Blass, and D.J. Wuebbles, 1997: Radiative forcing calculations for CH_3Cl and CH_3Br. *J. Geophys. Res.*, **102**, 13651-13656.

Gupta, M., L., R.J. Cicerone, and S. Elliott, 1998: Perturbation to global tropospheric oxidizing capacity due to latitudinal redistribution of surface sources of NO_x, CH_4, and CO. *Geophys. Res. Lett.*, **25**, 3931-3934.

Haigh, J.D., 1994: The role of stratospheric ozone in modulating the solar radiative forcing of climate. *Nature*, **370**, 544-546.

Haigh, J.D., 1996: The impact of solar variability on climate. *Science*, **272**, 981-984.

Haigh, J.D., 1999: A GCM study of climate change in response to the 11-year solar cycle. *Quart. J. Roy. Meteorol. Soc.*, **125**, 871-892.

Han, Q., W.B. Rossow, and A.A. Lacis, 1994: Near-global survey of effective droplet radii in liquid water clouds using ISCCP data. *J. Climate*, **7**, 465-497.

Han, Q., W.B. Rossow, J. Chou, and R.M. Welch, 1998: Global survey of the relationships of cloud albedo and liquid water path with droplet size using ISCCP. *J. Climate*, **7**, 1516-1528.

Hansen, J., D. Johnson, A. Lacis, S. Lebedeff, P. Lee, D. Rind, and G. Russell, 1981: Climate impact of increasing atmospheric carbon dioxide. *Science*, **213**, 957-966.

Hansen, J., A. Lacis, D. Rind, G. Russell, P. Stone, I. Fung, R. Ruedy, and J. Lerner, 1984: Climate sensitivity: analysis of feedback mechanisms, in Climate Processes and Climate Sensitivity Geophys. Monogr. Ser., New York, **29**, pp. 130-163.

Hansen, J., I. Fung, A. Lacis, D. Rind, S. Lebedeff, R. Ruedy, G. Russell, and P. Stone, 1988: Global climate changes as forecast by Goddard Institute for Space Studies 3-dimensional model. *J. Geophys. Res.*, **93**, 9341-9364.

Hansen, J., A. Lacis, R. Ruedy, M. Sato, and H. Wilson, 1993: How sensitive is the world's climate? *National Geographic Research and Exploration*, **9**, 142-158.

Hansen, J.E., M. Sato, and R. Ruedy, 1995: Long-term changes of the diurnal temperature cycle: implications about mechanisms of global climate change. *Atmos. Res.*, **37**, 175-209.

Hansen, J., M. Sato, and R. Ruedy, 1997a: Radiative forcing and climate response. *J. Geophys. Res.*, **102**, 6831-6864.

Hansen, J., M. Sato, A. Lacis, and R. Ruedy, 1997b: The missing climate forcing. *Phil. Trans. R. Soc. London B*, **352**, 231-240.

Hansen, J., M. Sato, A. Lacis, R. Ruedy, I. Tegen, and E. Matthews, 1998: Climate forcings in the Industrial Era. *Proc. Natl. Acad. Sci.*, **95**, 12753-12758.

Harris, N., R. Hudson, and C. Philips (eds.), 1998: Assessment of Trends in the Vertical Distribution of Ozone. WMO Ozone Research and Monitoring Project report number **43**, SPARC report number **1**, SPARC Office, Verrières le Buisson, France.

Harrison, R.G. and K.P. Shine, 1999: A review of recent studies of the influence of solar changes on the Earth's climate. Hadley Centre Technical Note **6**, available from: Hadley Centre for Climate Prediction and Research, The Met Office, London Road, Bracknell, Berks, RG12 2SY, UK.

Hauglustaine, D.A. and C. Granier, 1995: Radiative forcing by tropospheric ozone changes due to increased emissions of CH_4, CO and NO_x. In: Atmospheric Ozone as a Climate Gas [Wang, W.-C. and I. Isaksen (eds.)]. NATO-ASI Series, Springer-Verlag, Berlin, Germany, pp. 189-203.

Hauglustaine, D.A., C. Granier, and G.P. Brasseur, 1994: Impact of increased methane emissions on the atmospheric composition and related radiative forcing on the climate system. In: Non CO_2 Greenhouse Gases [van Ham, J. *et al.* (eds.)]. Kluwer Academic Publishers, pp. 253-259.

Hauglustaine, D. A., G. P. Brasseur, S. Walters, P. J. Rasch, J.-F. Muller, L. K. Emmons, and M. A. Carroll, J., 1998: MOZART, a global chemical transport model for ozone and related chemical tracers. Part 2: Model results and evaluation. *J. Geophys. Res.*, **103**, 28291-28335.

Hauglustaine, D.A., G.P. Brasseur, and J.S. Levine, 1999: A sensitivity simulation of tropospheric ozone changes due to the l997 Indonesian fire emissions. *Geophys. Res. Lett*, **26**, 3305-3308.

Hauglustaine, D.A., L.K. Emmons, M. Newchurch, G.P. Brasseur, T. Takao, K. Matsubara, J. Johnson, B. Ridley, J. Stith, and J. Dye, 2001: On the role of lightning NO_x in the formation of tropospheric ozone plumes: a global model perspective. *J. Atmos. Chem.*, **38**, 277-294.

Haywood, J.M. and K.P. Shine, 1995: The effect of anthropogenic sulfate and soot on the clear sky planetary radiation budget. *Geophys. Res. Lett.*, **22**, 603-606.

Haywood, J.M. and K.P. Shine, 1997: Multi-spectral calculations of the radiative forcing of tropospheric sulphate and soot aerosols using a column model. *Quart. J. R. Met. Soc.*, **123**, 1907-1930.

Haywood, J.M. and V. Ramaswamy, 1998: Global sensitivity studies of the direct radiative forcing due to anthropogenic sulfate and black carbon aerosols. *J. Geophys. Res.*, **103**, 6043-6058.

Haywood, J.M. and Boucher, O., 2000: Estimates of the direct and indirect radiative forcing due to tropospheric aerosols: a review. *Revs. Geophys.*, **38**, 513-543.

Haywood, J.M., D.L. Roberts, A. Slingo, J.M. Edwards, and K.P. Shine, 1997a: General circulation model calculations of the direct radiative forcing by anthropogenic sulphate and fossil-fuel soot aerosol. *J. Clim.*, **10**, 1562-1577.

Haywood, J.M., V. Ramaswamy, and L.J. Donner, 1997b: A limited-area-model case study of the effects of sub-grid scale variations in relative humidity and cloud upon the direct radiative forcing of sulfate aerosol. *Geophys. Res. Lett.*, **24**, 143-146.

Haywood, J., R. Stouffer, R. Wetherald, S. Manabe, and V. Ramaswamy, 1997c: Transient response of a coupled model to estimated changes in greenhouse gas and sulfate concentrations. *Geophys. Res. Lett.*, **24**, 1335-1338.

Haywood, J.M., M.D. Schwarzkopf, and V. Ramaswamy, 1998a: Estimates of radiative forcing due to modeled increases in tropospheric ozone. *J. Geophys. Res.*, **103**, 16999-17007.

Haywood, J.M., V. Ramaswamy, and L.J. Donner, 1998b: Reply: A limited-area-model case study of the effects of sub-grid scale variations in relative humidity and cloud upon the direct radiative forcing of sulfate aerosol. *Geophys. Res. Lett.*, **25**, 1041.

Haywood, J.M., V. Ramaswamy, and B.J. Soden, 1999: Tropospheric aerosol climate forcing in clear-sky satellite observations over the oceans. *Science*, **283**, 1299-1303.

Heathfield, A.E., C. Anastasi, P. Pagsberg, and A. McCulloch, 1998a: Atmospheric lifetimes of selected fluorinated ether compounds. *Atmos. Env.*, **32**, 711-717.

Heathfield, A.E., C. Anastasi, A. McCulloch, and F.M. Nicolaisen, 1998b: Integrated infrared absorption coefficients of several partially fluorinated ether compounds: CF_3OCF_2H, CF_2HOCF_2H, $CH_3OCF_2CF_2H$, CH_3OCF_2CFClH, $CH_3CH_2OCF_2CF_2H$, $CF_3CH_2OCF_2CF_2H$, and $CH_2=CHCH_2OCF_2CF_2H$. *Atmos. Env.*, **32**, 2825-2833.

Hegg, D.A., 1994: The cloud condensation nucleus-sulfate mass relationship and cloud albedo. *J. Geophys. Res.*, **99**, 25903-25907.

Heintzenberg, J. and M. Wendisch, 1996: On the sensitivity of cloud albedo to the partitioning of particulate absorbers in cloudy air. *Contrib. Atmos. Phys.*, **69**, 491-499.

Henderson-Sellers, A., 1995: Human effects on climate through the large-scale impacts of land-use change. In: Future Climates of the World: A Modeling Perspective. World Survey of Climatology, Vol. 16, Elsevier, Amsterdam.

Hewitt, C.D. and J.F.B. Mitchell, 1997: Radiative forcing and response of a GCM to ice age boundary conditions: cloud feedback and climate sensitivity. *Clim. Dyn.*, **13**, 821-834.

Highwood, E.J. and K.P. Shine, 2000: Radiative forcing and global warming potentials of 11 halogenated compounds. *J. Quant. Spectrosc. Radiat. Transf.*, **66**, 169-183.

Highwood, E.J., K.P. Shine, M.D. Hurley, and T.L. Wallington, 1999: Estimation of direct radiative forcing due to non-methane hydrocarbons. *Atmos. Env.*, **33**, 759-767.

Hobbs, P.V., J.S. Reid, R.A. Kotchenruther, R.J. Ferek, and R. Weiss, 1997: Direct radiative forcing by smoke from biomass burning. *Science*, **275**, 1776-1778.

von Hoyningen-Huene, W., K. Wenzel, and S. Schienbein, 1999: Radiative properties of desert dust and its effect on radiative balance. *J. Aerosol Sci.*, **30**, 489-502.

Hoyt, D.V. and K.H. Schatten, 1993: A discussion of plausible solar irradiance variations, 1700-1992. *J. Geophys. Res.*, **98**, 18895-18906.

Hoyt, D.V. and K.H. Schatten, 1998: Group sunspot numbers: A new solar activity reconstruction. *Solar Phys.*, **181**, 491-512.

Hudson, R.D. and A.M. Thompson, 1998: Tropical tropospheric ozone from total ozone mapping spectrometer by a modified residual method. *J. Geophys. Res.*, **103**, 22129-22145.

Huntrieser, H., H. Schlager, C. Feigl, and H. Höller, 1998: Transport and production of NO_x in electrified thunderstorms: Survey of previous studies and new observations at midlatitudes. *J. Geophys. Res.*, **103**, 28247-28264.

Iacobellis, S.F., R. Frouin, and R.C.J. Somerville, 1999: Direct climate forcing by biomass-burning aerosols: impact of correlations between controlling variables. *J. Geophys. Res.*, **104**, 12031-12045.

Imasu, R., A. Suga, and T. Matsuno, 1995: Radiative effects and halocarbon global warming potentials of replacement compounds for chlorofluorocarbons. *J. Meteorol. Soc. Japan.*, **73**, 1123-1136.

IPCC, 1990: Climate Change 1990: The Intergovernmental Panel on Climate Change Scientific Assessment [Houghton, J.T., B.A. Callander, and S.K. Varney (eds)]. Cambridge University Press, Cambridge, United Kingdom and New York, NY, USA.

IPCC, 1992: Climate Change 1992: The Supplementary Report to the Intergovernmental Panel on Climate Change Scientific Assessment [Houghton, J.T., B.A. Callander, and S.K. Varney (eds.)]. Cambridge University Press, Cambridge, United Kingdom and New York, NY, USA, 100 pp.

IPCC, 1994: Climate Change 1994: Radiative Forcing of Climate Change and an Evaluation of the IPCC IS92 Emission Scenarios [Houghton, J.T., L.G. Meira Filho, J. Bruce, H. Lee, B.A. Callander, E.F. Haites, N. Harris, and K. Maskell (eds.)]. Cambridge University Press, Cambridge, United Kingdom and New York, NY, USA.

IPCC, 1996a: Climate Change 1995: The Science of Climate Change. Contribution of Working Group I to the Second Assessment Report of the Intergovernmental Panel on Climate Change [Houghton, J.T., L.G. Meira Filho, B.A. Callander, N. Harris, A. Kattenberg, and K. Maskell (eds.)]. Cambridge University Press, Cambridge, United Kingdom and New York, NY, USA, 572 pp.

IPCC, 1996b, Climate change 1995: Impacts, Adaptations and Mitigation of Climate Change: Scientific-Technical Analyses. Contribution of Working Group II to the Second Assessment Report of the Intergovernmental Panel on Climate Change [Watson, R.T., M.C. Zinyowera, and R.H. Moss (eds.)]. Cambridge University Press, Cambridge, United Kingdom and New York, NY, USA, 880 pp.

IPCC, 1999: Intergovernmental Panel on Climate Change Special Report on Aviation and the Global Atmosphere [Penner, J.E., D.H. Lister, D.J. Griggs, D.J. Dokken, and M. McFarland (eds.)]. Cambridge University Press, Cambridge, United Kingdom and New York, NY, USA, 373 pp.

Iversen, T., A. Kirkevåg, J.E. Kristjansson, and Ø. Seland, 2000: Climate effects of sulphate and black carbon estimated in a global climate model. In: Air pollution and its Application XIV [Gryning S.E. and F.A. Schiermeier (eds.)], Kluwer/Plenum, in press.

Jacobson, M.Z., 2000: A physically-based treatment of elemental carbon optics: Implications for global direct forcing of aerosols. *Geophys. Res. Lett.*, **27**, 217-220.

Jacobson, M.Z., 2001: Global direct radiative forcing due to multicomponent anthropogenic and natural aerosols. *J. Geophys. Res.*, **106**, 1551-1568.

Jacquinet-Husson, N., E. Arie, J. Ballard, A. Barbe, G. Bjoraker, B. Bonnet, L.R. Brown, C. Camy-Peyret, J.P. Champion, A. Chédin, A. Chursin, C. Clerbaux, G. Duxbury, J.-M. Flaud, N. Fourrie, A. Fayt, G. Graner, R. Gamache, A. Goldman, V. Golovko, G. Guelachvili,

J.M. Hartmann, J.C. Hilico, J. Hillman, G. Lefevre, E. Lellouch, S.N. Mikhailenko, O.V. Naumenko, V. Nemtchinov, D.A. Newnham, A. Nikitin, J. Orphal, A. Perrin, D. C. Reuter, C.P. Rinsland, L. Rosenmann, L.S. Rothman, N.A. Scott, J. Selby, L.N. Sinitsa, J.M. Sirota, A.M. Smith, K.M. Smith, V.G. Tyuterev, R.H. Tipping, S. Urban, P. Varanasi, and M. Weber, 1999: The 1997 spectroscopic GEISA databank. *J. Quant. Spectrosc. Radiat. Transfer*, **62**, 205-254.

Jain, A.K., B.P. Briegleb, K. Minschwaner, and D.J. Wuebbles, 2000: Radiative forcings and global warming potentials of 39 greenhouse gases. *J. Geophys. Res.*, **105**, 20773-20790.

Jensen, E.J. and O.B. Toon, 1997: The potential impact of soot from aircraft exhaust on cirrus clouds. *Geophys. Res. Lett.*, **24**, 249-252.

Johns, T.C., J.M. Gregory, W.J. Ingram, C.E. Johnson, A. Jones, J.A. Lowe, J.F.B. Mitchell, D.L. Roberts, D.M.H. Sexton, D.S. Stevenson, S.F.B. Tett, and M.J. Woodage, 2001: Anthropogenic climate change for 1860 to 2100 simulated with the HadCM3 model under updated emissions scenarios. Hadley Centre Technical Note **22**, Hadley Centre, Met Office, Bracknell, Berks, UK.

Johnson, C.E. and R.G. Derwent, 1996: Relative radiative forcing consequences of global emissions of hydrocarbons, carbon monoxide, and NO_x from human activities estimated with a zonally-averaged two-dimensional model. *Clim. Change*, **34**, 439-462.

Jones, A. and A. Slingo, 1996: Predicting cloud-droplet effective radius and indirect sulphate aerosol forcing using a general circulation model. *Q. J. R. Meteorol. Soc.*, **122**, 1573-1595.

Jones, A. and A. Slingo, 1997: Climate model studies of sulphate aerosols and clouds. *Phil. Trans. R. Soc. London B*, **352**, 221-229.

Jones, A., D.L. Roberts, and A. Slingo, 1994: A climate model study of the indirect radiative forcing by anthropogenic sulphate aerosols. *Nature*, **370**, 450-453.

Jones, A., D.L. Roberts, and M.J. Woodage, 1999: The indirect effects of anthropogenic sulphate aerosol simulated using a climate model with an interactive sulphur cycle. Hadley Centre Technical Note no. **14**, 38 pp., available from: Hadley Centre for Climate Prediction and Research, The Met Office, London Road, Bracknell, Berks, RG12 2SY, UK.

Joos, F., M. Bruno, R. Fink, T.F. Stocker, U. Siegenthaler, C. Le Quéré, and J.L. Sarmiento, 1996: An efficient and accurate representation of complex oceanic and biospheric models of anthropogenic carbon uptake. *Tellus*, **48B**, 397-417.

Jørgensen, T.S. and A.W. Hansen, 2000: Comment on "Variation of cosmic ray flux and global cloud coverage - a missing link in solar-climate relationships" by H. Svensmark and E. Friis-Christensen. *J. Atmos. Solar-Terrest. Phys.*, **62**, 73-77.

Karl, T.R., P.D. Jones, R.W. Knight, G. Kukla, N. Plummer, V. Razuvayev, K.P. Gallo, J. Lindseay, R.J. Charlson, and T.C. Peterson, 1993: A new perspective on recent global warming: Asymmetric trends of daily maximum and minimum temperature. *Bull. Am. Meteorol. Soc.*, **74**, 1007-1023.

Kasibhatla, P., W.L. Chameides, and J. St. John, 1997: A three-dimensional global model investigation of the seasonal variation in the atmospheric burden of anthropogenic sulfate aerosols. *J. Geophys. Res.*, **102**, 3737-3759.

Kattenberg, A., F. Giorgi, H. Grassl, G.A. Meehl, J.F.B. Mitchell, R.J. Stouffer, T. Tokioka, A.J. Weaver, and T.M.L. Wigley, 1996: Climate Models – Projections of Future Climate. In: Climate Change 1995: The Science of Climate Change. The Contribution of Working Group I to the Second Assessment Report of the International Panel on Climate Change [Houghton, J.T., L.G. Meira Filho, B.A. Callender, N. Harris, A. Kattenberg, and K. Maskell (eds.)]. Cambridge University Press, Cambridge, United Kingdom and New York, NY, USA, pp. 285-357.

Kaufman, Y.J. and T. Nakajima, 1993: Effect of Amazon smoke on cloud microphysics and albedo - Analysis from satellite imagery. J. Appl. Meteorol., **32**, 729-744.

Kaufman, Y.J. and R.S. Fraser, 1997: Control of the effect of smoke particles on clouds and climate by water vapor. *Science*, **277**, 1636-1639.

Kaufman, Y.J., D. Tanré, A. Karnieli, and L.A. Remer, 2001: Absorption of sunlight by dust as inferred from satellite and ground-based remote sensing. *Geophys. Res. Lett.*, **28**, 1479-1482.

Kernthaler, S.C., R. Toumi, and J.D. Haigh, 1999: Some doubts concerning a link between cosmic ray fluxes and global cloudiness. *Geophys. Res. Lett.*, **26**, 863-866.

Kiehl, J.T. and B.P. Briegleb, 1993: The relative roles of sulfate aerosols and greenhouse gases in climate forcing. *Science* **260**, 311-314.

Kiehl, J.T. and H. Rodhe, 1995: Modelling geographical and seasonal forcing due to aerosols. In: Aerosol Forcing of Climate [Charlson, R.J. and J. Heintzenberg (eds.)]. J. Wiley and Sons Ltd, pp. 281-296.

Kiehl, J. T., T. L. Schneider, R. W. Portmann, and S. Solomon, 1999: Climate forcing due to tropospheric and stratospheric ozone. *J. Geophys. Res.*, **104**, 31239-31254.

Kiehl, J.T., T.L. Schneider, P.J. Rasch, M.C. Barth, and J. Wong, 2000: Radiative forcing due to sulfate aerosols from simulations with the National Center for Atmospheric Research Community Climate Model, Version 3. *J. Geophys. Res.*, **105**, 1441-1457.

Kleinman, L.I., 1994: Low and high NO_x tropospheric photochemistry. *J. Geophys. Res.*, **99**, 16831-16838.

Klonecki, A. and H. Levy II, 1997: Tropospheric chemical ozone tendencies in $CO-CH_4-NO_y-H_2O$ system: Their sensitivities to variations in environmental parameters and their application to global chemistry transport model study. *J. Geophys. Res.*, **102**, 21221-21237.

Ko, M., R.-L. Shia, N.-D. Sze, H. Magid, and R.G. Bray, 1999: Atmospheric lifetime and global warming potential of HFC-245fa. *J. Geophys. Res.*, **104**, 8173-8181.

Koch, D., D. Jacob, I. Tegen, D. Rind, and M. Chin, 1999: Tropospheric sulfur simulation and sulfate direct radiative forcing in the Goddard Institute for Space Studies general circulation model. *J. Geophys. Res.*, **104**, 23799-23822.

Kodera, K., 1995: On the origin and nature of the interannual variability of the winter stratospheric circulation in the northern hemisphere. *J. Geophys. Res.*, **100**, 14077-14087.

Kogan, Z.N., Y.L. Kogan, and D.K. Lilly, 1996: Evaluation of sulfate aerosols indirect effect in marine stratocumulus clouds using observation-derived cloud climatology. *Geophys. Res. Lett.*, **23**, 1937-1940.

Kogan, Z.N., Y.L. Kogan, and D.K. Lilly, 1997: Cloud factor and seasonality of the indirect effect of anthropogenic sulfate aerosols. *J. Geophys. Res.*, **102**, 25927-25939.

Kondratyev, K.Y., 1999: Climatic Effects of Aerosols and Clouds. Springer-Verlag, Berlin, 264 pp.

Kristjánsson, J.E. and J. Kristiansen, 2000: Is there a cosmic ray signal in recent variations in global cloudiness and cloud radiative forcing? *J. Geophys. Res.*, **105**, 11851-11863.

Kristensson, A., J.-F. Gayet, J. Ström, and F. Auriol, 2000: In situ observations of a reduction in effective crystal diameter in cirrus clouds near flight corridors. *Geophys. Res. Lett.*, **27**, 681-684.

Krol, M., P.J. van Leeuwen, and J. Lelieveld, 1998: Global OH trend inferred from methylchloroform measurements. *J. Geophys. Res.*, **103**, 10697-10711.

Kuang, Z., Y. Jiang, and Y.L. Yung, 1998: Cloud optical thickness variations during 1983-1991: Solar cycle or ENSO? *Geophys. Res. Lett.*, **25**, 1415-1417.

Kulmala, M., A. Laaksonen, P. Korhonen, T. Vesala, and T. Ahonen, 1993: The effect of atmospheric nitric acid vapor on cloud condensation nucleus activation. *J. Geophys. Res.*, **98**, 22949-22958.

Kulmala, M., P. Korhonen, A. Laaksonen, and T. Vesala, 1995: Changes in cloud properties due to NO_x emissions. *Geophys. Res. Lett.*, **22**, 239-242.

Kulmala, M., A. Toivonen, T. Mattila, and P. Korhonen, 1998: Variations of cloud droplet concentrations and the optical properties of clouds due to changing hygroscopicity: A model study. *J. Geophys. Res.*, **103**, 16183-16195.

Laaksonen, A., J. Hienola, M. Kulmala, and F. Arnold, 1997: Supercooled cirrus cloud formation modified by nitric acid pollution of the upper troposphere. *Geophys. Res. Lett.*, **24**, 3009-3012.

Labitzke, K. and H. van Loon, 1997: The signal of the 11-year sunspot cycle in the upper troposphere-lower stratosphere. *Space Sci. Rev.*, **80**, 393-410.

Lacis, A., J. Hansen, and M. Sato, 1992: Climate forcing by stratospheric aerosols. *Geophys. Res. Lett.*, **19**, 1607-1610.

Lacis, A.A., D.J. Wuebbles, and J.A. Logan, 1990: Radiative forcing by changes in the vertical distribution of ozone. *J. Geophys. Res.*, **95**, 9971-9981.

Lamarque, J.-F., G.P. Brasseur, P.G. Hess, and J.-F. Muller, 1996: Three-dimensional study of the relative contributions of the different nitrogen sources in the troposphere. *J. Geophys. Res.*, **101**, 22955-22968.

Langner, J. and H. Rodhe, 1991: A global three-dimensional model of the tropospheric sulfur cycle. *J. Atmos. Chem.*, **13**, 225-263.

Larkin, A., J.D. Haigh, and S. Djavidnia, 2000: The effect of solar UV irradiance variations on the Earth's atmosphere. *Space Science Reviews*, **94**, 199-214.

Lashof, D.A. and D.R. Ahuja, 1990: Relative contributions of greenhouse gas emissions to global warming. *Nature*, **344**, 529-531.

Lean, J. and D. Rind, 1998: Climate forcing by changing solar radiation. *J. Clim.*, **11**, 3069-3094.

Lean, J., A. Skumanitch, and O. White, 1992: Estimating the Sun's radiative output during the Maunder Minimum. *Geophys. Res. Lett.*, **19**, 1591-1594.

Lean, J., J. Beer and R.S. Bradley, 1995: Reconstruction of solar irradiance since 1610: Implications for climate change. *Geophys. Res. Lett.*, **22**, 3195-3198.

Lean, J.L., G.J. Rottman, H.L. Kyle, T.N. Woods, J.R. Hickey, and L.C. Puga, 1997: Detection and parameterization of variations in solar mid- and near-ultraviolet radiation (200-400 nm). *J. Geophys. Res.*, **102**, 29939-29956.

Lelieveld, J. and P.J. Crutzen, 1992: Indirect chemical effects of methane on climate warming. *Nature*, **355**, 339-342.

Lelieveld, J. and P.J. Crutzen, 1994: Role of deep cloud convection in the ozone budget of the troposphere. *Science*, **264**, 1759-1761.

Lelieveld, J. and F. Dentener, 2000: What controls tropospheric ozone? *J. Geophys. Res.*, **105**, 3531-3551.

Lelieveld, J., P.J. Crutzen, and C. Brühl, 1993: Climate effects of atmospheric methane. *Chemosphere*, **26**, 739-768.

Lelieveld, J., P. Crutzen, and F.J. Dentener, 1998: Changing concentration, lifetime and climate forcing of atmospheric methane. *Tellus*, **50B**, 128-150.

Le Treut, H., M. Forichon, O. Boucher, and Z-X. Li, 1998: Sulfate aerosol, indirect effect and CO_2 greenhouse forcing: Equilibrium response of the LMD GCM and associated cloud feedbacks. *J. Climate*, **11**, 1673-1684.

Levin, Z., E. Ganor, and V. Gladstein, 1996: The effects of desert particles coated with sulfate on rain formation in the Eastern Mediterranean. *J. Appl. Meteor.*, **35**, 1511-1523.

Levy, II, H., W.J. Moxim, and P.S. Kasibhatla, 1996: A global three-dimensional time-dependent lightning source of tropospheric NO_x. *J. Geophys. Res.*, **101**, 22911-22922.

Li, Z., Z. Tao, V. Naik, D.A. Good, J. Hansen, G.-R. Jeong, J.S. Francisco, A.K. Jain, and D.J. Wuebbles, 2000: Global warming potential assessment for $CF_3OCF=CF_2$. *J. Geophys. Res.*, **105**, 4019-4029.

Liao, H. and J.H. Seinfeld, 1998: Effect of clouds on direct radiative forcing of climate. *J. Geophys. Res.*, **103**, 3781-3788.

Lindberg, J.D., J.B. Gillespie, and B.D. Hinds, 1976: Measurements of the refractive indices of atmospheric particulate matter from a variety of geographic locations. In: Proceedings of the International Symposium on Radiation in the Atmosphere, Garmisch-Partenkirchen [Bolle, H.J., (ed.)]. Science Press, New York.

Liousse, C., J.E. Penner, C.C. Chuang, J.J. Walton, and H. Eddleman, 1996. A global three dimensional model study of carbonaceous aerosols. *J. Geophys. Res.*, **101**, 19411-19432.

Lockwood, M. and R. Stamper, 1999: Long-term drift of the coronal source magnetic flux and the total solar irradiance. *Geophys. Res. Lett.*, **26**, 2461-2464.

Lohmann, U. and J. Feichter, 1997: Impact of sulfate aerosols on albedo and lifetime of clouds: A sensitivity study with the ECHAM4 GCM. *J. Geophys. Res.*, **102**, 13685-13700.

Lohmann, U., J. Feichter, C.C. Chuang, and J.E. Penner, 1999a: Prediction of the number of cloud droplets in the ECHAM GCM. *J. Geophys. Res.*, **104**, 9169-9198.

Lohmann, U., J. Feichter, C.C. Chuang, and J.E. Penner, 1999b: Correction. *J. Geophys. Res.*, **104**, 24557-24563.

Lohmann, U., J. Feichter, J. Penner, and R. Leaitch, 2000: Indirect effect of sulfate and carbonaceous aerosols: A mechanistic treatment. *J. Geophys. Res.*, **105**, 12193-12206.

Ma, J. and M. van Weele, 2000: Effects of stratospheric ozone depletion on the net production of ozone in polluted rural areas. *Chemosphere-Global Change Science*, **2**, 23-37.

Madronich, S. and C. Granier, 1994: Tropospheric chemistry changes due to increased UV-B radiation. In: Stratospheric Ozone Depletion/UV-B Radiation in the Biosphere [Biggs, R.H. and M.E.B. Joyner (eds.)]. NATO ASI Series, Springer-Verlag, Berlin, Germany.

Manabe, S. and R. Wetherald, 1967: Thermal equilibrium of the atmosphere with a given distribution of relative humidity. *J. Atmos. Sci.*, **24**, 241-259.

Manabe, S. and R. Wetherald, 1980: On the distribution of climate change resulting from an increase in CO_2 content of the atmosphere. *J. Atmos. Sci.*, **37**, 99-118.

Manabe, S. and A.J. Broccoli, 1985: The influence of continental ice sheets on the climate of an ice age. *J. Geophys. Res.*, **90**, 2167-2190.

Marenco, A., H. Gouget, P. Nédélec, J.P. Pagés, and F. Karcher, 1994: Evidence of a long-term increase in tropospheric ozone from Pic du Midi data series: Consequences: positive radiative forcing. *J. Geophys. Res.*, **99**, 16617-16632.

McCormack, J.P. and L.L. Hood, 1996: Apparent solar cycle variations of upper stratospheric ozone and temperature: Latitude and seasonal dependencies. *J. Geophys. Res.*, **101**, 20933-20944.

Meerkötter, R., U. Schumann, D.R. Doelling, P. Minnis, T. Nakajima, and Y. Tsushima, 1999: Radiative forcing by contrails. *Ann. Geophys.*, **17**, 1080-1094.

Mendoza, B., 1997: Estimates of Maunder Minimum solar irradiance and Ca II H and K fluxes using rotation rates and diameters. *Astrophys. J.*, **483**, 523-526.

Mickley, L.J., P.P. Murti, D.J. Jacob, J.A. Logan, D.M. Koch, and D. Rind, 1999: Radiative forcing from tropospheric ozone calculated with a unified chemistry-climate model. *J. Geophys. Res.*, **104**, 30153-30172.

Miller, R. and I. Tegen, 1998: Climate response to soil dust aerosols. *J. Clim.*, **11**, 3247-3267.

Miller, R. and I. Tegen, 1999: Radiative forcing of a tropical direct circulation by soil dust aerosols. *J. Atmos. Sci*, **56**, 2403-2433.

Minnis, P., U. Schumann, D.R. Doelling, K.M. Gierens, and D.W. Fahey, 1999: Global distribution of contrail radiative forcing. *Geophys. Res. Lett.*, **26**, 1853-1856.

Minschwaner, K., R.W. Carver, B.P. Briegleb, and A.E. Roche, 1998: Infrared radiative forcing and atmospheric lifetimes of trace species based on observations from UARS. *J. Geophys. Res.*, **103**, 23243-23253.

Mishchenko, M.I., L.D. Travis, R.A. Kahn, and R.A. West, 1997:

Modeling phase functions for dustlike tropospheric aerosols using a shape mixture of randomly orientated polydisperse spheroids. *J. Geophys. Res.*, **102**, 16831-16847.

Mitchell, J.F.B. and T.C. Johns, 1997: On modification of global warming by sulfate aerosols. *J. Clim.*, **10**, 245-267.

Molina, L.T., P.J. Wooldridge, and M.J. Molina, 1995: Atmospheric reactions and ultraviolet and infrared absorptivities of nitrogen trifluoride. *Geophys. Res. Lett.*, **22**, 1873-1876.

Murphy, D.M., J.R Anderson, P.K. Quinn, L.M. McInnes, M. Posfai, D.S. Thomson, and P.R. Buseck, 1998: Influence of sea-salt on aerosol radiative properties in the Southern Ocean marine boundary layer. *Nature*, **392**, 62-65.

Myhre, G. and F. Stordal, 1997: Role of spatial and temporal variations in the computation of radiative forcing and GWP. *J. Geophys. Res.*, **102**, 11181-11200.

Myhre, G., E.J. Highwood, K.P. Shine, and F. Stordal, 1998b: New estimates of radiative forcing due to well mixed greenhouse gases. *Geophys. Res. Lett.*, **25**, 2715-2718.

Myhre, G., F. Stordal, B. Rognerud, and I.S.A. Isaksen, 1998a: Radiative forcing due to stratospheric ozone. In: Atmospheric Ozone: Proceedings of the XVIII Quadrennial Ozone Symposium [Bojkov, R.D. and G. Visconti (eds.)]. L'Aquila, Italy, Parco Scientifico e Technologico d'Abruzzo, pp. 813-816.

Myhre, G., F. Stordal, K. Restad, and I. Isaksen, 1998c: Estimates of the direct radiative forcing due to sulfate and soot aerosols. *Tellus*, **50B**, 463-477.

Myhre, G., C.J. Nielsen, D.L. Powell, and F. Stordal, 1999: Infrared absorption cross section, radiative forcing, and GWP of four hydrofluoro(poly)ethers. *Atmos. Env.*, **33**, 4447-4458.

Myhre, G., S. Karlsdóttir, I.S.A. Isaksen, and F. Stordal, 2000: Radiative forcing due to changes in tropospheric ozone in the period 1980 to 1996. *J. Geophys. Res.*, **105**, 28935-28942.

Myhre, G, A. Myhre, and F. Stordal, 2001: Historical evolution of radiative forcing of climate. *Atmos. Env.*, **35**, 2361-2373.

Naik, V., A. Jain, K.O. Patten, and D.J. Wuebbles, 2000: Consistent sets of atmospheric lifetimes and radiative forcings on climate for CFC replacements: HCFCs and HFCs. *J. Geophys. Res.*, **105**, 6903-6914.

Nakićenović, N., J. Alcamo, G. Davis, B. de Vries, J. Fenhann, S. Gaffin, K. Gregory, A. Grübler, T.Y. Jung, T. Kram, E.L. La Rovere, L. Michaelis, S. Mori, T. Morita, W. Pepper, H. Pitcher, L. Price, K. Raihi, A. Roehrl, H.-H. Rogner, A. Sankovski, M. Schlesinger, P. Shukla, S. Smith, R. Swart, S. van Rooijen, N. Victor, and Z. Dadi, 2000: IPCC Special Report on Emissions Scenarios, Cambridge University Press, Cambridge, United Kingdom and New York, NY, USA, 599 pp.

Nedoluha, G.E., R. Bevilacqua, R.M. Gomez, D.E. Siskind, B. Hicks, J. Russell, and B.J. Connor, 1998: Increases in middle atmospheric water vapor as observed by the Halogen Occultation Experiment and the ground-based Water Vapor Millimeter-wave Spectrometer from 1991 to 1997. *J. Geophys. Res.*, **103**, 3531-3543.

Nemesure, S., R. Wagener, and S.E. Schwartz, 1995: Direct shortwave forcing of climate by anthropogenic sulfate aerosol: sensitivity to particle size, composition, and relative humidity. *J. Geophys. Res.*, **100**, 26105-26116.

Nesme-Ribes, E., E.N. Ferreira, R. Sadourny, H. Le Treut, and Z.X. Li, 1993: Solar dynamics and its impact on solar irradiance and the terrestrial climate. *J. Geophys. Res.*, **98**, 18923-18935.

Novakov, T. and J.E. Penner, 1993: Large contribution of organic aerosols to cloud-condensation-nuclei concentrations. *Nature*, **365**, 823-826.

Novakov, T. and C.E. Corrigan, 1996: Cloud condensation nucleus activity of the organic component of biomass smoke particles. *Geophys. Res. Lett.*, **23**, 2141-2144.

O'Dowd, D., J.A. Lowe, and M.H. Smith, 1999: Coupling sea-salt and sulphate interactions and its impact on cloud droplet concentration

predictions. *Geophys. Res. Lett.*, **26**, 1311-1314.

Oltmans, S.J. and D.J. Hofmann, 1995: Increase in lower-stratospheric water vapour at a mid-latitude Northern hemisphere site from 1981 to 1994. *Nature*, **374**, 146-149.

Oltmans, S.J., A.S. Lefohn, H.E. Sheel, J.M. Harris, H. Levy, I.E. Galbally, E.-G. Brunke, C.P. Meyer, J.A. Lathrop, B.J. Johnson, D.S. Shadwick, E. Cuevas, F.J. Schmidlin, D.W. Tarasick, H. Claude, J.B. Kerr, O. Uchino, and V. Mohnen, 1998: Trends of ozone in the troposphere. *Geophys. Res. Lett.*, **25**, 139-142.

Pan, W., M.A. Tatang, G.J. McRae, and R.G. Prinn, 1997: Uncertainty analysis of the direct radiative forcing by anthropogenic sulfate aerosols. *J. Geophys. Res.*, **102**, 21915-21924.

Papasavva, S., S. Tai, K.H. Illinger, and J.E. Kenny, 1997: Infrared radiative forcing of CFC substitutes and their atmospheric reaction products. *J. Geophys. Res.*, **102**, 13643-13650.

Penner, J.E., R.E. Dickinson, and C.A. O'Neill, 1992: Effects of aerosol from biomass burning on the global radiation budget. *Science*, **256**, 1432-1434.

Penner, J.E, R.J. Charlson, J.M. Hales, N.S. Laulainen, R. Leifer, T. Novakov, J. Ogren, L.F. Radke, S.E. Schwartz, and L. Travis, 1994: Quantifying and minimising uncertainty of climate forcing by anthropogenic aerosols. *Bull. Am. Meterol. Soc.*, **75**, 375-400.

Penner, J.E., C.C. Chuang, and C. Liousse, 1996: The contribution of carbonaceous aerosols to climate change. In: Nucleation and Atmospheric Aerosols [Kulmala, M. and P.E. Wagner (eds.)]. Elsevier Science, pp. 759-769.

Penner, J.E., T.M. Wigley, P. Jaumann, B.D. Santer, and K.E. Taylor, 1997: Anthropogenic aerosols and climate change: A method for calibrating forcing. In: Assessing Climate Change: Results from the Model Evaluation Consortium for Climate Assessment [Howe, W. and A. Henderson-Sellers (eds.)]. Gordon and Breach Publishers, Sydney, Australia, pp. 91-111.

Penner, J.E., D.J. Bergmann, J. Walton, D. Kinnison, M.J. Prather, D. Rotman, C. Price, K. Pickering, and S.L. Baughcum, 1998a: An evaluation of upper tropospheric NO$_x$ in two models. *J. Geophys. Res.*, **103**, 22097-22113.

Penner, J. E., C.C. Chuang, and K. Grant, 1998b: Climate forcing by carbonaceous and sulfate aerosols. *Clim. Dyn.*, **14**, 839-851.

Petit, J.R., 1990. Paleoclimatological and chronological implications of the Vostok core dust record. *Nature*, **343**, 56-58.

Pham, M., J.F. Muller, G. Brasseur, C. Granier, and G. Mégie, 1995: A three-dimensional study of the tropospheric sulfur cycle. *J. Geophys. Res.*, **100**, 26061-26092.

Pilinis, C., S.N. Pandis, and J.H. Seinfeld, 1995: Sensitivity of direct climate forcing by atmospheric aerosols to aerosol size and composition. *J. Geophys. Res.*, **100**, 18739-18754.

Pincus, R. and M. Baker, 1994: Precipitation, solar absorption, and albedo susceptibility in marine boundary layer clouds. *Nature*, **372**, 250-252.

Pinnock, S. and K.P. Shine, 1998: The effects of changes in HITRAN and uncertainties in the spectroscopy on infrared irradiance calculations. *J. Atmos. Sci.*, **55**, 1950-1964.

Pinnock, S., M.D. Hurley, K.P. Shine, T.J. Wallington, and T.J. Smyth, 1995: Radiative forcing of climate by hydrochlorofluorocarbons and hydrofluorocarbons. *J. Geophys. Res.*, **100**, 23277-23238.

Platnick, S. and S. Twomey, 1994: Determining the susceptibility of cloud albedo to changes in droplet concentration with the Advanced Very High Resolution Radiometer. *J. Appl. Meteorol.*, **33**, 334-347.

Pope, V.D., M.L. Gallani, P.R. Rowntree, and R.A. Stratton, 2000: The impact of new physical parametrizations in the Hadley Centre climate model: HadAM3. *Clim. Dyn.*, **16**, 123-146.

Portmann, R.W., S. Solomon, J. Fishman, J.R. Olson, J.T. Kiehl, and B. BrieGleb, 1997: Radiative forcing of the Earth-climate system due to tropical tropospheric ozone production. *J. Geophys. Res.*, **102**, 9409-9417.

Prather, M. J., 1994: Lifetimes and eigenstates in atmospheric chemistry. *Geophys. Res. Lett.*, **21**, 801-804.

Prather, M. J., 1996: Times scales in atmospheric chemistry: theory, GWPs for CH_4 and CO, and runaway growth. *Geophys. Res. Lett.*, **23**, 2597-2600.

Pruppacher, H.R., 1973: Electrofreezing of supercooled water. *Pure Appl. Geophys.*, **104**, 623-634.

Pruppacher, H.R. and J.D. Klett, 1997: Microphysics of clouds and precipitation. 2nd Ed., Kluwer.

Ramanathan, V., 1981: The role of ocean-atmosphere interactions in the CO_2-climate problem. *J. Atmos. Sci.*, **38**, 918-930.

Ramanathan, V. and J. Coakley, 1978: Climate modeling through radiative-convective models. *Rev. Geophys. Space Phys.*, **16**, 465-490.

Ramanathan, V. and R. Dickinson, 1979: The role of stratospheric ozone in the zonal and seasonal radiative energy balance of the earth-troposphere system. *J. Atmos. Sci.*, **36**, 1084-1104.

Ramanathan, V., R. Cicerone, H. Singh, and J. Kiehl, 1985: Trace gas trends and their potential role in climate change. *J. Geophys. Res*, **90**, 5547-5566.

Ramanathan,V., L. Callis, R. Cess, J. Hansen, I. Isaksen, W. Kuhn, A. Lacis, F. Luther, J. Mahlman, R. Reck, and M. Schlesinger, 1987: Climate-chemical interactions and effects of changing atmospheric trace gases. *Rev. Geophys.*, **25**, 1441-1482.

Ramanathan, V., R.D. Cess, E.F. Harrison, P. Minnis, B. Barkstrom, E. Ahmed, and D. Hartmann, 1989: Cloud radiative forcing and climate: Results from the Earth Radiation Budget Experiment. *Science*, **243**, 57-63.

Ramaswamy, V. and C.-T. Chen, 1997a: Linear additivity of climate response for combined albedo and greenhouse perturbations. *Geophys. Res. Lett.*, **24**, 567-570.

Ramaswamy, V. and C.-T. Chen, 1997b: Climate forcing-response relationships for greenhouse and shortwave radiative perturbations. *Geophys. Res. Lett.*, **24**, 667-670.

Randel, W. J., F. Wu, J. R. Russell III, and J. Waters, 1999: Space-time patterns of trends in stratospheric constituents derived from UARS measurements. *J. Geophys. Res.*, **104**, 3711-3727.

Reid, G., 1997: Solar forcing of global climate change since the mid-17th century. *Clim. Change*, **37**, 391-405.

Reilly, J., R. Prinn, J. Harnisch, J. Fitzmaurice, H. Jacoby, D. Kicklighter, J. Mellilo, P. Stone, A. Sokolov, and C. Wang, 1999: Multi-gas assessment of the Kyoto Protocol. *Nature*, **401**, 549-554.

Remer, L.A., Y.J. Kaufman, D. Tanré, Z. Levin, and D.A. Chu, 1999: Principles in remote sensing of aerosol from MODIS over land and ocean, Proceedings of the ALPS99 International Conference, 18-22 January 1999, Méribel, France, pp. WK1/O/05/1-4.

Restad, K., I. Isaksen, and T.K. Berntsen, 1998: Global distribution of sulfate in the troposphere: A 3-dimensional model. *Atmos Env.*, **32**, 3593-3609.

Rind, D. and N.K. Balachandran, 1995: Modelling the effects of UV variability and the QBO on the troposphere-stratosphere system. Part II: The troposphere. *J. Clim.*, **8**, 2080-2095.

Rind, D., D. Peteet, and G. Kukla, 1989: Can Milankovitch orbital variations initiate the growth of ice sheets in a GCM? *J. Geophys. Res.*, **94**, 12851-12871.

Robock, A. and M.P. Free, 1995: Ice cores as an index of global volcanism from 1850 to the present. *J. Geophys. Res.*, **100**, 11549-11567.

Robock, A. and M.P. Free, 1996: The volcanic record in ice cores for the past 2000 years. In: Climatic Variations and Forcing Mechanisms of the Last 2000 Years [Jones, P., R. Bradley, and J. Jouzel (eds.)]. Springer-Verlag, Berlin, Germany, pp. 533-546.

Rodhe, H. and P.J. Crutzen, 1995: Climate and CCN. *Nature*, **375**, 111.

Rodhe, H., R.J. Charlson, and T.L. Anderson, 2000: Avoiding circular logic in climate modeling. *Clim. Change*, **44**, 419-422.

Roeckner, E., T. Siebert, and J. Feichter, 1994: Climatic response to anthropogenic sulfate forcing simulated with a general circulation model. In: Aerosol Forcing of Climate [Charlson, R. and J. Heintzenberg (eds.)]. John Wiley and Sons, pp. 349-362.

Roehl, C.M., D. Boglu, C. Brühl, and G.K. Moortgat, 1995: Infrared band intensities and global warming potentials of CF_4, C_2F_6, C_3F_8, C_4F_{10}, C_5F_{12}, and C_6F_{14}. *Geophys. Res. Lett.*, **22**, 815-818.

Roelofs, G.-J., J. Lelieveld, and R. van Dorland, 1997: A three dimensional chemistry/general circulation model simulation of anthropogenically derived ozone in the troposphere and its radiative climate forcing. *J. Geophys. Res.*, **102**, 23389-23401.

Roelofs, G.-J., J. Lelieveld, and L. Ganzeveld, 1998: Simulation of global sulfate distribution and the influence on effective cloud drop radii with a coupled photochemistry-sulfur cycle model. *Tellus*, **50B**, 224-242.

Ross, J.L., P.V. Hobbs, and B. Holben, 1998: Radiative characteristics of regional hazes dominated by smoke from biomass burning in Brazil: Closure tests and direct radiative forcing. *J. Geophys. Res.*, **103**, 31925-31941.

Rothman, L.S., R.R. Gamache, R.H. Tipping, C.P. Rinsland, M.A.H. Smith, D.C. Benner, V.M. Devi, J.-M. Flaud, C. Camy-Peyret, A. Perrin, A. Goldman, S.T. Massie, L.R. Brown, and R.A. Toth, 1992: The HITRAN molecular database: Editions of 1991 and 1992. *J. Quant. Spectrosc. Radiat. Transf.*, **48**, 469-507.

Rothman, L.S., C.P. Rinsland, A. Goldman, S.T. Massie, D.P. Edwards, J.-M. Flaud, A. Perrin, C. Camy-Peyret, V. Dana, J.-Y. Mandin, J. Schroeder, A. McCann, R.R. Gamache, R.B. Wattson, K. Yoshino, K.V. Chance, K.W. Jucks, L.R. Brown, V. Nemtchinov, and P. Varanasi, 1998: The HITRAN molecular spectroscopic database and HAWKS (HITRAN atmospheric workstation): 1996 edition. *J. Quant. Spectrosc. Radiat. Transfer*, **60**, 665-710.

Rotstayn, L.D., 1999: Indirect forcing by anthropogenic aerosols: A GCM calculation of the effective-radius and cloud lifetime effects. *J. Geophys. Res.*, **104**, 9369-9380.

Rotstayn, L.D., 2000: On the "tuning" of autoconversion parameterizations in climate models. *J. Geophys. Res.*, **105**, 15495-15507.

Rotstayn, L.D. and J.E. Penner, 2001: Indirect aerosol forcing, quasi-forcing, and climate response. *J. Climate*, in press.

Rottman, G.J, T.N. Woods, and T.P. Sparn, 1993: Solar Stellar Irradiance Comparison Experiment: Instrument design and operation. *J. Geophys. Res.*, **98**, 10667-10678.

Rowntree, P., 1998: Global average climate forcing and temperature response since 1750. *Int. J. Climatol.*, **18**, 355-377.

SAR, see IPCC, 1996a.

Sato, M., J. Hansen, M.P. McCormick, and J.B. Pollack, 1993: Stratospheric aerosol optical depths, 1850-1990. *J. Geophys. Res.*, **98**, 22987-22994.

Sausen, R., K. Gierens, M. Ponater, and U. Schumann, 1998: A diagnostic study of the global distribution of contrails: Part I: Present day climate. *Theor. Appl. Climatol.*, **61**, 127-141.

Schlesinger, M.E., X. Jiang, and R.J. Charlson, 1992: Implication of anthropogenic atmospheric sulphate for the sensitivity of the climate system. In: Climate Change and Energy Policy: Proceedings of the International Conference on Global Climate Change: Its Mitigation through Improved Production and Use of Energy [Rosen, L. and R. Glasser (eds.)]. Amer. Inst. Phys., New York, NY, USA, pp. 75-108.

Shi, G., 1992: Radiative forcing and greenhouse effect due to the atmospheric trace gases. Science in China (Series B), **35**, 217-229.

Shindell, D., D. Rind, N. Balachandran, J. Lean, and P. Lonergan, 1999: Solar cycle variability, ozone, and climate. *Science*, **284**, 305-308.

Shine, K.P. and P.M. de F. Forster, 1999: The effects of human activity on radiative forcing of climate change: a review of recent developments. *Global and Planetary Change*, **20**, 205-225.

Shine, K.P., B.P. Briegleb, A. Grossman, D. Hauglustaine, H. Mao, V. Ramaswamy, M.D. Schwarzkopf, R. van Dorland, and W-C. Wang,

1995: Radiative forcing due to changes in ozone: a comparison of different codes. In Atmospheric Ozone as a Climate Gas, NATO ASI Series, Springer-Verlag, Berlin, **32**, 373-396.

Shine, K.P., Y. Fouquart, V. Ramaswamy, S. Solomon, and J. Srinivasan, 1996: Radiative Forcing of Climate Change, Chapter 2.4. In: Climate Change 1995: The Science of Climate Change. Contribution of Working Group I to the Second Assessment Report of the Intergovernmental Panel on Climate Change [Houghton, J.T., L.G. Meira Filho, B.A. Callender, N. Harris, A. Kattenberg, and K. Maskell (eds.)]. Cambridge University Press, Cambridge, United Kingdom and New York, NY, USA, pp. 108-118.

Shine, K.P., R.S. Freckleton, and P.M. de F. Forster, 1998: Comment on "Climate forcing by stratospheric ozone depletion calculated from observed temperature trends" by Zhong *et al. Geophys. Res. Lett.*, **25**, 663-664.

Shira, K., M.D. Hurley, K.P. Shine, and T.J. Wallington, 2001: Updated radiative forcing estimates of sixty-five halocarbons and non-methane hydrocarbons. *J. Geophys. Res.*, in press.

Siegenthaler, U. and F. Joos, 1992: Use of a simple model for studying oceanic tracer distributions and the global carbon cycle. *Tellus*, **44B**, 186-207.

Sinha, A. and R. Toumi, 1996: A comparison of climate forcings due to chlorofluorocarbons and carbon monoxide. *Geophys. Res. Lett.*, **23**, 65-68.

Sinha, A. and J.E. Harries, 1997: Possible change in climate parameters with zero net radiative forcing. *Geophys. Res. Lett.*, **24**, 2355-2358.

Sinha, A. and R. Toumi, 1997: Tropospheric ozone, lightning and climate change. *J. Geophys. Res.*, **102**, 10667-10672.

Skodvin, T. and J.S. Fuglestvedt, 1997: A comprehensive approach to climate change: Political and scientific considerations. *Ambio*, **26**, 351-358.

Smith, S.J. and T.M.L. Wigley, 2000a: Global warming potentials, 1, Climatic implications of emissions reductions. *Clim. Change*, **44**, 445-457.

Smith, S.J. and T.M.L. Wigley, 2000b: Global warming potentials, 2, Accuracy. *Clim. Change*, **44**, 459-469.

Smith, W.L., S. Ackerman, H. Revercomb, H. Huang, D.H. DeSlover, W. Feltz, L. Gumley and A. Collard, 1998: Infrared spectral absorption of nearly invisible cirrus clouds. *Geophys. Res. Lett.*, **25**, 1137-1140.

Sokolik, I.N. and O.B. Toon, 1996: Direct radiative forcing by anthropogenic airborne mineral aerosols. *Nature*, **381**, 681-683.

Sokolik, I.N., A. Andronova, and T.C. Johnson, 1993: Complex refractive index of atmospheric dust aerosols. *Atmos. Env.*, **27A**, 2495-2502.

Sokolik, I.N., O.B. Toon, and R.W. Bergstrom, 1998: Modeling the direct radiative characteristics of airborne mineral aerosols at infrared wavelengths. *J. Geophys. Res.*, **103**, 8813-8826.

Solanki, S. K. and M. Fligge, 1998: Solar irradiance since 1874 revisited. *Geophys. Res. Lett.*, **25**, 341-344.

Solomon, S. and J.S. Daniel, 1996: Impact of the Montreal Protocol and its Amendments on the rate of change of global radiative forcing. *Clim. Change*, **32**, 7-17.

Solomon, S., R.W. Portmann, R.R. Garcia, L.W. Thomason, L.R. Poole, and M.P. McCormick, 1996: The role of aerosol variations in anthropogenic ozone depletion at northern midlatitudes. *J. Geophys. Res.*, **101**, 6713-6727.

Solomon, S., R.W. Portmann, R.W. Sanders, J.S. Daniel, W. Madsen, B. Bartram, and E.G. Dutton, 1999: On the rôle of nitrogen dioxide in the absorption of solar radiation. *J. Geophys. Res.*, **104**, 12047-12058.

Soon, W.H., E.S. Posmentier, and S.L. Baliunas, 1996: Inference of solar irradiance variability from terrestrial temperature changes, 1880-1993: An astrophysical application of the Sun-climate connection. *Astrophys. J.*, **472**, 891-982.

Stenchikov, G., I. Kirchner, A. Robock, H.-F. Graf, J.C. Antuna, R.G. Grainger, A. Lambert, and L. Thomason, 1998: Radiative forcing

from the 1991 Mount Pinatubo volcanic eruption. *J. Geophys. Res.*, **103**, 13837-13857.

Stern, D.I. and R.K. Kaufmann, 1996: Estimates of global anthropogenic sulfate emissions 1860-1993, Center for Energy and Environmental Studies, Working Papers Series, Number 9602, Boston University.

Stevens, M.J. and G.R. North, 1996: Detection of the climate response to the solar cycle. *J. Atmos. Sci.*, **53**, 2594-2608.

Stevenson, D.S., C.E. Johnson, W.J. Collins, R.G. Derwent, K.P. Shine, and J.M. Edwards, 1998: Evolution of tropospheric ozone radiative forcing. *Geophys. Res. Lett.*, **25**, 3819-3822.

Ström, J. and S. Ohlsson, 1998: In situ measurements of enhanced crystal number densities in cirrus clouds caused by aircraft exhaust. *J. Geophys. Res.*, **103**, 11355-11361.

Stuiver, M. and P.J. Reimer, 1993: Extended ^{14}C data base and revised CALIB 3.0 ^{14}C age calibration program. *Radiocarbon*, **35**, 215-230.

Sturges, W.T., T.J. Wallington, M.D. Hurley, K.P. Shine, K. Sihra, A. Engel, D.E. Oram, S.A. Penkett, R. Mulvaney, and C.A.M. Brenninkmeijer, 2000: A potent greenhouse gas identified in the atmosphere: SF_5CF_3. *Science*, **289**, 611-613.

Svensmark, H., 1998: Influence of cosmic rays on Earth's climate. *Phys. Rev. Lett.*, **81**, 5027-5030.

Svensmark, H. and E. Friis-Christensen, 1997: Variation of cosmic ray flux and global cloud coverage - A missing link in solar-climate relationships. *J. Atmos. Solar Terrest. Phys.*, **59**, 1225-1232.

Taylor, J.P. and A. McHaffie, 1994: Measurements of cloud susceptibility. *J. Atmos. Sci.*, **51**, 1298-1306.

Taylor, K.E. and J.E. Penner, 1994: Response of the climate system to atmospheric aerosols and greenhouse gases. *Nature*, **369**, 734-737.

Tegen, I. and I. Fung, 1995: Contribution to the atmospheric mineral aerosol load from land surface modification. *J. Geophys. Res.*, **100**, 18707-18726.

Tegen, I. and A. Lacis, 1996: Modeling of particle size distribution and its influence on the radiative properties of mineral dust. *J. Geophys, Res.*, **101**, 19237-19244.

Tegen, I., A. Lacis, and I. Fung, 1996: The influence of mineral aerosols from disturbed soils on climate forcing. *Nature*, **380**, 419-422.

Tegen, I., P. Hollrigl, M. Chin, I. Fung, D. Jacob, and J.E. Penner, 1997: Contribution of different aerosol species to the global aerosol extinction optical thickness: Estimates from model results. *J. Geophys. Res.*, **102**, 23895-23915.

ten Brink, H.M., C. Kruisz, G.P.A. Kos, and A. Berner, 1997: Composition/size of the light-scattering aerosol in the Netherlands. *Atmos. Env.*, **31**, 3955-3962.

Tinsley, B.A., 1996: Correlations of atmospheric dynamics with solar wind-induced changes of air-earth current density into cloud top. *J. Geophys. Res.*, **101**, 29701-29714.

Tokuhashi, K, A. Takahashi, M. Kaise, S. Kondo, A. Sekiya, S. Yamashita, and H. Ito, 1999a: Rate constants for the reactions of OH radicals with $CH_3OCF_2CF_3$, $CH_3OCF_2CF_2CF_3$, and $CH_3OCF(CF_3)_2$, *International Journal of Chemical Kinetics*, **31**, 846-853.

Tokuhashi, K., H. Nagai, A. Takahashi, M. Kaise, S. Kondo, A. Sekiya, M. Takahashi, Y. Gotoh, and A. Suga, 1999b: Measurement of the OH reaction rate constants for CF_3CH_2OH, $CF_3CF_2CH_2OH$, and $CF_3CH(OH)CF_3$, *J. Phys. Chem.* A, **103**, 2664-2672.

Tokuhashi, K., A. Takahashi, M. Kaise, S. Kondo, A. Sekiya, S. Yamashita, and H. Ito, 2000: Rate constants for the reactions of OH radicals with $CH_3OCF_2CHF_2$, $CHF_2OCH_2CF_2CHF_2$, $CHF_2OCH_2CF_2CF_3$, and $CF_3CH_2OCF_2CHF_2$ over the temperature range 250-430K, *J. Phys. Chem.* A, **104**, 1165-1170.

Toumi, R., S. Bekki, and K. Law, 1994: Indirect influence of ozone depletion on climate forcing by clouds. *Nature*, **372**, 348-351.

Toumi, R., S. Bekki, and K. Law, 1995: Response to Climate and CCN by Rodhe and Crutzen. *Nature*, **375**, 111.

Toumi, R., J.D. Haigh, and K. Law, 1996: A tropospheric ozone-lightning climate feedback. *Geophys. Res. Lett.*, **23**, 1037-1040.

Twomey, S., 1974: Pollution and the planetary albedo. *Atmos. Env.*, **8**, 1251-1256.

Twohy, C.H., A.D. Clarke, S.G. Warren, L.F. Radke, and R.J. Charlson, 1989: Light-absorbing material extracted from cloud droplets and its effect on cloud albedo. *J. Geophys. Res.*, **94**, 8623-8631.

Van Dop, H. and M. Krol, 1996: Changing trends in tropospheric methane and carbon monoxide: a sensitivity analysis of the OH-radical. *J. Atmos. Chem.*, **25**, 271-288.

Veefkind, J.P., J.C.H. van der Hage, H.M. ten Brink, 1996: Nephelometer derived and directly measured aerosol optical depth of the atmospheric boundary layer. *Atmos. Res.*, **41**, 217-228.

Wallington, T.J., W.F. Schneider, J. Sehested, M. Bilde, J. Platz, O.J. Nielsen, L.K. Christensen, M.J. Molina, L.T. Molina, and P.W. Wooldridge, 1997: Atmospheric chemistry of HFE-7100 ($C_4F_9OCH_3$): Reaction with OH radicals, UV spectra, and kinetic data for $C_4F_9OCH_2$ and $C_4F_9OCH_2O_2$ radicals and the atmospheric fate of $C_4F_9OCH_2O$ radicals. *J. Phys. Chem.* A., **101**, 8264-8274.

Wang, H.J., D.M. Cunnold, and X. Bao, 1996: A critical analysis of stratospheric aerosol and gas experiment ozone trends. *J. Geophys. Res.*, **101**, 12495-12514.

Wang, W.-C., J.P. Pinto, and Y.L. Yung, 1980: Climatic effects due to halogenated compounds in the Earth's atmosphere. *J. Atmos. Sci.*, **37**, 333-338.

Wang, W.-C., M. Dudek, X-Z. Liang, and J. Kiehl, 1991: Inadequacy of effective CO_2 as a proxy in simulating the greenhouse effect of other radiatively active gases. *Nature*, **350**, 573-577.

Wang, W.-C., M. Dudek, and X-Z. Liang, 1992: Inadequacy of effective CO_2 as a proxy in assessing the regional climate change due to other radiatively active gases. *Geophys. Res. Lett.*, **19**, 1375-1378.

Wang, Y. and D.J. Jacob, 1998: Anthropogenic forcing on tropospheric ozone and OH since preindustrial times. *J. Geophys. Res.*, **103**, 31123-31135.

Wang, Y., A. W. DeSilva, G. C. Goldenbaum, and R. R. Dickerson, 1998: Nitric oxide production by simulated lightning: dependence on current, energy, and pressure, *J. Geophys. Res.*, **103**, 19,149-19,159, 1998.

Warren, S.G., C.J. Hahn, J. London, R.M. Chervin, and R.L. Jenne, 1988: Global distribution of total cloud cover and cloud type amounts over the ocean. NCAR Technical Note TN-317 + STR, Boulder, CO, 42 pp. + 170 maps.

Wetherald, R. and S. Manabe, 1988: Cloud feedback processes in a general circulation model. *J. Atmos. Sci.*, **45**, 1397-1415.

White, O.R., A. Skumanich, J. Lean, W.C. Livingston, and S. Keil, 1992: The sun in a non-cycling state. *Public. Astron. Soc. Pacif.*, **104**, 1139-1143.

Wigley, T.M.L., P.D. Jones, and S.C.B. Raper, 1997: The observed global warming record: What does it tell us? *Proc. Natl. Acad. Sci.* USA, **94**, 8314-8320.

Willson, R.C., 1997: Total solar irradiance trend during solar cycles 21 and 22. *Science*, **277**, 1963-1965.

WMO, 1986: Atmospheric Ozone: 1985, Global Ozone Research and Monitoring Project, World Meteorological Organization, Report No. **16**, Chapter 15, Geneva, Switzerland.

WMO, 1992: Scientific Assessment of Ozone Depletion: 1991, Global Ozone Research and Monitoring Project, World Meteorological Organization, Report No. **25**, Geneva, Switzerland.

WMO, 1995: Scientific Assessment of Ozone Depletion: 1994, Global Ozone Research and Monitoring Project, World Meteorological Organization, Report No. **37**, Geneva, Switzerland.

WMO, 1999: Scientific Assessment of Ozone Depletion: 1998, Global Ozone Research and Monitoring Project, World Meteorological Organization, Report No. **44**, Geneva, Switzerland.

Wuebbles, D.J. and J.M. Calm, 1997: An environmental rationale for retention of endangered chemicals. *Science*, **278**, 1090-1091.

Wuebbles, D.J., C.F. Wei, and K.O. Patten, 1998: Effects on stratospheric ozone and temperature during the Maunder Minimum. *Geophys. Res. Lett.*, **25**, 523-526.

Wyser, K. and J. Ström, 1998: A possible change in cloud radiative forcing due to aircraft exhaust. *Geophys. Res. Lett.*, **25**, 1673-1676.

Zhong, W., R. Toumi, and J.D. Haigh, 1996: Climate forcing by stratospheric ozone depletion calculated from observed temperature trends. *Geophys. Res. Lett.*, **23**, 3183-3186.

Zhong, W., R. Toumi, and J.D. Haigh 1998: Reply to comment on "Climate forcing by stratospheric ozone depletion calculated from observed temperature trends" by K.P. Shine, R.S. Freckleton, and P.M. de F. Forster. *Geophys. Res. Lett.*, **25**, 665.

7

Physical Climate Processes and Feedbacks

Co-ordinating Lead Author
T.F. Stocker

Lead Authors
G.K.C. Clarke, H. Le Treut, R.S. Lindzen, V.P. Meleshko, R.K. Mugara, T.N. Palmer,
R.T. Pierrehumbert, P.J. Sellers, K.E. Trenberth, J. Willebrand

Contributing Authors
R.B. Alley, O.E. Anisimov, C. Appenzeller, R.G. Barry, J.J. Bates, R. Bindschadler, G.B. Bonan,
C.W. Böning, S. Bony, H. Bryden, M.A. Cane, J.A. Curry, T. Delworth, A.S. Denning, R.E. Dickinson,
K. Echelmeyer, K. Emanuel, G. Flato, I. Fung, M. Geller, P.R. Gent, S.M. Griffies, I. Held,
A. Henderson-Sellers, A.A.M. Holtslag, F. Hourdin, J.W. Hurrell, V.M. Kattsov, P.D. Killworth,
Y. Kushnir, W.G. Large, M. Latif, P. Lemke, M.E. Mann, G. Meehl, U. Mikolajewicz, W. O'Hirok,
C.L. Parkinson, A. Payne, A. Pitman, J. Polcher, I. Polyakov, V. Ramaswamy, P.J. Rasch, E.P. Salathe,
C. Schär, R.W. Schmitt, T.G. Shepherd, B.J. Soden, R.W. Spencer, P. Taylor, A. Timmermann,
K.Y. Vinnikov, M. Visbeck, S.E. Wijffels, M. Wild

Review Editors
S. Manabe, P. Mason

Contents

Executive Summary

Considerable advances have been made in the understanding of processes and feedbacks in the climate system. This has led to a better representation of processes and feedbacks in numerical climate models, which have become much more comprehensive. Because of the presence of non-linear processes in the climate system, deterministic projections of changes are potentially subject to uncertainties arising from sensitivity to initial conditions or to parameter settings. Such uncertainties can be partially quantified from ensembles of climate change integrations, made using different models starting from different initial conditions. They necessarily give rise to probabilistic estimates of climate change. This results in more quantitative estimates of uncertainties and more reliable projections of anthropogenic climate change. While improved parametrizations have built confidence in some areas, recognition of the complexity in other areas has not indicated an overall reduction or shift in the current range of uncertainty of model response to changes in atmospheric composition.

Atmospheric feedbacks largely control climate sensitivity. Important progress has been made in the understanding of those processes, partly by utilising new data against which models can be compared. Since the Second Assessment Report (IPCC, 1996) (hereafter SAR), there has been a better appreciation of the complexity of the mechanisms controlling water vapour distribution. Within the boundary layer, water vapour increases with increasing temperatures. In the free troposphere above the boundary layer, where the greenhouse effect of water vapour is most important, the situation is less amenable to straightforward thermodynamic arguments. In models, increases in water vapour in this region are the most important reason for large responses to increased greenhouse gases.

Water vapour feedback, as derived from current models, approximately doubles the warming from what it would be for fixed water vapour. Since the SAR, major improvements have occurred in the treatment of water vapour in models, although detrainment of moisture from clouds remains quite uncertain and discrepancies exist between model water vapour distributions and those observed. It is likely that some of the apparent discrepancy is due to observational error and shortcomings in intercomparison methodology. Models are capable of simulating the moist and very dry regions observed in the tropics and sub-tropics and how they evolve with the seasons and from year to year, indicating that the models have successfully incorporated the basic processes governing water vapour distribution. While reassuring, this does not provide a definitive check of the feedbacks, though the balance of evidence favours a positive clear-sky water vapour feedback of a magnitude comparable to that found in simulations.

Probably the greatest uncertainty in future projections of climate arises from clouds and their interactions with radiation. Cloud feedbacks depend upon changes in cloud height, amount, and radiative properties, including short-wave absorption. The radiative properties depend upon cloud thickness, particle size, shape, and distribution and on aerosol effects. The evolution of clouds depends upon a host of processes, mainly those governing the distribution of water vapour. The physical basis of the cloud parametrizations included into the models has also been greatly improved. However, this increased physical veracity has not reduced the uncertainty attached to cloud feedbacks: even the sign of this feedback remains unknown. A key issue, which also has large implications for changes in precipitation, is the sensitivity of sub-grid scale dynamical processes, turbulent and convective, to climate change. It depends on sub-grid features of surface conditions such as orography. Equally important are microphysical processes, which have only recently been introduced explicitly in the models, and carry major uncertainties. The possibility that models underestimate solar absorption in clouds remains controversial, as does the effect of such an underestimate on climate sensitivity. The importance of the structure of the stratosphere and both radiative and dynamical processes have been recognised, and limitations in representing stratospheric processes add some uncertainty to model results.

Considerable improvements have taken place in modelling ocean processes. In conjunction with an increase in resolution, these improvements have, in some models, allowed a more realistic simulation of the transports and air-sea fluxes of heat and fresh water, thereby reducing the need for flux adjustments in coupled models. These improvements have also contributed to better simulations of natural large-scale circulation patterns such as El Niño-Southern Oscillation (ENSO) and the oceanic response to atmospheric variability associated with the North Atlantic Oscillation (NAO). However, significant deficiencies in ocean models remain. Boundary currents in climate simulations are much weaker and wider than in nature, though the consequences of this fact for the global climate sensitivity are not clear. Improved parametrizations of important sub-grid scale processes, such as mesoscale eddies, have increased the realism of simulations but important details are still under debate. Major uncertainties still exist with the representation of small-scale processes, such as overflows and flow through narrow channels (e.g., between Greenland and Iceland), western boundary currents (i.e., large-scale narrow currents along coastlines), convection, and mixing.

In the Atlantic, the thermohaline circulation (THC) is responsible for the major part of the ocean meridional heat transport associated with warm and saline surface waters flowing northward and cold and fresh waters from the North Atlantic returning at depth. The interplay between the large-scale atmospheric forcing, with warming and evaporation in low latitudes, and cooling and net precipitation at high latitudes, forms the basis of a potential instability of the present Atlantic THC. Changes in ENSO may also influence the Atlantic THC by altering the fresh water balance of the tropical Atlantic, therefore providing a coupling between low and high latitudes. Uncertainty resides with the relative importance of feedbacks associated with processes influencing changes in high latitude sea surface temperatures and salinities, such as atmosphere-ocean heat and fresh water fluxes, formation and transport of sea ice, continental runoff and the large-scale transports in ocean and atmosphere. The Atlantic THC is likely to change over the coming century but its evolution continues to be an unresolved issue. While some recent calculations find little changes in the THC, most projections suggest a gradual and significant decline of the THC. A complete shut-down of the THC is simulated in a number of models if the warming continues, but knowledge about the locations of thresholds for such a shut-down is very limited. Models with reduced THC appear to be more susceptible for a shut-down. Although a shut-down during the next 100 years is unlikely, it cannot be ruled out.

Recent advances in our understanding of vegetation photosynthesis and water use have been used to couple the terrestrial energy, water and carbon cycles within a new generation of physiologically based land-surface parametrizations. These have been tested against field observations and implemented in General Circulation Models (GCMs) with demonstrable improvements in the simulation of land-atmosphere fluxes. There has also been significant progress in specifying land parameters, especially the type and density of vegetation. Importantly, these data sets are globally consistent in that they are primarily based on one type of satellite sensor and one set of interpretative algorithms. Satellite observations have also been shown to provide a powerful diagnostic capability for tracking climatic impacts on surface conditions; e.g., droughts and the recently observed lengthening of the boreal growing season; and direct anthropogenic impacts such as deforestation. The direct effects of increased carbon dioxide (CO_2) on vegetation physiology could lead to a relative reduction in evapotranspiration over the tropical continents, with associated regional warming over that predicted for conventional greenhouse warming effects. On time-scales of decades these effects could significantly influence the rate of atmospheric CO_2 increase, the nature and extent of the physical climate system response, and ultimately, the response of the biosphere itself to global change. In addition, such models must be used to account for the climatic effects of land-use change which can be very significant at local and regional scales. However, realistic land-use change scenarios for the next 50 to 100 years are not expected to give rise to global scale climate changes comparable to those resulting from greenhouse gas warming. Significant modelling problems remain to be solved in the areas of soil moisture processes, runoff prediction, land-use change, and the treatment of snow and sub-grid scale heterogeneity.

Increasingly complex snow schemes are being used in some climate models. These schemes include parametrizations of the metamorphic changes in snow albedo arising from age dependence or temperature dependence. Recent modelling studies of the effects of warming on permafrost predict a 12 to 15% reduction in near-surface area and a 15 to 30% increase in thickness of the seasonally thawing active layer by the mid-21st century. The representation of sea-ice processes continues to improve with several climate models now incorporating physically based treatments of ice dynamics. The effects of sub-grid scale variability in ice cover and thickness, which can significantly influence albedo and atmosphere-ocean exchanges, are being introduced. Understanding of fast-flow processes for land ice and the role of these processes in past climate events is growing rapidly. Representation of ice stream and grounding line physics in land-ice dynamics models remains rudimentary but, in global climate models, ice dynamics and thermodynamics are ignored entirely.

Climate change may manifest itself both as shifting means as well as changing preference of specific regimes, as evidenced by the observed trend toward positive values for the last 30 years in the NAO index and the climate "shift" in the tropical Pacific in about 1976. While coupled models simulate features of observed natural climate variability such as the NAO and ENSO, suggesting that many of the relevant processes are included in the models, further progress is needed to depict these natural modes accurately. Moreover, because ENSO and NAO are key determinants of regional climate change, and they can possibly result in abrupt and counter-intuitive changes, there has been an increase in uncertainty in those aspects of climate change that critically depend on regional changes.

Possible non-linear changes in the climate system as a result of anthropogenic climate forcing have received considerable attention in the last few years. Employing the entire climate model hierarchy, and combining these results with palaeoclimatic evidence and instrumental observation, has shown that mode changes in all components of the climate system have occurred in the past and may also take place in the future. Such changes may be associated with thresholds in the climate system. Such thresholds have been identified in many climate models and there is an increasing understanding of the underlying processes. Current model simulations indicate that the thresholds may lie within the reaches of projected changes. However, it is not yet possible to give reliable values of such thresholds.

7.1 Introduction

The key problem to be addressed is the response of the climate system to changes in forcing. In many cases there is a fairly direct and linear response and many of the simulated changes fall into that class (Chapter 9). These concern the large-scale general circulations of the atmosphere and ocean, and they are in principle represented in current comprehensive coupled climate models. Such possible large-scale dynamical feedbacks are also influenced by small-scale processes within the climate system. The various feedbacks in the climate system may amplify (positive feedbacks), or diminish (negative feedbacks) the original response. We often approximate the response of a particular variable to a small forcing by a scaling number. A prominent example of such a quantification is the equilibrium global mean temperature increase per Wm^{-2} change in the global mean atmospheric radiative forcing (see Chapter 9, Section 9.2.1).

While many aspects of the response of the climate system to greenhouse gas forcing appear to be linear, regime or mode transitions cannot be quantified by a simple number because responses do not scale with the amplitude of the forcing: small perturbations can induce large changes in certain variables of the climate system. This implies the existence of thresholds in the climate system which can be crossed for a sufficiently large perturbation. While such behaviour has long been known and studied in the context of simple models of the climate system, such thresholds are now also found in the comprehensive coupled climate models currently available. Within this framework, the possibility for irreversible changes in the climate system exists. This insight, backed by the palaeoclimatic record (see Chapter 2, Section 2.4), is a new challenge for global change science because now thresholds have to be identified and their values need to be estimated using the entire hierarchy of climate models.

To estimate the response properly, we must represent faithfully the physical processes in models. Not only must the whole system model perform reasonably well in comparison with observation (both spatial and temporal), but so too must the component models and the processes that are involved in the models. It is possible to tune a model so that some variable appears consistent with that observed, but we must also ask whether it comes out that way for the right reason and with the right variability. By examining how well individual processes are known and can be modelled, we can comment on the capabilities and usefulness of the models, and whether they are likely to be able to properly represent possible non-linear responses of the climate system.

7.1.1 Issues of Continuing Interest

Examples of some processes in the climate system and its components that have been dealt with in the Second Assessment Report (IPCC, 1996) (hereafter SAR) and still are important topics of progress:

- *The hydrological cycle.* Progress has been made in all aspects of the hydrological cycle in the atmosphere, involving evaporation, atmospheric moisture, clouds, convection, and precipitation. Because many facets of these phenomena are sub-grid scale, they must be parametrized. While models reasonably simulate gross aspects of the observed behaviour on several time-scales, it is easy to point out shortcomings, although it is unclear as to how much these affect the sensitivity of simulated climate to changes in climate forcing.

- *Water vapour feedback.* An increase in the temperature of the atmosphere increases its water-holding capacity; however, since most of the atmosphere is undersaturated, this does not automatically mean that water vapour, itself, must increase. Within the boundary layer (roughly the lowest 1 to 2 km of the atmosphere), relative humidity tends to remain fixed, and water vapour does increase with increasing temperature. In the free troposphere above the boundary layer, where the water vapour greenhouse effect is most important, the behaviour of water vapour cannot be inferred from simple thermodynamic arguments. Free tropospheric water vapour is governed by a variety of dynamical and microphysical influences which are represented with varying degrees of fidelity in general circulation models. Since water vapour is a powerful greenhouse gas, increasing water vapour in the free troposphere would lead to a further enhancement of the greenhouse effect and act as a positive feedback; within current models, this is the most important reason for large responses to increased anthropogenic greenhouse gases.

- *Cloud radiation feedback.* Clouds can both absorb and reflect solar radiation (thereby cooling the surface) and absorb and emit long-wave radiation (thereby warming the surface). The compensation between those effects depends on cloud height, thickness and cloud radiative properties. The radiative properties of clouds depend on the evolution of atmospheric water vapour, water drops, ice particles and atmospheric aerosols. These cloud processes are most important for determining radiative, and hence temperature, changes in models. Although their representation is greatly improved in models, the added complexity may explain why considerable uncertainty remains; this represents a significant source of potential error in climate simulations. The range in estimated climate sensitivity of 1.5 to 4.5°C for a CO_2 doubling is largely dictated by the interaction of model water vapour feedbacks with the variations in cloud behaviour among existing models.

- *Sub-grid scale processes in ocean models.* Although improved parametrizations in coarse resolution models are available to represent mixing processes, they do not, in their present form, induce sufficient variability on the broad range of time-scales exhibited by eddy-resolving ocean models. Thus present generations of coarse resolution ocean models may not be able to decide questions about those types of natural variability that are associated with sub-grid scale processes, and increased resolution is highly desirable.

7.1.2 New Results since the SAR

Improved atmospheric and oceanic modules of coupled climate models, especially improved representation of clouds, parametrizations of boundary layer and ocean mixing, and increased grid resolution, have helped reduce and often eliminate the need for flux adjustment in some coupled climate models. This has not reduced the range of sensitivities in projection experiments.

There is a growing appreciation of the importance of the stratosphere, particularly the lower stratosphere, in the climate system. Since the mass of the stratosphere represents only about 10 to 20% of the atmospheric mass, the traditional view has been that the stratosphere can play only a limited role in climate change. However, this view is changing. The transport and distribution of radiatively active constituents, especially water vapour and ozone, are important for radiative forcing. Moreover, waves generated in the troposphere propagate into the stratosphere and are absorbed, so that stratospheric changes alter where and how they are absorbed, and effects can extend downward into the troposphere.

Observational records suggest that the atmosphere may exhibit specific regimes which characterise the climate on a regional to hemispheric scale. Climate change may thus manifest itself both as shifting means as well as changing preference of specific regimes. Examples are the North Atlantic Oscillation (NAO) index, which shows a bias toward positive values for the last 30 years, and the climate "shift" in the tropical Pacific at around 1976.

While considerable advances have been made in improving feedbacks and coupled processes and their depiction in models, the emergence of the role of natural modes of the climate system such as the El Niño-Southern Oscillation (ENSO) and NAO as key determinants of regional climate change, and possibly also shifts, has led to an increase in uncertainty in those aspects of climate change that critically depend on regional changes. It is encouraging that the most advanced models exhibit natural variability that resembles the most important modes such as ENSO and NAO.

The coupled ocean-atmosphere system contains important non-linearities which give rise to a multiplicity of states of the Atlantic thermohaline circulation (THC). Most climate models respond to global warming by a reduction of the Atlantic THC. A complete shut-down of the THC in response to continued warming cannot be excluded and would occur if certain thresholds are crossed. Models have identified the maximum strength of greenhouse gas induced forcing and the rate of increase as thresholds for the maintenance of the THC in the Atlantic ocean, an important process influencing the climate of the Northern Hemisphere. While such thresholds have been found in a variety of fundamentally different models, suggesting that their existence in the climate system is a robust result, we cannot yet determine with accuracy the values of these thresholds, because they crucially depend on the response of the atmospheric hydrological cycle to climate change.

The representation of sea-ice dynamics and sub-grid scale processes in coupled models has improved significantly, which is an important prerequisite for a better understanding of, the current variability in, and a more accurate prediction of future changes in polar sea-ice cover and atmosphere-ocean interaction in areas of deep water formation.

Recent model simulations, including new land-surface parametrizations and field observations, strongly indicate that large-scale changes in land use can lead to significant impacts on the regional climate. The terrestrial carbon and water cycles are also linked through vegetation physiology, which regulates the ratio of carbon dioxide (CO_2) uptake (photosynthesis) to water loss (evapotranspiration). As a result, vegetation water-use efficiency is likely to change with increasing atmospheric CO_2, leading to a reduction in evapotranspiration over densely vegetated areas. Tropical deforestation, in particular, is associated with local warming and drying. However, realistic land-use change scenarios for the next 50 to 100 years are not expected to give rise to global scale climate changes comparable to those resulting from greenhouse gas warming.

7.1.3 Predictability of the Climate System

The Earth's atmosphere-ocean dynamics is chaotic: its evolution is sensitive to small perturbations in initial conditions. This sensitivity limits our ability to predict the detailed evolution of weather; inevitable errors and uncertainties in the starting conditions of a weather forecast amplify through the forecast (Palmer, 2000). As well as uncertainty in initial conditions, such predictions are also degraded by errors and uncertainties in our ability to represent accurately the significant climate processes. In practice, detailed weather prediction is limited to about two weeks.

However, because of the more slowly varying components of the climate system, predictability of climate is not limited to the two week time-scale. Perhaps the most well-known and clear cut example of longer-term predictability is El Niño (see Section 7.6.5) which is predictable at least six months in advance. There is some evidence that aspects of the physical climate system are predictable on even longer time-scales (see Section 7.6). In practice, if natural decadal variability is partially sensitive to initial conditions, then projections of climate change for the 21st century will exhibit a similar sensitivity.

In order to be able to make reliable forecasts in the presence of both initial condition and model uncertainty, it is now becoming common to repeat the prediction many times from different perturbed initial states, and using different global models (from the stock of global models that exist in the world weather and climate modelling community). These so-called multi-model, multi-initial-condition ensembles are the optimal basis of probability forecasts (e.g., of a weather event, El Niño, or the state of the THC).

Estimating anthropogenic climate change on times much longer than the predictability time-scale of natural climate fluctuations does not, by definition, depend on the initial state. On these time-scales, the problem of predicting climate change is one of estimating changes in the probability distribution of climatic states (e.g., cyclonic/anticyclonic weather, El Niño, the THC, global mean temperature) as atmospheric composition is altered in some prescribed manner. Like the initial value problems mentioned above, estimates of such changes to the probability distribution of climate states must be evaluated using ensemble prediction techniques.

The number of ensemble members required to estimate reliably changes to the probability distribution of a given climatic phenomenon depends on the phenomenon in question. Estimating changes in the probability distribution of localised extreme weather events, which by their nature occur infrequently, may require very large ensembles with hundreds of members. Estimating changes in the probability distribution of large scale frequent events (e.g., the probability of above-average hemispheric mean temperature) requires much smaller ensembles.

An important question is whether a multi-model ensemble made by pooling the world climate community's stock of global models adequately spans the uncertainty in our ability to represent faithfully the evolution of climate. Since the members of this stock of models were not developed independently of one another, such an ensemble does not constitute an independent unbiased sampling of possible model formulations.

7.2 Atmospheric Processes and Feedbacks

7.2.1 Physics of the Water Vapour and Cloud Feedbacks

Any increase in the amount of a greenhouse gas contained in the Earth's atmosphere would reduce the emission of outgoing long-wave radiation (OLR) if the temperature of the atmosphere and surface were held fixed. The climate achieves a new equilibrium by warming until the OLR increases enough to balance the incoming solar radiation. Addition of greenhouse gases affects the OLR primarily because the tropospheric temperature decreases with height. With a fixed temperature profile, increasing the greenhouse gas content makes the higher parts of the atmosphere more opaque to infrared radiation upwelling from below, replacing this radiation with OLR emitted from the colder regions.

Determination of the new equilibrium is complicated by the fact that water vapour is itself a potent greenhouse gas, and the amount and distribution of water vapour will generally change as the climate changes. The atmospheric water vapour content responds to changes in temperature, microphysical processes and the atmospheric circulation. An overarching consideration is that the maximum amount of water vapour air can hold increases rapidly with temperature, in accord with the Clausius-Clapeyron relation. This affects all aspects of the hydrological cycle. Unlike CO_2, water vapour concentration varies substantially in both the vertical and horizontal. An increase in water vapour reduces the OLR only if it occurs at an altitude where the temperature is lower than the ground temperature, and the impact grows sharply as the temperature difference increases. If water vapour at such places increases as the climate warms, then the additional reduction in OLR requires the new equilibrium to be warmer than it would have been if water vapour content had remained fixed. This is referred to as a *positive water vapour feedback*.

Clouds are intimately connected to the water vapour pattern, as clouds occur in connection with high relative humidity, and cloud processes in turn affect the moisture distribution. Clouds

affect OLR in the same way as a greenhouse gas, but their net effect on the radiation budget is complicated by the fact that clouds also reflect incoming solar radiation. As clouds form the condensation releases latent heat, which is a central influence in many atmospheric circulations.

The boundary layer is the turbulent, well-mixed shallow layer near the ground, which can be regarded as being directly moistened by evaporation from the surface. In the boundary layer, the increase in water vapour with temperature in proportion with the Clausius-Clapeyron relation is uncontroversial. Observations (e.g., Wentz and Schabel, 2000) clearly show a very strong relation of total column water vapour (precipitable water) with surface and tropospheric temperature. Because the boundary-layer temperature is similar to that of the ground, however, boundary-layer water vapour is not of direct significance to the water vapour feedback. Furthermore, half of the atmospheric water vapour is below 850 mb, so measurements of total column water have limited utility in understanding water vapour feedback. The part of the troposphere above the boundary layer is referred to as the "free troposphere". Water vapour is brought to the free troposphere by a variety of mixing and transport processes, and water vapour feedback is determined by the aggregate effects of changes in the transport and in the rate at which water is removed by precipitation occurring when air parcels are cooled, usually by rising motions.

The complexity of water vapour radiative impact is reflected in the intricate and strongly inhomogeneous patterns of the day-to-day water vapour distribution (Figure 7.1a). The very dry and very moist regions reveal a strong influence of the large-scale dynamical transport. Model simulations exhibit similar patterns (Figure 7.1b), with a notable qualitative improvement at higher resolution (Figure 7.1c). Understanding the dominant transport processes that set up those patterns, and how they can be affected by a modified climate, should help assess the representation of the water vapour feedbacks in the corresponding region found in model simulations.

(i) In the tropical free troposphere, ascent is generally concentrated in intense, narrow, and nearly vertical cumulonimbus towers, with descent occurring slowly over the broad remaining area. The moist regions are broader than the narrow ascending towers and are mostly descending. The vicinity of convective towers is moistened by evaporation of precipitation from the cirrus outflow of the cumulus towers above about 5 km and by the dissipation and mixing of cumulonimbus towers below about 5 km and above the trade wind boundary layer. Cloud coverage thus depends on the detrainment of ice from cumulus towers (Gamache and Houze, 1983). The ice content of the detrained air depends in turn on the efficiency of precipitation within the tower, i.e., condensed water which falls as rain within the towers is no longer available for detrainment. The association of high stratiform cloud cover with high relative humidity was clearly illustrated by Udelhofen and Hartmann (1995), and in the tropics upper-tropospheric humidity (UTH) is tied to the frequency of deep convection (Soden and Fu, 1995).

The tropical dry regions shown in Figure 7.1 are associated with large-scale descent. A number of studies (Sherwood, 1996; Salathe and Hartmann, 1997; Pierrehumbert and Roca, 1998;

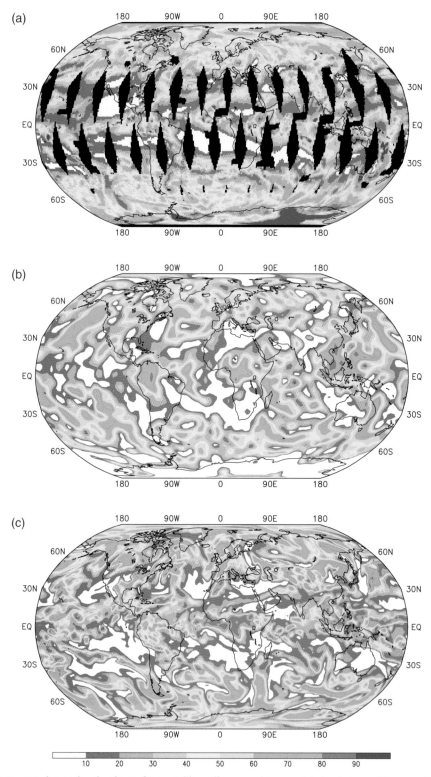

Figure 7.1: Comparison between an observational estimate from satellite radiances and two model simulations of the complex structure of mid-tropospheric water vapour distribution for the date May 5. At any instant, water vapour is unevenly distributed in the atmosphere with very dry areas adjacent to very moist areas. Any modification in the statistics of those areas participates in the atmospheric feedback. The observed small-scale structure of the strong and variable gradients (a) is not well resolved in a simulation with a climate model of the spatial resolution currently used for climate projections (b), but simulated with much better fidelity in models with significantly higher resolution (c). (a): Distribution of mean relative humidity in layer 250 to 600 mb on May 5, 1998, as retrieved from observations on SSM/T-2 satellite (Spencer and Braswell, 1997). Missing data are indicated by black areas and the retrieval is most reliable in the latitude band 30°S to 30°N. (b): Relative humidity at about 400 mb for May 5 of an arbitrary year from a simulation with the GFDL R30L14 atmospheric general circulation model used in the AMIP I simulation (Gates *et al.*, 1999; Lau and Nath, 1999). In small polar areas (about 5% of the globe) some relative humidities are negative (set to zero) due to numerical spectral effects. (c): Relative humidity at 400 mb from the ECHAM4 T106 simulation for May 5 of an arbitrary year (Roeckner *et al.*, 1996; Wild *et al.*, 1998).

Soden, 1998; Dessler and Sherwood, 2000) have shown that the predominant source of water vapour here is large-scale advection from the convective region. Some of the transient moist filaments which account for much of the moisture of the climatologically dry zone can be seen in Figure 7.1a,c. Dry-zone humidity depends on saturation vapour pressure at remote points along the relevant trajectories. The way the water vapour feedback operates in the dry regions is dependent on the manner in which these points change as the climate warms.

The moist regions are also cloudy. In the tropics, water vapour feedback could involve changes in the water vapour content of either the convective or dry regions, or changes in the relative area of the two regions.

(ii) In the extra-tropical free troposphere, convection still plays a role (Hu and Liu, 1998; Kirk-Davidoff and Lindzen, 2000); however, ascent and descent are primarily associated with large-scale wave motions with comparable ascending and descending areas (Yang and Pierrehumbert, 1994; Hu and Liu, 1998). Moisture in the troposphere is directly affected by evaporation, which is a maximum in the sub-tropics, and the boundary-layer reservoir of water vapour that is tapped by extra-tropical waves. Observations confirm that extra-tropical free tropospheric water vapour is mostly not associated with deep convection (Soden and Fu, 1995), though there may still be some influence from the extra-tropical surface via local convection. Condensation in the broad ascending areas leads directly to stratiform clouds.

7.2.1.1 Water vapour feedback

Water vapour feedback continues to be the most consistently important feedback accounting for the large warming predicted by general circulation models in response to a doubling of CO_2. Water vapour feedback acting alone approximately doubles the warming from what it would be for fixed water vapour (Cess *et al.*, 1990; Hall and Manabe, 1999; Schneider *et al.*, 1999; Held and Soden, 2000). Furthermore, water vapour feedback acts to amplify other feedbacks in models, such as cloud feedback and ice albedo feedback. If cloud feedback is strongly positive, the water vapour feedback can lead to 3.5 times as much warming as would be the case if water vapour concentration were held fixed (Hall and Manabe, 1999).

As noted by Held and Soden (2000), the relative sensitivity of OLR to water vapour changes at various locations depends on how one perturbs the water vapour profile; the appropriate choice depends entirely on the nature of the water vapour perturbation anticipated in a changing climate. The sensitivity is also affected by cloud radiative effects, which tend to mask the influence of sub-cloud water vapour on OLR. Incorporating cloud radiative effects and a fixed relative humidity perturbation (argued to be most appropriate to diagnosing GCM water vapour feedback), Held and Soden suggest that OLR is almost uniformly sensitive to water vapour perturbations throughout the tropics. Roughly 55% of the total is due to the free troposphere in the "tropics" (30°N to 30°S) with 35% from the extra-tropics. Allowing for polar amplification of warming increases the proportion of water vapour feedback attributable to the extra-tropics. Of the tropical contribution, about two thirds, or 35% of the global total, is due to the

upper half of the troposphere, from 100 to 500 mb. The boundary layer itself accounts for only 10% of the water vapour feedback globally. Simulations incorporating cloud radiative effects in a doubled CO_2 experiment (Schneider *et al.*, 1999) and a clear-sky analysis based on 15 years of global data (Allan *et al.*, 1999) yield maximum sensitivity to water vapour fluctuations in the 400 to 700 mb layer (see also Le Treut *et al.*, 1994). In a simulation analysed by Schneider *et al.* (1999) extra-tropical water vapour feedback affected warming 50% more than did tropical feedback.

Most of the free troposphere is highly undersaturated with respect to water, so that local water-holding capacity is not the limiting factor determining atmospheric water vapour. Within the constraints imposed by Clausius-Clapeyron alone there is ample scope for water vapour feedbacks either stronger or weaker than those implied by constant relative humidity, especially in connection with changes in the area of the moist tropical convective region (Pierrehumbert, 1999). It has been estimated that, without changes in the relative area of convective and dry regions, a shift of water vapour to lower levels in the dry regions could, at the extreme, lead to a halving of the currently estimated water vapour feedback, but could not actually cause it to become a negative, stabilising feedback (Harvey, 2000).

Attempts to directly confirm the water vapour feedback by correlating spatial surface fluctuations with spatial OLR fluctuations were carried out by Raval and Ramanathan (1989). Their results are difficult to interpret, as they involve the effects of circulation changes as well as direct thermodynamic control (Bony *et al.*, 1995). Inamdar and Ramanathan (1998) showed that a positive correlation between water vapour, greenhouse effect and SST holds for the entire tropics at seasonal time-scales. This is consistent with a positive water vapour feedback, but it still cannot be taken as a direct test of the feedback as the circulation fluctuates in a different way over the seasonal cycle than it does in response to doubling of CO_2.

7.2.1.2 Representation of water vapour in models

General circulation models do not impose a fixed relative humidity. Assumptions built into the models directly govern the relative humidity only in the sparse set of grid boxes that are actively convecting, where the choice of convection scheme determines the humidity. In the case of moist convective adjustment, the relative humidity after convection is explicitly set to a predetermined profile, whereas the mass flux schemes compute a humidity profile based on microphysical assumptions of varying complexity. The relative humidity elsewhere is determined by explicitly resolved dynamical processes, and in fact undergoes marked spatial (e.g., Figure 7.1b,c) and temporal fluctuations. Nonetheless, all models studied to date produce a positive water vapour feedback consistent with the supposition that water vapour increases in such a way as to keep the relative humidity approximately unchanged at all levels (Held and Soden, 2000). The strength of the water vapour feedback is consistent amongst models, despite considerable differences in the treatment of convection and microphysics (Cess *et al.*, 1990).

The "convective region" is potentially a source of modelling errors, since the evaporation of detrained precipitation and other poorly characterised and heavily parametrized processes are essential to the moistening of the atmosphere. Tompkins and Emanuel (2000) showed that in a single-column model, numerical convergence of the simulated water vapour profile requires a vertical resolution better than 25 mb in pressure (see also Emanuel and Zivkovic-Rothman, 1999). Thus, the apparent lack of sensitivity of water vapour feedback in current GCMs to the way convection is treated may be an artefact of insufficient vertical resolution. An alternative view is that the water vapour flux is vigorous enough to keep the convective region at a substantial fraction of saturation, so that the details of the moistening process do not matter much. In this case the convection essentially extends the boundary layer into the free troposphere, making condensed water abundantly available to evaporation, but in the form of droplets or crystals. Indeed, Dessler and Sherwood (2000) find that satellite observations of tropical water vapour can be understood without detailed microphysical considerations, even in the convective region. Radiosonde observations (Kley *et al.*, 1997) and satellite UTH data (Soden and Fu, 1995) reveal free troposphere relative humidities of 50 to 70% throughout the convective region. Udelhofen and Hartmann (1995) find that moisture decays away from convective systems with a characteristic scale of 500 km, so that convection need not be too closely spaced to maintain a uniformly moistened region. As noted by Held and Soden (2000), the evidence for the opposite view, that convective region air becomes drier as temperature increases, is weak.

This moistening of large-scale tropical subsiding zones by large-scale advection is a process explicitly resolved in models, and in which one can have reasonably high confidence. Care must still be taken with the design of numerical advection methods to avoid spurious transport. In this regard, there has been notable progress since the SAR as models are increasingly using semi-Lagrangian and other advanced advection methods in place of spectral advection of water vapour (Williamson and Rasch, 1994). These methods virtually eliminate spurious transport arising from the need to adjust negative moisture regions, as commonly happens in spectral advection.

In the extra-tropics the ubiquitous large-scale synoptic eddies which dominate moisture transport are explicitly represented in GCMs, and there is reasonably high confidence in simulated mid-latitude water vapour feedback. The eddies maintain a fairly uniform monthly mean relative humidity of 30 to 50% throughout the year (Soden and Fu, 1995; Bates and Jackson, 1997).

Simulation of water vapour variations with natural climate fluctuations such as the annual cycle and El Niño can help provide tests of the verisimilitude of models but may not be sufficient for assessing climate change due to increases in greenhouse gases (Bony *et al.*, 1995). Models are quite successful at reproducing the climatological free tropospheric humidity pattern (e.g., Figure 7.1b,c). Model-satellite comparisons of UTH are still at a rather early stage of development; the AMIP simulations did not archive enough fields to provide an adequate understanding of UTH, and there are problems in determining which model level should be compared to the satellite data. Also there are uncertainties in UTH retrievals (see Chapter 2). Caveats notwithstanding, the ensemble mean of the AMIP simulations shows a moist bias in UTH compared to satellite data, but this result is rendered uncertain by apparent inconsistencies in the computation of model relative humidities for intercomparison purposes. The models studied by AMIP (Bates and Jackson, 1997) appear to show little skill in reproducing the seasonal or interannual variations of convective region UTH, though many show significant correlations between simulated and observed sub-tropical and extra-tropical UTH fluctuations. The latter lends some confidence to the simulated water vapour feedback, as the dry sub-tropics and extra-tropics account for a large part of the feedback. Del Genio *et al.* (1994) found good agreement between the observed seasonal cycle of zonal mean UTH, and that simulated by the GISS model.

Humidity is important to water vapour feedback only to the extent that it alters OLR. Because the radiative effects of water vapour are logarithmic in water vapour concentration, rather large errors in humidity can lead to small errors in OLR, and systematic underestimations in the contrast between moist and dry air can have little effect on climate sensitivity (Held and Soden, 2000). Most GCMs reproduce the climatological pattern of clear-sky OLR very accurately (Duvel *et al.*, 1997). In addition, it has been shown that the CCM3 model tracks the observed seasonal cycle of zonal mean clear-sky OLR to within 5 Wm^{-2} (Kiehl *et al.*, 1998), the GFDL model reproduces tropical mean OLR fluctuations over the course of an El Niño event (Soden, 1997), and the collection of models studied under the AMIP project reproduces interannual variability of the clear-sky greenhouse parameter with errors generally under 25% for sea surface temperatures (SSTs) under 25°C (Duvel *et al.*, 1997). Errors become larger over warmer waters, owing to inaccuracies in the way the simulated atmospheric circulation responds to the imposed SSTs (which may also not be accurate). The LMD model shows close agreement with the observed seasonal cycle of greenhouse trapping (Bony *et al.*, 1995), but models vary considerably in their ability to track this cycle (Duvel *et al.*, 1997).

Indirect evidence for validity of the positive water vapour feedback in present models can be found in studies of interannual variability of global mean temperature: Hall and Manabe (1999) found that suppression of positive water vapour feedback in the GFDL model led to unrealistically low variability. An important development since the SAR is the proliferation of evidence that the tropics during the Last Glacial Maximum were 2 to 5°C colder than the present tropics. Much of the cooling can be simulated in GCMs which incorporate the observed reduction in CO_2 at the Last Glacial Maximum (see Chapter 8). Without water vapour feedback comparable to that currently yielded by GCMs, there would be a problem accounting for the magnitude of the observed cooling in the tropics and Southern Hemisphere.

7.2.1.3 Summary on water vapour feedbacks

The decade since the First IPCC Assessment Report (IPCC, 1990) has seen progressive evolution in sophistication of thinking about water vapour feedback. Concern about the role of upper-tropospheric humidity has stimulated much theoretical, model diagnostic and observational study. The period since the SAR has seen continued improvement in the analysis of observations of water vapour from sondes and satellite instrumentation. Theoretical understanding of the atmospheric hydrological cycle has also increased. As a result, observational tests of how well models represent the processes governing water vapour content have become more sophisticated and more meaningful. Since the SAR, appraisal of the confidence in simulated water vapour feedback has shifted from a diffuse concern about upper-tropospheric humidity to a more focused concern about the role of microphysical processes in the convection parametrizations, and particularly those affecting tropical deep convection. Further progress will almost certainly require abandoning the artificial diagnostic separation between water vapour and cloud feedbacks.

In the SAR, a crude distinction was made between the effect of "upper-tropospheric" and "lower-tropospheric" water vapour, and it was implied that lower-tropospheric water vapour feedback was a straightforward consequence of the Clausius-Clapeyron relation. It is now appreciated that it is only in the boundary layer that the control of water vapour by Clausius-Clapeyron can be regarded as straightforward, so that "lower-tropospheric" feedback is no less subtle than "upper-tropospheric". It is more meaningful to separate the problem instead into "boundary layer" and "free tropospheric" water vapour, with the former contributing little to the feedback.

The successes of the current models lend some confidence to their results. For a challenge to the current view of water vapour feedback to succeed, relevant processes would have to be incorporated into a GCM, and it would have to be shown that the resulting GCM accounted for observations at least as well as the current generation. A challenge that meets this test has not yet emerged. Therefore, the balance of evidence favours a positive clear-sky water vapour feedback of magnitude comparable to that found in simulations.

7.2.2 Cloud Processes and Feedbacks

7.2.2.1 General design of cloud schemes within climate models
The potential complexity of the response of clouds to climate change was identified in the SAR as a major source of uncertainty for climate models. Although there has been clear progress in the physical content of the models, clouds remain a dominant source of uncertainty, because of the large variety of interactive processes which contribute to cloud formation or cloud-radiation interaction: dynamical forcing – large-scale or sub-grid scale, microphysical processes controlling the growth and phase of the various hydrometeors, complex geometry with possible overlapping of cloud layers. Most of these processes are sub-grid scale, and need to be parametrized in climate models.

As can be inferred from the description of the current climate models gathered by AMIP (AMIP, 1995; Gates *et al.*, 1999) the cloud schemes presently in use in the different modelling centres vary greatly in terms of complexity, consistency and comprehensiveness. However, there is a definite tendency toward a more consistent treatment of the clouds in climate models. The more widespread use of a prognostic equation for cloud water serves as a unifying framework coupling together the different aspects of the cloud physics, as noted in the

SAR. The evolution of the cloud schemes in the different climate models has continued since then.

The main model improvements can be summarised as follows:

(i) Inclusion of additional conservation equations representing different types of hydrometeors
A first generation of so-called prognostic cloud schemes (Le Treut and Li, 1991; Roeckner *et al.*, 1991; Senior and Mitchell, 1993; Del Genio *et al.*, 1996), has used a budget equation for cloud water, defined as the sum of all liquid and solid cloud water species that have negligible vertical fall velocities. The method allows for a temperature-dependent partitioning of the liquid and ice phases, and thereby enables a bulk formulation of the microphysical processes. By providing a time-scale for the residence of condensed water in the atmosphere, it provides an added physical consistency between the respective simulations of condensation, precipitation and cloudiness. The realisation that the transition between ice and liquid phase clouds was a key to some potentially important feedbacks has prompted the use of two or more explicit cloud and precipitation variables, thereby allowing for a more physically based distinction between cloud water and cloud ice (Fowler *et al.*, 1996; Lohmann and Roeckner, 1996).

(ii) Representation of sub-grid scale processes
The conservation equations to determine the cloud water concentration are written at the scale explicitly resolved by the model, whereas a large part of the atmospheric dynamics generating clouds is sub-grid scale. This is still an inconsistency in many models, as clouds generated by large-scale or convective motions are very often treated in a completely separate manner, with obvious consequences on the treatment of anvils for example. Several approaches help to reconcile these contradictions. Most models using a prognostic approach of cloudiness use probability density functions to describe the distribution of water vapour within a grid box, and hence derive a consistent fractional cover (Smith, 1990; Rotstayn, 1997). An alternative approach, initially proposed by Tiedtke (1993), is to use a conservation equation for cloud air mass as a way of integrating the many small-scale processes which determine cloud cover (Randall, 1995). Representations of sub-grid scale cloud features also require assumptions about the vertical overlapping of cloud layers, which in turn affect the determination of cloud radiative forcing (Jacob and Klein, 1999; Morcrette and Jakob, 2000; Weare, 2000a).

(iii) Inclusion of microphysical processes
Incorporating a cloud budget equation into the models has opened the way for a more explicit representation of the complex microphysical processes by which cloud droplets (or crystals) form, grow and precipitate (Houze, 1993). This is necessary to maintain a full consistency between the simulated changes of cloud droplet (crystal) size distribution, cloud water content, and cloud cover, since the nature, shape, number and size distribution of the cloud particles influence cloud formation and lifetime, the onset of precipitation (Albrecht, 1989), as well as cloud inter-

action with radiation, in both the solar and long-wave bands (Twomey, 1974). Some parametrizations of sub-grid scale condensation, such as convective schemes, are also complemented by a consistent treatment of the microphysical processes (Sud and Walker, 1999).

In warm clouds these microphysical processes include the collection of water molecules on a foreign substance (heterogeneous nucleation on a cloud condensation nucleus), diffusion, collection of smaller drops when falling through a cloud (coalescence), break-up of drops when achieving a certain threshold size, and re-evaporation of drops when falling through a layer of unsaturated air. In cold clouds, ice particles may be nucleated from either the liquid or vapour phase, and spontaneous homogeneous freezing of supercooled liquid drops is also relevant at temperatures below approximately –40°C. At higher temperatures the formation of ice particles is dominated by heterogeneous nucleation of water vapour on ice condensation nuclei. Subsequent growth of ice particles is then due to diffusion of vapour toward the particle (deposition), collection of other ice particles (aggregation), and collection of supercooled drops which freeze on contact (riming). An increase in ice particles may occur by fragmentation. Falling ice particles may melt when they come into contact with air or liquid particles with temperatures above 0°C.

Heterogeneous nucleation of soluble particles and their subsequent incorporation into precipitation is also an important mechanism for their removal, and is the main reason for the indirect aerosol effect. The inclusion of microphysical processes in GCMs has produced an impact on the simulation of the mean climate (Hahmann and Dickinson, 1997).

Measurements of cloud drop size distribution indicate a significant difference in the total number of drops and drop effective radius in the continental and maritime atmosphere, and some studies indicate that inclusion of more realistic drop size distribution may have a significant impact on the simulation of the present climate (Hahmann and Dickinson, 1997).

7.2.2.2 Convective processes

The interplay of buoyancy, moisture and condensation on scales ranging from millimetres to tens of kilometres is the defining physical feature of atmospheric convection, and is the source of much of the challenge in representing convection in climate models. Deep convection is in large measure responsible for the very existence of the troposphere. Air typically receives its buoyancy through being heated by contact with a warm, solar-heated underlying surface, and convection redistributes the energy received by the surface upwards throughout the troposphere. Shallow convection also figures importantly in the structure of the atmospheric boundary layer and will be addressed in Section 7.2.2.3.

Latent heat release in convection drives many of the important atmospheric circulations, and is a key link in the cycle of atmosphere-ocean feedbacks leading to the ENSO phenomenon. Convection is a principal means of transporting moisture vertically, which implies a role of convection in the radiative feedback due to both water vapour and clouds. Convection also in large measure determines the vertical temperature lapse rate of the atmosphere, and particularly so in the tropics. A strong

decrease of temperature with height enhances the greenhouse effect, whereas a weaker temperature decrease ameliorates it. The effect of lapse rate changes on clear-sky water vapour feedback has been studied by Zhang et al. (1994), but the significance of the lapse rate contribution (cf. item (5) of the SAR, Technical Summary) has been somewhat exaggerated through a misinterpretation of the paper. In fact the variation in lapse rate effects among the models studied alters the water vapour feedback factor by only 0.25 W/(m²K), or about 10% of the total (Table 1 of Zhang et al., 1994).

There is ample theoretical and observational evidence that deep moist convection locally establishes a "moist adiabatic" temperature profile that, loosely speaking, is neutrally buoyant with respect to ascending, condensing parcels (Betts, 1982; Xu and Emanuel, 1989). This adjustment happens directly at the scale of individual convective clouds, but dynamical processes plausibly extend the radius of influence of the adjustment to the scale of a typical GCM grid box, and probably much further in the tropics, where the lapse rate adjusts close to the moist adiabat almost everywhere. All convective schemes, from the most simple Moist Adiabatic Adjustment to those which attempt a representation of cloud-scale motions (Arakawa and Schubert, 1974; Emanuel, 1991), therefore agree in that they maintain the temperature at a nearly moist adiabatic profile. Moist Adiabatic Adjustment explicitly resets the temperature to the desired profile, whereas mass flux schemes achieve the adjustment to a near-adiabat as a consequence of equations governing the parametrized convective heating field. The constraint on temperature, however, places only a limited constraint on the moisture profile remaining after adjustment, and the performance in terms of moisture, clouds and precipitation may be very variable.

Since the SAR, a variety of simulations of response to CO_2 doubling accounting for combinations of different parametrizations have been realised with different models (Colman and McAvaney, 1995; Yao and Del Genio, 1999; Meleshko et al., 2000). The general effects of the convection parametrization on climate sensitivity are difficult to assess because the way a model responds to changes in convection depends on a range of other parametrizations, so results are somewhat inconsistent between models (Colman and McAvaney, 1995; Thompson and Pollard, 1995; Zhang and McFarlane, 1995). There is some indication that the climate sensitivity in models with strong negative cloud feedback is insensitive to convective parametrization whereas models with strong positive cloud feedback show more sensitivity (Meleshko et al., 2000).

7.2.2.3 Boundary-layer mixing and cloudiness

Turbulent motions affect all exchanges of heat, water, momentum and chemical constituents between the surface and the atmosphere, and these motions are also responsible for the mixing processes inside the atmospheric boundary layer. Consequently, the turbulent motions impact on the formation and existence of fog and boundary-layer clouds such as cumulus, stratocumulus and stratus. Cumulus is typically found in fair weather conditions over land and sea, while layers of stratocumulus and stratus can be found in subsidence areas such as the anticyclonic areas in the eastern part of the sub-tropical oceans, or the polar regions.

The atmospheric boundary layer with clouds is typically characterised by a well-mixed sub-cloud layer of order 500 metres, and by a more extended conditionally unstable layer with boundary-layer clouds up to 2 km. The latter layer is very often capped by a temperature inversion. If the clouds are of the stratocumulus or stratus type, then conservative quantities are approximately well mixed in the cloud layer. The lowest part (say 10%) of the sub-cloud layer is known as the surface layer. In this layer the vertical gradients of variables are normally significant, even with strong turbulent mixing. Physical problems associated with the surface layer depend strongly on the type of surface considered (such as vegetation, snow, ice, steep orography) and are treated in the corresponding sub-sections. We note that the surface characteristics do impact on the formation of boundary-layer clouds, because of the turbulent mixing inside the boundary layer.

Atmospheric models have great difficulty in the proper representation of turbulent mixing processes in general. This also impacts on the representation of boundary-layer clouds. At present, the underprediction of boundary-layer clouds is still one of the most distinctive and permanent errors of AGCMs. This has been demonstrated through AMIP intercomparisons by Weare and Mokhov (1995) and also by Weare (2000b). It has a very great importance, because the albedo effect of these clouds is not compensated for by a significant greenhouse effect in both clear-sky and cloudy conditions.

The persisting difficulty in simulation of observed boundary layer cloud properties is a clear testimony of the still inadequate representation of boundary-layer processes. A variety of approaches is followed, ranging from bulk schemes in which the assumption of a well-mixed layer is made *a priori*, to discretised approaches considering diffusion between a number of vertical layers. Here the corresponding diffusion coefficients are being computed from dimensional analysis and observational data fitted to it. The use of algorithms based on the prognostic computation of turbulent kinetic energy and higher-order closure hypotheses is also becoming more common and new schemes continue to be proposed (Abdella and McFarlane, 1997). A critical review and evaluation of boundary-layer schemes was recently made by Ayotte *et al.* (1996). They found that all schemes have difficulty with representing the entrainment processes at the top of even the clear boundary layer.

Important and still open problems include the decoupling between the turbulence at the surface and that within clouds, the non-local treatment of semi-convective cells (thermals) that can transport heat and substances upward, the role of moist physics, and microphysical aspects (Ricard and Royer, 1993; Moeng *et al.*, 1995; Cuijpers and Holtslag, 1998; Grenier and Bretherton, 2001). Several studies (Moeng *et al.*, 1996; Bretherton *et al.*, 1999; Duynkerke *et al.*, 1999) show that column versions of the climate models may predict a reasonable cloud cover in response to observed initial and boundary conditions, but have more difficulty in maintaining realistic turbulent fluxes. The results are also very dependent on vertical resolution and numerical aspects (Lenderink and Holtslag, 2000). This points to the need for new approaches for boundary-layer turbulence, both for clear-sky and cloudy conditions which are not so sensitive to vertical resolution (see also contributions in Holtslag and Duynkerke, 1998).

7.2.2.4 Cloud-radiative feedback processes

Clouds affect radiation both through their three-dimensional geometry and the amount, size and nature of the hydrometeors which they contain. In climate models these properties translate into cloud cover at different levels, cloud water content (for liquid water and ice) and cloud droplet (or crystal) equivalent radius. The interaction of clouds and radiation also involves other parameters (asymmetry factor of the Mie diffusion) which depend on cloud composition, and most notably on their phase. The subtle balance between cloud impact on the solar short-wave (SW) and terrestrial long-wave (LW) radiation may be altered by a change in any of those parameters. In response to any climate perturbation the response of cloudiness thereby introduces feedbacks whose sign and amplitude are largely unknown. While the SAR noted some convergence in the cloud radiative feedback simulated by different models between two successive intercomparisons (Cess *et al.*, 1990, 1995, 1996), this convergence was not confirmed by a separate consideration of the SW and LW components.

Schemes predicting cloudiness as a function of relative humidity generally show an upward displacement of the higher troposphere cloud cover in response to a greenhouse warming, resulting in a positive feedback (Manabe and Wetherald, 1987). While this effect still appears in more sophisticated models, and even cloud resolving models (Wu and Moncrieff, 1999; Tompkins and Emanuel, 2000), the introduction of cloud water content as a prognostic variable, by decoupling cloud and water vapour, has added new features (Senior and Mitchell, 1993; Lee *et al.*, 1997). As noted in the SAR, a negative feedback corresponding to an increase in cloud cover, and hence cloud albedo, at the transition between ice and liquid clouds occurs in some models, but is crucially dependent on the definition of the phase transition within models. The sign of the cloud cover feedback is still a matter of uncertainty and generally depends on other related cloud properties (Yao and Del Genio, 1999; Meleshko *et al.*, 2000).

Most GCMs used for climate simulations now include interactive cloud optical properties. Cloud optical feedbacks produced by these GCMs, however, differ both in sign and strength. The transition between water and ice may be a source of error, but even for a given water phase, the sign of the variation of cloud optical properties with temperature can be a matter of controversy. Analysis from the ISCCP data set, for example, revealed a decrease of low cloud optical thickness with cloud temperature in the sub-tropical and tropical latitudes and an increase at middle latitudes in winter (Tselioudis *et al.*, 1992; Tselioudis and Rossow, 1994). A similar relationship between cloud liquid water path and cloud temperature was found in an analysis of microwave satellite observations (Greenwald *et al.*, 1995). This is opposite to the assumptions on adiabatic increase of cloud liquid water content with temperature, adopted in early studies, and still present in many models. Changes in cloud water path reflect

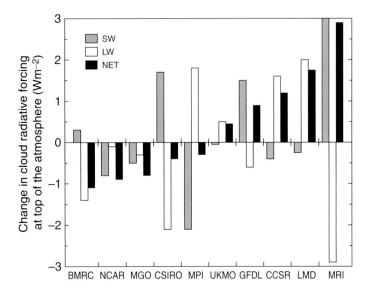

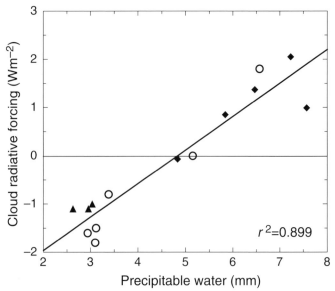

Figure 7.2: Change in the Top of the Atmosphere (TOA) Cloud Radiative Forcing (CRF) associated with a CO_2 doubling (from a review by Le Treut and McAvaney, 2000). The models are coupled to a slab ocean mixed layer and are brought to equilibrium for present climatic conditions and for a double CO_2 climate. The sign is positive when an increase of the CRF (from present to double CO_2 conditions) increases the warming, negative when it reduces it. The contribution of the short-wave (SW, solar) and long-wave (LW, terrestrial) components are first distinguished, and then added to provide a net effect (black bars). Results presented in the diagram are bounded by a 3 Wm^{-2} limit. As in Chapter 8, Table 8.1, the acronyms refer to the atmospheric models of the following institutions: BMRC is the Bureau of Meteorology Research Center (Australia); NCAR is the National Center for Atmospheric Research (USA); CSIRO is the Commonwealth Scientific and Industrial Research Organization (Australia); MPI the Max Planck Institute for Meteorology (Germany); UKMO refers to the model of the Hadley Centre (UK); GFDL is the Geophysical Fluid Dynamics Laboratory (USA); CCSR is the Center for Climate System Research (Japan); LMD is the Laboratoire de Météorologie Dynamique (the corresponding coupled model being referenced as IPSL, France); MRI is the Meteorological Research Institute (Japan). The MGO model appears only in this intercomparison and is the model of the Main Geophysical Observatory (Russia) (see reference in Meleshko *et al.*, 2000).

Figure 7.3: Relationship between simulated global annually averaged variation of net cloud radiative forcing at the top of the atmosphere and precipitable water due to CO_2 doubling produced in simulations with different parametrizations of cloud related processes. Results are from Colman and McAvaney (1995) denoted by triangles, Meleshko *et al.* (2000) denoted by open circles, and Yao and Del Genio (1999) denoted by diamonds.

different effects which may partially compensate, such as changes in cloud vertical extension or cloud water content. The role of low cloud optical thickness dependence on climate was tested in $2\times CO_2$ experiments using the GISS GCM, in which simulations with fixed or simulated cloud optical properties were compared (Tselioudis *et al.*, 1998). In spite of a low impact on global sensitivity, these results showed a strong cloud impact on the latitudinal distribution of the warming. High latitude warming decreased while low latitude warming increased, resulting in a large decrease in the latitudinal amplification of the warming (by 20% in the Northern Hemisphere and by 40% in the Southern Hemisphere).

Since the SAR, there has been much progress in the use of simplified tropical models to understand the impact of cloud

feedbacks, and to address the issue of whether cloud feedbacks impose an upper limit on SST. This problem has been addressed in simplified models that maintain consistency with the whole-tropics energy budget (Pierrehumbert, 1995; Clement and Seager, 1999; Larson *et al.*, 1999). Diagnostic studies have suggested both destabilising (Chou and Neelin, 1999) and stabilising (Lindzen *et al.*, 2001) cloud feedbacks. None of this supports the existence of a strict limitation of maximum tropical SST of the sort proposed by Ramanathan and Collins (1991), which has been criticised on the grounds that it does not respect the whole-tropics energy budget, and that it employs an incorrect means of determining the threshold temperature for convection. It is beyond question that the increased cloudiness prevailing over the warmer portions of the Pacific has a strong effect on the surface energy budget, which is fully competitive with the importance of evaporation. Determination of SST, however, requires a consistent treatment of the top-of-atmosphere energy budget, and cannot be effected with reference to the surface budget alone. This does not preclude the possibility that other cloud feedback mechanisms could have a profound effect on tropical SST, and in no way implies that cloud representation is inconsequential in the tropics. Meehl *et al.* (2000) illustrate this point when they show how a change from a diagnostic prescription of clouds in the NCAR atmospheric GCM to a prognostic cloud liquid water formulation changes the sign of net cloud forcing in the eastern tropical Pacific and completely alters the nature of the coupled model response to increased greenhouse gases.

7.2.2.5 Representation of cloud processes in models

Over the last 10 years, the generalised availability of satellite retrievals (OLR, water path, cloud cover, cloud top temperature) has strongly increased the possibility of assessing clouds simulated by climate models, in spite of remaining uncertainties concerning the vertical structure of the cloud systems, of large importance for all feedback effects. Yet this remains a very difficult task, which may be illustrated by a few figures. Recent simulations of equilibrium climate with a doubling of CO_2 indicate that the induced variation of net cloud radiative forcing short-wave or long-wave, ranges within ± 3 Wm^{-2}, which is a small fraction of the cloud mean long-wave warming (30 to 35 Wm^{-2}) or cloud mean short-wave cooling (45 to 50 Wm^{-2}) (Le Treut and McAvaney, 2000; Figure 7.2). Although most models simulate CRF values within the uncertainty range of the observed values, the discrepancy in terms of response to CO_2 increase is large, both in sign, amplitude, and share between long-wave and short-wave. This disagreement (amplified by the water vapour feedback) reflects the sensitivity of the simulated feedbacks to model formulation (Watterson *et al.*, 1999; Yao and Del Genio, 1999; Meleshko *et al.*, 2000) and is the cause of the large spread in model climate sensitivity (see Figure 7.3; and Chapters 8 and 9). The correct simulation of the mean distribution of cloud cover and radiative fluxes is therefore a necessary but by no means sufficient test of a model's ability to handle realistically the cloud feedback processes relevant for climate change.

Satellite records provide some access to the study of natural climate fluctuations, such as the seasonal or ENSO cycles, which can be used to test the ability of the models to represent different feedbacks. At these time-scales, however, cloud structures are characterised by large shifts in their latitudinal or longitudinal position, as well as changes in their vertical distribution in response to SST changes. Only those latter effects are really relevant to test climate models in the context of climate change. Bony *et al.* (1997) have shown that in the inter-tropical regions, it was possible to isolate them by looking at the dependence of cloud properties on SST for specified dynamical regimes. These regressions are remarkably consistent when carried out with different sets of independent data (ISCCP, ERBE, TOVS, SSMI) and offer a constraining test for models. Similar methodologies are being developed for the mid-latitudes (Tselioudis *et al.*, 2000). It is encouraging, though, that CCM3 has demonstrated considerable accuracy in the reproduction of regional all-sky long-wave and short-wave radiation budgets (Kiehl *et al.*, 1998).

Since the SAR, there has been progress in the qualitative understanding of the complexity of cloud/climate relation. There is also an important and ongoing evolution of the cloud parametrizations included into the models, which is characterised by a greater physical consistency and a greatly enhanced physical content: in particular many models now include a more comprehensive representation of sub-grid scales, convective or turbulent, and explicit microphysics, liquid or ice. Simulation of current climate with those advanced models demonstrate capability to reproduce realistically many features of the cloud radiative forcing and its seasonal variations, for both the solar and terrestrial components.

In spite of these improvements, there has been no apparent narrowing of the uncertainty range associated with cloud feedbacks in current climate change simulations. A straightforward approach of model validation is not sufficient to constrain the models efficiently and a more dedicated approach is needed. This should be favoured by a greater availability of satellite measurements.

7.2.3 Precipitation

Precipitation drives the continental hydrology, and influences the salinity of the ocean. The distribution of the precipitation in space and time, is therefore of key importance. The amount of precipitation also constitutes a measure of the latent heat release within the atmosphere. The long-term global mean precipitation of 984 mm/yr implies a vertically integrated mean heating rate of 78 Wm^{-2} (Arkin and Xie, 1994; Kiehl and Trenberth, 1997). The local precipitation rate on horizontal spatial scales of 1 to 10 km are 10 to 1,000 times the global number so that the associated heating rates often dominate all other effects and strongly influence local and global circulations. Precipitation also acts to remove and transport aerosols and soluble gases both within and below clouds, and thus strongly affects the chemical composition and aerosol distribution. The suspended and precipitating condensates provide sites for aqueous and surface chemical reactions.

The highly variable rain rates and enormous spatial variability makes determination of mean precipitation difficult, let alone how it will change as the climate changes. For instance, a detailed examination of spatial structure of daily precipitation amounts by Osborn and Hulme (1997) shows that in Europe the average separation distance between climate stations where the correlation falls to 0.5 is about 150 km in summer and 200 km in winter. This complexity makes it difficult to model precipitation reliably, as many of the processes of importance cannot be resolved by the model grid (typically 200 km) and so sub-grid scale processes have to be parametrized.

7.2.3.1 Precipitation processes

Precipitation is usually considered to be of stratiform or convective nature, or a mixture of the two. In stratiform precipitation the vertical velocity of air, usually forced by developing low pressure systems, monsoonal circulations, or underlying orography, is comparable to or smaller than the fall speed of snow and ice crystals. Stratiform precipitation dominates in the extra-tropics except over continents in summer, and there is substantial spatial coherency on scales up to and beyond about 100 km. In contrast, convective precipitation systems are associated with vigorous latent-heat-driven vertical circulations on horizontal scales of a few kilometres. Convection is responsible for most of the precipitation in the tropics and middle latitude continents in summer. In many cases, convective and stratiform precipitation interact or occur together, for instance as convective cells are embedded within areas of stratiform precipitation.

7.2.3.2 Precipitation modelling

There is increasing evidence that the use of one or more prognostic cloud water variables for the modelling of stratiform precipitation is fairly successful in simulating continental and sub-continental scale precipitation distributions, including orographic precipitation, provided the synoptic-scale circulation is properly accounted for. This is the case for numerical weather predictions in the short-term range of 1 to 2 days (Petroliagis *et al.*, 1996), and regional climate models driven by observed lateral boundary conditions (e.g., Jones *et al.*, 1995; Lüthi *et al.*, 1996; Christensen *et al.*, 1998; Giorgi *et al.*, 1998). Cloud schemes that include an explicit cloud water variable and some parametrization of the ice phase (either by carrying an explicit cloud ice variable or by including a temperature-dependent formulation of microphysical conversion rates) appear able to credibly reproduce some of the major features of the observed intensity-frequency relations (Frei *et al.*, 1998; Murphy, 1999).

In contrast, the simulation of convective precipitation, which is fully parametrized, is of substantially poorer quality over continental regions. In a recent intercomparison study that attempted to simulate dry and wet summers over the continental US, large inter-model and model-observation differences were found (Takle *et al.*, 1999), although the larger-scale atmospheric circulation was prescribed. One specific difficulty is the strong diurnal cycle of convective precipitation over land, which is accompanied by the build-up of a well-mixed boundary layer in response to solar heating prior to the onset of convection. Recent studies (Yang and Slingo, 1998; Dai *et al.*, 1999) find that moist convection schemes tend to initiate convection prematurely as compared with the real world, and instability does not build up adequately. Premature cloud formation prevents the correct solar heating from occurring, impacting the development of the well-mixed boundary layer and continental-scale convergence at the surface, which in turn affects the triggering of convection. Scale interactions between convection that is organised by somewhat larger scales also seem to underlie the difficulties all GCMs have in simulating the Madden-Julian Oscillation of intra-seasonal variations in the deep tropics (Slingo *et al.*, 1996).

7.2.3.3 The temperature-moisture feedback and implications for precipitation and extremes

With increasing temperature, the surface energy budget tends to become increasingly dominated by evaporation, owing to the increase in the water holding capacity of the boundary layer. The increase of evaporation is not strictly inevitable (Pierrehumbert, 1999), but it occurs in all general circulation models, though with varying strength. Simulated evapotranspiration and net atmospheric moisture content is also found to increase (Del Genio *et al.*, 1991; Trenberth, 1998), as is observed to be happening in many places (Hense *et al.*, 1988; Gaffen *et al.*, 1992; Ross and Elliot, 1996; Zhai and Eskridge, 1997). Globally there must be an increase in precipitation to balance the enhanced evaporation but the processes by which precipitation is altered locally are not well understood. Over land, enhanced evaporation can occur only to the extent that there is sufficient soil moisture in the unperturbed state. Naturally occurring droughts are likely to be exacerbated by enhanced potential evapotranspiration, which quickly robs soil of its moisture.

Because moisture convergence is likely to be proportionately enhanced as the moisture content increases, it should lead to

similarly enhanced precipitation rates. Moreover, the latent heat released feeds back on the intensity of the storms. These factors suggest that, while global precipitation exhibits a small increase with modest surface warming, it becomes increasingly concentrated in intense events, as is observed to be happening in many parts of the world (Karl *et al.*, 1995), including the USA (Karl and Knight, 1998), Japan (Iwashima and Yamamoto, 1993) and Australia (Suppiah and Hennessy, 1998), thus increasing risk of flooding. However, the overall changes in precipitation must equal evaporation changes, and this is smaller percentage-wise than the typical change in moisture content in most model simulations (e.g., Mitchell *et al.*, 1987; Roads *et al.*, 1996). Thus there are implications for the frequency of storms or other factors (duration, efficiency, etc.) that must come into play to restrict the total precipitation. One possibility is that individual storms could be more intense from the latent heat enhancement, but are fewer and farther between (Trenberth, 1998, 1999).

These aspects have been explored only to a limited extent in climate models. No studies deal with true intensity of rainfall, which requires hourly (or higher resolution) data, and the analysis is typically of daily rainfall amounts. Increases in rain intensity and dry periods are simulated along with a general decrease in the probability of moderate precipitation events (Whetton *et al.*, 1993; Cubasch *et al.*, 1995; Gregory and Mitchell, 1995; Mearns *et al.*, 1995; Jones *et al.*, 1997; Zwiers and Kharin, 1998; McGuffie *et al.*, 1999). For a given precipitation intensity of 20 to 40 mm/day, the return periods become shorter by a factor of 2 to 5 (Hennessy *et al.*, 1997). This effect increases with the strength of the event (Fowler and Hennessy, 1995; Frei *et al.*, 1998). However, estimates of precipitation and surface long-wave radiation suggest that the sensitivity of the hydrological cycle in climate models to changes in SST may be systematically too weak (Soden, 2000). Accordingly, it is important that much more attention should be devoted to precipitation rates and frequency, and the physical processes which govern these quantities.

7.2.4 Radiative Processes

7.2.4.1 Radiative processes in the troposphere

Radiative processes constitute the ultimate source and sink of energy in the climate system. They are generally well known, and particularly the clear-sky long-wave transfer, including the absorption properties of most greenhouse gases. Their treatment in atmospheric general circulation models relies on several approximations: the fluxes are computed as averaged quantities over a few spectral intervals, the propagation is limited to the vertical upwelling or downwelling directions, and the role of sub-grid scale features of the clouds or aerosols is essentially neglected. Although this methodology is believed to have only a marginal impact on the accuracy of computed long-wave fluxes, the analysis of satellite, ground-based or aircraft measurements (Cess *et al.*, 1995; Pilewskie and Valero, 1995; Ramanathan *et al.*, 1995) has generated a concern that the radiative algorithms used in climate models could significantly underestimate the atmospheric shortwave absorption. Similar results have been obtained as part of the ARM/ARESE experiment (Valero *et al.*,

1997; Zender *et al.*, 1997). The excess, or "anomalous" observed absorption may reach typical values of about 30 to 40 Wm^{-2}. Its very existence, however, remains controversial and is at odds with other investigations. Li *et al.* (1995) analysed large global data sets from satellite (ERBE) and ground observations (GEBA). They did not find the anomalous absorption except for a few tropical sites over a short period of time. The exception appears to be induced by enhanced absorption due to biomass burning aerosols (Li, 1998). Limited accuracy of the measurements (Imre *et al.*, 1996) and the methodology of analysis (Arking *et al.*, 1996; Stephens, 1996; Barker and Li, 1997) have been raised as possible contributing factors to the finding of the anomalous absorption.

A comparison of the ARM-ARESE measurements with a state of the art radiation code, which uses measured atmospheric quantities as an input, and relies on the same physical assumptions as used in climate models, indicated that the anomalous absorption is larger for cloudy than for clear conditions, and also tends to be larger for visible rather than for near-infrared fluxes (Zender *et al.*, 1997). These two characteristics are consistent with the results of a comparison between the output of the CCM3 general circulation model and observations from Nimbus 7 (Collins, 1998). However, Li *et al.* (1999) analysed all the data sets collected during the ARM/ARESE experiment made by space-borne, air-borne and ground-based instruments and did not find any significant absorption anomaly. They traced the source of controversial findings to be associated with inconsistent measurements made by some air-borne radiometers as used in the studies of Valero *et al.* (1997) and Zender *et al.* (1997) with those from all other instruments. However, other studies (Cess *et al.*, 1999; Pope and Valero, 2000; Valero *et al.*, 2000), that do find a significant absorption anomaly, have further demonstrated that the ARESE data do indeed satisfy a number of consistency tests as well as being in agreement with measurements made by other instruments. Meanwhile, evidence of enhanced cloud absorption has been found from measurements of the MGO aircraft laboratory (Kondratyev *et al.*, 1998). Inclusion of anomalous absorption has been found to improve the representation of atmospheric tides (Braswell and Lindzen, 1998). There finally exists evidence of an effect associated with clear-sky conditions – in the presence of aerosols: models tend to overpredict clear-sky diffusion to the surface (Kato *et al.*, 1997).

The three dimensional nature of the solar radiation diffusion by cloud (Breon, 1992; Cahalan *et al.*, 1994; Li and Moreau, 1996) is unaccounted for by present climate models and may play some role in anomalous absorption. In a recent study using a Monte-Carlo approach to simulate three-dimensional radiative transfer, O'Hirok and Gautier (1998a; 1998b) show that cloud inhomogeneities can increase both the near-infrared gaseous absorption and cloud droplet absorption in morphologically complex cloud fields. The enhancement is caused when photons diffused from cloud edges more easily reach water-rich low levels of the atmosphere and when photons entering the sides of clouds become trapped within the cloud cores. The amplitude of those effects remains limited to an average range of 6 to 15 Wm^{-2}, depending on the solar angle. The inhomogeneities significantly affect the vertical distribution of the heating, though, with

potential consequences on cloud development. Incorporating the effect of cloud inhomogeneities in radiative algorithms may become a necessity, and recent efforts have been made along that path (Oreopoulos and Barker, 1999).

If anomalous absorption turns out to be real, it is an effect that will need to be incorporated into radiation schemes. Evaluation of its importance is hampered by lack of knowledge of a physical mechanism responsible for the absorption, and hence lack of a physical basis for any parametrization. Modelling studies by Kiehl *et al.* (1995) have demonstrated the sensitivity of the simulated climate to changes in the atmospheric absorption. As a radiative forcing, anomalous absorption is fundamentally different from water vapour or CO_2 in that it does not significantly alter the Earth's net radiation budget. Instead, it shifts some of the deposition of solar energy from the ground to the atmosphere (Li *et al.*, 1997), with implications for the hydrological cycle and vertical temperature profile of the atmosphere. Anomalous absorption may not, however, appreciably affect climate sensitivity (Cess *et al.*, 1996).

Model validation is also affected. Many of the data which are used to tune or validate model parametrizations, such as the Liquid Water Path (LWP) or the droplet equivalent radius, are obtained from space measurements by the inversion of radiative algorithms, which ignore cloud inhomogeneities and anomalous absorption. This gives a strong importance to satellite instruments such as POLDER which provide measurements of the same scene at a variety of viewing angles, and provides a good test of the plane-parallel hypothesis in retrievals of cloud quantities.

7.2.4.2 Radiative processes in the stratosphere
The stratosphere lies immediately above the troposphere, with the height of the bounding tropopause varying from about 15 km in the tropics to about 7 km at high latitudes. The mass of the stratosphere represents only about 10 to 20% of the total atmospheric mass, but changes in stratospheric climate are important because of their effect on stratospheric chemistry, and because they enter into the climate change detection problem (Randel and Wu, 1999; Shine and Forster, 1999). In addition there is a growing realisation that stratospheric effects can have a detectable and perhaps significant influence on tropospheric climate.

Solar radiative heating of the stratosphere is mainly from absorption of ultraviolet (UV) and visible radiation by ozone, along with contributions due to the near-infrared absorption by carbon dioxide and water vapour. Depletion of the direct and diffuse solar beams arises from scattering by molecules, aerosols, clouds and surface (Lacis and Hansen, 1974).

The long-wave process consists of absorption and emission of infrared radiation, principally by carbon dioxide, methane, nitrous oxide, ozone, water vapour and halocarbons (CFCs, HFCs, HCFCs, PFCs etc.). The time-scales for the radiative adjustment of stratospheric temperatures is less than about 50 to 100 days.

For CO_2, part of the main 15 micron band is saturated over quite short vertical distances, so that some of the upwelling radiation reaching the lower stratosphere originates from the cold upper troposphere. When the CO_2 concentration is increased, the increase in absorbed radiation is quite small and increased emission leads to a cooling at all heights in the stratosphere. But for gases such as the CFCs, whose absorption bands are generally in the 8 to 13 micron "atmospheric window", much of the upwelling radiation originates from the warm lower troposphere, and a warming of the lower stratosphere results, although there are exceptions (see Pinnock *et al.*, 1995). Methane and nitrous oxide are in between. In the upper stratosphere, increases in all well-mixed gases lead to a cooling as the increased emission becomes greater than the increased absorption. Equivalent CO_2 is the amount of CO_2 used in a model calculation that results in the same radiative forcing of the surface-troposphere system as a mixture of greenhouse gases (see e.g., IPCC, 1996) but does not work well for stratospheric temperature changes (Wang *et al.*, 1991; Shine, 1993; WMO, 1999).

An ozone loss leads to a reduction in the solar heating, while the major long-wave radiative effects from the 9.6 and 14 micron bands (Shine *et al.*, 1995) produce a cooling tendency in the lower stratosphere and a positive radiative change above (Ramaswamy *et al.*, 1996; Forster *et al.*, 1997). Large transient loadings of aerosols in the stratosphere follow volcanic eruptions (IPCC, 1996) which leads to an increase of the heating in the long-wave. For the solar beam, aerosols enhance the planetary albedo while the interactions in the near-infrared spectrum yield a heating which is about one third of the total solar plus long-wave heating (IPCC, 1996; WMO, 1999). In addition, ozone losses can result from heterogeneous chemistry occurring on or within sulphate aerosols, and those changes produce a radiative cooling (Solomon *et al.*, 1996).

The Antarctic ozone hole is a stratospheric phenomenon with a documented impact on temperature and, during the period 1979 to 1994, ozone decreases very likely contributed a negative radiative forcing of the troposphere-surface that offset perhaps as much as one half of the positive radiative forcing attributable to the increases in CO_2 and other greenhouse gases (Hansen *et al.*, 1997; Shine and Forster, 1999). It appears that most of the observed decreases in upper-tropospheric and lower-stratospheric temperatures were due to ozone decreases rather than increased CO_2 (Ramaswamy *et al.*, 1996; Tett *et al.*, 1996; Bengtsson *et al.*, 1999).

The subject of solar effects on climate and weather (see Section 6.10) has enjoyed a recent resurgence, in part because of observational studies (Labitzke and van Loon, 1997), but more so because of modelling studies that suggest viable mechanisms involving the stratosphere. As solar irradiance changes, proportionally much greater changes are found in the ultraviolet which leads to photochemically induced ozone changes, and the altered UV radiation changes the stratospheric heating rates per amount of ozone present (Haigh, 1996; Shindell *et al.*, 1999a). Including the altered ozone concentrations gave an enhanced tropospheric response provided the stratosphere was adequately resolved.

7.2.5 Stratospheric Dynamics

Waves generated in the troposphere propagate into the stratosphere and are absorbed, so that stratospheric changes alter where and how they are absorbed, and effects can extend downward into the troposphere through a mechanism called "downward control"

(Haynes *et al.*, 1991). The downward propagation of zonal-mean anomalies provides a purely dynamical stratosphere-troposphere link, which may account for the well-documented troposphere-stratosphere anomaly correlations seen in observations (Baldwin *et al.*, 1994; Perlwitz and Graf, 1995). The North Atlantic Oscillation (Section 7.6.4) thus could be coupled with the strength of the wintertime Arctic vortex (Thompson and Wallace, 1998).

The dominant wave-induced forcing in the stratosphere is believed to come from tropospherically generated planetary scale Rossby waves in wintertime. These waves are explicitly resolved in models and data. Thus the meridional mass circulation, although two-celled, is predominantly directed towards the winter pole (e.g., Eluszkiewicz *et al.*, 1996), and leads to a significant warming and weakening of the polar night vortex relative to its radiatively determined state (Andrews *et al.*, 1987). Variability and changes in planetary wave forcing thus lead directly to variability and changes in wintertime polar temperatures, which modulate chemical ozone loss (WMO, 1999).

The principal uncertainties in wave-induced forcing come from gravity waves, which are undetected in analyses, but whose role is inferred from systematic errors in climate models. The most notable such error is the tendency of all atmospheric GCMs to suffer from excessively cold polar temperatures in the winter stratosphere, together with an excessively strong polar night jet, especially in the Southern Hemisphere (Boville, 1995). Enhanced Rayleigh friction improves the results (Manzini and Bengtsson, 1996; Butchart and Austin, 1998), but its physical basis is unclear (Shepherd *et al.*, 1996). The principal forcing of gravity waves arises from unresolved sub-grid scale processes, such as convection, and more physically based gravity wave parametrizations are being developed.

A dominant factor determining the interannual variability of the stratosphere is the quasi-biennial oscillation (QBO). It is driven by wave drag (momentum transport), but it remains unclear exactly which waves are involved (Dunkerton, 1997). Most current atmospheric GCMs do not simulate the QBO and are therefore incomplete in terms of observed phenomena. It appears that QBO-type oscillations are found in models with higher vertical resolution (better than 1 km, Takahashi, 1996; Horinouchi and Yoden, 1998). It is still not clear what aspects of vertical resolution, energy dissipation, and wave spectrum are necessary to generate the QBO in climate models in a self-consistent way.

The meridional mass circulation, known as the Brewer-Dobson circulation, transports chemical species poleward in the stratosphere (Andrews *et al.*, 1987). Air entering the stratosphere in the tropics returns to the troposphere in the extratropics with a time-scale of about five years (Rosenlof, 1995). Current models indicate shorter time-scales (Waugh *et al.*, 1997), but the reasons are not currently well understood. The variation of the height of the troposphere with latitude is also important for meridional transport and troposphere-stratosphere exchange. This is because mid-latitude cross-tropopause mixing is preferentially along isentropic surfaces which are in the troposphere in the tropics but in the stratosphere at higher latitudes. Transport processes in the lowermost stratosphere are important factors affecting tropospheric climate (Pan *et al.*, 1997).

The mean climate and variability of the stratosphere are not well simulated in current models. Because there is increasing evidence of effects of the stratosphere on the troposphere, this increases uncertainty in model results for tropospheric climate change. A key concern is how well mixing on small scales is done in the lowermost stratosphere. While overdue attention is being given to the stratosphere as more resolution in the vertical is added to models, further increases in resolution are desirable.

7.2.6 Atmospheric Circulation Regimes

There is growing evidence that patterns of atmospheric intra-seasonal and interannual variability have preferred states (i.e., local maxima in the probability density function of atmospheric variables in phase space) corresponding to circulation regimes. Using atmospheric data, Corti *et al.* (1999) showed four distinct circulation patterns in the wintertime Northern Hemisphere. These geographical patterns correspond to conventional patterns of low-frequency atmospheric variability which include the so-called "Cold Ocean Warm Land" (COWL; Section 7.6) pattern, the negative Pacific North American pattern, and the so-called negative Arctic Oscillation pattern.

Simplified dynamical models, which represent fundamental aspects of atmospheric circulation, react to external forcing initially by changes in the recurrence frequency of the patterns rather than by changes in the patterns themselves (Palmer, 1999). Therefore, anthropogenically forced climate changes may also be expressed in an altered pattern frequency. It appears that the observed northern hemispheric changes can be associated, to some extent, with a more frequent occurrence of the COWL pattern (Corti *et al.*, 1999), i.e., the horizontal structure of recent climate change is correlated with the horizontal pattern of natural variability.

These results indicate that detailed predictions of anthropogenic climate change require models which can simulate accurately natural circulation patterns and their associated variability, even though the dominant time-scale of such variability may be much shorter than the climate change signal itself. This regime view of climate change has recently been shown in GCM simulations (Monahan *et al.*, 2000).

7.2.7 Processes Involving Orography

The major mountain ranges of the world play an important role in determining the strength and location of the atmospheric jet streams, mainly by generating planetary-scale Rossby waves and through surface drag. Orography acts both through large-scale resolved lifting and diversion of the flow over and around major mountain ranges, and through sub-grid scale momentum transport due to vertically propagating gravity waves at horizontal scales between 10 and 100 km. The limited horizontal resolution of climate GCMs implies a smoothing of the underlying topography which has sometimes been counteracted by enhancing the terrain by using an envelope orography, but this has adverse effects by displacing other surface physical processes. The sub-grid scale momentum transport acts to decelerate the upper-level flow and is included by gravity-wave drag parametrization schemes (e.g., Palmer *et al.*, 1986; Kim, 1996; Lott and Miller, 1997).

No systematic studies are available to assess the impacts of these procedures and schemes on climate sensitivity and variability. The orographic impact is most pronounced in the Northern Hemisphere winter. Gravity wave drag implies a reduction in the strength of the mid-latitude jet stream by almost 20 ms^{-1} (Kim, 1996), and alters the amplitude and location of the planetary-scale wave structure (e.g., Zhou *et al.*, 1996). Convective precipitation in models is often spuriously locked onto high topography. Thus the numerical simulation of many key climatic elements, such as rainfall and cloud cover, strongly depends upon orography and is strongly sensitive to the horizontal resolution employed. As such, it is possible that phenomena of climate variability are sensitive to orographic effects and their parametrization (Palmer and Mansfield, 1986). These issues may have potentially important consequences for the planetary-scale distribution of climate change.

7.3 Oceanic Processes and Feedbacks

The ocean influences climate and climate change in various ways. Ocean currents transport a significant amount of heat, usually directed poleward and thus contributing to a reduction of the pole-to-equator temperature gradient; a remarkable exception exists however in the South Atlantic where heat is transported equator-ward, i.e. up-gradient, into the North Atlantic. Because of its large heat capacity, ocean heat storage largely controls the time-scales of variability to changes in the ocean-atmosphere system, including the time-scales of adjustment to anthropogenic radiative forcing. The ocean is coupled to the atmosphere primarily through the fluxes of heat and fresh water which are strongly tied to the sea surface temperature (see Section 7.6.1), and also through the fluxes of radiatively active trace gases such as CO_2 (see Chapter 3) which can directly affect the atmospheric radiation balance. All ocean processes which ultimately can influence these fluxes are relevant for climate change. Processes in the ocean surface layer which are associated with seasonal time-scales hence are of obvious relevance. As the budgets in the surface layer depend on the exchange with deeper layers in the ocean, it is also necessary to consider the processes which affect the circulation and water mass distribution in the deep ocean, in particular when the response of the climate system at decadal and longer time-scales is considered. Moreover, processes governing vertical mixing are important in determining the time-scales on which changes of, for example, deep ocean temperature and sea level evolve.

Since the SAR, the assessment of the status of ocean processes in climate models has changed in two ways. On the one hand, advances in model resolution and in the representation of sub-grid scale processes have led to a somewhat improved realism in many model simulations. On the other hand, however, growing evidence for a very high sensitivity of model results to the representation of certain small-scale processes, in particular those associated with the THC, has been found. As a consequence, considerable uncertainties still exist concerning the extent to which present climate models correctly describe the oceanic response to changes in the forcing.

7.3.1 Surface Mixed Layer

The surface mixed layer is directly influenced by the atmospheric fluxes which are connected to the ocean interior by vigorous three-dimensional turbulence. That turbulence is driven primarily by the surface wind stress and convective buoyancy flux, and includes the wave driven Langmuir circulation (e.g., Weller and Price, 1988; McWilliams *et al.*, 1997). As a result, the upper ocean often becomes well mixed.

The heat budget of the mixed layer is determined by horizontal advection, surface heating, entrainment and the vertical heat flux at the mixed-layer base. Entrainment occurs when there is sufficient turbulent energy to deepen the mixed layer and can result in rapid cooling when accompanied by upwelling from Ekman pumping. This pumping velocity at the mixed-layer base results from divergent mixed-layer flow driven by the wind stress. The shallowing of the mixed layer leads to a transfer of water from the mixed layer to the interior of the ocean. The water that passes the deepest mixed-layer depth will not be re-entrained within a seasonal cycle and is subducted. Large subduction rates are found where horizontal gradients in mixed layer depth are large. Thus variations in mixed-layer depth are of primary importance in setting the structure of the interior of the ocean. This process can temporarily shield heat anomalies generated in the mixed layer from the atmosphere. The subduction process itself is relatively well understood (Spall *et al.*, 2000), although the role of sub-grid processes in modifying the subduction process needs further clarification (Hazeleger and Drijfhout, 2000).

The surface buoyancy flux (combined net heat and fresh water flux) effectively drives a cross-isopycnal mass flux by converting mixed-layer water from one density class to another. Waters of intermediate density are transformed into both lighter waters and heavier waters. In general, but especially in the Indian and Pacific Oceans, the thermal and haline contributions are additive in forming light tropical waters, but opposed in forming heavy polar waters. Despite this cancellation about 15 Sv (1 Sverdrup = 1 Sv = $10^6 m^3 s^{-1}$) of North Atlantic Deep Water (NADW) is formed thermally. With more cancellation, less than a few Sverdrups of Antarctic Bottom Water (AABW) is formed in the Southern Ocean. However, nearly 30 Sv of Antarctic Intermediate Water is formed mostly by the haline effect.

In summary, proper parametrization of turbulence in the surface mixed layer is crucial to correctly simulate air-sea exchange, SST and sea ice (e.g., Large *et al.*, 1997; Goosse *et al.*, 1999), and thereby reduce the need for flux adjustments in coupled models. While several schemes are in use (Large and Gent, 1999), a systematic intercomparison of the properties, behaviour and accuracy of these parametrizations is still lacking.

7.3.2 Convection

Open ocean convection occurs every winter at high latitudes when buoyancy loss at the sea surface causes the surface layer to become denser than the water below, and results in highly variable mixing depth as a function of space and time (see recent review by Marshall and Schott, 1999). Convection directly affects the SST locally, and on larger scales indirectly through its effect on water mass properties and circulation. The maximum depth of convection occurs at the end of the cooling season, and depends on the balance between the cumulative air-sea fluxes, including ice-melt and precipitation, and the oceanic advection of buoyancy. During the summer, shallow warm surface mixed layers isolate the newly formed deep water from the atmosphere and mean currents and mesoscale eddies steadily transfer the newly formed deep water into the abyssal ocean.

Deep convective mixing is an essential ingredient of the THC, in particular in the North Atlantic, and is thus important for climate problems. It constitutes a very efficient vertical transfer process, and only a few small regions are needed to offset the slow diffusive buoyancy gain due to vertical (diapycnal) mixing (see also Section 7.3.3; e.g., Winton, 1995). Major sites of known open ocean deep convection are the centre of the Greenland Sea (Schott *et al.*, 1993; Visbeck *et al.*, 1995), Labrador Sea (LabSea Group, 1998) and a small region in the north-western Mediterranean Sea (Schott *et al.*, 1996). However, only the Labrador Sea is in direct contact with the NADW which replenishes the deep waters of the Atlantic, Pacific and Indian Oceans. In the Greenland Sea deep and bottom waters remain local to the Arctic Ocean and deep basins north of Iceland and have no direct influence on the Denmark Strait Overflow (Mauritzen, 1996). Deep water formed in the Mediterranean Sea also never directly outflows through the Strait of Gibraltar. Dense water also forms on the shelf where convection can reach to the sea floor producing a well-mixed layer of dense water. Several mechanisms, such as eddies, flow over canyons and time-varying shelf break fronts, allow the dense water to enter into the deep ocean via descending plumes. This 'shelf convection' is believed to be fairly widespread around the Antarctic continent and is probably the primary mechanism by which AABW is formed (Orsi *et al.*, 1999). AABW is the densest bottom water and penetrates into all of the three major oceans.

The overall effect of open ocean convection is usually parametrized through simple convective-adjustment schemes (Marotzke, 1991) which have been found to work well (Klinger *et al.*, 1996). More advanced schemes have a somewhat increased performance (Paluszkiewicz and Romea, 1997). Lateral exchange between the deep convective centres and the surrounding boundary currents can significantly alter the convective process (Maxworthy, 1997). Within those small regions, deep mixing is affected by mesoscale eddies in two ways: cyclonic eddies provide an additional preconditioning (Legg *et al.*, 1998), and collectively exchange fluid with the periphery of the deep mixed region. Coarse and medium resolution ocean models have shown significantly improved simulations of the deep convective regions when more sophisticated parametrizations of mesoscale eddies were employed (Danabasoglu and McWilliams, 1995; Visbeck *et al.*, 1997). In particular the unrealistic widespread convective mixing (>500 m) over much of the Southern Ocean was significantly reduced.

Shelf plume convection is more difficult to represent in coarse resolution climate models. The problem is challenging because shelf convection is heavily influenced by the details of the bathymetry, coastal fronts and mesoscale eddies

(Gwarkiewicz and Chapman, 1995), as well as entrainment of the ambient water (Baringer and Price, 1997). Several different attempts have been made to parametrize its overall effect (e.g., Beckmann and Döscher, 1997; Killworth and Edwards, 1999). Most ocean models have, however, not yet implemented such schemes and ventilate, e.g., the Southern Ocean by means of extensive open ocean (polynya) convection. The effect on the sensitivity of the current coupled climate models to forcings which involve changes in the convective processes is not known.

The convective activity in the Greenland and Labrador Seas varies inversely on decadal time-scales (see Chapter 2, Section 2.2.2.5). The switch of the convective activity from the Greenland Sea to the Labrador Sea has been attributed to changes in the index of the North Atlantic Oscillation (Dickson *et al.*, 1996). The magnitude of the corresponding change in THC intensity is, however, controversial. While model results suggest moderate fluctuations (10 to 15%), it has been claimed from analysis of hydrographic observations that these changes might be much larger, with heat transport changes of more than 0.3 PW at 48°N, and in excess of 0.5 PW at 36°N (Koltermann *et al.*, 1999).

In summary, the representation of oceanic convection in current climate models is satisfactory to simulate the convection changes observed over the last decades. It is, however, not certain that current schemes will work equally well in situations that involve substantial changes in the THC.

7.3.3 Interior Ocean Mixing

Diapycnal mixing across density surfaces provides the main way to warm the cold deep waters which are formed by convection and sink at high latitudes, thus allowing them to rise through the thermocline and complete the lower limb of the "conveyor belt". Diapycnal mixing is an essential part of the ocean circulation, in particular the THC, and can affect surface conditions and climate change on decadal to centennial and longer time-scales.

The processes leading to mixing in the main thermocline involve random internal wave breaking (Gregg, 1989; Polzin *et al.*, 1995), and to some degree also double-diffusive mixing (St. Laurent and Schmitt, 1999). Diffusivities in most of the main thermocline are typically of the order 10^{-5} m^{2}s^{-1} or less (Gregg, 1989; Ledwell *et al.*, 1993), whereas diffusivities of an order of magnitude larger are required to close the THC (Munk, 1966). Recent observations in the deep ocean have, however, found that turbulence is greatly enhanced 1,000 to 2,000 m above the bottom in regions of rough bottom topography (Polzin *et al.*, 1997), with mixing rates reaching values of 1 to 3 × 10^{-4} m^{2}s^{-1} (Ledwell *et al.*, 2000). The most likely cause is internal waves generated by tidal flows over kilometre-scale bathymetric features. In addition to tidal-driven mixing, the calculations of Munk and Wunsch (1998) indicate that wind-driven mixing in the Antarctic Circumpolar Current contributes substantially. Preliminary estimates of mixing rates based on the internal wave parametrization suggest that the wind-driven mixing is more uniformly distributed throughout the water column (Polzin and Firing, 1997). Whereas the classical assumption of uniform mixing and

upwelling led to an expectation of poleward interior flow in the deep ocean, the new data suggest downwelling in the interior, with upwelling confined to the many ocean bottom fracture zone valleys perpendicular to the ridge axis (Ledwell *et al.*, 2000).

Evidence for a significant role for double diffusion in ocean mixing has emerged. The "salt finger" and "diffusive convection" processes transfer heat and salt at different rates resulting in a density transport against its mean gradient. New data suggest that salt fingers may have a somewhat subtle role over widespread areas of the sub-tropical gyres (St. Laurent and Schmitt, 1999). The implications of widespread double diffusion for the general circulation are, however, controversial. Gargett and Holloway (1992) found in a model study with a simplistic representation of double-diffusive mixing that there were dramatic changes in the circulation and water masses. Recent studies using a more conservative parametrization of double-diffusion suggest modest though still significant changes in the circulation (Zhang *et al.*, 1998).

Diapycnal mixing in the ocean is usually associated with energy conversions, and parametrizations are often based on energy arguments. In their most simple form, these suggest formulating mixing coefficients in terms of stability frequency and/or Richardson number. Double-diffusive mixing has been included in a few models only to very a limited extent. The higher mixing rates found in the abyssal ocean over rough topography have so far not been included in the vertical mixing scheme of coupled climate models, but have been used with some effect on the deep circulation in ocean models (e.g., Hasumi and Suginohara, 1999).

The sensitivity of climate model results to the mixing parametrization is not fully clear. In a state of thermohaline equilibrium, the intensity of the meridional overturning circulation should be strongly dependent on the average internal mixing rate, and in ocean-only models the meridional overturning circulation often varies with a certain power of the average mixing rate (Zhang *et al.*, 1999a). Some models indicate, however, that significant transports may be involved in wind-driven flows along isopycnals which outcrop in different hemispheres. The buoyancy changes are confined to the surface mixed layers, and little or no interior mixing is required. The scaling laws and dynamics of such flows have yet to be clarified, but they do raise the possibility of a meridional heat transport independent of the interior mixing rate.

In summary, the uncertainties associated with interior ocean mixing parametrizations are likely to be small for climate projections over a few decades but could be considerable over longer time scales.

7.3.4 Mesoscale Eddies

Mesoscale eddies in the ocean have a scale of 50 to 100 km, and correspond dynamically to high and low pressure systems in the atmosphere. The role of eddies for climate change arises from their influence on the circulation by (i) transporting and mixing temperature and salinity, (ii) exchanging (usually extracting) potential energy from the mean flow, and (iii) exchanging momentum with the mean flow (in both directions). Eddy processes are of primary importance for the dynamics of intense western boundary currents, through the exchange of momentum and energy via instability and/or rectification processes, and also influence the dynamics of the Southern Ocean. The eddy contribution to the meridional transport of heat and fresh water is small in many regions but cannot be ignored on the global scale. Long lived, propagating eddies such as Agulhas rings determine a major part of the inter-basin exchange between Indian and South Atlantic Oceans. The decaying rings provide a source of warm salty water important for the global thermohaline circulation (de Ruijter *et al.*, 1999), and variations in that exchange may generate THC variations.

Considerable progress has been made in recent years with the parametrization of eddies in climate models of coarse and medium resolution (cf., Chapter 8, Table 8.1) which do not explicitly represent eddies. New schemes are based on eddy dynamics and the physics of baroclinic instability. Lateral eddy mixing of tracers such as potential temperature and salinity is mainly directed along isopycnals, and has a small effect on the dynamics but is important for water mass properties. The traditional mixing along horizontal surfaces which is still used in several climate models leads to unrealistic upwelling in the western boundary current (Boening *et al.*, 1995) and strongly resolution-dependent simulations of meridional overturning and heat transport (Roberts and Marshall, 1998). Isopycnal mixing is natural in models with an isopycnal vertical co-ordinate, for other models a stable and conservative algorithm for isopycnal rotation of the diffusion tensors is now available (Griffies *et al.*, 1998).

The dynamical effects of baroclinic instability are now frequently parametrized as an additional eddy advection of tracers (Gent *et al.*, 1995). With this parametrization, the northward heat transport in the North Atlantic Ocean is less dependent on resolution, and matches the observational estimates much better. Aspects of the Southern Hemisphere circulation are also improved by this parametrization. Theoretical studies (e.g., Killworth, 1997) suggest somewhat different formulations based on down-gradient mixing of potential vorticity. The strengths of isopycnal mixing and eddy-induced advection are usually described empirically through coefficients which are constant or dependent on grid size. For modelling climate change, it is, however, imperative to relate these coefficients to properties of the mean flow (Visbeck *et al.*, 1997).

The parametrization of exchange of eddy momentum with the mean flow remains a challenge. Some studies suggest a substantial eddy influence on the mean barotropic flow based on the interaction of a statistical eddy field with bottom topography (Eby and Holloway, 1994; Merryfield and Holloway, 1997). However, due to the near-geostrophy of ocean currents, it is possible that these processes are not as critical as the sub-grid scale effects on tracers for the overall quality of ocean model solutions.

In summary, while the effects of ocean eddies for climate change are likely to be moderately small, a quantitative assessment will require coupled simulations with eddy-permitting ocean models.

7.3.5 Flows over Sills and through Straits

The water mass structure in the deep ocean is largely dominated by the flows across a few shallow sills and through straits. The role of such flows for climate change arises from their influence on the THC which in turn can affect surface conditions. Once water has crossed a sill, it descends the continental slope as a dense gravity current and provides the water source for the downstream basin. The water mass properties are determined by the entrainment of and mixing with ambient water (Baringer and Price, 1997). For the NADW which is at the heart of the sinking branch of the global conveyor, the overflow of cold water across the Greenland-Iceland-Scotland ridge is the principal source. Observations show intense flow with speeds up to 1 m/s, in a layer of less than 100 m thickness above the bottom and within 20 km of the continental slope. Model calculations suggest that an interruption of this overflow would lead to a breakdown of the Atlantic THC and the associated heat transport within less than a decade (Döscher *et al.*, 1994). In the past decades the overflow appears to have been fairly steady. Another prominent feature is the Indonesian Throughflow which has substantial contributions to the variability of the THC and heat transport in the Indian Ocean (Godfrey, 1996).

In coarse and medium resolution climate models, the overflow across the Greenland-Iceland-Scotland ridge has been found to be highly sensitive to the precise geometry used. For example, changes in topography by as little as one grid cell resulted in gross changes not only to the amount of cross-ridge flux but also to its location, and to the composition of the water mass actually crossing the sill. Thus, a 50% change in heat flux at the Greenland-Iceland-Scotland ridge latitude could be achieved in a model by the addition or subtraction of a single grid box (Roberts and Wood, 1997). Even eddy-permitting simulations give flows through sills which are very sensitive to model details (Willebrand *et al.*, 2001). The physical processes involve hydraulic control and are not properly represented in climate models, and it is unclear whether a situation with substantial overflow changes can be modelled correctly. Parametrizations for the flux across a sill are only available for the simplest process models (Killworth, 1994; Pratt and Chechelnitsky, 1997) that do not include mixing or unsteadiness, both of which are known to be important at sills (Spall and Price, 1998).

Climate models which are based on depth co-ordinates obtain far too much mixing near a sill. This is caused by both poor mixing parametrizations in such models and by excessive diapycnal mixing resulting from the 'staircase'-like representation of bottom topography. As a result, water mass structure downstream of a sill is poorly represented in climate models. Isopycnic models are free of this erroneous mixing, and addition of Richardson number dependent mixing to such models results in realistic tongues of dense water moving downslope, mixing at a rate consistent with observations (Hallberg, 2000). A promising new development is the explicit description of the turbulent bottom boundary layer in climate models which yields more realistic flow of dense water down slopes (Beckmann and Döscher, 1997; Killworth and Edwards, 1999).

In summary, the uncertainties in the representation of the flows across the Greenland-Iceland-Scotland ridge limit the ability of models to simulate situations that involve a rapid change in the THC.

7.3.6 Horizontal Circulation and Boundary Currents

The horizontal circulation in ocean gyres contributes to the meridional transports of heat and fresh water in the climate system (see Section 7.6.2), and therefore is of immediate relevance for climate change. Much of that transport occurs through Western Boundary Currents (WBCs) such as the Gulf Stream, Kuroshio, Agulhas Current and others which are prominent elements of the ocean circulation. These currents are mainly driven by the wind, and have a typical width of 50 km. Once a WBC has left the continental slope, it is characterised by recirculation regimes and strong mesoscale variability. The dynamical influence of the WBCs on the less vigorous interior circulation is not fully understood. Their pathways are, however, crucial in determining the location of the sub-polar front which separates the warm waters of the sub-tropics from the colder sub-polar waters. The lower branch of the THC is also dominated by deep WBCs. In the Atlantic, these are concentrated within 30 km of the continental slope, and are accompanied by substantial recirculation and variability (Lee *et al.*, 1996).

The width and strength of boundary currents in climate models are very sensitive to resolution, they become stronger and narrower as the resolution is increased, provided that at the same time the sub-grid scale transports are also reduced. For example, a recent simulation for the North Atlantic using a grid resolution of 1/10° (equivalent to 11 km or better) indicates that features such as the width, location and variability of boundary currents, the eddy field and its statistics, as well as regional current features are in rather good agreement with observations (Smith *et al.*, 2000). On the other hand, in non-eddy-resolving climate models, boundary currents are quite unrealistic, they lack the observed sharp fronts and recirculation regimes and hence miss the associated air-sea heat fluxes and their dynamical influence on mid-ocean circulation. The effect of having weaker and wider boundary currents for the model's climate and climate sensitivity has so far not been systematically evaluated.

7.3.7 Thermohaline Circulation and Ocean Reorganisations

In the Atlantic Ocean, the THC is responsible for the relatively mild climate in Western Europe (see Section 7.6.1). While palaeo-oceanographic analyses suggest that the Atlantic THC has been relatively stable for the last 8,000 years, a series of large and rapid climatic changes during the last Ice Age has been reconstructed from numerous palaeoclimatic archives (see Chapter 2, Section 2.4). Based on the presently available evidence, these changes are best explained by major reorganisations involving the THC (Broecker, 1997; Stocker and Marchal, 2000). Changes of the THC, due to natural variability or slowly changing surface forcing, thus have an important effect on the climate on a regional to hemispheric scale, and numerous model studies since the SAR have investigated potential changes in

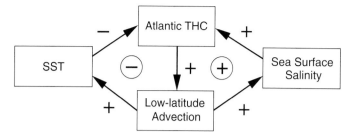

Figure 7.4: Idealised schematics of two advective feedback loops involving the thermohaline circulation (THC). The signs attached to the arrows indicate the correlation between changes in the quantity of the outgoing box with that of the ingoing box, e.g., warmer sea surface temperatures (SST) lead to weaker THC. Resulting correlations of a loop are circled and they indicate whether a process is self-reinforced (positive sign) or damped (negative sign). A stabilising loop (left) is associated with changes in SST due to changes in the advection of heat. This loop may give rise to oscillations. The second loop (right) is due to the influence of advection of low latitude salty waters into the areas of deep water formation. The resulting correlation is positive and the loop may therefore cause instabilities.

THC and elucidated the underlying mechanisms and their impact on the climate system.

Generally, in high latitudes the ocean loses heat and gains fresh water (precipitation and continental runoff) which has opposite effects on the density of ocean water. In addition, the density of sea water is influenced by the supply of warm and salty water from the low latitudes which constitutes the positive feedback maintaining the Atlantic THC (Figure 7.4). This balance is influenced by the surface fluxes of heat and fresh water, i.e., precipitation, evaporation, continental runoff and sea-ice formation, all processes that are likely to change in the future. Some models also suggest an influence of Southern Ocean wind on the Atlantic THC (Toggweiler and Samuels, 1995).

The response of the THC to a perturbation depends on the relative strength of further feedbacks. Reduced convection and advective heat transport into a region lead to colder SSTs which counteracts the effect of salinity on density and thus limits the strength of the destabilising oceanic feedbacks (Figure 7.4). The atmospheric heat transport compensates for parts of the changes in ocean heat transport and is a key factor in determining the stability of the THC (Zhang *et al.*, 1993; Mikolajewicz and Maier-Reimer, 1994). Atmospheric moisture transport among basins provides another feedback (Schiller *et al.*, 1997). Model results suggest that in the presence of overflow between Greenland and Iceland the THC is not very sensitive to changes in the atmosphere-ocean fluxes (Lohmann and Gerdes, 1998).

It is likely that sea ice also has an effect on the stability of the THC: decreased THC and hence oceanic heat transport leads to more sea ice formation (Schiller *et al.*, 1997). Formation of sea-ice tends to increase the density of sea water both through brine rejection and cooling of the overlying air via increased surface albedo and enhanced insulation. Increased export of sea ice from these areas represents a significant fresh water transport over long distances which can decrease deep water formation, thus representing a positive feedback. A reliable estimate of the net effect of sea ice on the stability of the THC is hampered by the still crude representation of sea ice in most current climate models and the resulting unrealistic simulation of sea-ice distributions. A negative feedback contributing to a stabilisation of the THC was recently proposed by Latif *et al.* (2000). They suggested that during El Niño conditions fresh water export from the tropical Atlantic via the atmosphere is enhanced; this tendency is also found in observations (Schmittner *et al.*, 2000). This implies an increased supply of saltier waters towards the northern North Atlantic which facilitates deep water formation, hence a stabilising process for the THC; the opposite effect occurs for La Niña conditions.

A special concern are possible reductions of the Atlantic THC caused by warming and freshening of high latitude surface water associated with global warming (Manabe and Stouffer, 1993; Stocker and Schmittner, 1997). Both the high latitude warming and an enhanced poleward transport of moisture in the atmosphere contribute to the reduction of water density in the formation regions of NADW. However, it is not clear which of these two processes is the more important (Dixon *et al.*, 1999; Mikolajewicz and Voss, 2000). Most coupled climate models indicate a reduction in the meridional overturning circulation, a measure of the THC, from 10 to 50% in response to increasing CO_2 concentration in the atmosphere for the next 100 years (see Chapter 9). A notable exception is one coupled climate model (Latif *et al.*, 2000), but this result must be corroborated with other comprehensive climate models that are capable of resolving well both ENSO and the THC processes.

A less likely, but not impossible, scenario is a complete shut-down of the THC, which would have a dramatic impact on the climate around the North Atlantic. A complete shut-down requires the passing of a critical threshold and may be an irreversible process because of multiple equilibria of the THC. Multiple equilibria have been reported by the entire ocean and climate model hierarchy ranging from simple models (Stommel, 1961; Stocker and Wright, 1991), uncoupled OGCMs (Bryan, 1986; Marotzke and Willebrand, 1991; Weaver and Hughes, 1994) to fully coupled ocean-atmosphere GCMs (Manabe and Stouffer, 1988). The structure of the oceanic reorganisation beyond the threshold, simulated in models of different complexity, is robust: the Atlantic THC ceases (Manabe and Stouffer, 1993; Schmittner and Stocker, 1999), which leads to a reduction in the meridional heat transport in the Atlantic, and hence a regional cooling counteracts the temperature increase. It depends on the model's climate sensitivity whether the combined effect results in a net warming or net cooling in the regions most affected by the meridional heat transport of the Atlantic THC. In addition, such reorganisations would have a profound impact on the north-south distribution of the warming and precipitation (Schiller *et al.*, 1997), on sea level rise (see Chapter 11, Section 11.5.4.1; Knutti and Stocker, 2000), and on the biogeochemical cycles (see Chapter 3, Section 3.7.2; Joos *et al.*, 1999).

Model simulations with different climate models have found a complete shut-down of the THC when the global atmospheric temperature increases by an amount between 3.7 and 7.4°C (Manabe and Stouffer, 1993; Stocker and Schmittner, 1997; Dixon *et al.*, 1999). It is however not clear whether the climate system possesses a threshold in this range, as the values depend on the response of the hydrological cycle to the warming and the parametrizations of mixing processes in the ocean. Furthermore, the rate of temperature increase also determines stability: the THC is less stable under faster perturbations (Stocker and Schmittner, 1997; Stouffer and Manabe, 1999). Model simulations further suggest that the proximity to the threshold depends on the strength of the THC: a weaker THC is more likely to shut down completely (Tziperman, 2000).

In summary, a reduction of the Atlantic THC is a likely response to increased greenhouse gas forcing based on currently available model simulations. Uncertainties in the model simulation of the hydrological cycle, the vertical transport of heat in the ocean interior and the parametrization of deep water formation processes, however, translate directly to uncertainties regarding the stability and future evolution of the THC. While none of the current projections with coupled models exhibit a complete shut-down of the THC during the next 100 years (see Chapter 9, Figure 9.21), one cannot exclude the possibility that such thresholds lie in the range of projected climate changes. Furthermore, since natural variability in the climate system is not fully predictable, it follows that there are inherent limitations to predicting transitions and thresholds.

7.4 Land-Surface Processes and Feedbacks

The net radiation absorbed by the continents is partitioned mainly into sensible and latent (evapotranspiration) heat fluxes whose release back into the atmosphere directly influences local air temperature and humidity and thence other climate system variables. In any given locale, soil moisture availability and vegetation state largely determine the fraction of net radiation that is used for evapotranspiration, as well as the photosynthetic and respiration rates. Thus, realistic modelling of land surface-atmosphere interactions is essential to realistic prediction of continental climate and hydrology. In doing this, attention must be paid to the links between vegetation and the terrestrial energy, water and carbon cycles, and how these might change due to eco-physiological responses to elevated CO_2 and changes in land use.

7.4.1 Land-Surface Parametrization (LSP) Development

The exchanges of energy, momentum, water, heat and carbon between the land surface and the atmosphere must be more realistically and accurately calculated in the next generation of coupled models (Sellers *et al.*, 1997). Fluxes of the first four quantities, traditionally defined as physical climate system variables, are routinely parametrized in Numerical Weather Prediction (NWP) models and climate models as functions of the surface albedo, aerodynamic roughness length and surface "moisture availability" (Betts *et al.*, 1998; Viterbo *et al.*, 1999). These land-surface properties can all be defined as functions of the type and density of the local vegetation, and the depth and physical properties of the soil. The first generation of land-surface parametrizations (LSPs) developed in the 1970s took little account of these relationships, and were replaced by biophysically realistic models, complete with supporting vegetation and soil databases, which led

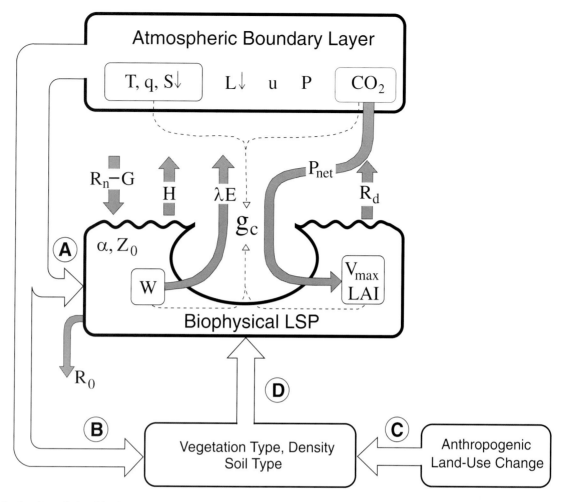

Figure 7.5: Schematic showing relationships between a simulation of the Atmospheric Boundary Layer (ABL), a Land-Surface Parametrization (LSP), vegetation and soil properties, and anthropogenic change. Interactions are shown by broad white arrows marked with capital letters, fluxes by grey arrows, and dependencies by dotted lines. (A) Diurnal-seasonal interactions between the ABL and the LSP; the ABL variables of air temperature, humidity, downward short-wave radiation, downward long-wave radiation, wind speed and precipitation (T, q, $S\downarrow$, $L\downarrow$, u, P) are used to force the LSP which calculates net radiation minus ground heat, sensible heat, and latent heat fluxes (R_n–G, H, λE), which in turn feed back to the atmosphere. Three surface parameters in the LSP are critical to these calculations: Albedo and surface roughness (α, Z_0) determine the radiative balance and turbulent exchange regime, and in third generation LSPs, the canopy conductance term, g_c (equivalent to the summation of all the leaf stomatal conductances) determines the vegetation evapotranspiration rate (λE) and net photosynthetic rate (P_{net}). On time-scales of minutes to hours, g_c is a direct function of T, q, $S\downarrow$, CO_2 concentration and soil moisture (W). Increasing CO_2 concentration can act to significantly reduce g_c and hence limit λE. The maximum value of g_c is determined by parameters related to vegetation density or leaf area index (LAI), and biochemical capacity (V_{max}). Long-term climatic forcing (B) and land-use change (C) can alter the vegetation type and density, soil properties and ecosystem respiration rates, R_d, by which carbon is returned to the atmosphere from the vegetation and soil. (D) Changes in vegetation properties affect V_{max} and LAI, and changes in soil properties affect soil moisture (W) and runoff (R_0).

to significant improvements in NWP and climate model performance in the 1980s and early 1990s. However, these second generation models incorporated only empirical descriptions of the evapotranspiration process, by which water is taken up from the soil by plant roots and released into the atmosphere through tiny pores in leaf surfaces called stomata, while CO_2 is drawn from the atmosphere into leaf interiors for photosynthesis through these same stomata. Research has shown that living plants appear to actively control stomatal widths (conductance) in response to changes in water vapour and CO_2 concentration to optimise the ratio of water vapour losses to CO_2 uptake, and simple, robust

models of the photosynthesis-conductance system in plant leaves have been constructed based on this idea, see Figure 7.5 (Farquhar *et al.*, 1980; Collatz *et al.*, 1991; Sellers *et al.*, 1992a). These models have been parametrized and verified at the leaf level, and can also be scaled up to describe vegetation canopy processes at regional scales using satellite data. These third generation LSPs, published in the late 1990s, thus combine consistent descriptions of the physical climate system transfer processes for energy, momentum, water and heat, with the biophysics of photosynthesis (Bonan, 1995; Sellers *et al.*, 1996c; Dickinson *et al.*, 1998). Why is this important?

Photosynthesis and respiration are climatically sensitive and exhibit interannual variations following climate variations (Dai and Fung, 1993; Francey *et al.*, 1995; Goulden *et al.*, 1996; Myneni *et al.*, 1997; Randerson *et al.*, 1997; Xiao *et al.*, 1998; Randerson *et al.*, 1999; Tian *et al.*, 1999). Furthermore, there appears to have been a net enhancement of terrestrial photosynthesis over respiration over the last two decades, according to inverse modelling results (Tans *et al.*, 1990; Ciais *et al.*, 1995; Keeling *et al.*, 1995; Denning *et al.*, 1996a,b; Randerson *et al.*, 1999); and isotopic analyses (Fung *et al.*, 1997). While disagreements remain about the longitudinal distribution (Rayner and Law, 1999) and the processes responsible for this uptake (Holland *et al.*, 1997a; Field and Fung, 1999; Houghton and Hackler, 1999), these results indicate that changes in the terrestrial carbon balance could be a significant factor in determining future trajectories of atmospheric CO_2 concentration and thus the rate and extent of global warming; see further discussion in Chapter 3. The third generation LSPs have incorporated advances in our understanding of how green plants function, how they alter isotopic fractions of gases they come into contact with, and how they interact with radiation to produce distinctive signatures that can be observed by remote sensing satellites (Tucker *et al.*, 1986; Sellers *et al.*, 1992a; Myneni *et al.*, 1995). As a result, LSPs can now be used to calculate mutually consistent land-atmosphere fluxes of energy, heat, water and carbon (Denning *et al.*, 1996a,b; Randall *et al.*, 1996).

Photosynthesis and stomatal conductance also exhibit strong diurnal variation, and improved representation of this has led to better simulation of the diurnal variation of surface heat and water fluxes and hence more realistic forcings for boundary-layer dynamics and convection (Denning *et al.*, 1996a; Randall *et al.*, 1996). These physiologically driven variations in the surface fluxes have a direct influence on the diurnal surface air temperature range in continental interiors and are directly sensitive to changes in atmospheric CO_2 concentration (Collatz *et al.*, 2000). Furthermore, increasing atmospheric CO_2 is likely to have a direct effect on vegetation stomatal function through feedbacks in the photosynthesis-conductance system. Increased CO_2 concentrations allow vegetation to maintain the same photosynthetic rate with a lower evapotranspiration rate. In a recent GCM study, tropical photosynthesis and transpiration rates were calculated to change only slightly under a CO_2 concentration of 700 ppm, while the additional surface net radiation due to global warming was mainly returned to the atmosphere as sensible heat flux, boosting warming over the tropical continents by 0.4 to 0.9°C above the direct greenhouse warming of 1.7°C (Sellers *et al.*, 1996a). It has been hypothesised that this effect may be partially countered by increased vegetation growth (Betts *et al.*, 1997), but it is not clear to what extent this would be significant in already densely vegetated areas such as the tropical forests. To what extent can we trust these new models and their predictions?

The third generation LSPs combine biology and atmospheric physics in an economical and plausible way and require fewer parameters than their empirical predecessors (Bonan, 1995; Sellers *et al.*, 1996c; Dickinson *et al.*, 1998). Equally important, these models provide more opportunities for parameter calibration, constraint, process sub-model validation and "upscaling" algorithm verification through comparison with observations. This is because a powerful range of carbon-related and physiological fluxes and variables may be added to the physical climate system parameters conventionally used for LSP validation (Colello *et al.*, 1998; Delire and Foley, 1999). It has also been found that the leaf-level physiological photosynthesis-conductance models developed and tested under laboratory and field conditions can be spatially integrated to describe processes over large areas using routine satellite observations. Over the last few years, satellite measurements have been used to construct global data sets of vegetation type and other surface parameters at monthly time resolution and one degree spatial resolution or better (Sellers *et al.*, 1996b). Most importantly, these data sets are globally consistent in that they are primarily based on one type of satellite sensor and one set of interpretative algorithms. Satellite observations also provide a powerful diagnostic capability for tracking climatic impacts on surface conditions; e.g., droughts and the recently observed lengthening of the boreal growing season; and direct anthropogenic impacts, e.g., deforestation. Validation of the process models, upscaling methodologies and the satellite algorithms used to define global parameter fields, has been achieved through a series of large-scale field experiments (Andre *et al.*, 1986; Sellers *et al.*, 1992b; Bolle *et al.*, 1993; Goutorbe, 1994; Sellers *et al.*, 1995). In every environment studied so far (mid-latitude grassland and forest, boreal forest, arid zones) these methods have been used to calculate regional land-atmosphere flux fields that concur with satellite observations and fluxes measured by surface rigs and low flying aircraft (Cihlar *et al.*, 1992; Desjardins *et al.*, 1992; Sellers *et al.*, 1992c).

The field experiments have provided other significant benefits to the modelling community: LSPs in NWP models have been enhanced, leading to direct improvements in precipitation and cloudiness prediction over the continents (Beljaars *et al.*, 1996; Betts *et al.*, 1996), and large biases in some NWP surface parameter fields, for example winter albedo in the boreal forest, have been corrected (Betts *et al.*, 1998). Climate modellers have similarly used the results from field experiments to improve process sub-models (Shao and Henderson-Sellers, 1996), global parameter sets (Sellers *et al.*, 1996b), and to develop better methods for dealing with land-surface spatial heterogeneity (Avissar, 1998; Avissar *et al.*, 1998). Model validation and intercomparison tests have used field experiment data sets to baseline the performance of LSPs within a rigorous intercomparison framework (see Chapter 8, Section 8.5.4).

While there has been considerable success in upscaling process sub-models and parameter sets to describe continental-scale fluxes and states, there remain some difficult areas, notably the treatment of heterogeneous landscapes, soil water transport and catchment hydrology (see Section 7.4.3); the treatment of snow (see Section 7.5.1); and the coupling of LSPs with atmospheric boundary-layer models (see Section 7.2). Sophisticated scaling methodologies are needed to deal with the many non-linear components in these systems and accurate, realistic modelling of these processes on a large scale is likely to remain a challenge for some time to come.

Global biogeochemical models have been developed independently of LSPs to investigate carbon cycling by the terrestrial biosphere (e.g., Field *et al.*, 1995; Foley *et al.*, 1996, 1998; Cao and Woodward, 1998; Kicklighter *et al.*, 1998; see also Chapter 3). Biogeochemical models are now being coupled with LSPs so that a complete, internally consistent carbon balance can be performed along with conventional surface energy balance calculations (see also Section 7.4.2). Changes in the isotopic fractions of key gases as a result of interactions with the vegetation physiology can also be calculated (Ciais *et al.*, 1997; Fung *et al.*, 1997; Randerson and Thompson, 1999) and preliminary work is under way to calculate dry deposition rates and thus refine calculations of the global tropospheric budget of O_3 and reactive N-species within LSPs.

The eventual result should be powerful, internally consistent, model combinations that will be capable of simulating the influence of the physical climate system on the terrestrial carbon cycle, and *vice versa*. This kind of model will be important for realistic simulation of the global climate out beyond a few decades, during which carbon cycle-climate interactions could significantly influence the rate of atmospheric CO_2 increase, the nature and extent of the physical climate system response, and ultimately, the response of the biosphere itself to global change. Furthermore, the performance of such models can be checked, or constrained, by a combination of conventional physical climate system observations, atmospheric gas and isotope fraction concentrations, and satellite data.

Many scientific and technical problems must be solved to achieve these goals. Sections 7.4.2, 7.4.3 and 7.5.1 cover some particular issues that will be important for the construction of realistic, accurate LSPs in the future.

7.4.2 Land-Surface Change

Climate and carbon cycle simulations extending over more than a few decades must take account of land-surface change for two main reasons. First, changes in the physical character of the land surface can affect land-atmosphere exchanges of radiation, momentum, heat and water (see Figure 7.5 and the simulation studies discussed below). These effects must be allowed for within climate simulations or analyses to avoid confusion with the effects of global warming. Second, changes in vegetation type, density and associated soil properties usually lead to changes in terrestrial carbon stocks and fluxes that can then directly contribute to the evolution of atmospheric CO_2 concentration. Therefore, any historical analysis of the atmospheric CO_2 record must estimate these contributions to avoid inaccurate attribution of carbon sinks or sources. Similarly, model simulations extending over the next 50 to 100 years should allow for significant perturbations to the atmospheric carbon budget from changes in terrestrial ecosystems (Woodwell *et al.*, 1998; see also Chapter 3).

There are two types of land-surface change; direct anthropogenic change, such as deforestation and agriculture; and indirect change, where changes in climate or CO_2 concentration force changes in vegetation structure and function within biomes, or the migration of biomes themselves. With respect to direct

anthropogenic change, population growth in the developing countries and the demand for economic development worldwide has led to regional scale changes in vegetation type, vegetation fraction and soil properties (Henderson-Sellers *et al.*, 1996; Ramankutty and Foley, 1998). Such changes can now be continuously monitored from space, and the satellite data record extends back to 1973. Large-scale deforestation in the humid tropics (South America, Africa and Southeast Asia) has been identified as an important ongoing process, and its possible impact on climate has been the topic of several field campaigns (Gash *et al.*, 1996), and modelling studies (for example, Nobre *et al.*, 1991; Lean *et al.*, 1996; Xue and Shukla, 1996; Zhang *et al.*, 1996a; Hahmann and Dickinson, 1997; Lean and Rowntree, 1997). Some significant extra-tropical impacts have also been identified in several model experiments (e.g., Sud *et al.*, 1996; Zhang *et al.*, 1996b). Replacement of tropical forest by degraded pasture has been observed to reduce evaporation, and increase surface temperature; these effects are qualitatively reproduced by most models. However, large uncertainties still persist about the impact of large-scale deforestation on the hydrological cycle over the Amazon in particular. Some numerical studies point to a reduction of moisture convergence while others tend to increase the inflow of moisture into the region. This lack of agreement occurs during the rainy season and reflects our poor understanding of the interaction of convection and land-surface processes (Polcher, 1995; Zhang *et al.*, 1997a), in addition to the effects of differences between the formulations in the land-surface schemes, their parameter fields, and the host GCMs used in the studies.

Other simulation work has indicated that the progressive cultivation of large areas in the East and Midwest USA over the last century may have induced a regional cooling of the order of 1 to 2°C due to enhanced evapotranspiration rates and increased winter albedo (Bonan, 1999). Snow-vegetation albedo effects significantly influence the near-surface climate; assignment of an open snow albedo value to the winter boreal forest in an NWP led to the prediction of air temperatures that were 5 to 10°C too low over large areas of Canada (Betts *et al.*, 1998). Work has also been done on the interaction between Sahelian vegetation and rainfall that suggests that the persistent rainfall anomaly observed there in the 1970s and 1980s could be related to land-surface changes (Claussen, 1997; Xue, 1997). All these studies indicate that large-scale land-use changes can lead to significant regional climatic impacts. However, it is unlikely that the aggregate of realistic land-use changes over the next 50 to 100 years will contribute to global scale climate changes comparable to those resulting from the warming associated with the continuing increase in greenhouse gases.

Changes to the land surface resulting from climate change or increased CO_2 concentration are likely to become important over the mid- to long term. For example, the extension of the growing season in high latitudes (Myneni *et al.*, 1997) will probably result in increases in biomass density, biogeochemical cycling rates, photosynthesis, respiration and fire frequency in the northern forests, leading to significant changes in albedo, evapotranspiration, hydrology and the carbon balance of the zone (Bonan *et al.*, 1992; Thomas and Rowntree, 1992; Levis *et al.*, 1999). There have been several attempts to calculate patterns of vegetation type and density as a function of climate (e.g., Zeng *et al.*, 1999); most of these have made use of climate predictions to calculate the future steady-state distribution of terrestrial biomes but some have attempted to model transitional cases (Ciret and Henderson-Sellers, 1998).

However, over the next 50 to 100 years, it is more likely that changes in vegetation density and soil properties within existing biome borders will make a greater contribution to modifying physical climate system and carbon cycle processes than any large-scale biogeographical shifts. In some cases, soil physical and chemical properties will limit the rate at which biomes can "migrate"; for example, colonisation of the tundra by boreal forest species is likely to be slowed by the lack of soil. Climate-vegetation relations are discussed further in Chapter 8, Section 8.5.5 with respect to past climates.

At present, only limited global data sets for LSPs are available and these need to be further improved. A comprehensive land-use/land cover data set, providing a global time-series of vegetation and soil parameters over the last two centuries at GCM resolution, would be a very useful tool to separate land-use change impacts on regional climate from global scale warming effects. Additionally, for both historical analyses and future projections, there is a need for interactive vegetation models that can simulate changes in vegetation parameters and carbon cycle variables in response to climate change. These proposed fourth generation models are just beginning to be designed and implemented within climate models.

7.4.3 Land Hydrology, Runoff and Surface-Atmosphere Exchange

Soil moisture conditions directly influence the net surface energy balance and determine the partitioning of the surface heat flux into sensible and latent contributions, which in turn control the evolution of the soil moisture distribution. There have been studies of the importance of soil moisture anomalies for episodes of drought (Atlas and Wolfson, 1993) and flooding (Beljaars *et al.*, 1996; Giorgi *et al.*, 1996), and the impact of initial soil moisture conditions on mid-latitude weather (Betts *et al.*, 1996; Schär *et al.*, 1999). Results from other GCM studies (e.g., Milly and Dunne, 1994; Bonan, 1996; Ducharne *et al.*, 1996) and regional and global water budgets analyses (e.g., Brubaker *et al.*, 1993; Brubaker and Entekhabi, 1996) have deepened our appreciation of the importance of land-surface hydrology in the regional and global energy and water exchanges. In relation to climate change, such mechanisms are relevant since they might lead to, or intensify, a reduction in summer soil moisture in mid- and high latitude semi-arid regions under doubled CO_2 conditions (Wetherald and Manabe, 1999). Most of these studies reported some impact of soil conditions upon land precipitation during episodes of convective activity, and there is observational evidence from lagged correlation analysis between soil moisture conditions and subsequent precipitation over Illinois that this mechanism is active in mid-latitudes (Findell and Eltahir, 1997). The formulation of surface runoff and baseflow has been calculated to have an indirect but strong impact on the surface energy balance (Koster and Milly, 1997).

The feedback mechanisms between soil moisture conditions and precipitation are particularly relevant to climate change studies since they may interact with, and determine the response to, larger-scale changes in atmospheric circulation, precipitation and soil moisture anomalies. The modelling of soil moisture-climate interactions is complicated by the range of time-scales involved, as soil moisture profiles can have a "memory" of many months, and the interaction of vertical soil moisture transfers with the larger-scale horizontal hydrology. Work is continuing to improve the realism of vertical water transfers, the effect of soil water on evapotranspiration rates, and the parametrization of sub-grid scale variability in land hydrological components (e.g., Avissar and Schmidt, 1998; Wood *et al.*, 1998). To date, there have been few attempts to describe the effects of within-grid horizontal transfers of water, but there has been success in connecting river routing schemes to GCMs (Dümenil *et al.*, 1997; see also Chapter 8, Section 8.5.4.2). Development in this area has lagged significantly behind that of vegetation canopy processes, despite the fact that the former are critical to a land-surface scheme's overall performance.

7.5 Cryosphere Processes and Feedbacks

The cryosphere, comprising snow and ice within the Earth system, introduces forcings that can affect oceanic deep water formation and feedbacks that can amplify climate variability and change. Important feedbacks involve: (i) the dependence of surface albedo on the temperature, depth and age of ice and snow; (ii) the influence of melt/freeze processes on sea surface salinity and deep-water formation. Palaeoclimatic evidence (Chapter 2, Section 2.4), indicating that extreme climate excursions have been induced by cryospheric processes, as well as recent observations that Arctic sea ice has decreased significantly in both extent and thickness (see Box 7.1), motivates the addition of this new section to the TAR.

7.5.1 Snow Cover and Permafrost

The presence of snow and ice adds complexity to surface energy and water balance calculations due to changes in surface albedo and roughness and the energies involved in phase changes and heat transfer within the snow/soil profile (Slater *et al.*, 1998a; Viterbo *et al.*, 1999). The parametrizations of snow processes have received significant attention since the SAR and more complex snow schemes are now used in some climate models (Loth *et al.*, 1993; Verseghy *et al.*, 1993; Lynch-Stieglitz, 1994). These models include advanced albedo calculations based on snow age or temperature and may explicitly model the metamorphism of snow as well as representing liquid water storage and wind-blown snow. Douville *et al.* (1995), Yang *et al.* (1997), Loth *et al.* (1993) and Slater *et al.* (1998b) examined the ability of snow modules within specific land-surface schemes to simulate snow cover. An offline evaluation of many schemes by Schlosser *et al.* (2000) focused on how successfully current land-surface schemes simulated snow over an 18 year period; they found considerable scatter in the simulation of snow and no evidence that the ability to simulate cold climate hydrology was related to scheme complexity.

Permafrost, defined as any soil/rock material that remains frozen throughout two or more consecutive years, underlies almost 25% of the exposed land surface in the Northern Hemisphere (Zhang *et al.*, 1999b). The uppermost layer of ground above permafrost, which experiences seasonal thawing, is called the active layer. The most distinct feature of land-atmosphere interactions in permafrost regions is that mass exchange is usually limited to this relatively shallow active layer, with complex transfers of heat by conduction and percolation across the ice/water interface. Recent modelling studies indicate that by the middle of the 21st century, climatic warming may result in a 12 to 15% reduction of the near-surface permafrost area and a 15 to 30% increase of the active layer thickness (Anisimov and Nelson, 1996, 1997; Anisimov *et al.*, 1997). Because of the latent heat involved, thawing of ice-rich permafrost under the changing climatic conditions will be slow, while the reaction of the active layer will be very fast.

There are two major longer term feedbacks between climate and permafrost: release of greenhouse gases from thawing permafrost (Goulden *et al.*, 1998; see also Chapter 4, Section 4.2) and changes in the vegetation associated with the thickening of the active layer. The first contributes directly to the global radiative forcing, while the second alters parameters of the radiation balance and surface hydrology.

7.5.2 Sea Ice

Sea ice plays an important role in moderating heat exchange between the ocean and atmosphere at high latitudes, especially by controlling the heat flux through openings in the ice. Sea ice also interacts with the broader climate system via the ice albedo feedback, which amplifies projected climate warming at high latitudes, and by oceanic feedbacks involving ice growth and melt and the fresh water balance at the ocean surface (Curry and

Webster, 1999; Lewis, 2000) Two feedbacks associated with sea ice are illustrated in Figure 7.6.

The Arctic Ocean sea-ice cover evolves from a highly reflective snow covered surface with few openings in May to a decaying sea-ice cover, mottled with melt ponds and interrupted by frequent openings in July. The seasonal changes in mean albedo in the central Arctic, from roughly 0.8 in May to 0.5 in mid-August, are known to within about ± 0.06 to 0.08. The mean spatial pattern in each summer month features lower albedos in the central Arctic and values 0.1 to 0.2 higher along the ice margins, but the spatial evolution of the pattern in any given year is variable. Averages observed at Soviet North Pole drifting stations, from 1950 to 1991, tend to be higher than satellite-based estimates (Marshunova and Mishin, 1994). Albedo depends on wavelength of radiation and on type and thickness of ice; however, ice type and thickness become unimportant if the snow cover exceeds 3 cm water equivalent. The representation of sea-ice albedo in AGCMs may take account of fractional snow cover, specified or predicted sea-ice thickness, ice surface temperature and the fraction of openings and puddles (Barry, 1996). Most models treat visible and near-infrared spectral ranges but there is still a wide variety of snow and ice albedo parametrizations among atmospheric and ocean GCMs. Important new data sets on puddle albedos (Morassutti and LeDrew, 1996) and the temporal evolution of melt pond coverage (Fetterer and Untersteiner, 1998) will enable more realistic albedo formulations to be developed.

Since the SAR, several coupled climate models have incorporated an explicit treatment of openings in sea ice, often in conjunction with ice dynamics. This is typically effected by partitioning a model grid cell into ice-free and ice-covered fractions. However, sub-grid scale variability in ice thickness, not represented in these schemes, can have a potentially important influence on sea-ice mass balance (Schramm *et al.*, 1997), ice/ocean fluxes of heat and fresh water (Holland *et al.*, 1997b) and the sensitivity of sea ice to thermodynamic perturbations (Holland and Curry, 1999). Recent advances in modelling the thickness distribution function make representing sub-grid scale variability, and the accompanying effects, feasible in global climate simulations (Bitz *et al.*, 2001). Other developments since the SAR include updated parametrizations of snow ageing and associated albedo changes and the implementation in some models of a multi-layer formulation of heat conduction through the ice. Snow plays a particularly important role in sea-ice thermodynamics by modifying the surface albedo, reducing thermal conductivity (and hence ice growth rates), and in some locations causing submergence and surface flooding of ice. Considerable effort continues to be devoted to the development and testing of improved physically based parametrizations suitable for use in such models. Recent field experiments, most notably SHEBA (Randall *et al.*, 1998; Perovich *et al.*, 1999), have provided observational data particularly suited to evaluating climate model parametrizations of sea-ice thermodynamic processes and initial attempts at this are underway. Although sea-ice thermodynamic processes are crudely approximated in many coupled climate models (see Chapter 8, Section 8.5.3), it is unclear how these approximations contribute to errors in climate model simulations.

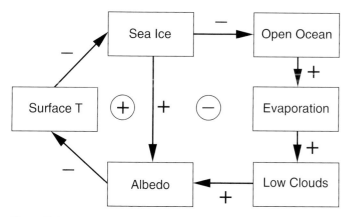

Figure 7.6: Idealised feedback loops involving sea ice. The signs attached to the arrows indicate the correlation between changes in the quantity of the outgoing box with that of the ingoing box, e.g., more sea ice leads to larger albedo. Resulting correlations of a loop are circled and they indicate whether a process is self-reinforced (positive sign) or damped (negative sign). Generally, resulting negative correlations can lead to oscillations, whereas resulting positive correlations may lead to instabilities. The classical ice-albedo effect is shown in the left loop, a feedback involving the overlying atmosphere is on the right.

Box 7.1: Sea ice and climate change.

Sea-ice processes:
Sea ice in the Arctic and around Antarctica responds directly to climate change and may, if properly monitored, become increasingly important for detecting climate change. Although sea ice covers only about 5% of the Earth's surface, its extent and thickness have important influences in the coupled atmosphere-ocean system. Increasing the understanding of these processes and representing them more realistically in climate models is important for making more reliable climate change projections. Several processes associated with sea ice are climatically relevant. The sea-ice albedo effect is an important contributor to the amplification of projected warming at high latitudes. Albedo decreases if the extent of sea ice is reduced and more ocean surface is exposed, resulting in increased heat absorption and hence warming. Melting of snow and the formation of melt-water ponds also reduces albedo and alters the radiation balance. Changes in sea-ice thickness and lead (open water) fraction modify the heat transfer from the ocean: thinner sea ice and more leads result in enhanced heat loss from the exposed ocean thus further warming the atmosphere. Changes in cloud cover may influence how large this effect really is. A principal mechanism for dense water formation in the ocean around Antarctica and in shelf regions of the Arctic is the rejection of brine as sea water freezes. Changes in sea-ice formation alter the properties and formation rates of ocean deep water and therefore have an influence on the water mass structure that reaches far beyond the area of sea ice. Finally, ice export from the Arctic represents an important southward flux of fresh water which influences the density structure of the upper ocean in the Nordic, Labrador and Irminger Seas.

Observations of sea ice:
Observations of sea-ice extent and concentration (the fraction of local area covered by ice) are based primarily on satellite data available since the late 1970s. Sea-ice thickness is also important in assessing possible changes in the amount of sea ice; however, thickness observations are more difficult to make. For the Arctic, thickness data come primarily from sonar measurements from submarines and a few oceanographic moorings. Although limited, the observations indicate statistically significant decreases in ice extent and thickness over the past few decades, with Arctic sea-ice extent declining at a rate of about 3% per decade since the late 1970s. Sea-ice retreat in the Arctic spring and summer over the last few decades is consistent with an increase in spring temperature, and to a lesser extent, summer temperatures at high latitudes. Thickness data show a near 40% decrease in the summer minimum thickness of Arctic sea ice over approximately the last 30 years. Estimates using independent methods for the winter, but over a much shorter period, also suggest thickness reductions, but at a markedly slower rate. However, due to limited sampling, uncertainties are difficult to estimate, and the influence of decadal to multi-decadal variability cannot yet be assessed.

While Arctic sea-ice extent and thickness have clearly decreased in the last 20 years, changes in Antarctic sea-ice extent have been insignificant. The earlier part of the data set indicates somewhat greater ice extents in the early 1970s, and indirect evidence from historical records also points to more northerly sea-ice margins in the 1930s and 1940s. Warming over much of Antarctica has only been about 0.5°C over the last 50 years with the notable exception of the Antarctic Peninsula where temperatures have increased by about 2°C for reasons that remain unclear.

Sea-ice modelling and projection:
Sea ice is particularly difficult to simulate in climate models because it is influenced directly by both the atmosphere (temperature, radiation, wind) and the ocean (heat transport and mixing, and surface currents), and because many of the relevant processes require high grid resolution or must be parametrized. Recent coupled climate models include a sea-ice component that incorporates openings in the ice, often in conjunction with ice dynamics (motion and deformation). Furthermore, updated parametrizations of snow ageing and associated albedo changes and multi-layer formulations of heat conduction through the ice and overlying snow cover are being implemented in some models. Although many thermodynamic processes are crudely approximated, it is unclear how these approximations contribute to errors in climate model simulations. Sea-ice dynamics is important in determining local ice thickness and export of sea ice from the formation areas, but despite the rather mature status of physically based sea-ice dynamics models, only a few of the current coupled climate models include such a component. Coupled model simulations of the seasonal cycle of sea-ice coverage in both hemispheres exhibit large deviations from the limited observational data base, as illustrated in Chapter 8, and current research is aimed at improving model performance.

    Coupled model projections of the future distribution of sea ice differ quantitatively from one to another as shown in Chapter 9. However, they agree that sea-ice extent and thickness will decline over the 21st century as the climate warms. Box 7.1, Figure 1 illustrates this with annual mean Arctic ice extent results from two coupled models. The simulations of ice extent decline over the past 30 years are in good agreement with the observations, lending confidence to the subsequent projections which show a substantial decrease of Arctic sea-ice cover leading to roughly 20% reduction in annual mean Arctic sea-ice extent by the year 2050.

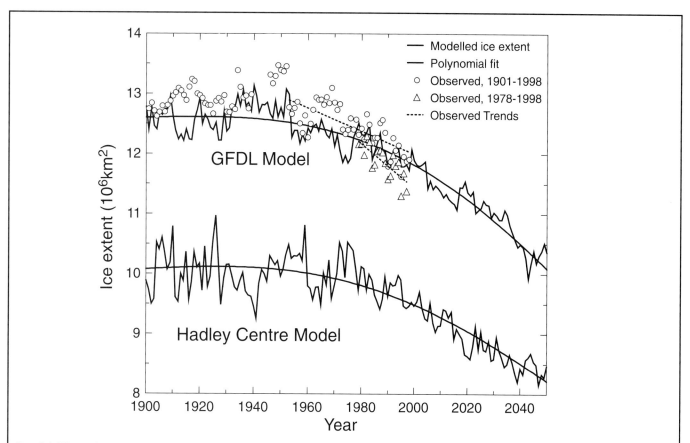

Box 7.1, Figure 1: Observed and modelled variations of annual averages of Northern Hemisphere sea-ice extent (10^6 km^2). Observed data for 1901 to 1998 are denoted by open circles (Chapman and Walsh, 1993, revised and updated) and for 1978 to 1998 by open triangles (Parkinson *et al.*, 1999, updated). The modelled sea-ice extents are from the GFDL and Hadley Centre climate model runs forced by observed CO$_2$ and aerosols. Modelled data are smoothed by a polynomial fit. Sea-ice extent in these models was determined as the area which had a thickness exceeding 2 cm. This criterion was determined to yield the best agreement with the observed mean during 1953 to 1998; this choice also reproduces the seasonal cycle realistically. Figure from Vinnikov *et al.* (1999).

Ice motion is driven by wind and ocean currents and resisted by ice-ocean drag and internal ice stresses. The representation of internal stresses is the primary distinguishing feature among ice dynamics models. Sensitivity experiments with stand alone sea-ice models (Hibler, 1984; Pollard and Thompson, 1994; Arbetter *et al.*, 1997) indicate that inclusion of ice dynamics can alter the modelled sensitivity of the ice cover to climatic perturbations. An assessment of sea-ice dynamic schemes suited to use in global climate models has been undertaken by the ACSYS Sea-Ice Model Intercomparison Project (SIMIP) (Lemke *et al.*, 1997), which has initially focused on the evaluation of sea-ice rheologies (the relationship between internal stresses and deformation). Results are summarised in Kreyscher *et al.* (1997) and indicate that the elliptical yield curve, viscous-plastic scheme of Hibler (1979) generally outperforms the other schemes evaluated in terms of comparisons to observed ice drift statistics, ice thickness and Fram Strait outflow. Clearly, the effects of ice-related fresh water transports and of other potentially important processes influenced by ice dynamics are not included in climate models which ignore ice motion (see Chapter 8, Section 8.5.3). The

effect on climate sensitivity remains to be assessed. Since the SAR, progress has been made at improving the efficiency of numerical sea-ice dynamic models, making them more attractive for use in coupled climate models (Hunke and Dukowicz, 1997; Zhang and Hibler, 1997).

The high latitude ocean fresh water budget is dominated by cryospheric processes. These include growth, melt and transport of sea ice, glacial melt, snowfall directly on ice and its subsequent melt and the runoff of snow melt water from adjacent land. Decadal-scale oscillations in atmospheric circulation may result in changes of the distribution of precipitation over watersheds that empty into the Arctic Ocean as well as of the snow cover on sea ice itself. The various components of the Arctic fresh water budget are reviewed in detail in Lewis (2000). Fresh water transport by sea ice is becoming increasingly important in coupled climate models as the trend toward eliminating flux adjustments requires explicit representation of this type of redistribution of the fresh water entering the ocean's surface.

Transport of sea ice through Fram Strait has long been implicated in modulating sea surface salinity and deep water formation in the North Atlantic (Dickson *et al.*, 1988; Belkin *et al.*, 1998). Recent modelling studies show that wind forcing dominates the variability in ice outflow (Hakkinen, 1993; Harder *et al.*, 1998; Hilmer *et al.*, 1998) and suggest that decadal-scale oscillations in atmospheric circulation are reflected in overall ice transport patterns (Proshutinsky and Johnson, 1997; Polyakov *et al.*, 1999). Net export of ice from the Arctic Ocean to the North Atlantic amounts to an annual loss of roughly 0.4 m of fresh water (Vinje *et al.*, 1998), which is offset by net ice growth. The salt released by this ice growth is in turn largely offset by inflow of fresher Pacific water, via the Bering Strait, and river runoff, maintaining a stable stratification. Changes in ice outflow, such as might arise under a changing climate, could alter the distribution of fresh water input to the Arctic and North Atlantic, with consequences for ocean circulation, heat transport and carbon cycling.

Around Antarctica, ice transport primarily exports fresh water from coastal regions where it is replaced by net ice growth. The salt released by this growth contributes directly to Antarctic deep water production (Goosse *et al.*, 1997; Legutke *et al.*, 1997; Stössel *et al.*, 1998; Goosse and Fichefet, 1999). However, the details of how salt released by freezing sea ice is distributed in the water column can have a substantial impact on water mass formation and circulation (Duffy and Caldeira, 1997; Legutke *et al.*, 1997; Duffy *et al.*, 1999) and hence on heat and carbon sequestration in a changing climate.

7.5.3 Land Ice

Ice stream instability, ice shelf break-up and switches in routing and discharge of glacial melt water present themselves as mechanisms for altering the surface salinity of oceans and inducing changes in the pattern and strength of the THC. There are positive and negative feedbacks associated with changing land ice masses (Figure 7.7). In coupled climate models, all these cryospheric processes are represented simply as surface fresh water inputs to the ocean (see Chapter 8, Section 8.5.4.2).

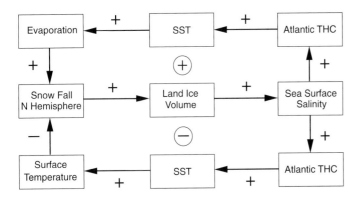

Figure 7.7: Feedback loops associated with land ice masses. Changes in their volume affect the salt balance at the sea surface and may influence the thermohaline circulation (THC). While there is palaeoclimatic evidence that this has happened often during the last Ice Age, the above processes are unlikely to play a major role for future climate change.

Ice sheets are continental scale masses of fresh water ice formed by the burial and densification of snow. They can be divided into areas that are grounded on the land surface, either below or above sea level, and areas that are afloat. Grounded parts of ice sheets exhibit slow and fast modes of flow. The slow sheet-flowing component is the more prevalent but the fast stream-flowing component can account for most of the ice discharge. Switching between slow and fast modes of flow may occur thermally, mechanically or hydrologically and is suggestive of "surging", a known cyclic instability of certain glaciers (Kamb *et al.*, 1985; Bindschadler, 1997). The slow-flow process for ice sheets is internal creep and the fast-flow processes relevant to ice streams are bottom sliding, enhanced creep and sub-glacial sediment deformation (Alley, 1989; Iken *et al.*, 1993; Engelhardt and Kamb, 1998). The factors controlling onset, discharge and width of ice streams are subject to poorly understood geological, topographic, thermal and hydrological controls (Clarke and Echelmeyer, 1996; Anandakrishnan and Alley, 1997; Bindschadler, 1997; Anandakrishnan *et al.*, 1998; Bell *et al.*, 1998; Jacobson and Raymond, 1998).

Although in GCMs the albedo of land ice is typically fixed, satellite observations could be used to remove this limitation; the melt area, as identified from satellite passive microwave observations (Mote *et al.*, 1993; Abdalati and Steffen, 1995), and AVHRR-derived albedo estimates can now be mapped (Stroeve *et al.*, 1997). Land ice dynamics and thermodynamics, ignored in current coupled GCMs, respond to changes in the temperature and balance of accumulation and melt at upper and lower surfaces. These boundary conditions involve couplings to the atmosphere, ocean and lithosphere. Ice sheet models typically represent atmospheric boundary conditions using simple elevation-based parametrizations. However, results obtained for Antarctica using this approach (Huybrechts and Oerlemans, 1990; Fastook and Prentice, 1994) compare favourably with those using more comprehensive AGCM-derived boundary conditions (Thompson and Pollard, 1997). Predictive models of future evolution must incorporate past ice-mass variations because ice sheets continue to respond to climate change for several thousand years. Time-scales complicate the inclusion of land ice in existing coupled climate models, and it is usual in these models to regard land ice as passive, providing only a static boundary for the atmosphere. Models of ocean circulation seldom incorporate explicit treatments of melt water inputs from land ice.

Models simulate the coupled evolution of ice sheet flow, form and temperature. The flow law in Glen (1955) is commonly adopted but such models must employ a poorly justified flow enhancement factor to correctly capture the height-to-width ratios of ice sheets. In general, these models perform well in intercomparison exercises (Huybrechts *et al.*, 1996; Payne *et al.*, 2000), but there is uncertainty in the predicted thermal structure and bottom melting conditions. This constitutes a potentially serious shortcoming because several models of ice sheet instability invoke thermal trigger mechanisms (MacAyeal, 1992). Ice stream flow models are limited by the current knowledge of the controlling physical processes. Outstanding issues involve the relative importance of bottom drag and lateral drag (Whillans and van der Veen, 1993; MacAyeal *et al.*, 1995) and the representation of fast-flow and water-transport processes (Fowler and Schiavi, 1998; Hindmarsh, 1998; Tulaczyk *et al.*, 1998). Recently, coupled models of the evolution of ice streams within ice sheets have been developed (Marshall and Clarke, 1997) which incorporate the crucial processes of ice stream onset (Bell *et al.*, 1998) and margin migration (van der Veen and Whillans, 1996).

Floating margins of ice sheets are called ice shelves, the largest of which are found in West Antarctica and mediate the transfer of ice between fast-flowing ice streams and the Southern Ocean. Break-up of one or both of the large West Antarctic ice shelves, either associated with ice stream instability or independent of it, cannot be discounted and would be considerably more probable than complete disintegration of the West Antarctic ice sheet (Bentley, 1998; see also Chapter 11, Section 11.3.3). Such an event would be accompanied by a large increase in iceberg flux to the Southern Ocean and uncertain effects on the production of AABW and ocean circulation. Different ice shelf models compare well (MacAyeal *et al.*, 1996); however, the stability of grounding lines remains an issue and could significantly influence the behaviour of numerical models (Hindmarsh, 1993). The suggestion that the grounding line is inherently unstable (Thomas and Bentley, 1978) is not supported by more recent modelling studies (Muszynski and Birchfield, 1987; Huybrechts and Oerlemans, 1990; Hindmarsh, 1993). However, this issue has yet to be resolved.

Megafloods of water stored beneath ice sheets have been postulated (Shaw *et al.*, 1996) but the subject is controversial. The estimated total volume of lake water beneath Antarctica is equivalent to 10 to 35 mm of sea level rise (Dowdeswell and Siegert, 1999) but the simultaneous release of all this stored water is unlikely. The timing would be unpredictable and climate impacts, if any, would be associated with the effect of the fresh water pulse on AABW formation. Modelling such phenomena is beyond the scope of existing ice dynamics models. Although progress has been made in the understanding of ice stream processes, there are still many unanswered questions and representation of ice stream and grounding line physics in land-ice dynamics models remains rudimentary.

7.6 Processes, Feedbacks and Phenomena in the Coupled System

This section deals with some of the processes and feedbacks in the coupled atmosphere-ocean system; those involving land and the cryosphere are dealt with in Sections 7.4 and 7.5, respectively. Increasingly, natural modes of variability of the atmosphere and the atmosphere-ocean system are also understood to play key roles in the climate system and how it changes, and thus the teleconnections which are believed to be most important for climate change are briefly discussed.

7.6.1 Surface Fluxes and Transport of Heat and Fresh Water

Fundamental issues in the global climate system are the relative amounts of heat and fresh water carried by the oceans and the atmosphere to balance the global heat and water budgets.

Because the top-of-the-atmosphere radiation largely determines the distribution of net energy on the Earth, the fluid components of the climate system, the atmosphere and ocean, transport heat and energy from regions of surplus to regions of deficit, with the total heat transport constrained. In the SAR a major problem with atmospheric models was that their surface fluxes with the ocean and the implied oceanic heat transports did not agree with the quite uncertain observational estimates and the biggest problem was clearly with clouds. The result was that all coupled models of the current climate either had a bias in their simulation of the mean state or that a "flux adjustment" was employed to keep the mean state close to that observed. Although some recent coupled ocean-atmosphere models (Boville and Gent, 1998; Gordon *et al.*, 2000) do not have adjustments to the modelled surface heat, water and momentum fluxes (see Chapter 8, Section 8.4.2), the drift in surface climate is small. This implies that the surface air-sea heat and water fluxes calculated by the model are reasonably consistent with the simulated ocean heat and fresh water transports. However, only comparison with flux values of known accuracy can verify the predicted fluxes, and thereby increase confidence in the modelled physical processes.

The role of the oceans in maintaining the global heat and water budgets may be assessed using bulk formulae, residual and direct (hydrographic) methods. Direct measurements of the sensible and latent heat fluxes have led to improved parametrizations (Fairall *et al.*, 1996; Weller and Anderson, 1996; Godfrey *et al.*, 1998), but large 20 to 30% uncertainties amounting to several tens of Wm^{-2} remain (DeCosmo *et al.*, 1996; Gleckler and Weare, 1997). Recent global air-sea flux climatologies based on ship data and bulk formulae (da Silva *et al.*, 1994; Josey *et al.*, 1999) exhibit an overall global imbalance; on average the ocean gains heat at a rate of about 30 Wm^{-2}. This was adjusted by globally scaling the flux estimates (da Silva *et al.*, 1994), but spatially uniform corrections are not appropriate (Josey *et al.*, 1999). While satellites give data over all ocean regions (e.g., Schulz and Jost, 1998; Curry *et al.*, 1999), and precipitation estimates over the oceans are only viable using satellites (Xie and Arkin, 1997), their accuracy is still unknown. Overall, surface evaporation minus precipitation (E–P) estimates are probably accurate to no better than about 30%.

In the residual method the atmospheric energy transport is subtracted from the top-of-the-atmosphere radiation budget to derive the ocean heat transport. Until fairly recently, the atmospheric poleward transport estimates were too small and led to overestimates of the ocean transports. This has improved with local applications of global numerical atmospheric analyses based on four dimensional data assimilation (Trenberth and Solomon, 1994; Keith, 1995). With reanalysis of atmospheric observations, more reliable atmospheric transports are becoming available (Table 7.1). Sun and Trenberth (1998) show that large interannual changes in poleward ocean heat transports are inferred with El Niño events in the Pacific. The moisture budget estimates of E–P are generally superior to those computed directly from E and P estimates from the model used in the assimilation (e.g., Trenberth and Guillemot, 1998) and are probably to be preferred to bulk flux estimates, given the inaccuracy of bulk estimates of P.

In the direct method the product of ocean velocity and

temperature measured over the boundaries is integrated to determine the ocean heat transport divergence for the volume. While it deals with ocean circulation and the mechanisms of ocean heat transport, estimates could only be made at a few locations where high quality observations were available and assumptions, such as geostrophic velocity estimates, are reasonable. Most of the recent estimates are for the Atlantic Ocean (see Table 7.1), for which there is general agreement that there is northward ocean heat transport at all latitudes. It is generally understood that this northward heat transport is associated with the THC in which NADW is formed in the polar and sub-polar North Atlantic and subsequently flows southward as a deep western boundary current into the Southern Ocean. In the North Pacific Ocean at 10°N (Wijffels *et al.*, 1996), the northward heat flux of 0.7 ± 0.5 PW (1 PetaWatt = 1 PW = 10^{15} W) is due primarily to a shallow Ekman upper thermocline cell. It is estimated that a 6 Sv change in Ekman transport, which is comparable to the differences among existing climatologies, would cause a 0.4 PW change in meridional heat transport. For the South Pacific (Koshlyakov and Sazhina, 1996; Tsimplis *et al.*, 1998) and Indian oceans (Robbins and Toole, 1997) the determination of ocean heat transport is hampered by uncertainties in the size of the Indonesian Throughflow and in the Agulhas Current transports (Beal and Bryden, 1997), and the estimates of northward heat transport across 30°S in the Indian Ocean range from –0.4 to –1.3 PW (Robbins and Toole, 1997; Macdonald, 1998). Macdonald (1998) and Macdonald and Wunsch (1996) have made a global inverse analysis of selected high quality hydrographic sections covering all ocean basins taken prior to the World Ocean Circulation Experiment (WOCE) observational period to produce meridional heat transports (Table 7.1). For the heat fluxes, results from the methods are beginning to converge. While some new direct ocean fresh water transport measure-

ments (Saunders and King, 1995) indicate that the South Atlantic receives adequate excess precipitation to supply the cross equatorial flux of fresh water for North Atlantic excess evaporation, this runs contrary to previous estimates (Wijffels *et al.*, 1992) and the recent atmospheric moisture budget results (Trenberth and Guillemot, 1998).

The latest results of the zonal mean ocean heat transports computed from (i) observational data as residuals, (ii) models run without flux adjustments and (iii) direct measurements agree within error estimates (Table 7.1). This suggests that the models are now converging on the correct values for the zonally averaged heat fluxes. However, significant regional biases and compensating errors in the radiative and turbulent fluxes still exist (Doney *et al.*, 1998; Gordon *et al.*, 2000). The coupled model fresh water flux estimates are more problematic. For example, the inter-tropical convergence zone may become skewed and spuriously migrate from one hemisphere to the other, seriously distorting the precipitation fields (e.g., Boville and Gent, 1998).

The compatibility and improvements of the ocean and atmospheric heat transports in models is evidently the primary reason why coupled runs can now be made without flux adjustments (Gregory and Mitchell, 1997). This is due to a better representation of key processes in both atmosphere and ocean components of climate models, and improved spatial resolution. These key processes include, in the atmosphere, convection, boundary-layer physics, clouds, and surface latent heat fluxes (Hack, 1998), and in the ocean, boundary layer and mesoscale eddy mixing processes (Doney *et al.*, 1998; Large and Gent, 1999). However, while simulated SSTs generally agree well with observations (deviations of less than 2°C), there are large areas where consistent errors of SST occur in the models (e.g., too cold in the North Pacific; too warm in coastal upwelling regions, see Chapter 8, Figure 8.1a). These point to processes that must be

Table 7.1: Ocean heat transport estimates, positive northwards in PetaWatts (1 PetaWatt = 1 PW = 10^{15}W), from analyses of individual hydrographic sections from pre WOCE sections (Macdonald, 1998), indirect methods (Trenberth et al., 2001), and years 81 to 120 of the HadCM3 (UKMO) coupled model (Gordon et al., 2000) and from the CSM 1.0 (NCAR) (Boville and Gent, 1998). Typical error bars are ±0.3 PW. For the Atlantic the sections are: 55°N (Bacon, 1997), 24°N (Lavin et al., 1998), 14°N (Klein et al., 1995), 11°S (Speer et al., 1996), and at 45°S (Saunders and King, 1995). For the Pacific: 47°N (Roemmich and McCallister, 1989), 24°N (Bryden et al., 1991), 10°N (Wijffels et al., 1996). Because of the Indonesian throughflow, South Pacific and Indian Ocean transports make most sense if combined.

Atlantic	Sections	Macdonald	Trenberth	HadCM3	CSM 1.0
55°N	0.28	–	0.29	–	0.63
48°N	–	0.65	0.41	0.54	0.81
24°N	1.27	1.07	1.15	1.14	1.31
14°N	1.22	–	1.18	–	1.27
11°N	–	1.39	1.15	1.12	1.21
11°S	0.60	0.89	0.63	–	0.65
23°S	–	0.33	0.51	0.67	0.61
45°S	0.53	–	0.62	0.64	–
Pacific					
47°N	−0.09	−0.08	−0.06	0.17	0.11
24°N	0.76	0.45	0.73	0.50	0.69
10°N	0.70	0.44	0.85	0.68	0.87
Pacific + Indian					
32°S	–	−1.34	−1.14	−1.19	−1.13

improved in future models. For instance, stratocumulus decks are not well simulated in coupled models, resulting in significant deviations of SST from the observed (see Chapter 8, Figure 8.1a and Section 8.4.2 for a discussion).

7.6.2 Ocean-atmosphere Interactions

There is no clear separation between the wind-driven circulation and the THC (see Section 7.3.6) because they interact with each other on several time-scales. While there is a great deal of empirical evidence that the ocean and sea surface temperatures co-vary with the atmosphere, this may only indicate that the atmosphere forces the ocean, and it does not necessarily signify a feedback or a truly coupled process that contributes to the variability. Moreover, it is very difficult to establish such coupling from observational studies. This topic was not dealt with thoroughly by the SAR. In the tropics there is clear evidence of the ocean forcing the atmosphere, such as in El Niño (see Section 7.6.5). In the extra-tropics much of what can be seen is accountable through fairly random wind variations; essentially stochastic forcing of the ocean is converted into low frequency ocean variability and gives a red spectrum in oceanic temperatures and currents up to the decadal time-scale (Hasselmann, 1976; Hall and Manabe, 1997). Feedback to the atmosphere is not involved. In the spatial resonance concept (Frankignoul and Reynolds, 1983) there is still no feedback from the ocean to the atmosphere, but oceanic quantities may exhibit a spectral peak through an advective time-scale (Saravanan and McWilliams, 1998) or Rossby wave dynamics time-scale (Weng and Neelin, 1998). In coupled air-sea modes, such as those proposed by Latif and Barnett (1996) for the North Pacific and by Groetzner *et al.* (1998) for the North Atlantic, there is a feedback from the ocean to the atmosphere. Spectral peaks are found in both the ocean and the atmosphere, and the period of the oscillation is basically determined by the adjustment time of the sub-tropical gyre to changes in the wind stress curl.

Coupled models indicate that, in mid-latitudes, the predominant process is the atmosphere driving the ocean as seen by the surface fluxes and as observed, yet when an atmospheric model is run with specified SSTs, the fluxes are reversed in sign, showing the forcing of the atmosphere from the now infinite heat capacity of the ocean (implied by specified SSTs). Recent ensemble results (Rodwell *et al.*, 1999; Mehta *et al.*, 2000) have been able to reproduce the decadal North Atlantic atmospheric variations from observed SSTs but with much reduced amplitude. Bretherton and Battisti (2000) suggest that this is consistent with a predominant stochastic driving of the ocean by the atmosphere with some modest feedback on the atmosphere, and that the signal only emerges through ensemble averaging.

In the extra-tropics, a key question remains the sensitivity of the mid-latitude atmosphere to surface forcing from sea ice and sea surface temperature anomalies. Different modelling studies with similar surface conditions yield contradictory results (e.g., Robertson *et al.*, 2000a,b). The crude treatment of processes involving sea ice, oceanic convection, internal ocean mixing and eddy-induced transports and the coarse resolution of most coupled climate models, adds considerably to the uncertainty.

7.6.3 Monsoons and Teleconnections

The weather and climate around the world in one place is generally strongly linked to that in other places through atmospheric linkages. In the tropics and sub-tropics, large-scale overturning in the atmosphere, which is manifested as the seasonal monsoon variations, links the wet summer monsoons to the dry subsiding regions usually in the tropics and sub-tropics of the winter hemisphere. Throughout the world, teleconnections link neighbouring regions mainly through large-scale, quasi-stationary atmospheric Rossby waves. A direct consequence of these linkages is that some regions are wetter and/or hotter than the prevailing global scale changes, while half a wavelength away the regions may be dryer and/or cooler than the global pattern. Moreover, because of the way these patterns set up relative to land and ocean, they can alter the global mean changes. In addition, errors in models (such as in convection in the tropical Pacific) can be manifested non-locally through teleconnections, e.g., in the North Pacific SSTs (see Figure 8.1a), although other processes are also involved.

The term "monsoon" is now generally applied to tropical and sub-tropical seasonal reversals in both the atmospheric circulation and associated precipitation. These changes arise from reversals in temperature gradients between continental regions and the adjacent oceans with the progression of the seasons. The dominant monsoon systems in the world are the Asian-Australian, African and the American monsoons. As land heats faster than ocean in summer, heated air rises and draws moist low-level maritime air inland where convection and release of latent heat fuel the monsoon circulation. For the Asian monsoon, a regional meridional temperature gradient extending from the tropical Indian Ocean north to mid-latitude Asia develops prior to the monsoon through a considerable depth of the troposphere (Webster *et al.*, 1998). To a first order, the stronger this meridional temperature gradient, the stronger the monsoon. Thus land-surface processes, such as soil moisture and snow cover in Asia can influence the monsoon and, along with SST variations, may induce quasi-biennial variability (Meehl, 1997). Additionally, large-scale forcing associated with tropical Pacific SSTs influences monsoon strength through the large-scale east-west overturning in the atmosphere. Anomalously cold (warm) Pacific SSTs often are associated with a strong (weak) monsoon, though these connections are somewhat intermittent.

Some teleconnections arise simply from natural preferred modes of the atmosphere associated with the mean climate state and the land-sea distribution. Several are directly linked to SST changes (Trenberth *et al.*, 1998). The most prominent are the Pacific-North American (PNA) and the North Atlantic Oscillation (NAO; see Section 7.6.4) in the Northern Hemisphere, and both account for a substantial part of the pattern of northern hemispheric temperature change, especially in winter (Hurrell, 1996), in part through the "cold ocean warm land" (COWL) pattern (Wallace *et al.*, 1995; Hurrell and Trenberth, 1996) (see

Chapter 2). Although evidently a prominent mode of the atmosphere alone, the PNA is also influenced by changes in ENSO (see Section 7.6.5). Thompson and Wallace (1998, 2000) suggest that the NAO may be the regional manifestation of an annular (zonally symmetric) hemispheric mode of variability characterised by a seesaw of atmospheric mass between the polar cap and the middle latitudes in both the Atlantic and Pacific Ocean basins and they call this the Arctic Oscillation (AO). A similar, even more zonal structure is dominant in the Southern Hemisphere (Trenberth *et al.*, 1998) (the Southern Annular Mode, sometimes called the Antarctic Oscillation, AAO). The vertical structure of both AO and AAO extends well into the stratosphere (Perlwitz and Graf, 1995; Thompson and Wallace, 1998).

In the Atlantic, an important emerging coupled mode of variability is the so-called tropical Atlantic dipole, which involves variations of opposite sign in the sea level pressure field across the equatorial Atlantic and corresponding variations in the ITCZ location. Given the background SST and wind fields, anomalous SSTs of opposite sign across the equatorial region are apt to alter the surface winds in such a way as to enhance or reduce evaporative cooling of the ocean and reinforce the original SST pattern (Carton *et al.*, 1996; Chang *et al.*, 1997). The ocean provides a decadal time-scale to the coupled interactions.

Dominant large-scale patterns of ocean-atmosphere interactions are also found in the tropical Indian Ocean (Saji *et al.*, 1999; Webster *et al.*, 1999). They have characteristics similar to El Niño and are associated with large east-west SST changes and a switch of the major tropical convection areas from Africa to Indonesia. There are indications that this atmosphere-ocean process is somewhat independent of ENSO and represents a natural mode of the tropical Indian Ocean.

The vital processes for improved monsoon simulation in models are those associated with the hydrological cycle, especially in the tropics. These include convection, precipitation and other atmospheric processes (see Section 7.2) as well as land surface processes (see Section 7.4), and interactions of the atmosphere with complex topography and with the ocean. The difficulties in assembling all of these elements together has led to problems in simulating mean precipitation as well as interannual monsoon variability, although improvements are evident (Webster *et al.*, 1998, and see Chapter 8, Section 8.7.3).

For teleconnections and regional climate patterns, not only are there demanding requirements on simulating the variations in tropical SSTs that drive many of the interannual and longer-term fluctuations through latent heating in the associated tropical precipitation, but results also depend on the mean state of the atmosphere through which Rossby waves propagate. Feedbacks from changes in storm tracks, and thus momentum and heat transports by transient atmospheric disturbances, as well as interactions with changed land-surface soil moisture from precipitation changes and interactions with extra-tropical oceans are critical. While there is scope for further improvements, great strides have been made in modelling all these aspects in the recent years.

7.6.4 North Atlantic Oscillation and Decadal Variability

The NAO is the dominant pattern of wintertime atmospheric circulation variability over the extra-tropical North Atlantic (Hurrell, 1995), and has exhibited decadal variability and trends (see Chapter 2, Section 2.6). There is strong evidence indicating that much atmospheric circulation variability in the form of the NAO arises from internal atmospheric processes (Saravanan, 1998; Osborn *et al.*, 1999). During winters when the stratospheric vortex is stronger than normal, the NAO (and AO) tends to be in a positive phase suggesting an interaction and perhaps even a downward influence from the stratosphere to the troposphere (see Sections 7.2.5 and 7.6.3; Baldwin and Dunkerton, 1999). The recent trend in the NAO/AO could possibly thus be related to processes which are known to affect the strength of the stratospheric polar vortex such as tropical volcanic eruptions (Kodera, 1994; Kelly *et al.*, 1996), ozone depletion, and changes in greenhouse gas concentrations resulting from anthropogenic forcing (Shindell *et al.*, 1999b).

It has long been recognised that fluctuations in SST are related to the strength of the NAO, and Dickson *et al.* (1996, 2000) have shown a link to the ocean gyre and thermohaline circulations. The leading mode of SST variability over the North Atlantic during winter is associated with the NAO. During high NAO years anomalous SSTs form a tri-polar pattern with a cold anomaly in the sub-polar region, a warm anomaly in the middle latitudes, and a cold sub-tropical anomaly (e.g., Deser and Blackmon, 1993), consistent with the spatial form of the anomalous surface fluxes associated with the NAO pattern (Cayan, 1992). This indicates that SST is responding to atmospheric forcing on seasonal time-scales (Deser and Timlin, 1997). However, GCM simulations suggest that SST in the North Atlantic can, in turn, have a marked effect on NAO (see Section 7.6.2). Winter SST anomalies were observed to spread eastward along the path of the Gulf Stream and North Atlantic Current with a transit time-scale of a decade (Sutton and Allen, 1997). These SST anomalies reflect anomalies in the heat content of the deep winter mixed layers that when exposed to the atmosphere in winter (Alexander and Deser, 1995) could affect the NAO, imprinting the advective time-scale of the gyre on the atmosphere (McCartney *et al.*, 1996). Moreover, similar processes were identified in coupled GCM integrations (Groetzner *et al.*, 1998; Timmermann *et al.*, 1998; Delworth and Mann, 2000) where changes in SST due to oceanic processes (gyre advection or thermohaline circulation) affected the NAO. This, in turn, leads to changes in heat and fresh water fluxes, and in wind stress forcing of the oceanic circulation. Oceanic response to such changes in the forcing produced a negative feedback loop, leading to decadal oscillations. However, the role of these mechanisms is yet to be established.

Watanabe and Nitta (1999) have suggested that high latitude snow cover on land is responsible for decadal changes in the NAO. Changes in sea-ice cover in both the Labrador and Greenland Seas as well as over the Arctic also appear to be well correlated with the NAO (Deser *et al.*, 2000). Such changes may also affect the atmosphere because of the large changes in sensible and latent heat fluxes along the ice edge.

Box 7.2: Changes in natural modes of the climate system.

Observed changes in climate over the Northern Hemisphere in winter reveal large warming over the main continental areas and cooling over the North Pacific and North Atlantic. This "cold ocean – warm land" pattern has been shown to be linked to changes in the atmospheric circulation, and, in particular, to the tendency in the past few decades for the North Atlantic Oscillation (NAO) to be in its positive phase. Similarly, the Pacific-North American (PNA) teleconnection pattern has been in a positive phase in association with a negative Southern Oscillation index or, equivalently, the tendency for El Niño-Southern Oscillation (ENSO) to prefer the warm El Niño phase following the 1976 climate shift (Chapter 2). Because of the differing heat capacities of land and ocean, the "cold ocean-warm land" pattern has amplified the Northern Hemisphere warming. A fingerprint of global warming from climate models run with increasing greenhouse gases indicates greater temperature increases over land than over the oceans, mainly from thermodynamic (heat capacity and moisture) effects. This anthropogenic signal is therefore very similar to that observed, although an in-depth analysis of the processes involved shows that the dynamical effects from atmospheric circulation changes are also important. In other words, the detection of the anthropogenic signal is potentially masked or modified by the nature of the observed circulation changes, at least in the northern winter season. The detection question can be better resolved if other seasons are also analysed (Chapter 12). Attribution of the cause of the observed changes requires improved understanding of the origin of the changes in atmospheric circulation. In particular, are the observed changes in ENSO and the NAO (and other modes) perhaps a consequence of global warming itself?

There is no simple answer to this question at present. Because the natural response of the atmosphere to warming (or indeed to any forcing) is to change large-scale waves, some regions will warm while others cool more than the hemispheric average, and counterintuitive changes can be experienced locally. Indeed, there are preferred modes of behaviour of the atmospheric circulation, sometimes manifested as preferred teleconnection patterns (see this chapter) that arise from the planetary waves in the atmosphere and the distribution of land, high topography, and ocean. Often these modes are demonstrably natural modes of either the atmosphere alone or the coupled atmosphere-ocean system. As such, it is also natural for modest changes in atmospheric forcing to project onto changes in these modes, through changes in their frequency and preferred sign, and the evidence suggests that changes can occur fairly abruptly. This is consistent with known behaviour of non-linear systems, where a slow change in forcing or internal mechanisms may not evoke much change in behaviour until some threshold is crossed at which time an abrupt switch occurs. The best known example is the evidence for a series of abrupt climate changes in the palaeoclimate record apparently partly in response to slow changes in sea level and the orbit of the Earth around the Sun (Milankovitch changes, see Chapter 2). There is increasing evidence that the observed changes in the NAO may well be, at least in part, a response of the system to observed changes in sea surface temperatures, and there are some indications that the warming of tropical oceans is a key part of this (see this chapter for more detail). ENSO is not simulated well enough in global climate models to have confidence in projected changes with global warming (Chapter 8). It is likely that changes in ENSO will occur, but their nature, how large and rapid they will be, and their implications for regional climate change around the world are quite uncertain and vary from model to model (see this chapter and Chapter 9). On time-scales of centuries, the continuing increase of greenhouse gases in the atmosphere may cause the climate system to cross a threshold associated with the Atlantic thermohaline circulation: beyond this threshold a permanent shutdown of the thermohaline circulation results (see this chapter and Chapter 9).

Therefore, climate change may manifest itself both as shifting means as well as changing preference of specific regimes, as evidenced by the observed trend toward positive values for the last 30 years in the NAO index and the climate "shift" in the tropical Pacific about 1976. While coupled models simulate features of observed natural climate variability such as the NAO and ENSO, suggesting that many of the relevant processes are included in the models, further progress is needed to depict these natural modes accurately. Moreover, because ENSO and NAO are key determinants of regional climate change, and they can possibly result in abrupt changes, there has been an increase in uncertainty in those aspects of climate change that critically depend on regional changes.

7.6.5 El Niño-Southern Oscillation (ENSO)

The strongest natural fluctuation of climate on interannual time-scales is the El Niño-Southern Oscillation (ENSO) phenomenon, and ENSO-like fluctuations also dominate decadal time-scales (sometimes referred to as the Pacific decadal oscillation). ENSO originates in the tropical Pacific but affects climate conditions globally. The importance of changes in ENSO as the climate changes and its potential role in possible abrupt shifts have only recently been appreciated. Observations and modelling of ENSO are addressed in Chapters 2, 8 and 9; here the underlying processes are discussed. Observational and modelling results suggest that more frequent or stronger ENSO events are possible in the future. Because social and ecological systems are particularly vulnerable to rapid changes in climate, for the next decades, these may prove of greater consequence than a gradual rise in mean temperature.

7.6.5.1 ENSO processes

ENSO is generated by ocean-atmosphere interactions internal to the tropical Pacific and overlying atmosphere. Positive temperature anomalies in the eastern equatorial Pacific (characteristic of an El Niño event) reduce the normally large sea surface temperature difference across the tropical Pacific. As a consequence, the trade winds weaken, the Southern Oscillation index (defined as the sea level pressure difference between Tahiti and Darwin) becomes anomalously negative, and sea level falls in the west and rises in the east by as much as 25 cm as warm waters extend eastward along the equator. At the same time, these weakened trades reduce the upwelling of cold water in the eastern equatorial Pacific, thereby strengthening the initial positive temperature anomaly. The weakened trades also cause negative off-equatorial thermocline depth anomalies in the central and western Pacific. These anomalies propagate westward to Indonesia, where they are reflected and propagate eastward along the equator. Thus some time after their generation, these negative anomalies cause the temperature anomaly in the east to decrease and change sign. The combination of the tropical air-sea instability and the delayed negative feedback due to sub-surface ocean dynamics can give rise to oscillations (for a summary of theories see Neelin *et al.*, 1998). Two of these feedbacks are schematically illustrated in Figure 7.8. Beyond influencing tropical climate, ENSO seems to have a global influence: during and following El Niño, the global mean surface temperature increases as the ocean transfers heat to the atmosphere (Sun and Trenberth, 1998).

The shifts in the location of the organised rainfall in the tropics and the associated latent heat release alters the heating patterns of the atmosphere which forces large-scale waves in the atmosphere. These establish teleconnections, especially the PNA and the southern equivalent, the Pacific South American (PSA) pattern, that extend into mid-latitudes altering the winds and changing the jet stream and storm tracks (Trenberth *et al.*, 1998), with ramifications for weather patterns and societal impacts around the world.

Another related feedback occurs in the sub-tropics. The normally cold waters off the western coasts of continents (such as California and Peru) encourage the development of extensive low stratocumulus cloud decks which block the Sun, and this helps keep the ocean cold. A warming of the waters, such as during El Niño, eliminates the cloud deck and leads to further sea surface warming through solar radiation. Kitoh *et al.* (1999) found that this mechanism could lead to interannual variations in the Pacific Ocean without involving equatorial ocean dynamics. Currently, stratocumulus decks are not well simulated in coupled models, resulting in significant deviations of SST from the observed (see Chapter 8, Figure 8.1).

Indices of ENSO for the past 120 years (Figure 7.9), indicate that there is considerable variability in the ENSO cycle in the modern record. This variability has been variously attributed to: (i) stochastic forcing due to weather and other high-frequency "noise", and the Madden-Julian intra-seasonal oscillation in particular; (ii) deterministic chaos arising from internal non-linearities of the tropical Pacific ENSO system; (iii) forcing within the climate system but external to the tropical Pacific, and (iv) changes in exogenous forcing (see Neelin *et al.*, 1998 and references therein). Palaeo-proxies, archaeological evidence, and instrumental data (see Chapter 2) all indicate variations in ENSO behaviour over the past centuries, and throughout the Holocene. Much of this variability appears to be internal to the Earth's climate system, but there is evidence that the rather weak forcing due to orbital variations may be responsible for a systematic change to weaker ENSO cycles in the mid-Holocene (Sandweiss *et al.*, 1996; Clement *et al.*, 1999; Rodbell *et al.*, 1999). However, it appears that the character of ENSO can change on a much faster time-scale than that of small amplitude insolation change imposed by the Earth's varying orbit. The inference to be drawn from observed ENSO variability is that small forcings are able to cause large alterations in the behaviour of this non-linear system.

Following the apparent climate "shift" in the tropical Pacific around 1976 (Graham, 1994; Trenberth and Hurrell, 1994) (Figure 7.9), which is part of the recent tropical Pacific warming, the last two decades are characterised by relatively more El Niño variability, including the two strongest El Niño events (1982/83 and 1997/98) in the 130 years of instrumental records and a long-lasting warm spell in the early 1990s (Trenberth and Hoar, 1997; see Wunsch, 1999, and Trenberth and Hurrell, 1999, for a discussion on statistical significance). The tropical Pacific warming may be linked to anthropogenic forcing (Knutson and Manabe, 1998), but attribution is uncertain in view of the strong natural variability observed (Latif *et al.*, 1997; Zhang *et al.*, 1997b) and the inability of models to fully simulate ENSO realistically.

Whereas the above discussion has dealt with processes on multiple time-scales, separation of the time-scales into inter-annual (periods less than a decade) and inter-decadal (Zhang *et al.*, 1997b) show quite similar patterns. More focused equatorial signals occur on interannual time-scales, and are identifiable with the equatorial dynamically active waves in the ocean, while the lower-frequency variations, called the Pacific Decadal Oscillation (PDO), have relatively a somewhat larger expression in mid-latitudes of both hemispheres, but with an "El Niño-like" pattern. Because the patterns are highly spatially correlated and

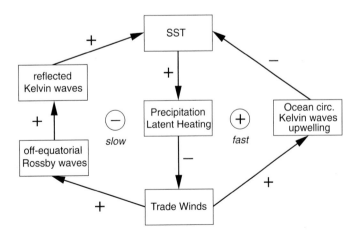

Figure 7.8: Simplified principal feedback loops active in El Niño-Southern Oscillation (ENSO). The fast loop (right) gives rise to an instability responsible for the development of an El Niño, the slow loop (left) tends to dampen and reverse the anomalies, so that together, these processes excite oscillations.

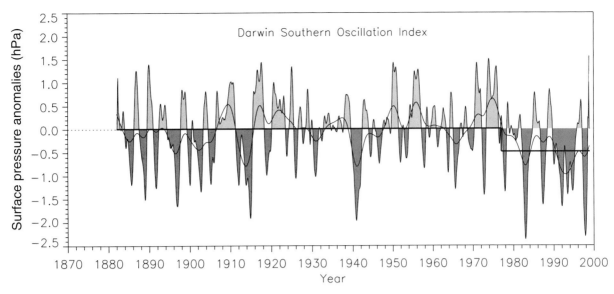

Figure 7.9: Darwin Southern Oscillation Index (SOI) represented as monthly surface pressure anomalies in hPa. Data cover the period from January 1882 to December 1998. Base period climatology computed from the period January 1882 to December 1981. The step function fit is illustrative only, to highlight a possible shift around 1976 to 1977.

thus not orthogonal, there is no unique projection onto them at any one time and they should not be considered independently.

In the tropics, it has been difficult for models to simulate all of the important processes correctly, and errors exist in the mean state of the tropical Pacific and the variability (see Chapter 8, Section 8.7.1). It is unlikely that it is possible to simulate the mean annual cycle correctly without simulating realistic ENSO events, as ENSO alters the mean state through rectification effects. For example, it is known that some regions only receive rainfall during an El Niño event. Hence it would be impossible to simulate correct rainfall in these regions without also simulating El Niño. Moreover, ENSO events are phase-locked to the annual cycle and changes in the mean state alter ENSO variability. In the Zebiak and Cane (1991) model the tropical east Pacific SST is systematically cooler during the "weak ENSO" regime than during the "strong ENSO" regime. Cloud radiative forcing feedbacks, such as the interactions between stratus regimes and ocean temperature, are poorly simulated, and the greenhouse effects of clouds versus their possible brightening with climate change are very uncertain.

Model studies have shown that greenhouse gas forcing is likely to change the statistics of ENSO variability, but the character of the change is model dependent. A comprehensive simulation suggests that greenhouse warming may induce greater ENSO activity marked by larger interannual variations relative to the warmer mean state (Timmermann *et al.*, 1999). More El Niños would increase the probability of certain weather regimes which favour COWL patterns associated with increased hemispheric mean surface temperatures (Hurrell, 1996; Corti *et al.*, 1999). The positive feedbacks involved in ENSO imply that small errors in simulating the relevant processes can be amplified. Yet increasing evidence suggests that ENSO plays a fundamental role in global climate and its interannual variability, and increased credibility in both regional and global climate

projections will be gained once realistic ENSOs and their changes are simulated.

7.6.5.2 ENSO and tropical storms
Any changes in ENSO as the climate changes will impact the distribution and tracks of tropical cyclones (including intense storms such as hurricanes and typhoons). During an El Niño, for example, the incidence of hurricanes typically decreases in the Atlantic and far western Pacific and Australian regions, but increases in the central and eastern Pacific (Lander, 1994). Thus it should be recognised that decreases in one area may be offset by increases in another area because of the global connectivity of the tropical atmospheric circulation. Global warming from increasing greenhouse gases in the atmosphere suggests increased convective activity but there is a possible trade-off between individual versus organised convection. While increases in sea surface temperatures favour more and stronger tropical cyclones, increased isolated convection stabilises the tropical troposphere and this in turn suppresses organised convection making it less favourable for vigorous tropical cyclones to develop (Yoshimura *et al.*, 1999). Thus changes in atmospheric stability (Bengtsson *et al.*, 1996) and circulation may produce offsetting tendencies (e.g., Royer *et al.*, 1998). General circulation models of the atmosphere (see Chapter 8, Section 8.8.4) do not resolve the scales required to properly address this issue; for instance, moist convection and hurricanes are not resolved adequately.

7.7 Rapid Changes in the Climate System

Small changes in the climate system can be sufficiently understood by assuming linear relationships between variables. However, many climate processes are non-linear by nature, and conclusions based on linear models and processes may in these cases no longer be valid. Non-linearity is a prerequisite for the existence of thresholds in the climate system: small perturbations or changes in the forcing can trigger large reorganisations if thresholds are passed. The result is that atmospheric and oceanic circulations may change from one regime to another. This could possibly be manifested as rapid climate change.

There is no clear definition of "rapid climate change". In general, this notion is used to describe climate changes that are of significant magnitude (relative to the natural variability) and occur as a shift in the mean or variability from one level to another. In order to distinguish such changes from "extreme events", a certain persistence of the change is required. Among the classical cases are spontaneous transitions from one preferred mode to another or transitions triggered by slowly varying forcing. This occurs in non-linear systems which have multiple equilibria (Lorenz, 1993). Evidence for the possibility of such transitions can be found in palaeoclimatic records (see Chapter 2, Section 2.4; and Stocker, 2000), in observations of changes in large-scale circulation patterns from the instrumental record (see Section 7.6.5.1), and contemporary observations of regional weather patterns (e.g., Corti *et al.*, 1999).

Here, we briefly summarise non-linear changes that have captured attention in the recent literature and that have been assessed in this chapter:

- The Northern Hemispheric atmospheric circulation exhibits different regimes that are associated with the North Atlantic Oscillation. Recent analyses of observations suggest that relatively rapid regime changes are possible (see Section 7.2.6) and that they may have happened frequently in the past (Appenzeller *et al.*, 1998). Several studies suggest that the recent changes may be a response to anthropogenic forcing, but our understanding of the processes generating NAO is not sufficient to have confidence in whether this is the case.

- Observed variability of ENSO suggests that a transition to more frequent El Niño occurred around 1976 (see Section 7.6.5.1). Current understanding of ENSO processes does not yet permit a distinction as to the extent this is a response to anthropogenic forcing versus part of the long-term natural variability of the tropical atmosphere-ocean system, or both (see also Chapter 2, Section 2.6.2).

- Results from most climate models suggest that the Atlantic thermohaline circulation slows down in response to global warming; some models simulate a complete shut-down if certain thresholds are passed (see Section 7.3.6 and Chapter 9,

Section 9.3.4.3). Such a shut-down is in general not abrupt but evolves on a time-scale which is determined by the warming, i.e., a few decades to centuries. Processes for such an evolution are increasingly understood. As model resolution increases and high latitude processes are better represented in these models (sea ice, topography), additional feedbacks influencing the Atlantic THC will be investigated and their relative importance must be explored. This will lead to a better quantification of the overall stability of the THC.

- Large polar ice masses, ice shelves or even complete ice sheets may be destabilised by sea level rise (see Section 7.5 and Chapter 11, Section 11.3.3), thereby contributing to further sea level rise.

- Warming in the high latitudes may lead to significant reductions in sea ice and associated feedbacks may accelerate this development (see Box 7.1).

- Large-scale and possibly irreversible changes in the terrestrial biosphere and vegetation cover are thought to have occurred in the past when anthropogenic perturbation was negligible (e.g., the development of the Saharan desert, Claussen *et al.*, 1999). These changes may be interpreted as non-linear changes triggered by slow changes in external forcing and thus cannot be excluded to occur in the future. Knowledge on these phenomena, however, is not advanced yet.

Reducing uncertainty in climate projections also requires a better understanding of these non-linear processes which give rise to thresholds that are present in the climate system. Observations, palaeoclimatic data, and models suggest that such thresholds exist and that transitions have occurred in the past. The occurrence of such transitions can clearly not be excluded in a climate that is changing. On the contrary, model simulations indicate that such transitions lie within the range of changes that are projected for the next few centuries if greenhouse gas concentrations continue to increase. A particular concern is the fact that some of these changes may even be irreversible due to the existence of multiple equilibrium states in the climate system.

Comprehensive climate models in conjunction with sustained observational systems, both *in situ* and remote, are the only tool to decide whether the evolving climate system is approaching such thresholds. Our knowledge about the processes, and feedback mechanisms determining them, must be significantly improved in order to extract early signs of such changes from model simulations and observations.

References

Abdalati, W. and K. Steffen, 1995: Passive microwave-derived snow-melt regions on the Greenland ice sheet. *Geophys. Res. Lett.,* **22**, 787-790.

Abdella, K. and N. McFarlane, 1997: A new second-order turbulence closure scheme for the planetary boundary layer. *J. Atm. Sci.,* **54**, 1850-1867.

Albrecht, B., 1989: Aerosols, cloud microphysics and fractional cloudiness. *Science,* **245**, 1227-1230.

Alexander, M.A. and C. Deser, 1995: A mechanism for the recurrence of winter time midlatitude SST anomalies. *J. Phys. Oceanogr.,* **25**, 122-137.

Allan, R.P., K.P. Shine, A. Slingo and J.A. Pamment, 1999: The dependence of clear-sky outgoing long-wave radiation on surface temperature and relative humidity. *Quart. J. Roy. Met. Soc.,* **125**, 2103-2126.

Alley, R.B., 1989: Water-pressure coupling of sliding and bed deformation: II. Velocity-depth profiles. *J. Glaciol.,* **35**, 119-129.

AMIP, 1995: *Proceedings of the First International AMIP Scientific Conference* [Gates, W. L. (ed.)]. World Meteorological Organization, Geneva.

Anandakrishnan, S. and R.B. Alley, 1997: Stagnation of ice stream C, West Antarctica, by water piracy. *Geophys. Res. Lett.,* **24**, 265-268.

Anandakrishnan, S., D.D. Blankenship, R.B. Alley and P.L. Stoffa, 1998: Influence of subglacial geology on the position of a West Antarctic ice stream from seismic observations. *Nature,* **394**, 62-65.

Andre, J.C., J.P. Goutorbe and A. Perrier, 1986: HAPEX/MOBILHY: A hydrologic atmospheric experiment for the study of water budget and evaporation flux at the climate scales. *Bull. Am. Met. Soc.,* **67**, 138-144.

Andrews, D.G., J.R. Holton and C.B. Leovy, 1987: *Middle Atmosphere Dynamics.* Academic Press, Florida. Chapter 2.

Anisimov, O.A. and F.E. Nelson, 1996: Permafrost distribution in the northern hemisphere under scenarios of climatic change. *Glob. Planet. Change,* **14**, 59-72.

Anisimov, O.A. and F.E. Nelson, 1997: Permafrost zonation and climate change: results from transient general circulation models. *Clim. Change,* **35**, 241-258.

Anisimov, O.A., N.I. Shiklomanov and F.E. Nelson, 1997: Effects of global warming on permafrost and active-layer thickness: results from transient general circulation models. *Glob. Planet. Change,* **61**, 61-77.

Appenzeller, C., T.F. Stocker and M. Anklin, 1998: North Atlantic oscillation dynamics recorded in Greenland ice cores. *Science,* **282**, 446-449.

Arakawa, A. and W.H. Schubert, 1974: Interaction of a cumulus cloud ensemble with the large-scale environment. Part I. *J. Atm. Sci.,* **31**, 674-701.

Arbetter, T.E., J.A. Curry, M.M. Holland and J.A. Maslanik, 1997: Response of sea-ice models to perturbations in surface heat flux. *Ann. Glaciol.,* **25**, 193-197.

Arkin, P.A. and P. Xie, 1994: The Global Precipitation Climatology Project: First algorithm intercomparison project. *Bull. Am. Met. Soc.,* **75**, 401-420.

Arking, A., M.-D. Chou and W.L. Ridgway, 1996: On estimating the effects of clouds on atmospheric absorption based on flux observations above and below cloud level. *Geophys. Res. Lett.,* **23**, 829-832.

Atlas, R.M. and N. Wolfson, 1993: The effect of SST and soil moisture anomalies on GLA model simulations of the 1988 U.S. summer drought. *J. Clim.,* **6**, 2034-2048.

Avissar, R., 1998: Which type of soil-vegetation-atmosphere transfer scheme is needed for general circulation models: a proposal for a higher-order scheme. *J. Hydrol.,* **212**, 136-154.

Avissar, R. and T. Schmidt, 1998: An evaluation of the scale at which ground-surface heat flux patchiness affects the convective boundary layer using large-eddy simulations. *J. Atm. Sci.,* **55**, 2666-2689.

Avissar, R., E.W. Eloranta, K. Gurer and G.J. Tripoli, 1998: An evaluation of the large-eddy simulation option of the regional atmospheric modeling system in simulating a convective boundary layer: a FIFE case study. *J. Atm. Sci.,* **55**, 1109-1130.

Ayotte, K.W., P.P. Sullivan, A. Andren, S.C. Doney, A.A.M. Holtslag, W.G. Large, J.C. McWilliams, C.-H. Moeng, M. Otte, J.J. Tribbia and J.C. Wyngaard, 1996: An evaluation of neutral and convective planetary boundary-layer parameterization relative to large eddy simulations. *Bound.-Lay. Meteorol.,* **79**, 131-175.

Bacon, S., 1997: Circulation and fluxes in the North Atlantic between Greenland and Ireland. *J. Phys. Oceanogr.,* **27**, 1420-1435.

Baldwin, M.P. and T.J. Dunkerton, 1999: Propagation of the Arctic Oscillation from the stratosphere to the troposphere. *J. Geophys. Res.,* **104**, 30937-30946.

Baldwin, M.P., X. Cheng and T.J. Dunkerton, 1994: Observed correlations between winter-mean tropospheric and stratospheric circulation anomalies. *Geophys. Res. Lett.,* **21**, 1141-1144.

Baringer, M.O. and J.F. Price, 1997: Mixing and spreading of the Mediterranean outflow. *J. Phys. Oceanogr.,* **27**, 1654-1677.

Barker, H.W. and Z. Li, 1997: Interpreting shortwave albedo-transmittance plots: True or apparent anomalous absorption. *Geophys. Res. Lett.,* **24**, 2023-2026.

Barry, R.G., 1996: The parameterization of surface albedo for sea ice and its snow cover. *Prog. Phys. Geogr.,* **20**, 63-79.

Bates, J.J. and D.L. Jackson, 1997: A comparison of water vapor observations with AMIP 1 simulations. *J. Geophys. Res.,* **102**, 21837-21852.

Beal, L.M. and H.L. Bryden, 1997: Observations of an Agulhas Undercurrent. *Deep-Sea Res. I,* **44**, 1715-1724.

Beckmann, A. and R. Döscher, 1997: A method for improved representation of dense water spreading over topography in geopotential-coordinate models. *J. Phys. Oceanogr.,* **27**, 581-591.

Beljaars, A.C.M., P. Viterbo, M.J. Miller and A.K. Betts, 1996: The anomalous rainfall over the United States during July 1993: Sensitivity to land-surface parameterization and soil-moisture anomalies. *Mon. Wea. Rev.,* **124**, 362-383.

Belkin, I.M., S. Levitus, J. Antonov and S.-A. Malmberg, 1998: "Great Salinity Anomalies" in the North Atlantic. *Prog. Oceanogr.,* **41**, 1-68.

Bell, R.E., D.D. Blankenship, C.A. Finn, D.L. Morse, T.A. Scambos, J.M. Brozena and S.M. Hodge, 1998: Influence of subglacial geology on the onset of a West Antarctic ice stream from aerogeophysical observations. *Nature,* **394**, 58-62.

Bengtsson, L., M. Botzet and M. Esch, 1996: Will greenhouse gas-induced warming over the next 50 years lead to higher frequency and greater intensity of hurricanes? *Tellus,* **48A**, 57-73.

Bengtsson, L., E. Roeckner and M. Stendel, 1999: Why is the global warming proceeding much slower than expected. *J. Geophys. Res.,* **104**, 3865-3876.

Bentley, C.R., 1998: Rapid sea-level rise from a West Antarctic ice-sheet collapse: a short-term perspective. *J. Glaciol.,* **44**, 157-163.

Betts, A.K., 1982: Saturation point analysis of moist convective overturning. *J. Atm. Sci.,* **39**, 1484-1505.

Betts, A.K., J.H. Ball, A.C.M. Beljaars, M.J. Miller and P.A. Viterbo, 1996: The land-surface atmosphere interaction: A review based on observational and global modeling perspectives. *J. Geophys. Res.,* **101**, 7209-7225.

Betts, A.K., P. Viterbo, A.C.M. Beljaars, H.L. Pan, S.Y. Hong, M. Goulden and S. Wofsy, 1998: Evaluation of land-surface interaction in ECMWF and NCEP/NCAR reanalysis models over grassland (FIFE) and boreal forest (BOREAS). *J. Geophys. Res.,* **103**, 23079-23085.

Betts, R.A., P.M. Cox, S.E. Lee and F.I. Woodward, 1997: Contrasting physiological and structural vegetation feedbacks in climate change simulations. *Nature,* **387**, 796-799.

Bindschadler, R., 1997: Actively surging West Antarctic ice streams and their response characteristics. *Ann. Glaciol.,* **24**, 409-414.

Bitz, C.M., M.M. Holland, A.J. Weaver and M. Eby, 2001: Simulating the ice-thickness distribution in a coupled climate model. *J. Geophys. Res.,* **106**, 2441-2464.

Boening, C.W., W.R. Holland, F.O. Bryan, G. Danabasoglu and J.C. McWilliams, 1995: An overlooked problem in model simulations of the thermohaline circulation and heat transport in the Atlantic Ocean. *J. Clim.,* **8**, 515-523.

Bolle, H.-J., J.-C. Andre, J.L. Arrue, H.K. Barth, P. Bessemoulin, A. Brasa, H.A.R. de Bruin, J. Cruces, G. Dugdale, E.T. Engman, D.L. Evans, R. Fantechi, F. Fiedler, A. van de Griend, A.C. Imeson, A. Jochum, P. Kabat, T. Kratzsch, J.-P. Lagouarde, I. Langer, R. Llamas, E. Lopez-Baeza, J. Melia Miralles, L.S. Muniosguren, F. Nerry, J. Noilhan, H.R. Oliver, R. Roth, S.S. Saatchi, J. Sanchez Diaz, M. de Santa Olalla, W.J. Shuttleworth, H. Sogaard, H. Stricker, J. Thornes, M. Vauclin and D. Wickland, 1993: EFEDA: European field experiment in a desertification threatened area. *Ann. Geophys.,* **11**, 173-189.

Bonan, G.B., 1995: Land-atmosphere CO_2 exchange simulated by a land surface process model coupled to an atmospheric general circulation model. *J. Geophys. Res.,* **100**, 2817-2831.

Bonan, G.B., 1996: Sensitivity of a GCM simulation to subgrid infiltration and surface runoff. *Clim. Dyn.,* **12**, 279-285.

Bonan, G.B., 1999: Frost followed the plow: Impacts of deforestation on the climate of the United States. *Ecol. Appl.,* **9**, 1305-1315.

Bonan, G.B., D. Pollard and S.L. Thompson, 1992: Effects of boreal forest vegetation on global climate. *Nature,* **359**, 716-718.

Bony, S., J.-P. Duvel and H. Le Treut, 1995: Observed dependence of the water vapor and clear-sky greenhouse effect on sea surface temperature: comparison with climate warming experiments. *Clim. Dyn.,* **11**, 307-320.

Bony, S., K.M. Lau and Y.C. Sud, 1997: Sea surface temperature and large-scale circulation influences on tropical greenhouse effect and cloud radiative forcing. *J. Clim.,* **10**, 2055-2077.

Boville, B.A., 1995: Middle atmosphere version of CCM2 (MACCM2): annual cycle and interannual variability. *J. Geophys. Res.,* **100**, 9017-9039.

Boville, B.A. and P.R. Gent, 1998: The NCAR climate system model, version one. *J. Clim.,* **11**, 1115-1130.

Braswell, W.D. and R.S. Lindzen, 1998: Anomalous short wave absorption and atmospheric tides. *Geophys. Res. Lett.,* **25**, 1293-1296.

Breon, F., 1992: Reflectance of broken cloud fields: Simulation and parameterization. *J. Atm. Sci.,* **49**, 1221-1232.

Bretherton, C.S. and D.S. Battisti, 2000: An interpretation of the results from atmospheric general circulation models forced by the time history of the observed sea surface temperature distribution. *Geophys. Res. Lett.,* **27**, 767-770.

Bretherton, C.S., M.K. McVean, P. Bechtold, A. Chlond, W.R. Cotton, J. Cuxart, H. Cuijpers, M. Khairoutdinov, B. Kosovic, D. Lewellen, C.-H. Moeng, P. Siebesma, B. Stevens, D.E. Stevens, I. Sykes and M.C. Wyant, 1999: An intercomparison of radiatively-driven entrainment and turbulence in a smoke cloud, as simulated by different numerical models. *Quart. J. Roy. Met. Soc.,* **125**, 391-423.

Broecker, W.S., 1997: Thermohaline circulation, the Achilles heel of our climate system: will man-made CO_2 upset the current balance? *Science,* **278**, 1582-1588.

Brubaker, K.L. and D. Entekhabi, 1996: Analysis of feedback mechanisms in land-atmosphere interaction. *Water Resour. Res.,* **32**, 1343-1357.

Brubaker, K.L., D. Entekhabi and P.S. Eagleson, 1993: Estimation of continental precipitation recycling. *J. Clim.,* **6**, 1077-1089.

Bryan, F., 1986: High-latitude salinity effects and interhemispheric thermohaline circulations. *Nature,* **323**, 301-304.

Bryden, H.L., D.H. Roemmich and J.A. Church, 1991: Ocean heat transport across 24N in the Pacific. *Deep-Sea Res.,* **38**, 297-324.

Butchart, N. and J. Austin, 1998: Middle atmosphere climatologies from

the troposphere-stratosphere configuration of the UKMO's unified model. *J. Atm. Sci.,* **55**, 2782-2809.

Cahalan, R.F., W. Ridgeway, W.J. Wiscombe, S. Gollmer and Harshvardhan, 1994: Independent pixel and Monte Carlo estimates of stratocumulus albedo. *J. Atm. Sci.,* **51**, 3776-3790.

Cao, K.K. and F.I. Woodward, 1998: Dynamic responses of terrestrial ecosystem carbon cycling to global climate change. *Nature,* **393**, 249-252.

Carton, J.A., X. Cao, B.S. Giese and A.M. da Silva, 1996: Decadal and interannual SST variability in the tropical Atlantic. *J. Phys. Oceanogr.,* **26**, 1165-1175.

Cayan, D.R., 1992: Latent and sensible heat flux anomalies over the northern oceans: the connection to monthly atmospheric circulation. *J. Clim.,* **5**, 354-369.

Cess, R.D., G.L. Potter, J.P. Blanchet, G.J. Boer, A.D.D. Genio, M. Deque, V. Dymnikov, V. Galin, W.L. Gates, S.J. Ghan, J.T. Kiehl, A.A. Lacis, H. Le Treut, Z.X. Li, X.-Z. Liang, B.J. McAvaney, V.P. Meleshko, J.F.B. Mitchell, J.-J. Morcrette, D.A. Randall, L. Rikus, E. Roeckner, J.F. Roer, U. Schlese, D.A. Sheinin, A. Slingo, A.P. Sokolov, K.E. Taylor, W.M. Washington, R.T. Wetherald, I. Yagai and M.-H. Zhang, 1990: Intercomparison and interpretation of climate feedback processes in 19 atmospheric general circulation models. *J. Geophys. Res.,* **95**, 16601-16615.

Cess, R.D., M. Zhang, P. Minnis, L. Corsetti, E.G. Dutton, B.W. Forgan, D.P. Garber, W.L. Gates, J.J. Hack, E.F. Harrisson, X. Jing, J.T. Kiehl, C.N. Long, J.-J. Morcrette, G.L. Potter, V. Ramanathan, B. Subasilar, C.H. Whitlock, D.F. Young and Y. Zhou, 1995: Absorption of solar radiation by clouds: Observations versus models. *Science,* **267**, 496-499.

Cess, R.D., M.H. Zhang, W.J. Ingram, G.L. Potter, V. Alekseev, H.W. Barker, E. Cohen-Solal, R.A. Colman, D.A. Dazlich, A.D. Del Genio, M.R. Dix, V. Dymnikov, M. Esch, L.D. Fowler, J.R. Fraser, V. Galin, W.L. Gates, J.J. Hack, J.T. Kiehl, H. LeTreut, K.K.-W. Lo, B.J. McAvaney, V.P. Meleshko, J.-J. Morcrette, D.A. Randall, E. Roeckner, J.-F. Royer, M.E. Schlesinger, P.V. Sporyshev, B. Timbal, E.M. Volodin, K.E. Taylor, W. Wang and R.T. Wetherald, 1996: Cloud feedback in atmospheric general circulation models: An update. *J. Geophys. Res.,* **101**, 12791-12794.

Cess, R.D., M.H. Zhang, F.P.J. Valero, S.K. Pope, A. Bucholtz, B. Bush, C.S. Zender and J. Vitko, 1999: Absorption of solar radiation by the cloudy atmosphere: Further interpretations of collocated aircraft measurements. *J. Geophys. Res.,* **104**, 2059-2066.

Chang, P., L. Ji and H. Li, 1997: A decadal climate variation in the tropical ocean from thermodynamic air-sea interactions. *Nature,* **385**, 516-518.

Chapman, W.L. and J.E. Walsh, 1993: Recent variations of sea ice and air temperature in high latitudes. *Bull. Am. Met. Soc.,* **74**, 33-48.

Chou, C. and J.D. Neelin, 1999: Cirrus detrainment-temperature feedback. *Geophys. Res. Lett.,* **26**, 1295-1298.

Christensen, O.B., J.H. Christensen, B. Machenhauer and M. Botzet, 1998: Very high-resolution regional climate simulations over Scandinavia - Present climate. *J. Clim.,* **11**, 3204-3229.

Ciais, P., P.P. Tans, J.W.C. White, M. Trolier, R.J. Francey, J.A. Berry, D.R. Randall, P.J. Sellers, J.G. Collatz and D.S. Schimel, 1995: Partitioning of ocean and land uptake of CO_2 as inferred by $\delta^{13}C$ measurements from the NOAA Climate Monitoring and Diagnostics Laboratory Global Air Sampling Network. *J. Geophys. Res.,* **100**, 5051-5070.

Ciais, P., A.S. Denning, P.P. Tans, J.A. Berry, D.A. Randall, G.J. Collatz, P.J. Sellers, J.W.C. White, M. Trolier, H.A.J. Meijer, R.J. Francey, P. Monfray and M. Heimann, 1997: A three-dimensional synthesis study of delta O-18 in atmospheric CO_2. 1. Surface fluxes. *J. Geophys. Res.,* **102**, 5857-5872.

Cihlar, J., P.H. Carmori, P.H. Schuepp, R.L. Desjardins and J.I. Macpherson, 1992: Relationship between satellite-derived vegetation indices and aircraft-based CO_2 measurements. *J. Geophys. Res.,* **97,** 18515-18521.

Ciret, C. and A. Henderson-Sellers, 1998: Sensitivity of global vegetation models to present-day climate grassland simulated by global climate models. *Glob. Biogeochem. Cyc.,* **11,** 1141-1169.

Clarke, T.S. and K. Echelmeyer, 1996: Seismic-reflection evidence for a deep subglacial trough beneath Jakobshavns Isbrae, West Greenland. *J. Glaciol.,* **43,** 219-232.

Claussen, M., 1997: Modeling bio-geophysical feedbacks in the African and Indian monsoon region. *Clim. Dyn.,* **13,** 247-257.

Claussen, M., C. Kubatzki, V. Brovkin, A. Ganopolski, P. Hoelzmann and H.-J. Pachur, 1999: Simulation of an abrupt change in Saharan vegetation in the mid-Holocene. *Geophys. Res. Lett.,* **26,** 2037-2040.

Clement, A. and R. Seager, 1999: Climate and the tropical oceans. *J. Clim.,* **12,** 3383-3401.

Clement, A., R. Seager and M.A. Cane, 1999: Orbital controls on tropical climate. *Paleoceanogr.,* **14,** 441-455.

Colello, G.D., C. Grivet, P.J. Sellers and J.A. Berry, 1998: Modeling of energy, water and CO_2 flux in a temperate grassland ecosystem with SiB2: May-October 1987. *J. Atm. Sci.,* **55,** 1141-1169.

Collatz, G.J., J.T. Ball, C. Grivet and J.A. Berry, 1991: Physiological and environmental regulation of stomatal conductance, photosynthesis and transpiration: A model that includes a laminar boundary layer. *Agric. For. Meteorol.,* **54,** 107-136.

Collatz, G.J., L. Bounoua, S.O. Los, D.A. Randall, I.Y. Fung and P.J. Sellers, 2000: A mechanism for the influence of vegetation on the response of the diurnal temperature range to a changing climate. *Geophys. Res. Lett.,* **27,** 3381-3384.

Collins, W.G., 1998: Complex quality control of significant level rawinsonde temperature. *J. Atmos. Oc. Tech.,* **15,** 69-79.

Colman, R.A. and B.J. McAvaney, 1995: Sensitivity of the climate response of the atmospheric general circulation model to changes in convective parameterization and horizontal resolution. *J. Geophys. Res.,* **100,** 3155-3172.

Corti, S., F. Molteni and T.N. Palmer, 1999: Signature of recent climate change in frequencies of natural atmospheric circulation regimes. *Nature,* **398,** 799-802.

Cubasch, U., J. Waszkewitz, G. Hegerl and J. Perlwitz, 1995: Regional climate changes as simulated in time-slice experiments. *Clim. Change,* **31,** 273-304.

Cuijpers, J.W.M. and A.A.M. Holtslag, 1998: Impact of skewness and nonlocal effects on scalar and buoyancy fluxes in convective boundary layers. *J. Atm. Sci.,* **55,** 151-162.

Curry, J., C.A. Clayson, W.B. Rossow, R. Reeder, Y.-C. Zhang, P.J. Webster, G. Liu and R.-S. Sheu, 1999: High-resolution satellite-derived dataset of the surface fluxes of heat, freshwater, and momentum for the TOGA COARE IOP. *Bull. Am. Met. Soc.,* **80,** 2059-2080.

Curry, J.A. and P.J. Webster, 1999: *Thermodynamics of Atmospheres and Oceans.* Academic Press, 465.

da Silva, A.M., C.C. Young and S. Levitus, 1994: Atlas of Surface Marine Data 1994 Volume 1: Algorithms and Procedures. In: *NOAA Atlas NESDIS (6)* . U.S. Department of Commerce, Washington D. C., 83.

Dai, A., F. Giorgi and K.E. Trenberth, 1999: Observed and model-simulated diurnal cycles of precipitation over the contiguous United States. *J. Geophys. Res.,* **104,** 6377-6402.

Dai, A.G. and I.Y. Fung, 1993: Can climate variability contribute to the "missing" CO_2 sink? *Global Biogeochemical Cycles,* **7,** 543-567.

Danabasoglu, G. and J.C. McWilliams, 1995: Sensitivity of the global ocean circulation to parameterizations of mesoscale tracer transports. *J. Clim.,* **8,** 2967-2987.

de Ruijter, W.P.M., A. Biastoch, S.S. Drijfhout, J.R.E. Lutjeharms, R.P. Matano, T. Pichevin, P.J. van Leeuwen and W. Weijer, 1999: Indian-Atlantic interocean exchange: dynamics, estimation and transport. *J. Geophys. Res.,* **104,** 20885-20910.

DeCosmo, J., K.B. Katsaros, S.D. Smith, R.J. Anderson, W. Oost, K. Bumke and H. Chadwick, 1996: Air-sea exchange of water vapor and sensible heat: The humidity exchange over the sea (HEXOS) results. *J. Geophys. Res.,* **101,** 12001-12016.

Del Genio, A.D., A.A. Lacis and R.A. Ruedy, 1991: Simulations of the effect of a warmer climate on atmospheric humidity. *Nature,* **251,** 382-385.

Del Genio, A.D., W. Kovari and M.-S. Yao, 1994: Climatic implications of the seasonal variation of upper tropospheric water vapor. *Geophys. Res. Lett.,* **21,** 2701-2704.

Del Genio, A.D., M.-S. Yao, W. Kovari and K.K.W. Lo, 1996: A prognostic cloud water parameterization for global climate models. *J. Clim.,* **9,** 270-304.

Delire, C. and J.A. Foley, 1999: Evaluating the performance of a land surface/ecosystem model with biophysical measurements from contrasting environments. *J. Geophys. Res.,* **104,** 16895-16909.

Delworth, T.L. and M.E. Mann, 2000: Observed and simulated multidecadal variability in the Northern Hemisphere. *Clim. Dyn.,* **16,** 661-676.

Denning, A.S., G.J. Collatz, C. Zhang, D.A. Randall, J.A. Berry, P.J. Sellers, G.D. Colello and D.A. Dazlich, 1996a: Simulations of terrestrial carbon metabolism and atmospheric CO_2 in a general circulation model. Part I: Surface carbon fluxes. *Tellus,* **48B,** 521-542.

Denning, A.S., D.A. Randall, G.J. Collatz and P.J. Sellers, 1996b: Simulations of terrestrial carbon metabolism and atmospheric CO_2 in a general circulation model. Part II: simulated CO_2 concentrations. *Tellus,* **48B,** 543-567.

Deser, C. and M.L. Blackmon, 1993: Surface climate variations over the North Atlantic Ocean during winter: 1900-1993. *J. Clim.,* **6,** 1743-1753.

Deser, C. and M.S. Timlin, 1997: Atmosphere-ocean interaction on weekly time scales in the North Atlantic and Pacific. *J. Clim.,* **10,** 393-408.

Deser, C., J.E. Walsh and M.S. Timlin, 2000: Arctic sea ice variability in the context of recent wintertime atmospheric circulation trends. *J. Clim.,* **13,** 617-633.

Desjardins, J., R.J. Hart, J.I. MacPherson, P.H. Schuepp and S.B. Verma, 1992: Aircraft- and tower-based fluxes of carbon dioxide, latent, and sensible heat. *J. Geophys. Res.,* **97,** 18477-18485.

Dessler, A.E. and S.C. Sherwood, 2000: Simulations of tropical upper tropospheric humidity. *J. Geophys. Res.,* **105,** 20155-20163.

Dickinson, R.E., M. Shaikh, R. Bryant and L. Graumlich, 1998: Interactive canopies for a climate model. *J. Clim.,* **11,** 2823-2836.

Dickson, R.R., J. Lazier, J. Meincke, P. Rhines and J. Swift, 1996: Long-term coordinated changes in the convective activity of the North Atlantic. *Prog. Oceanogr.,* **38,** 241-295.

Dickson, R.R., J. Meincke, S.-A. Malmberg and A.J. Lee, 1988: The "Great Salinity Anomaly" in the northern North Atlantic 1968-1982. *Prog. Oceanogr.,* **20,** 103-151.

Dickson, R.R., T.J. Osborn, J.W. Hurrell, J. Meincke, J. Blindheim, B. Adlandsvik, T. Vinje, G. Alekseev and W. Maslowski, 2000: The Arctic Ocean response to the North Atlantic Oscillation. *J. Clim.,* **13,** 2671-2696.

Dixon, K.W., T.L. Delworth, M.J. Spelman and R.J. Stouffer, 1999: The influence of transient surface fluxes on North Atlantic overturning in a coupled GCM climate change experiment. *Geophys. Res. Lett.,* **26,** 2749-2752.

Doney, S.C., W.G. Large and F.O. Bryan, 1998: Surface ocean fluxes and water-mass transformation rates in the coupled NCAR Climate System Model. *J. Clim.,* **11,** 1420-1441.

Döscher, R., C.W. Boening and P. Herrmann, 1994: Response of circulation and heat transport in the North Atlantic to changes in thermohaline forcing in northern latitudes: A model study. *J. Phys. Oceanogr.,* **24,** 2303-2320.

Douville, H., J.F. Royer and J.F. Mahfouf, 1995: A new snow parameterization for the Meteo-France climate model. 1. Validation in stand-alone experiments. *Clim. Dyn.,* **12**, 21-35.

Dowdeswell, J.A. and M.J. Siegert, 1999: The dimensions and topographic setting of Antarctic subglacial lakes and implications for large-scale water storage beneath continental ice sheets. *Geology,* **111**, 254-263.

Ducharne, A., K. Laval and J. Polcher, 1996: Sensitivity of the hydrological cycle to the parameterization of soil hydrology in a GCM. *Clim. Dyn.,* **14**, 307-327.

Duffy, P.B. and K. Caldeira, 1997: Sensitivity of simulated salinity in a three-dimensional ocean model to upper-ocean transport of salt from sea-ice formation. *Geophys. Res. Lett.,* **24**, 1323-1326.

Duffy, P.B., M. Eby and A.J. Weaver, 1999: Effects of sinking of salt rejected during formation of sea ice on results of an ocean-atmosphere-sea ice climate model. *Geophys. Res. Lett.,* **26**, 1739-1742.

Dümenil, L., S. Hagemann and K. Arpe (eds.), 1997: *Validation of the hydrological cycle in the Arctic using river discharge data*, Workshop on Polar Processes in Global Climate, Cancun, Mexico.

Dunkerton, T.J., 1997: The role of gravity waves in the quasi-biennial oscillation. *J. Geophys. Res.,* **102**, 26053-26076.

Duvel, J.-P., S. Bony and H. Le Treut, 1997: Clear-sky greenhouse effect sensitivity to sea surface temperature changes: an evaluation of AMIP simulations. *Clim. Dyn.,* **13**, 259-273.

Duynkerke, P.G., P.J. Jonker, A. Chlond, M.C. van Zanten, J. Cuxart, P. Clark, E. Sanchez, G. Martin, G. Lenderink and J. Teixeira, 1999: Intercomparison of three- and one-dimensional model simulations and aircraft observations of stratocumulus. *Bound.-Lay. Meteorol.,* **92**, 453-488.

Eby, M. and G. Holloway, 1994: Sensitivity of a large-scale ocean model to a parameterization of topographic stress. *J. Phys. Oceanogr.,* **24**, 2577-2588.

Eluszkiewicz, J.E., R. Zurek, L. Elson, E. Fishbein, L. Froidevaux, J. Waters, R. Grainger, A. Lambert, R. Harwood and G. Peckham, 1996: Residual circulation in the stratosphere and lower mesosphere as diagnosed from Microwave Limb Sounder data. *J. Atm. Sci.,* **53**, 217-240.

Emanuel, K., 1991: A scheme for representing cumulus convection in large scale models. *J. Atm. Sci.,* **48**, 2313-2333.

Emanuel, K.A. and M. Zivkovic-Rothman, 1999: Development and evaluation of a convection scheme for use in climate models. *J. Atm. Sci.,* **56**, 1766-1782.

Engelhardt, H. and B. Kamb, 1998: Basal sliding of Ice Stream B, West Antarctica. *J. Glaciol.,* **44**, 223-230.

Fairall, C.W., E.F. Bradley, D.P. Rogers, J.B. Edson and G.S. Young, 1996: Bulk parametrization of air-sea fluxes for TOGA COARE. *J. Geophys. Res.,* **101**, 1295-1308.

Farquhar, G.D., S. von Caemmerer and J.A. Berry, 1980: A biochemical model of photosynthetic CO_2 fixation in leaves of C3 species. *Planta,* **149**, 78-90.

Fastook, J.L. and M. Prentice, 1994: A finite-element model of Antarctica: sensitivity test for meteorological mass-balance relationship. *J. Glaciol.,* **40**, 167-175.

Fetterer, F. and N. Untersteiner, 1998: Observations of melt ponds on Arctic sea ice. *J. Geophys. Res.,* **103**, 24821-24835.

Field, C.B. and I.Y. Fung, 1999: The not-so-big US carbon sink. *Science,* **285**, 544-545.

Field, C.B., J.T. Randerson and C.M. Malmstrong, 1995: Global net primary production: combining ecology and remote sensing. *Rem. Sens. Environ.,* **51**, 74-88.

Findell, K.L. and E.A.B. Eltahir, 1997: An analysis of the soil moisture-rainfall feedback, based on direct observations from Illinois. *Water Resour. Res.,* **33**, 725-735.

Foley, J.A., I.C. Prentice, N. Ramankutty, S. Levis, D. Pollard, S. Sitch and A. Haxeltine, 1996: An integrated biosphere model of land surface processes, terrestrial carbon balance and vegetation dynamics. *Glob. Biogeochem. Cyc.,* **10**, 603-628.

Foley, J.A., S. Levis, I.C. Prentice, D. Pollard and S.L. Thompson, 1998: Coupling dynamic models of climate and vegetation. *Glob. Change Biol.,* **4**, 561-579.

Forster, P., R.S. Freckleton and K.P. Shine, 1997: On aspects of the concept of radiative forcing. *Clim. Dyn.,* **13**, 547-560.

Fowler, A.C. and E. Schiavi, 1998: A theory of ice-sheet surges. *J. Glaciol.,* **44**, 104-118.

Fowler, A.M. and K.J. Hennessy, 1995: Potential impacts of global warming on the frequency and magnitude of heavy precipitation. *Natural Hazards,* **11**, 283-304.

Fowler, L.D., D.A. Randall and S.A. Rutledge, 1996: Liquid and ice cloud microphysics in the CSU general circulation model. Part I: Model description and simulated microphysical processes. *J. Clim.,* **9**, 489-529.

Francey, R.J., P.P. Tans, C.E. Allison, I.G. Enting, J.W.C. White and M. Troller, 1995: Changes in oceanic and terrestrial carbon uptake since 1982. *Nature,* **373**, 326-330.

Frankignoul, C. and R.W. Reynolds, 1983: Testing a dynamical model for mid-latitude sea surface temperature anomalies. *J. Phys. Oceanogr.,* **13**, 1131-1145.

Frei, C., C. Schär, D. Lüthi and H.C. Davies, 1998: Heavy precipitation processes in a warmer climate. *Geophys. Res. Lett.,* **25**, 1431-1434.

Fung, I.Y., C.B. Field, J.A. Berry, M.V. Thompson, J.T. Randerson, C.M. Malmstrom, P.M. Vitousek, G.J. Collatz, P.J. Sellers, D.A. Randall, A.S. Denning, F. Badeck and J. John, 1997: Carbon-13 exchanges between the atmosphere and biosphere. *Glob. Biogeochem. Cyc.,* **11**, 507-533.

Gaffen, D.J., A. Robock and W.P. Elliott, 1992: Annual cycles of tropospheric water vapor. *J. Geophys. Res.,* **97**, 18185-18193.

Gamache, J.F. and R.A. Houze, 1983: Water budget of a meso-scale convective system in the tropics. *J. Atm. Sci.,* **40**, 1835-1850.

Gargett, A.E. and G. Holloway, 1992: Sensitivity of the GFDL ocean model to different diffusivities for heat and salt. *J. Phys. Oceanogr.,* **22**, 1158-1177.

Gash, J.H.C., C.A. Nobre, J.M. Robert and R.L. Victoria, 1996: *Amazonian Deforestation and Climate.* Wiley, Chichester, 595.

Gates, W.L., J.S. Boyle, C. Covey, C.G. Dease, C.M. Doutriaux, R.S. Drach, M. Fiorino, P.J. Gleckler, J.J. Hnilo, S.M. Marlais, T.J. Phillips, G.L. Potter, B.D. Santer, K.R. Sperber, K.E. Taylor and D.N. Williams, 1999: An overview of the results of the Atmospheric Model Intercomparison Project (AMIP I). *Bull. Am. Met. Soc.,* **80**, 29-55.

Gent, P.R., J. Willebrand, T.J. McDougall and J.C. McWilliams, 1995: Parameterizing eddy-induced transports in ocean circulation models. *J. Phys. Oceanogr.,* **25**, 463-474.

Giorgi, F., L.O. Mearns, C. Shields and L. Mayer, 1996: A regional model study of the importance of local versus remote controls of the 1988 drought and the 1993 flood over the Central United States. *J. Clim.,* **9**, 1150-1162.

Giorgi, F., L.O. Mearns, C. Shields and L. McDaniel, 1998: Regional nested model simulations of present day and 2 x CO_2 climate over the central plains of the US. *Clim. Change,* **40**, 457-493.

Gleckler, P.J. and B.C. Weare, 1997: Uncertainties in global ocean surface heat flux climatologies derived from ship observations. *J. Clim.,* **10**, 2764-2781.

Glen, J.W., 1955: The creep of polycrystalline ice. *Proc. Roy. Soc. Lond. A,* **228**, 519-538.

Godfrey, J.S., 1996: The effect of the Indonesian throughflow on ocean circulation and heat exchange with the atmosphere: a review. *J. Geophys. Res.,* **101**, 12217-12238.

Godfrey, J.S., R.A. Houze, R.H. Johnson, R. Lukas, J.-L. Redelsperger, A. Sumi and R. Weller, 1998: Coupled Ocean-Atmosphere Response Experiment (COARE): An interim report. *J. Geophys. Res.,* **103**,

14395-14450.

Goosse, H. and T. Fichefet, 1999: Importance of ice-ocean interactions for the global ocean circulation: a model study. *J. Geophys. Res.*, **104**, 23337-23355.

Goosse, H., J.M. Campin, T. Fichefet and E. Deleersnijder, 1997: The impact of sea-ice formation on the properties of Antarctic Bottom Water. *Ann. Glaciol.*, **25**, 276-281.

Goosse, H., E. Deleersnijder, T. Fichefet and M. England, 1999: Sensitivity of a global coupled ocean-sea ice model to the parameterization of vertical mixing. *J. Geophys. Res.*, **104**, 13681-13695.

Gordon, C., C. Cooper, C.A. Senior, H. Banks, J.M. Gregory, T.C. Johns, J.F.B. Mitchell and R.A. Wood, 2000: The simulation of SST, sea ice extents and ocean heat transports in a version of the Hadley Centre coupled model without flux adjustments. *Clim. Dyn.*, **16**, 147-168.

Goulden, M., J. Munger, S.-M. Fan, B. Daube and S. Wofsy, 1996: Exchange of carbon dioxide by a deciduous forest: response to interannual climate variability. *Science,* **271**, 1576-1578.

Goulden, M., S. Wofsy, J. Harden, S. Trumbore, P. Crill, S. Gower, T. Fries, B. Daube, S. Fan, D. Sutton, A. Bazzaz and J. Munger, 1998: Sensitivity of boreal forest carbon balance to soil thaw. *Science,* **279**, 214-217.

Goutorbe, J.-P., 1994: HAPEX-Sahel: A large-scale study of land-atmosphere interactions in the semi-arid tropics. *Ann. Geophys.*, **12**, 53-64.

Graham, N.E., 1994: Decadal-scale climate variability in the tropical and North Pacific during the 1970s and 1980s: Observations and model results. *Clim. Dyn.*, **10**, 135-162.

Greenwald, T.J., G.L. Stephens, S.A. Christopher and T.H. Vonder Haar, 1995: Observations of the global characteristics and regional radiative effects of marine cloud liquid water. *J. Clim.*, **8**, 2928-2946.

Gregg, M.C., 1989: Scaling turbulent dissipation in the thermocline. *J. Geophys. Res.*, **94**, 9686-9698.

Gregory, J.M. and J.F.B. Mitchell, 1995: Simulation of daily variability of surface temperature and precipitation over Europe in the current and 2xCO$_2$ climates of the UKMO climate model. *Quart. J. Roy. Met. Soc.*, **121**, 1451-1476.

Gregory, J.M. and J.F.B. Mitchell, 1997: The climate response to CO$_2$ of the Hadley Centre coupled AOGCM with and without flux adjustment. *Geophys. Res. Lett.*, **24**, 1943-1946.

Grenier, H. and C.S. Bretherton, 2001: A moist PBL parameterization for large-scale models and its application to subtropical cloud-topped marine boundary layers. *Mon. Wea. Rev.*, **129**, 357-377.

Griffies, S.M., A. Gnanadesikan, R.C. Pacanowski, V.D. Larichev, J.K. Dukowicz and R.D. Smith, 1998: Isoneutral diffusion in a z-coordinate ocean model. *J. Phys. Oceanogr.*, **28**, 805-830.

Groetzner, A., M. Latif and T.P. Barnett, 1998: A decadal climate cycle in the North Atlantic Ocean as simulated by the ECHO coupled GCM. *J. Clim.*, **11**, 831-847.

Gwarkiewicz, G. and D.C. Chapman, 1995: A numerical study of dense water formation and transport on a shallow, sloping continental shelf. *J. Geophys. Res.*, **100**, 4489-4507.

Hack, J.J., 1998: Analysis of the improvement in implied meridional ocean energy transport as simulated by the NCAR CCM3. *J. Clim.*, **11**, 1237-1244.

Hahmann, A.N. and R. Dickinson, 1997: RCCM2-BATS model over tropical South America: Application to tropical deforestation. *J. Clim.*, **10**, 1944-1964.

Haigh, J.D., 1996: The impact of solar variability on climate. *Science,* **272**, 981-983.

Hakkinen, S., 1993: An Arctic source for the Great Salinity Anomaly: A simulation of the Arctic ice-ocean system for 1955-1975. *J. Geophys. Res.*, **98**, 16397-16410.

Hall, A. and S. Manabe, 1997: Can local linear stochastic theory explain sea surface temperature and salinity variability? *Clim. Dyn.*, **13**, 167-180.

Hall, A. and S. Manabe, 1999: The role of water vapor feedback in unperturbed climate variability and global warming. *J. Clim.*, **12**, 2327-2346.

Hallberg, R., 2000: Time integration of diapycnal diffusion and Richardson number-dependent mixing in isopycnal coordinate ocean models. *Mon. Wea. Rev.*, **128**, 1402-1419.

Hansen, J.E., M. Sato and R. Ruedy, 1997: Radiative forcing and climate response. *J. Geophys. Res.*, **102**, 6831-6864.

Harder, M., P. Lemke and M. Hilmer, 1998: Simulation of sea ice transport through Fram Strait: Natural variability and sensitivity to forcing. *J. Geophys. Res.*, **103**, 5595-5606.

Harvey, L.D.D., 2000: An assessment of the potential impact of a downward shift of tropospheric water vapor on climate sensitivity. *Clim. Dyn.*, **16**, 491-500.

Hasselmann, K., 1976: Stochastic climate models. Part I: Theory. *Tellus,* **28**, 473-485.

Hasumi, H. and N. Suginohara, 1999: Effects of locally enhanced vertical diffusivity over rough bathymetry on the world ocean circulation. *J. Geophys. Res.*, **104**, 23367-23374.

Haynes, P.H., C.J. Marks, M.E. McIntyre, T.G. Shepherd and K.P. Shine, 1991: On the downward control of extratropical diabatic circulations by eddy-induced mean forces. *J. Atm. Sci.*, **48**, 651-678.

Hazeleger, W. and S.S. Drijfhout, 2000: Eddy subduction in a model of the subtropical gyre. *J. Phys. Oceanogr.*, **30**, 677-695.

Held, I.M. and B.J. Soden, 2000: Water vapor feedback and global warming. *Ann. Rev. Energy Env.*, **25**, 441-475.

Henderson-Sellers, A., H. Zhang and W. Howe, 1996: Human and physical aspects of tropical deforestation. In: *Climate Change: Developing Southern Hemisphere Perspectives* [Giambelluca, T. W. and A. Henderson-Sellers (eds.)]. John Wiley & Sons, Chichester, 259-292.

Hennessy, K.J., J.M. Gregory and J.F.B. Mitchell, 1997: Changes in daily precipitation under enhanced greenhouse conditions. *Clim. Dyn.*, **13**, 667-680.

Hense, A., P. Krahe and H. Flohn, 1988: Recent fluctuations of tropospheric temperature and water vapour content in the tropics. *Meteorology and Atmospheric Physics,* **38**, 215-227.

Hibler, W.D., 1979: A dynamic thermodynamic sea ice model. *J. Phys. Oceanogr.*, **9**, 815-846.

Hibler, W.D., 1984: The role of sea ice dynamics in modeling CO$_2$ increases. In: *Climate Processes and Climate Sensitivity* [Hansen, J. E. and T. Takahashi (eds.)]. American Geophysical Union, Washington DC, 238-253.

Hilmer, M., M. Harder and P. Lemke, 1998: Sea ice transport: a highly variable link between Arctic and North Atlantic. *Geophys. Res. Lett.*, **25**, 3359-3362.

Hindmarsh, R.C.A., 1993: Qualitative dynamics of marine ice sheets. In: *NATO ASI, I 12: Ice in the Climate System* [Peltier, W. R. (ed.)]., 67-99.

Hindmarsh, R.C.A., 1998: The stability of a viscous till sheet coupled with ice flow, considered at wavelengths less than the ice thickness. *J. Glaciol.*, **44**, 285-292.

Holland, E.A., B.H. Braswell, J.-F. Lamarque, A. Townsend, J. Sulzman, J.-F. Muller, F. Dentener, G. Brasseur, H. Levy, J.E. Penner and G.-U. Roelofs, 1997a: Variations in the predicted spatial distribution of atmospheric nitrogen deposition and their impact on carbon uptake by terrestrial ecosystems. *J. Geophys. Res.*, **102**, 15849-15866.

Holland, M.M. and J.A. Curry, 1999: The role of different physical processes in determining the interdecadal variability of Central Arctic sea ice. *J. Clim.*, **12**, 3319-3330.

Holland, M.M., J.A. Curry and J.L. Schramm, 1997b: Modeling the thermodynamics of a sea ice thickness distribution. 2. Sea ice/ocean interactions. *J. Geophys. Res.*, **102**, 23093-23107.

Holtslag, A.A.M. and P.G. Duynkerke, 1998: *Clear and cloudy boundary layers.* Royal Netherlands Academy of Arts and Sciences, 372 pp.

Horinouchi, T. and S. Yoden, 1998: Wave-mean flow interaction associated with a QBO like oscillation in a simplified GCM. *J. Atm. Sci.,* **55**, 502-526.

Houghton, R.A. and J.L. Hackler, 1999: Emissions of carbon from forestry and land-use change in tropical Asia. *Glob. Change Biol.,* **5**, 481-492.

Houze, R.A., 1993: *Cloud Dynamics.* Academic Press, San Diego, 570.

Hu, H. and W.T. Liu, 1998: The impact of upper tropospheric humidity from microwave limb sounder on the midlatitude greenhouse effect. *Geophys. Res. Lett.,* **25**, 3151-3154.

Hunke, E.C. and J.K. Dukowicz, 1997: An elastic-viscous-plastic model for sea ice dynamics. *J. Phys. Oceanogr.,* **27**, 1849-1867.

Hurrell, J.W., 1995: Decadal trends in the North Atlantic Oscillation regional temperatures and precipitation. *Science,* **269**, 676-679.

Hurrell, J.W., 1996: Influence of variations in extratropical wintertime teleconnections on Northern Hemisphere temperatures. *Geophys. Res. Lett.,* **23**, 665-668.

Hurrell, J.W. and K.E. Trenberth, 1996: Satellite versus surface estimates of air temperature since 1979. *J. Clim.,* **9**, 2222-2232.

Huybrechts, P. and J. Oerlemans, 1990: Response of the Antarctic Ice Sheet to future greenhouse warming. *Clim. Dyn.,* **5**, 93-102.

Huybrechts, P., A.J. Payne and EISMINT Intercomparison Group, 1996: The EISMINT benchmarks for testing ice-sheet models. *Ann. Glaciol.,* **23**, 1-12.

Iken, A., K. Echelmeyer, W. Harrison and M. Funk, 1993: Mechanism of fast flow in Jakobshavns Isbrae, West Greenland: Part I. Measurements of temperature and water level in deep boreholes. *J. Glaciol.,* **39**, 15-25.

Imre, D.G., E.H. Abramson and P.H. Daum, 1996: Quantifying cloud-induced shortwave absorption: An examination of uncertainties and of recent arguments for large excess absorption. *J. Appl. Met.,* **35**, 1991-2010.

Inamdar, A.K. and V. Ramanathan, 1998: Tropical and global scale interactions among water vapor, atmospheric greenhouse effect, and surface temperature. *J. Geophys. Res.,* **103**, 32177-32194.

IPCC, 1990: *Climate Change – The IPCC Scientific Assessment* [Houghton, J.T., G.J. Jenkins and J.J. Ephraums (eds.)]. Cambridge University Press, 365 pp.

IPCC, 1996: *Climate Change 1995: The Science of Climate Change. Contribution of Working Group I to the Second Assessment Report of the Intergovernmental Panel on Climate Change* [Houghton, J. T., L. G. Meira Filho, B. A. Callander, N. Harris, A. Kattenberg and K. Maskell (eds.)]. Cambridge University Press, 572 pp.

Iwashima, T. and R. Yamamoto, 1993: A statistical analysis of the extreme events: Long-term trend of heavy daily precipitation. *J. Met. Soc. Japan,* **71**, 637-640.

Jacob, C. and S.A. Klein, 1999: The role of vertically varying cloud fraction in the parameterization of microphysical processes in the ECMWF model. *Quart. J. Roy. Met. Soc.,* **125**, 941-965.

Jacobson, H.P. and C.F. Raymond, 1998: Thermal effects on the location of ice stream margins. *J. Geophys. Res.,* **103**, 12111-12122.

Jones, R.G., J.M. Murphy and M. Noguer, 1995: Simulation of climate change over Europe using a nested regional climate model. Part I: Assessment of control climate including sensitivity to location of lateral boundaries. *Quart. J. Roy. Met. Soc.,* **121**, 1413-1449.

Jones, R.G., J.M. Murphy, M. Noguer and A.B. Keen, 1997: Simulation of climate change over Europe using a nested regional climate model. Part II: Comparison of driving and regional model responses to a doubling of carbon dioxide. *Quart. J. Roy. Met. Soc.,* **123**, 265-292.

Joos, F., G.-K. Plattner, T.F. Stocker, O. Marchal and A. Schmittner, 1999: Global warming and marine carbon cycle feedbacks on future atmospheric CO_2. *Science,* **284**, 464-467.

Josey, S.A., E.C. Kent and P.K. Taylor, 1999: New insights into the ocean heat budget closure problem from analysis of the SOC air-sea flux climatology. *J. Clim.,* **12**, 2685-2718.

Kamb, B., C.F. Raymond, W.D. Harrison, H. Engelhardt, K.A. Echelmeyer, N. Humphrey, M.M. Brugman and T. Pfeffer, 1985: Glacier surge mechanism: 1982-1983 surge of Variegated Glacier, Alaska. *Science,* **227**, 469-479.

Karl, T.R. and R.W. Knight, 1998: Secular trends of precipitation amount, frequency and intensity in the USA. *Bull. Am. Met. Soc.,* **79**, 231-242.

Karl, T.R., R.W. Knight and N. Plummer, 1995: Trends in high-frequency climate variability in the twentieth century. *Nature,* **377**, 217-220.

Kato, S., T.P. Ackerman, E.E. Clothiaux and J.H. Mather, 1997: Uncertainties in modeled and measured clear-sky surface shortwave irradiances. *J. Geophys. Res.,* **102**, 25881-25898.

Keeling, C.D., T.P. Whorf, M. Wahlen and M. van der Plicht, 1995: Interannual extremes in the rate of rise of atmospheric carbon dioxide since 1980. *Nature,* **375**, 666-670.

Keith, D.A., 1995: Meridional energy transport: Uncertainty in zonal means. *Tellus,* **47A**, 30-44.

Kelly, P.M., P. Jia and P.D. Jones, 1996: The spatial response of the climate system to explosive volcanic eruptions. *Int. J. Clim.,* **16**, 537-550.

Kicklighter, D.W., M. Bruno, S. Donges, G. Esser, M. Heimann, J. Helfrich, F. Ift, F. Joos, J. Kaduk, G.H. Kohlmaier, A.D. McGuire, J.M. Melillo, R. Meyer, B. Moore, A. Nadler, I.C. Prentice, W. Sauf, A.L. Schloss, S. Sitch, U. Wittenberg and G. Wurth, 1998: A first-order analysis of the potential role of CO_2 fertilization to affect the global carbon budget: a comparison of four terrestrial biosphere models. *Tellus,* **51B**, 343-366.

Kiehl, J.T. and K.E. Trenberth, 1997: Earth's annual global mean energy budget. *Bull. Am. Met. Soc.,* **78**, 197-208.

Kiehl, J.T., J.J. Hack, M.H. Zhang and R.D. Cess, 1995: Sensitivity of a GCM climate to enhanced shortwave absorption. *J. Atm. Sci.,* **8**, 2200-2212.

Kiehl, J.T., J.J. Hack and J.W. Hurrell, 1998: The energy budget of the NCAR Community Climate Model CCM3. *J. Clim.,* **11**, 1151-1178.

Killworth, P.D., 1994: On reduced-gravity flow through sills. *Geophys. Astrophys. Fluid Dyn.,* **75**, 91-106.

Killworth, P.D., 1997: On the parametrization of eddy transfer. Part I: Theory. *J. Mar. Res.,* **55**, 1171-1197.

Killworth, P.D. and N.R. Edwards, 1999: A turbulent bottom boundary layer code for use in numerical ocean models. *J. Phys. Oceanogr.,* **29**, 1221-1238.

Kim, Y.J., 1996: Representation of subgrid-scale orographic effects in a general circulation model, 1: Impact on the dynamics of simulated January climate. *J. Clim.,* **9**, 2698-2717.

Kirk-Davidoff, D.B. and R.S. Lindzen, 2000: An energy balance model based on potential vorticity homogenization. *J. Clim.,* **13**, 431-448.

Kitoh, A., T. Motoi and H. Koide, 1999: SST variability and its mechanism in a coupled atmosphere-mixed layer ocean model. *J. Clim.,* **12**, 1221-1239.

Klein, B., R.L. Molinari, T.J. Muller and G. Siedler, 1995: A transatlantic section at 14.5N: Meridional volume and heat fluxes. *J. Mar. Res.,* **53**, 929-957.

Kley, D., H.G.J. Smit, H. Vömel, H. Grassl, V. Ramanathan, P.J. Crutzen, S. Williams, J. Meywerk and S.J. Oltmans, 1997: Tropospheric water-vapour and ozone cross-sections in a zonal plane over the central equatorial Pacific Ocean. *Quart. J. Roy. Met. Soc.,* **123**, 2009-2040.

Klinger, B., J. Marshall and U. Send, 1996: Representation of convective plumes by vertical adjustment. *J. Geophys. Res.,* **101**, 18175-18182.

Knutson, T.R. and S. Manabe, 1998: Model assessment of decadal variability and trends in the Tropical Pacific Ocean. *J. Clim.,* **11**, 2273-2296.

Knutti, R. and T.F. Stocker, 2000: Influence of the thermohaline circulation on projected sea level rise. *J. Clim.,* **13**, 1997-2001.

Kodera, K., 1994: Influence of volcanic eruptions on the troposphere through stratospheric dynamical processes in the Northern Hemisphere winter. *J. Geophys. Res., 99*, 1273-1282.

Koltermann, K.P., A.V. Sokov, V.P. Tereschenkov, S.A. Dobroliubov, K. Lorbacher and A. Sy, 1999: Decadal changes in the thermohaline circulation of the North Atlantic. *Deep-Sea Res. II, 46*, 109-138.

Kondratyev, K.Y., V.I. Bieneko and I.N. Melnikova, 1998: Absorption of solar radiation by clouds and aerosols in the visible wavelength region. *Meteorology and Atmospheric Physics, 65*, 1-10.

Koshlyakov, M.N. and T.G. Sazhina, 1996: Meridional volume and heat transport by large-scale geostrophic currents in the Pacific sector of the Antarctic. *Oceanology, 35*, 767-777.

Koster, R. and P.C.D. Milly, 1997: The interplay between transpiration and runoff formulations in land surface schemes used with atmospheric models. *J. Clim., 10*, 1578-1591.

Kreyscher, M., M. Harder and P. Lemke, 1997: First results of the Sea-Ice Model Intercomparison Project (SIMIP). *Ann. Glaciol., 25*, 8-11.

Labitzke, K. and H. van Loon, 1997: The signal of the 11-year sunspot cycle in the upper troposphere-lower stratosphere. *Space Sciences Reviews, 80*, 393-410.

LabSea Group, 1998: The Labrador Sea deep convection experiment. *Bull. Am. Met. Soc., 79*, 2033-2058.

Lacis, A.A. and J.E. Hansen, 1974: A parameterization for the absorption of solar radiation in the earth's atmosphere. *J. Atm. Sci., 31*, 118-133.

Lander, M., 1994: An exploratory analysis of the relationship between tropical storm formation in the Western North Pacific and ENSO. *Mon. Wea. Rev., 122*, 636-651.

Large, W.G. and P.R. Gent, 1999: Validation of vertical mixing in an equatorial ocean model using large eddy simulations and observations. *J. Phys. Oceanogr., 29*, 449-464.

Large, W.G., G. Danabasoglu, S.C. Doney and J.C. McWilliams, 1997: Sensitivity to surface forcing and boundary layer mixing in a global ocean model: Annual-mean climatology. *J. Phys. Oceanogr., 27*, 2418-2447.

Larson, K., D.L. Hartmann and S.A. Klein, 1999: The role of clouds, water vapor, circulation, and boundary layer structure in the sensitivity of the tropical climate. *J. Clim., 12*, 2359-2374.

Latif, M. and T.P. Barnett, 1996: Decadal climate variability over the North Pacific and North America: Dynamics and predictability. *J. Clim., 9*, 2407-2423.

Latif, M., R. Kleeman and C. Eckert, 1997: Greenhouse warming, decadal variability or El Niño: An attempt to understand the anomalous 1990s. *J. Clim., 10*, 2221-2239.

Latif, M., E. Roeckner, U. Mikolajewicz and R. Voss, 2000: Tropical stabilization of the thermohaline circulation in a greenhouse warming simulation. *J. Clim., 13*, 1809-1813.

Lau, N.-C. and M.J. Nath, 1999: Observed and GCM-simulated westward-propagating, planetary-scale fluctuation with approximately three-week periods. *Mon. Wea. Rev., 127*, 2324-2345.

Lavin, A., H.L. Bryden and G. Parrilla, 1998: Meridional transport and heat flux variations in the subtropical North Atlantic. *Glob. Atm. Oc. Sys., 6*, 269-293.

Le Treut, H. and Z.X. Li, 1991: The sensitivity of an atmospheric general circulation model to prescribed SST changes: Feedback effects associated with the simulation of cloud optical properties. *Clim. Dyn., 5*, 175-187.

Le Treut, H., Z.X. Li and M. Forichon, 1994: Sensitivity of the LMD general circulation model to greenhouse forcing associated with 2 different cloud-water parameterizations. *J. Clim., 7*, 1827-1841.

Le Treut, H. and B. McAvaney, 2000: Equilibrium climate change in response to a CO_2 doubling: an intercomparison of AGCM simulations coupled to slab oceans. *Technical Report, Institut Pierre Simon Laplace, 18*, 20 pp.

Lean, J., C.B. Bunton, C.A. Nobre and P.R. Rowntree, 1996: The simulated impacts of Amazonian deforestation on climate using

measured ABRACOS vegetation characteristics. In: *Amazonian Deforestation and Climate* [Gash, J. H. C., C. A. Nobre, J. M. Robert and R. L. Victoria (eds.)]. Wiley, Chichester, 549-576.

Lean, J. and P. Rowntree, 1997: Understanding the sensitivity of a GCM simulation of Amazonian deforestation to the specification of vegetation and soil characteristics. *J. Clim., 10*, 1216-1235.

Ledwell, J.R., E.T. Montgomery, K.L. Polzin, L.C. St. Laurent, R.W. Schmitt and J.M. Toole, 2000: Evidence for enhanced mixing over rough topography in the abyssal ocean. *Nature, 403*, 179-181.

Ledwell, J.R., A.J. Watson and C.S. Law, 1993: Evidence for slow mixing across the pycnocline from an open-ocean tracer-release experiment. *Nature, 364*, 701-703.

Lee, T.N., W. Johns, R. Zantopp and E. Fillenbaum, 1996: Moored observations of western boundary current variability and thermohaline circulation at 26.5°N in the subtropical North Atlantic. *J. Phys. Oceanogr., 26*, 962-983.

Lee, W.-H., S.F. Iacobellis and R.C.J. Somerville, 1997: Cloud radiation forcings and feedbacks: general circulation model test and observational validation. *J. Clim., 10*, 2479-2496.

Legg, S., J. McWilliams and G. Jianbo, 1998: Localization of deep ocean convection by a mesoscale eddy. *J. Phys. Oceanogr., 28*, 944-970.

Legutke, S., E. Maier-Reimer, A. Stössel and A. Hellbach, 1997: Ocean-sea-ice coupling in a global ocean general circulation model. *Ann. Glaciol., 25*, 116-120.

Lemke, P., W.D. Hibler, G. Flato, M. Harder and M. Kreyscher, 1997: On the improvement of sea-ice models for climate simulations: the Sea-Ice Model Intercomparison Project. *Ann. Glaciol., 25*, 183-187.

Lenderink, G. and A.A.M. Holtslag, 2000: Evaluation of the kinetic energy approach for modelling fluxes in stratocumulus. *Mon. Wea. Rev., 128*, 244-258.

Levis, S., J.A. Foley and D. Pollard, 1999: Potential high-latitude vegetation feedbacks on CO_2-induced climate change. *Geophys. Res. Lett., 26*, 747-750.

Lewis, E.L., E.P. Jones, P. Lemke, T.D. Prowse, P. Wadhams (eds.), 2000: *The Freshwater Budget of the Arctic Ocean.* NATO Science Series 2-70. Kluwer Academic Publishers, 625.

Li, Z., 1998: Influence of absorbing aerosols on the inference of solar surface radiation budget and cloud absorption. *J. Clim., 11*, 5-17.

Li, Z. and L. Moreau, 1996: Alteration of atmospheric solar absorption by clouds: simulation and observation. *J. Appl. Met., 35*, 653-670.

Li, Z., H.W. Barker and L. Moreau, 1995: The variable effect of clouds on atmospheric absorption of solar radiation. *Nature, 376*, 486-490.

Li, Z., L. Moreau and A. Arking, 1997: On solar energy disposition: A perspective from observation and modeling. *Bull. Am. Met. Soc., 78*, 53-70.

Li, Z., A. Trishchenko, H.W. Barker, G.L. Stephens and P.T. Partain, 1999: Analysis of Atmospheric Radiation Measurement (ARM) programs's Enhanced Shortwave Experiment (ARESE) multiple data sets for studying cloud absorption. *J. Geophys. Res., 104*, 19127-19134.

Lindzen, R.S., M.-D. Chou and A. Hou, 2001: Does the earth have an adaptive iris? *Bull. Am. Met. Soc., 82*, 417-432.

Lohmann, U. and E. Roeckner, 1996: Design and performance of a new cloud microphysics scheme developed for the ECHAM general circulation model. *Clim. Dyn., 12*, 557-572.

Lohmann, G. and R. Gerdes, 1998: Sea ice effects on the sensitivity of the thermohaline circulation. *J. Clim., 11*, 2789-2803.

Lorenz, E.N., 1993: *The Essence of Chaos.* University of Washington Press, Seattle, USA, 227.

Loth, B., H.-F. Graf and J.M. Oberhuber, 1993: Snow cover model for global climate simulations. *J. Geophys. Res., 98*, 10451-10464.

Lott, F. and M.J. Miller, 1997: A new subgrid-scale orographic drag parametrization: Its formulation and testing. *Quart. J. Roy. Met. Soc., 123*, 101-127.

Lüthi, D., A. Cress, H.C. Davies, C. Frei and C. Schär, 1996: Interannual variability and regional climate simulations. *Theor. Appl. Clim.,* **53**, 185-209.

Lynch-Stieglitz, M., 1994: The development and validation of a simple snow model for the GISS GCM. *J. Clim.,* **7**, 1842-1855.

MacAyeal, D.R., 1992: Irregular oscillations of the West Antarctic Ice Sheet. *Nature,* **359**, 29-32.

MacAyeal, D.R., R.A. Bindschadler and T.A. Scambos, 1995: Basal friction of Ice Stream E, West Antarctica. *J. Glaciol.,* **41**, 247-262.

MacAyeal, D.R., V. Rommelaere, C.L. Hulbe, J. Determann and C. Ritz, 1996: An ice-shelf model test based on the Ross Ice Shelf, Antarctica. *Ann. Glaciol.,* **23**, 46-51.

Macdonald, A.M., 1998: The global ocean circulation: a hydrographic estimate and regional analysis. *Prog. Oceanogr.,* **41**, 281-382.

Macdonald, A.M. and C. Wunsch, 1996: An estimate of global ocean circulation and heat fluxes. *Nature,* **382**, 436-439.

Manabe, S. and R.J. Stouffer, 1988: Two stable equilibria of a coupled ocean atmosphere model. *J. Clim.,* **1**, 841-866.

Manabe, S. and R.J. Stouffer, 1993: Century-scale effects of increased atmospheric CO_2 on the ocean-atmosphere system. *Nature,* **364**, 215-218.

Manabe, S. and R.T. Wetherald, 1987: Large-scale changes of soil wetness induced by an increase in atmospheric carbon dioxide. *J. Atm. Sci.,* **44**, 1211-1235.

Manzini, E. and L. Bengtsson, 1996: Stratospheric climate and variability from a general circulation model and observations. *Clim. Dyn.,* **12**, 615-639.

Marotzke, J., 1991: Influence of convective adjustment on the stability of the thermohaline circulation. *J. Phys. Oceanogr.,* **21**, 903-907.

Marotzke, J. and J. Willebrand, 1991: Multiple equilibria of the global thermohaline circulation. *J. Phys. Oceanogr.,* **21**, 1372-1385.

Marshall, J. and F. Schott, 1999: Open-ocean convection: observations, theory and models. *Rev. Geophys.,* **37**, 1-64.

Marshall, S.J. and G.K.C. Clarke, 1997: A continuum mixture model of ice stream thermodynamics in the Laurentide Ice Sheet 1. Theory. *J. Geophys. Res.,* **102**, 20599-20613.

Marshunova, M.S. and A.A. Mishin, 1994: Handbook on the Radiation Regime of the Arctic Basin: Results from the Drift Stations. In: *Arctic Ocean Snow and Meteorological Observations from Drifting Stations: University of Washington, Applied Physics Laboratory, Tech. Rep. APL-URW TR 9413* [Radionov, V. F. and R. Colony (eds.)]., 52. (Available on: Arctic Ocean Snow and Meteorological Observations from Drifting Stations, CD ROM, NSIDC, University of Colorado, 1996).

Mauritzen, C., 1996: Production of dense overflow waters feeding the North Atlantic across the Greenland-Scotland Ridge: Evidence for a revised circulation scheme. *Deep-Sea Res.,* **43**, 769-806.

Maxworthy, T., 1997: Convection into domains with open boundaries. *Ann. Rev. Fluid Mech.,* **29**, 327-371.

McCartney, M.S., R.G. Curry and H.F. Bezdek, 1996: North Atlantic's transformation pipeline chills and redistributed subtropical water. *Oceanus,* **39**, 19-23.

McGuffie, K., A. Henderson-Sellers, N. Holbrook, Z. Kothavala, O. Balachova and J. Hoekstra, 1999: Assessing simulations of daily temperature and precipitation variability with global climate models for present and enhanced greenhouse climates. *Int. J. Clim.,* **19**, 1-26.

McWilliams, J.C., P.P. Sullivan and C.-H. Moeng, 1997: Langmuir turbulence in the ocean. *J. Fluid Mech.,* **334**, 1-30.

Mearns, L.O., F. Giorgi, L. McDaniel and C. Shields, 1995: Analysis of daily variability of precipitation in a nested regional climate model: comparison with observations and doubled CO_2 results. *Glob. Planet. Change,* **10**, 55-78.

Meehl, G.A., 1997: The south Asian monsoon and the tropospheric biennial oscillation. *J. Clim.,* **10**, 1921-1943.

Meehl, G.A., W.D. Collins, B.A. Boville, J.T. Kiehl, T.M.L. Wigley and J.M. Arblaster, 2000: Response of the NCAR Climate System Model to increased CO_2 and the role of physical processes. *J. Clim.,* **13**, 1879-1898.

Mehta, V.M., M.J. Suarez, J.V. Manganello and T.L. Delworth, 2000: Oceanic influence of the North Atlantic Oscillation and associated Northern Hemisphere climate variations: 1959-1993. *Geophys. Res. Lett.,* **27**, 121-124.

Meleshko, V.M., Kattsov, P.V. Sporyshev, S.V. Vavulin and V.A. Govorkova, 2000: Feedback processes in climate system: cloud radiation and water vapour feedbacks interaction. *Meteorologia i Gidrologia (Russian Meteorology and Hydrology),* **2**, 22-45.

Merryfield, W.J. and G. Holloway, 1997: Topographic stress parameterization in a quasi-geostrophic barotropic model. *J. Fluid Mech.,* **341**, 1-18.

Mikolajewicz, U. and E. Maier-Reimer, 1994: Mixed boundary conditions in ocean general circulation models and their influence on the stability of the model's conveyor belt. *J. Geophys. Res.,* **99**, 22633-22644.

Mikolajewicz, U. and R. Voss, 2000: The role of the individual air-sea flux components in CO_2-induced changes of the ocean's circulation and climate. *Clim. Dyn.,* **16**, 627-642.

Milly, P.C.D. and K.A. Dunne, 1994: Sensitivity of the global water cycle to the water-holding capacity of land. *J. Clim.,* **7**, 506-526.

Mitchell, J.F.B., C.A. Wilson and W.M. Cunnington, 1987: On CO_2 climate sensitivity and model dependence of results. *Quart. J. Roy. Met. Soc.,* **113**, 293-322.

Moeng, C.-H., D.H. Lenschow and D.A. Randall, 1995: Numerical investigations of the roles of radiative and evaporative feedbacks in stratocumulus entrainment and breakup. *J. Atm. Sci.,* **52**, 2869-2883.

Moeng, C.-H., W.R. Cotton, C.S. Bretherton, A. Chlond, M. Khairoutdinov, S. Krueger, W.S. Lewellen, M.K.M. Vean, J.R.M. Pasquier, H.A. Rand, A.P. Siebesma, R.I. Sykes and B. Stevens, 1996: Simulations of a stratocumulus-topped PBL: Intercomparison among different numerical codes. *Bull. Am. Met. Soc.,* **77**, 261-278.

Monahan, A.H., J.C. Fyfe and G.M. Flato, 2000: A regime view of northern hemisphere atmospheric variability and change under global warming. *Geophys. Res. Lett.,* **27**, 1139-1142.

Morassutti, M.P. and E.F. LeDrew, 1996: Albedo and depth of melt ponds on sea ice. *Int. J. Clim.,* **16**, 817-838.

Morcrette, J.-J. and C. Jakob, 2000: The response of the ECMWF model to changes in cloud overlap assumption. *Mon. Wea. Rev.,* **128**, 1707-1732.

Mote, T.L., M.R. Anderson, K.C. Kuivinen and C.M. Rowe, 1993: Passive microwave-derived spatial and temporal variations of summer melt on the Greenland ice sheet. *Ann. Glaciol.,* **17**, 233-238.

Munk, W., 1966: Abyssal recipes. *Deep-Sea Res.,* **13**, 707-730.

Munk, W. and C. Wunsch, 1998: Abyssal recipes II, energetics of tidal and wind mixing. *Deep-Sea Res.,* **45**, 1977-2010.

Murphy, J., 1999: An evaluation of statistical and dynamical techniques for downscaling local climate. *J. Clim.,* **12**, 2256-2284.

Muszynski, I. and G.E. Birchfield, 1987: A coupled marine ice-stream ice-shelf model. *J. Glaciol.,* **33**, 3-15.

Myneni, R.B., F.G. Hall, P.J. Sellers and A.L. Marshak, 1995: The interpretation of spectral vegetation indexes. *IEEE Trans. Geosci. Rem. Sens.,* **33**, 481-486.

Myneni, R.B., C.D. Keeling, C.J. Tucker, G. Asrar and R.R. Nemani, 1997: Increased plant growth in the northern high latitudes from 1981-1991. *Nature,* **386**, 698-701.

Neelin, J.D., D.S. Battisti, A.C. Hirst, F.-F. Jin, Y. Wakata, T. Yamagata and S.E. Zebiak, 1998: ENSO theory. *J. Geophys. Res.,* **103**, 14,261-14,290.

Nobre, C.A., P.J. Sellers and J. Shukla, 1991: Amazonian deforestation and regional climate change. *J. Clim.,* **4**, 957-988.

O'Hirok, W. and C. Gautier, 1998a: A three-dimensional radiative transfer model to investigate the solar radiation within a cloudy atmosphere. Part I: Spatial effects. *J. Atm. Sci.,* **55**, 2162-2179.

O'Hirok, W. and C. Gautier, 1998b: A three-dimensional radiative transfer model to investigate the solar radiation within a cloudy atmosphere. Part II: Spectral effects. *J. Atm. Sci.,* **55**, 3065-3076.

Oreopoulos, L. and H.W. Barker, 1999: Accounting for subgrid-scale cloud variability in a multi-layer 1D solar radiative transfer radiative algorithm. *Quart. J. Roy. Met. Soc.,* **125**, 301-330.

Orsi, A., G. Johnson and J. Bullister, 1999: Circulation, mixing and production of Antarctic Bottom Water. *Prog. Oceanogr.,* **43**, 55-109.

Osborn, T.J., K.R. Briffa, S.F.B. Tett, P.D. Jones and R.M. Trigo, 1999: Evaluation of the North Atlantic oscillation as simulated by a coupled climate model. *Clim. Dyn.,* **15**, 685-702.

Osborn, T.J. and M. Hulme, 1997: Development of a relationship between station and grid-box rainday frequencies for climate model evaluation. *J. Clim.,* **10**, 1885-1908.

Palmer, T.N., 1999: A nonlinear dynamical perspective on climate prediction. *J. Clim.,* **12**, 575-591.

Palmer, T.N., 2000: Predicting uncertainty in forecasts of weather and climate. *Rep. Prog. Phys.,* **63**, 71-116.

Palmer, T.N. and D.A. Mansfield, 1986: A study of the wintertime circulation anomalies during past El Niño events, using a high resolution general circulation model. I. Influence of model climatology. *Quart. J. Roy. Met. Soc.,* **112**, 613-638.

Palmer, T.N., G.J. Shutts and R. Swinbank, 1986: Alleviation of a systematic westerly bias in general circulation and numerical weather prediction models through an orographic gravity wave drag parametrization. *Quart. J. Roy. Met. Soc.,* **112**, 1001-1039.

Paluszkiewicz, T. and R.D. Romea, 1997: A one-dimensional model for the parameterization of deep convection in the ocean. *Dyn. Atm. Ocean,* **26**, 95-130.

Pan, L., S. Solomon, W. Randel, J.F. Lamarque, P. Hess, J. Gille, E.W. Chiou and M.P. McCormick, 1997: Hemispheric asymmetries and seasonal variations of the lowermost stratospheric water vapor and ozone derived from SAGE II data. *J. Geophys. Res.,* **102**, 28177-28184.

Parkinson, C.L., D.J. Cavalieri, P. Gloersen, H.J. Zwally and J.C. Comiso, 1999: Arctic sea ice extent, areas, and trends, 1978-1996. *J. Geophys. Res.,* **104**, 20837-20856.

Payne, A.J., P. Huybrechts, A. Abe-Ouchi, R. Calov, J.L. Fastook, R. Greve, S.J. Marshall, I. Marsiat, C. Ritz, L. Tarasov and M.P.A. Thomassen, 2000: Results from the EISMINT model intercomparison: the effects of thermomechanical coupling. *J. Glaciol.,* **46**, 227-238

Perlwitz, J. and H.-F. Graf, 1995: The statistical connection between tropospheric and stratospheric circulation of the Northern Hemisphere in winter. *J. Clim.,* **8**, 2281-2295.

Perovich, D.K., E.L. Andreas, J.A. Curry, H. Eiken, C.W. Fairall, T.C. Grenfell, P.S. Guest, J. Intrieri, D. Kadko, R.W. Lindsay, M.G. McPhee, J. Morison, R.E. Moritz, C.A. Paulson, W.S. Pegau, P.O.G. Persson, R. Pinkel, J.A. Richter-Menge, T. Stanton, H. Stern, M. Sturm, W.B. Tucker III and T. Uttal, 1999: Year on ice gives climate insights. *EOS; Trans. Am. Geophys. Un.,* **80**, 481-486.

Petroliagis, T., R. Buizza, A. Lanzinger and T.N. Palmer, 1996: Extreme rainfall prediction using the European centre for medium-range weather forecasts ensemble prediction system. *J. Geophys. Res.,* **101**, 26227-26236.

Pierrehumbert, R.T., 1995: Thermostats, radiator fins and the local runaway greenhouse. *J. Atm. Sci.,* **52**, 1784-1806.

Pierrehumbert, R.T., 1999: Subtropical water vapor as a mediator of rapid climate change. In: *Geophysical Monograph: Mechanisms of global climate change at millennial time scales (112)* [Clark, P. U., R. S. Webb and L. D. Keigwin (eds.)]. American Geophysical Union, Washington, 339-361.

Pierrehumbert, R.T. and R. Roca, 1998: Evidence for control of Atlantic subtropical humidity by large scale advection. *Geophys. Res. Lett.,* **25**, 4537-4540.

Pilewskie, P. and F.P.J. Valero, 1995: Direct observations of excess absorption by clouds. *Science,* **267**, 1626-1629.

Pinnock, S., M.D. Hurley, K.P. Shine, T.J. Wallington and T.J. Smyth, 1995: Radiative forcing by hydrochlorofluorocarbons and hydrofluorocarbons. *J. Geophys. Res.,* **100**, 23227-23238.

Polcher, J., 1995: Sensitivity of tropical convection to land surface processes. *J. Atm. Sci.,* **52**, 3143-3161.

Pollard, D. and S.L. Thompson, 1994: Sea-ice dynamics and CO_2 sensitivity in a global climate model. *Atm.-Oce.,* **32**, 449-467.

Polyakov, I.V., A.Y. Proshutinsky and M.A. Johnson, 1999: Seasonal cycles in two regimes of Arctic climate. *J. Geophys. Res.,* **104**, 25761-25788.

Polzin, K., J.M. Toole and R.W. Schmitt, 1995: Finescale parameterizations of turbulent dissipation. *J. Phys. Oceanogr.,* **25**, 306-328.

Polzin, K.L. and E. Firing, 1997: Estimates of diapycnal mixing using LADCP and CTD data from 18S. *International WOCE Newsletter,* **29**, 39-42.

Polzin, K.L., J.M. Toole, J.R. Ledwell and R.W. Schmitt, 1997: Spatial variability of turbulent mixing in the abyssal ocean. *Science,* **276**, 93-96.

Pope, S.K. and F.P.J. Valero, 2000: Observations and models of irradiance profiles, column transmittance, and column reflectance during the Atmospheric Radiation Measurements Enhanced Shortwave Experiment. *J. Geophys. Res.,* **105**, 12521-12528.

Pratt, L.J. and M. Chechelnitsky, 1997: Principles for capturing the upstream effects of deep sills in low resolution ocean models. *Dyn. Atm. Ocean,* **26**, 1-25.

Proshutinsky, A.Y. and M.A. Johnson, 1997: Two circulation regimes of the wind-driven Arctic Ocean. *J. Geophys. Res.,* **102**, 12493-12514.

Ramanathan, V. and W. Collins, 1991: Thermodynamic regulation of ocean warming by cirrus clouds deduced from observations of the 1987 El Niño. *Nature,* **351**, 27-32.

Ramanathan, V., B. Subasilar, G. Zhang, W. Conant, R.D. Cess, J.T. Kiehl, H. Grassl and L. Shi, 1995: Warm Pool heat budget and shortwave cloud forcing - A missing physics. *Science,* **267**, 499-503.

Ramankutty, N. and J.A. Foley, 1998: Characterizing patterns of global land use: an analysis of global croplands data. *Glob. Biogeochem. Cyc.,* **12**, 667-685.

Ramaswamy, V., M.D. Schwarzkopf and W. Randel, 1996: Fingerprint of ozone depletion in the spatial and temporal pattern of recent lower-stratospheric cooling. *Nature,* **382**, 616-618.

Randall, D., J. Curry, D. Battisti, G. Flato, R. Grumbine, S. Hakkinen, D. Martinson, R. Preller, J. Walsh and J. Weatherly, 1998: Status and outlook for large-scale modelling of atmosphere-ice-ocean interactions in the Arctic. *Bull. Am. Met. Soc.,* **79**, 197-219.

Randall, D.A., 1995: Parameterizing fractional cloudiness produced by cumulus detrainment. In: *Technical Document: Workshop on Cloud Microphysics Parameterizations in Global Atmospheric General Circulation Models (713)*. World Meteorological Organization, 1-16.

Randall, D.A., P.J. Sellers, J.A. Berry, D.A. Dazlich, C. Zhang, C.J. Collatz, A.S. Denning, S.O. Los, C.B. Field, I. Fung, C.O. Justice and C.J. Tucker, 1996: A revised land surface parameterization (SiB2) for atmospheric GCMs. Part 3: The greening of the CSU GCM. *J. Clim.,* **9**, 738-763.

Randel, W.J. and F. Wu, 1999: Cooling of the Arctic and Antarctic polar stratosphere due to ozone depletion. *J. Clim.,* **12**, 1467-1479.

Randerson, J.T. and M.V. Thompson, 1999: Linking C-13 based estimates of land and ocean sinks with predictions of carbon storage from CO_2 fertilization of plant growth. *Tellus,* **51B**, 668-678.

Randerson, J.T., M.V. Thompson, T.J. Conway, I. Fung and C.B. Field, 1997: The contribution of terrestrial sources and sinks to trends in the seasonal cycle of atmospheric carbon dioxide. *Glob. Biogeochem. Cyc.,* **11**, 535-560.

Randerson, J.T., C.B. Field, I.Y. Fung and P.P. Tans, 1999: Increases in early season ecosystem uptake explain recent changes in the seasonal cycle of atmospheric CO_2 at high northern latitudes. *Geophys. Res. Lett.,* **26**, 2765-2768.

Raval, A. and V. Ramanathan, 1989: Observational determination of the greenhouse effect. *Nature,* **342**, 758-761.

Rayner, P. and R. Law, 1999: The interannual variability of the global carbon cycle. *Tellus,* **51B**, 210-212.

Ricard, J.-L. and J.-F. Royer, 1993: A statistical cloud scheme for use in an AGCM. *Ann. Geophys.,* **11**, 1095-1115.

Roads, J.O., S. Marshall, R. Oglesby and S.-C. Chen, 1996: Sensitivity of the CCM1 hydrological cycle to CO_2. *J. Geophys. Res.,* **101**, 7321-7339.

Robbins, P.E. and J.M. Toole, 1997: The dissolved silica budget as a constraint on the meridional overturning circulation of the Indian Ocean. *Deep-Sea Res. I,* **44**, 879-906.

Roberts, M. and D. Marshall, 1998: Do we require adiabatic dissipation schemes in eddy-resolving ocean models? *J. Phys. Oceanogr.,* **28**, 2050-2063.

Roberts, M.J. and R.A. Wood, 1997: Topography sensitivity studies with a Bryan-Cox type ocean model. *J. Phys. Oceanogr.,* **27**, 823-836.

Robertson, A.W., M. Ghil and M. Latif, 2000a: Interdecadal changes in atmospheric low-frequency variability with and without boundary forcing. *J. Atm. Sci.,* **57**, 1132-1140.

Robertson, A.W., C.R. Mechoso and Y.-J. Kim, 2000b: The influence of Atlantic sea surface temperature anomalies on the North Atlantic Oscillation. *J. Clim.,* **13**, 122-138.

Rodbell, D., G.O. Seltzer, D.M. Anderson, D.B. Enfield, M.B. Abbott and J.H. Newman, 1999: A high-resolution 15,000 year record of El Niño driven alluviation in southwestern Ecuador. *Science,* **283**, 516-520.

Rodwell, M.J., D.P. Rowell and C.K. Folland, 1999: Oceanic forcing of the wintertime North Atlantic oscillation and European climate. *Nature,* **398**, 320-323.

Roeckner, E., M. Rieland and E. Keup, 1991: Modeling of cloud and radiation in the ECHAM model. In: *Procs. ECMWF/WCRP Workshop on "Clouds, radiative transfer and the hydrological cycle" 12-15 November 1990* . ECMWF, Reading, UK, 199-222.

Roeckner, E., K. Arpe, L. Bengtsson, M. Christoph, M. Claussen, L. Dümenil, M. Esch, M. Giorgetta, U. Schlese and U. Schulzweida, 1996: The atmospheric general circulation model ECHAM-4: Model description and simulation of present-day climate. Report **218**, *Max-Planck Institut für Meteorologie,* Hamburg, 90 pp.

Roemmich, D. and T. McCallister, 1989: Large scale circulation of the North Pacific Ocean. *Prog. Oceanogr.,* **22**, 171-204.

Rosenlof, K.H., 1995: Seasonal cycle of the residual mean meridional circulation in the stratosphere. *J. Geophys. Res.,* **100**, 5173-5191.

Ross, R.J. and W.P. Elliot, 1996: Tropospheric water vapor climatology and trends over North America: 1973-93. *J. Clim.,* **9**, 3561-3574.

Rotstayn, L.D., 1997: A physically based scheme for the treatment of clouds and precipitation in large-scale models. I: Description and evaluation of the microphysical processes. *Quart. J. Roy. Met. Soc.,* **123**, 1227-1282.

Royer, J.-F., F. Chauvain, P. Timbal, P. Araspin and D. Grimal, 1998: A GCM study of the impact of greenhouse gas increase on the frequency of occurrence of tropical cyclones. *Clim. Change,* **38**, 307-343.

Saji, N.H., B.N. Goswami, P.N. Vinayachandran and T. Yamagata, 1999: A dipole mode in the tropical Indian Ocean. *Nature,* **401**, 360-363.

Salathe, E.P. and D.L. Hartmann, 1997: A trajectory analysis of tropical upper-tropospheric moisture and convection. *J. Clim.,* **10**, 2533-2547.

Sandweiss, D.H., J.B. Richardson, E.J. Reitz, H.B. Rollins and K.A. Maasch, 1996: Geoarchaeological evidence from Peru for a 5,000 BP onset of El Niño. *Science,* **273**, 1531-1533.

SAR, see IPCC, 1996.

Saravanan, R., 1998: Atmospheric low frequency variability and its relationship to midlatitude SST variability: studies using the NCAR Climate System Model. *J. Clim.,* **11**, 1386-1404.

Saravanan, R. and J.C. McWilliams, 1998: Advective ocean-atmosphere interaction: An analytical stochastic model with implications for decadal variability. *J. Clim.,* **11**, 165-188.

Saunders, P.M. and B.A. King, 1995: Oceanic fluxes on the WOCE A11 section. *J. Phys. Oceanogr.,* **25**, 1942-1958.

Schär, C., D. Lüthi, U. Beyerle and E. Heise, 1999: The soil-precipitation feedback: a process study with a regional climate model. *J. Clim.,* **12**, 722-741.

Schiller, A., U. Mikolajewicz and R. Voss, 1997: The stability of the North Atlantic thermohaline circulation in a coupled ocean-atmosphere general circulation model. *Clim. Dyn.,* **13**, 325-347.

Schlosser, C.A., A.G. Slater, A.J. Pitman, A. Robock, K.Y. Vinnikov, A. Henderson-Sellers, N.A. Speranskaya, K. Mitchell, A. Boone, H. Braden, F. Chen, P. Cox, P. de Rosnay, C.E. Desborough, R.E. Dickinson, Y.-J. Dai, Q. Duan, J. Entin, P. Etchevers, N. Gedney, Y.M. Gusev, F. Habets, J. Kim, V. Koren, E. Kowalczyk, O.N. Nasonova, J. Noilhan, J. Schaake, A.B. Shmakin, T.G. Smirnova, D. Verseghy, P. Wetzel, Y. Xue and Z.-L. Yang, 2000: Simulations of a boreal grassland hydrology at Valdai, Russia: PILPS Phase 2(d). *Mon. Wea. Rev.,* **128**, 301-321.

Schmittner, A., C. Appenzeller and T.F. Stocker, 2000: Enhanced Atlantic freshwater export during El Niño. *Geophys. Res. Lett.,* **27**, 1163-1166.

Schmittner, A. and T.F. Stocker, 1999: The stability of the thermohaline circulation in global warming experiments. *J. Clim.,* **12**, 1117-1133.

Schneider, E.K., B.P. Kirtman and R.S. Lindzen, 1999: Tropospheric water vapor and climate sensitivity. *J. Atm. Sci.,* **36**, 1649-1658.

Schott, F., M. Visbeck and J. Fischer, 1993: Observations of vertical currents and convection in the central Greenland Sea during winter 1988/89. *J. Geophys. Res.,* **98**, 14401-14421.

Schott, F., M. Visbeck, U. Send, J. Fischer, L. Stramma and Y. Desaubies, 1996: Observations of deep convection in the Gulf of Lions, northern Mediterranean, during the winter of 1991/92. *J. Phys. Oceanogr.,* **26**, 505-524.

Schramm, J.L., M.M. Holland, J.A. Curry and E.E. Ebert, 1997: Modeling the thermodynamics of a sea ice thickness distribution. 1. Sensitivity to ice thickness resolution. *J. Geophys. Res.,* **102**, 23079-23091.

Schulz, J. and V. Jost (eds.), 1998: *HOAPS: A Satellite-Derived Water Balance Climatology,* WOCE Conference, May, Halifax.

Sellers, P.J., J.A. Berry, G.J. Collatz, C.B. Field and F.G. Hall, 1992a: Canopy reflectance, photosynthesis and transpiration. Part III: A reanalysis using enzyme kinetics-electron transport models of leaf physiology. *Rem. Sens. Environ.,* **42**, 187-216.

Sellers, P.J., F.G. Hall, G. Asrar, D.E. Strebel and R.E. Murphy, 1992b: An overview of the First International Satellite Land Surface Climatology Project (ISLSCP) Field Experiment (FIFE). *J. Geophys. Res.,* **97**, 18345-18731.

Sellers, P.J., M.D. Heiser and F.G. Hall, 1992c: Relationship between surface conductance and spectral vegetation indices at intermediate ($100 \, m^2$-15 km^2) length scales. *J. Geophys. Res.,* **97**, 19033-19060.

Sellers, P.J., F.G. Hall, H. Margolis, B. Kelly, D. Baldocchi, J. denHartog, J. Cihlar, M. Ryan, B. Goodison, P. Crill, J. Ranson, D. Lettenmaier and D.E. Wickland, 1995: The Boreal Ecosystem-Atmosphere Study (BOREAS): an overview and early results from the 1994 field year. *Bull. Am. Met. Soc.,* **76**, 1549-1577.

Sellers, P.J., L. Bounoua, G.J. Collatz, D.A. Randall, D.A. Dazlich, S.O. Los, J.A. Berry, I. Fung, C.J. Tucker, C.B. Field and T.G. Jensen, 1996a: Comparison of radiative and physiological effects of doubled atmospheric CO_2 on climate. *Science,* **271**, 1402-1406.

Sellers, P.J., B.W. Meeson, J. Closs, F. Corprew, D. Dazlich, F.G. Hall, Y. Kerr, R. Koster, S. Los, K. Mitchell, J. McManus, D. Myers, K.-J. Sun and P. Try, 1996b: The ISLSCP Initiative I Global Data Sets: Surface boundary conditions and atmospheric forcings for land-atmosphere studies. *Bull. Am. Met. Soc.,* **77**, 1987-2005.

Sellers, P.J., D.A. Randall, G.J. Collatz, J.A. Berry, C.B. Field, D.A. Dazlich, C. Zhang, G.D. Collelo and L. Bounoua, 1996c: A revised land surface parameterization (SiB2) for atmospheric GCMs. Part I: Model formulation. *J. Clim.,* **9**, 676-705.

Sellers, P.J., R.E. Dickinson, D.A. Randall, A.K. Betts, F.G. Hall, J.A. Berry, G.J. Collatz, A.S. Denning, H.A. Mooney, C.A. Nobre, N. Sato, C.B. Field and A. Henderson-Sellers, 1997: Modeling the exchange of energy, water and carbon between continents and the atmosphere. *Science, 275,* 502-509.

Senior, C.A. and J.F.B. Mitchell, 1993: Carbon dioxide and climate: the impact of cloud parameterization. *J. Clim.,* **6**, 393-418.

Shao, Y. and A. Henderson-Sellers, 1996: Validation of soil moisture simulation in land surface parameterization schemes with HAPEX data. *Glob. Planet. Change,* **13**, 11-46.

Shaw, J., B. Rains, R. Eyton and L. Weissling, 1996: Laurentide subglacial outburst floods: landform evidence from digital elevation models. *Can. J. Earth Sci.,* **33**, 1154-1168.

Shepherd, T.G., K. Semeniuk and J.N. Koshyk, 1996: Sponge layer feedbacks in middle atmosphere models. *J. Geophys. Res.,* **101**, 23447-23464.

Sherwood, S.C., 1996: Maintenance of the free-troposphere tropical water vapor distribution. Part II: Simulation by large-scale advection. *J. Clim.,* **9**, 2919-2934.

Shindell, D., D. Rind, N. Balachandran, J. Lean and P. Lonergan, 1999a: Solar cycle variability, ozone, and climate. *Science, 284,* 305-308.

Shindell, D.T., R.L. Miller, G.A. Schmidt and L. Pandolfo, 1999b: Simulation of recent northern winter climate trends by greenhouse-gas forcing. *Nature, 399,* 452-455.

Shine, K.P., 1993: The greenhouse effect and stratospheric change. In: *NATO ASI I8: The role of the stratosphere in global change* [Chanin, M.-L. (ed.)]. Springer-Verlag, Berlin, 285-300.

Shine, K.P. and P.M.F. Forster, 1999: The effect of human activity on radiative forcing of climate change: a review of recent developments. *Glob. Planet. Change,* **20**, 205-225.

Shine, K.P., B. Briegleb, A. Grossman, D. Hauglustaine, H. Mao, V. Ramaswamy, D. Schwarzkopf, R.V. Dorland and W.-C. Wang, 1995: Radiative forcing due to changes in ozone: A comparison of different codes. In: *NATO ASI I32: Atmospheric Ozone as a Climate Gas* [Wang, W.-C. and I. Isaksen (eds.)]., 373-396.

Slater, A.G., A.J. Pitman and C.E. Desborough, 1998a: The simulation of freeze-thaw cycles in a GCM land surface scheme. *J. Geophys. Res.,* **103**, 11303-11312.

Slater, A.G., A.J. Pitman and C.E. Desborough, 1998b: The validation of a snow parameterization designed for use in General Circulation Models. *Int. J. Clim.,* **18**, 595-617.

Slingo, J.M., K.R. Sperber, J.S. Boyle, J.P. Ceron, M. Dix, B. Dugas, W. Ebisuzaki, J. Fyfe, D. Gregory, J.F. Gueremy, J. Hack, A. Harzallah, P. Inness, A. Kitoh, W.K.M. Lau, B. McAvaney, R. Madden, A. Matthews, T.N. Palmer and C.K. Park, 1996: Intraseasonal oscillations in 15 atmospheric general circulation models: results from an AMIP diagnostic subproject. *Clim. Dyn.,* **12**, 325-357.

Smith, R.D., M.E. Maltrud, F.O. Bryan and M.W. Hecht, 2000: Numerical simulation of the North Atlantic at 1/10°. *J. Phys. Oceanogr.,* **30**, 1532-1561.

Smith, R.N.B., 1990: A scheme for predicting layer clouds and their water content in a general circulation model. *Quart. J. Roy. Met. Soc.,* **116**, 435-460.

Soden, B.J., 1997: Variations in the tropical greenhouse effect during El Niño. *J. Clim.,* **10**, 1050-1055.

Soden, B.J., 1998: Tracking upper tropospheric water vapor radiances: A satellite perspective. *J. Geophys. Res.,* **103**, 17069-17081.

Soden, B.J., 2000: The sensitivity of the tropical hydrological cycle to ENSO. *J. Clim.,* **13**, 538-549.

Soden, B.J. and R. Fu, 1995: A satellite analysis of deep convection, upper-tropospheric humidity, and the greenhouse effect. *J. Clim.,* **8**, 2333-2351.

Solomon, S., R.W. Portmann, R.R. Garcia, L.W. Thomason, L.R. Poole and M.P. McCormick, 1996: The role of aerosol variations in anthropogenic ozone depletion at northern midlatitudes. *J. Geophys. Res.,* **101**, 6713-6727.

Spall, M.A. and J.F. Price, 1998: Mesoscale variability in Denmark Strait: the PV outflow hypothesis. *J. Phys. Oceanogr.,* **28**, 1598-1623.

Spall, M.A., R.A. Weller and P. Furey, 2000: Modelling the three-dimensional upper ocean heat budget and subduction rate during the Subduction Experiment. *J. Geophys. Res.,* **105**, 26151-26166.

Speer, K.G., J. Holfort, T. Reynard and G. Siedler, 1996: South Atlantic heat transport at 11°S. In: *The South Atlantic: Present and Past Circulation* [Wefer, G., W. H. Berger, G. Siedler and D. J. Webb (eds.)]. Springer, 105-120.

Spencer, R.W. and W.D. Braswell, 1997: How dry is the tropical free troposphere? Implications for global warming theory. *Bull. Am. Met. Soc.,* **78**, 1097-1106.

St. Laurent, L. and R.W. Schmitt, 1999: The contribution of salt fingers to vertical mixing in the North Atlantic Tracer Release Experiment. *J. Phys. Oceanogr.,* **29**, 1404-1424.

Stephens, G.L., 1996: How much solar radiation do clouds absorb? *Science, 271,* 1131-1133.

Stocker, T.F., 2000: Past and future reorganization in the climate system. *Quat. Sci. Rev.,* **19**, 301-319.

Stocker, T.F. and D.G. Wright, 1991: Rapid transitions of the ocean's deep circulation induced by changes in surface water fluxes. *Nature,* **351**, 729-732.

Stocker, T.F. and A. Schmittner, 1997: Influence of CO_2 emission rates on the stability of the thermohaline circulation. *Nature,* **388**, 862-865.

Stocker, T.F. and O. Marchal, 2000: Abrupt climate change in the computer: is it real? *PNAS,* **97**, 1362-1365.

Stommel, H., 1961: Thermohaline convection with two stable regimes of flow. *Tellus,* **13**, 224-230.

Stössel, A., S.-J. Kim and S.S. Drijfhout, 1998: The impact of Southern Ocean sea ice in a global ocean model. *J. Phys. Oceanogr.,* **28**, 1999-2018.

Stouffer, R.J. and S. Manabe, 1999: Response of a coupled ocean-atmosphere model to increasing atmospheric carbon dioxide: sensitivity to the rate of increase. *J. Clim.,* **12**, 2224-2237.

Stroeve, J., A.W. Nolin and K. Steffen, 1997: Comparison of AVHRR-derived and in situ surface albedo over the Greenland ice sheet. *Rem. Sens. Environ.,* **62**, 262-276.

Sud, Y.C. and G.K. Walker, 1999: Microphysics of clouds with the relaxed Arakawa-Schubert scheme (McRAS). Part I: design and evaluation with GATE phase III data. *J. Atm. Sci.,* **56**, 3196-3220.

Sud, Y.C., G.K. Walker, J.-H. Kim, G.E. Liston, P.J. Sellers and W.K.-M. Lau, 1996: Biogeophysical consequences of a tropical deforestation scenario: a GCM simulation study. *J. Clim.,* **9**, 3225-3247.

Sun, D.-Z. and K.E. Trenberth, 1998: Coordinated heat removal from the tropical Pacific during the 1986-87 El Niño. *Geophys. Res. Lett.,* **25**, 2659-2662.

Suppiah, R. and K.J. Hennessy, 1998: Trends in the intensity and frequency of heavy rainfall in tropical Australia and links with the Southern Oscillation. *Austr. Meteorol. Mag.,* **45**, 1-17.

Sutton, R.T. and M.R. Allen, 1997: Decadal predictability of North Atlantic sea surface temperature and climate. *Nature, 388,* 563-567.

Takahashi, M., 1996: Simulation of the stratospheric quasi-biennial oscillation using a general circulation model. *Geophys. Res. Lett.,* **23**, 661-664.

Takle, E.S., J. Gutowski, W. J., R.W. Arritt, Z. Pan, C.J. Anderson, R.R. da Silva, D. Caya, S.-C. Chen, F. Giorgi, J.H. Christensen, S.-Y. Hong, H.-M.H. Juang, J. Katzfey, W.M. Lapenta, R. Laprise, G.E. Liston, P. Lopez, J. McGregor, R.A. Pielke Sr. and J.O. Roads, 1999: GEWEX Continental-scale International Project (GCIP), Part 2: Project to intercompare regional climate simulations (PIRCS):

Description and initial results. *J. Geophys. Res.,* **104**, 19443-19462.

Tans, P.P., I.Y. Fung and T. Takahashi, 1990: Observational constraints on the global atmospheric CO_2 budget. *Science,* **247**, 1431-1438.

Tett, S.F.B., J.F.B. Mitchell, D.E. Parker and M.R. Allen, 1996: Human influence on the atmospheric vertical temperature structure: detection and observations. *Science,* **274**, 1170-1173.

Thomas, G. and P.R. Rowntree, 1992: The boreal forest and climate. *Quart. J. Roy. Met. Soc.,* **118**, 469-497.

Thomas, R.H. and C.R. Bentley, 1978: A model for Holocene retreat of the West Antarctic Ice Sheet. *Quat. Res.,* **10**, 150-170.

Thompson, D.W.J. and J.M. Wallace, 1998: The Arctic Oscillation signature in the wintertime geopotential height and temperature fields. *Geophys. Res. Lett.,* **25**, 1297-1300.

Thompson, D.W.J. and J.M. Wallace, 2000: Annular modes in the extra-tropical circulation Part I: month-to-month variability. *J. Clim.,* **13**, 1000-1016.

Thompson, S.L. and D. Pollard, 1995: A global climate model (GENESIS) with a land-surface transfer scheme (LSX). Part II CO_2 sensitivity. *J. Clim.,* **8**, 1104-1121.

Thompson, S.L. and D. Pollard, 1997: Greenland and Antarctic mass balances for present and doubled atmospheric CO_2 from the GENESIS version-2 global climate model. *J. Clim.,* **10**, 871-900.

Tian, H., J.M. Melillo, D.W. Kicklighter, A.D. McGuire and J. Helfrich, 1999: The sensitivity of terrestrial carbon storage to historical climate variability and atmospheric CO_2 in the United States. *Tellus,* **51B**, 414-452.

Tiedtke, M., 1993: Representation of clouds in large-scale models. *Mon. Wea. Rev.,* **121**, 3040-3061.

Timmermann, A., M. Latif, R. Voss and R.A. Grotzner, 1998: Northern hemisphere interdecadal variability: a coupled air-sea mode. *J. Clim.,* **11**, 1906-1931.

Timmermann, A., J.M. Oberhuber, A. Bacher, M. Esch, M. Latif and E. Roeckner, 1999: Increased El Niño frequency in a climate model forced by future greenhouse warming. *Nature,* **398**, 694-696.

Toggweiler, J.R. and B. Samuels, 1995: Effect of Drake Passage on the global thermohaline circulation. *Deep-Sea Res. I,* **42**, 477-500.

Tompkins, A.M. and K.A. Emanuel, 2000: The vertical resolution sensitivity of simulated equilibrium temperature and water vapour profiles. *Quart. J. Roy. Met. Soc.,* **126**, 1219-1238.

Trenberth, K.E., 1998: Atmospheric moisture residence times and cycling: Implications for rainfall rates and climate change. *Clim. Change,* **39**, 667-694.

Trenberth, K.E., 1999: Conceptual framework for changes of extremes of the hydrological cycle with climate change. *Clim. Change,* **42**, 327-339.

Trenberth, K.E. and J.W. Hurrell, 1994: Decadal atmosphere-ocean variations in the Pacific. *Clim. Dyn.,* **9**, 303-319.

Trenberth, K.E. and A. Solomon, 1994: The global heat balance: heat transports in the atmosphere and ocean. *Clim. Dyn.,* **10**, 107-134.

Trenberth, K.E. and T.J. Hoar, 1997: El Niño and climate change. *Geophys. Res. Lett.,* **24**, 3057-3060.

Trenberth, K.E. and C.J. Guillemot, 1998: Evaluation of the atmospheric moisture and hydrological cycle in the NCEP/NCAR reanalyses. *Clim. Dyn.,* **14**, 213-231.

Trenberth, K.E. and J.W. Hurrell, 1999: Comments on "The interpretation of short climate records with comments on the North Atlantic and Southern Oscillations". *Bull. Am. Met. Soc.,* **80**, 2721-2722.

Trenberth, K.E., G.W. Branstator, D. Karoly, A. Kumar, N.-C. Lau and C. Ropelewski, 1998: Progress during TOGA in understanding and modeling global teleconnections associated with tropical sea surface temperatures. *J. Geophys. Res.,* **103**, 14291-14324.

Trenberth, K.E., J.M. Caron and D.P. Stepaniak, 2001: The atmospheric energy budget and implications for surface fluxes and ocean heat transports. *Clim. Dyn.,* **17**, 259-276.

Tselioudis, G. and W.B. Rossow, 1994: Global, multiyear variations of optical thickness with temperature in low and cirrus clouds. *Geophys. Res. Lett.,* **21**, 2211-2214.

Tselioudis, G., W. Rossow and D. Rind, 1992: Global patterns of cloud optical thickness variation with temperature. *J. Clim.,* **5**, 1484-1495.

Tselioudis, G., A.D. Del Genio, W. Kowari Jr. and M.-S. Yao, 1998: Temperature dependence of low cloud optical thickness in the GISS GCM: Contributing mechanisms and climate implication. *J. Clim.,* **11**, 3268-3281.

Tselioudis, G., Y. Zhang and W.B. Rossow, 2000: Cloud and radiation variations associated with northern midlatitude low and high sea level pressure regimes. *J. Clim.,* **13**, 312-327.

Tsimplis, M.N., S. Bacon and H.L. Bryden, 1998: The circulation of the sub-tropical South Pacific derived from hydrographic data. *J. Geophys. Res.,* **103**, 21443-21468.

Tucker, C.J.B., I.Y. Fung, C.D. Keeling and R.H. Gammon, 1986: Relationship between atmospheric CO_2 variation and a satellite-derived vegetation index. *Nature,* **319**, 195-199.

Tulaczyk, S., B. Kamb, R.P. Scherer and H. Engelhardt, 1998: Sedimentary processes at the base of a West Antarctic ice stream: Constraints from textural and compositional properties of subglacial debris. *J. Sed. Res.,* **68**, 487-496.

Twomey, S.A., 1974: Pollution and the planetary albedo. *Atmos. Env.,* **8**, 1251-1256.

Tziperman, E., 2000: Proximity of the present-day thermohaline circulation to an instability threshold. *J. Phys. Oceanogr.,* **30**, 90-104.

Udelhofen, P.M. and D.L. Hartmann, 1995: Influence of tropical cloud systems on the relative humidity in the upper troposphere. *J. Geophys. Res.,* **100**, 7423-7440.

Valero, F.P.J., R.D. Cess, M. Zhang, S.K. Pope, A. Bucholtz, B. Bush and J. Vitko Jr., 1997: Absorption of solar radiation by clouds. *J. Geophys. Res.,* **102**, 29917-29927.

Valero, F.P.J., P. Minnis, S.K. Pope, A. Bucholtz, D.R. Doelling, W.L. Smith and X.Q. Dong, 2000: Absorption of solar radiation by the atmosphere as determined using satellite, aircraft, and surface data during the Atmospheric Radiation Measurement Enhanced Shortwave Experiment (ARESE). *J. Geophys. Res.,* **105**, 4743-4758.

van der Veen, C.J. and I.M. Whillans, 1996: Model experiments on the evolution and stability of ice streams. *Ann. Glaciol.,* **23**, 129-137.

Verseghy, D.L., N.A. McFarlane and M. Lazare, 1993: CLASS - A Canadian land surface scheme for GCMs. II: Vegetation model and coupled runs. *Int. J. Clim.,* **13**, 347-370.

Vinje, T., N. Nordlund and A. Kvambekk, 1998: Monitoring ice thickness in Fram Strait. *J. Geophys. Res.,* **103**, 10437-10449.

Vinnikov, K.Y., A. Robock, R.J. Stouffer, J.E. Walsh, C.L. Parkinson, D.J. Cavalieri, J.F.B. Mitchell, D. Garrett and V.F. Zakharov, 1999: Global warming and Northern Hemisphere sea ice extent. *Science,* **286**, 1934-1937.

Visbeck, M., J. Fischer and F. Schott, 1995: Preconditioning the Greenland Sea for deep convection: Ice formation and ice drift. *J. Geophys. Res.,* **100**, 18489-18502.

Visbeck, M., J. Marshall, T. Haine and M. Spall, 1997: Specification of eddy transfer coefficients in coarse-resolution ocean circulation models. *J. Phys. Oceanogr.,* **27**, 381-402.

Viterbo, P., A.C.M. Beljaars, J.-F. Mahfouf and J. Teixeira, 1999: The representation of soil moisture and freezing and its impact on the stable boundary layer. *Quart. J. Roy. Met. Soc.,* **125**, 2401-2426.

Wallace, J.M., Y. Zhang and J.A. Renwick, 1995: Dynamic contribution to hemispheric mean temperature trends. *Science,* **270**, 780-783.

Wang, W.-C., M.P. Dudek, X.-Z. Liang and J.T. Kiehl, 1991: Inadequacy of effective CO_2 as a proxy in simulating the greenhouse effect of other radiatively active gases. *Nature,* **350**, 573-577.

Watanabe, M. and T. Nitta, 1999: Decadal changes in the atmospheric circulation and associated surface climate variations in the Northern Hemisphere winter. *J. Clim.,* **12**, 494-510.

Watterson, I.G., M.R. Dix and R.A. Colman, 1999: A comparison of present and doubled CO_2 climates and feedbacks simulated by three general circulation models. *J. Geophys. Res.,* **104**, 1943-1956.

Waugh, D.W., T.M. Hall, W.J. Randel, K.A. Boering, S.C. Wofsy, B.C. Daube, J.W. Elkins, D.W. Fahey, G.S. Dutton, C.M. Volk and P. Vohralik, 1997: Three-dimensional simulations of long-lived tracers using winds from MACCM2. *J. Geophys. Res.,* **102**, 21493-21513.

Weare, B.C., 2000a: Insights into the importance of cloud vertical structure in climate. *Geophys. Res. Lett.,* **27**, 907-910.

Weare, B.C., 2000b: Near-global observations of low clouds. *J. Clim.,* **13**, 1255-1268.

Weare, B.C. and I.I. Mokhov, 1995: Evaluation of total cloudiness and its variability in the Atmospheric Model Intercomparison Project. *J. Clim.,* **8**, 2224-2238.

Weaver, A.J. and T.M.C. Hughes, 1994: Rapid interglacial climate fluctuations driven by North Atlantic ocean circulation. *Nature,* **367**, 447-450.

Webster, P.J., V.O. Magaña, T.N. Palmer, J. Shukla, R.A. Tomas, M. Yanai and T. Yasunari, 1998: Monsoons: Processes, predictability and the prospects for prediction. *J. Geophys. Res.,* **103**, 14451-14510.

Webster, P.J., A.M. Moore, J.P. Loschnigg and R.R. Leben, 1999: Coupled ocean-atmosphere dynamics in the Indian Ocean during 1997-98. *Nature,* **401**, 356-360.

Weller, R.A. and S.P. Anderson, 1996: Surface Meteorology and air-sea fluxes in the western equatorial Pacific warm pool during the TOGA Coupled Ocean-Atmosphere Experiment. *J. Clim.,* **9**, 1959-1990.

Weller, R.A. and J.F. Price, 1988: Langmuir circulation within the oceanic mixed layer. *Deep-Sea Res.,* **35**, 711-747.

Weng, W.J. and J.D. Neelin, 1998: On the role of ocean-atmosphere interaction in midlatitude interdecadal variability. *Geophys. Res. Lett.,* **25**, 167-170.

Wentz, F.J. and M. Schabel, 2000: Precise climate monitoring using complementary satellite data sets. *Nature,* **403**, 414-416.

Wetherald, R.T. and S. Manabe, 1999: Detectability of summer dryness caused by greenhouse warming. *Clim. Change,* **43**, 495-511.

Whetton, P.H., A.M. Fowler and M.R. Haylock, 1993: Implications of climate change due to the enhanced greenhouse effect on floods and draughts in Australia. *Clim. Change,* **25**, 289-317.

Whillans, I.M. and C.J. van der Veen, 1993: New and improved determinations of velocity of Ice Streams B and C, West Antarctica. *J. Glaciol.,* **39**, 483-490.

Wijffels, S.E., R.W. Schmitt, H.L. Bryden and A. Stigebrandt, 1992: Transport of freshwater by the oceans. *J. Phys. Oceanogr.,* **22**, 155-162.

Wijffels, S.E., J.M. Toole, H.L. Bryden, R.A. Fine, W.J. Jenkins and J.L. Bullister, 1996: The water masses and circulation at 10N in the Pacific. *Deep-Sea Res.,* **43**, 501-544.

Wild, M., A. Ohmura, H. Gilgen, E. Roeckner, M. Giorgetta and J.-J. Morcrette, 1998: The disposition of radiative energy in the global climate system: GCM-calculated versus observational estimates. *Clim. Dyn.,* **14**, 853-869.

Willebrand, J., B. Barnier, C. Böning, C. Dieterich, P. Killworth, C. LeProvost, Y. Jia, J.-M. Molines and A.L. New, 2001: Circulation characteristics in three eddy-permitting models of the North Atlantic. *Prog. Oceanogr.,* **in press**,

Williamson, D.L. and P.J. Rasch, 1994: Water-vapor transport in the NCAR-CCM2. *Tellus,* **46A**, 34-51.

Winton, M., 1995: Why is the deep sinking narrow? *J. Phys. Oceanogr.,* **25**, 997-1005.

WMO, 1999: Scientific Assessment of Ozone Depletion: 1998, Global Ozone Research and Monitoring Project. Report *World Meteorological Organization,* Geneva, Chapter 5

Wood, E.F., D.P. Lettenmaier, X. Liang, D. Lohmann, A. Boone, S. Chang, F. Chen, Y. Dai, R.E. Dickinson, Q. Duan, M. Ek, Y.M. Gusev, F. Habets, P. Irannejad, R. Koster, K.E. Mitchel, O.N. Nasonova, J. Noilhan, J. Schaake, A. Schlosser, Y. Shao, A.B. Shmakin, D.

Verseghy, K. Warrach, P. Wetzel, Y. Xue, Z.-L. Yang and Q.-C. Zeng, 1998: The Project for Intercomparison of Land-surface Parameterization Schemes (PILPS) Phase 2(c) Red-Arkansas River basin experiment: 1. Experiment description and summary intercomparisons. *Glob. Planet. Change,* **19**, 115-135.

Woodwell, G.M., F.T. Mackenzie, R.A. Houghton, M. Apps, E. Gorham and E. Davidson, 1998: Biotic feedbacks in the warming of the Earth. *Clim. Change,* **40**, 495-518.

Wu, X. and M.W. Moncrieff, 1999: Effects of sea surface temperature and large-scale dynamics on the thermodynamic equilibrium state and convection over the tropical western Pacific. *J. Geophys. Res.,* **104**, 6093-6100.

Wunsch, C., 1999: The interpretation of short climate records, with comments on the North Atlantic and Southern Oscillations. *Bull. Am. Met. Soc.,* **80**, 245-256.

Xiao, X., J.M. Melillo, D.W. Kicklighter, A.D. McGuire, R.G. Prinn, C. Wang, P.H. Stone and A. Sokolov, 1998: Transient climate change and net ecosystem production of the terrestrial biosphere. *Glob. Biogeochem. Cyc.,* **12**, 345-360.

Xie, P.P. and P.A. Arkin, 1997: Global precipitation: A 17-year monthly analysis based on gauge observations, satellite estimates, and numerical model outputs. *Bull. Am. Met. Soc.,* **78**, 2539-2558.

Xu, K.M. and K.A. Emanuel, 1989: Is the tropical atmosphere conditionally unstable? *Mon. Wea. Rev.,* **117**, 1471-1479.

Xue, Y., 1997: Biosphere feedback on regional climate in tropical north Africa. *Quart. J. Roy. Met. Soc.,* **123**, 1483-1515.

Xue, Y. and J. Shukla, 1996: The influence of land surface properties on Sahel climate. Part II: Afforestation. *J. Clim.,* **9**, 3260-3275.

Yang, G. and J. Slingo, 1998: The seasonal mean and diurnal cycle of tropical convection as inferred from CLAUS data and the Unified Model. *UGAMP Tech. Rep.,* **47**,(unpublished)

Yang, H. and R.T. Pierrehumbert, 1994: Production of dry air by isentropic mixing. *J. Atm. Sci.,* **51**, 3437-3454.

Yang, Z.-L., R.E. Dickinson, A. Robock and K.Y. Vinnikov, 1997: Validation of the snow sub-model of the biosphere-atmosphere transfer scheme with Russian snow cover and meteorological observational data. *J. Clim.,* **10**, 353-373.

Yao, M.-S. and A.D. Del Genio, 1999: Effect of cloud parameterization on the simulation of climate changes in the GISS GCM. *J. Clim.,* **12**, 761-779.

Yoshimura, J., M. Sugi and A. Noda (eds.), 1999: *Influence of greenhouse warming on tropical cyclone frequency simulated by a high-resolution AGCM.,* 23rd Conference on Hurricanes and Tropical Cyclones, Dallas, TX, 10-15.

Zebiak, S.E. and M.A. Cane, 1991: Natural climate variability in a coupled model. In: *Greenhouse Gas-Induced Climatic Change: Critical Appraisal of Simulations and Observations* [Schlesinger, M. E. (ed.)]. Elsevier, 457-470.

Zender, C.S., B. Bush, S.K. Pope, A. Bucholtz, W.D. Collins, J.T. Kiehl, F.P.J. Valero and J. Vitko Jr., 1997: Atmospheric absorption during the Atmospheric Radiation Measurement (ARM) Enhanced Shortwave Experiment (ARESE). *J. Geophys. Res.,* **102**, 29901-29915.

Zeng, N., J.D. Neelin, W.K.-M. Lau and C.J. Tucker, 1999: Enhancement of interdecadal climate variability in the Sahel by vegetation interaction. *Science,* **286**, 1537-1539.

Zhai, P. and R.E. Eskridge, 1997: Atmospheric water vapor over China. *J. Clim.,* **10**, 2643-2652.

Zhang, G.J. and N.A. McFarlane, 1995: Sensitivity of climate simulations to the parameterization of cumulus convection in the Canadian Climate Centre general circulation model. *Atm.-Oce.,* **33**, 407-446.

Zhang, H., A. Henderson-Sellers, B. McAvaney and A. Pitman, 1997a: Uncertainties in GCM evaluations of tropical deforestation: A comparison of two model simulations. In: *Assessing Climate Change: Results from the Model Evaluation Consortium for Climate Assessment* [Howe, W. and A. Henderson-Sellers (eds.)]. Gordon and

Breach Science Publisher, Sydney, 418.

Zhang, H., A. Henderson-Sellers and K. McGuffie, 1996a: Impacts of tropical deforestation I: Process analysis of local climate change. *J. Clim.,* **9**, 1497-1517.

Zhang, H., K. McGuffie and A. Henderson-Sellers, 1996b: Impacts of tropical deforestation II: The role of large-scale dynamics. *J. Clim.,* **9**, 2498-2521.

Zhang, J. and W.D. Hibler, 1997: On an efficient numerical method for modeling sea ice dynamics. *J. Geophys. Res.,* **102**, 8691-8702.

Zhang, J., R.W. Schmitt and R.X. Huang, 1998: Sensitivity of the GFDL Modular Ocean Model to the parameterization of double-diffusive processes. *J. Phys. Oceanogr.,* **28**, 589-605.

Zhang, J., R.W. Schmitt and R.X. Huang, 1999a: The relative influence of diapycnal mixing and hydrologic forcing on the stability of the thermohaline circulation. *J. Phys. Oceanogr.,* **29**, 1096-1108.

Zhang, M.H., J.J. Hack, J.T. Kiehl and R.D. Cess, 1994: Diagnostic study of climate feedback processes in atmospheric general circulation models. *J. Geophys. Res.,* **99**, 5525-5537.

Zhang, S., R.J. Greatbatch and C.A. Lin, 1993: A reexamination of the polar halocline catastrophe and implications for coupled ocean-atmosphere modelling. *J. Phys. Oceanogr.,* **23**, 287-299.

Zhang, T., R.G. Barry, K. Knowles, J.A. Heginbottom and J. Brown, 1999b: Statistics and characteristics of permafrost and ground ice distribution in the Northern Hemisphere. *Polar Geogr.,* **23**, 147-169.

Zhang, Y., J.M. Wallace and D.S. Battisti, 1997b: ENSO-like interdecadal variability: 1900-93. *J. Clim.,* **10**, 1004-1020.

Zhou, J.Y., Y.C. Sud and K.M. Lau, 1996: Impact of orographically induced gravity-wave drag in the GLA GCM. *Quart. J. Roy. Met. Soc.,* **122**, 903-927.

Zwiers, F.W. and V.V. Kharin, 1998: Changes in the extremes of the climate simulated by CCC GCM2 under CO_2 doubling. *J. Clim.,* **11**, 2200-2222.

8

Model Evaluation

Co-ordinating Lead Author
B.J. McAvaney

Lead Authors
C. Covey, S. Joussaume, V. Kattsov, A. Kitoh, W. Ogana, A.J. Pitman, A.J. Weaver, R.A. Wood, Z.-C. Zhao

Contributing Authors
K. AchutaRao, A. Arking, A. Barnston, R. Betts, C. Bitz, G. Boer, P. Braconnot, A. Broccoli, F. Bryan,
M. Claussen, R. Colman, P. Delecluse, A. Del Genio, K. Dixon, P. Duffy, L. Dümenil, M. England,
T. Fichefet, G. Flato, J.C. Fyfe, N. Gedney, P. Gent, C. Genthon, J. Gregory, E. Guilyardi, S. Harrison,
N. Hasegawa, G. Holland, M. Holland, Y. Jia, P.D. Jones, M. Kageyama, D. Keith, K. Kodera, J. Kutzbach,
S. Lambert, S. Legutke, G. Madec, S. Maeda, M.E. Mann, G. Meehl, I. Mokhov, T. Motoi, T. Phillips,
J. Polcher, G.L. Potter, V. Pope, C. Prentice, G. Roff, F. Semazzi, P. Sellers, D.J. Stensrud, T. Stockdale,
R. Stouffer, K.E. Taylor, K. Trenberth, R. Tol, J. Walsh, M. Wild, D. Williamson, S.-P. Xie, X.-H. Zhang,
F. Zwiers

Review Editors
Y. Qian, J. Stone

Contents

Executive Summary

This chapter evaluates the suitability of models (in particular coupled atmosphere-ocean general circulation models) for use in climate change projection and in detection and attribution studies. We concentrate on the variables and time-scales that are important for this task. Models are evaluated against observations and differences between models are explored using information from a number of systematic model intercomparisons. Even if a model is assessed as performing credibly when simulating the present climate, this does not necessarily guarantee that the response to a perturbation remains credible. Therefore, we also assess the performance of the models in simulating the climate over the 20th century and for selected palaeoclimates. Incremental improvements in the performance of coupled models have occurred since the IPCC WGI Second Assessment Report (IPCC, 1996) (hereafter SAR) resulting from advances in the modelling of the atmosphere, ocean, sea ice and land surface as well as improvements in the coupling of these components.

Highlights include:

• Coupled models can provide credible simulations of both the present annual mean climate and the climatological seasonal cycle over broad continental scales for most variables of interest for climate change. Clouds and humidity remain sources of significant uncertainty but there have been incremental improvements in simulations of these quantities.

• Confidence in model projections is increased by the improved performance of several models that do not use flux adjustment. These models now maintain stable, multi-century simulations of surface climate that are considered to be of sufficient quality to allow their use for climate change projections.

• There is no systematic difference between flux adjusted and non-flux adjusted models in the simulation of internal climate variability. This supports the use of both types of model in detection and attribution of climate change.

• Confidence in the ability of models to project future climates is increased by the ability of several models to reproduce the warming trend in 20th century surface air temperature when driven by radiative forcing due to increasing greenhouse gases and sulphate aerosols. However, only idealised scenarios of only sulphate aerosols have been used.

• Some modelling studies suggest that inclusion of additional forcings such as solar variability and volcanic aerosols may improve some aspects of the simulated climate variability of the 20th century.

• Confidence in simulating future climates has been enhanced following a systematic evaluation of models under a limited number of past climates.

• The performance of coupled models in simulating the El Niño-Southern Oscillation (ENSO) has improved; however, the region of maximum sea surface temperature variability associated with El Niño events is displaced westward and its strength is generally underestimated. When suitably initialised with an ocean data assimilation system, some coupled models have had a degree of success in predicting El Niño events.

• Other phenomena previously not well simulated in coupled models are now handled reasonably well, including monsoons and the North Atlantic Oscillation.

• Some palaeoclimate modelling studies, and some land-surface experiments (including deforestation, desertification and land cover change), have revealed the importance of vegetation feedbacks at sub-continental scales. Whether or not vegetation changes are important for future climate projections should be investigated.

• Analysis of, and confidence in, extreme events simulated within climate models is emerging, particularly for storm tracks and storm frequency. "Tropical cyclone-like" vortices are being simulated in climate models, although enough uncertainty remains over their interpretation to warrant caution in projections of tropical cyclone changes.

Final Assessment

Coupled models have evolved and improved significantly since the SAR. In general, they provide credible simulations of climate, at least down to sub-continental scales and over temporal scales from seasonal to decadal. The varying sets of strengths and weaknesses that models display lead us to conclude that no single model can be considered "best" and it is important to utilise results from a range of coupled models. We consider coupled models, as a class, to be suitable tools to provide useful projections of future climates.

8.1 Summary of Second Assessment Report

The systematic evaluation of coupled climate models was only beginning to emerge at the time of the IPCC WGI Second Assessment Report (IPCC, 1996) (hereafter SAR). Suitable formalisms for evaluating fully coupled models were in very early stages of development whereas considerable progress had been made in the evaluation of the performance of individual components (atmosphere, ocean, land surface and sea ice and their interactions).

The need for flux adjustment and the widely varying spin-up methodologies in coupled models were areas of concern and the need for a more systematic evaluation of these on the simulated climate was expressed. It was noted that, although flux adjustments were generally large in those models that used them, the absence of flux adjustment generally affected the realism of the simulated climate and could adversely affect the associated feedback processes. It was hoped that the need for flux adjustment would diminish as components were improved.

A new feature of coupled model evaluation was an analysis of the variability of the coupled system over a range of time-scales. The new opportunities that provided for more comprehensive evaluation were an important component of the overall assessment of coupled model capabilities.

Evaluation of the performance of individual components of the coupled system (especially for atmosphere-only models) was much more advanced than previously. Results from the first phase of the Atmospheric Model Intercomparison Project (AMIP) demonstrated that *"current atmospheric models generally provide a realistic portrayal of the phase and amplitude of the seasonal march of the large-scale distribution of temperature, pressure and circulation"*. The simulation of clouds and their seasonal variation was noted as the major source of uncertainty in atmospheric models. In general it was found that *"atmospheric models respond realistically to large-scale sea surface temperature (SST) patterns"* and hence can reproduce many facets of interannual variability. In the case of the land-surface component, however, it was noted that *"the general agreement found among the results of relatively simple land-surface schemes in 1990 has been reduced by the introduction of more complex parametrizations"*. Ocean and sea-ice models were found to *"portray the observed large-scale distribution of temperature, salinity and sea ice more accurately than in 1990"*, but some reservations were expressed regarding the possible adverse effects of the relatively coarse resolution of the ocean components of current coupled models.

The overall assessment of coupled models was that *"current models are now able to simulate many aspects of the observed climate with a useful level of skill"* and *"model simulations are most accurate at large space scales (e.g., hemispheric or continental); at regional scales skill is lower"*.

8.2 What is Meant by Evaluation?

8.2.1 The Approach: Mean State and Variability in Climate Models

In this chapter, we (as the authors of this chapter) have attempted a two-pronged approach to evaluation. As is traditional, we discuss how well models simulate the mean seasonal climate for a number of variables (i.e., the average for a given season taken over many simulated years). Since the characterisation of a climate state includes its variability, we also describe simulated climate variability over a range of time-scales. In addition, we discuss aspects of the variability in the behaviour of specific phenomena. Evaluation of the performance of global models in specific geographical regions is the subject of Chapter 10.

We use a wide range of "observations" in order to evaluate models. However, often the most useful source for a particular variable is a product of one of the reanalysis projects (most commonly that of the National Centers for Environmental Prediction (NCEP) (Kalnay *et al.*, 1996) or from the European Centre for Medium-Range Weather Forecasts (ECMWF) (Gibson *et al.*, 1997)). Although products from a data assimilation system are not direct "observations" (they are the outcome of a combination of observed and model data), the global grided nature and high time resolution of these products makes them extremely useful when their accuracy is not in question. Some additional useful products from reanalysis are not, in fact, the result of a direct combination of observed and model data but are in fact the outcome of model integration and hence must be used with caution. It is important to note that the various variables available are not all of the same quality and, especially for data-sparse regions, implicitly contain contributions from the errors in the underlying model (see also Chapter 2). The overall quality of reanalysis products is continually assessed at regular International Reanalysis Workshops.

8.2.2 The Basis

Recent discussions by Randall and Wielicki (1997), Shackley *et al.* (1998 and 1999), Henderson-Sellers and McGuffie (1999) and Petersen (2000) illustrate many of the confusions and uncertainties that accompany attempts to evaluate climate models especially when such models become very complex. We recognise that, unlike the classic concept of Popper (1982), our evaluation process is not as clear-cut as a simple search for "falsification". While we do not consider that the complexity of a climate model makes it impossible to ever prove such a model "false" in any absolute sense, it does make the task of evaluation extremely difficult and leaves room for a subjective component in any assessment. The very complexity of climate models means that there are severe limits placed on our ability to analyse and understand the model processes, interactions and uncertainties (Rind, 1999). It is always possible to find errors in simulations of particular variables or processes in a climate model. What is important to establish is whether such errors make a given model "unusable" in answering specific questions.

Two fundamentally different ways are followed to evaluate models. In the first, the important issues are the degree to which a model is physically based and the degree of realism with which essential physical and dynamical processes and their interactions have been modelled. This first type of evaluation is undertaken in Chapter 7. (We discuss the related aspects of the numerical formulation and numerical resolution in Section 8.9.) In the second, there are attempts to quantify model errors, to consider the causes for those errors (where possible) and attempts to understand the nature of interactions within the model. We fully recognise that many of the evaluation statements we make contain a degree of subjective scientific perception and may contain much "community" or "personal" knowledge (Polanyi, 1958). For example, the very choice of model variables and model processes that are investigated are often based upon the subjective judgement and experience of the modelling community.

The aim of our evaluation process is to assess the ability of climate models to simulate the climate of the present and the past. Wherever possible we will be concentrating on coupled models, however, where necessary we will examine the individual model components. This assessment then acts as a guide to the capabilities of models used for projections of future climate.

8.2.3 Figures of Merit

There have been many attempts to obtain a "figure of merit" for climate models. Usually such quantification is only attempted for well-observed atmospheric variables and range from calculation of simple root mean square errors (r.m.s.) between a model variable and an observation, to more complex multi-variate calculations. Among the most promising attempts at generating skill scores deemed more suitable for climate models are: the normalised mean square error approach of Williamson (1995) that follows on, in part, from Murphy (1988); and the categorisation of models in terms of combination of the error in the time mean and the error in temporal variability along the lines suggested by Wigley and Santer (1990) (see Chapter 5, Section 5.3.1.1. of the SAR for an example). Other less widely used non-dimensional measures have also been devised (e.g., Watterson, 1996). Although a number of skill scoring methods have been devised and used for the seasonal prediction problem (e.g., Potts *et al.*, 1996; linear error in probability space score – LEPS) these have not found general application in climate models. Attempts to derive measures of the goodness of fit between model results and data containing large uncertainties have been partially successful in the oceanographic community for a limited number of variables (Frankignoul *et al.*, 1989; Braconnot and Frankignoul, 1993). Fuzzy logic techniques have been trialled by the palaeoclimatology community (Guiot *et al.*, 1999). It is important to remember that the types of error measurement that have been discussed are restricted to relatively few variables. It has proved elusive to derive a fully comprehensive multi-dimensional "figure of merit" for climate models.

Since the SAR, Taylor (2000) has devised a very useful diagrammatic form (termed a "Taylor diagram" – see Section 8.5.1.2 for description) for conveying information about the pattern similarity between a model and observations. This same type of diagram can be used to illustrate the relative accuracy amongst a number of model variables or different observational data sets (see Section 8.5.1). One additional advantage of the "Taylor diagram" is that there is no restriction placed on the time or space domain considered.

While at times we use a figure of merit to intercompare models for some selected variables, we usually apply more subjective assessments in our overall evaluations; we do not believe it is objectively possible to state which model is "best overall" for climate projection, since models differ amongst themselves (and with available observations) in many different ways. Even if a model is assessed as performing credibly when simulating the present climate, we cannot be sure that the response of such a model to a perturbation remains credible. Hence we also rely on evaluating models in their performance with individual processes (see Chapter 7) as well as past climates as in Section 8.5.5.

8.3 Model Hierarchy

8.3.1 Why is a Hierarchy of Models Important?

The impact of anthropogenic perturbation on the climate system can be projected by calculating all the key processes operating in the climate system through a mathematical formulation which, due to its complexity, can only be implemented in a computer program, referred to as a climate model. If all our current understanding of the climate system were explicitly included, the model would be too complex to run on any existing computer; hence, for practical purposes, simplifications are made so that the system has reduced complexity and computing requirements. Since different levels of simplifications are possible, a hierarchy of models tends to develop (see Chapter 1 and Harvey *et al.*, 1997).

The need to balance scientific understanding against computational efficiency and model realism often guides the choice of the particular class of models used. In addition, it is usually necessary to balance the relative level of detail in the representation, and the level of parametrization, within each component of the climate system.

8.3.2 Three-dimensional Climate Models

The most complex climate models, termed coupled atmosphere-ocean general circulation models (and abbreviated as AOGCM in this report), involve coupling comprehensive three-dimensional atmospheric general circulation models (AGCMs), with ocean general circulation models (OGCMs), with sea-ice models, and with models of land-surface processes, all of which are extensively reviewed in the SAR (Chapters 4 and 5). For AOGCMs, information about the state of the atmosphere and the ocean adjacent to, or at the sea surface, is used to compute exchanges of heat, moisture and momentum between the two components. Computational limitations mean that the majority of sub-grid scale processes are parametrized (see Randall and Wielicki, 1997 and Chapter 7). Occasionally atmospheric models with simple mixed-layer ocean models (much discussed and utilised in the SAR) are still used.

8.3.3 Simple Climate Models

Simplifications can be made so that the climate model has reduced complexity (e.g., a reduction in dimensionality to two or even zero). Simple models allow one to explore the potential sensitivity of the climate to a particular process over a wide range of parameters. For example, Wigley (1998) used a modified version of the Wigley and Raper (1987, 1992) upwelling diffusion-energy climate model (see Kattenberg *et al.*, 1996; Raper *et al.*, 1996) to evaluate Kyoto Protocol implications for increases in global mean temperatures and sea level. While such a simple climate model relies on climate sensitivity and ocean heat uptake parameters based on coupled atmosphere-ocean models and ice-melt parameters based upon more complex ice sheet and glacier models, it nevertheless allows for a first-order analysis of various post-Kyoto emission reductions. Simple climate models are also used within larger integrated assessment models to analyse the costs of emission reduction (Peck and Teisberg, 1996; Manne and Richels, 1999) and impacts of climate change (Nordhaus, 1994; Tol, 1999).

8.3.4 Earth System Models of Intermediate Complexity

Recently, significant advances have occurred in the development of Earth System Models of Intermediate Complexity (EMIC), which are designed to bridge the gap between the three-dimensional comprehensive models and simple models. The main characteristic of EMICs is that they describe most of the processes implicit in comprehensive models, albeit in a more reduced (i.e., more parametrized) form. They also explicitly simulate the interactions among several components of the climate system including biogeochemical cycles. On the other hand, EMICs are computationally efficient enough to allow for long-term climate simulations over several tens of thousands of years or a broad range of sensitivity experiments over several millennia. As for AOGCMs, but in contrast to simple models, the number of degrees of freedom of an EMIC exceeds the number of adjustable parameters by several orders of magnitude. Currently, there are several EMICs in operation such as: two-dimensional, zonally averaged ocean models coupled to a simple atmospheric module (e.g., Stocker *et al.*, 1992; Marchal *et al.*, 1998) or geostrophic two-dimensional (e.g., Gallee *et al.*, 1991) or statistical-dynamical (e.g., Petoukhov *et al.*, 2000) atmospheric modules; three-dimensional models with a statistical-dynamical atmospheric and oceanic modules (Petoukhov *et al.*, 1998; Handorf *et al.*, 1999); reduced-form comprehensive models (e.g., Opsteegh *et al.*, 1998) and those that involve an energy-moisture balance model coupled to an OGCM and a sea-ice model (e.g., Fanning and Weaver, 1996). Some EMICs have been used to investigate both the climate of the last glacial maximum (see Section 8.5) as well as to investigate the cause of the collapse of the conveyor in global warming experiments (Stocker and Schmittner, 1997; Rahmstorf and Ganopolski, 1999) while others have been used to undertake a number of sensitivity studies including the role of sub-grid scale ocean mixing in global warming experiments (Wiebe and Weaver, 1999).

EMIC development involves the same evaluation procedure as AOGCMs use, albeit restricted due to the reduced complexity of some, or all, of the constituent sub-components. While EMIC evaluation is in its early stages, the nature of these models allows for a detailed comparison with both historical and proxy observa-

tional data. Initial analyses (referenced above) suggest that EMICs hold promise as exploratory tools to understand important processes, and their interactions and feedbacks within the climate system. However, they are not useful for assessing regional aspects of climate change.

8.4 Coupled Climate Models – Some Methodologies

8.4.1 Model Initialisation

In this chapter, we assess climate models on the basis of their ability to simulate present and past climates. What it means to simulate a particular climate state is linked to the question of model initialisation. Ideally, given a "perfect model", and "perfect knowledge" of the present climate state, one could simply initialise a climate model with the present state. Then, given perpetual present day forcing (from trace substances and solar radiation), one might expect the model to remain close to the present state, perhaps with some level of variability. However, in practice, this ideal is not achieved and a model initialised in this way adjusts from the initial state. This adjustment has been characterised by two time-scales (Bryan, 1998): in the initial "fast" adjustment, the atmosphere, land surface, ocean mixed layer and sea ice reach a state of near equilibrium, typically taking 5 to 50 years. Surface temperature error patterns (e.g., Figure 8.1a) are generally established on this time-scale, and persist over many centuries of integration, but there typically remain slight imbalances in the surface heat and fresh water fluxes. These imbalances drive a second, slower, adjustment phase (often called "climate drift") which takes place over centuries to millennia and involves adjustment of the deep ocean to the surface imbalances. This "climate drift" can make interpretation of transient climate change simulations difficult, so models are generally allowed to adjust to a state where such drifts have become acceptably slow, before starting climate change simulations. A number of techniques have been used to achieve this (see Stouffer and Dixon, 1998), but it is not possible, in general, to say which of these procedures gives "better" initial conditions for a climate change projection run.

A number of model runs have now been made which are forced by the historical record of natural and anthropogenic forcing from the mid-19th century to the present. As well as avoiding the "cold start" problem in climate projections (SAR Chapter 6, Section 6.2.4 and Chapter 9), these runs can be compared against historical observations and form a valuable model evaluation tool (Section 8.6.1).

8.4.2 Flux Adjustment and Energy Transports

Given present day greenhouse gas concentrations, most coupled models at the time of the SAR had difficulty in obtaining a stable climate near to the present day state. Therefore "flux adjustment" terms were often added to the surface fluxes of heat, water and (sometimes) momentum which were passed from the atmosphere to the ocean model. Flux adjustments are non-physical in that they cannot be related to any physical process in the climate system and do not *a priori* conserve heat and water across the atmosphere-ocean interface. The flux adjustments were specifically chosen to give a stable and realistic simulation of present surface climate (especially the sea surface temperature and sea-ice cover), and were often as large as the annual mean model fluxes themselves. The need to use such adjustments was clearly a source of uncertainty: the approach inherently disguises sources of systematic error in the models, and may distort their sensitivity to changed radiative forcing. Models which did not use flux adjustment produced unrealistic simulations of fundamental aspects of the climate system such as the strength of the North Atlantic thermohaline circulation (SAR Chapter 5, Table 5.5).

Recently a number of coupled models have emerged with greatly improved surface climatologies without using flux adjustments. Figure 8.1a shows SST errors from one such model, about 100 years after initialisation. Errors are generally less than 2°C, and the error pattern shown is stable over several centuries of integration. However some larger errors are seen, in particular a cooling in the North Pacific, a warming in the eastern tropical ocean basins (probably due to a lack of stratocumulus cloud there), and a warming in the southern ocean. These errors appear to be common to a number of more recent non-flux adjusted models (Action de Recherche Petite Echelle Grand Echelle/Océan Parallélisé (ARPEGE/OPA1), Guilyardi and Madec 1997; Climate System Model (CSM 1.0), Boville and Gent 1998; Hadley Centre Coupled Model (HadCM3), Gordon *et al.*, 2000), but other models show different error patterns (ARPEGE/OPA2, Institut Pierre Simon Laplace/Coupled Atmosphere-Ocean-Vegetation Model (IPSL-CM2), Barthelet *et al.*, 1998a,b). The surface climatologies in several non-flux adjusted models are now considered good enough and stable enough to use those models for climate change projections. Typical flux adjusted models do show smaller SST errors, because the flux adjustments are chosen specifically to minimise those errors (Figure 8.1b). For comparison, Figure 8.1c shows the SST errors when the older, flux adjusted model was run in non-flux adjusted mode.

It appears that the success of the recent models which do not require heat flux adjustments is related to an improved ability to simulate the large-scale heat balances described in Chapter 7 (Weaver and Hughes, 1996; Guilyardi and Madec, 1997; Johns *et al.*, 1997; Bryan, 1998; Gordon *et al.*, 2000). Improvements to both atmospheric (Section 8.5.1.2.2) and oceanic (Section 8.5.2.2) components of the models have played a part in this advance. Such models have the advantage over flux-adjusted models that, provided the large-scale balances are obtained using a physically justifiable choice of model parameters, these models are physically self-consistent representations of the climate system. However, in some cases only very loose physical constraints can be placed on the model parameters.

The fresh water budget is more complex than the heat budget because of the effects of land surface processes, rivers and sea ice. Water budget errors are potentially far reaching because there is no direct feedback between surface salinity errors and the surface fresh water flux, for example, persistent freshening at high latitudes could lead to a collapse of the ocean thermohaline circulation (Manabe and Stouffer, 1997; see also Chapter 7, Section 7.6.2). Some aspects of the large-scale hydrological cycle

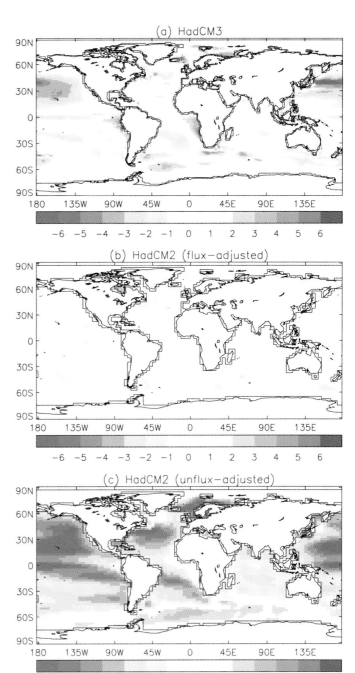

Figure 8.1: Decadal mean SST errors relative to the GISST climatology for (a) the non-flux adjusted model HadCM3 (Gordon *et al.*, 2000), (b) the previous generation, flux adjusted model HadCM2 (Johns *et al.* 1997), (c) the HadCM2 model when run without flux adjustments (Gregory and Mitchell, 1997). The figures are from representative periods after at least 100 years of each control run. Multi-century drifts in each run are much smaller than the differences between the runs. The errors are smallest in (b), because the flux adjustments were chosen specifically to minimise the errors. The errors in (c) result from a number of complex feedbacks. The model was designed to work in flux adjusted mode and it is possible that the non-flux adjusted SST errors could have been reduced by relatively minor "tuning" of the model.

Table 8.1: *Model control runs: a consolidated list of coupled AOGCMs that are assessed in Chapter 8 and used in other Chapters. The naming convention for the models is as agreed by all modelling groups involved. Under the heading CMIP: 1,2 indicate that the model control run is included in the Coupled Model Intercomparison Phase 1 and 2 (CMIP1 and 2) databases, respectively.*

	MODEL NAME	CENTRE	REFERENCE	CMIP	Ch 9	Ch 11	Ch 12	ATMOSPHERIC RESOLUTION	OCEAN RESOLUTION	LAND SURFACE	SEA ICE	FLUX ADJUST
1	ARPEGE/OPA1	CERFACS	Guilyardi and Madec, 1997	1	– – –			T21 (5.6 × 5.6) L30	2.0×2.0 L31*	C	(d)	–
2	ARPEGE/OPA2	CERFACS	Barthelet *et al.*, 1998a,b	2	C – –			T31(3.9 × 3.9) L19	2.0×2.0 L31*	C	T	–
3	BMRCa	BMRC	Power *et al.*, 1993	1	C – –			R21 (3.2 × 5.6) L9	3.2×5.6 L12	M,B	T	–
4	BMRCb	BMRC	Power *et al.*, 1998	2	– – –			R21 (3.2 × 5.6) L17	3.2×5.6 L12*	M,B	T	H,W
5	CCSR/NIES	CCSR/NIES	Emori *et al.*, 1999	1,2	C – D			T21 (5.6 × 5.6) L20	2.8×2.8 L17	M,BB	T	H,W
6	CGCM1	CCCma	Boer *et al.*, 2000; Flato *et al.*, 2000	1,2	C – D	*	*	T32 (3.8 × 3.8) L10	1.8×1.8 L29	M,BB	T	H,W
7	CGCM2	CCCma	Flato and Boer, 2001	–	– S –	*	*	T32 (3.8 × 3.8) L10	1.8×1.8 L29	M,BB	T,R	H,W
8	COLA1	COLA	Schneider *et al.*, 1997; Schneider and Zhu, 1998	1	– – –			R15 (4.5 × 7.5) L9	1.5×1.5 L20*	C	T	–
9	COLA2	COLA	Dewitt and Schneider, 1999	1	– – –			T30 (4 × 4) L18	3.0×3.0 L20*	C	T	–
10	CSIRO Mk2	CSIRO	Gordon and O'Farrell, 1997	1,2	C – D			R21 (3.2 x 5.6) L9	3.2×5.6 L21	C	T,R	H,W,M
11	CSM 1.0	NCAR	Boville and Gent, 1998	1,2	C – –			T42 (2.8 × 2.8) L18	2.0×2.4 L45*	C	T,R	–
12	CSM 1.3	NCAR	Boville *et al.*, 2001	–	– S D			T42 (2.8 × 2.8) L18	2.0×2.4 L45*	C	T,R	–
13	ECHAM1/LSG	DKRZ	Cubasch *et al.*, 1992; von Storch, 1994; von Storch *et al.*, 1997	1	– – –		*	T21 (5.6 × 5.6) L19	4.0×4.0 L11	C	T	H,W,M
14	ECHAM3/LSG	DKRZ	Cubasch et al 1997; Voss *et al.*, 1998	1,2	C – D		*	T21 (5.6 × 5.6) L19	4.0×4.0 L11	C	T	H,W,M
15	ECHAM4/OPYC3	DKRZ	Roeckner *et al.*, 1996	1	C – D	*	*	T42 (2.8 × 2.8) L19	2.8×2.8 L11*	C	T,R	H,W(*)
16	GFDL_R15_a	GFDL	Manabe *et al.*, 1991; Manabe and Stouffer1996	1,2	C – D		*	R15 (4.5 × 7.5) L9	4.5×3.7 L12	B	T,F	H,W
17	GFDL_R15_b	GFDL	Dixon and Lanzante, 1999	–	C – –	*		R15 (4.5 × 7.5) L9	4.5×3.7 L12	B	T,F	H,W
18	GFDL_R30_c	GFDL	Knutson *et al.*, 1999	–	C S –	*	*	R30 (2.25 × 3.75)L14	1.875×2.25 L18	B	T,F	H,W
19	GISS1	GISS	Miller and Jiang, 1996	1	– – –			4.0 × 5.0 L9	4.0×5.0 L16	C	T	–
20	GISS2	GISS	Russell *et al.*, 1995	1,2	C – –			4.0 × 5.0 L9	4.0×5.0 L13	C	T	–
21	GOALS	IAP/LASG	Wu *et al.*, 1997; Zhang *et al.*, 2000	1,2	C – –			R15 (4.5 × 7.5) L9	4.0×5.0 L20	C	T	H,W,M
22	HadCM2	UKMO	Johns 1996; Johns *et al.*, 1997	1,2	C – D		*	2.5 × 3.75 L19	2.5×3.75 L20	C	T,F	H,W
23	HadCM3	UKMO	Gordon *et al.*, 2000	2	C S D		*	2.5 × 3.75 L19	1.25 × 1.25 L20	C	T,F	–
24	IPSL-CM1	IPSL/LMD	Braconnot *et al.*, 2000	1	– – –			5.6 × 3.8 L15	2.0×2.0 L31*	C	(d)	–
25	IPSL-CM2	IPSL/LMD	Laurent *et al.*, 1998;	2	C – –			5.6 × 3.8 L15	2.0×2.0 L31*	C	T	–
26	MRI1[a]	MRI	Tokioka *et al.*, 1996	1,(2)[a]	C – –			4.0 × 5.0 L15	2.0×2.5 L21(23)[a]*	M,B	T,F	H,W
27	MRI2	MRI	Yukimoto *et al.*, 2000	–	C S –	*		T42(2.8 × 2.8) L30	2.0×2.5 L23*	C	T,F	H,W,M
28	NCAR1	NCAR	Meehl and Washington, 1995; Washington and Meehl, 1996	1,2	– – –			R15 (4.5 × 7.5) L9	1.0×1.0 L20	B	T,R	–
29	NRL	NRL	Hogan and Li, 1997; Li and Hogan, 1999	1,2	– – –			T47 (2.5 × 2.5) L18	1.0 × 2.0 L25*	BB	T(p)	H,W(*)
30	DOE PCM	NCAR	Washington *et al.*, 2000	2	C S D			T42 (2.8 × 2.8) L18	0.67 × 0.67 L32	C	T,R	–
31	CCSR/NIES2	CCSR/NIES	Nozawa *et al.*, 2000	–	C S –			T21 (5.6 × 5.6) L20	2.8 × 3.8 L17	M,BB	T	H,W
I1	BERN2D	PIUB	Stocker *et al.*, 1992; Schmittner & Stocker, 1999	–	– – –	*		10* × ZA L1	10* × ZA L15	–	T	–
I2	UVIC	UVIC	Fanning and Weaver, 1996; Weaver *et al.*, 1998	–	– – –	*		1.8 × 3.6 L1	1.8 × 3.6 L19	–	T,R	–
I3	CLIMBER	PIK	Petoukhov *et al.*, 2000	–	– – –	*		10 × 51 L2	10 × ZA L11	C	T,F	–

[a] Model MRI1 exists in two versions. At the time of writing, more complete assessment data was available for the earlier version, whose control run is in the CMIP1 database. This model is used in Chapter 8. The model used in Chapter 9 has two extra ocean levels and a modified ocean mixing scheme. Its control run is in the CMIP2 database. The equilibrium climate sensitivities and Transient Climate Responses (Chapter 9, Table 9.1) of the two models are the same.

CMIP: 1,2 indicate that the model control run is included in the CMIP1 and CMIP2 databases, respectively.

Ch 9: C indicates that a run or runs with the CMIP2 1% p.a. CO_2 increase scenario is used in Chapter 9 (irrespective of whether the data is included in the CMIP database). S indicates that SRES scenario runs (including at least A2 and B2) are used in Chapter 9. D indicates that model output is lodged at the IPCC Data Distribution Centre.

Ch 11, Ch 12: An asterisk indicates that the model has been used to make sea level projections (Chapter 11) or in detection/attribution studies (Chapter 12).

Atmospheric resolution: Horizontal and vertical resolution. The former is expressed either as degrees latitude × longitude or as a spectral truncation with a rough translation to degrees latitude × longitude. An asterisk indicates enhanced meridional resolution in midlatitudes. ZA indicates a zonally averaged model (360° zonal resolution). Vertical resolution is expressed as "Lmm", where mm is the number of vertical levels.

Ocean resolution: Horizontal and vertical resolution. The former is expressed as degrees latitude × longitude, while the latter is expressed as "Lmm", where mm is the number of vertical levels. An asterisk indicates enhanced horizontal resolution near the Equator. ZA indicates a zonally averaged model for each ocean basin. The following classification of ocean horizontal resolution is used throughout Chapters 7 to 9: Coarse: >2°, Medium: 2/3° to 2°, Eddy-permitting: 1/6° to 2/3°, Eddy-resolving: <1/6°.

Land surface scheme: B = standard bucket hydrology scheme; BB = modified bucket scheme with spatially varying soil moisture capacity and/or a surface resistance; M = multi-layer temperature scheme; C = a complex land surface scheme usually including multi-soil layers for temperature and soil moisture, and an explicit representation of canopy processes.

Sea ice model: T = thermodynamic ice model only; F = 'free drift' dynamics; R = ice rheology included; (d) = ice extent/thickness determined diagnostically from ocean surface temperature; (p) = ice extent prescribed.

Flux adjustment: H = heat flux; W = fresh water flux; M = momentum flux. An asterisk indicates annual mean flux adjustment only.

are subject to large observational uncertainty (Wijffels *et al.*, 1992), and this has inhibited evaluation and improvements in the water budget. Nonetheless, some models are now able to produce stable multi-century runs without water flux adjustments

8.4.2.1 Does the use of flux adjustments in a model have a significant impact on climate change projections?

Marotzke and Stone (1995) show that using flux adjustment to correct surface errors in the control climate does not necessarily correct errors in processes which control the climate change response. Flux adjustments can also result in spurious multiple equilibrium states of the tropical (Neelin and Dijkstra, 1995) and thermohaline (Dijkstra and Neelin, 1999) ocean circulation. On the other hand, a good representation of, say, sea-ice extent may be important to produce the correct magnitude of ice-albedo feedback under climate change, and it may be preferable to use flux adjustments to give a good sea-ice distribution than to omit the flux adjustments but to have a poorer sea-ice extent. Overall, differences have been seen in the climate change response of flux adjusted and equivalent non-flux adjusted models (Fanning and Weaver, 1997b; Gregory and Mitchell, 1997), but it is not clear whether the differences are due to the flux adjustment itself, or to the systematic errors in the non-flux adjusted model. The only practical way to resolve this issue may be to continue the progress which has been made towards models which achieve good surface climatology without flux adjustment, whereupon the effect of flux adjustments will cease to be of concern.

8.5 Coupled Climate Models – Means

8.5.1 Atmospheric Component

8.5.1.1 Development since the SAR

The model evaluation chapter of the IPCC Second Assessment Report (Gates *et al.*, 1996) found that "large-scale features of the current climate are well simulated on average by current coupled models." However, two major points of concern were noted.

Firstly, the SAR found that simulation of clouds and related processes "remains a major source of uncertainty in atmospheric models". As discussed in Chapter 7, these processes continue to account for most of the uncertainty in predicting human-induced climate change. Secondly, the SAR noted an unsatisfactory situation involving flux adjustments (Section 8.4.2): they "are relatively large in the models that use them, but their absence affects the realism of the control climate and the associated feedback processes". Improvements in coupled climate models since the SAR have addressed both points of concern. For the atmospheric (as well as the oceanic) component, these improvements have included higher horizontal resolution (which means less numerical diffusion and better representation of topography), and advances in parametrizations. In addition, the advent of the Coupled Model Intercomparison Program (CMIP; see Meehl *et al.*, 2000a) since the SAR has provided an additional database for evaluating AOGCMs. Some basic details of models evaluated in this chapter and used elsewhere in this report are presented in Table 8.1.

8.5.1.2 Tropospheric climate

8.5.1.2.1 Surface quantities

The SAR's evaluation of coupled-model simulations focused on surface air temperature, sea level pressure and precipitation. The SAR concluded that model simulations of surface air temperature were "very similar" to observations. Simulations of the other two quantities were found to be less accurate but nevertheless reasonable: the SAR concluded that coupled models represented "the observed large-scale geographical distribution" of sea level pressure "rather well", and that they were "generally successful in simulating the broad-scale structure of the observed precipitation".

Figures 8.2 and 8.3 (reproduced from Lambert and Boer, 2001) update this assessment using coupled model output from the CMIP1 database. For each quantity, the figures show both a map of the average over all fifteen models ("model mean") and

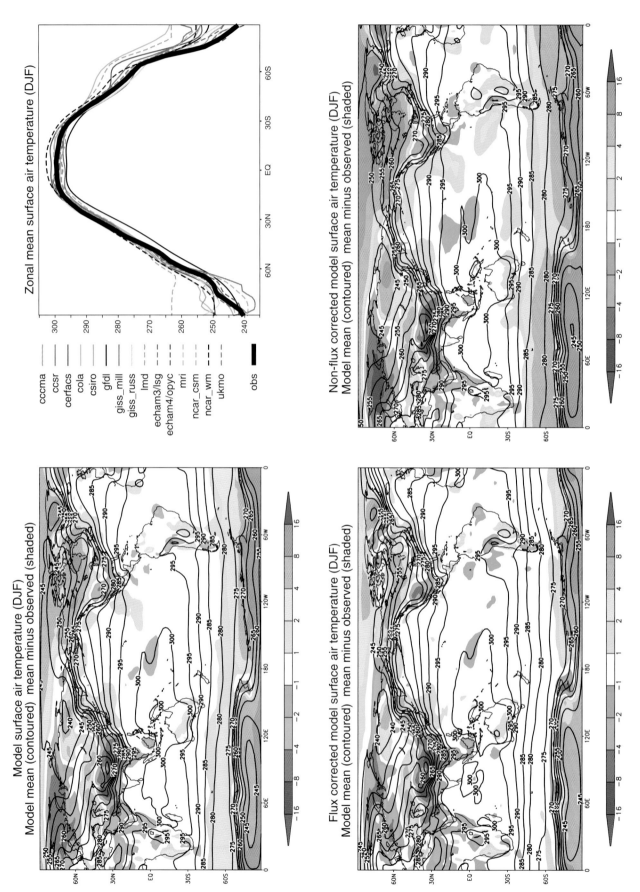

Figure 8.2: December-January-February climatological surface air temperature in K simulated by the CMIP1 model control runs. Averages over all models (upper left), over all flux adjusted models (lower left) and over all non-flux adjusted models (lower right) are shown together with zonal mean values for individual models (upper right). Observed value is shown in the zonal mean plot (thick solid line), and the difference between "average model" and observation is shown on the longitude-latitude maps. From Lambert and Boer (2001).

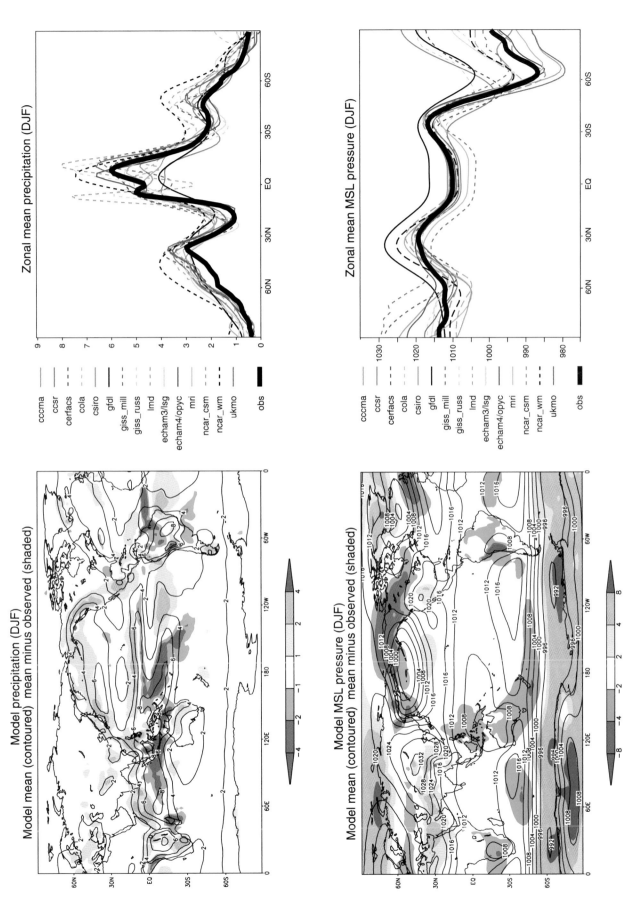

Figure 8.3: December-January-February climatological precipitation in mm per day (top) and mean sea level pressure in hPa (bottom) simulated by CMIP1 model control runs. Averages over all models are shown at left and zonal mean values for individual models are shown at right. Observed values are shown on the zonal mean plot (thick solid line) and the difference between an "average model" and the observations is shown by shading on the longitude-latitude maps. From Lambert and Boer (2001).

zonal means for all individual fifteen models. Lambert and Boer (2001) demonstrate that the model mean exhibits good agreement with observations, often better than any of the individual models. Inspection of these portions of the figures reaffirms the SAR conclusions summarised above. The errors in model-mean surface air temperature rarely exceed 1°C over the oceans and 5°C over the continents; precipitation and sea level pressure errors are relatively greater but the magnitudes and patterns of these quantities are recognisably similar to observations. The bottom portion of Figure 8.2 shows maps of the model mean taken separately over all flux adjusted models (lower left) and all non-flux adjusted models (lower right). Flux adjusted models are generally more similar to the observations – and to each other – than are non-flux adjusted models. However, errors in the non-flux adjusted model mean are not grossly larger than errors in the flux adjusted model mean (except in polar regions). This result from the "inter-model" CMIP database suggests that the SAR was correct in anticipating that the need for flux adjustments would diminish as coupled models improve. It is reinforced by "intra-model ensembles", i.e., by the experience that improvements to individual models can reduce the need for flux adjustments (e.g., Boville and Gent, 1998).

The foregoing points are made in a more quantitative fashion by Figures 8.4 to 8.6 (reproduced from Covey *et al.*, 2000b). Figure 8.4 gives the standard deviation and correlation with observations of the total spatial and temporal variability (including the seasonal cycle, but omitting the global mean) for surface air temperature, sea level pressure and precipitation in the CMIP2 simulations. The standard deviation is normalised to its observed value and the correlation ranges from zero along an upward vertical line to unity along a line pointing to the right. Consequently, the observed behaviour of the climate is represented by a point on the horizontal axis a unit distance from the origin. In this coordinate system, the linear distance between each model's point and the "observed" point is proportional to the r.m.s. model error (Taylor, 2000; see also Box 8.1). Surface air temperature is particularly well simulated, with nearly all models closely matching the observed magnitude of variance and exhibiting a correlation > 0.95 with the observations. Sea level pressure and precipitation are simulated less well, but the simulated variance is still within ±25% of observed and the correlation with observations is noticeably positive (about 0.7 to 0.8 for sea level pressure and 0.4 to 0.6 for precipitation).

Observational uncertainties are indicated in Figure 8.4 by including extra observational data sets as additional points, as if they were models. These additional points exhibit greater agreement with the baseline observations as expected. It is noteworthy, however, that the differences between alternate sets of observations are not much smaller than the differences between models and the baseline observations. This result implies that in terms of variance and space-time pattern correlation, the models nearly agree with observations to within the observational uncertainty.

Figures 8.5 and 8.6 show global mean errors ("bias") and root mean square (r.m.s.) errors normalised by standard deviations for surface air temperature and precipitation in

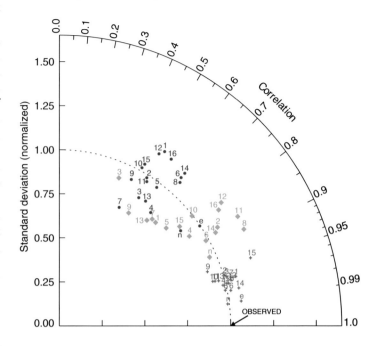

Total space-time component AOGCM control runs

- • Precipitation compared against Xie-Arkin.
- + Surface Air Temperature compared against Jones/Parker.
- ◆ Sea Level Pressure compared against ERA15.

Figure 8.4: Second-order statistics of surface air temperature, sea level pressure and precipitation simulated by CMIP2 model control runs. The radial co-ordinate gives the magnitude of total standard deviation, normalised by the observed value, and the angular co-ordinate gives the correlation with observations. It follows that the distance between the OBSERVED point and any model's point is proportional to the r.m.s model error (see Section 8.2). Numbers indicate models counting from left to right in the following two figures. Letters indicate alternate observationally based data sets compared with the baseline observations: e = 15-year ECMWF reanalysis ("ERA"); n = NCAR/NCEP reanalysis. From Covey *et al.* (2000b).

CMIP2 model simulations. Both the r.m.s. errors and the background standard deviations are calculated from the full spatial and temporal variability of the fields. The r.m.s. errors are divided into a number of components such as zonal mean vs. deviations and annual mean vs. seasonal cycle. For nearly all models the r.m.s. error in zonal- and annual-mean surface air temperature is small compared with its natural variability. The errors in the other components of surface air temperature, and in zonal mean precipitation, are relatively larger but generally not excessive compared with natural variability.

In Figures 8.5 and 8.6, models are divided into flux adjusted and non-flux adjusted classes. Slight differences between the two may be discerned, but it is not obvious that these are statistically significant if the two classes are considered as random samples from a large population of potential climate models. The same conclusion was reached in a detailed study of the seasonal cycle of surface air temperature in the CMIP models (Covey *et al.*,

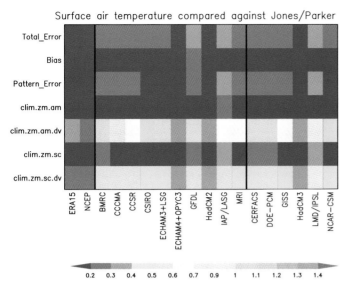

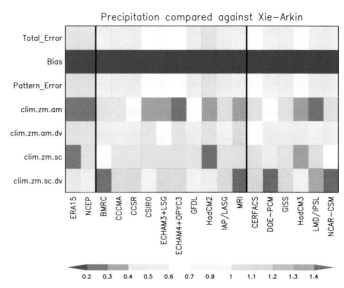

Figure 8.5: Components of space-time errors of surface air temperature (climatological annual cycle) simulated by CMIP2 model control runs. Shown are the total errors, the global and annual mean error ("bias"), the total r.m.s ("pattern") error, and the following components of the climatological r.m.s. error: zonal and annual mean ("clim.zm.am"); annual mean deviations from the zonal mean ("clim.zm.am.dv"), seasonal cycle of the zonal mean ("clim.zm.sc"); and seasonal cycle of deviations from the zonal mean ("clim.zm.sc.dv"). For each component, errors are normalised by the component's observed standard deviation. The two left-most columns represent alternate observationally based data sets, ECMWF and NCAR/NCEP reanalyses, compared with the baseline observations (Jones *et al.*, 1999). Remaining columns give model results: the ten models to the left of the second thick vertical line are flux adjusted and the six models to the right are not. From Covey *et al.* (2000b).

Figure 8.6: As in the Figure 8.5, for precipitation. The two left-most columns represent alternate observationally based data sets, 15-year ECMWF reanalysis ("ECMWF") and NCAR/NCEP reanalysis ("NCEP"), compared with the baseline observations (Xie and Arkin, 1996). From Covey *et al.* (2000b).

Box 8.1: Taylor diagrams

To quantify how well models simulate an observed climate field, it is useful to rely on three non-dimensional statistics: the ratio of the variances of the two fields:

$$\gamma^2 = \delta^2_{mod} / \delta^2_{obs}$$

the correlation between the two fields (R, which is computed after removing the overall means), and the r.m.s difference between the two fields (E, which is normalised by the standard deviation of the observed field). The ratio of variance indicates the relative amplitude of the simulated and observed variations, whereas the correlation indicates whether the fields have similar patterns of variation, regardless of amplitude. The normalised r.m.s error can be resolved into a part due to differences in the overall means (E_0), and a part due to errors in the pattern of variations (E').

These statistics provide complementary, but not completely independent, information. Often the overall differences in means (E_0) is reported separately from the three pattern statistics (E', γ, and R), but they are in fact related by the following equation:

$$E'^2 = E^2 - E_0^2 = 1 + \gamma^2 - 2\gamma R$$

This relationship makes it possible to display the three pattern statistics on a two-dimensional plot like that in Figure 8.4. The plot is constructed based on the Law of Cosines. The observed field is represented by a point at unit distance from the origin along the abscissa. All other points, which represent simulated fields, are positioned such that γ is the radial distance from the origin, R is the cosine of the azimuthal angle, and E' is the distance to the observed point. When the distance to the point representing the observed field is relatively short, good agreement is found between the simulated and observed fields. In the limit of perfect agreement (which is, however, generally not achievable because there are fundamental limits to the predictability of climate), E' would approach zero, and γ and R would approach unity.

2000a). (That study, however, also noted that many of the non-flux adjusted models suffered from unrealistic "climate drift" up to about 1°C / century in global mean surface temperature.) The relatively small differences between flux adjusted and non-flux adjusted models noted above suggest that flux adjustments could be – indeed, are being – dispensed with at acceptable cost in many climate models, at least for the century time-scale integrations of interest in detecting and predicting anthropogenic climate change. In recent models that omit flux adjustment, the representation of atmospheric fields has in some cases actually improved, compared with older, flux-adjusted versions of the models. Examples include the HadCM3 model and the CSM 1.0 model. In CSM 1.0, atmospheric temperature, precipitation and atmospheric circulation are close to values simulated when the atmospheric component of the CSM 1.0 model is driven by observed sea surface temperatures (Boville and Hurrell, 1998).

8.5.1.2.2 Surface and top of atmosphere (TOA) fluxes
In this and the following two sub-sections we discuss simulations by AGCMs that are provided observed sea surface temperatures and sea-ice distributions as input boundary conditions. AOGCM control runs have not yet been thoroughly examined in studies of surface boundary fluxes or mid-tropospheric and stratospheric quantities.

Satellite observations over the past quarter of a century have provided estimates of top of atmosphere (TOA) flux that are considered reliable. Any discrepancies between models and observations are usually attributed to the inadequate modelling of clouds, since they are difficult to specify and accurately model, and account for most of the variability.

Unfortunately, there are no global estimates of surface flux that do not rely heavily on models. The best model-independent estimates come from the Global Energy Balance Archive (GEBA), a compilation of observations from more than 1,000 stations (Gilgen et al., 1998). Compared with GEBA observations, surface solar insolation is overestimated in most AGCMs (Betts et al., 1993; Garratt, 1994; Wild et al., 1997, 1998; Garratt et al., 1998). Downwelling long wave radiation, on the other hand, is underestimated (Garratt and Prata, 1996; Wild et al., 1997). The shortwave discrepancy is of more concern: it is more than a factor of two larger than the long-wave discrepancy, and could be due to missing absorption processes in the atmosphere.

The observations indicate that about 25% of the incident solar flux at the TOA is absorbed in the atmosphere, but most models underestimate this quantity by 5 to 8% of the of the incident solar flux (Arking, 1996, 1999; Li et al., 1997). The extent and the source (or sources) of this discrepancy have been intensely debated over the past five years, with investigations yielding contradictory results on whether the discrepancy is associated with clouds, aerosols, water vapour, or is an artefact of the instrumentation and/or the methods by which sensors are calibrated and deployed.

This discrepancy is important for climate modelling because it affects the partitioning of solar energy between the atmosphere and the surface. If the observations are correct, then improving the models will reduce the energy available for surface evaporation by 10 to 20% with a corresponding reduction in precipitation (Kiehl et al., 1995) and a general weakening of the hydrological cycle.

8.5.1.2.3 Mid-tropospheric variables
The SAR concluded that although atmospheric models adequately simulate the three-dimensional temperature distribution and wind patterns, "current models portray the large-scale latitudinal structure and seasonal change of the observed total cloud cover with only fair accuracy". Subsequent studies have confirmed both the good and bad aspects of model simulations. Throughout most of the troposphere, errors in AMIP1 ensemble simulations of temperature and zonal wind are small compared with either inter-model scatter or the observed spatial standard deviation (Gates et al., 1999). (See Section 8.8 for brief discussion of storm tracks.) On the other hand, discrepancies between models and observations that substantially exceed the observational uncertainty are evident for both clouds (Mokhov and Love, 1995; Weare et al., 1995, 1996; Weare, 2000a, 2000b) and upper tropospheric humidity (see Chapter 7).

Although solutions to these problems have proved elusive, incremental improvements have been noted since publication of the SAR. For total cloudiness, a revised subset of AMIP models exhibits noticeably less inter-model variation and significantly less average r.m.s error (Gates et al., 1999; Figure 8.7), compared with the original versions of the models. Several models adequately simulate seasonal changes in cloud radiative forcing (Cess et al., 1997). Model intercomparisons organised under the Global Energy and Water cycle Experiment (GEWEX) Cloud System Study (Stewart et al., 1998) will provide further information for improving cloud simulation. For tropospheric humidity, improved agreement with observations may result from improved numerical techniques (Section 8.9). Furthermore, even though the seasonal mean amounts of clouds and upper tropospheric water vapour are not well simulated in current climate models, *variations* of these quantities may be more important than absolute amounts for predicting climate changes. For example, Del Genio et al. (1994) noted that, in mid-latitudes, the seasonal cycle of upper tropospheric humidity can be simulated reasonably well by climate models. They argued that this variation provides a surrogate for decadal climate change in mid-latitudes because both are characterised by combined temperature increase and latitudinal temperature-gradient decrease, and thus both have similar effects on storms.

Examination of monsoons in climate models provides another measure of their ability to simulate hydrologic variations. Developments since publication of the SAR have been encouraging. Sperber and Palmer (1996) found that about half the original AMIP models obtained a realistic dependence of monsoon circulation on location and season. A follow-up study reveals that nearly all the revised AMIP models do so (Sperber et al., 1999; see Section 8.7.3).

8.5.1.3 Stratospheric climate
Simulation of the stratosphere in coupled climate models is advancing rapidly as the atmospheric components of these models enhance their vertical resolution in the upper part of their domain. Since publication of the SAR, it has become increas-

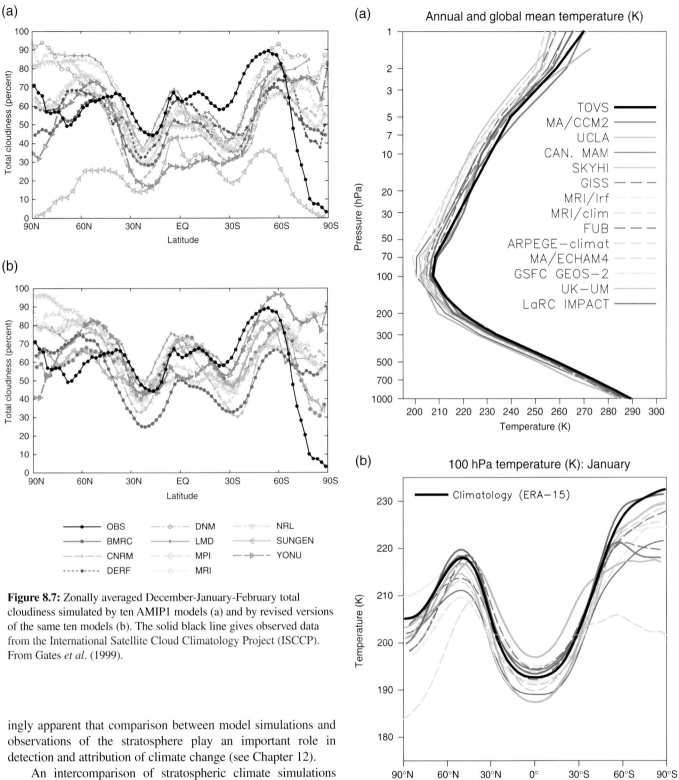

(a)

(b)

Figure 8.7: Zonally averaged December-January-February total cloudiness simulated by ten AMIP1 models (a) and by revised versions of the same ten models (b). The solid black line gives observed data from the International Satellite Cloud Climatology Project (ISCCP). From Gates *et al.* (1999).

(a) Annual and global mean temperature (K)

(b) 100 hPa temperature (K): January

Figure 8.8: (a) Vertical structure of the annual-mean, globally averaged temperature (K) from observations and the thirteen models. (b) Latitudinal structure of January temperature at 100 hPa. From Pawson *et al.* (2000).

ingly apparent that comparison between model simulations and observations of the stratosphere play an important role in detection and attribution of climate change (see Chapter 12).

An intercomparison of stratospheric climate simulations (Pawson *et al.*, 2000) shows that all models reproduce to some extent the zonally averaged latitudinal and vertical structure of the observed atmosphere, although several deficiencies are apparent. There is a tendency for the models to show a global mean cold bias at all levels (Figure 8.8a). The latitudinal distribution shows that almost all models are too cold in both hemispheres of the extra-tropical lower stratosphere (Figure 8.8b). There also is a large scatter in the tropical temperatures. Another common model deficiency is in the strengths and locations of the jets. The polar night jets in most models are inclined poleward with height, in noticeable contrast to an equatorward inclination of the observed jet. There is also a differing degree of separation in the models between the winter sub-tropical jet and the polar night jet.

Proper accounting of the role of gravity waves can improve stratospheric modelling. An orographic gravity wave scheme has been shown to improve the cold pole problem. Recent work with non-orographic gravity wave schemes show that waves of non-zero phase speed result in equatorward inclined jet through larger deceleration of westerly winds in polar regions (Manzini and McFarlane, 1998; Medvedev *et al.*, 1998). Nevertheless, all models have shortcomings in their simulations of the present day climate of the stratosphere, which might limit the accuracy of predictions of future climate change (Shindell *et al.*, 1999).

8.5.1.4 Summary

Coupled climate models simulate mean atmospheric fields with reasonable accuracy, with the exception of clouds and some related hydrological processes (in particular those involving upper tropospheric humidity). Since publication of the SAR, the models have continued to simulate most fields reasonably well while relying less on arbitrary flux adjustments. Problems in the simulation of clouds and upper tropospheric humidity, however, remain worrisome because the associated processes account for most of the uncertainty in climate model simulations of anthropogenic change. Incremental improvements in these aspects of model simulation are being made.

8.5.2 Ocean Component

8.5.2.1 Developments since the SAR

There have been a number of important developments in the ocean components of climate models since the SAR. Many climate models now being used for climate projections have ocean resolution of order 1 to 2° (Table 8.1), whereas at the time of the SAR most models used in projections had ocean resolution of order 3 to 5° (SAR Tables 5.1 and 6.3). The improved resolution may contribute to better representation of poleward heat transport (Section 8.5.2.2.2), although some key processes are still not resolved (see Sections 8.5.2.3, 8.9.2). Coupled models with even finer resolution are under development at the time of writing, but their computational expense makes their use in climate change projections impractical at present. Advances in the parametrization of sub-grid scale mixing (Chapter 7, Section 7.3.4) have also led to improved heat transports (Section 8.5.2.2.2). Some models have also adopted more advanced parametrizations of the surface mixed layer (Guilyardi and Madec, 1997; Gent *et al.*, 1998; see Chapter 7, Section 7.3.1)

A formal comparison project of a wide range of ocean-climate models has not yet been set up. This is largely because the specification of surface forcing for the ocean, and the long spinup time-scale, make a co-ordinated experimental design more difficult to achieve than for the atmosphere. Nonetheless, a number of smaller, focused projects have provided valuable information about the performance of different model types and the importance of specific processes (Chassignet *et al.*, 1996;

Roberts *et al.*, 1996; DYNAMO group 1997). Also, the Ocean Carbon Cycle Intercomparison Project (OCMIP) has compared the ocean circulation in a number of models (Sarmiento *et al.*, 2000; Orr *et al.*, 2001), and some comparisons have been made of the ocean components of coupled models under CMIP (see, e.g., Table 8.2; Jia, 2000)

The observational phase of the World Ocean Circulation Experiment (WOCE) was completed in 1997. Much analysis of the data to date has concentrated on individual sections or regions, and some of this analysis has been used in the assessment of climate models (e.g., Banks, 2000). Some initial attempts to put sections together into a consistent global picture also appear promising (MacDonald, 1998; de las Heras and Schlitzer, 1999). Such a global picture is an important baseline against which models can be tested (Gent *et al.*, 1998; Gordon *et al.*, 2000; see also Chapter 7, Section 7.6).

8.5.2.2 Present climate

8.5.2.2.1 Wind driven circulation

The wind-driven dynamics of the interior of the ocean basins are largely a linear response to the wind, and are generally well represented in current models, although there is still some observational debate over the reality of the classical Sverdrup balance (Wunsch and Roemmich, 1985). The main errors can usually be traced back to errors in the driving winds from the atmospheric model. The same can be said of surface Ekman transport, which makes an important contribution to poleward heat transport in the tropics (Danabasoglu, 1998). However, the western boundary currents and inertial recirculations which close the wind-driven gyres are generally poorly resolved by current models, and this may lead to an underestimate of the heat transport by this component of the system at higher latitudes (Fanning and Weaver, 1997a; Bryan and Smith 1998).

8.5.2.2.2 Heat transport and thermohaline circulation

Section 8.4.2 and Chapter 7, Section 7.6 discuss the fundamental importance of poleward heat transport in modelling the climate system. Ocean heat transport is greatly improved in some more recent models, compared with the models in use at the time of the SAR (see, e.g., Table 8.2; Chapter 7, Section 7.6). Increased horizontal resolution (Fanning and Weaver, 1997a; Gordon *et al.*, 2000) and improved parametrization of sub-grid scale mixing (Danabasoglu and McWilliams, 1996; Visbeck *et al.*, 1997; Gordon *et al.*, 2000) have been important factors in this. The fresh water transports of coupled models have not been widely evaluated (see Section 8.4.2 and Chapter 7, Section 7.6). Bryan (1998) shows how fresh water imbalances can lead to long-term drifts in deep ocean properties.

The thermohaline circulation (THC) plays an important role in poleward heat transport, especially in the North Atlantic. Table 8.2 shows the strength of North Atlantic THC for various models. In contrast to the SAR, some non-flux adjusted models are now able to produce a THC with a realistic strength of around 20 Sv, which is stable for many centuries. A common systematic error at the time of the SAR was a model thermocline that was too deep and diffusive, resulting in deficient heat transport because the

Table 8.2: *Diagnostics of the ocean circulation from a number of coupled model control runs (see Table 8.1 for the specification of models).*

MODEL NAME	OVERTURNING ATLANTIC 25°N (Sv)	TEMPERATURE CONTRAST ATLANTIC 25°N (°C)	HEAT TRANSPORT ATLANTIC 25°N (PW)	HEAT TRANSPORT PACIFIC 25°N (PW)	HEAT TRANSPORT ATLANTIC 30°S (PW)	HEAT TRANSPORT INDO-PACIFIC 30°S (PW)	ACC TRANSPORT (Sv)	NIÑO 3 SST STD DEV (°C)	NIÑO 4 SST STD DEV (°C)
ARPEGE/OPA1	16.0	12.4	0.74	0.83	0.20	−0.79	60		
ARPEGE/OPA2	12.6	15.9	0.77	0.64	0.33	−0.69	143		
CCSR/NIES*	20.8	10.0	0.73	0.32	−0.06	−2.44	200	0.8	0.6
CGCM1*	18.3	9.0	0.72	0.40	0.22	−0.97	62	0.2	0.3
COLA1	16.4	11.0	0.38	0.55	−0.13	−0.48	10		
CSIRO Mk 2*	13.0	12.1	0.81	0.43	0.21	−1.31	103		
CSM 1.0	24.6	14.0	1.30	0.74	0.54	−1.14	236	0.5	0.5
CSM 1.3	22		1.15	0.8	0.45	−1.10	178		
ECHAM1/LSG*	30						100		
ECHAM3/LSG*	28.1	9.9	0.76	0.26	0.09	−1.98	112	0.2	0.3
ECHAM4/OPYC3*	22.9	7.3	0.59	0.49	−0.19	−2.31	122	0.8	0.5
GFDL_R15_a*	15.0	9.9	0.59	0.34	0.08	−1.09	70	0.5	0.4
GISS1	18.6	12.3	1.01	0.76	0.37	−0.72	75		
GISS2	7.9	21.5	0.56	0.67	0.07	−0.60			
GOALS			0.68	0.44	0.05	−1.56	74		
HadCM2*	16.8	11.9	0.82	0.34	0.40	−0.93	216		
HadCM3	18.1	14.8	1.10	0.51	0.55	−1.25	204	1.1	1.0
IPSL-CM1	11.2	15.2	0.61	0.61	0.27	−0.81	66	0.3	0.2
IPSL-CM2	22.7	11.1	0.90	0.66	0.26	−1.48	164		
MRI1*	1.6		−0.29	0.82	−1.00	−1.08	50	0.4	0.7
MRI2*	17.0		0.86	0.78	0.26	−1.21	83	1.9	1.7
NCAR1	35.8	10.3	0.58	0.32	0.80	1.00	79	0.5	0.4
NRL*	3.1		0.43	0.72	-0.10	−0.77	66		
DOE PCM	27.2	12.7	1.13	0.77	0.40	−0.73			
BMRCa							39	0.4	0.4
COLA2								0.7	0.5
GFDL_R30_c*								0.4	0.6
OBSERVED	19.3[a]	14.1[a]	1.15[b]	0.76[c]	0.50[d]	−1.34[d]	123[e]	0.7[f]	0.5[f]

Positive heat transport values indicate northward transport. An asterisk indicates flux adjusted models. Cells are left blank where particular data items are unavailable.

Temperature contrast Atlantic 25°N: The difference between the mean temperatures of the northward flowing surface water and the southward flowing North Atlantic Deep Water at 25°N (Jia 2000).

NIÑO3/4 SST Std Dev: standard deviation of the sea surface temperature in the NIÑO3 and NIÑO4 regions of the tropical Pacific.

References:

[a] Hall and Bryden, 1982

[b] Trenberth, 1998a (see Chapter 7, Section 7.6)

[c] Bryden *et al.*, 1991

[d] MacDonald, 1998

[e] Whitworth and Petersen, 1985

[f] Parker *et al.*, 1995

temperature contrast between cold, southward and warm, northward flows was too weak. The models with realistic North Atlantic heat transports generally maintain a realistic temperature contrast (Table 8.2). Some models also show improved realism in the spatial structure of the THC, with separate deep water sources in the Nordic Seas and in the Labrador Sea (Wood *et al.*, 1999).

Interior diapycnal mixing plays a critical role in the thermohaline circulation. Recent process studies (part of WOCE) have confirmed that such mixing is highly localised in the deep ocean (Polzin *et al.*, 1997; Munk and Wunsch, 1998). This mixing is very crudely represented in climate models, and it is not known whether this deficiency has a significant effect on the model thermohaline circulations (Marotzke, 1997).

Although overall heat transports are now better represented in some models, the partition of the heat transport between different components of the circulation may not agree so well with observational estimates (Gordon *et al.*, 2000). How important such discrepancies may be in modelling the transient climate change response is not well understood.

The deep western boundary current, which carries much of the deep branch of the North Atlantic THC, contains a number of strong recirculating gyres (e.g., Hogg, 1983). These recirculations may act as a buffer, delaying the response of the THC to climate anomalies. The representation of these recirculations in climate models, and their importance in transient climate response, have not been evaluated.

8.5.2.2.3 Antarctic circumpolar current (ACC)

Many current models produce rather poor estimates of the volume transport of the Antarctic circumpolar current (ACC) (Table 8.2). The reason for this is not fully understood. Thermohaline as well as wind-driven processes are believed to be important (Cox 1989; Bryan 1998). The problem is shared by some eddy-permitting ocean models, so insufficient horizontal resolution does not seem to be the only factor. The path of the ACC is largely controlled by topography, and errors in the path can lead to significant local sea surface temperature errors. The Atlantic and Indian sectors of the southern ocean appear to be particularly susceptible (e.g., Figure 8.1b; Gent *et al.*, 1998; Gordon *et al.*, 2000). However, it is not clear how, or whether, the transport and SST errors impact on the atmospheric climate or on the climate change response of the models.

8.5.2.2.4 Water mass formation

At high latitudes, deep convection and subsequent spreading of dense water form the deep water masses that fill most of the volume of the ocean. At mid-latitudes, the processes of mode water formation and thermocline ventilation are the means by which surface changes are propagated into the thermocline (Chapter 7, Section 7.3.1). These processes play an important role in determining the effective rate of heat uptake by the ocean in response to climate change (see Chapter 9, Section 9.3.4.2), and in the response of the THC (see Chapter 9, Section 9.3.4.3). Water mass formation processes can be evaluated directly from model fields (Guilyardi, 1997; Doney *et al.*, 1998), or indirectly using model simulations of the ocean uptake of anthropogenic tracers such as CFCs and carbon 14 (Robitaille and Weaver,

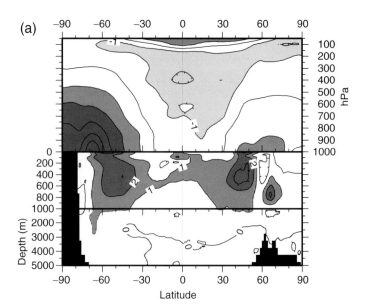

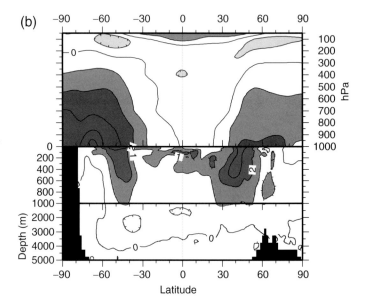

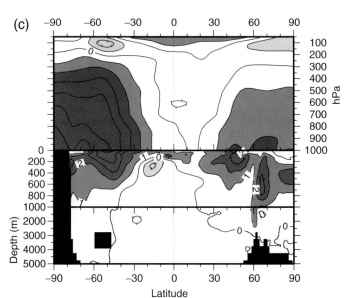

Figure 8.9: Zonal mean air and sea temperature "errors" in °C (defined here as the difference from the initial model state, which was derived from observations), for three different coupled models. The models are all versions of the ARPEGE/OPA model, with T31 atmospheric resolution, and differ only in the parametrization of lateral mixing used in the ocean component ((a) lateral diffusion, (b) isopycnal diffusion, (c) the scheme of Gent and McWilliams (1990)). The different mixing schemes produce different rates of heat transport between middle and high latitudes, especially in the Southern Hemisphere. The atmosphere must adjust in order to radiate the correct amount of heat to space at high latitudes (Chapter 7, Section 7.6 and Section 8.4.1), and this adjustment results in temperature differences at all levels of the atmosphere. From Guilyardi (1997).

Table 8.3: *Coupled model simulations (CMIP1) for December, January, February (DJF) and June, July, August (JJA) of sea-ice cover (columns 2 to 5) and snow cover (10^6 km^2) columns 6 and 7). Model names (column 1) are supplemented with ordinal numbers (in brackets) which refers to the models listed in Table 8.1. The observed sea-ice extent is from Gloersen et al. (1992) and the climatological observed snow is from Foster and Davy (1988).*

Model name	Sea-ice cover (10^6 km^2)				Snow cover (10^6 km^2)	
	Northern Hemisphere		Southern Hemisphere		Northern Hemisphere	
	DJF (winter)	JJA (summer)	JJA (winter)	DJF (summer)	DJF (winter)	JJA (summer)
ARPEGE/OPA1 (1)	10.1	8.8	2.5	1.9	50.6	19.2
BMRCa (3)	13.7	12.0	0.0	0.0	42.4	2.2
CCSR/NIES (5)	13.0	9.3	16.7	8.6	46.2	12.0
CGCM1 (6)	8.6	7.0	12.3	8.2	47.5	13.9
COLA1 (8)	9.4	5.9	0.0	0.0	58.7	2.5
CSIRO Mk2 (10)	14.3	14.1	14.2	13.6	48.8	18.9
CSM 1.0 (11)	18.6	13.1	22.8	10.0	43.7	4.7
ECHAM3/LSG (14)	12.5	10.4	11.1	7.3	35.8	9.1
ECHAM4/OPYC3 (15)	10.5	9.1	21.0	13.4		
GFDL_R15_a (16)	10.6	8.8	13.2	6.5	56.9	2.4
GISS1 (19)	15.3	14.6	8.7	7.1		
GISS2 (20)	15.7	15.2	10.9	9.5	43.2	9.3
HadCM2 (22)	12.0	10.1	24.7	11.8	45.0	8.2
IPSL-CM1 (24)					44.2	11.2
MRI1 (26)	19.4	18.3	14.5	4.1	60.2	11.6
NCAR1 (28)	11.6	10.6	20.8	16.4	38.9	3.6
Observed	14.5	11.5	16.0	7.0	49.3	3.7

1995; Dixon *et al.*, 1996; England and Rahmstorf, 1999; Goosse *et al.*, 1999; England and Maier-Reimer, 2001). A conclusion from many of these studies is that, while the models clearly show some skill in this area, ventilation processes are sensitive to the details of the ocean mixing parametrization used. Wiebe and Weaver (1999) show that the efficiency of ocean heat uptake is also sensitive to these parametrizations.

8.5.2.3 Summary

Considerable progress has been made since the SAR in the realism of the ocean component of climate models. Models now exist which simultaneously maintain realistic poleward heat transports, surface temperatures and thermocline structure, and this has been a vital contributor to the improvement in non-flux adjusted models. However, there are still a number of processes which are poorly resolved or represented, for example western boundary currents (see Chapter 7, Section 7.3.6), convection (Chapter 7, Section 7.3.2), overflows (Chapter 7, Section 7.3.5), Indonesian through flow, eddies (including Agulhas eddies which travel long distances and may be hard to treat by a local parametrization; Chapter 7, Section 7.3.4), Antarctic Bottom Water formation (Chapter 7, Section 7.3.2) and interior diapycnal mixing (Chapter 7, Section 7.3.3). In many cases, the importance of these processes in controlling transient climate change has not been evaluated. Over the next few years there is likely to be a further move to finer resolution models, and a wider range of model types; these developments are likely to reduce further some of these uncertainties. Finally, there is still only patchy understanding of the effects of sub-grid scale parametrizations in

the context of coupled models. Valuable understanding can be gained from sensitivity studies using ocean or atmosphere models alone, but Figure 8.9 shows the inherently coupled nature of the climate system – changes in ocean parametrizations can have a significant impact throughout the depth of the atmosphere (the reverse is also true). Further sensitivity studies in the coupled model context will help to quantify and reduce uncertainty in this area.

8.5.3 Sea Ice Component

While the important role of sea ice in projections of future climate has been widely recognised (Chapter 7, Section 7.5.2), results of systematic intercomparisons or sensitivity studies of AOGCM sea-ice components remain very limited. The sea-ice simulations of fifteen global coupled models contributed to CMIP1 are summarised in Table 8.3. (All these models are also presented in Table 8.1, where the last two columns indicate whether an ice dynamics scheme is included, and whether the model is flux adjusted.) Sea-ice thermodynamic formulations of the coupled models are mostly based on simplified schemes: few employ a multi-layer representation of heat transfer through the ice, while the rest assume a linear temperature profile. In addition, roughly half of the models ignore leads and polynyas in the ice although these account for principal thermodynamic coupling of the atmosphere and ocean. Some models also ignore the thermodynamic effects of snow on sea ice. Despite the rather mature status of sea-ice dynamics modelling (e.g., the Sea Ice Model Intercomparison Project (SIMIP), Lemke *et al.*, 1997),

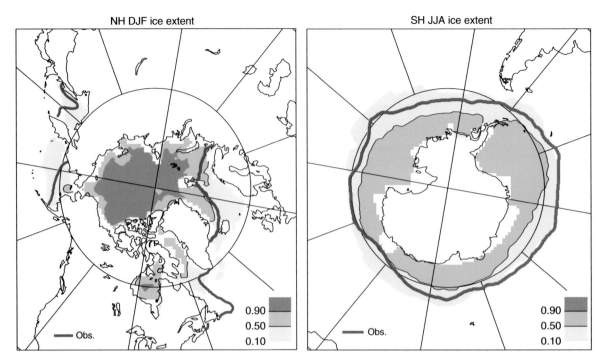

NH DJF ice extent   SH JJA ice extent

0.90
0.50
0.10
— Obs.

0.90
0.50
0.10
— Obs.

Figure 8.10: Illustration of the range of sea-ice extent in CMIP1 model simulations listed in Table 8.3: Northern Hemisphere, DJF (left) and Southern Hemisphere, JJA (right). For each model listed in Table 8.3, a 1/0 mask is produced to indicate presence or absence of ice. The fifteen masks were averaged for each hemisphere and season. The 0.5 contour therefore delineates the region for which at least half of the models produced sea ice. The 0.1 contour indicates the region outside of which only 10% of models produced ice, while the 0.9 contour indicates that region inside of which only 10% of models did not produce ice. The observed boundaries are based on GISST_2.2 (Rayner *et al.*, 1996) averaged over 1961 to 1990.

only two of the fifteen models include a physically based ice dynamics component. Three of the fifteen models allow ice to be advected with the ocean currents (the so-called 'free drift' scheme), and the remainder assume a motionless ice cover. Overall, this highlights the slow adoption, within coupled climate models, of advances in stand-alone sea ice and coupled sea-ice/ocean models (Chapter 7, Section 7.5.2).

Table 8.3 provides a comparison of ice extent, defined as the area enclosed by the ice edge (which is in turn defined as the 0.1 m thickness contour or the 15% concentration contour, depending on the data provided), for winter and summer seasons in each hemisphere. The last row of the table provides an observed estimate based on satellite data (Gloersen *et al.*, 1992) covering the period 1978 to 1987. It should be noted that assessment of sea-ice model performance continues to be hampered by observational problems. For the satellite period (1970s onward) the accuracy of observations of sea-ice concentration and extent is fair, however observational estimates of sea-ice thickness and velocity are far from satisfactory.

Figure 8.10 provides a visual presentation of the range in simulated ice extent, and was constructed as follows. For each model listed in Table 8.3, a 1/0 mask was produced to indicate presence or absence of ice. The fifteen masks were averaged for each hemisphere and season and the percentage of models that had sea ice at each grid point was calculated.

There is a large range in the ability of models to simulate the position of the ice edge and its seasonal cycle, particularly in the Southern Hemisphere. Models that employ flux adjustment tend,

on average, to produce smaller ice extent errors, but there is no obvious connection between fidelity of simulated ice extent and the inclusion of an ice dynamics scheme. The latter finding probably reflects the additional impact of errors in the simulated wind field and surface heat fluxes that offset, to a great extent, any improvements due to including more realistic parametrizations of the physics of ice motion. In turn this partially explains the relative slowness in the inclusion of sophisticated sea ice models with AOGCMs. However, even with quite simple formulations of sea ice, in transient simulations, some AOGCMs demonstrate ability to realistically reproduce observed annual trend in the Arctic sea ice extent during several past decades of the 20th century (see Chapter 2, Section 2.2.5.2), which adds some more confidence in the use of AOGCM for future climate projections (Vinnikov *et al.*, 1999)

8.5.4 Land Surface Component (including the Terrestrial Cryosphere)

8.5.4.1 Introduction
The role of the land surface (soil, vegetation, snow, permafrost and land ice) was discussed in detail in the SAR. The SAR noted that improvements had occurred in our ability to model land-surface processes but that there was a wide disparity among current land-surface schemes when forced by observed meteorology. Our physical understanding of the role of land-surface processes within the climate system was discussed in Chapter 7, Section 7.4.1.

8.5.4.2 Developments since the SAR

Most of the effort in trying to reduce the disparity in land-surface scheme performance has been performed in offline intercomparisons under the auspices of International Geosphere Biosphere Programme (IGBP) and the World Climate Research Programme (WCRP). Due to the difficulties in coupling multiple land-surface schemes into climate models (see Section 8.5.4.3) specific endeavours have been: the Project for the Intercomparison of Landsurface Parametrization Schemes (PILPS), the Global Soil Wetness Project (GSWP), the International Satellite Land Surface Climatology Project (ISLSCP), and the Biological Aspects of the Hydrological Cycle (BAHC) (e.g., Henderson-Sellers *et al.*, 1995; Polcher *et al.*, 1996; Dirmeyer *et al.*, 1999; Schlosser *et al.*, 2000). In comparisons between offline simulation results and observations, difficulties in partitioning available energy between sensible and latent heat and partitioning of available water between evaporation and runoff were highlighted (e.g., Chen *et al.*, 1997; Schlosser *et al.*, 2000). While we are far from a complete understanding of why land surface models differ by such a large degree, some progress has been made (e.g., Koster and Milly, 1997; Desborough, 1999). Significant progress has also been made in adding physical processes into land-surface models (see Chapter 7, Section 7.4). Where some observations exist (e.g., for incoming solar radiation, net radiation and soil moisture), an evaluation of the ability of current climate models to simulate these quantities suggests that significant problems remain (Wild *et al.*, 1997; Garratt *et al.*, 1998; Robock *et al.*, 1998). The evaluation of surface processes in climate models tends to focus on monthly and, less commonly, daily quantities. An evaluation of the ability of climate models to simulate land-surface quantities at the diurnal scale has yet to be performed systematically, although some efforts have been initiated since the SAR (e.g., Watterson, 1997).

Work since the SAR has also focused on trying to identify the relative significance of land-surface processes in comparison with other components of climate models. The sophistication in the representation of the land surface in coupled models is varied (see Chapter 7 and Table 8.1). The addition of more realistic plant physiology in some land-surface models (e.g., Bonan, 1995; Sellers *et al.*, 1996 (See Chapter 7, Section 7.4)) which permits the simulation of carbon dioxide (CO_2) and gas isotope fluxes, provides the opportunity to compare these quantities with local scale, regional scale and global scale observations from flux towers and satellites.

Snow, and the snow albedo feedback, are important components of the land surface. Current climate models incorporate snow in varying degrees of sophistication and there is currently major uncertainty in the ability of land-surface schemes to simulate snow mass or cover (see Chapter 7, Section 7.5). Frei and Robinson (1998) evaluated the simulation of monthly mean snow extent from 27 AMIP AGCMs and found weaknesses in the simulation of the seasonal cycle of snow extent and a general underestimation in interannual variability. These weaknesses limit confidence in the simulation of mid- and high latitude changes simulated by current climate models, since a failure to simulate snow accurately tends to impact significantly on albedo, surface roughness length and soil moisture (and therefore precipitation on subsequent seasons).

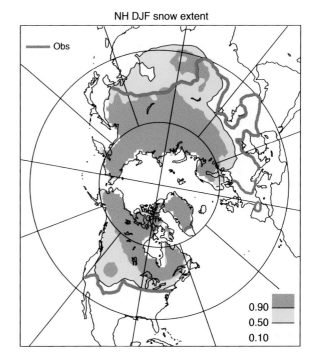

NH DJF snow extent

Obs

0.90
0.50
0.10

Figure 8.11: Illustration of the range of snow cover extent in CMIP1 model simulations listed in Table 8.3: Northern Hemisphere, DJF. The figure is constructed similarly to Figure 8.10 based on the prescribed 1 cm cutoff. The observed boundary is based on Foster and Davy (1988).

The Northern Hemisphere snow simulations of fourteen global coupled models contributed to CMIP are summarised in Figure 8.11 and Table 8.3. Figure 8.11 (constructed similarly to Figure 8.10) provides a visual presentation of the range in simulated (land only) snow extent. The relative error in simulated snow extent is larger in summer than in winter. There is no obvious connection between either flux adjustment or land-surface scheme and the quality of the simulated snow extent.

Other components of the land surface potentially important to climate change include lateral water flows from the continents into the ocean, permafrost, land-based ice and ice sheets. Some land-surface schemes now include river routing (e.g., Sausen *et al.*, 1994; Hagemann and Dümenil, 1998) in order to simulate the annual cycle of river discharge into the ocean. This appears to improve the modelling of runoff from some large drainage basins (Dümenil *et al.*, 1997), although water storage and runoff in regions of frozen soil moisture remain outstanding problems (Arpe *et al.*, 1997; Pitman *et al.*, 1999). River routing is also useful in diagnosing the representation of the hydrological cycle in models (e.g., Kattsov *et al.*, 2000). There has been limited progress towards developing a permafrost model for use in climate models (e.g., Malevsky-Malevich *et al.*, 1999) although existing simple models appear to approximate the observed range in permafrost reasonably well (e.g., Volodin and Lykosov, 1998). The dynamics of ice sheets and calving are not presently represented in coupled climate models. To close the fresh water budget for the coupled system, fresh water which accumulates on Greenland and Antarctica is usually uniformly distributed either over the entire ocean or just in the vicinity of the ice sheets (e.g., Legutke and Voss, 1999; Gordon *et al.*, 2000). The impact of these limitations has yet to be investigated.

8.5.4.3 Does uncertainty in land surface models contribute to uncertainties in climate prediction?

Uncertainty in climate simulations resulting from the land surface has traditionally been deduced from offline experiments (see Chapter 7 and Section 8.5.4.2) due to difficulties associated with comparing land-surface schemes when forced by different climate models (e.g., Polcher *et al.*, 1998b). Some work since the SAR has focused on the sensitivity of land-surface schemes to uncertainties in parameters (e.g., Milly, 1997) and on whether different land-surface schemes, coupled to climate models, lead to different climate simulations or different sensitivity to increasing CO_2 (Henderson-Sellers *et al.*, 1995).

Polcher *et al.* (1998a) used four different climate models, each coupled to two different land-surface schemes to explore the role of the land surface under $1 \times CO_2$ and $2 \times CO_2$. The modification to the land-surface scheme tended to focus on aspects of the soil-hydrology and vegetation/soil-moisture interactions (Gedney *et al.*, 2000). To measure the uncertainty associated with surface processes, the variance of anomalies caused by the changes to the surface scheme in the four climate models was computed. The uncertainty in climate models was computed by using the variance of annual anomalies relative to a consensus (the average of all models). This measure takes into account the differences between climate models, as well as the internal variance of the atmosphere. With these two variances, a ratio was constructed to evaluate the relative importance of the uncertainty linked to surface processes in comparison to the uncertainty linked to other aspects of the climate model (Crossley *et al.*, 2000). In Figure 8.12a this diagnostic is applied to zonal mean values over land for the $1 \times CO_2$ experiments. The highest values, indicating a large contribution of surface processes to the uncertainty, were obtained for evaporation, the variable most affected by surface processes (the asterisk indicates significance at the 95% of this measure). Surface air temperature was strongly dependent on surface processes in the tropics but at high latitudes its uncertainty was dominated by atmospheric processes. In the high latitudes, the hydrological cycle (characterised by precipitation and cloud cover) was partly controlled by the surface as indicated by high values of the ratio. Overall, Figure 8.12 shows that the contribution to total uncertainty in the simulation of climate resulting from the land surface may be large and varies geographically.

Figure 8.12b displays the same diagnostic but for the anomalies resulting from a doubling of CO_2. The maximum uncertainty is concentrated in the tropics and the variables most affected are cloud cover and temperature. The uncertainty in evaporation changes is large but does not dominate as in the control climate. In the Northern Hemisphere a secondary peak was found for cloud cover and to some extent for precipitation, indicating that a significant part of the uncertainties in the impact of climate change on the hydrological cycle originates from land-surface processes. The shapes of these curves are very different in the two figures, indicating that different processes are respon-

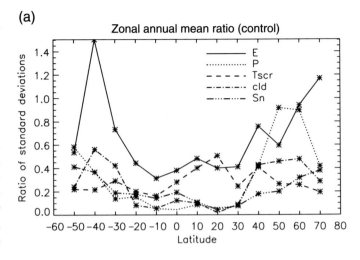

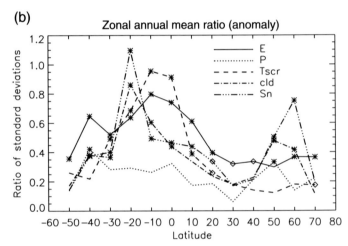

Figure 8.12: An uncertainty ratio for 10-degree latitude bands; (a) control simulations; (b) difference between the control and doubled greenhouse gas simulations. *E* is evaporation, *P* is precipitation, *Tscr* is screen temperature, *cld* is the percentage cloud cover and *Sn* is the net short-wave radiation at the surface. The units on the Y-axis are dimensionless. An asterisk means the value is statistically significant at 95% and a diamond at 90% (see Crossley *et al.*, 2000).

sible for the uncertainties in the control simulation and the climate change anomalies. This implies that the sensitivity of land-surface schemes to climate change needs to be evaluated and that it can not be deduced from results obtained for present day conditions. Gedney *et al.* (2000) analysed these results regionally and found that the simulations differ markedly in terms of their predicted changes in evapotranspiration and soil moisture. They conclude that uncertainty in the predicted changes in surface hydrology is more dependent on gross features of the runoff versus soil moisture relationship than on the detailed treatment of evapotranspiration. The importance of hydrology was also demonstrated by Ducharne *et al.* (1998) and Milly (1997).

Other work has involved using global climate fields provided by GCM analyses to force land-surface models offline (see Dirmeyer *et al.*, 1999). In this study, different land-surface models use the same atmospheric forcings, and the same soil and vegetation data sets. Model outputs are compared with regional runoff and soil moisture data sets and, where available, to observations from large-scale field experiments. To date, results highlight the differences between land-surface model treatments of large-scale hydrology and snow processes; it is anticipated that these and other trials will lead to significant improvements in these problem areas in the near future.

Uncertainty in land-surface processes, coupled with uncertainty in parameter data combines, at this time, to limit the confidence we have in the simulated regional impacts of increasing CO_2. In general, the evidence suggests that the uncertainty is largely restricted to surface quantities (i.e., the large-scale climate changes simulated by coupled climate models are probably relatively insensitive to land-surface processes). Our uncertainty derives from difficulties in the modelling of snow, evapotranspiration and below-ground processes. Overall, at regional scales, and if land-surface quantities are considered (soil moisture, evaporation, runoff, etc.), uncertainties in our understanding and simulation of land-surface processes limit the reliability of predicted changes in surface quantities.

8.5.5 Past Climates

Accurate simulation of current climate does not guarantee the ability of a model to simulate climate change correctly. Climate models now have some skill in simulating changes in climate since 1850 (see Section 8.6.1), but these changes are fairly small compared with many projections of climate change into the 21st century. An important motivation for attempting to simulate the climatic conditions of the past is that such experiments provide opportunities for evaluating how models respond to large changes in forcing. Following the pioneering work of the Co-operative Holocene Mapping Project (COHMAP-Members, 1988), the Paleoclimate Modeling Intercomparison Project (PMIP) (Joussaume and Taylor, 1995; PMIP, 2000) has fostered the more systematic evaluation of climate models under conditions during the relatively well-documented past 20,000 years. The mid-Holocene (6,000 years BP) was chosen to test the response of climate models to orbital forcing with CO_2 at pre-industrial concentration and present ice sheets. The orbital configuration intensifies (weakens) the seasonal distribution of the incoming solar radiation in the Northern (Southern) Hemisphere, by about 5% (e.g., 20 Wm^{-2} in the boreal summer). The last glacial maximum (21,000 years BP) was chosen to test the response to extreme cold conditions

8.5.5.1 Mid-Holocene

Atmosphere alone simulations
Within PMIP, eighteen different atmospheric general circulation models using different resolutions and parametrizations have been run under the same mid-Holocene conditions, assuming present-day conditions over the oceans (Joussaume *et al.*, 1999). In summer, all of the models simulate an increase and northward expansion of the African monsoon; conditions warmer than present in high northern latitudes, and drier than present in the interior of the northern continents. Palaeo-data do not support

drying in interior Eurasia (Harrison *et al.*, 1996; Yu and Harrison, 1996; Tarasov *et al.*, 1998), but they clearly show an expanded monsoon in northern Africa (Street-Perrott and Perrott 1993; Hoelzmann *et al.*, 1998; Jolly *et al.*, 1998a, 1998b), warming in the Arctic (Texier *et al.*, 1997), and drying in interior North America (Webb *et al.*, 1993).

Vegetation changes reconstructed from pollen data in the BIOME 6000 project (Jolly *et al.*, 1998b; Prentice and Webb III, 1998) provide a quantitative model-data comparison in northern Africa. The PMIP simulations produce a northward displacement of the desert-steppe transition, qualitatively consistent with biomes, but strongly underestimated in extent (Harrison *et al.*, 1998). At least an additional 100 mm/yr of precipitation would be required for most models to sustain grassland at 23°N, i.e., more than twice as much as simulated in this area (Joussaume *et al.*, 1999) (Figure 8.13). The increased area of lakes in the Sahara has also been quantified (Hoelzmann *et al.*, 1998) and, although the PMIP simulations do produce an increase, this latter is not large enough (Coe and Harrison, 2000). A similar underestimation is obtained at high latitudes over northern Eurasia, where PMIP simulations produce a northward shift of the Arctic tree-line in agreement with observed shifts (Tarasov *et al.*, 1998) but strongly underestimated in extent (Kutzbach *et al.*, 1996b; Texier *et al.*, 1997; Harrison *et al.*, 1998). Model-data discrepancies may, however, be due to missing feedbacks in the simplified PMIP experimental design.

Ocean feedbacks
Recent experiments with asynchronous (Kutzbach and Liu 1997; Liu *et al.*, 1999) and synchronous (Hewitt and Mitchell, 1998; Braconnot *et al.*, 2000) coupling of atmospheric models to full dynamical ocean models have been performed for the mid-Holocene. They all produce a larger enhancement of the African monsoon than shown in their PMIP atmosphere only experiments, resulting from the ocean thermal inertia and changes in the meridional ocean heat transport (Braconnot *et al.*, 2000). However, the changes are not sufficient to reproduce the observed changes in biome shifts over northern Africa.

Coupled models are also beginning to address the issue of changes in interannual to inter-decadal variability under conditions of large differences in the basic climate. Some palaeo-environmental evidence has suggested that short-term climate variability associated with El Niño-Southern Oscillation (ENSO) was reduced during the early to mid-Holocene (Sandweiss *et al.*, 1996; Rodbell, 1999). Up to now, ENSO variability has only been analysed in the CSM simulation, exhibiting no significant change at the mid-Holocene (Otto-Bliesner, 1999).

Land-surface feedbacks
Land-surface changes also provide an additional important feedback. During the mid-Holocene, vegetation changes over northern Africa have indeed favoured a larger increase in monsoon precipitation as shown through sensitivity experiments (Kutzbach *et al.*, 1996a; Brostrom *et al.*, 1998; Texier *et al.*, 2000) as well as through coupled atmosphere-vegetation experiments (Claussen and Gayler, 1997; Texier *et al.*, 1997; Pollard *et al.*, 1998; Doherty *et al.*, 2000; de Noblet *et al.*, 2000). Including

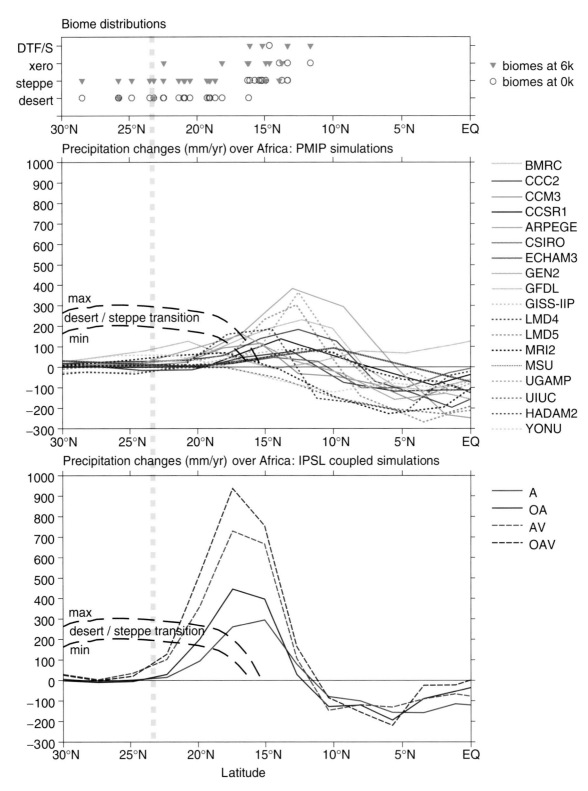

Figure 8.13: Annual mean precipitation changes (mm/yr) over Africa (20°W to 30°E) for the mid-Holocene climate: (upper panel) Biome distributions (desert, steppe, xerophytic and dry tropical forest/savannah (DTF/S)) as a function of latitude for present (red circles) and 6,000 yr BP (green triangles), showing that steppe vegetation replaces desert at 6,000 yr BP as far north as 23°N (vertical blue dashed line); (middle panel) 6000 yr BP minus present changes as simulated by the PMIP models. The black hatched lines are estimated upper and lower bounds for the excess precipitation required to support grassland, based on present climatic limits of desert and grassland taxa in palaeo-ecological records, the intersection with the blue vertical line indicates that an increase of 200 to 300 mm/yr is required to sustain steppe vegetation at 23°N at 6,000 yr BP (redrawn from Joussaume *et al.*, 1999); (lower panel) same changes for the IPSL atmosphere-alone (A), i.e., PMIP simulation, the coupled atmosphere-ocean (OA), the atmosphere-alone with vegetation changes from OA (AV) and the coupled atmosphere-ocean-vegetation (OAV) simulations performed with the IPSL coupled climate model. The comparison between AV and OAV emphasises the synergism between ocean and land feedbacks (redrawn from Braconnot *et al.*, 1999).

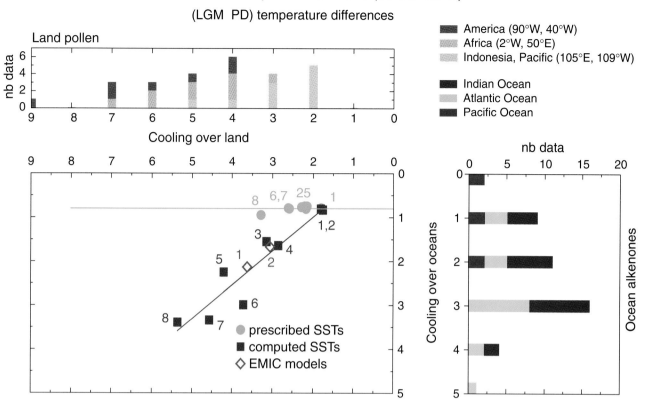

Figure 8.14: Annual mean tropical cooling at the last glacial maximum: comparison between model results and palaeo-data. (Centre panel) simulated surface air temperature changes over land are displayed as a function of surface temperature changes over the oceans, both averaged in the 30°S to 30°N latitudinal band, for all the PMIP simulations: models with prescribed CLIMAP SSTs (circles) and coupled atmosphere-mixed layer ocean models (squares) (from Pinot *et al.*, 1999). Numbers refer to different models: circles, 1: LMD4, 2-5: MRI2, ECHAM3, UGAMP, LMD5 (higher resolution), 6-7: CCSR/NIES1, LMD5, 8: GEN2. Squares : 1: LMD4,2: UGAMP, 3: GEN2, 4: GFDL, 5: HADAM2, 6: MRI2, 7: CCM1, 8: CCC2 (names refer to Tables 8.1 and 8.5). Results from two EMIC models including a dynamical ocean model have also been displayed (diamonds): 1-UVIC (Weaver *et al.*, 1998), 2-CLIMBER-2 (Petoukhov *et al.*, 2000).

The comparison with palaeo-data: (upper panel) over land is with estimates from various pollen data for altitudes below 1,500m (the label *"nb data"* refers to the number of data points in three different regions corresponding to the temperature change estimate plotted in the abscissa) from (Farrera, *et al.*, 1999); (right panel) the distribution of SST changes estimated from alkenones in the tropics from the Sea Surface Temperature Evolution Mapping Project based on Alkenone Stratigraphy (TEMPUS) (Rosell-Melé, *et al.*, 1998) (*nb data*: same as upper panel, number of data points for each temperature change). Caution: in this figure, model results are averaged over the whole tropical domain and not over proxy-data locations, which may bias the comparison (e.g., Broccoli and Marciniak, 1996). For example, for the pollen data, extreme values are obtained for specific regions: weakest values over the Indonesia-Pacific region and coldest values over South America.

the observed occurrence of large lakes and wetlands (Coe and Bonan, 1997; Brostrom *et al.*, 1998) also intensifies monsoon rains. Vegetation feedbacks also amplify the effects of orbital forcing at high latitudes where they led to greater and more realistic shifts of vegetation cover over northern Eurasia (Foley *et al.*, 1994; Kutzbach *et al.*, 1996b; Texier *et al.*, 1997). The importance of land-surface feedbacks has further been emphasised in the IPSL AOGCM coupled to a vegetation model (Braconnot *et al.*, 1999). The IPSL simulation shows that combined feedbacks between land and ocean lead to a closer agreement with palaeo-data (Figure 8.13). The ocean feedback increases the supply of water vapour, while the vegetation feedback increases local moisture recycling and the length of the monsoon season. The importance of land-surface feedbacks has also been shown by an EMIC (Ganopolski

et al., 1998a), further emphasising that vegetation feedbacks may explain abrupt changes in Saharan vegetation in the mid-Holocene (Claussen *et al.*, 1999).

8.5.5.2 *The last glacial maximum*

Results from the PMIP experiments

The Last Glacial Maximum (LGM) climate involves large changes in ice sheet extent and height, SSTs, albedo, sea level and CO_2 (200 ppm), but only minor changes in solar radiation. Over the oceans, two sets of experiments have been performed within PMIP using several atmospheric models, either prescribing SSTs estimated from macrofossil transfer functions (CLIMAP, 1981) or computing SSTs from a mixed-layer ocean model. An annual mean global cooling of about −4°C is obtained by all models forced by the Climate: Long-range Investigation, Mapping and Prediction (CLIMAP) SSTs, whereas the range of cooling is larger when using computed SSTs, from −6 to −2°C. This range of 4°C arises both from differences in the simulated radiative forcing associated primarily with different ice albedo values, and from differences in model climate sensitivity (see Chapter 9, Section 9.2.1).

Evaluating the consistency between the simulated climate and that reconstructed from palaeo-data can potentially provide an independent check that model sensitivity is neither too large nor too small. A detailed analysis of a subset of the PMIP models (Taylor *et al.*, 2000), shows that their forcing estimates for the LGM vary from about −4 to −6 Wm^{-2} and that their global climate sensitivity, given relatively to a doubling of CO_2, ranges from 3.2 to 3.9°C (assuming that climate sensitivity is independent of the type of forcing, although one model study shows a slightly stronger sensitivity at the LGM than for a CO_2 doubling (Hewitt and Mitchell, 1997)). A direct evaluation of climate sensitivity is, however, very difficult since global temperature changes are poorly known. Hoffert and Covey (1992) estimated a global cooling of −3 ± 0.6°C from CLIMAP (1981) SST data which would, using the simulated range of forcing, yield to a global climate sensitivity for a doubling of CO_2 ranging from 1.4 to 3.2°C, which probably gives a lower estimate of climate sensitivity since CLIMAP SSTs tend to be relatively too warm in the tropics (see below).

An alternative approach to evaluating climate sensitivity is provided by the detailed comparison of model results with proxy-data over different regions. The amplitude of the tropical cooling at LGM has long been disputed (Rind and Peteet, 1985; Guilderson *et al.*, 1994). Compared to a new synthesis of terrestrial data (Farrera *et al.*, 1999), PMIP simulations with prescribed CLIMAP sea-surface conditions produce land temperatures that are too warm (Pinot *et al.*, 1999) (Figure 8.14), which may be due to too-warm prescribed SSTs, as indicated by new marine data based on alkenone palaeo-thermometry (Rosell-Melé *et al.*, 1998) (Figure 8.14). Some mixed-layer ocean models have produced more realistic sea and land temperature cooling (Pinot *et al.*, 1999), enhancing our confidence in using such models to estimate climate sensitivity (see Chapter 9, Section 9.3.4) (Figure 8.14). The same conclusion is derived by Broccoli (2000) when accounting for uncertainties in both the forcing and reconstructed climate from various proxy data. Over Eurasia, all the models simulate a cooling in fairly good agreement with proxy data estimates, except over western Europe (Kageyama *et al.*, 2001), where they all underestimate the winter cooling shown from pollen data (Peyron *et al.*, 1998). However, such simulations have an important caveat since they prescribe present day ocean heat transport whereas changes in the North Atlantic deep water circulation shown by two EMIC models (Ganopolski *et al.*, 1998b; Weaver *et al.*, 1998) and also inferred by palaeo-oceanographic data (e.g., Duplessy *et al.*, 1988) may further decrease temperatures over Europe.

Land-surface feedbacks
Vegetation feedbacks at the LGM could have been due to

climate-induced shifts in biomes, CO_2-induced changes in vegetation structure (Jolly and Haxeltine, 1997; Street-Perrot *et al.*, 1997; Cowling, 1999), and CO_2-induced changes in leaf conductance (see Chapter 7, Section 7.4.2). Sensitivity experiments (Crowley and Baum, 1997; Levis *et al.*, 1999) suggest that the first two types dominated. Over much of Eurasia, forests were replaced by tundra or steppe (Prentice *et al.*, 1998) which may have contributed to the observed cooling over Europe (Crowley and Baum, 1997; Kubatzki and Claussen, 1998; Levis *et al.*, 1999). Permafrost may also have to be accounted for (Renssen *et al.*, 2000). In the tropics though, there is yet no systematic improvement of the simulated cooling, since the models find large areas of warming due to the simulated deforestation (Crowley and Baum, 1997; Levis *et al.*, 1999). However, land-surface feedbacks may also have affected climate through mineral aerosol (dust) concentrations (Mahowald *et al.*, 1999).

8.5.5.3 Summary

Mid-Holocene
Through PMIP experiments, it is now well-established that all atmospheric models are able to simulate several robust large-scale features of the Holocene climate but also that they all underestimate these changes. Several complementary simulations have shown that ocean and vegetation processes introduce important feedbacks which are necessary to explain the observed monsoon changes. These results urge for a systematic evaluation of coupled atmosphere-ocean-vegetation models for the mid-Holocene and for an investigation of the impact of vegetation changes, such as climate-induced density and land-use cover changes (see Chapter 7, Section 7.4.2), on future climate change projections.

Last Glacial Maximum
The more systematic evaluation of atmosphere alone models conducted within PMIP confirms that the LGM SST as estimated by CLIMAP (1981) need to be revised. Some simulations with atmospheric models coupled to mixed-layer models produce realistic results, especially in the tropics, and enhance our confidence in the estimates of climate sensitivity used in future climate change studies. However, such models neglect changes in ocean heat transport as well as land-surface feedbacks. Moreover, an evaluation of coupled AOGCMs is still needed at the LGM.

8.6 20th Century Climate and Climate Variability

8.6.1 20th Century Coupled Model Integrations Including Greenhouse Gases and Sulphate Aerosols

Since the pioneer experiments conducted at the Hadley Centre for Climate Prediction and Research (Mitchell *et al.* 1995) and at the Deutsche Klimarechenzentrum (DKRZ) (Hasselmann *et al.* 1995), reported in the SAR, a number of other groups internationally have reproduced the trend in the surface air temperature instrumental record over the 20th century. These include the Canadian Center for Climate Modelling and Analysis (CCCma) (Boer *et al.* 2000), Centre for Climate System Research/National Institute for

Observed and simulated global mean temperature

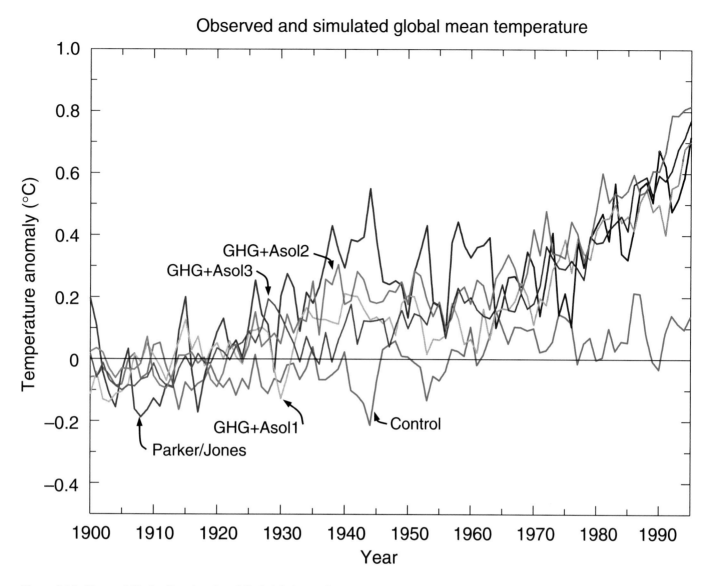

Figure 8.15: Observed (Parker/Jones) and modelled global annual mean temperature anomalies (°C) from the 1901 to 1930 climatological average. The control and three independent simulations with the same greenhouse gas plus aerosol forcing and slightly different initial conditions are shown from CGCM1 (Boer *et al.*, 2000). The greenhouse gas alone simulation is labelled GHG. The three greenhouse gas plus aerosol simulations are labelled GHG+Asol1, GHG+Asol2 and GHG+Asol3, respectively.

Environmental Studies (CCSR/NIES) (Emori *et al.* 1999), Commonwealth Scientific and Industrial Research Organization (CSIRO), Geophysical Fluid Dynamics Laboratory (GFDL) (Haywood *et al.* 1997) and the National Center for Atmospheric Research (NCAR) (Meehl *et al.* 2000b). Many of these new contributions, including recent experiments at the Hadley Centre and DKRZ, include an ensemble of projections over the 20th century (e.g., Figure 8.15). Such an ensemble allows for an estimate of intra-model variability, which in the case of the CCCma model (Figure 8.15), is larger than the possible anthropogenic signal through the early part of the 20th century (cf., inter-model variability shown in Figure 9.3).

Coupled models that have been used to simulate changes over the 20th century have all started with "control model" levels of atmospheric CO_2 (typically 330 ppm). This initial condition is

then referred to as the "pre-industrial" initial condition. Changes in radiative forcing are then calculated by taking the observed atmospheric equivalent CO_2 level over the 20th century as a difference relative to the actual pre-industrial level (280 ppm), and adding this as a perturbation to the control model levels. Implicit in this approach is the assumption that the climate system responds linearly to small perturbations away from the present climate. Haywood *et al.* (1997) demonstrated the near linear response of the GFDL-coupled model to changes in radiative forcing associated with increases in atmospheric greenhouse gases and sulphate aerosols. When added together, experiments which included aerosol and greenhouse gas increases separately over the 20th century yielded a similar transient response (in terms of globally averaged and geographical distribution of surface air temperature and precipitation) to

Linear trend in precipitation rate (%/100 yr)

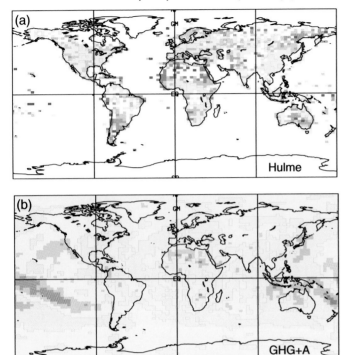

Figure 8.16: The geographical distribution of the linear trend in annual-mean precipitation (%/100 yr) for (a) the observations of Hulme (1992, 1994); (b) the ensemble of three greenhouse gas + aerosol integrations using the CCCma model. Taken from Boer *et al.* (2000).

an experiment which included both aerosol and greenhouse gas increases. This analysis is particularly important as it validates the methodological approach used in coupled model simulations of the 20th century climate.

As noted in the SAR, the inclusion of the direct effect of sulphate aerosols is important since the radiative forcing associated with 20th century greenhouse gas increase alone tends to overestimate the 20th century warming in most models. Groups that have included a representation of the direct effects of sulphate aerosols have found that their model generally reproduces the observed trend in the instrumental surface air temperature warming, thereby suggesting that their combination of model climate sensitivity and oceanic heat uptake is not unrealistic (see Chapter 9, Section 9.2.1 and Figure 9.7). These same models have more difficulty representing variability observed within the 20th century instrumental record (Sections 8.6.2, 8.6.3). As mentioned in Section 8.6.3, some modelling studies suggest that the inclusion of additional forcings from solar variability and volcanic aerosols may improve aspects of this simulated variability. Delworth and Knutson (2000), on the other hand, note that one of their six 20th century integrations (using GFDL_R30_c) bears a striking resemblance to the observed 20th century warming which occurs primarily in two distinct periods (from 1925 to 1944 and from 1978 to the present), without the

need for additional external forcing. In addition, all coupled models have shown a trend towards increasing global precipitation, with an intensification of the signal at the high northern latitudes, consistent with the observational record (Figure 8.16). Nevertheless, AOGCM simulations have yet to be systematically analysed for the occurrence of other key observed trends, such as the reduction in diurnal temperature range over the 20th century and the associated increase in cloud coverage.

The aforementioned studies all prescribed the temporal and geographical distribution of sulphate aerosols and included their radiative effects by perturbing the surface albedo according to the amount of sulphate loading in the atmospheric column above the surface (see Chapter 6, Sections 6.7, 6.8 and 6.14). This approach both ignores the indirect effect of these aerosols (i.e., their effects on cloud formation) as well as weather affects on aerosol redistribution and removal. Roeckner *et al.* (1999) made a major step forward by incorporating a sulphur cycle model into the ECHAM4(ECMWF/MPI AGCM)/OPYC3(Ocean isoPYCnal GCM) AOGCM to eliminate these shortcomings. In addition, they included the radiative forcing due to anthropogenic changes in tropospheric ozone by prescribing ozone levels obtained from an offline tropospheric chemistry model coupled to ECHAM4. The simulation of the 20th century climate obtained from this model (Bengtsson *et al.* 1999), which includes the indirect effect of aerosols, shows a good agreement with the general 20th century trend in warming (see Chapter 12, Section 12.4.3.3). The results of this study also suggest that the agreement between model and observed 20th century warming trends, achieved without the inclusion of the indirect aerosol effect, was probably accomplished with an overestimated direct effect or an overestimated transient oceanic heat uptake. Alternatively, since these studies include only idealised scenarios of sulphate radiative forcing alone (direct and/or indirect) that do not include the apparent effects of other aerosol types (Chapter 6), one might view the sulphate treatment as a surrogate, albeit with large uncertainty, for the radiative forcing associated with all anthropogenic aerosols.

As noted in Chapter 2, Sections 2.2.2 and 2.2.3, land surface temperatures show a greater rate of warming than do lower tropospheric air temperatures over the last 20 years (see also discussion in Chapter 2, Section 2.2.4). While noting uncertainties in the observational records (Chapter 2, Sections 2.2.2 and 2.2.3), the National Research Council (NRC) (2000) pointed out that models, which tend not to show such a differential trend, need to better capture the vertical and temporal profiles of the radiative forcing especially associated with water vapour and tropospheric and stratospheric ozone and aerosols, and the effects of the latter on clouds. Santer *et al.* (2000) provide further evidence to support this notion from integrations conducted with the ECHAM4/OPYC3 AOGCM (Bengtsson *et al.* 1999; Roeckner *et al.*, 1999) that includes a representation of the direct and indirect effects of sulphate aerosols, as well as changes in tropospheric ozone. They showed that the further inclusion of stratospheric ozone depletion and stratospheric aerosols associated with the Pinatubo eruption lead to a better agreement with observed tropospheric temperature changes since 1979, although discrepancies still remain (see Chapter 12, Section 12.3.2 and Figure 12.4).

8.6.2 Coupled Model Variability

8.6.2.1 Comparison with the instrumental record

Barnett (1999) concatenated the annual mean near-surface temperature anomaly fields from the first 100 years of integration of eleven CMIP control experiments to produce common empirical orthogonal functions (EOFs) for the eleven AOGCMs. By projecting the Global Sea Ice and Sea Surface Temperature (GISST) annual mean temperature anomaly data set (Rayner *et al.* 1996) onto these common EOFs, he was able to estimate to what extent model variability represented the observed variability. An analysis of the partial eigenvalue spectrum for the different models (Figure 8.17a) suggests that there is considerable disparity between the estimates of variability within the coupled models. Some of this disparity arises from model drift and other low-frequency variability. Intra-model disparity was much lower than inter-model disparity as demonstrated by a similar common EOF analysis obtained from ten 100 year segments in the 1,000 year GFDL_R15_a control run (Figure 8.17b). While the highest two modes were substantially underestimated in the GFDL_R15_a model, the higher modes agreed better with observations. Error bars on the observational data are large and when this is taken into account, model disagreement with observations may not be significant. As Barnett (1999) did not remove the trend over the 20th century in the GISST data set, the observations also contain responses to both natural and anthropogenic forcing. One would therefore expect control integrations from coupled climate models to underestimate the observed spectrum at low frequencies (e.g., Folland *et al.*, 1999).

An analogous study by Stouffer *et al.* (2000) compared the surface air temperature variability from three long 1,000-year CMIP integrations (GFDL_R15_a, HadCM2, ECHAM3/LSG (Large-Scale Geostrophic ocean model)) to the variability found in the same observational data set (Jones and Briffa, 1992; Jones 1994). They argued that, over the instrumental period, the simulated variability on annual to decadal time-scales was fairly realistic both in terms of the geographical distribution and the global mean values with a notable exception of the poor simulation of observed tropical Pacific variability (Figure 8.18). The HadCM2 model substantially overestimated tropical Pacific variability, whereas it was underestimated in the GFDL_R15_a and ECHAM3/LSG models. They also noticed that on the inter-decadal time-scale, the greatest variance in the models was generally located near sea-ice margins close to regions of deep oceanic convection and associated with low-frequency variations of the thermohaline circulation. While the three models generally agreed on the dominant modes of variability, there was substantial inter-model disparity in the magnitude of each mode. The analysis of Stouffer *et al.* (2000) can easily be reconciled with Barnett (1999) by realising that Stouffer *et al.* (2000) examined a subset (GFDL_R15_a, HadCM2, ECHAM3/LSG models) of those CMIP models considered by Barnett (1999) that did not experience climate drift, and that less emphasis was placed on the poor resolution of tropical Pacific variability.

As an extension to the above analysis, Bell *et al.* (2000) compared annual mean surface air temperature variability in sixteen CMIP control simulations to the thermometer record, on time-

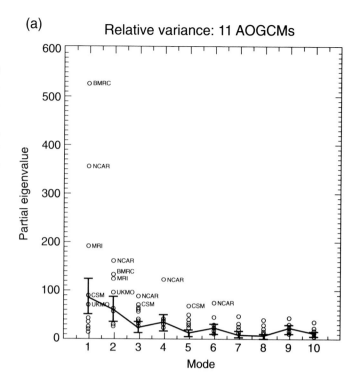

(a)

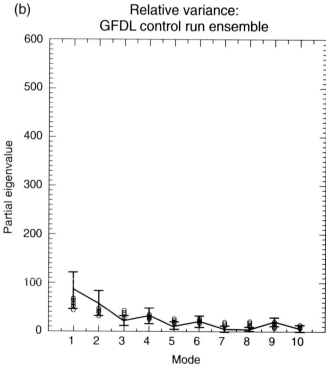

(b)

Figure 8.17: (a) Partial eigenvalue spectrum for the EOFs from eleven CMIP AOGCMs. The observations (heavy line), together with their 95% confidence limits (vertical bars), are obtained by projecting the observed surface air temperature (Jones and Briffa, 1992; Rayner *et al.*, 1996) onto the common EOFs obtained from eleven CMIP AOGCMs. (b) as in (a) except for ten separate 100-year segments of the GFDL control run. Taken from Barnett (1999).

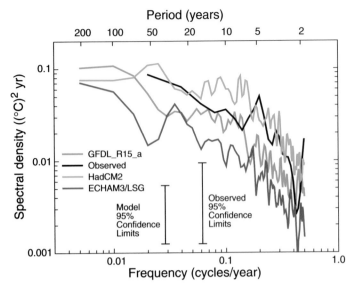

Figure 8.18: Power spectra of the detrended globally averaged annual mean surface air temperature (SAT) anomaly. The curves represent the estimates obtained from HadCM2 (blue), GFDL_R15_a (green) and ECHAM3/LSG (red). The observed (black line) is from the globally averaged annual mean SAT anomalies compiled by Jones and Briffa (1992). The spectra are smoothed Fourier transforms of the autocovariance function using a Tukey window of 100 lags for the models and 30 lags for the observations. The two vertical lines represent the range of 95% confidence in the spectral estimates for the model and the observations. Taken from Stouffer *et al.* (2000).

scales of 1 year to 40 years (Figure 8.19). The authors found that: (1) thirteen of the sixteen CMIP models underestimate variability in surface air temperatures over the global oceans; (2) twelve of the sixteen models overestimate variability over land; (3) all the models overestimate the ratio of air temperature variability over land to variability over oceans. These results likely reflect problems in both the ocean and land-surface components of climate models. In particular, underestimation of variability over oceans may be due, at least in part, to weak or absent representations of El Niño in the models; overestimation of variability over land may be due to poor land surface parametrizations including insufficient soil moisture.

Duffy *et al.* (2000) also partitioned the CMIP models into those that are flux adjusted and those that are not. They defined two measures of temperature variability and applied them to the CMIP control simulations. The simulations differed substantially in the amount of temperature variability they showed. However, on time-scales of 1 year to 20 years, the flux adjusted simulations did not have significantly less variability than the non-flux adjusted simulations; there is some suggestion that they may have *more* variability. Thus it cannot be argued, for example, that the use of flux adjusted models in studies of detection of anthropogenic climate change tends to make observed temperature changes seem more significant than they should be, compared to natural internal climate variability. Nevertheless, it is still an open question as to how coupled model variability depends on internal model parameters and resolution.

8.6.2.2 *Comparison with palaeo-data*

There have been relatively few studies which have undertaken a systematic comparison of AOGCM variability with variability found in the Holocene proxy temperature record. However, three studies (from the MPI (Max Planck Institute), Hadley Centre and GFDL) that focus on the analysis of long control integrations are available. Barnett *et al.* (1996) demonstrated that the GFDL_R15_a (Stouffer *et al.*, 1994) and ECHAM1/LSG (Cubasch *et al.*, 1994) models underestimate the levels of decadal-scale variability in summer palaeo-temperature proxies from 1600 to 1950 (expanded version of Bradley and Jones, 1993), with increasing disparity with observations at lower frequencies.

Using annual and decadal mean near-surface palaeo-temperature reconstructions at seventeen locations Jones *et al.* (1998) demonstrated, through a principal component analysis, that the standard deviation of the GFDL_R15_a model (Stouffer *et al.*, 1994) and the HadCM2 model (Johns *et al.*, 1997; Tett *et al.*, 1997) principal component time-series compared favourably with both proxy and observed data. Time-series of the top seven principal components did, however, show much less century time-scale variability than in the proxy time-series. This was especially true in the HadCM2 model that was dominated by tropic-wide decadal variability. Through cross-spectral analysis they concluded that the "GFDL control integration bears a remarkable similarity in its statistical properties to that obtained from the proxy data. In view of this similarity it appears the spatial structures from the control integration can be used to represent the spatial structures of naturally occurring variations in near-surface air temperature". This conclusion was also highlighted by Delworth and Mann (2000) who noted that both palaeo-temperature reconstructions (Mann *et al.*, 1998) and the GFDL_R15_a coupled model suggest a distinct oscillatory mode of climate variability (with an approximate time-scale of about 70 years) of hemispheric scale and centred around the North Atlantic.

8.6.3 *The Role of Volcanic and Solar Forcing and Changes in Land Use*

Coupled model control runs have a general tendency to underestimate the variability found in both the instrumental and palaeo-proxy record, especially over the oceans (the converse is true over land when compared to the instrumental record). With the exception of Santer *et al.* (2000), none of the aforementioned simulations examined the potential climatic effects of stratospheric aerosols associated with volcanic emissions. On the longer time-scales, none of these studies included variability in solar forcing. Crowley (2000) has estimated that changes in solar irradiance and volcanism may account for between 41 and 64% of pre-industrial, decadal-scale surface air temperature variations.

Cubasch *et al.* (1997) demonstrated that when solar variations were included in the ECHAM3/LSG model, their simulation from 1700 through the 20th century showed enhanced low-frequency variability associated with variability in solar irradiance (see also Lean and Rind, 1998). The implication of climate change detection and attribution studies (Chapter 12; Hegerl *et al.*, 1997; Tett *et al.*, 1999) for the reproduction of 20th century climate by AOGCMs, is that changes in solar irradiance may be important to include if one

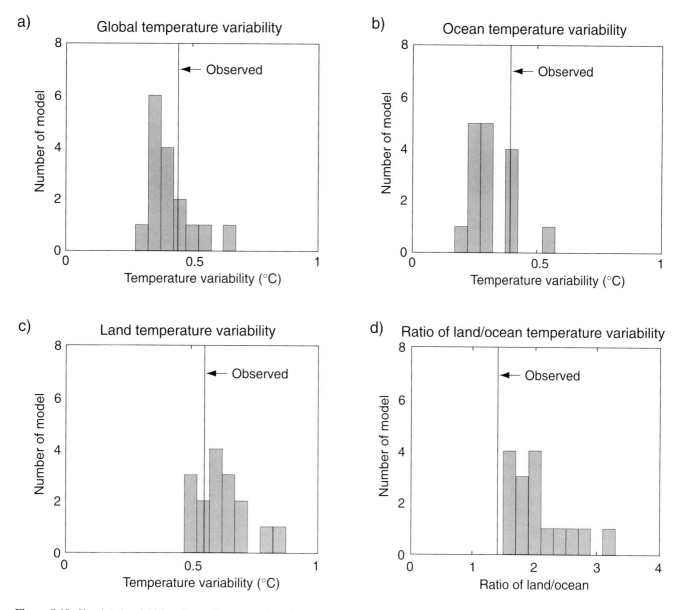

Figure 8.19: Simulated variability of annual mean surface air temperatures over the last 40 years, 1959 to 1998 in sixteen CMIP simulations and in observations (Jones, 1994; Parker *et al.*, 1995). (a) Global-mean temperature variability; four models show higher than observed amounts of variability. (b) Mean over-ocean temperature variability; three models show higher than observed amounts of variability. (c) Mean over-land temperature variability; four models show less than observed amounts of variability. (d) Ratio of over-land to over-ocean temperature variability; all models show higher than observed ratios. Taken from Bell *et al.* (2000).

wants to reproduce the warming in the early part of the century. As noted earlier, it is conceivable that this early warming may also be solely a result of natural internal climate variability (Delworth and Knutson, 2000). Energy balance/upwelling diffusion climate models and Earth system models of intermediate complexity, when forced with volcanic and solar variations for the past 400 years, capture the cooling associated with the Little Ice Age (Betrand *et al.*, 1999; Crowley and Kim, 1999; Free and Robock, 1999), although they are not capable of assessing regional climatic anomalies associated with local feedbacks or changes in atmospheric dynamics. These same models produce the observed warming of the past century when additionally forced with anthropogenic greenhouse gases and aerosols.

Hansen *et al.* (1997) conducted a systematic study of the climate system response to various radiative forcings for the period 1979 to 1995 (over which period Nimbus 7 satellite data were available) using the Goddard Institute for Space Studies (GISS) AGCM coupled to a mixed-layer ocean model. A series of ensemble simulations, with each ensemble consisting of five experiments, were conducted by cumulatively adding, one-by-one, radiative forcing effects due to stratospheric aerosols (associated with volcanic emissions), decreases in upper level ozone, increases in anthropogenic greenhouse gases, and changes in solar irradiance (Figure 8.20). While changes in tropospheric aerosols, either via direct or indirect effects, were not included in their calculations, over the short record a reasonable agreement

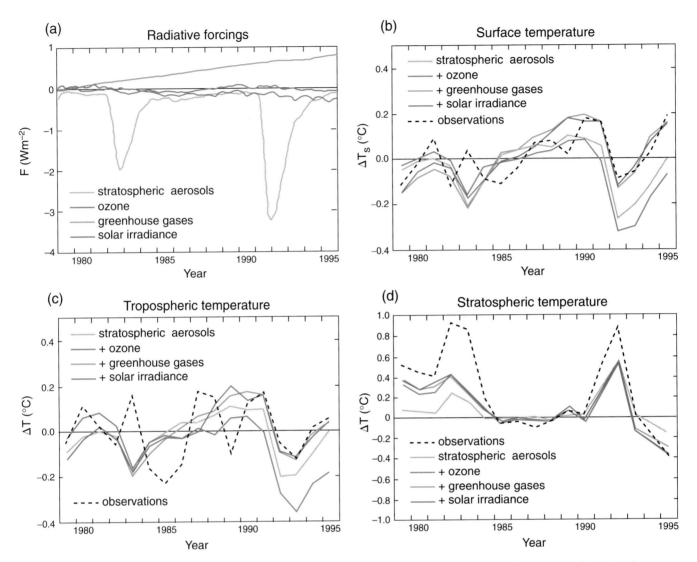

Figure 8.20: (a) Radiative forcing (Wm^{-2}) since 1979 due to changes in stratospheric aerosols, ozone, greenhouse gases and solar irradiance. (b) to (d) Observed global annual mean surface, tropospheric and stratospheric temperature changes and GISS GCM simulations as the successive radiative forcings are cumulatively added one by one. An additional constant 0.65 Wm^{-2} forcing was included in all their simulations, representing the estimated disequilibrium forcing for 1979. The base period defining the zero mean observed temperature was 1979 to 1990 for the surface and the troposphere and 1984 to 1990 for the stratosphere. Taken from Hansen *et al.* (1997).

with observations was obtained. Internal climate variability (e.g., the warming associated with the El Niño of 1983 and the cooling associated with the La Niña of 1989 in Figure 8.20b,c) is not well resolved in the model. These experiments point out that while solar irradiance changes caused minimal changes over the period (consistent with the analysis of Hegerl *et al.*, 1997; Tett *et al.*, 1999), stratospheric aerosols associated with volcanic emissions and changes in upper level ozone are important components which need to be included if one hopes to accurately reproduce the variations in the instrumental record.

Changing land-use patterns affect climate in several ways (see Chapter 6, Section 6.11). While the impact of land-use changes on radiative forcing is small (e.g., Hansen *et al.*, 1998) changes in roughness, soil properties and other quantities may be important (see Chapter 7, Section 7.4). Brovkin *et al.* (1999)

demonstrated that CLIMBER (Climate Biosphere Model) was able to capture the long-term trends and slow modulation of the Mann *et al.* (1998) reconstruction of Northern Hemisphere temperatures over the past 300 years provided changes in land-use patterns (as well as changes in atmospheric CO_2 and changing solar forcing) were taken into account. Some recent model results suggest that land cover changes during this century may have caused regional scale warming (Chase *et al.*, 2000; Zhao *et al.*, 2001) but this remains to be examined with a range of climate models.

8.6.4 *Climate of 20th Century: Summary*

Several coupled models are able to reproduce the major trend in 20th century surface air temperature, when driven by historical radiative forcing scenarios corresponding to the 20th century.

However, in these studies idealised scenarios of only sulphate radiative forcing have been used. One study using the ECHAM4/OPYC model that includes both the indirect and direct effects of sulphate aerosols, as well as changes in tropospheric ozone, suggests that the observed surface and tropospheric air temperature discrepancies since 1979 are reduced when stratospheric ozone depletion and stratospheric aerosols associated with the Pinatubo eruption are included. Systematic evaluation of 20th century AOGCM simulations for other trends found in observational fields, such as the reduction in diurnal temperature range over the 20th century and the associated increase in cloud coverage, have yet to be conducted.

The inclusion of changes in solar irradiance and volcanic aerosols has improved the simulated variability found in several AOGCMs. In addition, some evaluation studies aimed at the reproduction of 20th century climate have suggested that changes in solar irradiance may be important to include in order to reproduce the warming in the early part of the century. Another study has suggested that this early warming can be solely explained as a consequence of natural internal climate variability.

Taken together, we consider that there is an urgent need for a systematic 20th century climate intercomparison project with a standard set of forcings, including volcanic aerosols, changes in solar irradiance and land use, as well as a more realistic treatment of both the direct and indirect effects of a range of aerosols.

8.6.5 Commentary on Land Cover Change

Land-cover change can occur through human intervention (land clearance), via direct effects of changes in CO_2 on vegetation physiology and structure, and via climate changes (Chapter 7). Evidence from observational studies (see Chapter 7, Section 7.4.2) and modelling studies (e.g., Betts *et al.*, 1997, 2000; Chase *et al.*, 2000; Zhao *et al.*, 2001) demonstrate that changes in land cover can have a significant impact on the regional scale climate but suggestions that land clearance has an impact on the global scale climate is currently speculative. Evidence from palaeoclimate (Section 8.5) and modelling work (Section 8.5 and Chapter 7, Section 7.4) indicates that these changes in vegetation may lead to very significant local and regional scale climate changes which, in some cases, may be equivalent to those due to increasing CO_2 (Pitman and Zhao, 2000).

On time-scales of decades the impact of land cover change could significantly influence the rate of atmospheric CO_2 increase (Chapter 3), the nature and extent of the physical climate system response, and ultimately, the response of the biosphere to global change (Chapter 8). Models currently under development that can represent changes in land cover resulting from changes in climate and CO_2 should enable the simulations of these processes in the future. If these models can be coupled with scenarios representing human-induced changes in land cover over the next 50 to 100 years, the important effects of land-cover change can be included in climate models. While the inclusion of these models of the biosphere is not expected to change the global scale response to increasing CO_2, they may significantly effect the simulations of local and regional scale change.

8.7 Coupled Model: Phenomena

The atmosphere-ocean coupled system shows various modes of variability that range widely from intra-seasonal to inter-decadal time-scales (see Chapters 2 and 7). Since the SAR, considerable progress has been achieved in characterising the decadal to inter-decadal variability of the ocean-atmosphere system (Latif, 1998; Navarra, 1999). Successful evaluation of models over a wide range of phenomena increases our confidence.

8.7.1 El Niño-Southern Oscillation (ENSO)

ENSO is a phenomenon resulting from large-scale air-sea inter-actions (see Chapter 7, Section 7.6.5). ENSO modelling has advanced considerably since the SAR (e.g., Yukimoto *et al.*, 1996; Kimoto and Shen, 1997; Knutson *et al.*, 1997; Timmermann *et al.*, 1998). Some models now use enhanced horizontal resolution in the tropics to better resolve equatorial ocean dynamics. Models show SST variability in the tropical Pacific, which has some similarity to observed ENSO as is shown in upper panels of Figure 8.21. However, some aspects of ENSO are still not well captured by present day coupled models (Delecluse *et al.*, 1998). Latif *et al.* (1999) analysed the SST climatology and interannual variability simulated by twenty four models in the equatorial Pacific. When compared with observations, the models have flaws in reproducing the annual cycle. About half of the models are characterised by too weak interannual variability in the eastern equatorial Pacific, while models generally have larger variability in the central equatorial Pacific (Table 8.2). It was found that the majority of the models show the observed ENSO-monsoon relationship, that is, a weak Indian summer monsoon tends to be associated with El Niño.

Seasonal forecasting with coupled global models has just begun (Barnston *et al.*, 1999; McPhaden, 1999), although few of the models discussed in Chapter 9 are used. While forecast skill of coupled global models is still lower than statistical models (Landsea and Knaff, 2000), coupled global models have better skill than simple models. The 1997 to 1998 El Niño event (Trenberth, 1998b) is a good test of coupled model forecast systems. Figure 8.22 plots the SST anomaly during the 1997 to 1998 El Niño for predictions made with various initial conditions by prediction systems at ECMWF (Stockdale *et al.*, 1998), the Japan Meteorological Agency (JMA) (Ishii *et al.*, 1998), NCEP (Barnston *et al.*, 1999) and for the hindcast made at the Bureau of Meteorology Research Centre (BMRC) (Wang *et al.*, 2000). Those comprehensive models predicted unusually warm tropical Pacific SST for 1997, albeit with underestimation of the strength of the event and the warming speed. A similar conclusion is reached with other global climate models (Oberhuber *et al.*, 1998; Zhou *et al.*, 1998). Unusually strong Madden-Julian Oscillation (MJO, see Section 8.7.4) and westerly wind bursts may have affected not only the timing but also the amplitude of the 1997 to 1998 El Niño (McPhaden, 1999; Moore and Kleeman, 1999), and, in this respect, models may fail to forecast the onset of an El Niño in some circumstances. However, these results suggest an improved ability of coupled models to forecast El Niño if sufficient data to initialise the model are available from a good ocean data assimilation system.

(a) Observed

Time-scales < 12 years

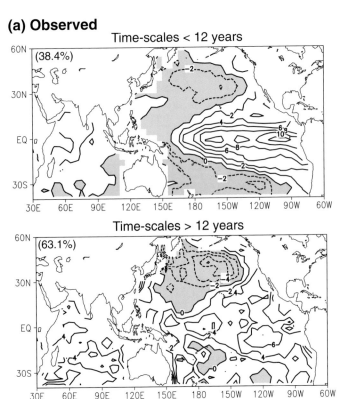

Time-scales > 12 years

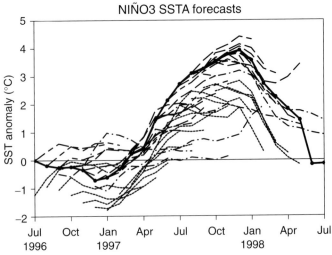

Figure 8.22: Niño-3 SST anomaly predictions and hindcast made at various times during the 1997 to 1998 El Niño event together with the subsequent observed SST anomaly (solid). Predictions made at ECMWF (Stockdale *et al.*, 1998, long dash), JMA (Ishii *et al.*, 1998, short dash), NCEP (Barnston *et al.*, 1999, long short dash) and the hindcast made at BMRC (Wang *et al.*, 2000, dot dash) are shown.

(b) Modelled

Time-scales < 12 years

Time-scales > 12 years

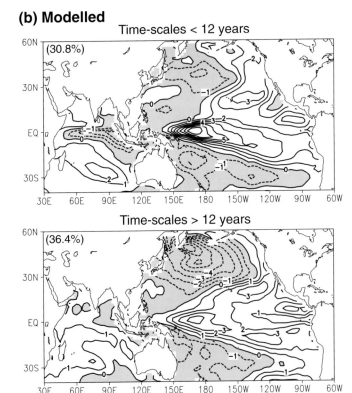

Figure 8.21: Comparison of eigenvectors for the leading EOFs of the SSTs between the ENSO time-scale (<12 years) (upper panels) and the decadal time scale (>12 years) (lower panels) for (a) observation, and (b) the MRI coupled climate model, respectively (Yukimoto, 1999). Numbers in bracket at the upper left show explained variance in each mode.

In summary, the higher resolution coupled climate models employed since the SAR are better able to simulate El Niño-like SST variability in the tropical Pacific. However, there still remain common model errors such as weaker amplitude of SST anomalies and westward shift of the variability maximum compared to the observations. Current models can predict major El Niño events with some accuracy, suggesting that, as the resolution increases and the model physics improves, El Niño simulation will also improve.

8.7.2 Pacific Decadal Oscillation (PDO)

The leading mode in the Pacific with decadal time-scale is usually called the Pacific Decadal Oscillation (PDO, see Chapter 2, Section 2.6.3). Unlike the well-documented interannual mode (ENSO), the decadal pattern does not have a distinctive equatorial maximum. Several coupled climate models are able to reproduce a pattern of this decadal variability broadly similar to the observed pattern (Latif and Barnett, 1996; Robertson, 1996; Yukimoto *et al.*, 1996, 2000; Knutson and Manabe, 1998; Yu *et al.*, 2000a). These modelling groups have proposed different mechanisms to explain the observed Pacific decadal variability based on analysis of large samples of simulated decadal variability in their coupled models. An example is shown in Figure 8.21b, where larger SST variability in the North Pacific than in the equatorial region is captured (Yukimoto, 1999). While the geographical location of the mid-latitude poles and the amplitude ratio between the tropical and mid-latitude poles vary slightly from one model to another, pattern correlation between the observed and model leading decadal EOFs are quite high.

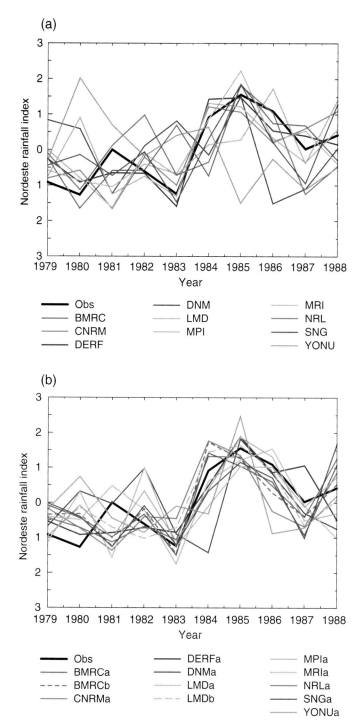

Figure 8.23: Simulated (March-May) and observed (March-April) averaged Nordeste (north-eastern Brazil) rainfall indices for (a) the original AMIP simulations, (b) the revised AMIP simulations (Sperber *et al.*, 1999).

8.7.3 Monsoons

Monsoon constitutes an essential phenomenon for a tropical climate (see Chapter 7, Section 7.6.3). The monsoon precipitation simulated by AGCMs has been evaluated in AMIP (Sperber and Palmer, 1996; Zhang *et al.*, 1997; Gadgil and Sajani, 1998). The seasonal migration of the major rain belt over the West African region is well simulated by almost all models. However coarse resolution climate models generally fail to give satisfactory simulations of the East Asian, East African and North American monsoons (Stensrud *et al.*, 1995; Lau and Yang, 1996; Semazzi and Sun, 1997; Yu *et al.*, 2000b). For example, models have excessive precipitation in the eastern periphery of the Tibetan Plateau. Increase of horizontal resolution can improve the precipitation details, but may not be sufficient to remove large-scale model biases (Kar *et al.*, 1996; Lal *et al.*, 1997; Stephenson *et al.*, 1998; Chandrasekar *et al.*, 1999; Martin, 1999; also see Section 8.9.1).

Interannual variations of Nordeste (north-eastern Brazil) rainfall are well captured with atmospheric models with prescribed interannually varying SST (Potts *et al.*, 1996; Sperber and Palmer, 1996). This is also the case for the South American monsoon (Robertson *et al.*, 1999) and the West African monsoon (Rowell *et al.*, 1995; Semazzi *et al.*, 1996; Rocha and Simmonds, 1997; Goddard and Graham, 1999). The precipitation variation over India is less well simulated. However the models show better skill in reproducing the interannual variability of a wind shear index over the Indian summer monsoon region, indicating that the models exhibit greater fidelity in capturing the large-scale dynamic fluctuations than the regional scale rainfall variations.

More recent atmospheric models with revised physical parametrizations show improved interannual variability of the all-India rainfall, Indian/Asian monsoon wind shear, Sahel and Nordeste rainfall (Figure 8.23) (Sperber *et al.*, 1999). Improvement in the simulation of interannual variability is associated with a better simulation of the observed climate by the models (Sperber and Palmer, 1996; Ferranti *et al.*, 1999; Martin and Soman, 2000). The observed rainfall/ENSO SST correlation pattern is better simulated by those models that have a rainfall climatology in closer agreement with observations (Gadgil and Sajani, 1998).

Coupled climate models that simulate El Niño-like SST variability in the tropical Pacific indicate a strong connection between ENSO and the strength of the Indian summer monsoon in qualitative agreement with observations (Meehl and Arblaster, 1998; Kitoh *et al.*, 1999; Latif *et al.*, 1999). Besides the ENSO time-scale, the South Asian monsoon reveals a strong biennial oscillation. Coupled models can reproduce this tropospheric biennial oscillation (TBO) (Meehl, 1997; Ogasawara *et al.*, 1999).

8.7.4 Madden and Julian Oscillation (MJO)

MJO is a 30 to 60 day oscillation that moves eastward in the tropical large-scale circulation, and affects both mid-latitude atmospheric circulation and the Asian-Australian monsoon. Slingo *et al.* (1996) showed that nearly all of the AMIP models have power in the intra-seasonal time-scale of equatorial upper troposphere zonal wind at higher frequencies than the observation. They also show that most models underestimated the strength of the MJO. Slingo *et al.* (1999) show that the HadAM3 model forced by the observed SST displays a decadal time-scale variability of MJO activity as observed, implying a possible link between long-term changes of tropical SST and MJO activity, and also the ability of a current atmospheric model to simulate it.

Recent studies suggest an important role of air-sea interaction on the intra-seasonal time-scale phenomena (Flatau *et al.*, 1997; Waliser *et al.*, 1999; Li and Yu, 2001), thus a possible improvement in reproducing the MJO by coupled climate models. This warrants a need to evaluate the MJO in coupled climate models, but this is yet to be undertaken.

8.7.5 *The North Atlantic Oscillation (NAO) and the Arctic Oscillation (AO)*

The North Atlantic Oscillation (NAO) is a regional mode of variability over the North Atlantic, while the Arctic Oscillation (AO) is a hemispheric mode of variability which resembles in many respects the NAO (see Chapter 7, Section 7.6.4). Coupled climate models simulate the NAO quite well, although there are some differences in its amplitude (Delworth, 1996; Laurent *et al.*, 1998; Saravanan, 1998; Osborn *et al.*, 1999). Atmospheric models with prescribed SST also simulate the spatial pattern of NAO variability fairly well (Rodwell *et al.*, 1999), although coupling to an interactive ocean does seem to produce the most realistic NAO pattern. A realistic AO is simulated in the CCCma (Fyfe *et al.*, 1999), GISS (Shindell *et al.*, 1999) and GFDL (Broccoli *et al.*, 1998) climate models. The AO extends into the mid-troposphere to lower stratosphere where it is associated with variations in westerly wind speed (see Chapter 2, Section 2.6.5 and Chapter 7, Section 7.6.4). This coupled troposphere-stratosphere mode of internal variability has been reproduced in the Meteorological Research Institute (MRI) coupled climate model (Kitoh *et al.*, 1996; Kodera *et al.*, 1996).

8.7.6 *Pacific-North American (PNA) and Western Pacific (WP) Patterns*

The Pacific-North American (PNA) and Western Pacific (WP) patterns are low-frequency teleconnection patterns (Wallace and Gutzler, 1981). Observations show that the PNA and WP are sensitive to the frequency distribution of the SST anomalies associated with ENSO. The HadAM3 model correctly reproduces the changes in frequency distribution of the PNA pattern between the El Niño years and the La Niña years (Renshaw *et al.*, 1998). However, this model fails to reproduce the WP mode distribution. On the other hand, the JMA atmospheric model showed an ability to simulate the WP with reasonable intensity, responding to SST anomalies (Kobayashi *et al.*, 2000). How extra-tropical air-sea interactions affect such weather regimes is not yet clear and an evaluation of the ability of coupled climate models to simulate these modes is yet to be undertaken.

8.7.7 *Blocking*

Blocking affects the large-scale flow and storm tracks and thus is important for mid-latitude climate. D'Andrea *et al.* (1998) evaluated the statistical behaviour of fifteen AMIP AGCMs in simulating Northern Hemisphere mid-latitude blocking. The AMIP models simulate reasonably well the seasonality and geographical location of blocking, but have a general tendency to underestimate both blocking frequency and the average duration of blocks. Using the ECMWF model, Brankovic and Molteni (1997) obtained a more realistic representation of Pacific blocking. This was due to reduced systematic error of zonal flow over the north-eastern Pacific. However, model deficiencies still remain in the Atlantic region. A link between the mean flow error in a model and blocking was also shown by Stratton (1999). In general, more recent atmospheric models show an improvement in ability to reproduce atmospheric blocking, but a corresponding evaluation of coupled climate models has not yet been undertaken.

8.7.8 *Summary*

Recent atmospheric models show improved performance in simulating many of the important phenomena, compared with those at the time of the SAR, by using better physical parametrizations and using higher resolutions both in the horizontal and in the vertical domain. A systematic evaluation of the ability of coupled climate models to simulate a full range of the phenomena referred in this section is yet to be undertaken. However, an intercomparison of El Niño simulations, one of the most important phenomena, has revealed the ability of coupled climate models to simulate the El Niño-like SST variability in the tropical Pacific and its associated changes in precipitation in the tropical monsoon regions, although the region of maximum SST variability is displaced further westward than in the observations.

8.8 Extreme Events

Since the SAR, there has been more attention paid to the analysis of extreme events in climate models. Unfortunately, none of the major intercomparison projects such as AMIP and CMIP have had diagnostic sub-projects that concentrated on analyses of extreme events. Very few coupled models have been subjected to any form of systematic extreme event analysis. Intercomparison of extreme events between models is also made very difficult due to the lack of consistent methodologies amongst the various analyses and also to the lack of access to high-frequency (at least daily) model data. Analysis has also been limited by the comparatively low resolution at which most models are run, this presents difficulties since most extreme events are envisaged to occur at the regional scale and have comparatively short lifetimes. However other forms of extreme event analysis have been developed which use the large-scale fields produced by a climate model and produce various indices of extreme events; such indices include maximum potential intensity of tropical cyclones (Holland, 1997) or maps of 20-year return values of variables such as precipitation or maximum temperature (Zwiers and

Table 8.4: *Analyses of extreme events in GCMs since the SAR. Wherever possible the model names have been made consistent with Table 8.1; however, since much of the analysis has been done with AGCMs alone (and often with comparatively old model versions) there often is no correspondence between these two tables. The references given refer to the particular analysis used, and are not necessarily tied to a specific model description.*

Names	References	Characteristics		Extreme events			
		AGCM	**OGCM**	**T**	**Pr**	**ETC**	**TC**
ARPEGE-C	Royer *et al.*,1998	T42, L30	no				F,G
CCC2	Zwiers and Kharin, 1998	T32, L10	no	D,L,R	D,L,R	P	P
CCM	Tsutsui and Kasahara, 1996	T42, L18	no				F,G
	Zhang and Wang, 1997	T42, L18	no				F
	Kothavala, 1997	T42L18	no		E		
CGCM1	Kharin and Zwiers, 2000	T32, L10	T64, L29	D,L.R	D,L,R	P	P
CSIRO	Watterson *et al.*, 1995	R21, L9	no				F,G
	Walsh and Pittock, 1998	R21, L9	no	E			I,T,W
	Schubert *et al.*, 1998	R21, L9	no	D,E			
ECHAM	Bengtsson *et al.*, 1995, 1996, 1999	T106, L19	no				F,I,N
	Lunkeit *et al.*, 1996	ECHAM2	OPYC			M	
	Beersma *et al.*, 1997	ECHAM3	no				F,S
	Christoph *et al.*1997	T42, L19	no				S
	Schubert *et al.*, 1998	T42, L19	LSG				F,I,S
FSU	Krishnamurti *et al.*, 1998	T42, L16	no				F
GFDL	Vitart *et al.*, 1997	T42, L18	no				F
	Haywood *et al.*, 1997	R15, L9	GFDL_R15_a		D		
	Knutson *et al.*, 1998	R30, L14	no				I
	Delworth *et al.*, 1999	R15L9	no	H			
	Wetherald and Manabe, 1999	R15L9	no		D		
HadCM2	Carnell and Senior, 1998	2.5×3.75, L19	2.5×3.75, L20				N,S
HadCM2b	Bhaskaran and Mitchell, 1998	2.5×3.75, L19	2.5×3.75, L20		E		
HadAM2	Thorncroft and Rowell, 1998	2.5×3.75,L19	no				L,W
	Durman *et al.*, 2001	2.5×3.75,L19	no		D,E		
JMA	Sugi *et al.*, 1997	T106, L21	no				C
	Yoshimura *et al.*, 1999	T106, L21	no				C
JMA/NIED	Matsuura *et al.*, 1999	T106, L21	0.5×1.0, L37				F
PMIP	Kageyama *et al.*, 1999	ECHAM3, LMD, UGAMP, UKMO	no		D		S
UKMO	Gregory and Mitchell1, 1995	2.5×3.75, L11	no	D	D		
	Hulme and Viner, 1998	UKTR	no				T
AGCMs	Hennessy *et al.*, 1997	CSIRO, UKHI	no		D		
	Henderson-Sellers *et al.*, 1998	GFDL,ECHAM3	no				F,G,I,N,U
	McGuffie *et al.*, 1999	BMRC, CCM	no		E,L,R		E,L,R
	Zhao *et al.*, 2000	IAP, NCC	IAP				

Under "Extreme events", column T denotes extremes in temperature, Pr denotes extremes in precipitation, ETC denotes extra-tropical cyclone, TC denotes tropical cyclone. The model names and characteristics are further explained (where possible) in Table 8.1.

GCM analyses have been with different techniques and methods designated as: C for cyclone centres; D for daily variability of temperature or precipitation; E for extreme temperature or precipitation; F for frequency of cyclones; G for Gray's yearly genesis parameter; H for heat index; I for intensity of cyclone; L for dry/wet spells or hot/cold spells; M for maximum eddy growth rate; N for numbers of cyclones; P for wind speed; R for return value or return period; S for storm track; T for sea surface temperature; U for maximum potential intensity; W for wave activity.

Kharin, 1998) (a 20-year return value implies that the value given is reached once in every 20 years).

In this chapter we assess the following types of extreme events that can be presented in terms of global patterns; frequency of tropical cyclones, daily maximum and minimum temperature, length of hot or cold spells, and precipitation intensity and frequency (floods and droughts). While it is arguable that extra-tropical cyclones belong to the class of "extreme events" we choose to include them here for consistency with other chapters. Table 8.4 summarises the climate models and the types of extreme events that have been analysed since the SAR. Assessments of extreme events that are purely local or regional are discussed in Chapter 10.

8.8.1 Extreme Temperature

Analysis of extreme temperature in climate model simulations has concentrated on the surface daily maximum and minimum temperature, or on the duration of hot/cold spells on the global scale (Schubert, 1998; Zwiers and Kharin, 1998; McGuffie *et al.*, 1999; Kharin and Zwiers, 2000).

Zwiers and Kharin (1998) and Kharin and Zwiers (2000) analysed the 20-year return values for daily maximum and minimum screen temperature simulated by both CCC GCM2 and CGCM1. Comparison with the NCEP reanalyses shows that the model reproduced the return values of both maximum and minimum temperature and warm/cold spells reasonably well.

Intercomparisons among five AGCMs for the return values of extreme temperature of <$-20°C$ and >$40°C$ over the globe show a reasonable level of agreement between the models in terms of global scale variability (McGuffie *et al.*, 1999).

8.8.2 Extreme Precipitation

Analysis of extreme precipitation simulated by climate models has included the daily variability of anomalous precipitation (Zwiers and Kharin, 1998; McGuffie *et al.*, 1999; Kharin and Zwiers, 2000), patterns of heavy rainfall (Bhaskran and Mitchell, 1998; Zhao *et al.*, 2000b), as well as wet and dry spells (Thorncroft and Rowell, 1998; McGuffie *et al.*, 1999). The results show some agreement with the available observations but the comparatively low model resolution is an inhibiting factor.

Ideally the simulated extreme rainfall should be compared with grided data calculated from the observed station data; however, observed grided data comparable to those produced by the models are scarce. Therefore, often the NCEP/NCAR reanalysis data are used as an "observed" data set despite the fact that this data set does not appear to reproduce daily variability well (Zwiers and Kharin, 1998). Another issue is the interpretation of precipitation simulated by a climate model, some authors treat simulated precipitation as grid-box averages; others argue that it should be treated as grid-point values (Zwiers and Kharin, 1998). Hennessy *et al.* (1997) compared the daily precipitation by both CSIRO and UKHI (United Kingdom High-Resolution) AGCMs coupled to mixed-layer ocean models. They found that simulated frequencies of daily precipitation were close to those for grid-box average observations.

In summary, in contrast with the simulations of extreme temperature by climate models, extreme precipitation is difficult to reproduce, especially for the intensities and patterns of heavy rainfall which are heavily affected by the local scale (see Chapter 10).

8.8.3 Extra-tropical Storms

Analyses of occurrences and tracks of extra-tropical storms have been performed for some climate models (Lunkeit *et al.*, 1996; Beersma *et al.*, 1997; Carnell and Senior, 1998; Schubert *et al.*, 1998; Zwiers and Kharin, 1998; Kharin and Zwiers, 2000). However, very different methods are used to characterise extra-tropical storms, among the methods used are: mid-latitude storm tracks defined by 1,000 hPa wind speed (Zwiers and Kharin, 1998), maximum eddy growth rate at 350 hPa and 775 hPa (Lunkeit *et al.*, 1996), index of storm tracks (such as 500 hPa height variability, sea level pressure, surface wind) (Beersma *et al.*, 1997), frequency, intensity and track of 500 hPa transient eddies (Schubert *et al.*, 1998), as well as low centres at 500 hPa (Carnell and Senior, 1998).

Kaurola (1997) compared the numbers of the observed extra-tropical storms north of 30°N for five winter seasons and from a 30-year simulation of the ECHAM3 atmospheric model for two periods during the control run. The comparisons indicated that the ratios of total numbers between the simulations and observations were 0.96 and 0.97 for two respective periods. It appears that the ECHAM3 model is able to simulate the numbers of storms north of 30°N in wintertime. The mid-latitude storm tracks over the North Pacific and Atlantic Oceans and over the southern circumpolar ocean were also well simulated by the CCC GCM2 (Zwiers and Kharin, 1998).

Kageyama *et al.* (1999) focused on the storm tracks of the Northern Hemisphere as simulated by several AGCMs. Intercomparisons of the nine AGCMs show that the models reproduce reasonably the storm tracks defined with high-pass second-order transient eddy quantities. These results also indicated that higher resolution models tend to be better at reproducing the storm tracks.

Schubert *et al.* (1998) analysed North Atlantic storms in the ECHAM3/LSG model. Their analysis indicated that the storm frequency, position and density agreed with the observations. Lunkeit *et al.* (1996) analysed storm activity in the ECHAM2/OPYC model. They found that the mean eddy activity and storm tracks in that simulation were in reasonable agreement with observations.

The general ability of models to simulate extra-tropical storms and storm tracks is most encouraging.

8.8.4 Tropical Cyclones

Tropical cyclones can be characterised in models by several measures such as their intensity, track, frequency and location of occurrence (Bengtsson *et al.*, 1996; Sugi *et al.*, 1997; Tsutsui *et al.*, 1999). Other broad-scale fields such as maximum wind speed, maximum potential intensity (Holland, 1997) and high sea surface temperature (Hulme and Viner, 1998) are also used as indicators of tropical cyclones. Thus it is important to consider the particular characteristics that are used to describe tropical cyclones in a given analysis when results from models are compared (Henderson-Sellers *et al.*, 1998; Krishnamurti *et al.*, 1998; Royer *et al.*, 1998; Walsh and Pittock, 1998).

Many analyses have been based on the physical parameters favourable for cyclogenesis as summarised by Gray (1981). Gray relates the climatological frequency of tropical cyclone genesis to six environmental factors: (1) large values of low-level relative vorticity, (2) Coriolis parameter (at least a few degrees poleward of the equator), (3) weak vertical shear of the horizontal winds, (4) high SST's exceeding 26°C and a deep thermocline, (5) conditional instability through a deep atmospheric layer, and (6) large values of relative humidity in the lower and middle troposphere (Gray, 1981; Henderson-Sellers *et al.*, 1998). Following the general concepts outlined by Emanuel (1987), Holland (1997) has derived an alternative thermodynamic approach to estimate maximum potential intensity of tropical cyclones. The approach requires an atmosphere sounding, SST, and surface pressure; it includes the oceanic feedback of increasing moist entropy associated with falling surface pressure over a steady SST, and explicitly incorporates a representation of the cloudy eye wall and a clear eye.

Several climate model simulations in Table 8.4 have been analysed using a variety of the above techniques to determine the frequency of tropical cyclones (Bengtsson *et al.*, 1995; Watterson *et al.*, 1995; Vitart *et al.*, 1997; Royer *et al.*, 1998). The ECHAM3 model has by far the highest horizontal resolution amongst these models. The numbers of simulated tropical cyclones are between 70 and 141 per year. The numbers of observed tropical cyclones per year are quite variable; 80 for the period 1958 to 1977 (Gray, 1979), 99 for the period 1952 to 1971 (Gray, 1975) and 86 for the period of 1970 to 1995 (Henderson-Sellers *et al.*, 1998). Despite the differing definitions of tropical cyclones used in the different analyses, the range of tropical cyclones numbers simulated by the models are similar to the observed data.

Bengtsson *et al.* (1995; 1996; 1999) have analysed a five-year simulation with ECHAM3 at T106 (100 km) horizontal resolution. They conclude that the model could reproduce some aspects of the characteristic structure of tropical cyclones and some aspects of their geographical distribution and seasonal variability. They also found that in certain areas, in particular in the north-east Pacific, a realistic number of tropical cyclones was only generated by the model when the horizontal resolution was finer than 100 km. Their results showed a reasonably good agreement with the observed distribution, tracks and annual variability of tropical cyclones (Bengtsson *et al.*, 1995, 1999). Sugi *et al.* (1997) and Yoshimura *et al.* (1999) used a 100 km version of the JMA AGCM and compared the simulated geographical distribution of tropical cyclones with observations. They obtained reasonably realistic geographical patterns. However, in contrast to the observations, they did not find a significant difference in tropical cyclone frequency when they used the SSTs representing El Niño and La Niña years.

Henderson-Sellers *et al.* (1998) suggested that AOGCMs could provide useful information of the frequency of tropical cyclones, but the models they studied all had coarse resolution

(about 500 km), climate drift (or flux adjustment) and unproven skill for present day tropical cyclones. A first attempt of tropical cyclones simulations with a high-resolution coupled climate model was performed by Matsuura *et al.* (1999) with a 100 km JMA atmospheric model coupled with the GFDL modular ocean model (MOM2) ($0.5° \times 1.0°$) model (but without sea ice). This model reproduced some aspects of the structure of observed tropical cyclones, although the simulated "tropical cyclones" are weaker and larger in scale than the observed. The model also shows the observed tendency of less (more) frequent tropical cyclones and an eastward (westward) shift of their locations over the northwestern equatorial Pacific during El Niño (La Niña) years. This result gives us some confidence in using a high-resolution coupled climate model in the future to explore the relationship between global warming and the frequency and intensity of tropical cyclones .

In summary, high horizontal resolution AGCMs (or AOGCMs) are able to simulate some aspects of "tropical cyclone-like vortices" with some degree of success, but it is still too computationally expensive to use such models for long experiments. The type of tropical cyclone index chosen in the analysis of low-resolution climate models is important, the use of maximum potential intensity may provide the most robust estimate, but analyses using this index remain infrequent.

8.8.5 Summary and Discussion

Since the SAR, more attention has been paid to the analysis of extreme events in climate model simulations. Evaluations indicate that climate models are more capable of reproducing the variability in maximum and minimum temperature in the global scale than the daily precipitation variability. The ability of climate models to simulate extra-tropical storm tracks and storm frequency is encouraging. When tropical cyclones were analysed, high-resolution models generally produced better results. It is worth noting that some high-resolution operational numerical weather prediction models have demonstrated reasonable ability in forecasting tropical cyclones. This increases our confidence that they may be better reproduced by high-resolution climate models in the future.

The lack of consistent methodologies used in analyses of extreme events prevents a ready intercomparison of results between models; future IPCC assessments would be greatly assisted if common approaches were adopted.

8.9 Coupled Models – Dependence on Resolution

The importance of numerical aspects of climate models continues to be well recognised and new numerical techniques are beginning to be tested for use in climate simulation. However, there has been very little systematic investigation of the impact of improved numerics for climate simulation and many important questions remain unanswered. The degree of interaction between horizontal and vertical resolution in climate models and the interaction of physical parametrizations at differing resolutions has made it extremely difficult to make general statements about the convergence of model solutions and hence the optimum resolution that should be used. An important question regarding the adequacy of resolution is deciding whether the information produced at finer scales at higher resolution feeds back on the larger scales or do the finer scales simply add to local effects (Williamson, 1999). Insufficient systematic work has been done with coupled models to answer this question. As well as improving numerical accuracy in advection, improved horizontal resolution can also improve the representation of the lower boundary of a model (the mountains) and the land-sea mask; this may improve the regional climate of a model but little systematic work has been carried out to assess this aspect.

8.9.1 Resolution in Atmospheric Models

A series of experiments that explores convergence characteristics has been conducted with the NCAR Community Climate Model (CCM) by Williamson (1999). In these experiments the grid and scale of the physical parametrizations was held fixed while the horizontal resolution of the dynamical core was increased. As the dynamical resolution was increased, but the parametrization resolution held fixed, the local Hadley circulation in the dual-resolution model simulations converged to a state close to that produced by a standard model at the fixed parametrization resolution. The mid-latitude transient aspects did not converge with increasing resolution when the scale of the physics was held fixed. Williamson (1999) concludes that the physical parametrizations used in climate models should explicitly take into account the scale of the grid on which it is applied. That does not seem to be common in parametrizations for global climate models today.

Pope *et al.* (1999) have also illustrated the positive impact of increased horizontal resolution on the climate of HADAM3. A number of systematic errors evident at low resolution are reduced as horizontal resolution is increased from 300 to 100 km. Improvements are considered to be mainly associated with better representation of storms. It is apparent that, for some models at least, neither the regional aspects of a climate simulation nor the processes that produce them converge over the range of horizontal resolutions commonly used (e.g., Déqué and Piedelievre, 1995; Stephenson and Royer, 1995; Williamson *et al.*, 1995; Stephenson *et al.*, 1998). As part of a European project (High Resolution Ten-Year Climate Simulations, HIRETYCS, 1998), it was found that increases in horizontal resolution did not produce systematic improvements in model simulations and any improvements found were of modest amplitude.

The need for consistency between horizontal and vertical resolution in atmospheric models was first outlined by Lindzen and Fox-Rabinovitz (1989) but little systematic study has been followed. Experiments with the NCAR CCM3 showed that increased vertical resolution (up to 26 levels) above the standard 18 levels typical of the modest vertical resolutions of climate models is beneficial to the simulations (Williamson *et al.*, 1998). Pope *et al.* (2000) also considered the impact of increased (up to 30 levels) vertical resolution on simulations with HADAM3. In both cases a number of improvements were noted due mostly to the improved representation of the tropopause as the resolution was increased. However, Bossuet *et al.* (1998) reached a somewhat different conclusion when they increased the vertical resolution in the ARPEGE model; they concluded that increasing vertical resolution produced little impact on the simulated mean climate of their model. They also found that the physical parametrizations they employed were resolution independent. Increased vertical resolution in the upper troposphere and stratosphere has generally reduced model systematic errors in that region (Pawson *et al.*, 2000).

Enhanced regional resolution within an AGCM is possible through the global variable-resolution stretched-grid approach that has been further developed since the SAR (e.g., Déqué and Piedelievre, 1995; Fox-Rabinovitz *et al.*, 1997); this is discussed in more detail in Chapter 10.

8.9.2 Resolution in Ocean Models

A number of important oceanic processes are not resolved by the current generation of coupled models, e.g., boundary currents, mesoscale eddy fluxes, sill through flows. Two model studies show an explicit dependence of ocean heat transport on resolution, ranging between $4°$ and $0.1°$ (Fanning and Weaver, 1997a; Bryan and Smith, 1998). However, this dependence appears to be much weaker when more advanced sub-grid scale mixing parametrizations are used, at least at resolutions of $0.4°$ or less (Gent *et al.*, 1999). As previously noted, a number of recent non-flux adjusted models produce acceptable large-scale heat transports. The need for ocean resolution finer than $1°$ is a matter of continuing scientific debate.

Some ocean models have been configured with increased horizontal resolution (usually specifically in the meridional direction) in the tropics in order to provide a better numerical framework to handle tropical ocean dynamics. Unfortunately at this time, there has been little systematic intercomparison of such model configurations.

8.9.3 Summary

The lack of carefully designed systematic intercomparison experiments exploring impacts of resolution is restricting our ability to draw firm conclusions. However, while the horizontal resolution of $2.5°$ (T42) or better in the atmospheric component of many coupled models is probably adequate to resolve most important features, the typical vertical resolution of around 20 levels is probably too low, particularly in the atmospheric boundary layer and near the tropopause. The potential exists for spurious numerical dispersion, when combined with errors in parametrizations and incompletely modelled processes, to produce erroneous entropy sources. This suggests that further careful investigation of model numerics is required as part of a continuing overall programme of model improvement. The vertical resolution required in the ocean component is still a matter of judgement and tends to be governed by available computing resources. There is still considerable debate on the adequacy of the horizontal resolution in the ocean component of coupled models and it is suggested that some results (those that are reliant on meridional heat transport) from coupled models with coarse ($>1°$) resolution ocean components should be treated cautiously.

8.10 Sources of Uncertainty and Levels of Confidence in Coupled Models

8.10.1 Uncertainties in Evaluating Coupled Models

Our attempts to evaluate coupled models have been limited by the lack of a more comprehensive and systematic approach to the collection and analysis of model output from well co-ordinated and well designed experiments. Important gaps still remain in our ability to evaluate the natural variability of models over the last several centuries. There are gaps in the specification of the radiative forcing (especially the vertical profile) as well as gaps in proxy palaeo-data necessary for the production of long time series of important variables such as surface air temperature and precipitation.

In order to assist future coupled model evaluation exercises, we would strongly encourage substantially expanded international programmes of systematic evaluation and intercomparison of coupled models under standardised experimental conditions. Such programmes should include a much more comprehensive and systematic system of model analysis and diagnosis, and a Monte Carlo approach to model uncertainties associated with parametrizations and initial conditions. The computing power now available to most major modelling centres is such that an ambitious programme that explores the differing direct responses of parametrizations (as well as some indirect effects) is now quite feasible.

Further systematic and co-ordinated intercomparison of the impact of physical parametrizations both on the ability to simulate the present climate (and its variability) and on the transient climate response (and its variability) is urgently needed.

The systematic analysis of extremes in coupled models remains considerably underdeveloped. Use of systematic analysis techniques would greatly assist future assessments.

It is important that in future model intercomparison projects the experimental design and data management takes heed of the detailed requirements of diagnosticians and the impacts community to ensure the widest possible participation in analysing the performance of coupled models.

8.10.2 Levels of Confidence

We have chosen to use the following process in assigning confidence to our assessment statements; the level of confidence we place in a particular finding reflects both the degree of consensus amongst modellers and the quantity of evidence that is available to support the finding. We prefer to use a qualitative three-level classification system following a proposal by Moss and Schneider (1999), where a finding can be considered:

"*well established*" – nearly all models behave the same way; observations are consistent with nearly all models; systematic experiments conducted with many models support the finding;

"*evolving*" – some models support the finding; different models account for different aspects of the observations; different aspects of key processes can be invoked to support the finding;

"*speculative*" – conceptually plausible idea that has only been tried in one model or has very large uncertainties associated with it.

8.10.3 Assessment

In this chapter, we have evaluated a number of climate models of the types used in Chapter 9. The information we have collected gives an indication of the capability of coupled models in general and some details of how individual coupled models have performed.

We regard the following as "well established":

- Incremental improvements in the performance of coupled models have occurred since the SAR, resulting from advances in the modelling of the oceans, atmosphere and land surface, as well as improvements in the coupling of these components.

- Coupled models can provide credible simulations of both the annual mean climate and the climatological seasonal cycle over broad continental scales for most variables of interest for climate change. Clouds and humidity remain sources of significant uncertainty but there have been incremental improvements in simulations of these quantities.

- Some non-flux adjusted models are now able to maintain stable climatologies of comparable quality to flux adjusted models.

- There is no systematic difference between flux adjusted and non-flux adjusted models in the simulation of internal climate variability. This supports the use of both types of model in detection and attribution of climate change.

- Several coupled models are able to reproduce the major trend in surface air temperature, when driven by radiative forcing scenarios corresponding to the 20th century. However, in these studies only idealised scenarios of only sulphate radiative forcing have been used.

- Many atmospheric models are able to simulate an increase of the African summer monsoon in response to insolation forcing for the Holocene but they all underestimate this increase if vegetation feedbacks are ignored.

We regard the following as "evolving":

- Coupled model simulation of phenomena such as monsoons and the NAO has improved since the SAR.

- Analysis of, and confidence in, extreme events simulated within climate models is emerging, particularly for storm tracks and storm frequency.

• The performance of coupled models in simulating ENSO has improved; however, the region of maximum SST variability is displaced westward and its strength is generally underestimated. When suitably initialised, some coupled models have had a degree of success in predicting ENSO events.

• Models tend to underestimate natural climate variability derived from proxy data over the last few centuries. This may be due to missing forcings, but this needs to be explored more systematically, with a wider range of more recent models.

• A reasonable simulation of a limited set of past climate states (over the past 20,000 years) has been achieved using a range of climate models, enhancing our confidence in using models to simulate climates different from the present day.

• Our ability to increase confidence in the simulation of land surface quantities in coupled models is limited by the need for significant advances in the simulation of snow, liquid and frozen soil moisture (and their associated water and energy fluxes).

• Coupled model simulations of the palaeo-monsoons produce better agreement with proxy palaeo-data when vegetation feedbacks are taken into account; this suggests that vegetation changes, both natural and anthropogenic, may need to be incorporated into coupled models used for climate projections.

• Models have some skill in simulating ocean ventilation rates, which are important in transient ocean heat uptake. However these processes are sensitive to choice of ocean mixing parametrizations.

• Some coupled models now include improved sea-ice components, but they do not yield systematic improvements in the sea-ice distributions. This may reflect the impact of errors in the simulated near surface wind fields, which offsets any improvement due to including sea-ice motion.

• Some coupled models produce good simulations of the large-scale heat transport in the coupled atmosphere-ocean system. This appears to be an important factor in achieving good model climatology without flux adjustment.

• The relative importance of increased resolution in coupled models remains to be evaluated systematically but many models show benefits from increased resolution.

• Our ability to make firmer statements regarding the minimum resolution (both horizontal and vertical) required in the components of coupled models is limited by the lack of systematic modelling studies.

We regard the following as "speculative":

• Tropical vortices with some of the characteristics of "tropical cyclones" may be simulated in high resolution atmospheric models but not yet in coupled climate models. Considerable debate remains over their detailed interpretation and behaviour.

• Some modelling studies suggest that adding forcings such as solar variability and volcanic aerosols to greenhouse gases and the direct sulphate aerosol effect improves the simulation of climate variability of the 20th century.

• Emerging modelling studies that add the indirect effect of aerosols and of ozone changes to greenhouse gases and the direct sulphate aerosol effect suggest that the direct aerosol effect may previously have been overestimated.

• Lack of knowledge of the vertical distribution of radiative forcing (especially aerosol and ozone) is contributing to the discrepancies between models and observations of the surface-troposphere temperature record.

Our overall assessment
Coupled models have evolved and improved significantly since the SAR. In general, they provide credible simulations of climate, at least down to sub-continental scales and over temporal scales from seasonal to decadal. The varying sets of strengths and weaknesses that models display lead us to conclude that no single model can be considered "best" and it is important to utilise results from a range of coupled models. We consider coupled models, as a class, to be suitable tools to provide useful projections of future climates.

References

Arking, A., 1996: Absorption of solar energy in the atmosphere: Discrepancy between model and observations. *Science*, **273**, 779-782.

Arking, A., 1999: The influence of clouds and water vapor on atmospheric absorption. *Geophys. Res. Lett.*, **26**, 2729-2732.

Arpe, K., H. Behr and L. Dümenil, 1997: Validation of the ECHAM4 climate model and re-analyses data in the Arctic region. Proc. Workshop on the Implementation of the Arctic Precipitation data Archive (APDA) at the Global Precipitation Climatology Centre (GPCC), Offenbach, Germany, World Climate Research Programme WCRP-98, WMO/TD No.804, 31-40.

Banks, H.T., 2000: Ocean heat transport in the South Atlantic in a coupled climate model. *J. Geophys. Res.* **105**, 1071-1092.

Barnett, T.P., 1999: Comparison of near-surface air temperature variability in 11 coupled global climate models. *J. Climate*, **12**, 511-518.

Barnett, T.P., B.D. Santer, P.D. Jones, R.S. Bradley and K.R. Briffa, 1996: Estimates of low frequency natural variability in near-surface air temperature. *The Holocene*, **6**, 255-263.

Barnston, A.G., Y. He and M.H. Glantz, 1999: Predictive skill of statistical and dynamical climate models in SST forecasts during the 1997-98 El Niño episode and the 1998 La Niña onset. *Bull. Am. Met. Soc.*, **80**, 217-243.

Barthelet, P., S. Bony, P. Braconnot, A. Braun, D. Cariolle, E. Cohen-Solal, J.-L. Dufresne, P. Delecluse, M. Déqué, L. Fairhead, M.-A. Filiberti, M. Forichon, J.-Y. Grandpeix, E. Guilyardi, M.-N. Houssais, M. Imbard, H. LeTreut, C. Lévy, Z.-X. Li, G. Madec, P. Marquet, O. Marti, S. Planton, L. Terray, O. Thual and S. Valcke, 1998a: Simulations couplées globales de changements climatiques associés à une augmentation de la teneur atmosphérique en CO₂. C. R. Acad. Sci. Paris, Sciences de la terre et des planètes, 326, 677-684 (in French with English summary).

Barthelet, P., L. Terray and S. Valcke, 1998b: Transient CO₂ experiment using the ARPEGE/OPAICE nonflux corrected coupled model. *Geophys. Res. Lett.*, **25**, 2277-2280.

Beersma, J.J., K.M. Rider, G.J. Komen, E. Kaas and V.V. Kharin, 1997: An analysis of extra-tropical storms in the North Atlantic region as simulated in a control and 2×CO₂ time-slice experiment with a high-resolution atmospheric model. *Tellus*, **49A**, 347-361.

Bell, J., P.B. Duffy, C. Covey, L. Sloan and the CMIP investigators, 2000: Comparison of temperature variability in observations and sixteen climate model simulations. *Geophys. Res. Lett.*, **27**, 261-264.

Bengtsson, L., M. Botzet and M. Esch, 1995: Hurricane-type vortices in a general circulation model. *Tellus*, **47A**, 175-196.

Bengtsson, L., M. Botzet and M. Esch, 1996: Will greenhouse gas-induced warming over the next 50 years lead to higher frequency and greater intensity of hurricanes? *Tellus*, **48A**, 57-73.

Bengtsson, L., E. Roeckner and M. Stendel, 1999: Why is global warming proceeding much slower than expected?. *J. Geophys., Res.*, **104**, 3865-3876.

Betrand, C., J.-P. Van Ypersele and A. Berger, 1999: Volcanic and solar impacts on climate since 1700. *Clim. Dyn.*, **15**, 355-367.

Betts, A.K., J.H. Ball and A.C.M. Beljaars, 1993: Comparison between the land surface response of the ECMWF model and the FIFE-1987 data. *Quart. J. R. Met. Soc.*, **119**, 975-1001.

Betts, R.A., P.M. Cox, S.E. Lee and F.I. Woodward, 1997: Contrasting physiological structural vegetation feedbacks in climate change simulations. *Nature*, **387**, 796-799.

Betts, R.A., P.M. Cox and F.I. Woodward, 2000: Simulated responses of potential vegetation to doubled-CO₂ climate change and feedbacks on near-surface temperature. *Global Ecology and Biogeography*, **9**, 171-180.

Bhaskaran, B. and J.F.B. Mitchell, 1998: Simulated changes in

Southeast Asian monsoon precipitation resulting from anthropogenic emissions. *Int. J. Climatol.*, **18**, 1455-1462.

Boer, G.J., G. Flato, M.C. Reader and D. Ramsden, 2000: A transient climate change simulation with greenhouse gas and aerosol forcing: experimental design and comparison with the instrumental record for the 20th century. *Clim. Dyn.*, **16**, 405-426.

Bonan, G.B., 1995: Land-atmosphere CO₂ exchange simulated by a land surface process model coupled to an atmospheric general circulation model. *J. Geophys. Res.*, **100**, 2817-2831.

Bossuet, C., M. Déué and D. Cariolle, 1998: Impact of a simple parameterization of convective gravity-wave drag in a stratosphere-troposphere general circulation model and its sensitivity to vertical resolution. *Ann. Geophysicae*, **16**, 238-249.

Boville, B.A. and P.R. Gent, 1998: The NCAR Climate System Model, Version One. *J. Climate*, **11**, 1115-1130.

Boville, B.A. and J.W. Hurrell, 1998: A comparison of the atmospheric circulations simulated by the CCM3 and CSM1. *J. Climate*, **11**, 1327-1341.

Boville, B.A., J.T. Kiehl, P.J. Rasch and F.O. Bryan, 2001: Improvements to the NCAR CSM-1 for transient climate simulations. *J. Climate*, **14**, 164-179.

Braconnot, P. and C. Frankignoul, 1993: Testing model simulations of the thermocline depth variability in the tropical Atlantic from 1982 to 1984. *J. Phys. Oceanogr.*, **23**, 626-647.

Braconnot, P., S. Joussaume, O. Marti and N. de Noblet, 1999: Synergistic feedbacks from ocean and vegetation on the African monsoon response to mid-Holocene insolation. *Geophys. Res. Lett.*, **26**, 2481-2484.

Braconnot, P., O. Marti, S. Joussaume and Y. Leclainche, 2000: Ocean feedback in response to 6 kyr BP insolation. *J. Climate*, **13**, 1537-1553.

Bradley, R.S. and P.D. Jones, 1993: 'Little Ice Age' summer temperature variations: their nature and relevance to recent global warming trends. *Holocene*, **3**, 367-376.

Brankovic, C. and F. Molteni, 1997: Sensitivity of the ECMWF model northern winter climate to model formulation. *Clim. Dyn.*, **13**, 75-101.

Broccoli, A.J., 2000: Tropical cooling at the LGM: An atmosphere-mixed layer ocean model simulation, *J. Climate*, **13**, 951-976.

Broccoli, A.J. and E.P. Marciniak, 1996: Comparing simulated glacial climate and paleodata: a reexamination. *Paleoceanogr.*, **11**, 3-14.

Broccoli, A.J., N.-C. Lau and M.J. Nath, 1998: The cold ocean-warm land pattern: Model simulation and relevance to climate change detection. *J. Climate*, **11**, 2743-2763.

Brostrom, A., M. Coe, S.P. Harrison, R. Gallimore, J.E. Kutzbach, J. Foley, I.C. Prentice and P. Behling, 1998: Land Surface feedbacks and palaeomonsoons in Northern Africa. *Geophys. Res. Lett.*, **25**, 3615-3618.

Brovkin, V., A. Ganopolski, M. Claussen, C. Kubatzki and V. Petoukhov, 1999: Modelling climate response to historical land cover change. *Global Ecology and Biogeography*, **8**, 509-517.

Bryan, F.O., 1998: Climate drift in a multi century integration of the NCAR Climate System Model. *J. Climate*, **11**, 1455-1471.

Bryan, F.O. and R.D. Smith, 1998: Modelling the North Atlantic circulation: from eddy-resolving to eddy-permitting. International WOCE Newsletter, **33**, 12-14.

Bryden, H.L., D.H. Roemmich and J.A. Church, 1991: Ocean heat transport across 24N in the Pacific. *Deep Sea Res.*, **38**, 297-324.

Carnell, R.E. and C.A. Senior, 1998: Changes in mid-latitude variability due to increasing greenhouse gases and sulphate aerosols. *Clim. Dyn.*, **14**, 369-383.

Cess, R., M.H. Zhang, G.L. Potter, V. Alekseev, H.W. Barker, S. Bony, R.A. Colman, D.A. Dazlich, A.D. Del Genio, M. Deque, M.R. Dix, V. Dymnikov, M. Esch, L.D. Fowler, J.R. Fraser, V. Galin, W.L. Gates, J.J. Hack, W.J. Ingram, J.T. Kiehl, Y. Kim, H. Le Treut, K.K.-

W. Lo, B.J. McAvaney, V.P. Meleshko, J.-J. Morcrette, D.A. Randall, E. Roeckner, J.-F. Royer, M.E. Schlesinger, P.V. Sporyshev, B. Timbal, E.M. Volodin, K.E. Taylor, W. Wang, W.C. Wang and R.T. Wetherald, 1997: Comparison of the seasonal change in cloud-radiative forcing from atmospheric general circulation models and satellite observations. *J. Geophys. Res.*, **102**, 16593-16603.

Chandrasekar, A., D.V.B. Rao and A. Kitoh, 1999: Effect of horizontal resolution on the simulation of Asian summer monsoon using the MRI GCM-II. *Pap. Met. Geophys.*, **50**, 65-80.

Chase, T.N., R.A. Pielke, T.G.F. Kittel, R. Nemani and S.W. Running, 2000: Simulated impacts of historical land cover changes on global climate. *Clim. Dyn.*, **16**, 93-105.

Chassignet, E.P., L.T. Smith, R. Bleck and F.O. Bryan, 1996: A model comparison: numerical simulations of the north and equatorial Atlantic oceanic circulation in depth and isopycnic coordinates. *J. Phys. Oceanogr.*, **26**, 1849-1867.

Chen, T.H., A. Henderson-Sellers, P.C.D. Milly, A.J. Pitman, A.C.M. Beljaars, J. Polcher, F. Abramopoulos, A. Boone, S. Chang, F. Chen, Y. Dai, C.E. Desborough, R.E. Dickenson, L. Dumenil, M. Ek, J.R. Garratt, N. Gedney, Y.M. Gusev, J. Kim, R. Koster, E.A. Kowalczyk, K. Laval, J. Lean, D. Lettenmaier, X. Liang, J.-F. Mahfouf, H.-T. Mengelkamp, K. Mitchell, O.N. Nasonova, J. Noilhan, J. Polcher, A. Robock, C. Rosenzweig, J. Schaake, C.A. Schlosser, J.-P. Schulz, A.B. Shmakin, D.L. Verseghy, P. Wetzel, E.F. Wood, Y. Xue, Z.-L. Yang and Q. Zeng, 1997: Cabauw experimental results from the project for intercomparison of land surface parameterization schemes. *J. Climate*, **10**, 1144-1215.

Christoph, M., U. Ulbrich and P. Speth, 1997: Midwinter suppression of Northern Hemisphere storm track activity in the real atmosphere and in GCM experiments. *J. Atmos. Sci.*, **54**, 1589-1599.

Claussen, M. and V. Gayler, 1997: The greening of Sahara during the mid-Holocene: results of an interactive atmosphere-biome model. *Global Ecol Biogeography Letters*, **6**, 369-377.

Claussen, M., C. Kubatzki, V. Brovkin and A. Ganopolski, 1999: Simulation of an abrupt change in Saharan vegetation in the mid-Holocene. *Geophys. Res. Lett.*, **26**, 2037-2040.

CLIMAP, 1981: Seasonal reconstructions of the Earth's surface at the last glacial maximum. Map Series, Technical Report MC-36, Geological Society of America, Boulder, Colorado.

Coe, M.T. and G.B. Bonan, 1997: Feedbacks between climate and surface water in northern Africa during the middle Holocene. *J. Geophys. Res.*, **102**, 11087-11101.

Coe, M.T. and S. Harrison, 2000: A comparison of the simulated surface water area in Northern Africa for the 6000 yr. BP experiments, In "Paleoclimate Modeling Intercomparison Project (PMIP) : proceedings of the third PMIP workshop, Canada, 4-8 October 1999" P. Braconnot (Ed), WCRP-111, WMO/TD-1007, 65-68.

COHMAP-Members, 1988: Climatic changes of the last 18,000 years: observations and model simulations. *Science*, **241**, 1043-1052.

Covey, C., A. Abe-Ouchi, G.J. Boer, G.M. Flato, B.A. Boville, G.A. Meehl, U. Cubasch, E. Roeckner, H. Gordon, E. Guilyardi, L. Terray, X. Jiang, R. Miller, G. Russell, T.C. Johns, H. Le Treut, L. Fairhead, G. Madec, A. Noda, S.B. Power, E.K. Schneider, R.J. Stouffer and J.-S. von Storch, 2000a: The Seasonal Cycle in Coupled Ocean-Atmosphere General Circulation Models, *Clim. Dyn.*, **16**, 775-787.

Covey, C., K.M. AchutaRao, S.J. Lambert and K.E. Taylor, 2000b: Intercomparison of Present and Future Climates Simulated by Coupled Ocean-Atmosphere GCMs. *PCMDI Report No 66*. Program for Climate Model Diagnosis and Intercomparison, Lawrence Livermore National Laboratory, University of California, Livermore, CA.

Cowling, S.A., 1999: Simulated effects of low atmospheric CO_2 on structure and composition of North American vegetation at the Last Glacial Maximum. *Global Ecology and Biogeography*, **8**, 81-93.

Cox, M.D., 1989: An idealised model of the world ocean. Part I: the global scale water masses. *J. Phys. Oceanogr.*, **19**, 1730-1752.

Crossley, J.F., J. Polcher, P.M. Cox, N. Gedney and S. Planton, 2000: Uncertainties linked to land surface processes in climate change simulations. *Clim. Dyn.* **16**, 949-961.

Crowley, T.J., 2000: Causes of climate change over the past 1000 years. *Science*, **289**, 270–277.

Crowley, T.J. and S.K. Baum, 1997: Effect of vegetation on an ice-age climate model simulation. *J. Geophys. Res.*, **102**, 16463-16480.

Crowley, T.J. and K.-Y. Kim, 1999: Modeling the temperature response to forced climate change over the last six centuries. *Geophys. Res. Lett.*, **26**, 1901-1904.

Cubasch, U., K. Hasselmann, H. Höck, E. Maier-Reimer, U. Mikolajewicz, B.D. Santer and R. Sausen, 1992: Time-dependent greenhouse warming computations with a coupled ocean-atmosphere model. *Clim. Dyn.*, **8**, 55-69.

Cubasch, U., B.D. Santer, A. Hellbach, G.C. Hegerl, H. Hock, E. Meir Reimer, U. Mikolajewicz , A. Stossel and R. Voss, 1994: Monte Carlo forecasts with a coupled ocean-atmosphere model. *Clim. Dyn.*, **10**, 1-19.

Cubasch, U., R. Voss, G.C. Hegerl, J. Waszkewitz and T.J. Crowley, 1997: Simulation of the influence of solar radiation variations on the global climate with an ocean-atmosphere general circulation model. *Clim. Dyn.*, **13**, 757-767.

Danabasoglu, G., 1998: On the wind-driven circulation of the uncoupled and coupled NCAR climate system ocean model. *J. Climate*, **11**, 1442-1454.

Danabasoglu, G. and J.C. McWilliams, 1996: Sensitivity of the global ocean circulation to parameterizations of mesoscale tracer transports. *J. Climate*, **8**, 2967-2987.

D'Andrea, F., S. Tibaldi and Co-authors, 1998: Northern Hemisphere atmospheric blocking as simulated by 15 atmospheric general circulation models in the period 1979-1988. *Clim. Dyn.*, **14**, 385-407.

de las Heras, M.M. and R. Schlitzer, 1999: On the importance of intermediate water flows for the global ocean overturning. *J. Geophys. Res.*, **104**, 15515-15536.

Delecluse, P., M.K. Davey, Y. Kitamura, S.G.H. Philander, M. Suarez and L. Bengtsson, 1998: Coupled general circulation modeling of the tropical Pacific. *J. Geophys. Res.*, **103**, 14357-14373.

Del Genio, A.D., W. Kovari Jr. and M.-S. Yao, 1994:Climatic implications of the seasonal variation of upper troposphere water vapor, *Geophys. Res. Lett.*, **21**, 2701-2704.

Delworth, T.L., 1996: North Atlantic interannual variability in a coupled ocean-atmosphere model. *J. Climate*, **9**, 2356-2375.

Delworth, T.L. and M.E. Mann, 2000: Observed and simulated multidecadal variability in the North Atlantic. *Clim. Dyn.*, **16**, 661-676.

Delworth, T.L., J.D. Mahlman and T.R. Knutson, 1999, Changes in heat index associated with CO_2-induced global warming. *Clim. Change*, **43**, 369-396.

Delworth, T.L. and T.R. Knutson, 2000: Simulation of the early 20th century global warming. *Science*, **287**, 2246–2250.

De Noblet, N., M. Claussen and C. Prentice, 2000: Mid-Holocene greening of the Sahara: first results of the GAIM 6000 yr BP experiment with two asynchronously coupled atmosphere/biome models. *Clim. Dyn.*, **16**, 643-659.

Déqué, M. and J.P. Piedelievre, 1995: High resolution climate simulation over Europe. *Clim. Dyn.*, **11**, 321-339.

Desborough, C.E, 1999: Surface energy balance complexity in GCM land surface models. *Clim. Dyn.*, **15**, 389-403.

Dewitt, D.G. and E.K. Schneider, 1999: The processes determining the annual cycle of equatorial sea surface temperature: A coupled general circulation model perspective. *Mon. Wea. Rev.*, **127**, 381-395.

Dijkstra , H.A. and J.D. Neelin, 1999: Imperfections of the thermoha-line circulation: multiple equilibria and flux correction. *J. Climate*,

12, 1382-1392.

Dirmeyer, P.A., A.J. Dolman and N. Sato, 1999: The pilot phase of the global soil wetness project. *Bull. Am. Met. Soc.,* **80**, 851-878.

Dixon, K.W. and J.R. Lanzante, 1999: Global mean surface air temperature and North Atlantic overturning in a suite of coupled GCM climate change experiments. *Geophys. Res. Lett.,* **26**, 1885-1888.

Dixon, K.W., J.L. Bullister, R.H. Gammon and R.J. Stouffer, 1996: Examining a coupled climate model using CFC-11 as an ocean tracer. *Geophys Res. Lett.,* **23**, 1957-1960.

Doherty, R., J. Kutzbach, J. Foley and D. Pollard, 2000: Fully-coupled climate/dynamical vegetation model simulations over Northern Africa during the mid-Holocene. *Clim. Dyn.,* **16**, 561-573.

Doney, S.C., W.G. Large and F.O. Bryan, 1998: Surface ocean fluxes and water-mass transformation rates in the coupled NCAR Climate System Model. *J. Climate,* **11**, 1420-1441.

Ducharne, A., K. Laval and J. Polcher, 1998: Sensitivity of the hydrological cycle to the parameterization of soil hydrology in a GCM. *Clim. Dyn.,* **14**, 307-327.

Duffy, P.B., J. Bell, C. Covey and L. Sloan, 2000: Effect of flux adjustments on temperature variability in climate models. *Geophys. Res. Lett.,* **27**, 763-766.

Dümenil, L., S. Hagemann, and K. Arpe, 1997: Validation of the hydrological cycle in the Arctic using river discharge data. Proc. Workshop on Polar Processes in Global Climate (13-15 November 1996, Cancun, Mexico), AMS, Boston, USA.

Duplessy, J.C., N.J. Shackleton, R.G. Fairbancks, L. Labeyrie, D. Oppo and N. Kallel, 1988: Deepwater source variations during the last climatic cycle and their impact on the global deepwater circulation. *Paleoceanography,* **3**, 343-360.

Durman, C.F., J.M. Gregory, D.C. Hassell, R.G. Jones and J.M. Murphy, 2001: A comparison of extreme European daily precipitation simulated by a global and regional climate model for present and future climates. *Quart. J. R. Met. Soc.,* in press.

DYNAMO Group, 1997: Dynamics of North Atlantic Models: simulation and assimilation with high resolution models. Report no. 294, Institut fuer Meereskunde, University of Kiel, Kiel, Germany, 334pp.

Emanuel, K.A., 1987: The dependence of hurricane intensity on climate. *Nature,* **326**, 483-485.

Emori, S., T. Nozawa, A. Abe-Ouchi, A. Numaguti, M. Kimoto and T. Nakajima, 1999: Coupled ocean-atmosphere model experiments of future climate change with an explicit representation of sulfate aerosol scattering. *J. Met. Soc. Japan,* **77**, 1299-1307.

England, M.H. and S. Rahmstorf, 1999: Sensitivity of ventilation rates and radiocarbon uptake to subsurface mixing parameterisation in ocean models. *J. Phys. Oceanogr.,* **29**, 2802-2827.

England, M.H. and E. Maier-Reimer, 2001: Using Chemical tracers to Assess Ocean models. *Rev. Geophys.,* **39**, 29-70.

Fanning, A.F. and A.J. Weaver, 1996: An Atmospheric energy-moisture balance model: Climatology, interpentadal climate change, and coupling to an ocean general circulation model. *J. Geophys. Res.,* **101**, 15111-15128.

Fanning, A.F. and A.J. Weaver, 1997a: A horizontal resolution and parameter sensitivity analysis of heat transport in an idealized coupled model. *J. Climate,* **10**, 2469-2478.

Fanning, A.F. and A.J. Weaver, 1997b: On the role of flux adjustments in an idealised coupled climate model. *Clim. Dyn.,* **13**, 691-701.

Farrera, I., S.P. Harrison, I.C. Prentice, G. Ramstein, J. Guiot, P.J. Bartlein, R. Bonnefille, M. Bush, W. Cramer, U. von Grafenstein, K. Holmgren, H. Hooghiemstra, G. Hope, D. Jolly, S.-E. Lauritzen, Y. Ono, S. Pinot, M. Stute and G. Yu, 1999: Tropical climates at the last glacial maximum: a new synthesis of terrestrial palaeoclimate data. I. Vegetation, lake-levels and geochemistry. *Clim. Dyn.,* **15**, 823-856.

Ferranti, L., J.M. Slingo, T.N. Palmer, B.J. Hoskins, 1999: The effect of land surface feedbacks on the monsoon circulation. *Quart. J. R. Met. Soc.,* **125**, 1527-1550.

Flatau, M., P.J. Flatau, P. Phoebus and P.P. Niiler, 1997: The feedback between equatorial convection and local radiative and evaporative processes: The implications for intra seasonal oscillations. *J. Atmos. Sci.,* **54**, 2373-2386.

Flato, G.M. and G.J. Boer, 2001: Warming Asymmetry in Climate Change Experiments. *Geophys. Res. Lett.,* **28**, 195-198.

Flato, G., G.J. Boer, W.G. Lee, N.A. McFarlane, D. Ramsden, M.C. Reader and A.J. Weaver, 2000: The Canadian Centre for Climate Modelling and Analysis Global Coupled Model and its Climate. *Clim. Dyn.,* **16**, 451-468.

Foley, J.A., J.E. Kutzbach, M.T. Coe and S. Levis, 1994: Feedbacks between climate and boreal forests during the Holocene epoch. *Nature,* **371**, 52-54.

Folland, C.K., D.E. Parker, A. Colman and R. Washington, 1999: Large scale modes of ocean surface temperature since the late nineteenth century. In: *Beyond El Niño: Decadal and Interdecadal Climate Variability.* Navarra, A., Ed., Springer-Verlag, Berlin, pp 73-102.

Foster, D.J., Jr. and R.D. Davy, 1988: Global snow depth climatology. USAF publication USAFETAC/TN-88/006, Scott Air Force Base, Illinois, 48 pp.

Fox-Rabinovitz, M.S., G.L. Stenchikov, M.J. Suarez and L.L. Takacs, 1997: A Finite-Difference GCM Dynamical Core with a Variable-Resolution Stretched Grid. *Mon. Wea. Rev.,* **125**, 2943–2968.

Frankignoul, C., C. Duchêne and M.A. Cane, 1989: A statistical approach to testing equatorial ocean models with observed data. *Journal of Physical Oceanography,* **19**, 1191-1209.

Free, M. and A. Robock, 1999: Global warming in the context of the Little Ice Age. *J. Geophys. Res.,* **104**, 19057-19070.

Frei, A. and D. Robinson, 1998: Evaluation of snow extent and its variability in the Atmospheric Model Intercomparison Project. *J. Geophys. Res.,* **103**, 8859-8871.

Fyfe, J.C., G.J. Boer and G.M. Flato, 1999: The Arctic and Antarctic oscillations and their projected changes under global warming. *Geophys. Res. Lett.,* **26**, 1601-1604.

Gadgil, S. and S. Sajani, 1998: Monsoon precipitation in the AMIP runs. *Clim. Dyn.,* **14**, 659-689.

Gallee, H., J.P. van Ypersele, T. Fichefet, C. Tricot and A. Berger, 1991: Simulation of the last glacial cycle by a coupled, sectorially averaged climate-ice sheet model. I. The climate model. *J. Geophys Res.,* **96**, 13139-13161.

Ganopolski, A., C. Kubatzki, M. Claussen, V. Brovkin and V. Petoukhov, 1998a: The influence of vegetation-atmosphere-ocean interaction on climate during the mid-Holocene. *Science,* **280**, 1916-1919.

Ganopolski, A., S. Rahmstorf, V. Petoukhov and M. Claussen, 1998b: Simulation of modern and glacial climates with a coupled global model of intermediate complexity. *Nature,* **391**, 351-356.

Garratt, J.R., 1994: Incoming shortwave fluxes at the surface - a comparison of GCM results with observations. *J. Climate,* **7**, 72-80.

Garratt, J.R. and A.J. Prata, 1996: Downwelling Longwave Fluxes at Continental Surfaces - A Comparison of Observations with GCM Simulations and Implications for the Global Land-Surface Radiation Budget. *J. Climate,* **9**, 646-655.

Garratt, J.R., A.J. Prata, L.D. Rotstayn, B.J. McAvaney and S. Cusack, 1998: The Surface Radiation Budget over Oceans and Continents. *J. Climate,* **11**, 951-1968.

Gates, W.L., A. Henderson-Sellers, G.J. Boer, C.K. Folland, A. Kitoh, B.J. McAvaney, F. Semazzi, N. Smith, A.J. Weaver and Q.-C. Zeng, 1996: Climate Models - Evaluation. In: *Climate Change 1995: The Science of Climate Change. Contribution of Working Group I to the Second Assessment Report of the Intergovernmental Panel on Climate Change* [Houghton, J.T., L.G. Meira Filho, B.A. Callander, N. Harris, A. Kattenberg, and K. Maskell (eds.)]. Cambridge University Press, Cambridge, United Kingdom and New York, NY, USA, pp 228-284.

Gates, W.L., J. Boyle, C. Covey, C. Dease, C. Doutriaux, R. Drach, M.

Fiorino, P. Gleckler, J. Hnilo, S. Marlais, T. Phillips, G. Potter, B.D. Santer, K.R. Sperber, K. Taylor and D. Williams, 1999: An Overview of the Results of the Atmospheric Model Intercomparison Project (AMIP I). *Bull. Am. Met. Soc.*, **80**, 29-55.

Gedney, N., P.M. Cox, H. Douville, J. Polcher and P.J. Valdes, 2000: Characterising GCM land surface schemes to understand their response to climate change. *J. Climate, 13*, 3066-3079.

Gent, P.R and J.C. McWilliams, 1990: Isopycnal mixing in ocean circulation models. *J. Phys.Oceanogr.*, **20**, 150-155.

Gent, P.R., F.O. Bryan, G. Danabasoglu, S.C. Doney, W.R. Holland, W.G. Large and J.C. McWilliams, 1998: The NCAR Climate System Model Global Ocean Component. *J. Climate, 11*, 1287-1306.

Gent, P.R., F.O. Bryan, S.C. Doney and W.G. Large, 1999: A perspective on the ocean component of climate models. *CLIVAR Exchanges* **4**(4).

Gibson, J.K., P. Kallberg, S. Uppala, A. Hernandez, A. Nomura and E. Serrano, 1997: ERA description. ECMWF Reanalysis Project Report Series 1, European Centre for Medium Range Weather Forecasts, Reading, UK, 66 pp.

Gilgen, H., M. Wild and A. Ohmura, 1998: Means and trends of shortwave irradiance at the surface estimated from Global Energy Balance Archive data. *J. Climate*, **11**, 2042-2061.

Gloersen, P., W.J. Campbell, D.J. Cavalieri, J.C. Comiso, C.L. Parkinson and H.J. Zwally, 1992: Arctic and Antarctic sea ice, 1978-1987: Satellite passive-microwave observations and analysis. NASA SP-511, National Aeronautics and Space Administration, Washington, 290 pp.

Goddard, L. and N.E. Graham, 1999: The importance of the Indian Ocean for GCM-based climate forecasts over eastern and southern Africa. *J. Geophys. Res.*, **104**, 19099-19116.

Goosse, H., E. Deleersnijder, T. Fichefet and M. England, 1999: Sensitivity of a global coupled ocean-sea ice model to the parameterization of vertical mixing. *J. Geophys. Res.*, **104**, 13681-13695.

Gordon, H.B. and S.P. O'Farrell, 1997: Transient climate change in the CSIRO coupled model with dynamic sea ice. *Mon. Wea. Rev.*, **125**, 875-907.

Gordon, C., C. Cooper, C.A. Senior, H.T. Banks, J.M. Gregory, T.C. Johns, J.F.B. Mitchell and R.A. Wood, 2000: The simulation of SST, sea ice extents and ocean heat transports in a version of the Hadley Centre coupled model without flux adjustments. *Clim. Dyn.*, **16**, 147-168.

Gray, W.M., 1975: Tropical cyclone genesis. CSU Report No.234, 121 pp.

Gray, W.M., 1979: Hurricanes: their formation, structure and likely role in the tropical circulation, in Shaw, D.B. (ed.), *Meteorology over the tropical oceans*, Royal Meteorological Society, J. Glaisher House, Grenville place, Bracknell, Berks, pp 155-218.

Gray, W.M. 1981: Recent advances in Tropical Cyclone Research from rawinsonde composite analysis. World Meteorological Organisation, 407 pp.

Gregory, J.M. and J.F.B. Mitchell, 1995: Simulation of daily variability of surface temperature and precipitation over Europe in the current and $2\times CO_2$ climate using the UKMO climate model. *Quart. J. R. Met. Soc.*, **121**, 1451-1476.

Gregory, J.M. and J.F.B. Mitchell, 1997: The climate response to CO_2 of the Hadley Centre coupled AOGCM with and without flux adjustment. *Geophys. Res. Lett.*, **24**, 1943-1946.

Guilderson, T., R. Fairbanks and J. Rubenstone, 1994: Tropical temperature variations since 20,000 years ago: modulating interhemispheric climate change. *Science*, **263**, 663-665.

Guilyardi, E., 1997: Role de la physique oceaniqe sur la formation/consummation des masses d'eau dans un modle couplé-atmosphè. Ph.D. thesis, Université Paul Sabatier, 195 pp.

Guilyardi, E. and G. Madec, 1997: Performance of the OPA/ARPEGE-T21 global ocean-atmosphere coupled model. *Clim. Dyn.*, **13**, 149-165.

Guiot, J., J.J. Boreux, P. Braconnot, F. Torre and PMIP participants, 1999: Data-model comparison using fuzzy logic in palaeoclimatology. *Clim. Dyn.*, **15**, 569-581.

Hagemann, S. and L. Dümenil, 1998: A parameterization of the lateral water flow for the global scale. *Clim. Dyn.*, **14**, 17-31.

Hall, M.M. and H.L. Bryden, 1982: Direct estimates of ocean heat transport. *Deep Sea Res.*, **29**, 339-359.

Handorf, D., V.K. Petoukhov, K. Dethloff, A.V. Eliseev, A. Weisheimer and I.I. Mokhov, 1999: Decadal climate variability in a coupled atmosphere-ocean climate model of moderate complexity. *J. Geophys. Res.*, **104**, 27253-27275.

Hansen, J., M. Sato, A. Lacis and R. Rueby, 1997: The missing climate forcing. *Phil. Trans. Roy. Soc. Lond.* B, **352**, 231-240.

Hansen, J., M. Sato, A. Lacis, R. Rueby, I. Tegen and E. Matthews, 1998: Climate forcing in the Industrial Era. *Proc. Nat. Acad. Sci.*, **95**, 12753-12758.

Harrison, S.P., G. Yu and P.E. Tarasov, 1996: The Late Quaternary lake-level record from northern Eurasia. *Quat. Res.*, **45**, 138-159.

Harrison, S.P., D. Jolly, F. Laarif, A. Abe-Ouchi, B. Dong, K. Herterich, C. Hewitt, S. Joussaume, J.E. Kutzbach, J. Mitchell, N. de Noblet and P. Valdes, 1998: Intercomparison of simulated global vegetation distributions in response to 6 kyr BP orbital forcing. *J. Climate*, **11**, 2721-2742.

Harvey, L.D.D., J. Gregory, M. Hoffert, A. Jain, M. Lal, R. Leemans, S.C.B. Raper, T.M.L. Wigley and J.R. de Wolde, 1997: An introduction to simple climate models used in the IPCC Second Assessment Report. *IPCC Technical Paper 2*, J.T. Houghton, L.G. Meira Filho, D.J. Griggs and K. Maskell (Eds.). IPCC, Geneva, Switzerland, 51 pp.

Hasselmann, K., L. Bengtsson, U. Cubasch, G.C. Hegerl, H. Rodhe, E. Roeckner, H. von Storch, R. Voss and J. Waszkewitz, 1995: Detection of anthropogenic climate change using a fingerprint method. In: *Proceedings of "Modern Dynamical Meteorology", Symposium in honor of Aksel Wiin-Nielsen, 1995*, P. Ditlevsen (ed.), ECMWF press, 1995.

Haywood, J.M., R.J. Stouffer, R.T. Wetherald, S. Manabe and V. Ramaswamy, 1997: Transient response of a coupled model to estimated changes in greenhouse gas and sulfate concentrations. *Geophys. Res. Lett.*, **24**, 1335-1338.

Hegerl, G.C., K. Hasselmann, U. Cubasch, J.F.B. Mitchell, E. Roeckner, R. Voss and J. Waszkewitz, 1997: Multi-fingerprint detection and attribution analysis of greenhouse gases, greenhouse gas-plus-aerosol and solar forced climate change. *Clim. Dyn.*, **13**, 613-634.

Henderson-Sellers, A., A.J. Pitman, P.K. Love, P. Irannejad and T. Chen, 1995: The project for Intercomparison of land surface parameterisation schemes PILPS) Phases 2 and 3. *Bull. Am. Met. Soc.*, **76**, 489-503.

Henderson-Sellers, A., H. Zhang, G. Berz, K. Emanuel, W. Gray, C. Landsea, G. Holland, J. Lighthill, S.-L. Shieh, P. Webster and K. McGuffie, 1998: Tropical cyclones and global climate change: a post-IPCC assessment. *Bull. Am. Met. Soc.*, **79**, 19-38.

Henderson-Sellers, A. and K. McGuffie, 1999: Concepts of good science in climate change modelling: Comments on S. Shackley *et al.* Climatic Change 38, 1998. *Clim. Change*, **42**, 597-610.

Hennessy, K.J., J.M. Gregory and J.F.B. Mitchell, 1997: Changes in daily precipitation under enhanced greenhouse conditions. *Clim. Dyn.*, **13**, 667-680.

Hewitt, C.D. and J.F.B. Mitchell, 1997: Radiative forcing and response of a GCM to ice age boundary conditions: cloud feedbacks and climate sensitivity. *Clim. Dyn.*, **13**, 821-834.

Hewitt, C.D. and J.F.B. Mitchell, 1998: A fully coupled GCM simulation of the climate of the mid-Holocene. *Geophys. Res. Lett.*, **25**, 361-364.

HIRETYCS (High Resolution Ten-Year Climate Simulations), 1998: Final Report, Contract No. ENV4-CT95-0184, European

Commission Environment and Climate Program, Brussels.

Hoelzmann, P., D. Jolly, S.P. Harrison, F. Laarif, R. Bonnefille and H.-J. Pachur, 1998: Mid-Holocene land-surface conditions in northern Africa and the Arabian peninsula: a data set for AGCM simulations. *Global Biogeochemical Cycles, 12*, 35-52.

Hoffert, M.I. and C. Covey, 1992 : Deriving global climate sensitivity from paleoclimate reconstructions. *Nature, 360*, 573-576.

Hogan, T.F. and T. Li, 1997: Long-term simulations with a coupled global atmosphere-ocean prediction system. *NRL Review*, 183-185.

Hogg, N.G., 1983: A note on the deep circulation of the western North Atlantic: its nature and causes. *Deep Sea Res., 30*, 945-961.

Holland, G.J., 1997: The maximum potential intensity of tropical cyclones. *J. Atmos. Sci., 54*, 2519-2541.

Hulme, M., 1992: Global land precipitation climatology for the evaluation of general circulation models. *Clim. Dyn., 7*, 57–72.

Hulme, M., 1994: Validation of large-scale precipitation fields in general circulation models. In*: Global Precipitation and Climate Change*. Desbois, M., and F. Desalmand, Eds., NATO ASI Series, Springer Verlag, Berlin.

Hulme, M. and D. Viner, 1998: A climate change scenario for the tropics. *Clim. Change, 39*, 145-176.

IPCC, 1996: Climate Change 1995: The Science of Climate Change. Contribution of Working Group I to the Second Assessment Report of the Intergovernmental Panel on Climate Change [Houghton, J.T., L.G. Meira Filho, B.A. Callander, N. Harris, A. Kattenberg, and K. Maskell (eds.)]. Cambridge University Press, Cambridge, United Kingdom and New York, NY, USA, 572 pp.

Ishii, M., N. Hasegawa, S. Sugimoto, I. Ishikawa, I. Yoshikawa and M. Kimoto, 1998: An El Niño prediction experiment with a JMA ocean-atmosphere coupled model, "Kookai". Proceedings of WMO International Workshop on Dynamical Extended Range Forecasting, Toulouse, France, 17-21 November 1997, WMO/TD-No. 881, 105-108.

Jia, Y., 2000: The ocean heat transport and meridional overturning near 25N in the Atlantic in the CMIP models. *CLIVAR Exchanges, 5*(3), 23-26.

Johns, T.C., 1996: A description of the Second Hadley Centre Coupled Model (HadCM2). *Climate Research Technical Note 71*, Hadley Centre, United Kingdom Meteorological Office, Bracknell Berkshire RG12 2SY, United Kingdom, 19 pp.

Johns, T.C., R.E. Carnell, J.F. Crossley, J.M. Gregory, J.F.B. Mitchell, C.A. Senior, S.F.B. Tett and R.A. Wood, 1997: The second Hadley Centre coupled atmosphere-ocean GCM: model description, spinup and validation. *Clim. Dyn., 13*, 103-134.

Jolly, D. and A. Haxeltine, 1997: Effect of low glacial atmospheric CO_2 on tropical African montane vegetation. *Science, 276*, 786-788.

Jolly, D., S.P. Harrison, B. Damnati and R. Bonnefille, 1998a: Simulated climate and biomes of Africa during the Late Quaternary: comparison with pollen and lake status data. *Quat. Sci. Rev., 17*, 629-657.

Jolly, D., I.C. Prentice, R. Bonnefille, A. Ballouche, M. Bengo, P. Brenac, G. Buchet, D. Burney, J.-P. Cazet, R. Cheddadi, T. Edorh, H. Elenga, S. Elmoutaki, J. Guiot, F. Laarif, H. Lamb, A.-M. Lezine, J. Maley, M. Mbenza, O. Peyron, M. Reille, I. Reynaud-Ferrera, G. Riollet, J. C. Ritchie, E. Roche, L. Scott, I. Ssemmanda, H. Straka, M. Umer, E. Van Campo, S. Vilimumbala, A. Vincens and M. Waller, 1998b: Biome reconstruction from pollen and plant macrofossil data for Africa and the Arabian peninsula at 0 and 6 ka. *J. Biogeogr., 25*, 1007-1027.

Jones, P.D., 1994: Hemispheric surface air temperature variations: a reanalysis and an update to 1993. *J. Climate, 7*, 1794-1802.

Jones, P.D. and K. Briffa, 1992: Global surface air temperature variations over the twentieth century. Part 1: Spatial, temporal, and seasonal details. *Holocene, 2*, 165–179.

Jones, P.D., K.R. Briffa, T.P. Barnett and S.F.B. Tett, 1998: High-resolution palaeoclimatic records for the last millennium: interpretation, integration and comparison with general circulation model control-run temperatures. *The Holocene, 8*, 455-471.

Jones, P.D., N. New, D.E. Parker, S. Martin and I.G. Rigor, 1999: Surface air temperature and its change over the past 150 years. *Rev. Geophys., 37*, 173-199.

Joussaume, S. and K.E. Taylor, 1995: Status of the Paleoclimate Modeling Intercomparison Project (PMIP). *Proceedings of the first international AMIP scientific conference*. WCRP Report, 425-430.

Joussaume, S., K.E. Taylor, P. Braconnot, J.F.B. Mitchell, J.E. Kutzbach, S.P. Harrison, I.C. Prentice, A.J. Broccoli, A. Abe-Ouchi, P.J. Bartlein, C. Bonfils, B. Dong, J. Guiot, K. Herterich, C.D. Hewitt, D. Jolly, J.W. Kim, A. Kislov, A. Kitoh, M.F. Loutre, V. Masson, B. McAvaney, N. McFarlane, N. de Noblet, W.R. Peltier, J.Y. Peterschmitt, D. Pollard, D. Rind, J.F. Royer, M.E. Schlesinger, J. Syktus, S. Thompson, P. Valdes, G. Vettoretti, R.S. Webb and U. Wyputta, 1999: Monsoon changes for 6000 years ago: results of 18 simulations from the Paleoclimate Modeling Intercomparison Project (PMIP). *Geophys. Res. Lett., 26*, 859-862.

Kageyama, M., P.J. Valdes, G. Ramstein, C. Hewitt and U. Wyputta, 1999: Northern Hemisphere storm tracks in present day and last glacial maximum climate simulations: a comparison of the European PMIP models. *J. Climate, 12*, 742-760.

Kageyama, M., O. Peyron, S. Pinot, P. Tarasov, J. Guiot, S. Joussaume and G. Ramstein, 2001: The Last Glacial Maximum climate over Europe and western Siberia: a PMIP comparison between models and data. *Clim. Dyn., 17*, 23-43.

Kalnay, E. and Coauthors, 1996: The NCEP/NCAR 40-year Reanalysis Project. *Bull. Am. Met. Soc., 77*, 437-471.

Kar, S.C., M. Sugi and N. Sato, 1996: Simulation of the Indian summer monsoon and its variability using the JMA Global Model. *Pap. Met. Geophys., 47*, 65-101.

Kattenberg, A., F. Giorgi, H. Grassl, G.A. Meehl, J.F.B. Mitchell, R.J. Stouffer, T. Tokioka, A.J. Weaver and T.M.L. Wigley, 1996: Climate models — projections of future climate. In: *Climate Change 1995 – The Science of Climate Change: Contribution of Working Group I to the Second Assessment Report of the Intergovernmental Panel on Climate Change.* Houghton, J.T., L.G. Meira Filho, B.A. Callander, N. Harris, A. Kattenburg and K. Maskell, Eds., Cambridge University Press, Cambridge, England, pp. 285-357.

Kattsov, V.M., J.E. Walsh, A. Rinke, K. Dethloff, 2000: Atmospheric climate models: simulation of the Arctic Ocean fresh water budget components. In *The Freshwater Budget of the Arctic Ocean* (E.L. Lewis, ed.) Kluwer Academic Publishers, Dordrecht, The Netherlands, pp 209-247.

Kaurola, J., 1997: Some diagnostics of the northern wintertime climate simulated by the ECHAM3 model. *J. Climate, 10*, 201-222.

Kharin, V.V. and F.W. Zwiers, 2000: Changes in the extremes in an ensemble of transient climate simulations with a coupled atmosphere-ocean GCM. *J. Climate, 13*, 3760-3780.

Kiehl, J.T., J.J. Hack, M.H. Zhang and R.D. Cess, 1995: Sensitivity of a GCM climate to enhanced shortwave cloud absorption. *J. Climate, 8*, 2200-2212.

Kimoto, M. and X. Shen, 1997: Climate variability study by GCM - monsoon and El Niño - Center for Climate System Research Series, No. 2, 91-116, CCSR, University of Tokyo (in Japanese).

Kitoh, A., H. Koide, K. Kodera, S. Yukimoto and A. Noda, 1996: Interannual variability in the stratosphere-troposphere circulation in a coupled ocean-atmosphere GCM. *Geophys. Res. Lett., 23*, 543-546.

Kitoh, A., S. Yukimoto and A. Noda, 1999: ENSO-monsoon relationship in the MRI coupled GCM. *J. Met. Soc. Japan, 77*, 1221-1245.

Knutson, T.R. and S. Manabe, 1998: Model assessment of decadal variability and trends in the tropical Pacific Ocean. *J. Climate, 11*, 2273-2296.

Knutson, T.R., S. Manabe and D. Gu, 1997: Simulated ENSO in a global coupled ocean-atmosphere model: Multidecadal amplitude modulation and CO_2 sensitivity. *J. Climate*, **10**, 138-161.

Knutson, R., R.E. Tuleya and Y. Kurihara, 1998: Simulated increase of hurricane intensities in a CO_2-warmed climate. *Science*, **279**, 1018-1020.

Knutson, T.R., T.L. Delworh, K.W. Dixon and R.J. Stouffer, 1999: Model assessment of regional surface temperature trends (1949-1997). *J. Geophys. Res.*, **104**, 30981-30996.

Kobayashi, C., K. Takano, S. Kusunoki, M. Sugi and A. Kitoh, 2000: Seasonal predictability in winter over eastern Asia using the JMA global model. *Quart. J. R. Met. Soc.*, **126**, 2111-2123.

Kodera, K., M. Chiba, H. Koide, A. Kitoh and Y. Nikaidou, 1996: Interannual variability of the winter stratosphere and troposphere in the Northern Hemisphere. *J. Met. Soc. Japan*, **74**, 365-382.

Koster, R.D. and P.C. Milly, 1997: The interplay between transpiration and runoff formulations in land surface schemes used with atmospheric models. *J. Climate*, **10**, 1578-1591.

Kothavala, Z., 1997: Extreme precipitation events and the applicability of global climate models to study floods and droughts. *Math. and Comp. in Simulation*, **43**, 261-268.

Krishnamurti, T.N., R. Correa-Torres, M. Latif and G. Daughenbaugh, 1998: The impact of current and possibly future sea surface temperature anomalies on the frequency of Atlantic hurricanes. *Tellus*, **50A**, 186-210.

Kubatzki, C. and M. Claussen, 1998: Simulation of the global biogeophysical interactions during the last glacial maximum. *Clim. Dyn.*, **14**, 461-471.

Kutzbach, J.E. and Z. Liu, 1997: Response of the African monsoon to orbital forcing and ocean feedbacks in the Middle Holocene. *Science*, **278**, 440-443.

Kutzbach, J.E., G. Bonan, J. Foley and S. Harrison , 1996a: Vegetation and soil feedbacks on the response of the African monsoon to forcing in the early to middle Holocene. *Nature*, **384**, 623-626.

Kutzbach, J.E., P.J. Bartlein, J.A. Foley, S.P. Harrison, S.W. Hostetler, Z. Liu, I.C. Prentice and T. Webb III, 1996b: Potential role of vegetation feedback in the climate sensitivity of high-latitude regions: A case study at 6000 years B.P. *Global Biogeochemical Cycles*, **10**, 727-736.

Lal, M., U. Cubasch, J. Perlwitz and J. Waszkewitz, 1997: Simulation of the Indian monsoon climatology in ECHAM3 climate model: Sensitivity to horizontal resolution. *Int. J. Climatol.*, **17**, 847-858.

Lambert, S.J. and G.J. Boer, 2001: CMIP1 evaluation and intercomparison of coupled climate models. *Clim. Dyn.*, **17**, 2/3, 83-106.

Landsea, C.W. and J.A. Knaff, 2000: How much skill was there in forecasting the very strong 1997-98 El Niño? *Bull. Am. Met. Soc.*, **81**, 2107-2120.

Latif, M., 1998: Dynamics of interdecadal variability in coupled ocean-atmosphere models. *J. Climate*, **11**, 602-624.

Latif, M. and T.P. Barnett, 1996: Decadal climate variability over the North Pacific and North America: Dynamics and predictability. *J. Climate*, **9**, 2407-2423.

Latif, M., K. Sperber and Co-authors, 1999: ENSIP: The El Niño simulation intercomparison project. CLIVAR Report.

Lau, W.K.-M. and S. Yang, 1996: Seasonal variation, abrupt transition, and intra seasonal variability associated with the Asian summer monsoon in the GLA GCM. *J. Climate*, **9**, 965-985.

Laurent, C., H. Le Treut, Z.X. Li, L. Fairhead and J.L. Dufresne, 1998: The influence of resolution in simulating inter-annual and inter-decadal variability in a coupled ocean-atmosphere GCM with emphasis over the North Atlantic. IPSL report N8.

Lean, J. and D. Rind, 1998: Climate forcing by changes in solar radiation. *J. Climate*, **11**, 3069-3094.

Legutke, S. and R. Voss, 1999: The Hamburg Atmosphere-Ocean Coupled Circulation Model ECHO-G. Deutches Klimarechenzentrum Tech. Rep. No.18, Hamburg.

Lemke, P., W.D. Hibler, G. Flato, M. Harder and M. Kreyscher, 1997: On the improvement of sea ice models for climate simulations: the Sea

Ice Model Intercomparison Project. *Ann. Glaciol.*, **25**, 183-187.

Levis, S., J.A. Foley and D. Pollard, 1999: Climate-vegetation feedbacks at the Last Glacial Maximum. *J. Geophys. Res.*, **104**, 31191-31198.

Li, T. and T.F. Hogan, 1999: The role of the annual mean climate on seasonal and interannual variability of the tropical Pacific in a coupled GCM. *J. Climate*, **12**, 780-792.

Li, W. and Y.-Q. Yu, 2001: Intraseasonal oscillation in coupled general circulation model. *Scientia Atmospherica Sinica* (in Chinese with English abstract), **25**, 118-132.

Li, Z., L. Moreau and A. Arking, 1997: On solar energy disposition: A perspective from observation and modeling. *Bull. Am. Met. Soc.*, **78**, 53-70.

Lindzen, R.S. and M.S. Fox-Rabinovitz, 1989: Consistent Vertical and Horizontal Resolution. *Mon. Wea. Rev.*, **117**, 2575-2583.

Liu, Z., R.G. Gallimore, J.E. Kutzbach, W. Xu, Y. Golubev, P. Behling and R. Selin, 1999: Modeling long-term climate changes with equilibrium asynchronous coupling. *Clim. Dyn.*, **15**, 324-340.

Lunkeit, F., M. Ponater, R. Sausen, M. Sogalla, U. Ulbrich and M. Windelband, 1996: Cyclonic activity in a warmer climate. *Contribution to Atmospheric Physics*, **69**, 393-407.

Macdonald, A.M., 1998: The global ocean circulation: a hydrographic estimate and regional analysis. *Progr. Oceanogr.*, **41**, 281-382.

Mahowald, N., K.E. Kohfeld, M. Hansson, Y. Balkanski, S.P. Harrison, I.C. Prentice, M. Schulz and H. Rodhe, 1999: Dust sources and deposition in the Last Glacial Maximum and current climate. *J. Geophys. Res.*, **104**(D13), 15895-15916.

Malevsky-Malevich, S.P., E.D. Nadyozhina, V.V. Simonov, O.B. Shklyarevich and E.K. Molkentin, 1999: The evaluation of climate change influence on the permafrost season soil thawing regime. *Contemporary Investigation at Main Geophysical Observatory*, **1**, 33-50 (in Russian).

Manabe, S.J. and R.J. Stouffer, 1996: Low-frequency variability of surface air temperature in a 1000-year integration of a coupled atmosphere-ocean-land model. *J. Climate*, **9**, 376-393.

Manabe , S. and R.J. Stouffer, 1997: Coupled ocean-atmosphere response to freshwater input: comparison to Younger Dryas event. *Palaeooceanography*, **12**, 321-336.

Manabe, S., R.J. Stouffer, M.J. Spelman and K. Bryan, 1991: Transient responses of a coupled ocean-atmosphere model to gradual changes of atmospheric CO_2. Part I: Annual mean response. *J. Climate*, **4**, 785-818.

Mann, M.E., R.S. Bradley and M.K. Hughes, 1998: Global-scale temperature patterns and climate forcing over the past six centuries. *Nature*, **392**, 779-787.

Manne, A.S. and R.G. Richels, 1999: The Kyoto Protocol: A Cost-Effective Strategy for Meeting Environmental Objectives? *Energy Journal Special Issue on the Costs of the Kyoto Protocol: A Multi-Model Evaluation*, 1-24.

Manzini , E. and N.A. McFarlane, 1998, The effect of varying the source spectrum of a gravity wave parameterization in a middle atmosphere general circulation model, *J. Geophys. Res.* **103**, 31523-31539.

Manzini and McFarlane, 1998: The effect of varying the source spectrum of a gravity wave parametrization in a middle atmosphere general circulation model. *J. Geophys. Res.*, **103**, 31523-31539.

Marchal, O., T.F. Stocker and F. Joos, 1998: A latitude-depth, circulation-biogeochemical ocean model for paleoclimate studies: Model development and sensitivities. *Tellus*, **50B**, 290-316.

Marotzke, J., 1997: Boundary mixing and the dynamics of three-dimensional thermohaline circulations. *J. Phys. Oceanogr.*, **27**, 1713-1728.

Marotzke, J. and P.H. Stone, 1995: Atmospheric transports, the thermohaline circulation, and flux adjustments in a simple coupled model. *J. Phys. Oceanogr.*, **25**, 1350-1364

Martin, G.M., 1999: The simulation of the Asian summer monsoon, and its sensitivity to horizontal resolution, in the UK Meteorological

Office Unified Model. *Quart. J. R. Met. Soc.*, **125**, 1499-1525.

Martin, G.M. and M.K. Soman, 2000: Effects of changing physical parametrisations on the simulation of the Asian summer monsoon in the UK Meteorological Office Unified Model. *Hadley Centre Technical Note No. 17*, Met Office, London Road, Bracknell, RG12 2SY, UK.

Matsuura, T., M. Yumoto, S. Lizuka and R. Kawamura, 1999: Typhoon and ENSO simulation using a high-resolution coupled GCM. *Geophys. Res. Lett.,* **26**, 1755-1758.

McGuffie, K., A. Henderson-Sellers, N. Holbrook, Z. Kothavala, O. Balachova and J. Hoestra, 1999: Assessing simulations of daily temperature and precipitation variability with global climate models for present and enhanced greenhouse climates. *Int. J. Climatol.*, **19**, 1-26.

McPhaden, M., 1999: Genesis and evolution of the 1997-98 El Niño. *Science*, **283**, 950-954.

Medvedev, A.S., G.P. Klassen and S.R. Beagley, 1998: On the role of an anisotropic gravity wave spectrum in maintaining the circulation of the middle atmosphere. *Geophys. Res. Lett.*, **25**, 509-512.

Meehl, G.A., 1997: The south Asian monsoon and the tropospheric biennial oscillation. *J. Climate*, **10**, 1921-1943.

Meehl, G.A. and W.M. Washington, 1995: Cloud albedo feedback and the super greenhouse effect in a global coupled GCM. *Clim. Dyn.*, **11**, 399-411.

Meehl, G.A. and J.M. Arblaster, 1998: The Asian-Australian monsoon and El Niño-Southern Oscillation in the NCAR Climate System Model. *J. Climate,* **11**, 1356-1385.

Meehl, G.A., G.J. Boer, C. Covey, M. Latif and R.J. Stouffer, 2000a: The Coupled Model Intercomparison Project (CMIP). *Bull. Am. Met. Soc.,* **81**, 313-318.

Meehl, G.A., W.M. Washington, J.M. Arblaster, T.W. Bettge and W.G. Strand Jr., 2000b: Anthropogenic forcing and decadal climate variability in sensitivity experiments of 20th and 21st century climate. *J. Climate*, **13**, (in press).

Miller, R.L. and X. Jiang, 1996: Surface energy fluxes and coupled variability in the Tropics of a coupled general circulation model. *J. Climate*, **9**, 1599-1620.

Milly, P.C.D., 1997: Sensitivity of greenhouse summer dryness to changes in plan rooting characteristics. *Geophys. Res. Lett.*, **24**, 269-271

Mitchell, J.F.B., T.J. Johns, J.M. Gregory and S.B.F Tett, 1995: Climate response to increasing levels of greenhouse gases and sulphate aerosols. *Nature*, **376**, 501-504.

Mokhov, I.I. and P.K. Love, 1995: Diagnostics of cloudiness evolution in the annual cycle and interannual variability in the AMIP. *Proc. First Int. AMIP Sci. Conf.*, WMO/TD-No.732, 49-53.

Moore, A.M. and R. Kleeman, 1999: Stochastic forcing of ENSO by the intra seasonal oscillation. *J. Climate*, **12**, 1199-1220.

Moss, D. and Schneider, 1999: Uncertainties in the IPCC TAR: Recommendations to lead authors for more consistent assessment and Reporting. (Available from IPCC WGI Technical Support Unit)

Munk, W. and C. Wunsch, 1998: Abyssal Recipes II. *Deep Sea Res.*, **45**, 1976-2009.

Murphy, A.H., 1988: Skill scores based on the mean square error and their relationships to the correlation coefficient. *Mon. Wea. Rev.*, **116**, 2417-2424.

National Research Council, 2000: Reconciling Observations of Global Temperature Change. National Academy Press, Washington, D.C., 85 pp.

Navarra, A. (ed.), 1999: Beyond El Niño: Decadal and Interdecadal Climate Variability. Springer-Verlag, Berlin, 374 pp.

Neelin, J.D. and H.A. Dijkstra, 1995: Ocean-atmosphere interaction and the tropical climatology. Part I: the dangers of flux correction. *J. Climate,* **8**, 1343-1359.

Nordhaus, W.D., 1994: Managing the Global Commons: The Economics of Climate Change. The MIT Press, Cambridge.

Nozawa, T., S. Emori, T. Takemura, T. Nakajima, A. Numaguti, A. Abe-Ouchi and M. Kimoto, 2000: Coupled ocean-atmosphere model experiments of future climate change based on IPCC SRES scenarios. *Preprints of the 11th Symposium on Global Change Studies*, 9-14 January 2000, Long Beach, USA, 352-355.

Oberhuber, J.M., E. Roeckner, M. Christoph, M. Esch and M. Latif, 1998: Predicting the '97 El Niño event with a global climate model. *Geophys. Res. Lett.*, **25**, 2273-2276.

Ogasawara, N., A. Kitoh, T. Yasunari and A. Noda, 1999: Tropospheric biennial oscillation of ENSO monsoon system in the MRI coupled GCM. *J. Met. Soc. Japan*, **77**, 1247-1270.

Opsteegh, J.D., R.J. Haarsma, F.M. Selten and A. Kattenberg, 1998: ECBILT: A dynamic alternative to mixed boundary conditions in ocean models. *Tellus*, **50A**, 348-367.

Orr, J.C., E. Maier-Reimer, U. Mikolajewicz, P. Monfray, J.L. Sarmiento, J.R. Toggweiler, N.K. Taylor, J. Palmer, N. Gruber, C.L. Sabine, C. Le Quéré, R.M. Key and J. Boutin, 2001: Estimates of anthropogenic carbon uptake from four 3-D global ocean models. *Global Biogeochem. Cycles*, **15**(1), 43-60.

Osborn, T.J., K.R. Briffa, S.F.B. Tett, P.D. Jones and R.M. Trigo, 1999: Evaluation of the North Atlantic Oscillation as simulated by a coupled climate model. *Clim. Dyn.*, **15**, 685-702.

Otto-Bliesner, B.L., 1999: El Niño/La Niña and Sahel precipitation during the middle Holocene. *Geophys. Res. Lett.*, **26**, 87-90.

Parker, D.E., C.K . Folland and M. Jackson, 1995: Marine surface temperature: observed variations and data requirements. *Clim. Change*, **31**, 559-600.

Pawson, S., K. Kodera and Coauthors, 2000: The GCM-Reality Intercomparison Project for SPARC (GRIPS): Scientific issues and initial results. *Bull. Am. Met. Soc.*, **81**, 781-796.

Peck, S.C. and T.J. Teisberg, 1996: International CO_2 Emissions Targets and Timetables: An Analysis of the AOSIS Proposal. *Environmental Modeling and Assessment,* **1**(4), 219-227.

Petersen, A.C., 2000: Philosophy of Climate Science. *Bull. Am. Met. Soc.*, **81**, 265-271.

Petoukhov, V.K., I.I. Mokhov, A.V. Eliseev and V.A. Semenov, 1998: The IAP RAS global climate model. *Dialogue-MSU*, Moscow, 110 pp.

Petoukhov, V., A. Ganapolski, V. Brovkin, M. Claussen, A. Eliseev, C. Kubatzki and S. Rahmstorf, 2000: CLIMBER-2: A Climate system model of intermediate complexity, Part I: Model description and performance for present climate. *Clim. Dyn.*, **16**, 1-17.

Peyron, O., J. Guiot, R. Cheddadi, P. Tarasov, M. Reille, J.L. de Beaulieu, S. Bottema and V. Andrieu, 1998: Climatic reconstruction in Europe for 18,000 years B.P. from pollen data. *Quat. Res.,* **49**, 183-196.

Pinot, S., G. Ramstein, S.P. Harrison, I.C. Prentice, J. Guiot, M. Stute, S. Joussaume and PMIP-participating-groups, 1999: Tropical paleoclimates at the Last Glacial Maximum: comparison of Paleoclimate Modeling Intercomparison Project (PMIP) simulations and paleodata. *Clim. Dyn.*, **15**, 857-874.

Pitman, A.J. and M. Zhao, 2000: The relative Impact of observed change in land cover and carbon dioxide as simulated by a climate model. *Geophys. Res. Lett.*, **27**, 1267-1270.

Pitman, A.J., A.G. Slater, C.E. Desborough and M. Zhao, 1999: Uncertainty in the simulation of runoff due to the parameterization of frozen soil moisture using the GSWP methodology. *J. Geophys. Res.*, **104**, 16879-16888.

PMIP, 2000: Paleoclimate Modeling Intercomparison Project (PMIP): *Proceedings of the third PMIP workshop*, Canada, 4-8 October 1999, P. Braconnot (Ed), WCRP-111, WMO/TD-1007, 271 pp.

Polanyi, M, 1958: Personal Knowledge, Routledge and Kegan Paul, London, 428 pp.

Polcher, J., K. Laval, L. Dümenil, J. Lean and P.R. Rowntree, 1996: Comparing three land surface schemes used in GCMs. *J. Hydrology*,

180, 373-394.

Polcher, J., J. Crossley, C. Bunton, H. Douville, N. Gedney, K. Laval, S. Planton, P.R. Rowntree and P. Valdes, 1998a: Importance of land-surface processes for the uncertainties of climate change: A European Project. *GEWEX News*, **8**(2), 11-13.

Polcher, J., B. McAvaney, P. Viterbo, M.-A. Gaertner, A. Hahmann, J.-F. Mahfouf, J. Noilhan, T. Phillips, A.J. Pitman, C.A. Schlosser, J.-P. Schulz, B. Timbal, D. Verseghy and Y. Xue, 1998b: A proposal for a general interface between land-surface schemes and general circulation models. *Global Planet. Change*, **19**, 263-278.

Pollard, D., J.C. Bergengren, L.M. Stillwell-Soller, B. Felzer and S.L. Thompson, 1998: Climate simulations for 10000 and 6000 years BP using the GENESIS global climate model. *Paleoclimates - Data and Modelling*, **2**, 183-218.

Polzin, K.L., J.M. Toole, J.R. Ledwell and R.W. Schmitt, 1997: Spatial variability of turbulent mixing in the ocean. *Science*, **276**, 93-96.

Pope, V.D., A. Pamment and R.A. Stratton, 1999: Resolution sensitivity of the UKMO climate model. *Research Activities in Atmospheric and Oceanic Modelling, No. 28*, WCRP CAS/JSC working group on numerical experimentation, WMO, Geneva.

Pope, V.D., M. Gallani, P.R. Rowntree and R.A. Stratton, 2000: The impact of new physical parametrisations in the Hadley Centre climate model - HadAM3. *Clim. Dyn.*, **16**, 123-146.

Popper, K., 1982: The Open Universe, Hutchinson, London.

Potts, J.M., C.K. Folland, I.T. Jolliffe and D. Sexton, 1996: Revised "LEPS" scores for assessing climate model simulations and long-range forecasts. *J. Climate*, **9**, 34-53.

Power, S.B., R.A. Colman, B.J. McAvaney, R.R. Dahni, A.M. Moore and N.R. Smith, 1993: The BMRC Coupled atmosphere/ocean/sea-ice model. *BMRC Research Report No. 37*, Bureau of Meteorology Research Centre, Melbourne, Australia, 58 pp.

Power, S.B., F. Tseitkin, R.A. Colman and A. Sulaiman, 1998: A coupled general circulation model for seasonal prediction and climate change research. *BMRC Research Report No 66*, Bureau of Meteorology, Australia.

Prentice, I.C. and T. Webb III, 1998: BIOME 6000: reconstructing global mid-Holocene vegetation patterns from palaeoecological records. *J. Biogeogr.*, **25**, 997-1005.

Prentice, I.C., D. Jolly and BIOME-6000-participants, 1998: Mid-Holocene and glacial-maximum vegetation geography of the northern continents and Africa. *J. Biogeogr.*, **25**, 997-1005.

Rahmstorf, S. and Ganopolski, 1999: Long-term warming scenarios computed with an efficient coupled climate model. *Clim. Change*, **43**, 353-367.

Randall, D.A. and B.A. Wielicki, 1997: Measurements, models and hypotheses in the atmospheric sciences. *Bull. Am. Met. Soc.*, **78**, 399-406.

Raper, S.C.B., T.M.L Wigley and R.A. Warrick, 1996: Global sea-level rise: Past and Future. In: Sea-Level rise and Coastal subsidence, Causes, Consequences and Strategies. Kluwer Academic Publishers, Dordrecht, 369 pp.

Rayner, N.A., E.B. Horton, D.E. Parker, C.K. Folland and R.B. Hackett, 1996: Version 2.2 of the Global Sea-Ice and Sea Surface Temperature dataset, 1903–1994. *CRTN Rep. 74*. HCCPR, Bracknell, United Kingdom.

Renshaw, A.C., D.P. Rowell and C.K. Folland, 1998: Wintertime low-frequency weather variability in the North Pacific-American sector 1949-93. *J. Climate*, **11**, 1073-1093.

Renssen H., R.F.B. Isarin, J. Vandenberghe, M. Lautenschlager and U. Schlese, 2000: Permafrost as a critical factor in palaeoclimate modelling: the Younger Dryas case in Europe. *Earth and Planetary Science Letters*, **176**, 1-5.

Rind, D., 1999: Complexity and climate. *Science*, **284**, 105-107.

Rind, D. and D. Peteet, 1985: Terrestrial conditions at the Last Glacial Maximum and CLIMAP sea-surface temperature estimates: Are they consistent? *Quat. Res.*, **24**, 1-22.

Roberts, M.J., R. Marsh, A.L. New and R.A. Wood, 1996: An intercomparison of a Bryan-Cox-type ocean model and an isopycnic ocean model. Part I: The subpolar gyre and high latitude processes. *J. Phys. Oceanogr.*, **26**, 1495-1527.

Robertson, A.W., 1996: Interdecadal variability over the North Pacific in a multi-century climate simulation. *Clim. Dyn.*, **12**, 227-241.

Robertson, A.W., C.R. Mechoso and Y.-J. Kim, 1999: Interannual variations in the South American monsoon and their teleconnection with the North Atlantic Oscillation. Preprints of 10th Symposium on Global Change Studies, 10-15 January 1999, Dallas, Texas, American Meteorological Society, 434-437.

Robitaille, D.Y. and A.J. Weaver, 1995: Validation of sub-grid-scale mixing schemes using CFCs in a global ocean model. *Geophys. Res. Lett.*, **22**, 2917-2920.

Robock, A., C.A. Schlosser, K.Y. Vinnikov, N.A. Speranskaya, J.K. Entin and S. Qiu, 1998: Evaluation of the AMIP soil moisture simulations. *Global Planet. Change*, **19**, 181-208.

Rocha, A. and I. Simmonds, 1997: Interannual variability of south-eastern African summer rainfall. Part II: Modelling the impact of sea-surface temperatures on rainfall and circulation. *Int. J. Climatol.*, **17**, 267-290.

Rodbell, D.T., 1999: An 15,000 year record of El Niño-driven alluviation in southwestern Ecuador. *Science*, **283**, 516-520.

Rodwell, M.J., D.P. Rowell and C.K. Folland, 1999: Oceanic forcing of the wintertime North Atlantic Oscillation and European climate. *Nature*, **398**, 320-323.

Roeckner, E., J.M. Oberhuber, A. Bacher, M. Christoph and I. Kirchner, 1996: ENSO variability and atmospheric response in a global coupled atmosphere-ocean GCM. *Clim. Dyn.*, **12**, 737-754.

Roeckner, E., L. Bengtsson, J. Feichter, J. Lelieveld and H. Rodhe, 1999: Transient climate change simulations with a coupled atmosphere-ocean GCM including the tropospheric sulfur cycle. *J. Climate*, **12**, 3004-3012.

Rosell-Melé, A., E. Bard, K.C. Emeis, P. Farrimond, J. Grimalt, P.J. MŸller and R.R. Schneider, 1998: Project takes a new look at past sea surface temperatures. *EOS*, **79**, 393-394.

Rowell, D.P., C.K. Folland, K. Maskell and M.N. Ward, 1995: Variability of summer rainfall over tropical North Africa 1906-1992, Observations and modelling. *Quart. J. R. Met. Soc.*, **121**, 669-704.

Royer J.-F., F. Chauvin, B. Timbal, P. Araspin and D. Grimal, 1998: A GCM study of the impact of greenhouse gas increase on the frequency of occurrence of tropical cyclones. *Clim. Change*, **38**, 307-343.

Russell, G.L., J.R. Miller and D. Rind, 1995: A coupled atmosphere-ocean model for transient climate change studies. *Atmos.-Ocean*, **33**, 683-730.

Sandweiss, D., J.B. Richardson, E.J. Rieitz, H.B. Rollins and K.A. Maasch, 1996: Geoarchaeological evidence from Peru for a 5000 years B.P. onset of El Niño. *Science*, **273**, 1531-1533.

Santer, B.D., T.M.L. Wigley, D.J. Gaffen, L. Bengtsson, C. Doutriaux, J.S. Boyle, M. Esch, J.J. Hnilo, P.D. Jones, G. A. Meehl, E. Roeckner, K.E. Taylor and M.F. Wehner, 2000: Interpreting differential temperature trends at the surface and in the lower troposphere. *Science*, **287**, 1227-1232.

SAR, see IPCC, 1996.

Saravanan, R., 1998: Atmospheric low-frequency variability and its relationship to midlatitude SST variability: Studies using the NCAR climate system model. *J. Climate*, **11**, 1386-1404.

Sarmiento, J.L., P. Monfray, E. Maier-Reimer, O. Aumont, R. Murnane and J.C. Orr, 2000: Sea-air CO_2 fluxes and carbon transport: a comparison of three ocean general circulation models. *Global Biogeochem. Cycles*, 14(4), 1267-1281.

Sausen, R., S. Schubert and L. Dumenil, 1994: A model of the river-runoff for use in coupled atmosphere-ocean models. *Journal of*

Hydrology, **155**, 337-352.

Schlosser, C.A., A.G. Slater, A. Robock, A.J. Pitman, K.Y. Vinnikov, A. Henderson-Sellers, N.A. Speranskaya, K. Mitchell, A. Boone, H. Braden, F. Chen, P. Cox, P. de Rosnay, C.E. Desborough, R.E. Dickinson, Y.-J. Dai, Q. Duan, J. Entin, P. Etchevers, N. Gedney, Y.M. Gusev, F. Habets, J. Kim, V. Koren, E. Kowalczyk, O.N. Nasonova, J. Noilhan, J. Schaake, A.B. Shmakin, T.G. Smirnova, D. Verseghy, P. Wetzel, Y. Xue and Z.-L. Yang, 2000: Simulations of a boreal grassland hydrology at Valdai, Russia: PILPS Phase 2(d). *Mon. Wea. Rev.,* **128**, 301-321.

Schmittner, A. and T.F. Stocker, 1999: The stability of the thermohaline circulation in global warming experiments. *J. Climate,* **12**, 1117-1133.

Schneider, E.K. and Z. Zhu, 1998: Sensitivity of the simulated annual cycle of sea surface temperature in the Equatorial Pacific to sunlight penetration. *J. Climate*, **11**, 1932-1950.

Schneider, E.K., Z. Zhu, B.S. Giese, B. Huang, B.P. Kirtman, J. Shukla and J.A. Carton, 1997: Annual cycle and ENSO in a coupled ocean-atmosphere general circulation model. *Mon. Wea. Rev.*, **125**, 680-702.

Schubert, S., 1998: Downscaling local extreme temperature changes in South-Eastern Australia from the CSIRO MARK2 GCM. *Int. J. Climatol.*, **18**, 1419-1438.

Schubert, M., J. Perlwitz, R. Blender, K. Fraedrich and F. Lunkeit, 1998: North Atlantic cyclones in CO_2-induced warm climate simulations: frequency, intensity, and tracks. *Clim. Dyn.*, **14**, 827-837.

Sellers, P.J., D.A. Randall, C.J. Collatz, J.A. Berry, C.B. Field, D.A. Dazlich, C. Zhang, G. Collelo and L. Bounoua, 1996: A revised land-surface parameterization (SiB2) for atmospheric GCMs. Part 1: Model formulation. *J. Climate*, **9**, 676-705.

Semazzi, F.H.M. and L. Sun, 1997: The role of orography in determining the Sahelian climate. *Int. J. Climatol.*, **17**, 581-596.

Semazzi, F.H.M., B. Burns, N.-H. Lin and J.-K. Schemm, 1996: A GCM study of the teleconnections between the continental climate of Africa and global sea surface temperature anomalies. *J. Climate*, **9**, 2480-2497.

Shackley, S., P. Young, S. Parkinson and B. Wynne, 1998: Uncertainty, complexity and concepts of good science in climate change modelling: Are GCMs the best tools? *Clim. Change*, **38**, 159-205.

Shackley, S., P. Young and S. Parkinson, 1999: Response to A. Henderson-Sellers and K. McGuffie. *Clim. Change*, **42**, 611-617.

Shindell, D.T., R.L. Miller, G. Schmidt and L. Pandolfo, 1999: Simulation of recent northern winter climate trends by greenhouse-gas forcing. *Nature*, **399**, 452-455.

Slingo, J.M., K.R. Sperber and Co-authors, 1996: Intra seasonal oscillations in 15 atmospheric general circulation models: results from an AMIP diagnostic subproject. *Clim. Dyn.*, **12,** 325-357.

Slingo, J.M., D.P. Rowell, K.R. Sperber and F. Nortley, 1999: On the predictability of the interannual behaviour of the Madden-Julian Oscillation and its relationship with El Niño. *Quart. J. R. Met. Soc.*, **125**, 583-609.

Sperber, K.R. and T. Palmer, 1996: Interannual tropical rainfall variability in general circulation model simulations associated with the Atmospheric Model Intercomparison Project. *J. Climate*, **9**, 2727-2750.

Sperber, K.R. and Participating AMIP Modelling Groups, 1999: Are revised models better models? A skill score assessment of regional interannual variability. *Geophys. Res. Lett.*, **26**, 1267-1270.

Stensrud, D.J., R.L. Gall, S.L. Mullen and K.W. Howard, 1995: Model climatology of the Mexican monsoon. *J. Climate*, **8**, 1775-1794.

Stephenson, D.B. and J.-F. Royer, 1995: GCM simulation of the Southern Oscillation from 1979-1988. *Clim. Dyn.*, **11**, 115-128.

Stephenson, D.B., F. Chauvin and J.-F. Royer, 1998: Simulation of the Asian summer monsoon and its dependence on model horizontal resolution. *J. Met. Soc. Japan*, **76**, 237-265.

Stewart, R.E., K.K. Szeto, R.F. Reinking, S.A. Clough and S.P. Ballard, 1998: Midlatitude cyclonic cloud systems and their features affecting large scales and climate. *Rev. Geophys.*, **36**, 245-268.

Stockdale, T.N., D.L.T. Anderson, J.O.S. Alves and M.A. Balmaseda, 1998: Global seasonal rainfall forecasts using a coupled ocean-atmosphere model. *Nature*, **392**, 370-373.

Stocker, T.F., D.G. Wright and L.A. Mysak, 1992: A zonally-averaged, coupled ocean-atmosphere model for paleoclimatic studies. *J. Climate*, **5**, 773-797.

Stocker, T.F. and A. Schmittner, 1997: Influence of CO_2 emission rates on the stability of the thermohaline circulation. *Nature,* **388**, 862-865.

Stouffer , R.J. and K.W. Dixon, 1998: Initialization of coupled models for use in climate studies: A review. In Research Activities in Atmospheric and Oceanic Modelling. Report No. 27, WMO/TD-No. 865, World Meteorological Organization, Geneva, Switzerland, I.1-I.8.

Stouffer, R.J., S. Manabe and K.Y. Yinnikov, 1994: Model assessment of the role of natural variability in recent global warming. *Nature*, **367**, 634–636.

Stouffer, R.J., G. Hegerl and S. Tett, 2000: A comparison of surface air temperature variability in three 1000-year coupled ocean-atmosphere model integrations. *J. Climate*, **13**, 513-537.

Stratton, R.A., 1999: A high resolution AMIP integration using the Hadley Centre model HadAM2b. *Clim. Dyn.*, **15**, 9-28.

Street-Perrott, F.A. and R.A. Perrott, 1993: Holocene vegetation, lake levels and climate of Africa. *Global Climates since the Last Glacial Maximum*. H.E.J. Wright, J.E. Kutzbach, T. Webb III, W.F. Ruddiman, F.A. Street-Perrott and P.J. Bartlein, Eds., University of Minnesota Press, 318-356.

Street-Perrot, F.A. *et al.*, 1997: Impact of lower atmospheric carbon dioxide on tropical mountain ecosystems. *Science,* **278**, 1422-1426.

Sugi, M., A. Noda and N. Sato, 1997: Influence of global warming on tropical cyclone climatology- an experiment with the JMA global model, Research Activities in Atmospheric and Oceanic Modelling. Report No. 25, WMO/TD-No.792, 7.69-7.70.

Tarasov, P.E., T.I. Webb, A.A. Andreev, N.B. Afanas'eva, N.A. Berezina, L.G. Bezusko, T.A. Blyakharchuk, N.S. Bolikhovskaya, R. Cheddadi, M.M. Chernavskaya, G.M. Chernova, N.I. Dorofeyuk, V.G. Dirksen, G.A. Elina, L.V. Filimonova, F.Z. Glebov, J. Guiot, V.S. Gunova, S.P. Harrison, D. Jolly, V.I. Khomutova, E.V. Kvavadze, I.R. Osipova, N.K. Panova, I.C. Prentice, L. Saarse, D.V. Sevastyanov, V.S. Volkova and V.P. Zernitskaya, 1998: Present-day and mid-Holocene biomes reconstructed from pollen and plant macrofossil data from the former Soviet Union and Mongolia. *J. Biogeogr.,* **25**, 1029-1053.

Taylor, K.E., 2000: Summarizing Multiple aspects of model performance in a single diagram. *PCMDI Report No 65.*, Program for Climate Model Diagnosis and Intercomparison, Lawrence Livermore National Laboratory, University of California, Livermore CA, 24 pp.

Taylor, K.E., C.D. Hewitt, P. Braconnot, A.J. Broccoli, C. Doutriaux, J.F.B. Mitchell and PMIP-Participating-Groups, 2000: Analysis of forcing, response and feedbacks in a paleoclimate modeling experiment. In "Paleoclimate Modeling Intercomparison Project (PMIP) : proceedings of the third PMIP workshop, Canada, 4-8 October 1999", P. Braconnot (Ed), WCRP-111, WMO/TD-1007, pp 43-50.

Tett, S.F.B., T.C. Johns and J.F.B. Mitchell, 1997: Global and regional variability in a coupled AOGCM. *Clim. Dyn.*, **13**, 303–323.

Tett, S.F.B., P.A. Stott, M.R. Allen, W.J. Ingram and J.F.B. Mitchell, 1999: Causes of twentieth century temperature change. *Nature*, **399**, 569-572.

Texier, D., N. de Noblet, S.P. Harrison, A. Haxeltine, D. Jolly, S. Joussaume, F. Laarif, I.C. Prentice and P. Tarasov, 1997: Quantifying the role of biosphere-atmosphere feedbacks in climate change: coupled model simulations for 6000 years BP and comparison with paleodata for northern Eurasia and northern Africa. *Clim. Dyn.*, **13**, 865-882.

Texier, D., N. de Noblet and P. Braconnot, 2000: Sensitivity of the African and Asian monsoons to mid-Holocene insolation and data-inferred surface changes. *J. Climate*, **13**, 164-191.

Thorncroft C.D. and D.P. Rowell, 1998: Interannual variability of African wave activities in a general circulation model. *Int. J. Climatol.*, **18**, 1305-1323.

Timmermann, A., J. Oberhuber, A. Bacher, M. Esch, M. Latif and E. Roeckner, 1998: ENSO response to greenhouse warming. Max-Planck-Institut fur Meteorologie, Report No. 251, 13 pp.

Tokioka, T., A. Noda, A. Kitoh, Y. Nikaidou, S. Nakagawa, T. Motoi, S. Yukimoto and K. Takata, 1996: A transient CO_2 experiment with the MRI CGCM: Annual mean response. *CGER's Supercomputer Monograph Report Vol. 2*, CGER-IO22-96, ISSN 1341-4356, Center for Global Environmental Research, National Institute for Environmental Studies, Environment Agency of Japan, Ibaraki, Japan, 86 pp.

Tol, R.S.J., 1999: 'Spatial and Temporal Efficiency in Climate Change: Applications of FUND'. *Environmental and Resource Economics,* **14** (1), 33-49.

Trenberth, K.E, 1998a: The heat budget of the atmosphere and ocean. *Proceedings of the First International Conference on Reanalysis.* pp 17-20. WCRP 104, WMO/TD-No. 876.

Trenberth, K.E., 1998b: Development and forecasts of the 1997/98 El Niño: CLIVAR scientific issues. *CLIVAR – Exchanges*, **3**(2/3), 4-14.

Tsutsui, J.I. and A. Kasahara, 1996: Simulated tropical cyclones using the National Center for Atmospheric Research Community climate model. *J. Geophys. Res.*, **101**(D10), 15013-15032.

Tsutsui, J., A. Kasahara and H. Hirakuchi, 1999: The impacts of global warming on tropical cyclones- a numerical experiment with the T42 version of NCAR CCM2, Preprint volume of the 10th Symposium on Global Change Studies, J16B.310-15 January 1999, Dallas, Texas.

Vinnikov, K.Y., A. Robock, R.J. Stouffer, J.E. Walsh, C.L. Parkinson, D.J. Cavalieri, J.F.B. Mitchell, D. Garrett and V.F. Zakharov, 1999: Global Warming and Northern Hemisphere Sea Ice Extent. *Science,* **286**, 1934-1937.

Visbeck, M., J. Marshall, T. Haine and M. Spall, 1997: On the specification of eddy transfer coefficients in coarse resolution ocean circulation models. *J. Phys. Oceanogr.*, **27**, 381-402.

Vitart, F., J.L. Anderson and W.E. Stern, 1997: Simulation of interannual variability of tropical storm frequency in an ensemble of GCM integrations, *J. Climate*, **10**, 745-760.

Volodin, E.M. and V.N. Lykosov, 1998: Parameterization of heat and moisture transfer in the soil-vegetation system for use in atmospheric general circulation models: 2. Numerical experiments on climate modelling. *Izv. RAS, Physics of Atmosphere and Ocean*, **34**, 622-633.

von Storch, J-S., 1994: Interdecadal variability in a global coupled model. *Tellus*, **46A**, 419-432.

von Storch, J-S., V.V. Kharin, U. Cubasch, G.C. Hegerl, D. Schriever, H. von Storch and E. Zorita, 1997: A description of a 1260-year control integration with the coupled ECHAM1/LSG general circulation model. *J. Climate*, **10**, 1525-1543.

Voss, R., R. Sausen, and U. Cubasch, 1998: Periodically synchronously coupled integrations with the atmosphere-ocean general circulation model ECHAM3/LSG. *Climate Dyn.* **14**, 249-266.

Waliser, D.E., K.M. Lau and J.-H. Kim, 1999: The influence of coupled sea surface temperatures on the Madden-Julian oscillation: a model perturbation experiment. *J. Atmos. Sci.*, **56**, 333-358.

Wallace, J.M. and D.S. Gutzler, 1981: Teleconnections in the geopotential height field during the Northern Hemisphere winter. *Mon. Wea. Rev.*, **109**, 784-812.

Walsh, K. and A.B. Pittock, 1998: Potential changes in tropical storms, hurricanes, and extreme rainfall events as a result of climate change. *Clim. Change*, **39**, 199-213.

Wang, G., R. Kleeman, N. Smith and F. Tseitkin, 2000: Seasonal predic-

tions with a coupled global ocean-atmosphere model. *BMRC Res. Rep. No.77.*

Washington, W.M. and G.A. Meehl, 1996: High-latitude climate change in a global coupled ocean-atmosphere-sea ice model with increased atmospheric CO_2. *J. Geophys. Res.*, **101**(D8), 12795-12801.

Washington, W.M. and 10 others, 2000: Parallel Climate Model (PCM): Control and Transient simulations. *Clim. Dyn.*, **16**, 755-774.

Watterson, I.G. 1996: Non-dimensional measures of climate model performance. *Int. J. Climatol.*, **16**, 379-391.

Watterson, I.G., 1997, The diurnal cycle of surface air temperature in simulated present and doubled CO_2 climates, *Climate Dynamics*, **13**, 533-545.

Watterson, I.G., J.L. Evans and B.F. Ryan, 1995: Seasonal and interannual variability of tropical cyclogenesis: Diagnostics from large-scale fields. *J. Climate*, **8**, 3052-3066.

Weare, B.C., 2000a: Insights into the importance of cloud vertical structure in climate. *Geophys. Res. Lett.*, **27**, 907-910.

Weare, B.C., 2000b: Near-global observations of low clouds. *J. Climate*, **13**, 1255-1268.

Weare, B.C., I.I. Mokhov and Project Members, 1995: Evaluation of cloudiness and its variability in global climate models. *J. Climate*, **8**, 2224-2238.

Weare, B.C. and AMIP Modeling Groups, 1996: Evaluation of the vertical structure of zonally averaged cloudiness and its variability in the Atmospheric Model Intercomparison Project. *J. Climate*, **9**, 3419-3431.

Weaver, A.J. and T.M.C. Hughes, 1996: On the incompatibility of ocean and atmosphere models and the need for flux adjustments. *Clim. Dyn.*, **12**, 141-170.

Weaver, A.J., M. Eby, A.F. Fanning and E.C.C Wiebe, 1998: Simulated influence of carbon dioxide, orbital forcing and ice sheets on the climate of the Last Glacial Maximum. *Nature*, **394**, 847-853.

Webb, T.I., P.J. Bartlein, S.P. Harrison and K.H. Anderson, 1993: Vegetation lake-levels, and climate in Western North America since 12,000 years. *Global Climates since the Last Glacial Maximum.* (H.E.J. Wright, J.E. Kutzbach, T. Webb III, W.F. Ruddiman, F.A. Street-Perrott and P.J. Bartlein, Eds.), University of Minnesota Press, 415-467.

Wetherald, R.T. and S. Manabe, 1999: Detectability of summer dryness caused by greenhouse warming. *Clim. Change*, **43**, 495-511.

Whitworth, T. and R.G. Petersen, 1985: Volume transport of the Antarctic Circumpolar Current from bottom pressure measurements. *J. Phys. Oeanogr.*, **15**, 810-816.

Wiebe, E.C. and A.J. Weaver, 1999: On the sensitivity of global warming experiments to the parameterisation of sub-grid-scale ocean mixing. *Clim. Dyn.*, **15**, 875-893.

Wigley, T.M.L., 1998: The Kyoto Protocol: CO_2, CH_4, and climate implications. *Geophys. Res. Lett.*, **25**, 2285-2288.

Wigley, T.M.L. and S.C.B. Raper, 1987: Thermal expansion of sea water associated with global warming. *Nature*, **330**, 127-131.

Wigley, T.M.L. and B.D. Santer, 1990: Statistical comparison of spatial fields in model validation, perturbation and predictability experiments. *J. Geophys. Res.*, **95**, 851-865.

Wigley, T.M.L. and S.C.B. Raper, 1992: Implication for climate and sea level of revised IPCC emissions scenarios. *Nature*, **357**, 293-300.

Wijffels, S.E., R.W. Schmitt, H.L. Bryden and A. Stigebrandt, 1992: Transport of freshwater by the oceans. *J. Phys. Oceanogr.,* **22**, 155-162.

Wild, M., A. Ohmura and U. Cubasch, 1997: GCM-simulated surface energy fluxes in climate change experiments. *J. Climate*, **10**, 3093-3110.

Wild, M., A. Ohmura, H. Gilgen, E. Roeckner, M. Giorgetta and J.-J. Morcrette, 1998: The disposition of radiative energy in the global climate system: GCM-calculated versus observational estimates. *Clim. Dyn.*, **14**, 853-869.

Williamson, D.L., 1995: Skill scores from the AMIP Simulations. *Proc AMIP Scientific Conference* Monterey CA. World Climate Research Program, 253-258.

Williamson, D.L. 1999: Convergence of atmospheric simulations with increasing horizontal resolution and fixed forcing scales. *Tellus,* **51A**, 663-673.

Williamson, D.L. J.J. Hack and J.T. Kiehl, 1995: Climate sensitivity of the NCAR Community Climate Model (CCM2) to horizontal resolution. *Clim. Dyn.*, **11**, 377-397.

Williamson, D.L., J.G. Olson and B.A. Boville, 1998: A comparison of semi-Lagrangian and Eulerian tropical climate simulations. *Mon. Wea. Rev.*, **126**, 1001-1012.

Wood, R.A., A.B. Keen, J.F.B. Mitchell and J.M. Gregory, 1999: Changing spatial structure of the thermohaline circulation in response to atmospheric CO_2 forcing in a climate model. *Nature,* **399**, 572-575.

Wu, G.-X., X.-H. Zhang, H. Liu, Y.-Q. Yu, X.-Z. Jin, Y.-F. Guo, S.-F. Sun and W.-P. Li, 1997: Global ocean-atmosphere-land system model of LASG (GOALS/LASG) and its performance in simulation study. *Quart. J. of Appl. Meteor.,* **8**, Supplement, 15-28 (in Chinese).

Wunsch, C. and D. Roemmich, 1985: Is the North Atlantic in Sverdrup Balance? *J. Phys. Oceanogr.*, **15**, 1876-1883.

Xie, P. and P.A. Arkin, 1996: Analyses of Global Monthly Precipitation Using Gauge Observations, Satellite Estimates, and Numerical Model Predictions. *J. Climate,* **9**, 840-858.

Yoshimura, J., M. Sugi and A. Noda, 1999: Influence of greenhouse warming on tropical cyclone frequency simulated by a high-resolution AGCM, Preprints of 10th Symposium on Global Change Studies, 10-15 January 1999, Dallas, USA, 555-558.

Yu, G. and S.P. Harrison, 1996: An evaluation of the simulated water balance of northern Eurasia at 6000 yr B.P. using lake status data. *Clim. Dyn.*, **12**, 723-735.

Yu, Y.-Q., Y.-F. Guo and X.-H. Zhang, 2000a: Interdecadal climate variability. In*: IAP Global Ocean-Atmosphere-Land System Model* [Zhang, X.-H. (ed)]. Science Press, Beijing, New York, pp 155-170.

Yu, R., W. Li, X. Zhang, Y. Liu, Y. Yu, H. Liu and T. Zhou, 2000b: Climatic features related to Eastern China summer rainfalls in the NCAR CCM3. *Advances in Atmospheric Sciences*, **17**, 503-518.

Yukimoto, S., 1999: The decadal variability of the Pacific with the MRI coupled models. In: Beyond El Niño: Decadal and Interdecadal Climate Variability [Navarra, A. (ed)]. Springer-Verlag, Berlin, pp 205-220.

Yukimoto, S., M. Endoh, Y. Kitamura, A. Kitoh, T. Motoi, A. Noda and T. Tokioka, 1996: Interannual and interdecadal variabilities in the Pacific in a MRI coupled GCM. *Clim. Dyn.*, **12**, 667-683.

Yukimoto, S., M. Endoh, Y. Kitamura, A. Kitoh, T. Motoi and A. Noda, 2000: ENSO-like interdecadal variability in the Pacific Ocean as simulated in a coupled GCM. *J. Geophys. Res.*, **105**, 13945-13963.

Zhang, X.-H., G.-Y. Shi, H. Liu and Y.-Q. Yu (eds), 2000: IAP Global Atmosphere-Land System Model. Science Press, Beijing, China, 259 pp.

Zhang, Y. and W.-C. Wang, 1997: Model simulated northern winter cyclone and anti-cyclone activity under a greenhouse warming scenario. *J. Climate*, **10**, 1616-1634.

Zhang, Y., K.R. Sperber, J.S. Boyle, M. Dix, L. Ferranti, A. Kitoh, K.M. Lau, K. Miyakoda, D. Randall, L. Takacs and R. Wetherald, 1997: East Asian winter monsoon: results from eight AMIP models. *Clim. Dyn.*, **13**, 797-820.

Zhao, Z.-C., X. Gao and Y. Luo, 2000: Investigations on short-term climate prediction by GCMs in China. *Acta Meteorologica Sinica -English Version*, **14**, 108-119.

Zhao, M., A.J. Pitman and T. Chase, 2001: The Impact of Land Cover Change on the Atmospheric Circulation. *Clim. Dyn.*, **17**, 467-477.

Zhou, G.Q., X. Li and Q.C. Zeng, 1998: A coupled ocean-atmosphere general circulation model for ENSO prediction and 1997/98 ENSO forecast. *Climatic and Environmental Research*, **3**, 349-357 (In Chinese with English abstract).

Zwiers, F.W. and V.V. Kharin, 1998: Changes in the extremes of the climate simulated by CCC GCM2 under CO_2 doubling. *J. Climate*, **11**, 2200-2222.

9

Projections of Future Climate Change

Co-ordinating Lead Authors
U. Cubasch, G.A. Meehl

Lead Authors
G.J. Boer, R.J. Stouffer, M. Dix, A. Noda, C.A. Senior, S. Raper, K.S. Yap

Contributing Authors
A. Abe-Ouchi, S. Brinkop, M. Claussen, M. Collins, J. Evans, I. Fischer-Bruns, G. Flato, J.C. Fyfe,
A. Ganopolski, J.M. Gregory, Z.-Z. Hu, F. Joos, T. Knutson, R. Knutti, C. Landsea, L. Mearns, C. Milly,
J.F.B. Mitchell, T. Nozawa, H. Paeth, J. Räisänen, R. Sausen, S. Smith, T. Stocker, A. Timmermann,
U. Ulbrich, A. Weaver, J. Wegner, P. Whetton, T. Wigley, M. Winton, F. Zwiers

Review Editors
J.-W. Kim, J. Stone

Contents

Executive Summary

The results presented in this chapter are based on simulations made with global climate models and apply to spacial scales of hundreds of kilometres and larger. Chapter 10 presents results for regional models which operate on smaller spatial scales. Climate change simulations are assessed for the period 1990 to 2100 and are based on a range of scenarios for projected changes in greenhouse gas concentrations and sulphate aerosol loadings (direct effect). A few Atmosphere-Ocean General Circulation Model (AOGCM) simulations include the effects of ozone and/or indirect effects of aerosols (see Table 9.1 for details). Most integrations[1] do not include the less dominant or less well understood forcings such as land-use changes, mineral dust, black carbon, etc. (see Chapter 6). No AOGCM simulations include estimates of future changes in solar forcing or in volcanic aerosol concentrations.

There are many more AOGCM projections of future climate available than was the case for the IPCC Second Assessment Report (IPCC, 1996) (hereafter SAR). We concentrate on the IS92a and draft SRES A2 and B2 scenarios. Some indication of uncertainty in the projections can be obtained by comparing the responses among models. The range and ensemble standard deviation are used as a measure of uncertainty in modelled response. The simulations are a combination of a forced climate change component together with internally generated natural variability. A number of modelling groups have produced ensembles of simulations where the projected forcing is the same but where variations in initial conditions result in different evolutions of the natural variability. Averaging these integrations preserves the forced climate change signal while averaging out the natural variability noise, and so gives a better estimate of the models' projected climate change.

For the AOGCM experiments, the mean change and the range in global average surface air temperature (SAT) for the 1961 to 1990 average to the mid-21st century (2021 to 2050) for IS92a is +1.3°C with a range from +0.8 to +1.7°C for greenhouse gas plus sulphates (GS) as opposed to +1.6°C with a range from +1.0 to +2.1°C for greenhouse gas only (G). For SRES A2 the mean is +1.1°C with a range from +0.5 to +1.4°C, and for B2, the mean is +1.2°C with a range from +0.5 to +1.7°C.

For the end of the 21st century (2071 to 2100), for the draft SRES marker scenario A2, the global average SAT change from AOGCMs compared with 1961 to 1990 is +3.0°C and the range is +1.3 to +4.5°C, and for B2 the mean SAT change is +2.2°C and the range is +0.9 to +3.4°C.

AOGCMs can only be integrated for a limited number of scenarios due to computational expense. Therefore, a simple climate model is used here for the projections of climate change for the next century. The simple model is tuned to simulate the response found in several of the AOGCMs used here. The forcings for the simple model are based on the radiative forcing estimates from Chapter 6, and are slightly different to the forcings used by the AOGCMs. The indirect aerosol forcing is

scaled assuming a value of −0.8 Wm^{-2} for 1990. Using the IS92 scenarios, the SAR gives a range for the global mean temperature change for 2100, relative to 1990, of +1 to +3.5°C. The estimated range for the six final illustrative SRES scenarios using updated methods is +1.4 to +5.6°C. The range for the full set of SRES scenarios is +1.4 to +5.8°C.

These estimates are larger than in the SAR, partly as a result of increases in the radiative forcing, especially the reduced estimated effects of sulphate aerosols in the second half of the 21st century. By construction, the new range of temperature responses given above includes the climate model response uncertainty and the uncertainty of the various future scenarios, but not the uncertainty associated with the radiative forcings, particularly aerosol. Note the AOGCM ranges above are 30-year averages for a period ending at the year 2100 compared to the average for the period 1961 to 1990, while the results for the simple model are for temperature changes at the year 2100 compared with the year 1990.

A traditional measure of climate response is equilibrium climate sensitivity derived from 2×CO$_2$ experiments with mixed-layer models, i.e., Atmospheric General Circulation Models (AGCMs) coupled to non-dynamic slab oceans, run to equilibrium. It has been cited historically to provide a calibration for models used in climate change experiments. The mean and standard deviation of this quantity from seventeen mixed-layer models used in the SAR are +3.8 and +0.8°C, respectively. The same quantities from fifteen models in active use are +3.5 and +0.9°C, not significantly different from the values in the SAR. These quantities are model dependent, and the previous estimated range for this quantity, widely cited as +1.5 to +4.5°C, still encompasses the more recent model sensitivity estimates.

A more relevant measure of transient climate change is the transient climate response (TCR). It is defined as the globally averaged surface air temperature change for AOGCMs at the time of CO$_2$ doubling in 1%/yr CO$_2$ increase experiments. The TCR combines elements of model sensitivity and factors that affect response (e.g., ocean heat uptake). It provides a useful measure for understanding climate system response and allows direct comparison of global coupled models. The range of TCR for current AOGCMs is +1.1 to +3.1°C with an average of 1.8°C. The 1%/yr CO$_2$ increase represents the changes in radiative forcing due to all greenhouse gases, hence this is a higher rate than is projected for CO$_2$ alone. This increase of radiative forcing lies on the high side of the SRES scenarios (note also that CO$_2$ doubles around mid-21st century in most of the scenarios). However these experiments are valuable for promoting the understanding of differences in the model responses.

The following findings from the models analysed in this chapter corroborate results from the SAR (projections of regional climate change are given in Chapter 10) for all scenarios considered. We assign these to be virtually certain to very likely (defined as agreement among most models, or, where only a small number of models have been analysed and their results are physically plausible, these have been assessed to characterise those from a larger number of models). The more recent results are generally obtained from models with improved parametrizations (e.g., better land-surface process schemes).

---

[1] In this report, the term "integration" is used to mean a climate model run.

- The troposphere warms, stratosphere cools, and near surface temperature warms.

- Generally, the land warms faster than the ocean, the land warms more than the ocean after forcing stabilises, and there is greater relative warming at high latitudes.

- The cooling effect of tropospheric aerosols moderates warming both globally and locally, which mitigates the increase in SAT.

- The SAT increase is smaller in the North Atlantic and circumpolar Southern Ocean regions relative to the global mean.

- As the climate warms, Northern Hemisphere snow cover and sea-ice extent decrease.

- The globally averaged mean water vapour, evaporation and precipitation increase.

- Most tropical areas have increased mean precipitation, most of the sub-tropical areas have decreased mean precipitation, and in the high latitudes the mean precipitation increases.

- Intensity of rainfall events increases.

- There is a general drying of the mid-continental areas during summer (decreases in soil moisture). This is ascribed to a combination of increased temperature and potential evaporation that is not balanced by increases in precipitation.

- A majority of models show a mean El Niño-like response in the tropical Pacific, with the central and eastern equatorial Pacific sea surface temperatures warming more than the western equatorial Pacific, with a corresponding mean eastward shift of precipitation.

- Available studies indicate enhanced interannual variability of northern summer monsoon precipitation.

- With an increase in the mean surface air temperature, there are more frequent extreme high maximum temperatures and less frequent extreme low minimum temperatures. There is a decrease in diurnal temperature range in many areas, with night-time lows increasing more than daytime highs. A number of models show a general decrease in daily variability of surface air temperature in winter, and increased daily variability in summer in the Northern Hemisphere land areas.

- The multi-model ensemble signal to noise ratio is greater for surface air temperature than for precipitation.

- Most models show weakening of the Northern Hemisphere thermohaline circulation (THC), which contributes to a reduction in the surface warming in the northern North Atlantic. Even in models where the THC weakens, there is still a warming over Europe due to increased greenhouse gases.

- The deep ocean has a very long thermodynamic response time to any changes in radiative forcing; over the next century, heat anomalies penetrate to depth mainly at high latitudes where mixing is greatest.

A second category of results assessed here are those that are new since the SAR, and we ascribe these to be very likely (as defined above):

- The range of the TCR is limited by the compensation between the effective climate sensitivity (ECS) and ocean heat uptake. For instance, a large ECS, implying a large temperature change, is offset by a comparatively large heat flux into the ocean.

- Including the direct effect of sulphate aerosols (IS92a or similar) reduces global mean mid-21st century warming (though there are uncertainties involved with sulphate aerosol forcing – see Chapter 6).

- Projections of climate for the next 100 years have a large range due both to the differences of model responses and the range of emission scenarios. Choice of model makes a difference comparable to choice of scenario considered here.

- In experiments where the atmospheric greenhouse gas concentration is stabilised at twice its present day value, the North Atlantic THC recovers from initial weakening within one to several centuries.

- The increases in surface air temperature and surface absolute humidity result in even larger increases in the heat index (a measure of the combined effects of temperature and moisture). The increases in surface air temperature also result in an increase in the annual cooling degree days and a decrease in heating degree days.

Additional new results since the SAR; these are assessed to be likely due to many (but not most) models showing a given result, or a small number of models showing a physically plausible result.

- Areas of increased 20 year return values of daily maximum temperature events are largest mainly in areas where soil moisture decreases; increases in return values of daily minimum temperature especially occur over most land areas and are generally larger where snow and sea ice retreat.

- Precipitation extremes increase more than does the mean and the return period for extreme precipitation events decreases almost everywhere.

Another category includes results from a limited number of studies which are new, less certain, or unresolved, and we assess these to have medium likelihood, though they remain physically plausible:

- Although the North Atlantic THC weakens in most models, the relative roles of surface heat and freshwater fluxes vary from model to model. Wind stress changes appear to play only a minor role.

- It appears that a collapse in the THC by the year 2100 is less likely than previously discussed in the SAR, based on the AOGCM results to date.

- Beyond 2100, the THC could completely shut-down, possibly irreversibly, in either hemisphere if the rate of change of radiative forcing was large enough and applied long enough. The implications of a complete shut-down of the THC have not been fully explored.

- Although many models show an El Niño-like change in the mean state of tropical Pacific SSTs, the cause is uncertain. It has been related to changes in the cloud radiative forcing and/or evaporative damping of the east-west SST gradient in some models.

- Future changes in El Niño-Southern Oscillation (ENSO) interannual variability differ from model to model. In models that show increases, this is related to an increase in thermocline intensity, but other models show no significant change and there are considerable uncertainties due to model limitations of simulating ENSO in the current generation of AOGCMs (Chapter 8).

- Several models produce less of the weak but more of the deeper mid-latitude lows, meaning a reduced total number of storms. Techniques are being pioneered to study the mechanisms of the changes and of variability, but general agreement among models has not been reached.

- There is some evidence that shows only small changes in the frequency of tropical cyclones derived from large-scale parameters related to tropical cyclone genesis, though some measures of intensities show increases, and some theoretical and modelling studies suggest that upper limit intensities could increase (for further discussion see Chapter 10).

- There is no clear agreement concerning the changes in frequency or structure of naturally occurring modes of variability such as the North Atlantic Oscillation.

9.1 Introduction

The purpose of this chapter is to assess and quantify projections of possible future climate change from climate models. A background of concepts used to assess climate change experiments is presented in Section 9.2, followed by Section 9.3 which includes results from ensembles of several categories of future climate change experiments, factors that contribute to the response of those models, changes in variability and changes in extremes. Section 9.4 is a synthesis of our assessment of model projections of climate change.

In a departure from the organisation of the SAR, the assessment of regional information derived in some way from global models (including results from embedded regional high resolution models, downscaling, etc.) now appears in Chapter 10.

9.1.1 Background and Recap of Previous Reports

Studies of projections of future climate change use a hierarchy of coupled ocean/atmosphere/sea-ice/land-surface models to provide indicators of global response as well as possible regional patterns of climate change. One type of configuration in this climate model hierarchy is an Atmospheric General Circulation Model (AGCM), with equations describing the time evolution of temperature, winds, precipitation, water vapour and pressure, coupled to a simple non-dynamic "slab" upper ocean, a layer of water usually around 50 m thick that calculates only temperature (sometimes referred to as a "mixed-layer model"). Such air-sea coupling allows those models to include a seasonal cycle of solar radiation. The sea surface temperatures (SSTs) respond to increases in carbon dioxide (CO_2), but there is no ocean dynamical response to the changing climate. Since the full depth of the ocean is not included, computing requirements are relatively modest so these models can be run to equilibrium with a doubling of atmospheric CO_2. This model design was prevalent through the 1980s, and results from such equilibrium simulations were an early basis of societal concern about the consequences of increasing CO_2.

However, such equilibrium (steady-state) experiments provide no information on time-dependent climate change and no information on rates of climate change. In the late 1980s, more comprehensive fully coupled global ocean/atmosphere/sea-ice/land-surface climate models (also referred to as Atmosphere-Ocean Global Climate Models, Atmosphere-Ocean General Circulation Models or simply AOGCMs) began to be run with slowly increasing CO_2, and preliminary results from two such models appeared in the 1990 IPCC Assessment (IPCC, 1990).

In the 1992 IPCC update prior to the Earth Summit in Rio de Janeiro (IPCC, 1992), there were results from four AOGCMs run with CO_2 increasing at 1%/yr to doubling around year 70 of the simulations (these were standardised sensitivity experiments, and consequently no actual dates were attached). Inclusion of the full ocean meant that warming at high latitudes was not as uniform as from the non-dynamic mixed-layer models. In regions of deep ocean mixing in the North Atlantic and Southern Oceans, warming was less than at other high latitude locations. Three of those four models used some form of flux adjustment

whereby the fluxes of heat, fresh water and momentum were either singly or in some combination adjusted at the air-sea interface to account for incompatibilities in the component models. However, the assessment of those models suggested that the main results concerning the patterns and magnitudes of the climate changes in the model without flux adjustment were essentially the same as in the flux-adjusted models.

The most recent IPCC Second Assessment Report (IPCC, 1996) (hereafter SAR) included a much more extensive collection of global coupled climate model results from models run with what became a standard 1%/yr CO_2-increase experiment. These models corroborated the results in the earlier assessment regarding the time evolution of warming and the reduced warming in regions of deep ocean mixing. There were additional studies of changes in variability in the models in addition to changes in the mean, and there were more results concerning possible changes in climate extremes. Information on possible future changes of regional climate was included as well.

The SAR also included results from the first two global coupled models run with a combination of increasing CO_2 and sulphate aerosols for the 20th and 21st centuries. Thus, for the first time, models were run with a more realistic forcing history for the 20th century and allowed the direct comparison of the model's response to the observations. The combination of the warming effects on a global scale from increasing CO_2 and the regional cooling from the direct effect of sulphate aerosols produced a better agreement with observations of the time evolution of the globally averaged warming and the patterns of 20th century climate change. Subsequent experiments have attempted to quantify and include additional forcings for 20th century climate (Chapter 8), with projected outcomes for those forcings in scenario integrations into the 21st century discussed below.

In the SAR, the two global coupled model runs with the combination of CO_2 and direct effect of sulphate aerosols both gave a warming at mid-21st century relative to 1990 of around 1.5°C. To investigate more fully the range of forcing scenarios and uncertainty in climate sensitivity (defined as equilibrium globally averaged surface air temperature increase due to a doubling of CO_2, see discussion in Section 9.2 below) a simpler climate model was used. Combining low emissions with low sensitivity and high emissions with high sensitivity gave an extreme range of 1 to 4.5°C for the warming in the simple model at the year 2100 (assuming aerosol concentrations constant at 1990-levels). These projections were generally lower than corresponding projections in IPCC (1990) because of the inclusion of aerosols in the pre-1990 radiative forcing history. When the possible effects of future changes of anthropogenic aerosol as prescribed in the IS92 scenarios were incorporated this led to lower projections of temperature change of between 1°C and 3.5°C with the simple model.

Spatial patterns of climate change simulated by the global coupled models in the SAR corroborated the IPCC (1990) results. With increasing greenhouse gases the land was projected to warm generally more than the oceans, with a maximum annual mean warming in high latitudes associated with reduced snow cover and increased runoff in winter, with greatest warming at high northern latitudes. Including the effects of aerosols led to a somewhat reduced warming in middle latitudes of the Northern Hemisphere and the maximum warming in northern high latitudes was less extensive since most sulphate aerosols are produced in the Northern Hemisphere. All models produced an increase in global mean precipitation but at that time there was little agreement among models on changes in storminess in a warmer world and conclusions regarding extreme storm events were even more uncertain.

9.1.2 New Types of Model Experiments since 1995

The progression of experiments including additional forcings has continued and new experiments with additional greenhouse gases (such as ozone, CFCs, etc., as well as CO_2) will be assessed in this chapter.

In contrast to the two global coupled climate models in the 1990 Assessment, the Coupled Model Intercomparison Project (CMIP) (Meehl *et al.*, 2000a) includes output from about twenty AOGCMs worldwide, with roughly half of them using flux adjustment. Nineteen of them have been used to perform idealised 1%/yr CO_2-increase climate change experiments suitable for direct intercomparison and these are analysed here. Roughly half that number have also been used in more detailed scenario experiments with time evolutions of forcings including at least CO_2 and sulphate aerosols for 20th and 21st century climate. Since there are some differences in the climate changes simulated by various models even if the same forcing scenario is used, the models are compared to assess the uncertainties in the responses. The comparison of 20th century climate simulations with observations (see Chapter 8) has given us more confidence in the abilities of the models to simulate possible future climate changes in the 21st century and reduced the uncertainty in the model projections (see Chapter 14). The newer model integrations without flux adjustment give us indications of how far we have come in removing biases in the model components. The results from CMIP confirm what was noted in the SAR in that the basic patterns of climate system response to external forcing are relatively robust in models with and without flux adjustment (Gregory and Mitchell, 1997; Fanning and Weaver, 1997; Meehl *et al.*, 2000a). This also gives us more confidence in the results from the models still using flux adjustment.

The IPCC data distribution centre (DDC) has collected results from a number of transient scenario experiments. They start at an early time of industrialisation and most have been run with and without the inclusion of the direct effect of sulphate aerosols. Note that most models do not use other forcings described in Chapter 6 such as soot, the indirect effect of sulphate aerosols, or land-use changes. Forcing estimates for the direct effect of sulphate aerosols and other trace gases included in the DDC models are given in Chapter 6. Several models also include effects of tropospheric and stratospheric ozone changes.

Additionally, multi-member ensemble integrations have been run with single models with the same forcing. So-called "stabilisation" experiments have also been run with the atmospheric greenhouse gas concentrations increasing by 1%/yr or following an IPCC scenario, until CO_2-doubling, tripling or quadrupling. The greenhouse gas concentration is then kept fixed and the model

integrations continue for several hundred years in order to study the commitment to climate change. The 1%/yr rate of increase for future climate, although larger than the actual CO_2 increase observed to date, is meant to account for the radiative effects of CO_2 and other trace gases in the future and is often referred to as "equivalent CO_2" (see discussion in Section 9.2.1). This rate of increase in radiative forcing is often used in model intercomparison studies to assess general features of model response to such forcing.

In 1996, the IPCC began the development of a new set of emissions scenarios, effectively to update and replace the well-known IS92 scenarios. The approved new set of scenarios is described in the IPCC Special Report on Emission Scenarios (SRES) (Nakićenović *et al.*, 2000; see more complete discussion of SRES scenarios and forcing in Chapters 3, 4, 5 and 6). Four different narrative storylines were developed to describe consistently the relationships between emission driving forces and their evolution and to add context for the scenario quantification (see Box 9.1). The resulting set of forty scenarios (thirty-five of which contain data on the full range of gases required for climate modelling) cover a wide range of the main demographic, economic and technological driving forces of future greenhouse gas and sulphur emissions. Each scenario represents a specific quantification of one of the four storylines. All the scenarios based on the same storyline constitute a scenario "family". (See Box 9.1, which briefly describes the main characteristics of the four SRES storylines and scenario families.) The SRES scenarios do not include additional climate initiatives, which means that no scenarios are included that explicitly assume implementation of the UNFCCC or the emissions targets of the Kyoto Protocol. However, greenhouse gas emissions are directly affected by non-climate change policies designed for a wide range of other purposes. Furthermore, government policies can, to varying degrees, influence the greenhouse gas emission drivers and this influence is broadly reflected in the storylines and resulting scenarios.

Because SRES was not approved until 15 March 2000, it was too late for the modelling community to incorporate the scenarios into their models and have the results available in time for this Third Assessment Report. Therefore, in accordance with a decision of the IPCC Bureau in 1998 to release draft scenarios to climate modellers (for their input to the Third Assessment Report) one marker scenario was chosen from each of four of the scenario groups based on the storylines (A1B, A2, B1 and B2) (Box 9.1). The choice of the markers was based on which initial quantification best reflected the storyline, and features of specific models. Marker scenarios are no more or less likely than any other scenarios but these scenarios have received the closest scrutiny. Scenarios were also selected later to illustrate the other two scenario groups (A1FI and A1T), hence there is an illustrative scenario for each of the six scenario groups. These latter two illustrative scenarios were not selected in time for AOGCM models to utilise them in this report. In fact, time and computer resource limitations dictated that most modelling groups could run only A2 and B2, and results from those integrations are evaluated in this chapter. However, results for all six illustrative scenarios are shown here using a simple climate model discussed below. The IS92a scenario is also used in a number of the results presented in this chapter in order to provide direct comparison with the results in the SAR.

Box 9.1: The Emissions Scenarios of the Special Report on Emissions Scenarios (SRES)

A1. The A1 storyline and scenario family describe a future world of very rapid economic growth, global population that peaks in mid-century and declines thereafter, and the rapid introduction of new and more efficient technologies. Major underlying themes are convergence among regions, capacity building and increased cultural and social interactions, with a substantial reduction in regional differences in per capita income. The A1 scenario family develops into three groups that describe alternative directions of techno-logical change in the energy system. The three A1 groups are distinguished by their technological emphasis: fossil intensive (A1FI), non-fossil energy sources (A1T), or a balance across all sources (A1B) (where balanced is defined as not relying too heavily on one particular energy source, on the assumption that similar improvement rates apply to all energy supply and end use technologies).

A2. The A2 storyline and scenario family describe a very heterogeneous world. The underlying theme is self-reliance and preser-vation of local identities. Fertility patterns across regions converge very slowly, which results in continuously increasing popula-tion. Economic development is primarily regionally oriented and *per capita* economic growth and technological change are more fragmented and slower than in other storylines.

B1. The B1 storyline and scenario family describe a convergent world with the same global population, that peaks in mid-century and declines thereafter, as in the A1 storyline, but with rapid change in economic structures toward a service and information economy, with reductions in material intensity and the introduction of clean and resource-efficient technologies. The emphasis is on global solutions to economic, social and environmental sustainability, including improved equity, but without additional climate initiatives.

B2. The B2 storyline and scenario family describe a world in which the emphasis is on local solutions to economic, social and environmental sustainability. It is a world with continuously increasing global population, at a rate lower than A2, intermediate levels of economic development, and less rapid and more diverse technological change than in the B1 and A1 storylines. While the scenario is also oriented towards environmental protection and social equity, it focuses on local and regional levels.

The final four marker scenarios contained in SRES differ in minor ways from the draft scenarios used for the AOGCM experiments described in this report. In order to ascertain the likely effect of differences in the draft and final SRES scenarios each of the four draft and final marker scenarios were studied using a simple climate model tuned to the AOGCMs used in this report. For three of the four marker scenarios (A1B, A2 and B2) temperature change from the draft and final scenarios are very similar. The primary difference is a change to the standardised values for 1990 to 2000, which is common to all these scenarios. This results in a higher forcing early in the period. There are further small differences in net forcing, but these decrease until, by 2100, differences in temperature change in the two versions of these scenarios are in the range 1 to 2%. For the B1 scenario, however, temperature change is significantly lower in the final version, leading to a difference in the temperature change in 2100 of almost 20%, as a result of generally lower emissions across the full range of greenhouse gases. For descriptions of the simula-tions, see Section 9.3.1.

9.2 Climate and Climate Change

Chapter 1 discusses the nature of the climate system and the climate variability and change it may undergo, both naturally and as a consequence of human activity. The projections of future climate change discussed in this chapter are obtained using climate models in which changes in atmospheric composition are specified. The models "translate" these changes in composition into changes in climate based on the physical processes governing the climate system as represented in the models. The simulated climate change depends, therefore, on projected changes in emissions, the changes in atmospheric greenhouse gas and particulate (aerosol) concentrations that result, and the manner in which the models respond to these changes. The response of the climate system to a given change in forcing is broadly characterised by its "climate sensitivity". Since the climate system requires many years to come into equilibrium with a change in forcing, there remains a "commitment" to further climate change even if the forcing itself ceases to change.

Observations of the climate system and the output of models are a combination of a forced climate change "signal" and internally generated natural variability which, because it is random and unpredictable on long climate time-scales, is charac-terised as climate "noise". The availability of multiple simula-tions from a given model with the same forcing, and of simula-tions from many models with similar forcing, allows ensemble methods to be used to better characterise projected climate change and the agreement or disagreement (a measure of reliability) of model results.

9.2.1 Climate Forcing and Climate Response

The heat balance
Broad aspects of global mean temperature change may be illustrated using a simple representation of the heat budget of the climate system expressed as:

$$dH/dt = F - \alpha T.$$

Here F is the radiative forcing change as discussed in Chapter 6; αT represents the net effect of processes acting to counteract changes in mean surface temperature, and dH/dt is the rate of heat storage in the system. All terms are differences from unperturbed equilibrium climate values. A positive forcing will act to increase the surface temperature and the magnitude of the resulting increase will depend on the strength of the feedbacks measured by αT. If α is large, the temperature change needed to balance a given change in forcing is small and vice versa. The result will also depend on the rate of heat storage which is dominated by the ocean so that $dH/dt = dH_o/dt = F_o$ where H_o is the ocean heat content and F_o is the flux of heat into the ocean. With this approximation the heat budget becomes $F = \alpha T + F_o$, indicating that both the feedback term and the flux into the ocean act to balance the radiative forcing for non-equilibrium conditions.

Radiative forcing in climate models

A radiative forcing change, symbolised by F above, can result from changes in greenhouse gas concentrations and aerosol loading in the atmosphere. The calculation of F is discussed in Chapter 6 where a new estimate of CO_2 radiative forcing is given which is smaller than the value in the SAR. According to Section 6.3.1, the lower value is due mainly to the fact that stratospheric temperature adjustment was not included in the (previous) estimates given for the forcing change. It is important to note that this new radiative forcing estimate does not affect the climate change and equilibrium climate sensitivity calculations made with general circulation models. The effect of a change in greenhouse gas concentration and/or aerosol loading in a general circulation model (GCM) is calculated internally and interactively based on, and in turn affecting, the three dimensional state of the atmosphere. In particular, the stratospheric temperature responds to changes in radiative fluxes due to changes in CO_2 concentration and the GCM calculation includes this effect.

Equivalent CO₂

The radiative effects of the major greenhouse gases which are well-mixed throughout the atmosphere are often represented in GCMs by an "equivalent" CO_2 concentration, namely the CO_2 concentration that gives a radiative forcing equal to the sum of the forcings for the individual greenhouse gases. When used in simulations of forced climate change, the increase in "equivalent CO_2" will be larger than that of CO_2 by itself, since it also accounts for the radiative effects of other gases.

1%/yr increasing CO₂

A common standardised forcing scenario specifies atmospheric CO_2 to increase at a rate of 1%/year compound until the concentration doubles (or quadruples) and is then held constant. The CO_2 content of the atmosphere has not, and likely will not, increase at this rate (let alone suddenly remain constant at twice or four times an initial value). If regarded as a proxy for all greenhouse gases, however, an "equivalent CO_2" increase of 1%/yr does give a forcing within the range of the SRES scenarios.

This forcing prescription is used to illustrate and to quantify aspects of AOGCM behaviour and provides the basis for the analysis and intercomparison of modelled responses to a specified forcing change (e.g., in the SAR and the CMIP2 intercomparison). The resulting information is also used to calibrate simpler models which may then be employed to investigate a broad range of forcing scenarios as is done in Section 9.3.3. Figure 9.1 illustrates the global mean temperature evolution for this standardised forcing in a simple illustrative example with no exchange with the deep ocean (the green curves) and for a full coupled AOGCM (the red curves). The diagram also illustrates the transient climate response, climate sensitivity and warming commitment.

TCR – Transient climate response

The temperature change at any time during a climate change integration depends on the competing effects of all of the processes that affect energy input, output, and storage in the ocean. In particular, the global mean temperature change which occurs at the time of CO_2 doubling for the specific case of a 1%/yr increase of CO_2 is termed the "transient climate response" (TCR) of the system. This temperature change, indicated in Figure 9.1, integrates all processes operating in the system, including the strength of the feedbacks and the rate of heat storage in the ocean, to give a straightforward measure of model response to a change in forcing. The range of TCR values serves to illustrate and calibrate differences in model response to the same standardised forcing. Analogous TCR measures may be used, and compared among models, for other forcing scenarios.

Equilibrium climate sensitivity

The "equilibrium climate sensitivity" (IPCC 1990, 1996) is defined as the change in global mean temperature, T_{2x}, that results when the climate system, or a climate model, attains a new equilibrium with the forcing change F_{2x} resulting from a doubling of the atmospheric CO_2 concentration. For this new equilibrium $dH/dt = 0$ in the simple heat budget equation and $F_{2x} = \alpha T_{2x}$ indicating a balance between energy input and output. The equilibrium climate sensitivity

$$T_{2x} = F_{2x} / \alpha$$

is inversely proportional to α, which measures the strength of the feedback processes in the system that act to counter a change in forcing. The equilibrium climate sensitivity is a straightforward, although averaged, measure of how the system responds to a specific forcing change and may be used to compare model responses, calibrate simple climate models, and to scale temperature changes in other circumstances.

In earlier assessments, the climate sensitivity was obtained from calculations made with AGCMs coupled to mixed-layer upper ocean models (referred to as mixed-layer models). In that case there is no exchange of heat with the deep ocean and a model can be integrated to a new equilibrium in a few tens of years. For a full coupled atmosphere/ocean GCM, however, the heat exchange with the deep ocean delays equilibration and several millennia, rather than several decades, are required to attain it. This difference is illustrated in Figure 9.1 where the smooth green curve illustrates the rapid approach to a new climate equilibrium in an idealised mixed-layer case while the red curve is the result of a coupled model integration and indicates the much longer time needed to attain equilibrium when there is interaction with the deep ocean.

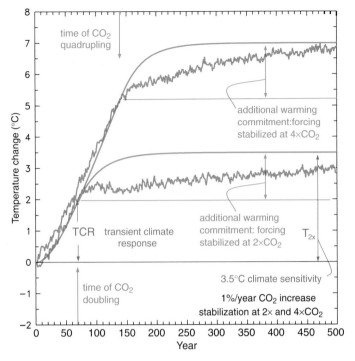

Figure 9.1: Global mean temperature change for 1%/yr CO_2 increase with subsequent stabilisation at $2\times CO_2$ and $4\times CO_2$. The red curves are from a coupled AOGCM simulation (GFDL_R15_a) while the green curves are from a simple illustrative model with no exchange of energy with the deep ocean. The "transient climate response", TCR, is the temperature change at the time of CO_2 doubling and the "equilibrium climate sensitivity", T_{2x}, is the temperature change after the system has reached a new equilibrium for doubled CO_2, i.e., after the "additional warming commitment" has been realised.

Effective climate sensitivity

Although the definition of equilibrium climate sensitivity is straightforward, it applies to the special case of equilibrium climate change for doubled CO_2 and requires very long simulations to evaluate with a coupled model. The "effective climate sensitivity" is a related measure that circumvents this requirement. The inverse of the feedback term α is evaluated from model output for evolving non-equilibrium conditions as

$$1/\alpha_e = T / (F - dH_o/dt) = T / (F - F_o)$$

and the effective climate sensitivity is calculated as

$$T_e = F_{2x} / \alpha_e$$

with units and magnitudes directly comparable to the equilibrium sensitivity. The effective sensitivity becomes the equilibrium sensitivity under equilibrium conditions with $2\times CO_2$ forcing. The effective climate sensitivity is a measure of the strength of the feedbacks at a particular time and it may vary with forcing history and climate state.

Warming commitment

An increase in forcing implies a "commitment" to future warming even if the forcing stops increasing and is held at a constant value. At any time, the "additional warming commitment" is the further increase in temperature, over and above the

increase that has already been experienced, that will occur before the system reaches a new equilibrium with radiative forcing stabilised at the current value. This behaviour is illustrated in Figure 9.1 for the idealised case of instantaneous stabilisation at $2\times$ and $4\times CO_2$. Analogous behaviour would be seen for more realistic stabilisation scenarios.

9.2.2 Simulating Forced Climate Change

9.2.2.1 Signal versus noise

A climate change simulation produces a time evolving three dimensional distribution of temperature and other climate variables. For the real system or for a model, and taking temperature as an example, this is expressed as $T = T_0 + T_0'$ for pre-industrial equilibrium conditions. T is now the full temperature field rather than the global mean temperature change of Section 9.2.1. T_0 represents the temperature structure of the mean climate, which is determined by the (pre-industrial) forcing, and T_0' the internally generated random natural variability with zero mean. For climate which is changing as a consequence of increasing atmospheric greenhouse gas concentrations or other forcing changes, $T = T_0 + T_f + T'$ where T_f is the deterministic climate change caused by the changing forcing, and T' is the natural variability under these changing conditions. Changes in the statistics of the natural variability, that is in the statistics of T_0' vs T', are of considerable interest and are discussed in Sections 9.3.5 and 9.3.6 which treat changes in variability and extremes.

The difference in temperature between the control and climate change simulations is written as $\Delta T = T_f + (T' - T_0') = T_f + T''$, and is a combination of the deterministic signal T_f and a random component $T'' = T' - T_0'$ which has contributions from the natural variability of both simulations. A similar expression arises when calculating climate change as the difference between an earlier and a later period in the observations or a simulation. Observed and simulated climate change are the sum of the forced "signal" and the natural variability "noise" and it is important to be able to separate the two. The natural variability that obscures the forced signal may be at least partially reduced by averaging.

9.2.2.2 Ensembles and averaging

An ensemble consists of a number of simulations undertaken with the same forcing scenario, so that the forced change T_f is the same for each, but where small perturbations to remote initial conditions result in internally generated climate variability that is different for each ensemble member. Small ensembles of simulations have been performed with a number of models as indicated in the "number of simulations" column in Table 9.1. Averaging over the ensemble of results, indicated by braces, gives the ensemble mean climate change as $\{\Delta T\} = T_f + \{T''\}$. For independent realisations, the natural variability noise is reduced by the ensemble averaging (averaging to zero for a large enough ensemble) so that $\{\Delta T\}$ is an improved estimate of the model's forced climate change T_f. This is illustrated in Figure 9.2, which shows the simulated temperature differences from 1975 to 1995 to the first decade in the 21st century for three climate change simulations made with the same model and the same forcing scenario but starting from

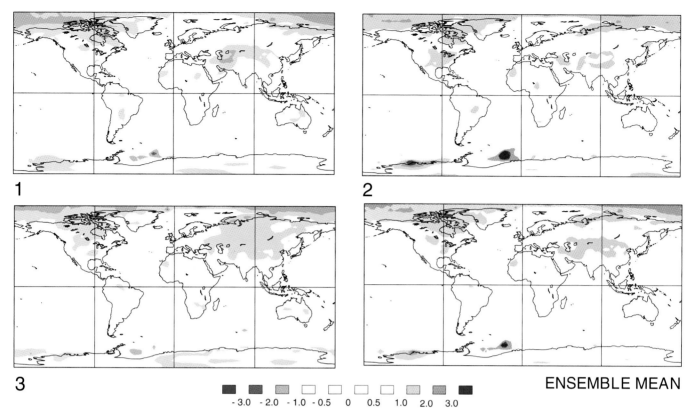

Figure 9.2: Three realisations of the geographical distribution of temperature differences from 1975 to 1995 to the first decade in the 21st century made with the same model (CCCma CGCM1) and the same IS92a greenhouse gas and aerosol forcing but with slightly different initial conditions a century earlier. The ensemble mean is the average of the three realisations. (Unit: °C).

slightly different initial conditions more than a century earlier. The differences between the simulations reflect differences in the natural variability. The ensemble average over the three realisations, also shown in the diagram, is an estimate of the model's forced climate change where some of this natural variability has been averaged out.

The ensemble variance for a particular model, assuming there is no correlation between the forced component and the variability, is $\sigma^2_{\Delta T} = \{(\Delta T - \{\Delta T\})^2\} = \{(T'' - \{T''\})^2\} = \sigma^2_N$ which gives a measure of the natural variability noise. The "signal to noise ratio", $\{\Delta T\}/\sigma_{\Delta T}$, compares the strength of the climate change signal to this natural variability noise. The signal stands out against the noise when and where this ratio is large. The signal will be better represented by the ensemble mean as the size of the ensemble grows and the noise is averaged out over more independent realisations. This is indicated by the width, $\{\Delta T\} \pm 2\sigma_{\Delta T}/\sqrt{n}$, of the approximate 95% confidence interval which decreases as the ensemble size n increases.

The natural variability may be further reduced by averaging over more realisations, over longer time intervals, and by averaging in space, although averaging also affects the information content of the result. In what follows, the geographical distributions ΔT, zonal averages $[\Delta T]$, and global averages $<\Delta T>$ of temperature and other variables are discussed. As the amount of averaging increases, the climate change signal is better defined, since the noise is increasingly averaged out, but the geographical information content is reduced.

9.2.2.3 Multi-model ensembles

The collection of coupled climate model results that is available for this report permits a multi-model ensemble approach to the synthesis of projected climate change. Multi-model ensemble approaches are already used in short-range climate forecasting (e.g., Graham *et al.*, 1999; Krishnamurti *et al.*, 1999; Brankovic and Palmer, 2000; Doblas-Reyes *et al.*, 2000; Derome *et al.*, 2001). When applied to climate change, each model in the ensemble produces a somewhat different projection and, if these represent plausible solutions to the governing equations, they may be considered as different realisations of the climate change drawn from the set of models in active use and produced with current climate knowledge. In this case, temperature is represented as $T = T_0 + T_F + T_m + T'$ where T_F is the deterministic forced climate change for the real system and $T_m = T_f - T_F$ is the error in the model's simulation of this forced response. T' now also includes errors in the statistical behaviour of the simulated natural variability. The multi-model ensemble mean estimate of forced climate change is $\{\Delta T\} = T_F + \{T_m\} + \{T''\}$ where the natural variability again averages to zero for a large enough ensemble. To the extent that unrelated model errors tend to average out, the ensemble mean or systematic error $\{T_m\}$ will be small, $\{\Delta T\}$ will approach T_F and the multi-model ensemble average will be a better estimate of the forced climate change of the real system than the result from a particular model.

As noted in Chapter 8, no one model can be chosen as "best" and it is important to use results from a range of models. Lambert and Boer (2001) show that for the CMIP1 ensemble of simulations of current climate, the multi-model ensemble means of temperature, pressure, and precipitation are generally closer to the observed distributions, as measured by mean squared differences, correlations, and variance ratios, than are the results of any particular model. The multi-model ensemble mean represents those features of projected climate change that survive ensemble averaging and so are common to models as a group. The multi-model ensemble variance, assuming no correlation between the forced and variability components, is $\sigma^2_{\Delta T} = \sigma^2_M + \sigma^2_N$, where $\sigma^2_M = \{(T_m - \{T_m\})^2\}$ measures the inter-model scatter of the forced component and σ^2_N the natural variability. The common signal is again best discerned where the signal to noise ratio $\{\Delta T\} / \sigma_{\Delta T}$ is largest.

Figure 9.3 illustrates some basic aspects of the multi-model ensemble approach for global mean temperature and precipitation. Each model result is the sum of a smooth forced signal, T_f, and the accompanying natural variability noise. The natural variability is different for each model and tends to average out so that the ensemble mean estimates the smooth forced signal. The scatter of results about the ensemble mean (measured by the ensemble variance) is an indication of uncertainty in the results and is seen to increase with time. Global mean temperature is seen to be a more robust climate change variable than precipitation in the sense that $\{\Delta T\} / \sigma_{\Delta T}$ is larger than $\{\Delta P\} / \sigma_{\Delta P}$. These results are discussed further in Section 9.3.2.

9.2.2.4 Uncertainty

Projections of climate change are affected by a range of uncertainties (see also Chapter 14) and there is a need to discuss and to quantify uncertainty in so far as is possible. Uncertainty in projected climate change arises from three main sources; uncertainty in forcing scenarios, uncertainty in modelled responses to given forcing scenarios, and uncertainty due to missing or misrepresented physical processes in models. These are discussed in turn below.

Forcing scenarios: The use of a range of forcing scenarios reflects uncertainties in future emissions and in the resulting greenhouse gas concentrations and aerosol loadings in the atmosphere. The complexity and cost of full AOGCM simulations has restricted these calculations to a subset of scenarios; these are listed in Table 9.1 and discussed in Section 9.3.1. Climate projections for the remaining scenarios are made with less general models and this introduces a further level of uncertainty. Section 9.3.2 discusses global mean warming for a broad range of scenarios obtained with simple models calibrated with AOGCMs. Chapter 13 discusses a number of techniques for scaling AOGCM results from a particular forcing scenario to apply to other scenarios.

Model response: The ensemble standard deviation and the range are used as available indications of uncertainty in model results for a given forcing, although they are by no means a complete characterisation of the uncertainty. There are a number of caveats associated with the ensemble approach. Common or systematic errors in the simulation of current climate (e.g., Gates *et al.*, 1999; Lambert and Boer, 2001; Chapter 8) survive ensemble averaging and contribute error to the ensemble mean while not contributing to the standard deviation. A tendency for

models to under-simulate the level of natural variability would result in an underestimate of ensemble variance. There is also the possibility of seriously flawed outliers in the ensemble corrupting the results. The ensemble approach nevertheless represents one of the few methods currently available for deriving information from the array of model results and it is used in this chapter to characterise projections of future climate.

Missing or misrepresented physics: No attempt has been made to quantify the uncertainty in model projections of climate change due to missing or misrepresented physics. Current models attempt to include the dominant physical processes that govern the behaviour and the response of the climate system to specified forcing scenarios. Studies of "missing" processes are often carried out, for instance of the effect of aerosols on cloud lifetimes, but until the results are well-founded, of appreciable magnitude, and robust in a range of models, they are considered to be studies of sensitivity rather than projections of climate change. Physical processes which are misrepresented in one or more, but not all, models will give rise to differences which will be reflected in the ensemble standard deviation.

The impact of uncertainty due to missing or misrepresented processes can, however, be limited by requiring model simulations to reproduce recent observed climate change. To the extent that errors are linear (i.e., they have proportionally the same impact on the past and future changes), it is argued in Chapter 12, Section 12.4.3.3 that the observed record provides a constraint on forecast anthropogenic warming rates over the coming decades that does not depend on any specific model's climate sensitivity, rate of ocean heat uptake and (under some scenarios) magnitude of sulphate forcing and response.

9.3 Projections of Climate Change

9.3.1 Global Mean Response

Since the SAR, there have been a number of new AOGCM climate simulations with various forcings that can provide estimates of possible future climate change as discussed in Section 9.1.2. For the first time we now have a reasonable number of climate simulations with different forcings so we can begin to quantify a mean climate response along with a range of possible outcomes. Here each model's simulation of a future climate state is treated as a possible outcome for future climate as discussed in the previous section.

These simulations fall into three categories (Table 9.1):

- The first are integrations with idealised forcing, namely, a 1%/yr compound increase of CO_2. This 1% increase represents equivalent CO_2, which includes other greenhouse gases like methane, NO_x etc. as discussed in Section 9.2.1. These runs extend at least to the time of effective CO_2 doubling at year 70, and are useful for direct model intercomparisons since they use exactly the same forcing and thus are valuable to calibrate model response. These experiments are collected in the CMIP exercise (Meehl *et al.*, 2000a) and referred to as "CMIP2" (Table 9.1).

• A second category of AOGCM climate model simulations uses specified time-evolving future forcing where the simulations start sometime in the 19th century, and are run with estimates of observed forcing through the 20th century (see Chapter 8). That state is subsequently used to begin simulations of the future climate with estimated forcings of greenhouse gases ("G") or with the additional contribution from the direct effect of sulphate aerosols ("GS") according to various scenarios, such as IS92a (see Chapter 1). These simulations avoid the cold start problem (see SAR) present in the CMIP experiments. They allow evaluation of the model climate and response to forcing changes that could be experienced over the 21st century. The experiments are collected in the IPCC-DDC. These experiments are assessed for the mid-21st century when most of the DDC experiments with sulphate aerosols finished.

• A third category are AOGCM simulations using as an initial state the end of the 20th century integrations, and then following the A2 and B2 (denoted as such in Table 9.1) draft marker SRES forcing scenarios to the year 2100 (see Section 9.1.2). These simulations are assessed to quantify possible future climate change at the end of the 21st century, and also are treated as members of an ensemble to better assess and quantify consistent climate changes. A simple model is also used to provide estimates of global temperature change for the end of the 21st century from a greater number of the SRES forcing scenarios.

Table 9.1 gives a detailed overview of all experiments assessed in this report.

9.3.1.1 1%/yr CO_2 increase (CMIP2) experiments

Figure 9.3 shows the global average temperature and precipitation changes for the nineteen CMIP2 simulations. At the time of CO_2 doubling at year 70, the 20-year average (years 61 to 80) global mean temperature change (the transient climate response TCR; see Section 9.2) for these models is 1.1 to 3.1°C with an average of 1.8°C and a standard deviation of 0.4°C (Figure 9.7). This is similar to the SAR results (Figure 6.4 in Kattenberg *et al.*, 1996).

At the time of CO_2 doubling at year 70, the 20-year average (years 61 to 80) percentage change of the global mean precipitation for these models ranges from −0.2 to 5.6% with an average of 2.5% and a standard deviation of 1.5%. This is similar to the SAR results.

For a hypothetical, infinite ensemble of experiments, in which T_m and T'' are uncorrelated and both have zero means,

$$\{\Delta T^2\} = T_f^2 + \{T_m^2\} + \{T''^2\} = T_f^2 + \sigma_M^2 + \sigma_N^2.$$

The ensemble mean square climate change is thus the sum of contributions from the common forced component (T_f^2), model differences (σ_M^2), and internal variability (σ_N^2). This framework is applied to the CMIP2 experiments in Figure 9.4. These components of the total change are estimated for each grid box separately, using formulas that allow for unbiased estimates of these when a limited number of experiments are available (Räisänen 2000, 2001). The variance associated with internal

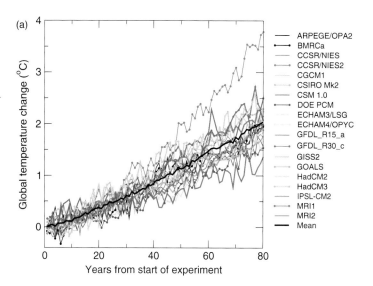

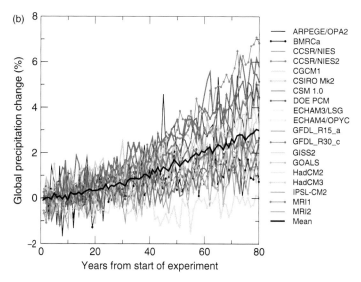

Figure 9.3: The time evolution of the globally averaged (a) temperature change relative to the control run of the CMIP2 simulations (Unit: °C). (b) ditto. for precipitation. (Unit: %). See Table 9.1 for more information on the individual models used here.

variability σ_N^2 is inferred from the temporal variability of detrended CO_2 run minus control run differences and the model-related variance σ_M^2 as a residual. Averaging the local statistics over the world, the relative agreement between the CMIP2 experiments is much higher for annual mean temperature changes (common signal makes up 86% of the total squared amplitude) than for precipitation (24%) (Figure 9.4).

The relative agreement on seasonal climate changes is slightly lower, even though the absolute magnitude of the common signal is in some cases larger in the individual seasons than in the annual mean. Only 10 to 20% of the inter-experiment variance in temperature changes is attributable to internal variability, which indicates that most of this variance arises from differences between the models themselves. The estimated contribution of internal variability to the inter-experiment variance in precipitation changes is larger, from about a third in

Table 9.1: *The climate change experiments assessed in this report.*

Model Number (see Chapter 8, Table 8.1)	Model Name and centre in italics (see Chapter 8, Table 8.1)	Scenario name	Scenario description	Number of simulations	Length of simulation or starting and final year	Transient Climate Response (TCR) (Section 9.2.1)	Equilibrium climate sensitivity (Section 9.2.1) (in bold used in Figure. 9.18 / Table 9.4)	Effective climate sensitivity (Section 9.2.1) (from CMIP2 yrs 61-80) in bold used in Table A1	References	Remarks
2	ARPEGE/OPA2 *CERFACS*	CMIP2	1% CO_2	1	80	1.64			Barthelet *et al.*, 1998a	
3	BMRCa *BMRC*	ML	Equilibrium 2×CO_2 in mixed-layer experiment	2	60		**2.2**		Colman and McAvaney, 1995; Colman, 2001	
		CMIP2	1% CO_2	1	100	1.63				
5	CCSR/NIES *CCSR/NIES*	ML	Equilibrium 2×CO_2 in mixed-layer experiment	1	40		**3.6**		Emori *et al.*, 1999	
		CMIP2	1% CO_2	1	80	1.8				
		G	Historical equivalent CO_2 to 1990 then 1% CO_2 (approx. IS92a)	1	1890-2099					
		GS	As G but including direct effect of sulphate aerosols	1	1890-2099					
		GS2	1% CO_2 + direct effect of sulphate aerosols but with explicit representation	1	1890-2099					
31	CCSR/NIES2 *CCSR/NIES*	ML	Equilibrium 2×CO_2 in mixed-layer experiment	1	40		**5.1**		Nozawa *et al.*, 2001	
		CMIP2	1% CO_2	1	80	3.1		11.6		
		A1	SRES A1 scenario	1	1890-2100					
		A2	SRES A2 scenario	1	1890-2100					
		B1	SRES B1 scenario	1	1890-2100					
		B2	SRES B2 scenario	1	1890-2100					
6	CGCM1 *CCCma*	ML	Equilibrium 2×CO_2 in mixed-layer experiment	1	30		3.5		Boer *et al.*, 1992	
		CMIP2	1% CO_2	1	80	1.96		**3.6**	Boer *et al.*, 2000a,b	1,000 yr control
		G	Historical equivalent CO_2 to 1990 then 1% CO_2 (approx. IS92a)	1	1900-2100					
		GS	As G but including direct effect of sulphate aerosols	3	1900-2100					
		GS2050	As GS but all forcings stabilised in year 2050	1	1000 after stability					
		GS2100	As GS but all forcings stabilised in year 2100	1	1000 after stability					
7	CGCM2 *CCCma*	GS	Historical equivalent CO_2 to 1990 then 1% CO_2 (approx. IS92a) and direct effect of sulphate aerosols	3	1900-2100				Flato and Boer, 2001	1,000 yr control
		A2	SRES A2 scenario	3	1990-2100					
		B2	SRES B2 scenario	3	1990-2100					
10	CSIRO Mk2 *CSIRO*	ML	Equilibrium 2×CO_2 in mixed-layer experiment	1	60		**4.3**		Watterson *et al.*, 1998	
		CMIP2	1% CO_2	1	80	2.00		**3.7**	Gordon and O'Farrell, 1997	
		G	Historical equivalent CO_2 to 1990 then 1% CO_2 (approx. IS92a)	1	1881-2100					
		G2080	As G but forcing stabilised at 2080 (3× initial CO_2)	1	700 after stability				Hirst, 1999	
		GS	As G + direct effect of sulphate aerosols	1	1881-2100				Gordon and O'Farrell, 1997	
		A2	SRES A2 scenario	1	1990-2100					
		B2	SRES B2 scenario	1	1990-2100					
11	CSM 1.0 *NCAR*	ML	Equilibrium 2×CO_2 in mixed-layer experiment	1	50		**2.1**		Meehl *et al.*, 2000a	
		CMIP2	1% CO_2	1	80	1.43		**1.9**		
12	CSM 1.3[a] *NCAR*	GS	Historical GHGs + direct effect of sulphate aerosols to 1990 then BAU CO_2 + direct effect of sulphate aerosols including effects of pollution control policies	1	1870-2100				Boville *et al.*, 2001; Dai *et al.*, 2001	
		GS2150	Historical GHGs + direct effect of sulphate to aerosols to 1990 then as GS except WRE550 scenario for CO_2 until it reaches 550 ppm in 2150	1	1870-2100					
		A1	SRES A1 scenario	1	1870-2100					
		A2	SRES A2 scenario	1	1870-2100					
		B2	SRES B2 scenario	1	1870-2100					
		CMIP2	1% CO_2	1	100	1.58		**2.2**		
14	ECHAM3/LSG *DKRZ*	G	Historical equiv CO_2 to 1990 then 1% CO_2 (approx. IS92a)	1	1881-2085				Cubasch *et al.*, 1992, 1994, 1996	
		G2050	As G but forcing stabilised at 2050 (2× initial CO_2)	1	850 after stability					
		G2110	As G but forcing stabilised at 2110 (4× initial CO_2)	2	850 after stability				Voss and Mikolajewicz, 2001	Periodically synchronous coupling
		GS	As G + direct effect of sulphate aerosols	2	1881-2050					
		ML	Equilibrium 2×CO_2 in mixed-layer experiment	1	60		**3.2**		Cubasch *et al.*, 1992, 1994, 1996b	

[a] CSM 1.3 was at the time of the printing of this report not archived completely in the DDC. It is therefore not considered in calculations and diagrams refering to the DDC experiments with the exception of Figure 9.5.

Table 9.1: *Continuation.*

Model Number (see Chapter 8, Table 8.1)	Model Name and centre in italics (see Chapter 8, Table 8.1)	Scenario name	Scenario description	Number of simulations	Length of simulation or starting and final year	Transient Climate Response (TCR) (Section 9.2.1)	Equilibrium climate sensitivity (Section 9.2.1) (in bold used in Figure. 9.18 / Table 9.4	Effective climate sensitivity (Section 9.2.1) (from CMIP2 yrs 61-80) in bold used in Table A1	References	Remarks
15	ECHAM4/OPYC *MPI*	CMIP2	1% CO_2	1	80	1.4		**2.6**	Roeckner *et al.*, 1999	
		G	Historical GHGs to 1990 then IS92a	1	1860-2099					
		GS	As G + direct effect of sulphate aerosol interactively calculated	1	1860-2049					
		GSIO	As GS + indirect effect of sulphate aerosol + ozone	1	1860-2049					
		A2	SRES A2 scenario	1	1990-2100				Stendel *et al.*, 2000	
		B2	SRES B2 scenario	1	1990-2100					
16	GFDL_R15_a *GFDL*	ML	Equilibrium 2×CO_2 in mixed-layer experiment	2	40		**3.7** (3.9)[b]		Manabe *et al.*, 1991	15,000 year control
		CMIP2	1% CO_2	2	80	2.15		**4.2**	Stouffer and Manabe, 1999	
		CMIP270	As CMIP2 but forcing stabilised at year 70 (2 × initial CO_2)	1	4000		(4.5)[c]			
		CMIP2140	As CMIP2 but forcing stabilised at year 140 (4 × initial CO_2)	1	5000					
		G	Historical equivalent CO_2 to 1990 then 1% CO_2 (approximate IS92a)	1	1766-2065				Haywood *et al.*, 1997; Sarmiento *et al.*, 1998	
		GS	As G + direct effect of sulphate aerosols	2	1766-2065					
17	GFDL_R15_b *GFDL*	CMIP2	1% CO_2	1	80	Data unavailable				
		GS	Historical equivalent CO_2 to 1990 then 1% CO_2 (approximate IS92a) + direct effect of sulphate aerosols	3 3 3	1766-2065 1866-2065 1916-2065				Dixon and Lanzante, 1999	
18	GFDL_R30_c *GFDL*	ML	Equilibrium 2×CO_2 in mixed-layer experiment	1	40		**3.4**			2×1,000 year control runs with different oceanic diapycnal mixing
		CMIP2	1% CO_2	2	80	1.96				Different oceanic diapycnal mixing
		CMIP270	As CMIP2 but forcing stabilised at year 70 (2 × initial CO_2)	1	140 after stability					
		CMIP2140	As CMIP2 but forcing stabilised at year 140 (4 × initial CO_2)	1	160 after stability					
		GS	Historical equivalent CO_2 to 1990 then 1% CO (approximate IS92a) + direct effect of sulphate aerosols	9	1866-2090				Knutson *et al.*, 1999	
		A2	SRES A2 scenario	1	1960-2090					
		B2	SRES B2 scenario	1	1960-2090					
20	GISS2 *GISS*	ML	Equilibrium 2×CO_2 in mixed-layer experiment	1	40		(**3.1**)[d]		Yao and Del Genio, 1999	
		CMIP2	1% CO_2	1	80	1.45			Russell *et al.*, 1995; Russell and Rind, 1999	
21	GOALS *IAP/LASG*	CMIP2	1% CO_2	1	80	1.65				
22	HadCM2 *UKMO*	ML	Equilibrium 2× CO_2 in mixed-layer experiment	1	40		**4.1**		Senior and Mitchell, 2000	
		CMIP2	1% CO_2	1	80	1.7		**2.5**	Keen and Murphy, 1997	1,000 year control run
		CMIP270	As CMIP2 but forcing stabilised at year 70 (2 × initial CO_2)	1	900 after stability				Senior and Mitchell, 2000	
		G	Historical equivalent CO_2 to 1990 then 1% CO_2 (approximate IS92a)	4	1881-2085				Mitchell *et al.*, 1995; Mitchell and Johns, 1997	
		G2150	As G but all forcings stabilised in year 2150	1	110 after stability				Mitchell *et al.*, 2000	
		GS	As G + direct effect of sulphate aerosols	4	1860-2100				Mitchell *et al.*, 1995; Mitchell and Johns, 1997	
23	HadCM3 *UKMO*	ML	Equilibrium 2×CO_2 in mixed-layer experiment	1	30		**3.3**		Williams *et al.*, 2001	1,800 year control run
		CMIP2	1% CO_2	1	80	2.0		3.0		
		G	Historical GHGs to 1990 then IS95a	1	1860-2100				Mitchell *et al.*, 1998; Gregory and Lowe, 2000 Johns *et al.*, 2001	
		GSIO	As G + direct and indirect effect of sulphate aerosols + ozone changes	1	1860-2100					
		A2	SRES A2 scenario	1	1990-2100					
		B2	SRES B2 scenario	1	1990-2100					

[b] The equilibrium climate sensitivity if the control SSTs from the coupled model are used.
[c] The equilibrium climate sensitivity calculated from the coupled model.
[d] The ML experiment used in Table 9.2 for the GISS model were performed with a different atmospheric model to that used in the coupled model listed here.

Table 9.1: *Continuation.*

Model Number (see Chapter 8, Table 8.1)	Model Name and centre in italics (see Chapter 8, Table 8.1)	Scenario name	Scenario description	Number of simulations	Length of simulation or starting and final year	Transient Climate Response (TCR) (Section 9.2.1)	Equilibrium climate sensitivity (Section 9.2.1) (in bold used in Figure. 9.18 / Table 9.4	Effective climate sensitivity (Section 9.2.1) (from CMIP2 yrs 61-80) in bold used in Table A1	References	Remarks
25	IPSL-CM2 *IPSL/LMD*	ML	Equilibrium 2×CO$_2$ in mixed-layer experiment	1	25		(3.6)[e]		Ramstein *et al.*, 1998	
		CMIP2	1% CO$_2$	1	140	1.96			Barthelet *et al.*, 1998b	
		CMIP270	As CMIP2 but forcing stabilised at year 70 (2 × initial CO$_2$)	1	50 after stability					
		CMIP2140	As CMIP2 but forcing stabilised at year 140 (4 × initial CO$_2$)	1	60 after stability					
26	MRI1[f] *MRI*	ML	Equilibrium 2×CO$_2$ in mixed-layer experiment	1	60		**4.8**		Noda *et al.*, 1999a	
		CMIP2	1% CO$_2$	1	150	1.6		2.5	Tokioka *et al.*, 1995, 1996	
		CMIP2S	As CMIP2 + direct effect of sulphate aerosols	1	100				Japan Met. Agency, 1999	
27	MRI2 *MRI*	ML	Equilibrium 2×CO$_2$ in mixed-layer experiment	1	50		**2.0**		Yukimoto *et al.*, 2001; Noda *et al.*, 2001	
		CMIP2	1% CO$_2$	1	150	1.1		1.5		
		G	Historical equivalent CO$_2$ to 1990 then 1% CO$_2$ (approx IS92a)	1	1900-2100					
		GS	As G + explicit representation of direct effect of sulphate aerosols	1	1900-2100					
		A2	SRES A2 scenario	1	1990-2100					
		B2	SRES B2 scenario	1	1990-2100					
30	DOE PCM *NCAR*	ML	Equilibrium 2×CO$_2$ in mixed-layer exp.	1	50		**2.1**		Washington *et al.*, 2000 Meehl *et al.*, 2001	
		CMIP2	1% CO$_2$	5	80	1.27		**1.7**		
		G	Historical GHGs + direct effect of sulphate aerosols to 1990 then BAU CO$_2$ + direct effect of sulphate aerosols including effects of pollution control policies		1870-2100					
		GS	Historical GHGs + direct effect of sulphate to 1990 then as GS except WRE550 scenario for CO$_2$ until it reaches 550 ppm in 2150	5	1870-2100					
		GS2150	Historical GHGs to 1990 then as GS except WRE550 scenario for CO$_2$ until it reaches 550 ppm in 2150.	5	1870-2100					
		A2	SRES A2 scenario	1	1870-2100					
		B2	SRES B2 scenario	1	1870-2100					

[e] The ML experiment used in Table 9.2 for the IPSL-CM2 model were performed with a slightly earlier version of the atmospheric model than that used in the coupled model, but tests have suggested the changes would not affect the equilibrium climate sensitivity.

[f] Model MRI1 exists in two versions. At the time of writing, more complete assessment data was available for the earlier version, whose control run is in the CMIP1 database. This model is used in Chapter 8. The model used in Chapter 9 has two extra ocean levels and a modified ocean mixing scheme. Its control run is in the CMIP2 database. The equilibrium climate sensitivities and Transient Climate Responses (shown in this table) of the two models are the same.

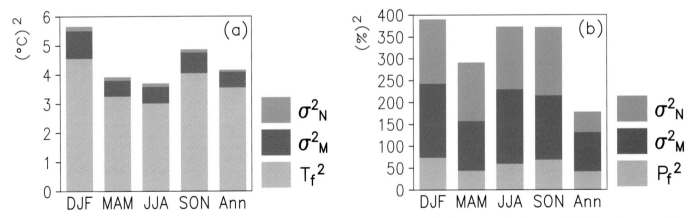

Figure 9.4: Intercomparison statistics for seasonal and annual (a) temperature and (b) precipitation changes in nineteen CMIP2 experiments at the doubling of CO$_2$ (years 61 to 80). The total length of the bars shows the mean squared amplitude of the simulated local temperature and precipitation changes averaged over all experiments and over the whole world. The lowermost part of each bar represents a nominally unbiased "common signal", the mid-part directly model-related variance and the top part the inter-experiment variance attributed to internal variability. Precipitation changes are defined as $100\% \times (P_G - P_{CTRL}) / Max(P_{CTRL}, 0.25 \text{ mm/day})$, where the lower limit of 0.25 mm/day is used to reduce the sensitivity of the global statistics to areas with very little control run precipitation.

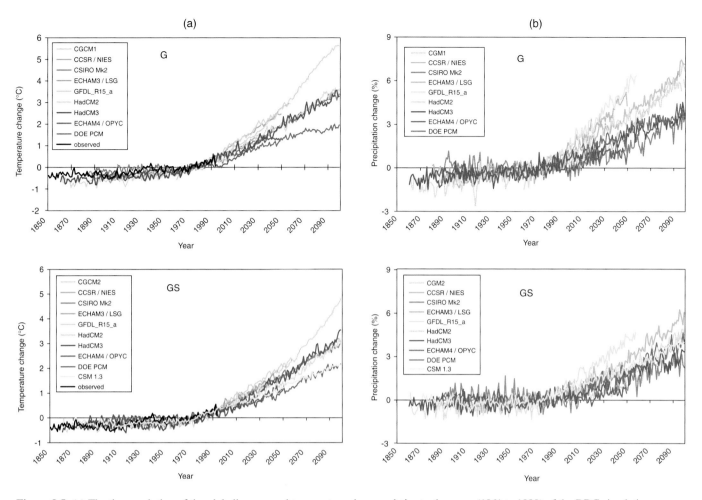

Figure 9.5: (a) The time evolution of the globally averaged temperature change relative to the years (1961 to 1990) of the DDC simulations (IS92a). G: greenhouse gas only (top), GS: greenhouse gas and sulphate aerosols (bottom). The observed temperature change (Jones, 1994) is indicated by the black line. (Unit: °C). See Table 9.1 for more information on the individual models used here. (b) The time evolution of the globally averaged precipitation change relative to the years (1961 to 1990) of the DDC simulations. GHG: greenhouse gas only (top), GS: greenhouse gas and sulphate aerosols (bottom). (Unit: %). See Table 9.1 for more information on the individual models used here.

the annual mean to about 50% in individual seasons. Thus there is more internal variability and model differences and less common signal indicating lower reliability in the changes of precipitation compared to temperature.

9.3.1.2 Projections of future climate from forcing scenario experiments (IS92a)

Please note that the use of projections for forming climate scenarios to study the impacts of climate change is discussed in Chapter 13.

These experiments include changes in greenhouse gases plus the direct effect of sulphate aerosol using IS92a type forcing (see Chapter 6 for a complete discussion of direct and indirect effect forcing from sulphate aerosols). The temperature change (Figures 9.5a and 9.7a, top) for the 30-year average 2021 to 2050 compared with 1961 to 1990 is +1.3°C with a range of +0.8 to +1.7°C as opposed to +1.6°C with a range of +1.0 to +2.1°C for greenhouse gases only (Cubasch and Fischer-Bruns, 2000). The experiments including sulphate aerosols show a smaller temperature rise compared to experiments without sulphate aerosols

due to the negative radiative forcing of these aerosols. Additionally, in these simulations CO_2 would double around year 2060. Thus for the averaging period being considered, years 2021 to 2050, the models are still short of the CO_2 doubling point seen in the idealised 1%/yr CO_2 increase simulations. These sensitivity ranges could be somewhat higher (about 30%) if the positive feedback effects from the carbon cycle are included interactively but the magnitude of these feedbacks is uncertain (Cox *et al.*, 2000; Friedlingstein, 2001). The globally averaged precipitation response for 2021 to 2050 for greenhouse gases plus sulphates is +1.5% with a range of +0.5 to +3.3% as opposed to +2.3% with a range of +0.9 to +4.4% for greenhouse gases only (Figures 9.5b and 9.7a, bottom).

9.3.1.3 Marker scenario experiments (SRES)

As discussed in Section 9.1.2, only the draft marker SRES scenarios A2 and B2 have been integrated with more than one AOGCM, because the scenarios were defined too late to have experiments ready from all the modelling groups in time for this report. Additionally, some new versions of models have been

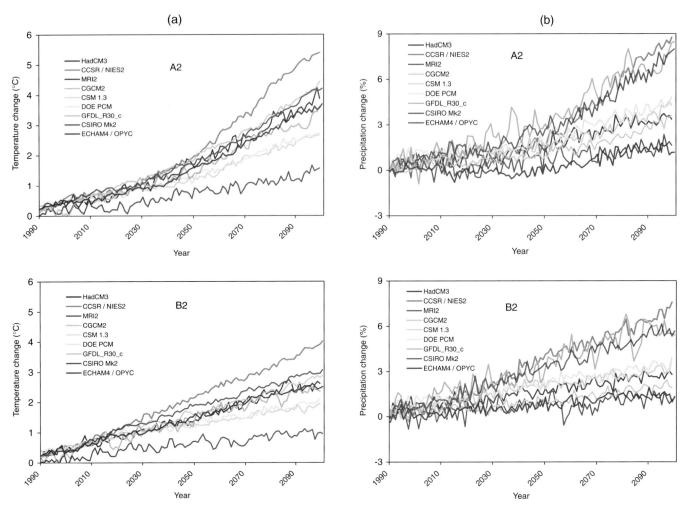

Figure 9.6: (a) The time evolution of the globally averaged temperature change relative to the years (1961 to 1990) of the SRES simulations A2 (top) and B2 (bottom) (Unit: °C). See Table 9.1 for more information on the individual models used here. (b) The time evolution of the globally averaged precipitation change relative to the years (1961 to 1990) of the SRES simulations A2 (top) and B2 (bottom) (Unit: %). See Table 9.1 for more information on the individual models used here.

used to run the A2 and B2 scenarios that have not had time to be evaluated by Chapter 8. Therefore, we present results from all the model simulations and consider them all as possible realisations of future climate change, but their ranges are not directly comparable to the simple model results in Section 9.3.3 (range: 1.4 to 5.8°C), because in the simple model analysis seven somewhat different versions of the nine models have been considered. Additionally, for the AOGCMs the temperature changes are evaluated for an average of years 2071 to 2100 compared with 1961 to 1990, while the simple model results are differences of the year 2100 minus 1990.

The average temperature response from nine AOGCMs using the SRES A2 forcing (Figures 9.6a and 9.7b, top) for the 30-year average 2071 to 2100 relative to 1961 to 1990 is +3.0°C with a range of +1.3 to +4.5°C, while using the SRES B2 scenarios it amounts to +2.2°C with a range of +0.9 to +3.4°C. The B2 scenario produces a smaller warming which is consistent with its lower positive radiative forcing at the end of the 21st century. For the 30-year average 2021 to 2050 using the A2

scenario, the globally averaged surface air temperature increase compared to 1961 to 1990 is +1.1°C with a range of +0.5 to +1.4°C, while using the SRES B2 scenarios it amounts to +1.2°C with a range of +0.5 to +1.7°C. The values for the SRES scenarios for the mid-21st century are lower than for the IS92a scenarios for the corresponding period due to differences in the forcing.

The average precipitation response using the SRES A2 forcing (Figures 9.6b and 9.7b, bottom) for the 30-year average 2071 to 2100 compared with 1961 to 1990 is an increase of 3.9% with a range of 1.3 to 6.8% , while using the SRES B2 scenarios it amounts to an increase of 3.3% with a range of 1.2 to 6.1%. The lower precipitation increase values for the B2 scenario are consistent with less globally averaged warming for that scenario at the end of the 21st century compared with A2. For the 30-year average 2021 to 2050 the globally averaged precipitation increases 1.2% for the A2 scenario, and 1.6% for B2 which is again consistent with the slightly greater global warming in B2 for mid-21st century compared with A2. Globally averaged

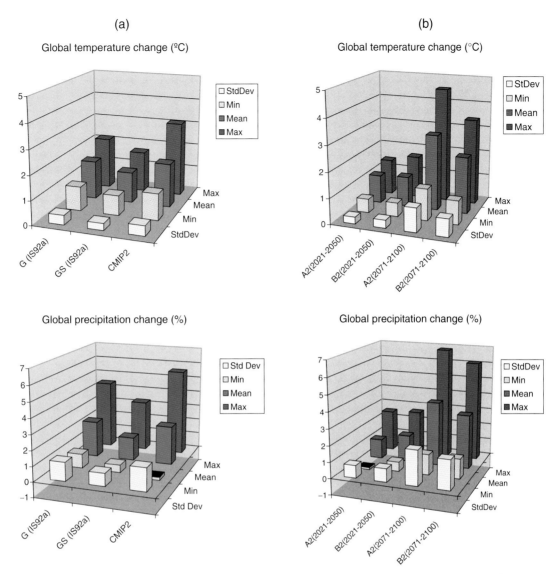

Figure 9.7: (a) The global mean, the maximum and minimum simulated by the respective models and the standard deviation for the CMIP2 experiments at the time of CO_2-doubling and for the DDC experiments during the years 2021 to 2050 relative to the years 1961 to 1990 for temperature (top) (Unit: °C) and precipitation (bottom) (Unit: %). G: greenhouse gases only, GS: greenhouse gases and sulphate aerosols. See Table 9.1 for more information on the individual models used here. (b) The global mean, the maximum and minimum simulated by the respective models and the standard deviation for the SRES scenario experiments A2 and B2 performed by the AOGCMs, for the years 2021 to 2050 and 2071 to 2100 relative to the years 1961 to 1990 for temperature (top) (Unit: °C) and precipitation (bottom) (Unit: %). See Table 9.1 for more information on the individual models used here.

changes of temperature and precipitation are summarised in Figure 9.7b. A more extensive analysis of globally averaged temperature changes for a wider range of SRES forcing scenarios using a simple climate model is given in Section 9.3.3.

9.3.2 Patterns of Future Climate Change

For the change in annual mean surface air temperature in the various cases, the model experiments show the familiar pattern documented in the SAR with a maximum warming in the high latitudes of the Northern Hemisphere and a minimum in the Southern Ocean (due to ocean heat uptake) evident in the zonal mean for the CMIP2 models (Figure 9.8) and the geographical

patterns for all categories of models (Figure 9.10). For the zonal means in Figure 9.8 there is consistent mid-tropospheric tropical warming and stratospheric cooling. The range tends to increase with height (Figure 9.8, middle) partly due to the variation in the level of the tropopause among the models. Ocean heat uptake also contributes to a minimum of warming in the North Atlantic, while land warms more rapidly than ocean almost everywhere (Figure 9.10). The large warming in high latitudes of the Northern Hemisphere is connected with a reduction in the snow (not shown) and sea-ice cover (Figure 9.9).

The ensemble mean temperature divided by its standard deviation $\{\Delta T\} / \sigma_{\{\Delta T\}}$ provides a measure of the consistency of the climate change patterns (Section 9.2). Different types and

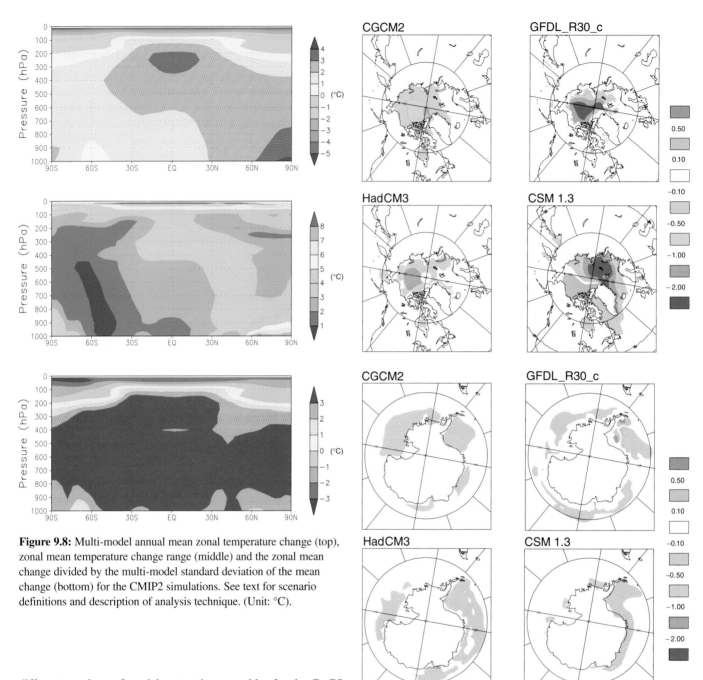

Figure 9.8: Multi-model annual mean zonal temperature change (top), zonal mean temperature change range (middle) and the zonal mean change divided by the multi-model standard deviation of the mean change (bottom) for the CMIP2 simulations. See text for scenario definitions and description of analysis technique. (Unit: °C).

different numbers of models enter the ensembles for the G, GS and SRES A2 and B2 cases and results will depend both on this and on the difference in forcing. Values greater than 1.0 are a conservative estimate of areas of consistent model response, as noted in Section 9.2.2 above.

There is relatively good agreement between the models for the lower latitude response, with larger range and less certain response at higher latitudes (Figure 9.10). For example, most models show a minimum of warming somewhere in the North Atlantic but the location is quite variable. There is a tendency for more warming (roughly a degree) in the tropical central and east Pacific than in the west, though this east-west difference in warming is generally less than a degree in the multi-model ensemble and is not evident with the contour interval in Figure 9.10 except in the B2 experiment in Figure 9.10e. This El Nino-like response is discussed further in Section 9.3.5.2.

Figure 9.9: Change in annual mean sea-ice thickness between the periods 1971 to 1990 and 2041 to 2060 as simulated by four of the most recent coupled models. The upper panels show thickness changes in the Northern Hemisphere, the lower panels show changes in the Southern Hemisphere. All models were run with similar forcing scenarios: historical greenhouse gas and aerosol loading, then future forcing as per the IS92a scenario. The colour bar indicates thickness change in metres – negative values indicate a decrease in future ice thickness.

The biggest difference between the CMIP2 G (Figure 9.10a,b) and GS experiments (Figure 9.10c) is the regional moderating of the warming mainly over industrialised areas in GS where the negative forcing from sulphate aerosols is greatest at mid-21st century (note the regional changes

Table 9.2: *The pattern correlation of temperature and precipitation change for the years (2021 to 2050) relative to the years (1961 to 1990) for the simulations in the IPCC DDC. Above the diagonal: G experiments, below the diagonal: GS experiments. The diagonal is the correlation between G and GS patterns from the same model.*

Temperature	CGCM1	CCSR/NIES	CSIRO Mk2	ECHAM3/LSG	GFDL_R15_a	HadCM2	HadCM3	ECHAM4/OPYC	DOE PCM
CGCM1	**0.96**	0.74	0.65	0.47	0.65	0.72	0.67	0.65	0.31
CCSR/NIES	0.75	**0.97**	0.77	0.45	0.72	0.77	0.73	0.80	0.49
CSIRO Mk2	0.61	0.71	**0.96**	0.40	0.75	0.72	0.67	0.75	0.63
ECHAM3/LSG	0.58	0.50	0.44	**0.46**	0.40	0.53	0.60	0.53	0.35
GFDL_R15_a	0.65	0.76	0.69	0.42	**0.73**	0.58	0.61	0.69	0.55
HadCM2	0.65	0.69	0.59	0.52	0.50	**0.85**	0.79	0.79	0.43
HadCM3	0.60	0.65	0.60	0.49	0.47	0.63	**0.90**	0.75	0.47
ECHAM4/OPYC	0.67	0.78	0.66	0.37	0.71	0.61	0.69	**0.89**	0.41
DOE PCM	0.30	0.38	0.63	0.24	0.36	0.40	0.44	0.37	**0.91**

Precipitation	CGCM1	CCSR/NIES	CSIRO Mk2	ECHAM3/LSG	GFDL_R15_a	HadCM2	HadCM3	ECHAM4/OPYC	DOE PCM
CGCM1	**0.88**	0.14	0.08	0.05	0.05	0.23	−0.16	−0.03	0.02
CCSR/NIES	0.14	**0.91**	0.13	0.21	0.34	0.36	0.29	0.33	0.18
CSIRO Mk2	0.15	0.14	**0.73**	0.13	0.29	0.32	0.31	0.07	0.11
ECHAM3/LSG	0.20	0.23	0.13	**0.39**	0.28	0.19	0.11	0.11	0.29
GFDL_R15_a	0.18	0.20	0.28	0.14	**0.41**	0.28	0.20	0.22	0.21
HadCM2	0.34	0.34	0.23	0.37	0.24	**0.73**	0.19	0.24	0.17
HadCM3	−0.20	0.06	0.31	−0.05	0.11	−0.01	**0.81**	0.25	0.09
ECHAM4/OPYC	0.13	0.30	0.09	0.07	0.04	0.23	0.20	**0.79**	0.01
DOE PCM	0.02	0.08	0.12	−0.09	0.06	0.13	−0.06	−0.07	**0.43**

discussed in Chapter 10). This regional effect was noted in the SAR for only two models, but Figure 9.10c shows this is a consistent response across the greater number of more recent models. The GS experiments only include the direct effect of sulphate aerosols, but two model studies have included the direct and indirect effect of sulphate aerosols and show roughly the same pattern (Meehl *et al.,* 1996; Roeckner *et al.,* 1999). The simulations performed with and without the direct sulphate effect (GS and G, respectively) with the same model are more similar to each other than to the other models, indicating that the individual response characteristics of the various models are dominating the response pattern rather than differences in the forcing. With greater CO_2 forcing, the simulated patterns are more highly correlated in the G simulations than in the GS simulations (Table 9.2, 26 of 36 possible model combinations for temperature, 22 of 36 for precipitation).

The SRES A2 and B2 integrations (Figure 9.10d,e) show a similar pattern of temperature change as the CMIP2 and G experiments. Since the positive radiative forcing from greenhouse gases overwhelms the sulphate aerosol forcing at the end of the 21st century in A2 and B2 compared to the GS experiments at mid-21st century, the patterns resemble more closely the G simulations in Figure 9.10a,b. The amplitude of the climate change patterns is weaker for the B2 than for the A2 simulations at the end of the 21st century (Figure 9.10d,e).

The relative change in the mean precipitation (Figure 9.11) for all models in all categories shows a general increase in the tropics (particularly the tropical oceans and parts of northern

Africa and south Asia) and the mid- and high latitudes, while the rainfall generally decreases in the sub-tropical belts. These changes are more evident for larger positive radiative forcing in the A2 and B2 scenario runs at the end of the 21st century (Figure 9.11d,e). This also applies to the areas of decrease that show a high inter-model variability and therefore little consistency among models, while in the tropics the change can exceed the variability of the signal by a factor of 2. This is particularly evident over the central and eastern tropical Pacific where the El Niño-like surface temperature warming is associated with an eastward shift of positive precipitation anomalies. The A2 and B2 scenario experiments exhibit a relatively large increase in precipitation over the Sahara and Arabia, but with large inter-model variability. This is partly an artefact of using percentage change rather than absolute values, since in these regions the absolute precipitation amount is very small.

Other manifestations of the changes in precipitation are reported by Noda and Tokioka (1989), Murphy and Mitchell (1995) and Royer *et al.* (1998) who found an increase in the global mean convective rain rate in the $2 \times CO_2$ climate compared with the $1 \times CO_2$ climate. Results from another model (Brinkop, 2001; see also Cubasch *et al.,* 1999) indicate a decrease in global mean convective precipitation. Essentially the results of Brinkop are consistent with Murphy and Mitchell, because in both transient climate simulations the strongest reduction in convective rain is found in the sub-tropics, and is most pronounced in the Southern Hemisphere. The increase in convective rain rate in the Northern Hemisphere is less strong in Brinkop compared to

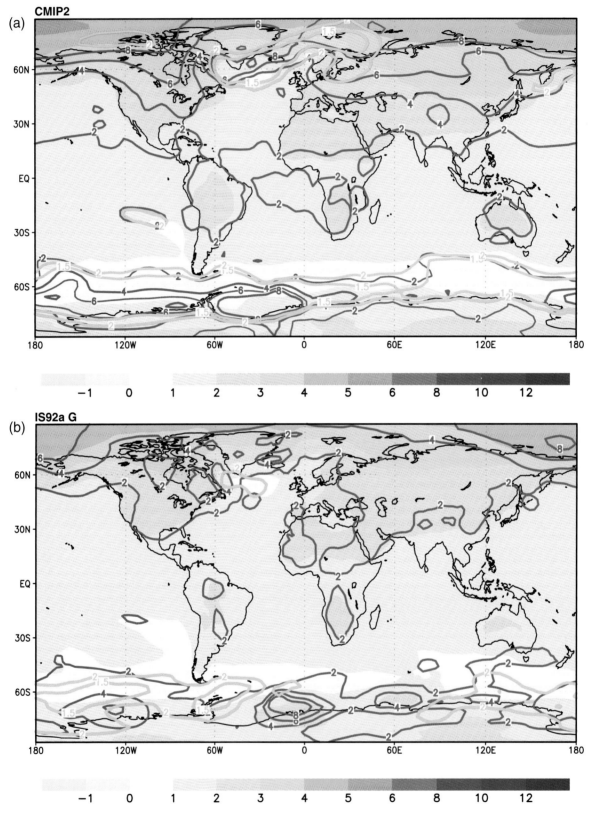

Figure 9.10: The multi-model ensemble annual mean change of the temperature (colour shading), its range (thin blue isolines) (Unit: °C) and the multi-model mean change divided by the multi-model standard deviation (solid green isolines, absolute values) for (a) the CMIP2 scenarios at the time of CO_2-doubling; (b) the IPCC-DDC scenario IS92a (G: greenhouse gases only) for the years 2021 to 2050 relative to the period 1961 to 1990; (c) the IPCC-DDC scenario IS92a (GS: greenhouse gases and sulphate aerosols) for the years 2021 to 2050 relative to the period 1961 to 1990; (d) the SRES scenario A2 and (e) the SRES scenario B2. Both SRES scenarios show the period 2071 to 2100 relative to the period 1961 to 1990. See text for scenario definitions and description of analysis technique. In (b) and (d) the ratio mean change/standard deviation is increasing towards the low latitudes as well as in (a), (c) and (e), while the high latitudes around Antarctica show a minimum.

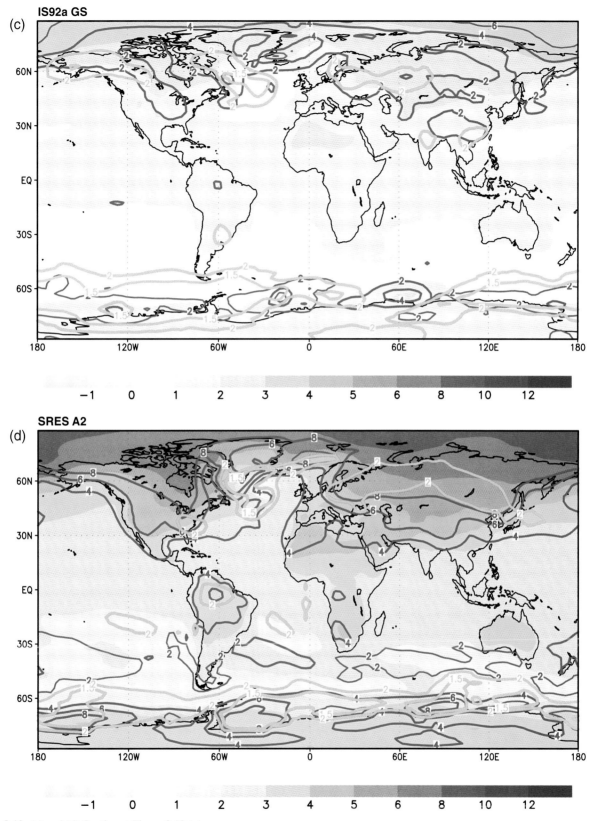

Figure 9.10: (c) and (d) Caption at Figure 9.10 (a).

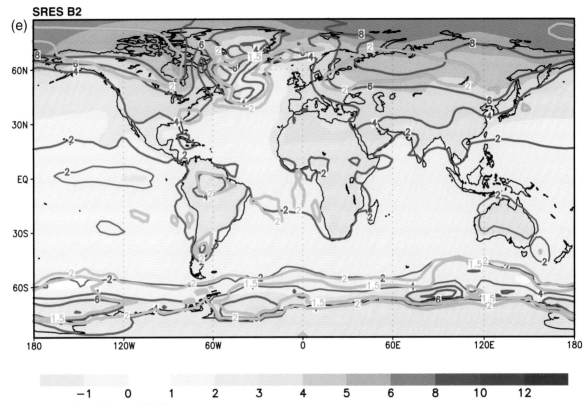

Figure 9.10: (e) Caption at Figure 9.10 (a).

Murphy and Mitchell, resulting in the decrease in global mean convective rain rate. In both models the origin of the decrease in convective precipitation is an increase in stability in the troposphere in the warmer climate. In accordance with the reduction in convective precipitation, Brinkop analysed a strong decrease (11% for JJA and 7.5% for DJF) of the global mean frequency of deep convection in the warmer climate. However, the frequency of shallow convection slightly increases.

The most consistent feature in the ensemble mean sea level pressure difference (Figure 9.12) is a decrease in the sea level pressure at high latitudes and an increase at mid-latitudes. In studies of the the Southern Hemisphere, this is related to a combination of changes in surface and mid-tropospheric temperature gradients (Räisänen, 1997; Fyfe *et al.*, 1999; Kushner *et al.*, 2001). Over wide regions of the Southern Hemisphere and Northern Hemisphere high latitudes, the ensemble mean signal generally exceeds the ensemble standard deviation indicating a consistent response across the models. For the A2 and B2 scenarios this is also found. Additionally a lowering of pressure can be found over the Sahara, probably due to thermal effects. The lowering of pressure is consistent across the A2 and B2 simulations.

9.3.2.1 Summary

First we note results assessed here that reconfirm results from the SAR:

- As the climate warms, Northern Hemisphere snow cover and sea-ice extent decrease. The globally averaged precipitation increases.

- As the radiative forcing of the climate system changes, the land warms faster than the ocean. The cooling effect of tropospheric aerosols moderates warming both globally and locally.
- The surface air temperature increase is smaller in the North Atlantic and circumpolar Southern Ocean regions.
- Most tropical areas, particularly over ocean, have increased precipitation, with decreases in most of the sub-tropics, and relatively smaller precipitation increases in high latitudes.
- The signal to noise ratio (from the multi-model ensemble) is greater for surface air temperature than for precipitation.

A second category of results assessed here are those that are new since the SAR:

- There are many more model projections for a given scenario, and more scenarios. The greater number of model simulations allows us to better quantify patterns of climate change for a given forcing and develop a measure of consistency among the models.
- Including the direct effect of sulphate aerosols according to an IS92a type estimate reduces global mean mid-21st century warming. The indirect effect, not included in most AOGCM experiments to date, is acknowledged to be uncertain, as discussed in Chapter 6.
- The geographic details of various forcing patterns are less important than differences among the models' responses for the scenarios considered here. This is the case for the global mean as well as for patterns of climate response. Thus, the choice of model and the choice of scenario are both important.

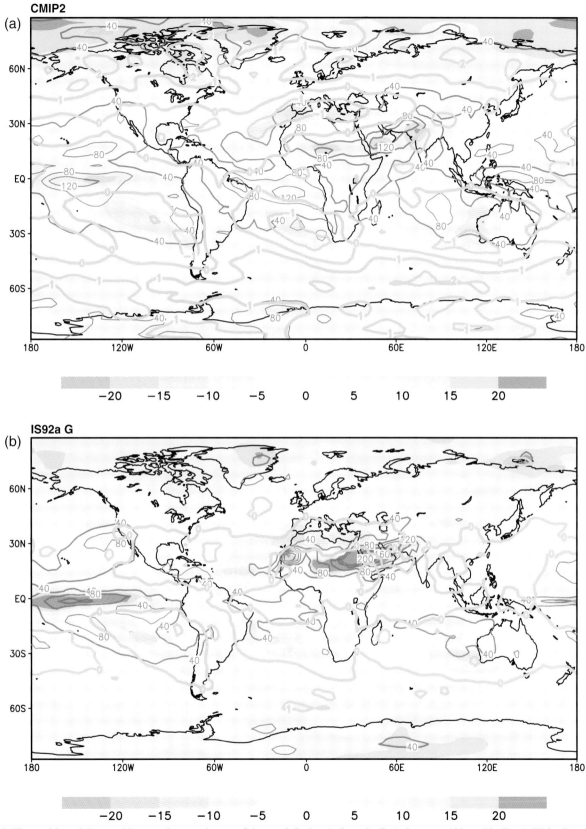

Figure 9.11: The multi-model ensemble annual mean change of the precipitation (colour shading), its range (thin red isolines) (Unit: %) and the multi-model mean change divided by the multi-model standard deviation (solid green isolines, absolute values) for (a) the CMIP2 scenarios at the time of CO_2-doubling; (b) the IPCC-DDC scenario IS92a (G: greenhouse gases only) for the years 2021 to 2050 relative to the period 1961 to 1990; (c) the IPCC-DDC scenario IS92a (GS: greenhouse gases and sulphate aerosols) for the years 2021 to 2050 relative to the period 1961 to 1990; (d) the SRES scenario A2; and (e) the SRES scenario B2. Both SRES-scenarios show the period 2071 to 2100 relative to the period 1961 to 1990. See text for scenario definitions and description of analysis technique.

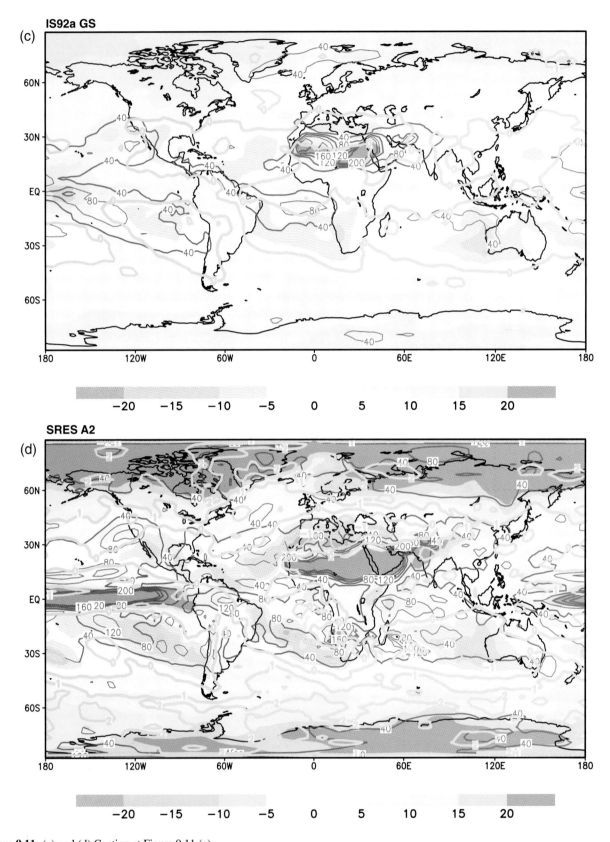

Figure 9.11: (c) and (d) Caption at Figure 9.11 (a).

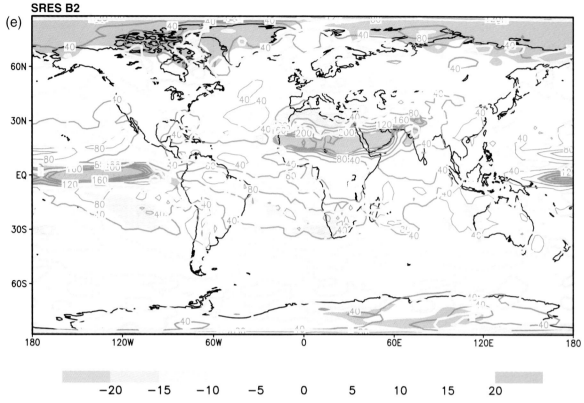

Figure 9.11: (e) Caption at Figure 9.11 (a).

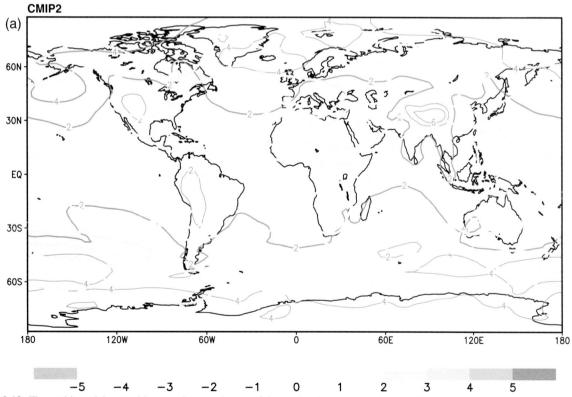

Figure 9.12: The multi-model ensemble annual mean change of the sea level pressure (colour shading), its range (thin red isolines) (Unit: hPa) and the multi-model mean change divided by the multi-model standard deviation (solid green isolines, absolute values) for (a) the CMIP2 scenarios at the time of CO_2-doubling; (b) the IPCC-DDC scenario IS92a (G: greenhouse gases only) for the years 2021 to 2050 relative to the period 1961 to 1990; (c) the IPCC-DDC scenario IS92a (GS: greenhouse gases and sulphate aerosols) for the years 2021 to 2050 relative to the period 1961 to 1990; (d) the SRES scenario A2 and (e) the SRES scenario B2. Both SRES-scenarios show the period 2071 to 2100 relative to the period 1961 to 1990. See text for scenario definitions and description of analysis technique.

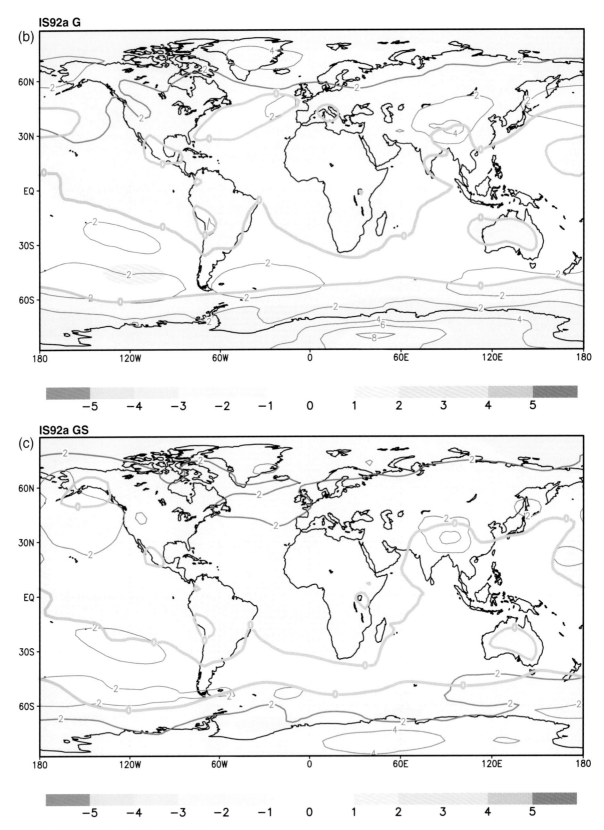

Figure 9.12: (b) and (c) Caption at Figure 9.12 (a).

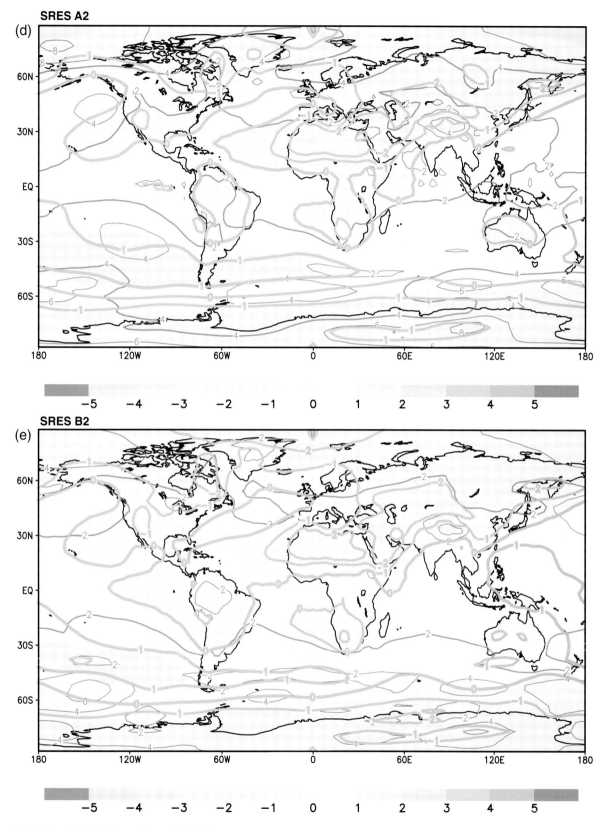

Figure 9.12: (d) and (e) Caption at Figure 9.12 (a).

9.3.3 Range of Temperature Response to SRES Emission Scenarios

This section investigates the range of future global mean temperature changes resulting from the thirty-five final SRES emissions scenarios with complete greenhouse gas emissions (Nakićenović *et al.*, 2000). This range is compared to the expected range of uncertainty due to the differences in the response of several AOGCMs. Forcing uncertainties are not considered in these calculations. As well as envelope results that incorporate all the SRES scenarios, six specific SRES scenarios are considered. These are the four illustrative marker scenarios A1B, A2, B1 and B2 and two further illustrative scenarios from the A1 family representing different energy technology options; A1FI and A1T (see Section 9.3.1.3 and Box 9.1). For comparison, results are also shown for some of the IS92 scenarios. As discussed in Section 9.3.1.3 some AOGCMs have run experiments with some or all of the four draft marker scenarios. In order to investigate the temperature change implications of the full range of the final SRES scenarios, a simple climate model is used as a tool to simulate the AOGCM results (Wigley and Raper, 1992; Raper *et al.*, 1996, 2001a). The tuning of the simple model to emulate the different AOGCM results is described in Appendix 9.1. The original SRES MiniCAM (Mini Climate Assessment Model from the Pacific Northwest National Laboratory, USA) scenarios did not contain emissions for the reactive gases CO, NMVOCs, and NO_x (Nakićenović *et al.*, 2000). To facilitate the calculations, the MiniCAM modelling team provided emissions paths for these gases.

For the six illustrative SRES scenarios, anthropogenic emissions are shown for CO_2 in Chapter 3, Figure 3.12, tabulated for CH_4 and N_2O in Appendix II and shown in Nakićenović *et al.* (2000), and shown for SO_2 in Chapter 5, Figure 5.13. It is evident that these scenarios encompass a wide range of emissions. Note in particular the much lower future sulphur dioxide emissions for the six SRES scenarios compared with the IS92a scenario.

The calculation of radiative forcing from the SRES emission scenarios for the temperature projections presented here follows closely that described in Chapters 3, 4, 5 and 6, with some exceptions as described below. Further details of the forcing for the collective procedures (MAGICC model) are given by Wigley (2000). Atmospheric concentrations of the greenhouse gases are calculated from the emissions using gas cycle models. For CO_2, the model of Wigley (1993) is used and as described therein, the CO_2 fertilisation factor is adjusted to give a balanced 1980s mean budget. To be consistent with Chapter 3, climate feedbacks are included and the model has been tuned to give results that are similar to those of the Bern-CC and ISAM models for a climate sensitivity of 2.5°C (Chapter 3, Figure 3.12). The strength of the climate feedbacks on the carbon cycle are very uncertain, but models show they are in the direction of greater temperature change giving greater atmospheric CO_2 concentration. The climate feedbacks in the Bern-CC model are greater than those of the ISAM model and the feedback strength used here is about half as big as that in the ISAM model. The gas cycle models for CH_4 and N_2O and the other trace gases are identical to those used in Chapter 4. The concentrations for the main greenhouse gases for the six SRES scenarios are shown in Chapter 4, Figure 4.14.

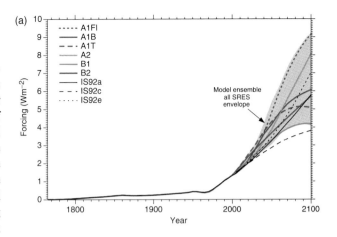

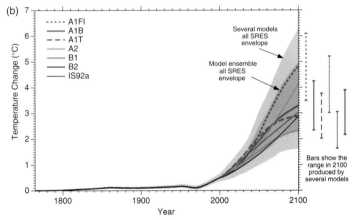

Figure 9.13: Simple model results. (a) Estimated historical anthropogenic radiative forcing followed by radiative forcing for the four illustrative SRES marker scenarios and for two additional scenarios from the A1 family illustrating different energy technology options. The blue shading shows the envelope of forcing that encompasses the full set of thirty-five SRES scenarios. The method of calculation closely follows Chapter 6 except where explained in the text. The values are based on the radiative forcing for a doubling of CO_2 from seven AOGCMs as given in Appendix 9.1, Table 9.A1. The IS92a, IS92c and IS92e forcing is also shown following the same method of calculation. (b) Historical anthropogenic global mean temperature change and future changes for the six illustrative SRES scenarios using a simple climate model tuned to seven AOGCMs. Also for comparison, following the same method, results are shown for IS92a. The dark blue shading represents the envelope of the full set of thirty-five SRES scenarios using the simple model ensemble mean results. The light blue envelope is based on the GFDL_R15_a and DOE PCM parameter settings. The bars show the range of simple model results in 2100 for the seven AOGCM model tunings.

Except for the treatment of organic carbon (OC), black carbon (BC) and indirect aerosol forcing, the method of calculation for the radiative forcing follows closely that described in Chapter 6 and includes tropospheric ozone, halocarbons, and stratospheric ozone. For OC and BC this report's best estimate forcing values for the present day given in Chapter 6, Table 6.11 are used. As pointed out in Chapter 5, past and future emissions of OC and BC are uncertain. Here fossil OC and BC direct aerosol

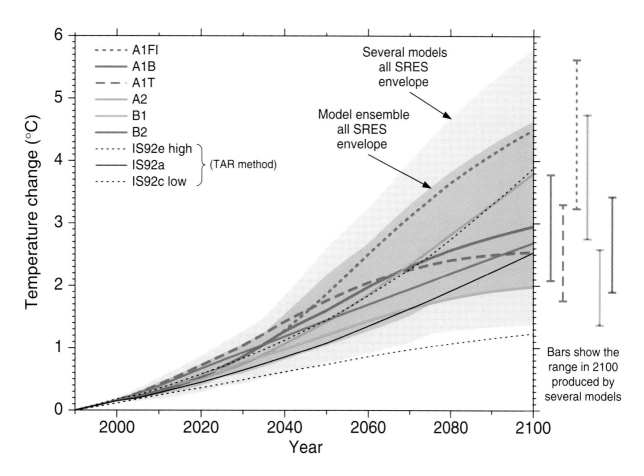

Figure 9.14: As for Figure 9.13b but results are relative to 1990 and shown for 1990 to 2100.

forcings are considered together and are scaled linearly with SO_2 emissions. Biomass burning OC and BC aerosol direct forcings are both scaled with gross deforestation. First (cloud albedo) indirect sulphate aerosol forcing components are included and scaled non-linearly with SO_2 emissions as derived by Wigley (1991). A present day indirect sulphate aerosol forcing of –0.8 Wm^{-2} is assumed. This is the same value as that employed in the SAR. It is well within the range of values recommended by Chapter 6, and is also consistent with that deduced from model simulations and the observed temperature record (Chapter 12).

Estimated total historical anthropogenic radiative forcing from 1765 to 1990 followed by forcing resulting from the six illustrative SRES scenarios are both shown in Figure 9.13a. It is evident that the six SRES scenarios considered cover nearly the full range of forcing that results from the full set of SRES scenarios. The latter is shown on figure 9.13a as an envelope since the forcing resulting from individual scenarios cross with time. For comparison, radiative forcing is also shown for the IS92a, IS92c and IS92e scenarios. It is evident that the range in forcing for the new SRES scenarios is wider and higher than in the IS92 scenarios. The range is wider due to more variation in emissions of non-CO_2 greenhouse gases. The shift to higher forcing is mainly due to the reduced future sulphur dioxide emissions of the SRES scenarios compared to the IS92 scenarios. Secondary factors include generally greater tropospheric ozone

forcing, the inclusion of climate feedbacks in the carbon cycle and slightly larger cumulative carbon emissions featured in some SRES scenarios.

Figure 9.13b shows the simple climate model simulations representing AOGCM-calibrated global mean temperature change results for the six illustrative SRES scenarios and for the full SRES scenario envelopes. The individual scenario time-series and inner envelope (darker shading) are the average results obtained from simulating the results of seven AOGCMs, denoted "ensemble". The average of the effective climate sensitivity of these AOGCMs is 2.8°C (see Appendix 9.1). The range of global mean temperature change from 1990 to 2100 given by the six illustrative scenarios for the ensemble is 2.0 to 4.5°C (see Figure 9.14). The range for the six illustrative scenarios encompassing the results calibrated to the DOE PCM and GFDL_R15_a AOGCM parameter settings is 1.4 to 5.6°C. These two AOGCMs have effective climate sensitivities of 1.7 and 4.2°C, respectively (see Table 9.1). The range for these two parameter settings for the full set of SRES scenarios is 1.4 to 5.8°C. Note that this is not the extreme range of possibilities, for two reasons. First, forcing uncertainties have not been considered. Second, some AOGCMs have effective climate sensitivities outside the range considered (see Table 9.1). For example, inclusion of the simple model's representation of the CCSR/NIES2 AOGCM would increase the high end of the range by several degrees C.

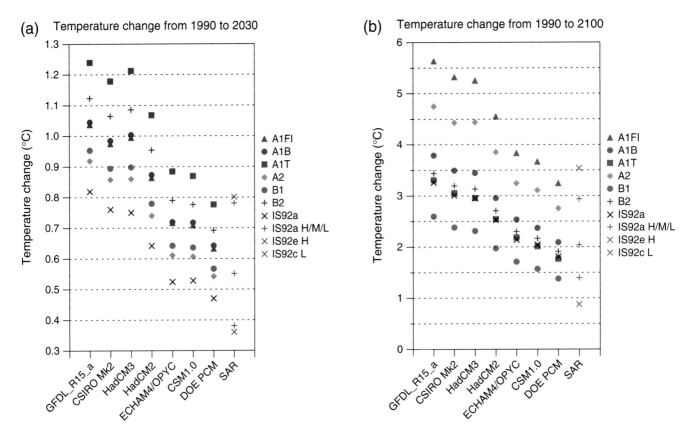

Figure 9.15: Simple model results: Temperature changes from (a) 1990 to 2030 and from (b) 1990 to 2100 for the six illustrative SRES scenarios and IS92a. The bottom axis indicates the AOGCM to which the simple model is tuned. For comparison results are also shown for the SAR version of the simple climate model using SAR forcing with some of the IS92 scenarios (see Kattenberg *et al.*, 1996). IS92a H/M/L refers to the IS92a scenario with climate sensitivity of 1.5, 2.5 and 4.5°C respectively. Also shown are the IS92e scenario with a sensitivity of 4.5°C and the IS92c scenario with a sensitivity of 1.5°C.

Since the AOGCM SRES results discussed in Section 9.3.1.3 are based on the *draft* marker SRES scenarios, it is important to note differences that would result from the use of the final SRES scenarios. Based on a comparison using the simple climate model, the final scenarios for the three markers A1B, A2 and B2 give temperature changes that are slightly smaller than those of the draft scenarios (Smith *et al.*, 2001). The main difference is a change in the standardised values for 1990 through 2000, which are common to all these scenarios. This results in higher forcing early in the period. There are further small differences in net forcing, but these decrease until, by 2100, differences in temperature change in the two versions of these scenarios are only 1 to 2%. For the B1 scenario, however, temperature changes are significantly lower in the final version. The difference is almost 20% in 2100, as a result of generally lower emissions across the whole range of greenhouse gases.

Temperature change results from the simple climate model tuned to individual AOGCMs using the six illustrative SRES scenarios are shown in Figure 9.15. For comparison, analogous results are shown for the IS92a scenario. For direct comparison with the SAR, results are also shown for some of the IS92 scenarios using the SAR forcing and the SAR version of the simple climate model (Kattenberg *et al.*, 1996). The results give rise to conclusions similar to those of Wigley (1999) and Smith

et al. (2001), which were drawn from sensitivity studies using the SAR version of the simple climate model. First, note that the range of temperature change for the SRES scenarios is shifted higher than the range for the IS92 scenarios, primarily because of the higher forcing as described above.

A second feature of the illustrative SRES scenarios is that their relative ranking in terms of global mean temperature changes varies with time (Wigley, 1999; Smith *et al.* 2001). The temperature-change values of the scenarios cross in about mid-century because of links between the emissions of different gases. In particular, for scenarios with higher fossil fuel use, and therefore carbon dioxide emissions (for example A2), sulphur dioxide emissions are also higher. In the near term (to around 2050) the cooling effect of higher sulphur dioxide emissions more than offsets the warming caused by increased emissions of greenhouse gases in scenarios such as A2. The effect of the high sulphur dioxide emissions in the IS92a scenario is similar. It causes IS92a to give rise to a lower 2030 temperature than any of the specific SRES scenarios considered (Figure 9.15a). The opposite effect is seen for scenarios B1 and B2, which have lower fossil fuel emissions, but also lower sulphur dioxide emissions. This leads to a larger near-term warming. In the longer term, however, the level of emissions of long-lived greenhouse gases such as carbon dioxide and nitrous oxide becomes the dominant determinant of the resulting global mean temperature changes. For example, by the latter part of the 21st century, the higher emissions of greenhouse gases in scenario A2 result in larger climate changes than in the other three marker scenarios (A1B, B1 and B2) even though this scenario also has higher sulphur dioxide emissions.

Considering the six illustrative scenarios, the bars on the right-hand side of Figure 9.14 show that scenarios A1FI and B1 alone, define the top and bottom of the range of projected temperature changes, respectively. Towards the middle of the range the scenario bars overlap, indicating that most of the projections fall within this region. In the corresponding sea level rise figure, because of the greater intertia in the ocean response, there is a greater overlap in the projected response to the various scenarios (see Chapter 11, Figure 11.12). In addition, the sea level range for a given scenario is broadened by inclusion of uncertainty in land ice estimates.

By 2100, the differences in the surface air temperature response across the group of climate models forced with a given scenario is as large as the range obtained by a single model forced with the different SRES scenarios (Figure 9.15). Given the quasi-linear nature of the simple model, projections which go outside the range as yet explored by AOGCMs must be treated with caution, since non-linear effects may come into play. Further uncertainties arise due to uncertainties in the radiative forcing. The uncertainty in sulphate aerosol forcing is generally characterised in terms of the 1990 radiative forcing. Wigley and Smith (1998) and Smith *et al.* (2001) examined the effect of this uncertainty on future temperature change by varying the assumed 1990 sulphate radiative forcing by 0.6 Wm^{-2} above and below a central value of -1.1 Wm^{-2}. Reducing the sulphate forcing increased the 1990 to 2100 warming by 0 to 7% (depending on the scenario), while increasing the sulphate forcing decreased warming over the next century by a similar amount. The sensitivity to the uncertainty in sulphate forcing was found to be significantly less in the new scenarios than in the IS92a scenario; in the latter the sensitivity to sulphate forcing was twice as large as the largest value for the SRES marker scenarios. Therefore, the smaller future emissions of sulphur dioxide in the new scenarios significantly lowers the uncertainty in future global mean temperature change due to the uncertain value of present day sulphate aerosol forcing. The climate effects described here use the SRES scenarios as contained in Nakićenović *et al.* (2000). Any feedbacks on the socio-economic development path, and hence on emissions, as a result of these climate changes have not been included.

9.3.3.1 Implications for temperature of stabilisation of greenhouse gases

The objective of Article 2 of the United Nations Framework Convention on Climate Change (United Nations, 1992) is "to achieve stabilisation of greenhouse gas concentrations in the atmosphere at a level that would prevent dangerous anthropogenic interference with the climate system." This section gives an example of the possible effect on future temperature change of the stabilisation of greenhouse gases at different levels using carbon dioxide stabilisation as a specific example.

The carbon dioxide concentration stabilisation profiles developed by Wigley *et al.* (1996) (see also Wigley, 2000) commonly referred to as the WRE profiles, are used. These profiles indirectly incorporate economic considerations. They are also in good agreement with observed carbon dioxide concentrations up to 1999. Corresponding stabilisation profiles for the other greenhouses gases have not yet been produced. To illustrate the effect on temperature of earlier reductions in carbon dioxide emissions, results are also presented for the original stabilisation profiles referred to as the S profiles (Enting *et al.*, 1994). The S profiles are, however, unrealistic because, for example, they require emissions and concentration values during the 1990s below those actually observed.

In order to define future radiative forcings fully, it is necessary to make assumptions about how the emissions or concentrations of the other gases may change in the future. In addition, it is necessary to have a base scenario against which the effect of the different stabilisation pathways may be assessed. The state of the science at present is such that it is only possible to give illustrative examples of possible outcomes (Wigley *et al.*, 1996; Schimel *et al.*, 1997; Mitchell *et al.*, 2000).

To produce these examples, the SRES scenario A1B is used as the base scenario. CO_2 concentrations for this scenario are close to the WRE CO_2 profiles in terms of their implied past and near-future values, so our choice satisfies the underlying WRE assumption that emissions should initially follow a baseline trajectory. This is not the case for the S profiles, however, because as pointed out above, present day CO_2 concentrations already exceed the values assumed for the S profiles. Note that the baseline scenario (A1B) is specified only out to 2100. For stabilisation cases, emissions of non-CO_2 gases are assumed to follow the A1B scenario out to 2100 and are thereafter held constant at their year 2100 level. For scenario A1B, this assumption of constant emissions from 2100 leads to stabilisation of the other gas concentrations at values close to their 2100 values. For gases with long lifetimes (such as N_2O) it takes centuries to reach stabilisation. In all cases, however, the net radiative forcing changes for the non-CO_2 gases are small after 2100 and negligible after about 2200. Note that, in comparing the baseline case with the various stabilisation cases, the only gas that changes is CO_2.

The models used to calculate the other gas concentrations and to convert concentrations and sulphur dioxide emissions to radiative forcing are the same as those used in Section 9.3.3. The simple climate model used is again that based on Wigley and Raper (1992) and Raper *et al.* (1996), tuned to the different AOGCMs using the CMIP2 data set (see Appendix 9.1).

The temperature consequences of the five WRE stabilisation profiles used, based on the assumptions described above and using the simple model ensemble (the average results from tuning the simple model to several AOGCMs), are shown in Figure 9.16. The temperature results for the S profiles are also included for comparison. The simple climate model can be expected to give results in good agreement to those that would be produced by the AOGCMs up to 2100. Thereafter the agreement becomes increasingly less certain and this increasing

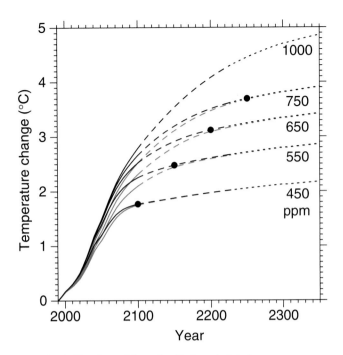

Figure 9.16: Simple model results: Projected global mean temperature changes when the concentration of CO_2 is stabilised following the WRE profiles. For comparison, results with the original S profiles are also shown in blue (S1000 not available). The results are ensemble means produced by a simple climate model tuned to seven AOGCMs (see Appendix 9.1). The baseline scenario is scenario A1B, this is specified only to 2100. After 2100, the emissions of gases other than CO_2 are assumed to remain constant at their A1B 2100 values. The projections are labelled according to the level of CO_2 stabilisation (in ppm). The broken lines after 2100 indicate increased uncertainty in the simple climate model results beyond 2100. The black dots indicate the time of CO_2 stabilisation. The stabilisation year for the WRE1000 profile is 2375.

uncertainty is indicated on the graph by the graduated broken lines. Indeed it has been shown in a comparison of results from the simple model and HadCM2 that the simple model underestimates the temperature change compared to HadCM2 on longer time-scales (Raper *et al.*, 2001a). This is at least in part due to the fact that the HadCM2 effective climate sensitivity increases with time (see Section 9.3.4.1). The results in Figure 9.16 are consistent with the assumption of time-constant climate sensitivities, the average value being 2.8°C.

Since sulphur dioxide emissions stabilise at 2100, the forcing from sulphate aerosols is constant thereafter. CH_4 concentrations stabilise before 2200, and the forcing change from N_2O concentration changes after 2200 is less than 0.1 Wm^{-2}. The continued increase in temperature after the time of CO_2 stabilisation (Figure 9.16) is in part due to the later stabilisation of the other gases but is primarily due to the inertia in the climate system which requires several centuries to come into equilibrium with a particular forcing.

Temperature changes from 1990 to 2100 and from 1990 to 2350, for the simple climate model tuned to seven AOGCMs, are shown in Figure 9.17. These Figures give some indication

Table 9.3: *Reduction in 1990 to 2100 temperature change, relative to the A1B scenario, achieved by five WRE profiles across all seven simple model AOGCM model tunings. WRE1000 refers to stabilisation at a CO_2 concentration of 1,000 ppm, etc.*

Profile	WRE1000	WRE750	WRE650	WRE550	WRE450
Percentage reduction in temperature change relative to A1B	4 – 6%	9 – 10%	14 – 15%	23 – 25%	39 – 41%

of the range of uncertainty in the results due to differences in AOGCM response. Figure 9.17a also shows the temperature change for the baseline scenario, A1B. The percentage reductions in temperature change relative to the baseline scenario that the WRE profiles achieve by 2100 are given in Table 9.3. These range from 4 to 6% for the WRE1000 profile to 39 to 41% for the WRE450 profile. Note that these reductions are for stabilisation of CO_2 concentrations alone.

Although only CO_2 stabilisation is explicitly considered here, it is important to note that the other gases also eventually stabilise in these illustrations. The potential for further reductions in warming, both up to 2100 and beyond, through non-CO_2 gases, depends on whether, in more comprehensive scenarios (when such become available), their stabilisation levels are less than the levels assumed here.

Only one AOGCM study has considered the regional effects of stabilising CO_2 concentrations (Mitchell *et al.*, 2000). HadCM2, which has an effective climate sensitivity in the middle of the IPCC range (Table 9.1), was run with the S550 ppm and S750 ppm stabilisation profiles ("S profiles"; Enting *et al.*, 1994; Schimel *et al.*, 1997). Simulations with a simple climate model (Schimel *et al.*, 1997) indicate that the global mean temperature response in these profiles is likely to differ by no more than about 0.2°C from the equivalent WRE profiles (Wigley *et al.*, 1996; see Figure 9.16), though the maximum rate of temperature change is likely to be lower with the S profiles. Global mean changes in the AOGCM experiments are similar to those in Schimel *et al.* (1994). Note that the AOGCM experiments consider stabilisation of CO_2 concentrations only, and do not take into account changes in other gases, effectively assuming that concentrations of other gases are stabilised immediately. To allow for ongoing increases in other greenhouse gases, one would have in practice to reduce CO_2 to even lower levels to obtain the same level of climate change. For example, in the IS92a scenario, other trace gases contribute 1.3 Wm^{-2} to the radiative forcing by 2100. If the emissions of these gases were to continue to increase as in the IS92a scenario, then CO_2 levels would have to be reduced by about 95 ppm to maintain the same level of climate change in these experiments.

Changes in temperature and precipitation averaged over five sub-continental regions at 2100 were compared to those in a baseline scenario based on 1%/yr increase in CO_2 concentrations from 1990. With both stabilisation profiles, there were significant reductions in the regional temperature changes but the significance of the regional precipitation changes depended on location and season. The response of AOGCMs to idealised stabilisation profiles is discussed in Section 9.3.4.4.

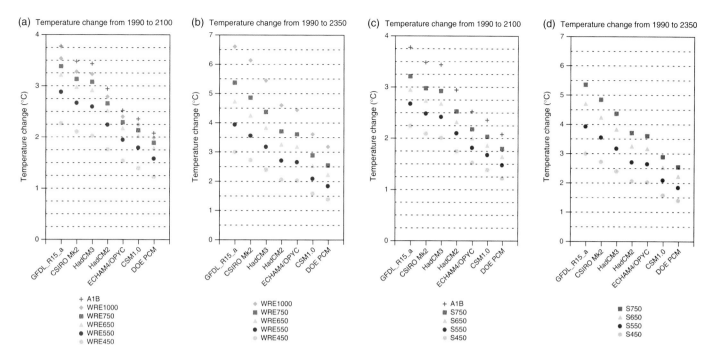

Figure 9.17: Simple model results: Temperature change (a) from 1990 to 2100 and (b) from 1990 to 2350, resulting from the five WRE stabilisation profiles, using the simple climate model tuned to different AOGCMs as indicated on the bottom axis. (c) and (d) show the corresponding results for the four S profiles. The underlying assumptions are the same as those for Figure 9.15. For comparison, temperature changes for the base scenario A1B are also shown in (a) and (c).

9.3.4 Factors that Contribute to the Response

9.3.4.1 Climate sensititivity

A variety of feedback processes operate in the climate system (Chapter 7) to determine the response to changes in radiative forcing. The climate sensitivity (see Section 9.2.1) is a broad measure of this response. Ideally, a coupled AOGCM's climate sensitivity would be obtained by integrating the model to a new climate equilibrium after doubling the CO_2 concentration in the model atmosphere. Since this requires a lengthy integration, climate sensitivities are usually estimated with atmospheric GCMs coupled to mixed-layer upper ocean models, for which the new equilibrium is obtained in decades rather than millennia. Equilibrium climate sensitivities for models in current use are compared with the results reported in the SAR. A related measure, the effective climate sensitivity, is obtained from non-equilibrium transient climate change experiments.

Equilibrium climate sensitivity from AGCMs coupled to mixed-layer upper ocean models
The blue diamonds in Figure 9.18 give the equilibrium climate sensitivity and the associated percentage change in global mean precipitation rate (sometimes termed the hydrological sensitivity) for seventeen equilibrium mixed-layer model calculations documented in Table 6.3 of the SAR (Kattenberg *et al.*, 1996). Table 9.4 gives the average sensitivity of the seventeen models as 3.8°C for temperature and 8.4% for precipitation, with a standard deviation or "inter-model scatter" of 0.78°C and 2.9%, respectively. LeTreut and McAvaney (2000) provide a recent compilation of climate sensitivities for mixed-layer models and this

information has been updated in Table 9.1 under the column headed "equilibrium climate sensitivity". These results, from fifteen models in active use, are represented by the red triangles in Figure 9.18. The associated statistics are given in Table 9.4 where the mean and standard deviation for temperature are 3.5 and 0.92°C, and for precipitation are 6.6 and 3.7%.

According to Table 9.4 and Figure 9.18, the average climate sensitivity, as estimated from AGCMs coupled to mixed-layer ocean models, has decreased slightly from about 3.8 to 3.5°C since the SAR. The inter-model standard deviation has increased and the range has remained essentially the same. The associated hydrological sensitivity has decreased from 8.4 to 6.6% but the inter-model standard deviation has increased. As explained in Section 9.2, these climate sensitivity values are not altered by the lower value for the radiative forcing change for doubled CO_2 discussed in Chapter 6.

These results indicate slightly lower average values of sensitivity in models in current use compared with the SAR. Although more recent models attempt to incorporate improvements in our ability to simulate the climate system, these mean results do not in themselves provide a clear indication that modelled climate sensitivity has decreased. In particular, the inter-model scatter has increased slightly, the range of results is not much changed, the differences are not statistically significant, and the reasons for the modest decrease in average sensitivity have not been identified.

Climate sensitivity from AGCMs coupled to full OGCMs
Because of the long time-scales associated with deep ocean equilibration, the direct calculation of coupled model equilibrium

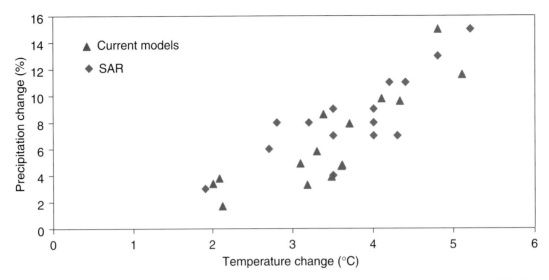

Figure 9.18: Equilibrium climate and hydrological senstivities from AGCMs coupled to mixed-layer ocean components; blue diamonds from the SAR, red triangles from models in current use (LeTreut and McAvaney, 2000 and Table 9.1).

Table 9.4: *Statistics of climate and hydrological sensitivity for mixed-layer models*

Source	No. of models	Temperature (°C)			Precipitation (%)		
		Mean	Standard deviation	Range	Mean	Standard deviation	Range
SAR	17	3.8	0.78	1.9 / 5.2	8.4	2.9	3 / 15
Current models	15	3.5	0.92	2.0 / 5.1	6.6	3.7	2 / 15

temperature change for doubled CO_2 requires an extended simulation and a considerable commitment of computer resources. One such calculation has been performed (Stouffer and Manabe, 1999, Table 9.1); a 4,000 year simulation with stabilisation at $2\times CO_2$, and a 5,000 year simulation with stabilisation at $4\times CO_2$. Figure 9.19 displays the temperature for the first 500 years of these simulations (the red curves) together with stabilisation results from other models discussed further below. The calculation shows that: (1) some 15 to 20 centuries are required for the coupled model to attain a new equilibrium after the forcing is stabilised, (2) for the $2\times CO_2$ case, the temperature change ultimately increases to 4.5°C for the GFDL_R15_a model, which exceeds the 3.9°C value obtained when a mixed-layer ocean is used to estimate the climate sensitivity and, (3) the $4\times CO_2$ equilibrium temperature change is very nearly twice that of the $2\times CO_2$ equilibrium temperature change for this model. In this case the mixed-layer value of climate sensitivity is resonably close to the full climate model value. The difference between coupled model and mixed-layer sensitivities for other models is unknown.

Effective climate sensitivity
The term effective climate sensitivity (Murphy, 1995) as defined in Section 9.2.1 is a measure of the strength of the feedbacks at a particular time in a transient experiment. It is a function of climate state and may vary with time. Watterson (2000) calculates the effective climate sensitivity from several experiments with different versions of an AOGCM. The results show considerable variability, particularly near the beginning of the integrations when the temperature change is small. That study nevertheless concludes that the effective climate sensitivity is approximately constant and close to the appropriate equilibrium result. However, estimates of effective climate sensitivity obtained from the HadCM2 model range from about 2.7°C at the time of stabilisation at $2\times CO_2$ to about 3.8°C after 900 years (Raper *et al.*, 2001a). Senior and Mitchell (2000) implicate time-dependent cloud-feedbacks associated with the slower warming of the Southern Ocean in that model as the cause for this variation in time. The effective climate sensitivity of this climate model is initially considerably smaller than the equilibrium sensitivity obtained with a mixed-layer ocean. As the coupled model integration approaches a new equilibrium, the effective climate sensitivity increases and appears to be approaching the equilibrium climate sensitivity.

If effective climate sensitivity varies with climate state, estimates of climate sensitivity made from a transient simulation may not reflect the ultimate warming the system will undergo. The use of a constant climate sensitivity in simple models will lead to inconsistencies which depend on the value of sensitivity chosen. This feature deserves further study.

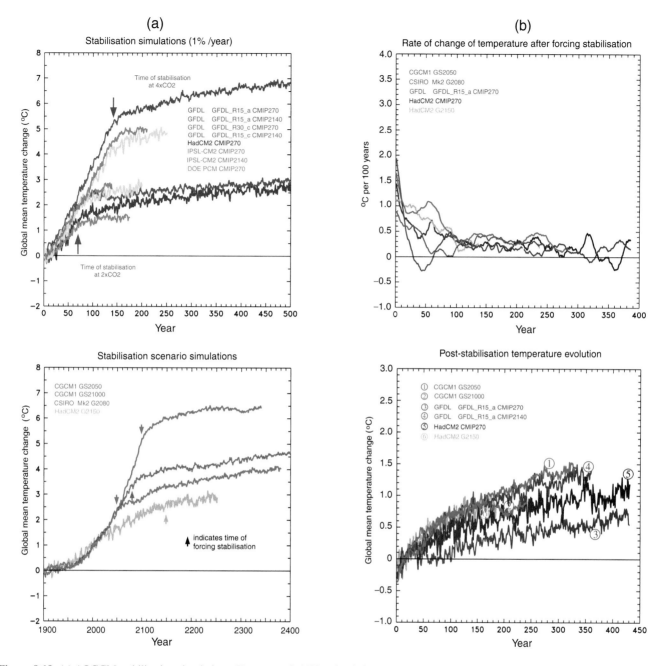

Figure 9.19: (a) AOGCM stabilisation simulations. Upper panel: 1%/yr simulations to stabilisation at and near 2× and 3×CO₂. Lower panel: stabilisation experiments starting with historical greenhouse gas forcing up to present day, then going up to 2× or 3×CO₂ concentrations following the IS92a scenario, then stabilising in the years 2050 and 2100. (b) Post-stabilisation temperature evolution. Upper panel: Rate of change of the temperature. Lower panel: Temperature evolution after the stabilisation of the greenhouse gas concentration.

Summary
The climate sensitivity is a basic measure of the response of the climate system to a change in forcing. It may be measured in several ways as discussed above. The equilibrium climate sensitivity, that is, the range of the surface air temperature response to a doubling of the atmospheric CO_2 concentration, was estimated to be between 1.5 and 4.5°C in the SAR (Kattenberg *et al.*, 1996). That range still encompasses the estimates from the current models in active use.

9.3.4.2 The role of climate sensitivity and ocean heat uptake
Earlier (Section 9.3.1), it was noted that the climate response varies from model to model even when the radiative forcing used to drive the models is similar. This difference in the climate models' response is mainly the result of differing climate sensitivities and differing rates of heat uptake by the oceans in each model, although differences in the AOGCM radiative forcing for a given CO_2 concentration also have a small effect (see Chapter 6, Section 6.3).

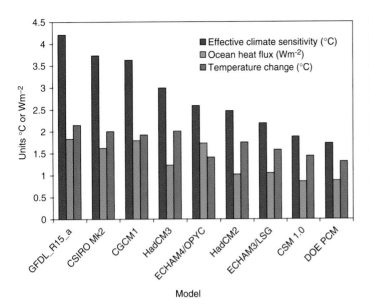

Figure 9.20: Comparison of CMIP2 model results for 20-year average values centred on year 70, the time of CO_2 doubling. Values are shown for the effective climate sensitivity, the net heat flux across the ocean surface multiplied by the ocean fraction and the global mean temperature change (TCR).

The effective climate sensitivity and ocean heat uptake are compared by Raper *et al.* (2001b) using the CMIP2 data set (1%/yr CO_2 increase to doubling). The effective climate sensitivities around the time of CO_2 doubling (average for the years 61 to 80), when the signal is strongest, agree reasonably well with the mixed-layer equilibrium climate sensitivities given in Figure 9.20. Results are shown for various models in Figure 9.20 It is evident that the models with high effective climate sensitivity also tend to have a large net heat flux into the ocean. This oceanic heat flux causes a delay in the climate response. The relationship between the effective climate sensitivity and the oceanic heat uptake was first described by Hansen *et al.* (1984, 1985) using a box diffusion model. Raper *et al.* (2001b) show that an additional ocean-feedback is possibly associated with the warming and freshening of the high latitude surface waters that enhances this relationship. Details of the individual model's sub-grid scale parametrizations also affect both the effective climate sensitivity and the oceanic heat uptake (Weaver and Wiebe, 1999). The evident relationship between effective climate sensitivity and ocean heat uptake leads to the transient climate response (TCR) having a smaller spread among the model results than the climate models' climate sensitivity alone would suggest (see Section 9.3.1). Since the oceanic heat uptake is directly related to the thermal expansion, the range for thermal expansion is correspondingly increased due to the compensation noted above (see Chapter 11 for a complete discussion of sea level rise).

9.3.4.3 Thermohaline circulation changes
In the SAR, it was noted that the thermohaline circulation (THC) weakens as CO_2 increases in the atmosphere in most coupled climate model integrations. The weakening of the THC is found in both the Northern and Southern Hemispheres. The amount of

weakening varied from model to model, but in some cases it was noted that the THC in the North Atlantic stopped completely (Manabe and Stouffer, 1994; Hirst 1999). The weakening of the THC in the Atlantic Ocean results in a reduction of the poleward heat transport that in turn leads to a minimum in the surface warming in the northern North Atlantic Ocean and/or in the circumpolar Ocean (see Section 9.3.2). The reduction in the warming in the North Atlantic region touches the extreme north-eastern part of North America and north-west Europe. The shutting off of the THC in either hemisphere could have long-term implications for climate. However, even in models where the THC weakens, there is still a warming over Europe. For example, in all AOGCM integrations where the radiative forcing is increasing, the sign of the temperature change over north-west Europe is positive (see Figure 9.10).

Figure 9.21 shows a comparison of the strength of the THC through a number of transient experiments with various models and warming scenarios over the 21st century. The initial (control state) absolute strength of the Atlantic thermohaline circulation (THC) varies by more than a factor of 2 between the models, ranging from 10 to 30 Sv (1 Sv = 10^6 m^3s^{-1}). The cause of this wide variation is unclear, but it must involve the sub-grid parametrization schemes used for mixing in the oceans (Bryan, 1987) and differences in the changes of the surface fluxes. The sensitivity of the THC to changes in the radiative forcing is also quite different between the models. Generally as the radiative forcing increases, most models show a reduction of THC. However, some models show only a small weakening of the THC and one model (ECHAM4/OPYC; Latif *et al.*, 2000) has no weakening in response to increasing greenhouse gases, as does the NCAR CSM as documented by Gent (2001). The exact reasons for the difference in the THC responses are unknown, but the role of the surface fluxes is certainly part of the reasons for the differences in the response (see below).

Stocker and Schmittner (1997), using an intermediate complexity model, found that the North Atlantic THC shut-down when the rate of 1%/yr of CO_2 increase was held fixed for approximately 100 years. This is in agreement with the earlier AOGCM study of Manabe and Stouffer (1994), where the THC shut-down in an integration where the CO_2 concentration increased by 1%/yr to four times its initial value. In integrations where the CO_2 stabilised at doubling, the THC did not shut-down in either study (Stocker and Schmittner 1997; Manabe and Stouffer 1994). Furthermore, in the Manabe and Stouffer (1994) AOGCM where the CO_2 is stabilised at four times its normal value, the THC recovers to the control integration value around model year 2300. A recent study (Stouffer and Manabe, 1999) found that the amount of weakening of the THC by the time of CO_2 doubling is a function of the rate of CO_2 increase and not the absolute increase in the radiative forcing. They found the slower the rate of increase, the more the weakening of the THC by the time of CO_2 doubling.

The evolution of the THC in response to future forcing scenarios is a topic requiring further study. It should be noted in particular that these climate model experiments do not currently include the possible effects of significant freshwater input arising from changes in land ice sheets (Greenland and Antarctic ice caps) and mountain glaciers, which might well lead to bigger reductions

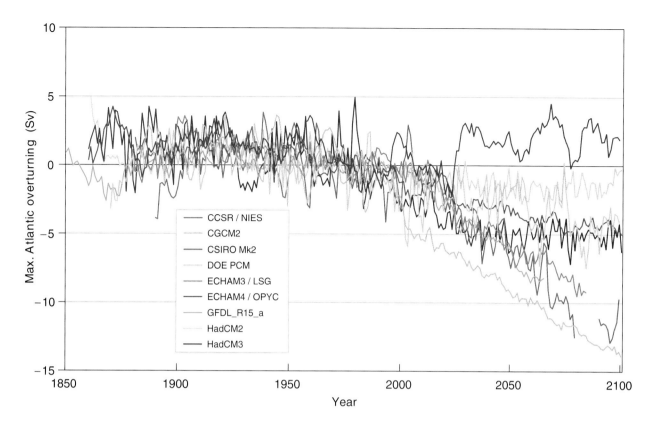

Figure 9.21: Simulated water-volume transport change of the Atlantic "conveyor belt" (Atlantic overturning) in a range of global warming scenarios computed by different climate research centres. Shown is the annual mean relative to the mean of the years (1961 to 1990) (Unit: SV, 10^6 m^{3}s^{-1}). The past forcings are only due to greenhouse gases and aerosols. The future-forcing scenario is the IS92a scenario. See Table 9.1 for more information on the individual models used here.

in the THC. It is too early to say with confidence whether irreversible shut-down of the THC is likely or not, or at what threshold it might occur. Though no AOGCM to date has shown a shut-down of the THC by the year 2100, climate changes over that period may increase the likelihood during subsequent centuries, though this is scenario-dependent. The realism of the representation of oceanic mechanisms involved in the THC changes also needs to be carefully evaluated in the models.

Role of the surface fluxes
The role of heat, fresh water and momentum fluxes in weakening the North Atlantic THC as a consequence of increasing atmospheric CO_2 concentration has been studied in two different AOGCMs (ECHAM3/LSG, Mikolajewicz and Voss, 2000; and GFDL_R15_b, Dixon *et al.*, 1999). In both these studies (Figure 9.22), two baseline integrations are performed; a control integration in which the CO_2 is held fixed, and a perturbation integration in which the CO_2 is increasing. The water fluxes from both of these integrations are archived and used as input in two new integrations.

In the first integration, the atmospheric CO_2 concentration is held fixed and the fresh water fluxes into the ocean are prescribed as those obtained from the perturbation integration. In the second integration, the CO_2 increases as in the perturbation integration and the water fluxes are prescribed to be the fluxes from the control integration (see Table 9.5). In this way, the relative roles of the fresh water and heat fluxes can be evaluated (Figure 9.22).

Table 9.5: *The THC-sensitivity experiments.*

Experiment	CO₂ concentration	Freshwater flux	Wind stress
FSS	fixed present day	simulated	simulated
ISS	increasing	simulated	simulated
IFS	increasing	from FSS	simulated
FIS	fixed present day	from ISS	simulated
FSI	fixed present day	simulated	from ISS
IFF	increasing	from FSS	from FSS

9.3.4.4 Time-scales of response
As mentioned earlier, the basis of the experiments discussed in the SAR is a transient increase of greenhouse gases throughout the integration. In the model integrations presented in this section, the CO_2 concentration increases up to a certain value (e.g., a doubling of the CO_2 concentration) and then remains constant for the remainder of the integration. Since this type of integration involves integrating the model for very long time periods (at least several centuries) only a few integrations have been performed using AOGCMs. Furthermore, no standard emission scenarios have been used for forcing these model runs and most have used idealised stabilisation values ($2\times CO_2$ or $3\times CO_2$ or $4\times CO_2$ for example). Again, in these integrations, the CO_2 changes represent the radiative forcing changes of all the greenhouse gases. Results from the models of intermediate complexity are used to help understand the coupled model results, or in some cases, to explore areas where

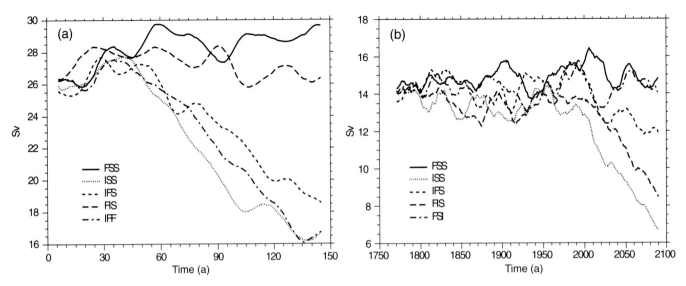

Figure 9.22: Time-series of the zonally integrated Atlantic mass transport stream function at 30°N and 1500 m depth, close to the maximum of the stream function simulated by the (a) ECHAM3/LSG model and the (b) GFDL_R15_b model. For a description of the experiments see Table 9.5.

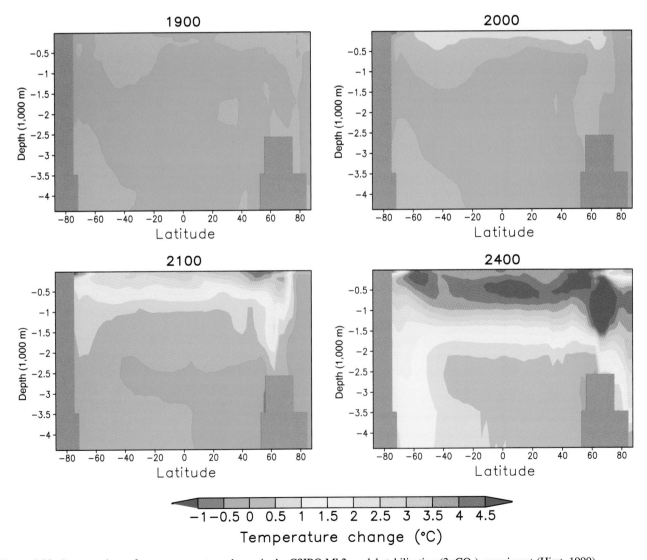

Figure 9.23: Cross-sections of ocean temperature change in the CSIRO Mk2 model stabilisation ($3\times CO_2$) experiment (Hirst, 1999).

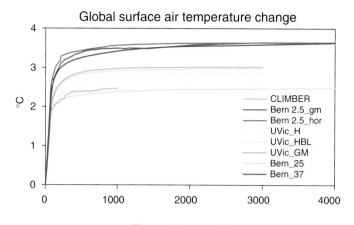

Global surface air temperature change

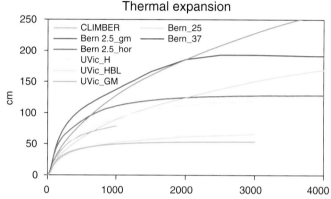

Thermal expansion

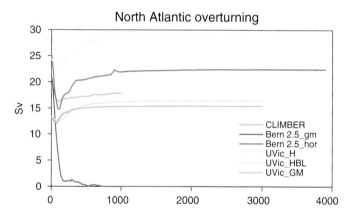

North Atlantic overturning

Figure 9.24: Global mean temperature change, thermal expansion and North Atlantic overturning for a number of models of intermediate complexity. The models have been forced by 1% increase of CO_2 until doubling, then the CO_2 concentration has been kept constant.

AOGCM integrations do not exist. Experiments where the coupled system is allowed time to reach equilibrium with the radiative forcing clearly show the response times of its various components.

Even after the radiative forcing becomes constant, the surface air temperature continues to increase for many centuries (Figure 9.19) as noted in Section 9.3.4.1. The rate of warming after stabilisation is relatively small (<0.3°C per century, Figure 9.19b); however, the total warming after the radiative forcing stabilises can be significant (more than 1°C) because the warming continues for a long time period (Figure 9.19b). From Figure 9.19, one notes that the rate of warming after stabilisation varies from model to model.

The slow rate of surface air temperature increase occurs as the heat anomaly slowly penetrates to depth in the ocean (Figure 9.23). The rate of penetration is dependent on the model's vertical mixing both resolved by the model's grid and by the sub-grid scale parametrizations. The effect of the oceanic mixing parametrizations on the coupled model response has been investigated using climate models of intermediate complexity (Figure 9.24): *CLIMBER* – Ganopolski *et al.* (2001); *Bern 2.5_gm, Bern 2.5_hor* – Stocker *et al.* (1992); *Uvic* – Fanning and Weaver (1996), Weaver *et al.* (1998), Wiebe and Weaver (2000); *Bern_25, Bern_37* – Siegenthaler and Joos (1992), Joos *et al.* (1996). The effect on the response of the global mean surface air temperature, thermal expansion (see Chapter 11, Section 11.5.4.1 for a more complete discussion) and THC can be seen by comparing the results obtained from Stocker's (Bern 2.5) models and the Uvic models (Weaver and Wiebe, 1999). The sub-grid scale mixing parametrizations vary in the AOGCMs, accounting for much of the difference in the rate of surface warming (as seen in Figure 9.19b).

The thermohaline circulation (THC) response is more complex than that of the surface air temperature in the stablisation integrations (compare Figure 9.19a with Figure 9.25, for example). Typically the THC weakens as the radiative forcing increases (Section 9.3.4.3). After the radiative forcing stabilises, the THC recovers to its control integration value. The initial weakening is caused by the warming of the mixed layer in the ocean and the increase in the freshwater flux in high latitudes. As the radiative forcing stablises, the tendency for the surface fluxes to weaken the THC is balanced by the changes in the ocean heat and water transports and vertical structure. It is found that the time-scale for this recovery varies from model to model (about a century to multi-centuries). Again it is likely that differences in the oceanic mixing are the cause for the differences in the recovery time.

The time rate of change in the radiative forcing also affects both the weakening and recovery of the THC (Figure 9.25). In the GFDL_R15_a model when the CO_2 increased at a rate of 1%/yr to doubling, the THC continued to weaken for 70 years after the point at which the CO_2 was held constant at the doubled value (year 70). In a second integration, the CO_2 increased at a rate of 0.25%/yr to doubling. In this integration, the THC does not weaken after the doubling point (Manabe and Stouffer, 1994), indicating that the behaviour of the THC response is highly dependent on the rate that the radiative forcing changes (Figure 9.25).

Finally, it is important to note that the transient THC response (i.e., the weakening) is quite different from the equilibrium response of the THC (i.e., little change). This fact makes the interpretation of comparisons between palaeo-proxy data and coupled model results presented here difficult, since one needs to know the details of the changes in the radiative forcing and resolve relatively small time-scales in the proxy record.

9.3.5 Changes in Variability

The capability of models to simulate the large-scale variability of climate, such as the El Niño-Southern Oscillation (ENSO) (a major source of global interannual variability) has improved substantially in recent years, with an increase in the number and quality of coupled ocean-atmosphere models (Chapter 8) and

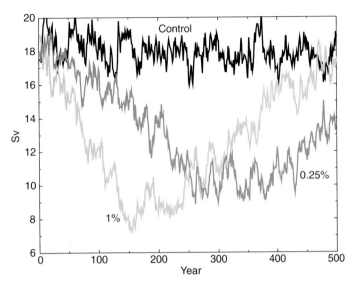

Figure 9.25: Time-series of the maximum value of the stream function (Sv) between 40°N and 60°N in the Atlantic Ocean for the control (black line), 1% (green line) and 0.25% (red line) integrations with the GFDL_R15_a model. See text for integration definitions.

with the running of multi-century experiments and multi-member ensembles of integrations for a given climate forcing (Section 9.2). There have been a number of studies that have considered changes in interannual variability under climate change (e.g., Knutson and Manabe, 1994; Knutson *et al.*, 1997; Tett *et al.* 1997; Timmermann *et al.* 1999; Boer *et al.* 2000b; Collins, 2000a,b). Other studies have looked at intra-seasonal variability in coupled models and the simulation of changes in mid-latitude storm tracks (e.g., Carnell *et al.* 1996; Lunkeit *et al.*, 1996; Carnell and Senior, 1998; Ulbrich and Christoph, 1999), tropical cyclones (Bengtsson *et al.*, 1996; Henderson-Sellers *et al.*, 1998; Knutson *et al.*, 1998; Krishnamurti *et al.*, 1998; Royer *et al.*, 1998) or blocking anticyclones (Lupo *et al.*, 1997; Zhang and Wang, 1997; Carnell and Senior, 1998). The results from these models must still be treated with caution as they cannot capture the full complexity of these structures, due in part to the coarse resolution in both the atmosphere and oceans of the majority of the models used (Chapter 8).

An expanding area of research since the SAR is the consideration of whether climate change may be realised as preferred modes of non-linear naturally occurring atmospheric circulation patterns, or so-called weather regimes as proposed by Palmer (1999). Recent work (e.g., Hurrell 1995, 1996; Thompson and Wallace 1998; Corti *et al.*, 1999) has suggested that the observed warming over the last few decades may be manifest as a change in frequency of these naturally preferred patterns (Chapters 2 and 7) and there is now considerable interest in testing the ability of climate models to simulate such weather regimes (Chapter 8) and to see whether the greenhouse gas forced runs suggest shifts in the residence time or transitions between such regimes on long time-scales. There are now several multi-ensemble simulations using scenarios of time-evolving forcing and multi-century experiments with stabilised forcing, which may help to separate the noise of decadal variability from the signal of climate change.

In this section, changes in variability (defined as the deviation from some mean value) will be considered on different time-scales (intra-seasonal, interannual, and decadal and longer). Particular attention will be given to changes in naturally occurring modes of variability such as ENSO, the Arctic Oscillation (AO; and its more spatially restricted counterpart, the North Atlantic Oscillation, NAO) and the Antarctic Oscillation (AAO) etc.

9.3.5.1 Intra-seasonal variability
Daily precipitation variability
Changes in daily variability of temperature and rainfall are most obviously manifest in changes in extreme events and much of the work in this area will be discussed in the extreme events section (Section 9.3.6). However, changes in short time-scale variability do not necessarily only imply changes in extreme weather. More subtle changes in daily variability, when integrated over time, could still have important socio-economic impacts. Hennessey *et al.* (1997) found that the simulated number of wet days (days where the rainfall is non-zero) in two mixed-layer models went down in mid-latitudes and up in high latitudes when CO_2 was doubled, whilst the mean precipitation increased in both areas. The global mean precipitation also increased, by around 10% in both models, typical of the changes in many mixed-layer models on doubling CO_2. An analysis of changes in daily precipitation variability in a coupled model (Durman *et al.*, 2001) suggests a similar reduction in wet days over Europe where the increase in precipitation efficiency exceeds the increase in mean precipitation.

Circulation patterns
Kattenberg *et al.* (1996) reported research on changes in inter-monthly temperatures and precipitation variability from two coupled models (Meehl *et al*, 1994; Parey, 1994). More recently, there have been several studies looking at changes in intra-seasonal circulation patterns using higher resolution atmosphere-only models with projected SSTs taken from coupled models at given time periods in the future (e.g., Beersma *et al.*, 1997; Schubert *et al.*, 1998). The effects of changes in extra-tropical storms on extreme wind and precipitation events are described in Section 9.3.6, but there has also been work on changes in lower-frequency variability such as persistent or "blocking" anti-cyclones. As discussed in the SAR, there still seems to be little consensus on the methodology for looking at changes in storms and blocks and it is likely that this is partly the reason for the lack of consistency in results. In new studies, Lupo *et al.* (1997) looked at the effect of doubled CO_2 on several of the characteristics of blocking. They found an increase in the number of continental blocks and a general increase in the persistence of blocks, but with weakened amplitude. In contrast, Carnell and Senior (1998) found the largest change was a decrease in blocking in the North Pacific Ocean in winter in their model. Earlier studies have pointed to the possible model dependency of results (Bates and Meehl, 1986) and Carnell and Senior (2000) suggest that the changes in blocking found in their earlier study (Carnell and Senior, 1998) may depend on the meridional gradient of temperature change in the model, which may in turn depend on the simulation of cloud feedback in their model.

Zhang and Wang (1997) found a decrease in the total number of Northern Hemisphere winter anticyclones under increased greenhouse gases, although they did not specifically look at blocking anticyclones.

Fyfe (1999) has looked at changes in African easterly waves due to a doubling of CO_2 in one model. Significant low-level warming and increases in atmospheric humidity over the Northern Sahara lead to an increase in the easterly wave activity. Again, these results must be considered speculative given the relatively low resolution of the model (T32, about 3.5° resolution), which leads to substantial systematic biases in the present day simulation of the low-level storm track in the region.

9.3.5.2 Interannual variability
ENSO

ENSO is associated with some of the most pronounced year-to-year variability in climate features in many parts of the world (Chapters 2 and 7). Since global climate models simulate some aspects of ENSO-like phenomena (Chapter 8), there have been a number of studies that have attempted to use climate models to assess the changes that might occur in ENSO in connection with future climate warming and in particular, those aspects of ENSO that may affect future climate extremes.

Firstly, will the long-term mean Pacific SSTs shift toward a more El Niño-like or La Niña-like regime? Since 1995, the analyses of several global climate models indicate that as global temperatures increase due to increased greenhouse gases, the Pacific climate will tend to resemble a more El Niño-like state (Knutson and Manabe, 1995; Mitchell *et al.*, 1995; Meehl and Washington, 1996; Timmermann *et al.*, 1999; Boer *et al.*, 2000b). However, the reasons for such a response are varied, and could depend on the model representation of cloud feedback (Senior, 1999; Meehl *et al.*, 2000b); the quality of the unperturbed El Niño state in the models (Chapter 8) or the stronger evaporative damping of the warming in the warm pool region, relative to the eastern Pacific due to the non-linear Clausius-Clapeyron relationship between temperature and saturation mixing ratios (e.g., Knutson and Manabe, 1995). Additionally, a different coupled model (Noda *et al.*, 1999b) shows a La Niña-like response and yet another model shows an initial La Niña-like pattern which becomes an El Niño-like pattern due to subducted warmed extra-tropical water that penetrates through the sub-tropics into the tropics (Cai and Whetton, 2000). A possible reason for the La Niña-like response has been suggested in a simple coupled model study where the dominant role of ocean dynamics in the heat balance over the tropical Pacific is seen for a specified uniform positive forcing across the Pacific basin (Cane *et al.*, 1997).

Secondly, will El Niño variability (the amplitude and/or the frequency of temperature swings in the equatorial Pacific) increase or decrease? Attempts to address this question using climate models have again shown conflicting results, varying from slight decreases or little change in amplitude (Tett 1995; Knutson *et al.*, 1997; Noda *et al.*, 1999b; Collins, 2000b; Washington *et al.*, 2001; Figure 9.26b) to a small increase in amplitude (Timmermann *et al.*, 1999; Collins, 2000a; Figure 9.26a), which has been attributed to an increase in the intensity of

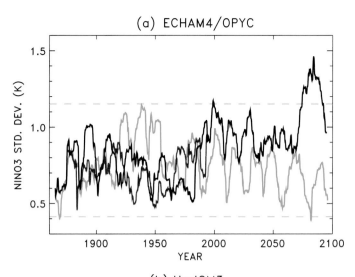

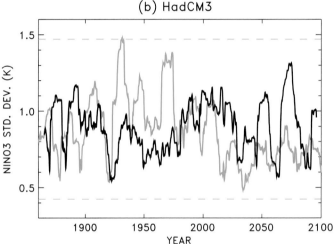

Figure 9.26: Standard deviations of Niño-3 SST anomalies (Unit: °C) as a function of time during transient greenhouse warming simulations (black line) from 1860 to 2100 and for the same period of the control run (green line). Minimum and maximum standard deviations derived from the control run are denoted by the dashed green lines. A low-pass filter in the form of a sliding window of 10 years width was used to compute the standard deviations. (a) ECHAM4/OPYC model. Also shown is the time evolution of the standard deviation of the observed from 1860 to 1990 (red line). Both the simulated and observed SST anomalies exhibit trends towards stronger interannual variability, with pronounced inter-decadal variability superimposed, (reproduced from Timmermann *et al.*, 1999), (b) HadCM3 (Collins, 2000b).

the thermocline in the tropical Pacific. Knutson *et al.* (1997) and Hu *et al.* (2001) find that the largest changes in the amplitude of ENSO occur on decadal time-scales with increased multi-decadal modulation of the ENSO amplitude. Several authors have also found changes in other statistics of variability related to ENSO. Timmermann *et al.* (1999) find that the interannual variability of their model becomes more skewed towards strong cold (La Niña type) events relative to the warmer mean climate. Collins (2000a) finds an increased frequency of ENSO events and a shift in the seasonal cycle, so that the maximum occurs between August and October rather than around January as in the unperturbed model and the observations. Some recent coupled models have achieved a stable climate without the use of flux adjustments and an important question to ask is what is the effect of flux adjustment on changes in variability. Collins (2000b) finds different responses in ENSO in two models, one of which has been run without the use of flux-adjustments. However, he concludes that differences in response are most likely to be due to differences in the response of the meridional temperature gradient in the two models arising from different cloud feedbacks (Williams *et al.*, 2001) rather than due to the presence or absence of flux adjustment.

Finally, how will ENSO's impact on weather in the Pacific Basin and other parts of the world change? Meehl *et al.* (1993) and Meehl and Washington (1996) indicate that future seasonal precipitation extremes associated with a given ENSO event are likely to be more intense due to the warmer, more El Niño-like, mean base state in a future climate. That is, for the tropical Pacific and Indian Ocean regions, anomalously wet areas could become wetter and anomalously dry areas become drier during future ENSO events. Also, in association with changes in the extra-tropical base state in a future warmer climate, the teleconnections to mid-latitudes, particularly over North America, may shift somewhat with an associated shift of precipitation and drought conditions in future ENSO events (Meehl *et al.*, 1993).

When assessing changes in ENSO, it must be recognised that an "El Niño-like" pattern can apparently occur at a variety of time-scales ranging from interannual to inter-decadal (Zhang *et al.*, 1997), either without any change in forcing or as a response to external forcings such as increased CO_2 (Meehl and Washington, 1996; Knutson and Manabe, 1998; Noda *et al.*, 1999a,b; Boer *et al.*, 2000b; Meehl *et al.*, 2000b). Making conclusions about "changes" in future ENSO events will be complicated by these factors. Additionally, since substantial internally generated variability of ENSO statistics on multi-decadal to century time-scales occurs in long unforced climate model simulations (Knutson *et al.*, 1997), the attribution of past and future changes in ENSO amplitude and frequency to external forcing may be quite difficult, perhaps requiring extensive use of ensemble climate experiments or long experiments with stabilised forcing (e.g., Knutson *et al.*, 1997).

Although there are now better ENSO simulations in global coupled climate models (Chapter 8), further model improvements are needed to simulate a more realistic Pacific climatology and seasonal cycle as well as more realistic ENSO variability (e.g., Noda *et al.*, 1999b). It is likely that such things as increased ocean resolution, atmospheric physics and possibly flux correction can have an important effect on the response of the ENSO in models. Improvements in these areas will be necessary to gain further confidence in climate model projections.

Monsoon
One of the most significant aspects of regional interannual variability is the Asian Monsoon. Several recent studies (Kitoh *et al.*, 1997; Hu *et al.*, 2000a; Lal *et al.*, 2000) have corroborated earlier results (Mitchell *et al.*, 1990; Kattenberg *et al.*, 1996) of an increase in the interannual variability of daily precipitation in the Asian summer monsoon with increased greenhouse gases. Lal *et al.* (2000) find that there is also an increase in intra-seasonal precipitation variability and that both intra-seasonal and inter-annual increases are associated with increased intra-seasonal convective activity during the summer. Less well studied is the Asian winter monsoon, although Hu *et al.* (2000b) find reductions in its intensity with a systematic weakening of the north-easterlies along the Pacific coast of the Eurasian continent. However, they find no change in the interannual or inter-decadal variability.

The effect of sulphate aerosols on Indian summer monsoon precipitation is to dampen the strength of the monsoon compared to that seen with greenhouse gases only (Lal *et al.*, 1995; Cubasch *et al.*, 1996; Meehl *et al.*, 1996; Mitchell and Johns 1997; Roeckner *et al.*, 1999), reinforcing preliminary findings in the SAR. The pattern of response to the combined forcing is at least partly dependent on the land-sea distribution of the aerosol forcing, which in turn may depend upon the relative size of the direct and indirect effects (e.g., Meehl *et al.*, 1996; Roeckner *et al.*, 1999). There is still considerable uncertainty in these forcings (Chapter 6). To date, the effect of aerosol forcing (direct and indirect) on the variability of the monsoon has not been investigated.

In summary, an intensification of the Asian summer monsoon and an enhancement of summer monsoon precipitation variability with increased greenhouse gases that was reported in the SAR has been corroborated by new studies. The effect of sulphate aerosols is to weaken the intensification of the mean precipitation found with increases in greenhouse gases, but the magnitude of the change depends on the size and distribution of the forcing.

9.3.5.3 Decadal and longer time-scale variability
A few studies have attempted to look at model-simulated changes in modes of low-frequency variability due to anthropogenic climate change. Particular attention has focused on changes in ENSO as reported in the SAR and in Section 9.3.5.2, and the AO or NAO and AAO which are prominent features of low-frequency variability in the Northern and Southern Hemispheres, respectively (e.g., Fyfe *et al.*, 1999; Osborn *et al.*, 1999; Paeth *et al.*, 1999; Shindell *et al.*, 1999; Ulbrich and Christoph, 1999; Zorita and González-Rouco, 2000; Monahan *et al.*, 2000). It should be noted that these studies have used a variety of methods for analysing trends in these modes of variability, including indices based on pressure differentials and principal components (PCs) of hemispheric sea level pressure (SLP). In addition, these indices are sensitive to changes in the SLP patterns with time or forcing and so trends must be treated with some caution. Wallace (2000) finds that in both observations and modelling studies of increased greenhouse gases, the trends are larger in the PC of SLP than in the pressure differential indices. Meehl *et al.* (2000c) show that the changed base climate state in a future warmer climate could affect the period of global ENSO-like decadal (10 to 20-year period) variability such that there could be a shift to longer periods.

Ulbrich and Christoph (1999) find that the NAO index, based on SLP fluctuations over the North Atlantic in the 300-year control run of their model, shows only a moderate increase over the length of a 240-year scenario run with increasing greenhouse gases. The long-term trend exceeds the variability of the control climate only at the end of the simulation in 2100. In contrast, the steadily growing storm track activity over north-west Europe

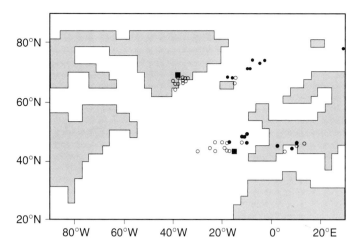

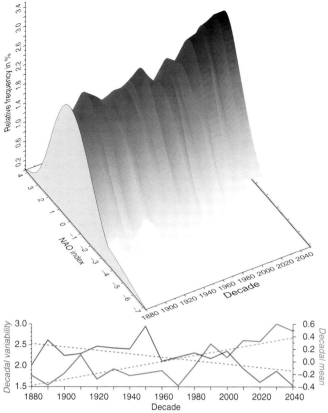

Figure 9.27: Locations of NAO centres (taken to be the position of maximum variance as computed from an EOF analysis of sea level pressure fields) of ECHAM4/OPYC (Ulbrich and Christoph, 1999). The average positions from the entire control run (using winter means) are marked by black squares, those of consecutive decades in the scenario run (using all individual months) are marked by open circles before year 2020 and by black dots thereafter.

already surpasses the standard deviation defined from the control run after about 160 years. This effect is associated with a change of the NAO pattern. During the length of the scenario experiment, empirical orthogonal functions for sequential 10-year periods show a systematic north-eastward shift of the NAO's northern variability centre from a position close to the east coast of Greenland, where it is also located in the control run, to the Norwegian Sea (Figure 9.27)

Osborn *et al.* (1999) show an initial small increase followed by a decrease in the NAO index in one model when forced with increases in greenhouse gases or with greenhouse gases and sulphate aerosols. Paeth *et al.* (1999) have assessed changes in both the mean and variance of the NAO on decadal time-scales at quadrupled CO_2-concentrations using an ensemble of four integrations of a single model. They find a statistically signifi-cant increase in the mean NAO index (at the 95% confidence level), especially during late summer/autumn and in winter, suggesting more westerly and typically milder weather over Europe during the cold season. However, the increase in the mean NAO index is accompanied by a reduction in the low-frequency variability of the NAO (Figure 9.28) (significant at the 5% significance level after 1910) suggesting that the NAO stabilises in the positive phase. Shindell *et al.* (1999) found a trend towards more positive values of the AO index with increased greenhouse gases in a model which included a representation of the stratosphere, but not in troposphere-only versions of the same model. They attribute this to the high correlation of the stratospheric circulation with SLP in the Arctic (e.g., Kitoh *et al.*, 1996; Kodera *et al.*, 1996).

In an ensemble of 1900 to 2100 transient integrations with greenhouse gas and aerosol forcing changes, Fyfe *et al.* (1999) find a positive trend in the mean AO and AAO indices. They argue that in their model this is as a result of essentially

Figure 9.28: Decadal probability density functions (PDF) of the ECHAM3/LSG transient greenhouse gas ensemble: each PDF (greyscale bars) consists of 160 NAO index realisations including the monthly means of November to February of the four simulations over one decade. The greyscale background indicates the relative frequency of the classified NAO indices based on a kernel function (Matyasovszky, 1998). The PDF's width indicates the decadal variability. At the bottom, the time-series of the decadal mean (solid green line) and the variability (solid red line) of each PDF as well as the corresponding linear trends (dashed lines) are shown (from Paeth *et al.*, 1999).

unchanged AO/AAO patterns superimposed onto a forced climate change. The result of Fyfe *et al.* (1999) suggests that since the mean AO/AAO increases, it might imply a change to higher-frequency variability, as the positive AO phase has enhanced westerlies and is typically correlated with above-average storminess. In a subsequent non-linear analysis by Monahan *et al.* (2000) of a 1,000-year control and 500-year stabilisation integration (with greenhouse gas and aerosol forcing fixed at their year 2100 levels) it is found that (1) in the control integration the AO is part of a more general non-linear mode of tropospheric variability which is strongly bimodal and partitions the variability into two distinct regimes, and (2) in the stabilisa-tion integration the occupancy statistics of these regimes change rather than the modes themselves.

In summary, there is not yet a consistent picture emerging from coupled models as to their ability to reproduce trends in climate regimes such as the recently observed upward trend in the NAO/AO index (Chapters 2 and 12). In addition, whilst several models show an increase in the NAO/AO index with increased greenhouse gases, this is not true for all models, and the magnitude and character of the changes vary across models. Such results do not necessarily suggest that the forced climate change is manifest as a change in the occurrence of only one phase of these modes of variability.

9.3.5.4 Summary

There are now a greater number of global coupled atmosphere-ocean models and a number of them have been run for multi-century time-scales. This has substantially improved the basis for estimating long time-scale natural unforced variability. There are still severe limitations in the ability of such models to represent the full complexity of observed variability and the conclusions drawn here about changes in variability must be viewed in the light of these shortcomings (Chapter 8).

Some new studies have reinforced results reported in the SAR. These are:

- The future mean Pacific climate base state could more resemble an El Niño-like state (i.e., a slackened west to east SST gradient with associated eastward shifts of precipitation). Whilst this is shown in several studies, it is not true of all.

- Enhanced interannual variability of daily precipitation in the Asian summer monsoon. The changes in monsoon strength depend on the details of the forcing scenario and model.

Some new results have challenged the conclusions drawn in earlier reports, such as:

- Little change or a decrease in ENSO variability. More recently, increases in ENSO variability have been found in some models where it has been attributed to increases in the strength of the thermocline. Decadal and longer time-scale variability complicates assessment of future changes in individual ENSO event amplitude and frequency. Assessment of such possible changes remains quite difficult. The changes in both the mean and variability of ENSO are still model dependent.

Finally there are areas where there is no clear indication of possible changes or no consensus on model predictions:

- Although many models show an El Niño-like change in the mean state of tropical Pacific SSTs, the cause is uncertain. In some models it has been related to changes in cloud forcing and/or changes in the evaporative damping of the east-west SST gradient, but the result remains model-dependent. For such an El Niño-like climate change, future seasonal precipitation extremes associated with a given ENSO would be more intense due to the warmer mean base state.

- There is still a lack of consistency in the analysis techniques used for studying circulation statistics (such as the AO, NAO and AAO) and it is likely that this is part of the reason for the lack of consensus from the models in predictions of changes in such events.

- The possibility that climate change may be expressed as a change in the frequency or structure of naturally occuring modes of low-frequency variability has been raised. If true, this implies that GCMs must be able to simulate such regime transitions to accurately predict the response of the system to climate forcing. This capability has not yet been widely tested in climate models. A few studies have shown increasingly positive trends in the indices of the NAO/AO or the AAO in simulations with increased greenhouse gases, although this is not true in all models, and the magnitude and character of the changes varies across models.

9.3.6 Changes of Extreme Events

In this section, possible future changes in extreme weather and climate phenomena or events (discussed in Chapter 2) will be assessed from global models. Regional information derived from global models concerning extremes will be discussed in Chapter 10.

Although the global models have improved over time (Chapter 8), they still have limitations that affect the simulation of extreme events in terms of spatial resolution, simulation errors, and parametrizations that must represent processes that cannot yet be included explicitly in the models, particularly dealing with clouds and precipitation (Meehl *et al.*, 2000d). Yet we have confidence in many of the qualitative aspects of the model simulations since they are able to reproduce reasonably well many of the features of the observed climate system not only in terms of means but also of variability associated with extremes (Chapter 8). Simulations of 20th century climate have shown that including known climate forcings (e.g., greenhouse gases, aerosols, solar) leads to improved simulations of the climate conditions we have already observed. Ensembles of climate change experiments are now being performed to enable us to better quantify changes of extremes.

9.3.6.1 Temperature

Models described in the IPCC First Assessment Report (Mitchell *et al.*, 1990) showed that a warmer mean temperature increases the probability of extreme warm days and decreases the probability of extreme cold days. This result has appeared consistently in a number of more recent different climate model configurations (Dai *et al.*, 2001; Yonetani and Gordon, 2001). There is also a decrease in diurnal temperature range (DTR) since the night-time temperature minima warm faster than the daytime maxima in many locations (e.g., Dai *et al.*, 2001). Although there is some regional variation as noted in Chapter 10, some of these changes in DTR have also been seen over a number of areas of the world in observations (see Chapter 2). In general, the pattern of change in return values for 20-year extreme temperature events from an equilibrium simulation for doubled CO_2 with a global atmospheric model coupled to a non-dynamic slab ocean shows moderate increases over oceans and larger increases over land masses (Zwiers and Kharin, 1998; Figure 9.29). This result from a slab ocean configuration without ocean currents is illustrative and could vary from model to model, though it is similar to results from the fully coupled version in a subsequent study (Kharin and Zwiers, 2000).

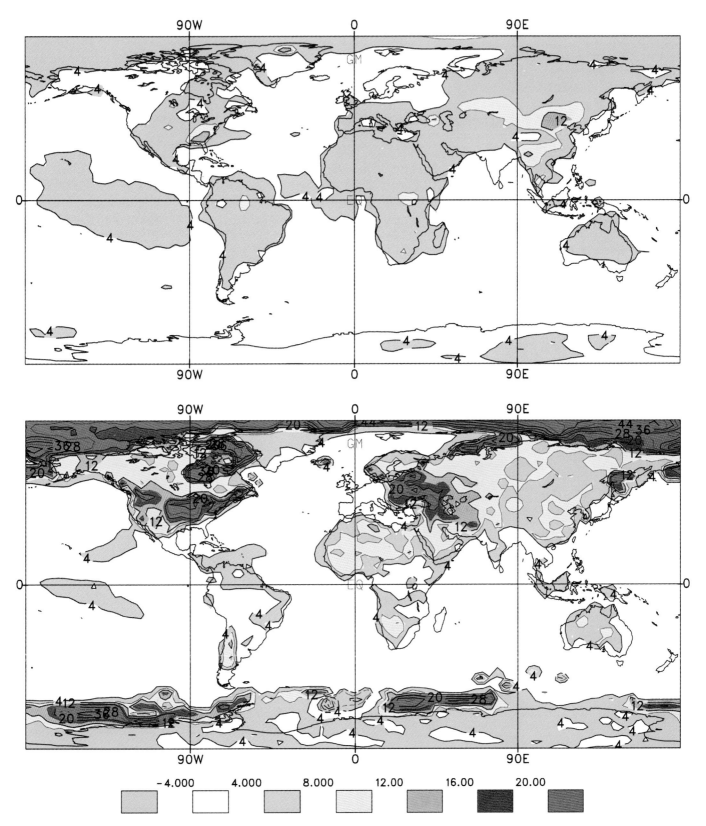

Figure 9.29: The change in 20-year return values for daily maximum (upper panel) and minimum (lower panel) surface air temperature (or screen temperature) simulated in a global coupled atmosphere-ocean model (CGCM1) in 2080 to 2100 relative to the reference period 1975 to 1995 (from Kharin and Zwiers, 2000). Contour interval is 4°C. Zero line is omitted.

The greatest increase in the 20-year return values of daily maximum temperature (Figure 9.29, top) is found in central and southeast North America, central and south-east Asia and tropical Africa, where there is a decrease in soil moisture content. Large extreme temperature increases are also seen over the dry surface of North Africa. In contrast, the west coast of North America is affected by increased precipitation resulting in moister soil and more moderate increases in extreme temperature. There are small areas of decrease in the Labrador Sea and Southern Ocean that are associated with changes in ocean temperature. The changes in the return values of daily minimum temperature (Figure 9.29, bottom) are larger than those of daily maximum temperature over land areas and high latitude oceans where snow and ice retreat. Somewhat larger changes are found over land masses and the Arctic while smaller increases in extreme minimum temperatures occur at the margins of the polar oceans. Thus, there is some asymmetry between the change in the extremes of minimum and maximum temperature (with a bigger increase for minima than maxima). This has to do with the change in the nature of the contact between atmosphere and the surface (e.g., minima increase sharply where ice and snow cover have retreated exposing either ocean or land, maxima increase more where the land surface has dried). Consequently there is a seasonal dependence related to changes in underlying surface conditions, which indroduces uncertainties in some regions in some models (Chapter 10).

Simulations suggest that both the mean and standard deviation of temperature are likely to change with a changed climate, and the relative contribution of the mean and standard deviation changes depends on how much each moment changes. Increased temperature variance adds to the probability of extreme high temperature events over and above what could be expected simply from increases in the mean alone. The increased variance of daily temperature in summer in northern mid-continental areas noted above has also been seen in other global models (Gregory and Mitchell, 1995). However, as noted in Chapter 10, such changes can vary from region to region and model to model (e.g., Buishand and Beersma (1996), who showed some small decreases over an area of Europe). The change in the mean is usually larger than the change in variance for most climate change simulations. Climate models have also projected decreased variability of daily temperature in winter over mid-continental Europe (Gregory and Mitchell, 1995). Such a decrease is partly related to a reduction of cold extremes, which are primarily associated with the increased mean of the daily minimum temperature. The detrimental effect of extreme summer heat is likely to be further exacerbated by increased atmospheric moisture. One model scenario shows an increase of about 5°C in July mean "heat index" (a measure which includes both the effects of temperature and moisture, leading to changes in the heat index which are larger than changes in temperature alone; it measures effects on human comfort; see further discussion in Chapter 10) over the southeastern USA by the year 2050 (Delworth *et al.*, 2000). Changes in the heating and cooling degree days are another likely extreme temperature-related effect of future greenhouse warming. For example, analysis of these measures

shows a decrease in heating degree days for Canada and an increase in cooling degree days in the southwest USA in model simulations of future climate with increased greenhouse gases (Zwiers and Kharin, 1998; Kharin and Zwiers, 2000), though this can be considered a general feature associated with an increase in temperature.

9.3.6.2 *Precipitation and convection*

Increased intensity of precipitation events in a future climate with increased greenhouse gases was one of the earliest model results regarding precipitation extremes, and remains a consistent result in a number of regions with improved, more detailed models (Hennessy *et al.*, 1997; Kothavala, 1997; Durman *et al.*, 2001; Yonetani and Gordon, 2001). There have been questions regarding the relatively coarse spatial scale resolution in climate models being able to represent essentially mesoscale and smaller precipitation processes. However, the increase in the ability of the atmosphere to hold more moisture, as well as associated increased radiative cooling of the upper troposphere that contributes to destabilisation of the atmosphere in some models, is physically consistent with increases in precipitation and, potentially, with increases in precipitation rate.

As with other changes, it is recognised that changes in precipitation intensity have a geographical dependence. For example, Bhaskharan and Mitchell (1998) note that the range of precipitation intensity over the south Asian monsoon region broadens in a future climate experiment with increased greenhouse gases, with decreases prevalent in the west and increases more widespread in the east (see further discussion in Chapter 10). Another model experiment (Brinkop, 2001) shows that extreme values of the convective rain rate and the maximum convective height occur more frequently during the 2071 to 2080 period than during the 1981 to 1990 period. The frequency of highest-reaching convective events increases, and the same holds for events with low cloud-top heights. In contrast, the frequency of events with moderate-top heights decreases. On days when it rains, the frequency of the daily rates of convective rainfall larger than 40 mm/day in JJA and greater than 50 mm/day for DJF, increases. Generally, one finds a strong increase in the rain rate per convective event over most of the land areas on the summer hemispheres and in the inter-tropical convergence zone (ITCZ). Between 10 and 30°S there are decreases in rain rate per event over the ocean and parts of the continents.

In global simulations for future climate, the percentage increase in extreme (high) rainfall is greater than the percentage increase in mean rainfall (Kharin and Zwiers, 2000). The return period of extreme precipitation events is shortened almost everywhere (Zwiers and Kharin, 1998). For example, they show that over North America the 20-year return periods are reduced by a factor of 2 indicating that extreme precipitation of that order occurs twice as often.

Another long-standing model result related to drought (a reduction in soil moisture and general drying of the mid-continental areas during summer with increasing CO_2) has been reproduced with the latest generation of global coupled climate models (Gregory *et al.*, 1997; Haywood *et al.*, 1997; Kothavala, 1999; Wetherald and Manabe, 1999). This summer drying is generally ascribed to a combination of increased temperature and potential evaporation not being balanced by precipitation. To address this problem more quantitatively, a global climate model with increased CO_2 was analysed to show large increases in frequency of low summer precipitation, the probability of dry soil, and the occurrence of long dry spells (Gregory *et al.*, 1997). The latter was ascribed to the reduction of rainfall events in the model rather than to decreases in mean precipitation. However, the magnitude of this summer drying response may be related to the model's simulation of net solar radiation at the surface, and more accurate simulation of surface fluxes over land will increase confidence in the GCM climate changes.

Alhough of great importance to society for their potential for causing destruction, as well as their human and economic impacts, there is little guidance from AOGCMs concerning the future behaviour of tornadoes, hail or lightning. This is because these phenomena are not explicitly resolved in AOGCMs, and any studies that have been done have had to rely on empirical relationships between model features and the phenomenon of interest. For example, Price and Rind (1994a) derive a relationship between lightning activity and convective cloud-top height to infer an increase of lightning with increasing CO_2. They take that relationship one step further to suggest a future increase in lightning-caused fires due to the increased lightning activity and decreased effective precipitation (Price and Rind, 1994b). Using another empirical relationship between daily minimum temperature and severe convective storm frequency for France, Dessens (1995) connects an increase in daily minimum temperature with greater convective storm frequency and more hail damage in a future climate with increased CO_2. However, there have been no recent studies examining this problem with the current generation of global climate models. Due to the fact that these severe weather phenomena are sub-grid scale (even more so than discussed below for tropical cyclones), and that second and third order linkages between model output and empirical relationships for limited regions must be used to derive results, we cannot reach any definitive conclusions concerning possible future increases in hail and lightning, and there is no information from AOGCMs concerning future changes in tornado activity.

9.3.6.3 Extra-tropical storms

Storms not only have obvious effects on extremes of temperature and precipitation, but also have severe impacts associated with wind, ocean waves, etc. Due to model limitations in previous generations of global climate models, until recently there have been few studies examining changes in extra-tropical cyclones in a future climate. With the improved recent generation of global climate models (see Chapter 8), such studies are now becoming more credible. An analysis of an ensemble of four future climate change experiments using a global coupled model with increased CO_2 and sulphate aerosols showed an increase in the number of deep low pressure systems in Northern Hemisphere winter, while the number of weaker storms was reduced (Carnell and Senior, 1998). Studies using different models show a similar change for both hemispheres (Sinclair and Watterson, 1999) or for a study region limited to the North Atlantic (Knippertz *et al.*, 2000).

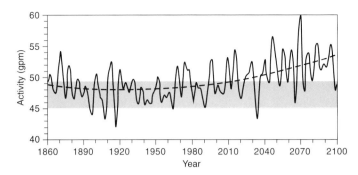

Figure 9.30: Storm track activity averaged over north-west Europe (6°W to 20°E, 40° to 70°N) in the ECHAM4/OPYC greenhouse gas scenario run (Unit: gpm). A 4-year running mean is shown for smoother display. The grey band indicates the variability of this index in the control run as measured by one standard deviation. The non-linear climate trend optimally obtained from quadratic curve fitting is marked by the dashed line; y-axis is activity in gpm (geopotential metres) and x-axis is time in calendar years. From Ulbrich and Christoph (1999).

The reasons given for this common result are still under discussion. Carnell and Senior (1998) ascribe it to a decrease in the mean meridional temperature gradient in the future climate, with high latitudes warming more than low latitudes (producing fewer storms), and greater latent heating in the moister atmosphere (resulting in deeper lows). Sinclair and Watterson (1999) point to the reduced mean sea level pressure and emphasise that vorticity as a measure of cyclone strength does not increase. Knippertz *et al.*(2000) consider the increasing upper tropospheric baroclinicity to be an important indicator of the change in surface cyclone activity. They also detect an increasing number of strong wind events in their simulation that can be assigned to the increasing number of deep lows. Upper air storm track activity (defined as the standard deviation of the band pass filtered 500 hPa height and related to the surface lows) has been found to increase over the East Atlantic and Western Europe with rising greenhouse gas forcing (such as seen in Figure 9.30 from Ulbrich and Christoph, 1999).

They related this increase to a change in the NAO (see discussion of possible NAO changes in Section 9.3.5.3). Several studies have tried to look at mechanisms of changes (e.g., Lunkeit *et al.,* 1998). For example, Christoph *et al.* (1997) identify a mid-winter suppression of the North Pacific storm track in present day climate which they attribute to very strong upper level winds at that time of year. In a 3×CO_2 climate model experiment, they note that very intense upper level winds occur more often, thus producing a more pronounced mid-winter suppression of the Pacific storm track.

Longer time-series from models have made the statistics more robust (e.g., Carnell and Senior, 1998). High-resolution models may improve the representation of storms, but the present experiments are mainly too short to provide indications of significant changes (e.g., Beersma *et al.*, 1997). As can be seen, there are now a growing number of studies addressing possible changes in storm activity, but in spite of an emerging common signal there remains uncertainty with respect to the governing mechanisms.

9.3.6.4 Tropical cyclones

Here we assess only AOGCM-related results pertaining to tropical cyclones. For further discussion of results from embedded and mesoscale models regarding possible future changes in tropical cyclone activity, see Chapter 10 (also refer to Box 10.2 for a summary). The ability of global models to accurately represent tropical cyclone phenomena, and their present limitations in this regard, is important for understanding their projection of possible future changes. These capabilities are discussed in detail in Chapter 8.

Some of the global climate models suggest an increase in tropical storm intensities with CO_2-induced warming (Krishnamurti *et al.*, 1998), though a limitation of that study is the short two year model run. However, the highest resolution global climate model experiment reported to date (Bengtsson *et al.*, 1996; see Chapter 10) still has a resolution too coarse (about 1°) to simulate the most intense storms or realistically simulate structures such as the hurricane eye.

Indices of tropical cyclone activity (Gray, 1979) summarise the necessary large-scale conditions for tropical cyclone activity from coarse resolution GCMs (Evans and Kempisty, 1998; Royer *et al.*, 1998). The latter study examined large-scale atmospheric and oceanic conditions (vertical shear, vorticity and thermo-dynamic stability), and suggested that only small changes in the tropical cyclone frequencies would occur (up to a 10% increase in the Northern Hemisphere primarily in the north-west Pacific, and up to a 5% decrease in the Southern Hemisphere). Climate change studies to date show a great sensitivity to the measure of convective activity chosen, and depend less on the model produced fields. Additionally, the broad geographic regions of cyclogenesis, and therefore also the regions affected by tropical cyclones, are not expected to change significantly (Henderson-Sellers *et al.,* 1998). This is because results from Holland's (1997) Maximum Potential Intensity model show that even with substantial (1 to 2°C) SST increases in the tropics from global warming, one would also get a correspondingly much bigger warming in the upper troposphere leading to very little change in the moist static stability (Holland, 1997). Another study shows areas of deep convection that can be associated with tropical cyclone formation would not expand with increases in CO_2 due to an increase of the SST threshold for occurrence of deep convection (Dutton *et al.*, 2000). Additionally, since tropical storm activity in most basins is modulated by El Niño/La Niña conditions in the tropical Pacific, projections of future regional changes in tropical storm frequencies may depend on accurate projections of future El Niño conditions, an area of considerable uncertainty for climate models (as noted in Section 9.3.5.2).

9.3.6.5 Commentary on changes in extremes of weather and climate

Although changes in weather and climate extremes are important to society, ecosystems, and wildlife, it is only recently that evidence for changes we have observed to date has been able to be compared to similar changes that we see in model simulations for future climate (generally taken to be the end of the 21st century as shown in this chapter). Though several simulations of 20th century climate with various estimates of observed forcings

now exist (see Chapter 8), few of these have been analysed for changes in extremes over the 20th century. So far, virtually all studies of simulated changes in extremes have been performed for future climate. A number of studies are now under way for simulated 20th century climate, but are not yet available for assessment. Additionally, in the 20th century climate integrations there is usually a significant signal/noise problem (especially for changes in phenomena like storms). Therefore, here we assess changes in extremes that have been observed during the 20th century (see Chapter 2), and compare these to simulated changes of extremes for the end of the 21st century from AOGCMs run with increases in greenhouse gases and other constituents. Agreement between the observations and model results would suggest that the changes in extremes we have already observed are qualitatively consistent in a very general way with those changes in climate model simulations of future climate, indicating these changes in extremes would be likely to continue into the future.

The assessment of extremes here relies on very large-scale changes that are physically plausible or representative of changes over many areas. There are some regions where the changes of certain extremes may not agree with the larger-scale changes (see Chapters 2 and 10). Therefore, the assessment here is a general one where observed and model changes appear to be representative and physically consistent with a majority of changes globally. Additionally, certain changes in observed extremes may not have been specifically itemised from model simulations, but are physically consistent with changes of related extremes in the future climate experiments and are denoted as such. Also note that the information for tropical cyclones is drawn from Chapter 10, and diurnal temperature range from Chapter 12. A further discussion of the synthesis of observed and modelled changes of extremes, along with results on how extremes can affect human society, ecosystems and wildlife, appears in Easterling *et al.* (2000).

The qualitative consistency between the observations from the latter half of the 20th century and the models for the end of the 21st century in Table 9.6 suggests that at least some of the changes we have observed to date are likely to be associated with changes in forcing we have already experienced over the 20th century. The implication is that these could continue to increase into the 21st century with the ongoing rise in forcing from ever greater amounts of greenhouse gases in the atmosphere.

Table 9.6 depicts an assessment of confidence in observed changes in extremes of weather and climate during the latter half of the 20th century (left column) and in projected changes during the 21st century (right column). As noted above, this assessment relies on observational and modelling studies, as well as the physical plausibility of future projections across all commonly used scenarios and is based on expert judgement. For more details, see Chapter 2 (observations) and Chapter 10 (regional projections).

For the projected changes in the right-hand column, "very likely" indicates that a number of models have been analysed for such a change, all those analysed show it in most regions, and it is physically plausible. No models have been analysed to show fewer frost days, but it is physically plausible, since most models

Table 9.6: *Estimates of confidence in observed and projected changes in extreme weather and climate events.*

Confidence in observed changes (latter half of the 20th century)	Changes in Phenomenon	Confidence in projected changes (during the 21st century)
Likely	**Higher maximum temperatures and more hot days[a] over nearly all land areas**	Very likely
Very likely	**Higher minimum temperatures, fewer cold days and frost days over nearly all land areas**	Very likely
Very likely	**Reduced diurnal temperature range over most land areas**	Very likely
Likely, over many areas	**Increase of heat index[b] over land areas**	Very likely, over most areas
Likely, over many Northern Hemisphere mid- to high latitude land areas	**More intense precipitation events[c]**	Very likely, over many areas
Likely, in a few areas	**Increased summer continental drying and associated risk of drought**	Likely, over most mid-latitude continental interiors. (Lack of consistent projections in other areas)
Not observed in the few analyses available	**Increase in tropical cyclone peak wind intensities[d]**	Likely, over some areas
Insufficient data for assessment	**Increase in tropical cyclone mean and peak precipitation intensities[d]**	Likely, over some areas

[a] Hot days refers to a day whose maximum temperature reaches or exceeds some temperature that is considered a critical threshold for impacts on human and natural systems. Actual thresholds vary regionally, but typical values include 32°C, 35°C or 40°C.

[b] Heat index refers to a combination of temperature and humidity that measures effects on human comfort.

[c] For other areas, there are either insufficient data or conflicting analyses.

[d] Past and future changes in tropical cyclone location and frequency are uncertain.

show an increase in night-time minimum temperatures, which would result in fewer frost days. The category "likely" indicates that theoretical studies and those models analysed show such a change, but only a few current climate models are configured in such a way as to reasonably represent such changes. "Hot days" refers to a day whose maximum temperature reaches or exceeds some temperature that is considered a critical threshold for impacts on human and natural systems. Actual thresholds vary regionally, but typical values include 32°C, 35°C or 40°C.

For some other extreme phenomena, many of which may have important impacts on the environment and society, there is currently insufficient information to assess recent trends, and climate models currently lack the spatial detail required to make confident projections. For example, very small-scale phenomena, such as thunderstorms, tornadoes, hail and lightning, are not simulated in climate models at present.

9.3.6.6 Conclusions

Much of what climate model studies show could happen to weather and climate extremes in a future climate with increased greenhouse gases is what we would intuitively expect from our understanding of how the climate system works. For example, a warming of the surface supplies more water vapour to the atmosphere, which is a greater source of moisture in storms and thus we would expect an increase in intense precipitation and more rainfall from a given rainfall event, both results seen in climate model simulations. There are competing effects of

decreased baroclinicity in some regions due to greater surface warming at high latitudes, and increasing mid-tropospheric baroclinicity due to greater mid-tropospheric low latitude warming (Kushner *et al.*, 2001). Additionally, a number of changes in weather and climate extremes from climate models have been seen in observations in various parts of the world (decreased diurnal temperature range, warmer mean temperatures associated with increased extreme warm days and decreased extreme cold days, increased rainfall intensity, etc.). Though the climate models can simulate many aspects of climate variability and extremes, they are still characterised by systematic simulation errors and limitations in accurately simulating regional climate such that appropriate caveats must accompany any discussion of future changes in weather and climate extremes.

Recent studies have reproduced previous results in the SAR and this gives us increased confidence in their credibility (although agreement between models does not guarantee that those changes will occur in the real climate system):

- An increase in mean temperatures leads to more frequent extreme high temperatures and less frequent extreme low temperatures.

- Night-time low temperatures in many regions increase more than daytime highs, thus reducing the diurnal temperature range.

• Decreased daily variability of temperature in winter and increased variability in summer in Northern Hemisphere mid-latitude areas.

• There is a general drying of the mid-continental areas during summer in terms of decreases in soil moisture, and this is ascribed to a combination of increased temperature and potential evaporation not being balanced by precipitation.

• Intensity of precipitation events increases.

Additional results since 1995 include:

• Changes in temperature extremes noted above have been related to an increase in a heat index (leading to increased discomfort and stress on the human body), an increase in cooling degree days and a decrease in heating degree days.

• Additional statistics relating to extremes are now being produced. For example, in one model the greatest increase in the 20-year return values of daily maximum temperature is found in central and Southeast North America, central and Southeast Asia and tropical Africa where there is a decrease in soil moisture content, and also over the dry surface of North Africa. The west coast of North America is affected by increased precipitation, resulting in moister soil and more moderate increases in extreme temperature. The increases in the return values of daily minimum temperature are larger than those of daily maximum temperature mainly over land areas and where snow and sea ice retreat.

• Precipitation extremes increase more than the mean and that means a decrease in return period for the extreme precipitation events almost everywhere (e.g., 20 to 10 years over North America).

Aspects which have been addressed but remain unresolved at this time include:

• There is no general agreement yet among models concerning future changes in mid-latitude storms (intensity, frequency and variability), though there are now a number of studies that have looked at such possible changes and some show fewer weak but greater numbers of deeper mid-latitude lows, meaning a reduced total number of cyclones.

• Due to the limitations of spatial resolution in current AOGCMs, climate models do not provide any direct information at present regarding lightning, hail, and tornadoes. Results derived from earlier models used empirical relationships to infer a possible future increase in lightning and hail, though there have been no recent studies to corroborate those results.

• There is some evidence that shows only small changes in the frequency of tropical cyclones derived from large-scale parameters related to tropical cyclone genesis, though some

measures of intensities show increases, and some theoretical and modelling studies suggest that upper limit intensities could increase.

9.4 General Summary

Figure 9.31 summarises some of the model results for projections of future climate change for the end of the 21st century. This figure can be compared to one for observations from the 20th century in Chapter 2 (Figure 2.37). A number of the observed changes are qualitatively consistent with those projected for future climate changes from climate models. A confidence scale is provided for the model projections in Figure 9.31, and is the same as the one used in the Executive Summary. Since there is considerable agreement between the observations in Figure 2.37 and the model results listed in Figure 9.31, we conclude that many of the larger observed climate changes to date are qualitatively consistent with those changes in climate models for future climate with increases of greenhouse gases.

(a) Temperature indicators

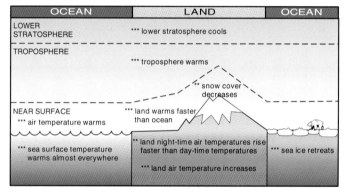

(b) Hydrological and storm-related indicators

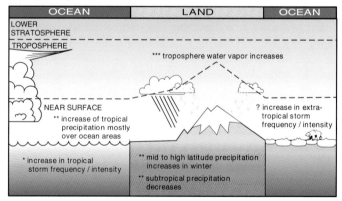

*** virtually certain
      (many models analysed and all show it)
** very likely
      (a number of models analysed show it, or change is physically
      plausible and could readily be shown for other models)
* likely
      (some models analysed show it, or change is physically plausible
      and could be shown for other models)
? medium likelihood
      (a few models show it, or results mixed)

Figure 9.31: Schematic of changes in the temperature and hydrological indicators from projections of future climate changes with AOGCMs. This figure can be compared with Figure 2.37 to note climate changes already observed, to provide a measure of qualitative consistency with what is projected from climate models.

Appendix 9.1: Tuning of a Simple Climate Model to AOGCM Results

The simple climate model MAGICC (Wigley and Raper; 1987, 1992; updated in Raper *et al.*, 1996) was used in the SAR to make temperature projections for various forcing scenarios and for sensitivity analyses. The justification for using the simple model for this purpose was the model's ability to simulate AOGCM results in controlled comparisons spanning a wide range of forcing cases (for example SAR Figure 6.13). The approach used in this report differs from that in the SAR. Thus the upwelling diffusion-energy balance model (UD/EB) model is not used here as a stand-alone model in its own right but instead it is tuned to individual AOGCMs and is used only as a tool to emulate and extend their results. In this way, a range of results is produced reflecting the range of AOGCM results. The tuning is based on the CMIP2 data analysis of Raper *et al.* (2001b). The validity of the tuning is tested by comparisons with AOGCM results in the DDC data set and, where available, with recent AOGCM results using the SRES scenarios. By using such simple models, differences between different scenarios can easily be seen without the obscuring effects of natural variability, or the similar variability that occurs in coupled AOGCMs (Harvey *et al.*, 1997). Simple models also allow the effect of uncertainties in the climate sensitivity and the ocean heat uptake to be quantified. Potentially, other simple models (for example, Watterson (2000), Visser *et al.* (2000)) could be used in a similar way.

The first step in the tuning process is to select appropriate values for the radiative forcing for a CO_2 doubling parameter, F_{2x}, and the climate sensitivity parameter, T_{2x}. In the SAR, $F_{2x}= 4.37$ Wm^{-2} was used, as given in the 1990 IPCC Assessment (Shine *et al.*, 1990). This value, which did not account for stratospheric adjustment and solar absorption by CO_2, is now considered to be too high (Myhre *et al.*, 1998). These authors suggest a best estimate of 3.71 Wm^{-2}; model-specific values are used here (see Table 9.A1). The effect on global mean temperature and sea level change of using lower values of F_{2x} has been investigated by Wigley and Smith (1998). The lower F_{2x} values result in slightly lower temperature projections. Different definitions and methods of calculation of model climate sensitivity are discussed in Section 9.3.4.1. Here the effective climate sensitivities based on the last twenty years of the CMIP2 data are used.

Having selected the value of F_{2x} and T_{2x} appropriate to a specific AOGCM, the simple model tuning process consists of matching the AOGCM net heat flux across the ocean surface by adjusting the simple model ocean parameters following Raper *et al.* (2001a), using the CMIP2 results analysed in Raper *et al.* (2001b). Sokolov and Stone (1998) show that when using a pure diffusion model to match the behaviour of different AOGCMs a wide range of diffusion coefficients is needed. The range here is much smaller because a 1-D upwelling diffusion model is used and changes in the strength of the thermohaline circulation are also accounted for. A decrease in the strength of the thermohaline circulation leads to an increased heat flux into the ocean. In the UD/EB model a weakening of the thermohaline circulation is represented by a decline in the upwelling rate (see SAR). The rate of sea level rise from thermal expansion for a collapse in the

thermohaline circulation in the UD/EB model is tuned to match that which occurs for an induced collapse in the GFDL model (GFDL_R15_a) control run. An instantaneous 30% decline in the UD/EB model upwelling rate gives rates of sea level rise comparable to that seen in the GFDL model over a period of 500 years. Thus a 30% decline in the UD/EB model upwelling rate represents a collapse in the thermohaline circulation. For the individual models the rate of decline in the strength of the thermohaline circulation relative to the global mean temperature change is based on the CIMP2 data and is specified by the parameter ΔT^+. It should be pointed out that the processes in the UD/EB model that determine the heat flux into the ocean are not necessarily physically realistic. Raper and Cubasch (1996) as well as Raper *et al.* (2001a) show that the net heat flux into the ocean in the UD/EB model can be tuned to match that in an AOGCM in several ways, using different sets of parameter values. Nevertheless, if the UD/EB model is carefully tuned to match the results of an AOGCM, and provided the extrapolations are not too far removed from the results used for tuning, the UD/EB model can be used to give reasonably reliable estimates of AOGCM temperature changes for different forcing scenarios. The thermal expansion results are less reliably reproduced because thermal expansion is related to the integrated heat flux into the ocean. Errors therefore tend to accumulate. In addition, the expansion depends on the distribution of warming in the ocean. Nonetheless, the simulation is adequate for comparison of scenarios.

Other parameters in the UD/EB model are adjusted in order to correctly simulate the greater surface temperature change over the land relative to the ocean as shown to a varying degree in different AOGCM results. The land-ocean, Northern-Southern Hemisphere temperature change contrasts are adjusted by parameters that govern the contrast in the land-ocean climate sensitivity and the land-ocean exchange coefficients. The specific parameter values used for the different AOGCMs are given in Table 9.A1.

Table 9.A1: Simple climate model parameter values used to simulate AOGCM results. In all cases the mixed-layer depth hm=60m, the sea ice parameter CICE=1.25 and the proportion of the upwelling that is scaled for a collapse of the thermohaline circulation is 0.3, otherwise parameters are as used in the SAR (Kattenberg et al., 1996; Raper et al., 1996).

AOGCM	F_{2x} (Wm^{-2})	T_{2x} (°C)	ΔT^+ (°C)	k (cm^2s^{-1})	RLO	LO and NS $(Wm^{-2}$ °$C^{-1})$
GFDL_R15_a	3.71[*]	4.2	8	2.3	1.2	1.0
CSIRO Mk2	3.45	3.7	5	1.6	1.2	1.0
HadCM3	3.74	3.0	25	1.9	1.4	0.5
HadCM2	3.47	2.5	12	1.7	1.4	0.5
ECHAM4/OPYC	3.8	2.6	20	9.0	1.4	0.5
CSM 1.0	3.60	1.9	-	2.3	1.4	0.5
DOE PCM	3.60	1.7	14	2.3	1.4	0.5

[*] Here the best estimate from Myhre *et al.* (1998) is used.
F_{2x} – the radiative forcing for double CO_2 concentration
T_{2x} – climate sensitivity
hm – mixed-layer depth
CICE – sea ice parameter (see Raper*et al.,* 2001a)
ΔT^+ – magnitude of warming that would result in a collapse of the THC
k – vertical diffusivity
RLO – ratio of the equilibrium temperature changes over land versus ocean
LO and NS – land/ocean and Northern Hemisphere/Southern Hemisphere exchange coefficients

References

Barthelet, P., L. Terray and S. Valcke, 1998a: Transient CO_2 experimets using the ARPEGE/OPAICE non-flux corrected coupled model. *Geophys. Res. Lett.*, **25**, 2277-2280.

Barthelet, P., S. Bony, P. Braconnot, A. Braun, D. Cariolle, E. Cohen-Solal, J.-L. Dufresne, P. Delecluse, M. Déqué, L.Fairhead, M.-A., Filiberti, M. Forichon, J.-Y. Grandpeix, E. Guilyardi, M.-N. Houssais, M. Imbard, H. LeTreut, C. Lévy, Z. X. Li, G. Madec, P. Marquet, O. Marti, S. Planton, L. Terray, O. Thual and S. Valcke, 1998b: Simulations couplées globales de changements climatiques associés à une augmentation de la teneur atmosphérique en CO_2. *C. R. Acad. Sci. Paris, Sciences de la terre et des planètes*, **326**, 677-684 (in French with English summary).

Bates, G.T. and G.A. Meehl, 1986: Effect of CO_2 concentration on the frequency of blocking in a general circulation model coupled to a simple mixed layer ocean model. *Mon. Wea. Rev.*, **114**, 687-701.

Beersma, J.J., K.M. Rider, G.J. Komen, E. Kaas and V.V. Kharin, 1997: An analysis of extratropical storms in the North Atlantic region as simulated in a control and $2\times CO_2$ time-slice experiment with a high-resolution atmospheric model. *Tellus*, **49A**, 347-361.

Bengtsson, L., M. Botzet and M. Esch, 1996. Will greenhouse gas-induced warming over the next 50 years lead to a higher frequency and greater intensity of hurricanes? *Tellus*, **48A**, 175-196.

Bhaskharan B. and Mitchell J.F.B., 1998: Simulated changes in the intensity and variability of the southeast Asian monsoon in the twenty first century resulting from anthropogenic emissions scenarios. *Int. J. Climatol.*, **18**, 1455-1462.

Boer, G.J., K. Arpe, M. Blackburn, M. Deque, W.L. Gates, T.L. Hart, H. le Treut, H. E. Roeckner, D.A. Sheinin, I. Simmonds, R.N.B. Smith, T. Tokioka, R.T. Wetherald and D. Williamson, 1992: Some results from an intercomparison of climates simulated by 14 atmospheric general circulation models. *J. Geophys. Res.*, **97**, 12,771-12,786.

Boer, G.J., G. Flato, M. C. Reader and D. Ramsden, 2000a: A transient climate change simulation with greenhouse gas and aerosol forcing: experimental design and comparison with the instrumental record for the 20th century. *Clim. Dyn.* **16**, 405-425.

Boer, G.J., G. Flato, and D. Ramsden, 2000b: A transient climate change simulation with greenhouse gas and aerosol forcing: projected climate for the 21st century. *Clim. Dyn.* **16**, 427-450.

Boville, B.A., J.T. Kiehl, P.J. Rasch and F.O. Bryan, 2001: Improvements to the NCAR-CSM-1 for transient climate simulations. *J. Climate, 14, 164-179.*

Brankovic, C. and T. Palmer, 2000: Seasonal skill and predictability of ECMWF PROVOST ensemble, *Quart. J. R. Met. Soc., 126, 2035-2069.*

Brinkop, S., 2001: Change of convective activity and extreme events in a transient climate change simulation, DLR-Institut fuer Physik der Atmosphaere, Report No. 142, [Available from DLR-Oberpfaffenhofen, Institut fuer Physik der Atmosphaere, D-82234 Wessling, Germany].

Bryan, F. 1987: Parameter sensitivity of primitive equation ocean general circulation model. *J. Phys. Oceanogr.*, **17**. 970-985.

Buishand, T.A. and J.J. Beersma, 1996: Statistical tests for comparison of daily variability in observed and simulated climates. *J. Climate*, **9**, 2538-2550.

Cai, W. and P.H. Whetton, 2000: Evidence for a time-varying pattern of greenhouse warming in the Pacific Ocean. *Geophys. Res. Lett.*, **27**, 2577-2580.

Cane M.A., A.C. Clement, A. Kaplan, Y. Kushnir, D. Pozdnyakov, R. Seager, S.E. Zebiak and R. Murtugudde, 1997: Twentieth century sea surface temperature trends. *Science*, **275**, 957-960.

Carnell, R.E., C.A. Senior and J.F.B. Mitchell, 1996: An assessment of measures of storminess: simulated changes in Northern Hemisphere winter due to increasing CO_2. *Clim. Dyn.*, **12**, 467-476.

Carnell, R.E. and C.A. Senior, 1998. Changes in mid-latitude variability due to increasing greenhouse gases and sulphate aerosols. *Clim. Dyn.*, **14**, 369-383.

Carnell, R.E. and C.A. Senior, 2000: Mechanisms of changes in storm tracks with increased greenhouse gases. Hadley Centre Technical Note 18. Available from Met Office, London Road Bracknell, RG12 2SZ, UK.

Christoph, M., U. Ulbrich and P. Speth, 1997: Midwinter suppression of Northern Hemisphere storm track activity in the real atmosphere and in GCM experiments. *J. Atmos. Sci.*, **54**, 1589-1599.

Collins, M., 2000a: The El-Niño Southern Oscillation in the second Hadley Centre coupled model and its response to greenhouse warming. *J. Climate*, **13**, 1299-1312.

Collins, M., 2000b: Understanding uncertainties in the response of ENSO to greenhouse warming. *Geophys. Res. Lett.*, **27**, 3509-3512.

Colman, R.A. and B.J. McAvaney, 1995: Sensitivity of the climate response of an atmospheric general circulation model to changes in convective parameterisation and horizontal resolution. *J. Geophys. Res.* **100**, 3155-3172.

Colman, R.A., 2001: On the vertical extent of GCM feedbacks. *Clim. Dyn.*, in press.

Corti, S., F. Molteni, T.N. Palmer, 1999: Signature of recent climate change in frequencies of natural atmospheric circulation regimes. *Nature*, **398**, 799-802.

Cox, P.M., R.A. Betts, C.D. Jones, S.A. Spall, and I.J. Totterdell, 2000: Acceleration of global warming by carbon cycle feedbacks in a 3D coupled model. *Nature*, **408**, 184-187.

Cubasch, U., K. Hasselmann, H. Höck, E. Maier-Reimer, U. Mikolajewicz, B. D. Santer and R. Sausen, 1992: Time-dependent greenhouse warming - computations with a coupled ocean-atmosphere model. *Clim. Dyn.*, **8**, 55-69.

Cubasch, U., B. D. Santer, A. Hellbach, G. Hegerl, H. Höck, E. Maier-Reimer, U. Mikolajewicz, A. Stössel and R. Voss, 1994: Monte Carlo climate change forecasts with a global coupled ocean-atmosphere model. *Clim. Dyn.*, **10**, 1-19.

Cubasch, U., G. C. Hegerl and J. Waszkewitz, 1996: Prediction, detection and regional assessment of anthropogenic climate change. *Geophysica*, **32**, 77-96.

Cubasch, U., M. Allen, P. Barthelet, M. Beniston, C.Bertrand, S. Brinkop, J.-Y.Caneill, J.-L. Dufresne, L. Fairhead, M.-A. Filiberti, J. Gregory, G. Hegerl, G. Hoffmann, T. Johns, G. Jones, C. Laurent, R. McDonald, J. Mitchell, D. Parker, J. Oberhuber, C. Poncin, R. Sausen, U. Schlese, P. Stott, L. Terray, S. Tett, H. leTreut, U. Ulbrich, S. Valcke, R. Voss, M. Wild, J.-P. van Ypersele, 1999: Summary Report of the Project Simulation, Diagnosis and Detection of the Anthropogenic Climate Change (SIDDACLICH), EU-Commission, Brussels, EUR 19310, ISBN 92-828-8864-9.

Cubasch, U. and I. Fischer-Bruns, 2000: An intercomparison of scenario simulations performed with different AOGCMs, in: RegClim, General Technical Report **No. 4**, DNMI (Norwegian Meteorological Institute), eds. T. Iversen and B.A.K. Hoiskar.

Dai, A., T.M.L. Wigley, B. A. Boville, J.T. Kiehl, and L.E. Buja, 2001: Climates of the 20th and 21st centuries simulated by the NCAR climate system model. *J. Climate*, **14**, 485-519.

Delworth, T.L., J.D. Mahlman, and T.R. Knutson: 2000: Changes in heat index associated with CO_2-induced global warming. *Clim. Change*, **43**, 369-386.

Derome, J., G. Brunet, A. Plante, N. Gagnon, G.J. Boer, F. Zwiers, S.Lambert, J. Sheng and H. Ritchie, 2001: Seasonal prediction based on two dynamical models, *Atmos.-Ocean*, in press.

Dessens, J., 1995: Severe convective weather in the context of a night-time global warming. *Geophys. Res. Lett.*, **22**, 1241-1244.

Dixon, K.W. and J.R. Lanzante, 1999: Global mean surface air temperature and North Atlantic overturning in a suite of coupled GCM climate change experiments. *Geophys. Res. Lett.*, **26**, 1885-1888.

Dixon, K. W., T. L. Delworth, M. J. Spelman and R. J. Stouffer, 1999: The influence of transient surface fluxes on North Atlantic overturning in a coupled GCM climate change experiment, *Geophys. Res. Lett.*, **26**, 2749-2752.

Doblas-Reyes, J., M. Deque and J.-P. Piedelievre, 2000: Multi-model spread and probabilistic seasonal forecasts in PROVOST, *Quart. J. R.*

Met. Soc., **126**, 2069 - 2089.

Durman, C.F., J.M. Gregory, D.C. Hassell, R.G. Jones and J.M. Murphy, 2001: A comparison of extreme European daily precipitation simulated by a global and a regional model for present and future climates. Quart. J. R. Met. Soc., in press.

Dutton, J.F., C.J. Poulsen, and J.L. Evans, 2000: The effect of global climate change on the regions of tropical convection in CSM1. *Geophys. Res. Lett.*, **27**, 3049-3052.

Easterling, D.R., G A. Meehl, C. Parmesan, S.A. Changnon, T.R. Karl and L.O. Mearns, 2000: Climate extremes: observations, modelling and impacts. *Science*, **289**, 2068-2074.

Emori, S., T. Nozawa, A. Abe-Ouchi, A. Numaguti and M. Kimoto, 1999: Coupled ocean-atmosphere model experiments of future climate change with an explicit representation of sulphate aerosol scattering. *J. Met. Soc. Japan*, **77**, 1299-1307.

Enting, I.G., T.M.L. Wigley and M. Heimann, 1994: Future emissions and concentrations of carbon dioxide: key ocean/atmosphere/land analyses, CSIRO Division of Atmospheric Research Technical Paper No. 31.

Evans, J.L. and T. Kempisty, 1998: Tropical cyclone signatures in the climate. *AMS Symposium on Tropical Cyclone Intensity Change*, 11-16 January, 1998, Phoenix AZ.

Fanning, A.F. and A.J. Weaver, 1996: An atmospheric energy-moisture balance model: climatology, interpentadal climate change, and coupling to an ocean general circulation model. *J. Geophys. Res.*, **101**, 15,111-15,128.

Fanning, A.F. and A.J. Weaver, 1997: On the role of flux adjustments in an idealised coupled climate model. *Clim. Dyn.*, **13**, 691-701.

Flato, G.M. and G.J. Boer, 2001: Warming asymmetry in climate change simulations. *Geophys. Res. Lett.*, **28**, 195-198.

Friedlingstein, P., L. Bopp, P. Ciais, J.-L. Dufresne, L. Fairhead, H. LeTreut, P. Monfray, and J. Orr, 2001: Positive feedback of the carbon cycle on future climate change. *Geophys. Res. Lett.*, in press.

Fyfe, J.C., 1999: On climate simulations of African easterly waves. *J. Climate*, **12**, 1747-1769.

Fyfe, J.C., G.J. Boer, and G.M. Flato, 1999: The Arctic and Antarctic oscillations and their projected changes under global warming, *Geophys. Res. Lett.*, **26**, 1601-1604.

Ganopolski, A., V. Petoukhov, S. Rahmstorf, V. Brovkin, M. Claussen, A. Eliseev and C. Kubatzki, 2001: CLIMBER-2: A climate system model of intermediate complexity. Part II: validation and sensitivity tests. *Clim. Dyn.*, in press.

Gates, W.L., J.S. Boyle, C.Covey, C.G. Dease, C.M. Doutriaux, R.S. Drach, M. Fiorino, P.J. Gleckler, J.J. Hnilo, S.M. Marlais, T.J. Phillips, G.L. Potter, B.D. Santer, K.R. Sperber, K.E.Taylor and D.N. Williams, 1999: An overview of the results of the atmospheric model intercomparison project (AMIP I). *Bull. Am. Met. Soc.*, **80**, 29-55.

Gent, P.R., 2001: Will the North Atlantic Ocean thermohaline circulation weaken during the 21st century? *Geophys. Res. Lett.*, in press.

Gordon, H.B. and S.P. O´Farrell, 1997: Transient climate change in the CSIRO coupled model with dynamic sea ice. *Mon. Wea. Rev.*, **125**, 875-907.

Graham, R.J., A.D.L. Evans, K.R. Mylen, M.S.J. Harrison and K.B. Robertson, 1999: An assessment of seasonal predictability using atmospheric general circulation models. *Forecasting Research Scientific Paper No. 54*. UK Met Office, Bracknell Berkshire RG12 2SY, UK, 22 pp

Gray, W. M., 1979: Hurricanes: their formation, structure and likely role in the tropical circulation, In: *Meteorology Over the Tropical Oceans*, D.B. Shaw (ed), Royal Meteorological Society, J. Glaisher House, Grenville Place, Bracknell, Berks, pp. 155-218.

Gregory, J. M. and J. F. B. Mitchell, 1995: Simulation of daily variability of surface temperature and precipitation over Europe in the current and $2 \times CO_2$ climate using the UKMO high-resolution climate model. *Quart. J. R. Met. Soc.*, **121**, 1451-1476.

Gregory, J.M. and J.F.B. Mitchell, 1997: The climate response to CO_2 of the Hadley Centre coupled AOGCM with and without flux adjustment. *Geophys. Res. Lett.*, **24**, 1943-1946.

Gregory, J.M., J.F.B. Mitchell and A.J. Brady, 1997: Summer drought in

Northern midlatitudes in a time-dependent CO_2 climate experiment. *J. Climate*, **10**, 662-686.

Gregory, J.M. and J.A. Lowe, 2000: Predictions of global and regional sea-level rise using AOGCMs with and without flux adjustment. *Geophys. Res. Lett.*, **27**, 3069-3072.

Hansen, J., A. Lacis, D. Rind, G. Russell, P. Stone, I. Fung, R. Ruedy and J.Lerner, 1984: Climate sensitivity: analysis of feedback mechanisms, *Met. Monograph*, **29**, 130-163.

Hansen, J., G. Russell, A. Lacis, I. Fung, D. Rind and P. Stone, 1985: Climate response times: dependence on climate sensitivity and ocean mixing. *Science*, **299**, 857-859.

Harvey, D., J. Gregory, M. Hoffert, A. Jain, M. Lal, R. Leemans, S. Raper T. Wigley and J. de Wolde, 1997: An introduction to simple climate models used in the IPCC Second Assessment Report. [J.T. Houghton, L. G. Meira Filho, D. J. Griggs and K. Maskell (eds.)] IPCC Technical Paper II.

Haywood, J.M., R.J. Stouffer, R.T. Wetherald, S. Manabe and V. Ramaswamy, 1997: Transient response of a coupled model to estimated changes in greenhouse gas and sulphate concentrations. *Geophys. Res. Lett.*, **24**, 1335-1338.

Henderson-Sellers, A., H. Zhang, G. Berz, K. Emanuel, W. Gray, C. Landsea, G. Holland, J. Lighthill, S.-L. Shieh, P. Webster and K. McGuffie, 1998. Tropical cyclones and global climate change: a post IPCC assessment. *Bull. Am. Met. Soc.*, **79**, 19-38.

Hennessy, K.J., J.M. Gregory and J.F.B. Mitchell, 1997: Changes in daily precipitation under enhanced greenhouse conditions: comparison of UKHI and CSIRO9 GCM. *Clim. Dyn.*, **13**, 667-680.

Hirst, A.C. 1999: The Southern Ocean response to global warming in the CSIRO coupled ocean-atmosphere model. *Environmental Modelling and Software*, **14**, 227-241.

Holland, G.J., 1997: Maximum potential intensity of tropical cyclones. *J. Atmos. Sci.*, **54**, 2519-2541.

Hu, Z.-Z., M. Latif, E. Roeckner and L. Bengtsson, 2000a: Intensified Asian summer monsoon and its variability in a coupled model forced by increasing greenhouse gas concentrations. *Geophys. Res. Lett.*, **27**, 2681-2684.

Hu, Z.-Z., L. Bengtsson and K. Arpe, 2000b: Impact of the global warming on the Asian winter monsoon in a coupled GCM. *J. Geophys. Res.*, **105**, 4607-4624.

Hu, Z.-Z., L. Bengtsson, E. Roeckner, M. Christoph, A. Bacher and J. Oberhuber, 2001: Impact of global warming on the interannual and interdecadal climate modes in a coupled GCM. *Clim. Dyn.*, in press.

Hurrell J.W., 1995: Decadal trends in the North Atlantic Oscillation: regional temperatures and precipitation. *Science*, **269**, 676-679.

Hurrell J.W., 1996: Influence of variations in extratropical wintertime teleconnections on Northern Hemisphere temperature. *Geophys. Res. Lett.*, **23**, 1665-1668.

IPCC, 1990: *Climate Change: The IPCC Scientific Assessment. Contribution of Working Group I to the First Assessment Report of the Intergovernmental Panel on Climate Change*. [Houghton, J.T., G.J. Jenkins and J.J. Ephraums (eds.)]. Cambridge University Press, Cambridge, United Kingdom and New York, NY, USA, 365 pp.

IPCC, 1992: *Climate Change 1992: The Supplementary Report to the IPCC Scientific Assessment. Report prepared for IPCC by Working Group I*. [Houghton, J.T., B.A.Callander and S.K.Varney (eds.)]. Cambridge University Press, Cambridge, United Kingdom and New York, NY, USA, 200 pp.

IPCC, 1996: *Climate Change 1995: The Science of Climate Change. Contribution of Working Group I to the Second Assessment Report of the Intergovernmental Panel on Climate Change*. [Houghton, J.T., L.G. Meira Filho, B.A. Callander, N. Harris, A. Kattenberg, and K. Maskell (eds.)]. Cambridge University Press, Cambridge, United Kingdom and New York, NY, USA, 572 pp.

Japan Meteorological Agency, 1999: *Information of Global Warming, Vol. 3 -Climate change due to increase of CO_2 and sulphate aerosol projected with a coupled atmosphere ocean model* (in Japanese). 70pp. (CD-ROM data are available from JMA.)

Jones, P.D., 1994: Hemispheric surface air temperature variations: a reanalysis and an update to 1993. *J. Climate,* **7,** 1794-1802.

Johns, T.C., J.M. Gregory, W.J. Ingram, C.E. Johnson, A. Jones, J.A. Lowe, J.F.B. Mitchell, D.L. Roberts, D.M.H. Sexton, D.S. Stevenson, S.F.B. Tett and M.J. Woodge, 2001: Anthropogenic climate change for 1860 to 2100 simulated with the HadCM3 model under updated emissions scenarios. Hadley Centre Technical Note No. **22,** available from The Hadley Centre for Climate Prediction and Research, The Met Office, London Road, Bracknell, RG12 2SY, UK.

Joos, F., M. Bruno, R. Fink, T.F.Stocker, U. Siegenthaler, C. Le Quéré and J.L. Sarmiento, 1996: An efficient and accurate representation of complex oceanic and biospheric models of anthropogenic carbon uptake. *Tellus,* **48B,** 397-417.

Kattenberg, A., F. Giorgi , H. Grassl, G.A. Meehl, J.F.B. Mitchell, R.J. Stouffer, T. Tokioka, A.J. Weaver and T.M.L.Wigley, 1996. In: *Climate Change 1995: The Science of Climate Change. Contribution of Working Group I to the Second Assessment Report of the Intergovernmental Panel on Climate Change* [Houghton, J.T., L.G. Meira Filho, B.A. Callander, N. Harris, A. Kattenberg, and K. Maskell (eds.)]. Cambridge University Press, Cambridge, United Kingdom and New York, NY, USA, 572 pp.

Keen, A.B. and J.M. Murphy, 1997: Influence of natural variability and the cold start problem on the simulated transient response to increasing CO_2. *Clim. Dyn.* , **13,** 847-864.

Kharin, V.V. and F.W. Zwiers, 2000: Changes in the extremes in an ensemble of transient climate simulations with a coupled atmosphere-ocean GCM. *J. Climate,* **13,** 3760–3788.

Kitoh, A., H. Koide, K. Kodera, S. Yukimoto and A. Noda, 1996: Interannual variability in the stratospheric-tropospheric circulation in an ocean-atmosphere coupled GCM. *Geophys. Res. Lett.,* **23,** 543-546.

Kitoh, A., S. Yukimoto, A. Noda and T. Motoi, 1997. Simulated changes in the Asian summer monsoon at times of increased atmospheric CO_2. *Journal of the Meteorological Society of Japan,* **75,** 1019-1031.

Knippertz, P., U. Ulbrich and P. Speth, 2000: Changing cyclones and surface wind speeds over the North Atlantic and Europe in a transient GHG experiment. *Clim. Res.,* **15,** 109-122.

Knutson T.R. and S. Manabe, 1994: Impact of increased CO_2 on simulated ENSO-like phenomena. *Geophys. Res. Lett.,* **21,** 2295-2298.

Knutson, T.R., and S. Manabe, 1995: Time-mean response over the tropical Pacific to increased CO_2 in a coupled ocean-atmosphere model. *J. Climate,* **8,** 2181-2199.

Knutson, T.R., S. Manabe and D. Gu, 1997: Simulated ENSO in a global coupled ocean-atmosphere model: multidecadal amplitude modulation and CO_2-sensitivity. *J. Climate,* **10,** 138-161.

Knutson, T.R. and S. Manabe, 1998: Model assessment of decadal variability and trends in the tropical Pacific ocean. *J. Climate,* **11,** 2273-2296.

Knutson, T.R., R.E. Tuleya and Y. Kurihara, 1998: Simulated increase of hurricane intensities in a CO_2-warmed climate. *Science,* **279,** 1018-1020.

Knutson, T.R., T.L. Delworth, K.W. Dixon and R.J. Stouffer, 1999: Model assessment of regional surface temperature trends (1949-97). *J. Geophys. Res.,* **104,** 30,981-30,996.

Kodera, K., M. Chiba, H. Koide, A. Kitoh and Y. Nikaidou, 1996. Interannual variability of the winter stratosphere and troposphere in the Northern Hemisphere. *Journal of the Meteorological Society of Japan,* **74,** 365-382.

Kothavala, Z., 1997: Extreme precipitation events and the applicability of global climate models to study floods and droughts. *Math. and Comp. in Simulation,* **43,** 261-268.

Kothavala, Z., 1999: The duration and severity of drought over eastern Australia simulated by a coupled ocean-atmosphere GCM with a transient increase in CO_2. *Environmental Modelling Software,* **14,** 243-252.

Krishnamurti, T.N., R. Correa-Torres, M. Latif and G. Daughenbaugh, 1998. The impact of current and possibly future SST anomalies on the frequency of Atlantic hurricanes. *Tellus,* **50A,** 186-210.

Krishnamurti, T.N., C.M. Kishtawal, T.E. LaRow, D.R. Bachiochi, Z. Zhang, C.E. Williford, S. Gadgil and S. Surendran, 1999: Improved

weather and seasonal climate forecasts from multimodel superensemble. *Science,* **285,** 1548-1550.

Kushner, P.J., I.M. Held and T.L. Delworth, 2001: Southern-hemisphere atmospheric circulation response to global warming. *J. Climate,* in press.

Lal, M., U. Cubasch, R. Voss and J. Waszkewitz, 1995: The effect of transient increase of greenhouse gases and sulphate aerosols on monsoon climate. *Curr. Sci.,* **69,** 752-763.

Lal, M., G.A. Meehl and J.M. Arblaster, 2000: Simulation of Indian summer monsoon rainfall and its intraseasonal variability. *Regional Environmental Change,* in press.

Lambert, S.J. and G.J. Boer, 2001: CMIP1 evaluation and intercomparison of coupled climate models. *Clim. Dyn.,* **17,** 83-106.

Latif, M., E. Roeckner, U. Mikolajewicz and R. Voss, 2000: Tropical stabilisation of the thermohaline circulation in a greenhouse warming simulation. *J.Climate,* **13,** 1809-1813.

LeTreut, H. and B.J. McAvaney, 2000: A model intercomparison of equilibrium climate change in response to CO_2 doubling. Note du Pole de Modelisation de l'IPSL, Number 18, Institut Pierre Simon LaPlace, Paris, France.

Lunkeit, F., M. Ponater, R. Sausen, M. Sogalla, U. Ulbrich and M. Windelband, 1996: Cyclonic activity in a warmer climate. *Contrib. Atmos. Phys.,* **69,** 393-407.

Lunkeit, F., S.E. Bauer and K. Fraedrich, 1998: Storm tracks in a warmer climate: sensitivity studies with a simplified global circulation model. *Clim. Dyn.,* **14,** 813-826.

Lupo, A.R., R.J. Oglesby and I.I. Mokhov, 1997: Climatological features of blocking anticyclones: a study of Northern Hemisphere CCM1 model blocking events in present-day and double CO_2 concentrations. *Clim. Dyn.,* **13,** 181-195.

Manabe, S., R.J. Stouffer, M.J. Spelman and K. Bryan, 1991: Transient responses of a coupled ocean-atmosphere model to gradual changes of atmospheric CO_2. Part I: annual mean response. *J. Climate,* **4,** 785-818.

Manabe, S. and R.J. Stouffer, 1994: Multiple-century response of a coupled ocean-atmosphere model to an increase of the atmospheric carbon dioxide. *J. Climate,* **7,** 5-23.

Matyasovszky,I., 1998: Non-parametric estimation of climate trends. *Quarterly Journal of the Hungarian Meteorolgical Service,* **102,** 149-158.

Meehl, G.A., G.W. Branstator and W.M. Washington, 1993: Tropical Pacific interannual variability and CO_2 climate change. *J. Climate,* **6,** 42-63.

Meehl, G. A., M. Wheeler and W.M. Washington, 1994: Low-frequency variability and CO_2 transient climate change. Part 3. Intermonthly and interannual variability. *Clim. Dyn.* **10,** 277-303.

Meehl, G.A., W.M. Washington, D.J. Erickson III, B.P. Briegleb and P.J. Jaumann, 1996: Climate change from increased CO_2 and direct and indirect effects of sulphate aerosols. *Geophys. Res. Lett.,* **23,** 3755-3758.

Meehl, G.A. and W.M. Washington, 1996: El Nino-like climate change in a model with increased atmospheric CO_2-concentrations. *Nature,* **382,** 56-60.

Meehl, G.A., G.J. Boer, C. Covey, M. Latif and R.J. Stouffer, 2000a: The Coupled Model Intercomparison Project (CMIP). *Bull. Am. Met. Soc.,* **81,** 313-318.

Meehl, G.A., W. Collins, B. Boville, J.T. Kiehl, T.M.L. Wigley and J.M. Arblaster, 2000b: Response of the NCAR Climate System Model to increased CO_2 and the role of physical processes. *J. Climate,* **13,** 1879-1898.

Meehl, G.A., W.M. Washington, J.M. Arblaster, T.W. Bettge and W.G. Strand Jr., 2000c: Anthropogenic forcing and decadal climate variability in sensitivity experiments of 20th and 21st century climate. *J. Climate,* **13,** 3728-3744.

Meehl, G.A., F. Zwiers, J. Evans, T. Knutson, L. Mearns and P. Whetton, 2000d: Trends in extreme weather and climate events: issues related to modelling extremes in projections of future climate change. *Bull. Am. Met. Soc.,* **81,** 427-436.

Meehl, G.A., P. Gent, J.M. Arblaster, B. Otto-Bliesner, E. Brady and A. Craig, 2001: Factors that affect amplitude of El Nino in global coupled climate models. *Clim. Dyn.*, **17**, 515-526.

Mikolajewicz, U. and R. Voss, 2000: The role of the individual air-sea flux components in CO$_2$-induced changes of the ocean's circulation and climate. *Clim. Dyn.* **16**, 627-642.

Mitchell, J.F.B., S. Manabe, V. Meleshko and T. Tokioka, 1990. Equilibrium climate change – and its implications for the future. In *Climate Change. The IPCC Scientific Assessment. Contribution of Working Group 1 to the first assessment report of the Intergovernmental Panel on Climate Change,* [Houghton, J. L, G. J. Jenkins and J. J. Ephraums (eds)], Cambridge University Press, Cambridge, pp. 137-164.

Mitchell, J.F.B., T.C. Johns, J.M. Gregory and S.F.B. Tett, 1995: Climate response to increasing levels of greenhouse gases and sulphate aerosols. *Nature*, **376**, 501-504.

Mitchell, J.F.B. and T.C. Johns, 1997: On the modification of global warming by sulphate aerosols. *J. Climate*, **10**, 245-267.

Mitchell J.F.B., T.C. Johns and C.A. Senior, 1998: Transient response to increasing greenhouse gases using models with and without flux adjustment. *Hadley Centre Technical Note 2*. Available from Met Office, London Road Bracknell, RG12 2SZ, UK.

Mitchell, J.F.B., T.C. Johns, W.J. Ingram and J.A. Lowe, 2000: The effect of stabilising atmospheric carbon dioxide concentrations on global and regional climate change. *Geophys. Res. Lett.* **27**, 2977-2930.

Monahan, A.H., J.C. Fyfe and G.M. Flato, 2000: A regime view of Northern Hemisphere atmospheric variability and change under global warming. *Geophys. Res. Lett*, **27**, 1139-1142.

Murphy, J.M., 1995: Transient response of the Hadley Centre coupled ocean-atmosphere model to increasing carbon dioxide. Part III: analysis of global-mean response using simple models. *J. Climate*, **8**, 496-514.

Murphy, J.M. and J.F.B. Mitchell, 1995: Transient response of the Hadley Centre coupled ocean-atmosphere model to increasing carbon dioxide. Part II: spatial and temporal structure of response. *J. Climate*, **8**, 57-80.

Myhre, G., E.J. Highwood, K.P. Shine and F. Stordal, 1998: New estimates of radiative forcing due to well mixed greenhouse gases. *Geophys. Res. Lett.*, **25**, 2715-2718.

Nakićenović, N., J. Alcamo, G. Davis, B. de Vries, J. Fenhann, S. Gaffin, K. Gregory, A. Grübler, T. Y. Jung, T. Kram, E. L. La Rovere, L. Michaelis, S. Mori, T. Morita, W. Pepper, H. Pitcher, L. Price, K. Raihi, A. Roehrl, H.-H. Rogner, A. Sankovski, M. Schlesinger, P. Shukla, S. Smith, R. Swart, S. van Rooijen, N. Victor, Z. Dadi, 2000: IPCC Special Report on Emissions Scenarios, Cambridge University Press, Cambridge, United Kingdom and New York, NY, USA, 599 pp.

Noda, A. and T. Tokioka, 1989: The effect of doubling the CO$_2$ concentration on convective and non-convective precipitation in a general circulation model coupled with a simple mixed layer ocean model. *J. Met. Soc. Japan*, **67**, 1057-1069.

Noda, A., K. Yoshimatsu, A. Kitoh and H. Koide, 1999a: Relationship between natural variability and CO$_2$-induced warming pattern: MRI coupled atmosphere/mixed-layer (slab) ocean GCM (SGCM) Experiment. *10th Symposium on Global Change Studies* , 10-15 January 1999, Dallas, Texas. pp. 355-358, American Meteorological Society, Boston. Mass.

Noda, A., K. Yoshimatsu, S. Yukimoto, K. Yamaguchi and S. Yamaki, 1999b: Relationship between natural variability and CO$_2$-induced warming pattern: MRI AOGCM Experiment. *10th Symposium on Global Change Studies*, 10-15 January 1999, Dallas, Texas. American Meteorological Society, Boston. Mass. pp. 359-362

Noda, A., S. Yukimoto, S. Maeda, T. Uchiyama, K. Shibata and S. Yamaki, 2001: A new meteorological research institute coupled GCM (MRI-CGCM2): Transient response to greenhouse gas and aerosol scenarios. CGER's supercomputer monograph report Vol. 7, National Institute for Environmental Studies, Tsukuba, Japan, 66pp (in press).

Nozawa, T., S. Emori, A. Numaguti, Y. Tsushima, T. Takemura, T. Nakajima, A. Abe-Ouchi and M. Kimoto, 2001: Projections of future climate change in the 21st century simulated by the CCSR/NIES CGCM under the IPCC SRES scenarios, In: *Present and Future of Modelling Global Environmental Change – Toward Integrated Modelling*, T. Matsuno (ed), Terra Scientific Publishing Company, Tokyo (in press).

Osborn, T.J., K.R. Briffa, S.F.B. Tett, P.D. Jones and R.M. Trigo, 1999. Evaluation of the North Atlantic Oscillation as simulated by a coupled climate model. *Clim. Dyn.*, **15**, 685-702.

Paeth, H., A. Hense, R. Glowienka-Hense, R. Voss and U. Cubasch, 1999: The North Atlantic Oscillation as an indicator for greenhouse-gas induced climate change. *Clim. Dyn.*, **15**, 953-960.

Palmer, T. N., 1999. A nonlinear dynamical perspective on climate prediction. *J. Climate*, **12**, 575-591.

Parey, S, 1994: Simulations de Trente ans 1xCO$_2$, 2xCO$_2$, 3xCO$_2$, avec le modele du LMD (64x50x11) premiers resultata. *EDF (Electricité de France), Direction des études et Recherches*, HE-33/94/008.

Price, C., and D. Rind, 1994a: Possible implications of global climate change on global lightning distributions and frequencies. *J. Geophys. Res.*, **99**, 10,823-10,831.

Price, C., and D. Rind, 1994b: The impact of a 2xCO$_2$ climate on lightning-caused fires. *J. Climate*, **7**, 1484-1494.

Räisänen, J., 1997: Objective comparison of patterns of CO$_2$-induced climate change in coupled GCM experiments. *Clim. Dyn.*, **13**, 197-221.

Räisänen, J., 2000: CO$_2$-induced climate change in Northern Europe: comparison of 12 CMIP2 experiments. Reports Meteorology and Climatology No. 87, *SMHI*, 59 pp.

Räisänen, J., 2001: CO$_2$-induced climate change in CMIP2 experiments. Quantification of agreement and role of internal variability. *J. Climate*, in press.

Ramstein, G., Y. Serafini-Le Treut, H. Le Treut, M. Forichon and S. Joussaume, 1998. Cloud processes associated with past and future climate changes. *Clim. Dyn.*, **14**, 233-247.

Raper, S.C.B, T.M.L. Wigley and R.A. Warrick, 1996: Global sea-level rise: past and future, In: *Sea-Level rise and Coastal Subsidence*, J.D. Milliman and B.U. Haq (eds), Kluwer Academic Publishers, 11-46.

Raper, S.C.B. and U. Cubasch, 1996: Emulation of the results from a coupled general circulation model using a simple climate model. *Geophys. Res. Lett.*, **23**, 1107-1110.

Raper, S.C.B., J.M. Gregory and T.J. Osborn, 2001a: Use of an upwelling-diffusion energy balance climate model to simulate and diagnose AOGCM results. *Clim. Dyn.*, in press.

Raper, S.C.B., J.M. Gregory and R.J. Stouffer, 2001b: The role of climate sensitivity and ocean heat uptake on AOGCM transient temperature and thermal expansion response. *J. Climate*, in press.

Roeckner, E., L. Bengtsson, J. Feichter, J. Lelieveld and H. Rodhe, 1999: Transient climate change with a coupled atmosphere-ocean GCM including the tropospheric sulfur cycle. *J. Climate*, **12**, 3004-3032.

Royer, J.-F., F. Chauvin, B. Timbal, P. Araspin and D. Grimal, 1998. A GCM study of the impact of greenhouse gas increase on the frequency of occurrence of tropical cyclones. *Clim. Change*, **38**, 307-343.

Russell, G.L., J.R. Miller and D. Rind, 1995: A coupled atmosphere-ocean model for transient climate change studies. *Atmosphere-Ocean*, **33**, 683-730.

Russell, G.L. and D. Rind, 1999. Response to CO$_2$ transient increase in the GISS coupled model. Regional cooling in a warming climate. *J. Climate*, **12**, 531-539.

Sarmiento, J.L., T.M.C. Hughes, R.J. Stouffer and S. Manabe, 1998: Simulated response of the ocean carbon cycle to anthropogenic climate warming. *Nature*, **393**, 245-249.

Schimel, D., I.G. Enting, M. Heimann, T.M.L. Wigley, D. Raynaud, D. Alves and U. Siegenthaler, 1994. CO$_2$ and the carbon cycle. In: *Climate Change, 1994: Radiative Forcing of Climate Change and an Evaluation of the IPCC IS92 Emission Scenarios*. [J.T. Houghton, L.G. Meira Filho, J. Bruce, H Lee, B.A. Callander, E.F. Haites, N. Harris and K. Maskell (eds.)] Cambridge University Press, Cambridge, UK, 35-71.

Schimel, D., M. Grubb, F. Joos, R. Kufmann, R. Moss, W. Ogana, R. Richels, T. Wigley , 1997: Stabilisation of Atmospheric Greenhouse Gases: Physical, Biological and Socio-economic Implications. [J.T. Houghton, L. Gylvan Meira Filho, D.J. Griggs and K. Maskell (eds.)] IPCC Technical Paper III.

Schubert, M., J. Perlwitz, R. Blender, K. Fraedrich, and F. Lunkeit, 1998: North Atlantic cyclones in CO$_2$-induced warm climate simulations: Frequency, intensity, and tracks. *Clim. Dyn.* , **14**, 827-837.

Senior, C.A., 1999: Comparison of mechanisms of cloud-climate feedbacks in a GCM. *J. Clim.*, **12**, 1480-1489.

Senior, C.A. and J.F.B. Mitchell, 2000: The time-dependence of climate sensitivity. *Geophys. Res. Lett.*, **27**, 2685-2688.

Shindell, D.T., R.L. Miller, G.A. Schmidt and L. Pandolfo, 1999. Simulation of recent northern winter climate trends by greenhouse gas forcing. *Nature*, **399**, 452-455.

Shine, K.P., R.G. Derwent, D.J. Wuebbles and J.-J.Morcrette, 1990: Radiative forcing of climate. In: *Climate Change: The IPCC Scientific Assessment.* [Houghton, J.T., G.J. Jenkins and J.J. Ephraums (eds.)]. Cambridge University Press, Cambridge, United Kingdom and New York, NY, USA, pp. 41-68.

Siegenthaler, U. and Joos, F., 1992: Use of a simple model for studying oceanic tracer distributions and the global carbon cycle. *Tellus*, **44B**, 186-207.

Sinclair, M.R. and I.G. Watterson, 1999: Objective assessment of extratropical weather systems in simulated climates. *J. Climate*, **12**, 3467-3485.

Smith, S. J., T.M.L.Wigley, N. Nakićenović and S.C.B. Raper, 2001: Climate implications of greenhouse gas emission scenarios. *Technological Forecasting and Social Change,* **65**, 195-204.

Sokolov, A.P. and P.H. Stone 1998: A flexible climate model for use in integrated assessments. *Clim. Dyn.*, **14**, 291-303.

Stendel, M., T. Schmith, E. Roeckner and U. Cubasch, 2000: The climate of the 21st century: transient simulations with a coupled atmosphere-ocean general circulation model. Danmarks Klimacenter Report 00-6, Danish Meteorological Institute, Lyngbyvej 100, DK-2100 Copenhagen, Denmark, ISBN: 87-7478-427-7.

Stocker T.F., D.G. Wright and L.A. Mysak, 1992: A zonally averaged, coupled ocean-atmosphere model for paleoclimate studies. *J. Climate*, **5**, 773-797.

Stocker, T.F. and A. Schmittner, 1997: Influence of CO$_2$ emission rates on the stability of the thermohaline circulation. *Nature*, **388**, 862-865.

Stouffer, R.J. and S. Manabe, 1999: Response of a coupled ocean-atmosphere model to increasing atmospheric carbon dioxide: Sensitivity to the rate of increase. *J. Climate*, **12**, 2224-2237.

Tett, S.F.B., 1995: Simulation of El Niño-Southern Oscillation-like variability in a global coupled AOGCM and its response to CO$_2$-increase. *J. Climate*, **8**, 1473-1502.

Tett, S.F.B., T.C. Johns and J.F.B. Mitchell, 1997: Global and regional variability in a coupled AOGCM. *Clim. Dyn.*, **13**, 303 –323.

Thompson, D.W.J. and J.M. Wallace, 1998: The Arctic Oscillation signature in the wintertime geopotential height and temperature fields. *Geophys. Res. Lett.*, **25**, 1297-1300.

Timmermann, A., J. Oberhuber, A. Bacher, M. Esch, M. Latif and E. Roeckner, 1999: Increased El Niño frequency in a climate model forced by future greenhouse warming. *Nature*, **398**, 694-696.

Tokioka, T., A. Noda, A. Kitoh, Y. Nikaidou, S. Nakagawa, T. Motoi, Y. Yukimoto and K. Takata, 1995: Transient CO$_2$ experiment with the MRI CGCM - Quick report. *J. Met. Soc. Japan*, **73**, 817-826.

Tokioka, T., A. Noda, A. Kitoh, Y. Nikaidou, S. Nakagawa, T. Motoi, Y. Yukimoto and K. Takata, 1996: Transient CO$_2$ experiment with the MRI CGCM - Annual mean response -. *CGER's Supercomputer Monograph Report Vol. 2*, National Institute for Environmental Studies, Tsukuba, Japan, 86 pp.

Ulbrich, U., and M. Christoph, 1999: A shift of the NAO and increasing storm track activity over Europe due to anthropogenic greenhouse gas forcing. *Clim. Dyn.*, **15**, 551-559.

United Nations, 1992: United Nations framework convention on climate change, UNFCCC. http://www.unfccc.int/

Visser, H., R.J.M. Folkert, J. Hoekstra and J.J.de Wolff, 2000: Identifying key sources of uncertainty in climate change projections. *Clim. Change*, **45**, 421-457.

Voss, R. and U. Mikolajewicz, 2001: Long-term climate changes due to increased CO$_2$ concentration in the coupled atmosphere-ocean general circulation model ECHAM3/LSG. *Clim. Dyn.*, **17**, 45-60.

Wallace, J. M., 2000. North Atlantic Oscillation / annular mode: Two

paradigms – one phenomenon. *Quart. J. R. Met. Soc*, **126**, 791-806.

Washington, W.M., J.W. Weatherly, G.A. Meehl, A.J. Semtner Jr., T.W. Bettge, A.P. Craig, W.G. Strand Jr., J.M. Arblaster, V.B. Wayland, R. James and Y. Zhang, 2000: Parallel climate model (PCM) control and transient simulations. *Clim. Dyn.*, **16**, *755-774.*

Watterson, I. G., M.R. Dix and R.A. Colman, 1998: A comparison of present and doubled CO$_2$ climates and feedbacks simulated by three general circulation models. *J. Geophys. Res.*, **104**, 1943-1956.

Watterson, I. G. 2000: Interpretation of simulated global warming using a simple model. *J. Climate*, **13**, 202-215.

Weaver, A.J., M. Eby, A.F. Fanning and E.C. Wiebe, 1998: Simulated influence of carbon dioxide, orbital forcing and ice sheets on the climate of the last glacial maximum. *Nature*, **394**, 847-853.

Weaver, A.J. and E.C. Wiebe, 1999: On the sensitivity of projected oceanic thermal expansion to the parameterisation of sub-grid scale ocean mixing. *Geophys. Res. Lett.*, **26**, 3461-3464.

Wetherald, R. T. and S. Manabe, 1999: Detectability of summer dryness caused by greenhouse warming. *Climatic Change*, **43**, 495-511.

Wiebe, E.C. and A.J. Weaver, 2000: On the sensitivity of global warming experiments to the parameterisation of sub-grid scale ocean mixing. *Clim. Dyn.*, **15**, 875-893.

Wigley T.M.L. and Raper S.C.B., 1987: Thermal expansion of sea water associated with global warming. *Nature*, **330**, 127-131.

Wigley, T.M.L., 1991: Could reducing fossil-fuel emissions cause global warming? *Nature,* **349**, 503-506.

Wigley T.M.L. and S.C.B. Raper,1992: Implications for climate and sea level of revised IPCC emissions scenarios. *Nature*, **357**, 293-300.

Wigley, T.M.L. 1993: Balancing the carbon budget. Implications for projections of future carbon dioxide concentration changes. *Tellus*, **45B**, 409-425.

Wigley, T.M.L, R. Richels and J.A. Edmonds, 1996: Economic and environmental choices in the stabilisation of atmospheric CO$_2$ concentrations. *Nature*, **379**, 242-245.

Wigley, T.M.L. and S.J. Smith, 1998: Uncertainties in projections of future global-mean temperature change. In: *Do We Understand Global Climate change?* Norwegian Academy of Technological Sciences (NTVA), Trondheim, Norway, 185-195.

Wigley, T.M.L., 1999: The science of climate change: global and U.S. perspectives. Pew Centre, 2101 Wilson Blvd., Arlington, VA, USA, 48pp.

Wigley, T.M.L., 2000: TAR version of MAGICC forcing. http://www.acacia.ucar.edu

Williams, K. D., C.A. Senior and J.F.B. Mitchell, 2001. Transient climate change in the Hadley Centre models: The roles of physical processes. *J. Climate*, in press.

Yao, M-S. and Del Genio, A., 1999. Effects of parameterisation on the simulation of climate changes in the GISS GCM. *J. Climate.*, **12**, 761-779.

Yonetani, T. and H.B. Gordon, 2001: Simulated changes in the frequency of extremes and regional features of seasonal/annual temperature and precipitation when atmospheric CO$_2$ is doubled. *J.Climate*, in press.

Yukimoto, S., A. Noda, A. Kitoh, M. Sugi, Y. Kitamura, M. Hosaka, K. Shibata, S. Maeda and T. Uchiyama, 2001: A new meteorological research institute coupled GCM (MRI-CGCM2) – model climate and its variability. *Pap. Meteor. Geophys.*, **51**, 47-88.

Zhang, Y. and W.-Ch. Wang, 1997: Model simulated northern winter cyclone and anti-cyclone activity under a greenhouse warming scenario. *J. Climate,* **10,** 1616-1634.

Zhang, Y., J.M. Wallace and D.S. Battisti, 1997: ENSO-like interdecadal variability: 1900-93. *J. Climate*, **10**, 1004-1020.

Zorita, E. and F. González-Rouco, 2000. Disagreement between predictions of the future Atrctic Oscillation as simulated in two different climate models: Implications for global warming. *Geophys. Res. Lett.*, **27**, 1755-1758

Zwiers, F.W. and V. V. Kharin, 1998: Changes in the extremes of the climate simulated by CCC GCM2 under CO$_2$-doubling. *J. Climate*, **11**, 2200-2222.

10

Regional Climate Information – Evaluation and Projections

Co-ordinating Lead Authors
F. Giorgi, B. Hewitson

Lead Authors
J. Christensen, M. Hulme, H. Von Storch, P. Whetton, R. Jones, L. Mearns, C. Fu

Contributing Authors
R. Arritt, B. Bates, R. Benestad, G. Boer, A. Buishand, M. Castro, D. Chen, W. Cramer, R. Crane,
J. F. Crossley, M. Dehn, K. Dethloff, J. Dippner, S. Emori, R. Francisco, J. Fyfe, F.W. Gerstengarbe,
W. Gutowski, D. Gyalistras, I. Hanssen-Bauer, M. Hantel, D.C. Hassell, D. Heimann, C. Jack, J. Jacobeit,
H. Kato, R. Katz, F. Kauker, T. Knutson, M. Lal, C. Landsea, R. Laprise, L.R. Leung, A.H. Lynch,
W. May, J.L. McGregor, N.L. Miller, J. Murphy, J. Ribalaygua, A. Rinke, M. Rummukainen, F. Semazzi,
K. Walsh, P. Werner, M. Widmann, R. Wilby, M. Wild, Y. Xue

Review Editors
M. Mietus, J. Zillman

Contents

Executive Summary

Introduction

This chapter assesses regional climate information from Atmosphere-Ocean General Circulation Models (AOGCMs) and techniques used to enhance regional detail. These techniques have been substantially improved since the IPCC WGI Second Assessment Report (IPCC, 1996) (hereafter SAR) and have become more widely applied. They fall into three categories: high and variable resolution Atmosphere General Circulation Models (AGCMs); regional (or nested limited area) climate models (RCMs); and empirical/statistical and statistical/dynamical methods. The techniques exhibit different strengths and weaknesses and their use depends on the needs of specific applications.

Simulations of present day climate

Coarse resolution AOGCMs simulate atmospheric general circulation features well in general. At the regional scale the models display area-average biases that are highly variable from region-to-region and among models, with sub-continental area-averaged seasonal temperature biases typically within 4°C and precipitation biases mostly between −40 and +80% of observations. In most cases, these represent an improvement compared to the AOGCM results evaluated in the SAR.

The development of high resolution/variable resolution AGCMs since the SAR shows that the models' dynamics and large-scale flow improve as resolution increases. In some cases, however, systematic errors are worsened compared with coarser resolution models although only very few results have been documented.

RCMs consistently improve the spatial detail of simulated climate compared to General Circulation Models (GCMs). RCMs driven by observed boundary conditions show area-averaged temperature biases (regional scales of 10^5 to 10^6 km^2) generally within 2°C and precipitation biases within 50% of observations. Statistical downscaling demonstrates similar performance, although greatly depending on the methodological implementation and application.

Simulation of climate change for the late decades of the 21st century

Climate means

The following conclusions are based on seasonal mean patterns at sub-continental scales emerging from current AOGCM simulations. Based on considerations of consistency of changes from two IS92a-type emission scenarios and preliminary results from two SRES emission scenarios, within the range of these four scenarios:

• It is very likely that: nearly all land areas will warm more rapidly than the global average, particularly those at high latitudes in the cold season; in Alaska, northern Canada, Greenland, northern Asia, and Tibet in winter and central Asia and Tibet in summer the warming will exceed the global mean warming in each model by more than 40% (1.3 to 6.9°C

for the range of models and scenarios considered). In contrast, the warming will be less than the global mean in south and Southeast Asia in June-July-August (JJA), and in southern South America in winter.

• It is likely that: precipitation will increase over northern mid-latitude regions in winter and over northern high latitude regions and Antarctica in both summer and winter. In December-January-February (DJF), rainfall will increase in tropical Africa, show little change in Southeast Asia and decrease in central America. There will be increase or little change in JJA over South Asia. Precipitation will decrease over Australia in winter and over the Mediterranean region in summer. Change of precipitation will be largest over the high northern latitudes.

Results from regional studies indicate that at finer scales the changes can be substantially different in magnitude or sign from the large area average results. A relatively large spread exists between models, although attribution is unclear.

Climate variability and extremes

The following conclusions are based on patterns emerging from a limited number of studies with current AOGCMs, older GCMs and regionalisation studies.

• Daily to interannual variability of temperature will likely decrease in winter and increase in summer in mid-latitude Northern Hemisphere land areas.

• Daily high temperature extremes will likely increase in frequency as a function of the increase in mean temperature, but this increase is modified by changes in daily variability of temperature. There is a corresponding decrease in the frequency of daily low temperature extremes.

• There is a strong correlation between precipitation interannual variability and mean precipitation. Future increases in mean precipitation will very likely lead to increases in variability. Conversely, precipitation variability will likely decrease only in areas of reduced mean precipitation.

• For regions where daily precipitation intensities have been analysed (e.g., Europe, North America, South Asia, Sahel, southern Africa, Australia and the South Pacific) extreme precipitation intensity may increase.

• Increases in the occurrence of drought or dry spells are indicated in studies for Europe, North America and Australia.

Tropical cyclones

Despite no clear trends in the observations, a series of theoretical and model-based studies, including the use of a high resolution hurricane prediction model, suggest:

• It is likely that peak wind intensities will increase by 5 to 10% and mean and peak precipitation intensities by 20 to 30% in some regions;

- There is no direct evidence of changes in the frequency or areas of formation.

Recommendations
The material assessed identifies key priorities for future work:

GCMs:
- Continued improvement in GCMs, as their use is fundamental to deriving regional climate information.
- GCM simulations with a greater range of forcing scenarios and an increased ensemble size to assess the spread of regional predictions.
- More assessment of GCM regional attributes and climate change simulations.
- A much greater effort in the evaluation of variability (daily to interannual) and extreme events.

RCMs:
- A more systematic and wide application of RCMs to adequately assess their performance and to provide information for regional scenarios.
- Ensemble RCM simulations with a range of regional models driven by different AOGCM simulations.

- A much greater effort in the evaluation of variability (daily to interannual) and extreme events.

Empirical/statistical and statistical/dynamical methods:
- More regional observations to provide for more comprehensive statistical downscaling functions.
- Much further work to identify the important climate change predictors for statistical downscaling.
- Application of different techniques to a range of AOGCM simulations.

Tropical cyclones:
- A greater range of models and techniques for a comprehensive assessment of the future behaviour of tropical cyclones.

Cross-cutting:
- Systematic comparisons of the relative strengths and weaknesses of techniques to derive regional climate information.
- The development of high-resolution observed climatologies, especially for remote and physiographically complex regions.
- A systematic evaluation of uncertainties in regional climate information.

10.1 Introduction

This chapter is a new addition compared with previous IPCC assessment reports. It stems from the increasing need to better understand the processes that determine regional climate and to evaluate regional climate change information for use in impact studies and policy planning. To date, a relatively high level of uncertainty has characterised regional climate change information. This is due to the complexity of the processes that determine regional climate change, which span a wide range of spatial and temporal scales, and to the difficulty in extracting fine-scale regional information from coarse resolution coupled Atmosphere-Ocean General Circulation Models (AOGCMs).

Coupled AOGCMs are the modelling tools traditionally used for generating projections of climatic changes due to anthropogenic forcings. The horizontal atmospheric resolution of present day AOGCMs is still relatively coarse, of the order of 300 to 500 km, due to the centennial to millennial time-scales associated with the ocean circulation and the computing requirements that these imply. However, regional climate is often affected by forcings and circulations that occur at the sub-AOGCM horizontal grid scale (e.g., Giorgi and Mearns, 1991). Consequently, AOGCMs cannot explicitly capture the fine-scale structure that characterises climatic variables in many regions of the world and that is needed for impact assessment studies (see Chapter 13).

Therefore, a number of techniques have been developed with the goal of enhancing the regional information provided by coupled AOGCMs and providing fine-scale climate information. Here these are referred to as "regionalisation" techniques and are classified into three categories:

- high resolution and variable resolution Atmosphere GCM (AGCM) experiments;
- nested limited area (or regional) climate models (RCMs);
- empirical/statistical and statistical/dynamical methods.

Since the IPCC WGI Second Assessment Report (IPCC, 1996) (hereafter SAR), a substantial development has been achieved in all these areas of research. This chapter has two fundamental objectives. The first is to assess whether the scientific community has been able to increase the confidence that can be placed in the projection of regional climate change caused by anthropogenic forcings since the SAR. The second is to evaluate progress in regional climate research. It is not the purpose of this chapter to provide actual regional climate change information for use in impact work, although the material discussed in this chapter serves most often for the formation of climate change scenarios (see Chapter 13).

The assessment is based on all the different modelling tools that are currently available to obtain regional climate information, and includes: (a) an evaluation of the performance, strengths and weaknesses of different techniques in reproducing present day climate characteristics and in simulating processes of importance for regional climate; and (b) an evaluation of simulations of climate change at the regional scale and associated uncertainties.

Evaluation of present day climate simulations is important because, even though a good simulation of present day climate does not necessarily imply a more accurate simulation of future climate change (see also Chapter 13), confidence in the realism of a model's response to an anomalous climate forcing can be expected to be higher when the model is capable of reproducing observed climate. In addition, interpretation of the response is often facilitated by understanding the behaviour of the model in simulating the current climate. When possible, the capability of models to simulate climates different from the present, such as palaeoclimates, may also provide additional confidence in the predicted climatic changes.

The chapter is organised as follows. In the remainder of this section a summary is first presented of the conclusions reached in the SAR concerning regional climate change. This is followed by a brief discussion of the regional climate problem. Section 10.2 examines the principles behind different approaches to the generation of regional climate information. Regional attributes of coupled AOGCM simulations are discussed in Section 10.3. This discussion is important for different reasons: first, because AOGCMs are the starting point in the generation of regional climate change scenarios; second, because many climate impact assessment studies still make use of output from coupled AOGCM experiments without utilising any regionalisation tool; and third because AOGCMs provide the baseline against which to assess the added value of regionalisation techniques. Sections 10.4, 10.5 and 10.6 are devoted to the analysis of experiments using high resolution and variable resolution AGCMs, RCMs and empirical/statistical and statistical/dynamical methods, respectively. Section 10.7 analyses studies in which different regionalisation techniques have been intercompared and Section 10.8 presents a summary assessment.

10.1.1 Summary of SAR

The analysis of regional climate information in the SAR (Section 6.6 of Kattenberg *et al.*, 1996) consisted of two primary segments. In the first, results were analysed from an intercomparison of AOGCM experiments over seven large (sub-continental) regions of the world. The intercomparison included AOGCMs with and without ocean flux correction and focused on summer and winter precipitation and surface air temperature. Biases in the simulation of present day climate with respect to observations and sensitivities at time of greenhouse gas (GHG) doubling were analysed. A broad inter-model range of regionally averaged biases and sensitivities was found, with marked inter-regional variability. Temperature biases were mostly in the range of ±5°C, with several instances of larger biases (even in excess of 10°C). Precipitation biases were mostly in the range of ±50%, but with a few instances of biases exceeding 100%. The range of sensitivities was lower for both variables.

The second segment of the analysis focused on results from nested RCMs and statistical downscaling experiments. Both these techniques were still at the early stages of their development and application, so that only a limited set of studies was available. The primary conclusions from these studies were that (a) both RCMs and downscaling techniques showed a promising performance in reproducing the regional detail in surface climate characteristics as forced by topography, lake, coastlines and land use distribution; and (b) high resolution surface forcings can modify the surface climate change signal at the sub-AOGCM grid scale.

Overall, the SAR placed low confidence in the simulation of regional climate change produced by available modelling tools, primarily because of three factors:

- errors in the reproduction of present day regional climate characteristics;
- wide range in the simulated regional climatic changes by different models;
- need to more comprehensively use regionalisation techniques to study the sub-AOGCM grid scale structure of the climate change signal.

Other points raised in the SAR were the need for better observational data sets for model validation at the regional scale and the need to examine higher order climate statistics.

10.1.2 The Regional Climate Problem

A definition of regional scale is difficult, as different definitions are often implied in different contexts. For example, definitions can be based on geographical, political or physiographic considerations, considerations of climate homogeneity, or considerations of model resolution. Because of this difficulty, an operational definition is adopted in this chapter based on the range of "regional scale" found in the available literature. From this perspective, regional scale is here defined as describing the range of 10^4 to 10^7 km^2. The upper end of the range (10^7 km^2) is also often referred to as sub-continental scale, and marked climatic inhomogeneity can occur within sub-continental scale regions in many areas of the globe. Circulations occurring at scales greater than 10^7 km^2 (here referred to as "planetary scales") are clearly dominated by general circulation processes and interactions. The lower end of the range (10^4 km^2) is representative of the smallest scales resolved by current regional climate models. Scales smaller than 10^4 km^2 are referred to as "local scale".

Given these definitions, the climate of a given region is determined by the interaction of forcings and circulations that occur at the planetary, regional and local spatial scales, and at a wide range of temporal scales, from sub-daily to multi-decadal. Planetary scale forcings regulate the general circulation of the global atmosphere. This in turn determines the sequence and characteristics of weather events and weather regimes that characterise the climate of a region. Embedded within the planetary scale circulation regimes, regional and local forcings and mesoscale circulations modulate the spatial and temporal structure of the regional climate signal, with an effect that can in turn influence planetary scale circulation features. Examples of regional and local scale forcings are those due to complex topography, land-use characteristics, inland bodies of water, land-ocean contrasts, atmospheric aerosols, radiatively active gases, snow, sea ice, and ocean current distribution. Moreover, climatic

variability of a region can be strongly influenced through teleconnection patterns originated by forcing anomalies in distant regions, such as in the El Niño-Southern Oscillation (ENSO) and North Atlantic Oscillation (NAO) phenomena.

The difficulty of simulating regional climate change is therefore evident. The effects of forcings and circulations at the planetary, regional and local scale need to be properly represented, along with the teleconnection effects of regional forcing anomalies. These processes are characterised by a range of temporal variability scales, and can be highly non-linear. In addition, similarly to what happens for the global Earth system, regional climate is also modulated by interactions among different components of the climate system, such as the atmosphere, hydrosphere, cryosphere, biosphere and chemosphere, which may require coupling of these components at the regional scale.

Therefore, a cross-disciplinary and multi-scale approach is necessary for a full understanding of regional climate change processes. This is based on the use of AOGCMs to simulate the global climate system response to planetary scale forcings and the variability patterns associated with broad regional forcing anomalies (see Chapter 9). The information provided by the AOGCMs can then be enhanced to account for regional and local processes via a suitable use of the regionalisation techniques discussed in this chapter.

10.2 Deriving Regional Information: Principles, Objectives and Assumptions

For some applications, the regional information provided by AOGCMs might suffice (Section 10.2.1), while in other cases regionalisation techniques are needed in order to enhance the regional information provided by coupled AOGCMs. The "added value" expected of a regionalisation technique essentially depends on the specific problem of interest. Examples in which regionalisation tools can enhance the AOGCM information include the simulation of the spatial structure of temperature and precipitation in areas of complex topography and land-use distribution, the description of regional and local atmospheric circulations (e.g., narrow jet cores, mesoscale convective systems, sea-breeze type circulations, tropical storms) and the representation of processes at high frequency temporal scales (e.g., precipitation frequency and intensity distributions, surface wind variability, monsoon front onset and transition times).

The basic principles behind the regionalisation methods identified here are discussed in Section 10.2.2, high resolution and variable resolution AGCM experiments; Section 10.2.3, RCMs; and Section 10.2.4, empirical/statistical and statistical/dynamical models. The general philosophy behind regionalisation techniques is to use input data from AOGCMs to produce more detailed regional information. By design, many of these techniques are not intended to strongly modify the planetary scale circulations produced by the forcing AOGCM. This ensures consistency with the AOGCM simulation and facilitates the interpretation of the additional detail as due to the increase in resolution. However, high and variable resolution AGCMs, as well as RCMs with sufficiently large domains, can yield significant modification of the large-scale circulations, often leading to an improved simulation of them. This would tend to increase confidence in the simulations, but the implications of inconsistencies with the AOGCM forcing fields would need to be considered carefully in the interpretation of the climate change information.

Note that RCMs and statistical models are often referred to as "downscaling" tools of AOGCM information. The concept of "downscaling" implies that the regional climate is conditioned but not completely determined by the larger scale state. In fact, regional states associated with similar larger scale states may vary substantially (e.g., Starr, 1942; Roebber and Bosart, 1998).

10.2.1 Coupled AOGCMs

The majority of climate change impact studies have made use of climate change information provided by transient runs with coupled AOGCMs without any further regionalisation processing. The primary reason for this is the ready availability of this information, which is global in nature and is routinely stored by major laboratories. Data can easily be drawn from the full range of currently available AOGCM experiments of the various modelling centres for any region of the world and uncertainty due to inter-model (or inter-run) differences can thus be evaluated (e.g., Hulme and Brown, 1998). In addition, data can be obtained for a large range of variables down to short (sub-daily) time-scales.

From the theoretical viewpoint, the main advantage of obtaining regional climate information directly from AOGCMs is the knowledge that internal physical consistency is maintained. The feedback resulting from climate change in a particular region on planetary scale climate and the climate of other regions is allowed for through physical and dynamical processes in the model. This may be an important consideration when the simulation of regional climate or climate change is compared across regions.

The limitations of AOGCM regional information are, however, well known. By definition, coupled AOGCMs cannot provide direct information at scales smaller than their resolution (order of several hundred kilometres), neither can AOGCMs capture the detailed effects of forcings acting at sub-grid scales (unless parametrized). Biases in the climate simulation at the AOGCM resolution can thus be introduced by the absence of sub-grid scale variations in forcing. As an example, a narrow (sub-grid scale) mountain range can be responsible for rain shadow effects at the broader scale. Many important aspects of the climate of a region (e.g., climatic means in areas of complex topography or extreme weather systems such as tropical cyclones) can only be directly simulated at much finer resolution than that of current AOGCMs. Analysis relevant to these aspects is undertaken with AOGCM output, but various qualifications need to be considered in the interpretation of the results. Past analyses have indicated that even at their smallest resolvable scales, which still fall under our definition of regional, AOGCMs have substantial problems in reproducing present day climate characteristics. The minimum skilful scale of a model is of several grid lengths, since these are necessary to describe the smallest wavelengths in the model and

since numerical truncation errors are most severe for the smallest resolved spatial scales. Furthermore, non-linear interactions are poorly represented for those scales closest to the truncation of a model because of the damping by dissipation terms and because only the contribution of larger scale (and not smaller scale) eddies is accounted for (e.g., von Storch, 1995).

Advantages and disadvantages of using AOGCM information in impact studies can weigh-up differently depending on the region and variables of interest. For example, in instances for which sub-grid scale variation is weak (e.g., for mean sea level pressure) the practical advantages of using direct AOGCM data may predominate (see also Chapter 13). However, even if resolution factors limit the feasibility of using regional information from coupled AOGCMs for impact work, AOGCMs are the starting point of any regionalisation technique presently used. Therefore, it is of utmost importance that AOGCMs show good performance in simulating circulation and climatic features affecting regional climates, such as jet streams and storm tracks. Indeed, most indications are that, in this regard, the AOGCM performance is generally improving, because of both increased resolution and improvements in the representation of physical processes (see Chapter 8).

10.2.2 High Resolution and Variable Resolution AGCM Experiments

Though simulations of many centuries are required to fully integrate the global climate system, for many applications regional information on climate or climate change is required for at most several decades. Over these time-scales AGCM, simulations are feasible at resolutions of the order of 100 km globally, or 50 km locally, with variable resolution models. This suggests identifying periods of interest within AOGCM transient simulations and modelling these with a higher resolution or variable resolution AGCM to provide additional spatial detail (e.g., Bengtsson *et. al.*, 1995; Cubasch *et al.*, 1995; Dèquè and Piedelievre, 1995).

Here the AGCM is used to provide a reinterpretation of the atmospheric response to the anomalous atmospheric forcing (from GHG and aerosols) experienced in a transient AOGCM simulation. Hence, both this forcing and its accumulated effect on the ocean surface have to be provided to the AGCM. In a typical experiment (e.g., May and Roeckner, 2001), two time slices, say 1961 to 1990 and 2071 to 2100, are selected from a transient AOGCM simulation. The simulations include prescribed time-dependent GHG and aerosol concentrations as in the corresponding periods of the AOGCM run. Also prescribed as lower boundary conditions are the time-dependent Sea Surface Temperature (SST) and sea-ice distributions simulated by the AOGCM. The AGCM simulations are initialised using atmospheric and land-surface conditions interpolated from the corresponding AOGCM fields.

Alternative experimental designs may be more appropriate. Large systematic errors in the AOGCM simulation of SST and sea ice may induce significant biases in the climatology of the AGCM. In this case, observed SSTs and sea-ice distributions could be used for the present day simulation and changes derived from the AOGCM experiment can be added to provide the forcing for the anomaly simulation. If the AOGCM calculates the aerosol concentrations from prescribed sources then the AGCM may use the same method. This has the advantage of providing aerosol concentrations consistent with the AGCM circulations, although its global and regional effects may be different from those in the AOGCM.

The philosophy behind the use of high or variable resolution AGCM simulations is that, given the SST, sea ice, trace gas and aerosol forcing, relatively high-resolution information can be obtained globally or regionally without having to perform the whole transient simulation with high resolution models. The main theoretical advantage of this approach is that the resulting simulations are globally consistent, capturing remote responses to the impact of higher resolution. The use of higher resolution can lead to improved simulation of the general circulation in addition to providing regional detail (e.g., HIRETYCS, 1998; Stratton, 1999a).

In general, AGCMs will evolve their own planetary scale climatology. Therefore, in a climate change simulation they are providing a reinterpretation of the impact on the atmosphere of the sea surface and radiative forcings compared to that given by the driving AOGCM. This may lead to inconsistency with the AOGCM-derived forcing. This issue has yet to be explored but should be considered carefully when interpreting AGCM responses. It would be of less concern if a model simulation of the resolved planetary scale variables were asymptoting to a solution as resolution increased, i.e., if the solution would not change fundamentally in character with resolution but just add extra detail at the finer scales. Evidence shows that this is not the case at the current resolution of AOGCMs (Williamson, 1999).

A current weakness of high resolution AGCMs is that they generally use the same formulations as at the coarse resolution for which these have been optimised to reproduce current climate. Some processes may be represented less accurately when finer scales are resolved and so the model formulations would need to be optimised for use at higher resolution. Experience with high resolution GCMs is still limited, so that, at present, increasing the resolution of an AGCM generally both enhances and degrades different aspects of the simulations. With global variable resolution models, this issue is further complicated as the model physics parametrizations have to be designed in such a way that they can be valid, and function correctly, over the range of resolutions covered by the model.

Another issue concerning the use of variable resolution models is that feedback effects from fine scales to larger scales are represented only as generated by the region of interest. Conversely, in the real atmosphere, feedbacks derive from different regions and interact with each other so that a variable resolution model, based on a single high resolution region, might give an improper description of fine-to-coarse scale feedbacks. In addition, a sufficient minimal resolution must be retained outside the high resolution area of interest in order to prevent a degradation of the simulation of the whole global system.

Use of high resolution and variable resolution global models is computationally very demanding, which poses limits to the increase in resolution obtainable with this method. However, it has been suggested that high-resolution AGCMs could be used to obtain forcing fields for higher resolution RCMs or statistical downscaling, thus effectively providing an intermediate step between AOGCMs and regional and empirical models.

10.2.3 Regional Climate Models (RCMs)

The nested regional climate modelling technique consists of using initial conditions, time-dependent lateral meteorological conditions and surface boundary conditions to drive high-resolution RCMs. The driving data is derived from GCMs (or analyses of observations) and can include GHG and aerosol forcing. A variation of this technique is to also force the large-scale component of the RCM solution throughout the entire domain (e.g., Kida *et al.*, 1991; Cocke and LaRow, 2000; von Storch *et al.*, 2000)

To date, this technique has been used only in one-way mode, i.e., with no feedback from the RCM simulation to the driving GCM. The basic strategy is, thus, to use the global model to simulate the response of the global circulation to large-scale forcings and the RCM to (a) account for sub-GCM grid scale forcings (e.g., complex topographical features and land cover inhomogeneity) in a physically-based way; and (b) enhance the simulation of atmospheric circulations and climatic variables at fine spatial scales.

The nested regional modelling technique essentially originated from numerical weather prediction, and the use of RCMs for climate application was pioneered by Dickinson *et al.* (1989) and Giorgi (1990). RCMs are now used in a wide range of climate applications, from palaeoclimate (Hostetler *et al.*, 1994, 2000) to anthropogenic climate change studies (Section 10.5). They can provide high resolution (up to 10 to 20 km or less) and multi-decadal simulations and are capable of describing climate feedback mechanisms acting at the regional scale. A number of widely used limited area modelling systems have been adapted to, or developed for, climate application. More recently, RCMs have begun to couple atmospheric models with other climate process models, such as hydrology, ocean, sea-ice, chemistry/aerosol and land-biosphere models.

Two main theoretical limitations of this technique are the effects of systematic errors in the driving fields provided by global models; and lack of two-way interactions between regional and global climate (with the caveats discussed in Section 10.2.2 for variable resolution models). Practically, for a given application, consideration needs to be given to the choice of physics parametrizations, model domain size and resolution, technique for assimilation of large-scale meteorological conditions, and internal variability due to non-linear dynamics not associated with the boundary forcing (e.g., Giorgi and Mearns, 1991, 1999; Ji and Vernekar 1997). Depending on the domain size and resolution, RCM simulations can be computationally demanding, which has limited the length of many experiments to date. Finally, GCM fields are not routinely stored at high temporal frequency (6-hourly or higher), as required for RCM boundary conditions, and thus careful co-ordination between global and regional modellers is needed in order to perform RCM experiments.

10.2.4 Empirical/Statistical and Statistical/Dynamical Downscaling

Statistical downscaling is based on the view that regional climate may be thought of as being conditioned by two factors: the large-scale climatic state, and regional/local physiographic features (e.g., topography, land-sea distribution and land use; von Storch, 1995, 1999a). From this viewpoint, regional or local climate information is derived by first determining a statistical model which relates large-scale climate variables (or "predictors") to regional and local variables (or "predictands"). Then the predictors from an AOGCM simulation are fed into this statistical model to estimate the corresponding local and regional climate characteristics.

A range of statistical downscaling models, from regressions to neural networks and analogues, has been developed for regions where sufficiently good data sets are available for model calibration. In a particular type of statistical downscaling method, called statistical-dynamical downscaling (see Section 10.6.2.3), output of atmospheric mesoscale models is used in statistical relationships. Statistical downscaling techniques have their roots in synoptic climatology (*Growetterlagen*; e.g., Baur *et al.*, 1944; Lamb, 1972) and numerical weather prediction (Klein and Glahn, 1974), but they are also currently used for a wide range of climate applications, from historical reconstruction (e.g., Appenzeller *et al.*, 1998; Luterbacher *et al.*, 1999), to regional climate change problems (see Section 10.6). A number of review papers have dealt with downscaling concepts, prospects and limitations: von Storch (1995); Hewitson and Crane (1996); Wilby and Wigley (1997); Zorita and von Storch (1997); Gyalistras *et al.* (1998); Murphy (1999,2000).

One of the primary advantages of these techniques is that they are computationally inexpensive, and thus can easily be applied to output from different GCM experiments. Another advantage is that they can be used to provide local information, which can be most needed in many climate change impact applications. The applications of downscaling techniques vary widely with respect to regions, spatial and temporal scales, type of predictors and predictands, and climate statistics (see Section 10.6). In addition, empirical downscaling methods often offer a framework for testing the ability of physical models to simulate the empirically found links between large-scale and small-scale climate (Busuioc *et al.*, 1999; Murphy, 1999; Osborn *et al.*, 1999; von Storch *et al.*, 1993; Noguer, 1994).

The major theoretical weakness of statistical downscaling methods is that their basic assumption is not verifiable, i.e., that the statistical relationships developed for present day climate also hold under the different forcing conditions of possible future climates. In addition, data with which to develop relationships may not be readily available in remote regions or regions with complex topography. Another caveat is that these empirically-based techniques cannot account for possible systematic changes in regional forcing conditions or feedback processes. The possibility of tailoring the statistical model to the requested regional or local information is a distinct advantage. However, it has the drawback that a systematic assessment of the uncertainty of this type of technique, as well as a comparison with other techniques, is difficult and may need to be carried out on a case-by-case basis.

10.2.5 Sources of Uncertainty in the Generation of Regional Climate Change Information

There are several levels of uncertainty in the generation of regional climate change information. The first level, which is not dealt with in this chapter, is associated with alternative scenarios of future emissions, their conversion to atmospheric concentrations and the radiative effects of these (see Chapter 13). The second level is related to the simulation of the transient climate response by AOGCMs for a given emission scenario (see also Chapters 8 and 9). This uncertainty has a global aspect, related to the model global sensitivity to forcing, as well as a regional aspect, more tied to the model simulation of general circulation features. This uncertainty is important both, when AOGCM information is used for impact work without the intermediate step of a regionalisation tool, and when AOGCM fields are used to drive a regionalisation technique. The final level of uncertainty occurs when the AOGCM data are processed through a regionalisation method.

Sources of uncertainty in producing regional climate information are of different nature. On the modelling and statistical downscaling side, uncertainties are associated with imperfect knowledge and/or representation of physical processes, limitations due to the numerical approximation of the model's equations, simplifications and assumptions in the models and/or approaches, internal model variability, and inter-model or inter-method differences in the simulation of climate response to given forcings. It is also important to recognise that the observed regional climate is sometimes characterised by a high level of uncertainty due to measurement errors and sparseness of stations, especially in remote regions and in regions of complex topography. Finally, the internal variability of the global and regional climate system adds a further level of uncertainty in the evaluation of a climate change simulation.

Criteria to evaluate the level of confidence in a regional climate change simulation can be based on how well the models reproduce present day climate or past climates and how well the climate change simulations converge across models and methods (see Chapters 8 and 9). These criteria will be drawn upon in evaluating available simulations. We add that the emerging activity of seasonal to interannual climate forecasting, particularly at the regional scale, may give valuable insights into the capability of models to simulate climatic changes and may provide objective methodologies for evaluating the long-term prediction performance of climate models at the regional scale.

10.3 Regional Attributes of AOGCMs

10.3.1 Simulations of Current Climate

10.3.1.1 Mean climate
Although current AOGCMs simulate well the observed global pattern of surface temperature (see Chapter 8), at the regional scale substantial biases are evident. To give an overview of the regional performance of current models, results are presented of Giorgi and Francisco (2000b), who compared model and observed seasonal mean temperature and precipitation averaged

for the regions indicated in Figure 10.1. The AOGCM experiments they considered were a selection of those available through the IPCC Data Distribution Centre (DDC) and included single simulations using the CSIRO Mk2, CCSR/NIES and ECHAM/OPYC models, a three-member ensemble of CGCM1 simulations and a four-member ensemble of HadCM2 simulations (see Table 9.1 for further model details). Figure 10.2 shows the biases in regionally averaged seasonal mean surface temperature and precipitation for 1961 to 1990 using as reference the gridded analysis of New *et al.* (1999). Nearly all regional temperature biases are within the range of ±4°C. The main exceptions to this are negative biases of more than 5°C in some models over Asia in DJF. Precipitation biases are mostly between −40 and +80%, with the exception of positive biases in DJF in excess of 100% over central America (CAM), northern Africa (WAF and SAH), Alaska (ALA), and some parts of Asia (EAS, SAS, TIB and NAS). The regional biases of Figure 10.2 are, in general terms, smaller than those of a similar analysis presented in the SAR (see also Kittel *et al.*, 1998) which, for example, showed regional temperature biases as high as 10 to 15°C in some models and regions. Given that the current analysis also includes many more regions, this difference in general performance strongly suggests that simulation of surface climate at the sub-continental scale is improved in current generation AOGCMs.

Current generation AOGCM simulations in which historical changes in climate forcing over the 20th century are used enable simulated regional climatic trends to be assessed against observations. This was done by Boer *et al.* (2000a) for temperature and precipitation for the regions of southern Europe, North America, Southeast Asia, Sahel and Australia (defined as in the SAR) using the CGCM1 model. Simulated and observed regional linear temperature trends agreed for all regions, except the Sahel, when sulphate forcing was included. Little could be said about agreement in observed and model precipitation trends as these trends were weak over the period in both the model and the observations.

It should be stressed that assessments of model regional performance based on area-averaging of AOGCM output over broad regularly-shaped regions should not be assumed to apply to all areas within these regions. Many of the regions considered contain a number of distinct climate regimes, and model performance may vary considerably from regime to regime. For the purpose of assessing model performance in a particular region, more detailed analysis is appropriate.

Where studies have examined spatial patterns within regions (e.g., Joubert and Tyson, 1996; Labraga and Lopez, 1997; Lal *et al.*, 1998a), reasonable correspondence with observations was found, especially for temperature and mean sea level pressure (MSLP). Most studies focus on seasonal mean conditions, but models can be analysed to focus on simulation of specific climate features. For example, Arritt *et al.* (2000) examined circulation and precipitation patterns associated with the onset of the North American monsoon in simulations with the HadCM2 model, and found this feature to be well simulated. Some studies have identified important errors in current simulations of regional MSLP, such as the tendency for pressure to be too low over Europe and too high north and south of this area (Machenhauer *et al.*, 1998). Such errors contribute significantly to local temperature and precipitation biases both in the global climate model and in nested high-resolution RCM simulations (Risbey and Stone, 1996; Machenhauer *et al.*, 1998; Noguer *et al.*, 1998).

As would be expected, GCM simulations of current climate are often poor at the local scale (e.g., Schubert, 1998). However, in areas without complex topography, it is possible for the model results at individual grid points to compare well with observations (Osborn *et al.*, 1999).

10.3.1.2 Climate variability and extreme events
Analysis of global climate model performance in reproducing observed regional climate variability has given widely varying results depending on model and region. Interannual variability in temperature was assessed regionally, as well as globally, in a long control simulation with the HadCM2 model (Tett *et al.*, 1997). Many aspects of model variability compared well against observations, although there was a tendency for temperature variability to be too high over land. In the multi-regional study of Giorgi and Francisco (2000a), both regional temperature and precipitation interannual variability of HadCM2 were found to be generally overestimated. Similar results were obtained in the European study of Machenhauer *et al.* (1998) using the ECHAM/OPYC3 model. However, in a 200-year control simulation with the CGCM1 model (see Table 9.1), Flato *et al.* (2000) noted that simulated interannual variability in seasonal temperature and precipitation compared well with observations both globally and in five selected study regions (Sahel, North America, Australia, southern Europe and Southeast Asia).

Comparison against observations of daily precipitation variability as simulated at grid boxes in GCMs is problematic because the corresponding variability in the real world operates at much finer spatial scale (see Hennessy *et al.*, 1997). A significant development in this area has been the work of Osborn and Hulme (1997) who devised a method of calculating grid box average observed daily precipitation that corrects for biases commonly introduced by insufficient station density. Using this correction, agreement between observations and the results of the CSIRO GCM were significantly improved. In an analysis of different AGCMs over Europe, Osborn and Hulme (1998) found that the models commonly simulated precipitation in winter to be more frequent and less intense than observed. Daily temperature variability over Europe was found to be too high in winter in the Hadley Centre model (Gregory and Mitchell, 1995) and in winter and spring in the ECHAM3/LSG model (Buishand and Beersma, 1996).

Synoptic circulation variability at daily and longer time-scales operates at a spatial scale which GCMs can simulate directly and work has focused on GCM performance in this area (e.g., Katzfey and McInnes, 1996; Huth, 1997; Schubert, 1998; Wilby *et al.*, 1998a; Osborn *et al.*, 1999; Fyfe, 1999). Regions studied include North America, Europe, southern Africa, Australia and East Asia. Although in many respects model performance is good, some studies have noted synoptic variability to be less than in the observations and the more

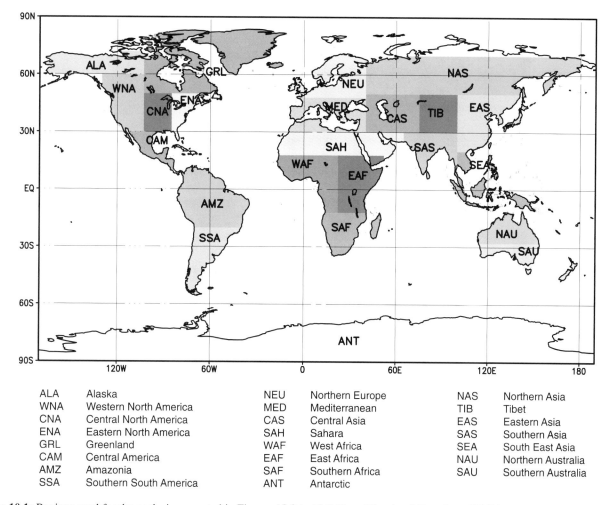

Figure 10.1: Regions used for the analysis presented in Figures 10.2 to 10.5 (from Giorgi and Francisco, 2000b).

ALA	Alaska	NEU	Northern Europe	NAS	Northern Asia
WNA	Western North America	MED	Mediterranean	TIB	Tibet
CNA	Central North America	CAS	Central Asia	EAS	Eastern Asia
ENA	Eastern North America	SAH	Sahara	SAS	Southern Asia
GRL	Greenland	WAF	West Africa	SEA	South East Asia
CAM	Central America	EAF	East Africa	NAU	Northern Australia
AMZ	Amazonia	SAF	Southern Africa	SAU	Southern Australia
SSA	Southern South America	ANT	Antarctic		

extreme deviations from the mean flow to be less intense or less frequent than observed (e.g., Osborn *et al.*, 1999).

Simulated climatic variability has also been examined as part of assessing model representation of the link between atmospheric circulation and local climate. Results have shown considerable regional differences. Osborn *et al.* (1999) examined the relationship between the circulation anomalies and grid-box average temperature and precipitation anomalies and found this to be well represented by the HadCM2 model. However, Wilby and Wigley (2000) found HadCM2 less satisfactory in reproducing the observed correlations between daily precipitation over six regions in the United States and a variety of different atmospheric predictor variables. In a similar investigation over Europe, Busuioc *et al.* (1999) found the performance of the ECHAM3 AGCM to be good in some seasons.

Widmann and Bretherton (2000) examined precipitation variability from atmospheric reanalyses as an alternative method of validating GCMs under historic flow conditions. By virtue of this approach, the atmospheric circulation is constrained to be unbiased, but the precipitation is calculated according to model physics and parametrizations. Results based on the GCM used in the reanalysis were found to be in good agreement with observations over Oregon and Washington.

10.3.2 Simulations of Climate Change

10.3.2.1 Mean climate

Giorgi and Francisco (2000b) analysed regional temperature change in five AOGCMs under a range of forcing scenarios. In all regions warming depended strongly on the forcing scenario used and inter-model differences in simulated warming were large compared to differences between ensemble members from a single model. Figure 10.3a presents some results from Giorgi and Francisco (2000b) relative to scenarios of 1%/yr increase in GHG concentration without sulphate aerosol effects. Most regional warmings for 2071 to 2100 compared to 1961 to 1990 are in the range of 2 to 8°C. Exceptions are the high northern latitudes in DJF (5 to 11°C in GRL, NAS and ALA) and central and eastern Asia in DJF (3 to 11°C in EAS, CAS and TIB). In many regions, the warming is 2 to 3°C higher in the CCC simulation than in the other models (e.g., in the African regions of WAF, EAF, SAF and SAH). Giorgi and Francisco (2000b) also considered corresponding simulations in which large increases in sulphate aerosols (consistent with IS92a emission scenarios) were included in addition to the GHG changes and found significantly reduced regional warming (the range for most regions is 1.5 to 7°C) (Figure 10.3b). Nearly all the temperature changes in

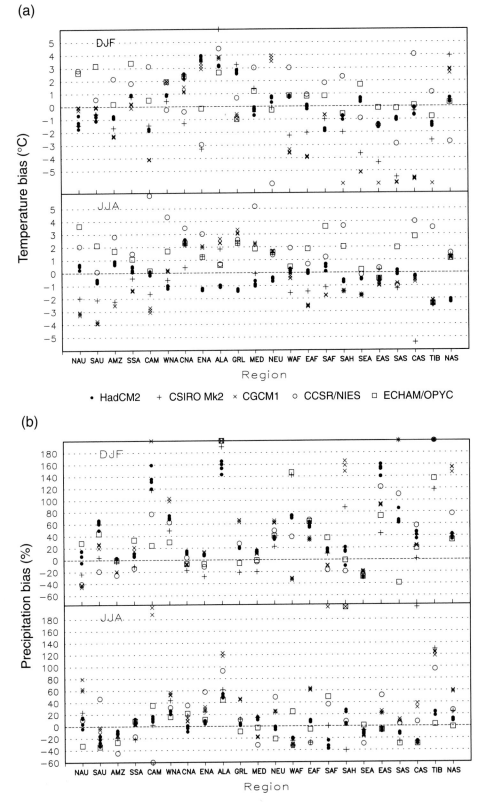

Figure 10.2: Surface temperature biases (in °C) and precipitation biases (% of observed) for 1961 to 1990 for experiments using the AOGCMs of CSIRO Mk2, CCSR/NIES, ECHAM/OPYC, CGCM1 (a three-member ensemble) and HadCM2 (a four-member ensemble) with historical forcing including sulphates (further experimental details are in Table 9.1). Regions are as indicated in Figure 10.1 and observations are from New *et al.* (1999a,b). (a) surface air temperature, (b) precipitation (from Giorgi and Francisco, 2000b).

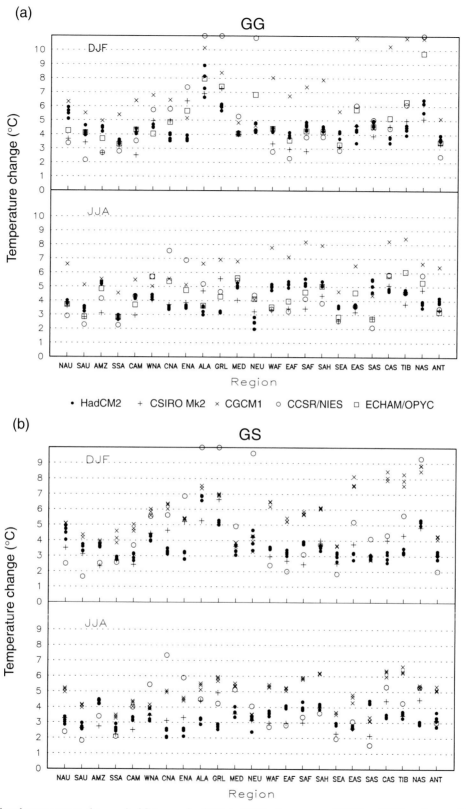

Figure 10.3: Simulated temperature changes in °C (mean for 2071 to 2100 minus mean of 1961 to 1990) under conditions of 1%/yr increasing CO_2 without and with sulphate forcing using experiments undertaken with the AOGCMs of CSIRO Mk2, CCSR/NIES, ECHAM/OPYC, CGCM1 and Hadley Centre (further experimental details are in Table 9.1). Under both forcing scenarios a four-member ensemble is included of the Hadley Centre model, and under the CO_2 plus sulphate scenario a three-member ensemble is included for the CGCM1 model. (a) increased CO_2 only (GG), (b) increased CO_2 and sulphate aerosols (GS). Global model warming values in the CO_2 increase-only experiments are 3.07°C for HadCM2 (ensemble average), 3.06°C for CSIRO Mk2, 4.91°C for CGCM1, 3.00°C for CCSR/NIES and 3.02°C for ECHAM/OPYC. Global model warming values for the experiments including sulphate forcing are 2.52°C for HadCM2 (ensemble average), 2.72°C for CSIRO Mk2, 3.80°C for CGCM1 (ensemble average) and 2.64°C for CCSR/NIES (from Giorgi and Francisco, 2000b).

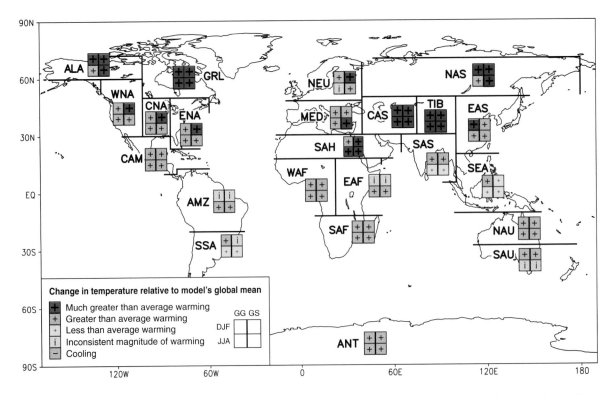

Figure 10.4: Analysis of inter-model consistency in regional warming relative to each model's global warming, based on the results presented in Figure 10.3. Regions are classified as showing either agreement on warming in excess of 40% above the global average ("Much greater than average warming"), agreement on warming greater than the global average ("Greater than average warming"), agreement on warming less than the global average ("Less than average warming"), or disagreement amongst models on the magnitude of regional relative warming ("Inconsistent magnitude of warming"). There is also a category for agreement on cooling (which is not used). GG is the greenhouse gas only case (see Figure 10.3a), and, GS, the greenhouse gas with increased sulphate case (see Figure 10.3b). In constructing the figure, ensemble results were averaged to a single case, and "agreement" was defined as having at least four of the five GG models agreeing or three of the four GS models agreeing. The global annual average warming of the models used span 3.0 to 4.9°C for GG and 2.5 to 3.8°C for GS, and therefore a regional 40% amplification represents warming ranges of 4.2 to 6.9°C for GG and 3.5 to 5.3°C for GS.

Figure 10.3a were statistically significant at the 5% confidence level (Giorgi and Francisco, 2000b).

Inter-model differences in regional warming partially reflect differences in the global climate sensitivities of the models concerned. This effect may be set aside by comparing the regional warmings given in Giorgi and Francisco (2000b) with the corresponding global average warmings of the simulations used (Figure 10.4). Nearly all land areas warm more rapidly than the global average, particularly those at high latitudes in the cold season. For both the non-sulphate and sulphate cases, in the northern high latitudes, central Asia and Tibet (ALA, GRL, NAS, CAS and TIB) in DJF and in northern Canada, Greenland and central Asia and Tibet (GRL, CAS and TIB) in JJA, the warming is in excess of 40% above the global average. In both cases, warming is less than the global average in South and Southeast Asia, and southern South America (SAS, SEA and SSA) in JJA. In this analysis, differences between the non-sulphate and sulphate cases are minor. A strong contribution to the enhancement of warming over cold climate regions is given by the snow and sea ice albedo feedback mechanism (Giorgi and Francisco, 2000b). The snow albedo feedback also tends to enhance warming over high elevation regions (Fyfe and Flato, 1999).

In line with the globally averaged precipitation increase given by all models (see Chapter 9), precipitation is also simulated to increase regionally in the majority of cases. However, regions of precipitation decrease are also simulated. Precipitation reduction can be due to changes in large and synoptic scale features (e.g., changes in storm track characteristics) and/or to local feedback processes (e.g., between soil moisture and precipitation). The results of the regional analysis of Giorgi and Francisco (2000b) are presented in Figure 10.5 (as percentage changes for each model, region and forcing scenario) and are used in an analysis of inter-model consistency which is presented in Figure 10.6. In both the non-sulphate and sulphate cases for DJF, most simulations show increased precipitation for regions in the mid- to high latitudes of the Northern Hemisphere (ALA, GRL, WNA, ENA, CNA, NEU, NAS, CAS and TIB) and over Antarctica (ANT). In the tropics, models consistently show increase in Africa (EAF and WAF), increase or little change in South America (AMZ) and little change in Southeast Asia (SEA). Simulated regional precipitation decreases are common in subtropical latitudes, but only for central America (CAM) and northern Australia (NAU) are decreases indicated by most models in both cases. The pattern is broadly similar in JJA,

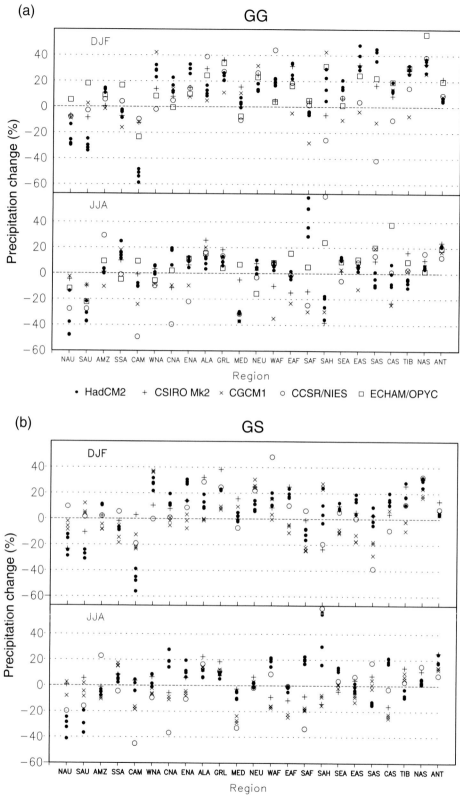

Figure 10.5: As Figure 10.3, but for percentage precipitation change (from Giorgi and Francisco, 2000b).

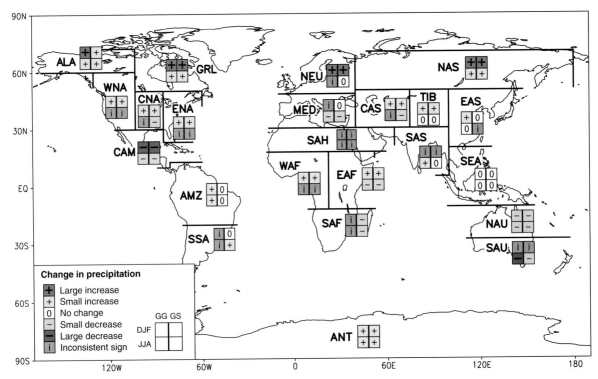

Figure 10.6: Analysis of inter-model consistency in regional precipitation change based on the results presented in Figure 10.5. Regions are classified as showing either agreement on increase with an average change of greater than 20% ("Large increase"), agreement on increase with an average change between 5 and 20% ("Small increase"), agreement on a change between –5 and +5% or agreement with an average change between –5 and 5% ("No change"), agreement on decrease with an average change between –5 and –20% ("Small decrease"), agreement on decrease with an average change of less than –20% ("Large decrease"), or disagreement ("Inconsistent sign"). GG is the greenhouse gas only case (see Figure 10.5a), and, GS, the greenhouse gas with increased sulphate case (see Figure 10.5b). In constructing the figure, ensemble results were averaged to a single case, and "agreement" was defined as having at least of four the five GG models agreeing or three of the four GS models agreeing.

although with some features shifting northwards. Only the high-latitude regions (ANT, ALA, GRL and NAS) show consistent increase. There is disagreement on the direction of change in a number of regions in the northern mid-latitudes and the sub-tropics, although consistent decrease is evident in the Mediterranean Basin (MED) and central American (CAM) regions. Some regions along the Inter-Tropical Convergence Zone (ITCZ) show consistent increase or little change (AMZ , SEA and SAS), but eastern Africa (EAF) shows a consistent decrease. In the Southern Hemisphere, only Australia (SAU and NAU) shows a consistent pattern of change in both cases (decrease). When the non-sulphate and sulphate cases are contrasted, more frequent simulated precipitation decrease may be noted in parts of North America, Africa and Asia for the case with increased sulphate aerosols (see results in Figure 10.6 for SAF, CNA, CAS and EAS).

The magnitude of regional precipitation change varies considerably amongst models, with the typical range being around 0 to 50% where the direction of change is strongly indicated and around –30 to +30% where it is not. Larger ranges occur in some regions (e.g., –30 to +60% in southern Africa in JJA for GHG only forcing), but this occurs mainly in regions of low seasonal precipitation where the implied range in absolute terms would not be large. Changes are consistently

large (greater than 20% averaged across models) in both the sulphate and non-sulphate cases in northern high latitude regions (GRL, NEU and NAS, positive change) in DJF and central America (CAM, negative change) in DJF. The number of precipitation changes statistically significant at the 5% confidence level varied widely across regions and seasons.

A number of new transient AOGCM simulations for the SRES A2 and B2 scenarios have recently become available and a preliminary analysis was conducted by the lead authors. This follows the procedure similar to that described in this section in relation to Figures 10.3 to 10.6. The results are presented in Box 10.1.

The analysis described above is for broad area-averages only and the results described should not be assumed to apply to all areas within these regions. More focused regional studies have examined within-region spatial patterns of change (Joubert and Tyson, 1996; Machenhauer *et al.*, 1996, 1998; Pittock *et al.*, 1995; Whetton *et al.*, 1996b; Carril *et al.*, 1997; Labraga and Lopez, 1997). Such studies can reveal important features which are consistent amongst models but are not apparent in area-average regional results. For example, Labraga and Lopez (1997) noted a tendency for simulated rainfall to decrease in northern Amazonia and to increase in southern parts of this region. Jones R.N. *et al.* (2000) noted a

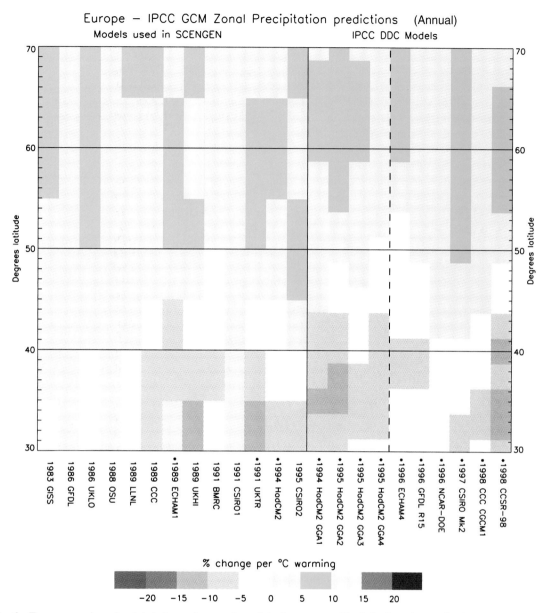

Figure 10.7: For the European region, simulated change in annual precipitation, averaged by latitude and normalised to % change per °C of global warming. Results are given for twenty-three enhanced GHG simulations (forced by CO_2 change only) produced between the years 1983 and 1998. The earlier experiments are those used in the SCENGEN climate scenario generator (Hulme *et al.*, 1995) and include some mixed-layer $1\times$ and $2\times CO_2$ equilibrium experiments; the later ones are the AOGCM experiments available through the DDC. From Hulme *et al.* (2000).

predominance of rainfall increase in the central equatorial Pacific (northern Polynesia), but in the areas to the west and south-west the direction of rainfall change was not clearly indicated.

To illustrate further inter-model variations in simulated regional precipitation change, results obtained in model inter-comparison studies for the Australian, Indian, North American and European regions are examined. All of these regions have been extensively studied over the years using equilibrium $2\times CO_2$ experiments (such as those featured in IPCC, 1990), first generation transient coupled AOGCMs (as in the SAR), and more recent AOGCMs available in the DDC (Table 9.1).

This comparison also enables an assessment of how the regional precipitation projections have changed as the models evolved.

In the Australian region, the pattern of simulated precipitation change in winter (JJA) has remained broadly similar across these three groups of experiments and consists of rainfall decrease in sub-tropical latitudes and rainfall increase south of 35 to 40°S (Whetton *et al.*, 1996a, 2001). However, as the latitude of the boundary between these two zones varied between models, southernmost parts of Australia lay in the zone where the direction of precipitation change was inconsistent amongst models. In summer (DJF) the equilibrium $2\times CO_2$ experiments showed a

Box 10.1: Regional climate change in AOGCMs which use SRES emission scenarios

Introduction

This box summarises results on regional climate change obtained from a set of nine AOGCM simulations undertaken using SRES preliminary marker emission scenarios A2 and B2. The models are CGCM2, CSIRO Mk2, CSM 1.3, ECHAM4/OPYC, GFDL_R30_c, HadCM3, MRI2, CCSR/NIES2, DOE PCM, (numbered 7, 10, 12, 15, 18, 23, 27, 31 and 30 in Chapter 9, Table 9.1). The results are based on data for 2071 to 2100 and 1961 to 1990 that have been directly analysed and assessed by the lead authors. These results should be treated as preliminary only.

Analysis

Regional changes in precipitation and temperature were calculated using the same methodology as that of Giorgi and Francisco (2000b) (see Figures 10.1, 10.3 and 10.5). The results were then assessed for inter-model consistency using the same method as that used in Figures 10.4 and 10.6 for the earlier set of simulations. The results for temperature are in Box 10.1, Figure 1 and for precipitation in Box 10.2, Figure 2.

The SRES results may be compared with the earlier results summarised in Figures 10.4 and 10.6 (which will be referred to here as the IS92a results). However, it should be noted that these two sets of results differ in the set of models used (both in the model versions and in the total number of simulations), and in the scenarios contrasted in each case (for IS92a it is GHG-only versus GHG+sulphate and for SRES it is A2 versus B2). Also, due to differences in the number of models, thresholds for agreement are not the same in each case (although they have been chosen to be as nearly equivalent as possible).

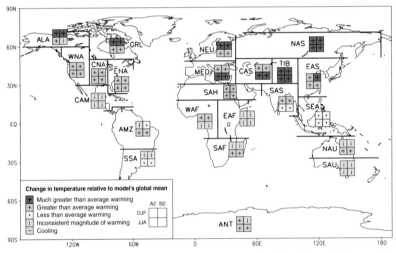

Box 10.1, Figure 1: Analysis of inter-model consistency in regional relative warming (warming relative to each model's global warming). Regions are classified as showing either agreement on warming in excess of 40% above the global average ('Much greater than average warming'), agreement on warming greater than the global average ('Greater than average warming'), agreement on warming less than the global average ('Less than average warming'), or disagreement amongst models on the magnitude of regional relative warming ('Inconsistent magnitude of warming'). There is also a category for agreement on cooling (which never occurs). A consistent result from at least seven of the nine models is deemed necessary for agreement. The global annual average warming (DJF and JJA combined) of the models used span 1.2 to 4.5°C for A2 and 0.9 to 3.4°C for B2, and therefore a regional 40% amplification represents warming ranges of 1.7 to 6.3°C for A2 and 1.3 to 4.7°C for B2.

Results

SRES

- Under both SRES cases, most land areas warm more rapidly than the global average. The warming is in excess of 40% above the global average in all high northern latitude regions and Tibet (ALA, GRL, NEU, NAS and TIB) in DJF, and in the Mediterranean basin, central and northern Asia and Tibet (MED, CAS, NAS, and TIB) in JJA. Only in South Asia and southern South America (SAS and SSA) in JJA and southeast Asia (SEA) in both seasons do the models consistently show warming less than the global average.
- For precipitation, consistent increase is evident in both SRES scenarios over high latitude regions (ALA, GRL, NAS and ANT) in both seasons, northern mid-latitude regions and tropical Africa (WNA, ENA, NEU, CAS, TIB, WAF and EAF) in DJF, and South Asia, East Asia and Tibet (SAS, EAS and TIB) in JJA. Consistent precipitation decrease is present over Central America (CAM) in DJF and over Australia and southern Africa (NAU, SAU and SAF) in JJA.
- Differences between the A2 and B2 results are minor and are mainly evident for precipitation. In the B2 scenario there are fewer regions showing consistently large precipitation changes, and there is a slight increase in the frequency of regions showing "inconsistent" and "no change" results. As the climate forcing is smaller in the B2 case and the climate response correspondingly weaker, some differences of this nature are to be expected.

SRES versus IS92a

- In broad terms, the temperature results from SRES are similar to the IS92a results. In each of the two SRES and IS92a cases, warming is in excess of 40% above the global average in Alaska, northern Canada, Greenland, northern Asia, and Tibet (ALA, GRL, NAS and TIB) in DJF and in central Asia and Tibet (CAS and TIB) in JJA. All four cases also show warming less than the global average in South and Southeast Asia, and southern South America (SAS, SEA and SSA) in JJA.

- The main difference in the results is that there are substantially more instances for the SRES cases where there is disagreement on the magnitude of the relative regional warming. This difference is mainly evident in tropical and Southern Hemisphere regions.

- The precipitation results from SRES are also broadly similar to the corresponding IS92a results. There are many regions where the direction of precipitation change (although not necessarily the magnitude of this change) is consistent across all four cases. In DJF this is true for increase in northern mid- to high latitude regions, Antarctica and tropical Africa (ALA, GRL, WNA, ENA, NEU, NAS, TIB, CAS, WAF, EAF and ANT) and decrease in Central America (CAM). In JJA it is true for increase in high latitude regions (ALA, GRL, NAS and ANT) and for decrease in southern and northern Australia (SAU and NAU). Little change in Southeast Asia in DJF and little change or increase over South Asia in JJA are also consistent results.

- Although there are no cases where the SRES and IS92a results indicate precipitation changes of opposite direction, there are some notable differences. In the Sahara and in East Asia (SAH and EAS) in JJA, the results for both SRES scenarios show consistent increase whereas this was not true in either of the IS92a cases. On the other hand, in central North America and northern Australia (CNA and NAU) in DJF, and in East Africa (EAF) in JJA, the results for both SRES scenarios show model disagreement whereas the IS92a scenarios showed a consistent direction of change (increase in CNA, and decrease in EAF and NAU). It is also notable that the consistent decrease in JJA precipitation over the Mediterranean basin (MED) seen for both IS92a cases is present for SRES only for the A2 scenario (for which the decrease is large).

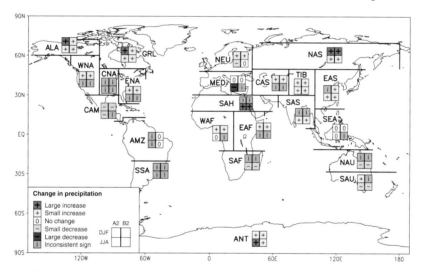

Box 10.1, Figure 2: Analysis of inter-model consistency in regional precipitation change. Regions are classified as showing either agreement on increase with an average change of greater than 20% ('Large increase'), agreement on increase with an average change between 5 and 20% ('Small increase'), agreement on a change between −5 and +5% or agreement with an average change between −5 and 5% ('No change'), agreement on decrease with an average change between −5 and −20% ('Small decrease'), agreement on decrease with an average change of less than −20% ('Large decrease'), or disagreement ('Inconsistent sign'). A consistent result from at least seven of the nine models is deemed necessary for agreement.

Uncertainty

The above comparisons concern the quantification of two different sources of uncertainty represented in the cascade of uncertainty described in Chapter 13, Section 13.5.1 (Figure 13.2). These include uncertainties in future emissions (IS92a GG and GS; SRES A2 and B2), and uncertainties in modelling the response of the climate system to a given forcing (samples of up to nine AOGCMs). Agreement across the different scenarios and climate models suggests, relatively speaking, less uncertainty about the nature of regional climate change than where there is disagreement. For example, the agreement for northern latitude winter precipitation extends across all emission scenarios and all models, whereas there is considerable disagreement (greater uncertainty) for tropical areas in JJA. Note that these measures of uncertainty are qualitative and applied on a relatively coarse spatial scale. It should also be noted that the range of uncertainty covered by the four emissions scenarios does not encompass the entire envelope of uncertainty of emissions (see Chapter 9, Section 9.2.2.4, and Chapter 13, Section 13.5.1). The range of models (representing the uncertainties in modelling the response to a given forcing) is somewhat more complete than in earlier analyses, but also limited.

strong tendency for precipitation to increase, particularly in the north-west of the continent. This tendency was replaced in the first coupled AOGCMs by one of little change or precipitation decrease, which has remained when the most recent coupled models are considered. Whetton *et al.* (1996a) partly attributed the contrast in the regional precipitation response of the two types of experiments to contrasts in their hemispheric patterns of warming.

Lal *et al.* (1998b) surveyed the results for the Indian subcontinent of seventeen climate change experiments including both equilibrium $2\times CO_2$ and transient AOGCM simulations with and without sulphate aerosol forcing. In the simulations forced only by GHG increases, most models show wet season (JJA) rainfall increases over the region of less than 5% per degree of global warming. A minority of experiments show rainfall decreases. The experiments which included scenarios of increasing sulphate forcing all showed reduced rainfall increases, or stronger rainfall decreases, than their corresponding GHG-only experiments.

For the central plains of North America, IPCC (1990) noted a good deal of similarity in the response of equilibrium $2\times CO_2$ experiments, with precipitation decreases prevailing in the summer and increases in the winter of less than 10%. In the second group of experiments (nine transient runs with AOGCMs) a wider range of responses was found (in the SAR). In winter, changes in precipitation ranged from about -12 to $+20\%$ for the time of CO_2 doubling, and most of the models (six out of nine) exhibited increases. In summer, the range of change was narrower, within $\pm 10\%$, but there was no clear majority response towards increases or decreases. Doherty and Mearns (1999) found that the CGCM1 and HadCM2 models simulated opposite changes in precipitation in both seasons over North America. While overall there is a tendency for more decreases to be simulated in the summer and more increases in the winter, there does not seem to be a reduction in the uncertainty for this region through the progression of climate models.

Many studies have considered GCM-simulated patterns of climate change in the European region (e.g., Barrow *et al.*, 1996; Hulme and Brown, 1998; Osborn and Hulme, 1998; Räisänen, 1998; Benestad *et al.* 1999; Osborn *et al.*, 1999). Hulme *et al.* (2000) provide an overview of simulated changes in the region by considering the results of twenty-three climate change simulations (forced by GHG change only) produced between the years 1983 and 1998 and including mixed-layer $1\times$ and $2\times CO_2$ equilibrium experiments as well as transient experiments. Figure 10.7 shows their results for simulated change in annual precipitation, averaged by latitude and normalised to percentage change per degree of global warming. It may be seen that the consensus amongst current models for drying in southern Europe and wetter conditions in northern Europe represents a continuation of a pattern established amongst the earlier simulations. The effect of model development has primarily been to intensify this pattern of response.

Variations across simulations in the regional enhanced GHG results of AOGCMs, which are particularly evident for precipitation, represent a major uncertainty in any assessment of regional climate change. Such variation may arise due to differences in forcing, systematic model-to-model differences in the regional

response to a given forcing or differences due to natural decadal to inter-decadal scale variability in the models. Giorgi and Francisco (2000a,b) analysed AOGCM simulations including different models, forcing scenarios and ensembles of simulations, and found that the greatest source of uncertainty in regional climate change simulation was due to inter-model differences, with intra-ensemble and inter-scenario differences being less important (see Figures 10.3 and 10.5). However, it should be noted that Giorgi and Francisco (2000a,b) used long (thirty year) means and large (sub-continental scale) regions and that the uncertainty due to simulated natural variability would be larger when shorter averaging periods, or smaller regions, are used. The results of Hulme *et al.* (1999) also suggest that low-frequency natural climatic variability is important at the sub-regional scale in Europe and can mask the enhanced GHG signal.

Regional changes in the mean pattern of atmospheric circulation have been noted in various studies, although typically the changes are not marked (e.g., Huth, 1997; Schubert, 1998). Indeed, the work of Conway (1998) and Wilby *et al.* (1998b) suggests that the contribution of changes in synoptic circulation to regional climate change may be relatively small compared to that of sub-synoptic processes.

10.3.2.2 Climate variability and extreme events

Gregory and Mitchell (1995) identified in an equilibrium $2\times CO_2$ simulation with the Hadley Centre model a tendency for daily temperature variability over Europe to increase in JJA and to decrease in DJF. Subsequent work on temperature variability at daily to monthly and seasonal time-scales has tended to confirm this pattern, as found by Buishand and Beersma (1996) over Europe, Beersma and Buishand (1999) over southern Europe, northern Europe and central North America and Boer *et al.* (2000b) throughout the northern mid-latitudes. This tendency can also be seen in the results of Giorgi and Francisco (2000a) for a set of transient HadCM2 simulations over different regions of the globe.

Daily high temperature extremes are likely to increase in frequency as a function of the increase in mean temperature, but this increase is modified by changes in daily variability of temperature. There is a corresponding decrease in the frequency of daily low temperature extremes. Kharin and Zwiers (2000) and Zwiers and Kharin (1997) found that in all regions of the globe the CGCM1 model simulated substantial increases in the magnitude of extreme daily maximum and minimum temperatures, with an average frequency of occurrence of once per twenty years. Delworth *et al.* (1999) considered simulated changes of a 'heat index' (a measure which combines the effect temperature and moisture) in the GFDL R15a model. Their results indicated that seasonally warm and humid areas such as the south-eastern United States, India, Southeast Asia and northern Australia can experience increases in the heat index substantially greater than that expected due to warming alone.

There is a strong correlation between precipitation inter-annual variability and mean precipitation. Increases in mean precipitation are likely to be associated with increases in variability, and precipitation variability is likely to decrease in areas of reduced mean precipitation. In general, where simulated changes in regional precipitation variability have been examined, increases are more commonly noted. Giorgi and Francisco (2000a) found a tendency for regional interannual variability of seasonal mean precipitation to increase in HadCM2 simulations in many of the regions they considered. Increases in interannual variability also predominated in the CGCM1 simulation (Boer *et al.*, 2000b) although there were areas of decrease, particularly in areas where mean rainfall decreased. Beersma and Buishand (1999) mostly found increases in monthly precipitation variance over southern Europe, northern Europe and central North America. A number of studies have reported a tendency for interannual rainfall variability to increase over South Asia (SAR; Lal *et al.*, 2000). McGuffie *et al.* (1999) identified a tendency for increased daily rainfall variability in two models over the Sahel, North America, South Asia, southern Europe and Australia. It should also be noted that in many regions interannual climatic variability is strongly related to ENSO, and thus will be affected by changes in ENSO behaviour (see Chapter 9).

The tendency for increased rainfall variability in enhanced GHG simulations is reflected in a tendency for increases in the intensity and frequency of extreme heavy rainfall events. Such increases have been documented in regionally focused studies for Europe, North America, South Asia, the Sahel, southern Africa, Australia and the South Pacific (Hennessy *et al.*, 1997; Bhaskaran and Mitchell, 1998; McGuffie et al 1999; Jones, R.N. *et al.*, 2000) as well as in the global studies of Kharin and Zwiers (2000) and Zwiers and Kharin (1998). For example, Hennessy *et al.* (1997) found that under $2\times CO_2$ conditions the one-year return period events in Europe, Australia, India and the USA increased in intensity by 10 to 25% in two models.

Changes in the occurrence of dry spells or droughts have been assessed for some regions using recent model results. Joubert *et al.* (1996) examined drought occurrence over southern Africa in an equilibrium $2\times CO_2$ CSIRO simulation and noted areas of both substantial increase and decrease. Gregory *et al.* (1997) looked at drought occurrence over Europe and North America in a transient simulation using both rainfall-based and soil moisture-based measures of drought. In all cases, marked increases were obtained. This was attributed primarily to a reduction in the number of rainfall events rather than a reduction in mean rainfall. Marked increases in the frequency and intensity of drought were found also by Kothavala (1997) over Australia using the Palmer drought severity index.

Fewer studies have considered changes in variability and extremes of synoptic circulation under enhanced GHG conditions. Huth (1997) noted little change in synoptic circulation variability under equilibrium $2\times CO_2$ conditions over North America and Europe. Katzfey and McInnes (1996) found that the intense cut-off lows off the Australian east coast became less common under equilibrium $2\times CO_2$ conditions in the CSIRO model, although they had limited confidence in this result.

10.3.3 Summary and Recommendations

Analysis of transient simulations with AOGCMs indicates that average climatic features are generally well simulated at the planetary and continental scale. At the regional scale, area-average

biases in the simulation of present day climate are highly variable from region to region and across models. Seasonal temperature biases are typically within the range of $\pm 4°C$ but exceed $\pm 5°C$ in some regions, particularly in DJF. Precipitation biases are mostly between -40 and $+80\%$, but exceed 100% in some regions. These regional biases are, in general terms, smaller than those of a similar analysis presented in the SAR. When it has been assessed, many aspects of model variability have compared well against observations, although significant model-dependent biases have been noted. Model performance was poorer at the finer scales, particularly in areas of strong topographical variation. This highlights the need for finer resolution regionalisation techniques.

Simulated changes in mean climatic conditions for the last decades of the 21st century (compared to present day climate) vary substantially among models and among regions. All land regions undergo warming in all seasons, with the warming being generally more pronounced over cold climate regions and seasons. Average precipitation increases over most regions, especially in the cold season, due to an intensified hydrological cycle. However, some exceptions occur in which most models concur in simulating decreases in precipitation. The magnitude of regional precipitation change varies considerably among models with the typical range being around 0 to 50%, where the direction of change is strongly indicated, and around -30 to $+30\%$ where it is not. There is strong tendency for models to simulate regional increases in precipitation variability with associated increases in the frequency of extreme rainfall events. Increased interannual precipitation variability is also commonly simulated and, in some regions, increases in drought or dry-spell occurrence have been noted. Daily to inter-annual variability of temperature is simulated to decrease in winter and increase in summer in mid-latitude Northern Hemisphere land areas.

10.4 GCMs with Variable and Increased Horizontal Resolution

This section deals with the relatively new idea of deriving regional climate information from AGCMs with variable and increased horizontal resolution. Although the basic methodology is suggested in the work of Bengtsson *et al.* (1995), where a high resolution GCM was used to simulate changes in tropical cyclones in a warmer climate, it is only in the last few years that such models have been used more widely to predict regional aspects of climate change. Even so, only a limited number of experiments have been conducted to date (see Table 10.1) and hence what follows is not a definitive evaluation of the technique but an initial exploration of its potential.

10.4.1 Simulations of Current Climate

Analysis of current climate simulations has considered both deviations from the observed climate and effects of changes in resolution on the model's climatology. Most studies have considered just the mean climate and some measures of variability, either globally or for a particular region of interest. The only extreme behaviour studied in any detail was the simulation of tropical cyclones. Even for mean climate, no comprehensive assessment

Table 10.1: *Enhanced and variable resolution GCM control and anomaly simulations. Resolution is given as either the spectral truncation or grid-point spacing depending on the model's formulation (and with a range for variable resolution models). The equivalent grid-point resolution of spectral truncation T42 is 2.8°32.8° (scaling linearly).*

Institution	Model	Horizontal Resolution	Control Forcing	Anomaly Forcing	Region of interest
MPI	ECHAM3	T42	ECHAM/LSG	ECHAM/LSG	Euro/Global
MPI	ECHAM3/4	T106	Obs	ECHAM/OPYC	Euro/Global
UKMO	HadAM2b	0.83°×1.25°	Obs		Global
UKMO	HadAM3a	0.83°×1.25°	Obs	HadCM3	Euro/Global
MRI	JMA	T106	Obs	MRI/GFDL/+2°C	Tropics
CNRM	ARPEGE	T213–T21, T106	Obs/HadCM2	HadCM2	Euro/Global
LGGE	LMDZ	100 to 700km	Obs	CLIMAP	Polar regions

of the surface climatology of variable or high resolution models has been attempted. Europe has been the most common area of study to date, although southern Asia and the polar regions have also received attention

10.4.1.1 Mean climate

The mean circulation is generally well simulated by AGCMs, though relatively large regional-scale biases can still be present. Many features of the large-scale climate of AGCMs are retained at higher resolution (Dèquè and Piedelievre, 1995; Stendel and Roeckner, 1998; May, 1999; Stratton, 1999a). A common change is a poleward shift of the extra-tropical storm track regions. It has been suggested that this is linked to a general deepening of cyclones, noted as a common feature in high-resolution atmospheric models (Machenhauer *et al.*, 1996; Stratton 1999a). More intense activity is also seen at higher resolution in the tropics. For example, a stronger Hadley circulation was observed in ECHAM4 and HadAM3a that worsened agreement with observations (Stendel and Roeckner, 1998; Stratton, 1999b).

The repositioning of the storm tracks generally improves the simulations in the Northern Hemisphere, as it reduces a positive polar surface pressure bias which is present in the models at standard resolution. In the case of HadAM3a, this leads to substantial improvements in Northern Hemisphere low level flow in winter (Figure 10.8). In the Southern Hemisphere, the impact on the circumpolar flow is not consistently positive across models (Figure 10.8; Krinner *et al.*, 1997). In ECHAM4 and HadAM3a, increased resolution has little impact on the negative surface pressure bias over the tropics but improves the low-level South Asian monsoon flow (Lal *et al.*, 1997; Stratton, 1999b).

The existence of these common responses to increased resolution suggests that they result from improved representation of the resolved variables. In contrast, an increase in the intensity of sub-tropical anticyclones observed in ECHAM4 results from a tropospheric warming promoted by excessive cirrus clouds attributed to a scale-dependent response in the relevant parametrization (Stendel and Roeckner, 1998).

The aim of increasing resolution in AGCMs is generally to improve the simulation of surface climatology compared to coarser resolution models (Cubash *et al.*, 1995). Early experience shows a much more mixed response. ECHAM3 at T42 improved the seasonal cycle of surface temperature in seven regions, compared

to the driving AOGCM, but overall surface temperature was too high (by 2 to 5°C). Increasing the resolution to T106 did not improve winter temperatures and, in summer, the spatial patterns were better but the regional biases worse (Cubasch *et al.*, 1996). For precipitation, spatial patterns were improved in summer but degraded in winter. The summer warming was due to excessive insolation from reduced cloud cover and overly transparent clear skies (Wild *et al.*, 1995). Improved physics in ECHAM4 reduced some of the radiation errors but the precipitation and temperature biases remained (Wild et al, 1996; Stendel and Roeckner, 1998). In simulations of European climate with ARPEGE (Dèquè and Piedelievre, 1995) and HadAM2b/3a (Jones, 1999; Stratton, 1999a), improved flow at higher resolution generally led to better surface temperatures and precipitation. However, over south-eastern Europe, precipitation biases increased in both models, as did the warm temperature bias in HadAM3a.

The increased summer temperatures in Europe in HadAM3a were caused by reduced cloud cover at higher resolution (Jones 1999) and warming and drying, in summer, was seen over all extra-tropical continents (Stratton, 1999b). This clearly demonstrates a potential drawback of increasing the resolution of a model without comprehensively retuning the physics. Krinner *et al.* (1997) showed that, to obtain a reasonable simulation of the surface climatology of the Antarctic with the LMD variable resolution AGCM, many modifications to the model physics were required. The model was then able to simulate surface temperatures to within 2 to 4°C of observations and to provide a good simulation of the ice mass balance (snow accumulation), with both aspects being better than at standard resolution.

10.4.1.2 Climate variability and extreme events

Enhanced resolution improves many aspects of the AGCMs' intra-seasonal variability of circulation at low and intermediate frequencies (Stendel and Roeckner, 1998). However, in some cases values underestimated at standard resolution are overestimated at enhanced resolution (Dèquè and Piedelievre, 1995; Stratton, 1999a,b). Martin (1999) found little sensitivity to resolution in either the interannual or intra-seasonal variability of circulation and precipitation of the South Asian monsoon in HadAM3a. Extreme events have not been studied, with the exception of tropical cyclones. This subject cuts across various sections and chapters and thus is dealt with in Box 10.2.

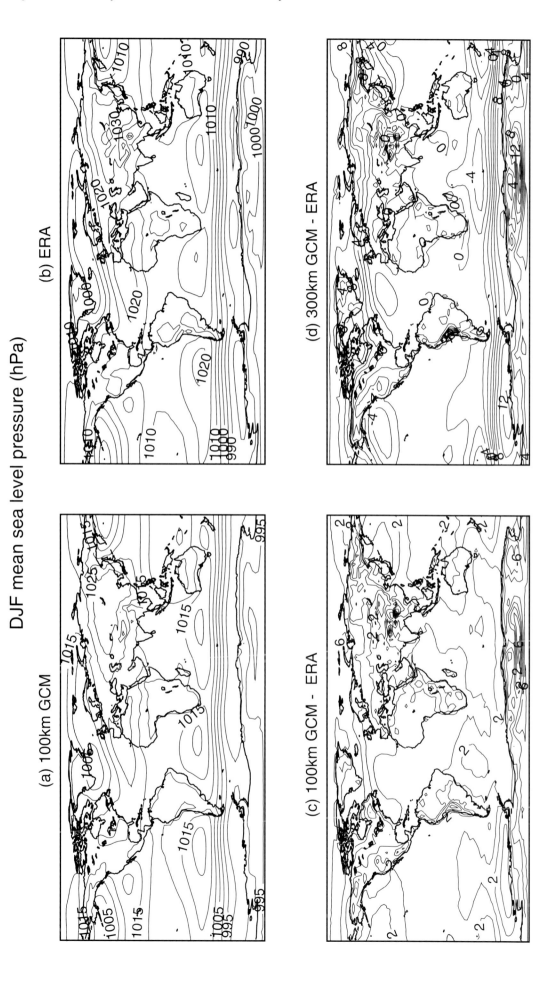

Figure 10.8: Mean sea level pressure for DJF in: (a) HadAM3a at high resolution (100 km), (b) ECMWF reanalysis (ERA), (c) HadAM3a high resolution minus ERA, (d) HadAM3a at standard resolution (300 km) minus ERA. Adapted from Stratton (1999b).

Box 10.2: Tropical cyclones in current and future climates

Simulating a climatology of tropical cyclones

Tropical cyclones can have devastating human and economic impacts (e.g., Pielke and Landsea, 1998) and therefore accurate estimates of future changes in their frequency, intensity and location would be of great value. However, because of their relatively small extent (in global modelling terms) and intense nature, detailed simulation of tropical cyclones for this purpose is difficult. Atmospheric GCMs can simulate tropical cyclone-like disturbances which increase in realism at higher resolution though the intense central core is not resolved (e.g., Bengtsson *et al.*, 1995; McDonald, 1999). Further increases of resolution, by the use of RCMs, provide greater realism (e.g., Walsh and Watterson, 1997) with a very high resolution regional hurricane prediction model giving a reasonable simulation of the magnitude and location of maximum surface wind intensities for the north-west Pacific basin (Knutson *et al.*, 1998). GCMs generally provide realistic simulation of the location and frequency of tropical cyclones (e.g., Tsutsui and Kasahara, 1996; Yoshimura *et al.*, 1999). See also Chapter 8 for more details on tropical cyclones in GCMs.

Tropical cyclones in a warmer climate

Much effort has gone into obtaining and analysing good statistics on tropical cyclones in the recent past. The main conclusion is that there is large decadal variability in the frequency and no significant trend during the last century. One study looking at the century time-scale has shown an increase in the frequency of North Atlantic cyclones from 1851 to 1890 and 1951 to 1990 (Fernandez-Partagas and Diaz, 1996). See Chapter 2 for more details on observed tropical cyclones.

Most assessments of changes in tropical cyclone behaviour in a future climate have been derived from GCM or RCM studies of the climate response to anthropogenically-derived atmospheric forcings (e.g., Bengtsson *et al.*, 1996, 1997; Walsh and Katzfey, 2000). Recently, more focused approaches have been used: nesting a hurricane prediction model in a GCM climate change simulation (Knutson *et al.*, 1998); inserting idealised tropical cyclones into an RCM climate change simulation (Walsh and Ryan, 2000).

In an early use of a high-resolution AGCM, a T106 ECHAM3 experiment simulated a decrease in tropical cyclones in the Northern Hemisphere and a reduction of 50% in the Southern Hemisphere (Bengtsson *et al.*, 1996, 1997). However, the different hemispheric responses raised questions about the model's ability to properly represent tropical cyclones and methodological concerns about the experimental design were raised (Landsea, 1997). In a similar experiment, the JMA model also simulated fewer tropical cyclone-like vortices in both hemispheres (Yoshimura *et al.*, 1999). Other GCM studies have shown consistent basin-dependent changes in tropical cyclone formation under $2 \times CO_2$ conditions (Royer *et al.*, 1998; Tsutsui *et al.*, 1999). Frequencies increased in the north-west Pacific, decreased in the North Atlantic, and changed little in the south-west Pacific. A high resolution HadAM3a simulation reproduced the latter changes, giving changes in timing in the north-west Pacific and increases in frequency in the north-east Pacific and the north Indian basin (McDonald, 1999). Some GCM studies show increases in tropical storm intensity in a warmer climate (Krishnamurti *et al.*, 1998) though these results can probably not be extrapolated to tropical cyclones as the horizontal resolution of these models is insufficient to resolve the cyclone eye. The likely mean response of tropical Pacific sea surface warming having an El Niño-like structure suggests that the pattern of tropical cyclone frequency may become more like that observed in El Niño years (see Chapter 9).

An indication of the likely changes in maximum intensity of cyclones will be better provided by models able to simulate realistic tropical cyclone intensities. A sample of GCM-generated tropical cyclone cases nested in a hurricane prediction model gave increases in maximum intensity (of wind speed) of 5 to 11% in strong cyclones over the north-west Pacific for a 2.2°C SST warming (Knutson and Tuleya, 1999). The RCM study of idealised tropical cyclones (in the South Pacific) showed a small, but not statistically significant, increase in maximum intensity (Walsh and Ryan, 2000). These results are supported by the theory of the maximum potential intensity (MPI) of huricanes (Emanuel, 1987). A calculation using the MPI framework of Holland (1997) suggested increases of 10 to 20% for a $2 \times CO_2$ climate (Henderson-Sellers *et al.*, 1998). This study also acknowledges physical omissions that would reduce this estimate though Emanuel (2000) suggests there is a linear relationship betweeen MPI and the wind speed of real events. Published modelling studies to date neglect the possible feedback of sea surface cooling induced by the cyclone. However, a recently submitted study using a hurricane model with ocean coupling indicates that the increased maximum intensity by CO_2 warming would still occur even when the sea surface cooling feedback is included (Knutson *et al.*, 2000).

The extreme precipitation associated with tropical cyclones can also be very damaging. The very high resolution studies discussed above suggest that increases in the intensity of tropical cyclones will be accompanied by increases in mean and maximum precipitation rates. In the cases studied, precipitation in the vicinity of the storm centre increased by 20% whereas peak rates increased by 30%. Part of these increases may be due to the increased moisture-holding capacity of a warmer atmosphere but nevertheless point to substantially increasing destructive capacity of tropical cyclones in a warmer climate.

In conclusion, there is some evidence that regional frequencies of tropical cyclones may change but none that their locations will change. There is also evidence that the peak intensity may increase by 5% to 10% and precipitation rates may increase by 20% to 30%. There is a need for much more work in this area to provide more robust results.

10.4.2 Simulations of Climate Change

10.4.2.1 Mean climate

Climate change simulations using ECHAM3 at T42 and T106 resolutions predicted substantially different responses for southern Europe (Cubash *et al.*, 1996). For example, surface temperature response of less than +2°C in summer at T42 increased by over 4°C for much of the region at T106 and winter precipitation increased more at T106 than at T42. An important factor in generating the different responses was the substantial difference in the control simulations. Wild *et al.* (1997) showed a large positive summer surface temperature bias in the T106 control derived from a positive feedback between excessive surface insolation and summer dryness. This mechanism provided a large increase in the insolation, and thus the surface temperature, in the anomaly experiment. As this process was handled poorly in the control simulation, little confidence can be placed in the warming amplification simulated at T106.

A variable grid AGCM climate change experiment using the ARPEGE model and sea surface forcing from HadCM2 predicted moderate warmings over Europe, 1.5°C (northern) to 2.5°C (southern) in winter and 1°C to 3.5°C in summer (Dèquè *et al.*, 1998). In contrast, HadCM2 predicted greater warming and a larger north-south gradient in winter (Figure 10.9). These differences result mainly from the ARPEGE large-scale flow being too zonal and too strong over mainland Europe, which enhances the moderating influence of the SSTs. The precipitation responses are more similar, especially in summer, when both models predict a decrease over most of Europe, maximum –30% in the south. Differences in the control simulations suggest that little confidence should be placed in this result.

In a similar experiment, HadAM3a at 1.25°×0.83° resolution used observed sea surface forcing and anomalies from a HadCM3 GHG simulation and produced a response at the largest scales in the annual mean similar to the AOGCM (Johns, 1999). However, regionally or seasonally, many differences were evident in the two models, notably in land sea contrasts, monsoon precipitation and some circulation features. Over Europe, large-scale responses in surface temperature and precipitation were similar except for a larger winter surface warming in northern Europe in HadCM3. This was due to a greater melting back of Arctic sea ice which was too extensive in the HadCM3 control (Jones, 1999). In a 30-year ECHAM4 T106 experiment driven by ECHAM4/OPYC simulations for 1970 to 1999 and 2060 to 2089, the simulations of future climate were more similar to each other than those for the present day (May, 1999). This implies that the differences in the control simulations would determine a proportion of the difference in the responses. In these cases better control simulations at high resolution increase the confidence in their responses.

10.4.2.2 Climate variability and extreme events

Due to the limited number and length of simulations and a lack of comprehensive analyses, this subject has been almost completely ignored. The only response in variability or extremes that has received any attention is that of tropical cyclones (Box 10.2).

10.4.3 Summary and Recommendations

Since the SAR, several variable and high-resolution GCMs have been used to provide high-resolution simulations of climate change. Clearly the technique is still in its infancy with only a few modelling studies carried out and for only a limited number of regions. Also, there is little in-depth analysis of the performance of the models and only preliminary conclusions can be drawn.

Many aspects of the models' dynamics and large-scale flow are improved at higher resolution, though this is not uniformly so geographically or across models. Some models also demonstrate improvements in their surface climatologies at higher resolution. However, substantial underlying errors are often still present in high-resolution versions of current AGCMs. In addition, the direct use of high-resolution versions of current AGCMs, without some allowance of the dependence of models physical parametrizations on resolution, leads to some deterioration in the performance of the models.

Regional responses currently appear more sensitive to the AGCM than the SST forcing used. This result is partially due to some of the model responses being dependent on their control simulations and systematic errors within them. These factors and the small number of studies carried out imply that little confidence can be attached to any of the regional projections provided by high and variable resolution AGCM simulations. The improvements seen with this technique are encouraging, but more effort should be put in analysing, and possibly improving the performance of current models at high resolution. This is particularly important in view of the fact that future AOGCMs will likely use models approaching the resolution considered here in the next 5 to 10 years.

10.5 Regional Climate Models

Since the SAR, much insight has been provided into fundamental issues concerning the nested regional modelling technique.

Multi-year to multi-decadal simulations must be used for climate change studies to provide meaningful climate statistics, to identify significant systematic model errors and climate changes relative to internal model and observed climate variability, and to allow the atmospheric model to equilibrate with the land surface conditions (e.g., Jones *et al.*, 1997; Machenhauer *et al.*, 1998; Christensen 1999; McGregor *et al.*, 1999; Kato *et al.*, 2001).

The choice of an appropriate domain is not trivial. The influence of the boundary forcing can reduce as region size increases (Jones *et al.*, 1995; Jacob and Podzun, 1997) and may be dominated by the internal model physics for certain variables and seasons (Noguer *et al.*, 1998). This can lead to the RCM solution significantly departing from the driving data, which can make the interpretation of down-scaled regional climate changes more difficult (Jones *et al.*, 1997). The domain size has to be large enough so that relevant local forcings and effects of enhanced resolution are not damped or contaminated by the application of the boundary conditions (Warner *et al.*, 1997). The exact location of the lateral boundaries can influence the sensitivity to internal parameters (Seth and Giorgi, 1998) or may

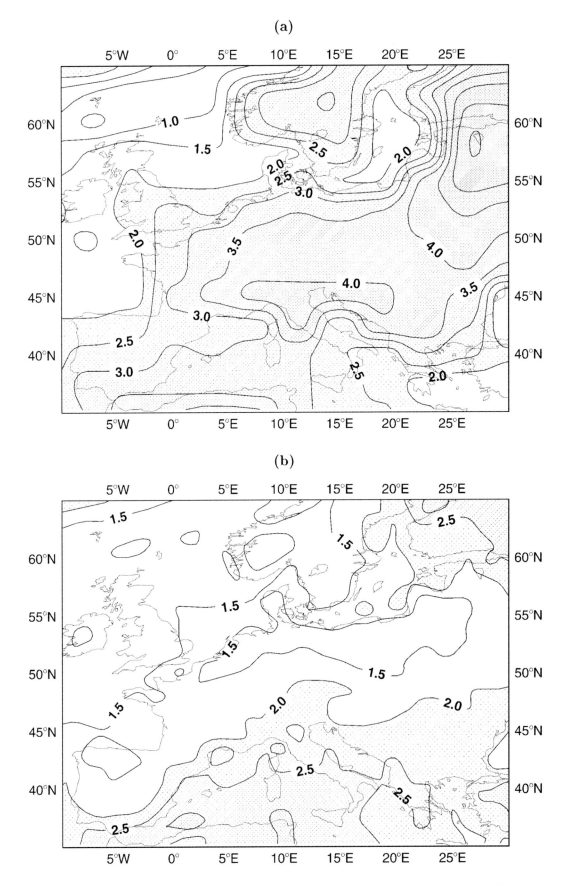

Figure 10.9: Winter surface air temperature change (°C) over Europe at the time of CO_2 doubling in (a) a transient climate change experiment with the AOGCM HadCM2 and (b) the stretched grid AGCM ARPEGE driven by SSTs and sea-ice from the HadCM2 integration. From Dèquè *et al.* (1998).

have no significant impact (Bhaskaran *et al.*, 1996). Finally, location of boundaries over areas with significant topography may lead to inconsistencies and noise generation (e.g., Hong and Juang, 1998).

Surface forcing due to land, ocean and sea ice greatly affects regional climate simulation (e.g., Giorgi *et al.*, 1996; Seth and Giorgi, 1998; Wei and Fu, 1998; Christensen, 1999; Pan *et al.*, 1999; Pielke *et al.*, 1999; Rinke and Dethloff, 1999; Chase *et al.*, 2000; Maslanik *et al.*, 2000, Rummukainen *et al.*, 2000). In particular, RCM experiments do not start with equilibrium conditions and therefore the initialisation of surface variables, such as soil moisture and temperature, is important. For example, to reach equilibrium it can require a few seasons for the rooting zone (about 1 m depth) and years for the deep soils (Christensen, 1999).

The choice of RCM resolution can modulate the effects of physical forcings and parametrizations (Giorgi and Marinucci, 1996a; Laprise *et al.*, 1998). The description of the hydrologic cycle generally improves with increasing resolution due to the better topographical representation (Christensen *et al.*, 1998; Leung and Ghan, 1998). Resolving more of the spectrum of atmospheric motions at high resolution improves the representation of cyclonic systems and vertical velocities, but can sometimes worsen aspects of the model climatology (Machenhauer *et al.*, 1998; Kato *et al.*, 1999). Different resolutions may be required to capture relevant forcings in different sub-regions, which can be achieved via multiple one-way nesting (Christensen *et al.*, 1998; McGregor *et al.*, 1999), two-way nesting (Liston *et al.*, 1999) or smoothly varying horizontal grids (Qian and Giorgi, 1999). Only limited studies of the effects of changing vertical resolution have been published (Kato *et al.*, 1999).

RCM model physics configurations are derived either from a pre-existing (and well tested) limited area model system with modifications suitable for climate application (Pielke *et al.*, 1992; Giorgi *et al.*, 1993b,c; Leung and Ghan, 1995, 1998; Copeland *et al.*, 1996; Miller and Kim, 1997; Liston and Pielke 2000; Rummukainen *et al.*, 2000) or are implemented directly from a GCM (McGregor and Walsh, 1993; Jones *et al.*, 1995; Christensen *et al.*, 1996; Laprise *et al.*, 1998). In the first approach, each set of parametrizations is developed and optimised for the respective model resolutions. However, this makes interpreting differences between nested model and driving GCM more difficult, as these will not result only from changes in resolution. Also, the different model physics schemes may result in inconsistencies near the boundaries (Machenhauer *et al.*, 1998; Rummukainen *et al.*, 2000). The second approach maximises compatibility between the models. However, physics schemes developed for coarse resolution GCMs may not be adequate for the high resolutions used in nested regional models and may, at least, require recalibration (Giorgi and Marinucci, 1996a; Laprise *et al.*, 1998; see also Section 10.4). Overall, both strategies have shown performance of similar quality (e.g., IPCC, 1996), and either one may be preferable (Giorgi and Mearns, 1999). In the context of climate change simulations, if there is no resolution dependence, the second approach may be preferable to maximise consistency between RCM and GCM responses to the radiative forcing.

Ocean RCMs have been developed during the last decades for a broad variety of applications. To date, the specific use of these models, in a context similar to the use of nested atmospheric RCMs for climate change studies, is very limited (Kauker, 1998). Although the performance of ocean RCMs has yet to be assessed, it is known that a very high resolution, few tens of kilometres or less, is needed for accurate ocean simulations.

The construction of coupled RCMs is a very recent development. They comprise atmospheric RCMs coupled to other models of climate system components, such as lake, ocean/sea ice, chemistry/aerosol, and land biosphere/hydrology models (Hostetler *et al.*, 1994; Lynch *et al.*, 1995, 1997a,b, 1998; Leung *et al.*, 1996; Bailey *et al.*, 1997; Kim *et al.*, 1998; Qian and Giorgi 1999; Small *et al.*, 1999a,b; Bailey and Lynch, 2000a,b; Mabuchi *et al.*, 2000; Maslanik *et al.*, 2000; Rummukainen *et al.*, 2000; Tsvetsinskaya *et al.*, 2000; Weisse *et al.*, 2000). This promises the development of coupled "regional climate system models".

10.5.1 Simulations of Current Climate

Simulations of current climate conditions serve to evaluate the performance of RCMs. Since the SAR, a vast number of such simulations have been conducted (McGregor, 1997; Appendices 10.1 to 10.3). These fall into two categories, RCMs driven by observed (or "perfect") boundary conditions and RCMs driven by GCM boundary conditions. Observed boundary conditions are derived from Numerical Weather Prediction (NWP) analyses (e.g., European Centre for Medium Range Weather Forecast (ECMWF) reanalysis, Gibson *et al.* 1997; or National Center for Environmental Prediction (NCEP) reanalysis, Kalnay *et al.*, 1996). Over most regions they give accurate representation of the large-scale flow and tropospheric temperature structure (Gibson *et al.*, 1997), although errors are still present due to poor data coverage and to observational uncertainty. The analyses may be used to drive RCM simulations for short periods, for comparison with individual episodes, or over long periods to allow statistical evaluation of the model climatology. Comparison with climatologies is the only available evaluation tool for RCMs driven by GCM fields, with the caveats applied to GCM validation concerning the influence of sample size and decadal variability (see Sections 10.2, 10.3, and 10.4). Despite these, relatively short simulations (several years) can identify major systematic RCM biases if they yield departures from observations significantly greater than the observed natural variability (Machenhauer *et al.*, 1996, 1998; Christensen *et al.*, 1997; Jones *et al.*, 1999).

Often a serious problem in RCM evaluation is the lack of good quality high-resolution observed data. In many regions, observations are extremely sparse or not readily available. In addition, only little work has been carried out on how to use point measurements to evaluate the grid-box mean values from a climate model, especially when using sparse station networks or stations in complex topographical terrain (e.g., Osborn and Hulme, 1997). Most of the observational data available at typical RCM resolution (order of 50 km) is for precipitation and daily minimum and maximum temperature. While these fields have been shown to be useful for evaluating model performance, they are also the end product of a series of complex processes, so that the evaluation of individual model dynamical and physical processes is necessarily limited. Additional fields need to be examined in model evaluation to broaden the perspective on model performance and to help delineate sources of model error. Examples are the surface energy and water fluxes.

Despite these problems, the situation is steadily improving in terms of grid-cell climatologies (Daly *et al.*, 1994; New *et al.*, 1999, 2000; Widman and Bretherton, 2000), with various groups developing high-resolution regional climatologies (e.g., Christensen *et al.*, 1998; Frei and Schär, 1998). In addition, regional programs such as the Global Energy and Water Cycle Experiment (GEWEX) Continental-Scale International Program (GCIP) have been designed with the purpose of developing sets of observation databases at the regional scale for model evaluation (GCIP, 1998).

10.5.1.1 Mean climate: Simulations using analyses of observations

Ideally, experiments using analyses of observations to drive the RCMs should precede any attempt to simulate climate change. The model behaviour, with realistic forcing, should be as close as possible to that of the real atmosphere and experiments driven by analyses of observations can reveal systematic model biases primarily due to the internal model dynamics and physics.

A list of published RCM simulations driven by analyses of observations is given in Appendix 10.1. Many of these studies present regional differences (or biases) of seasonally or monthly-averaged surface air temperature and precipitation from observed values. They indicate that current RCMs can reproduce average observations over regions of size 10^5 to 10^6 km^2 with errors generally below 2°C and within 5 to 50% of observed precipitation, respectively (Giorgi and Shields, 1999; Small *et al.*, 1999a,b; van Lipzig, 1999; Pan *et al.*, 2000). Uncertainties in the analysis fields, used to drive the models, and, in the observed station data sets, should be considered in the interpretation of these biases.

Various RCM intercomparison studies have been carried out to identify different or common model strengths and weaknesses, over Europe by Christensen *et al.* (1997), over the USA by Takle *et al.* (1999), and over East Asia by Leung *et al.* (1999a). For Europe a wide range of performance was reported, with the better models exhibiting a good simulation of surface air temperature (sub-regional monthly bias in the range ±2°C), except over south-eastern Europe during summer. For the USA, a major finding was that the model ability to simulate precipitation episodes varied depending on the scale of the relevant dynamical forcing. Organised synoptic-scale precipitation systems were well simulated deterministically, while episodes of mesoscale and convective precipitation were represented in a more stochastic sense, with less degree of agreement with the observed events and among models. Over East Asia, a major factor in determining the model performance was found to be the simulation of cloud radiative processes.

10.5.1.2 Mean climate: Simulations using GCM boundary conditions

Since the SAR, evaluation of RCMs driven by GCM simulations of current climate has gained much attention (Appendix 10.2), as this is the context in which many RCMs are used (e.g., for climate change experiments). Errors introduced by the GCM representation of large-scale circulations are transmitted to the RCM as, for example, clearly shown by Noguer *et al.* (1998). However, since the SAR, regional biases of seasonal surface air temperature and precipitation have been reduced and are mostly within 2°C, and 50 to 60% of observations (with exceptions in all seasons), respectively (Giorgi and Marinucci, 1996b; Noguer *et al.*, 1998; Jones *et al.*, 1999 for Europe; Giorgi *et al.*, 1998 for the continental USA; McGregor *et al.*, 1998 for Southeast Asia; Kato *et al.*, 2001 for East Asia). The reduction of biases is due to both better large-scale boundary condition fields and improved aspects of internal physics and dynamics in the RCMs.

The regionally averaged biases in the nested RCMs are not necessarily smaller than those in the driving GCMs. However, all the experiments mentioned above, along with those of Leung *et al.* (1999a,b), Laprise *et al.* (1998), Christensen *et al.* (1998) and Machenhauer *et al.* (1998) clearly show that the spatial patterns produced by the nested RCMs are in better agreement with observations because of the better representation of high-resolution topographical forcings and improved land/sea contrasts. For example, in simulations over Europe and central USA, Giorgi and Marinucci (1996a) and Giorgi *et al.* (1998) find correlation coefficients between simulated and observed seasonally averaged precipitation in the range of +0.53 to +0.87 in a nested RCM and –0.69 to +0.85 in the corresponding driving GCM.

The role of the high-resolution forcing was clearly demonstrated in the study of Noguer *et al.* (1998), which showed that the skill in simulating the mesoscale component of the climate signal (Giorgi *et al.*, 1994; Jones *et al.*, 1995) was little sensitive to the quality of the driving data (Noguer *et al.*, 1998). On the other hand, interactions between the large-scale driving data and high resolution RCM forcings can have negative effects. In simulations over the European region of Machenauer *et al.* (1998), the increased shelter due to the better-resolved mountains in the RCMs caused an intensification of the GCM-simulated excessively dry and warm summer conditions over south-eastern Europe.

Horizontal resolution is especially important for the simulation of the hydrologic cycle. Christensen *et al.* (1998) showed that only at a very high resolution do the mountain chains in Norway and Sweden become sufficiently well resolved to yield a realistic simulation of the surface hydrology (Figure 10.10). An alternative strategy is to utilise a sub-grid scale scheme capable of resolving complex topographical features (Leung *et al.*, 1999a).

10.5.1.3 Climate variability and extreme events

A number of studies have investigated the interannual variability in RCM simulations driven by analyses of observations over different regions (e.g., Lüthi *et al.*, 1996 for Europe; Giorgi *et al.*, 1996 and Giorgi and Shields 1999 for the continental USA; Sun *et al.*, 1999 for East Africa; Small *et al.*, 1999a for central Asia; Rinke *et al.*, 1999 for the Arctic; van Lipzig, 1999 for Antarctica). These show that RCMs can reproduce well interannual anomalies of precipitation and surface air temperature, both in sign and magnitude, over sub-regions varying in size from a few hundred kilometres to about 1,000 km (Figure 10.11).

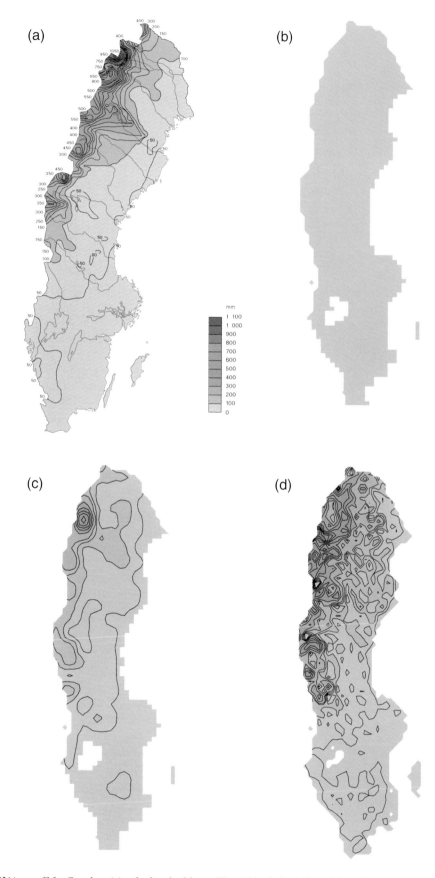

Figure 10.10: Summer (JJA) runoff for Sweden. (a) calculated with a calibrated hydrological model, using daily meteorological station observations and stream gauging stations (Raab and Vedin, 1995); (b) GCM simulation; (c) 55 km RCM simulation; (d) 18 km resolution RCM. Units are mm (from Christensen *et al.*, 1998).

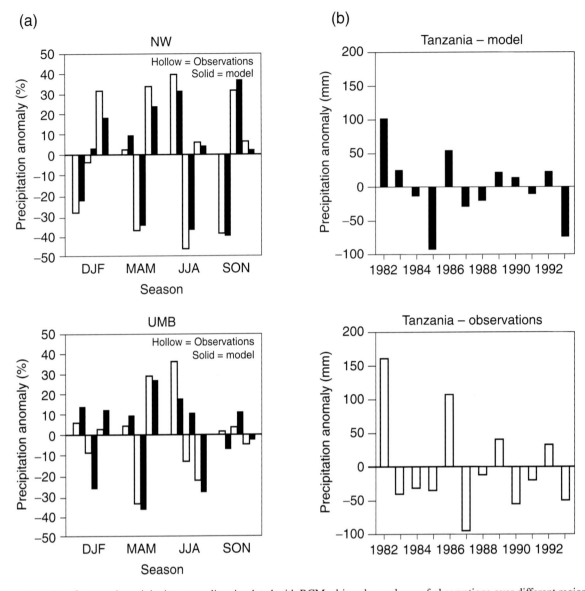

Figure 10.11: Examples of seasonal precipitation anomalies simulated with RCMs driven by analyses of observations over different regions. In all cases the anomalies are calculated as the difference between the precipitation of an individual season and the average for the seasonal value for the entire simulation. (a) (top) Northwestern USA (NW), and (bottom) Upper Mississippi Basin (UMB) for a three year simulation (1993 to 1996) over the continental USA. The three pairs of observed (hollow bars) and simulated (solid bars) anomalies for each season are grouped in sequential order from 1993 to 1996. Units are percentage of the three-year seasonal average (from Giorgi and Shields, 1999, Figure 9). (b) Precipitation anomalies for twelve short-rains periods over Tanzania for the October-December season: (top) model simulation, and (bottom) observations. Units are mm. (From Sun *et al.*, 1999).

At the intra-seasonal scale, the timing and positioning of regional climatological features such as the East Asia rain belt and the Baiu front can be reproduced with a high degree of realism with an RCM (Fu *et al.*, 1998). A good simulation of the intra-seasonal evolution of precipitation during the short rain season of East Africa has also been documented (Sun *et al.*, 1999). However, at shorter time-scales, Dai *et al.* (1999) found that, despite a good simulation of average precipitation, significant problems were exhibited by an RCM simulation of the observed diurnal cycle of precipitation over different regions of the USA.

Only a few examples are available of analysis of variability in RCMs driven by GCMs. At the intra-seasonal scale, Bhaskaran

et al. (1998) showed that the leading mode of sub-seasonal variability of the South Asia monsoon, a 30 to 50 day oscillation of circulation and precipitation anomalies, was more realistically captured by an RCM than the driving GCM. Hassell and Jones (1999) then showed that a nested RCM captured observed precipitation anomalies in the active break phases of the South Asia monsoon (5 to 10 periods of anomalous circulations and precipitation) that were absent from the driving GCM (Figure 10.12).

At the daily time-scale, some studies have shown that nested RCMs tend to simulate too many light precipitation events compared with station data (Christensen *et al.* 1998; Kato *et al.*, 2001). However, RCMs produce more realistic statistics of heavy precipitation events than the driving GCMs, sometimes capturing

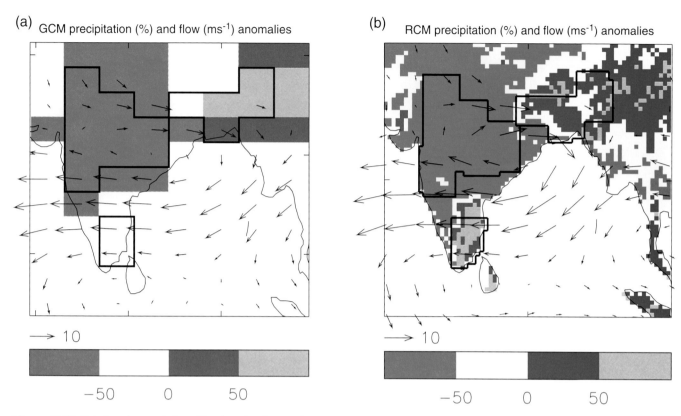

Figure 10.12: Relative characteristics of break and active precipitation composites of the Indian monsoon as simulated by (a) GCM and (b) RCM. Each field is the difference in the break and active composite precipitation as a percentage of the full mean. Overlaid are the 850 hPa wind anomalies (break composite minus active composite, units ms^{-1}). Regions marked where observed ratios are <−50% (central India) and >+50% (Tamil Nadu and north-eastern India) according to Hamilton (1977). From Hassel and Jones (1999).

extreme events entirely absent in the GCMs (Christensen *et al.*, 1998; Jones, 1999). Part of this is due to the inherent disaggregation of grid-box mean values resulting from the RCM's higher horizontal resolution. However, in one study, even when aggregated to the GCM grid scale, the RCM was closer to observations than the driving GCM (Durman *et al.*, 2001).

10.5.2 Simulations of Climate Change

Since the SAR, several multi-year RCM simulations of anthropogenic climate change, either from equilibrium experiments or for time slices of transient simulations, have become available (Appendix 10.3).

10.5.2.1 Mean climate
An important issue when analysing RCM simulations of climate change is the significance of the modelled responses. To date RCM simulations have been mostly aimed at evaluating models and processes rather than producing projections and, as such, they have been relatively short (10 years or less). At short time-scales, natural climate variability may mask all but the largest responses. For example, in an analysis of 10-year RCM simulations over Europe, Machenhauer *et al.* (1998) concluded that generally only the full area averaged seasonal mean surface temperature responses were statistically significant, and in only a few cases were sub-domain deviations from the mean response

significant. The changes in precipitation were highly variable in space, and, in each season, they were only significant in those few sub-areas having the largest changes. Similar results were documented by Pan *et al.* (2000) and Kato *et al.* (2001) for the USA and East Asia, respectively. Hence, 30-year samples may be required to confidently assess the mesoscale response of a RCM (Jones *et al.*, 1997). Partly to improve signal to noise definition, a transient RCM simulation of 140 years duration was recently conducted (Hennessy *et al.*, 1998; McGregor *et al.*, 1999).

Despite the limitations in simulation length, most RCM experiments clearly indicate that, while the large-scale patterns of surface climate change in the nested and driving models are similar, the mesoscale details of the simulated changes can be quite different. For example, significantly different patterns of temperature and rainfall changes were found in a regional climate change simulation for Australia (Whetton *et al.*, 2001). This was most clearly seen in mountainous areas (Figure 10.13). Winter rainfall in southern Victoria increased in the RCM simulation, but decreased in the driving GCM. High resolution topographical modification of the regional precipitation change signal in nested RCM simulations has been documented in other studies (Jones *et al.*, 1997; Giorgi *et al.*, 1998; Machenhauer *et al.*, 1998; Kato *et al.*, 2001).

The response in an RCM can also be modified by changes in regional feedbacks. In a 20 year nested climate change experiment for the Indian monsoon region, Hassell and Jones (1999)

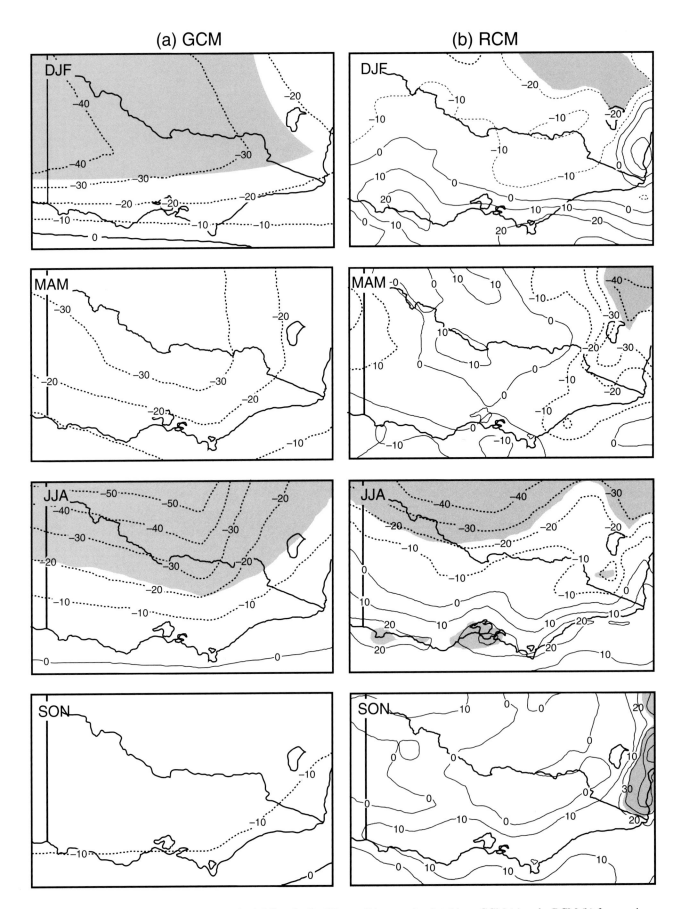

Figure 10.13: Percentage change in mean seasonal rainfall under $2\times CO_2$ conditions as simulated by a GCM (a) and a RCM (b) for a region around Victoria, Australia. Areas of change statistically significant at the 5% confidence level are shaded. Whetton *et al.* (2001).

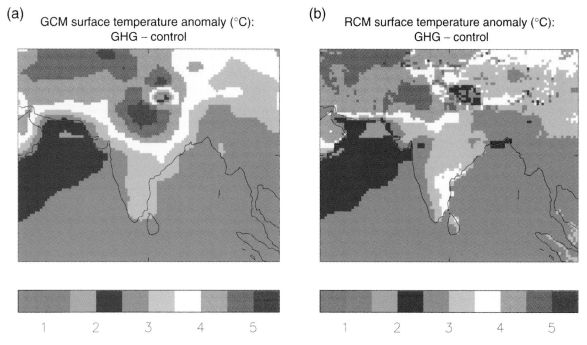

Figure 10.14: Simulated surface air temperature anomaly (°C) for JJA, Indian monsoon region. GHG (2040 to 2060) minus control 20 year average for (a) GCM and (b) RCM. From Hassel and Jones (1999).

showed that a maximum anomaly of 5°C seen in central northern India in the GCM simulation was reduced and moved to the north-west in the nested RCM, with a secondary maximum appearing to the south-east (Figure 10.14). The shift of the main maximum was attributed to deficiencies in the GCM control climate that promoted excessive drying of the soil in North-west India. The secondary maximum was attributed to a complex response involving the RCM's better representation of the flow patterns in southern India resulting from an improved representation of the Western Ghats mountains. In this instance, it was argued that the improved realism of the RCM's control simulation increases confidence in its response.

The high resolution representation of mountainous areas in an RCM has made it possible to show that the simulated surface air temperature change signal due to $2 \times CO_2$ concentration could have a marked elevation dependency, resulting in more pronounced warming at high elevations than low elevations as shown in Figure 10.15 (Giorgi *et al.*, 1997). This is primarily caused by a depletion of the snow pack in enhanced GHG conditions and the associated snow albedo feedback mechanism, and it is consistent with observed temperature trends for anomalous warm winters over the alpine region. A similar elevation modulation of the climate change signal has been confirmed in later studies utilising both RCMs and GCMs (e.g., Leung and Ghan, 1999b; Fyfe and Flato, 1999).

The impact of land-use changes on regional climate has been addressed in RCM simulations (e.g., Wei and Fu, 1998; Pan *et al.*, 1999; Pielke *et al.*, 1999; Chase *et al.*, 2000). Land-use changes due to human activities could induce climate modifications, at the regional and local scale, of magnitude similar to the observed climatic changes during the last century (Pielke *et al.*, 1999; Chase *et al.*, 2000). The issue of regional climate modification by

land-use change has been little explored within the context of the global change debate and, because of its potential importance, is in need of further examination.

10.5.2.2 Climate variability and extreme events

Changes in climate variability between control and $2 \times CO_2$ simulations with a nested RCM for the Great Plains of the USA have been reported (Mearns, 1999; Mearns *et al.*, 1999). There is indication of significant decreases in daily temperature variability in winter and increases in temperature variability in summer. These changes are very similar to those of the driving GCM, while changes in variability of precipitation are quite different in the nested and driving models, particularly in summer, with increases being more pronounced in the RCM. Similar results have been documented over the Iberian Peninsula (Gallardo *et al.*, 1999).

Different studies have analysed changes in the frequency of heavy precipitation events in enhanced GHG climate conditions over the European region (Schär *et al.*, 1996; Frei *et al.*, 1998; Durman *et al.*, 2001). They all indicate an increase of up to several tens of percentage points in the frequency of occurrence of precipitation events exceeding 30 mm/day, with these increases being less than those simulated by the driving GCMs (see also Jones *et al.*, 1997). In a transient RCM simulation for 1961 to 2100 over south-eastern Australia, substantial increases were found in the frequency of extreme daily rainfall and days of extreme high maximum temperature (Hennessy *et al.*, 1998), In this long simulation, changes in the frequency of long-duration extreme events (such as droughts) were identified. Finally, increases in the number of typhoons reaching mainland China and in the number of heavy rain days were reported for enhanced GHG conditions in RCM simulations over East Asia (Gao *et al.*, 2001).

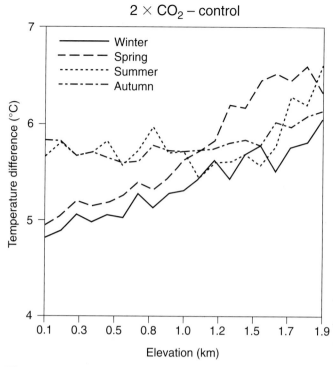

Figure 10.15: Difference between 2×CO$_2$ and control run surface air temperature as a function of elevation over the Alpine sub-region for the four seasons. Units are °C. From Giorgi *et al.* (1997).

10.5.3 Summary and Recommendations

Since the SAR, significant improvements have been achieved in the areas of development and understanding of the nested regional climate modelling technique. These include many new RCM systems, multiple nesting, coupling with different components of the climate system and research into the effects of domain size, resolution, boundary forcing and internal model variability. As a result, a number of RCM systems are currently available with the capability of high-resolution, multi-decadal simulations in a variety of regional settings. Nested RCMs have shown marked improvements in their ability to reproduce present day average climate, with some of this improvement due to better quality driving fields provided by GCMs. Seasonal temperature and precipitation biases in state-of-the-art RCMs are generally less than 1 to 2°C and a few percent to 50 to 60 % of observed precipitation, respectively, over regions of size 10^5 to 10^6 km^2. However, it is imperative for the effective use of RCMs in climate change work that the quality of GCM large-scale driving fields continues to improve. Research aiming at reducing systematic errors in both GCMs and RCMs should be carried out. With significantly improved model systems the evidence, so far, indicates that improved regional climate change simulations can be produced in the near future.

The analysis of RCM simulations has extended beyond simple averages to include higher-order climate statistics, and has indicated that RCMs can effectively reproduce interannual variability when driven by good quality forcing fields. However, more anlysis and improvements are needed of the model

performance in simulating climate variability at short time-scales (daily to sub-daily).

A serious problem concerning RCM evaluation is a general lack of good quality high-resolution observed data. In many areas, observations are extremely sparse due to complex geography or remoteness of settings. In addition, only a little work has been carried out on how to use point measurements to evaluate the grid-box mean values from a climate model, especially when using sparse station networks. This limits the ability to assess model skill in complex terrain and remote regions. It is essential for the advancement of regional climate understanding and modelling, that more research aiming at improving the quality of data for model evaluation is performed.

Overall, the evidence is strong that regional models consistently improve the spatial detail of simulated climate compared to GCMs because of their better representation of sub-GCM grid scale forcings, especially in regard to the surface hydrologic budget. This is not necessarily the case for region-averaged climate. The increased resolution of RCMs also allows the simulation of a broader spectrum of weather events, in particular concerning higher order climate statistics such as daily precipitation intensity distributions. Analysis of some RCM experiments indicate that this is in the direction of increased agreement with observations.

Several RCM studies have been important for understanding climate change processes, such as the elevation signature of the climate change signal or the effect of climate change at the river catchment level. However, a consistent set of RCM simulations of climate change for different regions which can be used as climate change scenarios for impact work is still not available. Most RCM climate change simulations have been sensitivity and process studies aimed at specific goals. The need is there to co-ordinate RCM simulation efforts and to extend studies to more regions so that ensemble simulations with different models and scenarios can be developed to provide useful information for impact assessments. This will need to be achieved under the auspices of international or large national programmes. Within this context, an important issue is to provide RCM simulations of increasing length to minimise limitations due to sampling problems.

10.6 Empirical/Statistical and Statistical/Dynamical Methods

10.6.1 Introduction

As with the dynamical downscaling of RCMs, the methods described in this section rely on the concept that regional climates are largely a function of the large-scale atmospheric state. In empirical downscaling the cross-scale relationship is expressed as a stochastic and/or deterministic function between a set of large-scale atmospheric variables (predictors) and local/regional climate variables (predictands). Predictor and predictand can be the same variables on different spatial scales (e.g., Bürger, 1997; Wilks, 1999b; Widmann and Bretherton, 2000), but more commonly are different.

When using downscaling for assessing regional climate change, three implicit assumptions are made:

- The predictors are variables of relevance to the local climate variable being derived, and are realistically modelled by the GCM. Tropospheric quantities such as temperature or geopotential height are more skilfully represented than derived variables such as precipitation at the regional or grid scale (e.g., Osborn and Hulme, 1997; Trigo and Palutikof, 1999). Furthermore, there is no theoretical level of spatial aggregation at which GCMs can be considered skilful, though there is evidence that this is several grid lengths (Widmann and Bretherton, 2000).

- The transfer function is valid under altered climatic conditions (see Section 10.6.2.2). This cannot be proven in advance, as it would require the observational record to span all possible future realisations of the predictors. However, it could be evaluated with nested AOGCM/RCM simulations of present and future climate, using the simulation of present climate to determine the downscaling function and testing the function against the future time slice.

- The predictors fully represent the climate change signal. Most downscaling approaches to date have relied entirely on circulation-based predictors and, therefore, can only capture this component of the climate change. More recently other important predictors, e.g., atmospheric humidity, have been considered (e.g., Charles *et al.*, 1999b; Hewitson, 1999).

A diverse range of downscaling methods has been developed, but, in principle, these models are based on three techniques:

- Weather generators, which are random number generators of realistic looking sequences of local climate variables, and may be conditioned upon the large-scale atmospheric state (Section 10.6.2.1);

- Transfer functions, where a direct quantitative relationship is derived through, for example, regression (Section 10.6.2.2);

- Weather typing schemes based on the more traditional synoptic climatology concept (including analogues and phase space partitioning) and which relate a particular atmospheric state to a set of local climate variables (Section 10.6.2.3).

Each of these approaches has relative strengths and weaknesses in representing the range of temporal variance of the local climate predictand. Consequently, the above approaches are often used in conjunction with one another in order to compensate for the relative deficiencies in one method.

Most downscaling applications have dealt with temperature and precipitation. However, a diverse array of studies exists in which other variables have been investigated. Appendix 10.4 provides a non-exhaustive list of past studies indicating predictands, geographical domain, and technique category. In light of the diversity in the literature, we concentrate on references to applications since 1995 and based on recent global climate change projections.

10.6.2 Methodological Options

10.6.2.1 Weather generators
Weather generators are statistical models of observed sequences of weather variables (Wilks and Wilby, 1999). Most of them focus on the daily time-scale, as required by many impact models, but sub-daily models are also available (e.g., Katz and Parlange, 1995). Various types of daily weather generators are available, based on the approach to modelling daily precipitation occurrence, and usually these rely on stochastic processes. Two of the more common are the Markov chain approach (e.g., Richardson, 1981; Hughes *et al.*, 1993, Lettenmaier, 1995; Hughes *et al.*, 1999; Bellone *et al.*, 2000) and the spell length approach (Roldan and Woolhiser, 1982; Racksko *et al.*, 1991; Wilks, 1999a). The adequacy of the stochastic models analysed in these studies varied with the climate characteristics of the locations. For example, Wilks (1999a) found the first-order Markov model to be adequate for the central and eastern USA, but that spell length models performed better in the western USA. An alternative approach would include stochastic mechanisms of storm arrivals able to produce the clustering found in observed sequences (e.g., Smith and Karr, 1985; Foufoula-Georgiou and Lettenmeier, 1986; Gupta and Waymrie, 1991; Cowpertwait and O'Connel, 1997; O'Connell, 1999).

In addition to statistical models of precipitation frequency and intensity, weather generators usually produce time-series of other variables, most commonly maximum and minimum temperature, and solar radiation. Others also include additional variables such as relative humidity and wind speed (Wallis and Griffiths, 1997; Parlange and Katz, 2000.) The most common means of including variables other than precipitation is to condition them on the occurrence of precipitation (Richardson, 1981), most often via a multiple variable first-order autoregressive process (Perica and Foufoula-Georgiou, 1996a,b; Wilks, 1999b). The parameters of the weather generator can be conditioned upon a large-scale state (see Katz and Parlange, 1996; Wilby, 1998; Charles *et al.*, 1999a), or relationships between large-scale parameter sets and local-scale parameters can be developed (Wilks, 1999b).

10.6.2.2 Transfer functions
The more common transfer functions are derived from regression-like techniques or piecewise linear or non-linear interpolations. The simplest approach is to build multiple regression models with free atmosphere grid-cell values as predictors for surface variables such as local temperatures (e.g., Sailor and Li, 1999). Other regression models have used fields of spatially distributed variables (e.g., D. Chen *et al.*, 1999), principal components of geopotential height fields (e.g., Hewitson and Crane, 1992), Canonical Correlation Analysis (CCA) and a variant termed redundancy analysis (WASA, 1998) and Singular Value Decomposition (e.g., von Storch and Zwiers, 1999).

Most applications have dealt with monthly or seasonal rainfall (e.g., Busuioc and von Storch, 1996; Dehn and Buma, 1999); local pressure tendencies (a proxy for local storminess; Kaas *et al.*, 1996); climate impact variables such as salinity and oxygen (Heyen and Dippner, 1998; Zorita and Laine, 1999); sea level (e.g., Cui *et al.*, 1996); and ecological variables such as abundance of species (e.g., Kröncke *et al.*, 1998). In addition statistics of extreme events such as storm surge levels (e.g., von Storch and Reichardt, 1997) and ocean wave heights (WASA, 1998) have been simulated.

An alternative to linear regression is piecewise linear or non-linear interpolation (Brandsma and Buishand, 1997; Buishand and Brandsma, 1999), for example, the "kriging" tools from geostatistics (Biau *et al.*, 1999). One application of this approach is a non-linear model of snow cover duration in Austria derived from European mean temperature and altitude (Hantel *et al.*, 1999). An alternative approach is based on Artificial Neural Networks (ANNs) that allow the fit of a more general class of statistical model (Hewitson and Crane, 1996; Trigo and Palutikof, 1999). For example, Crane and Hewitson (1998) apply ANN downscaling to GCM data in a climate change application over the west coast of the USA using atmospheric circulation and humidity as predictors to represent the climate change signal. The approach was shown to accurately capture the local climate as a function of atmospheric forcing. In application to GCM data, the regional results revealed significant differences from the co-located GCM grid cell, e.g., a significant summer increase in precipitation in the downscaled data (Figure 10.16).

10.6.2.3 Weather typing

This synoptic downscaling approach relates "weather classes" to local and regional climate variations. The weather classes may be defined synoptically or fitted specifically for downscaling purposes by constructing indices of airflow (Conway *et al.*, 1996). The frequency distributions of local or regional climate are then derived by weighting the local climate states with the relative frequencies of the weather classes. Climate change is then estimated by determining the change of the frequency of weather classes. However, typing procedures contain a potentially critical weakness in assuming that the characteristics of the weather classes do not change.

In many cases, the local and regional climate states are derived by sampling the observational record. For example, Wanner *et al.* (1997) and Widmann and Schär (1997) used changing global to continental scale synoptic structures to understand and reconstruct Alpine climate variations. The technique was applied similarly for New Zealand (Kidson and Watterson, 1995) and to a study of changing air pollution mechanisms (Jones and Davies, 2000).

An extreme form of weather typing is the analogue method (Zorita *et al.*, 1995). A similar concept, although mathematically more demanding, is Classification And Tree Analysis (CART) which uses a randomised design for picking regional distributions (Hughes *et al.*, 1993; Lettenmaier, 1995). Both analogue and CART approaches return approximately the right level of variance and correct spatial correlation structures.

Weather typing is also used in statistical-dynamical downscaling (SDD), a hybrid approach with dynamical elements (Frey-Buness *et al.*, 1995 and see references in Appendix 10.4). GCM results of a multi-year climate period are disaggregated into non-overlapping multi-day episodes of quasi-stationary large-scale flow patterns. Similar episodes are then grouped in classes of different weather types, and, members of these classes are simulated with an RCM. The RCM results are statistically evaluated, and the frequency of occurrence of the respective classes determines their statistical weight. An advantage of the SDD technique over other empirical downscaling techniques is that it specifies a complete three-dimensional climate state. The advantage over continuous RCM simulations is the reduction in computing time, as demonstrated in Figure 10.17.

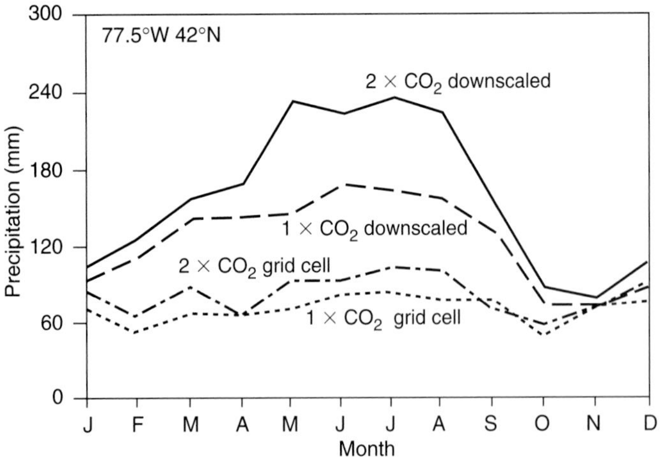

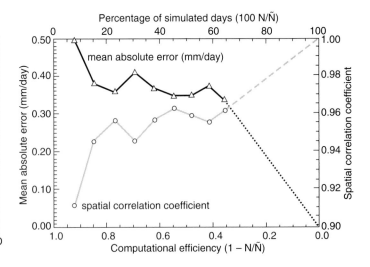

Figure 10.16: Climate change scenario of monthly mean precipitation (mm) over the Susquehanna river basin, USA. Monthly means derived using daily down-scaled precipitation generated with an Artificial Neural Network (ANN) and atmospheric predictors from 1xCO_2 and 2xCO_2 GCM simulations. Also shown are the GCM grid cell precipitation values from the co-located grid cell. From Crane and Hewitson (1998).

Figure 10.17: Similarity of time mean precipitation distributions obtained in a continuous RCM simulation and through statistical-dynamical downscaling (SDD) for different levels of disaggregation. Black line: mean absolute error (mm/day), grey line: spatial correlation coefficient. Horizontal axis: computational load of SDD. N is the number of days simulated in SDD, Ñ the number of days simulated with the continuous RCM simulation.

10.6.3 Issues in Statistical Downscaling

10.6.3.1 Temporal variance

Transfer function approaches and some weather typing methods suffer from an under prediction of temporal variability, as this is related only in part to the large-scale climate variations (see Katz and Parlange, 1996). Two approaches have been used to restore the level of variability: inflation and randomisation. In the inflation approach the variation is increased by the multiplication of a suitable factor (Karl *et al.*, 1990). A more sophisticated version is "expanded downscaling", a variant of Canonical Correlation Analysis that ensures the right level of variability (Bürger, 1996; Huth, 1999; Dehn *et al.*, 2000). In the randomisation approachs, the unrepresented variability is added as noise, possibly conditioned on synoptic state (Buma and Dehn, 1998; Dehn and Buma 1999; Hewitson, 1999; von Storch, 1999b).

Often weather generators have difficulty in representing low frequency variance, and conditioning the generator parameters on the large-scale state may alleviate this problem (see Katz and Parlange, 1996; Wilby, 1998; Charles *et al.*, 1999a). For example, Katz and Parlange (1993, 1996) modelled daily time-series of precipitation as a chain-dependent process, conditioned on a discrete circulation index. The results demonstrated that the mean and standard deviation of intensity and the probability of precipitation varied significantly with the circulation, and reproduced the precipitation variance statistics of the observations better than the unconditioned model. The method describes the mean precipitation as a linear function of the circulation state, and the standard deviation as a non-linear function (Figure 10.18).

10.6.3.2 Evaluation

The evaluation of downscaling techniques is essential but problematic. It requires that the validity of the downscaling functions under future climates be demonstrated, and that the predictors represent the climate change signal. It is not possible to achieve this rigorously as the empirical knowledge available is insufficient. The analysis of historical developments, e.g., by comparing downscaling models between recent and historical periods (Jacobeit *et al.*, 1998), as well as simulations with GCMs can provide support for these assumptions. However, the success of a statistical downscaling technique for representing present day conditions does not necessarily imply that it would give skilful results under changed climate conditions, and may need independent confirmation from climate model simulations (Charles *et al.*, 1999b).

The classical validation approach is to specify the downscaling technique from a segment of available observational evidence and then assess the performance of the empirical model by comparing its predictions with independent observed values. This approach is particularly valuable when the observational record is long and documents significant changes (greater than 50 years in some cases; Hanssen-Bauer and Førland (1998, 2000)). An example is the analysis of absolute pressure tendencies in the North Atlantic (Kaas *et al.*, 1996). As another example, Wilks (1999b) developed a downscaling function on dry years and found it functioned well in wet years.

An alternative approach is to use a series of comparisons between models and transfer functions (e.g., González-Rouco *et al.*, 1999, 2000). For instance, empirically derived links were shown to be incorporated in a GCM (Busuioc *et al.*, 1999) and a RCM (Charles *et al.*, 1999b). Then a climatic change due to doubling of CO_2 was estimated through the empirical link and compared with the result of the dynamical models. In both cases, the dynamical response was found consistent for the winter season, indicating the validity of the empirical approach, although less robust results were noted in the other seasons.

10.6.3.3 Choice of predictors

There is little systematic work explicitly evaluating the relative skill of different atmospheric predictors (Winkler *et al.*, 1997). This is despite the availability of disparate studies that evaluate a broad range of predictors, predictands and techniques (see Appendix 10.4). Useful summaries of downscaling techniques and the predictors used are also presented in Rummukainen (1997), Wilby (1998), and Wilby and Wigley (2000).

The choice of the predictor variables is of utmost importance. For example, Hewitson and Crane (1996) and Hewitson (1999) have demonstrated how the down-scaled projection of future change in mean precipitation and extreme events may alter significantly depending on whether or not humidity is included as a predictor. The downscaled results can also depend on whether absolute or relative humidity is used as a predictor (Charles *et al.*, 1999b). The implication here is that while a predictor may or may not appear as the most significant when developing the downscaling function under present climates, the changes in that predictor under a future climate may be critical for determining the climate change. Some estimation procedures, for example stepwise regression, are not able to recognise this and exclude variables that may be vital for climate change.

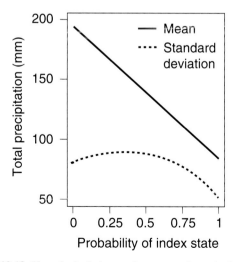

Figure 10.18: Hypothetical changes in mean and standard deviation of January total precipitation at Chico, California, as a function of changing probability that January mean sea level pressure is above normal.

A similar issue exists with respect to downscaling temperature. Werner and von Storch (1993), Hanssen-Bauer and Førland (2000) and Mietus (1999) noted that low-frequency changes in local temperature during the 20th century could only partly be related to changes in circulation. Schubert (1998) makes a vital point in noting that changes of local temperature under doubled atmospheric CO_2 may be dominated by changes in the radiative properties of the atmosphere rather than circulation changes. These can be accounted for by incorporating the large-scale temperature field from the GCM as a surrogate indicator of the changed radiative properties of the atmosphere (Dehn and Buma, 1999) or by using several large-scale predictors, such as gridded temperature and circulation fields (e.g., Gyalistras *et al.*, 1998; Huth, 1999).

With the recent availability of global reanalyses (Kalnay *et al.*, 1996; Gibson *et al.*, 1997), the number of candidate predictor fields has been greatly enhanced (Solman and Nuñez, 1999). Prior to this the empirical evidence about the co-variability of regional/local predictands and large-scale predictors was limited mostly to gridded near surface temperature and/or air pressure. These "new" data sets allow significant improvements in the design of empirical downscaling techniques, in particular by incorporating knowledge about detailed meteorological processes. Taking advantage of these new data sets have allowed systematic evaluation of a broad range of possible predictors for daily precipitation. It has been found that indicators of mid-tropospheric circulation and humidity to be the most critical predictors, with surface flow and humidity information being important under orographic rainfall.

10.6.4 Intercomparison of Statistical Downscaling Methodologies

An increasing number of studies comparing different downscaling studies have emerged since the SAR. However, there is a paucity of systematic studies that use common data sets applied to different procedures and over the same geographic region. A number of articles discussing different empirical and dynamical downscaling approaches present summaries of the relative merits and shortcomings of different procedures (Giorgi and Mearns, 1991; Hewitson and Crane, 1996; Rummukainen, 1997; Wilby and Wigley, 1997; Gyalistras *et al.*, 1998; Kidson and Thompson, 1998; Biau *et al.*, 1999; Murphy, 1999; ; von Storch, 1999b; Zorita and von Storch, 1999; Murphy, 2000). However, these inter-comparisons vary widely with respect to predictors, predictands and measures of skill. Consequently, a systematic, internationally co-ordinated inter-comparison project would be particularly helpful in addressing this issue.

The most systematic and comprehensive study so far compared empirical transfer functions, weather generators, and circulation classification schemes over the same geographical region using climate change simulations and observational data (Wilby and Wigley, 1997; Wilby, 1998). This considered a demanding task to downscale daily precipitation for six locations over North America, spanning arid, moist tropical, maritime, mid-latitude, and continental climate regimes. Fourteen measures of skill were used, strongly emphasising daily statistics, and included

wet and dry spell length, 95th percentile values, wet-wet day probabilities, and several measures of standard deviation. Downscaling procedures in the study included two different weather generators, two variants of an ANN-based technique, and two stochastic/circulation classification schemes based on vorticity classes.

The results require careful evaluation as they indicate relative merits and shortcoming of the different procedures rather than recommending one over another. Overall, the weather generators captured the wet-day occurrence and the amount distributions in the data well, but were less successful at capturing the interannual variability, while the opposite results was found for the ANN procedures. The stochastic/circulation typing schemes, as something of a combination of the principles underlying the other methods, were a better all-round performer.

A factor not yet fully evaluated in any comparative study is that of the temporal evolution of daily events which may be critical for some applications, e.g., hydrological modelling. While a downscaling procedure may correctly represent, for example, the number of rain days, the temporal sequencing of these may be as important. Zorita *et al.* (1995) and Zorita and von Storch (1997) compared a CART technique, a CCA and an ANN technique with the analogue technique, and found the simpler analogue technique performed as well as the more complicated methods.

A number of analyses have dealt with the relative merits of non-linear and linear approaches. For example, the relationships between daily precipitation and circulation indicators are often non-linear (Conway *et al.*, 1996; Brandsma and Buishand, 1997). Similarly, Corte-Real *et al.* (1995) applied multivariate adaptive regression splines (MARS) to approximate the non-linear relationships between large-scale circulation and monthly mean precipitation. In a comparison of kriging and analogues, Biau *et al.* (1999) and von Storch (1999c) show that kriging resulted in better specifications of averaged quantities but too low variance, whereas analogues returned the right variance but lower correlation. In general, it appears that downscaling of the short-term climate variance benefits from the use of non-linear models.

Most of the comparative studies mentioned above come to the conclusion that techniques differ in their success of specifying regional climate, and the relative merits and shortcomings emerge differently in different studies and regions. This is not surprising, as there is considerable flexibility in setting up a downscaling procedure, and the suitability of a technique and the adaptation to the problem at hand varies. This flexibility is a distinct advantage of empirical methods.

10.6.5 Summary and Recommendations

A broad range of statistical downscaling techniques has been developed in the past few years. Users of GCM-based climate information may choose from a large variety of methods conditional upon their needs. Weather generators provide realistic sequences of high temporal resolution events. With transfer functions, statistics of regional and local climate, such as conditional means or quantiles, may consistently be derived from GCM generated data. Techniques based on weather typing serve both purposes, but are less adapted to specific applications.

Downscaling means post-processing GCM data; it cannot account for insufficiencies in the driving GCM. As statistical techniques combine the existing empirical knowledge, statistical downscaling can describe only those links that have been observed in the past. Thus, it is based on the assumption that presently found links will prevail under different climate conditions. It may be, in particular, that under present conditions some predictors appear less relevant, but become significant in describing climate change. It is recommended to test statistical downscaling methods by comparing their estimates with high resolution dynamical model simulations. The advent of decades-long atmospheric reanalyses has offered the community many more atmospheric large-scale variables to incorporate as predictors.

Statistical downscaling requires the availability of long and homogeneous data series spanning the range of observed variance, while the computational resources needed are small. Therefore, statistical downscaling techniques are suitable tools for scientific communities without access to supercomputers and with little experience in process-based climate modelling. Furthermore, statistical techniques may relate directly GCM-derived data to impact relevant variables, such as ecological variables or ocean wave heights, which are not simulated by contemporary climate models.

It is concluded that statistical downscaling techniques are a viable complement to process-based dynamical modelling in many cases, and will remain so in the future.

10.7 Intercomparison of Methods

Few formal comparative studies of different regionalisation techniques have been carried out. To date, published work has mostly focused on the comparison between RCMs and statistical downscaling techniques. Early applications of RCMs for climate change simulations (Giorgi and Mearns, 1991; Giorgi *et al.*, 1994) compared the models against observations or against the driving GCMs, but not against statistical/empirical techniques.

Kidson and Thompson (1998) compared the RAMS (Regional Atmospheric Modelling System) dynamical model and a statistical regression-based technique. Both approaches were applied to downscale reanalysis data (ECMWF) over New Zealand to a grid resolution of 50 km. The statistical downscaling used a screening regression technique to predict local minimum and maximum temperature and daily precipitation, at both monthly and daily time-scales. The regression technique limits each regression equation to five predictors (selected from Empirical Orthogonal Functions (EOFs) of atmospheric fields). Both monthly and daily results indicated little difference in skill between the two techniques, and Kidson and Thompson (1998) suggested that, subject to the assumption of statistical relationships remaining viable under a future climate, the computational requirements do not favour the use of the dynamical model. They also noted, however, that the dynamical model performed better with the convective components of precipitation.

Bates *et al.* (1998) compared a south-western Australia simulation using the DARLAM (CSIRO Division of Atmospheric Research Limited Area Model) model with a down-scaled DARLAM simulation where the downscaling model had been fitted independently to observational data. The downscaling reproduced observed precipitation probabilities and wet and dry spell frequencies while the DARLAM simulation underestimated the frequency of dry spells and over estimated the probability of precipitation and the frequency of wet spells. In a climate change follow-on experiment, again using both methods, Charles *et al.* (1999b) found a small decrease in probability of precipitation under future climate conditions.

Murphy (1999) evaluated the UK Meteorological Office Unified Model (UM) RCM over Europe against a statistical downscaling model based on regression. Monthly mean surface temperature and precipitation anomalies were down-scaled using predictor sets chosen from a range of candidate variables similar to those used by Kidson and Thompson (1998) (EOFs of atmospheric fields). The results showed similar levels of skill for the dynamical and statistical methods, in line with the Kidson and Thompson (1998) study. The statistical method was nominally better for summertime estimates of temperature, while the dynamical model gave better estimates of wintertime precipitation. Again, the conclusion was drawn that the sophistication of the dynamical model shows little advantage over statistical techniques, at least for present day climates.

Murphy (2000) continued the comparative study by deriving climate change projections for 2080 to 2100 from a simulation with the HadCM2 AOGCM. The dynamical and statistical downscaling techniques were the same regional and statistical models as used by Murphy (1999). The statistical and dynamical techniques produced significantly different predictions of climate change, despite exhibiting similar skill when validated against present day observations. The study identifies two main sources of divergence between the dynamical and statistical techniques: firstly, differences between the strength of the observed and simulated predictor/predictand relationships, and secondly, omission from the regression equations of variables which represent climate change feedbacks, but are weak predictors of natural variability. In particular, the exclusion of specific humidity led to differences between the dynamical and statistical simulations of precipitation change. This point would seem to confirm the humidity issue raised in Section 10.6.3 (Hewitson and Crane 1996, Crane and Hewitson, 1998, Charles *et al.*, 1999b; Hewitson 1999).

Mearns *et al.* (1999) compared RCM simulations and statistical downscaling using a regional model and a semi-empirical technique based on stochastic procedures conditioned on weather types which were classified from circulation fields (700hPa geopotential heights). While Mearns *et al.* suggest that the semi-empirical approach incorporates more physical meaning into the relationships than a pure statistical approach does, this approach does impose the assumption that the circulation patterns are robust into a future climate in addition to the normal assumption that the cross-scale relationships are stationary in time. For both techniques, the driving fields were from the CSIRO AOGCM (Watterson *et al.*, 1995). The variables of interest were maximum and minimum daily temperature and precipitation over central-northern USA (Nebraska). As with the preceding studies, the validation under present climate conditions indicated similar skill levels for the dynamical and statistical approaches, with some advantage by the statistical technique.

In line with the Murphy (2000) study, larger differences were also noted by Mearns *et al.* (1999) when climate change projections were produced. Notably for temperature, the statistical technique produced an amplified seasonal cycle compared to both the RCM and CSIRO data, although similar changes in daily temperature variances were found in both the RCM and the statistical technique (with the statistical approach producing mostly decreases). The spatial patterns of change showed greater variability in the RCM compared with the statistical technique. Mearns *et al.* (1999) suggested that some of the differences found in the results were due to the climate change simulation exceeding the range of data used to develop the statistical model, while the decreases in variance were likely to be a true reflection of changes in the circulation controls. The precipitation results from Mearns *et al.* (1999) are different from earlier studies with the same RCM (e.g., Giorgi *et al.*, 1998) that produced few statistically significant changes.

Extending the comparison beyond simple methodological performance, Wilby *et al.* (2000) compared hydrological responses using data from dynamically and statistically downscaled climate model output for the Animas River basin in Colorado, USA. While not a climate change projection, the use of output from an RCM and a statistical downscaling approach to drive a distributed hydrological model exemplify the objective of the downscaling. The results indicate that both the statistical and dynamical methods had greater skill (in terms of modelling hydrology) than the coarse resolution reanalysis output used to drive the downscaling. The statistical method had the advantage of requiring very few parameters, an attribute making the procedure attractive for many hydrological applications. The dynmical model output, once elevation-corrected, provided better water balance estimates than raw or elevation-corrected reanalysis output.

Overall, the above comparative studies indicate that for present climate both techniques have similar skill. Since statistical models are based on observed relationships between predictands and predictors, this result may represent a further validation of the performance of RCMs. Under future climate conditions more differences are found between the techniques, and the question arises as to which is "more correct". While the dynamical model should clearly provide a better physical basis for change, it is still unclear whether different regional models generate similar downscaled changes. With regard to statistical/empirical techniques, it would seem that careful attention must be given to the choice of predictors, and that methodologies which internally select predictors based on explanatory power under present climates may exclude predictors important for determining change under future climate modes.

10.8 Summary Assessment

Today different modelling tools are available to provide climate change information at the regional scale. Coupled AOGCMs are the fundamental models used to simulate the climatic response to anthropogenic forcings and, to date, results from AOGCM simulations have provided the climate information for the vast majority of impact studies. On the other hand, resolution limita-

tions pose severe constraints on the usefulness of AOGCM information, especially in regions characterised by complex physiographic settings. Three classes of regionalisation techniques have been developed to enhance the regional information of coupled AOGCMs: high resolution and variable resolution time-slice AGCM experiments, regional climate modelling, and empirical/statistical and statistical/dynamical approaches.

Since the SAR, substantial progress has been achieved in all regionalisation methods, including better understanding of the techniques, development of a wide variety of modelling systems and methods, application of the techniques to a wide range of studies and regional settings, and reduction of model biases. Modelling work has indicated that regionalisation techniques enhance some aspects of AOGCM regional information, such as the high resolution spatial detail of precipitation and temperature, and the statistics of daily precipitation events. It is important to stress that AOGCM information is the starting point for the application of all regionalisation techniques, so that a foremost requirement in the simulation of regional climate change is that the AOGCMs simulate well the circulation features that affect regional climates. In this respect, indications are that the performance of current AOGCMs is generally improving.

Analysis of AOGCM simulations for broad (sub-continental scale) regions indicates that biases in the simulation of present day regionally and seasonally averaged surface climate variables, although highly variable across regions and models, are generally improved compared with the previous generation models. This implies increased confidence in simulated climatic changes. The performance of models in reproducing observed interannual variability varies across regions and models.

Regional analysis of AOGCM transient simulations extending to 2100, for different scenarios of GHG increase and sulphate aerosol effects, and with a number of modelling systems (some simulations include ensembles of realisations) indicate that the average climatic changes for the late decades of the 21st century compared to present day climate vary substantially across regions and models. The primary source of uncertainty in the simulated changes is associated with inter-model range of changes, with inter-scenario and intra-ensemble range of simulated changes being less pronounced. Despite the range of inter-model results, some common patterns of sub-continental scale climatic changes are emerging, and thus providing increased confidence in the simulation of these changes.

Work performed with all regionalisation techniques indicates that sub-GCM grid scale structure in the regional climate change signal can occur in response to regional and local forcings, although more work is needed to assess the statistical significance of the sub-GCM grid scale signal. In particular, modelling evidence clearly indicates that topography, land use and the surface hydrologic cycle strongly affect the surface climate change signal at the regional to local scale. This implies that the use of AOGCM information for impact studies needs to be taken cautiously, especially in regions characterised by pronounced sub-GCM grid scale variability in forcings, and that suitable regionalisation techniques should be used to enhance the AOGCM results over these regions.

Considerations of various types may enter the choice of the regionalisation technique, as different techniques may be most suitable for different applications and different working environments. High resolution AGCMs offer the primary advantage of global coverage and two-way interactions between regional and global climate. However, due to their computational cost, the resolution increase that can be expected from these models is limited. Variable resolution and RCMs yield a greater increase in resolution, with current RCMs reaching resolutions as fine as a few tens of kilometres or less. RCMs can capture physical processes and feedbacks occurring at the regional scale, but they are affected by the errors of the AOGCM driving fields, and they do not represent regional-to-global climate feedbacks. The effects of regional-to-global feedback processes depend on the specific problem and in many cases may not be important. Two-way GCM-RCM nesting would allow the description of such effects, and some research efforts in that direction are currently under way. Statistical downscaling techniques offer the advantages of being computationally inexpensive, of providing local information which is needed in many impact applications, and of offering the possibility of being tailored to specific applica-

tions. However, these techniques have limitations inherent in their empirical nature.

The combined use of different techniques may provide the most suitable approach in many instances. For example, a high-resolution AGCM simulation could represent an important intermediate step between AOGCM information and RCM or statistical downscaling models. The convergence of results from different approaches applied to the same problem can increase the confidence in the results and differences between approaches can help to understand the behaviour of the models.

Despite recent improvements and developments, regionalisation research is still a maturing process and the related uncertainties are still rather poorly known. One of the reasons for this is that most regionalisation research activities have been carried out independently of each other and aimed at specific objectives. Therefore a coherent picture of regional climate change via available regionalisation techniques cannot yet be drawn. More co-ordinated efforts are thus necessary to improve the integrated hierarchy of models, evaluate the different methodologies, intercompare methods and models and apply these methods to climate change research in a comprehensive strategy.

Appendix 10.1:

List of regional climate model simulations of duration longer than 3 months nested within analyses; also including oceanic RCMs (O-RCM).

References	Grid size	Duration	Region
a) Individual January/July present-day simulations			
Walsh and McGregor (1996)	125 km	7 × 1 month	Antarctica
Rinke *et al.* (1999)	55 km	11 × 1 month	Arctic
Takle *et al.* (1999)	50 km	7 × 2 months	USA
Katzfey (1999)	125 km	8 × 1 month	Australia
b) Seasonally-varying present-day simulations			
Giorgi *et al.* (1993a)	60 km	2 years	USA
Christensen *et al.* (1995)	56 km	20 months	Europe
Leung and Ghan (1995)	30 and 90 km	1 year	North-west USA
Kim (1997)	20 km	6 months	Western USA
Christensen *et al.* (1997)	26 to 57 km	11 months to 10 years	Europe
Jenkins (1997)	110 km	2 × 4 months	West Africa
Kidson and Thompson (1998)	50 km	5 years	New Zealand
McGregor *et al.* (1998)	44 km	1 year	Southeast Asia
Noguer *et al.* (1998)	50 km	10 years	Europe
Ruti *et al.* (1998)	30 km	19 months	Europe
Seth and Giorgi (1998)	60 km	2 × 4 months	USA
Leung and Ghan (1998)	90 km	3 years	North-west USA
Kauker (1998)	15 km	15 years	North Sea (O-RCM)
Christensen (1999)	55 km	7 × 1 year	Mediterranean area
Giorgi and Shields (1999)	60 km	3 years	USA
Giorgi *et al.* (1999)	60 km	13 month	East Asia
Small *et al.* (1999a)	60 km	5.5 years	Central Asia
van Lipzig (1999)	55 km	10 years	Antarctica
Liston and Pielke (1999)	50 km	1 year	USA
Hong and Leetmaa (1999)	50 km	4 × 3 months	USA
Christensen and Kuhry (2000)	16 km	15 years	Arctic Russia
Pan *et al.* (2000)	55 km	2 × 10 years	USA
Mabuchi *et al.* (2000)	30 km	6.5 years	Japanese Islands
Jacob and Podzun (2000)	55 km	10 years	Northern Europe
c) Seasonal tropical or monsoon simulations			
Bhaskaran *et al.* (1996)	50 km	4 months	Indian monsoon
Ji and Vernekar (1997)	80 km	3 × 5.5 months	Indian monsoon
Wei *et al.* (1998)	60 km	4 months	Temperate East Asia
Sun *et al.* (1999)	60 km	10 × 3 month	East Africa
Leung *et al.* (1999a)	60 km	3 × 3 month	East Asia
Chen and Fu (2000)	60 km	3 years	East Asia

[1995, 1] Third International Conference on Modelling of Global Climate Change and Variability, Hamburg, Germany, 4 to 8 September 1995.

[2000, 2] Submitted to Research Activities in Atmospheric and Oceanic Modelling. (CAS/JSC Working Group on Numerical Experimentation Report) [Geneva]: WMO.

[2000, 3] 80th AMS Annual Meeting, Long Beach, California, 9 to 14 January 2000.

Appendix 10.2:

List of regional climate model simulations of duration longer than 3 months nested within a GCM present day simulation; also including oceanic RCMs (O-RCM) and variable resolution GCMs (var.res.GCM).

References	Grid size	Duration	Region
a) Perpetual January simulation			
McGregor and Walsh (1993)	250 km	10 months	Australia
b) Individual January/July simulations			
Giorgi (1990)	60 km	6 × 1 month	USA
Marinucci and Giorgi (1992)	70 km	5 × 1 month	Europe
McGregor and Walsh (1994)	125 km/60 km	10 × 1 month	Tasmania
Marinucci *et al.* (1995)	20 km	5 × 1 month	Europe (Alps)
Walsh and McGregor (1995)	125 km	10 × 1 month	Australasia
Podzun *et al.* (1995)	55 km	5 × 1 month	Europe
Rotach *et al.* (1997)	20 km	5 × 1 month	Europe (Alps)
Joubert *et al.* (1999)	125 km	20 × 1 month	South Africa
c) Seasonally-varying simulations			
Giorgi *et al.* (1994)	60 km	3.5 years	USA
Dèquè and Piedelievre (1995)	T21-T200	10 years	Europe (var.res.GCM)
Hirakuchi and Giorgi (1995)	50 km	5 years	East Asia
Jones *et al.* (1995)	50 km	10 years	Europe
McGregor *et al.* (1995)	125 km	10 years	Australasia
Giorgi and Marinucci (1996b)	50 km	5 years	Europe
Giorgi *et al.* (1997)	50 km	5 years	Europe
Krinner *et al.* (1997)	~100 km	5 years	Antarctica (var.res.GCM)
Jenkins and Barron (1997)	108 km	7 months	USA – AMIP
Jacob and Podzun (1997)	55 km	4 years	Europe
Walsh and McGregor (1997)	125 km	5 × 18 months	Australasia – AMIP
Christensen *et al.* (1998)	57 and 19 km	9 years	Scandinavia
Krinner and Genthon (1998)	~100 km	3 years	Greenland (var.res.GCM)
Dèquè *et al.* (1998)	~60 km	10 years	Europe
Giorgi *et al.* (1998)	50 km	5 years	USA
Katzfey *et al.* (1998)	60 and 125 km	20 years	Australia
Laprise *et al.* (1998)	45 km	5 years	West Canada
Machenhauer *et al.* (1998)	19 to 70 km	5 to 30 years	Europe
McGregor *et al.* (1998)	44 km	10 years	Southeast Asia
Noguer *et al.* (1998)	50 km	10 years	Europe
Renwick *et al.* (1998)	50 km	10 years	New Zealand
Böhm *et al.* (1998)	55 km	13 month	Northern South America
Kauker (1998)	15 km	5 years	North Sea (O-RCM)
Leung and Ghan (1999a)	90 km	7 years	North-west USA
Gallardo *et al.* (1999)	50 km	10 years	Iberian Peninsula
Leung *et al.* (1999b)	90 km	2 years	North-west USA
Haugen *et al.* (1999)	55 km	20 years	North-west Europe
Jacob and Podzun (2000)	55 km	10 years	Northern Europe
Pan *et al.* (2000)	55 km	2 × 10 years	USA
Rummukainen *et al.* (2000)	44 km	10 years	Europe
Kato *et al.* (2001)	50 km	10 years	East Asia
Gao *et al.* (2000)	60 km	5 year	China
Chen and Fu (2000)	60 km	3 years	East Asia
c) Seasonal tropical or monsoon simulations			
Jacob *et al.* (1995)	55 km	6 months	Indian monsoon
Bhaskaran *et al.* (1998)	50 km	10 years	India – AMIP
Hassel and Jones (1999)	50 km	20 years	Indian monsoon

[1995, 1] Third International Conference on Modelling of Global Climate Change and Variability, Hamburg, Germany, 4 to 8 September 1995.

[1998, 2] International Conference on The Role of Topography in Modelling Weather and Climate. International Centre for Theoretical Physics, Trieste, Italy, 22 to 26 June 1998.

[2000, 3] Submitted to Research Activities in Atmospheric and Oceanic Modelling. (CAS/JSC Working Group on Numerical Experimentation Report) [Geneva]: WMO.

Appendix 10.3:

List of regional climate model simulations of duration longer than 3 months nested within a GCM climate change simulation; also including oceanic RCMs (O-RCM) and variable resolution GCMs (var.res.GCM).

References	Grid size	Duration	Region
a) Individual January/July 2×CO₂ simulations			
Giorgi *et al.* (1992)	70 km	5 × 1 month	Europe
McGregor and Walsh (1994)	60 km	10 × 1 month	Tasmania
Rotach *et al.* (1997)	20 km	5 × 1 month	Europe (Alps)
b) Seasonally-varying 2×CO₂ time-slice simulations			
Giorgi *et al.* (1994)	60 km	3.5 years	USA
Hirakuchi and Giorgi (1995)	50 km	5 years	East Asia
McGregor *et al.* (1995)	125 km	10 years	Australasia
Giorgi *et al.* (1997)	50 km	3 years	Europe
Jones *et al.* (1997)	50 km	10 years	Europe
Dèquè *et al.* (1998)	About 60 km	10 years	Europe (var.res.GCM)
Giorgi *et al.* (1998)	50 km	5 years	USA
Joubert *et al.* (1998)	125 km	10 years	Southern Africa
Laprise *et al.* (1998)	45 km	5 years	West Canada
Machenhauer *et al.* (1998)	19 to 70 km	5 to 30 years	Europe
McGregor *et al.* (1998)	44 km	10 years	South-east Asia
Renwick *et al.* (1998)	50 km	10 years	New Zealand
Kauker (1998)	15 km	5 years	North Sea (O-RCM)
Räisänen *et al.* (1999)	44 km	10 years	Europe
Hassel and Jones (1999)	50 km	20 years	Indian monsoon
Gallardo *et al.* (1999)	50 km	10 years	Iberian Peninsula
Haugen *et al.* (1999)	55 km	20 years	North-west Europe
Leung and Ghan (1999b)	90 km	8 years	North-west USA
Pan *et al.* (2000)	55 km	2 × 10 years	USA
Kato *et al.* (2001)	50 km	10 years	East Asia
Gao *et al.* (2000)	60 km	5 year	China
c) Seasonally-varying fully transient CO₂ simulations			
McGregor *et al.* (1999)	125 km	140 years	Australasia
McGregor *et al.* (1999)	60 km	140 years	South-east Australia

[1995, 1] Third International Conference on Modelling of Global Climate Change and Variability, Hamburg, Germany, 4 to 8 September 1995.

[1998, 2] IntInternational Conference on The Role of Topography in Modelling Weather and Climate. IntInternational Centre for Theoretical Physics,Trieste, Italy, 22 to 26 June 1998.

Appendix 10.4: Examples of downscaling studies.

Technique (utilised in the above categories):

- WG = weather generators (e.g.: Markov-type procedures, conditional probability).
- TF = transfer functions (e.g.: Regression, canonical correlation analysis, and artificial neural networks).
- WT = weather typing (e.g.: cluster analysis, self-organising map, and extreme value distribution).

Predictor variables: C = circulation based (e.g.: sea level pressure fields and geopotential height fields). T = temperature (at surface or on one or more atmospheric levels). TH = thickness between pressure levels. VOR = vorticity. W = wind related. Q = specific humidity (at surface or on one or more atmospheric levels). RH = relative humidity (at surface or on one or more atmospheric levels). Cld = cloud cover. ZG = spatial gradients of the predictors. O = other.

Predictands: T (temperature); Tmax (maximum temperature); Tmin (minimum temperature); P (precipitation).

Region is the geographic domain.

Time is the time-scale of the predictor and predictand: H (hourly), D (daily), M (monthly), S (seasonal), and A (annual).

Region	Technique	Predictor	Predictand	Time	Author (s)
Africa					
South Africa	TF	C	P	D	Hewitson and Crane, 1996
America					
USA	WT	T	Tmax, Tmin	D	Brown and Katz, 1995
USA	WG	C	P	D	Zorita *et al.*, 1995
USA	WG, TF	C, T, VOR	P	D	Wilby and Wigley, 1997
USA	TF	C, Q	P	D	Crane and Hewitson, 1998
USA	WG, TF	C, T, VOR	T, P	D	Wilby *et al.*, 1998a, b
USA	WG, WT	C	T, P	D	Mearns *et al.*, 1999
USA	TF	C, T, RH, W	T	D	Sailor and Li, 1999
USA	WG		P	D	Bellone *et al.*, 1999
Mexico and USA	TF	C, TH, O	P	D	Cavazos, 1997
Mexico and USA	TF, WT	C, TH, Q	P	D	Cavazos, 1999
Central Argentina	TF	C, W	T, Tmax, Tmin	M	Solman and Nuñez, 1999
Asia					
Japanese coast	TF	C	Sea level	M	Cui *et al.*, 1995, 1996
Chinese coast	TF		Sea level variability	M	Cui and Zorita, 1998
Oceania					
New Zealand	WT	C	Tmax, Tmin, P	D	Kidson and Watterson, 1995
New Zealand	TF	C, TH, VOR, W	T, P	D	Kidson and Thompson, 1998
Australia	TF	C	Tmax, Tmin	D	Schubert and Henderson-Sellers, 1997
Australia	TF	C	Tmax, Tmin	D	Schubert, 1998
Australia	WT	C, T	P		Timbal and McAvaney, 1999
Australia	WT				Schnur and Lettenmaier, 1999
Europe					
Europe	WG	VOR, W			Conoway *et al.*, 1996
Europe	WG, TF	C, P, Tmax, Tmin, O	T, P	D	Semenov and Barrow, 1996
Europe	TF	C, W, VOR, T, Q, O	T, P	M	Murphy, 1998a, b
Europe	TF	C	T, P, vapour pressure	D	Weichert and Bürger, 1998
Germany	TF	T	Phenological event		Maak and van Storch, 1997
Germany	TF	C	Storm surge	M	Von Storch and Reichardt, 1997
Germany	TF		Salinity		Heyen and Dippner, 1998
Germany	WT		Thunderstorms	D	Sept, 1998

Region	Technique	Predictor	Predictand	Time	Author (s)
Germany	TF		Ecological variables		Krönke *et al.*, 1998
Iberian Peninsula	WG	C	P	D	Cubash *et al.*, 1996
Iberian Peninsula	TF	C	Tmax, Tmin	D	Trigo and Palutikof, 1998
Iberian Peninsula	TF		T, P		Boren *et al.*, 1999
Iberian Peninsula	TF		T, P		Ribalaygua *et al.*, 1999
Spain (and USA)	TF	C	Tmax, Tmin	D	Palutikof *et al.*, 1997
Spain (and USA)	TF	C	Tmax, Tmin	D	Winkler *et al.*, 1997
Spain	WT			D	Goodess and Palutikof, 1998
Portugal	TF	C	P	M	Corte-Real *et al.*, 1995
Portugal	WT	C		D	Corte-Real *et al.*, 1999
The Netherlands	WT	C, VOR, W	T, P	D,M	Buishand and Brandsma, 1997
Norway	TF	C, O	T, P and others	M	Benestad, 1999a, b
Norway (glaciers)	TF	C, O	Local weather	D	Reichert *et al.*, 1999
Romania	TF	C	P	M	Busuioc and von Storch, 1996
Romania	TF	C	P	M	Busuioc *et al*, 1999
Switzerland	TF		P		Buishand and Klein Tank, 1996
Switzerland	TF		P		Brandsma and Buishand, 1997
Switzerland	TF			D	Widmann and Schär, 1997
Switzerland	WG	C	Local Weather	H	Gyalistras *et al.*, 1997
Switzerland	TF		P		Buishand and Brandsma, 1999
Poland	TF	C	T, sea level, wave height, salinity, wind, run-off	D,M	Mietus, 1999
Alps	WT				Fuentes and Heimann, 1996
Alps	TF	C, T	T, P	M	Fischlin and Gylistras, 1997
Alps	WT	C	Snow		Martin *et al.*, 1997
Alps	WT				Fuentes *et al.*, 1998
Alps	TF	C, T	T, P,		Gyalistras *et al.*, 1998
Alps,	TF	C, T	Snow cover		Hantel *et al.*, 1998
Alps	WT	C, T	Landslide activity		Dehn, 1999a, b
Alps	WT		T, P	D	Heimann and Sept, 1999
Alps	WT		P	D	Fuentes and Heimann, 1999
Alps	TF, WG	C, T	Weather statistics	M	Riedo *et al.*, 1999
Alps	TF	C	P	M	Burkhardt, 1999
Mediterranean	TF	C, P	T		Palutikof and Wigley, 1995
Mediterranean	TF	C	P	S	Jacobeit, 1996
North Atlantic	TF	C	Pressure tendencies	M	Kaas *et al.*, 1996
North Atlantic	TF	C	Wave height	M	WASA, 1998
North Sea	TF		Ecological variables		Dippner, 1997a, b
North Sea coast	TF	C	Sea level	M	Langenberg *et al.*, 1999
Baltic Sea	TF	SLP	Sea level	M	Heyen *et al.*, 1996
Region not specified					
	WT				Frey-Buness *et al.*, 1995
	WT	C			Matyasovszky and Bogardi, 1996
	WT				Enke and Spekat, 1997
	TF	C, VOR, W			Kilsby *et al.*, 1998
	TF		Ecological variables		Heyen *et al.*, 1998
	TF		P		Biau *et al.*, 1999
	WG	P	P	D	Wilks, 1999
	WT		P	D	Zorita and von Storch, 1999

References

Appenzeller, C., T.F. Stocker and M. Anklin, 1998: North Atlantic Oscillation dynamics recorded in Greenland ice cores. *Science,* **282**, 446–449.

Arritt, R., D.C. Goering and C.J. Anderson, 2000: The North American monsoon system in the Hadley Centre coupled ocean-atmosphere GCM. *Geophys. Res. Lett.,* **27**, 565-568.

Bailey, D.A., A.H. Lynch and K.S. Hedström, 1997: The impact of ocean circulation on regional polar simulations using the Arctic regional climate system model. *Ann. Glaciol.,* **25**, 203-207.

Bailey, D.A. and A.H. Lynch, 2000a: Development of an Antarctic regional climate system model: Part I: Sea ice and large-scale circulation, *J. Climate,* **13**, 1337-1350.

Bailey, D.A. and A.H. Lynch, 2000b: Development of an Antarctic regional climate system model: Part II Station validation and surface energy ballance, *J. Climate,* **13**, 1351-1361.

Barrow, E., M. Hulme and M. Semenov, 1996: Effect of using difference methods in the construction of climate change scenarios: examples from Europe. *Clim. Res.,* **7**, 195–211.

Bates, B., S.P. Charles and J.P. Hughes, 1998: Stochastic downscaling of numerical climate model simulations. *Env. Modelling and Software,* **13**, 325–331.

Baur, F., P. Hess and H. Nagel, 1944: Kalender der Grosswetterlagen Europas 1881-1939. Bad Homburg, 35 pp.

Beersma, J.J. and T.A. Buishand, 1999: A simple test for equality of variances in monthly climate data. *J. Climate,* **12**, 1770-1779.

Bellone, E., J.P. Hughes and P. Guttrop, 2000: A hidden Markov model for downcaling synoptic atmospheric patterns to precipitation amounts. *Clim. Res.,* **15**, 1-12.

Benestad, R.E., 1999: Pilot Studies on Enhanced Greenhouse Gas Scenarios for Norwegian Temperature and Precipitation from Empirical Downscaling, DNMI-Klima report 18/99, Norwegian Met. Inst., PO Box 43 Blindern, N0313 Oslo, Norway.

Benestad, R.E., I. Hanssen-Bauer, E.J. Førland, K.A. Iden and O.E. Tveito, 1999: Evaluation of monthly mean data fields from the ECHAM4/OPYC3 control integration, DNMI-Klima report 14/99, Norwegian Met. Inst., P.O. Box 43 Blindern, N0313 Oslo, Norway.

Bengtsson, L., M. Botzet and M. Esch, 1995: Hurricane-type vortices in a general circulation model. *Tellus,* **47A**, 175–196.

Bengtsson, L., M. Botzet and M. Esch, 1997: Numerical simulation of intense tropical storms. Hurricanes: Climate and Socioeconomics Impacts, Springer, 67–92.

Bengtsson, L., M. Botzet and M. Esch, 1996: Will greenhouse gas induced warming over the next 50 years lead to higher frequency and greater intensity of hurricanes? *Tellus,* **48A**, 57–73.

Bhaskaran B., R.G. Jones, J.M. Murphy and M. Noguer, 1996: Simulations of the Indian summer monsoon using a nested regional climate model: Domain size experiments. *Clim. Dyn.,* **12**, 573–587.

Bhaskaran, B. and J.F.B. Mitchell, 1998: Simulated changes in southeast Asian monsoon precipitation resulting from anthropogenic emissions. *Int. J. Climatology,* **18**, 1455-1462.

Bhaskaran, B., J.M. Murphy and R.G. Jones, 1998: Intraseasonal Oscillation in the Indian Summer Monsoon Simulated by Global and nested Regional Climate Models. *Mon. Wea. Rev.,* **126**, 3124–3134.

Biau, G., E. Zorita, H. von Storch and H. Wackernagel, 1999: Estimation of precipitation by kriging in EOF space. *J. Climate,* **12**, 1070–1085.

Boer, G.J., G. Flato and D. Ramsden, 2000b: A transient climate change simulation with greenhouse gas and aerosol forcing: projected climate for the 21st century. *Clim. Dyn.,* **16**, 427-450.

Boer, G.J., G. Flato, M.C. Reader and D. Ramsden, 2000a: A transient climate change simulation with greenhouse gas and aerosol forcing: experimental design and comparison with the instrumental record for the 20th century. *Clim. Dyn.,* **16**, 405-426.

Böhm, U., R. Podzun and D. Jacob, 1998: Surface Water Balance Estimation for a Semi-Arid Region using a Regional Climate Model and Comparison of Water Balance Components with Global Circulation Model Output and Analysis Data. *Phys. Chem. Earth,* **23**, 405-411.

Brandsma, T. and T. A. Buishand, 1997: Statistical linkage of daily precipitation in Switzerland to atmospheric circulation and temperature. *J. Hydrology,* **198**, 98–123.

Brown, B. G. and R. W. Katz, 1995: Regional analysis of temperature extremes: Spatial analog for climate change? *J. Climate,* **8**, 108–119.

Buishand, T.A. and J.J. Beersma, 1996: Statistical tests for comparison of daily variability in observed and simulated climates. *J. Climate,* **10**, 2538–2550.

Buishand, T.A. and A.M.G. Klein Tank, 1996: Regression model for generating time series of daily precipitation amounts for climate change impact studies. *Stochastic Hydrology and Hydraulics,* **10**, 87–106.

Buishand, T.A. and T. Brandsma, 1997: Comparison of circulation classification schemes for predicting temperature and precipitation in the Netherlands, *Int. J. Climatol.,* **17**, 875–889.

Buishand, T.A. and T. Brandsma, 1999: The dependence of precipitation on temperature at Florence and Livorno. *Clim. Res.,* **12**, 53–63.

Buma, J. and M. Dehn, 1998: A method for predicting the impact of climate change on slope stability. *Environmental Geology,* **35**, 190–196.

Buma, J. and M. Dehn, 2000: The impact of climate change on a landslide in South East France, simulated using different GCM-scenarios and downscaling methods for local precipitation. *Clim. Res.,* **15**, 69-81.

Bürger, G., 1996: Expanded downscaling for generating local weather scenarios *Clim. Res.,* **7**, 111–128.

Bürger, G., 1997: On the disaggregation of climatological means and anomalies. *Clim. Res.,* **8**, 183–194.

Burkhardt, U., 1999: Alpine precipitation in a tripled CO_2 climate. *Tellus,* **51A**, 289–303.

Busuioc, A. and H. von Storch, 1996: Changes in the winter precipitation in Romania and its relation to the large scale circulation. *Tellus,* **48A**, 538–552.

Busuioc, A., H. von Storch and R. Schnur, 1999: Verification of GCM generated regional precipitation and of statistical downscaling estimates. *J. Climate,* **12**, 258–272.

Carril, A. F.,C.G. Menéndez and M.N. Nuñez, 1997: Climate Change Scenarios over the South American Region: An Intercomparison of Coupled General Atmosphere-Ocean Circulation Models. *Int. J. Climatol.,* **17**, 1613–1633.

Cavazos, T., 1997: Downscaling large-scale circulation to local rainfall in North-Eastern Mexico. *Int.. J. Climatology,* **17**, 1069-1082.

Cavazos, T., 1999: Large-Scale Circulation Anomalies Conducive to Extreme Precipitation Events and Derivation of Daily Rainfall in Northeastern Mexico and Southeastern Texas. *J. Climate,* **12**, 1506-1523.

Charles, S.P., B.C. Bates, and J.P. Hughes, 1999a: A spatio-temporal model for downscaling precipitation occurrence and amounts. *J. Geophys. Res.,* **104**, 31657-31669.

Charles, S.P., B.C. Bates, P.H. Whetton and J.P. Hughes, 1999b: Validation of downscaling models for changed climate conditions: Case study of southwestern Australia. *Clim. Res.,* **12**, 1-14.

Chase T.N., R.A. Pielke, T.G.F. Kittel, R.R. Nemani and S.W. Running, 2000: Simulated impacts of historical land cover changes on global climate in northern winter. *Clim. Dyn. ,* **16**, 93-106.

Chen, D, C. Hellstroem and Y. Chen, 1999: Preliminary analysis and statistical downscaling of monthly temperature in Sweden. Department of Physical Geography, Göteborg, C 16 1999, 76 pp.

Chen, M., and C. Fu, 2000: A nest procedure between regional and global climate models and its application in long term regional climate simulation. *Chinese J. Atmos. Sci.,* **24**, 233-262.

Christensen, J.H., O.B. Christensen and B. Machenhauer, 1995: On the performance of the HIRHAM limited area regional climate model over Europe, Third Int. Conference on Modelling of Global Climate Change and Variability, Hamburg, Germany, 4-8 September 1995.

Christensen J.H., O.B. Christensen, P. Lopez, E. van Meijgaard and M. Botzet, 1996: The HIRHAM4 Regional Atmospheric Climate Model. *DMI Sci. Rep.*, 96-4., 51pp. DMI, Copenhagen.

Christensen J.H., B. Machenhauer, R.G. Jones, C. Schär, P.M. Ruti, M. Castro and G. Visconti, 1997: Validation of present-day regional climate simulations over Europe: LAM simulations with observed boundary conditions. *Clim. Dyn.*, **13**, 489–506.

Christensen O.B., J.H. Christensen, B. Machenhauer and M. Botzet, 1998: Very high-resolution regional climate simulations over Scandinavia – Present climate. *J. Climate*, **11**, 3204-3229.

Christensen O.B., 1999: Relaxation of soil variables in a regional climate model. *Tellus*, **51A**, 674-685.

Christensen J.H. and P. Kuhry, 2000: High resolution regional climate model validation and permafrost simulation for the East-European Russian Arctic. *J. Geophys. Res.*, **105**, 29647-29658.

Cocke, S.D. and T.E. LaRow, 2000: Seasonal Prediction using a Regional Spectral Model embedded within a Coupled Ocean-Atmosphere Model. *Mon. Wea. Rev.*, **128**, 689-708.

Conway, D., 1998: Recent climate variability and future climate change scenarios for the British Isles. *Prog. Phys. Geog.*, **22**, 350-374.

Conway, D., R.L. Wilby and P.D. Jones, 1996: Precipitation and air flow indices over the British Isles, *Clim. Res.*, **7**, 169–183.

Copeland, J.H., R.A. Pielke and T.G.F. Kittel, 1996: Potential climatic impacts of vegetation change: A regional modeling study. *J. Geophys. Res.*, **101**, 7409–7418.

Corte-Real, J., H. Xu and B. Qian, 1999: A weather generator for obtaining daily precipitation scenarios based on circulation patterns. *Clim. Res.*, **12**, 61–75.

Corte-Real, J., X. Zhang and X. Wang, 1995: Downscaling GCM information to regional scales: a non-parametric multivariate regression approach, *Clim. Dyn.*, **11**, 413–424.

Cowpertwait, P.S.P. and P.E. O'Connel, 1997: A regionalized Neymann-Scott model of rainfall with convective and stratiform cells, *Hydrology and Earth Sciences*, **1**, 71–80.

Crane, R.G. and B.C. Hewitson, 1998: Doubled CO_2 climate change scenarios for the Susquehanna Basin: Precipitation. *Int. J. Climatology*, **18**, 65–76.

Cubasch, U., H. von Storch, J. Waszkewitz and E. Zorita, 1996: Estimates of climate change in southern Europe using different downscaling techniques. *Clim. Res.*, **7**, 129–149.

Cubasch, U., J. Waszkewitz, G. Hegerl and J. Perlwitz, 1995: Regional climate changes as simulated in time-slice experiments. *Clim. Change*, **31**, 273–304.

Cui, M. and E. Zorita, 1998: Analysis of the sea-level variability along the Chinese coast and estimation of the impact of a CO_2-perturbed atmospheric circulation. *Tellus*, **50A**, 333–347.

Cui, M., H. von Storch and E. Zorita, 1995: Coastal sea level and the large-scale climate state: A downscaling exercise for the Japanese Islands. *Tellus*, **47A**, 132–144.

Cui, M., H. von Storch and E. Zorita, 1996: Coast sea level and the large-scale climate state: a downscaling exercise for the Japanese Islands. *Studia Marina Sinica* **36**, 13–32 (in Chinese).

Dai, A., F. Giorgi and K. Trenberth, 1999: Observed and model simulated precipitation diurnal cycles over the continental United States. *J. Geophys. Res.*, **104**, 6377–6402.

Daly, C., R.P. Neilson and D.L. Phillips, 1994: A statistical-topographic model for mapping climatological precipitation over mountainous terrain. *J. Appl. Met.*, **33**, 140-158.

Dehn, M., 1999a: Szenarien der klimatischen Auslösung alpiner Hangrutschungen. Simulation durch Downscaling allgemeiner Zirkulationsmodelle der Atmosphare. PhD thesis, Bonn. Bonner Geographische Abhandlungen, 99. Bonn, 99pp.

Dehn, M., 1999b: Application of an analog downscaling technique for assessment of future landslide activity - a case study in the Italian Alps. *Clim. Res.*, **13**, 103-113.

Dehn, M., and J. Buma, 1999: Modelling future landslide activity based on general circulation models. *Geomorphology*, **30**, 175-187.

Dehn, M., G. Bürger, J. Buma and P. Gasparetto, 2000: Impact of climate change on slope stability using expanded downscaling. *Engineering Geology*, **55**, 193-204.

Delworth, T.L., J.D. Mahlman and T.R. Knutson, 1999: Changes in heat index associated with CO_2-induced global warming. *Clim. Change*, **43**, 369-386.

Déqué, M. and J. P. Piedelievre, 1995: High resolution climate simulation over Europe. *Clim. Dyn.*, **11**, 321–339.

Déqué, M., P. Marquet and R. G. Jones, 1998: Simulation of climate change over Europe using a global variable resolution general circulation model. *Clim. Dyn.*, **14**, 173-189.

Dickinson R.E., R.M. Errico, F. Giorgi and G.T. Bates, 1989: A regional climate model for western United States. *Clim. Change*, 15, 383–422.

Dippner, J. W., 1997a: Recruitment success of different fish stocks in the North Sea in relation to climate variability. *Dtsche Hydrogr. Z.*, **49**, 277-293

Dippner, J.W., 1997b: SST anomalies in the North Sea in relation to the North Atlantic Oscillation and the influence on the theoretical spawning time of fish. *Dtsche Hydrogr. Z.*, **49**, 267-275.

Doherty, R. and Mearns, L.O., 1999: A Comparison of Simulations of Current Climate from Two Coupled Atmosphere - Ocean GCMs Against Observations and Evaluation of Their Future Climates. Report to the NIGEC National Office. National Center for Atmospheric Research, Boulder, Colorado, 47pp.

Durman, C.F., J.M. Gregory, D.H. Hassell and R.G. Jones, 2001. The comparison of extreme European daily precipitation simulated by a global and a regional climate model for present and future climates. *Quart. J. R.. Met. Soc.*, (in press).

Emanuel, K.A., 1987: The dependence of hurricane intensity on climate. *Nature*, **326**, 483-485.

Emanuel, K., 2000: A statistical analysis of tropical cyclone intensity. *Mon. Wea. Rev.*, **128**, 1139-1152.

Enke, W and A. Spekat, 1997: Downscaling climate model outputs into local and regional weather elements by classification and regression. *Clim. Res.*, **8**, 195–207.

Fernandez-Partagas, J. and H.F. Diaz, 1996: Atlantic hurricanes in the second half of the 19th Century. *Bull. Am. Met. Soc.*, **77**, 2899-2906.

Fischlin, A. and D. Gyalistras, 1997: Assessing impacts of climate change on forests in the Alps. *Glob. Ecol. Biogeogr. Lett.*, 6, 19–37.

Flato, G.M., Boer, G.J., Lee, W.G., McFarlane, N.A., Ramsden, D., Reader, M.C. and Weaver, A.J., 2000: The Canadian Centre for Climate Modelling and Analysis global coupled model and its climate. *Clim. Dyn.*, **16**, 451-467.

Foufoula-Georgiou, E., and D.P. Lattenmaier, 1987: A Markov renewal model for rainfall occurrences. *Water. Resour. Res.*, **23**, 875-884.

Frei C., C. Schär, D. Lüthi and H.C. Davies, 1998: Heavy precipitation processes in a warmer climate. *Geophys. Res. Lett.*, **25**, 1431–1434.

Frei, C. and C. Schär, 1998: A precipitation climatology for the Alps from high resolution rain guage observations. *Int. J. Climatol.*, **18**, 873–900.

Frey-Buness, F., D. Heimann and R. Sausen, 1995: A statistical-dynamical downscaling procedure for global climate simulations. *Theor. Appl. Climatol.*, **50**, 117–131.

Fu, C.B., H.L.Wei, M.Chen, B.K.Su, M.Zhao and W.Z. Zheng, 1998: Simulation of the evolution of summer monsoon rainbelts over Eastern China from a regional climate model, *Scientia Atmospherica Sinica*, **22**, 522–534.

Fuentes, U. and D. Heimann, 1996: Verification of statistical-dynamical downscaling in the Alpine region. *Clim. Res.*, **7**, 151–186.

Fuentes, U., and D. Heimann, 1999: An improved statistical-dynamical downscaling scheme and its application to the Alpine precipitation climatology. *Theor. Appl. Climatol.*, **65**, 119-135.

Fyfe, J.C., 1999: On climate simulations of African Easterly waves. *J. Climate*, **12**, 6, 1747-1769.

Fyfe, J.C. and G.M. Flato, 1999: Enhanced Climate Change and its Detection over the Rocky Mountains. *J. Climate*, **12**, 230–243.

Gallardo, C., M.A. Gärtner, J.A. Prego and M. Castro, 1999: Multi-year simulations with a high resolution Regional Climate Model over the Iberian Peninsula: Current climate and $2 \times CO_2$ scenario. Internal Report, Dept. Geofísica y Meteorología, 45 pp. (Available in English on request from: Facultad de Física, Universidad Complutense, 28040 Madrid, Spain).

Gao, X., Z. Zhao Y. Ding, R. Huang and F. Giorgi, 2001: Climate change due to greenhouse effects in China as simulated by a regional climate model, *Acta Meteorologica Sinica*, in press.

GCIP, 1998: Global Energy and Water Cycle Experiment (GEWEX) Continental-Scale International Project. A Review of Progress and Opportunities. National Academy Press, Washington, D. C., 99 pp.

Gibson J. K., P. Källberg, S. Uppala, A. Hernández, A Nomura and E. Serrano, 1997, ERA description. ECMWF Reanalysis Project Report Series 1, ECMWF, Reading, 66 pp.

Giorgi F., 1990: Simulation of regional climate using a limited area model nested in a general circulation model. *J. Climate*, **3**, 941–963.

Giorgi, F., G.T Bates and S.J. Nieman, 1993a: The multiyear surface climatology of a regional atmospheric model over the western United States. *J. Climate*, **6**, 75–95.

Giorgi, F., C.S. Brodeur and G.T. Bates, 1994: Regional climate change scenarios over the United States produced with a nested regional climate model. *J. Climate*, **7**, 375–399.

Giorgi, F. and R. Francisco, 2000a: Uncertainties in regional climate change predictions. A regional analysis of ensemble simulations with the HadCM2 GCM. *Clim. Dyn.*, **16**, 169-182.

Giorgi, F. and R. Francisco, 2000b: Evaluating uncertainties in the prediction of regional climate change. *Geophys. Res. Lett*, **27**, 1295-1298.

Giorgi, F., J.W. Hurrell, M.R. Marinucci and M. Beniston, 1997: Elevation signal in surface climate change: A model study. *J. Climate*, **10**, 288–296.

Giorgi, F.,Y. Huang, K. Nishizawa and C. Fu, 1999: A seasonal cycle simulation over eastern Asia and its sensitivity to radiative transfer and surface processes. *J. Geophys. Res.*, **104**, 6403-6424.

Giorgi, F. and M.R. Marinucci, 1996a: An investigation of the sensitivity of simulated precipitation to model resolution and its implications for climate studies. *Mon. Wea. Rev.*, **124**, 148-166.

Giorgi, F. and M.R. Marinucci, 1996b: Improvements in the simulation of surface climatology over the European region with a nested modeling system. *Geophys. Res. Lett.*, **23**, 273–276.

Giorgi, F., M.R. Marinucci, G.T. Bates and G. DeCanio, 1993b: Development of a second-generation regional climate model (RegCM2). Part II: Convective processes and assimilation of lateral boundary conditions. *Mon. Wea. Rev.*, **121**, 2814–2832.

Giorgi, F., M.R. Marinucci, G.T. Bates, 1993c: Development of a second-generation regional climate model (RegCM2). Part I: Boundary-layer and radiative transfer processes. *Mon. Weath. Rev.*, **121**, 2794–2813.

Giorgi, F., M.R Marinucci and G. Visconti, 1992: A $2 \times CO_2$ climate change scenario over Europe generated using a limited area model nested in a general circulation model. 2. Climate change scenario. *J. Geophys. Res.*, **97**, 10011–10028.

Giorgi, F. and L.O. Mearns, 1991: Approaches to the simulation of regional climate change: a review, *Rev. Geophys.*, **29**, 191–216.

Giorgi, F. and L.O. Mearns, 1999: Regional climate modeling revisited. An introduction to the special issue. *J. Geophys. Res.*, **104**, 6335–6352.

Giorgi, F., L.O. Mearns, C. Shields and L. Meyer, 1996: A regional model study of the importance of local vs. remote controls in maintaining the drought 1988 and flood 1993 conditions in the central U.S. *J. Climate*, **9**, 1150–1162.

Giorgi, F., L.O. Mearns, C. Shields and L. McDaniel, 1998: Regional nested model simulations of present day and $2 \times CO_2$ climate over the Central Plains of the U.S. *Clim. Change*, **40**, 457–493.

Giorgi, F. and C. Shields, 1999: Tests of precipitation parameterizations available in the latest version of the NCAR regional climate model (RegCM) over the continental U.S. *J. Geophys. Res.*, **104**, 6353–6376.

Goodess, C. and J. Palutikof, 1998: Development of daily rainfall scenarios for southeast Spain using a circulation-type approach to downscaling. *Int.. J. Climatol.*, **18**, 1051-1083.

Gregory, J.M. and J.F.B. Mitchell, 1995: Simulation of daily variability of surface temperature and precipitation over Europe in the current and $2 \times CO_2$ climates using the UKMO climate model. *Quart. J. R. Met. Soc.*, **121**, 1451-1476.

Gregory, J.M., J.F.B. Mitchell and A.J. Brady, 1997: Summer drought in northern midlatitudes in a time-dependent CO_2 climate experiment. *J. Climate*, **10**, 662-686.

González-Rouco, F., H. Heyen, E. Zorita and F. Valero, 1999: Testing the validity of a statistical downscaling method in simulations with global climate models. GKSS Report 99/E/29, GKSS Research Centre, Geesthacht, Germany.

González-Rouco, F., H. Heyen, E. Zorita and F. Valero, 2000: Agreement between observed rainfall trends and climate change simulations in the southwest of Europe. *J. Climate*, **13**, 3057-3065.

Gupta,V.K. and E. Waymire, 1991: On lognormality and scaling in spatial rainfall averages? *Non-linear variability in geophysics*, 175-183.

Gyalistras, D., A. Fischlin and M. Riedo, 1997: Herleitung stündlicher Wetterszenarien unter zuküftigen Klimabedingungen. In: Fuhrer, J. (ed), Klimaänderungen und Grünland - eine Modellstudie über die Auswirkungen zukünftiger Klimaveränderungen auf das Dauergrünland in der Schweiz. vdf, Hochschulverlag AG an der ETH Zürich, 207-276.

Gyalistras, D., C. Schär, H.C. Davies and H. Wanner, 1998: Future Alpine climate. In: Cebon, P., U. Dahinden, H.C. Davies, D. Imboden, and C.C. Jaeger (eds) Views from the Alps. Regional Perspectives on Climate Change. Cambridge, Massachusetts: MIT Press, 171–223.

Hanssen-Bauer, I. and E.J. Førland 1998: Long-term trends in precipitation and temperature in the Norwegian Arctic: can they be explained by changes in atmospheric circulation patterns? *Clim. Res.*, **10**, 143–153.

Hanssen-Bauer I. and E.J. Førland 2000: Temperature and precipitation variations in Norway and their links to atmospheric circulation. *Int. J. Climatology.*, **20**, 1693-1708.

Hantel, M., M. Ehrendorfer and A. Haslinger, 1999: Climate sensitivity of snow cover duration in Austria. *Int.. J. Climatology.*, **20**, 615-640.

Hassell, D. and R.G. Jones, 1999: Simulating climatic change of the southern Asian monsoon using a nested regional climate model (HadRM2), HCTN 8, Hadley Centre for Climate Prediction and Research, London Road, Bracknell, UK.

Haugen, J.E., D. Bjørge and T.E. Nordeng, 1999: A 20-year climate change experiment with HIRHAM, using MPI boundary data. RegClim Techn. Rep. No. 3, 37–44. Available from NILU, P.O. Box 100, N-2007 Kjeller, Norway.

Heimann, D. and V. Sept, 2000: Climatic change of summer temperature and precipitation in the Alpine region - a statistical-dynamical assessment. *Theor. Appl. Climatol.*, **66**, 1-12.

Henderson-Sellers A, H. Zhang, G. Berz, K. Emanuel, W. Gray, C. Landsea, G. Holland, J. Lighthill, S.-L. Shieh, P. Webster and K. McGuffie, 1998: Tropical cyclones and global climate change: a post-IPCC assessment. *Bull. Am. Met. Soc.*, **79**, 19-38.

Hennessy, K.J., J.M. Gregory and J.F.B. Mitchell, 1997: Changes in daily precipitation under enhanced greenhouse conditions. *Clim. Dyn.*, **13**, 667–680.

Hennessy, K.J., P.H. Whetton, J.J. Katzfey, J.L. McGregor, R.N. Jones, C.M. Page and K.C. Nguyen, 1998: Fine-resolution climate change scenarios for New South Wales. Annual report 1997-1998, research undertaken for the New South Wales Environmental Protection Authority. Chatswood, N.S.W., NSW Environment protection Authority, 48 pp.

Hewitson, B.C., 1999: Deriving regional precipitation scenarios from general circulation models, Report K751/1/99, Water Research Commission, Pretoria, South Africa.

Hewitson, B.C., and Crane, R.G., 1992: Large-scale atmospheric controls on local precipitation in tropical Mexico. *Geophys. Res. Lett.*, **19**(18), 1835-1838.

Hewitson, B.C., and Crane, R.G., 1996: Climate downscaling: Techniques and application, *Clim. Res.*, **7**, 85-95.

Heyen, H., and J. W. Dippner, 1998: Salinity in the southern German Bight estimated from large-scale climate data. *Tellus*, **50A**, 545–556.

Heyen, H., E. Zorita and H. von Storch, 1996: Statistical downscaling of monthly mean North Atlantic air-pressure to sea-level anomalies in the Baltic Sea. *Tellus*, **48A**, 312–323.

Heyen, H., H. Fock and W. Greve, 1998: Detecting relationships between the interannual variability in ecological time series and climate using a multivariate statistical approach - a case study on Helgoland Roads zooplankton. *Clim. Res.*, **10**, 179–191.

Hirakuchi, H., Giorgi, F., 1995: Multiyear present-day and $2\times CO_2$ simulations of monsoon climate over eastern Asia and Japan with a regional climate model nested in a general circulation model. *J. Geophys. Res.*, **100**, 21105–21125.

HIRETYCS, 1998. Final report of the HIRECTYCS project, European Commission Climate and Environment Programme, Brussels.

Holland, G.J., 1997: The maximum potential intensity of tropical cyclones. *J. Atmos. Sci.* **54**, 2519-2541.

Hong, S.-Y., and H. - M. H. Juang, 1998: Orography blending in the lateral boundary of a regional model, *Mon. Wea. Rev.*, **126**, 1714-1718.

Hong, S.Y. and A. Leetmaa, 1999: An evaluation of the NCEP RCM for Regional Climate Modeling. *J. Climate*, **12**, 592-609.

Hostetler, S.W., F. Giorgi, G.T. Bates and P.J. Bartlein, 1994: Lake-atmosphere feedbacks associated with paleolakes Bonneville and Lahontan. *Science*, **263**, 665–668.

Hostetler, S.W., P.J. Bartlein, P.U. Clark, E.E. Small and A.M. Solomon, 2000: Simulated influence of Lake Agassiz on the climate of Central North America 11,000 years ago. *Nature*, **405**, 334-337.

Hughes, J. P., D.P. Lettenmaier, and P. Guttorp, 1993: A stochastic approach for assessing the effect of changes in synoptic circulation patterns on gauge precipitation. *Water Resour. Res.*, **29** (10), 3303–3315.

Hughes, J.P., P. Guttorpi and S.P. Charles, 1999: A non-homogeneous hidden Markov model for precipitation occurrence. *Appl. Statist.*, **48**, 15–30.

Hulme, M. and O. Brown, 1998: Portraying climate scenario uncertainties in relation to tolerable regional climate change. *Clim. Res.*, **10**, 1–14.

Hulme, M., E.M. Barrow, N. Arnell, P.A. Harrison, T.E. Downing and T.C. Johns, 1999: Relative impacts of human-induced climate change and natural climate variability. *Nature*, **397**, 688–691.

Hulme, M., J. Crossley, D. Lister, K.R. Briffa and P.D. Jones, 2000: Climate observations and GCM validation Interim Annual Report to the Department of the Environment, Transport and the Regions, April 1999 to March 2000, CRU, Norwich, UK.

Huth, R., 1997: Continental-scale circulation in the UKHI GCM. *J. Climate*, **10**, 1545–1561.

Huth, R., 1999: Statistical downscaling in central Europe: Evaluation of methods and potential predictors. *Clim. Res.*, **13**, 91-101.

IPCC, 1990: Climate Change: The IPCC Scientific Assessment. [J.T. Houghton , G.J. Jenkins and J.J. Ephraums, (Eds.)], Cambridge University Press, Cambridge, 365 pp.

IPCC, 1996:Climate change 1995: The Science of Climate Change. Contribution of Working Group I to the Second Assessment Report of the Intergovernmental Panel on Climate Change [J.T. Houghton, L.G. Meira Filho, B.A. Callander, N. Harris, A. Kattenberg and K Maskell (eds.)], Cambridge University Press, Cambridge, 572 pp.

Jacob, D. and R. Podzun, 1997: Sensitivity study with the Regional Climate Model REMO, *Meteorol. Atmos. Phys.*, **63**, 119-129.

Jacob, D. and R. Podzun, 2000: In: Research Activities in Atmospheric and Oceanic Modelling. (CAS/JSC Working Group on Numerical Experimentation Report) [Geneva]: WMO, February 2000, **30**, 7.10-7.11

Jacob, D., R. Podzun, M. Lal and U. Cubasch, 1995: Summer monsoon climatology as simulated in a regional climate model nested in a genearal circulation model, Third Int. Conference on Modelling of Global Climate Change and Variability, Hamburg, Germany, 4-8 September 1995.

Jacobeit, J., 1996: Atmospheric circulation changes due to increased greenhouse warming and its impact on seasonal rainfall in the Mediterranean area. In: Climate Variability and Climate Change - Vulnerability and Adaptation, I. Nemesová (Ed.), Proceedings, Prague, Czech Republic, September 11-15, 1995, 71–80

Jacobeit, J. C. Beck and A. Philipp, 1998 : Annual to decadal variability in climate in Europe - objectives and results of the German contribution to the European climate research project ADVICE. - Würzburger Geographische Manuskripte 43, 163 pp.

Jenkins, G. S., 1997: The 1988 and 1990 summer season simulations for West Africa. *J. Climate*, **10**, 51255–1272.

Jenkins, G. S. and E.J. Barron, 1997: Global climate model and coupled regional climate model. *Global Planet. Change*, **15**, 3–32.

Ji, Y. and A.D. Vernekar, 1997: Simulation of the Asian summer monsoons of 1987 and 1988 with a regional model nested in a global GCM. *J. Climate*, **10**, 1965–1979.

Jóhannesson, T., T. Jónsson, E. Källén and E. Kaas, 1995: Climate change scenarios for the Nordic countries. *Clim. Res.*, **5**: 181–195.

Johns, T.C. , 1999: Initial assessment of global response to greenhouse gas increases in a high resolution atmospheric timeslice experiment compared to the HadCM3 coupled model. DETR report, March 1999, Hadley Centre for Climate Prediction and Research, London Road, Bracknell, UK.

Jones, J.M. and T.D. Davies, 2000: Investigation of the climatic influence of air and precipitation chemistry over Europe, and applications to a downscaling methodology to assess future acidic deposition. *Clim. Res.* **14**, 7-24.

Jones, R.G., 1999: The response of European climate to increased greenhouse gases simulated by standard and high resolution GCMs. DETR report, March 1999, Hadley Centre for Climate Prediction and Research, London Road, Bracknell, UK.

Jones, R.G., D. Hassel and K. Ward, 1999: First results from the Hadley Centre's new regional climate model including effects of enhanced resolution. DETR report, March 1999, Hadley Centre for Climate Prediction and Research, London Road, Bracknell, UK.

Jones, R.G., J.M. Murphy and M. Noguer, 1995: Simulations of climate change over Europe using a nested regional climate model. I: Assessment of control climate, including sensitivity to location of lateral boundaries. *Quart. J. R. Met. Soc.*, **121**, 1413–1449.

Jones, R.G., J.M. Murphy, M. Noguer and A.B. Keen, 1997: Simulation of climate change over Europe using a nested regional climate model. II: Comparison of driving and regional model responses to a doubling of carbon dioxide. *Quart. J. R.. Met. Soc.*, **123**, 265–292.

Jones, R.N., K.J. Hennessy, C.M. Page, A.B. Pittock, R. Suppiah, K.J.E.

Walsh and P.H. Whetton, 2000: An analysis of the effects of the Kyoto protocol on Pacific Island countries, part Two: Regional climate change scenarios and risk assessment methods. CSIRO/SPREP, in press.

Joubert, A.M. and P.D. Tyson, 1996: Equilibrium and fully coupled GCM simulations of future Southern African climates. *South African Journal of Science*, **92**, 471–484.

Joubert, A.M., S.J. Mason and J.S. Galpin, 1996: Droughts over southern Africa in a doubled-CO_2 climate. *Int. J. Climatology*, **16**, 1149-1156.

Joubert, A., J.J. Katzfey and J. McGregor, 1998: Int. Conference on The Role of Topography in Modelling Weather and Climate. Int. Centre for Theoretical Physics, Trieste, Italy, 22-26 June 1998.

Joubert, A.M., J.J. Katzfey, J.L. McGregor, and K.C. Nguyen, 1999: Simulating mid-summer climate over Southern Africa using a nested regional climate model. *J. Geophys. Res.*, **104**, 19015-19025.

Kaas, E., T.-S. Li and T. Schmith, 1996: Statistical hindcast of wind climatology in the North Atlantic and Northwestern European region. *Clim. Res.*, **7**, 97–110.

Kalnay, E., M. Kanamitsu, R. Kistler, W. Collins, D. Deaven, L. Gandin, M. Iredell, S. Saha, G. White, J. Woollen, Y. Zhu, M. Chelliah, W. Ebisuzaki, W. Higgins, J. Janowiak, K.C. Mo, C. Ropelewski, J. Wang, A. Leetmaa, R. Reynolds, R. Jenne and D. Joseph, 1996: The NCEP/NCAR 40-Year Reanalysis Project. *Bull. Am. Met. Soc.*, **77**, 437–471.

Karl, T.R., W.C. Wang, M.E. Schlesinger, R.W. Knight, D. Portman, 1990: A method of relating general circulation model simulated climate to observed local climate. Part I: Seasonal statistics. *J. Climate*, **3**, 1053-1079.

Kattenberg, A, F. Giorgi, H. Grassl, G. A. Meehl, J. F. B. Mitchell, R. J. Stouffer, T. Tokioka, A. J. Weaver, T. M. L. Wigley, 1996. Climate Models – Projections of Future Climate. In : *Climate Change 1995: The Science of Climate Changs. Contribution of Working Group I to the Second Assessment Report of the IPCC.* [Houghton, J.T., Meira Filho, L.G., Callander, B.A., Harris, N., Kattenberg, A. and Varney, S. K. (eds.)]. Cambridge University Press, Cambridge, 285-357.

Kato, H., H. Hirakuchi, K. Nishizawa, and F. Giorgi, 1999: Performance of the NCAR RegCM in the simulations of June and January climates over eastern Asia and the high-resolution effect of the model, *J. Geophys. Res.*, **104**, 6455-6476.

Kato, H., K. Nishizawa, H. Hirakuchi, S. Kadokura, N. Oshima and F. Giorgi. 2001: Performance of the RegCM2.5/NCAR-CSM nested system for the simulation of climate change in East Asia caused by global warming. *J. Met. Soc. Japan* (in press*)*.

Katz, R. W. and M. B. Parlange, 1993: Effects of an index of atmospheric circulation on stochastic properties of precipitation. *Water Resour. Res.*, **29**, 2335–2344.

Katz, R.W. and M.B. Parlange, 1995: Generalizations of chain-dependent processes: Application to hourly precipitation. *Water Resour. Res.*, **31**, 1331-1341.

Katz, R.W. and M. B. Parlange, 1996: Mixtures of stochastic processes: applications to statistical downscaling. *Clim. Res.*, **7**, 185–193.

Katzfey, J.J., 1999: Storm tracks in regional climate simulations: Verification and changes with doubled CO_2. *Tellus*, **51A**, 803-814.

Katzfey, J.J. and K.L. McInnes, 1996: GCM simulations of eastern Australian cutoff lows. *J. Climate*, **9**, 2337–2355.

Katzfey, J.J., J. McGregor and P.H. Whetton, 1998: Int. Conference on The Role of Topography in Modelling Weather and Climate. Int. Centre for Theoretical Physics, Trieste, Italy, 22-26 June 1998.

Kauker, F.,1998: Regionalization of climate model results for the North sea. PhD thesis University of Hamburg, 109 pp.

Kharin, V.V. and F.W. Zwiers, 2000: Changes in the extremes in an ensemble of transient climate simulations with a coupled atmosphere-ocean GCM. *J. Climate*, **13**, 3760-3788.

Kida, H., T. Koide, H. Sasaki and M. Chiba, 1991: A new approach to coupling a limited area model with a GCM for regional climate simulations. *J. Met. Soc. Japan.*, **69**, 723-728.

Kidson, J. W. and I.G. Watterson, 1995: A synoptic climatological evaluation of the changes in the CSIRO nine-level model with doubled CO_2 in the New Zealand region. *Int. J. Climatology.*, **15**, 1179–1194.

Kidson, J.W. and C.S. Thompson, 1998: Comparison of statistical and model-based downscaling techniques for estimating local climate variations. *J. Climate*, **11**, 735–753.

Kilsby, C.G., P.S.P. Cowpertwait, P.E. O'Connell and P.D. Jones, 1998: Predicting rainfall statistics in England and Wales using atmospheric circulation variables. *Int. J. Climatology*, **18**, 523-539.

Kim, J., 1997: Precipitation and snow budget over the southwestern United States during the 1994-1995 winter season in a mesoscale model simulation. *Water. Resour. Res.*, **33**, 2831-2839.

Kim, J., N.L. Miller, A.K. Guetter and K.P. Georgakakos, 1998: River flow response to precipitation and snow budget in California during the 1994-95 winter. *J. Climate*, **11**, 2376–2386.

Kittel, T.G.F., F. Giorgi and G.A. Meehl, 1998: Intercomparison of region al biases and doubled CO_2-sensitivity of coupled atmospheric-ocean general circulation model experiments. *Clim. Dyn.*, **14**, 1–15.

Klein, W.H. and H.R. Glahn, 1974: Forecasting local weather by means of model output statistics, *Bull. Amer. Met. Soc.*, **55**, 1217-1227.

Knutson, T.R. and R.E. Tuleya, 1999: Increased hurricane intensities with CO_2-induced warmings as simulated using the GFDL hurricane prediction system. *Clim. Dyn.*, **15**, 503-519.

Knutson, T.R., R.E. Tuleya and Y. Kurihara, 1998: Simulated increase of hurricane intensities in a CO_2-warmed world. *Science*, **279**, 1018–1020.

Knutson, T.R., R.E. Tuleya, W. Shen and I. Ginis, 2000: Impact of CO_2-induced warming on hurricane intensities as simulated in a hurricane model with ocean coupling. *J. Climate*, **14**, 2458-2468.

Kothavala, Z., 1997: Extreme precipitation events and the applicability of global climate models to study floods and droughts. *Math. and Comp. in Simulation*, **43**, 261-268.

Krinner G, C. Genthon, Z.-X. Li and P. Le Van, 1997: Studies of the Antarctic climate with a stretched grid GCM, *J. Geophys. Res.*, **102**, 13731-13745.

Krinner G. and C. Genthon, 1998: GCM simulations of the Last Glacial Maximum surface climate of Greenland and Antarctica, *Clim. Dyn.*, **14**, 741-758.

Krishnamurti, T.N., R. Correa-Torres, M. Latif and G. Daughenbaugh, 1998: The impact of current and posssibly future sea surface temperature anomalies on the frequency of Atlantic hurricanes. *Tellus*, **50A**, 186-210.

Kröncke, I., J.W. Dippner, H. Heyen and B. Zeiss, 1998: Long-term changes in macrofauna communities off Norderney (East Frisia, Germany) in relation to climate variability. *Mar. Ecol. Prog. Series*, **167**, 25–36.

Labraga, L.C. and M. Lopez, 1997: A comparison of the climate response to increased carbon dioxide simulated by general circulation models with mixed-layer and dynamic ocean representations in the region of South America. *Int. J. Climatology*, **17**, 1635–1650.

Lal, M., U. Cubasch, J. Perlwitz and J. Waszkewitz, 1997: Simulation of the Indian monsoon climatology in ECHAM3 climate model: sensitivity to horizontal resolution. *Int. J. Climatology*, **17**, 847-858.

Lal M., G.A. Meehl and J.M. Arblaster, 2000: Simulation of Indian summer monsoon rainfall and its intraseasonal variability in the NCAR climate system model. *J. Climate*, in press.

Lal, M., P.H. Whetton, A.B. Pittock, and B. Chakraborty, 1998a: Simulation of present-day climate over the Indian subcontinent by general circulation models. *Terrestrial Atmospheric and Oceanic Sciences*, **9**, 69–96.

Lal, M., P.H. Whetton, A.B. Pittock, and B. Chakraborty, 1998b: The greenhouse gas-induced climate change over the Indian subcontinent as projected by general circulation model experiments. *Terrestrial Atmospheric and Oceanic Sciences*, **9**, 673–690.

Lamb, H.H., 1972: British Isles weather types and a register of daily sequence of circulation patterns, 1861–1971. Geophysical Memoir 116, HMSO, London, 85 pp.

Landsea, C.W., 1997: Comments on "Will greenhouse gas induced warming over the next 50 years lead to higher frequency and greater intensity of hurricanes?", *Tellus*, **49A**, 622–623.

Langenberg, H, A. Pfizenmayer, H. von Storch and J. Sndermann, 1999: Storm related sea level variations along the North Sea coast: natural variability and anthropogenic change. *Cont. Shelf Res.* **19**, 821–842.

Laprise R., D. Caya, M. Giguère, G. Bergeron, H. Côte, J.-P. Blanchet, G.J. Boer and N.A. McFarlane, 1998: Climate and climate change in western Canada as simulated by the Canadian regional climate model. *Atmosphere-Ocean*, **36**, 119–167.

Lettenmaier, D., 1995: Stochastic Modeling of precipitation with applications to climate model downscaling. In: H. von Storch and A. Navarra (eds) "Analysis of Climate Variability: Applications of Statistical Techniques", Springer Verlag, 197–212 (ISBN 3-540-58918-X).

Leung, L.R. and S.J. Ghan, 1995: A subgrid parameterization of orographic precipitation. *Theor. Appl. Climatol.*, **52**, 95-118.

Leung, L.R. and S.J. Ghan, 1998: Parameterizing subgrid orographic precipitation and surface cover in climate models. *Mon. Wea. Rev.*, **126**, 3271-3291.

Leung, L.R. and S.J. Ghan, 1999a: Pacific Northwest climate sensitivity simulated by a regional climate model driven by a GCM. Part I: control simulations. *J. Climate*, **12**, 2010-2030.

Leung, L.R. and S.J. Ghan, 1999b: Pacific Northwest climate sensitivity simulated by a regional climate model driven by a GCM. Part II: $2\times CO_2$ simulations. *J. Climate*, **12**, 2031-2053.

Leung, L.R., Wigmosta, M. S., Ghan, S.J., Epstein, D.J., Vail, L.W., 1996: Application of a subgrid orographic precipitation/surface hydrology scheme to a mountain watershed. *J. Geophys. Res.*, **101**, 12803–12817.

Leung, L.R., S.J. Ghan, Z.C. Zhao, Y. Luo, W.-C. Wang, and H.-L. Wei, 1999a: Intercomparison of regional climate simulations of the 1991 summer monsoon in eastern Asia. *J. Geophys. Res.*, 104, 6425–6454.

Leung, L.R., A.F. Hamlet, D.P. Lettenmaier and A. Kumar, 1999b: Simulations of the ENSO hydroclimate signals in the Pacific Northwest Columbia River Basin. *Bull. Am. Met. Soc.*, **80**(11), 2313-2329.

Liston, G.E. and R.A. Pielke, 2000: A climate version of the regional atmospheric modeling system. *Theor.l Appl. Climatol.*, **66**, 29-47..

Liston, G.E., R.A. Pielke, Sr., and E.M. Greene, 1999: Improving first-order snow-related deficiencies in a regional climate model. *J. Geophysical Res.*, **104** (D16), 19559-19567.

Luterbacher, J., C. Schmutz, D. Gyalistras, E. Xoplaki and H. Wanner, 1999: Reconstruction of monthly NAO and EU indices back to AD 1675. *Geophys. Res. Lett.*, **26**, 2745-2748.

Lüthi, D., A. Cress, H.C. Davies, C. Frei and C. Schär, 1996: Interannual variability and regional climate simulations. *Theor. Appl. Climatol.*, **53**, 185–209.

Lynch, A. H., W.L. Chapman, J.E. Walsh. and G. Weller, 1995: Development of a regional climate model of the western Arctic. *J. Climate*, **8**, 1555–1570.

Lynch, A.H., M.F. Glück, W.L. Chapman, D.A. Bailey,= and J.E. Walsh, 1997a: Remote sensing and climate modeling of the St. Lawrence Is. Polynya, *Tellus*, **49A**, 277-297.

Lynch, A.H., D.L. McGinnis, W.L. Chapman and J.S. Tilley, 1997b: A Multivariate comparison of two land surface models integrated into an Arctic regional climate system model. *Ann. Glaciol.*, **25**, 127-131.

Lynch, A.H., D.L. McGinnis, and D.A. Bailey, 1998: Snow-albedo feedback and the spring transition in a regional climate system model: Influence of land surface model, *J. Geophys. Res.*, **103**, 29037-29049.

Maak, K. and H. von Storch, 1997: Statistical downscaling of monthly mean temperature to the beginning of flowering of Galanthus nivalis L. in Northern Germany. *Int. J. Biometeor.*, **41**, 5–12.

Mabuchi, K., Y. Sato and H. Kida, 2000: Numerical study of the relationships between climate and the carbon dioxide cycle on a regional scale. *J. Met. Soc. Japan,* **78**, 25-46.

Machenhauer, B., M. Windelband, M. Botzet, R. Jones and M. Déqué, 1996: Validation of present-day regional climate simulations over Europe: Nested LAM and variable resolution global model simulations with observed or mixed layer ocean boundary conditions. MPI Reprot No. 191, MPI, Hamburg, Germany.

Machenhauer, B., M. Windelband, M. Botzet, J.H. Christensen, M. Deque, R. Jones, P.M. Ruti and G. Visconti, 1998: Validation and analysis of regional present-day climate and climate change simulations over Europe. MPI Report No.275, MPI, Hamburg, Germany.

Marinucci, M. R. and F. Giorgi, 1992: A $2\times CO_2$ climate change scenario over Europe generated using a limited area model nested in a general circulation model 1. Present-day seasonal climate simulation. *J. Geophys. Res.*, **97**, 9989–10009.

Marinucci, M. R., F. Giorgi, M. Beniston, M. Wild, P. Tschuck, A. Ohmura and A. Bernasconi, 1995: High resolution simulations of January and July climate over the western alpine region with a nested regional modeling system. *Theor. Appl. Climatol.*, **51**, 119–138.

Martin, E., B. Timbal and E. Brun, 1997: Downscaling of general circulation model outputs: simulation of the snow climatology of the French Alps and sensitivity to climate change. *Clim. Dyn.*, **13**, 45–56.

Martin, G.M., 1999: The simulation of the Asian summer monsoon, and its sensitivity to horizontal resolution, in the UK Meteorological Office Unified Model. *Quart. J. R. Met. Soc.*, **125**, 1499-1525.

Maslanik, J.A., A.H. Lynch and M. Serreze, 2000: A case study simulation of Arctic regional climate in a coupled model. *J. Climate.* **13**, 383-401.

May, W., 1999: A time-slice experiment with the ECHAM4 AGCM at high resolution: The experimental design and the assessment of climate change as compared to a greenhouse gas experiment with ECHAM4/OPYC at low resolution, DMI Scientific Report 99-2, DMI, Copenhagen.

May, W. and E. Roeckner, 2001: A time-slice experiment with the ECHAM4 AGCM at high resolution: The impact of resolution on the assessment of anthropogenic climate change, *Clim. Dyn.*,(in press).

McDonald, R. 1999: Changes in tropical cyclones as greenhouse gases are increased. PMSR report, March 1999, Hadley Centre for Climate Prediction and Research, London Road, Bracknell, UK.

McGregor, J.L., 1997: Regional climate modelling. *Meteorology and Atmospheric Physics.* **63**,105-117.

McGregor, J.L., J.J. Katzfey and K.C. Nguyen, 1995: Seasonally-varying nested climate simulations over the Australian region, Third Int. Conference on Modelling of Global Climate Change and Variability, Hamburg, Germany, 4-8 September 1995.

McGregor, J.L., J.J. Katzfey and K.C. Nguyen, 1998: Fine resolution simulations of climate change for southeast Asia. Final report for a Research Project commissioned by Southeast Asian Regional Committee for START (SARCS), Aspendale, Vic.: CSIRO Atmospheric Research. VI, 15, 35pp.

McGregor, J.L., J.J. Katzfey and K.C. Nguyen, 1999: Recent regional climate modelling experiments at CSIRO. In: Research Activities in Atmospheric and Oceanic Modelling. H. Ritchie (ed). (CAS/JSC Working Group on Numerical Experimentation Report; 28; WMO/TD – no. 942) [Geneva]: WMO. P. 7.37–7.38.

McGregor, J.L., K. Walsh, 1993: Nested simulations of perpetual January climate over the Australian region. *J. Geophys. Res.*, **98**, 23283–23290.

McGregor, J. L., K. Walsh, 1994: Climate change simulations of Tasmanian precipitation using multiple nesting. *J. Geophys. Res.*, **99**,

20889–20905.

McGuffie, K., A. Henderson-Sellers, N. Holbrook, Z. Kothavala, O. Balachova and J. Hoekstra, 1999: Assessing simulations of daily temperature and precipitation variability with global climate models for present and enhanced greenhouse conditions, *Int. J. Climatol.* **19**, 1–26.

Mearns, L.O., I. Bogardi, F. Giorgi, I. Matayasovszky and M. Palecki, 1999: Comparison of climate change scenarios generated daily temperature and precipitation from regional climate model experiments and statistical downscaling, *J. Geophys. Res.*, **104**, 6603–6621.

Mearns, L. O., 1999: The effect of spatial and temporal resolution of climate change scenarios on changes in frequencies of temperature and precipitation extremes. In: S. Hassol and J. Katzenberger, Elements of Change 1998, Session 2: Climate Extremes: Changes, Impacts, and Projections. Aspen: Aspen Global Change Institute, 316-323.

Mietus M., 1999. The role of regional atmospheric circulation over Europe and North Atlantic in formation of climatic and oceanographic condition in the Polish coastal zone, Research Materials of Institute of Meteorology and Water Management, Meteorological Series, 29, 147pp (in Polish with English summary).

Miller, N.L. and J. Kim, 1997: The Regional Climate System Model. In: 'Mission Earth: Modeling and simulation for a sustainable global system', M.G. Clymer and C.R. Mechoso (eds.), Society for Computer Simulation International, 55-60.

Murphy, J.M., 1999: An evaluation of statistical and dynamical techniques for downscaling local climate. *J. Climate*, **12**, 2256-2284.

Murphy, J.M., 2000: Predictions of climate change over Europe using statistical and dynamical downscaling techniques. *Int. J. Climatology.*, **20**, 489-501.

New, M., M. Hulme, and P. Jones, 1999: Representing Twentieth-century space-time climate variability. Part 1: development of a 1961-1990 mean monthly terrestrial climatology. *J. Climate*, **12**, 829–856.

New, M., M. Hulme, and P. Jones, 2000: Representing Twentieth-century space-time climate variability. Part 2: development of a 1901-1996 mean monthly terrestrial climatology. *J. Climate*, **13**, 2217-2238.

Noguer, M., 1994: Using statistical techniques to deduce local climate distributions. An application for model validation. *Met. Apps.*, **1**, 277–287.

Noguer M., R.G. Jones and J.M. Murphy, 1998: Sources of systematic errors in the climatology of a nested regional climate model (RCM) over Europe. *Clim. Dyn.*, **14**, 691–712.

O'Connell, P. E. 1999: Producing rainfall scenarios for hydrologic impact modelling. In: G. Kiely et al., (editors). Abstract Volume from the EU Environmental and Climate Programme Advanced Study Course on European Water Resources and Climate Change Processes, 53pp.

Osborn, T.J. and M. Hulme, 1997: Development of a relationship between station and grid-box rain-day frequencies for climate model evaluation. *J. Climate*, **10**, 1885–1908.

Osborn, T.J. and M. Hulme, 1998: Evaluation of the European daily precipitation characteristics from the Atmospheric Model Intercomparison Project, *Int. J. Climatology*, **18**, 505–522.

Osborn, T.J., D. Conway, M. Hulme, J.M. Gregory and P.D. Jones, 1999: Air flow influences on local climate: observed and simulated mean relationships for the UK. *Clim. Res.*, **13**, 173-191.

Palutikof, J.P., J.A. Winkler, C.M. Goodess and J.A. Andresen, 1997: The simulation of daily temperature time series from GCM output. Part 1: Comparison of model data with observations, *J. Climate*, **10**, 2497–2513.

Pan, Z., E. S. Takle, M. Segal and R. Arritt, 1999: Simulation of potential impacts of man-made land use changes on U.S. summer climate under various synoptic regimes, *J. Geophys. Res.*, **104**, 6515-6528.

Pan, Z., J.H. Christensen, R.W. Arritt., W.J. Gutowski, E.S. Takle and F. Otieno, 2000: Evaluation of uncertainties in regional climate change

simulations. *J. Geophys, Res.*, in press.

Parlange, M.B. and R.W. Katz, 2000: An extended version of the Richardson model for simulationg daily weather variables. *J. Appl. Meteorol.*, **39**, 610-622.

Perica, S. and E. Foufoula-Georgiou, 1996a: Linkage of scaling and thermodynamic parameters of rainfall: Results from midlatitude mesoscale convective systems, *J. Geophys. Res.*, **101**, 7431-7448.

Perica, S. and E. Foufoula-Georgiou, 1996b: Model for multiscale dissagregation of spatial rainfall based on coupling meteorological and scaling descriptions, *J. Geophys. Res.*, **101(D21)**, 26347-26361.

Pielke, R.A., Jr., and C.W. Landsea, 1998: Normalized hurricane damages in the United States: 1925-95. *Weath. Forecast.*, **13**, 621-631.

Pielke, R.A., Sr., W.R. Cotton, R.L. Walko, C.J. Tremback, W.A. Lyons, L.D. Grasso, M.E. Nicholls, M.D. Moran, D.A. Wesley, T.J. Lee and J.H. Copeland, 1992: A comprehensive meteorological modeling system—RAMS, *Meteor. Atmos. Phys.*, **49**, 69-91.

Pielke, R.A., Sr., R.L. Walko, L. Steyaert, P.L. Vidale, G.E. Liston and W.A. Lyons, 1999: The influence of anthropogenic landscape changes on weather in south Florida. *Mon. Wea. Rev.*, **127**, 1663-1673.

Pittock, A.B., M.R. Dix, K.J. Hennessy, D.R. Jackett, J.J. Katzfey, T.J. McDougall, K.L. McInnes, S.P. O'Farrell, I.N. Smith, R. Suppiah, K.J.E. Walsh, P.H. Whetton and S.G. Wilson, 1995: Progress towards climate change scenario and impact studies for the southwest Pacific. In: The science and impacts of climate change in the Pacific islands, Apia, Western Samoa. Apia, Western Samoa: Conference on the Science and Impacts of Climate Change in the Pacific Islands. 15-2, 21-46.

Podzun, R., A. Cress, D. Majewski and V. Renner, 1995: Simulation of European climate with a limited area model. Part II: AGCM boundary conditions. *Contrib. Atmos. Phys.*, **68**, 205–225.

Qian, Y. and F. Giorgi, 1999: Interactive coupling of regional climate and sulfate aerosol models over East Asia. *J. Geophys. Res.*, **104**, 6501–6514.

Raab, B. and H. Vedin (Eds.), 1995: Climate, Lakes and Rivers: National Atlas for Sweden. SNA Publishing, 176pp and 198 plates.

Racksko, P., L. Szeidl and M. Semenov, 1991: A serial approach to local stochastic weather models. *Ecological Modeling*, **57**, 27–41.

Räisänen, J., 1998: Intercomparison studies of general circulation model simulations of anthropogenic climate change. Academic dissertation presented at the Faculty of Science, Department of Meterorology, Report 57, University of Helsinki, 50pp.

Räisänen, J., M. Rummukainen, A. Ullerstig, B. Bringfelt, U. Hansson and U. Willén, 1999: The first Rossby Centre regional climate scenario - dynamical downscaling of CO$_2$-induced climate change in the HadCM2 GCM. SMHI Reports Meteorology and Climatology No. 85, 56 pp. Swedish Meteorological and Hydrological Institute, SE-601 76 Norrköping, Sweden.

Reichert, B.K., L. Bengtsson and Ove Åkesson, 1999: A statistical modeling approach for the simulation of local paleoclimatic proxy records using GCM output. *J. Geophys. Res.* **104**(D16), 19071–19083.

Renwick, J. A., J.J. Katzfey, K.C. Nguyen and J.L. McGregor, 1998: Regional model simulations of New Zealand climate. *J. Geophys. Res.*, **103**, 5973–5982.

Richardson, C.W., 1981: Stochastic simulation of daily precipitation, temperature, and solar radiation. *Water Resour. Res.*, **17**(1),182–190.

Riedo, M., D. Gyalistras, A Fischlin and J. Fuhrer, 1999: Using an ecosystem model linked to GCM derived local weather scenarios to analyse effects of climate change and elevated CO$_2$ on dry matter production and partitioning, and water use in temperate managed graslands. *Global Change Biol.*, **5**, 213–223.

Rinke A. and K. Dethloff, 1999: Sensitivity studies concerning initial and boundary conditions in a regional climate model of the Arctic, *Clim.*

Res., **14**, 101-113.

Rinke, A., K. Dethloff and J.H. Christensen, 1999: Arctic winter climate and its interannual variations simulated by a regional climate model. *J. Geophys. Res.*, **104**, 19027–19038.

Risbey, J. and P. Stone, 1996: A case study of the adequacy of GCM simulations for input to regional climate change assessments. *J. Climate*, **9**, 1441–1467.

Roebber, P.J. and L.F. Bosart, 1998: The sensitivity of precipitation to circulation details. Part I: An analysis of regional analogs. *Mon. Wea. Rev.*, **126**, 437–455.

Roldan, J. and D.A. Woolhiser, 1982, Stochastic daily precipitation models. 1. A comparison of occurrence processes. *Water Resour. Res.*, **18**, 1451-1459.

Rotach, M.W., M.R.M. Marinucci, M. Wild, P. Tschuck, A. Ohmura, M. Beniston, 1997: Nested regional simulation of climate change over the Alps for the scenario of doubled greenhouse forcing. *Theor. Appl. Climatol.*, **57**, 209–227.

Royer, J.-F., F. Chauvin, B. Timbal, P. Araspin and D. Grimal, 1998: A GCM study of the impact of greenhouse gas increase on the frequency of occurrence of tropical cyclones. *Clim. Change*, **38**, 307-343.

Rummukainen, M., 1997: Methods for statistical downscaling of GCM simulations, RMK No 80, Swedish Meteorological and Hydrological Institute, Norrköping, Sweden.

Rummukainen, M., J. Räisänen, B. Bringfelt, A. Ullerstig, A. Omstedt, U. Willen, U. Hansson and C. Jones, 2000: RCA1 Regional Climate Model for the Nordic Region – Model Description and Results from the Control Run Downscaling of Two GCMs. *Clim. Dyn.*, (in press).

Ruti P.M., C. Cacciamani, T. Paccagnella, A. Bargagli and C. Cassardo, 1998: LAM multi-seasonal numerical integrations: performance analysis with different surface schemes. *Contrib. Atmos. Phys.*, **71**, 321–346.

SAR, see IPCC (1996a).

Sailor, D.J. and X. Li, 1999: A semiempirical downscaling approach for predicting regional temperature impacts associated with climatic change. *J. Climate*, **12**, 103–114.

Schär C., C. Frei, D. Lüthi and H.C. Davies, 1996: Surrogate climate change scenarios for regional climate models. *Geophys. Res. Lett.*, **23**, 669–672.

Schnur, R., and D. Lettenmaier, 1998: A case study of statistical downscaling in Australia using weather classification by recursive partitioning. *J. Hydrol.*, **212**–123, 362–379.

Schubert, S. 1998: Downscaling local extreme temperature changes in south-eastern Australia from the CSIRO Mark2 GCM. *Int. J. Climatol.*, **18**, 1419–1438.

Schubert, S. and A. Henderson-Sellers, 1997: A statistical model to downscale local daily temperature extremes from synoptic-scale atmospheric circulation patterns in the Australian region. *Clim. Dyn.*, **13**, 223-234.

Semenov, M.A. and E.M. Barrow, 1997: Use of a stochastic weather generator in the development of climate change scenarios. *Clim. Change*, **35**, 397–414.

Sept, V., 1998: Untersuchungen der Gewitteraktivitt im sddeustchen Raum mitteln statistisch-dynamischer Regionalisierung. Deutsches Zentrum für Luft- und Raumfahrt e.V., Forschungsbericht 98–27, 113 pp.

Seth, A. and F. Giorgi, 1998: The effects of domain choice on summer precipitation simulation and sensitivity in a regional climate model, *J. Climate*, **11**, 2698–2712.

Small, E.E., F. Giorgi and L.C. Sloan, 1999a: Regional climate model simulation of precipitation in central Asia: Mean and interannual variability. *J. Geophys. Res.*, **104**, 6563–6582.

Small, E.E., L.C. Sloan, S. Hostetler and F. Giorgi, 1999b: Simulating the water balance of the Aral Sea with a coupled regional climate-lake model. *J. Geophys. Res.*, **104**, 6583–6602.

Smith, J. A. and A. F. Karr, 1985: Statistical inference for point process models of rainfall, *Water Resour. Res.*, **21**(1), 73-80.

Solman, S.A. and M.N. Nuñez, 1999: Local estimates of global change: a statistical downscaling approach. *Int. J. Climatology,.* **19**, 835–861.

Starr, V.P., 1942: Basic principles of weather forecasting. Harper Brothers Publishers, New York, London, 299pp.

Stendel, M. and E. Röckner 1998: Impacts of horizontal resolution on simulated climate statistics in ECHAM4. MPI report 253, MPI, Hamburg, Germany.

Stratton, R.A. 1999a: A high resolution AMIP integration using the Hadley Centre model HadAM2b. *Clim. Dyn.*, **15**, 9–28.

Stratton, R.A. 1999b: The impact of increasing resolution on the HadAM3 climate simulation. HCTN 13, Hadley Centre, The Met. Office, Bracknell, UK.

Sun, L., F.H.M. Semazzi, F. Giorgi and L. Ogallo, 1999: Application of the NCAR regional climate model to eastern Africa. 2. Simulations of interannual variability of short rains. *J. Geophys. Res.*, **104**, 6549–6562.

Takle, E.S., W.J. Gutowski, R.W. Arritt, Z. Pan, C.J. Anderson, R.S. da Silva, D. Caya, S.-C. Chen, F. Giorgi, J.H. Christensen, S.-Y. Hong, H.-M. H. Juang, J. Katzfey, W.M. Lapenta, R. Laprise, P. Lopez, G.E. Liston, J. McGregor, A. Pielke and J.O. Roads, 1999: Project to intercompare regional climate simulation (PIRCS)): Description and initial results. *J. Geophys. Res.*, **104**, 19443-19461.

Tett, S.F.B., T.C. Johns and J.F.B. Mitchell, 1997: Global and Regional Variability in a coupled AOGCM. *Clim. Dyn.*, **13**, 303–323.

Timbal, B. and B.J. McAvaney, 1999: A downscaling procedure for Australia. BMRC scientific report 74.

Trigo, R.M. and J.P. Palutikof, 1999: Simulation of daily temperatures for climate change scenarios over Portugal: a neural network model approach. *Clim. Res.* **13**, 61-75.

Tsutsui, J., A. Kasahara and H. Hirakuchi, 1999: The impacts of global warming on tropical cyclones - a numerical experiment with the T42 version of NCAR CCM2. Proceedings of the 23rd Conference on Hurricanes and tropical meteorology, 10-15 January 1999, Dallas, American Meteorological Society, 1077–1080.

Tsutsui, J.-I. and A. Kasahara, 1996: Simulated tropical cyclones using the National Center for Atmospheric Research community climate model. *J. Geophys. Res.*, **101**, 15013–15032.

Tsvetsinskaya, E., L.O. Mearns and W.E. Easterling, 2000: Investigating the effect of seasonal plant growth and development in 3-dimensional atmospheric simulations. *J. Climate*, in press.

van Lipzig, 1999: The surface mass balance of the Antarctic ice sheet, a study with a regional atmospheric model, Ph. D. thesis, Utrecht University, 154 pp.

von Storch, H., 1995: Inconsistencies at the interface of climate impact studies and global climate research. Meteorol. Zeitschrift 4 NF, 72–80.

von Storch, H., 1999a: The global and regional climate system. In: H. von Storch and G. Flöser: Anthropogenic Climate Change, Springer Verlag, ISBN 3-540-65033-4, 3–36.

von Storch., H., 1999b: Representation of conditional random distributions as a problem of "spatial" interpolation. In. J. Gómez-Hernández, A. Soares and R. Froidevaux (Eds.): geoENV II - Geostatistics for Environmental Applications. Kluwer Adacemic Publishers, Dordrecht, Boston, London, ISBN 0-7923-5783-3, 13–23.

von Storch, H., 1999c: On the use of "inflation" in downscaling. *J. Climate*, **12**, 3505-3506.

von Storch, H., H. Langenberg and F. Feser, 2000: A spectral nudging technique for dynamical downscaling purposes, *Mon. Wea. Rev.*, **128**, 3664-3673.

von Storch, H. and H. Reichardt, 1997: A scenario of storm surge statistics for the German Bight at the expected time of doubled atmospheric carbon dioxide concentration. *J. Climate*, **10**,

2653–2662.

von Storch, H., E. Zorita and U. Cubasch, 1993: Downscaling of global climate change estimates to regional scales: An application to Iberian rainfall in wintertime. *J. Climate*, **6**, 1161–1171.

von Storch, H., and F.W. Zwiers, 1999: Statistical Analysis in Climate Research, Cambridge University Press, ISBN 0 521 45071 3, 494 pp.

Wallis, T.W.R. and J.F. Griffiths, 1997. Simulated meteorological input for agricultural models. *Agricultural and Forest Meteorology*, **88**, 241-258.

Walsh, K. and J.L McGregor, 1995: January and July climate simulations over the Australian region using a limited area model. *J. Climate*, **8**, 2387–2403.

Walsh, K. and J.L McGregor, 1996: Simulations of Antarctic climate using a limited area model. *J. Geophys. Res.*, **101**, 19093–19108.

Walsh, K. and J.L. McGregor, 1997: An assessment of simulations of climate variability over Australia with a limited area model. *Int. J. Climatology*, **17**, 201–233.

Walsh J.E and J.J. Katzfey, 2000: The impact of climate change on the poleward movement of tropical cyclone-like vortices in a regional model. *J. Climate*, in press.

Walsh J.E. and B.F. Ryan, 2000: Tropical cyclone intensity increase near Australia as a result of climate change. *J. Climate*, **13**, 3029-3036.

Walsh, K. and I.G. Watterson, 1997: Tropical cyclone-like vortices in a limited area model: comparison with observed climatology. *J. Climate*, **10**, 2240-2259.

Wanner, H., R. Rickli, E. Salvisberg, C. Schmutz and M. Schepp, 1997: Global climate change and variability and its influence on Alpine climate - concepts and observations. *Theor. Appl. Climatol.*, **58**, 221–243.

Warner, T. T., R.A. Peterson and R.E. Treadon, 1997: A tutorial on lateral conditions as a basic and potentially serious limitation to regional numerical weather prediction. *Bull. Am. Met.. Soc.*, **78**, 2599-2617.

WASA, 1998: Changing waves and storms in the Northeast Atlantic? *Bull. Am. Met. Soc.* **79**, 741–760.

Watterson, I.G., M.R. Dix, H.B. Gordon and J.L McGregor, 1995: The CSIRO nine-level atmospheric general circulation model and its equilibrium present and doubled CO_2 climate, *Australian Meteorological Magazine*, **44**, 111–125.

Wei, H. and C. Fu, 1998: Study of the sensitivity of a regional model in response to land cover change over Northern China. *Hydrological processes*, **12**, 2249-2265.

Wei, H., Fu, C., Wang, W.-C., 1998: The effect of lateral boundary treatment of regional climate model on the East Asian summer monsoon rainfall simulation. *Chinese J. Atmos. Sci.*, **22**, 231–243.

Weichert, A. and G. Bürger, 1998: Linear versus nonlinear techniques in downscaling. *Clim. Res.*, **10**, 83–93.

Weisse, R., H. Heyen and H. von Storch, 2000: Sensitivity of a regional atmospheric model to a state dependent roughness and the need of ensemble calculations. *Mon. Wea. Rev.*, **128**, 3631-3642.

Werner, P.C. and F.-W. Gerstengarbe, 1997: A proposal for the development of climate scenarios. *Clim. Res.*, **8**, 171-182.

Werner, P.C. and H. von Storch, 1993: Interannual variability of Central European mean temperature in January/February and its relation to the large-scale circulation. *Clim. Res.*, **3**, 195–207.

Whetton, P.H., M.H. England, S.P. O'Farrell, I.G. Watterson and A.B. Pittock, 1996a: Global comparison of the regional rainfall results of enhanced greenhouse coupled and mixed layer ocean experiments: implications for climate change scenario development. *Clim. Change*, **33**, 497–519.

Whetton P.H., J.J. Katzfey, K.J. Hennesey, X. Wu, J.L. McGregor and K. Nguyen, 2001: Using regional climate models to develop fine resolution scenarios of climate change: An example for Victoria,

Australia. *Clim. Res.*, **16**, 181-201.

Whetton, P. H., A.B. Pittock, J.C. Labraga, A.B. Mullan and A.M. Joubert, 1996b: Southern Hemisphere climate: comparing models with reality. In: Climate change: developing southern hemisphere perspectives. T. W. Giambelluca, and A. Henderson-Sellers (Eds.). (Research and Developments in Climate and Climatology) Chichester: Wiley, 89–130.

Widmann, M. and C. Schär, 1997: A principal component and long-term trend analysis of daily precipitation in Switzerland. *Int. J. Climatology*, **17**, 1333–1356.

Widmann, M. and C.S. Bretherton, 2000: Validation of mesoscale precipitation in the NCEP reanalysis using a new gridcell dataset for the northwestern United States. *J. Climate*, **13**, 1936-1950.

Wilby, R.L., 1998: Statistical downscaling of daily precipitation using daily airflow and seasonal teleconnection indices. *Clim. Res.* **10**:163–178.

Wilby, R.L., H. Hassan and K. Hanaki, 1998a: Statistical downscaling of hydrometeorological variables using general circulation model output. *J. Hydrology*, **205**, 1–19.

Wilby, R.L., T.M.L. Wigley, D. Conway, P.D. Jones, B.C. Hewitson, J. Main and D.S.Wilks, 1998b: Statistical downscaling of general circulation model output: A comparison of methods, *Water Resou. Res.*, **34**, 2995–3008.

Wilby, R.L. and T.M.L. Wigley, 1997: Downscaling general circulation model output: a review of methods and limitations. *Prog. Phys. Geography*, **21**, 530–548.

Wilby, R.L. and Wigley, T.M.L., 2000: Precipitation predictors for downscaling: observed and General Circulation Model relationships. *Int. J. Climatology*, **20**, No. 6, 641-661.

Wilby, R.L., L.E. Hay, W.J. Gutowski, R.W. Arritt, E.S. Takle, Z. Pan, G.H. Leavesley and M.P. Clark, 2000: Hydrological responses to dynamically and statistically downscaled climate model output. *Geophys. Res. Lett.*, **27**, No. 8, p1199.

Wild, M., A. Ohmura, H. Gilgen and E. Roeckner, 1995: Regional climate simulation with a high resolution GCM: surface radiative fluxes. *Clim. Dyn.*, **11**, 469-486.

Wild, M., L. Dümenil, and J.P. Schulz, 1996: Regional climate simulation with a high resolution GCM: surface hydrology. Climate Dynamics, **12**, 755-774.

Wild M., A. Ohmura and U. Cubasch, 1997: GCM simulated surface energy fluxes in climate change experiments. *J. Climate*, **10**, 3093-3110.

Wilks, D. S., 1999a: Interannual variability and extreme-value characteristics of several stochastic daily precipitation models. *Agric. For. Meteorol.*, **93**, 153–169.

Wilks, D., 1999b: Multisite downscaling of daily precipitation with a stochastic weather generator. *Clim. Res.*, **11**, 125–136.

Wilks, D. S. and R. L. Wilby, 1999: The weather generator game: A review of stochastic weather models. *Prog. Phys. Geography*, **23**, 329-358.

Williamson, D.L., 1999: Convergence of atmospheric simulations with increasing horizontal resolution and fixed forcing scales. *Tellus*, **51A**, 663-673.

Winkler, J.A., J.P. Palutikof, J.A. Andresen and C.M. Goodess, 1997: The simulation of daily temperature time series from GCM output: Part II: Sensitivity analysis of an empirical transfer function methodology. *J. Climate*, **10**, 2514–2535.

Yoshimura, J., M. Sugi and A. Noda, 1999: Influence of greenhouse warming on tropical cyclone frequency simulated by a high-resolution AGCM. Proceedings of the 23rd Conference on Hurricanes and tropical meteorology, 10-15 January 1999, Dallas, American Meteorological Society, 1081–1084.

Zorita, E., J. Hughes, D. Lettenmaier, and H. von Storch, 1995: Stochastic characterisation of regional circulation patterns for climate model diagnosis and estimation of local precipitation. *J.*

Climate, **8**, 1023–1042.

Zorita, E., and A. Laine, 1999: Dependence of salinity and oxygen concentrations in the Baltic Sea on the large-scale atmospheric circulation. *Clim. Res.*, **14**, 25-34.

Zorita, E. and H. von Storch, 1997: A survey of statistical downscaling techniques. GKSS report 97/E/20.

Zorita, E. and H. von Storch, 1999: The analog method - a simple statistical downscaling technique: comparison with more complicated methods. *J. Climate*, **12**, 2474-2489.

Zwiers, F.W. and V.V. Kharin, 1997: Changes in the extremes of climate simulated by CCC GCM2 under CO_2 doubling. *J. Climate*, **11**, 2200-2222.

11

Changes in Sea Level

Co-ordinating Lead Authors
J.A. Church, J.M. Gregory

Lead Authors
P. Huybrechts, M. Kuhn, K. Lambeck, M.T. Nhuan, D. Qin, P.L. Woodworth

Contributing Authors
O.A. Anisimov, F.O. Bryan, A. Cazenave, K.W. Dixon, B.B. Fitzharris, G.M. Flato, A. Ganopolski,
V. Gornitz, J.A. Lowe, A. Noda, J.M. Oberhuber, S.P. O'Farrell, A. Ohmura, M. Oppenheimer,
W.R. Peltier, S.C.B. Raper, C. Ritz, G.L. Russell, E. Schlosser, C.K. Shum, T.F. Stocker, R.J. Stouffer,
R.S.W. van de Wal, R. Voss, E.C. Wiebe, M. Wild, D.J. Wingham, H.J. Zwally

Review Editors
B.C. Douglas, A. Ramirez

Contents

Executive Summary

This chapter assesses the current state of knowledge of the rate of change of global average and regional sea level in relation to climate change. We focus on the 20th and 21st centuries. However, because of the slow response to past conditions of the oceans and ice sheets and the consequent land movements, we consider changes in sea level prior to the historical record, and we also look over a thousand years into the future.

Past changes in sea level
From recent analyses, our conclusions are as follows:

• Since the Last Glacial Maximum about 20,000 years ago, sea level has risen by over 120 m at locations far from present and former ice sheets, as a result of loss of mass from these ice sheets. There was a rapid rise between 15,000 and 6,000 years ago at an average rate of 10 mm/yr.

• Based on geological data, global average sea level may have risen at an average rate of about 0.5 mm/yr over the last 6,000 years and at an average rate of 0.1 to 0.2 mm/yr over the last 3,000 years.

• Vertical land movements are still occurring today as a result of these large transfers of mass from the ice sheets to the ocean.

• During the last 6,000 years, global average sea level variations on time-scales of a few hundred years and longer are likely to have been less than 0.3 to 0.5 m.

• Based on tide gauge data, the rate of global average sea level rise during the 20th century is in the range 1.0 to 2.0 mm/yr, with a central value of 1.5 mm/yr (as with other ranges of uncertainty, it is not implied that the central value is the best estimate).

• Based on the few very long tide gauge records, the average rate of sea level rise has been larger during the 20th century than the 19th century.

• No significant acceleration in the rate of sea level rise during the 20th century has been detected.

• There is decadal variability in extreme sea levels but no evidence of widespread increases in extremes other than that associated with a change in the mean.

Factors affecting present day sea level change
Global average sea level is affected by many factors. Our assessment of the most important is as follows.

• Ocean thermal expansion leads to an increase in ocean volume at constant mass. Observational estimates of about 1 mm/yr over recent decades are similar to values of 0.7 to 1.1 mm/yr obtained from Atmosphere-Ocean General Circulation Models (AOGCMs) over a comparable period. Averaged over the 20th

century, AOGCM simulations result in rates of thermal expansion of 0.3 to 0.7 mm/yr.

• The mass of the ocean, and thus sea level, changes as water is exchanged with glaciers and ice caps. Observational and modelling studies of glaciers and ice caps indicate a contribution to sea level rise of 0.2 to 0.4 mm/yr averaged over the 20th century.

• Climate changes during the 20th century are estimated from modelling studies to have led to contributions of between –0.2 and 0.0 mm/yr from Antarctica (the results of increasing precipitation) and 0.0 to 0.1 mm/yr from Greenland (from changes in both precipitation and runoff).

• Greenland and Antarctica have contributed 0.0 to 0.5 mm/yr over the 20th century as a result of long-term adjustment to past climate changes.

• Changes in terrestrial storage of water over the period 1910 to 1990 are estimated to have contributed from –1.1 to +0.4 mm/yr of sea level rise.

The sum of these components indicates a rate of eustatic sea level rise (corresponding to a change in ocean volume) from 1910 to 1990 ranging from –0.8 to 2.2 mm/yr, with a central value of 0.7 mm/yr. The upper bound is close to the observational upper bound (2.0 mm/yr), but the central value is less than the observational lower bound (1.0 mm/yr), i.e., the sum of components is biased low compared to the observational estimates. The sum of components indicates an acceleration of only 0.2 mm/yr/century, with a range from –1.1 to +0.7 mm/yr/century, consistent with observational finding of no acceleration in sea level rise during the 20th century. The estimated rate of sea level rise from anthropogenic climate change from 1910 to 1990 (from modelling studies of thermal expansion, glaciers and ice sheets) ranges from 0.3 to 0.8 mm/yr. It is very likely that 20th century warming has contributed significantly to the observed sea level rise, through thermal expansion of sea water and widespread loss of land ice.

Projected sea level changes from 1990 to 2100
Projections of components contributing to sea level change from 1990 to 2100 (this period is chosen for consistency with the IPCC Second Assessment Report), using a range of AOGCMs following the IS92a scenario (including the direct effect of sulphate aerosol emissions) give:

• thermal expansion of 0.11 to 0.43 m, accelerating through the 21st century;
• a glacier contribution of 0.01 to 0.23 m;
• a Greenland contribution of –0.02 to 0.09 m;
• an Antarctic contribution of –0.17 to 0.02 m.

Including thawing of permafrost, deposition of sediment, and the ongoing contributions from ice sheets as a result of climate change since the Last Glacial Maximum, we obtain a range of global-average sea level rise from 0.11 to 0.77 m. This range reflects systematic uncertainties in modelling.

For the 35 SRES scenarios, we project a sea level rise of 0.09 to 0.88 m for 1990 to 2100, with a central value of 0.48 m. The central value gives an average rate of 2.2 to 4.4 times the rate over the 20th century. If terrestrial storage continued at its present rates, the projections could be changed by –0.21 to +0.11 m. For an average AOGCM, the SRES scenarios give results which differ by 0.02 m or less for the first half of the 21st century. By 2100, they vary over a range amounting to about 50% of the central value. Beyond the 21st century, sea level rise will depend strongly on the emissions scenario.

The West Antarctic ice sheet (WAIS) has attracted special attention because it contains enough ice to raise sea level by 6 m and because of suggestions that instabilities associated with its being grounded below sea level may result in rapid ice discharge when the surrounding ice shelves are weakened. The range of projections given above makes no allowance for ice-dynamic instability of the WAIS. It is now widely agreed that major loss of grounded ice and accelerated sea level rise are very unlikely during the 21st century.

Our confidence in the regional distribution of sea level change from AOGCMs is low because there is little similarity between models. However, models agree on the qualitative conclusion that the range of regional variation is substantial compared with the global average sea level rise. Nearly all models project greater than average rise in the Arctic Ocean and less than average rise in the Southern Ocean.

Land movements, both isostatic and tectonic, will continue through the 21st century at rates which are unaffected by climate change. It can be expected that by 2100 many regions currently experiencing relative sea level fall will instead have a rising relative sea level.

Extreme high water levels will occur with increasing frequency (i.e. with reducing return period) as a result of mean sea level rise. Their frequency may be further increased if storms become more frequent or severe as a result of climate change.

Longer term changes

If greenhouse gas concentrations were stabilised, sea level would nonetheless continue to rise for hundreds of years. After 500 years, sea level rise from thermal expansion may have reached only half of its eventual level, which models suggest may lie within ranges of 0.5 to 2.0 m and 1 to 4 m for CO_2 levels of twice and four times pre-industrial, respectively.

Glacier retreat will continue and the loss of a substantial fraction of the total glacier mass is likely. Areas that are currently marginally glaciated are most likely to become ice-free.

Ice sheets will continue to react to climate change during the next several thousand years even if the climate is stabilised. Models project that a local annual-average warming of larger than 3°C sustained for millennia would lead to virtually a complete melting of the Greenland ice sheet. For a warming over Greenland of 5.5°C, consistent with mid-range stabilisation scenarios, the Greenland ice sheet contributes about 3 m in 1,000 years. For a warming of 8°C, the contribution is about 6 m, the ice sheet being largely eliminated. For smaller warmings, the decay of the ice sheet would be substantially slower.

Current ice dynamic models project that the WAIS will contribute no more than 3 mm/yr to sea level rise over the next thousand years, even if significant changes were to occur in the ice shelves. However, we note that its dynamics are still inadequately understood to make firm projections, especially on the longer time-scales.

Apart from the possibility of an internal ice dynamic instability, surface melting will affect the long-term viability of the Antarctic ice sheet. For warmings of more than 10°C, simple runoff models predict that a zone of net mass loss would develop on the ice sheet surface. Irreversible disintegration of the WAIS would result because the WAIS cannot retreat to higher ground once its margins are subjected to surface melting and begin to recede. Such a disintegration would take at least a few millennia. Thresholds for total disintegration of the East Antarctic Ice Sheet by surface melting involve warmings above 20°C, a situation that has not occurred for at least 15 million years and which is far more than predicted by any scenario of climate change currently under consideration.

11.1 Introduction

Sea level change is an important consequence of climate change, both for societies and for the environment. In this chapter, we deal with the measurement and physical causes of sea level change, and with predictions for global-average and regional changes over the next century and further into the future. We reach qualitatively similar conclusions to those of Warrick *et al.* (1996) in the IPCC WGI Second Assessment Report (IPCC, 1996) (hereafter SAR). However, improved measurements and advances in modelling have given more detailed information and greater confidence in several areas. The impacts of sea level change on the populations and eco-systems of coastal zones are discussed in the IPCC WGII TAR (IPCC, 2001).

The level of the sea varies as a result of processes operating on a great range of time-scales, from seconds to millions of years. Our concern in this report is with climate-related processes that have an effect on the time-scale of decades to centuries. In order to establish whether there is a significant anthropogenic influence on sea level, the longer-term and non-climate-related processes have to be evaluated as well.

"Mean sea level" at the coast is defined as the height of the sea with respect to a local land benchmark, averaged over a period of time, such as a month or a year, long enough that fluctuations caused by waves and tides are largely removed. Changes in mean sea level as measured by coastal tide gauges are called "relative sea level changes", because they can come about either by movement of the land on which the tide gauge is situated or by changes in the height of the adjacent sea surface (both considered with respect to the centre of the Earth as a fixed reference). These two terms can have similar rates (several mm/yr) on time-scales greater than decades. To infer sea level changes arising from changes in the ocean, the movement of the land needs to be subtracted from the records of tide gauges and geological indicators of past sea level. Widespread land movements are caused by the isostatic adjustment resulting from the slow viscous response of the mantle to the melting of large ice sheets and the addition of their mass to the oceans since the end of the most recent glacial period ("Ice Age") (Section 11.2.4.1). Tectonic land movements, both rapid displacements (earthquakes) and slow movements (associated with mantle convection and sediment transport), can also have an important effect on local sea level (Section 11.2.6).

We estimate that global average eustatic sea level change over the last hundred years is within the range 0.10 to 0.20 m (Section 11.3.2). ("Eustatic" change is that which is caused by an alteration to the volume of water in the world ocean.) These values are somewhat higher than the sum of the predictions of the contributions to sea level rise (Section 11.4). The discrepancy reflects the imperfect state of current scientific knowledge. In an attempt to quantify the processes and their associated rates of sea level change, we have critically evaluated the error estimates (Box 11.1). However, the uncertainties remain substantial, although some have narrowed since the SAR on account of improved observations and modelling.

Box 11.1: Accuracy

For indicating the uncertainty of data (measurements or model results), two options have been used in this chapter.

1. For data fulfilling the usual statistical requirements, the uncertainty is indicated as ± 1 standard deviation (± 1σ).
2. For limited data sets or model results, the full range is shown by quoting either all available data or the two extremes. In these cases, outliers may be included in the data set and the use of an arithmetic mean or central value might be misleading.

To combine uncertainties when adding quantities, we used the following procedures:

- Following the usual practice for independent uncertainties, the variances were added (i.e. the standard deviations were combined in quadrature).
- Ranges were combined by adding their extreme values, because in these cases the true value is very likely to lie within the overall range.
- To combine a standard deviation with a range, the standard deviation was first used to derive a range by taking ± 2 standard deviations about the mean, and then the ranges were combined.

Eustatic sea level change results from changes to the density or to the total mass of water. Both of these relate to climate. Density is reduced by thermal expansion occurring as the ocean warms. Observational estimates of interior temperature changes in the ocean reported by Warrick *et al.* (1996) were limited, and estimates of thermal expansion were made from simple ocean models. Since the SAR, more observational analyses have been made and estimates from several Atmosphere-Ocean General Circulation Models (AOGCMs) have become available (Section 11.2.1). Thermal expansion is expected to contribute the largest component to sea level rise over the next hundred years (Section 11.5.1.1). Because of the large heat capacity of the ocean, thermal expansion would continue for many centuries after climate had been stabilised (Section 11.5.4.1).

Exchanges with water stored on land will alter the mass of the ocean. (Note that sea level would be unaffected by the melting of sea ice, whose weight is already supported by the ocean.) Groundwater extraction and impounding water in reservoirs result in a direct influence on sea level (Section 11.2.5). Climate change is projected to reduce the amount of water frozen in glaciers and ice caps (Sections 11.2.2, 11.5.1.1) because of increased melting and evaporation. Greater melting and evaporation on the Greenland and Antarctic ice sheets (Sections 11.2.3, 11.5.1.1) is also projected, but might be outweighed by increased precipitation. Increased discharge of ice from the ice sheets into the ocean is also possible. The ice sheets react to climate change by adjusting their shape and size on time-scales of up to millennia, so they could still be gaining or losing mass as a result of climate variations over a history extending as far back as the last glacial period, and they would continue to change for thousands of years after climate had been stabilised (Section 11.5.4.3).

Sea level change is not expected to be geographically uniform (Section 11.5.2), so information about its distribution is needed to inform assessments of the impacts on coastal regions. Since the SAR, such information has been calculated from several AOGCMs. The pattern depends on ocean surface fluxes, interior conditions and circulation. The most serious impacts are caused not only by changes in mean sea level but by changes to extreme sea levels (Section 11.5.3.2), especially storm surges and exceptionally high waves, which are forced by meteorological conditions. Climate-related changes in these therefore also have to be considered.

11.2 Factors Contributing to Sea Level Change

11.2.1 Ocean Processes

The pattern of sea level in ocean basins is maintained by atmospheric pressure and air-sea fluxes of momentum (surface wind stress), heat and fresh water (precipitation, evaporation, and fresh-water runoff from the land). The ocean is strongly density stratified with motion preferentially along density surfaces (e.g. Ledwell *et al.*, 1993, 1998). This allows properties of water masses, set by interaction with the atmosphere or sea ice, to be carried thousands of kilometres into the ocean interior and thus provides a pathway for warming of surface waters to enter the ocean interior.

As the ocean warms, the density decreases and thus even at constant mass the volume of the ocean increases. This thermal expansion (or steric sea level rise) occurs at all ocean temperatures and is one of the major contributors to sea level changes during the 20th and 21st centuries. Water at higher temperature or under greater pressure (i.e., at greater depth) expands more for a given heat input, so the global average expansion is affected by the distribution of heat within the ocean. Salinity changes within the ocean also have a significant impact on the local density and thus local sea level, but have little effect on global average sea level change.

The rate of climate change depends strongly on the rate at which heat is removed from the ocean surface layers into the ocean interior; if heat is taken up more readily, climate change is retarded but sea level rises more rapidly. Climate change simulation requires a model which represents the sequestration of heat in the ocean and the evolution of temperature as a function of depth.

The large heat capacity of the ocean means that there will be considerable delay before the full effects of surface warming are felt throughout the depth of the ocean. As a result, the ocean will not be in equilibrium and global average sea level will continue to rise for centuries after atmospheric greenhouse gas concentrations have stabilised.

The geographical distribution of sea level change is principally determined by alterations to the ocean density structure, with consequent effects on ocean circulation, caused by the modified surface momentum, heat and water fluxes. Hsieh and Bryan (1996) have demonstrated how the first signals of sea level rise are propagated rapidly from a source region (for instance, a region of heat input) but that full adjustment takes place more slowly. As a result, the geographical distribution of sea level change may take many decades to centuries to arrive at its final state.

11.2.1.1 *Observational estimates of ocean warming and ocean thermal expansion*

Previous IPCC sea level change assessments (Warrick and Oerlemans, 1990; Warrick *et al.*, 1996) noted that there were a number of time-series which indicate warming of the ocean and a resultant thermal expansion (i.e. a steric sea level rise) but there was limited geographical coverage. Comparison of recent ocean temperature data sets (particularly those collected during the World Ocean Circulation Experiment) with historical data is beginning to reveal large-scale changes in the ocean interior. (Section 2.2.2.5 includes additional material on ocean warming, including studies for which there are no estimates of ocean thermal expansion.) However, the absence of comprehensive long ocean time-series data makes detection of trends difficult and prone to contamination by decadal and interannual variability. While there has been some work on interannual variability in the North Atlantic (e.g. Levitus, 1989a,b, 1990) and North Pacific (e.g. Yasuda and Hanawa, 1997; Zhang and Levitus, 1997), few studies have focused on long-term trends.

The most convincing evidence of ocean warming is for the North Atlantic. An almost constant rate of interior warming, with implied steric sea level rise, is found over 73 years at Ocean Station S (south-east of Bermuda). Comparisons of trans-ocean sections show that these changes are widespread (Table 11.1). On decadal time-scales, variations in surface steric height from station S compare well with sea level at Bermuda (Roemmich, 1990) and appear to be driven by changes in the wind stress curl (Sturges and Hong, 1995; Sturges *et al.*, 1998). Variability in the western North Atlantic (Curry *et al.*, 1998) is related to changes in convective activity in the Labrador Sea (Dickson *et al.*, 1996). Over the 20 years up to the early 1990s there has been a cooling of the Labrador Sea Water (as in the Irminger Sea, Read and Gould, 1992), and more recently in the western North Atlantic (Koltermann *et al.*, 1999). For the South Atlantic, changes are more uncertain, particularly those early in the 20th century.

A warming of the Atlantic layer in the Arctic Ocean is deduced by comparison of modern oceanographic sections collected on board ice-breakers (e.g., Quadfasel *et al.*, 1991; Carmack *et al.*, 1997; Swift *et al.*, 1997) and submarines (e.g. Morison *et al.*, 1998; Steele and Boyd, 1998) with Russian Arctic Ocean atlases compiled from decades of earlier data (Treshnikov, 1977; Gorshkov, 1983). It is not yet clear whether these changes result from a climate trend or, as argued by Grotefendt *et al.* (1998), from decadal variability. The published studies do not report estimates of steric sea level changes; we note that a warming of 1°C over the central 200 m of the Atlantic layer would result in a local rise of steric sea level of 10 to 20 mm.

Observations from the Pacific and Indian Oceans cover a relatively short period, so any changes seen may be a result of decadal variability. Wong (1999), Wong *et al.* (1999), Bindoff and McDougall (1994) and Johnson and Orsi (1997) studied changes in the South Pacific. Bindoff and McDougall (2000) studied changes in the southern Indian Ocean. These authors

Table 11.1: *Summary of observations of interior ocean temperature changes and steric sea level rise during the 20th century.*

Reference	Dates of data	Location, section or region	Depth range (m)	Temperature change (°C/century)	Steric rise (mm/yr) (and heat uptake)	
North Atlantic Ocean						
Read and Gould (1992)	1962–1991	55°N, 40°–10°W	50–3000	−0.3		
Joyce and Robbins (1996)	1922–1995	Ocean Station S 32.17°N, 64.50°W	1500–2500	0.5	0.9 (0.7 W/m^{-2})	
Joyce et al. (1999)	1958, 1985, 1997	20°N–35°N 52°W and 66°W		0.57	1.0	
Parrilla et al. (1994), Bryden et al. (1996)	1957, 1981, 1992	24°N	800–2500	Peak of 1 at 1100 m	0.9 (1 W/m^{-2})	
Roemmich and Wunsch (1984)	1959, 1981	36°N	700–3000	Peak of 0.8 at 1500 m	0.9	
Arhan et al. (1998)	1957, 1993	8°N	1000–2500	Peak of 0.45 at 1700 m	0.6	
Antonov (1993)	1957–1981	45°N–70°N	0–500	Cooling		
			800–2500	0.4		
South Atlantic Ocean						
Dickson et al. (2001), Arbic and Owens (2001)	1926, 1957	8°S, 33.5°W–12.5°W	1000–2000 (Steric expansion for 100 m to bottom is shown in the right-hand half of the last column)	0.30	−0.1	0.0
	1926, 1957	8°S, 12°W–10.5°E		0.23	0.2	0.2
	1983, 1994	11°S, 34°W–13°W		0.30	1.1	4.4
	1983, 1994	11°S, 12.5°W–12°E		0.08	0.3	2.2
	1926, 1957	16°S, 37°W–14°W		0.10	−0.8	−2.5
	1926, 1957	16°S, 13.5°W–10.5°E		0.05	−0.2	−0.7
	1958, 1983	24°S, 40.5°W–14°W		0.41	0.1	1.0
	1958, 1983	24°S, 13.5°W–12.5°E		0.46	0.6	1.0
	1925, 1959	32°S, 48.5°W–14°W		0.13	−0.4	−0.2
Arctic Ocean						
See text			200–1500	Peak of >1 at 300 m		
North Pacific Ocean						
Thomson and Tabata (1989)	1956-1986	Ocean station Papa 50°N, 145°W			1.1	
Roemmich (1992)	1950-1991	32°N (off the coast of California)	0–300		0.9 ± 0.2	
Wong (1999), Wong et al. (1999, 2001)	1970s, 1990s	3.5°S–60°N			1.4	
		31.5°S–60°N			0.85	
Antonov (1993)	1957-1981	North of 30°N	0–500	Cooling		
South Pacific Ocean						
Holbrook and Bindoff (1997)	1955-1988	S. Tasman Sea	0–100		0.3	
Ridgway and Godfrey (1996), Holbrook and Bindoff (1997)	1955, mid-1970s	Coral and Tasman Seas	0–100	Warming		
	Since mid-1970s		0–450	Cooling		
Bindoff and Church (1992)	1967, 1989-1990	Australia-170°E 43°S			0.9	
		28°S			1.4	
Shaffer et al. (2000)	1967-1995	Eastern S Pacific 43°S			0.5	
		28°S			1.1	
Indian Ocean						
Bindoff and McDougall (1999)	1959-1966, 1987	30°S –35°S	0–900		1.6	
Atlantic, Pacific and Indian Oceans						
Levitus et al. (2000), Antonov et al. (2000)	1955-1995	Global average	0–300	0.7	(0.3 Wm^{-2})	
			0–3000		0.55 mm/yr (0.5 Wm^{-2})	

found changes in temperature and salinity in the upper hundreds of metres of the ocean which are consistent with a model of surface warming and freshening in the formation regions of the water masses and their subsequent subduction into the upper ocean. Such basin-scale changes are not merely a result of vertical thermocline heave, as might result from variability in surface winds.

In the only global analysis to date, Levitus *et al.* (2000) finds the ocean has stored 20×10^{22} J of heat between 1955 and 1995 (an average of 0.5 Wm^{-2}), with over half of this occurring in the upper 300 m for a rate of warming of 0.7°C/century. The steric sea level rise equivalent is 0.55 mm/yr, with maxima in the sub-tropical gyre of the North Atlantic and the tropical eastern Pacific.

In summary, while the evidence is still incomplete, there are widespread indications of thermal expansion, particularly in the sub-tropical gyres, of the order 1 mm/yr (Table 11.1). The evidence is most convincing for the North Atlantic but it also extends into the Pacific and Indian Oceans. The only area where cooling has been observed is in the sub-polar gyre of the North Atlantic and perhaps the North Pacific sub-polar gyre.

11.2.1.2 Models of thermal expansion

A variety of ocean models have been employed for estimates of ocean thermal expansion. The simplest and most frequently quoted is the one-dimensional (depth) upwelling-diffusion (UD) model (Hoffert *et al.*, 1980; Wigley and Raper, 1987, 1992, 1993; Schlesinger and Jiang, 1990; Raper *et al.*, 1996), which represents the variation of temperature with depth. Kattenberg *et al.* (1996) demonstrated that results from the GFDL AOGCM could be reproduced by the UD model of Raper *et al.* (1996). Using this model, the best estimate of thermal expansion from 1880 to 1990 was 43 mm (with a range of 31 to 57 mm) (Warrick *et al.*, 1996). Raper and Cubasch (1996) and Raper *et al.* (2001) discuss ways in which the UD model requires modification to reproduce the results of other AOGCMs. The latter work shows that a UD model of the type used in the SAR may be inadequate to represent heat uptake into the deep ocean on the time-scale of centuries. De Wolde *et al.* (1995, 1997) developed a two dimensional (latitude-depth, zonally averaged) ocean model, with similar physics to the UD model. Their best estimate of ocean thermal expansion in a model forced by observed sea surface temperatures over the last 100 years was 35 mm (with a range of 22 to 51 mm). Church *et al.* (1991) developed a subduction model in which heat is carried into the ocean interior through an advective process, which they argued better represented the oceans with movement of water along density surfaces and little vertical mixing. Jackett *et al.* (2000) developed this model further and tuned it

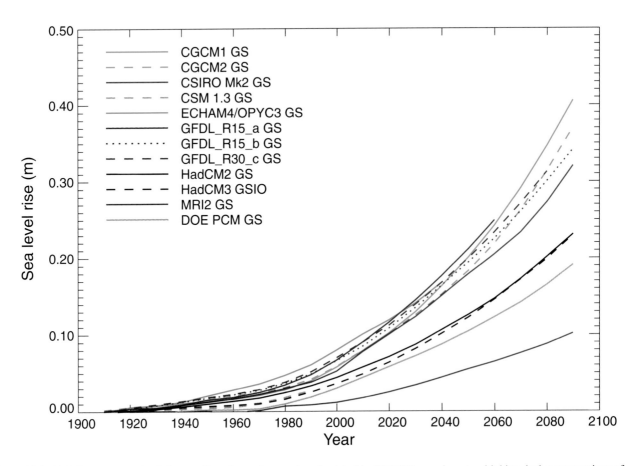

Figure 11.1: Global average sea level changes from thermal expansion simulated in AOGCM experiments with historical concentrations of greenhouse gases in the 20th century, then following the IS92a scenario for the 21st century, including the direct effect of sulphate aerosols. See Tables 8.1 and 9.1 for further details of models and experiments.

by comparison with an AOGCM, obtaining an estimate of 50 mm of thermal expansion over the last 100 years.

The advantage of these simple models is that they require less computing power than AOGCMs and so the sensitivity of results to a range of uncertainties can easily be examined. However, the simplifications imply that important processes controlling the penetration of heat from the surface into the ocean interior are not reproduced and they cannot provide information on the regional distribution of sea level rise. The most satisfactory way of estimating ocean thermal expansion is through the use of AOGCMs (Chapter 8, Section 8.3) (Gregory, 1993; Cubasch *et al.*, 1994; Bryan, 1996; Jackett *et al.*, 2000; Russell *et al.*, 2000; Gregory and Lowe, 2000). Improvements over the last decade relate particularly to the representation of the effect on mixing by processes which operate on scales too small to be resolved in global models, but which may have an important influence on heat uptake (see Section 8.5.2.2.4). The geographical distribution of sea level change due to density and circulation changes can be obtained from AOGCM results (various methods are used; see Gregory *et al.*, 2001). The ability of AOGCMs to simulate decadal variability in the ocean interior has not yet been demonstrated adequately, partly because of the scarcity of observations of decadal variability in the ocean for testing these models. This is not only an issue of evaluation of model performance; it is also relevant for deciding whether observed trends in sea level and interior ocean temperatures represent a change which is significantly larger than the natural internal variability of the climate system.

A number of model simulations of the 20th century (Table 9.1) have recently been completed using realistic greenhouse gas and aerosol forcings. Results for global average thermal expansion over periods during the 20th century are given in Figure 11.1 and Table 11.2. They suggest that over the last hundred years the average rate of sea level rise due to thermal expansion was of the order of 0.3 to 0.7 mm/yr, a range which encompasses the simple model estimates, rising to 0.6 to 1.1 mm/yr in recent decades, similar to the observational estimates (Section 11.2.1.1).

11.2.2 Glaciers and Ice Caps

> **Box 11.2:** Mass balance terms for glaciers, ice caps and ice sheets
>
> A glacier, ice cap or ice sheet gains mass by accumulation of snow (snowfall and deposition by wind-drift), which is gradually transformed to ice, and loses mass (ablation) mainly by melting at the surface or base with subsequent runoff or evaporation of the melt water. Some melt water may refreeze within the snow instead of being lost, and some snow may sublime or be blown off the surface. Ice may also be removed by discharge into a floating ice shelf or glacier tongue, from which it is lost by basal melting and calving of icebergs. Net accumulation occurs at higher altitude, net ablation at lower altitude; to compensate for net accumulation and ablation, ice flows downhill by internal deformation of the ice and sliding and bed deformation at the base. The rate at which this occurs is mainly controlled by the surface slope, the ice thickness, the effective ice viscosity, and basal thermal and physical conditions. The mass balance for an individual body of ice is usually expressed as the rate of change of the equivalent volume of liquid water, in m^3/yr; the mass balance is zero for a steady state. Mass balances are computed for both the whole year and individual seasons; the winter mass balance mostly measures accumulation, the summer, surface melting. The specific mass balance is the mass balance averaged over the surface area, in m/yr. A mass balance sensitivity is the derivative of the specific mass balance with respect to a climate parameter which affects it. For instance, a mass balance sensitivity to temperature is in m/yr/°C.

11.2.2.1 Mass balance studies

The water contained in glaciers and ice caps (excluding the ice sheets of Antarctica and Greenland) is equivalent to about 0.5 m of global sea level (Table 11.3). Glaciers and ice caps are rather sensitive to climate change; rapid changes in their mass are possible, and are capable of producing an important contribution to the rate of sea level rise. To evaluate this contribution, we need to know the rate of change of total glacier mass. Unfortunately sufficient measurements exist to determine the mass balance (see Box 11.2 for definition) for only a small minority of the world's 10^5 glaciers.

Table 11.2: Rate and acceleration of global-average sea level rise due to thermal expansion during the 20th century from AOGCM experiments with historical concentrations of greenhouse gases, including the direct effect of sulphate aerosols. See Tables 8.1 and 9.1 for further details of models and experiments. The rates are means over the periods indicated, while a quadratic fit is used to obtain the acceleration, assumed constant. Under this assumption, the rates apply to the midpoints (1950 and 1975) of the periods. Since the midpoints are 25 years apart, the difference between the rates is 25 times the acceleration. This relation is not exact because of interannual variability and non-constant acceleration.

	Rate of sea level rise (mm/yr)		Acceleration (mm/yr/century)
	1910 [a] to 1990 [b]	1960 to 1990 [b]	1910 [a] to 1990 [b]
CGCM1 GS	0.48	0.79	0.7 ± 0.2
CGCM2 GS	0.50	0.71	0.5 ± 0.3
CSIRO Mk2 GS	0.47	0.72	1.1 ± 0.2
CSM 1.3 GS	0.34	0.70	1.2 ± 0.3
ECHAM4/OPYC3 GS	0.75	1.09	1.0 ± 0.5
GFDL_R15_a GS	0.59	0.97	1.4 ± 0.4
GFDL_R15_b GS	0.60	0.88	1.1 ± 0.3
GFDL_R30_c GS	0.64	0.97	1.2 ± 0.3
HadCM2 GS	0.42	0.60	0.8 ± 0.2
HadCM3 GSIO	0.32	0.64	1.3 ± 0.4
DOE PCM GS	0.25	0.63	0.8 ± 0.4

[a] The choice of 1910 (rather than 1900) is made to accommodate the start date of some of the model integrations.

[b] The choice of 1990 (rather than 2000) is made because observational estimates referred to here do not generally include much data from the 1990s.

Table 11.3: Some physical characteristics of ice on Earth.

	Glaciers	**Ice caps**	**Glaciers and ice caps** [a]	**Greenland ice sheet** [b]	**Antarctic ice sheet** [b]
Number	>160 000	70			
Area (10^6 km^2)	0.43	0.24	0.68	1.71	12.37
Volume (10^6 km^3)	0.08	0.10	0.18 ± 0.04	2.85	25.71
Sea-level rise equivalent [d]	0.24	0.27	0.50 ± 0.10	7.2 [c]	61.1 [c]
Accumulation (sea-level equivalent, mm/yr) [d]			1.9 ± 0.3	1.4 ± 0.1	5.1 ± 0.2

Data sources: Meier and Bahr (1996), Warrick *et al.* (1996), Reeh *et al.* (1999), Huybrechts *et al.* (2000), Tables 11.5 and 11.6.

[a] Including glaciers and ice caps on the margins of Greenland and the Antarctic Peninsula, which have a total area of 0.14×10^6 km^2 (Weideck and Morris, 1996). The total area of glaciers and ice-caps outside Greenland and Antarctica is 0.54×10^6 km^2 (Dyurgerov and Meier, 1997a). The glaciers and ice caps of Greenland and Antarctica are included again in the next two columns.

[b] Grounded ice only, including glaciers and small ice caps.

[c] For the ice sheets, sea level rise equivalent is calculated with allowance for isostatic rebound and sea water replacing grounded ice, and this therefore is less than the sea level equivalent of the ice volume.

[d] Assuming an oceanic area of 3.62×10^8 km^2.

A possible approximate approach to this problem is to group glaciers into climatic regions, assuming glaciers in the same region to have a similar specific mass balance. With this method, we need to know only the specific mass balance for a typical glacier in each region (Kuhn *et al.*, 1999) and the total glacier area of the region. Multiplying these together gives the rate of change of glacier mass in the region. We then sum over all regions.

In the past decade, estimates of the regional totals of area and volume have been improved by the application of high resolution remote sensing and, to a lesser extent, by radio-echo-sounding. New glacier inventories have been published for central Asia and the former Soviet Union (Dolgushin and Osipova, 1989; Liu *et al.*, 1992; Kuzmichenok, 1993; Shi *et al.*, 1994; Liu and Xie, 2000; Qin *et al.*, 2000), New Zealand (Chinn, 1991), India (Kaul, 1999) South America (Casassa, 1995; Hastenrath and Ames, 1995; Skvarca *et al.*, Aniya *et al*, 1997; Kaser, 1999; 1995; Kaser *et al.*, 1996; Rott *et al.*, 1998), and new estimates made for glaciers in Antarctica and Greenland apart from the ice sheets (Weidick and Morris, 1996).

By contrast, specific mass balance is poorly known. Continuous mass balance records longer than 20 years exist for about forty glaciers worldwide, and about 100 have records of more than five years (Dyurgerov and Meier, 1997a). Very few have both winter and summer balances; these data are critical to relating glacier change to climatic elements (Dyurgerov and Meier, 1999). Although mass balance is being monitored on several dozen glaciers worldwide, these are mostly small (<20 km^2) and not representative of the size class that contains the majority of the mass (>100 km^2). The geographical coverage is also seriously deficient; in particular, we are lacking information on the most important maritime glacier areas. Specific mass balance exhibits wide variation geographically and over time (Figure 11.2). While glaciers in most parts of the world have had negative mass balance in the

past 20 years, glaciers in New Zealand (Chinn, 1999; Lamont *et al.*, 1999) and southern Scandinavia (Tvede and Laumann, 1997) have been advancing, presumably following changes in the regional climate.

Estimates of the historical global glacier contribution to sea level rise are shown in Table 11.4. Dyurgerov and Meier (1997a) obtained their estimate by dividing a large sample of measured glaciers into seven major regions and finding the mass balance for each region, including the glaciers around the ice sheets. Their area-weighted average for 1961 to 1990 was equivalent to 0.25 ± 0.10 mm/yr of sea level rise. Cogley and Adams (1998) estimated a lower rate for 1961 to 1990. However, their results may be not be representative of the global average because they do not make a correction for the regional biases in the sample of well investigated glaciers (Oerlemans, 1999). When evaluating data based on observed mass balance, one should note a worldwide glacier retreat

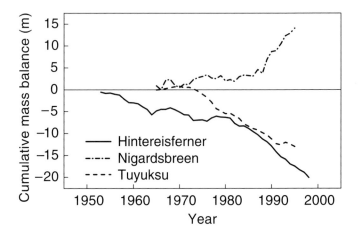

Figure 11.2: Cumulative mass balance for 1952-1998 for three glaciers in different climatic regimes: Hintereisferner (Austrian Alps), Nigardsbreen (Norway), Tuyuksu (Tien Shan, Kazakhstan).

Table 11.4: *Estimates of historical contribution of glaciers to global average sea level rise.*

Reference	Period	Rate of sea-level rise (mm/yr)	Remarks
Meier (1984)	1900 to 1961	0.46 ± 0.26	
Trupin *et al.* (1992)	1965 to 1984	0.18	
Meier (1993)	1900 to 1961	0.40	
Zuo and Oerlemans (1997), Oerlemans (1999)	1865 to 1990	0.22 ± 0.07[a]	Observed temperature changes with mass balance sensitivities estimated from precipitation in 100 regions
	1961 to 1990	0.3[a]	
Dyurgerov and Meier (1997b)	1961 to 1990	0.25 ± 0.10	Area-weighted mean of observed mass balance for seven regions
Dowdeswell *et al.* (1997)	1945 to 1995 approx.	0.13	Observed mass balance, Arctic only
Gregory and Oerlemans (1998)	1860 to 1990	0.15[a]	General Circulation Model (GCM) temperature changes with mass balance sensitivities from Zuo and Oerlemans (1997)
	1960 to 1990	0.26[a]	

[a] These papers give the change in sea level over the period indicated, from which we have calculated the rate of sea level rise.

following the high stand of the middle 19th century and subsequent small regional readvances around 1920 and 1980.

11.2.2.2 Sensitivity to temperature change

A method of dealing with the lack of mass balance measurements is to estimate the changes in mass balance as a function of climate, using mass balance sensitivities (see Box 11.2 for definition) and observed or modelled climate change for glacier covered regions. Mass-balance modelling of all glaciers individually is not practical because no detailed description exists for the great majority of them, and because local climate data are not available; even regional climate models do not have sufficient resolution, while downscaling methods cannot generally be used because local climate measurements have not been made (see Section 10.7). A number of authors have estimated past glacier net mass loss using past temperature change with present day glacier covered areas and mass balance sensitivities (Table 11.4). In this report, we project future mass balance changes using regional mass balance sensitivities which take account of regional and seasonal climatic information, instead of using the heuristic model of Wigley and Raper (1995) employed by Warrick *et al.* (1996).

Meier (1984) intuitively scaled specific mass balance according to mass balance amplitude (half the difference between winter and summer specific mass balance). Braithwaite and Zhang (1999) demonstrated a dependence of mass balance sensitivity on mass balance amplitude. Oerlemans and Fortuin (1992) derived an empirical relationship between the mass balance sensitivity of a glacier to temperature change and the local average precipitation, which is the principal factor determining its mass turnover rate. Zuo and Oerlemans (1997) extended this idea by distinguishing the effects of temperature changes in summer and outside summer; the former have a stronger influence on mass loss, in general. They made a calculation of glacier net mass loss since 1865. For 1961 to 1990, they obtained a rate of 0.3 mm/yr of sea level rise (i.e., a total of 8 mm, Oerlemans, 1999), very similar to the result of Dyurgerov and

Meier (1997b). Gregory and Oerlemans (1998) applied local seasonal temperature changes over 1860 to 1990 calculated by the HadCM2 AOGCM forced by changing greenhouse gases and aerosols (HadCM2 GS in Table 9.1) to the glacier model of Zuo and Oerlemans.

Zuo and Oerlemans (1997), Gregory and Oerlemans (1998) and Van de Wal and Wild (2001) all stress that the global average glacier mass balance depends markedly on the regional and seasonal distribution of temperature change. For instance, Gregory and Oerlemans (1998) find that projected future glacier net mass loss is 20% greater if local seasonal variation is neglected, and 20% less if regional variation is not included. The first difference arises because annual average temperature change is greater than summer temperature change at high latitudes, but the mass balance sensitivity is greater to summer change. The second is because the global average temperature change is less than the change at high latitudes, where most glaciers are found (Section 9.3.2).

Both the observations of mass balance and the estimates based on temperature changes (Table 11.4) indicate a reduction of mass of glaciers and ice caps in the recent past, giving a contribution to global-average sea level of 0.2 to 0.4 mm/yr over the last hundred years.

11.2.2.3 Sensitivity to precipitation change

Precipitation and accumulation changes also influence glacier mass balance, and may sometimes be dominant (e.g. Raper *et al.*, 1996). Generally, glaciers in maritime climates are more sensitive to winter accumulation than to summer conditions (Kuhn, 1984). AOGCM experiments suggest that global-average annual mean precipitation will increase on average by 1 to 3%/°C under the enhanced greenhouse effect (Figure 9.18). Glacier mass balance modelling indicates that to compensate for the increased ablation from a temperature rise of 1°C a precipitation increase of 20% (Oerlemans, 1981) or 35% (Raper *et al.*, 2000) would be required. Van de Wal and Wild (2001) find that the effect of precipitation changes on calculated global-average glacier mass changes in the 21st century is only 5% of the temperature effect. Such results suggest that the evolution of the global glacier mass is controlled principally by temperature changes rather than precipitation changes. Precipitation changes might be significant in particular localities, especially where precipitation is affected by atmospheric circulation changes, as seems recently to have been the case with southern Scandinavian glaciers (Oerlemans, 1999).

11.2.2.4 Evolution of area

The above calculations all neglect the change of area that will accompany loss of volume. Hence they are inaccurate because reduction of area will restrict the rate of melting. A detailed computation of transient response with dynamic adjustment to decreasing glacier sizes is not feasible at present, since the required information is not available for most glaciers. Oerlemans *et al.* (1998) undertook such detailed modelling of twelve individual glaciers and ice caps with an assumed rate of temperature change for the next hundred years. They found that neglecting the contraction of glacier area could lead to an overestimate of net mass loss of about 25% by 2100.

Dynamic adjustment of glaciers to a new climate occurs over tens to hundreds of years (Jóhannesson *et al.*, 1989), the time-scale being proportional to the mean glacier thickness divided by the specific mass balance at the terminus. Since both quantities are related to the size of the glacier, the time-scale is not necessarily longer for larger glaciers (Raper *et al.*, 1996; Bahr *et al.*, 1998), but it tends to be longer for glaciers in continental climates with low mass turnover (Jóhannesson *et al.*, 1989; Raper *et al.*, 2000).

Meier and Bahr (1996) and Bahr *et al.* (1997), following previous workers, proposed that for a glacier or an ice sheet in a steady state there may exist scaling relationships of the form $V \propto A^c$ between the volume V and area A, where c is a constant. Such relationships seem well supported by the increasing sample of glacier volumes measured by radio-echo-sounding and other techniques, despite the fact that climate change may be occurring on time-scales similar to those of dynamic adjustment. If one assumes that the volume-area relationship always holds, one can use it to deduce the area as the volume decreases. This idea can be extended to a glacier covered region if one knows the distribution of total glacier area among individual glaciers, which can be estimated using empirical functions (Meier and Bahr, 1996; Bahr, 1997). Using these methods, Van de Wal and Wild (2001) found that contraction of area reduces the estimated glacier net mass loss over the next 70 years by 15 to 20% (see also Section 11.5.1.1).

11.2.3 Greenland and Antarctic Ice Sheets

Together, the present Greenland and Antarctic ice sheets contain enough water to raise sea level by almost 70 m (Table 11.3), so that only a small fractional change in their volume would have a significant effect. The average annual solid precipitation falling onto the ice sheets is equivalent to 6.5 mm of sea level, this input being approximately balanced by loss from melting and iceberg calving. The balance of these processes is not the same for the two ice sheets, on account of their different climatic regimes. Antarctic temperatures are so low that there is virtually no surface runoff; the ice sheet mainly loses mass by ice discharge into floating ice shelves, which experience melting and freezing at their underside and eventually break up to form icebergs. On the other hand, summer temperatures on the Greenland ice sheet are high enough to cause widespread melting, which accounts for about half of the ice loss, the remainder being discharged as icebergs or into small ice-shelves.

Changes in ice discharge generally involve response times of the order of 10^2 to 10^4 years. The time-scales are determined by isostasy, the ratio of ice thickness to yearly mass turnover, processes affecting ice viscosity, and physical and thermal processes at the bed. Hence it is likely that the ice sheets are still adjusting to their past history, in particular the transition to interglacial conditions. Their future contribution to sea level change therefore has a component resulting from past climate changes as well as one relating to present and future climate changes.

For the 21st century, we expect that surface mass balance changes will dominate the volume response of both ice sheets. A key question is whether ice-dynamical mechanisms could operate which would enhance ice discharge sufficiently to have an appreciable additional effect on sea level rise.

11.2.3.1 Mass balance studies

Traditionally, the state of balance of the polar ice sheets has been assessed by estimating the individual mass balance terms, and making the budget. Only the mass balance of the ice sheet resting on bedrock (the grounded ice sheet) needs to be considered, because changes in the ice shelves do not affect sea level as they are already afloat. Recent mass balance estimates for Greenland and Antarctica are shown in Tables 11.5 and 11.6. Most progress since the SAR has been made in the assessment of accumulation, where the major obstacle is poor coverage by *in situ* measurements. New methods have made use of atmospheric moisture convergence analysis based on meteorological data, remotely sensed brightness temperatures of dry snow, and GCMs (see references in the tables). Recent accumulation estimates display a tendency for convergence towards a common value, suggesting a remaining error of less than 10% for both ice sheets.

For Greenland (Table 11.5), runoff is an important term but net ablation has only been measured directly at a few locations and therefore has to be calculated from models, which have considerable sensitivity to the surface elevation data set and the parameters of the melt and refreezing methods used (Reeh and Starzer, 1996; Van de Wal, 1996; Van de Wal and Ekholm, 1996; Janssens and Huybrechts, 2000). Summing best estimates of the various mass balance components for Greenland gives a balance of $-8.5 \pm 10.2\%$ of the input, or $+0.12 \pm 0.15$ mm/yr of global sea level change, not significantly different from zero.

During the last five years, some mass balance estimates have been made for individual Greenland sectors. A detailed comparison of the ice flux across the 2,000 m contour with total accumulation revealed most of the accumulation zone to be near to equilibrium, albeit with somewhat larger positive and negative local imbalances (Thomas *et al.*, 1998, 2000). These results are

Table 11.5: *Current state of balance of the Greenland ice sheet (10^{12} kg/yr).*

Source and remarks	A Accumulation	B Runoff	C Net accumulation	D Iceberg production	E Bottom melting	F Balance
Benson (1962)	500	272	228	215		+13
Bauer (1968)	500	330	170	280		−110
Weidick (1984)	500	295	205	205		± 0
Ohmura and Reeh (1991): New accumulation map	535					
Huybrechts *et al.* (1991): Degree-day model on 20 km grid	539	256	283			
Robasky and Bromwich (1994): Atmospheric moisture budget analysis from radiosonde data, 1963-1989	545					
Giovinetto and Zwally (1995a): Passive microwave data of dry snow	461[a]					
Van de Wal (1996): Energy-Balance model on 20 km grid	539	316	223			
Jung-Rothenhäusler (1998): Updated accumulation map	510					
Reeh *et al.* (1999)	547	276	271	239	32	± 0
Ohmura *et al.* (1999): Updated accumulation map with GCM data; runoff from ablation-summer temperature parametrization	516	347	169			
Janssens and Huybrechts (2000): recalibrated degree-day model on 5 km grid; updated precipitation and surface elevation maps	542	281	261			
Zwally and Giovinetto (2000): Updated calculation on 50 km grid			216[b]			
Mean and standard deviation	520 ± 26	297 ± 32	225 ± 41	235 ± 33	32 ± 3[c]	−44 ± 53[d]

[a] Normalised to ice sheet area of 1.676×10^6 km^2 (Ohmura and Reeh, 1991).

[b] Difference between net accumulation above the equilibrium line and net ablation below the equilibrium line.

[c] Melting below the fringing ice shelves in north and northeast Greenland (Rignot *et al.*, 1997).

[d] Including the ice shelves, but nearly identical to the grounded ice sheet balance because the absolute magnitudes of the other ice-shelf balance terms (accumulation, runoff, ice-dynamic imbalance) are very small compared to those of the ice sheet *(F=A–B–D–E).*

Table 11.6: *Current state of balance of the Antarctic ice sheet (10^{12} kg/yr).*

Source and remarks	A Accumulation over grounded ice	B Accumulation over all ice sheet	C Ice shelf melting	D Runoff	E Iceberg production	F Flux across grounding line
Kotlyakov *et al.* (1978)		2000	320	60	2400	
Budd and Smith (1985)	1800	2000			1800	1620
Jacobs *et al.* (1992). Ice shelf melting from observations of melt water outflow, glaciological field studies and ocean modelling.	1528	2144	544	53	2016	
Giovinetto and Zwally (1995a). Passive microwave data of dry snow.	1752[a]	2279[a]				
Budd *et al.* (1995). Atmospheric moisture budget analysis from GASP data, 1989 to 1992.		2190[b]				
Jacobs *et al.* (1996). Updated ice-shelf melting assessment.			756			
Bromwich *et al.* (1998). Atmospheric moisture budget analysis from ECMWF reanalysis and evaporation/ sublimation forecasts, 1985 to 1993.		2190[b]				
Turner *et al.* (1999). Atmospheric moisture budget analysis from ECMWF reanalysis, 1979 to 1993.		2106				
Vaughan *et al.* (1999). 1800 *in situ* measurements interpolated using passive microwave control field.	1811	2288				
Huybrechts *et al.* (2000). Updated accumulation map.	1924	2344				
Giovinetto and Zwally (2000). Updated map on 50 km grid.	1883[c]	2326[c]				
Mean and standard deviation.	1843 ± 76[d]	2246 ± 86[d]	540 ± 218	10 ± 10[e]	2072 ± 304	

[a] Normalised to include the Antarctic Peninsula.

[b] Specific net accumulation multiplied by total area of 13.95×10^6 km^2 (Fox and Cooper, 1994).

[c] Normalised to include the Antarctic Peninsula, and without applying a combined deflation and ablation adjustment.

[d] Mean and standard deviation based only on accumulation studies published since 1995.

[e] Estimate by the authors.

The mass balance of the ice sheet including ice shelves can be estimated as $B–C–D–E=–376 \pm 384 \times 10^{12}$ kg/yr, which is $–16.7 \pm 17.1\%$ of the total input B.

Assuming the ice shelves are in balance (and noting that the runoff derives from the grounded ice, not the ice shelves) would imply that $0=F+(B–A)–C–E$, in which case the flux across the grounding line would be $F=A–B+C+E=2209 \pm 391 \times 10^{12}$ kg/yr.

likely to be only little influenced by short-term variations, because in the ice sheet interior, quantities that determine ice flow show little variation on a century time-scale. Recent studies have suggested a loss of mass in the ablation zone (Rignot *et al.*, 1997; Ohmura *et al.*, 1999), and have brought to light the important role played by bottom melting below floating glaciers (Reeh *et al.*, 1997, 1999; Rignot *et al.*, 1997); neglect of this term led to erroneous results in earlier analyses.

For Antarctica (Table 11.6), the ice discharge dominates the uncertainty in the mass balance of the grounded ice sheet, because of the difficulty of determining the position and thickness of ice at the grounding line and the need for assumptions about the vertical distribution of velocity. The figure of Budd and Smith (1985) of $1,620 \times 10^{12}$ kg/yr is the only available estimate. Comparing this with an average value of recent accumulation estimates for the grounded ice sheet would suggest a positive mass balance of around +10% of the total input, equivalent to -0.5 mm/yr of sea level. Alternatively, the flux across the grounding line can be obtained by assuming the ice shelves to be in balance and using estimates of the calving rate (production of icebergs), the rate of melting on the (submerged) underside of the ice shelves, and accumulation on the ice shelves. This results in a flux of $2,209 \pm 391 \times 10^{12}$ kg/yr across the grounding line and a mass balance for the grounded ice equivalent to $+1.04 \pm 1.06$ mm/yr of sea level (Table 11.6). However, the ice shelves may not be in balance, so that the error estimate probably understates the true uncertainty.

11.2.3.2 Direct monitoring of surface elevation changes

Provided that changes in ice and snow density and bedrock elevation are small or can be determined, elevation changes can be used to estimate changes of mass of the ice sheets. Using satellite altimetry, Davis *et al.* (1998) reported a small average thickening between 1978 and 1988 of 15 ± 20 mm/yr of the Greenland ice sheet above 2,000 m at latitudes up to 72°N. Krabill *et al.* (1999) observed a similar pattern above 2,000 m from 1993 to 1998 using satellite referenced, repeat aircraft laser altimetry. Together, these results indicate that this area of the Greenland ice sheet has been nearly in balance for two decades, in agreement with the mass budget studies mentioned above (Thomas *et al.*, 2000). Krabill *et al.* (1999) observed markedly different behaviour at lower altitudes, with thinning rates in excess of 2 m/yr in the south and east, which they attributed in part to excess flow, although a series of warmer-than-average summers may also have had an influence. In a recent update, Krabill *et al.* (2000) find the total ice sheet balance to be -46×10^{12} kg/yr or 0.13 mm/yr of sea level rise between 1993 and 1999, but could not provide an error bar. Incidentally, this value is very close to the century time-scale imbalance derived from the mass budget studies (Table 11.5), although the time periods are different and the laser altimetry results do not allow us to distinguish between accumulation, ablation, and discharge.

Small changes of ± 11 mm/yr were reported by Lingle and Covey (1998) in a region of East Antarctica between 20° and 160°E for the period 1978 to 1988. Wingham *et al.* (1998) examined the Antarctic ice sheet north of 82°S from 1992 to 1996, excluding the marginal zone. They observed no change in

East Antarctica to within ± 5 mm/yr, but reported a negative trend in West Antarctica of -53 ± 9 mm/yr, largely located in the Pine Island and Thwaites Glacier basins. They estimated a century-scale mass imbalance of $-6\% \pm 8\%$ of accumulation for 63% of the Antarctic ice sheet, concluding that the thinning in West Antarctica is likely to result from a recent accumulation deficit. However, the measurements of Rignot (1998a), showing a 1.2 ± 0.3 km/yr retreat of the grounding line of Pine Island Glacier between 1992 and 1996, suggest an ice-dynamic explanation for the observed thinning. Altimetry records are at present too short to confidently distinguish between a short-term surface mass-balance variation and the longer-term ice-sheet dynamic imbalance. Van der Veen and Bolzan (1999) suggest that at least five years of data are needed on the central Greenland ice sheet.

11.2.3.3 Numerical modelling

Modelling of the past history of the ice sheets and their underlying beds over a glacial cycle is a way to obtain an estimate of the present ice-dynamic evolution unaffected by short-term (annual to decadal) mass-balance effects. The simulation requires time-dependent boundary conditions (surface mass balance, surface temperature, and sea level, the latter being needed to model grounding-line changes). Current glaciological models employ grids of 20 to 40 km horizontal spacing with 10 to 30 vertical layers and include ice shelves, basal sliding and bedrock adjustment.

Huybrechts and De Wolde (1999) and Huybrechts and Le Meur (1999) carried out long integrations over two glacial cycles using 3-D models of Greenland and Antarctica, with forcing derived from the Vostok (Antarctica) and Greenland Ice Core Project (GRIP) ice cores. The retreat history of the ice sheet along a transect in central west Greenland in particular was found to be in good agreement with a succession of dated moraines (Van Tatenhove *et al.*, 1995), but similar validation elsewhere is limited by the availability of well-dated material. Similar experiments were conducted as part of the European Ice Sheet Modelling Initiative (EISMINT) intercomparison exercise (Huybrechts *et al.*, 1998). These model simulations suggest that the average Greenland contribution to global sea level rise has been between -0.1 and 0.0 mm/yr in the last 500 years, while the Antarctic contribution has been positive. Four different Antarctic models yield a sea level contribution of between $+0.1$ and $+0.5$ mm/yr averaged over the last 500 years, mainly due to incomplete grounding-line retreat of the West Antarctic ice sheet (WAIS) since the Last Glacial Maximum (LGM) (Figure 11.3). However, substantial uncertainties remain, especially for the WAIS, where small phase shifts in the input sea level time-series and inadequate representation of ice-stream dynamics may have a significant effect on the model outcome. Glacio-isostatic modelling of the solid earth beneath the Antarctic ice sheet with prescribed ice sheet evolution (James and Ivins, 1998) gave similar uplift rates as those presented in Huybrechts and Le Meur (1999), indicating that the underlying ice sheet scenarios and bedrock models were similar, but observations are lacking to validate the generated uplift rates. By contrast, Budd *et al.* (1998) find that Antarctic ice volume is currently increasing at a rate of about 0.08 mm/yr of sea level lowering because in their modelling the Antarctic ice sheet was actually smaller during the LGM than today (for which there

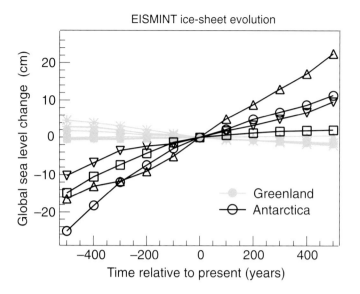

EISMINT ice-sheet evolution

Figure 11.3: Modelled evolution of ice sheet volume (represented as sea level equivalent) centred at the present time resulting from ongoing adjustment to climate changes over the last glacial cycle. Data are from all Antarctic and Greenland models that participated in the EISMINT intercomparison exercise (From Huybrechts *et al.*, 1998).

is, however, little independent evidence) and the effect of the higher accumulation rates during the Holocene dominates over the effects of grounding line changes.

Model simulations of this kind have not included the possible effects of changes in climate during the 20th century. The simulations described later (Section 11.5.1.1), in which an ice sheet model is integrated using changes in temperature and precipitation derived from AOGCM simulations, suggest that anthropogenic climate change could have produced an additional contribution of between –0.2 to 0.0 mm/yr of sea level from increased accumulation in Antarctica over the last 100 years, and between 0.0 and 0.1 mm/yr from Greenland, from both increased accumulation and ablation. The model results for Greenland exhibit substantial interannual variability. Furthermore, because of rising temperatures during the 20th century, the contribution for recent decades is larger than the average for the century. These points must be borne in mind when comparing with results of the direct observation methods for short periods in recent decades (Sections 11.2.3.1 and 11.2.3.2). Note also that the observational results include the ongoing response to past climate change as well as the effect of 20th century climate change.

11.2.3.4 Sensitivity to climatic change

The sensitivity of the ice sheet's surface mass balance has been studied with multiple regression analyses, simple meteorological models, and GCMs (Table 11.7). Most progress since the SAR has been made with several coupled AOGCMs, especially in the "time-slice" mode in which a high-resolution AGCM (Atmospheric General Circulation Model) is driven by output from a low-resolution transient AOGCM experiment for a limited duration of time. Model resolution of typically 100 km allows for a more realistic topography crucial to better

resolving temperature gradients and orographic forcing of precipitation along the steep margins of the polar ice sheets. Even then, GCMs do not yet perform well in reproducing melting directly from the surface energy fluxes. The ablation zone around the Greenland ice sheet is mostly narrower than 100 km, and the important role played by topography therefore requires the use of downscaling techniques to transfer information to local and even finer grids (Glover, 1999). An additional complication is that not all melt water runs off to the ocean and can be partly retained on or in the ice sheet (Pfeffer *et al.*, 1991; Janssens and Huybrechts, 2000).

For Greenland, estimates of the sensitivity to a 1°C local warming over the ice sheet are close to 0.3 mm/yr (with a total range of +0.1 to +0.4 mm/yr) of global sea level equivalent. This range mainly reflects differences in the predicted precipitation changes and the yearly distribution of the temperature increase, which is predicted to be larger in winter than in summer in the GCMs, but is assumed uniform in the studies of Van de Wal (1996) and Janssens and Huybrechts (2000). Another difference amongst the GCM results concerns the time window over which the sensitivities are assessed. The CSIRO9/T63 sensitivities are estimated from high-resolution runs forced with observed SSTs for the recent past (Smith *et al.*, 1998; Smith, 1999), whereas the ECHAM data are given as specific mass balance changes for doubled minus present atmospheric CO_2. Thompson and Pollard (1997) report similar results to the ECHAM studies but the corresponding sensitivity value could not be calculated because the associated temperature information is not provided. Some palaeoclimatic data from central Greenland ice cores indicate that variations in precipitation during the Holocene are related to changes in atmospheric circulation rather than directly to local temperature (Kapsner *et al.*, 1995; Cuffey and Clow, 1997), such that precipitation might not increase with temperature (in contrast with Clausen *et al.*, 1988). For glacial-interglacial transitions, the ice cores do exhibit a strong positive correlation between temperature and precipitation (Dansgaard *et al.*, 1993; Dahl-Jensen *et al.*, 1993; Kapsner *et al.*, 1995; Cuffey and Marshall, 2000), as simulated by AOGCMs for anthropogenic warming. Although other changes took place at the glacial-interglacial transition, this large climate shift could be argued to be a better analogue for anthropogenic climate change than the smaller fluctuations of the Holocene. To allow for changes in circulation patterns and associated temperature and precipitation patterns, we have used time-dependent AOGCM experiments to calculate the Greenland contribution (Section 11.5.1).

For Antarctica, mass-balance sensitivities for a 1°C local warming are close to –0.4 mm/yr (with one outlier of –0.8 mm/yr) of global sea level equivalent. A common feature of all methods is the insignificant role of melting, even for summer temperature increases of a few degrees, so that only accumulation changes need to be considered. The sensitivity for the case that the change in accumulation is set proportional to the relative change in saturation vapour pressure is at the lower end of the sensitivity range, suggesting that in a warmer climate changes in atmospheric circulation and increased moisture advection can become equally important, in particular close to the ice sheet margin (Bromwich, 1995; Steig, 1997). Both ECHAM3 and

Table 11.7: Mass balance sensitivity of the Greenland and Antarctic ice sheets to a 1°C local climatic warming.

Source	dB/dT (mm/yr/°C)	Method
Greenland ice sheet		
Van de Wal (1996)	+0.31 [a]	Energy balance model calculation on 20 km grid
Ohmura et al. (1996)	+0.41 [c] [+0.04] [cd]	ECHAM3/T106 time slice [2×CO_2 – 1×CO_2]
Smith (1999)	[–0.306] [d]	CSIRO9/T63 GCM forced with SSTs 1950-1999
Janssens and Huybrechts (2000)	+0.35 [a] [+0.26] [b]	Recalibrated degree-day model on 5 km grid with new precipitation and surface elevation maps
Wild and Ohmura (2000)	+0.09 [c] [–0.13] [cd]	ECHAM4-OPYC3/T106 GCM time slice [2×CO_2 – 1×CO_2]
Antarctic ice sheet		
Huybrechts and Oerlemans (1990)	–0.36	Change in accumulation proportional to saturation vapour pressure
Giovinetto and Zwally (1995b)	–0.80 [e]	Multiple regression of accumulation to sea-ice extent and temperature
Ohmura et al. (1996)	–0.41 [c]	ECHAM3/T106 time slice [2×CO_2 – 1×CO_2]
Smith et al. (1998)	–0.40	CSIRO9/T63 GCM forced with SSTs 1950-1999
Wild and Ohmura (2000)	–0.48 [c]	ECHAM4-OPYC3/T106 time slice [2×CO_2 – 1×CO_2]

dB/dT Mass balance sensitivity to local temperature change expressed as sea level equivalent. Note that this is not a sensitivity to global average temperature change.

[a] Constant precipitation.

[b] Including 5% increase in precipitation.

[c] Estimated from published data and the original time slice results.

[d] Accumulation changes only.

[e] Assuming sea-ice edge retreat of 150 km per °C.

ECHAM4/OPYC3 give a similar specific balance change over the ice sheet for doubled versus present atmospheric CO_2 to that found by Thompson and Pollard (1997).

In summary, the static sensitivity values suggest a larger role for Antarctica than for Greenland for an identical local temperature increase, meaning that the polar ice sheets combined would produce a sea level lowering, but the spread of the individual estimates includes the possibility that both ice sheets may also balance one another for doubled atmospheric CO_2 conditions (Ohmura et al., 1996; Thompson and Pollard, 1997). For CO_2 increasing according to the IS92a scenario (without aerosol), studies by Van de Wal and Oerlemans (1997) and Huybrechts and De Wolde (1999) calculated sea level contributions for 1990 to 2100 of +80 to +100 mm from the Greenland ice sheet and about –80 mm from the Antarctic ice sheet. On this hundred year time-scale, ice-dynamics on the Greenland ice sheet was found to counteract the mass-balance-only effect by between 10 and 20%. Changes in both the area-elevation distribution and iceberg discharge played a role, although the physics controlling the latter are poorly known and therefore not well represented in the models. Because of its longer response time-scales, the Antarctic ice sheet hardly exhibits any dynamic response on a century time-scale, except when melting rates below the ice shelves were prescribed to rise by in excess of 1 m/yr (O'Farrell et al., 1997; Warner and Budd, 1998; Huybrechts and De Wolde, 1999; see also Section 11.5.4.3).

11.2.4 Interaction of Ice Sheets, Sea Level and the Solid Earth

11.2.4.1 Eustasy, isostasy and glacial-interglacial cycles

On time-scales of 10^3 to 10^5 years, the most important processes affecting sea level are those associated with the growth and decay of the ice sheets through glacial-interglacial cycles. These contributions include the effect of changes in ocean volume and the response of the earth to these changes. The latter are the glacio-hydro-isostatic effects: the vertical land movements induced by varying surface loads of ice and water and by the concomitant redistribution of mass within the Earth and oceans. While major melting of the ice sheets ceased by about 6,000 years ago, the isostatic movements remain and will continue into the future because the Earth's viscous response to the load has a time-constant of thousands of years. Observational evidence indicates a complex spatial and temporal pattern of the resulting isostatic sea level change for the past 20,000 years. As the geological record is incomplete for most parts of the world, models (constrained by the reliable sea level observations) are required to describe and predict the vertical land movements and changes in ocean area and volume. Relative sea level changes caused by lithospheric processes, associated for example with tectonics and mantle convection, are discussed in Section 11.2.6.

Figure 11.4 illustrates global-average sea level change over the last 140,000 years. This is a composite record based on oxygen isotope data from Shackleton (1987) and Linsley

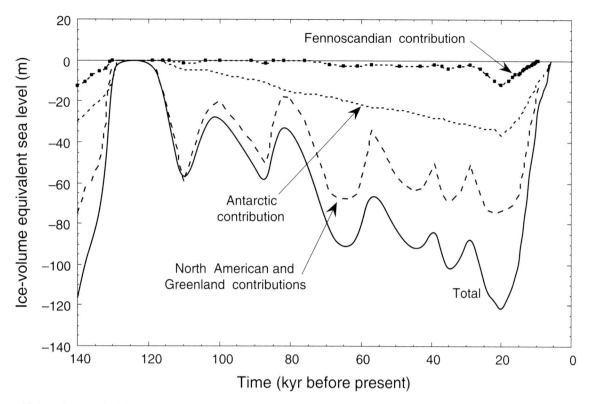

Figure 11.4: Estimates of global sea level change over the last 140,000 years (continuous line) and contributions to this change from the major ice sheets: (i) North America, including Laurentia, Cordilleran ice, and Greenland, (ii) Northern Europe (Fennoscandia), including the Barents region, (iii) Antarctica. (From Lambeck, 1999.)

(1996), constrained by the Huon terrace age-height relationships of Chappell *et al.* (1996a), the estimate of the LGM sea level by Yokoyama *et al.* (2000), the late-glacial eustatic sea level function of Fleming *et al.* (1998), and the timing of the Last Interglacial by Stirling *et al.* (1998). These fluctuations demonstrate the occurrence of sea level oscillations during a glacial-interglacial cycle that exceed 100 m in magnitude at average rates of up to 10 mm/year and more during periods of decay of the ice sheets and sometimes reaching rates as high as 40 mm/year (Bard *et al.*, 1996) for periods of very rapid ice sheet decay. Current best estimates indicate that the total LGM land-based ice volume exceeded present ice volume by 50 to 53×10^6 km^3 (Yokoyama *et al.*, 2000).

Local sea level changes can depart significantly from this average signal because of the isostatic effects. Figure 11.5 illustrates typical observational results for sea level change since the LGM in regions with no significant land movements other than of a glacio-hydro-isostatic nature. Also shown are model predictions for these localities, illustrating the importance of the isostatic effects. Geophysical models of these isostatic effects are well developed (see reviews by Lambeck and Johnston, 1998; Peltier, 1998). Recent modelling advances have been the development of high-resolution models of the spatial variability of the change including the detailed description of ice loads and of the melt-water load distribution (Mitrovica and Peltier, 1991; Johnston, 1993) and the examination of different assumptions about the physics of the earth (Peltier and Jiang, 1996; Johnston *et al.*, 1997; Kaufmann and Wolf, 1999; Tromp and Mitrovica, 1999).

Information about the changes in ice sheets come from field observations, glaciological modelling, and from the sea level observations themselves. Much of the emphasis of recent work on glacial rebound has focused on improved calculations of ice sheet parameters from sea level data (Peltier, 1998; Johnston and Lambeck, 2000; see also Section 11.3.1) but discrepancies between glaciologically-based ice sheet models and models inferred from rebound data remain, particularly for the time of, and before, the LGM. The majority of ice at this time was contained in the ice sheets of Laurentia and Fennoscandia but their combined estimated volume inferred from the rebound data for these regions (e.g., Nakada and Lambeck, 1988; Tushingham and Peltier, 1991, 1992; Lambeck *et al.*, 1998) is less than the total volume required to explain the sea level change of about 120 to 125 m recorded at low latitude sites (Fairbanks, 1989; Yokoyama *et al.*, 2000). It is currently uncertain how the remainder of the ice was distributed. For instance, estimates of the contribution of Antarctic ice to sea level rise since the time of the LGM range from as much as 37 m (Nakada and Lambeck, 1988) to 6 to 13 m (Bentley, 1999; Huybrechts and Le Meur, 1999). Rebound evidence from the coast of Antarctica indicates that ice volumes have changed substantially since the LGM (Zwartz *et al.*, 1997; Bentley, 1999) but these observations, mostly extending back only to 8,000 years ago, do not provide good constraints on the LGM volumes. New evidence from exposure age dating of moraines and rock surfaces is beginning to provide new constraints on ice thickness in Antarctica (e.g., Stone *et al.*, 1998) but the evidence is not yet sufficient to constrain past volumes of the entire ice sheet.

Figure 11.5: Examples of observed relative sea level change (with error bars, right-hand side) and model predictions for four different locations. The model predictions (left-hand side) are for the glacio-hydro-eustatic contributions to the total change (solid line, right hand side). (a) Angermann River, Sweden, near the centre of the former ice sheet over Scandinavia. The principal contribution to the sea level change is the crustal rebound from the ice unloading (curve marked ice, left-hand side) and from the change in ocean volume due to the melting of all Late Pleistocene ice sheets (curve marked esl). The combined predicted effect, including a small water loading term (not shown), is shown by the solid line (right-hand side), together with the observed values. (b) A location near Stirling, Scotland. Here the ice and esl contributions are of comparable magnitude but opposite sign (left-hand side) such that the rate of change of the total contribution changes sign (right-hand side). This result is typical for locations near former ice margins or from near the centres of small ice sheets. (c) The south of England where the isostatic contributions from the water (curve marked water) and ice loads are of similar amplitude but opposite sign. The dominant contribution to sea level change is now the eustatic contribution. This behaviour is characteristic of localities that lie well beyond the ice margins where a peripheral bulge created by the ice load is subsiding as mantle material flows towards the region formerly beneath the ice. (d) A location in Australia where the melt-water load is the dominant cause of isostatic adjustment. Here sea level has been falling for the past 6,000 years. This result is characteristic of continental margin sites far from the former areas of glaciation. (From Lambeck, 1996.)

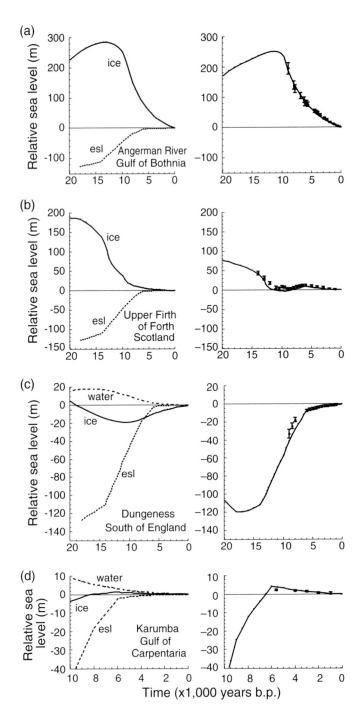

11.2.4.2 Earth rotation constraints on recent sea level rise

Changes in the Earth's ice sheets introduce a time-dependency in the Earth's inertia tensor whose consequences are observed both in the planet's rotation (as an acceleration in its rotation rate and as a shift in the position of the rotation axis) and in the acceleration of the rotation of satellite orbits about the Earth (Wu and Peltier, 1984; Lambeck, 1988). Model estimates of these changes are functions of mass shifts within and on the Earth and are dependent, therefore, on the past ice sheet geometries, on the Earth's rheology, and on the recent past and present rates of melting of the residual ice sheets. Other geophysical processes also contribute to the time-dependence

of the rotational and dynamical parameters (e.g. Steinberger and O'Connell, 1997). Hence, unique estimates of recent melting cannot be inferred from the observations.

Some constraints on the present rates of change of the ice sheets have, nevertheless, been obtained, in particular through a combination of the rotational observations with geological and tide gauge estimates of sea level change (Wahr *et al.*, 1993; Mitrovica and Milne, 1998; Peltier, 1998; Johnston and Lambeck, 1999). Results obtained so far are preliminary because observational records of the change in satellite orbits are relatively short (Nerem and Klosko, 1996; Cheng *et al.*, 1997) but they will become important as the length of the record increases. Peltier

(1998) has argued that if the polar ice sheets contributed, for example, 0.5 mm/yr to the global sea level rise, then the rotational constraints would require that most of this melting derived from Greenland. Johnston and Lambeck (1999) concluded that a solution consistent with geological evidence, including constraints on sea level for the past 6,000 years (Section 11.3.1), is for a non-steric sea level rise (i.e., not resulting from ocean density changes) of 1.0 ± 0.5 mm/yr for the past 100 years, with 5 to 30% originating from Greenland melting. However, all such estimates are based on a number of still uncertain assumptions such that the inferences are more indicative of the potential of the methodology than of actual quantitative conclusions.

11.2.5 Surface and Ground Water Storage and Permafrost

An important contribution to present day sea level rise could result from changes in the amount of water stored in the ground, on the surface in lakes and reservoirs, and by modifications to surface characteristics affecting runoff or evapotranspiration rates. Changing practices in the use of land and water could make these terms larger in future. However, very little quantitative information is available. For some of the components of the terrestrial water budget, Gornitz *et al.* (1997), updated by Gornitz (2000), give net results which differ substantially from those of Sahagian (2000) and Vörösmarty and Sahagian (2000), and also from those of Sahagian *et al.* (1994) used by Warrick *et al.* (1996). The largest positive contribution to sea level probably comes from ground water mining, which means the extraction of ground water from storage in aquifers in excess of the rate of natural recharge. Gornitz *et al.* (1997) estimate that ground water is mined at a rate that has been increasing in time, currently equivalent to 0.2 to 1.0 mm/yr of sea level, but they assume that much of this infiltrates back into aquifers so the contribution to sea level rise is only 0.1 to 0.4 mm/yr. Sahagian (2000) considers fewer aquifers; consequently he obtains a smaller total of 0.17 mm/yr from mining, but assumes that all of this water eventually reaches the ocean through the atmosphere or runoff. If Sahagian's assumption were applied to the inventory of Gornitz *et al.* it would imply a sea level contribution of 0.2 to 1.0 mm/yr.

Volumes of many of the world's large lakes have been reduced in recent decades through increased irrigation and other water use. Sahagian *et al.* (1994) and Sahagian (2000) estimate that the reduced volumes of the Caspian and Aral Seas (and associated ground water) contribute 0.03 and 0.18 mm/year to sea level rise, on the assumption that the extracted water reaches the world ocean by evapotranspiration. Recent *in situ* records and satellite altimetry data indicate that substantial fluctuations in the level of the Caspian Sea can occur on decadal time-scales (Cazenave *et al.*, 1997) which suggests that short records may not give a good indication of the long-term average. The reduction of lake volumes in China may contribute a further 0.005 mm/yr (Shi and Zhou, 1992). Assuming there are no other large sources, we take 0.2 mm/yr as the upper limit of the present contribution to sea level from lakes. Gornitz *et al.* (1997) do not include a term from lake volume changes, because they assume the water extracted for irrigation largely enters the ground water rather than the world ocean, so we take zero as the lower limit.

Gornitz *et al.* (1997) estimate there is 13.6 mm of sea level equivalent impounded in reservoirs. Most of this capacity was created, at roughly a constant rate, from 1950 to 1990. This rate of storage represents a reduction in sea level of 0.34 mm/yr. They assume that annually $5 \pm 0.5\%$ of the water impounded seeps into deep aquifers, giving a 1990 rate of seepage of 0.61 to 0.75 mm/yr, and a total volume of 15 mm sea level equivalent. We consider that this represents an upper bound, because it is likely that the rate of seepage from any reservoir will decrease with time as the surrounding water table rises, as assumed by Sahagian (2000). On the basis of a typical porosity and area affected, he estimates that the volume trapped as ground water surrounding reservoirs is 1.2 times the volume impounded in reservoirs. His estimate of the storage in reservoirs is 14 to 28 mm sea level equivalent; hence the ground water storage is an additional 17 to 34 mm sea level equivalent. Lack of global inventories means that these estimates of storage may well be too small because of the many small reservoirs not taken into account (rice paddies, ponds, etc., provided they impound water above the water table) (Vörösmarty and Sahagian, 2000). The total stored could be up to 50% larger (Sahagian, 2000).

Gornitz *et al.* (1997) estimate that evapotranspiration of water from irrigated land leads to an increase in atmospheric water content and hence a fall in sea level of 0.14 to 0.15 mm/yr. We consider this to be an overestimate, because it implies a 20th century increase in global tropospheric water content which substantially exceeds observations (Section 2.5.3.2). They further suggest that irrigation water derived from surface sources may infiltrate into aquifers, removing 0.40 to 0.48 mm/yr of sea level equivalent, based on the same assumption as for seepage from reservoirs. Urbanisation leads to reduced infiltration and increased surface runoff, which Gornitz *et al.* (1997) estimate may contribute 0.35 to 0.41 mm/yr of sea level rise. We consider these two terms to be upper bounds because, as with infiltration from reservoirs, a new steady state will be achieved after a period of years, with no further change in storage.

Estimates of the water contributed by deforestation are 0.1 mm/yr (Gornitz *et al.*, 1997) and 0.14 mm/yr (Sahagian, 2000) of sea level rise. Water released by oxidation of fossil fuels, vegetation and wetlands loss is negligible (Gornitz *et al.*, 1997).

Gornitz *et al.* (1997) estimate the total contribution to the 1990 rate of sea level change as -1.2 to -0.5 mm/yr. Integrating their estimates over 1910 to 1990 gives between -32 and -11 mm of sea level rise. In contrast, the estimate of Vörösmarty and Sahagian (2000) for the rate of sea level change from terrestrial storage is 0.06 mm/yr, equivalent to 5.4 mm over 80 years. The estimate of Sahagian *et al.* (1994), quoted by Warrick *et al.* (1996), was 12 mm during the 20th century. These discrepancies emphasise again the unsatisfactory knowledge of these contributions to sea level change.

Table 11.8 shows the ranges we have adopted for the various terms, based on the foregoing discussion. We integrate these terms over 1910 to 1990. (We use the time profiles of Gornitz *et al.* (1997) except that the infiltration from reservoirs is based on the approach of Sahagian (2000), and the rate of

Table 11.8: *Estimates of terrestrial storage terms. The values given are those of Gornitz et al. (1997) and Sahagian (2000). The estimates used in this report are the maximum and minimum values from these two studies. The average rates over the period 1910 to 1990 are obtained by integrating the decadal averages using the time history of contributions estimated by Gornitz et al. (1997).*

	Rate of sea level rise for 1990 (mm/yr)						Average rate 1910 to 1990 (mm/yr)	
	Gornitz *et al.* (1997)		Sahagian (2000)		This assessment			
	min	max	min	max	min	max	min	max
Groundwater mining	+0.1	+0.4	+0.17		+0.1	+1.0	0.0	0.5
Lakes	0.0		+0.2		0.0	+0.2	0.0	0.1
Impoundment in reservoirs	−0.38	−0.30	−0.70	−0.35	−0.7	−0.3	−0.4	−0.2
Infiltration from reservoirs	−0.75	−0.61	−0.84	−0.42	−0.8	−0.4	−0.5	−0.2
Evapotranspiration	−0.15	−0.14	0.0		−0.1	0.0	−0.1	0.0
Infiltration from irrigation	−0.48	−0.40	0.0		−0.5	0.0	−0.2	0.0
Runoff from urbanisation	+0.35	+0.41	0.0		0.0	+0.4	0.0	0.1
Deforestation	+0.1		+0.14		+0.1	0.14	0.1	0.1
Total					−1.9	+1.0	−1.1	0.4

withdrawal from lakes is assumed constant over the last five decades.) This gives a range of −83 to +30 mm of sea level equivalent, or −1.1 to +0.4 mm/yr averaged over the period. However note that the rate of each of the terms increases during the 20th century.

This discussion suggests three important conclusions: (i) the effect of changes in terrestrial water storage on sea level may be considerable; (ii) the net effect on sea level could be of either sign, and (iii) the rate has increased over the last few decades (in the assessment of Gornitz *et al.* (1997) from near zero at the start of the century to −0.8 mm/yr in 1990).

Estimates of ice volume in northern hemisphere permafrost range from 1.1 to 3.7×10^{13} m³ (Zhang *et al.*, 1999), equivalent to 0.03 to 0.10 m of global-average sea level. It occupies 25% of land area in the northern hemisphere. The major effects of global warming in presently unglaciated cold regions will be changes in the area of permafrost and a thickening of the active layer (the layer of seasonally thawed ground above permafrost). Both of these factors result in conversion of ground ice to liquid water, and hence in principle could contribute to the sea level change. Anisimov and Nelson (1997) estimated that a 10 to 20% reduction of area could occur by 2050 under a moderate climate-change scenario. In the absence of information about the vertical distribution of the ice, we make the assumption that the volume change is proportional to the area change. By 2100, the upper limit for the conversion of permafrost to soil water is thus about 50% of the total, or 50 mm sea level equivalent.

A thickening active layer will result in additional water storage capacity in the soil and thawing of ground ice will not necessarily make water available for runoff. What water is released could be mainly captured in ponds, thermokarst lakes, and marshes, rather than running off. Since the soil moisture in permafrost regions in the warm period is already very high, evaporation would not necessarily increase. We know of no quantitative estimates for these storage terms. We assume that the fraction which runs off lies within 0 and 50% of the available water. Hence we estimate the contribution of

permafrost to sea level 1990 to 2100 as 0 to 25 mm (0 to 0.23 mm/yr). For the 20th century, during which the temperature change has been about five times less than assumed by Anisimov and Nelson for the next hundred years, our estimate is 0 to 5 mm (0 to 0.05 mm/yr).

11.2.6 Tectonic Land Movements

We define tectonic land movement as that part of the vertical displacement of the crust that is of non-glacio-hydro-isostatic origin. It includes rapid displacements associated with earthquake events and also slow movements within (e.g., mantle convection) and on (e.g., sediment transport) the Earth. Large parts of the earth are subject to active tectonics which continue to shape the planet's surface. Where the tectonics occur in coastal areas, one of its consequences is the changing relationship between the land and sea surfaces as shorelines retreat or advance in response to the vertical land movements. Examples include the Huon Peninsula of Papua New Guinea (Chappell *et al.*, 1996b), parts of the Mediterranean (e.g. Pirazzoli *et al.*, 1994; Antonioli and Oliverio, 1996), Japan (Ota *et al.*, 1992) and New Zealand (Ota *et al.*, 1995). The Huon Peninsula provides a particularly good example (Figure 11.6) with 125,000 year old coral terraces at up to 400 m above present sea level. The intermediate terraces illustrated in Figure 11.6 formed at times when the tectonic uplift rates and sea level rise were about equal. Detailed analyses of these reef sequences have indicated that long-term average uplift rates vary between about 2 and 4 mm/yr, but that large episodic (and unpredictable) displacements of 1 m or more occur at repeat times of about 1,000 years (Chappell *et al.*, 1996b). Comparable average rates and episodic displacements have been inferred from Greek shoreline evidence (Stiros *et al.*, 1994). With major tectonic activity occurring at the plate boundaries, which in many instances are also continental or island margins, many of the world's tide gauge records are likely to contain both tectonic and eustatic signals. One value of the geological data is that it permits evaluations to be made of tectonic stability of the tide gauge locality.

Figure 11.6: The raised 125,000 year old coral terraces of the Huon Peninsula of Papua New Guinea up to 400 m above present sea level (Chappell *et al.*, 1996b).

Over very long time-scales (greater than 10^6 years), mantle dynamic processes lead to changes in the shape and volume of the ocean basins, while deposition of sediment reduces basin volume. These affect sea level but at very low rates (less than 0.01 mm/yr and 0.05 mm/yr, respectively; e.g., Open University, 1989; Harrison, 1990).

Coastal subsidence in river delta regions can be an important contributing factor to sea level change, with a typical magnitude of 10 mm/yr, although the phenomenon will usually be of a local character. Regions of documented subsidence include the Mediterranean deltas (Stanley, 1997), the Mississippi delta (Day *et al.*, 1993) and the Asian deltas. In the South China Sea, for example, the LGM shoreline is reported to occur at a depth of about 165 m below present level (Wang *et al.*, 1990), suggesting that some 40 m of subsidence may have occurred in 20,000 years at an average rate of about 2 mm/yr. Changes in relative sea level also arise through accretion and erosion along the coast; again, such effects may be locally significant.

11.2.7 Atmospheric Pressure

Through the inverse barometer effect, a local increase in surface air pressure over the ocean produces a depression in the sea surface of 1 cm per hPa (1 hPa = 1 mbar). Since water is practically incompressible, this cannot lead to a global-average sea level rise, but a long-term trend in surface air pressure patterns could influence observed local sea level trends. This has been investigated using two data sets: (i) monthly mean values of surface air pressure on a $10° \times 5°$ grid for the period 1873 to 1995 for the Northern Hemisphere north of 15°N obtained from the University of East Anglia Climatic Research Unit, and (ii) monthly mean values on a global $5° \times 5°$ grid for the period 1871 to 1994 obtained from the UK Met Office (see Basnett and Parker, 1997, for a discussion of the various data sets). The two data sets present similar spatial pattens of trends for their geographical and temporal overlaps. Both yield small trends of the order 0.02 hPa/yr; values of –0.03 hPa/yr occur in limited

regions of the high Arctic and equatorial Pacific. As found by Woodworth (1987), trends are only of the order of 0.01 hPa/yr in northern Europe, where most of the longest historical tide gauges are located. We conclude that long-term sea level trends could have been modified to the extent of ± 0.2 mm/yr, considerably less than the average eustatic rate of rise. Over a shorter period larger trends can be found. For example, Schönwiese *et al.* (1994) and Schönwiese and Rapp (1997) report changes in surface pressure for the period 1960 to 1990 that could have modified sea level trends in the Mediterranean and around Scandinavia by –0.05 and +0.04 mm/yr respectively.

11.3 Past Sea Level Changes

11.3.1 Global Average Sea Level over the Last 6,000 Years

The geological evidence for the past 10,000 to 20,000 years indicates that major temporal and spatial variation occurs in relative sea level change (e.g., Pirazzoli, 1991) on time-scales of the order of a few thousand years (Figure 11.5). The change observed at locations near the former centres of glaciation is primarily the result of the glacio-isostatic effect, whereas the change observed at tectonically stable localities far from the former ice sheets approximate the global average sea level change (for geologically recent times this is primarily eustatic change relating to changes in land-based ice volume). Glacio-hydro-isostatic effects (the Earth's response to the past changes in ice and water loads) remain important and result in a spatial variability in sea level over the past 6,000 years for localities far from the former ice margins. Analysis of data from such sites indicate that the ocean volume may have increased to add 2.5 to 3.5 m to global average sea level over the past 6,000 years (e.g., Fleming *et al.*, 1998), with a decreasing contribution in the last few thousand years. If this occurred uniformly over the past 6,000 years it would raise sea level by 0.4 to 0.6 mm/year. However, a few high resolution sea level records from the French Mediterranean coast indicate that much of this increase occurred between about 6,000 and 3,000 years ago and that the rate over the past 3,000 years was only about 0.1 to 0.2 mm/yr (Lambeck and Bard, 2000). These inferences do not constrain the source of the added water but likely sources are the Antarctic and Greenland ice sheets with possible contributions from glaciers and thermal expansion.

In these analyses of Late Holocene observations, the relative sea level change is attributed to both a contribution from any change in ocean volume and a contribution from the glacio-hydro-isostatic effect, where the former is a function of time only and the latter is a function of both time and position. It is possible to use the record of sea level changes to estimate parameters for a model of isostatic rebound. In doing this, the spatial variability of sea level change determines the mantle rheology, whereas the time dependence determines any correction that may be required to the assumed history of volume change. Solutions from different geographic regions may lead to variations in the rheology due to lateral variations in mantle temperature, for example, but the eustatic term should be the same, within observational and model uncertainties, in each case (Nakada and Lambeck, 1988). If it is assumed that no eustatic change has occurred in the past 6,000 years or so, but in fact eustatic change actually has occurred, the solution for Earth-model parameters will require a somewhat stiffer mantle than a solution in which eustatic change is included. The two solutions may, however, be equally satisfactory for interpolating between observations. For example, both approaches lead to mid-Holocene highstands at island and continental margin sites far from the former ice sheets of amplitudes 1 to 3 m. The occurrence of such sea level maxima places a upper limit on the magnitude of glacial melt in recent millennia (e.g., Peltier, 2000), but it would be inconsistent to combine estimates of ongoing glacial melt with results of calculations of isostatic rebound in which the rheological parameters have been inferred assuming there is no ongoing melt.

The geological indicators of past sea level are usually not sufficiently precise to enable fluctuations of sub-metre amplitude to be observed. In some circumstances high quality records do exist. These are from tectonically stable areas where the tidal range is small and has remained little changed through time, where no barriers or other shoreline features formed to change the local conditions, and where there are biological indicators that bear a precise and consistent relationship to sea level. Such areas include the micro-atoll coral formations of Queensland, Australia (Chappell, 1982; Woodroffe and McLean, 1990); the coralline algae and gastropod vermetid data of the Mediterranean (Laborel *et al.*, 1994; Morhange *et al.*, 1996), and the fresh-to-marine transitions in the Baltic Sea (Eronen *et al.*, 1995; Hyvarinen, 1999). These results all indicate that for the past 3,000 to 5,000 years oscillations in global sea level on time-scales of 100 to 1,000 years are unlikely to have exceeded 0.3 to 0.5 m. Archaeological evidence for this interval places similar constraints on sea level oscillations (Flemming and Webb, 1986). Some detailed local studies have indicated that fluctuations of the order of 1 m can occur (e.g., Van de Plassche *et al.*, 1998) but no globally consistent pattern has yet emerged, suggesting that these may be local rather than global variations.

Estimates of current ice sheet mass balance (Section 11.2.3.1) have improved since the SAR. However, these results indicate only that the ice sheets are not far from balance. Earth rotational constraints (Section 11.2.4.2) and ice sheet altimetry (Section 11.2.3.2) offer the prospect of resolving the ice sheet mass balance in the future, but at present the most accurate estimates of the long-term imbalance (period of several hundred years) follows from the comparison of the geological sea level data with the ice sheet modelling results (Section 11.2.3.3). The above geological estimates of the recent sea level rates may include a component from thermal expansion and glacier mass changes which, from the long-term temperature record in Chapter 2 (Section 2.3.2), could contribute to a sea level lowering by as much as 0.1 mm/yr. These results

Table 11.9: *Recent estimates of sea level rise from tide gauges. The standard error for these estimates is also given along with the method used to correct for vertical land movement (VLM).*

	Region	**VLM** [a]	**Rate ± s.e.** [b] **(mm/yr)**
Gornitz and Lebedeff (1987)	Global	Geological	1.2 ± 0.3
Peltier and Tushingham (1989, 1991)	Global	ICE-3G/M1	2.4 ± 0.9 [c]
Trupin and Wahr (1990)	Global	ICE-3G/M1	1.7 ± 0.13
Nakiboglu and Lambeck (1991)	Global	Spatial decomposition	1.2 ± 0.4
Douglas (1991)	Global	ICE-3G/M1	1.8 ± 0.1
Shennan and Woodworth (1992)	NW Europe	Geological	1.0 ± 0.15
Gornitz (1995) [d]	N America E Coast	Geological	1.5 ± 0.7 [c]
Mitrovica and Davis (1995), Davis and Mitrovica (1996)	Global far field (far from former ice sheets)	PGR Model	1.4 ± 0.4 [c]
Davis and Mitrovica (1996)	N America E Coast	PGR Model	1.5 ± 0.3 [c]
Peltier (1996)	N America E Coast	ICE-4G/M2	1.9 ± 0.6 [c]
Peltier and Jiang (1997)	N America E Coast	Geological	2.0 ± 0.6 [c]
Peltier and Jiang (1997)	Global	ICE-4G/M2	1.8 ± 0.6 [c]
Douglas (1997) [d]	Global	ICE-3G/M1	1.8 ± 0.1
Lambeck *et al.* (1998)	Fennoscandia	PGR Model	1.1 ± 0.2
Woodworth *et al.* (1999)	British Isles	Geological	1.0

[a] This column shows the method used to correct for vertical land motion. ICE-3G/M1 is the Post Glacial Rebound (PGR) model of Tushingham and Peltier (1991). ICE-4G/M2 is a more recent PGR model based on the deglaciation history of Peltier (1994) and the mantle viscosity model of Peltier and Jiang (1996). Nakiboglu and Lambeck (1991) performed a spherical harmonic decomposition of the tide-gauge trends and took the the zero-degree term as the global-average rate. They indicated that a PGR signal would make little contribution to this term. The use of geological data is discussed in the text.

[b] The uncertainty is the standard error of the estimate of the global average rate.

[c] This uncertainty is the standard deviation of the rates at individual sites.

[d] See references in these papers for estimates of sea level rise for various other regions.

suggest that the combined long-term ice sheet imbalance lies within the range 0.1 to 0.3 mm/yr. Results from ice sheet models for the last 500 years indicate an ongoing adjustment to the glacial-interglacial transition of Greenland and Antarctica together of 0.0 to 0.5 mm/yr. These ranges are consistent. We therefore take the ongoing contribution of the ice sheets to sea level rise in the 20th and 21st centuries in response to earlier climate change as 0.0 to 0.5 mm/yr. This is additional to the effect of 20th century and future climate change.

11.3.2 Mean Sea Level Changes over the Past 100 to 200 Years

11.3.2.1 Mean sea level trends

The primary source of information on secular trends in global sea level during the past century is the tide gauge data set of the Permanent Service for Mean Sea Level (PSMSL) (Spencer and Woodworth, 1993). The tide gauge measurement is of the level of the sea surface relative to that of the land upon which the gauge is located and contains information on both the displacement of the land and on changes in ocean volume (eustatic changes). The land displacement may be of two types: that caused by active tectonics and that caused by glacial rebound. Corrections for these effects are required if the change in ocean volume is to be extracted from the tide gauge record. Both corrections are imperfectly known and are based on sea level observations themselves, usually from long geological records. Different strategies have been developed for dealing with these corrections but differences remain that are not inconsequential (see Table 11.9).

The sea level records contain significant interannual and decadal variability and long records are required in order to estimate reliable secular rates that will be representative of the last century. In addition, sea level change is spatially variable because of land movements and of changes in the ocean circulation. Therefore, a good geographic distribution of observations is required. Neither requirement is satisfied with the current tide gauge network and different strategies have been developed to take these differences into consideration. Warrick *et al.* (1996), Douglas (1995) and Smith *et al.* (2000) give recent reviews of the subject, including discussions of the Northern Hemisphere geographical bias in the historical data set.

In the absence of independent measurements of vertical land movements by advanced geodetic techniques (Section 11.6.1), corrections for movements are based on either geological data or geophysical modelling. The former method uses geological evidence from locations adjacent to the gauges to estimate the long-term relative sea level change which is assumed to be caused primarily by land movements, from whatever cause. This is subtracted from the gauge records to estimate the eustatic change for the past century. However, this procedure may underestimate the real current eustatic change because the observed geological data may themselves contain a long-term component of eustatic sea level rise (Section 11.3.1). The latter method, glacial rebound modelling, is also constrained by geological observations to estimate earth response functions or ice load parameters, which may therefore themselves contain a component of long-term eustatic sea level

change unless this component is specifically solved for (Section 11.3.1).

A further underestimate of the rate of sea level rise from the geological approach, compared to that from glacial rebound models, will pertain in forebulge areas, and especially the North American east coast, where the linear extrapolation of geological data could result in an underestimate of the corrected rate of sea level change for the past century typically by 0.3 mm/yr because the glacial rebound signal is diminishing with time (Peltier, 2000). However, in areas remote from the former ice sheets this bias will be considerably smaller.

Also, in adding recent mass into the oceans, most studies have assumed that it is distributed uniformly and have neglected the Earth's elastic and gravitational response to the changed water loading (analogous to glacio-hydro-isostatic effect). This will have the effect of reducing the observed rise at continental margin sites from ongoing mass contributions by as much as 30% (cf. Nakiboglu and Lambeck, 1991).

Table 11.9 summarises estimates of the corrected sea level trends for the past century. Estimates cover a wide range as a result of different assumptions and methods for calculating the rate of vertical land movement, of different selections of gauge records for analysis, and of different requirements for minimum record length.

There have been several more studies since the SAR of trends observed in particular regions. Woodworth *et al.* (1999) provided a partial update to Shennan and Woodworth (1992), suggesting that sea level change in the North Sea region has been about 1 mm/yr during the past century. Lambeck *et al.* (1998) combined coastal tide gauge data from Fennoscandinavia together with lake level records and postglacial rebound models to estimate an average regional rise for the past century of 1.1 ± 0.2 mm/yr. Studies of the North American east coast have been particularly concerned with the spatial dependence of trends associated with the Laurentian forebulge. Peltier (1996) concluded a current rate of order 1.9 ± 0.6 mm/yr, larger than the 1.5 mm/yr obtained by Gornitz (1995), who used the geological data approach, and Mitrovica and Davis (1995), who employed Post Glacial Rebound (PGR) modelling. Note that the observations of thermal expansion (Section 11.2.1.1) indicate a higher rate of sea level rise over recent decades in the sub-tropical gyres of the North Atlantic (i.e., off the North American east coast) than the higher latitude sub-polar gyre. Thus the differences between three lower European values compared with the higher North American values may reflect a real regional difference (with spatial variations in regional sea level change being perhaps several tenths of a millimetre per year – see also Section 11.5.2). In China, relative sea level is rising at about 2 mm/yr in the south but less than 0.5 mm/yr in the north (National Bureau of Marine Management, 1992), with an estimated average of the whole coastline of 1.6 mm/yr (Zhen and Wu, 1993) and with attempts to remove the spatially dependent component of vertical land movement yielding an average of 2.0 mm/yr (Shi, 1996). The two longest records from Australia (both in excess of 80 years in length and not included in Douglas, 1997) are from Sydney and Fremantle, on opposite sides of the continent. They show observed rates of relative sea level rise of 0.86 ± 0.12 mm/yr

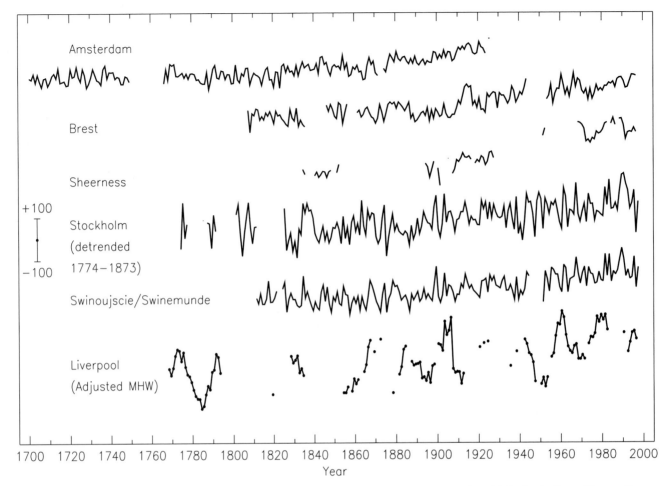

Figure 11.7: Time-series of relative sea level for the past 300 years from Northern Europe: Amsterdam, Netherlands; Brest, France; Sheerness, UK; Stockholm, Sweden (detrended over the period 1774 to 1873 to remove to first order the contribution of postglacial rebound); Swinoujscie, Poland (formerly Swinemunde, Germany); and Liverpool, UK. Data for the latter are of "Adjusted Mean High Water" rather than Mean Sea Level and include a nodal (18.6 year) term. The scale bar indicates ±100 mm. (Adapted from Woodworth, 1999a.)

and 1.38 ± 0.18 mm/yr over the periods 1915 to 1998 and 1897 to 1998 (Mitchell *et al.*, 2000), corresponding to approximately 1.26 mm/yr and 1.73 mm/yr after glacial rebound correction using the Peltier ICE-4G/M2 model, or 1.07 mm/yr and 1.55 mm/yr using the corrections of Lambeck and Nakada (1990).

There have been only two analyses of global sea level change based on the PSMSL data set published since the SAR. Douglas (1997) provided an update to Douglas (1991) and applied the PGR model of Tushingham and Peltier (1991) to a selected set of twenty-four long tide gauge records, grouped into nine geographical areas, with minimum record length 60 years and average length 83 years. However, the only Southern Hemisphere sites included in this solution were from Argentina and New Zealand. The overall global average of 1.8 ± 0.1 mm/yr agreed with the 1991 analysis, with considerable consistency between area-average trends. The standard error of the global rate was derived from the standard deviation of regional trends, assuming that temporal and spatial variability is uncorrelated between regions. Peltier and Jiang (1997) used essentially the same set of stations as Douglas and a new model for postglacial rebound.

From Table 11.9 one can see that there are six global estimates determined with the use of PGR corrections derived from global models of isostatic adjustment, spanning a range from 1.4 mm/yr (Mitrovica and Davis, 1995; Davis and Mitrovica, 1996) to 2.4 mm/yr (Peltier and Tushingham, 1989, 1991). We consider that these five are consistent within the systematic uncertainty of the PGR models, which may have a range of uncertainty of 0.5 mm/yr depending on earth structure parametrization employed (Mitrovica and Davis, 1995). The average rate of the five estimates is 1.8 mm/yr. There are two other global analyses, of Gornitz and Lebedeff (1987) and Nakiboglu and Lambeck (1991), which yield estimates of 1.2 mm/yr, lower than the first group. Because of the issues raised above with regard to the geological data method for land movement correction, the value of Gornitz and Lebedeff may be underestimated by up to a few tenths of a millimetre per year, although such considerations do not affect the method of Nakiboglu and Lambeck. The differences between the former five and latter two analyses reflect the analysis methods, in particular the differences in corrections for land movements and in selections of tide gauges used, including the effect of any spatial variation in thermal expansion. However, all the discrepancies which could arise as a consequence of different analysis methods remain to be more thoroughly investigated. On the basis of the published literature, we therefore cannot rule out an average rate of sea level rise of as little as 1.0 mm/yr during the 20th century. For the upper bound, we adopt a limit of 2.0 mm/yr, which includes all recent global estimates with some allowance for systematic uncertainty. As with other ranges (see Box 11.1), we do not imply that the central value is the best estimate.

11.3.2.2 Long-term mean sea level accelerations

Comparison of the rate of sea level rise over the last 100 years (1.0 to 2.0 mm/yr) with the geological rate over the last two millennia (0.1 to 0.2 mm/yr; Section 11.3.1) implies a comparatively recent acceleration in the rate of sea level rise. The few very long tide gauge records are especially important in the search for "accelerations" in sea level rise. Using four of the longest (about two centuries) records from north-west Europe (Amsterdam, Brest, Sheerness, Stockholm), Woodworth (1990) found long-term accelerations of 0.4 to 0.9 mm/yr/century (Figure 11.7). Woodworth (1999a) found an acceleration of order 0.3 mm/yr/century in the very long quasi-mean sea level (or 'Adjusted Mean High Water') record from Liverpool. From these records, one can infer that the onset of the acceleration occurred during the 19th century, a suggestion consistent with separate analysis of the long Stockholm record (Ekman, 1988, 1999; see also Mörner, 1973). It is also consistent with some geological evidence from north-west Europe (e.g., Allen and Rae, 1988). In North America, the longest records are from Key West, Florida, which commenced in 1846 and which suggest an acceleration of order 0.4 mm/year/century (Maul and Martin, 1993), and from New York which commenced in 1856 and which has a similar acceleration. Coastal evolution evidence from parts of eastern North America suggest an increased rate of rise between one and two centuries before the 20th century (Kearney and Stevenson, 1991; Varekamp *et al.*, 1992; Kearney, 1996; Van de Plassche *et al.*, 1998; Varekamp and Thomas, 1998; Shaw and Ceman, 2000).

There is no evidence for any acceleration of sea level rise in data from the 20th century data alone (Woodworth, 1990; Gornitz and Solow, 1991; Douglas, 1992). Mediterranean records show decelerations, and even decreases in sea level in the latter part of the 20th century, which may be caused by increases in the density of Mediterranean Deep Water and air pressure changes connected to the North Atlantic Oscillation (NAO) (Tsimplis and Baker, 2000), suggesting the Mediterranean might not be the best area for monitoring secular trends. Models of ocean thermal expansion indicate an acceleration through the 20th century but when the model is subsampled at the locations of the tide gauges no significant acceleration can be detected because of the greater level of variability (Gregory *et al.*, 2001). Thus the absence of an acceleration in the observations is not necessarily inconsistent with the model results.

11.3.2.3 Mean sea level change from satellite altimeter observations

In contrast to the sparse network of coastal and mid-ocean island tide gauges, measurements of sea level from space by satellite

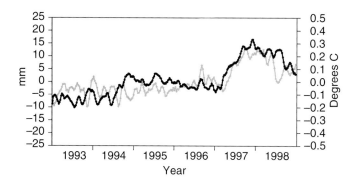

Figure 11.8: Global mean sea level variations (light line) computed from the TOPEX/POSEIDON satellite altimeter data compared with the global averaged sea surface temperature variations (dark line) for 1993 to 1998 (Cazenave *et al.*, 1998, updated). The seasonal components have been removed from both time-series.

radar altimetry provide near global and homogenous coverage of the world's oceans, thereby allowing the determination of regional sea level change. Satellite altimeters also measure sea level with respect to the centre of the earth. While the results must be corrected for isostatic adjustment (Peltier, 1998), satellite altimetry avoids other vertical land movements (tectonic motions, subsidence) that affect local determinations of sea level trends measured by tide gauges. However, achieving the required sub-millimetre accuracy is demanding and requires satellite orbit information, geophysical and environmental corrections and altimeter range measurements of the highest accuracy. It also requires continuous satellite operations over many years and careful control of biases.

To date, the TOPEX/POSEIDON satellite-altimeter mission, with its (near) global coverage from 66°N to 66°S (almost all of the ice-free oceans) from late 1992 to the present, has proved to be of most value to direct estimates of sea level change. The current accuracy of TOPEX/POSEIDON data allows global average sea level to be estimated to a precision of several millimetres every 10 days, with the absolute accuracy limited by systematic errors.

Careful comparison of TOPEX/POSEIDON data with tide gauge data reveals a difference in the rate of change of local sea level of −2.3 ± 1.2 mm/yr (Mitchum, 1998) or −2 ± 1.5 mm/yr (Cazenave *et al.*, 1999). This discrepancy is caused by a combination of instrumental drift, especially in the TOPEX Microwave Radiometer (TMR) (Haines and Bar-Sever, 1998), and vertical land motions which have not been allowed for in the tide gauge data. The most recent estimates of global average sea level rise from the six years of TOPEX/POSEIDON data (using corrections from tide gauge comparisons) are 2.1 ± 1.2 mm/yr (Nerem *et al.*, 1997), 1.4 ± 0.2 mm/yr (Cazenave *et al.*, 1998; Figure 11.8), 3.1 ± 1.3 mm/yr (Nerem, 1999) and 2.5 ± 1.3 mm/yr (Nerem, 1999), of which the last assumes that all instrumental drift can be attributed to the TMR. When Cazenave *et al.* allow for the TMR drift, they compute a sea level rise of 2.6 mm/yr. Their uncertainty of

± 0.2 mm/yr does not include allowance for uncertainty in instrumental drift, but only reflects the variations in measured global sea level. Such variations correlate with global average sea surface temperature, perhaps indicating the importance of steric effects through ocean heat storage. Cazenave *et al.* (1998) and Nerem *et al.* (1999) argue that ENSO events cause a rise and a subsequent fall in global averaged sea level of about 20 mm (Figure 11.8). These findings indicate that the major 1997/98 El Niño-Southern Oscillation (ENSO) event could bias the above estimates of sea level rise and also indicate the difficulty of separating long-term trends from climatic variability.

After upgrading many of the geophysical corrections on the original European Remote Sensing (ERS) data stream, Cazenave *et al.* (1998) find little evidence of sea level rise over the period April 1992 to May 1996. However, over the time span of overlap between the ERS-1 and TOPEX/POSEIDON data, similar rates of sea level change (about 0.5 mm/yr) are calculated. For the period April 1992 to April 1995, Anzenhofer and Gruber (1998) find a sea level rise of 2.2 ± 1.6 mm/yr.

In summary, analysis of TOPEX/POSEIDON data suggest a rate of sea level rise during the 1990s greater than the mean rate of rise for much of the 20th century. It is not yet clear whether this is the result of a recent acceleration, of systematic differences between the two measurement techniques, or of the shortness of the record (6 years).

11.3.3 Changes in Extreme Sea Levels: Storm Surges and Waves

Lack of adequate data sets means we can not ascertain whether there have been changes in the magnitude and/or frequency of storm surges (aperiodic changes associated with major meteorological disturbances resulting in sea level changes of up to several metres and lasting a few hours to days) in many regions of the world. Zhang *et al.* (1997, 2000) performed a comprehensive analysis of hourly tide gauge data from the east coast of North America, and concluded that there had been no discernible secular trend in storm activity or severity during the past century. European analyses include that of Woodworth (1999b), who found no significant increase in extreme high water level distributions from Liverpool from 1768 to 1993 to those from later epochs, other than what can be explained in terms of changes in local tidal amplitudes, mean sea level and vertical land movement. Vassie (reported in Pugh and Maul, 1999) and Bijl *et al.* (1999) concluded that there was no discernible trend over the last century in the statistics of non-tidal sea level variability around the UK and the eastern North Sea (Denmark, Germany and the Netherlands), above the considerable natural sea level variability on decadal time-scales. In South America, D'Onofrio *et al.* (1999) observed a trend of extreme levels at Buenos Aires of 2.8 mm/yr over 1905 to 1993. On the basis of available statistics, the South American result is consistent with the local mean sea level trend.

Variations in surge statistics can also be inferred from analysis of meteorological data. Kass *et al.* (1996) and the WASA Group (1998) showed that there are no significant overall trends in windiness and cyclonic activity over the North Atlantic

and north-west Europe during the past century, although major variations on decadal times-scales exist. An increase in storminess in the north-east Atlantic in the last few decades (Schmith *et al.*, 1998) and a recent trend towards higher storm surge levels on the German and Danish coasts (Langenberg *et al.*, 1999) is consistent with natural variability evident over the last 150 years. Pirazzoli (2000) detected evidence for a slight decrease in the main factors contributing to surge development on the French Atlantic coast in the last 50 years. Correlation between the frequency of Atlantic storms and ENSO was demonstrated by Van der Vink *et al.* (1998).

Increases in wave heights of approximately 2 to 3 m over the period 1962 to 1985 off Land's End, south-west England (Carter and Draper, 1988), increases in wave height over a neighbouring area at about 2%/yr since 1950 (Bacon and Carter, 1991, 1993) and wave height variations simulated by the Wave Action Model (WAM) (Günther *et al.*, 1998) are all consistent with decadal variations over most of the north-east Atlantic and North Sea. This variability could be related to the NAO (Chapter 2, Section 2.6.5).

11.4 Can 20th Century Sea Level Changes be Explained?

In order to have confidence in our ability to predict future changes in sea level, we need to confirm that the relevant processes (Section 11.2) have been correctly identified and evaluated. We attempt this by seeing how well we can account for the current rate of change (Section 11.3). We note that:

- some processes affecting sea level have long (centuries and longer) time-scales, so that current sea level change is also related to past climate change,

- some relevant processes are not determined solely by climate,

- fairly long records (at least 50 years according to Douglas, 1992) are needed to detect a significant trend in local sea level, because of the influence of natural variability in the climate system, and

- the network of tide gauges with records of this length gives only a limited coverage of the world's continental coastline and almost no coverage of the mid-ocean.

The estimated contributions from the various components of sea level rise during the 20th century (Table 11.10, Figure 11.9) were constructed using the results from Section 11.2. The sum of these contributions for the 20th century ranges from –0.8 mm/yr to 2.2 mm/yr, with a central value of 0.7 mm/yr. The upper bound is close to the observational upper bound (2.0 mm/yr), but the central value is less than the observational lower bound (1.0 mm/yr), and the lower bound is negative i.e. the sum of components is biased low compared to the observational estimates. Nonetheless, the range is narrower than the range given by Warrick *et al.* (1996), as a result of greater constraints on all the contributions, with the exception of the terrestrial storage terms. In particular, the long-term contribution from the

Table 11.10: *Estimated rates of sea level rise components from observations and models (mm/yr) averaged over the period 1910 to 1990. (Note that the model uncertainties may be underestimates because of possible systematic errors in the models.) The 20th century terms for Greenland and Antarctica are derived from ice sheet models because observations cannot distinguish between 20th century and long-term effects. See Section 11.2.3.3.*

	Minimum (mm/yr)	Central value (mm/yr)	Maximum (mm/yr)
Thermal expansion	0.3	0.5	0.7
Glaciers and ice caps	0.2	0.3	0.4
Greenland – 20th century effects	0.0	0.05	0.1
Antarctica – 20th century effects	− 0.2	− 0.1	0.0
Ice sheets – adjustment since LGM	0.0	0.25	0.5
Permafrost	0.00	0.025	0.05
Sediment deposition	0.00	0.025	0.05
Terrestrial storage (not directly from climate change)	− 1.1	− 0.35	0.4
Total	− 0.8	0.7	2.2
Estimated from observations	1.0	1.5	2.0

ice sheets has been narrowed substantially from those given in Warrick *et al.* (1996) by the use of additional constraints (geological data and models of the ice sheets) (Section 11.3.1).

The reason for the remaining discrepancy is not clear. However, the largest uncertainty (by a factor of more than two) is in the terrestrial storage terms. Several of the components of the terrestrial storage term are poorly determined and the quoted limits require several of the contributions simultaneously to lie at the extremes of their ranges. This coincidence is improbable unless the systematic errors affecting the estimates are correlated. Furthermore, while coupled models have improved considerably in recent years, and there is general agreement between the observed and modelled thermal expansion contribution, the models' ability to quantitatively simulate decadal changes in ocean temperatures and thus thermal expansion has not been adequately tested. Given the poor global coverage of high quality tide gauge records and the uncertainty in the corrections for land motions, the observationally based rate of sea level rise this century should also be questioned.

In the models, at least a third of 20th century anthropogenic eustatic sea level rise is caused by thermal expansion, which has a geographically non-uniform signal in sea level change. AOGCMs do not agree in detail about the patterns of geographical variation (see Section 11.5.2). They all give a geographical spread of 20th century trends at individual grid points which is characterised by a standard deviation of 0.2 to 0.5 mm/yr (Gregory *et al.*, 2001). This spread is a result of a combination of spatial non-uniformity of trends and the uncertainty in local trend estimates arising from temporal variability. As yet no published study has revealed a stable pattern of observed non-uniform sea level change. Such a pattern would provide a critical test of models. If there is significant non-uniformity, a trend from a single location would be an inaccurate estimate of the global average. For example, Douglas (1997) averaged nine regions and found a standard deviation of about 0.3 mm/yr (quoted by Douglas as a standard error), similar to the range expected from AOGCMs.

A common perception is that the rate of sea level rise should have accelerated during the latter half of the 20th century. The tide gauge data for the 20th century show no significant acceleration (e.g., Douglas, 1992). We have obtained estimates based on AOGCMs for the terms directly related to anthropogenic climate change in the 20th century, i.e., thermal expansion (Section 11.2.1.2), ice sheets (Section 11.2.3.3), glaciers and ice caps (Section 11.5.1.1) (Figure 11.10a). The estimated rate of sea level rise from anthropogenic climate change ranges from 0.3 to 0.8 mm/yr (Figure 11.10b). These terms do show an acceleration through the 20th century (Figure 11.10a,b). If the terrestrial storage terms have a negative sum (Section 11.2.5), they may offset some of the acceleration in recent decades. The total computed rise (Figure 11.10c) indicates an acceleration of only 0.2 mm/yr/century, with a range from −1.1 to +0.7 mm/yr/century,

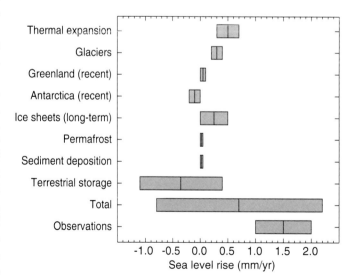

Figure 11.9: Ranges of uncertainty for the average rate of sea level rise from 1910 to 1990 and the estimated contributions from different processes.

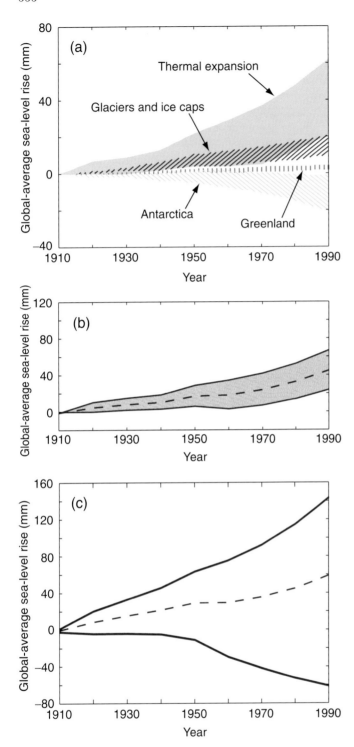

Figure 11.10: Estimated sea level rise from 1910 to 1990. (a) The thermal expansion, glacier and ice cap, Greenland and Antarctic contributions resulting from climate change in the 20th century calculated from a range of AOGCMs. Note that uncertainties in land ice calculations have not been included. (b) The mid-range and upper and lower bounds for the computed response of sea level to climate change (the sum of the terms in (a) plus the contribution from permafrost). These curves represent our estimate of the impact of anthropogenic climate change on sea level during the 20th century. (c) The mid-range and upper and lower bounds for the computed sea level change (the sum of all terms in (a) with the addition of changes in permafrost, the effect of sediment deposition, the long-term adjustment of the ice-sheets to past climate change and the terrestrial storage terms).

consistent with observational finding of no acceleration in sea level rise during the 20th century (Section 11.3.2.2). The sum of terms not related to recent climate change is −1.1 to +0.9 mm/yr (i.e., excluding thermal expansion, glaciers and ice caps, and changes in the ice sheets due to 20th century climate change). This range is less than the observational lower bound of sea level rise. Hence it is very likely that these terms alone are an insufficient explanation, implying that 20th century climate change has made a contribution to 20th century sea level rise.

Recent studies (see Sections 2.3.3, 2.3.4) suggest that the 19th century was unusually cold on the global average, and that an increase in solar output may have had a moderate influence on warming in the early 20th century (Section 12.4.3.3). This warming might have produced some thermal expansion and could have been responsible for the onset of glacier recession in the early 20th century (e.g., Dowdeswell *et al.*, 1997), thus providing a possible explanation of an acceleration in sea level rise commencing before major industrialisation.

11.5 Future Sea Level Changes

11.5.1 Global Average Sea Level Change 1990 to 2100

Warrick *et al.* (1996) made projections of thermal expansion and of loss of mass from glaciers and ice-sheets for the 21st century for the IS92 scenarios using two alternative simple climate models. Since the SAR, time-dependent experiments have been run with several AOGCMs (Chapter 9.1.2, Table 9.1) following the IS92a scenario (Leggett *et al.*, 1992) for future concentrations of greenhouse gases, including the direct effect of sulphate aerosols. In Section 11.5.1.1, we use the AOGCM IS92a results to derive estimates of thermal expansion and land ice melt, employing methods from the literature as described in Section 11.2, and we add contributions from thawing of permafrost, sediment deposition, and the continuing adjustment of the ice sheets to climate changes since the LGM. The choice of scenario is not the principal consideration; the main point is that the AOGCMs all follow the same scenario, so the range of results reflects the systematic uncertainty inherent in the modelling of sea level changes. The use of IS92a also facilitates comparison with the result of Warrick *et al.* (1996).

To quantify the uncertainty resulting from the uncertainty in future emissions, and to obtain results consistent with the global-average temperature change projections of Section 9.3.3, in Section 11.5.1.2 we derive projections for thermal expansion and land ice melt for the scenarios of the IPCC Special Report on Emissions Scenarios (SRES) (Nakićenović *et al.*, 2000) (see also Box 9.1 in Chapter 9, Section 9.1). The results are given as sea level change relative to 1990 in order to facilitate comparison with previous IPCC reports, which used 1990 as their base date.

11.5.1.1 Projections for a single scenario based on a range of AOGCMs
Thermal expansion
Over the hundred years 1990 to 2090, the AOGCM experiments for IS92a including sulphate aerosols (GS experiments – see Chapter 9, Table 9.1) show global-average sea level rise from

Table 11.11: Sea level rise from thermal expansion from AOGCM experiments following the IS92a scenario for the 21st century, including the direct effect of sulphate aerosols. See Chapter 8, Table 8.1 and Chapter 9, Table 9.1 for further details of models and experiments. See Table 11.2 for thermal expansion from AOGCM experiments for the 20th century.

Experiment	ΔT_g (°C)	Sea level rise (m)	
	1990 to 2090[a]	1990 to 2040	1990 to 2090[a]
CGCM1 GS	3.7	0.12	0.37
CGCM2 GS	3.6	0.11	0.33
CSIRO Mk2 GS	2.7	0.11	0.28
ECHAM4/OPYC3 GS[b]	—	0.11	—
GFDL_R15_a GS[c]	—	0.13	—
GFDL_R15_b GS	3.2	0.12	0.29
GFDL_R30_c GS	2.8	0.12	0.31
HadCM2 GS	2.5	0.07	0.20
HadCM3 GSIO	2.8	0.07	0.20
MRI2 GS	1.5	0.04	0.09
DOE PCM GS	1.9	0.07	0.17

[a] An end date of 2090, rather than 2100, is chosen to match the last available date in some of the experiments.

[b] This experiment ends at 2050.

[c] This experiment ends at 2065.

ΔT_g Global average surface air temperature change.

thermal expansion in the range 0.09 to 0.37 m (Figure 11.1, Table 11.11). There is an acceleration through the 21st century; expansion for 2040 to 2090 is greater than for 1990 to 2040 by a factor of 1.4 to 2.1. Since the models experience the same forcing, the differences in the thermal expansion derive from differences in the physical behaviour of the models. Broadly speaking, the range of results reflects the systematic uncertainty of modelling in three factors: the size of the surface warming, the effectiveness of heat uptake by the ocean for a given warming (Gregory and Mitchell, 1997) and the expansion resulting from a given heat uptake (Russell *et al.*, 2000). The separation of the first two factors parallels the distinction made in Section 9.3.4.2 between the effects of climate feedback and heat uptake on the rate of climate change. Since models differ in regard to the second and third factors, experiments with a similar temperature change do not necessarily have a similar thermal expansion, as the results demonstrate.

Glaciers and ice caps

To make projections for future loss of mass from glaciers and ice caps, we have applied the methods of Gregory and Oerlemans (1998) and Van de Wal and Wild (2001) (Sections 11.2.2.2, 11.2.2.4) to the seasonally and geographically dependent temperature changes given by a range of AOGCM IS92a experiments including sulphate aerosols (Table 11.12). We adjust the results to be consistent with the assumption that the climate of 1865 to 1895 was 0.15 K warmer than the steady state for glaciers, following Zuo and Oerlemans (1997) (see also Section 11.4). Precipitation changes are not included, as they are not expected to have a strong influence on the global average (Section 11.2.2.3).

Table 11.12: Calculations of glacier melt from AOGCM experiments following the IS92a scenario for the 21st century, including the direct effect of sulphate aerosols. See Tables 8.1 and 9.1 for further details of models and experiments.

Experiment	B (mm/yr) 1990	ΔT_g (°C) 1990 to 2090	Sea level rise (m) 1990 to 2090		$\partial B/\partial T_g$ (mm/yr/°C)
			Constant area	Changing area	
CGCM1 GS	0.43	3.7	0.15	0.11	0.65
CSIRO Mk2 GS	0.52	2.7	0.15	0.11	0.73
CSM 1.3 GS	0.45	1.8	0.10	0.07	0.61
ECHAM4/OPYC3 GS[a]	0.56	–	–	–	0.64
GFDL_R15_a GS[b]	0.42	–	–	–	0.58
GFDL_R15_b GS	0.44	3.2	0.13	0.09	0.54
GFDL_R30_c GS	0.33	2.8	0.12	0.08	0.53
HadCM2 GS[c]	0.44	2.5	0.11	0.08	0.61
HadCM3 GSIO	0.31	2.8	0.11	0.08	0.62
MRI2 GS	0.22	1.5	0.06	0.05	0.60
DOE PCM GS	0.42	1.9	0.09	0.06	0.59

[a] This experiment ends at 2050.

[b] This experiment ends at 2065.

[c] Similar results for constant area were obtained for an ensemble of HadCM2 GS experiments by Gregory and Oerlemans (1998).

B Global glacier mass balance for constant glacier area, expressed as sea level equivalent.

$\partial B/\partial T_g$ Sensitivity of global glacier mass balance for constant glacier area, expressed as sea level equivalent, to global average surface air temperature change.

ΔT_g Global average surface air temperature change.

Table 11.13: *Calculations of ice sheet mass changes using temperature and precipitation changes from AOGCM experiments following the IS92a scenario for the 21st century, including the direct effect of sulphate aerosols, to derive boundary conditions for an ice sheet model. See Tables 8.1 and 9.1 for further details of models and experiments.*

Experiment	Greenland					Antarctica			
	Sea level rise (m) 1990 to 2090	Sensitivity (mm/yr/°C)		$\Delta T/\Delta T_g$	1/P dP/dT (%/°C)	Sea level rise (m) 1990 to 2090	Sensitivity (mm/yr/°C)		$\Delta T/\Delta T_g$
		dB/dT_g	dB/dT				dB/dT_g	dB/dT	
CGCM1 GS	0.03	0.13	0.10	1.3	2.7	− 0.02	− 0.12	− 0.11	1.1
CSIRO Mk2 GS	0.02	0.16	0.08	2.0	5.9	− 0.07	− 0.37	− 0.33	1.1
CSM 1.3 GS	0.02	0.15	0.05	3.1	7.8	− 0.04	− 0.31	− 0.27	1.1
ECHAM4/OPYC3 GS [a]	–	0.03	0.03	1.2	6.5	–	− 0.48	− 0.32	1.5
GFDL_R15_a GS [b]	–	0.12	0.06	1.9	4.1	–	− 0.18	− 0.22	0.8
HadCM2 GS	0.02	0.10	0.07	1.4	4.0	− 0.04	− 0.21	− 0.17	1.2
HadCM3 GSIO	0.02	0.09	0.06	1.4	4.5	− 0.07	− 0.35	− 0.28	1.3
MRI2 GS	0.01	0.08	0.05	1.6	4.4	− 0.01	− 0.14	− 0.12	1.2
DOE PCM GS	0.02	0.14	0.06	2.2	5.6	− 0.07	− 0.48	− 0.30	1.6

[a] This experiment ends at 2050.
[b] This experiment ends at 2065.
dB/dT_g Ice-sheet mass balance sensitivity to global-average surface air temperature change, expressed as sea level equivalent.
dB/dT Ice-sheet mass balance sensitivity to surface air temperature change averaged over the ice sheet, expressed as sea level equivalent.
$\Delta T/\Delta T_g$ Slope of the regression of surface air temperature change averaged over the ice sheet against global-average change.
1/P dP/dT Fractional change in ice-sheet average precipitation as a function of temperature change.

For constant glacier area, from the AOGCM IS92a experiments including sulphate aerosol, predicted sea level rise from glacier melt over the hundred years 1990 to 2090 lies in the range 0.06 to 0.15 m. The variation is due to three factors. First, the global average temperature change varies between models. A larger temperature rise tends to give more melting, but they are not linearly related, since the total melt depends on the time-integrated temperature change. Second, the global mass balance sensitivity to temperature change varies among AOGCMs because of their different seasonal and regional distribution of temperature change. Third, the glaciers are already adjusting to climate change during the 20th century, and any such imbalance will persist during the 21st century, in addition to the further imbalance due to future climate change. The global average temperature change and glacier mass balance sensitivity may not be independent factors, since both are affected by regional climate feedbacks. The sensitivity and the present imbalance are related factors, because a larger sensitivity implies a greater present imbalance.

With glacier area contracting as the volume reduces, the estimated sea level rise contribution is in the range 0.05 to 0.11 m, about 25% less than if constant area is assumed, similar to the findings of Oerlemans *et al.* (1998) and Van de Wal and Wild (2001). The time-dependence of glacier area means the results can no longer be represented by a global glacier mass balance sensitivity.

Glaciers and ice caps on the margins of the Greenland and Antarctic ice sheets are omitted from these calculations, because they are included in the ice sheet projections below. These ice masses have a large area (Table 11.3), but experience little ablation on account of being in very cold climates. Van de Wal

and Wild (2001) find that the Greenland marginal glaciers contribute an additional 6% to glacier melt in a scenario of CO_2 doubling over 70 years. Similar calculations using the AOGCM IS92a results give a maximum contribution of 14 mm for 1990 to 2100. For the Antarctic marginal glaciers, the ambient temperatures are too low for there to be any significant surface runoff. Increasing temperatures will increase the runoff and enlarge the area experiencing ablation, but their contribution is very likely to remain small. For instance, Drewry and Morris (1992) calculate a contribution of 0.012 mm/yr/°C to the global glacier mass balance sensitivity from the glacier area of 20,000 km² which currently experiences some melting on the Antarctic Peninsula.

Lack of information concerning glacier areas and precipitation over glaciers, together with uncertainty over the projected changes in glacier area, lead to uncertainty in the results. This is assessed as ± 40%, matching the uncertainty of the observed mass balance estimate of Dyurgerov and Meier (1997b).

Greenland and Antarctic ice sheets

To make projections of Greenland and Antarctic ice sheet mass changes consistent with the IS92a AOGCM experiments including sulphate aerosols, we have integrated the ice-sheet model of Huybrechts and De Wolde (1999) using boundary conditions of temperature and precipitation derived by perturbing present day climatology according to the geographically and seasonally dependent pattern changes predicted by the T106 ECHAM4 model (Wild and Ohmura, 2000) for a doubling of CO_2. To generate time-dependent boundary conditions, these patterns were scaled with the area average changes over the ice sheets as a function of time for each AOGCM experiment using a method similar to that described by Huybrechts *et al.* (1999).

Table 11.14: *Sea level rise 1990 to 2100 due to climate change derived from AOGCM experiments following the IS92a scenario, including the direct effect of sulphate aerosols. See Tables 8.1 and 9.1 for further details of models and experiments. Results were extrapolated to 2100 for experiments ending at earlier dates. The uncertainties shown in the land ice terms are those discussed in this section. For comparison the projection of Warrick et al. (1996) (in the SAR) is also included. Note that the minimum of the sum of the components is not identical with the sum of the minima because the smallest values of the components do not all come from the same AOGCM, and because for each model the land ice uncertainties have been combined in quadrature; similarly for the maxima, which also include non-zero contributions from smaller terms.*

Experiment		Sea level rise (m) 1990 to 2100								
		Expansion		Glaciers		Greenland		Antarctica[a]		Sum[b]	
		min	max	min	max	min	max	min	max	min	max
CGCM1 GS		0.43		0.03	0.23	0.00	0.07	−0.07	0.02	0.45	0.77
CSIRO Mk2 GS		0.33		0.02	0.22	−0.01	0.08	−0.12	−0.04	0.29	0.60
ECHAM4/OPYC3 GS		0.30		0.02	0.18	−0.02	0.03	−0.17	−0.06	0.19	0.48
GFDL_R15_a GS		0.38		0.02	0.19	−0.01	0.09	−0.09	−0.01	0.37	0.67
HadCM2 GS		0.23		0.02	0.17	−0.01	0.05	−0.09	0.00	0.21	0.48
HadCM3 GSIO		0.24		0.02	0.18	0.00	0.05	−0.13	−0.03	0.18	0.46
MRI2 GS		0.11		0.01	0.11	0.00	0.03	−0.04	0.00	0.11	0.31
DOE PCM GS		0.19		0.01	0.13	−0.01	0.06	−0.13	−0.04	0.12	0.37
Range		0.11	0.43	0.01	0.23	−0.02	0.09	−0.17	0.02	0.11	0.77
Central value		0.27		0.12		+0.04		−0.08		0.44	
SAR 7.5.2.4	Best estimate	0.28		0.16		+0.06		−0.01		0.49	
	Range									0.20	0.86

[a] Note that this range does not allow for uncertainty relating to ice-dynamical changes in the West Antarctic ice sheet. See Section 11.5.4.3 for a full discussion.

[b] Including contributions from permafrost, sedimentation, and adjustment of ice sheets to past climate change.

The marginal glaciers and ice caps on Greenland and Antarctica were included in the ice sheet area. The calculated contributions from these small ice masses have some uncertainty resulting from the limited spatial resolution of the ice sheet model.

For 1990 to 2090 in the AOGCM GS experiments, Greenland contributes 0.01 to 0.03 m and Antarctica −0.07 to −0.01 m to global average sea level (Table 11.13). Note that these sea level contributions result solely from recent and projected future climate change; they do not include the response to past climate change (discussed in Sections 11.2.3.3 and 11.3.1).

Mass balance sensitivities are derived by regressing rate of change of mass against global or local temperature change (note that they include the effect of precipitation changes) (Table 11.13). The Greenland local sensitivities are smaller than some of the values reported previously from other methods (Section 11.2.3.4 and Table 11.7) and by Warrick *et al.* (1996) because of the larger precipitation increases and the seasonality of temperature changes (less increase in summer) predicted by AOGCMs, and the smaller temperature rise in the ablation zone (as compared to the ice-sheet average) projected by the T106 ECHAM4 time slice results. The Antarctic sensitivities are less negative than those in Table 11.7 because the AOGCMs predict smaller precipitation increases.

The use of a range of AOGCMs represents the uncertainty in modelling changing circulation patterns, which lead to both changes in temperature and precipitation, as noted by Kapsner *et al.* (1995) and Cuffey and Clow (1997) from the results from Greenland ice cores. The range of AOGCM thermodynamic and circulation responses gives a range of 4 to 8%/°C for Greenland precipitation increases, generally less than indicated by ice-cores

for the glacial-interglacial transition, but more than for Holocene variability (Section 11.2.3.4). If precipitation did not increase at all with greenhouse warming, Greenland local sensitivities would be larger, by 0.05 to 0.1 mm/yr/°C (see also Table 11.7). Given that all AOGCMs agree on an increase, but differ on the strength of the relationship, we include an uncertainty of ±0.02 mm/yr/°C in Table 11.13 on the Greenland local sensitivities, being the product of the standard deviation of precipitation increase (1.5%/°C) and the current Greenland accumulation (1.4 mm/yr sea level equivalent, Table 11.5).

Estimates of Greenland runoff (Table 11.5) have a standard error of about ± 10%. This reflects uncertainty in the degree-day method (Braithwaite, 1995) and refreezing parametrization (Janssens and Huybrechts, 2000) used to calculate Greenland ablation. Given that a typical size of the sensitivity of ablation to temperature change is 0.3 mm/yr/°C (Table 11.7), we adopt an additional uncertainty of ± 0.03 mm/yr/°C for the local Greenland sensitivities in Table 11.13. We include a separate uncertainty of the same size to reflect the possible sensitivity to use of different high-resolution geographical patterns of temperature and precipitation change (the T106 ECHAM4 pattern was the only one available). As an estimate of the uncertainty related to changes in iceberg discharge and area-elevation distribution, we ascribe an uncertainty of ± 10% to the net mass change, on the basis of the magnitude of the dynamic response for Greenland described in Section 11.2.3.4.

For Antarctica, uncertainty introduced by ablation model parameters need not be considered because melting remains very small for the temperature scenarios considered for the 21st century. Ice-dynamical uncertainties are much more difficult to

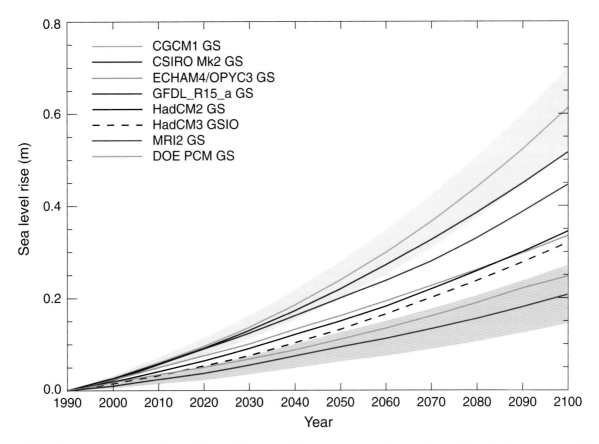

Figure 11.11: Global average sea level rise 1990 to 2100 for the IS92a scenario, including the direct effect of sulphate aerosols. Thermal expansion and land ice changes were calculated from AOGCM experiments, and contributions from changes in permafrost, the effect of sediment deposition and the long-term adjustment of the ice sheets to past climate change were added. For the models that project the largest (CGCM1) and the smallest (MRI2) sea level change, the shaded region shows the bounds of uncertainty associated with land ice changes, permafrost changes and sediment deposition. Uncertainties are not shown for the other models, but can be found in Table 11.14. The outermost limits of the shaded regions indicate our range of uncertainty in projecting sea level change for the IS92a scenario.

determine. See Section 11.5.4.3 for a detailed discussion. We include an uncertainty of 0.08 mm/yr/°C on the local sensitivity, which is its inter-model standard deviation, to reflect the spread of precipitation changes as a function of temperature.

Total

To obtain predictions of global average sea level rise for 1990-2100 for the IS92a scenario with sulphate aerosols, we calculate the sum of the contributions from thermal expansion, glaciers and ice sheets for each AOGCM, and add the 0 to 0.5 mm/yr from the continuing evolution of the ice sheets in response to past climate change (Section 11.2.3.1) and smaller terms from thawing of permafrost (Section 11.2.5) and the effect of sedimentation (Section 11.2.6). The range of our results is 0.11 to 0.77 m (Table 11.14, Figure 11.11), which should be compared with the range of 0.20 to 0.86 m given by Warrick *et al.* (1996) (SAR Section 7.5.2.4, Figure 7.7) for the same scenario. The AOGCMs have a range of effective climate sensitivities from 1.4 to 4.2°C (Table 9.1), similar to the range of 1.5 to 4.5°C used by Warrick *et al.* The AOGCM thermal expansion values are generally larger than those of Warrick *et al.* (SAR Section 7.5.2.4, Figure 7.8), but the other terms are mostly smaller (i.e., more negative in the case of Antarctica).

Warrick *et al.* included a positive term to allow for the possible instability of the WAIS. We have omitted this because it is now widely agreed that major loss of grounded ice and accelerated sea level rise are very unlikely during the 21st century (Section 11.5.4.3). The size of our range is an indication of the systematic uncertainty in modelling radiative forcing, climate and sea level changes. Uncertainties in modelling the carbon cycle and atmospheric chemistry are not covered by this range, because the AOGCMs are all given similar atmospheric concentrations as input.

11.5.1.2 Projections for SRES scenarios

Few AOGCM experiments have been done with any of the SRES emissions scenarios. Therefore to establish the range of sea level rise resulting from the choice of different SRES scenarios, we use results for thermal expansion and global-average temperature change from a simple climate model based on that of Raper *et al.* (1996) and calibrated individually for seven AOGCMs (CSIRO Mk2, CSM 1.3, ECHAM4/OPYC3, GFDL_R15_a, HadCM2, HadCM3, DOE PCM). The calibration is discussed in Chapter 9, Section 9.3.3 and the Appendix to Chapter 9. The AOGCMs used have a range of effective climate sensitivity of 1.7 to 4.2°C (Table 9.1). We

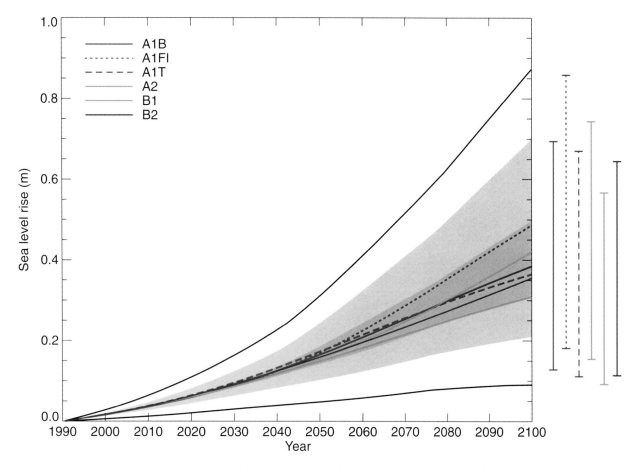

Figure 11.12: Global average sea level rise 1990 to 2100 for the SRES scenarios. Thermal expansion and land ice changes were calculated using a simple climate model calibrated separately for each of seven AOGCMs, and contributions from changes in permafrost, the effect of sediment deposition and the long-term adjustment of the ice sheets to past climate change were added. Each of the six lines appearing in the key is the average of AOGCMs for one of the six illustrative scenarios. The region in dark shading shows the range of the average of AOGCMs for all 35 SRES scenarios. The region in light shading shows the range of all AOGCMs for all 35 scenarios. The region delimited by the outermost lines shows the range of all AOGCMs and scenarios including uncertainty in land-ice changes, permafrost changes and sediment deposition. Note that this range does not allow for uncertainty relating to ice-dynamical changes in the West Antarctic ice sheet. See 11.5.4.3 for a full discussion. The bars show the range in 2100 of all AOGCMs for the six illustrative scenarios.

calculate land-ice changes using the global average temperature change from the simple model and the global average mass balance sensitivities estimated from the AOGCM IS92a experiments in Section 11.5.1.1 (Tables 11.12 and 11.13). We add contributions from the continuing evolution of the ice sheets in response to past climate change, thawing of permafrost, and the effect of sedimentation (the same as in Section 11.5.1.1). The methods used to make the sea level projections are documented in detail in the Appendix to this chapter.

For the complete range of AOGCMs and SRES scenarios and including uncertainties in land-ice changes, permafrost changes and sediment deposition, global average sea level is projected to rise by 0.09 to 0.88 m over 1990 to 2100, with a central value of 0.48 m (Figure 11.12). The central value gives an average rate of 2.2 to 4.4 times the rate over the 20th century.

The corresponding range reported by Warrick *et al.* (1996) (representing scenario uncertainty by using all the IS92 scenarios with time-dependent sulphate aerosol) was 0.13 to 0.94 m, obtained using a simple model with climate sensitivities

of 1.5 to 4.5°C. Their upper bound is larger than ours. Ice sheet mass balance sensitivities derived from AOGCMs (see Section 11.5.1.1) are smaller (less positive or more negative) than those used by Warrick *et al.*, while the method we have employed for calculating glacier mass loss (Sections 11.2.2 and 11.5.1.1) gives a smaller sea level contribution for similar scenarios than the heuristic model of Wigley and Raper (1995) employed by Warrick *et al.*

In addition, Warrick *et al.* included an allowance for ice-dynamical changes in the WAIS. The range we have given does not include such changes. The contribution of the WAIS is potentially important on the longer term, but it is now widely agreed that major loss of grounded ice from the WAIS and consequent accelerated sea-level rise are very unlikely during the 21st century. Allowing for the possible effects of processes not adequately represented in present models, two risk assessment studies involving panels of experts concluded that there was a 5% chance that by 2100 the WAIS could make a substantial contribution to sea level rise, of 0.16 m (Titus and

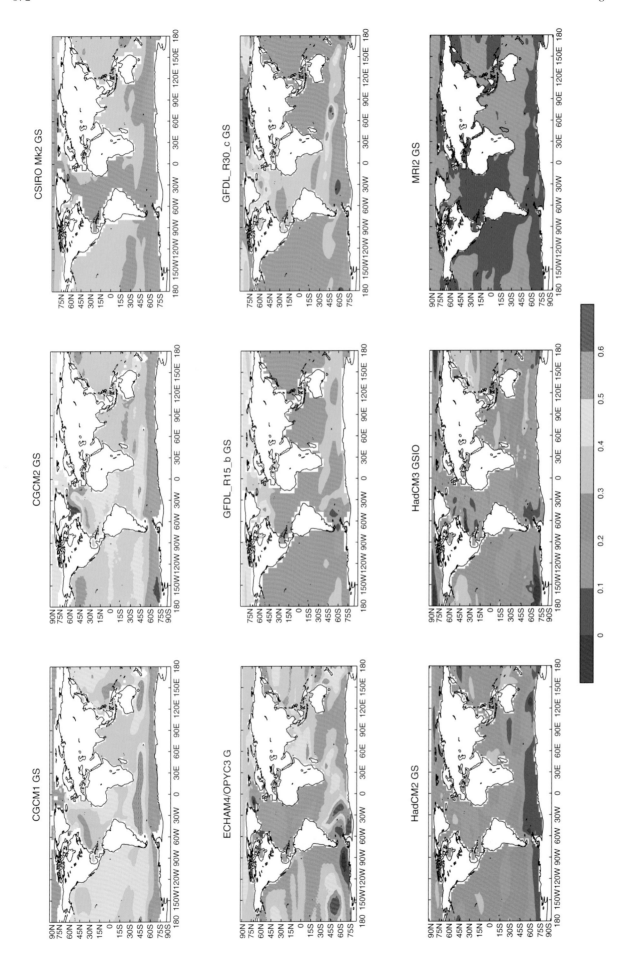

Figure 11.13: Sea level change in metres over the 21st century resulting from thermal expansion and ocean circulation changes calculated from AOGCM experiments following the IS92a scenario and including the direct effect of sulphate aerosol (except that ECHAM4/OPYC3 G is shown instead of GS, because GS ends in 2050). Each field is the difference in sea level change between the last decade of the experiment and the decade 100 years earlier. See Tables 8.1 and 9.1 for further details of models and experiments.

Table 11.15: *Spatial standard deviation, local minimum and local maximum of sea level change during the 21st century due to ocean processes, from AOGCM experiments following the IS92a scenario for greenhouse gases, including the direct effect of sulphate aerosols. See Tables 8.1 and 9.1 for further details of models and experiments. Sea level change was calculated as the difference between the final decade of each experiment and the decade 100 years earlier. Sea level changes due to land ice and water storage are not included.*

Experiment	Std deviation (m)	Divided by global average		
		Std deviation	Minimum	Maximum
CGCM1 GS	0.07	0.19	0.3	1.6
CGCM2 GS	0.08	0.23	0.2	2.2
CSIRO Mk2 GS	0.05	0.15	0.5	1.3
ECHAM4/OPYC3 G[a]	0.10	0.34	−1.2	2.3
GFDL_R15_b GS	0.05	0.18	0.3	1.8
GFDL_R30_c GS	0.07	0.25	0.2	2.5
HadCM2 GS	0.06	0.29	−0.1	1.7
HadCM3 GSIO	0.07	0.32	−0.5	2.2
MRI2 GS	0.04	0.35	−1.2	2.2

[a] This experiment does not include sulphate aerosols. The ECHAM4/OPYC3 experiment including sulphates extends only to 2050.

Narayanan, 1996) or 0.5 m (Vaughan and Spouge, 2001). These studies also noted a 5% chance of WAIS contributing a sea level fall of 0.18 m or 0.4 m respectively. (See Section 11.5.4.3 for a full discussion.)

The range we have given also does not take account of uncertainty in modelling of radiative forcing, the carbon cycle, atmospheric chemistry, or storage of water in the terrestrial environment. The recent publications by Gornitz *et al.* (1997) and Sahagian (2000) indicate that this last term could be significant (Section 11.2.5). Future changes in terrestrial storage depend on societal decisions on the use of ground water, the building of reservoirs and other factors. We are not currently in a position to make projections incorporating future changes in these factors, although we note that the assumptions behind the construction of the SRES scenarios imply increasing water consumption, which may entail both more ground water extraction and more reservoir capacity. Continued anthropogenic water storage on land at its current rate could change the projected sea level rise 1990 to 2100 by between −0.21 and +0.11 m. We emphasise that estimates of the relevant factors are highly uncertain (see Sections 11.2.5 and 11.4).

The evolution of sea level rise for the average of the seven AOGCMs for each of the six illustrative SRES scenarios is shown in Figure 11.12, and the shading shows the range for all 35 SRES scenarios. It is apparent that the variation due to the choice of scenario alone is relatively small over the next few decades. The range spanned by the SRES scenarios by 2040 is only 0.02 m or less. By 2100, the scenario range has increased to 0.18 m, about 50% of the central value. All the AOGCMs have a similar range at 2100 expressed as a fraction of their central value. Of the six illustrative scenarios, A1FI gives the largest sea level rise and B1 the smallest.

The average-AOGCM range for all 35 scenarios (dark shading in Figure 11.12) covers about one third of the all-AOGCM range (light shading). That is, for sea level rise 1990 to 2100, the uncertainty in climate sensitivity and heat uptake, represented by the spread of AOGCMs, is more important than the uncertainty from choice of emissions scenario. This is

different for three reasons from the case of global average temperature change (Section 9.3.2.1), where the scenario and modelling uncertainties are comparable. First, the compensation between climate sensitivity and heat uptake does not apply to thermal expansion. Second, models with large climate sensitivity and temperature change consequently have a large land-ice melt contribution to sea level. Third, both thermal expansion and land-ice melt depend on past climate change, being approximately proportional to the time-integral of temperature change; the SRES scenarios differ by less in respect of the time-integral of temperature change over the interval 1990 to 2100 than they do in respect of the temperature change at 2100.

11.5.2 Regional Sea Level Change

The geographical distribution of sea level change caused by ocean processes can be calculated from AOGCM results (see Gregory *et al.*, 2001, for methods). This was not possible with the simple climate models used by Warrick *et al.* (1996). Results for sea level change from ocean processes in the 21st century are shown in Figure 11.13 for AOGCM experiments used in Section 11.5.1.1. Some regions show a sea level rise substantially more than the global average (in many cases of more than twice the average), and others a sea level fall (Table 11.15) (note that these figures do not include sea level rise due to land ice changes). The standard deviation of sea level change is 15 to 35% of the global average sea level rise from thermal expansion.

In each of these experiments, a non-uniform pattern of sea level rise emerges above the background of temporal variability in the latter part of the 21st century. However, the patterns given by the different models (Figure 11.13) are not similar in detail. The largest correlations are between models which are similar in formulation: 0.65 between CGCM1 and CGCM2, 0.63 between GFDL_R15_b and GFDL_R30_c. The largest correlations between models from different centres are 0.60 between CSIRO Mk2 and HadCM2, 0.58 between CGCM2 and GFDL_R30_c. The majority of correlations are less than 0.4, indicating no significant similarity (Gregory *et al.*, 2001). The disagreement between models is partly a reflection of the differences in ocean model formulation that are also responsible for the spread in the global average heat uptake and thermal expansion (Sections 11.2.1.2, 11.5.1.1). In addition, the models predict different changes in surface windstress, with consequences for changes in ocean circulation and subduction. More detailed analysis is needed to elucidate the reasons for the differences in patterns. The lack of similarity means that our confidence in predictions of local sea level changes is low. However, we can identify a few common features on the regional and basin scale (see also Gregory *et al.*, 2001).

Seven of the nine models in Table 11.14 (also Bryan, 1996; Russell *et al.*, 2000) exhibit a maximum sea level rise in the Arctic Ocean. A possible reason for this is a freshening of the Arctic due to increased river runoff or precipitation over the ocean (Bryan, 1996; Miller and Russell, 2000). The fall in salinity leads to a reduction of density, which requires a compensating sea level rise in order to maintain the pressure gradient at depth.

Seven of the models (also Gregory, 1993; Bryan, 1996) show a minimum of sea level rise in the circumpolar Southern Ocean south of 60°S. This occurs despite the fact that the Southern Ocean is a region of pronounced heat uptake (e.g., Murphy and Mitchell, 1995; Hirst *et al.*, 1996). The low thermal expansion coefficient at the cold temperatures of the high southern latitudes, changes in wind patterns and transport of the heat taken up to lower latitudes are all possible explanations.

Bryan (1996) draws attention to a dipole pattern in sea level change in the north-west Atlantic; there is a reduced rise south of the Gulf Stream extension and enhanced rise to the north, which corresponds to a weakening of the sea surface gradient across the current. This would be consistent with a weakening of the upper

branch of the North Atlantic circulation, which is a response to greenhouse warming observed in many AOGCM experiments (e.g., Manabe and Stouffer, 1993, 1994; Hirst, 1998). This can be seen in all the models considered here except ECHAM4/OPYC3, in which the Atlantic thermohaline circulation does not weaken (Latif and Roeckner, 2000).

Local land movements, both isostatic and tectonic (Sections 11.2.4.1, 11.2.6), will continue in the 21st century at rates which are unaffected by climate change, and should be added to the regional variation described in this section. On account of the increased eustatic rate of rise in the 21st century (Section 11.5.1) it can be expected that by 2100 many regions currently experiencing relative sea level fall owing to isostatic rebound will instead have a rising relative sea level.

All the global models discussed here have a spatial resolution of 1 to 3°. To obtain information about mean sea level changes at higher resolution is currently not practical; a regional model such as that of Kauker (1998) would be needed.

11.5.3 Implications for Coastal Regions

To determine the practical consequences of projections of global sea level rise in particular coastal regions, it is necessary to understand the various components leading to relative sea level changes. These components include local land movements, global eustatic sea level rise, any spatial variability from that global average, local meteorological changes and changes in the frequency of extreme events.

11.5.3.1 Mean sea level

Titus and Narayanan (1996) propose a simple method for computing local projections of mean sea level rise given historical observations at a site and projections of global average sea level (such as Figure 11.12). To allow for local land movements and our current inability to model sea level change accurately (Section 11.4), they propose linearly extrapolating the historical record and adding to this a globally averaged projection. However, they point out that to avoid double counting, it is necessary to correct the global projection for the corresponding modelled trend of sea level rise during the period of the historical observations. Caution is required in applying this method directly to the projections of this chapter for several reasons. First, current model projections indicate substantial spatial variability in sea level rise. This variability has a standard deviation of up to 0.1 m by 2100; some locations experience a sea level rise of more than twice the global-average thermal expansion, while others may have a fall in sea level (Section 11.5.2; Table 11.15). Second, there are uncertainties in the accuracy of the trend from the historical record and in modelling of past sea level changes (Sections 11.3.2.1 and 11.4). Third, as well as changes in mean sea level there may be changes in the local meteorological regime resulting in modified storm surge statistics (Section 11.5.3.2). If the method of Titus and Narayanan is not applied, it is nonetheless important to recognise that in all models and scenarios the rate of local sea level rise in the 21st century is projected to be greater than in the 20th century at the great majority of coastal locations.

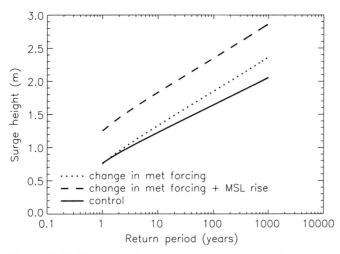

Figure 11.14: Frequency of extreme water level, expressed as return period, from a storm surge model for present day conditions (control) and the projected climate around 2100 for Immingham on the east coast of England, showing changes resulting from mean sea level rise and changes in meteorological forcing. The water level is relative to the sum of present day mean sea level and the tide at the time of the surge. (From Lowe *et al.*, 2001.)

11.5.3.2 Extremes of sea level: storm surges and waves

The probability of flood risk in coastal areas is generally expressed in terms of extreme sea level distributions. Such distributions are usually computed from observed annual maximum sea levels from several decades of tide gauge data, or from numerical models. While such distributions are readily available for many locations, a worldwide set has never been computed to common standards for studies of impacts of global sea level change.

Changes in the highest sea levels at a given locality could result mainly from two effects. First, if mean sea level rises, the present extreme levels will be attained more frequently, all else being equal. This may imply a significant increase in the area threatened with inundation (e.g., Hubbert and McInnes, 1999) and an increased risk within the existing flood plain. The effect can be estimated from a knowledge of the present day frequency of occurrence of extreme levels (e.g., Flather and Khandker, 1993; Lowe *et al.*, 2001; Figure 11.14).

Second, changes in storm surge heights would result from alterations to the occurrence of strong winds and low pressures. At low-latitude locations, such as the Bay of Bengal, northern Australia and the southern USA, tropical cyclones are the primary cause of storm surges. Changes in frequency and intensity of tropical cyclones could result from alterations to sea surface temperature, large-scale atmospheric circulation and the characteristics of ENSO (Pittock *et al.*, 1996) but no consensus has yet emerged (see Box 10.2). In other places, such as southern Australia and north-west Europe, storm surges are associated with mid-latitude low-pressure systems. For instance, Hubbert and McInnes (1999) showed that increasing the wind speeds in historical storm surge events associated with the passage of cold fronts could lead to greater flooding in Port Phillip Bay, Victoria, Australia. Changes in extratropical storms also cannot be predicted with confidence (Section 9.3.6.3).

Several studies have attempted to quantify the consequences of changes in storm climatology for the north-west European continental shelf using regional models of the atmosphere and ocean. Using five-year integrations of the ECHAM T106 model for present and doubled CO_2, Von Storch and Reichardt (1997) and Flather and Smith (1998) did not find any significant changes in extreme events compared with the variability of the control climate (see also WASA Group, 1998). However, Langenberg *et al.* (1999) reported increases of 0.05 to 0.10 m in five-winter-mean high-water levels around all North Sea coasts, judged to be significant compared with observed natural variability. Lowe *et al.* (2001) undertook a similar study using multi-decadal integrations of the Hadley Centre regional climate model for the present climate and the end of the 21st century (Figure 11.14), finding statistically significant changes of up to 0.2 m in five-year extremes in the English Channel. Differences between these various results relate to the length of model integration and to systematic uncertainty in the modelling of both the atmospheric forcing and the ocean response.

Changes in wind forcing could result in changes to wave heights, but with the short integrations available, the WASA Group (Rider *et al.*, 1996) were not able to identify any significant changes for the North Atlantic and North Sea for a doubling of CO_2. Günther *et al.* (1998) noted that changes in future wave climate were similar to patterns of past variation.

11.5.4 Longer Term Changes

Anthropogenic emissions beyond 2100 are very uncertain, and we can only indicate a range of possibilities for sea level change. On the time-scale of centuries, thermal expansion and ice sheet changes are likely to be the most important processes.

11.5.4.1 Thermal expansion

The most important conclusion for thermal expansion is that it would continue to raise sea level for many centuries after stabilisation of greenhouse gas concentrations, so that its eventual contribution would be much larger than at the time of stabilisation.

Table 11.16: *Sea level rise due to thermal expansion in $2 \times CO_2$ and $4 \times CO_2$ experiments. See Chapter 8, Table 8.1 for further details of models.*

	Sea level rise (m) in $2 \times CO_2$ experiment		$\Delta h_2 / \Delta h_1$	Final sea level rise (m)	
	at $2 \times CO_2$ (Δh_1)	500 yr later (Δh_2)		$2 \times CO_2$ experiment	$4 \times CO_2$ experiment
CLIMBER	0.16	0.67	4.2	0.78	1.44
ECHAM3/LSG	0.06	0.57	9.2	1.53[a]	2.56[a]
GFDL_R15_a	0.13	1.10	8.5	1.96	3.46
HadCM2	0.09	0.70	7.8	–	–
BERN2D GM	0.23	1.12	4.9	1.93	3.73
BERN2D HOR	0.22	0.92	4.2	1.28	4.30
UVic GM	0.11	0.44	3.9	0.53	1.24
UVic H	0.13	0.71	5.6	1.19	2.62
UVic HBL	0.10	0.44	4.3	0.65	1.78

[a] Estimated from the ECHAM3/LSG experiments by fitting the time-series with exponential impulse response functions (Voss and Mikolajewicz, 2001).

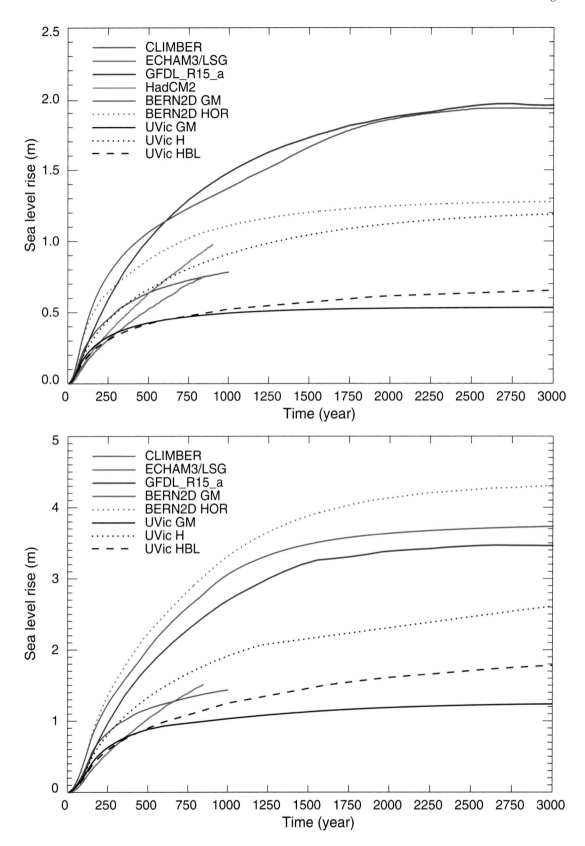

Figure 11.15: Global average sea level rise from thermal expansion in model experiments with CO_2 (a) increasing at 1%/yr for 70 years and then held constant at 2× its initial (preindustrial) concentration; (b) increasing at 1%/yr for 140 years and then held constant at 4× its initial (preindustrial) concentration.

A number of investigations have aimed to quantify this delayed but inevitable consequence of the enhanced greenhouse effect, using a simple scenario in which carbon dioxide concentration increases rapidly (at 1%/yr, not intended as a realistic historical scenario) up to double or four times its initial value (referred to as $2\times CO_2$ and $4\times CO_2$), and thereafter remains constant. ($2\times CO_2$ is about 540 ppm by volume, and $4\times CO_2$ about 1080 ppm.) Long experiments of this kind have been run with three AOGCMs (Chapter 8, Table 8.1): GFDL (Manabe and Stouffer, 1994; Stouffer and Manabe, 1999), ECHAM3/LSG (Voss and Mikolajewicz, 2001) and HadCM2 (Senior and Mitchell, 2000), but owing to the computational requirement, only one of these (GFDL) has been continued until a steady state is reached. Models of intermediate complexity (Chapter 8, Table 8.1) have also been employed: CLIMBER, BERN2D and UVic. These models have a less detailed representation of some important processes, but are less expensive to run for millennia.

Thermal expansion could be greater in one model than another either because the surface warming is larger, or because the warming penetrates more deeply (Figure 11.15, Table 11.16; the suffixes to the BERN2D and UVic model names indicate versions of the models with different parametrizations of heat transport processes). For instance, UVic H and UVic GM show markedly different expansion, although they have similar surface warming (Weaver and Wiebe, 1999). The $4\times CO_2$ experiment with each model generally has around twice the expansion of the $2\times CO_2$ experiment, but the BERN2D HOR $4\times CO_2$ experiment has more than three times, because the Atlantic thermohaline circulation collapses, permitting greater warming and adding about 0.5 m to thermal expansion (Knutti and Stocker, 2000). (See also Section 9.3.4.4.) In all the models reported here, a vertical temperature gradient is maintained, although in some cases weakened. If the whole depth of the ocean warmed to match the surface temperature, thermal expansion would be considerably larger.

The long time-scale (of the order of 1,000 years) on which thermal expansion approaches its eventual level is characteristic of the weak diffusion and slow circulation processes that transport heat to the deep ocean. On account of the time-scale, the thermal expansion in the $2\times CO_2$ experiments after 500 years of constant CO_2 is 4 to 9 times greater than at the time when the concentration stabilises. Even by this time, it may only have reached half of its eventual level, which models suggest may lie within a range of 0.5 to 2.0 m for $2\times CO_2$ and 1 to 4 m for $4\times CO_2$. For the first 1,000 years, the $4\times CO_2$ models give 1 to 3 m.

11.5.4.2 Glaciers and ice caps

Melting of all existing glaciers and ice caps would raise sea level by 0.5 m (Table 11.3). For 1990 to 2100 in IS92a, the projected loss from land-ice outside Greenland and Antarctica is 0.05 to 0.11 m (Table 11.12). Further contraction of glacier area and retreat to high altitude will restrict ablation, so we cannot use the 21st century rates to deduce that there is a time by which all glacier mass will have disappeared. However, the loss of a substantial fraction of the total glacier mass is likely. The viability of any particular glacier or ice cap will depend on whether there remains any part of it, at high altitude, where ablation does not exceed accumulation over the annual cycle. Areas which are currently marginally glaciated are most likely to become ice-free.

11.5.4.3 Greenland and Antarctic ice sheets

Several modelling studies have been conducted for time periods of several centuries to millennia (Van de Wal and Oerlemans, 1997; Warner and Budd, 1998; Huybrechts and De Wolde, 1999; Greve, 2000). A main conclusion is that the ice sheets would continue to react to the imposed climatic change during the next millennium, even if the warming stabilised early in the 22nd century. Whereas Greenland and Antarctica may largely counteract one another for most of the 21st century (Section 11.2.3.4), this situation would no longer hold after that and their combined contribution would be a rise in sea level.

Greenland ice sheet

The Greenland ice sheet is the most vulnerable to climatic warming. As the temperature rises, ablation will increase. For moderate warming, the ice sheet can be retained with reduced extent and modified shape if this results in less ablation and/or a decrease in the rate of ice discharge into the sea, each of which currently account for about half the accumulation (Section 11.2.3). The discharge can be reduced by thinning of the ice sheet near the grounding line. Ablation can be reduced by a change in the area-elevation distribution. However, once ablation has increased enough to equal accumulation, the ice sheet cannot survive, since discharge cannot be less than zero. This situation occurs for an annual-average warming of 2.7°C for the present ice-sheet topography, and for a slightly larger warming for a retreating ice sheet (Huybrechts *et al.*, 1991; see also Oerlemans,

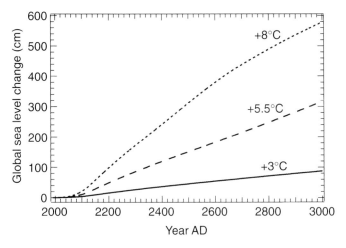

Figure 11.16: Response of the Greenland Ice Sheet to three climatic warming scenarios during the third millennium expressed in equivalent changes of global sea level. The curve labels refer to the mean annual temperature rise over Greenland by 3000 AD as predicted by a two-dimensional climate and ocean model forced by greenhouse gas concentration rises until 2130 AD and kept constant after that. (From Huybrechts and De Wolde, 1999.) Note that projected temperatures over Greenland are generally greater than globally averaged temperatures (by a factor of 1.2 to 3.1 for the range of AOGCMs used in this chapter). See Table 11.13 and Chapter 9, Fig 9.10c.

1991; Van de Wal and Oerlemans, 1994). Models show that under these circumstances the Greenland ice sheet eventually disappears, except for residual glaciers at high altitudes. By using the AOGCM ratios of the Greenland temperature to the global average (Table 11.13) with the results of the calibrated simple model (Section 11.5.1.2 and Chapter 9, Section 9.3.3) we project increases in Greenland temperatures by 2100 of more than 2.7°C for nearly all combinations of SRES scenarios and AOGCMs. The maximum by 2100 is 9°C.

Huybrechts and De Wolde (1999) (Figure 11.16) (see also Letreguilly *et al.*, 1991) find the Greenland ice sheet contributes about 3 m of sea level rise equivalent over a thousand years under their mid-range scenario, in which the Greenland temperature change passes through 4°C in 2100 before stabilising at 5.5°C after 2130. Taking into account the high-latitude amplification of warming, this temperature change is consistent with mid-range stabilisation scenarios (Chapter 9, Section 9.3.3.1 and Figure 9.17(b)). For a warming of 8°C, they calculate a contribution of about 6 m. Their experiments take into account the effect of concomitant increases in precipitation (which reduces sensitivity) but also of the precipitation fraction falling as rain (which strongly enhances sensitivity for the larger temperature increases). Disregarding the effects of accumulation changes and rainfall, Greve (2000) reports that loss of mass would occur at a rate giving a sea level rise of between 1 mm/yr for a year-round temperature perturbation of 3°C to as much as 7 mm/yr for a sustained warming of 12°C, the latter being an extreme scenario in which the ice sheet would be largely eliminated within 1,000 years.

West Antarctic ice sheet

The WAIS contains enough ice to raise sea level by 6 m. It has received particular attention because it has been the most dynamic part of the Antarctic ice sheet in the recent geological past, and because most of it is grounded below sea level – a situation that according to models proposed in the 1970s could lead to flow instabilities and rapid ice discharge into the ocean when the surrounding ice shelves are weakened (Thomas, 1973; Weertman, 1974; Thomas *et al.*, 1979). Geological evidence suggests that WAIS may have been smaller than today at least once during the last million years (Scherer *et al.*, 1998). The potential of WAIS to collapse in response to future climate change is still a subject of debate and controversy.

The discharge of the WAIS is dominated by fast-flowing ice streams, dynamically constrained at four boundaries: the transition zone where grounded ice joins the floating ice shelf (Van der Veen, 1985; Herterich, 1987), the interface of ice with bedrock that is lubricated by sediment and water (Blankenship *et al.*, 1986; Anandakrishnan *et al.*, 1998; Bell *et al.*, 1998), the shear zone where fast-moving ice meets relatively static ice at the transverse margins of ice streams (Echelmeyer *et al.*, 1994; Jacobson and Raymond, 1998), and the ice-stream onset regions where slowly flowing inland ice accelerates into the ice streams. Mechanisms have been proposed for dynamic changes at each of these boundaries.

Early studies emphasised the role of the ice shelf boundary in ice discharge by introducing the concept of a "back-stress"

believed to buttress the grounded ice sheet and prevent it from collapsing. Recent work, however, both modelling and measurement, places greater emphasis on the other ice stream boundaries. Force balance studies on Ice Stream B show no evidence of stresses generated by the Ross ice shelf, and mechanical control emanates almost entirely from the lateral margins (Whillans and Van der Veen, 1997). If confirmed for the other Siple Coast ice streams, this suggests ice stream flow fields to be little influenced by conditions in the ice shelf, similar to the situation elsewhere at the Antarctic margin (Mayer and Huybrechts, 1999).

Nonetheless, there is a considerable body of evidence for ice stream variability, and the above analyses may not apply to dynamic situations involving large thinning or grounding line change. Ice Stream C largely stopped about a century ago (Retzlaff and Bentley, 1993) and Ice Stream B decelerated by 20% within a decade (Stephenson and Bindschadler, 1988). The mechanisms for these oscillations are not well understood and have been ascribed to processes such as basal water diversion to a neighbouring ice stream (Anandakrishnan and Alley, 1997) or thermomechanical interactions between competing catchment areas (Payne, 1998). Despite ice stream variability in the latter model on the millenial time-scale, the overall volume of the WAIS hardly changed, supporting the suggestion that ice streams may act to remove the imbalance of individual drainage basins and to stabilise rather than destabilise WAIS (Hindmarsh, 1993).

The WAIS was much larger during the LGM and has probably lost up to two thirds of its volume since then (Bindschadler, 1998). The largest losses have been from grounding line retreat below the present WAIS ice shelves (Ross and Filchner-Ronne), most likely as a gradual response to rising sea levels subsequent to melting of the Northern Hemisphere ice sheets (Ingolfsson *et al.*, 1998). Local imbalances, both positive and negative, are presently occurring (Shabtaie and Bentley, 1987; Whillans and Bindschadler, 1988; Bindschadler and Vornberger, 1998; Hamilton *et al.*, 1998; Rignot, 1998a,b; Wingham, *et al.*, 1998), but there is no conclusive observational evidence (from monitoring of surface elevation, see Section 11.2.3.2) that WAIS overall is making a significant contribution to global average sea level change (Bentley, 1997, 1998a,b; Bindschadler, 1998; Oppenheimer, 1998; Wingham *et al.* 1998). Conway *et al.* (1999) suggest that grounding line retreat since the LGM may still be ongoing, giving an average rate of recession corresponding to a rate of sea level rise of 0.9 mm/yr (Bindschadler, 1998). If projected into the future, this would imply disappearance of WAIS in 4,000 to 7,000 years (Bindschadler, 1998). However, geological estimates of ocean volume increase over the last 6,000 years place an upper limit of about half this amount on global sea level rise (Section 11.3.1).

Recent spectacular break-ups of the Larsen ice shelves in the Antarctic Peninsula (Vaughan and Doake, 1996; Doake *et al.*, 1998) demonstrate the existence of an abrupt thermal limit on ice shelf viability associated with regional atmospheric warming (Skvarca *et al.*, 1998). However, the WAIS ice shelves are not immediately threatened by this mechanism, which would require a further warming of 10°C before the −5°C mean annual isotherm reached their ice fronts (Vaughan and Doake, 1996). Although atmospheric warming would increase the rate of deformation of the ice, causing the ice shelf to thin, response time-scales are of the order of several hundred years (Rommelaere and MacAyeal, 1997; Huybrechts and de Wolde, 1999).

In view of these considerations, it is now widely agreed that major loss of grounded ice, and accelerated sea level rise, is very unlikely during the 21st century. An interdisciplinary panel of international experts applying the techniques of risk assessment to the future evolution of WAIS concluded that there is a 98% chance that WAIS will not collapse in the next 100 years, defined as a change that contributes at least 10 mm/yr to global sea level change (Vaughan and Spouge, 2001). The probability of a contribution to sea level (exceeding 0.5 m) by the year 2100 was 5%. These results are broadly consistent with an earlier assessment by Titus and Narayanan (1996) based on a US-only panel, who found a 5% chance of a 0.16 m contribution and 1% chance of a 0.3 m contribution to sea level rise from WAIS by 2100. We note that Vaughan and Spouge also report a probability of 5% for WAIS giving a sea level fall exceeding 0.4 m within the same time frame, while Titus and Narayanan give 0.18 m.

Nonetheless, on a longer time-scale, changes in ice dynamics could result in significantly increased outflow of ice into the ice shelves and a grounding line retreat. Large-scale models show both of these phenomena to be sensitive to basal melting below the ice shelves (Warner and Budd, 1998; Huybrechts and De Wolde, 1999). Model studies do not agree on the sensitivity of the basal melting to an oceanic warming: for instance, one shows a quadrupling of the basal melting rate below the Amery ice shelf in East Antarctica for an adjacent sea warming of 1°C (Williams *et al.*, 1998), while another claims that warmer sea temperatures would reduce melting rates below the Ronne-Filchner ice shelves through alteration to sea-ice formation and the thermohaline circulation (Nicholls, 1997). Changes in open ocean circulation may also play a role. Warner and Budd (1998) suggest that even for moderate climatic warmings of a few degrees, a large increase in bottom melting of 5 m/yr becomes the dominant factor in the longer-term response of the Antarctic ice sheet. In their model, this causes the demise of WAIS ice shelves in a few hundred years and would float a large part of the WAIS (and marine portions of East Antarctica) after 1,000 years. Predicted rates of sea level rise are between 1.5 and 3.0 mm/yr depending on whether accumulation rates increase together with the warming. Allowing for runoff in addition to increased accumulation, Huybrechts and De Wolde (1999) find a maximum Antarctic contribution to global sea level rise of 2.5 mm/yr for an extreme scenario involving a warming of 8°C and a bottom melting rate of 10 m/yr. These figures are upper limits based on results currently available from numerical models, which do not resolve ice streams explicitly and which may not adequately predict the effect of ice shelf thinning on grounding line retreat owing to physical uncertainties.

Based on a wide-ranging review, Oppenheimer (1998) argues that WAIS could disintegrate within five to seven centuries following a warming of only a few degrees. Such a collapse implies a rate of sea level rise of 10 mm/yr and an average speed-up of the total outflow by at least a factor of 10 (Bentley, 1997, 1998a,b). However, the majority opinion of a recent expert panel reported by Vaughan and Spouge (2001) is that such outflow

rates are not attainable. It is, therefore, also plausible that WAIS may not make a significant contribution to sea level rise over time-scales less than a millennium. Vaughan and Spouge (2001) attribute a 50% probability to the latter scenario, but retained an equally large probability that the sea level rise will be larger than 2 mm/yr after 1,000 years, emphasising the inadequacy of our current understanding of the dynamics of WAIS, especially for predictions on the longer time-scales.

Independent of bottom melting below the ice shelves and the possibility of an ice-dynamic instability, surface melting sets an upper temperature limit on the viability of the Antarctic ice sheet, because runoff would eventually become the dominant wastage mechanism (as would be the case for Greenland in a climate several degrees warmer than today). For warmings of more than 10°C, simple runoff models predict that an ablation zone would develop around the Antarctic coast, making the mass balance at sea level sufficiently negative that the grounded ice would no longer be able to feed an ice shelf. Also the WAIS ice shelves would disintegrate to near to their inland limits as summer temperatures rise above the thermal limit of ice shelf viability believed to be responsible for the recent collapse of ice shelves at the northern tip of the Antarctic Peninsula. Disintegration of WAIS would in that case result, because the WAIS cannot retreat to higher ground once its margins are subjected to surface melting and begin to recede (Huybrechts, 1994). Depending on the strength of the warming, such a disintegration would take at least a few millennia.

East Antarctic ice sheet

Thresholds for disintegration of the East Antarctic ice sheet by surface melting involve warmings above 20°C, a situation that has not occurred for at least the last 15 million years (Barker *et al.*, 1999), and which is far more than thought possible under any scenario of climatic change currently under consideration. In that case, the ice sheet would decay over a period of at least 10,000 years. However, the recent inference of complex flow patterns in the interior of the East Antarctic ice sheet demonstrates the existence of ice-streaming features penetrating far inland, which may be indicative of a more dynamic regime than believed so far (Bamber *et al.*, 2000; Huybrechts *et al.*, 2000).

11.6 Reducing the Uncertainties in Future Estimates of Sea Level Change

It is valuable to note that the reduction in the uncertainty of estimation of the long-term ice sheet imbalance reported in Sections 11.3.1 and 11.4 came from indirect constraints and the synthesis of information of different types. Such syntheses offer promise for further progress.

11.6.1 Observations of Current Rates of Global-averaged and Regional Sea Level Change

Sections 11.3.2.1 and 11.4 reveal significant uncertainty in the analysis of 20th century sea level change. Also, we have little knowledge of the regional pattern of sea level change. Observational determination of such a pattern would be a powerful test of the coupled models required for projections of globally averaged and regional sea level rise. Requirements for reducing uncertainties include:

- A global tide gauge network (the 'GLOSS Core Network') (IOC, 1997) for measuring relative change.
- A programme of measurements of vertical land movements at gauge sites by means of GPS (Global Positioning System), DORIS Beacons and/or absolute gravity meters (Neilan *et al.*, 1998).
- Improved models of postglacial rebound.
- A reanalysis of the historical record, including allowing for the impact of variable atmospheric forcing.
- A subset of mostly island tide gauge stations devoted to ongoing calibration of altimetric sea level measurements (Mitchum, 1998).
- An ongoing high-quality (TOPEX/POSEIDON class) satellite radar altimeter mission (Koblinsky *et al.*, 1992) and careful control of biases within a mission and between missions.
- Space gravity missions to estimate the absolute sea surface topography (Balmino *et al.*, 1996, 1999) and its temporal changes, to separate thermal expansion from an increase in ocean mass from melting of glaciers and ice sheets (NASA, 1996, NRC, 1997) and changes in terrestrial storage.

For assessment of possible changes to the severity of storm surges, analyses of historical storm surge data in conjunction with meteorological analyses are needed for the world's coastlines, including especially vulnerable regions.

11.6.2   Ocean Processes

Requirements for improved projections of ocean thermal expansion include:

- Global estimates of ocean thermal expansion through analysis of the historical data archive of ocean observations and a programme of new observations, including profiling floats measuring temperature and salinity and limited sets of full-depth repeat oceanographic sections and time-series stations.
- Testing of the ability of AOGCMs to reproduce the observed three-dimensional and time-varying patterns of ocean thermal expansion.
- An active programme of ocean and atmosphere model improvement, with a particular focus on the representation of processes which transport heat into and within the interior of the ocean.

11.6.3   Glaciers and Ice Caps

Requirements for improved projections of glacier contributions include (see also Haeberli *et al.*, 1998):

- A strategy of worldwide glacier monitoring, including the application of remote sensing techniques (laser altimetry, aerial photography, high-resolution satellite visible and infrared imagery e.g. from ASTER and Landsat).
- A limited number of detailed and long-term mass balance measurements on glaciers in different climatic regions of the world, with an emphasis on winter and summer balances in order to provide a more direct link with meteorological observations.

- Development of energy balance and dynamical models for more detailed quantitative analysis of glacier geometry changes with respect to mass balance and climate change.
- Glacier inventory data to determine the distribution of glacier parameters such as area and area-altitude relations, so that mass balance, glacier dynamics and runoff/sea level rise models can be more realistically framed.

11.6.4   Greenland and Antarctic Ice Sheets

- Continued observations with satellite altimeters, including the upcoming satellite laser altimeter on ICESat and the radar interferometer on CRYOSAT. Measurements should be continued for at least 15 years (with intercalibration between missions) to establish the climate sensitivities of the mass balance and decadal-scale trends.
- Satellite radar altimetry and synthetic aperture radar interferometry (ERS-1, ERS-2 and Radarsat) for detailed topography, changes in ice sheet volume and surface velocity of the ice sheets (Mohr *et al.*, 1998; Joughin *et al.*, 1999), as well as short-term variability in their flow (Joughin *et al.*, 1996a,b) and grounding line position (Rignot, 1998a,b,c).
- Determination of the Earth's time-variant gravity field by the Gravity Recovery and Climate Experiment (GRACE) satellite flown concurrently with ICESat to provide an additional constraint on the contemporary mass imbalances (Bentley and Wahr, 1998). This could provide estimates of sea level change to an accuracy of ±0.35 mm/yr.
- Geological observations of sea level change during recent millennia combined with improved postglacial rebound models and palaeoclimatic and palaeoglaciological studies to learn what changes have occurred in the past.
- Further analysis of Earth rotational parameters in combination with sea level measurements.
- Improved estimates of surface mass balance (including its spatial and temporal variability) from *in situ* observations, accumulation rates inferred from atmospheric moisture budgets and improved estimates of the rate of iceberg calving and the melt-water flux.
- Improved calculation of the surface mass balance within ice sheet models or by atmospheric models, with attention to modelling of changes in sea-ice concentration because of the consequent effect on moisture transports and accumulation.
- Improved understanding and modelling of the dynamics of ice sheets, ice streams and ice shelves (requiring combined studies using glaciological, oceanographic and satellite observations), including the physics of iceberg calving.

11.6.5   Surface and Ground Water Storage

Surface and ground water storage changes are thought to be having a significant impact on sea level, but their contribution is very uncertain (Table 11.10, Figure 11.9), and could be either positive or negative. They may become more important in the future, as a result of changes related not only to climate, but also to societal decisions that are beyond the scope of this scientific assessment. There are several general issues in climate-related aspects:

- A more thorough investigative search of historical records could provide addition information on ground water mining, and storage in reservoirs.
- Accurate satellite measurements of variations in the Earth's gravity (Herring, 1998) to detect changes in land water storage due to water-table variations and impoundments.
- A better understanding of seepage losses beneath reservoirs and in irrigation is required.
- A unified systems approach is needed to trace the path of water more accurately through the atmosphere, hydrologic, and biosphere sub-systems, and to account for various feedbacks (including the use of GCMs and improved hydrologic models).
- Satellite remote sensing offers useful technology for monitoring the global hydrologic budget. A cumulative volume estimate for the many small reservoirs might be possible using high-resolution radar data, targeted ground studies and a classification of land use classes from satellite data and also of changes in deforestation and other land-use transformations (Koster *et al.*, 1999).

11.6.6 Summary

Sea level change involves many components of the climate system and thus requires a broad range of research activities. A more detailed discussion of the requirements is given in the report of the recent IGBP/GAIM Workshop on sea level changes (Sahagian and Zerbini, 1999). We recognise that it is important to assign probabilities to projections, but this requires a more critical and quantitative assessment of model uncertainties than is possible at present.

Appendix 11.1: *Methods for projections of global-average sea level rise*

This Appendix describes the methods used in this report to make sea level rise projections for the SRES scenarios for the 21st century. The results are discussed in Section 11.5.1.2 and shown in Figure 11.12 and Appendix II.

Global-average sea-level rise $\Delta h(t)$ is a function of time t and is expressed relative to the level in 1990. It comprises several components, which are all zero at 1990:

$$\Delta h(t) = X(t) + g(t) + G(t) + A(t) + I(t) + p(t) + s(t)$$

The components are sea-level rise due to:

X	thermal expansion.
g	loss of mass of glaciers and ice caps.
G	loss of mass of the Greenland ice sheet due to projected and recent climate change.
A	loss of mass of the Antarctic ice sheet due to projected and recent climate change.
I	loss of mass of the Greenland and Antarctic ice sheets due to the ongoing adjustment to past climate change.
p	runoff from thawing of permafrost.
s	deposition of sediment on the ocean floor.

The components X, g, G and A are estimated for each of 35 SRES scenarios using the projections of an upwelling-diffusion energy-balance (UD/EB) model calibrated separately for each of seven AOGCMs (Appendix 9.1).

Thermal expansion X is obtained directly from the thermal expansion $X_m(t)$ projected by the UD/EB model:

$$X(t) = X_m(t) - X_m(1990)$$

No uncertainty is included in this term, because the uncertainty is sufficiently represented by the use of a range of AOGCMs. The term g from glaciers and ice caps is estimated using the global average temperature change $T_m(t)$ projected by the UD/EB model. First, we obtain the loss of mass g_u with respect to the glacier steady state without taking contraction of glacier area into account.

$$g_u(t) = g_{1990} + \int_{1990}^{t} (T_{1990} + \Delta T_b + T_m(t') - T_m(1990)) \frac{\partial B_g}{\partial T_g} \, dt'$$

where g_{1990} is the sea-level rise from glaciers and ice caps up to 1990 calculated from AOGCM results without contraction of glacier area, T_{1990} is the AOGCM global average temperature change at 1990 with respect to the climate of the late 19th century, $\Delta T_b = 0.15$ K the difference in the global average temperature between the late 19th century and the glacier steady state (see 11.5.1.1) and $\partial B_g/\partial T_g$ is the sensivity of global glacier mass balance for constant glacier area to global-average temperature change, expressed as sea level equivalent (from Table 11.11). Second, we estimate the loss of mass g_s with respect to the glacier steady state taking into account contraction of glacier area. This is done by using an empirical relationship between the loss of mass for changing and for constant area. The relationship was obtained by a quadratic fit to the AOGCM IS92a results of Section 11.5.1.1.

$$g_s(t) = 0.934 g_u(t) - 1.165 g_u^2(t)$$

for g_u and g_s in metres. Third, we calculate the change since 1990.

$$g(t) = g_s(t) - g_s(1990)$$

The uncertainty $\delta g(t)$ on this term is calculated assuming an uncertainty of ±40% (standard deviation) in the mass balance sensitivities, as discussed in Section 11.5.1.1.

$$\delta g(t) = 0.40 g(t)$$

The term G from the Greenland ice sheet is calculated according to

$$G(t) = \int_{1990}^{t} (T_{1990} + T_m(t') - T_m(1990)) \frac{dB_G}{dT_g} \, dt'$$

where dB_G/dT_g is the sensitivity of the Greenland mass balance to global-average temperature change, expressed as sea level equivalent (from Table 11.12). The uncertainty on this term comprises two components, as discussed in Section 11.5.1.1. The first uncertainty is a mass balance uncertainty

$$\delta G_1(t) = \int_{1990}^{t} (T_{1990} + T_m(t') - T_m(1990)) \frac{\Delta T_G}{\Delta T_g} \delta m_G \, dt'$$

where $\delta m_G = 0.05$ mm/yr/°C and $\Delta T_G/\Delta T_g$ is the ratio of Greenland average temperature change to global average temperature change (from Table 11.12). The first uncertainty is the combination in quadrature of 0.03 mm/yr/°C from ablation parametrization,

0.03 mm/yr/°C from high-resolution patterns, and 0.02 mm/yr/°C from precipitation changes, as discussed in section 11.5.1.1. The second uncertainty is an ice-dynamic uncertainty.

$$\delta G_2(t) = 0.1G(t)$$

The term A from the Antarctic ice sheet is calculated according to

$$A(t) = \int_{1990}^{t} (T_{1990} + T_m(t') - T_m(1990)) \frac{dB_A}{dT_g} dt'$$

where dB_A/dT_g is the sensitivity of the Antarctic mass balance to global-average temperature change, expressed as sea level equivalent (from Table 11.12). Ice-dynamical uncertainty for the Antarctic is not included and is discussed in Section 11.5.4.3. There is no uncertainty for ablation. Precipitation change uncertainty is calculated as discussed in Section 11.5.1.1 according to

$$\delta A(t) = \int_{1990}^{t} (T_{1990} + T_m(t') - T_m(1990)) \frac{\Delta T_A}{\Delta T_g} \delta m_A \, dt'$$

where $\delta m_A = 0.08$ mm/yr/°C and $\Delta T_A/\Delta T_g$ is the ratio of Antarctic average temperature change to global average temperature change (from Table 11.12).

The uncertainties on the above terms are combined in quadrature:

$$\delta h_v = \sqrt{(\delta g)^2 + (\delta G_1)^2 + (\delta G_1)^2 + (\delta A)^2}$$

The remaining terms are calculated assuming they contribute to sea-level rise at a constant rate, independent of AOGCM and scenario, thus:

$$I(t) = \int_{1990}^{t} \frac{dI}{dt'} dt' \qquad p(t) = \int_{1990}^{t} \frac{dp}{dt'} dt' \qquad s(t) = \int_{1990}^{t} \frac{ds}{dt'} dt'$$

The rates each have a range of uncertainty. For dI/dt, this is 0.0 to 0.5 mm/yr (Section 11.3.1, Table 11.9), for dp/dt 0 to 0.23 mm/yr (the upper bound is more precisely 25 mm divided by 110 years, section 11.2.5), for ds/dt 0 to 0.05 mm/yr (Section 11.2.6, Table 11.9). The central rates are 0.25, 0.11 and 0.025 mm/yr for the three terms. We denote I calculated at the minimum rate by I_{min} and at the maximum rate by I_{max}; similarly for p and s. The minimum projected sea-level rise $\Delta h_{min}(t)$ for a given AOGCM and SRES scenario is given by

$$\Delta h_{min}(t) = X(t) + g(t) + G(t) + A(t) - 2\delta h_v(t) + I_{min}(t) + P_{min}(t) + s_{min}(t)$$

and the maximum is

$$\Delta h_{max}(t) = X(t) + g(t) + G(t) + A(t) + 2\delta h_v(t) + I_{max}(t) + P_{max}(t) + s_{max}(t)$$

In these formulae, δh_v has been doubled to convert from an uncertainty to a range, following Box 11.1.

Table 11.17: *Parameters used in sea-level projections to simulate AOGCM results.*

AOGCM	T_{1990} (°C)	g_{1990} (m)	$\partial B_g/\partial T_g$ (mm/yr°C)	dB_G/dT_g (mm/yr/°C)	dB_A/dT_g (mm/yr/°C)	$\Delta T_G/\Delta T_g$	$\Delta T_A/\Delta T_g$
CSIRO Mk2	0.593	0.022	0.733	0.157	− 0.373	2.042	1.120
CSM 1.3	0.567	0.021	0.608	0.146	− 0.305	3.147	1.143
ECHAM4/OPYC3	0.780	0.027	0.637	0.029	− 0.478	1.153	1.484
GFDL_R15_a	0.635	0.015	0.576	0.121	− 0.177	1.879	0.799
HadCM2	0.603	0.027	0.613	0.096	− 0.214	1.441	1.239
HadCM3	0.562	0.021	0.622	0.085	− 0.354	1.443	1.288
DOE PCM	0.510	0.017	0.587	0.136	− 0.484	2.165	1.618

References

Allen, J.R.L. and J.E. Rae, 1988: Vertical salt marsh accretion since the Roman period in the Severn Estuary, southwest Britain. *Marine Geology*, **83**, 225-235.

Anandakrishnan, S., and R.B. Alley, 1997: Stagnation of ice stream C, West Antarctica by water piracy. *Geophys. Res. Lett.*, **24**, 265-268.

Anandakrishnan, S., D.D. Blankenship, R.B. Alley, and P.L. Stoffa, 1998: Influence of subglacial geology on the position of a West Antarctic ice stream. *Nature*, **394** , 62-65.

Anisimov, O.A. and F.E. Nelson, 1997: Permafrost zonation and climate change in the northern hemisphere: results from transient general circulation models. *Clim. Change*, **35**, 241-258.

Aniya, M., H. Sato, R. Naruse, P. Skvarca, G. Casassa, 1997: Recent glacier variations in the Southern Patagonian icefield, South America. *Arctic and Alpine Research*, **29 (1)**, 1-12.

Antonioli, F. and M. Oliverio, 1996: Holocene sea level rise recorded by a radiocarbon-dated mussel in a submerged speleotherm beneath the Mediterranean Sea, *Quat. Res.*, **45**, 241-244.

Antonov, J.I., 1993: Linear trends of temperature at intermediate and deep layers of the North Atlantic and the Pacific Oceans: 1957-1981. *J. Climate*, **6**, 1928-42.

Antonov, J.I., S. Levitus and T. P. Boyer, 2000: Ocean temperature and salinity impact on sea level rise duriing 1957-1994. *EOS Trans.*, *AGU*, **81**, Fall Meet Suppl., Abstract, U61B-03-2000.

Anzenhofer, T. G, 1998: Fully reprocessed ERS-1 altimeter data from 1992 to 1995: Feasibility of the detection of long term sea level change. *J. Geophys. Res.*, **103(C4)**, 8089-112.

Arbic, B.K. and B. Owens, 2001: Climatic Warming of Atlantic Intermediate Waters, *J. Climate*, in press.

Arhan, M., H. Mercier, B. Bourles, and Y. Gouriou, 1998: Hydrographic sections across the Atlantic at 7 30N and 4 30S. *Deep-Sea Res., I*, **45**, 829-72.

Bacon, S. and D.J.T. Carter, 1991: Wave climate changes in the North Atlantic and North Sea. *Int. J. Climatol.*, **11**, 545-558.

Bacon, S. and D.J.T. Carter, 1993: A connection between mean wave height and atmospheric pressure gradient in the North Atlantic. *Int. J. Climatol.*, **13**, 423-436.

Bahr, D.B., 1997: Global distributions of glacier properties: a stochastic scaling paradigm. *Water Resour. Res.*, **33**, 1669-1679.

Bahr, D.B., M.F. Meier, and S.D. Peckham, 1997: The physical basis of glacier volume-area scaling. *J. Geophys. Res.*, **102(B9)**, 20355-20362.

Bahr, D.B., W.T. Pfeffer, C. Sassolas, and M.F. Meier, 1998: Response time of glaciers as a function of size and mass balance: 1. Theory. *Journal of Geophysical Research*, **103**, 9777-9782.

Balmino, G., R. Sabadini, C. Tscherning, and P.L. Woodworth, 1996: Gravity Field and Steady-State Ocean Circulation Mission. Reports for assessment: the nine candidate Earth Explorer Missions. European Space Agency Report SP-1196(1). 77pp.

Balmino, G., R. Rummel, P. Visser, and P. Woodworth, 1999. Gravity Field and Steady-State Ocean Circulation Mission. Reports for assessment: the four candidate Earth Explorer Core Missions. European Space Agency Report SP-1233(1). 217pp.

Bamber, J.L., D.G.Vaughan, and I. Joughin, 2000: Widespread complex flow in the interior of the Antarctic ice sheet, *Science*, **287**, 1248-1250.

Bard, E., B. Hamelin, M. Arnold, L.F. Montaggioni, G. Cabioch, G. Faure and F. Rougerie, 1996: Deglacial sea level record from Tahiti corals and the timing of global meltwater discharge. *Nature*, **382**, 241-244.

Barker, P. F., P.J. Barrett, A.K. Cooper, and P. Huybrechts, 1999: Antarctic glacial history from numerical models and continental margin sediments. *Palaeogeography, Palaeoecology, Palaeoclimatology*, **150**, 247-267.

Basnett, T.A. and D.E. Parker, 1997: Development of the global mean sea level pressure data set GMSLP2. Climatic Research Technical Note No.79, Hadley Centre, Meteorological Office, Bracknell, 16pp. & Appendices.

Bauer, A., 1968: Nouvelle estimation du bilan de masse de l'Inlandsis du Groenland. *Deep Sea Res.*, **14**, 13-17.

Bell, R. E., D.D. Blankenship, C.A. Finn, D.L. Morse, T.A. Scambos, J.M. Brozena, and S.M. Hodge, 1998: Influence of subglacial geology on the onset of a West Antarctic ice stream from aerogeophysical observations. *Nature*, **394**, 58-62.

Benson, C.S., 1962: Stratigraphic studies in the snow and firn of the Greenland Ice Sheet. *US Army SIPRE (now CRREL) Research Report*, **70**, 93

Bentley, C.R., 1997: Rapid sea level rise soon from West Antarctic ice sheet collapse? *Science*, **275**, 1077-1078.

Bentley, C.R., 1998a: Ice on the fast track. *Nature*, **394**, 21-22.

Bentley, C.R., 1998b: Rapid sea level rise from a West Antarctic ice-sheet collapse: a short-term perspective. *Journal of Glaciology*, **44** (146) 157-163.

Bentley, C. R., and J. Wahr, 1998: Satellite gravimetry and the mass balance of the Antarctic ice sheet. *Journal of Glaciology*, **44** (147), 207-213.

Bentley, M.J., 1999: Volume of Antarctic Ice at the Last Glacial Maximum, and its impact on global sea level change [Review]. *Quaternary Science Reviews*, **18**, 1569-1595.

Bijl, W., R. Flather, J.G. de Ronde and T. Schmith, 1999: Changing storminess? An analysis of long-term sea level data sets. *Clim. Res.*, **11**, 161-172.

Bindoff, N.L. and J.A. Church, 1992: Warming of the water column in the southwest Pacific Ocean. *Nature*, **357**, 59-62.

Bindoff, N.L., and T.J. McDougall, 1994: Diagnosing Climate Change and Ocean Ventilation using Hydrographic Data. *Journal of Physical Oceanography*, **24**, 1137-1152.

Bindoff, N.L., and T.J. McDougall, 2000: Decadal changes along an Indian Ocean section at 32S and their interpretation. *Journal of Physical Oceanography*, **30** (June), 1207-1222.

Bindschadler, R., and P. Vornberger, 1998: Changes in the West Antarctic ice sheet since 1963 from declassified satellite photography. *Science*, **279**, 689-692.

Bindschadler, R., 1998: Future of the West Antarctic Ice Sheet. *Science*, **282**, 428-429.

Blankenship, D. D., C.R. Bentley, S.T. Rooney, and R.B. Alley, 1986: Seismic measurements reveal a saturated porous layer beneath an active Antarctic ice stream. *Nature*, **322**, 54-57.

Braithwaite, R.J., 1995: Positive degree-day factors for ablation on the Greenland Ice Sheet studied by energy-balance modeling, *Journal of Glaciology*, **41(137)**, 153-160.

Braithwaite, R.J. and Y. Zhang, 1999: Modelling changes in glacier mass balance that may occur as a result of climate changes. *Geografiska Annaler*, **81A**, 489-496.

Bromwich, D., 1995: Ice sheets and sea level. *Nature*, **373**, 18-19.

Bromwich, D.H., R.I.Cullather and M.L. van Woert, 1998: Antarctic precipitation and its contribution to the global sea level budget. *Annals of Glaciology*, **27**, 220-226.

Bryan, K., 1996: The steric component of sea level rise associated with enhanced greenhouse warming: a model study. *Climate Dynamics*, **12**, 545-55.

Bryden, H.L., M.J. Griffiths, A.M. Lavin, R.C. Milliard, G. Parilla, and W.M. Smethie, 1996: Decadal changes in water mass characteristics at 24°N in the Subtropical North Atlantic Ocean. *Journal of Climate*, **9(12)**, 3162-86.

Budd, W. F., and I.N. Smith, 1985: The state of balance of the Antarctic ice sheet, an updated assessment. In: *Glaciers, ice sheets and sea level: effects of a CO2-induced climatic change*. National Academy Press, Washington, pp. 172-177.

Budd, W. F., P.A. Reid, and L.J. Minty, 1995: Antarctic moisture flux and net accumulation from global atmospheric analyses. *Annals of Glaciology*, **21**, 149-156.

Budd, W.F., B. Coutts, and R. Warner, 1998: Modelling the Antarctic and northern hemisphere ice-sheet changes with global climate through the glacial cycle. Ann. Glaciol. **27**, 153-160.

Carmack, E.C., K. Aargaard, J.H. Swift, R.W. MacDonald, F.A. McLaughlin, E.P. Jones, R.G. Perkin, J.N. Smith, K.M. Ellis, and L.R. Killius, 1997: Changes in temperature and tracer distributions within the Arctic Ocean: Results from the 1994 Arctic Ocean section. *Deep-Sea Research II,* **44(8)**, 1487-502.

Carter, D.J.T. and L. Draper, 1988: Has the north-east Atlantic become rougher? *Nature*, **332**, 494.

Casassa, G., 1995: Glacier inventory in Chile: current status and recent glacier variations. *Annals of Glaciology,* **21**, 317-322.

Cazenave, A., P. Bonnefond, K. Dominh, and P. Schaeffer, 1997: Caspian sea level from Topex-Poseidon altimetry: Level now falling, *Geophysical Research Letters,* **24**, 881-884.

Cazenave, A., K. Dominh, M.C. Gennero, and B. Ferret, 1998: Global mean sea level changes observed by Topex-Poseidon and ERS-1. *Physical Chemical Earth,* **23**, 1069-75.

Cazenave, A., K. Dominh, L. Soudrarin, F. Ponchaut, and C. Le Provost, 1999: Sea level changes from TOPEX-POSEIDON altimetry and tide gauges and vertical crustal motions from DORIS. *Geophysical Research Letters,* **26**, 2077-2080.

Chappell, J., 1982: Evidence for smoothly falling sea level relative to north Queensland, Australia, during the past 6000 years. *Nature,* **302**, 406-408.

Chappell, J., A. Ohmura, T. Esat, M. McCulloch, J. Pandolfi, Y. Ota, and B. Pillans, 1996a: Reconciliation of late Quaternary sea levels derived from coral terraces at Huon Peninsula with deep sea oxygen isotope records. *Earth Planetary Science Letters,* **141**: 227-236.

Chappell, J., Y. Ota, and K. Berryman, 1996b: Late Quaternary coseismic uplift history of Huon Peninsula, Papua New Guinea, *Quaternary Science Review,* **15**, 7-22.

Cheng, M.K., C.K. Shum and B.D. Tapley, 1997: Determination of long-term changes in the Earth's gravity field from satellite laser ranging observations. *Journal Geophysical Research* **102**, 22,377-22,390.

Chinn, T.J., 1991: Glacier inventory of New Zealand. Institute of Geological and Nuclear Sciences, Dunedin, New Zealand.

Chinn, T.J., 1999: New Zealand glacier response to climate change of the past 2 decades. *Global and Planetary Change,* **22**, 155-168.

Church, J.A., J.S. Godfrey, D.R. Jackett, and T.J. McDougall, 1991: A model of sealevel rise caused by ocean thermal expansion. *Journal of Climate,* **4(4)**, 438-56.

Clausen, H. B., N. S. Gundestrup, S. J. Johnsen, R. A. Bindschadler and H. J. Zwally, 1988: Glaciological investigations in the Crete area, Central Greenland. A search for a new drilling site. *Annals of Glaciology*, **10**, 10-15.

Cogley, J.G., W.P. Adams, 1998: Mass balance of glaciers other than ice sheets. *Journal of Glaciology*, **44**, 315-325.

Conway, H.W., B.L. Hall, G.H. Denton, A.M. Gades, and E.D.Waddington, 1999: Past and future grounding-line retreat of the West Antarctic ice sheet, *Science*, **286**, 280-286.

Cubasch, U., B.D. Santer, A. Hellbach, G. Hegerl, H. Höck, E. Meier-Reimer, U. Mikolajewicz, A. Stössel, R. Voss, 1994: Monte Carlo climate change forecasts with a global coupled ocean-atmosphere model. *Climate Dynamics*, **10**, 1-19.

Cuffey, K.M. and G.D. Clow, 1997: Temperature, accumulation, and ice sheet elevation in central Greenland through the last deglacial transition, *J. Geophys. Res.*, **102**, 26383-26396.

Cuffey, K.M. and S.J. Marshall, 2000: Substantial contribution to sea-level rise during the last interglacial from the Grennland ice sheet, *Nature.*, **404**, 591-594.

Curry, R.G., M.S. McCartney, and T.M. Joyce, 1998: Oceanic transport of subpolar climate signals to mid-depth subtropical waters. *Nature,* **391**, 575-7.

Dahl-Jensen, D., S.J. Johnsen, C.U. Hammer, H.B. Clausen and J. Jouzel, 1993: Past accumulation rates derived from observed annual layers in the GRIP ice core from Summit, Central Greenland. In: Ice in the climate system, W.R. Peltier (ed), NATO ASI Series, **I 12**, pp. 517-532.

Dansgaard, W., S.J. Johnsen, H.B. Clausen, D. Dahl-Jensen, N.S. Gundestrup, C.U. Hammer, C.S. Hvidberg, J.P. Steffensen, A.E. Sveinbjoernsdottir, J. Jouzel and G.C. Bond, 1993: Evidence for general instability of past climate from a 250-kyr ice-core record. *Nature*, **364**, 218-220.

Davis, J.L. and J.X. Mitrovica, 1996. Glacial isostatic adjustment and the anomalous tide gauge record of eastern North America. *Nature*, **379**, 331-333.

Davis, C. H., C.A. Kluever, and B.J. Haines, 1998: Elevation change of the southern Greenland Ice Sheet. *Science*, **279**, 2086-2088.

Day, J.W., W.H. Conner, R. Costanza, G.P. Kemp, and I.A. Mendelssohn, 1993: Impacts of sea level rise on coastal systems with special emphasis on the Mississippi River deltaic plain. In: *Climate and sea level change: observations, projections and implications*, R.A. Warrick, E.M. Barrow and T.M.L. Wigley (eds), Cambridge University Press, Cambridge, UK, pp 276-296.

De Wolde, J.R., R. Bintanja, and J. Oerlemans, 1995: On thermal expansion over the last hundred years. *Climate Dynamics*, **11**, 2881-2891.

De Wolde, J. R., P. Huybrechts, J. Oerlemans, and R.S.W. van de Wal, 1997: Projections of global mean sea level rise calculated with a 2D energy-balance climate model and dynamic ice sheet models. *Tellus*, **49A**, 486-502.

Dickson, R., J. Lazier, J. Meincke, P. Rhines and J. Swift, 1996: Long-term coordinated changes in the convective activity of the North Atlantic. *Progress in Oceanography*, **38**, 241-295.

Dickson, R., Nathan Bindoff, Annie Wong, Brian Arbic, Breck Owens, Shiro Imawacki and Jim Hurrell, 2001: The World during WOCE. In: Ocean Circulation and Climate, G. Siedler and J.A. Church (eds), Academic Press, London, pp 557-583.

Doake, C.S.M., H.F.J. Corr, H. Rott, P. Skvarca, and N.W. Young, 1998: Breakup and conditions for stability of the northern Larsen Ice Shelf, Antarctica. *Nature*, **391**, 778-780.

Dolgushin, L.D. and G.B. Osipova, 1989: *Glaciers.* Mysl, Moscow, 444pp.

D'Onofrio, E.E., M.M.E. Fiore and S.I. Romero, 1999: Return periods of extreme water levels estimated for some vulnerable areas of Buenos Aires. *Continental Shelf Research*, **19**, 1681-1693.

Douglas, B.C. 1991: Global sea level rise. *Journal of Geophysical Research*, **96**, 6981-6992.

Douglas, B.C., 1992: Global sea level acceleration. *Journal of Geophysical Research*, **97**, 12699-12706.

Douglas, B.C., 1995: Global sea level change: determination and interpretation. Reviews of Geophysics, Supplement, 1425-1432. (U.S. National Report to the International Union of Geodesy and Geophysics 1991-1994).

Douglas, B.C., 1997: Global sea rise: a redetermination. *Surveys in Geophysics,* **18**, 279-292.

Dowdeswell, J.A., J.O. Hagen, H. Björnsson, A.F. Glazovsky, W.D. Harrison, P. Holmlund, J. Jania, R.M. Koerner, B. Lefauconnier, C.S.L. Ommanney, and R.H. Thomas, 1997: The mass balance of circum-Arctic glaciers and recent climate change. *Quaternary Research*, **48**, 1-14.

Drewry, D.J. and E.M. Morris, 1992: The response of large ice sheets to climatic change. *Philosophical Transactions of the Royal Society of London* B, **338**, 235-242.

Dyurgerov, M.B., and M.F. Meier, 1997a: Mass balance of mountain and subpolar glaciers: a new assessment for 1961–1990. *Arctic and*

Alpine Research, **29**, 379-391.

Dyurgerov, M.B., and M.F. Meier, 1997b: Year-to-year fluctuations of global mass balance of small glaciers and their contribution to sea level changes. *Arctic and Alpine Research*, **29**, 392-402.

Dyurgerov, M.B., and M.F. Meier, 1999: Analysis of winter and summer glacier mass balances. *Geografiska Annaler*, **81A**, 541-554.

Echelmeyer, K. A., W.D. Harrison, C. Larsen, and J.E. Mitchell, 1994: The role of the margins in the dynamics of an active ice stream. *Journal of Glaciology*, **40**, 527-538.

Ekman, M., 1988: The world's longest continued series of sea level observations. *Pure and Applied Gephysics*, **127**, 73-77.

Ekman, M., 1999: Climate changes detected through the world's longest sea level series. *Global and Planetary Change*, **21**, 215-224.

Eronen, M., G. Glückert, O. van de Plassche, J. van der Plicht, and P. Rantala, 1995: Land uplift in the Olkilvoto-Pyhäjärvi area, southwestern Finland, during the last 8000 years, In: Nuclear Waste Commission of Finnish Power Companies, Helsinki, 26 pp.

Fairbanks, R.G., 1989: A 17,000-year glacio-eustatic sea level record: influence of glacial melting dates on the Younger Dryas event and deep ocean circulation. *Nature*, **342**, 637-642.

Flather, R.A. and H. Khandker, 1993: The storm surge problem and possible effects of sea level changes on coastal flooding in the Bay of Bengal, in *Climate and sea level change: observations, projections and implications*, R.A. Warrick, E.M. Barrow and T.M.L.Wigley (eds), Cambridge, Cambridge University Press, 424pp.

Flather, R.A., and J.A. Smith, 1998: First estimates of changes in extreme storm surge elevations due to the doubling of CO_2. *Global Atmospheric Ocean Systems*, **6**, 193-208.

Fleming, K., P. Johnston, D. Zwartz, Y. Yokoyama, K. Lambeck, and J. Chappell, 1998: Refining the eustatic sea level curve since the Last Glacial Maximum using far- and intermediate-field sites, *Earth Planetary Science Letters*, **163**, 327-342.

Flemming, N.C. and C.O. Webb, 1986: Regional patterns of coastal tectonics and eustatic change of sea level in the Mediterranean during the last 10,000 years derived from archaeological data. Zeitschrift für Geomorphologie. December, Suppl - Bd62, p.1-29.

Fox, A. J., and A.P.R. Cooper, 1994: Measured properties of the Antarctic ice sheet derived from the SCAR Antarctic digital database. *Polar Record*, **30**, 201-206.

Giovinetto, M. B., and H.J. Zwally, 1995a: An assessment of the mass budgets of Antarctica and Greenland using accumulation derived from remotely sensed data in areas of dry snow. *Zeitschrift für Gletscherkunde und Glazialgeologie*, **31**, 25-37.

Giovinetto, M. B., and H.J. Zwally, 1995b: Annual changes in sea ice extent and of accumulation on ice sheets: implications for sea level variability. *Zeitschrift für Gletscherkunde und Glazialgeologie*, **31**, 39-49.

Giovinetto, M.B., and H.J. Zwally, 2000: Spatial distribution of net surface accumulation on the Antarctic ice sheet, *Annals of Glaciology*, **31**, 171-178.

Glover, R.W., 1999: Influence of spatial resolution and treatment of orography on GCM estimates of the surface mass balance of the Greenland Ice Sheet, *Journal of Climate*, **12**, 551-563.

Gornitz, V., 1995: A comparison of differences between recent and late Holocene sea level trends from eastern North America and other selected regions. *Journal of Coastal Research*, Special Issue 17, Holocene Cycles: Climate, Sea Levels and Sedimentation, C.W. Finkl, Jr. (ed.), pp.287-297.

Gornitz, V., 2000: Impoundment, groundwater mining, and other hydrologic transformations: impacts on global sea level rise. In *Sea level rise: history and consequences*, B.C. Douglas, M.S. Kearney and S.P. Leatherman (eds.), Academic Press, 97-119.

Gornitz, V., and S. Lebedeff, 1987: Global sea level changes during the past century. In: *sea level Fluctuation and Coastal Evolution*, D. Nummedal, O.H. Pilkey and J.D. Howard (eds.), Society for

Economic Paleontologists and Mineralogists, pp. 3-16 (SEPM Special Publication No. 41).

Gornitz, V. and Solow, A. 1991: Observations of long-term tide-gauge records for indications of accelerated sea level rise. In, *Greenhouse-gas-induced climatic change: a critical appraisal of simulations and observations*, M.E. Schlesinger (ed.), Elsevier, Amsterdam, pp.347-367.

Gornitz, V., C. Rosenzweig, and D. Hillel, 1997: Effects of anthropogenic intervention in the land hydrological cycle on global sea level rise. *Global and Planetary Change*, **14**, 147-161.

Gorshkov, S.G. (Ed.), 1983: Arctic Ocean Vol. 3, In The World Ocean Atlas Series (in Russian), Pergamon, Oxford.

Gregory, J.M., 1993: Sea level changes under increasing atmospheric CO_2 in a transient coupled ocean-atmophere GCM Experiment. *Journal of Climate*, **6**, 2247-62.

Gregory, J.M. and J.F.B. Mitchell, 1997: The climate response to CO2 of the Hadley Centre coupled AOGCM with and without flux adjustment, *Geophys Res Lett*, **24**, 1943-1946.

Gregory, J.M. and J. Oerlemans, 1998: Simulated future sea level rise due to glacier melt based on regionally and seasonally resolved temperature changes. *Nature*, **391**, 474-6.

Gregory, J.M. and J.A. Lowe, 2000: Predictions of global and regional sea level rise using AOGCMs with and without flux adjustment. *Geophysical Research Letters*, **27**, 3069-3072.

Gregory, J.M., J.A. Church, G.J. Boer, K.W. Dixon, G.M. Flato, D. R. Jackett, J. A. Lowe, S. P. O'Farrell, E. Roeckner, G.L. Russell, R. J. Stouffer, M. Wintern, 2001: Comparison of results from several AOGCMs for global and regional sea level change 1900-2100. *Clim. Dynam.*, in press..

Greve, R., 2000: On the response of the Greenland Ice Sheet to greenhouse climate change. *Climatic Change*, **46**, 289-283.

Grotefendt, K., K. Logemann, D. Quadfasel, and S. Ronski, 1998: Is the Arctic Ocean warming? *Journal of Geophysical Research*, **103**, 27,679-27,687.

Günther, H., W. Rosenthal, M. Stawarz, J.C. Carretero, M. Gomez, L. Lozano, O. Serrano, and M. Reistad, 1998: The wave climate of the Northeast Atlantic over the period 1955-1994: the WASA wave hindcast. The Global Atmosphere and Ocean System, **6**, 121-163.

Haeberli, W., M. Hoelzle and S. Suter (Eds.), 1998: Into the second century of worldwide glacier monitoring: prospects and strategies. A contribution to the International Hydrological Programme (IHP) and the Global Environment Monitoring System (GEMS). UNESCO - Studies and Reports in Hydrology, 56.

Haines, B.J., Y.E. Bar-Sever, 1998: Monitoring the TOPEX microwave radiometer with GPS: Stability of columnar water vapor measurements. *Geophysical Research Letters*, **25**, 3563-6.

Hamilton, G. S., I.M. Whillans, and P.J. Morgan, 1998: First point measurements of ice-sheet thickness change in Antarctica. *Annals of Glaciology*, **27**, 125-129.

Harrison, C.G.A., 1990: Long-term eustasy and epeirogeny in continents. pp.141-158 of *sea level Change*, National Research Council Surveys in Geophysics.

Hastenrath, S. and A. Ames 1995: Diagnosing the imbalance of Yanamarey Glacier in the Cordillera Blanca of Perú. *J. Geoph. Res.*, **100**, 5105-5112.

Herring, T.A., 1998: Appreciate the gravity. *Nature*, **391**, 434-435.

Herterich, K., 1987: On the flow within the transition zone between ice sheet and ice shelf. In: *Dynamics of the West Antarctic ice sheet*, C.J. van der Veen, and J. Oerlemans (eds.). D. Reidel, Dordrecht, pp. 185-202.

Hindmarsh, R. C. A., 1993: Qualitative dynamics of marine ice sheets. In: *Ice in the Climate System*, W.R. Peltier (eds.). NATO ASI Series I12, pp. 68-99.

Hirst, A.C., 1998: The Southern Ocean response to global warming in the CSIRO coupled ocean-atmosphere model. *Environmental Modeling*

and Software, **14**, 227-241.

Hirst, A.C., H.B. Gordon, and S.P. O'Farrell, 1996: Response of a coupled ocean-atmosphere model including oceanic eddy-induced advection to anthropogenic CO$_2$ increase. *Geophysical Research Letters*, **23**, 3361-3364.

Hoffert, M.I., A.J. Callegari and C.T. Hsieh, 1980: The role of deep sea heat storage in the secular response to climate forcing. *J. Geophys. Res.*, **85**, 6667-6679.

Holbrook, N.J., and N.L. Bindoff, 1997: Interannual and decadal temperature variability in the Southwest Pacific Ocean between 1955 and 1988. *Journal of Climate*, **10**, 1035-49.

Hsieh, W.W., and K. Bryan, 1996: Redistribution of sea level rise associated with enhanced greenhouse warming: a simple model study. *Climate Dynamics*, **12**, 535-544.

Hubbert, G.D., and K. McInnes, 1999: A storm surge inundation model for coastal planning and impact studies. *Journal of Coastal Research*, **15**, 168-185.

Huybrechts, P., and J. Oerlemans, 1990: Response of the Antarctic Ice Sheet to future greenhouse warming. *Climate Dynamics*, **5**, 93-102.

Huybrechts, P., 1994: Formation and disintegration of the Antarctic ice sheet. *Annals of Glaciology*, **20**, 336-340.

Huybrechts, P., A. Letreguilly, and N. Reeh, 1991: The Greenland Ice Sheet and greenhouse warming. *Paleogeography, Paleoclimatology, Paleoecology (Global and Planetary Change Section)*, **89**, 399-412.

Huybrechts, P., and J. De Wolde, 1999: The dynamic response of the Greenland and Antarctic ice sheets to multiple-century climatic warming. *Journal of Climate*, **12**, 2169-2188.

Huybrechts, P., and E. Le Meur, 1999: Predicted present-day evolution patterns of ice thickness and bedrock elevation over Greenland and Antarctica. *Polar Research*, **18**, 299-308.

Huybrechts, P., A. Abe-Ouchi, I. Marsiat, F. Pattyn, T. Payne, C. Ritz, and V. Rommelaere, 1998: Report of the Third EISMINT Workshop on Model Intercomparison, European Science Foundation (Strasbourg), 140 p.

Huybrechts, P., C. Mayer, H. Oerter and F. Jung-Rothenhäusler, 1999: Climate change and sea level: ice-dynamics and mass-balance studies on the Greenland Ice Sheet, Report on the Contribution of the Alfred Wegener Institute to EU Contract No. ENV4-CT95-0124, European Commision, DG XII, 18 pp.

Huybrechts, Ph., D. Steinhage, F. Wilhelms, and J.L. Bamber, 2000: Balance velocities and measured properties of the Antarctic ice sheet from a new compilation of gridded data for modelling. *Annals of Glaciology*, **30**, 52-60.

Hyvarinen, H., 1999: Shore displacement and stone age dwelling sites near Helsinki, southern coast of Finland. In: *Digitall, Papers dedicated to Ari Siiriainen*. Finnish Antiquarian Society, 79-86p.

Ingolfsson, O., C. Hjort, P.A. Berkman, S. Björck, E. Colhoun, I.D. Goodwin, B. Hall, K. Hirakawa, M. Melles, P. Möller, and M.L. Prentice, 1998: Antarctic glacial history since the Last Glacial Maximum: an overview of the record on land. *Antarctic Science*, **10**, 326-344.

IOC (Intergovernmental Oceanographic Commission), 1997: Global Sea Level Observing System (GLOSS) implementation plan-1997. Intergovernmental Oceanographic Commission, Technical Series, No. 50, 91pp. & Annexes.

IPCC, 1996: *Climate Change 1995: The Science of Climate Change. Contribution of Working Group I to the Second Assessment Report of the Intergovernmental Panel on Climate Change*. Houghton, J.T., L.G. Meira Filho, B.A. Callander, N. Harris, A. Kattenberg, and K. Maskell (eds.). Cambridge University Press, Cambridge, United Kingdom and New York, NY, USA, 572 pp.

IPCC, 2001: *Climate Change 2001: Impacts, Adaptations and Vulnerability: Contribution of Working Group II to the Third Assessment Report of the Intergovernmental Panel on Climate Change*. J.J. McCarthy, O.F. Canziani, N.A. Leary, D.J. Dokken, K.S.

White (eds.). Cambridge University Press, Cambridge, United Kingdom and New York, NY, USA.

Jackett, D.R., T.J. McDougall, M.H. England, and A.C. Hirst, 2000: Thermal expansion in ocean and coupled general circulation models. *Journal of Climate*. **13**, 1384-1405.

Jacobs, S.S., H.H. Hellmer, C.S.M. Doake, A. Jenkins, and R. Frolich, 1992: Melting of ice shelves and the mass balance of Antarctica. *Journal of Glaciology*, **38**, 375-387

Jacobs, S.S., H.H. Hellmer, and A. Jenkins, 1996: Antarctic ice sheet melting in the Southeast Pacific. *Geophysical Research Letters*, **23**, 957-960.

Jacobson, H. P., and C.F. Raymond, 1998: Thermal effects on the location of ice stream margins. *Journal of Geophysical Research*, **103**, 12111-12122.

James, T.S., and E.R. Ivins. 1998. Predictions of Antarctic crustal motions driven by present-day ice sheet evolution and by isostatic memory of the Last Glacial Maximum. *J. Geophys. Res.*, **103**, 4993-5017.

Janssens, I., and P. Huybrechts, 2000: The treatment of meltwater retention in mass-balance parameterisations of the Greenland Ice Sheet, *Annals of Glaciology*, **31**, 133-140.

Jóhannesson, T., C. Raymond and E. Waddington, 1989: Timescale for adjustment of glaciers to changes in mass balance. *J. Glaciol.*, **35**, 355-369.

Johnson, G.C., A. Orsi, 1997: Southwest Pacific Ocean Water-Mass Changes between 1968/69 and 1990/91. *Journal of Climate*, **10**, 306-16.

Johnston, P., 1993: The effect of spatially non-uniform water loads on the prediction of sea level change. *Geophysical Journal International*, **114**, 615-634.

Johnston, P., K. Lambeck, and D. Wolf, 1997: Material versus isobaric internal boundaries in the Earth and their influence on postglacial rebound, *Geophysical Journal International*, **129**, 252-268.

Johnston, P.J. and K. Lambeck, 1999: Postglacial rebound and sea level contributions to changes in the geoid and the Earth's rotation axis. *Geophysical Journal International*, **136**, 537-558.

Johnston, P. and K. Lambeck, 2000: Automatic inference of ice models from postglacial sea level observations: Theory and application to the British Isles. *Journal of Geophysical Research*, 105, 13179-13194.

Joughin, I., R. Kwok, M. Fahnestock, 1996a: Estimation of ice-sheet motion using satellite radar interferometry: method and error analysis with application to Humboldt Glacier, Greenland. *Journal of Glaciology*, **42**, 564-575.

Joughin, I., S. Tulaczyk, M. Fahnestock, R. Kwok, 1996b: A mini-surge on the Ryder Glacier, Greenland, observed by satellite radar interferometry. *Science*, **274**, 228-230.

Joughin, I., L. Gray, R.A. Bindschadler, S. Price, D.L. Morse, C.L. Hulbe, K. Mattar, and C. Werner, 1999: Tributaries of West Antarctic ice streams revealed by RADARSAT interferometry, *Science*, **286**, 283-286.

Joyce, T.M. and P. Robbins, 1996: The long-term hydrographic record at Bermuda. *Journal of Climate*, **9**, 3121-3131.

Joyce, T.M., R.S. Pickart, and R.C. Millard, 1999: Long-term hydrographic changes at 52 and 66W in the North Atlantic subtropical gyre and Caribbean. *Deep-Sea Research II*, **46**, 245-78.

Jung-Rothenhäusler, F., 1998: Remote sensing and GIS studies in North-East Greenland, Berichte zur Polarforschung (Alfred-Wegener-Institut, Bremerhaven), **280**, 161 p.

Kapsner, W.R., R.B. Alley, C.A. Shuman, S. Anandakrishnan and P.M. Grootes, 1995: Dominant influence of atmospheric circulation on snow accumulation in central Greenland, *Nature*, **373**, 52-54.

Kaser, G. 1999: A review of modern fluctuations of tropical glaciers. *Glob. and Plan. Change*, **22**, 93-104.

Kaser, G., Ch. Georges and A. Ames 1996: Modern glacier fluctuations in the Huascrán-Chopiqualqui Massif of the Cordillera Blanca, Perú.

Z. Gletscherkunde Glazialgeol., **32**, 91-99.

Kass, E., T.-S. Li, and T. Schmith, 1996: Statistical hindcast of wind climatology in the North Atlantic and northwestern European region. *Climate Research*, **7**, 97-110.

Kattenberg, A., F. Giorgi, H. Grassl, G.A. Meehl, J.F.B. Mitchell, R.J. Stouffer, T. Tokioka, A.J. Weaver and T.M.L. Wigley, 1996: Climate models - projections of future climate, in Houghton, J.T., L.G. Meira Filho, B.A. Callander, N. Harris, A. Kattenberg and K. Maskell (eds), *Climate Change 1995. The Science of Climate Change*, Cambridge.

Kaufmann, G. and D. Wolf, 1999: Effects of lateral viscosity variations on postglacial rebound: an analytical approach. *Geophysical Journal International*, **137**, 489-500.

Kauker, F., 1998: Regionalization of climate model results for the North Sea. PhD thesis, University of Hamburg, 109pp.

Kaul, M.K., 1999: Inventory of the Himalayan glaciers. Special publication of the Geological Survey of India.

Kearney, M.S., 1996: sea level change during the last thousand years in Chesapeake Bay. *Journal of Coastal Research*, **12**, 977-983.

Kearney, M.S., and J.C. Stevenson, 1991: Island land loss and marsh vertical accretion rate evidence for historical sea level changes in Chesapeake Bay. *Journal of Coastal Resources*, **7**, no. 2, pp 403-415.

Knutti, R. and T.F. Stocker, 2000: Influence of the thermohaline circulation on projected sea level rise. *Journal of Climate*, **13**, 1997-2001.

Koblinsky, C.J., P. Gaspar and G. Lagerloef, 1992: The future of spaceborne altimetry: oceans and climate change. A long-term strategy. Joint Ocenographic Institutions Inc., Washington DC, 75pp.

Koltermann, K.P., A.V. Sokov, V.P. Tereschenkov, S.A. Dobroliubov, K. Lorbacher, and A. Sy, 1999: Decadal changes in the thermohaline circulation of the North Atlantic. *Deep-Sea Research II*, **46**, 109-38.

Koster, R.D., P.R. Houser, E.T. Engman, and W.P. Kustas. 1999: Remote sensing may provide unprecedented hydrological data. *EOS*, **80**, 156.

Kotlyakov, V. M., K.S. Losev, and I.A. Loseva, 1978: The ice budget of Antarctica. *Polar Geogr. Geol.*, **2**, 251-262.

Krabill, W., E. Frederick, S. Manizade, C. Martin, J. Sonntag, R. Swift, R. Thomas, W. Wright, and J. Yungel, 1999: Rapid thinning of parts of the southern Greenland Ice Sheet. *Science*, **283**, 1522-1524.

Krabill, W., W. Abdalati, E. Frederick, S. Manizade, C. Martin, J. Sonntag, R. Swift, R. Thomas, W. Wright, and J. Yungel, 2000: Greenland Ice Sheet: high-elevation balance and peripheral thinning, *Science*, **289**, 428-430.

Kuhn, M. 1984. Mass budget imbalances as criterion for a climatic classification of glaciers. *Geogr. Annaler*, **66A**, 229-238.

Kuhn, M., E. Dreiseitl, S. Hofinger, G. Markl, N. Span, G. Kaser 1999: Measurements and models of the mass balance of Hintereisferner. *Geogr. Annaler*, **81**A, 659-670.

Kuzmichenok, V.A., 1993: Glaciers of the Tien Shan. Computerised analysis of the Inventory. *Materialy Glyatsiologicheskikh Isledovaniy*, **77**, 29-40.

Laborel, J., C. Morhange, R. Lafont, J. Le Campion, F. Laborel-Deguen, and S. Sartoretto, 1994: Biological evidence of sea level rise during the last 4500 years on the rocky coasts of continental southwestern France and Corsica, *Marine Geology*, **120**, 203-223.

Lambeck, K., 1988: *Geophysical Geodesy: The Slow Deformations of the Earth*. Oxford University Press, 718 pp.

Lambeck, K., 1996: Limits on the areal extent of the Barents Sea ice sheet in Late Weischelian time, *Palaeogeography Palaeoclimatology Palaeoecology*, **12**, 41-51.

Lambeck, K., and M. Nakada, 1990: Late Pleistocene and Holocene sea level change along the Australian coast. *Palaeogeography Palaeoclimatology Palaeoecology*, **89**, 143-176.

Lambeck, K. and P. Johnston, 1998: The viscosity of the mantle: Evidence from analyses of glacial rebound phenomena. In: *The Earth's mantle*, Jackson, I. (ed), Cambridge University Press, Cambridge, 461-502pp.

Lambeck, K., C. Smither, and M. Ekman, 1998: Tests of glacial rebound models for Fennoscandinavia based on instrumented sea- and lake-level records. *Geophysical Journal International*, **135**, 375-387.

Lambeck, K. and E. Bard, 2000: Sea level change along the French Mediterranean coast since the time of the Last Glacial Maximum. *Earth Planetary Science Letters*, **175**, 203–222.

Lamont, G.N., T.J.Chinn, and B.B. Fitzharris, 1999: Slopes of glacier ELAs in the Southern Alps of New Zealand in relation to atmospheric circulation patterns. *Global and Planetary Change*, **22**, 209-219.

Langenberg, H., A. Pfizenmayer, H. von Storch and J. Suendermann, 1999: Storm-related sea level variations along the North Sea coast: natural variability and anthropogenic change, *Continental Shelf Research*, **19**, 821-842.

Latif, M. and E. Roeckner, 2000: Tropical stabilisation of the thermohaline circulation in a greenhouse warming simulation, *J. Climate*, **13**, 1809-1813.

Ledwell, J.R., A.J.Watson and C.S. Law, 1993: Evidence for slow mixing across the pycnocline from an open-ocean tracer-release experiment. *Nature*, **364**, 701-703.

Ledwell, J.R., A.J. Watson and C.S. Law, 1998: Mixing of a tracer in the pycnocline. *Journal of Geophysical Research*, **103**, 21,499-21,529.

Leggett, J., W.J. Pepper and R.J. Swart, 1992: Emissions scenarios for the IPCC: an update. In *Climate change 1992: the supplementary report to the IPCC scientific assessment*, Houghton, J.T., B.A. Callander and S.K. Varney (eds.), Cambridge University Press, Cambridge

Letreguilly, A., P. Huybrechts, N. Reeh, 1991: Steady-state characteristics of the Greenland ice sheet under different climates. *J. Glaciology*, **37**, 149-157.

Levitus, S., 1989a: Interpentadal variability of temperature and salinity at intermediate depths of the North Atlantic, 1970-1974 versus 1955-1959. *Journal of Geophysical Research*, **94**, 6091-131.

Levitus, S., 1989b: Interpentadal variability of temperature and salinity in the deep North Atlantic, 1970-1974 versus 1955-1959. *Journal of Geophysical Research*, **94**, 16,125-16,131.

Levitus, S., 1990: Interpentadal variability of steric sea level and geopotential thickness of the North Atlantic, 1970-1974 versus 1955-1959. *Journal of Geophysical Research*, **95**, 5233-8.

Levitus, S., J.I. Antonov, T.P. Boyer, and C. Stephens, 2000: Warming of the World Ocean. *Science*, **287**, 2225-2229.

Lingle, C., and D.N. Covey, 1998: Elevation changes on the East Antarctic ice sheet, 1978-93, from satellite radar altimetry: a preliminary assessment. *Annals of Glaciology*, **27**, 7-18.

Linsley, B.K., 1996: Oxygen-isotope record of sea level and climate variations in the Sulu Sea over the past 150,000 years, *Nature*, **380**, 234-237.

Liu, S. Y. and Z. C. Xie, 2000: Glacier mass balance and fluctuations. In: *Glaciers and Environment in China*, Y. F. Shi (ed.), Science Press, Beijing, 101-103.

Liu, C.H., G.P. Song, M.X. Jin, 1992: Recent change and trend prediction of glaciers in Qilian Mountains. In Memoirs of Lanzhou Institute of Glaciology and Geocryology, Chinese Academy of Sciences, No. 7 (The monitoring of glacier, climate runoff changes and the research of cold region hydrology in Qilian Mountains), Beijing, Science Press.

Lowe, J.A., J.M. Gregory and R.A.Flather, 2001: Changes in the occurrence of storm surges around the United Kingdom under a future climate scenario using a dynamic storm surge model driven by the Hadley Centre climate model. *J. Climate*, in press.

Manabe, S., R.J. Stouffer, 1993: Century-scale effects of increased atmospheric CO_2 on the ocean-atmosphere system. *Nature*, **364**, 215-8.

Manabe, S., R. Stouffer, 1994: Multiple-century response of a coupled ocean-atmosphere model to an increase of atmospheric carbon dioxide. *Journal of Climate*, **7**, 5-23.

Maul, G.A. and Martin, D.M. 1993: Sea level rise at Key West, Florida, 1846-1992: America's longest instrument record? *Geophysical*

Research Letters, **20**, 1955-1958.

Mayer, C., and P. Huybrechts, 1999: Ice-dynamic conditions across the grounding zone, Ekströmisen, East Antarctica. *Journal of Glaciology*, **45**, 384-393.

Meier, M.F., 1984: Contribution of small glaciers to global sea level. *Science*, **226**, 1418-21.

Meier, M.F., 1993: Ice, climate and sea level: do we know what is happening? in: *Ice in the climate system*, Peltier, W.R. (ed.), NATO ASI Series I, Springer-Verlag, Heidelberg, 141-160.

Meier, M.F., and D.B. Bahr, 1996: Counting glaciers: use of scaling methods to estimate the number and size distribution of the glaciers of the world. In: *Glaciers, Ice Sheets and Volcanoes: A Tribute to Mark F. Meier*, S.C. Colbeck (ed.), CRREL Special Report 96-27, 89-94. U. S. Army Hanover, New Hampshire.

Miller, J.R. and G.L. Russell, 2000: Projected impact of climate change on the freshwater and salt budgets of the Arctic Ocean by a global climate model. *Geophysical Research Letters*, **27**, 1183-1186.

Mitchell, W., J.Chittleborough, B.Ronai and G.W.Lennon, 2000: Sea Level Rise in Australia and the Pacific. *The South Pacific Sea Level and Climate Change Newsletter, Quarterly Newsletter*, **5**, 10-19.

Mitchum, G.T., 1998: Monitoring the stability of satellite altimeters with tide gauges. *Journal of Atmospheric and Oceanic Technology*, **15**, 721-730.

Mitrovica J.X. and J.L. Davis, 1995: Present-day post-glacial sea level change far from the Late Pleistocene ice sheets: Implications for recent analyses of tide gauge records. *Geophys. Res. Lett.*, **22**, 2529-32.

Mitrovica, J. X. and G.A. Milne, 1998: Glaciation-induced perturbations in the Earth's rotation: a new appraisal. *Journal Geophysical Research*, **103**, 985-1005.

Mitrovica, J.X. and W.R. Peltier, 1991: On postglacial geoid subsidence over the equatorial oceans, *Journal of Geophysical Research*, **96**, 20053-20071.

Mohr, J. C., N. Reeh, and S.N. Madsen, 1998: Three-dimensional glacial flow and surface elevation measured with radar interferometry. *Nature*, **391**, 273-276.

Morhange, C., J. Laborel, A. Hesnard, and A. Prone, 1996: Variation of relative mean sea level during the last 4000 years on the northern shores of Lacydon, the ancient harbour of Marseille (Chantier J. Verne). *Journal of Coastal Research*, **12**, 841–849.

Morison, J., M. Steele, and R. Andersen, 1998: Hydrography of the upper Arctic Ocean measured from the nuclear submarine U.S.S. Pargo. *Deep-Sea Research I*, **45**, 15-38.

Mörner, N.A. 1973. Eustatic changes during the last 300 years. Palaeogeography, Palaeoclimatology, Palaeoecology, **13**, 1-14.

Murphy, J.M. and J.F.B. Mitchell, 1995: Transient response of the Hadley Centre coupled ocean-atmosphere model to increasing carbon dioxide. Part II: Spatial and temporal structure of response. *J. Climate*, **8**, 57-80.

Nakada, M. and K. Lambeck, 1988: The melting history of the Late Pleistocene Antarctic ice sheet, *Nature*, **333**, 36-40.

Nakiboglu, S.M. and K. Lambeck, 1991: Secular sea level change. In: *Glacial Isostasy, Sea Level and Mantle Rheology*, R. Sabadini, K. Lambeck and E. Boschi, (eds.), Kluwer Academic Publ., 237-258.

Nakićenović, N., J. Alcamo, G. Davis, B. de Vries, J. Fenhann, S. Gaffin, K. Gregory, A. Grubler, T. Y. Jung, T. Kram, E. L. La Rovere, L. Michaelis, S. Mori, T. Morita, W. Pepper, H. Pitcher, L. Price, K. Raihi, A. Roehrl, H-H. Rogner, A. Sankovski, M. Schlesinger, P. Shukla, S. Smith, R. Swart, S. van Rooijen, N. Victor, Z. Dadi: 2000: *IPCC Special Report on Emissions Scenarios*, Cambridge University Press, Cambridge, United Kingdom and New York, NY, USA, 599 pp.

National Bureau of Marine Management, 1992: *Bulletin of China's sea level*. National Bureau of Marine Management, Beijing, 26pp.

NASA (National Aeronautics and Space Administration), 1996: Gravity Recovery and Climate Experiment. Proposal to NASA's Earth System Science Pathfinder Program.

Neilan, R., P.A. Van Scoy, and P.L. Woodworth, (eds)., 1998: Proceedings of the workshop on methods for monitoring sea level: GPS and tide gauge benchmark monitoring and GPS altimeter calibration. Workshop organised by the IGS and PSMSL, Jet Propulsion Laboratory, Pasadena, California, 17-18 March 1997. 202pp.

Nerem, R. S. and S.M. Klosko, 1996: Secular variations of the zonal harmonics and polar motion as geophysical constraints. In: *Global gravity field and its temporal variations*, Rapp, R. H., Cazenave, A. A. and Nerem, R. S. (eds.)Springer, New York, 152-163.

Nerem, R.S., B.J. Haines, J. Hendricks, J.F. Minster, G.T. Mitchum, and W.B. White, 1997: Improved determination of global mean sea level variations using TOPEX/POSEIDON altimeter data. *Geophysical Research Letters.*, **24**, 1331-4.

Nerem, R.S., 1999: Measuring very low frequency sea level variations using satellite altimeter data. *Global and Planetary Change*, **20**, 157-171.

Nerem, R.S., D.P. Chambers E.W. Leuliette, G.T. Mitchum, and B.S. Giese, 1999: Variations in global mea sea level associated with the 1997-98 ENSO event: implications for measuring long term sea level change. *Geophysical Research Letters*, **26**, 3005-3008.

Nicholls, K. W., 1997: Predicted reduction in basal met rates of an Antarctic ice shelf in a warmer climate. *Nature*, **388**, 460-462.

NRC (National Research Council), 1997: Satellite gravity and the geosphere. National Academy Press: Washington, D.C.

Oerlemans, J., 1981: Effect of irregular fluctuation in Antarctic precipitation on global sea level. *Nature*, **290**, 770-772.

Oerlemans, J., 1991: The mass balance of the Greenland ice sheet: sensitivity to climate change as revealed by energy balance modelling. *Holocene*, **1**, 40-49.

Oerlemans, J., 1999: Comments on "Mass balance of glaciers other than the ice sheets", by J. Graham Cogley and W.P. Adams. *J Glaciology*, **45**, 397-398.

Oerlemans, J., and J.P.F. Fortuin, 1992: Sensitivity of glaciers and small ice caps to greenhouse warming. *Science*, **258**, 115-117.

Oerlemans, J., B. Anderson, A. Hubbard, P. Huybrechts, T. Jóhanneson, W.H. Knap, M. Schmeits, A.P. Stroeven, R.S.W. van de Wal, J. Wallinga, and Z. Zuo, 1998: Modelling the response of glaciers to climate warming. *Climate Dynamics*, **14**, 267-274.

O'Farrell, S.P., J.L. McGregor, L.D. Rotstayn, W.F. Budd, C. Zweck, and R. Warner, 1997: Impact of transient increases in atmospheric CO_2 on the accumulation and mass balance of the Antarctic ice sheet. *Annals of Glaciology*, **25**, 137-144.

Ohmura, A., and N. Reeh, 1991: New precipitation and accumulation maps for Greenland. *J. Glaciol.*, **37**, 140-148.

Ohmura, A., M. Wild, and L. Bengtsson, 1996: A possible change in mass balance of Greenland and Antarctic ice sheets in the coming century. *Journal of Climate*, **9**, 2124-2135.

Ohmura, A., P. Calanca, M. Wild, and M. Anklin, 1999: Precipitation, accumulation and mass balance of the Greenland Ice Sheet. *Zeitschrift für Gletscherkunde und Glazialgeologie*, **35**, 1-20.

Open University, 1989. Waves, tides and shallow-water processes. Open University Oceanography Series Vol.4. Pergamon Press, Oxford, in association with the Open University, 187pp.

Oppenheimer, M., 1998: Global warming and the stability of the West Antarctic ice sheet. *Nature*, **393**, 325-331.

Ota, Y., A. Omura, and T. Miyauchi, 1992: Last interglacial shoreline map of Japan. Japanese Working Group for IGCP Project 274.

Ota, Y., L.J. Brown, K.R. Berryman, T. Fujimori, and T. Miyauchi, 1995: Vertical tectonic movement in northeastern Marlborough: stratigraphic, radiocarbon, and paleoecological data from Holocene estuaries, *New Zealand Journal of Geological and Geophysical*, **38**, 269-282.

Parrilla, G., A. Lavin, H. Bryden, M. Garcia, and R. Millard, 1994: Rising temperatures in the subtropical North Atlantic Ocean over the past 35 years. *Nature,* **369,** 48-51.

Payne, A.J., 1998: Dynamics of the Siple Coast ice streams, West Antarctica: results from a thermomechanical ice sheet model. *Geophysical Research Letters,* **25,** 3173-3176.

Peltier, W.R., 1994: Ice age paleotopography. *Science,* **265,** 195-201.

Peltier, W.R., 1996: Global sea level rise and glacial isostatic adjustment: an analysis of data from the east coast of North America. *Geophysical Research Letters,* **23,** 717-720.

Peltier, W.R., 1998: Postglacial variations in the level of the Sea: implications for climate dynamics and solid-earth geophysics, *Review of Geophysics,* **36,** 603-689.

Peltier, W.R., 2000: Global glacial isostatic adjustment. In *Sea level rise: history and consequences,* B.C. Douglas, M.S. Kearney and S.P. Leatherman (eds.), Academic Press, 65-95.

Peltier, W.R. and A.M. Tushingham, 1989: Global sea level rise and the greenhouse effect: Might they be connected? *Science,* **244,** 806-10.

Peltier, W.R. and A.M. Tushingham, 1991: Influence of glacial isostatic adjustment on tide gauge measurements of secular sea level change. *J Geophys Res,* **96,** 6779-6796.

Peltier, W.R. and X. Jiang, 1996: Mantle viscosity from the simultaneous inversion of multiple datasets pertaining to postglacial rebound. *Geophys. Res. Lett.,* **33,** 503-506.

Peltier, W.R., and X. Jiang, 1997: Mantle viscosity, glacial isostatic adjustment and the eustatic level of the sea. *Surveys in Geophysics,* **18,** 239-277.

Pfeffer, W. T., M.F. Meier, and T.H. Illangasekare, 1991: Retention of Greenland runoff by refreezing: implications for projected future sea level change. *J. Geophys. Res.,* **96,** 22117-22124.

Pirazzoli, P.A., 1991: *World Atlas of Holocene sea level changes.* Elsevier, Amsterdam, 300p.

Pirazzoli, P.A., 2000: Surges, atmospheric pressure and wind change, and flooding probability on the Atlantic coast of France. *Oceanologica Acta,* **23,** 643-661.

Pirazzoli, P.A., S.C. Stiros, M. Arnold, J. Laborel, F. Laborel-Deguen, and S. Papageorgiou, 1994: Episodic uplift deduced from Holocene shorelines in the Perachora Peninsula, Corinth area, Greece, *Tectonophysics,* **229,** 201-209.

Pittock, A.B., K. Walsh, and K. McInnes, 1996: Tropical cyclones and coastal inundation under enhanced greenhouse conditions. *Water, Air and Soil Pollution,* **92,** 159-169.

Pugh, D.T. and G.A. Maul, 1999: Coastal sea level prediction for climate change. In: *Coastal Ocean Prediction. Coastal and Estuarine Studies,* American Geophysical Union, Washington, D.C., **56,** 377-404.

Quadfasel, D., A. Sy, and D. Wells, 1991: Warming in the Arctic. *Nature,* **350,** 385.

Qin, D., P.A. Mayewski, C.P. Wake, S.C. Kang, J.W. Ren, S.G. Hou, T.D. Yao, Q.Z. Yang, Z.F. Jin, D.S. Mi, 2000: Evidence for recent climate change from ice cores in the central Himalaya. *Annals of Glaciology,* **31,** 153-158.

Raper, S.C.B., and U. Cubasch, 1996: Emulation of the results from a coupled general circulation model using a simple climate model. *Geophysical Research Letters,* **23,** 1107-1110.

Raper, S.C.B., T.M.L Wigley, and R.A. Warrick, 1996: Global sea level rise: Past and Future. In: *Sea Level Rise and Coastal Subsidence, Causes, Consequences and Strategies,* Kluwer Academic Publishers, Dordrecht, 369pp.

Raper, S.C.B., J.M. Gregory and T.J. Osborn, 2001: Use of an upwelling-diffusion energy balance climate model to simulate and diagnose A/OGCM results. *Clim. Dyn.,* in press.

Raper, S.C.B., O. Brown and R.J. Braithwaite, 2000: A geometric glacier model suitable for sea level change calculations, *Journal of Glaciology.,* **46,** 357-368.

Read, J.F., W.J. Gould, 1992: Cooling and freshening of the subpolar North Atlantic Ocean since the 1960s. *Nature,* **360,** 55-7.

Reeh, N., and W. Starzer, 1996: Spatial resolution of ice-sheet topography: influence on Greenland mass-balance modelling. *GGU rapport ,* **1996/53,** 85-94.

Reeh, N., H.H. Thomsen, O.B. Olesen, and W. Starzer, 1997: Mass balance of north Greenland. *Science,* **278,** 207-209.

Reeh, N., C. Mayer, H. Miller, H.H. Thomson, and A. Weidick, 1999: Present and past climate control on fjord glaciations in Greenland: implications for IRD-deposition in the sea. *Geophysical Research Letters,* **26,** 1039-1042.

Retzlaff, R., and C.R. Bentley, 1993: Timing of stagnation of ice stream C, West Antarctica, from short-pulse radar studies of buried surface crevasses. *Journal of Glaciology,* **39,** 553-561.

Rider, K.M., G.J. Komen, and J.J. Beersma, 1996: Simulations of the response of the ocean waves in the North Atlantic and North Sea to CO_2 doubling in the atmosphere. KNMI Scientific Report WR 96-95, De Bilt, Netherlands.

Ridgway, K.R. and J.S. Godfrey, 1996: Long-term temperature and circulation changes off eastern Australia. *Journal of Geophysical Research,* **101,** 3615-27.

Rignot, E.J., 1998a: Fast recession of a West Antarctic glacier. *Science,* **281,** 549-551.

Rignot, E., 1998b: Radar interferometry detection of hinge-line migration on Rutford Ice Stream and Carlson Inlet, Antarctica. *Annals of Glaciology,* **27,** 25-32.

Rignot, E.J., 1998c: Hinge-line migration of Petermann Gletscher, north Greenland, detected using satellite-radar interferometry. *Journal of Glaciology,* **44,** 469-476.

Rignot, E.J., S.P. Gogineni, W.B. Krabill, and S. Ekholm, 1997: North and Northeast Greenland Ice Discharge from Satellite Radar Interferometry. *Science,* **276,** 934-937.

Robasky, F. M., and D.H. Bromwich, 1994: Greenland precipitation estimates from the atmospheric moisture budget. *Geophysical Research Letters,* **21,** 2485-2498.

Roemmich, D. and C. Wunsch, 1984: Apparent changes in the climatic state of the deep North Atlantic Ocean. *Nature,* **307,** 447-450.

Roemmich, D., 1990: Sea level and the thermal variability of the oceans. In: *Sea Level Change,* National Academy Press, Washington DC, 1990, 208-229.

Roemmich, D., 1992: Ocean warming and sea level rise along the southwest U.S. coast. *Science,* **257,** 373-5.

Rommelaere, V., and D.R. MacAyeal, 1997: Large-scale rheology of the Ross Ice Shelf, Antarctica, computed by a control method. *Annals of Glaciology,* **24,** 43-48.

Rott, H., M. Stuefer, and A. Siegel, 1998: Mass fluxes and dynamics of Moreno Glacier, Southern Patagonian Icefield. *Geoph. Res. Letters,* **25,** 1407-1410.

Russell, G.L., V. Gornitz and J.R. Miller, 2000: Regional sea level changes projected by the NASA/GISS atmosphere-ocean model. *Climate Dynamics,* 16, 789-797.

Sahagian, D., 2000: Global physical effects of anthropogenic hydrological alterations: sea level and water redistribution. *Global and Planetary Change,* **25,** 39-48.

Sahagian, D.L., and S. Zerbini, 1999: Global and regional sea level changes and the hydrological cycle. IGBP/GAIM Report Series, No. 8.

Sahagian, D.L., F.W. Schwartz, and D.K. Jacobs, 1994: Direct anthropogenic contributions to sea level rise in the twentieth century. *Nature,* **367,** 54-7.

SAR, see IPCC, 1996.

Scherer, R. P., A. Aldahan, S. Tulaczyk, G. Possnert, H. Engelhardt, and B. Kamb, 1998: Pleistocene collapse of the West Antarctic ice sheet. *Science,* **281,** 82-85.

Schlesinger, M.E. and X. Jiang, 1990: Simple model representation of atmosphere-ocean GCMs and estimation of the timescale of CO_2-induced climate change. *J Climate*, **3**, 1297-1315.

Schmith, T., E. Kaas, and T.-S. Li, 1998: Northeast Atlantic winter storminess 1875-1995 re-analysed. *Climate Dynamics*, **14**, 529-536.

Schönwiese, C.-D., J. Rapp, T. Fuchs and M. Denhard, 1994: Observed climate trends in Europe 1891-1990. *Meteorol. Zeitschrift*, **3**, 22-28.

Schönwiese, C.-D. and J. Rapp, 1997: *Climate Trend Atlas of Europe Based on Observations 1891-1990*. Kluwer Academic Publishers, Dordrecht, 228 p.

Senior, C.A. and J.F.B. Mitchell, 2000: The time dependence of climate sensitivity. *Geophys Res. Lett.*, **27**, 2685-2688.

Shabtaie, S., and C.R. Bentley, 1987: West Antarctic ice streams draining into the Ross ice shelf: configuration and mass balance. *J. Geophys. Res.*, **92**, 1311-1336.

Shackleton, N.J., 1987: Oxygen isotopes, ice volume and sea level, *Quaternary Science Review*, **6**, 183-190.

Shaffer, G., O. Leth, O. Ulloa, J. Bendtsen, G. Daneri, V. Dellarossa, S. Hormazabal, and P. Sehstedt, 2000: Warming and circulation change in the eastern South Pacific Ocean. *Geophys. Res. Lett.*, GRL **27(9)**, 1247-1250.

Shaw, J. and J. Ceman, 2000: Salt-marsh aggradation in response to late-Holocene sea level rise at Amherst Point, Nova Scotia, Canada. *The Holocene*, **9**, 439-451.

Shennan, I. and P.L. Woodworth, 1992: A comparison of late Holocene and twentieth-century sea level trends from the UK and North Sea region. *Geophysical Journal International*, **109**, 96-105.

Shi, Y.F., 1996 (ed.): Sea level change in China. In: *Climatic and sea level changes and their trends and effects in China*. Shangdong Science And Technology Press, Jinan, pp464.

Shi, Y.F. and K.J. Zhou, 1992: Characteristics and recent variation of surface water in China, and its effect on the environment. In: *Preliminary research on global change in China*, Yie, N.Z. and B.Q. Chen, (eds.), Beijing, Seismological Press, pp85-158.

Shi, Y. F., Z.C. Kong and S.M. Wang, L.Y. Tang, F.B. Wang, T.D. Yao, X.T. Zhao, P.Y. Zhang, S.H. Shi 1994: Climates and environments of the Holocene megathermal maximum in China. *Science in China (series B)*, **37**, 481-493.

Skvarca, P., H. Rott, T. Nagler, 1995: Drastic retreat of Upsala Glacier, Southern Patagonia, revealed by ESR-1/SAR images and field survey. Revista Selper, 11, 51-55.

Skvarca, P., W. Rack, H. Rott, T. Ibarzabal y Donangelo, 1998: Evidence of recent climatic warming on the eastern Antarctic Peninsula. *Annals of Glaciology*, **27**, 628-632.

Smith, I. N., W.F. Budd, and P. Reid, 1998: Model estimates of Antarctic accumulation rates and their relationship to temperature changes. *Annals of Glaciology*, **27**, 246-250.

Smith, I., 1999: Estimating mass balance components of the Greenland Ice Sheet from a long-term GCM simulation. *Global and Planetary Change*, **20**, 19-32.

Smith, D., S. Raper, S. Zerbini, A. Sanchez-Arcilla, and R. Nicholls, (eds.) 2000: Sea Level Change and Coastal Processes: Implications for Europe, Office for official Publications of the European Union Community, Luxembourg, 247p.

Spencer, N.E. and P.L. Woodworth, 1993: Data holdings of the Permanent Service for Mean Sea Level (November 1993). Bidston, Birkenhead: Permanent Service for Mean Sea Level. 81pp.

Stanley, D.J., 1997: Mediterranean deltas: Subsidence as a major control of relative sea level rise, Monaco, *Bulletin de l'Institut océanographique*, **Special no. 18**, 35-62.

Steele, M. and T. Boyd, 1998: Retreat of the cold halocline layer in the Arctic Ocean. *Journal of Geophysical Research*, **103**, 10,419-10,435.

Steig, E. J., 1997: How well can we parameterize past accumulation rates in polar ice sheets? *Annals of Glaciology*, **25**, 418-422.

Steinberger, B. and R.J. O'Connell, 1997: Changes of the Earth's rotation axis owing to advection of mantle density heterogeneities.

Nature, **387**, 169-173.

Stephenson, S. N., and R.A. Bindschadler, 1988: Observed velocity fluctuations on a major Antarctic ice stream. *Nature*, **334**, 695-697.

Stirling, C.H., T.M. Esat, K. Lambeck and M.T. McCulloch, 1998: Timing and duration of the Last Interglacial: evidence for a restricted interval of widespread coral reef growth, *Earth Planetary Science Letters*, **160**, 745-762.

Stiros, S.C., Marangou, L. and Arnold, M., 1994: Quaternary uplift and tilting of Amorgos Island (southern Aegean) and the 1956 earthquake, *Earth Planetary Science Letters*, **128**, 65-76.

Stouffer, R.J. and S. Manabe, 1999: Response of a coupled ocean-atmosphere model to increasing carbon dioxide: sensitivity to the rate of increase. *J. Climate*, **12**, 2224-2237.

Stone, J.O., D. Zwartz, M.C.G. Mabin, K. Lambeck, D. Fabel, and L.K. Fifield, 1998: Exposure dating constraints on ice volume and retreat history in East Antarctica, and prospects in West Antarctica. R. Bindschadler and H. Borns (eds.), *Chapman Conference on West Antarctic Ice Sheet*. 13-18 September 1998, American Geophysical Union, Orano, Maine, USA.

Sturges, W., B.D. Hong, 1995: Wind forcing of the Atlantic thermocline along 32N at low frequencies. *Journal of Physical Oceanographer*, **25(July)**, 1706-15.

Sturges, W., B.G. Hong, and A.J. Clarke, 1998: Decadal wind forcing of the North Atlantic subtropical gyre. *Journal of Physical Oceanographer*, **28(April)**, 659-68.

Swift, J.H., E.P. Jones, K. Aagaard, E.C. Carmack, M. Hingston, R.W. MacDonald, F.A. McLaughlin, and R.G. Perkin, 1997: Waters of the Makarov and Canada Basins. *Deep-Sea Research II*, **44**, 1503-29.

Thomas, R. H., 1973: The creep of ice shelves: theory. *J. Glaciol.*, **12**, 45-53.

Thomas, R. H., T.J.O. Sanderson, and K.E. Rose, 1979: Effect of climatic warming on the West Antarctic ice sheet. *Nature*, **277**, 355-358.

Thomas, R. H., B.M. Csatho, S. Gogineni, K.C. Jezek, K. Kuivinen, 1998: Thickening of the western part of the Greenland Ice Sheet. *Journal of Glaciology*, **44**, 653-658.

Thomas, R., T. Akins, B. Csatho, M Fahnestock, P. Gogineni, C. Kim, and J. Sonntag, 2000: Mass balance of the Greenland Ice Sheet at high elevations, *Science*, **289**, 426-428.

Thompson, S.L., and D. Pollard, 1997: Greenland and Antarctic mass balances for present and doubled atmospheric CO_2 from the GENESIS version-2 global climate model. *Journal of Climate*, **10**, 871-900.

Thomson, R.E. and S. Tabata, 1989: Steric sea level trends in the northeast Pacific Ocean: Possible evidence of global sea level rise. *Journal of Climate*, **2**, 542-53.

Titus, J.G., and V. Narayanan, 1996: The risk of sea level rise, *Climatic Change*, **33**, 151-212.

Treshnikov, A.F., 1977: Water masses of the Arctic Basin, in Polar Oceans, pp.17-31, edited by M. Dunbar.

Tromp, J. and J.X. Mitrovica, 1999: Surface loading of a viscoelastic earth - I. General theory. *Geophysical Journal International*, **137**, 847-855

Trupin, A. and J. Wahr, 1990: Spectroscopic analysis of global tide gauge sea level data. *Geophysical Journal International*, **100**, 441-53.

Trupin, A.S., M.F. Meier and J.M. Wahr, 1992: Effects of melting glaciers on the Earth's rotation and gravitational field: 1965-1984. *Geophys. J. Intern.*, **108**, 1-15.

Tsimplis, M.N. and Baker, T.F. 2000. Sea level drop in the Mediterranean Sea: an indicator of deep water salinity and temperature changes? *Geophysical Research Letters*, **27**, 1731-1734.

Turner, J., W.M. Connolley, S. Leonard, G.J. Marshall, and D.G. Vaughan, 1999: Spatial and temporal variability of net snow accumulation over the Antarctic from ECMWF re-analysis project data. *International Journal of Climatology*, **19**, 697-724.

Tushingham, A.M. and W.R. Peltier, 1991: Ice-3G: a new global model of Late Pleistocene deglaciation based upon geophysical predictions of post-glacial relative sea level change. *Journal of Geophysical Research*, **96**, 4497-4523.

Tushingham, A.M. and W.R. Peltier, 1992: Validation of the ICE-3G model of Würm-Wisconsin deglaciation using a global data base of relative sea level histories. *Journal of Geophysical Research*, **97**, 3285-3304.

Tvede, A.M. and T. Laumann, 1997: Glacial variations on a meso-scale example from glaciers in the Aurland Mountains, southern Norway. *Annals of Glaciology*, **24**, 130-134.

Van de Plassche, O., K. Van der Borg and A.F.M. De Jong, 1998: Sea level-climate correlation during the past 1400 yr. *Geology*, **26**, 319-322.

Van de Wal, R. S. W., 1996: Mass balance modelling of the Greenland Ice Sheet: a comparison of an energy balance and a degree-day model. *Annals of Glaciology*, **23**, 36-45.

Van de Wal, R.S.W. and J. Oerlemans, 1994: An energy balance model for the Greenland ice sheet. *Glob. Planetary Change*, **9**, 115-131.

Van de Wal, R. S. W. and J. Oerlemans, 1997: Modelling the short term response of the Greenland Ice Sheet to global warming. *Climate Dynamics*, **13**, 733-744.

Van de Wal, R. S. W. and S. Ekholm, 1996: On elevation models as input for mass balance calculations of the Greenland Ice Sheet. *Annals of Glaciology*, **23**, 181-186.

Van de Wal, R.S.W. and M. Wild, 2001: Modelling the response of glaciers to climate change, applying volume area scaling in combination with a high-resolution GCM. IMAU Report R-01-06, Utrecht University, Netherlands. Also *Climate Dynamics* in press.

Van der Veen, C. J., 1985: Response of a marine ice sheet to changes at the grounding line. *Quat. Res.*, **24**, 257-267.

Van der Veen, C. J. and J.F. Bolzan, 1999: Interannual variability in net accumulation on the Greenland Ice Sheet: observations and implications for mass balance measurements. *Journal of Geophysical Research*, **104**, 2009-2014.

Van der Vink, G. *et al.*, 1998: Why the United States is becoming more vulnerable to natural disasters. *EOS, Transactions of the American Geophysical Union*, **79**, 533-537.

Van Tatenhove, F.G.M., J.J.M. van der Meer and P. Huybrechts, 1995: Glacial-geological/geomorphological research in West Greenland used to test an ice-sheet model. *Quaternary Research*, **44**, 317-327.

Varekamp, J.C. and E. Thomas, 1998: Climate change and the rise and fall of sea level over the millennium. EOS, Transactions of the American Geophysical Union, 79, 69 and 74-75.

Varekamp, J.C., E. Thomas, and O. Van de Plassche, 1992: Relative sea level rise and climate change over the last 1500 years. *Terra Nova*, **4**, pp. 293-304. (R12689 POL library).

Vaughan, D. G., and C.S.M. Doake, 1996: Recent atmospheric warming and retreat of ice shelves on the Antarctic Peninsula. *Nature*, **379**, 328-331.

Vaughan, D. G., J.L. Bamber, M. Giovinetto, J. Russell, and A.P.R. Cooper, 1999: Reassessment of net surface mass balance in Antarctica. *Journal of Climate*, **12**, 933-946.

Vaughan, D.G. and J.R. Spouge, 2001: Risk estimation of collapse of the West Antarctic ice sheet, *Climatic Change*, in press.

Von Storch, H. and H. Reichardt, 1997: A scenario of storm surge statistics for the German Bight at the expected time of doubled atmospheric carbon dioxide concentration. *Journal of Climate*, **10**, 2653-2662.

Vörösmarty, C.J. and D. Sahagian, 2000: Anthropogenic disturbance of the terrestrial water cycle. *Bioscience*, **50**, 753-765.

Voss, R. and U. Mikolajewicz, 2001: Long-term climate changes due to increased CO_2 concentration in the coupled atmosphere-ocean general circulation model ECHAM3/LSG. *Clim. Dyn.*, **17**, 45-60.

Wahr, J., H. Dazhong, A. Trupin and V. Lindqvist, 1993: Secular changes in rotation and gravity: evidence of post-glacial rebound or of changes in polar ice? *Advanced Space Research*, **13**, 257-269.

Wang, S.H., J.M. Yang, X.L. Sun, C.S. Ceng, M.T. Yu, X.Z. Wu, 1990: Sea level changes since deglaciation in the downstream area of Min Jiang and surroundings. *Acta Oceanologica Sinica*, **12**, 64-74.

Warner, R. C., and W.F. Budd, 1998: Modelling the long-term response of the Antarctic ice sheet to global warming. *Annals of Glaciology*, **27**, 161-168.

Warrick, R.A., J. Oerlemans, 1990: Sea Level Rise. In: *Climate Change, The IPCC Scientific Assessment*. J.T. Houghton, G.J. Jenkins, J.J. Ephraums (eds.), pp 260-281.

Warrick, R.A., C. Le Provost, M.F. Meier, J. Oerlemans, P.L. Woodworth, 1996: Changes in Sea Level. In: *Climate Change 1995, The Science of Climate Change*, J.T. Houghton, L.G. Meira Filho, B.A. Callander, N. Harris, A. Klattenberg, K. Maskell (eds.). Cambridge University Press, 359-405.

WASA Group, 1998: Changing waves and storms in the northeast Atlantic. *Bulletin American Meteorogical Society*, **79**, 741-760.

Weaver, A.J. and E.C. Wiebe, 1999: On the sensitivity of projected oceanic thermal expansion to the parameterisation of sub-grid scale ocean mixing. *Geophysical Research Letters*, **26**, 3461-3464.

Weertman, J., 1974: Stability of the junction of an ice sheet and an ice shelf. *Journal of Glaciology*, **13**, 3-11.

Weidick, A., 1984: Review of glacier changes in West Greenland. *Zeitschrift für Gletscherkunde und Glazialgeologie*, **21**, 301-309.

Weidick, A. and E. Morris, 1996: Local glaciers surrounding continental ice sheets. In Haeberli, W., M. Hoezle and S. Suter (eds.), *Into the second century of world glacier monitoring?prospects and strategies*. A contribution to the IHP and the GEMS. Prepared by the World Glacier Monitoring Service.

Whillans, I. M., and R.A. Bindschadler, 1988: Mass balance of ice stream B, West Antarctica. *Ann. Glaciol.*, **11**, 187-193.

Whillans, I. M., and C.J. van der Veen, 1997: The role of lateral drag in the dynamics of Ice Stream B, Antarctica. *Journal of Glaciology*, **43**, 231-237.

Wigley, T.M.L and S.C.B. Raper, 1987: Thermal expansion of sea water associated with global warming. *Nature*, **330**, 127-31.

Wigley, T.M.L and S.C.B. Raper, 1992: Implications for climate and sea level of revised IPCC emissions scenarios. *Nature*, **357** (May 28), 293-300.

Wigley, T.M.L. and S.C.B. Raper, 1993: Future changes in global mean temperature and sea level, in Warrick, R.A., E.M. Barrow and T.M.L.Wigley (eds), *Climate and sea level change: observations, projections and implications*, Cambridge University Press, Cambridge, UK, 424pp.

Wigley, T.M.L and S.C.B. Raper, 1995: An heuristic model for sea level rise due to the melting of small glaciers. *Geophysical Research Letters*, **22**, 2749-2752.

Wild, M., and A. Ohmura, 2000: Changes in Mass balance of the polar ice sheets and sea level under greenhouse warming as projected in high resolution GCM Simulations. *Ann.Glaciol.*, **30**, 197-203.

Williams, M. J. M., R.C. Warner, and W.F. Budd, 1998: The effects of ocean warming on melting and ocean circulation under the Amery ice shelf, East Antarctica. *Annals of Glaciology*, **27**, 75-80.

Wingham, D., A.J. Ridout, R. Scharroo, R.J. Arthern, and C.K. Shum, 1998: Antarctic elevation change from 1992 to 1996. *Science*, **282**, 456-458.

Wong, A.P.S., 1999: Water mass changes in the North and South Pacific Oceans between the 1960s and 1985-94, PhD Thesis, University of Tasmania, 249pp.

Wong, A.P.S., N.L. Bindoff, and J.A. Church, 1999: Coherent large-scale freshening of intermediate waters in the Pacific and Indian Oceans. *Nature*, **400**, 440-443.

Wong, A.P.S., N.L. Bindoff, and J.A. Church, 2001: Freshwater and heat changes in the North and South Pacific Oceans between the 1960s and 1985-94. *J. Climate*, **14**(7), 1613-1633.

Woodroffe, C. and R. McLean, 1990: Microatolls and recent sea level change on coral atolls. *Nature,* **334**, 531-534.

Woodworth, P.L., 1987: Trends in U.K. mean sea level. *Marine Geodesy,* **11**, 57-87.

Woodworth, P.L. 1990: A search for accelerations in records of European mean sea level. *International Journal of Climatology,* **10**, 129-143.

Woodworth, P.L., 1999a: High waters at Liverpool since 1768: the UK's longest sea level record. *Geophysical Research Letters,* **26**, 1589-1592.

Woodworth, P.L. 1999b: A study of changes in high water levels and tides at Liverpool during the last two hundred and thirty years with some historical background. Proudman Oceanographic Laboratory Report No.56, 62 pp.

Woodworth, P.L., M.N. Tsimplis, R.A. Flather and I. Shennan, 1999: A review of the trends observed in British Isles mean sea level data measured by tide gauges. *Geophysical Journal International,* **136**, 651-670.

Wu, P. and Peltier W. R., 1984: Pleistocene deglaciation and the Earth's rotation: A new analysis. *Geophys. J.,* **76**, 202-242.

Yasuda, T. and K. Hanawa, 1997: Decadal changes in the mode waters in the midlatitude North Pacific. *Journal of Physical Oceanography,* **27**, 858-70.

Yokoyama, Y., K. Lambeck, P. de Dekker, P. Johnston, and K. Fifield, 2000: Timing of the last glacial maximum from observed sea level minima. *Nature,* **406**, 713-716.

Zhang, R.-H. and S. Levitus, 1997: Structure and cycle of decadal variability of upper-ocean temperature in the North Pacific. *Journal of Climate,* **10**, 710-27.

Zhang, K., B.C. Douglas, and S.P. Leatherman, 1997: East coast storm surges provide unique climate record. *EOS, Transactions of the American Geophysical Union,* **78(37)**, 389-397.

Zhang, K., B.C. Douglas and S.P. Leatherman, 2000: Twentieth-century storm activity along the U.S. east coast. *J Climate,* **13**, 1748-1761.

Zhang, T., R.G. Barry, K. Knowles, J.A. Heginbottom, and J. Brown, 1999. Statistics and characteristics of permafrost and ground ice distribution in the Northern Hemisphere. *Polar Geography,* **23**, 147-169.

Zhen, W.Z. and R.H. Wu, 1993: Sea level changes of the World and China. *Marine Science Bulletin,* **12**, 95-99.

Zuo, Z., and J. Oerlemans, 1997: Contribution of glacier melt to sea level rise since AD 1865: a regionally differentiated calculation. *Climate Dynamics,* **13**, 835-845.

Zwally, H.J., and M.B. Giovinetto, 2000: Spatial distribution of surface mass balance on Greenland, *Annals of Glaciology,* **31**, 126-132.

Zwartz, D., K. Lambeck, M. Bird, and J. Stone, 1997: Constraints on the former Antarctic Ice Sheet from sea level observations and geodynamics modelling. In: *The Antarctic Region: Geological Evolution and Processes,* C. Ricci (ed.), Terra Antarctica Publications, Siena, 861-868.

12

Detection of Climate Change and Attribution of Causes

Co-ordinating Lead Authors
J.F.B. Mitchell, D.J. Karoly

Lead Authors
G.C. Hegerl, F.W. Zwiers, M.R. Allen, J. Marengo

Contributing Authors
V. Barros, M. Berliner, G. Boer, T. Crowley, C. Folland, M. Free, N. Gillett, P. Groisman, J. Haigh,
K. Hasselmann, P. Jones, M. Kandlikar, V. Kharin, H. Kheshgi, T. Knutson, M. MacCracken, M. Mann,
G. North, J. Risbey, A. Robock, B. Santer, R. Schnur, C. Schönwiese, D. Sexton, P. Stott, S. Tett,
K. Vinnikov, T. Wigley

Review Editors
F. Semazzi, J. Zillman

Contents

Executive Summary

The IPCC WG1 Second Assessment Report (IPCC, 1996) (hereafter SAR) concluded, "the balance of evidence suggests that there is a discernible human influence on global climate". It noted that the detection and attribution of anthropogenic climate change signals can only be accomplished through a gradual accumulation of evidence. The SAR authors also noted uncertainties in a number of factors, including the magnitude and patterns of internal climate variability, external forcing and climate system response, which prevented them from drawing a stronger conclusion. The results of the research carried out since 1995 on these uncertainties and other aspects of detection and attribution are summarised below.

A longer and more closely scrutinised observational record

Three of the five years (1995, 1996 and 1998) added to the instrumental record since the SAR are the warmest in the instrumental record of global temperatures, consistent with the expectation that increases in greenhouse gases will lead to continued long-term warming. The impact of observational sampling errors has been estimated for the global and hemispheric mean surface temperature record and found to be small relative to the warming observed over the 20th century. Some sources of error and uncertainty in both the Microwave Sounding Unit (MSU) and radiosonde observations have been identified that largely resolve discrepancies between the two data sets. However, current climate models cannot fully account for the observed difference in the trend between the surface and lower-tropospheric temperatures over the last twenty years even when all known external influences are included. New reconstructions of the surface temperature record of the last 1,000 years indicate that the temperature changes over the last 100 years are unlikely to be entirely natural in origin, even taking into account the large uncertainties in palaeo-reconstructions.

New model estimates of internal variability

Since the SAR, more models have been used to estimate the magnitude of internal climate variability. Several of the models used for detection show similar or larger variability than observed on interannual to decadal time-scales, even in the absence of external forcing. The warming over the past 100 years is very unlikely to be due to internal variability alone as estimated by current models. Estimates of variability on the longer time-scales relevant to detection and attribution studies are uncertain. Nonetheless, conclusions on the detection of an anthropogenic signal are insensitive to the model used to estimate internal variability and recent changes cannot be accounted for as pure internal variability even if the amplitude of simulated internal variations is increased by a factor of two or more. In most recent studies, the residual variability that remains in the observations after removal of the estimated anthropogenic signals is consistent with model-simulated variability on the space- and time-scales used for detection and attribution. Note, however, that the power of the consistency test is limited. Detection studies to date have shown that the observed large-scale changes in surface temperature in recent decades are unlikely (bordering on very unlikely) to be entirely the result of internal variability.

New estimates of responses to natural forcing

Fully coupled ocean-atmosphere models have used reconstructions of solar and volcanic forcings over the last one to three centuries to estimate the contribution of natural forcing to climate variability and change. Including their effects produces an increase in variance on all time-scales and brings the low-frequency variability simulated by models closer to that deduced from palaeo-reconstructions. Assessments based on physical principles and model simulations indicate that natural forcing alone is unlikely to explain the increased rate of global warming since the middle of the 20th century or changes in vertical temperature structure. The reasons are that the trend in natural forcing has likely been negative over the last two decades and natural forcing alone is unlikely to account for the observed cooling of the stratosphere. However, there is evidence for a detectable volcanic influence on climate. The available evidence also suggests a solar influence in proxy records of the last few hundred years and also in the instrumental record of the early 20th century. Statistical assessments confirm that natural variability (the combination of internal and naturally forced) is unlikely to explain the warming in the latter half of the 20th century.

Improved representation of anthropogenic forcing

Several studies since the SAR have included an explicit representation of greenhouse gases (as opposed to an equivalent increase in carbon dioxide (CO_2)). Some have also included tropospheric ozone changes, an interactive sulphur cycle, an explicit radiative treatment of the scattering of sulphate aerosols, and improved estimates of the changes in stratospheric ozone. While detection of the climate response to these other anthropogenic factors is often ambiguous, detection of the influence of greenhouse gas increases on the surface temperature changes over the past 50 years is robust.

Sensitivity to estimates of climate change signals

Since the SAR, more simulations with increases in greenhouse gases and some representation of aerosol effects have become available. In some cases, ensembles of simulations have been run to reduce noise in the estimates of the time-dependent response. Some studies have evaluated seasonal variation of the response. Uncertainties in the estimated climate change signals have made it difficult to attribute the observed climate change to one specific combination of anthropogenic and natural influences. Nevertheless, all studies since the SAR have found a significant anthropogenic contribution is required to account for surface and tropospheric trends over at least the last 30 years.

Qualitative consistencies between observed and modelled climate changes

There is a wide range of evidence of qualitative consistencies between observed climate changes and model responses to anthropogenic forcing, including global warming, increasing land-ocean temperature contrast, diminishing Arctic sea-ice extent, glacial retreat and increases in precipitation in Northern Hemisphere high latitudes. Some qualitative inconsistencies remain, including the fact that models predict a faster rate of warming in the mid- to upper troposphere which is not observed in either satellite or radiosonde tropospheric temperature records.

A wider range of detection techniques

A major advance since the SAR is the increase in the range of techniques used, and the evaluation of the degree to which the results are independent of the assumptions made in applying those techniques. There have been studies using pattern correlations, optimal detection studies using one or more fixed patterns and time-varying patterns, and a number of other techniques. Evidence of a human influence on climate is obtained using all these techniques.

Results are sensitive to the range of temporal and spatial scales that are considered. Several decades of data are necessary to separate the forced response from internal variability. Idealised studies have demonstrated that surface temperature changes are detectable only on scales greater than 5,000 km. Studies also show that the level of agreement found between simulations and observations in pattern correlation studies is close to what one would expect in theory.

Attribution studies have applied multi-signal techniques to address whether or not the magnitude of the observed response to a particular forcing agent is consistent with the modelled response and separable from the influence of other forcing agents. The inclusion of time-dependent signals has helped to distinguish between natural and anthropogenic forcing agents. As more response patterns are included, the problem of degeneracy (different combinations of patterns yielding near identical fits to the observations) inevitably arises. Nevertheless, even with the responses to all the major forcing factors included in the analysis, a distinct greenhouse gas signal remains detectable. Overall, the magnitude of the model-simulated temperature response to greenhouse gases is found to be consistent with the observed greenhouse response on the scales considered. However, there remain discrepancies between the modelled and observed responses to other natural and anthropogenic factors, and estimates of signal amplitudes are model-dependent. Most studies find that, over the last 50 years, the estimated rate and magnitude of warming due to increasing concentrations of greenhouse gases alone are comparable with, or larger than, the observed warming. Furthermore, most model estimates that take into account both greenhouse gases and sulphate aerosols are consistent with observations over this period.

The increase in the number of studies, the breadth of techniques, increased rigour in the assessment of the role of anthropogenic forcing in climate, the robustness of results to the assumptions made using those techniques, and consistency of results lead to increased confidence in these results. Moreover, to be consistent with the signal observed to date, the rate of anthropogenic warming is likely to lie in the range 0.1 to 0.2°C/decade over the first half of the 21st century under the IS92a (IPCC, 1992) emission scenario.

Remaining uncertainties

A number of important uncertainties remain. These include:

- Discrepancies between the vertical profile of temperature change in the troposphere seen in observations and models. These have been reduced as more realistic forcing histories have been used in models, although not fully resolved. Also, the difference between observed surface and lower-tropospheric trends over the last two decades cannot be fully reproduced by model simulations.

- Large uncertainties in estimates of internal climate variability from models and observations, though as noted above, these are unlikely (bordering on very unlikely) to be large enough to nullify the claim that a detectable climate change has taken place.

- Considerable uncertainty in the reconstructions of solar and volcanic forcing which are based on proxy or limited observational data for all but the last two decades. Detection of the influence of greenhouse gases on climate appears to be robust to possible amplification of the solar forcing by ozone/solar or solar/cloud interactions, provided these do not alter the pattern or time dependence of the response to solar forcing. Amplification of the solar signal by these processes, which are not yet included in models, remains speculative.

- Large uncertainties in anthropogenic forcing are associated with the effects of aerosols. The effects of some anthropogenic factors, including organic carbon, black carbon, biomass aerosols, and changes in land use, have not been included in detection and attribution studies. Estimates of the size and geographic pattern of the effects of these forcings vary considerably, although individually their global effects are estimated to be relatively small.

- Large differences in the response of different models to the same forcing. These differences, which are often greater than the difference in response in the same model with and without aerosol effects, highlight the large uncertainties in climate change prediction and the need to quantify uncertainty and reduce it through better observational data sets and model improvement.

Synopsis

The SAR concluded: "The balance of evidence suggests a discernible human influence on global climate". That report also noted that the anthropogenic signal was still emerging from the background of natural climate variability. Since the SAR, progress has been made in reducing uncertainty, particularly with respect to distinguishing and quantifying the magnitude of responses to different external influences. Although many of the sources of uncertainty identified in the SAR still remain to some degree, new evidence and improved understanding support an updated conclusion.

- There is a longer and more closely scrutinised temperature record and new model estimates of variability. The warming over the past 100 years is very unlikely to be due to internal variability alone, as estimated by current models. Reconstructions of climate data for the past 1,000 years also indicate that this warming was unusual and is unlikely to be entirely natural in origin.

- There are new estimates of the climate response to natural and anthropogenic forcing, and new detection techniques have been applied. Detection and attribution studies consistently find evidence for an anthropogenic signal in the climate record of the last 35 to 50 years.

- Simulations of the response to natural forcings alone (i.e., the response to variability in solar irradiance and volcanic eruptions) do not explain the warming in the second half of the 20th century. However, they indicate that natural forcings may have contributed to the observed warming in the first half of the 20th century.

- The warming over the last 50 years due to anthropogenic greenhouse gases can be identified despite uncertainties in forcing due to anthropogenic sulphate aerosol and natural factors (volcanoes and solar irradiance). The anthropogenic sulphate aerosol forcing, while uncertain, is negative over this period and therefore cannot explain the warming. Changes in natural forcing during most of this period are also estimated to be negative and are unlikely to explain the warming.

- Detection and attribution studies comparing model simulated changes with the observed record can now take into account uncertainty in the magnitude of modelled response to external forcing, in particular that due to uncertainty in climate sensitivity.

- Most of these studies find that, over the last 50 years, the estimated rate and magnitude of warming due to increasing concentrations of greenhouse gases alone are comparable with, or larger than, the observed warming. Furthermore, most model estimates that take into account both greenhouse gases and sulphate aerosols are consistent with observations over this period.

- The best agreement between model simulations and observations over the last 140 years has been found when all the above anthropogenic and natural forcing factors are combined. These results show that the forcings included are sufficient to explain the observed changes, but do not exclude the possibility that other forcings may also have contributed.

In the light of new evidence and taking into account the remaining uncertainties, most of the observed warming over the last 50 years is likely to have been due to the increase in greenhouse gas concentrations.

Furthermore, it is very likely that the 20th century warming has contributed significantly to the observed sea level rise, through thermal expansion of sea water and widespread loss of land ice. Within present uncertainties, observations and models are both consistent with a lack of significant acceleration of sea level rise during the 20th century.

12.1 Introduction

12.1.1 The Meaning of Detection and Attribution

The response to anthropogenic changes in climate forcing occurs against a backdrop of natural internal and externally forced climate variability that can occur on similar temporal and spatial scales. Internal climate variability, by which we mean climate variability not forced by external agents, occurs on all time-scales from weeks to centuries and millennia. Slow climate components, such as the ocean, have particularly important roles on decadal and century time-scales because they integrate high-frequency weather variability (Hasselmann, 1976) and interact with faster components. Thus the climate is capable of producing long time-scale internal variations of considerable magnitude without any external influences. Externally forced climate variations may be due to changes in natural forcing factors, such as solar radiation or volcanic aerosols, or to changes in anthropogenic forcing factors, such as increasing concentrations of greenhouse gases or sulphate aerosols.

Definitions

The presence of this natural climate variability means that the detection and attribution of anthropogenic climate change is a statistical "signal-in-noise" problem. *Detection* is the process of demonstrating that an observed change is significantly different (in a statistical sense) than can be explained by natural internal variability. However, the detection of a change in climate does not necessarily imply that its causes are understood. As noted in the SAR, the unequivocal *attribution* of climate change to anthropogenic causes (i.e., the isolation of cause and effect) would require controlled experimentation with the climate system in which the hypothesised agents of change are systematically varied in order to determine the climate's sensitivity to these agents. Such an approach to attribution is clearly not possible. Thus, from a practical perspective, attribution of observed climate change to a given combination of human activity and natural influences requires another approach. This involves statistical analysis and the careful assessment of multiple lines of evidence to demonstrate, within a pre-specified margin of error, that the observed changes are:

- unlikely to be due entirely to internal variability;

- consistent with the estimated responses to the given combination of anthropogenic and natural forcing; and

- not consistent with alternative, physically plausible explanations of recent climate change that exclude important elements of the given combination of forcings.

Limitations

It is impossible, even in principle, to distinguish formally between all conceivable explanations with a finite amount of data. Nevertheless, studies have now been performed that include all the main natural and anthropogenic forcing agents that are generally accepted (on physical grounds) to have had a substan-

tial impact on near-surface temperature changes over the 20th century. Any statement that a model simulation is consistent with observed changes can only apply to a subset of model-simulated variables, such as large-scale near-surface temperature trends: no numerical model will ever be perfect in every respect. To attribute all or part of recent climate change to human activity, therefore, we need to demonstrate that alternative explanations, such as pure internal variability or purely naturally forced climate change, are unlikely to account for a set of observed changes that can be accounted for by human influence. Detection (ruling out that observed changes are only an instance of internal variability) is thus one component of the more complex and demanding process of attribution. In addition to this general usage of the term detection (that some climate change has taken place), we shall also discuss the detection of the influence of individual forcings (see Section 12.4).

Detection and estimation

The basic elements of this approach to detection and attribution were recognised in the SAR. However, detection and attribution studies have advanced beyond addressing the simple question "have we detected a human influence on climate?" to such questions as "how large is the anthropogenic change?" and "is the magnitude of the response to greenhouse gas forcing as estimated in the observed record consistent with the response simulated by climate models?" The task of detection and attribution can thus be rephrased as an estimation problem, with the quantities to be estimated being the factor(s) by which we have to scale the model-simulated response(s) to external forcing to be consistent with the observed change. The estimation approach uses essentially the same tools as earlier studies that considered the problem as one of hypothesis testing, but is potentially more informative in that it allows us to quantify, with associated estimates of uncertainty, how much different factors have contributed to recent observed climate changes. This interpretation only makes sense, however, if it can be assumed that important sources of model error, such as missing or incorrectly represented atmospheric feedbacks, affect primarily the amplitude and not the structure of the response to external forcing. The majority of relevant studies suggest that this is the case for the relatively small-amplitude changes observed to date, but the possibility of model errors changing both the amplitude and structure of the response remains an important caveat. Sampling error in model-derived signals that originates from the model's own internal variability also becomes an issue if detection and attribution is considered as an estimation problem – some investigations have begun to allow for this, and one study has estimated the contribution to uncertainty from observational sampling and instrumental error. The robustness of detection and attribution findings obtained with different climate models has been assessed.

Extensions

It is important to stress that the attribution process is inherently open-ended, since we have no way of predicting what alternative explanations for observed climate change may be proposed, and be accepted as plausible, in the future. This problem is not unique to the climate change issue, but applies to any problem of establishing cause and effect given a limited sample of observations. The possibility of a confounding explanation can never be ruled out completely, but as successive alternatives are tested and found to be inadequate, it can be seen to become progressively more unlikely. There is growing interest in the use of Bayesian methods (Dempster, 1998; Hasselmann, 1998; Leroy, 1998; Tol and de Vos, 1998; Barnett *et al.*, 1999; Levine and Berliner, 1999; Berliner *et al.*, 2000). These provide a means of formalising the process of incorporating additional information and evaluating a range of alternative explanations in detection and attribution studies. Existing studies can be rephrased in a Bayesian formalism without any change in their conclusions, as demonstrated by Leroy (1998). However, a number of statisticians (e.g., Berliner *et al.*, 2000) argue that a more explicitly Bayesian approach would allow greater flexibility and rigour in the treatment of different sources of uncertainty.

12.1.2 Summary of the First and Second Assessment Reports

The first IPCC Scientific Assessment in 1990 (IPCC, 1990) concluded that the global mean surface temperature had increased by 0.3 to 0.6°C over the previous 100 years and that the magnitude of this warming was broadly consistent with the predictions of climate models forced by increasing concentrations of greenhouse gases. However, it remained to be established that the observed warming (or part of it) could be attributed to the enhanced greenhouse effect. Some of the reasons for this were that there was only limited agreement between model predictions and observations, because climate models were still in the early stages of development; there was inadequate knowledge of natural variability and other possible anthropogenic effects on climate and there was a scarcity of suitable observational data, particularly long, reliable time-series.

By the time of the SAR in 1995, considerable progress had been made in attempts to identify an anthropogenic effect on climate. The first area of significant advance was that climate models were beginning to incorporate the possible climatic effects of human-induced changes in sulphate aerosols and stratospheric ozone. The second area of progress was in better defining the background variability of the climate system through multi-century model experiments that assumed no changes in forcing. These provided important information about the possible characteristics of the internal component of natural climate variability. The third area of progress was in the application of pattern-based methods that attempted to attribute some part of the observed changes in climate to human activities, although these studies were still in their infancy at that time.

The SAR judged that the observed trend in global climate over the previous 100 years was unlikely to be entirely natural in origin. This led to the following, now well-known, conclusion: "Our ability to quantify the human influence on global climate is currently limited because the expected signal is still emerging from the noise of natural variability, and because there are uncertainties in key factors. Nevertheless, the balance of evidence suggests that there is a discernible human influence on global climate". It also noted that the magnitude of the influence was uncertain.

12.1.3 Developments since the Second Assessment Report

In the following sections, we assess research developments since the SAR in areas crucial to the detection of climate change and the attribution of its causes. First, in Section 12.2, we review advances in the different elements that are needed in any detection and attribution study, including observational data, estimates of internal climate variability, natural and anthropogenic climate forcings and their simulated responses, and statistical methods for comparing observed and modelled climate change. We draw heavily on the assessments in earlier chapters of this report, particularly Chapter 2 – Observed Climate Variability and Change, Chapter 6 – Radiative Forcing of Climate Change, Chapter 8 – Model Evaluation, and Chapter 9 – Projections of Future Climate Change.

In Section 12.3, a qualitative assessment is made of observed and modelled climate change, identifying general areas of agreement and difference. This is based on the observed climate changes identified with most confidence in Chapter 2 and the model projections of climate change from Chapter 9.

Next, in Section 12.4, advances obtained with quantitative methods for climate change detection and attribution are assessed. These include results obtained with time-series methods, pattern correlation methods, and optimal fingerprint methods. The interpretation of optimal fingerprinting as an estimation problem, finding the scaling factors required to bring the amplitude of model-simulated changes into agreement with observed changes, is discussed. Some remaining uncertainties are discussed in Section 12.5 and the key findings are drawn together in Section 12.6.

12.2 The Elements of Detection and Attribution

12.2.1 Observed Data

Ideally, a detection and attribution study requires long records of observed data for climate elements that have the potential to show large climate change signals relative to natural variability. It is also necessary that the observing system has sufficient coverage so that the main features of natural variability and climate change can be identified and monitored. A thorough assessment of observed climate change, climate variability and data quality was presented in Chapter 2. Most detection and attribution studies have used near-surface air temperature, sea surface temperature or upper air temperature data, as these best fit the requirement above.

The quality of observed data is a vital factor. Homogeneous data series are required with careful adjustments to account for changes in observing system technologies and observing practices. Estimates of observed data uncertainties due to instrument errors or variations in data coverage (assessed in Chapter 2) are included in some recent detection and attribution studies.

There have been five more years of observations since the SAR. Improvements in historical data coverage and processing are described in Chapter 2. Confidence limits for observational sampling error have been estimated for the global and hemispheric mean temperature record. Applications of improved pre-instrumental proxy data reconstructions are described in the next two sections.

12.2.2 Internal Climate Variability

Detection and attribution of climate change is a statistical "signal-in-noise" problem, it requires an accurate knowledge of the properties of the "noise". Ideally, internal climate variability would be estimated from instrumental observations, but a number of problems make this difficult. The instrumental record is short relative to the 30 to 50 year time-scales that are of interest for detection and attribution of climate change, particularly for variables in the free atmosphere. The longest records that are available are those for surface air temperature and sea surface temperature. Relatively long records are also available for precipitation and surface pressure, but coverage is incomplete and varies in time (see Chapter 2). The instrumental record also contains the influences of external anthropogenic and natural forcing. A record of natural internal variability can be reconstructed by removing estimates of the response to external forcing (for example, Jones and Hegerl, 1998; Wigley *et al.*, 1998a). However, the accuracy of this record is limited by incomplete knowledge of the forcings and by the accuracy of the climate model used to estimate the response.

Estimates using palaeoclimatic data
Palaeo-reconstructions provide an additional source of information on climate variability that strengthens our qualitative assessment of recent climate change. There has been considerable progress in the reconstruction of past temperatures. New reconstructions with annual or seasonal resolution, back to 1000 AD, and some spatial resolution have become available (Briffa *et al.*, 1998; Jones *et al.*, 1998; Mann *et al.*, 1998, 2000; Briffa *et al.*, 2000; Crowley and Lowery, 2000; see also Chapter 2, Figure 2.21). However, a number of difficulties, including limited coverage, temporal inhomogeneity, possible biases due to the palaeo-reconstruction process, uncertainty regarding the strength of the relationships between climatic and proxy indices, and the likely but unknown influence of external forcings inhibit the estimation of internal climate variability directly from palaeo-climate data. We expect, however, that the reconstructions will continue to improve and that palaeo-data will become increasingly important for assessing natural variability of the climate system. One of the most important applications of this palaeo-climate data is as a check on the estimates of internal variability from coupled climate models, to ensure that the latter are not underestimating the level of internal variability on 50 to 100 year time-scales (see below). The limitations of the instrumental and palaeo-records leave few alternatives to using long "control" simulations with coupled models (see Figure 12.1) to estimate the detailed structure of internal climate variability.

Estimates of the variability of global mean surface temperature
Stouffer *et al.* (2000) assess variability simulated in three 1,000-year control simulations (see Figure 12.1). The models are found to simulate reasonably well the spatial distribution of variability and the spatial correlation between regional and global mean variability, although there is more disagreement between models at long time-scales (>50 years) than at short time-scales. None of the long model simulations produces a secular trend which is

comparable to that observed. Chapter 8, Section 8.6.2. assesses model-simulated variability in detail. Here we assess the aspects that are particularly relevant to climate change detection. The power spectrum of global mean temperatures simulated by the most recent coupled climate models (shown in Figure 12.2) compares reasonably well with that of detrended observations (solid black line) on interannual to decadal time-scales. However, uncertainty of the spectral estimates is large and some models are clearly underestimating variability (indicated by the asterisks). Detailed comparison on inter-decadal time-scales is difficult because observations are likely to contain a response to external forcings that will not be entirely removed by a simple linear trend. At the same time, the detrending procedure itself introduces a negative bias in the observed low-frequency spectrum.

Both of these problems can be avoided by removing an independent estimate of the externally forced response from the observations before computing the power spectrum. This independent estimate is provided by the ensemble mean of a coupled model simulation of the response to the combination of natural and anthropogenic forcing (see Figure 12.7c). The resulting spectrum of observed variability (dotted line in Figure 12.2) will not be subject to a negative bias because the observed data have not been used in estimating the forced response. It will, however, be inflated by uncertainty in the model-simulated forced response and by noise due to observation error and due to incomplete coverage (particularly the bias towards relatively noisy Northern Hemisphere land temperatures in the early part of the observed series). This estimate of the observed spectrum is therefore likely to overestimate power at all frequencies. Even so, the more variable models display similar variance on the decadal to inter-decadal time-scales important for detection and attribution.

Estimates of spatial patterns of variability
Several studies have used common empirical orthogonal function (EOF) analysis to compare the spatial modes of climate variability between different models. Stouffer *et al.* (2000) analysed the variability of 5-year means of surface temperature in 500-year or longer simulations of the three models most commonly used to estimate internal variability in formal detection studies. The distribution of the variance between the EOFs was similar between the models and the observations. HadCM2 tended to overestimate the variability in the main modes, whereas GFDL and ECHAM3 underestimated the variability of the first mode. The standard deviations of the dominant modes of variability in the three models differ from observations by less than a factor of two, and one model (HadCM2) has similar or more variability than the observations in all leading modes. In general, one would expect to obtain conservative detection and attribution results when natural variability is estimated with such a model. One should also expect control simulations to be less variable than observations because they do not contain externally forced variability. Hegerl *et al.* (2000) used common EOFS to compare 50-year June-July-August (JJA) trends of surface temperature in ECHAM3 and HadCM2. Standard deviation differences between models

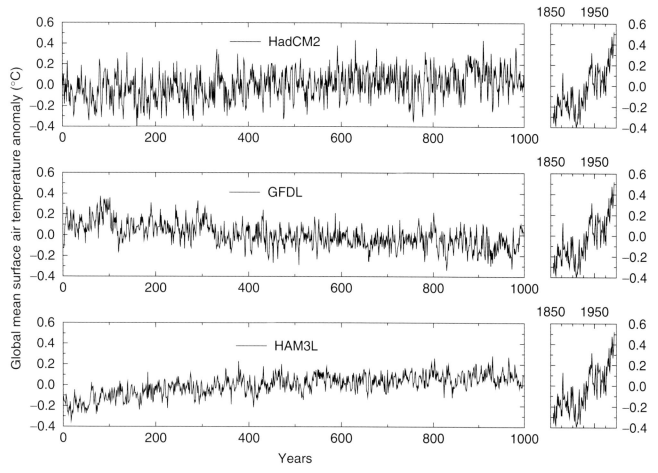

Figure 12.1: Global mean surface air temperature anomalies from 1,000-year control simulations with three different climate models, HadCM2, GFDL R15 and ECHAM3/LSG (labelled HAM3L), compared to the recent instrumental record (Stouffer *et al.*, 2000). No model control simulation shows a trend in surface air temperature as large as the observed trend. If internal variability is correct in these models, the recent warming is likely not due to variability produced within the climate system alone.

were marginally larger on the 50-year time-scale (less than a factor of 2.5). Comparison with direct observations cannot be made on this time-scale because the instrumental record is too short.

Variability of the free atmosphere
Gillett *et al.* (2000a) compared model-simulated variability in the free atmosphere with that of detrended radiosonde data. They found general agreement except in the stratosphere, where present climate models tend to underestimate variability on all time-scales and, in particular, do not reproduce modes of variability such as the quasi-biennial oscillation (QBO). On decadal time-scales, the model simulated less variability than observed in some aspects of the vertical patterns important for the detection of anthropogenic climate change. The discrepancy is partially resolved by the inclusion of anthropogenic (greenhouse gas, sulphate and stratospheric ozone) forcing in the model. However, the authors also find evidence that solar forcing plays a significant role on decadal time-scales, indicating that this should be taken into account in future detection studies based on changes in the free atmosphere (see also discussion in Chapter 6 and Section 12.2.3.1 below).

Comparison of model and palaeoclimatic estimates of variability
Comparisons between the variability in palaeo-reconstructions and climate model data have shown mixed results to date. Barnett *et al.* (1996) compared the spatial structure of climate variability of coupled climate models and proxy time-series for (mostly summer) decadal temperature (Jones *et al.*, 1998). They found that the model-simulated amplitude of the dominant proxy mode of variation is substantially less than that estimated from the proxy data. However, choosing the EOFs of the palaeo-data as the basis for comparison will maximise the variance in the palaeo-data and not the models, and so bias the model amplitudes downwards. The neglect of naturally forced climate variability in the models might also be responsible for part of the discrepancy noted in Barnett *et al.* (1996) (see also Jones *et al.*, 1998). The limitations of the temperature reconstructions (see Chapter 2, Figure 2.21), including for example the issue of how to relate site-specific palaeo-data to large-scale variations, may also contribute to this discrepancy. Collins *et al.* (2000) compared the standard deviation of large-scale Northern Hemisphere averages in a model control simulation and in tree-ring-based proxy data for the last 600 years on decadal time-scales. They found a

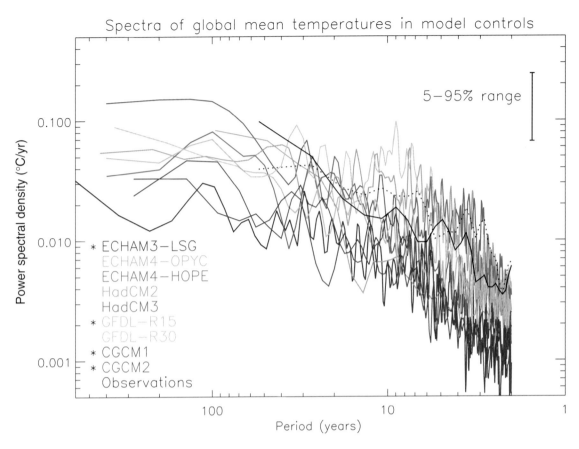

Figure 12.2: Coloured lines: power spectra of global mean temperatures in the unforced control integrations that are used to provide estimates of internal climate variability in Figure 12.12. All series were linearly detrended prior to analysis, and spectra computed using a standard Tukey window with the window width (maximum lag used in the estimate) set to one-fifth of the series length, giving each spectral estimate the same uncertainty range, as shown (see, e.g., Priestley, 1981). The first 300 years were omitted from ECHAM3-LSG, CGCM1 and CGCM2 models as potentially trend-contaminated. Solid black line: spectrum of observed global mean temperatures (Jones *et al.*, 2001) over the period 1861 to 1998 after removing a best-fit linear trend. This estimate is unreliable on inter-decadal time-scales because of the likely impact of external forcing on the observed series and the negative bias introduced by the detrending. Dotted black line: spectrum of observed global mean temperatures after removing an independent estimate of the externally forced response provided by the ensemble mean of a coupled model simulation (Stott *et al.*, 2000b, and Figure 12.7c). This estimate will be contaminated by uncertainty in the model-simulated forced response, together with observation noise and sampling error. However, unlike the detrending procedure, all of these introduce a positive (upward) bias in the resulting estimate of the observed spectrum. The dotted line therefore provides a conservative (high) estimate of observed internal variability at all frequencies. Asterisks indicate models whose variability is significantly less than observed variability on 10 to 60 year time-scales after removing either a best-fit linear trend or an independent estimate of the forced response from the observed series. Significance is based on an F-test on the ratio observed/model mean power over this frequency interval and quoted at the 5% level. Power spectral density (PSD) is defined such that unit-variance uncorrelated noise would have an expected PSD of unity (see Allen *et al.*, 2000a, for details). Note that different normalisation conventions can lead to different values, which appear as a constant offset up or down on the logarithmic vertical scale used here. Differences between the spectra shown here and the corresponding figure in Stouffer *et al.* (2000) shown in Chapter 8, Figure 8.18 are due to the use here of a longer (1861 to 2000) observational record, as opposed to 1881 to 1991 in Figure 8.18. That figure also shows 2.5 to 97.5% uncertainty ranges, while for consistency with other figures in this chapter, the 5 to 95% range is displayed here.

factor of less than two difference between model and data if the tree-ring data are calibrated such that low-frequency variability is better retained than in standard methods (Briffa *et al.*, 2000). It is likely that at least part of this discrepancy can be resolved if natural forcings are included in the model simulation. Crowley (2000) found that 41 to 69% of the variance in decadally smoothed Northern Hemisphere mean surface temperature reconstructions could be externally forced (using data from Mann *et al.* (1998) and Crowley and

Lowery (2000)). The residual variability in the reconstructions, after subtracting estimates of volcanic and solar-forced signals, showed no significant difference in variability on decadal and multi-decadal time-scales from three long coupled model control simulations. In summary, while there is substantial uncertainty in comparisons between long-term palaeo-records of surface temperature and model estimates of multi-decadal variability, there is no clear evidence of a serious discrepancy.

Summary

These findings emphasise that there is still considerable uncertainty in the magnitude of internal climate variability. Various approaches are used in detection and attribution studies to account for this uncertainty. Some studies use data from a number of coupled climate model control simulations (Santer *et al.*, 1995; Hegerl *et al.*, 1996, 1997, North and Stevens, 1998) and choose the most conservative result. In other studies, the estimate of internal variance is inflated to assess the sensitivity of detection and attribution results to the level of internal variance (Santer *et al.*, 1996a, Tett *et al.*, 1999; Stott *et al.*, 2001). Some authors also augment model-derived estimates of natural variability with estimates from observations (Hegerl *et al.*, 1996). A method for checking the consistency between the residual variability in the observations after removal of externally forced signals (see equation A12.1.1, Appendix 12.1) and the natural internal variability estimated from control simulations is also available (e.g., Allen and Tett, 1999). Results indicate that, on the scales considered, there is no evidence for a serious inconsistency between the variability in models used for optimal fingerprint studies and observations (Allen and Tett, 1999; Tett *et al.*, 1999; Hegerl *et al.*, 2000, 2001; Stott *et al.*, 2001). The use of this test and the use of internal variability from the models with the greatest variability increases confidence in conclusions derived from optimal detection studies.

12.2.3 Climate Forcings and Responses

The global mean change in radiative forcing (see Chapter 6) since the pre-industrial period may give an indication of the relative importance of the different external factors influencing climate over the last century. The temporal and spatial variation of the forcing from different sources may help to identify the effects of individual factors that have contributed to recent climate change.

The need for climate models

To detect the response to anthropogenic or natural climate forcing in observations, we require estimates of the expected space-time pattern of the response. The influences of natural and anthropogenic forcing on the observed climate can be separated only if the spatial and temporal variation of each component is known. These patterns cannot be determined from the observed instrumental record because variations due to different external forcings are superimposed on each other and on internal climate variations. Hence climate models are usually used to estimate the contribution from each factor. The models range from simpler energy balance models to the most complex coupled atmosphere-ocean general circulation models that simulate the spatial and temporal variations of many climatic parameters (Chapter 8).

The models used

Energy balance models (EBMs) simulate the effect of radiative climate forcing on surface temperature. Climate sensitivity is included as an adjustable parameter. These models are computationally inexpensive and produce noise-free estimates of the climate signal. However, EBMs cannot represent dynamical components of the climate signal, generally cannot simulate

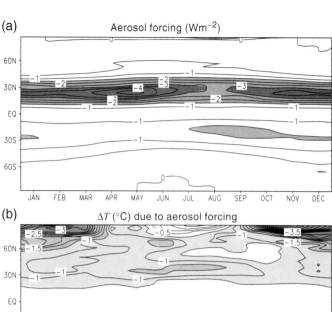

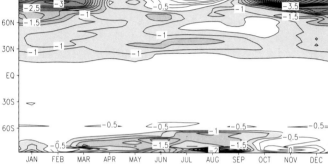

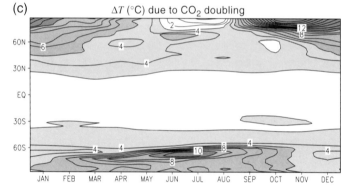

Figure 12.3: Latitude-month plot of radiative forcing and model equilibrium response for surface temperature. (a) Radiative forcing (Wm^{-2}) due to increased sulphate aerosol loading at the time of CO_2 doubling. (b) Change in temperature due to the increase in aerosol loading. (c) Change in temperature due to CO_2 doubling. Note that the patterns of radiative forcing and temperature response are quite different in (a) and (b), but that the patterns of large-scale temperature responses to different forcings are similar in (b) and (c). The experiments used to compute these fields are described by Reader and Boer (1998).

variables other than surface temperature, and may omit some of the important feedback processes that are accounted for in more complex models. Most detection and attribution approaches therefore apply signals estimated from coupled Atmosphere Ocean General Circulation Models (AOGCMs) or atmospheric General Circulation Models (GCMs) coupled to mixed-layer ocean models. Forced simulations with such models contain both the climate response to external forcing and superimposed internal climate variability. Estimates of the climate response computed from model output will necessarily contain at least some noise from this source, although this can be reduced by the use of ensemble simulations. Note that different models can produce quite different patterns of response to a given forcing due to differences in the representation of feedbacks arising from changes in cloud (in particular), sea ice and land surface processes.

The relationship between patterns of forcing and response
There are several reasons why one should not expect a simple relationship between the patterns of radiative forcing and temperature response. First, strong feedbacks such as those due to water vapour and sea ice tend to reduce the difference in the temperature response due to different forcings. This is illustrated graphically by the response to the simplified aerosol forcing used in early studies. The magnitude of the model response is largest over the Arctic in winter even though the forcing is small, largely due to ice-albedo feedback. The large-scale patterns of change and their temporal variations are similar, but of opposite sign, to that obtained in greenhouse gas experiments (Figure 12.3, see also Mitchell *et al.*, 1995a). Second, atmospheric circulation tends to smooth out temperature gradients and reduce the differences in response patterns. Similarly, the thermal inertia of the climate system tends to reduce the amplitude of short-term fluctuations in forcing. Third, changes in radiative forcing are more effective if they act near the surface, where cooling to space is restricted, than at upper levels, and in high latitudes, where there are stronger positive feedbacks than at low latitudes (Hansen *et al.*, 1997a).

In practice, the response of a given model to different forcing patterns can be quite similar (Hegerl *et al.*, 1997; North and Stevens, 1998; Tett *et al.*, 1999). Similar signal patterns (a condition often referred to as "degeneracy") can be difficult to distinguish from one another. Tett *et al.* (1999) find substantial degeneracy between greenhouse gas, sulphate, volcanic and solar patterns they used in their detection study using HadCM2. On the other hand, the greenhouse gas and aerosol patterns generated by ECHAM3 LSG (Hegerl *et al.*, 2000) are more clearly separable, in part because the patterns are more distinct, and in part because the aerosol response pattern correlates less well with ECHAM3 LSG's patterns of internal variability. The vertical patterns of temperature change due to greenhouse gas and stratospheric ozone forcing are less degenerate than the horizontal patterns.

Summary
Different models may give quite different patterns of response for the same forcing, but an individual model may give a surprisingly similar response for different forcings. The first point means that attribution studies may give different results when using signals generated from different models. The second point means that it may be more difficult to distinguish between the response to different factors than one might expect, given the differences in radiative forcing.

12.2.3.1 Natural climate forcing
Since the SAR, there has been much progress in attempting to understand the climate response to fluctuations in solar

luminosity and to volcanism. These appear to be the most important among a broad range of natural external climate forcings at decadal and centennial time-scales. The mechanisms of these forcings, their reconstruction and associated uncertainties are described in Chapter 6, and further details of the simulated responses are given in Chapter 8, Section 8.6.3.

Volcanic forcing
The radiative forcing due to volcanic aerosols from the recent El Chichon and Mt. Pinatubo eruptions has been estimated from satellite and other data to be -3 Wm^{-2} (peak forcing; after Hansen *et al.*, 1998). The forcing associated with historic eruptions before the satellite era is more uncertain. Sato *et al.* (1993) estimated aerosol optical depth from ground-based observations over the last century (see also Stothers, 1996; Grieser and Schoenwiese, 1999). Prior to that, reconstructions have been based on various sources of data (ice cores, historic documents etc.; see Lamb, 1970; Simkin *et al.*, 1981; Robock and Free, 1995; Crowley and Kim, 1999; Free and Robock, 1999). There is uncertainty of about a factor of two in the peak forcing in reconstructions of historic volcanic forcing in the pre-satellite era (see Chapter 6).

Solar forcing
The variation of solar irradiance with the 11-year sunspot cycle has been assessed with some accuracy over more than 20 years, although measurements of the magnitude of modulations of solar irradiance between solar cycles are less certain (see Chapter 6). The estimation of earlier solar irradiance fluctuations, although based on physical mechanisms, is indirect. Hence our confidence in the range of solar radiation on century time-scales is low, and confidence in the details of the time-history is even lower (Harrison and Shine, 1999; Chapter 6). Several recent reconstructions estimate that variations in solar irradiance give rise to a forcing at the Earth's surface of about 0.6 to 0.7 Wm^{-2} since the Maunder Minimum and about half this over the 20th century (see Chapter 6, Figure 6.5; Hoyt and Schatten, 1993; Lean *et al.*, 1995; Lean, 1997; Froehlich and Lean, 1998; Lockwood and Stamper, 1999). This is larger than the 0.2 Wm^{-2} modulation of the 11-year solar cycle measured from satellites. (Note that we discuss here the forcing at the Earth's surface, which is smaller than that at the top of the atmosphere, due to the Earth's geometry and albedo.) The reconstructions of Lean *et al.* (1995) and Hoyt and Schatten (1993), which have been used in GCM detection studies, vary in amplitude and phase. Chapter 6, Figure 6.8 shows time-series of reconstructed solar and volcanic forcing since the late 18th century. All reconstructions indicate that the direct effect of variations in solar forcing over the 20th century was about 20 to 25% of the change in forcing due to increases in the well-mixed greenhouse gases (see Chapter 6).

Reconstructions of climate forcing in the 20th century indicate that the net natural climate forcing probably increased during the first half of the 20th century, due to a period of low volcanism coinciding with a small increase in solar forcing. Recent decades show negative natural forcing due to increasing volcanism, which overwhelms the direct effect, if real, of a small increase in solar radiation (see Chapter 6, Table 6.13).

(a)

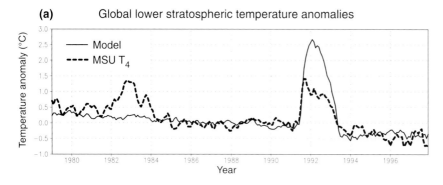

Global lower stratospheric temperature anomalies

(b)

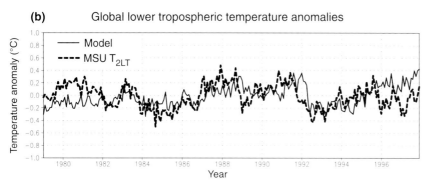

Global lower tropospheric temperature anomalies

Figure 12.4: (a) Observed microwave sounding unit (MSU) global mean temperature in the lower stratosphere, shown as dashed line, for channel 4 for the period 1979 to 97 compared with the average of several atmosphere-ocean GCM simulations starting with different atmospheric conditions in 1979 (solid line). The simulations have been forced with increasing greenhouse gases, direct and indirect forcing by sulphate aerosols and tropospheric ozone forcing, and Mt. Pinatubo volcanic aerosol and stratospheric ozone variations. The model simulation does not include volcanic forcing due to El Chichon in 1982, so it does not show stratospheric warming then. (b) As for (a), except for 2LT temperature retrievals in the lower troposphere. Note the steady response in the stratosphere, apart from the volcanic warm periods, and the large variability in the lower troposphere (from Bengtsson *et al.*, 1999).

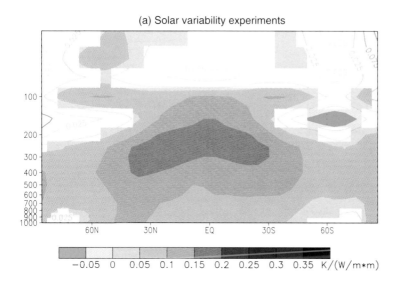

(a) Solar variability experiments

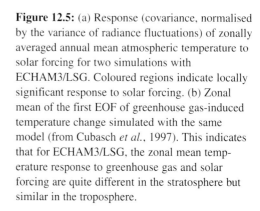

Figure 12.5: (a) Response (covariance, normalised by the variance of radiance fluctuations) of zonally averaged annual mean atmospheric temperature to solar forcing for two simulations with ECHAM3/LSG. Coloured regions indicate locally significant response to solar forcing. (b) Zonal mean of the first EOF of greenhouse gas-induced temperature change simulated with the same model (from Cubasch *et al.*, 1997). This indicates that for ECHAM3/LSG, the zonal mean temperature response to greenhouse gas and solar forcing are quite different in the stratosphere but similar in the troposphere.

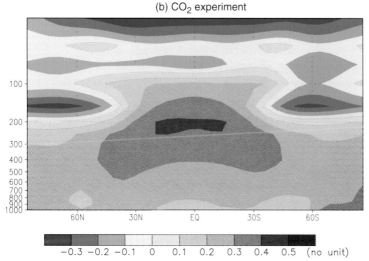

(b) CO_2 experiment

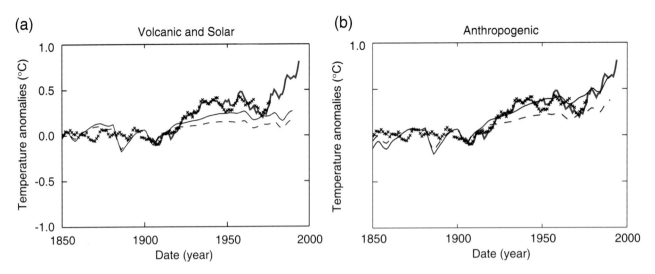

Figure 12.6: (a) Five-year running mean Northern Hemisphere temperature anomalies since 1850 (relative to the 1880 to 1920 mean) from an energy-balance model forced by Dust Veil volcanic index and Lean *et al.* (1995) solar index (see Free and Robock, 1999). Two values of climate sensitivity to doubling CO_2 were used; 3.0°C (thin solid line), and 1.5°C (dashed line). Also shown are the instrumental record (thick red line) and a reconstruction of temperatures from proxy records (crosses, from Mann *et al.*, 1998). The size of both the forcings and the proxy temperature variations are subject to large uncertainties. Note that the Mann temperatures do not include data after 1980 and do not show the large observed warming then. (b) As for (a) but for simulations with volcanic, solar and anthropogenic forcing (greenhouse gases and direct and indirect effects of tropospheric aerosols). The net anthropogenic forcing at 1990 relative to 1760 was 1.3 Wm^{-2}, including a net cooling of 1.3 Wm^{-2} due to aerosol effects.

12.2.3.2 Climatic response to natural forcing

Response to volcanic forcing

The climate response to several recent volcanic eruptions has been studied in observations and simulations with atmospheric GCMs (e.g., Robock and Mao, 1992, 1995; Graf *et al.*, 1996; Hansen *et al.*, 1996; Kelly *et al.*, 1996; Mao and Robock, 1998; Kirchner *et al.*, 1999). The stratosphere warms and the annual mean surface and tropospheric temperature decreases during the two to three years following a major volcanic eruption. A simulation incorporating the effects of the Mt. Pinatubo eruption and observed changes in stratospheric ozone in addition to anthropogenic forcing approximately reproduces the observed stratospheric variations (Figure 12.4; Bengtsson *et al.*, 1999). It shows stratospheric warming after the volcanic eruption, superimposed on a long-term cooling trend. Although the surface temperature response in the Northern Hemisphere warm season following a volcanic eruption is dominated by global scale radiative cooling, some models simulate local warming over Eurasia and North America in the cold season due to changes in circulation (e.g., Graf *et al.*, 1996; Kirchner *et al.*, 1999). Variability from other sources makes assessment of the observed climate response difficult, particularly as the two most recent volcanic eruptions (Mt. Pinatubo and El Chichon) occurred in El Niño-Southern Oscillation (ENSO) warm years. Simulations with simple models (Bertrand *et al.*, 1999; Crowley and Kim, 1999; Grieser and Schoenwiese, 2001) and AOGCMs (Tett *et al.*, 1999; Stott *et al.*, 2001) produce a small decadal mean cooling in the 1980s and 1990s due to several volcanic eruptions in those decades. Some simulations also produce global warming in the early 20th century as a recovery from a series of strong eruptions around the turn of the 20th century. It is unclear whether such a long-term response is realistic.

Response to solar forcing

Since the SAR, there have been new modelling and observational studies on the climate effects of variations in solar irradiance. The surface temperature response to the 11-year cycle is found to be small (e.g., Cubasch *et al.*, 1997; White *et al.*, 1997; North and Stevens, 1998; Crowley and Kim, 1999; Free and Robock, 1999). Low-frequency solar variability over the last few hundred years gives a stronger surface temperature response (Cubasch *et al.*, 1997; Drijfhout *et al.*, 1999; Rind *et al.*, 1999; Tett *et al.*, 1999; Stott *et al.*, 2001). Model results show cooling circa 1800 due to the hypothesised solar forcing minimum and some warming in the 20th century, particularly in the early 20th century. Time-dependent experiments produce a global mean warming of 0.2 to 0.5°C in response to the estimated 0.7 Wm^{-2} change of solar radiative forcing from the Maunder Minimum to the present (e.g., Lean and Rind, 1998, Crowley and Kim, 1999).

Ozone changes in the Earth's atmosphere caused by the 11-year solar cycle could affect the temperature response in the free atmosphere. A relation between 30 hPa geopotential and a solar index has been shown over nearly four solar cycles by Labitzke and van Loon (1997). Van Loon and Shea (1999, 2000) found a related connection between upper to middle tropospheric temperature and a solar index over the last 40 years, which is particularly strong in July and August. Variations in ozone forcing related to the solar cycle may also affect surface temperature via radiative and dynamical processes (see discussion in Chapter 6; Haigh, 1999; Shindell *et al.*, 1999, 2001), but observational evidence remains ambiguous (e.g., van Loon and Shea, 2000). The assessment of ozone-related Sun-climate interactions is uncertain as a result of the lack of long-term, reliable observations. This makes it difficult to separate effects of volcanic eruptions and solar forcing on ozone. There has also been speculation that the solar cycle might influence cloudiness and hence surface temperature through cosmic rays (e.g., Svensmark and Friis-Christensen, 1997; Svensmark, 1998). The latter effect is difficult to assess due to limitations in observed data and the shortness of the correlated time-series.

As discussed earlier in Section 12.2.3, differences between the response to solar and greenhouse gas forcings would make it easier to distinguish the climate response to either forcing. However, the spatial response pattern of surface air temperature to an increase in solar forcing was found to be quite similar to that in response to increases in greenhouse gas forcing (e.g., Cubasch *et al.*, 1997). The vertical response to solar forcing (Figure 12.5) includes warming throughout most of the troposphere. The response in the stratosphere is small and possibly locally negative, but less so than with greenhouse gas forcing, which gives tropospheric warming and strong stratospheric cooling. The dependence of solar forcing on wavelength and the effect of solar fluctuations on ozone were generally omitted in these simulations. Hence, the conclusion that changes in solar forcing have little effect on large-scale stratospheric temperatures remains tentative.

The different time-histories of the solar and anthropogenic forcing should help to distinguish between the responses. All reconstructions suggest a rise in solar forcing during the early decades of the 20th century with little change on inter-decadal time-scales in the second half. Such a forcing history is unlikely to explain the recent acceleration in surface warming, even if amplified by some unknown feedback mechanism.

Studies linking forcing and response through correlation techniques

A number of authors have correlated solar forcing and volcanic forcing with hemispheric and global mean temperature time-series from instrumental and palaeo-data (Lean *et al.*, 1995; Briffa *et al.*, 1998; Lean and Rind, 1998; Mann *et al.*, 1998) and found statistically significant correlations. Others have compared the simulated response, rather than the forcing, with observations and found qualitative evidence for the influence of natural forcing on climate (e.g., Crowley and Kim, 1996; Overpeck *et al.*, 1997; Wigley *et al.*, 1997; Bertrand *et al.*, 1999) or significant correlations (e.g., Schönwiese *et al.*, 1997; Free and Robock, 1999; Grieser and Schönwiese, 2001). Such a comparison is preferable as the climate response may differ substantially from the forcing. The results suggest that global scale low-frequency temperature variations are influenced by variations in known natural forcings. However, these results show that the late 20th century surface warming cannot be well represented by natural forcing (solar and volcanic individually or in combination) alone (for example Figures 12.6, 12.7; Lean and Rind, 1998; Free and Robock, 1999; Crowley, 2000; Tett *et al.*, 2000; Thejll and Lassen, 2000).

Mann *et al.* (1998, 2000) used a multi-correlation technique and found significant correlations with solar and, less so, with the volcanic forcing over parts of the palaeo-record. The authors concluded that natural forcings have been important on decadal-to-century time-scales, but that the dramatic warming of the 20th century correlates best and very significantly with greenhouse gas forcing. The use of multiple correlations avoids the possibility of spuriously high correlations due to the common trend in the solar and temperature time-series (Laut and Gunderman, 1998). Attempts to estimate the contributions of natural and anthropogenic forcing to 20th century temperature evolution simultaneously are discussed in Section 12.4.

Summary

We conclude that climate forcing by changes in solar irradiance and volcanism have likely caused fluctuations in global and hemispheric mean temperatures. Qualitative comparisons suggest that natural forcings produce too little warming to fully explain the 20th century warming (see Figure 12.7). The indication that the trend in net solar plus volcanic forcing has been negative in recent decades (see Chapter 6) makes it unlikely that natural forcing can explain the increased rate of global warming since the middle of the 20th century. This question will be revisited in a more quantitative manner in Section 12.4.

12.2.3.3 Anthropogenic forcing

In the SAR (Santer *et al.*, 1996c), pattern-based detection studies took into account changes in well-mixed greenhouse gases (often represented by an equivalent increase in CO_2), the direct effect of sulphate aerosols (usually represented by a seasonally constant change in surface albedo) and the influence of changes in stratospheric ozone. Recent studies have also included the effect of increases in tropospheric ozone and a representation of the indirect effect of sulphate aerosols on cloud albedo. Many models now include the individual greenhouse gases (as opposed to a CO_2 equivalent) and include an interactive sulphur cycle and an explicit treatment of scattering by aerosols (as opposed to using prescribed changes in surface albedo). Note that representation of the sulphur cycle in climate models is not as detailed as in the offline sulphur cycle models reported in Chapter 5. Detection and attribution studies to date have not taken into account other forcing agents discussed in Chapter 6, including biogenic aerosols, black carbon, mineral dust and changes in land use. Estimates of the spatial and temporal variation of these factors have not been available long enough to have been included in model simulations suitable for detection studies. In general, the neglected forcings are estimated to be small globally and there may be a large degree of cancellation in their global mean effect (see Chapter 6, Figure 6.8). It is less clear that the individual forcings will cancel regionally. As discussed in Section 12.4, this will add further uncertainty in the attribution of the response to individual forcing agents, although we believe it is unlikely to affect our conclusions about the effects of increases in well-mixed greenhouse gases on very large spatial scales.

Global mean anthropogenic forcing

The largest and most certain change in radiative forcing since the pre-industrial period is an increase of about 2.3 Wm^{-2} due to an increase in well-mixed greenhouse gases (Chapter 6, Figure 6.8 and Table 6.1). Radiative forcing here is taken to be the net downward radiative flux at the tropopause (see Chapter 6). Smaller, less certain contributions have come from increases in tropospheric ozone (about 0.3 Wm^{-2}), the direct effect of increases in sulphate aerosols (about $-0.4\,Wm^{-2}$) and decreases in

(a)

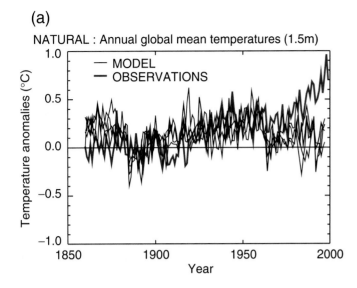

(b)

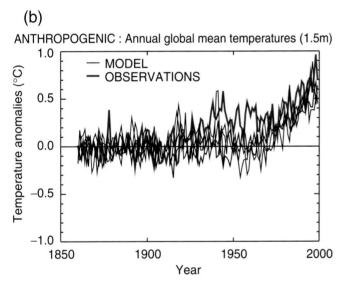

(c)

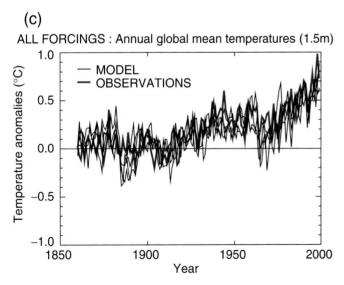

Figure 12.7: Global mean surface temperature anomalies relative to the 1880 to 1920 mean from the instrumental record compared with ensembles of four simulations with a coupled ocean-atmosphere climate model (from Stott *et al.*, 2000b; Tett *et al.*, 2000) forced (a) with solar and volcanic forcing only, (b) with anthropogenic forcing including well mixed greenhouse gases, changes in stratospheric and tropospheric ozone and the direct and indirect effects of sulphate aerosols, and (c) with all forcings, both natural and anthropogenic. The thick line shows the instrumental data while the thin lines show the individual model simulations in the ensemble of four members. Note that the data are annual mean values. The model data are only sampled at the locations where there are observations. The changes in sulphate aerosol are calculated interactively, and changes in tropospheric ozone were calculated offline using a chemical transport model. Changes in cloud brightness (the first indirect effect of sulphate aerosols) were calculated by an offline simulation (Jones *et al.*, 1999) and included in the model. The changes in stratospheric ozone were based on observations. The volcanic forcing was based on the data of Sato *et al.* (1993) and the solar forcing on Lean *et al.* (1995), updated to 1997. The net anthropogenic forcing at 1990 was 1.0 Wm^{-2} including a net cooling of 1.0 Wm^{-2} due to sulphate aerosols. The net natural forcing for 1990 relative to 1860 was 0.5 Wm^{-2}, and for 1992 was a net cooling of 2.0 Wm^{-2} due to Mt. Pinatubo. Other models forced with anthropogenic forcing give similar results to those shown in b (see Chapter 8, Section 8.6.1, Figure 8.15; Hasselmann *et al.*, 1995; Mitchell *et al.*, 1995b; Haywood *et al.*, 1997; Boer *et al.*, 2000a; Knutson *et al.*, 2000).

stratospheric ozone (about -0.2 Wm^{-2}). There is a very uncertain and possibly large negative contribution from the indirect effects of aerosols. Other factors such as that due to increases in fossil fuel organic carbon, aviation, changes in land use and mineral dust are very poorly known and not yet incorporated into simulations used in formal detection studies. Their contribution is generally believed to be small relative to well-mixed greenhouse gases, though they could be of importance on regional scales.

In order to assess temperature changes over the last two decades, Hansen *et al.* (1997b) estimated the net radiative forcing due to changes in greenhouse gases (including ozone), solar variations and stratospheric aerosols from 1979 to 1995 from the best available measurements of the forcing agents. The negative forcing due to volcanoes and decreases in stratospheric ozone compensated for a substantial fraction of the increase in greenhouse gas forcing in this period (see Chapter 6, Table 6.13).

Patterns of anthropogenic forcing
Many of the new detection studies take into account the spatial variation of climate response, which will depend to some extent on the pattern of forcing (see also Section 12.2.3). The patterns of forcing vary considerably (see Chapter 6, Figure 6.7). The magnitude of the overall forcing due to increases in well-mixed greenhouse gases varies from almost 3 Wm^{-2} in the sub-tropics to about 1 Wm^{-2} around the poles. The warming due to increases in tropospheric ozone is mainly in the tropics and northern sub-tropics. Decreases in stratospheric ozone observed over the last couple of decades have produced negative forcing of up to about 0.5 Wm^{-2} around Antarctica. The direct effect of sulphate aerosols predominates in the Northern Hemisphere industrial regions where the negative forcing may exceed 2 Wm^{-2} locally.

Temporal variations in forcing
Some of the new detection studies take into account the temporal as well as spatial variations in climate response (see Section 12.4.3.3). Hence the temporal variation of forcing is also important. The forcing due to well-mixed greenhouse gases (and tropospheric ozone) has increased slowly in the first half of the century, and much more rapidly in recent decades (Chapter 6, Figure 6.8). Contributions from other factors are smaller and more uncertain. Sulphur emissions increased steadily until World War I, then levelled off, and increased more rapidly in the 1950s, though not as fast as greenhouse gas emissions. This is reflected in estimates of the direct radiative effect of increases in sulphate aerosols. Given the almost monotonic increase in greenhouse gas forcing in recent decades, this means the ratio of sulphate to greenhouse gas forcing has probably been decreasing since about 1960 (see Chapter 6, Figure 6.8). This should be borne in mind when considering studies that attempt to detect a response to sulphate aerosols. The decreases in stratospheric ozone have been confined to the last two to three decades.

Uncertainties in aerosol forcing
Some recent studies have incorporated the indirect effect of increases in tropospheric aerosols. This is very poorly understood (see Chapter 6), but contributes a negative forcing which could be negligible or exceed 2 Wm^{-2}. The upper limit would imply very little change in net global mean anthropogenic forcing over the last century although there would still be a quite strong spatial pattern of heating and cooling which may be incompatible with recent observed changes (see, for example, Mitchell *et al.*, 1995a). A negligible indirect sulphate effect would imply a large increase in anthropogenic forcing in the last few decades. There is also a large range in the inter-hemispheric asymmetry in the different estimates of forcing (see Chapter 6, Table 6.4). Given this high level of uncertainty, studies using simulations including estimates of indirect sulphate forcing should be regarded as preliminary.

Summary
Well-mixed greenhouse gases make the largest and best-known contribution to changes in radiative forcing over the last century or so. There remains a large uncertainty in the magnitude and patterns of other factors, particularly those associated with the indirect effects of sulphate aerosol.

12.2.3.4 Climatic response to anthropogenic forcing
We now consider the simulated response to anthropogenic forcing. Models run with increases in greenhouse gases alone give a warming which accelerates in the latter half of the century. When a simple representation of aerosol effects is included (Mitchell *et al.*, 1995b; Cubasch *et al.*, 1996; Haywood *et al.*, 1997; Boer *et al.*, 2000a,b) the rate of warming is reduced (see also Chapter 8, Section 8.6.1). The global mean response is similar when additional forcings due to ozone and the indirect effect of sulphates are included. GCM simulations (Tett *et al.*, 1996; Hansen *et al.*, 1997b) indicate that changes in stratospheric ozone observed over the last two decades yield a global mean surface temperature cooling of about 0.1 to 0.2°C. This may be too small to be distinguishable from the model's internal variability and is also smaller than the warming effects due to the changes in the well-mixed greenhouse gases over the same time period (about 0.2 to 0.3°C). The lack of a statistically significant surface temperature change is in contrast to the large ozone-induced cooling in the lower stratosphere (WMO, 1999; Bengtsson *et al.* 1999).

The response of the vertical distribution of temperature to anthropogenic forcing
Increases in greenhouse gases lead to a warming of the troposphere and a cooling of the stratosphere due to CO_2 (IPCC, 1996). Reductions in stratospheric ozone lead to a further cooling, particularly in the stratosphere at high latitudes. Anthropogenic sulphate aerosols cool the troposphere with little effect on the stratosphere. When these three forcings are included in a climate model (e.g., Tett *et al.*, 1996, 2000) albeit in a simplified way, the simulated changes show tropospheric warming and stratospheric cooling, as observed and as expected on physical principles (Figure 12.8). Note that this structure is distinct from that expected from natural (internal and external) influences.

The response of surface temperature to anthropogenic forcing
The spatial pattern of the simulated surface temperature response to a steady increase in greenhouse gases is well documented (e.g., Kattenberg *et al.*, 1996; Chapter 10). The warming is greater over

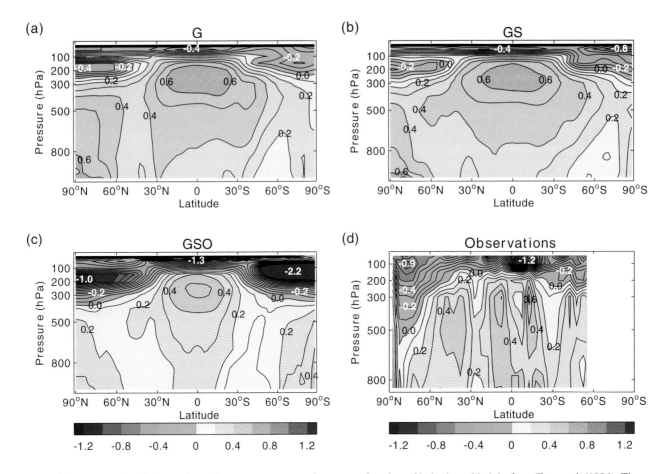

Figure 12.8: Simulated and observed zonal mean temperature change as a function of latitude and height from Tett *et al.* (1996). The contour interval is 0.1°C. All signals are defined to be the difference between the 1986 to 1995 decadal mean and the 20 year 1961 to 1980 mean. (a), increases in CO_2 only (G); (b), as (a), but with a simple representation of sulphate aerosols added (GS); (c), as (b), with observed changes in stratospheric ozone (GSO); (d), observed changes.

land than over ocean and generally small during the 20th century over the Southern Ocean and northern North Atlantic where mixing extends to considerable depth. The warming is amplified in high latitudes in winter by the recession of sea ice and snow, and is close to zero over sea ice in summer.

Despite the qualitative consistency of these general features, there is considerable variation from model to model. In Chapter 9, it was noted that the spatial correlation between the transient response to increasing CO_2 in *different* models in scenarios to the middle of the 21st century was typically 0.65. In contrast, the spatial correlation between the temperature response to greenhouses gases only, and greenhouse gases and aerosols in the *same* model was typically 0.85 (see Chapter 9, Table 9.2). Hence, attempts to detect separate greenhouse gas and aerosol patterns in different models may not give consistent results (see Section 12.4.3.2).

12.2.4 Some Important Statistical Considerations

Most recent studies (Hegerl *et al.*, 1996, 1997, 2000, 2001; North and Stevens, 1998; Allen and Tett, 1999; Tett *et al.*, 1999, 2000; Berliner *et al.*, 2000; North and Wu, 2001; Stott *et al.*, 2001) have used a regression approach in which it is assumed that observa-

tions can be represented as a linear combination of candidate signals plus noise (see Appendices 12.1 and 12.2). Other approaches, such as pattern correlation (Santer *et al.*, 1995, 1996a; see also Appendix 12.3), complement the regression approach, being particularly valuable in cases where model-simulated response patterns are particularly uncertain. In all cases, the signal patterns are obtained from climate models. In the regression approach, the unknown signal amplitudes are estimated from observations. The uncertainty of these estimates that is caused by natural variability in the observations is expressed with confidence intervals. Detection of an individual signal is achieved when the confidence interval for its amplitude does not include zero. Overall detection (that some climate change has taken place) is achieved when the joint confidence interval on the signals considered does not encompass the origin.

Attribution and consistency

Detecting that some climate change has taken place does not immediately imply that we know the cause of the detected change. The practical approach to attribution that has been taken by climatologists includes a demand for consistency between the signal amplitudes projected by climate models and estimated from observations (Hasselmann, 1997). Consequently, several studies, including Hegerl *et al.* (1997, 2000) and Tett *et al.* (1999, 2000) have performed an "attribution" consistency test that is designed to detect inconsistency between observed and model projected signal amplitudes. This test is a useful adjunct to detection because it provides an objective means of identifying model-simulated signal amplitudes that are significantly different from those estimated from observations. However, the test does not give the final word on attribution because it is designed to identify evidence of inconsistency rather than evidence for consistency between modelled and observed estimates of signal strength. A further refinement (e.g., Stott *et al.*, 2001) is to consider the full range of signals believed, on physical grounds, to be likely to have had a significant impact on recent climate change and to identify those subsets of these signals that are consistent with recent observations. If all these subsets contain an anthropogenic component, for example, then at least part of the observed change can be attributed to anthropogenic influence. Levine and Berliner (1999) point out that a test that searches for consistency is available (Brown *et al.*, 1995), but it has not yet been used in attribution studies. Bayesian statisticians approach the problem more directly by estimating the posterior probability that the signal amplitudes projected by climate models are close to those in the observed climate. Berliner *et al.* (2000) provides a demonstration.

The use of climate models to estimate natural internal variability
Climate models play a critical role in these studies because they provide estimates of natural internal variability as well as the signals. In most studies an estimate of natural internal variability is needed to optimise the search for the signal and this is usually obtained from a long control simulation. In addition, a separate estimate of natural variability is required to determine the uncertainty of the amplitude estimates. Unfortunately, the short instrumental record gives only uncertain estimates of variability on the 30 to 50 year time-scales that are important for detection and attribution and palaeo-data presently lacks the necessary spatial coverage (see Section 12.2.2). Thus a second control integration is generally used to estimate the uncertainty of the amplitude estimates that arises from natural climate variability (e.g., Hegerl *et al.*, 1996; Tett *et al.*, 1999).

Temporal and spatial scales used in detection studies
While a growing number of long control simulations are becoming available, there remain limitations on the spatial scales that can be included in global scale detection and attribution studies. Present day control simulations, which range from 300 to about 2,000 years in length, are not long enough to simultaneously estimate internal variability on the 30 to 50 year time-scale over a broad range of spatial scales. Consequently, detection and attribution studies are conducted in a reduced space that includes only large spatial scales. This space is selected so that it represents the signals well and allows reliable estimation of internal variability on the scales retained (see Appendix 12.4). Recently, the scale selection process has been augmented with a statistical procedure that checks for consistency between model simulated and observed variability on the scales that are retained (Allen and Tett, 1999).

Fixed and temporally-varying response patterns
Detection and attribution studies performed up to the SAR used fixed signal patterns that did not evolve with time. These studies were hampered because the mean large-scale response of climate to different types of anomalous forcing tends to be similar (e.g., Mitchell *et al.*, 1995a; Reader and Boer, 1998; see also Figure 12.3). Recent studies have been able to distinguish more clearly between signals from anthropogenic and other sources by including information from climate models about their temporal evolution. Tett *et al.* (1999, 2000) and Stott *et al.* (2001) in related studies have used a *space-time* approach in which the signal pattern evolves on the decadal time-scale over a 50-year period. North and Wu (2001) also use a space-time approach. North and Stevens (1998) used a related *space-frequency* approach (see Appendix 12.2).

Allowance for noise in signal patterns
Most studies have assumed that signal patterns are noise free. This is a reasonable assumption for fixed pattern studies (see Appendix 12.2) but space-time estimates of the 20th century climate change obtained from small ensembles of forced climate simulations are contaminated by the model's internal variability. Allen and Tett (1999) point out that noise in the signal patterns will tend to make the standard detection algorithm (e.g., Hasselmann, 1993, 1997) somewhat conservative. Methods for accommodating this source of noise have been available for more than a century (Adcock, 1878; see also Ripley and Thompson, 1987). Allen and Stott (2000) recently applied such a method and found that, while the question of which signals could be detected was generally unaffected, the estimated amplitude of individual signals was sensitive to this modification of the procedure. Another source of uncertainty concerns differences in signal patterns between different models. Recent studies (Allen *et al.*, 2000a,b; Barnett *et al.*, 2000; Hegerl *et al.*, 2000) consider the sensitivity of detection and attribution results to these differences.

12.3 Qualitative Comparison of Observed and Modelled Climate Change

12.3.1 Introduction

This section presents a qualitative assessment of consistencies and inconsistencies between the observed climate changes identified in Chapter 2 and model projections of anthropogenic climate change described in Chapter 9.

Most formal detection and attribution studies concentrate on variables with high climate change signal-to-noise ratios, good observational data coverage, and consistent signals from different model simulations, mainly using mean surface air temperatures or zonal mean upper-air temperatures. To enhance the signal-to-noise ratio, they generally consider variations on large spatial scales and time-scales of several decades or longer.

There are many studies that have identified areas of qualitative consistency and inconsistency between observed and modelled climate change. While the evidence for an anthropogenic influence on climate from such studies is less compelling than from formal attribution studies, a broad range of evidence of qualitative consistency between observed and modelled climate change is also required. In addition, areas of qualitative consistency may suggest the possibility for further formal detection and attribution study.

12.3.2 Thermal Indicators

Surface temperature
Global mean surface air temperature has been used in many climate change detection studies. The warming shown in the instrumental observations over the last 140 years is larger than that over a comparable period in any of the multi-century control simulations carried out to date (e.g., Figure 12.1; Stouffer *et al.*, 2000). If the real world internal variability on this time-scale is no greater than that of the models, then the temperature change over the last 140 years has been unusual and therefore likely to be externally forced. This is supported by palaeo-reconstructions of the last six centuries (Mann *et al.*, 1998) and the last 1,000 years (Briffa *et al.*, 1998; 2000; Jones *et al.*, 1998; Crowley, 2000; Crowley and Lowery, 2000; Mann *et al.*, 2000), which show that the 20th century warming is highly unusual. Three of the five years (1995, 1996 and 1998) added to the instrumental record since the SAR are the warmest globally in the instrumental record, consistent with the expectation that increases in greenhouse gases will lead to sustained long-term warming.

When anthropogenic factors are included, models provide a plausible explanation of the changes in global mean temperature over the last hundred years (Figure 12.7). It is conceivable that this agreement between models and observations is spurious. For example, if a model's response to greenhouse gas increases is too large (small) and the sulphate aerosol forcing too large (small), these errors could compensate. Differences in the spatio-temporal patterns of response to greenhouse gases and sulphate forcing nevertheless allow some discrimination between them, so this compensation is not complete. On the other hand, when forced with known natural forcings, models produce a cooling over the second half of the 20th century (see Figure 12.7) rather than the warming trend shown in the observed record. The discrepancy is too large to be explained through model estimates of internal variability and unlikely to be explained through uncertainty in forcing history (Tett *et al.*, 2000). Schneider and Held (2001) applied a technique to isolate those spatial patterns of decadal climate change in observed surface temperature data over the 20th century which are most distinct from interannual variability. They find a spatial pattern which is similar to model-simulated greenhouse gas and sulphate aerosol fingerprints in both July and December. The time evolution of this pattern shows a strong trend with little influence of interannual variability. (Note that this technique is related to optimal fingerprinting, but does not use prior information on the pattern of expected climate change.)

Other thermal indicators
While most attention in formal detection and attribution studies has been paid to mean surface air temperatures, a number of other thermal indicators of climate variations are also discussed in Chapter 2. Many of these, including warming in sub-surface land temperatures measured in bore holes, warming indicators in ice

cores and corresponding bore holes, warming in sub-surface ocean temperatures, retreat of glaciers, and reductions in Arctic sea-ice extent and in snow cover, are consistent with the recent observed warming in surface air temperatures and with model projections of the response to increasing greenhouse gases. Other observed changes in thermal indicators include a reduction in the mean annual cycle (winters warming faster than summers) and in the mean diurnal temperature range (nights warming faster than days) over land (see Chapter 2). While the changes in annual cycle are consistent with most model projections, the observed changes in diurnal temperature range are larger than simulated in most models for forcings due to increasing greenhouse gases and sulphate aerosols this century (see Chapters 2 and 8). However, the spatial and temporal coverage of data for changes in observed diurnal temperature range is less than for changes in mean temperatures, leading to greater uncertainty in the observed global changes (Karoly and Braganza, 2001; Schnur, 2001). Also, the observed reductions in diurnal temperature range are associated with increases in cloudiness (see Chapter 2), which are not simulated well by models. Few models include the indirect effects of sulphate aerosols on clouds.

Changes in sea-ice cover and snow cover in the transition seasons in the Northern Hemisphere are consistent with the observed and simulated high latitude warming. The observed trends in Northern Hemisphere sea-ice cover (Parkinson *et al.*, 1999) are consistent with those found in climate model simulations of the last century including anthropogenic forcing (Vinnikov *et al.*, 1999). Sea-ice extent in the Southern Hemisphere does not show any consistent trends.

Compatibility of surface and free atmosphere temperature trends
There is an overall consistency in the patterns of upper air temperature changes with those expected from increasing greenhouse gases and decreasing stratospheric ozone (tropospheric warming and stratospheric cooling). It is hard to explain the observed changes in the vertical in terms of natural forcings alone, as discussed in Section 12.2.3.2 (see Figure 12.8). However, there are some inconsistencies between the observed and modelled vertical patterns of temperature change. Observations indicate that, over the last three to four decades, the tropical atmosphere has warmed in the layer up to about 300 hPa and cooled above (Parker *et al.*, 1997; Gaffen *et al.*, 2000). Model simulations of the recent past produce a warming of the tropical atmosphere to about 200 hPa, with a maximum at around 300 hPa not seen in the observations. This discrepancy is less evident when co-located model and radiosonde data are used (Santer *et al.*, 2000), or if volcanic forcing is taken into account, but does not go away entirely (Bengtsson *et al.*, 1999; Brown *et al.*, 2000b). The MSU satellite temperature record is too short and too poorly resolved in the vertical to be of use here.

Comparison of upper air and surface temperature data in Chapter 2 shows that the lower to mid-troposphere has warmed less than the surface since 1979. The satellite-measured temperature over a broad layer in the lower troposphere around 750 hPa since 1979 shows no significant trend, in contrast to the warming trend measured over the same time period at the surface. This disparity has been assessed recently by a panel of experts

(National Academy of Sciences, 2000). They concluded that "the troposphere actually may have warmed much less rapidly than the surface from 1979 to the late 1990s, due both to natural causes (e.g., the sequence of volcanic eruptions that occurred within this particular 20-year period) and human activities (e.g., the cooling in the upper troposphere resulting from ozone depletion in the stratosphere)" (see also Santer *et al.*, 2000). They also concluded that "it is not currently possible to determine whether or not there exists a fundamental discrepancy between modelled and observed atmospheric temperature changes since the advent of satellite data in 1979". Over the last 40 years, observed warming trends in the lower troposphere and at the surface are similar, indicating that the lower troposphere warmed faster than the surface for about two decades prior to 1979 (Brown *et al.*, 2000a; Gaffen *et al.*, 2000). However, in the extra-tropical Eurasian winter some additional warming of the surface relative to the lower or mid-troposphere might be expected since 1979. This is due to an overall trend towards an enhanced positive phase of the Arctic Oscillation (Thompson *et al.*, 2000) which has this signature.

Model simulations of large-scale changes in tropospheric and surface temperatures are generally statistically consistent with the observed changes (see Section 12.4). However, models generally predict an enhanced rate of warming in the mid- to upper troposphere over that at the surface (i.e., a negative lapse-rate feedback on the surface temperature change) whereas observations show mid-tropospheric temperatures warming no faster than surface temperatures. It is not clear whether this discrepancy arises because the lapse-rate feedback is consistently over-represented in climate models or because of other factors such as observational error or neglected forcings (Santer *et al.*, 2000). Note that if models do simulate too large a negative lapse-rate feedback, they will tend to underestimate the sensitivity of climate to a global radiative forcing perturbation.

Stratospheric trends

A recent assessment of temperature trends in the stratosphere (Chanin and Ramaswamy, 1999) discussed the cooling trends in the lower stratosphere described in Chapter 2. It also identified large cooling trends in the middle and upper stratosphere, which are consistent with anthropogenic forcing due to stratospheric ozone depletion and increasing greenhouse gas concentrations. An increase in water vapour, possibly due to increasing methane oxidation, is another plausible explanation for the lower strato-spheric cooling (Forster and Shine, 1999) but global stratospheric water vapour trends are poorly understood.

12.3.3 Hydrological Indicators

As discussed in Chapter 2, there is less confidence in observed variations in hydrological indicators than for surface temperature, because of the difficulties in taking such measurements and the small-scale variations of precipitation. There is general consistency between the changes in mean precipitation in the tropics over the last few decades and changes in ENSO. There is no general consistency between observed changes in mean tropical precipitation and model simulations. In middle and high latitudes in the Northern Hemisphere, the observed increase in precipita-tion is consistent with most model simulations. Observed changes in ocean salinity in the Southern Ocean appear to be consistent with increased precipitation there, as expected from model simulations (Wong *et al.*, 1999; Banks *et al.*, 2000).

The observed increases in the intensity of heavy precipitation in the tropics and in convective weather systems described in Chapter 2 are consistent with moist thermodynamics in a warmer atmosphere and model simulations. Observed increases of water vapour in the lower troposphere in regions where there is adequate data coverage are also consistent with model simulations. As discussed in Chapter 7, different theories suggest opposite variations of water vapour in the upper troposphere associated with an increased greenhouse effect and surface warming. The quality, amount and coverage of water vapour data in the upper troposphere do not appear to be sufficient to resolve this issue.

12.3.4 Circulation

In middle and high latitudes of both hemispheres, there has been a trend over the last few decades towards one phase of the North Atlantic Oscillation/Arctic Oscillation and of the Antarctic high latitude mode, sometimes also referred to as "annular modes", (Chapter 2; Thompson *et al.*, 2000). These are approximately zonally symmetric modes of variability of the atmospheric circulation. Both trends have been associated with reduced surface pressure at high latitudes, stronger high latitude jets, a stronger polar vortex in the winter lower stratosphere and, in the Northern Hemisphere, winter warming over the western parts of the continents associated with increased warm advection from ocean regions. The trend is significant and cannot be explained by internal variability in some models (Gillett *et al.*, 2000b). These dynamical changes explain only part of the observed Northern Hemisphere warming (Gillett *et al.*, 2000b; Thompson *et al.*, 2000). Modelling studies suggest a number of possible causes of these circulation changes, including greenhouse gas increases (Fyfe *et al.*, 1999; Paeth *et al.*, 1999; Shindell *et al.*, 1999) and stratospheric ozone decreases (Graf *et al.*, 1998; Volodin and Galin, 1999). Some studies have also shown that volcanic eruptions (Graf *et al.*, 1998; Mao and Robock, 1998; Kirchner *et al.*, 1999) can induce such changes in circulation on interannual time-scales. Shindell *et al.* (2001) show that both solar and volcanic forcing are unlikely to explain the recent trends in the annular modes.

The majority of models simulate the correct sign of the observed trend in the North Atlantic or Arctic Oscillation when forced with anthropogenic increases in greenhouse gases and sulphate aerosols, but almost all underestimate the magnitude of the trend (e.g., Osborn *et al.*, 1999; Gillett *et al.*, 2000b; Shindell *et al.*, 1999). Some studies suggest that a better resolved stratosphere is necessary to simulate the correct magnitude of changes in dynamics involving the annular modes (e.g., Shindell *et al.*, 2001).

12.3.5 Combined Evidence

The combination of independent but consistent evidence should strengthen our confidence in identifying a human influence on climate. The physical and dynamical consistency of most of the thermal and hydrological changes described above supports this conclusion. However, it is important to bear in mind that much of this evidence is associated with a global and regional pattern of warming and therefore cannot be considered to be completely independent evidence.

An elicitation of individual experts' subjective assessment of evidence for climate change detection and attribution is being carried out (Risbey *et al.*, 2000). This will help to better understand the nature of the consensus amongst experts on the subject of climate change attribution.

12.4 Quantitative Comparison of Observed and Modelled Climate Change

A major advance since the SAR has been the increase in the range of techniques used to assess the quantitative agreement between observed and modelled climate change, and the evaluation of the degree to which the results are independent of the assumptions made in applying those techniques (Table 12.1). Also, some studies have based their conclusions on estimates of the amplitude of anthropogenic signals in the observations and consideration of their consistency with model projections. Estimates of the changes in forcing up to 1990 used in these studies, where available, are given in Table 12.2. In this section we assess new studies using a number of techniques, ranging from descriptive analyses of simple indices to sophisticated optimal detection techniques that incorporate the time and space-dependence of signals over the 20th century.

We begin in Section 12.4.1 with a brief discussion of detection studies that use simple indices and time-series analyses. In Section 12.4.2 we discuss recent pattern correlation studies (see Table 12.1) that assess the similarity between observed and modelled climate changes. Pattern correlation studies were discussed extensively in the SAR, although subsequently they received some criticism. We therefore also consider the criticism and studies that have evaluated the performance of pattern correlation techniques. Optimal detection studies of various kinds are assessed in Section 12.4.3. We consider first studies that use a single fixed spatial signal pattern (Section 12.4.3.1) and then studies that simultaneously incorporate more than one fixed signal pattern (Section 12.4.3.2). Finally, optimal detection studies that take into account temporal as well as spatial variations (so-called space-time techniques) are assessed in Section 12.4.3.3.

We provide various aids to the reader to clarify the distinction between the various detection and attribution techniques that have been used. Box 12.1 in Section 12.4.3 provides a simple intuitive description of optimal detection. Appendix 12.1 provides a more technical description and relates optimal detection to general linear regression. The differences between fixed pattern, space-time and space-frequency optimal detection methods are detailed in Appendix 12.2 and the relationship between pattern correlation and optimal detection methods is discussed in Appendix 12.3. Dimension reduction, a necessary part of optimal detection studies, is discussed in Appendix 12.4.

12.4.1 Simple Indices and Time-series Methods

An index used in many climate change detection studies is global mean surface temperature, either as estimated from the instrumental record of the last 140 years, or from palaeo-reconstructions. Some studies of the characteristics of the global mean and its relationship to forcing indices are assessed in Section 12.2.3. Here we consider briefly some additional studies that examine the spatial structure of observed trends or use more sophisticated time-series analysis techniques to characterise the behaviour of global, hemispheric and zonal mean temperatures.

Spatial patterns of trends in surface temperature
An extension of the analysis of global mean temperature is to compare the spatial structure of observed trends (see Chapter 2, Section 2.2.2.4) with those simulated by models in coupled control simulations. Knutson *et al.* (2000) examined observed 1949 to 1997 surface temperature trends and found that over about half the globe they are significantly larger than expected from natural low-frequency internal variability as simulated in long control simulations with the GFDL model (Figure 12.9). A similar result was obtained by Boer *et al.* (2000a) using 1900 to 1995 trends. The level of agreement between observed and simulated trends increases substantially in both studies when observations are compared with simulations that incorporate transient greenhouse gases and sulphate aerosol forcing (compare Figure 12.9c with Figure 12.9d, see also Chapter 8, Figure 8.18). While there are areas, such as the extra-tropical Pacific and North Atlantic Ocean, where the GFDL model warms significantly more than has been observed, the anthropogenic climate change simulations do provide a plausible explanation of temperature trends over the last century over large areas of the globe. Delworth and Knutson (2000) find that one in five of their anthropogenic climate change simulations showed a similar evolution of global mean surface temperature over the 20th century to that observed, with strong warming, particularly in the high latitude North Atlantic, in the first half of the century. This would suggest that the combination of anthropogenic forcing and internal variability may be sufficient to account for the observed early-century warming (as suggested by, e.g., Hegerl *et al.*, 1996), although other recent studies have suggested that natural forcing may also have contributed to the early century warming (see Section 12.4.3).

Correlation structures in surface temperature
Another extension is to examine the lagged and cross-correlation structure of observed and simulated hemispheric mean temperature as in Wigley *et al.*, (1998a). They find large differences between the observed and model correlation structure that can be explained by accounting for the combined influences of anthropogenic and solar forcing and internal variability in the observations. Solar forcing alone is not found to be a satisfactory explanation for the discrepancy between the correlation structures of the observed and simulated temperatures. Karoly and Braganza (2001) also examined the correlation structure of surface air temperature variations. They used several simple indices, including the land-ocean contrast, the meridional

Table 12.1: *Summary of the main detection and attribution studies considered.*

Study	Signals	Signal source	Noise source	Method	S, V	Sources of uncertainty	Time-scale	No. of patterns	Detect
Santer et al., 1996	G, GS, O etc.	Equilibrium / future LLNL, GFDL R15, HadCM2	GFDL R15, HadCM2, ECHAM1	F, Corr	V	Internal variability	25 year Annual and seasonal	1	GSO
Hegerl, 1996, 1997	G, GS	Future ECHAM3, HadCM2	GFDL R15, ECHAM1, HadCM2; observation	F, Pattern	S	Internal variability	30, 50 years Annual and JJA	1, 2	G, GS, S
Tett et al., 1996	G, GS, GSO	Historical HadCM2	HadCM2	F, Corr	V	Internal variability	35 years	1	GSO
Hegerl et al., 2000	G, GS, Vol, Sol	Future, ECHAM3, HadCM2	ECHAM3, HadCM2	F, Pattern	S	Internal variability; model uncertainty	30, 50 years Annual and JJA	1, 2	GS, G, S (not all cases)
Allen and Tett, 1999	G, GS, GSO	Historical HadCM2	HadCM2	F, pattern	V	Internal variability	35 years Annual	1, 2	GSO and also G
Tett et al., 1999 Stott et al., 2001	G, GS, Sol, Vol	Historical HadCM2	HadCM2	Time-space	S	Internal variability, 2 solar signals	50 years decadal and seasonal	2 or more	G, GS, Sol (Vol)
North and Stevens, 1998 Leroy, 1998 North and Wu, 2001	G, GS, Sol, Vol	Historical EBM	GFDL ECHAM1, EBM Same+Had CM2	Freq-Space Time-space	S	Internal variability	Annual and hemispheric summer Annual	4	G, S, Vol G, Vol
Barnett et al., 1999	G, GS, GSIO Sol+vol	Future ECHAM3, ECHAM4, HadCM2, GFDL R15	ECHAM3, ECHAM4, HadCM2, GFDL R15	F, Pattern	S	Observed sampling error, model uncertainty, internal variability	50 years JJA trends	2	GS, G, S (S not all cases)
Hill et al., 2001	G, GSO, Sol	Historical HadCM2	HadCM2	F, pattern	V	Internal variability	35 years annual	3	G
Tett et al., 2000	G, GSTI, GSTIO, Nat	Historical HadCM3	HadCM3	Time-space	S	Internal variability	50, 100 years decadal	2 or more	G, SIT, GSTIO and Nat
				F, pattern	V	Internal variability	35 years, annual	2	GSTI

The columns contain the following information:

Study :   the main reference to the study.

Signals :   outlines the principal signals considered: G-greenhouse gases, S-sulphate aerosol direct effect, T-tropospheric ozone, I-sulphate aerosol indirect effect, O-stratospheric ozone, Sol-solar, Vol-volcanoes, Nat-solar and volcanoes.

Signal source :   "historical" indicates the signal is taken from a historical hindcast simulation, "future" indicates that the pattern is taken from a prediction.

Noise source :   origin of the noise estimates.

Method :   "F" means fixed spatial pattern, "corr" indicates a correlation study, "pattern" an optimal detection study.

S, V :   "V" indicates a vertical temperature pattern, "S" a horizontal temperature pattern.

Sources of uncertainty : any additional uncertainties allowed for are indicated. Modelled internal variability is allowed for in all studies.

Time-scale :   the lengths of time interval considered. (JJA= June-July-August)

No. of patterns :   the number of patterns considered simultaneously.

Detect :   signals detected.

Table 12.2: *Estimated forcing from pre-industrial period to 1990 in simulations used in detection studies (Wm⁻²). GS indicates only direct sulphate forcing included, GSI indicates both direct and indirect effects included. Other details of the detection studies are given in Table 12.1. Details of the models are given in Chapter 8, Table 8.1.*

Model	Aerosol	Baseline forcing	1990 aerosol forcing	1990 greenhouse gas forcing	Source of estimate
HadCM2	GS	1760	−0.6	1.9	Mitchell and Johns, 1997
HadCM3	GSI	1860	−1.0	2.0	Tett *et al.*, 2000
ECHAM3/LSG	GS	1880	−0.7	1.7	Roeckner
ECHAM4/OPYC	GSI	1760	−0.9	2.2	Roeckner *et al.*, 1999
GFDL_R30	GS	1760	−0.6	2.1	Stouffer
CGCM1,2	GS	1760	~ −1.0	~2.2	Boer *et al.*, 2000a,b

gradient, and the magnitude of the seasonal cycle, to describe global climate variations and showed that for natural variations, they contain information independent of the global mean temperature. They found that the observed trends in these indices over the last 40 years are unlikely to have occurred due to natural climate variations and that they are consistent with model simulations of anthropogenic climate change.

Statistical models of time-series
Further extensions involve the use of statistical "models" of global, hemispheric and regional temperature time-series. Note however, that the stochastic models used in these time-series studies are generally not built from physical principles and are thus not as strongly constrained by our knowledge of the physical climate system as climate models. All these studies depend on inferring the statistical properties of the time-series from an assumed noise model with parameters estimated from the residuals. As such, the conclusions depend on the appropriateness or otherwise of the noise model.

Tol and de Vos (1998), using a Bayesian approach, fit a hierarchy of time-series models to global mean near-surface temperature. They find that there is a robust statistical relationship between atmospheric CO_2 and global mean temperature and that natural variability is unlikely to be an explanation for the observed temperature change of the past century. Tol and Vellinga (1998) further conclude that solar variation is also an unlikely explanation. Zheng and Basher (1999) use similar time-series models and show that deterministic trends are detectable over a large part of the globe. Walter *et al.* (1998), using neural network models, estimate that the warming during the past century due to greenhouse gas increases is 0.9 to 1.3°C and that the counter-balancing cooling due to sulphate aerosols is 0.2 to 0.4°C. Similar results are obtained with a multiple regression model (Schönwiese *et al.*, 1997). Kaufmann and Stern (1997) examine the lagged-covariance structure of hemispheric mean temperature and find it consistent with unequal anthropogenic aerosol forcing in the two hemispheres. Smith *et al.* (2001), using similar bivariate time-series models, find that the evidence for causality becomes weak when the effects of ENSO are taken into account. Bivariate time-series models of hemispheric mean temperature that account for box–diffusion estimates of the response to anthropogenic and solar forcing are found to fit the observations

significantly better than competing statistical models. All of these studies draw conclusions that are consistent with those of earlier trend detection studies (as described in the SAR).

In summary, despite various caveats in each individual result, time-series studies suggest that natural signals and internal variability alone are unlikely to explain the instrumental record, and that an anthropogenic component is required to explain changes in the most recent four or five decades.

12.4.2 Pattern Correlation Methods

12.4.2.1 Horizontal patterns

Results from studies using pattern correlations were reported extensively in the SAR (for example, Santer *et al.*, 1995, 1996c; Mitchell *et al.*, 1995b). They found that the patterns of simulated surface temperature change due to the main anthropogenic factors in recent decades are significantly closer to those observed than expected by chance. Pattern correlations have been used because they are simple and are insensitive to errors in the amplitude of the spatial pattern of response and, if centred, to the global mean response. They are also less sensitive than regression-based optimal detection techniques to sampling error in the model-simulated response. The aim of pattern-correlation studies is to use the differences in the large-scale patterns of response, or "fingerprints", to distinguish between different causes of climate change.

Strengths and weaknesses of correlation methods
Pattern correlation statistics come in two types – centred and uncentred (see Appendix 12.3). The centred (uncentred) statistic measures the similarity of two patterns after (without) removal of the global mean. Legates and Davis (1997) criticised the use of centred correlation in detection studies. They argued that correlations could increase while observed and simulated global means diverge. This was precisely the reason centred correlations were introduced (e.g., Santer *et al.*, 1993): to provide an indicator that was statistically independent of global mean temperature changes. If both global mean changes and centred pattern correlations point towards the same explanation of observed temperature changes, it provides more compelling evidence than either of these indicators in isolation. An explicit analysis of the role of the global mean in correlation-based studies can be provided by the

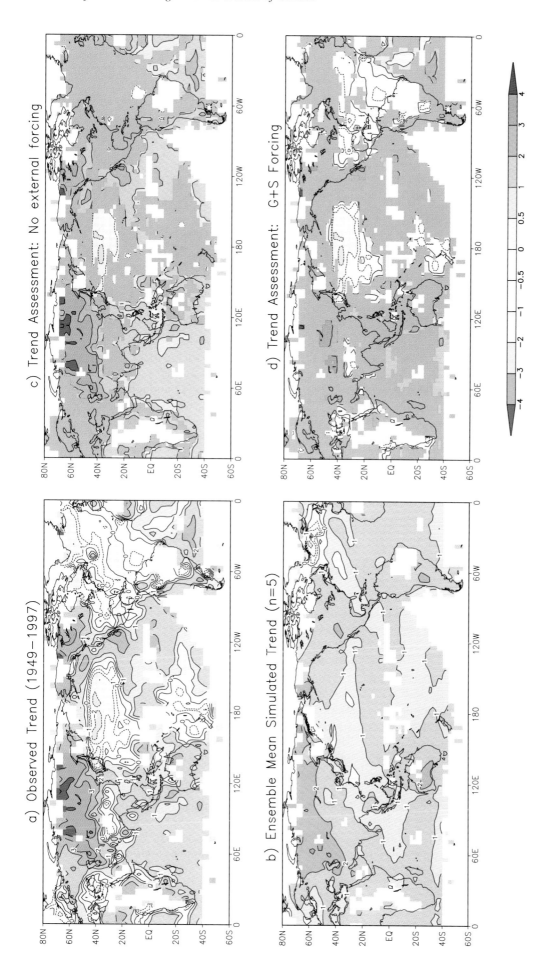

Figure 12.9: (a) Observed surface air temperature trends for 1949 to 1997. (b) Simulated surface air temperature trends for the same period as estimated from a five-member greenhouse gas plus sulphate ensemble run with the GFDL R30 model. (c) Observed trends (in colour) that lie outside the 90% natural variability confidence bounds as estimated from the GFDL R30 control run. Grey areas show regions where the observed trends are consistent with the local 49-year temperature trends in the control run. (d) As for (c) but showing observed 1949 to 1997 trends (in colour) that are significantly different (as determined with a t-test at the 10% level) from those simulated by the greenhouse gas plus aerosol simulations performed with the GFDL R30 model (from Knutson *et al.*, 2000). The larger grey areas in (d) than (c) indicate that the observed trends are consistent with the anthropogenic forced simulations over larger regions than the control simulations.

use of both centred and uncentred statistics. Pattern correlation-based detection studies account for spatial auto-correlation implicitly by comparing the observed pattern correlation with values that are realised in long control simulations (see Wigley *et al.*, 2000). These studies do not consider the amplitude of anthropogenic signals, and thus centred correlations alone are not sufficient for the attribution of climate change.

Wigley *et al.* (1998b) studied the performance of correlation statistics in an idealised study in which known spatial signal patterns were combined with realistic levels of internal variability. The statistics were found to perform well even when the signal is contaminated with noise. They found, in agreement with Johns *et al.* (2001), that using an earlier base period can enhance detectability, but that much of this advantage is lost when the reduced data coverage of earlier base periods is taken into account. They also found that reasonable combinations of greenhouse gas and aerosol patterns are more easily detected than the greenhouse gas pattern on its own. This last result indicates the importance of reducing the uncertainty in the estimate of aerosol forcing, particularly the indirect effects. In summary, we have a better understanding of the behaviour of pattern correlation statistics and reasons for the discrepancies between different studies.

12.4.2.2 Vertical patterns

As noted in Section 12.3.2, increases in greenhouse gases produce a distinctive change in the vertical profile of temperature. Santer *et al.* (1996c) assessed the significance of the observed changes in recent decades using equilibrium GCM simulations with changes in greenhouse gases, sulphate aerosols and stratospheric ozone. This study has been extended to include results from the transient AOGCM simulations, additional sensitivity studies and estimates of internal variability from three different models (Santer *et al.*, 1996a). Results from this study are consistent with the earlier results – the 25-year trend from 1963 to 1988 in the centred correlation statistic between the observed and simulated patterns for the full atmosphere was significantly different from the population of 25-year trends in the control simulations. The results were robust even if the estimates of noise levels were almost doubled, or the aerosol response (assumed linear and additive) was halved. The aerosol forcing leads to a smaller warming in the Northern Hemisphere than in the Southern Hemisphere.

Tett *et al.* (1996) refined Santer *et al.*'s (1996a) study by using ensembles of transient simulations which included increases in CO_2, and sulphate aerosols, and reductions in stratospheric ozone, as well as using an extended record of observations (see Figure 12.8). They found that the best and most significant agreement with observations was found when all three factors were included[1]. Allen and Tett (1999) find that the effect of greenhouse gases can be detected with these signal patterns using optimal detection (see Appendix 12.1).

Folland *et al.* (1998) and Sexton *et al.* (2001) take a complementary approach using an atmospheric model forced with sea

surface temperatures (SST) and ice extents prescribed from observations. The correlation between the observed and simulated temperature changes in the vertical relative to the base period from 1961 to 1975 was computed. The experiments with anthropogenic forcing (including some with tropospheric ozone changes), give significantly higher correlations than when only SST changes are included.

Interpretation of results

Weber (1996) and Michaels and Knappenburger (1996) both criticised the Santer *et al.* (1996a) results, quoting upper air measurements analysed by Angell (1994). Weber argued that the increasing pattern similarity over the full atmosphere (850 to 50 hPa) resulted mainly from a Southern Hemisphere cooling associated with stratospheric ozone depletion. Santer *et al.* (1996b) pointed out that when known biases in the radiosonde data are removed (e.g., Parker *et al.*, 1997), or satellite or operationally analysed data are used, the greater stratospheric cooling in the Southern Hemisphere all but disappears. Weber (1996) is correct that stratospheric cooling due to ozone will contribute to the pattern similarity over the full atmosphere, but decreases in stratospheric ozone alone would be expected to produce a tropospheric cooling, not a warming as observed. This point should be born in mind when considering a later criticism of the pattern correlation approach. Both Weber (1996) and Michaels and Knappenburger (1996) note that the greater warming of the Southern Hemisphere relative to the Northern Hemisphere from 1963 to 1988 has since reversed. They attribute the Southern Hemisphere warming from 1963 to the recovery from the cooling following the eruption of Mount Agung. Santer *et al.* (1996b) claim that this change in asymmetry is to be expected, because the heating due to increases in greenhouse gases over the most recent years has probably been growing faster than the estimated cooling due to increases in aerosols (see Section 12.2.3.3). Calculations of the difference in the rate of warming between the Northern and Southern Hemispheres vary between different climate models and as a function of time, depending on the relative forcing due to greenhouse gases and sulphate aerosols, and on the simulated rate of oceanic heat uptake in the Southern Hemisphere (Santer *et al.*, 1996b; Karoly and Braganza, 2001).

Assessing statistical significance of changes in the vertical patterns of temperature

There are some difficulties in assessing the statistical significance in detection studies based on changes in the vertical temperature profile. First, the observational record is short, and subject to error, particularly at upper levels (Chapter 2). Second, the model estimates of variability may not be realistic (Section 12.2.2), particularly in the stratosphere. Third, because of data and model limitations, the number of levels used to represent the stratosphere in detection studies to date is small, and hence may not be adequate to allow an accurate representation of the stratospheric response. Fourth, all models produce a maximum warming in the upper tropical troposphere that is not apparent in the observations and whose impact on detection results is difficult to quantify. Nevertheless, all the studies indicate that

---

[1] Correction of an error in a data mask (Allen and Tett, 1999) did not affect these conclusions, though the additional improvement due to adding sulphate and ozone forcing was no longer significant.

Box 12.1: Optimal detection

Optimal detection is a technique that may help to provide a clearer separation of a climate change fingerprint from natural internal climate variations. The principle is sketched in Figure 12.B1, below (after Hasselmann, 1976).

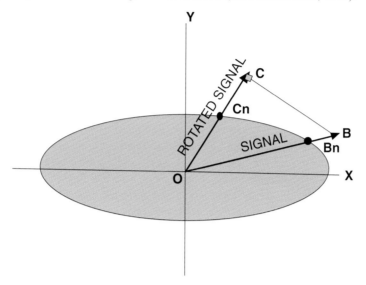

Suppose for simplicity that most of the natural variability can be described in terms of two modes (well-defined spatial patterns) of variability. In the absence of climate change, the amplitudes of these two modes, plotted on a 2D diagram along OX and OY will vary with time, and for a given fraction of occasions (usually chosen as 95 %), the amplitude of each mode will lie within the shaded ellipse. Suppose we are attempting to detect a fingerprint that can be made up of a linear combination of the two patterns such that it lies along OB. The signal to noise ratio is given by OB/OBn. Because our signal lies close to the direction of the main component of variability, the signal to noise ratio is small. On the other hand, we can choose a direction OC that overlaps less with the main component of natural variability such that the signal to noise ratio OC/OCn for the component of the signal that lies in direction OC is larger even though the projected signal OC is smaller then the full signal OB. Optimal detection techniques merely choose the direction OC that maximises the signal to noise ratio. This is equivalent to general linear regression (see Appendix 12.1). A good estimate of natural internal variability is required to optimise effectively.

anthropogenic factors account for a significant part of recent observed changes, whereas internal and naturally forced variations alone, at least as simulated by current models, cannot explain the observed changes. In addition, there are physical arguments for attributing the changes in the vertical profile of temperature to anthropogenic influence (Section 12.3.2).

12.4.3 Optimal Fingerprint Methods

The use of "optimal" techniques can increase the detectability of forced climate changes. These techniques increase the signal-to-noise ratio by looking at the component of the response away from the direction of highest internal variability (see, e.g., Hasselmann, 1979, 1997, 1993; North *et al.*, 1995; see also Box 12.1 on optimal detection and Appendix 12.1). Several new approaches to the optimal detection of anthropogenic climate change have been undertaken since the SAR. We focus on optimal detection studies that use a single pattern of climate change in the following section. Attribution (see Section 12.1.1), which requires us to consider several signals simultaneously, will be considered in Sections 12.4.3.2 and 12.4.3.3.

12.4.3.1 Single pattern studies
Since the SAR, optimal detection studies of surface temperature have been extended (Hegerl *et al.*, 1997, 2000; Barnett *et al.*, 1999) and new studies of data other than surface air temperature have been conducted (Allen and Tett, 1999; Paeth and Hense, 2001; Tett *et al.*, 2000).

Surface temperature patterns
The Hegerl *et al.* (1996) optimal detection study was extended to include more recent estimates of internal variability and simulations with a representation of sulphate aerosols (Hegerl *et al.*, 1997). As in the previous study, different control simulations were used to determine the optimal fingerprint and the significance level of recent temperature change. The authors find significant evidence for a "greenhouse gas plus sulphate aerosol" (GS) fingerprint in the most recent observed 30-year temperature trends regardless of whether internal variability is estimated from models or observations. The 30-year trend ending in the 1940s was found to be significantly larger than expected from internal variability, but less so than the more recent trends. This work has been extended to include other models (Figure 12.10a; see also Barnett *et al.*, 1999; Hegerl *et*

al., 2000), examining whether the amplitude of the 50-year summer surface temperature trends in the GS simulations is consistent with that estimated in the observations. In eleven out of fourteen cases (seven models each evaluated using the fingerprints from the two original models), the model trends are consistent with observations. The greenhouse gas only simulations are generally not consistent with observations, as their warming trends are too large. Berliner *et al.* (2000) detect a combined greenhouse gas and sulphate signal in a fixed pattern detection study of temperature changes using Bayesian techniques.

Vertical patterns of temperature

Allen and Tett (1999) use optimal detection methods to study the change in the vertical profile of zonal mean temperature between 1961 to 1980 and 1986 to 1995. Estimated signals from ensemble AOGCM simulations with greenhouse gas alone (G), greenhouse gas plus direct sulphate (GS), and also including stratospheric ozone forcing (GSO; Tett *et al.*, 1996) are considered. The G and GSO signals are detected separately. The amplitude of the GSO fingerprint estimated from observations is found to be consistent with that simulated by the model, while the model-simulated response to greenhouse gases alone was found to be unrealistically strong. The variance of the residuals that remain after the estimated signal is removed from the observations is consistent with internal variability estimated from a control run.

Other climatic variables

Schnur (2001) applied the optimal detection technique to trends in a variety of climate diagnostics. Changes in the annual mean surface temperature were found to be highly significant (in agreement with previous results from Hegerl *et al.*, 1996, 1997). The predicted change in the annual cycle of temperature as well as winter means of diurnal temperature range can also be detected in most recent observations. The changes are most consistent with those expected from increasing greenhouse gases and aerosols. However, changes in the annual mean and annual cycle of precipitation were small and not significant.

Paeth and Hense (2001) applied a correlation method related to the optimal fingerprint method to 20-year trends of lower tropospheric mean temperature (between 500 and 1,000 hPa) in the summer half of the year in the Northern Hemisphere north of 55°N. Greenhouse gas fingerprints from two models were detected. The combined greenhouse gas plus (direct) sulphate (GS) fingerprints from the two models were not detected.

Summary

All new single-pattern studies published since the SAR detect anthropogenic fingerprints in the global temperature observations, both at the surface and aloft. The signal amplitudes estimated from observations and modelled amplitudes are consistent at the surface if greenhouse gas and sulphate aerosol forcing are taken into account, and in the free atmosphere if ozone forcing is also included. Fingerprints based on smaller areas or on other variables yield more ambiguous results at present.

12.4.3.2 Optimal detection studies that use multiple fixed signal patterns

Surface temperature patterns

Hegerl *et al.* (1997) applied a two-fingerprint approach, using a greenhouse gas fingerprint and an additional sulphate aerosol fingerprint that is made spatially independent (orthogonalised) of the greenhouse fingerprint. They analysed 50-year trends in observed northern summer temperatures. The influence of greenhouse gas and sulphate aerosol signals were both detected simultaneously in the observed pattern of 50-year temperature trends, and the amplitudes of both signals were found to be consistent between model and observations. Simulations forced with greenhouse gases alone and solar irradiance changes alone were not consistent with observations.

Hegerl *et al.* (2000) repeated this analysis using parallel simulations from a different climate model. The combined effect of greenhouse gases and aerosols was still detectable and consistent with observations, but the separate influence of sulphate aerosol forcing, as simulated by this second model, was not detectable. This was because the sulphate response was weaker in the second model, and closely resembled one of the main modes of natural variability. Hence, the detection of the net anthropogenic signal is robust, but the detection of the sulphate aerosol component is very sensitive to differences in model-simulated responses.

As in the single-pattern case, this study has been extended to include seven model GS simulations and to take into account observational sampling error (Figure 12.10b,c, see also Barnett *et al.*, 1999; Hegerl *et al.* 2001). A simple linear transformation allows results to be displayed in terms of individual greenhouse and sulphate signal amplitudes, which assists comparison with other results (see Figure 12.10; Hegerl and Allen, 2000). The amplitudes of the greenhouse gas and sulphate components are simultaneously consistent with the observed amplitudes in 10 of the fourteen GS cases (seven models for two sets of fingerprints) displayed. This contrasts with eleven out of fourteen in the combined amplitude test described in Section 12.4.3.1. If the trends to 1995 are used (Figure 12.10c), the results are similar, though in this case, the ellipse just includes the origin and six out of the fourteen GS cases are consistent with observations. The inconsistency can be seen to be mainly due to large variations in the amplitudes of the model-simulated responses to sulphate aerosols (indicated by the vertical spread of results). Model-simulated responses to greenhouse gases are generally more consistent both with each other and with observations. Two of the cases of disagreement are based on a single simulation rather than an ensemble mean and should therefore be viewed with caution (see Barnett *et al.*, 2000). Barnett *et al.* (1999) found that the degree of agreement between the five models and observations they considered was similar, whether or not the global mean response was removed from the patterns. Signal amplitudes from simulations with greenhouse gas forcing only are generally inconsistent with those estimated from observations (Figure 12.10b,c).

In most of the cases presented here, the response to natural forcings was neglected. In a similar analysis to that just described, Hegerl *et al.* (2000); see also Barnett *et al.*, 1999) also assessed simulations of the response to volcanic and solar forcing. They find, in agreement with Tett *et al.* (1999), that there is better agreement between observations and simulations when these natural forcings are included, particularly in the early 20th century, but that natural forcings alone cannot account for the late-century warming.

In summary, the estimation of the contribution of individual factors to recent climate change is highly model dependent, primarily due to uncertainties in the forcing and response due to sulphate aerosols. However, although the estimated amplitude varies from study to study, all studies indicate a substantial contribution from anthropogenic greenhouse gases to the changes observed over the latter half of the 20th century.

Vertical patterns of temperature
Allen and Tett (1999) also used spatial fingerprints in the vertical derived from simulations with greenhouse gas forcing alone and simulations with greenhouse gas, sulphate aerosol and strato-spheric ozone forcing. These authors show that, even if both greenhouse and other anthropogenic signals are estimated simultaneously in the observed record, a significant response to greenhouse gases remains detectable. Hill *et al.* (2001) extended this analysis to include model-simulated responses to both solar and volcanic forcing, and again found that the response to greenhouse gases remains detectable. Results with non-optimised fingerprints are consistent with the optimised case, but the uncertainty range is larger.

In summary, the fixed pattern studies indicate that the recent warming is unlikely (bordering on very unlikely) to be due to internal climate variability. A substantial response to anthro-pogenic greenhouse gases appears to be necessary to account for recent temperature trends but the majority of studies indicate that greenhouse gases alone do not appear to be able to provide a full explanation. Inclusion of the response to the direct effect of sulphate aerosols usually leads to a more satisfactory explanation of the observed changes, although the amplitude of the sulphate signal depends on the model used. These studies also provide some evidence that solar variations may have contributed to the early century warming.

12.4.3.3 Space-time studies
Here we consider studies that incorporate the time evolution of forced signals into the optimal detection formalism. These studies use evolving patterns of historical climate change in the 20th century that are obtained from climate models forced with historical anthropogenic and natural forcing. Explicit representa-tion of the time dimension of the signals yields a more powerful approach for both detecting and attributing climate change (see Hasselmann, 1993; North *et al.*, 1995) since it helps to distin-guish between responses to external forcings with similar spatial patterns (e.g., solar and greenhouse gas forcing). The time variations of the signals can be represented either directly in the time domain or transformed to the frequency domain.

Surface temperature
Tett *et al.* (1999) and Stott *et al.* (2001) describe a detection and attribution study that uses the space-time approach (see Appendix 12.2). They estimate the magnitude of modelled 20th century

greenhouse gas, aerosol, solar and volcanic signals in decadal mean data. Signals are fitted by general linear regression to moving fifty-year intervals beginning with 1906 to 1956 and ending 1946 to 1996. The signals are obtained from four ensembles of transient change simulations, each using a different historical forcing scenario. Greenhouse gas, greenhouse gas plus direct sulphate aerosol, low frequency solar, and volcanic forcing scenarios were used. Each ensemble contains four independent simulations with the same transient forcing. Two estimates of natural variability, one used for optimisation and the other for the estimation of confidence intervals, are obtained from separate segments of a long control simulation.

Signal amplitudes estimated with multiple regression become uncertain when the signals are strongly correlated ("degenerate"). Despite the problem of degeneracy, positive and significant greenhouse gas and sulphate aerosol signals are consistently detected in the most recent fifty-year period (Figure 12.11) regardless of which or how many other signals are included in the analysis (Allen *et al.*, 2000a; Stott *et al.*, 2001). The residual variation that remains after removal of the signals is consistent with the model's internal variability. In contrast, recent decadal temperature changes are not consistent with the model's internal climate variability alone, nor with any combination of internal variability and naturally forced signals, even allowing for the possibility of unknown processes amplifying the response to natural forcing.

Tett *et al.* (2000) have completed a study using a model with no flux adjustments, an interactive sulphur cycle, an explicit representation of individual greenhouse gases and an explicit treatment of scattering by aerosols. Two ensembles of four simulations for the instrumental period were run, one with natural (solar and volcanic) forcing only and the other anthropogenic (well-mixed greenhouse gases, ozone and direct and indirect sulphate aerosol) forcing only (see Figure 12.4). They find a substantial response to anthropogenic forcing is needed to explain observed changes in recent decades, and that natural forcing may have contributed significantly to early 20th century climate change. The best agreement between model simulations and observations over the last 140 years has been found when all the above anthropogenic and natural forcing factors are included (Stott *et al.*, 2000b; Figure 12.7c). These results show that the forcings included are sufficient to explain the observed changes, but do not exclude the possibility that other forcings may also have contributed.

The detection of a response to solar forcing in the early part of the century (1906 to 1956) is less robust and depends on the details of the analysis. If seasonally stratified data are used (Stott *et al.*, 2001), the detection of a significant solar influence on climate in the first half of the century becomes clearer with the solar irradiance reconstruction of Hoyt and Schatten (1993), but weaker with that from Lean *et al.* (1995). Volcanism appears to show only a small signal in recent decadal temperature trends and could only be detected using either annual mean data or specifi-cally chosen decades (Stott *et al.*, 2001). The residual variability that remains after the naturally forced signals are removed from the observations of the most recent five decades are not consis-tent with model internal variability, suggesting that natural

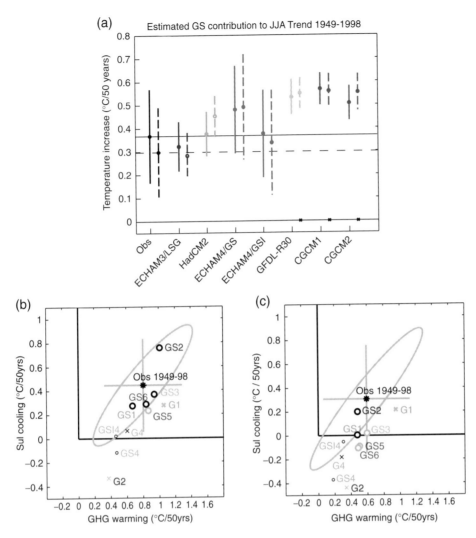

Figure 12.10: Comparison between the amplitude of anthropogenic signals from observed and modelled JJA trend patterns using fingerprints from two different climate models (ECHAM3/LSG and HadCM2) and data from five climate models. (a) Comparison of the amplitude of a single greenhouse gas + sulphate aerosol (GS) signal (expressed as change in global mean temperature [°C] over 50 years). Results show that a significant GS signal can be detected in observed trend patterns 1949 to 1998 at a 5% significance level (one-sided test), independent of which pair of fingerprints was used. The observed signal amplitude is consistent with contemporaneous GS amplitudes for most models' GS simulations. 90% confidence intervals are shown by solid lines for estimates using ECHAM3/LSG fingerprints and by dashed lines for estimates based on HadCM2 fingerprints. Cases where a model's and the observed amplitude disagree are marked by a cross on the axis. (b) and (c) show an estimate of the observed amplitude of a greenhouse gas signal (horizontal axis) and a sulphate aerosol signal (vertical axis) estimated simultaneously. Both signal amplitudes can be estimated as positive from observations based on ECHAM3/LSG fingerprints shown in (b) while only the greenhouse gas signal is detected based on HadCM2 fingerprints shown in panel (c). The amplitudes of both signals from the observations are compared with those from model simulations forced with various forcing histories and using different climate models (1: HadCM2; 2: ECHAM3/LSG; 3: GFDL; 4: ECHAM4/OPYC; 5: CCCma1; 6: CCCma2). Simulations with symbols shown in black are consistent with observations relative to the uncertainty in observations (grey ellipse) and that of the model simulations (not shown). Simulations which are inconsistent are shown in grey. Model simulations where only a single ensemble member is available are illustrated by thin symbols, those based on ensembles of simulations by fat symbols.

Results from consistency tests indicate that most greenhouse gas only simulations (G, shown by "×") are inconsistent with observations. Ten of the GS simulations in both panels are in agreement with observed trend patterns, discrepancies arise mostly from the magnitude of a sulphate signal (vertical axis). The failure to detect a sulphate signal as well as a greenhouse gas signal in panel (c) is due to the two signals being very highly correlated if only spatial patterns are used- this makes separation of the signals difficult. These results show that estimates of a sulphate aerosol signal from observations are model dependent and quite uncertain, while a single anthropogenic signal can be estimated with more confidence.

All units are in °C/50 year, values in the upper right quadrant refer to a physically meaningful greenhouse warming and sulphate aerosol cooling signal. The consistency test establishes whether the difference between a model's and the observed amplitude estimate is significantly larger than the combined uncertainty in the observations (internal variability + observational uncertainty) and the model simulation (internal variability). The figure is derived by updating the data used by Barnett *et al.* (1999) (for details of the analysis see Hegerl *et al.*, 2000) and then applying a simple linear transformation of the multi-regression results (Hegerl and Allen, 2000).

Results for 1946 to 1995 period used by Barnett *et al.* (1999) are similar, except fewer of the models in b and c agree with observations and the case of both signals being zero in c is not rejected. Simulations of natural forcing only ending before 1998 are also rejected in that case.

forcing alone cannot explain the observed 20th century temperature variations. Note that Delworth and Knutson (2000) find one out of five of their simulations with only anthropogenic forcing can reproduce the early century global mean warming, including the enhanced warming in Northern Hemisphere high latitudes. Hence a substantial response to anthropogenic (specifically greenhouse) forcing appears necessary to account for the warming over the past 50 years, but it remains unclear whether natural external forcings are necessary to explain the early 20th century warming.

Sensitivity of results

A variety of sensitivity tests confirm that the detection of anthropogenic signals is insensitive to differences between solar forcing reconstructions, the inclusion of additional forcing through the specification of observed stratospheric ozone concentrations, and to varying details of the analysis (including omitting the signal-to-noise optimisation). Tett *et al.* (1999, 2000) also found that detection of an anthropogenic signal continues to hold even when the standard deviation of the control simulation is inflated by a factor of two. Uncertainty in the signals is unavoidable when ensembles are small, as is the case in Tett *et al.* (1999), and biases the estimates of the signal amplitudes towards zero. Consistent results are obtained when this source of uncertainty is taken into account (Allen and Stott, 2000; Stott *et al.*, 2000a). However amplitude estimates become more uncertain, particularly if the underlying signal is small compared with internal climate variability. Accounting for sampling uncertainty in model-simulated signals indicates a greater degree of greenhouse warming and compensating aerosol cooling in the latter part of the century than shown by Tett *et al.* (1999). Gillett *et al.* (2000b) find that discounting the temperature changes associated with changes in the Arctic Oscillation (Thompson and Wallace, 1998; Thompson *et al.*, 2000), which are not simulated by the model, does not significantly alter the Tett *et al.* (1999) results.

Confidence intervals and scaling factors

Confidence intervals for the signal amplitudes that are obtained from the regression of modelled signals onto observations can be re-expressed as ranges of scaling factors that are required to make modelled signal amplitudes consistent with those estimated from observations (see, e.g., Allen and Tett, 1999). The results show that the range of scaling factors includes unity (i.e., model is consistent with observations) for both the greenhouse gas and the sulphate aerosol signal, and that the scaling factors vary only to a reasonable (and consistent) extent between 50-year intervals.

The scaling factors can also be used to estimate the contribution from anthropogenic factors other than well-mixed greenhouse gases. Using the methodology of Allen and Stott (2000) on the simulations described by Tett *et al.* (2000), the 5 to 95% uncertainty range for scaling the combined response changes in tropospheric ozone and direct and indirect sulphate forcing over the last fifty years is 0.6 to 1.6. The simulated indirect effect of aerosol forcing is by far the biggest contributor to this signal. Ignoring the possible effects of neglected forcings and assuming that the forcing can be scaled in the same way as the response, this translates to a -0.5 to -1.5 Wm^{-2} change in

forcing due to the indirect effect since pre-industrial times. This range lies well within that given in Chapter 6 but the limits obtained are sensitive to the model used. Note that large values of the indirect response are consistently associated with a greater sensitivity to greenhouse gases. This would increase this model's estimate of future warming: a large indirect effect coupled with decreases in sulphate emissions would further enhance future warming (Allen *et al.*, 2000b).

Allen *et al.* (2000a) have determined scaling factors from other model simulations (Figure 12.12) and found that the modelled response to the combination of greenhouse gas and sulphate aerosol forcing is consistent with that observed. The scaling factors ranging from 0.8 to 1.2 and the corresponding 95% confidence intervals cover the range 0.5 to 1.6. Scaling factors for 50-year JJA trends are also easily derived from the results published in Hegerl *et al.* (2000). The resulting range of factors is consistent with that of Allen *et al.* (2000a), but wider because the diagnostic used in Allen *et al.* (2000b) enhances the signal-to-noise ratio. If it is assumed that the combination of greenhouse warming and sulphate cooling simulated by these AOGCMs is the only significant external contributor to inter-decadal near-surface temperature changes over the latter half of the 20th century, then Allen *et al.* (2000a) estimate that the anthropogenic warming over the last 50 years is 0.05 to 0.11°C/decade. Making a similar assumption, Hegerl *et al.* (2000) estimate 0.02 to 0.12°C/decade with a best guess of 0.06 to 0.08°C/decade (model dependent, Figure 12.10). The smallness of the range of uncertainty compared with the observed change indicates that natural internal variability alone is unlikely (bordering on very unlikely) to account for the observed warming.

Given the uncertainties in sulphate aerosol and natural forcings and responses, these single-pattern confidence intervals give an incomplete picture. We cannot assume that the response to sulphate forcing (relative to the greenhouse signal) is as simulated in these greenhouse-plus-sulphate simulations; nor can we assume the net response to natural forcing is negligible even though observations of surface temperature changes over the past 30 to 50 years are generally consistent with both these assumptions. Hence we need also to consider uncertainty ranges based on estimating several signals simultaneously (Figure 12.12, right hand panels). These are generally larger than the single-signal estimates because we are attempting to estimate more information from the same amount of data (Tett *et al.*, 1999; Allen and Stott, 2000; Allen *et al.*, 2000a). Nevertheless, the conclusion of a substantial greenhouse contribution to the recent observed warming trend is unchanged.

Estimation of uncertainty in predictions

The scaling factors derived from optimal detection can also be used to constrain predictions of future climate change resulting from anthropogenic emissions (Allen *et al.*, 2000b). The best guess scaling and uncertainty limits for each component can be applied to the model predictions, providing objective uncertainty limits that are based on observations. These estimates are independent of possible errors in the individual model's climate sensitivity and time-scale of oceanic adjustment, provided these

G&S

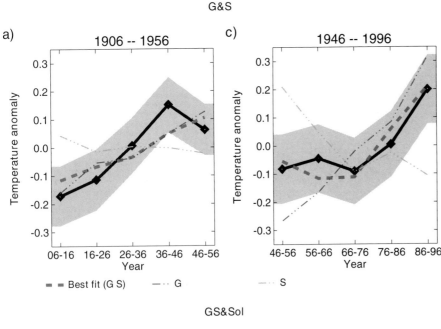

Figure 12.11: Best-estimate contributions to global mean temperature change. Reconstruction of temperature variations for 1906 to 1956 (a and b) and 1946 to 1995 (c and d) for G and S (a and c) and GS and SOL (b and d). (G denotes the estimated greenhouse gas signal, S the estimated sulphate aerosol signal, GS the greenhouse gas / aerosol signal obtained from simulations with combined forcing, SOL the solar signal). Observed (thick black), best fit (dark grey dashed), and the uncertainty range due to internal variability (grey shading) are shown in all plots. (a) and (c) show contributions from GS (orange) and SOL (blue). (b) and (d) show contributions from G (red) and S (green). All time-series were reconstructed with data in which the 50-year mean had first been removed. (Tett *et al.*, 1999).

GS&Sol

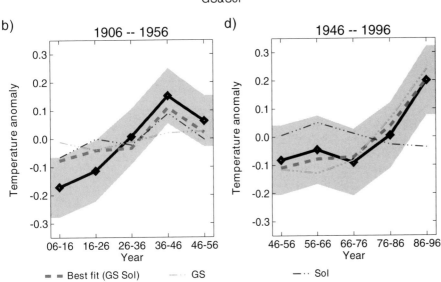

errors are persistent over time. An example based on the IS92a (IPCC, 1992) GS scenario (whose exact forcing varies between models, see Chapter 9, Table 9.1 for details) is shown in Figure 12.13 based on a limited number of model simulations. Note that in each case, the original warming predicted by the model lies in the range consistent with the observations. A rate of warming of 0.1 to 0.2°C/decade is likely over the first few decades of the 21st century under this scenario. Allen *et al.* (2000b) quote a 5 to 95% ("very likely") uncertainty range of 0.11 to 0.24°C/decade for the decades 1996 to 2046 under the IS92a scenario, but, given the uncertainties and assumptions behind their analysis, the more cautious "likely" qualifier is used here. For comparison, the simple model tuned to the results of seven AOGCMs used for projections in Chapter 9 gives a range of 0.12 to 0.22°C/decade under the IS92a scenario, although it should be noted that this similarity may reflect some cancellation of errors and equally good agreement between the two approaches should not be expected for all scenarios, nor for time-scales longer than the few

decades for which the Allen *et al.* (2000b) approach is valid. Figure 12.13 also shows that a similar range of uncertainty is obtained if the greenhouse gas and sulphate components are estimated separately, in which case the estimate of future warming for this particular scenario is independent of possible errors in the amplitude of the sulphate forcing and response. Most of the recent emission scenarios indicate that future sulphate emissions will decrease rather than increase in the near future. This would lead to a larger global warming since the greenhouse gas component would no longer be reduced by sulphate forcing at the same rate as in the past. The level of uncertainty also increases (see Allen *et al.*, 2000b). The final error bar in Figure 12.13 shows that including the model-simulated response to natural forcing over the 20th century into the analysis has little impact on the estimated anthropogenic warming in the 21st century.

It must be stressed that the approach illustrated in Figure 12.13 only addresses the issue of uncertainty in the large-scale climate response to a particular scenario of future greenhouse gas

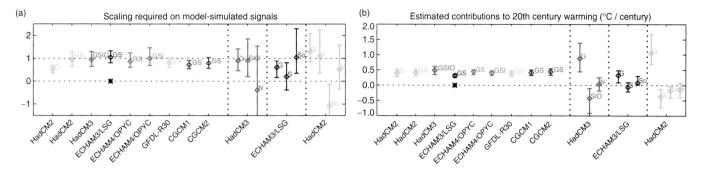

Figure 12.12: (a) Estimates of the "scaling factors" by which we have to multiply the amplitude of several model-simulated signals to reproduce the corresponding changes in the observed record. The vertical bars indicate the 5 to 95% uncertainty range due to internal variability. A range encompassing unity implies that this combination of forcing amplitude and model-simulated response is consistent with the corresponding observed change, while a range encompassing zero implies that this model-simulated signal is not detectable (Allen and Stott, 2000; Stott *et al.*, 2000a). Signals are defined as the ensemble mean response to external forcing expressed in large-scale (>5000 km) near-surface temperatures over the 1946 to 1996 period relative to the 1896 to 1996 mean. The first entry (G) shows the scaling factor and 5 to 95% confidence interval obtained if we assume the observations consist only of a response to greenhouse gases plus internal variability. The range is significantly less than one (consistent with results from other models), meaning that models forced with greenhouse gases alone significantly overpredict the observed warming signal. The next eight entries show scaling factors for model-simulated responses to greenhouse and sulphate forcing (GS), with two cases including indirect sulphate and tropospheric ozone forcing, one of these also including stratospheric ozone depletion (GSI and GSIO respectively). All but one (CGCM1) of these ranges is consistent with unity. Hence there is little evidence that models are systematically over- or under-predicting the amplitude of the observed response under the assumption that model-simulated GS signals and internal variability are an adequate representation (i.e. that natural forcing has had little net impact on this diagnostic). Observed residual variability is consistent with this assumption in all but one case (ECHAM3, indicated by the asterisk). We are obliged to make this assumption to include models for which only a simulation of the anthropogenic response is available, but uncertainty estimates in these single-signal cases are incomplete since they do not account for uncertainty in the naturally forced response. These ranges indicate, however, the high level of confidence with which we can reject internal variability as simulated by these various models as an explanation of recent near-surface temperature change.

A more complete uncertainty analysis is provided by the next three entries, which show corresponding scaling factors on individual greenhouse (G), sulphate (S), solar-plus-volcanic (N), solar-only (So) and volcanic-only (V) signals for those cases in which the relevant simulations have been performed. In these cases, we estimate multiple factors simultaneously to account for uncertainty in the amplitude of the naturally forced response. The uncertainties increase but the greenhouse signal remains consistently detectable. In one case (ECHAM3) the model appears to be overestimating the greenhouse response (scaling range in the G signal inconsistent with unity), but this result is sensitive to which component of the control is used to define the detection space. It is also not known how it would respond to the inclusion of a volcanic signal. In cases where both solar and volcanic forcing is included (HadCM2 and HadCM3), G and S signals remain detectable and consistent with unity independent of whether natural signals are estimated jointly or separately (allowing for different errors in S and V responses). (b) Estimated contributions to global mean warming over the 20th century, based on the results shown in (a), with 5 to 95% confidence intervals. Although the estimates vary depending on which model's signal and what forcing is assumed, and are less certain if more than one signal is estimated, all show a significant contribution from anthropogenic climate change to 20th century warming (from Allen *et al.*, 2000a).

concentrations. This is only one of many interlinked uncertainties in the climate projection problem, as illustrated in Chapter 13, Figure 13.2. Research efforts to attach probabilities to climate projections and scenarios are explored in Chapter 13, Section 13.5.2.3.

Forest *et al.* (2000) used simulations with an intermediate complexity climate model in a related approach. They used optimal detection results following the procedure of Allen and Tett (1999) to rule out combinations of model parameters that yield simulations that are not consistent with observations. They find that low values of the climate sensitivity (<1°C) are consistently ruled out, but the upper bound on climate sensitivity and the rate of ocean heat uptake remain very uncertain.

Other space-time approaches
North and Stevens (1998) use a space-frequency method that is closely related to the space-time approach used in the studies discussed above (see Appendix 12.2). They analyse 100-year

surface temperature time-series of grid box mean surface temperatures in a global network of thirty six large (10°×10°) grid boxes for greenhouse gas, sulphate aerosol, volcanic and solar cycle signals in the frequency band with periods between about 8 and 17 years. The signal patterns were derived from simulations with an EBM (see Section 12.2.3). The authors found highly significant responses to greenhouse gas, sulphate aerosol, and volcanic forcing in the observations. Some uncertainty in their conclusions arises from model uncertainty (see discussion in Section 12.2.3) and from the use of control simulations from older AOGCMs, which had relatively low variability, for the estimation of internal climate variability.

A number of papers extend and analyse the North and Stevens (1998) approach. Kim and Wu (2000) extend the methodology to data with higher (monthly) time resolution and demonstrate that this may improve the detectability of climate change signals. Leroy (1998) casts the results from North and Stevens (1998) in a Bayesian framework. North and

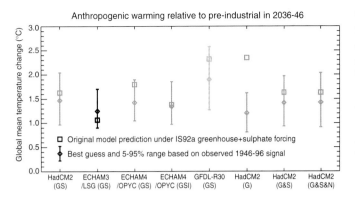

Figure 12.13: Global mean temperature in the decade 2036 to 2046 (relative to pre-industrial, in response to greenhouse gas and sulphate aerosol forcing following the IS92a (IPCC, 1992) scenario), based on original model simulations (squares) and after scaling to fit the observed signal as in Figure 12.12(a) (diamonds), with 5 to 95% confidence intervals. While the original projections vary (depending, for example, on each model's climate sensitivity), the scale should be independent of errors in both sensitivity and rate of oceanic heat uptake, provided these errors are persistent over time. GS indicates combined greenhouse and sulphate forcing. G shows the impact of setting the sulphate forcing to zero but correcting the response to be consistent with observed 20th century climate change. G&S indicates greenhouse and sulphate responses estimated separately (in which case the result is also approximately independent, under this forcing scenario, to persistent errors in the sulphate forcing and response) and G&S&N indicates greenhouse, sulphate and natural responses estimated separately (showing the small impact of natural forcing on the diagnostic used for this analysis). (From Allen *et al.*, 2000b.)

Wu (2001) modified the method to perform space-time (rather than space-frequency) detection in the 100-year record. Their results are broadly similar to those of Tett *et al.*, (1999), Stott *et al.* (2001) and North and Stevens (1998). However, their best guess includes a small sulphate aerosol signal countered by a relatively small, but highly significant, greenhouse gas signal.

All of the space-time and space-frequency optimal detection studies to date indicate a discernible human influence on global climate and yield better-constrained estimates of the magnitude of anthropogenic signals than approaches using spatial information alone. In particular, the inclusion of temporal information can reduce the degeneracy that may occur when more than one climate signal is included. Thus, results from time-space methods generally detect anthropogenic signals even if natural forcings are estimated simultaneously and show that the combination of natural signals and internal variability is inconsistent with the observed surface temperature record.

12.4.3.4 Summary of optimal fingerprinting studies
Results from optimal fingerprint methods indicate a discernible human influence on climate in temperature observations at the surface and aloft and over a range of applications. These methods can also provide a quantitative

estimate of the magnitude of this influence. The use of a number of forced climate signals, and the extensive treatment of various (but not all) sources of uncertainty increases our confidence that a considerable part of the recent warming can be attributed to anthropogenic influences. The estimated signals and scaling factors remain subject to the considerable uncertainty in our knowledge of historic climate forcing from sources other than greenhouse gases. While estimates of the amplitude of a single anthropogenic signal are quite consistent between different model signals (see Figures 12.10, 12.12) and different approaches, joint estimates of the amplitude of several signals vary between models and approaches. Thus quantitative separation of the observed warming into anthropogenic and naturally forced components requires considerable caution. Nonetheless, all recent studies reject natural forcing and internal variability alone as a possible explanation of recent climate change. Analyses based on a single anthropogenic signal focusing on continental and global scales indicate that:

- Changes over the past 30 to 50 years are very unlikely to be due to internal variability as simulated by current models.

- The combined response to greenhouse and sulphate forcing is more consistent with the observed record than the response to greenhouse gases alone.

- Inclusion of the simulated response to stratospheric ozone depletion improves the simulation of the vertical structure of the response.

Analyses based on multiple anthropogenic and natural signals indicate that:

- The combination of natural external forcing (solar and volcanic) and internal variability is unlikely to account for the spatio-temporal pattern of change over the past 30 to 50 years, even allowing for possible amplification of the amplitude of natural responses by unknown feedback processes.

- Anthropogenic greenhouse gases are likely to have made a significant and substantial contribution to the warming observed over the second half of the 20th century, possibly larger than the total observed warming.

- The contribution from anthropogenic sulphate aerosols is less clear, but appears to lie in a range broadly consistent with the spread of current model simulations. A high sulphate aerosol forcing is consistently associated with a stronger response to greenhouse forcing.

- Natural external forcing may have contributed to the warming that occurred in the early 20th century.

Results based on variables other than continental and global scale temperature are more ambiguous.

12.5 Remaining Uncertainties

The SAR identified a number of factors that limited the degree to which any human influence on climate could be quantified. It was noted that detection and attribution of anthropogenic climate change signals would be accomplished through a gradual accumulation of evidence, and that there were appreciable uncertainties in the magnitude and patterns of natural variability, and in the radiative forcing and climate response resulting from human activity.

The SAR predicted an increase in the anthropogenic contribution to global mean temperature of slightly over 0.1°C in the five years following the SAR, which is consistent with the observed change since the SAR (Chapter 2). The predicted increase in the anthropogenic signal (and the observed change) are small compared to natural variability, so it is not possible to distinguish an anthropogenic signal from natural variability on five year time-scales.

Differences in surface and free atmosphere temperature trends
There are unresolved differences between the observed and modelled temperature variations in the free atmosphere. These include apparent changes in the temperature difference between the surface and the lower atmosphere, and differences in the tropical upper troposphere. While model simulations of large-scale changes in free atmospheric and surface temperatures are generally consistent with the observed changes, simulated and observed trends in troposphere minus surface temperature differences are not consistent. It is not clear whether this is due to model or observational error, or neglected forcings in the models.

Internal climate variability
The precise magnitude of natural internal climate variability remains uncertain. The amplitude of internal variability in the models most often used in detection studies differs by up to a factor of two from that seen in the instrumental temperature record on annual to decadal time-scales, with some models showing similar or larger variability than observed (Section 12.2; Chapter 8). However, the instrumental record is only marginally useful for validating model estimates of variability on the multi-decadal time-scales that are relevant for detection. Some palaeo-climatic reconstructions of temperature suggest that multi-decadal variability in the pre-industrial era was higher than that generated internally by models (Section 12.2; Chapter 8). However, apart from the difficulties inherent in reconstructing temperature accurately from proxy data, the palaeoclimatic record also includes the climatic response to natural forcings arising, for example, from variations in solar output and volcanic activity. Including the estimated forcing due to natural factors increases the longer-term variability simulated by models, while eliminating the response to external forcing from the palaeo-record brings palaeo-variability estimates closer to model-based estimates (Crowley, 2000).

Natural forcing
Estimates of natural forcing have now been included in simulations over the period of the instrumental temperature record.

Natural climate variability (forced and/or internally generated) on its own is generally insufficient to explain the observed changes in temperature over the last few decades. However, for all but the most recent two decades, the accuracy of the estimates of forcing may be limited, being based entirely on proxy data for solar irradiance and on limited surface data for volcanoes. There are some indications that solar irradiance fluctuations have indirect effects in addition to direct radiative heating, for example due to the substantially stronger variation in the UV band and its effect on ozone, or hypothesised changes in cloud cover (see Chapter 6). These mechanisms remain particularly uncertain and currently are not incorporated in most efforts to simulate the climate effect of solar irradiance variations, as no quantitative estimates of their magnitude are currently available.

Anthropogenic forcing
The representation of greenhouse gases and the effect of sulphate aerosols has been improved in models. However, some of the smaller forcings, including those due to biomass burning and changes in land use, have not been taken into account in formal detection studies. The major uncertainty in anthropogenic forcing arises from the indirect effects of aerosols. The global mean forcing is highly uncertain (Chapter 6, Figure 6.8). The estimated forcing patterns vary from a predominantly Northern Hemisphere forcing similar to that due to direct aerosol effects (Tett *et al.*, 2000) to a more globally uniform distribution, similar but opposite in sign to that associated with changes in greenhouse gases (Roeckner *et al.*, 1999). If the response to indirect forcing has a component which can be represented as a linear combination of the response to greenhouse gases and to the direct forcing by aerosols, it will influence amplitudes of the responses to these two factors estimated through optimal detection.

Estimates of response patterns
Finally, there remains considerable uncertainty in the amplitude and pattern of the climate response to changes in radiative forcing. The large uncertainty in climate sensitivity, 1.5 to 4.5°C for a doubling of atmospheric carbon dioxide, has not been reduced since the SAR, nor is it likely to be reduced in the near future by the evidence provided by the surface temperature signal alone. In contrast, the emerging signal provides a relatively strong constraint on forecast transient climate change under some emission scenarios. Some techniques can allow for errors in the magnitude of the simulated global mean response in attribution studies. As noted in Section 12.2, there is greater pattern similarity between simulations of greenhouse gases alone, and of greenhouse gases and aerosols using the same model, than between simulations of the response to the same change in greenhouse gases using different models. This leads to some inconsistency in the estimation of the separate greenhouse gas and aerosol components using different models (see Section 12.4.3).

In summary, some progress has been made in reducing uncertainty, particularly with respect to distinguishing the responses to different external influences using multi-pattern techniques and in quantifying the magnitude of the modelled and observed responses. Nevertheless, many of the sources of uncertainty identified in the SAR still remain.

12.6 Concluding Remarks

In the previous sections, we have evaluated the different lines of evidence on the causes of recent climate change. Here, we summarise briefly the arguments that lead to our final assessment. The reader is referred to the earlier sections for more detail.

20th century climate was unusual.
Palaeoclimatic reconstructions for the last 1,000 years (e.g., Chapter 2, Figure 2.21) indicate that the 20th century warming is highly unusual, even taking into account the large uncertainties in these reconstructions.

The observed warming is inconsistent with model estimates of natural internal climate variability.
While these estimates vary substantially, on the annual to decadal time-scale they are similar, and in some cases larger, than obtained from observations. Estimates from models and observations are uncertain on the multi-decadal and longer time-scales required for detection. Nonetheless, conclusions on the detection of an anthropogenic signal are insensitive to the model used to estimate internal variability. Recent observed changes cannot be accounted for as pure internal variability even if the amplitude of simulated internal variations is increased by a factor of two or more. It is therefore unlikely (bordering on very unlikely) that natural internal variability alone can explain the changes in global climate over the 20th century (e.g., Figure 12.1).

The observed warming in the latter half of the 20th century appears to be inconsistent with natural external (solar and volcanic) forcing of the climate system.
Although there are measurements of these forcings over the last two decades, estimates prior to that are uncertain, as the volcanic forcing is based on limited measurements, and the solar forcing is based entirely on proxy data. However, the overall trend in natural forcing over the last two, and perhaps four, decades of the 20th century is likely to have been small or negative (Chapter 6, Table 6.13) and so is unlikely to explain the increased rate of global warming since the middle of the 20th century.

The observed change in patterns of atmospheric temperature in the vertical is inconsistent with natural forcing.
The increase in volcanic activity during the past two to four decades would, if anything, produce tropospheric cooling and stratospheric warming, the reverse to what has occurred over this period (e.g., Figure 12.8). Increases in solar irradiance could account for some of the observed tropospheric warming, but mechanisms by which this could cool the stratosphere (e.g., through changes in stratospheric ozone) remain speculative. Observed increases in stratospheric water vapour might also account for some of the observed stratospheric cooling. Estimated changes in solar radiative forcing over the 20th century are substantially smaller than those due to greenhouse gas forcing, unless mechanisms exist which enhance the effects of solar radiation changes at the ground. Palaeo-data show little evidence of such an enhancement at the surface in the past. Simulations based solely on the response to natural forcing (e.g.,

Figure 12.7a) are inconsistent with the observed climate record even if the model-simulated response is allowed to scale up or down to match the observations. It is therefore unlikely that natural forcing and internal variability together can explain the instrumental temperature record.

Anthropogenic factors do provide an explanation of 20th century temperature change.
All models produce a response pattern to combined greenhouse gas and sulphate aerosol forcing that is detectable in the 20th century surface temperature record (e.g., Figures 12.10, 12.12 (one model produces an estimate of internal variability which is not consistent with that observed)). Given that sulphate aerosol forcing is negative, and hence tends to reduce the response, detection of the response to the combined forcing indicates the presence of a greenhouse gas signal that is at least as large as the combined signal.

The effect of anthropogenic greenhouse gases is detected, despite uncertainties in sulphate aerosol forcing and response.
The analysis used to derive Figures 12.10a and 12.12, left box, assumes that the ratio of the greenhouse gas and sulphate aerosol responses in each model is correct. Given the uncertainty in sulphate aerosol forcing, this may not be the case. Hence one must also consider the separate responses to greenhouse gases and aerosols simultaneously. A greenhouse gas signal is consistently detected in the observations (e.g., Figure 12.10b,c, Figure 12.12 right hand boxes; North and Wu, 2001; Tett *et al.* 2000). The greenhouse gas responses are consistent with the observations in all but one case The two component studies all indicate a substantial detectable greenhouse gas signal, despite uncertainties in aerosol forcing. The spread of estimates of the sulphate signal emphasises the uncertainty in sulphate aerosol forcing and response.

It is unlikely that detection studies have mistaken a natural signal for an anthropogenic signal.
In order to demonstrate an anthropogenic contribution to climate, it is necessary to rule out the possibility that the detection procedure has mistaken part or all of a natural signal for an anthropogenic change. On physical grounds, natural forcing is unlikely to account completely for the observed warming over the last three to five decades, given that it is likely that the overall trend in natural forcing over most of the 20th century is small or negative. Several studies have involved three or more components – the responses to greenhouse gases, sulphate aerosols and natural (solar, volcanic or volcanic and solar) forcing. These studies all detect a substantial greenhouse gas contribution over the last fifty years, though in one case the estimated greenhouse gas amplitude is inconsistent with observations. Thus it is unlikely that we have misidentified the solar signal completely as a greenhouse gas response, but uncertainty in the amplitude of the response to natural forcing continues to contribute to uncertainty in the size of the anthropogenic signal.

The detection methods used should not be sensitive to errors in the amplitude of the global mean forcing or response.
Signal estimation methods (e.g., Figures 12.10, 12.11 and 12.12)

allow for errors in the amplitude of the response, so the results should not be sensitive to errors in the magnitude of the forcing or the magnitude of the simulated model response. This would reduce the impact of uncertainty in indirect sulphate forcing on the estimated greenhouse and net sulphate signal amplitudes, to the extent that the pattern of response to indirect sulphate forcing resembles the pattern of response to direct sulphate forcing. Some models indicate this is may be the case, others do not, so this remains an important source of uncertainty. Note that if the spatio-temporal pattern of response to indirect sulphate forcing were to resemble the greenhouse response, it would lead to the amplitude of the greenhouse response being underestimated in cases where indirect sulphate forcing has not been included in the model. Detection and attribution results are also expected to be insensitive to all but the largest scale details of radiative forcing patterns. Detection is only possible at the largest spatial scales (e.g., Stott and Tett, 1998). In addition, atmospheric motions and large-scale feedbacks smooth out the response. All these arguments tend to reduce the impact of the large uncertainty in the magnitude of the forcing due to indirect sulphate aerosols. The inclusion of forcing from additional aerosols (see Chapter 6) is unlikely to alter our conclusion concerning the detection of a substantial greenhouse gas signal, though it is likely to affect estimates of the sulphate aerosol response. This is because part of the response to sulphate aerosols can be considered as surrogate for other aerosols, even though the patterns of forcing and response may differ on smaller scales. In general, the estimates of global mean forcing for other neglected factors are small (see Chapter 6, Figure 6.6).

Studies of the changes in the vertical patterns of temperature also indicate that there has been an anthropogenic influence on climate over the last 35 years.

One study finds that even when changes in stratospheric ozone and solar irradiance are taken into account, there is a detectable greenhouse gas signal in the vertical temperature record.

Observed and simulated vertical lapse rate changes are inconsistent over the last two decades, but there is an anthropogenic influence on tropospheric temperatures over a longer period.

Over the last twenty years, the observed warming trend in the lower troposphere has been smaller than at the surface. This contrasts with model simulations of the response to anthropogenic greenhouse gases and sulphate aerosols. Natural climate variability and the influence of natural external forcing, such as volcanism, can explain part of this difference. However, a discrepancy remains that cannot be accounted for with current climate models. The reduced warming in the lower troposphere does not, however, call into question the fact that the surface temperature has been warming over the satellite period (e.g., National Academy of Sciences, 2000). Over the longer period for which radiosonde data are available, an anthropogenic influence due to increasing greenhouse gases and decreasing stratospheric ozone is detected in all studies.

Natural factors may have contributed to the early century warming.

Most of the discussion in this section has been concerned with evidence relating to a human effect on late 20th century climate. The observed global mean surface temperature record shows two main periods of warming. Some studies detect a solar influence on surface temperature over the first five decades of the century, with perhaps a small additional warming due to increases in greenhouse gases. One study suggests that the early warming could be due to a combination of anthropogenic effects and a highly unusual internal variation. Thus the early century warming could be due to some combination of natural internal variability, changes in solar irradiance and some anthropogenic influence. The additional warming in the second half-century is most likely to be due to a substantial warming due to increases in greenhouse gases, partially offset by cooling due to aerosols, and perhaps by cooling due to natural factors towards the end of the period.

Appendix 12.1: Optimal Detection is Regression

The detection technique that has been used in most "optimal detection" studies performed to date has several equivalent representations (Hegerl and North, 1997; Zwiers, 1999). It has recently been recognised that it can be cast as a multiple regression problem with respect to generalised least squares (Allen and Tett, 1999; see also Hasselmann, 1993, 1997) in which a field of *n* "observations" **y** is represented as a linear combination of signal patterns **g**$_1$,...,**g**$_m$ plus noise **u**

$$\mathbf{y} = \sum_{i=1}^{m} a_i \mathbf{g}_i + \mathbf{u} = \mathbf{Ga} + \mathbf{u} \qquad (A12.1.1)$$

where **G**=(**g**$_1$|...|**g**$_m$) is the matrix composed of the signal patterns and **a**=(a$_1$,...,a$_m$)^T is the vector composed of the unknown amplitudes. The field usually contains temperature observations, arrayed in space, either at the surface as grid box averages of surface temperature observations (typically 5×5 degrees; Santer *et al.*, 1995; Hegerl *et al.*, 1997; Tett *et al.*, 1999), or in the vertical as zonal averages of radiosonde observations (Karoly *et al.*, 1994; Santer *et al.*, 1996a; Allen and Tett, 1999). The fields are masked so that they represent only those regions with adequate data. The fields may also have a time dimension (Allen and Tett, 1999; North and Stevens; 1998; Stevens and North, 1996). Regardless of how the field is defined, its dimension *n* (the total number of observed values contained in any one single realisation of the field) is large. The signal patterns, which are obtained from climate models, and the residual noise field, have the same dimension. The procedure consists of efficiently estimating the unknown amplitudes **a** from observations and testing the null hypotheses that they are zero. In the event of rejection, testing the hypothesis that the amplitudes are unity for some combination of signals performs the attribution consistency test. This assumes, of course, that the climate model signal patterns have been normalised. When the signal is noise-free, estimates of the amplitudes are given by

$$\tilde{\mathbf{a}} = (\mathbf{G}^T \mathbf{C}_{uu}^{-1} \mathbf{G})^{-1} \mathbf{G}^T \mathbf{C}_{uu}^{-1} \mathbf{y} \qquad (A12.1.2)$$

where **C**$_{uu}$ is the *n*×*n* covariance matrix of the noise (Hasselmann, 1997, 1998; Allen and Tett, 1999; Levine and Berliner, 1999). Generalisations allow for the incorporation of signal uncertainties (see, for example, Allen *et al.*, 2000b). A schematic two-dimensional example is given in Box 12.1. In essence, the amplitudes are estimated by giving somewhat greater weight to information in the low variance parts of the field of observations. The uncertainty of this estimate, expressed as the *m*×*m* covariance matrix of **C**$_{aa}$ of **ã**, is given by

$$\mathbf{C}_{aa} = (\mathbf{G}^T \mathbf{C}_{uu}^{-1} \mathbf{G})^{-1} \qquad (A12.1.3)$$

This leads to a (1−α)×100% confidence ellipsoid for the unknown amplitudes when **u** is the multivariate Gaussian that is given by

$$(\tilde{\mathbf{a}} - \mathbf{a})^T \mathbf{G}^T \mathbf{C}_{uu}^{-1} \mathbf{G} (\tilde{\mathbf{a}} - \mathbf{a}) \leq \chi_{1-\alpha}^2 \qquad (A12.1.4)$$

where $\chi_{1-\alpha}^2$ is the (1−α) critical value of the chi-squared distribution with *m* degrees of freedom. Marginal confidence ellipsoids can be constructed for subsets of signals simply by removing the appropriate rows and columns from $\mathbf{G}^T\mathbf{C}_{uu}^{-1}\mathbf{G}$ and reducing the number of degrees of freedom. The marginal (1−α)×100% confidence interval for the amplitude of signal *i* (i.e., the confidence interval that would be obtained in the absence of information about the other signals) is given by

$$\tilde{a}_i - z_{1-\alpha/2}(\mathbf{G}^T\mathbf{C}_{uu}^{-1}\mathbf{G})_{ii} \leq a_i \leq \tilde{a}_i + z_{1-\alpha/2}(\mathbf{G}^T\mathbf{C}_{uu}^{-1}\mathbf{G})_{ii} \quad (A12.1.5)$$

where $Z_{1-\alpha/2}$ is the (1−α/2) critical value for the standard normal distribution. Signal *i* is said to be detected at the α/2×100% significance level if the lower limit confidence interval (A12.1.5) is greater than zero. However, "multiplicity" is a concern when making inferences in this way. For example, two signals that are detected at the α/2×100% significance level may not be jointly detectable at this level. The attribution consistency test is passed when the confidence ellipsoid contains the vector of units (1,...,1)^T.

Appendix 12.2: Three Approaches to Optimal Detection

Optimal detection studies come in several variants depending upon how the time evolution of signal amplitude and structure is treated.

Fixed pattern studies (Hegerl *et al.*, 1996, 1997, 2000a; Berliner *et al.*, 2000; Schnur, 2001) assume that the spatial structure of the signals does not change during the epoch covered by the instrumental record. This type of study searches for evidence that the amplitudes of fixed anthropogenic signals are increasing with time. The observed field $\mathbf{y}=\mathbf{y}(t)$ that appears on the left hand side of equation (A12.1.1) is typically a field of 30 to 50-year moving window trends computed from annual mean observations. The regression equation (A12.1.1) is solved repeatedly with a fixed signal matrix $\mathbf{G}$ as the moving 30 to 50-year window is stepped through the available record.

Studies with *time-varying patterns* allow the shape of the signals, as well as their amplitudes, to evolve with time. Such studies come in two flavours.

The *space-time* approach uses enlarged signal vectors that consist of a sequence of spatial patterns representing the evolution of the signal through a short epoch. For example, Tett *et al.* (1999) use signal vectors composed of five spatial patterns representing a sequence of decadal means. The enlarged signal matrix $\mathbf{G}=\mathbf{G}(t)$ evolves with time as the 5-decade window is moved one decade at a time. The observations are defined

similarly as extended vectors containing a sequence of observed decadal mean temperature patterns. As with the fixed pattern approach, a separate model is fitted for each 5-decade window so that the evolution of the signal amplitudes can be studied.

The *space-frequency* approach (North *et al.*, 1995) uses annual mean signal patterns that evolve throughout the analysis period. A Fourier transform is used to map the temporal variation of each signal into the frequency domain. Only the low-frequency Fourier coefficients representing decadal-scale variability are retained and gathered into a signal vector. The observations are similarly transformed. The selection of time-scales that is effected by retaining only certain Fourier coefficients is a form of dimension reduction (see Dimension Reduction, Appendix 12.4) in the time domain. This is coupled with spatial dimension reduction that must also be performed. The result approximates the dimension reduction that is obtained by projecting observations in space and time on low order space-time EOFs (North *et al.*, 1995). A further variation on this theme is obtained by increasing the time resolution of the signals and the data by using monthly rather than annual means. Climate statistics, including means, variances and covariances, have annual cycles at this time resolution, and thus dimension reduction must be performed with cyclo-stationary space-time EOFs (Kim and Wu, 2000).

Given the same amount of data to estimate covariance matrices, the space-time and space-frequency approaches will sacrifice spatial resolution for temporal resolution.

Appendix 12.3: Pattern Correlation Methods

The pattern correlation methods discussed in this section are closely related to optimal detection with one signal pattern. Pattern correlation studies use either a centred statistic, R, which correlates observed and signal anomalies in space relative to their respective spatial means, or an uncentred statistic, C (Barnett and Schlesinger, 1987), that correlates these fields without removing the spatial means. It has been argued that the latter is better suited for detection, because it includes the response in the global mean, while the former is more appropriate for attribution because it better measures the similarity between spatial patterns. The similarity between the statistics is emphasised by the fact that they can be given similar matrix-vector representations. In the one pattern case, the optimal (regression) estimate of signal amplitude is given by

$$\tilde{a} = \mathbf{g}_1^T \mathbf{C}_{uu}^{-1} \mathbf{y} / \mathbf{g}_1^T \mathbf{C}_{uu}^{-1} \mathbf{g}_1 \qquad (A12.3.1)$$

The uncentred statistics may be written similarly as

$$C = \mathbf{g}_1^T \mathbf{y} / \mathbf{g}_1^T \mathbf{g}_1 = \mathbf{g}_1^T \mathbf{I} \mathbf{y} / \mathbf{g}_1^T \mathbf{I} \mathbf{g}_1 \qquad (A12.3.2)$$

where $\mathbf{I}$ is the $n \times n$ identity matrix. Similarly, the centred statistic can be written (albeit with an extra term in the denominator) as

$$R = \mathbf{g}_1^T (\mathbf{I} - \mathbf{U}) \mathbf{y} / [(\mathbf{g}_1^T (\mathbf{I} - \mathbf{U}) \mathbf{g}_1)^{1/2} (\mathbf{y}^T (\mathbf{I} - \mathbf{U}) \mathbf{y})^{1/2}] \qquad (A12.3.3)$$

where $\mathbf{U}$ is the $n \times n$ matrix with elements $u_{i,j} = 1/n$. The matrix $\mathbf{U}$ removes the spatial means. Note that area, mass or volume weighting, as appropriate, is easily incorporated into these expressions. The main point is that each statistic is proportional to the inner product with respect to a matrix "kernel" between the signal pattern and the observations (Stephenson, 1997). In contrast with the pattern correlation statistics, the optimal signal amplitude estimate, which is proportional to a correlation coefficient using the so-called Mahalonobis kernel (Stephenson, 1997), maximises the signal-to-noise ratio.

Appendix 12.4: Dimension Reduction

Estimation of the signal amplitudes, as well as the detection and attribution consistency tests on the amplitudes, requires an estimate of the covariance matrix $\mathbf{C_{uu}}$ of the residual noise field. However, as $\mathbf{y}$ typically represents climate variation on time-scales similar to the length of the observed instrumental record, it is difficult to estimate the covariance matrix reliably. Thus the covariance matrix is often estimated from a long control simulation. Even so, the number of independent realisations of $\mathbf{u}$ that are available from a typical 1,000 to 2,000-year control simulation is substantially smaller than the dimension of the field, and thus it is not possible to estimate the full covariance matrix. The solution is to replace the full fields $\mathbf{y}$, $\mathbf{g_1},...,\mathbf{g_m}$ and $\mathbf{u}$ with vectors of dimension k, where $m<k<<n$, containing indices of their projections onto the dominant patterns of variability $\mathbf{f_1},...,\mathbf{f_k}$ of $\mathbf{u}$. These patterns are usually taken to be the k highest variance EOFs of a control run (North and Stevens, 1998; Allen and Tett, 1999; Tett *et al.*, 1999) or a forced simulation (Hegerl *et al.*, 1996, 1997; Schnur, 2001). Stott and Tett (1998) showed with a "perfect model" study that climate change in surface air temperature can only be detected at very large spatial scales. Thus Tett *et al.* (1999) reduce the spatial resolution to a few spherical harmonics prior to EOF truncation. Kim *et al.* (1996) and Zwiers and Shen (1997) examine the sampling properties of spherical harmonic coefficients when they are estimated from sparse observing networks.

An important decision, therefore, is the choice of k. A key consideration in the choice is that the variability of the residuals should be consistent with the variability of the control simulation in the dimensions that are retained. Allen and Tett (1999) describe a simple test on the residuals that makes this consistency check. Rejection implies that the model-simulated variability is significantly different from that of the residuals. This may happen when the number of retained dimensions, k, is too large because higher order EOFs may contain unrealistically low variance due to sampling deficiencies or scales that are not well represented. In this situation, the use of a smaller value of k can still provide consistent results: there is no need to require that model-simulated variability is perfect on all spatio-temporal scales for it to be adequate on the very large scales used for detection and attribution studies. However, failing the residual check of Allen and Tett (1999) could also indicate that the model does not have the correct timing or pattern of response (in which case the residuals will contain forced variability that is not present in the control regardless of the choice of k) or that the model does not simulate the correct amount of internal variability, even at the largest scales represented by the low order EOFs. In this case, there is no satisfactory choice of k. Previous authors (e.g., Hegerl *et al.*, 1996, 1997; Stevens and North, 1996; North and Stevens, 1998) have made this choice subjectively. Nonetheless, experience in recent studies (Tett *et al.* 1999; Hegerl *et al.* 2000, 2001; Stott *et al.*, 2001) indicates that their choices were appropriate.

Appendix 12.5: Determining the Likelihood of Outcomes (*p*-values)

Traditional statistical hypothesis tests are performed by comparing the value of a detection statistic with an estimate of its natural internal variability in the unperturbed climate. This estimate must be obtained from control climate simulations because detection statistics typically measure change on time-scales that are a substantial fraction of the length of the available instrumental record (see Appendix 12.4). Most "optimal" detection studies use two data sets from control climate simulations, one that is used to develop the optimal detection statistic and the other to independently estimate its natural variability. This is necessary to avoid underestimating natural variability. The *p*-value that is used in testing the no signal null hypothesis is often computed by assuming that both the observed and simulated projections on signal patterns are normally distributed. This is convenient, and is thought to be a reasonable assumption given the variables and the time and space scales used for detection and attribution. However, it leads to concern that very small *p*-values may be unreliable, because they correspond to events that have not been explored by the model in the available control integrations (Allen and Tett, 1999). They therefore recommend that *p*-values be limited to values that are consistent with the range visited in the available control integrations. A non-parametric approach is to estimate the *p*-value by comparing the value of the detection statistic with an empirical estimate of its distribution obtained from the second control simulation data set. If parametric methods are used to estimate the *p*-value, then very small values should be reported as being less than $1/n_p$ where n_p represents the equivalent number of independent realisations of the detection statistic that are contained in the second control integration.

References

Adcock,R.J., 1878: *The Analyst* (Des Moines, Iowa), **5**, 53.

Allen, M.R. and S.F.B.Tett, 1999: Checking for model consistency in optimal fingerprinting. *Clim. Dyn.*, **15**, 419-434.

Allen, M.R. and P.A. Stott, 2000: Interpreting the signal of anthropogenic climate change I: estimation theory. Tech. Report RAL-TR-2000-045, Rutherford Appleton Lab., Chilton, X11 0QX, UK.

Allen, M.R., N.P. Gillett, J.A. Kettleborough, R. Schnur, G.S. Jones, T. Delworth, F. Zwiers, G. Hegerl, T.P. Barnett, 2000a: Quantifying anthropogenic influence on recent near-surface temperature change", Tech. Report RAL-TR-2000-046, Rutherford Appleton Lab., Chilton, X11 0QX, UK. Also accepted in *Surveys in Geophysics.*

Allen, M.R., P.A. Stott, R. Schnur, T. Delworth and J.F.B. Mitchell, 2000b: Uncertainty in forecasts of anthropogenic climate change. *Nature*, **407**, 617-620.

Angell, J.K.,1994: Global, hemispheric and zonal temperature anomalies derived from radiosonde records. pp. 636-672. In "Trends '93: A compendium of data on global change" (Eds. T. A. Boden *et al.*), ORNL/CDIAC-65, Oak Ridge, TN.

Bankes, H.T., R.A. Wood, J.M. Gregory, T.C. Johns and G.S Jones, 2000: Are observed changes in intermediate water masses a signature of anthropogenic climate change? *Geophys. Res. Lett.* , **27**, 2961-2964.

Barnett, T.P. and M.E. Schlesinger, 1987: Detecting changes in global climate induced by greenhouse gases. *J. Geophys. Res.*, **92**, 14772-14780.

Barnett, T.P., B.D. Santer, P.D. Jones, R.S. Bradley and K.R. Briffa, 1996: Estimates of low Frequency Natural Variability in Near-Surface Air Temperature. *The Holocene*, **6**, 255-263.

Barnett, T.P., K. Hasselmann, M. Chelliah, T. Delworth, G. Hegerl, P. Jones, E. Rasmussen, E. Roeckner, C. Ropelewski, B. Santer and S. Tett, 1999: Detection and attribution of climate change: A Status report. *Bull. Am. Met. Soc.*, **12**, 2631-2659.

Barnett, T. P., G.C. Hegerl, T. Knutson and S.F.B. Tett, 2000: Uncertainty levels in predicted patterns of anthropogenic climate change. *J. Geophys.Res.*, **105**, 15525-15542.

Bengtsson, L., E.Roeckner, M.Stendel, 1999: Why is the global warming proceeding much slower than expected. *J. Geophys. Res.*, **104**, 3865-3876.

Berliner, L.M., R.A. Levine and D.J. Shea, 2000: Bayesian climate change assessment. *J.Climate,* **13**, 3805-3820.

Bertrand, C., J.-P. van Ypersele, A. Berger, 1999: Volcanic and solar impacts on climate since 1700. *Clim. Dyn.*, **15**, 355-367

Boer, G. J., G. Flato, M.C. Reader and D. Ramsden, 2000a: A transient climate change simulation with greenhouse gas and aerosol forcing: experimental design and comparison with the instrumental record for the 20th century. *Clim. Dyn.*, **16**, 405-426.

Boer, G.J., G. Flato, M.C. Reader and D. Ramsden, 2000b: A transient climate change simulation with greenhouse gas and aerosol forcing: projected for the 21st century. *Clim. Dyn.*, **16**, 427-450.

Briffa, K.R., P.D. Jones, F.H. Schweingruber and T.J. Osborn, 1998: Influence of volcanic eruptions on Northern Hemisphere summer temperature over the past 600 years. *Nature*, **393**, 450-454.

Briffa, K.R., T.J. Osborn, F.H. Schweingruber, I.C. Harris, P.D. Jones, S.G. Shiyatov and E.A. Vaganov, 2000: Low-frequency temperature variations from a northern tree-ring density network. *J.Geophys.Res.* **106**, 2929-2942.

Brown, L., G. Casella and T.G. Hwang, 1995: Optimal confidence sets, bioequivalence and the limacon of Pascal. *J. Amer. Statist. Soc.*, **90**, 880-889.

Brown, S.J., D.E. Parker, C.K. Folland and I. Macadam, 2000a: Decadal variability in the lower-tropospheric lapse rate. *Geophys. Res. Lett.*, **27**, 997-1000.

Brown, S.J., D.E. Parker and D.M.H. Sexton, 2000b: Differential changes in observed surface and atmospheric temperature since 1979. Hadley Centre Technical Note No. 12, The Met. Office Bracknell, UK, pp41.

Chanin, M.-L. and V. Ramaswamy, 1999: "Trends in Stratospheric Temperatures" in WMO (World Meteorological Organization), Scientific Assessment of Ozone Depletion: 1998, Global Ozone Research and Monitoring Project - Report No. 44, Geneva. 5.1-5.59

Collins, M., T.J. Osborn, S.F.B. Tett, K.R. Briffa and F.H. Schwingruber, 2000: A comparison of the variability of a climate model with a network of tree-ring densities. Hadley Centre Technical Note 16.

Crowley, T.J., 2000: Causes of climate change over the last 1000 years, *Science*, **289**, 270-277.

Crowley, T.J. and K.-Y. Kim, 1996: Comparison of proxy records of climate change and solar forcing. *Geophys. Res. Lett.* **23**: 359-362.

Crowley, T.J. and Kim K.-Y., 1999: Modelling the temperature response to forced climate change over the last six centuries. *Geophys. Res. Lett.*, **26**, 1901-1904.

Crowley, T.J. and T. Lowery, 2000: How warm was the Medieval warm period? *Ambio*, **29**, 51-54.

Cubasch, U., G.C. Hegerl and J. Waszkewitz,1996: Prediction, detection and regional assessment of anthropogenic climate change. *Geophysica*, **32**, 77-96.

Cubasch, U., G.C. Hegerl, R. Voss, J. Waszkewitz and T.J. Crowley, 1997: Simulation of the influence of solar radiation variations on the global climate with an ocean-atmosphere general circulation model. *Clim. Dyn.*, **13**, 757-767.

Delworth, T.L. and T.R. Knutson, 2000. Simulation of early 20th century global warming. *Science*, **287**, 2246-2250.

Dempster, A.P., 1998: Logicist statistics I. Models and modeling. *Statistical Science*, **13**, 248-276.

Drijfhout, S.S., R.J. Haarsma, J.D.Opsteegh, F.M. Selten, 1999: Solar-induced versus internal variability in a coupled climate model. *Geophys. Res. Lett.*, **26**, 205-208.

Folland, C.K., D.M.H. Sexton, D.J.K. Karoly, C.E. Johnston, D.P. Rowell and D.E. Parker, 1998: Influences of anthropogenic and oceanic forcing on recent climate change. *Geophys. Res. Lett.*, **25**, 353-356.

Forest, C.E., M.R. Allen, P.H. Stone, A.P. Sokolov, 2000: Constraining uncertainties in climate models using climate change detection techniques. *Geophys. Res. Lett.*, **27**, 569-572.

Forster, P.M. and K.P. Shine, 1999: Stratospheric water vapour changes as a possible contributor to observed stratospheric cooling. *Geophys. Res. Lett.*, **26**,3309-3312.

Free, M. and A. Robock, 1999: Global Warming in the Context of the little Ice Age. *J.Geophys Res*, **104**, 19057-19070.

Froehlich, K. and J. Lean, 1998: The Sun's Total Irradiance: Cycles, Trends and Related Climate Change Uncertainties since 1976. *Geophys. Res. Lett.*, **25**, 4377-4380.

Fyfe, J.C., G.J. Boer and G.M. Flato 1999: The Arctic and Antarctic Oscillations and their projected changes under global warming. *Geophys. Res. Lett.*, **26**, 1601-1604.

Gaffen, D.J., B.D. Santer, J.S. Boyle, J.R. Christy, N.E. Graham and R.J.Ross, 2000. Multi-decadal changes in vertical temperature structure of the tropical troposphere. *Science*, **287**, 1242-1245.

Gillett, N., M.R. Allen and S.F.B. Tett, 2000a: Modelled and observed variability in atmospheric vertical temperature structure. *Clim Dyn*, **16**, 49-61.

Gillett, N, G.C. Hegerl, M.R. Allen, P.A. Stott, 2000b: Implications of observed changes in the Northern Hemispheric winter circulation for the detection of anthropogenic climate change. *Geophys.Res.Lett.*, **27**, 993-996..

Graf, H.-F., I. Kirchner, I. Schult 1996: Modelling Mt. Pinatubo Climate Effects. NATO-ASI Series, Vol 142, The Mount Pinatubo Eruption, Fiocco G and Dua D. eds., Springer, 1996: 219-231.

Graf, H-F., I. Kirchner and J. Perlwitz,1998: Changing lower stratospheric circulation: The role of ozone and greenhouse gases. *J.*

Geophys. Res., **103**, 11251-11261.

Grieser, J.and C.-D. Schoenwiese, 1999: Parameterization of spatio-temporal patterns of volcanic aerosol induced stratospheric optical depth and its climate radiative forcing. *Atmosfera*, **12**, 111-133

Grieser, J. and C.-D. Schoenwiese, 2001: Process, forcing and signal analysis of global mean temperature variations by means of a three box energy balance model. *Clim. Change*, in press.

Haigh, J.D., 1999: A GCM study of climate change in response to the 11-year solar cycle. *Quart. J. R. Met. Soc.*, **125**, 871-892.

Hansen J., Mki.Sato, R. Ruedy, A. Lacis, K. Asamoah, S. Borenstein, E. Brown, B. Cairns, G. Caliri, M. Campbell, B. Curran, S. deCastro, L. Druyan , M. Fox, C. Johnson, J. Lerner, M.P. McCormick, R. Miller, P. Minnis, A. Morrison, L. Pandolfo, I. Ramberran, F. Zaucker, M. Robinson, P. Russell, K. Shah, P. Stone, I. Tegen, L. Thomason, J. Wilder and H. Wilson 1996: A Pinatubo climate modeling investgation. In "The Mount Pinatubo Eruption: Effects on the Atmosphere and Climate", Fiocco G and Fua Visconti G, Springer Verlag, Berlin, pp 233-272.

Hansen, J., M. Sato and R. Reudy, 1997a: Radiative forcing and climate response. *J. Geophys. Res.*, **102**, 6831-6864.

Hansen, J.E., M. Sato, A. Lacis and R. Reudy, 1997b: The missing climate forcing. *Roy. Soc. Phil. Trans. B*, **352**, 231-240.

Hansen, J.E., M.Sato, A. Lacis, R.Reudy, I .Tegen and E. Matthews, 1998. Climate forcings in the industrial era. *Proc. Nat. Acad. Sci. U.S.A.*, **95**, 12753-12758.

Harrison, R.G. and K.P. Shine, 1999: A review of recent studies of the influence of solar changes on the Earth's climate. Hadley Centre Tech. Note 6. Hadley Centre for Climate Prediction and Research, Meteorological Office, Bracknell RG12 2SY UK pp 65

Hasselmann, K., 1976: Stochastic climate models. Part 1. Theory. *Tellus*, **28**, 473-485.

Hasselmann, K., 1979. On the signal-to-noise problem in atmsopheric response studies. Meteorology over the tropical oceans, D.B. Shaw ed., Roy Meteorol. Soc., 251-259

Hasselmann, K., 1993: Optimal fingerprints for the Detection of Time dependent Climate Change. *J. Climate*, **6**: 1957-1971.

Hasselmann, K. ,1997: Multi-pattern fingerprint method for detection and attribution of climate change. *Clim. Dyn.*, **13**: 601-612.

Hasselmann, K., 1998: Conventional and Bayesian approach to climate-change detection and attribution. *Quart. J. R. Met. Soc.*, **124**, 2541-2565.

Hasselmann, K., L. Bengtsson , U. Cubasch, G.C. Hegerl, H. Rodhe, E. Roeckner, H. von Storch, R.Voss and J. Waskewitz,1995: Detection of an anthropogenic fingerprint. In:*Proceedings of Modern Meteorology" Symposium in honour of Aksle Wiin-Nielsen, 1995*. P. Ditvelson (ed) ECMWF.

Haywood, J.M., R. Stouffer, R. Wetherald, S. Manabe and V. Ramaswamy, 1997: Transient response of a coupled model to estimated changes in grenhouse gas and sulphate concentrations. *Geophys. Res. Lett.* , **24**, 1335- 1338.

Hegerl, G.C. and G.R. North 1997: Statistically optimal methods for detecting anthropogenic climate change. *J. Climate*, **10**: 1125-1133.

Hegerl, G.C. and M.R. Allen, 2000: Physical interpretation of optimal detection. Tech Report RAL-TR-2001-010, Rutherford Appleton Lab., Chilton, X11 0QX, U.K.

Hegerl, G.C., von Storch, K. Hasselmann, B.D. Santer, U. Cubasch and P.D. Jones 1996: Detecting Greenhouse Gas induced Climate Change with an Optimal Fingerprint Method. *J. Climate*, **9**, 2281-2306.

Hegerl, G.C., K Hasselmann, U. Cubasch, J.F.B. Mitchell, E. Roeckner, R. Voss and J. Waszkewitz 1997: Multi-fingerprint detection and attribution of greenhouse gas- and aerosol forced climate change. *Clim. Dyn.* **13**, 613-634

Hegerl, G.C., P. Stott, M. Allen, J.F.B.Mitchell, S.F.B.Tett and U.Cubasch, 2000: Detection and attribution of climate change: Sensitivity of results to climate model differences. *Clim. Dyn.*, **16**,

737-754

Hegerl, G.C., P.D. Jones and T.P. Barnett. 2001: Effect of observational sampling error on the detection of anthropogenic climate change. *J. Climate*, **14**, 198-207.

Hill, D.C., M.R. Allen, P.A. Stott, 2001: Allowing for solar forcing in the detection of human influence on atmospheric vertical temperature structures. *Geophys. Res. Lett.*, **28**, 1555-1558.

Hoyt, D.V. and K.H. Schatten,1993: A discussion of plausible solar irradiance variations, 1700-1992. *J. Geophys. Res.*, **98**, 18895-18906.

IPCC, 1990: Climate change: the IPCC scientific assessment [Houghton, J.T., G.J. Jenkins and J.J. Ephraums (eds.)] Cambridge University Press, Cambridge, United Kingdom, 365pp.

IPCC, 1992: Climate Change 1992: The Supplementary Report to the Intergovernmental Panel on Climate Change Scientific Assessment [Houghton, J.T., B.A. Callander and S.K. Varney (eds.)]. Cambridge University Press, Cambridge, United Kingdom and New York, NY, USA, 100 pp.

IPCC,1996: Climate Change 1995. The IPCC second scientific assessment. Houghton JT, L.G. Meira Filho, B.A. Callander, N. Harris, A. Kattenberg , K. Maskell (eds.). Cambridge University Press, Cambridge, 572 pp.

Johns, T.C., J.M. Gregory, P. Stott and J.F.B. Mitchell, 2001: Assessment of the similarity between modelled and observed near surface temperature changes using ensembles. *Geophys. Res. Lett.* 28, 1007-1010.

Jones, A., D.L. Roberts and M.J. Woodage, 1999: The indirect effects of anthropogenic aerosl simulated using a climate model with an interactive sulphur cycle . Hadley Centre Tech. Note 14. Hadley Centre for Climate Prediction and Research, Meteorological Office, Bracknell RG12 2SY UK, pp39.

Jones, P.D. and G.C. Hegerl, 1998: Comparisons of two methods of removing anthropogenically related variability from the near—surface observational temperature field. *J.Geophys.Res.*, **103**, D12: 1377-13786

Jones, P.D., K.R. Briffa, T.P. Barnett and S.F.B. Tett, 1998: High-resolution paleoclimatic records for the last millenium. *The Holocene* **8**: 467-483.

Jones, P.D., T.J. Osborn, K.R. Briffa, C.K. Folland, B. Horton, L.V. Alexander, D.E. Parker and N.A. Raynor, 2001:Adjusting for sampling density in grid-box land and ocean surface temperature time series, *J. Geophys. Res.*, **106**, 3371-3380

Karoly, D. J. and K. Braganza, 2001: Identifying global climate change using simple indices. *Geophys. Res. Lett.*, in press.

Karoly, D.J., J.A. Cohen, G.A. Meehl, J.F.B. Mitchell, A.H. Oort, R.J. Stouffer, R.T. Weatherald, 1994: An example of fingerprint detection of greenhouse climate change. *Clim. Dyn*, **10**, 97-105.

Kattenberg, A., F. Giorgi, H. Grassl, G.A. Meehl, J.F.B. Mitchell, R. Stouffer, T. Tokioka, A.J. Weaver and T.M.L. Wigley, 1996: Climate Models - Projections of Future Climate. Houghton *et al.* (eds.), The IPCC Second Scientific Assessment. Cambridge University Press, Cambridge, 285-357.

Kaufmann, R.K. and D.I. Stern, 1997: Evidence for human influence on climate from hemispheric temperature relations. *Nature*, **388**, 39-44.

Kelly, P.M., P.D. Jones, J. Pengqun, 1996: The spatial response of the climate system to explosive volcanic eruptions. *Int. J. Climatol.* **16**, 537-550.

Kim, K.-Y. and Q. Wu, 2000: Optimal detection using cyclostationary EOFs. *J. Climate*, **13**, 938-950.

Kim, K.-Y., G.R. North, S.S. Shen, 1996: Optimal estimation of spherical harmonic components from a sample of spatially nonuniform covariance statistics. *J. Climate*, **9**, 635-645.

Kirchner, I., G.L. Stenchikov, H.-F.Graf, A.Robock and J.C.Antuna, 1999: Climate model simulation of winter warming and summer cooling following the 1991 Mount Pinatubo volcanic eruption. *J. Geophys. Res.*, **104**, 19,039-19,055.

Knutson, T.R., T.L. Delworth, K. Dixon and R.J. Stouffer, 2000: Model assessment of regional surface temperature trends (1949-97), *J. Geophys. Res.*, **104**, 30981-30996.

Labitzke, K. and H.van Loon, 1997: The signal of the 11-year sunspot cycle in the upper troposphere-lower stratosphere. *Space Sci. Rev.*, **80**, 393-410.

Lamb, H. H., 1970: Volcanic dust in the atmosphere; with a chronology and assessment of its meteorological significance, Philosophical transactions of the Royal Society of Londen, A 266, 425-533

Laut, P. and J. Gunderman, 1998: Does the correlation between the solar cycle and hemispheric land temperature rule out any significant global warming from greenhouse gases? *J. Atmos. Terrest. Phys.*, **60**, 1-3.

Lean, J., 1997: The Sun's Variable Raditation and its relevance for earth. *Annu. Rev. Astrophys.* **35**, 33-67

Lean, J., J.Beer and R.Bradley, 1995: Reconstruction of solar irradiance since 1600: Implications For climate change. *Geophys. Res. Lett.*, **22**, 3195-3198.

Lean, J., D. Rind, 1998: Climate Forcing by Changing Solar Radiation. *J. Climate*, **11**, 3069-3094.

Legates, D. R. and R. E. Davis, 1997 The continuing search for an anthropogenic climate change signal: limitations of correlation based approaches. *Geophys. Res. Lett.*, **24**, 2319-2322

Leroy, S., 1998: Detecting Climate Signals: Some Bayesian Aspects. *J. Climate* **11**, 640-651.

Levine, R.A. and L.M. Berliner, 1999: Statistical principles for climate change studies. *J. Climate*, **12**, 564-574.

Lockwood, M. and R. Stamper 1999: Long-term drift of the coronal source magnetic flux and the total solar irradiance, *Geophys. Res. Lett.*, **26**, 2461-2464

Mann, M.E., R.S. Bradley, M.K. Hughes, 1998: Global scale tmperature patterns and climate forcing over the past six centuries. *Nature*, **392**, 779-787.

Mann, M.E., E. Gille, R.S. Bradley, M.K. Hughes, J.T. Overpeck, F.T. Keimig and W. Gross, 2000b: Global temperature patterns in past centuries: An interactive presentation. *Earth Interactions*, **4/4**, 1-29.

Mao, J. and A. Robock, 1998: Surface air temperature simulations by AMIP general circulation models: Volcanic and ENSO signals and systematic errors. *J. Climate*, **11**, 1538-1552.

Michaels, P.J. and P.C. Knappenburger, 1996: Human effect on global climate? *Nature*, **384**, 523-524.

Mitchell, J.F.B. and T.C. Johns, 1997: On the modification of greenhouse warming by sulphate aerosols. *J. Climate*, **10**, 245-267

Mitchell, J.F.B., R.A. Davis, W.J. Ingram and C.A. Senior, 1995a: On surface temperature, greenhouse Gases and aerosols, models and observations. *J. Climate*, **10**, 2364-3286.

Mitchell, J.F.B., T.J. Johns, J.M. Gregory and S.F.B. Tett, 1995b: Transient climate response to Increasing sulphate aerosols and greenhouse gases. *Nature*, **376**, 501-504.

National Academy of Sciences, 2000; Reconciling observations of global temperature change. 85pp. Nat. Acad. Press, Washington, DC.

North, G.R. and M. Stevens,1998: Detecting Climate Signals in the Surface Temperature Record. *J. Climate*, **11**,: 563-577.

North, G.R. and Q. Wu, 2001: Detecting Climate Signals Using Space Time EOFs. *J. Climate*, **14**, 1839-1863.

North, G.R., K-Y Kim, S. Shen and J.W. Hardin 1995: Detection of Forced Climate Signals, Part I: Filter Theory. *J. Climate*, **8**, 401-408.

Osborn, T.J., K.R. Briffa, S.F.B. Tett, P.D. Jones and R.M. Trigo, 1999: Evaluation of the North Atlantic Oscillation as simulated by a coupled climate model. *Clim. Dyn.* **15**, 685-702.

Overpeck, J., K. Hughen, D. Hardy, R. Bradley, R. Case, M. Douglas, B. Finney, K. Gajewski, G. Jacoby, A. Jennings, S. Lamoureux, A. Lasca, G. MacDonald, J. Moore, M. Retelle, S. Smith, A. Wolfe, G. Zielinksi,1997: Arctic Environmental Change of the Last Four Centuries. *Science*, *278*, 1251-1256.

Paeth, H. and H. Hense, 2001: Signal Analysis of the northern hemisphere atmospheric mean temperature 500/1000 hPa north of 55N between 1949 and 1994. *Clim. Dyn.*, in press.

Paeth, H., A. Hense, R. Glowenka-Hense, R.Voss and U.Cubasch, 1999: The North Atlantic Oscillation as an indicator for greenhouse gas induced climate change. *Clim. Dyn.*, **15**, 953-960.

Parker, D.E., M.Gordon, D.P.M.Cullum, D.M.H.Sexton, C.K.Folland and N.Rayner 1997: A new global gridded radiosonse temperature data base and recent temperature trends. *Geophys. Res. Lett.*, **24**, 1499-1502.

Parkinson, C. L., D. J. Cavalieri, P. Gloersen, H. J. Zwally and J. C.Comiso, 1999: Arctic sea ice extents, areas and trends, 1978-1996, *J. Geophys. Res.*, **104**, 20,837-20,856.

Priestley, M. B., "Spectral Analysis and Time-series", Volume 1, Academic Press, 1981.

Reader, M.C. and G.J. Boer, 1998: The modification of greenhouse gas warming by the direct effect of sulphate aerosols. *Clim. Dyn.*, **14**, 593-607.

Rind, D., J. Lean and R. Healy, 1999: Simulated time-depended climate response to solar radiative forcing since 1600. *J. Geophys. Res.* 104: 1973-1990.

Ripley, B.D. and M. Thompson, 1987: Regression techniques for the detection of analytical bias. *Analyst*, **112**, 377-383.

Risbey, J., M. Kandlikar and D. Karoly, 2000: A framework to articulate and quantify uncertainty in climate change detection and attribution. *Clim. Res.*, **16**, 61-78.

Robock, A. and J. Mao 1992: Winter warming from large volcanic eruptions. *Geophys. Res. Lett.* **19**, 2405-2408.

Robock, A. and M. Free,1995: Ice cores as an index of global volcanism from 1850 to the present, *J. Geophys. Res.*, **100**, 11567-11576.

Robock, A. and M. Jianping, 1995: The volcanic signal in surface temperature observations. *J. Climate*, **8**, 1086-1103.

Roeckner, E., L. Bengtsson, J. Feichter, J. Lelieveld and H. Rodhe, 1999: Transient climate change simulations with a coupled atmosphere-ocean GCM including the tropospheric sulfur cycle. *J. Climate*, **12**, 3004-3032.

Santer, B.D., T.M.L. Wigley and P.D. Jones, 1993:Correlation methods in fingerprint detection studies. *Clim Dyn.*, **8**, 265-276.

Santer, B.D., K.E. Taylor, J.E. Penner, T.M.L. Wigley, U. Cubasch and P.D. Jones, 1995: Towards the detection and attribution of an anthropogenic effect on climate. *Clim. Dyn.*, **12**, 77-100.

Santer, B.D., K.E. Taylor, T.M.L. Wigley, P.D. Jones, D.J. Karoly, J.F.B. Mitchell, A.H. Oort, J.E.Penner, V. Ramaswamy, M.D. Schwarzkopf, R.S. Stouffer and S.F.B. Tett, 1996a: A search for human influences on the thermal structure in the atmosphere. *Nature*, **382**, 39-46.

Santer, B.D., K.E. Taylor, T.M.L. Wigley, T.C. Johns, P.D. Jones, D.J. Karoly, J.F.B. Mitchell, A.H. Oort, J.E. Penner, V. Ramaswamy, M.D. Schwartzkopf, R.J. Stouffer and S.F.B. Tett, 1996b: Human effect on global climate? *Nature*, **384**, 522-524.

Santer, B.D., T.M.L. Wigley, T.P. Barnett, E. Anyamba, 1996c: Detection of climate change and attribution of causes. in: J. T. Houghton *et al.* (eds.) Climate change 1995. The IPCC Second Scientific Assessment, Cambridge University Press, Cambridge, pp. 407-444.

Santer, B.D., T.M.L. Wigley, D.J. Gaffen, C. Doutriaux, J.S. Boyle, M. Esch, J.J. Hnilo, P.D. Jones, G.A. Meehl, E. Roeckner, K.E. Taylor and M.F. Wehner 2000: Interpreting Differential Temperature Trends at the Surface and in the lower Troposphere, *Science*, **287**, 1227-1231.

Sato, M., J.E. Hansen, M.P. McCormick and J. Pollack, 1993: Statospheric aerosol optical depths (1850-1990). *J. Geophys. Res.* **98**, 22987-22994

Schönwiese, C.-D., M. Denhard, J. Grieser and A. Walter, 1997: Assessments of the global anthropogenic greenhouse and sulphate signal using different types of simplified climate models. *Theor. Appl. Climatol.*, **57**, 119-124.

Schneider T and I. Held, 2001: Discriminants of twentieth-century changes in Earth surface temperatures. *J. Climate*, **14**, 249-254.

Schnur, R., 2001: Detection of climate change using advanced climate diagnostics: seasonal and diurnal cycle. Max-Planck-Institut fuer Meteorologie, Report No. 312, Bundesstr. 55, 20146 Hamburg, Germany.

Sexton, D.M.H., D.P. Rowell, C.K. Folland and D.J. Karoly, 2001: Detection of anthropogenic climate change using an atmospheric GCM. *Clim. Dyn.*, in press.

Shindell, D. T., R.L. Miller, G. Schmidt and L. Pandolfo, 1999: Simulation of recent northern winter climate trends by greenhouse-gas forcing. *Nature*, **399**, 452-455.

Shindell, D. T., G.Schmidt, R.L. Miller and D. Rind, 2001: Northern Hemispheric climate response to greenhouse gas, ozone, solar and volcanic forcing. *J. Geophys. Res.* (Atmospheres), in press.

Simkin, T., L.Siebert, L.McClelland, D.Bridge, C.G.Newhall and J.H. Latter, 1981: Volcanoes of the World, 232 pp. Van Nostrand Reinhold, New York, 1981

Smith, R.L., T.M.L. Wigley and B.D. Santer, 2001: A bivariate time-series approach to anthropogenic trend detection in hemispheric mean temperatures. *J. Climate*, in press.

Stephenson, D.B., 1997: Correlation of spatial climate/weather maps and the advantages of using the Mahalonbis metric in predictions. *Tellus*, **49A**, 513-527.

Stevens, M.J. and G.R. North, 1996: Detection of the climate response to the solar cycle. *J. Atmos. Sci.*, **53**, 2594-2608.

Stothers, R.B., 1996. Major optical depth perturbations to the strato-sphere from volcanic eruptions: Pyrheliometric period 1881-1960. *J.Geophys. Res.*, **101**, 3901-3920.

Stott, P. A. and S.F.B. Tett,1998: Scale-dependent detection of climate change. *J. Climate*, **11**: 3282-3294.

Stott, P.A., M.R. Allen and G.S. Jones, 2000a: Estimating signal amplitudes in optimal fingerprinting II: Application to general circula-tion models. Hadley Centre Tech Note 20, Hadley Centre for Climate Prediction and Response, Meteorological Office, RG12 2SY UK.

Stott, P.A., S.F.B. Tett, G.S. Jones, M.R. Allen, J.F.B.Mitchell and G.J. Jenkins, 2000b: External control of twentieth century temperature variations by natural and anthropogenic forcings. *Science,* **15**, 2133-2137.

Stott, P.A., S.F.B. Tett, G.S. Jones, M.R. Allen, W.J. Ingram and J.F.B. Mitchell, 2001: Attribution of Twentieth Century Temperature Change to Natural and Anthropogenic Causes. *Clim. Dyn.* **17**,1-22.

Stouffer, R.J., G.C. Hegerl and S.F.B. Tett, 1999: A comparison of Surface Air Temperature Variability in Three 1000-Year Coupled Ocean-Atmosphere Model Integrations. *J. Climate*, **13**, 513-537.

Svensmark, H. and E. Friis-Christensen, 1997: Variations of cosmic ray flux and global cloud coverage – a missing link in solar-climate relationships. *J. Atmos. Solar-Terrestrial Phys.*, **59**, 1226-1232.

Svensmark, H., 1998: Influence of cosmic rays on Earth's climate. *Phys. Rev. Lett.*, **81**, 5027-5030.

Tett, S.F.B., J.F.B. Mitchell, D.E. Parker and M.R. Allen, 1996: Human Influence on the Atmospheric Vertical Temperature Structure: Detection and Observations. *Science,* **274**, 1170-1173.

Tett, S.F.B., P.A. Stott, M.A. Allen, W.J. Ingram and J.F.B. Mitchell, 1999: Causes of twentieth century temperature change. *Nature*, **399**, 569-572.

Tett, S.F.B., G.S. Jones, P.A. Stott, D.C. Hill, J.F.B. Mitchell, M.R. Allen, W.J. Ingram, T.C. Johns, C.E. Johnson, A. Jones, D.L. Roberts, D.M.H. Sexton and M.J. Woodage, 2000: Estimation of natural and anthropogenic contributions to 20th century. Hadley Centre Tech Note 19, Hadley Centre for Climate Prediction and Response, Meteorological Office, RG12 2SY, UK pp52.

Thejll, P. and K. Lassen, 2000: Solar forcing of the Northern Hemisphere land air temperature: New data. *J. Atmos. Solar-Terrestrial Phys,* **62**,1207-1213.

Thompson, D.W.J. and J.M. Wallace, 1998: The Arctic oscillation signature in the wintertime geopotential height and temperature fields. *Geophys. Res. Lett.*, **25**, 1297-1300.

Thompson, D.W.J., J.M. Wallace and G.C. Hegerl 2000: Annular Modes in the Extratropical Circulation. Part II Trends. *J. Climate*, **13**, 1018-1036 .

Tol, R.S.J. and P. Vellinga, 1998: 'Climate Change, the Enhanced Greenhouse Effect and the Influence of the Sun: A Statistical Analysis', *Theoretical and Applied Climatology,* **61**, 1-7.

Tol, R.S.J. and A.F. de Vos, 1998: A Bayesian statistical analysis of the enhanced greenhouse effect. *Clim. Change*, **38**, 87-112.

van Loon, H. and D. J. Shea, 1999: A probable signal of the 11-year solar cycle in the troposphere of the Northern Hemisphere, *Geophys. Res. Lett.*, **26**, 2893-2896.

van Loon, H. and D. J. Shea, 2000: The global 11-year solar signal in July-August. *J. Geophys. Res.*, **27**, 2965-2968.

Vinnikov, K. Y., A.Robock, R. J.Stouffer, J.E.Walsh, C. L. Parkinson, D.J.Cavalieri, J.F.B Mitchell, D.Garrett and V.F. Zakharov, 1999: Global warming and Northern Hemisphere sea ice extent. *Science,* **286**, 1934-1937.

Volodin, E.M. and V.Y. Galin, 1999: Interpretation of winter warming on Northern Hemisphere continents 1977-94. *J. Climate*, 12, 2947-2955.

Walter, A., M. Denhard, C.-D. Schoenwiese, 1998: Simulation of global and hemispheric temperature variation and signal detection studies using neural networks. Meteor. Zeitschrift, N.F. **7**, 171-180.

Weber, G.R., 1996: Human effect on global climate? *Nature*, **384**, 524-525.

White, W.B., J. Lean, D.R. Cayan and M.D. Dettinger, 1997: A response of global upper ocean temperature to changing solar irradiance. *J. Geophys. Res.*, **102**, 3255-3266.

Wigley, T.M.L., P.D. Jones and S.C.B. Raper, 1997: The observed global warming record: What does it tell us? *Proc. Natl. Acad. Sci. USA*, **94**, 8314-8320.

Wigley, T.M.L., R.L. Smith and B.D. Santer, 1998a: Anthropogenic influence on the autocorrelation structure of hemispheric-mean temperatures. *Science*, **282**, 1676-1679.

Wigley, T.M.L., P.J. Jaumann, B.D. Santer and K.E Taylor, 1998b: Relative detectablility of greenhouse gas and aerosol climate change signals *Clim Dyn*, **14**, 781-790.

Wigley, T.M.L., B.D.Santer and K.E.Taylor, 2000: Correlation approaches to detection. *Geophys. Res. Lett.* **27**, 2973-2976.

WMO, 1999: Scientific assessment of ozone depletion : 1998, Global Ozone Research and Monitoring Project, World Meteorological Organisation, Report No. 44, Geneva.

Wong, A.P.S., N.L. Bindoff and J.A. Church, 1999: Large-scale freshening of intermediate water masses in the Pacific and Indian Oceans. *Nature*, **400**, 440-443.

Zheng, X. and R.E. Basher, 1999: Structural time-series models and trend detection in global and regional temperature series. *J. Climate*, 12, 2347-2358.

Zwiers, F.W., 1999: The detection of climate change. In "Anthropogenic Climate Change", H. von Storch and G. Floeser, eds., Springer-Verlag, 161-206.

Zwiers, F.W. and S.S. Shen, 1997: Errors in estimating spherical harmonic coefficients from partially sampled GCM output. *Clim. Dyn.*, **13**, 703-716.

13

Climate Scenario Development

Co-ordinating Lead Authors
L.O. Mearns, M. Hulme

Lead Authors
T.R. Carter, R. Leemans, M. Lal, P. Whetton

Contributing Authors
L. Hay, R.N. Jones, R. Katz, T. Kittel, J. Smith, R. Wilby

Review Editors
L.J. Mata, J. Zillman

Contents

Executive Summary

The Purpose of Climate Scenarios

A climate scenario is a plausible representation of future climate that has been constructed for explicit use in investigating the potential impacts of anthropogenic climate change. Climate scenarios often make use of climate projections (descriptions of the modelled response of the climate system to scenarios of greenhouse gas and aerosol concentrations), by manipulating model outputs and combining them with observed climate data.

This new chapter for the IPCC assesses the methods used to develop climate scenarios. Impact assessments have a very wide range of scenario requirements, ranging from global mean estimates of temperature and sea level, through continental-scale descriptions of changes in mean monthly climate, to point or catchment-level detail about future changes in daily or even sub-daily climate.

The science of climate scenario development acts as an important bridge from the climate science of Working Group I to the science of impact, adaptation and vulnerability assessment, considered by Working Group II. It also has a close dependence on emissions scenarios, which are discussed by Working Group III.

Methods for Constructing Scenarios

Useful information about possible future climates and their impacts has been obtained using various scenario construction methods. These include climate model based approaches, temporal and spatial analogues, incremental scenarios for sensitivity studies, and expert judgement. This chapter identifies advantages and disadvantages of these different methods (see Table 13.1).

All these methods can continue to serve a useful role in the provision of scenarios for impact assessment, but it is likely that the major advances in climate scenario construction will be made through the refinement and extension of climate model based approaches.

Each new advance in climate model simulations of future climate has stimulated new techniques for climate scenario construction. There are now numerous techniques available for scenario construction, the majority of which ultimately depend upon results obtained from general circulation model (GCM) experiments.

Representing the Cascade of Uncertainty

Uncertainties will remain inherent in predicting future climate change, even though some uncertainties are likely to be narrowed with time. Consequently, a range of climate scenarios should usually be considered in conducting impact assessments.

There is a cascade of uncertainties in future climate predictions which includes unknown future emissions of greenhouse gases and aerosols, the conversion of emissions to atmospheric concentrations and to radiative forcing of the climate, modelling the response of the climate system to forcing, and methods for regionalising GCM results.

Scenario construction techniques can be usefully contrasted according to the sources of uncertainty that they address and those that they ignore. These techniques, however, do not always provide consistent results. For example, simple methods based on direct GCM changes often represent model-to-model differences in simulated climate change, but do not address the uncertainty associated with how these changes are expressed at fine spatial scales. With regionalisation approaches, the reverse is often true.

A number of methods have emerged to assist with the quantification and communication of uncertainty in climate scenarios. These include pattern-scaling techniques to interpolate/extrapolate between results of model experiments, climate scenario generators, risk assessment frameworks and the use of expert judgement. The development of new or refined scenario construction techniques that can account for multiple uncertainties merits further investigation.

Representing High Spatial and Temporal Resolution Information

The incorporation of climate changes at high spatial (e.g., tens of kilometres) and temporal (e.g., daily) resolution in climate scenarios currently remains largely within the research domain of climate scenario development. Scenarios containing such high resolution information have not yet been widely used in comprehensive policy relevant impact assessments.

Preliminary evidence suggests that coarse spatial resolution AOGCM (Atmosphere-Ocean General Circulation Model) information for impact studies needs to be used cautiously in regions characterised by pronounced sub-GCM grid scale variability in forcings. The use of suitable regionalisation techniques may be important to enhance the AOGCM results over such regions.

Incorporating higher resolution information in climate scenarios can substantially alter the assessment of impacts. The incorporation of such information in scenarios is likely to become increasingly common and further evaluation of the relevant methods and their added value in impact assessment is warranted.

Representing Extreme Events

Extreme climate/weather events are very important for most climate change impacts. Changes in the occurrence and intensity of extremes should be included in climate scenarios whenever possible.

Some extreme events are easily or implicitly incorporated in climate scenarios using conventional techniques. It is more difficult to produce scenarios of complex events, such as tropical cyclones and ice storms, which may require specialised techniques. This constitutes an important methodological gap in scenario development. The large uncertainty regarding future changes in some extreme events exacerbates the difficulty in incorporating such changes in climate scenarios.

Applying Climate Scenarios in Impact Assessments

There is no single "best" scenario construction method appropriate for all applications. In each case, the appropriate method is determined by the context and the application of the scenario.

The choice of method constrains the sources of uncertainty that can be addressed. Relatively simple techniques, such as those that rely on scaled or unscaled GCM changes, may well be the most appropriate for applications in integrated assessment modelling or for informing policy; more sophisticated techniques, such as regional climate modelling or conditioned stochastic weather generation, are often necessary for applications involving detailed regional modelling of climate change impacts.

Improving Information Required for Scenario Development
Improvements in global climate modelling will bring a variety of benefits to most climate scenario development methods. A more diverse set of model experiments, such as AOGCMs run under a broader range of forcings and at higher resolutions, and regional climate models run either in ensemble mode or for longer time periods, will allow a wider range of uncertainty to be represented in climate scenarios. In addition, incorporation of some of the physical, biological and socio-economic feedbacks not currently simulated in global models will improve the consistency of different scenario elements.

13.1 Introduction

13.1.1 Definition and Nature of Scenarios

For the purposes of this report, a climate scenario refers to a plausible future climate that has been constructed for explicit use in investigating the potential consequences of anthropogenic climate change. Such climate scenarios should represent future conditions that account for both human-induced climate change and natural climate variability. We distinguish a climate scenario from a climate projection (discussed in Chapters 9 and 10), which refers to a description of the response of the climate system to a scenario of greenhouse gas and aerosol emissions, as simulated by a climate model. Climate projections alone rarely provide sufficient information to estimate future impacts of climate change; model outputs commonly have to be manipulated and combined with observed climate data to be usable, for example, as inputs to impact models.

To further illustrate this point, Box 13.1 presents a simple example of climate scenario construction based on climate projections. The example also illustrates some other common considerations in performing an impact assessment that touch on issues discussed later in this chapter.

We also distinguish between a climate scenario and a climate change scenario. The latter term is sometimes used in the scientific literature to denote a plausible future climate. However, this term should strictly refer to a representation of the difference between some plausible future climate and the current or control climate (usually as represented in a climate model) (see Box 13.1, Figure 13.1a). A climate change scenario can be viewed as an interim step toward constructing a climate scenario. Usually a climate scenario requires combining the climate change scenario with a description of the current climate as represented by climate observations (Figure 13.1b). In a climate impacts context, it is the contrasting effects of these two climates – one current (the observed "baseline" climate), one

---

Box 13.1: Example of scenario construction.

Example of basic scenario construction for an impact study: the case of climate change and world food supply (Rosenzweig and Parry, 1994).

Aim of the study

The objective of this study was to estimate how global food supply might be affected by greenhouse gas induced climate change up to the year 2060. The method adopted involved estimating the change in yield of major crop staples under various scenarios using crop models at 112 representative sites distributed across the major agricultural regions of the world. Yield change estimates were assumed to be applicable to large regions to produce estimates of changes in total production which were then input to a global trade model. Using assumptions about future population, economic growth, trading conditions and technological progress, the trade model estimated plausible prices of food commodities on the international market given supply as defined by the production estimates. This information was then used to define the number of people at risk from hunger in developing countries.

Scenario information

Each of the stages of analysis required scenario information to be provided, including:
- scenarios of carbon dioxide (CO_2) concentration, affecting crop growth and water use, as an input to the crop models;
- climate observations and scenarios of future climate, for the crop model simulations;
- adaptation scenarios (e.g., new crop varieties, adjusted farm management) as inputs to the crop models;
- scenarios of regional population and global trading policy as an input to the trade model.

To the extent possible, the scenarios were mutually consistent, such that scenarios of population (United Nations medium range estimate) and Gross Domestic Product (GDP) (moderate growth) were broadly in line with the transient scenario of greenhouse gas emissions (based on the Goddard Institute for Space Studies (GISS) scenario A, see Hansen *et al.*, 1988), and hence CO_2 concentrations. Similarly, the climate scenarios were based on $2 \times CO_2$ equilibrium GCM projections from three models, where the radiative forcing of climate was interpreted as the combined concentrations of CO_2 (555 ppm) and other greenhouse gases (contributing about 15% of the change in forcing) equivalent to a doubling of CO_2, assumed to occur in about 2060.

Construction of the climate scenario

Since projections of current (and hence future) regional climate from the GCM simulations were not accurate enough to be used directly as an input to the crop model, modelled changes in climate were applied as adjustments to the observed climate at a location. Climate change by 2060 was computed as the difference (air temperature) or ratio (precipitation and solar radiation) of monthly mean climate between the GCM (unforced) control and $2 \times CO_2$ simulations at GCM grid boxes coinciding with the crop modelling sites (Figure 13.1b). These estimates were used to adjust observed time-series of daily climate for the baseline period (usually 1961 to 1990) at each site (Figure 13.1b,c). Crop model simulations were conducted for the baseline climate and for each of the three climate scenarios, with and without CO_2 enrichment (to estimate the relative contributions of CO_2 and climate to crop yield changes), and assuming different levels of adaptation capacity.

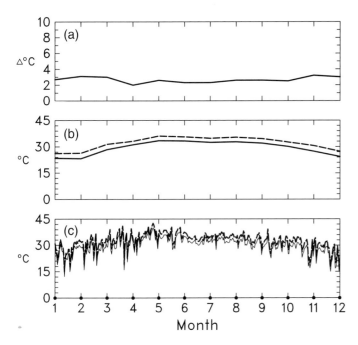

Figure 13.1: Example of the stages in the formation of a simple climate scenario for temperature using Poza Rica (20.3° N, 97.3° W) as a typical site used in the Mexican part of the Rosenzweig and Parry (1994) study.

(a) Mean monthly differences (Δ) (2×CO$_2$ minus control) of average temperature (°C) as calculated from the control and 2×CO$_2$ runs of the Geophysical Fluid Dynamics Laboratory (GFDL) GCM (Manabe and Wetherald, 1987) for the model grid box that includes the geographic location of Poza Rica. The climate model spatial resolution is 4.4° latitude by 7.5° longitude.

(b) The average 17-year (1973 to 1989) observed mean monthly maximum temperature for Poza Rica (solid line) and the 2×CO$_2$ mean monthly maximum temperature produced by adding the differences portrayed in (a) to this baseline (dashed line). The crop models, however, require daily climate data for input.

(c) A sample of one year's (1975) observed daily maximum temperature data (solid line) and the 2×CO$_2$ daily values created by adding the monthly differences in a) to the daily data (dashed line). Thus, the dashed line is the actual daily maximum temperature time-series describing future climate that was used as one of the weather inputs to the crop models for this study and for this location (see Liverman *et al.*, 1994 for further details).

future (the climate scenario) – on the exposure unit[1] that determines the impact of the climate change (Figure 13.1c).

A treatment of climate scenario development, in this specific sense, has been largely absent in the earlier IPCC Assessment Reports. The subject has been presented in independent IPCC Technical Guidelines documents (IPCC, 1992, 1994), which were briefly summarised in the Second Assessment Report of Working Group II (Carter *et al.*, 1996b). These documents, while serving a useful purpose in providing guidelines for scenario use,

---

[1] An exposure unit is an activity, group, region or resource exposed to significant climatic variations (IPCC, 1994).

did not fully address the science of climate scenario development. This may be, in part, because the field has been slow to develop and because only recently has a critical mass of important research issues coalesced and matured such that a full chapter is now warranted.

The chapter also serves as a bridge between this Report of Working Group I and the IPCC Third Assessment Report of Working Group II (IPCC, 2001) (hereafter TAR WG II) of climate change impacts, adaptation and vulnerability. As such it also embodies the maturation in the IPCC assessment process – that is, a recognition of the interconnections among the different segments of the assessment process and a desire to further integrate these segments. Chapter 3 performs a similar role in the TAR WG II (Carter and La Rovere, 2001) also discussing climate scenarios, but treating, in addition, all other scenarios (socio-economic, land use, environmental, etc.) needed for undertaking policy-relevant impact assessment. Chapter 3 serves in part as the other half of the bridge between the two Working Group Reports.

Scenarios are neither predictions nor forecasts of future conditions. Rather they describe alternative plausible futures that conform to sets of circumstances or constraints within which they occur (Hammond, 1996). The true purpose of scenarios is to illuminate uncertainty, as they help in determining the possible ramifications of an issue (in this case, climate change) along one or more plausible (but indeterminate) paths (Fisher, 1996).

Not all possibly imaginable futures can be considered viable scenarios of future climate. For example, most climate scenarios include the characteristic of increased lower tropospheric temperature (except in some isolated regions and physical circumstances), since most climatologists have very high confidence in that characteristic (Schneider *et al.*, 1990; Mahlman, 1997). Given our present state of knowledge, a scenario that portrayed global tropospheric cooling for the 21st century would not be viable. We shall see in this chapter that what constitutes a viable scenario of future climate has evolved along with our understanding of the climate system and how this understanding might develop in the future.

It is worth noting that the development of climate scenarios predates the issue of global warming. In the mid-1970s, for example, when a concern emerged regarding global cooling due to the possible effect of aircraft on the stratosphere, simple incremental scenarios of climate change were formulated to evaluate what the possible effects might be worldwide (CIAP, 1975).

The purpose of this chapter is to assess the current state of climate scenario development. It discusses research issues that are addressed by researchers who develop climate scenarios and that must be considered by impacts researchers when they select scenarios for use in impact assessments. This chapter is not concerned, however, with presenting a comprehensive set of climate scenarios for the IPCC Third Assessment Report.

13.1.2 Climate Scenario Needs of the Impacts Community

The specific climate scenario needs of the impacts community vary, depending on the geographic region considered, the type of impact, and the purpose of the study. For example, distinctions can be made between scenario needs for research in climate scenario development and in the methods of conducting impact assessment (e.g., Woo, 1992; Mearns *et al.*, 1997) and scenario needs for direct application in policy relevant impact and integrated assessments (e.g., Carter *et al.*, 1996a; Smith *et al.*, 1996; Hulme and Jenkins, 1998).

The types of climate variables needed for quantitative impacts studies vary widely (e.g., White, 1985). However, six "cardinal" variables can be identified as the most commonly requested: maximum and minimum temperature, precipitation, incident solar radiation, relative humidity, and wind speed. Nevertheless, this list is far from exhaustive. Other climate or climate-related variables of importance may include CO_2 concentration, sea-ice extent, mean sea level pressure, sea level, and storm surge frequencies. A central issue regarding any climate variable of importance for impact assessment is determining at what spatial and temporal scales the variable in question can sensibly be provided, in comparison to the scales most desired by the impacts community. From an impacts perspective, it is usually desirable to have a fair amount of regional detail of future climate and to have a sense of how climate variability (from short to long time-scales) may change. But the need for this sort of detail is very much a function of the scale and purpose of the particular impact assessment. Moreover, the availability of the output from climate models and the advisability of using climate model results at particular scales, from the point of view of the climate modellers, ultimately determines what scales can and should be used.

Scenarios should also provide adequate quantitative measures of uncertainty. The sources of uncertainty are many, including the

trajectory of greenhouse gas emissions in the future, their conversion into atmospheric concentrations, the range of responses of various climate models to a given radiative forcing and the method of constructing high resolution information from global climate model outputs (Pittock, 1995; see Figure 13.2). For many purposes, simply defining a single climate future is insufficient and unsatisfactory. Multiple climate scenarios that address at least one, or preferably several sources of uncertainty allow these uncertainties to be quantified and explicitly accounted for in impact assessments. Moreover, a further important requirement for impact assessments is to ensure consistency is achieved among various scenario components, such as between climate change, sea level rise and the concentration of actual (as opposed to equivalent) CO_2 implied by a particular emissions scenario.

As mentioned above, climate scenarios that are developed for impacts applications usually require that some estimate of climate change be combined with baseline observational climate data, and the demand for more complete and sophisticated observational data sets of climate has grown in recent years. The important considerations for the baseline include the time period adopted as well as the spatial and temporal resolution of the baseline data.

Much of this chapter is devoted to assessing how and how successfully these needs and requirements are currently met.

13.2 Types of Scenarios of Future Climate

Four types of climate scenario that have been applied in impact assessments are introduced in this section. The most common scenario type is based on outputs from climate models and receives most attention in this chapter. The other three types have usually been applied with reference to or in conjunction with model-based scenarios, namely: incremental scenarios for sensitivity studies, analogue scenarios, and a general category of "other scenarios". The origins of these scenarios and their mutual linkages are depicted in Figure 13.3.

The suitability of each type of scenario for use in policy-relevant impact assessment can be assessed according to five criteria adapted from Smith and Hulme (1998):

1. *Consistency* at regional level with global projections. Scenario changes in regional climate may lie outside the range of global mean changes but should be consistent with theory and model-based results.
2. *Physical plausibility and realism*. Changes in climate should be physically plausible, such that changes in different climatic variables are mutually consistent and credible.
3. *Appropriateness* of information for impact assessments. Scenarios should present climate changes at an appropriate temporal and spatial scale, for a sufficient number of variables, and over an adequate time horizon to allow for impact assessments.
4. *Representativeness* of the potential range of future regional climate change.
5. *Accessibility*. The information required for developing climate scenarios should be readily available and easily accessible for use in impact assessments.

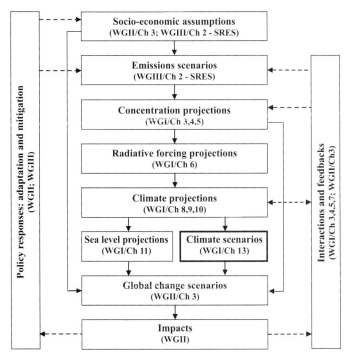

Figure 13.2: The cascade of uncertainties in projections to be considered in developing climate and related scenarios for climate change impact, adaptation and mitigation assessment.

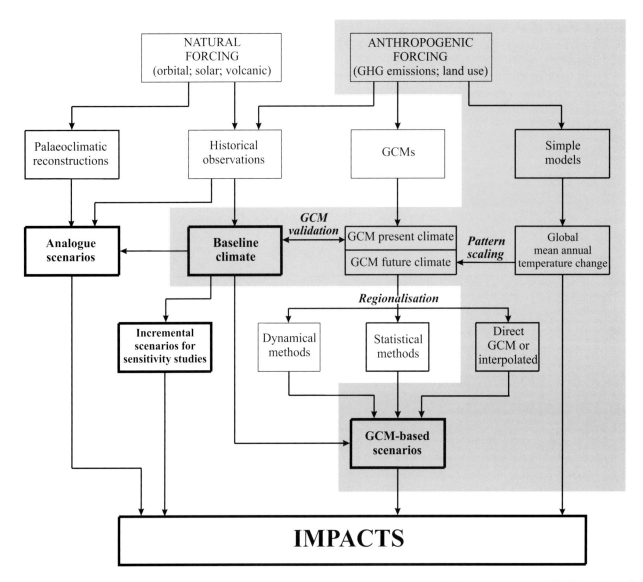

Figure 13.3: Some alternative data sources and procedures for constructing climate scenarios for use in impact assessment. Highlighted boxes indicate the baseline climate and common types of scenario (see text for details). Grey shading encloses the typical components of climate scenario generators.

A summary of the major advantages and disadvantages of different scenario development methods, based on these criteria, is presented in Table 13.1. The relative significance of the advantages and disadvantages is highly application dependent.

13.2.1 Incremental Scenarios for Sensitivity Studies

Incremental scenarios describe techniques where particular climatic (or related) elements are changed incrementally by plausible though arbitrary amounts (e.g., +1, +2, +3, +4°C change in temperature). Also referred to as synthetic scenarios (IPCC, 1994), they are commonly applied to study the sensitivity of an exposure unit to a wide range of variations in climate, often according to a qualitative interpretation of projections of future regional climate from climate model simulations ("guided sensitivity analysis", see IPCC-TGCIA, 1999). Incremental scenarios facilitate the construction of response surfaces – graphical devices for plotting changes in climate against some

measure of impact (for example see Figure 13.9b) which can assist in identifying critical thresholds or discontinuities of response to a changing climate. Other types of scenarios (e.g., based on model outputs) can be superimposed on a response surface and the significance of their impacts readily evaluated (e.g., Fowler, 1999). Most studies have adopted incremental scenarios of constant changes throughout the year (e.g., Terjung *et al.*, 1984; Rosenzweig *et al.*, 1996), but some have introduced seasonal and spatial variations in the changes (e.g., Whetton *et al.*, 1993; Rosenthal *et al.*, 1995) and others have examined arbitrary changes in interannual, within-month and diurnal variability as well as changes in the mean (e.g., Williams *et al.*, 1988; Mearns *et al.*, 1992; Semenov and Porter, 1995; Mearns *et al.*, 1996).

Incremental scenarios provide information on an ordered range of climate changes and can readily be applied in a consistent and replicable way in different studies and regions, allowing for direct intercomparison of results. However, such scenarios do

Table 13.1: *The role of various types of climate scenarios and an evaluation of their advantages and disadvantages according to the five criteria described in the text. Note that in some applications a combination of methods may be used (e.g., regional modelling and a weather generator).*

Scenario type or tool	Description/Use	Advantages[a]	Disadvantages[a]
Incremental	• Testing system sensitivity • Identifying key climate thresholds	• Easy to design and apply (5) • Allows impact response surfaces to be created (3)	• Potential for creating unrealistic scenarios (1, 2) • Not directly related to greenhouse gas forcing (1)
Analogue: Palaeoclimatic	• Characterising warmer periods in past	• A physically plausible changed climate that really did occur in the past of a magnitude similar to that predicted for ~2100 (2)	• Variables may be poorly resolved in space and time (3, 5) • Not related to greenhouse gas forcing (1)
Instrumental	• Exploring vulnerabilities and some adaptive capacities	• Physically realistic changes (2) • Can contain a rich mixture of well-resolved, internally consistent, variables (3) • Data readily available (5)	• Not necessarily related to greenhouse gas forcing (1) • Magnitude of the climate change usually quite small (1) • No appropriate analogues may be available (5)
Spatial	• Extrapolating climate/ecosystem relationships • Pedagogic	• May contain a rich mixture of well-resolved variables (3)	• Not related to greenhouse gas forcing (1, 4) • Often physically implausible (2) • No appropriate analogues may be available (5)
Climate model based: Direct AOGCM outputs	• Starting point for most climate scenarios • Large-scale response to anthropogenic forcing	• Information derived from the most comprehensive, physically-based models (1, 2) • Long integrations (1) • Data readily available (5) • Many variables (potentially) available (3)	• Spatial information is poorly resolved (3) • Daily characteristics may be unrealistic except for very large regions (3) • Computationally expensive to derive multiple scenarios (4, 5) • Large control run biases may be a concern for use in certain regions (2)
High resolution/stretched grid (AGCM)	• Providing high resolution information at global/continental scales	• Provides highly resolved information (3) • Information is derived from physically-based models (2) • Many variables available (3) • Globally consistent and allows for feedbacks (1,2)	• Computationally expensive to derive multiple scenarios (4, 5) • Problems in maintaining viable parametrizations across scales (1,2) • High resolution is dependent on SSTs and sea ice margins from driving model (AOGCM) (2) • Dependent on (usually biased) inputs from driving AOGCM (2)
Regional models	• Providing high spatial/temporal resolution information	• Provides very highly resolved information (spatial and temporal) (3) • Information is derived from physically-based models (2) • Many variables available (3) • Better representation of some weather extremes than in GCMs (2, 4)	• Computationally expensive, and thus few multiple scenarios (4, 5) • Lack of two-way nesting may raise concern regarding completeness (2) • Dependent on (usually biased) inputs from driving AOGCM (2)
Statistical downscaling	• Providing point/high spatial resolution information	• Can generate information on high resolution grids, or non-uniform regions (3) • Potential, for some techniques, to address a diverse range of variables (3) • Variables are (probably) internally consistent (2) • Computationally (relatively) inexpensive (5) • Suitable for locations with limited computational resources (5) • Rapid application to multiple GCMs (4)	• Assumes constancy of empirical relationships in the future (1, 2) • Demands access to daily observational surface and/or upper air data that spans range of variability (5) • Not many variables produced for some techniques (3, 5) • Dependent on (usually biased) inputs from driving AOGCM (2)
Climate scenario generators	• Integrated assessments • Exploring uncertainties • Pedagogic	• May allow for sequential quantification of uncertainty (4) • Provides 'integrated' scenarios (1) • Multiple scenarios easy to derive (4)	• Usually rely on linear pattern scaling methods (1) • Poor representation of temporal variability (3) • Low spatial resolution (3)
Weather generators	• Generating baseline climate time-series • Altering higher order moments of climate • Statistical downscaling	• Generates long sequences of daily or sub-daily climate (2, 3) • Variables are usually internally consistent (2) • Can incorporate altered frequency/intensity of ENSO events (3)	• Poor representation of low frequency climate variability (2, 4) • Limited representation of extremes (2, 3, 4) • Requires access to long observational weather series (5) • In the absence of conditioning, assumes constant statistical characteristics (1, 2)
Expert judgment	• Exploring probability and risk • Integrating current thinking on changes in climate	• May allow for a 'consensus' (4) • Has the potential to integrate a very broad range of relevant information (1, 3, 4) • Uncertainties can be readily represented (4)	• Subjectivity may introduce bias (2) • A representative survey of experts may be difficult to implement (5)

[a] Numbers in parentheses under Advantages and Disadvantages indicate that they are relevant to the criteria described. The five criteria are: (1) *Consistency* at regional level with global projections; (2) *Physical plausibility and realism*, such that changes in different climatic variables are mutually consistent and credible, and spatial and temporal patterns of change are realistic; (3) *Appropriateness* of information for impact assessments (i.e., resolution, time horizon, variables); (4) *Representativeness* of the potential range of future regional climate change; and (5) *Accessibility* for use in impact assessments.

not necessarily present a realistic set of changes that are physically plausible. They are usually adopted for exploring system sensitivity prior to the application of more credible, model-based scenarios (Rosenzweig and Iglesias, 1994; Smith and Hulme, 1998).

13.2.2 Analogue Scenarios

Analogue scenarios are constructed by identifying recorded climate regimes which may resemble the future climate in a given region. Both spatial and temporal analogues have been used in constructing climate scenarios.

13.2.2.1 Spatial analogues

Spatial analogues are regions which today have a climate analogous to that anticipated in the study region in the future. For example, to project future grass growth, Bergthórsson et al. (1988) used northern Britain as a spatial analogue for the potential future climate over Iceland. Similarly, Kalkstein and Greene (1997) used Atlanta as a spatial analogue of New York in a heat/mortality study for the future. Spatial analogues have also been exploited along altitudinal gradients to project vegetation composition, snow conditions for skiing, and avalanche risk (e.g., Beniston and Price, 1992; Holten and Carey, 1992; Gyalistras et al., 1997). However, the approach is severely restricted by the frequent lack of correspondence between other important features (both climatic and non-climatic) of a study region and its spatial analogue (Arnell et al., 1990). Thus, spatial analogues are seldom applied as scenarios, *per se*. Rather, they are valuable for validating the extrapolation of impact models by providing information on the response of systems to climatic conditions falling outside the range currently experienced at a study location.

13.2.2.2 Temporal analogues

Temporal analogues make use of climatic information from the past as an analogue for possible future climate (Webb and Wigley, 1985; Pittock, 1993). They are of two types: palaeo-climatic analogues and instrumentally based analogues.

Palaeoclimatic analogues are based on reconstructions of past climate from fossil evidence, such as plant or animal remains and sedimentary deposits. Two periods have received particular attention (Budyko, 1989; Shabalova and Können, 1995): the mid-Holocene (about 5 to 6 ky BP[2]) and the Last (Eemian) Interglacial (about 120 to 130 ky BP). During these periods, mean global temperatures were as warm as or warmer than today (see Chapter 2, Section 2.4.4), perhaps resembling temperatures anticipated during the 21st century. Palaeoclimatic analogues have been adopted extensively in the former Soviet Union (e.g., Frenzel et al., 1992; Velichko et al., 1995a,b; Anisimov and Nelson, 1996), as well as elsewhere (e.g., Kellogg and Schware, 1981; Pittock and Salinger, 1982). The major disadvantage of using palaeoclimatic analogues for climate scenarios is that the causes of past changes in climate (e.g., variations in the Earth's orbit about the Sun; continental configuration) are different from

---

[2] ky BP = thousand years before present.

those posited for the enhanced greenhouse effect, and the resulting regional and seasonal patterns of climate change may be quite different (Crowley, 1990; Mitchell, 1990). There are also large uncertainties about the quality of many palaeoclimatic reconstructions (Covey, 1995). However, these scenarios remain useful for providing insights about the vulnerability of systems to abrupt climate change (e.g., Severinghaus et al., 1998) and to past El Niño-Southern Oscillation (ENSO) extremes (e.g., Fagan, 1999; Rodbell et al., 1999). They also can provide valuable information for testing the ability of climate models to reproduce past climate fluctuations (see Chapter 8).

Periods of observed global scale warmth during the historical period have also been used as analogues of a greenhouse gas induced warmer world (Wigley et al., 1980). Such scenarios are usually constructed by estimating the difference between the regional climate during the warm period and that of the long-term average or a similarly selected cold period (e.g., Lough et al., 1983). An alternative approach is to select the past period on the basis not only of the observed climatic conditions but also of the recorded impacts (e.g., Warrick, 1984; Williams et al., 1988; Rosenberg et al., 1993; Lapin et al., 1995). A further method employs observed atmospheric circulation patterns as analogues (e.g., Wilby et al., 1994). The advantage of the analogue approach is that the changes in climate were actually observed and so, by definition, are internally consistent and physically plausible. Moreover, the approach can yield useful insights into past sensitivity and adaptation to climatic variations (Magalhães and Glantz, 1992). The major objection to these analogues is that climate anomalies during the past century have been fairly minor compared to anticipated future changes, and in many cases the anomalies were probably associated with naturally occurring changes in atmospheric circulation rather than changes in greenhouse gas concentrations (e.g., Glantz, 1988; Pittock, 1989).

13.2.3 Scenarios Based on Outputs from Climate Models

Climate models at different spatial scales and levels of complexity provide the major source of information for constructing scenarios. GCMs and a hierarchy of simple models produce information at the global scale. These are discussed further below and assessed in detail in Chapters 8 and 9. At the regional scale there are several methods for obtaining sub-GCM grid scale information. These are detailed in Chapter 10 and summarised in Section 13.4.

13.2.3.1 Scenarios from General Circulation Models

The most common method of developing climate scenarios for quantitative impact assessments is to use results from GCM experiments. GCMs are the most advanced tools currently available for simulating the response of the global climate system to changing atmospheric composition.

All of the earliest GCM-based scenarios developed for impact assessment in the 1980s were based on equilibrium-response experiments (e.g., Emanuel et al., 1985; Rosenzweig, 1985; Gleick, 1986; Parry et al., 1988). However, most of these scenarios contained no explicit information about the time of realisation of changes, although time-dependency was introduced in some studies using pattern-scaling techniques (e.g., Santer *et al.*, 1990; see Section 13.5).

The evolving (transient) pattern of climate response to gradual changes in atmospheric composition was introduced into climate scenarios using outputs from coupled AOGCMs from the early 1990s onwards. Recent AOGCM simulations (see Chapter 9, Table 9.1) begin by modelling historical forcing by greenhouse gases and aerosols from the late 19th or early 20th century onwards. Climate scenarios based on these simulations are being increasingly adopted in impact studies (e.g., Neilson *et al.*, 1997; Downing *et al.*, 2000) along with scenarios based on ensemble simulations (e.g., papers in Parry and Livermore, 1999) and scenarios accounting for multi-decadal natural climatic variability from long AOGCM control simulations (e.g., Hulme *et al.*, 1999a).

There are several limitations that restrict the usefulness of AOGCM outputs for impact assessment: (i) the large resources required to undertake GCM simulations and store their outputs, which have restricted the range of experiments that can be conducted (e.g., the range of radiative forcings assumed); (ii) their coarse spatial resolution compared to the scale of many impact assessments (see Section 13.4); (iii) the difficulty of distinguishing an anthropogenic signal from the noise of natural internal model variability (see Section 13.5); and (iv) the difference in climate sensitivity between models.

13.2.3.2 *Scenarios from simple climate models*
Simple climate models are simplified global models that attempt to reproduce the large-scale behaviour of AOGCMs (see Chapter 9). While they are seldom able to represent the non-linearities of some processes that are captured by more complex models, they have the advantage that multiple simulations can be conducted very rapidly, enabling an exploration of the climatic effects of alternative scenarios of radiative forcing, climate sensitivity and other parametrized uncertainties (IPCC, 1997). Outputs from these models have been used in conjunction with GCM information to develop scenarios using pattern-scaling techniques (see Section 13.5). They have also been used to construct regional greenhouse gas stabilisation scenarios (e.g., Gyalistras and Fischlin, 1995). Simple climate models are used in climate scenario generators (see Section 13.5.2) and in some integrated assessment models (see Section 13.6).

13.2.4 *Other Types of Scenarios*

Three additional types of climate scenarios have also been adopted in impact studies. The first type involves extrapolating ongoing trends in climate that have been observed in some regions and that appear to be consistent with model-based projections of climate change (e.g., Jones *et al.*, 1999). There are obvious dangers in relying on extrapolated trends, and especially in assuming that recent trends are due to anthropogenic forcing rather than natural variability (see Chapters 2 and 12). However, if current trends in climate are pointing strongly in one direction, it may be difficult to defend the credibility of scenarios that posit a trend in the opposite direction, especially over a short projection period.

A second type of scenario, which has some resemblance to the first, uses empirical relationships between regional climate and global mean temperature from the instrumental record to extrapolate future regional climate on the basis of projected global or hemispheric mean temperature change (e.g. Vinnikov and Groisman, 1979; Anisimov and Poljakov, 1999). Again, this method relies on the assumption that past relationships between local and broad-scale climate are also applicable to future conditions.

A third type of scenario is based on expert judgement, whereby estimates of future climate change are solicited from climate scientists, and the results are sampled to obtain probability density functions of future change (NDU, 1978; Morgan and Keith, 1995; Titus and Narayanan, 1996; Kuikka and Varis, 1997; Tol and de Vos, 1998). The main criticism of expert judgement is its inherent subjectivity, including problems of the representativeness of the scientists sampled and likely biases in questionnaire design and analysis of the responses (Stewart and Glantz, 1985). Nevertheless, since uncertainties in estimates of future climate are inevitable, any moves towards expressing future climate in probabilistic terms will necessarily embrace some elements of subjective judgement (see Section 13.5).

13.3 Defining the Baseline

A baseline period is needed to define the observed climate with which climate change information is usually combined to create a climate scenario. When using climate model results for scenario construction, the baseline also serves as the reference period from which the modelled future change in climate is calculated.

13.3.1 *The Choice of Baseline Period*

The choice of baseline period has often been governed by availability of the required climate data. Examples of adopted baseline periods include 1931 to 1960 (Leemans and Solomon, 1993), 1951 to 1980 (Smith and Pitts, 1997), or 1961 to 1990 (Kittel *et al.*, 1995; Hulme *et al.*, 1999b).

There may be climatological reasons to favour earlier baseline periods over later ones (IPCC, 1994). For example, later periods such as 1961 to 1990 are likely to have larger anthropogenic trends embedded in the climate data, especially the effects of sulphate aerosols over regions such as Europe and eastern USA (Karl *et al.*, 1996). In this regard, the "ideal" baseline period would be in the 19th century when anthropogenic effects on global climate were negligible. Most impact assessments, however, seek to determine the effect of climate change with respect to "the present", and therefore recent baseline periods such as 1961 to 1990 are usually favoured. A further attraction of using 1961 to 1990 is that observational climate data coverage and availability are generally better for this period compared to earlier ones.

Whatever baseline period is adopted, it is important to acknowledge that there are differences between climatological averages based on century-long data (e.g., Legates and Wilmott, 1990) and those based on sub-periods. Moreover, different 30-year periods have been shown to exhibit differences in regional annual mean baseline temperature and precipitation of up to

±0.5°C and ±15% respectively (Hulme and New, 1997; Visser *et al.*, 2000; see also Chapter 2).

13.3.2 *The Adequacy of Baseline Climatological Data*

The adequacy of observed baseline climate data sets can only be evaluated in the context of particular climate scenario construction methods, since different methods have differing demands for baseline climate data.

There are an increasing number of gridded global (e.g., Leemans and Cramer, 1991; New *et al.*, 1999) and national (e.g., Kittel *et al.*, 1995, 1997; Frei and Schär, 1998) climate data sets describing mean surface climate, although few describe inter-annual climate variability (see Kittel *et al.*, 1997; Xie and Arkin, 1997; New *et al.*, 2000). Differences between alternative gridded regional or global baseline climate data sets may be large, and these may induce non-trivial differences in climate change impacts that use climate scenarios incorporating different baseline climate data (e.g., Arnell, 1999). These differences may be as much a function of different interpolation methods and station densities as they are of errors in observations or the result of sampling different time periods (Hulme and New, 1997; New, 1999). A common problem that some methods endeavour to correct is systematic biases in station locations (e.g., towards low elevation sites). The adequacy of different techniques (e.g., Daly *et al.*, 1994; Hutchinson, 1995; New *et al.*, 1999) to interpolate station records under conditions of varying station density and/or different topography has not been systematically evaluated.

A growing number of climate scenarios require gridded daily baseline climatological data sets at continental or global scales yet, to date, the only observed data products that meet this criterion are experimental (e.g., Piper and Stewart, 1996; Widmann and Bretherton, 2000). For this and other reasons, attempts have been made to combine monthly observed climatologies with stochastic weather generators to allow "synthetic" daily observed baseline data to be generated for national (e.g., Carter *et al.*, 1996a; Semenov and Brooks, 1999), continental (e.g., Voet *et al.*, 1996; Kittel *et al.*, 1997), or even global (e.g., Friend, 1998) scales. Weather generators are statistical models of observed sequences of weather variables, whose outputs resemble weather data at individual or multi-site locations (Wilks and Wilby, 1999). Access to long observed daily weather series for many parts of the world (e.g., oceans, polar regions and some developing countries) is a problem for climate scenario developers who wish to calibrate and use weather generators.

A number of statistical downscaling techniques (see Section 13.4 and Chapter 10, Section 10.6, for definition) used in scenario development employ Numerical Weather Prediction (NWP) reanalysis data products as a source of upper air climate data (Kalnay *et al.*, 1996). These reanalysis data sets extend over periods up to 40 years and provide spatial and temporal resolution sometimes lacking in observed climate data sets. Relatively little detailed work has compared such reanalysis data with independent observed data sets (see Santer *et al.*, 1999, and Widmann and Bretherton, 2000, for two exceptions), but it is known that certain reanalysis variables – such as precipitation and some other hydrological variables – are unreliable.

13.3.3 *Combining Baseline and Modelled Data*

Climate scenarios based on model estimates of future climate can be constructed either by adopting the direct model outputs or by combining model estimates of the changed climate with observational climate data. Impact studies rarely use GCM outputs directly because GCM biases are too great and because the spatial resolution is generally too coarse to satisfy the data requirements for estimating impacts. Mearns *et al.* (1997) and Mavromatis and Jones (1999) provide two of the few examples of using climate model output directly as input into an impact assessment.

Model-based estimates of climate change should be calculated with respect to the chosen baseline. For example, it would be inappropriate to combine modelled changes in climate calculated with respect to model year 1990 with an observed baseline climate representing 1951 to 1980. Such an approach would "disregard" about 0.15°C of mean global warming occurring between the mid-1970s and 1990. It would be equally misleading to apply modelled changes in climate calculated with respect to an unforced (control) climate representing "pre-industrial" conditions (e.g., "forced" t_3 minus "unforced" t_1 in Figure 13.4) to an observed baseline climate representing some period in the 20th century. Such an approach would introduce an unwarranted amount of global climate change into the scenario. This latter definition of modelled climate change was originally used in transient climate change experiments to overcome problems associated with climate "drift" in the coupled AOGCM simulations (Cubasch *et al.*, 1992), but was not designed to be used in conjunction with observed climate data. It is more appropriate to define the modelled change in climate with respect to the same baseline period that the observed climate data set is representing (e.g., "forced" t_3 minus "forced" t_1 in Figure 13.4, added to a 1961 to 1990 baseline climate).

Whatever baseline period is selected, there are a number of ways in which changes in climate can be calculated from model results and applied to baseline data. For example, changes in

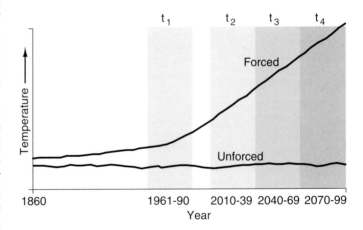

Figure 13.4: A schematic representation of different simulations and periods in a coupled AOGCM climate change experiment that may be used in the definition of modelled climate change. t_1 to t_4 define alternative 30-year periods from either forced or unforced experiments.

climate can be calculated either as the difference or as the ratio between the simulated future climate and the simulated baseline climate. These differences or ratios are then applied to the observed baseline climate – whether mean values, monthly or a daily time-series. Differences are commonly applied to temperature (as in Box 13.1), while ratios are usually used with those surface variables, such as precipitation, vapour pressure and radiation, that are either positive or zero. Climate scenarios have been constructed using both absolute and relative changes for precipitation. The effects of the two different approaches on the resulting climate change impacts depend on the types of impacts being studied and the region of application. Some studies report noticeable differences in impacts (e.g., Alcamo *et al.*, 1998), especially since applying ratio changes alters the standard deviation of the original series (Mearns *et al.*, 1996); in others, differences in impacts were negligible (e.g., Torn and Fried, 1992).

13.4 Scenarios with Enhanced Spatial and Temporal Resolution

The spatial and temporal scales of information from GCMs, from which climate scenarios have generally been produced, have not been ideal from an impacts point of view. The desire for information on climate change regarding changes in variability as well as changes in mean conditions and for information at high spatial resolutions has been consistent over a number of years (Smith and Tirpak, 1989).

The scale at which information can appropriately be taken from relatively coarse-scale GCMs has also been debated. For example, many climate scenarios constructed from GCM outputs have taken information from individual GCM grid boxes, whereas most climate modellers do not consider the outputs from their simulation experiments to be valid on a single grid box scale and usually examine the regional results from GCMs over a cluster of grid boxes (see Chapter 10, Section 10.3). Thus, the scale of information taken from coarse resolution GCMs for scenario development often exceeds the reasonable resolution of accuracy of the models themselves.

In this section we assess methods of incorporating high resolution information into climate scenarios. The issue of spatial and temporal scale embodies an important type of uncertainty in climate scenario development (see Section 13.5.1.5).

Since spatial and temporal scales in atmospheric phenomena are often related, approaches for increasing spatial resolution can also be expected to improve information at high-frequency temporal scales (e.g., Mearns *et al.*, 1997; Semenov and Barrow, 1997; Wang *et al.*, 1999; see also Chapter 10).

13.4.1 Spatial Scale of Scenarios

The climate change impacts community has long bemoaned the inadequate spatial scale of climate scenarios produced from coarse resolution GCM output (Gates, 1985; Lamb, 1987; Robinson and Finkelstein, 1989; Smith and Tirpak, 1989; Cohen, 1990). This dissatisfaction emanates from the perceived mismatch of scale between coarse resolution GCMs (hundreds of

kilometres) and the scale of interest for regional impacts (an order or two orders of magnitude finer scale) (Hostetler, 1994; IPCC, 1994). For example, many mechanistic models used to simulate the ecological effects of climate change operate at spatial resolutions varying from a single plant to a few hectares. Their results may be highly sensitive to fine-scale climate variations that may be embedded in coarse-scale climate variations, especially in regions of complex topography, along coastlines, and in regions with highly heterogeneous land-surface covers.

Conventionally, regional "detail" in climate scenarios has been incorporated by applying changes in climate from the coarse-scale GCM grid points to observation points that are distributed at varying resolutions, but often at resolutions higher than that of the GCMs (e.g., see Box 13.1; Whetton *et al.*, 1996; Arnell, 1999). Recently, high resolution gridded baseline climatologies have been developed with which coarse resolution GCM results have been combined (e.g., Saarikko and Carter, 1996; Kittel *et al.*, 1997). Such relatively simple techniques, however, cannot overcome the limitations imposed by the fundamental spatial coarseness of the simulated climate change information itself.

Three major techniques (referred to as regionalisation techniques) have been developed to produce higher resolution climate scenarios: (1) regional climate modelling (Giorgi and Mearns, 1991; McGregor, 1997; Giorgi and Mearns, 1999); (2) statistical downscaling (Wilby and Wigley, 1997; Murphy, 1999); and (3) high resolution and variable resolution Atmospheric General Circulation Model (AGCM) time-slice techniques (Cubasch *et al.*, 1995; Fox-Rabinovitz *et al.*, 1997). The two former methods are dependent on the large-scale circulation variables from GCMs, and their value as a viable means of increasing the spatial resolution of climate change information thus partially depends on the quality of the GCM simulations. The variable resolution and high resolution time-slice methods use the AGCMs directly, run at high or variable resolutions. The high resolution time-slice technique is also dependent on the sea surface temperature simulated by a coarser resolution AOGCM. There have been few completed experiments using these AGCM techniques, which essentially are still under development (see Chapter 10, Section 10.4). Moreover, they have rarely been applied to explicit scenario formation for impacts purposes (see Jendritzky and Tinz, 2000, for an exception) and are not discussed further in this chapter. See Chapter 10 for further details on all techniques.

13.4.1.1 Regional modelling
The basic strategy in regional modelling is to rely on the GCM to reproduce the large-scale circulation of the atmosphere and for the regional model to simulate sub-GCM scale regional distributions or patterns of climate, such as precipitation, temperature, and winds, over the region of interest (Giorgi and Mearns, 1991; McGregor, 1997; Giorgi and Mearns, 1999). The GCM provides the initial and lateral boundary conditions for driving the regional climate model (RCM). In general, the spatial resolution of the regional model is on the order of tens of kiilometres, whereas the GCM scale is an order of magnitude coarser. Further details on the techniques of regional climate modelling are covered in Chapter 10, Section 10.5.

13.4.1.2 Statistical downscaling

In statistical downscaling, a cross-scale statistical relationship is developed between large-scale variables of observed climate such as spatially averaged 500 hPa heights, or measure of vorticity, and local variables such as site-specific temperature and precipitation (von Storch, 1995; Wilby and Wigley, 1997; Murphy, 1999). These relationships are assumed to remain constant in the climate change context. Also, it is assumed that the predictors selected (i.e., the large-scale variables) adequately represent the climate change signal for the predictand (e.g., local-scale precipitation). The statistical relationship is used in conjunction with the change in the large-scale variables to determine the future local climate. Further details of these techniques are provided in Chapter 10, Section 10.6.

13.4.1.3 Applications of the methods to impacts

While the two major techniques described above have been available for about ten years, and proponents claim use in impact assessments as one of their important applications, it is only quite recently that scenarios developed using these techniques have actually been applied in a variety of impact assessments, such as temperature extremes (Hennessy *et al.*, 1998; Mearns, 1999); water resources (Hassall and Associates, 1998; Hay *et al.*, 1999; Wang *et al.*, 1999; Wilby *et al.*, 1999; Stone *et al.*, 2001); agriculture (Mearns *et al.*, 1998, 1999, 2000a, 2001; Brown *et al.*, 2000) and forest fires (Wotton *et al.*, 1998). Prior to the past couple of years, these techniques were mainly used in pilot studies focused on increasing the temporal (and spatial) scale of scenarios (e.g., Mearns *et al.*, 1997; Semenov and Barrow, 1997).

One of the most important aspects of this work is determining whether the high resolution scenario actually leads to significantly different calculations of impacts compared to that of the coarser resolution GCM from which the high resolution scenario was partially derived. This aspect is related to the issue of uncertainty in climate scenarios (see Section 13.5). We provide examples of such studies below.

Application of high resolution scenarios produced from a regional model (Giorgi *et al.*, 1998) over the Central Plains of the USA produced changes in simulated crop yields that were significantly different from those changes calculated from a coarser resolution GCM scenario (Mearns *et al.*, 1998; 1999, 2001). For simulated corn in Iowa, for example, the large-scale (GCM) scenario resulted in a statistically significant decrease in yield, but the high resolution scenario produced an insignificant increase (Figure 13.5). Substantial differences in regional economic impacts based on GCM and RCM scenarios were also found in a recent integrated assessment of agriculture in the south-eastern USA (Mearns *et al.*, 2000a,b). Hay *et al.* (1999), using a regression-based statistical downscaling technique, developed downscaled scenarios based on the Hadley Centre Coupled AOGCM (HadCM2) transient runs and applied them to a hydrologic model in three river basins in the USA. They found that the standard scenario from the GCM produced changes in surface runoff that were quite different from those produced from the downscaled scenario (Figure 13.6).

13.4.2 Temporal Variability

The climate change information most commonly taken from climate modelling experiments comprises mean monthly, seasonal, or annual changes in variables of importance to impact assessments. However, changes in climate will involve changes in variability as well as mean conditions. As mentioned in Section 13.3 on baseline climate, the interannual variability in climate scenarios constructed from mean changes in climate is most commonly inherited from the baseline climate, not from the climate change experiment. Yet, it is known that changes in variability could be very important to most areas of impact assessment (Mearns, 1995; Semenov and Porter, 1995). The most obvious way in which variability changes affect resource systems is through the effect of variability change on the frequency of extreme events. As Katz and Brown (1992) demonstrated, changes in standard deviation have a proportionately greater effect than changes in means on changes in the frequency of extremes. However, from a climate scenario point of view, it is the relative size of the change in the mean versus standard deviation of a variable that determines the final relative contribution of these statistical moments to a change in extremes. The construction of scenarios incorporating extremes is discussed in Section 13.4.2.2.

The conventional method of constructing mean change scenarios for precipitation using the ratio method (discussed in Section 13.3) results in a change in variability of daily precipitation intensity; that is, the variance of the intensity is changed by a factor of the square of the ratio (Mearns *et al.*, 1996). However, the frequency of precipitation is not changed. Using the difference method (as is common for temperature variables) the variance of the time-series is not changed. Hence, from the perspective of variability, application of the difference approach to precipitation produces a more straightforward scenario. However, it can also result in negative values of precipitation. Essentially neither approach is realistic in its effect on the daily characteristics of the time-series. As mean (monthly) precipitation changes, both the daily intensity and frequency are usually affected.

13.4.2.1 Incorporation of changes in variability: daily to interannual time-scales

Changes in variability have not been regularly incorporated in climate scenarios because: (1) less faith has been placed in climate model simulations of changes in variability than of changes in mean climate; (2) techniques for changing variability are more complex than those for incorporating mean changes; and (3) there may have been a perception that changes in means are more important for impacts than changes in variability (Mearns, 1995). Techniques for incorporating changes in variability emerged in the early 1990s (Mearns *et al.*, 1992; Wilks, 1992; Woo, 1992; Barrow and Semenov, 1995; Mearns, 1995).

Some relatively simple techniques have been used to incorporate changes in interannual variability alone into scenarios. Such techniques are adequate in cases where the impact models use monthly climate data for input. One approach is to calculate present day and future year-by-year anomalies relative to the modelled baseline period, and to apply these anomalies (at an annual, seasonal or monthly resolution) to the long-term mean

Crop model response to scenario resolution

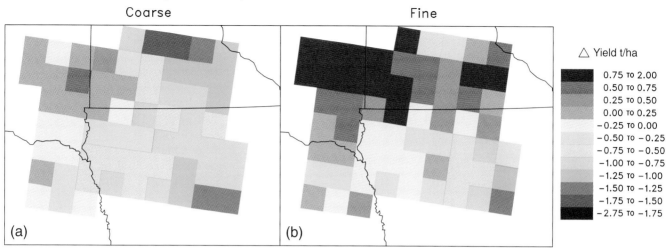

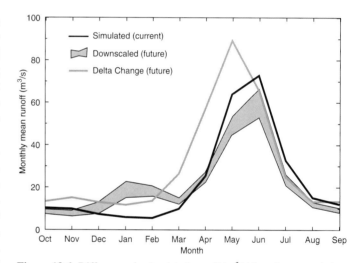

Figure 13.5: Spatial pattern of differences (future climate minus baseline) in simulated corn yields based on two different climate change scenarios for the region covering north-west Iowa and surrounding states (a) coarse spatial resolution GCM scenario (CSIRO); (b) high spatial resolution region climate model scenario (RegCM) (modified from Mearns *et al.*, 1999).

observed baseline climate. This produces climate time-series having an interannual variability equivalent to that modelled for the present day and future, both superimposed on the observed baseline climate. The approach was followed in evaluating impacts of variability change on crop yields in Finland (Carter *et al.*, 2000a), and in the formation of climate scenarios for the United States National Assessment, though in the latter case the observed variability was retained for the historical period.

Another approach is to calculate the change in modelled interannual variability between the baseline and future periods, and then to apply it as an inflator or deflator to the observed baseline interannual variability. In this way, modelled changes in interannual variability are carried forward into the climate scenario, but the observed baseline climate still provides the initial definition of variability. This approach was initially developed in Mearns *et al.* (1992) and has recently been experimented with by Arnell (1999). However, this approach can produce unrealistic features, such as negative precipitation or inaccurate autocorrelation structure of temperature, when applied to climate data on a daily time-scale (Mearns *et al.*, 1996).

The major, most complete technique for producing scenarios with changes in interannual and daily variability involves manipulation of the parameters of stochastic weather generators (defined in Section 13.3.2). These are commonly based either on a Markov chain approach (e.g., Richardson, 1981) or a spell length approach (e.g., Racksko *et al.*, 1991), and simulate changes in variability on daily to interannual time-scales (Wilks, 1992). More detailed information on weather generators is provided in Chapter 10, Section 10.6.2.

To bring about changes in variability, the parameters of the weather generator are manipulated in ways that alter the daily variance of the variable of concern (usually temperature or precipitation) (Katz, 1996). For precipitation, this usually involves changes in both the frequency and intensity of daily precipitation. By manipulating the parameters on a daily time-scale, changes in

Figure 13.6: Differences in simulated runoff (m³/s) based on a statistically downscaled climate scenario and a coarse resolution GCM scenario (labelled Delta Change) for the Animas River Basin in Colorado (modified from Hay *et al.*, 1999). The downscaled range (grey area) is based on twenty ensembles.

variability are also induced on the interannual time-scale (Wilks, 1992). Some weather generators operating at sub-daily time-scales have also been applied to climate scenario generation (e.g., Kilsby *et al.*, 1998).

A number of crop model simulations have been performed to determine the sensitivity of crop yields to incremental changes in daily and interannual variability (Barrow and Semenov, 1995; Mearns, 1995; Mearns *et al.*, 1996; Riha *et al.*, 1996; Wang and Erda, 1996; Vinocur et al., 2000). In most of these studies, changes in variability resulted in significant changes in crop yield. For example, Wang and Erda (1996) combined systematic incremental changes in daily variance of temperature and precipitation with mean climate scenarios in their study of climate change and corn yields in China. They found that increases in the variance of temperature and precipitation combined, further decreased crop yields compared to the effect of the mean change scenarios alone taken from several GCMs.

Studies using the variance changes in addition to mean changes from climate models to form climate scenarios also emerged in the past decade (Kaiser *et al.*, 1993; Bates *et al.* 1994). For example, Bates *et al.* (1994, 1996) adapted Wilks' (1992) method and applied it to changes in daily variability from doubled CO_2 runs of the Commonwealth Scientific and Industrial Research Organisation (CSIRO) climate model (CSIRO9). They then applied the changed time-series to a hydrological model. Combined changes in mean and variability are also evident in a broad suite of statistical downscaling methods (e.g., Katz and Parlange, 1996; Wilby *et al.*, 1998). See also Chapter 10, Section 10.6.3, for further discussion of statistical downscaling and changes in variability.

In recent years, more robust and physically meaningful changes in climatic variability on daily to interannual time scales have been found in runs of GCMs and RCMs for some regions (e.g., Gregory and Mitchell, 1995; Mearns *et al.*, 1995a,b; Whetton *et al.*, 1998a; Mearns, 1999; Boer *et al.*, 2000). For example, on both daily and interannual time-scales many models simulate temperature variability decreases in winter and increases in summer in northern mid-latitude land areas (see Chapter 9, Section 9.3). This result is likely to encourage the further application of model-derived variability changes in climate scenario construction.

The most useful studies, from the point of view of elucidating uncertainty in climate scenarios and impacts, are those that compare applying scenarios with only mean changes to those with mean and variability change. Semenov and Barrow (1997) and Mearns *et al.* (1997) used mean and variance changes from climate models, formed scenarios of climate change using weather generators and applied them to crop models. In both studies important differences in the impacts of climatic change on crop yields were calculated when including the effect of variance change, compared to only considering mean changes. They identified three key aspects of changed climate relevant to the role played by change in daily to interannual variability of climate: the marginality of the current climate for crop growth, the relative size of the mean and variance changes, and the timing of these changes.

It is difficult to generalise the importance of changes in variability to climate change impacts since significance of changes in variability is region, variable, and resource system specific. For example, based on results of equilibrium control and $2\times CO_2$ experiments of DARLAM (a regional model developed in Australia) nested within the CSIRO climate model over New South Wales, Whetton *et al.* (1998a) emphasised that most of the change in temperature extremes they calculated resulted from changes in the mean, not through change in the daily variance. In contrast, Mearns (1999) found large changes (e.g., decreases in winter) in daily variance of temperature in control and $2\times CO_2$ experiments with a regional climate model (RegCM2) over the Great Plains of the U.S. (Giorgi *et al.*, 1998). These changes were sufficient to make a significant difference in the frequency of daily

temperature extremes. Note, however, that these results are not contradictory since they concern two very different regions. More generalised statements may be made regarding the importance of change in the variability of precipitation from climate change experiments for determining changes in the frequency of droughts and floods (e.g., Gregory *et al.*, 1997; Kothavala, 1999). As noted in Chapters 9 and 10, high intensity rainfall events are expected to increase in general, and precipitation variability would be expected to increase where mean precipitation increases.

Other types of variance changes, on an interannual time-scale, based on changes in major atmospheric circulation oscillations, such as ENSO and North Atlantic Oscillation (NAO), are difficult to incorporate into impact assessments. The importance of the variability of climate associated with ENSO phases for resources systems such as agriculture and water resources have been well demonstrated (e.g., Cane *et al.*, 1994; Chiew *et al.*, 1998; Hansen *et al.*, 1998).

Where ENSO signals are strong, weather generators can be successfully conditioned on ENSO phases; and therein lies the potential for creating scenarios with changes in the frequency of ENSO events. By conditioning on the phases, either discretely (Wang and Connor, 1996) or continuously (Woolhiser *et al.*, 1993), a model can be formed for incorporating changes in the frequency and persistence of such events, which would then induce changes in the daily (and interannual) variability of the local climate sites. Weather generators can also be successfully conditioned using NAO signals (e.g., Wilby, 1998). However, it must be noted that there remains much uncertainty in how events such as ENSO might change with climate change (Knutson, *et al.*, 1997; Timmerman *et al.*, 1999; Walsh *et al.*, 1999; see also Chapter 9, Section 9.3.5, for further discussion on possible changes in ENSO events). While there is great potential for the use of conditioned stochastic models in creating scenarios of changed variability, to date, no such scenario has actually been applied to an impact model.

13.4.2.2 Other techniques for incorporating extremes into climate scenarios

While the changes in both the mean and higher order statistical moments (e.g., variance) of time-series of climate variables affect the frequency of relatively simple extremes (e.g., extreme high daily or monthly temperatures, damaging winds), changes in the frequency of more complex extremes are based on changes in the occurrence of complex atmospheric phenomena (e.g., hurricanes, tornadoes, ice storms). Given the sensitivity of many exposure units to the frequency of extreme climatic events (see Chapter 3 of TAR WG II, Table 3.10 (Carter and La Rovere, 2001)), it would be desirable to incorporate into climate scenarios the frequency and intensity of some composite atmospheric phenomena associated with impacts-relevant extremes.

More complex extremes are difficult to incorporate into scenarios for the following reasons: (1) high uncertainty on how they may change (e.g., tropical cyclones); (2) the extremes may not be represented directly in climate models (e.g., ice storms); and (3) straightforward techniques of how to incorporate changes at a particular location have not been developed (e.g., tropical cyclone intensity at Cairns, Australia).

The ability of climate models to adequately represent extremes partially depends on their spatial resolution (Skelly and Henderson-Sellers, 1996; Osborn, 1997; Mearns, 1999). This is particularly true for complex atmospheric phenomena such as hurricanes (see Chapter 10, Box 10.2). There is some very limited information on possible changes in the frequency and intensity of tropical cyclones (Bengtsson *et al.*, 1996; Henderson-Sellers *et al.*, 1998; Krishnamurti *et al.*, 1998; Knutson and Tuleya, 1999; Walsh and Ryan, 2000); and of mid-latitude cyclones (Schubert *et al.*, 1998), but these studies are far from definitive (see Chapter 9, Section 9.3.6, and Chapter 10 for discussion on changes of extremes with changes in climate).

In the case of extremes that are not represented at all in climate models, secondary variables may sometimes be used to derive them. For example, freezing rain, which results in ice storms, is not represented in climate models, but frequencies of daily minimum temperatures on wet days might serve as useful surrogate variables (Konrad, 1998).

An example of an attempt to incorporate such complex changes into climate scenarios is the study of McInnes *et al.* (2000), who developed an empirical/dynamical model that gives return period versus height for tropical cyclone-related storm surges for Cairns on the north Australian coast. To determine changes in the characteristics of cyclone intensity, they prepared a climatology of tropical cyclones based on data drawn from a much larger area than Cairns locally. They incorporated the effect of climate change by modifying the parameters of the Gumbel distribution of cyclone intensity based on increases in tropical cyclone intensity derived from climate model results over a broad region characteristic of the location in question. Estimates of sea level rise also contributed to the modelled changes in surge height. Other new techniques for incorporating such complex changes into quantitative climate scenarios are yet to be developed.

13.5 Representing Uncertainty in Climate Scenarios

13.5.1 Key Uncertainties in Climate Scenarios

Uncertainties about future climate arise from a number of different sources (see Figure 13.2) and are discussed extensively throughout this volume. Depending on the climate scenario construction method, some of these uncertainties will be explicitly represented in the resulting scenario(s), while others will be ignored (Jones, 2000a). For example, scenarios that rely on the results from GCM experiments alone may be able to represent some of the uncertainties that relate to the modelling of the climate response to a given radiative forcing, but might not embrace uncertainties caused by the modelling of atmospheric composition for a given emissions scenario, or those related to future land-use change. Section 13.5.2 therefore assesses different approaches for representing uncertainties in climate scenarios. First, however, five key sources of uncertainty, as they relate to climate scenario construction, are very briefly described. Readers are referred to the relevant IPCC chapters for a comprehensive discussion.

13.5.1.1 Specifying alternative emissions futures
In previous IPCC Assessments, a small number of future greenhouse gas and aerosol precursor emissions scenarios have been presented (e.g., Leggett *et al.*, 1992). In the current Assessment, a larger number of emissions scenarios have been constructed in the Special Report on Emissions Scenarios (SRES) (Nakićenović *et al.*, 2000), and the uncertain nature of these emissions paths have been well documented (Morita and Robinson, 2001). Climate scenarios constructed from equilibrium GCM experiments alone (e.g., Howe and Henderson-Sellers, 1997; Smith and Pitts, 1997) do not consider this uncertainty, but some assumption about the driving emissions scenario is required if climate scenarios are to describe the climate at one or more specified times in the future. This source of uncertainty is quite often represented in climate scenarios (e.g., Section 13.5.2.1).

13.5.1.2 Uncertainties in converting emissions to concentrations
It is uncertain how a given emissions path converts into atmospheric concentrations of the various radiatively active gases or aerosols. This is because of uncertainties in processes relating to the carbon cycle, to atmospheric trace gas chemistry and to aerosol physics (see Chapters 3, 4 and 5). For these uncertainties to be reflected in climate scenarios that rely solely on GCM outputs, AOGCMs that explicitly simulate the various gas cycles and aerosol physics are needed. At present, however, they are seldom, if ever, represented in climate scenarios.

13.5.1.3 Uncertainties in converting concentrations to radiative forcing
Even when presented with a given greenhouse gas concentration scenario, there are considerable uncertainties in the radiative forcing changes, especially aerosol forcing, associated with changes in atmospheric concentrations. These uncertainties are discussed in Chapters 5 and 6, but again usually remain unrepresented in climate scenarios.

13.5.1.4 Uncertainties in modelling the climate response to a given forcing
An additional set of modelling uncertainties is introduced into climate scenarios through differences in the global and regional climate responses simulated by different AOGCMs for the same forcing. Different models have different climate sensitivities (see Chapter 9, Section 9.3.4.1), and this remains a key source of uncertainty for climate scenario construction. Also important is the fact that different GCMs yield different regional climate change patterns, even for similar magnitudes of global warming (see Chapter 10). Furthermore, each AOGCM simulation includes not only the response (i.e., the "signal") to a specified forcing, but also an unpredictable component (i.e., the "noise") that is due to internal climate variability. This latter may itself be an imperfect replica of true climate variability (see Chapter 8). A fourth source of uncertainty concerns important processes that are missing from most model simulations. For instance AOGCM-based climate scenarios do not usually allow for the effect on climate of future land use and land cover change (which is itself, in part, climatically induced). Although the first two sources of model uncertainty − different climate sensitivities and regional climate

change patterns – are usually represented in climate scenarios, it is less common for the third and fourth sources of uncertainty – the variable signal-to-noise ratio and incomplete description of key processes and feedbacks – to be effectively treated.

13.5.1.5 Uncertainties in converting model response into inputs for impact studies

Most climate scenario construction methods combine model-based estimates of climate change with observed climate data (Section 13.3). Further uncertainties are therefore introduced into a climate scenario because observed data sets seldom capture the full range of natural decadal-scale climate variability, because of errors in gridded regional or global baseline climate data sets, and because different methods are used to combine model and observed climate data. These uncertainties relating to the use of observed climate data are usually ignored in climate scenarios. Furthermore, regionalisation techniques that make use of information from AOGCM and RCM experiments to enhance spatial and temporal scales introduce additional uncertainties into regional climate scenarios (their various advantages and disadvantages are assessed in Chapter 10 and in Section 13.4). These uncertainties could be quantified by employing a range of regionalisation techniques, but this is rarely done.

13.5.2 Approaches for Representing Uncertainties

There are different approaches for representing each of the above five generic sources of uncertainty when constructing climate scenarios. The cascade of uncertainties, and the options for representing them at each of the five stages, can result in a wide range of climate outcomes in the finally constructed scenarios (Henderson-Sellers, 1996; Wigley, 1999; Visser *et al.*, 2000). Choices are most commonly made at the stage of modelling the climate response to a given forcing, where it is common for a set of climate scenarios to include results from different GCMs. In practice, this sequential and conditional approach to representing uncertainty in climate scenarios has at least one severe limitation: at each stage of the cascade, only a limited number of the conditional outcomes have been explicitly modelled. For example, GCM experiments have used one, or only a small number, of the concentration scenarios that are plausible (for example, most transient AOGCM experiments that have been used for climate scenarios adopted by impacts assessments have been forced with a scenario of a 1% per annum growth in greenhouse gas concentration). Similarly, regionalisation techniques have been used with only a small number of the GCM experiments that have been conducted. These limitations restrict the choices that can be made in climate scenario construction and mean that climate scenarios do not fully represent the uncertainties inherent in climate prediction.

In order to overcome some of these limitations, a range of techniques has been developed to allow more flexible treatment of the entire cascade of uncertainty. These techniques manipulate or combine different modelling results in a variety of ways. If we are truly to assess the risk of climate change being dangerous, then impact and adaptation studies need scenarios that span a very substantial part of the possible range of future climates (Pittock,

1993; Parry *et al.*, 1996; Risbey, 1998; Jones, 1999; Hulme and Carter, 2000). The remainder of this section assesses four aspects of climate scenario development that originate from this concern about adequately representing uncertainty:

1. scaling climate response patterns across a range of forcing scenarios;
2. defining appropriate climate change signals;
3. risk assessment approaches;
4. annotation of climate scenarios to reflect more qualitative aspects of uncertainty.

13.5.2.1 Scaling climate model response patterns

Pattern-scaling methods allow a wider range of possible future forcings (e.g., the full range of IS92 (Leggett *et al.*, 1992) or SRES emissions scenarios) and climate sensitivities (e.g., the 1.5°C to 4.5°C IPCC range) to be represented in climate scenarios than if only the direct results from GCM experiments were used. The approach involves normalising GCM response patterns according to the global mean temperature change (although in some cases zonal mean temperature changes have been used). These normalised patterns are then rescaled using a scalar derived from simple climate models and representing the particular scenario under consideration.

This pattern-scaling method was first suggested by Santer *et al.* (1990) and was employed in the IPCC First Assessment Report to generate climate scenarios for the year 2030 (Mitchell *et al.*, 1990) using patterns from $2 \times CO_2$ GCM experiments. It has subsequently been widely adopted in climate scenario generators (CSGs), for example in ESCAPE (Rotmans *et al.*, 1994), IMAGE-2 (Alcamo *et al.*, 1994), SCENGEN (Hulme *et al.*, 1995a,b), SILMUSCEN (Carter *et al.*, 1995, 1996a), COSMIC (Schlesinger *et al.*, 1997) and CLIMPACTS (Kenny *et al.*, 2000). A climate scenario generator is an integrated suite of simple models that takes emissions or forcing scenarios as inputs and generates geographically distributed climate scenarios combining response patterns of different greenhouse gases from GCMs with observational climate data. CSGs allow multiple sources of uncertainty to be easily represented in the calculated scenarios, usually by using pattern-scaling methods.

Two fundamental assumptions of pattern-scaling are, first, that the defined GCM response patterns adequately depict the climate "signal" under anthropogenic forcing (see Section 13.5.2.2) and, second, that these response patterns are representative across a wide range of possible anthropogenic forcings. These assumptions have been explored by Mitchell *et al.* (1999) who examined the effect of scaling decadal, ensemble mean temperature and precipitation patterns in the suite of HadCM2 experiments. Although their response patterns were defined using only 10-year means, using four-member ensemble means improved the performance of the technique when applied to reconstructing climate response patterns in AOGCM experiments forced with alternative scenarios (see Figure 13.7). This confirmed earlier work by Oglesby and Saltzman (1992), among others, who demonstrated that temperature response patterns derived from equilibrium GCMs were fairly uniform over a wide range of concentrations, scaling linearly with global mean temperature. The main exception occurred in the

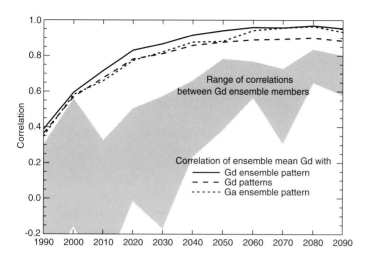

Figure 13.7: Pattern correlations between the decadal ensemble mean temperature (Northern Hemisphere only) from the HadCM2 experiment forced with a 0.5%/yr increase in greenhouse gas concentrations (Gd) and: the scaled ensemble mean pattern (solid line); the four scaled individual ensemble member patterns – average coefficient (dashed line); and the scaled ensemble mean pattern derived from the HadCM2 experiment forced with a 1%/yr increase in greenhouse gas concentrations (Ga) (dotted line). The correlations increase with time as the pattern of greenhouse gas response (the "signal") increasingly dominates the random effects of internal climate variability (the "noise"). The shaded area shows the spread of correlations between the pairs of the individual members of the Gd ensemble; these correlations are lower than those between the realised and scaled patterns above, indicating that the scaled pattern is not due to internal climate variability. (Source: Mitchell *et al.*, 1999.)

regions of enhanced response near sea ice and snow margins. Mitchell *et al.* (1999) concluded that the uncertainties introduced by scaling ensemble decadal mean temperature patterns across different forcing scenarios are smaller than those due to the model's internal variability, although this conclusion may not hold for variables with high spatial variability such as precipitation.

Two situations where the pattern-scaling techniques may need more cautious application are in the cases of stabilisation forcing scenarios and heterogenous aerosol forcing. Whetton *et al.* (1998b) have shown that for parts of the Southern Hemisphere a highly non-linear regional rainfall response was demonstrated in an AOGCM forced with a stabilisation scenario, a response that could not easily be handled using a linear pattern-scaling technique. In the case of heterogeneous forcing, similar global mean warmings can be associated with quite different regional patterns, depending on the magnitude and pattern of the aerosol forcing. Pattern-scaling using single global scalars is unlikely to work in such cases. There is some evidence, however, to suggest that separate greenhouse gas and aerosol response patterns can be assumed to be additive (Ramaswamy and Chen, 1997) and pattern-scaling methods have subsequently been adapted by Schlesinger *et al.* (1997, 2000) for the case of heterogeneously forced scenarios. This is an area, however, where poor signal-to-noise ratios hamper the application of the technique and caution is advised.

The above discussion demonstrates that pattern-scaling techniques provide a low cost alternative to expensive AOGCM and RCM experiments for creating a range of climate scenarios that embrace uncertainties relating to different emissions, concentration and forcing scenarios and to different climate model responses. The technique almost certainly performs best in the case of surface air temperature and in cases where the response pattern has been constructed so as to maximise the signal-to-noise ratio. When climate scenarios are needed that include the effects of sulphate aerosol forcing, regionally differentiated response patterns and scalars must be defined and signal-to-noise ratios should be quantified. It must be remembered, however, that while these techniques are a convenient way of handling several types of uncertainty simultaneously, they introduce an uncertainty of their own into climate scenarios that is difficult to quantify. Little work has been done on exploring whether patterns of change in inter-annual or inter-daily climate variability are amenable to scaling methods.

13.5.2.2 Defining climate change signals
The question of signal-to-noise ratios in climate model simulations was alluded to above, and has also been discussed in Chapters 9 and 12. The treatment of "signal" and "noise" in constructing climate scenarios is of great importance in interpreting the results of impact assessments that make use of these scenarios. If climate scenarios contain an unspecified combination of signal plus noise, then it is important to recognise that the impact response to such scenarios will only partly be a response to anthropogenic climate change; an unspecified part of the impact response will be related to natural internal climate variability. However, if the objective is to specify the impacts of the anthropogenic climate signal alone, then there are two possible strategies for climate scenario construction:

- attempt to maximise the signal and minimise the noise;

- do not try to disentangle signal from noise, but supply impact assessments with climate scenarios containing both elements and also companion descriptions of future climate that contain only noise, thus allowing impact assessors to generate their own impact signal-to-noise ratios (Hulme *et al.*, 1999a).

The relative strength of the signal-to-noise ratio can be demonstrated in a number of ways. Where response patterns are reasonably stable over time, this ratio can be maximised in a climate change scenario by using long (30-year or more) averaging periods. Alternatively, regression or principal component techniques may be used to extract the signal from the model response (Hennessy *et al.*, 1998). A third technique is to use results from multi-member ensemble simulations, as first performed by Cubasch *et al.* (1994). Sampling theory shows that in such simulations the noise is reduced by a factor of $\sqrt{(n)}$, where n is the ensemble size. Using results from the HadCM2 four-member ensemble experiments, Giorgi and Francisco (2000), for example, suggest that uncertainty in future regional climate change associated with internal climate variability at sub-continental scales (10^7 km^2), is generally smaller than the

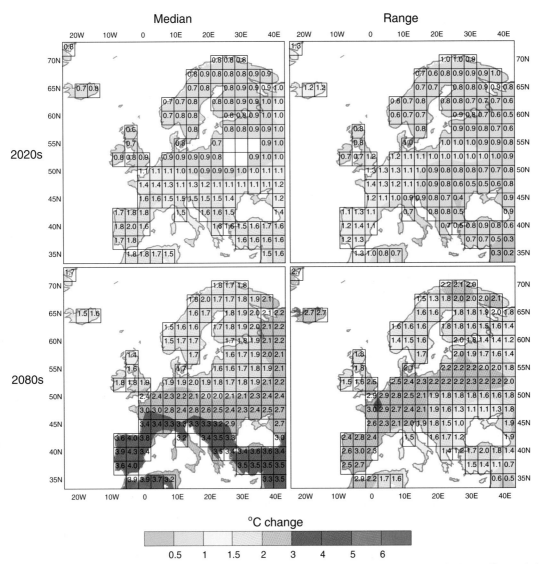

Figure 13.8: A summer (JJA) temperature change scenario for Europe for the 2020s and 2080s. Left panel is the median scaled response of five GCM experiments available on the IPCC Data Distribution Centre (http://ipcc-ddc.cru.uea.ac.uk/) and the right panel is the inter-model range (largest scaled response minus the smallest scaled response). (Source: Hulme and Carter, 2000.)

uncertainty associated with inter-model or forcing differences. This conclusion is scale- and variable-dependent, however (see Chapter 9, Figure 9.4; see also Räisänen, 1999), and the inverse may apply at the smaller scales (10^4 to 10^5 km^2) at which many impact assessments are conducted. Further work is needed on resolving this issue for climate scenario construction purposes.

A different way of maximising the climate change signal is to compare the responses of single realisations from experiments completed using different models. If the error for different models is random with zero mean, then sampling theory shows that this model average will yield a better estimate of the signal than any single model realisation. This approach was first suggested in the context of climate scenarios by Santer *et al.* (1990) and is illustrated further in Chapter 9, Section 9.2.2. Treating different GCM simulations in this way, i.e., as members of a multi-model ensemble, is one way of defining a more robust climate change signal, either for use in pattern-scaling techniques or directly in

constructing a climate scenario. The approach has been discussed by Räisänen (1997) and used recently by Wigley (1999), Hulme and Carter (2000; see Figure 13.8) and Carter *et al.* (2000b) in providing regional characterisations of the SRES emissions scenarios.

The second strategy requires that the noise component be defined explicitly. This can be done by relying either on observed climate data or on model-simulated natural climate variability (Hulme *et al.*, 1999a; Carter *et al.*, 2000b). Neither approach is ideal. Observed climate data may often be of short duration and therefore yield a biased estimate of the noise. Multi-decadal internal climate variability can be extracted from multi-century unforced climate simulations such as those performed by a number of modelling groups (e.g., Stouffer *et al.*, 1994; Tett *et al.*, 1997; von Storch *et al.*, 1997). In using AOGCM output in this way, it is important not only to demonstrate that these unforced simulations do not drift significantly (Osborn, 1996), but also to evaluate the extent to which model estimates of low-frequency variability are comparable to those estimated from measured climates (Osborn *et al.*, 2000) or reconstructed palaeoclimates (Jones *et al.*, 1998). Furthermore, anthropogenic forcing may alter the character of multi-decadal climate variability and therefore the noise defined from model control simulations may not apply in the future.

13.5.2.3 Risk assessment approaches

Uncertainty analysis is required to perform quantitative risk or decision analysis (see Toth and Mwandosya (2001) for discussion of decision analysis). By itself, scenario analysis is not equivalent to uncertainty analysis because not all possible scenarios are necessarily treated and, especially, because probabilities are not attached to each scenario (see Morgan and Henrion (1990) for a general treatment of uncertainty analysis; see Katz (2000) for a more recent overview focusing on climate change). Recognising this limitation, a few recent studies (Jones, 2000b; New and Hulme, 2000) have attempted to modify climate scenario analysis, grouping a range of scenarios together and attaching a probability to the resultant classes. Such an approach can be viewed as a first step in bridging the gap between scenario and uncertainty analysis. Single climate scenarios, by definition, are limited to plausibility with no degree of likelihood attached. Since risk analysis requires that probabilities be attached to each climate scenario, subjective probabilities can be applied to the input parameters that determine the climate outcomes (e.g., emissions scenarios, the climate sensitivity, regional climate response patterns), thus allowing distributions of outcomes to be formally quantified.

In formal risk analysis, the extremes of the probability distribution should encompass the full range of possible outcomes, although in climate change studies this remains hard to achieve. The ranges for global warming and sea level rise from the IPCC WGI Second Assessment Report (IPCC, 1996) (hereafter SAR), for example, deliberately did not encompass the full range of possible outcomes and made no reference to probability distributions. As a consequence, the bulk of impact assessments have treated these IPCC ranges as having a uniform probability, i.e., acting as if no information is available about what changes are more likely than others. As pointed out by Titus and Narayanan (1996), Jones (1998, 2000a), and Parkinson and Young (1998), however, where several sources of uncertainty are combined, the resulting probability distribution is not uniform but is a function of the component probability distributions and the relationship between the component elements. For example, descriptions of regional changes in temperature and rainfall over Australia constructed from regional response patterns have been used in a number of hydrological studies where the extreme outcomes have been considered as likely as outcomes in the centre of the range (e.g., Chiew *et al.*, 1995; Schreider *et al.*, 1996; Whetton, 1998). However, when the two component ranges – global warming and normalised local temperature and rainfall change – are randomly sampled and then multiplied together, they offer a distinctly non-uniform distribution (see Figure 13.9a). Further refinements of these approaches for quantifying the risk of climate change are needed (New and Hulme, 2000).

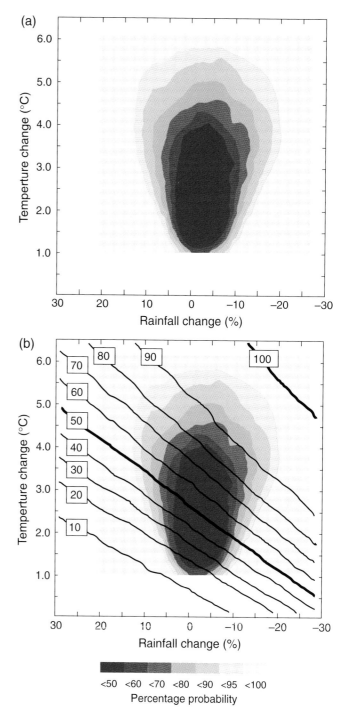

Figure 13.9: (a) Projected ranges of regional annual temperature and rainfall change for inland southern Australia in 2100 extrapolated from CSIRO (1996) with temperature sampled randomly across the projected ranges of both global and normalised regional warming and then multiplied together. Projected regional ranges for normalised seasonal rainfall change were randomly sampled, multiplied by the randomly sampled global warming as above, and then averaged. The resulting probability density surface reveals the likelihood of different climate change outcomes for this region; (b) Response surface of irrigation demand for the same region superimposed on projected climate changes as (a), showing the likelihood of exceeding an annual allocation of irrigation water supply. Risk can be calculated by summing the probabilities of all climates below a given level of annual exceedance of annual water supply; e.g., 50%, or exceedance of the annual water limit in at least one of every two years. (Source: Jones, 2000b.)

This approach to portraying uncertainty has potentially useful applications when combined with climate impact sensitivity response surfaces (see Section 13.2.1; see also Chapter 3 of TAR WG II (Carter and La Rovere, 2001)). The superimposed response surfaces allow the calculation of probabilities for exceeding particular impact thresholds (Figure 13.9b). Another method of assessing risk using quantified probability distributions is through a series of linked models such as those used for calculating sea level rise (Titus and Narayanan, 1996) and economic damage due to sea level rise (Yohe and Schlesinger, 1998), for quantifying climate uncertainty (Visser *et al.*, 2000), and in integrated assessments (Morgan and Dowlatabadi, 1996). Efforts to make explicit probabilistic forecasts of the climate response to a given emissions scenario for the near future have been made using the current observed climate trajectory to constrain the "forecasts" from several GCMs (Allen *et al.,* 2000). More details on this technique are given in Section 12.4.3.3.

13.5.2.4 Annotation of climate scenarios

Even if quantifiable uncertainties are represented, further uncertainties in climate scenarios may still need to be documented or explicitly treated. These include the possible impact on scenarios of errors in the unforced model simulation, the possibility that current models cannot adequately simulate the enhanced greenhouse response of a climatic feature of interest, or inconsistencies between results of model simulations and emerging observed climatic trends. For these reasons climate scenarios are often annotated with a list of caveats, along with some assessment as to their importance for the scenario user.

When choosing which GCM(s) to use as the basis for climate scenario construction, one of the criteria that has often been used is the ability of the GCM to simulate present day climate. Many climate scenarios have used this criterion to assist in their choice of GCM, arguing that GCMs that simulate present climate more faithfully are likely to simulate more plausible future climates (e.g., Whetton and Pittock, 1991; Robock *et al.*, 1993; Risbey and Stone, 1996; Gyalistras *et al.*, 1997; Smith and Pitts, 1997; Smith and Hulme, 1998; Lal and Harasawa, 2000). A good simulation of present day climate, however, is neither a necessary nor a sufficient condition for accurate simulation of climate change (see Chapter 8). It is possible, for example, that a model with a poor simulation of present day climate could provide a more accurate simulation of climate change than one which has a good simulation of present climate, if it contains a better representation of the dominant feedback processes that will be initiated by radiative forcing. While such uncertainties are difficult to test, useful insights into the ability of models to simulate long-term climate change can also be obtained by comparing model simulations of the climate response to past changes in radiative forcing against reconstructed paleoclimates.

This approach to GCM selection, however, raises a number of questions. Over which geographic domain should the GCM be evaluated – the global domain or only over the region of study? Which climate variables should be evaluated – upper air synoptic features that largely control the surface climate, or only those climate variables, mostly surface, that are used in impact studies? Recent AOGCMs simulate observed 1961 to 1990 mean climate

more faithfully than earlier GCMs (Kittel *et al.*, 1998; see also Chapter 8), but they still show large errors in simulating inter-annual climate variability in some regions (Giorgi and Francisco, 2000; Lal *et al.*, 2000) and in replicating ENSO-like behaviour in the tropics (Knutson *et al.*, 1997). These questions demonstrate that there is no easy formula to apply when choosing GCMs for climate scenario construction; there will always be a role for informed but, ultimately, individual judgement. This judgement, however, should be made not just on empirical grounds (for example, which model's present climate correlates best with observations) but also on the basis of understanding the reasons for good or bad model performance, particularly if those reasons are important for the particular scenario application.

Several examples of such annotations can be given. Lal and Giorgi (1997) suggested that GCMs that cannot simulate the observed interannual variability of the Indian monsoon correctly should not be used as the basis for climate scenarios. Giorgi *et al.* (1998) commented that model-simulated spring temperatures over the USA Central Plains were too cold in both the CSIRO GCM and in the CSIRO-driven RegCM2 control simulations and affected the credibility of the ensuing temperature climate scenarios. Finally, scenarios prepared for the Australian region have often been accompanied by the note that ENSO is an important component of Australian climate that may change in the future, but that is not yet adequately simulated in climate models (e.g., Hennessy *et al.*, 1998). Expert judgment can also be used to place confidence estimates on scenario ranges (Morgan and Keith, 1995). For example, Jones *et al.* (2000) placed "high confidence" on the temperature scenarios (incorporating quantifiable uncertainty) prepared for the South Pacific, but only "moderate to low confidence" in the corresponding rainfall scenarios.

13.6 Consistency of Scenario Components

This section discusses some of the caveats of climate scenario development and focuses on the need for consistency in representing different physical aspects of the climate system. It does not discuss the many possible inconsistencies with respect to socio-economic issues in scenario development. Chapter 3 of the TAR WG II (Carter and La Rovere, 2001) and Chapter 2 of the TAR WG III (Morita and Robinson, 2001) provide a detailed treatment of these issues. Three common inconsistencies in applying climate scenarios are discussed, concerning the representation of ambient versus equivalent CO_2 concentrations, biosphere-ocean-atmosphere interactions and time lags between sea level rise and temperature change.

The climate system consists of several components that interact with and influence each other at many different temporal and spatial scales (see Chapter 7). This complexity adds further constraints to the development of climate scenarios, though their relevance is strongly dependent on the objectives and scope of the studies that require scenarios. Most climate scenarios are based on readily available climate variables (e.g., from AOGCMs) and, where these are used in impact assessments, studies are often restricted to an analysis of the effects of changes in climate alone. However, other related environmental aspects may also change, and these are often neglected or inadequately represented, thus potentially reducing the comprehensiveness of the impact assessment. Furthermore, some feedback processes that are seldom considered in AOGCM simulations, may modify regional changes in climate (e.g., the effect of climate-induced shifts in vegetation on albedo and surface roughness).

Concurrent changes in atmospheric concentrations of gases such as CO_2, sulphur dioxide (SO_2) and ozone (O_3) can have important effects on biological systems. Studies of the response of biotic systems require climate scenarios that include consistent information on future levels of these species. For example, most published AOGCM simulations have used CO_2-equivalent concentrations to represent the combined effect of the various gases. Typically, only an annual 1% increase in CO_2-equivalent concentrations, which approximates changes in radiative forcing of the IS92a emission scenario (Leggett *et al.*, 1992), has been used. However, between 10 and 40% of this increase results from non-CO_2 greenhouse gases (Alcamo *et al.*, 1995). The assumption that CO_2 concentrations equal CO_2-equivalent concentrations (e.g., Schimel *et al.*, 1997; Walker *et al.*, 1999) has led to an exaggeration of direct CO_2 effects. If impacts are to be assessed more consistently, proper CO_2 concentration levels and CO_2-equivalent climate forcing must be used. Many recent impact assessments that recognise these important requirements (e.g., Leemans *et al.*, 1998; Prinn *et al.*, 1999; Downing *et al.*, 2000) make use of tools such as scenario generators (see Section 13.5.2.1) that explicitly treat atmospheric trace gas concentrations. Moreover, some recent AOGCM simulations now discriminate between the individual forcings of different greenhouse gases (see Chapter 9, Table 9.1)

The biosphere is an important control in defining changes in greenhouse gas concentrations. Its surface characteristics, such as albedo and surface roughness, further influence climate patterns. Biospheric processes, such as CO_2-sequestration and release, evapotranspiration and land-cover change, are in turn affected by climate. For example, warming is expected to result in a poleward expansion of forests (IPCC, 1996b). This would increase biospheric carbon storage, which lowers future CO_2 concentrations and change the surface albedo which would directly affect climate. A detailed discussion of the role of the biosphere on climate can be found elsewhere (Chapters 3 and 7), but there is a clear need for an improved treatment of biospheric responses in scenarios that are designed for regional impact assessment. Some integrated assessment models, which include simplifications of many key biospheric responses, are beginning to provide consistent information of this kind (e.g., Alcamo *et al.*, 1996, 1998; Harvey *et al.*, 1997; Xiao *et al.*, 1997; Goudriaan *et al.*, 1999).

Another important input to impact assessments is sea level rise. AOGCMs usually calculate the thermal expansion of the oceans directly, but this is only one component of sea level rise (see Chapter 11). Complete calculations of sea level rise, including changes in the mass balance of ice sheets and glaciers, can be made with simpler models (e.g., Raper *et al.*, 1996), and the transient dynamics of sea level rise should be explicitly calculated because the responses are delayed (Warrick *et al.*, 1996). However, the current decoupling of important dynamic processes in most simple models could generate undesirable inaccuracies in the resulting scenarios.

Climate scenario generators can comprehensively address some of these inconsistencies. Full consistency, however, can only be attained through the use of fully coupled global models (earth system models) that systematically account for all major processes and their interactions, but these are still under development.

References

Alcamo, J., G.J. van denBorn, A.F. Bouwman, B.J. deHaan, K. Klein Goldewijk, O. Klepper, J. Krabec, R. Leemans, J.G.J. Olivier, A.M.C. Toet, H.J.M. deVries, and H.J. van der Word, 1994: Modeling the global society-biosphere-climate system. *Water Air Soil Pollut.*, **76**, 1-78.

Alcamo, J., A. Bouwman, J. Edmonds, A. Grübler, T. Morita and A. Sugandhy, 1995: An evaluation of the IPCC IS92 emission scenarios. In: *Climate Change 1994: Radiative forcing of climate change and an evaluation of the IPCC IS92 emission scenarios. Reports of Working Groups I and III of the Intergovernmental Panel on Climate Change* [Houghton, J.T., L.G. Meira Filho, J. Bruce, H. Lee, B.A. Callander, E. Haites, N. Harris and K. Maskell (eds.)]. Cambridge University Press, Cambridge, United Kingdom and New York, NY, USA, pp. 247-304.

Alcamo, J., G.J.J. Kreileman, J.C. Bollen, G.J. van den Born, R. Gerlagh, M.S. Krol, A.M.C. Toet and H.J.M. de Vries, 1996: Baseline scenarios of global environmental change. *Global Environ. Change*, **6**, 261-303.

Alcamo, J., R. Leemans, and E. Kreileman (eds.), 1998: *Global Change Scenarios of the 21st Century*. Pergamon & Elseviers Science, London, UK, 296 pp.

Allen, M. R., P. A. Stott, R. Schnur, T. Delworth, and J. F. B. Mitchell, 2000: Uncertainty in forecasts of anthropogenic climate change. Nature, **407**, 617-620.

Anisimov, O.A. and F.E. Nelson, 1996: Permafrost distribution in the northern hemisphere under scenarios of climatic change. *Global Planet. Change*, **14**, 59-72.

Anisimov, O.A. and V.Yu. Poljakov, 1999: On the prediction of the air temperature in the first quarter of the XXI century. *Meteorology and Hydrology*, **2**, 25-31 (in Russian).

Arnell, N.W., 1999: A simple water balance model for the simulation of streamflow over a large geographic domain. *J. Hydrol.*, **217**, 314-335.

Arnell, N.W., R.P.C. Brown, and N.S. Reynard, 1990: *Impact of Climatic Variability and Change on River Flow Regimes in the UK*. IH Report No. 107, Institute of Hydrology, Wallingford, UK, 154 pp.

Barrow, E. M., and M. A. Semenov, 1995: Climate change scenarios with high resolution and temporal resolution and agricultural applications. *Forestry*, **68,** 349-360.

Bates, B.C., S.P. Charles, N.R. Sumner, and P.M. Fleming, 1994: Climate change and its hydrological implications for South Australia. *T. Roy. Soc. South. Aust.* **118(1)**, 35-43.

Bates, B.C., A.J. Jakeman, S.P. Charles, N.R. Sumner, and P.M. Fleming, 1996: Impact of climate change on Australia's surface water resources. In: *Greenhouse: Coping with Climate Change* [Bouma, W.J., G.I. Pearman, and M.R. Manning (eds.)]. CSIRO, Melbourne, pp. 248-262.

Bengtsson, L., M. Botzet, and M. Esch, 1996: Will greenhouse gas induced warming over the next 50 years lead to higher frequency and greater intensity of hurricanes? *Tellus*, **48A**, 57-73.

Beniston, M. and M. Price, 1992: Climate scenarios for the Alpine regions: a collaborative effort between the Swiss National Climate Program and the International Center for Alpine Environments. *Env. Conserv.*, **19**, 360-363.

Bergthórsson, P., H. Björnsson, O. Dórmundsson, B. Gudmundsson, A. Helgadóttir, and J.V. Jónmundsson, 1988: The effects of climatic variations on agriculture in Iceland. In: *The Impact of Climatic*

Variations on Agriculture, Volume 1, Assessments in Cool Temperate and Cold Regions [Parry, M.L., T.R. Carter, and N.T. Konijn (eds.)]. Kluwer, Dordrecht, The Netherlands, pp. 381-509.

Boer, G.J., G.M. Flato, and D. Ramsden, 2000: A transient climate change simulation with greenhouse gas and aerosol forcing: projected climate to the twenty-first century. *Clim. Dyn.*, **16(6)**, 427-450.

Brown, R.A., N. J. Rosenberg, W.E. Easterling, C. Hays, and L. O. Mearns, 2000: Potential production and environmental effects of switchgrass and traditional crops under current and greenhouse-altered climate in the MINK region of the central United States. *Ecol. Agric. Environ.*, **78**, 31-47.

Budyko, M.I., 1989: Empirical estimates of imminent climatic changes. *Soviet Meteorology and Hydrology,* **10**, 1-8.

Cane, M.A., G. Eshel, and R.W. Buckland, 1994: Forecasting Zimbabwean maize yield using eastern equatorial Pacific sea surface temperature. *Nature*, **370**, 204-205.

Carter, T., M. Posch, and H. Tuomenvirta, 1995: SILMUSCEN and CLIGEN user's guide. Guidelines for the construction of climatic scenarios and use of a stochastic weather generator in the Finnish Research Programme on Climate Change (SILMU). *Publications of the Academy of Finland, 1/95*, 62 pp.

Carter, T.R., M. Posch, and H. Tuomenvirta, 1996a: The SILMU scenarios: specifying Finland's future climate for use in impact assessment. *Geophysica*, **32**, 235-260.

Carter, T.R., M. Parry, S. Nishioka, and H. Harasawa, 1996b: Technical Guidelines for assessing climate change impacts and adaptations. In: *Climate Change 1995. Impacts, Adaptations and Mitigation of Climate Change: Scientific-Technical Analysis. Contribution of Working Group II to the Second Assessment Report of the Intergovernmental Panel on Climate Change* [Watson, R.T., M.C. Zinyowere and R.H. Moss (eds.)]. Cambridge University Press, Cambridge, United Kingdom and New York, NY, USA, pp. 823-833.

Carter, T.R., R.A. Saarikko, and S.K.H. Joukainen, 2000a: Modelling climate change impacts on wheat and potatoes in Finland. In: *Climate Change, Climate Variability and Agriculture in Europe: An Integrated Assessment* [Downing, T.E., P.E. Harrison, R.E. Butterfield and K.G. Lonsdale (eds.)]. Research Report No. 21, ECU, University of Oxford, Oxford, UK, pp. 289-312.

Carter, T.R., M. Hulme, J.F. Crossley, S. Malyshev, M.G. New, M.E. Schlesinger and H. Tuomenvirta, 2000b: *Climate change in the 21st century: interim characterizations based on the new IPCC SRES Emissions Scenarios.* Finnish Environment Institute Report, No. 433, Helsinki, Finland, 148pp.

Carter, T.R. and E.L. La Rovere, 2001: Developing and Applying Scenarios. In *Climate Change: Impacts, Adaptation and Vulnerability. Contribution of Working Group II to the IPCC Third Assessment Report.*

Chiew, F.H.S., P.H. Whetton, T.A. McMahon, and A.B. Pittock, 1995: Simulation of the impacts of climate change on runoff and soil moisture in Australian catchments. *J. Hydrol.*, **167**, 121-147.

Chiew, F.H.S., T.C. Piechota, J.A. Dracup, and T.A. McMahon, 1998: El Niño Southern Oscillation and Australian rainfall, streamflow and drought—links and potential for forecasting. *J. Hydrol.*, **204**, 138-149.

Climate Impact Assessment Program (CIAP), 1975: *Impacts of Climatic Change on the Biosphere.* CIAP Monograph 5, Part 2 - Climate Effects. Department of Transportation, CIAP, Washington, D.C.

Cohen, S. J., 1990: Bringing the global warming issue closer to home: the challenge of regional impact studies. *Bull. Am. Met. Soc.*, **71**, 520-526.

Covey, C., 1995: Using paleoclimates to predict future climate: how far can analogy go? *Clim. Change*, **29**, 403-407.

Crowley, T.J., 1990: Are there any satisfactory geologic analogs for a future greenhouse warming? *J. Climate*, **3**, 1282-1292.

CSIRO, 1996: OzClim: A climate scenario generator and impacts package for Australia. CSIRO, Canberra, November 1996, 4pp. [Available on line from http://www.dar.csiro.au/publications/ozclim.htm]

Cubasch, U., K. Hasselmann, H. Hock, E. Maier-Reimer, U. Mikolajewicz, B.D. Santer, and R. Sausen, 1992: Time-dependent greenhouse warming computations with a coupled ocean-atmosphere model. *Clim. Dyn.*, **8**, 55-69.

Cubasch, U., B.D. Santer, A. Hellbach, G. Hegerl, H. Höck, E. Maier-Reimer, U. Mikolajewicz, A. Stössel, and R. Voss, 1994: Monte Carlo climate change forecasts with a global coupled ocean-atmosphere model. *Clim. Dyn.*, **10**, 1-19.

Cubasch, U., J. Waszkewitz, G. Hegerl, and J. Perlwitz, 1995: Regional climate changes as simulated in time-slice experiments. *Clim. Change*, **31**, 273-304.

Daly, C., R.P. Neilson, and D.L. Phillips, 1994: A statistical-topographic model for maping climatological precipitation over mountainous terrain. *J. Appl. Met.*, **33**, 140-158.

Downing, T.E., P.A. Harrison, R.E. Butterfield, and K.G. Lonsdale (eds.), 2000: *Climate Change, Climatic Variability and Agriculture in Europe: An Integrated Assessment.* Research Report No. 21, Environmental Change Unit, University of Oxford, Oxford, UK, 445 pp.

Emanuel, W.R., H.H. Shugart, and M.P. Stevenson, 1985: Climatic change and the broad-scale distribution of terrestrial ecosystem complexes. *Clim. Change*, **7**, 29-43.

Fagan, B. 1999: *Floods, Famines and Emperors; El Nino and the Fate of Civilizations.* Basic Books, NY, 284 pp.

Fisher, R. W., 1996: Future Energy Use. *Future Research Quarterly*, **31**, 43-47.

Fowler, A., 1999: Potential climate change impacts on water resources in the Auckland Region (New Zealand). *Clim. Res.*, **11**, 221-245.

Fox-Rabinovitz, M. S., G. Stenchikov, and L.L.Takacs, 1997: A finite-difference GCM dynamical core with a variable resolution stretched grid. *Mon. Wea. Rev.*, **125**, 2943-2961.

Frei, C. and C. Schär, 1998: A precipitation climatology of the Alps from high-resolution rain-gauge observations. *Int. J. Climatol.*, **18**, 873-900.

Frenzel, B., B. Pecsi, and A.A. Velichko (eds.), 1992: *Atlas of Palaeoclimates and Palaeoenvironments of the Northern Hemisphere.* INQUA/Hungarian Academy of Sciences, Budapest.

Friend, A.D., 1998: Parameterisation of a global daily weather generator for terrestrial ecosystem modelling. *Ecol. Modelling*, **109**, 121-140.

Gates, W. L., 1985: The use of general circulation models in the analysis of the ecosystem impacts of climatic change. *Clim. Change*, **7**, 267-284.

Giorgi, F. and L.O. Mearns, 1991: Approaches to regional climate change simulation: A review. *Rev. Geophys.*, **29**, 191-216.

Giorgi, F. and L.O. Mearns, 1999: Regional climate modeling revisited: an introduction to the special issue. *J. Geophys. Res.*, **104(D6)**, 6335-6352.

Giorgi, F. and R. Francisco, 2000: Uncertainties in regional climate change prediction: a regional analysis of ensemble simulations with HadCM2 coupled AOGCM. *Clim. Dyn.*, **16**, 169-182.

Giorgi, F., L. Mearns, S. Shields, and L. McDaniel, 1998: Regional nested model simulations of present day and 2xCO$_2$ climate over the Central Great Plains of the United States. *Clim. Change*, **40**, 457-493.

Glantz, M., 1988: *Societal Responses to Regional Climatic Change: Forecasting by Analogy.* Westview Press, Boulder, Colorado, 77 pp.

Gleick, P.H., 1986: Methods for evaluating the regional hydrologic impacts of global climatic changes. *J. Hydrol.*, **88**, 97-116.

Goudriaan, J., H.H. Shugart, H. Bugmann, W. Cramer, A. Bondeau, R.H. Gardner, L.A. Hunt, W.K. Lauwenroth, J.J. Landberg, S. Linder, I.R. Noble, W. J. Parton, L.F. Pitelka, M. Staford Smith, R.W.

Sutherst, C. Valentin and F.I. Woodward, 1999: Use of models in global change studies. In: *The Terrestrial Biosphere and Global Change: Implications for Natural and Managed Ecosystems* [Walker, B., W. Steffen, J. Canadell and J. Ingram (eds.)]. Cambridge University Press, Cambridge, pp. 106-140.

Gregory, J.M. and J.F.B. Mitchell, 1995: Simulation of daily variability of surface temperature and precipitation over Europe in the current and 2xCO2 climates using the UKMO climate model. *Quart. J. R. Met. Soc.*, **121**, 1451-1476.

Gregory, J.M., J.F.B. Mitchell, and A.J. Brady, 1997: Summer drought in northern midlatitudes in a time-dependent CO2 climate experiment. *J. Climate*, **10**, 662-686.

Gyalistras, D. and A. Fischlin, 1995. Downscaling: applications to ecosystems modelling. In: *Proceedings of the 6th International Meeting on Statistical Climatology, Galway, Ireland, June 19-23, 1995* [Muirterchaigh, I. (ed.)]. University College, Galway, pp. 189-192.

Gyalistras, D., C. Schär, H.C. Davies, and H. Wanner, 1997: Future alpine climate. In: *Views from the Alps: Regional Perspectives on Climate Change* [Cebon, P., U. Dahinden, H. Davies, D.M. Imboden, and C.C. Jaeger (eds.)]. The MIT Press, Cambridge, MA, pp. 171-223.

Hammond, A., 1996: *Which World?* Island Press, Washington, D.C., 306 pp.

Hansen, J., Il Fung, A. Lacis, D. Rind, G. Russell, S. Lebedeff, R. Ruedy, and P. Stone, 1988: Global climate changes as forecast by the GISS 3-D model. *J. Geophys. Res.*, **93(D8),** 9341-9364.

Hansen, J.W., A.W. Hodges, and J.W. Jones, 1998: ENSO influences on agriculture in the southeastern US. *J. Climate*, **11**, 404-411.

Harvey, D., J. Gregory, M. Hoffert, A. Jain, M. Lal, R. Leemans, S. Raper, T. Wigley and J. de Wolde, 1997: *An Introduction to Simple Climate Models Used in the IPCC Second Assessment Report.* Intergovernmental Panel on Climate Change, Geneva, February 1977, IPCC Technical Paper II, 50 pp.

Hassall and Associates, 1998: *Climate Change Scenarios and Managing the Scarce Water Resources of the Macquarie River.* Australian Greenhouse Office, Canberra, 113 pp.

Hay, E.L., R. B. Wilby, and G.H. Leavesy, 1999: A comparison of delta change and downscaled GCM scenarios: implications for climate change scenarios in three mountainous basins in the United States. In: *Proceedings of the AWRA Specialty conference on Potential Consequences of Climatic Variability and Change to Water Resources of the United States, May, 1999, Atlanta, GA* [Adams, D.B. (ed.)]. AWRA, Middleburg, VA, 424 pp.

Henderson-Sellers, A., 1996: Can we integrate climatic modelling and assessment? *Environ. Modeling & Assessment*, **1**, 59-70.

Henderson-Sellers, A., H. Zhang, G. Berz, K. Emanuel, W. Gray, C. Landsea, G. Holland, J. Lighthill, S.L. Shieh, P. Webster, and K. McGuffie, 1998: Tropical cyclones and global climate change: A post-IPCC assessment. *Bull. Am. Met. Soc.*, **79**, 19-38.

Hennessy, K.J., P.H. Whetton, J.J. Katzfey, J.L. McGregor, R.N. Jones, C.M. Page, and K.C. Nguyen, 1998: *Fine Resolution Climate Change Scenarios for New South Wales.* Annual Report, 1997/98, CSIRO, Australia, 48 pp.

Holten, J.I. and P.D. Carey, 1992: *Responses of Climate Change on Natural Terrestrial Ecosystems in Norway.* NINA Forskningsrapport 29, Norwegian Institute for Nature Research, Trondheim, Norway, 59 pp.

Hostetler, S., 1994: Hydrologic and atmospheric models: the problem of discordant scales: an editorial comment. *Clim. Change*, **27**, 345-350.

Howe, W. and A. Henderson-Sellers (eds.), 1997: *Assessing Climate Change. Results from the Model Evaluation Consortium for Climate Assessment.* Gordon Breach Science Publishers, Amsterdam, 418 pp.

Hulme, M. and M. New, 1997: The dependence of large-scale precipitation climatologies on temporal and spatial gauge sampling. *J. Climate*, **10**, 1099-1113.

Hulme, M. and G.J. Jenkins, 1998: *Climate change scenarios for the United Kingdom.* UKCIP Technical Note No.1, Climatic Research Unit, Norwich, UK, 80 pp.

Hulme, M. and T.R. Carter, 2000: The changing climate of Europe. Chapter 3 in: *Assessment of the Potential Effects of Climate Change in Europe* [Parry, M.L. (ed.)]. Report of the ACACIA Concerted Action, November 2000, UEA, Norwick, UK, 350 pp.

Hulme, M., T. Jiang, and T.M.L. Wigley, 1995a: *SCENGEN: A Climate Change SCENario GENerator, Software User Manual, Version 1.0.* Climatic Research Unit, University of East Anglia, Norwich, 38 pp.

Hulme, M., S.C.B. Raper, and T.M.L. Wigley, 1995b: An integrated framework to address climate change (ESCAPE) and further developments of the global and regional climate modules (MAGICC). *Energy Policy*, **23**, 347-355.

Hulme, M., E.M. Barrow, N. Arnell, P.A. Harrison, T.E. Downing, and T.C. Johns, 1999a: Relative impacts of human-induced climate change and natural climate variability. *Nature*, **397**, 688-691.

Hulme, M., J.F.B. Mitchell, W. Ingram, T.C. Johns, J.A. Lowe, M.G. New and D. Viner, 1999b: Climate change scenarios for global impacts studies. *Global Environ. Change,* **9**, S3-S19.

Hutchinson, M.F., 1995: Interpolating mean rainfall using thin-plate smoothing splines. *Int. J. Geogr. Inf. Systems*, **9**, 385-403.

IPCC, 1992: *Preliminary Guidelines for Assessing Impacts of Climate Change. Prepared by Intergovernmental Panel on Climate Change,* Working Group II for WMO and UNEP [Carter, T.R., M.L. Parry, S. Nishioka, and H. Harasawa (eds.)]. Cambridge University Press, Cambridge, United Kingdom.

IPCC, 1994: *IPCC Technical Guidelines for Assessing Climate Change Impacts and Adaptations.* Prepared by Working Group II [Carter, T.R., M.L. Parry, H.Harasawa, and S. Nishioka] and WMO/UNEP. CGER-IO15-'94. University College London, UK and Center for Global Environmental Research, National Institute for Environmental Studies, Tsukuba, Japan, 59 pp.

IPCC, 1996a: *Climate Change 1995: The Science of Climate Change.* Contribution of Working Group I to the Second Assessment Report of the Intergovernmental Panel on Climate Change [Houghton, J.T., L.G. Meira Filho, B.A. Callander, N. Harris, A. Kattenberg, and K. Maskell (eds.)]. Cambridge University Press, Cambridge, United Kingdom and New York, NY, USA, 572 pp.

IPCC, 1996b: *Climate Change 1995: Impacts, Adaptations and Mitigation of Climate Change: Scientific-Technical Analyses.* Contribution of Working Group II to the Second Assessment Report of the Intergovernmental Panel on Climate Change [Watson, R.T., M.C. Zinyowera, and R.H. Moss (eds.)]. Cambridge University Press, Cambridge, United Kingdom and New York, NY, USA, 880 pp.

IPCC, 1997: *An introduction to simple climate models used in the IPCC Second Assessment Report* [Harvey, L.D., J. Gregory, M.Hoffert, A. Jain, M. Lal, R. Leemans, S. Raper, T. Wigley and J. de Wolde]. IPCC Technical Paper 2, Intergovernmental Panel on Climate Change, Geneva, Switzerland, 50 pp.

IPCC, 2001: Climate Change: Impacts, Adaptation and Vulnerability. Contribution of Working Group II to the Third Assessment Report of the Intergovernmental Panel on Climate Change.

IPCC-TGCIA, 1999: *Guidelines on the Use of Scenario Data for Climate Impact and Adaptation Assessment.* Version 1. Prepared by Carter, T.R., M. Hulme and M. Lal, Intergovernmental Panel on Climate Change, Task Group on Scenarios for Climate Impact Assessment. Download from: http://ipcc-ddc.cru.uea.ac.uk/cru_data/support.html

Jendritzky, G. and B. Tinz, 2000: Human bioclimate maps for climate impact research. In: *Proceedings of Int. Conf. Biomet and Int. Conf. Urban Climate.* ICB-ICUC '99, Sidney, 8-12 November 1999 (in press).

Jones, R.N., 1998: Climate change scenarios, impact thresholds and risk. In: *Proceedings of the Workshop on Impacts of Global Change on Australian Temperate Forests* [Gorman, J. and S.M. Howden (eds.)]. February 1998, CSIRO Wildlife and Ecology, Canberra, Australia, pp. 14-28.

Jones, R.N., 1999: Climate change scenarios, impact thresholds and risk. In: *Impacts of Global Change on Australian Temperate Forests* [Howden, S.M. and J. Gorman (eds.)]. Working Paper Series 99/08, Resource Future Program, CSIRO Wildlife and Ecology, Canberra, Australia, pp. 40-52.

Jones, R.N., 2000a: Managing uncertainty in climate change projections-issues for impact assessment. *Clim. Change,* **45(3/4)**, 403-419.

Jones, R.N., 2000b: Analyzing the risk of climate change using an irrigation demand model. *Clim. Res.*, **14**, 89-100.

Jones, P.D., K.R. Briffa, T.P. Barnett, and S.F.B. Tett, 1998: High-resolution palaeoclimatic records for the last millennium: interpretation, integration and comparison with GCM control-run temperatures. *The Holocene*, **8**, 455-471.

Jones, R.N., K.J. Hennessy, and D.J. Abbs, 1999: *Climate Change Analysis Relevant to Jabiluka*, Attachment C. Assessment of the Jabiluka Project, Report of the Supervising Scientist to the World Heritage Committee. Environment Australia, Canberra.

Jones, R.N., K.J. Hennessy, C.M. Page, A.B. Pittock, R. Suppiah, K.J.E. Walsh, and P.H. Whetton, 2000: *An analysis of the effects of the Kyoto Protocol on Pacific Island countries. Part 2: regional climate change scenarios and risk assessment methods.* South Pacific Regional Environment Programme, Apia, Western Samoa (in press).

Kaiser, H.M., S.J. Riha, D.S. Wilks, D.G. Rossiter, and R. Sampath, 1993: A farm-level analysis of economic and agronomic impacts of gradual climate warming. *Am. J. Ag. Econ.*, **75**, 387-398.

Kalkstein, L.S. and J.S. Greene, 1997: An evaluation of climate/mortality relationships in large U.S. cities and possible impacts of a climate change. *Environmental Health Perspectives*, **105**, 84-93.

Kalnay, E., M. Kanamitsu, R. Kistler, W. Collins, D. Deaven, L. Gandin, M. Iredell, S. Saha, G. White, J. Woollen, Y. Zhu, A. Leetmaa, R. Reynolds, M. Chelliah, W. Ebisuzaki, W. Higgins, J. Janowiak, K. Mo, C. Ropelewski, J. Wang, R. Jenne, and D. Joseph, 1996: The NCEP/NCAR 40-year reanalysis project. *Bull. Am. Met. Soc.*, **77**, 437-472.

Karl, T.R., R.W. Knight, D.R. Easterling, and R.G. Quayle, 1996: Indices of climate change for the United States. *Bull. Am. Met. Soc.*, **77**, 279-292.

Katz, R.W., 1996: The use of stochastic models to generate climate change scenarios. *Clim. Change*, **32**, 237-255.

Katz, R.W., 2000: Techniques for estimating uncertainty in climate change scenarios and impact studies. *Clim. Res.* (accepted).

Katz, R.W. and B.G. Brown, 1992: Extreme events in a changing climate: variability is more important than averages. *Clim. Change*, **21**, 289-302.

Katz, R.W., and M.B. Parlange, 1996: Mixtures of stochastic processes: Application to statistical downscaling. *Clim. Res.*, **7**, 185-193.

Kellogg, W.W. and R. Schware, 1981: *Climate Change and Society: Consequences of Increasing Atmospheric Carbon Dioxide.* Westview Press, Boulder, Colorado, 178 pp.

Kenny, G.J., R.A. Warrick, B.D. Campbell, G.C. Sims, M. Camilleri, P.D. Jamieson, N.D. Mitchell, H.G. McPherson, and M.J. Salinger, 2000: Investigating climate change impacts and thresholds: an application of the CLIMPACTS integrated assessment model for New Zealand agriculture. *Clim. Change*, **46(1/2)**, 91-113.

Kilsby, C.G., P.E. O'Connell, and C.S. Fallows, 1998: Producing rainfall scenarios for hydrological impact modelling. In: *Hydrology in a Changing Environment* [Wheater, H. and C. Kirby (eds.)]. Wiley, Chichester, volume 1, pp. 33-42.

Kittel, T.G.F., N.A. Rosenbloom, T.H. Painter, D.S. Schimel, and VEMAP Modeling Participants, 1995: The VEMAP integrated database for modeling United States ecosystem/vegetation sensitivity to climate change. *J. Biogeography*, **22**, 857-862.

Kittel, T.G.F., J.A. Royle, C. Daly, N.A. Rosenbloom, W.P. Gibson, H.H. Fisher, D.S. Schimel, L.M. Berliner, and VEMAP2 Participants, 1997: A gridded historical (1895-1993) bioclimate dataset for the conterminous United States. In: *Proceedings of the 10th Conference on Applied Climatology*, 20-24 October 1997, Reno, NV, American Meteorological Society, Boston, pp. 219-222.

Kittel, T.G.F., F. Giorgi, and G.A.Meehl, 1998: Intercomparison of regional biases and doubled CO_2-sensitivity of coupled atmosphere-ocean general circulation model experiments. *Clim. Dyn.*, **14**, 1-15.

Knutson, T.R. and R.E. Tuleya, 1999: Increased hurricane intensities with CO2-induced warming as simulated using the GFDL hurricane prediction system. *Clim. Dyn.*, **15(7)**, 503-519.

Knutson, T.R., S. Manabe, and D. Gu, 1997: Simulated ENSO in a global coupled ocean-atmosphere model: Multidecadal amplitude modulation and CO2 sensitivity. *J. Climate*, **10**, 138-161.

Konrad, C.E., 1998: An empirical approach for delineating fine scaled spatial patterns of freezing rain in the Appalachian region of the USA. *Clim. Res.*, **10**, 217-227.

Kothavala, Z., 1999: Extreme precipitation events and the applicability of global climate models to study floods and droughts. *Math. and Comp. in Simulation*, **43**, 261-268.

Krishnamurti, T.N., R. Correa-Torres, M. Latif, and G. Daughenbaugh, 1998: The impact of current and possibly future sea surface temperature anomalies on the frequency of Atlantic hurricanes. *Tellus*, **50A**, 186-210.

Kuikka, S. and O. Varis, 1997: Uncertainties of climatic change impacts in Finnish watersheds: a Bayesian network analysis of expert knowledge. *Boreal Environment Research*, **2**, 109-128.

Lal, M. and F. Giorgi, 1997: *Regional Climate Modelling and Applications.* Paper presented at the World Climate Research Programme: achievements, benefits and challenges, Geneva, Switzerland, 26-28 August, 1997, WMO, Geneva.

Lal, M. and H. Harasawa, 2000: Comparison of the present-day climate simulation over Asia in selected coupled atmosphere-ocean global climate models. *J.Met.Soc.Japan* (in press).

Lal, M., G.A. Meehl and J.M. Arblaster, 2000: Simulation of Indian summer monsoon rainfall and its intraseasonal variability. *Regional Environmental Change* (in press).

Lamb, P., 1987: On the development of regional climatic scenarios for policy oriented climatic impact assessment. *Bull. Am. Met. Soc.*, **68**, 1116-1123.

Lapin, M., E. Nieplová, and P Fa?ko, 1995: Regional scenarios of temperature and precipitation changes for Slovakia (In Slovak with English abstract). *Monographs of the Slovak National Climate Program*, **3**, 17-57.

Leemans, R. and W. Cramer, 1991: *The IIASA Database for Mean Monthly Values of Temperature, Precipitation and Cloudiness on a Global Terrestrial Grid.* IIASA Report RR-91-18, Laxenburg, 63 pp.

Leemans, R. and A.M. Solomon, 1993: Modeling the potential change in yield and distribution of the earth's crops under a warmed climate. *Clim. Res.*, **3**, 79-96.

Leemans, R., E. Kreileman, G. Zuidema, J. Alcamo, M. Berk, G.J. van den Born, M. den Elzen, R. Hootsmans, M. Janssen, M. Schaeffer, A.M.C. Toet and H.J.M. de Vries, 1998: *The IMAGE User Support System: Global Change Scenarios from IMAGE 2.1.* National Institute of Public Health and the Environment, Bilthoven. October 1998. RIVM Publication, CD-rom.

Legates, D.R. and C.J. Willmott, 1990: Mean seasonal and spatial variability in gauge-corrected, global precipitation. *Int. J. Climatol.*, **10**, 111-128

Leggett, J., W.J. Pepper and R.J. Swart, 1992. Emissions Scenarios for the IPCC: an Update. In: *Climate Change 1992. The Supplementary Report to the IPCC Scientific Assessment* [Houghton, J.T., B.A.

Callander and S.K. Varney (eds.)]. Cambridge University Press, Cambridge, UK and New York, NY, USA, pp. 71-95.

Liverman, D., M. Dilley, and K. O'Brien, 1994: Possible impacts of climate change on maize yields in Mexico. In: *Implications of Climate Change for International Agriculture: Crop Modeling Study*. EPA Report 230-B-94-003, United States Environmental Protection Agency, Policy, Planning and Evaluation, Climate Change Division, Washington, DC.

Lough, J.M., T.M.L. Wigley, and J.P. Palutikof, 1983: Climate and climate impact scenarios for Europe in a warmer world. *J. Clim. Appl. Met.*, **22**, 1673-1684.

Magalhães, A.R. and M.H. Glantz, 1992: *Socioeconomic Impacts of Climate Variations and Policy Responses in Brazil*. United Nations Environment Programme/SEPLAN-CE/Esquel Foundation, Brasilia, Brazil.

Mahlman, J.D., 1997: Uncertainties in projections of human-caused climate warming. *Science*, **278**, 1416-1417 (in Policy Forum).

Manabe, W. and R.T. Wetherald, 1987: Large-scale changes of soil wetness induced by an increase in atmospheric carbon dioxide. *J. Atmos. Sci.*, **44**, 1211-1235.

Mavromatis, T. and P.D. Jones, 1999: Evaluation of HadCM2 and direct use of daily GCM data in impact assessment studies. *Clim. Change*, **41**, 583-614.

McGregor, J.J., 1997: Regional climate modeling. *Meteorological Atmospheric Physics*, **63**, 105-117.

McInnes, K.L., K.J.E. Walsh, and A.B. Pittock, 2000: *Impact of sea level rise and storm surges on coastal resorts*. A report for CSIRO Tourism Research: final report. CSIRO Atmospheric Research, Aspendale, Vic., 13 pp.

Mearns, L. O., 1995: Research issues in determining the effects of changing climatic variability on crop yields. In: *Climate Change and Agriculture: Analysis of Potential International Impacts* [Rosenzweig, C. (ed.)]. American Society of Agronomy Special Publication No. 58, Madison, ASA, Chapter 6, pp. 123-146.

Mearns, L. O., 1999: The effect of spatial and temporal resolution of climate change scenarios on changes in frequencies of temperature and precipitation extremes. In: *Elements of Change 1998* [Hassol, S.J. and J. Katzenberger (eds.)]. Aspen Global Change Institute, Aspen, CO, pp. 316-323.

Mearns, L.O., C. Rosenzweig, and R. Goldberg, 1992: Effect of changes in interannual climatic variability on CERES-Wheat yields: sensitivity and 2×CO$_2$ general circulation model studies. *Agric. For. Meteorol.*, **62**, 159-189.

Mearns, L.O., F. Giorgi, C. Shields, and L. McDaniel, 1995a: Analysis of the variability of daily precipitation in a nested modeling experiment: comparison with observations and 2xCO$_2$ results. *Global Planet. Change*, **10**, 55-78.

Mearns, L.O., F. Giorgi, C. Shields, and L. McDaniel, 1995b: Analysis of the diurnal range and variability of daily temperature in a nested model experiment: comparison with observations and 2xCO2 results. *Clim. Dyn.*, **11**, 193-209.

Mearns, L. O., C. Rosenzweig, and R. Goldberg, 1996: The effect of changes in daily and interannual climatic variability on CERES-wheat: a sensitivity study. *Clim. Change*, **32,** 257-292.

Mearns, L. O., C. Rosenzweig, and R. Goldberg, 1997: Mean and variance change in climate scenarios: methods, agricultural applications, and measures of uncertainty. *Clim. Change, ***35,** 367-396.

Mearns, L. O., W. Easterling, and C. Hays 1998: *The Effect of Spatial Scale of Climate Change Scenarios on the Determination of Impacts: An Example of Agricultural Impacts on the Great Plains*. Proceedings of the International Workshop on Regional Modeling of the General Monsoon System in Asia, Beijing, October 20—23, START Regional Center for Temperate East Asia, TEACOM Report No. 4, pp. 70-73.

Mearns, L.O., T. Mavromatis, E. Tsvetsinskaya, C. Hays, and W.

Easterling, 1999: Comparative responses of EPIC and CERES crop models to high and low resolution climate change scenario. *J. Geophys. Res.*, **104(D6)**, 6623-6646.

Mearns, L.O., G. Carbone, W. Gao, L. McDaniel, E. Tsvetsinskaya, B. McCarl, and R. Adams, 2000a: The issue of spatial scale in integrated assessments: An example of agriculture in the Southeastern U.S. In: *Preprints of the American Meteorological Society of America Annual Meeting*, 9-14 January, Long Beach, CA. AMS, Boston, MA, pp. 38-41.

Mearns, L.O., G. Carbone, L. McDaniel, W. Gao, B. McCarl, R. Adams, and W. Easterling, 2000b: The importance of spatial scale of climate scenarios for regional climate change impacts analysis: implications for regional climate modeling activities. In: *Preprints of the Tenth PSU/NCAR Mesoscale Model Users' Workshop*, June 21-22 2000, National Center for Atmospheric Research, Mesoscale and Microscale Meteorology Division, Boulder, Colorado, pp. 127-130.

Mearns, L. O., W. Easterling, and C. Hays, 2001: Comparison of agricultural impacts of climate change calculated from high and low resolution climate model scenarios: part I: the uncertainty due to spatial scale. *Clim. Change* (in press).

Mitchell, J.F.B., 1990: Greenhouse warming: is the mid-Holocene a good analogue? *J. Climate*, **3**, 1177-1192.

Mitchell, J.F.B., S. Manabe, V. Meleshko, and T. Tokioka, 1990: Equilibrium climate change - and its implications for the future. In: *Climate Change: The IPCC Scientific Assessment. Report prepared by Working Group I* [Houghton, J.T., G.J. Jenkins, and J.J. Ephraums (eds.)]. Cambridge University Press, Cambridge, United Kingdom, and New York, NY, USA, pp. 131-164.

Mitchell, J.F.B., T.C. Johns, M. Eagles, W.J. Ingram, and R.A. Davis, 1999: Towards the construction of climate change scenarios. *Clim. Change*, **41**, 547-581.

Morgan, M.G. and M. Henrion, 1990: *Uncertainty: A Guide to Dealing with Uncertainty in Quantitative Risk and Policy Analysis*. Cambridge University Press, Cambridge, UK.

Morgan, M.G. and D. Keith, 1995: Subjective judgements by climate experts. *Envir. Sci. Tech.*, **29**, 468-476.

Morgan, M.G. and H. Dowlatabadi, 1996: Learning from integrated assessment of climate change. *Clim. Change*, **34**, 337-368.

Morita, T. and J. Robinson, 2001: Greenhouse Gas Emission Mitigation Scenarios and Implications. In *Climate Change 2001: Mitigation*. Contribution of Working Group III to the IPCC Third Assessment Report.

Murphy, J.M., 1999: An evaluation of statistical and dynamical techniques for downscaling local climate. *J. Climate*, **12(8)**, 2256-2284.

Nakićenović, N., J. Alcamo, G. Davis, B. de Vries, J. Fenhann, S. Gaffin, K. Gregory, A. Grübler, T. Y. Jung, T. Kram, E. L. La Rovere, L. Michaelis, S. Mori, T. Morita, W. Pepper, H. Pitcher, L. Price, K. Raihi, A. Roehrl, H-H. Rogner, A. Sankovski, M. Schlesinger, P. Shukla, S. Smith, R. Swart, S. van Rooijen, N. Victor, Z. Dadi, 2000: IPCC Special Report on Emissions Scenarios, Cambridge University Press, Cambridge, United Kingdom and New York, NY, USA, 599 pp.

NDU, 1978: *Climate Change to the Year 2000*. Dept. of Defense, National Defense University, US Government Printing Office, Washington, D.C., 109 pp.

Neilson, R.P., I.C. Prentice, B. Smith, 1997: Simulated changes in vegetation distribution under global warming. In: *The Regional Impacts of Climate Change: An Assessment of Vulnerability. A Special Report of IPCC Working Group II* [Watson, R.T., M.C. Zinyowera, and R.H. Moss (eds.)]. Cambridge University Press, Cambridge, United Kingdom and New York, NY, USA, pp. 439-456.

New, M., 1999: Uncertainty in representing the observed climate. In: *Proceedings of the ECLAT-2 Helsinki Workshop, 14-16 April, 1999* [Carter, T.R., M. Hulme, and D. Viner (eds.)]. Climatic Research

Unit, Norwich, UK, pp. 59-66.

New, M. and M.Hulme, 2000: Representing uncertainties in climate change scenarios: a Monte Carlo approach. *Integrated Assessment,* **1**, 203-213.

New, M., M. Hulme, M. and P.D. Jones, 1999: Representing twentieth century space-time climate variability. Part 1: development of a 1961-90 mean monthly terrestrial climatology. *J. Climate,* **12**, 829-856.

New, M., M. Hulme, and P.D. Jones, 2000: Representing twentieth century space-time climate variability. Part 2: development of 1901-96 monthly grids of terrestrial surface climate. *J. Climate,* **13(13)**, 2217-2238.

Oglesby, R.J. and B. Saltzman, 1992: Equilibrium climate statistics of a general circulation model as a function of atmospheric carbon dioxide, part I. *J. Climate,* **5**, 66-92.

Osborn, T.J., 1996: Comment on Climate drift in a global ocean general circulation model. *J. Phys. Oceanogr.,* **26**, 166a-1663.

Osborn, T.J., 1997: Areal and point precipitation intensity changes: Implications for the application of climate models. *Geopohys. Res. Lett.,* **24**, 2829-2832.

Osborn, T.J., K.R. Briffa, S.F.B. Tett, P.D. Jones, and R.M. Trigo, 2000: Evaluation of the North Atlantic Oscillation as simulated by a coupled climate model. *Clim. Dyn.,* **15(9)**, 685-702.

Parkinson, S.D. and P.C. Young, 1998: Uncertainty and sensitivity in global carbon cycle monitoring. *Clim. Res.,* **9**, 157–174.

Parry, M.L. and M. Livermore (eds.), 1999: A new assessment of the global effects of climate change. *Global Environ.l Change,* **9**, S1-S107, Supplementary Issue.

Parry, M.L., T.R. Carter, and M. Hulme, 1996: What is a dangerous climate change? *Global Environ. Change,* **6**, 1-6.

Parry, M.L., T.R. Carter, and N.T. Konijn (eds.), 1988*: The Impact of Climatic Variations on Agriculture. Volume 1. Assessments in Cool Temperate and Cold Regions.* Kluwer, Dordrecht, The Netherlands, 876 pp.

Piper, S.C. and E.F. Stewart, 1996: A gridded global data set of daily temperature and precipitation for terrestrial biospheric modelling. *Global Biogeochem. Cycles,* **10**, 757-782.

Pittock, A.B., 1989: Book review: Societal Responses to Regional Climatic Change [Glantz, M.] *Bull. Am. Met. Soc.,* **70**, 1150-1152.

Pittock, A.B., 1993: Climate scenario development. In: *Modelling Change in Environmental Systems* [Jakeman, A.J., M.B. Beck, and M.J. McAleer (eds.)]. John Wiley, Chichester, New York, pp. 481-503.

Pittock, A.B., 1995: Comments on climate change scenario development. *Math. Comput. Modelling,* **21**, 1-4.

Pittock , A.B. and M.J. Salinger, 1982: Toward regional scenarios for a CO_2-warmed earth. *Clim. Change,* **4**, 23-40.

Prinn, R. H. Jacoby, A. Sokolov, C. Wang, X. Xiao, Z. Yang, R. Eckaus, P. Stone, D. Ellerman, J. Melillo, J. Fitzmaurice, D. Kicklighter, Y. Liu, and G. Holian, 1999: Integrated global system model for climate policy analysis: Feedbacks and sensitivity studies. *Clim. Change,* **41**, 469-546.

Racksko, P., L. Szeidl, and M. Semenov, 1991: A serial approach to local stochastic weather models. *Ecological Modeling,* **57**, 27-41.

Räisänen, J., 1997: Objective comparison of patterns of CO_2 induced climate change in coupled GCM experiments. *Clim. Dyn.,* **13**, 197-212.

Räisänen, J., 1999: Internal variability as a cause of qualitative inter-model disagreement on anthropogenic climate changes *Theor. Appl. Climatol.,* **64**, 1-13.

Ramaswamy, V. and T.C. Chen, 1997: Linear additivity of climate response for combined albedo and greenhouse perturbations. *Geophys. Res. Lett.,* **24**, 567-570.

Raper, S.C.B., T.M.L. Wigley, and R.A. Warrick, 1996: Global sea level rise: Past and future. In: *Sea-Level Rise and Coastal Subsidence: Causes, Consequences and Strategies* [Milliman, J. and B.U. Haq

(eds.)]. Kluwer Academic Publishers, Dordrecht, The Netherlands, pp. 11-45.

Richardson, C.W., 1981: Stochastic simulation of daily precipitation, temperature, and solar radiation. *Water Resour. Res.,* **17**, 182-190.

Riha, S. J., D. W. Wilks, and P. Simoens, 1996: Impact of temperature and precipitation variability on crop model predictions. *Clim. Change,* **32(3)**, 293-311.

Risbey, J.S., 1998: Sensitivities of water supply planning decisions to streamflow and climate scenario uncertainties. *Water Policy,* **1**, 321-340.

Risbey, J.S. and P.H. Stone, 1996: A case study of the adequacy of GCM simulations for input to regional climate change assessments. *J. Climate,* **9**, 1441-1467.

Robinson, P. J., and P. L. Finkelstein, 1989: *Strategies for Development of Dlimate Scenarios.* Final Report to the U. S. Environmental Protection Agency. Atmosphere Research and Exposure Assessment Laboratory, Office of Research and Development, USEPA, Research Triangle Park, NC, 73 pp.

Robock, A., R.P. Turco, M.A. Harwell, T.P. Ackerman, R. Andressen, H.S. Chang, and M.V.K. Sivakumar, 1993: Use of general circulation model output in the creation of climate change scenarios for impact analysis. *Clim. Change,* **23**, 293-355.

Rodbell, D., G. O. Seltzer, D. M. Anderson, D. B. Enfield, M. B. Abbott, and J. H. Newman, 1999: A high-resolution 15000 year record of El Nino driven alluviation in southwestern Ecuador. *Science,* **283**, 516-520.

Rosenberg, N.J., P.R. Crosson, K.D. Frederick, W.E. Easterling, M.S. McKenney, M.D. Bowes, R.A. Sedjo, J. Darmstadter, L.A. Katz, and K.M. Lemon, 1993: Paper 1. The MINK methodology: background and baseline. *Clim. Change,* **24**, 7-22.

Rosenthal, D.H., H.K. Gruenspecht, and E.A. Moran, 1995: Effects of global warming on energy use for space heating and cooling in the United States. *Energy J.,* **16**, 41-54.

Rosenzweig, C. 1985: Potential CO_2-induced climate effects on North American wheat-producing regions. *Clim. Change,* **4**, 239-254.

Rosenzweig, C. and A. Igesias (eds.), 1994: *Implications of Climate Change for International Agriculture: Crop Modeling Study.* EPA Report 230-B-94-003, United States Environmental Protection Agency, Policy, Planning and Evaluation, Climate Change Division, Washington, DC.

Rosenzweig, C. and M. L. Parry, 1994: Potential impact of climate change on world food supply. *Nature,* **367**, 133-137.

Rosenzweig, C., J. Phillips, R. Goldberg, J. Carroll, and T. Hodges, 1996: Potential impacts of climate change on citrus and potato production in the US. *Agricultural Systems,* **52**, 455-479.

Rotmans, J., M. Hulme, and T.E. Downing, 1994: Climate change implications for Europe: an application of the ESCAPE model. *Global Env. Change,* **4**, 97-124.

Saarikko, R.A., and T.R. Carter, 1996: Estimating regional spring wheat development and suitability in Finland under climatic warming. *Clim. Res.,* **7**, 243-252.

Santer, B.D., T.M.L. Wigley, M.E. Schlesinger, and J.F.B.Mitchell, 1990: *Developing Climate Scenarios from Equilibrium GCM results.* Report No. 47, Max-Planck-Institut-für-Meteorologie, Hamburg, 29 pp.

Santer, B.D., J.J. Hnilo, T.M.L. Wigley, J.S. Boyle, C. Doutriaux, M. Fiorino, D.E. Parker, and K.E. Taylor, 1999: Uncertainties in observationally based estimates of temperature change in the free atmosphere. *J. Geophys. Res.,* **104**, 6305-6333.

Schimel, D.S., W. Emanuel, B. Rizzo, T. Smith, F.I. Woodward, H. Fisher, T.G.F. Kittel, R. Mckeown, T. Painter, N. Rosenbloom, D.S. Ojima, W.J. Parton, D.W. Kicklighter, A.D. Mcguire, J.M. Melillo, Y. Pan, A. Haxeltine, C. Prentice, S. Sitch, K. Hibbard, R. Nemani, L. Pierce, S. Running, J. Borchers, J. Chaney, R. Neilson, and B.H. Braswell, 1997: Continental scale variability in ecosystem processes:

models, data, and the role of disturbance. *Ecological Monographs*, **67(2)**, 251-271.

Schlesinger, M.E., N. Andronova, A. Ghanem, S. Malyshev, T. Reichler, E. Rozanov, W. Wang, and F. Yang, 1997: *Geographical Scenarios of Greenhouse Gas and Anthropogenic-Sulfate-Aerosol Induced Climate Changes*. CRG manuscript, July 1997, University of Illinois, Department of Atmospheric Sciences, Urbana, IL, 85 pp.

Schlesinger, M.E., S. Malyshev, E.V. Rozanov, F. Yang, N.G. Andronova, B. de Vries, A. Grübler, K. Jiang, T. Masui, T. Morita, J. Penner, W. Pepper, A. Sankovski and Y. Zhang, 2000: Geographical distributions of temperature change for scenarios of greenhouse gas and sulfur dioxide emissions. *Tech. Forecast. Soc. Change*, **65**, 167-193.

Schneider, S. H., P. H. Gleick and L. O. Mearns, 1990: Prospects for Climate Change. In: *Climate and Water: Climate Change, Climatic Variability, and the Planning and Management of U. S. Water Resources*. Wiley Press, New York, Chapter 3, pp. 41-73.

Schreider, S. Yu, A.J. Jakeman, A.B. Pittock, and P.H. Whetton, 1996: Estimation of possible climate change impacts on water availability, extreme flow events and soil moisture in the Goulburn and Ovens Basins, Victoria. *Clim. Change*, **34**, 513–546.

Schubert, M., J. Perlwitz, R. Blender, K. Fraedrich, and F. Lunkeit, 1998: North Atlantic cyclones in CO2-induced warm climate simulations: Frequency, intensity, and tracks. *Clim. Dyn.*, **14**, 827-837.

Semenov, M.A. and J.R. Porter, 1995: Climatic variability and the modelling of crop yields. *Agric. For. Meteorol.*, **73**, 265-283.

Semenov, M. A. and E. Barrow, 1997: Use of a stochastic weather generator in the development of climate change scenarios. *Clim. Change*, **35**, 397-414.

Semenov, M.A. and R.J. Brooks, 1999: Spatial interpolation of the LARS-WG stochastic weather generator in Great Britain. *Clim. Res.*, **11**, 137-148.

Severinghaus, J.P., T. Sowers, E. Brook, R.B. Alley, and M.L. Bender, 1998: Timing of abrupt climate change at the end of the Younger Dryas interval from thermally fractionated gases in polar ice. *Nature*, **391**, 141-146.

Shabalova, M.V., and G.P. Können, 1995: Climate change scenarios: comparisons of paleo-reconstructions with recent temperature changes. *Clim. Change*, **29**, 409-428.

Skelly, W.C. and A. Henderson-Sellers, 1996: Grid box or grid point: what type of data do GCMs deliver to climate impacts researchers? *Int. J. Climatol.*, **16**, 1079-1086.

Smith, J.B. and G.J. Pitts, 1997: Regional climate change scenarios for vulnerability and adaptation assessments. *Clim. Change*, **36**, 3-21.

Smith, J.B. and M. Hulme, 1998: Climate change scenarios. In: *Handbook on Methods of Climate Change Impacts Assessment and Adaptation Strategies* [Feenstra, J., I. Burton, J.B. Smith, and R.S.J. Tol (eds.)]. UNEP/IES, Version 2.0, October, Amsterdam, Chapter 3.

Smith, J.B. and D.A. Tirpak (eds.) 1989: *The Potential Effects of Global Climate Change on the United States*. Report to Congress, United States Environmental Protection Agency, EPA-230-05-89-050, Washington, D.C., 409 pp.

Smith, J.B., S. Huq, S. Lenhart, L.J. Mata, I. Neme?ová, and S. Toure (eds.), 1996: *Vulnerability and Adaptation to Climate Change: Interim Results from the U.S. Country Studies Program*. Kluwer, Dordrecht, The Netherlands, 366 pp.

Stewart, T.R. and M.H.Glantz, 1985: Expert judgment and climate forecasting: a methodological critique of Climate Change to the Year 2000. *Clim. Change*, **7**, 159-183.

Stone, M.C., R.H. Hotchkiss, C.M. Hubbard, T.A. Fontaine, L.O. Mearns, and J.G. Arnold, 2001: Impacts of climate change on the water yield of the Missouri River. *J. Am. Water Resour. As.* (accepted).

Stouffer, R.J., S. Manabe, and K.Ya. Vinnikov, 1994: Model assessment of the role of natural variability in recent global warming. *Nature*, **367**, 634-636.

Terjung, W.H., D.M. Liverman, J.T. Hayes, and collaborators, 1984: Climatic change and water requirements for grain corn in the North American Great Plains. *Clim. Change*, **6**, 193-220.

Tett, S.F.B., T.C. Johns, and J.F.B. Mitchell, 1997: Global and regional variability in a coupled AOGCM. *Clim. Dyn.*, **13**, 303-323.

Timmermann, A., J. Oberhuber, A. Bacher, M. Esch, M. Latif, and E. Roeckner, 1999: Increased El Nino frequency in a climate model forced by future greenhouse warming. *Nature*, **398**, 694-696.

Titus, J.G. and V. Narayanan, 1996: The risk of sea level rise. *Clim. Change*, **33**, 151-212.

Tol, R.S.J. and A.F. de Vos, 1998. A Bayesian statistical analysis of the enhanced greenhouse effect. *Clim. Change*, **38**, 87-112.

Torn, M.S and J.S. Fried, 1992: Predicting the impacts of global warming on wildland fire. *Clim. Change*, **21**, 257-274.

Toth, L. and M. Mwandosya, 2001: Decision-making Frameworks. In *Climate Change 2001: Mitigation*. Contribution of Working Group III to the IPCC Third Assessment Report, Cambridge University Press.

Velichko, A. A., O. K. Borisova, E. M. Zelikson, and V. P Nechaev, 1995a: Permafrost and vegetation response to global warming in North Eurasia. In: *Biotic Feedbacks in the Global Climatic System*. New York, Oxford, Oxford University Press, pp. 134-156.

Velichko, A.A., L.O. Karpachevsky, and T.D. Morozova, 1995b: Water reserves in soils under global climate warming: an approach to forecasting for Eastern Europe as an example. *Pochvovedenie*, **8**, 933-942 (in Russian).

Vinnikov, K. Ya. and P. Ya. Groisman, 1979: An empirical model of present-day climatic changes. *Meteor. Gidrol.*, **3**, 25-28 (in Russian).

Vinocur, M.G., R.A. Seiler, and L.O. Mearns, 2000: Forecasting the impact of climate variability on peanut crop production in Argentina. In: *Proceedings of the International Forum on Climate Prediction, Agriculture and Development*. International Research Institute for Climate Prediction (IRI), Palisades, NY (in press).

Visser, H., R.J.M. Folkert, J. Hoekstra, and J.J. deWolff, 2000: Identifying key sources of uncertainty in climate change projections. *Clim. Change*, **45(3/4)**, 421-457.

Voet, P. van der, K. Kramer, and C.A. van Diepen, 1996: *Parametrization of the Richardson Weather Generator within the European Union*. DLO Winand Staring Centre Report 92, Wageningen, The Netherlands, 73 pp.

von Storch, H., 1995: Inconsistencies at the interface of climate impact studies and global climate research. *Meteor. Z.*, **4 NF**, 72-80.

von Storch, J.S., V. Kharin, U. Cubasch, G.C. Hergel, D. Schriever, H. von Storch, and E. Zorita, 1997: A description of a 1260-year control integration with the coupled ECHAM1/LSG general circulation model. *J. Climate*, **10**, 1525-1543.

Walker, B., W. Steffen, J. Canadell and J. Ingram (eds.), 1999: *The Terrestrial Biosphere and Global Change: Implications for Natural and Managed Ecosystems*. Cambridge University Press, Cambridge, 439 pp.

Walsh, K.J.E. and B.F. Ryan, 2000: Tropical cyclone intensity increase near Australia as a result of climate change. *J. Climate*, **13**, 3029-3036.

Walsh, K., R. Allan, R. Jones, B. Pittock, R. Suppiah, and P. Whetton, 1999: *Climate Change in Queensland under Enhanced Greenhouse Conditions*. First Annual Report, CSIRO Atmospheric Branch, Aspendale, Victoria, Australia, 84 pp.

Wang, Y. P., and D. J. Connor, 1996: Simulation of optimal development for spring wheat at two locations in southern Australia under present and changed climate conditions. *Agric. For. Meteorol.*, **79**, 9-28.

Wang, J. and L. Erda, 1996: The Impacts of potential climate change and climate variability on simulated maize production in China. *Water Air Soil Pollut.*, **92**, 75-85.

Wang, Q. J., R. J. Nathan, R. J. Moran, and B. James, 1999: *Impact of climate Changes on the Security of the Water Supply of the Campaspe System*. Proceedings of 25th Hydrology and Water Resources

Symposium, Vol. 1, 6-8 July 1999, Brisbane, Institution of Engineers, Australia, Water 99 Joint Congress, pp. 135-140.

Warrick, R.A., 1984: The possible impacts on wheat production of a recurrence of the 1930s drought in the U.S. Great Plains. *Clim. Change*, **6**, 5-26

Warrick, R.A., C. Leprovost, M.F. Meier, J. Oerlemans, P.L. Woodworth, R.B. Alley, R.A. Bindschadler, C.R. Bentley, R.J. Braithwaite, J.R. de Wolde, B.C. Douglas, M. Dyurgerov, N.C. Flemming, C. Genthon, V. Gornitz, J. Gregory, W. Haeberli, P. Huybrechts, T. Jóhannesson, U. Mikolajewicz, S.C.B. Raper, D.L. Sahagian, R.S.W. van de Wal and T.M.L. Wigley, 1996: Changes in sea level. In: *Climate Change 1995: The Science of Climate Change. Contribution of Working Group I to the Second Assesesment Report of the Intergovernmental Panel on Climate Change* [Houghton, J.T., L.G.M. Filho, B.A. Callander, N. Harris, A. Kattenberg and K. Maskell (eds.)]. Cambridge University Press, Cambridge, United Kingdom and New York, NY, USA, pp. 359-405.

Webb, T., III, and T.M.L. Wigley, 1985: What past climates can indicate about a warmer world. In: *Projecting the Climatic Effects of Increasing Carbon Dioxide* [MacCracken, M.C. and F.M. Luther (eds.)]. United States Department of Energy, Office of Energy Research, DOE/ER-0237, Washington, D.C., pp. 237-257.

Whetton, P. H., 1998: Climate change impacts on the spatial extent of snow-cover in the Australian Alps. In: *Snow: A Natural History, An Uncertain Future* [K. Green (ed.)]. Australian Alps Liaison Committee, Canberra, pp. 195-206.

Whetton, P.H. and A.B. Pittock, 1991: *Australian Region Intercomparison of the Results of some Greenhouse General Circulation Modelling Experiments*. Tech. Paper No. 21, CSIRO Div. of Atmospheric Research, Melbourne, Australia, 73 pp.

Whetton, P.H., A.M. Fowler, M.R. Haylock, and A.B. Pittock, 1993: Implications of climate change due to the enhanced greenhouse effect on floods and droughts in Australia. *Climatic Change*, **25(3-4)**, 289-317.

Whetton, P.H., M.R. Haylock, and R. Galloway, 1996: Climate change and snow-cover duration in the Australian Alps. *Clim. Change*, **32(4)**, 447-479.

Whetton, P.H., J.J. Katzfey, K.C. Nguyen, J.L. McGregor, C.M. Page, T.I. Elliott, and K.J. Hennessy, 1998a: *Fine Resolution Climate Change Scenarios for New South Wales. Part 2: Climatic Variability.* Annual Report, 1996/7, CSIRO, Australia, 51 pp.

Whetton, P. H., Z. Long, and I.N. Smith, 1998b: Comparison of simulated climate change under transient and stabilised increased CO2 conditions. In: *Coupled Climate Modelling: Abstracts of Presentations at the Tenth Annual BMRC Modelling Workshop, BMRC, Melbourne* [P. J. Meighen (ed.)]. BMRC Research Report, 69, Melbourne, Bureau of Meteorology Research Centre, pp. 93-96.

White, M.R., 1985: *Characterization of Information Requirements for Studies of CO₂ Effects: Water Resources, Agriculture, Fisheries, Forests, and Human Health.* US Dept. of Energy, Office of Energy Research, DOE/ER-0236, Washington, DC.

Widmann, M. and C.S. Bretherton, 2000: Validation of mesoscale precipitation in the NCEP reanalysis using a new grid-cell data set for the northwestern United States. *J.Climate*, **13(11)**, 936-950.

Wigley, T.M.L., 1999: *The Science of Climate Change: Global and US Perspectives*. Pew Center on Global Climate Change, Arlington, VA, 48 pp.

Wigley, T.M.L., P.D. Jones, and P.M. Kelly, 1980: Scenarios for a warm, high-CO₂ world. *Nature*, **288**, 17-21.

Wilby, R.L., 1998: Statistical downscaling of daily precipitation using daily airflow and seasonal teleconnection indices. *Clim. Res.*, **10**, 163-178.

Wilby, R.L. and T.M.L. Wigley, 1997: Downscaling general circulation model output: a review of methods and limitations. *Progress in Physical Geography*, **21**, 530-548.

Wilby, R.L., B. Greenfield, and C. Glenny, 1994: A coupled synoptic-hydrological model for climate change impact assessment. *Journal of Hydrology*, **153**, 265-290.

Wilby, R.L., T.M.L. Wigley, D. Conway, P.D. Jones, B.C. Hewitson, J. Main, and D.S. Wilks, 1998: Statistical downscaling of General Circulation Model output: a comparison of methods. *Water Resources Research*, **34**, 2995-3008.

Wilby, R.L., L.E. Hay, and G.H. Leavesley, 1999: A comparison of downscaled and raw GCM output: implications for climate change scenarios in the San Juan River Basin, Colorado. *J. Hydrol.* **225**, 67-91.

Wilks, D. S., 1992: Adapting stochastic weather generation algorithms for climate change studies. *Clim.Change,* **22**, 67-84.

Wilks, D. S. and R.L. Wilby, 1999: The weather generation game: A review of stochastic weather models. *Progress in Physical Geography*, **23**, 329-358.

Williams, G.D.V., R.A. Fautley, K.H. Jones, R.B. Stewart, and E.E. Wheaton, 1988: Estimating effects of climatic change on agriculture in Saskatchewan, Canada. In: *The Impact of Climatic Variations on Agriculture, Volume 1, Assessments in Cool Temperate and Cold Regions* [Parry, M.L., T.R. Carter, and N.T. Konijn (eds.)]. Kluwer, Dordrecht, The Netherlands, pp. 219-379.

Woo, M.K., 1992: Application of stochastic simulation to climatic change studies. *Clim.Change,* **20**, 313-330.

Woolhiser, D.A., T.O. Keefer, and K.T. Redmond, 1993: Southern Oscillation effects on daily precipitation in the southwestern United States. *Water Resour. Res.,* **29**, 1287-1295.

Wotton, B.M., B. J. Stocks, M.D. Flannigan, R. LaPrise, and J-P Blanchet, 1998: Estimating future 2xCO₂ fire climates in the boreal forest of Canada using a regional climate model. In: *Proceedings of Third International Conference on Forest Fire Research and the 14th Conference in Fire and Forest Meteorology*. University of Coimbra, Portugal, pp. 1207-1221.

Xiao, X., D.W. Kicklighter, J.M. Melillo, A.D. McGuire, P.H. Stone and A. P. Sokolov, 1997: Linking a global terrestrial biogeochemical model with a 2-dimensional climate model: Implications for global carbon budget. *Tellus*, **49B**, 18-37.

Xie, P. and P.A. Arkin, 1997: Global precipitation: a 17-year monthly analysis based on gauge observations, satellite estimates, and numerical model outputs. *Bull. Am. Met. Soc.*, **78**, 2539-2558.

Yohe, G.W. and M.E. Schlesinger, 1998: Sea-level change: the expected economic cost of protection or abandonment in the United States. *Clim. Change*, **38**, 447-472.

14

Advancing Our Understanding

Co-ordinating Lead Author
B. Moore III

Lead Authors
W.L. Gates, L.J. Mata, A. Underdal

Contributing Author
R.J. Stouffer

Review Editors
B. Bolin, A. Ramirez Rojas

Contents

Executive Summary

Further work is required to improve the ability to detect, attribute, and understand climate change, to reduce uncertainties, and to project future climate changes. In particular, there is a need for additional systematic observations, modelling and process studies. A serious concern is the decline of observational networks. Further work is needed in eight broad areas:

• *Reverse the decline of observational networks in many parts of the world.* Unless networks are significantly improved, it may be difficult or impossible to detect climate change over large parts of the globe.

• *Sustain and expand the observational foundation for climate studies by providing accurate, long-term, consistent data including implementation of a strategy for integrated global observations.* Given the complexity of the climate system and the inherent multi-decadal time-scale, there is a need for long-term consistent data to support climate and environmental change investigations and projections. Data from the present and recent past, climate-relevant data for the last few centuries, and for the last several millennia are all needed. There is a particular shortage of data in polar regions and data for the quantitative assessment of extremes on the global scale.

• *Understand better the mechanisms and factors leading to changes in radiative forcing; in particular, improve the observations of the spatial distribution of greenhouse gases and aerosols.* It is particularly important that improvements are realised in deriving concentrations from emissions of gases and particularly aerosols, and in addressing biogeochemical sequestration and cycling, and specifically, in determining the spatial-temporal distribution of carbon dioxide (CO_2) sources and sinks, currently and in the future. Observations are needed that would decisively improve our ability to model the carbon cycle; in addition, a dense and well-calibrated network of stations for monitoring CO_2 and oxygen (O_2) concentrations will also be required for international verification of carbon sinks. Improvements in deriving concentrations from emissions of gases and in the prediction and assessment of direct and indirect aerosol forcing will require an integrated effort involving *in situ* observations, satellite remote sensing, field campaigns and modelling.

• *Understand and characterise the important unresolved processes and feedbacks, both physical and biogeochemical, in the climate system.* Increased understanding is needed to improve prognostic capabilities generally. The interplay of observation and models will be the key for progress. The rapid forcing of a non-linear system has a high prospect of producing surprises.

• *Address more completely patterns of long-term climate variability including the occurrence of extreme events.* This topic arises both in model calculations and in the climate system. In simulations, the issue of climate drift within model calculations needs to be clarified better in part because it compounds the difficulty of distinguishing signal and noise. With respect to the long-term natural variability in the climate system *per se*, it is important to understand this variability and to expand the emerging capability of predicting patterns of organised variability such as El Niño-Southern Oscillation (ENSO). This predictive capability is both a valuable test of model performance and a useful contribution in natural resource and economic management.

• *Improve methods to quantify uncertainties of climate projections and scenarios, including development and exploration of long-term ensemble simulations using complex models.* The climate system is a coupled non-linear chaotic system, and therefore the long-term prediction of future climate states is not possible. Rather the focus must be upon the prediction of the probability distribution of the system's future possible states by the generation of ensembles of model solutions. Addressing adequately the statistical nature of climate is computationally intensive and requires the application of new methods of model diagnosis, but such statistical information is essential.

• *Improve the integrated hierarchy of global and regional climate models with a focus on the simulation of climate variability, regional climate changes, and extreme events.* There is the potential for increased understanding of extremes events by employing regional climate models; however, there are also challenges in realising this potential. It will require improvements in the understanding of the coupling between the major atmospheric, oceanic, and terrestrial systems, and extensive diagnostic modelling and observational studies that evaluate and improve simulation performance. A particularly important issue is the adequacy of data needed to attack the question of changes in extreme events.

• *Link models of the physical climate and the biogeochemical system more effectively, and in turn improve coupling with descriptions of human activities.* At present, human influences generally are treated only through emission scenarios that provide external forcings to the climate system. In future more comprehensive models, human activities need to begin to interact with the dynamics of physical, chemical, and biological sub-systems through a diverse set of contributing activities, feedbacks, and responses.

Cutting across these foci are crucial needs associated with strengthening international co-operation and co-ordination in order to utilise better scientific, computational, and observational resources. This should also promote the free exchange of data among scientists. A special need is to increase the observational and research capacities in many regions, particularly in developing countries. Finally, as is the goal of this assessment, there is a continuing imperative to communicate research advances in terms that are relevant to decision making.

The challenges to understanding the Earth system, including the human component, are daunting, but these challenges simply must be met.

14.1 Introduction

There has been encouraging progress over this first decade of the IPCC process. We understand better the coupling of the atmosphere and ocean. Significant steps have been taken in linking the atmosphere and the terrestrial systems although the focus tends to be on water-energy and the biosphere with fixed vegetation patterns. Even so, revealing and unexpected teleconnections are being discovered; moreover, progress is being made towards model structures and data sets that will allow implementation of coupled atmosphere-ocean-terrestrial models that include key biological-biogeochemical feedbacks. There is also encouraging progress in developing integrated assessment models that couple economic activity, with associated emissions and impacts, with models of the biogeochemical and climate systems. This work has yielded preliminary insights into system behaviour and key policy-relevant uncertainties.

The challenges are significant, but the record of progress suggests that within the next decade the scientific community will develop fully coupled dynamical (prognostic) models of the full Earth system (e.g., the coupled physical climate, biogeochemical, human sub-systems) that can be employed on multidecadal time-scales and at spatial scales relevant to strategic impact assessment. Future models should certainly advance in completeness and sophistication; however, the key will be to demonstrate some degree of prognostic skill. The strategy will be to couple the biogeochemical-physical climate system to representations of key aspects of the human system, and then to develop more coherent scenarios of human actions in the context of feedbacks from the biogeochemical-physical climate system.

Developing these coupled models is an important step. From the perspective of understanding the Earth system, determining the nature of the link between the biogeochemical system and the physical climate system represents a fundamental scientific goal. Present understanding is incomplete, and a successful attack will require extensive interdisciplinary collaboration. It will also require global data that clearly document the state of the system and how that state is changing as well as observations to illuminate important processes more clearly.

14.2 The Climate System

14.2.1 Overview

Models of physical processes in the ocean and atmosphere provide much of our current understanding of future climate change. They incorporate the contributions of atmospheric dynamics and thermodynamics through the methods of computational fluid dynamics. This approach was initially developed in the 1950s to provide an objective numerical approach to weather prediction. It is sometimes forgotten that the early development of "supercomputers" at that time was motivated in large part by the need to solve this problem. In the 1960s, versions of these weather prediction models were developed to study the "general circulation" of the atmosphere, i.e., the physical statistics of weather systems satisfying requirements of conservation of mass, momentum, and energy. To obtain realistic simulations, it was found necessary to include additional energy sources and sinks: in particular, energy exchanges with the surface and moist atmospheric processes with the attendant latent heat release and radiative heat inputs.

Development of models for the general circulation of the ocean started later, but has proceeded in a similar manner. Models that deal with the physics of the oceans have been developed and linked to models of the atmospheric system. Within ocean models, the inclusion of geochemical and biological interactions has begun, with a focus upon the carbon cycle. Since the late 1960s, the geochemical aspects of the carbon cycle have been included in low-dimensional box models. More recently, including the carbon chemistry system in general circulation models has simply been a question of allocation of computing resources. Modelling of the biological system, however, has been more challenging, and it has only been recently that primitive ecosystem models have been incorporated in global general circulation ocean models. Even though progress has been significant, much remains to be done. Eddy-resolving ocean models with chemistry and biology need to be tested and validated in a transient mode, and the prognostic aspects of marine ecosystems including nutrient dynamics need greater attention at basin and global scales.

Model development for the ocean and atmosphere has had a fundamental theoretical advantage: it is based on firmly established hydrodynamic equations. At present there is less theoretical basis for a "first principles" development of the dynamical behaviour of the terrestrial system. There is a need to develop a fundamental methodology to describe this very heterogeneous and complex system. For the moment, it is necessary to rely heavily upon parametrizations and empirical relationships. Such reliance is data intensive and hence independent validation of terrestrial system models is problematical. In spite of these difficulties, a co-ordinated strategy has been developed to improve estimates of terrestrial primary productivity and respiration by means of measurement and modelling. The strategy has begun to yield dividends. Techniques from statistical mechanics have been wedded to biogeochemistry and population ecology, yielding new vegetation dynamic models. Global terrestrial models at meso-spatial scales (roughly 50 km grids) now exist which capture complex ecophysiological processes and ecosystem dynamics.

Expanded efforts are needed in these domain-specific models. In the ocean, we need to consider better the controls on thermohaline circulation, on potential changes in biological productivity, and on the overall stability of the ocean circulation system. Within terrestrial systems the question of the carbon sink-source pattern is central: what is it and how might it change? Connected to this question is the continued development of dynamic vegetation models, which treat competitive processes within terrestrial ecosystems and their response to multiple stresses. And for the atmosphere, a central question has been, is, and likely will be the role of clouds. Also, there is a corresponding non-linearity associated with change in the distribution and extent of sea ice. Further increased efforts will be needed in linking terrestrial ecosystems with the atmosphere, the land with the ocean, the ocean (and its ecosystems) with the atmosphere, the chemistry of the atmosphere with the physics of the atmosphere, and finally linking the human system to them all. Such models will also need to be able to highlight different regions with increased spatial and temporal detail.

Models, however, depend upon high quality data. Data allow hypotheses about processes and their linkages to be rejected or to be given increased consideration. Giving formal (e.g., quantitative) expression to processes is at the heart of the scientific enterprise. Such expressions reflect our knowledge and form the basis for models. Models are simply formal expressions of processes and how they fit together. And all rest upon data. Models are of limited use without observations; the value of observations increases by interaction with models. Systematic global observations are an essential underpinning of research to improve understanding of the climate system. For numerous applications in climate-impact research, information about the complex nature of the system is needed. Unfortunately, there continue to be justifiable concerns about the loss of some monitoring of climate parameters and deterioration of coverage. There is a basic need for more observations with better coverage, higher accuracy, and with increased availability. This overriding importance of data has been recognised repeatedly in the past and in this volume (e.g., Chapter 2, Section 2.8; Chapter 3, Section 3.5; Chapter 4, Section 4.2; Chapter 6, Section 6.14; Chapter 11, Section 11.6.1 and Chapter 12, Section 12.4), and there are reasons for guarded optimism on the issue of data even though there are also significant reasons for concern. One such reason for tempered optimism is the plan for and beginning implementation of global observing systems such as the Global Climate Observing System (GCOS), Global Ocean Observing System (GOOS), and Global Terrestrial Observing System (GTOS). However plans in themselves do not produce data, and data that are not accessible are of limited value. The issue of data remains central for progress.

14.2.2 Predictability in a Chaotic System

The climate system is particularly challenging since it is known that components in the system are inherently chaotic; there are feedbacks that could potentially switch sign, and there are central processes that affect the system in a complicated, non-linear manner. These complex, chaotic, non-linear dynamics are an inherent aspect of the climate system. As the IPCC WGI Second Assessment Report (IPCC, 1996) (hereafter SAR) has previously noted, "future unexpected, large and rapid climate system changes (as have occurred in the past) are, by their nature, difficult to predict. This implies that future climate changes may also involve 'surprises'. In particular, these arise from the non-linear, chaotic nature of the climate system … Progress can be made by investigating non-linear processes and sub-components of the climatic system." These thoughts are expanded upon in this report: "Reducing uncertainty in climate projections also requires a better understanding of these non-linear processes which give rise to thresholds that are present in the climate system. Observations, palaeoclimatic data, and models suggest that such thresholds exist and that transitions have occurred in the past … Comprehensive climate models in conjunction with sustained observational systems, both *in situ* and remote, are the only tool to decide whether the evolving climate system is approaching such thresholds. Our

knowledge about the processes, and feedback mechanisms determining them, must be significantly improved in order to extract early signs of such changes from model simulations and observations." (See Chapter 7, Section 7.7).

14.2.2.1 Initialisation and flux adjustments

Integrations of models over long time-spans are prone to error as small discrepancies from reality compound. Models, by definition, are reduced descriptions of reality and hence incomplete and with error. Missing pieces and small errors can pose difficulties when models of sub-systems such as the ocean and the atmosphere are coupled. As noted in Chapter 8, Section 8.4.2, at the time of the SAR most coupled models had difficulty in reproducing a stable climate with current atmospheric concentrations of greenhouse gases, and therefore non-physical "flux adjustment terms" were added. In the past few years significant progress has been achieved, but difficulties posed by the problem of flux adjustment, while reduced, remain problematic and continued investigations are needed to reach the objective of avoiding dependence on flux adjustment (see Chapter 8, Section 8.4.2; see also Section 8.5.1.1).

Another important (and related) challenge is the initialisation of the models so that the entire system is in balance, i.e., in statistical equilibrium with respect to the fluxes of heat, water, and momentum between the various components of the system. The problem of determining appropriate initial conditions in which fluxes are dynamically and thermodynamically balanced throughout a coupled stiff system, such as the ocean-atmosphere system, is particularly difficult because of the wide range of adjustment times ranging from days to thousands of years. This can lead to a "climate drift", making interpretation of transient climate calculations difficult (see Chapter 8, Section 8.4.1).

The initialisation of coupled models is important because it produces the climate base state or "starting point" for climate change experiments. Climate model initialisation continues to be an area of active research and refinement of techniques (see Chapter 8, Section 8.4). Most groups use long integrations of the sub-component models to provide a dynamically and thermodynamically balanced initial state for the coupled model integration. However, there are at least as many different methods used to initialise coupled models as there are modelling groups. See Stouffer and Dixon (1998) for a more complete discussion of the various issues and methods used to initialise coupled models.

Since the SAR, improvements in developing better initialisation techniques for coupled models have been realised. For instance, starting with observed oceanic conditions has yielded improved simulations with reduced climate drift (Gordon *et al.*, 1999). Earlier attempts with this technique usually resulted in relatively large trends in the surface variables (Meehl and Washington, 1995; Washington and Meehl, 1996). Successfully starting long coupled integrations from observations is important for a number of reasons: it simplifies the initialisation procedure, saves time and effort, and reduces the overhead for starting new coupled model integrations.

Such progress is important, but again further work is needed. We simply do not fully understand the causes of climate drift in coupled models (see Chapter 8, Section 8.4.2).

14.2.2.2 Balancing the need for finer scales and the need for ensembles

There is a natural tendency to produce models at finer spatial scales that include both a wider array of processes and more refined descriptions. Higher resolution can lead to better simulations of atmospheric dynamics and hydrology (Chapter 8, Section 8.9.1), less diffusive oceanic simulations, and improved representation of topography. In the atmosphere, fine-scale topography is particularly important for resolving small-scale precipitation patterns (see Chapter 8, Section 8.9.1). In the ocean, bottom topography is very important for the various boundary flows (see Chapter 7, Section 7.3.4). The use of higher oceanic resolution also improves the simulation of internal variability such as ENSO (see Chapter 8, Section 8.7.1). However, in spite of the use of higher resolution, important climatic processes are still not resolved by the model's grid, necessitating the continued use of sub-grid scale parametrizations.

It is anticipated that the grids used in the ocean sub-components of the coupled climate models will begin to resolve eddies by the next report. As the oceanic eddies become resolved by the grid, the need for large diffusion coefficients and various mixing schemes should be reduced (see Chapter 8, Section 8.9.3; see also, however, the discussion in Section 8.9.2). In addition, the amount of diapycnal mixing, which is used for numerical stability in this class of ocean models, will also be reduced as the grid spacing becomes smaller. This reduction in the sub-grid scale oceanic mixing should reduce the uncertainty associated with the mixing schemes and coefficients currently being used.

Underlying this issue of scale and detail is an important tension. As the spatial and process detail in a model is increased, the required computing resources increase, often significantly; models with less detail may miss important non-linear dynamics and feedbacks that affect model results significantly, and yet simpler models may be more appropriate to generating the needed statistics. The issue of spatial detail is intertwined with the representation of the physical (and other) processes, and hence the need for a balance between level of process detail and spatial detail. These tensions must be recognised forthrightly, and strategies must be devised to use the available computing resources wisely. Analyses to determine the benefits of finer scale and increased resolution need to be carefully considered. These considerations must also recognise that the potential predictive capability will be unavoidably statistical, and hence it must be produced with statistically relevant information. This implies that a variety of integrations (and models) must be used to produce an ensemble of climate states. Climate states are defined in terms of averages and statistical quantities applying over a period typically of decades (see Chapter 7, Section 7.1.3 and Chapter 9, Section 9.2.2).

Fortunately, many groups have performed ensemble integrations, that is, multiple integrations with a single model using identical radiative forcing scenarios but different initial conditions. Ensemble integrations yield estimates of the variability of the response for a given model. They are also useful in determining to what extent the initial conditions affect the magnitude and pattern of the response. Furthermore, many groups have now performed model integrations using similar radiative forcings. This allows ensembles of model results to be constructed (see Chapter 9, Section 9.3; see also the end of Chapter 7, Section 7.1.3 for an interesting question about ensemble formation).

In sum, a strategy must recognise what is possible. In climate research and modelling, we should recognise that we are dealing with a coupled non-linear chaotic system, and therefore that the long-term prediction of future climate states is not possible. The most we can expect to achieve is the prediction of the probability distribution of the system's future possible states by the generation of ensembles of model solutions. This reduces climate change to the discernment of significant differences in the statistics of such ensembles. The generation of such model ensembles will require the dedication of greatly increased computer resources and the application of new methods of model diagnosis. Addressing adequately the statistical nature of climate is computationally intensive, but such statistical information is essential.

14.2.2.3 Extreme events

Extreme events are, almost by definition, of particular importance to human society. Consequently, the importance of understanding potential extreme events is first order. The evidence is mixed, and data continue to be lacking to make conclusive cases. Chapter 9, Sections 9.3.5 and 9.3.6 consider projections of changes in patterns of variability (discussed in the next section) and changes in extreme events (see also Chapters 2 and 10). Though the conclusions are mixed in both of these topical areas, certain results begin to appear robust. There appear to be some consistent patterns with increased CO_2 with respect to changes in variability: (a) the Pacific climate base state could be a more El Niño-like state and (b) an enhanced variability in the daily precipitation in the Asian summer monsoon with increased precipitation intensity (Chapter 9, Section 9.3.5). More generally, the intensification of the hydrological cycle with increased CO_2 is a robust conclusion. For possible changes in extreme weather and climate events, the most robust conclusions appear to be: (a) an increased probability of extreme warm days and decreased probability of extreme cold days and (b) an increased chance of drought for mid-continental areas during summer with increasing CO_2 (see Chapter 9, Section 9.3.6).

The evaluation of many types of extreme events is made difficult because of issues of scale. Damaging extreme events are often at small temporal and spatial scales. Intense, short-duration events are not well-represented (or not represented at all) in model-simulated climates. In addition, there is often a basic mismatch between the scales resolved in models and those of the validating data. A promising approach is to use multi-fractal models of rainfall events in that they naturally generate extreme events. Reanalysis has also helped in this regard, but reanalysis *per se* is not the sole answer because the models used for reanalysis rely on sub-grid scale parametrizations almost as heavily as climate models do.

One area that is possibly ripe for a direct attack on improving the modelling of extreme events is tropical cyclones (see Section Chapter 2, 2.7.3.1; Chapter 8, Section 8.8.4; Chapter 9, Section 9.3.6.4, and Chapter 10, Box 10.2). Also, there is the potential for increased understanding of extreme events by employing regional climate models (RCMs); however, there are also challenges to realising this potential (see Chapter 10). It must be established that RCMs produce more realistic extremes than general circulation models (GCMs). Most RCM simulations to date are not long enough (typically 5 or 10 years for nested climate change simulations) to evaluate extremes well (see Chapter 10, Section 10.5.2).

Another area in which developments are needed is that of extremes associated with the land surface (flood and drought). There is still a mismatch between the scale of climate models and the finer scales appropriate for surface hydrology. This is particularly problematical for impact studies. For droughts there is a basic issue of predictability; drought prediction is difficult regardless of scale.

A particularly important issue is the adequacy of data needed to attack the question of changes in extreme events. There have been recent advances in our understanding of extremes in simulated climates (see, for example, Meehl *et al.*, 2000), but thus far the approach has not been very systematic. Atmospheric Model Intercomparison Project 2 (AMIP2) provides an opportunity for a more systematic approach: AMIP2 will be collecting and organising some of the high-frequency data that are needed to study extremes. However, it must be recognised that we are still unfortunately short of data for the quantitative assessment of extremes on the global scale in the observed climate.

Finally, it is often stated that the impacts of climate change will be felt through changes in extremes because they stress our present day adaptations to climate variability. What does this imply for the research agenda for the human dimension side of climate studies?

14.2.2.4 Organised variability

An overriding challenge to modelling and to the IPCC is prediction. This challenge is particularly acute when predictive capability is sought for a system that is chaotic, that has significant non-linearities, and that is inherently stiff (i.e., widely varying time constants). And within prognostic investigations of such a complex system, the issue of predicting extreme events presents a particularly vexing yet important problem. However, there appear to be coherent modes of behaviour that not only support a sense of optimism in attacking the prediction problem, but also these modes may offer measurable prediction targets that can be used as benchmarks for evaluating our understanding of the climate system. In addition, predictions of these modes represent valuable contributions in themselves.

Evaluating the prognostic skill of a model and understanding the characteristics of this skill are clearly important objectives. In the case of weather prediction, one can estimate the range of predictability by evaluating the change of the system from groups of initial states that are close to each other. The differences in these time-evolving states give a measure of the predictive utility of the model. In addition, one has the near-term reality of the evolving weather as a constant source of performance metrics. For the climate issue, the question of predictability is wrapped up with understanding the physics behind the low-frequency variability of climate and distinguishing the signal of climate change (see Chapter 9, Section 9.2.2.1). In other words, there are the paired challenges of capturing (predicting) "natural" variability of climate as well as the emerging anthropogenically forced climate signal. This dual challenge is distinctively climatic in nature, and whereas the longer-term character of climate projections is unavoidable and problematic, the intra-seasonal to inter-decadal modes of climate variability (e.g., ENSO, Pacific Decadal Oscillation (PDO), and North Atlantic Oscillation (NAO) – see also Chapter 7, Box 7.2) offer opportunities to test prognostic climate skill. Here, some predictive skill for the climate system appears to exist on longer time-scales. One example is the ocean-atmosphere phenomenon of ENSO. This skill has been advanced and more clearly demonstrated since the SAR, and this progress and demonstration are important (see Chapter 7, Section 7.6; Chapter 8, Section 8.7 and Chapter 9, Section 9.3.5). Such demonstrations and the insights gained in developing and making prognostic statements on climate modes frame an important area for further work.

This opportunity is well summarised in Chapter 8 (in particular, Section 8.7), "The atmosphere-ocean coupled system shows various modes of variability that range widely from intra-seasonal to inter-decadal time-scales (see Chapters 2 and 7). Since the SAR, considerable progress has been achieved in characterising the decadal to inter-decadal variability of the ocean-atmosphere system. Successful evaluation of models over a wide range of phenomena increases our confidence."

14.2.3 Key Sub-systems and Phenomena in the Physical-Climate System

Central to the climate system are the coupled dynamics of the atmosphere-ocean-terrestrial system, the physical processes associated with the energy and water cycles and the associated biological and chemical processes controlling the biogeochemical cycles, particularly carbon, nitrogen, phosphorus, sulphur, iron, and silicon. The atmosphere plays a unique role in the climate system since on a zeroth order basis it sets the radiative forcing. Specific sub-systems that are important and yet still poorly understood are clouds and sea ice; the thermohaline ocean circulation is a fundamentally important phenomenon that needs to be known better, and underlying these sub-systems and phenomena are the still ill-understood non-linear processes of advection (large-scale) and convection (small-scale) of dynamical and thermodynamical oceanic and atmospheric quantities. These sub-systems, phenomena, and processes are important and merit increased attention to improve prognostic capabilities generally.

14.2.3.1 Clouds

The role of clouds in the climate system continues to challenge the modelling of climate (e.g., Chapter 7, Section 7.2.2). It is generally accepted that the net effect of clouds on the radiative balance of the planet is negative and has an average magnitude of about 10 to 20 Wm^{-2}. This balance consists of a short-wave cooling (the albedo effect) of about 40 to 50 Wm^{-2} and a long-wave warming of about 30 Wm^{-2}. Unfortunately, the size of the uncertainties in this budget is large when compared to the expected anthropogenic greenhouse forcing. Although we know that the overall net effect of clouds on the radiative balance is slightly negative, we do not know the sign of cloud feedback with respect to the increase of greenhouse gases, and it may vary with the region. In fact, the basic issue of the nature of the future cloud feedback is not clear. Will it remain negative? If the planet warms, then it is plausible that evaporation will increase, which probably implies that liquid water content will increase but the volume of clouds may not. What will be the effect and how will the effects be distributed in time and space? Finally, the issue of cloud feedbacks is also coupled to the very difficult issue of indirect aerosol forcing (see Chapter 5, Section 5.3).

The importance of clouds was summarised in the SAR: "The single largest uncertainty in determining the climate sensitivity to either natural or anthropogenic changes are clouds and their effects on radiation and their role in the hydrological cycle" (Kattenberg *et al.*, 1996, p.345). And yet, the single greatest source of uncertainty in the estimates of the climate sensitivity continues to be clouds (see also Chapter 7, Section 7.2). Since the SAR, there have been a number of improvements in the simulation of both the cloud distribution and in the radiative properties of clouds (Chapter 7, Section 7.2.2). The simulation of cloud distribution has improved as the overall simulation of the atmospheric models has improved. In addition, the cloud sub-component models used in the coupled models have become more realistic. Also, our understanding of the radiative properties of clouds and their effects on climate sensitivity have improved. And yet in Chapter 7, Section 7.2.2 we find that, "In spite of these improvements, there has been no apparent narrowing of the uncertainty range associated with cloud feedbacks in current climate change simulations."

Handling the physics and/or the parametrization of clouds in climate models remains a central difficulty. There is a need for increased observations. J. Mitchell highlighted the challenge in a recent paper at the World Climate Research Programme (WCRP) Workshop on Cloud Properties and Cloud Feedbacks in Large-scale Models where he stated that "Reducing the uncertainty in cloud-climate feedbacks is one of the toughest challenges facing atmospheric physicists" (Mitchell, 2000).

Cloud modelling is a particularly challenging scientific problem because it involves processes covering a very wide range of space- and time-scales. For example, cloud systems extending over thousands of kilometres to cloud droplets and aerosols of microscopic size are all important components of the climate system. The time-scales of interest can range from hundreds of years (e.g., future equilibrium climates) to fractions of a second (e.g., droplet collisions). This is not to say that all cloud micro-physics must be included in modelling cloud formation and cloud properties, but the demarcation between what must be included and what can be parametrized remains unclear. Clarifying this demarcation and improving both the resulting phenomenological characterisations and parametrizations will depend critically on improved global observations of clouds (see Chapter 2, Section 2.5.5; see also Senior, 1999). Of particular importance are observations of cloud structure and distribution against natural patterns of climate variability (e.g., ENSO). Complementing the broad climatologies will be important observations of cloud ice-

water and liquid-water content, radiative heating and optical depth profiles, and precipitation occurrence and cloud geometry.

The recently approved CloudSat and PICASSO missions, which will fly in formation with the National Aeronautics and Space Administration (USA) (NASA) Earth Observing System (EOS) PM (the Aqua Mission), will provide valuable profiles of cloud ice and liquid content, optical depth, cloud type, and aerosol properties. These observations, combined with wider swath radiometric data from EOS PM sensors, will provide a rich new source of information about the properties of clouds (Stephens *et al.*, 2000).

And yet, this question of cloud feedback remains open, and it is not clear how it will be answered. Given that the current generation of global climate models represents the Earth in terms of grid-points spaced roughly 200 km apart, many features observed on smaller scales, such as individual cloud systems and cloud geometry, are not explicitly resolved. Without question, the strategy for attacking the feedback question will involve comparison of model simulations with appropriate observations on global or local scales. The interplay of observation and models, again, will be the key for progress. Mitchell (Mitchell, 2000) states this clearly, "Unless there are stronger links between those making observations and those using climate models, then there is little chance of a reduction in the uncertainty in cloud feedback in the next twenty years." This is echoed in this report (see Chapter 7, Section 7.2.2), "A straightforward approach of model validation is not sufficient to constrain efficiently the models and a more dedicated approach is needed. It should be favoured by a larger availability of satellite measurements."

14.2.3.2 Thermohaline circulation

In the oceanic component of climate models, ocean current patterns are represented significantly better in models of higher resolution in large part because ocean current systems (including mesoscale eddies), ocean variability (including ENSO events), and the thermohaline circulation (and other vertical mixing processes) and topography which greatly influence the ocean circulation, can be better represented. Improved resolution and understanding of the important facets of coupling in both atmosphere and ocean components of global climate models have also been proven to reduce flux imbalance problems arising in the coupling of the oceanic and the atmospheric components. However, it must still be noted that uncertainties associated with clouds still cause problems in the computation of surface fluxes. With the availability of computer power, a central impediment to the gain in model accuracy is being reduced; however, there is still a long way to go before many of the important processes are explicitly resolved by the numerical grid. In addition there continues to be a necessary "concomitant" increase in resources for process studies and for diagnosis as computer power increases. It must still be remembered that the system presents chaotic characteristics that can only be evaluated through an analysis of ensembles statistics, and these ensembles must be generated by running suites of models under varied initial and forcing conditions.

In a few model calculations, a large rate of increase in the radiative forcing of the planet is enough to cause the ocean's global thermohaline circulation almost to disappear, though in some experiments it reappears given sufficiently long integration times (see Chapter 7, Section 7.3.7 and Chapter 9, 9.3.4.3). This circulation is important because in the present climate it is responsible for a large portion of the heat transport from the tropics to higher latitudes, and it plays an important role in the oceanic uptake of CO_2. Palaeo-oceanographic investigations suggest that aspects of longer-term climate change are associated with changes in the ocean's thermohaline circulation. We need appropriate observations of the thermohaline circulation, and its natural variations, to compare with model simulations (see Chapter 9, Section 9.3.4.3; see also Chapter 7, Section 7.6 and Chapter 8, Section 8.5.2.2).

The coming decade will be important for ocean circulation in the context of climate. A particularly exciting development is the potential for assimilating synoptic ocean observations (e.g., the US/French ocean TOPography satellite altimeter EXperiment (TOPEX-POSEIDON) and Argo) into ocean general circulation models. Key questions, such as how well do the ocean models capture the inferred heat flux or tracer distributions, are central to the use of these models in climate studies. The effort of comparing models with data, as the direct path for model rejection and model improvement, is central to increasing our understanding of the system.

14.2.3.3 Arctic sea ice

There is increasing evidence that there is a decline in the extent and thickness of Arctic sea ice in the summer that appears to be connected with the observed recent Arctic warming (see Chapter 2, Section 2.2.5.2; Chapter 7, Box 7.1, and Chapter 8, Section 8.5.3; see also Chapter 7, Section 7.5.2 for a general discussion on the role of sea ice in the climate system as well as recent advances in modelling sea ice).

It is not known whether these changes reflect anthropogenic warming transmitted either from the atmosphere or the ocean or whether they mostly reflect a major mode of multi-decadal variability. Some of this pattern of warming has been attributed to recent trends in the Arctic Oscillation (see Section 2.6); however, how the anthropogenic signal is imprinted on the natural patterns of climate variability remains a central question. What does seem clear is that the changes in Arctic sea ice are significant, and there is a positive feedback that could be triggered by declines in sea-ice extent through changes in the planetary albedo. If the Arctic shifted from being a bright summer object to a less bright summer object, then this would be an important positive feedback on a warming pattern (see the "left loop" in Chapter 7, Figure 7.6).

In addition to these recently available observations, there have been several models (Commonwealth Scientific and Industrial Research Organisation (Australia) (CSIRO) – Gordon and O'Farrell, 1997; Department of Energy (USA) Parallel Climate Model (DOE PCM) – Washington *et al.*, 2000; National Center for Atmospheric Research (USA) Climate System Model (NCAR CSM) – Weatherly *et al.*, 1998; see also Chapter 7, Section 7.5.2 and Chapter 8, Section 8.5.3) that have improved their sea ice representation since the SAR. These improvements include simulation of open water within the ice pack, snow cover upon the ice, and sea ice dynamics. The incorporation of sophisticated sea ice components in climate models provides a framework for testing and calibrating these models with observations. Further, as the formulation of sea ice dynamics becomes more realistic, the validity of spatial patterns of the simulated wind stress over the polar oceans is becoming an issue in Atmosphere-Ocean General Circulation Model (AOGCM) simulations. Hence, improvements, such as the above-mentioned data, in the observational database will become increasingly relevant to climate model development. In addition, satellite observations have recently been used to determine sea-ice velocity (Emery *et al.*, 1997) and melt season (Smith, 1998).

New field programmes are under way with the explicit goal of improving the accuracy of model simulations of sea ice and polar climate (see Randall *et al.*, 1998, for a review). In order to improve model representations and validation, it will be essential to enhance the observations over the Arctic including ocean, atmosphere, and sea ice state variables. This will help provide more reliable projections for a region of the world where significant changes are expected.

The refinement of sea-ice models along with enhanced observations reduces the uncertainty associated with ice processes. (See Chapter 7, Section 7.5 and Chapter 8, Section 8.5.3 for more discussion and evaluation of model performance; for some open issues see Chapter 9, Section 9.4.) This progress is important, and efforts are needed to expand upon it and, as stated, to improve the observational basis significantly.

14.2.4 The Global Carbon Cycle

From measurements of air trapped in ice cores and from direct measurements of the atmosphere, we know that in the past 200 years the abundance of CO_2 in the atmosphere has increased by over 30% (i.e., from a concentration of 280 ppm by volume (pppmv) in 1700 to nearly 370 ppmv in 2000). We also know that the concentration was relatively constant (roughly within ±10 ppmv of 275) for more than 1,000 years prior to the human-induced rapid increase in atmospheric CO_2 (see Chapter 3, Figures 3.2a and 3.2b).

Looking further back in time, we find an extraordinarily regular record of change. The Vostok core (Figure 3.2d) captures a remarkable and intriguing signal of the periodicity of inter-glacial and glacial climate periods in step with the transfer of significant pools of carbon from the land (most likely through the atmosphere) to the ocean and then the recovery of terrestrial carbon back from the ocean. The repeated pattern of a 100 to 120 ppmv decline in atmospheric CO_2 from an inter-glacial value of 280 to 300 ppmv to a 180 ppmv floor and then the rapid recovery as the planet exits glaciation suggests a tightly governed control system. There is a similar methane (CH_4) cycle between 320 to 350 ppbv (parts per billion by volume) and 650 to 770 ppbv. What begs explanation is not just the linked periodicity of carbon and glaciation, but also the apparent consistent limits on the cycles over the period. See Chapter 3, Box 3.4.

Today's atmosphere, imprinted with the fossil fuel CO_2 signal, stands at nearly 90 to 70 ppmv above the previous inter-glacial maximum of 280 to 300 ppmv. The current methane value is even further (percentage-wise) from its previous inter-glacial high values. In essence, carbon has been moved from a relatively immobile pool (in fossil fuel reserves) in the slow carbon cycle to the relatively mobile pool (the atmosphere) in the fast carbon cycle, and the ocean, terrestrial vegetation and soils have yet to equilibrate with this "rapidly" changing concentration of CO_2 in the atmosphere.

Given this remarkable and unprecedented history one cannot help but wonder about the characteristics of the carbon cycle in the future (Chapter 3). To understand better the global carbon cycle, two themes are clear: (1) there is a need for global observations that can contribute significantly to determining the sources and sinks of carbon and (2) there is a need for fundamental work on critical biological processes and their interaction with the physical system. Two observational needs must be highlighted:

- Observations that would decisively improve our ability to model the carbon cycle. For example, a dense and well-calibrated network for monitoring CO_2 and O_2 concentrations that will also be required for international verification of carbon sources and sinks is central.

- "Benchmarks" data sets that allow model intercomparison activities to move in the direction of becoming data-model comparisons and not just model-model comparisons.

We note that the Subsidiary Body for Scientific and Technological Advice (SBSTA) of the United Nations Framework Convention on Climate Change (UNFCCC) recognised the importance of an Integrated Global Observing Strategy Partnership in developing observing systems for the oceans and terrestrial carbon sources and sinks in the global carbon cycle and in promoting systematic observations.

There is also a range of areas where present day biogeochemistry modelling is not only in need of additional data, but is also crucially limited by insufficient understanding at the level of physical or biological processes. Clarifying these processes and their controls is central to a better understanding of the global carbon cycle.

14.2.4.1 The marine carbon system

The marine carbon cycle plays an important role in the partitioning of CO_2 between the atmosphere and the ocean (Chapter 3, Section 3.2.3). The primary controls are the circulation of the ocean (a function of the climate system), and two important biogeochemical processes: the solubility pump and the biological pump, both of which act to create a global mean increase of dissolved inorganic carbon with depth.

The physical circulation and the interplay of the circulation and the biogeochemical processes are central to understanding the ocean carbon system and future concentrations of CO_2 in the atmosphere. In the ocean, the prevailing focus on surface conditions and heat transport has led to a comparative neglect of transport processes below about 800 m depth. For carbon cycle modelling, however, vertical transports and deep horizontal transports assume fundamental importance. The importance of the thermohaline circulation is obviously

important (and insufficiently well understood; see Section 14.2.3.2) in moving carbon from the surface to deeper layers. Similarly, the regional distribution of upwelling, which brings carbon- and nutrient-rich water to surface layers, is poorly known and inconsistently simulated in models. The ventilation of the Southern Ocean provides an extreme, though not unique, example.

It has been pointed out by a number of modelling studies that if there were no marine biological system, then the pre-industrial atmospheric CO_2 concentration would have been 450 ppmv instead of 280 ppmv (Sarmiento and Toggweiler 1984; Maier-Raimer *et al.*, 1996). Any complete model of the natural ocean carbon cycle should therefore include the biological system; however, most recent assessments of the oceanic uptake of anthropogenic CO_2 have assumed that the biological system would not be affected by climate change and have therefore only modelled the chemical solubility in addition to the physical circulation. This was based on the understanding that nitrate or other nutrients limit marine phytoplankton growth. There would therefore be no CO_2 fertilisation effect as has been suggested for terrestrial plants and that, unless there was a large change in the nutrient supply to the upper ocean because of a climate-induced shift in circulation, then no extra anthropogenic CO_2 could be sequestered to the deep ocean by the organic matter pump. More recently, a number of studies have suggested possible ways in which the organic matter pump might be affected by climate change over a 200-year timescale (see Chapter 3, Sections 3.2.3.2 and 3.2.3.3). The main conclusion was that, because of the complexity of biological systems, it was not yet possible to say whether some of the likely feedbacks would be positive or negative. However, it is clear that our understanding of these issues needs to be improved.

Simulating the calcium carbonate system with a process-oriented model presents another level of complexity beyond simulating the organic matter formation-decomposition: the distribution of particular phytoplankton species (mainly coccolithophorids) must be simulated. The calcium carbonate pump, however, contributes relatively little to the vertical dissolved inorganic carbon (DIC) gradient compared to the organic matter and solubility pumps. The importance of this pump needs careful evaluation and its past (palaeo) role in the carbon cycle needs to be considered (see end of Chapter 3, Section 3.2.3.3).

In the ocean, models incorporating biology are relatively underdeveloped and incorporate empirical assumptions (such as fixed Redfield (nutrient) ratios) rather than explicitly modelling the underlying processes. As a result, present models may be unduly constrained in the range of responses they can show to changes in climate and ocean dynamics. A better understanding is required concerning the workings of nutrient constraints on productivity, the controls of nitrogen fixation, and the controls on the geographical distribution of biogeochemically important species and functional types in the ocean. To develop this understanding it will be necessary to combine remotely sensed information with a greatly expanded network of continuous biogeochemical monitoring sites, and to gather data on the space-time patterns of variability in species composition of marine ecosystems in relation to climate variability phenomena such as ENSO and NAO. (See Chapter 3, Sections 3.6.3 and 3.7).

14.2.4.2 The terrestrial system

The metabolic processes that are responsible for plant growth and maintenance and the microbial turnover, which is associated with dead organic matter decomposition, control the cycle of carbon, nutrients, and water through plants and soil on both rapid and intermediate time-scales. Moreover, these cycles affect the energy balance and provide key controls over biogenic trace gas production. Looking at the carbon fixation-organic material decomposition as a linked process, one sees that some of the carbon fixed by photosynthesis and incorporated into plant tissue is perhaps delayed from returning to the atmosphere until it is oxidised by decomposition or fire. This slower carbon loop through the terrestrial component of the carbon cycle affects the rate of growth of atmospheric CO_2 concentration and, in its shorter term expression, imposes a seasonal cycle on that trend (Chapter 3, Figure 3.2a). The structure of terrestrial ecosystems, which respond on even longer time-scales, is determined by the integrated response to changes in climate and to the intermediate time-scale carbon-nutrient machinery. The loop is closed back to the climate system, since it is the structure of ecosystems, including species composition, that largely sets the terrestrial boundary condition of the climate in terms of surface roughness, albedo, and latent heat exchange (see Chapter 3, Section 3.2.2).

Modelling interactions between terrestrial and atmospheric systems requires coupling successional models to biogeochemical models and physiological models that describe the exchange of water and energy between vegetation and the atmosphere at fine time-scales. At each step toward longer time-scales, the climate system integrates the more fine-scaled processes and applies feedbacks onto the terrestrial biome. At the finest time-scales, the influence of temperature, radiation, humidity and winds has a dramatic effect on the ability of plants to transpire. On longer time-scales, integrated weather patterns regulate biological processes such as the timing of leaf emergence or excision, uptake of nitrogen by autotrophs, and rates of organic soil decay and turnover of inorganic nitrogen. The effect of climate at the annual or interannual scale defines the net gain or loss of carbon by the biota, its water status for the subsequent growing season, and even its ability to survive.

As the temporal scale is extended, the development of dynamic vegetation models, which respond to climate and human land use as well as other changes, is a central issue. These models must not only treat successional dynamics, but also ecosystem redistribution. The recovery of natural vegetation in abandoned areas depends upon the intensity and length of the agricultural activity and the amount of soil organic matter on the site at the time of abandonment. To simulate the biogeochemistry of secondary vegetation, models must capture patterns of plant growth during secondary succession. These patterns depend substantially on the nutrient pools inherited from the previous stage. The changes in hydrology need also to be considered, since plants that experience water stress will alter the allocation of carbon to allocate more carbon to roots. Processes such as reproduction, establishment, and light competition have been added to such models, interactively with the carbon, nitrogen, and water cycles. Disturbance regimes such as fire are also incorporated into the models, and these disturbances are essential

in order to treat successfully competitive dynamics and hence future patterns of ecosystem. It should be noted also that these forcing terms themselves might be altered by the changes that result from changes in the terrestrial system.

This coupling across time-scales represents a significant challenge. Immediate challenges that confront models of the terrestrial-atmosphere system include exchanges of carbon and water between the atmosphere and land, and the terrestrial sources and sinks of trace gases.

Prognostic models of terrestrial carbon cycle and terrestrial ecosystem processes are central for any consideration of the effects of environmental change and analysis of mitigation strategies; moreover, these demands will become even more significant as countries begin to adopt carbon emission targets. At present, several rather complex models are being developed to account for the ecophysiological and biophysical processes, which determine the spatial and temporal features of primary production and respiration (see Chapter 3, Sections 3.6.2 and 3.7.1). Despite recent progress in developing and evaluating terrestrial biosphere models, several crucial questions remain open. For example, current models are highly inconsistent in the way they treat the response of Net Primary Production (NPP) to climate variability and climate change – even though this response is fundamental to predictions of the total terrestrial carbon balance in a changing climate. Models also differ significantly in the degree of CO_2 fertilisation they allow, and the extent to which CO_2 responses are constrained by nutrient availability; the extent to which CO_2 concentrations affect the global distribution of C_3 and C_4 photosynthetic pathways; and the impacts of climate, CO_2 and land management on the tree-grass balance. These are all areas where modelling capability is limited by lack of knowledge, thus making it crucially important to expand observational and experimental research. Important areas are interannual variability in terrestrial fluxes and the interplay of warming, management, and CO_2 enrichment responses at the ecosystem scale. Moreover, these issues must be far better resolved if there is to be an adequate verification scheme to confirm national performance in meeting targets for CO_2 emissions. (See Chapter 3, Sections 3.6.2 and 3.7.1.)

Finally, while progress will be made on modelling terrestrial processes, more integrative studies are also needed wherein terrestrial systems are coupled with models of the physical atmosphere and eventually with the chemical atmosphere as well.

14.2.5 Precipitation, Soil Moisture, and River Flow: Elements of the Hydrological Cycle

Changes in precipitation could have significant impacts on society. Precipitation is an essential element in determining the availability of drinking water and the level of soil moisture. Improved treatment of precipitation (see Section Chapter 7, 7.2.3) is an essential step.

Soil moisture is a key component in the land surface schemes in climate models, since it is closely related to evapotranspiration and thus to the apportioning of sensible and latent heat fluxes. It is primary in the formation of runoff and hence river-flow. Further, soil moisture is an important determinant of ecosystem structure and therein a primary means by which climate regulates (and is partially regulated by) ecosystem distribution. Soil moisture is an important regulator of plant productivity and sustainability of natural ecosystems. In turn terrestrial ecosystems recycle water vapour at the land-surface/atmosphere boundary, exchange numerous important trace gases with the atmosphere, and transfer water and biogeochemical compounds to river systems (see also the discussion in Chapter 7, Section 7.4.3 and Chapter 8, Section 8.5.4). New efforts are needed in the development of models, which successfully represent the space-time dynamics interaction between soil, climate and vegetation. If water is a central controlling aspect, then the interaction necessarily passes all the way through the space-time dynamics of soil moisture. Finally, adequate soil moisture is an essential resource for human activity. Consequently, accurate prediction of soil moisture is crucial for simulation of the hydrological cycle, of soil and vegetation biochemistry, including the cycling of carbon and nutrients, and of ecosystem structure and distribution as well as climate.

River systems are linked to regional and continental-scale hydrology through interactions among precipitation, evapotranspiration, soil water, and runoff in terrestrial ecosystems. River systems, and more generally the entire global water cycle, control the movement of constituents over vast distances, from the continental land-masses to the world's oceans and to the atmosphere. Rivers are also central features of human settlement and development.

It appears, however, that a significant level of variance exists among land models, associated with unresolved differences among parametrization details (particularly difficulties in the modelling of soil hydrology) and parameter sets. In fact, many of the changes in land-surface models since the SAR fall within this range of model diversity. It is not known to what extent these differences in land-surface response translate into differences in global climate sensitivity (see Chapter 8, Section 8.5.4.3) although the uncertainty associated with the land-surface response must be smaller than the uncertainty associated with clouds (Lofgren, 1995). There is model-based evidence indicating that these differences in the land-surface response may be significant for the simulation of the local land-surface climate and regional atmospheric climate changes (see Chapter 7, Section 7.4).

Much attention in the land-surface modelling community has been directed toward the diversity of parametrizations of water and energy fluxes (see Chapter 7, Sections 7.4, 7.5, and Chapter 8, Section 8.5). Intercomparison experiments (see Chapter 8, Section 8.5.4) have quantified the inter-model differences in response to prescribed atmospheric forcing, and have demonstrated that the most significant outliers can be understood in terms of unrealistic physical approximations in their formulation, particularly the neglect of stomatal resistance. Some coupled models now employ some form of stomatal resistance to evaporation.

Climate-induced changes in vegetation have potentially large climatic implications, but are still generally neglected in the coupled-model experiments used to estimate future changes in climate (see Chapter 8).

There is, obviously, a direct coupling between predicted soil moisture and predicted river flows and the availability of water for human use. Complex patterns of locally generated runoff are transformed into horizontal transport as rivers through the drainage basin. Moreover, any global perspective on surface hydrology must explicitly recognise the impact of human intervention in the water cycle, not only through climate and land-use change, but also through the operation of impoundments, inter-basin transfers, and consumptive use.

Recognition of the importance of land hydrology for the salinity distribution of the oceans is one reason for seeking improvements in models for routing runoff to the oceans (see more precise cites here and in Chapter 7). Most coupled models now return land runoff to the ocean as fresh water (see Chapter 8). Runoff is collected over geographically realistic river basins and mixed into the ocean at the appropriate river mouths. Although this routing is performed instantaneously in some models, the trend is toward model representation of the significant time-lag (order of a month) in runoff production to river-ocean discharge. What is needed for a variety of reasons, however, is for river flow itself to be treated in models of the climate system. (See Chapter 7, Section 7.4.3.)

On land, surface processes have until very recently been treated summarily in Atmospheric General Circulation Models (AGCMs). The focus of evaluating AGCMs has been on large-scale dynamics and certain meteorological variables; far less so on the partitioning of sensible and latent heat flux, or the moisture content of the planetary boundary layer. When the goals of climate modelling are expanded to include terrestrial biosphere function, such aspects become of central importance as regulators of the interaction between the carbon and water cycles. Terrestrial flux and boundary-layer measurements represent a new, expanding and potentially hugely important resource for improving our understanding of these processes and their representation in models of the climate system. (See Chapter 7, Section 7.4.1.)

The spatial resolution of current global climate models, roughly 200 km, is too coarse to simulate the impact of global change on most individual river basins. To verify the transport models will require budgets of water and other biogeochemical constituents for large basins of the world. This requires ground-based meteorology in tandem with remotely sensed data for a series of variables, including information on precipitation, soils, land cover, surface radiation, status of the vegetative canopy, topography, floodplain extent, and inundation. Model results can be constrained by using a database of observed discharge and constituent fluxes at key locations within the drainage basins analysed. Climate time-series and monthly discharge data for the past several decades at selected locations provide the opportunity for important tests of models, including appraisal of the impact of episodic events, such as El Niño, on surface water balance and river discharge. It will be necessary to inventory, document, and make available such data sets to identify gaps in our knowledge, and where it is necessary to collect additional data. Even in the best-represented regions of the globe coherent time-series are available for only the last 30 years or less. This lack of data constrains our ability to construct and test riverine flux models. Standardised protocols, in terms of sampling frequency, spatial distribution of sampling networks, and chemical analyses are needed to ensure the production of comparable data sets in disparate parts of the globe. Upgrades of the basic monitoring system for discharge and riverborne constituents at the large scale are therefore required.

In sum, hydrological processes and energy exchange, especially those involving clouds, surface exchanges, and interactions of these with radiation are crucial for further progress in modelling the atmosphere. Feedbacks with land require careful attention to the treatments of evapotranspiration, soil moisture storage, and runoff. All of these occur on spatial scales which are fine compared with the model meshes, so the question of scaling must be addressed. These improvements must be paralleled by the acquisition of global data sets for validation of these treatments. Validation of models against global and regional requirements for conservation of energy is especially important in this regard. As noted in Chapter 8 (Section 8.5.4.3), "Uncertainty in land surface processes, coupled with uncertainty in parameter data combines, at this time, to limit the confidence we have in the simulated regional impacts of increasing CO_2."

14.2.6 Trace Gases, Aerosols, and the Climate System

The goal is a completely interactive simulation of the dynamical, radiative, and chemical processes in the atmosphere-ocean-land system with a central theme of characterising adequately the radiative forcing in the past, in the present, and into the future (See Chapter 6, Sections 6.1 and 6.2; see also Chapter 9, Section 9.1). Such a model will be essential in future studies of the broad question on the role of the oceans, terrestrial ecosystems, and human activities in the regulation of atmospheric concentrations of CO_2 and other radiatively active atmospheric constituents. It will be required for understanding tropospheric trace constituents such as nitrogen oxides, ozone, and sulphate aerosols. Nitrogen oxides are believed to control the production and destruction of tropospheric ozone, which controls the chemical reactivity of the lower atmosphere and is itself a significant greenhouse gas. Tropospheric sulphate aerosols, carbonaceous aerosols from both natural and anthropogenic processes, dust, and sea salt, on the other hand, are believed to affect the Earth's radiation budget significantly, by scattering solar radiation and through their effects on clouds. Systematic observations of different terrestrial ecosystems and surface marine systems under variable meteorological conditions are needed along with the development of ecosystem and surface models that will provide parametrizations of these exchanges.

Models that incorporate atmospheric chemical processes provide the basis for much of our current understanding in such critical problem areas as acid rain, photochemical smog production in the troposphere, and depletion of the ozone layer in the stratosphere. These formidable problems require models that include chemical, dynamical, and radiative processes, which through their mutual interactions determine the circulation, thermal structure, and distribution of constituents in the atmosphere. That is, the problems require a coupling of the physics and chemistry of the atmosphere. Furthermore, the models must be applicable on a variety of spatial (regional-to-global) and temporal (days-to-decades) scales (see Chapter 6). A particularly important and challenging issue is the need to reduce the uncertainty on the size and spatial pattern of the indirect aerosol effects (see Chapter 6, Section 6.8).

Most of the effort in three-dimensional atmospheric chemistry models over the last decade has been in the use of transport models in the analysis of certain chemically active species, e.g., long-lived gases such as nitrous oxide (N_2O) or the chlorofluorocarbons (CFCs). In part, the purpose of these studies was not to improve our understanding of the chemistry of the atmosphere, but rather to improve the transport formulation associated with general circulation models and, in association with this improvement, to understand sources and sinks of CO_2. The additional burden imposed by incorporating detailed chemistry into a comprehensive general circulation model has made long-term simulations and transient experiments with existing computing resources challenging. Current three-dimensional atmospheric chemistry models which focus on the stratosphere seek a compromise solution by employing coarse resolution (both vertical and horizontal dimensions); incorporating constituents by families (similar to the practice used in most two-dimensional models); omitting or simplifying parametrizations for tropospheric physical processes; or conducting "off line" transport simulations in which previously calculated wind and temperature fields are used as known input to continuity equations including chemical source/sink terms. This last approach renders the problem tractable and has produced much progress towards understanding the transport of chemically reacting species in the atmosphere. The corresponding disadvantage is the lack of the interactive feedback between the evolving species distributions and the atmospheric circulation. Better descriptions of the complex relationship between hydrogen, nitrogen, and oxygen species as well as hydrocarbons and other organic species are needed in order to establish simplified chemical schemes that will be implemented in chemical/transport models. In parallel, better descriptions of how advection, turbulence, and convection affect the chemical composition of the atmosphere are needed. (See Chapter 4, Section 4.5.2.)

We also need improved understanding of the processes involving clouds, surface exchanges, and their interactions with radiation. The coupling of aerosols with both the energy and water cycles as well as with the chemistry components of the system is of increasing importance. Determining feedbacks between the land surface and other elements of the climate system will require careful attention to the treatments of evapotranspiration, soil moisture storage and runoff. All of these occur on spatial scales that are small compared with the model meshes, so the question of scaling must be addressed. These improvements must be paralleled by the acquisition of global data sets for validation of these treatments. Validation of models against global and regional requirements for conservation of energy is especially important in this regard. (See Chapter 4, Section 4.5.1.)

The problems associated with how to treat clouds within the climate system are linked to problems associated with aerosols. Current model treatments of climate forcing from aerosols predict effects that are not easily consistent with the past climate record. A major challenge is to develop and validate the treatments of the microphysics of clouds and their interactions with aerosols on the scale of a general circulation model grid. A second major challenge is to develop an understanding of the carbon components of the aerosol system. Meeting this challenge requires that we develop data for a mechanistic understanding of carbonaceous aerosol effects on clouds as well as developing an understanding of the magnitude of the anthropogenic and natural components of the carbonaceous aerosol system. (See Chapter 6, Sections 6.7 and 6.8; see also Chapter 4, Section 4.5.1.2.)

As attention is turned toward the troposphere, the experimental strategy simply cannot adopt the stratospheric simplifications. The uneven distribution of emission sources at the surface of the Earth and the role of meteorological processes at various scales must be addressed directly. Fine-scaled, three-dimensional models of chemically active trace gases in the troposphere are needed to resolve transport processes at the highest possible resolution. These models should be designed to simulate the chemistry and transport of atmospheric tracers on global and regional scales, with accurate parametrizations of sub-scale processes that affect the chemical composition of the troposphere. It is therefore necessary to pursue an ambitious long-term programme to develop comprehensive models of the troposphere system, including chemical, dynamical, radiative, and eventually biological components. (See Chapter 4, Sections 4.4 to 4.6.)

The short-lived radiatively important species pose an observational challenge. The fact that they are short-lived implies that observations of the concentrations are needed over wide spatial regions and over long periods of time. This is particularly important for aerosols. The current uncertainties are non-trivial (see again Chapter 6, Figure 6.7) and need to be reduced.

In sum, there needs to be an expanded attack on the key contributors to uncertainty about the behaviour of the climate system today and in the future. As stated in Chapter 13, Section 13.1.2, "Scenarios should also provide adequate quantitative measures of uncertainty. The sources of uncertainty are many, including the trajectory of greenhouse gas emissions in the future, their conversion into atmospheric concentrations, the range of responses of various climate models to a given radiative forcing and the method of constructing high resolution information from global climate model outputs (see Chapter 13, Figure 13.2). For many purposes, simply defining a single climate future is insufficient and unsatisfactory. Multiple climate scenarios that address at least one, or preferably several, sources of uncertainty allow these uncertainties to be quantified and explicitly accounted for in impact assessments."

In addition to this needed expansion in the attack on uncertainties in the climate system, there is an important new challenge that should now be addressed more aggressively. It is time to link more formally physical climate-biogeochemical

models with models of the human system. At present, human influences generally are treated only through emission scenarios that provide external forcings to the climate system. In future comprehensive models, human activities will interact with the dynamics of physical, chemical, and biological sub-systems through a diverse set of contributing activities, feedbacks, and responses. This does not mean that it is necessary or even logical to attempt to develop prognostic models of human actions since much will remain inherently unpredictable; however, the scenarios analysis could and should be more fully coupled to the coupled physical climate-biogeo-chemical system.

As part of the foundation-building to meet this challenge, we turn attention now to the human system.

14.3 The Human System

14.3.1 Overview

Human processes are critically linked to the climate system as contributing causes of global change, as determinants of impacts, and through responses. Representing these linkages poses perhaps the greatest challenge to modelling the total Earth system. But understanding them is essential to understanding the behaviour of the whole system and to providing useful advice to inform policy and response. Significant progress has been made, but formidable challenges remain.

Human activities have altered the Earth system, and many such influences are accelerating with population growth and technological development. The use of fossil fuels and chemical fertilisers are major influences, as is the human transformation of much of the Earth's surface in the past 300 years.

Land-use change illustrates the potential complexity of linkages between human activity and major non-human components of the Earth system. The terrestrial biosphere is fundamentally modified by land clearing for agriculture, industrialisation, urbanisation, and by forest and rangeland management practices. These changes affect the atmosphere through an altered energy balance over the more intensively managed parts of the land surface, as well as through changed fluxes of water vapour, CO_2, CH_4 and other trace gases between soils, vegetation, and the atmosphere. Changed land use also greatly alters the fluxes of carbon, nutrients, and inorganic sediments into river systems, and consequently into oceanic coastal zones.

The response of the total Earth system to these changes in anthropogenic forcing is currently not known. Sensitivity studies with altered land cover distributions in general circulation models have shown that drastic changes, such as total deforestation of all tropical or boreal forests, may lead to feedbacks in atmospheric circulation and a changed climate that would not support the original vegetation (e.g., Claussen, 1996). Regional climate simulations, on the other hand, have shown that at the continental scale, important teleconnections may exist through which more modest tropical forest clearing may cause a change in climate in undisturbed areas. Coupling the global to the local is a key challenge; regional studies may prove to be uniquely valuable.

Human land-use change will continue and probably accelerate due to increasing demands for food and fibre, changes in forest and water management practices, and possibly large-scale projects to sequester carbon in forests or to produce biomass fuels. In addition, anthropogenic changes in material and energy fluxes, resulting from such activities as fossil fuel combustion and chemical fertiliser use, are expected to increase in the coming decades. Predictions of changes in the carbon and nitrogen cycles are sensitive to estimates of human activity and predictions of the impacts of these global changes must take into account human vulnerability, adaptation, and response. Predicting the future response of the Earth system to changes in climate and in parallel to changes in land use and land cover will require projections of trends in the human contributions to these global changes; this sort of modelling presents difficult challenges because of the multiple factors operating at local, regional, continental, and global levels to influence local land-use decisions.

In sum, the human element probably represents the most important aspect both of the causes and effects of climate change and environmental impacts. Any policy intervention will have human activities as its immediate target.

14.3.2 Humans: Drivers of Global Change: Recipients of Global Change

The provision of useful guidance to inform policy requires observation and description of human contributions to global change, as well as theoretical studies of the underlying social processes that shape them. We also need to understand how global change affects human welfare. This requires not merely studies of direct exposure but also of the capacity to respond.

Causal models of social processes have large uncertainties, and pose problems that are of a qualitatively different character than those encountered in modelling non-human components of the Earth system. This is due, first and foremost, to the inherent reflexivity of human behaviour; i.e., the fact that human beings have intellectual capabilities and emotional endowments enabling them to invent new solutions and transcend established "laws" in ways that no other species can do. As a consequence, predictive models may well alter the behaviour that they seek to predict and explain – indeed, such models are sometimes deliberately used exactly for that purpose. Moreover, the diversity of societies, cultures, and political and economic systems often frustrates attempts to generalise findings and propositions from one setting to another. Representation of human behaviour at the micro (individual) and macro (collective) scale may require fundamentally different approaches (see Gibson *et al.*, 1998).

These kinds of difficulties intrinsically limit the predictive power that can be ascribed to models of social processes. As a consequence, research on human drivers and responses to climate change cannot be expected to produce conventional predictions beyond a very short time horizon. This does not imply, however, that research on human behaviour and social processes cannot provide knowledge and insight that can inform policy deliberations. A considerable amount of basic knowledge and insight exist, and this knowledge can be used, *inter alia*, for constructing

scenarios showing plausible trajectories and identifying the critical factors that will have to be targeted in order to switch from one trajectory to another. From the perspective of policy-makers, this can indeed be an important contribution.

To make the most of this potential, further progress is required along two main frontiers. One challenge is to develop a more integrated understanding of social systems and human behaviour. With some exceptions, the first generation of models in this area represented "the human system" by a few key variables. For example, resource use was often conceived of as a function of population size and income level. The performance of such simplistic models was by-and-large poor. It is abundantly clear that the impact of human activities as drivers of climate change depends upon a complex set of interrelated factors, including also technologies in use, social institutions, and individual beliefs, attitudes, and values. At present, it seems fair to say that we have a reasonably good theoretical grasp of important types of institutions, such as markets and hierarchies, in ideal-type form. What we need to understand better is how their impure real-world counterparts work, and to improve our understanding of the intricate interplay of institutional complexes, i.e., how markets, governments and other social institutions interact to shape human behaviour. Research in political economy clearly indicates that phenomena such as economic growth are to a significant extent affected by the functioning of interlocking networks of institutional arrangements.

Similarly, we have a fairly good grasp on particular kinds of intellectual processes – in particular, the logic of rational choice – but we are doing less well when it comes to understanding how beliefs, attitudes and values change and how change in these factors in turn affects manifest human behaviour, such as consumption patterns. To address these challenges we need more interdisciplinary research designed to integrate knowledge from different fields and sub-fields into a more holistic understanding of "the human system". The intellectual and organisational problems involved should not be underestimated, but we are confident that the prospects for making progress along this frontier are better now than ever before.

The other main challenge is to find better ways of integrating models of the biogeophysical Earth system with models of social systems and human behaviour. Some encouraging progress has been made at this interface, particularly over the last decade. For example, there has been a rapid increase in attempts to integrate representations of human activities in models with explicit formal linkages to other components of the Earth system. Such integrated assessment models have offered preliminary characterisations of human-climate linkages, particularly through models of multiple linked human and climate stresses on land cover. Moreover, they have provided preliminary characterisation of broad classes of policy responses, and have been employed to characterise and prioritise policy-relevant uncertainties.

Yet, effective integration is frustrated by at least two main obstacles. One is incongruity of temporal and spatial scales. Social science research cannot match the long time horizons of much natural science research. On the other hand, in studying consequences for human welfare and responses to these consequences, social scientists need estimates of biophysical impacts of climate change differentiated by political units or even smaller social systems. Aggregate global-scale estimates are of limited use in this context; human sensitivity to climate change varies significantly across regions and social groups, and so does response capacity. We can expect to see some progress in alleviating the spatial resolution problem, as regional-scale models of climate change are further developed, but we have to recognise that the scale problems are fundamental and that no quick fixes are in sight. The other problem pertains to the interface between different methodological approaches. In particular, concerted efforts are required to develop better tools for coupling approaches relying on numerical modelling with "softer" approaches using interpretative frameworks and qualitative methods. Some of these differences are too profound to be eliminated, but that does not imply that bridges cannot be built. Learning how to work more effectively across these methodological divides is essential to the further development of integrated global change research. Again, some encouraging progress is being made.

14.4 Outlook

There is a growing recognition in the scientific community and more broadly that:

- The Earth functions as a system, with properties and behaviour that are characteristic of the system as a whole. These include critical thresholds, "switch" or "control" points, strong non-linearities, teleconnections, chaotic elements, and unresolvable uncertainties. Understanding the components of the Earth system is critically important, but is insufficient on its own to understand the functioning of the Earth system as a whole.

- Humans are now a significant force in the Earth system, altering key process rates and absorbing the impacts of global environmental changes. The environmental significance of human activities is now so profound that the current geological era can be called the "Anthropocene" (Crutzen and Stoermer, 2000).

A scientific understanding of the Earth system is required to help human societies develop in ways that sustain the global life support system. The clear challenge of understanding climate variability and change and the associated consequences and feedbacks is a specific and important example of the need for a scientific understanding of the Earth as a system. It is also clear that the scientific study of the whole Earth system, taking account of its full functional and geographical complexity over time, requires an unprecedented effort of international collaboration. It is well beyond the scope of individual countries and regions.

The world's scientific community, working in part through the three global environmental change programmes (the International Geosphere-Biosphere Programme (IGBP), the International Human Dimensions Programme on Global Environmental Change (IHDP), and the World Climate Research Programme (WCRP)), has built a solid base for understanding the Earth system. The IGBP, IHDP and WCRP have also developed effective and efficient strategies for implementing global environmental change research at the international level. The challenge to

IGBP, IHDP and WCRP is to build an international programme of Earth system science, driven by a common mission and common questions, employing visionary and creative scientific approaches, and based on an ever closer collaboration across disciplines, research themes, programmes, nations and regions.

We need to build on our existing understanding of the Earth system and its interactive human and non-human processes through time in order to:

- improve evaluation and understanding of current and future global change; and

- place on an increasingly firm scientific basis the challenge of sustaining the global environment for future human societies.

The climate system is particularly challenging since it is known that components in the system are inherently chaotic, and there are central components which affect the system in a non-linear manner and potentially could switch the sign of critical feedbacks. The non-linear processes include the basic dynamical response of the climate system and the interactions between the different components. These complex, non-linear dynamics are an inherent aspect of the climate system. Amongst the important non-linear processes are the role of clouds, the thermohaline circulation, and sea ice. There are other broad non-linear components, the biogeochemical system and, in particular, the carbon system, the hydrological cycle, and the chemistry of the atmosphere.

Given the complexity of the climate system and the inherent multi-decadal time-scale, there is a central and unavoidable need for long-term consistent data to support climate and environmental change investigations. Data from the present and recent past, credible global climate-relevant data for the last few centuries, along with lower frequency data for the last several millennia, are all needed. Research observational data sets that span significant temporal and spatial scales are needed so that models can be refined, validated, or perhaps, most importantly, rejected. The elimination of models because they are in conflict with climate-relevant data is particularly important. Running unrealistic models will consume scarce computing resources, and the results may add unrealistic information to the needed distribution functions. Such data must be adequate in temporal and spatial coverage, in parameters measured, and in precision, to permit meaningful validation. We are still unfortunately short of data for the quantitative assessment of extremes on the global scale in the observed climate.

In sum, there is a need for:

- more comprehensive data, contemporary, historical, and palaeological, relevant to the climate system;

- expanded process studies that more clearly elucidate the structure of fundamental components of the Earth system and the potential for changes in these central components;

- greater effort in testing and developing increasingly comprehensive and sophisticated Earth system models;

- increased emphasis upon producing ensemble calculations of Earth system models that yield descriptions of the likelihood of a broad range of different possibilities, and finally;

- new efforts in understanding the fundamental behaviour of large-scale non-linear systems.

These are significant challenges, but they are not insurmountable. The challenges to understanding the Earth system including the human component are daunting, and the pressing needs are significant. However, the opportunity for progress exists, and, in fact, this opportunity simply must be realised. The issues are too important, and they will not vanish. The challenges simply must be met.

References

Claussen, M., 1996: Variability of global biome patterns as a function of initial and boundary conditions in a climate model, *Clim. Dyn.,* **12**, 371-379.

Crutzen, P., and E. Stoermer, 2000: *International Geosphere Biosphere Programme (IGBP) Newsletter,* **41**.

Emery, W. J., C. W. Fowler, and J. A. Maslanik, 1997: Satellite-derived maps of Arctic and Antarctic sea-ice motion: 1988-1994. *Geophys. Res. Lett.,* **24**, 897-900.

Gibson, C., E. Ostrom, and Toh-Kyeong Ahn. 1998: Scaling Issues in the Social Sciences. IHDP Working Paper No. 1. International Human dimensions Programme on Global Environmental Change. Bonn, Germany. See also http://www.uni-bonn.de/ihdp.

Gordon, H. B., and S. P. O´Farrell, 1997: Transient climate change in the CSIRO coupled model with dynamic sea ice. *Mont. Weath. Rev.,* **125(5),** 875-907.

Gordon, C., C. Cooper, C.A. Senior, H. Banks, J.M. Gregory, T.C. Johns, J.F.B. Mitchell, and R.A. Wood, 1999: The simulation of SST, sea-ice extents and ocean heat transport in a version of the Hadley Centre coupled model without flux adjustments. Accepted, *Clim. Dyn.*

IPCC, 1996: Climate Change 1995: The Science of Climate Change. Contribution of Working Group 1 to the Second Assessment Report of the Intergovernmental Panel on Climate Change. Houghton, J.T., L.G.M. Filho, B.A. Callandar, N. Harris, A. Kattenberg, and K. Maskell (eds). Cambridge University Press, New York, 572 pp.

Kattenberg, A., F. Giorgi, H. Grassl, G.A. Meehl, J.F.B. Mitchell, R.J. Stouffer, T. Tokioka, A.J. Weaver, and T.M.L. Wigley. 1996: Climate Models - Projections of Future Climate. In: Houghton, J.T., L.G.M. Filho, B.A. Callandar, N. Harris, A. Kattenberg, and K. Maskell (eds). 1996. Climate Change 1995: The Science of Climate Change. Contribution of Working Group 1 to the Second Assessment Report of the Intergovernmental Panel on Climate Change. p. 285-357. Cambridge University Press, New York, 572 pp.

Lofgren, B.M. 1995: Sensitivity of the land-ocean circulations, precipitation and soil moisture to perturbed land surface albedo. *J. Climate,* **8**, 2521-2542.

Maier-Reimer, E., U. Mikolajewicz, and A. Winguth, 1996: Future ocean uptake of CO2 - Interaction between ocean circulation and biology. *Clim. Dyn.,* **12**, 711-721.

Meehl, G.A and W.M. Washington, 1995: Cloud albedo feedback and the super greenhouse effect in a global coupled GCM. *Clim. Dyn.,* **11**, 399-411.

Meehl, G. A., G.J. Boer, C. Covey, M. Latif, and R.J. Stouffer, 2000: The Coupled Model Intercomparison Project (CMIP). *Bull. Am. Met. Soc.,* **81(2)**, 313-318.

Mitchell, J. 2000. Modelling cloud-climate feedbacks in predictions of human-induced climate change. In: Workshop on Cloud Processes and Cloud Feedbacks in Large-scale Models. World Climate Research Programme. WCRP-110; WMO/TD-No.993. Geneva.

Randall, S.D., J. Curry, D. Battisti, G. Flato, R. Grumbine, S. Hakkinen, D. Martinson, R. Preller, J. Walsh, J. Weatherly, 1998: Status of and outlook for large-scale modelling of atmosphere-ice-ocean interactions in the Arctic. *Bull. Am. Met. Soc.,* **79**, 197-219.

SAR, see IPCC, 1996.

Sarmiento, J.L. and J.R. Toggweiler, 1984: A new model for the role of the oceans in determining atmospheric CO_2. *Nature,* **308**, 621-624.

Senior, C.A., 1999. Comparison of mechanisms of cloud-climate feedbacks in a GCM. *J. Climate,* **12**, 1480-1489.

Smith, D. M., 1998: Recent increase in the length of the melt season of perennial Arctic sea ice. *Geophys. Res. Lett.,* **25**, 655-658.

Stephens, G., D. Varne, S. Walker. 2000: The CLOUDSAT mission: A new dimension to space-based observations of cloud in the coming millenium. In: Workshop on Cloud Processes and Cloud Feedbacks in Large-scale Models. World Climate Research Programme. WCRP-110; WMO/TD-No.993. Geneva.

Stouffer, R. J., and K. W. Dixon, 1998: Initialization of coupled models for use in climate studies: A review. In: *Research Activities in Atmospheric and Oceanic Modelling,* Report No. **27**, WMO/TD-No. 865, World Meteorological Organization, Geneva, Switzerland, I.1-I.8.

Washington, W.M., and G.A. Meehl, 1996: High-latitude climate change in a global coupled ocean-atmosphere-sea ice model with increased atmospheric CO_2. *J. Geophys. Res.,* **101**, 12795-12801.

Washington, W.M., J.W. Weatherly, G.A. Meehl, A.J. Semtner Jr., T.W. Bettge, A.P. Craig, W.G. Strand Jr., J.M. Arblaster, V.B. Wayland, R. James, Y. Zhang, 2000: Parallel climate model (PCM) control and 1% per year CO_2 simulations with a 2/3 degree ocean model and a 27 km dynamical sea ice model. *Clim. Dyn.,* **16 (10,11)**, 755-774.

Weatherly, J. W., B. P. Briegleb, W. G. Large, J. A. Maslanik, 1998: Sea ice and polar climate in the NCAR CSM. *J. Climate,* **11**, 1472-1486.

Appendix I

Glossary

Editor: A.P.M. Baede

A → indicates that the following term is also contained in this Glossary.

Adjustment time
See: →Lifetime; see also: →Response time.

Aerosols
A collection of airborne solid or liquid particles, with a typical size between 0.01 and 10 µm and residing in the atmosphere for at least several hours. Aerosols may be of either natural or anthropogenic origin. Aerosols may influence climate in two ways: directly through scattering and absorbing radiation, and indirectly through acting as condensation nuclei for cloud formation or modifying the optical properties and lifetime of clouds. See: →Indirect aerosol effect.

The term has also come to be associated, erroneously, with the propellant used in "aerosol sprays".

Afforestation
Planting of new forests on lands that historically have not contained forests. For a discussion of the term →forest and related terms such as afforestation, →reforestation, and →deforestation: see the IPCC Report on Land Use, Land-Use Change and Forestry (IPCC, 2000).

Albedo
The fraction of solar radiation reflected by a surface or object, often expressed as a percentage. Snow covered surfaces have a high albedo; the albedo of soils ranges from high to low; vegetation covered surfaces and oceans have a low albedo. The Earth's albedo varies mainly through varying cloudiness, snow, ice, leaf area and land cover changes.

Altimetry
A technique for the measurement of the elevation of the sea, land or ice surface. For example, the height of the sea surface (with respect to the centre of the Earth or, more conventionally, with respect to a standard "ellipsoid of revolution") can be measured from space by current state-of-the-art radar altimetry with centrimetric precision. Altimetry has the advantage of being a measurement relative to a geocentric reference frame, rather than relative to land level as for a →tide gauge, and of affording quasi-global coverage.

Anthropogenic
Resulting from or produced by human beings.

Atmosphere
The gaseous envelope surrounding the Earth. The dry atmosphere consists almost entirely of nitrogen (78.1% volume mixing ratio) and oxygen (20.9% volume mixing ratio), together with a number of trace gases, such as argon (0.93% volume mixing ratio), helium, and radiatively active →greenhouse gases such as →carbon dioxide (0.035% volume mixing ratio), and ozone. In addition the atmosphere contains water vapour, whose amount is highly variable but typically 1% volume mixing ratio. The atmosphere also contains clouds and →aerosols.

Attribution
See: →Detection and attribution.

Autotrophic respiration
→Respiration by photosynthetic organisms (plants).

Biomass
The total mass of living organisms in a given area or volume; recently dead plant material is often included as dead biomass.

Biosphere (terrestrial and marine)
The part of the Earth system comprising all →ecosystems and living organisms, in the atmosphere, on land (terrestrial biosphere) or in the oceans (marine biosphere), including derived dead organic matter, such as litter, soil organic matter and oceanic detritus.

Black carbon
Operationally defined species based on measurement of light absorption and chemical reactivity and/or thermal stability; consists of soot, charcoal, and/or possible light-absorbing refractory organic matter. (Source: Charlson and Heintzenberg, 1995, p. 401.)

Burden
The total mass of a gaseous substance of concern in the atmosphere.

Carbonaceous aerosol
Aerosol consisting predominantly of organic substances and various forms of →black carbon. (Source: Charlson and Heintzenberg, 1995, p. 401.)

Carbon cycle
The term used to describe the flow of carbon (in various forms, e.g. as carbon dioxide) through the atmosphere, ocean, terrestrial →biosphere and lithosphere.

Carbon dioxide (CO_2)
A naturally occurring gas, also a by-product of burning fossil fuels and →biomass, as well as →land-use changes and other industrial processes. It is the principal anthropogenic →greenhouse gas that affects the earth's radiative balance. It is the reference gas against which other greenhouse gases are measured and therefore has a →Global Warming Potential of 1.

Carbon dioxide (CO_2) fertilisation
The enhancement of the growth of plants as a result of increased atmospheric CO_2 concentration. Depending on their mechanism of →photosynthesis, certain types of plants are more sensitive to changes in atmospheric CO_2 concentratioin. In particular, →C_3 plants generally show a larger response to CO_2 than →C_4 plants.

Charcoal
Material resulting from charring of biomass, usually retaining some of the microscopic texture typical of plant tissues; chemically it consists mainly of carbon with a disturbed graphitic structure, with lesser amounts of oxygen and hydrogen. See: →Black carbon; Soot particles. (Source: Charlson and Heintzenberg, 1995, p. 402.)

Climate
Climate in a narrow sense is usually defined as the "average weather", or more rigorously, as the statistical description in terms of the mean and variability of relevant quantities over a period of time ranging from months to thousands or millions of years. The classical period is 30 years, as defined by the World Meteorological Organization (WMO). These quantities are most often surface variables such as temperature, precipitation, and wind. Climate in a wider sense is the state, including a statistical description, of the →climate system.

Climate change
Climate change refers to a statistically significant variation in either the mean state of the climate or in its variability, persisting for an extended period (typically decades or longer). Climate change may be due to natural internal processes or external forcings, or to persistent anthropogenic changes in the composition of the atmosphere or in land use.
Note that the →Framework Convention on Climate Change (UNFCCC), in its Article 1, defines "climate change" as: "a change of climate which is attributed directly or indirectly to human activity that alters the composition of the global atmosphere and which is in addition to natural climate variability observed over comparable time periods". The UNFCCC thus makes a distinction between "climate change" attributable to human activities altering the atmospheric composition, and "climate variability" attributable to natural causes.
See also: →Climate variability.

Climate feedback
An interaction mechanism between processes in the →climate system is called a climate feedback, when the result of an initial process triggers changes in a second process that in turn influences the initial one. A positive feedback intensifies the original process, and a negative feedback reduces it.

Climate model (hierarchy)
A numerical representation of the →climate system based on the physical, chemical and biological properties of its components, their interactions and feedback processes, and accounting for all or some of its known properties. The climate system can be represented by models of varying complexity, i.e. for any one component or combination of components a *hierarchy* of models can be identified, differing in such aspects as the number of spatial dimensions, the extent to which physical, chemical or biological processes are explicitly represented, or the level at which empirical →parametrizations are involved. Coupled atmosphere/ocean/sea-ice General Circulation Models (AOGCMs) provide a comprehensive representation of the climate system. There is an evolution towards more complex models with active chemistry and biology.

Climate models are applied, as a research tool, to study and simulate the climate, but also for operational purposes, including monthly, seasonal and interannual →climate predictions.

Climate prediction
A climate prediction or climate forecast is the result of an attempt to produce a most likely description or estimate of the actual evolution of the climate in the future, e.g. at seasonal, interannual or long-term time scales. See also: →Climate projection and →Climate (change) scenario.

Climate projection
A →projection of the response of the climate system to →emission or concentration scenarios of greenhouse gases and aerosols, or →radiative forcing scenarios, often based upon simulations by →climate models. Climate projections are distinguished from →climate predictions in order to emphasise that climate projections depend upon the emission/concentration/

radiative forcing scenario used, which are based on assumptions, concerning, e.g., future socio-economic and technological developments, that may or may not be realised, and are therefore subject to substantial uncertainty.

Climate scenario

A plausible and often simplified representation of the future climate, based on an internally consistent set of climatological relationships, that has been constructed for explicit use in investigating the potential consequences of anthropogenic →climate change, often serving as input to impact models. →Climate projections often serve as the raw material for constructing climate scenarios, but climate scenarios usually require additional information such as about the observed current climate. A *climate change scenario* is the difference between a climate scenario and the current climate.

Climate sensitivity

In IPCC Reports, *equilibrium climate sensitivity* refers to the equilibrium change in global mean surface temperature following a doubling of the atmospheric (→equivalent) CO_2 concentration. More generally, equilibrium climate sensitivity refers to the equilibrium change in surface air temperature following a unit change in →radiative forcing ($°C/Wm^{-2}$). In practice, the evaluation of the equilibrium climate sensitivity requires very long simulations with Coupled General Circulation Models (→Climate model).

The *effective climate sensitivity* is a related measure that circumvents this requirement. It is evaluated from model output for evolving non-equilibrium conditions. It is a measure of the strengths of the →feedbacks at a particular time and may vary with forcing history and climate state. Details are discussed in Section 9.2.1 of Chapter 9 in this Report.

Climate system

The climate system is the highly complex system consisting of five major components: the →atmosphere, the →hydrosphere, the →cryosphere, the land surface and the →biosphere, and the interactions between them. The climate system evolves in time under the influence of its own internal dynamics and because of external forcings such as volcanic eruptions, solar variations and human-induced forcings such as the changing composition of the atmosphere and →land-use change.

Climate variability

Climate variability refers to variations in the mean state and other statistics (such as standard deviations, the occurrence of extremes, etc.) of the climate on all temporal and spatial scales beyond that of individual weather events. Variability may be due to natural internal processes within the climate system (*internal variability*), or to variations in natural or anthropogenic external forcing (*external variability*). See also: →Climate change.

Cloud condensation nuclei

Airborne particles that serve as an initial site for the condensation of liquid water and which can lead to the formation of cloud droplets. See also: →Aerosols.

CO_2 fertilisation

See →Carbon dioxide (CO_2) fertilisation

Cooling degree days

The integral over a day of the temperature above 18°C (e.g. a day with an average temperature of 20°C counts as 2 cooling degree days). See also: →Heating degree days.

Cryosphere

The component of the →climate system consisting of all snow, ice and permafrost on and beneath the surface of the earth and ocean. See: →Glacier; →Ice sheet.

C_3 plants

Plants that produce a three-carbon compound during photosynthesis; including most trees and agricultural crops such as rice, wheat, soyabeans, potatoes and vegetables.

C_4 plants

Plants that produce a four-carbon compound during photosynthesis; mainly of tropical origin, including grasses and the agriculturally important crops maize, sugar cane, millet and sorghum.

Deforestation

Conversion of forest to non-forest. For a discussion of the term →forest and related terms such as →afforestation, →reforestation, and deforestation: see the IPCC Report on Land Use, Land-Use Change and Forestry (IPCC, 2000).

Desertification

Land degradation in arid, semi-arid, and dry sub-humid areas resulting from various factors, including climatic variations and human activities. Further, the UNCCD (The United Nations Convention to Combat Desertification) defines land degradation as a reduction or loss, in arid, semi-arid, and dry sub-humid areas, of the biological or economic productivity and complexity of rain-fed cropland, irrigated cropland, or range, pasture, forest, and woodlands resulting from land uses or from a process or combination of processes, including processes arising from human activities and habitation patterns, such as: (i) soil erosion caused by wind and/or water; (ii) deterioration of the physical, chemical and biological or economic properties of soil; and (iii) long-term loss of natural vegetation.

Detection and attribution

Climate varies continually on all time scales. *Detection* of →climate change is the process of demonstrating that climate has changed in some defined statistical sense, without providing a reason for that change. *Attribution* of causes of climate change is the process of establishing the most likely causes for the detected change with some defined level of confidence.

Diurnal temperature range

The difference between the maximum and minimum temperature during a day.

Dobson Unit (DU)
A unit to measure the total amount of ozone in a vertical column above the Earth's surface. The number of Dobson Units is the thickness in units of 10^{-5} m, that the ozone column would occupy if compressed into a layer of uniform density at a pressure of 1013 hPa, and a temperature of 0°C. One DU corresponds to a column of ozone containing 2.69×10^{20} molecules per square meter. A typical value for the amount of ozone in a column of the Earth's atmosphere, although very variable, is 300 DU.

Ecosystem
A system of interacting living organisms together with their physical environment. The boundaries of what could be called an ecosystem are somewhat arbitrary, depending on the focus of interest or study. Thus the extent of an ecosystem may range from very small spatial scales to, ultimately, the entire Earth.

El Niño-Southern Oscillation (ENSO)
El Niño, in its original sense, is a warm water current which periodically flows along the coast of Ecuador and Peru, disrupting the local fishery. This oceanic event is associated with a fluctuation of the intertropical surface pressure pattern and circulation in the Indian and Pacific oceans, called the Southern Oscillation. This coupled atmosphere-ocean phenomenon is collectively known as El Niño-Southern Oscillation, or ENSO. During an El Niño event, the prevailing trade winds weaken and the equatorial countercurrent strengthens, causing warm surface waters in the Indonesian area to flow eastward to overlie the cold waters of the Peru current. This event has great impact on the wind, sea surface temperature and precipitation patterns in the tropical Pacific. It has climatic effects throughout the Pacific region and in many other parts of the world. The opposite of an El Niño event is called *La Niña*.

Emission scenario
A plausible representation of the future development of emissions of substances that are potentially radiatively active (e.g. →greenhouse gases, →aerosols), based on a coherent and internally consistent set of assumptions about driving forces (such as demographic and socio-economic development, technological change) and their key relationships.

Concentration scenarios, derived from emission scenarios, are used as input into a climate model to compute →climate projections.

In IPCC (1992) a set of emission scenarios was presented which were used as a basis for the →climate projections in IPCC (1996). These emission scenarios are referred to as the IS92 scenarios. In the IPCC Special Report on Emission Scenarios (Nakićenović *et al.,* 2000) new emission scenarios, the so called →SRES scenarios, were published some of which were used, among others, as a basis for the climate projections presented in Chapter 9 of this Report. For the meaning of some terms related to these scenarios, see →SRES scenarios.

Energy balance
Averaged over the globe and over longer time periods, the energy budget of the →climate system must be in balance. Because the climate system derives all its energy from the Sun, this balance implies that, globally, the amount of incoming →solar radiation must on average be equal to the sum of the outgoing reflected solar radiation and the outgoing →infrared radiation emitted by the climate system. A perturbation of this global radiation balance, be it human induced or natural, is called →radiative forcing.

Equilibrium and transient climate experiment
An *equilibrium climate experiment* is an experiment in which a →climate model is allowed to fully adjust to a change in →radiative forcing. Such experiments provide information on the difference between the initial and final states of the model, but not on the time-dependent response. If the forcing is allowed to evolve gradually according to a prescribed →emission scenario, the time dependent response of a climate model may be analysed. Such experiment is called a *transient climate experiment.* See: →Climate projection.

Equivalent CO$_2$ (carbon dioxide)
The concentration of →CO$_2$ that would cause the same amount of →radiative forcing as a given mixture of CO$_2$ and other →greenhouse gases.

Eustatic sea-level change
A change in global average sea level brought about by an alteration to the volume of the world ocean. This may be caused by changes in water density or in the total mass of water. In discussions of changes on geological time-scales, this term sometimes also includes changes in global average sea level caused by an alteration to the shape of the ocean basins. In this Report the term is not used with that sense.

Evapotranspiration
The combined process of evaporation from the Earth's surface and transpiration from vegetation.

External forcing
See: →Climate system.

Extreme weather event
An extreme weather event is an event that is rare within its statistical reference distribution at a particular place. Definitions of "rare" vary, but an extreme weather event would normally be as rare as or rarer than the 10th or 90th percentile. By definition, the characteristics of what is called *extreme weather* may vary from place to place.

An *extreme climate event* is an average of a number of weather events over a certain period of time, an average which is itself extreme (e.g. rainfall over a season).

Faculae
Bright patches on the Sun. The area covered by faculae is greater during periods of high →solar activity.

Feedback
See: →Climate feedback.

Flux adjustment

To avoid the problem of coupled atmosphere-ocean general circulation models drifting into some unrealistic climate state, adjustment terms can be applied to the atmosphere-ocean fluxes of heat and moisture (and sometimes the surface stresses resulting from the effect of the wind on the ocean surface) before these fluxes are imposed on the model ocean and atmosphere. Because these adjustments are precomputed and therefore independent of the coupled model integration, they are uncorrelated to the anomalies which develop during the integration. In Chapter 8 of this Report it is concluded that present models have a reduced need for flux adjustment.

Forest

A vegetation type dominated by trees. Many definitions of the term forest are in use throughout the world, reflecting wide differences in bio-geophysical conditions, social structure, and economics. For a discussion of the term forest and related terms such as →afforestation, →reforestation, and →deforestation: see the IPCC Report on Land Use, Land-Use Change and Forestry (IPCC, 2000).

Fossil CO_2 (carbon dioxide) emissions

Emissions of CO_2 resulting from the combustion of fuels from fossil carbon deposits such as oil, gas and coal.

Framework Convention on Climate Change See: →United Nations Framework Convention on Climate Change (UNFCCC).

General Circulation

The large scale motions of the atmosphere and the ocean as a consequence of differential heating on a rotating Earth, aiming to restore the →energy balance of the system through transport of heat and momentum.

General Circulation Model (GCM)

See: →Climate model.

Geoid

The surface which an ocean of uniform density would assume if it were in steady state and at rest (i.e. no ocean circulation and no applied forces other than the gravity of the Earth). This implies that the geoid will be a surface of constant gravitational potential, which can serve as a reference surface to which all surfaces (e.g., the Mean Sea Surface) can be referred. The geoid (and surfaces parallel to the geoid) are what we refer to in common experience as "level surfaces".

Glacier

A mass of land ice flowing downhill (by internal deformation and sliding at the base) and constrained by the surrounding topography e.g. the sides of a valley or surrounding peaks; the bedrock topography is the major influence on the dynamics and surface slope of a glacier. A glacier is maintained by accumulation of snow at high altitudes, balanced by melting at low altitudes or discharge into the sea.

Global surface temperature

The global surface temperature is the area-weighted global average of (i) the sea-surface temperature over the oceans (i.e. the subsurface bulk temperature in the first few meters of the ocean), and (ii) the surface-air temperature over land at 1.5 m above the ground.

Global Warming Potential (GWP)

An index, describing the radiative characteristics of well mixed →greenhouse gases, that represents the combined effect of the differing times these gases remain in the atmosphere and their relative effectiveness in absorbing outgoing →infrared radiation. This index approximates the time-integrated warming effect of a unit mass of a given greenhouse gas in today's atmosphere, relative to that of →carbon dioxide.

Greenhouse effect

→Greenhouse gases effectively absorb →infrared radiation, emitted by the Earth's surface, by the atmosphere itself due to the same gases, and by clouds. Atmospheric radiation is emitted to all sides, including downward to the Earth's surface. Thus greenhouse gases trap heat within the surface-troposphere system. This is called the *natural greenhouse effect*.

Atmospheric radiation is strongly coupled to the temperature of the level at which it is emitted. In the →troposphere the temperature generally decreases with height. Effectively, infrared radiation emitted to space originates from an altitude with a temperature of, on average, −19°C, in balance with the net incoming solar radiation, whereas the Earth's surface is kept at a much higher temperature of, on average, +14°C.

An increase in the concentration of greenhouse gases leads to an increased infrared opacity of the atmosphere, and therefore to an effective radiation into space from a higher altitude at a lower temperature. This causes a →radiative forcing, an imbalance that can only be compensated for by an increase of the temperature of the surface-troposphere system. This is the *enhanced greenhouse effect*.

Greenhouse gas

Greenhouse gases are those gaseous constituents of the atmosphere, both natural and anthropogenic, that absorb and emit radiation at specific wavelengths within the spectrum of infrared radiation emitted by the Earth's surface, the atmosphere and clouds. This property causes the →greenhouse effect. Water vapour (H_2O), carbon dioxide (CO_2), nitrous oxide (N_2O), methane (CH_4) and ozone (O_3) are the primary greenhouse gases in the Earth's atmosphere. Moreover there are a number of entirely human-made greenhouse gases in the atmosphere, such as the →halocarbons and other chlorine and bromine containing substances, dealt with under the →Montreal Protocol. Beside CO_2, N_2O and CH_4, the →Kyoto Protocol deals with the greenhouse gases sulphur hexafluoride (SF_6), hydrofluorocarbons (HFCs) and perfluorocarbons (PFCs).

Gross Primary Production (GPP)

The amount of carbon fixed from the atmosphere through →photosynthesis.

Grounding line/zone
The junction between →ice sheet and →ice shelf or the place where the ice starts to float.

Halocarbons
Compounds containing either chlorine, bromine or fluorine and carbon. Such compounds can act as powerful →greenhouse gases in the atmosphere. The chlorine and bromine containing halocarbons are also involved in the depletion of the →ozone layer.

Heating degree days
The integral over a day of the temperature below 18°C (e.g. a day with an average temperature of 16°C counts as 2 heating degree days). See also: →Cooling degree days.

Heterotrophic respiration
The conversion of organic matter to CO_2 by organisms other than plants.

Hydrosphere
The component of the climate system comprising liquid surface and subterranean water, such as: oceans, seas, rivers, fresh water lakes, underground water etc.

Ice cap
A dome shaped ice mass covering a highland area that is considerably smaller in extent than an→ice sheet.

Ice sheet
A mass of land ice which is sufficiently deep to cover most of the underlying bedrock topography, so that its shape is mainly determined by its internal dynamics (the flow of the ice as it deforms internally and slides at its base). An ice sheet flows outwards from a high central plateau with a small average surface slope. The margins slope steeply, and the ice is discharged through fast-flowing ice streams or outlet glaciers, in some cases into the sea or into ice-shelves floating on the sea. There are only two large ice sheets in the modern world, on Greenland and Antarctica, the Antarctic ice sheet being divided into East and West by the Transantarctic Mountains; during glacial periods there were others.

Ice shelf
A floating →ice sheet of considerable thickness attached to a coast (usually of great horizontal extent with a level or gently undulating surface); often a seaward extension of ice sheets.

Indirect aerosol effect
→Aerosols may lead to an indirect →radiative forcing of the →climate system through acting as condensation nuclei or modifying the optical properties and lifetime of clouds. Two indirect effects are distinguished:
First indirect effect
A radiative forcing induced by an increase in anthropogenic aerosols which cause an initial increase in droplet concentration and a decrease in droplet size for fixed liquid water content, leading to an increase of cloud →albedo. This effect is also known as the *Twomey effect*. This is sometimes referred to as the *cloud albedo effect*. However this is highly misleading since the second indirect effect also alters cloud albedo.
Second indirect effect
A radiative forcing induced by an increase in anthropogenic aerosols which cause a decrease in droplet size, reducing the precipitation efficiency, thereby modifying the liquid water content, cloud thickness, and cloud life time. This effect is also known as the *cloud life time effect* or *Albrecht effect*.

Industrial revolution
A period of rapid industrial growth with far-reaching social and economic consequences, beginning in England during the second half of the eighteenth century and spreading to Europe and later to other countries including the United States. The invention of the steam engine was an important trigger of this development. The industrial revolution marks the beginning of a strong increase in the use of fossil fuels and emission of, in particular, fossil carbon dioxide. In this Report the terms *pre-industrial* and *industrial* refer, somewhat arbitrarily, to the periods before and after 1750, respectively.

Infrared radiation
Radiation emitted by the earth's surface, the atmosphere and the clouds. It is also known as terrestrial or long-wave radiation. Infrared radiation has a distinctive range of wavelengths ("spectrum") longer than the wavelength of the red colour in the visible part of the spectrum. The spectrum of infrared radiation is practically distinct from that of →solar or short-wave radiation because of the difference in temperature between the Sun and the Earth-atmosphere system.

Integrated assessment
A method of analysis that combines results and models from the physical, biological, economic and social sciences, and the interactions between these components, in a consistent framework, to evaluate the status and the consequences of environmental change and the policy responses to it.

Internal variability
See: →Climate variability.

Inverse modelling
A mathematical procedure by which the input to a model is estimated from the observed outcome, rather than *vice versa*. It is, for instance, used to estimate the location and strength of sources and sinks of CO_2 from measurements of the distribution of the CO_2 concentration in the atmosphere, given models of the global →carbon cycle and for computing atmospheric transport.

Isostatic land movements
Isostasy refers to the way in which the →lithosphere and mantle respond to changes in surface loads. When the loading of the lithosphere is changed by alterations in land ice mass, ocean mass, sedimentation, erosion or mountain building, vertical isostatic adjustment results, in order to balance the new load.

Kyoto Protocol

The Kyoto Protocol to the United Nations →Framework Convention on Climate Change (UNFCCC) was adopted at the Third Session of the Conference of the Parties (COP) to the United Nations →Framework Convention on Climate Change, in 1997 in Kyoto, Japan. It contains legally binding commitments, in addition to those included in the UNFCCC. Countries included in Annex B of the Protocol (most OECD countries and countries with economies in transition) agreed to reduce their anthropogenic →greenhouse gas emissions (CO_2, CH_4, N_2O, HFCs, PFCs, and SF_6) by at least 5% below 1990 levels in the commitment period 2008 to 2012. The Kyoto Protocol has not yet entered into force (April 2001).

Land use

The total of arrangements, activities and inputs undertaken in a certain land cover type (a set of human actions). The social and economic purposes for which land is managed (e.g., grazing, timber extraction, and conservation).

Land-use change

A change in the use or management of land by humans, which may lead to a change in land cover. Land cover and land-use change may have an impact on the →albedo, →evapotranspiration, →sources and →sinks of →greenhouse gases, or other properties of the →climate system and may thus have an impact on climate, locally or globally. See also: the IPCC Report on Land Use, Land-Use Change, and Forestry (IPCC, 2000).

La Niña

See: →El Niño-Southern Oscillation.

Lifetime

Lifetime is a general term used for various time-scales characterising the rate of processes affecting the concentration of trace gases. The following lifetimes may be distinguished:

Turnover time (T) is the ratio of the mass M of a reservoir (e.g., a gaseous compound in the atmosphere) and the total rate of removal S from the reservoir: T = M/S. For each removal process separate turnover times can be defined. In soil carbon biology this is referred to as *Mean Residence Time (MRT)*.

Adjustment time or *response time* (T_a) is the time-scale characterising the decay of an instantaneous pulse input into the reservoir. The term *adjustment time* is also used to characterise the adjustment of the mass of a reservoir following a step change in the source strength. *Half-life* or *decay constant* is used to quantify a first-order exponential decay process. See: →Response time, for a different definition pertinent to climate variations. The term *lifetime* is sometimes used, for simplicity, as a surrogate for *adjustment time.*

In simple cases, where the global removal of the compound is directly proportional to the total mass of the reservoir, the adjustment time equals the turnover time: $T = T_a$. An example is CFC-11 which is removed from the atmosphere only by photochemical processes in the stratosphere. In more complicated cases, where several reservoirs are involved or where the removal is not proportional to the total mass, the equality $T = T_a$ no longer holds.

→Carbon dioxide (CO_2) is an extreme example. Its turnover time is only about 4 years because of the rapid exchange between atmosphere and the ocean and terrestrial biota. However, a large part of that CO_2 is returned to the atmosphere within a few years. Thus, the adjustment time of CO_2 in the atmosphere is actually determined by the rate of removal of carbon from the surface layer of the oceans into its deeper layers. Although an approximate value of 100 years may be given for the adjustment time of CO_2 in the atmosphere, the actual adjustment is faster initially and slower later on. In the case of methane (CH_4) the adjustment time is different from the turnover time, because the removal is mainly through a chemical reaction with the hydroxyl radical OH, the concentration of which itself depends on the CH_4 concentration. Therefore the CH_4 removal S is not proportional to its total mass M.

Lithosphere

The upper layer of the solid Earth, both continental and oceanic, which comprises all crustal rocks and the cold, mainly elastic, part of the uppermost mantle. Volcanic activity, although part of the lithosphere, is not considered as part of the →climate system, but acts as an external forcing factor. See: →Isostatic land movements.

LOSU (Level of Scientific Understanding)

This is an index on a 4-step scale (High, Medium, Low and Very Low) designed to characterise the degree of scientific understanding of the radiative forcing agents that affect climate change. For each agent, the index represents a subjective judgement about the reliability of the estimate of its forcing, involving such factors as the assumptions necessary to evaluate the forcing, the degree of knowledge of the physical/ chemical mechanisms determining the forcing and the uncertainties surrounding the quantitative estimate.

Mean Sea Level

See: →Relative Sea Level.

Mitigation

A human intervention to reduce the →sources or enhance the →sinks of →greenhouse gases.

Mixing ratio

See: →Mole fraction.

Model hierarchy

See: →Climate model.

Mole fraction

Mole fraction, or *mixing ratio*, is the ratio of the number of moles of a constituent in a given volume to the total number of moles of all constituents in that volume. It is usually reported for dry air. Typical values for long-lived →greenhouse gases are in the order of μmol/mol (parts per million: ppm), nmol/mol (parts per billion: ppb), and fmol/mol (parts per trillion: ppt). Mole fraction differs from *volume mixing ratio*, often expressed in ppmv etc., by the corrections for non-ideality of gases. This correction is significant relative to measurement precision for many greenhouse gases. (Source: Schwartz and Warneck, 1995).

Montreal Protocol

The Montreal Protocol on Substances that Deplete the Ozone Layer was adopted in Montreal in 1987, and subsequently adjusted and amended in London (1990), Copenhagen (1992), Vienna (1995), Montreal (1997) and Beijing (1999). It controls the consumption and production of chlorine- and bromine-containing chemicals that destroy stratospheric ozone, such as CFCs, methyl chloroform, carbon tetrachloride, and many others.

Net Biome Production (NBP)

Net gain or loss of carbon from a region. NBP is equal to the →Net Ecosystem Production minus the carbon lost due to a disturbance, e.g. a forest fire or a forest harvest.

Net Ecosystem Production (NEP)

Net gain or loss of carbon from an →ecosystem. NEP is equal to the →Net Primary Production minus the carbon lost through →heterotrophic respiration.

Net Primary Production (NPP)

The increase in plant →biomass or carbon of a unit of a landscape. NPP is equal to the →Gross Primary Production minus carbon lost through →autotrophic respiration.

Nitrogen fertilisation

Enhancement of plant growth through the addition of nitrogen compounds. In IPCC Reports, this typically refers to fertilisation from anthropogenic sources of nitrogen such as human-made fertilisers and nitrogen oxides released from burning fossil fuels.

Non-linearity

A process is called "non-linear" when there is no simple proportional relation between cause and effect. The →climate system contains many such non-linear processes, resulting in a system with a potentially very complex behaviour. Such complexity may lead to →rapid climate change.

North Atlantic Oscillation (NAO)

The North Atlantic Oscillation consists of opposing variations of barometric pressure near Iceland and near the Azores. On average, a westerly current, between the Icelandic low pressure area and the Azores high pressure area, carries cyclones with their associated frontal systems towards Europe. However, the pressure difference between Iceland and the Azores fluctuates on time-scales of days to decades, and can be reversed at times.

Organic aerosol

→Aerosol particles consisting predominantly of organic compounds, mainly C, H, O, and lesser amounts of other elements. (Source: Charlson and Heintzenberg, 1995, p. 405.) See: →Carbonaceous aerosol.

Ozone

Ozone, the triatomic form of oxygen (O_3), is a gaseous atmospheric constituent. In the →troposphere it is created both naturally and by photochemical reactions involving gases resulting from human activities ("smog"). Tropospheric ozone acts as a →greenhouse gas. In the →stratosphere it is created by the interaction between solar ultraviolet radiation and molecular oxygen (O_2). Stratospheric ozone plays a decisive role in the stratospheric radiative balance. Its concentration is highest in the →ozone layer.

Ozone hole

See: →Ozone layer.

Ozone layer

The →stratosphere contains a layer in which the concentration of ozone is greatest, the so called ozone layer. The layer extends from about 12 to 40 km. The ozone concentration reaches a maximum between about 20 and 25 km. This layer is being depleted by human emissions of chlorine and bromine compounds. Every year, during the Southern Hemisphere spring, a very strong depletion of the ozone layer takes place over the Antarctic region, also caused by human-made chlorine and bromine compounds in combination with the specific meteorological conditions of that region. This phenomenon is called the *ozone hole*.

Parametrization

In →climate models, this term refers to the technique of representing processes, that cannot be explicitly resolved at the spatial or temporal resolution of the model (sub-grid scale processes), by relationships between the area or time averaged effect of such sub-grid scale processes and the larger scale flow.

Patterns of climate variability

Natural variability of the →climate system, in particular on seasonal and longer time-scales, predominantly occurs in preferred spatial patterns, through the dynamical non-linear characteristics of the atmospheric circulation and through interactions with the land and ocean surfaces. Such spatial patterns are also called "regimes" or "modes". Examples are the →North Atlantic Oscillation (NAO), the Pacific-North American pattern (PNA), the →El Niño-Southern Oscillation (ENSO), and the Antarctic Oscillation (AO).

Photosynthesis

The process by which plants take CO_2 from the air (or bicarbonate in water) to build carbohydrates, releasing O_2 in the process. There are several pathways of photosynthesis with different responses to atmospheric CO_2 concentrations. See: →Carbon dioxide fertilisation.

Pool

See: →Reservoir.

Post-glacial rebound

The vertical movement of the continents and sea floor following the disappearance and shrinking of →ice sheets, e.g. since the Last Glacial Maximum (21 ky BP). The rebound is an →isostatic land movement.

Ppm, ppb, ppt

See: → Mole fraction.

Precursors

Atmospheric compounds which themselves are not →greenhouse gases or →aerosols, but which have an effect on greenhouse gas or aerosol concentrations by taking part in physical or chemical processes regulating their production or destruction rates.

Pre-industrial

See: →Industrial revolution.

Projection (generic)

A projection is a potential future evolution of a quantity or set of quantities, often computed with the aid of a model. Projections are distinguished from *predictions* in order to emphasise that projections involve assumptions concerning, e.g., future socio-economic and technological developments that may or may not be realised, and are therefore subject to substantial uncertainty. See also →Climate projection; →Climate prediction.

Proxy

A proxy climate indicator is a local record that is interpreted, using physical and biophysical principles, to represent some combination of climate-related variations back in time. Climate related data derived in this way are referred to as proxy data. Examples of proxies are: tree ring records, characteristics of corals, and various data derived from ice cores.

Radiative forcing

Radiative forcing is the change in the net vertical irradiance (expressed in Watts per square metre: Wm^{-2}) at the →tropopause due to an internal change or a change in the external forcing of the →climate system, such as, for example, a change in the concentration of →carbon dioxide or the output of the Sun. Usually radiative forcing is computed after allowing for stratospheric temperatures to readjust to radiative equilibrium, but with all tropospheric properties held fixed at their unperturbed values. Radiative forcing is called *instantaneous* if no change in stratospheric temperature is accounted for. Practical problems with this definition, in particular with respect to radiative forcing associated with changes, by aerosols, of the precipitation formation by clouds, are discussed in Chapter 6 of this Report.

Radiative forcing scenario

A plausible representation of the future development of →radiative forcing associated, for example, with changes in atmospheric composition or land-use change, or with external factors such as variations in →solar activity. Radiative forcing scenarios can be used as input into simplified →climate models to compute →climate projections.

Radio-echosounding

The surface and bedrock, and hence the thickness, of a glacier can be mapped by radar; signals penetrating the ice are reflected at the lower boundary with rock (or water, for a floating glacier tongue).

Rapid climate change

The →non-linearity of the →climate system may lead to rapid climate change, sometimes called *abrupt events* or even *surprises*. Some such abrupt events may be imaginable, such as a dramatic reorganisation of the →thermohaline circulation, rapid deglaciation, or massive melting of permafrost leading to fast changes in the →carbon cycle. Others may be truly unexpected, as a consequence of a strong, rapidly changing, forcing of a non-linear system.

Reforestation

Planting of forests on lands that have previously contained forests but that have been converted to some other use. For a discussion of the term →forest and related terms such as →afforestation, reforestation, and →deforestation: see the IPCC Report on Land Use, Land-Use Change and Forestry (IPCC, 2000).

Regimes

Preferred →patterns of climate variability.

Relative Sea Level

Sea level measured by a →tide gauge with respect to the land upon which it is situated. Mean Sea Level (MSL) is normally defined as the average Relative Sea Level over a period, such as a month or a year, long enough to average out transients such as waves.

(Relative) Sea Level Secular Change

Long term changes in relative sea level caused by either →eustatic changes, e.g. brought about by →thermal expansion, or changes in vertical land movements.

Reservoir

A component of the →climate system, other than the atmosphere, which has the capacity to store, accumulate or release a substance of concern, e.g. carbon, a →greenhouse gas or a →precursor. Oceans, soils, and →forests are examples of reservoirs of carbon. *Pool* is an equivalent term (note that the definition of pool often includes the atmosphere). The absolute quantity of substance of concerns, held within a reservoir at a specified time, is called the *stock*.

Respiration

The process whereby living organisms convert organic matter to CO_2, releasing energy and consuming O_2.

Response time

The response time or *adjustment time* is the time needed for the →climate system or its components to re-equilibrate to a new state, following a forcing resulting from external and internal processes or →feedbacks. It is very different for various components of the climate system. The response time of the →troposphere is relatively short, from days to weeks, whereas the →stratosphere comes into equilibrium on a time-scale of typically a few months. Due to their large heat capacity, the oceans have a much longer response time, typically decades, but up to centuries or millennia. The response time of the strongly coupled surface-troposphere system is, therefore, slow compared to that of the stratosphere, and mainly determined by the oceans. The →biosphere may respond fast, e.g. to droughts, but also very slowly to imposed changes.

See: →Lifetime, for a different definition of response time pertinent to the rate of processes affecting the concentration of trace gases.

Scenario (generic)
A plausible and often simplified description of how the future may develop, based on a coherent and internally consistent set of assumptions about driving forces and key relationships. Scenarios may be derived from →projections, but are often based on additional information from other sources, sometimes combined with a "narrative storyline". See also: →SRES scenarios; →Climate scenario; →Emission scenarios.

Sea level rise
See: →Relative Sea Level Secular Change; →Thermal expansion.

Sequestration
See: →Uptake.

Significant wave height
The average height of the highest one-third of all sea waves occurring in a particular time period. This serves as an indicator of the characteristic size of the highest waves.

Sink
Any process, activity or mechanism which removes a →greenhouse gas, an →aerosol or a precursor of a greenhouse gas or aerosol from the atmosphere.

Soil moisture
Water stored in or at the land surface and available for evaporation.

Solar activity
The Sun exhibits periods of high activity observed in numbers of →sunspots, as well as radiative output, magnetic activity, and emission of high energy particles. These variations take place on a range of time-scales from millions of years to minutes. See: →Solar cycle.

Solar ("11 year") cycle
A quasi-regular modulation of →solar activity with varying amplitude and a period of between 9 and 13 years.

Solar radiation
Radiation emitted by the Sun. It is also referred to as short-wave radiation. Solar radiation has a distinctive range of wavelengths (spectrum) determined by the temperature of the Sun. See also: →Infrared radiation.

Soot particles
Particles formed during the quenching of gases at the outer edge of flames of organic vapours, consisting predominantly of carbon, with lesser amounts of oxygen and hydrogen present as carboxyl and phenolic groups and exhibiting an imperfect graphitic structure. See: →Black carbon; Charcoal. (Source: Charlson and Heintzenberg, 1995, p. 406.)

Source
Any process, activity or mechanism which releases a greenhouse gas, an aerosol or a precursor of a greenhouse gas or aerosol into the atmosphere.

Spatial and temporal scales
Climate may vary on a large range of spatial and temporal scales. Spatial scales may range from local (less than 100,000 km²), through regional (100,000 to 10 million km²) to continental (10 to 100 million km²). Temporal scales may range from seasonal to geological (up to hundreds of millions of years).

SRES scenarios
SRES scenarios are →emission scenarios developed by Nakićenović *et al.* (2000) and used, among others, as a basis for the climate projections in Chapter 9 of this Report. The following terms are relevant for a better understanding of the structure and use of the set of SRES scenarios:
(Scenario) Family
Scenarios that have a similar demographic, societal, economic and technical-change storyline. Four scenario families comprise the SRES scenario set: A1, A2, B1 and B2.
(Scenario) Group
Scenarios within a family that reflect a consistent variation of the storyline. The A1 scenario family includes four groups designated as A1T, A1C, A1G and A1B that explore alternative structures of future energy systems. In the Summary for Policymakers of Nakićenović *et al.* (2000), the A1C and A1G groups have been combined into one 'Fossil Intensive' A1FI scenario group. The other three scenario families consist of one group each. The SRES scenario set reflected in the Summary for Policymakers of Nakićenović *et al.* (2000) thus consist of six distinct scenario groups, all of which are equally sound and together capture the range of uncertainties associated with driving forces and emissions.
Illustrative Scenario
A scenario that is illustrative for each of the six scenario groups reflected in the Summary for Policymakers of Nakićenović *et al.* (2000). They include four revised 'scenario markers' for the scenario groups A1B, A2, B1, B2, and two additional scenarios for the A1FI and A1T groups. All scenario groups are equally sound.
(Scenario) Marker
A scenario that was originally posted in draft form on the SRES website to represent a given scenario family. The choice of markers was based on which of the initial quantifications best

reflected the storyline, and the features of specific models. Markers are no more likely than other scenarios, but are considered by the SRES writing team as illustrative of a particular storyline. They are included in revised form in Nakićenović *et al.* (2000). These scenarios have received the closest scrutiny of the entire writing team and via the SRES open process. Scenarios have also been selected to illustrate the other two scenario groups (see also 'Scenario Group' and 'Illustrative Scenario').

(Scenario) Storyline
A narrative description of a scenario (or family of scenarios) highlighting the main scenario characteristics, relationships between key driving forces and the dynamics of their evolution.

Stock
See: →Reservoir.

Storm surge
The temporary increase, at a particular locality, in the height of the sea due to extreme meteorological conditions (low atmospheric pressure and/or strong winds). The storm surge is defined as being the excess above the level expected from the tidal variation alone at that time and place.

Stratosphere
The highly stratified region of the atmosphere above the →troposphere extending from about 10 km (ranging from 9 km in high latitudes to 16 km in the tropics on average) to about 50 km.

Sunspots
Small dark areas on the Sun. The number of sunspots is higher during periods of high →solar activity, and varies in particular with the →solar cycle.

Thermal expansion
In connection with sea level, this refers to the increase in volume (and decrease in density) that results from warming water. A warming of the ocean leads to an expansion of the ocean volume and hence an increase in sea level.

Thermohaline circulation
Large-scale density-driven circulation in the ocean, caused by differences in temperature and salinity. In the North Atlantic the thermohaline circulation consists of warm surface water flowing northward and cold deep water flowing southward, resulting in a net poleward transport of heat. The surface water sinks in highly restricted sinking regions located in high latitudes.

Tide gauge
A device at a coastal location (and some deep sea locations) which continuously measures the level of the sea with respect to the adjacent land. Time-averaging of the sea level so recorded gives the observed →Relative Sea Level Secular Changes.

Transient climate response
The globally averaged surface air temperature increase, averaged over a 20 year period, centred at the time of CO_2 doubling, i.e., at year 70 in a 1% per year compound CO_2 increase experiment with a global coupled →climate model.

Tropopause
The boundary between the →troposphere and the →stratosphere.

Troposphere
The lowest part of the atmosphere from the surface to about 10 km in altitude in mid-latitudes (ranging from 9 km in high latitudes to 16 km in the tropics on average) where clouds and "weather" phenomena occur. In the troposphere temperatures generally decrease with height.

Turnover time
See: →Lifetime.

Uncertainty
An expression of the degree to which a value (e.g. the future state of the climate system) is unknown. Uncertainty can result from lack of information or from disagreement about what is known or even knowable. It may have many types of sources, from quantifiable errors in the data to ambiguously defined concepts or terminology, or uncertain projections of human behaviour. Uncertainty can therefore be represented by quantitative measures (e.g. a range of values calculated by various models) or by qualitative statements (e.g., reflecting the judgement of a team of experts). See Moss and Schneider (2000).

United Nations Framework Convention on Climate Change (UNFCC)
The Convention was adopted on 9 May 1992 in New York and signed at the 1992 Earth Summit in Rio de Janeiro by more than 150 countries and the European Community. Its ultimate objective is the "stabilisation of greenhouse gas concentrations in the atmosphere at a level that would prevent dangerous anthropogenic interference with the climate system". It contains commitments for all Parties. Under the Convention, Parties included in Annex I aim to return greenhouse gas emissions not controlled by the Montreal Protocol to 1990 levels by the year 2000. The convention entered into force in March 1994. See: →Kyoto Protocol.

Uptake
The addition of a substance of concern to a →reservoir. The uptake of carbon containing substances, in particular carbon dioxide, is often called (carbon) *sequestration*.

Volume mixing ratio
See: →Mole fraction.

Sources:

Charlson, R. J., and J. Heintzenberg (Eds.): *Aerosol Forcing of Climate*, pp. 91-108, copyright 1995 ©John Wiley and Sons Limited. Reproduced with permission.

IPCC, 1992: Climate Change 1992: *The Supplementary Report to the IPCC Scientific Assessment* [J. T. Houghton, B. A. Callander and S. K. Varney (eds.)]. Cambridge University Press, Cambridge, UK, xi + 116 pp.

IPCC, 1994: *Climate Change 1994: Radiative Forcing of Climate Change and an Evaluation of the IPCC IS92 Emission Scenarios,* [J. T. Houghton, L. G. Meira Filho, J. Bruce, Hoesung Lee, B. A. Callander, E. Haites, N. Harris and K. Maskell (eds.)]. Cambridge University Press, Cambridge, UK and New York, NY, USA, 339 pp.

IPCC, 1996: *Climate Change 1995: The Science of Climate Change. Contribution of Working Group I to the Second Assessment Report of the Intergovernmental Panel on Climate Change* [J. T. Houghton., L.G. Meira Filho, B. A. Callander, N. Harris, A. Kattenberg, and K. Maskell (eds.)]. Cambridge University Press, Cambridge, United Kingdom and New York, NY, USA, 572 pp.

IPCC, 1997a: *IPCC Technical Paper 2: An introduction to simple climate models used in the IPCC Second Assessment Report,* [J. T. Houghton, L.G. Meira Filho, D. J. Griggs and K. Maskell (eds.)]. 51 pp.

IPCC, 1997b: *Revised 1996 IPCC Guidelines for National Greenhouse Gas Inventories* (3 volumes) [J. T. Houghton, L. G. Meira Filho, B. Lim, K. Tréanton, I. Mamaty, Y. Bonduki, D. J. Griggs and B. A. Callander (eds.)].

IPCC, 1997c: *IPCC technical Paper 4: Implications of proposed CO$_2$ emissions limitations.* [J. T. Houghton, L.G. Meira Filho, D. J. Griggs and M Noguer (eds.)]. 41 pp.

IPCC, 2000:*Land Use, Land-Use Change, and Forestry. Special Report of the IPCC.* [R.T. Watson, I.R. Noble, B. Bolin, N.H. Ravindranath and D. J. Verardo, D. J. Dokken, , (eds.)] Cambridge University Press, Cambridge, United Kingdom and New York, NY, USA, 377 pp.

Maunder, W. John , 1992: *Dictionary of Global Climate Change,* UCL Press Ltd.

Moss, R. and S. Schneider, 2000: *IPCC Supporting Material, pp. 33-51:Uncertainties in the IPCC TAR: Recommendations to Lead Authors for more consistent Assessment and Reporting,* [R. Pachauri, T. Taniguchi and K. Tanaka (eds.)]

Nakićenović, N., J. Alcamo, G. Davis, B. de Vries, J. Fenhann, S. Gaffin, K. Gregory, A. Grübler, T. Y. Jung, T. Kram, E. L. La Rovere, L. Michaelis, S. Mori, T. Morita, W. Pepper, H. Pitcher, L. Price, K. Raihi, A. Roehrl, H-H. Rogner, A. Sankovski, M. Schlesinger, P. Shukla, S. Smith, R. Swart, S. van Rooijen, N. Victor, Z. Dadi, 2000: *Emissions Scenarios, A Special Report of Working Group III of the Intergovernmental Panel on Climate Change.* Cambridge University Press, Cambridge, United Kingdom and New York, NY, USA, 599 pp.

Schwartz, S. E. and P. Warneck, 1995: *Units for use in atmospheric chemistry,* Pure & Appl. Chem., 67, pp. 1377-1406.

Appendix II

SRES Tables

Contents

Introduction

Appendix II gives, in tabulated form, the values for emissions, abundances and burdens, and, radiative forcing of major greenhouse gases and aerosols based on the SRES[1] scenarios (Nakićenović *et. al.*, 2000). The Appendix also presents global projections of changes in surface air temperature and sea level using these SRES emission scenarios.

The emission values are only anthropogenic emissions and are the ones published in Appendix VII of the SRES Report. Apart from the CO_2 emissions, for which deforestation and land use values are given in the SRES Report, the SRES scenarios for the rest of the gases define only the changes in direct anthropogenic emissions and do not specify the current magnitude of the natural emissions nor the concurrent changes in natural emissions due either to direct human activities such as land-use change or to the indirect impacts of climate change. Emissions for black carbon (BC) aerosols and organic matter carbonaceous (OC) aerosols species not covered in the SRES Report, are calculated by scaling to the SRES anthropogenic CO emissions.

The abundances and burdens for each of the species are calculated with the latest climate chemistry and climate carbon models (see Chapters 3, 4 and 5 for details).

The radiative forcings due to well-mixed greenhouse gases are computed using each of the simplified expressions given in

Chapter 6, Table 6.2. The radiative forcings associated with future tropospheric O_3 increase are calculated on the basis of the O_3 changes presented in Chapter 4 for the various SRES scenarios. The mean forcing per DU estimated from the various models, and given in Chapter 6, Table 6.3 (i.e., 0.042 Wm^{-2}/DU), is used to derive these future forcings. For each aerosol species, the ratio of the column burdens for the particular scenario to that of the year 2000 is multiplied by the "best estimate" of the present day radiative forcing (see Chapter 6 for more details). The radiative forcings for all the species have been calculated since pre-industrial time.

The global mean surface air temperature and sea level projections, based on the SRES scenarios, have been calculated using Simple Climate models which have been "tuned" to get similar responses to the AOGCMs in the global mean (see Chapters 9 and 11 for details).

The results presented are global mean values, every ten years from 2000 to 2100, for a range of scenarios. These scenarios are the final approved Illustrative Marker Scenarios (A1B, A1T, A1FI, A2, B1, and B2); the preliminary marker scenarios (A1p, A2p, B1p, B2p, approved by the IPCC Bureau in June 1998) and, for comparison and for some species, results based on a previous scenario used by IPCC (IS92a) have also been added. For some gases, the values tabulated in the IPCC Second Assessment Report (IPCC, 1996; hereafter SAR), for that IS92a scenario using the previous generation of chemistry and climate models, are also given.

---

[1] IPCC Special Report on Emission Scenarios (Nakićenović *et. al.*, 2000), herafter SRES.

Main Chemical Symbols used in this Appendix:

CO_2	carbon dioxide	O_3	ozone
CH_4	methane	OH	hydroxyl
CFC	chlorofluorocarbon	PFC	perfluorocarbon
CO	carbon monoxide	SO_2	sulphur dioxide
HFC	hydrofluorocarbon	SO_4^{2-}	sulphate ion
N_2O	nitrous oxide	SF_6	sulphur hexafluoride
NO_x	the sum of NO (nitric oxide) and NO_2 (nitrogen dioxide)	VOC	volatile organic compound

II.1: Anthropogenic Emissions

II.1.1: CO_2 emissions (PgC/yr)

CO_2 emissions from fossil fuel and industrial processes (PgC/yr)

Year	A1B	A1T	A1FI	A2	B1	B2	A1p	A2p	B1p	B2p	IS92a
2000	6.90	6.90	6.90	6.90	6.90	6.90	6.8	6.8	6.8	6.8	7.1
2010	9.68	8.33	8.65	8.46	8.50	7.99	9.7	8.4	7.7	7.9	8.68
2020	12.12	10.00	11.19	11.01	10.00	9.02	12.2	10.9	8.3	8.9	10.26
2030	14.01	12.26	14.61	13.53	11.20	10.15	14.2	13.3	8.4	10.0	11.62
2040	14.95	12.60	18.66	15.01	12.20	10.93	15.2	14.7	9.1	10.8	12.66
2050	16.01	12.29	23.10	16.49	11.70	11.23	16.2	16.4	9.8	11.1	13.7
2060	15.70	11.41	25.14	18.49	10.20	11.74	15.9	18.2	10.4	11.6	14.68
2070	15.43	9.91	27.12	20.49	8.60	11.87	15.6	20.2	10.1	11.8	15.66
2080	14.83	8.05	29.04	22.97	7.30	12.46	15.0	22.7	8.7	12.4	17.0
2090	13.94	6.27	29.64	25.94	6.10	13.20	14.1	25.6	7.5	13.1	18.7
2100	13.10	4.31	30.32	28.91	5.20	13.82	13.2	28.8	6.5	13.7	20.4

CO_2 emissions from deforestation and land use (PgC/yr)

Year	A1B	A1T	A1FI	A2	B1	B2	A1p	A2p	B1p	B2p	IS92a
2000	1.07	1.07	1.07	1.07	1.07	1.07	1.6	1.6	1.6	1.6	1.3
2010	1.20	1.04	1.08	1.12	0.78	0.80	1.5	1.6	0.8	1.8	1.22
2020	0.52	0.26	1.55	1.25	0.63	0.03	1.6	1.7	1.3	1.6	1.14
2030	0.47	0.12	1.57	1.19	−0.09	−0.25	0.7	1.5	0.7	0.3	1.04
2040	0.40	0.05	1.31	1.06	−0.48	−0.24	0.3	1.3	0.6	0.0	0.92
2050	0.37	−0.02	0.80	0.93	−0.41	−0.23	−0.2	1.2	0.5	−0.3	0.8
2060	0.30	−0.03	0.55	0.67	−0.46	−0.24	−0.3	0.7	0.7	−0.2	0.54
2070	0.30	−0.03	0.16	0.40	−0.42	−0.25	−0.3	0.4	0.8	−0.2	0.28
2080	0.35	−0.03	−0.36	0.25	−0.60	−0.31	−0.4	0.3	1.0	−0.2	0.12
2090	0.36	−0.01	−1.22	0.21	−0.78	−0.41	−0.5	0.2	1.2	−0.2	0.06
2100	0.39	0.00	−2.08	0.18	−0.97	−0.50	−0.6	0.2	1.4	−0.2	−0.1

CO_2 emissions – total (PgC/yr)

Year	A1B	A1T	A1FI	A2	B1	B2	A1p	A2p	B1p	B2p	IS92a
2000	7.97	7.97	7.97	7.97	7.97	7.97	8.4	8.4	8.4	8.4	8.4
2010	10.88	9.38	9.73	9.58	9.28	8.78	11.2	10.0	8.5	9.7	9.9
2020	12.64	10.26	12.73	12.25	10.63	9.05	13.8	12.6	9.6	10.5	11.4
2030	14.48	12.38	16.19	14.72	11.11	9.90	14.9	14.8	9.1	10.3	12.66
2040	15.35	12.65	19.97	16.07	11.72	10.69	15.5	16.0	9.7	10.8	13.58
2050	16.38	12.26	23.90	17.43	11.29	11.01	16.0	17.6	10.3	10.8	14.5
2060	16.00	11.38	25.69	19.16	9.74	11.49	15.6	18.9	11.1	11.4	15.22
2070	15.73	9.87	27.28	20.89	8.18	11.62	15.3	20.6	10.9	11.6	15.94
2080	15.18	8.02	28.68	23.22	6.70	12.15	14.6	23.0	9.7	12.2	17.12
2090	14.30	6.26	28.42	26.15	5.32	12.79	13.6	25.8	8.7	12.9	18.76
2100	13.49	4.32	28.24	29.09	4.23	13.32	12.6	29.0	7.9	13.5	20.3

II.1.2: CH_4 emissions ($Tg(CH_4)$/yr)

Year	A1B	A1T	A1FI	A2	B1	B2	A1p	A2p	B1p	B2p	IS92a
2000	323	323	323	323	323	323	347	347	347	347	390
2010	373	362	359	370	349	349	417	394	367	389	433
2020	421	415	416	424	377	384	484	448	396	448	477
2030	466	483	489	486	385	426	547	506	403	501	529
2040	458	495	567	542	381	466	531	560	423	528	580
2050	452	500	630	598	359	504	514	621	444	538	630
2060	410	459	655	654	342	522	464	674	445	544	654
2070	373	404	677	711	324	544	413	732	446	542	678
2080	341	359	695	770	293	566	370	790	447	529	704
2090	314	317	715	829	266	579	336	848	413	508	733
2100	289	274	735	889	236	597	301	913	379	508	762

II.1.3: N₂O emissions (TgN/yr)

Year	A1B	A1T	A1FI	A2	B1	B2	A1p	A2p	B1p	B2p	IS92a
2000	7.0	7.0	7.0	7.0	7.0	7.0	6.9	6.9	6.9	6.9	5.5
2010	7.0	6.1	8.0	8.1	7.5	6.2	7.3	7.9	7.4	7.1	6.2
2020	7.2	6.1	9.3	9.6	8.1	6.1	7.7	9.4	8.1	7.1	7.1
2030	7.3	6.2	10.9	10.7	8.2	6.1	7.5	10.5	8.3	6.7	7.7
2040	7.4	6.2	12.8	11.3	8.3	6.2	7.1	11.1	8.6	6.4	8.0
2050	7.4	6.1	14.5	12.0	8.3	6.3	6.8	11.8	8.9	6.0	8.3
2060	7.3	6.0	15.0	12.9	7.7	6.4	6.3	12.7	8.8	5.8	8.3
2070	7.2	5.7	15.4	13.9	7.4	6.6	5.9	13.7	8.7	5.5	8.4
2080	7.1	5.6	15.7	14.8	7.0	6.7	5.5	14.6	8.6	5.4	8.5
2090	7.1	5.5	16.1	15.7	6.4	6.8	5.2	15.5	8.3	5.2	8.6
2100	7.0	5.4	16.6	16.5	5.7	6.9	4.9	16.4	8.0	5.1	8.7

II.1.4: PFCs, SF₆ and HFCs emissions (Gg/yr)

CF₄ emissions (Gg/yr)

Year	A1B	A1T	A1FI	A2	B1	B2	A1p	A2p	B1p	B2p
2000	12.6	12.6	12.6	12.6	12.6	12.6	26.7	26.7	26.7	26.7
2010	15.3	15.3	15.3	20.3	14.5	21.0	28.4	28.9	27.0	29.9
2020	21.1	21.1	21.1	25.2	15.7	27.1	41.0	35.2	29.6	37.7
2030	30.1	30.1	30.1	31.4	16.6	34.6	59.4	43.0	31.4	47.4
2040	38.2	38.2	38.2	37.9	18.5	43.6	71.7	50.9	33.1	58.9
2050	43.8	43.8	43.8	45.6	20.9	52.7	77.3	60.0	35.5	70.5
2060	48.1	48.1	48.1	56.0	23.1	59.2	76.7	72.6	36.1	78.5
2070	52.1	52.1	52.1	63.6	22.5	63.1	64.2	84.7	29.6	85.1
2080	56.1	56.1	56.1	73.2	21.3	64.2	40.6	97.9	19.7	86.6
2090	58.9	58.9	58.9	82.8	22.5	62.9	46.8	110.9	20.8	84.7
2100	57.0	57.0	57.0	88.2	22.2	59.9	53.0	117.9	20.5	80.6

C₂F₆ emissions (Gg/yr)

Year	A1B	A1T	A1FI	A2	B1	B2	A1p	A2p	B1p	B2p
2000	1.3	1.3	1.3	1.3	1.3	1.3	2.7	2.7	2.7	2.7
2010	1.5	1.5	1.5	2.0	1.5	2.1	2.8	2.9	2.7	3.0
2020	2.1	2.1	2.1	2.5	1.6	2.7	4.1	3.5	3.0	3.8
2030	3.0	3.0	3.0	3.1	1.7	3.5	5.9	4.3	3.1	4.7
2040	3.8	3.8	3.8	3.8	1.8	4.4	7.2	5.1	3.3	5.9
2050	4.4	4.4	4.4	4.6	2.1	5.3	7.7	6.0	3.6	7.1
2060	4.8	4.8	4.8	5.6	2.3	5.9	7.7	7.3	3.6	7.9
2070	5.2	5.2	5.2	6.4	2.2	6.3	6.4	8.5	3.0	8.5
2080	5.6	5.6	5.6	7.3	2.1	6.4	4.1	9.8	2.0	8.7
2090	5.9	5.9	5.9	8.3	2.2	6.3	4.7	11.1	2.1	8.5
2100	5.7	5.7	5.7	8.8	2.2	6.0	5.3	11.8	2.1	8.1

SF$_6$ emissions (Gg/yr)

Year	A1B	A1T	A1FI	A2	B1	B2	A1p	A2p	B1p	B2p
2000	6.2	6.2	6.2	6.2	6.2	6.2	6.2	6.2	6.2	6.2
2010	6.7	6.7	6.7	7.6	5.6	7.4	7.2	8.0	6.4	7.7
2020	7.3	7.3	7.3	9.7	5.7	8.4	7.9	10.2	6.5	9.9
2030	10.2	10.2	10.2	11.6	7.2	9.2	10.7	12.0	8.0	12.5
2040	15.2	15.2	15.2	13.7	8.9	11.7	15.8	14.0	9.7	15.8
2050	18.3	18.3	18.3	16.0	10.4	12.1	18.8	16.8	11.2	18.6
2060	19.5	19.5	19.5	18.8	10.9	12.2	20.0	18.7	11.6	20.4
2070	17.3	17.3	17.3	19.8	9.5	11.4	17.8	19.7	10.2	22.0
2080	13.5	13.5	13.5	20.7	7.1	9.6	12.0	20.6	6.8	22.8
2090	13.0	13.0	13.0	23.4	6.5	10.0	13.5	23.3	7.2	23.9
2100	14.5	14.5	14.5	25.2	6.5	10.6	15.0	25.1	7.2	24.4

HFC−23 emissions (Gg/yr)

Year	A1B	A1T	A1FI	A2	B1	B2	A1p	A2p	B1p	B2p
2000	13	13	13	13	13	13	13	13	13	13
2010	15	15	15	15	15	15	15	15	15	15
2020	5	5	5	5	5	5	5	5	5	5
2030	2	2	2	2	2	2	2	2	2	2
2040	2	2	2	2	2	2	2	2	2	2
2050	1	1	1	1	1	1	0	0	0	0
2060	1	1	1	1	1	1	0	0	0	0
2070	1	1	1	1	1	1	0	0	0	0
2080	1	1	1	1	1	1	0	0	0	0
2090	1	1	1	1	1	1	0	0	0	0
2100	1	1	1	1	1	1	0	0	0	0

HFC−32 emissions (Gg/yr)

Year	A1B	A1T	A1FI	A2	B1	B2	A1p	A2p	B1p	B2p
2000	0	0	0	0	0	0	2	2	2	2
2010	4	4	4	4	3	3	3	3	3	3
2020	8	8	8	6	6	6	8	6	6	7
2030	14	14	14	9	8	9	14	9	8	10
2040	19	19	19	11	10	11	19	10	10	12
2050	24	24	24	14	14	14	24	13	14	16
2060	28	28	28	17	14	17	26	16	14	19
2070	29	29	29	20	14	20	27	19	14	21
2080	30	30	30	24	14	22	28	23	14	23
2090	30	30	30	29	14	24	28	28	13	24
2100	30	30	30	33	13	26	28	33	13	25

HFC−125 emissions (Gg/yr)

Year	A1B	A1T	A1FI	A2	B1	B2	A1p	A2p	B1p	B2p	IS92a
2000	0	0	0	0	0	0	7	7	7	7	0
2010	12	12	12	11	11	11	11	10	10	10	1
2020	27	27	27	21	21	22	26	19	20	22	9
2030	45	45	45	29	29	30	44	27	28	32	46
2040	62	62	62	35	36	38	62	33	35	40	111
2050	80	80	80	46	48	49	78	43	47	52	175
2060	94	94	94	56	48	58	84	53	48	62	185
2070	98	98	98	66	48	67	88	62	47	70	194
2080	100	100	100	79	48	76	91	74	46	75	199
2090	101	101	101	94	46	83	92	89	45	79	199
2100	101	101	101	106	44	89	93	104	43	83	199

HFC–134a emissions (Gg/yr)

Year	A1B	A1T	A1FI	A2	B1	B2	A1p	A2p	B1p	B2p	IS92a
2000	80	80	80	80	80	80	147	147	147	147	148
2010	176	176	176	166	163	166	220	204	206	216	290
2020	326	326	326	252	249	262	427	315	319	359	396
2030	515	515	515	330	326	352	693	412	422	496	557
2040	725	725	725	405	414	443	997	508	545	638	738
2050	931	931	931	506	547	561	1215	635	734	816	918
2060	1076	1076	1076	633	550	679	1264	800	732	991	969
2070	1078	1078	1078	758	544	799	1272	962	718	1133	1020
2080	1061	1061	1061	915	533	910	1247	1169	698	1202	1047
2090	1029	1029	1029	1107	513	1002	1204	1422	667	1261	1051
2100	980	980	980	1260	486	1079	1142	1671	627	1317	1055

HFC–143a emissions (Gg/yr)

Year	A1B	A1T	A1FI	A2	B1	B2	A1p	A2p	B1p	B2p
2000	0	0	0	0	0	0	6	6	6	6
2010	9	9	9	9	8	8	8	8	8	8
2020	21	21	21	16	15	16	20	15	15	17
2030	34	34	34	22	21	22	34	21	21	24
2040	47	47	47	27	26	27	48	26	26	30
2050	61	61	61	35	35	35	60	33	35	39
2060	70	70	70	43	35	42	64	41	35	47
2070	74	74	74	51	35	49	67	48	35	53
2080	75	75	75	61	35	55	69	58	35	57
2090	76	76	76	73	34	60	70	70	33	60
2100	76	76	76	82	32	65	70	81	32	63

HFC–152a emissions (Gg/yr)

Year	A1B	A1T	A1FI	A2	B1	B2	A1p	A2p	B1p	B2p	IS92a
2000	0	0	0	0	0	0	0	0	0	0	0
2010	0	0	0	0	0	0	0	0	0	0	0
2020	0	0	0	0	0	0	0	0	0	0	18
2030	0	0	0	0	0	0	0	0	0	0	114
2040	0	0	0	0	0	0	0	0	0	0	281
2050	0	0	0	0	0	0	0	0	0	0	448
2060	0	0	0	0	0	0	0	0	0	0	495
2070	0	0	0	0	0	0	0	0	0	0	542
2080	0	0	0	0	0	0	0	0	0	0	567
2090	0	0	0	0	0	0	0	0	0	0	568
2100	0	0	0	0	0	0	0	0	0	0	570

HFC–227ea emissions (Gg/yr)

Year	A1B	A1T	A1FI	A2	B1	B2	A1p	A2p	B1p	B2p
2000	0	0	0	0	0	0	8	8	8	8
2010	13	13	13	12	13	14	12	11	11	12
2020	22	22	22	17	18	20	21	16	17	18
2030	34	34	34	21	24	26	33	19	22	25
2040	48	48	48	26	30	33	48	24	28	32
2050	62	62	62	32	39	41	57	29	38	41
2060	72	72	72	40	40	50	60	37	37	49
2070	71	71	71	48	39	59	60	44	37	57
2080	68	68	68	58	38	67	59	53	36	60
2090	65	65	65	70	36	74	56	64	34	63
2100	61	61	61	80	34	80	53	76	32	66

HFC–245ca emissions (Gg/yr)

Year	A1B	A1T	A1FI	A2	B1	B2	A1p	A2p	B1p	B2p
2000	0	0	0	0	0	0	38	38	38	38
2010	62	62	62	59	60	61	56	52	53	55
2020	100	100	100	79	80	85	98	73	75	84
2030	158	158	158	98	102	112	159	92	97	114
2040	222	222	222	121	131	144	229	113	128	149
2050	292	292	292	149	173	178	281	140	173	188
2060	350	350	350	190	173	216	298	179	172	229
2070	343	343	343	228	170	255	299	216	168	266
2080	330	330	330	276	166	290	287	262	163	280
2090	312	312	312	334	159	323	271	319	155	291
2100	288	288	288	388	150	353	251	376	145	302

HFC43–10mee emissions (Gg/yr)

Year	A1B	A1T	A1FI	A2	B1	B2	A1p	A2p	B1p	B2p
2000	0	0	0	0	0	0	5	5	5	5
2010	7	7	7	7	6	6	6	6	6	6
2020	9	9	9	8	7	7	8	7	7	7
2030	12	12	12	8	8	8	10	7	7	8
2040	15	15	15	9	9	10	13	8	9	9
2050	18	18	18	11	11	11	15	9	10	11
2060	22	22	22	12	11	12	17	11	10	12
2070	24	24	24	14	11	14	20	12	10	13
2080	27	27	27	16	11	15	22	14	10	14
2090	29	29	29	19	11	17	24	17	10	15
2100	30	30	30	22	10	18	26	19	10	15

Note: Table II.1.4 contains supplementary data to the SRES Report (Nakićenović *et. al.*, 2000): The data contained in the SRES Report was insufficient to break down the individual contributions to HFCs, PFCs and SF_6, these emissions were supplied by Lead Authors of the SRES Report and are also available at the CIESIN (Center for International Earth Science Information Network) Website (http://sres.ciesin.org). The sample scenario IS92a is only included for HFC–125, HFC–134a, and HFC–152a.
All PFCs, SF_6 and HFCs emissions are the same for family A1 (A1B, A1T and A1FI).

II.1.5: NO_x emissions (TgN/yr)

Year	A1B	A1T	A1FI	A2	B1	B2	A1p	A2p	B1p	B2p	IS92a
2000	32.0	32.0	32.0	32.0	32.0	32.0	32.5	32.5	32.5	32.5	37.0
2010	39.3	38.8	39.7	39.2	36.1	36.7	41.0	39.6	34.8	37.6	43.4
2020	46.1	46.4	50.4	50.3	39.9	42.7	48.9	50.7	39.3	43.4	49.8
2030	50.2	55.9	62.8	60.7	42.0	48.9	52.5	60.8	40.7	48.4	55.2
2040	48.9	59.7	77.1	65.9	42.6	53.4	50.9	65.8	44.8	52.8	59.6
2050	47.9	61.0	94.9	71.1	38.8	54.5	49.3	71.5	48.9	53.7	64.0
2060	46.0	59.6	102.1	75.5	34.3	56.1	47.2	75.6	48.9	55.4	67.8
2070	44.2	51.7	108.5	79.8	29.6	56.3	45.1	80.1	48.9	55.6	71.6
2080	42.7	42.8	115.4	87.5	25.7	59.2	43.3	87.3	48.9	58.5	75.4
2090	41.4	34.8	111.5	98.3	22.2	60.9	41.8	97.9	41.2	60.1	79.2
2100	40.2	28.1	109.6	109.2	18.7	61.2	40.3	109.7	33.6	60.4	83.0

Note: NO_x is the sum of NO and NO_2

II.1.6: CO emissions (Tg(CO)/yr)

Year	A1B	A1T	A1FI	A2	B1	B2	A1p	A2p	B1p	B2p	IS92a
2000	877	877	877	877	877	877	1036	1036	1036	1036	1048
2010	1002	1003	1020	977	789	935	1273	1136	849	1138	1096
2020	1032	1147	1204	1075	751	1022	1531	1234	985	1211	1145
2030	1109	1362	1436	1259	603	1111	1641	1413	864	1175	1207
2040	1160	1555	1726	1344	531	1220	1815	1494	903	1268	1282
2050	1214	1770	2159	1428	471	1319	1990	1586	942	1351	1358
2060	1245	1944	2270	1545	459	1423	2174	1696	984	1466	1431
2070	1276	2078	2483	1662	456	1570	2359	1816	1026	1625	1504
2080	1357	2164	2776	1842	426	1742	2455	1985	1068	1803	1576
2090	1499	2156	2685	2084	399	1886	2463	2218	1009	1948	1649
2100	1663	2077	2570	2326	363	2002	2471	2484	950	2067	1722

II.1.7: Total VOC emissions (Tg/yr)

Year	A1B	A1T	A1FI	A2	B1	B2	A1p	A2p	B1p	B2p	IS92a
2000	141	141	141	141	141	141	151	151	151	151	126
2010	178	164	166	155	141	159	178	164	143	172	142
2020	222	190	192	179	140	180	207	188	151	192	158
2030	266	212	214	202	131	199	229	210	144	202	173
2040	272	229	256	214	123	214	255	221	147	215	188
2050	279	241	322	225	116	217	285	235	150	217	202
2060	284	242	361	238	111	214	324	246	155	214	218
2070	289	229	405	251	103	202	301	260	160	202	234
2080	269	199	449	275	99	192	263	282	165	192	251
2090	228	167	435	309	96	178	223	315	159	178	267
2100	193	128	420	342	87	170	174	352	154	170	283

Note: Volatile Organic Compounds (VOC) include non–methane hydrocarbons (NMHC) and oxygenated NMHC (e.g., alcohols, aldehydes and organic acids).

II.1.8: SO$_2$ emissions (TgS/yr)

Year	A1B	A1T	A1FI	A2	B1	B2	A1p	A2p	B1p	B2p	IS92a
2000	69.0	69.0	69.0	69.0	69.0	69.0	69.0	69.0	69.0	69.0	79.0
2010	87.1	64.7	80.8	74.7	73.9	65.9	87.4	74.7	59.8	68.2	95.0
2020	100.2	59.9	86.9	99.5	74.6	61.3	100.8	99.5	56.2	65.0	111.0
2030	91.0	59.6	96.1	112.5	78.2	60.3	91.4	111.9	53.5	59.9	125.8
2040	68.9	45.9	94.0	109.0	78.5	59.0	77.9	108.1	53.3	58.8	139.4
2050	64.1	40.2	80.5	105.4	68.9	55.7	64.3	105.4	51.4	57.2	153.0
2060	46.9	34.4	56.3	89.6	55.8	53.8	51.2	86.3	51.2	53.7	151.8
2070	35.7	30.1	42.6	73.7	44.3	50.9	44.9	71.7	49.2	51.9	150.6
2080	30.7	25.2	39.4	64.7	36.1	50.0	30.7	64.2	42.2	49.1	149.4
2090	29.1	23.3	39.8	62.5	29.8	49.0	29.1	61.9	33.9	48.0	148.2
2100	27.6	20.2	40.1	60.3	24.9	47.9	27.4	60.3	28.6	47.3	147.0

Note: The SRES emissions for SO$_2$ are used with a linear offset in all scenarios to 69.0 TgS/yr in year 2000.

II.1.9: BC aerosol emissions (Tg/yr)

Year	A1B	A1T	A1FI	A2	B1	B2	A1p	A2p	B1p	B2p	IS92a
2000	12.4	12.4	12.4	12.4	12.4	12.4	12.4	12.4	12.4	12.4	12.4
2010	13.9	13.9	14.1	13.6	11.3	13.1	15.2	13.6	10.2	13.6	13.0
2020	14.3	15.6	16.3	14.8	10.9	14.1	18.3	14.8	11.8	14.5	13.6
2030	15.2	18.2	19.1	17.0	9.1	15.2	19.6	16.9	10.3	14.1	14.3
2040	15.8	20.5	22.6	18.0	8.3	16.5	21.7	17.9	10.8	15.2	15.2
2050	16.4	23.1	27.7	19.0	7.5	17.7	23.8	19.0	11.3	16.2	16.1
2060	16.8	25.2	29.1	20.4	7.4	18.9	26.0	20.3	11.8	17.5	17.0
2070	17.2	26.8	31.6	21.8	7.4	20.7	28.2	21.7	12.3	19.4	17.9
2080	18.1	27.8	35.1	24.0	7.0	22.8	29.4	23.8	12.8	21.6	18.7
2090	19.8	27.7	34.0	26.8	6.7	24.5	29.5	26.5	12.1	23.3	19.6
2100	21.8	26.8	32.7	29.7	6.2	25.9	29.6	29.7	11.4	24.7	20.5

Note: Emissions for BC are scaled to SRES anthropogenic CO emissions offset to year 2000.

II.1.10: OC aerosol emissions (Tg/yr)

Year	A1B	A1T	A1FI	A2	B1	B2	A1p	A2p	B1p	B2p	IS92a
2000	81.4	81.4	81.4	81.4	81.4	81.4	81.4	81.4	81.4	81.4	81.4
2010	91.2	91.3	92.6	89.3	74.5	86.0	100.0	89.3	66.7	89.4	85.2
2020	93.6	102.6	107.1	97.0	71.5	92.8	120.3	97.0	77.4	95.2	89.0
2030	99.6	119.5	125.3	111.4	59.9	99.8	128.9	111.0	67.9	92.3	93.9
2040	103.6	134.7	148.1	118.1	54.2	108.3	142.6	117.4	71.0	99.6	99.8
2050	107.9	151.6	182.1	124.7	49.5	116.1	156.4	124.6	74.0	106.2	105.8
2060	110.3	165.2	190.9	133.9	48.6	124.3	170.8	133.3	77.3	115.2	111.5
2070	112.8	175.8	207.6	143.1	48.3	135.9	185.4	142.7	80.6	127.7	117.2
2080	119.1	182.5	230.6	157.2	46.0	149.4	192.9	156.0	83.9	141.7	122.9
2090	130.3	181.9	223.5	176.2	43.8	160.7	193.5	174.3	79.3	153.1	128.6
2100	143.2	175.7	214.4	195.2	41.0	169.8	194.2	195.2	74.6	162.4	134.4

Note: Emissions for OC are scaled to SRES anthropogenic CO emissions offset to year 2000.

II.2: Abundances and burdens

II.2.1: CO$_2$ abundances (ppm)

ISAM model (reference) – CO$_2$ abundances (ppm)

Year	A1B	A1T	A1FI	A2	B1	B2	A1p	A2p	B1p	B2p	IS92a	IS92a/ SAR
1970	325	325	325	325	325	325	325	325	325	325	325	326
1980	337	337	337	337	337	337	337	337	337	337	337	338
1990	353	353	353	353	353	353	353	353	353	353	353	354
2000	369	369	369	369	369	369	369	369	369	369	369	372
2010	391	389	389	390	388	388	393	391	388	390	390	393
2020	420	412	417	417	412	408	425	419	409	414	415	418
2030	454	440	455	451	437	429	461	453	429	438	444	446
2040	491	471	504	490	463	453	499	492	450	462	475	476
2050	532	501	567	532	488	478	538	535	472	486	508	509
2060	572	528	638	580	509	504	577	583	497	512	543	544
2070	611	550	716	635	525	531	615	637	522	539	582	580
2080	649	567	799	698	537	559	652	699	544	567	623	620
2090	685	577	885	771	545	589	685	771	563	597	670	664
2100	717	582	970	856	549	621	715	856	578	630	723	715

ISAM model (low) – CO$_2$ abundances (ppm)

Year	A1B	A1T	A1FI	A2	B1	B2	A1p	A2p	B1p	B2p	IS92a
2000	368	368	368	368	368	368	368	368	368	368	368
2010	383	381	381	382	380	380	385	383	380	382	382
2020	405	398	403	402	398	394	409	404	395	400	401
2030	432	419	433	429	416	410	438	431	410	417	423
2040	461	443	473	460	436	427	467	461	425	435	446
2050	493	466	525	493	455	446	498	495	442	454	472
2060	524	486	584	532	470	466	528	534	460	473	499
2070	554	501	647	576	480	486	557	577	479	492	529
2080	582	511	715	626	486	507	583	627	495	513	561
2090	607	516	783	686	490	530	607	686	507	536	598
2100	630	516	851	755	490	554	627	755	517	561	640

ISAM model (high) – CO$_2$ abundances (ppm)

Year	A1B	A1T	A1FI	A2	B1	B2	A1p	A2p	B1p	B2p	IS92a
2000	369	369	369	369	369	369	369	369	369	369	369
2010	397	394	394	395	394	393	398	396	393	396	396
2020	431	422	427	427	422	417	435	429	418	424	426
2030	470	455	471	466	452	443	477	469	444	453	460
2040	513	491	527	511	483	472	521	514	469	482	498
2050	560	527	597	561	514	502	568	564	496	512	539
2060	609	560	678	617	541	534	615	620	527	543	583
2070	656	590	767	681	563	567	661	682	558	577	631
2080	703	613	863	754	581	602	706	755	586	612	682
2090	748	631	962	838	594	640	749	838	611	650	739
2100	790	642	1062	936	603	680	789	936	634	691	804

Note: A "reference" case was defined with climate sensitivity 2.5°C, ocean uptake corresponding to the mean of the ocean model results in Chapter 3, Figure 3.10, and terrestrial uptake corresponding to the mean of the responses of mid–range models, LPJ, IBIS and SDGM (Chapter 3, Figure 3.10). A "low CO$_2$" parametrization was chosen with climate sensitivity 1.5°C and maximal CO$_2$ uptake by oceans and land. A "high CO$_2$" parametrization was defined with climate sensitivity 4.5°C and minimal CO$_2$ uptake by oceans and land. See Chapter 3, Box 3.7, and Jain *et al.* (1994) for more details on the ISAM model.

The IS92a column values are calculated using the ISAM parametrization noted above with IS92a emissions starting in the year 2000; whereas the IS92a/SAR column refers to values as reported in the SAR using IS92a emissions starting in 1990, using the SAR parametrization of ISAM.

Bern–CC model (reference) – CO$_2$ abundances (ppm)

Year	A1B	A1T	A1FI	A2	B1	B2	A1p	A2p	B1p	B2p	IS92a	IS92a/ SAR
1970	325	325	325	325	325	325	325	325	325	325	325	325
1980	337	337	337	337	337	337	337	337	337	337	337	337
1990	352	352	352	352	352	352	352	352	352	352	352	353
2000	367	367	367	367	367	367	367	367	367	367	367	370
2010	388	386	386	386	386	385	390	388	385	387	387	391
2020	418	410	415	414	410	406	421	416	407	412	413	416
2030	447	435	449	444	432	425	454	447	425	433	439	444
2040	483	466	495	481	457	448	490	484	445	457	468	475
2050	522	496	555	522	482	473	529	525	467	481	499	507
2060	563	523	625	568	503	499	569	571	492	506	533	541
2070	601	545	702	620	518	524	606	622	515	532	568	577
2080	639	563	786	682	530	552	642	683	537	559	607	616
2090	674	572	872	754	538	581	674	754	555	588	653	660
2100	703	575	958	836	540	611	702	836	569	618	703	709

Bern–CC model (low) – CO$_2$ abundances (ppm)

Year	A1B	A1T	A1FI	A2	B1	B2	A1p	A2p	B1p	B2p	IS92a
2000	367	367	367	367	367	367	367	367	367	367	367
2010	383	381	381	381	381	380	384	383	380	382	383
2020	407	400	405	404	400	396	411	406	397	402	403
2030	432	419	432	428	417	410	437	431	410	417	424
2040	460	442	472	459	436	427	466	461	425	434	448
2050	491	464	521	492	455	445	496	495	440	452	473
2060	522	483	577	529	470	464	524	531	458	470	500
2070	548	496	636	569	479	482	550	569	475	487	527
2080	575	505	700	617	485	502	575	616	490	507	559
2090	598	508	763	671	487	522	596	670	501	528	593
2100	617	506	824	735	486	544	613	734	509	550	632

Bern–CC model (high) – CO$_2$ abundances (ppm)

Year	A1B	A1T	A1FI	A2	B1	B2	A1p	A2p	B1p	B2p	IS92a
2000	367	367	367	367	367	367	367	367	367	367	367
2010	395	393	393	393	392	392	397	395	392	394	395
2020	436	427	433	431	426	422	441	434	424	430	431
2030	483	467	484	477	463	454	491	482	455	465	471
2040	538	514	552	533	503	491	548	538	488	504	517
2050	599	562	638	597	544	531	609	602	524	544	568
2060	666	610	743	670	584	575	675	675	566	588	624
2070	732	653	859	753	617	620	738	757	608	632	684
2080	797	689	985	848	645	668	802	851	648	680	750
2090	860	717	1118	957	666	718	863	959	682	730	822
2100	918	735	1248	1080	681	769	918	1082	713	782	902

Note: A "reference" case was defined with an average ocean uptake for the 1980s of 2.0 PgC/yr. A "low CO$_2$" parameterisation was obtained by combining a "fast ocean" (ocean uptake of 2.54 PgC/yr for the 1980s) and no response of heterotrophic respiration to temperature. A "high CO$_2$" parameterisation was obtained by combining a "slow ocean " (ocean uptake of 1.46 PgC/yr for the 1980s) and capping CO$_2$ fertilisation. Climate sensitivity was set to 2.5°C for a doubling of CO$_2$. See Chapter 3, Box 3.7 for more details on the Bern–CC model.
The IS92a/SAR column refers to values as reported in the SAR using IS92a emissions; whereas the IS92a column is calculated using IS92a emissions but with year 2000 starting values and the BERN-CC model as described in Chapter 3.
The Bern-CC model was initialised for observed atmospheric CO$_2$ which was prescribed for the period 1765 to 1999. The CO$_2$ data were smoothed by a spline. Scenario calculations started at the begining of the year 2000. This explains the difference in the values given for the years upto 2000. Values shown are for the beginning of each year. Annual-mean values are generally higher (up to 7ppm) depending on the scenario and the year.

II.2.2: CH$_4$ abundances (ppb)

Year	A1B	A1T	A1FI	A2	B1	B2	A1p	A2p	B1p	B2p	IS92a	IS92a/ SAR
1970	1420	1420	1420	1420	1420	1420	1420	1420	1420	1420	1420	1420
1980	1570	1570	1570	1570	1570	1570	1570	1570	1570	1570	1570	1570
1990	1700	1700	1700	1700	1700	1700	1700	1700	1700	1700	1700	1700
2000	1760	1760	1760	1760	1760	1760	1760	1760	1760	1760	1760	1810
2010	1871	1856	1851	1861	1827	1839	1899	1861	1816	1862	1855	1964
2020	2026	1998	1986	1997	1891	1936	2126	1997	1878	2020	1979	2145
2030	2202	2194	2175	2163	1927	2058	2392	2159	1931	2201	2129	2343
2040	2337	2377	2413	2357	1919	2201	2598	2344	1963	2358	2306	2561
2050	2400	2503	2668	2562	1881	2363	2709	2549	2009	2473	2497	2793
2060	2386	2552	2875	2779	1836	2510	2736	2768	2049	2552	2663	3003
2070	2301	2507	3030	3011	1797	2639	2669	2998	2077	2606	2791	3175
2080	2191	2420	3175	3252	1741	2765	2533	3238	2100	2625	2905	3328
2090	2078	2310	3307	3493	1663	2872	2367	3475	2091	2597	3019	3474
2100	1974	2169	3413	3731	1574	2973	2187	3717	2039	2569	3136	3616

Note: The IS92a/SAR column refers to values as reported in the SAR using IS92a emissions; whereas the IS92a column is calculated using IS92a emissions but with year 2000 starting values and the new feedbacks on the lifetime. See Chapter 4 for details.

II.2.3: N₂O abundances (ppb)

Year	A1B	A1T	A1FI	A2	B1	B2	A1p	A2p	B1p	B2p	IS92a	IS92a/ SAR
1970	295	295	295	295	295	295	295	295	295	295	295	295
1980	301	301	301	301	301	301	301	301	301	301	301	301
1990	308	308	308	308	308	308	308	308	308	308	308	308
2000	316	316	316	316	316	316	316	316	316	316	316	319
2010	324	323	325	325	324	323	324	325	324	324	324	328
2020	331	328	335	335	333	328	332	335	333	331	333	339
2030	338	333	347	347	341	333	340	347	341	338	343	350
2040	344	338	361	360	349	338	346	360	350	343	353	361
2050	350	342	378	373	357	342	351	373	358	347	363	371
2060	356	345	396	387	363	346	355	386	366	350	372	382
2070	360	348	413	401	368	350	358	400	373	352	381	391
2080	365	350	429	416	371	354	360	415	380	354	389	400
2090	368	352	445	432	374	358	361	430	385	355	396	409
2100	372	354	460	447	375	362	361	446	389	356	403	417

Note: The IS92a/SAR column refers to values as reported in the SAR using IS92a emissions; whereas the IS92a column is calculated using IS92a emissions but with year 2000 starting values and the new feedbacks on the lifetime. See Chapter 4 for details.

II.2.4: PFCs, SF₆ and HFCs abundances (ppt)

CF₄ abundances (ppt)

Year	A1B	A1T	A1FI	A2	B1	B2	A1p	A2p	B1p	B2p
1990	70	70	70	70	70	70	70	70	70	70
2000	82	82	82	82	82	82	82	82	82	82
2010	91	91	91	92	91	93	100	100	100	100
2020	103	103	103	107	101	108	122	121	118	122
2030	119	119	119	125	111	128	154	146	138	150
2040	141	141	141	148	122	153	197	176	159	184
2050	168	168	168	175	135	184	245	212	181	226
2060	198	198	198	208	150	221	296	255	204	274
2070	230	230	230	246	164	261	342	306	226	327
2080	265	265	265	291	179	302	377	365	242	383
2090	303	303	303	341	193	344	405	433	256	439
2100	341	341	341	397	208	384	437	508	269	493

C₂F₆ abundances (ppt)

Year	A1B	A1T	A1FI	A2	B1	B2	A1p	A2p	B1p	B2p
1990	2	2	2	2	2	2	2	2	2	2
2000	3	3	3	3	3	3	3	3	3	3
2010	4	4	4	4	4	4	4	4	4	4
2020	5	5	5	5	4	5	6	6	6	6
2030	6	6	6	6	5	6	8	7	7	8
2040	7	7	7	7	6	8	11	9	8	10
2050	9	9	9	9	7	10	14	12	10	12
2060	11	11	11	11	8	12	17	14	11	16
2070	13	13	13	14	8	15	20	18	12	19
2080	15	15	15	17	9	17	22	21	13	22
2090	17	17	17	20	10	20	24	26	14	26
2100	20	20	20	23	11	22	26	30	15	30

SF$_6$ abundances (ppt)

Year	A1B	A1T	A1FI	A2	B1	B2	A1p	A2p	B1p	B2p
1990	3	3	3	3	3	3	3	3	3	3
2000	5	5	5	5	5	5	5	5	5	5
2010	7	7	7	7	7	7	7	7	7	7
2020	10	10	10	11	9	10	10	11	10	11
2030	13	13	13	15	12	14	14	15	12	15
2040	18	18	18	20	15	18	19	20	16	21
2050	25	25	25	26	19	23	26	26	20	27
2060	32	32	32	32	23	27	33	33	24	35
2070	39	39	39	40	27	32	41	41	29	43
2080	45	45	45	48	30	36	46	48	32	52
2090	50	50	50	56	33	40	51	57	35	61
2100	56	56	56	65	35	44	57	66	37	70

HFC–23 abundances (ppt)

Year	A1B	A1T	A1FI	A2	B1	B2	A1p	A2p	B1p	B2p
1990	8	8	8	8	8	8	8	8	8	8
2000	15	15	15	15	15	15	15	15	15	15
2010	26	26	26	26	26	26	26	26	26	26
2020	33	33	33	33	33	33	33	33	33	33
2030	35	35	35	35	35	35	35	35	35	35
2040	35	35	35	35	35	35	36	35	35	35
2050	35	35	35	35	35	35	35	35	35	35
2060	35	35	35	35	34	35	34	34	33	34
2070	35	35	34	34	34	34	33	32	32	33
2080	34	34	34	34	33	34	32	31	31	31
2090	34	34	34	34	33	34	31	30	30	30
2100	34	34	34	33	32	34	30	29	29	29

HFC–32 abundance (ppt)

Year	A1B	A1T	A1FI	A2	B1	B2	A1p	A2p	B1p	B2p
1990	0	0	0	0	0	0	0	0	0	0
2000	0	0	0	0	0	0	0	0	0	0
2010	1	1	1	1	1	1	1	1	1	1
2020	3	3	3	3	3	3	3	3	3	3
2030	7	7	6	4	4	4	7	4	4	5
2040	10	10	10	6	5	6	11	5	5	7
2050	14	14	13	7	7	8	15	7	7	9
2060	17	17	16	9	8	10	18	9	8	11
2070	19	19	18	11	8	12	20	11	8	13
2080	19	21	19	14	8	14	21	13	8	14
2090	20	22	20	17	8	15	21	16	8	15
2100	19	22	20	20	8	17	20	20	8	16

HFC–125 abundance (ppt)

Year	A1B	A1T	A1FI	A2	B1	B2	A1p	A2p	B1p	B2p	IS92a
1990	0	0	0	0	0	0	0	0	0	0	0
2000	0	0	0	0	0	0	0	0	0	0	0
2010	2	2	2	2	2	2	4	3	3	3	0
2020	9	9	9	8	8	8	10	8	8	9	2
2030	21	21	21	16	16	16	22	15	16	17	12
2040	37	37	37	24	24	26	38	23	24	27	40
2050	57	56	55	34	33	36	57	32	33	38	87
2060	77	78	76	45	43	48	78	43	42	51	137
2070	97	98	95	58	49	61	96	54	49	65	177
2080	112	115	111	72	54	75	111	68	54	77	210
2090	124	129	124	89	57	88	123	83	57	89	236
2100	133	140	134	107	58	102	132	101	58	99	255

HFC–134a abundance (ppt)

Year	A1B	A1T	A1FI	A2	B1	B2	A1p	A2p	B1p	B2p	IS92a
1990	0	0	0	0	0	0	0	0	0	0	0
2000	12	12	12	12	12	12	12	12	12	12	12
2010	58	58	58	55	55	56	80	76	76	79	94
2020	130	130	129	111	108	113	172	141	142	155	183
2030	236	235	233	170	165	179	319	214	215	250	281
2040	375	373	366	231	223	250	522	290	294	356	401
2050	537	535	521	299	293	330	754	375	393	477	537
2060	698	701	675	382	352	424	954	480	476	615	657
2070	814	832	791	480	380	526	1092	606	515	756	743
2080	871	912	859	594	391	633	1167	753	530	878	807
2090	887	952	893	729	390	737	1185	930	531	968	850
2100	875	956	899	877	379	835	1157	1132	522	1041	878

HFC–143a abundance (ppt)

Year	A1B	A1T	A1FI	A2	B1	B2	A1p	A2p	B1p	B2p
1990	0	0	0	0	0	0	0	0	0	0
2000	0	0	0	0	0	0	0	0	0	0
2010	3	3	3	3	2	2	4	4	4	4
2020	11	11	11	10	9	9	12	11	11	11
2030	26	26	26	20	18	19	27	20	20	22
2040	47	47	47	32	29	31	48	31	31	35
2050	73	73	72	45	43	45	75	44	44	51
2060	103	103	101	62	57	62	104	60	58	69
2070	132	133	130	81	68	81	131	78	69	89
2080	158	161	157	103	77	101	156	98	79	110
2090	181	185	180	129	85	121	179	123	86	129
2100	200	207	201	157	90	142	197	151	92	147

HFC–152a abundance (ppt)

Year	A1B	A1T	A1FI	A2	B1	B2	A1p	A2p	B1p	B2p	IS92a
1990	0	0	0	0	0	0	0	0	0	0	0
2000	0	0	0	0	0	0	0	0	0	0	0
2010	0	0	0	0	0	0	0	0	0	0	0
2020	0	0	0	0	0	0	0	0	0	0	2
2030	0	0	0	0	0	0	0	0	0	0	12
2040	0	0	0	0	0	0	0	0	0	0	33
2050	0	0	0	0	0	0	0	0	0	0	56
2060	0	0	0	0	0	0	0	0	0	0	67
2070	0	0	0	0	0	0	0	0	0	0	74
2080	0	0	0	0	0	0	0	0	0	0	79
2090	0	0	0	0	0	0	0	0	0	0	81
2100	0	0	0	0	0	0	0	0	0	0	82

HFC–227ea abundance (ppt)

Year	A1B	A1T	A1FI	A2	B1	B2	A1p	A2p	B1p	B2p
1990	0	0	0	0	0	0	0	0	0	0
2000	0	0	0	0	0	0	0	0	0	0
2010	2	2	2	2	2	2	3	3	3	3
2020	6	6	6	5	6	6	7	6	6	7
2030	13	13	13	10	10	11	13	9	10	11
2040	22	22	22	14	15	17	22	13	15	17
2050	33	33	32	19	21	24	33	18	20	23
2060	45	45	44	25	27	31	43	23	26	31
2070	56	56	55	32	31	40	52	29	30	39
2080	63	65	62	40	34	49	60	36	33	47
2090	68	71	68	49	35	59	64	45	34	54
2100	70	74	71	60	36	68	67	55	35	60

HFC–245ca abundance (ppt)

Year	A1B	A1T	A1FI	A2	B1	B2	A1p	A2p	B1p	B2p
1990	0	0	0	0	0	0	0	0	0	0
2000	0	0	0	0	0	0	0	0	0	0
2010	8	8	8	8	8	8	11	10	10	10
2020	20	20	20	17	17	18	20	16	16	18
2030	34	34	33	23	23	26	35	21	22	26
2040	52	51	50	29	29	34	55	27	28	35
2050	72	72	69	36	38	44	76	34	38	46
2060	92	93	88	46	43	55	92	43	44	58
2070	102	105	99	58	44	67	101	55	44	70
2080	101	108	101	72	43	80	101	68	44	79
2090	97	107	99	88	42	92	96	84	43	84
2100	90	101	94	105	40	103	88	101	41	88

HFC–43–10mee abundance (ppt)

Year	A1B	A1T	A1FI	A2	B1	B2	A1p	A2p	B1p	B2p
1990	0	0	0	0	0	0	0	0	0	0
2000	0	0	0	0	0	0	0	0	0	0
2010	1	1	1	1	1	1	1	1	1	1
2020	2	2	2	2	1	1	2	2	2	2
2030	3	3	3	2	2	2	3	2	2	2
2040	4	4	4	3	2	3	4	2	2	3
2050	5	5	5	3	3	3	5	3	3	3
2060	7	7	6	4	3	4	6	3	3	4
2070	8	8	8	4	4	5	7	4	3	4
2080	9	9	9	5	4	5	8	4	4	5
2090	10	11	10	6	4	6	9	5	4	5
2100	11	12	11	7	4	7	10	6	4	6

Note: Even though all PFCs, SF6 and HFCs emissions are the same for family A1 (A1B, A1T and A1FI), the OH changes due to CH_4, NO_x, CO and VOC (affecting only HFCs burdens). Hence the burden for HFCs can diverge for each of these scenarios within familiy A1. See Chapter 4 for details.

II.2.5: Tropospheric O₃ burden (global mean column in DU)

Year	A1B	A1T	A1FI	A2	B1	B2	A1p	A2p	B1p	B2p	IS92a	IS92a/ SAR
1990	34.0	34.0	34.0	34.0	34.0	34.0	34.0	34.0	34.0	34.0	34.0	34.0
2000	34.0	34.0	34.0	34.0	34.0	34.0	34.0	34.0	34.0	34.0	34.0	34.3
2010	35.8	35.6	35.8	35.7	34.8	35.2	36.2	35.6	34.3	35.4	35.5	34.8
2020	37.8	37.7	38.4	38.2	35.6	36.7	38.8	38.2	35.4	37.1	37.1	35.3
2030	39.3	40.3	41.5	40.8	35.9	38.4	40.5	40.7	35.7	38.5	38.7	35.8
2040	39.7	41.9	45.1	42.6	35.8	39.8	41.3	42.4	36.5	39.9	40.1	36.5
2050	39.8	42.9	49.6	44.2	35.0	40.7	41.6	44.1	37.5	40.6	41.6	37.1
2060	39.6	43.1	51.9	45.7	34.0	41.5	41.8	45.6	37.7	41.2	42.9	37.7
2070	39.1	41.9	53.8	47.2	33.1	42.1	41.4	47.1	37.9	41.6	44.0	38.2
2080	38.5	40.2	55.9	49.3	32.1	43.0	40.8	49.1	38.1	42.3	45.1	38.7
2090	38.0	38.4	55.6	52.0	31.2	43.7	39.9	51.8	36.8	42.6	46.1	39.1
2100	37.5	36.5	55.2	54.8	30.1	44.2	38.9	54.7	35.2	42.8	47.2	39.5

Note: IS92a/SAR column refers to IS92a emissions as reported in the SAR which estimated this O₃ change only as an indirect feedback effect from CH₄ increases; whereas IS92a column uses the latest models (see Chapter 4) which include also changes in emissions of NO_x, CO and VOC. A mean tropospheric O₃ content of 34 DU in 1990 is adopted; and 1 ppb of tropospheric O₃ = 0.65 DU.

These projected increases in troposheric O₃ are likely to be 25% too large, see note to Table 4.11 of Chapter 4 describing corrections made after government review.

II.2.6: Tropospheric OH (as a factor relative to year 2000)

Year	A1B	A1T	A1FI	A2	B1	B2	A1p	A2p	B1p	B2p	IS92a
2000	1.00	1.00	1.00	1.00	1.00	1.00	1.00	1.00	1.00	1.00	1.00
2010	0.99	0.99	0.99	1.00	1.01	0.99	0.98	1.00	1.02	0.99	1.00
2020	0.97	0.98	0.99	1.00	1.02	0.99	0.94	1.00	1.01	0.97	0.99
2030	0.94	0.96	0.98	0.99	1.04	0.98	0.90	0.99	1.02	0.96	0.98
2040	0.91	0.93	0.96	0.98	1.06	0.96	0.85	0.98	1.03	0.95	0.96
2050	0.90	0.89	0.94	0.96	1.06	0.93	0.81	0.96	1.04	0.93	0.95
2060	0.89	0.87	0.92	0.94	1.05	0.91	0.78	0.94	1.03	0.92	0.93
2070	0.89	0.84	0.90	0.92	1.04	0.89	0.77	0.92	1.01	0.90	0.92
2080	0.89	0.81	0.88	0.90	1.04	0.87	0.77	0.90	1.01	0.89	0.91
2090	0.90	0.81	0.86	0.89	1.04	0.86	0.80	0.89	0.98	0.89	0.90
2100	0.90	0.82	0.86	0.88	1.05	0.84	0.82	0.88	0.97	0.89	0.89

II.2.7: SO₄²⁻ aerosol burden (TgS)

Year	A1B	A1T	A1FI	A2	B1	B2	A1p	A2p	B1p	B2p	IS92a
2000	0.52	0.52	0.52	0.52	0.52	0.52	0.52	0.52	0.52	0.52	0.52
2010	0.66	0.49	0.61	0.56	0.56	0.50	0.66	0.56	0.45	0.51	0.64
2020	0.76	0.45	0.65	0.75	0.56	0.46	0.76	0.75	0.42	0.49	0.76
2030	0.69	0.45	0.72	0.85	0.59	0.45	0.69	0.84	0.40	0.45	0.87
2040	0.52	0.35	0.71	0.82	0.59	0.44	0.59	0.81	0.40	0.44	0.98
2050	0.48	0.30	0.61	0.79	0.52	0.42	0.48	0.79	0.39	0.43	1.08
2060	0.35	0.26	0.42	0.68	0.42	0.41	0.39	0.65	0.39	0.40	1.07
2070	0.27	0.23	0.32	0.56	0.33	0.38	0.34	0.54	0.37	0.39	1.06
2080	0.23	0.19	0.30	0.49	0.27	0.38	0.23	0.48	0.32	0.37	1.05
2090	0.22	0.18	0.30	0.47	0.22	0.37	0.22	0.47	0.26	0.36	1.04
2100	0.21	0.15	0.30	0.45	0.19	0.36	0.21	0.45	0.22	0.36	1.03

Note: Global burden is scaled to emissions: 0.52 Tg burden for 69.0 TgS/yr emissions.

II.2.8: BC aerosol burden (Tg)

Year	A1B	A1T	A1FI	A2	B1	B2	A1p	A2p	B1p	B2p	IS92a
2000	0.26	0.26	0.26	0.26	0.26	0.26	0.26	0.26	0.26	0.26	0.26
2010	0.29	0.29	0.30	0.29	0.24	0.27	0.32	0.29	0.21	0.29	0.27
2020	0.30	0.33	0.34	0.31	0.23	0.30	0.38	0.31	0.25	0.30	0.28
2030	0.32	0.38	0.40	0.36	0.19	0.32	0.41	0.35	0.22	0.29	0.30
2040	0.33	0.43	0.47	0.38	0.17	0.35	0.46	0.37	0.23	0.32	0.32
2050	0.34	0.48	0.58	0.40	0.16	0.37	0.50	0.40	0.24	0.34	0.34
2060	0.35	0.53	0.61	0.43	0.16	0.40	0.55	0.43	0.25	0.37	0.36
2070	0.36	0.56	0.66	0.46	0.15	0.43	0.59	0.46	0.26	0.41	0.37
2080	0.38	0.58	0.74	0.50	0.15	0.48	0.62	0.50	0.27	0.45	0.39
2090	0.42	0.58	0.71	0.56	0.14	0.51	0.62	0.56	0.25	0.49	0.41
2100	0.46	0.56	0.68	0.62	0.13	0.54	0.62	0.62	0.24	0.52	0.43

Note: Global burden is scaled to emissions: 0.26 Tg burden for 12.4 Tg/yr emsissions.

II.2.9: OC aerosol burden (Tg)

Year	A1B	A1T	A1FI	A2	B1	B2	A1p	A2p	B1p	B2p	IS92a
2000	1.52	1.52	1.52	1.52	1.52	1.52	1.52	1.52	1.52	1.52	1.52
2010	1.70	1.70	1.73	1.67	1.39	1.61	1.87	1.67	1.25	1.67	1.59
2020	1.75	1.92	2.00	1.81	1.34	1.73	2.25	1.81	1.45	1.78	1.66
2030	1.86	2.23	2.34	2.08	1.12	1.86	2.41	2.07	1.27	1.72	1.75
2040	1.94	2.51	2.77	2.21	1.01	2.02	2.66	2.19	1.32	1.86	1.86
2050	2.01	2.83	3.40	2.33	0.92	2.17	2.92	2.33	1.38	1.98	1.97
2060	2.06	3.09	3.56	2.50	0.91	2.32	3.19	2.49	1.44	2.15	2.08
2070	2.11	3.28	3.88	2.67	0.90	2.54	3.46	2.66	1.51	2.38	2.19
2080	2.22	3.41	4.31	2.94	0.86	2.79	3.60	2.91	1.57	2.65	2.29
2090	2.43	3.40	4.17	3.29	0.82	3.00	3.61	3.25	1.48	2.86	2.40
2100	2.67	3.28	4.00	3.65	0.77	3.17	3.63	3.64	1.39	3.03	2.51

Note: Global burden is scaled to emissions: 1.52 Tg burden for 81.4 Tg/yr emissions.

II.2.10: CFCs and HFCs abundances from WMO98 Scenario A1(baseline) following the Montreal (1997) Amendments (ppt)

Year	CFC–11	CFC–12	CFC–113	CFC–114	CFC–115	CCl₄	CH₃CCl₃	HCFC–22	HCFC–141b	HCFC–142b	HCFC–123	CF₂BrCl	CF₃Br	EESCl
1970	50	109	4	6	0	56	13	13	0	0	0	0	0	1.25
1975	106	199	9	8	1	77	36	25	0	0	0	0	0	1.54
1980	164	290	18	10	1	92	75	41	0	0	0	1	0	1.99
1985	207	373	34	12	3	100	102	64	0	0	0	2	1	2.44
1990	258	467	67	15	5	102	125	90	0	1	0	3	2	2.87
1995	271	520	86	16	7	100	110	112	3	7	0	4	2	3.30
2000	267	535	85	16	9	92	44	145	13	15	0	4	3	3.28
2010	246	527	81	16	9	75	6	257	22	33	2	4	3	3.03
2020	214	486	72	15	9	59	1	229	16	32	3	3	3	2.74
2030	180	441	64	15	9	47	0	137	9	23	2	2	3	2.42
2040	149	400	57	14	9	37	0	88	6	17	2	1	3	2.16
2050	123	362	51	14	9	29	0	46	2	11	1	1	3	1.94
2060	101	328	45	13	9	23	0	20	1	6	1	0	2	1.76
2070	83	298	40	13	9	18	0	9	0	4	0	0	2	1.62
2080	68	270	36	12	8	14	0	4	0	2	0	0	2	1.51
2090	56	245	32	12	8	11	0	2	0	1	0	0	2	1.41
2100	45	222	28	12	8	9	0	1	0	1	0	0	1	1.33

Notes: Only significant greenhouse halocarbons shown (ppt).
EESCl = Equivalent Effective Stratospheric Chlorine in ppb (includes Br).
[Source: UNEP/WMO Scientific Assessment of Ozone Depletion: 1998 (Chapter 11), Version 5, June 3, 1998, Calculations by John Daniel and Guus Velders – guus.velders@rivm.nl & jdaniel@al.noaa.gov]

II.3: Radiative Forcing (Wm^{-2}) (relative to pre-industrial period, 1750)

The concentrations of CO_2 and CH_4 considered here correspond to the year 2000 and differ slightly from those considered in Chapter 6 which used the values corresponding to the year 1998 (as appropriate for the time frame when Chapter 6 began its preparation). The resulting difference in the computed present day forcings is about 3% in the case of CO_2 and about 2% in the case of CH_4. For N_2O, the difference in the computed forcings is negligible. In the case of tropospheric ozone, the forcing for the year 2000 given here and that in Chapter 6 are the results of slightly different scenarios employed which leads to about a 9% difference in the forcings. For the halogen containing compounds, the absolute differences in concentrations between here and Chapter 6 lead to a difference in present day forcing of less than 0.002 Wm^{-2} for any species.

II.3.1: CO_2 radiative forcing (Wm^{-2})

ISAM model (reference) – CO_2 radiative forcing (Wm^{-2})

Year	A1B	A1T	A1FI	A2	B1	B2	A1p	A2p	B1p	B2p	IS92a	IS92a/ SAR
2000	1.51	1.51	1.51	1.51	1.51	1.51	1.51	1.51	1.51	1.51	1.51	1.56
2010	1.82	1.80	1.80	1.81	1.78	1.78	1.85	1.82	1.78	1.81	1.81	1.85
2020	2.21	2.10	2.17	2.17	2.10	2.05	2.27	2.19	2.07	2.13	2.14	2.18
2030	2.62	2.46	2.64	2.59	2.42	2.32	2.71	2.61	2.32	2.43	2.50	2.53
2040	3.04	2.82	3.18	3.03	2.73	2.61	3.13	3.05	2.58	2.72	2.87	2.88
2050	3.47	3.15	3.81	3.47	3.01	2.90	3.53	3.50	2.83	2.99	3.23	3.24
2060	3.86	3.43	4.44	3.93	3.24	3.18	3.91	3.96	3.11	3.27	3.58	3.59
2070	4.21	3.65	5.06	4.42	3.40	3.46	4.25	4.44	3.37	3.54	3.95	3.93
2080	4.54	3.81	5.65	4.93	3.52	3.74	4.56	4.93	3.59	3.81	4.32	4.29
2090	4.82	3.91	6.20	5.46	3.60	4.02	4.82	5.46	3.78	4.09	4.71	4.66
2100	5.07	3.95	6.69	6.02	3.64	4.30	5.05	6.02	3.92	4.38	5.11	5.05

ISAM model (low) – CO_2 radiative forcing (Wm^{-2})

Year	A1B	A1T	A1FI	A2	B1	B2	A1p	A2p	B1p	B2p	IS92a
2000	1.50	1.50	1.50	1.50	1.50	1.50	1.50	1.50	1.50	1.50	1.50
2010	1.71	1.69	1.69	1.70	1.67	1.67	1.74	1.71	1.67	1.70	1.70
2020	2.01	1.92	1.99	1.97	1.92	1.87	2.07	2.00	1.88	1.95	1.96
2030	2.36	2.19	2.37	2.32	2.16	2.08	2.43	2.35	2.08	2.17	2.25
2040	2.71	2.49	2.84	2.69	2.41	2.30	2.78	2.71	2.27	2.40	2.53
2050	3.06	2.76	3.40	3.06	2.64	2.53	3.12	3.09	2.48	2.62	2.83
2060	3.39	2.99	3.97	3.47	2.81	2.76	3.43	3.49	2.69	2.84	3.13
2070	3.69	3.15	4.52	3.90	2.92	2.99	3.72	3.91	2.91	3.05	3.44
2080	3.95	3.26	5.05	4.34	2.99	3.21	3.96	4.35	3.09	3.28	3.76
2090	4.18	3.31	5.54	4.83	3.03	3.45	4.18	4.83	3.21	3.51	4.10
2100	4.38	3.31	5.99	5.35	3.03	3.69	4.35	5.35	3.32	3.76	4.46

ISAM model (high) – CO_2 radiative forcing (Wm^{-2})

Year	A1B	A1T	A1FI	A2	B1	B2	A1p	A2p	B1p	B2p	IS92a
2000	1.51	1.51	1.51	1.51	1.51	1.51	1.51	1.51	1.51	1.51	1.51
2010	1.91	1.87	1.87	1.88	1.87	1.85	1.92	1.89	1.85	1.89	1.89
2020	2.35	2.23	2.30	2.30	2.23	2.17	2.40	2.32	2.18	2.26	2.28
2030	2.81	2.64	2.82	2.76	2.60	2.49	2.89	2.80	2.50	2.61	2.69
2040	3.28	3.04	3.42	3.26	2.96	2.83	3.36	3.29	2.80	2.94	3.12
2050	3.75	3.42	4.09	3.76	3.29	3.16	3.82	3.78	3.10	3.27	3.54
2060	4.20	3.75	4.77	4.27	3.56	3.49	4.25	4.29	3.42	3.58	3.96
2070	4.59	4.03	5.43	4.79	3.78	3.81	4.63	4.80	3.73	3.91	4.39
2080	4.96	4.23	6.06	5.34	3.94	4.13	4.99	5.35	3.99	4.22	4.80
2090	5.30	4.39	6.64	5.90	4.06	4.46	5.30	5.90	4.21	4.54	5.23
2100	5.59	4.48	7.17	6.49	4.14	4.79	5.58	6.49	4.41	4.87	5.68

Bern–CC model (reference) – CO_2 radiative forcing (Wm^{-2})

Year	A1B	A1T	A1FI	A2	B1	B2	A1p	A2p	B1p	B2p	IS92a	IS92a/ SAR
2000	1.49	1.49	1.49	1.49	1.49	1.49	1.49	1.49	1.49	1.49	1.49	1.53
2010	1.78	1.76	1.76	1.76	1.76	1.74	1.81	1.78	1.74	1.77	1.77	1.82
2020	2.18	2.08	2.14	2.13	2.08	2.03	2.22	2.16	2.04	2.10	2.12	2.16
2030	2.54	2.40	2.56	2.50	2.36	2.27	2.62	2.54	2.27	2.37	2.44	2.50
2040	2.96	2.76	3.09	2.93	2.66	2.55	3.03	2.97	2.52	2.66	2.79	2.87
2050	3.37	3.10	3.70	3.37	2.94	2.84	3.44	3.40	2.78	2.93	3.13	3.21
2060	3.78	3.38	4.33	3.82	3.17	3.13	3.83	3.85	3.05	3.20	3.48	3.56
2070	4.12	3.60	4.96	4.29	3.33	3.39	4.17	4.31	3.30	3.47	3.82	3.91
2080	4.45	3.78	5.56	4.80	3.45	3.67	4.48	4.81	3.52	3.74	4.18	4.26
2090	4.74	3.86	6.12	5.34	3.53	3.94	4.74	5.34	3.70	4.01	4.57	4.63
2100	4.96	3.89	6.62	5.89	3.55	4.21	4.96	5.89	3.83	4.27	4.96	5.01

Bern–CC model (low) – CO$_2$ radiative forcing (Wm^{-2})

Year	A1B	A1T	A1FI	A2	B1	B2	A1p	A2p	B1p	B2p	IS92a
2000	1.49	1.49	1.49	1.49	1.49	1.49	1.49	1.49	1.49	1.49	1.49
2010	1.71	1.69	1.69	1.69	1.69	1.67	1.73	1.71	1.67	1.70	1.71
2020	2.04	1.95	2.01	2.00	1.95	1.89	2.09	2.03	1.91	1.97	1.99
2030	2.36	2.19	2.36	2.31	2.17	2.08	2.42	2.35	2.08	2.17	2.26
2040	2.69	2.48	2.83	2.68	2.41	2.30	2.76	2.71	2.27	2.38	2.55
2050	3.04	2.74	3.36	3.05	2.64	2.52	3.10	3.09	2.46	2.60	2.84
2060	3.37	2.96	3.91	3.44	2.81	2.74	3.39	3.46	2.67	2.81	3.14
2070	3.63	3.10	4.43	3.83	2.91	2.94	3.65	3.83	2.87	3.00	3.42
2080	3.89	3.19	4.94	4.27	2.98	3.16	3.89	4.26	3.03	3.21	3.74
2090	4.10	3.23	5.40	4.71	3.00	3.37	4.08	4.71	3.15	3.43	4.05
2100	4.27	3.20	5.81	5.20	2.99	3.59	4.23	5.19	3.24	3.65	4.39

Bern–CC model (high) – CO$_2$ radiative forcing (Wm^{-2})

Year	A1B	A1T	A1FI	A2	B1	B2	A1p	A2p	B1p	B2p	IS92a
2000	1.49	1.49	1.49	1.49	1.49	1.49	1.49	1.49	1.49	1.49	1.49
2010	1.88	1.85	1.85	1.85	1.84	1.84	1.91	1.88	1.84	1.87	1.88
2020	2.41	2.30	2.37	2.35	2.28	2.23	2.47	2.38	2.26	2.33	2.35
2030	2.96	2.78	2.97	2.89	2.73	2.62	3.04	2.94	2.64	2.75	2.82
2040	3.53	3.29	3.67	3.48	3.17	3.04	3.63	3.53	3.01	3.18	3.32
2050	4.11	3.77	4.44	4.09	3.59	3.46	4.20	4.13	3.39	3.59	3.82
2060	4.67	4.20	5.26	4.71	3.97	3.89	4.75	4.75	3.80	4.01	4.33
2070	5.18	4.57	6.04	5.33	4.27	4.29	5.23	5.36	4.19	4.39	4.82
2080	5.63	4.86	6.77	5.97	4.50	4.69	5.67	5.99	4.53	4.79	5.31
2090	6.04	5.07	7.45	6.61	4.67	5.08	6.06	6.62	4.80	5.17	5.80
2100	6.39	5.20	8.03	7.26	4.79	5.44	6.39	7.27	5.04	5.53	6.30

II.3.2: CH$_4$ radiative forcing (Wm^{-2})

Year	A1B	A1T	A1FI	A2	B1	B2	A1p	A2p	B1p	B2p	IS92a	IS92a/ SAR
2000	0.49	0.49	0.49	0.49	0.49	0.49	0.49	0.49	0.49	0.49	0.49	0.51
2010	0.53	0.52	0.52	0.53	0.51	0.52	0.54	0.53	0.51	0.53	0.52	0.56
2020	0.59	0.58	0.57	0.58	0.54	0.55	0.62	0.58	0.53	0.58	0.57	0.63
2030	0.65	0.64	0.64	0.63	0.55	0.60	0.71	0.63	0.55	0.64	0.62	0.69
2040	0.69	0.70	0.71	0.70	0.55	0.64	0.77	0.69	0.56	0.70	0.68	0.76
2050	0.71	0.74	0.79	0.76	0.53	0.70	0.80	0.76	0.58	0.73	0.74	0.83
2060	0.71	0.76	0.85	0.83	0.52	0.74	0.81	0.82	0.59	0.76	0.79	0.89
2070	0.68	0.74	0.90	0.89	0.50	0.78	0.79	0.89	0.60	0.77	0.83	0.94
2080	0.64	0.72	0.94	0.96	0.48	0.82	0.75	0.96	0.61	0.78	0.86	0.98
2090	0.60	0.68	0.97	1.02	0.45	0.85	0.70	1.02	0.61	0.77	0.90	1.02
2100	0.57	0.63	1.00	1.09	0.42	0.88	0.64	1.08	0.59	0.76	0.93	1.06

II.3.3: N$_2$O radiative forcing (Wm^{-2})

Year	A1B	A1T	A1FI	A2	B1	B2	A1p	A2p	B1p	B2p	IS92a	IS92a/ SAR
2000	0.15	0.15	0.15	0.15	0.15	0.15	0.15	0.15	0.15	0.15	0.15	0.16
2010	0.18	0.17	0.18	0.18	0.18	0.17	0.18	0.18	0.18	0.18	0.18	0.19
2020	0.20	0.19	0.21	0.21	0.21	0.19	0.20	0.21	0.21	0.20	0.21	0.22
2030	0.22	0.21	0.25	0.25	0.23	0.21	0.23	0.25	0.23	0.22	0.24	0.26
2040	0.24	0.22	0.29	0.29	0.25	0.22	0.25	0.29	0.26	0.24	0.27	0.29
2050	0.26	0.23	0.34	0.33	0.28	0.23	0.26	0.33	0.28	0.25	0.30	0.32
2060	0.28	0.24	0.39	0.37	0.30	0.25	0.27	0.36	0.31	0.26	0.32	0.35
2070	0.29	0.25	0.44	0.41	0.31	0.26	0.28	0.40	0.33	0.26	0.35	0.38
2080	0.30	0.26	0.48	0.45	0.32	0.27	0.29	0.45	0.35	0.27	0.37	0.40
2090	0.31	0.26	0.53	0.49	0.33	0.28	0.29	0.49	0.36	0.27	0.39	0.43
2100	0.32	0.27	0.57	0.53	0.33	0.29	0.29	0.53	0.37	0.28	0.41	0.45

II.3.4: PFCs, SF$_6$ and HFCs radiative forcing (Wm^{-2})

CF$_4$ radiative forcing (Wm^{-2})

Year	A1B	A1T	A1FI	A2	B1	B2	A1p	A2p	B1p	B2p
2000	0.003	0.003	0.003	0.003	0.003	0.003	0.003	0.003	0.003	0.003
2010	0.004	0.004	0.004	0.004	0.004	0.004	0.005	0.005	0.005	0.005
2020	0.005	0.005	0.005	0.005	0.005	0.005	0.007	0.006	0.006	0.007
2030	0.006	0.006	0.006	0.007	0.006	0.007	0.009	0.008	0.008	0.009
2040	0.008	0.008	0.008	0.009	0.007	0.009	0.013	0.011	0.010	0.012
2050	0.010	0.010	0.010	0.011	0.008	0.012	0.016	0.014	0.011	0.015
2060	0.013	0.013	0.013	0.013	0.009	0.014	0.020	0.017	0.013	0.019
2070	0.015	0.015	0.015	0.016	0.010	0.018	0.024	0.021	0.015	0.023
2080	0.018	0.018	0.018	0.020	0.011	0.021	0.027	0.026	0.016	0.027
2090	0.021	0.021	0.021	0.024	0.012	0.024	0.029	0.031	0.017	0.032
2100	0.024	0.024	0.024	0.029	0.013	0.028	0.032	0.037	0.018	0.036

C$_2$F$_6$ radiative forcing (Wm^{-2})

Year	A1B	A1T	A1FI	A2	B1	B2	A1p	A2p	B1p	B2p
2000	0.001	0.001	0.001	0.001	0.001	0.001	0.001	0.001	0.001	0.001
2010	0.001	0.001	0.001	0.001	0.001	0.001	0.001	0.001	0.001	0.001
2020	0.001	0.001	0.001	0.001	0.001	0.001	0.002	0.002	0.002	0.002
2030	0.002	0.002	0.002	0.002	0.001	0.002	0.002	0.002	0.002	0.002
2040	0.002	0.002	0.002	0.002	0.002	0.002	0.003	0.002	0.002	0.003
2050	0.002	0.002	0.002	0.002	0.002	0.003	0.004	0.003	0.003	0.003
2060	0.003	0.003	0.003	0.003	0.002	0.003	0.004	0.004	0.003	0.004
2070	0.003	0.003	0.003	0.004	0.002	0.004	0.005	0.005	0.003	0.005
2080	0.004	0.004	0.004	0.004	0.002	0.004	0.006	0.005	0.003	0.006
2090	0.004	0.004	0.004	0.005	0.003	0.005	0.006	0.007	0.004	0.007
2100	0.005	0.005	0.005	0.006	0.003	0.006	0.007	0.008	0.004	0.008

SF$_6$ radiative forcing (Wm^{-2})

Year	A1B	A1T	A1FI	A2	B1	B2	A1p	A2p	B1p	B2p
2000	0.003	0.003	0.003	0.003	0.003	0.003	0.003	0.003	0.003	0.003
2010	0.004	0.004	0.004	0.004	0.004	0.004	0.004	0.004	0.004	0.004
2020	0.005	0.005	0.005	0.006	0.005	0.005	0.005	0.006	0.005	0.006
2030	0.007	0.007	0.007	0.008	0.006	0.007	0.007	0.008	0.006	0.008
2040	0.009	0.009	0.009	0.010	0.008	0.009	0.010	0.010	0.008	0.011
2050	0.013	0.013	0.013	0.014	0.010	0.012	0.014	0.014	0.010	0.014
2060	0.017	0.017	0.017	0.017	0.012	0.014	0.017	0.017	0.012	0.018
2070	0.020	0.020	0.020	0.021	0.014	0.017	0.021	0.021	0.015	0.022
2080	0.023	0.023	0.023	0.025	0.016	0.019	0.024	0.025	0.017	0.027
2090	0.026	0.026	0.026	0.029	0.017	0.021	0.027	0.030	0.018	0.032
2100	0.029	0.029	0.029	0.034	0.018	0.023	0.030	0.034	0.019	0.036

HFC–23 radiative forcing (Wm^{-2})

Year	A1B	A1T	A1FI	A2	B1	B2	A1p	A2p	B1p	B2p
2000	0.002	0.002	0.002	0.002	0.002	0.002	0.002	0.002	0.002	0.002
2010	0.004	0.004	0.004	0.004	0.004	0.004	0.004	0.004	0.004	0.004
2020	0.005	0.005	0.005	0.005	0.005	0.005	0.005	0.005	0.005	0.005
2030	0.006	0.006	0.006	0.006	0.006	0.006	0.006	0.006	0.006	0.006
2040	0.006	0.006	0.006	0.006	0.006	0.006	0.006	0.006	0.006	0.006
2050	0.006	0.006	0.006	0.006	0.006	0.006	0.006	0.006	0.006	0.006
2060	0.006	0.006	0.006	0.006	0.005	0.006	0.005	0.005	0.005	0.005
2070	0.006	0.006	0.005	0.005	0.005	0.005	0.005	0.005	0.005	0.005
2080	0.005	0.005	0.005	0.005	0.005	0.005	0.005	0.005	0.005	0.005
2090	0.005	0.005	0.005	0.005	0.005	0.005	0.005	0.005	0.005	0.005
2100	0.005	0.005	0.005	0.005	0.005	0.005	0.005	0.005	0.005	0.005

HFC–32 radiative forcing (Wm^{-2})

Year	A1B	A1T	A1FI	A2	B1	B2	A1p	A2p	B1p	B2p
2000	0.000	0.000	0.000	0.000	0.000	0.000	0.000	0.000	0.000	0.000
2010	0.000	0.000	0.000	0.000	0.000	0.000	0.000	0.000	0.000	0.000
2020	0.000	0.000	0.000	0.000	0.000	0.000	0.000	0.000	0.000	0.000
2030	0.001	0.001	0.001	0.000	0.000	0.000	0.001	0.000	0.000	0.000
2040	0.001	0.001	0.001	0.001	0.000	0.001	0.001	0.000	0.000	0.001
2050	0.001	0.001	0.001	0.001	0.001	0.001	0.001	0.001	0.001	0.001
2060	0.002	0.002	0.001	0.001	0.001	0.001	0.002	0.001	0.001	0.001
2070	0.002	0.002	0.002	0.001	0.001	0.001	0.002	0.001	0.001	0.001
2080	0.002	0.002	0.002	0.001	0.001	0.001	0.002	0.001	0.001	0.001
2090	0.002	0.002	0.002	0.002	0.001	0.001	0.002	0.001	0.001	0.001
2100	0.002	0.002	0.002	0.002	0.001	0.002	0.002	0.002	0.001	0.001

HFC–125 radiative forcing (Wm^{-2})

Year	A1B	A1T	A1FI	A2	B1	B2	A1p	A2p	B1p	B2p	IS92a
2000	0.000	0.000	0.000	0.000	0.000	0.000	0.000	0.000	0.000	0.000	0.000
2010	0.000	0.000	0.000	0.000	0.000	0.000	0.001	0.001	0.001	0.001	0.000
2020	0.002	0.002	0.002	0.002	0.002	0.002	0.002	0.002	0.002	0.002	0.000
2030	0.005	0.005	0.005	0.004	0.004	0.004	0.005	0.003	0.004	0.004	0.003
2040	0.009	0.009	0.009	0.006	0.006	0.006	0.009	0.005	0.006	0.006	0.009
2050	0.013	0.013	0.013	0.008	0.008	0.008	0.013	0.007	0.008	0.009	0.020
2060	0.018	0.018	0.017	0.010	0.010	0.011	0.018	0.010	0.010	0.012	0.032
2070	0.022	0.023	0.022	0.013	0.011	0.014	0.022	0.012	0.011	0.015	0.041
2080	0.026	0.026	0.026	0.017	0.012	0.017	0.026	0.016	0.012	0.018	0.048
2090	0.029	0.030	0.029	0.020	0.013	0.020	0.028	0.019	0.013	0.020	0.054
2100	0.031	0.032	0.031	0.025	0.013	0.023	0.030	0.023	0.013	0.023	0.059

HFC–134a radiative forcing (Wm^{-2})

Year	A1B	A1T	A1FI	A2	B1	B2	A1p	A2p	B1p	B2p	IS92a
2000	0.002	0.002	0.002	0.002	0.002	0.002	0.002	0.002	0.002	0.002	0.002
2010	0.009	0.009	0.009	0.008	0.008	0.008	0.012	0.011	0.011	0.012	0.014
2020	0.020	0.020	0.019	0.017	0.016	0.017	0.026	0.021	0.021	0.023	0.027
2030	0.035	0.035	0.035	0.026	0.025	0.027	0.048	0.032	0.032	0.038	0.042
2040	0.056	0.056	0.055	0.035	0.033	0.038	0.078	0.043	0.044	0.053	0.060
2050	0.081	0.080	0.078	0.045	0.044	0.050	0.113	0.056	0.059	0.072	0.081
2060	0.105	0.105	0.101	0.057	0.053	0.064	0.143	0.072	0.071	0.092	0.099
2070	0.122	0.125	0.119	0.072	0.057	0.079	0.164	0.091	0.077	0.113	0.111
2080	0.131	0.137	0.129	0.089	0.059	0.095	0.175	0.113	0.079	0.132	0.121
2090	0.133	0.143	0.134	0.109	0.059	0.111	0.178	0.140	0.080	0.145	0.128
2100	0.131	0.143	0.135	0.132	0.057	0.125	0.174	0.170	0.078	0.156	0.132

HFC–143a radiative forcing (Wm^{-2})

Year	A1B	A1T	A1FI	A2	B1	B2	A1p	A2p	B1p	B2p
2000	0.000	0.000	0.000	0.000	0.000	0.000	0.000	0.000	0.000	0.000
2010	0.000	0.000	0.000	0.000	0.000	0.000	0.001	0.001	0.001	0.001
2020	0.001	0.001	0.001	0.001	0.001	0.001	0.002	0.001	0.001	0.001
2030	0.003	0.003	0.003	0.003	0.002	0.002	0.004	0.003	0.003	0.003
2040	0.006	0.006	0.006	0.004	0.004	0.004	0.006	0.004	0.004	0.005
2050	0.009	0.009	0.009	0.006	0.006	0.006	0.010	0.006	0.006	0.007
2060	0.013	0.013	0.013	0.008	0.007	0.008	0.014	0.008	0.008	0.009
2070	0.017	0.017	0.017	0.011	0.009	0.011	0.017	0.010	0.009	0.012
2080	0.021	0.021	0.020	0.013	0.010	0.013	0.020	0.013	0.010	0.014
2090	0.024	0.024	0.023	0.017	0.011	0.016	0.023	0.016	0.011	0.017
2100	0.026	0.027	0.026	0.020	0.012	0.018	0.026	0.020	0.012	0.019

HFC–152a radiative forcing (Wm^{-2})

Year	A1B	A1T	A1FI	A2	B1	B2	A1p	A2p	B1p	B2p	IS92a
2000	0.000	0.000	0.000	0.000	0.000	0.000	0.000	0.000	0.000	0.000	0.000
2010	0.000	0.000	0.000	0.000	0.000	0.000	0.000	0.000	0.000	0.000	0.000
2020	0.000	0.000	0.000	0.000	0.000	0.000	0.000	0.000	0.000	0.000	0.000
2030	0.000	0.000	0.000	0.000	0.000	0.000	0.000	0.000	0.000	0.000	0.001
2040	0.000	0.000	0.000	0.000	0.000	0.000	0.000	0.000	0.000	0.000	0.003
2050	0.000	0.000	0.000	0.000	0.000	0.000	0.000	0.000	0.000	0.000	0.005
2060	0.000	0.000	0.000	0.000	0.000	0.000	0.000	0.000	0.000	0.000	0.006
2070	0.000	0.000	0.000	0.000	0.000	0.000	0.000	0.000	0.000	0.000	0.007
2080	0.000	0.000	0.000	0.000	0.000	0.000	0.000	0.000	0.000	0.000	0.007
2090	0.000	0.000	0.000	0.000	0.000	0.000	0.000	0.000	0.000	0.000	0.007
2100	0.000	0.000	0.000	0.000	0.000	0.000	0.000	0.000	0.000	0.000	0.007

HFC–227ea radiative forcing (Wm^{-2})

Year	A1B	A1T	A1FI	A2	B1	B2	A1p	A2p	B1p	B2p
2000	0.000	0.000	0.000	0.000	0.000	0.000	0.000	0.000	0.000	0.000
2010	0.001	0.001	0.001	0.001	0.001	0.001	0.001	0.001	0.001	0.001
2020	0.002	0.002	0.002	0.002	0.002	0.002	0.002	0.002	0.002	0.002
2030	0.004	0.004	0.004	0.003	0.003	0.003	0.004	0.003	0.003	0.003
2040	0.007	0.007	0.007	0.004	0.004	0.005	0.007	0.004	0.004	0.005
2050	0.010	0.010	0.010	0.006	0.006	0.007	0.010	0.005	0.006	0.007
2060	0.014	0.014	0.013	0.008	0.008	0.009	0.013	0.007	0.008	0.009
2070	0.017	0.017	0.016	0.010	0.009	0.012	0.016	0.009	0.009	0.012
2080	0.019	0.020	0.019	0.012	0.010	0.015	0.018	0.011	0.010	0.014
2090	0.020	0.021	0.020	0.015	0.010	0.018	0.019	0.014	0.010	0.016
2100	0.021	0.022	0.021	0.018	0.011	0.020	0.020	0.016	0.010	0.018

HFC–245ca radiative forcing (Wm^{-2})

Year	A1B	A1T	A1FI	A2	B1	B2	A1p	A2p	B1p	B2p
2000	0.000	0.000	0.000	0.000	0.000	0.000	0.000	0.000	0.000	0.000
2010	0.002	0.002	0.002	0.002	0.002	0.002	0.003	0.002	0.002	0.002
2020	0.005	0.005	0.005	0.004	0.004	0.004	0.005	0.004	0.004	0.004
2030	0.008	0.008	0.008	0.005	0.005	0.006	0.008	0.005	0.005	0.006
2040	0.012	0.012	0.012	0.007	0.007	0.008	0.013	0.006	0.006	0.008
2050	0.017	0.017	0.016	0.008	0.009	0.010	0.017	0.008	0.009	0.011
2060	0.021	0.021	0.020	0.011	0.010	0.013	0.021	0.010	0.010	0.013
2070	0.023	0.024	0.023	0.013	0.010	0.015	0.023	0.013	0.010	0.016
2080	0.023	0.025	0.023	0.017	0.010	0.018	0.023	0.016	0.010	0.018
2090	0.022	0.025	0.023	0.020	0.010	0.021	0.022	0.019	0.010	0.019
2100	0.021	0.023	0.022	0.024	0.009	0.024	0.020	0.023	0.009	0.020

HFC–43–10mee radiative forcing (Wm^{-2})

Year	A1B	A1T	A1FI	A2	B1	B2	A1p	A2p	B1p	B2p
2000	0.000	0.000	0.000	0.000	0.000	0.000	0.000	0.000	0.000	0.000
2010	0.000	0.000	0.000	0.000	0.000	0.000	0.000	0.000	0.000	0.000
2020	0.001	0.001	0.001	0.001	0.000	0.000	0.001	0.001	0.001	0.001
2030	0.001	0.001	0.001	0.001	0.001	0.001	0.001	0.001	0.001	0.001
2040	0.002	0.002	0.002	0.001	0.001	0.001	0.002	0.001	0.001	0.001
2050	0.002	0.002	0.002	0.001	0.001	0.001	0.002	0.001	0.001	0.001
2060	0.003	0.003	0.002	0.002	0.001	0.002	0.002	0.001	0.001	0.002
2070	0.003	0.003	0.003	0.002	0.002	0.002	0.003	0.002	0.001	0.002
2080	0.004	0.004	0.004	0.002	0.002	0.002	0.003	0.002	0.002	0.002
2090	0.004	0.004	0.004	0.002	0.002	0.002	0.004	0.002	0.002	0.002
2100	0.004	0.005	0.004	0.003	0.002	0.003	0.004	0.002	0.002	0.002

II.3.5: Tropospheric O$_3$ radiative forcing (Wm^{-2})

Year	A1B	A1T	A1FI	A2	B1	B2	A1p	A2p	B1p	B2p	IS92a	IS92a/SAR
2000	0.38	0.38	0.38	0.38	0.38	0.38	0.38	0.38	0.38	0.38	0.38	0.39
2010	0.45	0.45	0.45	0.45	0.41	0.43	0.47	0.45	0.39	0.44	0.44	0.41
2020	0.54	0.53	0.56	0.55	0.45	0.49	0.58	0.55	0.44	0.51	0.51	0.43
2030	0.60	0.64	0.69	0.66	0.46	0.56	0.65	0.66	0.45	0.57	0.58	0.45
2040	0.62	0.71	0.84	0.74	0.45	0.62	0.68	0.73	0.48	0.63	0.63	0.48
2050	0.62	0.75	1.03	0.81	0.42	0.66	0.70	0.80	0.52	0.66	0.70	0.51
2060	0.61	0.76	1.13	0.87	0.38	0.69	0.71	0.87	0.53	0.68	0.75	0.53
2070	0.59	0.71	1.21	0.93	0.34	0.72	0.69	0.93	0.54	0.70	0.80	0.55
2080	0.57	0.64	1.30	1.02	0.30	0.76	0.66	1.01	0.55	0.73	0.84	0.58
2090	0.55	0.56	1.29	1.13	0.26	0.79	0.63	1.13	0.50	0.74	0.89	0.59
2100	0.52	0.48	1.27	1.25	0.21	0.81	0.58	1.25	0.43	0.75	0.93	0.61

II.3.6: SO$_4^{2-}$ aerosols (direct effect) radiative forcing (Wm^{-2})

Year	A1B	A1T	A1FI	A2	B1	B2	A1p	A2p	B1p	B2p	IS92a
2000	−0.40	−0.40	−0.40	−0.40	−0.40	−0.40	−0.40	−0.40	−0.40	−0.40	−0.40
2010	−0.51	−0.38	−0.47	−0.43	−0.43	−0.38	−0.51	−0.43	−0.35	−0.39	−0.49
2020	−0.58	−0.35	−0.50	−0.58	−0.43	−0.35	−0.58	−0.58	−0.32	−0.38	−0.58
2030	−0.53	−0.35	−0.55	−0.65	−0.45	−0.35	−0.53	−0.65	−0.31	−0.35	−0.67
2040	−0.40	−0.27	−0.55	−0.63	−0.45	−0.34	−0.45	−0.62	−0.31	−0.34	−0.75
2050	−0.37	−0.23	−0.47	−0.61	−0.40	−0.32	−0.37	−0.61	−0.30	−0.33	−0.83
2060	−0.27	−0.20	−0.32	−0.52	−0.32	−0.32	−0.30	−0.50	−0.30	−0.31	−0.82
2070	−0.21	−0.18	−0.25	−0.43	−0.25	−0.29	−0.26	−0.42	−0.28	−0.30	−0.82
2080	−0.18	−0.15	−0.23	−0.38	−0.21	−0.29	−0.18	−0.37	−0.25	−0.28	−0.81
2090	−0.17	−0.14	−0.23	−0.36	−0.17	−0.28	−0.17	−0.36	−0.20	−0.28	−0.80
2100	−0.16	−0.12	−0.23	−0.35	−0.15	−0.28	−0.16	−0.35	−0.17	−0.28	−0.79

II.3.7: BC aerosols radiative forcing (Wm^{-2})

Year	A1B	A1T	A1FI	A2	B1	B2	A1p	A2p	B1p	B2p	IS92a
2000	0.40	0.40	0.40	0.40	0.40	0.40	0.40	0.40	0.40	0.40	0.40
2010	0.45	0.45	0.46	0.45	0.37	0.42	0.49	0.45	0.32	0.45	0.42
2020	0.46	0.51	0.52	0.48	0.35	0.46	0.58	0.48	0.38	0.46	0.43
2030	0.49	0.58	0.62	0.55	0.29	0.49	0.63	0.54	0.34	0.45	0.46
2040	0.51	0.66	0.72	0.58	0.26	0.54	0.71	0.57	0.35	0.49	0.49
2050	0.52	0.74	0.89	0.62	0.25	0.57	0.77	0.62	0.37	0.52	0.52
2060	0.54	0.82	0.94	0.66	0.25	0.62	0.85	0.66	0.38	0.57	0.55
2070	0.55	0.86	1.02	0.71	0.23	0.66	0.91	0.71	0.40	0.63	0.57
2080	0.58	0.89	1.14	0.77	0.23	0.74	0.95	0.77	0.42	0.69	0.60
2090	0.65	0.89	1.09	0.86	0.22	0.78	0.95	0.86	0.38	0.75	0.63
2100	0.71	0.86	1.05	0.95	0.20	0.83	0.95	0.95	0.37	0.80	0.66

II.3.8: OC aerosols radiative forcing (Wm^{-2})

Year	A1B	A1T	A1FI	A2	B1	B2	A1p	A2p	B1p	B2p	IS92a
2000	−0.50	−0.50	−0.50	−0.50	−0.50	−0.50	−0.50	−0.50	−0.50	−0.50	−0.50
2010	−0.56	−0.56	−0.57	−0.55	−0.46	−0.53	−0.62	−0.55	−0.41	−0.55	−0.52
2020	−0.58	−0.63	−0.66	−0.60	−0.44	−0.57	−0.74	−0.60	−0.48	−0.59	−0.55
2030	−0.61	−0.73	−0.77	−0.68	−0.37	−0.61	−0.79	−0.68	−0.42	−0.57	−0.58
2040	−0.64	−0.83	−0.91	−0.73	−0.33	−0.66	−0.88	−0.72	−0.43	−0.61	−0.61
2050	−0.66	−0.93	−1.12	−0.77	−0.30	−0.71	−0.96	−0.77	−0.45	−0.65	−0.65
2060	−0.68	−1.02	−1.17	−0.82	−0.30	−0.76	−1.05	−0.82	−0.47	−0.71	−0.68
2070	−0.69	−1.08	−1.28	−0.88	−0.30	−0.84	−1.14	−0.88	−0.50	−0.78	−0.72
2080	−0.73	−1.12	−1.42	−0.97	−0.28	−0.92	−1.18	−0.96	−0.52	−0.87	−0.75
2090	−0.80	−1.12	−1.37	−1.08	−0.27	−0.99	−1.19	−1.07	−0.49	−0.94	−0.79
2100	−0.88	−1.08	−1.32	−1.20	−0.25	−1.04	−1.19	−1.20	−0.46	−1.00	−0.83

II.3.9: Radiative forcing (Wm⁻²) from CFCs and HCFCs following the Montreal (1997) Amendments

Year	CFC-11	CFC-12	CFC-113	CFC-114	CFC-115	CCl₄	CH₃CCl₃	HCFC-22	HCFC-141b	HCFC-142b	HCFC-123	CF₂BrCl	CF₃Br	SUM
2000	0.0668	0.1712	0.0255	0.0050	0.0016	0.0120	0.0026	0.0290	0.0018	0.0030	0.0000	0.0012	0.0010	0.3206
2010	0.0615	0.1686	0.0243	0.0050	0.0016	0.0098	0.0004	0.0514	0.0031	0.0066	0.0004	0.0012	0.0010	0.3348
2020	0.0535	0.1555	0.0216	0.0047	0.0016	0.0077	0.0001	0.0458	0.0022	0.0064	0.0006	0.0009	0.0010	0.3015
2030	0.0450	0.1411	0.0192	0.0047	0.0016	0.0061	0.0000	0.0274	0.0013	0.0046	0.0004	0.0006	0.0010	0.2529
2040	0.0373	0.1280	0.0171	0.0043	0.0016	0.0048	0.0000	0.0176	0.0008	0.0034	0.0004	0.0003	0.0010	0.2166
2050	0.0308	0.1158	0.0153	0.0043	0.0016	0.0038	0.0000	0.0092	0.0003	0.0022	0.0002	0.0003	0.0010	0.1848
2060	0.0253	0.1050	0.0135	0.0040	0.0016	0.0030	0.0000	0.0040	0.0001	0.0012	0.0002	0.0000	0.0006	0.1585
2070	0.0208	0.0954	0.0120	0.0040	0.0016	0.0023	0.0000	0.0018	0.0000	0.0008	0.0000	0.0000	0.0006	0.1393
2080	0.0170	0.0864	0.0108	0.0037	0.0014	0.0018	0.0000	0.0008	0.0000	0.0004	0.0000	0.0000	0.0006	0.1230
2090	0.0140	0.0784	0.0096	0.0037	0.0014	0.0014	0.0000	0.0004	0.0000	0.0002	0.0000	0.0000	0.0006	0.1098
2100	0.0113	0.0710	0.0084	0.0037	0.0014	0.0012	0.0000	0.0002	0.0000	0.0002	0.0000	0.0000	0.0003	0.0977

II.3.10: Radiative Forcing (Wm⁻²) from fosil fuel plus biomass Organic and Black Carbon as used in the Chapter 9 Simple Model SRES Projections

Year	A1B	A1T	A1FI	A2	B1	B2	IS92a
1990	−0.0997	−0.0997	−0.0997	−0.0997	−0.0997	−0.0997	−0.0998
2000	−0.1361	−0.1361	−0.1361	−0.1361	−0.1361	−0.1361	−0.1586
2010	−0.1308	−0.1468	−0.1280	−0.1392	−0.1081	−0.1203	−0.1357
2020	−0.0524	−0.0799	−0.1714	−0.1248	−0.0926	−0.0516	−0.1103
2030	−0.0562	−0.0598	−0.1745	−0.1088	−0.0154	−0.0148	−0.0872
2040	−0.0780	−0.0644	−0.1614	−0.1064	0.0349	−0.0075	−0.0610
2050	−0.0804	−0.0603	−0.1351	−0.1029	0.0280	−0.0049	−0.0339
2060	−0.0948	−0.0615	−0.1417	−0.1002	0.0241	0.0015	−0.0190
2070	−0.1071	−0.0613	−0.1193	−0.0939	0.0147	0.0064	−0.0026
2080	−0.1161	−0.0629	−0.0644	−0.0871	0.0300	0.0180	0.0166
2090	−0.1178	−0.0619	0.0365	−0.0816	0.0421	0.0341	0.0390
2100	−0.1208	−0.0629	0.0565	−0.0762	0.0351	0.0510	0.0635

II.3.11: Total Radiative Forcing (Wm⁻²) from GHG plus direct and indirect aerosol effects as used in the Chapter 9 Simple Model SRES Projections

Year	A1B	A1T	A1FI	A2	B1	B2	IS92a
1990	1.03	1.03	1.03	1.03	1.03	1.03	1.03
2000	1.33	1.33	1.33	1.33	1.33	1.33	1.31
2010	1.65	1.85	1.69	1.74	1.73	1.82	1.63
2020	2.16	2.48	2.17	2.04	2.15	2.36	2.00
2030	2.84	3.07	2.78	2.56	2.56	2.81	2.40
2040	3.61	3.76	3.67	3.22	2.93	3.26	2.82
2050	4.16	4.31	4.83	3.89	3.30	3.70	3.25
2060	4.79	4.73	5.99	4.71	3.65	4.11	3.76
2070	5.28	4.97	7.02	5.56	3.92	4.52	4.24
2080	5.62	5.11	7.89	6.40	4.09	4.92	4.74
2090	5.86	5.12	8.59	7.22	4.18	5.32	5.26
2100	6.05	5.07	9.14	8.07	4.19	5.71	5.79

II.4: Model Average Surface Air Temperature Change (°C)

Year	A1B	A1T	A1FI	A2	B1	B2	IS92a
1750 to 1990	0.33	0.33	0.33	0.33	0.33	0.33	0.34
1990	0.00	0.00	0.00	0.00	0.00	0.00	0.00
2000	0.16	0.16	0.16	0.16	0.16	0.16	0.15
2010	0.30	0.40	0.32	0.35	0.34	0.39	0.27
2020	0.52	0.71	0.55	0.50	0.55	0.66	0.43
2030	0.85	1.03	0.85	0.73	0.77	0.93	0.61
2040	1.26	1.41	1.27	1.06	0.98	1.18	0.80
2050	1.59	1.75	1.86	1.42	1.21	1.44	1.00
2060	1.97	2.04	2.50	1.85	1.44	1.69	1.26
2070	2.30	2.25	3.10	2.33	1.63	1.94	1.52
2080	2.56	2.41	3.64	2.81	1.79	2.20	1.79
2090	2.77	2.49	4.09	3.29	1.91	2.44	2.08
2100	2.95	2.54	4.49	3.79	1.98	2.69	2.38

Note: See Chapter 9 for details.

II.5: Sea Level Change (mm)

Note: Values are for the middle of the year..

II.5.1: Total sea level change (mm)

Models average – Total sea level change (mm)

Year	A1B	A1T	A1FI	A2	B1	B2
1990	0	0	0	0	0	0
2000	17	17	17	17	17	17
2010	37	39	37	38	38	38
2020	61	66	61	61	62	64
2030	91	97	90	88	89	94
2040	127	134	126	120	118	126
2050	167	175	172	157	150	160
2060	210	217	228	201	183	197
2070	256	258	290	250	216	235
2080	301	298	356	304	249	275
2090	345	334	424	362	281	316
2100	387	367	491	424	310	358

Note: The sum of the components listed in Appendix II.5.2 to II.5.5 does not equal the values shown above owing to the addition of other terms. See Chapter 11, Section 11.5.1 for details.

Models minimum – Total sea level change (mm)

Year	A1B	A1T	A1FI	A2	B1	B2
1990	0	0	0	0	0	0
2000	6	6	6	6	6	6
2010	13	13	13	13	13	13
2020	22	22	24	21	22	23
2030	34	33	36	31	32	34
2040	48	47	49	44	42	45
2050	63	66	64	58	52	56
2060	78	89	77	75	63	68
2070	93	113	89	93	72	79
2080	107	137	99	113	80	91
2090	119	160	106	133	87	103
2100	129	182	111	155	92	114

Note: The final values of these timeseries correspond to the lower limit of the coloured bars on the right–hand side of Chapter 11, Figure 11.12.

Model maximum – Total sea level change (mm)

Year	A1B	A1T	A1FI	A2	B1	B2
1990	0	0	0	0	0	0
2000	29	29	29	29	29	29
2010	63	63	65	64	64	65
2020	103	104	110	104	105	109
2030	153	153	164	149	151	159
2040	214	214	228	204	203	216
2050	284	291	299	269	259	277
2060	360	386	375	343	319	344
2070	442	494	453	430	381	414
2080	527	612	529	526	444	488
2090	611	735	602	631	507	566
2100	694	859	671	743	567	646

Note: The final values of these timeseries correspond to the upper limit of the coloured bars on the right–hand side of Chapter 11, Figure 11.12.

II.5.2: Sea level change due to thermal expansion (mm)

Year	A1B	A1T	A1FI	A2	B1	B2
1990	0	0	0	0	0	0
2000	10	10	10	10	10	10
2010	23	24	23	23	23	24
2020	39	43	39	39	39	42
2030	60	66	60	57	58	62
2040	87	93	86	81	79	85
2050	117	123	122	109	101	110
2060	150	155	166	142	125	137
2070	185	186	217	180	149	165
2080	220	216	272	224	173	196
2090	255	243	329	272	195	227
2010	288	267	388	325	216	260

II.5.3: Sea level change due to glaciers and ice caps (mm)

Year	A1B	A1T	A1FI	A2	B1	B2
1990	0	0	0	0	0	0
2000	4	4	4	4	4	4
2010	9	10	9	10	10	10
2020	16	17	16	16	16	16
2030	23	25	23	23	23	24
2040	32	35	32	31	31	34
2050	43	46	44	41	41	44
2060	55	58	57	52	50	54
2070	67	71	72	65	61	66
2080	80	83	89	79	71	77
2090	93	95	105	93	82	89
2100	106	106	120	108	92	101

II.5.4: Sea level change due to Greenland (mm)

Year	A1B	A1T	A1FI	A2	B1	B2
1990	0	0	0	0	0	0
2000	0	0	0	0	0	0
2010	1	1	1	1	1	1
2020	2	2	2	2	2	2
2030	4	4	4	4	4	4
2040	5	6	5	5	5	6
2050	8	8	8	7	7	8
2060	10	11	11	10	9	10
2070	13	14	15	13	12	13
2080	17	17	19	16	14	16
2090	20	21	24	20	17	19
2100	24	24	29	25	20	22

II.5.5: Sea level change due to Antarctica (mm)

Year	A1B	A1T	A1FI	A2	B1	B2
1990	0	0	0	0	0	0
2000	−2	−2	−2	−2	−2	−2
2010	−5	−5	−5	−5	−5	−5
2020	−8	−9	−8	−8	−8	−9
2030	−12	−14	−13	−12	−13	−13
2040	−18	−20	−18	−17	−17	−19
2050	−25	−27	−25	−23	−23	−25
2060	−33	−35	−35	−31	−30	−32
2070	−42	−45	−46	−40	−37	−41
2080	−52	−54	−59	−50	−44	−49
2090	−63	−64	−74	−62	−53	−59
2100	−74	−75	−90	−76	−61	−70

References

IPCC, 1996: *Climate Change 1995: The Science of Climate Change. Contribution of Working Group I to the Second Assessment Report of the Intergovernmental Panel on Climate Change* [Houghton, J.T., L.G. Meira Filho, B.A. Callander, N. Harris, A. Kattenberg, and K. Maskell (eds.)]. Cambridge University Press, Cambridge, United Kingdom and New York, NY, USA, 572 pp.

Jain, A.K., H.S. Kheshgi, and D.J. Wuebbles, 1994: Integrated Science Model for Assessment of Climate Change. Lawrence Livermore National Laboratory, UCRL-JC-116526.

Nakićenović, N., J. Alcamo, G. Davis, B. de Vries, J. Fenhann, S. Gaffin, K. Gregory, A. Grübler, T. Y. Jung, T. Kram, E. L. La Rovere, L. Michaelis, S. Mori, T. Morita, W. Pepper, H. Pitcher, L. Price, K. Raihi, A. Roehrl, H-H. Rogner, A. Sankovski, M. Schlesinger, P. Shukla, S. Smith, R. Swart, S. van Rooijen, N. Victor, Z. Dadi, 2000: *IPCC Special Report on Emissions Scenarios*, Cambridge University Press, Cambridge, United Kingdom and New York, NY, USA, 599 pp.

WMO, 1999: Scientific Assessment of Ozone Depletion: 1998. Global Ozone Research and Monitoring Project - Report No. 44, World Meteorological Organization, Geneva, Switzerland, 732 pp.

Appendix III

Contributors

to the IPCC WGI Third Assessment Report

Technical Summary

Co-ordinating Lead Authors

D.L. Albritton	NOAA Aeronomy Laboratory, USA
L.G. Meira Filho	Agência Espacial Brasileira, Brazil

Lead Authors

U. Cubasch	Max-Planck Institute for Meteorology, Germany
X. Dai	IPCC WGI Technical Support Unit, UK/National Climate Center, China
Y. Ding	IPCC WGI Co-Chairman, National Climate Center, China
D.J. Griggs	IPCC WGI Technical Support Unit, UK
B. Hewitson	University of Capetown, South Africa
J.T. Houghton	IPCC WGI Co-Chairman, UK
I. Isaksen	University of Oslo, Norway
T. Karl	NOAA National Climatic Data Center, USA
M. McFarland	Dupont Fluoroproducts, USA
V.P. Meleshko	Voeikov Main Geophysical Observatory, Russia
J.F.B. Mitchell	Hadley Centre for Climate Prediction and Research, Met Office, UK
M. Noguer	IPCC WGI Technical Support Unit, UK
B.S. Nyenzi	Zimbabwe Drought Monitoring Centre, Tanzania
M. Oppenheimer	Environmental Defense, USA
J.E. Penner	University of Michigan, USA
S. Pollonais	Environment Management Authority, Trinidad and Tobago
T. Stocker	University of Bern, Switzerland
K.E. Trenberth	National Center for Atmospheric Research, USA

Contributing Authors

M.R. Allen	Rutherford Appleton Laboratory, UK
A.P.M. Baede	Koninklijk Nederlands Meteorologisch Instituut, Netherlands
J.A. Church	CSIRO Division of Marine Research, Australia
D.H. Ehhalt	Institut für Chemie der KFA Jülich GmbH, Germany
C.K. Folland	Hadley Centre for Climate Prediction and Research, Met Office, UK
F. Giorgi	Abdus Salam International Centre for Theoretical Physics, Italy
J.M. Gregory	Hadley Centre for Climate Prediction and Research, Met Office, UK
J.M. Haywood	Hadley Centre for Climate Prediction and Research, Met Office, UK
J.I. House	Max-Plank Institute for Biogeochemistry, Germany
M. Hulme	University of East Anglia, UK
V.J. Jaramillo	Instituto de Ecologia, UNAM, Mexico

A. Jayaraman	Physical Research Laboratory, India
C.A. Johnson	IPCC WGI Technical Support Unit, UK
S. Joussaume	Institut Pierre Simon Laplace, Laboratoire des Sciences du Climat et de l'Environnement, France
D.J. Karoly	Monash University, Australia
H. Kheshgi	Exxon Mobil Research and Engineering Company, USA
C. Le Quéré	Max Plank Institute for Biogeochemistry, France
K. Maskell	IPCC WGI Technical Support Unit, UK
L.J. Mata	Universitaet Bonn, Germany
B.J. McAvaney	Bureau of Meteorology Research Centre, Australia
L.O. Mearns	National Center for Atmospheric Research, USA
G.A. Meehl	National Center for Atmospheric Research, USA
B. Moore III	University of New Hampshire, USA
R.K. Mugara	Zambia Meteorological Department, Zambia
M. Prather	University of California, USA
C. Prentice	Max-Planck Institute for Biogeochemistry, Germany
V. Ramaswamy	NOAA Geophysical Fluid Dynamics Laboratory, USA
S.C.B. Raper	University of East Anglia, UK
M.J. Salinger	National Institute of Water & Atmospheric Research, New Zealand
R. Scholes	Division of Water, Environment and Forest Technology, South Africa
S. Solomon	NOAA Aeronomy Laboratory, USA
R. Stouffer	NOAA Geophysical Fluid Dynamics Laboratory, USA
M.-X. Wang	Institute of Atmospheric Physics, Chinese Academy of Sciences, China
R.T. Watson	Chairman IPCC, The World Bank, USA
K.-S. Yap	Malaysian Meteorological Service, Malaysia

Review Editors

F. Joos	University of Bern, Switzerland
A. Ramirez-Rojas	Universidad Central Venezuela, Venezuela
J.M.R. Stone	Environment Canada, Canada
J. Zillman	Bureau of Meteorology, Australia

Chapter 1. The Climate System: an Overview

Co-ordinating Lead Author

A.P.M. Baede	Koninklijk Nederlands Meteorologisch Instituut, Netherlands

Lead Authors

E. Ahlonsou	National Meteorological Service, Benin
Y. Ding	IPCC WG1 Co-Chairman, National Climate Center, China
D. Schimel	Max-Planck Institute for Biogeochemistry, Germany/NCAR, USA

Review Editors

B. Bolin	Retired, Sweden
S. Pollonais	Environment Management Authority, Trinidad and Tobago

Chapter 2. Observed Climate Variability and Change

Co-ordinating Lead Authors

C.K. Folland	Hadley Centre for Climate Prediction and Research, Met Office, UK
T.R. Karl	NOAA National Climatic Data Center, USA

Lead Authors

J.R. Christy	University of Alabama, USA
R.A. Clarke	Bedford Institute of Oceanography, Canada
G.V. Gruza	Institute for Global Climate and Ecology, Russia

J. Jouzel	Institut Pierre Simon Laplace, Laboratoire des Sciences du Climat et de l'Environment, France
M.E. Mann	University of Virginia, USA
J. Oerlemans	University of Utrecht, Netherlands
M.J. Salinger	National Institute of Water & Atmospheric Research, New Zealand
S.-W. Wang	Peking University, China

Contributing Authors

J. Bates	NOAA Environmental Research Laboratories, USA
M. Crowe	NOAA National Climatic Data Center, USA
P. Frich	Hadley Centre for Climate Prediction and Research, Met Office, UK
P. Groissman	NOAA National Climatic Data Center, USA
J. Hurrell	National Center for Atmospheric Research, USA
P. Jones	University of East Anglia, UK
D. Parker	Hadley Centre for Climate Prediction and Research, Met Office, UK
T. Peterson	NOAA National Climatic Data Center, USA
D. Robinson	Rutgers University, USA
J. Walsh	University of Illinois at Urbana-Champaign, USA
M. Abbott	Oregon State University, USA
L. Alexander	Hadley Centre for Climate Prediction and Research, Met Office, UK
H. Alexanderson	Swedish Meteorological and Hydrological Institute, Sweden
R. Allan	CSIRO Division of Atmospheric Research, Australia
R. Alley	Pennsylvania State University, USA
P. Ambenjie	Department of Meteorology, Kenya
P. Arkin	Lamont-Doherty Earth Observatory of Columbia University, USA
L. Bajuk	Mathsoft Data Analysis Products Division, USA
R. Balling	Arizona State University, USA
M.Y. Bardin	Institute for Global Climate and Ecology, Russia
R. Bradley	University of Massachusetts, USA
R. Brázdil	Masaryk University, Czech Republic
K.R. Briffa	University of East Anglia, UK
H. Brooks	NOAA National Severe Storms Laboratory, USA
R.D. Brown	Atmospheric Environment Service, Canada
S. Brown	Hadley Centre for Climate Prediction and Research, Met Office, UK
M. Brunet-India	University Rovira I Virgili, Spain
M. Cane	Lamont-Doherty Earth Observatory of Columbia University, USA
D. Changnon	Northern Illinois University, USA
S. Changnon	University of Illinois at Urbana-Champaign, USA
J. Cole	University of Colorado, USA
D. Collins	Bureau of Meteorology, Australia
E. Cook	Lamont-Doherty Earth Observatory of Columbia University, USA
A. Dai	National Center for Atmospheric Research, USA
A. Douglas	Creighton University, USA
B. Douglas	University of Maryland, USA
J.C. Duplessy	Institut Pierre Simon Laplace, Laboratoire des Sciences du Climat et de l'Environnement, France
D. Easterling	NOAA National Climatic Data Center, USA
P. Englehart	USA
R.E. Eskridge	NOAA National Climatic Data Center, USA
D. Etheridge	CSIRO Division of Atmospheric Research, Australia
D. Fisher	Geological Survey of Canada, Canada
D. Gaffen	NOAA Air Resources Laboratory, USA
K. Gallo	National Environmental Satellite, Data and Information Service, USA
E. Genikhovich	Main Geophysical Observatory, Russia
D. Gong	Peking University, China
G. Gutman	National Environmental Satellite, Data and Information Service, USA
W. Haeberli	University of Zurich, Switzerland
J. Haigh	Imperial College, UK
J. Hansen	Goddard Institute for Space Studies, USA

D. Hardy	University of Massachusetts, USA
S. Harrison	Max-Planck Institute for Biogeochemistry, Germany
R. Heino	Finnish Meteorological Institute, Finland
K. Hennessy	CSIRO Division of Atmospheric Research, Australia
W. Hogg	Atmospheric Environment Service, Canada
S. Huang	University of Michigan, USA
K. Hughen	Woods Hole Oceanographic Institute, USA
M.K. Hughes	University of Arizona, USA
M. Hulme	University of East Angelia, UK
H. Iskenderian	Atmospheric and Environmental Research, Inc., USA
O.M. Johannessen	Nasen Environmental and Remote Sensing Center, Norway
D. Kaiser	Oak Ridge National Laboratory, USA
D. Karoly	Monash University, Australia
D. Kley	Institut fuer Chemie und Dynamik der Geosphaere, Germany
R. Knight	NOAA National Climatic Data Center, USA
K.R. Kumar	Indian Institute of Tropical Meteorology, India
K. Kunkel	Illinois State Water Survey, USA
M. Lal	Indian Institute of Technology, India
C. Landsea	NOAA Atlantic Oceanographic & Meteorological Laboratory, USA
J. Lawrimore	NOAA National Climatic Data Center, USA
J. Lean	Naval Research Laboratory, USA
C. Leovy	University of Washington, USA
H. Lins	US Geological Survey, USA
R. Livezey	NOAA National Weather Service, USA
K.M. Lugina	St Petersburg University, Russia
I. Macadam	Hadley Centre for Climate Prediction and Research, Met Office, UK
J.A. Majorowicz	Northern Geothermal, Canada
B. Manighetti	National Institute of Water & Atmospheric Research, New Zealand
J. Marengo	Instituto Nacional de Pesquisas Espaciais, Brazil
E. Mekis	Environment Canada, Canada
M.W. Miles	Nasen Environmental and Remote Sensing Center, Norway
A. Moberg	Stockholm University, Sweden
I. Mokhov	Institute of Atmospheric Physics, Russia
V. Morgan	University of Tasmania, Australia
L. Mysak	McGill University, Canada
M. New	Oxford University, UK
J. Norris	NOAA Geophysical Fluid Dynamics Laboratory, USA
L. Ogallo	University of Nairobi, Kenya
J. Overpeck	NOAA National Geophysical Data Center, USA
T. Owen	NOAA National Climatic Data Center, USA
D. Paillard	Institut Pierre Simon Laplace, Laboratoire des Sciences du Climat et de l'Environnement, France
T. Palmer	European Centre for Medium-range Weather Forecasting, UK
C. Parkinson	NASA Goddard Space Flight Center, USA
C.R. Pfister	Unitobler, Switzerland
N. Plummer	Bureau of Meteorology, Australia
H. Pollack	University of Michigan, USA
C. Prentice	Max-Planck Institute for Biogeochemistry, Germany
R. Quayle	NOAA National Climatic Data Center, USA
E.Ya. Rankova	Institute for Global Climate and Ecology, Russia
N. Rayner	Hadley Centre for Climate Prediction and Research, Met Office, UK
V.N. Razuvaev	Chief Climatology Department, Russia
G. Ren	National Climate Center, China
J. Renwick	National Institute of Water & Atmospheric Research, New Zealand
R. Reynolds	NOAA National Centers for Environmental Prediction, USA
D. Rind	Goddard Institute of Space Studies, USA
A. Robock	Rutgers University, USA
R. Rosen	Atmospheric and Environmental Research, Inc., USA

S. Rösner	Department Climate and Environment, Deutscher Wetterdienst, Germany
R. Ross	NOAA Air Resources Laboratory, USA
D. Rothrock	Applied Physics Laboratory, USA
J.M. Russell	Hampton University, USA
M. Serreze	University of Colorado, USA
W.R. Skinner	Environment Canada, Canada
J. Slack	US Geological Survey, USA
D.M. Smith	Hadley Centre for Climate Prediction and Research, Met Office, UK
D. Stahle	University of Arkansas, USA
M. Stendel	Danish Meteorological Institute, Denmark
A. Sterin	RIHMI-WDCB, Russia
T. Stocker	University of Bern, Switzerland
B. Sun	University of Massachusetts, USA
V. Swail	Environment Canada, Canada
V. Thapliyal	India Meteorological Department, India
L. Thompson	Ohio State University, USA
W.J. Thompson	University of Washington, USA
A. Timmermann	Koninklijk Nederlands Meteorologisch Instituut, Netherlands
R. Toumi	Imperial College, UK
K. Trenberth	National Center for Atmospheric Research, USA
H. Tuomenvirta	Finnish Meteorological Institute, Finland
T. van Ommen	University of Tasmania, Australia
D. Vaughan	British Antarctic Survey, UK
K.Y. Vinnikov	University of Maryland, USA
U. von Grafenstein	Institut Pierre Simon Laplace, Laboratoire des Sciences du Climat et de l'Environnement, France
H. von Storch	GKSS Research Center, Germany
M. Vuille	University of Massachusetts, USA
P. Wadhams	Scott Polar Research Institute, UK
J.M. Wallace	University of Washington, USA
S. Warren	University of Washington, USA
W. White	Scripps Institution of Oceanography, USA
P. Xie	NOAA National Centers for Environmental Prediction, USA
P. Zhai	National Climate Center, China

Review Editors

R. Hallgren	American Meteorological Society, USA
B. Nyenzi	Zimbabwe Drought Monitoring Centre, Tanzania

Chapter 3. The Carbon Cycle and Atmospheric Carbon Dioxide

Co-ordinating Lead Author

I.C. Prentice	Max-Planck Institute for Biogeochemistry, Germany

Lead Authors

G.D. Farquhar	Australian National University, Australia
M.J.R. Fasham	Southampton Oceanography Centre, UK
M.L. Goulden	University of California, USA
M. Heimann	Max-Planck Institute for Biogeochemistry, Germany
V.J. Jaramillo	Instituto de Ecologia, UNAM, Mexico
H.S. Kheshgi	Exxon Mobil Research and Engineering Company, USA
C. Le Quéré	Max-Planck Institute for Biogeochemistry, Germany
R.J. Scholes	Division of Water, Environment and Forest Technology, South Africa
D.W.R. Wallace	Universitat Kiel, Germany

Contributing Authors

D. Archer	University of Chicago, USA

M.R. Ashmore	University of Bradford, UK
O. Aumont	Institut Pierre Simon Laplace, Laboratoire des Sciences du Climat et de l'Environnement, France
D. Baker	Princeton University, USA
M. Battle	Bowdoin College, USA
M. Bender	Princeton University, USA
L.P. Bopp	Institut Pierre Simon Laplace, Laboratoire des Sciences du Climat et de l'Environnement, France
P. Bousquet	Institut Pierre Simon Laplace, Laboratoire des Sciences du Climat et de l'Environnement, France
K. Caldeira	Lawrence Livermore National Laboratory, USA
P. Ciais	CEA, LMCE/DSM, France
P.M. Cox	Hadley Centre for Climate Prediction and Research, Met Office, UK
W. Cramer	Potsdam Institute for Climate Impact Research, Germany
F. Dentener	Environment Institute, Italy
I.G. Enting	CSIRO Division of Atmospheric Research, Australia
C.B. Field	Carnegie Institute of Washington, USA
P. Friedlingstein	Institut Pierre Simon Laplace, Laboratoire des Sciences du Climat et de l'Environnement, France
E.A. Holland	Max-Planck Institute for Biochemistry, Germany
R.A. Houghton	Woods Hole Research Center, USA
J.I. House	Max-Planck Institute for Biogeochemistry, Germany
A. Ishida	Institute for Global Change Research, Japan
A.K. Jain	University of Illinois, USA
I.A. Janssens	Universiteit Antwerpen, Belgium
F. Joos	University of Bern, Switzerland
T. Kaminski	Max-Planck Institute for Meteorology, Germany
C.D. Keeling	University of California at San Diego, USA
R.F. Keeling	University of California at San Diego, USA
D.W. Kicklighter	Marine Biological Laboratory, USA
K.E. Kohfeld	Max-Planck Institute for Biogeochemistry, Germany
W. Knorr	Max-Planck Institute for Biogeochemistry, Germany
R. Law	Monash University, Australia
T. Lenton	Institute of Terrestrial Ecology, UK
K. Lindsay	National Center for Atmospheric Research, USA
E. Maier-Reimer	Max-Planck Institute for Meteorology, Germany
A.C. Manning	University of California at San Diego, USA
R.J. Matear	CSIRO Division of Marine Research, Australia
A.D. McGuire	University of Alaska at Fairbanks, USA
J.M. Melillo	Woods Hole Oceanographic Institution, USA
R. Meyer	University of Bern, Switzerland
M. Mund	Max-Planck Institute for Biogeochemistry, Germany
J.C. Orr	Institut Pierre Simon Laplace, Laboratoire des Sciences du Climat et de l'Environnement, France
S. Piper	Scripps Institution of Oceanography, USA
K. Plattner	University of Bern, Switzerland
P.J. Rayner	CSIRO Division of Atmospheric Research, Australia
S. Sitch	Institut für Klimafolgenforschung, Germany
R. Slater	Princeton University Atmospheric and Oceanic Sciences Program, USA
S. Taguchi	National Institute for Research & Environment, Japan
P.P. Tans	NOAA Climate Monitoring & Diagnostics Laboratory, USA
H.Q. Tian	Marine Biological Laboratory, USA
M.F. Weirig	Alfred Wegener Institute for Polar and Marine Research, Germany
T. Whorf	University of California at San Diego, USA
A. Yool	Southampton Oceanography Centre, UK

Review Editors

L. Pitelka	University of Maryland, USA
A. Ramirez Rojas	Universidad Central Venezuela, Venezuela

Chapter 4. Atmospheric Chemistry and Greenhouse Gases

Co-ordinating Lead Authors

D. Ehhalt   Institut für Chemie der KFA Jülich GmbH, Germany
M. Prather   University of California, USA

Lead Authors

F. Dentener   Institute for Marine and Atmospheric Research, Netherlands
R. Derwent   Met Office, UK
E. Dlugokencky   NOAA Climate Monitoring & Diagnostics Laboratory, USA
E. Holland   Max-Planck Institute for Biogeochemistry, Germany
I. Isaksen   University of Oslo, Norway
J. Katima   University of Dar-Es-Salaam, Tanzania
V. Kirchhoff   Instituto Nacional de Pesquisas Espaciais, Brazil
P. Matson   Stanford University, USA
P. Midgley   M&D Consulting, Germany
M. Wang   Institute of Atmospheric Physics, China

Contributing Authors

T. Berntsen   Centre for International Climate and Environmental Research, Norway
I. Bey   Harvard University, USA/France
G. Brasseur   Max-Planck Institute for Meteorology, Germany
L. Buja   National Center for Atmospheric Research, USA
W.J. Collins   Hadley Centre for Climate Prediction and Research, Met Office, UK
J. Daniel   NOAA Aeronomy Laboratory, USA
W.B. DeMore   Jet Propulsion Laboratory, USA
N. Derek   CSIRO Division of Atmospheric Research, Australia
R. Dickerson   University of Maryland, USA
D. Etheridge   CSIRO Division of Atmospheric Research, Australia
J. Feichter   Max-Planck Institute for Meteorology, Germany
P. Fraser   CSIRO Division of Atmospheric Research, Australia
R. Friedl   Jet Propulsion Laboratory, USA
J. Fuglestvedt   University of Oslo, Norway
M. Gauss   University of Oslo, Norway
L. Grenfell   NASA Goddard Institute for Space Studies, USA
A. Grübler   International Institute for Applied Systems Analysis, Austria
N. Harris   European Ozone Research Coordinating Unit, UK
D. Hauglustaine   Center National de la Recherche Scientifique, Service Aeronomie, France
L. Horowitz   National Center for Atmospheric Research, USA
C. Jackman   NASA Goddard Space Flight Center, USA
D. Jacob   Harvard University, USA
L. Jaeglé   Harvard University, USA
A. Jain   University of Illinois, USA
M. Kanakidou   Environmental Chemical Processes Laboratory, Greece
S. Karlsdottir   University of Oslo, Norway
M. Ko   Atmospheric & Environmental Research Inc., USA
M. Kurylo   NASA Headquarters, USA
M. Lawrence   Max-Planck Institute for Chemistry, Germany
J.A. Logan   Harvard University, USA
M. Manning   National Institute of Water & Atmospheric Research, New Zealand
D. Mauzerall   Princeton University, USA
J. McConnell   York University, Canada
L. Mickley   Harvard University, USA
S. Montzka   NOAA Climate Monitoring & Diagnostics Laboratory, USA
J.F. Muller   Belgian Institute for Space Aeronomy, Belgium
J. Olivier   National Institute of Public Health and the Environment, Netherlands
K. Pickering   University of Maryland, USA

G. Pitari	Università Degli Studi dell' Aquila, Italy
G.J. Roelofs	University of Utrecht, Netherlands
H. Rogers	University of Cambridge, UK
B. Rognerud	University of Oslo, Norway
S. Smith	Pacific Northwest National Laboratory, USA
S. Solomon	NOAA Aeronomy Laboratory, USA
J. Staehelin	Federal Institute of Technology, Switzerland
P. Steele	CSIRO Division of Atmospheric Research, Australia
D. S. Stevenson	Met Office, UK
J. Sundet	University of Oslo, Norway
A. Thompson	NASA Goddard Space Flight Center, USA
M. van Weele	Konjnklijk Nederlands Meteorologisch Instituut, Netherlands
R. von Kuhlmann	Max-Planck Institute for Chemistry, Germany
Y. Wang	Georgia Institute of Technology, USA
D. Weisenstein	Atmospheric & Envrionmental Research Inc., USA
T. Wigley	National Center for Atmospheric Research, USA
O. Wild	Frontier Research System for Global Change, Japan
D. Wuebbles	University of Illinois, USA
R. Yantosca	Harvard University, USA

Review Editors

| F. Joos | University of Bern, Switzerland |
| M. McFarland | Dupont Fluoroproducts, USA |

Chapter 5. Aerosols, their Direct and Indirect Effects

Co-ordinating Lead Author

| J.E. Penner | University of Michigan, USA |

Lead Authors

M. Andreae	Max-Planck Institute for Chemistry, Germany
H. Annegarn	University of the Witwatersrand, South Africa
L. Barrie	Atmospheric Environment Service, Canada
J. Feichter	Max-Planck Institute for Meteorology, Germany
D. Hegg	University of Washington, USA
A. Jayaraman	Physical Research Laboratory, India
R. Leaitch	Atmospheric Environment Service, Canada
D. Murphy	NOAA Aeronomy Laboratory, USA
J. Nganga	University of Nairobi, Kenya
G. Pitari	Università Degli Studi dell' Aquil, Italy

Contributing Authors

A. Ackerman	NASA Ames Research Center, USA
P. Adams	Caltech, USA
P. Austin	University of British Columbia, Canada
R. Boers	CSIRO Division of Atmospheric Research, Australia
O. Boucher	Laboratoire d'Optique Atmospherique, France
M. Chin	Goddard Space Flight Center, USA
C. Chuang	Lawrence Livermore National Laboratory, USA
W. Collins	Met Office, UK
W. Cooke	NOAA Geophysical Fluid Dynamics Laboratory, USA
P. DeMott	Colorado State University, USA
Y. Feng	University of Michigan, USA
H. Fischer	Scripps Institution of Oceanography, Germany
I. Fung	University of California, USA
S. Ghan	Pacific Northwest National Laboratory, USA

P. Ginoux	NASA Goddard Space Flight Center, USA
S.-L. Gong	Atmospheric Environment Service, Canada
A. Guenther	National Center for Atmospheric Research, USA
M. Herzog	University of Michigan, USA
A. Higurashi	National Institute for Environmental Studies, Japan
Y. Kaufman	NASA Goddard Space Flight Center, USA
A. Kettle	Max-Planck Institute for Chemistry, Germany
J. Kiehl	National Center for Atmospheric Research, USA
D. Koch	National Center for Atmospheric Research, USA
G. Lammel	Max-Planck Institute for Meteorology, Germany
C. Land	Max-Planck Institute for Meteorology, Germany
U. Lohmann	Dalhousie University, Canada
S. Madronich	National Center for Atmospheric Research, USA
E. Mancini	Università Degli Studi dell' Aquila, Italy
M. Mishchenko	NASA Goddard Institute for Space Studies, USA
T. Nakajima	University of Tokyo, Japan
P. Quinn	National Oceanographic and Atmospheric Administration, USA
P. Rasch	National Center for Atmospheric Research, USA
D.L. Roberts	Hadley Centre for Climate Prediction and Research, Met Office, UK
D. Savoie	University of Miami, USA
S. Schwartz	Brookhaven National Laboratory, USA
J. Seinfeld	California Institute of Technology, USA
B. Soden	Princeton University, USA
D. Tanré	Laboratoire d'Optique Atmospherique, France
K. Taylor	Lawrence Livermore National Laboratory, USA
I. Tegen	Max-Planck Institute for Biogeochemistry, Germany
X. Tie	National Center for Atmospheric Research, USA
G. Vali	University of Wyoming, USA
R. Van Dingenen	Enviroment Institute of European Commission, Italy
M. van Weele	Koninklijk Nederlands Meteorologisch Instituut, The Netherlands
Y. Zhang	University of Michigan, USA

Review Editors
| B. Nyenzi | Zimbabwe Drought Monitoring Centre, Tanzania |
| J. Prospero | University of Miami, USA |

Chapter 6. Radiative Forcing of Climate Change

Co-ordinating Lead Author
| V. Ramaswamy | NOAA Geophysical Fluid Dynamics Laboratory, USA |

Lead Authors
O. Boucher	Max-Planck Institute for Chemistry, Germany/Laboratoire d'Optique Atmospherique, France
J. Haigh	Imperial College, UK
D. Hauglustaine	Center National de la Recherche Scientifique, France
J. Haywood	Meteorological Research Flight, Met Office, UK
G. Myhre	University of Oslo, Norway
T. Nakajima	University of Tokyo, Japan
G.Y. Shi	Institute of Atmospheric Physics, China
S. Solomon	NOAA Aeronomy Laboratory, USA

Contributing Authors
R. Betts	Hadley Centre for Climate Prediction and Research, Met Office, UK
R. Charlson	Stockholm University, Sweden
C. Chuang	Lawrence Livermore National Laboratory, USA
J.S. Daniel	NOAA Aeronomy Laboratory, USA

A. Del Genio	NASA Goddard Institute for Space Studies, USA
J. Feichter	Max-Planck Institute for Meteorology, Germany
J. Fuglestvedt	University of Oslo, Norway
P.M. Forster	Monash University, Australia
S.J. Ghan	Pacific Northwest National Laboratory, USA
A. Jones	Hadley Centre for Climate Prediction and Research, Met Office, UK
J.T. Kiehl	National Center for Atmospheric Research, USA
D. Koch	Yale University, USA
C. Land	Max-Planck Institute for Meteorology, Germany
J. Lean	Naval Research Laboratory, USA
U. Lohmann	Dalhousie University, Canada
K. Minschwaner	New Mexico Institute of Mining and Technology, USA
J.E. Penner	University of Michigan, USA
D.L. Roberts	Hadley Centre for Climate Prediction and Research, Met Office, UK
H. Rodhe	University of Stockholm, Sweden
G.J. Roelofs	University of Utrecht, Netherlands
L.D. Rotstayn	CSIRO, Australia
T.L. Schneider	Institute for World Forestry and Ecology, Germany
U. Schumann	Institut für Physik der Atmosphäre, Germany
S.E. Schwartz	Brookhaven National Laboratory, USA
M.D. Schwartzkopf	NOAA Geophysical Fluid Dynamics Laboratory, USA
K.P. Shine	University of Reading, UK
S. Smith	Pacific Northwest National Laboratory, USA
D.S. Stevenson	Met Office, UK
F. Stordal	Norwegian Institute for Air Research, Norway
I. Tegen	Max-Planck Institute for Biogeochemistry, Germany
R. van Dorland	Knoinklijk Nederlands Meteorologisch Instituut, The Netherlands
Y. Zhang	University of Michigan, USA

Review Editors

| J. Srinivasan | Indian Institute of Science, India |
| F. Joos | University of Bern, Switzerland |

Chapter 7. Physical Climate Processes and Feedbacks

Co-ordinating Lead Author

| T.F. Stocker | University of Bern, Switzerland |

Lead Authors

G.K.C. Clarke	University of British Columbia, Canada
H. Le Treut	Laboratoire de Météorologie Dynamique du Center National de la Recherche Scientifique, France
R.S. Lindzen	Massachusetts Institute of Technology, USA
V.P. Meleshko	Voeikov Main Geophysical Observatory, Russia
R.K. Mugara	Zambia Meteorological Department, Zambia
T.N. Palmer	European Centre for Medium-range Weather Forecasting, UK
R.T. Pierrehumbert	University of Chicago, USA
P.J. Sellers	NASA Johnson Space Center, USA
K.E. Trenberth	National Center for Atmospheric Research, USA
J. Willebrand	Institut für Meereskunde an der Universität Kiel, Germany

Contributing Authors

R.B. Alley	Pennsylvania State University, USA
O.E. Anisimov	State Hydrological Institute, Russia
C. Appenzeller	University of Bern, Switzerland
R.G. Barry	University of Colorado, USA

J.J. Bates	NOAA Environmental Research Laboratories, USA
R. Bindschadler	NASA Goddard Space Flight Center, USA
G.B. Bonan	National Center for Atmospheric Research, USA
C.W. Böning	Universtat Kiel, Germany
S. Bony	Laboratoire de Météorologie Dynamique du Center National de la Recherche Scientifique, France
H. Bryden	Southampton Oceanography Centre, UK
M.A. Cane	Lamont-Doherty Earth Observatory of Columbia Univeristy, USA
J.A. Curry	Aerospace Engineering, USA
T. Delworth	NOAA Geophysical Fluid Dynamics Laboratory, USA
A.S. Denning	Colorado State University, USA
R.E. Dickinson	University of Arizona, USA
K. Echelmeyer	University of Alaska, USA
K. Emanuel	Massachusetts Institute of Technology, USA
G. Flato	Canadian Centre for Climate Modelling & Analysis, Canada
I. Fung	University of California, USA
M. Geller	New York State University, USA
P.R. Gent	National Center for Atmospheric Research, USA
S.M. Griffies	NOAA Princeton University, USA
I. Held	NOAA Geophysical Fluid Dynamics Laboratory, USA
A. Henderson-Sellers	Australian Nuclear Science and Technology Organisation, Australia
A.A.M. Holtslag	Royal Netherlands Meteorological Institute, Netherlands
F. Hourdin	Center National de la Recherche Scientifique, Laboratoire de Météorologie Dynamique, France
J.W. Hurrell	National Center for Atmospheric Research, USA
V.M. Kattsov	Voeikov Main Geophysical Observatory, Russia
P.D. Killworth	Southampton Oceanography Centre, UK
Y. Kushnir	Lamont-Doherty Earth Observatory of Columbia Univeristy, USA
W.G. Large	National Center for Atmospheric Research, USA
M. Latif	Max-Planck Institute for Meteorology, Germany
P. Lemke	Alfred-Wegener Institute for Polar & Marine Research, Germany
M.E. Mann	University of Virginia, USA
G. Meehl	National Center for Atmospheric Research, USA
U. Mikolajewicz	Max-Planck Institute for Meteorology, Germany
W. O'Hirok	Institute for Computational Earth System Science, USA
C.L. Parkinson	NASA Goddard Space Flight Center, USA
A. Payne	University of Southampton, UK
A. Pitman	Macquarie University, Australia
J. Polcher	Center National de la Recherche Scientifique, Laboratoire de Météorologie Dynamique, France
I. Polyakov	Princeton University, USA
V. Ramaswamy	NOAA Geophysical Fluid Dynamics Laboratory, USA
P.J. Rasch	National Center for Atmospheric Research, USA
E.P. Salathe	University of Washington, USA
C. Schär	Institut fur Klimaforschung ETH, Switzerland
R.W. Schmitt	Woods Hole Oceanographic Institution, USA
T.G. Shepherd	University of Toronto, Canada
B.J. Soden	Princeton University, USA
R.W. Spencer	Marshall Space Flight Center, USA
P. Taylor	Southampton Oceanography Centre, UK
A. Timmermann	Koninklijk Nederlands Meteorologisch Instituut, Netherlands
K.Y. Vinnikov	University of Maryland, USA
M. Visbeck	Lamont Doherty Earth Observatory of Columbia University, USA
S.E. Wijffels	CSIRO Division of Marine Research, Australia
M. Wild	Swiss Federal Institute of Technology, Switzerland

Review Editors

S. Manabe	Institute for Global Change, Japan
P. Mason	Met Office, UK

Chapter 8. Model Evaluation

Co-ordinating Lead Author
B.J. McAvaney   Bureau of Meteorology Research Centre, Australia

Lead Authors

C. Covey	Lawrence Livermore National Laboratory, USA
S. Joussaume	Institut Pierre Simon Laplace, Laboratoire des Sciences du Climat et de l'Environment, France
V. Kattsov	Voeikov Main Geophysical Observatory, Russia
A. Kitoh	Meteorological Research Institute, Japan
W. Ogana	University of Nairobi, Kenya
A.J. Pitman	Macquarie University, Australia
A.J. Weaver	University of Victoria, Canada
R.A. Wood	Hadley Centre for Climate Prediction and Research, Met Office, UK
Z.-C. Zhao	National Climate Center, China

Contributing Authors

K. AchutaRao	Lawrence Livermore National Laboratory, USA
A. Arking	NASA Goddard Space Flight Center, USA
A. Barnston	NOAA Climate Prediction Center, USA
R. Betts	Hadley Centre for Climate Prediction and Research, Met Office, UK
C. Bitz	Quaternary Research, USA
G. Boer	Canadian Center for Climate Modelling & Analysis, Canada
P. Braconnot	Institut Pierre Simon Laplace, Laboratoire des Sciences du Climat et de l'Environment, France
A. Broccoli	NOAA Geophysical Fluid Dynamics Laboratory, USA
F. Bryan	Programe in Atmospheric and Oceanic Sciences, USA
M. Claussen	Potsdam Institute for Climate Impact Research, Germany
R. Colman	Bureau of Meteorology Research Centre, Australia
P. Delecluse	Institut Pierre Simon Laplace, Laboratoire d'Oceanographie Dynamique et Climatologie, France
A. Del Genio	NASA Goddard Institute for Space Studies, USA
K. Dixon	NOAA Geophysical Fluid Dynamics Laboratory, USA
P. Duffy	Lawrence Livermore National Laboratory, USA
L. Dümenil	Max-Planck Institute for Meteorology, Germany
M. England	University of New South Wales, Australia
T. Fichefet	Universite Catholique de Louvain, Belgium
G. Flato	Canadian Centre for Climate Modelling & Analysis, Canada
J.C. Fyfe	Canadian Centre for Climate Modelling & Analysis, Canada
N. Gedney	Hadley Centre for Climate Prediction and Research, Met Office, UK
P. Gent	National Center for Atmospheric Research, USA
C. Genthon	Laboratoire de Glaciologie et Geophysique de l'Environment, France
J. Gregory	Hadley Centre for Climate Prediction and Research, Met Office, UK
E. Guilyardi	Institut Pierre Simon Laplace, Laboratoire d'Oceanographie Dynamique et Climatologie, France
S. Harrison	Max-Planck Institute for Biogeochemistry, Germany
N. Hasegawa	Japan Environment Agency, Japan
G. Holland	Bureau of Meteorology Research Centre, Australia
M. Holland	National Center for Atmospheric Research, USA
Y. Jia	Southampton Oceanography Centre, UK
P.D. Jones	University of East Angelia, UK
M. Kageyama	Institut Pierre Simon Laplace, Laboratoire Sciences du Climat et de l'Environment, France
D. Keith	Harvard University, USA
K. Kodera	Meteorological Research Institute, Japan
J. Kutzbach	University of Wisconsin at Madison, USA
S. Lambert	University of Victoria, Canada
S. Legutke	Deutsches Klimarechenzentrum GmbH, Germany
G. Madec	Institut Pierre Simon Laplace, Laboratoire d'Oceanographie Dynamique et Climatologie, France
S. Maeda	Meteorological Research Institute, Japan

M.E. Mann	University of Virginia, USA
G. Meehl	National Center for Atmospheric Research, USA
I. Mokhov	Institute of Atmospheric Physics, Russia
T. Motoi	Frontier Research System for Global Change, Japan
T. Phillips	Lawrence Livermore National Laboratory, USA
J. Polcher	Center National de la Recherche Scientifique, Laboratoire de Météorologie Dynamique, France
G.L. Potter	Lawrence Livermore National Laboratory, USA
V. Pope	Hadley Centre for Climate Prediction and Research, Met Office, UK
C. Prentice	Max-Planck Institute for Biogeochemistry, Germany
G. Roff	Bureau of Meteorology Research Centre, Australia
P. Sellers	NASA Johnson Space Center, USA
F. Semazzi	Southampton Oceanography Centre, UK
D.J. Stensrud	NOAA National Severe Storms Laboratory, USA
T. Stockdale	European Centre for Medium-range Weather Forecasting, UK
R. Stouffer	NOAA Geophysical Fluid Dynamics Laboratory, USA
K.E. Taylor	Lawrence Livermore National Laboratory, USA
R. Tol	Vrije Universitiet, Netherlands
K. Trenberth	National Center for Atmospheric Research, USA
J. Walsh	University of Illinois at Urbana-Champaign, USA
M. Wild	Swiss Federal Institute of Technology, Switzerland
D. Williamson	National Center for Atmospheric Research, USA
S.-P. Xie	University of Hawaii at Manoa, USA
X.-H. Zhang	Chinese Academy of Sciences, China
F. Zwiers	Canadian Centre for Climate Modelling and Analysis, Canada

Review Editors
| Y. Qian | Nanjing University, China |
| J. Stone | Environment Canada, Canada |

Chapter 9. Projections of Future Climate Change

Co-ordinating Lead Authors
| U. Cubasch | Max-Planck Institute for Meteorology, Germany |
| G.A. Meehl | National Center for Atmospheric Research, USA |

Lead Authors
G.J. Boer	University of Victoria, Canada
R.J. Stouffer	NOAA Geophysical Fluid Dynamics Laboratory, USA
M. Dix	CSIRO Division of Atmospheric Research, Australia
A. Noda	Meteorological Research Institute, Japan
C.A. Senior	Hadley Centre for Climate Prediction and Research, Met Office, UK
S. Raper	University of East Anglia, UK
K.S. Yap	Malaysian Meteorological Service, Malaysia

Contributing Authors
A. Abe-Ouchi	University of Tokyo, Japan
S. Brinkop	Institute für Physik der Atmosphäre, Germany
M. Claussen	Potsdam Institute for Climate Impact Research, Germany
M. Collins	Hadley Centre for Climate Prediction and Research, Met Office, UK
J. Evans	Pennsylvania State University, USA
I. Fischer-Bruns	Max-Planck Institute for Meteorology, Germany
G. Flato	Canadian Centre for Climate Modelling & Analysis, Canada
J.C. Fyfe	Canadian Centre for Climate Modelling & Analysis, Canada
A. Ganopolski	Potsdam Institute for Climate Impact Research, Germany
J.M. Gregory	Hadley Centre for Climate Prediction and Research, Met Office, UK
Z.-Z. Hu	Center for Ocean-Land-Atmosphere Studies, USA

F. Joos	University of Bern, Switzerland
T. Knutson	NOAA Geophysical Fluid Dynamics Laboratory, USA
C. Landsea	NOAA Atlantic Oceanographic & Meteorological Laboratory, USA
L. Mearns	National Center for Atmospheric Research, USA
C. Milly	US Geological Survey, USA
J.F.B. Mitchell	Hadley Centre for Climate Prediction and Research, Met Office, UK
T. Nozawa	National Institute for Environmental Studies, Japan
H. Paeth	Universität Bonn, Germany
J. Räisänen	Swedish Meteorological and Hydrological Institute, Sweden
R. Sausen	Institute für Physik der Atmosphäre, Germany
S. Smith	Pacific Northwest National Laboratory, USA
T. Stocker	University of Bern, Switzerland
A. Timmermann	Royal Netherlands Meteorological Institute, Netherlands
U. Ulbrich	Institut fuer Geophysik und Meteorolgie, Germany
A. Weaver	University of Victoria, Canada
J. Wegner	Deutsches Klimarechenzentrum, Germany
P. Whetton	CSIRO Division of Atmospheric Research, Australia
T. Wigley	National Center for Atmospheric Research, USA
M. Winton	NOAA Geophysical Fluid Dynamics Laboratory, USA
F. Zwiers	Canadian Centre for Climate Modelling and Analysis, Canada

Review Editors

| J. Stone | Environment Canada, Canada |
| J.-W. Kim | Yonsei University, South Korea |

Chapter 10. Regional Climate Information - Evaluation and Projections

Co-ordinating Lead Authors

| F. Giorgi | Abdus Salam International Centre for Theoretical Physics, Italy |
| B. Hewitson | University of Capetown, South Africa |

Lead Authors

J. Christensen	Danish Meteorological Institute, Denmark
M. Hulme	University of East Anglia, UK
H. Von Storch	GKSS, Germany
P. Whetton	CSIRO Division of Atmospheric Research, Australia
R. Jones	Hadley Centre for Climate Prediction and Research, Met Office, UK
L. Mearns	National Center for Atmospheric Research, USA
C. Fu	Institute of Atmospheric Physics, China

Contributing Authors

R. Arritt	Iowa State University, USA
B. Bates	CSIRO Land and Water, Australia
R. Benestad	Det Norske Meteorologiske Institutt, Norway
G. Boer	Canadian Centre for Climate Modelling & Analysis, Canada
A. Buishand	Koninklijk Nederlands Meteorologisch Instituut, Netherlands
M. Castro	Universidad Complutense de Madrid, Spain
D. Chen	Göteborg University, Sweden
W. Cramer	Potsdam Institute for Climate Impact Research, Germany
R. Crane	The Pennsylvania State University, USA
J.F. Crossley	University of East Anglia, UK
M. Dehn	University of Bonn, Germany
K. Dethloff	Alfred Wegener Institute for Polar and Marine Research, Germany
J. Dippner	Institute for Baltic Research, Germany
S. Emori	National Institute for Environmental Studies, Japan
R. Francisco	Weather Bureau, Philippines

J. Fyfe	Canadian Centre for climate modelling and analysis, Canada
F.W. Gerstengarbe	Potsdam Institute for Climate Impact Research, Germany
W. Gutowski	Iowa State University, USA
D. Gyalistras	University of Berne, Switzerland
I. Hanssen-Bauer	The Norwegian Meteorological Institute, Norway
M. Hantel	University of Vienna, Austria
D.C. Hassell	Hadley Centre for Climate Prediction and Research, Met Office, UK
D. Heimann	Institute of Atmospheric Physics, Germany
C. Jack	University of Cape Town, South Africa
J. Jacobeit	Universitaet Wuerzburg, Germany
H. Kato	Central Research Institute of Electric Power Industry, Japan
R. Katz	National Center for Atmospheric Research, USA
F. Kauker	Alfred Wegener Institute for Polar and Marine Research, Germany
T. Knutson	NOAA Geophysical Fluid Dynamics Laboratory, USA
M. Lal	Indian Institute of Technology, India
C. Landsea	NOAA Atlantic Oceanographic & Meteorological Laboratory, USA
R. Laprise	University of Quebec at Montreal, Canada
L.R. Leung	Pacific Northwest National Laboratory, USA
A.H. Lynch	University of Colorado, USA
W. May	Danish Meteorological Institute, Denmark
J.L. McGregor	CSIRO Division of Atmospheric Research, Australia
N.L. Miller	Lawrence Berkeley National Laboratory, USA
J. Murphy	Hadley Centre for Climate Prediction and Research, Met Office, UK
J. Ribalaygua	Fundación para la Investigación del Clima, Spain
A. Rinke	Alfred Wegener Institute for Polar and Marine Research, Germany
M. Rummukainen	Swedish Meteorological and Hydrological Institute, Sweden
F. Semazzi	Southampton Oceanography Centre, UK
K. Walsh	CSIRO Division of Atmospheric Research, Australia
P. Werner	Potsdam Institute for Climate Impact Research, Germany
M. Widmann	GKSS Research Centre, Germany
R. Wilby	University of Derby, UK
M. Wild	Swiss Federal Institute of Technology, Switzerland
Y. Xue	University of California at Los Angeles, USA

Review Editors

M. Mietus	Institute of Meteorology & Water Management, Poland
J. Zillman	Bureau of Meteorology, Australia

Chapter 11. Changes in Sea Level

Co-ordinating Lead Authors

J.A. Church	CSIRO Division of Marine Research, Australia
J.M. Gregory	Hadley Centre for Climate Prediction and Research, Met Office, UK

Lead Authors

P. Huybrechts	Vrije Universiteit Brussel, Belgium
M. Kuhn	Innsbruck University, Austria
K. Lambeck	Australian National University, Australia
M.T. Nhuan	Hanoi University of Sciences, Vietnam
D. Qin	Chinese Academy of Sciences, China
P.L. Woodworth	Bidston Observatory, UK

Contributing Authors

O.A. Anisimov	State Hydrological Institute, Russia
F.O. Bryan	Programe in Atmospheric and Oceanic Sciences, USA
A. Cazenave	Groupe de Recherche de Geodesie Spatiale CNES, France

K.W. Dixon	NOAA Geophysical Fluid Dynamics Laboratory, USA
B.B. Fitzharris	University of Otago, New Zealand
G.M. Flato	Canadian Centre for Climate Modelling & Analysis, Canada
A. Ganopolski	Potsdam Institute for Climate Impact Research, Germany
V. Gornitz	Goddard Institute for Space Studies, USA
J.A. Lowe	Hadley Centre for Climate Prediction and Research, Met Office, UK
A. Noda	Japan Meteorological Agency, Japan
J.M. Oberhuber	German Climate Computing Centre, Germany
S.P. O'Farrell	CSIRO Division of Atmospheric Research, Australia
A. Ohmura	Geographisches Institute ETH, Switzerland
M. Oppenheimer	Environmental Defense, USA
W.R. Peltier	University of Toronto, Canada
S.C.B. Raper	University of East Anglia, UK
C. Ritz	Laboratoire de Glaciologie et Geophysique de l'Environment, France
G.L. Russell	NASA Goddard Institute for Space Studies, USA
E. Schlosser	Innsbruck University, Austria
C.K. Shum	Ohio State University, USA
T.F. Stocker	University of Bern, Switzerland
R.J. Stouffer	NOAA Geophysical Fluid Dynamics Laboratory, USA
R.S.W. van de Wal	Institute for Marine and Atmospheric Research, Netherlands
R. Voss	Deutsches Klimarechenzentrum, Germany
E.C. Wiebe	University of Victoria, Canada
M. Wild	Swiss Federal Institute of Technology, Switzerland
D.J. Wingham	University College London, UK
H.J. Zwally	NASA Goddard Space Flight Center, USA

Review Editors

| B.C. Douglas | University of Maryland, USA |
| A. Ramirez | Universidad Central Venezuela, Venezuela |

Chapter 12. Detection of Climate Change and Attribution of Causes

Co-ordinating Lead Authors

| J.F.B. Mitchell | Hadley Centre for Climate Prediction and Research, Met Office, UK |
| D.J. Karoly | Monash University, Australia |

Lead Authors

G.C. Hegerl	Texas A&M University, USA/Germany
F.W. Zwiers	University of Victoria, Canada
M.R. Allen	Rutherford Appleton Laboratory, UK
J. Marengo	Instituto Nacional de Pesquisas Espaciais, Brazil

Contributing Authors

V. Barros	Ciudad Universitaria, Argentina
M. Berliner	Ohio State University, USA
G. Boer	Canadian Centre for Climate Modelling & Analysis, Canada
T. Crowley	Texas A&M University, USA
C. Folland	Hadley Centre for Climate Prediction and Research, Met Office, UK
M. Free	NOAA Air Resources Laboratory, USA
N. Gillett	University of Oxford, UK
P. Groissman	NOAA National Climatic Data Center, USA
J. Haigh	Imperial College, UK
K. Hasselmann	Max-Planck Institute for Meteorology, Germany
P. Jones	University of East Anglia, UK
M. Kandlikar	Carnegie-Mellon University, USA
V. Kharin	Canadian Centre for Climate Modelling and Analysis, Canada

H. Khesghi	Exxon Mobil Research & Engineering Company, USA
T. Knutson	NOAA Geophysical Fluid Dynamics Laboratory, USA
M. MacCracken	Office of the US Global Change Research Program, USA
M. Mann	University of Virginia, USA
G. North	Texas A&M University, USA
J. Risbey	Carnegie-Mellon University, USA
A. Robock	Rutgers University, USA
B. Santer	Lawrence Livermore National Laboratory, USA
R. Schnur	Max-Planck Institute for Meteorology, Germany
C. Schönwiese	J.W. Goethe University, Germany
D. Sexton	Hadley Centre for Climate Prediction and Research, Met Office, UK
P. Stott	Hadley Centre for Climate Prediction and Research, Met Office, UK
S. Tett	Hadley Centre for Climate Prediction and Research, Met Office, UK
K. Vinnikov	University of Maryland, USA
T. Wigley	National Center for Atmospheric Research, USA

Review Editors

F. Semazzi	Southampton Oceanography Centre, UK
J. Zillman	Bureau of Meteorology, Australia

Chapter 13. Climate Scenario Development

Co-ordinating Lead Authors

L.O. Mearns	National Center for Atmospheric Research, USA
M. Hulme	University of East Anglia, UK

Lead Authors

T.R. Carter	Finnish Environment Institute, Finland
R. Leemans	Rijksinstituut voor Volksgezondheid en Milieu, Netherlands
M. Lal	Indian Institute of Technology, India
P. Whetton	CSIRO Division of Atmospheric Research, Australia

Contributing Authors

L. Hay	US Geological Survey, USA
R.N. Jones	CSIRO Division of Atmospheric Research, Australia
R. Katz	National Center for Atmospheric Research, USA
T. Kittel	National Center for Atmospheric Research, USA
J. Smith	Stratus Consulting Inc., USA
R. Wilby	University of Derby, UK

Review Editors

L.J. Mata	Universidad Central Venezuela, Venezuela
J. Zillman	Bureau of Meteorology, Australia

Chapter 14. Advancing our Understanding

Co-ordinating Lead Author

B. Moore III	University of New Hampshire, USA

Lead Authors

W.L. Gates	Lawrence Livermore National Laboratory, USA
L.J. Mata	Universidad Central Venezuela, Venezuela
A. Underdal	University of Oslo, Norway

Contributing Author
R.J. Stouffer   NOAA Geophysical Fluid Dynamics Laboratory, USA

Review Editors
B. Bolin   Retired, Sweden
A. Ramirez Rojas   Universidad Central Venezuela, Venezuela

Appendix IV

Reviewers
of the IPCC WGI Third Assessment Report

Argentina

M. Nuñez   Ciudad Universitaria

Australia

K. Abel   Australian Greenhouse Office
G. Ayers   CSIRO Division of Atmospheric Research
S. Barrell   Bureau of Meteorology
P. Bate   Bureau of Meteorology
B. Bates   CSIRO Division of Land and Water
T. Beer   CSIRO Division of Atmospheric Research
R. Boers   CSIRO Division of Atmospheric Research
W. Budd   University of Tasmania
I. Carruthers   Australian Greenhouse Office
S. Charles   CSIRO Division of Atmospheric Research
J. Church   CSIRO Division of Marine Research
D. Collins   Bureau of Meteorology
R. Colman   Bureau of Meteorology Research Centre
D. Cosgrove   Bureau of Transport Economics
S. Crimp   Department of Natural Resources
B. Curran   Bureau of Meteorology
M. Davison   Australian Industry Greenhouse Network
M. Dix   CSIRO Division of Atmospheric Research
B. Dixon   Bureau of Meteorology
M. England   University of New South Wales
I. Enting   CSIRO Division of Atmospheric Research
D. Etheridge   CSIRO Division of Atmospheric Research
G. Farquhar   Australian National University
P. Forster   Monash University
R. Francey   CSIRO Division of Atmospheric Research
P. Fraser   CSIRO Division of Atmospheric Research
R. Gifford   CSIRO Division of Plant Industry
I. Goodwin   University of Tasmania
J. Gras   CSIRO Division of Atmospheric Research
G. Hassall   Australian Greenhouse Office
A. Henderson-Sellers   Australian Nuclear Science and Technology Organisation

K. Hennessy	CSIRO Division of Atmospheric Research
A. Ivanovici	Australian Greenhouse Office
J. Jacka	Australian Antarctic Division
I. Jones	University of Sydney
R. Jones	CSIRO Division of Atmospheric Research
D. Karoly	Monash University
J. Katzfey	CSIRO Division of Atmospheric Research
B. Kininmonth	Australasian Climate Research
J. Lough	Australian Institute of Marine Science
G. Love	Bureau of Meteorology
M. Manton	Bureau of Meteorology Research Centre
B. McAvaney	Bureau of Meteorology Research Centre
T. McDougall	CSIRO Division of Marine Research
A. McEwan	Bureau of Meteorology
J. McGregor	CSIRO Division of Atmospheric Research
L. Minty	Bureau of Meteorology
B. Mitchell	Flinders University of South Australia
N. Plummer	Bureau of Meteorology
L. Powell	Australian Greenhouse Office
L. Quick	Australian Greenhouse Office
P. Rayner	CSIRO Division of Atmospheric Research
L. Rikus	Bureau of Meteorology Research Centre
L. Rotstayn	CSIRO Division of Atmospheric Research
W. Scherer	Flinders University of South Australia
I. Smith	CSIRO Division of Atmospheric Research
P. Steele	CSIRO Division of Atmospheric Research
K. Walsh	CSIRO Division of Atmospheric Research
I. Watterson	CSIRO Division of Atmospheric Research
P. Whetton	CSIRO Division of Atmospheric Research
J. Zillman	Bureau of Meteorology

Austria

| M. Hantel | University of Vienna |
| K. Radunsky | Federal Environment Agency |

Belgium

T. Fichefet	Université Catholique de Louvain
J. Franklin	Solvay Research and Technology
A. Mouchet	Astrophysics and Geophysics Institute
J. van Ypersele	Université Catholique de Louvain
R. Zander	University of Liege

Benin

| E. Ahlonsou | National Meteorological Service |

Brazil

| P. Fearnside | National Institute for Research in the Amazon |
| J. Marengo | Instituto Nacional de Pesquisas Espaciais |

Canada

P. Austin	University of British Columbia
E. Barrow	Atmospheric and Hydrologic Science Division
J. Bourgeois	Geological Survey of Canada
R. Brown	Atmospheric Environment Service
E. Bush	Environment Canada
M. Demuth	Geological Survey of Canada
K Denman	Department of Fisheries and Oceans
P. Edwards	Environment Canada
W. Evans	Trent University
D. Fisher	Geological Survey of Canada
G. Flato	University of Victoria
W. Gough	University of Toronto at Scarbrough
D. Harvey	University of Toronto
H. Hengeveld	Environment Canada
W. Hogg	Atmospheric Environment Service
P. Kertland	Natural Resources Canada
R. Koerner	Geological Survey of Canada
R. Laprise	University of Quebec at Montreal
Z. Li	Natural Resources Canada
U. Lohmann	Dalhousie University
J. Majorowicz	Northern Geothermal
L. Malone	Environment Canada
N. McFarlane	University of Victoria
L. Mysak	McGill University
W. Peltier	University of Toronto
I. Perry	Fisheries and Oceans Canada
J. Rudolph	York University
P. Samson	Natural Resources Canada
J. Sargent	Finance Canada
J. Shaw	Geological Survey of Canada
S. Smith	Natural Resources Canada
J. Stone	Environment Canada
R. Street	Environment Canada
D. Whelpdale	Environment Canada
R. Wong	Government of Alberta
F. Zwiers	University of Victoria

China

D. Gong	Peking University
W. Li	Institute of Atmospheric Physics
G. Ren	National Climate Center
S. Sun	Institute of Atmospheric Physics
R. Yu	Institute of Atmospheric Physics
P. Zhai	National Climate Center
X. Zhang	Institute of Atmospheric Physics
G. Zhou	Institute of Atmospheric Physics
T. Zhou	Institute of Atmospheric Physics

Czech Republic

R. Brazdil	Masaryk University

Denmark

J. Bates	University of Copenhagen
B. Christiansen	Danish Meteorological Institute
P. Frich	Danmarks Miljøundersøgelser (DMU)
A. Hansen	University of Copenhagen
A. Jørgensen	Danish Meteorological Institute
T. Jørgensen	Danish Meteorological Institute
E. Kaas	Danish Meteorological Institute
P. Laut	Technical University of Denmark
B. Machenhauer	Danish Meteorological Institute
L. Prahm	Danish Meteorological Institute
M. Stendel	Danish Meteorological Institute
P. Thejll	Danish Meteorological Institute

Finland

T. Carter	Finnish Environment Institute
E. Holopainen	University of Helsinki
R. Korhonen	Technical Research Centre of Finland (VTT)
M. Kulmala	University of Helsinki
J. Launiainen	Finnish Institute of Marine Research
H. Tuomenvirta	Finnish Meteorological Institute

France

A. Alexiou	Intergovernmental Oceanographic Commission
P. Braconnot	Institut Pierre Simon Laplace, Laboratoire des Sciences du Climat et de l'Environment
J. Brenguier	Meteo France
N. Chaumerliac	Université Blaisi Pascal
M. Deque	Meteo France
Y. Fouquart	Université des Science & Techn de Lille
C. Genthon	Laboratoire de Glaciologie et Geophysique de l'Environment du CNRS
M. Gillet	Mission Interministerielle de l'Effet de Serre
S. Joussaume	Institut Pierre Simon Laplace, Laboratoire des Sciences du Climat et de l'Environment
J. Jouzel	Institut Pierre Simon Laplace, Laboratoire des Sciences du Climat et de l'Environment
R. Juvanon du Vachat	Mission Interministerielle de l'Effet de Serre
H. Le Treut	Center National de la Recherche Scientifique, Laboratoire de Météorologie Dynamique
M. Petit	Ecole Polytechnique
P. Pirazzoli	Center National de la Recherche Scientifique, Laboratoire de Géographie Physique
S. Planton	Meteo France
J. Polcher	Center National de la Recherche Scientifique, Laboratoire de Météorologie Dynamique
A. Riedacker	INRA
J. Salmon	Ministère de l'Aménagement du Territoire et de l'Environnement
D. Tanre	Laboratoire d'Optigue Atmospherique

Germany

H. Ahlgrimm	Federal Agricultural Research Center
M. Andreae	Max-Planck Institut für Biochemistry
R. Benndorf	Federal Environmental Agency
U. Boehm	Universität Potsdam
O. Boucher	Max-Planck Institut für Chemie
S. Brinkop	Institut für Physik der Atmosphäre

M. Claussen   Potsdam Institute for Climate Impact Research
M. Dehn   Universität Bonn
P. Dietze   Private
E. Holland   Max-Planck Institut für Biochemistry
J. Jacobeit   Universität Wuerzburg
K. Kartschall   Federal Environmental Agency
B. Kärcher   Institut für Physik der Atmosphäre
K. Lange   Federal Ministry for Environment, Nature Conservation and Nuclear Safety
P. Mahrenholz   Federal Environmental Agency
J. Oberhuber   German Climate Computing Centre
R. Sartorius   Federal Environmental Agency
C. Schoenwiese   J.W. Goethe University
U. Schumann   Institut für Physik der Atmosphäre
U. Ulbrich   Institut für Geophysik und Meteorolgie
T. Voigt   Federal Environment Agency
A. Volz-Thomas   Forschungsezentrum Juelich
G. Weber   Gesamtverband Steinkohlenbergbau (GVST)
G. Wefer   Universität Bremen
M. Widmann   GKSS-Forschungszentrum

Hungary

G. Koppány   University of Szeged

Iceland

T. Johannesson   Icelandic Meteorological Office

Israel

P. Alpert   Tel Aviv University
S. Krichark   Tel Aviv University
C. Price   Tel Aviv University
Z. Levin   Tel Aviv University

Italy

W. Dragoni   Perugia Universita
A. Mariotti   National Agency for New Technology, Energy and Environment (ENEA)
T. Nanni   ISAO National Research Council
P. Ruti   National Agency for New Technology, Energy and Environment (ENEA)
R. van Dingenen   Enviroment Institute of European Commission
G. Visconti   Università Degli Studi dell' Aquila

Japan

M. Amino   Japan Meteorological Agency
T. Asoh   Japan Meteorological Agency
H. Isobe   Japan Meteorological Agency
H. Kanzawa   Environment Agency
H. Kato   Central Research Institute of Electric Power Industry
M. Kimoto   University of Tokyo

K. Kurihara	Japan Meteorological Agency
S. Kusunoki	Meteorological Research Institute
S. Manabe	Institute for Global Change
S. Nagata	Environment Agency
Y. Nikaidou	Japan Meteorological Agency
J. Ohyama	Japan Meteorological Agency
Y. Sato	Meteorological Research Institute
A. Sekiya	National Institute of Materials and Chemical Research
M. Shinoda	Tokyo Metropolitan University
S. Taguchi	National Institute for Research & Environment
T. Tokioka	Japan Meteorological Agency
Y. Tsutsumi	Japan Meteorological Agency
O. Wild	Frontier Research System for Global Change
R. Yamamoto	Kyoto University

Kenya

| J. Ng'ang'a | University of Nairobi |
| N. Sabogal | United Nations Environment Programme |

Malaysia

| A. Chan | Malaysian Meteorological Service |

Morocco

A. Allali	Ministry of Agriculture & Moroccan Association for Environment Protection
S. Khatri	Meteorological Office of Morocco
A. Mokssit	Meteorological Office of Morocco
A. Sbaibi	Universite Hassan II - Mohammedia

Netherlands

A.P.M. Baede	Koninklijk Nederlands Meteorologisch Instituut
J. Beersma	Koninklijk Nederlands Meteorologisch Instituut
L. Bijlsma	Rijksinstituut voor Kust en Zee
T. Buishand	Koninklijk Nederlands Meteorologisch Instituut
G. Burgers	Koninklijk Nederlands Meteorologisch Instituut
H. Dijkstra	University of Utrecht
S. Drijfhout	Koninklijk Nederlands Meteorologisch Instituut
W. Hazeleger	Koninklijk Nederlands Meteorologisch Instituut
B. Holtslag	Wageningen University
C. Jacobs	Koninklijk Nederlands Meteorologisch Instituut
A. Jeuken	Koninklijk Nederlands Meteorologisch Instituut
H. Kelder	Koninklijk Nederlands Meteorologisch Instituut
G. Komen	Koninklijk Nederlands Meteorologisch Instituut and University of Utrecht
N. Maat	Koninklijk Nederlands Meteorologisch Instituut
L. Meyer	Ministry of Housing, Spatial Planning & the Environment
J. Olivier	Rijksinstituut voor Volksgezondheid en Milieu
J. Opsteegh	Koninklijk Nederlands Meteorologisch Instituut
A. Petersen	Vrije Universiteit
H. Radder	Vrije Universiteit
H. Renssen	Vrije Universiteit

J. Ronde	Rijksinstituut voor Kust en Zee
M. Scheffers	Rijksinstituut voor Kust en Zee
C. Schuurmans	University of Utrecht
P. Siegmund	Koninklijk Nederlands Meteorologisch Instituut
A. Sterl	Koninklijk Nederlands Meteorologisch Instituut
H. ten Brink	Energieonderzoek Centrum Nederland
R. Tol	Vrije Universiteit
S. van de Geijn	Plant Research International
R. van Dorland	Koninklijk Nederlands Meteorologisch Instituut
G. van Tol	Expertisecentrum LNV
A. van Ulden	Koninklijk Nederlands Meteorologisch Instituut
M. van Weele	Koninklijk Nederlands Meteorologisch Instituut
P. Veefkind	Koninklijk Nederlands Meteorologisch Instituut
G. Velders	Rijksinstituut voor Volksgezondheid en Milieu
J. Verbeek	Koninklijk Nederlands Meteorologisch Instituut
H. Visser	KEMA

New Zealand

C. de Freitas	University of Auckland
B. Fitzharris	University of Otago
V. Gray	Climate Consultant, New Zealand
J. Kidson	National Institute of Water & Atmospheric Research
H. Larsen	National Institute of Water & Atmospheric Research
P. Maclaren	University of Canterbury
M. Manning	National Institute of Water & Atmospheric Research
J. Renwick	National Institute of Water & Atmospheric Research

Norway

T. Asphjell	Norwegian State Pollution Control Authority
R. Benestad	Norwegian Meteorological Institute
O. Christophersen	Ministry of Environment
E. Forland	Norwegian Meteorological Institute
J. Fuglestvedt	University of Oslo
O. Godal	University of Oslo
S. Grønås	University of Bergen
I. Hanssen-Bauer	Norwegian Meteorological Institute
E. Jansen	University of Bergen
N. Koc	Norsk Polarinstitutt
H. Loeng	Institute of Marine Research
S. Mylona	Norwegian State Pollution Control Authority
M. Pettersen	Norwegian State Pollution Control Authority
A. Rosland	Norwegian State Pollution Control Authority
T. Segalstad	University of Oslo
J. Winther	Norwegian Polar Institute

Peru

| N. Gamboa | Pontificia Universidad Catolica del Peru |

Poland

M. Mietus   Institute of Meteorology & Water Management

Portugal

C. Borrego   Universidade de Aveiro

Russian Federation

O. E. Anisimov   State Hydrological Institute
R. Burlutsky   Hydrometeorological Research Centre of Russia
N. Datsenko   Hydrometeorological Research Centre of Russia
G. Golitsyn   Institute of Atmospheric Physics
N. Ivachtchenko   Hydrometeorological Research Centre of Russia
I. Karol   Main Geophysical Observatory
K. Kondratyev   Research Centre for Ecological Safety
V. P. Meleshko   Main Geophysical Observatory
I. Mokhov   Institute of Atmospheric Physics
D. Sonechkin   Hydrometeorological Research Centre of Russia

Saudi Arabia

M. Al-Sabban   Ministry of Petroleum

Slovak Republic

M. Lapin   Comenius University
K. Mareckova   Slovak Hydrometeorological Institute

Slovenia

A. Kranjc   Hydrometeorological Institute of Slovenia

Spain

S. Alonso   Universitat de les Illes Balears
L. Balairon   National Institute of Meteorology
Y. Castro-Diez   Universidad de Granada
J. Cortina   Universitat d'Alacant
M. de Luis   Universitat d'Alacant
E. Fanjul   Clima Maritimo - Puertos del Estado
B. Gomez   Clima Maritimo - Puertos del Estado
M. Gomez-Lahoz   Puertos del Estado
J. Gonzalez-Hidalgo   University of Zaragoza
A. Lavin   Instituto Español de Oceanografía
J. Peñuelas   Universitat Autònoma de Barcelona
J. Raventos   Universitat d'Alacant
J. Sanchez   Universitat d'Alacant
I. Sanchez-Arevalo   Clima Maritimo - Puertos del Estado
M. Vazquez   Instituto de Astrofísica de Canarias

Sudan

N. Awad	Higher Council for Environment & Natural Resources
I. Elgizouli	Higher Council for Environment & Natural Resources
N. Goutbi	Higher Council for Environment & Natural Resources

Sweden

R. Charlson	Stockholm University
E. Källén	Stockholm University
A. Moberg	Stockholm University
N. Morner	Stockholm University
J. Raisanen	Swedish Meteorological and Hydrological Institute
H. Rodhe	Stockholm University
M. Rummukainen	Swedish Meteorological and Hydrological Institute

Switzerland

U. Baltensperger	Paul Scherrer Institute
D. Gyalistras	University of Bern
W. Haeberli	University of Zurich
F. Joos	University of Bern
H. Lang	Swiss Federal Institute of Technology
C. Pfister	Unitobler
J. Romero	Federal Office of Environment, Forests and Landscape
C. Schaer	Swiss Federal Institute of Technology
J. Staehelin	Swiss Federal Institute of Technology
H. Wanner	University of Bern
M. Wild	Swiss Federal Institute of Technology

Thailand

J. Boonjawat	Chulalongkorn University

Togo

A. Ajavon	Universite du Benin

Turkey

A. Danchev	Fatih University
M. Turkes	Turkish State Meteorological Service

United Kingdom

M. Allen	Rutherford Appleton Laboratory
S. Allison	Southampton Oceanography Centre
R. Betts	Hadley Centre for Climate Prediction and Research, Met Office
S. Boehmer-Christiansen	Sussex University
R. Braithwaite	University of Manchester
K. Briffa	University of East Anglia

S. Brown	Hadley Centre for Climate Prediction and Research, Met Office
I. Colbeck	University of Essex
R. Courtney	European Science and Environment Forum
M. Crompton	Department of the Environment, Transport and the Regions
X. Dai	IPCC WGI Technical Support Unit
C. Doake	British Antarctic Survey
C. Folland	Hadley Centre for Climate Prediction and Research, Met Office
N. Gedney	Hadley Centre for Climate Prediction and Research, Met Office
N. Gillett	University of Oxford
W. Gould	Southampton Oceanography Centre
J. Gregory	Hadley Centre for Climate Prediction and Research, Met Office
S. Gregory	University of Sheffield
D. J Griggs	IPCC WGI Technical Support Unit
J. Grove	University of Cambridge
J. Haigh	Imperial College
R. Harding	Centre for Ecology and Hydrology
M. Harley	English Nature
J. Haywood	Meteorological Research Flight, Met Office
J. Houghton	IPCC WGI Co-Chairman
W. Ingram	Hadley Centre for Climate Prediction and Research, Met Office
T. Iversen	European Centre for Medium-range Weather Forecasting
J. Lovelock	Retired, United Kingdom
K. Maskell	IPCC WGI Technical Support Unit
A. McCulloch	Marbury Technical Consulting, United Kingdom
G. McFadyen	Department of the Environment, Transport and the Regions
J. Mitchell	Hadley Centre for Climate Prediction and Research, Met Office
J. Murphy	Hadley Centre for Climate Prediction and Research, Met Office
C. Newton	Environment Agency
M. Noguer	IPCC WGI Technical Support Unit
T. Osborn	University of East Anglia
D. Parker	Hadley Centre for Climate Prediction and Research, Met Office
D. Pugh	Southampton Oceanography Centre
S. Raper	University of East Anglia
D. Roberts	Hadley Centre for Climate Prediction and Research, Met Office
D. Sexton	Hadley Centre for Climate Prediction and Research, Met Office
K. Shine	University of Reading
K. Smith	University of Edinburgh
P. Smithson	University of Sheffield
P. Stott	Hadley Centre for Climate Prediction and Research, Met Office
S. Tett	Hadley Centre for Climate Prediction and Research, Met Office
P. Thorne	University of East Anglia
R. Toumi	Imperial College
P. Viterbo	European Centre for Medium-range Weather Forecasting
D. Warrilow	Department of the Environment, Transport and the Regions
R. Wilby	University of Derby
P. Williamson	Plymouth Marine Laboratory
P. Woodworth	Bidston Observatory

United States of America

M. Abbott	Oregon State University
W. Abdalati	NASA Goddard Space Flight Center
D. Adamec	NASA Goddard Space Flight Center
R. B. Alley	Pennsylvania State University
R. Andres	University of Alaska at Fairbanks
J. Angel	Illinois State Water Survey

P. Arkin   Columbia University
R. Arritt   Iowa State University
E. Atlas   National Center for Atmospheric Research
D. Bader   Department of Energy
T. Baerwald   National Science Foundation
R. Bales   University of Arizona
R. Barber   Duke University
T. Barnett   Scripps Institute of Oceanography
P. Bartlein   University of Oregon
J. J. Bates   NOAA Environmental Technology Laboratory
T. Bates   NOAA Pacific Marine Environmental Laboratory
M. Bender   Princeton University
C. Bentley   University of Wisconsin at Madison
K. Bergman   NASA Global Modeling and Analysis Program
C. Berkowitz   Pacific Northwest National Laboratory
M. Berliner   Ohio State University
J. Berry   Carnegie Institution of Washington
R. Bindschadler   NASA Goddard Space Flight Center
D. Blake   University of California at Irvine
T. Bond   University of Washington
A. Broccoli   Princeton University
W. Broecker   Lamont Doherty Earth Observatory of Columbia University
L. Bruhwiler   NOAA Climate Monitoring and Diagnostics Laboratory
K. Bryan   Princeton University
K. Caldeira   Lawrence Livermore National Laboratory
M. A. Cane   Lamont Doherty Earth Observatory of Columbia University
A. Carleton   Pennsylvania State University
R. Cess   State University of New York
W. Chameides   Georgia Institute of Technology
T. Charlock   NASA Langley Research Center
M. Chin   NASA Goddard Space Flight Center
K. Cook   Cornell University
W. Cooke   Princeton University
C. Covey   Lawrence Livermore National Laboratory
T. Crowley   Texas A&M University
D. Cunnold   Georgia Institute of Technology
J. A. Curry   University of Colorado
R. Dahlman   Department of Energy
A. Dai   National Center for Atmospheric Research
B. DeAngelo   Environmental Protection Agency
P. DeCola   NASA
P. DeMott   Colorado State University
A. S. Denning   Colorado State University
W. Dewar   Florida State University
R. E. Dickerson   University of Maryland
R. Dickinson   Georgia Institute of Technology
L. Dilling   NOAA Office of Global Programs
E. Dlugokencky   NOAA Climate Monitoring & Diagnostics Laboratory
S. Doney   National Center for Atmospheric Research
S. Drobot   University of Nebraska
H. Ducklow   Virginia Institute of Marine Sciences
W. Easterling   Pennsylvania State University
J. Elkins   NOAA Climate Monitoring & Diagnostics Laboratory
E. Elliott   National Science Foundation
W. Elliott   NOAA Air Resources Laboratory
H. Ellsaesser   Atmospheric Consultant
S. Esbensen   Oregon State University

C. Fairall	NOAA Environmental Technology Laboratory
Y. Fan	Center for Ocean-Land-Atmosphere Studies
P. Farrar	Naval Oceanographic Office
R. Feely	NOAA Pacific Marine Environmental Laboratory
F. Fehsenfeld	NOAA Environmental Research Laboratories
G. Feingold	NOAA Environmental Technology Laboratory
R. Fleagle	University of Washington
R. Forte	Environmental Protection Agency
M. Fox-Rabinovitz	University of Maryland
J. Francis	Rutgers University
M. Free	NOAA Air Resources Laboratory
R. Friedl	Jet Propulsion laboratory
I. Fung	University of California
D. Gaffen	NOAA Air Resources Laboratory
W. Gates	Lawrence Livermore National Laboratory
C. Gautier	University of California at Santa Barbara
P. Geckler	Lawrence Livermore National Laboratory
L. Gerhard	University of Kansas
S. Ghan	Pacific Northwest National Laboratory
M. Ghil	University of California at Los Angeles
P. Gleckler	Lawrence Livermore National Laboratory
V. Gornitz	NASA Goddard Institute for Space Studies
V. Grewe	NASA Goddard Institute for Space Studies
W. Gutowski	Iowa State University
P. Guttorp	University of Washington
R. Hallgren	American Meteorological Society
D. Hardy	University of Massachusetts
E. Harrison	NOAA Pacific Marine Environmental Laboratory
G. Hegerl	Texas A&M University
B. Hicks	NOAA Air Resources Laboratory
W. Higgins	NOAA Climate Protection Center
D. Houghton	University of Wisconsin at Madison
R. Houghton	Woods Hole Research Center
Z. Hu	Center for Ocean-Land-Atmosphere Studies
B. Huang	Center for Ocean-Land-Atmosphere Studies
J. Hudson	Desert Research Institute
M. Hughes	University of Arizona
C. Hulbe	NASA Goddard Space Flight Center
D. Jacob	Harvard University
S. Jacobs	Columbia University
M. Jacobson	Stanford University
A. Jain	University of Illinois
D. James	National Science Foundation
G. Johnson	NOAA Pacific Marine Environmental Laboratory
R. Johnson	Colorado State University
T. Joyce	Woods Hole Oceanographic Institution
R. Katz	National Center for Atmospheric Research
R. Keeling	Scripps Institute of Oceanography
J. Kiehl	National Center for Atmospheric Research
J. Kim	Lawrence Berkeley National Laboratory
J. Kinter	Center for Ocean-Land-Atmosphere Studies
B. Kirtman	Center for Ocean-Land-Atmosphere Studies
T. Knutson	NOAA Geophysical Fluid Dynamics Laboratory
D. Koch	National Center for Atmospheric Research
S. Kreidenweis	Colorado State University
V. Krishnamurthy	Center for Ocean-Land-Atmosphere Studies
D. Kruger	Environmental Protection Agency

J. Kutzbach	University of Wisconsin at Madison
C. Landsea	NOAA Atlantic Oceanographic & Meteorological Laboratory
N. Laulainen	Pacific Northwest National Laboratory
J. Lean	Naval Research Laboratory
M. Ledbetter	National Science Foundation
T. Ledley	TERC
A. Leetmaa	NOAA National Weather Service
C. Leith	Lawrence Livermore National Laboratory
S. Levitus	NOAA National Oceanographic Data Center
J. Levy	NOAA Office of Global Programs
L. Leung	Pacific Northwest National Laboratory
R. Lindzen	Massachusetts Institute of Technology
C. Lingle	University of Alaska at Fairbanks
J. Logan	Harvard University
A. Lupo	University of Missouri
M. MacCracken	Office of the US Global Change Research Program
G. Magnusdottir	University of California
J. Mahlman	Princeton University
T. Malone	Connecticut Academy of Science and Engineering
M. E. Mann	University of Virginia
P. Matrai	Bigelow Laboratory for Ocean Sciences
D. Mauzerall	Princeton University
M. McFarland	Dupont Fluoroproducts
A. McGuire	University of Alaska at Fairbanks
S. Meacham	National Science Foundation
M. Meier	Institute of Arctic & Alpine Research
P. Michaels	University of Virginia
N. Miller	Lawrence Berkeley National Laboratory
M. Mishchenko	NASA Goddard Institute for Space Studies
V. Misra	Center for Ocean-Land-Atmosphere Studies
R. Molinari	NOAA Atlantic Oceanographic and Meteorological Laboratory
S. Montzka	NOAA Climate Monitoring & Diagnostics Laboratory
K. Mooney	NOAA Office of Global Programs
A. Mosier	Department of Agriculture
D. Neelin	University of California at Los Angeles
R. Neilson	Oregon State University
J. Norris	Princeton University
G. North	Texas A & M University
T. Novakov	Lawrence Berkeley National Laboratory
W. O'Hirok	Institute for Computational Earth System Science
M. Palecki	Illinois State Water Survey
S. Pandis	Carnegie Mellon University
C. L. Parkinson	NASA Goddard Space Flight Center
J. Penner	University of Michigan
K. Pickering	University of Maryland
R. Pielke	Colorado State University
S. Piper	Scripps Institution of Oceanography
H. Pollack	University of Michigan
G. Potter	Lawrence Livermore National Laboratory
M. Prather	University of California at Irvine
R. Prinn	Massachusetts Institute of Technology
N. Psuty	State University of New Jersey
V. Ramanathan	Scripps Institute of Oceanography
V. Ramaswamy	Princeton University
R. Randall	The Rainforest Regeneration Institution
J. Randerson	California Institute of Technology
C. Raymond	University of Washington

P. Rhines	University of Washington
C. Rinsland	NASA Langley Research Center
D. Ritson	Stanford University
A. Robock	Rutgers University
B. Rock	University of New Hampshire
J. Rodriguez	University of Miami
R. Ross	NOAA Air Resources Laboratory
D. Rotman	Lawrence Livermore National Laboratory
C. Sabine	University of Washington
D. Sahagian	University of New Hampshire
E. Saltzman	National Science Foundation
S. Sander	NASA Jet Propulsion Laboratory
E. Sarachik	University of Washington
V. Saxena	North Carolina State University
S. Schauffler	National Center for Atmospheric Research
E. Scheehle	Environmental Protection Agency
W. Schlesinger	Duke University
C. Schlosser	Center for Ocean-Land-Atmosphere Studies
R. W. Schmitt	Woods Hole Oceanographic Institution
E. Schneider	Center for Ocean-Land-Atmosphere Studies
S. Schneider	Stanford University
S. Schwartz	Brookhaven National Laboratory
M. Schwartzkopf	Princeton University
J. Seinfeld	California Institute of Technology
A. Semtner	Naval Postgraduate School
J. Severinghaus	University of California
D. Shindell	NASA Goddard Institute for Space Studies
H. Sievering	University of Colorado
J. Simpson	University of California
H. Singh	NASA Ames Research Center
D. Skole	Michigan State University
S. Smith	Pacific Northwest National Laboratory
B. J. Soden	Princeton University
R. Somerville	University of California
M. Spector	Lehigh University
T. Spence	National Science Foundation
P. Stephens	National Science Foundation
P. Stone	Massachusetts Institute of Technology
R. Stouffer	Princeton University
D. Straus	Center for Ocean-Land-Atmosphere Studies
C. Sucher	NOAA Office of Global Programs
Y. Sud	NASA Goddard Space Flight Center
B. Sun	University of Massachusetts
P. Tans	NOAA Climate Monitoring & Diagnostics Laboratory
R. Thomas	NASA Wallops Flight Facility
D. Thompson	University of Washington
J. Titus	Environmental Protection Agency
K. E. Trenberth	National Center for Atmospheric Research
S. Trumbore	University of California at Irvine
G. Tselioudis	NASA Goddard Institute for Space Studies
C. van der Veen	Ohio State University
M. Visbeck	Lamont Doherty Earth Observatory of Columbia University
M. Vuille	University of Massachusetts
M. Wahlen	University of California
J. Wallace	University of Washington
J. Walsh	University of Illinois at Urbana-Champaign
J. Wang	NOAA Air Resources Laboratory

W. Wang	State University of New York at Albany
Y. Wang	Georgia Institute of Technology
M. Ward	Lamont Doherty Earth Observatory of Columbia University
S. Warren	University of Washington
W. Washington	National Center for Atmospheric Research
B. Weare	University of California at Davis
T. Webb	Brown University
M. Wehner	Lawrence Livermore National Laboratory
R. Weller	Woods Hole Oceanographic Institution
P. Wennberg	California Institute of Technology
H. Weosky	Federal Aviation Administration
D. Williamson	National Center for Atmospheric Research
D. Winstanley	Illinois State Water Survey
S. Wofsy	Harvard University
J. Wong	NOAA Air Resources Laboratory
C. Woodhouse	NOAA National Geophysical Data Center
Z. Wu	Centre for Ocean-Land-Atmosphere Studies
X. Xiao	University of New Hampshire
Z. Yang	University of Arizona
S. Yvon-Lewis	NOAA Atlantic Oceanographic & Meteorological Laboratory
C. Zender	University of California at Irvine

United Nations Organisations and Specialised Agencies

N. Harris	European Ozone Research Coordinating Unit, United Kingdom
F. Raes	Enviroment Institute of European Commission, Italy

Non-Governmental Organisations

J. Owens	3M Company
C. Kolb	Aerodyne Research Inc.
H. Feldman	American Petroleum Institute
J. Martín-Vide	Asociación Española de Climatología, Spain
M. Ko	Atmospheric & Environmental Research Inc.
S. Baughcum	Boeing Company
C. Field	Carnegie Institute of Washington
K. Gregory	Centre for Business and the Environment, United Kingdom
W. Hennessy	CRL Energy Ltd., New Zealand
E. Olaguer	The Dow Chemical Company
D. Fisher	DuPont Company
A. Salamanca	ECO Justicia, Spain
C. Hakkarinen	Electric Power Research Institute, USA
M. Oppenheimer	Environmental Defense, USA
H. Kheshgi	Exxon Mobil Research & Engineering Company, USA
S. Japar	Ford Motor Company
W. Hare	Greenpeace International, Netherlands
L. Bishop	Honeywell International Inc.
J. Neumann	Industrial Economics, Incorporated
I. Smith	International Energy Agency Coal Research, United Kingdom
L. Bernstein	International Petroleum Industry Environmental Conservation Association
J. Grant	International Petroleum Industry Environmental Conservation Association
D. Hoyt	Raytheon
K. Green	Reason Public Policy Institute
S. Singer	Science & Environmental Policy Project, USA
J. Le Cornu	SHELL Australia Ltd.

Appendix V

Acronyms and Abbreviations

AABW	Antarctic Bottom Water
AAO	Antarctic Oscillation
ABL	Atmospheric Boundary Layer
ACC	Antarctic Circumpolar Current
ACE	Aerosol Characterisation Experiment
ACRIM	Active Cavity Radiometer Irradiance Monitor
ACSYS	Arctic Climate System Study
ACW	Antarctic Circumpolar Wave
AEROCE	Atmosphere Ocean Chemistry Experiment
AGAGE	Advanced Global Atmospheric Gases Experiment
AGCM	Atmospheric General Circulation Model
AGWP	Absolute Global Warming Potential
AMIP	Atmospheric Model Intercomparison Project
ANN	Artificial Neural Networks
AO	Arctic Oscillation
AOGCM	Atmosphere-Ocean General Circulation Model
ARESE	Atmospheric Radiation Measurement Enhanced Shortwave Experiment
ARGO	Part of the Integrated Global Observation Strategy
ARM	Atmospheric Radiation Measurement
ARPEGE/OPA	Action de Recherche Petite Echelle Grande Echelle/Océan Parallélisé
ASHOE/MAESA	Airborne Southern Hemisphere Ozone Experiment/Measurement for Assessing the Effects of Stratospheric Aircraft
AVHRR	Advanced Very High Resolution Radiometer
AWI	Alfred Wegener Institute (Germany)
BAHC	Biospheric Aspects of the Hydrological Cycle
BC	Black Carbon
BERN2D	Two-dimensional Climate Model of University of Bern
BIOME 6000	Global Palaeo-vegetation Mapping Project
BMRC	Bureau of Meteorology Research Centre (Australia)
CART	Classification and Tree Analysis
CCA	Canonical Correlation Analysis
CCC(ma)	Canadian Centre for Climate (Modelling and Analysis) (Canada)
CCM	Community Climate Model
CCMLP	Carbon Cycle Model Linkage Project
CCN	Cloud Condensation Nuclei
CCSR	Centre for Climate System Research (Japan)
CERFACS	European Centre for Research and Advanced Training in Scientific Computation (France)
CIAP	Climate Impact Assessment Program

CLIMAP	Climate: Long-range Investigation, Mapping and Prediction
CLIMBER	Climate-Biosphere Model
CLIMPACTS	Integrated Model for Assessment of the Effects of Climate Change on the New Zealand Environment
CMAP	CPC Merged Analysis of Precipitation
CMDL	Climate Monitoring and Diagnostics Laboratory of NOAA (USA)
CMIP	Coupled Model Intercomparison Project
CNRM	Centre National de Recherches Météorologiques (France)
CNRS	Centre National de la Recherche Scientifique (France)
COADS	Comprehensive Ocean Atmosphere Data Set
COHMAP	Co-operative Holocene Mapping Project
COLA	Center for Ocean-Land-Atmosphere Studies (USA)
COSAM	Comparison of Large-scale Atmospheric Sulphate Aerosol Model
COSMIC	Country Specific Model for Intertemporal Climate
COWL	Cold Ocean Warm Land
CPC	Climate Prediction Center of NOAA (USA)
CRF	Cloud Radiative Forcing
CRU	Climatic Research Unit of UEA (UK)
CRYOSat	Cryosphere Satellite
CSG	Climate Scenario Generator
CSIRO	Commonwealth Scientific and Industrial Research Organisation (Australia)
CSM	Climate System Model
CTM	Chemistry Transport Model
DARLAM	CSIRO Division of Atmospheric Research Limited Area Model
DDC	Data Distribution Centre of IPCC
DGVM	Dynamic Global Vegetation Model
DERF	Dynamical Extended Range Forecasting group of GFDL (USA)
DIC	Dissolved Inorganic Carbon
DJF	December, January, February
DKRZ	Deutsche KlimaRechenZentrum (Germany)
DMS	Dimethylsulfide
DMSP	Defense Meteorological Satellite Program
DNM	Department of Numerical Mathematics (Russia)
DOC	Dissolved Organic Carbon
DOE	Department of Energy (USA)
DORIS	Determination d'Orbite et Radiopositionnement Intégrés par Satellite
DRF	Direct Radiative Forcing
DTR	Diurnal Temperature Range
DYNAMO	Dynamics of North Atlantic Models
EBM	Energy Balance Model
ECHAM	ECMWF/MPI AGCM
ECMWF	European Centre for Medium-range Weather Forecasting
ECS	Effective Climate Sensitivity
EDGAR	Emission Database for Global Atmospheric Research
EISMINT	European Ice Sheet Modelling initiative
EMDI	Ecosystem Model/Data Intercomparison
EMIC	Earth system Models of Intermediate Complexity
ENSO	El Niño-Southern Oscillation
EOF	Empirical Orthogonal Function
EOS	Earth Observing System
ERA	ECMWF Reanalysis
ERB	Earth Radiation Budget
ERBE	Earth Radiation Budget Experiment
ERBS	Earth Radiation Budget Satellite
ESCAPE	Evaluation of Strategies to Address Climate Change by Adapting to and Preventing Emissions
ESMR	Electrically Scanning Microwave Radiometer
EURECA	European Retrievable Carrier
FACE	Free Air Carbon-dioxide Enrichment

FAO	Food and Agriculture Organisation (UN)
FCCC	Framework Convention on Climate Change
FDH	Fixed Dynamical Heating
FF	Fossil Fuel
FPAR	Plant-absorbed Fraction of Incoming Photosynthetically Active Radiation
FSU	Former Soviet Union
GASP	Global Assimilation and Prediction
GCIP	GEWEX Continental-scale International Program
GCM	General Circulation Model
GCOS	Global Climate Observing System
GCR	Galactic Cosmic Ray
GDP	Gross Domestic Product
GEBA	Global Energy Balance Archive
GEIA	Global Emissions Inventory Activity
GEISA	Gestion et Etude des Informations Spectroscopiques Atmosphériques
GEWEX	Global Energy and Water cycle Experiment
GFDL	Geophysical Fluid Dynamics Laboratory (USA)
GHCN	Global Historical Climate Network
GHG	Greenhouse Gas
GIM	Global Integration and Modelling
GISP	Greenland Ice Sheet Project
GISS	Goddard Institute for Space Studies (USA)
GISST	Global Sea Ice and Sea Surface Temperature
GLOSS	Global Sea Level Observing System
GOALS	Global Ocean-Atmosphere-Land System
GPCP	Global Precipitation Climatology Project
GPP	Gross Primary Production
GPS	Global Positioning System
GRACE	Gravity Recovery and Climate Experiment
GRIP	Greenland Ice Core Project
GSFC	Goddard Space Flight Center (USA)
GSWP	Global Soil Wetness Project
GUAN	GCOS Upper Air Network
GWP	Global Warming Potential
HadCM	Hadley Centre Coupled Model
HIRETYCS	High Resolution Ten-Year Climate Simulations
HITRAN	High Resolution Transmission Molecular Absorption Database
HLM	High Latitude Mode
HNLC	High Nutrient-Low Chlorophyll
HRBM	High Resolution Biosphere Model
IAHS	International Association of Hydrological Science
IAP	Institute of Atmospheric Physics (China)
IASB	Institut d'Aéronomie Spatiale de Belgique (Belgium)
IBIS	Integrated Biosphere Simulator
ICESat	Ice, Cloud and Land Elevation Satellite
ICSI	International Commission on Snow and Ice
ICSU	International Council of Scientific Unions
IGAC	International Global Atmospheric Chemistry
IGBP	International Geosphere Biosphere Programme
IGCR	Institute for Global Change Research (Japan)
IHDP	International Human Dimensions Programme on Global Environmental Change
IMAGE	Integrated Model to Assess the Global Environment
IN	Ice Nuclei
INDOEX	Indian Ocean Experiment
IOC	Intergovernmental Oceanographic Commission
IPCC	Intergovernmental Panel on Climate Change
IPO	Interdecadal Pacific Oscillation

IPSL-CM	Institut Pierre Simon Laplace/Coupled Atmosphere-Ocean-Vegetation Model
ISAM	Integrated Science Assessment Model
ISCCP	International Satellite Cloud Climatology Project
ISLSCP	International Satellite Land Surface Climatology Project
ITCZ	Inter-Tropical Convergence Zone
IUPAC	International Union of Pure and Applied Chemistry
JGOFS	Joint Global Ocean Flux Study
JJA	June, July, August
JMA	Japan Meteorological Agency (Japan)
JPL	Jet Propulsion Laboratory of NASA (USA)
KNMI	Koninklijk Nederlands Meteorologisch Instituut (Netherlands)
LAI	Leaf Area Index
LASG	State Key Laboratory of Numerical Modelling for Atmospheric Sciences and Geophysical Fluid Dynamics (China)
LBA	Large-scale Biosphere-atmosphere Experiment in Amazonia
LGGE	Laboratoire de Glaciologie et Géophysique de l'Environnement (France)
LGM	Last Glacial Maximum
LLNL	Lawrence Livermore National Laboratory (USA)
LMD	Laboratoire de Météorologie Dynamique (France)
LOSU	Level of Scientific Understanding
LPJ	Land-Potsdam-Jena Terrestrial Carbon Model
LSAT	Land Surface Air Temperature
LSG	Large-Scale Geostrophic Ocean Model
LSP	Land Surface Parameterisation
LT	Lifetime
LWP	Liquid Water Path
MAGICC	Model for the Assessment of Greenhouse-gas Induced Climate Change
MAM	March, April, May
MARS	Multivariate Adaptive Regression Splines
MGO	Main Geophysical Observatory (Russia)
MJO	Madden-Julian Oscillation
ML	Mixed Layer
MLOPEX	Mauna Loa Observatory Photochemistry Experiment
MODIS	Moderate Resoluting Imaging Spectroradiometer
MOGUNTIA	Model of the General Universal Tracer Transport in the Atmosphere
MOM	Modular Ocean Model
MOZART	Model for Ozone and Related Chemical Tracers
MPI	Max-Plank Institute for Meteorology (Germany)
MRI	Meteorological Research Institute (Japan)
MSLP	Mean Sea Level Pressure
MSU	Microwave Sounding Unit
NADW	North Atlantic Deep Water
NAO	North Atlantic Oscillation
NARE	North Atlantic Regional Experiment
NASA	National Aeronautics and Space Administration (USA)
NBP	Net Biome Production
NCAR	National Center for Atmospheric Research (USA)
NCC	National Climate Centre (China)
NCDC	National Climatic Data Center of NOAA (USA)
NCEP	National Centers for Environmental Prediction of NOAA (USA)
NDVI	Normalised Difference Vegetation Index
NEP	Net Ecosystem Production
NESDIS	National Environmental Satellite, Data and Information Service of NOAA (USA)
NIC	National Ice Center of NOAA (USA)
NIED	National Research Institute for Earth Science and Disaster Prevention (Japan)
NIES	National Institute for Environmental Studies (Japan)
NMAT	Night Marine Air Temperature

NMHC	Non-Methane Hydrocarbon
NOAA	National Oceanic and Atmospheric Administration (USA)
NPP	Net Primary Production
NPZD	Nutrients, Phytoplankton, Zooplankton and Detritus
NRC	National Research Council (USA)
NRL	Naval Research Laboratory (USA)
NWP	Numerical Weather Prediction
OC	Organic Carbon
OCMIP	Ocean Carbon-cycle Model Intercomparison Project
OCS	Organic Carbonyl Sulphide
OGCM	Ocean General Circulation Model
OLR	Outgoing Long-wave Radiation
OPYC	Ocean Isopycnal GCM
OxComp	Tropospheric Oxidant Model Comparison
PC	Principal Component
PCM	Parallel Climate Model
PDF	Probability Density Function
PDO	Pacific Decadal Oscillation
PEM	Pacific Exploratory Missions
PFT	Plant Functional Type
PGR	Post-Glacial Rebound
PhotoComp	Ozone Photochemistry Model Comparison
PICASSO	Pathfinder Instruments for Cloud and Aerosol Spaceborne Observations
PIK	Potsdam Institute for Climate Impact Research (Germany)
PILPS	Project for the Intercomparison of Land-surface Parameterisation Schemes
PIUB	Physics Institute University of Bern (Switzerland)
PMIP	Palaeoclimate Model Intercomparison Project
PNA	Pacific-North American
PNNL	Pacific Northwest National Laboratory (USA)
POC	Particulate Organic Carbon
POLDER	Polarisation and Directionality of the Earth's Reflectances
POPCORN	Photo-Oxidant Formation by Plant Emitted Compounds and OH Radicals in North-eastern Germany
PSMSL	Permanent Service for Mean Sea Level
PT	Perturbation Lifetime
QBO	Quasi-Biennial Oscillation
RAMS	Regional Atmospheric Modelling System
RCM	Regional Climate Model
RIHMI	Research Institute for Hydrometeorological Information
SAGE	Stratospheric Aerosol & Gas Experiment
SAR	IPCC Second Assessment Report
SAT	Surface Air Temperature
SBUV	Solar Backscatter Ultra Violet
SCAR-B	Smoke Cloud and Radiation-Brazil
SCE	Snow Cover Extent
SCENGEN	Scenario Generator
SCSWP	Small-scale Severe Weather Phenomena
SDD	Statistical-Dynamical Downscaling
SDGVM	Sheffield Dynamic Global Vegetation Model
SEFDH	Seasonally Evolving Fixed Dynamical Heating
SHEBA	Surface Heat Balance of the Arctic Ocean
SHI	State Hydrological Institute (Russia)
SIMIP	Sea Ice Model Intercomparison Project
SIO	Scripps Institution of Oceanography (USA)
SLP	Sea Level Pressure
SMMR	Scanning Multichannel Microwave Radiometer
SOA	Secondary Organic Aerosol
SOC	Southampton Oceanography Centre (UK)

SOHO	Solar Heliospheric Observatory
SOI	Southern Oscillation Index
SOLSTICE	Solar Stellar Irradiance Comparison Experiment
SON	September, October, November
SONEX	Subsonic Assessment Program Ozone and Nitrogen Oxide Experiment
SOS	Southern Oxidant Study
SPADE	Stratospheric Photochemistry, Aerosols, and Dynamics Expedition
SPARC	Stratospheric Processes and Their Role in Climate
SPCZ	South Pacific Convergence Zone
SRES	IPCC Special Report on Emission Scenarios
SSM/T-2	Special Sensor Microwave Water Vapour Sounder
SSM/I	Special Sensor Microwave/Imager
SST	Sea Surface Temperature
SSU	Stratospheric Sounding Unit
STRAT	Stratospheric Tracers of Atmospheric Transport
SUCCESS	Subsonic Aircraft Contrail and Cloud Effects Special Study
SUNGEN	State University of New York at Albany/NCAR Global Environmental and Ecological Simulation of Interactive Systems
SUSIM	Solar Ultraviolet Spectral Irradiance Monitor
TAR	IPCC Third Assessment Report
TARFOX	Tropospheric Aerosol Radiative Forcing Observational Experiment
TBFRA	Temperate and Boreal Forest Resource Assessment
TBO	Tropospheric Biennial Oscillation
TCR	Transient Climate Response
TEM	Terrestrial Ecosystem Model
TEMPUS	Sea Surface Temperature Evolution Mapping Project based on Alkenone Stratigraphy
THC	Thermohaline Circulation
TMR	TOPEX Microwave Radiometer
TOA	Top of the Atmosphere
TOMS	Total Ozone Mapping Spectrometer
TOPEX/POSEIDON	US/French Ocean Topography Satellite Altimeter Experiment
TOVS	Television Infrared Observation Satellite Operational Vertical Sounder
TPI	Trans Polar Index
TRIFFID	Top-down Representation of Interactive Foliage and Flora Including Dynamics
TSI	Total Solar Irradiance
UARS	Upper Atmosphere Research Satellite
UCAM	University of Cambridge (UK)
UCI	University of California at Irvine (USA)
UD/EB	Upwelling Diffusion-Energy Balance
UEA	University of East Anglia (UK)
UGAMP	University Global Atmospheric Modelling Project
UIO	Universitetet I Oslo (Norway)
UIUC	University of Illinois at Urbana-Champaign (USA)
UKHI	United Kingdom High-resolution climate model
UKMO	United Kingdom Met Office (UK)
UKTR	United Kingdom Transient climate experiment
ULAQ	Università degli studi dell'Aquila (Italy)
UM	Unified Model
UNEP	United Nations Environment Programme
UNESCO	United Nations Education, Scientific and Cultural Organisation
UNFCCC	United Nations Framework Convention on Climate Change
USSR	Union of Soviet Socialist Republics
UTH	Upper Tropospheric Humidity
UV	Ultraviolet radiation
UVic	University of Victoria (Canada)
VIRGO	Variability of Solar Irradiance and Gravity Oscillations
VLM	Vertical Land Movement

VOC	Volatile Organic Compounds
WAIS	West Antarctic Ice Sheet
WASA	Waves and Storms in the North Atlantic
WAVAS	Water Vapour Assessment
WBCs	Western Boundary Currents
WCRP	World Climate Research Programme
WMGGs	Well-Mixed Greenhouse Gases
WMO	World Meteorological Organization
WOCE	World Ocean Circulation Experiment
WP	Western Pacific
WRE	Wigley, Richels and Edmonds
YONU	Yonsei University (Korea)

Appendix VI

Units

SI (Systeme Internationale) Units:

Physical Quantity	Name of Unit	Symbol
length	metre	m
mass	kilogram	kg
time	second	s
thermodynamic temperature	kelvin	K
amount of substance	mole	mol

Fraction	Prefix	Symbol	Multiple	Prefix	Symbol
10^{-1}	deci	d	10	deca	da
10^{-2}	centi	c	10^2	hecto	h
10^{-3}	milli	m	10^3	kilo	k
10^{-6}	micro	μ	10^6	mega	M
10^{-9}	nano	n	10^9	giga	G
10^{-12}	pico	p	10^{12}	tera	T
10^{-15}	femto	f	10^{15}	peta	P

Special Names and Symbols for Certain SI-Derived Units:

Physical Quantity	Name of SI Unit	Symbol for SI Unit	Definition of Unit
force	newton	N	$kg\ m\ s^{-2}$
pressure	pascal	Pa	$kg\ m^{-1}\ s^{-2}\ (=N\ m^{-2})$
energy	joule	J	$kg\ m^2\ s^{-2}$
power	watt	W	$kg\ m^2\ s^{-3}\ (=J\ s^{-1})$
frequency	hertz	Hz	s^{-1} (cycles per second)

Decimal Fractions and Multiples of SI Units Having Special Names:

Physical Quantity	Name of Unit	Symbol for Unit	Definition of Unit
length	Ångstrom	Å	10^{-10} m = 10^{-8} cm
length	micron	μm	10^{-6} m
area	hectare	ha	10^4 m^2
force	dyne	dyn	10^{-5} N
pressure	bar	bar	10^5 N m^{-2} = 10^5 Pa
pressure	millibar	mb	10^2 N m^{-2} = 1 hPa
mass	tonne	t	10^3 kg
mass	gram	g	10^{-3} kg
column density	Dobson units	DU	2.687×10^{16} molecules cm^{-2}
streamfunction	Sverdrup	Sv	10^6 m^3 s^{-1}

Non-SI Units:

°C	degree Celsius (0 °C = 273 K approximately)
	Temperature differences are also given in °C (=K) rather than the more correct form of "Celsius degrees".
ppmv	parts per million (10^6) by volume
ppbv	parts per billion (10^9) by volume
pptv	parts per trillion (10^{12}) by volume
yr	year
ky	thousands of years
bp	before present

The units of mass adopted in this report are generally those which have come into common usage and have deliberately not been harmonised, e.g.,

GtC	gigatonnes of carbon (1 GtC = 3.7 Gt carbon dioxide)
PgC	petagrams of carbon (1 PgC = 1 GtC)
MtN	megatonnes of nitrogen
TgC	teragrams of carbon (1 TgC = 1 MtC)
Tg(CH$_4$)	teragrams of methane
TgN	teragrams of nitrogen
TgS	teragrams of sulphur

Appendix VII

Some chemical symbols used in this report

C	carbon (there are three isotopes: ^{12}C, ^{13}C, ^{14}C)	DOC	dissolved organic carbon
Ca	calcium	H_2	hydrogen
$CaCO_3$	calcium carbonate	halon-1211	CF_2ClBr
CCl_4	carbon tetrachloride	halon-1301	CF_3Br
CF_4	perfluoromethane	halon-2402	CF_2BrCF_2Br
C_2F_6	perfluoroethane	HCFC	hydrochlorofluorocarbon
C_3F_8	perfluoropropane	HCFC-21	$CHCl_2F$
C_4F_8	perfluorocyclobutane	HCFC-22	CHF_2Cl
C_4F_{10}	perfluorobutane	HCFC-123	$C_2F_3HCl_2$
C_5F_{12}	perfluoropentane	HCFC-124	CF_3CHClF
C_6F_{14}	perfluorohexane	HCFC-141b	CH_3CFCl_2
CFC	chlorofluorocarbon	HCFC-142b	CH_3CF_2Cl
CFC-11	$CFCl_3$ (trichlorofluoromethane)	HCFC-225ca	$CF_3CF_2CHCl_2$
CFC-12	CF_2Cl_2 (dichlorodifluoromethane)	HCFC-225cb	$CClF_2CF_2CHClF$
CFC-13	CF_3Cl (chlorotrifluoromethane)	HCFE-235da2	$CF_3CHClOCHF_2$
CFC-113	$CF_2ClCFCl_2$ (trichlorotrifluoroethane)	HCO_3^-	bicarbonate ion
CFC-114	CF_2ClCF_2Cl (dichlorotetrafluoroethane)	HFC	hydrofluorocarbon
CFC-115	CF_3CF_2Cl (chloropentafluoroethane)	HFC-23	CHF_3
CF_3I	trifluoroiodomethane	HFC-32	CH_2F_2
CH_4	methane	HFC-41	CH_3F
C_2H_6	ethane	HFC-125	CHF_2CF_3
C_5H_8	isoprene	HFC-134	CHF_2CHF_2
C_6H_6	benzene	HFC-134a	CF_3CH_2F
C_7H_8	toluene	HFC-143	$CH_2F\,CHF_2$
$C_{10}H_{16}$	terpene	HFC-143a	CH_3CF_3
CH_3Br	methylbromide	HFC-152	CH_2FCH_2F
CH_3CCl_3	methyl chloroform	HFC-152a	CH_3CHF_2
$CHCl_3$	chloroform/trichloromethane	HFC-161	CH_3CH_2F
CH_2Cl_2	dichloromethane/methylene chloride	HFC-227ea	CF_3CHFCF_3
CH_3Cl	methylchloride	HFC-236cb	$CF_3CF_2CH_2F$
CH_3OCH_3	dimethyl ether	HFC-236ea	$CF_3CHFCHF_2$
CO	carbon monoxide	HFC-236fa	$CF_3CH_2CF_3$
CO_2	carbon dioxide	HFC-245ca	$CH_2FCF_2CHF_2$
CO_3^{2-}	carbonate ion	HFC-245ea	$CHF_2CHFCHF_2$
DIC	dissolved inorganic carbon	HFC-245eb	CF_3CHFCH_2F

HFC-245fa	$CHF_2CH_2CF_3$	**HFOC-134**	CF_2HOCF_2H
HFC-263fb	$CF_3CH_2CH_3$	**HFOC-143a**	CF_3OCH_3
HFC-338pcc	$CHF_2CF_2CF_2CF_2H$	**HFOC-152a**	CH_3OCHF_2
HFC-356mcf	$CF_3CF_2CH_2CH_2F$	**HFOC-245fa**	$CHF_2OCH_2CF_3$
HFC-356mff	$CF_3CH_2CH_2CF_3$	**HFOC-356mmf**	$CF_3CH_2OCH_2CF_3$
HFC-365mfc	$CF_3CH_2CF_2CH_3$	**HG-01**	$CHF_2OCF_2CF_2OCHF_2$
HFC-43-10mee	$CF_3CHFCHFCF_2CF_3$	**HG-10**	$CHF_2OCF_2OCHF_2$
HFC-458mfcf	$CF_3CH_2CF_2CH_2CF_3$	**H-Galden 1040x**	$CHF_2OCF_2OC_2F_4OCHF_2$
HFC-55-10mcff	$CF_3CF_2CH_2CH_2CF_2CF_3$	**HNO₃**	nitric acid
HFE-125	CF_3OCHF_2	**HO₂**	hydroperoxyl
HFE-134	CF_2HOCF_2H	**HOₓ**	the sum of OH and HO_2
HFE-143a	CF_3OCH_3	**H₂O**	water vapour
HFE-152a	CH_3OCHF_2	**H₂SO₄**	sulphuric acid
HFE-227ea	$CF_3CHFOCF_3$	**N₂**	molecular nitrogen
HFE-236ea2	$CF_3CHFOCHF_2$	**NF₃**	nitrogen trifluoride
HFE-236fa	$CF_3CH_2OCF_3$	**NH₃**	ammonia
HFE-245cb2	$CF_3CF_2OCH_3$	**NH₄⁺**	ammonium ion
HFE-245fa1	$CHF_2CH_2OCF_3$	**NMHC**	non-methane hydrocarbon
HFE-245fa2	$CHF_2OCH_2CF_3$	**NO**	nitric oxide
HFE-254cb2	$CHF_2CF_2OCH_3$	**NO₂**	nitrogen dioxide
HFE-263fb2	$CF_3CH_2OCH_3$	**NOₓ**	nitrogen oxides (the sum of NO and NO_2)
HFE-329mcc2	$CF_3CF_2OCF_2CHF_2$	**NO₃**	nitrate radical
HFE-338mcf2	$CF_3CF_2OCH_2CF_3$	**NO₃⁻**	nitrate ion
HFE-347mcc3	$CF_3CF_2CF_2OCH_3$	**N₂O**	nitrous oxide
HFE-347mcf2	$CF_3CF_2OCH_2CHF_2$	**O₂**	molecular oxygen
HFE-356mec3	$CF_3CHFCF_2OCH_3$	**O₃**	ozone
HFE-356mff2	$CF_3CH_2OCH_2CF_3$	**OCS**	organic carbonyl sulphide
HFE-356pcc3	$CHF_2CF_2CF_2OCH_3$	**OH**	hydroxyl radical
HFE-356pcf2	$CHF_2CF_2OCH_2CHF_2$	**PAN**	peroxyacetyl nitrate
HFE-356pcf3	$CHF_2CF_2CH_2OCHF_2$	**PFC**	perfluorocarbon
HFE-365mcf3	$CF_3CF_2CH_2OCH_3$	**SF₆**	sulphur hexafluoride
HFE-374pc2	$CHF_2CF_2OCH_2CH_3$	**SF₅CF₃**	trifluoromethyl sulphur pentafluoride
HFE-7100	$C_4F_9OCH_3$	**SO₂**	sulphur dioxide
HFE-7200	$C_4F_9OC_2H_5$	**SO₄²⁻**	sulphate ion
HFOC-125	CF_3OCHF_2	**VOC**	volatile organic compounds

Appendix VIII

Index